Environmental Microbiology: Fundamentals and Applications

Jean-Claude Bertrand • Pierre Caumette
Philippe Lebaron • Robert Matheron
Philippe Normand • Télesphore Sime-Ngando
Editors

Environmental Microbiology: Fundamentals and Applications

Microbial Ecology

Editors

Jean-Claude Bertrand
Institut Méditerranéen
 d'Océanologie (MIO)
UM 110, CNRS 7294 IRD 235,
 Université de Toulon, Aix-Marseille
 Université
Marseille Cedex 9, France

Philippe Lebaron
Observatoire Océanologique de Banyuls,
 Laboratoire de Biodiversité et
 Biotechnologie Microbiennes (LBBM)
Sorbonne Universités, UPMC Univ
 Paris 06, USR CNRS 3579
Banyuls-sur-Mer, France

Philippe Normand
Microbial Ecology Center
UMR CNRS 5557/USC INRA 1364
Villeurbanne, France

Pierre Caumette
Institut des Sciences Analytiques et de Physico-chimie pour
 l'Environnement et les Matériaux (IPREM)
UMR CNRS 5254, Université de Pau et des Pays de l'Adour
Pau Cedex, France

Robert Matheron
Institut Méditerranéen de Biodiversité et
 d'Ecologie marine et continentale (IMBE)
UMR-CNRS-IRD 7263, Aix-Marseille Université
Marseille Cedex 20, France

Télesphore Sime-Ngando
Laboratoire Microorganismes: Génome
 et Environnement (LMGE)
UMR CNRS 6023, Université Blaise Pascal,
 Clermont Université
Aubère Cedex, France

Editors
Jean-Claude Bertrand
Pierre Caumette
Philippe Lebaron
Robert Matheron
Philippe Normand

Title
Écologie microbienne
Microbiologie des milieux naturels et anthropisés

Publisher/Location
Presses Universitaires de Pau et des Pays de l'Adour
www.presses-univ-pau.fr

Copyright year : 2011

Translation
The contributing authors in their role as translators of the chapters from French to English.

Content
Updated (previously published content—in the French title—but now includes
updated material and/or sections that have been added)

ISBN 978-94-017-9117-5   ISBN 978-94-017-9118-2 (eBook)
DOI 10.1007/978-94-017-9118-2
Springer Dordrecht Heidelberg New York London

Library of Congress Control Number: 2014956023

© Springer Science+Business Media Dordrecht 2015
This work is subject to copyright. All rights are reserved by the Publisher, whether the whole or part of the material is concerned, specifically the rights of translation, reprinting, reuse of illustrations, recitation, broadcasting, reproduction on microfilms or in any other physical way, and transmission or information storage and retrieval, electronic adaptation, computer software, or by similar or dissimilar methodology now known or hereafter developed. Exempted from this legal reservation are brief excerpts in connection with reviews or scholarly analysis or material supplied specifically for the purpose of being entered and executed on a computer system, for exclusive use by the purchaser of the work. Duplication of this publication or parts thereof is permitted only under the provisions of the Copyright Law of the Publisher's location, in its current version, and permission for use must always be obtained from Springer. Permissions for use may be obtained through RightsLink at the Copyright Clearance Center. Violations are liable to prosecution under the respective Copyright Law.
The use of general descriptive names, registered names, trademarks, service marks, etc. in this publication does not imply, even in the absence of a specific statement, that such names are exempt from the relevant protective laws and regulations and therefore free for general use.
While the advice and information in this book are believed to be true and accurate at the date of publication, neither the authors nor the editors nor the publisher can accept any legal responsibility for any errors or omissions that may be made. The publisher makes no warranty, express or implied, with respect to the material contained herein.

Printed on acid-free paper

Springer is part of Springer Science+Business Media (www.springer.com)

Preface

The average citizen thinks of microbes as dangerous or at least as nasty enemies to her/his health that she/he cannot see. The invisibility to the naked eye is the reason why microbes are so poorly known, their wide action and importance in nature almost fully neglected. And this also has the consequence that they play no or only a subordinate role in the international conventions for the protection of species. The French born English writer Hilaire Belloc (1870–1953) put this dilemma into a simple poem that starts: "The microbe is so very small, you cannot make him out at all, but many sanguine people hope to see him through the microscope ...". There is no doubt, that only a minority of people have the opportunity to do so and thus the majority stays ignorant and continues to harbor only the negative image of microbes.

This book about microbial ecology will open the ignorant eyes to understand and appreciate the role and importance of the ubiquitous "invisible microbes" in nature as well as in daily life. In the twenty-first century, with mankind facing new threats like contamination of soils, rivers, the oceans, air pollution, global warming, climate change, desertification, newly evolving diseases, etc., it is necessary – more than ever before – to assess the importance of microbes, of their ecological behavior and implication in these processes. A prerequisite for that is a multi-facetted knowledge of microbes, their diversity, their metabolism, their interactions, the ecological rules controlling them and the biotopes they inhabit. Tools and methods for the study of ecological interactions are as necessary as the physicochemical axioms and mathematics behind them. Ecology is predominantly seen as an outdoor science, but has by principle two quite different localities of research: Laboratory work is as important as the study of the sometimes rather inaccessible natural habitats. It is essential for the modern ecologist to be as competent and experienced in the field as at the lab bench.

The six editors of this book, Jean Claude Bertrand, Pierre Caumette, Philippe Lebaron, Robert Matheron, Philippe Normand and Télesphore Sime-Ngando are to be congratulated for finding such a perfect list of authorial teams, all very well known experts in the field. The book is divided in five parts: "general chapters", "taxonomy and evolution", "microbial habitats", "role and functioning of microbial ecosystems", and "tools and microbial genetics applied to the environment", each containing 3–5 chapters, altogether written by 59 authors, of whom many were involved in more than one chapter. Thirteen chapters are co-authored by one or more of the editors! Thus and by means of multiple authorship the editors made sure that the book comes out as a single consistent effort, not a collection of more or less divergent chapters. As a comprehensive treatise it definitely has a chance to become at least an important reference and textbook for graduate and postgraduate students in microbial ecology, but most probably in general microbiology, too. Textbooks in general microbiology (with exceptions) usually put emphasis on metabolism, systematics, structure etc., not on ecology. This book, however, positions ecology at the center and the other fields around it. Quite often one misses the microbial eukaryotes or viruses in microbial ecology books. Not in this one. This is an all microbes embracing effort and therefore constitutes a complete treatise of the field. What is welcomed and emphasizes the completeness of the book is that it devotes one chapter to the history of microbial ecology, depicting the development of theories, increase of knowledge, changes in thinking over the centuries towards more modern and global views. In other chapters, historical aspects are also mentioned where necessary. Divergent ideas and conceptions are not suppressed, but openly discussed, as scientific dispute demands.

Therefore it will gain the full attention of not only microbial ecologists, but of all microbiologists who need to modernize their views of ecology. Also plant and animal ecologists in general should brush up their knowledge on the often neglected part that microbes play in the global environment. For all who are involved in contamination and remediation of soils and waters, in waste treatment, in landscaping, in maintenance of natural resorts, food microbiology, biotechnology, phytopathology, medical and veterinarian microbiology, this book contains valuable information. This treatise is relevant for every aspect of modern microbiology and belongs into the hand of every microbiologist.

Persons who have studied this book thoroughly will be able to appreciate the profound truth of Louis Pasteur's sentence: *"Ce sont les microbes qui auront le dernier mot"* ("It is the microbes that will have the last say").

Bonn, Germany   Hans G. Trüper

Acknowledgements

We are grateful to all those who helped us to achieve this book whose publication has required the assistance and comments of many contributors.

Most of the illustrations are due to the talent of Marie-Josée Bodiou and Valérie Domien, graphic designers at Arago Laboratory in Banyuls-sur-mer, whose contribution was indispensable to the realization of the work; we wish to thank both very much for all the work accomplished with skill and good humor.

Our thanks also go to the following persons who were responsible for proofreading parts of the book, who expressed constructive criticism, or helped us with their skills: Michel Guiliano, Fortunat Joos, Dominique Lefevre, Louis Legendre, Roland Marmeisse, Danielle Marty, Gilbert Mille, Fereidoun Rassoulzadegan, Marc Troussellier, Francesca Vidussi, and all members of the GDR "Food Webs."

Contents

Part I General Chapters

1. **The Thematic Fields of Microbial Ecology** 3
   Jean-Claude Bertrand, Pierre Caumette, Philippe Lebaron,
   and Philippe Normand

2. **Some Historical Elements of Microbial Ecology** 9
   Pierre Caumette, Jean-Claude Bertrand, and Philippe Normand

3. **Structure and Functions of Microorganisms: Production
   and Use of Material and Energy** 25
   Robert Matheron and Pierre Caumette

Part II Taxonomy and Evolution

4. **For Three Billion Years, Microorganisms Were the Only
   Inhabitants of the Earth** 75
   Jean-Claude Bertrand, Céline Brochier-Armanet,
   Manolo Gouy, and Frances Westall

5. **Systematic and Evolution of Microorganisms: General Concepts** 107
   Charles-François Boudouresque, Pierre Caumette, Jean-Claude Bertrand,
   Philippe Normand, and Télesphore Sime-Ngando

6. **Taxonomy and Phylogeny of Prokaryotes** 145
   Pierre Caumette, Céline Brochier-Armanet, and Philippe Normand

7. **Taxonomy and Phylogeny of Unicellular Eukaryotes** 191
   Charles-François Boudouresque

Part III Microbial Habitats: Diversity, Adaptation and Interactions

8. **Biodiversity and Microbial Ecosystems Functioning** 261
   Philippe Normand, Robert Duran, Xavier Le Roux, Cindy Morris,
   and Jean-Christophe Poggiale

9. **Adaptations of Prokaryotes to Their Biotopes and to Physicochemical
   Conditions in Natural or Anthropized Environments** 293
   Philippe Normand, Pierre Caumette, Philippe Goulas,
   Petar Pujic, and Florence Wisniewski-Dyé

10. **The Extreme Conditions of Life on the Planet and Exobiology** 353
    Jean-Luc Cayol, Bernard Ollivier, Didier Alazard, Ricardo Amils,
    Anne Godfroy, Florence Piette, and Daniel Prieur

11	**Microorganisms and Biotic Interactions**...........................	395
	Yvan Moënne-Loccoz, Patrick Mavingui, Claude Combes, Philippe Normand, and Christian Steinberg	
12	**Horizontal Gene Transfer in Microbial Ecosystems**.................	445
	Céline Brochier-Armanet and David Moreira	

Part IV Role and Functioning of Microbial Ecosystems

13	**Microbial Food Webs in Aquatic and Terrestrial Ecosystems**..........	485
	Behzad Mostajir, Christian Amblard, Evelyne Buffan-Dubau, Rutger De Wit, Robert Lensi, and Télesphore Sime-Ngando	
14	**Biogeochemical Cycles**..	511
	Jean-Claude Bertrand, Patricia Bonin, Pierre Caumette, Jean-Pierre Gattuso, Gérald Grégori, Rémy Guyoneaud, Xavier Le Roux, Robert Matheron, and Franck Poly	
15	**Environmental and Human Pathogenic Microorganisms**..............	619
	Philippe Lebaron, Benoit Cournoyer, Karine Lemarchand, Sylvie Nazaret, and Pierre Servais	
16	**Applied Microbial Ecology and Bioremediation**.....................	659
	Jean-Claude Bertrand, Pierre Doumenq, Rémy Guyoneaud, Benoit Marrot, Fabrice Martin-Laurent, Robert Matheron, Philippe Moulin, and Guy Soulas	

Part V Tools of Microbial Ecology

17	**Methods for Studying Microorganisms in the Environment**...........	757
	Fabien Joux, Jean-Claude Bertrand, Rutger De Wit, Vincent Grossi, Laurent Intertaglia, Philippe Lebaron, Valérie Michotey, Philippe Normand, Pierre Peyret, Patrick Raimbault, Christian Tamburini, and Laurent Urios	
18	**Contributions of Descriptive and Functional Genomics to Microbial Ecology**...	831
	Philippe N. Bertin, Valérie Michotey, and Philippe Normand	
19	**Modeling in Microbial Ecology**.....................................	847
	Jean-Christophe Poggiale, Philippe Dantigny, Rutger DeWit, and Christian Steinberg	

Glossary.. 883

Subject Index.. 903

Taxonomy Index.. 917

Contributors

Chantal Abergel CNRS, Villeurbanne Cedex, France

Wafa Achouak UMR 7265 CNRS-CEA-Aix Marseille University, Institute of Environmental Biology and Biotechnology IBEB/DSV/CEA, CEA Cadarache, 13108 Saint-Paul-lez-Durance, France

Didier Alazard Institut Méditerranéen d'Océanologie (MIO), UM 110, CNRS 7294 IRD 235, Université de Toulon, Aix-Marseille Université, Campus de Luminy, 13288 Marseille Cedex 9, France

Christian Amblard Laboratoire Microorganismes: Génome et Environnement (LMGE), UMR CNRS 6023, Université Blaise Pascal, Clermont Université, 63171 Aubière Cedex, France

Ricardo Amils Centro de Biología Molecular Severo Ochoa (CSIC-UAM), Universidad Autónoma de Madrid, Madrid, Spain

Centro de Astrobiología (CSIC-INTA), Torrejón de Ardoz, Madrid, Spain

Sibel Berger Environ. Microb. Genom. Group, École Centrale de Lyon, UMR CNRS 5005, Laboratoire Ampère, Écully, France

Philippe N. Bertin Génétique Moléculaire, Génomique, Microbiologie (GMGM), UMR 7156, Université de Strasbourg, 67084 Strasbourg Cedex, France

Jean-Claude Bertrand Institut Méditerranéen d'Océanologie (MIO), UM 110, CNRS 7294 IRD 235, Université de Toulon, Aix-Marseille Université, Campus de Luminy, 13288 Marseille Cedex 9, France

Stephane Blain Laboratoire d'Océanographie Microbienne, LOMIC UMR CNRS-UPMC, Observatoire Océanologique de Banyuls, 66650 Banyuls-sur-Mer, France

Eric Blanchart IRD, UMR 210 ECO&SOLS, Montpellier, France

Patricia Bonin Institut Méditerranéen d'Océanologie (MIO), UM 110, CNRS 7294 IRD 235, Université de Toulon, Aix-Marseille Université, Campus de Luminy, 13288 Marseille Cedex 9, France

Charles-François Boudouresque Institut Méditerranéen d'Océanologie (MIO), UM 110, CNRS 7294 IRD 235, Université de Toulon, Aix-Marseille Université, Campus de Luminy, 13288 Marseille Cedex 9, France

Alain Brauman IRD, UMR 210 ECO&SOLS, Montpellier, France

Céline Brochier-Armanet Laboratoire de Biométrie et Biologie Évolutive, UMR CNRS 5558, Université Claude Bernard Lyon 1, 69622 Villeurbanne Cedex, France

Evelyne Buffan-Dubau Laboratoire d'Écologie Fonctionnelle et Environnement (ECOLAB), UMR CNRS 5245, Université Paul Sabatier, 31062 Toulouse Cedex 9, France

Pierre Caumette Institut des Sciences Analytiques et de Physico-chimie pour l'Environnement et les Matériaux (IPREM), UMR CNRS 5254, Université de Pau et des Pays de l'Adour, B.P. 1155, 64013 Pau Cedex, France

Jean-Luc Cayol Institut Méditerranéen d'Océanologie (MIO), UM 110, CNRS 7294 IRD 235, Université de Toulon, Aix-Marseille Université, Campus de Luminy, 13288 Marseille Cedex 9, France

Jean-Michel Claverie Aix-Marseille University, Marseille, France

Claude Combes Member of the Académie Française des Sciences (French Academy of Sciences), 16, rue du Vallon, Perpignan, Paris, France

Benoit Cournoyer Microbial Ecology Center, UMR CNRS 5557 / USC INRA 1364, Université Lyon 1, 69622 Villeurbanne, France

Philippe Cuny Institut Méditerranéen d'Océanologie (MIO), UM 110, CNRS 7294 IRD 235, Université de Toulon, Aix-Marseille Université, Campus de Luminy, 13288 Marseille Cedex 9, France

Philippe Dantigny Laboratoire des procédés alimentaires et microbiologiques, UMR PAM, AgroSup Dijon et Université de Bourgogne, 21000 Dijon, France

Maude M. David Environ. Microb. Genom. Group, UMR CNRS 5005, Laboratoire Ampère, École Centrale de Lyon, Écully, France

Rutger De Wit Écologie des systèmes marins côtiers (ECOSYM, UMR5119), Universités Montpellier 2 et 1, CNRS-Ifremer-IRD, 34095 Montpellier Cedex 05, France

Joël Doré Unité Micalis (INRA, UMR 1319), Paris, France

Pierre Doumenq Laboratoire de Chimie de l'Environnement, Aix-Marseille Université, FRE CNRS LCE 3416, Batiment Villemin, Europôle Environnement, 13545 Aix-en-Provence Cedex, France

Robert Duran Institut des Sciences Analytiques et de Physico-chimie pour l'Environnement et les Matériaux (IPREM), UMR CNRS 5254, Université de Pau et des Pays de l'Adour, B.P. 1155, 64013 Pau Cedex, France

Gerard Fonty Laboratoire des Micro-organismes: Génome et Environnement (LMGE), UMR CNRS 6023, Université Blaise Pascal, Clermont Université, 63171 Aubère Cedex, France

Jean-Pierre Gattuso Observatoire Océanologique, Laboratoire d'Océanographie, CNRS-UPMC, 06234 Villefranche-sur-Mer Cedex, France

Anne Godfroy Laboratoire de Microbiologie des Environnements Extrêmes, UMR 6197 (UMR CNRS/IFREMER/Université de Bretagne Occidentale), Plouzané, France

Philippe Goulas Institut des Sciences Analytiques et de Physico-chimie pour l'Environnement et les Matériaux (IPREM), UMR CNRS 5254, Université de Pau et des Pays de l'Adour, B.P. 1155, 64013 Pau Cedex, France

Manolo Gouy Laboratoire de Biométrie et Biologie Évolutive, UMR CNRS 5558, Université Claude Bernard Lyon 1, 69622 Villeurbanne Cedex, France

Gérald Grégori Institut Méditerranéen d'Océanologie (MIO), UM 110, CNRS 7294 IRD 235, Université de Toulon, Aix-Marseille Université, Campus de Luminy, 13288 Marseille Cedex 9, France

Regis Grimaud Environ. Microb. Genom. Group, Ecole Centrale de Lyon, UMR CNRS 5005, Laboratoire Ampère, Écully, France

Vincent Grossi Laboratoire de Géologie de Lyon: Terre, Planètes, Environnement, UMR CNRS 5276, Université Claude Bernard Lyon 1, 69622 Villeurbanne Cedex, France

Rémy Guyoneaud Institut des Sciences Analytiques et de Physico-chimie pour l'Environnement et les Matériaux (IPREM), UMR CNRS 5254, Université de Pau et des Pays de l'Adour, B.P. 1155, 64013 Pau Cedex, France

Thierry Heulin UMR 7265 CNRS-CEA-Aix Marseille University, Institute of Environmental Biology and Biotechnology IBEB/DSV/CEA, CEA Cadarache, 13108 Saint-Paul-lez-Durance, France

Jean-François Humbert INRA, UMR BIOEMCO, ENS, Paris, France

Laurent Intertaglia Observatoire Océanologique de Banyuls sur Mer UPMC/CNRS – UMS 2348, Plateforme BIO2MAR, 66650 Banyuls-sur-Mer, France

Fabien Joux Laboratoire d'Océanographie Microbienne (LOMIC), UMR 7621 CNRS-UPMC, Observatoire Océanologique de Banyuls, 66650 Banyuls-sur-Mer, France

Xavier Le Roux Microbial Ecology Center, UMR CNRS 5557 / USC INRA 1364, Université Lyon 1, 69622 Villeurbanne, France

Philippe Lebaron Observatoire Océanologique de Banyuls, Laboratoire de Biodiversité et Biotechnologie Microbiennes (LBBM), Sorbonne Universités, UPMC Univ Paris 06, USR CNRS 3579, 66650 Banyuls-sur-Mer, France

Franck Lejzerowicz Microbial Ecology Center, UMR CNRS 5557 / USC INRA 1364, Université Lyon 1, 69622 Villeurbanne, France

Karine Lemarchand Laboratoire de Bactériologie marine et Écotoxicologie microbienne, Institut des sciences de la mer de Rimouski, Rimouski, QC, Canada

Robert Lensi Centre d'Écologie Fonctionnelle et Évolutive (CEFE), Département d'Écologie Fonctionnelle, UMR 5175, 34293 Montpellier Cedex 5, France

Corinne Leyval Laboratoire Interdisciplinaire des Environnements Continentaux (LIEC, UMR 7360), Université de Lorraine, CNRS, Vandoeuvre-les-Nancy, France

Michel Magot Institut des Sciences Analytiques et de Physico-chimie pour l'Environnement et les Matériaux (IPREM), UMR CNRS 5254, Université de Pau et des Pays de l'Adour, B.P. 1155, 64013 Pau Cedex, France

Roland Marmeisse Microbial Ecology Center, UMR CNRS 5557 / USC INRA 1364, Université Lyon 1, 69622 Villeurbanne, France

Benoit Marrot Laboratoire de Mécanique, Modélisation et Procédés Propres, Aix-Marseille Université, M2P2-UMR 6181, Europole de l'Arbois, 13545 Aix en Provence Cedex, France

Fabrice Martin-Laurent Laboratoire Microbiologie du Sol et de l'Environnement, INRA, UMR 1347 Agroécologie, 21065 Dijon Cedex, France

Danielle Marty Institut Méditerranéen d'Océanologie (MIO), UM 110, CNRS 7294 IRD 235, Université de Toulon, Aix-Marseille Université, Campus de Luminy, 13288 Marseille Cedex 9, France

Robert Matheron Institut Méditerranéen de Biodiversité et d'Ecologie marine et continentale (IMBE), UMR-CNRS-IRD 7263, Aix-Marseille Université, 13397 Marseille Cedex 20, France

Patrick Mavingui Microbial Ecology Center, UMR CNRS 5557 / USC INRA 1364, Université Lyon 1, 69622 Villeurbanne, France

Valérie Michotey Institut Méditerranéen d'Océanologie (MIO), UM 110, CNRS 7294 IRD 235, Université de Toulon, Aix-Marseille Université, Campus de Luminy, 13288 Marseille Cedex 9, France

Cécile Militon Institut Méditerranéen d'Océanologie (MIO), UM 110, CNRS 7294 IRD 235, Université de Toulon, Aix-Marseille Université, Campus de Luminy, 13288 Marseille Cedex 9, France

Yvan Moënne-Loccoz Microbial Ecology Center, UMR CNRS 5557 / USC INRA 1364, Université Lyon 1, 69622 Villeurbanne, France

Hervé Moreau UMR 7232 LOBB, Observatoire Océanographique de Banyuls, 66650 Banyuls-sur-Mer, France

David Moreira Unité d'Écologie, Systématique et Évolution, UMR CNRS 8079, Université Paris-Sud, 91405 Orsay Cedex, France

Cindy Morris Unité de Recherches de Pathologie Végétale, INRA, 81413 Montfavet Cedex, France

Behzad Mostajir Écologie des systèmes marins côtiers (ECOSYM, UMR5119), Universités Montpellier 2 et 1, CNRS-Ifremer-IRD, 34095 Montpellier Cedex 05, France

Philippe Moulin Laboratoire de Mécanique, Modélisation et Procédés Propres, Aix-Marseille Université, M2P2-UMR 6181, Europole de l'Arbois, 13545 Aix en Provence Cedex, France

Thierry Moutin Institut Méditerranéen d'Océanologie (MIO), UM 110, CNRS 7294 IRD 235, Université de Toulon, Aix-Marseille Université, Campus de Luminy, 13288 Marseille Cedex 9, France

Sylvie Nazaret Microbial Ecology Center, UMR CNRS 5557 / USC INRA 1364, Université Lyon 1, 69622 Villeurbanne, France

Xavier Nesme Laboratoire d'écologie microbienne, UMR 5557 Université Claude Bernard Lyon 1, 69622 Villeurbanne Cedex, France

Philippe Normand Microbial Ecology Center, UMR CNRS 5557 / USC INRA 1364, Université Lyon 1, 69622 Villeurbanne, France

Bernard Ollivier Institut Méditerranéen d'Océanologie (MIO), UM 110, CNRS 7294 IRD 235, Université de Toulon, Aix-Marseille Université, Campus de Luminy, 13288 Marseille Cedex 9, France

Pierre Peyret EA CIDAM 4678, CBRV, Université d'Auvergne, Clermont-Ferrand, France

Florence Piette Laboratoire de Biochimie, Université de Liège, Liège, Belgique

Jean-Christophe Poggiale Institut Méditerranéen d'Océanologie (MIO), UM 110, CNRS 7294 IRD 235, Université de Toulon, Aix-Marseille Université, Campus de Luminy, 13288 Marseille Cedex 9, France

Franck Poly Microbial Ecology Center, UMR CNRS 5557 / USC INRA 1364, Université Lyon 1, 69622 Villeurbanne, France

Daniel Prieur Laboratoire de Microbiologie des Environnements Extrêmes, UMR 6197 (UMR CNRS/IFREMER/Université de Bretagne Occidentale), Plouzané, France

Petar Pujic Microbial Ecology Center, UMR CNRS 5557 / USC INRA 1364, Université Lyon 1, 69622 Villeurbanne, France

Bernard Quéguiner Institut Méditerranéen d'Océanologie (MIO), UM 110, CNRS 7294 IRD 235, Université de Toulon, Aix-Marseille Université, Campus de Luminy, 13288 Marseille Cedex 9, France

Patrick Raimbault Institut Méditerranéen d'Océanologie (MIO), UM 110, CNRS 7294 IRD 235, Université de Toulon, Aix-Marseille Université, Campus de Luminy, 13288 Marseille Cedex 9, France

Pierre Servais Laboratoire d'Écologie des Systèmes Aquatiques (ESA), Université Libre de Bruxelles, 1050 Bruxelles, Belgium

Télesphore Sime-Ngando Laboratoire Microorganismes: Génome et Environnement (LMGE), UMR CNRS 6023, Université Blaise Pascal, Clermont Université, 63171 Aubère Cedex, France

Pascal Simonet Environ. Microb. Genom. Group, École Centrale de Lyon, UMR CNRS 5005, Laboratoire Ampère, Écully, France

Guy Soulas Œnologie INRA, 33405 Talence Cedex, France

Christian Steinberg Pôle des Interactions Plante-Microorganismes, INRA UMR 1347 Agroécologie, AgroSup-INRA-Université de Bourgogne, 21065 Dijon Cedex, France

Georges Stora Institut Méditerranéen d'Océanologie (MIO), UM 110, CNRS 7294 IRD 235, Université de Toulon, Aix-Marseille Université, Campus de Luminy, 13288 Marseille Cedex 9, France

Christian Tamburini Institut Méditerranéen d'Océanologie (MIO), UM 110, CNRS 7294 IRD 235, Université de Toulon, Aix-Marseille Université, Campus de Luminy, 13288 Marseille Cedex 9, France

Laurent Urios Institut des Sciences Analytiques et de Physico-chimie pour l'Environnement et les Matériaux (IPREM), UMR CNRS 5254, Université de Pau et des Pays de l'Adour, B.P. 1155, 64013 Pau Cedex, France

France Van Wambeke Institut Méditerranéen d'Océanologie (MIO), UM 110, CNRS 7294 IRD 235, Université de Toulon, Aix-Marseille Université, Campus de Luminy, 13288 Marseille Cedex 9, France

Timothy M. Vogel Environ. Microb. Genom. Group, École Centrale de Lyon, UMR CNRS 5005, Laboratoire Ampère, Écully, France

Frances Westall Centre de Biophysique Moléculaire, CNRS, 45071 Orléans Cedex 2, France

Florence Wisniewski-Dyé Microbial Ecology Center, UMR CNRS 5557 / USC INRA 1364, Université Lyon 1, 69622 Villeurbanne, France

Part I
General Chapters

The Thematic Fields of Microbial Ecology

Jean-Claude Bertrand, Pierre Caumette, Philippe Lebaron, and Philippe Normand

Abstract

The microbial world, generally invisible to the naked eye, has largely shaped our environment and has been instrumental in the emergence and evolution of all other living organisms on Earth. These microscopic unicellular organisms were for 3 billion years the only forms of life on our planet.

Their most spectacular action was the modification of the primitive atmosphere: the dioxygen certainly not present initially reached its present concentration (21 % of the gas content of the atmosphere) through the action of microorganisms that are able of oxygenic photosynthesis. For the evolution of life, it is now widely accepted that multicellular life forms extremely complex have emerged from eukaryote microorganisms classified in the kingdom Plantae and in the Stramenopiles and Opisthokonta (especially metazoans which includes humans). These life forms are still dependent on the activity of microorganisms.

If a disaster, whether natural or caused by humans, should annihilate all nonmicrobial living species, it is likely that some microorganisms that have colonized all oceans (from the surface to the abyssal domain) and the earth's crust (to a depth of hundreds of meters) would be spared and would allow the initiation of a new evolution process, whatever the new environmental conditions at the end of this disaster, except in the absence of liquid water.

Keywords

Biogeochemical cycle • Distribution • Diversity • Ecosystems • Evolution • Interactions • Origin • Taxonomy • Xenobiotics

---

All authors contributed equally to this chapter.

J.-C. Bertrand
Institut Méditerranéen d'Océanologie (MIO), UM 110, CNRS 7294 IRD 235, Université de Toulon, Aix-Marseille Université, Campus de Luminy, 13288 Marseille Cedex 9, France
e-mail: Jean-Claude.bertrand@univ-amu.fr

P. Caumette
Institut des Sciences Analytiques et de Physico-chimie pour l'Environnement et les Matériaux (IPREM), UMR CNRS 5254, Université de Pau et des Pays de l'Adour, B.P. 1155, 64013 Pau Cedex, France
e-mail: pierre.caumette@univ-pau.fr

P. Lebaron
Observatoire Océanologique de Banyuls, Laboratoire de Biodiversité et Biotechnologie Microbiennes (LBBM), Sorbonne Universités, UPMC Univ Paris 06, USR CNRS 3579, 66650 Banyuls-sur-Mer, France
e-mail: lebaron@obs-banyuls.fr

P. Normand
Microbial Ecology Center, UMR CNRS 5557 / USC INRA 1364, Université Lyon 1, 69622 Villeurbanne, France
e-mail: philippe.normand@univ-lyon1.fr

The microbial world, generally invisible to the naked eye, has largely shaped our environment and has been instrumental in the emergence and evolution of all other living organisms on Earth. These microscopic unicellular organisms were for 3 billion years the only forms of life on our planet.

Their most spectacular action was the modification of the primitive atmosphere: the dioxygen certainly not present initially reached its present concentration (21 % of the gas content of the atmosphere) through the action of microorganisms that are able of oxygenic photosynthesis. For the evolution of life, it is now widely accepted that extremely complex multicellular life forms have emerged from eukaryote microorganisms classified in the kingdom *Plantae* and in the Stramenopiles and Opisthokonta (especially metazoans which includes humans). These life forms are still dependent on the activity of microorganisms.

If a disaster, whether natural or caused by humans, should annihilate all nonmicrobial living species, it is likely that some microorganisms that have colonized all oceans (from the surface to the abyssal domain) and the earth's crust (to a depth of hundreds of meters) would be spared and would allow the initiation of a new evolution process, whatever the new environmental conditions at the end of this disaster, except in the absence of liquid water.

The activity of the biosphere as a whole is totally dependent on the action of microorganisms. The major goal of microbial ecology (which, in general, can be defined as the study of microorganisms in natural and anthropized ecosystems) is to improve our understanding of:

1. *Their origin and evolution*

The first inhabitants of the Earth, microorganisms have developed mechanisms that allowed them to adapt to all climatic and geological changes that have occurred on the Earth since the origin of life. They are actually present in all habitats and are able to adapt to the most extreme conditions of life in terms of temperature, pressure, pH, salinity, radiation, etc.

To explain the longevity and ubiquity of microorganisms, it is important to take into account the very short duration of their cell cycle, their high mutation rates and adaptability to environmental changes, their extraordinary metabolic plasticity that allows them to adapt to a large diversity of carbon and energy sources that are available in their environment (including xenobiotics, which are molecules produced by humans), the diversity as well as the frequency of genetic information transfer between populations and communities, and their ability to survive under starvation conditions and all resistance strategies to face environmental stresses. This resistance can involve morphological and metabolic changes such as the production of detoxifying enzymes, DNA repair mechanisms, and the production of resistance forms (endospores, cysts, etc.). In addition, their small size results in a very high surface/volume ratio and therefore in a high exchange capacity with the external environment. This capacity promotes their ubiquity in natural environments (Woese 1987).

2. *Their taxonomic and functional diversities and their abundance and distribution in ecosystems*

The microbial ecologist is led to inventory microorganisms in the environment, which means all prokaryotic and eukaryotic unicellular microorganisms but also viruses in their functional dimension. The microbial compartment is a component of all ecosystems, similar to the animal or plant components. In addition to this taxonomic diversity, microbial ecologists have to describe the functions that microorganisms are in charge of, which means understanding which genes are involved and expressed in situ (Brock 1966).

3. *Their role in the functioning of ecosystems and in particular that of biogeochemical cycles*

Due to their important biomass, metabolic diversity, and capacity to adapt to the environment, microorganisms play a major role in the organization, functioning, and evolution of ecosystems. They are both producers, consumers, and decomposers, and they participate in all stages of the organic and mineral matter transformation, and they are alone to perform some transformation processes. Their role is also essential in the functioning of food webs. In the absence of plants and animals, biogeochemical cycles could continue to operate based solely on the activity of microorganisms. The study of their role in the different cycles requires the description, location, and quantification of microbial populations and communities that are involved, as well as the characterization of all organic matter degradation and synthesis pathways and the measurement of their activities. On this last point, it is important to note that if the action of microorganisms proceed at the microenvironment scale, the impact of their activities apply at the earth scale (Fenchel et al. 2000).

4. *The interactions between microorganisms and between microorganisms and plants, animals and humans*

All together, these interactions which are in most cases positive (beneficial) or negative (adverse) are classified as biotic interactions. In the environment, microorganisms do not live alone: they form populations often grouped into communities and can form highly structured assemblages such as biofilms or microbial mats. From the first stages of life, microorganisms have assembled to form multilayer structures called stromatolites which are microbial mats made with autotrophic and heterotrophic prokaryotes and fossilized for millions of years. The success of these associations is quite exceptional: if they have appeared more than 3 billion years ago, the stromatolites are still present. Studies of biotic interactions must take into account not only the interactions between microorganisms but also their interactions with multicellular organisms such as plants, animals, and humans (Margulis 1981).

5. *Interactions between microorganisms and their environment, considered as abiotic interactions*

Studies of these abiotic interactions should consider, on the one hand, how both physical and chemical properties of the environment may affect the presence, spatial distribution, and also the growth and activity of microorganisms and, on the other hand, the feedback of microorganisms on their environment. Among the environmental parameters interacting with microorganisms, the most important are pH, pressure, light intensity and wavelength, and the electron donor and acceptor concentration. These donors and acceptors are most often heterogeneously distributed in the natural environment and/or present at low concentration. As a consequence, they do not allow the maximum development of microorganisms and their survival under environmental conditions that are not adapted to their physiology. In most environments, microorganisms are most often living under nutrient starvation, leading to long periods of dormancy. They adapt to these starvation conditions by developing nutrient storage systems or other mechanisms (chemotactic, phototactic, magnetotactic) enabling them an easier access to nutrients. More generally, microorganisms must continuously adapt to survive to regular and sometimes fast changes of the environmental parameters.

The interaction of microorganisms with natural surfaces (soil or sediment particles, leaves and roots, digestive tract of humans and animals, skin, etc.) must also be considered. In the natural environment, many microorganisms are not free-living organisms, but they attach to most surfaces through the secretion of extracellular compounds such as polysaccharides that allow them to form biofilms or microbial mats (Fenchel et al. 2000).

6. *Their extraordinary capacity to degrade pollutants, especially xenobiotics, as a service to ecosystems and to the bioremediation of contaminated sites*

Microbial ecology can provide solutions to many environmental problems currently faced by human societies whose relationships with the environment have changed dramatically during the last centuries due to an important increase of the world population, to the settlement of populations over an extended range of biotopes, and more recently to the industrial development and, therefore, to the production of more and more domestic and industrial wastes, leading to an increasing pollution of environmental sites. Nowadays, pollution has reached such levels that it endangers the survival of a large number of species on Earth. An important goal in microbial ecology is to enhance the metabolic properties of microorganisms to degrade most pollutants that affect our planet. Microbial ecology is also concerned by public health problems such as an increasing need to produce drinking water, food contamination, dispersion, and changes in the behavior of pathogens in the environment (emerging infectious diseases, viable but nonculturable state of pathogens, antibiotics resistance, etc.).

These are the objectives of microbial ecology, at the interface between microbiology and ecology. To achieve these goals, it is important to develop appropriate and efficient methods, sometimes specific to the study of microorganisms in natural environments. In this matter, real-time methods are more and more requested to develop early warning detection systems. Sergei Winogradsky, the first microbiologist that has introduced the expression "environmental microbiology," pointed out the limitations and drawbacks of the methods that are commonly used in medical and industrial microbiology. Based on the study of microorganisms that have been isolated and grown under controlled conditions, most of these methods cannot be used to investigate the life and physiology of environmental microorganisms. This vision was especially appropriate since it is now well established that many microorganisms that are present in the environment are unknown and only 1–10 % of microbes present in natural biotopes have been grown in pure culture (Singh 2011).

Techniques have been developed that allow a direct access (without any cultivation step) to microbial populations and communities in the environment thanks to the increased performances of microscopy instruments and analytical chemistry techniques (development of fluorescence techniques and **biomarkers***, identification of metabolic pathways), development of microelectrodes to assess the distribution and activity of microorganisms at the microscale, use of isotopic techniques, development of techniques to determine the activity of microorganisms under conditions very close to those encountered in the natural environment (e.g., hyperbaric instruments that generate high pressure, temperature, and oligotrophic conditions that exist in the deep ocean, etc.), and development and improvement of flow cytometry instruments to access the properties of individual cells (numbers, physiological state, presence or absence of certain metabolic activities, etc.).

Nevertheless, the main technological revolution in this field was the implementation, development, and use of molecular biology techniques: DNA fingerprinting, DNA chips (taxonomic and functional), metagenomic libraries, transcriptomic and proteomic studies, full genome sequencing, etc. Environmental genomics has allowed the detection, identification, and quantification of microorganisms in the natural environment after DNA extraction, with access to non-cultivable species. They also provided access to the physiological state of cells and were a considerable contribution to our knowledge of the diversity of microorganisms (taxonomic, functional, and genetic diversity), their phylogeny, activities, and interactions.

This book aims to address all issues related to this discipline. It consists of 19 chapters that are organized into 5 parts.

– The first part is devoted to "General Chapters." After an introduction (Chap. 1, "The Thematic Fields of Microbial Ecology"), the next chapters present a brief history of the discipline, including an introduction to the work of Winogradsky (Chap. 2, "Some Historical Elements of

Microbial Ecology") and a description of the structure and microbial metabolisms (Chap. 3 "Structure and Functions of Microorganisms; Production and Use of Material and Energy"). Indeed, understanding the role of microorganisms in the environment requires that their metabolism be known to understand their action in the transformation of organic and inorganic compounds within biogeochemical cycles or wastewater treatment processes (catabolism pathways, energy metabolism), as well as their action in interactions and food webs (anabolism pathways, biosynthesis). The remarkable metabolic diversity of prokaryotes, which is developed in this chapter (different aerobic and anaerobic respirations, fermentations, and photosynthesis), has allowed them to colonize and to be very active in all habitats of our planet and perhaps other planets.

- The second part "Taxonomy and Evolution" starts with a chapter entitled "For Three Billion Years, Microorganisms Were the Only Inhabitants of the Earth" (Chap. 4), where a description is presented of the last universal common ancestor (LUCA), different hypotheses of the emergence of three kingdoms that make up the living world today (Bacteria, Archaea, Eukarya), and the main stages of the evolution of microorganisms, in particular the main evolutionary scenarios that can explain the emergence of eukaryotic organization.

Chapter 5, "Systematic and Evolution of Microorganisms: General Concepts" presents general concepts related to this topic. In eukaryotes, it addresses primary endosymbioses at the origin of mitochondria and chloroplasts (in the kingdom of plants) and perhaps of the kinetic processes and secondary and tertiary endosymbioses causing photosynthesis in other eukaryotic kingdoms.

The "Taxonomy and Phylogeny of Prokaryotes" are presented in Chap. 6. The description of criteria used for phenotypic and genotypic characterization leads to a discussion of the problems associated to the definition of species in prokaryotes. The characteristics of the two prokaryotic kingdoms Bacteria and Archaea are defined. The current phylogeny of their major taxa is also presented.

Chapter 7, entitled "Taxonomy and Phylogeny of Unicellular Eukaryotes," suggests a unified terminology for the description of the cytology, morphology, reproduction, and biological cycles of eukaryotes. Major taxa of unicellular eukaryotes are presented with a focus on specific biochemical and cytological markers for each of them. Their position in the eukaryotic phylogenetic tree is specified.

- The third part "Microbial Habitats: Diversity, Adaptation, and Interactions" focuses on the study of "Biodiversity and Microbial Ecosystems Functioning" (Chap. 8).

Microbial biodiversity is a concept that emerged a few years ago to identify all the players in the microbial biotopes. This concept is first discussed in a historical perspective in relation to the paradigms in ecology. Then, mathematical approaches that are used to characterize all microbial actors are presented as well as the technical tools developed to quantify microbial biodiversity. Approaches to explore the relationship between biodiversity and taxonomic functional biodiversity are described for major ecosystems (soil, water, etc.) and the main functions (respiration, photosynthesis, fixation, and use of nitrogen). Finally, in the conclusion, there is a discussion of the prospects offered by the tools under development.

Chapter 9, "Adaptations of Prokaryotes to Their Biotopes and to Physicochemical Conditions in Natural or Anthropized Environments," concerns the adaptive processes developed by prokaryotic cells (two components regulatory systems, chemotaxis, etc.) and their metabolic and genetic responses to adapt to physicochemical conditions prevailing in the terrestrial and aquatic habitats.

Chapter 10, "The Extreme Conditions of Life on the Planet and Exobiology," deals with the microbial life in ecosystems of the planet that are characterized by extreme physicochemical conditions. In the first part, the great diversity of prokaryotes is described that colonize such environments, which were most often seen in the past as hostile to life. The ecological, physiological, and taxonomic properties of these indigenous extremophiles microorganisms are then exposed with particular emphasis on psychrophilic, thermophilic, acidophilic, alkalophilic, halophilic, and piezophilic microorganisms.

Chapter 11, "Microorganisms and Biotic Interactions," describes the interactions (cooperation, commensalism, competition, mutualism or symbiosis, parasitism, predation) involving microorganisms and their significance in microbe/microbe, microorganism/plant, and microorganism/animal or man interactions. It describes molecular and evolutionary mechanisms of bipartite and/or multi-interactions. The biological functions involved in these interactions are addressed in terms of impact on matter and energy fluxes in the environment, biotechnology, agronomy, and public health.

Chapter 12, "Horizontal Gene Transfer in Microbial Ecosystems," describes general features of horizontal gene transfer (HGT) (mechanisms, discovery, etc.). HGT are discussed from a qualitative point of view in relation to the adaptive response of microorganisms in the environmental conditions but also from a quantitative point of view, as a major evolutionary mechanism in microorganisms.

- The fourth part, "Role and Functioning of Microbial Ecosystems," discusses the intervention of microorganisms in "Microbial Food Webs in Aquatic and Terrestrial Ecosystems" (Chap. 13). The organisms involved in microbial food webs (MFWs), including viruses, Archaea, Bacteria, and many eukaryotic taxa

(Fungi, Alveolates, Mycetobiontes, etc.), play an essential role in the functioning of aquatic and terrestrial ecosystems. From this point of view, pelagic aquatic ecosystems are particularly original since primary production is only due to microbial communities who provide most of the biomass in these ecosystems. After an overview of MFWs and their main players, the various bottom-up (resources) and top-down (predators) factors controlling microorganisms are presented. Differences and similarities between aquatic and terrestrial MFWs are described.

Chapter 14, "Biogeochemical Cycles," concerns the role of microorganisms in the functioning of natural and anthropogenic ecosystems by studying detailed processes and mechanisms that are involved in the main biogeochemical cycles (carbon, nitrogen, sulfur, phosphorus, silicate, metals), in soils and in freshwater and marine ecosystems. The exchanges and biotransformations of organic and mineral components between the oxic and anoxic zones of the different biotopes are presented.

Chapter 15 presents "Environmental and Human Pathogenic Microorganisms" involved in humans or animal diseases through a large diversity of infectious mechanisms. The diversity of these pathogenic microorganisms, their mode of infection, and dissemination but also their behavior in the environment as well as the specific methods used for their detection are reported.

"Applied Microbial Ecology and Bioremediation" (Chap. 16) concerns the use microorganisms for preventive remediation (wastewater treatment plants, treatment of gaseous effluents) or bioremediation in contaminated sites (bioaugmentation, biostimulation, rhizostimulation, bioleaching). Most of the processes used to recover contaminated soils, sediments, and coastal effluents of treatment plants are reported. An important part of this chapter is devoted to the description of main pollutants in the environment as well as natural attenuation processes due to microbial activities (biodegradation and/or biotransformation).

- The fifth part "Tools the Microbial Ecology" describes the "Methods to Study Microorganisms in the Environment" (Chap. 17): sampling, microbial biomass and activity measurements, structure and diversity of microbial populations, and communities using cultural and noncultural techniques and laboratory studies. These methods are described for different types of microorganisms (prokaryotes and eukaryotes, heterotrophs and autotrophs) and different types of biotopes (water, soil, sediment, biofilms, etc.) providing informations on their advantages and limitations and using varied approaches and instruments such as cytometry, molecular biology, biochemistry, and isotopic and molecular electrochemistry.

Chapter 18 focuses on "Contributions of Genomics and Proteomics in Microbial Ecology" for studying the organization and functioning of complex microbial communities as a whole. Genomics and related methods (transcriptomics, proteomics, metabolomics) are addressed from historical and technical points of view. Some examples of the contribution of these techniques to the knowledge on microorganisms are reported, addressing their physiology and ecology in different environments.

Actually, the growing interest in microbial ecology also requires the quantification of microbial activities and biotic and abiotic interactions.

Chapter 19 on "The Modeling Microbial Ecology" aims to provide informations on the development of models in various fields of microbial ecology through some examples but also to highlight the current limitations of modeling in this field. The second objective of this chapter is to provide the reader the necessary bases to understand the scientific literature related to microbial ecology and for which mathematical modeling should be seen as a tool that complements more traditional methods of investigation (molecular biology, culture, genomics, etc.).

At the end of each chapter, a list of references including general books and major articles related to the different topics may help the reader to complete the scientific information presented in the chapter. Words in bold with an asterisk in the text are defined in a glossary at the end of the book, before the index.

References

Brock TD (1966) Principles of microbial ecology. Prentice-Hall, Englewood Cliffs
Fenchel T, King GM, Blackburn TH (2000) Bacterial biogeochemistry. Academic Press, San Diego
Margulis L (1981) Symbiosis in cell evolution. W.H. Freeman, San Francisco
Singh SN (2011) Microbial degradation of xenobiotics. Environmental science and engineering. Springer, Heidelberg
Woese CR (1987) Bacterial evolution. Microbiol Rev 51:221–271

2. Some Historical Elements of Microbial Ecology

Pierre Caumette, Jean-Claude Bertrand, and Philippe Normand

Abstract

We present briefly, first, the history of the discovery of microorganisms and particularly bacteria with the pioneering works of Antoni van Leeuwenhoek, Louis Pasteur, and Robert Koch, essentially. In a second and more detailed part, the history of microbial ecology is presented with particularly the very important work of Sergei Winogradsky and his discoveries of the main bacterial groups active in biogeochemical cycles. It is followed by a description of the major microbial ecologists who have been very active in promoting and developing microbial ecology throughout the world. Their role in the advances of microbial ecology is presented and discussed.

Keywords

Antoni van Leeuwenhoek • History of microbiology • Louis Pasteur • Microbial ecology • Microorganisms discovery • Robert Koch • Sergei Winogradsky

2.1 Introduction

Ecology, from the Greek words "oikos" (the house and its operation) and "logos" (knowledge, discourse, laws), can be defined as "knowledge of the laws governing the operation of the 'house,'" i.e., the science that studies the relationships between organisms and their biotic and abiotic environments.

Thus, the German biologist Ernst Haëckel (1866) was the first to propose this word and to use it in that sense. By extension, microbial ecology is the science that studies the interrelationships between microorganisms and their biotic and abiotic environments. Since the early 1960s, this term has been commonly used and this discipline has developed considerably. These studies have been mainly focused on determining the role of living organisms and especially microorganisms in maintaining the ecological equilibrium of ecosystems and the wider environment. Since the 1970s, ecology has become popular as it developed not only into a science but also turned into a major social and political issue. Later, microbial ecology became a full-fledged discipline of microbiology. Today, it is particularly necessary in view of the marked deterioration of our environment and the need to "make the house livable" for the survival of humanity and the wider living world (biodiversity), that is to say, for the maintenance of quality and ecosystem balance in our environment. This discipline requires knowledge of microorganisms, their biodiversity, and their role in their immediate surroundings but also, to know at the cellular level, their metabolism and functional abilities. Thus, in this discipline, it is necessary to analyze closely the links between field approaches that study

---

*Chapter Coordinator

P. Caumette* (✉)
Institut des Sciences Analytiques et de Physico-chimie pour l'Environnement et les Matériaux (IPREM), UMR CNRS 5254, Université de Pau et des Pays de l'Adour, B.P. 1155, 64013 Pau Cedex, France
e-mail: pierre.caumette@univ-pau.fr

J.-C. Bertrand
Institut Méditerranéen d'Océanologie (MIO), UM 110, CNRS 7294 IRD 235, Université de Toulon, Aix-Marseille Université, Campus de Luminy, 13288 Marseille Cedex 9, France
e-mail: Jean-Claude.bertrand@univ-amu.fr

P. Normand
Microbial Ecology Center, UMR CNRS 5557 / USC INRA 1364, Université Lyon 1, 69622 Villeurbanne, France
e-mail: philippe.normand@univ-lyon1.fr

microorganisms in their environment and in pure culture, which, although microorganisms have been isolated from their environment, can permit to understand them from a functional point of view and thus their metabolic potential. It is through these global studies that scientists have been able to demonstrate the key role of microorganisms in the balance of biogeochemical cycles, in the restoration of environments polluted by various chemicals, in the biodegradation or biotransformation of persistent organic pollutants, of heavy metals, etc. Thus, microbial ecology is now becoming an important discipline whose practical implications are obvious and at the service of human activities that aim at maintaining our environment livable and make development sustainable.

2.2 On the Path to Discovery of Microorganisms

Although microorganisms are present all over the earth where they are the most abundant living beings, it was not until the mid-seventeenth century with the invention of the microscope that they were actually seen. An English philosopher, Robert Hooke, was the first to observe and to describe in 1655, using a microscope, various types of fungi (molds) and protozoa. However, it appears he was not able to observe bacteria, given the low quality of the lens he used. Soon after, a French scientist Louis Joblot (1645–1723) constructed different types of microscopes with which he observed the morphologies of many microorganisms. Born in Bar-le-Duc, he was appointed "Royal" professor of mathematics and taught mathematics and geometry at the Royal Academy of Painting and Sculpture in Paris; he devoted much of his time to develop microscopes with which he described many "animalcules" and other microscopic "snakes, fish and eels" he observed in infused preparations of various plants or vinegar, and "the largest part is invisible to the ordinary scope of our eyes." He published his descriptions and drawings in a book published in 1718: "Descriptions and uses of several new microscopes and New observations on many insects, and on Animalcules which are present in prepared liquors and those are not so" (Lechevalier 1976; Le Coustumier 2010). Among unicellular microorganisms, he observed probably bacteria ("tiny eels and snakes of vinegar").

At the same time, Antoni van Leeuwenhoek (1674–1723, Fig. 2.1a), a cloth merchant and amateur scientist from Delft (Holland), created a very simple type of microscope with higher magnification (Leeuwenhoek microscope, Fig. 2.1b) to control the quality of the threads of sheets and other woven wares for commercial purposes (Dobell 1923). By observing different samples (pond water, maceration of plants, biological fluids, etc.), he was surprised to find through his microscopic lenses tiny organisms with particular forms "apparently moving with a purpose." He thus watched many types of single-celled eukaryotes and even bacteria. In a series of letters to the Royal Society in London, he described and drew different bacterial forms (cocci, rods, spirilla), indicating their movement and behavior when they were placed under different physicochemical conditions, paving the way for the first microbial ecology observations. With his observations and discoveries of microbes in different biotopes, he firmly opposed the supporters of spontaneous generation, who at that time, believed that microorganisms resulted from the decay of organic matter and not the reverse. For example, in that perspective, meat "engendered" microorganisms through its own decay. A century later, the Italian naturalist Lazzaro Spallanzani (1729–1799), in order to disprove this theory, showed that the decomposition of organic substances was caused by microorganisms that were not spontaneous but that these multiplied by cell division and could be eliminated by heating. He tried to explain to proponents of spontaneous generation that substances heated and separated from ambient air in a tight manner could not decompose and yield microorganisms, but could be kept without microbial attacks. However, his works were not totally conclusive because some samples contained bacteria capable of withstanding the heat treatments he used (spore-forming or heat-resistant bacteria). These results have provided arguments for the theory of spontaneous generation which persisted in spite of that work. Charles Cagniard-Latour (1838), Theodor Schwann (1837), and Friedrich Kützing (1837) have, in turn, attempted to disprove this theory by highlighting the role of yeasts in the fermentation of sugars to alcohol: fermentation was shown not to occur in the absence of yeast. They described the yeasts present in fermenting wine and beer and showed their role in the fermentation of sugars. In addition, Schwann (1837) demonstrated the role of airborne microorganisms in the breakdown of sugars by an experimental setup (vials with a narrow tube heated to incandescence), which allowed to leave sugars in contact with air while preventing the penetration of microbes in the heated and sterilized tubing. In a control flask open to air, organic matter was decomposed while in the bottle with the heat-sterilized narrow tube, organic matter in contact with sterilized air did not decompose. However, these experiences were not enough to convince supporters of the spontaneous generation theory who persisted in maintaining that heating of the tubes to incandescence modified the air quality in the bottles and that "disturbed air prevented the phenomenon of decomposition and spontaneous generation." It was not until the work of Louis Pasteur, who confirmed in a series of experiments (Pasteur 1857, 1860) the role of microorganisms in fermentation, that the theory of spontaneous generation was definitely disproved. Using narrow tubing vials, curved (swan

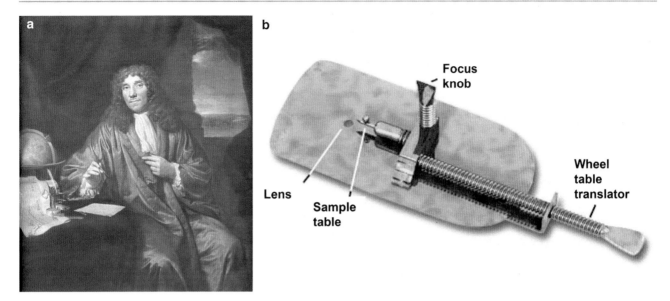

Fig. 2.1 (**a**) Antoni van Leeuwenhoek (1632–1723) (Copyright: courtesy of the Archives of the School of Microbiology of Delft, Technological University of Delft). (**b**) Microscope of A. van Leeuwenhoek (magnification × 200) (Modified from Wikipedia)

neck flasks, Fig. 2.2a) and unheated so as "not to disturb the air," he brought to a boil organic fluids where he could show, through microscopic observations and methods of bacterial cultures, that sterilized liquids were not recontaminated. These sealed and still sterile vials can still be seen in the Pasteur Museum at the Pasteur Institute in Paris. In parallel, the British physicist John Tyndall showed that air free of "particles" and optically clear did not cause degradation of organic substances, and resolved the problem of heat resistance of bacterial endospores by discontinuous heat treatments (Tyndall 1877). He was able to eliminate endospores that are resistant to boiling by successive heating cycles. The time between two heating cycles (24 h) allowed the germination of spores that had resisted the first heat treatment; such germinated bacteria were then killed by a second heating to a boil. This sterilization technique has been called **tyndallization***. His works were complementary to those of the German botanist Ferdinand Julius Cohn (1876) who described the endospores of *Bacillus subtilis* and their heat resistance properties. The studies of Pasteur, Tyndall, and Cohn have shown among other things that bacteria could survive in hostile environments by producing endospores and that these spores were airborne and carried by currents and could be scattered from one ecosystem to another.

A pioneer of the theory of microbial diseases was Ignaz Philipp Semmelweis (1818–1865) of Hungarian origin, a doctor at the obstetrics ward of the Vienna hospital. He had noticed that the death rate from puerperal fever in recently delivered women varied in different wards, ranging from 1 to 2 % in a service run by midwives to 30 % in another service staffed by physicians practicing dissections that were then not followed by hand washings. In 1847, he proposed to make hand washing mandatory when moving between services, which permitted to reduce quickly the death rate (Semmelweis 1861) but also caused deep resentment from upset department heads who greatly doubted the existence of *germs* as shown from the following quotation from the department head involved: "Herr Semmelweis claims that we carry on our hands little things that would be the cause of puerperal fever. What are these little things, those particles that no eye can see? This is ridiculous! The little things exist only in the imagination of Herr Semmelweis!"

2.3 The Beginnings of Microbiology

The major developments of microbiology from 1860 until the mid-twentieth century were largely influenced by the work of Louis Pasteur (Fig. 2.2b) and Robert Koch (Fig. 2.3).

Louis Pasteur was born in Dole (France) in 1822, made parts of his studies at the "Ecole Normale Supérieure" (ENS) in Paris. He was appointed professor of physics and chemistry, successively in Dijon, at the Faculty of Sciences of Strasbourg, at the Faculty of Sciences of Lille, and finally to the ENS in Paris. Strongly interested in the socioeconomic problems of his time, Pasteur, who had been trained as an organic chemist, has very quickly begun work on fermentation because of the problems encountered by producers of wine and beer and distillers and manufacturers of vinegar. His studies on fermentation between 1857 and 1876 have

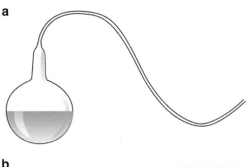

Fig. 2.2 (**a**) Gooseneck bottle used by Pasteur to study fermentation and to disprove the theory of "spontaneous generation" (see text) (Source: Wikipedia, Pasteur Institute, http://commons.wikimedia.org/wiki/file:col_de_cygne.png; free of rights). (**b**) Louis Pasteur (1822–1895) (Copyright: Institut Pasteur, Paris)

clearly established the role of anaerobic microorganisms in various fermentation processes, "fermentation is life without air" (Louis Pasteur). Indeed, during his work on fermentation, Pasteur discovered the ability of certain organisms to live in the absence of free oxygen (that is to say, in the absence of air). He called these anaerobes. (The words "aerobic" and "anaerobic" were coined by him.) From 1876 onward, Pasteur worked on different bacterial sterilization techniques: on porcelain porous filters and on the autoclave developed by Charles Chamberland (1851–1908) and the sterilization by high-temperature dry heating of the glass equipment used in microbiology (bottles, pipettes, etc.) in ovens that now carry his name (Pasteur oven).

Thereafter, he devoted much of his life to the role of pathogenic microorganisms in various infectious processes and in the genesis of infectious diseases. He thus took part in the progress of the germ theory of infectious diseases, mainly developed by Robert Koch. In 1880, Pasteur discovered the staphylococcus, which he identified as responsible for boils and osteomyelitis. Thus, for 6 years, Pasteur and Koch, scientific competitors and rivals, independently but in an intense emulation fanned by the Franco-Prussian war (1870–1871) studied infectious diseases and their causal microbes.

The last part of Pasteur's career was devoted to vaccines, that is to say, methods to prevent and suppress infectious diseases. This technique had been developed by the English physician Edward Jenner (1749–1823) who had noticed that farmworkers exposed to vaccinia, a cow disease caused by a virus similar to smallpox, which is also the origin of the name "vaccine," were subsequently protected against smallpox. Jenner had devised a standardized protocol whereby he took vaccinia blister exudates from infected workers, suspended it in water and inoculated it to children, and obtained a high rate of protection against smallpox. In the summer of 1879, Pasteur and his collaborators, Emile Roux and Emile Duclaux, discovered that chickens inoculated with old cultures of the fowl cholera microbe not only did not die but were resistant to new infections; it was the discovery of a new type of "vaccine." Contrary to what had been the case in vaccination against smallpox, it did not use as a vaccine a benign "virus"[1] (vaccinia) supplied by nature as a mild disease that immunizes against a severe disease, but the artificial attenuation of a highly virulent strain is provoked on purpose, and it is the resulting attenuated strain that is used as a vaccine. The theory of vaccination by attenuated "viruses" was confirmed by Pasteur's work on the "anthrax bacillus." After several successful results with various pathogens, prevention of rabies was the last great work of Pasteur and the only vaccination that he applied to humans. Pasteur having published his first success, his rabies vaccine soon became famous. He died in 1895 and became a national and world celebrity with streets and institutes named after him all over the world.

Robert Koch (Fig. 2.3) was born in Clausthal-Zellerfeld in Germany in 1843 and studied medicine, botany, physics, and mathematics at the University of Göttingen in 1862–1866. He was a doctor in Wollstein (Silesia) in 1876 when he published a memoir of microbiology, where he showed the role of the "anthrax bacillus" in the infectious disease. He isolated the bacterium in pure culture for the first time, demonstrated its ability to produce spores, and demonstrated its ability to induce the infectious disease. Therefore, in parallel to Pasteur, he founded in Germany

---

[1] At the time of Pasteur, the name "virus" was given to all infectious agents, parasites, bacteria, or real viruses; the word virus derives from Latin and means poison.

Fig. 2.3 Robert Koch (1843–1910) (Copyright: Institut Pasteur, Paris)

the discipline of microbiology. He did set up a makeshift laboratory, which allowed him to study infectious germs. Due to his discoveries, he was recruited by the Institute of Hygiene in Berlin. Koch has mainly developed general bacteriological techniques (isolation of pure strains, cultures, etc.) and has actively pursued the identification of pathogens. He isolated and identified a number of pathogens including the killer of the time that came to bear his name, Koch's bacillus (*Mycobacterium tuberculosis*), the agent of tuberculosis, and the "*Komma-Bacillus*" or *Vibrio cholerae*, the cholera agent. The announcement of pure culture of the slow-growing (15 days) tubercle bacillus was hailed at the time as a major achievement heralding the possibility of a cure. His procedure to identify the causative agent of a given infectious disease, called Koch's postulates, has four stages:
1. The suspected agent is present in sick hosts.
2. The agent must be cultivated in pure culture.
3. A pure culture can infect a healthy host, and the inoculated microorganism produces the classic symptoms of the disease.
4. The "same" microorganism can be isolated from the new hosts.

In 1887, Julius Richard Petri (1852–1921), a German bacteriologist assistant to R. Koch, invented a box that became well known in all microbiological laboratories. Robert Koch, with his assistant and his colleagues (Koch 1882, 1883), thus developed methods for the isolation and maintenance of pure cultures of many bacteria on solid media. The first tests on boiled potatoes were not very conclusive. Subsequently, Koch used gelatin in the Petri dish, and then agar, a seaweed extract, which allowed him to solidify any liquid medium regardless of its composition. This method is still used today and allows microbiologists to isolate microorganisms in pure cultures and to study them in detail. This discovery was very important because it helped to understand bacterial cells, their metabolism, behavioral changes following modifications in growth conditions, and so on. It was initially implemented mainly for the study of pathogenic bacteria but also contributed to the isolation of many environmental bacteria. However, this great discovery, which has permitted to obtain detailed knowledge of bacterial life at the cellular level through the study of pure strains, has contributed to isolate the bacteria from their environment and to study them in synthetic media that do not necessarily reflect the reality of their ecosystem. In addition, it does not permit to observe and evaluate the processes of interaction between microorganisms and their biotic or abiotic environment. Despite these remarks that highlight the limitations of this cultivation method in particular in microbial ecology, the technique proposed by Koch has deeply influenced the approaches used by most microbiologists. Actually, since the early days of microbiology and for most the twentieth century, most microbiologists influenced by Pasteur, Koch, and their students have mainly studied isolated microorganisms at the cellular or subcellular level, in order to understand their metabolic and physiological capabilities and their genetic potential. This enables today to reconstitute, at the molecular level, the metabolic and adaptive processes that govern the operation of these unicellular organisms. This knowledge is essential to understand how they function in their environment. However, it considers only microorganisms isolated and adapted to growing conditions imposed on them. Despites these limitations, his work has strongly influenced the history of microbiology, and Robert Koch was awarded the Nobel Prize for medicine and physiology in 1905. He died in 1910 in the German spa town of Baden-Baden.

The first to observe interactions between microorganisms was Sir Alexander Fleming who, in 1929, observed the inhibition of a culture of staphylococci by a contaminating mold, the fungus *Penicillium notatum* (Fleming 1929). He observed that an inhibition of the culture of staphylococci occurred, yielding a clear halo around the colony of *Penicillium*, which he hypothesized was due to diffusion of a substance secreted by the fungus. This substance, penicillin, was responsible for the great discovery of antibiotics and of the phenomenon of antibiosis in many soil bacteria that produce these molecules, giving them a competitive advantage in the fight between bacterial communities to occupy ecological niches. The field of antibiotics research has consistently expanded from thereon, for example, encompassing actinobacteria such as *Streptomyces* that the Russian-born American Selman A. Waksman showed able to synthesize a whole array of antibiotics like the aminoglycoside streptomycin that proved active against the tuberculosis agent *Mycobacterium tuberculosis*, which earned him the Nobel Prize for medicine and physiology in 1952. Microbial ecology is based on this type of observation and thus depends on the development of new methods to understand precisely such interactions. The idea of wanting to know the role and behavior of microorganisms in their biotic and abiotic environments developed from the work of Sergei Winogradsky (1856–1953, *cf.* Sect. 2.4) and has emerged from the 1970s and was instrumental in the emergence of microbial ecology. This discipline has developed alongside mainstream microbiology disciplines such as medical microbiology that studies the role of pathogenic bacteria in the genesis of infectious diseases and food and industrial microbiology that studies the metabolic role of certain bacteria for biotechnology.

Parallel to the discovery of bacteria and their role in infectious diseases, the Russian microbiologist Sergei Ivanovski (Ivanovski and Polovtsev 1890) was the first to demonstrate infectious particles of a size much smaller than that of bacteria. Ivanovski observed in the mosaic disease of tobacco leaves an infectious process caused by a pathogen unable to grow using the classical methods of microbiology and not retained on the candle filters (Chamberland candle filters made up of porous porcelain) with which one sterilized liquid by filtration and retention of bacteria on the filters. He called these particles "filterable viruses" to refer to infectious agents not retained on the filters of porous porcelain candles. He was the first to describe the tobacco mosaic virus (TMV) followed by the American virologist Rous who discovered the first animal virus in the early twentieth century (Rous 1911): the retrovirus causing leukemia in chickens. Later, the Canadian Felix d'Herelle described, in 1917, the first viruses of bacteria, called bacteriophages (d'Hérelle 1921). It was not until 1939 and the construction of the electron microscope that these viral particles could be directly observed.

2.4 The Beginnings of Microbial Ecology

The work of microbiologists mainly focused on pure cultures have partly contributed to a long-time neglect for ecological research on microorganisms. In addition, microbial ecology has had to face major methodological difficulties due to the small size of microorganisms. Thus, for a long time, mainstream macro-ecology ecologists either considered microorganisms as static particles or have generally ignored by considering their communities as "black boxes" in which their roles in biogeochemical processes are estimated by flux measurements of organic or inorganic material.

However, pioneering work in the nineteenth century has permitted to show the importance of knowledge of microorganisms in relation to their environments. In the early nineteenth century, the Swiss Nicolas-Théodore de Saussure (1767–1845) brought to light the capacity of soils to oxidize hydrogen and worked on various aspects of soil chemistry (de Saussure 1804). As this hydrogen-oxidizing capacity was inhibited by heating the soil or following the incorporation of sulfuric acid, he concluded that the oxidation activity was due to microorganisms. Similarly, the French Jacques-Théophile Schloesing and Achilles Muntz (1877) showed the oxidation of ammonium nitrate present in wastewater occurred when flowing through a sand column. The fact that this activity was destroyed by chloroform vapors and restored upon addition of a soil inoculum enabled them to conclude that it was due to the activity of microorganisms. At the same time, Pasteur was clearly establishing the role of microorganisms in the biodegradation of organic substances. He tried unsuccessfully to prove the role of microorganisms in the transformation of mineral substances, including the oxidation of ammonium nitrate, but he could not succeed. All these pioneering studies were performed on natural samples without the use of isolated bacteria.

It was not until the great discoveries of the Russian microbiologist Sergei Winogradsky (1856–1953, Fig. 2.4) from 1887 to really demonstrate the fundamental role of microorganisms in pathways of transformation of mineral compounds. He was the first to speak of the "microbiology of natural biotopes" and devoted 50 years of his life to microbiological research. He worked in several European universities.

In 1885, he left Russia and began his work in the botany laboratory of the University of Strasbourg. From 1887, he demonstrated the autotrophic ability of filamentous sulfide-oxidizing bacteria of genus *Beggiatoa*, which constituted one of his main microbial models (Winogradsky 1887). He discovered in 1988 (Winogradsky 1888) purple photosynthetic bacteria and sulfide-producing bacteria. He described taxonomically many anoxygenic phototrophic bacteria (Fig. 2.5), sulfate-reducing bacteria, and iron-oxidizing, nitrifying, denitrifying, and nitrogen-fixing bacteria. He thus discovered many metabolic pathways and showed the great diversity of microbial metabolisms. To elucidate the interactions between bacteria of the sulfur cycle, he designed an experimental device, called a "Winogradsky column" in which he was able to control the flow of nutrients and light (*cf.* Box 14.13). He could thus follow the appearance of sulfide production and development of sulfide-oxidizing purple and colorless bacteria and establish conditions for development of these microorganisms (*cf.* Sect. 14.4.3). These experiments allowed him to grow simultaneously communities of

Fig. 2.4 Sergei Winogradsky (1856–1953) (Copyright: Institut Pasteur, Paris)

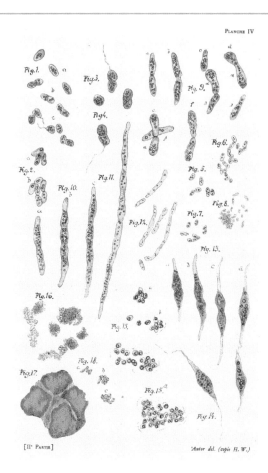

Fig. 2.5 Hand drawings by Sergei Winogradsky describing forms of phototrophic purple bacteria that appeared in his major book *Microbiologie du Sol* or (*Soil Microbiology*) published in 1949 (Winogradsky 1949, "planche IV") (Courtesy of Editions Elsevier Masson)

anaerobic, aerobic, microaerophilic, and photosynthetic bacteria along gradients of oxygen, sulfide, and light.

Then he went to Zurich in 1889 to work in a university laboratory on the problem of nitrification and proved that this process occurs in two steps, involving two groups of microorganisms. The first group performs the formation of nitrite (from the oxidation of ammonium) and the latter group the formation of nitrate (from the oxidation of nitrite to nitrate) (Winogradsky 1890).

Following these discoveries, in 1890, he was invited by Pasteur to settle in Paris but had to return to Russia where he was appointed in 1891 Head of the Department of General Microbiology at the Imperial Institute of Experimental Medicine in St. Petersburg. He remained there for 15 years, and he continued his research on nitrification, focusing on the study of the anaerobic decomposition of cellulose, used in the retting of flax. There was an especially important finding highlighting atmospheric nitrogen fixation by fermentative bacteria: he isolated an anaerobic bacillus, *Clostridium pasteurianum*, able to perform this function.

In 1906, he had to interrupt his scientific work and was appointed university professor in Belgrade in a laboratory with little means (1921).

Then he accepted the invitation of the student of Pasteur, Dr. Roux, who asked him to move to Paris to continue his research (1922). He settled in 1925 and created in Brie-Comte-Robert, close to Paris, a laboratory for the study of the natural environment, specifically soil, where he remained until 1950. He was interested especially in bacteria metabolizing nitrogen, iron, and manganese. He developed his early concepts of "ecological microbiology" and presented the summary of his work on the microbiology of natural environments, soil and water, in a book published in 1949, entitled *Soil Microbiology: Problems and Methods* which to this day is considered a classic in microbial ecology (Winogradsky 1949). Given the means of observation and analysis available to him – archaic compared to those currently available in laboratories – one can only admire the width and modernity of the microbial ecology concepts he developed. His extensive work is very difficult to summarize, but from all the results obtained from the isolation (by enrichment culture; see below Martinus Beijerinck) and the study of bacteria involved in the cycles of sulfur, nitrogen, and iron, three major concepts have emerged:

1. Chemolithotrophic prokaryotes base their metabolism on the oxidation of inorganic compounds such as ammonia

or nitrite, coupling it to the release of energy (*cf.* Sect. 3.3. 2).
2. Chemolithotrophic prokaryotes are autotrophic.
3. The demonstration of the physiological process of bacterial nitrogen fixation previously discovered by Beijerinck in 1888.

Before concluding this summary of the work of S. Winogradsky, it is necessary to present his thoughts on the main methods to study microorganisms in the environment; based on the recommendations stated in his monograph, he initiated a number of research themes in contemporary microbial ecology.

Without underestimating the value of pure cultures, he insisted on the limitations of axenic cultures, "absolutely impossible in nature," that moreover remove a microorganism from its biotic and abiotic environment and place it under artificial conditions, "sometimes bordering on pathology". Another disadvantage of this approach is that the study of population dynamics is impossible. Moreover, "protected as it is in its jar," the organism is not subject to competition, and more generally it eliminates interactions with other organisms. From these remarks, he stressed the difficulty of extrapolating the results obtained in pure culture to the natural environment.

He was also aware of the diversity of the microflora of the natural environment "that provides a habitat for a swarming mass of microscopic organisms, a variety that defies imagination." He showed that few organisms can be isolated in axenic culture: the culture, on "conventional" media, thus only providing an overview of diversity. For example, he found that direct counting by microscopic examination yielded counts of billions of cells per gram of soil, while the counting of bacterial colonies on standard culture media did not exceed several tens of millions. He concluded that "therefore only a small part, comprising only 10 to 5 %, sometimes less, of the total population." It is now well established (thanks to molecular biology techniques) that only 0.1–1 % of bacteria are grown on media conventionally used in microbiology.

He also drew attention to the fact that in the natural environment, namely, the soil, "the vast majority of germs, at some point, are in a state of latent life, only a minority being in an active state." This question led to the notion of the physiological state of a microorganism that will determine its activity level and survival, in particular, its ability to grow on a culture medium, some microorganisms being "viable" but "nonculturable," unable to divide while retaining some of their physiological functions (*cf.* Sect. 15.8). About survival, especially under conditions of dietary deficiency, he felt that "the activity of the soil microbial population seems resistant to adverse conditions."

He concluded that it was necessary to use culture media that allow the development of microorganisms capable of performing a specific function and the elimination of the majority of microorganisms unable to fulfill this function; culture by enrichment is the most effective because it "reveals the diversity of functions of microbes, and allows to isolate the agents capable of fulfilling most functions including in soils: oxidation of nitrite and nitrate, oxidation of sulfur and ferrous salts, fixing gaseous nitrogen, anaerobic decomposition of cellulose and pectic substances, to name only the most interesting."

Despite reservations about classic techniques of microbiology, Winogradsky was aware that "research in Biochemistry provides us with notions of great importance in showing us the intimate ways and the chemical mechanism of life processes." Indeed, the cultivation of a microorganism fulfilling a specific function, free from competition, is essential to know its physiology (e.g., its optimum growth conditions), biochemistry (e.g., understanding the metabolic pathways of degradation of an organic compound), and genetics (e.g., presence of a gene responsible for an activity). Culture under controlled artificial conditions provides valuable clues about the ecological role of a microorganism. He also questioned the activity levels measured in the laboratory: these measures provide an estimate of the potential activity, but not the actual activity that is expressed in situ. For him, "the method of analysis of actual activities of microorganisms in nature must be based not on the behavior of isolated species outside the natural environment, but on the reactions of the entire microbial community, in this environment." It was almost impossible to achieve this at the time of Winogradsky. Technological developments presented in Chap. 17 show that serious answers are made to what remains a major concern for microbial ecologists.

Another school of soil microbiology developed in parallel in the Netherlands. Dutch microbiologist Martinus Beijerinck (1851–1931, Fig. 2.6) developed from 1905 a laboratory of microbiology at the University of Delft, the birthplace of Antoni van Leeuwenhoek. Long before the emergence of microbial ecology, he wrote that his "approach of the microbial world was within microbial ecology, that is to say the relationship between environmental conditions and forms of life that exist." He was led to the discovery of nitrogen-fixing symbiotic and nonsymbiotic bacteria (Beijerinck 1888) and was the first to isolate sulfate-reducing bacteria. He demonstrated the important processes of recycling of sulfur and nitrogen compounds in soil, highlighting the importance of biotransformations in terrestrial ecosystems and their roles in soil fertility (Beijerinck 1895). His work contributed greatly to our understanding of biogeochemical cycles and of microbial biotransformations on a global scale. Associated with the work of Winogradsky, Beijerinck's work showed the significant role of microorganisms in the recycling of elements and the balance of ecosystems necessary for the maintenance of environmental

Fig. 2.6 Martinus Beijerinck (1851–1931) (Copyright: courtesy of the Archives of the School of Microbiology of Delft, Technological University of Delft)

Fig. 2.7 Albert Jan Kluyver (1888–1956). "Rector Magnificus" of Delft University (Copyright: courtesy of the Archives of the School of Microbiology of Delft, Technological University of Delft)

quality and the maintenance of life on Earth. As a result of numerous studies, he was able to show the ubiquity of microorganisms in soils and the emergence of a community based on the selective influence of the environment. On the basis of the famous often-cited sentence "all microorganisms are everywhere, the environment selects," expressed by the Dutch microbiologist Lourens Baas Becking (see below), he developed methods to obtain and select by enrichment and isolate targeted microorganisms even if they were in low numbers in the original samples.

At the same time, other studies completed the work of Beijerinck and Winogradsky on the role of bacteria in biogeochemical cycles and confirmed the importance of emerging microbial ecology, particularly with regard to the reduction of nitrates (Deherain 1897), methanogenesis and methanotrophy (Söhngen 1906), and the isolation of hydrogenotrophic bacteria (Kaserer 1906).

2.5 Microbial Ecology During the Twentieth Century

Fundamental discoveries about the role of microorganisms in the recycling of elements in ecosystems, initiated in the early twentieth century, were pursued by a few microbiologists who continued the pioneering work of Winogradsky and Beijerinck. At the University of Delft, Albert Jan Kluyver (1888–1956, Fig. 2.7) succeeded Beijerinck as the team leader of microbiology. His team developed mainly physiological studies of soil bacteria and studied different metabolic pathways in oxidative, fermentative, and chemolithotrophic bacteria. He discovered new metabolic types and demonstrated through comparative studies, many common metabolic pathways, illustrating the diversity of the microbial world.

One of the famous students of Kluyver, Cornelius Bernardus Van Niel (1897–1985, Fig. 2.8) began his work at the University of Delft but soon joined the United States, invited by Lourens Baas Becking (1895–1963) who was a professor of physiology at Stanford University in 1928. Van Niel began a career as a microbiologist at the Hopkins Marine Station of Stanford University in California in 1929. He pursued there the tradition of the Delft School of Microbiology on comparative microbial ecophysiology. He showed in particular the existence of physiological similarities between the use of hydrogen sulfide and of water in the photosystem apparatus of plants and photosynthetic bacteria. He became interested in the ecological role, diversity, and versatility of microorganisms in soils and in different types of aquatic environments.

An excellent teacher, Van Niel was the initiator of the famous microbiology courses held there to study microorganisms in their environments. Today, annual microbial ecology courses carried out at different locations around the world continue the tradition. Through his classes, he enthused

Fig. 2.8 Cornelius Bernardus Van Niel (1897–1985) (Copyright: courtesy of the Archives of the Center for History of Microbiology, American Society of Microbiology)

many students including Nobel laureates who continued his work on the physiology of bacteria, their role, and their interactions with their biotic and abiotic environments.

Another microbiologist who continued the tradition of Beijerinck and Kluyver was Hans Gunter Schlegel (1924–) in Germany. After studying under the direction of Buder in Halle, he accepted the chair of microbiology at the University of Göttingen in 1958. He contributed to knowledge on the development of purple bacteria in light gradients of lakes, as well as hydrogen-oxidizing bacteria. He was editor of the journal *Archives of Microbiology*, which has long been a journal of reference in the fields of microbiology and microbial ecology.

The work of Claude ZoBell Ephraim (1904–1989, Fig. 2.9) must also be mentioned. The scope of his investigations in the field of environmental microbiology has been very wide and has yielded over 300 publications. In the summary presentation of his results, the reader should pay particular attention to the date of publication of ZoBell's results. Indeed, he was a scientific pioneer in the study of the microbial role in corrosion from fundamental and applied points of view, the adhesion of microorganisms to solid surfaces (where he describes the different stages of development of a biofilm), and the microbiology of petroleum. On the latter topic, he particularly highlighted the role of microorganisms in the degradation of petroleum products in the environment and isolated and characterized a significant number of hydrocarbonoclastic bacteria providing extensive information on their physiology (ZoBell 1950). He also showed the ability of bacteria (especially sulfate-reducing bacteria) to synthesize certain hydrocarbons. But the most remarkable results he obtained were in marine

Fig. 2.9 Claude Ephraim ZoBell (1904–1989) (Copyright: Archives of Scripps Institution of Oceanography, University of California, San Diego)

microbiology, and he is considered the "father" of this discipline. The book he published in 1946 *Marine Microbiology* (ZoBell 1946), by the richness of the concepts presented, is a landmark document in the development of this discipline, and its reading is always recommended to researchers who wish to study the microbiology of the oceans. In this document, descriptions of the physiology and the role of microorganisms in the water column and sediment and factors controlling their distribution and activity (degradation of organic matter, nitrogen, sulfur, and phosphorus biogeochemical cycles) can be found. He brought particular attention to the effect of pressure on marine microorganisms and was the first to demonstrate the existence of bacteria capable of growing in the deep ocean. For example, during his participation to the research ship Galathea expedition, he proved with his student Richard Morita (1923–) that bacterial life was possible at depths exceeding 10,000 m and temperatures approaching 2.5 °C (ZoBell 1952; Morita 1975).

In the field of microbiology of deep-sea ecosystems, an important step was then passed with the work of Holder Windekilde Jannasch (1927–1998) and colleagues. This author who took part in numerous oceanographic cruises (about 30 in the Pacific and Atlantic Oceans, the Mediterranean and Black Sea) did develop a device for the harvesting of samples in the conditions prevailing in situ (Jannasch et al. 1973). This technological breakthrough enabled

H. W. Jannasch to isolate bacteria from their deep-sea environments by avoiding the decompression shock (Jannasch et al. 1982). Moreover, through the construction of a continuous culture system under pressure, he was able to study the kinetics of growth of barophilic bacteria at temperatures between 3 and 8 °C in the presence of low concentrations of carbon sources, both of which being characteristic of deep environments; he demonstrated that barophily, psychrophily, and oligotrophy were closely related (Jannasch et al. 1996). Through its expertise in the field of microbiology of deep environments (Jannasch and Taylor 1984), he also approached the study of prokaryotes living near the sites of deep oceanic hydrothermal vents. He thus contributed significantly to the microbiological study of these ecosystems based on chemosynthesis (*cf.* Sect. 10.3.1), in particular the role of sulfur-reducing bacteria living in symbiosis with *Riftia pachyptila*, the giant worm, devoid of mouth, digestive tract, and anus that harbors in its trophosome (*cf.* Sect. 10.3.1) endosymbiotic chemolithoautotrophic bacteria (Cananaugh et al. 1981).

With H. Jannasch many microbiologists were inspired by the teachings of Cornelius Van Niel. Among them:

- Roger Stanier (1916–1982) did his graduate studies at the University of Vancouver (Canada) and then at Stanford University (USA). In 1938, he met Prof. Van Niel at the Hopkins Marine Station of Stanford University in California during a summer course of microbiology; he then worked on the taxonomy of bacteria in Van Niel's laboratory (Pacific Grove). In 1947, he became professor of microbiology at the University of Berkeley (California, USA) and participated in the development of the *Bergey's Manual of Systematic Bacteriology*, which is the reference work for identification of bacterial species. In 1957, he published with his colleagues the first edition of *The Microbial World* that has been reedited five times (Stanier et al. 1976). He contributed greatly to knowledge on photosynthetic bacteria, including the positioning of Cyanophyceae (blue-green algae) in the "kingdom of prokaryotes" and thus renamed Cyanobacteria. In 1971, he accepted the position of department head of the microbial physiology unit offered him by the Pasteur Institute; he was appointed professor at the Pasteur Institute and developed the reference cyanobacteria laboratory. He initiated with Norbert Pfennig and Moshe Shilo a series of triennial symposia on photosynthetic prokaryotes that continues to the present days, with the 14th edition in 2012.
- Norbert Pfennig (1925–2008) studied microbiology at the University of Göttingen. After his doctorate in 1957, he was appointed assistant at the same university. Stimulated by Schlegel, he devoted himself to the study of purple and green phototrophic bacteria. He then went on to work in the laboratory of Cornelius Van Niel at the Hopkins Marine Station (California, United States) where he met Roger Stanier. He then returned to the University of Göttingen as professor and, with his collaborators, developed the physiological, systematic, and ecological study of phototrophic bacteria involved in the sulfur cycle. His work brought him to isolate hundreds of strains constituting many new species and genera not only of phototrophic bacteria but also of sulfate-reducing bacteria and fermentative bacteria. He became a world reference for photosynthetic (Stanier et al. 1981) and sulfate-reducing bacteria (Pfennig et al. 1981). He was responsible for new concepts on the role of these bacteria in the biogeochemical cycles of carbon and sulfur. He showed their important role in the biodegradation of organic compounds and even some xenobiotics under anoxic conditions. He helped with the committee of the *Bergey's Manual*, to reorganize the whole systematics of phototrophic bacteria with his colleague, Hans G. Trüper, professor at the University of Bonn from 1972 to 2001, founder of the Institute of Microbiology and Biotechnology of Bonn and FEMS president from 2000 to 2005. In 1980, Norbert Pfennig was appointed professor of microbiology and limnology at the University of Konstanz where he was still professor emeritus at the end of his days. Throughout his teaching career, he was renowned for his qualities as a thinker and humanist with a very open mind; he stimulated many students such as Bernhard Schink, Friedrich Widdel, Heribert Cypionka, and Jörg Overmann, who became professors of microbiology, and continued his leadership in these areas of research in microbial ecology.
- Moshe Shilo (1920–1990), born in Moscow, emigrated to Israel in 1934. He studied at the University of Jerusalem. After his appointment as professor at the University Givat Ram in Jerusalem, he founded the institute for the study of fish diseases in 1949 and the Marine Station of Eilat in 1964. He was the initiator of microbial ecology in Israel and spent his life as a researcher in the study of biotic interactions between organisms or between hosts and parasites and to the study of cyanobacteria in aquatic or marine coastal lakes. Inspired by the "Delft School" after the pioneering work of Beijerinck and after working in the laboratory of Cornelius Van Niel at the Hopkins Marine Station (United States), he became interested in the biology and blooms of cyanobacteria in aquatic environments. He initiated studies of cyanobacteria in microbial mats on which he worked with colleagues in many aquatic environments such as freshwater, marine, and hypersaline biotopes. Thus, he and his collaborators could demonstrate the existence of anoxygenic photosynthesis in cyanobacteria that are probably the origin of stromatolites, the diurnal regulatory pathways of photosynthesis and sulfate reduction in the vertical gradients established in microbial mats, and the interrelationships between microorganisms in

microbial mats. His qualities of enthusiastic researcher stimulated many students and researchers who pursued his study of microbial mats including extreme conditions of salinity (Cohen et al. 1977).

- Robert E. Hungate (1906–2004) was a pioneer of anaerobic microbial ecology. After studying at the University of Washington and then Stanford, he was the first doctoral student of Cornelius Van Niel. In 1935, he joined the laboratory of zoology, University of Texas, as a teacher and devoted himself to the study of fermentative bacteria that degrade cellulose. He initiated sophisticated methods and practices (Hungate tubes; see Sect. 17.8.2) to successfully isolate oxygen-sensitive anaerobic bacteria, fermentative bacteria, and especially methanogenic bacteria. He became interested in the termite gut and rumen of cattle. He was able to demonstrate the operation of anaerobic rumen bacterial communities and isolated syntrophic bacteria in total interdependence in the rumen. He could well describe the fermentative microbial communities and rumen methanogens which thus became the first fully characterized microbial ecosystem. His works on the rumen microbial ecology have been assembled into a book (*The Rumen and Its Microbes*: Hungate 1966) which brought a significant understanding of this unique ecosystem. With many students and collaborators, he developed the concept of the complete study of an ecosystem that requires not only knowledge of bacteria and their interactions but also a quantification of their synergistic activities.
- Ralph S. Wolfe (1921–) was educated at the University of Pennsylvania (United States) until his doctorate in 1953. After following the teachings of Cornelius Van Niel, he joined the University of Illinois as a professor where he has been professor emeritus since 1991. It was at the origin of physiological and biochemical studies of methanogenic bacteria and their role in anoxic environments and fermenters.

In France, following the work of Sergei Winogradsky, a school of soil microbiology developed at the Annex Brie-Comte-Robert of the Pasteur Institute. Professors J. Pochon, H. Barjac, and P. Tardieux developed their studies on soil microbiology including bacteria involved in the nitrogen cycle. They contributed to a better understanding of soil bacteria including their isolation from soil in two books: *Traité de microbiologie du sol* (*Treaty of Soil Microbiology*) published by Dunod in 1958 (Pochon and Barjac 1958) and *Techniques d'analyses en microbiologie des sols* (*Technical Analysis in Soil Microbiology*) (Pochon and Tardieux 1962).

In parallel, a school of soil science and soil ecology developed in Strasbourg and Nancy. In these laboratories, Yvon Dommergues (director of research at CNRS) developed studies on the physiology and ecology of soil bacteria including nitrogen-fixing bacteria. With his colleague François Mangenot, professor at the Faculty of Nancy, they wrote an important book in soil microbiology: *Ecologie microbienne du sol* (*Soil Microbial Ecology*) published by Masson in 1970 (Dommergues and Mangenot 1970). Under ORSTOM sponsorship, Yvon Dommergues created and directed the microbiology laboratory soil in Dakar (Senegal) in which he resumed the study of soil microorganisms, including studying the symbiotic nitrogen fixation and isolated first *Frankia* of *Casuarina* (Diem et al. 1983). He was very involved in the practical applications of his discoveries and participated in the installation of a green barrier of several hundred kilometers in Africa, using symbiotic nitrogen fixation, to stop erosion and expansion of the desert. Soil microbiology was then strongly developed in France by René Bardin who has worked on various aspects of nitrogen cycling in particular nitrification (Degrange and Bardin 1995).

John Postgate has made a great contribution to the study of sulfate-reducing bacteria for many years. After his doctoral studies at Oxford in the 1950s, he was employed by the Department of Chemical Research DSIR in London where he studied the physiological role of sulfate-reducing bacteria in corrosion. He later became professor at the University of Sussex in England, in the Department of Microbiology, where he studied nitrogen fixation at a biochemical level (Postgate 1998) and demonstrated the role of leghemoglobin in the legume-*Rhizobium* symbiosis and established in Brighton, United Kingdom (later moved to Norwich), an institute specialized on nitrogen fixation. He was president of the English Society for Microbiology (SGM) from 1984 to 1987.

Julian Davies, former chairman of the ASM worked on the physiology of many microorganisms, especially the secondary metabolism of *Streptomyces* and demonstrated the feasibility of the metagenomic approach for the synthesis of new metabolites (Seow et al. 1997).

David Hopwood has been the driving force behind the study of antibiotic synthesizing *Streptomyces* at the John Innes Institute in Norwich. In 1985, his group described production of the first hybrid antibiotic by genetic engineering. In 1994, he was made Knight Bachelor in the Queen's Birthday Honors for services to genetics and thus called Sir. In 2002, it was his group together with the Wellcome Trust Sanger Institute in Cambridge that published the complete genome sequence of *Streptomyces coelicolor*.

In the field of microbial ecology of extreme hyper-hot environments, we should mention the pioneering work of Thomas D. Brock (1926–). After studies at the University of Ohio (United States) and a Ph.D. in botany in 1952, Thomas Brock was formed in microbiology and molecular ecology in the research department of a company producing antibiotics. In 1960, he was appointed professor of bacteriology at Indiana University in 1971 and then as a professor of bacteriology at the University of Wisconsin-Madison, where he became director of the department of bacteriology in 1979. He devoted much of his career to the ecology of microorganisms including hyperthermophiles in the hot springs of Yellowstone Park since 1965. He discovered many bacteria and archaea and hyperthermophiles including the thermophilic bacterium *Thermus aquaticus*, which he named and from which Taq polymerase was purified, an enzyme which is widely used in molecular biology protocols that use PCR. He became interested in the boundary conditions of life on the planet and with his students he initiated research on microbial life in extreme conditions. He published the famous book *Biology of Microorganisms* which has been and still remains a basic book for many microbiology students over the world. This book has been updated regularly and is now in its 13th edition; it continues to be the reference book for many microbiologists. Thomas Brock is now a professor emeritus, and one of his students who became professor of microbiology, Michael Madigan, continues these lines of research on life under extreme conditions of high and low temperatures and on the biogeochemistry of bacteria living under these extreme conditions and continues the edition of the famous book *Biology of Microorganisms* (Madigan et al. 2010).

In the 1970s, studies of microbial ecology of biogeochemical cycles were developed under the impetus of the work of Tom Fenchel, Henry Blackburn, and Bo Barker Jørgensen at the University of Aarhus in Denmark where new methods using microelectrodes were developed to study the sedimentary gradients that develop between the oxic and anoxic conditions and microbial activities that determine in biogeochemical cycles (Jørgensen and Fenchel 1974). These micromethods were subsequently used for the study of microphysical and chemical gradients that characterize the microbial mats (Jørgensen et al. 1979; Van Gemerden et al. 1989). Fenchel and Blackburn published an important book in microbial ecology, *Bacteria and Mineral Cycling*, in which they laid the foundations of the microbial physiology of biogeochemical cycles (Fenchel and Blackburn 1979).

Regarding the environmental microbiology and applied microbial ecology, the pioneering work of Richard Bartha opened the way to studies of the role of microorganisms in the biodegradation of polluting compounds and xenobiotics. Born in Hungary, where he made some of his university studies, he participated in the uprising of 1956, and then he fled to Germany where he could prepare his microbiology doctorate with Hans Schlegel. In 1964, he joined the Department of Biochemistry and Microbiology, School of Biological Sciences and the Environment of Rutgers University (United States), where he remained professor until his retirement in 1998. He devoted his time to research on the role of microbes in the detoxication of environments contaminated with pesticides, hydrocarbons, and heavy metals, either in soil or in the marine environment. With his students including Ronald Atlas, he was responsible for the discovery of the biodegradation of petroleum compounds by bacteria thus opening the way to an important field of microbial ecology on biodegradation and biotransformation of pollutants. He showed the ability of bacteria to degrade or biotransform not only hydrocarbons including polyaromatics but also organochlorinated pesticides and heavy metals. It was at the origin of bioremediations methods to rehabilitate sites contaminated with pollutants (soils, oil spills, etc.). He became aware of the weakness of the teaching of microbial ecology, and with his student Ronald Atlas, professor of biology at the University of Louisville (United States) and former president of the American Society for Microbiology, they published the first book on *Microbial Ecology* in 1981 following a report published in 1975 in the Bulletin of the American Society for Microbiology "the teaching of microbial ecology unfortunately demands a substantial contribution with a well-documented book." They thus decided to write the book *Microbial Ecology: Fundamentals and Applications* which appeared for the first time in 1981 and, together with its fourth updated edition (Atlas and Bartha 1998), remains a basic book in microbial ecology.

Robert Mah, who was a professor of environmental microbiology at the University of California (UCLA, United States) from 1970 to 1995 and is now professor emeritus, brought a great contribution to the knowledge of methanogenic bacteria and also anaerobic halophilic bacteria.

In the field of marine microbial ecology, the work of Rita Colwell (1934–) is important. After studying microbiology and oceanography, she became professor of microbiology at the University of Maryland where she studied with many students marine microbiology in general, and especially *Vibrio* pathogens, and has shown the importance of culturability fluctuations of bacteria in the marine environment. Her pioneering work on viable nonculturable bacteria has opened the way for a whole area of research on viable nonculturable bacteria (Chaiyanan et al. 2001), which led to her being appointed to head the US NSF where she promoted work on microbial ecology. She has held many responsibilities in various American institutions where she has always promoted microbial ecology and the problem of survival of bacteria, including pathogens in aquatic environments. With Richard Morita, she contributed to an important book on marine microbiology: *Effect of the Ocean Environment on Microbial Activities* (Colwell and Morita 1974).

With regard to marine microbiology and aquatic environments, the School of Microbiology of Marseille has developed following the creation of the CNRS laboratory of bacterial chemistry by Jacques Sénez (1915–1999). After working at the Pasteur Institute in the laboratory of anaerobes with Professor Prevost, Jacques Sénez returned to Marseille where he created the laboratory of bacterial chemistry and biological corrosion in 1947; in this laboratory, he began his basic research with technological applications on sulfate-reducing bacteria and hydrocarbon biodegradation and corrosion of concrete by sulfur-oxidizing bacteria. He created, at the University of Marseille, the second Chair of Microbiology in France after Paris, in 1950. He developed his research team in 1962 and created the great laboratory of bacterial chemistry which became a reference laboratory for environmental bacteria and biotechnology, which he directed until 1983. He wrote a standard reference book in microbiology, *Microbiologie Générale* (*General Microbiology*) published by DOIN (Sénez 1968). He inspired many researchers at CNRS, ORSTOM, and at the university who continued his research paths on sulfate-reducing bacteria (Le Gall and Fauque 1988) and allowed the development of several research laboratories on marine microbiology devoted to the role of bacteria in biogeochemical cycling or in the biodegradation of hydrocarbons in the marine environment. Researchers in these laboratories contributed to the first French book for marine microbiology, *Microorganismes dans les écosystemes océaniques* (*Microorganisms in Ocean Ecosystems*) (Bianchi et al. 1989).

It has long been recognized that animal organisms harbor a large number of microbes on their surface and the Russian Ilya Ilyich Mechnikov, born in 1845, was one of the first to study them. He was formed at the University of Kharkoff in natural sciences and went on to study marine fauna at Heligoland and then to various universities (Göttingen, Munich, Giessen) where he studied intracellular digestion in the flatworm and phagocytosis. In addition to his work on immunity that earned him the Nobel Prize in Physiology and Medicine in 1908, he started the study of the flora of the human intestine and developed a theory that gut microbes' metabolites caused poisoning, which could be prevented through a diet containing fermented milk, containing large amounts of lactic acid.

The French René Dubos, born in Val d'Oise near Paris, in 1901 was recruited in the National Agronomical Institute (INRA) as an agronomist in 1921. He soon moved to work with O. T. Avery at Rockefeller University in New York where he studied enzymes that break down the capsule of pneumococci and developed gramicidin, the first commercialized antibiotic. He was elected to the US National Academy of Sciences in 1941 as a consequence of this discovery but was always an advocate that biological equilibrium was a better way to fight diseases than antibiotics. He thus coined the well-known motto "think globally, act locally" to suggest that global environmental problems can be solved only by considering ecological, economic, and cultural aspects of our local surroundings.

His most popular work *Mirage of Health* (1959) is a reflection on the balance between hosts and pathogens where health is never definite because disease results from the dynamic process of life.

Microbial ecology of the digestive tract, animal and human, has experienced strong development in the group of P. Raibaud in the laboratory of INRA at Jouy-en-Josas, by using several techniques such as axenic animals and antibiotherapy, allowing to know the dominant flora at different stages of life and the changes associated with certain diseases (Ducluzeau and Raibaud 1979).

2.6 Microbial Ecology Today

Since 1970, microbial ecology has developed considerably, and many research laboratories around the world are involved in this field of research. Microbial ecologists study interactions between microorganisms and their environment or between microorganisms and other biological components of ecosystems. Therefore, they require a wide knowledge in the fields of physics, chemistry, and biology. The study of microbial ecology therefore requires close links with various disciplines for studies of microorganisms and their roles in their environments, whether in geophysics, geochemistry, soil science, limnology and oceanography, climatology, general biology, botany and zoology, biochemistry and molecular biology, and statistics and biostatistics.

The methods of molecular biology, which have greatly expanded since the 1980s, have led to major advances in all biological sciences and especially in microbial ecology. They, among others, contributed greatly to the knowledge of the diversity and adaptations of microbial communities in ecosystems. They have allowed understanding microbial interactions by gene flow and response capacities of microbial communities to environmental stresses and increasingly important inputs of toxic compounds in ecosystems. Considering these different subject areas, the community of microbial ecologists is very large today, as shown by the number of scientific articles in constant increase in the flagship journals of the discipline such as *Microbial Ecology* (quarterly since 1974), *FEMS Microbiology Ecology* (quarterly since 1985), *Applied and Environmental Microbiology* (monthly since 1976), *Environmental Microbiology* (monthly since 1999), *Geomicrobiology Journal* (quarterly since 1983), *Aquatic Microbial Ecology* (monthly since 1995), and *ISME Journal* (monthly since 2007) and in numerous journals of ecology, soil science, and aquatic or ocean studies.

Given these important developments, microbial ecologists have decided to establish an international association, *ISME* (*International Society for Microbial Ecology*), created in 1977. This society aims to enable global exchanges between microbial ecologists using Internet forums, with a scientific periodical (*ISME Journal*, which is a monthly magazine with articles of microbial ecology and information on *ISME*) and especially with symposia that have taken place every third year from 1977 to 2004 and have been held every second year since 2004. Thus, the 14 symposia held since 1977 have brought together about 1,500–2,000 participants who have found opportunities for exchange, interaction, or cooperation in the field of microbial ecology at the global level.

Many national associations have as a function to encourage exchanges between members and organize meetings. This is the case of the *ASM* (*American Society for Microbiology*) founded in 1899, the *SGM* (*Society for General Microbiology*) founded in 1945, the *CSM* (*Canadian Society for Microbiology*) founded in 1952, the *CSM* (*Chinese Society for Microbiology*) founded in 1952, the *SFM* (*French Society of Microbiology*) founded in 1937, and the *AFEM* (*Association Francophone d'Écologie Microbienne*) founded in 2004.

References

Atlas R, Bartha R (1998) Microbial ecology, fundamental and applications, 4th edn. Addison Wesley Longman, Reading
Beijerinck MW (1888) Die Bacterien der Papilionaceen-Knöllchen. Bot Ztg 46:725–735
Beijerinck WM (1895) Über *Spirillum desulfuricans* als Ursache von Sulfat-reduktion. Zentralbl Bakteriol Abt L 1–9:104–114
Bianchi M, Marty D, Bertrand J-C, Caumette P, Gauthier M (1989) Micro-organismes dans les écosystèmes océaniques. Masson, Paris
Cagniard de La Tour C (1838) Mémoire sur la fermentation vineuse. Ann Chim Phys 68:206–222
Cananaugh CM, Gardiner SL, Jones ML, Jannasch HW, Waterbury JB (1981) Procaryotics cells in the hydrothermal vent tube worm *Riftia pachyptila* Jones: possible chemoautotrophic symbionts. Science 213:340–341
Chaiyanan S, Huq A, Maugel T, Colwell RR (2001) Viability of the nonculturable *Vibrio cholerae* O1 and O139. Syst Appl Microbiol 24:331–341
Cohen Y, Krumbein WE, Shilo M (1977) Solar Lake (Sinai). Distribution of photosynthetic microorganisms. Limnol Oceanogr 22:609–620
Cohn F (1876) Untersuchungen uber Bakterien IV. Beitrage zur Biologie der Bacillen. Beitr Biol Pflanzen 2:249–276
Colwell RR, Morita RY (1974) The effects of the ocean environment on microbial activities. University Park Press, Baltimore
d'Hérelle F (1921) Le Bactériophage: Son rôle dans l'Immunité. Masson, Paris, 227 p
de Saussure NT (1804) Recherches chimiques sur la végétation. Vve Nyon, Paris, 327 p
Degrange V, Bardin R (1995) Detection and counting of *Nitrobacter* populations in soil by PCR. Appl Environ Microbiol 61:2093–2098
Deherain PP (1897) La réduction des nitrates dans la terre arable. C R Acad Sci Paris 124:269–273
Diem H, Gauthier D, Dommergues Y (1983) An effective strain of *Frankia* from *Casuarina* sp. Can J Bot 61:2815–2821
Dobell C (1923) A protozoological bicentenary: Antony van Leeuwenhoek (1632–1723) and Louis Joblot (1645–1723). Parasitology 15:308–319
Dommergues Y, Mangenot F (1970) Écologie microbienne du sol. Masson, Paris
Ducluzeau R, Raibaud P (1979) Écologie microbienne du tube digestif. INRA and Masson, Paris
Fenchel T, Blackburn H (1979) Bacteria and mineral cycling. Academic Press, London
Fleming A (1929) On the antibacterial action of cultures of a *Penicillium*, with special reference to their use in the isolation of *B. influenzae*. Br J Exp Pathol 10:226–236
Haeckel E (1866) Generelle Morphologie. I: Allgemeine Anatomie der Organismen. II: Allgemeine Entwickelungsgeschichte der Organismen. Reimer, Berlin
Hungate R (1966) The rumen and its microbes. Academic Press, New York
Ivanovski D, Polovtsev VV (1890) Die Pockenkrankheit der Tabakspflanze. Mem Acad Sci St Petersbourg Ser 7(37):1–24
Jannasch HW, Taylor CD (1984) Deep sea microbiology. Annu Rev Microbiol 38:487–514
Jannasch HW, Wirsen CO, Winget CL (1973) A bacteriological pressure-retaining deep-sea sampler and culture vessel. Deep-Sea Res 20:661–664
Jannasch HW, Wirsen CO, Taylor CD (1982) Deep sea bacteria: isolation in the absence of decompression. Science 216:1315–1317
Jannash HW, Wirsen CO, Doherty KM (1996) A pressurized chemostat for the study of marine barophile and oligotrophic bacteria. Appl Environ Microbiol 62:1593–1596
Jørgensen BB, Fenchel T (1974) The sulfur cycle of a marine sediment model system. Mar Biol 24:189–201
Jørgensen BB, Revsbech NP, Blackburn TH, Cohen Y (1979) Diurnal cycles of oxygen and sulphide microgradients and microbial photosynthesis in a cyanobacterial mat. Appl Environ Microbiol 38:46–58
Kaserer H (1906) Die oxydation des Wasserstoffs durch Mikroorganismen. Zentralbl Bakteriol II Abt 16:681–696
Koch R (1882) Die Aetiologie der Tuberculose. Berl Klin Wochenschr 19:221–230
Koch R (1883) Ueber die neuen Untersuchungsmethoden zum Nachweis der Mikroorganismen in Boden, Luft und Wasser. Aerztliches Vereinsblatt für Dtschl 237:244–250
Kützing FT (1837) Microscopische Untersuchungen über die Hefe und Essigmutter, nebst mehreren andern dazu gehörigen vegetabilischen Gebilden. J Prakt Chem 11:385–409
Le Coustumier A (2010) Louis Joblot et ses microscopes. Bull Soc F Microbiol 25:89–100
Le Gall J, Fauque G (1988) Dissimilatory reduction of sulfur compounds. In: Zehnder AJB (ed) Biology of anaerobic microorganisms. Wiley, New York, pp 587–640
Lechevalier H (1976) Louis Joblot and his microscopes. Bacteriol Rev 40:241–258
Madigan M, Martinko JM, Stahl D, Clark DP (2010) Brock: biology of Microorganisms, 13th edn. Pearson Benjamin-Cummings, San Francisco
Morita R (1975) Psychrophilic bacteria. Bacteriol Rev 39:144–167
Pasteur L (1857) Mémoire sur la fermentation alcoolique. C R Acad Sci Paris 45:1032–1036
Pasteur L (1860) Mémoire sur la fermentation alcoolique. Ann Chim Phys 58:323–426
Pfennig N, Widdel F, Trüper HG (1981) The dissimilatory sulfate-reducing bacteria. In: Starr MP, Stolp H, Trüper HG, Balows A, Schlegel HG (eds) The prokaryotes. Springer, Berlin, pp 926–940
Pochon J, De Barjac H (1958) Traité de microbiologie du sol. Dunod, Paris

Pochon J, Tardieux P (1962) Techniques d'analyses en microbiologie du sol. éditions de la Tourelle, Saint Mandé

Postgate J (1998) Nitrogen fixation, 3rd edn. Cambridge University Press, Cambridge, UK

Rous P (1911) A sarcoma of the fowl transmissible by an agent separable from the tumor cells. J Exp Med 13:397–411

Schloesing J, Muntz A (1877) Sur la Nitrification par les Ferments Organisés. C R Acad Sci Paris 84:301–303

Schwann T (1837) Vorläufige Mittheilung, bettreffend Versuche über die Weingährung und Fäulniss. Ann Chim Phys 41:184–193

Semmelweis IP (1861) Die Aetiologie der Begriff und die Prophylaxis des Kindbettfiebers. Hartleben, Pest, Vienna/Leipzig

Senez J (1968) Microbiologie générale. Doin, Paris

Seow KT, Meurer G, Gerlitz M, Wendt-Pienkowski E, Hutchinson CR, Davies J (1997) A study of iterative type II polyketide synthases, using bacterial genes cloned from soil DNA: a means to access and use genes from uncultured microorganisms. J Bacteriol 179:7360–7368

Söhngen NL (1906) Über Bakterien, welche Methan als Kohlenstoffnahrung und Energiequelle gebrauchen. Zentralbl Bakteriol Parasitenkd Infektionskr Hyg Abt II 15:513–517

Stanier RY, Aldelberg EA, Ingraham J-L (1976) The microbial world. Prentice Hall, Englewood Cliffs

Stanier RY, Pfennig N, Trüper HG (1981) Introduction to the phototrophic prokaryotes. In: Starr MP, Stolp H, Trüper HG, Balows A, Schlegel HG (eds) The prokaryotes. Springer, Berlin, pp 197–211

Tyndall J (1877) On heat as a germicide when discontinuously applied. Proc R Soc Lond 25:569

van Gemerden H, Tughan R, de Wit R, Herbert A (1989) Laminated microbial ecosystems on sheltered beaches in Scapa Flow, Orkney Islands. FEMS Microbiol Ecol 62:87–102

Winogradsky SN (1887) Über Schwefelbakterien. Bot Ztg 45:489–600

Winogradsky SN (1888) Über Eisenbakterien. Bot Ztg 46:261–270

Winogradsky SN (1890) Sur les organismes de la nitrification. C R Acad Sci Paris 60:1013–1016

Winogradsky S (1949) Microbiologie du sol. Problèmes et méthodes. Masson, Paris

ZoBell CE (1946) Marine microbiology. Chronica Botanica Compagny, Waltman

ZoBell CE (1950) Assimilation of hydrocarbons by microorganisms. In: Nord FF (ed) Advances in enzymology. Interscience Publishers, New York/London, pp 443–486

ZoBell CE (1952) Bacterial life at the bottom of the Philippine Trench. Science 115:507–508

Structure and Functions of Microorganisms: Production and Use of Material and Energy

Robert Matheron and Pierre Caumette

Abstract

The cellular structures of prokaryotic and eukaryotic microorganisms and the characters distinguishing the three domains of life (*Archaea, Bacteria, Eukarya*) are first described. Then, the metabolic diversity of microorganisms is discussed, the knowledge of which is essential to understand the role of microorganisms in natural and anthropogenic environments. The different degradation pathways for mineral and organic compounds that provide cellular energy are described (aerobic and anaerobic respirations, fermentations) as well as the photosynthetic processes (aerobic and anaerobic photosynthesis). Finally, the mechanisms of biosynthesis are presented: autotrophy and heterotrophy, assimilation of C1, and assimilation of organic and inorganic compounds (mainly nitrogen and sulfur assimilation).

Keywords

Aerobic respirations • Anaerobic respirations • CO_2 assimilation • Eukaryotic cells • Fermentations • Heterotrophic biosynthesis • Inorganic compound assimilation • Photosynthesis • Prokaryotic cells

3.1 Structure and Functions of Prokaryotes and Eukaryotes: Major Features and Differences

The living world is now divided into three major areas (see the second part of the book, Chaps. 5, 6, and 7) in which the cellular structure is either of the prokaryotic or eukaryotic type. The *Bacteria* and *Archaea* domains consist of microorganisms of usually unicellular prokaryotic type, and the *Eukarya* domain includes microorganisms and multicellular organisms that are all of eukaryotic type. One major difference between prokaryotic and eukaryotic cells is the absence of a nuclear membrane in prokaryotes which is needed to define a true nucleus (chromosomes have been separated from the cytoplasm by the nuclear membrane) in eukaryotes (from the Greek *eu*: true, *caryon*: nucleus). In prokaryotes, the generally circular chromosome present in a single copy is directly in the cytoplasm (pro: that precedes, *caryon*: nucleus). In some prokaryotes, it may be linear and in multiple copies (*Rhodobacter sphaeroides*: two copies; *Halobacterium* sp.: three copies). Both types of structure are also differentiated by their size and cell contents including the presence of various organelles in eukaryotes (Fig. 3.1a; Table 3.1).

3.1.1 Prokaryotic Microorganisms (*Bacteria* and *Archaea*, cf. Chaps. 5 and 6)

Prokaryotes are unicellular microorganisms. However, some may associate to form clusters more or less regular, single filaments or branched filaments from a few cells to hundreds

---

* Chapter Coordinator

R. Matheron* (✉)
Institut Méditerranéen de Biodiversité et d'Ecologie marine et continentale (IMBE), UMR-CNRS-IRD 7263, Aix-Marseille Université, 13397 Marseille Cedex 20, France
e-mail: matheron.robert@gmail.com

P. Caumette
Institut des Sciences Analytiques et de Physico-chimie pour l'Environnement et les Matériaux (IPREM), UMR CNRS 5254, Université de Pau et des Pays de l'Adour, B.P. 1155, 64013 Pau Cedex, France
e-mail: pierre.caumette@univ-pau.fr

Fig. 3.1 Schematic representation of prokaryotic (**a**) and eukaryotic (**b**) cells (Drawing: M.-J. Bodiou)

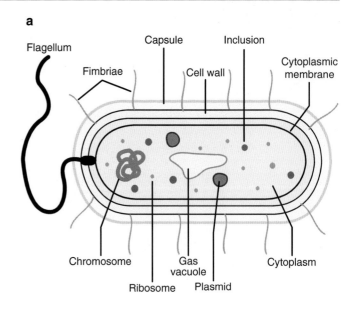

of cells, some with functional specificity, the first step toward multicellularity (*cf.* Sect. 5.2). The cells of prokaryotes are generally small, close to the micrometer, spherical (cocci) or elongated rod shaped, straight, curved, or spiral. Particular forms have stalks or protuberances (Fig. 3.2). Some prokaryotic cells may have large sizes, visible with the naked eye (such as *Thiomargarita namibiensis*, 300–750 µm in diameter) (*cf.* Chap. 5, Fig. 5.1c); in contrast, nanobacteria whose cells are less than 0.2 µm in diameter were isolated from different environments.

Exchanges between cells and the external environment are realized at the cell surface. The surface/volume ratio which is inversely proportional to cell size gives an advantage to prokaryotes compared to all other living organisms. However, this feature makes it more sensitive to changes in the environment. Whatever the form, prokaryotic cell organization remains the same: more or less complex cell envelopes are surrounding cytoplasm that contains a small number of inclusions (chromosome, plasmids, ribosomes, gas vacuoles, etc.). Some also contain a membrane

Table 3.1 Some characteristics of differentiation between the three domains of life

	Bacteria	Archaea	Eukarya
Organelles	Absent	Absent	Present
Nuclear membrane	Absent	Absent	Present
Cellular wall	Present muramic acid present	Present muramic acid absent	Present or absent
Membrane lipids	Linear chains	Branched aliphatic chains	Linear chains
Circular DNA	Present	Present	Absent
Introns	Absent	Absent	Present
Operons	Present	Present	Rare
mRNAs			
RNA polymerases	One	Several	Three
TATA sequence in the promoter	Absent	Present	Present
Addition of a polyA sequence	Rare	No	Yes
Addition of methylguanosine	No	No	Yes
Transfer RNA			
Thymine in the T Ψ C loop	In general present	Absent	In general present
Dihydrouracil	In general present	In general absent	In general present
AA carried by initiator tRNA	Formylmethionine	Methionine	Methionine
Ribosomes			
Size in Svedberg units	70 S	70 S	80 S
Size of two subunits in Svedberg units	30 S, 50 S	30 S, 50 S	40 S, 60 S
Size of RNA in Svedberg units	16 S	16 S	18 S
Functions			
Anaerobic respirations	Present	Present	Generally absent
Methanogenesis	Absent	Present	Absent
Chemolithotrophy	Present	Present	Absent
Chlorophyll photosynthesis	Present	Absent	Present

defining an inner cytoplasmic zone containing the chromosome (*cf.* Sect. 6.6.2: phylum *Planctomycetes*).

3.1.1.1 Cell Envelopes of Prokaryotes

Prokaryotic cells have generally two types of envelopes; on the outside, the rigid wall maintains the shape of the cells and covers the cytoplasmic membrane delimiting the cytoplasm. The composition of the wall defines three groups of prokaryotes: Gram-positive bacteria (Gram +), Gram-negative bacteria (Gram−), and archaea. *Mycoplasma* and some archaea do not have a wall.

Cytoplasmic Membrane

In bacteria, the cytoplasmic membrane is composed of proteins (*cf.* Sect. 4.1.7, Fig. 4.5) and a phospholipid bilayer (Fig. 3.3) that forms the basic structure. In this double layer, phospholipids are oriented such that the apolar chains are placed inside and polar ends located on the surface of the membrane. Many molecules are included in the membrane:

1. Terpenoid derivatives (hopanes) acting as stabilizers of membrane structure (*cf.* Sect. 4.1.5, Fig. 4.15; *cf.* Sect. 16. 8.2, Fig. 16.40).
2. Enzymes, pigments, and electron carriers involved in respiratory and photosynthesis activities. The cytoplasmic membrane of prokaryotes is the principal location of the production of cellular energy. This energy is derived from enzymes, pigments, and electron carriers involved in the respiratory and bacterial membrane proton-motive force (Δp) resulting from the formation of a transmembrane proton gradient (*cf.* Sect. 3.3.1).

In archaea, the cytoplasmic membrane plays the same role, but its bilayer or monolayer structure is different. Branched hydrocarbon chains associated with the glycerol by ether linkages – stronger than the ester linkages of bacteria – replace linear fatty acids. Branched chains have variable lengths in carbon number of C20 (bilayer) to C40 (monolayer). In the latter case, typical of many extreme thermophilic archaea, the monolayer consists of molecules having two glycerols linked by four ether linkages (tetraether) to the ends of the two branched chains, allowing for greater stability and rigidity of the membrane at high temperatures (*cf.* Sect. 4.1.7). Diethers and tetraethers may be mixed with lipids (phospholipids, sulfolipids, or glycolipids). The *Halobacteria* (extreme halophilic archaea) synthesize a modified membrane, the purple membrane, by inserting of a protein pigment, bacteriorhodopsin, close to the rhodopsin of the retina of the eye. The activation of this pigment by light allows the membrane energy production via the formation of a gradient of protons (*cf.* Sect. 3.3.4, Fig. 3.30).

Fig. 3.2 Some shapes and associations of prokaryotic cells (Drawing: M.-J. Bodiou)

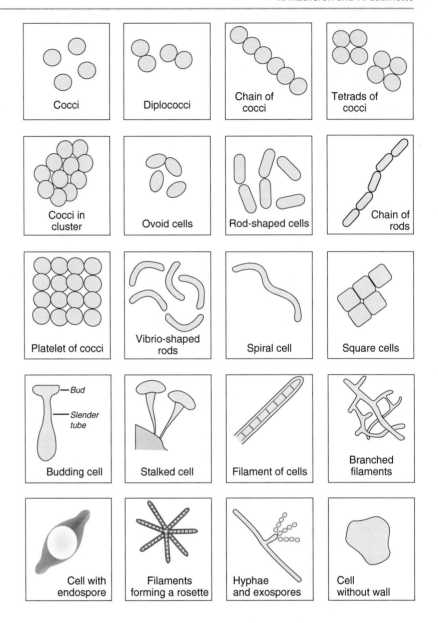

The membrane structure is a fluid mosaic in which phospholipids move quickly in their own layer and slowly from one layer to the other resulting in the displacement of other molecules. The membrane is a diffusion barrier that controls the exchanges of ions, solutes, metabolites, or macromolecules between the inside of the cell and its environment. The transport of substances through the membrane interacts with the cellular metabolism.

Simple diffusion transports concern small inorganic or organic molecules (O_2, CO_2, NH_3, H_2O, ethanol, etc.). They require a concentration gradient on both sides of the membrane and do not allow the intracellular accumulation of substances (passive diffusion). However, specific proteins may help the passage of solutes in the presence of a concentration gradient; it is the facilitated diffusion that does not involve energy and allows passage of molecules such as glycerol, fatty acids, aromatic acids, etc.

Active transports are used to transport substances when they are in low concentration in the environment of the cell. These transports, allowing accumulation of substances against concentration gradients, require energy. Several mechanisms are involved:

1. ABC transporters (ATP-binding cassette transporters) use the binding proteins located on the external face of the membrane that bind to the substance to be transported (mono- or disaccharides, amino acids, organic acids, nucleosides, some inorganic ions). These proteins then bind to an intramembranous specific carrier. Finally, energy from the hydrolysis of ATP releases the nutrient that is transferred into the cytoplasm.

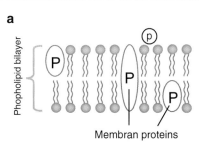

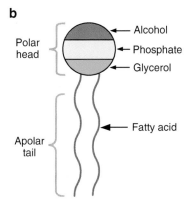

Fig. 3.3 The cytoplasmic membrane, general organization. (**a**) General scheme of the cytoplasmic membrane. (**b**) Schematic organization of a phospholipid (alcohols = glycerol, ethanolamine, etc.) (Drawing: M.-J. Bodiou)

2. Other transporters, secondary active transporters, use the energy of a proton or concentration gradient. Some are symport transporters for which penetration of the substance (sugars, amino acids) is generally associated with a positive gradient of protons or Na^+, the transfer of protons or Na^+ resulting in the passage of the active substance (Fig. 3.4). Other carriers are antiport transporters in which the transport of the substance is associated with a negative gradient (against gradient) of Na^+ or proton or of another substance, for example, the penetration of malate is associated with the output of lactate during the malolactic fermentation in lactobacilli.
3. During active transport, the transported molecules can be chemically modified. The best known mechanism is the transport of sugars (glucose, fructose, mannose) involving the phosphotransferase system which phosphorylates the sugars at the level of carrier and releases the phosphorylated sugar in the cytoplasm; for example, in the case of glucose transport, the phosphate group of glucose-6-phosphate originates from the hydrolysis of the phosphate bond of phosphoenolpyruvate.

In phototrophic purple bacteria, the cytoplasmic membrane, which is folded on itself, is at the origin of complex structures in the form of tubules, lamellae, folds, or vesicles according to bacterial groups (Fig. 3.27). These invaginations, observed also in nitrifying bacteria, increase the surface-to-cell volume ratio and thus promote exchange and energy activities.

Cell Wall

The cell wall allows the maintenance of cell shape; it is less complex in Gram-positive than in Gram-negative bacteria.

In bacteria staining Gram positive, the wall is made as a thick and rigid peptidoglycan or murein (Fig. 3.5b). Peptidoglycan which represents 90 % of weight of the wall is a very large polymer formed by the sequence of two derivatives of amino sugars, N-acetylglucosamine (AG) and N-acetylmuramic acid (AM). Residues of AM are substituted by tetrapeptides (TP). Polypeptide bridges (e.g., pentaglycine bridge) connect the tetrapeptides between them and thus form a rigid entanglement (Fig. 3.5a). Teichoic acids consisting of phosphoglycerol or ribitol phosphate polymers, which are sometimes associated with sugars and alanine, are also involved in wall rigidity.

In bacteria staining Gram negative, the wall structure is more complex (Fig. 3.5b). The peptidoglycan is thinner and represents only 10 % of the wall; there is no teichoic acid. On the external side, the wall is limited by a membrane, the outer membrane (Fig. 3.5c). The outer layer of this membrane contains chains of lipopolysaccharides (LPS) which extend outside the membrane forming side chains (antigen O). The LPS is the endotoxin of Gram-negative bacteria. The outer membrane is a diffusion barrier less effective than the cytoplasmic membrane. Proteins form channels (porins) for nonspecific passage of small molecules. The larger molecules are transported specifically through the membrane. The space between the outer membrane and cytoplasmic membrane is called the periplasmic space or periplasm. It contains many enzymatic proteins and may represent up to 40 % of bacterial volume (Fig. 3.5b). Some Gram-negative bacteria secrete into the environment outer membrane nanovesicles formed by the bacterial outer membrane and containing components of the periplasm and cytoplasm (enzymes and toxins).

In archaea, the cell wall has a different composition and is more variable. Some archaea have walls with a single thick layer formed of various polymers such as pseudomurein in which N-acetylgalactosamine replaces N-acetylglucosamine. Others have walls consisting of polysaccharides or heteropolysaccharides which are sulfated or non-sulfated. The wall of other archaea is formed of a layer or double layer of protein or glycoprotein subunits (Madigan et al. 2010).

Some bacteria and archaea are devoid of wall; in this case, the cytoplasmic membrane is the unique cellular envelope.

The prokaryotic microorganisms are frequently surrounded by a mucous layer called the glycocalyx. This layer is composed of polysaccharides which are often associated with polypeptides and that are secreted and accumulated around the cell. The glycocalyx may be a thin mucous layer called EPS ("exopolysaccharides" or "exopolymeric substances") or a thick layer more or less rigid, called the capsule. These highly hydrated exopolymers allow cells to agglomerate into biofilms (*cf.* Sect. 9.7.3), to

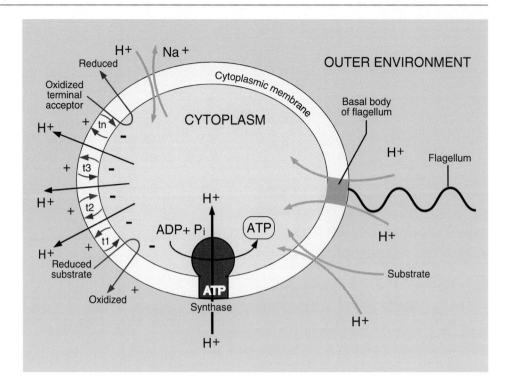

Fig. 3.4 Different modes of active transport and formation of a membrane proton gradient. t1 to tn: electron carriers (Drawing: M.-J. Bodiou)

adhere to surfaces, and to resist more or less effectively to environmental stresses. In some pathogenic bacteria, the capsule confers an additional pathogenicity, increasing the resistance to phagocytosis by the host organism.

Finally, some bacteria and archaea have a layer of proteins or glycoproteins on the outer surface of their walls, called S-layer that increases the protection of cells against environmental constraints.

3.1.1.2 Cytoplasm and Nucleoid

The cytoplasm is an aqueous solution of mineral salts and organic molecules necessary for the metabolic activity. It contains ribosomes, small particles of 15 nm in diameter, composed of proteins and ribosomal RNAs. Ribosomes are the site of the cellular protein synthesis and are composed of two subunits: the 50S subunit formed by proteins and 23S and 5S rRNAs and 30S subunit that consists of protein and 16S rRNA. The cytoplasm may contain inclusions of organic substances of reserves (glycogen, poly-β-hydroxyalkanoates, lipids) or inorganic compounds (polyphosphate, sulfur, iron) depending on the microbial type and metabolic activity. Other inclusions are more specific to some bacterial groups:

1. Cyanophycin granules, polypeptides constituting reserves of nitrogen in cyanobacteria.
2. Gas vacuoles for buoyancy of some prokaryotes.
3. Bacterial microstructures as carboxysomes, small particles containing ribulose 1,5-bisphosphate carboxylase necessary to the CO_2 fixation in some autotrophic prokaryotes.
4. Magnetosomes (*cf.* Sect. 14.5.2, Fig. 14.46) (containing magnetite). These structures provide magnetotactic bacteria the ability to move in a magnetic field.
5. Thylakoids, membrane structures forming flattened vesicles, location of the photosynthetic activity in cyanobacteria.
6. Chlorosomes, ovoid structures adjacent to the cytoplasmic membrane in phototrophic green bacteria, containing photosynthetic pigments (Fig. 3.27).

In some bacteria, protein filaments forming filamentous structures (cytoskeleton) in the cytoplasm under the cytoplasmic membrane may be involved in cell shape (Carballido-López 2006).

In a more or less central area of the cytoplasm, there is the nucleoid, consisting of a single double-stranded DNA molecule. It is the bacterial chromosome, usually single, circular, folded in on itself, and associated with proteins. Besides the chromosome, plasmids, small DNA molecules, can be present and transferred between cells in prokaryotes. Chromosome and extrachromosomal elements such as plasmids are part of the bacterial genome. Plasmids have few genes that are associated with specific properties of resistance (antibiotics, pollutants), virulence, or metabolism (biodegradation of xenobiotics). In bacteria belonging to the order *Planctomycetales*, an intracytoplasmic membrane divides the cytoplasm into two compartments, the peripheral paryphoplasm containing no ribosome and the central riboplasm containing ribosomes and the nucleoid (Fig. 14.32). In some of them, a

Fig. 3.5 The cell wall of bacteria. (**a**) Structure of peptidoglycan. (**b**) Cell wall structure of Gram-positive and Gram-negative bacteria. (**c**) Structure of the outer membrane of Gram-negative bacteria. (*1*) Pentaglycine bridges (Gram-positive bacteria) or direct bond between two amino acids of tetrapeptides (Gram-negative bacteria) (Drawing: M.-J. Bodiou)

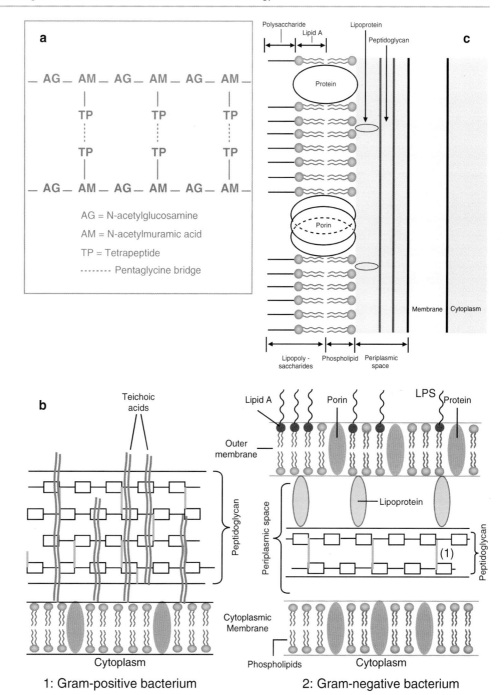

3.1.1.3 Spores and Cellular Appendages

nuclear envelope consisting of a double membrane surrounds the nucleoid inside the riboplasm, surprising discovery in prokaryotes, although this structure is not a true nucleus in the eukaryotic sense (Ward et al. 2006).

Some bacteria are able to produce endospores of resistance when their environmental conditions become unfavorable. The endospores are resistant structures to extreme conditions (temperature, UV, pH, redox potential, drought, etc.). They are a means of survival and can stay alive for very long periods estimated at thousands of years. The endospore contains genetic material and a very dehydrated cytoplasm that are protected by a complex system consisting of several resistant envelopes (Fig. 3.6a). Other structures of resistance exist in prokaryotes such as **akinetes*** of C*yanobacteria*, cysts of *Azotobacter* and *Myxobacteria*, or exospores of *Actinobacteria*.

Flagella are filamentous structures 15–20 nm in diameter in the form of rigid spirals, which are composed of proteins, principally flagellin (Fig. 3.6b). They are inserted on a basal body formed of protein rings included in the envelopes of the

without flagellum, by sliding on their support ("gliding bacteria," cf. Sect. 9.7.2).

Some pathogenic bacteria (*Escherichia coli*, *Pseudomonas aeruginosa*, *Erwinia* spp., etc.) possess secretion systems (e.g., type III or bacterial injectisome) in their envelopes, allowing them to inject cytoplasmic proteins in the cytosol of infected cells.

Intercellular nanotubes (diameter between 30 and 130 nm) have been described as forming conduits for exchanges of molecules between bacterial cells of the same species or different species growing on a solid surface (Dubey and Ben-Yehuda 2011).

Fimbriae or pili are present in bacteria. These are rigid filaments, thinner than flagella (3–7 nm in diameter), that play a role in cell adhesion to surfaces.

Finally, sex pili (10 nm in diameter) are produced by some bacteria to allow binding between bacteria and the formation of a cytoplasmic bridge used to transfer genes in conjugation between two bacterial cells. They provide the transfer of genetic material from one donor bacterium producing pilus to a receiver bacterium (*cf.* Chap. 12).

3.1.2 Eukaryotic Microorganisms (*cf.* Chaps. 5 and 7)

The cellular organization of eukaryotic microorganisms is much more complex than prokaryotic microorganisms. Eukaryotic microorganisms include photosynthetic and heterotrophic microorganisms that have important functional and morphological differences, but the basic cellular organization remains the same. Electronic micrograph of a cross section of eukaryote cell shows a central nucleus bounded by a nuclear membrane and surrounded by a cytoplasm containing structured membrane systems (Fig. 3.1b). The cytoplasm is surrounded by a membrane sometimes covered by a rigid wall in various taxa. The rigid wall is of chitinous, siliceous, or cellulosic nature. In some taxa, the cell is protected by a shell (theca) composed of protein, siliceous, or calcium substances. The phospholipid bilayer of the cytoplasmic membrane is asymmetrical. In the outer layer, glycolipids are inserted together with glycoproteins that serve as cellular receptors.

Sterols stabilize the membrane. In animal cells and many other taxa, the macromolecules and small particles penetrate through invaginations of the membrane (endocytosis) and are released into the cytoplasm as small vesicles or larger vacuoles (phagocytic vacuoles). The opposite (exocytosis) allows the release of cytoplasmic substances by fusion of intracellular vesicles with the cytoplasmic membrane. The nucleus is surrounded by a double phospholipid membrane (nuclear membrane) equipped with pores that allow exchanges with the cytoplasm. The interior of the nucleus

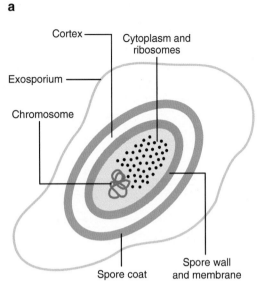

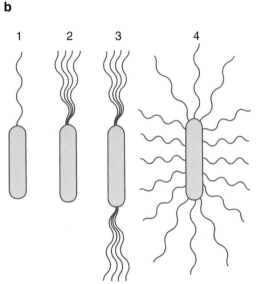

Fig. 3.6 Schematic representation of a bacterial endospore (**a**) and flagella and their location (**b**) (Drawing: M.-J. Bodiou)

cell. The activation of the basal body with membrane energy from the proton gradient allows rotation flagellum that drives the movement of the cell (Fig. 3.4). According to the bacteria, the implantation of flagella can be on one end (polar monotrichous or lophotrichous), at both ends (bipolar or amphitrichous), or on the whole bacterial cell (peritrichous). Spirochetes, spiral bacteria, do not possess a real flagellum but one axial filament resulting from the agglomeration of the flagella in the periplasm. Some bacteria move

Table 3.2 Characteristic organelles of eukaryotic cells

Organelles	Description	Function
Endoplasmic reticulum (ER)	Flattened cavities (limited by smooth membranes or membranes covered with ribosomes) and vesicles	Synthesis of glycoproteins, lipids, sterols; transport of various substances
Ribosomes	80 S particles formed of two subunits of 40S and 60S	Synthesis of proteins, forming polysomes if arranged in chains fixed to mRNA
Golgi apparatus (GA)	Pile of flattened sacs formed from vesicles, derived from ER	Finishing synthesis (proteins), packaging, and secreting of various cellular substances (lysosomes)
Lysosomes	Set of vesicles from the ER and GA	Containing hydrolytic enzymes, digestion of substrates
Vacuoles	Origin: ER and GA; large size in plant cells	Contractile vacuoles, digestive vacuoles; sequestration and storage of substances; hydric balance of plant cells
Peroxisomes	Vesicles formed from the ER	Containing redox enzymes
Proteasome	Enzymatic complex	Digestion of endogenous proteins; regulation of enzyme activity
Mitochondria	Rod about 2–3 μm long, double membrane with folded inner membrane; containing DNA of prokaryotic type	Respiratory center of the cell, respiratory transporters, and enzymes included in the inner membrane
Chloroplasts	Various lamellar structures (thylakoids); containing DNA of prokaryotic type	Containing the photosynthetic pigments, photosynthesis
Centrioles	Formed by the association of microtubules	Formation of the centrosome; participate in the formation of undulipodia
Centrosome	Formed from centrioles	Intracellular movements
Undulipodia	Cytoplasmic expansions containing microtubules inserted on basal granules (kinetosomes)	Cellular movements

is occupied by chromatin, entanglement of double-stranded DNA molecules associated with proteins, histones. The chromosomes become visible at the time of cell division. In the middle of the chromatin, usually a single nucleolus is the location of the synthesis of ribosomal RNA.

The soluble portion of cytoplasm or cytosol contains a set of microfilaments and microtubules (cytoskeleton), involved in maintaining cell shape, intracellular movements, cell locomotion by forming pseudopods, and cellular exchanges (endocytosis and exocytosis). Many inclusions (droplets of lipids, polysaccharide reserve) and many organelles are located in the cytosol (Table 3.2). Some organelles are formed by double-layer phospholipid membrane, delineating compartments of various shapes (vesicles, tubules, lamellae, etc.). Among these organelles, some are limited by a single membrane (endoplasmic reticulum, Golgi apparatus, lysosomes, peroxisomes), others by a double membrane (mitochondria and chloroplasts). There are also organelles composed of proteins: centriole, centrosome, ribosomes (RNAs and proteins), cilia, flagella (undulipodia), and proteasomes. Mitochondria and chloroplasts (plants), organelles of the same size as bacteria, have the particularity to contain DNA bacterial type (*cf.* Sects. 4.3.1 and 5.4.2). They multiply independently of the host cell. The amyloplasts of plants are specialized in synthesis and accumulation of starch. The presence of organelles and their importance in eukaryotic cells depends on the types of microorganisms and cellular activity.

Some eukaryotic microorganisms have a filamentous vegetative apparatus (Fungi). According to the taxa, these filaments are consisting of cells or a mass of multinucleated cytoplasm (syncytium). It may also be alternating between a single-cell generation and a filamentous generation (*cf.* Sect. 7.14.3).

3.2 Concept of Metabolism

The maintenance of cellular integrity, growth, and reproduction of living organisms requires the synthesis of cellular material, which depends on nutrients entering the interior of cells and subjected to a series of chemical changes. The sum of these changes is responsible for the production of energy (catabolism) and synthesis of biomolecules (anabolism); it is the metabolism of the cell (Fig. 3.7). Metabolic reactions are catalyzed by enzymes that, by combining with the biological molecules, decrease the activation energy of reactions and determine the reaction pathway to be used for the transformation of substances.

As sources of energy, the nutrient is oxidized, and the energy produced during the oxidation is transferred as energy-rich compounds, mostly in the form of ATP. These reactions of oxidation or degradation are the cellular catabolism:

1. In general, energy sources are organic compounds (**chemoorganotrophic microorganisms***).

   The oxidation of organic compounds can be partial or complete. When it is partial, the organic molecules of small molecular weight are produced, more oxidized than organic sources given as nutrients. The complete

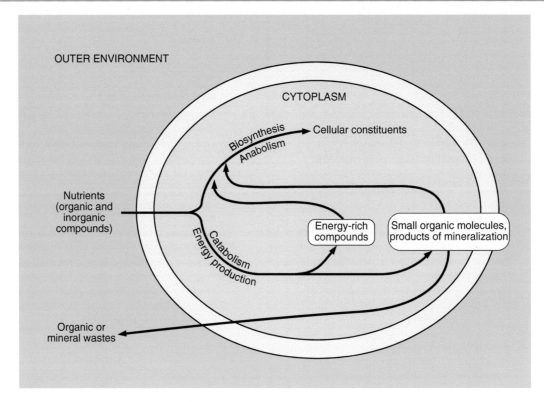

Fig. 3.7 Role of nutrients in cell metabolism (Drawing: M.-J. Bodiou)

oxidation results in formation of mineral compounds (CO_2, H_2O, NH_3, etc.). In this case, it is called a process of mineralization of organic sources.

Products of oxidation may remain inside the cell or be excreted out in the form of wastes. Thus, the chemoorganotrophic microorganisms are the decomposers of organic matter and ensure the progressive mineralization of organic matter, releasing mineral compounds in external environment (CO_2, NH_3, NO_3^-, PO_4^{3-}, SO_4^{2-}, HS^-).

2. The source of energy can be inorganic (**chemolithotrophic microorganisms***), consisting of reduced inorganic compounds such as dihydrogen, nitrogen or sulfur compounds, metals, etc.
3. The energy source can also be photonic (light) in the case of **photosynthetic*** or **phototrophic microorganisms*** which have pigments and photosynthetic systems able to react under the light action, thus converting light energy into chemical energy.

As sources of cellular constituents, the simple organic molecules produced in the cell or from the external environment are used as the basis of biosynthetic activities:

1. For most microorganisms, carbon organic compounds are required; these microorganisms are considered as **heterotrophs***. They can use of low-weight organic molecules produced in energy reactions of degradation or taken from the environment.
2. Other microorganisms use CO_2 as sole carbon source to synthesize all their cellular components; these are **autotrophic microorganisms*** using a mineral source of energy (chemolithotrophic microorganisms) or light (phototrophic microorganisms).

In summary, several nutritional types are defined in microorganisms:

1. The energy source is:
   - *Chemical*: chemotrophic microorganisms using an organic source (**chemoorganotrophs***) or an inorganic source (**chemolithotrophs***)
   - *Photonic*: phototrophic microorganisms using an organic compound (**photoorganotrophs***) or a mineral compound (**photolithotrophs***) as a source of electrons
2. The carbon source is a compound:
   - *Organic*: heterotrophs
   - *Mineral* (CO_2): autotrophs

Cellular metabolism is formed by means of the redox reactions in catabolism for energy production (energy metabolism) and the reactions of biosynthesis necessary for anabolism (production of cellular components) (Fig. 3.8). To be functional and viable, a cell should maintain its internal environment in a reduced state in order to ensure cohesion and structure of macromolecules. For this, it must constantly produce energy and **reducing power*** during chemical or

Fig. 3.8 General scheme of cellular metabolism (Drawing: M.-J. Bodiou)

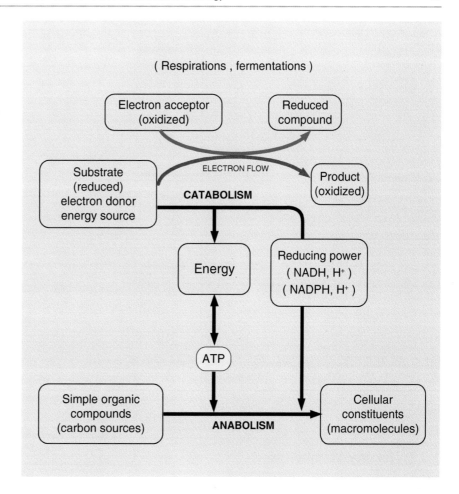

photonic redox reactions. Too important processes of internal super-oxidation (oxidative stress, *cf.* Sect. 9.6.2) cause cell death.

For some microorganisms, all the needs of biosynthesis are covered by a single carbon source; these microorganisms are described as **prototrophs***. Others require, in addition to the main carbon source, organic compounds that cannot be synthesized such as amino acids, fatty acids, vitamins, etc. These compounds are considered essential for microorganisms and are called growth factors; microorganisms are then defined as **auxotrophs*** for a given essential compound.

At the level of a community of microorganisms in an ecosystem, the cellular catabolism corresponds to the role of decomposers or mineralizers, and anabolism corresponds to the role of producers of biomass for the food chain. Photosynthetic production of organic material by photosynthetic eukaryotes and *Cyanobacteria* (oxygenic photosynthesis) from CO_2 and H_2O is called **primary production*** because these organisms use an external source of energy to terrestrial ecosystems (light) and inorganic compounds which are not necessarily derived from the metabolism of other microorganisms. Primary producers synthesize organic matter from which all other biological activities depend. Similarly, anoxygenic phototrophic bacteria that use for their photosynthesis reduced mineral electron donors (sulfur compounds, reduced iron, dihydrogen) from geochemical activities are also considered as primary producers. In contrast, phototrophic or chemolithotrophic microorganisms using compounds from the degradation of organic matter cannot be defined as primary producers and are therefore referred to as **paraprimary producers***. This is the case of phototrophic chemoorganotrophic and chemolithotrophic bacteria using electron donors derived from microbial metabolism and chemolithotrophic autotrophic microorganisms that respire dioxygen resulting from photosynthesis and thus are depending on the actual primary producers. In natural environments, the degradation of organic matter from primary or paraprimary producers requires the intervention of a succession of mineralizing microorganisms exchanging produced and usable substrates (*cf.* Chap. 13).

Some molecules, especially xenobiotic products, can be degraded or transformed only in the presence of easily usable substrates as sources of carbon and energy. This phenomenon is described as the **cometabolism***.

When the growth of microorganisms is limited by another factor that the sources of carbon and energy, for example, the source of nitrogen, sulfur, or phosphorus, growth slows or stops, but the energy production rate is unaffected. This

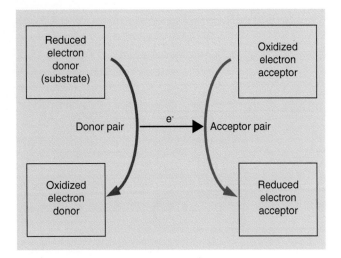

Fig. 3.9 Electron transfer between a donor and an acceptor in a redox reaction (Drawing: M.-J. Bodiou)

decoupling between growth and energy production is described as energy decoupling. The excess energy can be diverted to the production of glycogen and polyhydroxyalkanoates among microorganisms capable to accumulate these substances in their cells as carbon reserve.

3.3 Energy Metabolism

3.3.1 General Principles

In all cases (respiration, fermentation, photosynthesis), energy metabolism is based on redox processes between electron donor couple at low redox potential and electron acceptor couple at more high potential (Fig. 3.9).

During redox reactions that produce energy, the available energy for microorganisms appears mainly in the form of high-energy phosphate bonds, mainly in the form of adenosine triphosphate (ATP). ATP synthesis is an endergonic reaction that requires input of energy from the catabolism (potential phosphorylation: $\Delta G'p = +44$ kJ.mol^{-1}). The released energy ($\Delta G^{\circ\prime}$) during the hydrolysis of ATP in ADP, which is -32 kJ.mol^{-1}, is used for biosynthesis and other cellular functions. ATP is not the only energy-rich molecule. Other molecules can be used such as phosphoenolpyruvate (PEP, $\Delta G^{\circ\prime} = -52$ kJ.mol^{-1}), acetyl phosphate ($\Delta G^{\circ\prime} = -45$ kJ.mol^{-1}), or acetyl-CoA ($\Delta G^{\circ\prime} = -36$ kJ.mol^{-1}).

In microorganisms, there are two types of mechanism for the synthesis of ATP, the **substrate-level phosphorylation*** and the phosphorylation during electron transfer by a carrier chain (**oxidative phosphorylation*** and **photophosphorylation***); these mechanisms involving the redox reactions of catabolism are associated often to transfer protons during dehydrogenation:

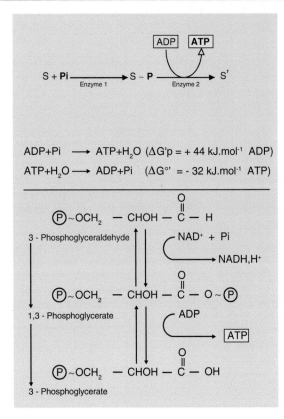

Fig. 3.10 Substrate-level phosphorylation. *Top*: Synthesis of ATP by phosphorylation of ADP and energy value of the ADP phosphorylation and dephosphorylation of ATP. *Bottom*: Example of substrate-level phosphorylation, phosphorylation glyceraldehyde in glycolysis. *Pi* inorganic phosphate, ~ energy-rich bond, *P* phosphate group (Drawing: M.-J. Bodiou)

1. Phosphorylation at the substrate level. The synthesis of ATP is directly coupled to the enzymatic oxidation of an organic substance. During the oxidation reaction, the organic substrate S is phosphorylated using an enzyme 1, and a phosphate ester with a high-energy bond (P ~ S) is produced. The high-energy phosphate is then transferred by means of an enzyme 2 to ADP to form ATP with an additional high-energy phosphate bond and releases a product S′ (Fig. 3.10). The enzymes involved in the phosphorylation at the substrate level are present in the cytoplasm and are soluble. The phosphorylation at the substrate level is the main mechanism of ATP production in fermentations.

2. Phosphorylation during electron transfer by a chain of membrane carriers. This is the mode of formation of ATP in the respiration (oxidative phosphorylation) and in the photosynthesis (photophosphorylation). During the oxidation–reduction reactions, electrons are transferred by a series of carriers from an initial electron donor with a low redox potential to a terminal electron acceptor at higher redox potential. Each carrier is characterized by a couple of redox (redox potential) between the oxidized and reduced forms (Fig. 3.11). During this transfer, electrons lose energy

Fig. 3.11 General scheme of a chain of membrane electron carriers (Drawing: M.-J. Bodiou)

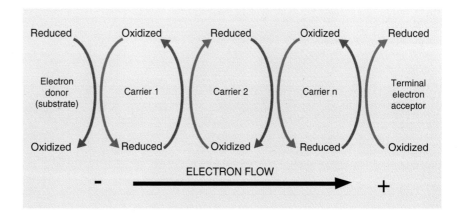

Table 3.3 Different types of respirations

Respiration in microorganisms	Electron donors	Final electron acceptors	Microorganisms
Aerobic chemoorganotrophs	Organic	Dioxygen	Prokaryotic and eukaryotic
Anaerobic chemoorganotrophs	Organic	Nitrate, sulfate, ferric iron, fumarate, etc.	Prokaryotic (sometimes eukaryotic)
Aerobic chemolithotrophs	Inorganic (ammonium, sulfide, ferrous iron, etc.)	Dioxygen	Prokaryotic
Anaerobic chemolithotrophs	Inorganic (ammonium, sulfide, hydrogen, etc.)	Nitrate, sulfate, etc.	Prokaryotic

which is recovered to synthesize ATP by phosphorylation of ADP. This synthesis is catalyzed by ATP synthases, **insoluble membrane enzymes***. During respiration, electron transfer between carriers takes place in the cytoplasmic membrane among prokaryotic microorganisms and in the internal mitochondrial membrane among eukaryotic microorganisms; the resulting ATP production is obtained by oxidative phosphorylation. For photosynthesis, electron transfer occurs in specialized membranes (thylakoids in cyanobacteria and chloroplasts in photosynthetic eukaryotes or photosynthetic cytoplasmic membrane in anoxygenic phototrophic bacteria), and ATP is produced by photophosphorylation. Almost all of enzymes associated with electron transport are included in the membrane and are insoluble. The transfer of electrons and protons generates a proton-motive force used as energy source or for the phosphorylation of ADP (Fig. 3.4).

3.3.2 Respirations in Microorganisms

The respiratory metabolism is more diverse in prokaryotic microorganisms than in eukaryotic microorganisms:
1. The energy source (electron donor) is necessarily organic in eukaryotes, while it may be organic or inorganic in prokaryotes (chemoorganotrophs or chemolithotrophs).
2. The terminal electron acceptor is usually dioxygen in eukaryotes, while in prokaryotes there are a variety of terminal electron acceptors in addition to dioxygen, such as nitrate, sulfate, iron, or manganese, thus defining a variety of anaerobic respirations.
3. The composition of respiratory chain (electron carriers) is variable in prokaryotes, while it is fairly constant in eukaryotes.

Thus, four types of respirations are known in microorganisms. Their characteristics are presented in Table 3.3.

Chemoorganotrophic microorganisms are generally heterotrophs, but some heterotrophs can use dihydrogen as donor of electrons. Chemolithotrophic microorganisms are mostly autotrophs, but some may use organic compounds as carbon sources. The microorganisms that use a mineral energy source and an organic carbon source are called **mixotrophic microorganisms***.

3.3.2.1 Aerobic Respiration in Chemoorganotrophic Microorganisms

Electrons or reducing equivalents (symbolized e^- or [H]) derived from the oxidation of a compound organic (electron donor or substrate) are transferred to dioxygen (terminal electron acceptor) by a chain of intermediate carriers (respiratory

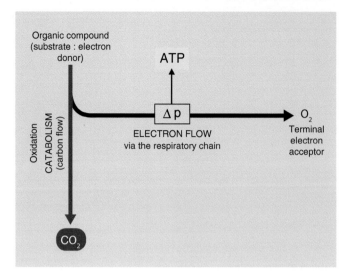

Fig. 3.12 Carbon flux and electron flow in aerobic respiration. Δp = Proton-motive force (Drawing: M.-J. Bodiou)

chain). During this transfer, energy is maintained particularly in the form of ATP produced by oxidative phosphorylation. The **reducing power*** or reducing equivalent sum that is produced during the oxidation of organic substrates is essentially reserved for energy production (Fig. 3.12). However, the reducing power operates in all cases where a reduction process is necessary: enzymatic reactions, biosynthesis, etc.

The oxidation reactions that take place in several steps are often reactions to dehydrogenation involving electrons and protons. The resulting reducing power is transferred to coenzymes (NAD^+, $NADP^+$, and FAD). There are many metabolic pathways of oxidation. For carbohydrates, the main pathway is glycolysis or the Embden–Meyerhof pathway associated with the tricarboxylic acid cycle (TAC), also called citric acid cycle, organic acid cycle to four carbon atoms (C4 cycle: succinic, fumaric, and malic acids), or the Krebs cycle.

Carbohydrate Oxidation

During glycolysis (Fig. 3.13a), glucose is oxidized to pyruvate in several enzymatic steps with concomitant formation of ATP by phosphorylation at the substrate level and reduced coenzymes:

$$Glucose + 2ADP + 2\,Pi + 2\,NAD^+$$
$$\rightarrow 2\;pyruvate + 2\;ATP + 2NADH, H^+$$

Pyruvate is then decarboxylated to acetyl-CoA and releases a molecule of CO_2 and reducing power ($NADH, H^+$):

$$2\;pyruvate + 2\;CoA + 2\;NAD^+$$
$$\rightarrow 2\;acetyl - CoA + 2\;CO_2 + 2\;NADH, H^+$$

The two molecules of acetyl-CoA are then oxidized via the Krebs cycle (Fig. 3.13b), by combining the acetyl group to a molecule of oxaloacetate to form a compound to six carbon atoms (citrate) that regenerates C4 compound (oxaloacetate) via a series of oxidation, decarboxylation, dehydration, and hydration reactions. During one cycle, the acetyl-CoA is oxidized to two molecules of CO_2 and allows the production of one molecule of ATP by phosphorylation at the substrate level, two molecules of $NADH, H^+$, one of $NADP\,H, H^+$, and one of $FADH_2$.

Thus, the total oxidation of glucose by glycolysis (Embden–Meyerhof pathway) and the Krebs cycle (Fig. 3.13c) produces six molecules of CO_2, four of ATP by phosphorylation at the substrate level, and 24 reducing equivalents [H] in the form of eight $NADH, H^+$, 2 $NADP\,H, H^+$, and two $FADH_2$. In prokaryotic microorganisms, glycolysis and the Krebs cycle take place in the cytoplasm, whereas in eukaryotes, glycolysis occurs in the cytoplasm and the Krebs cycle in the matrix of mitochondria.

Glucose can be degraded by other metabolic pathways such as the pentose phosphate methylglyoxal, phosphoketolase, and Entner–Doudoroff pathways.

Lipid Oxidation

Under the action of lipase, triglycerides are hydrolyzed to glycerol and fatty acids (Fig. 3.14a).

Glycerol enters into the Embden–Meyerhof pathway via the phosphoglycerate (Fig. 3.14b). The fatty acids are degraded by β-oxidation (Fig. 3.14c). During the β-oxidation, the fatty acid is esterified to acyl-CoA in the presence of coenzyme A. The acyl-CoA is then oxidized in position β in three steps (two dehydrogenations and one hydration). A new esterification in β releases a molecule of acetyl-CoA and a molecule of acyl-CoA that has lost two carbon atoms. A new round of β-oxidation begins. Finally, the fatty acid is cut into a succession of acetyl-CoA. These are metabolized through the glyoxylate cycle (Fig. 3.40).

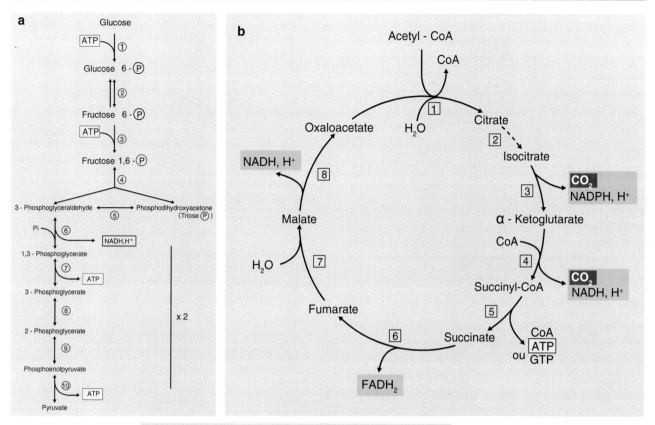
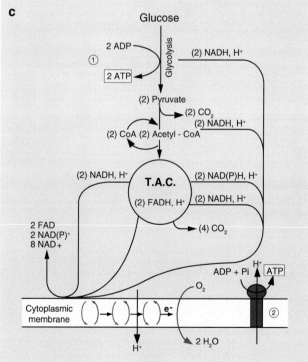

Fig. 3.13 Catabolism of glucose. (**a**) Glycolysis, a degradation pathway of glucose. *1* Glucokinase, *2* glucomutase, *3* phosphofructokinase, *4* aldolase, *5* isomerase, *6* 3-phosphoglyceraldehyde dehydrogenase, *7* 3-phosphoglycerate kinase, *8* phosphoglycerate mutase, *9* enolase, and *10* pyruvate kinase. *Pi* inorganic phosphate, *triose phosphates* 3-phosphoglyceraldehyde, and phosphodihydroxyacétone. (**b**) The Krebs cycle (tricarboxylic acid cycle, TAC). *1* Citrate synthase, *2* aconitate hydratase, *3* isocitrate dehydrogenase, *4* ketoglutarate dehydrogenase, *5* succinate thiokinase, *6* succinate dehydrogenase, *7* fumarase, and *8* malate dehydrogenase. (**c**) Scheme of glucose catabolism: glycolysis and Krebs cycle and electron transfer through the membrane respiratory chain. *1* Substrate-level phosphorylation, *2* oxidative phosphorylation via proton pores and ATP synthases (Drawing: M.-J. Bodiou)

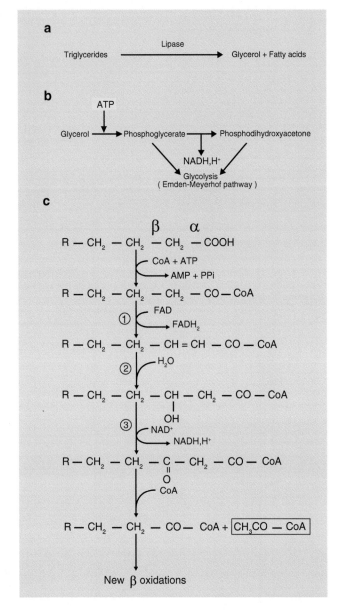

Fig. 3.14 Oxidation of lipids. (**a**) Enzymatic oxidation of triglycerides. (**b**) Incorporation of glycerol in glycolysis. (**c**) Fatty acid degradation by β-oxidation. *1* and *3* dehydrogenation reactions, *2* hydration reaction, *PPi* pyrophosphate (Drawing: M.-J. Bodiou)

With peer fatty acids, the last β-oxidation releases two acetyl-CoA, with odd fatty acids, one acetyl-CoA and one propionyl-CoA.

Protein Oxidation

Proteins are hydrolyzed by proteases to amino acids. Amino acids are then deaminated to organic acids that are oxidized via the Krebs cycle (Table 3.4).

Energy Synthesis

The synthesis of the energy is mainly produced at the level of the respiratory chain that catalyzes the oxidation of electron donors

Table 3.4 Organic acids resulting from deamination of amino acids

Amino acids	Metabolic intermediates
Alanine, glycine, cysteine, serine, threonine	Pyruvate
Asparagine, aspartate	Oxaloacetate
Tyrosine, phenylalanine, aspartate	Fumarate
Isoleucine, methionine, threonine, valine	Succinate
Glutamate, glutamine, histidine, proline, arginine	α-Ketoglutarate
Isoleucine, leucine, tryptophan, lysine, phenylalanine, tyrosine	Acetyl-CoA

by transferring electrons to dioxygen (Fig. 3.15). The respiratory chain is located in the cytoplasmic membrane in prokaryotes and in the inner membrane of the mitochondria in eukaryotes. The respiratory chain consists of two types of carriers: electron and proton carriers (flavoproteins and quinones) and electron carriers only (protein Fe/S and cytochromes).

The respiratory chain receives electrons from reduced coenzymes (NADH, H^+, NADP H, H^+, and $FADH_2$) but also from some organic molecules (lactate, succinate, etc.). The reduced coenzymes are reoxidized by dehydrogenases, an enzymatic complex of the chain formed by the flavoproteins and Fe/S proteins. The electrons then pass through the quinones or ubiquinones, followed by a more or less complex pathway depending on the microorganisms, which comprises cytochromes, and finally by the cytochrome oxidase that transfers electrons to dioxygen which is reduced to H_2O. During this transport, protons are expelled ("proton translocation") outside of the cytoplasmic membrane in prokaryotes or in the intermembrane space of mitochondria in eukaryotes. The translocation of protons is due to the alternation of the two types of carriers in the respiratory chain and the role of proton pumps attributed to dehydrogenases and cytochromes (often cytochrome oxidases). The membranes are impermeable to protons; consequently, protons cannot return naturally in the cytoplasm, while electrons can return to the inner face of the membrane. It results in a charge distribution on both sides of the membrane causing a double gradient, pH and electric charge gradients, constituting the proton-motive force Δp (chemiosmotic theory of Mitchell) which is the main source of energy among aerobic chemoorganotrophic heterotrophic microorganisms. This energy will be used to produce ATP, the active transport of nutrients, cell movements, etc. The synthesis of ATP is due to return protons through pores associated with proton transmembrane enzyme complexes, the ATP synthases, which catalyze the phosphorylation of ADP to ATP.

If the organization of the mitochondrial respiratory chain is relatively constant, structure changes occur in aerobic prokaryotes mainly at the level of prokaryotic cytochromes and with environmental conditions (Fig. 3.16).

Differences in redox potential between electron donors and the final acceptor (dioxygen) define the quantity of free energy generated by electron transfer. For example, the calculated

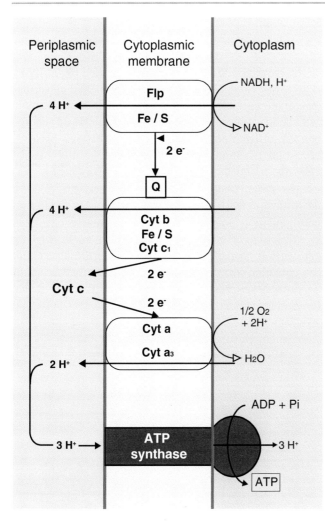

energy produced per mole of NADH is -220 kJ. The potential phosphorylation of ADP to ATP ($\Delta Gp'$) is 44 kJ.mol^{-1} ATP. Potentially, five moles of ATP could be synthesized in the transfer of two electrons to dioxygen from a mole of NADH, H$^+$. In fact according to the microorganisms, the actual energy yields are lower and, for example, can vary in general from 24 to 36 ATP per mole of glucose.

In general, the organic electron donor is also the source of carbon among chemoorganotrophic microorganisms that are also heterotrophs. In the case of *Escherichia coli*, about 50 % of the substrate is used for energy production, the rest acting as carbon source for biosynthesis; biosynthesis consumes the majority of produced cellular energy.

3.3.2.2 Aerobic Respiration in Chemolithotrophic Microorganisms

Chemolithotrophic microorganisms are all prokaryotes (*Bacteria* and *Archaea*) belonging to relatively small groups. They use oxidation of reduced inorganic compounds to produce energy by transferring electrons to the dioxygen via a membrane respiratory chain in the same way as that of chemoorganotrophic microorganisms (Figs. 3.17 and 3.18a–d).

Some bacteria are obligate chemolithotrophs, and others are facultative chemolithotrophs (can also use organic compounds). Many of chemolithotrophic bacteria use CO_2 as a carbon source and are chemolithoautotrophs, but some chemolithotrophs are heterotrophs for source of carbon and considered as chemolithoheterotrophs and are therefore mixotrophic microorganisms.

Main Electron Donors in Chemolithotrophic Microorganisms

The main reduced mineral compounds used as electron donors for chemolithotrophic prokaryotes are compounds of nitrogen and sulfur, iron, dihydrogen, and carbon monoxide.

Fig. 3.15 Respiratory chain of aerobic chemoorganotrophic bacteria (*Paracoccus denitrificans*). *Flp* flavoprotein, *Fe/S* iron–sulfur protein, *Q* quinone, *Cyt* cytochrome, *complex [Flp-Fe/S]* NADH dehydrogenase, *complex [Q-Cyt b, Fe/S, Cyt c1]* catalyzes the recycling of electrons (quinone cycle), *Cyt c* cytochrome c oxidoreductase, *complex [Cyt a, Cyt a3]* cytochrome oxidase (Drawing: M.-J. Bodiou)

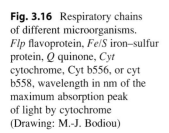

Fig. 3.16 Respiratory chains of different microorganisms. *Flp* flavoprotein, *Fe/S* iron–sulfur protein, *Q* quinone, *Cyt* cytochrome, Cyt b556, or cyt b558, wavelength in nm of the maximum absorption peak of light by cytochrome (Drawing: M.-J. Bodiou)

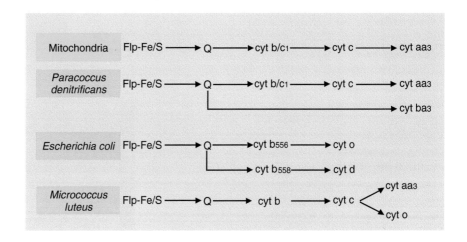

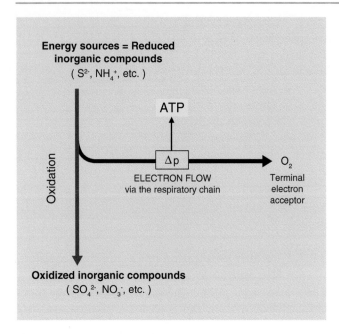

Fig. 3.17 General scheme of electron transfer in chemolithotrophic prokaryotes (Drawing: M.-J. Bodiou)

They can originate from aerobic or anaerobic degradation of organic matter or geochemical and biogeochemical processes (*cf.* Chap. 14).

The reduced nitrogen compounds are used by nitrifying bacteria that form a heterogeneous group including various genera and species all highly specialized in nitrification. Two large groups can be distinguished based on the reduced nitrogen compounds used by nitrifying bacteria: the group of ammonia-oxidizing bacteria that oxidize ammonia to nitrite (*Nitrosomonas*, *Nitrosococcus*, etc.) and the group of nitrite-oxidizing bacteria that oxidize nitrite to nitrate (*Nitrobacter*, *Nitrospira*, etc.). Most of these microorganisms are chemolithoautotrophs with the exception of some mixotrophs such as *Nitrobacter* that can use acetate as carbon source. In the absence of dioxygen or in dioxygen-limiting conditions, *Nitrosomonas europaea* can express a denitrifying activity. It was shown that ammonia-oxidizing archaea were abundant in natural environments.

Reduced sulfur compounds, mainly sulfide (S^{2-} or HS^- or H_2S), elemental sulfur ($S°$), thiosulfate ($S_2O_3^{2-}$),

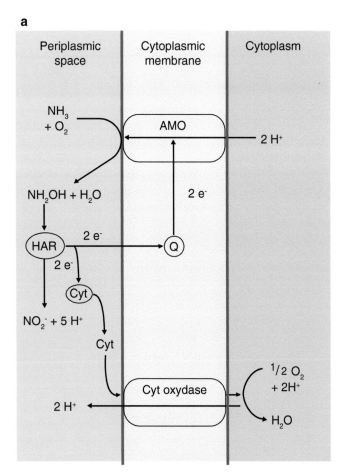
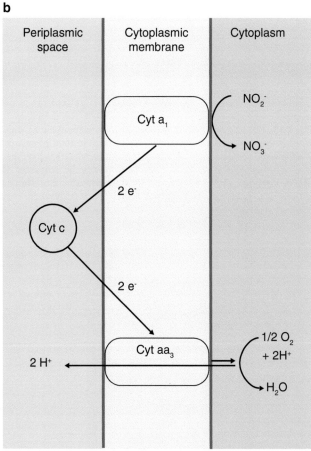

Fig. 3.18 Schemes of the respiratory chains of chemolithotrophic bacteria: ammonia-oxidizing bacteria, nitrite-oxidizing bacteria, sulfur-oxidizing bacteria, and iron-oxidizing bacteria. (**a**) Respiratory chain of *Nitrosomonas* (Modified and redrawn from Hooper et al. 1997). *AMO* ammonia monooxygenase, *NH₂OH* hydroxylamine, and *HAR* hydroxylamine oxidoreductase. (Note: Energy is necessary for the first stage, which is provided by the return of two electrons to AMO). (**b**) Respiratory chain of *Nitrobacter*.

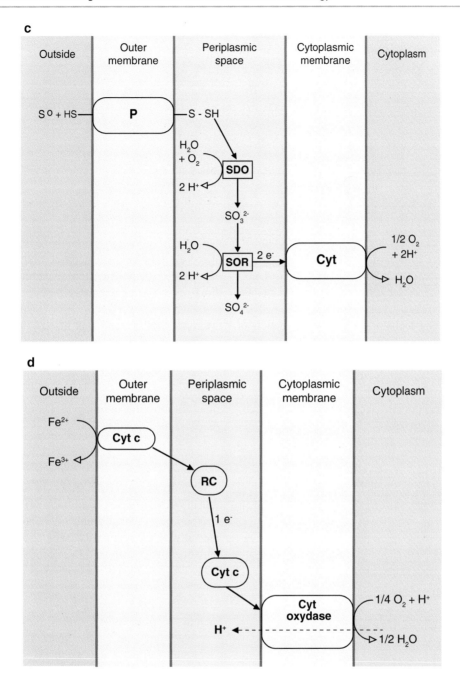

Fig. 3.18 (continued) The values of the redox potentials do not allow the transfer of electrons between the cytochrome a1 ($Eo' = + 0.35$ V) and c ($Eo' = + 0.27$ V). However, high oxidase activity of cytochrome aa3 maintains cytochrome c in a highly oxidized state that renders possible the transfer. (**c**) Respiratory chain of *Acidithiobacillus* (Modified and redrawn from Rohwerder and Sand 2003). The sulfur atom (S °) is mobilized in the form of persulfide by reaction with a thiol group of a protein (P) of the periplasmic membrane. Only two of six electrons involved are transferred by the respiratory chain to dioxygen. *SDO* sulfur dioxygenase, *SOR* sulfite oxidoreductase, and *Cyt* cytochrome(s). (**d**) Respiratory chain of *Acidithiobacillus ferrooxidans* (Modified and redrawn from Valdés et al. 2008). Cytochrome oxidase: cytochrome oxidase, *Cyt c* cytochrome c, and *RC* rusticyanin (copper protein) (Drawing: M.-J. Bodiou)

tetrathionate ($S_4O_6^{2-}$), and sulfite (SO_3^{2-}), are used by aerobic sulfur-oxidizing bacteria (or colorless sulfur-oxidizing bacteria) described as colorless compared to the phototrophic purple and green sulfur-oxidizing bacteria. Among aerobic sulfur-oxidizing bacteria, some are obligate chemolithotrophs, others are facultative chemolithotrophs, others are autotrophs, others are heterotrophs, and even some are obligate chemoorganotrophic heterotrophs (Table 3.5). Two major

Table 3.5 Different metabolisms of colorless sulfur bacteria and archaea

Trophic types	Energy source		Carbon sources		Microorganisms[a]	Growth pH	
	Inorganic	Organic	Inorganic	Organic		Neutral	Acidic
Obligatory autotrophic chemolithotrophs	X		X		*Thiobacillus*[b], *Acidithiobacillus*[b], *Acidianus*[c], *Sulfolobus*[c], *Hydrogenobacter*[b], *Thiomicrospira*[b]	X	X
Facultative autotrophic chemolithotrophs	X	X	X	X	*Thiobacillus*[b], *Sulfolobus*[c], *Acidianus*[c], *Thiosphaera*[b] *Paracoccus*[b], *Metallosphaera*[c], *Beggiatoa*[b]	X	X
Chemolithoheterotrophs	X	X		X	*Thiobacillus*[b], *Beggiatoa*[b]	X	
Chemoorganoheterotrophs		X		X	*Beggiatoa*[b], *Macromonas*[b], *Thiobacterium*[b], *Thiothrix*[b]	X	

[a]At least one species of the mentioned genus is concerned with the physiological characteristics
[b]Bacteria
[c]Achaea

groups are distinguished morphologically: the group of unicellular bacteria (*Thiobacillus*, *Thiospira*, etc.) and the group of filamentous bacteria (*Beggiatoa*, *Thiothrix*, etc.). Most bacteria are neutrophiles and some are extreme acidophiles (*Acidithiobacillus*). *Sulfolobus* is one of thermophilic and acidophilic sulfur-oxidizing archaea. During the oxidation of sulfur compounds, sulfur-oxidizing bacteria form globules of sulfur accumulated outside (*Thiobacillus*, *Acidithiobacillus*) or inside the cells (*Beggiatoa*, *Sulfolobus*, etc.). *Acidithiobacillus ferrooxidans* can also use iron (Fe^{2+}) as electron donor in extreme acidophilic conditions. The accumulated sulfur by aerobic sulfur-oxidizing bacteria can form large deposits.

The reduced iron or ferrous iron (Fe^{2+}) is abundant in nature, but only few microorganisms are capable to use mainly because of its low solubility in water and because of its rapid oxidation by the dioxygen at neutral pH. It is used by aerobic bacteria as an electron donor and is oxidized to ferric iron (Fe^{3+}) via a respiratory chain which transfers electrons to dioxygen (Fig. 3.18d). At neutral pH, few iron bacteria are able to use ferrous iron (*Gallionella*, *Mariprofundus*) and are obligate chemolithotrophic autotrophs. At acidic pH, ferrous iron is more soluble and chemically stable. It is a source of usable energy for acidophilic iron-oxidizing bacteria such as *Acidithiobacillus ferrooxidans*, *Leptospirillum ferrooxidans* which are obligate chemolithoautotrophs, or *Acidimicrobium* which is facultative chemolithoautotroph. These bacteria live at a pH between pH 1 and pH 3. The thermoacidophilic archaeal *Sulfolobus* is also able to use ferrous iron.

Aerobic iron-oxidizing bacteria and aerobic sulfur-oxidizing bacteria are restricted to aerobic interfaces between the **oxic*** and **anoxic zones*** because the used reduced compounds are generally spontaneously oxidized in the presence of dioxygen and can only be maintained in the reduced state in anoxic areas.

Biotic or abiotic (hydrothermal) dihydrogen is used as a donor of electrons by a large number of aerobic or microaerophilic microorganisms which have a membrane hydrogenase acting in oxidation of dihydrogen and release of protons. The protons excreted contribute to the formation of a proton gradient, and electrons are transferred to dioxygen via a membrane respiratory chain. The hydrogenase is sensitive to dioxygen; also many bacteria are **microaerophiles*** and live at only low dioxygen tension (1–5 %). These microorganisms are facultative chemolithoautotrophic bacteria belonging to various genera such as *Pseudomonas*, *Alcaligenes*, *Nocardia*, *Gordonia*, *Hydrogenophaga*, etc. Some of them are obligate chemolithotrophic autotrophs (*Hydrogenovibrio*, *Hydrogenothermus*, *Aquifex*):

$$H_2 \rightarrow 2\,H^+ + 2\,e^-$$

Carbon monoxide (CO) present in large amounts in habitats rich in organic matter and low in O_2 (e.g., rice paddies) is metabolized by a small number of microorganisms which also generally use dihydrogen as donor electrons and possess a CO dehydrogenase transferring the electrons from the oxidation of CO to CO_2 to a membrane respiratory chain. These are facultative chemolithotrophic bacteria such as *Oligotropha carboxidovorans*, *Pseudomonas carboxydohydrogena*, *Mycobacterium* spp., or some *Bacillus* (*Bacillus schlegelii*).

Other compounds may serve as electron donors for chemolithotrophic microorganisms in order to produce energy, such as the reduced compounds of copper, manganese, antimony, selenium, or arsenic. These compounds are often toxic, and bacteria that use them must be able to resist to their toxicity.

Energy Production in Chemolithotrophic Microorganisms

The electrons from the oxidation of reduced mineral compounds are transferred to dioxygen via a respiratory chain located in the cytoplasmic membrane. The redox potential of these mineral compounds is generally too high; the respiratory chain described previously in chemoorganotrophic microorganisms is not used in its entirety but only

Table 3.6 Redox potentials of redox couples and $\Delta G^{\circ\prime}$ of the oxidation reactions of various energy sources utilized by chemolithotrophic microorganisms

Redox pair	E_0' in volts	$\Delta G^{\circ\prime}$ in kJ/2 electrons
CO_2/CO	−0.52	−258.6
H^+/H_2	−0.42	−239.3
SO_4^{2-}/HS^-	−0.22	−200.7
NO_2^-/NH_4^+	+0.34	−92.6
NO_3^-/NO_2^-	+0.43	−75.3
Fe^{3+}/Fe^{2+}	+0.77	−9.6[a]
O_2/H_2O	+0.82	
$NAD^+/NADH$	−0.32	−219.9

[a] $\Delta G^{\circ\prime}$ to pH 7; ΔG° is −65.8 kJ/2e$^-$ at physiological pH of *Acidithiobacillus ferrooxidans* (pH 2)

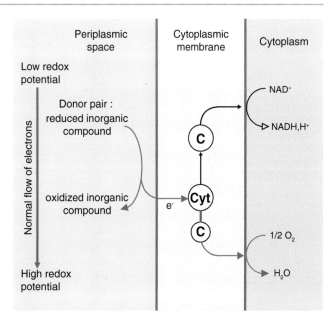

Fig. 3.19 General scheme of reverse electron flow in chemolithotrophic bacteria. C = electron carriers. The reverse flow corresponds to the electron transport from cytochromes (cyt) to NAD$^+$ via membrane carriers (C) (Drawing: M.-J. Bodiou)

in its terminal part (cytochromes, Fig. 3.18a–d). Thus, the potential difference between the donor couple (reduced mineral compound) and acceptor couple (O_2/H_2O) (Table 3.6) is relatively low, so the amount of energy will always be low except for paths using dihydrogen or CO as electron donors. As a result, the chemolithotrophic bacteria oxidize a large amount of reduced compounds (electron donors) for a relatively low energy and therefore biomass. In the case of sulfur-oxidizing bacteria and nitrifying bacteria, this great oxidation activity can lead a significant production of acid compounds (H_2SO_4, HNO_3) which, in environments poorly buffered, are responsible for the phenomena of corrosion.

The translocation of protons resulting from activity of the respiratory chain leads to a proton-motive force that can be used as energy or allow the synthesis of ATP by oxidative phosphorylation. In most chemolithotrophic bacteria, ATP production is exclusively obtained by oxidative phosphorylation, with the exception of sulfur-oxidizing bacteria in which a small amount of ATP is produced by substrate-level phosphorylation in the oxidation of sulfite to sulfate.

In chemolithotrophic autotrophic microorganisms, the reduction of CO_2 into organic compounds used for biosynthesis requires reducing power in the form of large amounts of reduced coenzymes (NADH, H^+ or NADPH, H^+). The electrons required for reduction of these coenzymes are derived from the oxidation of electron donor (reduced inorganic compound). However, apart from the case of dihydrogen and CO, the redox potentials of redox couples of electron donors (energy sources) are higher than that of redox couple of $NAD^+/NADH$, H^+, so electrons cannot be transferred spontaneously to NAD$^+$. The result is a transfer by an **electron reverse flow*** through a portion of the respiratory chain that consumes energy (Fig. 3.19). The role of energy is to place the electrons in an energy level sufficient to reduce the coenzymes. Thus, the small amount of energy produced by respiration in chemolithotrophic bacteria is largely used to produce reducing power (via the intermediary of the reverse electron flow) required for biosynthesis.

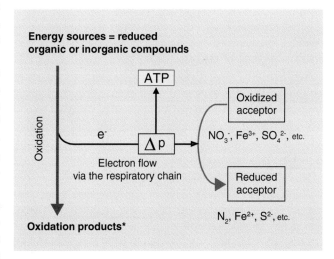

Fig. 3.20 General scheme of anaerobic respirations. Oxidation of organic energy sources can be complete (CO_2) or partial (Drawing: M.-J. Bodiou)

3.3.2.3 Anaerobic Respirations

In anoxic environments and in the absence of light, microorganisms can use two systems that produce energy, fermentations and anaerobic respirations. During anaerobic respirations, mechanisms of energy conservation are very similar to aerobic respiration (Fig. 3.20). However, during these respirations, the electrons from the oxidation of electron donor (substrate) are not transferred through the respiratory chain to dioxygen but to other oxidized compounds which act as terminal electron acceptor (Table 3.7). These inorganic (nitrate, sulfate, ferric iron, CO_2, chlorate, etc.) or sometimes organic

Table 3.7 Anaerobic respirations in prokaryotes and eukaryotes

Main anaerobic respirations[a] of prokaryotes[b]	Terminal electron acceptors
Iron respiration (b)	$Fe^{3+} + e^- \rightarrow Fe^{2+}$
Nitrate respiration (e.g., denitrification) (b, a)	$NO_3^- + 5\ e^- \rightarrow N_2$
Sulfate respiration (sulfate reducing) (b, a)	$SO_4^{2-} + 8\ e^- \rightarrow S^{2-}$
Sulfur respiration (sulfur reducing) (b, a)	$S° + 2\ e^- \rightarrow S^{2-}$
CO_2 respiration (acetogenesis) (b)	$CO_2 + 2\ e^- \rightarrow$ acetate
CO_2 respiration (methanogenesis) (a)	$CO_2 + 4\ e^- \rightarrow CH_4$
Fumarate respiration (b)	Fumarate $+ 2\ e^- \rightarrow$ succinate
Anaerobic respirations of eukaryotes	Main microorganisms
Fumarate respiration	*Euglena*, *Leishmania*
Nitrate respiration	*Fusarium*, *Cylindrocarpon* (Fungi)

[a]*Bacteria* (b), *Archaea* (a)
[b]Other electron acceptors for anaerobic respirations in prokaryotes: manganese oxide (MnO_2), arsenate (AsO_4^{3-}), selenate (SeO_4^{2-}), chromate (CrO_4^{2-}), vanadium oxide (V_2O_5), dimethyl sulfoxide, organic chloride compounds, trimethylamine oxide, and chlorate

Table 3.8 Some genera of microorganisms containing denitrifiers

Eukaryotes	Examples of microorganisms
	Globobulimina[a], *Cylindrocarpon*[b], *Fusarium*[b]
Prokaryotes	Examples of microorganisms
Archaea	*Ferroglobus, Haloarcula, Halobacterium, Haloferax, Pyrococcus*
Bacteria	*Nitrobacter, Paracoccus, Pseudomonas, Rhizobium, Rhodopseudomonas, Thiobacillus, Alcaligenes, Neisseria, Nitrosomonas, Zoogloea, Acinetobacter, Marinobacter, Shewanella, Wolinella*
	Bacillus, Frankia, Nocardia
	"*Candidatus* Kuenenia stuttgartiensis" (anammox)

[a]Foraminifera
[b]Fungi

terminal acceptors (fumarate, dimethyl sulfoxide, chloride organic compounds) are reduced. The reduction of the terminal acceptor is catalyzed by different reductases associated with a respiratory chain. When the electron donor is an organic compound, its oxidation may be complete, releasing CO_2, or partial. The product of the reduction of the terminal acceptor is excreted in the external environment and is not used to be assimilated in the biosynthesis, and thus anaerobic respirations are **dissimilatory reductions*** of terminal electron acceptors. However, all dissimilatory reductions are not anaerobic respirations. Some bacteria eliminate an excess of reducing power by transferring electrons to mineral acceptors.

According to the redox potentials of terminal electron acceptor couple (Table 3.6), the energy generated by electron transfer during anaerobic respiration will be different. Indeed, the energy production will be all larger than the terminal electron acceptor has a redox couple with a high potential. This is the case for nitrate and iron respirations, whose redox couples are close to the O_2/H_2O couple. Thus, for most conventional anaerobic respiration, a series of electron acceptors in order of decreasing energy production is as follows:

$$O_2 > Fe^{3+} > NO_3^- > Mn_4^+ > SO_4^{2-} > CO_2$$

In an anoxic environment, according to the available electron acceptors, it is always the anaerobic respiration which produces the most available energy that dominates in the anaerobic microbial community. This rule is always verified if the terminal electron acceptor is present under non-limiting conditions. Anaerobic respirations are present in all three domains of life, but they are especially important and common in prokaryotic domains (*Bacteria* and *Archaea*).

Dissimilatory Nitrate Reduction

The use of nitrate (the most oxidized nitrogen, oxidation state + V) as terminal electron acceptor is widespread in prokaryotes (Table 3.8). The first reduction step leading to NO_2^- (+III) produces energy via the respiratory chain linked to proton translocation (Fig. 3.21). Subsequently, there are two possible futures for NO_2^-. It can accumulate in the environment, but it is often reduced to NH_3 ($-$III) especially among *Enterobacteriaceae*, *Staphylococcus*, and *Fusarium* (ammonia dissimilative reduction). NO_2^- may also be reduced to N_2 (oxidation state 0) in several steps via the respiratory chain. This is the denitrification (reduction of NO_3^- to N_2) in which each step is producing energy and forms gaseous compounds of nitrogen that can be released into the atmosphere (nitrogen loss) (*cf*. Sects. 14.3.3 and 14.3.5).

The different reduction steps are catalyzed by reductases: nitrate reductase Nar (3.1), NADPH-dependent nitrite reductase (3.2), nitrite reductase (3.3), nitric oxide reductase (3.4), and nitrous oxide reductase (3.5):

$$NO_3^- + 2\ e^- + 2\ H^+ \rightarrow NO_2^- + H_2O \qquad (3.1)$$

$$NO_2^- + 3\ NADPH, H^+ + H^+ \rightarrow NH_3 + 2H_2O + 3\ NADP^+ \qquad (3.2)$$

$$NO_2^- + e^- + 2\ H^+ \rightarrow NO + H_2O \qquad (3.3)$$

$$NO + e^- + H^+ \rightarrow ½\ N_2O + ½\ H_2O \qquad (3.4)$$

$$½\ N_2O + e^- + H^+ \rightarrow ½\ N_2 + ½\ H_2O \qquad (3.5)$$

The electrons necessary for reducing the nitrogen compounds in reactions (3.1), (3.3), (3.4), and (3.5) are produced during the oxidation of the substrate and transferred through the respiratory chain (Fig. 3.21). The amount of energy obtained by the nitrate respiration is very high,

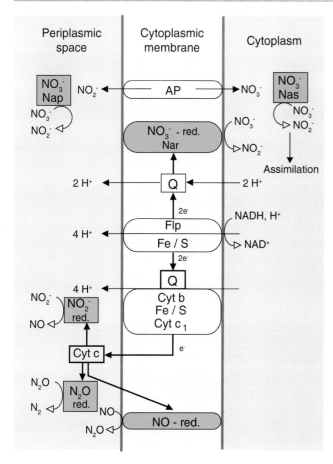

Fig. 3.21 Scheme of the respiratory chain of *Paracoccus denitrificans*, chemoorganotrophic denitrifying bacterium. NO_3^- *red Nar* dissimilatory nitrate reductase, NO_3^- *Nap* periplasmic nitrate reductase (aerobic denitrification), NO_3^- *Nas* assimilatory nitrate reductase, NO_2^- *red* nitrite reductase, *NO red* nitric oxide reductase, N_2O *red* nitrous oxide reductase, *AP* antiport transport system NO_3^-–NO_2^- (Drawing: M.-J. Bodiou)

close to that obtained by aerobic respiration. For example, with glucose as substrate,

$C_6H_{12}O_6 + 6\,O_2 \rightarrow 6\,H_2O + 6\,CO_2 \qquad \Delta G^{\circ'} = -2,868\,kJ$
$C_6H_{12}O_6 + 12\,NO_3^- \rightarrow 12\,NO_2^- + 6\,CO_2 + 6\,H_2O \quad \Delta G^{\circ'} = -1,764\,kJ$
$C_6H_{12}O_6 + 4\,NO_3^- \rightarrow 2\,N_2 + 6\,CO_2 + 6\,H_2O \qquad \Delta G^{\circ'} = -2,425\,kJ$

Prokaryotes that perform nitrate respiration are facultative anaerobic microorganisms, able to use the dioxygen in oxic conditions and nitrate under anoxic conditions. Enzymes that reduce nitrogen compounds are sensitive to dioxygen, and thus nitrate respiration will begin only when dioxygen is in very low concentration in the cell environment. Denitrifying enzymes have a sensitivity to dioxygen that is increasing from nitrate reductase which is tolerant to dioxygen, to nitrous oxide reductase, which is the most sensitive; so, when dioxygen is in low concentration, nitrate reduction can start but not the reduction of nitrous oxide whose reductase is inhibited at low-concentration dioxygen,

thereby releasing nitrous oxide in the atmosphere (*cf.* Sect. 14.3.5, Fig. 14.31). In most bacteria that reduce nitrate, these different enzymes are inducible under anaerobic conditions and in the presence of nitrogen compounds, with the exception of some bacteria where they are constitutive (*Thiomicrospira denitrificans*). Like other carriers of the respiratory chain, the reductases are included into the cytoplasmic membrane with the exception of nitrite reductase and N_2O reductase which are periplasmic. In some bacteria, there is a constitutive periplasmic nitrate reductase (nitrate reductase Nap) that allows for a real aerobic denitrification. Prokaryotes that perform nitrate respiration are predominantly chemoorganotrophic heterotrophic microorganisms. Some are obligate or facultative chemolithotrophs. Among the denitrifying microorganisms, some do not have complete enzymatic equipment and cannot perform the full steps of denitrification.

Ettwig et al. (2010) described an "intra-aerobic" pathway for the reduction of nitrite and the oxidation of methane. In the bacterium "*Candidatus* Methylomirabilis oxyfera," the final reduction of nitrite to dinitrogen involves the conversion of two molecules of nitric oxide in dinitrogen and dioxygen, the latter being used by the microorganism to oxidize methane.

A Particular Case, the Anammox (*cf.* Sect. 14.3.5)

A new mode of use of nitrogen compounds was discovered among *Planctomycetes*, a group of bacteria with complex cell structure. The anammox process is an energy conservation based on the anaerobic oxidation of ammonium with nitrite as electron acceptor:

$$NH_4^+ + NO_2^- \rightarrow N_2 + 2\,H_2O \quad \Delta G^{\circ'} = -357\,kJ$$

Dissimilatory Sulfate Reduction

The sulfur compounds are important terminal electron acceptor for anaerobic respiration. Among them, sulfate, the most oxidized form of sulfur, is used as an electron acceptor by bacteria or archaea during a respiration called sulfate reduction. Sulfate-reducing microorganisms are generally strict anaerobes and produce sulfide as a toxic end product of their respiration. Other less oxidized sulfur compounds (sulfite, thiosulfate, sulfur, etc.) can also play the role of electron acceptors. Seeing the variety of electron donors and metabolic pathways that respiration of sulfur compounds involves, the development of a unified model of electron transfer is impossible with the exception of sulfate reduction steps. The reduction of sulfate to sulfide involves the transfer of eight electrons and the oxidation state of sulfur changes from + VI to − II:

$$SO_4^{2-} + 8\,H^+ + 8\,e^- \rightarrow S^{2-} + 4\,H_2O$$

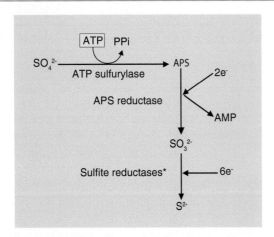

Fig. 3.22 Pathway of sulfate reduction. *APS* adenosine phosphosulfate, * 4 sulfite reductases known (desulfoviridin, desulforubidin, pigment P582, and desulfofuscidin), and *AMP* adenosine monophosphate (Drawing: M.-J. Bodiou)

Table 3.9 Different genera of sulfate-reducing and sulfur-reducing microorganisms

Sulfate-reducing bacteria	1. Complete oxidation	Gram +	*Desulfotomaculum*
		Gram −	*Desulfobacter*
			Desulfococcus
			Desulfobacterium
			Desulfosarcina
			Desulfomonile
			Desulfonema
			Desulfoarculus
			Desulfacinum
			Desulforhabdus
	2. Incomplete oxidation	Gram +	*Desulfotomaculum*
		Gram −	*Desulfovibrio*
			Desulfomicrobium
			Desulfobulbus
			Desulfobotulus
			Desulfobacula
			Desulfofustis
Sulfate-reducing archaea			*Archaeoglobus*
Sulfur-reducing bacteria, non-sulfate reducing			*Desulfuromonas*
			Desulfurella
			Sulfurospirillum
			Wolinella
Sulfur-reducing archaea, non-sulfate reducing			*Pyrobaculum*
			Pyrodictium
			Pyrococcus
			Thermoproteus
			Thermophilum
			Thermococcus
			Desulfurococcus

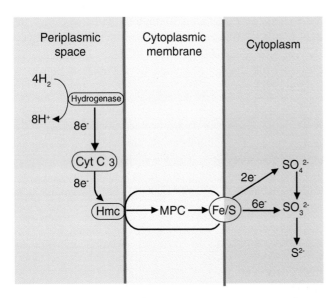

Fig. 3.23 Respiratory chain and formation of an H$^+$ gradient in *Desulfovibrio* from hydrogen as electron donor. *Cyt c3* cytochrome c3, *Hmc* high molecular weight cytochrome, *MPC* multiprotein complex, *Fe/S* iron–sulfur protein (Modified and redrawn from Voordouw 1995, Heidelberg et al. 2004, and Mathias et al. 2005) (Drawing: M.-J. Bodiou)

During this reduction, the sulfate must be activated by ATP. The activation reaction catalyzed by ATP sulfurylase leads to the formation of adenosine phosphosulfate (APS) (Fig. 3.22). The APS and sulfite accept electrons supplied by the respiratory chain.

The flow of electrons along the respiratory chain and the creation of a gradient of H$^+$ from molecular hydrogen oxidized by a periplasmic hydrogenase are shown in Fig. 3.23. The electrons are transferred to sulfur compounds by the APS reductase and sulfite reductase.

With lactate as electron donor, the hydrogenases are also involved in the formation of a proton gradient.

Sulfate reducers are usually chemoorganotrophic microorganisms, but they are often chemolithotrophic microorganisms using dihydrogen as electron donor. The organic substrates used are mostly of small molecules (lactate, acetate, pyruvate, ethanol, propionate, butyrate, etc.) from fermentations of organic matter. Sulfate reducers are separated into two groups according to their ability to oxidize organic substrates, those which partially oxidize and those which completely oxidize to CO_2 (Table 3.9). Some are also able to oxidize some more complex substrates such as hydrocarbons (*cf.* Sect. 16.8.1), benzoate, phenol, starch, peptides, indole, sugars, amino acids, or glycerol.

During a reaction of syntrophy (*cf.* Sect. 14.3.3), sulfate reduction by sulfate-reducing bacteria is coupled to the oxidation of methane by anaerobic methanotrophic archaea (reverse methanogenesis). Milucka et al. (2012) revealed the presence of methanotrophic archaea capable of coupling the anaerobic oxidation of methane to sulfate reduction; sulfate-reducing bacteria associated with methanotrophs get their energy from the **disproportionation*** of reduced sulfur compound (disulfide) released by methanotrophs.

The energy production is low, regardless of the substrates:

$$4H_2 + SO_4^{2-} \rightarrow S^{2-} + 4 H_2O \qquad \Delta G^{\circ'} = -152 \text{ kJ}$$
$$2 \text{ lactate} + SO_4^{2-} \rightarrow S^{2-} + 2 \text{ acetate} + 2 CO_2 + 2 H_2O \qquad \Delta G^{\circ'} = -160 \text{ kJ}$$

ATP synthesis is the result of oxidative phosphorylation during operation of the respiratory chain; however, during the degradation of lactate by *Desulfovibrio*, an additional ATP synthesis occurs by phosphorylation at the substrate level:

$$\text{Lactate} \rightarrow \text{pyruvate} + 2 e^- + 2 H^+$$
$$\text{Pyruvate} + \text{ADP} + \text{Pi} \rightarrow \text{acetate} + CO_2 + \text{ATP} + 2 e^- + 2 H^+$$

During the oxidation of acetate to CO_2, the majority of sulfate reducers use the **inverse pathway of acetyl-CoA*** (inverse pathway of the acetogenesis). The pathway of the citric acid is specific to some sulfate-reducing bacteria (*Desulfobacter, Desulfuromonas, Desulfurella*). The sulfate reducers are heterotrophs, using small organic molecules, but some are facultative autotrophs and can fix CO_2 via the acetyl-CoA or via the reverse tricarboxylic acid cycle (*cf.* Sect. 3.4.1). The majority of sulfate reducers can also use sulfite, thiosulfate, and sulfur as electron acceptors. In addition, some are capable of thiosulfate and sulfite disproportionation, releasing sulfate and sulfide:

$$S_2O_3^{2-} + H_2O \rightarrow SO_4^{2-} + H_2S \qquad \Delta G^{\circ'} = -22 \text{ kJ}$$
$$4 SO_3^{2-} + 2 H^+ \rightarrow 3 SO_4^{2-} + H_2S \qquad \Delta G^{\circ'} = -60 \text{ kJ}$$

The disproportionation of sulfur is thermodynamically unfavorable:

$$4 S^\circ + 4 H_2O \rightarrow SO_4^{2-} + 3 H_2S + 2H^+ \qquad \Delta G^{\circ'} = +10 \text{ kJ}$$

However, if the sulfide is oxidized with the help of a metal (iron or manganese), the reaction is favorable:

$$3 H_2S + 2 \text{FeOOH} \rightarrow S^\circ + 2 \text{FeS} + 4 H_2O \qquad \Delta G^{\circ'} = -144 \text{ kJ}$$

Indeed, the sum of the two reactions above becomes

$$3S^\circ + 2 \text{FeOOH} \rightarrow SO_4^{2-} + 2 \text{FeS} + 2 H^+ \qquad \Delta G^{\circ'} = -134 \text{ kJ}$$

In addition to sulfate reducers, there are sulfur-reducing microorganisms, reducing sulfur but unable to reduce sulfate. They are mostly found in *Archaea* (Table 3.9).

In the absence of sulfur compounds, sulfate-reducing bacteria can use other compounds as terminal electron acceptors. These are organic (fumarate, malate, organochlorines) or inorganic compounds (nitrate, derivatives of uranium, iron, selenium or arsenic, etc.) and more surprising the dioxygen for "anaerobic" bacteria. Research has shown that microorganisms belonging to the genus *Desulfovibrio* were particularly resistant to dioxygen (Le Gall and Xavier 1996). This is particularly true for those isolated from environments containing anoxic microniches or subjected to conditions of very fluctuating redox (interface zones, biofilms, microbial mats, etc.). Resistance enzymes to oxidative stress (catalase, superoxide dismutase, superoxide reductase) were found in some sulfate reducers. In addition, an oxygenase reductase has even been discovered, responsible for the dioxygen reduction by aerobic respiratory chains incompletely described in sulfate reducers (Santana 2008). In most cases, energy production by oxidative phosphorylation induced by the presence of dioxygen allows only the survival of sulfate reducers. However, the possibility of growth of *Desulfovibrio* has been partially demonstrated in partial pressure of dioxygen, close to that of the atmosphere (Lobo et al. 2007).

Ferric Iron Respiration or Ferric Reduction

The redox potential of the Fe^{3+}/Fe^{2+} couple (+770 mV) close to the potential of the O_2/H_2O couple (+820 mV) suggests a significant energy production during the respiration of iron (III). But, the very low solubility of Fe^{3+} ion at pH 7 (lower than 10^{-16} M) makes this process inefficient. However, anaerobic growth of *Shewanella oneidensis* and *Geobacter metallireducens* depends on the reduction of iron (III). This property is found in other bacteria (*Thermotoga, Thermus, Geothrix, Ferribacterium, Acidiphilum*) and archaea (*Pyrobaculum*). The link between microorganisms and insoluble iron oxides can be established by chelators (*Geothrix*), by direct contact with particles of iron oxides (*Geobacter*), or by the intermediate carriers called "shuttle" responsible for transfer electrons to insoluble iron (*Geobacter*). Humic substances widespread in natural environments can play the role of shuttles. Indeed, humic substances contain quinone groups that can undergo oxidation–reduction cycles; microorganisms transfer electrons to humic substances that oxidize again in contact with particles of iron (III) (*cf.* Sect. 14.5.2). Bacteria that use iron as a terminal electron acceptor is usually facultative anaerobic chemoorganotrophic bacteria except of some strict anaerobes (*Geobacter metallireducens*). They use various organic substrates as electron donors and carbon sources. Their activity is limited to interfaces between oxic and anoxic environments where anaerobiosis and the presence of iron (III) can coexist (*cf.* Sect. 14.5.2).

Fumarate Respiration

Fumarate is not abundant in natural environments. However, fumarate respiration is widespread among microorganisms, probably due to the fact that fumarate is a common metabolite, formed from the catabolism of carbohydrates and proteins. This type of respiration is known in *Wolinella succinogenes*, *Enterobacteriaceae*, *Clostridia*, *Paenibacillus macerans*, sulfate-reducing bacteria, and *Propionibacteria*.

The reducing power resulting from the oxidation electron donors, mainly hydrogen, formate, or NADH, is transferred by the respiratory chain to a fumarate reductase which reduces the fumarate to succinate:

$$H_2 + \text{fumarate} \rightarrow \text{succinate}$$
$$\text{Formiate} + \text{fumarate} + H^+ \rightarrow CO_2 + \text{succinate}$$

With a redox potential of +30 mV, the fumarate/succinate couple is not the source of important production of energy. For example, the first reaction above produces only -43 kJ.mol^{-1} dihydrogen and therefore will require more than one mole of dihydrogen to synthesize one mole of ATP.

In addition to the fumarate, other organic molecules can play the role of electron acceptors during anaerobic respirations such as glycine, dimethyl sulfoxide, and trimethylamine oxide reduced, respectively, in acetate, dimethyl sulfide, or trimethylamine.

CO₂ Respirations (Acetogenesis and Methanogenesis)

CO_2 from metabolism of chemoorganotrophic microorganisms is an abundant compound in natural environments and serves as a terminal electron acceptor for two groups of strict anaerobic microorganisms, acetogenic bacteria and methanogenic archaea. In anoxic environments, these microorganisms have at their disposal electron donors from the decomposition of organic matter, in particular dihydrogen. Thus, some of them are chemolithotrophic autotrophs. CO_2 reduction leads to the formation of acetate in acetogens and methane in methanogens. Redox couples are very electronegative (CO_2/CH_4, -0.24 V and CO_2/acetate, -0.29 V); consequently, the two respirations are low in energy:

$$4 H_2 + CO_2 \rightarrow CH_4 + 2 H_2O \quad \Delta G^{o'} = -131 \text{ kJ}$$
$$4 H_2 + 2 CO_2 \rightarrow CH_3COOH + 2 H_2O \quad \Delta G^{o'} = -95 \text{ kJ}$$

– *Acetogenesis*

Acetogenic bacteria form a very heterogeneous group of Gram-positive bacteria essentially (*Acetobacterium*, *Butyribacterium*, *Clostridium*, *Eubacterium*, *Moorella*, *Sporomusa*). They use CO_2, CO, or formate as electron acceptors and produce acetate as an end product of their respiration. The electron donor is mostly dihydrogen. However, molecules such as sugars, alcohols, organic acids, or aromatic compounds can serve as electron donors and carbon sources. In all cases, the reduction of CO_2 passes through acetyl-CoA (Wood–Ljungdahl pathway) (*cf.* Sect. 3.4.1).

With dihydrogen as electron donor, the reduction of one molecule of CO_2 as a methyl group and the reduction of a second molecule of CO_2 in the form of carbonyl function lead to the formation of acetyl-CoA in the presence of coenzyme A; acetyl-CoA is subsequently phosphorylated to acetyl phosphate which releases a molecule acetate and an ATP molecule (substrate-level phosphorylation).

During this process, energy can also be produced by the formation of a Na^+ gradient, instead of H^+ gradient, at the origin of a sodium-motive force (Muller 2003; Detkova and Pusheva 2006).

– *Methanogenesis*

All organisms are methanogenic archaea. Methanogenesis with dihydrogen as substrate and CO_2 as electron acceptor and carbon source is very widespread among the methanogenic microorganisms. However, other compounds involving other pathways could be used by some methanogens. Methanogenic reactions can be divided into two groups (Table 3.10). In the first group, methanogenesis involves the reduction of C1 molecules (CO, CO_2, and methanol) with dihydrogen or alcohols having more than one carbon atom as electron donors. The second group relates to microorganisms carrying out disproportionation reactions of compounds in C1 (CO, formate, formaldehyde, and methanol), methylamines, and methylsulfide (different genera of methanogens) or acetate (*Methanosarcina* and *Methanothrix*). During these reactions, a fraction of the compounds is oxidized to CO_2 and the other is reduced to CH_4 (*cf.* Sect. 14.2.6).

Of all the ways, the reduction of CO_2 with dihydrogen as electron donor is the best known. This reaction requires a series of specific coenzymes, carriers of carbon groups, particularly the coenzyme M (CoM-SH).

Conservation of energy is coupled to the reduction of the disulfide bridge of heterodisulfide (CoB-SS-CoM). Figure 3.24 shows a representation of mechanism of energy conservation in a methanogen leading to the formation of a proton gradient allowing energy production. The electrons

Table 3.10 Main metabolic pathways of methanogenic archaea

Metabolic reactions	$\Delta G^{o'}$ in kJ/mole of CH_4
$4H_2 + CO_2 \rightarrow CH_4 + 2H_2O$	-131
$H_2 + CH_3OH \rightarrow CH_4 + H_2O$	-112
$2CH_3CH_2OH + CO_2 \rightarrow CH_4 + 2CH_3COO^- + 2H^+$	-116
$4CO + 2H_2O \rightarrow CH_4 + 3CO_2$	-450
$4CH_3OH \rightarrow 3CH_4 + CO_2 + 2H_2O$	-130
$4(CH_3)_3NH^+ + 6H_2O \rightarrow 9CH_4 + 3CO_2 + 4NH_4^+$	-76
$CH_3COOH \rightarrow CH_4 + CO_2$	-36

Fig. 3.24 Pathway of CO_2 reduction to CH_4 (CO_2 respiration) and formation of the membrane potential in a methanogenic (Modified and redrawn from Deppenmeier et al. 1999). CO_2 is successively reduced in formyl, methenyl, methylene, and methyl groups which are linked to coenzymes acting as carriers of carbon groups, and finally CH_4 released during the formation of heterodisulfide (CoB-SS-CoM). *[CH₃-X]* CH_3 bound, *CoM-SH* coenzyme M, *HS-CoB* coenzyme B, *Cyt* cytochrome, *Mp* oxidized methanophenazine, *MpH₂* reduced methanophenazine, and *Hsr* heterodisulfide reductase (Drawing: M.-J. Bodiou)

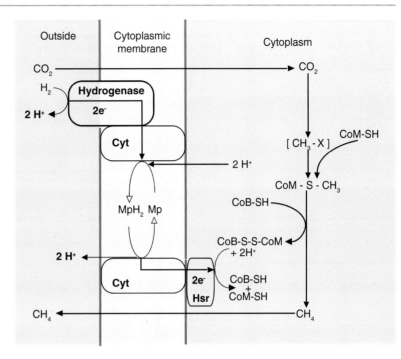

from the oxidation of dihydrogen are transferred by the respiratory chain to a reductase which reduces the disulfide bridge, thus releasing the CoM-SH available once again for the final reduction of CO_2. H^+ gradient is created by the oxidation–reduction of an intermediary carrier, the methanophenazine. In addition, during the CO_2 reduction cycle to CH_4, a sodium pump is activated and generates a sodium gradient allowing a sodium-motive force (Deppenmeier et al. 1999; Muller et al. 2008).

Methanogenic pathways using other compounds (methanol, acetate) are different, but some carriers are common.

3.3.3 Fermentations

In the absence of light, dioxygen, and other extracellular acceptors of electrons, the energy required for various cellular activities can be provided by fermentations. Pasteur has defined fermentation as "la vie sans air" ("life without air"). This definition is now not accurate. Mechanisms of energy producers that are not fermentations occur in the absence of air (anaerobic respirations, anoxygenic photosynthesis), and certain fermentations occur in the presence of air (e.g., lactic fermentation). A more accurate definition might be as follows: fermentations are energy producer mechanisms occurring usually under anaerobic conditions, energy being conserved mainly by substrate-level phosphorylation. Electron donors are organic compounds. Electron acceptors are formed by endogenous organic compounds obtained from cellular metabolism and derived from the partial oxidation of electron donors. Unlike process of aerobic and anaerobic respirations, in fermentations there is no exogenous electron acceptor, with the exception of the fermentation of some amino acids which require an amino acid that acts as a donor of electrons and an amino acid that plays the role of electron acceptor (Stickland reaction). In the majority of fermentations, energy production does not involve a chain of membrane electron carriers; redox processes take place into the cytoplasm.

While in most respirations, the electron donor is completely oxidized to CO_2, in fermentations, the electron donor is only partially oxidized and thus provides fermentation products (organic acids, alcohols, etc.) that are released into the external environment and that often characterize the type of fermentation. A practical consequence of this excretion of products is the use of fermentations in the manufacture of foods and products for food, pharmaceutical, or chemical industries. The reducing power is temporarily transferred to coenzymes (NAD^+ in general) which oxidize again by transferring electrons to an organic compound (which serves as a final electron acceptor) originating from the oxidation pathway of the electron donor (Fig. 3.25). The main fermentation products are CO_2, dihydrogen, formate, acetate, lactate, and short-chain fatty acids. Ammonium, sulfide, methyl mercaptans, and aromatic compounds come from the fermentation of amino acids.

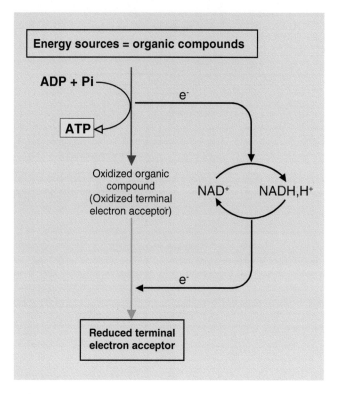

Fig. 3.25 General scheme of fermentation (Drawing: M.-J. Bodiou)

3.3.3.1 Dihydrogen Production by Fermentations

Many fermentation processes produce dihydrogen to maintain their redox balance. Hydrogenase or a formate hydrogen lyase catalyzes the formation of dihydrogen:

Pyruvate $\rightarrow e^- \rightarrow$ ferredoxin $\rightarrow 2H^+ \rightarrow H_2$(hydrogenase)
Formate $\rightarrow CO_2 + H_2$(formate hydrogen lyase)
$NADH, H^+ \rightarrow e^- \rightarrow$ ferredoxin $\rightarrow 2H^+ \rightarrow H_2$(hydrogenase)

The third reaction thermodynamically unfavorable can take place only during interspecies dihydrogen transfer.

3.3.3.2 Mechanism of Energy Conservation

ATP is synthesized by substrate-level phosphorylation. For the same substrate, the fermentation is much less efficient than respiration. For example, for one mole of glucose, respiration in yeast produces 2,872 kJ and alcoholic fermentation only 236 kJ:

$C_6H_{12}O_6 + 6O_2 \rightarrow 6CO_2 + 6H_2O \quad \Delta G^{\circ'} = -2,872$ kJ.mole^{-1}
$C_6H_{12}O_6 \rightarrow 2C_2H_6O + 2CO_2 \quad \Delta G^{\circ'} = -236$ kJ.mole^{-1}

With a free energy of −236 kJ per mole of glucose fermented, several moles of ATP should be produced. In fact, the alcoholic fermentation releases only two moles of ATP.

This low efficiency, which characterizes all fermentations, obliges the fermentative microorganisms to degrade large amounts of substrate and thus produce large quantities of products used in biotechnology.

Among the many reactions of substrate-level phosphorylation, three are frequent during the fermentation of carbohydrates:

1, 3 − diphosphoglycerate + ADP → 3 − phosphoglycerate + ATP
Phosphoenolpyruvate + ADP → pyruvate + ATP
Acetyl phosphate + ADP → acetate + ATP

In addition to phosphorylation at the substrate level, there are, in some fermentative bacteria, other mechanisms of energy conservation:

1. During the propionic fermentation by *Propionibacterium*, in addition to the phosphorylation at the substrate level, the bacterium uses the formation of membrane gradients of H^+ and Na^+ (proton- and sodium-motive forces). The fermentative pathway is complex and involves the formation of C4 dicarboxylic acids, leading to the reduction of fumarate to succinate via a chain of electron carriers. This transfer is coupled to proton translocation. Then the succinate is decarboxylated under the action of a membrane decarboxylase resulting in Na^+ excretion.
2. During the malolactic fermentation, malate enters into the cell as a di-anion via a permease which exchanges it with a lactate mono-anion. This ionic imbalance which is equivalent to an output of H^+ generates a proton-motive force.

Table 3.11 The major fermentations

Fermented substrates	Names of fermentations	Fermentative products	Examples of microorganisms
Sugars	Alcoholic	Ethanol, CO_2	Yeasts, *Zymomonas*
	Lactic	Lactic acid and sometimes ethanol, acetic acid, CO_2	*Lactobacillus, Lactococcus, Leuconostoc*
	Butyric	Butyric acid, acetic acid, CO_2, H_2	*Clostridium, Butyribacterium*
	Acetonobutylic	Acetone, butanol, CO_2, H_2	*Clostridium acetobutylicum*
	Acetic	Acetic acid	*Clostridium thermoaceticum*
	Mixed acids	Formic acid, acetic acid, lactic acid, succinic acid, ethanol, CO_2, H_2	*Escherichia, Salmonella, Shigella, Proteus*
	2,3-Butanediol	2,3-Butanediol, lactic acid, formic acid, ethanol, CO_2, H_2	*Enterobacter, Serratia, Erwinia*
	Propionic	Propionic acid, acetic acid, CO_2	*Propionibacterium, Corynebacterium*
Organic acids			
Lactic acid	Propionic	Propionic acid, CO_2	*Clostridium propionicum*
Malic acid	Malolactic	Lactic acid, CO_2	*Leuconostoc oenos*
Citric acid		Acetoin, diacetyl, acetic acid, lactic acid, CO_2	*Lactococcus cremoris, Leuconostoc cremoris*
Amino acids			
Alanine		Propionic acid, acetic acid, NH_3, CO_2	*Clostridium propionicum*
Glycine		Acetic acid, NH_3, CO_2	*Peptococcus anaerobius*
Threonine		Propionic acid, NH_3, H_2, CO_2	*Clostridium propionicum*
Arginine		Ornithine, CO_2, NH_3	*Clostridium, Streptococcus*
Cysteine		H_2S, NH_3, pyruvic acid	*Proteus, Escherichia, Propionibacterium*
Tryptophan		Indole propionic acid, indole pyruvic acid, NH_3	*Clostridium sporogenes*
		Indole, pyruvate, NH_3	*Escherichia coli*
Heterocyclic compounds			
Guanine, xanthine		Glycine, formic acid, NH_3, acetic acid, CO_2	*Clostridium cylindrosporum*
Urate		Acetic acid, CO_2, NH_3	*Clostridium acidurici*

3. In homolactic bacteria, excretion of lactate takes place by symport with H^+ forming a proton gradient.

These different processes which concern only special cases in fermentations allow an additional production of energy.

3.3.3.3 Diversity of Fermentations and Their Metabolic Pathways

Fermentations are classified according to the formed products and consumed substrates (Table 3.11). Very widespread among microorganisms, the fermentative processes are very diverse. Microorganisms may be facultative fermentative microorganisms (also capable of aerobic or anaerobic respirations) or obligate fermentative microorganisms, and in this case either anaerobic or **air tolerant***.

The group of prokaryotes contains the majority of fermentative microorganisms. However, fermentative pathways are known in some eukaryotes. Besides yeasts, some fungi (*Anaeromyces, Neocallimastix, Orpinomyces,* etc., present in the digestive tract of herbivorous) possess a fermentative metabolism. Some protozoa, parasites (*Giardia, Entamoeba, Trichomonas,* etc.), as well as commensal or symbiotic (*Dasytricha, Isotricha, Trichonympha*) and free protozoa (*Hexamita, Trimyema*) can obtain the energy necessary to their activity during fermentation. Many of these fungi and protozoa lack mitochondria but have particular organelles, hydrogenosomes, for the origin of dihydrogen and ATP productions (*cf.* Sects. 4.3.4 and 5.4.2).

Many metabolic pathways have been described in prokaryotes, particularly in bacteria. During fermentation, sugars are generally oxidized to pyruvate which can be considered the main key to fermentative metabolism of sugars (Fig. 3.26). It is the source of various fermentation products resulting from the use of different pathways. Pyruvate or its degradation products become electron acceptors to allow oxidation of reduced coenzymes. Pyruvate is also a fermentative intermediate for other substrates but is not obligatory intermediate for all fermentations. Table 3.12 shows the equations and the productions of free energy of some fermentations.

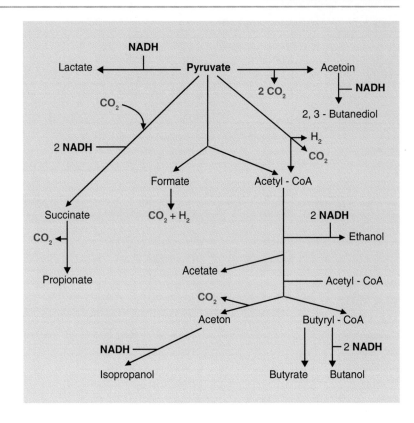

Fig. 3.26 Major pathways of reduction of pyruvate in the fermentation (Modified and redrawn from Stanier et al. 1986. Drawing: M.-J. Bodiou)

Table 3.12 Examples of metabolic reactions in some fermentations

Fermentations	Equations of fermentations	$\Delta G^{\circ\prime}$ in kJ/reaction	Examples of microorganisms
Alcoholic	Glucose → 2 ethanol + 2 CO_2	−236	Yeasts, *Zymomonas*
Homolactic	Glucose → 2 lactate	−198	*Lactobacillus*
Heterolactic	Glucose → lactate + ethanol + CO_2	−177	*Lactobacillus*
	Ribose → lactate + acetate	−210	*Leuconostoc*
Butyric acid	Glucose → butyrate + 2 CO_2 + 2 H_2	−224	*Clostridium*
Malolactic	Malate → lactate + CO_2	−67.3	*Leuconostoc*
Propionic acid	3 lactate → 2 propionate + acetate + CO_2	−170	*Propionibacterium*
Alanine	3 alanine + 2H_2O → 3 NH_3 + CO_2 + 3 acetate + 2 propionate	−135	*Clostridium*
Alanine (Stickland reaction)	Alanine + 2 glycine + 2 H_2O → 3 acetate + CO_2 + 3 NH_3	−107	*Clostridium*
Glutamate	5 glutamate + 6 H_2O → 5 NH_3 + 5 CO_2 + 6 acetate + 2 butyrate + H_2	−300	*Clostridium*
Glycine	4 glycine + 2 H_2O → 4 NH_3 + 2 CO_2 + 3 acetate	−217	*Eubacterium*

3.3.3.4 Syntrophy Interspecies Hydrogen Transfer

During fermentation, the reaction of **syntrophy*** or interspecies hydrogen transfer is a mechanism involving two microorganisms in anoxic conditions: a first microorganism that ferments a low fermentable substrate (propionate, butyrate, ethanol) producing acetate and dihydrogen (syntrophic microorganism: *Syntrophobacter*, *Syntrophomonas*), only in the presence of a second microorganism that consumes the dihydrogen as it is produced. The direct reaction of fermentation of these poorly fermentable substrates by the syntrophic microorganism is impossible because it is thermodynamically unfavorable. The reaction requires energy and therefore cannot be coupled to a mechanism that generates energy. This reaction is only possible if the dihydrogen produced during fermentation is maintained at a very low level (partial pressure of about 10^{-4} atm or less) by the activity of a second microorganism that consumes dihydrogen as soon as it is produced; the main

Table 3.13 Fermentative reactions in syntrophic microorganisms

Reactions of syntrophic metabolism	$\Delta G^{\circ\prime}$ in kJ	Syntrophic microorganisms
Butyrate + 2H$_2$O → 2 acetate + H$^+$ + 2H$_2$	+48.1	*Syntrophomonas*
Propionate + 2H$_2$O → acetate + 3H$_2$ + CO$_2$	+71.3	*Syntrophobacter*
Lactate + 2H$_2$O → acetate + 2 H$_2$ + CO$_2$	−9	*Desulfovibrio*
CH$_4$ + 2 H$_2$O → CO$_2$ + 4 H$_2$	+130	Methanotrophic archaea (reverse methanogenesis)

microorganisms that consume dihydrogen are sulfate-reducing bacteria, acetogenic bacteria, or methanogenic archaea. This interspecies hydrogen transfer reaction is called syntrophic reaction because both microorganisms involved acquire a benefit from their interrelationship (Stams and Plugge 2009).

A well-known case of syntrophy is the fermentation of ethanol to acetate and methane by a coculture of a syntrophic bacterium and a methanogenic archaea. The syntrophic bacterium ferments ethanol to acetate and dihydrogen:

$$2\text{ ethanol} + 2\text{ H}_2\text{O} \rightarrow 2\text{ acetate} + 4\text{ H}_2 + 2\text{ H}^+ \quad \Delta G^{\circ\prime} = +20 \text{ kJ}$$

The endergonic reaction cannot allow the development of this bacterium unless the dihydrogen partial pressure is lowered to 10^{-4} atm. Only in this case, the fermentation by the syntrophic bacteria becomes exergonic and releases energy (−44 kJ) for its growth (Zehnder and Stumm 1988). The methanogenic archaea consume dihydrogen and thus the syntrophic bacteria produces energy and can grow. The resultant of the two equations is in favor of energy production:
Syntrophic bacterium:

$$2\text{ ethanol} + 2\text{ H}_2\text{O} \rightarrow 2\text{ acetate} + 4\text{ H}_2 + 2\text{ H}^+ \quad \Delta G^{\circ\prime} = +20 \text{ kJ}$$

Methanogenic archaea:

$$4\text{H}_2 + \text{CO}_2 \rightarrow \text{CH}_4 + 2\text{ H}_2\text{O} \quad \Delta G^{\circ\prime} = -131 \text{ kJ}$$

Results:

$$2\text{ ethanol} + \text{CO}_2 \rightarrow 2\text{ acetate} + \text{CH}_4 + 2\text{H}^+ \quad \Delta G^{\circ\prime} = -111 \text{ kJ}$$

In Table 3.13, some examples of possible syntrophic fermentations by interspecies hydrogen transfers are presented.

3.3.4 Photosynthesis

Phototrophic microorganisms (also called photosynthetic microorganisms) use solar energy as an energy source that they capture and convert into chemical energy. Solar energy becomes thus available for the different cellular activities and particularly for the synthesis of ATP and the reduction of coenzymes (NAD$^+$ and NADP$^+$). Phototrophic microorganisms in their great majority autotrophs (photoautotrophic microorganisms) use these coenzymes to reduce CO$_2$ into organic compounds. This synthesis of organic compounds with light as an energy source is called photosynthesis. Photosynthesis occurs in two phases: a light phase during which light energy is converted into chemical energy producing ATP and reduced coenzymes, followed by a dark phase of CO$_2$ reduction and synthesis of organic compounds that consumes ATP and reduced coenzymes. Photosynthesis depends on the presence of photosynthetic pigments, in particular of chlorophylls. However, there are specific cases where the mechanism of transformation of light energy into chemical energy is independent of chlorophylls and depends on totally different types of pigments (e.g., bacteriorhodopsin) especially in some extreme halophilic archaea. Photosynthesis is the most important metabolic processes for life on our planet at the base of most food chains; it allows the introduction of the energy which is necessary for the preservation of life into the terrestrial biosystem.

In chlorophyll microorganisms, the presence or absence of a release of dioxygen during photosynthesis subdivides microorganisms into oxygenic phototrophs and anoxygenic phototrophs.

In oxygenic phototrophic microorganisms that include oxygenic phototrophic bacteria (cyanobacteria) and photosynthetic eukaryotes, electron donor for the reduction of coenzymes is water, and photosynthesis is accompanied by dioxygen evolution. In anoxygenic phototrophic microorganisms that are all of the bacterial domain, the electron donors are more reduced than water. They can be organic compounds of low molecular weight which also serve as carbon source (anoxygenic photoorganotrophic and photoheterotrophic bacteria) or inorganic compounds such as reduced sulfur compounds, dihydrogen, or sometimes the ferrous iron (anoxygenic photolithotrophic bacteria) (*cf.* Sects. 14.4.3 and 14.5.2). These photolithotrophic bacteria use CO$_2$ as a carbon source and thus are photoautotrophic bacteria (Table 3.14).

3.3.4.1 Photosynthetic Pigments of Chlorophyll-Containing Phototrophic Microorganisms

The chlorophylls (Chls) of oxygenic phototrophic microorganisms and bacteriochlorophylls (BChls) of anoxygenic phototrophic bacteria consist of a porphyrin core containing in its center a magnesium atom. Five chlorophylls and six bacteriochlorophylls are known. They differ in the nature of the substituents connected to the

Table 3.14 Main characteristics of chlorophyll-containing phototrophic microorganisms

	Electron donors	Carbon source	Phototrophic types
Oxygenic phototrophic microorganisms			
Photosynthetic microeukaryotes	H_2O	CO_2	Photolithotroph, photoautotroph
Cyanobacteria	H_2O	CO_2	Photolithotroph, photoautotroph
Anoxygenic phototrophic bacteria			
Purple and green sulfur bacteria	H_2S, $S°$, $S_2O_3^{2-}$, Fe^{2+}, H_2	CO_2	Photolithotroph, photoautotroph
Purple and green nonsulfur bacteria	Organic compounds	Organic compounds	Photoorganotroph, photoheterotroph

Table 3.15 Wavelengths of light absorption of different chlorophylls (Chl) and bacteriochlorophylls (Bchl) in whole cells

Chlorophylls	Wavelengths of light absorption
Chl a	430–440, 670–675
Chl b	480, 650
Chl c (c1, c2, c3)	450, 630
Chl d	440, 670, 718
Divinyl-chlorophyll a[a]	440, 660
BChl a	375, 590, 805, 830–920
BChl b	400, 605, 835–850, 1020–1040
BChl c	460, 745–755
BChl d	450, 715–745
BChl e	460, 710–725
BChl g	375, 419, 575, 670, 788

[a]Divinyl-chlorophyll is present in prochlorophytes
Chl chlorophyll, *BChl* bacteriochlorophyll

periphery of the porphyrin. In vivo, Chls and BChls form complexes with proteins that absorb light at different characteristic wavelengths (λ) (Table 3.15).

The absorption maxima are complementary: photosynthesis in microorganisms is active in a wide band of the solar spectrum from 400 to 1,040 nm. Other pigments associated with chlorophylls also trap light in phototrophic microorganisms. The most common are phycobiliproteins (phycocyanin and phycoerythrin) of cyanobacteria and carotenoid pigments. They absorb light radiations between 450 and 650 nm, region where Chls and BChls absorb little or no light. They allow better utilization of wavelengths and protect the cell and the Chls and BChls from the photooxidation reactions. Abundant carotenoids mask the color of the BChls. In anoxygenic phototrophic bacteria called "purple bacteria" (e.g., *Chromatium*, *Rhodospirillum*, etc.) that have BChl a or b, carotenoid pigments give them colors ranging from yellow-orange, pink, red, brown to purple-violet. Anoxygenic phototrophic bacteria that are qualified of phototrophic green bacteria (e.g., *Chlorobium*, *Chloroflexus*, etc.) which possess BChl c, d, or e and specific carotenoids in small amounts reveal cell suspension colors from yellow-green to green-brown (*cf.* Sect. 14.4.3; Fig. 14.39). The combination of chlorophyll and phycocyanin gives a blue-green color to cells of cyanobacteria. This color is the origin of the old name given to these prokaryotic microorganisms (Cyanophyceae, Cyanophyta, or blue-green algae). Some cyanobacteria that contain phycoerythrin have cell colors ranging from pink to red. According to the microorganisms, the pigments are combined in different antenna or light-harvesting molecules that are responsible for transferring the light energy to a reaction center (RC) where the transformation of the light energy into chemical energy occurs. Only a very small number of Chl a molecules are present in the reaction centers of oxygenic phototrophic microorganisms and, usually, Bchl a molecules in the reaction centers of phototrophic bacteria, either purple or green. These molecules associated with electron transport chains are inserted into specialized membranes, generating a proton-motive force and allowing the reduction of coenzymes.

The intracellular location of antenna and RC depends on the type of phototrophic microorganism (Fig. 3.27). In photosynthetic microeukaryotes, all pigments are included in the specialized membrane systems (thylakoids) within chloroplasts. In cyanobacteria (prokaryotic microorganisms), the thylakoids are directly present into the cytoplasm and contain the reaction centers in their membrane, while the antennae pigments form aggregates on their surface. In anoxygenic phototrophic purple bacteria, all the pigments are inserted into the cytoplasmic membrane that develops and forms characteristic invaginations like lamellae, tubules, vesicles, etc. In anoxygenic phototrophic green bacteria, the antennae are located into ovoid structures (chlorosomes) attached to the inner surface of the cytoplasmic membrane, while the associated RCs are inserted directly in the cytoplasmic membrane.

3.3.4.2 Oxygenic Photosynthesis (Fig. 3.28a, b)

The reaction center (RC) of phototrophic eukaryotic microorganisms and cyanobacteria is incorporated into photosystems included in the thylakoid membrane. A photosystem comprises an antenna complex and a reaction center which is associated with a chain of electron carriers.

RCs are composed of proteins, Chl a, and initial electron acceptors that convert photon energy into chemical energy. Two coupled photosystems are required. In the photosystem I, the reaction center has a Chl a dimer which absorbs light at 700 nm (P700), whereas in the photosystem II, the dimer is active at a wavelength of 680 nm (P680). In general, the two

Fig. 3.27 Cellular localization of pigments in phototrophic microorganisms. *AP* antenna pigment; *RC* reaction center (Drawing: M.-J. Bodiou)

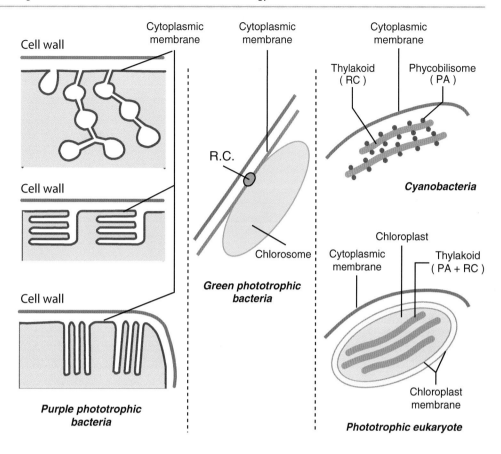

photosystems operate simultaneously; electrons are transferred from photosystem II to photosystem I.

By absorbing a photon, the P680 dimer of photosystem II passes from a ground state to an excited state which is strongly reduced, allowing the transfer of electron along the first chain of electron carriers to the RC of photosystem I (P 700). For this, P680 dimer gives its electron very quickly to a pheophytin (Chl molecule devoid of magnesium) at low redox potential which transfers it to the chain of electron carriers. A new light excitement transfers another electron and so on. The electrons are used for reducing the dimer P700 when excited by light. This transfer is accompanied by a translocation of protons in the center of the thylakoid to the origin of a proton-motive force that generates ATP by photophosphorylation via ATP synthases of the thylakoid. Chl of photosystem II, which has lost electrons during its excitation by light, recovers electrons from a water molecule. The $Chl^{oxdized}$/Chl couple having a redox potential higher than O_2/H_2O redox couple (+0.9 and +0.82 volts, respectively) gives this reaction thermodynamically possible. The oxidation of the water molecule by loss of electrons is accompanied by release of dioxygen. The protons released contribute to the formation of the proton gradient at the level of the thylakoid membrane:

$$H_2O \rightarrow 2\,e^- + 2\,H^+ + \tfrac{1}{2}\,O_2$$

In photosystem I, the electron of the dimer of Chl *a* (P700) excited by a photon is transferred to a first carrier A0 (Chl *a* modified) and then a second A1. Then the electrons pass through a chain of carriers containing iron sulfur protein and ferredoxin. Finally, the electrons are transferred to NADP reductase that reduces the coenzyme. The reduction of the coenzyme with water as electron donor is thermodynamically unfavorable:

$$NADP^+ + H_2O \rightarrow NADPH, H^+ + \tfrac{1}{2}\,O_2 \quad \Delta G^{\circ\prime} = +220\,kJ/NADP^+$$

Both activation steps are needed to bring the electrons at the top of photosystem I which represents the energy change required for efficient transfer of electrons from water to $NADP^+$. This flux of electrons in one direction is noncyclic transport or noncyclic photophosphorylation which is also called schema "Z."

When the reducing power (NADPH, H^+) is sufficiently large, cyclic electron transfer or cyclic photophosphorylation takes place involving only photosystem I. The electrons transit directly from Fd to complex b/f to reduce the P700 Chl. This transfer creates a proton gradient for ATP formation.

3.3.4.3 Anoxygenic Photosynthesis (Fig. 3.29a–c)

Anoxygenic phototrophic bacteria have only one photosystem. Photosystem of phototrophic purple bacteria is

Fig. 3.28 Schemes of oxygenic photosynthesis. (**a**) The Z scheme of electron flow. *P680 and P700* chlorophyll of reaction centers of photosystems II and I, respectively, *P680* and P700**: excited chlorophylls, *Ph* pheophytin, *Q* quinone, *PQ* plastoquinone (substituted quinones), *Cyt b/f* cytochromes b and f complex, *Pc* plastocyanin (copper protein), *A0 and A1* electron acceptors (A 0, chlorophyll amended), *Fe/S* iron–sulfur protein, and *Fd* ferredoxin (iron–sulfur protein). (**b**) Location of the photosynthetic apparatus in the thylakoid membrane. *PS I and PS II* photosystems I and II, respectively, *PQ* plastoquinones, *PC* plastocyanin, *Fd* ferredoxin, *NADP-red* NADP reductase, and *ATPs* ATP synthase. The complex Cyt b/f plays the role of proton pump (Drawing: M.-J. Bodiou)

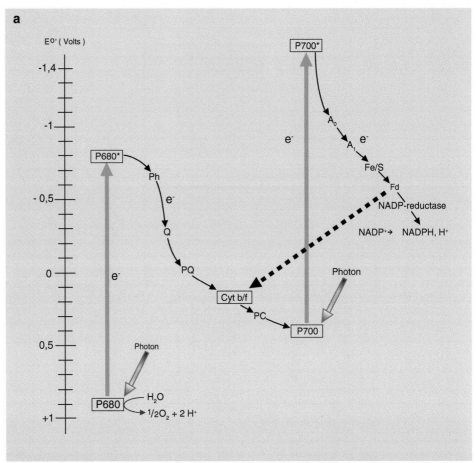

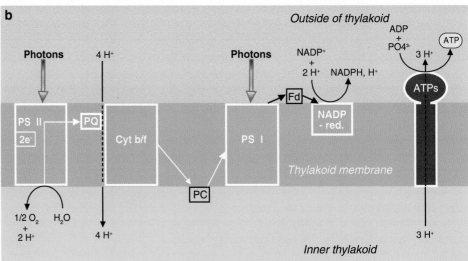

similar to photosystem II, while that of phototrophic green bacteria is similar to photosystem I. Reaction centers are included in the normal cytoplasmic membrane in green bacteria or in invaginated cytoplasmic membrane in purple bacteria. Because there is only one photosystem, the electron transfer is cyclic (cyclic photophosphorylation) and allows the production of chemical energy necessary for cellular activities. Noncyclic transfer is necessary for the reduction of coenzymes. This transfer requires energy and a reverse flux of electrons for purple bacteria.

The redox potential of BChl *a* which is involved in the reaction center is less electropositive than the O_2/H_2O redox couple. Thus, the reduction of coenzymes requires electron donors more reduced than water (reduced sulfur compounds,

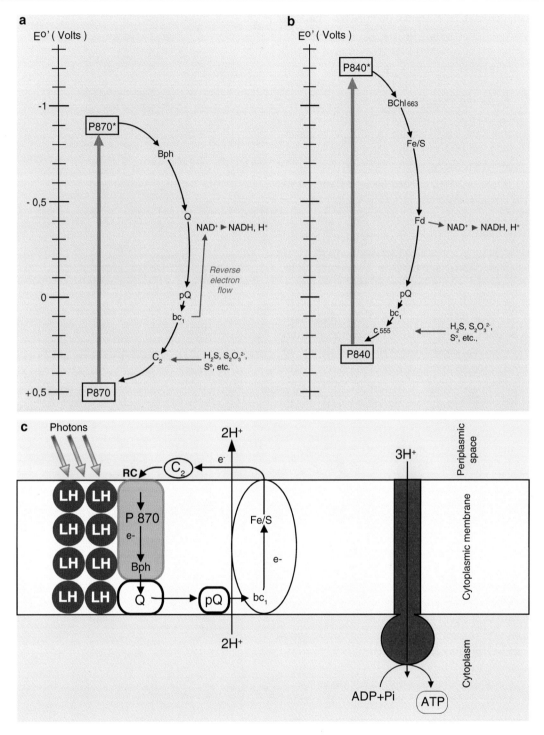

Fig. 3.29 Photosynthetic system of anoxygenic phototrophic bacteria. (**a**) Purple phototrophic bacteria. (**b**) Green sulfur phototrophic bacteria. (**c**) Membrane localization of the photosynthetic apparatus of purple bacteria, proton transfer, and photophosphorylation. *P840 and P870* bacteriochlorophylls of reaction centers (RC), *P870* and P840** excited bacteriochlorophylls, *Bph* bacteriopheophytin, *BChl* $_{663}$ amended bacteriochlorophyll a, *Q* quinone, *pQ* quinone pool, *Fe/S* iron–sulfur protein, *Fd* ferredoxin, *bc1* complex of cytochromes b and c1, *c2* cytochrome c2, and *c555* cytochrome c555. *LH* right harvesting pigments. *RC* reaction center (Drawing: M.-J. Bodiou)

H_2, organic compounds). Anoxygenic phototrophic bacteria so have an anaerobic photosynthetic activity.

Anoxygenic phototrophic activity (photosystem of type II) was demonstrated in planktonic aerobic heterotrophic bacteria or aerobic anoxygenic phototrophs (*Erythrobacter, Roseobacter, Erythromicrobium, Roseococcus, Porphyrobacter*, etc.). However, photophosphorylations only provide a supplementary supply of energy, unable to ensure by itself the growth of these microorganisms (Yurkov and Beatty 1998; Fuchs et al. 2007).

The BChl *a* (P870 or P840) excited by light transfers its electrons to an electron acceptor at low potential (Bph or BChl 663). The electrons are then transferred to a series of electron carriers. During this transfer, the electron energy level drops. Finally, these low-energy electrons return to the reaction centers and reduce again the Bchl. During the cycle, protons are transferred to the outside of the cytoplasmic membrane. H^+ gradient is thus created; it is at the origin of a proton-motive force that generates ATP by photophosphorylation. The reduction of NAD^+ is carried by electrons coming out of cyclical flow, thus causing a deficit of electrons. This is compensated by a contribution from inorganic external donors (reduced sulfur compounds, H_2) or organic (succinate, malate) via the cytochromes C555 or C2 in the case of sulfur compounds and plastoquinone (PQ) in that of the succinate. The electrons return to reaction centers where they make up the deficit of BChl which is consecutive to its excitement. In the case of phototrophic purple bacteria, the electrons which reduce coenzymes pass via the complex cytochrome b/c1 (Fig. 3.29a). The positive redox potential of this complex makes impossible the direct reduction of NAD^+ whose redox potential is very negative (−0.32 V). A reverse flow of electrons that consumes energy is needed to allow the reduction of coenzymes. In phototrophic green bacteria, the redox potential of ferredoxin is sufficiently electronegative (−0.42 V) to directly reduce coenzymes (Fig. 3.29b).

3.3.4.4 Use of Light Energy by the Archaea

Some extreme halophilic archaea, *Halobacterium salinarum* in particular, can use light as an energy source when dioxygen levels in their natural environment are too low. The membrane of archaea is largely invaded by dark red color that characterizes the purple membrane containing a colored protein, bacteriorhodopsin, near the rhodopsin of the retina of eyes. Bacteriorhodopsin has a carotenoid as prosthetic group, the all-trans-retinal bound by a **Schiff base*** which serves as photoreceptor. When the all-trans-retinal is activated by a photon, it isomerizes into 11-cis-retinal causing a change in the pattern of the protein, and the Schiff base loses a proton (Fig. 3.30) which is excreted to the outside. Then, the cis-retinal returns to its stable form of trans-retinal and

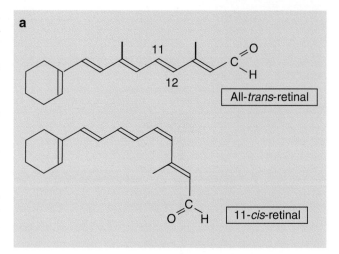

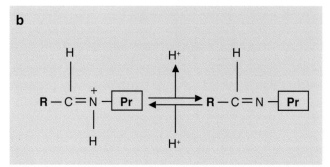

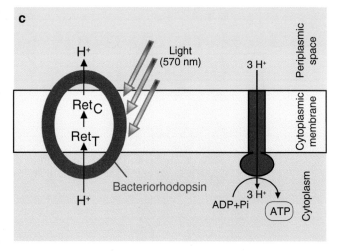

Fig. 3.30 Scheme of proton transfer by bacteriorhodopsin. (**a**) The two retinal isomers. (**b**) The mechanism of proton transfer by bacteriorhodopsin. Retinal (R) and protein (Pr) linked by a Schiff base. (**c**) Scheme of proton transfer and formation of proton-motive force in the purple membrane of halobacteria. *RetC* cis-retinal and *RetT* trans-retinal (Drawing: M.-J. Bodiou)

the Schiff base recovers a cytoplasmic proton. A new isomerization process can take place. Thus, bacteriorhodopsin acts as proton pump transferring protons from the inside to the outside of the cell under the influence of light. It establishes a proton gradient which results in a proton-motive force, an energy source for cells. Proteorhodopsin, a new rhodopsin similar to bacteriorhodopsin of the archaea, is widely distributed in marine bacteria and functions as a proton pump light dependent (Walter et al. 2007).

3.4 Production of Cellular Material and Biosyntheses

An important part of the energy produced by the cells is used to synthesize their cellular constituents (biosynthesis); this is anabolism. From inorganic and/or simple organic compounds, microorganisms produce more and more complex molecules at the origin of all structural and functional components of cells (Fig. 3.31).

The autotrophic microorganisms synthesize carbohydrates, lipids, proteins, nucleotides, and other constituents from inorganic molecules and ions: CO_2, ammonium, sulfate, phosphate, etc. The heterotrophic microorganisms depend on the organic molecules produced by the autotrophs. They are unable to synthesize their organic molecules from CO_2 and even sometimes prefer to use organic nitrogen of the amino acids. Several require growth factors, essential organic compounds they are unable to synthesize: vitamins, essential amino acids, purines, etc.

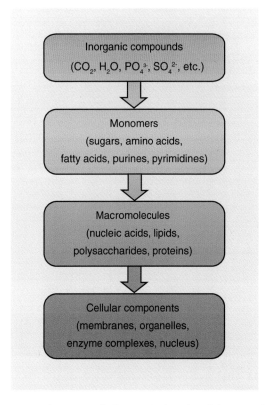

Fig. 3.31 Main steps of the synthesis of cellular constituents (Drawing: M.-J. Bodiou)

3.4.1 The Autotrophic Microorganisms: CO_2 Assimilation

Several metabolic pathways are known in microorganisms for assimilation (or fixation) of CO_2:
1. The pathway of ribulose 1,5-diphosphate or Calvin cycle
2. The reverse cycle of tricarboxylic acids
3. The reductive pathway of acetyl-CoA
4. The cycle of 3-hydroxypropionate
5. The pathway of C4

3.4.1.1 The Pathway of Ribulose 1,5-Diphosphate

This pathway or Calvin cycle (Calvin–Benson cycle, Calvin–Benson–Bassham cycle, or C3 cycle) is one of the most important biosynthetic processes of the biosphere. It is used to fix CO_2 by many autotrophic microorganisms (photosynthetic microeukaryotes, cyanobacteria, anoxygenic phototrophic purple bacteria, chemolithotrophic bacteria) and by plants.

The key enzyme of the Calvin cycle is ribulose 1,5-bisphosphate carboxylase or RuBisCo. It catalyzes the carboxylation of ribulose 1,5-diphosphate and thus allows fixing a CO_2 molecule in an organic form. The Calvin cycle can be divided into three phases (Fig. 3.32). Phase I corresponds to the phase of the fixation of three CO_2 molecules by reaction with three molecules of ribulose 1,5-diphosphate to form six molecules of 3-phosphoglycerate. During phase II or phase of reduction, the six molecules of 3-phosphoglycerate are reduced to 3-phosphoglyceraldehyde (PGA), one molecule being reserved for biosynthesis and the other five used to regenerate three molecules of ribulose 1,5-diphosphate in a series of complex biochemical reactions (phase III). Thus, one molecule of 3-phosphoglyceraldehyde (C3) is synthesized by the Calvin cycle from three molecules of CO_2:

$$3\ CO_2 + 9\ ATP + 6\ NADH, H^+ (or\ NADPH, H^+)$$
$$\rightarrow 1\ PGA + 9\ ADP + 8Pi + 6\ NAD^+ (or\ NADP^+)$$

To synthesize one molecule of PGA, nine ATP and six NADPH, H^+ are needed; CO_2 fixation is a process that requires considerable energy and reducing power. If the light is an inexhaustible source of energy available to phototrophic microorganisms, it is not the same for chemolithotrophic microorganisms where oxidation of the energy source usually generates little free energy. For example, in the case of *Nitrobacter* (nitrite-oxidizing bacteria), it is generally accepted that the oxidation of one ion-gram of NO_2^- allows the translocation of two protons that generate 0.6 mole of ATP. Moreover, the reduction of coenzymes involves a reverse electron flow that consumes six ATP by

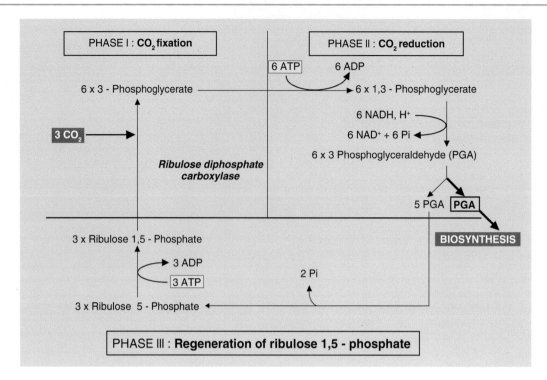

Fig. 3.32 The Calvin cycle (Drawing: M.-J. Bodiou)

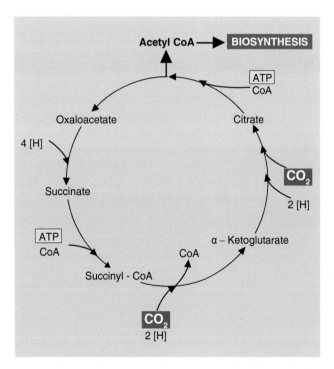

Fig. 3.33 The reverse tricarboxylic acid cycle (Drawing: M.-J. Bodiou)

reduced NAD^+. Thus, the reduction of three moles of CO_2 to form one mole of PGA requires 45 ATP or 75 ion-grams of NO_2^-. In these circumstances, it is clearly conceivable that these microorganisms oxidize a large amount of substrate and growth based on the biosynthesis is slow.

The Calvin cycle takes place in the stroma of chloroplasts in photosynthetic eukaryotes where RuBisCo is soluble, or is located in the pyrenoid in algae having this structure. In some prokaryotes, photosynthetic or non-photosynthetic (cyanobacteria, nitrifying bacteria, and aerobic sulfur-oxidizing bacteria), the Calvin cycle occurs in the cytoplasm where the RuBisCo is condensed as crystalline cytoplasmic inclusions, the carboxysomes. Studies have shown activity of RuBisCo in archaea (*Thermococcus*, *Archaeoglobus*, *Pyrococcus*, methanogens), although no Calvin cycle has been clearly identified (Mueller-Cajar and Badger 2007).

3.4.1.2 The Reverse Tricarboxylic Acid Cycle

Phototrophic green bacteria (*Chlorobium*), hydrogenotrophic bacteria (*Hydrogenobacter*), sulfate-reducing bacteria (*Desulfobacter*), and sulfur-reducing archaea (*Thermoproteus*, *Pyrobaculum*) fix CO_2 during the reverse cycle of tricarboxylic acids (Fig. 3.33). The steps from ketoglutarate to succinate and citrate to oxaloacetate are nonreversible steps of the normal cycle of Krebs. They are catalyzed by new enzymes and require energy. Each running cycle uses two ATP and eight reducing equivalents to reduce two CO_2 to obtain one molecule of acetyl-CoA reserved for cellular synthesis.

3.4.1.3 The Acetyl-CoA Reductive Pathway

Acetyl-CoA reductive pathway or Wood–Ljungdahl pathway described in acetogenic bacteria (*cf.* Sect. 3.3.2) allows

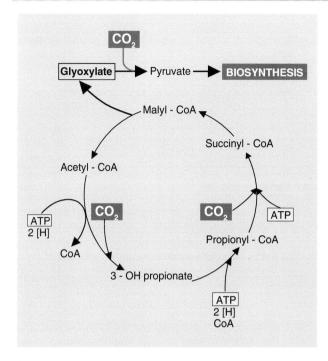

Fig. 3.34 The 3-hydroxypropionate cycle (Drawing: M.-J. Bodiou)

CO_2 fixation by most acetogenic bacteria, few sulfate-reducing bacteria (*Desulfobacterium*), or archaea (*Archaeoglobus*, methanogens).

3.4.1.4 The 3-Hydroxypropionate Cycle

3-Hydroxypropionate cycle is the way of CO_2 fixation in *Chloroflexus* (filamentous phototrophic green nonsulfur bacteria) and in some autotrophic archaea (*Sulfolobus*, *Acidianus*, *Metallosphaera*). During the cycle (Fig. 3.34), three ATP and four reducing equivalents are used to fix a first CO_2 in a C2 unit and a second CO_2 in a C3 unit to form a C4 compound (malyl-CoA). The enzymatic cleavage of malyl-CoA releases glyoxylate subsequently used in biosynthetic pathways and regenerates acetyl-CoA.

Zarzycki et al. (2009) proposed for *Chloroflexus aurantiacus*, a mechanism of CO_2 fixation involving the combination of two cycles: a previous cycle produces glyoxylate that is consumed by a second cycle which releases pyruvate for biosynthesis.

3.4.1.5 The C4 Pathway

Abundant in the marine environment, inorganic carbon is mainly in the form of bicarbonate. Normally, to provide a good functioning of the Calvin cycle or C3 pathway, the intracellular concentration of CO_2 must be high. In algae, carbonic anhydrase which interconverts CO_2 and bicarbonate is the main component of the mechanism of intracellular concentration of inorganic carbon that feeds the Calvin cycle:

$$HCO_3^- + H^+ \leftrightarrow CO_2 + H_2O$$

Another pathway optimizes the absorption and storage of CO_2 in the cells. This way or C4 pathway, known in terrestrial plants from hot and dry climate, has been described for diatoms which represent a significant fraction of the oceanic phytoplankton. The source of inorganic carbon is bicarbonate ion. This ion reacts with phosphoenolpyruvate to form a C4 compound, oxaloacetate, under the action of carboxylase. Oxaloacetate produces malate which is a form of CO_2 storage. Decarboxylation of malate produced pyruvate used in the biosynthesis and CO_2 which supplies the Calvin cycle.

In diatoms, the C4 pathway (Fig. 3.35) occurs in the cytoplasm and the Calvin cycle in the chloroplast. Diatoms in their natural environment are subject to significant fluctuations of light intensities depending on their position in the water column. At low light intensities, the C4 pathway allows storage of CO_2 in the form of malate. This CO_2 is released and used as a supplement during periods of high demand for high light intensities.

3.4.1.6 Other Ways of CO_2 Fixation

Two ways of CO_2 fixation, the hydroxypropionate–hydroxybutyrate and dicarboxylate–hydroxybutyrate cycles, have been described in archaea (Berg et al. 2010).

3.4.2 C1 Compound Assimilation and Related Compounds

The reduced C1 compounds (methane, methanol, formaldehyde, chloromethane), methylamines and methylated sulfur compounds, are assimilated by microorganisms, especially aerobic and anaerobic prokaryotes, microorganisms known as **methylotrophs***. The assimilation of these compounds is less costly in energy than CO_2 fixation because it requires less ATP and less reduced coenzymes.

3.4.2.1 Aerobic Methylotrophic Microorganisms

Some bacteria are obligate aerobic methylotrophs (*Methylobacter*, *Methylocystis*, *Methylobacillus*, etc.), and others may be facultative methylotrophs also able to use substrates containing more than one carbon atom (*Methylobacterium*, *Methylosulfonomonas*, *Paracoccus*, *Bacillus*, etc.). Among the obligate methylotrophs, only methanotrophic bacteria are capable of assimilating methane (*Methylobacter*, *Methylocystis*).

The C1 compounds are assimilated as formaldehyde (Fig. 3.36). The fixation of formaldehyde involves the

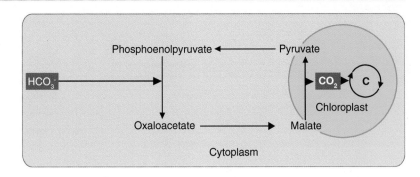

Fig. 3.35 The C4 pathway of diatoms (Modified and redrawn from Riebesell 2000). *c* Calvin cycle (Drawing: M.-J. Bodiou)

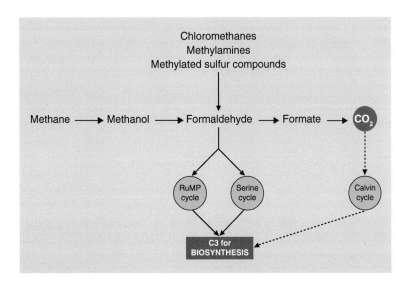

Fig. 3.36 Assimilation of C1 compounds (Modified and redrawn from Lidstrom 1991. Drawing: M.-J. Bodiou)

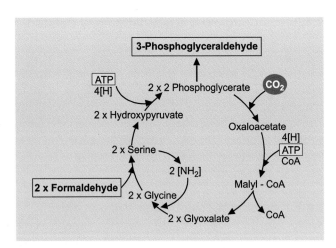

Fig. 3.37 The cycle of serine (Drawing: M.-J. Bodiou)

condensation with a compound containing more than one carbon atom and the production of a C3 molecule that feeds the biosynthesis. Two cycles are responsible for this fixation, the serine cycle (Fig. 3.37) and the ribulose monophosphate cycle (Fig. 3.38).

The formaldehyde is condensed with glycine (Fig. 3.37) or ribulose 5-phosphate (Fig. 3.38) to form glycerate 3-phosphate or dihydroxyacetone 3-phosphate, respectively. Bacteria known as pseudomethylotrophs can oxidize C1 to CO_2 which is then used by the Calvin cycle. Formaldehyde is also assimilated by eukaryotic microorganisms (yeasts) by condensation with xylulose 5-phosphate which produces glyceraldehyde 3-phosphate and dihydroxyacetone 3-phosphate.

3.4.2.2 Anaerobic Methylotrophic Microorganisms

Formation of acetyl-CoA described among acetogenic bacteria (*cf.* Sect. 3.3.2) is the usual way of fixing reduced C1 compounds among anaerobic microorganisms: methanogens and acetogens.

3.4.3 Heterotrophic Microorganisms

The heterotrophic microorganisms require organic compounds for growth. These compounds restore energy in various forms (proton-motive force, ATP, reduced coenzymes) and provide the carbon skeletons that are used in the biosynthesis.

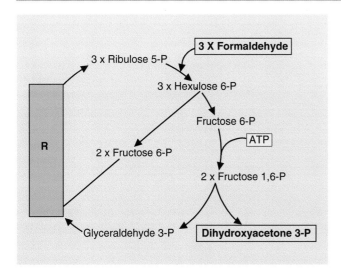

Fig. 3.38 The cycle of ribulose monophosphate (RUMP). *R* regeneration phase of the ribulose 5-P; *P* phosphate group (Drawing: M.-J. Bodiou)

In natural environments, the most abundant organic substances available to the microorganisms are macromolecules or biopolymers (polysaccharides, proteins, lipids, nucleic acids, etc.) that cannot penetrate into cells through the cytoplasmic membranes. Most heterotrophic microorganisms must excrete extracellular enzymes or exoenzymes that divide the polymer into small molecules (monomers) more easily transported inside the cells. The exoenzymes are mainly hydrolases that degrade polysaccharides (amylases, cellulases, pectinases, chitinases, xylanases, etc.), proteins (proteases), lipids (lipases), and nucleic acids (nucleases).

Some macromolecules are not degraded by hydrolases. This is, for example, the case of lignins that require the presence of polyphenoloxidases.

Unlike these enzymes that are released in the environment as true exoenzymes, certain enzymes involved in degradation of polymers can be fixed on the cell surfaces. Cellulolytic bacteria such as *Clostridium* and *Bacteroides* synthesize large protrusions attached on their surface, called cellulosomes. The cellulosomes are composed of multienzyme complexes containing endocellulases (enzymes hydrolyzing the bonds within the polysaccharide chain), xylanases, and other degradative enzymes. The attachment of cellulosomes to cellulose fibers provides intimate contact that allows the hydrolysis of the polymer.

Protozoa incorporate organic particles in their cells (organic debris, bacteria) which they feed by endocytosis. Monomers (amino acids, sugars) can also enter by endocytosis, by diffusion, or with the aid of carriers.

The hydrolysis of polysaccharides essentially provides hexoses. Lipases hydrolyze lipids into glycerol and fatty acids that penetrate the cells. During the β-oxidation, the fatty acids are cut into acetyl-CoA molecules. Proteases hydrolyze proteins into amino acids which are mostly deaminated inside cells into their corresponding keto acids. Nucleases hydrolyze nucleotides in nucleic acids that are absorbed by the cells after dephosphorylation and then degraded. All of these simple molecules resulting from enzymatic activities become the metabolic precursors of biosyntheses. They enter into the **central metabolic pathways*** to produce the monomers required for the synthesis of cellular macromolecules.

3.4.3.1 Assimilation of C2 Compounds

Some heterotrophic microorganisms can grow on C2 compounds as sole carbon source. To assimilate acetate, anaerobic bacteria activate the molecule in the form of acetyl-CoA, and then a following carboxylation produces pyruvate under reducing conditions. Aerobic microorganisms assimilate acetyl-CoA through the glyoxylate cycle (Fig. 3.40); other C2 compounds, glycine, glycolate, and oxalate, are first converted to glyoxylate and finally glycerate. Other metabolic pathways have been described for acetate assimilation such as citramalate cycle or acetoacetyl-CoA pathway.

3.4.3.2 Central Metabolic Pathways and Formation of the Carbon Skeleton of the Main Monomers

The central metabolic pathways (Fig. 3.39) are common to the vast majority of eukaryotic and prokaryotic microorganisms. Four major pathways are involved:

1. The glycolysis
2. The neoglucogenesis
3. The tricarboxylic acid cycle (TCA) or Krebs cycle
4. The pentose phosphate cycle

The carbon skeletons of the main monomers originate from metabolic intermediates of these pathways. The biosyntheses of purines and pyrimidines are complex reactions involving several sources of carbon and nitrogen.

Some steps of neoglucogenesis are nonreversible steps of glycolysis. The conversion of pyruvate to phosphoenolpyruvate occurs in two steps: a step of carboxylation to oxaloacetate and a decarboxylation step of oxaloacetate to phosphoenolpyruvate under the respective actions of a carboxylase and a carboxykinase with energy consumption (Fig. 3.39). The direct phosphorylation of pyruvate to phosphoenolpyruvate was described in Gram-negative bacteria. This reaction is catalyzed by phosphoenolpyruvate synthase and requires two energy-rich phosphate bonds. Two other steps of glycolysis are not reversible, dephosphorylation of fructose 1,6-P and of glucose-6-P which depend on the action of two specific phosphatases.

Biosynthesis pathways are less known in *Archaea*. The reverse path of glycolysis works even if the degradation of glucose does not always pass by this path. A complete cycle of oxidative citric acid, involving the same enzymes of bacteria, operates in aerobic archaea (archaeal halophiles,

Fig. 3.39 The central metabolic pathways, formation of carbon skeletons of the main monomers. *Gluconeogenesis* nonreversible steps of glycolysis are in *red*, *Pi* inorganic phosphate, *PPC* pentose phosphate cycle, *TAC* tricarboxylic acid cycle (Krebs cycle), *P* phosphate group, and *Frames* origin of the carbon skeletons of the monomers (Drawing: M.-J. Bodiou)

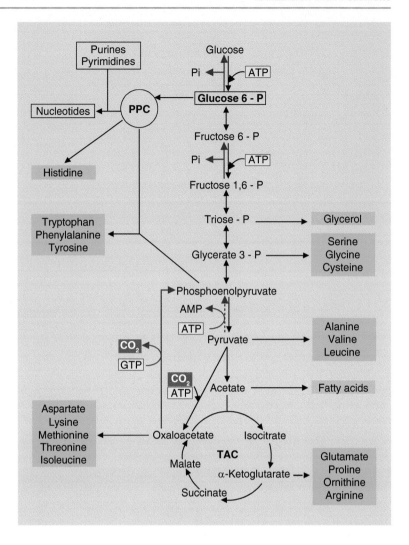

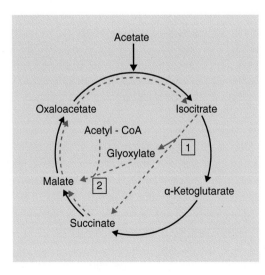

Fig. 3.40 Scheme of the glyoxylate cycle (in *red* in the Krebs cycle). *Black arrows*, normal citric acid; *1*, isocitrate lyase; and *2*, malate synthase (Drawing: M.-J. Bodiou)

Thermoplasma, *Sulfolobus*), while an incomplete cycle was often described in the anaerobic archaea (*Pyrococcus*, *Methanosarcina*, *Archaeoglobus*) (Danson et al. 2007). In some archaea, the whole of genes encoding enzymes of the cycle has been identified (*Halobacterium*, *Thermoplasma*, *Picrophilus*). In others, the cycle can run in the direction of reduction (reverse cycle) for CO_2 fixation during autotrophic growth (Hu and Holden 2006).

The oxaloacetate is required for operating the Krebs cycle. The removal of this compound for biosyntheses (neoglucogenesis, amino acid synthesis) must be compensated for the cycle to continue to operate. Reactions providing the compound or precursors are **anaplerotic sequences***. These reactions catalyzed by a carboxylase (3.6) and a carboxykinase (3.7) produce oxaloacetate:

$$\text{Pyruvate} + CO_2 + \text{ATP} \rightarrow \text{oxaloacetate} + \text{ADP} + \text{Pi} \quad (3.6)$$

$$\text{Phosphoenolpyruvate} + CO_2 + \text{GDP} \\ \leftrightarrow \text{oxaloacetate} + \text{GTP} \quad (3.7)$$

3 Structure and Functions of Microorganisms: Production and Use of Material and Energy

Table 3.16 Main monomers required for the biosynthesis of macromolecules

Monomers	Polymers
Glycerol and fatty acids	Lipids
Glucose 6-phosphate	Polysaccharides
Amino acids	Proteins
Pentose phosphates and purines or pyrimidines	Nucleic acids

Table 3.17 Properties of different nitrate reductases

	Assimilatory nitrate reductase (Nas)	Dissimilatory nitrate reductase (Nar)	Periplasmic nitrate reductase (Nap)
Location	Cytoplasm	Membrane	Periplasm
Inhibition by O_2	No	Yes	No
Inhibition by NH_3	Yes	No	No
Enzymatic induction	No[a]	Yes	No[a]
Function	NO_3^- assimilation	Anaerobic respiration of nitrate	Regulation of redox potential[b]

[a]Assimilative and periplasmic nitrate reductases are constitutive
[b]Nap removes excess reducing power by producing toxic NO_2^- for eventual competitors or providing NO_2^- to nitrite reductase, which allows microorganisms to adapt to rapid changes in oxygenation

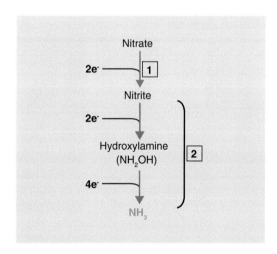

Fig. 3.41 The assimilatory reduction of nitrate. *1* nitrate reductase Nas; *2* nitrite reductase (Drawing: M.-J. Bodiou)

Another source of oxaloacetate is the formation of glyoxylate from isocitrate, which shortens the Krebs cycle avoiding the step of ketoglutarate (Fig. 3.40). The shorter Krebs cycle is called glyoxylate cycle.

The main monomers synthesized from different carbon skeletons are used to build macromolecules (Table 3.16).

3.4.4 Assimilation of Nitrogen, Sulfur, and Essential Elements

3.4.4.1 Assimilation of Nitrogen Compounds

The most abundant nitrogen sources for microorganisms are ammonia nitrogen, nitrate, and dinitrogen. Only ammonia nitrogen is directly assimilated by the microorganisms. Nitrate and dinitrogen must be reduced to ammonia to be assimilated. Protozoa are dependent to organic nitrogen (*cf.* Sect. 14.3.2).

Assimilatory Reduction of Nitrate

Nitrate is reduced to nitrite by assimilatory nitrate reductase (Nas) whose properties are different from dissimilative reductases previously described (Table 3.17) and used in energy metabolism. Nitrite is then reduced to ammonia in several steps by a nitrite assimilatory reductase (Fig. 3.41). The reduction of nitrate to ammonium nitrogen is coupled to a high consumption of ATP and reducing power (reduced coenzymes):

$$NO_3^- + 8e^- + 9H^+ \rightarrow NH_3 + 3H_2O$$

Microorganisms unable to reduce nitrate must find the ammonia nitrogen in their environment.

Dinitrogen Fixation

Many aerobic and anaerobic microorganisms can assimilate dinitrogen (Table 3.18). This assimilation or dinitrogen fixation is under the control of an enzyme complex called nitrogenase complex. Nitrogenase is very sensitive to dioxygen. Among aerobic heterotrophs, nitrogen fixation occurs when the respiration equals or exceeds the rate of dioxygen diffusion into the cells. In some filamentous cyanobacteria, nitrogenase is localized in specialized cells (heterocysts) devoid of photosystem II which produces dioxygen.

Nitrogenase is a metalloprotein complex. Some nitrogenases contain vanadium or iron, but the most common nitrogenases contain molybdenum. The functional enzyme is composed of two types of soluble proteins: one is molybdenum iron protein (MoFe) or dinitrogenase, and the other is known as protein iron (Fe) or dinitrogenase reductase. In a first step, a flavodoxin or ferredoxin is reduced (Fig. 3.42). Then the dinitrogenase reductase accepts electrons and transmits them to the dinitrogenase that reduces dinitrogen into ammonia. ATP needs are important to reduce the triple bond linking the two nitrogen atoms:

$$N_2 + 8\,e^- + 8\,H^+ + 16\,ATP \rightarrow 2\,NH_3 + H_2 + 16\,ADP + 16\,Pi$$

If microorganisms are provided with another source of assimilable nitrogen, they do not fix dinitrogen. Ammonia nitrogen also represses the synthesis of nitrogenase complex.

Table 3.18 Examples of microorganisms fixing dinitrogen under nonsymbiotic conditions

Chemotrophic microorganisms	Phototrophic microorganisms
Aerobic microorganisms	Aerobic microorganisms
Azotobacter, Azomonas, Klebsiella, Azospirillum, Gluconacetobacter, Bacillus, Thiobacillus, Alcaligenes, methylotrophic bacteria	Cyanobacteria
Anaerobic microorganisms	Anaerobic microorganisms
Bacteria: *Clostridium*, sulfate-reducing bacteria	Anoxygenic phototrophic bacteria
Archaea: methanogens	

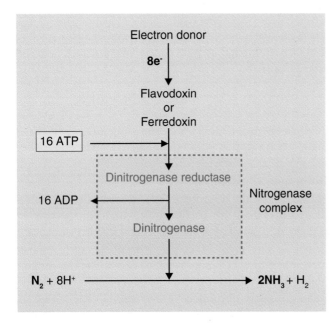

Fig. 3.42 Steps of dinitrogen fixation (Drawing: M.-J. Bodiou)

The bacteria mentioned above that fix dinitrogen without any symbiotic association are called free nitrogen-fixing bacteria. Bacteria known as symbiotic nitrogen-fixing bacteria fix dinitrogen only in association with plants.

Rhizobium and other genera (*Bradyrhizobium, Azorhizobium*), belonging to alphaproteobacteria, infect the roots of legumes, and these cause the formation of nodules. In the nodules, the cells of *Rhizobium* are isolated from the plant material by the surrounding membranes called sequestration membranes. Inside the sequestration membrane, the bacteria continue to multiply at the expense of plant metabolites and then differentiate into bacteroids fixing dinitrogen. In this form, the multiplication of bacteria stops. The leghemoglobin that is present in the nodules is synthesized in part by the plant (globin) and partly by the bacteroids (heme). This molecule regulates the supply of dioxygen to the microorganisms and thereby protects nitrogenase and allows proper operation. Bacteroids fix dinitrogen and release of ammonia nitrogen assimilated by the plant that provides organic energy sources for microorganisms.

The symbiotic nitrogen-fixing actinomycetes of the genus *Frankia* establish associations with different plants. They form root nodules with woody plants such as alders and various other plants belonging to the families Rhamnaceae, Myricaceae, Rosaceae, etc. In general, dinitrogen fixation by free *Frankia*, but also often by symbiotic *Frankia*, is correlated with the presence of vesicles containing nitrogenase. Thick envelopes of these structures limit diffusion of dioxygen and create favorable conditions for enzyme activity.

The presence of free-fixing bacteria associated with plants has been described. Nodules of *Psychotria* leaves (plant of Rubiaceae family) contain *Klebsiella*. Fixing cyanobacteria (*Anabaena*) are associated with aquatic ferns (*Azolla*). Some bacteria of the genus *Gluconacetobacter* are endophytes of tissues of sugarcane; others multiply in the rhizosphere of coffee plants. *Azotobacter* grows in the rhizosphere of corn. It is not certain that in these cases the bacteria that benefit from root exudates contribute to nitrogen nutrition of plants (*cf.* Sect. 11.3).

Incorporation of Ammonium

Microorganisms incorporate ammonia nitrogen in organic compounds by using one of the mechanisms described in the following equations:

$$\alpha\text{-ketoglutarate} + NH_3 + NADPH, H^+ \rightarrow \text{L-glutamate} + H_2O + NADP^+ \textit{glutamate dehydrogenase}$$

$$\text{Pyruvate} + NH_3 + NADPH, H^+ \rightarrow \text{L-alanine} + H_2O + NAD^+$$
alanine dehydrogenase

$$\text{Glutamate} + NH_3 + ATP \rightarrow \text{glutamine} + ADP + Pi$$
glutamine synthetase

$$\alpha\text{-ketoglutarate} + \text{glutamine} + 2\,H^+ + e^- \rightarrow 2\,\text{glutamate}$$
Glutamine α-ketoglutarate aminotransferase

Only a few microorganisms can incorporate ammonia nitrogen in the pyruvate. Glutamate and glutamine are the nitrogen source of the cellular amino acids. Glutamate is the major source of nitrogen.

The other amino acids result from changes to the molecular structure of glutamate to a small number of them (arginine, proline), but essentially from transaminations (transfer reactions of the amino group catalyzed by transaminase) with keto acids (aspartate, arginine, alanine, threonine,

Fig. 3.43 Assimilatory reduction of sulfate (**a**) and incorporation of the sulfide (**b**). (**a**) Steps in the reduction of assimilative sulfate. *1* ATP sulfurylase, *2* phosphokinase, *3* PAPS reductase, *4* sulfite reductase. *RSH* reduced thioredoxin, *RSSR* oxidized form of thioredoxin (RSH regeneration with NADPH, H⁺), *APS* adenosine 5′-phosphosulfate, *PAPS* phosphoadenosine 5′-phosphosulfate, and *PAP* phosphoadenosine 5′-phosphate, *PPi* pyrophosphate. (**b**) The incorporation of sulfide to O-acetylserine. *1* Serine transacetylase; *2* O-acetylserine sulfhydrylase (Drawing: M.-J. Bodiou)

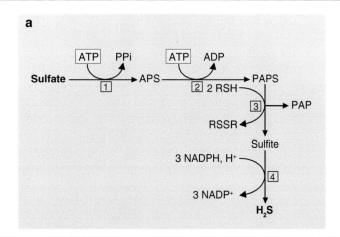

isoleucine, methionine, lysine, leucine, valine, serine, glycine, cysteine, phenylalanine, tyrosine). The synthesis of aspartate by transfer of amino group of glutamate to oxaloacetic acid catalyzed by glutamate oxaloacetate transaminase is an example of transamination (*cf.* Sect. 14.3.3, Fig. 14.29):

Oxaloacetate + L-glutamate → α-ketoglutarate
+ L-aspartate

Some transamidations (transfer of the amide group of glutamine) are involved in the synthesis of some amino acids (tryptophan, arginine).

The nitrogen of purines and pyrimidines, other important cellular nitrogen compounds, is derived from transamination and transamidation reactions and transfers of carbamyl group (NH_2–CO–) of carbamyl phosphate (NH_2–CO–O–PO_3^{2-}) whose synthesis is shown in the following equation:

Glutamine + CO_2 + 2 ATP → glutamate + carbamyl
phosphate + 2 ADP + 2 Pi *Carbamyl phosphate synthase*

The synthesis of arginine also implies a carbamyl group. That of histidine is a complex metabolic pathway where the nitrogen has several origins (purine, transamination, and transamidation reactions).

3.4.4.2 Assimilation of Sulfur Compounds

The most abundant form of sulfur in the biosphere is sulfate which is the most oxidized sulfur state (oxidation state + VI) (*cf.* Sect. 14.4.1, Fig. 14.34). Sulfate is the sulfur source used by most microorganisms; animal cells require sulfur in organic form.

In organic matter, sulfur is reduced, mainly in the form of sulfhydryl groups (oxidation state − II). Sulfate represents a too oxidized sulfur source to be included in organic compounds. To be assimilated, it must be reduced to oxidation state − II. The first step in the assimilatory sulfate reduction is the activation in the form adenosine 5′-phosphosulfate or APS (Fig. 3.43). Subsequent phosphorylation leads to the phosphoadenosine 5′-phosphosulfate or PAPS. PAPS is reduced to sulfide in two stages. The sulfide is incorporated into the O-acetylserine to form cysteine. This metabolic pathway has also been described among eukaryotic parasites (*Entamoeba*, *Leishmania*). Cysteine is the main precursor of other sulfur compounds from cells (methionine, coenzyme A, lipoic acid, thiamine, glutathione, iron–sulfur centers).

Microorganisms incapable of assimilatory sulfate reduction must find reduced sulfur compounds in their environment. They are often anaerobic microorganisms that grow in anoxic environments where reduced sulfur compounds are generally abundant.

3.4.4.3 Uptake of Other Inorganic Compounds, Trace Elements, and Essential Factors

Under the action of ATP synthase, phosphate that enters the cell by specific carrier reacts with ADP to give ATP. ATP is the source of phosphate for the synthesis of macromolecules, in particular nucleic acids, phospholipids, and nucleotides.

Microorganisms should find other mineral compounds in their environment such as the general anions and cations which penetrate cells by carrier systems. Potassium, calcium, magnesium, and iron are major elements essential to the cells for several reasons, including enzyme cofactors. Magnesium is present in cell walls, membranes, and phosphoric esters, calcium in coenzymes and bacterial spores in the form of dipicolinate, and iron in metalloproteins (cytochromes, ferredoxins, and iron–sulfur proteins). In oxic environments, extremely insoluble ferric salts are inaccessible to microorganisms. For that reason, microorganisms produce and excrete organic compounds or siderophores that chelate and trap iron (*cf.* Sect. 14.5.2). This chelated iron is absorbed by the cells. However, in anoxic environments, iron present in the form of soluble ferrous iron is readily available to cells.

In general the sodium requirements are low, except in halophilic microorganisms that require higher concentrations of ions Na^+ in their environment and sometimes accumulate in the cytoplasm to compensate osmotic pressure. Diatoms are important needs in silicon, an element essential to the formation of cell walls.

In addition to these compounds, many other elements or trace elements are needed in low concentrations for cellular activities. These are mainly manganese, zinc, molybdenum, selenium, cobalt, nickel, copper, and tungsten. They act as enzyme cofactors or integral part of enzymes.

3.5 Conclusion

Unlike their poorly differentiated structures compared to eukaryotes, prokaryotes have a very diverse metabolism especially in their energy sources (organic or inorganic substrates used through different aerobic and anaerobic respirations, fermentations, or photosynthesis) placing them as key actors with the major role in the transformation of organic or inorganic elements on the planet (*cf.* Chap. 14). Moreover, the metabolic versatility of many prokaryotes gives them a plasticity allowing their adaptation to changing environmental conditions. For example, members of the genus *Shewanella*, facultative anaerobic bacteria that grow in chemolithotrophic conditions with dihydrogen as an energy source or in chemoorganotrophic conditions, exhibit a very large diversity of respiration. In the absence of dioxygen, many of organic or inorganic compounds, including toxic elements and metals, are terminal electron acceptors (trimethylamine oxide, iron III, manganese IV, chromium VI, VI uranium, sulfur, polysulfide, sulfite, thiosulfate, dimethyl sulfoxide, arsenate, succinate, etc.). Similarly, some anoxygenic phototrophic bacteria show great adaptability to environmental conditions. This is the case of representatives of the genus *Thiocapsa* that, under anoxic conditions and in the light, use light radiations as energy source, reduced inorganic compounds of sulfur, dihydrogen, and small organic molecules as electron source, and organic molecules or carbon dioxide for their carbon source. Placed in the dark, under oxic conditions, these microorganisms exhibit an aerobic respiratory mechanism using organic compounds or inorganic sulfur, and under anoxic conditions, they ferment their intracellular reserves of polysaccharides using intracellular sulfur as electron trap. Our knowledge on the adaptability of microorganisms to environmental conditions has increased with the discovery of unexpected metabolic pathways such as aerobic respiration in bacteria long time considered as strictly anaerobic (sulfate-reducing bacteria) and anoxygenic photosynthesis in aerobic chemoorganotrophic heterotrophic bacteria (aerobic anoxygenic phototrophic bacteria).

Prokaryotes are the microorganisms most important in the biodegradation processes, and thus they are used either naturally or artificially in the processes of decontamination and biopurification of anthropized ecosystems (*cf.* Chap. 16). Eukaryotes have developed structures more and more complex with the recovery of both the most energetic prokaryotic metabolisms, both in relation to dioxygen: aerobic respiration and oxygenic photosynthesis.

References

Berg IA, Kockelkorn D, Ramos-Vera WH, Say RF, Zarzycki J, Hügler M, Alber BE, Fuchs G (2010) Autotrophic carbon fixation in *Archaea*. Nature rev 8:447–460

Carballido-López R (2006) The bacterial actin-like cytoskeleton. Microbiol Mol Biol Rev 70:888–909

Danson MJ, Lamble HJ, Hough DW (2007) Central metabolism. In: Cavicchioli R (ed) Archaea: molecular and cellular biology. ASM Press, Washington, DC, pp 260–287

Deppenmeier U, Lienard T, Gottschalk G (1999) Novel reactions involved in energy conservation by methanogenic archaea. FEBS Letters 457:291–297

Detkova EN, Pusheva MA (2006) Energy metabolism in halophilic and alkaliphilic acetogenic bacteria. Microbiology 75:1–11

Dubey GP, Ben-Yehuda S (2011) Intracellular nanotubes mediate bacterial communication. Cell 144:590–600

Ettwig FK et al (2010) Nitrite-driven anaerobic methane oxidation by oxygenic bacteria. Nature 464:543–547

Fuchs BM et al (2007) Characterization of a marine gammaproteobacterium capable of aerobic anoxygenic photosynthesis. Proc Natl Acad Sci USA 104:2891–2896

Heidelberg JF et al (2004) The genome sequence of the anaerobic sulfate-reducing bacterium *Desulfovibrio vulgaris* Hildenborough. Nature Biotechnol 22:554–559

Hooper AB, Vannelli T, Bergmann DJ, Arciero DM (1997) Enzymology of the oxidation of ammonia to nitrite by bacteria. Antonie van Leeuwenhoek 70:59–67

Hu Y, Holden JF (2006) Citric acid cycle in the hyperthermophilic archaeon *Pyrobaculum islandicum* grown autotrophically, heterotrophically, and mixotrophically with acetate. J Bacteriol 188:4350–4355

Le Gall J, Xavier AV (1996) Anaerobe response to oxygen: The sulfate-reducing bacteria. Anaerobe 2:1–9

Lidstrom ME (1991) Aerobic Methylotrophic bacteria. In: Balows A, Trüper HG, Dworkin M, Harder W, Schleifer KH (eds) The prokaryotes. Springer, New York, pp 431–445

Lobo SAL, Melo AMP, Carita JN, Teixeira M, Saraiva LM (2007) The anaerobe *Desulfovibrio desulfuricans* ATCC 27774 grows at nearly atmospheric oxygen levels. FEBS Letters 581:433–436

Madigan MT, Martinko JM, Stahl DA, Clark DP (2010) Brock: biology of microorganisms, 13th edn. Pearson Benjamin-Cummings, San-Francisco

Matias PM, Pereira IAC, Soares CM, Carrondo MA (2005) Sulphate respiration from hydrogen in *Desulfovibrio* bacteria: a structure biology overview. Progr Biophys Mol Biol 89:292–329

Milucka J et al (2012) Zero-valent sulphur is a key intermediate in marine methane oxidation. Nature 491:541–546

Mueller-Cajar O, Badger MR (2007) New roads lead to Rubisco in archaebacteria. Bioessays 29:722–724

Muller V (2003) Energy conservation in acetogenic bacteria. Appl Environm Microbiol 69:6345–6353

Muller V, Blaut M, Heise R, Winner C, Gottschalk G (2008) Sodium bioenergetics in methanogens and acetogens. FEMS Microbiol Letters 87:373–376

Riebesell U (2000) Carbon fix for diatom. Nature 407:959–960

Rohwerder T, Sand W (2003) The sulfane sulfur of persulfides is the actual substrate of the sulfur-oxidizing enzymes from *Acidithiobacillus* and *Acidiphilium* spp. Microbiology 149:1699–1709

Santana M (2008) Presence and expression of terminal oxygen reductases in strictly anaerobic sulfate reducing bacteria isolated from salt-marsh sediments. Anaerobe 14:145–156

Stams ASH, Plugge C (2009) Electron transfer in syntrophic communities of anaerobic bacteria and archaea. Nat Rev Microbiol 7:568–577

Stanier RV, Ingraham JL, Wheelis ML, Painter PR (1986) The microbial world. Prentice-Hall, Englewood cliffs

Valdés J, Pedroso I, Quatrini R, Dodson RJ, Tettelin H, Blake R, Eisen JA, Hokmes DS (2008) *Acidithiobacillus ferrooxidans* metabolism: from genome to industrial applications. BMC genomics 9:597–621

Voordouw G (1995) The genus *Desulfovibrio*: the centennial. Appl Environ Microbiol 61:2813–2819

Walter JM, Greenfield D, Bustamante C, Liphardt J (2007) Light-powering *Escherichia coli* with proteorhodopsin. Proc Nat Acad Sci 104:2408–2412

Ward N, Staley JT, Fuerst JA, Giovannovi S, Schlesner H, Stackbrandt E (2006) The order *Planctomycetales*, including the genera *Planctomyces*, *Pirellula*, *Gemmata* and *Isosphaera* and the *Candidatus* genera "Brocadia, Kuenenia and Scalindua". In: Dworkin M, Falkow S, Rosenberg E, Schleifer K, Stackebrandt E (eds) The prokaryotes. Springer, New York, pp 757–793

Yurkov VV, Beatty JT (1998) Aerobic anoxygenic phototrophic bacteria. Microbiol Mol Biol Rev 62:695–724

Zarzycki J, Brecht V, Müller M, Fuchs G (2009) Identifying the missing steps of the autotrophic 3-hydroxypropionate CO_2 fixation cycle in *Chloroflexus aurantiacus*. PNAS 106:21317–21322

Zehnder AJB, Stumm W (1988) Geochemistry and biogeochemistry of anaerobic habitats. In: Zehnder AJB (ed) Biology of anaerobic microorganisms. Wiley, New York, pp 1–38

Part II

Taxonomy and Evolution

4. For Three Billion Years, Microorganisms Were the Only Inhabitants of the Earth

Jean-Claude Bertrand, Céline Brochier-Armanet, Manolo Gouy, and Frances Westall

Abstract

Microorganisms were the sole inhabitants of our planet for almost 3 billion years. They have survived the intense geological upheavals that have marked the history of the Earth. They profoundly shaped their environment, thus participating in a true co-evolution between the biosphere and the geosphere. Through their activity, they also created favourable conditions for the emergence of multicellular aerobic organisms (particularly with an intense production of oxygen released into the atmosphere).

Among past microorganisms, LUCA occupied a central position in the evolutionary history of life. The possible origin and the large uncertainties about the nature of LUCA are discussed: where and when did LUCA live? Was it a hyperthermophilic, thermophilic or mesophilic organism? How did its genome look like?

Scenarios and hypotheses regarding the emergence and the relationships of the three domains of life – *Archaea*, *Bacteria* and *Eucarya* – as well as the transition from a prokaryotic to eukaryotic cell organisation are discussed in the light of the most recent data. Possible major steps in the evolution of microorganisms are deduced from genomic investigations and from the geological record (fossils, isotopic ratios, biomarkers). Although the early steps of microbial metabolic evolution are still hotly debated, it is possible to speculate on the occurrence of the first living entities, from the primordial metabolisms to the advent of photosynthesis.

Keywords

Early evolution • Evolution of the Earth • Evolutionary scenarios of life • Fossils • Geological record • History of life on Earth • LUCA • Tree of life

* Chapter Coordinator

J.-C. Bertrand*
Institut Méditerranéen d'Océanologie (MIO), UM 110, CNRS 7294 IRD 235, Université de Toulon, Aix-Marseille Université, Campus de Luminy, 13288 Marseille Cedex 9, France
e-mail: Jean-claude.bertrand@univ-amu.fr

C. Brochier-Armanet • M. Gouy (✉)
Laboratoire de Biométrie et Biologie Évolutive, UMR CNRS 5558, Université Claude Bernard Lyon 1, 69622 Villeurbanne Cedex, France

F. Westall
Centre de Biophysique Moléculaire, CNRS, 45071 ORLEANS Cedex 2, France

4.1 From The 'RNA World' to the Last Common Ancestor of All Living Beings: LUCA

This chapter will not present the various hypotheses that have been proposed to explain the terrestrial or extraterrestrial origin of life. The plot will begin with the appearance of the last common ancestor of all living beings. The Earth was formed about 4.5 billion years ago, and life must have emerged before 3.5 billion years (Ga) ago because the oldest fossil and isotopic traces of life have been found in South Africa and in Australia in particularly well-preserved rocks dated circa 3.5 Ga (that is to say, a billion years after the formation of the Earth) (see review in Westall 2011). The oldest traces indicative of possible biological activity (Box 4.1) (derived from the value of the $^{12}C/^{13}C$ ratio) were detected in rocks collected from the Isua site in West Greenland dating back to 3.7 billion years (Rosing 1999). However, given the state of preservation of these rocks, the exact date of the appearance of the first traces of life is still the subject of intense controversy.

In the history of the evolution of life, there is consensus about the initial appearance of an 'RNA world' (Fig. 4.1), supported by several arguments. Although RNA (like DNA) is a molecule present in all extant cells, the deoxyribonucleotides of DNA are synthesised from ribonucleotides which form the RNA, suggesting that RNA may have preceded DNA. According to Guy Ourisson and Yoichi Nakatani (1994), the first membrane structures were formed by self-assembly of polyprenyl phosphates organised in bilayers. The question of whether membranes or RNA came first is still under discussion. Nevertheless, a membrane structure containing RNA molecules is a plausible evolutionary step. The following step was protein synthesis by RNA. Indeed, Thomas Cech and Sidney Altman showed that some RNA molecules, called ribozymes, are capable of performing catalytic functions. Most noticeably, peptidyl transferase, responsible for the synthesis of the chemical bonds linking amino acids in proteins, is a ribozyme. This discovery therefore argues strongly in favour of the anteriority of RNA vis-à-vis proteins. DNA, the current carrier of genetic information, would have appeared later by replacement of the ribose of RNA with deoxyribose and by the substitution of uracil by thymine.

This transformation could have taken place through a mechanism such as 'reverse transcriptase' and would have resulted in the transfer of the information contained in RNA to DNA (Fig. 4.1a). This more stable molecule would have allowed the building of larger genomes and therefore increased storage of genetic information. An alternative hypothesis involves viruses. For Patrick Forterre, the transformation RNA → U-DNA → T-DNA took place in the viral world and was later transferred to the cellular world (Fig. 4.1b). The passage from RNA to DNA in viruses could have been an adaptation to escape the defence systems of their hosts directed against viral RNA genomes (Forterre 2002). The existence of viruses incorporating uracil instead of thymine in their DNA seems to support this hypothesis.

4.1.1 Definition(s) of LUCA

It is very likely that many cellular structures populated the early Earth. One of them led to the last universal common ancestor of all living beings, commonly known by the

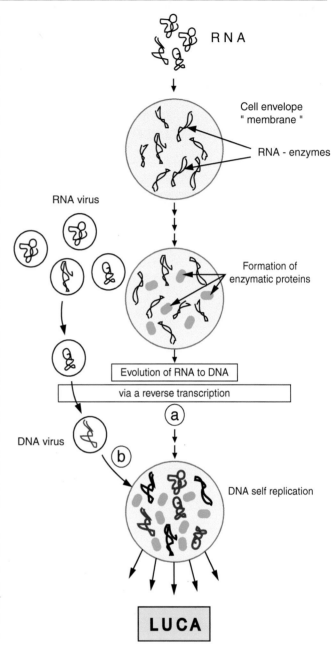

Fig. 4.1 Sketch explaining the possible passage from the 'RNA world' to the last common ancestor of all living organisms (Drawing: M.-J. Bodiou)

acronym LUCA. This term was coined in 1996 at an international conference on the initiative of Patrick Forterre. For a number of authors, LUCA was probably the result of a long evolutionary history and lived at a time when life was probably very abundant and diverse. This implies that LUCA lived contemporaneously with many organisms that did not leave any living descendants today (Fig. 4.2).

It is important to note that there is no consensus in the scientific community about the precise nature of LUCA and

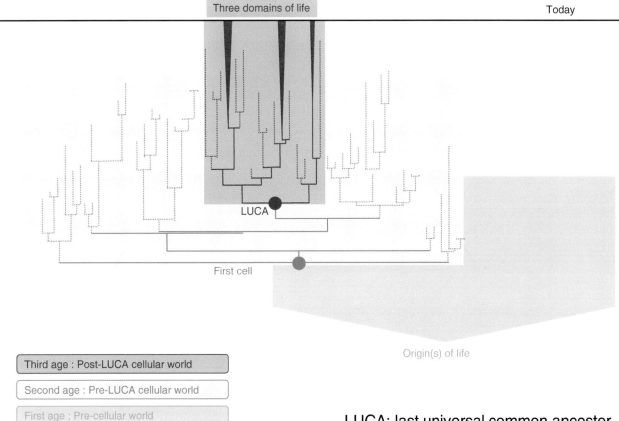

Fig. 4.2 Evolution of life on the Earth since its origin(s). Three evolutionary stages can be distinguished: (1) the first period including noncellular life forms (precellular world); (2) the second period, which includes descendants of the first cell (cellular pre-LUCA world) and ends with LUCA; and (3) the third period, which begins with LUCA and continues to this day (post-LUCA cellular world). It is important to note that the representatives of the different stages coexisted at certain periods of time. Thus, the first cells (representatives of the second period) very likely coexisted with their acellular cousins, descendants of the first age. Similarly, the descendants of LUCA probably coexisted for a time with their cousins in the second period. The duration of these three evolutionary stages and of possible periods of cohabitation remains completely unknown. It is probable that the majority of evolutionary lineages that came in existence became extinct without leaving any descendants today (Modified and redrawn after Forterre et al. 2005)

there are two opposing schools of thought. For some, LUCA is the organism (i.e. the cell) from which evolutionary lineages that are at the origin of the three domains of today's living world diverged *(Archaea, Bacteria* and *Eucarya)*. It is a genealogical vision of evolution, based on the research of ancestor-to-descendant relationships between organisms. Accordingly, LUCA was a single cell from which all existing cells descend (Fig. 4.3a). The other school of thought includes, in addition to the genealogy of organisms, the evolutionary history of all their constituents (i.e. genes). Because of the exchange of genetic material between organisms (**horizontal gene transfer***, HGTs), there would not be a 'single' LUCA but as many LUCAs as there are genes (Fig. 4.3b), each with its own evolutionary history that may differ from that of the other genes. For the purposes of this chapter, the term LUCA designates the common cellular ancestor of all extant cells (Fig. 4.3a).

4.1.2 When and Where Did LUCA Live

To try and describe LUCA, especially its physiology and metabolism, it is essential to understand its habitat and therefore at what time it could have lived. Palaeontological data suggest that from 3.2 to 3.5 Ga, there was an already diverse cellular life involving complex interactions between different populations, all of which are of the prokaryotic type (Westall 2011). However, the presence of diverse cellular life does not indicate whether these organisms were descendants of LUCA (red zone of Fig. 4.2) or its ancestors (yellow zone). Therefore, the uncertainty about the date of the occurrence of LUCA is considerable, although it is likely that LUCA lived during the Hadean-Archaean (the geological period prior to 4 and up to 2.5 Ga).

Despite all the uncertainties about the environmental conditions prevailing at this time, and taking into account

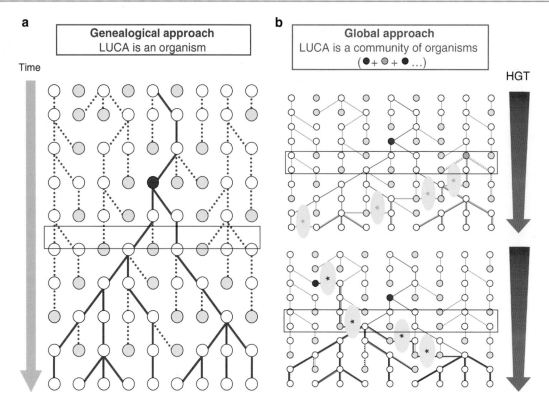

Fig. 4.3 Two views about LUCA. *Circles* represent organisms that inhabited the Earth at time t. Each *line of circles* represents a generation. For reasons of simplicity, it is assumed that (1) reproduction from one generation to the other is asexual and synchronous (i.e. no overlapping generations); (2) each organism gives rise to zero, one, two or three descendants; and (3) the number of organisms that inhabit the Earth at each generation is constant. (**a**) LUCA is the common ancestor to all currently living organisms. Starting from organisms inhabiting the Earth now, it is possible to go from generation to generation, until the so-called point of coalescence (most recent point from which all existing organisms are derived). This coalescence point necessarily exists and represents LUCA. This implies that none of the contemporaries of LUCA left descendants living now. (**b**) Here, the origin and the evolution of each gene of existing organisms is traced. The history of organisms is outlined in *red*. Other colours (*green* and *blue*) represent the history of two genes. Because of genetic material exchange between contemporary organisms (*asterisks*), the origin and evolutionary history of each gene can be distinct and differ from those of the organisms. There exist virtually as many LUCAs as genes (*coloured circles*). The *arrow* situated on the right of panel b shows the frequency of horizontal gene transfers (HGTs): initially high and gradually lower (*red*)

that these conditions most likely varied during the Archaean, it is possible to provide some clues about the climatic conditions of the primitive Earth that were markedly different from those known today (Nisbet and Fowler 2003). Volcanic and hydrothermal activity was more intense than at present. The amount of energy emitted by the sun that reached the surface of the Earth was 20–30 % lower than today. The atmosphere was probably richer in CO_2 but also contained one or more other greenhouse gases, such as methane (CH_4). The intensity of the greenhouse effect would have offset the weaker luminescence of the early sun and allowed water to remain on the surface in the liquid state. The presence of trace amounts of molecular nitrogen (N_2), ammonia (NH_3), nitric oxide (NO), nitrate (NO_3^-), carbon monoxide (CO), dihydrogen (H_2), sulphide (H_2S) and sulphate (SO_4^{2-}) in the atmosphere is also plausible. One important issue is the concentration of dioxygen (O_2). Most current hypotheses postulate that the atmosphere and the oceans were virtually anoxic. However, some authors believe that dioxygen generated by the photolysis of water was present in the atmosphere and the ocean surface at low concentrations in the range of 0.2–2 % of the current atmospheric level (21 %). Others argue instead that the concentration of dioxygen could have been much larger (Ohmoto 2004).

Regarding the temperature of the ocean, isotopic analysis of oxygen and silicon trapped in Archaean rocks suggests that the average temperature was higher than today, possibly about 50 °C (van den Boorn et al. 2007) or even 70–80 °C (Knauth and Lowe 2003). Under such conditions, the majority of available ecological niches would have been habitable by thermophilic organisms. These high temperatures would imply a very high level of CO_2 in the atmosphere (2–6 bar) and, therefore, extremely acidic rains (pH 3.7), which would have rapidly corroded exposed proto-continental rocks. However, there is no geological evidence for large scale

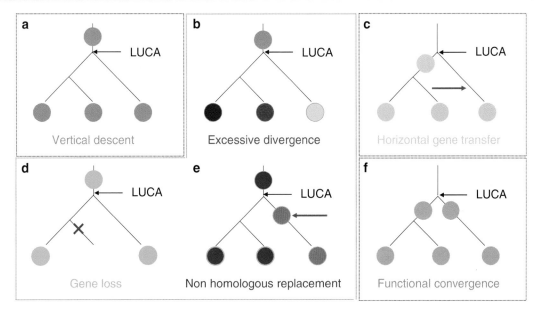

Fig. 4.4 Possible evolutionary scenarios explaining the presence or absence of a character in the three domains of life. (**a**) A gene appeared in an ancestor of LUCA (*pink circle*) and is present in LUCA and conserved in the three domains of life throughout evolution; (**b**) a gene appeared in an ancestor of LUCA (*pink circle*) and is present in LUVA and conserved in the three domains of life throughout evolution, but cannot be identified as a universal character because of excessive divergence; (**c**) a gene appeared in a descendant of LUCA (*blue dot*) and was transmitted by horizontal transfer (*red arrow*); (**d**) loss (*red cross*) of a gene present in an ancestor of LUCA (*green circle*); (**e**) replacement of a gene present in an ancestor of LUCA (*green circle*); (**f**) independent appearance (convergence) of the same function performed by nonhomologous genes; (**b, d, e**) a gene present in LUCA but not simultaneously detected in all three domains; (**c**) a universal gene that does not exist in LUCA, (**f**) a universal function but nonuniversal genes

corrosion of such magnitude, possibly because most continental surfaces were under water. The atmospheric CO_2 content was therefore probably lower and the ocean temperature lower than the higher estimates (Kasting and Howard 2006).

4.1.3 How to Describe LUCA?

The description of LUCA, a now defunct organism, may seem, a priori, a utopian enterprise. However, comparative studies based on the study of living organisms allow the formulation of a number of hypotheses. As some of them are very speculative, any dogmatic position should be avoided.

The starting point of these studies is based on the fact that the current members of the three domains of life have a number of common traits, such as a DNA genome, the same genetic code, a complex ribosome and a cytoplasmic membrane. The most parsimonious explanation for the presence of features common to all organisms is to consider that they were inherited from their last common ancestor, that is to say, LUCA. The development of techniques for large-scale DNA sequencing resulted in the determination of a large number of complete genome sequences. It has extended the study of the traits shared by the three domains of life to the information contained in the genome (e.g. genes). Such genes are called universal markers.

As with the other characters, a simple interpretation of the existence of **homologous genes*** (that is to say, having a common evolutionary origin) present in the genomes of all living beings is that they are the descendants of an ancestral gene that was already present in LUCA (Fig. 4.4a). The RNA molecules of the small and large subunits of the ribosome (LSU rRNA and SSU rRNA) are commonly cited as examples of universal markers. Indeed, all extant living organisms *(Archaea, Bacteria* and *Eucarya)* have one or more copies of these genes in their genome. The most parsimonious hypothesis is to assume that all existing genes coding for SSU rRNA and LSU rRNA are descendants of genes that were present in LUCA. But reality is not always as simple.

First of all, some genes present in all extant organisms (universal genes) that were inherited from LUCA may have diverged so much in sequence that we are no longer able to recognise their common ancestry (Fig. 4.4b).

A second scenario is that, during very early evolution, genes absent in LUCA may have appeared in one of three domains and were then horizontally transmitted to the other two domains (Fig. 4.4c). This implies that some universal genes may not have been inherited from LUCA. The role of the exchange of genetic information between the three domains, the importance of which was confirmed by comparative genomics, is discussed in Chap. 12.

Other scenarios may further complicate analyses. For example, some genes inherited from LUCA may have been lost in one or more domains. Thus, genes absent in one or more domains (or even in all extant organisms) may have existed in LUCA (Fig. 4.4d).

Similarly, genes inherited from LUCA may have been replaced in one domain. Thus, functions present in representatives of one or more domains, but due to nonhomologous genes, could have existed in LUCA (Fig. 4.4e). A final possibility is that nonhomologous genes encoding proteins with similar functions have appeared independently in the three domains of life. This is a phenomenon called **evolutionary convergence*** (Fig. 4.4f). In fact, very different molecules may have been independently recruited to perform similar functions. The flagellum that allows cell motility is an example of evolution by convergence (for another example, see the proteins carrying DNA replication, next section). A flagellum composed of several proteins exists in the three domains of life, suggesting that a flagellum was already present in LUCA. However, the detailed analysis of these flagella in the three domains revealed they have three very different organisations and that their constituents are not homologous. The existence of three types of flagella (one for each domain of life) can support the hypothesis that LUCA had no flagellum and that the latter was established independently in the three domains of life from different components. Alternatively, LUCA could have had one of three types of flagella that was replaced twice during evolution by new and different structures performing the same function, cell motility.

These examples illustrate that the research and analysis of characteristics common to the three domains is complex. On this point, the contribution of molecular phylogeny is crucial because, in reconstructing the evolutionary history of genes, it has identified which genes are more likely to have been present in the genome of LUCA. These approaches are limited in part by our still very fragmented knowledge of extant organisms and by the fact that the vast majority of past organisms became extinct without leaving any descendants or fossils.

4.1.4 The Main Features of LUCA

Despite the significant limitations set forth above, it is possible to sketch a portrait of LUCA. LUCA was a complex organism with probably:
1. A DNA-based genome. The assumption of a LUCA with a DNA genome is not universally accepted, though, because the majority of proteins carrying DNA replication in *Bacteria* are not homologous to those of *Archaea* and *Eucarya*. Three explanations have been advanced to explain the non-homology of these proteins: (a) LUCA had a DNA genome and proteins carrying DNA replication were replaced by other proteins in the lineage leading to *Bacteria* or in the *Archaea/Eucarya* ancestor (Fig. 4.4e); (b) LUCA had an RNA genome and DNA appeared twice, independently, in the lineages of *Bacteria* and of *Archaea/Eucarya* (Fig. 4.4f); (c) LUCA had an RNA genome and the transition to DNA from RNA occurred in one of the domains or in viruses (Fig. 4.1). Genes that carry out DNA replication and the synthesis of deoxyribonucleotides were secondarily transferred to other domains, which acquired the ability to make DNA (Fig. 4.4c).
2. A plasma membrane conforming to the 'fluid mosaic' model (*cf.* Sect. 3.3.1), which implies that LUCA was probably a cellular organism. The membrane of LUCA very likely integrated an ATP-generating system (because of the universal presence of ATPases) and a system of protein export. The nature of the lipids constituting the membrane of LUCA is still the subject of much controversy.
3. A genetic code controlling the translation of RNA into proteins.
4. Ribosomes, transfer RNAs, aminoacyl tRNA synthetases, and a number of translation factors involved in protein synthesis.

Many other issues, such as the presence or absence of introns and the type of metabolism, are currently being discussed, but the scientific community has not yet reached a consensus.

4.1.5 The Genome of LUCA

Another very hotly debated issue regards the number of genes in the genome of LUCA. There are two opposing views: the proponents of LUCA with a minimal genome (i.e. very small, containing few genes) and those arguing that LUCA had a large genome, containing thousands of genes. In the first view, the evolution of genomes after LUCA would be primarily by 'increasing complexity' (gene gain, through duplication and/or horizontal gene transfer), whereas in the second view, changes in gene stock would have been by gene loss. One of the key ideas upon which the supporters of LUCA with a small genome base their hypothesis is that the enzymes encoded by early genes could have had a broader spectrum than those of current enzymes. This very low specificity could result in a number of essential enzymes lower than that required for present microorganisms.

4.1.5.1 The Smallest Known Cellular Genome

The study of the smallest extant prokaryotic genomes has led to a few estimates of the minimum number of genes that the genome of a cellular organism may contain and could therefore have been present in the genome of LUCA.

One of the smallest cellular genome known as of today is that of the endosymbiotic bacterium *Carsonella ruddii* (Nakabachi et al. 2006). This is a gammaproteobacterium that infects all species of psyllid, Hemiptera insects feeding exclusively on plant sap. Its genome consists of 160 kilobases and contains 182 protein-coding genes. This genome is also extreme in its base composition: 84 % of A or T and only 16 % of C or G. *Carsonella ruddii* depends very closely on its host since it is completely devoid of genes involved in the biosynthesis of the membrane and in the metabolism of lipids and nucleotides. In contrast, the genome of this symbiont contains a large number of genes encoding proteins involved in the synthesis of essential amino acids (~18 % of genes). This symbiont is essential to the survival of the host insect, which does not find these amino acids in its single food source. Some authors interpret this genomic reduction as a process comparable to that experienced by mitochondrial and chloroplastic genomes. Others, on the contrary, think that this is an evolutionary dead end that will ultimately lead to the demise of this symbiont. More recently, even smaller genomes have been discovered such as the betaproteobacterium '*Candidatus* Tremblaya princeps' (139 kilobases, 121 protein-coding genes) or the alphaproteobacterium '*Candidatus* Hodgkinia cicadicola' (148 kilobases, 169 protein-coding genes) (McCutcheon and Moran 2012).

In general terms, the smallest sequenced prokaryotic genomes contain a few hundreds of genes. They come from symbionts or parasites whose highly specialised ways of life have led to very intense genome reduction: *Nanoarchaeum equitans* in *Archaea* and *Mycoplasma* and *Buchnera* in *Bacteria*. The small genome size of these extant cellular organisms is believed to be the result of recent and independent genomic reduction events rather than a feature inherited from a common ancestor.

4.1.5.2 Experimental Approaches

Systematic experiments of in vitro gene inactivation were used to identify genes essential for maintaining cellular life. The underlying assumption is that these essential genes could be inherited from LUCA. The results show that not all genes of a prokaryotic genome are necessary for the survival of a cell. For example, of the 480 genes in the genome of *Mycoplasma* genitalium, only two-thirds are actually needed for growth in laboratory. Similarly, in *Bacillus subtilis*, only 271 of the 4,100 genes present in the genome have proven essential to the maintenance of cellular life in the laboratory. In the yeast *Saccharomyces cerevisiae*, nearly 1,000 genes are detected as required. Despite significant variation according to the organisms studied and the experimental methods used to determine the number of essential genes, a 'minimal' cell seems to emerge from these studies: it is a simple prokaryotic cell consisting of a compartment defined by a membrane and performing typical cellular functions:

1. Replication
2. Isolation of the compartment from the outer medium
3. Protein synthesis
4. A rudimentary metabolism for energy production (glycolysis)

A major limitation of these experiments is that they involve the inactivation of a single gene at a time. Thus, the effect of the simultaneous inactivation of several individually dispensable genes is not taken into account (e.g. two genes can be individually dispensable without being so simultaneously). For example, the bacterium *Escherichia coli* has genes coding for two types of ribonucleoside diphosphate reductases that catalyse the reduction of ribonucleoside diphosphates into deoxyribonucleoside diphosphates under aerobic conditions and anaerobic conditions, respectively. Inactivation of the genes encoding the enzyme operating under aerobic conditions does not prevent the growth of *E. coli* under anaerobic conditions. They are therefore dispensable genes. Similarly, inactivation of genes encoding the enzyme operating anaerobically does not prevent the aerobic growth of *E. coli*. They are therefore also dispensable genes. However, the simultaneous inactivation of genes encoding the two types of enzymes is lethal. Finally, we must not forget that these experiments were conducted in laboratory conditions that are optimal for the microorganisms studied. The experimenters supplement all their nutritional requirements; they are neither subjected to stress nor are they in competition with other microorganisms. While most individual gene inactivation does not prevent strain growth in the laboratory, microorganisms that carry one such inactivation, however, are unable to thrive in a natural environment where the number of essential genes is much more important.

4.1.5.3 Comparative Genomics

Another approach to try and define the genome of LUCA is based on the comparison of the gene content of extant genomes to identify those genes inherited from the last common ancestor that were retained in all cells during the course of evolution. The first implementation of this approach dates back to 1996 with the search for **orthologous genes*** conserved in both pathogenic bacteria *Mycoplasma* *genitalium* and *Haemophilus influenzae* (genomes of 0.58 Mb for 468 genes and 1.83 Mb for 1,700 genes, respectively) that have diverged from a common ancestor probably more than 1.5 billion years ago (Mushegian and Koonin 1996). This approach identified 256 conserved genes in both bacteria, supposedly close to the minimum number necessary for the maintenance of cellular life. It is important to note that this number is similar to those obtained by the experimental methods described above. In particular, the minimal genome contains:

1. The DNA replication machinery
2. A rudimentary system for DNA repair and recombination
3. An almost complete transcription system but without regulatory factors
4. Chaperones
5. An anaerobic intermediary metabolism restricted to glycolysis
6. No system for the synthesis of amino acids and nucleotides
7. A limited biosynthetic pathway of lipids (no fatty acid synthesis)
8. A protein export system
9. A limited repertoire of proteins for the transport of metabolites

The nutritional needs of such a cell are quite large: it would need to import all of its amino acids, nucleotides, fatty acids and coenzyme complexes.

One limitation of this study is that the inferred genome is probably closer to that present in the last ancestor of bacteria than to the genome of LUCA. Since this analysis compares two bacterial genomes, it is very likely that this minimal genome contains bacterial solutions for some functions (such as a system of DNA replication of bacterial type; *cf*. Sects. 4.1.4 and 4.1.5). For example, the same study applied to 21 genomes of *Bacteria* and *Archaea* led to a list of only 52 genes encoding mainly for proteins involved in the formation of the ribosome and in protein biosynthesis.

Therefore, a less restrictive approach has been proposed. It is based on the search for universally conserved functions (which may have been present in LUCA) rather than on the search for universally conserved homologous genes. This search suggests a universal common ancestor with only 500–600 genes. The resulting cell could de novo synthesise its amino acids, nucleotides, complex carbohydrates and some coenzymes and would depend only on a small number of precursors taken from the environment (Koonin 2003).

All these studies show that the genome of LUCA probably contained at least a few hundred genes and corresponds to that of a relatively modern cell (i.e. ensuring functions similar to those currently observed).

4.1.6 LUCA: Hyperthermophilic, Thermophilic or Mesophilic?

Another discussion point concerns the optimal growth temperature of LUCA. The question is whether the ability to live at high temperatures is a character acquired early or late during evolution. The temperature at which LUCA lived is often confused with the temperature at which life emerged. These issues must clearly be separated because LUCA and the emergence of life are two separate events in time and space (Fig. 4.2). Currently, the majority of known hyperthermophilic organisms (living above 80–85 °C) are *Archaea*. There are also hyperthermophilic bacteria such as *Thermotogales* and *Aquificales*. However, no eukaryotic organism able to live above 60–65 °C is currently known.

The first school of thought, championed by Carl Woese, Karl Stetter and Norman Pace, considers that LUCA was a hyperthermophilic organism (Stetter 2006). This hypothesis is based mainly on the fact that hyperthermophilic organisms occupy a basal position within the bacterial and archaeal domains in the universal tree of life based on the small subunit ribosomal RNA and rooted in the bacterial branch (see Sect. 6.6.2). Moreover, these microorganisms are associated with shorter branches than those associated with mesophilic (living below 60 °C) or thermophilic (living between 60 and 80 °C) microorganisms. These two observations have been interpreted as indicating an ancient origin and a slower evolutionary rate of these microorganisms, which are therefore likely to have retained a large number of traits inherited from LUCA, in particular hyperthermophily. The ability to live in mesophilic environments would therefore have appeared secondarily and independently in the three domains of life.

Although the idea of a hyperthermophilic LUCA is generally taught, it is disputed by some scientists who believe that LUCA was not hyperthermophilic and that the adaptation to extremely hot environments observed in extant organisms appeared secondarily (Forterre et al. 2000). For example, the analysis of the most evolutionarily conserved parts of ribosomal RNA genes suggests that the basal emergence of both hyperthermophilic bacterial phyla could be an artefact of the tree reconstruction artefact rather than a reflection of their evolutionary origin (Brochier and Philippe 2002).

The hypothesis of a mesophilic LUCA is reinforced by results obtained by Galtier and colleagues (1999) who estimated the composition in G or C bases of the ribosomal RNA of LUCA by statistical methods. Knowing that stable RNAs (rRNA, tRNA) are rich in G and C in hyperthermophiles, these authors concluded that LUCA could not have survived temperatures above 70 °C. Further work on rRNAs and universal proteins has confirmed and expanded these results (Boussau et al. 2008). It seems that LUCA was mesophilic but that the ancestors of the bacterial and of the archaeal domains were (hyper)thermophilic organisms. Extant mesophilic microorganisms would therefore have had a thermophilic domain ancestor and a, yet more distant, mesophilic universal ancestor, LUCA.

Valuable information is also provided by experimental palaeobiochemistry, a discipline aimed at characterising macromolecules from extinct organisms. For example, Gaucher and colleagues (2003) conducted a statistical reconstruction of the sequence of the elongation factor EF-Tu (an essential protein in the translation process) in the last common ancestor of *Bacteria*. They then synthesised the corresponding gene in vitro and produced the enzyme by cloning its gene in an expression vector. Finally, they determined that the optimum temperature of the enzyme activity is 73 °C, suggesting that the last common ancestor of bacteria was thermophilic.

The various arguments that have been presented above concerning the temperature at which LUCA lived show that the debate on this issue is far from over.

4.1.7 Molecular Structure of Membrane Lipids in LUCA

The membranes of contemporary cells consist mainly of lipids arranged in a double layer or, in hyperthermophilic *Archaea*, in single layer. In *Bacteria* and in *Eucarya*, the lipid bilayer is formed by phospholipids, that is, glycerol-3-phosphate (G3P) molecules linked to fatty acid molecules by ester bonds (Fig. 4.5a). In *Archaea*, a glycerol-1-phosphate (G1P, a stereoisomer of glycerol 3-phosphate) is connected to isoprenoid chains by an ether bond (Fig. 4.5b).

The above description of lipids of the *Archaea* on the one hand and of the *Eucarya/Bacteria* on the other hand is not without exceptions. Indeed, lipids with an ether linkage have been described in *Eucarya* and in some thermophilic *Bacteria*, whereas side chains composed of fatty acids have been identified in *Archaea*. Homologues of genes involved in the biosynthesis of fatty acids were found in the genomes of archaea. However, to date no exception has been found in the glycerol-phosphate stereochemistry (G1P in archaea *vs.* G3P in bacteria and eukaryotes).

The enzymes responsible for the synthesis of G1P and G3P from dihydroxyacetone phosphate, G1PDH and G3PDH, are nonhomologous (i.e. have different evolutionary origins) and belong to different families of proteins showing no structural similarity. The important question is, what was the membrane lipid stereochemistry of LUCA? Several models have been proposed. One of them, for example, assumes that G1P and G3P were used interchangeably in the membranes of the first cells. The homochiral membrane would have appeared secondarily and independently in archaea, bacteria and eukaryotes following the acquisition of G1PDH and G3PDH (Lombard et al. 2012).

4.1.8 Metabolism in LUCA

It is generally accepted that the early atmosphere at the time when LUCA lived was largely anoxic. Its energy metabolism was thus either fermentative, or based on anaerobic respiration, or based on anoxygenic photosynthesis. If some oxygen was present in the environment, even in trace amounts, LUCA could have adopted aerobic respiration. This second hypothesis would imply the occurrence of respiration before oxygenic photosynthesis, an evolutionary scenario that, although surprising, cannot be completely ruled out (*see* Sect. 4.4).

4.1.9 LUCA Prokaryote, Eukaryote or Something Else?

The most widely defended (and taught) hypothesis is that LUCA had a cellular prokaryotic organisation, since the prokaryotic cell has a simpler organisation than the eukaryotic cell. This assumption is reflected in the names prokaryotes (the 'pro-' prefix means before; the 'caryos' root means nucleus) and eukaryotes ('eu-' means true), which imply that the latter derived from the former. Alternatively, it was suggested that both prokaryotic and eukaryotic modern organisms were derived from a proto-eukaryotic ancestral structure.

The prokaryotic organisation by its own qualities – simplicity and fast cell division, metabolic flexibility, gene exchange capacity and limited nutritional needs – has experienced extraordinary evolutionary success, irreversibly invading the primitive biosphere and resisting all the ecological upheavals that have marked the history of our planet. The success of the 'eukaryotic' approach would be equivalent, but using a different evolutionary strategy.

4.1.10 The Emergence of Three Domains

4.1.10.1 The Root of the Universal Tree of Life

Which among the *Archaea, Bacteria* and *Eucarya* came first? This issue is still hotly debated in the scientific community. In order to know in what order the three domains of life diverged, it would be necessary to:

1. Reconstruct the universal tree of life (i.e. based on a molecular marker present in all living organisms)
2. Root this tree (that is to say, define the direction of evolutionary time along tree branches)

Fig. 4.5 Sketch showing the organisation of polar membrane lipids in Bacteria and Archaea. (**a**) In *Bacteria*, the most abundant polar lipids are glycerophospholipids in which two molecules of fatty acids are linked by an ester linkage to glycerol at positions C-1 and C-2. Phospholipids form a bilayer. (**b**) Among non-hyperthermophilic *Archaea*, the basic unit of polar lipids is of the isoprenic glycerol ether type: two phytanyl molecules are linked by ether bonds to a glycerol molecule. In the cytoplasmic membrane, diethers are organised in a double layer (R = phosphate or aminophosphate with polar head). Diethers have also been identified in some thermophilic *Bacteria* (*Thermodesulfobacterium commune*, *Ammonifex degensii*, *Aquifex pyrophilus*). (**c**) In hyperthermophilic *Archaea*, the basic structure is glycerol tetraether: two identical, saturated, biphytanyl chains in $nC_{40}H_{82}$ are linked by an ether bond to two opposite glycerol molecules forming a monolayer (R1 = sugar residue). This type of molecular organisation withstands temperatures above 80 °C

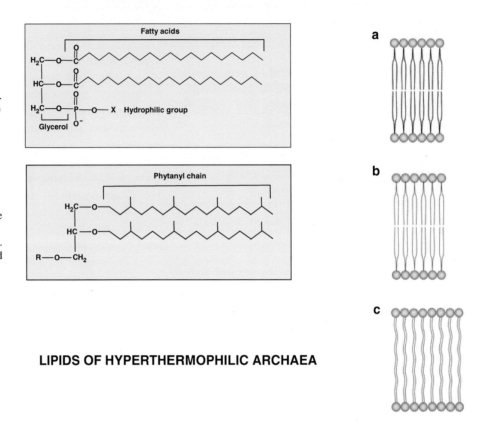

Rooting a phylogenetic tree requires inclusion in the analysis of what is called an **outgroup***, a set of genes that are homologous to the selected molecular marker, but branching outside the study group. For the universal tree of life, a major difficulty arises because all organisms are inside the tree. In 1978, Schwartz and Dayhoff proposed an elegant theoretical solution to this problem using the protein sequences of **paralogous genes*** that have undergone a duplication event prior to LUCA (Schwartz and Dayhoff 1978). If no gene loss subsequently occurred during the course of evolution, these genes should be present in two copies in LUCA and also in the members of the three domains of life. Inferring the phylogeny of these paralogues provides a symmetrical phylogenetic tree in which the phylogeny of one set of paralogues is rooted by the other paralogous sequences and vice versa (Fig. 4.6b). This approach was tested independently by two research groups (Gogarten et al. 1989; Iwabe et al. 1989) who used the sequences of elongation factors EF-1α and EF-2 and the sequences of the F and vacuolar H^+-ATPases, respectively. Both studies resulted in placing the root in the bacterial branch (Fig. 4.6a, tree 1).

The position of the root of the universal tree of life in the bacterial branch was quickly adopted by a majority of the scientific community because it is consistent with the hypothesis that LUCA was a prokaryotic organism and that bacteria form the most ancient evolutionary lineage. However, this result was also very controversial. Some believe that the amount of phylogenetic information contained in these markers is not sufficient to infer events as ancient as these. Indeed, the same markers used with different methods of analysis can place the universal root in the eukaryotic branch (Fig. 4.6a, tree 2).

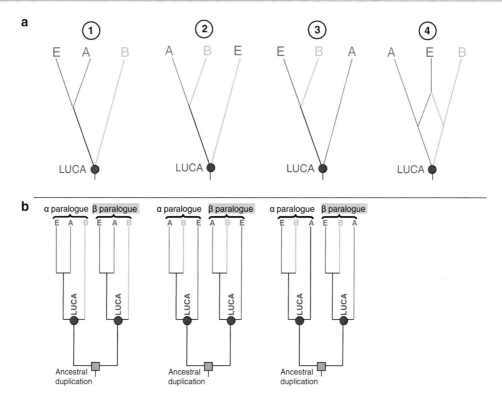

Fig. 4.6 Appearance of the three domains of life. (**a**) If the three domains of life, *Eucarya* (E), *Bacteria* (B) and *Archaea* (A) are monophyletic; three evolutionary relationships are possible between them (trees 1, 2 and 3). Tree 4 illustrates another possibility where the domain *Eucarya* results from the fusion of a bacterium and an archaeon. (**b**) The three theoretically expected phylogenetic trees in the analysis of paralogous genes resulting from a duplication predating the divergence between the three domains of life, whose descendants are still present in the genomes of extant species from all three domains. Each tree has a symmetrical structure and supports one of the trees 1–3 of part (**a**). The applicability of this method requires rebuilding the phylogenetic tree of compared genes without error

4.1.10.2 The Three Scenarios

Currently, three main scenarios are proposed to describe the evolutionary relationship between the three domains of life:

1. The tree is rooted in the branch of *Bacteria; Archaea* and *Eucarya* therefore appeared later (Fig. 4.6a, tree 1); LUCA was very probably of the prokaryotic type.
2. The tree is rooted between *Eucarya* and *Procarya (Archaea + Bacteria)*; from the origin, eukaryotes and prokaryotes are separated (Fig. 4.6a, tree 2); LUCA could have had a prokaryotic, eukaryotic or other type of cellular organisation.
3. A combination of one bacterium and one archaeon gives rise to *Eucarya* (Fig. 4.6a, tree 4); again, LUCA was of the prokaryotic type.

All of the three scenarios are based on solid arguments, but none can lead to a definitive conclusion. The timing of appearance of the three domains and the knowledge of the first steps that gave rise to the eukaryotic cell are still the subject of hot debate.

4.2 Geological Evidence of the Oldest Microbial Life Forms

The geological evidence of the oldest microbial life forms is in the form of biosignatures, i.e. signatures of life that are preserved in rocks. Even though microorganisms have no hard parts to relegate to the rocks, evidence of their former presence may be retained in the form of fossilised cells or bioconstructions, such as **stromatolites***, or of fossil organic molecules (Peters et al. 2007; Brocks et al. 1999; Summons et al. 1999) called **biomarkers***. There are other indications of microbial activity such as elemental isotope fractionation (Schidlowski 2001; Brocks and Summons 2003), the precipitation of biominerals or even microbial corrosion of rocks and minerals (see review in Westall and Cavalazzi 2011). Study of the most ancient traces of life is hampered by the lack of preservation of well-conserved rocks older than 3.5 billion years (Ga). Although older rocks exist, they have been significantly altered by high temperature and pressure deep in the crust, and any biosignatures that they may have contained are either unrecognisable or highly questionable. Nevertheless, there are a few locations, such as Pilbara (northwestern Australia) and Barberton (eastern South Africa) dating from the period 3.5 to 3.2 Ga containing exceptionally well-preserved rocks that can be used for studying the primitive Earth. This means that the record of early life is limited to these two ancient terranes.

4.2.1 How Do Microorganisms Fossilise?

Microorganisms rapidly degrade in the natural environment and, in order to be preserved as fossils, they need to be very rapidly encrusted and replaced by a mineral (Westall and Cavalazzi 2011). This happens where a particular mineral is supersaturated, the organic microorganisms acting as nuclei for the precipitation of the mineral. Polymerisation and dehydration of the mineral takes place, and, in anaerobic environments, such as the early Earth or present-day anoxic environments in the deep sea or lakes, the organic molecules can become trapped in the mineral matrix. In oxidising environments, for instance, in shallow water surface sediments or hot springs, the organic matter is completely oxidised and all that is left of the microorganism is the mineral crust (and sometimes mould or cast). Further processing is necessary to complete the preservation of the encrusted microorganisms in the form of lithification (cementation – again by another mineral) of the sedimentary or rocky microbial hosts. The lithified rocks then need to survive geological processing that includes plate tectonic recycling, metamorphism and erosion.

4.2.2 Microorganisms Perform Isotopic Discrimination

Extant living organisms preferentially use lighter isotopes: $^{12}C > ^{13}C$, $^{1}H > ^{2}H$, $^{14}N > ^{15}N$, $^{32}S > ^{34}S$, $^{54}Fe > ^{56}Fe$ (Box 4.1). Thus, a change in the ratio of the 'heavy' isotope to the 'light' isotope of an element suggests the presence of different metabolic activities in ancient sediments. For example, autotrophic activity will result in an enrichment of ^{12}C atoms compared to ^{13}C atoms; an enrichment of ^{32}S is evidence of sulphate-reducing activity, while ^{54}Fe enrichment indicates a capacity to respire iron. Note, however, that some caution is needed in the use of isotopic ratios because it has been shown that abiotic processes can also produce isotopic ratios that overlap with those generated by biological fractionation.

Box 4.1: Stable Isotopes and Isotopic Fractionation
Vincent Grossi

A chemical element X is characterised by its atomic number Z (equal to the number of protons in the nucleus) and its atomic weight A (equal to the number of protons and neutrons). It is represented as $^{A}_{Z}X$ (e.g. $^{12}_{6}C$). **Isotopes** are atoms of the same element with differing atomic masses (e.g. carbon-12, carbon-13 and carbon-14). That is, they have the same number of protons but different numbers of neutrons. Of the 92 existing natural chemical elements, 71 have multiple isotopes. Some isotopes are **stable** (e.g. carbon-13 or ^{13}C), while others are unstable and **radioactive** (e.g. carbon-14 or ^{14}C) and disintegrate over time into radiogenic isotopes (radionuclides).

Among the stable isotopes of an element, the lightest is the most abundant (e.g. $^{12}C = 98.9$ %, $^{13}C = 1.1$ %; $^{16}O = 99.76$ %, $^{17}O = 0.04$ %, $^{18}O = 0.20$ %) and tends to be 'used' (or reacts) more easily (or faster) than the heavier isotopes. So-called **isotopic fractionation** is the discrimination between the light and a heavy isotopes of an element that occurs during (bio)chemical reactions and physico-chemical processes. This fractionation leads to a very small (on the order of ‰) but significant change in the ratio of heavy to light isotopes (e.g. $^{13}C/^{12}C$, $^{18}O/^{16}O$, $^{2}H/^{1}H$ or D/H) between a reagent (e.g. CO_2 fixed by phototrophic organisms) and its transformation products (e.g. organic compounds formed from the CO_2 fixed by phototrophic organisms).

The heavy/light isotopic ratio of an element in a sample is determined using an **isotope ratio mass spectrometer** (**IRMS**), which measures this ratio in a purified gas (e.g. CO_2, H_2, N_2) produced from the sample (originally solid, liquid or gas), and compares it with that of a reference gas produced from an **international** (or **universal**) **standard**. These standards exist for each element (C, H, O, N, S) commonly measured by IRMS; for example, the standard for carbon is a belemnite (calcite) from the Pee Dee Formation (so-called PDB) with a $^{13}C/^{12}C$ ratio $= 0.01124$ and that for nitrogen is atmospheric

---

V. Grossi
Laboratoire de Géologie de Lyon: Terre, Planètes, Environnement, UMR CNRS 5276, Université Claude Bernard Lyon 1, 69622 Villeurbanne Cedex, France

(continued)

Box 4.1 (continued)

nitrogen with a $^{15}N/^{14}N$ ratio $= 0.00368$. The accuracy of an IRMS measurement is $\sim 0.1–0.2‰$ for $^{13}C/^{12}C$ (ca. 5–7‰ for D/H). The differences between the isotopic ratios of an element in a sample (R_{samp}) and in a standard (R_{std}) are very small; they are expressed as 'δ' values, which record the enrichment in units of ‰ relative to the standard according to the equation:

$$\delta = \left[\left(R_{samp} - R_{std}\right)/R_{std}\right] \times 1{,}000$$

Thus, for carbon, this equation is expressed as

$$\delta^{13}C_{samp} = \left[\left(^{13}C/^{12}C_{samp}/^{13}C/^{12}C_{PDB}\right) - 1\right] \times 1{,}000$$

If the isotopic ratio of the sample is lower than that of the standard (which is often the case, especially for carbon), δ is negative. A sample with a δ value lower than another sample is said to be 'depleted' in the heavy isotope of the element considered. Conversely, higher δ values are described as 'enriched'.

The measurement of δ values in geological samples represents a highly powerful tool (or 'proxy') that can provide information on the origin (e.g. biological vs abiotic) of the analysed element and can be used to reconstruct its geological or biogeochemical 'history'. For example, a $\delta^{13}C$ value of -50 to $-130‰$ in organic matter indicates formation through methanotrophic bacterial activity, while a $\delta^{34}S$ value of about $-20‰$ in diagenetic sulphides (e.g. pyrite) signifies sulphate-reducing bacterial activity.

It is important to remember that the δ value of a complex sample (or changes in δ values observed through time) is not necessarily representative, since multiple origins and processes can overlap and 'hide' (or modify) the original isotopic signature, thus limiting the interpretation of the measured bulk signal. This is particularly the case for ancient biological samples, whose stable isotopic composition could have been altered over time by abiotic processes such as thermal maturation, pressure and evaporation (Hoefs 2004). In this kind of study, a 'multi-proxy' approach is generally preferred.

4.2.3 Prokaryotic Fossils: Isolated or in Colonies

4.2.3.1 Criteria for the Identification of Microfossils

The identification of microbial microfossils is tricky, but taking great care and with well-preserved specimens, it is possible (Westall and Cavalazzi 2011). The first difficulty lies in the fact that non-biological structures can look strikingly like bacterial cells (Garcia-Ruiz et al. 2003) (Fig. 4.7) and, thus, it is necessary to determine the biogenicity (biological origin) of a particular biosignature. Moreover, recent contamination by **endolithic*** microorganisms is also possible. Thus, 8000 year-old cyanobacteria and filamentous fungi, were identified in geological formations dating back to 3.7 Ga (Westall and Folk 2003).

To reduce errors in the identification of microbial microfossils (some of the controversies caused by misinterpretation will be discussed below), it is imperative not to rely on a single biosignature, whether morphological or biochemical, because no individual signature is in itself sufficient evidence of biogenicity since many biosignatures can be mimicked by abiogenic features. To validate the biological origin of an observation, three types of studies should be undertaken (Westall and Cavalazzi 2011):

1. Study of the geological setting at the macroscopic, microscopic or elementary scales to determine whether the environment in which the rocks formed could have hosted life. It is also necessary to determine what changes the rock may have undergone since its formation (diagenesis, metamorphism), which could alter the biosignatures.
2. Geochemical study of the proposed biosignature by measurement of carbon and sulphur isotopes, determination of the presence of life-essential elements, such as carbon, hydrogen, oxygen, nitrogen, sulphur and phosphorus, as well as the presence of organic molecules showing compositional/structural complexity. For example, carbon isotopes with $\delta^{13}C$ values generally less than $-30‰$ are typical of methanogenic archaea and those around $-27‰$ of oxygenic photosynthetic bacteria.
3. Morphological study of the presumed microfossils. At the cellular level, features such as size, shape, division or lysis are sought. On a larger scale, colonies, biofilms or mats should exhibit: (a) associations of structures of the same size and shape that may represent cells of the same species (colonies) or of structures of different size and shape (formation of a consortium), (b) the presence of microbial exopolymers (extracellular polymeric substances, referred to as EPS), and (c) structures with a

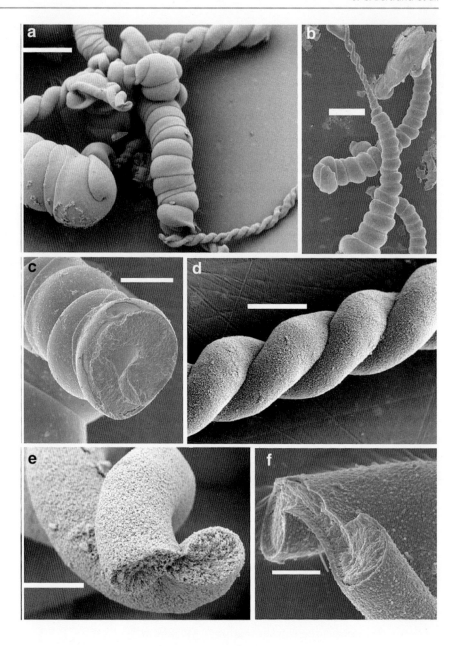

Fig. 4.7 Abiotic structures. Photomicrographs using scanning electron microscopy of silica structures resembling biological forms. These composite materials are formed of amorphous silica and crystalline barium carbonate. Both phases self-assemble creating structures that mimic biological forms. The formation of these inorganic materials is a plausible phenomenon in certain geochemical conditions and demonstrates that this type of morphology cannot be used as an unambiguous criterion of structures of biological origin (Photographs courtesy of Juan-Manuel García-Ruiz Bars: **a**: 40 µm; **b**: 50 µm, **c**, **d** and **e**: 10 µm; **f**: 5 µm)

relatively complex architecture, such as biofilms or microbial mats.

In addition to problems related to the interpretation of the observations, sample preparation itself can introduce artefacts and abiotic contaminants. It is not surprising that, given all these constraints, the study of microfossils is the subject of much controversy.

4.2.3.2 Stromatolites: A Definition

Stromatolites or stromatolitic structures have a particular place as a biosignature in the rock record and have been identified in all geological eras (from the Archaean (4.0–2.5 Ga) to the present). The oldest stromatolite-containing formations are located in the Archaean greenstone belts in the Pilbara in northwestern Australia (Hofmann et al. 1999) and at Barberton in eastern South Africa (Byerly et al. 1986).

A stromatolite is primarily a lithified microbial mat, whose growth is the result of the trapping of sediment particles and the precipitation of carbonate by microbial action. A stromatolite is thus essentially composed of a nonliving part consisting of superimposed organo-sedimentary strata, and on its surface, it is covered by a complex microbial mat community whose thickness may reach a few centimetres (*cf*. Sect. 9.7.3). Stromatolites can be either two-dimensional (tabular or stratiform) or three-dimensional (in the form of columns, domes, tepees) (Krumbein 1983). In the broad sense of this definition, any microbial mat is a potential stromatolite. Stromatolites are formed exclusively by photosynthetic microbial mats growing in shallow water habitats within the influence of sunlight. Other kinds of non-photosynthetic microbial mats may exist in other environments, for example, sulphate-reducing bacteria around deep water hydrothermal vents (see Sect. 9.7.3).

4.2.3.3 Microbial Mats Forming a Stromatolite

How can stromatolites that can reach one metre or more in height be formed from mats whose individual thicknesses vary from a few millimetres to a few centimetres (see Sect. 9.7.3)? Several mechanisms are involved that allow the vertical growth of microbial mats. Microorganisms forming the mat layers are embedded in a matrix of gelatinous, sticky exopolysaccharides. In today's oxygenic environment, exopolysaccharides are mainly produced by cyanobacteria and, because of their adhesive properties, contribute to the trapping of sedimentary particles in successive layers. Simultaneously, within the mat, and through the activity of microorganisms that increase alkalinity, calcium carbonate precipitates, thus cementing sedimentary particles trapped in the exopolysaccharide network. To gain access to sunlight, filamentous cyanobacteria glide through the matrix of exopolysaccharides and agglutinated sediments to form a new layer on the surface of the mat. Above this organo-sedimentary layer, another mat will develop that will give birth to another organo-sedimentary layer. This repeated process results in a finely laminated vertical structure.

4.2.3.4 The Stromatolites of the Early Archaean

The oldest known stromatolites occur in ~3.45-Ga-old strata in the Pilbara (Hofmann et al. 1999; Allwood et al. 2006). These features apparently formed on a shallow water carbonate platform and exhibit different morphologies related to the different microenvironments in which they formed: domical forms in the supra intertidal zone and conical or cuspate forms slightly in the intertidal zone (Fig. 4.9). These early photosynthesising microbial mats that formed both vertical and tabular stromatolites were anaerobic (Westall et al. 2011a).

4.2.3.5 Evolution of the Importance of Stromatolites Through the Geological Eras

One of the noticeable characteristics of the earliest stromatolites is their small size compared to stromatolites dating from the Later Archaean/Proterozoic (~2.7 Ga to ~542 Ga) to today (Figs. 4.8, 4.9, and 4.10). The difference in size appears to be related to the appearance of oxygenic photosynthesisers that could outcompete anaerobic metabolisms.

From ~ 2.7 Ga, stromatolites reach metric or plurimetric sizes. For example, stromatolites found at Belingwe in Zimbabwe (2.7 billion years) are several metres high (Grassineau et al. 2001).

The late Archaean and Proterozoic stromatolites form huge reefs, representing the dominant biosignature in the geological record until the end of the latter eon, about 560 million years ago (My). Their decline, which continued throughout the Phanerozoic, may be linked to the appearance of predators and to competition with other species (Fig. 4.11).

4.2.3.6 The Controversy over the Date of Appearance of Cyanobacteria

To complete this description of ancient and modern stromatolites, it is worth mentioning a controversy over whether or not cyanobacteria were among the oldest fossils of microbial life dating from the Early Archaean. For the last forty years, cyanobacteria-like structures have been described from these ancient rocks. However, several recent studies have shown that, even if the supposed microfossils contain carbon, they may be abiotic in origin.

The most striking of these controversies occurred following the work published by J. William Schopf of the University of California (Schopf 1993). Schopf describes microfossil remains in rocks dated at 3.465 Ga in the Pilbara terrane of Western Australia. These remains consist of long filaments of apparently tens of cells, surrounded by a 'wall', similar to that of extant filamentous cyanobacteria. He thus identified eleven different species of cyanobacteria. In a more recent publication, Schopf detected **kerogen*** in these same structures (Schopf et al. 2002). These organic molecules were interpreted as evidence of a bacterial origin. For this author, these microfossils are therefore the oldest traces of microbial life on our planet.

However, there are a number of flaws in Schopf's reasoning. Garcia-Ruiz et al. (2003) synthesised complex inorganic structures whose morphology is similar to that of microfossils described by Schopf (Fig. 4.7), and they consequently questioned some of Schopf's conclusions as to the biogenicity of his 'microfossils'.

In addition, Martin Brasier and colleagues from Oxford University also questioned the biogenicity of Schopf's microfossils on other grounds (Brasier et al. 2002). After having analysed rock samples taken from the same site, Brasier and colleagues conclude that the fossils described by Schopf are artefacts, produced by the reprecipitation of kerogen in a vein of hydrothermal silica. Although Brasier does not deny the presence of organic material in the samples, he believes that its origin is not biological, but produced by a chemical reaction called Fischer-Tropsch. This is a catalytic reaction producing paraffinic and olefinic hydrocarbons (and, to a lesser degree, oxygenated compounds: alcohols, aldehydes, ketones) from carbon monoxide and dihydrogen ($CO + H_2$):

$$n\,CO + (2n + 1)\,H_2 \rightarrow C_nH_{2n+2} + H_2O$$

However, the products formed have a simple chemical structure and the subsequent transformation of alkanes into kerogen remains is still to be explained.

Fig. 4.8 Modern stromatolite. Living dome-shaped stromatolites, observed at low tide in the Shark Bay lagoon, Australia (Photograph: courtesy of Daniele Marchesini)

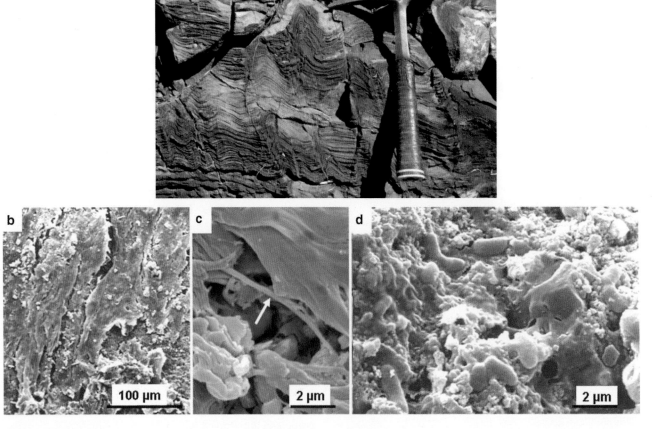

Fig. 4.9 Early Archaean stromatolite. (**a**) Planar view of a section of a dome-shaped stromatolite found in the Pilbara, Western Australia (Westall 2005) (Copyright: courtesy of University Press of Bordeaux); (**b**, **c**) Fossil microbial mat (tabular stromatolite) dated at 3.3 billion years (Westall et al. 2006a, 2011b). (**d**) Fossil structures interpreted as microorganisms occurring in the tabular stromatolite of figures **b** and **c** (Westall et al. 2006a) (Photographs **b**, **c** and **d**: copyright: courtesy of the Royal Society)

Fig. 4.10 Stromatolites from the Pilbara region (northwestern Australia). All of these rocks constitute a stromatolite. (**a**) The vertical cliff is about 4 m high. (**b**) The person indicates the size of the stromatolite (Photographs: courtesy David Moreira)

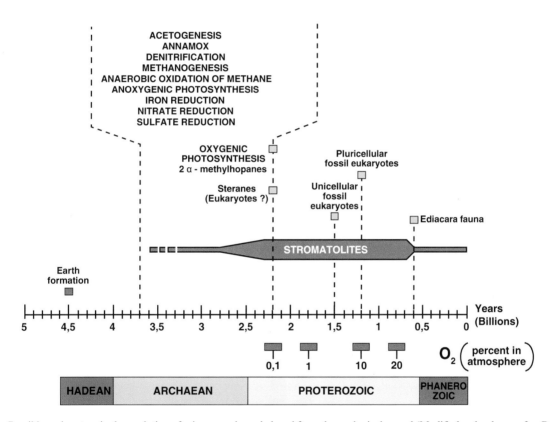

Fig. 4.11 Possible major steps in the evolution of microorganisms deduced from the geological record (Modified and redrawn after Brocks et al. 2003; Canfield et al. 2006; Nealson and Rye 2003; Schidlowski 1993; Rasmussen et al. 2008) (Drawing: M.-J. Bodiou)

Despite their divergences, both teams admit that life existed at that time. In particular, Martin Brasier and his team identified an enrichment of ^{12}C atoms compared to ^{13}C atoms which is a reflection of autotrophic activity. However, they interpret the ^{12}C enrichment as a reflection of the activity of methanogens and not as the activity of photosynthetic bacteria (Brasier et al. 2006).

4.2.3.7 Other Fossil Traces of Prokaryotic Cells

Stromatolites and photosynthetic microbial mats are not the only traces of ancient life. Sediments on the early Earth were mainly composed of volcanic ash and dust that could have provided habitats for chemolithotrophic microorganisms. The glassy surfaces of lava extruded underwater (pillow lava) are also a favoured habitat for chemolithotrophs.

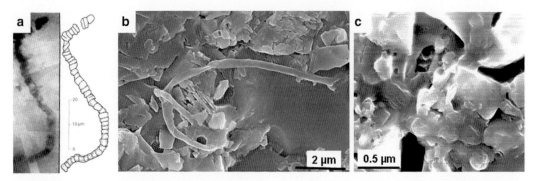

Fig. 4.12 Microscopic examination of microfossils (Pilbara 3.4 billion). (**a**) Filament-shaped microfossil that J. William Schopf (1993) had described as fossil cyanobacteria (Copyright: Science). (**b**) Filament-shaped fossil microorganism (Westall et al. 2006b) (Copyright: courtesy of the 'Geological Society of America'). (**c**) chained, coccoidal-shaped fossil microorganisms (Westall 2005) (Copyright: courtesy of University Press of Bordeaux)

Microfossils of likely biological origin have been described in some very well-preserved, 3.5–3.2-Ga-old, slightly metamorphosed sediments (Westall et al. 2006a, b, 2011a, b). The rocks containing these microfossils were collected in eastern South Africa (Barberton area) and in northwestern Australia (Pilbara area). They contain a wide variety of small, cellular structures, shaped like coccoids, vibrios, rods and filaments. Sizes range from ≤ 1 µm for the coccoids, while the filaments may range up to a few tens of micrometres in length and 0.5–2.5 µm in width (Fig. 4.12). None of these cell structures can be identified as a cyanobacterium. As noted above, identification of these structures as microbial cells requires the association of other biosignatures, such as carbon isotope ratios or molecular complexity.

Recently, some enigmatic, carbonaceous bacteriomorph structures have been identified in these ancient Early Archaean sediments. They are generally a couple of orders of magnitude larger than the tiny microfossils described above and are either spherical or platy in shape (Sugitani et al. 2009; Javaux et al. 2010). Their origin and evolutionary attribution is as yet uncertain, but their presence may indicate that early life was even more diverse than previously thought.

4.2.4 Fossil Eukaryotic Cells

Ideally, the characterisation of eukaryotic fossils should use cytological, genetic, biochemical and metabolic characteristics, for example, the presence of a nucleus, organelles, etc. Most of them rarely leave interpretable fossil remains and often only morphological trace fossils are accessible. It is therefore difficult to attribute a eukaryotic origin to a fossil with certainty. Prokaryotic cells today are generally small (<2 µm), while eukaryotic cells most often are larger

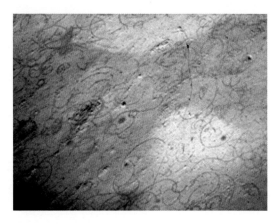

Fig. 4.13 *Grypania spiralis* observed in the Negaunee iron formations, Michigan, USA (Photographs: Xavie Vazquez, Wikipedia, GNU Free Documentation Licence)

than 10 µm. A large, cellular fossil is thus a good eukaryotic candidate. There are exceptions, however. Consider *Ostreococcus*, which is a eukaryotic micro-alga about 1 µm long and 0.7 µm thick (Courties et al. 1994). Conversely, Schulz et al. (1999) discovered a bacterium, *Thiomargarita namibiensis*, whose cells have a diameter of up to 750 µm that are visible to the naked eye (see Sect. 5.1.2). A very old fossil *Grypania spiralis* was discovered by Han and Runnegar (1992) in rocks from Michigan dated about 1.9 Ga. Similar forms are found in other parts of the world, at least until 1.3 Ga. *G. spiralis* is shaped like a filament and is several tens of centimetres in length, 1–2 mm wide, and coiled like a spring (Fig. 4.13). This fossil was originally considered as having a eukaryotic organisation. Some authors have speculated that this organism exhibits a **coenocytic*** organisation and have assigned it to Dasycladales, a group of Chlorobionta (Viridiplantae), that were common in the palaeozoic and mesozoic before declining. However, no solid argument reinforces this assignment. *G. spiralis* might actually be a eukaryotic fossil that cannot be assigned to any currently known lineage. Although some authors consider *Grypania* as a composite of filamentous prokaryotes, it is more likely that this fossil is eukaryotic in nature (Knoll et al. 2006).

The attribution of acritarchs, encountered from 1.9 to 1.8 Ga, to the eukaryotic domain is less controversial. They are unicellular organisms with very resistant cell walls, which may resemble cysts of extant dinoflagellates (see Sect. 7.8.3). The term 'acritarch' means a cellular, resistant organic structure. These cells are probably polyphyletic (i.e. consisting of various lineages that are not particularly related to each other) and may represent ancestors of extant unicellular eukaryotic lineages and/or extinct lineages (see Chap. 7).

Tappania plana, *Valeria lophostriata*, *Dictyosphaeria* sp. and *Satka favosa* were discovered in the formation of Roper, Northern Australia, and dated at 1.5 Ga (Fig. 4.14). These unicellular organisms were larger than 200 μm and lived in estuaries or deltas. The presence of protrusions in the cell wall for budding is difficult to explain without assuming the presence of a cytoskeleton in the cytoplasm, a distinctive eukaryotic feature (Javaux et al. 2001).

4.2.5 Fossils of the First Multicellular Organisms

The oldest clearly multicellular eukaryotic fossil, *Bangiomorpha pubescens*, was discovered by Butterfield (2000) in carbonates dated at 1.2 Ga in Somerset Island, Canada. Moreover, this is the earliest taxon that could be attached to an extant lineage, bangiophytes (a class of red algae). Indeed, *B. pubescens* has similar characters to those involved in the sexual reproduction of extant bangiophytes. It is, to date, the oldest fossil demonstrating sexuality, a hallmark of eukaryotes. Between 1.2 and 0.6 Ga, other fossils more or less credibly attributable to multicellular eukaryotes are known. Then suddenly, between 600 and 540 Ma, i.e. just before the beginning of the Palaeozoic era (its beginning is dated to 542 Ma, with the Cambrian), and in sites now very far apart from each other (Ediacara in South Australia, Doushantuo in South China, Spaniard's Bay in Newfoundland), most extant eukaryotic lineages (including metazoa) appear in the fossil record. The beginning of the Cambrian (542–500 Ma) is marked by an explosion in biodiversity, as documented in particular in the Burgess (British Columbia) and Chenjang (China) deposits.

The discovery in 2010 of macrofossils dated 2.1 Ga, interpreted as forms of multicellular life, would push the emergence of multicellular life back by a billion years (El Albani et al. 2010).

Was the emergence and diversification of eukaryotic lineages, initiated between 2.7 Ga (see Sect. 5.4.1) and 2.1 to 1.9 Ga, progressive or did it experience a sudden acceleration at 600 Ma? There is indeed a change in the rate of the diversification of life at the end of the Proterozoic but as yet no consensus as to the cause – rising dioxygen and nutrient levels and/or climatic forcing through a series of apparently global glaciations that occurred at the end of the Proterozoic between about 780 and 640 Ma (Lenton and Watson 2011).

4.2.6 The Analysis of Fossil Molecules: Molecular Palaeontology

4.2.6.1 The First Traces of Oxygenic Photosynthesis

After cell death, a very small portion of the organic matter escapes the mineralisation process and is buried in the sediment. Once buried, the organic material will sink gradually beneath layers of increasingly thick sediment, up to several hundred metres. Under the combined effects of temperature and pressure, the various constituents of the organic matter will undergo chemical transformations whose importance will depend, inter alia, of their molecular structure. Some of these molecules are sufficiently resistant to the processes of diagenesis and catagenesis to keep the carbon skeleton of their precursors and can be used as **biomarkers***. Thus, 2-α methylhopanes are considered as specific biomarkers of extant cyanobacteria that perform oxygenic photosynthesis. What is their origin? The plasma membranes of prokaryotes contain amphipathic lipid molecules called biohopanoids that are functionally equivalent to sterols in eukaryotes. Biohopanoids derive from a family of molecules, hopanes and pentacyclic triterpenoid compounds with 30 carbon atoms (Fig. 4.15a). The analysis of cyanobacteria, either in culture or in microbial mats, reveals the presence of particular hopanes called bacteriohopanepolyols with a methyl group in position 2 of ring A (Fig. 4.15b). After cell death and during burial in sediments, these molecules experience defunctionalisation processes (loss of the OH groups) and stereochemical changes, but their carbon backbone (5 rings) is retained and they occur in sediments in the form of 2-α methylhopanes (Fig. 4.15c). The discovery of 2-α methylhopanes in significant quantities in sediments dating back 2.7 Ga collected in Western Australia has been a strong argument for the existence of oxygenic photosynthesis-performing cyanobacteria at this period (Brocks et al. 1999). Since then, it has been shown that, although the rocks analysed were dated correctly, the 2-α methylhopanes came from younger (<2.2 Ga), contaminating carbon-rich fluids, i.e. a period during which oxygenic photosynthesis had already appeared (Rasmussen et al. 2008).

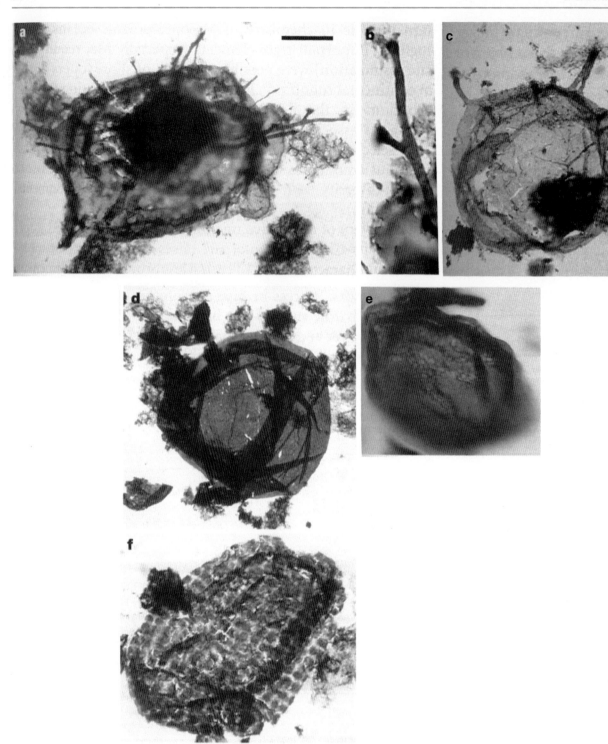

Fig. 4.14 Eukaryotic fossils from the Roper formation (northern Australia). (**a–c**) *Tappania plana* **b**: detail of the protuberance indicated by the *arrow* in a; (**d**) *Valeria lophostriata*; (**e**) Dictyosphaeria sp.; (**f**) *Satka favosa*. In *Satka favosa,* the cell wall determines the fragmented appearance. Scale: 35 μm for **a** and **c**, 10 μm for **b**, 100 μm for **d**, 15 μm for **e** and 40 μm for **f** (Javaux et al. 2001) (Photographs: Copyright courtesy of Nature Publishing Group)

4.2.6.2 Eukaryotic Biomarkers

The detection of steranes, fossil molecules formed from sterols (Fig. 4.16), in ancient sediments, is considered a proof of the existence of eukaryotic organisms. Their occurrence in the sediments mentioned above suggests the presence of such organisms 2.2 Ga ago. However, sterols have been detected in a small number of contemporary prokaryotes, weakening the argument coupling their

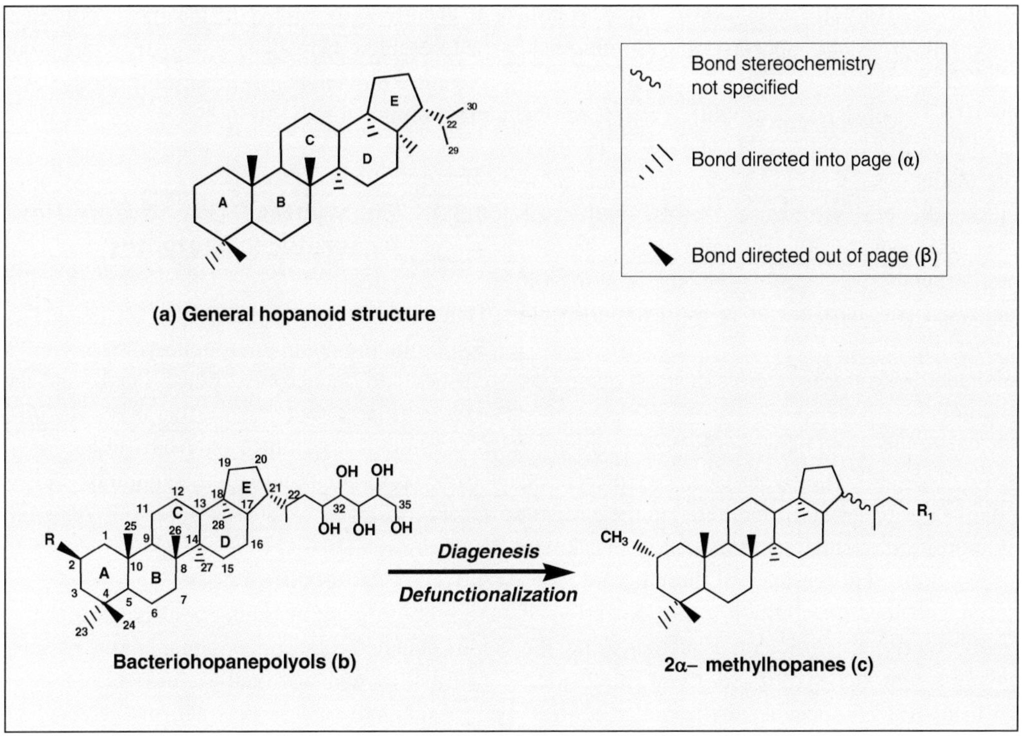

Fig. 4.15 Formation of 2-α methylhopanes. (**a**) General skeleton of hopanes; (**b**) R = H or CH$_3$; (**c**) R1 = H or CH$_3$ to nC$_5$H$_{11}$ (Modified and redrawn from Summons et al. 1999)

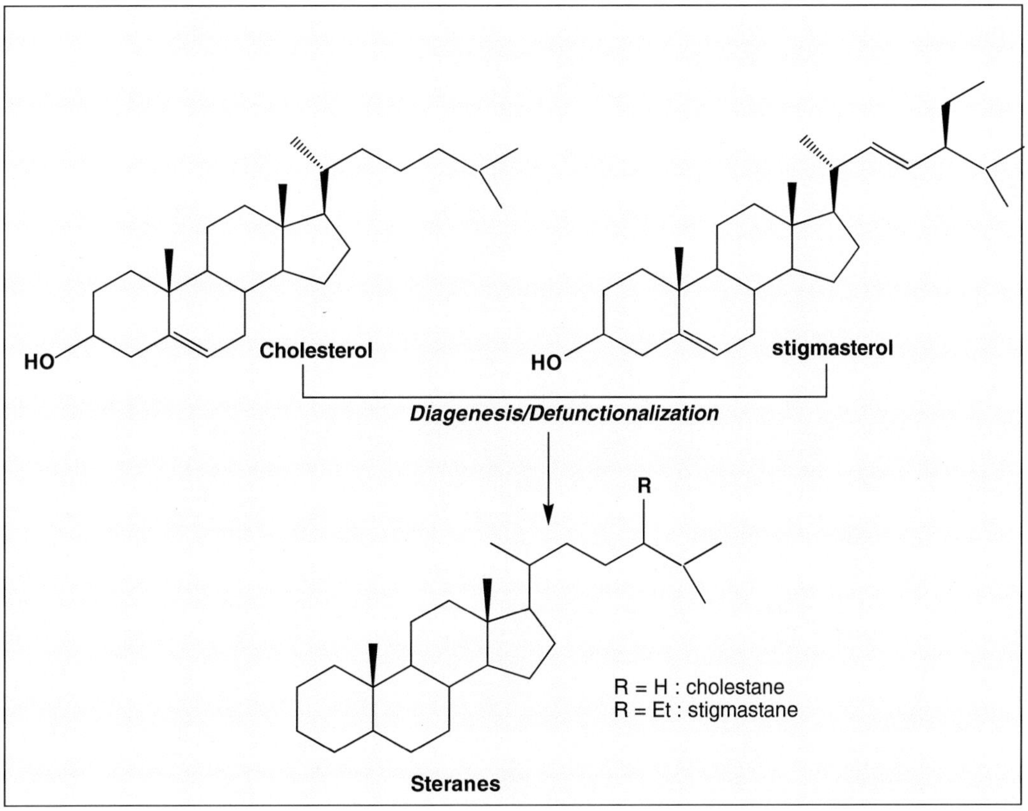

Fig. 4.16 Sterols and steranes

presence with that of eukaryotic cells. For example, sterols have been reported in a few species of cyanobacteria. However, in all cases, the sterols were found at very low concentrations (0.03 % of dry cell weight). To explain the presence of sterols in low quantities in cyanobacteria, three proposals were made:
1. Cyanobacteria do not synthesise sterols, and their presence resulted from sample contaminations.
2. Cyanobacteria synthesise sterols, but the amount of biomass analysed was insufficient in most experiments for their detection.
3. Some cyanobacteria synthesise sterols in small quantities and others do not.

This debate is not closed and has recently been revived by the identification of genes homologous to eukaryotic genes, coding for certain steps of sterol biosynthesis, in the genome of several cyanobacteria (such as *Anabaena*) (Volkman 2005). For some authors, the presence of these homologues in cyanobacterial genomes (in addition to the detection of sterols in cyanobacteria) is sufficient to demonstrate the ability of cyanobacteria to synthesise these molecules. For others however, it is not, because the presence of these homologous genes does not necessarily mean that they are functional and they encode for proteins involved in the biosynthesis of sterols (i.e. they may have completely different functions).

Sterols have also been discovered in several non-photosynthetic prokaryotes, such as *Methylococcus capsulatus* (Gammaproteobacteria), *Gemmata obscuriglobus* (Planctomycetales) and various myxobacteria (deltaproteobacteria).

The evidence therefore seems to indicate that sterols are present in some prokaryotes. Although the capacity to synthesise sterols in small amounts and in a limited number of bacteria is still unexplained, their existence requires caution in the use of these molecules as specific eukaryotic biomarkers.

Using all the information obtained from the study of the fossil record and of biogeochemical markers, it is possible to propose a diagram depicting the main possible stages of the evolution of microorganisms (Fig. 4.11).

4.3 The Passage from a Cellular Organisation of the Prokaryotic Type to a Cellular Organisation of the Eukaryotic Type

Intermediate forms between the 'prokaryotic' and the 'eukaryotic' cellular organisations have not been found, and there is almost no chance this will ever happen because the differences are of the kind that cannot be preserved as fossils. Many evolutionary scenarios have been proposed in the literature and only the most 'consensual' ones will be presented here. These scenarios are mainly based on current knowledge of the cell biology of contemporary microorganisms.

4.3.1 Eukaryotes Descend from Heterotrophic Anaerobic Prokaryotes

This scenario was proposed by Christian De Duve (1995, 2002). In the long evolutionary history of life, it distinguishes an anaerobic phase and an aerobic phase. Assumptions about events that have occurred during the first phase – the transition from prokaryotes to anaerobic eukaryotes – are speculative. In contrast, those that led to the emergence of aerobic eukaryotes rely on solid arguments.

4.3.1.1 Anaerobic Phase
Membrane invaginations could have appeared in an anaerobic and heterotrophic organism of the prokaryotic type that had lost its cell wall and increased in size, thus increasing the exchange surface between the cell and its surrounding environment (Fig. 4.17a–c). Some of these invaginations gave rise to vesicles, sequestering food and digestive enzymes (Fig. 4.17d). Digestion then became intracellular.

Subsequently, vesicles would have differentiated in several kinds, creating a real membranous network with the possible appearance of the smooth and rough endoplasmic reticulum, the Golgi apparatus and lysosomes (Fig. 4.17e). Later, cytoskeletal and motor organs (not shown in the figure) would have appeared. Finally, the membranous network would have wrapped and isolated DNA from the cytosol (Fig. 4.17f). The appearance of these membrane structures would have allowed functional specialisation of compartments within the cell. It is at this stage that DNA transcription and mRNA translation became decoupled.

4.3.1.2 Aerobic Phase
The emergence and development of cyanobacteria performing oxygenic photosynthesis radically transformed the chemical composition of the atmosphere, which gradually became aerobic. This transition had a great impact on the biodiversity of the time, in particular by fostering the emergence and expansion of organisms that had developed the ability to use dioxygen via aerobic respiration. It is very likely that many lineages of anaerobic organisms went extinct. However not all did, in particular, the ancestor of eukaryotes. Some may have found shelter in anoxic habitats, while others managed to develop defence mechanisms against dioxygen. Lynn Margulis, an American microbiologist who in the 1960s developed 'the endosymbiotic theory' (Margulis 1970), hypothesised that

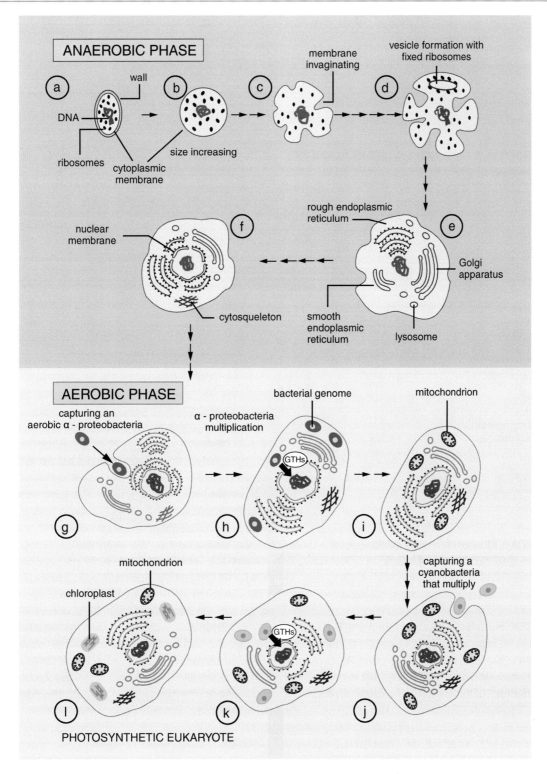

Fig. 4.17 The different stages of a potential scenario involving the theory of endosymbiosis proposed by Margulis (Modified and redrawn after de Duve 1995, 2002; Maynard Smith and Szathmáry 2000; Nelson and Cox 2000; Sadava et al. 2007) (Drawing: M.-J. Bodiou)

mitochondria and chloroplasts were former bacteria that, during evolution, had established a symbiotic relationship with a eukaryotic ancestor, probably heterotrophic and anaerobic.

According to this theory, an aerobic bacterium belonging to the group of alphaproteobacteria, perhaps a relative of the extant *Rickettsia,* would have been captured (probably by phagocytosis) by a heterotrophic anaerobic organism.

This bacterium would have multiplied inside its host (Fig. 4.17g, h). This association would have become permanent and, both partners finding mutual benefit, would have evolved into intracellular mutualistic symbiosis. The bacterium that consumes dioxygen found 'board and lodging' in its host, that is, protection and abundant food.

The exact nature of the immediate benefit to the host of the mitochondrial symbiosis is still debated today. The ability to synthesise ATP with maximum efficiency by oxidative phosphorylation could have been originally selected. Alternatively, this character appeared only secondarily and the primary selective effect was that the symbiotic dioxygen consumer was able to protect its host from damage caused by dioxygen, thus allowing it to 'conquer' aerobic ecological niches. In any event, a major consequence of this endosymbiosis for early eukaryotes would be the transition from an anaerobic to an aerobic lifestyle. The transition from the prokaryotic symbiont to the mitochondrion was accompanied by gene transfers from the prokaryotic genome to the nuclear genome of its host (Fig. 4.17i).

Following the acquisition of mitochondria, major extant eukaryotic groups diversified. Later, a new symbiotic association occurred in one lineage with the capture of a photosynthetic bacterium related to cyanobacteria (Fig. 4.17j). Similarly to mitochondrial endosymbiosis, gene transfers occurred from the cyanobacterium to the nucleus of the host (Fig. 4.17k), and the symbiotic photosynthetic bacterium evolved into the chloroplast (Fig. 4.17l).

A comprehensive study of processes involved in these endosymbioses is presented in Chap. 5.

4.3.1.3 Ample Evidence Supporting the Endosymbiotic Theory

1. Phylogenetic studies based on the comparison of sequences of rRNAs and protein-encoding genes from the genomes of endosymbionts have clearly demonstrated that mitochondria and chloroplasts are related to *Bacteria*, specifically alphaproteobacteria for mitochondria and cyanobacteria for chloroplasts. Mitochondrial genomes contain between 10 and 100 genes, which is much lower than the number of genes found in extant alphaproteobacteria. Two complementary explanations can be advanced: (a) many genes initially present in the symbiont were lost during evolution and (b) genes were transferred from the symbiont to the nucleus of the host. Supporting this point, molecular phylogenies have shown that some nuclear genes coding for proteins that sometimes act in the mitochondria are close to alphaproteobacteria. HGTs to the host nuclear genome are also frequent in the case of chloroplasts, whose genomes contain a higher number (100–200) of genes. The large size of the chloroplast genome in comparison with that of mitochondria has sometimes been interpreted as an indication that mitochondrial endosymbiosis predates the chloroplastic endosymbiosis. However, it may also simply indicate a different mode of evolution. A second argument in favour of the temporal priority of mitochondria is that the reception of a dioxygen producer is only conceivable for a host capable of consuming dioxygen. But the most decisive argument is that all current eukaryotic cells have or have had mitochondrion-harbouring ancestors, while the chloroplast endosymbiosis is specific to some eukaryotic lineages.
2. Mitochondria and chloroplasts contain their own ribosomes which are undoubtedly of the bacterial type. In addition, the same antibiotics as those acting on free-living bacteria, for example, streptomycin, inhibit their function.
3. Photosynthesis is virtually identical in chloroplasts and in extant cyanobacteria. Moreover, the respiratory chain process of prokaryotes (localised in the cytoplasmic membrane) is comparable to that of eukaryotes (located in the mitochondrial inner membrane).
4. The acquisition of photosynthesis continues today (see Sect. 5.4.4).

4.3.2 Eukaryotes Result of an Association Between a Bacterium and an Archaeon

Many other hypotheses apart from that proposed by De Duve have been put forward to explain the origin of eukaryotes. Two in particular suggest that the eukaryotic cell is the result of the establishment of a syntrophic association between bacteria and archaea. Such hypotheses have been advanced to explain the fact that the eukaryotic genome contains both genes that are close relatives of archaeal homologues and genes that are relatives of bacterial genes.

The hydrogen hypothesis was proposed in 1998 by Martin and Müller. It is based on the establishment of associations in an anoxic environment between H_2- and CO_2-producing fermentative alphaproteobacteria and methanogenic archaea that could use the dihydrogen and CO_2 produced by the alphaproteobacteria. Such associations, based on interspecies hydrogen transfer, have been observed among extant organisms. For these authors, an increase of the contact surface between the partners by expansion of the membranes of the methanogens around alphaproteobacteria would have resulted in total engulfment of the latter (Fig. 4.18a, b). In parallel, several alphaproteobacterial genes were transferred to the genome of the methanogenic archaea and this is how the genome of the primitive eukaryote (or proto-eukaryote) would have evolved. The other part of the alphaproteobacterial genome

Fig. 4.18 Sketch presenting the theory of proposed mechanisms (Modified and redrawn from Martin and Müller (1998) (**a**, **b**) and López-García and Moreira (2006) (**c–e**)) (Drawing: M.-J. Bodiou)

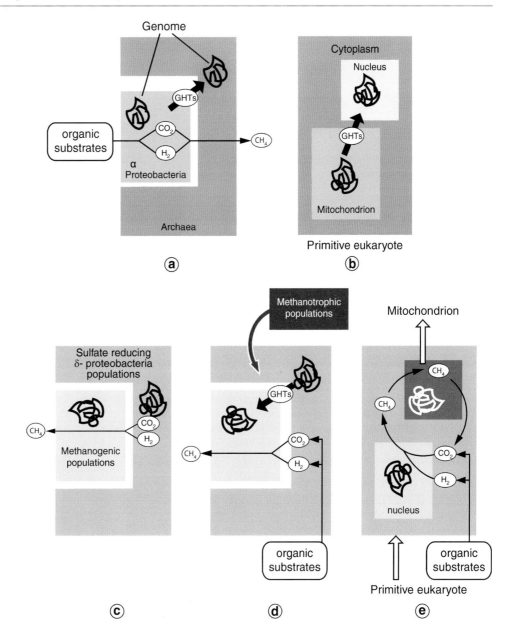

would have evolved into the mitochondrion. According to this hypothesis:

1. The ancestor of mitochondria would have initially had an anaerobic metabolism, and extant mitochondria are derived from it.
2. The origin of eukaryotes coincides with mitochondrial endosymbiosis, which implies that amitochondrial eukaryotes never existed.

The syntrophic hypothesis, proposed by López-García and Moreira (Moreira and López-García 1998; López-García and Moreira 2006), involves a symbiotic association between sulphate-reducing bacterial communities belonging to the group of deltaproteobacteria (such as myxobacteria) and methanogenic archaea that would use the dihydrogen produced by fermentation by the sulphate-reducing bacteria (Fig. 4.18c–e). As in the hydrogen hypothesis, this association would be accompanied by membrane extensions to maximise contact and exchange between partners. Gene transfer from deltaproteobacteria to archaea would have also occurred until total disappearance of the bacterial genome, thus giving rise to a chimerical eukaryotic genome. This hypothesis involves a third actor, a community of anaerobic, methanotrophic alphaproteobacteria that would consume the methane produced by methanogenic archaea and produce CO_2, thus stimulating methanogenic activity. In this scenario, these methanotrophic alphaproteobacteria would be the ancestors of mitochondria. As in the hydrogen hypothesis, the ancestor of mitochondria would initially have had an anaerobic metabolism.

The two hypotheses differ about the number and nature of partners involved in the association; but both support a chimerical eukaryotic genome and rely on metabolic interactions that are still widespread in nature. Both assume that the archaeal partners were methanogens and that the alphaproteobacteria at the origin of mitochondria were initially anaerobic. According to these two hypotheses, the acquisition of chloroplasts would have occurred as described above. Other associations between bacteria and archaea have been hypothesised, for example, by Lynn Margulis.

4.3.3 Eukaryotes and Archaea Evolved from Actinobacteria

Tom Cavalier-Smith (2002) suggested that *Archaea* and *Eucarya* have evolved from an actinobacterium. According to this author, this scenario is justified by a number of characteristics shared by actinobacteria, archaea and eukaryotes (a single membrane cell envelope instead of two as in most other bacteria, histone H1 and proteasome, for example). This scenario is attractive because it takes into account a number of observations. However, it is very controversial because it is:
1. Based on unproven assumptions. It assumes, for example, that the ancestral state is a two-membraned cell and that the loss of the outer membrane occurred only once.
2. Based on the subjective choice of a few characteristics to the exclusion of many others.
3. Not without exceptions (e.g. histone H1 is found today in some actinobacteria and eukaryotes only).

The scenario thus implies that H1 histones evolved in the common ancestor of actinobacteria, archaea and eukaryotes and were then lost in archaea.

4.3.4 Evolution by Simplification in Microorganisms

Although general trends towards simplification or complexity increase can be identified in some lineages, this does not mean that such trends can be generalised over time and over all structures of the lineages under consideration (Brinkmann and Philippe 2005). For example, if the eukaryotic cell is derived from a prokaryotic cell, this transition was probably accompanied by a phase of increased complexity in the eukaryotic evolutionary lineage (appearance of membrane compartments, transcription and translation uncoupling, appearance of eukaryotic-specific multiprotein structures, etc.). This does not mean that this trend has continued until today in all eukaryotic lineages. Similarly, changes in prokaryotic lineages should not be seen as a move towards simplification. Changes within whatever evolutionary lineage should be seen rather as alternating phases of simplification and complexity increase of structures. Simplifying a structure can even coexist with increase in complexity of another.

Formal evidence of evolution by simplification was provided by molecular phylogeny and confirmed by comparative genomics. A first example is given by mycoplasma, very small (2.2–0.8 mm) parasitic bacteria of eukaryotic cells that are devoid of a cell wall and have genomes consisting of 500–1,500 kilobases (kb). The genome of *Mycoplasma genitalium* (causing severe urinary disorders in humans) contains only 580 kb corresponding to 480 genes. Because of the small size of their genomes, mycoplasmas were initially considered as unsophisticated, 'primitive' prokaryotes. It is now clear that the small size of the genome of *Mycoplasma genitalium* is the result of a massive loss of genes in relation to its adaptation to a parasitic lifestyle including loss of the cell wall and a sharp reduction in its metabolic capacity. A similar phenomenon has been demonstrated for *Buchnera,* gammaproteobacterial symbionts of aphids. Different genes are lost in the case of *Mycoplasma* and in the case of *Buchnera*. This implies that there are different evolutionary pathways that may be associated with genomic reduction, each reflecting adaptation to different lifestyles (McCutcheon and Moran 2012).

Another example of evolution by simplification is given by Archezoa, a polyphyletic assemblage of unicellular eukaryotes (Diplomonads, Trichomonads and Microsporidia). Archezoa do not synthesise their ATP through mitochondrial respiration as most eukaryotic cells do because they do not have mitochondria. This feature, combined with their putative branching at the base of the eukaryotic tree (derived from analysis of the sequences of their rRNA in the 1980s), was initially interpreted as the fact that Archezoa were representatives of ancient eukaryotic lineages that diverged before mitochondrial endosymbiosis. However, several genes homologous to genes of alphaproteobacterial origin whose products function in mitochondria in most eukaryotes have been identified in the genomes of these organisms. One of them is the gene coding for the synthesis of proteins called 'heat shock protein 70' (Hsp70). These proteins have been visualised in the microsporidian *Trachipleistophora hominis* by using specific antibodies and are concentrated within intracellular vesicles measuring 50–90 nm, surrounded by a double membrane, called **mitosomes***(Williams et al. 2002). Similar vesicles were also identified in the Diplomonad *Giardia intestinalis* (pathogenic agent of an intestinal disease called giardiasis) and in other microsporidia. In Trichomonads such as *Trichomonas vaginalis*, genes of mitochondrial origin have been detected in other cell structures called **hydrogenosomes***. Hydrogenosomes are cellular compartments surrounded by a double membrane, which produce ATP and dihydrogen by fermentation and may contain DNA (see Sect. 5.4.2). It has been now clearly demonstrated that hydrogenosomes and mitosomes are highly derived mitochondria (Shiflett and Johnson 2010).

These examples show that the absence of mitochondria is not the mark of a poorly advanced cellular organisation but, rather, the result of evolution by simplification linked to a particular lifestyle, with loss of the mitochondrial genome as well as of all genes of the respiratory chain. Noticeably, a double-membraned intracellular organelle derived from mitochondria seems to have been conserved in all extant eukaryotic organisms.

The phenomenon of evolutionary simplification also applies to the evolutionary history of the agents of malaria, *Plasmodium falciparum,* and toxoplasmosis, *Toxoplasma gondii.* Sequencing the genome of these organisms has revealed the presence of extranuclear genomes sharing a common evolutionary origin with chloroplast genomes of photosynthetic eukaryotes. These genomes were localised in cell structures that were detected in microscopy but whose function was not previously understood. Apicomplexa, the phylum to which *Plasmodium* and *Toxoplasma* belong, are now seen as having a non-photosynthetic organelle, the apicoplast, evolved by simplification of an ancestral chloroplast (see Sect. 7.8.4).

4.4 Synthetic Approach of the Evolution of Metabolisms

What mechanisms of energy acquisition and carbon fixation did microorganisms use during evolution?

The present state of our knowledge does not allow determining precisely the different stages of development of these mechanisms. However, thanks to clues provided by geological, palaeontological, palaeoclimatic and phylogenetic studies and coupled with knowledge of the diversity of contemporary microbial metabolism, it is possible to speculate about the occurrence of different metabolisms during evolution (Nealson and Rye 2003). It is nevertheless necessary to bear in mind that this speculation says nothing (or very little) about the metabolism of LUCA.

4.4.1 The Primordial Metabolism: Heterotrophic or Autotrophic?

There are two hypotheses concerning the metabolism of the first living beings (Ehrlich 2002). In the first, the primitive metabolism would have been of the fermentative type and would thus be heterotrophic. Given the universality of glycolysis in today's living world, it is tempting to assume that this, probably very ancient, mechanism was present in LUCA and even perhaps in the first organisms. In this view, ATP, which is the storage form of energy, was produced by substrate-level phosphorylation. Fermented molecules would have had an abiotic origin. The fermentation products – organic acids – released into the environment would have decreased the extracellular pH. Because the first membranes were supposed to have been permeable to protons, the cell must have had a proton pump consuming a large part of the ATP produced by fermentation to maintain its internal pH (Fig. 4.19a). Subsequently, membrane proteins constituting rudimentary electron transfer chains would have appeared, fuelled by non-fermentable organic substrates present in the environment (Fig. 4.19b). The expulsion of a proton (H^+), possibly coupled to the activity of the respiratory chains, would have allowed the maintenance of intracellular pH compatible with the physiology of the cells and the creation of a membrane potential via a proton gradient between the inside and the outside of the cell. As the membrane became impermeable to protons, their return within the cell would have occurred by means of an ATP-synthesising enzyme, functionally comparable to the ATP synthase present in extant cells (Fig. 4.19c). This mechanism – the coupling between a proton gradient generated by electron transfer and ATP synthesis – corresponds to Peter Mitchell's chemiosmotic theory. Its presence in the three domains of life suggests that it appeared early during evolution. Some cells (the ancestors of future autotrophic organisms) would then have acquired the ability to use carbon dioxide as a carbon source.

It is also possible (second hypothesis) that the first living organisms were able, from the outset, to assimilate CO_2; in this case, autotrophy would have preceded heterotrophy. Organic molecules synthesised by autotrophic organisms could have been a source of food for the first heterotrophic organisms. If autotrophy appeared first, it is unclear if it was chemolitho-autotrophy or photoautotrophy (Ehrlich 2002).

Assuming chemolitho-autotrophy preceded phototrophy, energy needs would have been met by the use of mineral elements present on the early Earth (dihydrogen, iron, hydrogen sulphide, etc.). In the absence of dioxygen, electrons produced by redox reactions would have been transferred to a final acceptor, such as iron (Vargas et al. 1998) or elementary sulphur (Fig. 4.19d), and these organisms would have been chemolitho-autotrophic. Data obtained from isotopic fractionation suggest the appearance of autotrophic CO_2 fixation around 3.8 billion years. Similarly, sulphate-reducing organisms could have been active around 3.47 billion years (Shen and Buick 2004). Moreover, direct association of fossil microorganisms with a volcanic particle substrate in similarly aged rocks is strongly indicative of chemolithotrophy (Westall et al. 2011a, b). However, Philippot et al. (2007) suggested sulphide could be produced by disproportionation of elementary sulphur rather than sulphate reduction.

absence of solar energy in microorganisms living near hydrothermal vents that would have acquired the ability to use the infrared radiation emitted by these environments (Nisbet et al. 1995). Current black smokers formed at oceanic hydrothermal vents emit radiation with wavelengths between 700 and 1,000 nm. These wavelengths correspond to the absorption spectrum of bacteriochlorophylls of contemporary anoxygenic bacteria such as purple bacteria (absorbing at wavelengths around 900 nm and greater than 1,000 nm) or green sulphur bacteria (whose maximum absorption spectrum is around 750 nm (see Sect. 3.3.4)). Beatty and his team isolated from a black smoker an unknown green sulphur bacterium whose only source of light energy would be of geothermal origin (Beatty et al. 2005). This important result, although not yet confirmed by other studies, would demonstrate the existence of microorganisms capable of photosynthesis in the absence of solar energy. Even if this finding were validated, it would not constitute evidence that photosynthesis appeared in such environments. It might be a secondary colonisation of deep environments by initially solar energy-dependent photosynthetic organisms.

Whichever mechanism appeared first, phototrophy or chemolitho-autotrophy, the order of appearance of oxygenic and anoxygenic photosyntheses still remains an open issue.

Jin Xiong and Carl E. Bauer (2002) from Indiana University propose a model based on the study of the phylogeny of genes involved in the synthesis of photosynthetic pigments. The first photosynthetic lineage would be proteobacteria, which perform anoxygenic photosynthesis with photosystem II. Green nonsulphur bacteria (anoxygenic photosynthesis based on photosystem II), green sulphur bacteria, and heliobacteria (anoxygenic photosynthesis through photosystem I) would have appeared secondarily. Cyanobacteria (oxygenic photosynthesis based on photosystems I and II) would have acquired these two photosystems by HGTs from green sulphur bacteria and proteobacteria, respectively. While this study provides evidence that anoxygenic photosynthesis predated oxygenic photosynthesis, it does not allow any conclusions to be drawn about the ancestry of phototrophy compared to chemolitho-autotrophy.

In contrast, a study based on genomic analyses published by Mulkidjanian and colleagues (2006) suggests that oxygenic photosynthesis performed by cyanobacteria would have appeared before anoxygenic photosynthesis.

The rock record contains evidence for photosynthesis by 3.46–3.3 Ga, but there is no evidence that this was oxygenic because there are no signatures of oxygenic photosynthesis associated.

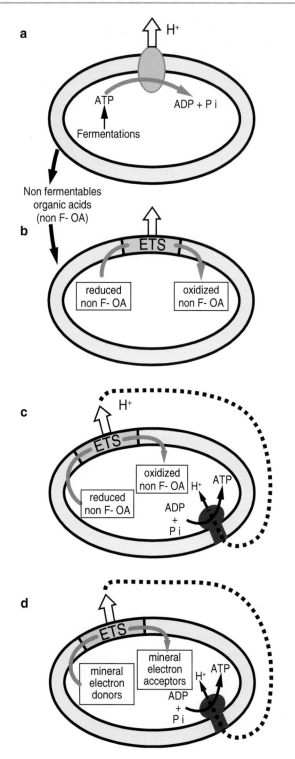

Fig. 4.19 Sketch presenting the simplified hypothetical steps in the appearance of chemolitho-autotrophy (Drawing: M.-J. Bodiou)

If, however, phototrophy preceded chemolitho-autotrophy, it would have harnessed solar energy. However, it was also suggested that phototrophy evolved in the

4.4.2 Appearance of Oxygenic Photosynthesis and Its Consequences

4.4.2.1 Modification of the Atmospheric Composition

During the evolution of the usage of solar energy as an energy source, a major step was taken with the appearance of cyanobacteria capable of oxygenic photosynthesis. They acquired the ability to use an energy source (solar energy) and an electron donor (water) that were available in inexhaustible quantities in the environment. In evolutionary terms, their appearance represents a major innovation and cyanobacteria experienced considerable expansion. Massively producing dioxygen through time, they profoundly changed the composition of the atmosphere of the planet, allowing the development of aerobic respiration. Without cyanobacteria, eukaryotes and especially multicellular organisms, as currently known, would have probably never appeared. The appearance of dioxygen upset the biogeochemical cycles of many elements including iron, which, spontaneously oxidised in the presence of dioxygen, precipitates as in mineral form and becomes a life-limiting factor in the oceans.

It is difficult to estimate with precision the evolution of the concentration of dioxygen in the atmosphere (Fig. 4.11). It seems clear that the activity of cyanobacteria in the oceans is the source of almost all atmospheric O_2, whose content greatly increased around 2.3 Ga ago to reach its current value more than a billion years later. However, the production of O_2 by cyanobacteria could have started much earlier. The time lag between dioxygen production by cyanobacteria and its accumulation in the atmosphere could be explained by the oxygenation of the multitude of reduced mineral species, such as iron, and by the oxygenation of the oceans. The transformation of ferrous iron to ferric iron led to the formation of deposits called 'banded iron formations' (BIF) that are today an important source of iron ore. After the complete conversion of ferrous iron to ferric iron precipitate, the oceans would have gradually become saturated in dioxygen, which subsequently diffused into the atmosphere to reach its present concentration.

The atmospheric dioxygen content before 2 billion years is hotly debated. For most scientists, the primitive atmosphere and oceans were predominantly anoxic. For others, the O_2 content would have been sufficient to allow the emergence of a form of aerobic life. It has also been suggested that the presence in the oceans and in shallow lagoon systems of 'oases', powered by relatively intense photosynthesis, could have promoted the emergence of an aerobic metabolism before the overall oxygenation of the atmosphere.

The dioxygen produced by cyanobacteria, which gradually invaded all terrestrial and aquatic habitats, could have been a dangerous poison for all forms of prokaryotic or eukaryotic anaerobic life. This is because, although dioxygen itself is not toxic, its radical derivatives called 'free radicals' (the superoxide ion hydrogen peroxide) are highly reactive, toxic chemical species. However, the increase of dioxygen in the atmosphere and in the oceans did not occur abruptly, as described above. Very likely, in parallel to the gradual increase of the O_2 concentration, some organisms developed defence systems to eliminate these free radicals (appearance of enzymes such as superoxide dismutase, catalase and peroxidase).

4.4.2.2 Appearance of a Protective Ozone Layer

Another consequence of increased dioxygen was the formation of an ozone (O_3) layer, which is a very effective barrier against ultraviolet radiation of short wavelength, lethal to all living organisms.

As this kind of radiation is efficiently attenuated by water (a few tens of centimetres to several metres of water suffice), ocean life could develop even in the absence of an ozone layer. The appearance of a protective ozone layer could allow the expansion of microorganisms to the land surface.

4.4.2.3 Emergence of New Pathways in Biogeochemical Cycles

The oxygenation of the atmosphere completely modified the functioning of biogeochemical cycles: aerobic biodegradation of organic matter, oxidation of ammonium into nitrate and of hydrogen sulphide to sulphate, etc. (See Chap. 14).

4.4.2.4 Dioxygen Is Used as a Terminal Electron Acceptor: The Appearance of Aerobic Respiration

The generally accepted paradigm, which was previously presented in this chapter, posits that oxygenic photosynthesis predated aerobic respiration. Aerobic respiration would have appeared either from the evolution of electron transport chains of the photosynthetic apparatus or from the respiratory chain of denitrifying microorganisms (Saraste and Castresana 1994). One possible explanation for the success of aerobic respiration is the higher energetic efficiency than possible with anaerobic respiration.

The emergence of oxygenic photosynthesis before that of aerobic respiration is however not unanimously accepted. Indeed, for some scientists, aerobic respiration predated oxygenic photosynthesis. Their hypothesis is based on phylogenetic analyses suggesting that LUCA could have possessed at least one cytochrome oxidase, an enzyme reducing dioxygen in H_2O and, therefore, could have been an aerobic organism (Castresana 2004). If aerobic respiration actually predated the onset of oxygenic photosynthesis, the question of the origin of respired dioxygen arises. One possibility is its production by abiotic reactions such as photolysis of water vapour in the upper atmosphere. Indeed, even if the amount of dioxygen produced by this mechanism is very small compared to that produced by oxygenic photosynthesis, high concentrations of dioxygen are not always necessary for aerobic respiration, as in the case of microaerophilic bacteria.

4.5 Conclusion

Data concerning very ancient life (fossils, isotopic ratios, biomarkers) are rare and sometimes difficult to interpret. Despite that, the existence of cellular life between 3.5 and 3.2 Ga is well established. The discovery of fossil biomarkers (steranes methylhopanes) and the use of stable isotopes (carbon, sulphur, nitrogen) have provided important information on the possible dates of occurrence of many metabolisms.

Since the appearance of the first cell, microorganisms were the only inhabitants of our planet for almost 3 billion years. They have survived the intense geological upheavals that have marked the history of the Earth. They profoundly modified their environment to such an extent that there was a true co-evolution between the biosphere and the geosphere. Through their activity, they also created favourable conditions for the emergence of multicellular aerobic organisms (particularly through oxygenation of the atmosphere).

Among all microorganisms that populated the Earth, LUCA occupies a central position in evolution and is the subject of much research because it is the ancestor of all extant living organisms.

It is possible to outline some of the major stages of the evolution of microorganisms (especially the most recent steps). The endosymbiotic origin of mitochondria and chloroplasts is a virtual certainty. It is now established that many bacterial genes have contributed, via mitochondrial endosymbiosis (and chloroplast endosymbiosis for photosynthetic eukaryotes), to build the eukaryotic genome. However, this chapter also highlights the large uncertainties about the nature of LUCA and the sequence of events that led to modern organisms. The main reason is due to the extremely old age of these events. Recall, for example, that the exact causes of the extinction of the dinosaurs that occurred only 65 million years ago are still debated. Compared to the several billion-year-old events discussed in this chapter, it was yesterday. It is thus no surprise that many questions remain about the history of microorganisms: was LUCA hyperthermophilic, thermophilic or mesophilic? Was it heterotrophic or autotrophic? How and when did eukaryotes appear? Was it 2.2 billion years ago, as suggested by the discovery of steranes, or 1.5 billion years ago, the age of some eukaryotic microfossils? What role did viruses play in the evolution of cellular organisms? Did oxygenic photosynthesis predate aerobic respiration or vice versa?

Despite huge gaps, our knowledge has progressed considerably on all these issues. Some of today's speculative hypotheses may become consensual tomorrow. Peter Mitchell's chemiosmotic hypothesis proposed in 1960 is a good example. It is currently considered a mainstream theory after having been initially judged unacceptable by specialists of bioenergetics. However, in 1975, Peter Mitchell received the Nobel Prize. He had explained mechanisms that most likely emerged during very early phases of cellular life.

While many questions remain unanswered, they offer many opportunities for research: improvement in the analytical techniques of ancient rocks samples, intensification of genome sequencing (especially protists), deepening of our understanding of the physiology and biochemistry of contemporary microorganisms, and improvement of techniques for the isolation and cultivation of microorganisms. Indeed, microbial biodiversity has not revealed all its secrets because only a very small part of the microorganisms that inhabit our planet is cultivable with current techniques (0.01–1 %).

Some answers might be brought in by exobiology (especially, the research of traces of extraterrestrial life) (see Chap. 10) or by a better understanding of incompletely or partially explored terrestrial habitats including subglacial Antarctic lakes, biota of the subsurface biosphere such as petroleum reservoirs, underground aquifers and deep rocks.

References

Allwood AC, Walter MR, Kamber BS, Marshall CP, Burch IW (2006) Stromatolite reef from the Early Archaean era of Australia. Nature 441:714–718

Beatty JT et al (2005) An obligately photosynthetic bacterial anaerobe from a deep-sea hydrothermal vent. Proc Natl Acad Sci U S A 102:9306–9310

Boussau B, Blanquart S, Necsulea A, Lartillot N, Gouy M (2008) Parallel adaptations to high temperature in the Archaean eon. Nature 456:942–945

Brasier MD et al (2002) Questioning the evidence for Earth's oldest fossils. Nature 416:76–81

Brasier MD, Mc Loughlin N, Green O, Wacey D (2006) A fresh look at the fossil evidence for early Archaean cellular life. Philos Trans R Soc Lond B Biol Sci 361:887–902

Brinkmann H, Philippe H (2005) The universal tree of life: from simple to complex or from complex to simple. In: Gargaud M, Barbier B, Martin H, Reisse J (eds) Lectures in astrobiology: vol I. Advances in astrobiology and biogeophysics. Springer, pp 617–656

Brochier C, Philippe H (2002) A non-hyperthermophilic ancestor for bacteria. Nature 417:244

Brocks JJ, Summons RE (2003) Sedimentary hydrocarbons, biomarkers for early life. In: Holland HD, Turekian KK (eds) Treatise on geochemistry. New-Haven, Elsevier, pp 63–115

Brocks JJ, Logan GA, Buick R, Summons RE (1999) Archean molecular fossils and the early rise of eukaryotes. Science 285:1033–1036

Brocks JJ, Buick R, Summons RE, Logan GA (2003) A reconstruction of Archean biological diversity based on molecular fossils from the 2.78 to 2.45 billion-year-old Mount Bruce Supergroup, Hamersley Basin, Western Australia. Geochim Cosmochim Acta 67:4321–4335

Butterfield NJ (2000) *Bangiomorpha pubescens* n.gen., n. sp.: implications for the evolution of sex, multicellularity, and the Mesoproterozoic/Neoproterozoic radiation of Eukaryotes. Paleobiology 26:386–404

Byerly GR, Lowe DR, Walsh MM (1986) Stromatolites from the 3300–3500 Myr Swaziland Supergroup, Barberton Mountain Land, South Africa. Nature 319:489–491

Canfield DE, Rosing MT, Bjerrum C (2006) Early anaerobic metabolisms. Philos Trans R Soc Lond B Biol Sci 361:1819–1834; discussion 1835–1816

Castresana J (2004) Evolution and phylogenetic analysis of respiration. In: Zannoni D (ed) Respiration in archaea and bacteria. Kluwer Academic Publishers, Dorderecht/Boston/London, pp 1–14

Cavalier-Smith T (2002) The neomuran origin of archaebacteria, the negibacterial root of the universal tree and bacterial megaclassification. Int J Syst Evol Microbiol 52:7–76

Courties C et al (1994) Smallest eukaryotic organism. Nature 370:255

De Duve C (1995) Vital dust: life as a cosmic imperative. Basic Books, New York

De Duve C (2002) À l'écoute du vivant. Odile Jacob, Paris

Ehrlich HL (2002) The origin of life and its early history. In: Geomicrobiology. M. Dekker, Inc, New York/Basel, pp 21–48

El Albani A et al (2010) Large colonial organisms with coordinated growth in oxygenated environments 2.1 Gyr ago. Nature 466:100–104

Forterre P (2002) The origin of DNA genomes and DNA replication proteins. Curr Opin Microbiol 5:525–532

Forterre P, Bouthier De La Tour C, Philippe H, Duguet M (2000) Reverse gyrase from hyperthermophiles: probable transfer of a thermoadaptation trait from archaea to bacteria. Trends Genet 16:152–154

Forterre P, Gribaldo S, Brochier C (2005) Luca: the last universal common ancestor. Med Sci 21:860–865

Galtier N, Tourasse N, Gouy M (1999) A nonhyperthermophilic common ancestor to extant life forms. Science 283:220–221

Garcia-Ruiz JM, Hyde ST, Carnerup AM, Christy AG, van Kranendonk MJ, Welham NJ (2003) Self-assembled silica-carbonate structures and detection of ancient microfossils. Science 302:1194–1197

Gaucher EA, Thomson JM, Burgan MF, Benner SA (2003) Inferring the palaeoenvironment of ancient bacteria on the basis of resurrected proteins. Nature 425:285–288

Gogarten JP et al (1989) Evolution of the vacuolar H + − ATPase: implications for the origin of eukaryotes. Proc Natl Acad Sci U S A 86:6661–6665

Grassineau NV et al (2001) Antiquity of the biological sulphur cycle: evidence from sulphur and carbon isotopes in 2700 million-year-old rocks of the Belingwe belt, Zimbabwe. Proc Roy Soc Lond B 268:113–119

Han TM, Runnegar B (1992) Megascopic eukaryotic algae from the 2.1-billion-year-old negaunee iron-formation, Michigan. Science 257:232–235

Hoefs J (2004) Stable isotopes geochemistry. 5th completely revised, updated, and enlarged edition. Springer, Berlin, 244 p

Hofmann HJ, Grey K, Hickman AH, Thorpe RI (1999) Origin of 3.45 Ga coniform stromatolites in Warrawoona Group, Western Australia. Geol Soc A Bull 111:1256–1262

Iwabe N, Kuma K, Hasegawa M, Osawa S, Miyata T (1989) Evolutionary relationship of archaebacteria, eubacteria, and eukaryotes inferred from phylogenetic trees of duplicated genes. Proc Natl Acad Sci U S A 86:9355–9359

Javaux EJ, Knoll AH, Walter MR (2001) Morphological and ecological complexity in early eukaryotic ecosystems. Nature 412:66–69

Javaux EJ, Marshall CP, Bekker A (2010) Organic-walled microfossils in 3.2-billion-year-old shallow-marine siliciclastic deposits. Nature 463:934–938

Kasting JF, Howard MT (2006) Atmospheric composition and climate on the early Earth. Philos Trans R Soc Lond B Biol Sci 361:1733–1741; discussion 1741–1732

Knauth LP, Lowe DR (2003) High Archean climatic temperature inferred from oxygen isotope geochemistry of cherts in the 3.5 Ga Swaziland Supergroup. S Afr Geol Soc Am Bull 115:566–580

Knoll AH, Javaux EJ, Hewitt D, Cohen P (2006) Eukaryotic organisms in Proterozoic oceans. Philos Trans R Soc Lond B Biol Sci 361:1023–1038

Koonin EV (2003) Comparative genomics, minimal gene-sets and the last universal common ancestor. Nat Rev Microbiol 1:127–136

Krumbein WE (1983) Stromatolites, the challenge of a term in space and time. Dev Precambrian Geol 7:385–423

Lenton T, Watson A (2011) Revolutions that made the Earth. Oxford University Press, Oxford

Lombard J, López-García P, Moreira D (2012) The early evolution of lipid membranes and the three domains of life. Nat Rev Microbiol 10:507–515

López-García P, Moreira D (2006) Selective forces for the origin of the eukaryotic nucleus. Bioessays 28:525–533

Margulis L (1970) Origin of eukaryotic cells. Yale University Press, Yale

Martin W, Muller M (1998) The hydrogen hypothesis for the first eukaryote. Nature 392:37–41

Maynard Smith J, Szathmáry E (2000) The origins of life. Oxford University Press, Oxford

McCutcheon JP, Moran NA (2012) Extreme genome reduction in symbiotic bacteria. Nat Rev Microbiol 10:13–26

Moreira D, López-García P (1998) Symbiosis between methanogenic archaea and delta-proteobacteria as the origin of eukaryotes: the syntrophic hypothesis. J Mol Evol 47:517–530

Mulkidjanian AY et al (2006) The cyanobacterial genome core and the origin of photosynthesis. Proc Natl Acad Sci U S A 103:13126–13131

Mushegian AR, Koonin EV (1996) A minimal gene set for cellular life derived by comparison of complete bacterial genomes. Proc Natl Acad Sci U S A 93:10268–10273

Nakabachi A, Yamashita A, Toh H, Ishikawa H, Dunbar HE, Moran NA, Hattori M (2006) The 160-kilobase genome of the bacterial endosymbiont Carsonella. Science 314:267

Nealson KH, Rye R (2003) Evolution of metabolism. In: Holland HD, Turekian KK (eds) Treatise on geochemistry. New-Haven, Elsevier, pp 41–61

Nelson DL, Cox MM (2000) Lehninger principles of biochemistry. Worth Publishers, New York

Nisbet EG, Fowler CMR (2003) The early history of life. In: Holland HD, Turekian KK (eds) Treatise on geochemistry. New-Haven, Elsevier, pp 1–39

Nisbet EG, Cann JR, Lee van Dover C (1995) Origins of photosynthesis. Nature 373:479–480

Ohmoto H (2004) The Archaean atmosphere, hydrosphere and biosphere. In: Eriksson PG, Altermann W, Nelson DR, Mueller WU, Catueanu O (eds) The Precambrian earth: tempos and events. Elsevier, Amsterdam, pp 361–388

Ourisson G, Nakatani Y (1994) The terpenoid theory of the origin of cellular life: the evolution of terpenoids to cholesterol. Curr Biol 1:11–23

Peters KE, Walters CC, Moldowan JM (2007) The biomarker guide, vol 1–2. Cambridge University Press, Cambridge

Philippot P, Van Zuilen M, Lepot K, Thomazo C, Farquhar J, Van Kranendonk MJ (2007) Early archaean microorganisms preferred elemental sulfur, not *sulfate*. Science 317:1534–1537

Rasmussen B, Fletcher IR, Brocks JJ, Kilburn MR (2008) Reassessing the first appearance of eukaryotes and cyanobacteria. Nature 455:1101–1104

Rosing MT (1999) 13C-Depleted carbon microparticles in >3700-Ma sea-floor sedimentary rocks from west Greenland. Science 283:674–676

Sadava D, Heller HC, Orians GH, Purves WK, Hillis DM (2007) Life: the science of biology, 8th edn. Sinauer Associates, Sunderland

Saraste M, Castresana J (1994) Cytochrome oxidase evolved by tinkering with denitrification enzymes. FEBS Lett 341:1–4

Schidlowski M (1993) The initiation of biological processes on Earth. In: Engel MH, Macko SA (eds) Organic geochemistry. Plenum Press, New York/London, pp 693–695

Schidlowski M (2001) Carbon isotopes as biogeochemical recorders of life over 3.8 Ga of Earth history: evolution of a concept. Precambrian Res 106:117–134

Schopf JW (1993) Microfossils of the Early Archean Apex chert: new evidence of the antiquity of life. Science 260:640–646

Schopf JW, Kudryavtsev AB, Agresti DG, Wdowiak TJ, Czaja AD (2002) Laser-Raman imagery of Earth's earliest fossils. Nature 416:73–76

Schulz HN, Brinkhoff T, Ferdelman TG, Hernández Mariné M, Teske A, Jørgensen BB (1999) Dense populations of a giant sulfur bacterium in Namibian shelf sediments. Science 284:493–495

Schwartz RM, Dayhoff MO (1978) Origins of prokaryotes, eukaryotes, mitochondria, and chloroplasts. Science 199:395–403

Shen Y, Buick R (2004) The antiquity of microbial sulfate reduction. Earth Sci Rev 64:243–272

Shiflett AM, Johnson PJ (2010) Mitochondrion-related organelles in eukaryotic protists. Annu Rev Microbiol 64:409–429

Stetter KO (2006) Hyperthermophiles in the history of life. Philos Trans R Soc Lond B Biol Sci 361:1837–1843

Sugitani K, Grey K, Nagaoka T, Mimura K, Walter M (2009) Taxonomy and biogenicity of Archaean spheroidal microfossils (ca. 3.0 Ga) from the Mount Goldsworthy-Mount Grant area in the northwestern Pilbara Craton, Western Australia. Precambrian Res 173:50–59

Summons RE, Jahnke LL, Hope JM, Logan GA (1999) 2-Methylhopanoids as biomarkers for cyanobacterial oxygenic photosynthesis. Nature 400:554–557

Van den Boorn S, Van Bergen MJ, Nijman W, Vroon PZ (2007) Dual role of seawater and hydrothermal fluids in Early Archean chert formation: evidence from silicon isotopes. Geology 35:939–942

Vargas M, Kashefi K, Blunt-Harris EL, Lovley DR (1998) Microbiological evidence for Fe(iii) reduction on early Earth. Nature 395:65–67

Volkman JK (2005) Sterols and other triterpenoids: source specificity and evolution of biosynthetic pathways. Org Geochem 36:139–159

Westall F (2005) The geological context for the origin of life and the mineral signatures of fossil life. In: Gargaud M, Barbier B, Martin H, Reisse J (eds) Lectures in astrobiology: vol I. Advances in astrobiology and biogeophysics. Springer, pp 195–226

Westall F (2011) Early life. In: Gargaud M et al (eds) Origins of life, an astrobiology perspective. Cambridge University Press, Cambridge, pp 391–413

Westall F, Cavalazzi B (2011) Biosignatures in rocks. In: Thiel V, Reitner J (eds) Encyclopedia of geobiology. Springer, Berlin, pp 189–201

Westall F, Folk RL (2003) Exogenous carbonaceous microstructures in Early Archaean cherts and BIFs from the Isua Greenstone Belt: implications for the search for life in ancient rocks. Precambrian Res 126:313–330

Westall F, de Ronde CEJ, Southam G, Grassineau N, Colas M, Cockell C, Lammer H (2006a) Implications of a 3.472-3.333 Gyr-old sub-aerial microbial mat from the Barberton greenstone belt, South Africa for the UV environmental conditions on the early Earth. Philos Trans R Soc Lond B Biol Sci 361:1857–1875

Westall F et al (2006b) The 3.466 Ga "Kitty's Gap Chert", an early Archean microbial ecosystem. Geol Soc America, special paper 405:105–131

Westall F, Foucher F, Cavalazzi B, de Vries ST, Nijman W, Pearson V, Watson J, Verchovsky A, Wright I, Rouzaud JN, Marchesini D, Anne S (2011a) Early life on Earth and Mars: a case study from ~3.5 Ga-old rocks from the Pilbara, Australia. Planet Space Sci 59:1093–1106

Westall F et al (2011b) Implications of in situ calcification for photosynthesis in a ~3.3 Ga-old microbial biofilm from the Barberton greenstone belt, South Africa. Earth Planet Sci Lett 310:468–479

Williams BAP, Hirt RP, Lucocq JM, Embley TM (2002) A mitochondrial remnant in the microsporidian trachipleistophora hominis. Nature 418:865–869

Xiong J, Bauer CE (2002) Complex evolution of photosynthesis. Annu Rev Plant Biol 53:503–521

Systematic and Evolution of Microorganisms: General Concepts

Charles-François Boudouresque, Pierre Caumette, Jean-Claude Bertrand, Philippe Normand, and Télesphore Sime-Ngando

Abstract

The diversity of metabolic activities is a characteristic of the microbial world. This enormous diversity needs to be structured in order to be understood, and as a result, taxonomy and systematics are constantly changing since the beginning of the history of microbiology and particularly today with the introduction in the last 20 years of phylogeny as the core of systematics. The history of concepts in systematics and classification is presented. Classification is the science of ordering microorganism groups (taxa) based on their interrelationships. Taxonomy is the discipline that defines the principles and laws of classification. Nomenclature is the science of defining and naming the taxonomic categories (species, genera, families, orders, classes, divisions, phyla, kingdoms, domains), according to their hierarchical rank. In this way, different schools of classification and bacterial systematics were developed in the twentieth century. Today, there is an international consensus based on the classification of the Bergey's Manual revisited with the concepts of phylogeny. Through this classification, the concept of the prokaryotic world organization has evolved. From the idea of a kingdom of prokaryotes, the concept of three domains in the organization of life supported by phylogenetic trees is fully accepted today. Among these three domains, two are prokaryotic: Bacteria and Archaea. In this chapter, the role of horizontal gene transfers in the evolution of life is discussed. The origin of eukaryotes with the primary, secondary, and tertiary endosymbioses is also presented. This allows to improve or to transform the concept of the tree of life from phylogeny to full genome study.

* Chapter Coordinator

C.-F. Boudouresque* (✉) • J.-C. Bertrand
Institut Méditerranéen d'Océanologie (MIO), UM 110, CNRS 7294 IRD 235, Université de Toulon, Aix-Marseille Université, Campus de Luminy, 13288 Marseille Cedex 9, France
e-mail: charles.boudouresque@mio.osupytheas.fr; Jean-Claude.bertrand@univ-amu.fr

P. Caumette (✉)
Institut des Sciences Analytiques et de Physico-chimie pour l'Environnement et les Matériaux (IPREM), UMR CNRS 5254, Université de Pau et des Pays de l'Adour, B.P. 1155, 64013 Pau Cedex, France
e-mail: pierre.caumette@univ-pau.fr

P. Normand
Microbial Ecology Center, UMR CNRS 5557 / USC INRA 1364, Université Lyon 1, 69622 Villeurbanne, France
e-mail: philippe.normand@univ-lyon1.fr

T. Sime-Ngando
Laboratoire Microorganismes: Génome et Environnement (LMGE), UMR CNRS 6023, Université Blaise Pascal, Clermont Université, B.P. 80026, 63171 Aubère Cedex, France
e-mail: telesphore.sime-ngando@univ-bpclermont.fr

> **Keywords**
> Endosymbiosis · Hierarchical classification · History of systematics · Life domains · Microbial classification · Microbial systematics · Nomenclature codes · Phylogeny · Tree of life

5.1 History and Development of Systematics

5.1.1 Interest of Systematics in Microbial Ecology

Each group of organisms should have a solid taxonomy to be studied in a structured way. It is of course possible to study the physiology and genetics of an organism without further justification beyond that it contributes to improving community's health and bioprocesses without a solid identification, but it will be impossible to replace the conclusions drawn in an evolutionary perspective, which is a crucial input in biology. If the concepts of taxonomy and phylogeny are necessary for the study of the ecology of multicellular eukaryotes (plants and animals in the traditional sense), these disciplines are even more essential in microbial ecology, since these organisms have few morphological characters to provide a reliable identification.

Microbial ecology aims at understanding the interactions of microorganisms with each other and with their environment, in particular the role of microorganisms in biogeochemical cycles. If such study must be analytical and predictive, it must be made in relation to specific taxa, and so a solid taxonomic framework should be proposed. This is the reason why microbial ecologists are the most interested in taxonomy and microbial evolution among microbiologists.

One of the important concepts regarding the microbial world is the enormous diversity of organisms on Earth. The diversity of metabolic activities is a characteristic and originality of the microbial world (*cf.* Chaps. 3 and 14). This enormous diversity needs to be structured and organized to be better understood. Man has always sought to understand and name the components of the world around him. If this attempt proved relatively easy with multicellular eukaryotes, it was much more difficult, if not artificial, with microorganisms, because of the change of scale and the need to extend our organs of perception with the instruments adapted such as techniques of microscopy, metabolic analysis, and tools of molecular biology. Thus, as and when there are changes in methods, our approach to design and organize the microbial world has changed. As a result, taxonomy and systematics are constantly changing since the beginning of the history of microbiology. They will continue to evolve and even be upset with the gradual addition of new techniques, becoming more sophisticated and powerful, such as genomics and proteomics. Such changes will not occur as rapidly in multicellular eukaryotes.

Thus, the systematics of microorganisms, which underlies all the studies directed towards the knowledge of their biodiversity, is an interesting central area for biotechnologists and epidemiologists as well as ecologists. They first try to isolate and study new microorganisms for large biotechnological or epidemiological potential. In parallel, microbial ecologists and biogeochemists study the role of microorganisms in ecosystems and their biodiversity in the history of life considering the environmental changes that result.

5.1.2 From the Discovery of the Microbial World to a Taxonomic Scheme of Its Diversity

Even before Pasteur, the cloth merchant Antonie van Leeuwenhoek in 1672 with his famous microscope could discern shapes in microorganisms. However, the number of categories that he could identify was limited, ranging from cocci (spherical shapes), bacilli (rod-shaped forms), spirilla (curved or spiral shapes), and filaments, and was much smaller than those that can be distinguished with present-day microscopes. The work of Pasteur (second half of the nineteenth century) and his students raised awareness that microorganisms were conducting more or less complex and diverse biochemical changes according to the types of microorganisms, including prokaryotes (*cf.* Sect. 2.3). Thus was born an attempt at bacterial classification based not only on their shape but also on their ability to perform specific biochemical transformations.

The microbial world concerns all microorganisms (visible under the microscope and usually unicellular), as single cells or groups of cells (colonies, filaments), whether their structure is prokaryotic (Bacteria and Archaea) or eukaryotic (which were named microalgae, protozoa, microscopic fungi that are polyphyletic groups; *cf.* Sects. 5.5.1, 5.5.2, and 5.5.3). Viruses require living prokaryotic (bacteriophages) or eukaryotic cells to multiply and cannot be considered as microorganisms in their own right. They are obligate intracellular parasites that have no cellular structure and no capacity for self-multiplication.[1] Regarding the *Mimivirus* (Raoult et al. 2004), the question of whether it is a giant virus or a new domain of the living world is not resolved (Box 5.1).

The size of microorganisms varies from a few tens to hundreds of micrometers for the larger microorganisms (especially eukaryotes) to a few micrometers or tenths of micrometers for the smallest prokaryotes such as *Nanoarchaeum equitans* (Fig. 5.1a). The diameter of eukaryotic cells is generally much larger than in prokaryotes: more than 10 µm. However, the findings of eukaryotic cells less than 2 µm in diameter (Box 5.2) have been successful; most of the taxa of higher eukaryotes have one or more lineages of pico-eukaryotes (Baldauf 2003). For instance, *Ostreococcus tauri* (Prasinophyta, Chlorobionta, Viridiplantae; *cf.* Sect. 7.5.5) measures 0.7–1.0 µm in diameter (Courties et al. 1994). Conversely, bacteria can be large, such as *Beggiatoa* (Fig. 5.1b) and *Thioploca*. Schulz et al. (1999) found a "giant" bacterium, *Thiomargarita namibiensis*, with a cell diameter that can reach 750 µm, which means that it is visible to the naked eye (Fig. 5.1c).

The cells of microorganisms differ from the cells of multicellular eukaryotic organisms by the fact that the latter are not able to live alone in nature but as components of an organism. In contrast, a single microbial cell is theoretically capable of performing its own vital processes of energy production, growth, and multiplication. Generations of microbial cells develop in rapid succession, usually within a few minutes to a few hours or even tens of hours. Thus, the microbial world, under optimal development conditions, will be constantly changing, modifying at any time the number and diversity of its cellular components. Microorganisms, multiplying rapidly, form sets of cells transmitting their cellular contents and their genetic information to their offsprings with a probability of change, genetic mutations, and thus higher evolution than in multicellular organisms. The consequence is an even bigger difficulty to assess their diversity and structure their systematic.

Box 5.1: Giant Viruses (Megaviridae)
Jean-Michel Claverie and Chantal Abergel

In the collective unconscious of the general public as well as that of biologists, the concept of "virus" (derived from the Latin word meaning "poison") seems to be frozen as it was at the end of the nineteenth century. Indeed, it is at that time that the "germ theory" (that is to say, the "bacterial" theory) of infectious diseases was established through the work of Robert Koch in Germany and Louis Pasteur in France. However, as soon as 1881, Pasteur himself already knew that the infectious agent of rabies was not a regular microbe as it was invisible to the optical microscope and could not be propagated on a culture medium but only in the brain of a live rabbit. Roughly at the same time, one of Pasteur's assistant was developing the porcelain filter (also called the "Chamberland" filter after his name) able to retain all bacteria. Soon after, it was then realized that many infectious diseases, including the famous tobacco mosaic disease, were caused by infectious agents capable of passing through the Chamberland filter and thus referred to as "ultrafiltrable." This original notion of a virus being too small to be contained ("contagium vivum fluidum") and unable to propagate without the support of a living system (an animal, a tissue, or a cell) is the one that durably remained in everybody's mind.

For 20 years, however, the world of viruses and that of cellular "microorganisms" (Bacteria and Archaea) continuously moved closer and closer to one another and then interpenetrated so much as to preclude discriminations based on size, genomic complexity, or even the recourse to an absolute parasitic lifestyle. On the one hand, many bacteria (e.g., *Rickettsia*, *Buchnera*, *Chlamydia*, etc.) were discovered that can only survive and multiply within a host cell due to their incomplete metabolism. On the other hand, viruses whose size (0.4–1.2 µm) and genomic complexity (number of genes) greatly exceed those of these parasitic bacteria are gradually discovered, most of them in aquatic environments or sediments.

Among these giant viruses (for which we coined the term "girus"; Claverie et al. 2006), a whole new family is emerging which is constituted of *Mimivirus* (Box Fig. 5.1) (Claverie and Abergel 2009) and its relatives *Megavirus* (Arslan et al. 2011) and *Moumouvirus* (Yoosuf et al. 2012). Their exceptional

---

[1] Viruses are functionally inactive when they are outside their host. However, a virus, as parasite of the archaea *Acidianus convivator*, was discovered in hydrothermal vents of Pozzuoli (Italy) and was able to generate outside of the host, a double tail protein (800 amino acids). It was named ATV (*Acidianus* two-tailed virus) (Häring et al. 2005).

J.M. Claverie (✉)
Aix-Marseille University, Marseille, France

C. Abergel
CNRS, Villeurbanne Cedex, France

(continued)

Box 5.1 (continued)

features reignited the debate on the origin of viruses and the role they might have played in both the emergence of the eukaryotes (i.e., of the nucleus) and the transition from RNA to DNA as the chemistry of choice for the storage of genetic information in cells (Claverie 2006; Claverie and Abergel 2010).

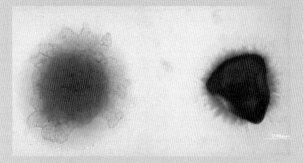

Box Fig. 5.1 A virus particle as big as a bacterium. A *Mimivirus* particle (*right*) is shown side by side with a mycoplasma cell (*left*) (transmission electron microscopy, negative staining of a Mimivirus – ureaplasma co-culture) (From La Scola et al. 2003)

These giant viruses (referred to as "Megaviridae") infect amoeba (from the genus *Acanthamoeba*). Their double-stranded DNA genomes are up to 1.3 million base pairs in length and could code for more than 1,000 genes. Most remarkably, they encode many central components of the protein translation apparatus, including up to seven different aminoacyl tRNA synthetases. The function of these enzymes is to load the right amino acid onto its cognate tRNA, thus using the universal genetic code. The presence of such a remnant translation apparatus, a hallmark of cellular organisms, strongly suggests that these viruses evolved from a cellular ancestor, most likely predating the emergence (or at least the radiation) of the eukaryotes. As shown in Box Fig. 5.2, such a scenario is consistent with the molecular phylogenetic analysis of the DNA polymerase amino acid sequences (as well as concatenations of other conserved protein sequences), where these giant viruses appear to define a new 4th domain in the tree of life or at least a

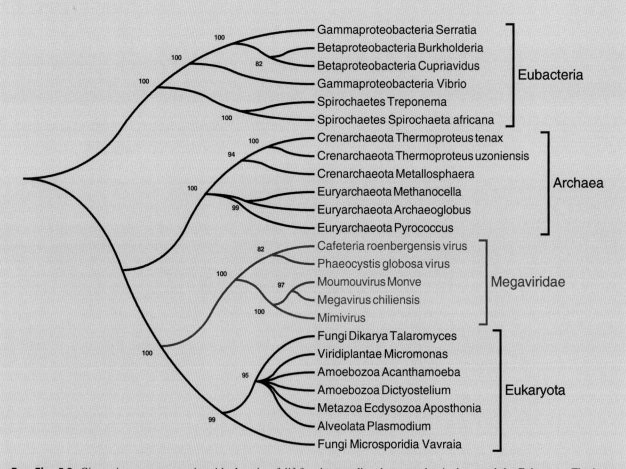

Box Fig. 5.2 Giant viruses: an emerging 4th domain of life? Phylogenetic tree built from the alignment of 24 DNA polymerase amino acid sequences (neighbor-joining method using 505 ungapped positions, JTT substation model, midpoint rooting). The giant viruses (corresponding to the five largest fully sequenced genomes) visibly cluster into a separate clade, intermediate between the Archaea and the Eukaryota. The interpretation of this clade as a *bona fide* new domain in the tree of life or as an early diverging branch of the Eukaryota domain is a matter of taste. This tree nevertheless suggests that the ancestors of today's giant viruses could have been involved in the emergence of eukaryotes

(continued)

Box 5.1 (continued)

subdomain that diverged very early on from the branch leading to the modern eukaryotes (Box Fig. 5.2).

The controversy remains strong: did the emergence of DNA viruses predate that of eukaryotes? Were these two phenomena related? Is it absurd to imagine that viruses might have emerged before their hosts? Or are these viruses the last representative of a 4th cellular domain now extinct and that only managed to survive as parasites within the extant eukaryotic domain?

In the latter case, these giant viruses might be the result of reductive evolution from ancestral microorganisms (somewhat like the one followed by *Rickettsia*). All these questions can only be answered by identifying more giant viruses whose genomes should have kept other traces (e.g., ribosomal proteins) of reductive evolution. In the present state of our knowledge, the presence/absence of functional protein of the translation system remains the most fundamental criterion for discriminating viruses from the most reduced parasitic bacteria. In the context of reductive evolution, the loss of one key component of the translation apparatus would constitute the initiating event irreversibly committing a cellular parasite to a viruslike evolutionary pathway, rather than becoming an organelle as it is the normal fate for obligate intracellular bacteria (Claverie 2006; Claverie and Abergel 2010).

Although *Mimivirus* was discovered fortuitously only recently, environmental DNA sequencing ("metagenomics") hinted soon afterward that many other members of the Megaviridae existed in the marine environment where they constitute a significant fraction of large DNA viruses (Monier et al. 2008). Ironically, this is probably because of their size and nonfilterability which delayed the discovery of these abundant viruses (as well as the fact that they do not produce a visible disease in humans or animals).

A handful of large DNA viruses related to the *Acanthamoeba*-infecting Megaviridae have been found to infect different species of unicellular eukaryotes such as *Cafeteria roenbergensis* (Fischer et al. 2010) and *Phaeocystis globosa* (Santini et al. 2013). These marine giruses are often isolated in the context of blooms of their host species, the population of which they regulate. Ironically, the latest discovered viruses are the largest, the most abundant, and possibly those whose geoclimatic influence is the most significant on our planet.

Box 5.2: Pico-eukaryotes: Highly Diversified Small Cells
Herve Moreau

Pico-eukaryotes are eukaryotic microorganisms that have a size below 2–3 µm. This definition based on the size of the cells has no phylogenetic significance but corresponds to operational considerations enabling to divide marine plankton in subcategories like pico-, nano-, and microplankton.

Probably because they are small and their morphology is very simple, the diversity of pico-eukaryotes is often underestimated. It is only in 2001 that several studies based on sequencing of the 18S ribosomal gene revealed this diversity in the sea (Moon-van der Staay et al. 2001; López-García et al. 2001). For practical reasons (easy detection of pigments), diversity of autotrophic pico-eukaryotes is the best known. In the sea, they belong essentially to the phylogenetic groups Prasinophyceae (Chlorobionta), Dinophyceae, Bacillariophyceae (Diatoms), Cryptophyceae, Prymnesiophyceae, and Bolidophyceae (Vaulot et al. 2008). In marine picoplankton, the respective contribution of prokaryotes (*Cyanobacteria*) and pico-eukaryotes to primary production remains difficult to determine. Prokaryotes are clearly more numerous in terms of cell number (around 80 %), but the bigger size of eukaryotes and their higher productivity allow these organisms, at least in coastal areas, to be responsible for the major part of biomass production in this cell size compartment (Worden et al. 2004).

Heterotrophic pico-eukaryotes are less known, although most of the ribosomal sequences found in metagenomic studies correspond to organisms belonging to heterotrophic lineages. It is, however, now clear that partial amplification of the 18S ribosomal gene (the most used gene marker) by "universal primers" introduces a bias towards detection of heterotrophs. Few of these organisms are cultivated, and it is sometimes impossible to determine the autotrophic or heterotrophic nature of the organisms corresponding to these 18S sequences. Cultivation of microorganisms of which only the sequence of one marker gene is known remains challenging, and a combination of techniques like in situ hybridization, cell sorting, and single-cell whole-genome amplification is more and more used to obtain functional information.

H. Moreau (✉)
UMR 7232 LOBB, Observatoire Océanographique de Banyuls, 66650 Banyuls-sur-Mer, France

(continued)

Box 5.2 (continued)

Pico-eukaryotes have been essentially described from aquatic environments although other ecosystems (e.g., soils or sediments) are not yet explored in detail for these organisms. It is, however, highly probable that many pico-eukaryotes are living in these ecosystems and that they can represent a distinct diversity compared to what is already described in fresh or marine waters.

5.1.3 Hierarchical Organization in Search of a Phylogeny

5.1.3.1 The Need for a Hierarchical Organization

It is important to emphasize here that all work on the hierarchical organization of prokaryotic microorganisms was done on pure culture isolates obtained in particular by following the work of Robert Koch (1880) on semisolid media, thus constituting of course an important step but one limited to culturable microorganisms on the media used. It is considered that in fact only 0.1–1 % of the prokaryotes present on our planet can be grown in synthetic media and maintained

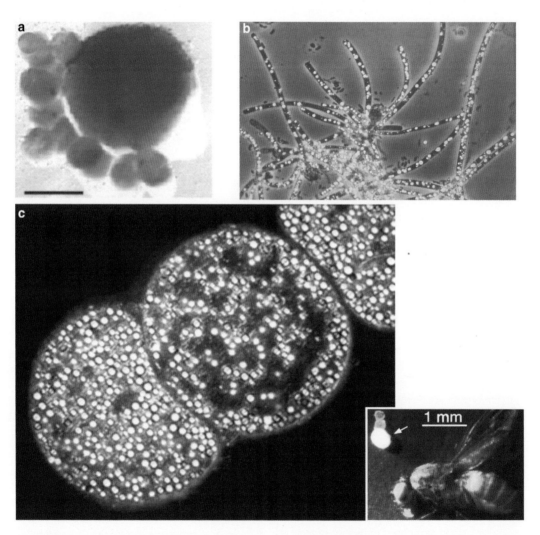

Fig. 5.1 (a) *Nanoarchaeum equitans* (**Archaea**) is one of the smallest known microorganisms (400-nm diameter) and can be isolated by filtration on 0.45-μm membrane, shown here surrounding a cell of *Ignicoccus* (Bacteria); scale = 1 μm (Huber et al. 2002) (Copyright: courtesy of Nature Publishing Group). (b) *Beggiatoa*, one of the largest known bacteria, with *Thioploca* and *Thiomargarita*, 10 to a few hundred micrometers in diameter, 100 μm to a few mm in length (Microphotograph: Pierre Caumette). (c) *Thiomargarita namibiensis*, the largest known bacterium; cells are up to 750 μm in diameter, compared to the size of a *Drosophila*, the bacteria is in the upper left corner (*bottom right*) (Schulz et al. 1999) (Copyright: The American Association for the Advancement of Science)

in collection. This approach which has been used for decades and applied to the study of pathogenic bacteria and bacteria from the environment has led to the definition of a set of rules. Prokaryotic taxonomy has been based on a series of morphological and increasingly of biochemical criteria in the different editions of books of bacterial systematics.

Early taxonomic schemes were based on dichotomous keys similar to those used for botanical books with a priority given to radiative autotrophy (photosynthesis), the existence of a thick wall capable of withholding a dye (Gram stain), the presence of survival structures ("spores"), etc. Morphological criteria have later been supplemented by physiological criteria.

Besides characterization of microorganisms, classification, nomenclature, and identification are the three main areas of systematics.

Classification is the science of ordering microorganism groups (taxa) based on their interrelationships. The term classification is used not only for the method but also for the resulting system. Taxonomy is the study which defines the principles and laws of classification.

Nomenclature is the science of defining and naming the taxonomic categories (species, genera, families, orders, classes, divisions, phyla, kingdoms, domains), according to their hierarchical rank. It helps to provide internationally recognized microorganism names correctly and place them in the classification as belonging to specific and defined groups (taxa). These names are controlled and regulated by international codes of nomenclature, whether for prokaryotic (bacterial nomenclature) or eukaryotic (plant nomenclature, which includes the "fungi," and animal nomenclature which includes the "protozoa"). When a microorganism is rigorously characterized and classified, it is assigned a name that corresponds to the genus and species in which it is included. Each isolated microorganism strain is defined by its genus name followed by the species name and finally the code of the isolated strain.

Identification is a set of procedures used to determine whether an unknown microorganism belongs or not to a taxonomic group already described. If it shows enough differences with taxa of the same hierarchical level, it can be described and named as a new taxon.

Classification and identification are dependent on data for characterization of microorganisms. They are constantly changing based on scientific advances in data acquisition and characterization methods.

Nomenclature, by giving the correct names for taxa (classification) or unknown strains (identification), connects the two approaches and, in the same way, is constantly evolving to maintain coherence. A Nomenclatural Committee ("International Committee on Systematics of Prokaryotes" ICSP: www.the-icsp.org/) was created to resolve conflicts, express opinions, and standardize practices internationally. ICSP is responsible for regularly updating the International Code of Nomenclature of Bacteria,[2] which concerns Bacteria and Archaea.

5.1.3.2 Different Schools of Classification of Prokaryotes

Meanwhile, schools of classification and bacterial systematic were developed in the twentieth century. The French school, at the Institut Pasteur, proposed a hierarchical classification taking into account not only the morphological and physiological criteria but also criteria of pathogenicity. It was essentially a classification of pathogenic bacteria with dichotomous identification keys to quickly orient the diagnosis of species of pathogenic bacteria. This classification led to the "Treaty of bacterial systematics of Prévot" (*Traité de Systématique bactérienne de Prévot*), the name of the French microbiologist who was the first to propose such a classification in 1961. At the same time, a second classification was proposed by the Russian school of microbiology. It took into account the environmental bacteria in a classification very close to the plant systematics.

Today, there is an international consensus based on the classification of Bergey. It was proposed in 1923 by the American microbiologist D. H. Bergey (1860–1937) who was the coordinator of a committee of classification of bacteria set up under the authority of the American Society of Microbiologists in 1917. This classification has been adopted internationally for greater consistency with the systematic and bacterial nomenclature. The committee ("Bergey's trust") was subsequently extended to the international community and consists today of microbial taxonomists from different countries. It manages the classification and systematics of prokaryotes and regularly provides an updated systematics book based on new knowledge about prokaryotes. This is the "Bergey's Manual for Systematic Bacteriology"; the latest edition (2002–2012) was used for the present work. In this latest edition, each chapter written by one or several experts contains the elements necessary for the identification of prokaryotes. In this edition, many concepts used in molecular biology from gene sequencing of 16S ribosomal RNA and genomic studies were introduced in order to try to combine the genotypic data with the phenotypic characteristics for each species described. The main change in recent years has been the need to change the taxa so as to be more consistent with the phylogeny, which led to many large changes at all taxonomic levels.

---

[2] http://www.ncbi.nlm.nih.gov/books/bv.fcgi?call=bv.View.ShowTOC&rid=icnb.TOC&depth=2)

5.1.3.3 Classification of Eukaryotes and Problems of Nomenclatural Codes

The nomenclature (legitimate names of species, genera, and taxa of lower or higher rank) of eukaryotes is governed by two codes. For "plants" in the traditional sense of the term, it is the "International Code of Botanical Nomenclature." Until 1975, the code of botanical nomenclature also included the nomenclature of bacteria (in the modern sense of prokaryotes), causing duplication and confusion with the "International Code of Bacterial Nomenclature."[3] For "animals" in the traditional sense, that is to say, including "protozoa" (cf. Sect. 5.5.3), it is the "International Code of Zoological Nomenclature."

These codes perpetuate an archaic division between animals and plants. Modern phylogenies show that the separation between what tradition has called "plants" and "animals" is totally artificial. The term "plant," after Linnaeus[4] in the eighteenth century, concerns such heterogeneous (polyphyletic) organisms, which is not even a functional set (photosynthetic organisms). But the most annoying is that these codes are independent. This means not only that the rules for determining the legitimate name of a taxon may be different but that the same genus name can be used by different codes to designate different taxa.

Although this constitutes an amazing archaism in the twenty-first century, it is important to be aware of it, not to run the risk of serious confusion. Thus, *Gracilaria* is a Rhodobionta (Archaeplastida) in the botanical code but a moth (Lepidoptera, winged with little colors, close to moths) in the zoological code. *Rissoella* is Rhodobionta in the botanical code but a gastropod mollusk in the zoological code. *Bonamia* is a Magnoliophyta (Convolvulaceae) and a unicellular Haplosporidia, *Bostrychia* is a Rhodobionta and a bird, *Crambe* is a Magnoliophyta and a Porifera (common in the Mediterranean), *Chondrilla* is also a Magnoliophyta and a Porifera, and *Eisenia* is a Phaeophyceae (Stramenopiles) and an earthworm. *Posidonia* is a present-day Magnoliophyta (Viridiplantae) and a fossil mollusk. *Turbinaria* is a Phaeophyceae and a Dendrophylliidae[5] coral. Very slight differences in spelling can have major meaning differences: *Archaeopteryx* is a kind of feathered "reptile" considered the ancestor of birds, while *Archaeopteris* is a tree fern that lived 370 Ma ago and that could be the closest kin of Magnoliophytes (Archaeplastida).

In addition, some groups that were traditionally placed on the border of the plant and animal worlds are claimed by both codes of botanical and zoological nomenclature. As a result, some genera and species may have a valid name that differs depending on the code considered. This is the case in the Dinobionta (dinoflagellates for zoologists, Dinophyta or Dinophyceae for botanists), the euglenoids (Euglenophyta or Euglena[6]), and Chlorarachniobionta (Chlorarachniophyta, Chlorarachnida[7]).

The independence of the nomenclature of prokaryotes, compared to that of eukaryotes, is logical, given the very specific concepts that it uses (cf. Chap. 6). However, one can only wonder at the independence of codes of zoological and botanical nomenclature. A unified code, based on phylogeny, has been proposed (phylogenetic or PhyloCode), but it is currently not used, although this may change in the future. For instance, the code of botanical nomenclature still continues to claim a portion of the prokaryotic Cyanophyceae that it calls "blue-green algae" while these are cyanobacteria for the ICSP. This fact indicates that this code is doomed to a comprehensive reform in the short term.

5.1.4 Hierarchical Organization of the Living World and Evolution of Concepts from Taxonomy to Phylogeny

The taxonomic approach based on back and forth movements between nomenclature, natural history, evolution, and embryogenesis has been developed over two centuries for what was then called the "higher organisms" (in the modern sense of multicellular eukaryotes). The works of Linnaeus (1753) have provided a formal framework for it. Those of Lamarck (1809), Darwin (1859), and many others have provided a historical and evolutionary perspective, completed in the twentieth century with the data of evolutionary biology (Table 5.1). Regarding the microorganisms, especially prokaryotes, the available tools have frozen the

---

[3] For prokaryotes, including Cyanobacteria and Planctomycetes, only the "International Code of Bacterial Nomenclature" should have the authority. Discussions are underway to harmonize the nomenclature of Cyanobacteria and Planctomycetes with the "International Code of Botanical Nomenclature."

[4] The name of Linnaeus is often translated as "Linné." This goes back to the tradition that was, until the eighteenth century, to change its name according to the languages and countries. This is also the choice of the Royal Swedish Museum of Natural History (www2.nrm.se/fbo/hist/linnaeus/linnaeus.html.se) website that talks about Carl von Linné, which is the French form he chose himself when he was knighted in 1762 by the King of Sweden. However, the name that appears on the cover of the book *Species Plantarum*, chosen as the starting point of the nomenclature, is "Caroli Linnaei" the declined Latin form of Carolus Linnaeus.

[5] The confusion between *Turbinaria* "plants" and "animals" is all the more disturbing given that they can coexist in the same coral reef ecosystem and even live one on the other.

[6] For example, *Paranema* (Dujardin 1841) is a valid name for a colorless Euglena species according to zoologists. For botanists, the valid name of the same species is *Pseudoparanema* (Christen 1962) as *Paranema* (Don 1825) is a fern and has the priority.

[7] Chlorarachniobionta are amoebae equipped with chloroplasts.

Table 5.1 Evolution of the classification systems

Traditional	Copeland (1956)	Curtis (1968)	Edwards (1976)	Margulis (1981)	Woese et al. (1990)
Plantae	**Plantae**	**Plantae**	**Chlorobionta**	**Plantae**	**Eukaryota**
Bacteria	Bryophyta	Bryophyta	Green algae	Bryophyta	Bryophyta
Blue algae	Tracheophyta	Tracheophyta	Bryophyta	Tracheophyta	Tracheophyta
Green algae			Tracheophyta		Metazoa
Red algae					Protozoa
Fungi					
Bryophyta					
Tracheophyta					
Animalia	**Animalia**	**Animalia**	**Erythrobionta**	**Animalia**	*Bacteria*
Protozoa	Metazoa	Metazoa		Metazoa	
Metazoa				Poriferae	
	Protista	**Protista**	**Cyanochlorobionta**	**Protista**	*Archaea*
	Protozoa	*Bacteria*	Blue algae	Protozoa	
	Green algae	Blue algae		Algae with a nucleus	
	Chrysophyta	Protozoa		Myxomycota	
		Green algae			
		Chrysophyta			
	Monera		**Myxobiota**	**Fungi**	
	Bacteria		Myxomycota	Basidiomycota	
	Blue algae			Ascomycota	
				Zygomycota	
				Lichens	
			Fungi 1	**Monera**	
			Chitridia	*Bacteria*	
			Nonflagellated	*Cyanobacteria*	
				Prochlorophyta	
			Fungi 2		
			Oomycota		
			Chromobionta		
			Brown algae		
			Chrysophyta		

Taxa corresponding to *Bacteria* and *Archaea* are in italics. The horizontal lines represent the higher-level taxa (kingdoms, domains) recognized by a given classification; there is therefore no correspondence between taxa that are placed on the same line (Modified and completed after Margulis 1981)

process in hierarchical organization with multiple revisions over the years based on the observation of morphological and physiological parameters of microorganisms, but historical and evolutionary perspectives have been provided with the explosion of tools and concepts of molecular biology, placing the microorganisms in a comprehensive evolutionary scheme.

Starting with Haeckel (1894) (Fig. 5.2), the microbial world was organized with two major kingdoms: the animal kingdom including protozoa and the plant kingdom including microalgae and microscopic fungi, which are all eukaryotic microorganisms. Starting with Pasteur in 1860 and its successors, the bacteria (prokaryotic microorganisms) were included in the classification of life, but with great difficulty. They were long included in the plant kingdom, in separate branches, with microscopic fungi and microalgae (including cyanobacteria, long known as blue-green algae or Cyanophyceae). Gradually, this pattern changed when cytology and biochemistry brought to understand that the prokaryotes were fundamentally different from eukaryotes. Several evolutionary schemes were published in the period 1956–1990.

The desire to take into account evolutionary relationships between bacterial taxa has strengthened over the years. It is the study of the structure of bacterial proteins that led scientists to realize that the macromolecules not only had a functional or physiological sense but were also a reflection of their development and could therefore be considered **semantides***, that is to say, they made sense and carried information pertaining to evolution (Zuckerkandl and Pauling 1965). Thus, proteins have a closely related sequence in organisms that are close either from a

Fig. 5.2 Haeckel (1894) was the first to represent the diversity of life in the form of a tree and its branches. It is to him that we owe the current term " phylogenetic tree." The original version of this tree (genome.imb jena.de/stammbaum.html), in German, is poorly readable and has been replaced here by a later version (Photography: free of copyright (expired))

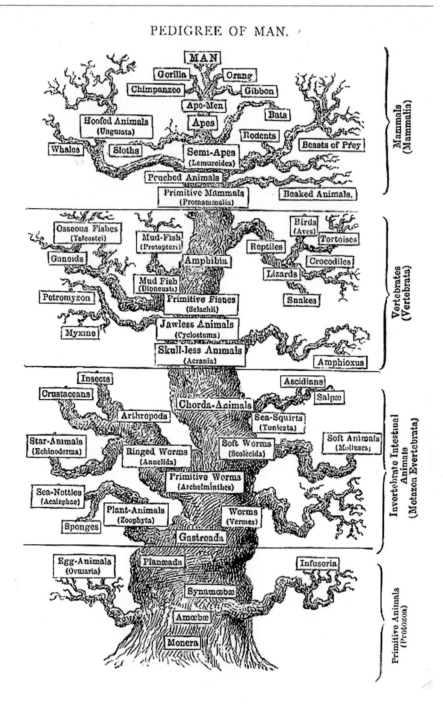

morphological or an evolutionary perspective (Fig. 5.3). The generality of this observation has profoundly changed the approach to microbial phylogeny.

This search for evolutionary relationships and a molecular clock (Box 5.3) has resulted in the use of the SSU (16S/18S) ribosomal RNA as a universal marker in microorganisms where morphological characters are not diversified and where paleontological data are rare (*cf.* Sect. 4.2). It is the work of Woese in the 1980s, which brought an end to the concept of a single taxon containing all prokaryotes ("kingdom of prokaryotes" Stanier 1974), by proposing a division of the living world in three domains, two of which are prokaryotic (Bacteria and Archaea) (Fig. 5.4, Table 5.1). The use of 16S ribosomal RNA has also clarified the concept of species showing that close bacterial strains have 16S ribosomal RNA gene sequences that are also close. It was thus found that strains with similarity between their sequences of the genes coding for 16S ribosomal RNA below 97 % did not belong to the same species.

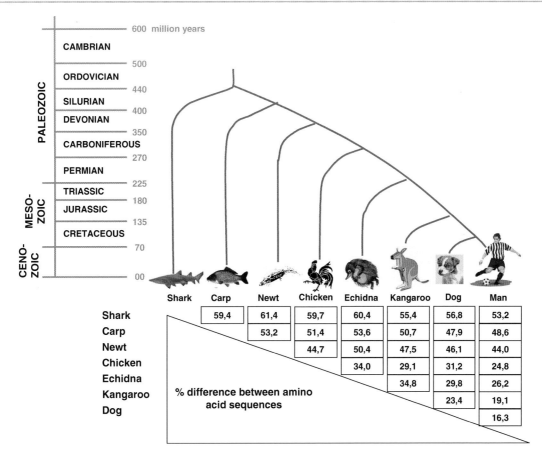

Fig. 5.3 Protein sequence alignment (partial) of the alpha hemoglobin of man, dog, kangaroo, echidna, chicken, newt, carp, and shark. The sequences of morphologically similar organisms are closer. From the calculation of genetic distances (% difference between the amino acid sequences in the protein), a distance matrix was generated and a phylogenetic tree was created, wherein the length of the branches corresponds to the number of differences. On the left side of the figure, a geological scale shows the probable age of the common ancestor of the different taxa (neighbor joining, distances according to Kimura 1968)

This proposal by Woese has profoundly changed the way we consider the evolution of prokaryotes. His choice of a conserved, universal, marker that nevertheless contained variations helped reshape the phylogeny and taxonomy of prokaryotes, characterize microbial habitats, identify non-culturable pathogens, and propose evolutionary scenarios that are regularly revisited in the light of developments in molecular biology. However, a weakness of this proposal is that it is based on one marker, the 16S ribosomal RNA, and a too limited number of phenotypic markers. It should also be noted that the term chosen for one of the two prokaryotic domains (*Archaea*) suggests a greater antiquity of these microorganisms. However, the current data cannot in any case be used to infer that the Archaea are the ancestral forms of life nor of course that would have had a hyperthermophilic archaeal origin from which the other prokaryotes would have derived. The characteristics of the last common ancestor ("last universal common

> **Box 5.3: Molecular Clock**
>
> Celine Brochier-Armanet
>
> This well-known hypothesis relies on the assumption that molecular sequences (DNA and proteins) evolve at a constant rate, meaning that their evolutionary rate (r) is constant over time. The molecular clock hypothesis was postulated in the early 1960s based on the first analyses of protein sequences (hemoglobin, cytochrome C, fibrinopeptides, etc.) which showed that the divergence observed between two sequences is proportional to the divergence time between the corresponding species. This is due to the fact that
>
> ---
>
> C. Brochier-Armanet (✉)
> Laboratoire de Biométrie et Biologie Évolutive,
> UMR CNRS 5558, Université Claude Bernard Lyon 1,
> 69622 Villeurbanne Cedex, France

(continued)

Box 5.3 (continued)

mutations occur randomly and mainly at neutral sites. Accordingly, mutations will be fixed (leading to substitutions) at a constant rate (Kumar 2005).

A constant rate r implies a linear relation between the divergence time (t) and the evolutionary distance (D) between two homologous sequences. More precisely, under a molecular clock assumption, $D = 2 * r * t$. Based on this relation, it is possible to perform molecular dating, meaning that knowing D, the evolutionary distance between two homologous sequences from A to B species, and r, their evolutionary rate r, it is possible to estimate t, the divergence time between A and B.

In the past few years, the molecular clock hypothesis has been called into question. Indeed, the analysis of numerous DNA and protein sequences revealed that there is no universal molecular clock, that is to say that there is no universal and constant evolutionary rate r. On the contrary, it was showed that:
1. Each molecular marker evolves at its own rate.
2. The evolutionary rate r of a given sequence can vary over time, and among lineages, due to selection pressure changes or functional shifts, for instance, following gene duplication events.
3. r may vary among sites within a given sequence because selective pressures are not uniform along a sequence.

For instance, the evolutionary rate of 18S rRNA sequences from Dasycladales (Chlorobionta) and Angiosperms (Viridiplantae) is about 0.01 substitutions per 25 million years, whereas it is much slower in *Cyanobacteria*. Similarly, the evolutionary rate of the superoxide dismutase is four time faster in mammals than in fungi. Finally, the four regions (A, B, C, and the signal peptide) of the proinsulin, the protein precursor of insulin, evolve at different rate, reflecting different selective pressures acting on these regions.

The concept of molecular clock must be used wisely and cautiously because recent advances in molecular evolution demonstrated that there is no universal molecular clock but a multitude of local molecular clocks, each having its own features. Far from being a problem, this offers the possibility to select accurate chronometers (i.e., genes, proteins, noncoding DNA, etc.) suitable for each question (e.g., fast evolving genes to study recent evolutionary events, slowly evolving genes to study ancient events, etc.).

ancestor" LUCA) are difficult to determine, and there is no consensus on its hyperthermophilic character (Galtier et al. 1999; Di Giulio 2003) or on its belonging to the bacterial or archaeal phyla, as proposed by Cavalier-Smith (2002a, b) (*cf.* Sect. 4.3.3).

Cavalier-Smith (2002a, b), who used different criteria to expand his vision of the phylogeny and taxonomy of prokaryotes, proposed taxa and a topology that are far from unanimously accepted in ICSP (Fig. 5.5). Despite the fact that they were validly published in the reference journal for bacterial taxonomy, many of these taxa are therefore not adopted in this book. The tree of life based on concepts and latest molecular tools will be presented later (*cf.* Sect. 5.5).

5.2 Microorganisms: Unicellular or Multicellular?

Microorganisms as a whole, including eukaryotes, do not constitute a monophyletic group of organisms (*cf.* Chap. 7). However, they can be considered as a well-defined set of organisms according to their morphology, particularly by their uni- or almost unicellularity. The notion of unicellularity (or almost unicellularity), as opposed to multicellularity, is actually much less clear as one may think at first. Between unicellularity and multicellularity, the boundary is in fact blurred, rather corresponding to a continuum.

In a unicellular organism, the cells are physically isolated and (at least theoretically) independent. A typical multicellular organism is (1) made by numerous cells; (2) its cells contain a single nucleus (in the case of a eukaryote); (3) its cells communicate, in one of several way (pores, synapses, etc.), allowing them to exchange morphogenetic informations and chemicals (products of metabolism) (Fig. 5.6f); and (4) a specialization appears between cells. A multicellular organism is therefore not an association of identical and totipotent cells but a set of specialized cells that can eventually organize themselves into tissues and organs.

As noted above, there are many intermediate cases between strict unicellularity and strict multicellularity: the main ones are those of colonial and filamentous organisms without cell specialization.

The closest organisms to unicellularity are colonial organisms which are single-celled organisms whose cells remain associated during their division by extracellular mucus; hydrodynamism can eventually dissociate them. Although there may be chemical interactions between cells, there is no physical communication between them. The colonial organisms are not multicellular. However, the question may arise for Labyrinthulobionta (Stramenopiles, eukaryotes) in which cells move in a common envelope (labyrinth) that they contribute to secrete (Fig. 5.6c). The

5 Systematic and Evolution of Microorganisms: General Concepts

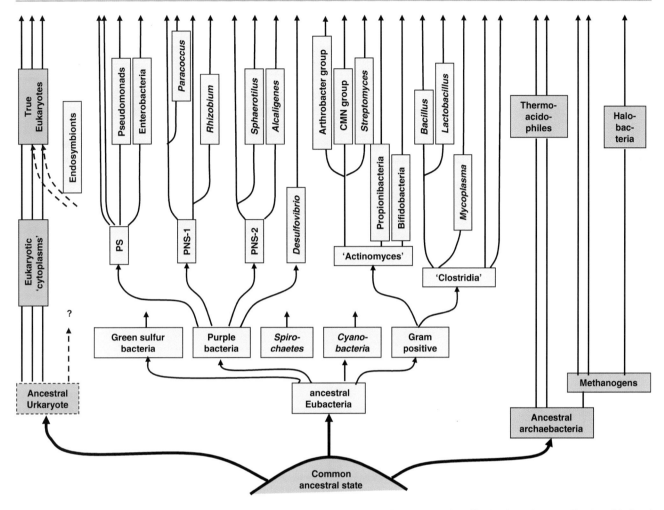

Fig. 5.4 The phylogenetic tree of Carl Woese. It establishes for the first time, based on the analysis of the gene 16S ribosomal RNA, the distinction between Bacteria and Archaea. The nomenclature has not been actualized. *PS* purple sulfur, *PNS* purple non sulfur (Modified and redrawn from Fox et al. 1980)

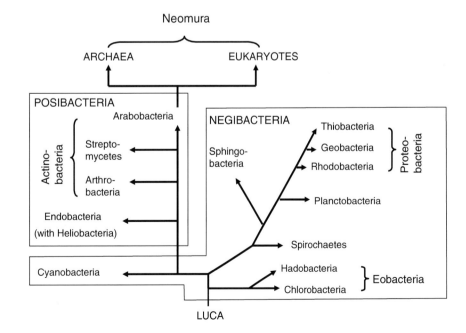

Fig. 5.5 Phylogeny of Bacteria and origin of Archaea and of eukaryotes (Modified, simplified, and redrawn from Cavalier-Smith 2002a)

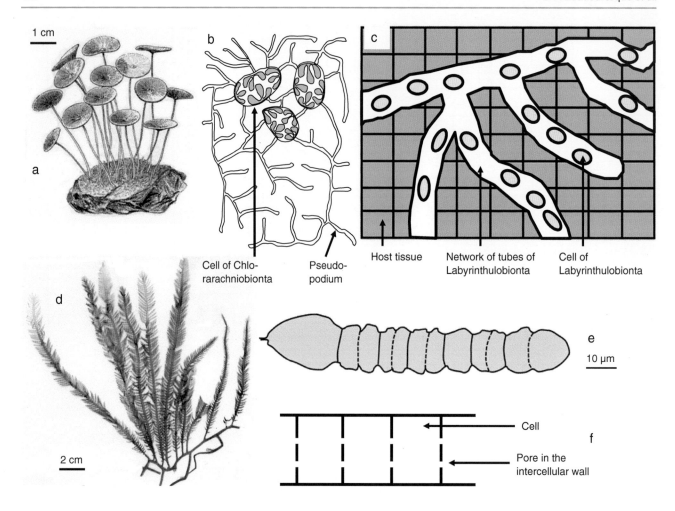

Fig. 5.6 Unicellularity and multicellularity. (**a**) A group of *Acetabularia acetabulum* (Chlorobionta, Viridiplantae); each individual consists in a single giant uninucleated cell (From Oltmanns 1904); (**b**) three cells of a Chlorarachniobionta (Rhizaria) linked by their network of pseudopodia (filopodia); (**c**) cells of Labyrinthulobionta (Stramenopile) gliding within the network of tubes; (**d**) *Caulerpa taxifolia* (Chlorobionta, Viridiplantae). The vegetative apparatus consists in a kind of supercell: a bag of cytoplasm containing millions of nuclei (coenocyte); (**e**) *Haplozoon axiothellae* (Dinobionta, Alveolata), a parasite thriving in the digestive tract of the annelid *Axiothella rubrocincta*. It is not really multicellular but consists in a syncytium compartmentalized by incomplete wall; (**f**) nonspecialized cells in a cyanobacterium filament, with pores between cells

case of Chlorarachniobionta (Rhizaria, eukaryotes), whose cells are joined by their pseudopodia, is a bit more complex (Fig. 5.6). The same goes for the soil bacterium species *Shewanella oneidensis* in which "nanowire" (100-nm diameter) connects the cells to each other, their role could be to move electrons up to the receptors located on the cell surface (Ball 2007). In both cases, there is no specialization between cells. Biofilms and microbial mats (*cf.* Sect. 9.7.3) may, however, be considered as an early stage of cell specialization, with external cells exposed to the fluctuating and "hostile" environment and internal cells trapped in an exopolymer matrix where they live in a different environment. We can also consider that the bacteria that form endospores, such as *Bacillus and Clostridium*, have structures (endospores) that are specialized in the resistance to environmental conditions (temperature, drought, etc.) to which vegetative cells cannot survive.

Most filamentous bacteria appear as filaments of identical cells which have no morphological, metabolic, and functional specialization. The cells are simply associated end to end or contained in a common sheath. They therefore cannot be considered multicellular.

Multicellularity in its strict sense has emerged repeatedly in bacteria: actinobacteria, cyanobacteria, and some other phylogenetically dispersed taxa. It allows a metabolic specialization with, in addition of vegetative cells, diazovesicles (cells that protect nitrogenase against oxygen) in *Frankia*, a N_2 fixing actinobacterium (Berry et al. 2003); heterocysts perform the same function in the cyanobacteria *Anabaena* and *Nostoc* (Stacey et al. 1979). In actinobacteria, there is production, at the end of filaments, of conidia, cells that allow them to resist and spread in the environment. Multicellularity has, in the same way, emerged six times in eukaryotes: in Metazoa and Fungi (Opisthokonta), in Amoebobionta (*Dictyostelium*), in Rhodobionta and Viridiplantae (Archaeplastida), and in Chromobionta (Stramenopiles). It should be noted that unicellular organisms may be derived from multicellular ancestors, which is the case of some *Saccharomyces* and *Candida* (Fungi, Opisthokonta).

There remains the question of where to place coenocytic eukaryotes such as Oobionta (Stramenopiles) and *Caulerpa* (Viridiplantae). Should they be considered as a single cell? Or should as multicellular because of the many nuclei (sometimes millions) in their cytoplasm? (Fig. 5.6d). Although Oobionta can be microscopic, some species are macroscopic. The vegetative apparatus of *Caulerpa* can measure several tens of centimeters or several meters. The Dinobionta *Haplozoon axiothellae*, who lives in the intestine of the annelid (*Axiothella rubrocincta*; Leander et al. 2002), is an intermediate situation between cellular and syncytial structures, a filament is outlined, with some unseparated cells and several nuclei (Fig. 5.6e).

There is also the case of giant cells, such as *Acetabularia* (Viridiplantae, eukaryotes). The organism consists of a single cell with a calcified wall, can measure up to 10 cm in length (Fig. 5.6a), exhibits a single nucleus, and is therefore undoubtedly unicellular, although, as for *Caulerpa*, it does not correspond to the idea usually associated with the notion of unicellularity: the microscopic size.

Finally, we must take into account the practices of microbiologists who regarded microorganisms as (1) all prokaryotes, even when they are clearly multicellular; (2) single-celled eukaryotes; (3) most taxa which lie on the borderline between unicellularity and multicellularity, primarily when they are microscopic or submicroscopic; and (4) a number of organisms traditionally grouped under the term "fungi" (customary meaning) (*cf.* Chap. 7). For the latter, the reason is understandable: they are unicellular or clearly multicellular, fungi represent, with some prokaryotes and unicellular eukaryotes, a significant risk to humans (human health, agriculture, aquaculture, etc.).

5.3 The Role of Gene Transfers in the Evolution of Life

Our vision of the evolution of the living world initially based solely on vertical gene transfers has been modified to take into account horizontal gene transfers (HGT; *cf.* Chap. 12). These transfers were demonstrated for the first time based on work on antibiotic resistance, which can spread rapidly between bacteria in hospitals. This spread has been shown to occur through the transfer of resistance factors, later identified as DNA. Thus, the transmission of genetic information and the evolution of the bacterial world are not only vertical but also horizontal. Bacteria exchange DNA fragments by various means such as cell-cell conjugation, transformation with naked DNA, or through transduction brought about by bacteriophages. These mechanisms allow the exchange of genes between very distant taxa, such as that between *Escherichia coli* and the yeast *Saccharomyces cerevisiae* (Nishikawa et al. 1992). The quantitative importance and frequency of such transfers in the evolution of life are increasingly taken into account. In *Escherichia coli*, 755 of its 4,288 genes could come from horizontal transfers (Lawrence and Ochman 1998), although this number is hotly debated (Daubin et al. 2001).

Environments where bacteria coexist in great abundance are particularly favorable to gene exchanges (e.g., rhizosphere, biofilms, and microbial mats).

The DNA molecule is intrinsically resistant, much more so than RNA, which explains that it can maintain itself in the environment (soil, sediment, etc.) long after cell death. It can bind to charged regions of solid components, e.g., to calcium ions on clays. It can then be incorporated into naturally transformable bacteria that will then express a new phenotype. The quantitative importance of this type of exchange remains difficult to assess at this time.

Extracellular DNA in the environment thus constitutes an important reservoir of genetic information for microorganisms that can have access to it. The conditions under which microorganisms can incorporate some of the "metagenome" remain little known. There may be enzyme systems that facilitate such a "fishing," or specific physiological or physicochemical conditions such as acidification of the environment, or a reduction in size which allows bacteria to penetrate crevices where the metagenomic DNA lies. In addition, meteorological phenomena such as lightning create strong electric fields which cause the incorporation of extracellular DNA by the phenomenon of transformative electroporation of bacteria (Cérémonie et al. 2004).

Insertion sequence (IS)* allows the insertion of a DNA fragment into the genome of its host cell. There are several types of IS, and they are present in almost all organisms. Each IS tends to increase its number of copies per genome. We thus find up to 26 copies of IS981 in *Lactococcus lactis* (Polzin and McKay 1991).

Overall, the tree of life, especially that of Bacteria and Archaea, is certainly far from the regularly connecting one that has dichotomous branchings found in traditional phylogenetic trees. It probably looks more like the one proposed by Doolittle (1999) called the "reticulated" or "cross-linked" tree (Fig. 5.7).

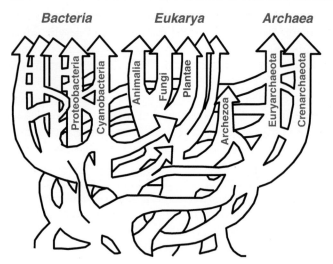

Fig. 5.7 The cross-linked tree of life. The names of taxa used by the author have been preserved, even if they do not match the nomenclatural choice of this book or if they correspond to a now abandoned concept (Archaezoa) (Modified and redrawn from Doolittle 1999)

5.4 The Origin of Eukaryotes

Eukaryotes differ from prokaryotes (Bacteria and Archaea) by a variety of cytological and biochemical characteristics (*cf.* Chap. 3), which confirm the robustness of molecular phylogenies. Although many of these characters can be observed in intermediate cases, there is no case of a species or higher taxon for which there is doubt as to its position within prokaryotes or eukaryotes.

According to Cavalier-Smith (2002b), it is the loss of the prokaryotic cell wall that made possible the acquisition of phagotrophy (ingestion of prey; secondarily lost in some taxa such as Fungi) which is the major event in the evolution of eukaryotes. Phagotrophy indeed has allowed endosymbiosis and therefore the acquisition of chloroplast, of mitochondria and perhaps of kinetic apparatus (*cf.* Sects. 4.3.1 and 5.4.2).

5.4.1 How Old Are the First Eukaryotes?

Forterre and Philippe (1999) do not exclude that eukaryotes are ancestral in the tree of life. Although this does not imply that they then possessed all the characteristics that characterize them today, this assumption is rarely accepted. It is generally accepted that eukaryotes are more recent than prokaryotes. Based on the discovery of steranes (degradation products of sterols and eukaryotic biomarkers[8]) in the 2.7-Ga-old rocks of the Pilbara Craton (Australia), Brocks et al. (1999) concluded that eukaryotes have the same age (i.e., 2.7 Ga) (*cf.* Chap. 4). The oldest fossil attributed to eukaryotes dates from 2.1 Ga (El Albani et al. 2010), but it is doubtful. Less questionable eukaryotic remains are 1.8 Ga old (Acritarchs, Buick 2010). *Bangiomorpha pubescens*, much like the contemporaneous Rhodobionta, is dated from 1.2 Ga (Butterfield 2000). Finally, Cavalier-Smith (2002a) argues for a more recent emergence of eukaryotes (about 850 Ma), considered a sister group of archaea. However, this hypothesis is rejected by a majority of authors. In the current state of our knowledge, the emergence of eukaryotes is therefore between 2.7 and 1.8 Ga.

5.4.2 The Discovery of Endosymbiosis

At the end of the nineteenth century, biologists have been intrigued by the brutality of the "jump" between the relative simplicity of prokaryotic cells and the extreme complexity of eukaryotic cells. Although prokaryotic cells are far from being as simple as we have expected, the separation between prokaryotes and eukaryotes is less clear-cut than previously thought, with exceptions. However, these exceptions concern few isolated criteria, so that, overall, the "jump" is not really questioned (*cf.* Sect. 5.4.1). For example, the fact that archaea (the hyperthermophilic genus *Thermoplasma*) have some eukaryotic characters do not undermine their archaeal phylogenetic position (Waggoner 2001).

The apparent resemblance between a eukaryotic chloroplast, most particularly of Rhodobionta, and a cyanobacterium has led a German botanist, A. F. W. Schimper in 1883, to suggest, in a context of a general indifference of scientists, that chloroplasts can derive from cyanobacteria. It then took more than 20 years, the time of one human generation, for the idea to resurface, due to Mereschkowsky (1905, 1910). The concepts developed by Mereschkowsky are insightful and even modern: Mereschkowsky was the first to compare chloroplasts to "little green slaves," despite the limited means of investigation at that time to buttress his hypothesis (McFadden 2001). Knowledge in biochemistry at that time was indeed modest, and the tools such as transmission or scanning electron microscopy and, of course, molecular phylogeny did not exist then.

Mereschkowsky knew that chloroplasts are "self-reproducing," that is to say, in a cell, a new chloroplast forms from the division of an existing chloroplast and, upon the division of the eukaryotic cell, not only the nucleus but also the chloroplast divide. There was no evidence of the autonomous nature of the chloroplast, known later from micromanipulation experiments: in a cell with a single chloroplast, if extracted, the cell is not able to form a new

---

[8] The use of steranes, as biomarkers of eukaryotes, is disputed (*cf.* Chap. 4).

Fig. 5.8 The chloroplast is a self-replicating organelle. (**a**) During the cell (symbolized by a *rectangle*) division, the nucleus (in *red*) and the chloroplast (in *green*) divide synchronously; (**b**) if the chloroplast is removed by micromanipulation, the cell is not able to regenerate a chloroplast. Further divisions of the cell give birth to a lineage deprived of chloroplasts

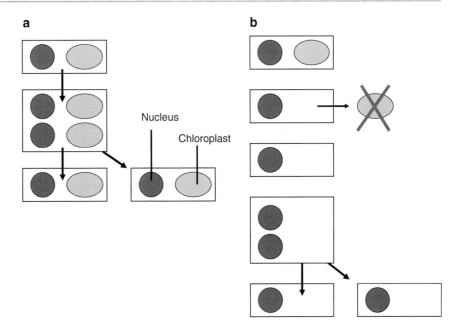

chloroplast. When a cell divides, if by accident, the two chloroplasts are segregated in one of the two daughter cells, the daughter cell without chloroplast permanently loses the photosynthetic function (Fig. 5.8). This means that the cell nucleus does not have the necessary genes to form chloroplast or does not have all the needed genes. But it is especially the discovery of DNA in the chloroplasts (and later on in the mitochondria), and the fact that this DNA is closest to that of prokaryotes, which marked the beginning of the "revolution." In order to understand the impact of this discovery, it is important to know that the nucleus was considered until then as the sole place of heredity and genes.

In 1970, Lynn Margulis and Peter H. Raven, separately and in two founding publications, have described the modern theory of endosymbiosis[9]: for these authors, chloroplasts are ancient cyanobacteria, while mitochondria are ancient non-photosynthetic bacteria. From that moment, despite some disputes, endosymbiosis has not been really contested. Carl Woese, especially quoted in this book for his work separating archaea from bacteria, is one of the many researchers who clearly rallied what was no longer a hypothesis but a reasonable certainty (Woese 1977). By the mid-1970s, gene comparison of chloroplasts and mitochondria with those of bacteria (e.g., Bonen and Doolittle 1975) indeed left no doubt on the origin of these organelles (Fig. 5.9).

However, a problem seemed to remain: if the chloroplasts of Rhodobionta are very close to those of cyanobacteria, namely, by the arrangement of thylakoids (isolated), by the presence of phycobilisomes containing accessory pigments (phycobilins) on the thylakoids, and by the presence of a single type of chlorophyll (chlorophyll *a*), this is not the case for other types of chloroplasts (in Viridiplantae, Chromobionta, etc.). In Viridiplantae, for example, the thylakoids are of two types (long and short), associated in large clusters, and chlorophyll *b* is present (in addition to chl *a*). In Chromobionta, the thylakoids are associated in clusters of 3 and chl *c* is present in addition to chl *a*. We therefore can hypothesize that other lines of oxygenic bacteria, with the chloroplast characteristics of today Viridiplantae and Chromobionta, had existed in the past along with cyanobacteria. The origin of the chloroplasts in each eukaryotic higher-order taxon would suggest the existence of different oxygenic bacteria, which means that the endosymbiosis was multiple and its origin was not from a single event. This hypothesis was acknowledged by Raven in 1970 (Raven 1970) and even earlier by Mereschkowsky in 1910. If such bacteria are not known in nature today, it is either that they still have to be discovered or that they have disappeared.

One hypothesis when confirmed usually constitutes a strong event in the history of sciences. This was the case in 1975, when Ralph Lewin discovered *Prochloron didemni*, an oxygenic bacteria mutualist of an ascidian Didemnidae from Baja California (Mexico), that corresponded well to sketch of the ancestor of Viridiplantae chloroplasts, with the following characteristics: presence of both chlorophyll *a* and *b*, absence of phycobilisomes and phycobilins, and even the stacking of thylakoids (Fig. 5.10). This discovery had a large

---

[9] In this chapter, we use the term symbiosis in its original and modern sense as relations between two taxonomically different organisms, and not in the sense of coexistence with mutual benefits, a meaning to which this has gradually derivated and continues to be accepted by current authors (*cf.* Chap. 10). Symbiosis therefore includes exploitation (predation, parasitism), competition, commensalism, amensalism, mutualism (mutually beneficial interaction), and helotism (servitude).

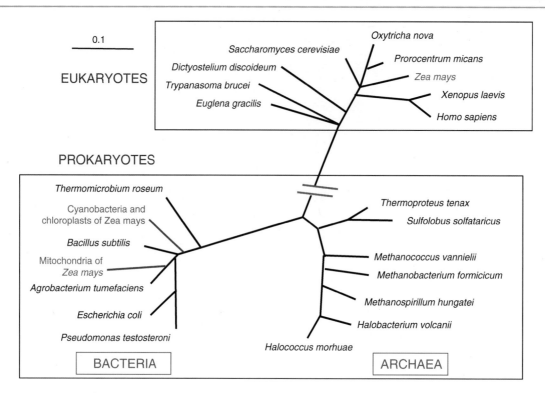

Fig. 5.9 Phylogenetic tree based on 16S-like rRNA genes. Mitochondria and chloroplasts of maize (*Zea mays*) are clearly positioned within the bacteria. The branch length is proportional to evolutionary distances (except the branch between archaea and eukaryotes). The scale bar corresponds to 0.1 per nucleotide mutation (10 %). The topology is based on an analysis of 920 nucleotides from complete sequences (Modified and redrawn from Woese and Olsen 1986) (The term 16S rRNA-like is used here to indicate that the 16S RNA (prokaryotes) and 18S RNA (eukaryotes) of the small subunits of ribosomes have been grouped to construct the same and common tree)

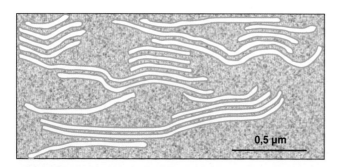

Fig. 5.10 The thylakoids of *Prochloron* seen in electron microscopy section. Note that they are arranged in clumps. *Prochloron* is an oxygenic phototrophic bacterium with chlorophylls *a* and *b*

echo and a new taxonomic division was created (the Prochlorophyta) (Lewin 1975; Lewin and Withers 1975; Lewin 1976).

But this division proved later to be wrong. Molecular phylogenies indeed showed that Prochlorophyta are a polyphyletic assemblage in which similarities are due to horizontal gene transfers (HGTs) and that they are not the direct ancestors of the chloroplasts of Viridiplantae (Turner et al. 1999; Chen et al. 2005). In the first approximation, all chloroplasts of eukaryotes therefore have a unique origin, and the founder event would have occurred in the ancestor of Archaeplastida. The possibility that several founding events have occurred however was mentioned for atypical cases (Stiller et al. 2003; Bodyl 2005; Nakayama and Ishida 2005; Burki et al. 2012). Apart from these cases, the uniqueness of the founding event of the chloroplast is now widely accepted.

The endosymbiotic origin of the mitochondria has been less debated than that of chloroplasts; the current bacterial taxon corresponding to the ancestors of mitochondria is indeed correctly identified: that of Alphaproteobacteria. The debate has focused more on other energy-converting organelles such as **hydrogenosomes***, peroxisomes, and mitosomes that, unlike mitochondria, generally do not have DNA, although DNA was discovered in hydrogenosomes of the ciliate *Nyctotherus ovalis* (Boxma et al. 2005). Hydrogenosomes are present, for example, in ciliates (Alveolata), in Parabasalia (Excavata), and in Chytridiomycota (Fungi). Are these organelles having an autogenous origin, that is to say, not endosymbiotic, by differentiation from the endo-membranes (endoplasmic reticulum) of the cytoplasm? Are they corresponding to endosymbioses distinct from those that have generated mitochondria, all the DNA being lost or transferred to the

Fig. 5.8 The chloroplast is a self-replicating organelle. (**a**) During the cell (symbolized by a *rectangle*) division, the nucleus (in *red*) and the chloroplast (in *green*) divide synchronously; (**b**) if the chloroplast is removed by micromanipulation, the cell is not able to regenerate a chloroplast. Further divisions of the cell give birth to a lineage deprived of chloroplasts

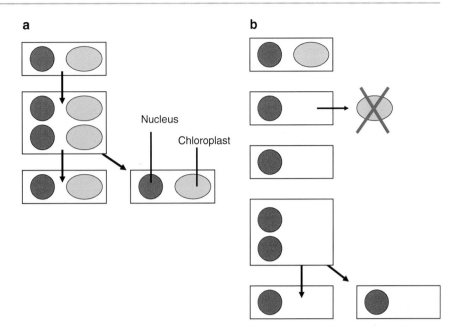

chloroplast. When a cell divides, if by accident, the two chloroplasts are segregated in one of the two daughter cells, the daughter cell without chloroplast permanently loses the photosynthetic function (Fig. 5.8). This means that the cell nucleus does not have the necessary genes to form chloroplast or does not have all the needed genes. But it is especially the discovery of DNA in the chloroplasts (and later on in the mitochondria), and the fact that this DNA is closest to that of prokaryotes, which marked the beginning of the "revolution." In order to understand the impact of this discovery, it is important to know that the nucleus was considered until then as the sole place of heredity and genes.

In 1970, Lynn Margulis and Peter H. Raven, separately and in two founding publications, have described the modern theory of endosymbiosis[9]: for these authors, chloroplasts are ancient cyanobacteria, while mitochondria are ancient non-photosynthetic bacteria. From that moment, despite some disputes, endosymbiosis has not been really contested. Carl Woese, especially quoted in this book for his work separating archaea from bacteria, is one of the many researchers who clearly rallied what was no longer a hypothesis but a reasonable certainty (Woese 1977). By the mid-1970s, gene comparison of chloroplasts and mitochondria with those of bacteria (e.g., Bonen and Doolittle 1975) indeed left no doubt on the origin of these organelles (Fig. 5.9).

However, a problem seemed to remain: if the chloroplasts of Rhodobionta are very close to those of cyanobacteria, namely, by the arrangement of thylakoids (isolated), by the presence of phycobilisomes containing accessory pigments (phycobilins) on the thylakoids, and by the presence of a single type of chlorophyll (chlorophyll *a*), this is not the case for other types of chloroplasts (in Viridiplantae, Chromobionta, etc.). In Viridiplantae, for example, the thylakoids are of two types (long and short), associated in large clusters, and chlorophyll *b* is present (in addition to chl *a*). In Chromobionta, the thylakoids are associated in clusters of 3 and chl *c* is present in addition to chl *a*. We therefore can hypothesize that other lines of oxygenic bacteria, with the chloroplast characteristics of today Viridiplantae and Chromobionta, had existed in the past along with cyanobacteria. The origin of the chloroplasts in each eukaryotic higher-order taxon would suggest the existence of different oxygenic bacteria, which means that the endosymbiosis was multiple and its origin was not from a single event. This hypothesis was acknowledged by Raven in 1970 (Raven 1970) and even earlier by Mereschkowsky in 1910. If such bacteria are not known in nature today, it is either that they still have to be discovered or that they have disappeared.

One hypothesis when confirmed usually constitutes a strong event in the history of sciences. This was the case in 1975, when Ralph Lewin discovered *Prochloron didemni*, an oxygenic bacteria mutualist of an ascidian Didemnidae from Baja California (Mexico), that corresponded well to sketch of the ancestor of Viridiplantae chloroplasts, with the following characteristics: presence of both chlorophyll *a* and *b*, absence of phycobilisomes and phycobilins, and even the stacking of thylakoids (Fig. 5.10). This discovery had a large

---

[9] In this chapter, we use the term symbiosis in its original and modern sense as relations between two taxonomically different organisms, and not in the sense of coexistence with mutual benefits, a meaning to which this has gradually derivated and continues to be accepted by current authors (*cf.* Chap. 10). Symbiosis therefore includes exploitation (predation, parasitism), competition, commensalism, amensalism, mutualism (mutually beneficial interaction), and helotism (servitude).

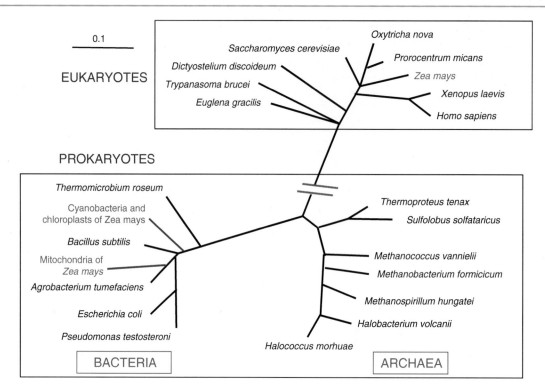

Fig. 5.9 Phylogenetic tree based on 16S-like rRNA genes. Mitochondria and chloroplasts of maize (*Zea mays*) are clearly positioned within the bacteria. The branch length is proportional to evolutionary distances (except the branch between archaea and eukaryotes). The scale bar corresponds to 0.1 per nucleotide mutation (10 %). The topology is based on an analysis of 920 nucleotides from complete sequences (Modified and redrawn from Woese and Olsen 1986) (The term 16S rRNA-like is used here to indicate that the 16S RNA (prokaryotes) and 18S RNA (eukaryotes) of the small subunits of ribosomes have been grouped to construct the same and common tree)

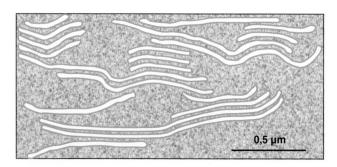

Fig. 5.10 The thylakoids of *Prochloron* seen in electron microscopy section. Note that they are arranged in clumps. *Prochloron* is an oxygenic phototrophic bacterium with chlorophylls *a* and *b*

echo and a new taxonomic division was created (the Prochlorophyta) (Lewin 1975; Lewin and Withers 1975; Lewin 1976).

But this division proved later to be wrong. Molecular phylogenies indeed showed that Prochlorophyta are a polyphyletic assemblage in which similarities are due to horizontal gene transfers (HGTs) and that they are not the direct ancestors of the chloroplasts of Viridiplantae (Turner et al. 1999; Chen et al. 2005). In the first approximation, all chloroplasts of eukaryotes therefore have a unique origin, and the founder event would have occurred in the ancestor of Archaeplastida. The possibility that several founding events have occurred however was mentioned for atypical cases (Stiller et al. 2003; Bodyl 2005; Nakayama and Ishida 2005; Burki et al. 2012). Apart from these cases, the uniqueness of the founding event of the chloroplast is now widely accepted.

The endosymbiotic origin of the mitochondria has been less debated than that of chloroplasts; the current bacterial taxon corresponding to the ancestors of mitochondria is indeed correctly identified: that of Alphaproteobacteria. The debate has focused more on other energy-converting organelles such as **hydrogenosomes***, peroxisomes, and mitosomes that, unlike mitochondria, generally do not have DNA, although DNA was discovered in hydrogenosomes of the ciliate *Nyctotherus ovalis* (Boxma et al. 2005). Hydrogenosomes are present, for example, in ciliates (Alveolata), in Parabasalia (Excavata), and in Chytridiomycota (Fungi). Are these organelles having an autogenous origin, that is to say, not endosymbiotic, by differentiation from the endo-membranes (endoplasmic reticulum) of the cytoplasm? Are they corresponding to endosymbioses distinct from those that have generated mitochondria, all the DNA being lost or transferred to the

Fig. 5.11 The five components of the eukaryotic kinetic apparatus. Other cell constituents (e.g., nucleus, mitochondria) are not presented. The shown example belongs to a photosynthetic Stramenopile. Anterior and posterior are in inverted commas because, as a matter of fact, both undulipodia are anterior (*cf.* Chap. 7)

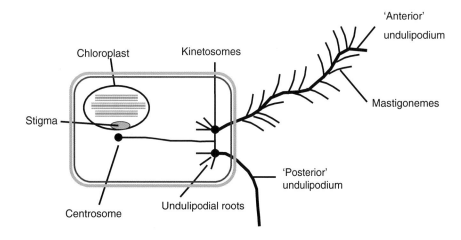

nucleus? Are they derived from the mitochondria, through loss of its genetic material? This last hypothesis is now the most accepted for hydrogenosomes (Hrdy et al. 2004; Carlton et al. 2007). In the case of peroxisomes present in all eukaryotic cells, it was thought that they could result from an endosymbiosis with prokaryotes, distinct from that which gave rise to the mitochondria, and have appeared very early in the history of eukaryotes. Peroxisomes could have served to protect the cell against the superoxidizing forms of oxygen, while the mitochondrial endosymbiosis had not yet occurred. However, the autogenous origin, by differentiation of the endoplasmic reticulum, is actually the most accepted hypothesis.

The debate also has focused on eukaryotes devoid of mitochondria (Microsporidia, Parabasalia, etc.): Did they appear before endosymbiosis as has long been thought? Or have they lost their mitochondria secondarily? The discovery, in the nucleus of these organisms, of genes whose origin is undoubtedly mitochondrial suggests that they have possessed mitochondria that were subsequently lost, as shown, for example, by Germot et al. (1997) for the Microsporidia *Nosema locustae*.

So far, the kinetic apparatus has not been addressed since its endosymbiotic origin is highly disputed or even refuted (Cavalier-Smith 2002b); the latter author considers that the kinetic apparatus was formed de novo, from the cytoskeleton. The term kinetic apparatus comprises a set of five parts: (1) one or more **undulipodia***; (2) one or more kinetosomes (basal granules); (3) undulipodium roots which fixes the kinetic apparatus on the cytoskeleton; (4) the centrosome (centriole), which plays a role in the division of the nucleus; and (5) the stigma, located in the chloroplast, a photosensitive structure capable of orienting the movements of the undulipodia as a function of light intensity, so somehow representing the "eye" of the cell (Fig. 5.11). Depending on taxa, these five components are not always present. For example, the stigma is present only in photosynthetic organisms. Moreover, in the same taxon, the undulipodia may be present (gametes, spores, etc.) or absent (vegetative cells).

The kinetic apparatus is present in almost all eukaryotes. When it is absent (Microsporidia, some Fungi, Rhodobionta), this generally concerns only part of a given taxa and can thus be interpreted as a secondary loss. In addition, some of its characteristics (organization of microfibrils 9 + 2 in undulipodium, 9 + 0 in kinetosome and centrosome, etc.) are very homogeneous in eukaryotes. This suggests a very ancient origin in the common ancestor of all eukaryotes (Bornens and Azimzadeh 2007). Whether its formation in eukaryotic cells is de novo as suggested by Cavalier-Smith (2002b) and Carvalho-Santos et al. (2011) or has originated from endosymbiosis as first suggested by Kozo-Polyansky (1924) is still questionable. As for chloroplasts and mitochondria, it is in the early 1970s that this hypothesis was given renewed interest, with a bacterial candidate for endosymbiosis: the group of spirochaetes (Margulis 1970, 1980; Margulis et al. 2000). However, it appeared unlikely that the spirochaetes are at the origin of the kinetic apparatus of eukaryotes, because of the importance of morphological and biochemical differences. The main argument in favor of the endosymbiotic origin is the absence of a credible alternative theory. How can we explain that a structure as complex as the kinetic apparatus did appear abruptly in the common ancestor of eukaryotes? Below, we consider the possibility that the kinetic apparatus is from an endosymbiosis for two reasons: (1) the absence of DNA in the kinetic apparatus can be due to the fact that it has been fully captured by the nucleus (*cf.* Sect. 5.4.3), and (2) it cannot be excluded that the bacterial ancestor of the kinetic apparatus has subsequently disappeared or has not yet been found in nature. Anyway, it must be clear that this is a very controversial hypothesis, which therefore cannot be put on the same level as the origin of chloroplasts and mitochondria.

Fig. 5.12 Stages of primary endosymbiosis at the origin of the kinetic apparatus (very hypothetical) of the mitochondrion and chloroplast. The genes of ancestral eukaryotic nucleus are symbolized by red crosses. Prokaryotic genes are shown in blue, including when they are captured by the nucleus. (**a**) Formation of the kinetic apparatus from a hypothetical bacterium with undulipodium that lost its entire genome; (**b**) an alphaproteobacteria becomes a mitochondrion after losing much of its genome; (**c**) a cyanobacterium becomes a chloroplast, after loss of the majority of its genome

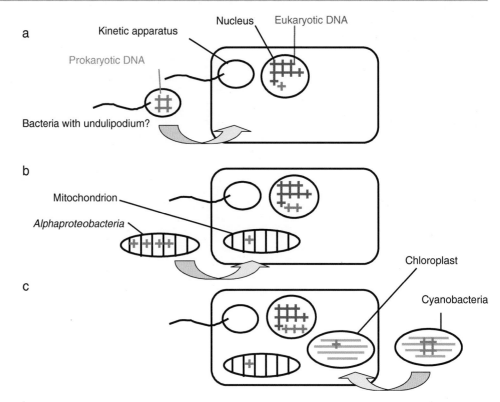

5.4.3 The Current Theory of Endosymbiosis

5.4.3.1 Primary Endosymbioses

The kinetic apparatus, either originating from endosymbiosis with a bacterium (which remains to be identified) (Fig. 5.12a) or formed de novo, remains ancestral in eukaryotes.

Mitochondria originated from an alphaproteobacterium (Boyen et al. 2001) (Fig. 5.12b). The current nearest relative of this bacterium is *Rickettsia prowazekii*, an obligate intracellular parasite. The endosymbiosis would have occurred before 1.6 Ga (Meyerowitz 2002), and the founding event would be unique (Cavalier-Smith 2002b). The common ancestor of all current eukaryotic kingdoms undoubtedly possessed these characteristics: a kinetic apparatus and one or more mitochondria.

Chloroplasts originate from a cyanobacterium (Fig. 5.12c). The founding event appears to be unique (*cf*. Sect. 5.4.2) and would have occurred in the kingdom of Archaeplastida, in the ancestor of Viridiplantae, Rhodobionta, and Glaucocystobionta. The cyanobacterium probably had chlorophylls *a* and *b* and phycobilin containing phycobilisomes as well (Bhattacharya et al. 2003); a cyanobacterium with these characteristics is not known in nature today; it may have disappeared. Thereafter, chlorophyll *b* was lost in the ancestor of Rhodobionta and Glaucocystobionta, while phycobilisomes and phycobilins (*cf*. Sect. 3.3.4) have been lost in the ancestor of Viridiplantae (Green 2005). There is a consensus that the endosymbiosis at the origin of chloroplasts occurred after the one at the origin of mitochondria, i.e., after 1.6 Ga (Meyerowitz 2002). The relatively "recent" date (580 Ma) proposed by Cavalier-Smith (2002b) is not consistent with the fossil register.

The genome of mitochondria and chloroplasts is 10 to 40 times smaller than that of a bacterium. For example, the genome of *Synechocystis* (Cyanobacteria) has about 3,300 genes (Kaneko et al. 1996), while that of the chloroplast *Porphyra purpurea* (Rhodobionta) only holds 200 genes (Reith and Munholland 1993). Genome reduction is generally more important for mitochondria than for chloroplasts. In the case of the kinetic apparatus, the loss of the bacterial genome is considered total. What happened to the missing genes? A part has been transferred to the nucleus, ensuring its control of the bacterium, but the major part, redundant with nuclear genes, has been lost (McFadden 2001; Dyall et al. 2004) (Fig. 5.12). The real "flood of genes," according to the words from Rivera and Lake (2004), from which the nucleus was submitted via organelles, has deeply influenced the structure of nucleus of eukaryotic cells. In *Arabidopsis* (Embryophyta, Viridiplantae), 18 % of nuclear genes came from cyanobacteria (Martin et al. 2002; Bhattacharya et al. 2003). Although more limited, gene transfers from the nucleus to the organelles also occurred, and even between organelles: chloroplastic DNA is present in the mitochondria of *Zea mays* (maize).

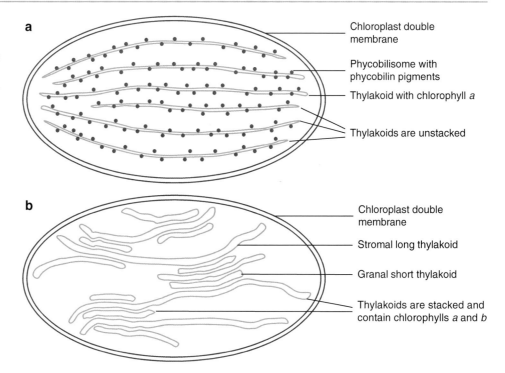

Fig. 5.13 Thylakoid structure and disposition in Archaeplastida chloroplasts. (**a**) Chloroplast of Rhodobionta. Note the absence of a true thylakoid lumen; (**b**) chloroplast of Chlorobionta. Note the presence of a lumen, i.e., a continuous aqueous phase enclosed by the thylakoid membrane

These genetic alterations are of great importance: in fact, from the moment bacteria hosted in the cytoplasm of eukaryotes did lose much of their genes, they were no longer autonomous. From this point of view, as rightly remarked by Cavalier-Smith (2002b), the term endosymbiosis, if used in the sense of mutualism, is not appropriate. Indeed, the mutualistic symbiosis involves an association with mutual benefit for two species, each of which is generally able to survive independently. From the moment most of the genes essential to their functioning, particularly those controlling their division, are present in the nucleus, these bacteria are no longer mutualistic symbionts but slaves, and the term **helotism*** is the most appropriate for the endosymbiotic-derived organelles.

5.4.3.2 Secondary and Tertiary Endosymbioses

The endosymbioses described previously (*cf.* Sect. 5.4.3) are primary endosymbioses: the hypothetical origin of the kinetic apparatus, the origin of the mitochondria, and the origin of chloroplasts in the kingdom of Archaeplastida. In the case of the first two types of organelles, only primary endosymbioses are known. In the case of the chloroplast, however, photosynthesis has developed, through secondary and tertiary endosymbioses, into up to four kingdoms of eukaryotes (Rhizaria, Alveolata, Stramenopiles, and Discicristata) and into two lineages whose phylogenetic position is not clear: the Haptobionta and the Cryptobionta (*cf.* Sect. 7.11).

But between the primary endosymbiosis that gave rise to the chloroplast and the secondary endosymbiosis in question, the chloroplast has changed considerably in Archaeplastida. In Glaucocystobionta and in Rhodobionta, chlorophyll *b* was lost while phycobilisomes and phycobilins were kept. Moreover, the thylakoids are isolated from each other (Fig. 5.13a), and as in most modern cyanobacteria, the possession or not of this character in the ancestor of the chloroplast is not established. In Viridiplantae, phycobilisomes and phycobilins were lost, and the thylakoids are arranged in packs (Fig. 5.13b).

The secondary endosymbiosis consists in the inclusion of unicellular Rhodobionta (called "red pathway") or a Viridiplantae ("green pathway") in a cell of another eukaryotic organism. This process has enabled the acquisition of a chloroplast and thus the photosynthesis. The nucleus of the secondary endosymbiont contained redundant genes: they were eliminated. The others will be captured by the nucleus of the host, so that the nucleus of the endosymbiont will eventually disappear. However, there are in the present environment "living fossils" that remain (McFadden 2001) in which this second nucleus has not yet been completely eliminated: the Cryptophyta and the Chlorarachniobionta. This second nucleus, very small compared to that of the host and which still contains some genes, is named **nucleomorph***. The nucleomorphs are valuable to scientists because they are proof of the secondary endosymbioses.

Another tracer of secondary endosymbiosis is the number of envelope layers around chloroplasts, which can form the nucleoplastidial complex (NPC) (Fig. 5.14). In organisms that have acquired the chloroplast by primary endosymbiosis (Archaeplastida), chloroplasts are surrounded by a double

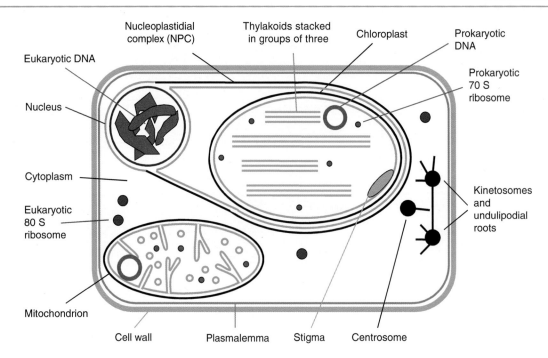

Fig. 5.14 The nucleoplastidial complex (NPC) within a cell of Chromobionta (Stramenopile). The NPC is a marker of the secondary endosymbiosis which brought the photosynthesis to this taxon. The cell structure and organelles are simplified, and the undulipodia originating from the two kinetosomes are not represented (for more details, see Fig. 7.43, Sect. 7.9.3). The plasmalemma and the membranes originating in the symbiotic bacterium and Rhodobionta plasmalemmas are in *blue*. The membranes originating in the host vacuole that embed the symbiont are in *black*

membrane. The inner membrane would be the plasmalemma (cytoplasmic membrane) of the cyanobacteria and the outer one the membrane of the vacuole in which the cyanobacterium was included. In secondary endosymbiosis, a second double membrane will surround the chloroplast; it comes from the plasmalemma of eukaryotic endosymbiont, surrounded by the membrane of the vacuole of the eukaryotic host. This results in a quadruple membrane, sometimes simplified into a triple membrane (e.g., in euglenoids). In Chromobionta, the second double membrane surrounding the chloroplast merged with the outer nuclear membrane to form the NPC (Fig. 5.14), a characteristic of this taxon (*cf.* Sect. 7.9.3) (although NPC has been observed in other taxa, including Viridiplantae; Selga et al. 2010).

Overall, Haptobionta, Cryptobionta, Chromobionta, and some Dinobionta have acquired photosynthesis by the "red pathway," that is to say, from Rhodobionta, while Chlorarachniobionta, Euglenoids, and other Dinobionta have acquired it through the "green pathway," that is to say, from Viridiplantae.

This system of endosymbioses, which can be likened to "Russian dolls," is in reality more complex: in Dinobionta, chloroplasts were acquired not only from Rhodobionta or from Viridiplantae (secondary endosymbiosis) but also from Haptobionta, Cryptobionta, or Chromobionta. In this case, the term tertiary endosymbiosis is used. Dinobionta constitutes the taxon in which the origin of photosynthesis is more complex and more diverse (Fig. 5.15); there are indeed species which:

1. Have acquired it from a Rhodobionta (by secondary endosymbiosis)
2. Have acquired it from a Viridiplantae (by secondary endosymbiosis), e.g., *Lepidodinium*
3. Have acquired it from a Cryptophyta (by tertiary endosymbiosis), e.g., *Rhodomonas*
4. Have acquired it from an Haptobionta (by tertiary endosymbiosis), e.g., *Gymnodinium*
5. Have acquired it from a diatom (Chromobionta) (by tertiary endosymbiosis), e.g., *Peridinium*
6. Have acquired it but have subsequently lost it
7. Have acquired it, have lost it, and have subsequently recovered it
8. Have ever owned it but practice kleptoplasty (see below), a process that will allow them perhaps to acquire the photosynthesis in the future
9. Or, perhaps, have never had photosynthesis (in the present state of our knowledge)

Secondary and tertiary endosymbioses are likely relatively recent, dating back to 285–260 Ma, at the historic minimum of CO_2 content of the atmosphere (Fig. 5.16) (Lee and Kugrens 2000). Indeed, given the pH of seawater, its CO_2 content is low, with mineral carbon present mainly in the form of HCO_3^-. Due to the fact that RuBisCo, the enzyme that incorporates inorganic carbon in an organic

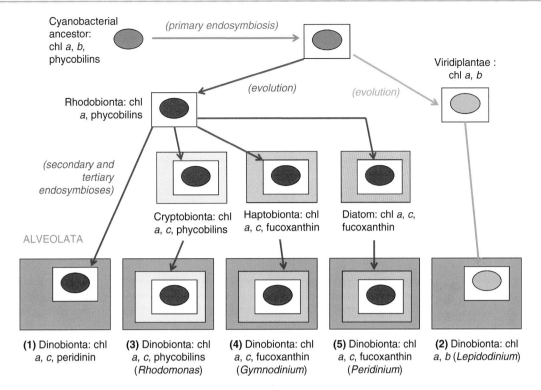

Fig. 5.15 Origin of the photosynthesis in Dinobionta (Alveolata, bottom), via secondary and tertiary endosymbioses. Acquisition pathways are numbered (1) through (5) (see text). The hypothetical cyanobacterial ancestor (*top left*) is in *blue*. Chloroplasts of the red and green pathway are in *red* and *green*, respectively. Peridinin and fucoxanthin are carotenoid pigments (Modified and redrawn from Falkowski et al. 2004)

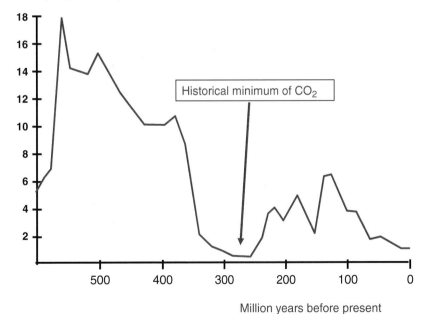

Fig. 5.16 Changes in the content of CO_2 in the atmosphere over the past 600 Ma. The value "1" corresponds to the current level of CO_2 (0.039 %) in the atmosphere (Modified and redrawn from Lee and Kugrens 2000)

molecule, exclusively uses CO_2 during photosynthesis, concentration mechanisms of CO_2 are necessary. The additional envelope around the chloroplast (especially the plastid-nuclear complex, Fig. 5.14), which characterizes the secondary and tertiary endosymbioses by creating around the chloroplast a micro-area of acidic pH, promotes the concentration of CO_2. According to Kugrens and Lee (2000), secondary and tertiary endosymbioses would therefore represent a key advantage at a time when the CO_2 content was unusually low and therefore have been selected by evolution. Paleontological data are consistent with this hypothesis: fossil taxa from secondary and tertiary endosymbioses, such as diatoms (Stramenopiles), are all younger than 260 Ma (Falkowski et al. 2004; Medlin and Kaczmarska 2004; Raven and Waite 2004).

5.4.4 Mechanisms and Processes of Endosymbiosis

Phagotrophy and/or parasitism may be the origin of primary endosymbiosis. **Parasitism*** of a eukaryotic cell by a non-photosynthetic bacteria constitutes a very reliable hypothesis for the origin of the mitochondria. Indeed, parasitism often precedes mutualism; in other words, mutualism is an advanced form of parasitism (Combes 1995, 2001). Moreover, K. W. Jeon has shown the relatively high speed of the shift from parasitism to mutualism-helotism in modern bacteria (Jeon 1972, 1991, 1995; Jeon and Jeon 1976, Box 5.4).

In the case of the origin of the chloroplast, predation (phagotrophy) of a cyanobacterium by a heterotrophic unicellular eukaryote can be the starting point of the endosymbiosis. The fact that cyanobacteria are retained in the cytoplasm instead of being digested is not an unrealistic assumption since in the present environment, non-photosynthetic eukaryotes host "wild" cyanobacteria, which are also found as free-living cells. This is the case, for example, of *Nostoc symbioticum* found in Glomeromycota (Fungi, Opisthokonta) *Geosiphon pyriformis* (Fig. 5.17).

In the case of secondary and tertiary endosymbioses which allowed non-photosynthetic taxa to acquire photosynthesis, phagotrophy and kleptoplasty are the most likely mechanism (Fig. 5.18). Normally, the prey is digested (Fig. 5.18a). However, there are many modern predators which are known to digest most of their prey but retain its chloroplasts (Fig. 5.18b); this phenomenon is called kleptoplasty. This is the case, for example, of the opisthobranch mollusk *Elysia viridis* (Rahat and Ben Ishac-Monselise 1979). This sea slug grazes on *Codium* (Chlorobionta, Viridiplantae) and sequesters its chloroplasts which are transferred, via the diverticula of the digestive tract, in its parapodia cells.[10] In the presence of light,

---

[10] The abundance of these chloroplasts gives a green color to *Elysia viridis*. Note that viridis means "green" in Latin.

the sea slug stops moving and displays its parapodia to allow the activity of chloroplasts, in order to benefit from the photosynthesis products.

However, the chloroplasts cannot divide because they are not autonomous; the controlling genes were located in the nucleus of the prey and were destroyed (digested) with the rest of the prey. The survival of chloroplasts in the parapodia of *Elysia* does not exceed a few days to a few weeks after which the sea slug must acquire new chloroplasts by grazing again on *Codium* (Williams and Walker 1999). Kleptoplasty is not uncommon; in addition to opisthobranchs, it has been observed particularly among ciliates (Alveolata) and foraminifera (Rhizaria).

To become truly photosynthetic, an organism that practices kleptoplasty must recover not only the chloroplasts but also the controlling genes located in the nucleus of its prey (Fig. 5.18d). This is undoubtedly a very improbable event, although it already happened in the course of evolution, at least once, as suggested by Cavalier-Smith (2002b), but more likely several times. An intermediate step is to recover not only the chloroplast (kleptoplasty) but also the nucleus (with chloroplast genes; karyoklepty) of the prey (Fig. 5.18c) (Hansen and Fenchel 2006; Johnson et al. 2007; Smith and Hansen 2007). In present-day nature, the beginning of the nuclear gene recovery process that controls chloroplast was observed. In the sea slug *Elysia chlorotica*, which feeds on *Vaucheria litorea* (Chromobionta, Stramenopile), if the juveniles do not encounter *Vaucheria*, they do not survive; chloroplasts survive as long as the life of the slug, 10 months (Fig. 5.19); kleptoplasty and photosynthesis have then became obligatory in *Elysia chlorotica* but not in *E. viridis*: in the absence of light, the latter survives via grazing (Rumpho et al. 2000; Pennisi 2006). Two large fragments of genes of photosynthetic organisms likely to participate in the survival of the chloroplast were discovered in the nucleus of *Elysia chlorotica* (Rumpho et al. 2000, 2008; Pennisi 2006). A similar discovery was made in *Elysia clarki* (Curtis et al. 2006).

Similarly, in the gammaproteobacterium *Carsonella ruddii*, an endosymbiont of psyllids, that has a tiny genome (160,000 bp), genes were lost and others transferred to the nucleus of the host: this bacterium is thus in a transitory step to becoming an organelle. Indeed, it is transmitted vertically by the host (Andersson 2006; Nakabachi et al. 2006). In endosymbiotic bacteria, gene capture by the host nucleus is not a general case; helotism can occur directly through loss of unnecessary genes (McCutcheon and Moran 2011).

It is therefore reasonable to think that kleptoplasty phenomena observed in nature today (mollusks, foraminifera, ciliates, etc.) are the beginnings of future secondary or tertiary endosymbioses, known as *endosymbiosis in progress* by McFadden (2001). This is really most likely because the period in which we live (Tertiary Age, Pleistocene) is characterized by

Box 5.4: A Very Fast Transition from Parasitism to Mutualism-Helotism

Charles-François Boudouresque

When free-living normal strains of amoebas (F) are infected by a pathogenic bacterium (P), they usually die (Box Fig. 5.4a). However, some infected amoebas survive in culture to harmful bacteria; they grow slower, are more sensitive to starvation, are smaller in size, and are more fragile than free-living strains. After 5 years of infection, the infective bacteria that were initially harmful to the host amoebas became harmless; the host amoeba (variant strain X) recovered a normal vitality (Box Fig. 5.4b).

Transplantation of the nucleus of the variant strain X into a normal free-living amoeba (microsurgical experiment) results in a low vitality amoeba (Box Fig. 5.4c). Transplantation of pathogenic bacteria P into the variant strain X of the amoeba results in its killing (Box Fig. 5.4d). In contrast, transplantation of bacteria from the strain X makes it recover its normal vitality.

The nucleus of the variant strain X therefore became dependent on the bacteria for its normal functions. Similarly, the bacteria established a stable symbiotic relationship with host amoebas that has resulted in mutual dependence for survival. Gene loss, exchange, and/or capture, between the amoeba and the bacterium, is probably involved in their mutual dependence.

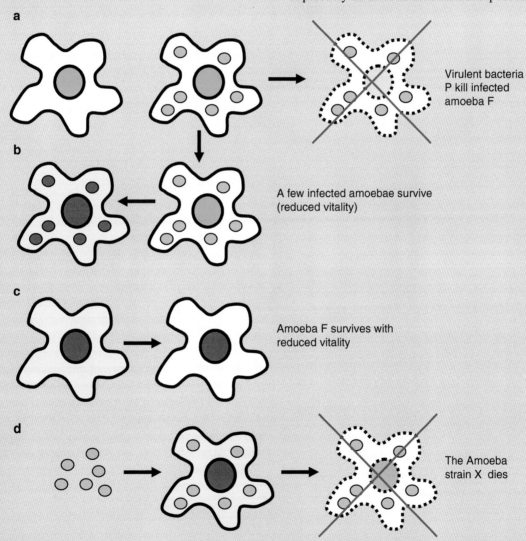

Box Fig. 5.4 (a) Free-living amoeba (F) infected by a pathogenic bacterium (P). (b) After 5 years, an amoeba strain (X) recovered a normal vitality despite the presence of bacteria. (c) Transplantation of the nucleus X into the amoeba F. (d) Transplantation of bacteria P into the amoeba strain X

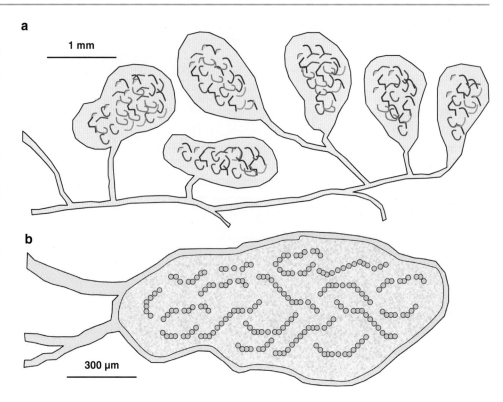

Fig. 5.17 *Geosiphon pyriformis* (Fungi, Glomeromycota). (**a**) filaments and vesicles hosting the mutualistic cyanobacterium *Nostoc symbioticum*; (**b**) detail of an enlarged vesicle. Capture of *Nostoc* by *Geosiphon* was observed experimentally: in contact with *Nostoc*, *Geosiphon* wall ruptures, the *Nostoc* enters the cytoplasm, and the cell wall is reconstituted. In the laboratory, both symbionts are able to survive separately; *Nostoc symbioticum* is not specific to *Geosiphon pyriformis*: it participates in other mutualistic symbioses (Modified and redrawn from Boullard 1990)

a new minimum CO_2 content in the atmosphere, unprecedented since the minimum of 285–260 Ma (Fig. 5.16) and, even beyond, since the formation of our planet.[11]

5.4.5 The "Travels" of the Chloroplast

The occurrence of chloroplast in Archaeplastida (primary endosymbiosis) corresponds to a massive horizontal gene transfer (HGT) from cyanobacteria. Thereafter, the chloroplast has evolved along with its host, Archaeplastida (vertical evolution), to differentiate into "red chloroplast" in Rhodobionta and into "green chloroplast" in Viridiplantae. Secondary endosymbioses (Fig. 5.20) that involve, on the one hand, the transfer of a portion of the green chloroplast of Viridiplantae to Euglenoids, Chlorarachniobionta, and some Dinobionta (the "green pathway") and, on the other hand, the transfer of red chloroplast of Rhodobionta to Haptobionta, Cryptobionta, Chromobionta and other Dinobionta (the "red pathway")[12] correspond, once again, to a massive series of HGT. In Dinobionta, in addition to direct green and red pathways, the chloroplast was also formed by tertiary endosymbioses (Fig. 5.21). Finally, the phenomena of kleptoplasty and karyoklepty, which are found in nature today, are perhaps future secondary or tertiary endosymbioses (Fig. 5.22).

Photosynthesis has not only been acquired: it could also have been lost secondarily. This is the case in Apicomplexa (Alveolata), particularly in *Plasmodium falciparum*, the causative agent of malaria. The apicoplast is the remnant of a chloroplast and would come from Rhodobionta[13] (Fig. 5.20; Köhler et al. 1997; McFadden and Waller 1997). The loss of photosynthesis has also been

---

[11] The reader will probably be surprised by the fact that the current CO_2 concentration was considered, in this chapter, as a historic minimum, while its increase, due to human activities, is so disturbing. Fig. 5.16, which shows the evolution of CO_2 in geological time, does not highlight, due to the scale, the current increase.

[12] Some authors (e.g., Embley and Martin 2006) hypothesized that in Chromalveolata a single secondary endosymbiotic event is the origin of photosynthesis; it would have occurred in the common ancestor of Chromalveolata. This hypothesis has been criticized by other authors (e.g., Bodyl et al. 2009; Burki et al. 2012).

[13] Some authors propose that the apicoplast of Apicomplexa is from a Viridiplantae (green pathway).

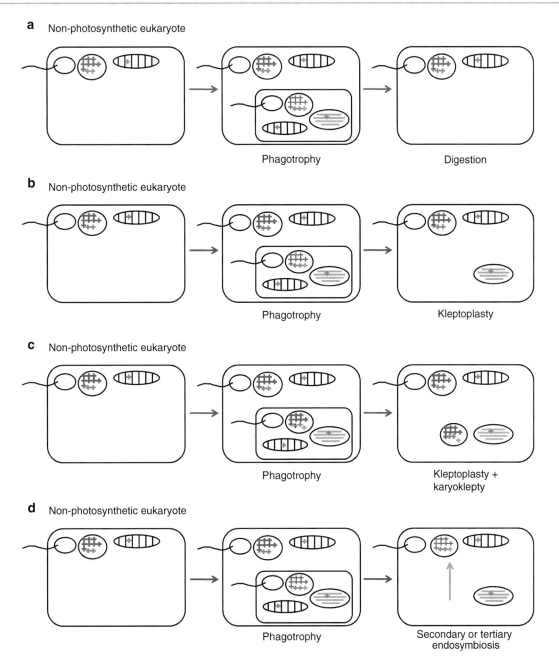

Fig. 5.18 The possible steps of secondary and tertiary endosymbioses. (**a**) The general case, the complete digestion of the prey following phagotrophy; (**b**) kleptoplasty, i.e., partial digestion of a photosynthetic prey, leaving the chloroplasts intact; (**c**) kleptoplasty and karyoklepty, i.e., partial digestion leaving the chloroplasts and the nuclei intact; (**d**) permanent acquisition of the chloroplasts by the predator via the transfer of the nuclear genes involved into photosynthesis from the prey nucleus to the predator nucleus. Green +: nuclear genes involved into photosynthesis. See legend of Fig. 5.12 for other symbols

demonstrated in some Dinobionta. The ancestors of Oobionta, one of the taxa belonging to the polyphyletic assemblage called "Fungi" (customary meaning *cf.* Sect. 5.5.3), have possessed photosynthesis ("green pathway") but subsequently have lost it (Tyler et al. 2006). Finally, in Chromobionta, photosynthesis appears to have been acquired in the first instance through the "green way," before being replaced through the "red pathway" (Dagan and Martin 2009; Moustafa et *al.* 2009).

Overall, the vertical phylogeny related to the evolution of nuclear genes overlaps with the phylogeny of chloroplasts, transmitted vertically (during reproduction) but also horizontally (during secondary and tertiary endosymbioses). Within a given taxon, such as Viridiplantae, phylogenies which are based on chloroplast and nuclear genes are congruent (Ané et al. 2005). However, they are not when taking into account the taxa involved in secondary and tertiary endosymbioses (Fig. 5.23).

Fig. 5.19 The sea slug *Elysia chlorotica* (*right*) is totally dependent on the kleptoplasty. The larvae are required to ingest chloroplasts of *Vaucheria litorea*, a Chromobionta Stramenopile (*left*), which are stored in its superficial cells and survive throughout its life (10 months). In the absence of light and therefore of photosynthesis, *E. chlorotica* cannot survive (From Rumpho et al. 2000) (Copyright: courtesy of Plant Physiology and Mary Rumpho)

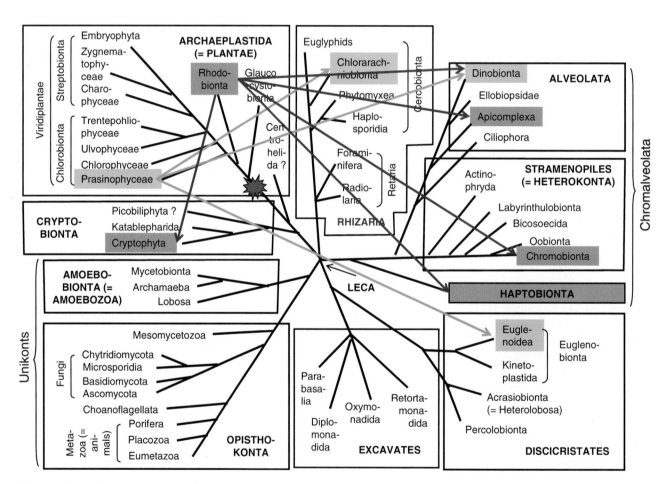

Fig. 5.20 Secondary endosymbioses. The *red star* indicates the place, in the phylogenetic tree, where the primary endosymbiosis with a cyanobacterium occurred, within the kingdom Archaeplastida; *red arrows*, the "red path" of secondary endosymbioses, originating from the Rhodobionta; *green arrows*, the "green path" of secondary endosymbioses, originating from Viridiplantae; *pink and green boxes*, the origin and the targets of the red and green paths, respectively; *blue box* (Dinobionta), target of both the red and the green paths. In Chromalveolata, a single endosymbiosis event, occurring in early evolutionary times, has been hypothesized (not presented). *LECA* last eukaryotic common ancestor. Eukaryotic tree modified, simplified, and redrawn from Baldauf (2003, 2008)

5 Systematic and Evolution of Microorganisms: General Concepts

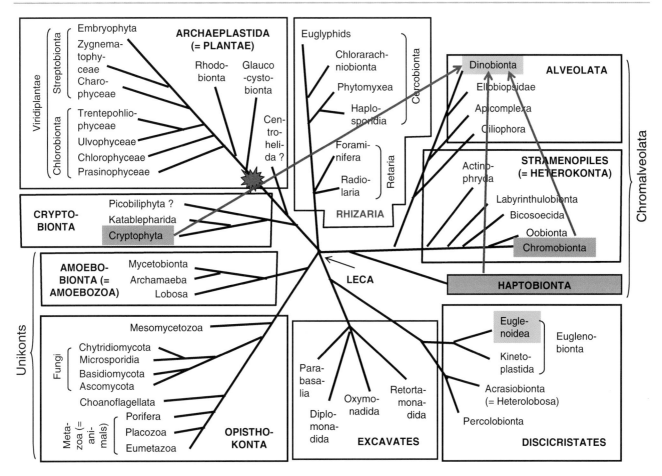

Fig. 5.21 Tertiary endosymbioses. Same legend as Fig. 5.20

5.5 The Current Tree of Life: From Phylogeny to Genome Study

5.5.1 The Current Tree of Life

Microorganisms exhibit a considerable diversity of species, although still very imperfectly known: thousands of new species or sequences are published annually. However, they are characterized more by their amazing phylogenetic diversity than by specific diversity. Even considering only the microorganisms in the strict sense (i.e., unicellular and nearly unicellular), they almost represent all prokaryotes and the vast majority of higher taxon levels of eukaryotes. Microorganisms are present in all kingdoms of eukaryotes, and many of these kingdoms are composed only of microorganisms. Thus, the vast majority of the tree of life is made up of microorganisms, while the higher-level taxa, which include multicellular organisms, are comparatively few in number (Fig. 5.24).

Regarding eukaryotes, the traditional representation, with animal and plant kingdoms, algae and "higher" plants, fungi, and protozoa, began to be questioned between the 1940s and 1970s. The consequences of the works of Feldmann and Feldmann (1946) and Hébant (1977), to cite but a few examples, although published in good journals, went unnoticed. Before the advent of molecular phylogenies, all elements existed to abandon the old system and various authors have drawn conclusions. Woese, in the 1980s, was the first to use molecular phylogeny in a systematic way and this was the trigger to decisively change our vision of the living world. The first books for students that initiated a revolution of taxonomy based on recent data, instead of perpetuating patterns partly known from Linnaeus in the eighteenth century, were those of Margulis (1970, 1981). More recently, Lecointre and Leguyader (2001, 2006) revised the taxonomy and phylogeny of the living world based on current data. In order to help the reader to identify the correspondence between the current taxonomy and the old classifications, two sections

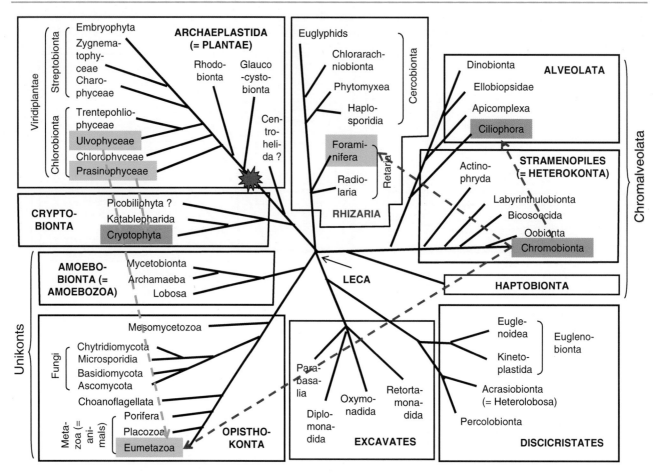

Fig. 5.22 Endosymbioses in progress? Kleptoplasty among present-day eukaryotes. See Fig. 5.20 for legend

Fig. 5.23 Phylogeny based on a number of chloroplast genes (rRNA, *psaA*, *psbA*, *tufA*). The names of the taxa have been brought into conformity with the nomenclature adopted here. The Rhodobionta are closer to Chromalveolata than Chlorobionta, while in the tree of eukaryotes based on nuclear genes (Fig. 5.20), it is the opposite (Modified and redrawn from Bhattacharya et al. 2003)

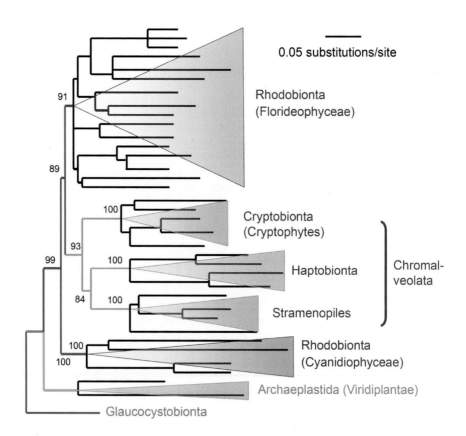

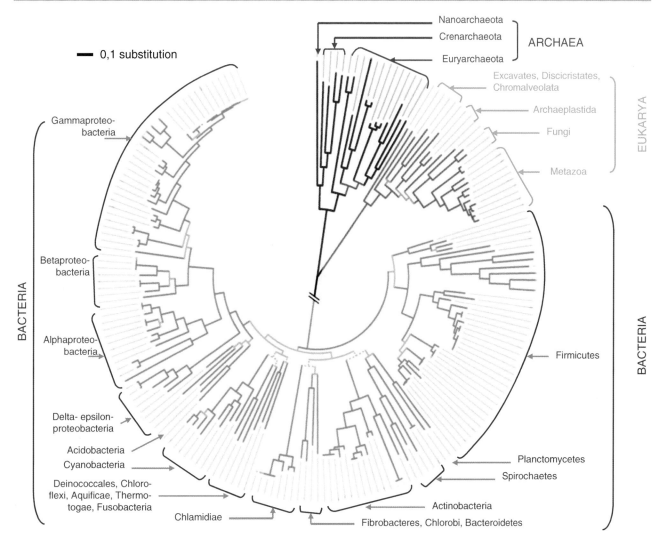

Fig. 5.24 The tree of life, based on 31 genes and nearly 200 completely sequenced genomes. The branch leading to Archaea and Eukarya has been shortened for graphical reasons. Although this tree based on sequenced genomes only gives a biased picture of the living world, it should be noted that the taxa that do not belong to microorganisms, that is to say, Metazoa and part of Archaeplastida, Chromalveolata, Fungi, and Amoebozoa, represent only a small part of the phylogenetic diversity of life (Modified and redrawn from Ciccarelli et al. 2006)

(i.e., "Where Are the Plants?" *cf*. Sect. 5.5.2 and "Where Are the Fungi, Algae, and Protozoa?"; *cf*. Sect. 5.5.3) are presented hereafter.

5.5.2 Where Are the Plants?

Intuitively, a plant is motionless while an animal is motile. Linnaeus, in the eighteenth century, has formalized these concepts by using the names of plant kingdom and animal kingdom. But in fact, things are much more complicated.

First of all, there are fixed, nonmobile animals (sponges, corals). In addition, many organisms that are placed more or less intuitively among plants can move from one place to another. Cyanobacteria crawl on the substrate, same for the diatoms *Navicula radiosa* and *Rhoicosphenia abbreviata* (Chromobionta), the latter species moving at a rate of 2 cm.h^{-1} with directional shifts (Bertrand 1991, 1992). *Spirogyra*, a filamentous Viridiplantae, propels itself into water at a speed of 3–5 cm.h^{-1} (Kim et al. 2005). Unicellulars equipped with undulipodia such as *Chlamydomonas* (Viridiplantae) and Dinobionta (Alveolata) are faster, reaching speeds of 72–180 cm.h^{-1}, which is considerable if one refers to their size (Raven and Richardson 1984).

Neither autotrophy nor the presence of chloroplasts constitutes shared characters among the living things traditionally known as plants. In all taxa where photosynthesis occurs, some taxa do not possess it (e.g., in Cyanobacteria and Dinobionta). In some taxa, this loss is secondary, but this is not the case of Fungi (modern meaning). Conversely, the case of non-photosynthetic organisms that practice

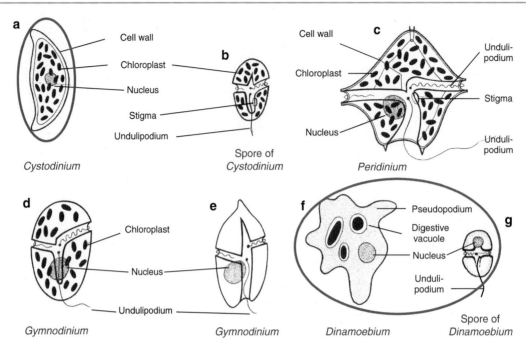

Fig. 5.25 Within a single taxon, Dinobionta (Alveolata), there is a continuum between features traditionally considered as characterizing "plants" (*top left*) and "animals" (*bottom right*). (**a**) *Cystodinium* exhibits chloroplasts, exhibits a cell wall, and is nonmotile and autotrophic; (**b**) however, the spores of *Cystodinium* are motile, thanks to two undulipodiums, whose movement is under the control of a photosensitive organelle, the stigma; (**c**) *Peridinium* also exhibits a cell wall and chloroplasts, but it is motile and engulfs preys (mixotrophy); (**d**) this species of *Gymnodinium* possesses chloroplasts and engulfs preys (mixotrophy) but lacks a cell wall; (**e**) this species of *Gymnodinium* lacks chloroplasts (heterotrophy); (**f**) *Dinamoebium* looks like an amoeba, lacks chloroplasts and cell wall, and engulfs preys (heterotrophy); (**g**) however, the spores of *Dinamoebium* exhibit the classical shape and undulipodiums of the Dinobionta (*cf.* Sect. 7.8.3) (Modified and redrawn from Chadefaud 1960; Gorenflot and Guern 1989; Boudouresque and Gómez 1995)

kleptoplasty is to be considered (*cf.* Sect. 5.4.4). It is not possible to review all the other characters, especially the cytological (e.g., cell wall) and biochemical (e.g., cellulose) ones that the tradition has associated with the notion of plant: none is discriminant enough. Overall, it is impossible to give an accurate definition to the word "plant."[14] Moreover, the continuum of characters that traditionally is attributed to a "typical" plant and to a "typical" animal can be observed within the same taxon, such as in Dinobionta (Fig. 5.25).

Knowledge of the acquisition mechanisms for photosynthesis (*cf.* Sects. 5.4.2, 5.4.3, and 5.4.4) allows us to understand why organisms classically classified as plants are in fact a polyphyletic assemblage, especially if we consider that, since Linnaeus, non-photosynthetic organisms, such as Fungi, are also included in it (Fig. 5.26). It should be noted that Fungi are closer to Metazoa (i.e., animals in the modern sense) in the kingdom of Opisthokonta, compared to Embryophyta in the kingdom of Archaeplastida, the latter organisms being the closest to the popular notion of plant (oak, daisy flower, and wheat are Embryophyta).

It should be noted that the kingdom of Archaeplastida is also named "Plantae," which can be confused with the popular notion of "plants" and therefore of "vegetal." Indeed, Archaeplastida (Plantae) could include both a portion of taxa traditionally considered as "plants" and a taxon (Centrohelida) traditionally considered as protozoa (Nicolaev et al. 2004), thus as an animal (*cf.* Sect. 5.5.3).

5.5.3 Where Are the Fungi, Algae, and Protozoa?

Organisms that since Linnaeus are called fungi have now proved to be polyphyletic (Fig. 5.27). Some of them, especially the "fungi with caps," which include mushrooms, milk caps, fly agaric, etc., are part of the Fungi (modern meaning) and are relatively close to Metazoa. Others, Oomycetes (here Oobionta) and Labyrinthulomycetes (here Labyrinthulobionta) are very close to Chromobionta ("kelp"), in the kingdom of Stramenopiles. The suffix "mycetes," which comes from the Greek *mykes*, means "fungi" in that language. It should be noted that long before the advent of molecular

---

[14] No single character can be considered characteristic of all organisms that tradition has gathered under the name of "plants." Similarly, no any combination of characters could define the set of "plants."

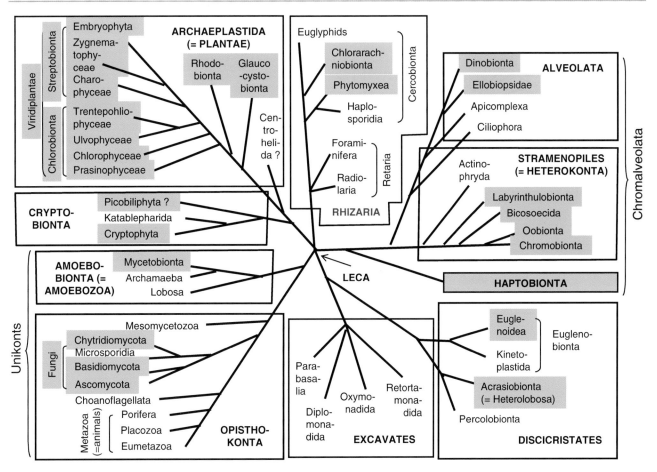

Fig. 5.26 Position, within the simplified tree of eukaryotes, of the taxa traditionally placed within the "plant kingdom" (*green boxes*). They represent a polyphyletic group, dispersed within 9 of the 10 kingdoms of eukaryotes. In addition, a number of prokaryotic taxa (Cyanobacteria, Actinobacteria, and Planctomycetes) were traditionally included within the "plant kingdom"

phylogeny, the great similarity between "brown algae" and "Oomycetes" had been established. Mereschkowsky (1910) had fully understood the consequences, in an earlier polyphyletic vision of algae and fungi that Cavalier-Smith (1981) recognized and reassessed in his kingdom of Chromista, which brought together Stramenopiles, Haptobionta, and Cryptobionta in this book. As in the case of the kingdom of Archaeplastida (*cf.* Sect. 5.5.2) often named Plantae, the adoption of the term Fungi in a restricted sense to designate only part of what was traditionally named "mushrooms" is a pity because it is a constant source of confusion.

Algae also represent a polyphyletic assemblage (Fig. 5.27). In the literature, the term has a somewhat confused sense, insofar as it refers to:

1. Sometimes photosynthetic eukaryotes, except embryophytes but with cyanobacteria (prokaryotes) photosynthetic or not
2. Sometimes photosynthetic eukaryotes, except embryophytes
3. Sometimes the unicellular eukaryotic taxa which are at least partly composed of photosynthetic species, as well as uni- or multicellular cyanobacteria
4. Sometimes unicellular photosynthetic eukaryotes and prokaryotes with oxygenic photosynthesis

The list is not exhaustive. In addition, the boundary between "algae" and "protozoa" is not very clear, a number of taxa being considered sometimes as algae, sometimes as protozoa (Fig. 5.28). In general, the reader must approximate the meaning of the terms "algae" depending on the context.

As plants and algae, protists are also a polyphyletic assemblage; however, the term is acceptable, as a morphologically defined assemblage, when reserved to the sole unicellular eukaryotes. But the term protozoa (Fig. 5.28) which refers to "animal" protists cannot be defined. It means (1) sometimes non-photosynthetic unicellular eukaryotes (whatever the taxon to which they belong), (2) sometimes unicellular eukaryotes belonging to predominantly non-photosynthetic taxa (whether or not they are in fact photosynthetic); and (3) sometimes all unicellular eukaryotes belonging to an unspecified, or arbitrary delineated, set of taxa. A number of these taxa are studied both by botanists who are specialists of "algae" and by zoologists who are specialists of "protozoa" (Fig. 5.28).

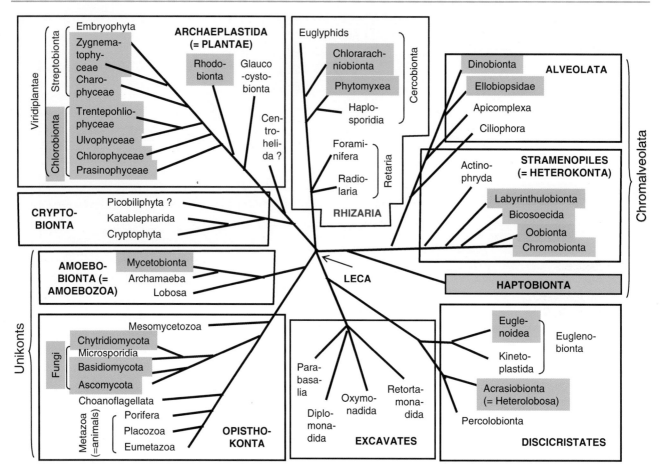

Fig. 5.27 Position, within the simplified tree of eukaryotes, of the taxa traditionally placed within the "algae" (*green boxes*) and the "fungi" (customary meaning) (*blue boxes*). They represent polyphyletic groups. In addition, a number of prokaryotic taxa were traditionally included within the "algae" and "fungi," *Cyanobacteria*, and *Actinobacteria* and *Planctomycetes*, respectively

This is particularly surprising in the case of the genus *Cafeteria* (Bicosoecids), classified by some authors in the algal class Bicosoecophyceae, while not only photosynthesis is absent, but it is perhaps the most abundant predator (number of individuals) of the planet (Baldauf 2008).

5.6 Conclusions and Perspectives

The quest for a universal, coherent, and robust phylogeny is an endeavor that has concerned man since the dawn of time, and the many myths of the origin have sought to fill our wish to answer the famous existential questions: "Where do we come from? Who are we? Where are we going?" Besides and in a more prosaic fashion, the constantly renewed success of the successive versions of the Bergey's Manual illustrates the general need for a solid, evolving, and consensual taxonomic framework.

Significant advances have been made in recent decades, mainly with the concept of biological sequences such as fruits and mirrors of evolution. The progress observed in recent years in sequencing techniques should continue with, for example, pyrosequencing, cheap genomes sequencing, and genomic counseling. The genomes of all major biological lineages will soon be available with the tools necessary to identify the evolutionary history of all genes in all lineages and thus trace the detailed history of the tree of life.

It will eventually be necessary to make the connection between the evolutionary history of all genes and taxa, on the one hand, and their forms and functions, on the other hand. This link will be complex to establish and will depend on expected progress in other areas, especially reverse genetics, mass spectrometry, and nuclear magnetic resonance to identify all metabolites of the cells in the functional exploration of genomes, in understanding the regulation of gene expression and the link between evolution and development.

DNA is a remarkably stable molecule, selected as long-term memory by the first forms of life. Nevertheless, it ages following, for instance, ionizing radiation and chemical attacks. It has been shown that DNA cannot be duplicated after a few hundred thousand years or else it accumulates too

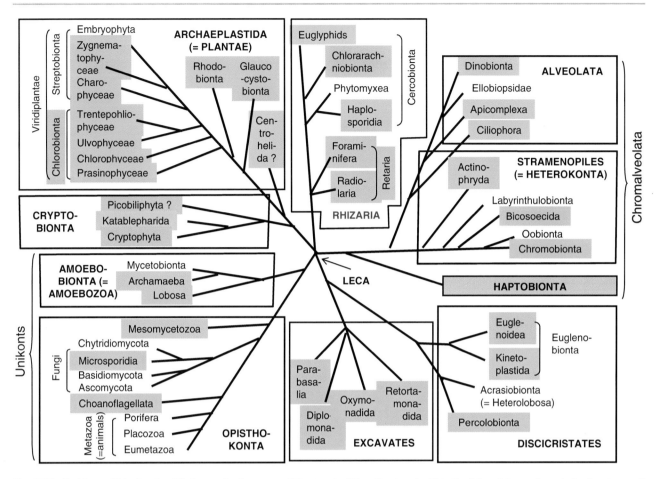

Fig. 5.28 Position, within the simplified tree of eukaryotes, of the taxa traditionally placed within the "algae" (*green boxes*), the "protozoans" (*ocher-colored boxes*), and claimed to be algae or protozoans by botanists and zoologists, respectively (*blue boxes*)

many artifactual mutations, so that fossils can hardly be studied by techniques such as PCR that involve polymerization. However, it is conceivable that the chemistry of ancient DNA progresses or else that it will become possible to determine the sequence of fossils other than through polymerization, making it possible to study the earliest fossils.

The extraterrestrial exploration of the moons of Saturn and Jupiter, as well as that of the subsurface of Mars, may also surprise us and may confirm the panspermia hypothesis developed by many researchers such as Anaxagoras, Benoit de Maillet, Hermann von Helmholtz, and closer to us Fred Hoyle and Chandra Wickramasinghe. Any microorganism found in these extraterrestrial environments will immediately be analyzed in every manner imaginable (and sequenced if of course it contains DNA) and positioned in the universal phylogeny. In addition, on our planet itself, it is likely that new lineages will be identified, particularly in extreme habitats that have been little explored (*cf.* Chap. 10) such as the deep earth. These lineage will complete the tree of living.

References

Andersson GE (2006) The bacterial world gets smaller. Science 314:259–260

Ané C, Burleigh JG, Mcmahon MM, Sanderson MJ (2005) Covarion structure in plastid genome evolution: a new statistical test. Mol Biol Evol 22(4):914–924

Arslan D et al (2011) Distant Mimivirus relative with a larger genome highlights the fundamental features of Megaviridae. Proc Natl Acad Sci U S A 108:17486–17491

Baldauf SL (2003) The deep roots of eucaryotes. Science 300:1703–1706

Baldauf SL (2008) An overview of the phylogeny and diversity of eukaryotes. J Syst Evol 46(3):263–273

Ball P (2007) Bacteria may be wiring up the soil. Nature 449:388

Berry AM, Harriott OT, Moreau RA, Osman SF, Benson DR, Jones AD (2003) Hopanoid lipids compose the *Frankia* vesicle envelope, presumptive barrier of oxygen diffusion to nitrogenase. Proc Natl Acad Sci U S A 90:6091–6094

Bertrand J (1991) Mouvement des diatomées. I – L'équilibre dynamique chez *Rhoicosphaenia abbreviata*. Cryptogamie Algol 12(1):11–29

Bertrand J (1992) Mouvement des diatomées. II – Synthèse des mouvements. Cryptogamie Algol 13(1):49–71

Bhattacharya D, Yoon HS, Hackett JD (2003) Photosynthetic eukaryotes unite: endosymbiosis connects the dots. Bioessays 25(1):50–60

Bodyl A (2005) Do plastid-related characters support the chromalveolate hypothesis? J Phycol 41:712–718

Bodyl A, Stiller JW, Mackiewicz P (2009) Chromalveolate plastids: direct descent or multiple endosymbioses? Trends Ecol Evol 24(3):119–121

Bonen L, Doolittle W (1975) On the prokaryotic nature of red algal chloroplasts. Proc Natl Acad Sci U S A 72:2310–2314

Bornens M, Azimzadeh J (2007) Origin and evolution of the centrosome. In: Jékeli G (ed) Eukaryotic membranes and cytoskeleton: origin and evolution. Land Bioscience, Austin (Tx), pp 119–129

Boudouresque CF, Gómez A (1995) Une approche moderne du monde végétal. Première Partie. GIS Posidonie Publishing, Marseille

Boullard B (1990) Guerre et paix dans le règne végétal. Edition Marketing, Paris, France

Boxma B, De Graaf RM, van der Staay GWM, van Alen TA, Ricard G, Gabaldon T, van Hoek AHAM, Moon-van der Staay SY, Koopman WJH, van Hellemond JJ, Tielens AGM, Friedrich T, Veenhuis M, Huynen MA, Hackstein JHP (2005) An anaerobic mitochondrion that produces hydrogen. Nature 434:74–79

Boyen C, Oudot MP, Loiseaux-De Goer S (2001) Origin and evolution of plastids and mitochondria: the phylogenetic diversity of algae. Cah Biol Mar 42:11–24

Brocks JJ, Logan GA, Buick R, Summons RE (1999) Archean molecular fossils and the early rise of eukaryotes. Science 285:1033–1036

Buick R (2010) Ancient acritarchs. Nature 463:885–886

Burki F, Okamoto N, Pombert JF, Keeling PJ (2012) The evolutionary history of haptophytes and cryptophytes: phylogenomic evidence for separate origins. Proc Royal Soc B 279:2246–2254

Butterfield NJ (2000) *Bangiomorpha pubescens* n.gen., n. sp.: implications for the evolution of sex, multicellularity, and the Mesoproterozoic/Neoproterozoic radiation of eukaryotes. Paleobiology 26(3):386–404

Carlton JM, Hirt RP, Silva JC, Delcher AL, Schatz M, Zhao Q, Wortman JR, Bidwell SL (2007) Draft genome sequence of the sexually transmitted pathogen *Trichomonas vaginalis*. Science 315:207–212

Carvalho-Santos Z, Azimzadeh J, Pereira-Leal JB, Bettencourt-Dias M (2011) Tracing the origin of centrioles, cilia and flagella. J Cell Biol 194(2):165–175

Cavalier-Smith T (1981) Eukaryote kingdoms: seven or nine? BioSystems 14:461–481

Cavalier-Smith T (2002a) The neomuran origin of archaebacteria, the negibacterial root of the universal tree and the bacterial megaclassification. Int J Syst Evol Microbiol 52:7–76

Cavalier-Smith T (2002b) The phagotrophic origin of eukaryotes and phylogenetic classification of protozoa. Int J Syst Evol Microbiol 52:295–354

Cérémonie H, Buret F, Simonet P, Vogel TM (2004) Isolation of lightning-competent soil bacteria. Appl Environ Microbiol 70:6342–6346

Chadefaud M (1960) Tome I: Les végétaux non vasculaires. Cryptogamie. In: Chadefaud M. et Emberger L (eds) Traité de botanique systématique. Masson et Cie Publishing, Paris, pp –i–xv + 1–1018

Chen M, Hiller RG, Howe CJ, Larkum AWD (2005) Unique origin and lateral transfer of prokaryotic chlorophyll-*b* and chlorophyll-*d* light harvesting systems. Mol Biol Evol 22(1):21–28

Ciccarelli FD, Doerks T, von Mering C, Creevey CJ, Snel B, Bork P (2006) Toward automatic reconstruction of a highly resolved tree of Life. Science 311:1283–1286

Claverie JM (2006) Viruses take center stage in cellular evolution. Genome Biol 7:110

Claverie JM, Abergel C (2009) Mimivirus and its virophage. Annu Rev Gen 43:49–66

Claverie JM, Abergel C (2010) Mimivirus: the emerging paradox of quasi-autonomous viruses. Trends Genet 26:431–437

Claverie JM, Otaga H, Audic S, Abergel C, Suhre K, Fournier PE (2006) Mimivirus and the emerging concept of "giant" virus. Virus Res 117:133–144

Combes C (1995) Les interactions durables. Ecologie et évolution du parasitisme. Masson Publishing, Paris

Combes C (2001) Les associations du vivant. L'art d'être parasite. Flammarion Publishing, Paris

Courties C, Vaquer A, Troussellier M, Lautier J, Chrétiennot-Dinet MJ, Neuveux J, Machado C, Claustre H (1994) Smallest eukaryotic organism. Nature 370:255

Curtis NE, Massey SE, Pierce SK (2006) The symbiotic chloroplasts in the Sacoglossan *Elysia clarki* are from several algal species. Invert Biol 125(4):336–345

Dagan T, Martin W (2009) Seeing green and red in diatom genomes. Science 324:1651–1652

Darwin C (1859) On the origin of species by means of natural selection, or the preservation of favoured races in the struggle for life. John Murray, London

Daubin V, Gouy M, Perriere G (2001) Bacterial molecular phylogeny using supertree approach. Genome Inform Ser Workshop Genome Inform 12:155–164

Di Giulio M (2003) The universal ancestor and the ancestor of bacteria were hyperthermophiles. J Mol Evol 57:721–730

Doolittle WF (1999) Phylogenetic classification and the universal tree. Science 284:2124–2128

Dyall SD, Brown MT, Johnson PJ (2004) Ancient invasions: from endosymbionts to organelles. Science 304:253–257

El Albani A, Bengston S, Canfield DE, Bekker A, Macchiarelli R, Mazurier A, Hammarlund EU, Boulvais P, Dupuy JJ, Fontaine C, Fürsich FT, Gauthier-Lafay F, Janvier P, Javaux E, Ossa Ossa F, Pierson-Wickmann AC, Riboulleau A, Sardini P, Vachard D, Whitehoute M, Meunier A (2010) Large colonial organisms with coordinated growth in oxygenated environments 2.1 Gyr ago. Nature 466:100–104

Embley TM, Martin W (2006) Eukaryotic evolution, changes and challenges. Nature 440:623–630

Falkowski PG, Katz ME, Knoll AH, Quigg A, Raven JA, Schofield O, Taylor FJR (2004) The evolution of modern eukaryotic phytoplankton. Science 305:354–360

Feldmann J, Feldmann G (1946) Recherches sur l'appareil conducteur des Floridées. Rev Cytol 8:159–209

Fischer MG et al (2010) Giant virus with a remarkable complement of genes infects marine zooplankton. Proc Natl Acad Sci U S A 107:19508–19513

Forterre P, Philippe H (1999) The last universal common ancestor (LUCA): simple or complex? Biol Bull 196:373–375; discussion: 375–377

Fox GE, Stackbrandt E, Hespell RB, Gibson J, Maniloff J, Dyer TA, Wolfe RS, Balch WE, Tanner RS, Magrum LJ, Zablen B, Blakemore R, Gupta R, Bonen L, Lewis BJ, Stahl DA, Luehrsen KR, Chen KN, Woese CR (1980) The phylogeny of prokaryotes. Science 209:457–463

Galtier N, Tourasse N, Gouy M (1999) A nonhyperthermophilic common ancestor to extant life forms. Science 283:220–221

Germot A, Philippe H, Le Guyader H (1997) Evidence for loss of mitochondria in Microsporidia from a mitochondrial-type HSP70 in *Nosema locustae*. Mol Biochem Parasit 87:159–168

Gorenflot R, Guern M (1989) Organisation et biologie des Thallophytes. Doin Publishing, Paris

Green BR (2005) Lateral gene transfer in the cyanobacteria: chlorophylls, proteins and scraps of ribosomal RNA. J Phycol 41:449–452

Haeckel E (1894) Systematische Phylogenie. Entwurf eines natürlichen Systems der Organismen auf Grund ihrer Stammesgeschichte.

Erster Theil, Systematische Phylogenie der Protisten und Pflanzen. Georg Reimer, Berlin/Allemagne

Hansen PJ, Fenchel T (2006) The bloom-forming ciliate *Mesodinium rubrum* harbours a single permanent endosymbiont. Mar Biol Res 2:169–177

Häring M, Vestergaard G, Rachel R, Chen L, Garrett RA, Prangishvili D (2005) Independent virus development outside a host. Nature 436:1101–1102

Hébant C (1977) The conducting tissues of bryophytes. Bryophytorum Bibl 10:1–157 + 80 pl. h.t

Hrdy I, Hirt RP, Dolezal P, Bardonová L, Foster PG, Tachezy J, Embley TM (2004) *Trichomonas* hydrogenosomes contain the NADH dehydrogenase module of mitochondrial complex I. Nature 432:618–622

Huber H, Hohn MJ, Rachel R, Fuchs T, Wimmer VC, Stetter KO (2002) A new phylum of Archaea represented by a nanosized hyperthermophilic symbiont. Nature 417:63–67

Jeon KW (1972) Development of cellular dependence on infective organisms: microsurgical studies in amoebas. Science 176:1122–1123

Jeon KW (1991) Amoeba and x-bacteria: symbiont acquisition and possible species change. In: Margulis L, et Fester R (eds) Symbiosis as a source of evolutionary innovation: speciation and morphogenesis. MIT Press, Cambridge, USA, pp 118–131

Jeon KW (1995) Bacterial endosymbiosis in amoebae. Trends Cell Biol 5:137–140

Jeon KW, Jeon MS (1976) Endosymbiosis in amoeba: recently established endosymbionts have become required cytoplasmic components. J Cell Physiol 89(2):337–344

Johnson MD, Oldach D, Delwiche CF, Stoecker DK (2007) Retention of transcriptionally active cryptophyte nuclei by the ciliate *Myrionecta rubra*. Nature 445:426–428

Kaneko T, Sato S, Kotani H, Tanaka A, Asamizu E, Nakamura Y, Miyajima N, Hirosawa M, Sasamoto S, Kimura T, Hosouchi T, Matsuno A, Muraki A, Nakazaki N, Naruo K, Okamura S, Shimpo S, Takeuchi C, Wada T, Watanabe A, Yamada M, Yasuda M, Tabata S (1996) Sequence analysis of genome of the unicellular cyanobacterium *Synechocystis* sp. strain PCC6803. II Sequence determination of the entire genome and assignment of potential protein-coding regions. DNA Res 3:109–116, 185–209

Kim GH, Yoon M, Klotchkova TA (2005) A moving mat: phototaxis in the filamentous green algae *Spirogyra* (Chlorophyta, Zygnemataceae). J Phycol 41:232–237

Kimura M (1968) Evolutionary rate at the molecular level. Nature 217:624–626

Köhler S, Delwiche CF, Denny PW, Tilney LG, Webster P, Wilson RJM, Palmer JD, Roos DS (1997) A plastid of probable green algal origin in Apicomplexan parasites. Science 275:1485–1489

Kozo-Polyanski BM (1924) A new principle of biology. Essay on the theory of symbiogenesis. Moscou, Russie (en russe)

Kumar S (2005) Molecular clocks: four decades of evolution. Nat Rev Genet 6(8):654–662

La Scola B, Audic S, Robert C, Jungang L, de Lamballerie X, Drancourt M, Birtles R, Claverie JM, Raoult D (2003) A giant virus in amoebae. Science 299:2033

Lamarck JB (1809) Philosophie zoologique. Dentu, Paris

Lawrence JG, Ochman H (1998) Molecular archaeology of the *Escherichia coli* genome. Proc Natl Acad Sci U S A 95(16):9413–9417

Leander BS, Saldarriaga JF, Keeling PJ (2002) Surface morphology of the marine parasite *Haplozoon axiothellae* Siebert (Dinoflagellata). Eur J Protistol 38:287–297

Lecointre G, Le Guyader H (2001) Classification phylogénétique du vivant. Belin Publishing, Paris

Lecointre G, Le Guyader H (2006) Classification phylogénétique du vivant, 3rd edn. Belin Publishing, Paris

Lee RE, Kugrens P (2000) Ancient atmospheric CO_2 and the timing of evolution of secondary endosymbioses. Phycologia 39(2):167–172

Lewin RA (1975) A marine *Synechocystis* (Cyanophyta, Chlorococcales) epizoic on ascidians. Phycologia 14:153–160

Lewin RA (1976) Prochlorophyta as a proposed new division of algae. Nature 261:687–698

Lewin RA, Withers N (1975) Extraordinary pigment composition of a prokaryotic alga. Nature 256:735–737

Linnaeus C (1753) Species plantarum, exhibentes plantas rite cognitas, ad genera relatas, cum differentiis specificis, nominibus trivialibus, synonymis selectis, locis natalibus, secundum systema sexuale digestas. Holmiae, Impensis Laurentii Salvii, Stockholm

López-García P, Rodríguez-Valera F, Pedrós-Alio C, Moreira D (2001) Unexpected diversity of small eukaryotes in deep-sea Antarctic plankton. Nature 409:603–607

Margulis L (1970) Origin of eukaryotic cells. Yale University Press, New Haven/London

Margulis L (1980) Undulipodium, flagella and cilia. BioSystems 12(1–2):105–108

Margulis L (1981) Symbiosis in cell evolution. W.H. Freeman, San Francisco

Margulis L, Dolan MF, Guerrero R (2000) The chimeric eukaryote: origin of the nucleus from the karyomastigont in amitochondriate protists. Proc Natl Acad Sci U S A 97:6954–6959

Martin W, Rujan T, Richly E, Hansen A, Cornelsen S, Lins T, Leister D, Stoebe B, Hasegawa M, Penny D (2002) Evolutionary analysis of *Arabidopsis*, cyanobacterial, and chloroplast genomes reveals plastid phylogeny and thousands of cyanobacterial genes in the nucleus. Proc Natl Acad Sci U S A 99:12246–12251

McCutcheon JP, Moran NA (2011) Extreme genome reduction in symbiotic bacteria. Nat Rev Microbiol, 14 pp

McFadden GI (2001) Primary and secondary endosymbiosis and the origin of plastids. J Phycol 37:951–959

McFadden GI, Waller RF (1997) Plastids in parasites of humans. Bioessays 19(11):1033–1040

Medlin LK, Kaczmarska I (2004) Evolution of the diatoms: V. Morphological and cytological support for the major clades and a taxonomic revision. Phycologia 43(3):245–270

Mereschkowsky C (1905) ÜberNatur und Ursprung der Chromatophoren im Pflanzenreiche. Biol Zentralbl 25:593–604

Mereschkowsky C (1910) Theorie der zwei Plasmaarten als Grundlage der Symbiogenesis, einer neuen Lehre von der Entstehung der Organismen. Biol Zentralbl 30:278–367

Meyerowitz EM (2002) Plants compared to animals: the broadest comparative study of development. Science 295:1482–1485

Monier A, Claverie JM, Ogata H (2008) Taxonomic distribution of large DNA viruses in the sea. Genome Biol 9:R106

Moon-van der Staay SY, De Wachter R, Vaulot D (2001) Oceanic 18S rDNA sequences from picoplankton reveal unsuspected eukaryotic diversity. Nature 409:607–610

Moustafa A, Beszteri B, Maier UW, Bowler C, Valentin K, Bhattacharya D (2009) Genomic footprints of a cryptic plastid endosymbiosis in Diatoms. Science 324:1724–1726

Nakabachi A, Yamashita A, Toh H, Ishikawa H, Dunbar HE, Moran NA, Hattori M (2006) The 160-kilobase genome of the bacterial endosymbiont *Carsonella*. Science 314:267

Nakayama T, Ishida K (2005) Another primary endosymbiosis? Origin of *Paulinella chromatophora* cyanelles. Phycologia 44(4):74

Nicolaev SI, Berney C, Fahrni JF, Bolivar I, Polet S, Mylnikov AP, Aleshin VV, Petrov NB, Pawlowski J (2004) The twilight of Heliozoa and rise of Rhizaria, an emerging supergroup of amoeboid eukaryotes. Proc Natl Acad Sci U S A 101(21):8066–8071

Nishikawa M, Suzuki K, Yoshida K (1992) DNA integration into recipient yeast chromosomes by trans-kingdom conjugation between *Escherichia coli* and *Saccharomyces cerevisiae*. Curr Genet 21:101–108

Oltmanns F (1904) Morphologie und Biologie der Algen. Erster Band. Gustav Fischer, Jena

Pennisi E (2006) Plant wannabes. Science 313:1229

Polzin KM, McKay LL (1991) Identification, DNA sequence, and distribution of IS981, a new, high-copy-number insertion sequence in lactococci. Appl Environ Microbiol 57:734–743

Rahat M, Ben Ishac-Monselise E (1979) Photobiology of the chloroplast hosting mollusc *Elysia timida* (Opisthobranchia). J Exp Mar Biol Ecol 79:225–233

Raoult D, Audio S, Robert C, Abergel C, Ernesto P, Ogata H, La Scola B, Suzan M, Claverie JM (2004) The 1.2-megabase genome sequence of Mimivirus. Science 306:1344–1350

Raven PH (1970) A multiple origin for plastids and mitochondria. Science 169:641–646

Raven JA, Richardson K (1984) Dinophyte flagella: a cost-benefit analysis. New Phytol 98:259–276

Raven JA, Waite AM (2004) The evolution of silicification in diatoms: inescapable sinking and sinking as escape? New Phytol 162:45–61

Reith M, Munholland J (1993) A high-resolution gene map of the chloroplast genome of the red alga *Porphyra purpurea*. Plant Cell 5:465–475

Rivera MC, Lake JA (2004) The ring of life provides evidence for a genome fusion origin of eukaryotes. Nature 431:152–155

Rumpho ME, Summer EJ, Manhart JR (2000) Solar-powered sea slugs. Mollusc/algal chloroplast symbiosis. Plant Physiol 123:29–38

Rumpho ME, Worful JM, Lee J, Kannan K, Tyler MS, Bhattacharya D, Moustafa A, Manhart JR (2008) Horizontal gene transfer of the algal nuclear gene *psbO* to the photosynthetic sea slug *Elysia chlorotica*. Proc Natl Acad Sci U S A 105(46):17867–17871

Santini S, Jeudy S, Bartoli J, Poirot O, Lescot M, Abergel C, Barbe V, Wommack KE, Noordeloos AAM, Brussaard CPD, Claverie JM (2013) The genome of *Phaeocystis globosa* virus PgV-16T highlights the common ancestry of the largest known DNA viruses infecting eukaryotes. Proc Natl Acad Sci U S A 110(26):10800–10805

Schulz HN, Brinkhoff T, Ferdelman TG, Hernández Mariné M, Teske A, Jørgensen BB (1999) Dense populations of a giant sulfur bacterium in Namibian shelf sediments. Science 284:493–495

Selga T, Selga M, Gobiņš V, Ozoliņa A (2010) Plastid-nuclear complexes: permanent structures in photosynthesizing tissues of vascular plants. Environ Exp Biol 8:85–92

Smith M, Hansen J (2007) Interaction between *Mesodinium rubrum* and its prey: importance of prey concentration, irradiance and pH. Mar Ecol Prog Ser 338:61–70

Stacey G, Bottomley PJ, Van Baalen C, Tabita FR (1979) Control of heterocyst and nitrogenase synthesis in cyanobacteria. J Bacteriol 137:321–326

Stanier RY (1974) Division I, the *Cyanobacteria*. In: Buchanan RE, et Gibbons NE (eds) Bergey's manual of determinative bacteriology. Williams & Wilkins Co, Baltimore

Stiller JW, Reel DC, Johnson JC (2003) A single origin of plastids revisited: convergent evolution in organellar genome content. J Phycol 39:95–105

Turner S, Pryer KM, Miao VP, Palmer JD (1999) Investigating deep phylogenetic relationships among cyanobacteria and plastids by small subunit rRNA sequence analysis. J Eukaryot Microbiol 46:327–338

Tyler BM, Tripathy S, Zhang X, Dehal P, Jiang RHY, Aerts A, Arredondo FD, Baxter L, Bensasson D, Beynon JL et al (2006) *Phytophthora* genome sequences uncover evolutionary origins and mechanisms of pathogenesis. Science 313:1261–1266

Vaulot D, Eikrem W, Viprey M, Moreau H (2008) The diversity of small eukaryotic phytoplankton (<or =3 micron) in marine Ecosystems. FEMS Microbiol Rev 32:795–820

Waggoner B (2001) Eukaryotes and multicells: origin. In: Encyclopedia of life sciences. Macmillan Publishing Ltd, Basingstoke, pp 1–9

Williams SI, Walker DI (1999) Mesoherbivore-macroalgal interactions: feeding ecology of sacoglossan sea slugs (Mollusca, Opisthobranchia) and their effects on their food algae. Oceanogr Mar Biol Annu Rev 37:87–128

Woese C (1977) Endosymbionts and mitochondrial origins. J Mol Evol 10:39–96

Woese CR, Olsen GJ (1986) Archaebacterial phylogeny: perspectives on the urkingdoms. Syst Appl Microbiol 7:161–177

Worden AZ, Nolan JK, Palenik B (2004) Assessing the dynamics and ecology of marine picophytoplankton: the importance of the eukaryotic component. Limnol Oceanogr 49:168–179

Yoosuf N et al (2012) Related giant viruses in distant locations and different habitats: Acanthamoeba polyphaga Moumouvirus represents a third lineage of the Mimiviridae that is close to the megavirus lineage. Gen Biol Evol 4:1324–1330

Zuckerkandl E, Pauling L (1965) Evolutionary divergence and convergence in proteins. In: Bryson V, et Vogel HJ (eds) Evolving genes and proteins. Academic, New York, pp 97–166

Taxonomy and Phylogeny of Prokaryotes

Pierre Caumette, Céline Brochier-Armanet, and Philippe Normand

Abstract

Classification of prokaryotes is hierarchically organized into seven levels: kingdoms, phyla, classes, orders, families, genera, and species. In prokaryotes, because they reproduce by clonal fission, the species, considered as the basic unit of the biological diversity, faces several problems such as the definition of an individual. A bacterial strain can be recognized as an individual belonging to a species. However, many inconsistencies exist between phenotypic similarity levels and evolutionary relationships deduced from molecular phylogenies. Most taxonomic groups have been reconsidered through phylogenetic analysis in the 1980s, and a consensus has been reached on the need for coherence between taxonomy and phylogeny. Thus, the multiple revisions of species, genera, or higher taxonomic levels pose many complex problems that are solved gradually. Prokaryotic microorganisms correspond to two of the three domains of life: *Archaea* and *Bacteria*. Their systematics is described in the "Bergey's Manual for Systematic Bacteriology, second edition" published in five volumes.

In the text, the Latin terms used are those accepted by the Nomenclature Committee, and the organization of the bacterial and archaeal domains is presented as they appear in the "Bergey's Manual for Systematic Bacteriology." They are discussed according to the recent data of the hierarchical classification of Prokaryotes.

Keywords

16S RNA homology • *Archaea* • *Bacteria* • Bacterial taxonomy • Dendrogram • DNA/DNA hybridization • Domains • G + C% • Genotypic criteria • Phenotypic criteria • Phylogenetic tree • Phyla • Systematics of prokaryotes

---

* Chapter Coordinator

P. Caumette* (✉)
Institut des Sciences Analytiques et de Physico-chimie pour l'Environnement et les Matériaux (IPREM), UMR CNRS 5254, Université de Pau et des Pays de l'Adour, B.P. 1155, 64013 Pau Cedex, France
e-mail: pierre.caumette@univ-pau.fr

C. Brochier-Armanet
Laboratoire de Biométrie et Biologie Évolutive, UMR CNRS 5558, Université Claude Bernard Lyon 1, 69622 Villeurbanne Cedex, France
e-mail: Celine.brochier-armanet@univ-lyon1.fr

P. Normand
Microbial Ecology Center, UMR CNRS 5557 / USC INRA 1364, Université Lyon 1, 69622 Villeurbanne, France
e-mail: philippe.normand@univ-lyon1.fr

6.1 The Concept of Species in Prokaryotes and Its Evolution

Classification of prokaryotes has long been based on the same hierarchical systems on which Linnaeus (Linnaeus 1753) based his nomenclatural system applied initially to plants and animals. These classifications are hierarchically organized into seven levels: **kingdoms*, phyla*, classes*, orders*, families*, genera*, and species***. In the kingdoms corresponding to the animal and the plant worlds, a robust classification has been established. Its success depends in part on the definition of a species that has been proposed for eukaryotes: a species is a group of individuals who have the ability to reproduce and yield fecund offspring and share many functional and morphological characteristics. This definition of species based on the "interfertility" criterion, which aims to be universal, is not applicable to prokaryotes because they reproduce by **clonal reproduction***. However, the species is considered as the basic unit of biological diversity, and its definition should be unambiguous and strong. In relation to this, the systematic of prokaryotes faces several problems such as the definition of an **individual*** and the definition of a species without being able to rely on the criterion of interfertility.

For years, the species concept in prokaryotes was based on morphological and physiological criteria (morphologic and physiologic are very seldom used) that have proven not to be really effective. Today molecular biology provides new tools to strengthen the concept of species replacing interfertility criteria and proposing quantitative criteria (Box 6.1). Thus, strains of prokaryotes for which hybridization of genomic DNA considered pairwise reaches 70 % are considered as belonging to the same species (Wayne et al. 1987). Other taxonomic levels are left to the discretion of the researchers and the consideration of a set of rules (priority, uniqueness, consistency, etc.). However, it is important to remember that the goal of this approach is to provide a solid framework in which the defined **taxa***, must be, inasmuch as possible, "natural" and "consistent" (i.e., reflecting relationships between microorganisms), which is not always possible. Also, Stackebrandt and Goebel (1994) have suggested an equivalence between the classical definition of a species based on the labor-intensive and delicate **genomic hybridization*** technique and an identification technique most commonly used in recent years. This approach is based on the comparison of genes encoding RNA of the small subunit of the ribosome (16S rRNA). By this method, which is being reevaluated (Stackebrandt and Ebers 2006), two strains are considered as not belonging to the same species if their 16S rRNA sequences have a similarity below a threshold currently set at 97 %.

Box 6.1: The Species Concept Paradox

Xavier Nesme

Species and speciation are constant topics of discussion in biology. In prokaryotes, this issue is particularly crucial at the time of metagenomics, which generates large volumes of nucleotide sequences from a variety of strains, species, genera, etc., that should be classified and ranked automatically with maximum biological sense. However, the bacterial species is paradoxically both well definable and lacking a consensus concept that would highlight the causes and biological consequences that this definition covers.

Bacterial Species Versus Eukaryotic Species

The bacterial species is defined simply on a technical basis by measuring the reassociation rate ("relative binding ratio" or RBR) of the genomic DNA of pairs of bacterial strains. For its part, the species in eukaryotes has benefited from profound reflections that led to the biological species concept ("biological species concept" or BSC), which highlights the role of the sexual isolation of species with, as a result, the containment of gene flow and therefore of genetic innovations to a given species.

The genomic definition of bacterial species (Wayne et al. 1987) is as follows:

"The bacterial species includes strains with both a relative DNA reassociation % superior to 70 % and 5 °C or less ΔTm; the two must be considered."

The biological species concept (BSC) in eukaryotes (Mayr 1942) is as follows:

"Species are groups of actually or potentially interbreeding populations, which are reproductively isolated from other such groups."

A simply operational definition of the species in bacteria corresponds to a biological species concept in eukaryotes: Are they the same thing? Can a biological concept be associated to the definition of bacterial species?

The genomic basis of the definition of a bacterial species was confirmed in 2002 by the International Committee of the definition of bacterial species, because since its publication in 1987, this definition proved operational in most lineages of prokaryotes (Stackebrandt et al. 2002). This definition, based on a rigorous physical measurement, has greatly reduced taxonomists' strifes about what should or not be grouped in the same species. This is unfortunately not the case for other taxonomic levels, especially the genus that do not currently have a definition as consensual, with as consequences technical and regulatory debates where purely scientific arguments are sometimes rare to find.

However, the same committee requested that alternative methods to RBR be proposed to define the

(continued)

X. Nesme
Laboratoire d'écologie microbienne, UMR 5557 Université Claude Bernard Lyon 1, Villeurbanne Cedex, France

Box 6.1 (continued)

"genomic" species. In addition, the committee strongly encourages research to find a biological basis for the bacterial species.

An Empirical but Not Arbitrary Definition

The canonical aspect of the 70 % RBR value is disturbing. In today's world of biology where Darwinian-derived concepts constitute a paradigm, the genomics definition of bacterial species has undeniably creationist's undertones. It seems that the RBR allows to capture the "essence" of the species and therefore suggests that such an "essence" exists. This Aristotelian view is reinforced by the declared justification of the polyphasic approach, in which different methodologies are used in conjunction with RBR to associate to it various phenotypic markers. This approach would give more highlights to this "essence" and yield more "shadows" on the wall of "Plato's cave."

In addition, setting a threshold at 70 % on RBR may seem arbitrary. In fact, this threshold was empirically obtained by comparing a large amount of RBR values. What Grimont (Grimont 1988) and others have shown is that there was no continuum in the distribution of RBR values between 100 and 0 %, but that there was a discontinuity around the value of 70 %. In addition, pairs of strains with RBR values above that threshold (Box Fig. 6.1) had all been previously classified in the same "species" on the basis of their biological similarities, using methods of numerical taxonomy applied to a large number of morpho-biochemical characters.

Finally, such a gap in the distribution of RBR obtained initially with Enterobacteriaceae has also been found with similar values in many if not in all bacterial taxa. This explains the operational success of this definition. One must recall, however, that the 70 % value is not an absolute one, but that it can and should be adjusted according to taxa. By virtue of its robustness, it is likely that this definition will persist because it guarantees the stability of nomenclature. This is however a regulatory argument that may not be totally "scientific." Therefore, the methodology for determining genomic species should be simplified to facilitate its implementation and understand the realities covered by such a definition.

Alternative Methodologies to RBR

Determining the RBR is quite tedious and requires equipment that is not available in all laboratories.

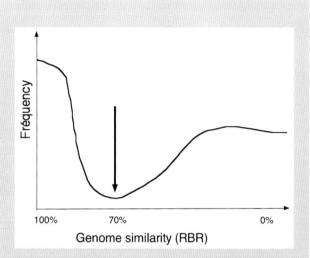

Box Fig. 6.1 Frequency distribution of reassociations measures between bacterial DNAs considered pairwise (RBR). There is a clear break in the distribution of values around 70% (*arrow*). This break may be observed with other methods for measuring the similarity or dissimilarity, with genetic or genomic sequences at 5°C for ΔTm, at 4–5 % of nucleotide divergence in MLSA, at 11 % of average genomic divergence (CGM) in AFLP, and at 93–94 nucleotide identity (ANI) between genomes. The left part of the curve corresponds to measurements involving pairs of strains belonging to the same species as they can be identified by morpho-biochemical criteria (Modified and redrawn from Grimont 1988)

Moreover, the method requires the availability of genomic DNA in large amounts for each strain analyzed and yields values for only pairs of strains. The method is not easily "portable" (that is to say used easily and without adaptation) from one laboratory to another and hardly lends itself to the analysis of many isolates as it is necessary in population genetics. Validated alternative methodologies are based on the molecular analysis of individual genomes. We can mention in that respect the amplified fragment length polymorphism or AFLP methodology, which proceeds by random sampling of different portions of the genome. AFLP allows to estimate the average mismatch rate or current genome mispairing (CGM). As it is perfectly correlated with RBR, it helps delineate genomic species with correspondence between the 11 % CGM threshold and the 70 % RBR threshold (Mougel et al. 2002). This method allows the phylogenomic analysis of numerous isolates and thus lends itself to the study of populations. It is now challenged by phylogenetic analysis of several housekeeping genes known as the multi-locus sequence analysis or MLSA. MLSA, which is completely portable, appears as the method of choice to determine the bacterial

(continued)

Box 6.1 (continued)

species. A value of 70 % RBR roughly corresponds to nucleotide mismatch rates on the order of 5 % in housekeeping genes. This is exactly what the ΔTm measures, with a drop of 1 °C per % genomics mismatch.

However, it is the significance of phylogenetic groupings measured, for instance, by the iterative "bootstrap" resampling method – and not an absolute threshold value – that must be used to delimit species in MLSA (Gevers et al. 2005). Other criteria, such as the average nucleotide identity or ANI (Goris et al. 2007) or other methods also based on the alignments of sequenced genomes, reveal not only discontinuities between species but also between genera. Because of easy access to complete genome sequencing, it can be expected that in the future, depositing the genomic reference sequence will be a prerequisite to the definition of new bacterial species.

As shown in Box Fig. 6.1, all these methods show that there is a discontinuity in the distribution of divergence measures (AFLP, MLSA, ΔTm) – or symmetrically of the similarity (RBR, ANI) – between genomes. This determination is empirical. What may be arbitrary is the choice of the degree of genomic difference to define entities called "species" about which one may wonder how they compare or not to eukaryotic species.

Speciation of Genomic Species Is "Fixed" in the Past

It must be understood that this definition implies that genomic species are bacterial lineages that have differentiated and have been isolated long enough that their genomic differences have led to the thresholds described above. An immediate consequence of this is that the events that led to this divergence have occurred long ago: speciation of genomic species was fixed in the past. In addition, since the infraspecific divergence is of the same order of magnitude for all species, speciation would have occurred roughly at the same time in various lineages! However, the phenomenon of speciation occurred and still occurs now, as is the case for many bacterial pathogens such as *Yersinia pestis*, *Mycobacterium tuberculosis*, or *Bacillus anthracis* that belong to wider "genomic" species. Since the genomic bacterial species definition does not integrate contemporary speciation, it is therefore irreconcilable with the BSC of eukaryotes.

Homologous Recombination and Sexual Isolation of Genomic Species

In eukaryotes, BSC is based on the sexual isolation of species. In prokaryotes, "sexuality" is both more promiscuous because it can involve very distantly related taxa (e.g., Firmicutes vs. Proteobacteria) and, very partial, because it concerns only a fraction of the genome. However, it has been suggested that the genetic divergence between species is such that it could result in a significant decrease in the frequency of homologous recombination (i.e., the mechanism by which foreign DNA is integrated into the bacterial genome) and thus lead to relative sexual isolation of species. Attempts have been made to explore this idea because it was attractive to be able to reconcile the two concepts. Briefly, it appears that in a model organism (the *Agrobacterium tumefaciens* species complex) and for a given marker gene, the decrease in the rate of homologous recombination was from 8 times, between very distant strains belonging to the same species, to 9 times, between strains belonging to different but closely related species (Costechareyre et al. 2009). This difference does not appear very significant, and homologous recombination is probably not sufficient to explain the genetic isolation of bacterial species. A sexual isolation, however, can have other causes. Studies of population genetics, for example, the work of Bailly et al. (2006), showed that sympatric species of *Sinorhizobium* are sufficiently isolated that gene exchanges occur significantly more between members of the same species than between species. The nature of the barrier, physiological, geographical, or otherwise, which leads to the sexual isolation of these species, is not known.

The Cause of Discontinuities Between Genomic Species

The existence of genomic groups is the result of forces that have swept or "purged" the diversity among these groups. Two types of models are available for these forces. One is based on selection. The other involves genetic drift with a predominant role of the founder effect.

Genomic Species Versus Ecological Species

For Cohan (2001), in the world of prokaryotes, each genomic group could correspond to an "ecotype" defined as a population of cells occupying the same

(continued)

Box 6.1 (continued)

ecological niche that are in intense competition with any adaptive mutant from this population. Well defined, the ecotypes share many properties attributed to eukaryotic species: the genetic diversity within ecotypes is limited by a cohesive force (here, the periodic selection), and the ecotypes are ecologically distinct. Therefore, ecotypes can be discovered and classified as genomic groups even when one remains ignorant of their ecology. This model is attractive because it is now possible to find the specific genome of a bacterial species and to then infer its specific ecological functions. This was done by Lassalle and collaborators (2011) who combined comparative genomics and reverse ecology to unmask the species-specific genes and then the species-specific ecological traits that differentiate *Agrobacterium fabrum* from its sister species within the *A. tumefaciens* complex. It is a great challenge for comparative genomic programs to try to find these features in the variable part of the genome. Conversely, the conserved genome part (the "core genome") has likely little effect on these specificities even if it bears the phylogenetic signatures used to characterize the species by MLSA. However, again for reasons of stability of the taxonomy, it may seem dangerous to raise each prokaryotic group, even a single clone, with proven ecological specialization and thus an "ecotype," to the status of a good and valid species. In connection with this model, the idea circulates that there may be a nesting of ecological specialization levels corresponding to nested taxonomic levels. Thus, genomic species could be ecological species even if ecotypes can also differentiate in these species (Box Fig. 6.2).

Metapopulations, Founder Effect, and Genetic Drift

The force that holds together the genomic cohesion of bacterial populations does not necessarily result from episodes of intense selection (Fraser et al. 2009). In models where populations go through bottlenecks that significantly reduce their effective size, the diversity is affected through genetic drift. In the island metapopulation model ("islands" in the sense of exploitable resources), islands of various sizes can be colonized by single founder genotypes coming at random from other islands. Strains may differentiate in an island, and one of these new genotypes can colonize at random another island. If some islands

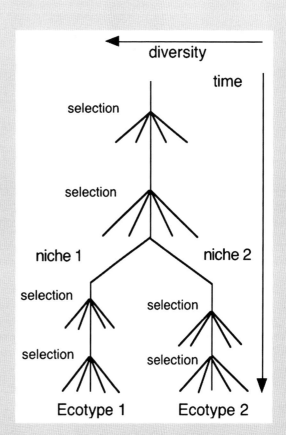

Box Fig. 6.2 Stable ecotype concept. One bacterial strain differentiates into two sublines which differ in certain aspects of their ecology. Selection periodically sweeps virtually all diversity occurred since the last episode of selection. As the two populations are ecologically distinct – that is, ecotypes – the periodic selection in a line does not influence the diversity of the other line. Ecotypes can then diverge to form separate species. In this model, the transfer of genes carrying genetic innovations plays an important role in adaptation to new ecological niches (Modified and redrawn from Cohan 2001)

become unable to support colonization, they lose their inhabitants and there follows a purge of diversity. This leads to a population differentiation between islands without selection of genotypes but just resulting from successive colonizations. Such a model could very well apply to microorganisms in the soil that undergo intense explosion of populations – for example, in contact with roots – followed by drastic reduction in disconnected soil microhabitats. It remains to be seen whether the model applies to all taxonomic levels such as from strains to species.

(continued)

> **Box 6.1** (continued)
> **Conclusion**
> The "genomic" definition of Eubacteria or Archaea species is efficient and confers stability to the taxonomic nomenclature. This is, however, a fixed version of the species concept that does not fit with the contemporary evolutionary concepts of speciation. Nevertheless, it may well be that it is within the "genomic species" that the majority of genetic exchanges would occur and thus that ecological innovations would be shared. Genomic species could well be "ecological species" adapted to specific ecological niches, at least at the time of speciation. However, it is also possible that genetic drift has played a major role in the individualization of genomic species with the result that their differentiations are purely contingent to the vagaries of the history of each species. The advantage of prokaryotes is that these alternatives are testable via comparative analysis of their genomes. In practice, it is possible to find the genes and functions that determine the specific niche of each species.

In addition to its use for the definition of species in prokaryotes, the 16S rDNA has also established itself as a reference via molecular phylogeny for the delimitation of higher taxonomic levels. However, the **gene*** for 16S ribosomal RNA is not the "yardstick" of bacterial taxonomy. It is sometimes too conserved to be able to distinguish between some close species such as in the case of the *Bacillus cereus* **species complex***. In addition, the **genomes*** of prokaryotes can contain multiple copies of this gene, reaching 10 (*Bacillus subtilis*) or even 15 copies (Rainey et al. 1996).

While in most cases, within a genome, the multiple copies of genes coding for 16S rRNA are identical or very similar, due to gene conversion (i.e., homologous recombination events leading to a homogenization of copies of a same gene in a genome), there are also taxa which are found to have markedly different copies. For example, in *Escherichia coli*, one of the seven copies present in the genome has up to 1 % difference (15 different bases/1,500 nt; Cilia et al. 1996), or in *Thermomonospora*, one of the six copies present in the genome has 10 % difference (Yap et al. 1999), showing thus the difficulty for phylogenetic studies depending on the selected 16S rRNA gene copy.

Finally, the use of a single **marker***, for the identification of new microorganisms or to study their position in a phylogeny, can lead to errors when there are gene transfers (*cf.* Sect. 12.2; Daubin et al. 2001) or when mutations in different lineages are convergent. This is why other genes are frequently used in addition to 16S rRNA. For example, for the identification of bacterial strains, additional markers are considered as the 23S ribosomal RNA gene, which is longer and more variable than 16S, allowing comparisons at finer scale. But a number of genes also known as **housekeeping gene**s* like *gyr*B, *rpo*B, etc., are more and more used through approaches called **multi-locus sequence typing** or **MLST***.

Despite the problems mentioned above, the use of the 16S rRNA gene has many advantages, the first of these being related to its abundance in public sequence databases. For example, the Ribosomal Database Project II entirely dedicated to rRNA contains 2,765,278 16S rRNA sequences aligned and annotated (release 10, update 32, May 14, 2013, Fig. 6.1). This abundance of 16S rRNA sequences is due to the fact that this molecule was used early as a reference marker. Indeed, the 16S rRNA genes have a number of advantages such as (1) ubiquity, that is, the presence in all living beings without exception; (2) the stability of the function of the gene product; (3) a low rate of mutation (allowing comparison across the living world) and making possible to design primers called universal, which allow amplification of almost all-known rRNA genes[1]; (4) a sufficient length; and (5) a low frequency of horizontal transfer.

6.2 Obtaining a Prokaryotic Strain: Strains Collection

Classical microbiology is based on obtaining prokaryotes **strains*** or isolates, that is to say on prokaryote cultures that may be kept for years, exchanged between laboratories, and compared with other isolates. However, isolates evolve over time so that subcultures can sometimes undergo major changes such as loss of plasmids, genomic recombination events, invasion by insertion elements (Polzin and McKay 1991), or have had point mutations with major phenotypic consequences, etc. It is therefore important to define the approach that will permit conservation of isolates as stable as possible. After being characterized, an isolate can become a reference strain which must be deposited in a reference **collection***, for example, ATCC in the USA (www.atcc.org), the DSMZ in Germany (www.dsmz.de), the NCIMB in England (http://www.ncimb.com/), the JCM in Japan (http://www.jcm.riken.go.jp/), or the collection of the Institute Pasteur in France (www.crbip.pasteur.fr). The mission of these reference collections is to maintain the different isolates and make them accessible to the entire international community of researchers in microbiology. Microorganisms are maintained as pure strains coded and referenced, maintained under freezing conditions (at −80 °C or in liquid nitrogen), or freeze-dried.

---

[1] This is an iterative process linked to the ongoing discovery of new taxa in new environments or better explored. In 2002, for instance, a bacteria was described in which 16S rRNA gene did not hybridize with the "universal" primers commonly used.

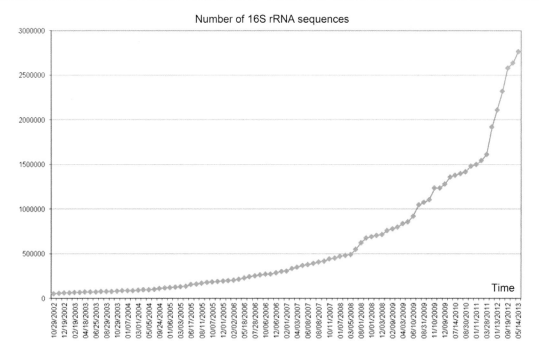

Fig. 6.1 Increase in the number of available 16S rRNA sequences. Graph showing the increasing number of 16S rRNA sequences in the databases, over the past 20 years (data extracted from the Ribosomal Database Project II http://rdp.cme.msu.edu/index.jsp)

If strain cultures are well preserved, microorganisms can be "revived" and subcultured when requested to be studied or used. Sometimes, however, strains with unusual physiological requirements may get lost, which is why it is so important that international collections of microorganisms should share their strains, and reference strains should be deposited in several international collections.

6.3 Characterization of a Prokaryotic Strain (Minimum Standards)

Strains of prokaryotes are characterized using two types of criteria: phenotypic and genotypic criteria. The "minimum standards" are the minimum phenotypic and genotypic characters requested by the International Committee on Systematics of Prokaryotes (ICSP) which guarantees the description of species, their nomenclature, and their taxonomic position. The International Committee on Systematics of Prokaryotes brings together researchers in microbiology from different countries chosen for their expertise in systematics of defined bacterial groups. The committee is assisted by subcommittees corresponding to different bacterial groups, for example, the subcommittee on photosynthetic bacteria, the subcommittee for sulfate-reducing bacteria, the subcommittee for Gram-positive bacteria, the subcommittee for thermophilic archaea, etc.

6.3.1 Phenotypic Criteria

These correspond to phenotypic traits expressed by microorganisms maintained in pure cultures. They must be carefully chosen to distinguish clearly between microorganisms (Table 6.1). They must be easy to determine and provide reproducible and reliable results. For a long time, these phenotypic criteria were limited to morphological characters: shape and size of the cells which constitute the pure strains, types of cell aggregates, response to Gram staining, presence of flagella, spore formation, the presence of capsules, cell particularities, etc. If morphological criteria have been used extensively (and successfully) for the classification of eukaryotes, the low morphological diversity of prokaryotes (cocci, rods, spirilla, *cf*. Sect. 3.1.1) did not allow developing a system for precise classification. However, this low morphological diversity is compensated by the high prokaryotic metabolic diversity especially in energy metabolism (*cf*. Sect. 3.3).

In prokaryotic microorganisms, the metabolic diversity, whether relating to catabolism (energy metabolism: donors and electron acceptors, respiratory, fermentative or photosynthetic types) or anabolism (sources of carbon, nitrogen, sulfur, etc.), has been used to characterize strains of prokaryotes relatively accurately. Many additional criteria were used in some cases to refine the characterization and allow a better classification. These are structural criteria (membrane lipids,

Table 6.1 Phenotypic criteria commonly used in taxonomy: classic minimum standards required by the International Committee for Bacterial Systematics

Category criteria	Major criteria
Morphology	Shape, size, aggregates, reaction to Gram stain, presence of spores, capsules, inclusions
Motility	Presence of flagella, association and insertion of flagella, gliding motility, gas vesicles
Nutrition and physiology	Mechanism of energy conservation: phototrophic, chemoorganotrophic, chemolithotrophic
	Ability to use a variety of sources of carbon, nitrogen, sulfur
	Relation to oxygen: aerobic and anaerobic respirations, fermentations
	Use of xenobiotics: by biodegradation or biotransformation
Ecophysiology	Optima and tolerances of salts, temperature, pH
Other factors	Pigments, types of membrane lipids, types of envelopes, antigens, pathogenicity, resistance to antibiotics, to heavy metals

coenzymes of the respiratory chains, etc.), ecophysiological criteria (adaptations to environmental conditions: pH optima, temperature, salinity, etc.), and antigenic criteria and resistance (presence of certain antigens, pathogenesis, antibiotic resistance, etc.) (Imhoff and Caumette 2004).

It is obvious that phenotypic properties have special meaning: not only do they allow classifying prokaryotes (called artificial classification), but they also tell something about the capabilities of microorganisms. They are therefore essential to understand the role of microorganisms in the environment where they live. For example, with respect to pathogenic bacteria, it is essential to know their metabolic and pathogenic features in order to develop sanitary applications and treatment of infectious diseases.

To identify a prokaryote, a taxonomist typically performs a series of phenotypic assays, some general, others specific to the type of microorganism studied. Identification keys can then be used. These dichotomous keys are constructed from the most general characteristics (cell shape, Gram stain, motility, etc.) to more specific characters (using specific substrates, the presence of antigens, etc.). The three major weaknesses of these methods are as follows:

1. The criteria tested may vary according to culture conditions.
2. A greater weight is given to taxonomic criteria used in the first part of the dichotomous key.
3. They allow the identification of organisms, but not the development of an evolutionary classification.

An attempt to resolve the first bias lies in the development of standardized methods for the determination provided by a single manufacturer and therefore still using the same components and the same culture conditions. This standardization of methods has to overcome the possible variations in the criteria used from one laboratory to another. Two methods are commonly used today:

1. The "API system" method using plastic plates comprising wells in which the substrates are in a lyophilized form to be assayed, the result is very often based on the change in pH during use of the substrate.
2. The "BIOLOG" method using microplates with wells in which the substrates tested can also be found in lyophilized form with a developer of redox potential, thereby indicating the respiratory activity of the microorganism following the use of the substrate.

These micromethods (API and BIOLOG) are mainly used in medical laboratories or hygiene control for rapid identification of pathogenic bacteria. However, additional tests (search for antigens, antibiotic resistance, etc.) are often needed to better characterize and confirm the taxon tested. These methods have been developed for specific groups of prokaryotes, including bacterial pathogens present in hospital environments (enterobacteria, clostridia, staphylococci, streptococci, etc.). They are thus not appropriate for all groups, including the prokaryotes grown from environments (sulfur-oxidizing or sulfate-reducing bacteria, methanogenic archaea, photosynthetic bacteria, etc.) which often have metabolic characteristics not covered by these tests.

6.3.2 Genetic Criteria

Generally for isolates, it is possible to match phenotypic criteria with genotypic criteria such as the percentage of bases G and C in the genomic DNA (% G + C), the gene sequence of the 16S ribosomal RNA (16S rDNA), MLST data, and DNA/DNA hybridization of genomic DNA between strains. These criteria are minimum standards required in the International Committee on Systematics of Prokaryotes. In the absence of phenotypic criteria distinguishing a group of isolates, it is possible to speak of **genomic species***, that is to say defined solely from molecular criteria. The genomic species will be unnamed, but listed.

6.3.2.1 The Percentage of G and C Bases in Genomes

The G + C content can be characterized by genomic DNA hydrolysis and chromatography or by measuring the melting temperature for denaturing DNA. Despite this criterion being not discriminating enough, it is still requested by scientific journals. This technique is based on the fact that the ATGC bases of DNA allow pairing of the two DNA strands with two hydrogen bonds between the A and T bases

Fig. 6.2 Composition range in DNA bases (G + C%)

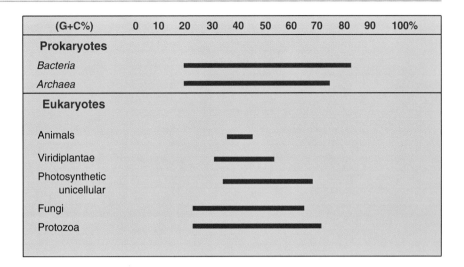

(T = A) and three hydrogen bonds between the G and C bases (G ≡ C). The calculation of G + C% is

$$G + C\% = \frac{G + C \text{ mol} \times 100}{\text{Mol}((G + C) + (A + T))}.$$

Due to the wide range of G + C percentages in genomic DNA in prokaryotes (13–75 %), this criterion has been proposed to differentiate groups of prokaryotes, assuming that bacterial strains or archaea belonging to the same species should have G + C percentages very similar if not identical (Fig. 6.2). Similarly, the variability within a genus was supposed to be very low. However, it appeared that the use of this criterion, although always required, allows differentiating neither species, genera, nor phyla (Fig. 6.3). Indeed, some important variations exist between organisms of the same genus. For example, within the genus *Mycobacterium*, the G + C% ranges from 57.8 % (*Mycobacterium leprae* TN) to 69.3 % (*Mycobacterium avium* K-10), or within the genus *Mycoplasma* rate varies from G + C 23.8 % (*Mycoplasma capricolum* ATCC 27343) to 40 % (*Mycoplasma pneumoniae* M129). The same applies to taxonomic levels of higher rank, such as **phyla***, for which G + C variations are too large to be used (Fig. 6.3). Other genetic criteria have therefore been subsequently proposed (DNA/DNA hybridization, 16S rRNA sequencing, etc.), whereas the G + C% is now mainly used to identify gene regions exchanged laterally in sequenced genomes (*cf*. Sect. 12.2).

6.3.2.2 The DNA/DNA Hybridization of Genomes

The gold standard for the identification of species is the DNA/DNA hybridization (Fig. 6.4). It is required by taxonomic flagship journals (e.g., International Journal of Systematic and Evolutionary Microbiology). This approach provides a 70 % threshold of DNA hybridized to define the membership of two strains of the same species. The two main problems of this approach are the need for culturing the strain in order to extract its DNA (thus excluding non-cultivated strains) and non-archivability, which therefore requires the cultivation of all the type strains of species with which a new strain should be compared.

6.3.2.3 The 16S rRNA Gene

The sequence of the gene coding for 16S ribosomal RNA is now provided by almost all the authors describing a new species. To be informative, it must be of good quality (less than 1 % of undetermined bases) and have a length of at least 1,000 nucleotides out of the 1,500 that make up the gene (*cf*. Sect. 17.7.4). The phylogenetic analysis of 16S rRNA sequences allows specifying the relationship of the studied microorganism compared to other microorganisms whose 16S rRNA sequences are known (see below). An advantage of this approach is that the sequencing of the gene coding for 16S rRNA can be done not only for isolates but also for complex communities (Stackebrandt et al. 1993) or in organs infected by a microorganism. Thus, the microorganisms are characterized by their 16S rRNA sequences and compared to other microorganisms. If the studied microorganism is not isolated, the name **Candidatus*** must be used (Murray and Stackebrandt 1995).

The sequences of other genes are sometimes used in order not to depend on the identification of a single marker. This approach called MLST was developed following the observation that the phylogenies of different genes are not always consistent with that of 16S rRNA. Indeed, the amount of information present in the 16S rRNA is sometimes not sufficient to reliably position strains. The use of alternative phylogenetic markers is therefore important to refine the phylogenetic position of studied microorganisms based on 16S rRNA gene alone. In addition, because of frequent

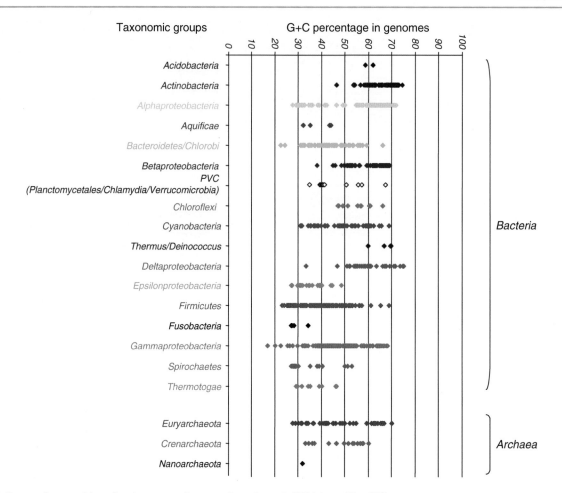

Fig. 6.3 Range of composition of major taxonomic groups in prokaryotic DNA base (G + C%)

genetic exchanges among prokaryotes, the use of several alternative markers is recommended. With the increase in sequencing capacity, it is quite possible that in a few years, the complete genome sequence will be required for the identification of an isolate.

6.4 Dendrograms and Phylogenetic Trees

6.4.1 Phenotypic Dendrograms, Numerical Taxonomy

Numerical taxonomy is associated with artificial classifications, as opposed to phylogenies that are associated with natural classifications (see below). Historically, numerical taxonomy appeared before molecular phylogenies, and today it can be used in addition to phylogenetic analyses for characterization at the species level. These dendrograms should not be confused with phylogenetic trees. The former represent phenotypic similarities, whereas the latter represent relations of kinship.

The principle of numerical taxonomy relies on comparisons of numerous phenotypic features for a set of strains in order to group them according to their degree of phenotypic similarity. Usually, more than 50 independent characters are used, but in some cases, a higher number may be required (e.g., up to 150 characters for the study of aerobic heterotrophic bacteria). Phenotypic characteristics tested must be independent, namely, overlapping characters must be avoided. The features of each strain are translated into "positive" or "negative" in a binary manner and gathered into a character matrix (Fig. 6.5a). The similarity between strains is subsequently quantified by a similarity coefficient $S_{(AB)}$, such as that of Jaccard, through pairwise comparisons of the strains. The $S_{(AB)}$ coefficient between two strains (Jaccard coefficient) is defined as follows:

$$S_{j(AB)} = a/(a + b + c)$$

In this formula, a represents the number of "positive" characters shared by the two trains, b corresponds to the number of "positive" features for strain A that are "negative" in strain B, whereas c is the number of "negative" characters

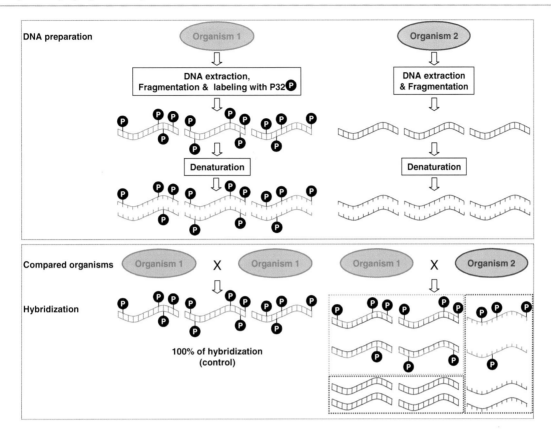

Fig. 6.4 Method of DNA/DNA hybridization; schematic and principle of the method. The DNA/DNA hybridization is done between the DNA of two organisms, 1 and 2. The DNA of one of the microorganisms is labeled with radioactive P^{32} phosphate. The DNA is then denatured, separated into single-stranded fragments. The denatured DNA of both organisms are then mixed by adding large excess in the unmarked DNA of organism 2, so as to limit the autohybridation between the labeled DNA of an organism. After hybridization (reconstitution of DNA double-stranded hybrids between the two microorganisms' DNA) and removal of remaining single strands by enzymatic digestion, the radioactivity of the double-stranded DNA hybrid is measured and compared to a control where 100 % of the labeled denatured DNA a microorganism are hybridized with each other. The radioactivity value obtained in the hybridization sample compared to that of the control 100 %, and thus gives the percentage hybridization between the two microorganisms

in strain A that are "positive" in strain B. The resulting coefficient is expressed as a percentage of similarity between two strains. It is important to note that the Jaccard coefficient considers only characters with a positive result in one or other of the two strains compared. There are other indices such as the Sokal and Michener index that take into account the sharing of negative characters. A $S_{(AB)}$ of 70 % is expected at the species level. The calculation of similarity coefficients among all pairs of strains provides a similarity matrix (Fig. 6.5b) that can be plotted as a dendrogram where the most similar strains are placed close together (Fig. 6.5c).

6.4.2 Phylogenetic Trees

In addition to phenotypic characterization, the identification and the classification of new isolates relies more and more frequently on a phylogenetic analysis. Besides deciphering of relationships among taxa, this approach allows studying the biodiversity and classifying prokaryotes via molecular techniques that do not require the direct cultivation of the corresponding microorganisms (*cf.* Sect. 8.4.2). A phylogenetic analysis is used to determine kinship relationships between taxa (strains, species, genera, etc.) and therefore allows identifying the closest known relatives of the strain of interest. The reference marker used for this analysis is the gene coding for 16S rRNA. Identification based on this gene opens the question of the similarity threshold above in which it is assumed that two closely related isolates belong to the same species which is classically defined by DNA/DNA hybridization (Wayne et al. 1987). Stackebrandt and Goebel (1994) suggested that a threshold of 70 % of genomic DNA hybridization between strains of the same species corresponded to a percent identity of at least 97 % of their genes encoding 16S rRNA.

One has to understand the 97 % level as a threshold below which there is no need to make a DNA/DNA hybridization because the two strains belong to different species, but this does not mean that two strains harboring more than 97 % identity at the 16S rDNA level always belong to the same

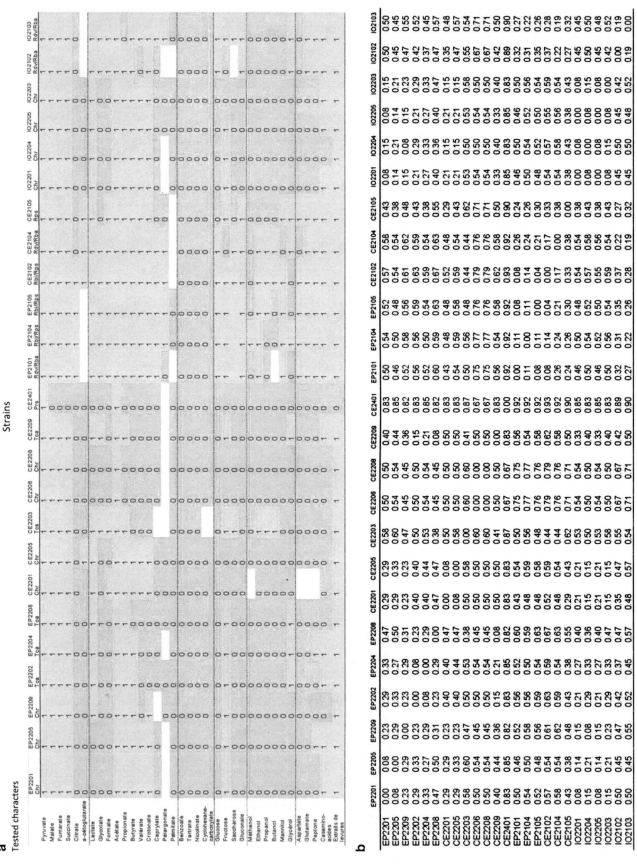

Fig. 6.5 (a) Matrix of characters used to classify 25 strains of sulfur-oxidizing photosynthetic bacteria isolated from three different lagoon environments. *1* positive test, *0* negative test. (b) Distance matrix derived from the matrix of $S_{(AB)}$ (1-$S_{(AB)}$). The $S_{(AB)}$ was calculated using the Jaccard coefficient. (c) Construction of the dendrogram from the distance matrix (b). (*Chr*) *Chromatium*, (*Tca*) *Thiocapsa*, (*Tcs*) *Thiocystis*, (*Rbi/Rps*) *Rhodobium/Rhodopseudomonas*, (*Rdv/Rba*) *Rhodovulum/Rhodobacter*, (*Rps*) *Rhodopseudomonas*, (*Prs*) *Prosthecochloris*. In most cases, strains belonging to the same genus are grouped together. The scale represents 10 % difference between strains. The difference is calculated as the sum of all horizontal distances between two strains. This dendrogram is constructed iteratively creating couples with the shortest distance and then recalculating a new matrix with this couple. This approach is called UPGMA

6 Taxonomy and Phylogeny of Prokaryotes

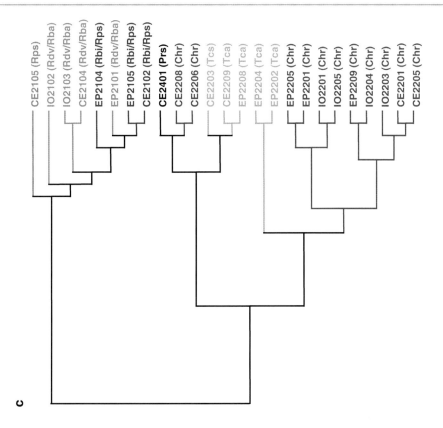

Fig. 6.5 (continued)

species. Therefore, in this latter case, DNA/DNA hybridization between the genomes of two strains is required (Fig. 6.6). The lack of equivalence agreed between the two techniques lies in the fact that 16S rDNA corresponds to one locus among the thousands of loci present in a genome. Therefore, it is recommended to use a multi-locus approach such as AFLP (amplified fragment length polymorphism) and correlate the results with data from DNA/DNA hybridization (Stackebrandt et al. 2002) (Fig. 6.7).

Phylogenetic analysis of **molecular markers*** (e.g., 16S rRNA) is a multistep process (Fig. 6.8), the first of which is to compare the sequence of interest to investigate whether homologous sequences are present in sequence public databases (e.g., general databases such as the nr database at the NCBI or more specialized databases such as the RDP II or silva in the case of 16S rRNA). This initial research is based on sequence similarity comparisons among the sequences of the database and the studied sequence. Indeed, the closer related the two strains are, the closer the sequences of their genes (including 16S rRNA) will be. One of the most frequently used software to perform these searches is the BLAST (Basic Local Alignment Search Tool). It helps to identify similar regions between two protein or nucleic acid sequences (Fig. 6.8a). The identification of most similar sequences provides only a crude taxonomic indication (genus level or higher). To get a finer taxonomic affiliation, it is thus important to perform a phylogenetic analysis. For the phylogenetic analysis, all homologous sequences from organisms close to the one analyzed are extracted from the database and aligned using multiple alignment software such as Clustal omega or Muscle (Fig. 6.8b, c). The resulting alignments are trimmed in order to remove regions where homology between sites is ambiguous (i.e., poorly conserved regions containing insertions/deletions that cannot be unambiguously positioned). The remaining sites are used to infer the phylogenetic trees (Fig. 6.8d). One of the tree reconstruction methods commonly used in taxonomy is the neighbor-joining (Saitou and Nei 1987) because of its simplicity and speed. The neighbor-joining method belongs to the family of distance methods because the first step of neighbor-joining requires the quantification of evolutionary distances among all pairs of sequences studied. This estimate of the evolutionary distances is based upon the use of an evolutionary model. The calculation of evolutionary distances between each pair of sequences allows the establishment of a pairwise distance matrix (Fig. 6.8e). These distances are then represented as phylogenetic trees where branch lengths are proportional to evolutionary distances among sequences (Fig. 6.8f). One of the advantages of the neighbor-joining is the ability to process a large number of sequences in a very short time (few seconds

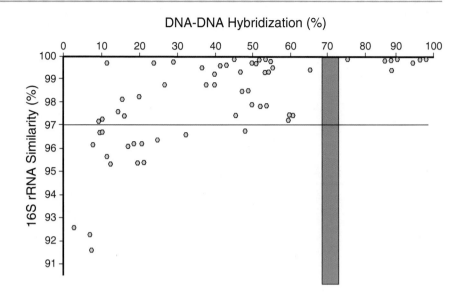

Fig. 6.6 Comparison of the DNA/DNA hybridization and the percentage of homology between 16S sequences. This meta-analysis yields a threshold of 97 % below in which it is considered that two bacterial strains are not likely to have a DNA/DNA hybridization greater than 70 % and thus belong to the same bacterial species. The converse is not true, because greater than 97 % similarity at the 16S rDNA sequences may correspond to a rate of DNA/DNA hybridization less than 70 % and thus to different species (redrawn from Stackebrandt and Goebel's work (1994) with permission of the authors)

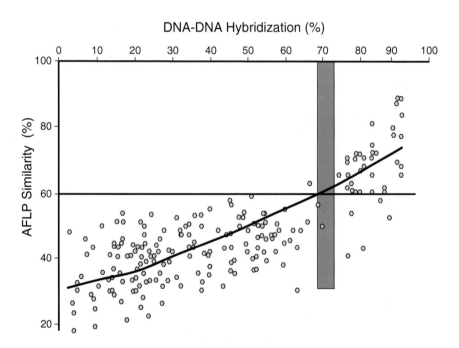

Fig. 6.7 AFLP method compared to DNA/DNA hybridization. According to work done on the genus *Vibrio*, valid also for many bacterial genera including *Xanthomonas*, *Burkholderia*, *Rhizobium*, and *Bacillus* (Modified from Thompson et al. (2004) with permission of the author and ASM)

to analyze hundreds of sequences). Its use is preferable to the **UPGMA*** that assumes that all sequences are evolving at the same rate (molecular clock hypothesis, Box 5.3; Ochman et al. 1999), but less satisfactory than the **Maximum Likelihood*** that is time-consuming.

6.5 Classification of Prokaryotic Microorganisms: Phylogeny Versus Phenotype

Classifications based on phenotypic criteria are more ancient than those based on molecular phylogeny. With regard to the prokaryotic microorganisms, many inconsistencies exist between phenotypic similarity levels and evolutionary relationships deduced from molecular phylogenies. Most taxonomic groups have been reconsidered through phylogenetic analysis in the 1980s, and a consensus was reached on the need for coherence between taxonomy and phylogeny. Reconciliation between the two schools of thought, similar in their basic goals but historically distinct, if it makes sense, is not easy. Thus, the multiple revisions of species, genera, or higher taxonomic levels pose many complex problems that are solved gradually. The difficulties are still present in some taxa where a tangle exists among species, leading to three possibilities all unsatisfactory: (1) accepting the existence of polymorphic taxa, with significant differences at the phenotypic level; (2) grouping all bodies taxon in a large catch-all using the rule of precedence; (3) subdividing heterogeneous taxa into smaller homogeneous subtaxa with the

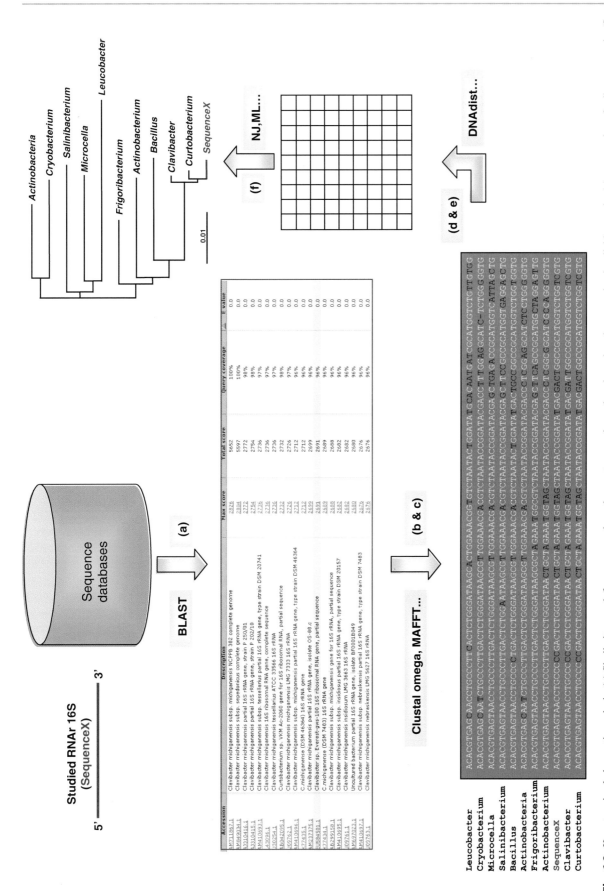

Fig. 6.8 Key steps in phylogenetic analysis. (**a**) Similarity search for sequences homologous to the sequence of interest in the sequence databases, (**b**) extraction of counterparts from closely related species, (**c**) multiple alignment of sequences, (**d**) filtering the alignment so that only regions where homology between sites is reliable remain, (**e**) construction of a distance matrix, (**f**) reconstruction of the phylogenetic tree and identification of the closest known relatives (see explanations in the text), the distance scale for the phylogenetic tree represents 1 % of difference

risk of having to cut into smaller and smaller units and thus increase beyond the reasonable the number of subgroups.

There is no solution that can be applied arbitrarily to all situations. For example, *Shigella flexneri* and *Escherichia coli*, two *Gammaproteobacteria* (family *Enterobacteriaceae*), exhibit a level of DNA/DNA hybridization above 70 %, which means that these two genera represent in fact the same species. However, there is a consensus to keep the two names because of the risks for human health due to their different pathogenicity characteristics. In this case, the phenotype has precedence over the genotype to maintain both species (and genera). However, the existence of pathogenic *Escherichia coli* strains may reduce the need to distinguish between the two genera and bring forward a name change in *Escherichia* and *Shigella*. Conversely, some broad and highly diversified genera such as *Bradyrhizobium* (*Alphaproteobacteria*), *Bacillus* (*Firmicutes*), or *Clostridium* (*Firmicutes*) gather groups of species which likely represent distinct genera. Finally, it should be remembered that any revision of the nomenclature associated with the reconsideration of some taxa raises the issue of names priority rules.

6.6 Systematic of Prokaryotic Microorganisms: Hierarchical Organization and Phylogenetics

Prokaryotic microorganisms correspond to two of the three **domains*** of life: *Archaea* and *Bacteria* (*cf.* Sect. 5.1.4, Fig. 5.9). Their systematics is described in the "Bergey's Manual for Systematic Bacteriology, second edition" published in five volumes (Boone and Castenholz 2001; Brenner et al. 2005; de Vos et al. 2008; Krieg et al. 2011; Goodfellow et al. 2012). For more information, the reader may consult the website http://www.bacterio.cict.fr/.

In the text of this book, the Latin terms used are those accepted by the Nomenclature Committee as they appear in the second edition of the "Bergey's Manual for Systematic Bacteriology" (Table 6.2).

The use of phylogenetic criteria for establishing a natural classification of prokaryotes (work of Carl Woese and George Fox in the late 1970s) (Woese and Fox 1977; Fox et al. 1980) revealed that no or very few phenotypic features can be used to define the highest taxonomic levels. For example, initial analysis of the 16S rDNA showed that there were not one but two groups of prokaryotes not distinguishable on the basis of morphological criteria, but as distant from each other in terms of their genetic distance as they are from eukaryotes. Similarly, these studies showed that no phenotypic feature can be specifically associated to a given phylum. For example, the phylum Proteobacteria was initially called purple bacteria because of their purple pigmentation resulting from carotenoids

Table 6.2 High taxonomic ranks of prokaryotic microorganisms proposed in Bergey's Manual of Systematic Bacteriology

Domain	Phyla (branches)	Classes
Archaea	A1: *Crenarchaeota 1:*	*Thermoprotei*
	A2: *Euryarchaeota*	1: *Methanobacteria*
		2: *Methanococci*
		3: *Halobacteria*
		4: *Thermoplasmata*
		5: *Thermococci*
		6: *Archaeoglobi*
		7: *Methanopyri*
	A3: *Korarchaeota*	
	A4: *Nanoarchaeota*	
Bacteria	B1: *Aquificae*	1: *Aquificae*
	B2: *Thermotogae*	1: *Thermotogae*
	B3: *Thermodesulfobacteria*	1: *Thermodesulfobacteria*
	B4: *Deinococcus-Thermus*	1: *Deinococci*
	B5: *Chrysiogenetes*	1: *Chrysiogenetes*
	B6: *Chloroflexi*	1: *Chloroflexi*
		2: *Anaerolineae*
		3: *Caldilineae*
	B7: *Thermomicrobia*	1: *Thermomicrobia*
	B8: *Nitrospirae*	1: *Nitrospirae*
	B9: *Deferribacteres*	1: *Deferribacteres*
	B10: *Synergistetes*	1: *Synergistia*
	B11: *Cyanobacteria*	1: *Cyanobacteria*
	B12: *Chlorobi*	1: *Chlorobia*
	B13: *Proteobacteria*	1: *Alphaproteobacteria*
		2: *Betaproteobacteria*
		3: *Gammaproteobacteria*
		4: *Deltaproteobacteria*
		5: *Epsilonproteobacteria*
		6: *Zetaproteobacteria*
	B14: *Firmicutes*	1: *Clostridia*
		2: *Bacilli*
	B15: *Tenericutes*	1: *Mollicutes*
	B16: *Actinobacteria*	1: *Actinobacteria*
	B17: *Planctomycetes*	1: *Planctomycetacia*
	B18: *Chlamydiae*	1: *Chlamydiae*
	B19: *Spirochaetes*	1: *Spirochaetes*
	B20: *Fibrobacteres*	1: *Fibrobacteres*
	B21: *Acidobacteria*	1: *Acidobacteria*
	B22: *Bacteroidetes*	1: *Bacteroides*
		2: *Flavobacteria*
		3: *Sphingobacteria*
	B23: *Fusobacteria*	1: *Fusobacteria*
	B24: *Verrucomicrobia*	1: *Verrucomicrobiae*
		2: *Opitutae*
	B25: *Gemmatimonadetes*	1: *Gemmatimonadetes*
	B26: *Lentisphaerae*	1: *Lentisphaeria*
	B27: *Dictyoglomi*	1: *Dictyoglomia*
	B28: *Caldiserica*	1: *Caldisericia*
	B29: *Elusimicrobia*	1: *Elusimicrobia*
	B30: *Armatimonadetes*	1: *Armatimonadia*
		2: *Chthonomonadetes*

associated with bacteriochlorophylls. In fact this phylum gathers a large number of lineages related to these photosynthetic pigments (hence, the purple pigmentation characteristic). Similarly, the so-called Gram-positive bacteria (testing positive for Gram stain) form two distinct bacterial phyla (i.e., the *Firmicutes* and *Actinobacteria*); this indicates that they are not specifically related to each other. In fact, only two phyla, the *Spirochetes* and the *Cyanobacteria*, were found to have been properly defined on the basis of phenotypic traits, meaning a spiral shape and the ability to achieve an oxygenic photosynthesis, respectively. This was a disappointment for those who believed that the use of molecular markers would allow discerning evolutionary traits associated with each phylum, which is not the case. An analogy can be made between the major phylogenetic groups and neighborhoods within large cities; knowing the large group to which a given microorganism belongs does not reveal much of its physiology and its characteristics, rather it may give some indications. An analogy can be made with the fact of belonging to a city neighborhood: it tells little about the personal characteristics of individuals who live there, and at most, it indicates membership in the area. This underlies the great plasticity of phenotypic and physiological features within high taxonomic groups.

In 1990, Woese and collaborators (1990) proposed a definition of each of the three domains of life:

1. The *Eucarya* from the Greek adjective "eu" meaning true and the Greek name "Karyon" signifying kernel include organisms with cells carrying genetic information in linear chromosomes enveloped by a nuclear membrane (core), containing eukaryotic-type ribosomes, and surrounded by a phospholipid bilayer (*cf.* Sect. 3.1.2) that consists of a glycerol core attached to two fatty acid molecules by ester bonds.
2. The *Bacteria* (from the Greek name "bakterios" meaning stick) denote cells with an prokaryotic-type organization (pro "Karyon" that does not have a true nucleus) surrounded by a phospholipids bilayer (*cf.* Sect. 3.1.1), consisting of a glycerol molecule attached to two molecules of fatty acid by ester bonds and which contain bacterial-type ribosomes.
3. The *Archaea* from the Greek adjective "arkeos" meaning ancient, primitive, group cells with archaeal-type ribosomes, having a prokaryotic-type organization, and surrounded by either a phospholipid bilayer mainly made of phospholipids consisting of a glycerol molecule attached to two molecules of isoprenoid by ether bonds or a phospholipid monolayer consisting of two glycerol molecules connected together by two isoprenoid chains (*cf.* Sect. 4.1.10, Fig. 4.5).

6.6.1 Domain *Archaea*

6.6.1.1 The Discovery of Archaea

The domain *Archaea* has been revealed by Carl R. Woese and George E. Fox who analyzed the oligonucleotide profiles derived from the digestion of the RNA component of small ribosomal subunits (16S for prokaryotes/18S for eukaryotes) (Woese 2007). Briefly, the 16S (or 18S) rRNAs of each organism are extracted and digested with different restriction enzymes (including the RNase T1). Digestion products are then subjected to bidirectional electrophoresis to build migration profiles (also called oligonucleotide catalogs) specific to each organism. The pairwise comparison of these catalogs (via Sokal and Michener index calculation) allows quantifying the similarities between rRNA profiles and thus studying and classifying the corresponding organisms. In their seminal study, Woese and Fox analyzed the rRNAs from the cytoplasm of various eukaryotes and prokaryotes (including methanogens), mitochondria and chloroplast. The results were surprising. As expected, the eukaryotes and prokaryotes rRNA sequences appeared very different from each other. However, a similar difference was observed between the prokaryotic 16S rRNA sequences, which form two distinct and distantly related groups. This means that at the genetic level, 16S rRNA sequences from two prokaryotes displaying strong phenotypic similarities can be more different than the 18S rRNA sequences from *Homo sapiens* and a plant or even than the 18S rRNA from *Homo sapiens* and the 16S rRNA from *E. coli*. These results have profoundly changed our view of the living world, by shifting from a eukaryote/prokaryote dichotomy to a tripartite divide. This divide was rapidly confirmed by subsequent phylogenetic analyses. At that time, the first group of prokaryotes gathered a wide variety of bacteria, while the second corresponded to methanogenic bacteria (i.e., carrying out the biosynthesis of methane). The former was named *Eubacteria* (eu = true) and the latter *Archaebacteria* (archaea = old) by Carl Woese. The name *Archaebacteria* reflects the widespread belief at the time that methanogenesis could have been one of the earliest metabolisms on Earth. Accordingly, present-day methanogens (and thus *Archaebacteria*) would have conserved this ancestral metabolism. Contradicting this hypothesis, Archaebacteria were rapidly enriched with new members, most of them being non-methanogens: the thermoacidophilic *Thermoplasma* that were formerly classified with *Mycoplasma* because like the latter they are devoid of cell wall; the *Halobacteria* which are extreme halophiles; the *Sulfolobales*, another group of thermoacidophiles; and various lineages of hyperthermophiles with optimal growth temperatures greater than 80 °C such as the *Thermococcales*.

A few years later, Carl Woese proposed to rename *Eubacteria* and *Archaebacteria* as *Bacteria* and *Archaea*, respectively, to remove the "bacteria" suffix which implies that *Archaea* are somehow bacteria and that both prokaryotic domains share a closer evolutionary link compared to Eucarya. In fact, the tree of life based on 16S/18S rRNA being unrooted (*cf.* Sect. 4.1.10), it is impossible to determine the relationships between the three domains.

6.6.1.2 Diversity of Archaea

While more than 10 bacterial phyla were described in the seminal works of Carl Woese, *Archaea* were divided into two phyla, *Crenarchaeota* and *Euryarchaeota*, corresponding to only three and nine cultivated orders (Woese 1987). *Euryarchaeota* (also referred to as euryotes or euryarchaeotes, from Greek "Euryos" meaning varied/diverse) bring together various lineages presenting very different lifestyle, such as hyperthermophilic or mesophilic methanogens (*Methanococcales*, *Methanobacteriales*, *Methanocellales*, *Methanomicrobiales*, *Methanosarcinales*, *Methanopyrales*, and the recently proposed "Methanoplasmatales"), extreme halophiles (*Halobacteriales* and the recently discovered "Nanohaloarchaea"), sulfate reducers (*Archaeoglobales*), thermoacidophiles (*Thermoplasmatales*), and some hyperthermophiles (*Thermococcales*). In contrast, the second phylum gathers exclusively thermophiles or hyperthermophiles. It was named *Crenarchaeota* (also referred to as crenotes or crenarchaeotes) from the Greek "Krenos" meaning source/origin because Carl Woese thought that the ancestor of Archaea was hyperthermophile, a feature retained by *Crenarchaeota*. For a long time, only two archaeal phyla were recognized, but new phyla (i.e., *Korarchaeota*, "Nanoarchaeota", *Thaumarchaeota*, and more recently the *Aigarchaeota*) have been proposed in the last few years.

Most archaea and bacteria look alike. They are similar in size, shape, and cellular organization. However a few atypical cell morphologies are found in archaea as polygonal, triangular, or ultrathin square cells or very irregular cells. As bacteria, archaea exhibit a variety of phenotypes and physiologies. In fact, except for methanogenesis, all metabolisms described in archaea exist also in bacteria. Conversely no photosynthesis involving chlorophyll, **spore production***, or pathogens have been reported in archaea so far. Archaea can be heterotrophic or autotrophic, can use various electron acceptors and electron donors, be aerobic, or anaerobic, etc. The main feature distinguishing bacteria from archaea is the nature of their cell envelope. In contrast to bacteria, the archaeal membranes contain lipids made of isoprene (and not fatty acid) chains, ether (and not ester) linkages, and L-glycerol (and not D-glycerol) moiety, in addition to the phosphate group. Moreover, archaeal envelope does not contain peptidoglycan or murein, resulting in insensitivity to the most common antibiotics targeting the bacterial cell wall (for a recent and complete review on the archaeal envelope, see Albers and Meyer 2011). Worth noting, the archaeal cell envelope is highly variable even among closely related lineages (Table 6.3), highlighting an unexpectedly very dynamic structure.

To the exception of *Ignicoccus hospitalis* (the host of *Nanoarchaeum equitans*), which harbors two membranes, the archaeal cell envelope is composed of a single membrane. In nearly all archaea characterized to date, this membrane is surrounded by a proteinaceous protein layer called the S-layer, which form a crystalline array (Table 6.3). S-layers are also found in bacteria. S-layer contributes to the shape, osmoprotection, and permeability of the cell. In most cases, the S-layer is composed of a single protein (or glycoprotein) aligned in lattices with oblique (p1 or p2), tetragonal (p4), or hexagonal (p3 or p6) symmetry. S-layer proteins have sizes ranging from 40 to 200 kDa and are also found in many bacteria. The composition and structure of the S-layer varies among archaea. While the hexagonal symmetry is predominant in archaeal S-layers, oblique or tetragonal lattices exist (Albers and Meyer 2011). In addition to S-layer, very atypical and stable proteinaceous structures have been reported in two unrelated species, namely, *Methanospirillum hungatei* (*Methanomicrobiales*) and *Methanosaeta concilii* (*Methanosarcinales*) (Table 6.3). It consists in tubular sheaths enclosing linear chains of cells, the individual cells within the chains being themselves surrounded either by an S-layer (*M. hungatei*) or by an amorphous granular protein layer (*M. concilli*) (Albers and Meyer 2011).

In many archaea, the S-layer is the only component of the cell wall, whereas in others various additional components can be found in the cell envelope. For instance, some methanogens (e.g., *Methanobacteriales* or *Methanopyrales*) harbor a cell wall composed of pseudopeptidoglycan (also called pseudomurein), which superficially resemble bacterial peptidoglycan. The comparison of the proteins involved in the biosynthesis of archaeal pseudomurein and bacterial murein showed no homology, suggesting that the two pathways emerged twice independently during evolution. A fibrillar polymer, called methanochondroitin, is found in the cell wall of members of the *Methanosarcina* genus when they form aggregates, but not in single cells. Strikingly a very similar polymer, the chondroitin, is part of the connective tissue matrix of vertebrates (Albers and Meyer 2011). Other polymers can be encountered in the cell envelope of some extreme halophiles belonging to *Halobacteriales* such as glutaminylglycan (*Natronococcus occultus*), highly sulfated heteropolysaccharides (*Halococcus morrhuae*), or halomucin (*Haloquadratum*

Table 6.3 Main features of the archaeal cell envelope for a subset of representative archaea

Phylum	Order	Species	S-layer	Second membrane	Pseudomurein	Glycocalyx (polysaccharide)	Sulfated heteropolysaccharide	Glutaminylglycan	Halomucin	Proteinaceous sheaths	Unspecified protein	Methanochondroitin
Thaumarchaeota		*Nitrosopumilus maritimus* SCM1	?	?	?	?	?	?	?	?	?	?
		Cenarchaeum symbiosum	?	?	?	?	?	?	?	?	?	?
Korarchaeota		'*Ca.* Korarchaeum cryptofilum'	X									
Crenarchaeota	Thermoproteales	*Thermofilum* sp.1505	X									
		Thermoproteus neutrophilus V24Sta	X									
		Thermoproteus tenax strain YS44	X									
		Pyrobaculum islandicum DSM 4184	X									
		Pyrobaculum organotrophum	X									
	Sulfolobales	*Acidianus brierleyi*	X									
		Metallosphaera prunae Ron12/II	X									
		Metallosphaera sedula DSM 5348	X									
		Sulfolobus solfataricus P2	X									
		Sulfolobus acidocaldarius DSM 639	X									
		Sulfolobus tokodaii str. 7	X									
	Desulfurococcales	*Pyrolobus fumarii* 1A	X									
		Hyperthermus butylicus DSM 5456		X								
		Ignicoccus hospitalis KIN4		X								
		Desulfurococcus mobilis	X									
		Desulfurococcus kamchatkensis	X									
		Staphylothermus marinus F1	X									
		Aeropyrum pernix K1	X									
Nanoarchaeota		*Nanoarchaeum equitans* Kin4−M	X									
Euryarchaeota	Methanopyrales	*Methanopyrus kandleri* AV19			X							
	Thermococcales	*Thermococcus acidaminovorans*	X									
		Pyrococcus furiosus DSM 3638	X									
	Methanobacteriales	*Methanothermus fervidus*	X		X							
		Methanobrevibacter ruminantium M1			X							
		Methanobrevibacter formicicum			X							
	Methanococcales	*Methanocaldococcus jannaschii*	X									
		Methanococcus voltae A3	X									
		Methanococcus vannielii SB	X									
		Methanococcus maripaludis C7	X									
	Archaeoglobales	*Ferroglobus placidus* DSM 10642	X									
		Archaeoglobus veneficus DSM11195	X									
		Archaeoglobus fulgidus DSM 4304	X									
	Thermoplasmatales	*Picrophilus torridus* DSM 9790	X									
		Thermoplasma acidophilum				X						
		Thermoplasma volcanium				X						
	Halobacteriales	*Halococcus morrhuae*					X					
		Natronococcus occultus SP4					X					
		Halobacterium salinarum R1	X									
		Haloarcula japonica TR−1	X									
		Haloferax volcanii DS2	X									
		Haloquadratum walsbyi C23	X						X			
		Haloquadratum walsbyi HBSQ001	X						X			
	Methanosarcinales	*Methanosaeta concilii* Opfikon								X		
		Methanosaeta concilii								X		
		Methanolobus tindarius DSM2278	X									
		Methanosarcina acetivorans C2A										X
		Methanosarcina mazei Go1										X
	Methanomicrobiales	*Methanocorpusculum sinense*	X									
		Methanocorpusculum parvum	X									
		Methanospirillum hungatei JF−1	X								X	
		Methanogenium tationis DSM 2702	X									
		Methanoculleus marisnigri JR1	X									
		Methanogenium marisnigri DSM 1498	X									
		Methanolacinia paynteri	X									
		Methanogenium cariaci DSM 1497	X									

"?" means: not known yet
The data have been extracted from Albers and Meyer (2011)

genus) (Table 6.3). Worth noting, halomucin is the largest archaeal protein (9,159 amino acids) known to date. Its amino acid composition and domain organization are similar to mammalian mucin, which acts as a shield against dehydration of various tissues, such as the bronchial epithelium and the eyes (Albers and Meyer 2011). Finally, a few archaea belonging to *Thermoplasmatales* harbor highly pleomorphic shapes, due to the lack of any cell wall (Table 6.3).

At the genetic level, *Archaea* are a mosaic. Their informational systems (i.e., systems involved in the transmission and expression of genetic information, namely, the machineries of transcription, translation, replication, and repair) are similar to eukaryotes (*cf.* Sect. 4.1.10), whereas their housekeeping and metabolic genes and their general cell organization are similar to bacteria. Since their discovery, *Archaea* have led to some of the most exciting discoveries in the field of biochemistry and biotechnologies, but archaeal genetics has been slow to get off the ground, until recently. In fact, the last past years have witnessed spectacular progress, and genetic tools and models are now available for the two major archaeal phyla, the *Euryarchaeota* (i.e., *Halobacteriales*, *Methanococcales*, *Methanosarcinales*, and *Thermococcales*) and the *Crenarchaeota* (i.e., *Sulfolobales*) (for more details on archaeal genetics, see the excellent review of Leigh et al. 2011). This will accelerate our understanding of the biology of *Archaea*.

Because the first described archaea inhabited some of the more inhospitable places on earth (from an anthropomorphic point of view), members of this domain have been considered for a long time as "exotic microbes" or "curiosities" by most microbiologists. Because archaea are the only living organisms able to grow optimally at temperatures above 100 °C, the dominance of archaea over bacteria in extremely hot environments was early recognized. In contrast, bacteria were considered as dominant over archaea in all other ecosystems. As a consequence, and to the exception of methanogens, the relevance of archaea in microbial ecosystems and in global biogeochemical cycles has been underestimated for years (Forterre et al. 2002; Gribaldo and Brochier-Armanet 2006). The situation has changed with the birth of molecular ecology at the end of the 1980s. The investigation of microbial ecosystems with molecular tools has uncovered the incredible genetic, physiological, and phenotypic diversity of *Archaea* (Schleper et al. 2005; Lopez-Garcia and Moreira 2008). A large number of new lineages, such as groups called I, II, III, IV, SA1, SA2, ARMAN, ANME-1, ANME-2, "Nanohaloarchaea," Miscellaneous Crenarchaeotal Group (MCG), etc., many of them representing likely high-level taxonomic groups were discovered. Importantly, these uncultured archaeal lineages could represent an important fraction of the biomass of some ecosystems (Narasingarao et al. 2011). Despite great advance in cultivation techniques, most of these lineages have resisted all cultivation attempts and remain poorly characterized. This underlines that our knowledge of *Archaea* based on cultured lineages is far from being representative of the real diversity of this domain. However, thanks to rapid progresses in DNA sequencing and metagenomics, complete genomes of representatives of some of these groups have been sequenced. Such culture-independent investigations will accelerate our understanding of these uncultured lineages and of *Archaea* in general.

6.6.1.3 Classification of Archaea
The *Crenarchaeota*

Based on 16S rRNA phylogenies, *Crenarchaeota* are divided into three orders (5 families and 22 genera) with cultured representatives: The *Sulfolobales*, the *Desulfurococcales*, and *Thermoproteales*, the first two being more closely related to each other (Fig. 6.9). Recently two additional orders have been proposed, namely, the *Acidilobales* and the *Fervidicoccales*, living in acidic hot springs. However, subsequent phylogenetic and genomic analyses suggested that they are rather *Desulfurococcales* (Brochier-Armanet et al. 2011). *Crenarchaeota* can be anaerobic, facultative anaerobic, or aerobic extreme thermophiles or hyperthermophiles. Their energetic metabolism is mainly based on sulfur, even if some of them are also able to use organic and other inorganic compounds or have lost the ability to use sulfur. Some of them are also acidophilic, being able to grow in pH ranging from 2 to 5.

The Sulfolobales

This order was proposed by Karl Stetter in 1989. *Sulfolobales* are extreme thermophilic or hyperthermophilic acidophiles thriving at temperatures ranging from 65 to 90 °C and at pH ranging from 1 to 5. Cells are regular to irregular cocci of about 1.0 up to 5 μm in diameter, occurring usually singly or in pairs. Most members of the *Sulfolobales* have been isolated from continental solfataric fields, from acidic hot soils, acidic hot springs, and smoldering slag heaps. In contrast, a few strains only have been isolated from submarine hydrothermal systems. This order contains a single family, the *Sulfolobaceae* divided into six genera: *Sulfolobus*, *Acidianus*, *Metallosphaera*, *Stygiolobus*, *Sulfurisphaera*, and *Sulfurococcus*. They can be aerobic, facultative anaerobic, or anaerobic. When growing autotrophically, they gain energy by oxidizing S^0, $S_2O_3^{2-}$, sulfidic ores, or H_2 and use CO_2 as a carbon source. In contrast, organotrophic growth occurs by aerobic respiration or anaerobic sulfur respiration or by fermentation of organic substrates. More precisely, *Sulfolobus* are obligate aerobes. Some of them can grow mixotrophically or heterotrophically by using complex organic compounds, sugar or amino acids,

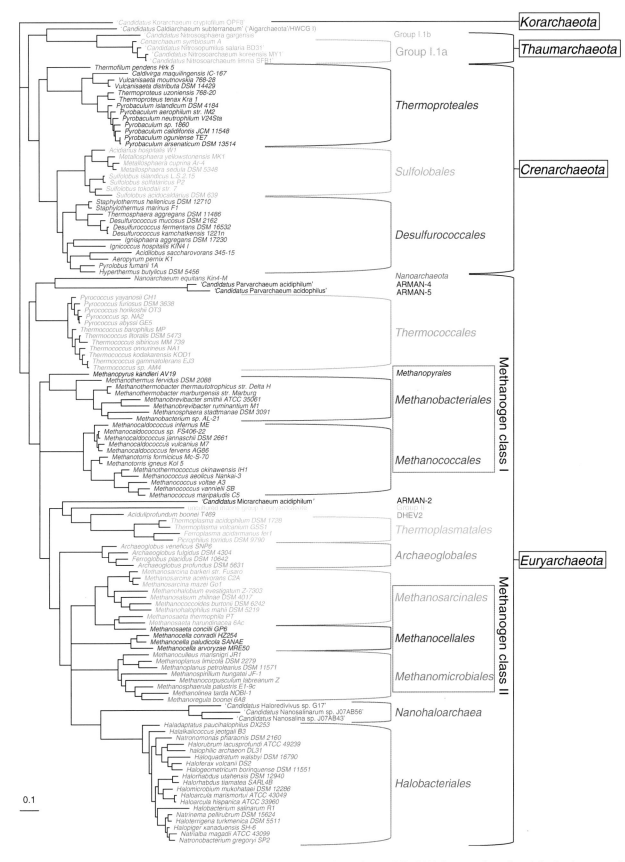

Fig. 6.9 Phylogeny of *Archaea*. Phylogeny of *Archaea* based on a concatenation of ribosomal proteins present in archaea for which complete genome sequences are available. These proteins are a good alternative to 16S rRNA because they allow inferring better resolved (and less biased) phylogenies. They are now commonly used to infer reference phylogenies of microorganisms

and oxidizing H_2 or S^0; lithoautotrophically by oxidizing sulfidic ores (pyrite, sphalerite, and chalcopyrite) or S^0; or chemolithotrophically to chemoheterotrophically by using complex organic compounds and amino acids and by oxidizing S^0. Finally some of them are also able to oxidize Fe^{2+} to Fe^{3+}. They are mainly found in terrestrial volcanic hot springs. *Sulfolobus* are very popular models for studies among others on informational processes (i.e., translation, transcription, replication, and repair), cell division, RNA processing, and metabolism and have been developed for genetic experiments (Leigh *et al.* 2011). A neighbor genus, *Acidianus*, harbors facultative aerobes that can be either obligate lithoautotrophs or facultative chemolithoautotrophs. Under aerobic conditions, they oxidize sulfur compounds as S^0, sulfidic ores, H_2, and Fe^{2+}, while under anaerobic conditions, they use H_2 as electron donor to reduce S^0, leading to the formation of H_2S as metabolic end product (S^0-H_2 autotrophy). *Metallosphaera* are aerobic facultative chemolithoautotrophs that oxidize sulfur compounds as S^0, sulfidic ores, metal sulfides (FeS), H_2S, $S_4O_6^{2-}$, H_2 and can also grow on complex organic substrates. In contrast, *Stygiolobus* are obligate chemolithoautotrophs growing under anaerobic conditions via the S^0-H_2 autotrophy. *Sulfurisphaera* are facultative anaerobes that grow mixotrophically or heterotrophically through S^0-H_2 autotrophy or the oxidation of complex organic compounds. *Sulfurococcus* are aerobic facultative chemolithoautotrophs which oxidize sulfidic ores or S^0 and use complex organic compounds, various sugars, and amino acids. Some of them are also able to oxidize Fe^{2+} to Fe^{3+}. For additional information on *Sulfolobales*, see Dworkin and collaborators (2006).

The *Desulfurococcales*

This order was defined by Harald Huber and Karl Stetter in 2001. *Desulfurococcales* cells are regular to irregular cocci (of about 0.5–15 μm), discs or dishes, which occur singly, in pairs, short chains, or aggregates. Some of them are flagellated. Most are neutrophilic or weakly acidophilic hyperthermophiles, living at temperatures ranging from 85 to 106 °C. *Desulfurococcales* occur mainly in hot marine environments, such as shallow marine sediments, springs, and venting waters (*Aeropyrum*, *Ignicoccus*, *Staphylothermus*, *Stetteria*, *Thermodiscus*, *Pyrodictium*, and *Hyperthermus*) or deep-sea hydrothermal systems and black smokers (*Ignicoccus*, *Staphylothermus*, *Pyrodictium*, and *Pyrolobus*). However, some representatives are also found in hot volcanic terrestrial ecosystems with low salinity and acidity to slightly alkaline pH values, such as hot springs, mud holes, and soils of continental solfataric fields (*Desulfurococcus*, *Sulfophobococcus*, *Thermosphaera*, and *Acidilobus*). *Desulfurococcales* gathers anaerobic, facultative anaerobic, or aerobic organisms. Under autotrophic conditions, they oxidize H_2 using S^0, $S_2O_3^{2-}$, NO_3^-, or NO_2^- as electron acceptor and CO_2 as a carbon source. Alternatively, they are also able to grow organotrophically through aerobic respiration, anaerobic sulfur respiration, or fermentation of organic substrates. The most hyperthermophilic organisms known to date belong to *Desulfurococcales*, among which *Pyrolobus fumarii* has an optimal growth temperature of 106 °C. It is able to grow at 113 °C but not below 90 °C. In contrast, even if some bacterial spores (e.g., *Morella*, Firmicutes) are able to resist 121 °C, no bacterial cells are able to survive above 100 °C. *Desulfurococcales* are divided into two main families: the *Pyrodictiaceae* and the *Desulfurococcaceae*. *Pyrodictiaceae* form a rather coherent cluster containing the genera *Geogemma* (which contain the famous strain 121 which was claimed to grow at 121 °C), *Hyperthermus*, *Pyrodictium*, and *Pyrolobus*, all having optimal growth temperature ranging from 95 to 106 °C. *Pyrodictium* forms networks of hollow cannulae, in which the disk- or dish-shapes cells are embedded. They are marine obligate anaerobes that can be chemolithoautotrophs to mixotrophs able to use S^0 or $S_2O_3^{2-}$ and H_2 under autotrophy with grow on complex organic compounds, or obligate heterotrophs fermenting complex organic compounds or peptides. *Hyperthermus* are obligate marine anaerobic heterotrophic cocci that ferment peptides products leading to the formation of organic acids, butanol, and CO_2. Finally *Pyrolobus* are marine anaerobic or microaerobic, obligately chemolithoautotrophic irregular cocci. They reduce NO_3^-, $S_2O_3^{2-}$, or O_2 with H_2. In contrast, *Desulfurococcaceae* grow optimally at 85–95 °C and are much more diverse than *Pyrodictiaceae*. They encompass the genera *Aeropyrum*, *Desulfurococcus*, *Ignicoccus* (the host of *Nanoarchaeum equitans*), *Ignisphaera*, *Staphylothermus*, *Stetteria*, *Sulfophobococcus*, *Thermodiscus*, *Thermogladius*, and *Thermosphaera*. *Desulfurococcus* are anaerobic cocci isolated from continental environments. They grow mixotrophically or heterotrophically by respiring S^0 or fermenting complex organic compounds, peptides, amino acids, starch, or glycogen. *Aeropyrum* are the only aerobic members of *Desulfurococcaceae*. These cocci are marine heterotrophs that respire complex organic compounds with O_2. *Ignicoccus* are marine anaerobic cocci which grow lithoautotrophically via S^0-H_2 autotrophy and produce exclusively H_2S as the metabolic end product. *Staphylothermus* are anaerobic marine coccoid organisms that grow in aggregates and gain energy by fermentation of complex organic substrates in the presence of S^0. *Stetteria* are coccoid marine anaerobic mixotrophs that respire S^0 or $S_2O_3^{2-}$ on complex organic compounds in presence of H_2. *Sulfophobococcus* are continental anaerobic obligate heterotrophic cocci in which growth is inhibited by S^0. *Thermodiscus* are dish- or disk-shaped anaerobic marine obligate heterotrophs, which carry out S^0 respiration and fermentation of complex organic compounds. *Thermosphaera* forms short chains or aggregates of cocci. They are anaerobic continental obligate heterotrophs which ferment complex organic compounds. Their growth is inhibited by elemental sulfur and also by H_2. Finally, *Desulfurococcales* include additional genera, namely, the *Acidilobus*, *Caldisphaera*, and *Fervidicoccus*. The former and the latter were recently proposed to represent new crenarchaeotal orders (*Acidilobales* and *Fervidicoccales*), whereas more recent analyses have confirmed that their membership of *Desulfurococcales* (see Brochier-Armanet et al. 2011), *Acidilobus* are acidophiles which grow optimally at pH close to 4 in terrestrial acidic hot springs. They are obligate heterotrophs growing via the fermentation of complex organic compounds. For additional information on *Desulfurococcales*, see Dworkin and collaborators (2006).

The *Thermoproteales*

This order was defined by Wolfram Zillig and collaborators in 1981. *Thermoproteales* are extreme thermophiles or hyperthermophiles with optimal growth temperatures ranging from 75 to 100 °C and from neutral to slightly acidic pH (from 3.7 to 7). Currently, the order is represented by two families: the *Thermoproteaceae* and the *Thermofilaceae*, which contain a single genus (*Thermofilum*) and five genera (*Caldivirga*, *Pyrobaculum*, *Thermocladium*, *Thermoproteus*, and *Vulcanisaeta*), respectively. The former are rods of at least 0.4 μm in diameter, whereas the latter are ultrathin filaments of only 0.15–0.35 μm in diameter and 1–100 μm in length. Some *Thermoproteales* bear spherical bodies ("golf clubs") at their terminals (e.g., *Vulcanisaeta*) and are flagellated (*Pyrobaculum*) or not (*Thermofilum*). *Thermofilaceae* and some *Thermoproteus* are obligate heterotrophic anaerobes that use peptides and S^0 as donor and accepter of electrons, respectively. Other *Thermoproteus* are facultative lithoautotrophic anaerobes that use H_2, simple or complex compounds as electron donors, and S^0 or malate as electron acceptors. In contrast, most *Pyrobaculum* are anaerobes, but some are also aerobes. They can be either obligate heterotrophs or facultative lithoautotrophs. They can use a wide range of substrates as electron donors (e.g., H2, $S_2O_3^{2-}$, and complex organic compounds) and electron acceptors (e.g., S^0, $S_2O_3^{2-}$, SO_3^-, O_2, NO_3^-, NO_2^-, Fe^{3+}, selenate, selenite, arsenate, L-cysteine, and oxidized glutathione). *Caldivirga* grows heterotrophically under anaerobic or microaerobic conditions at weakly acidic pH (3.7–4.2) by using complex organic compounds as electron donors and S^0, $S_2O_3^{2-}$, and SO_4^{2-} as electron acceptors. *Thermocladium* are obligate heterotrophs that use S^0, $S_2O_3^{2}$, SO_4^{2-}, and L-cystine as electron acceptor under anaerobic or microaerobic conditions. Finally, *Vulcanisaeta* grows optimally at pH 4.0–4.5 on proteinaceous substrates as carbon sources and use S^0 or $S_2O_3^{2-}$ as electron acceptor. To the exception of *Pyrobaculum aerophilum*, which was isolated from a marine hydrothermal system, all *Thermoproteales* have been obtained from terrestrial volcanic habitats (e.g., springs, water holes, mud holes, and soils of continental solfataric fields, etc.) with low salinity and acidity to neutral pH. In their ecosystem, *Thermoproteales* are important component of food webs. They can function as primary producers and/or as consumers of organic material. For additional information on *Thermoproteales*, see Dworkin and collaborators (2006).

The *Euryarchaeota*

Based on phylogenies of 16S rRNA, the *Euryarchaeota* have been divided initially into nine orders: *Methanobacteriales*, *Methanococcales*, *Methanopyrales*, *Methanomicrobiales*, *Methanosarcinales*, *Halobacteriales*, *Thermoplasmatales*, *Thermococcales*, and *Archaeoglobales*. Recently additional orders have been proposed: *Methanocellales*, *Methanoplasmatales*, etc. *Euryarchaeota* are much more diverse than *Crenarchaeota*. They present very different physiologies, live all types of environments: from temperate environments to the coldest or the hottest, from most alkaline to most acidic, and to most hypersaline ones. Large-scale phylogenetic analyses indicate that *Thermococcales* represent the earliest diverging lineage within *Euryarchaeota*; methanogens class I (*Methanobacteriales*, *Methanococcales*, *Methanopyrales*) and the large cluster containing *Thermoplasmatales*, "Methanoplasmatales," DHEV2, and group II, occupy an intermediate position, whereas *Archaeoglobales* and the large group gathering *Halobacteriales* and methanogens class II (i.e., *Methanosarcinales*, *Methanomicrobiales*, and *Methanocellales*) branch apically in the *Euryarchaeota* tree (Fig. 6.9).

The Methanogens (Fig. 6.10a–d)

Methanogens represent a large and highly diversified group of unrelated strictly anaerobic archaea, which produce large amount of methane (CH_4) as the major end product of their energy metabolism. Methanogens are key components of most anaerobic ecosystems due to their capacity to achieve the final step in the decomposition of organic matter. However, many methanogens are also autotrophs, being able to use CO_2 as sole source of carbon. Different types of reactions can lead to CH_4 production. The former ones are based on CO_2 reduction to CH_4. In this case, CO_2 is reduced by electrons provided by H_2 (hydrogenotrophic methanogens), formate, CO, or certain alcohols (e.g., 2-propanol, ethanol, etc.). Nearly all methanogens can use H_2 as electron donor, but many are also able to utilize formate, whereas very few use alcohols. The second type of reaction is based on methyl compounds. Indeed, some methanogens are able to use methyl-containing C-1 compounds (e.g., methanol, methylamine, dimethylamine, trimethylamine, dimethylsulfide, formate, etc.) as substrate

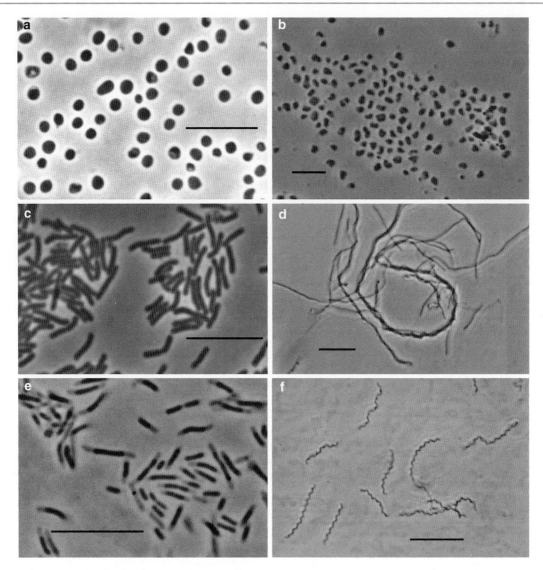

Fig. 6.10 Photomicrographs of methanogenic archaea and/or halophilic (**a–d**) bacteria. (**a**) *Methanosarcina mazei*, (**b**) *Methanocalculus halophilus*, (**c**) *Methanobacterium bryantii*, (**d**) *Methanobacterium oryzae*, (**e**) *Thermotoga hypogea*, (**f**) *Spirochaeta smaragdinae*. (**a** and **c**) Photographs: courtesy of Prof. Friedrich Widdel, MP Institute of Microbiology, Bremen, Germany; (**b, d, e**) Photographs: courtesy of Dr Jean-Luc Cayol, University of the Méditerranée, Marseille; (**f**) Photography: courtesy of Marie-Laure Fardeau, University of the Méditerranée, Marseille. The bars represent 10 μm

for methanogenesis. The methyl groups of these molecules are used as electron acceptors and are reduced directly to CH_4. A small amount of these molecules are used as electron donor through oxidation processes leading to the formation of CO_2. Finally, some *Methanosarcinales* species can catabolize acetate molecules by reducing their methyl carbon to CH_4 and by oxidizing their carboxyl carbon to CO_2. CH_4 production has been reported in non-methanogen organisms (e.g., aerobic bacteria, plants, mitochondria, etc.). However, this production results from side reactions of their normal metabolism and leads to very small amount of CH_4. Methanogens are currently divided into six orders, forming two distinct lineages. Methanogen class I gathers *Methanobacteriales*, *Methanococcales*, and *Methanopyrales*, whereas *Methanomicrobia* (i.e., methanogen class II) encompasses *Methanocellales*, *Methanomicrobiales*, and *Methanosarcinales*. However, the relationships among orders within each class are not resolved (Fig. 6.9). Most methanogens are mesophilic, although extremophiles living in hot (e.g., *Methanocaldococcus* and *Methanopyrus*), cold (e.g., *Methanogenium frigidum*), or hypersaline (e.g., *Methanothermobacter*, *Methanohalobium*) environments are known. Methanogens vary in shape (e.g., long or short rods, regular or irregular cocci, filaments, loops, etc.) and cell wall composition (see above). Methanogens have been developed as models to study archaeal informational processes, regulation, osmoregulation, protein structures, and microbial syntrophic associations (i.e., associations between organisms that facilitate the transfer of nutrients) (Leigh et al. 2011).

Well-developed genetic tools are now available for *Methanococcus* and *Methanosarcina* (Leigh et al. 2011). It has been suggested that methanogenesis was ancestral in *Archaea*, meaning that the last common ancestor to all present-day archaea was a methanogen. However, recent analyzes suggest rather that this metabolism appeared secondarily during the diversification of *Euryarchaeota*, namely, in the last common ancestor of all present-day methanogens (Bapteste et al. 2005). This implies that methanogenesis would have been secondarily lost in their non-methanogen relatives (e.g., *Archaeoglobales, Halobacteriales*, "Nanohaloarchaea," *Thermoplasmatales* and relatives, Fig. 6.9). Supporting this hypothesis, all the genes involved in methanogenesis are present in *Archaeaoglobales*, excepted those involved in the last step of this pathway, preventing the biosynthesis of CH_4 in these archaea (see below). A seventh order tentatively called "Methanoplasmatales," which likely correspond to the RC-III (Rice cluster III), has been recently proposed (Paul et al. 2012). These archaea have been detected in various habitats including marine environments, and soils, but also in the intestinal tracts of termites and mammals. This lineage is closely related to *Thermoplasmatales*, marine group II, and DHVE groups.

The order *Methanomicrobiales* has been proposed by William E. Balch and Ralph S. Wolfe in 1981. Members of this order are strictly anaerobic. They present very diverse morphologies: short rods, curved rods, plates, irregular cocci, filaments, etc., ranging from 0.4 to 2.6 μm in diameter and from 0.1 to 1 in width × 1.5–10 μm in length. Some of them are flagellated. All *Methanomicrobiales* are able to use H_2 and CO_2 as a substrate for methanogenesis, many of them can utilize formate, and some can also use alcohols. In contrast to their close relatives *Methanosarcinales*, they can use neither acetate nor methylated C-1 compounds (such as methanol, methylamines, or methyl sulfides) for methanogenesis, even if acetate can be used as a carbon source by some species. In contrast to *Methanobacteriales* and *Methanopyrales*, *Methanomicrobiales* cell wall does not contain pseudomurein. Some of them (*Methanospirillum hungatei*) are surrounded by proteinaceous sheaths (Table 6.3). Most *Methanobacteriales* are mesophilic, but psychrophilic (e.g., *Methanogenium frigidum*) and thermophilic (e.g., *Methanoculleus thermophilus*) are known. They have been reported in various anaerobic habitats (e.g., marine and freshwater sediments, swamps, anaerobic digesters, rumens of various animals, oil fields, underground waters, etc.). *Methanomicrobiales* are generally free-living, but some species (e.g., *Methanoplanus endosymbiosus*) are endosymbionts of anaerobic H_2 producer protists, such as ciliates living in freshwater sediments (e.g., *Metopus contortus*). *Methanomicrobiales* has been divided into four families. *Methanocorpusculaceae* and *Methanospirillaceae* each contain a single genus *Methanocorpusculum* and *Methanospirillum*. Both genera use H_2 with CO_2 and formate for methanogenesis, whereas some representatives can also use 2-propanol (or 2-butanol) with CO_2. *Methanoregulaceae* group three genera (*Methanolinea, Methanoregula*, and *Methanosphaerula*), whereas *Methanomicrobiaceae* gather six genera (*Methanoculleus, Methanofollis, Methanogenium, Methanolacinia, Methanomicrobium*, and *Methanoplanus*). Members of both families use H_2 with CO_2, but some are also able to use formate for CH_4 production. In addition *Methanomicrobiaceae* can produce CH_4 from 2-propanol and CO_2, from 2-butanol and CO_2, or from cyclopentanol and CO_2. From an evolutionary point of view, *Methanomicrobiales* belong to class II methanogens and are closely related to *Methanocellales* and *Methanosarcinales* and to *Halobacteriales* (Fig. 6.9). For additional information on *Methanomicrobiales*, see Dworkin and collaborators (2006).

The order *Methanosarcinales* has been proposed by David R. Boone and collaborators in 2002. Members of this order are strictly anaerobic. They present very diverse morphologies: coccoid, flat, polygonal, irregular cocci, spheroids, rods, pseudosarcinae, or sheathed rods, etc., ranging in size from 0.5 to 100 μm in diameter and from 0.8 wide × 7 μm long. They can form filaments and aggregates (that can be massive). Some of them are flagellated. Like *Methanomicrobiales*, *Methanosarcinales* cell wall does not contain pseudomurein, and some are surrounded by a proteinaceous sheath (*Methanosaeta concilii*) or methanochondroitin (*Methanosarcina mazei* and *Methanosarcina acetivorans*, Table 6.3). Most of them are mesophiles, even if some species are thermophilic (e.g., *Methanosaeta thermophila*). They live at pH neutral to weakly alkaline. Some *Methanosarcinales* are also halotolerant and halophilic (e.g., *Methanohalobium evestigatum*). They are able to use numerous substrates for methanogenesis. Contrarily to *Methanomicrobiales*, *Methanosarcinales* can grow by splitting acetate to CH_4 and CO_2. They can also dismutate methyl compounds (methanol, methyl amines, methyl sulfides, etc.) producing CO_2 and CH_4, or use H_2 to reduce methyl compounds. They are important components of ecosystems due to their capacity to achieve the terminal steps of the degradation of organic matter in anoxic environments where light and terminal electron acceptors other than CO_2 are limiting. Representatives of the *Methanosarcinales* are widespread and are found in very diverse anaerobic environments (e.g., freshwater, ocean, muds, sediments (even in extremely halophilic ones), gas industry pipelines, underground waters, sludge from anaerobic sewages, rumen and gastrointestinal tracts of metazoa, deep terrestrial subsurface, etc.). This order comprises three families: *Methanosaetaceae, Methanosarcinaceae*, and *Methermicoccaceae*. *Methanosaetaceae* are represented by a single genus (*Methanosaeta*). They are able to use acetate as sole energy source leading to production of CH_4 and CO_2, and acetate is the sole energy substrate. In contrast, *Methanosarcinaceae* encompass nine genera (*Halomethanococcus*, *Methanimicrococcus*, *Methanococcoides*, *Methanohalobium*, *Methanohalophilus*, *Methanolobus*, *Methanomethylovorans*, *Methanosalsum*, and *Methanosarcina* (Fig. 6.10a)) of coccoidal or pseudosarcinal bacteria. All representatives of this family can dismutate methyl compounds. Some are able to reduce acetate or CO_2 with H_2 but none catabolize formate. *Methermicoccaceae* contain a single genus *Methermicoccus* of small, thermophilic cocci, able to use methanol, methylamine, and trimethylamine as substrates for methanogenesis. From an evolutionary point of view, *Methanosarcinales* belong to class II methanogens and are closely related to *Methanomicrobiales* and *Methanocellales* and to *Halobacteriales* (Fig. 6.9). For additional information on *Methanosarcinales*, see Dworkin and collaborators (2006).

The *Methanocellales* order has been proposed by Sanae Sakai and colleagues in 2008. It corresponds to the taxon formerly designated RC-I (Rice Cluster I). This order contains a single family *Methanocellaceae* and a single genus *Methanocella*. Most cells are rod-shaped and occur singly; however, coccoid shaped appear in late-exponential culture. Described strains to date are nonmotile. They produce CH_4 from H_2 and formate. Optimal growth occurs at 35–37 °C, at neutral pH. Based on 16S rRNA environmental survey, *Methanocellales* appear to be widely distributed, especially in rice paddies that are one of the major sources of CH_4 on Earth, contributing about 10–25% of global CH_4 emission. From an evolutionary point of view, *Methanocellales* belong to class II methanogens and are closely related to *Methanomicrobiales* and *Methanosarcinales* and to *Halobacteriales* (Fig. 6.9).

The order *Methanobacteriales* has been proposed by William E. Balch and Ralph S. Wolfe in 1981. *Methanobacteriales* are strict anaerobic mesophilic, thermophilic, or hyperthermophilic microorganisms. They grow at temperature ranging from 15 to 97 °C. Optimal growth of most members of this order occurs at nearly neutral pH even if alkaliphilic (e.g., *Methanobacterium alcaliphilum*) or moderately acidophilic (e.g., *Methanobacterium espanolense*) strains have been characterized. They are found in anoxic habitats (e.g., freshwater and marine sediments, hot springs, groundwater, oil fields, peat bogs, rice paddies, terrestrial subsurface environments, anaerobic sewages, sludge, and gastrointestinal tracts of animals including humans, etc.). *Methanobacteriales* are generally free-living, but some species are endosymbionts of anaerobic H_2 producer protists, such as ciliates (e.g., *Nyctotherus ovalis*) thriving in the intestinal tracts of cockroaches, millipedes, and frogs. Cells are either rod-shaped or coccoid, often forming chains or aggregates. They can also form long filaments (up to 120 µm in length). Similarly to *Methanopyrales* but in contrast to other methanogens, the cell wall of *Methanobacteriales* contains pseudomurein. Most *Methanobacteriales* use H_2 as electron donor to reduce CO_2 leading to CH_4 formation. Some representatives of this order can also use formate, CO, or secondary alcohols as electron donors for CO_2 reduction, whereas members of a single genus (e.g., *Methanosphaera*) can form CH_4 by using H_2 to reduce methanol. *Methanobacteriales* are divided into two families: the *Methanobacteriaceae* and the *Methanothermaceae*. *Methanobacteriaceae* encompass rod-shaped mesophiles and thermophiles. Most of them are nonflagellated. This family is divided into four genera: *Methanobacterium*, *Methanobrevibacter*, *Methanosphaera*, and *Methanothermobacter*. They are H_2 oxidizers, although some species can also oxidize formate, CO, and/or secondary alcohols. *Methanosphaera* reduce methanol, whereas other *Methanobacteriaceae* use CO_2 as an electron acceptor. *Methanobrevibacter smithii* and *Methanosphaera stadtmanae* are two main methanogens found in human gut. The *Methanothermaceae* is represented by a single genus, *Methanothermus*, which grows by reducing CO_2 with H_2. Members of this family are hyperthermophilic, living at optimal temperature ranging from 80 to 85 °C, at pH 6.5. In contrast to *Methanobacteriaceae*, *Methanothermaceae* are flagellates. From an evolutionary point of view, *Methanobacteriales* belong to class I methanogens and are closely related to *Methanococcales* and *Methanopyrales* (Fig. 6.9). For additional information on *Methanobacteriales*, see Dworkin and collaborators (2006).

The order *Methanopyrales* has been proposed by Harald Huber and Karl Setter in 2001. It is represented by a single family *Methanopyraceae* and a single genus *Methanopyrus*, a single species *Methanopyrus kandleri* that was isolated from hydrothermally heated deep-sea sediment. *Methanopyrus* are the only methanogens known to date growing optimally at temperatures greater than 100 °C. *Methanopyrus* cells are rod-shaped and flagellated. Similarly to *Methanobacteriales* but in contrast to other methanogens, the cell wall of *Methanopyrales* contains pseudomurein. *Methanopyrus kandleri* uses H_2 as electron donor to reduce CO_2 leading to CH_4 formation. It is an obligate chemolithoautotroph that uses CO_2 as sole carbon source. From an evolutionary point of view, *Methanopyrales* belong to class I methanogens and are closely related to *Methanococcales* and *Methanobacteriales* (Fig. 6.9).

The order *Methanococcales* has been proposed by William E. Balch and Ralph S. Wolfe in 1981. Representatives of this order have been isolated from various anaerobic habitats (e.g., shores, estuary sediments, salt-marsh, coastal geothermally heated marine sediments, reservoir water, deep-sea hydrothermal vents, high-temperature oil reservoirs, etc.). *Methanococcales* have been divided into two families, each being represented by two genera. *Methanocaldococcaceae* gather *Methanocaldococcus* and *Methanotorris*, whereas *Methanococcaceae* encompass *Methanococcus* and

Methanothermococcus. All Methanococcales produce CH$_4$ by using H$_2$ to reduce CO$_2$, but *Methanococcaceae* are also able to use formate are electron donors. Cells occur as irregular flagellated cocci ranging from 0.9 to 3 μm in diameter occurring singly or in pairs. This order gathers mesophiles (e.g., *Methanococcus vannielii*), thermophiles (e.g., *Methanothermococcus thermolithotrophicus*), and hyperthermophiles (e.g., *Methanotorris igneus*, *Methanocaldococcus infernus*) at weakly acidic (e.g., *Methanotorris igneus*), neutral (e.g., *Methanothermococcus okinawensis*), or weakly alkaliphilic pH. Most members of this order are fast growing and require salt for growth. From an evolutionary point of view, *Methanococcales* belong to class I methanogens and are closely related to *Methanopyrales* and *Methanobacteriales* (Fig. 6.9). For additional information on *Methanococcales*, see Dworkin and collaborators (2006).

The *Halobacteriales*

This order was proposed by William D. Grant and Helge Larsen in 1989. *Halobacteriales* live in environments containing high salt levels exceeding 150–200 g/l. Most Halobacteriales cannot grow at salt concentrations below 2.5–3 M and are irreversibly damaged (or even lyse) when *suspended* in solutions containing less than 1–2 M salt. It comprises a single family the *Halobacteriaceae*, which gather 42 genera: *Haladaptatus*, *Halalkalicoccus*, *Halarchaeum*, *Haloarchaeobius*, *Haloarcula*, *Halobacterium*, *Halobaculum*, *Halobellus*, *Halobiforma*, *Halococcus*, *Haloferax*, *Halogeometricum*, *Halogranum*, *Halomarina*, *Halomicrobium*, *Halonotius*, *Halopelagius*, *Halopenitus*, *Halopiger*, *Haloplanus*, *Haloquadratum*, *Halorhabdus*, *Halorientalis*, *Halorubellus*, *Halorubrum*, *Halorussus*, *Halosarcina*, *Halosimplex*, *Halostagnicola*, *Haloterrigena*, *Halovivax*, *Natrialba*, *Natrinema*, *Natronoarchaeum*, *Natronobacterium*, *Natronococcus*, *Natronolimnobius*, *Natronomonas*, *Natronorubrum*, *Salarchaeum*, and *Salinarchaeum*. Most representatives of this order live in hypersaline marine biotopes or freshwaters, such as salt lakes (e.g., the Great Salt Lake, the Dead Sea, etc.) or saltern crystallizer ponds. These environments can be thalassohaline, meaning that they are dominated by Na$^+$ and Cl$^-$ ions. In contrast, athalassohaline environments present greatly different ionic compositions. Among them, the Dead Sea is dominated by Mg^{2+} and Ca^{2+}, in addition to Na$^+$ and K$^+$. *Halobacteriales* are also present in saline soils, mines, and arid areas (e.g., coasts, plains, mountains, deserts, etc.). Some *Halobacteriales* develop on products preserved by salt (e.g., food, hides, etc.) but contaminated through the use of crude solar salt. Indeed, during crystallization of halite, *Halobacteriales* cells can be trapped inside the growing crystals, and these may remain viable for a long time. The economic damages caused by these halophilic Archaea have triggered many of the early researches on *Halobacteriales*. In many hypersaline environments, *Halobacteriales* coexist with eukaryotes (such as green algae *Dunaliella*) and diverse bacteria such as *Salinibacter ruber* (*Cytophagales*). However, in the most extreme ones, *Halobacteriales* dominate microbial communities over bacteria. Some *Halobacteriales* (e.g., *Natronobacterium*) are haloalkaliphilic, living in alkaline hypersaline lakes characterized by salinity at (or close) to saturation and very high pH (9–11) due to high concentrations of carbonates. Many *Halobacteriales* are mesophilic to moderate thermophilic, having optimal growth temperatures ranging from 35 to 50 °C. This is not surprising given that many hypersaline environments inhabited by *Halobacteriales* are formed by evaporation processes occurring in warm areas. However, psychrotolerant members of *Halobacteriales* exist, such as those living in the very cold (but ice-free) hypersaline Deep Lake (Antarctica), which water temperature varies seasonally between below 0 and +11.5 °C. Some *Halobacteriales* cells are flagellated. *Halobacteriales* present many different morphotypes: rods, cocci, flat pleomorphic types, perfectly square flat cells, or even triangular and trapezoid cells. *Halobacteriales* are also known to carry the largest plasmids known to date, some of them being referred as to minichromosomes due to the presence of important or essential genes. For instance, 547 of the 2,674 (20 %) of the genes of *Halobacterium* sp. NRC1 are located on two megaplasmids, whereas 27 % of the *Haloferax volcanii* genes are carried on two megaplasmids and two small plasmids. In addition, many *Halobacteriales* are polyploid, for instance, there are 15–30 genome copies in *Haloferax volcanii* and *Halobacterium salinarum*.

The *Halobacteriales* are chemoorganotrophic. They oxidize various organic compounds under aerobic conditions and use O$_2$ as terminal electron acceptor. However, the availability of O$_2$ is often limited due to high microbial densities and the limited solubility of O$_2$ at high salt concentrations. Without surprise many *Halobacteriales* are able to use alternative pathways to produce their energy under microaerophilic or anaerobic conditions, including denitrification or fermentation of L-arginine. They are also able to use various compounds as electron acceptors, such as DMSO, TMAO, and fumarate. In addition, some *Halobacteriales* (e.g., *Halobacterium*) are phototrophic, meaning that they can use light to produce ATP. This process is carried out by a membrane protein called bacteriorhodopsin (*cf.* Sect. 3.3.4, Fig. 3.30) under anaerobic conditions. This protein is a light-driven H$^+$ pump that expulses H$^+$ from the cell, generating a H$^+$ gradient which in turn is converted into ATP by ATP synthases. Bacteriorhodopsin may be very abundant in cell membranes. It is responsible for the pink/red coloration of the cells. This phototrophic process is different from chlorophyll-based photosynthesis because it requires neither chlorophyll nor electron transport chains. Furthermore and contrarily to bacteriorhodopsin-based phototrophy, photosynthesis requires additional pigments known as "antennas" and is coupled to carbon fixation. In 2001, homologues of bacteriorhodopsins were discovered in uncultured planktonic marine SAR86 gammaproteobacteria (Beja et al. 2000). Accordingly, these were referred as proteorhodopsins. Since then, homologues of proteorhodopsin were shown to be widespread in the oceans and in many prokaryotic lineages including the ubiquitous and abundant SAR11 alphaproteobacteria, archaea from group II, etc. This suggests that proteorhodopsin-based phototrophy is a significant oceanic microbial process, and therefore could represent a significant source of energy for microbial communities living in the ocean photic zone.

To thrive in environments rich in salts, *Halobacteriales* must cope with high osmotic pressures. Most halophilic or halotolerant microorganisms use a "salt-out" strategy consisting in the expulsion of salt from the cytoplasm to maintain low intracellular ionic concentrations and the use of organic solutes (e.g., glycerol or glycine betaine) to maintain an osmotic balance with the extracellular medium. In contrast, *Halobacteriales*, a few anaerobic halophilic bacteria (*Haloanaerobiales*), and *Salinibacter ruber* (*Bacteroidetes*) have adopted a radically different strategy referred as to "salt-in," which consists in the molar accumulation of ions, especially K^+ and Cl^- ions (and often Na^+) in their cytoplasm. The import of K^+ allows balancing the high osmotic pressure generated by high environmental Na^+ levels. This equilibrium between intracellular K^+ and extracellular Na^+ is essential to prevent cells from dehydration. Some haloalkaliphilic archaea use organic osmotic solutes (e.g., 2-sulfotrehalose) in addition to high intracellular salt levels. The salt-out strategy is energetically costly due to the importance of organic-compatible solutes and less suitable at saturating salt levels, which probably explain why organisms using the salt-in strategy predominate under extreme hypersaline conditions. Due to the salt-in strategy, *halobacteriales* proteins must be adapted to function at molar salt levels. The side effect is that these proteins commonly denature in low-salt solutions. In addition, the cell wall of many *Halobacteriales* is composed of glycoproteins enriched in acidic amino acids (i.e., aspartate and glutamate). The negative charges provided by the carboxyl groups of these amino acids are surrounded by the Na^+ ions which stabilize these glycoproteins, ensuring the integrity of the wall. If Na^+ concentration becomes too low, the negatively charged glycoproteins repel each other, leading to destabilization of the cell wall and cell lysis. It has been early noticed that cytoplasmic proteins are also enriched in acidic amino acids. In addition they are depleted in hydrophobic amino acids, compared with their homologues found in nonhalophilic species. The replacement of large hydrophobic residues by small hydrophilic residues on cytoplasmic protein surface increases their overall polarity, preventing their aggregation and allowing them to remain functional. In parallel their increase in acidic residues creates a high density of negative charges coordinating a network of hydrated cations, which help to maintain the proteins in solution. *Halobacteriales* have been developed as genetic model because they are efficiently transformable. Moreover, they are simple to manipulate, in particular they are easy to culture, fast growing, and resistant to contamination by nonhalophilic microorganisms. They have been exploited to uncover genes involved in osmotic stress. Furthermore, they are also good models for structural biology because their proteins function under conditions of low water availability and biotechnology (Leigh et al. 2011). From an evolutionary point of view, *Halobacteriales* represent a relatively late diverging order within *Euryarchaeota*. They are related to "Nanohaloarchaea," a lineage of uncultured nanometric archaea (0.6 μm in diameter), which are prevalent in worldwide distributed hypersaline environments (Narasingarao et al. 2011). *Halobacteriales* are grouping with methanogens class II (i.e., *Methanocellales*, *Methanomicrobiales*, and *Methanosarcinales*) (Fig. 6.9). For additional information on *Halobacteriales*, see Dworkin and collaborators (2006).

The *Thermococcales*

This order has been proposed by Wolfram Zillig and collaborators in 1987. It gathers anaerobic heterotrophic hyperthermophiles which grow optimally at neutral pH (6.0–7.0), even if a few alkaliphilic strains able to grow at pH 9 have been reported. *Thermococcales* cells are spherical and some of them are flagellated. Their metabolism is based on fermentation. They use polymeric organic substrates like peptides and carbohydrates as energy and carbon sources. The fermentation process produces H_2 which in turn can be used to reduce S^0 to H_2S. Depending of the considered strains, S^0 can either be required for growth or to stimulate growth. In some strains (e.g., *Palaeococcus*), S^0 can be replaced by Fe^{2+}. From an evolutionary point of view, *Thermococcales* have diverged early within *Euryarchaeota* (Fig. 6.9). Members of this group are very abundant and commonly found within marine hot water environments. They represent therefore a major constituent of the biomass in these ecosystems. *Thermococcales* contain a single family *Thermococcaceae*, represented by three genera: *Pyrococcus*, *Thermococcus*, and *Paleococcus*, the two first being more closely related to each other. *Pyrococcus* live in marine hydrothermal vents, whereas *Thermococcus* are also found in terrestrial freshwater, marine solfataric ecosystems, deep-sea hydrothermal vents, and offshore oil wells. *Thermococcales* have been developed as genetic models for studying DNA replication and repair, transcription and its regulation, carbon and energy metabolism, CRISPR systems, and cellular responses to stress, such as oxidative, osmotic, temperature, and pressure (Leigh et al. 2011). *Thermococcales* are also used as models in various fields of biotechnology due to their capacity to efficiently utilize polymeric substrates and their capacity to produce a vast array of stable, polymer-degrading hydrolases. For additional information on *Thermococcales*, see Dworkin and collaborators (2006).

The *Thermoplasmatales*

This order was proposed by Reysenbach in 2001. It gathers aerobic and facultative anaerobic acidophiles which can be autotrophic or heterotrophic. The most acidophilic organisms known to date belong to this order. Most representatives of this order maintain their intracellular pH near neutrality, thanks to sophisticated mechanisms such as very effective H^+ pumps that expel H^+ outside of the cell and membranes that are highly impermeable to H^+. Some *Thermoplasmatales* are flagellated. The name is a reminder that the first described representatives of this order were thermophilic acidophiles, although currently mesophilic representatives are known. Thermoplasmatales are divided into three families: *Ferroplasmaceae*, *Picrophilaceae*, and *Thermoplasmataceae*, each containing a single genus, namely, *Ferroplasma*, *Picrophilus*, and *Thermoplasma*. *Thermoplasma* cells are of 0.2–5 μm in size. As they are devoid of cell wall or envelope (see above), the cells are pleomorphic, a feature shared with bacterial *Mycoplasma*. Because of this lack of cell wall and because the colonies of *Thermoplasma* resemble those of bacterial *Mycoplasma*, both were initially classified in the same taxonomic group. Molecular phylogeny studies have proven later that these features resulted from independent evolutionary processes (i.e., convergence) and that *Mycoplasma* and *Thermoplasma* are in fact unrelated. *Thermoplasma* are facultative aerobes growing at temperatures ranging from 33 to 67 °C (optimal near 60 °C) at pH 1–2. They are not able to survive at neutral pH. They are chemoorganotrophs. They can grow anaerobically through S^0 respiration which leads to the release of large amounts of H_2S. They are also able to grow anaerobically without S^0, suggesting that additional unidentified molecules can be used as electron acceptors. *Thermoplasma* can also grow aerobically. Hot springs, solfataric fields, warm acidic tropical swamps, marine hydrothermal systems, as well as aerobic and anaerobic zones of continental volcanic areas constitute natural habitats of *Thermoplasma*. They are also found in anthropized environments as coal refuse piles and associated water samples. *Ferroplasma* (as *Thermoplasma*) are pleomorphic. They are mesophilic chemolithoautotrophs. They are able to use CO_2 as a carbon source and Fe^{2+}, pyrite, or Mn^{2+} as energy source. Some of them are also able to grow heterotrophically. They occur in many sulfidic ore-containing mines and heaps on earth, as well as in acidic geothermal pool. *Picrophilus* are hyperacidophilic obligate aerobic heterotrophic irregular cocci of 1–1.5 μm. They grow optimally at temperatures near 60 °C and pH 0.7, but can still divide at pH 0. No growth occurs below 47 °C and above 65 °C, or at pH higher than 3.5. Unlike other *Thermoplasmatales*, intracellar pH of *Picrophilus* is low (approximately 4.6), suggesting specific adaptations allowing enzymatic activity in acidic medium. However, the nature of these modifications remains to be determined. They thrive in geothermal solfataric soils and springs, and terrestrial geothermal environments. *Thermoplasmatales* group with DHEV2, the recently proposed "Methanoplasmatales," and the uncultured group II (Fig. 6.9). For additional information on *Thermoplasmatales*, see Dworkin and collaborators (2006).

The *Archaeoglobales*

This order was proposed by Karl Stetter in 1989. It gathers regular to irregular lobe-shaped cocci of 0.3–1.3 μm in diameter occurring singly or in pairs that may be flagellated. *Archaeoglobales* grow anaerobically at temperatures ranging from 65 to 95 °C, at neutral or weakly acidic pH. This order contains a single family *Archaeoglobaceae*, which is divided into three genera *Archaeoglobus*, *Ferroglobus*, and *Geoglobus*. Most of them live in anoxic shallow and abyssal submarine hydrothermal vents, hot oil field waters but some have been identified in terrestrial environments including hot springs and terrestrial oil wells where SO_4^{2-} are abundant. *Archaeoglobus* are chemoorganotrophic sulfate reducers that oxidize H_2 or organic compounds (e.g., lactate, glucose, pyruvate, acetate, or formate) and reduce SO_4^{2-}, SO_3^{2-}, or $S_2O_3^{2-}$ (but not S^0) leading to H_2S formation. Some of them are also chemolithoautotrophic, being also able to use CO_2 as carbon source. In contrast, *Ferroglobus* are chemolithotrophic which grow anaerobically by using aromatic compounds such as benzoate as electron donors and Fe^{3+} as electron acceptor. They are also able to use H_2 and H_2S as energy sources and NO_3^- as a terminal electron acceptor leading to the formation of NO_2^-. $S_2O_3^{2-}$ can also be used as a terminal electron acceptor. *Geoglobus* are anaerobic chemoorganotrophs that grow by oxidizing acetate, pyruvate, palmitate, and stearate coupled to reduction of Fe^{3+}. They can also grow autotrophically with H_2 as electron donor and poorly crystalline Fe^{3+} oxide as electron acceptor. From an evolutionary point of view, *Archaeoglobales* branch after the methanogens class I (*Methanopyrales*, *Methanococcales*, *Methanobacteriales*) and *Thermoplasmatales*, but before the diversification of the large group containing *Halobacteriales*, *Methanomicrobiales*, *Methanosarcinales*, and *Methanocellales* (Fig. 6.9). Evolutionary and genomic studies have shown that *Archaeoglobales* are descendants of methanogens that have secondarily lost the capacity to produce methane. This hypothesis was strongly reinforced by the discovery in the genome of *Archaeoglobus fulgidus* of almost all genes encoding enzymes and cofactors involved in methanogenesis. The only missing genes are those encoding methyl CoM reductase (the enzyme involved in the final step of methanogenesis), which prevents the biosynthesis of methane by this organism. Instead these genes are used in the reverse direction during the oxidation of lactate. For additional information on *Archaeoglobales*, see Dworkin and collaborators (2006).

The *Nanoarchaeota*

This lineage was first described as a new phylum tentatively named "Nanoarchaeota" by Harald Huber and colleagues in 2002. The only representative of this group is "Nanoarchaeum equitans," a nanometric hyperthermophilic archaeon that grows attached to the surface of the surface of *Ignicoccus hospitalis* (*Desulfurococcales*). The nature of the association (i.e., symbiotic or parasitic) between these two archaea remains debated, even if most recent analyses favor the latter hypothesis. Environmental surveys suggest that members of the "Nanoarchaeota" are broadly distributed in hot biotopes, such as the deep-sea hydrothermal vents, shallow marine areas, and terrestrial solfataric fields. *Nanoarchaeota* are anaerobic cocci of only 400 nm in diameter. "Nanoarchaeum equitans" cells are not flagellated. They grow under strictly anaerobic conditions at temperatures ranging from 75 to 98 °C. Early phylogenetic analyses suggested that "Nanoarchaeota" branch deeply in archaeal tree, meaning before to the speciation of *Crenarchaeota* and *Euryarchaeota*. However, later analyses suggested that this position results from a tree reconstruction artifact called long branch attraction and that "Nanoarchaeota" represent in fact a fast-evolving lineage of *Euryarchaeota*, likely related to *Thermococcales* (Brochier et al. 2005) (Fig. 6.9). The relationship between "Nanoarchaeota" and *Euryarchaeota* has been strengthened by comparative genomic analyses. The 16S rRNA of "Nanoarchaeota" are highly divergent even in previously "universally" conserved regions, meaning that they cannot be easily amplified by PCR using universal 16S rRNA primers. The genome of "Nanoarchaeum equitans" is the smallest genome of archaea known to date (490 kb, corresponding to ~552 genes). It is G + C poor (31.6 %), has high gene density (95 %), and presents a number of unique genomic rearrangements (e.g., split genes such as tRNA, etc.) that were sometime interpreted as ancestral archaeal features in agreement with the deeply branching position of "Nanoarchaeota" observed in initial studies. Many biosynthetic pathways (e.g., lipids, cofactors, amino acids, nucleotides, etc.) and metabolic pathways (e.g., glycolysis, pentose phosphate pathway, carbon assimilation, etc.) are lacking. This suggests that "Nanoarchaeum equitans" is strictly dependent of its host and why all isolation attempts were unsuccessful so far. The genome sequence of a second representative of "Nanoarchaeota" (distantly related to "Nanoarchaeum equitans") has been obtained from an enrichment culture obtained from a terrestrial hot spring (Podar et al. 2013). In contrast to "Nanoarchaeum equitans," this new strain could be a symbiont of an uncultured *Sulfolobales*, but this remains to be formerly demonstrated. The genome of this new nanoarchaeon is larger than the genome of "Nanoarchaeum equitans," which contains less split genes, suggesting that the former has experienced less severe genome reduction than the latter. It encodes a complete gluconeogenesis pathway as well as a full set of archaeal flagellum proteins, suggesting that this new nanoarchaeon is motile. Altogether these findings indicate that the features of "Nanoarchaeum equitans" that were interpreted as ancestral are in fact derived characters linked to their particular lifestyle (Podar et al. 2013).

The *Korarchaeota*

Korarchaeota (from the Greek "koros" meaning young) was proposed by Susan M. Barns and colleagues in 1996 based on 16S rRNA environmental surveys of a hot spring in the Yellowstone National Park in the USA (Barns et al. 1996). This phylum is represented by a handful of 16S rRNA gene sequences from uncultured representatives. For a long time, these archaea remained elusive. This situation has changed following the report of the genome sequencing of the "*Candidatus* Korarchaeum cryptofilum" from an enrichment culture (Elkins et al. 2008). This ultrathin filamentous korarchaeaon is 0.16–0.18 µm wide × 26 µm long. The analysis of the genome suggests that "*Candidatus* Korarchaeum cryptofilum" ferments peptides to obtain carbon and energy. It lacks the ability to synthesize de novo many compounds and cofactors such as purines and CoA. Phylogenetic analyses based on 16S rRNA genes and on various conserved proteins support a closer relationship of *Korarchaeota* with *Crenarchaeota* and *Thaumarchaeota* (including "Aigarchaeota") than with *Euryarchaeota* (Fig. 6.9).

The *Thaumarchaeota*

Thaumarchaeota (from the Greek "thaumas" meaning wonder) was proposed in 2008 by Céline Brochier-Armanet and collaborators (Brochier-Armanet et al. 2008). *Thaumarchaeota* (formerly referred as group I or mesophilic crenarchaeota) have been discovered independently by the teams of Jed Fuhrman and Ed DeLong in 1992 through 16S rRNA surveys of environmental marine samples. This phylum constitutes one of the most abundant and diversified archaeal taxonomic groups (Brochier-Armanet et al. 2012). It gathers numerous sublineages (e.g., group I.1a, I.1b, I.1c, 1A/pSL12, ThAOA/HWCG III, SAGMCG-I, SCG, FSCG, etc.). Most Thaumarchaeota are uncultured free-living archaea occurring in very diverse environments (e.g., freshwaters, lakes, soils, sediments, oceans, hot springs, etc.), but some of them live in association with animals (e.g., *Cenarchaeum symbiosum*, a sponge symbiont): Worth noticing, Takuro Nunoura and colleagues have recently assembled a composite genome ("*Candidatus* Caldiarchaeum subterraneum") from a metagenomic library prepared from a geothermal water stream collected in a subsurface gold mine, which was proposed to represent an additional phylum tentatively called "Aigarchaeota" (corresponding to the formerly uncultured HWCG I group). However, subsequent phylogenetic and comparative genomics analyses suggest that "Aigarchaeota" rather represent an early branching lineage within *Thaumarchaeota* (Brochier-Armanet et al. 2011). *Thaumarchaeota* are the subject of much attention since environmental metagenomic surveys have shown that some members of this phylum carry ammonia monooxygenase genes, suggesting that they are nitrifiers (Treusch et al. 2005). This was confirmed after the isolation of *Nitrosopumilus maritimus*, a marine thaumarchaeon that grows chemolithoautotrophically by aerobically oxidizing ammonia to nitrite (Könneke et al. 2005). This key and limiting step of the nitrogen cycle was previously thought to be restricted to a few sublineages of autotrophic betaproteobacteria and gammaproteobacteria and a few heterotrophic nitrifiers. The discovery of nitrifying archaea extends the taxonomic range of microorganisms capable of nitrification and therefore opens up new perspectives on the origin and evolution of this key metabolism on Earth. Moreover, the widespread distribution of putative archaeal nitrifiers and their numerical dominance over their bacterial counterparts in most marine and terrestrial environments suggested that ammonia oxidizer thaumarchaeota play a major role in global nitrification (Pester et al. 2011). However, recent studies suggest that thaumarchaeotal nitrifiers dominate over bacteria only in environments containing low ammonium concentrations. Three orders of Thaumarchaeota have been proposed to date: *Cenarchaeales*, *Nitrosopumilales*, and *Nitrososphaerales*. The first two correspond to the former group I.1a, whereas the latter was previously referred as group I.1b. *Cenarchaeales* and *Nitrososphaerales* contain a single family and a single genus (*Cenarchaeaceae*/*Cenarchaeum* and *Nitrososphaeraceae*/*Nitrososphaera*, respectively), whereas *Nitrosopumilales* contains a single family (*Nitrosopumilaceae*) and two genera (*Candidatus* Nitrosoarchaeum and *Nitrosopumilus*). However, these lineages represent a small part of the real diversity of Thaumarchaeota, meaning that additional thaumarchaeota taxonomic groups should be proposed in the near future. Currently, it is not clear whether ammonia-oxidizing thaumarchaeota are strict autotrophs or also able to use other substrates (such as amino acids, oligopeptides, and glycerol) for their growth, indicating they could also be mixotrophs or even heterotrophs. Phylogenetic analyses of various conserved proteins support a closer relationship of *Thaumarchaeota* (including "Aigarchaeota") with *Crenarchaeota* and *Korarchaeota* than with *Euryarchaeota* (Fig. 6.9).

6.6.2 Domain *Bacteria*

In the second edition of "Bergey's Manual of Systematic Bacteriology," the domain is described with 30 bacterial phyla. Relationships between these phyla are not yet resolved and represent a major challenge in microbial evolution.

The 30 bacterial phyla are presented in Table 6.2. They bring together some 7,500 species characterized phenotypically from cultures of bacterial strains, and therefore they are defined from these bacterial isolates fully characterized.

6.6.2.1 Phyla B1 to B9

Phyla B1 to B9 correspond to bacteria generally positioned in the lower branches of the phylogenetic tree based on 16S rRNA bacteria gene sequences (Fig. 6.11) or bacteria having extremophilic characters and sometimes isolated from extreme environments. However, this position is not found in most phylogenetic analyses based on protein markers (Lopez-Garcia and Moreira 2008). Many species belonging to the phyla B1 to B9 are represented by bacteria capable of living in environments considered extreme, either hyper-hot or highly contaminated by metals or radiation. The phyla *Aquificae* and *Thermotogae*, and genus *Thermus* contain the most known thermophilic members among domain Bacteria.

The Phylum *Aquificae*
These Gram-negative aerobic autotrophic and hydrogenophilic bacteria can grow up to temperatures between 80 and 95 °C. This phylum includes a single class (*Aquificae*) containing a single order, *Aquificales*, in which there is a single family *Aquificaceae*. The main genera are *Aquifex*, *Calderobacterium*, *Hydrogenobacter*, and *Thermocrinis*. Representatives of these genera are hyperthermophilic bacteria isolated from marine and terrestrial hot springs, and developing either aerobic or microaerophilic or anaerobic and are chemoorganotrophs and some chemolithotrophs with hydrogen or thiosulfate as electron donors. These microorganisms occupy a basal position in the phylogenetic tree of Bacteria based on 16S rRNA. It was therefore suggested that these were very old lines having inherited from LUCA (*cf.* Sect. 4.1.1) the ability to live at high temperatures. However, this phylogenetic position is controversial, and these microorganisms may actually

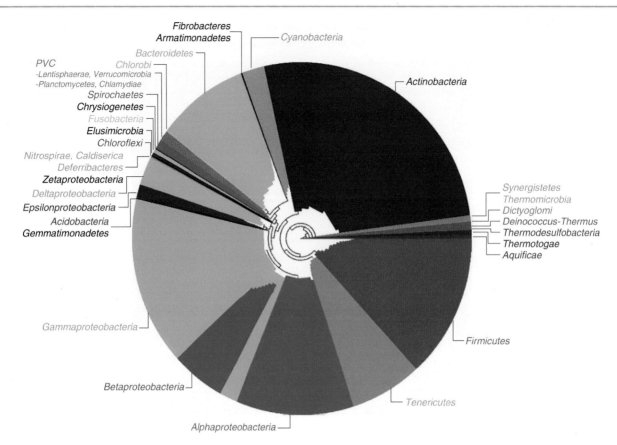

Fig. 6.11 Phylogeny of *Bacteria*. Phylogeny of *Bacteria* based on 8493 16S rRNA gene sequences of type strains (>1,200 bp) available at the Ribosomal Database Project II (release 10, http://rdp.cme.msu. edu/) in 2013. The tree was inferred with the neighbor-joining method. Bacterial phyla are shown in color. The width of areas is proportional to the number of sequences available for each phylum

be descendants of mesophilic microorganisms which are secondarily adapted to life at high temperature (*cf.* Sect. 4.1.6).

The Phylum *Thermotogae* (Fig. 6.10e)

Represented by a single class and a single order, *Thermotogales* includes hyperthermophilic bacteria, Gram-negative, anaerobic, and that has an outer membrane larger than the bacterial body and wrapping it as a "toga." One family (*Thermotogaceae*) includes several genera including types *Thermotoga*, *Geotoga*, *Petrotoga*, *Marinitoga*, and *Kosmotoga*, isolated for most of them from deep environments (aquifers, oil reservoirs) or volcanic sources in land or underwater. These bacteria are fermentative; some may use thiosulfate or sulfur as electron acceptors, and reduce H_2S.

Phyla *Thermodesulfobacteria*, *Thermomicrobia*, and *Deferribacteres*

They also include Gram-negative thermophiles or moderately thermophilic bacteria. They are also phyla represented by one class, one order, and one family. In *Thermodesulfobacteriaceae*, the genus *Thermodesulfobacterium* is strictly an anaerobic chemoorganotroph capable of sulfate-reducing and fermenting. In *Thermomicrobiaceae*, the genus *Thermomicrobium* is formed of irregular short rods that are aerobic and chemoorganotrophic. The *Deferribacteraceae* family includes several genera (among them *Deferribacter* and *Geovibrio*) which are constituted of straight or curved rods, anaerobes isolated from marine sediments or oil tanks, and are chemoorganotrophic using nitrate, iron, manganese, or cobalt as electron acceptors for anaerobic respirations.

The Phylum *Chrysiogenetes*

With one class, one order (*Chrysiogenales*), and one family (*Chrysiogenaceae*), this phylum is represented by a single genus *Chrysiogenes* formed of Gram-negative, curved, mesophilic, chemoorganotrophic performing anaerobic respiration with arsenate as electron donor.

The Phylum *Deinococcus/Thermus*

This phylum is a dream for biotechnologists. The only class includes two orders, *Deinococcales* and *Thermales*. The *Deinococcales* with a single family (*Deinococcaceae*) include Gram-positive chemoorganotrophs, aerobic mesophilic or marginally thermophilic, of the genus *Deinococcus* from which some representatives (*Deinococcus radiodurans*) are resistant to gamma irradiation of about 5,000 Gy, doses sufficient to eliminate most other microorganisms. The radiation resistance of *Deinococcus* could be linked to the presence of a mechanism specific (and very effective) for DNA repair (Cox and Battista 2005). These bacteria have a special envelope with a thick Gram-positive wall and in addition, an outer membrane, and an outer layer of S protein.

The *Thermales*, with one family, *Thermaceae*, combine several genera including genus *Thermus* isolated from hot springs (70–90 °C) which gave the thermostable DNA polymerase used for PCR, an enzyme that represents an annual market of 300 million dollars. This genus comprises bacteria in short filaments or rods, hyperthermophilic, chemoorganotrophic, and aerobic, with certain strains that are capable of using nitrate as electron acceptor under anaerobic conditions. The members of this phylum contain ornithine in their cell wall.

The Phylum *Chloroflexi*

This phylum contains Gram-negative photosynthetic anoxygenic bacteria (also called phototrophic green nonsulfur bacteria) formed of multicellular filaments living in microbial mats or in the water in hot environments anoxic and exposed to light, often in hot springs. Class *Chloroflexi* with order *Chloroflexales* contains *Chloroflexaceae* families (genera *Chloroflexus*, *Chloronema*) and *Oscillochloridaceae* (genus *Oscillochloris*). This class includes bacteria which contain bacteriochlorophyll c and carotenoids present in structures attached to the cytoplasmic membrane, the chlorosomes (*cf.* Sect. 3.3.4, Fig. 3.27), and the genus *Heliothrix* that contains bacteriochlorophyll a with no apparent membrane structures. These bacteria perform anoxygenic photosynthesis and use organic compounds in photoorganotrophic metabolism, and they can fix CO_2 by the pathway called hydroxypropionate cycle (*cf.* Sect. 3.4.1, Fig. 3.34). In this class, there are also many non-photosynthetic species grouped in order *Herpetosiphonales*. Classes of *Anaerolineae* (*Anaerolinea*) and *Caldilineae* (*Caldilinea*) include many genera and species of filamentous bacteria isolated from anaerobic fermenters. Other non-photosynthetic bacteria are also present in this group and have been detected so far by genetic analysis. These bacteria are in the majority inhabitants of environments contaminated by organic pollutants and metal, whether in soils or aquatic environments.

The Phylum *Nitrospirae*

With a single order (*Nitrospirales*) and one family (*Nitrospiraceae*), this phylum includes bacteria belonging to genera quite different, consisting of spiral or curved rods, Gram-negative. These microorganisms are aerobic chemolithotrophic nitrifiers (*Nitrospira*), iron-oxidizers (*Leptospirillum*), anaerobic chemoorganotrophic sulfate reducers, and thermophilic microbes (*Thermodesulfovibrio*) or microbes with magnetic iron inclusions ("*Magnetobacterium*" Fig. 14.46), mostly isolated from terrestrial and marine aquatic environments.

6.6.2.2 The Phylum B 10 Synergistetes

This phylum is Gram-negative, anaerobic, chemo- and fermentative organotrophic isolated from animals (rumen, genus *Synergists*) or man (genus *Jonquetella*). These bacteria are involved in anaerobic digestion operation in the rumen of animals and the human intestinal flora.

6.6.2.3 The Phylum Cyanobacteria B11 (Figs. 6.12 and 6.13)

This phylum consists of Gram-negative bacteria, performing oxygenic photosynthesis with photosystems I and II inserted in membrane systems (thylakoids) and producing oxygen (*cf.* Sect. 3.3.4, Fig. 3.28). This bacterial group would be considered as the ancestors of chloroplasts (*cf.* Sects. 3.3.4, 4.3.1, and 5.4.3). They form a group of bacteria of varied forms ranging from unicellular to multicellular filamentous and colonial types (Figs. 6.12 and 6.13). They contain chlorophyll, carotenoids, and phycobilins or biliproteins, giving them a dark blue-green pigmentation. Some cyanobacteria have the divinyl chlorophyll. They are present in surface waters as well as all kinds of wet surfaces exposed to light where their photosynthetic capacity gives them a competitive advantage. Many kinds of cyanobacteria fixing dinitrogen, including filamentous genera *Anabaena* and *Nostoc* and a proportion of them lives in symbiosis with Fungi (lichens) or with Viridiplantae (*Cycas*, *Azolla*). Some species of cyanobacteria in algal blooms can produce enterotoxic, hepatotoxic, or neurotoxic cyanotoxins that cause poisoning, which are sometimes fatal like microcystin in *Microcystis* spp. (Box 15.2). Cyanobacteria use the Calvin cycle for CO_2 fixation. For many years, cyanobacteria have been considered algae (hence the name blue-green algae) due to their ability to perform oxygenic photosynthesis. Their phylogenetic positioning has allowed classifying them as members of the domain *Bacteria* where they form a major phylum organized into five subsections. Classification within cyanobacteria is problematic because of their membership to the bacterial and botanical systematic codes. Numerous discussions between botanists and bacteriologists are still underway to harmonize the classification of cyanobacteria in both systematic codes.

Subsection 1 (*Chroococcales*): it relates unicellular cyanobacteria that can form clusters of cells such as *Microcystis*, *Synechococcus*, *Cyanobium*, *Cyanothece*, and *Gloeothece*.

Subsection 2 (*Pleurocapsales*): it includes cyanobacteria forming colonial structures and small spherical structures

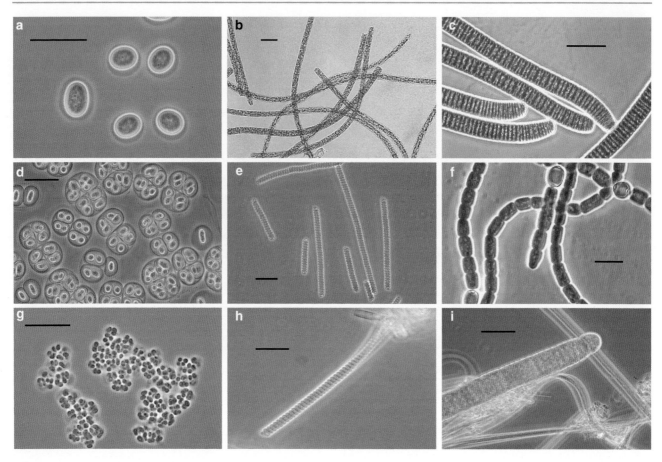

Fig. 6.12 Photomicrographs of *Cyanobacteria*. (**a**) *Cyanothece* sp., (**b**) *Planktothrix rubescens*, (**c**) *Oscillatoria sancta*, (**d**) *Gloeothece membranacea*, (**e**) *Spirulina subsala*, (**f**) *Anabaena cylindrical*, (**g**) *Gloeobacter violaceus*, (**h**) *Spirulina* sp., (**i**) *Oscillatoria* sp. (**a–g**) Photographs, courtesy of Thierry Laurent (cyanobacteria Unit, Institut Pasteur Paris) and Jean-François Humbert (Institute Pasteur, INRA); and (**h, i**) Photographs, Rémy Guyoneaud. The bars represent 10 μm

called beocytes for multiplication (such as genera *Pleurocapsa*, *Dermocarpa*).

Subsection 3 (*Oscillatoriales*): it is formed of filamentous cyanobacteria that divide by binary fission in a single plane constituting long gliding filaments of identical non-individualized cells as *Oscillatoria*, *Spirulina*, and *Microcoleus*.

Subsection 4 (*Nostocales*): it corresponds to the filamentous cyanobacteria with heterocysts – the filaments are formed of photosynthetic cells sometimes interspersed with differentiated cells (heterocysts). These do not make photosynthesis and do not produce oxygen but possess an enzyme (nitrogenase) sensitive to oxygen which enables them to fix atmospheric dinitrogen (N_2) by reducing it to ammonium used by other cells of the filament (Stacey et al. 1979). The best known are the genera *Nostoc*, *Anabaena*, and *Scytonema*.

Subsection 5 (*Stigonematales*): it includes branched filamentous cyanobacteria as genera *Fischerella* and *Stigonema*.

Many cyanobacteria even non-heterocystous ones are able to fix nitrogen from atmospheric dinitrogen. In this case, nitrogenase must be protected from oxygen produced during photosynthesis, either by being active especially in dark phase or by a spatial cell organization in clusters, allowing the cells at the center of the cluster to find an anoxic environment (*cf.* Sect. 3.4.4).

In the phylum, *Cyanobacteria* are also prochlorophytes with *Prochlorococcus*, *Prochloron*, and *Prochlorothrix* genera, which are unicellular or filamentous, abundant in marine phytoplankton. They contain chlorophyll a (as other cyanobacteria) and b (as Viridiplantae chloroplasts), but do not have phycobilins. These microorganisms thrive in the euphotic zone of oligotrophic oceanic waters in the wide open ocean. They are grouped within *Chroococcales* (Subsection 1).

6.6.2.4 Other Gram-Negative Bacteria

This set consists of phyla B12 to B30 except phyla B14, B15, and B16 (which include most of the Gram-positive bacteria), that is to say 16 phyla in total. It includes the majority of described species in the domain *Bacteria*. A phylum is

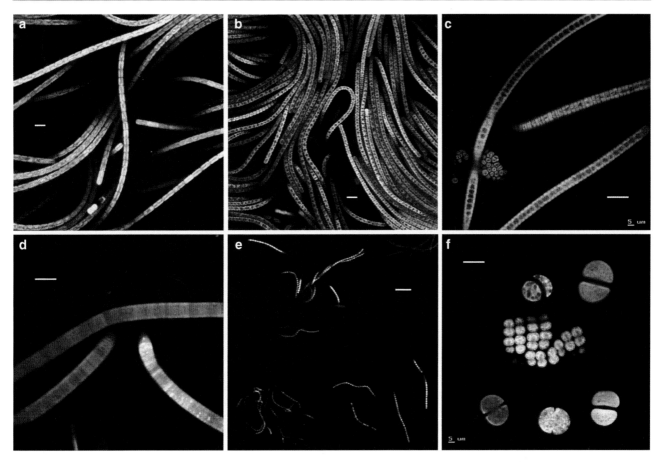

Fig. 6.13 Confocal photomicrographs of cyanobacteria observed in a microbial mat from the Ebro Delta in Spain. (**a**) *Microcoleus chtonoplastes*, (**b**) Morphology of *Phormidium*, (**c**) Morphology of *Lyngbia*, (**d**) Morphology of *Oscillatoria*, (**e**) Morphology of *Halomicronema*, (**f**) Representatives of the group "*Gloeocapsa.*" (**a–d**) Photographs: courtesy of Antonio Solé (Autonomous University of Barcelona); (**e**, **f**) Photographs, courtesy of Elia Diestra (Autonomous University of Barcelona). The bars represent 10 μm

especially important; it is the phylum *Proteobacteria* originally called the phylum of purple bacteria.

The Phylum *Proteobacteria*

This phylum was originally called "purple bacteria" because it includes pigmented photosynthetic bacteria. Anoxygenic photosynthesis is carried out by these microorganisms by using as an electron donor either an organic compound (phototrophic purple nonsulfur bacteria) or inorganic compounds such as sulfides, hydrogen, or iron (phototrophic purple sulfur bacteria) via bacteriochlorophylls a or b (*cf*. Sect. 3.3.4). Originally C. Woese had proposed the term "purple bacteria phylum" to emphasize the photosynthetic origin of this group; however, these photosynthetic bacteria are found only in three classes among the six classes that make up this phylum. In fact, in addition to the purple bacteria, this phylum contains a large number of nonphotosynthetic bacteria corresponding to the majority of Gram-negative bacterial genera currently known; it is a very large and very varied phylum with different types of bacterial metabolisms: phototrophy, chemoorganotrophy, and chemolithotrophy. Analyses suggest that the absence of photosynthesis observed in a majority of representatives of this phylum is the result of a loss of this metabolism initially present in their ancestor. Indeed, some nonphotosynthetic proteobacteria still have a portion of bacteriochlorophyll gene in their genome.

The Class *Alphaproteobacteria*

It includes many kinds of phototrophic purple nonsulfur bacteria such as *Rhodospirillum* (Fig. 6.14g), *Rhodobacter*, *Rhodovulum*, *Rhodobium*, and *Rhodopseudomonas* (Fig. 6.14d). These genera are grouped into three levels (*Rhodospirillales*, *Rhodobacterales*, and *Rhizobiales*). Many chemoorganotrophic bacteria with varied ecological niches (soil saprophytes, pathogens of plants or animals,

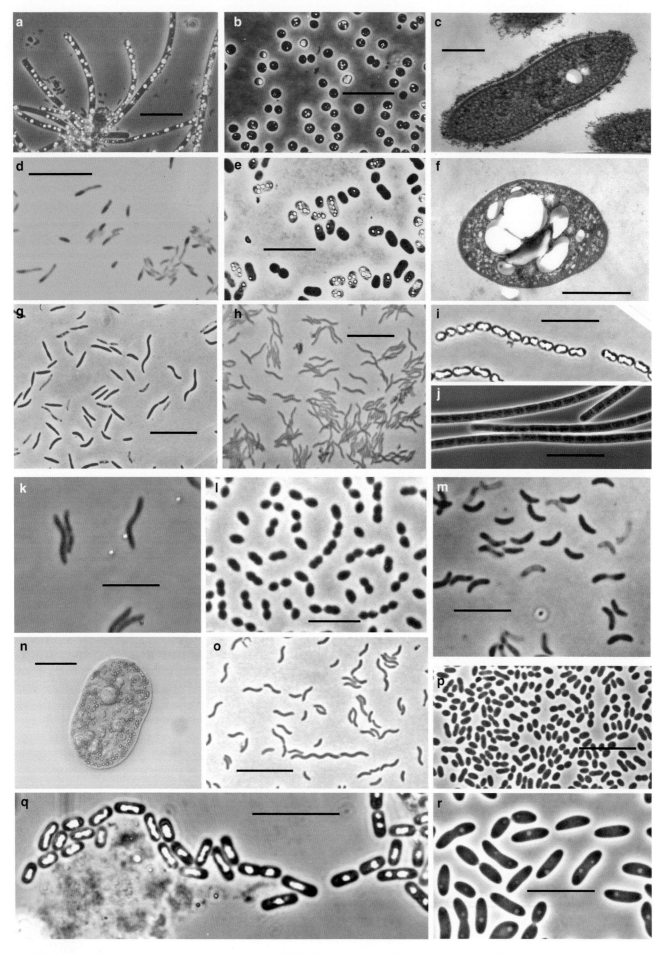

Fig. 6.14 Photomicrographs of Gram-negative bacteria of the phylum *Proteobacteria*. (**a**) *Beggiatoa* sp. (Photograph: Pierre Caumette); (**b**) *Halothiocapsa halophila* (Photograph: Pierre Caumette); (**c**) *Chromatium* sp., electron microscopy (Photograph: courtesy of

symbionts, etc.) are also present in this class with a large number of taxa in close interaction with host animals or plants. Among the best-known representatives, there are *Sinorhizobium meliloti*, the dinitrogen fixing root symbiont in alfalfa; *Agrobacterium tumefaciens*, the pathogen that transforms its host plant by injecting it with its own DNA; *Rickettsia conorii*, the agent of Q fever; and *Nitrobacter hamburgensis*, nitrifying soil bacteria.

In *Rhodospirillales*, an order consisting of two families (*Rhodospirillaceae* and *Acetobacteraceae*), the purple bacterium *Rhodospirillum* and other relatives (*Rhodovibrio, Rhodospira*, etc.) coexists with *Acetobacter, Gluconobacter* (bacteria oxidizing ethanol to acetic acid), and various other bacteria isolated from soil and aquatic continental or marine environments (*Azospirillum, Magnetospirillum*).

The *Rhodobacterales* with a single family (*Rhodobacteraceae*) contain many purple nonsulfur bacteria (*Rhodobacter, Rhodovulum*, etc.) and marine aerobic or denitrifying bacteria (*Amaricoccus, Antarctobacter, Paracoccus*, etc.) or containing bacteriochlorophyll *a* (*Roseobacter, Roseovivax*, etc.).

The *Rhizobiales* with 10 families contain some purple phototrophic bacteria in *Rhodobiaceae* (*Rhodobium*) or in *Bradyrhizobiaceae* (*Rhodopseudomonas*). Other families of this order include many soil bacteria, fixing dinitrogen. It is the case of *Rhizobiaceae* (*Rhizobium*), of *Phyllobacteriaceae* (*Phyllobacterium, Mesorhizobium*), of *Beijerinckiaceae* (*Beijerinckia*), and of *Bradyrhizobiaceae* (*Bradyrhizobium*) which also contain nitrifying bacteria (*Nitrobacter*). Other families concern methylotrophic bacteria as *Methylocystaceae, Methylobacteriaceae* (*Methylobacterium*). The *Hyphomicrobiaceae* (*Hyphomicrobium, Aquabacter*, and *Blastochloris*) isolated from soil or aquatic environments, *Brucellaceae* (*Brucella*) responsible for brucellosis, and *Bartonellaceae* (*Bartonella*) are also part of *Rhizobiales*.

The three other orders of this class are the *Rickettsiales* with three bacteria family symbionts or parasites, *Rickettsiaceae* (*Rickettsia, Wolbachia*) which are intracellular parasites or symbionts, the *Ehrlichiaceae* (*Ehrlichia*), and the *Holosporaceae*. The *Sphingomonadales* with family *Sphingomonadaceae* (*Sphingomonas*) and *Caulobacterales* with family *Caulobacteraceae* (*Caulobacter*) are the last two orders of this class which includes as a whole many bacteria which are diverse in their metabolism (phototrophic, chemolithotrophic, chemoorganotrophic) and multiply by budding or form stalks, enabling them to live on fixed supports (*Hyphomicrobium, Rhodopseudomonas, Rhodomicrobium, Blastochloris, Caulobacter, Ancalomicrobium*, and *Prosthecomicrobium*).

The Class *Betaproteobacteria*

It includes some phototrophic purple nonsulfur bacteria but also chemolithotrophic or chemoorganotrophic bacteria that live in varied ecological niches with a predominance of contaminated sites. It consists of six orders with two that contain some purple bacteria. These are *Burkholderiales* with family *Comamonadaceae* (*Rubrivivax, Rhodoferax*) which also contains many non-photosynthetic bacteria and *Rhodocyclales* with family *Rhodocyclaceae* (*Rhodocyclus*) which also includes the chemoorganotrophic genera *Propionibacter* or *Zooglea*.

The order *Burkholderiales* includes five families especially chemoorganotrophic using metals (*Ralstoniaceae* with *Ralstonia metallidurans*) or isolated from polluted soils (*Burkholderiaceae* with *Burkholderia, Alcaligenaceae* with *Alcaligenes*) capable of biodegrading hydrocarbons or xenobiotics (pesticides). The family *Comamonadaceae* includes many chemoorganotrophic bacteria (*Comamonas, Aquabacterium*) isolated from aquatic environments, or polluted environments including wastewaters (filamentous bacteria: *Leptothrix, Sphaerotilus*) and chemolithotrophic bacteria

Fig. 6.14 (continued) Michael Madigan, University of Illinois, Michigan, USA); (**d**) *Rhodopseudomonas palustris* (Photograph: Pierre Caumette); (**e**) *Allochromatium vinosum* (Photograph: Pierre Caumette); (**f**) *Allochromatium* sp. containing polyhydroxybutyrate globules, electron microscopy (Photograph: courtesy of Michael Madigan, University of Illinois, Michigan, USA); (**g**) *Rhodospirillum* sp. (Photograph: Pierre Caumette); (**h**) *Desulfovibrio halophilus* (Photograph: Pierre Caumette); (**i**) *Desulfobacterium vacuolatum* (Photograph: courtesy of Friedrich Widdel, MP Institute of Microbiology, Bremen, Germany); (**j**) *Desulfonema ishimotoi* (Photograph: courtesy of Friedrich Widdel, MP Institute of Microbiology, Bremen, Germany); (**k**) *Ectothiorhodospira* sp. (Photograph: courtesy of Remy Guyoneaud, IPREM, University of Pau, France); (**l**) *Desulfobulbus propionicus* (Photograph: courtesy of Friedrich Widdel, MP Institute of Microbiology, Bremen, Germany); (**m**) *Desulfovibrio* sp. (Photograph: courtesy of Friedrich Widdel, MP Institute of Microbiology, Bremen, Germany); (**n**) *Chromatium* sp. interference contrast (Photograph: courtesy of Heribert Cypionka, University of Oldenburg, Germany); (**o**) *Desulfovibrio desulfuricans* (Photograph: courtesy of Friedrich Widdel, MP Institute of Microbiology, Bremen, Germany; (**p**) *Desulfobacterium autotrophicum* (Photograph: courtesy of Friedrich Widdel, MP Institute of Microbiology, Bremen, Germany); (**q**) *Thiodictyon* sp., iron-oxidizing bacteria purple (Photograph: courtesy of Armin Ehrenreich, MP Institute of Microbiology, Bremen, Germany); (**r**) *Desulfobacter latus* (Photograph: courtesy of Friedrich Widdel, MP Institute of Microbiology, Bremen, Germany). The bars represent 10 μm except for figures **c** and **f** where they are 1 μm. With the exception of figures **c, f**, and **n**, all photographs correspond to phase contrast microscopy

using hydrogen (*Hydrogenophaga*) or sulfur (*Thiomonas*) as electron donors. Family *Oxalobacteraceae* (*Oxalobacter*) is part of this order. The order *Hydrogenophilales* with a single family (*Hydrogenophilaceae*) corresponds to aerobic chemolithotrophic bacteria using hydrogen (*Hydrogenophilus*) or sulfur (*Thiobacillus*) as electron donors.

The order *Methylophilales* concerns methylotrophic bacteria (*cf.* Sect. 3.4.2) grouped in a family (*Methylophilaceae*) with genera *Methylophilus*, *Methylobacillus*, etc. The order *Neisseriales* with a family (*Neisseriaceae*) contains very pathogenic Gram-negative cocci (*Neisseria gonorrhoeae*, *Neisseria pneumoniae*, *Neisseria meningitidis*) but also bacteria isolated from soil or water (*Aquaspirillum*, *Chromobacterium*).

The order *Nitrosomonadales* concerns chemolithotrophic bacteria with ammonium oxidizing family *Nitrosomonadaceae* (*Nitrosospira*, *Nitrosomonas*), ferro-oxidant family *Gallionellaceae* (*Gallionella*), and family *Spirillaceae* with chemoorganotrophic bacteria (*Spirillum*). Many filamentous bacteria are also present in this class with filamentous bacteria in sewage sludge (*Sphaerotilus*, *Zooglea*) or iron-oxidizing bacteria (*Leptothrix*).

The Class *Gammaproteobacteria*

It includes the largest number of *Proteobacteria* grouped in 13 orders. One (the *Chromatiales*) contains all the phototrophic purple sulfur bacteria oxidizing sulfide during anoxygenic photosynthesis. These bacteria accumulate intracellular sulfur globules (family *Chromatiaceae* with *Chromatium* as the type genus (Fig. 6.14c, e, f) and related genera, and *Thiocapsa*, *Halothiocapsa* (Fig. 6.14b), *Lamprobacter*, *Thiodictyon* (Fig. 6.14q) *Thiococcus*, *Thiocystis*, and *Thiopedia*) or extracellular (*Ectothiorhodospiraceae* family with the halophilic genus *Ectothiorhodospira* (Fig. 6.14k) or extreme halophilic *Halorhodospira*) and non-photosynthetic bacteria (*Nitrococcus*).

The *Xanthomonadales* with a family (*Xanthomonadaceae*) contain chemoorganotrophic aerobic bacteria often isolated from soil and plant pathogens (*Xanthomonas*, *Xylella*, etc.).

The *Pseudomonadales* include two families, *Pseudomonadaceae* with the genera *Pseudomonas*, common bacteria of soil and water, some plant pathogens, or nosocomial disease-causing (*Pseudomonas aeruginosa*), *Azomonas*, *Rhizobacter*, and *Moraxellaceae* with *Moraxella*, *Acinetobacter*, which are encountered in soils.

In the class *Gammaproteobacteria*, there is in particular the order *Enterobacteriales* for all *Enterobacteriaceae*, bacteria capable of chemoorganotrophic aerobic or anaerobic respiration (respiring nitrate) and fermentation (*cf.* Sect. 3.3.3), with coliforms (*Klebsiella*, *Hafnia*, *Citrobacter*, *Enterobacter*) and *Escherichia coli*, commensal or pathogenic bacteria being the microorganism most studied in terms of physiology and genetics, *Salmonella*, agents of typhoid and paratyphoid fever (*Salmonella typhi*, *Salmonella paratyphi*), *Proteus*, and *Shigella* (*Shigella sonnei*). In this order are also present *Yersinia pestis*, the agent of plague, and symbionts of insects or nematodes as *Photorhabdus*.

The order *Vibrionales* with a family *Vibrionaceae*, contains chemoorganotrophic bacteria isolated from waters, particularly coastal waters or estuaries as the genus *Vibrio* with *Vibrio cholerae* (cholera agent) and many marine *Vibrio* causing infections or poisoning of animal (fish) or human, and the genera *Salinivibrio* or *Photobacterium*, the latter containing luminescent bacteria with an enzyme, luciferase, also present in prokaryote light organs of certain deep marine organisms.

The *Aeromonadales* as *Vibrionales* and *Enterobacteriales* are also aerobic bacteria, facultative anaerobes (capable of respiring and fermentation) or strict anaerobes include two families: *Aeromonadaceae* with the genus *Aeromonas*, freshwater bacteria, some of which are pathogenic of fish (*Aeromonas salmonicida*), and *Succinivibrionaceae* which contain genera represented by strict fermentative bacteria (*Succinivibrio*, *Ruminobacter*, *Anaerobiospirillum*) isolated from fermenters or rumen.

The order *Pasteurellales* with a family (*Pasteurellaceae*) includes the genus *Pasteurella* chemoheterotrophic bacteria responsible for pulmonary infections in animals and *Haemophilus influenzae* pathogen of respiratory track, the first bacterium completely sequenced in 1995 (Fleischmann et al. 1995).

The order *Legionellales* with a single family (*Legionellaceae*) contains a single genus (*Legionella*) corresponding to *Legionella pneumophila*, the agent of Legionnaires' disease, surprise guest at the American Legionnaires Convention Philadelphia 1976, often isolated from ventilation systems, water circulation or refrigeration industries and communities, and responsible for serious lung infections.

The order *Oceanospirillales* includes aerobic chemoorganotrophic marine bacteria also capable of respiring nitrate, grouped into two families, *Oceanospirillaceae* with genera *Oceanospirillum*, *Marinospirillum*, and *Halomonadaceae* with genera *Halomonas* or *Alcanivorax*, bacteria capable of totally degrading hydrocarbons.

The *Alteromonadales* with the family *Alteromonadaceae* also comprise chemoorganotrophs marine halophilic or halotolerant bacteria (*Alteromonas*, *Marinobacter*, *Marinobacterium*) or bacteria living at very cold temperatures (*Psychromonas*).

The order *Thiotricales* is represented by chemolithotrophic sulfur-oxidizing bacteria filamentous and often gliding (*Thiothrix*, *Thiomargarita*, *Thiobacterium*, *Thioploca*,

Beggiatoa Fig. 6.14a, *Thiospira*) grouped in the family *Thiotricaceae*; two other families (*Piscirickettsiaceae* and *Francisellaceae*) are also part of this order.

The methylotrophic bacteria order (*Methylococcales*) is represented by family *Methylococcaceae* (*Methylococcus*, *Methylobacter*), and the *Cardiobacteriales* order with family *Cardiobacteriaceae* (*Cardiobacterium*) are also present in this *Gammaproteobacteria* class.

The Class *Deltaproteobacteria*

With seven orders, it includes among others the majority of sulfate-reducing bacteria which are chemoorganotrophic anaerobic bacteria for most, some may be chemolithotrophic, using sulfate as electron acceptor (*Desulfovibrio*, Fig. 6.14h, m, o; *Desulfobacter*, Fig. 6.14r; *Desulfobulbus*, Fig. 6.14l; *Desulfobacterium*, Fig. 6.14i, p) and anaerobic bacteria using ferric iron as electron acceptor (*Geobacter*, *Pelobacter*). These bacteria are grouped into four orders of this class.

The order *Desulfurellales* with family *Desulfurellaceae* contains the genus *Desulfurella*.

The order *Desulfovibrionales* includes three families, *Desulfovibrionaceae* (*Desulfovibrio*), especially encountered in coastal environments and capable of degrading partially simple organic compounds, *Desulfomicrobiaceae* (*Desulfomicrobium*), and *Desulfohalobiaceae* (*Desulfohalobium*, *Desulfonatronovibrio*, etc.), these latter including the halophilic sulfate-reducing bacteria isolated from hypersaline environments.

The *Desulfobacterales* with families *Desulfobacteraceae* (*Desulfobacter*, Fig. 6.14r; *Desulfobacterium*, Fig. 6.14i, p), *Desulfobulbaceae* (*Desulfobulbus*, Fig. 6.14l), and *Desulfoarculaceae* (*Desulfoarculus*, *Desulfomonile*) correspond to sulfate-reducing bacteria isolated from diverse environments, such as oceans, soils, deep earth, oil deposits, etc. They are able to degrade some complex organic compounds and of biotransformation of metals.

The order *Desulfomonadales* contains sulfate-reducing bacteria in the family *Desulfomonadaceae* (*Desulfomonas*) and anaerobic chemoorganotrophic iron-reducing bacteria in families *Geobacteraceae* (*Geobacter*) and *Pelobacteraceae* (*Pelobacter*).

In this class, there is also the order *Syntrophobacterales* with two families, *Syntrophobacteraceae* (*Syntrophobacter*) and *Syntrophaceae* (*Syntrophus*), which includes anaerobic bacteria living in syntrophy requiring interspecies hydrogen transfer but also some sulfate-reducing bacteria (*Desulfacinum*).

The order *Bdellovibrionales* (*Bdellovibrionaceae*) contains bacteriological predatory bacteria (*Bdellovibrio* or *Vampirovibrio*).

Finally, the order *Myxococcales* with the two families *Myxococcaceae* (*Myxococcus*, etc.) and *Polyangiaceae* (*Polyangium*, *Chondromyces*, etc.) is composed of bacteria with fruiting structures (myxobacteria).

The Class *Epsilonproteobacteria*

The fifth class of this great phylum contains some specific bacteria such as *Campylobacter*, sulfur-oxidizing bacteria (*Thiovulum*) grouped in the family *Campylobacteraceae*, and the pyloric gastric pathogen (*Helicobacter pylori*) in the family *Helicobacteraceae*.

The Class *Zetaproteobacteria*

This sixth and last class of *Proteobacteria* includes an order, the *Mariprofundales*, containing a single family, the *Mariprofundaceae* with a single genus *Mariprofundus* corresponding to microaerophilic chemolithotrophic curved rods, using iron as electron donor, with filamentous structures containing iron oxyhydroxide. One species *Mariprofundus ferrooxydans* was isolated from iron-rich microbial mats near deep hydrothermal vents of the Pacific Ocean.

The Phylum *Bacteroidetes*

It is also an important phylum of Gram-negative bacteria that includes aerobic and anaerobic heterotrophic bacteria living in soil or in water bodies and aquatic sediments. It is often called CFB group (*Cytophaga-Flavobacterium-Bacteroides*). It is divided into three classes.

Bacteroidia class with order *Bacteroidales* which contains four families corresponding to anaerobic bacteria encountered in environments rich in organic matter often from urban waste or fermenting materials, feces, or composted material. Families *Bacteroidaceae* (*Bacteroides*, etc.), *Rickenellaceae* (*Rickenella*, etc.), *Porphyromonadaceae* (*Porphyromonas*), and *Prevotellaceae* (*Prevotella*) constitute this order.

Flavobacteria class with *Flavobacteriales* order includes the families of *Flavobacteriaceae* (*Flavobacterium*, *Cellulophaga*, *Polaribacter*) of *Myroidaceae* (*Myroides*) and *Blattabacteriaceae* (*Blattabacterium*) including chemoorganotrophic aerobic or denitrifying bacteria which play an important role in the degradation of organic matter in soils and biodegradation of xenobiotics.

Sphingobacteria class in which the order *Sphingobacteriales* includes five families of chemoorganotrophic aerobic or denitrifying flexuous filamentous moving or gliding bacteria involved in the degradation of organic pollutants and xenobiotics: the *Sphingobacteriaceae* (*Sphingobacterium*), the *Saprospiraceae* (*Saprospira*), the *Flexibacteaceae* (*Flexibacter*, *Cytophaga*), the *Flammeovirgaceae* (*Flammeovirga*, *Flexithrix*, etc.), and *Crenotrichaceae* (*Crenothrix*, *Toxothrix*). From an evolutionary point of view, it would be close to the *Chlorobi* phylum.

The Phylum *Chlorobi*

Class *Chlorobia* contains one order (*Chlorobiales*) and a single family (*Chlorobiaceae*) with several genera (*Chlorobium, Chloroherpeton, Pelodyction*, etc.). This phylum corresponds to anoxygenic phototrophic green sulfur-oxidizing bacteria, containing bacteriochlorophyll *c*, *d*, or *e* in chlorosomes, structures attached to the inner layer of the cytoplasmic membrane. These bacteria use sulfur compounds as electron donors during anoxygenic photosynthesis (*cf.* Sect. 3.3.4, Fig. 3.29). From an evolutionary point of view, it would be close to the phylum *Bacteroidetes*.

6.6.2.5 The Major Phyla of Gram-Positive Bacteria
The Phylum *Acidobacteria*

It is composed of chemoorganotrophic bacteria commonly found in soils and sediments but of which very few representatives have been grown. Besides order *Acidobacteriales* represented by a family (*Acidobacteriaceae*), it contains, in addition to the genus *Acidobacterium* which corresponds to one of many taxa revealed as a result of direct in situ sequencing programs from soils, homoacetogenic bacteria (*Holophaga*), and iron-reducing bacteria (*Geothrix*).

The Phylum *Chlamydiae*

It contains chemoorganotrophic bacteria, obligatory intracellular parasites, and pathogens; it is constituted by one order (the *Chlamydiales*) containing four families: *Chlamydiaceae* (*Chlamydia*) responsible for lungs or urogenital infections, *Parachlamydiaceae* (*Parachlamydia*), *Simkaniaceae* (*Simkania*), and *Waddliaceae* (*Waddlia*). It is close to the phyla *Planctomycetes* and *Verrucomicrobia* [super phylum PVC (*Planctomycetes/Verrucomicrobia/Chlamydiae*)].

The Phylum *Planctomycetes*

This phylum, close to *Chlamydiae* and *Verrucomicrobia* (super phylum PVC), is formed by budding bacteria, unicellular or filamentous devoid of peptidoglycan, which was initially classified with Fungi. With the order *Planctomycetales* and one family *Planctomycetaceae* (*Planctomyces, Gemmata, Pirellula*, etc.), it consists of specific bacteria with internal membranes that keep DNA isolated from the cytoplasm. From the representatives of this family, the anaerobic oxidation of ammonia reaction (Anammox) was discovered, a reaction that takes place in a special membrane compartment (anammoxosome, Figs. 14.32 and 14.33).

The Phylum *Verrucomicrobia*

It includes two classes, *Verrumicrobiae* and *Opitutae*. The first class, with order *Verrucomicrobiales* and family *Verrucomicrobiaceae* (*Verrucomicrobium*), consists of unusual bacteria abundant in soil microbial communities but of which very few representatives have been isolated. The second class, with two orders, *Opitutales* and *Puniceicoccales*, includes Gram-negative chemoorganotrophic cocci, aerobic or fermentative, isolated from marine environments, hot springs, or paddy fields. This phylum would be close to phyla *Planctomycetes* and *Chlamydiae* (constituting together the PVC super phylum).

The Phylum *Spirochaetes*

It contains bacteria with a particular cell morphology: where cells are long, flexuous, and much spiraled. They are all motile by means of an axial filament. These are aerobic or anaerobic inhabitants of soils and aquatic environments. The spirochaetes are grouped in a single order, *Spirochaetales*, divided into three families. The family *Spirochaetaceae* contains pathogenic bacteria such as the well-known *Treponema pallidum* (agent of syphilis), *Borrelia* (relapsing agents transmitted by vectors such as ticks or lice and responsible for Lyme disease, relapsing fever or hemorrhagic fevers in animals and men), and many spirochetes (*Spirochaeta, Cristispira*) living under aerobic or anaerobic conditions in soils or aquatic sediments (Fig. 6.10f). *Serpulinaceae* family with genus *Serpulina* and the family *Leptospiraceae* with *Leptospira* consist of strict aerobic bacteria, the two families containing parasitic agents, pathogens of the oral cavity or causing kidney disease, and intestinal or liver infections in animals and man.

The Phylum *Fibrobacteres*

This phylum consists of one order, *Fibrobacterales* with the family *Fibrobacteraceae* containing one genus (*Fibrobacter*) corresponding to anaerobic chemoorganotrophic fermentative bacteria, some of which live in the digestive tract of animals and can degrade cellulose.

The Phylum *Fusobacteria*

Composed of one order, *Fusobacteriales*, with one family, *Fusobacteriaceae*, the phylum *Fusobacteria* contains obligate anaerobic bacteria commonly found in the oral cavity and in the intestines of animals, such as the genera *Fusobacterium, Leptotrichia*, and *Streptobacillus*, some of which are exclusively responsible for ENT infections and angina.

The Phylum *Gemmatimonadetes*

It includes a single order, *Gemmatimonadales*, with one family and one genus (*Gemmatimonadaceae: Gemmatimonas*) in which one species was isolated from a laboratory reactor, *Gemmatimonas aurantiaca*. It is a Gram-negative, motile, aerobic, and chemoorganotrophic bacterium that can divide by binary fission or budding.

The Phylum *Lentisphaerae*

In this phylum, the class *Lentisphaeria* includes two orders, order *Lentisphaerales* (*Lentisphaeraceae* family, genus *Lentisphaera*) with a single marine species isolated in Oregon (*Lentisphaera araneosa*) composed of Gram-negative spherical cells, chemoorganotrophs, and aerobic cocci, producing a transparent exopolymer, and order *Victivallales* with a single species corresponding to *Victivallis vadensis*, Gram-negative chemoorganotrophic anaerobic cocci isolated from feces.

The Phylum *Dictyoglomi*

This phylum (*Dictyoglomales* order, *Dictyoglomaceae* family) includes thermophilic bacteria and fermentative obligate anaerobes (genus *Dictyoglomus*).

The Phylum *Caldiserica*

With the order Caldisericales and the family Caldisericaceae, it comprises a single species, *Caldisericum exile*, isolated from a hot spring in Japan, consisting of multicellular filaments with a common envelope, motile by a polar flagellum, anaerobic and chemoheterotrophic, realizing anaerobic respirations with sulfur compounds such as thiosulfate, sulfite, and elemental sulfur as electron acceptors.

The Phylum *Elusimicrobia*

This phylum (order of *Elusimicrobiales*, family of *Elusimicrobiaceae*) comprises a single species in a single genus *Elusimicrobium* corresponding to pleomorphic rods, nonmotile, Gram-negative, anaerobic, fermentative, isolated from the intestinal tract of the larva *Pachnoda ephippiata* which feeds on humus.

The Phylum *Armatimonadetes*

This phylum includes two classes. Class *Armatimonadia* (order *Armatimonadales*, family *Armatimonadaceae*) consists of one genus, *Armatimonas*, formed of sticks to ovoid cells, nonmotile, Gram-negative, aerobic, chemoorganotrophic, isolated from the rhizoplane of an aquatic plant, *Phragmites australis*, in a freshwater lake in Japan. Class *Chthonomonadetes* (order *Chthonomonadales*, family *Chthonomonadaceae*) includes a genus, *Chthonomona*, corresponding to Gram-negative, aerobic, chemoorganotrophs, moderately thermophilic and acidophilic isolated from geothermal soils in New Zealand.

6.6.2.6 The Major Phyla of Gram-Positive Bacteria

The Phylum *Firmicutes*

It corresponds to the Gram-positive bacteria with low G + C % (GC% DNA in <50 %), but there are some exceptions with high G + C percentages (e.g., *Symbiobacterium* with a GC% as high as 69 %) (Fig. 6.3). The characteristic of having a thick peptidoglycan wall which gives its name to the taxon is shared with the *Actinobacteria*, which had resulted in many authors grouping into a category called "Gram-positive," a position which was reexamined on sequence analysis of many genes.

This group of microorganisms has a striking character, which is its ability to produce heat-resistant spores. Consisting of aerobic or anaerobic bacteria, it includes well-known taxa such as *Bacillus subtilis*, the model soil organism, or *Bacillus anthracis*, responsible of anthrax which has earned it a great reputation since biological weapons programs were announced in the 1990s. This group also contains bacteria that produce toxins such as the highly virulent *Clostridium botulinum*, the botulism agent that also contains microorganisms used in several biological weapons programs, or *Clostridium tetani*, the tetanus agent. In this group is also found *Desulfotomaculum*, a spore-forming sulfate-reducing bacterium. These spore-forming bacteria are found in two classes that constitute the phylum.

The class *Clostridia* contains three orders of chemoorganotrophic anaerobic fermentative bacteria. The order *Clostridiales* is composed of eight families. The family *Clostridiaceae* contains the genera *Clostridium* and *Acetonoma* (Fig. 6.15b, c), *Sporobacter* (Fig. 6.15i), *Anaerobacter*, etc. These are bacteria that live in soil and anoxic sediments, manures, and composts where they carry out fermentations. Among *Clostridium*, spore-forming bacteria, some are pathogens, producing toxins (botulism, tetanus). The families *Lachnospiraceae* (*Anaerophilum*, *Butyrivibrio*, *Coprococcus*), *Peptostreptococcaceae* (*Peptostreptococcus*), *Eubacteriaceae* (*Eubacterium*), and *Peptococcaceae* (*Peptococcus*, *Desulfotomaculum* (Fig. 6.15f, etc.) involve bacteria, cocci, or rods, obligately fermentative or sulfate-reducing bacteria, living in sediments, manure, anoxic sludges of organic-rich wastewater treatment plants, and feces. The family *Heliobacteriaceae* (*Heliobacterium*, *Heliobacillus*) corresponds to the phototrophic bacteria with bacteriochlorophyll g performing anoxygenic photosynthesis in photoorganotrophy (*cf.* Sect. 3.3.4). The families *Acidaminococcaceae* (*Acidaminococcus*, *Sporomusa*) and *Syntrophomonadaceae* (*Syntrophomonas*, *Anaerobaculum*) contain anaerobic cocci or rods developed by fermentation or syntrophy in the rumen or intestinal flora as well as in fermenters or anoxic organic sediments. The order *Thermoanaerobacteriales* with the family *Thermoanaerobacteriaceae* (*Thermoanaerobacterium*; *Thermoanaerobacter*, Fig. 6.15d) and the order *Haloanaerobiales* with the families *Haloanaerobiaceae* (*Haloanaerobium*, Fig. 6.15g; *Halothermothrix*, Fig. 6.15l) and *Halobacteroidaceae* (*Halobacteroides*, *Haloanaerobacter*) group fermentative bacteria living in sediments, or isolated from thermophilic or halophilic composts or fermenters, some of the bacteria being moderate halophiles or hyperhalophiles (*Haloanaerobiales*).

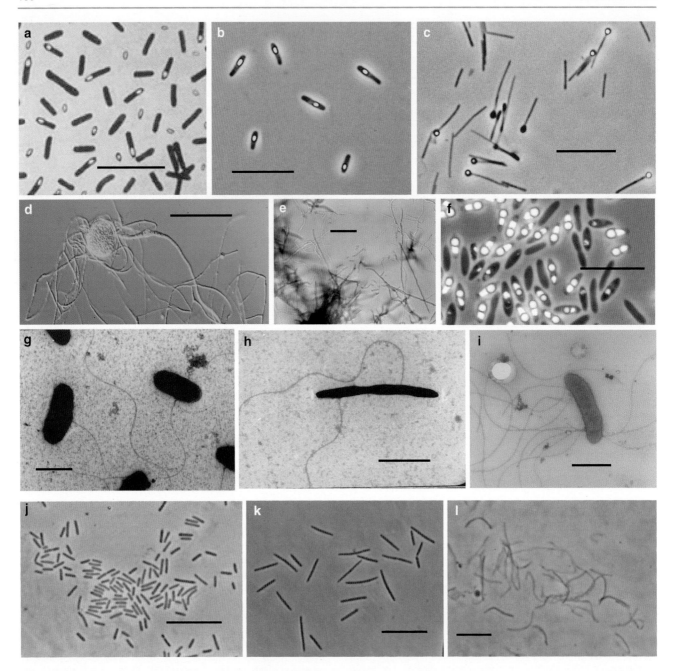

Fig. 6.15 Photomicrographs of Gram-positive bacteria. – Phylum *Firmicutes* (**a**) *Bacillus* sp. (Photograph: Pierre Caumette); (**b**) *Clostridium mayombei* (Photograph: courtesy of Alain Brauman); (**c**) *Acetonema longum* (Photograph: Jean-Luc Cayol); (**f**) *Desulfotomaculum acetoxidans* containing spores and gas vesicles (Photograph: courtesy of Friedrich Widdel, MP Institute of Microbiology, Bremen, Germany); (**g**) *Haloanaerobium* sp. (Photograph: Jean-Luc Cayol); (**h**) *Fusibacter paucivorans* (Photograph: Jean-Luc Cayol); (**i**) *Sporobacter termitidis* (Photograph: Jean-Luc Cayol); (**j**) *Thermoanaerobacter brockii* (Photograph: Jean-Luc Cayol); (**k**) *Thermohalobacter berrensis* (Photograph: Jean-Luc Cayol); (**l**) *Halothermothrix orenii* (Photograph: Jean-Luc Cayol). – Phylum *Actinobacteria* (**d**, **e**): *Frankia* sp. (Photograph: Philippe Normand). The bars represent 10 μm except for figures **g**, **h**, and **i** where they are 1 μm

The *Bacilli* class includes two orders.

The *Bacillales* order which is composed of nine families of chemoorganotrophic aerobic or facultative anaerobic cocci or rods with a fermentative and respiratory metabolism. Family *Bacillaceae* (genera *Bacillus*, Fig. 6.15, or *Saccharococcus*) comprises spore-forming bacilli abundant in soils and waters, ubiquitous due to the dissemination of their spores, and some being pathogens such as *Bacillus anthracis* (anthrax). Families *Planococcaceae* (*Planococcus*) and *Staphylococcaceae* (*Staphylococcus*, *Macrococcus*) involve aerobic or facultative anaerobic cocci isolated from soil, air, or water, some of which may be pathogenic of mucosa, skin, or responsible for poisoning (staphylococcal enteropathogenic *Staphylococcus aureus*). Families *Caryophanaceae* (*Caryophanon*), *Paenibacillaceae* (*Paenibacillus*), *Alicyclobacillaceae* (*Alicyclobacillus, Pasteuria*), *Thermoactinomycetaceae* (*Thermoactinomyces*), and *Sporolactobacillaceae* (*Sporolactobacillus*) group aerobic or fermentative bacteria isolated from soil and water sometimes thermophiles. The family *Listeriaceae* (*Listeria*) contains bacteria common in soil and water with a species (*Listeria monocytogenes*) which can be pathogenic and is responsible for food poisoning especially in dairy products.

The order *Lactobacillales* consisting of six family includes lactic acid bacteria and many cocci. The family *Lactobacillaceae* (*Lactobacillus, Pediococcus*) corresponds to chemoorganotrophic aerotolerant and fermentative lactic acid bacteria such as *Lactobacillus acidophilus* and *Lactobacillus bulgaricus* used in the manufacture of yogurt. The families *Aerococcaceae* (*Aerococcus*), *Carnobacteriaceae* (*Carnobacterium*), and *Enterococcaceae* (*Enterococcus*) include a majority of aerobic or anaerobic fermentative cocci, isolated from soil and water; some may reduce nitrates, iron, or manganese. The family *Leuconostocaceae* concerns aerotolerant fermentative or aerobic cocci used in the food industry of lactic fermentation such as *Leuconostoc* and *Oenococcus*, this latter being used in malolactic fermentation of grapes juice (winemaking) or cabbage (sauerkraut). The family *Streptococcaceae* corresponds to the genera *Streptococcus* with certain pathogens such as hemolytic streptococci (*Streptococcus pyogenes*) and *Lactococcus*, lactic acid fermentation bacteria.

The Phylum *Tenericutes*

This phylum, formerly grouped with *Firmicutes* was elevated to the rank of phylum in the second edition of Bergey's Manual of Systematic Bacteriology. It contains mycoplasma bacteria that are devoid of peptidoglycan but derive from Gram-positive bacteria while reacting negatively to this test. They are grouped into a single class, the class *Mollicutes*, including *Mycoplasma genitalium*, pathogen of animals and man, and *Spiroplasma citri*, citrus trees pathogen. This class *Mollicutes* is structured in five orders. The order *Mycoplasmatales* with the family *Mycoplasmataceae* (*Mycoplasma*), the order *Entomoplasmatales* with the families *Entomoplasmataceae* (*Entomoplasma*) and *Spiroplasmataceae* (*Spiroplasma*), the order *Acholeplasmatales* with the family *Acholeplasmataceae* (*Acholeplasma*), the order *Anaeroplasmatales* with the family *Anaeroplasmataceae* (*Anaeroplasma*), this family concerning anaerobic mycoplasma, and the order *Haloplasmatales* with the genus *Haloplasma* (halophilic mycoplasma). Mycoplasmas are obligate parasites of plants or animals and thus often pathogens in many infections (pulmonary or urogenital animals and man, leaf necrosis and degeneration in plants). A genus **Incertae sedis*** in the family *Erysipelotrichaceae* the *Erysipelothrix* genus is also present in this class *Mollicutes*, whereas it has a wall normal for Gram-positive bacteria. It is a bacterial pathogen of humans and animals (agent of swine erysipelas, zoonosis transmitted to man). This group has, in addition to unusual morphology (no wall), a high mutation rate (Dybvig and Voelker 1996).

The Phylum *Actinobacteria*

It concerns Gram-positive bacteria with high G + C% (% by GC in the DNA > usually 50 %) (Fig. 6.3). This group of microorganisms has as most striking character varied morphologies ranging from unicellular organisms to irregular shapes consisting of mycelial branched hyphae. This character, also found in Fungi, has been a source of confusion for years when taxonomy was not based on molecular sequences. It is now recognized unequivocally that this is an instance of convergent evolution with Fungi. *Actinobacteria*, especially the most famous of them, *Streptomyces coelicolor*, and related genera (*Actinomyces, Nocardia*) producing structures called conidiophores useful for dissemination. *Actinobacteria* are dominant in the soil where their dissemination structures (conidiophores and hyphal cells) give them a competitive advantage. They also have the ability to synthesize antibiotic compounds, some of which are used as important drugs in therapy of infectious diseases, and they also confer a competitive advantage vis-à-vis other soil bacteria (*cf.* Sect. 9.5.1). This phylum contains a class (*Actinobacteria*) which is composed of five subclasses (the *Acidimicrobidae*, the *Rubrobacteridae*, the *Coriobacteridae*, the *Sphaerobacteridae*, and the *Actinobacteridae*).

The first four subclasses contain very few species, often with only one representative and only one family. The *Acidimicrobidae* with the order *Acidimicrobiales* and the family *Acidimicrobiaceae* (*Acidimicrobium*), the *Rubrobacteridae* with the order *Rubrobacterales* and the family *Rubrobacteraceae* (*Rubrobacter*), the *Coriobacteridae* with the order *Coriobacteriales* which contains a single family and several *Coriobacteriaceae* genera (*Coriobacterium, Atopobium, Cryptobacterium*), and with the *Sphaerobacteridae* with the order *Sphaerobacterales* containing a family *Sphaerobacteraceae* (*Sphaerobacter*) correspond to aerobic Gram-positive facultative anaerobic or aerobic, with stick-shaped, more or less regular, isolated from soils or aquatic sediments, and which perform aerobic or nitrate respiration or fermentation.

The fifth subclass (*Actinobacteridae*) contains most of the bacteria of the phylum and is divided into two orders (*Actinomycetales* and *Bifidobacteriales*) containing numerous suborders and many families.

The order *Actinomycetales* comprises 10 suborders under which are grouped most of the bacteria of the phylum *Actinobacteria*. The suborder *Actinomycinae* with family *Actinomycetaceae* (*Actinomyces, Actinobaculum*) contains filamentous, branched bacteria, which are aerobic, chemoorganotrophic, producing antibiotics, colonizing the surface soil, sediment, or decaying leaves in the litter of forests. The suborder *Micrococcinae* includes several families with a Gram-positive cocci or irregularly shaped cells or sticks depending on the age of the culture: the family *Micrococcaceae* (*Micrococcus, Leucobacter, Arthrobacter*, etc.) consists of aerobic bacteria isolated from soil and freshwater or coastal or in the oral cavity (*Stomatococcus*), the *Brevibacteriaceae* (*Brevibacterium*) involve irregularly shaped bacteria, isolated from soils rich in organic matter, and the *Cellulomonadaceae* (*Cellulomonas*) include aerobic and fermentative bacteria capable of degrading cellulose in soils; families *Dermabacteraceae* (*Dermabacter*), *Dermatophilaceae* (*Dermatophilus*), *Intrasporangiaceae* (*Intrasporangium*), *Jonesiaceae* (*Jonesia*), *Microbacteriaceae* (*Microbacterium, Agromyces*, etc.), *Beutenbergiaceae* (*Beutenbergia*), and *Promicromonosporaceae* (*Promicromonospora*) correspond to different soil isolated bacteria, irregularly shaped or filamentous, important in the degradation of organic matter including macromolecules and polymers. The suborder *Corynebacterineae* includes several soil bacteria, many of which are pathogens: *Corynebacteriaceae* family (*Corynebacterium*) with *Corynebacterium diphtheriae* (diphtheria agent) and *Corynebacterium glutamicum*, a microorganism used in biotechnology for the synthesis of many molecules; it also contains other corynebacteria, organisms commonly found in soil and water, capable of degrading macromolecules of some xenobiotics; the family *Mycobacteriaceae* with the genus *Mycobacterium* consists of microorganisms of soil and water, some of which are pathogens such as *Mycobacterium tuberculosis*, the agent of tuberculosis, the bacterial disease that causes the greatest number of deaths worldwide, and *Mycobacterium leprae*, the agent of leprosy; these bacteria are characterized by their resistance to environmental factors that gives them their envelopes made of waxy mycolic acids. Family *Nocardiaceae* with *Nocardia* and *Rhodococcus* consists of chemoorganotrophic bacteria of soil and aquatic sediments, active in the biodegradation of organic pollutants in soils. Some *Nocardia* (*N. asteroides, N. cyriacigeorgica*) are pathogens of animals, and finally, families *Dietziaceae, Gordoniaceae, Tsukamurellaceae*, and *Williamsiaceae* contain heterotrophic and saprophytic irregularly shaped soil bacteria forming branches and buds. The suborder *Micromonosporineae* with a single family (*Micromonosporaceae*) includes nine types of bacteria in soil and aquatic environments, chemoorganotrophic, some of which are capable of producing spores (*Micromonospora*). They are irregularly shaped bacteria with buds that develop into filamentous mycelium (*Actinoplanes*) and produce spores inside a sporangium (*Dactylosporangium*). The suborder *Propionibacterineae* includes two families: *Propionibacteriaceae* (*Propionibacterium, Luteococcus, Propioniferax*) gathering aerotolerant or anaerobic bacteria, producing lactic acid fermentation, producing propionic acid (flavor of some cheeses), used in the dairy industry. Some species may be pathogenic, while others are inhabitants of the digestive tract and contribute to the intestinal flora. *Nocardioidaceae* (*Nocardioides*) bacteria are chemoorganotrophic, filamentous soil saprophytes, and some are able to degrade hydrocarbons. The suborder *Pseudonocardineae* with two families *Pseudonocardiaceae* (*Pseudonocardia*) and *Actinosynnemataceae* (*Actinosynnema*) group heterotrophic saprophytic filamentous soil bacteria. The suborder *Streptomycineae* with the family *Streptomycetaceae* including the genus *Streptomyces* contains chemoorganotrophic filamentous soil saprophytic bacteria. These bacteria form branched filaments, producing hyphae, and develop aerial mycelia with sporophores (bodies containing spores that are later released), which led to a confusion with the microscopic Fungi for a long time. These bacteria produce antibiotics and are the source of the majority of known antibiotics. There are also pathogenic species of plants, especially potato (*S. scabies*). The suborder *Streptosporangineae* with three families (*Streptosporangiaceae, Nocardiopsaceae, Thermomonosporaceae*) groups filamentous bacteria producing sporangia and containing spores which are close to the *Streptomycetaceae*. The suborder *Frankineae* is divided into six families. The family *Frankiaceae* with the genus *Frankia* (Fig. 6.15d, e) is a filamentous bacteria of the soil, often in symbiotic association with plants such as *Frankia alni* (symbiont of alder roots; Normand and Lalonde 1982). Families *Acidothermaceae, Cryptosporangiaceae, Geodermatophilaceae, Nakamurellaceae*, and *Sporichaetaceae* also contain heterotrophic saprophytic soil bacteria, forming mycelia. Some of these bacteria have special habitats such as *Acidothermus* found in hot springs in Yellowstone (the USA) or *Geodermatophilus* and *Modestobacter* found in irradiated soils. The suborder *Glycomycineae* with the genus *Glycomyces* also corresponds to mycelial saprophytic soil bacteria.

The order *Bifidobacteriales* with a family (*Bifidobacteriaceae*) includes *Bifidobacterium bifidus*, a bacterium of the digestive tract used as an additive in some yogurts.

6.7 Conclusion

Microorganisms have been on earth for billions of years (*cf.* Chap. 4) and have colonized almost all ecological niches, even those created by humans in recent decades. They have a very important adaptive potential. If a science fiction apocalyptic scenario such as that initiated by a meteorite impact or a massive volcanic episode of the kind that produced the Deccan Traps were to occur and eliminate a large proportion of taxa living on earth today, it is likely that life would not completely disappear because many microorganisms would survive and reinitiate evolutionary branches.

The emergence of new pathogens always surprises, either as a new genus *Legionella* in 1976 or as a particularly virulent new strain of a well-known species such as *E. coli* strain O157: H7 in 1985. Many environments are still poorly known, and probably only 1–10 % of prokaryotes have been described as of 2012. We must therefore expect many discoveries of new species in partly known environments including manmade and natural extreme environments, in particular, the deep subsurface environment and possibly in extraterrestrial environments that are being explored by metagenomic approaches (*cf.* Chap. 18). Moreover, it is clear that with many emerging pathogens and the accumulation of xenobiotic compounds in various environmental pollutants that cause the rapid evolution of new pathways of biodegradation, the world of prokaryotic microorganisms is in constant evolution. Consequently, their taxonomy cannot be frozen, and microbial taxonomists must describe a fluid world, with rapid modifications.

Websites http://www.bacterio.cict.fr/classificationgenera.html. This site is managed by Jean Euzéby of the ENV Toulouse; it maintains bacterial taxonomy entries based on the publications validly made in journals of bacterial taxonomy.

http://www.ncbi.nlm.nih.gov/sites/entrez?db=taxonomy. This site is managed by the American NCBI. It is complementary to the first; however, it incorporates several nonvalid taxa.

References

Albers SV, Meyer BH (2011) The archaeal cell envelope. Nat Rev Microbiol 9:414–426

Bailly X, Olivieri I, De Mita S, Cleyet-Marel J-C, Bena G (2006) Recombination and selection shape the molecular diversity pattern of nitrogen-fixing *Sinorhizobium* sp. associated to *Medicago*. Mol Ecol 15:2719–2734

Bapteste E, Brochier C, Boucher Y (2005) Higher level classification of the *Archaea*: evolution of methanogenesis and methanogens. Archaea 1:353–363

Barns SM, Delwiche CF, Palmer JD, Pace NR (1996) Perspectives on archaeal diversity, thermophily and monophyly from environmental rRNA sequences. Proc Natl Acad Sci U S A 93:9188–9193

Beja O et al (2000) Bacterial rhodopsin: evidence for a new type of phototrophy in the sea. Science 289:1902–1906

Boone DR, Castenholz RW (2001) Bergey's manual of systematic bacteriology, vol 1, 2nd edn. Springer, New York

Brenner DJ, Krieg NR, Staley JT (2005) Bergey's manual of systematic bacteriology: the proteobacteria, vol 2, 2nd edn. Springer, New York

Brochier C, Gribaldo S, Zivanovic Y, Confalonieri F, Forterre P (2005) *Nanoarchaea*: representatives of a novel archaeal phylum or a fast-evolving euryarchaeal lineage related to Thermococcales. Genome Biol 6:R42

Brochier-Aramanet C, Boussau B, Gribaldo S, Forterre P (2008) Mesophilic *Crenarchaeota*: proposal for a third archaeal phylum, the *Thaumarchaeota*. Nat Rev Microbiol 6:245–252

Brochier-Armanet C, Forterre P, Gribaldo S (2011) Phylogeny and evolution of the *Archaea*: one hundred genomes later. Curr Opin Microbiol 14:274–281

Brochier-Armanet C, Gribaldo S, Forterre P (2012) Spotlight on the *Thaumarchaeota*. ISME J 6:227–230

Cilia V, Lafay B, Christen R (1996) Sequence heterogeneities among 16S ribosomal RNA sequences, and their effect on phylogenetic analyses at the species level. Mol Biol Evol 13:451–461

Cohan FM (2001) Bacterial species and speciation. Syst Biol 50:513–524

Costechareyre D, Bertolla F, Nesme X (2009) Homologous recombination in *Agrobacterium*: potential implications for the genomic species concept in bacteria. Mol Biol Evol 26:167–176

Cox MM, Battista JR (2005) *Deinococcus radiodurans* the consummate survivor. Nat Rev Microbiol 3:882–892

Daubin V, Gouy M, Perriere G (2001) Bacterial molecular phylogeny using supertree approach. Genome Inform Ser Workshop 12:155–164

de Vos P, Garrity G, Jones D, Krieg NR, Ludwig W, Rainey FA, Schleifer KH, Whitman WD (2008) Bergey's manual of systematic bacteriology: the firmicutes, vol 3, 2nd edn. Springer, New York

Dworkin M, Falkow S, Rosenberg E, Schleifer KH, Stackebrandt E (2006) The prokaryotes – a handbook on the biology of bacteria, vol 3, 3rd edn. Springer, New York

Dybvig K, Voelker LL (1996) Molecular biology of mycoplasmas. Annu Rev Microbiol 50:25–57

Elkins JG et al (2008) Akorarchaeal genome reveals insights into the evolution of the *Archaea*. Proc Natl Acad Sci U S A 105:8102–8107

Fleischmann RD et al (1995) Whole-genome random sequencing and assembly of *Haemophilus influenzae* Rd. Science 269:496–512

Forterre P, Brochier C, Philippe H (2002) Evolution of the *Archaea*. Theor Popul Biol 61:409–422

Fox GE et al (1980) The phylogeny of prokaryotes. Science 209:457–463

Fraser C, Alm EJ, Polz MF, Spratt BG, Hanage WP (2009) The bacterial species challenge: making sense of genetic and ecological diversity. Science 323:741–746

Gevers D et al (2005) Opinion: re-evaluating prokaryotic species. Nat Rev Microbiol 3:733–739

Goodfellow M, Kämpfer P, Busse HJ, Trujillo ME, Suzuki K-I, Ludwig W, Whitman WB (2012) Bergey's manual of systematic bacteriology: the actinobacteria, vol 5, 2nd edn. Springer, New York

Goris J, Konstantinidis KT, Klappenbach JA, Coenye T, Vandamme P, Tiedje JM (2007) DNA-DNA hybridization values and their relationship to whole-genome sequence similarities. Int J Syst Evol Microbiol 57:81–91

Gribaldo S, Brochier-Armanet C (2006) The origin and evolution of *Archaea*: a state of the art. Philos Trans R Soc Lond B Biol Sci 361:1007–1022

Grimont PA (1988) Use of DNA reassociation in bacterial classification. Can J Microbiol 34:541–546

Huber H, Hohn MJ, Rachel R, Fuchs T, Wimmer VC, Stetter KO (2002) A new phylum of *Archaea* represented by a nanosized hyperthermophilic symbiont. Nature 417:63–67

Imhoff JF, Caumette P (2004) Recommended standards for the description of new species of anoxygenic phototrophic bacteria. Int J Syst Evol Microbiol 54:1415–1421

Könneke M, Bernhard AE, de la Torre JR, Walker CB, Waterbury JB, Stahl DA (2005) Isolation of an autotrophic ammonia-oxidizing marine archaeon. Nature 437:543–546

Krieg NR, Staley JT, Brown DR, Hedlund BP, Paster BJ, Ward NL, Ludwig W, Whitman WB (2011) Bergey's manual of systematic bacteriology: the *Bacteroidetes, Spirochaetes, Tenericutes (Mollicutes), Acidobacteria, Fibrobacteres, Fusobacteria, Dictyoglomi, Gemmatimonadetes, Lentisphaerae, Verrucomicrobia, Chlamydiae,* and *Planctomycete,* vol 4, 2nd edn. Springer, New York

Lassalle F et al (2011) Genomic species are ecological species as revealed by comparative genomics in *Agrobacterium tumefaciens*. Genome Biol Evol 3:762–781

Leigh JA, Albers SV, Atomi H, Allers T (2011) Model organisms for genetics in the domain Archaea: methanogens, halophiles, *Thermococcales* and *Sulfolobales*. FEMS Microbiol Rev 35:577–608

Linnaeus C (1753) Species plantarum. Stockholm. Holmiae: Impensis Laurentii Salvii

Lopez-Garcia P, Moreira D (2008) Tracking microbial biodiversity through molecular and genomic ecology. Res Microbiol 159:67–73

Mayr E (1942) Systematics and the origin of species from the viewpoint of a zoologist. Columbia University Press, New York

Mougel C, Thioulouse J, Perriere G, Nesme X (2002) A mathematical method for determining genome divergence and species delineation using AFLP. Int J Syst Evol Microbiol 52:573–586

Murray RG, Stackebrandt E (1995) Taxonomic note: implementation of the provisional status *Candidatus* for incompletely described procaryotes. Int J Syst Bacteriol 45:186–187

Narasingarao P, Podell S, Ugalde JA, Brochier-Armanet C, Emerson JB, Brocks JJ, Heidelberg KB, Banfield JF, Allen EE (2011) De novo metagenomic assembly reveals abundant novel major lineage of Archaea in hypersaline microbial communities. ISME J 6:81–93

Normand P, Lalonde M (1982) Evaluation of *Frankia* strains isolated from provenances of two *Alnus* species. J Can Microbiol 28:1133–1142

Ochman H, Elwyn S, Moran NA (1999) Calibrating bacterial evolution. Proc Natl Acad Sci U S A 96:12638–12643

Paul K, Nonoh JO, Mikulski L, Brune A (2012) "*Methanoplasmatales*", Thermoplasmatales-related archaea in termite guts and other environments, are the seventh order of methanogens. Appl Environ Microbiol 78:8245–8253

Pester M, Schleper C, Wagner M (2011) The *Thaumarchaeota*: an emerging view of their phylogeny and ecophysiology. Curr Opin Microbiol 14:300–306

Podar M, Makarova KS, Graham DE, Wolf YI, Koonin EV, Reysenbach AL (2013) Insights into archaeal evolution and symbiosis from the genomes of a nanoarchaeon and its inferred crenarchaeal host from Obsidian Pool, Yellowstone National Park. Biol direct, Apr 22; 8-9. doi:10.1186/1745-6150-8-9

Polzin KM, McKay LL (1991) Identification, DNA sequence, and distribution of IS981, a new, high-copynumber insertion sequence in lactococci. Appl Environ Microbiol 57:734–743

Rainey FA, Ward-Rainey NL, Janssen PH, Hippe H, Stackebrandt E (1996) *Clostridium paradoxum* DSM 7308T contains multiple 16S rRNA genes with heterogeneous intervening sequences. Microbiology 142:2087–2095

Saitou N, Nei M (1987) The neighbor-joining method: a new method for reconstructing phylogenetic trees. Mol Biol Evol 4:406–425

Schleper C, Jurgens G, Jonuscheit M (2005) Genomic studies of uncultivated *Archaea*. Nat Rev Microbiol 3:479–488

Stacey G, Bottomley PJ, van Baalen C, Tabita FR (1979) Control of heterocyst and nitrogenase synthesis in *Cyanobacteria*. J Bacteriol 137:321–326

Stackebrandt E, Ebers J (2006) Taxonomic parameters revisited: tarnished gold standards. Microbiol Today 33:152–155

Stackebrandt E, Goebel B (1994) Taxonomic note: a place for DNA-DNA reassociation and 16S rRNA sequence analysis in the present species definition in bacteriology. Int J Syst Bacteriol 44:846–849

Stackebrandt E, Liesack W, Goebel BM (1993) Bacterial diversity in a soil sample from a subtropical Australian environment as determined by 16S rDNA analysis. FASEB J 7:232–236

Stackebrandt E et al (2002) Report of the ad hoc committee for the re-evaluation of the species definition in bacteriology. Int J Syst Evol Microbiol 52:1043–1047

Thompson FL, Iida T, Swings J (2004) Biodiversity of vibrios. Microbiol Mol Biol Rev 68:403–431

Treusch AH, Leininger S, Kletzin A, Schuster SC, Klenk HP, Schleper C (2005) Novel genes for nitrite reductase and Amo-related proteins indicate a role of uncultivated mesophilic crenarchaeota in nitrogen cycling. Environ Microbiol 7:1985–1995

Wayne LG et al (1987) International Committee on Systematic Bacteriology. Report of the ad hoc committee on reconciliation of approaches to bacterial systematics. Int J Syst Bacteriol 37:463–464

Woese CR (1987) Bacterial evolution. Microbiol Rev 51:221–271

Woese CR (2007) The birth of the *Archaea*: a personal retrospective. In: Garrett RA, Klenk HP (eds) Archaea: evolution, physiology, and molecular biology. Blackwell publishing, Oxford, pp 1–15

Woese CR, Fox GE (1977) Phylogenetic structure of the prokaryotic domain: the primary kingdoms. Proc Natl Acad Sci U S A 74:5088–5090

Woese CR, Kandler O, Wheelis ML (1990) Towards a natural system of organisms: proposal for the domains *Archaea*, *Bacteria*, and *Eucarya*. Proc Natl Acad Sci U S A 87:4576–4579

Yap WH, Zhang Z, Wang Y (1999) Distinct types of rRNA operons exist in the genome of the actinomycete *Thermomonospora chromogena* and evidence for horizontal transfer of an entire rRNA operon. J Bacteriol 181:5201–5209

Taxonomy and Phylogeny of Unicellular Eukaryotes

Charles-François Boudouresque

Abstract

The notion of 'microbes' encompasses *Bacteria* (unicellular and multicellular) and *Archaea*, together with unicellular eukaryotes. In addition, microbiologists have traditionally included within their study field a number of eukaryotes named 'fungi' (customary meaning), e.g. Fungi (modern meaning), which are generally multicellular, and Oobionta, which are constituted by a giant multinucleate cell. Unicellular eukaryotes only represent ~10 % of all eukaryotic species; however, if only kingdoms, sub-kingdoms and phyla ('phyletic diversity') are taken into consideration, most eukaryotes are unicellular. Taking into consideration the huge phyletic diversity of unicellular eukaryotes and of affiliate taxa, which nearly fits the whole diversity of eukaryotes, it is impossible to present here a comprehensive description of the whole of these taxa. The choice was therefore to select a part of high-level taxa, likely to illustrate the amazing diversity of eukaryotes. For each selected taxon, some traits are recurrently tackled topics, e.g. the chloroplast structure and the photosynthetic pigments, the kinetic apparatus and the cell wall. Some derived characters, more or less specific to a taxon and prone to constitute a biomarker (genetic, biochemical, cytological and/or biological), are also emphasized. The biological life cycle of at least one species belonging to the taxon is illustrated in a standardized way. Finally, the role of the taxon in the functioning of the biosphere is described. Eukaryotes, one of the three domains of Life (together with *Bacteria* and *Archaea*), encompass a dozen or so high-level taxa (here kingdoms). Most of these kingdoms include taxa traditionally considered as belonging to the former polyphyletic plant kingdom together with taxa belonging to the former animal kingdom. Similarly, most of these kingdoms encompass ensembles formerly referred to as 'algae', 'fungi' (customary meaning) and protozoa, which modern phylogenies proved to be highly polyphyletic and therefore artificial. Here, eukaryotic taxa are placed within putatively monophyletic ensembles (kingdoms): Archaeplastida (=Plantae), Rhizaria, Alveolata, Stramenopiles (=Heterokonta), Haptobionta (=Haptophytes), Discicristates, Excavates, Opisthokonta (including Metazoa and Fungi, modern meaning), Amoebobionta (=Amoebozoa) and Cryptobionta.

---

* Chapter Coordinator

C.-F. Boudouresque* (✉)
Institut Méditerranéen d'Océanologie (MIO), UM 110, CNRS 7294
IRD 235, Université de Toulon, Aix-Marseille Université,
Campus de Luminy, 13288 Marseille Cedex 9, France
e-mail: charles.boudouresque@mio.osupytheas.fr

Keywords

Eukaryotes • Unicellular • Phylogeny • Archaeplastida • Rhizaria • Alveolata • Stramenopiles • Haptobionta • Discicristates • Excavates • Opisthokonta • Metazoa • Fungi • Amoebobionta • Cryptobionta

7.1 Introduction

In the vernacular language, the term "microbe" is usually regarded as a synonym of "bacteria," in fact both *Bacteria* and *Archaea* (prokaryotes), according to modern taxonomy.

The notion of "microbes" in fact encompasses *Bacteria* (unicellular and multicellular)[1] and *Archaea*, together with unicellular eukaryotes, which constitute the bulk of the eukaryote domain. In addition, microbiologists have traditionally included within their study field a number of eukaryotes named "fungi" (customary meaning), *e.g.* Fungi (modern meaning), which are generally multicellular, and Oobionta, which are constituted by a **coenocyte***, a giant multinucleate cell. The frontier between clearly unicellular and clearly multicellular organisms is often somewhat vague: between them, there is a continuum of intermediate situations (*cf*. Sect. 5.2).

The number of eukaryote species actually described, according to the criteria of the botanical and the zoological nomenclature codes, is not accurately known. This is due to the fact that many anciently described taxa are only known through very rough diagnoses, so that their taxonomic value is under debate. Some of them are taken into consideration in the lists, while they may just represent synonyms of better described taxa. Others were prematurely considered as possible synonyms of accepted taxa, while further studies could lead to reestablishing them as valid taxa. Overall, the number of eukaryote species currently recognized lies between 1,740,000 (Bouchet 2000; Lecointre and Le Guyader 2006) and 2,000,000 (Rokas 2006). The species diversity of eukaryotes would appear to be considerably higher than that of prokaryotes, with only 11,000 species or so (Lecointre and Le Guyader 2006). However, it is worth noting that the species concept is very different in eukaryotes and prokaryotes, so that the comparison is meaningless (*cf*. Sect. 6.1). The species that are definitely described represent a small part of the actual species diversity of the Earth. Estimates of this diversity lie between 3 and 50 million species (May 1997; Pedrós-Alió 2003); a statistical approach predicts ~8.7 million (±1.3 million SE), of which 2.2 million (±0.2 million SE) are marine (Mora et al. 2011).

The vast majority of known eukaryote species belong to the Metazoa (kingdom Opisthokonta; ~73 %) and Embryophyta (kingdom Archaeplastida; ~16 %), taxa which only encompass multicellular organisms. Unicellular eukaryotes only represent ~10 % of all eukaryotic species, even if coenocytic species, such as Oobionta (kingdom Stramenopiles) and species which are unicellular in only one phase of their life cycle, such as social amoeba, are taken into consideration. In contrast, if only taxa of high phylogenetic level, *e.g.* kingdoms, sub-kingdoms, and phyla (hereafter the "phyletic diversity"), are taken into consideration, most eukaryotes are unicellular (Fig. 7.1) or, though not sensu stricto unicellular, such as Phytomyxea, Oobionta, and Fungi (hereafter "affiliate taxa"), fall within the field of the microbiology. Moreover, because of the thorough exploration of the biosphere and molecular tools, new taxa of high phylogenetic level are continually being discovered; all of them are unicellular, which contributes to further enhancing the overwhelming importance of unicellular taxa within eukaryotes (Box 7.1).

Taking into consideration the huge phyletic diversity of unicellular eukaryotes and of affiliate taxa, which corresponds to nearly the whole diversity of eukaryotes (Fig. 7.1), it is impossible to present here a comprehensive description of the whole of these high-level taxa. Even a whole work would not be enough for such a purpose. In addition, some of these taxa are constituted by a single or a couple of poorly known species whose position within the eukaryote tree is uncertain. The choice was therefore to select one part, indeed a small part of these high-level taxa, but a part likely to illustrate the amazing diversity of eukaryotes. For each selected taxon, some traits will be recurrently tackled topics, *e.g.* the chloroplast structure and the photosynthetic pigments (photosynthetic species), the kinetic apparatus, and the cell wall. Some derived characters, more or less specific to a taxon and suitable to constitute a **biomarker*** (genetic, biochemical, cytological, and biological) will also be emphasized. The biological life cycle of at least one species belonging to the taxon will be analyzed and illustrated in a standardized way. Finally, the role of the taxon in the functioning of the biosphere will be described. Unfortunately, for some of the selected taxa, these aims were only partially achieved, due to the lack of knowledge.

The nomenclature of eukaryotes is ruled by two distinct codes: the International Code of Nomenclature for algae, fungi, and plants (ICN; "Botanical Code") and the International Code of Zoological Nomenclature (ICZN) (*cf*. Sect. 5.1.3). These codes are based upon the Linnean eighteenth century concept of a living world divided into two sharply differentiated kingdoms, namely the vegetable (or plant) kingdom and the animal kingdom. From the 1960s onward, this dualistic concept of Life was progressively demolished, though data challenging

---

[1] Many *Bacteria* are multicellular, such as. *Cyanobacteria*, *Actinobacteria* and Firmicutes (*e.g.* Candidatus *Arthromitus*).

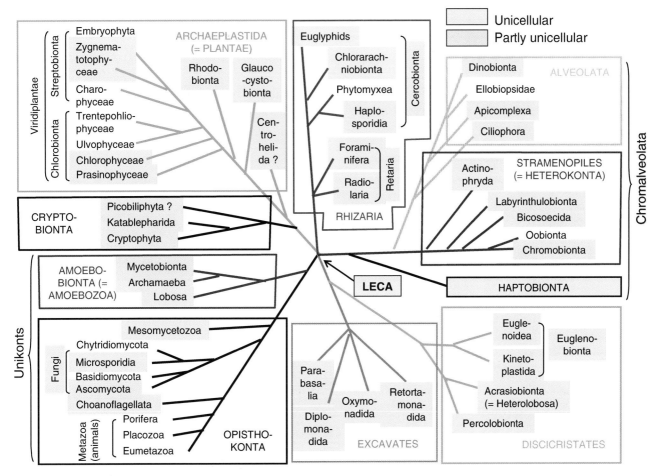

Fig. 7.1 *Unicellular eukaryotes.* The simplified phylogenetic tree of eukaryotes, based upon Baldauf (2003, 2008), Boudouresque et al. (2006) and Boudouresque (2011), modified and updated. *LECA* Last Eukaryotic Common Ancestor. Unicellular taxa are in *yellow*, partly unicellular taxa are in *light blue*. Coenocytic taxa (Oobionta and Phytomyxea), which are not unicellular sensu stricto, are nevertheless taken into consideration in the present work, together with Fungi, since they belong to the microbiology field. The figure illustrates the fact that most high-level taxa of eukaryotes are totally or partly unicellular

Box 7.1: Assessing Eukaryotic Diversity with Molecular Methods

Franck Lejzerowicz and Roland Marmeisse

The sequencing of taxonomic markers such as the ribosomal RNA genes (rDNA) amplified from environmental DNA, and referred to as "metagenetics," constitutes the gold standard approach for the characterization of microbial communities. This approach has revolutionized our understanding of micro-eukaryotic diversity, and with the advent of high-throughput sequencing technologies, has led to the discovery of micro-eukaryotic biodiversity patterns. Metagenetic analyses allowed to (i) identify novel phyla related to well-described taxa; (ii) re-evaluate the extant diversity of unicellular Eukaryotes toward higher species richness, (iii) describe surprisingly diverse micro-eukaryotic communities in environments traditionally considered as hostile to eukaryotic life.

New Eukaryotic Phyla

Environmental 18S rDNA sequences can be amplified using "universal" PCR primers and identified by phylogenetic inference. Some of the resulting phylotypes, belonging to closely-related species, form novel, well-supported clades distantly related to any described species. Two such clades might constitute novel lineages that

F. Lejzerowicz (✉) • R. Marmeisse
Microbial Ecology Center, UMR CNRS 5557 / USC INRA 1364, Université Lyon 1, 69622 Villeurbanne, France

(continued)

Box 7.1 (continued)

diverged early in the history of Eukaryota: the picobiliphytes (Not et al. 2007) and the rappemonades (Kim et al. 2011b). Usually, environmental lineages could be placed within known taxonomic groups such as Fungi, Archaeplastida, Rhizaria, Stramenopiles, or even associated to finer taxonomic levels. The discovery of widely distributed environmental lineages not only helps refine the Eukaryotic tree of Life but also provides a measure of the gap in our knowledge of eukaryotic biodiversity. Metagenetic surveys of micro-Eukaryotes revealed that many novel taxa are far from being marginal or rare and therefore may represent, a substantial fraction of the total microbial biomass. For example, diverse stramenopile phyla referred to as MAST (MArine STramenopiles) were identified all across the world's oceans, except for the MAST-4 group in polar areas (Massana et al. 2006; Box Fig. 7.1). Although MAST-4 is among the least diverse of all MASTs (Logares et al. 2012), it can represent up to 10 % of the heterotrophic eukaryotic cells thriving in a seawater sample. In the marine environment, such dominating phyla could greatly influence the primary production by regulating microbial population dynamics. However, the structural and functional characterization of the corresponding organisms remain particularly challenging. This can be achieved by fluorescent in situ hybridization (FISH) of specific probe sequences. In combination with the tracking of prokaryotic cells labeled with a different fluorescent dye, FISH proved useful to show the heterotrophic mode of life of some MAST phylotypes (Box Fig. 7.2).

Environmental phyla have long been overlooked possibly as a result of failures of cultivation attempt, but also because of their small cell sizes. Indeed, the majority of the pico-Eukaryotes (<2–3 µm) has not yet been morphologically characterized. This is the case in the Haptophytes where higher proportions of unknown phylotypes were found in the smaller size fractions (Bittner et al. 2013).

An Unexpected Species Richness

The number of formally described species is currently lower for unicellular than for multicellular Eukaryotes (mainly represented by animals and land plants). Turning to metagenetics could tip the scale toward unicellular taxa although sequences of multicellular organisms still dominate in 18S rDNA reference databases. The sequence diversity resulting from the exploration of micro-eukaryotic communities is

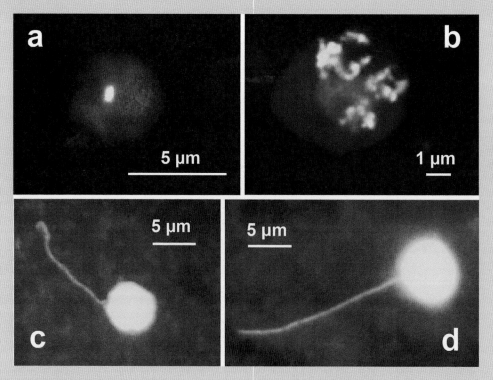

Box Fig. 7.1 Visualization of cells affiliated to a "Marine stramenopile" clade (MAST) and of their capacity to ingest bacteria. Epifluorescence microscopy images of cells belonging to the MAST clade 1B (**a** and **c**) and to the clade 1C (**b** and **d**). In (**a**) and (**b**), nuclei appear in *blue* (DAPI-staining), cytoplasms in *red* (following in situ hybridization (FISH) to a clade-specific oligonucleotide probe) and ingested bacteria in *yellow* (FITC labeling). (**c**) and (**d**), visualization of MAST cell flagella following whole cells FITC staining (According to Massana et al. 2006, courtesy of editions Wiley-Blackwell)

(continued)

Box 7.1 (continued)

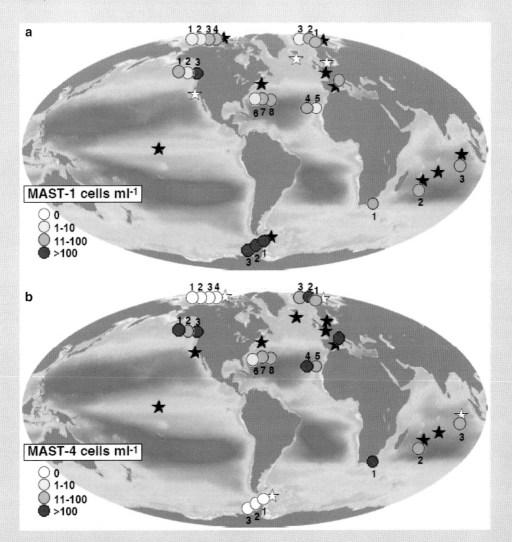

Box Fig. 7.2 Global distribution and abundance in oceans of cells belonging to clades MAST-1 (**a**) and MAST-4 (**b**). *Stars* indicate sampling sites from which 18S rDNA libraries were constructed and sequenced. *Black stars* indicate the presence of sequences affiliated to either MAST-1 or 4; *white stars* indicate their absence. *Circles* indicate sampling sites where plankton cells were counted following FISH staining with oligonucleotide-specific probes. Different colors indicate different abundance levels (According to Massana et al. 2006, courtesy of Wiley-Blackwell Editions)

prompting further inventory efforts in order to improve databases (for protists, see Pawlowski et al. 2012). Indeed, unexpected high levels of 18S rDNA sequence richness have been reported for diverse taxonomic groups targeted with specific PCR primers. For Diplonemids (Euglenozoa) (Lara et al. 2009) and Centroheliozoa (Cavalier-Smith and von der Heyden 2007), the number of phylotypes sequenced even exceeded the number of described species. For deep-sea Foraminifera, most of high-throughput sequencing reads corresponded to inconspicuous, poorly-described lineages (Lecroq et al. 2011).

These observations are not limited to "exotic" environments. Highly diverse and numerous, unknown taxa also live at our doorstep. For example, 83 Operational Taxonomic Units (OTUs) were delineated from only 377 sequences generated from a pond near Paris (France) and most could not be assigned to a known species, including the well-studied ciliates (Šlapeta et al. 2005b). Obviously, even among extensively studied taxa exist numerous genuine species that are difficult to observe (small sized, strict intracellular parasites, *etc.*) or that still need to be sequenced and deposed into databases. However, although the diversity stemming from morphologically similar but genetically distinct, cryptic species could be sequenced, metagenetic estimates are subject to multiple biases, especially at the sequencing scale provided by high-throughput sequencing technologies.

(continued)

> **Box 7.1** (continued)
>
> Group-specific PCR amplifications prior to sequencing are particularly suited for the characterization of highly diverse, albeit rare micro-eukaryotic taxa thriving in low abundance and low biomass-carrying capacity in some environments. In fact, targeted approaches are often the only way to detect taxa harboring highly divergent 18S rDNA sequences compared to those used for the design of so-called "universal" PCR primers. Using highly specific primers, foraminiferal SSU rDNA sequences could be enriched from degraded, ancient DNA preserved in subsurface deep-sea sediments as well as from soils-extracted DNA although foraminiferans are traditionally considered as strictly marine protists (Lejzerowicz et al. 2010, 2013).
>
> **Life in Extreme Environments**
>
> "Extreme environments" are usually defined as environments in which the values of one or several physicochemical parameters (*e.g.* temperature, pH, pressure, or salinity) are such that specific adaptations are required for the organisms to survive and multiply there. Life in extreme environments is not limited to prokaryotes and according to descriptive metagenetic analyses, is not limited to few taxonomic groups of eukaryotes. Indeed, diverse taxa were detected in extreme environments ranging from hyperacid (pH ±1) to hypersaline (6 M) waters, from anoxic basins to the deep-sea abysses (−3,000 m) including hydrothermal sediments (−2,200 m) as well as in polar areas. For example, eukaryotic microbial communities present in diverse micro-habitats of an hydrothermal vent system were dominated by 18S rDNA sequences belonging to kinetoplastids and ciliates although other sequences could be assigned to diverse groups of Alveolata, Radiolaria, or Opisthokonta (López-García et al. 2003). Conversely, the acid mine drainage waters surveyed by Baker et al. (2009) were found to host numerous sequences with no close relative in reference databases. These phylotypes might be the first representatives of a new phylum of high taxonomic rank named APC (Acidophilic Protist Clade). Given the high level of sequence divergence, specific probes designed for FISH experiments were successful at identifying the largest cells of the APC in environmental samples.
>
> The metagenetic exploration of micro-eukaryotic diversity is still in its infancy and whether the entire phyla are endemic to extreme environments or include cosmopolitan species remains to be tested. High throughput sequencing technologies are increasingly popular and rapidly satisfying the need for additional sequence data and the processing of numerous samples. However, although environmental sequences are informative for phylogenetic placement and taxonomic diversity description based on known, sequenced organisms, they constitute poor indicators of species physiology and ecological traits.

the Linnean system and announcing the phylogenetic revolution can be found as early as the late nineteenth century. Nowadays, eukaryotes, one of the three domains of Life (together with *Bacteria* and *Archaea*), encompass a dozen kingdoms.[2] Most of these kingdoms include taxa traditionally considered as belonging to the former plant kingdom and taxa belonging to the former animal kingdom (*cf.* Sects. 5.5.2 and 5.5.3). From the level of the species to that of the class, sometimes to phylum level, there is no major difficulty in applying the rules of either the botanical or the zoological code. However, beyond these levels, for high-level taxa, there is no reason for choosing one code, since the taxa include both former "plants" and "animals." For high-level taxa, the "rule" is therefore common practice or logic. This practice is very different from one author to another. Here, the practice of Lecointre and Le Guyader (2006) and/or of Boudouresque (2011) was adopted. The suffixes "phytes" (=plants), "mycota" (=fungi, in the customary meaning) and "zoa" (=animals, in the customary meaning) were therefore often changed into "bionta" (=living organism). This is especially necessary when *e.g.* Dinobionta are considered, a taxon claimed by both "botanists" (as Dinophyta or Dinophyceae) and "zoologists" (as Dinozoa and Dinoflagellates), though the situation is a bit more complicated than this rather facile shortcut would suggest. Within Dinobionta, some species are photosynthetic, some others are mixotrophic, while some others are heterotrophic, including formidable predators such as *Karlodinium armiger*. Classifying Dinobionta within a traditional Linnean kingdom, namely plants or animals, is therefore meaningless. Altogether, for every high-level taxon, the different names recently used in the literature (though they are not always exact synonyms) will be given.

7.2 General Characteristics of Eukaryotes

Eukaryotes differ from prokaryotes by a large number of characters. Overall, this set of differences is very robust and no organism appears to be intermediate between eukaryotes

---

[2] Within prokaryotes, the highest taxonomic rank is that of the phylum. In contrast, within eukaryotes, the highest taxonomic rank is constituted by kingdoms, though their use and definition are not ruled by the traditional nomenclature codes, constrained 'by nature' by the Linnean dichotomy plant-animals.

Table 7.1 *The pathways of methionine synthesis in eukaryotes.* The number of "+" is proportional to the relative importance of a pathway in the species of a given taxon (+ = ~20 %) (From Croft et al. (2005), modified)

Kingdom	Taxon	B_{12}-dependent (metH)	B_{12}-independent (metE)
Opisthokonta	Metazoa	+++++	–
Archaeplastida	Chlorobionta (except Charophyceae)	++	+++
	Charophyceae	–	+++++
	Embryophyta	–	+++++
	Glaucocystobionta	+++++	–
	Rhodobionta	+++++	–
Alveolata	Dinobionta	++++	+
Stramenopiles	Xanthophyceae	–	+++++
	Phaeophyceae	–	+++++
	Bacillariophyceae	+++	++
	Chrysophyceae	+++++	–
Cryptobionta	Cryptophyta	++++	+
Haptobionta		+++	++
Discicristates	Euglenoidea	++++	+

and prokaryotes. However, when a given eukaryotic character is considered alone, it can be absent from a taxon. This is considered to be a derived, not ancestral, character, corresponding to a secondary loss.

The differences between prokaryotes and eukaryotes are dealt with in Chap. 3. Here, we simply recall the characters common to all eukaryotes, or to most of them, characters of probable ancestral nature. These characters will not be recalled hereafter in the taxonomic section, except for their absence, probably due to a secondary loss.

First of all, the eukaryotic cell is characterized by the presence of a true nucleus. Most of the genetic material (DNA) is localized within this nucleus, separated from the cytoplasm by a double-layered membrane, the nuclear envelope. The nuclear DNA is of eukaryotic-type, with two strands wound around each other in a double helix; strands are linear, with open ends, unlike the prokaryotic-type DNA which is circular. The DNA is packaged by large proteic molecules, the histones. The genes are distributed into several chromosomes (at least two). The number of genes is generally high (up to 40,000). The major part of the genome is composed of repetitive DNA, with noncoding regions taking a large part (up to 90 %), within (introns) and between genes.

The enzymes involved in the vitamin B_{12} (cobalamin) synthesis have been characterized in some prokaryotic organisms only. In contrast, eukaryotes do not synthesize this vitamin, despite its role in the pathway that produces the amino acid methionine and in a variety of metabolic processes. Some eukaryotes, which possess only the vitamin-B12-dependent methionine synthase gene (*metH*), need vitamin B_{12} to synthesize methionine (Table 7.1); they either draw it from the environment or *via* a mutualistic symbiosis with a bacterium. Other eukaryotes, which possess a vitamin-B_{12}-independent methionine synthase gene (*metE*), use another pathway to synthesize methionine (Croft et al. 2005) (Table 7.1). Probably, the ancestors of eukaryotes had both genes; independent secondary losses of either the *metH* or *metE* gene subsequently occurred over evolutionary time. Only a few species, such as *Chlamydomonas reinhardtii* (Chlorobionta) and *Cyanidioschizon merolae* (Rhodobionta), have maintained the two genes (Andersen 2005; Croft et al. 2005). Overall, a large proportion of photosynthetic eukaryotes, usually considered as **autotrophs***, prove to be in fact **auxotrophs***.

Sterol biosynthesis is virtually ubiquitous among eukaryotes, where it likely constitutes an ancestral character. As a result, the presence of steranes in ancient rocks is used as evidence for eukaryotic evolution 2.7 Ga ago (*cf.* Sect. 5.4.1). The absence of sterols in Oobionta is a derived character. Sterols play essential roles in the physiology of eukaryotic organisms (*e.g.* the cellular membrane). Here, particular attention will be paid to sterols because of their diversity, making some of them good biochemical markers of the taxa they belong to.

Eukaryotic cells have a variety of internal membranes and structures, called **organelles***, and a cytoskeleton, which are absent in prokaryotes. Some of the organelles originate in ancient bacteria that became endosymbiotic, keeping only part of their prokaryotic-type DNA. This is the case of chloroplasts and mitochondria, together with **hydrogenosomes*** and **mitosomes***, which probably evolved from mitochondria (*cf.* Sect. 5.4). Other internal membranes and structures characteristic of the eukaryotic cells are the Golgi apparatus, the kinetic apparatus, the vacuole, and the endoplasmic reticulum (*cf.* Sect. 5.4.2 and Fig. 5.11). The kinetic apparatus and the chloroplast, because of the conspicuous cytological and biochemical differences they exhibit from one high-level taxon to another, provide efficient phyletic markers. Some organelles were secondarily lost (derived character) during evolution. Cytoplasmic ribosomes of eukaryotic cells are normally larger (80S) than prokaryotic ones (70S). Within chloroplasts and mitochondria, ribosomes are of the 70S-type, which is consistent with the prokaryotic origin of these organelles.

Sexual reproduction, which is absent in prokaryotes, was an early and pivotal evolutionary innovation of eukaryotic cells. It is characterized by two opposite events, fertilization (fusion of two gametes that doubles the number of chromosomes), a process where the conditions are highly taxon-dependent and varied, and meiosis, a special type of cell division (that divides the number of chromosomes by two). Fertilization, meiosis, and some associated processes constitute a series of transitions between generations and phases whose succession represents the life cycle (*cf.* Sect. 7.3). Patterns of life cycle are highly diverse and may constitute a biological marker characteristic of some high-level taxa. Sexual reproduction is however absent from some eukaryotic taxa, either because it never emerged (which is highly unlikely), or because it was secondarily lost, or finally because it is pending discovery.

Another pivotal evolutionary innovation in eukaryotes was the phagotrophy, the capability for an organism, either unicellular or multicellular, to engulf a particulate food, *e.g.* a prey, a process unknown in prokaryotes (Cavalier-Smith 1987a). Phagotrophy made possible a major process in early eukaryote evolution, endosymbiosis with *Bacteria* at the origin of mitochondria and chloroplasts (*cf.* Sect. 5.4). Its absence, *e.g.* in Fungi (kingdom Opisthokonta) is probably a loss and therefore a derived character.

7.3 Life Cycles

7.3.1 Why Is It Important to Use a Simple and Standardized Terminology?

Disciplinary boundaries, which were set in the eighteenth century, have been the source of a terminological proliferation. Botanists, zoologists, and bacteriologists created different terms to refer to the same notion while, at the same time, a given term was used with very different meanings. The same process of terminological proliferation occurred within each of these branches. Within botany, specialists in fungi (customary meaning), algae, mosses, ferns, and flowering plants created their own jargon. Sometimes, this confusing terminology can be found within a given family, even a given genus of the customary complexes called "fungi" (not the taxon named hereafter Fungi), "algae," and "protozoans" (*cf.* Sect. 5.5.3). Many teachers around the world are guilty of perpetuating old terminological customs and so spreading false similarities that mask the real similarities. In short, their teaching leaves much to be desired.

A given term, *e.g.* **spore***, refers to a kaleidoscope of meanings from one taxon to another. The specialists of a taxon (sometimes a particular species) have added a multitude of prefixes to account for highly redundant shades: aplanospore, ascospore, autospore, basidiospore, carpospore, chlamydospore, conchospore, conidiospore, ecidiospore, endospore, exospore, megaspore, meiospore, microspore, mitospore, monospore, oospore, paraspore, perispore, probasidiospore, seirospore, sporangiospore, teliospore, tetraspore, unispores, urediospore, zoospore, zygospore, *etc.* Conversely, the male gamete, a clear and simple notion, was named spermatozoon, antherozoid, or spermatium. Is it really useful or necessary to designate a filament consisting of cells, "trichome" in *Cyanobacteria* and "hypha" in Fungi?

Here it was decided to replace the customary terms stemming from the partitioning of the taxonomy by a single term, despite the risk of surprising, or even shocking, some experts. The choice of one term rather than another (there was often a choice) may seem arbitrary: the point is that this term has a precise definition, and that it facilitates the understanding of the real similarities and differences between taxa. When necessary, the correspondence to traditional terms is indicated. A little known book by a professor at the University of Montpellier (France), Jean Motte (1971), who tried to "groom" the customary terminology, served in part as the basis for the terminology used here.

Some figures which illustrate the present chapter are based upon multicellular organisms. The reasons are either that some multicellular eukaryotes are traditionally taken into account by microbiologists or because they help in understanding some of the concepts used here.

7.3.2 What Are the Differences Between Spores, Gametes, Conidia, Cuttings, Carpoconidia, and Zygotes?

A **gamete*** is a haploid (n) sexual cell that fuses with another gamete to produce a diploid (2n) zygote. Gametes contain only one set of dissimilar chromosomes, while zygotes contain two sets of (paired) chromosomes (Fig. 7.2). The fusion of gametes is called **fertilization*** and can be summed up in the form: $n + n \rightarrow 2n$. When more or less anthropomorphic criteria make the distinction possible, male and female gametes are distinguished: the male gamete is the smaller and/or more mobile, whereas the female is the larger and/or less mobile. When such criteria do not apply, gametes of opposing mating type are referred to by "+" and "−".

A **spore***[3] is a haploid (n) cell produced by meiosis. Meiosis begins with one diploid cell containing two copies of each chromosome. The cell divides twice (a reductional and an equational division), potentially producing up to

---

[3] Spore, from the ancient Greek word *spora* (meaning 'seed', 'sowing').

Fig. 7.2 *Simplified presentation of fertilization (top) and meiosis (bottom)*. In fact, divisions of meiosis are linked, so that the cells from meiosis I are usually not individualized. *Black circles*: diploid nuclei (2n). *Open circles*: haploid nuclei (n). Haploid nuclei of one of the sexes are distinguished by a central point. The association of male with (−) rather than (+) (top left) is arbitrary: by definition, the use of (−) and (+) means that no morphological or behavioral criteria offer a basis for distinguishing the sex (mating type)

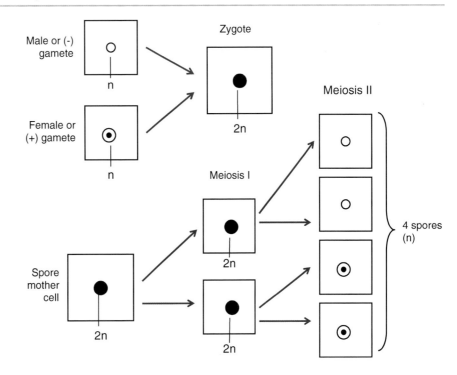

four haploid daughter cells (spores) containing one copy of each chromosome (Fig. 7.2). Meiosis can be summed up in the form: 2n → n. In some cases, only one of the four haploid cells produced by meiosis survives. In other cases (Fungi Ascomycota), the meiotic division is followed by a mitotic division, which results in eight haploid cells (spores). When the life cycle consists of a single diploid generation, the spores behave like gametes: they fuse together (fertilization) to form a zygote. It is then a rather facile shortcut to say that "meiosis produces gametes"; this is the case in Metazoa (kingdom Opisthokonta) and Fucales (an order in the brown algae, kingdom Stramenopiles). The production of spores, together with that of gametes and zygotes *via* fertilization, are characteristics of eukaryotes.[4] It should be noted that the term "spore" is often used erroneously, in the sense of what is defined below as conidium, carpoconidium, and cyst.

Conidia* are reproductive cells, the fate of which is to be disseminated, which give rise to individuals genetically identical to the one which produced them. They occur not only in eukaryotes but also in prokaryotes (Figs. 7.3 and 7.4). In eukaryotes, conidia are haploid when produced by a haploid individual, giving a new haploid individual (to sum up: n → n); they are diploid when produced by a diploid individual, giving a new diploid individual (to sum up: 2n → 2n). Conidia therefore constitute a cloning process. In addition, conidia do not interfere with the progress of the life cycle and are not necessary to the "closing of the circuit"; they are just optional reproductive cells, generating kinds of loops, the "conidial loops" (Fig. 7.9) (*cf.* Sect. 7.3.5).

Cuttings, like conidia, constitute a cloning process, in eukaryotes as in prokaryotes. The difference is that conidia are unicellular, while cuttings are multicellular. Cuttings can be non-specialized, *e.g.* a fragment of a filament which will give rise to a new individual filament (Fig. 7.3a). They can be specialized, *e.g.* a group of cells designed to become detached from the parent individual (Fig. 7.3b). In eukaryotes, as for conidia, the ploidy of the parent individual is preserved: n → n and 2n → 2n. Some botanists call the cuttings "propagules." In fact, "propagule" is more widely used with a different meaning: a non-specified element of dissemination, *e.g.* zygote, conidium, seed, cutting, zygote, and larva. Here, propagule will be used only with the latter meaning.

Carpoconidia* (most authors call them "carpospores") are reproductive cells similar to conidia in that, from a genetic point of view, they produce clones. The difference with conidia is that they do not constitute an optional loop in the life cycle; rather, they constitute an obligatory point of passage between two generations of the life cycle, a passage necessary for the progress of the life cycle and the "closing of the circuit." The generation which produces carpoconidia is generally, through its morphology and/or its biology, different from the generation stemming from carpoconidia. Though these two generations are genetically identical, the gene expression differs. Not all eukaryotes comprise carpoconidia in their life cycle.

---

[4] In prokaryotes, the term 'spore' is used with a different meaning from the one defined here for eukaryotes.

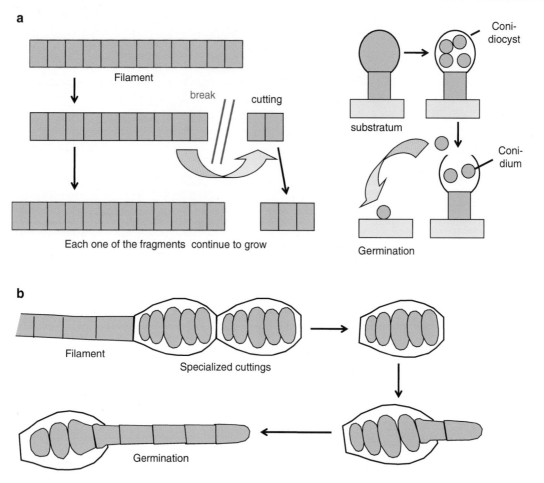

Fig. 7.3 *Conidia and cuttings in prokaryotes.* (**a**) *Top right:* Formation of conidia in a cyanobacterium of the genus *Dermocarpa*. The conidiocyst is the cell whose content turned into conidia. *Top left:* formation of a non-specialized cutting in a cyanobacterium of the genus *Oscillatoria*. (**b**) Formation and germination of specialized cuttings in a cyanobacterium of the genus *Westiella*

Fig. 7.4 *Formation of conidia in eukaryotes.* Formation of haploid (n) conidia by a haploid individual (*left*). Formation of diploid (2n) conidia by a diploid individual (*right*). *Open circles*: haploid nuclei (n). *Black circles*: diploid nuclei (2n). Conidia give rise to clones, *i.e.* individuals strictly identical (genome, morphology, ploidy) to the parents

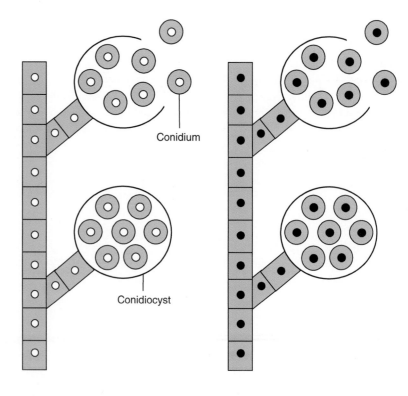

7.3.3 What Is a Cyst?

A cyst* is a non-motile resistant cell formed under adverse conditions (*e.g.*, low nutrient, predation), which may survive for months or years and is able to again become an active cell when conditions improve. A cyst is surrounded by a thick wall that can be almost indestructible, as is the case of cysts of Dinobionta (*cf.* Sect. 7.8.3). Cysts can originate in a variety of life stages (spores, conidia, zygotes).

7.3.4 Different Types of Fertilization

Fertilization is the fusion of two gametes, possibly of two vegetative cells which play the role of gametes. It gives rise to a zygote that initiates the development of a new individual organism. Fertilization can be decomposed into two steps, the fusion of cytoplasms (plasmogamy) and the fusion of nuclei (karyogamy). Generally, plasmogamy and karyogamy occur simultaneously or with only a slight delay. However, in some eukaryotes, such as Fungi, karyogamy is delayed and occurs a variable amount of time after plasmogamy. As a result, in addition to the classical haploid (n) and diploid (2n) phases, these organisms can present, at least during part of their life cycle, phases where nuclei of both sexes (male and female, or + and −), inherited from the two parents, coexist within the cytoplasm without merging (n + n), namely the **micthaploid*** and the **dikaryotic*** phases (Fig. 7.5). The phase is said to be micthaploid when numerous nuclei of both mating types coexist and divide within a single mass of cytoplasm, constituting a coenocyte. It is said to be dikaryotic when each cell harbors a couple of nuclei of both mating types.

A variety of types of fertilization occur in eukaryotes, mainly as a function of (i) the dissemination (or absence of dissemination) of the gametes of each sex, (ii) their mobility (or immobility) and (iii), when they are mobile, the way they move (Table 7.2). Some types of fertilization, such as planogamy and oogamy, can occur in several high-level taxa. Others, such as siphonogamy and protrichogamy, are specific to a taxon, Embryophyta and Rhodobionta, respectively.

Protrichogamy is specific to Bangiophyceae, one of the ancestral taxa in Rhodobionta (kingdom Archaeplastida). Considering that *Bangiomorpha pubescens*, a fossil discovered by Butterfield (2000), ascribed to Bangiophyceae and dated from 1.2 Ga (*cf.* Sect. 7.5.4), represents the oldest known evidence of sexuality, protrichogamy could testify to the ancestral forms of sexuality and tell us about the way it worked. In protrichogamy, female gametes are not scattered (Fig. 7.6a). Due to the absence of a kinetic apparatus in

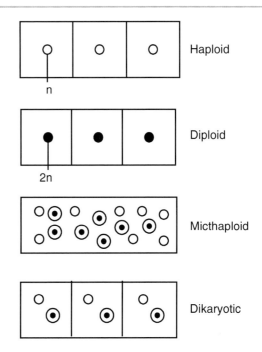

Fig. 7.5 *The four possible phases in the life cycle of eukaryotes. Open circles*: haploid nuclei (n). *Black circles*: diploid nuclei (2n). In the micthaploid and dikaryotic phases, the haploid nuclei of one of the two mating types are rendered with a central point, those of the other without point

Rhodobionta, the male gametes, though scattered, are motionless, *i.e.* unable to move. Consequently, the meeting of the male and the female gametes is unlikely and the efficiency of the fertilization probably extremely low. Trichogamy, which is specific to Florideophyceae (Rhodobionta) and Fungi (kingdom Opisthokonta), is characterized by the presence of a hairlike projection of the female gamete, called trichogyne, which receives (captures?) the male motionless gametes (Fig. 7.6b). The trichogyne, which can be relatively long, passively explores the surroundings and improves the likelihood of the meeting and fusion of gametes.

In oogamy, the female gamete is not scattered and remains fixed on the female "gametogen" (*i.e.* the gametogenous generation, which produces female gametes). In contrast, male gametes are spread and mobile by means of **undulipodiums***[5] (Fig. 7.6c). The male gamete localizes the female gamete through the pheromones produced by the latter and swims toward it. This class of pheromones,

---

[5] For the use of undulipodium rather than flagellum, in eukaryotes, see Sect. 5.4.2. Flagella are structures that characterize prokaryotes. Undulipodiums are much more complex structures, completely different from a biochemical and functioning point of view, which are specific to eukaryotes.

Table 7.2 *Different types of fertilization in eukaryotes.* X, in the columns "disseminated" or "non-disseminated," indicates the status which characterizes the considered type of fertilization. "Male" and "female" are to be replaced by "+" and "−", where gametes of both mating types cannot be distinguished, which occurs in cystogamy, somatogamy, and amoebogamy

Mobility	Male gamete		Female gamete		Name of the fertilization type
	Disseminated	Non-disseminated	Disseminated	Non-disseminated	
Both gametes are motionless	X		X		Protrichogamy
	X			X (+ trichogyne)	Trichogamy
		X (in fact, it is the male individual, namely the pollen grain, that is disseminated)		X (these are individuals, not gametes, that move closer)	Siphonogamy
		X (these are individuals, not gametes, that get close together)		X (these are individuals, not gametes, that move close together)	Cystogamy (=conjugation)
		X (Gametes are absent; these are individuals that move close together)		X (Gametes are absent; these are individuals that get close together)	Somatogamy
Male gamete mobile, female one non-mobile	X (undulipodium)		X		Fucogamy
	X (undulipodium)			X	Oogamy
Both gametes are mobile	X (undulipodium)		X (undulipodium)		Planogamy
	X (amoeboid movement)		X (amoeboid movement)		Amoebogamy

secreted by a partner to ellicit a sexual response in the other, is named sirenins. From the chemical point of view, sirenins are volatile compounds of low molecular mass, partly unsaturated olefin hydrocarbons. Sirenins are sometimes species-specific, more commonly specific to a group of taxonomically related species (Müller et al. 1985; Pitombo et al. 1989). Fucogamy is similar to oogamy, from which it differs by the fact that the motionless female gamete does not remain fixed on the female gametogenous individual and is disseminated (Fig. 7.6d).

Planogamy is characterized by the fact that both gametes are scattered and mobile, by means of undulipodiums (Fig. 7.6e). Gametes of both mating types can be of equal or unequal size. In the latter case, the larger gamete is considered as the female one. The female gamete swims and looks for a suitable substratum, where it fixes, loses its undulipodiums, and secretes a sirinin which attracts male gametes. According to Peters and Müller (1985), a sweet fragrance emanates from cultures of the female gametogenous generation ("gametogen") of *Dictyosiphon foeniculaceus* (Phaeophyceae, kingdom Stramenopiles), due to the secreted sirinin.

Cystogamy (=conjugation) is characterized by gametes of both mating types which are morphologically identical and are neither mobile nor disseminated (Fig. 7.6f). These are individuals that move close together. Two cells of the adjacent individuals form a conjugation tube, a kind of bridge, between them. In some species, once the tube is formed, both gametes crawl toward one another and fuse in the middle of the bridge, forming a zygote. In some other species, only one gamete crawls through the bridge, joins the other gamete on the other side and fuses with it. In the latter case, the gamete that crosses the bridge is considered, by convention, as the male one. Cystogamy occurs in *e.g.* diatoms (Bacillariophyceae, kingdom Stramenopiles).

Somatogamy (Fig. 7.7) is the most reduced form of sexual reproduction and fertilization. Gametes are not individualized. It involves the fusion of two somatic cells (acting as gametes) belonging to two cell filaments of different mating types. As a matter of fact, the fusion is incomplete as it only concerns cytoplasms (plasmogamy), not nuclei (karyogamy); the karyogamy is postponed to much later in the course of the life cycle. Somatogamy is very common in Basidiomycota but not prevalent in Ascomycota (Fungi, kingdom Opisthokonta). Some other types of fertilization, such as trichogamy, are also present in Fungi, though relatively uncommon.

7.3.5 What Is the Difference Between a Generation and a Phase, Within a Life Cycle?

A **generation*** is a step within the life cycle of an organism. It starts with a spore, a zygote, or a carpoconidium and ends, after pronounced vegetative activity (cell multiplication resulting, or not, in the edification of a multicellular body), with the production of spores, gametes, or carpoconidia.

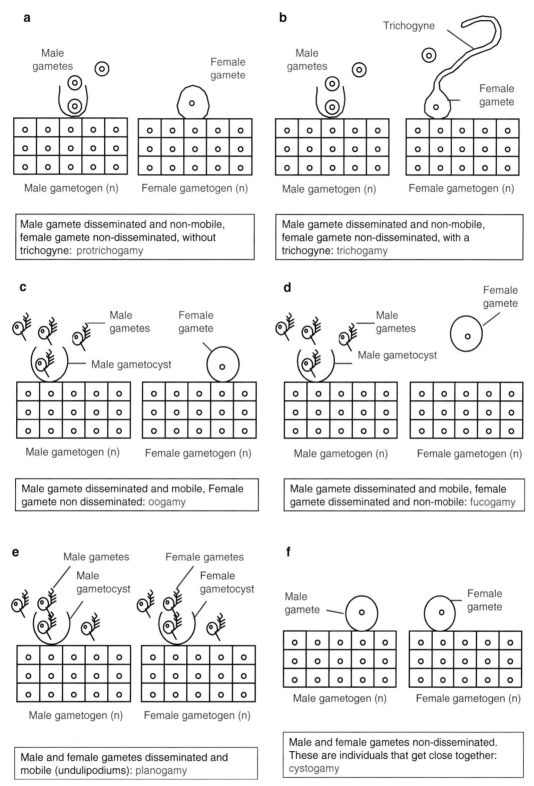

Fig. 7.6 *Some types of fertilization in eukaryotes*, protrichogamy (**a**), trichogamy (**b**), oogamy (**c**), fucogamy (**d**), planogamy (**e**), and cystogamy (**f**). The gametogenous generation ("gametogen"), which produces male and female gametes, is symbolized by a block of 15 cells. In some species, male and female gametes are not produced by distinct individuals (as represented here), but by the same individual (not represented). *Open circles* are haploid nuclei (n). Reproductive organs that may contain or carry the gametes are not represented (with the exception of some gametocysts). In the case of oogamy, fucogamy, and planogamy, the represented undulipodiums are those of Stramenopiles, though these types of fertilization can also occur in a variety of other taxa

Fig. 7.7 *Somatogamy in Fungi.* (**a**) Two growing filaments of the gametogenous generation (haploid phase, n), − and +. On the *right*, the apical cell. (**b**) The apical cells of both filaments orient their growth and divisions toward one another. (**c**) Apical cells fuse. They only merge their cytoplasm, not the nuclei, giving rise to a kind of binucleated zygote. (**d**) The binucleated "zygote" gives rise to the sporogenous generation (dikaryotic phase, n + n), with two nuclei within each cell. All nuclei (*open circles*) are haploid (n). *Open circles* with and without central point represent − and + mating types, respectively

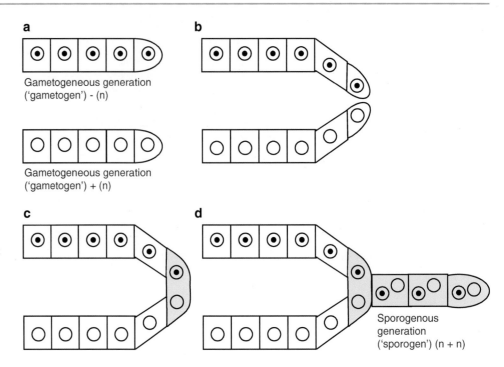

Table 7.3 *Terminology: reproductive cells, the structures that harbor them and the generation that produces them*. It is worth noting that no generation corresponds to the production of conidia. Conidia, such as cuttings, simply constitute a cloning process: they can be produced by the three generation types, namely gametogen, sporogen, and carpoconidiogen

The reproductive cell	The structure that harbors reproductive cells	The generation that produces these reproductive cells
Gamete	Gametocyst	Gametogen (=gametogenous generation)
Spore	Sporocyst	Sporogen (=sporogenous generation)
Conidium	Conidiocyst	
Carpoconidium	Carpoconidiocyst	Carpoconidiogen (=carpoconidiogenous generation)

Generations are named after the reproductive elements they produce (Table 7.3). The gametogenous generation (hereafter **"gametogen*"**) produces gametes, the sporogenous generation (hereafter **"sporogen*"**) produces spores and the carpoconidiogenous generation (hereafter **"carpoconidiogen*"**) produces carpoconidia.[6]

A **phase*** is a part of a life cycle characterized by a type of nuclear structure (ploidy), namely haploid (n), diploid (2n), micthaploid, and dikaryotic (n + n) (Fig. 7.5). A phase can encompass one or several generations. For example, in the life cycle of most Rhodobionta (kingdom Archaeplastida), the diploid phase comprises two generations, the carpoconidiogen and the sporogen (Fig. 7.9d).

Overall, a life cycle can encompass one, two, or three generations. Such life cycles are named monogenetic, digenetic, and trigenetic, respectively. Life cycles having a larger number of generations (up to 11) do exist, but they are anecdotal and do not apply to unicellular eukaryotes, with very few exceptions. Similarly, a life cycle can encompass one, two, or three of the ploidic phases, n, 2n, and n + n; they are named monophasic, diphasic, and triphasic, respectively. For example, the life cycle of Rhodobionta (kingdom Archaeplastida) is generally trigenetic (gametogen, carpoconidiogen, and sporogen) and diphasic (haploid and

---

[6] The gametogenous, sporogenous, and carpoconidiogenous generations (gametogen, sporogen, and carpoconidiogen) are traditionally named 'gametophyte', 'sporophyte', and 'carposporophyte', respectively, by most botanists. This terminology, stemming from the Linnean traditional dichotomy between a botanical and a zoological kingdoms, is particularly inappropriate in that similar generations and life cycles can be found within both these customary kingdoms. '-phyte' comes from the ancient Greek 'phuton' and means 'plant'.

Fig. 7.8 *Symbols used for eukaryote life cycles schematization.* See text (Sects. 7.3.2, 7.3.3, 7.3.4 and 7.3.5) for the meaning of the terms

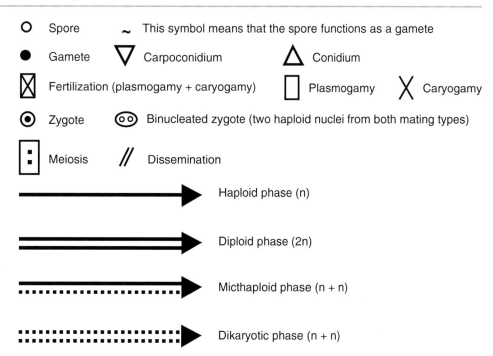

diploid phases). In order to represent, in a schematic and standardized form, life cycles, and to be able to compare them whatever the taxon considered, a set of symbols is proposed here (Fig. 7.8) and illustrated with some examples (Fig. 7.9).

7.4 Unikonts and Bikonts

Within eukaryotes, two sets of kingdoms can be distinguished: Unikonts (Amoebobionta and Opisthokonta) and Bikonts (the other kingdoms; Fig. 7.1). Their divergence probably constitutes an early event in eukaryote evolution. A lineage, Diphyllatia, particularly with the genus *Collodictyon*, is located close to the bifurcation between Unikonts and Bikonts (Zhao et al. 2012).

Unikonts (Unikonta) include the Opisthokonta and Amoebobionta kingdoms. Unicellular species, or mobile cells of pluricellular species, have a single emergent undulipodium. The presence of a second undulipodium, in some Amoebobionta, probably constitutes a derived character. In Opisthokonta, the undulipodium is situated at the rear relative to the direction of the stroke and acts as a propellant (Fig. 7.10). In Amoebobionta, it is located forward and plays the role of tractor.

Bikonts encompass all other eukaryote kingdoms (Cryptobionta, Archaeplastida, Rhizaria, Alveolata, Stramenopiles, Haptobionta, Discicritates, and Excavates). The unicellular species, or the swimming cells of multicellular species, are characterized by the presence of two undulipodiums. They are located at the front of the cell, relative to the direction of the stroke and act as tractors (Fig. 7.10). The occurrence of a larger number of undulipodiums (up to three dozen), or of only one undulipodium, in some taxa, probably constitutes derived characters.

The ensemble of kingdoms whose undulipodium(s) are located at the front of the cell (Amoebobionta and bikonts) is named Anterokonta. By this character, Anterokonta are opposite to Opisthokonta, the only kingdom where the undulipodium is located at the rear of the cell.

7.5 Kingdom Archaeplastida

7.5.1 General Remarks

Archaeplastida are also named Plantae and Primoplantae. The names Archaeplastida and Primoplantae are reminders of the fact that, within eukaryotes, the founder event of the photosynthesis, *via* an endosymbiosis with a cyanobacterium, occurred in this kingdom (*cf.* Sect. 5.4.3).

Despite its alternative name of Plantae, the modern kingdom Archaeplastida should not be confused either with the popular notion of "plant," or with the customary Linnean "vegetable kingdom" (plant kingdom). The reasons are that (i) Archaeplastida only include a small

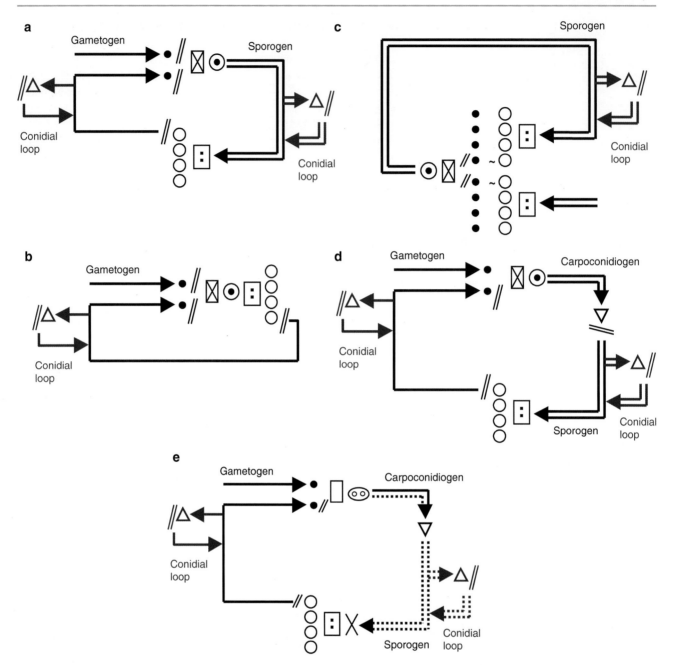

Fig. 7.9 *Some examples of life cycles in eukaryotes.* For the symbols used, see Fig. 7.8. Conidial loops are in blue. (**a**) Digenetic (gametogenous and sporogenous generations) diphasic (haploid and diploid phases) life cycle. (**b**) Monogenetic (gametogenous generation) monophasic (haploid phase) life cycle; it can result from the simplification of the previous cycle (a). (**c**) Monogenetic (gametogenous generation) monophasic (diploid phase) life cycle; it can also result from the simplification of the cycle (a); it is to be noted that the naming of the remaining generation (sporogen rather than gametogen) constitutes an arbitrary choice, though logical when compared with the cycle (a), where the diploid generation is the sporogen. (**d**) Trigenetic diphasic life cycle; the diploid phase encompasses two generations, the carpoconidiogen and the sporogen. This type of cycle is common in Rhodobionta (kingdom Archaeplastida). (**e**) Trigenetic triphasic (haploid, micthaploid, and dikaryotic phases) life cycle. This type of cycle occurs in some Fungi (kingdom Opisthokonta). Note the dissociation of the fertilization into plasmogamy and karyogamy. Many other types of life cycles occur in eukaryotes

part of the taxa that traditionalist botanists and the ICN (the code of botanical nomenclature) consider as plants, and (ii) a taxon that is traditionally considered as belonging to "protozoans" and therefore to the Linnean animal kingdom, namely Centrohelida, possibly falls within Archaeplastida (*cf.* Sect. 7.5.2).

Archaeplastida represent a kingdom whose importance is pivotal in the biosphere. They encompass ~300,000 species,

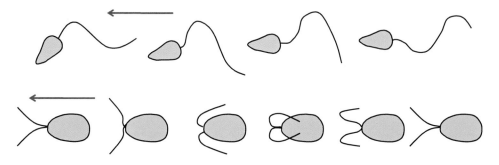

Fig. 7.10 *Swimming cells in eukaryotes. Top*: Opisthokonta. There is a single undulipodium, at the rear of the cell, which propels it. *Arrows* indicate the direction of swimming. *Below*: Bikonts. The two undulipodiums, located at the front of the cell, tow it (Modified and redrawn from Lecointre and Le Guyader (2006))

less than Opisthokonta but more than the other eukaryote kingdom and are dominant in biomass in the terrestrial realm where they constitute the basis for landscapes, because of Embryophyta, *e.g.* Magnoliophyta, Filicophyta (ferns), and Bryophyta (mosses). Some terrestrial Magnoliophyta returned to the marine realm ~100 Ma ago (*e.g.*, the ancestors of the modern genera *Posidonia, Thalassia,* and *Zostera*). Rhodobionta ("red algae") and Chlorobionta (part of the paraphyletic ensemble traditionally called "green algae") also belong to Archaeplastida. The inclusion of Glaucocystobionta is an issue under discussion. As far as Centrohelida are concerned (Nicolaev et al. 2004), they are of uncertain taxonomic position and their placement within Archaeplastida is just a working hypothesis.

Archaeplastida played a prominent role in eukaryote evolution. First, as mentioned above, it is the common ancestor of Viridiplantae and Rhodobionta, and possibly of Glaucocystobionta, which acquired photosynthesis, due to endosymbiosis with a cyanobacterium, the latter becoming a chloroplast. This common ancestor was the first photosynthetic eukaryote, which had so far only been heterotrophic organisms. Subsequently, some Archaeplastida, belonging to Rhodobionta and Chlorobionta, founded the photosynthesis in several other eukaryote kingdoms (*e.g.,* Rhizaria, Alveolata, and Stramenopiles), due to secondary and tertiary endosymbioses (*cf.* Sect. 5.4).

In the Archaeplastida configuration adopted here (including Glaucocystobionta and Centrohelida), there are very few shared characters other than those stemming from molecular phylogenies. The main shared character in Archaeplastida is the fine structure of mitochondria, with plate-like flat cristae.

7.5.2 Centrohelida

Centrohelida (also named Centrohelidia, Centroheliozoa, and Centrohelomonada) were once considered as Heliozoa, a polyphyletic ensemble whose taxa proved to belong to *e.g.* Rhizaria (Heliomonadida) and Stramenopiles (Actinophryda; *cf.* Sect. 7.9.1) (Nicolaev et al. 2004). The evolutionary position of Centrohelida is not clear. Depending upon the studied genes, phylogenies place Centrohelida close to Glaucocystobionta and, sometimes, Rhodobionta (Nicolaev et al. 2004; Šlapeta et al. 2005b; Moreira et al. 2007) or to Telonemia and Haptobionta (Yabuki et al. 2010; Burki et al. 2012). The best solution would probably be to consider Centrohelida as an *incertae sedis*[7] taxon (Okamoto and Inouye 2005; Sakaguchi et al. 2007). Several hundreds of species are known.

All Centrohelida are unicellular. Cells are spherical, around 30–80 μm in diameter, without cell wall and densely covered with pseudopods named **axopods*** (slender, raylike strands of cytoplasm supported internally by a microtubule). The axopod microtubules arise from a granule named centroplast, at the center of the cell. A few genera have no plasmalemma (cytoplasmic membrane) covering, but most have a coat of scales and spicules; these may be organic or siliceous and come in various shapes and sizes (Figs. 7.11 and 7.12) (Tregouboff 1953b). The nucleus is non-central. Mitochondria have plate-like flat cristae (Sakaguchi et al. 2007), a shared character among Archaeplastida. Undulipodiums never occur among Centrohelida (Cavalier-Smith and Chao 2002). Sexual reproduction is unknown.

Centrohelida are non-photosynthetic organisms. Axopods capture food, unicellular prey, and allow mobile forms to move about. However, some Centrohelida harbor within their cytoplasm photosynthetic unicellular symbionts. In *Acanthocystis turfacea*, the symbiont belongs to the genus *Chlorella* (Viridiplantae). Centrohelida thrive in the marine and freshwater plankton.

7.5.3 Glaucocystobionta

Glaucocystobionta (=Glaucophyta, Glaucophytes, Glaucocystophyta, Glaucocystophytes, glaucocystids) is a taxon of uncertain taxonomic placement. Depending upon the period and authors, they were included within Viridiplantae, Rhodobionta, Dinobionta (Alveolata), and

---

[7] *Incertae sedis* (Latin for 'of uncertain placement') is the term used to define a taxonomic group whose relationships are unknown or undefined.

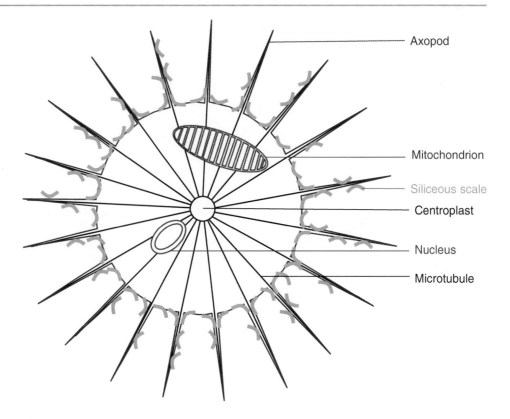

Fig. 7.11 *Simplified and theoretical scheme of a cell of Centrohelida.* Only part of the organelles are shown

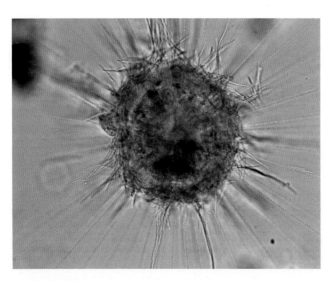

Fig. 7.12 *Raphidiophrys pallida (Centrohelida).* The axopods and siliceous spicules are visible (From Gazzaniga (2009). Photograph: courtesy of Maurizzio Gazzaniga)

even *Cyanobacteria* (Chadefaud 1960; Lecointre and Le Guyader 2006). According to Nozaki et al. (2009), they are closer to Chromalveolata and Viridiplantae than to Rhodobionta. Here, we have placed Glaucocystobionta within the kingdom Archaeplastida (Palmer et al. 2004; Yoon et al. 2006b; Leliaert et al. 2012), which is challenged by some authors (*e.g.*, Cuvelier et al. 2008). A dozen species are known, belonging to the genera *Cyanophora*, *Glaucocystis,* and *Gloeochaete*.

If the position of Glaucocystobionta in the phylogenetic tree of eukaryotes, adopted here, is correct, they are of considerable theoretical interest. Their chloroplasts would testify to the best preserved eukaryotic lineage which experienced the primary endosymbiosis with a cyanobacterium and allowed eukaryotes to acquire photosynthesis (*cf.* Sect. 5.4).

Glaucocystobionta are unicellular. The cell wall may be present, and in this case consists of cellulose (*Glaucocystis*), or absent (*Cyanophora*). Beneath the plasmalemma (cytoplasmic membrane), alveolae (alveolar sacs) similar to those of the kingdom Alveolata are present (Fig. 7.13). The chloroplast is much like a cyanobacterium; it has long been regarded as a mutual cyanobacterium, insofar as the peptidoglycan cell wall of *Cyanobacteria* is maintained between the two chloroplast envelopes, a unique feature in eukaryotes (Bhattacharya et al. 2003) (Fig. 7.13). If chloroplasts are experimentally removed, the cell dies, indicating that endosymbiosis is obligate. Within the chloroplast stroma, **thylakoids*** are isolated, as in Rhodobionta, not stacked; they contain chlorophyll *a*; they bear phycobilisomes, which contain two blue phycobilins (phycocyanine and allophycocyanin), photosynthetic pigments especially efficient at absorbing red, orange, yellow, and green light, wavelengths that are not well absorbed by chlorophyll *a*. The polysaccharides arising from photosynthesis, mainly true starch, a mixture of amylose (an

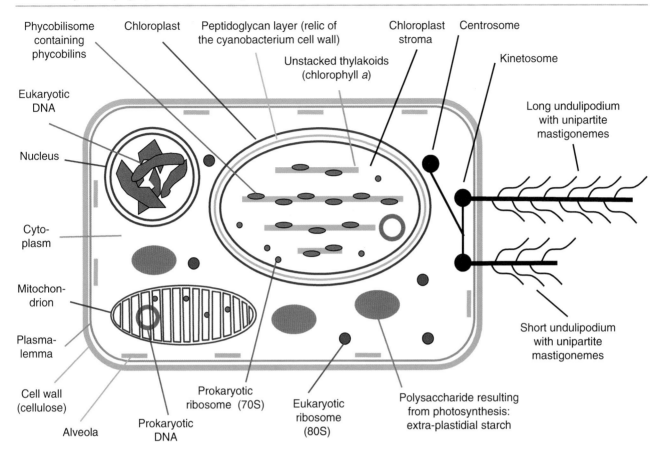

Fig. 7.13 *Theoretical scheme of a cell of Glaucocystobionta*. Vacuole, cytoskeleton and undulipodial roots are not represented. Some of the structures depicted herein may be absent in a given species

unbranched glucose polymer with α-1,4 bonds) and amylopectin (a branched glucose polymer with α-1,4 and a few α-1,6 bonds), are stored in the cytoplasm. Cells are motile due to two undulipodiums (*Cyanophora*), non-motile (*Glaucocystis*) or present at both motile and non-motile stages (*Gloeochaete*). When present, the two undulipodiums are unequal and bear two rows of unipartite mastigonemes (undulipodial hairs) (Fig. 7.13). Mitochondria have plate-like flat cristae, as have all other Archaeplastida taxa. Sexual reproduction is unknown.

Glaucocystobionta live in freshwater pools in temperate regions. They are also found in acidic lakes in northern regions (Lecointre and Le Guyader 2006). They are sometimes present in the soil.

7.5.4 Rhodobionta

Rhodobionta[8] (=Rhodophyceae, Rhodophycophyta, Rhodophyta, Rhodoplantae, red algae) constitute a relatively homogeneous taxon, compared with most other high-level taxa of eukaryotes. Cytological, biochemical, and biological (life cycle) similarities with Fungi (modern meaning), sometimes also with metazoans (Opisthokonta), proved to be homoplasies due to convergent evolution. Nearly 6,000 species of Rhodobionta were described; most of them belong to the class Florideophyceae and are multicellular. According to Lecointre and Le Guyader (2006), unicellularity in Rhodobionta is not ancestral, but a trait derived from multicellular ancestors. Unicellular species are present in Rhodellophyceae (*Rhodella*), Porphyridiophyceae (*Porphyridium*, *Erythrolobus*), Stylonematophyceae (*Rhodosorus* forms colonies), and Cyanidiophyceae (*Cyanidium, Cyanidoschyzon, Galdieria*) (Fig. 7.14).

The cell wall is mainly constituted by polymers of ester sulfated galactose (agar-agar, carrageenan, and porphyran). Cellulose is lacking or scarce (Gretz et al. 1982, 1984; Baldan et al. 2001). In most multicellular species (Florideophyceae and some Bangiophyceae), adjacent cells are linked by pit connections, a unique and distinctive feature of Rhodobionta. During cytokinesis, a small pore is left in the middle part of the newly formed wall; this pore is filled by a glycoproteic pit plug. Between the two daughter cells,

---

[8] Rhodobionta: from the ancient Greek words '*rhodon*' (pink) and '*biont*' (living thing).

the cytoplasmic continuity is blocked by the pit plug, while the plasmalemma is continuous (Fig. 7.15). Pit connections function as avenues for cell-to-cell communication and transport. Of course, they are absent in unicellular species.

Chloroplasts have a double membrane with an intermembrane space; they contain unstacked thylakoids, bearing phycobilisomes. As far as photosynthetic pigments are concerned, there is only one type of chlorophyll (chlorophyll *a*), within the thylakoids, and phycobilisomes are made of three phycobilin types (phycoerythrobilin, red, and phycocyanobilin and allophycocyanobilin, blue) (Fig. 7.15). Phycobilin pigments are especially efficient at absorbing red, yellow, and green light, wavelengths that are not absorbed by chlorophyll *a* while they are dominant at depth in the marine realm. Phycobilins are usually more abundant than chlorophyll *a*: they thus mask the green chlorophyll and give Rhodobionta their classic red color. The main storage polysaccharide is the floridean starch, a branched polymer of glucose with mainly α-1,4 and some α-1,6 bonds that, in several respects, resembles glycogen rather than true starch. It is stored outside the chloroplast, in the cytoplasm. Mitochondria have plate-like flat cristae, a shared character in Archaeplastida. Finally, the kinetic

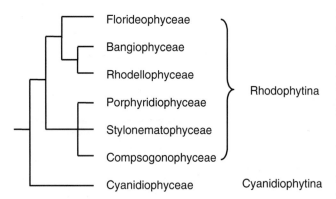

Fig. 7.14 *Phylogenetic tree of Rhodobionta*, based upon the genes PSI P700 chl *a*, *psa*A and *rbc*L, and a supermatrix containing data mined from GenBank (From Yoon et al. (2006b) and Verbruggen et al. (2010), modified and redrawn). The position of Bangiophyceae vs Rhodellophyceae is poorly resolved

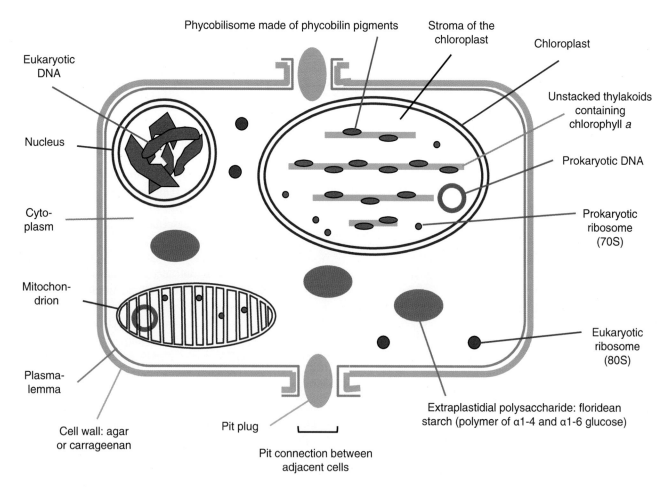

Fig. 7.15 *Theoretical scheme of a cell of Rhodobionta*. The vacuole, which can occupy most of the cell, and several other organelles (*e.g.* the endoplasmic reticulum) are not represented

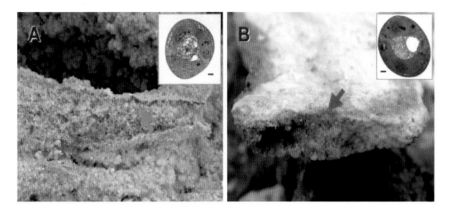

Fig. 7.16 Endolithic *Galdieria and Cyanidium* (Cyanidiophyceae, Rhodobionta) in Tuscany (A) and Naples (B), Italy. The *arrows* show the biomat of Cyanidiophyceae that thrives inside the rock at these sites. *Inserts*: *Galdieria* cells. Scale bars = 1 µm (From Yoon et al. (2006a). Copyright: with kind permission of BMC Evolutionary Biology, Biomed Central Ltd)

apparatus (undulipodiums, kinetosomes, undulipodial roots, centriole) is lacking; this may be due to a secondary loss (Fig. 7.15).

In Rhodobionta, the way of glycerol storage is unique. Unlike other taxa, in which glycerol is combined with fatty acids, producing glycerides, it is combined with sugars (→ heterosides) (Feldmann and Feldmann 1945). Floridoside (=α-D-galacto-pyranosyl-(1–2)-glycerol; combination of one galactose and one glycerol molecules) is the most common heteroside, often the main reserve product in the cell; it constitutes a reliable marker of Rhodobionta. Other heterosides can be present, *e.g.*, digeneaside (=α-D-manno-pyranosyl-(1–2)-glycérate) and isofloridoside. In *Erythrolobus coxiae* (Porphyridiophyceae), floridoside and digeneaside are present simultaneously (Scott et al. 2006). The low lipid content, in most Rhodobionta, may be a consequence of this particular pathway of glycerol storage. Halogenated compounds (chlorine and especially bromine) are relatively abundant, which constitutes another characteristic of Rhodobionta; Br+, sometimes Cl+, can take the place of H+ in the metabolism, resulting in *e.g.* bromoperoxydases and chloroperoxydases. Bromine is also present in defense compounds against herbivores. Finally and unexpectedly, the main sterol in Rhodobionta is cholesterol, a sterol that the general public associates more readily with animals rather than with "algae" (Bert et al. 1991).

In species with sexual reproduction, the typical life cycle is trigenetic and diphasic (Fig. 7.9d) and the fertilization is a trichogamy (Florideophyceae) or a protrichogamy (Bangiophyceae and Compsogonophyceae) (Fig. 7.6). In all cases, reproduction *via* conidia also occurs. In unicellular Rhodobionta, only the asexual reproduction has been observed, *via* binary fission and/or formation of conidia (often incorrectly named "endospores"); however, sexual reproduction does exist, as genetic recombination has been observed in *Galdieria*, Cyanidiophyceae (Yoon et al. 2006a, b).

Most Rhodobionta thrive in the marine realm. However, a few freshwater species are known. Multicellular species are benthic (living in close relationship with the bottom) while most unicellular species, such as *Rhodosorus, Porphyridium. Erythrolobus* and *Rhodella*, are pelagic (living in the water column). Species of the genera *Cyanidium, Cyanodioschyzon*, and *Galdieria* dwell in thermoacidic continental environments (pH = 0.5 to 6.0). *Galdieria* and *Cyanidium* can be endolithic (Fig. 7.16) (Yoon et al. 2006a). Because of their photosynthetic pigments, including phycobilins which utilize wavelengths not absorbed by chlorophylls, Rhodobionta are particularly well adapted to dim light; they are prevalent in sciaphilous and deep habitats. These are Rhodobionta which hold the depth record for photosynthetic organisms: 268 m under an illumination of only 0.0005 % of that reaching the surface of the ocean (Littler et al. 1985).

7.5.5 Viridiplantae

The Viridiplantae (Figs. 7.1 and 7.17) is the most important eukaryotic taxon, by the number of species (about 300,000), after Metazoa (Opisthokonta; about 850,000 species). The Chlorobionta match only part of what tradition has called "green algae." "Green algae" include most of the Viridiplantae, except Embryophyta (Fig. 7.17). "Green algae" are therefore a paraphyletic group, not a taxon.

The common ancestor of Chlorobionta and Streptobionta was a unicellular species that probably resembled the modern Prasinophyceae *Mesostigma viride* (Lemieux et al. 2000). These are Streptobionta, perhaps similar to the current Charophyceae, which conquered continents, about 475 Ma ago (Ordovician, Paleozoic era), and which are at the origin of much of the current terrestrial vegetation (Wellman et al. 2003).

A number of taxa of Viridiplantae only currently encompass multicellular or multinucleated (coenocytic) species: Trentepohliophyceae, Ulvophyceae, Dasycladophyceae, Bryopsidophyceae, Cladophorophyceae, Klebsormidiopyceae, Charophyceae, Coleochaetophyceae, and Embryophyta. Other taxa encompass both multicellular

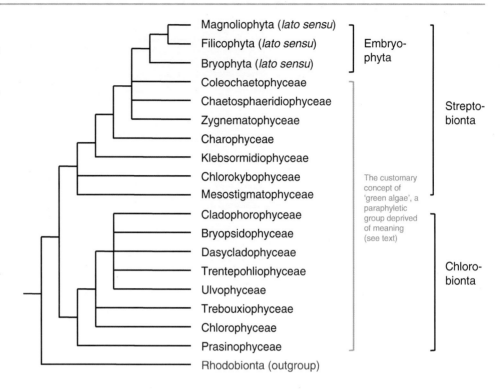

Fig. 7.17 *Phylogenetic tree of Viridiplantae*. The outgroup (Rhodobionta) is in *red*. This tree is a summary of the trees proposed by Hoek et al. (1998), Lecointre and Le Guyader (2006), Brodie et al. (2007) and Leliaert et al. (2012). Embryophyta, which are always multicellular, are not presented in detail. The Magnoliophyta encompass the flowering plants, the Filicophyta the ferns, and the Bryophyta the mosses

and unicellular species: Zygnematophyceae. Finally, Prasinophyceae, Chlorophyceae, Trebouxiophyceae, Mesostigmatophyceae, Chlorokybophyceae, and Chaetosphaeridiophyceae are only unicellular.

The cell wall of Viridiplantae (Chlorobionta and Streptobionta) is characterized by a wide variety of chemical compounds, which are often present simultaneously in the same taxon. It can comprise (i) cellulose, a glucose unbranched polymer with glycosidic β-1,4 bonds and with hydrogen bonds between neighbor chains, holding the chains firmly together side-by-side, (ii) mannane, a polymer of mannose, (iii) xylane, a polymer of xylose, (iv) hemicelluloses, polymers of galactose, or mannose close to the cellulose, (v) pectin, a polymer of α-1,4 galacturonic acid[9] (with carboxyl groups often esterified with methanol) with some α-1,2 rhamnose, (vi) and lignin, a very complex branched polymer of phenylpropanoids. The cellulose, though present in other eukaryotic taxa, and the lignin, characteristic of the Streptobionta,[10] are emblematic chemical compounds of the Viridiplantae. These two compounds, which are mainly (cellulose) or totally (lignin) synthesized by Viridiplantae, account for the major part of the Earth biomass, respectively 50 % and 30 % (Raven et al. 2000; Boerjan et al. 2003; Evert and Eichhorn 2013).

Chloroplasts of the Viridiplantae have a two-layer envelope. Thylakoids are stacked in lamellae (up to 30 thylakoids per lamella). There are two types of thylakoids, a unique feature in eukaryotes (Fig. 7.18): long thylakoids (also called intergranal thylakoids) and short thylakoid (granal thylakoids). Photosystem I and ATP synthase are mostly located in the long thylakoids, whereas Photosystem II is located mostly in the short thylakoids. In Chlorobionta, the stacks of short thylakoids are rather irregular in length; they are called pseudograna (*cf*. Chap. 5, Fig. 5.13). In contrast, in Streptobionta, stacks of short thylakoids are clear-cut; they are called grana. Two types of Chlorophyll are usually present simultaneously: chlorophyll *a* and *b*. A third type of chlorophyll, chlorophyll *c*, is present in some Prasinophyceae (Rodriguez et al. 2005; Six et al. 2005). The carotenoid pigments are represented by carotenes α and β and many xanthophylls: antheraxanthin, prasinoxanthin, siphonaxanthin, siphonein, violaxanthin, zeaxanthin, *etc*. Xanthophylls are less abundant than chlorophyll and therefore do not mask their color. The main storage polysaccharide is the starch ("true starch"), a mixture of amylose and amylopectin; amylose is an unbranched glucose polymer with α-1,4 bonds; amylopectin is a branched glucose polymer with α-1,4 and a few α-1,6 bonds (every 24–30 glucose units). Starch is stored within the chloroplasts, not in the cytoplasm, a unique feature among photosynthetic eukaryotes (Fig. 7.18). Plastids specialized in the storage of starch (amyloplasts) can be present in Steptobionta and

---

[9] Uronic acids (*e.g.* the galacturonic acid) are a class of sugar acids (hexoses) with both carbonyl and carboxylic acid functional groups. In the case of pectin, the hexose is galactose.

[10] Apart from Streptobionta, lignin is also present in a species of Rhodobionta living in the intertidal zone. According to Martone et al. (2009), the lignin biosynthetic pathways may have been present in the common unicellular ancestor of Viridiplantae and Rhodobionta.

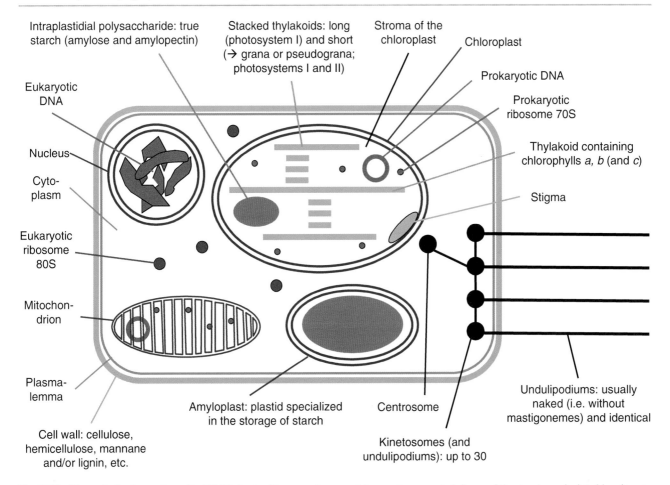

Fig. 7.18 *Theoretical scheme of a cell of Viridiplantae.* The vacuole, which can occupy most of the cell, and several other organelles and structures (*e.g.* endoplasmic reticulum, cytoskeleton, undulipodial roots) are not represented. Some of the structures depicted herein may be absent, or different, in a given species

in Ulvophyceae (Chlorobionta); they have not been observed in unicellular Chlorobionta. Sterols are extremely varied; the β-sitosterol is considered a marker of higher embryophytes (Puglisi et al. 2003); other sterols are not really specific to Viridiplantae.

Mitochondria have plate-like flat cristae, a shared character in Archaeplastida. The kinetic apparatus is present in Viridiplantae, except in Magnoliophytina (=angiosmerms; embryophytes) (Bornens and Azimzadeh 2007) and some unicellular Chlorobionta lacking of undulipodium stages, such as *Ostreococcus* (Prasinophyceae). As in other eukaryotes, where present, the fully developed kinetic apparatus comprises five parts: undulipodiums, kinetosomes, undulipodial roots, centrosome and, in photosynthetic cells, the stigma (Fig. 5.11). Motile cells of multicellular Viridiplantae, together with motile unicellular Viridiplantae, have 2–30 undulipodiums similar in form and length (they are said to be "isokont"). Undulipodiums are usually naked, *i.e.* do not bear mastigonemes (undulipodial hairs). However, in some species, unipartite (*e.g., Chlamydomonas*) or tripartite mastigonemes are present (*cf.* Sect. 7.9.3 for a typology of the mastigonemes). Undulipodial scales can cover the axis of the undulipodium (axoneme) (Inouye 1993). The axoneme is composed of 20 microtubules aligned in parallel; more specifically, on a cross section, the microtubules are arranged in a characteristic pattern known as the "9 + 2 pattern," nine sets of "doublet microtubules" (a specialized structure consisting of two linked microtubules) forming a ring around two central microtubules. Dynein arms are anchored to each doublet microtubule. This structure is a general feature in eukaryotes. However, in Viridiplantae, one of the two dynein arms (the outer one) of one of the doublet microtubules is lacking (Fig. 7.19); this feature is unique in eukaryotes and therefore constitutes a cytological marker of the Viridiplantae (Inouye 1993).

Sexual reproduction is known in all the taxa of Streptobionta, except Chlorokybophyceae and Mesostigmatophyceae. In Chlorobionta, sexual reproduction is probably present in all taxonomic classes, although it is poorly known in some unicellular taxa. Fertilization is usually a planogamy in Chlorobionta, while it is an oogamy in Streptobionta (Table 7.4). (i) In taxa of Chlorobionta and Streptobionta

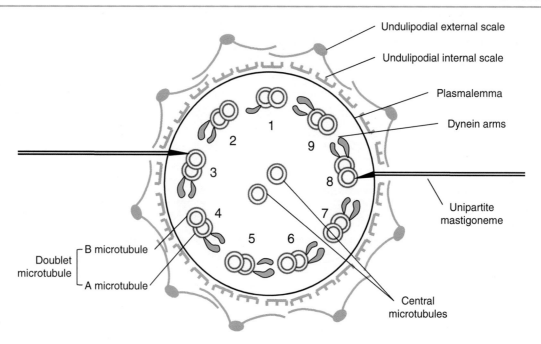

Fig. 7.19 *Theoretical cross section of an undulipodium of Chlorobionta.* Doublet microtubules are numbered *1–9*. One of the dynein arms of the first doublet is absent. The mastigonemes and undulipodial scales are not present in most Chlorobionta (From Inouye (1993), modified and redrawn)

Table 7.4 *Some morphological and life cycle traits in Viridiplantae.* The embryophytes, which are always multicellular, are not detailed here. The taxa are arranged in the same order as in the phylogenetic tree of Fig. 7.17

	Taxon	Uni- (1) or multi-cellular (X)	Sexual reproduction present (P) or unknown (?)	Fertilization	Life cycle mono- (1) or digenetic (2)	Ploidy n (haplophase), 2n (diplophase) or n + n (micthaplophase) and relative importance of the phases
Strepto-bionta	Magnoliophyta (*sensu lato*)	X	P	Siphonogamy	2	n <<< 2n
	Filicophyta (*sensu lato*)	X	P	Oogamy	2	n << 2n
	Bryophyta (*sensu lato*)	X	P	Oogamy	2	n > 2n
	Coleochaetophyceae	X	P	Oogamy	1	n
	Chaetosphaeridiophyceae	1	P	Oogamy	?	?
	Zygnematophyceae	1 or X	P	Cystogamy	1	n
	Charophyceae	X	P	Oogamy	1	n
	Klebsormidiophyceae	X	P	Oogamy	?	?
	Chlorokybophyceae	1	?	–	–	–
	Mesostigmatophyceae	1	?	–	–	–
Chloro-bionta	Cladophorophyceae	X	P	Planogamy	2	n = 2n
	Bryopsidophyceae	X	P	Planogamy	2	n > n + n
	Dasycladophyceae	X or 1	P	Planogamy	1	2n
	Trentepohliophyceae	X	P?	Planogamy	?	?
	Ulvophyceae	X	P	Planogamy	2	n = 2n
	Trebouxiophyceae	1	P	?	?	?
	Chlorophyceae	1	P	Planogamy, Oogamy	1	n
	Prasinophyceae	1	P	Planogamy	1	n

that branch near the base of the phylogenetic trees (="ancestral"), the life cycle has usually a single haploid generation (the gametogen). (ii) Higher up in the trees, there are digenetic diphasic life cycles, with alternation between a haploid gametogen and a diploid sporogen (gametogenous and sporogenous generations). (iii) Finally, in Streptobionta, approaching the top of the tree, there is a steady reduction of the importance of the gametogen, to only a few nuclei and cells in some Magnoliophyta, in favor of the sporogen (Table 7.4). This trend of reduction in the gametogenous generation results, in Dasycladophyceae (Chlorobionta), in its complete disappearance and therefore a monogenetic life cycle whose single generation is the diploid sporogen. This trend is also found in Chromobionta (*cf.* Sect. 7.9.3).

Unicellular Viridiplantae have colonized a wide variety of habitats, terrestrial, freshwater, and marine. They often play a major role in the functioning of ecosystems. (i) Most species are photosynthetic. In freshwater habitats, *e.g. Chlamydomonas, Volvox, Scenedesmus* (Chlorophyceae), *Micrasterias,* and *Closterium* (Zygnematophyceae) can be found. *Chaetosphaeridium minus* (Chaetosphaeridiophyceae) is epibiontic of freshwater MPOs.[11] In coastal lagoons and the open sea, the photosynthetic eukaryotes of the picoplankton (Box 5.2) mainly belong to Prasinophyceae: *Crustomastix, Dolichomastix, Halosphaera Mamiella, Micromonas, Ostreococcus,* and *Pyramimonas* (Rodriguez et al. 2005; Šlapeta et al. 2005a; Viprey et al. 2008). In the Thau Lagoon (France, Mediterranean Sea), a density of 90–250 millions cells per liter of *Ostreococcus tauri* was observed in summer (Courties et al. 1998). Photosynthetic eukaryotic picoplankton also consists of Trebouxiophyceae: *Chlorella, Nannochloris, Picochlorum* (Fuller et al. 2006; Viprey et al. 2008). *Trebouxia* (Trebouxiophyceae) is the green dust that covers tree trunks and walls. *Parachloroidium* thrives in corticolous biofilms (Neustupa et al. 2013). *Chlorokybus atmophyticus* (Chlorokybophyceae) lives in the soil surface of the terrestrial realm. (ii) A number of photosynthetic species are actually facultative heterotrophs; this is the case of *Chlamydomonas reinhardtii* (Chlorophyceae), common in fresh water and soil, which is able to grow in the dark with acetate as a carbon source. (iii) *Coccomyxa parasitica* (Chlorophyceae) is a unicellular parasite of the mantle of the sea scallop *Placopecten magellanicus* and the mussel *Mytilus edulis,* that has not lost its chloroplasts; *Coccomyxa ophiurae* and *C. astericola* are parasites of the brittle star *Ophiura texturata* and the sea star *Hippasteria phrygiana,* respectively; in the brittle star, *Coccomyxa ophiurae* is responsible for spine malformations (Stevenson and South 1974, 1976; Gray et al. 1999). *Prototheca* probably derived from *Chlorella* (Trebouxiophyceae) by loss of photosynthesis; members of this genus cause cutaneous infections in humans and cattle (Lass-Flörl and Mayr 2007). In Viridiplantae, the absence of photosynthesis results from a secondary loss; this is a very rare trait in unicellular taxa. (iv) Finally, some species have established mutualistic relations with other species. *Trebouxia* (Trebouxiophyceae) species are involved in the lichen mutualistic symbiosis, where they constitute the photosynthetic partner of the mushroom (Fungi), *e.g.* in *Xanthoria* and *Cladonia* (Feldmann 1978; Kroken and Taylor 2000). In the lichen family Verrucariaceae, a number of trebouxiophycean (mainly *Diplosphaera,* but also *Asterochloris, Myrmecia, Prasiola, etc.*), ulvophycean (*e.g., Dilabifilum*), and trentepohliophycean genera (*e.g., Trentepohlia*) are involved in the lichenic mutualistic symbiosis (Thüs et al. 2011). *Pedinomonas* and *Tetraselmis* (Prasinophyceae) are mutualistic symbionts of *Noctiluca* (Dinobionta) and the flatworms *Convoluta* (Platyhelminthes, Metazoa), respectively.

7.6 Kingdom Rhizaria

7.6.1 General Remarks

Rhizaria have sometimes been named Cercozoa. However, Cercozoa (here as Cercobionta) constitutes a subkingdom within the kingdom Rhizaria. The kingdom Rhizaria is fairly well characterized at the genetic level (Cavalier-Smith and Chao 2003). However, its limits are uncertain. From a cytological, biochemical, and biological perspective, it is difficult to identify characters shared by all Rhizaria. The phylogenetic diversity of Rhizaria is probably considerable, and this diversity is likely to increase when new taxa are sequenced. In addition, some taxa currently included within Rhizaria may prove to belong to other kingdoms (Lecointre and Le Guyader 2006).

The Rhizaria are mostly unicellular, though a multicellular form has been described (Brown et al. 2012). Most Rhizaria are amoeboid with peudopods. Many produce "shells," more or less complex in structure.

In this chapter, we have chosen to illustrate the phylogenetic diversity of Rhizaria by a small number of taxa particularly emblematic of this kingdom: Radiolaria, Chlorarachniobionta, Phytomyxea, and Foraminifera.

7.6.2 Radiolaria

Radiolaria are unicellular, but some species forming colonies are known. They usually measure between 0.1 and 2 mm in diameter. Colonial species can exceed 10 cm; a giant species, *Collozoum caudatum,* up to 2 m long, is known from equatorial and Gulf Stream waters in the Atlantic Ocean (Swanberg and Anderson 1981). Radiolaria, due to their siliceous skeleton, fossilize well; the oldest fossils date back to 600 Ma ago. Thousands of species have been described.

The cytoplasm consists of two parts. The inner part (endoplasm) contains the nucleus, most mitochondria, and the Golgi apparatus; it is limited by a capsular wall

---

[11] MPOs = Multicellular Photosynthetic Organisms.

Fig. 7.20 *Radiolaria with a spherical symmetry (left) and axial symmetry (right)* (From Wikipedia, Haeckel, 1904: *Kunstformen der Natur*)

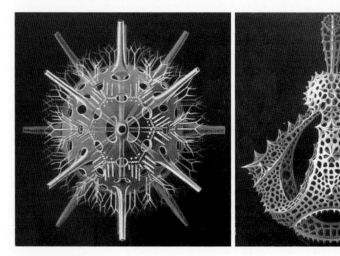

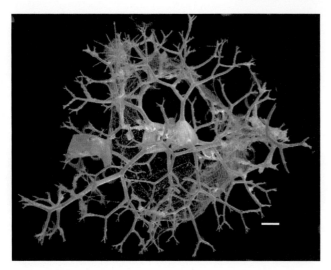

Fig. 7.21 *Siliceous skeleton of Coelodiceras spinosum (Phaeodaria, Radiolaria)*. Scale bar = 200 μm (Copyright: Deep Sea Research, Paterson et al. 2007)

perforated by pores, which is part of the skeleton. The outer part (ectoplasm) contains digestive vacuoles and lipid reserve droplets, keeping the cell buoyant (Tregouboff 1953a). At the periphery of the cell, not limited by a cell wall, the cytoplasm is extended by very narrow pseudopodia. Some pseudopodia are slender, raylike, and supported internally by a microtubule (axopods); others do not have microtubules and may anastomose (filopodes). The siliceous skeleton often has a radial symmetry (spherical), but symmetry may also be axial (Fig. 7.20). In some Radiolaria, the shape of the skeleton may even be irregular (Fig. 7.21).

Asexual reproduction is by binary fission. When the skeleton is compact, one of the offspring must reconstruct the skeleton (Tregouboff 1953a). Sexual reproduction, and thus life cycle are not known, although cells with two undulipodiums of unequal length were observed, perhaps corresponding to gametes or conidia (Kling and Boltovskoy 2002).

Radiolaria live in marine plankton, especially in the upper 100 m, but also in the greatest depths of the ocean. They capture small prey with their pseudopodia: bacteria, photosynthetic picoeukaryotes, diatoms, ciliates, dinobionts, haptobionts, even small copepods, and crustacean larvae. They thus constitute a trophic link between the photosynthetic unicellular plankton and the mesozooplankton. Many Radiolaria host, in their ectoplasm, photosynthetic unicellular organisms, *e.g.* *Cyanobacteria* close to *Prochlorococcus* and haptobionts. The transfer of carbon between the photosynthetic symbionts and the Radiolaria has been demonstrated, suggesting that we are dealing with a mutualistic symbiosis. The photosynthetic symbionts can move between the ectoplasm (at night) and pseudopodia (during the day) (Tregouboff 1953a; Kling and Boltovskoy 2002; Foster et al. 2006). In addition to their role in pelagic ecosystems, Radiolaria play an important role in the vertical transfer of silica between pelagos and sediments.

7.6.3 Chlorarachniobionta

Chlorarachniobionta[12] (=chlorarachniophyceae, Chlorarachniophyta, Chlorarachnida, Chlorarachnea) are unicellular. They have (or may have) a form of amoebae, with very thin pseudopodia (filopodes) that can anastomose between neighboring individuals (cells), achieving kinds of colonies named meroplasmodium (Fig. 5.6b). Six genera and a dozen species have been described (Ota et al. 2007, 2011).

---

[12] Chlorarachniobionta: from the Greek '*chloros*' (green) and '*arachne*' (spider).

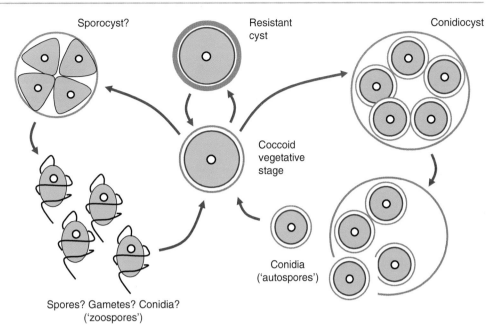

Fig. 7.22 *The life cycle without sexuality of Norrisiella sphaerica (Chlorarachniobionta)*. The "zoospores" that function as conidia could match former gametes or spores. The nuclei (*white circles*) are shown; it is not known if they are haploid (n) or diploid (2n)

In addition to the amoebae forms sensu stricto, *i.e.* with pseudopodia and without a cell wall, Chlorarachniobionta can be present in many different forms: amoebae with a cell wall, coccoid forms, with or without a cell wall and resistant cysts (Fig. 7.22). The cell wall, when present, does not contain any cellulose; it seems rather of pectic nature (Calderon-Saenz and Schnetter 1989).

Chlorarachniobionta are photosynthetic organisms. The cell has one or more chloroplasts, with thylakoids stacked into three-thylakoid lamellae. The chloroplasts are derived from a secondary endosymbiosis with a Chlorobionta close to the current genus *Tetraselmis* (Prasinophyceae) (Takahashi et al. 2007). In addition to its classical double membrane, the chloroplast is surrounded by a second double membrane, a remnant of the secondary endosymbiosis. A nucleomorph,[13] remnants of the nucleus of the Chlorobionta, is present in the periplatidial space, *i.e.* between the two inner and the two outer plastid membranes. Photosynthetic pigments are chlorophylls *a* and *b* and xanthophylls, neither of which appear to be specific. The main storage polysaccharide is the paramylon, a carbohydrate polymer of β-1,3 glucose with numerous β-1,6 branches. Paramylon is stored within cytoplasmic vesicles (de Reviers 2003). Outside Chlorarachniobionta, paramylon is only known in euglenoids (Excavates). The mitochondria have tubular cristae (Ota et al. 2007).

Chlorarachniobionta reproduce asexually by bipartition and conidia. Some species, such as *Cryptochlora perforans*, exhibit a kind of alternation of generations between a coccoid stage and several amoeboid stages, though sexuality is absent (Calderon-Saenz and Schnetter 1989). Cells equipped with undulipodiums ("zoospores"), which are present in all species (Fig. 7.22), function as conidia: they produce genetically identical individuals (clones) to the one which produced them. The fact that they are produced by four suggests that they came from meiosis, but in this case, a fertilization (never seen) would be needed to "close" the cycle. Either these "zoospores" are the remnants of a missing sex or the life cycle is still only partially known. The undulipodium has a unique character: it is wound helically around the cell; at its base, there are two kinetosomes, one of them being the remnant of a second undulipodium which has disappeared. The undulipodium is bordered by a row of tiny unilateral hairs.

While photosynthetic, Chlorarachniobionta capture and engulf, through their pseudopodia, small prey: bacteria, diatoms, *etc*. *Cryptochlora perforans* pierces and enters the dead filaments of Bryopsidophyceae (Chlorobionta) and ingests content (Calderon-Saenz and Schnetter 1989). The Chlorarachniobionta live in temperate and tropical marine environment in the pelagos (*Bigelowiella longifila*) or the benthos.

7.6.4 Phytomyxea

Phytomyxea is one of the taxa included within the customary concept of "mushrooms" (*cf*. Sect. 5.5.3). Phytomyxea comprises a dozen genera and about 50 species, divided into two sets, the Phagomyxida (with one genus: *Phagomyxa*) and Plasmodiophorida (*Plasmodiophora*, *Polymyxa*, *Spongospora*, *etc*.) (Braselton 2009).

---

[13] The nucleomorph of Chlorarachniobionta has lost most of its genes: it has only 380 kb.

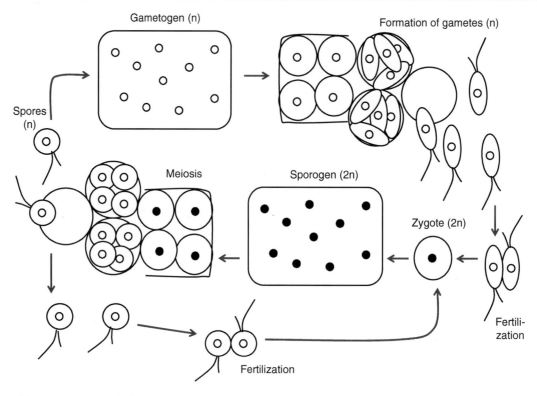

Fig. 7.23 *The life cycle of a Plasmodophora species (Phytomyxea). Open circles* = haploid nuclei, *solid circles* = diploid nuclei (Modified and redrawn from Chadefaud (1960))

The vegetative form of Phytomyxea is a coenocyte, *i.e.* a giant multinucleate cell (Fig. 7.23), located within the host organism. Plasmodiophorida possess a complex "extrusome" used for penetrating food cells (Braselton 1995; Bulman et al. 2001). The plasmodiophorid nuclei divide by an unusual and characteristic type of nuclear division, the cruciform nuclear division: an elongated nucleolus is arranged perpendicularly to the chromosomes at the equatorial plate (Fig. 7.24; Braselton 1995, 2009). The cell wall, where present, is composed of chitin.

The life cycle is digenetic diphasic, with alternation between haploid and diploid phases (Fig. 7.23), or monogenetic, with only a diploid phase (Chadefaud 1960). Spores and gametes have two undulipodiums usually naked, *i.e.* without mastigonemes, and of unequal length. Resistant spores (cysts) are present. The reproductive cells are spread when the host cell bursts.

The Plasmodiophorida are intracellular parasites of Embryophyta (Viridiplantae, Archaeplastida) and Chromobionta (Stramenopiles), in which they determine the formation of kinds of galls. For example, *Plasmodiophora brassicae* causes swellings or distortions of the roots of cabbages and related plants (the "clubroot disease"), *Spongospora subterranea* causes the "powdery scrab," a disease of potato tubers, and *Tetramyxa parasitica* is a parasite of *Potamogeton,* a genus of mostly freshwater Embryophyta (Chadefaud 1960, 1978). *Polymyxa betae* is a root parasite of

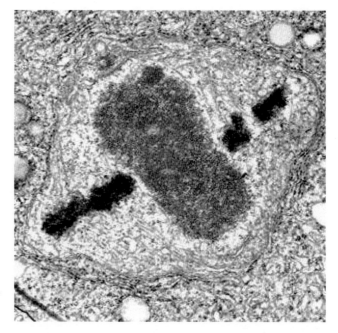

Fig. 7.24 The cruciform nuclear division. This type of division is one of the characteristic features of Plasmodiophorida (Photo: courtesy of Braselton (2009))

the sugar beet and vector of the beet necrotic yellow vein virus. *Polymyxa betae, P. graminis,* and *Spongospora subterranea* are obligate parasites of Magnoliophyta (Viridiplantae); they are responsible for the transmission of several very important

Fig. 7.25 *Ammonia tepida (Foraminifera) from San Francisco Bay (USA)*. The filopodes out of the apertures of the test are clearly visible (From English Wikipedia. Phase-contrast photomicrograph by Scott Fay, UC Berkeley, 2005)

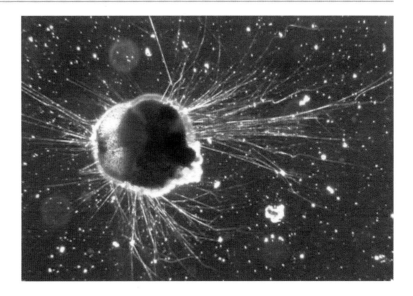

Fig. 7.26 *Heterostegina depressa (Foraminifera) from Sulawesi (Indonesia)*. Size: 4.4 mm (From English Wikipedia. Copyright: courtesy of Alain Couette (www.arenophile.fr))

viruses which cause serious diseases in cultivated plants (Braselton 2009). *Phagomyxa bellerocheae* and *P. odontellae* are parasites of marine diatoms (Schnepf et al. 2000).

7.6.5 Foraminifera

Foraminifera[14] are unicellular. They have the form of amoebae with thin pseudopodia (filopodes). They are protected by a test having openings through which the filopodes emerge (Fig. 7.25). Foraminifera have been known since the beginning of the Paleozoic era (about 540 Ma). The tests usually keep well in fossil form, so that the Foraminifera are excellent biostratigraphic markers. Between 4,000 and 6,000 living species are known.

Most Foraminifera produce a test (or "shell") which can have either one or multiple chambers. Chambers are arranged in single, double, or triple series, the most recent chamber being the largest (Fig. 7.26). Tests are commonly made of calcium carbonate or agglutinated sediment particles, and in one genus of silica. The test of *Spiculosiphon oceana* is made of agglutinated fragments of

---

[14] Foraminifera: from the Latin '*foramen*', an opening and '*ferre*', to bear. For short, 'hole bearers'.

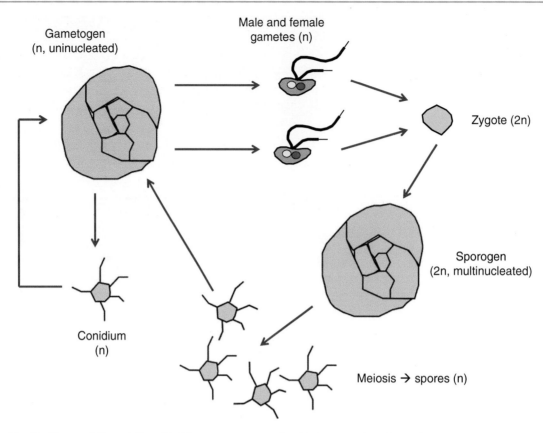

Fig. 7.27 *The life history of Foraminifera*. Nuclei are not represented, with the exception of haploid gamete nuclei (*red circles*); *yellow circles* = lipid droplets

sponge spicules; its body morphology strongly converges with that of the carnivorous sponges (Maldonado et al. 2013). The external surface of the test may bear spines, especially in planktonic species. Openings in the test, including those between chambers, are called apertures. The test diameter ranges between 20 μm and up to 7 cm (*Parafusulina*), 9 cm (*Loftusia*), 11 cm (*Camerina*), and 19 cm (*Neusina*) (Le Calvez 1953).

The life cycle is digenetic diphasic, with alternation between a uninucleated haploid gametogen, which produces gametes and therefore a zygote, and a multinucleated diploid sporogen, which produces spores giving rise to new gametogens (Fig. 7.27). Gametes typically have two undulipodiums which are unequal in length, naked, and ending in a single mastigoneme (Fig. 7.27). Spores have four undulipodiums. Gametogens also produce conidia which have four undulipodiums and account for asexual reproduction. Multiple rounds of asexual reproduction between sexual generations are not uncommon in benthic species. Fertilization is of planogamy type (Fig. 7.9d). In some species, the gametogen and the sporogen are morphologically different, so they have been described as a different species by ancient authors. Sexuality has been secondarily lost in some species, such as *Discorbis orbicularis*.

Foraminifera live mainly in the marine benthos, where they can reach densities of several thousand individuals per m^2. They can be found down to the greatest depths; Foraminifera were found in large numbers in the sediment of the Challenger deep, at 10,896 m depth, the deepest known point in the Earth's seabed hydrosphere (Todo et al. 2005). Some species are planktonic or live in coastal lagoons. Finally, Foraminifera are a widespread and diverse component of soil microbial communities (Lejzerowicz et al. 2010).

Pseudopodia (filopodes) are used for locomotion (a few centimeters per hour), anchoring and capturing food, which consists of organic debris and some organisms such as bacteria and diatoms. The engulfed food is held in digestive vacuoles. The giant (>4 cm) *Spiculosiphon oceana* captures prey probably larger than other predatory Foraminifera (Maldonado et al. 2013). A number of species host in their cytoplasm mutualistic symbionts that belong to various groups of photosynthetic unicellular eukaryotes: diatoms, Dinobionta, Chlorobionta, *etc.* (Lecointre and Le Guyader 2006). Other Foraminifera are kleptoplastic: they retrieve the chloroplasts of their prey, chloroplasts which continue to conduct photosynthesis for a few days (*cf.* Sect. 5.4.5, Fig. 5.22). A curious case is that of the benthic Foraminifera *Nonionella stella*, collected in the upper 3 cm of sediment off California at a depth of ~600 m (in the aphotic zone); it retains chloroplasts derived from diatoms closely related to *Skeletonema costatum* and *Odontella sinensis,* species which live in the photic pelagic zone; the chloroplasts are probably from fecal pellets of zooplanktonic organisms which fall to the bottom; they could be used for the assimilation of inorganic nitrogen, meeting the nitrogen requirements of the host (Grzymski et al. 2002).

7.7 Super-Kingdom Chromalveolata

The kingdoms Alveolata and Stramenopiles have a number of common characteristics. It seems that the monophyly of this super-group is strong. They are therefore referred to as Chromalveolata. Some *insertae sedis* taxa, such as Haptobionta, could belong to this super-group (Fig. 7.1).

Some authors, *e.g.* Palmer et al. (2004), consider that a single endosymbiosis founding event was at the origin of the chloroplast in Chromalveolata. This secondary endosymbiosis with a Rhodobionta would have occurred in the common ancestor of all Chromalveolata (*cf.* Sect. 5.4). According to this hypothesis, the absence of photosynthesis in taxa such as Oobionta, Labyrinthulobionta, and ciliates should be due to a secondary loss. The existence of a single founding event in the photosynthesis of Chromalveolata is however disputed (*e.g.*, Bodyl et al. 2009).

The Chromalveolata have a number of shared characters. They mainly concern the photosynthetic taxa. (i) Mitochondria have cristae[15] of the tubular type. (ii) Chlorophyll *a* is associated with chlorophyll *c* (c1, c2, and/or c3). (iii) The thylakoids are never isolated, but stacked into two- or three-layered lamellae. (iv) The chloroplasts (where present) are surrounded by a four-membrane envelope: the classical two-membrane envelope and an outer second two-membrane envelope. The latter constitutes the remains of the secondary endosymbiosis. This outer envelope (the periplastidial envelope) may (or not) be confluent with the nuclear envelope. It should be noted that these characters, taken in isolation, are not specific to Chromalveolata.

7.8 Kingdom Alveolata

7.8.1 General Remarks

The kingdom Alveolata (=alveolate, Alveolobionta) is a taxon strongly supported by molecular phylogenies. The most notable shared characteristic is the presence of flattened

---

[15] A crista is a fold in the inner membrane of a mitochondrion.

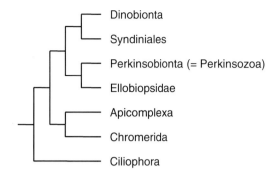

Fig. 7.28 *Simplified phylogenetic tree of the Alveolata.* This tree includes a greater number of taxa than in Fig. 7.1. It provides a possible synthesis of trees and data of Silberman et al. (2004), Groisillier et al. (2006) and Moore et al. (2008)

membrane-bound vesicles, the alveoli, just under the plasmalemma (cytoplasmic membrane). Compared to the simplified tree shown in Fig. 7.1, many other taxa are to be considered, although they are not the subject here of a detailed description (Fig. 7.28).

Alveolata bring together taxa that former taxonomists had scattered to place them either within the "vegetable kingdom" ("algae" and "fungi," customary meaning; plants) or within the "animal kingdom" (protozoa). The Ellobiopsidae were considered fungi (customary meaning). The Dinobionta, which often have photosynthesis, were considered as "algae," so as plants, although also claimed by zoologists. The Ciliates (Ciliophora) and Apicomplexa, which are normally not photosynthetic, were considered as protozoa, so as belonging to the animal kingdom. But the discovery that the apicoplast of Apicomplexa is actually an ancient chloroplast (McFadden and Waller 1997; McFadden et al. 2001), and that the ancestors of the present Apicomplexa were therefore photosynthetic would appear to demonstrate the irrationality of traditional classifications.

The Ellobiopsidae are parasites of crustaceans (Fig. 7.29), with the exception of the genus *Rhizellobiopsis* which parasites annelids. They are multinucleated, located inside their host, except for breeding filaments ending in a cell whose content is transformed into non-motile reproductive cells (*Ellobiopsis*) or reproductive cells provided with two undulipodiums (*Thalassomyces*). The life cycle is unknown (Silberman et al. 2004).

The Syndiniales (Syndiniophyceae) are intracellular parasites of Dinobionta (sensu stricto), radiolarians, ciliates, crustaceans, *etc. Amoebophrya* penetrates the nucleus of a Dinobionta and multiplies to fill the entire host cell. The cells of the parasite then meet in a vermiform cluster which leaves the host cell and swims. Subsequently, the cells of the vermiform stage disperse and attack new Dinobionta. DNA sequences of Syndiniales have a widespread distribution and are found in all samples of seawater

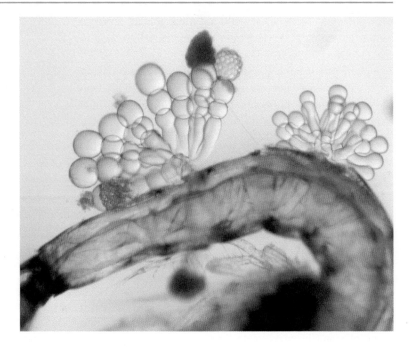

Fig. 7.29 Thalassomyces (Ellobiopsidae), parasite of a Mysidacea crustacean. It was observed in a sea cave near Marseilles (France, Mediterranean Sea) (Photo: courtesy of Christophe Lejeusne)

from the world ocean (Groisillier et al. 2006), suggesting that they play a major role in the control of unicellular plankton (Miller et al. 2012).

7.8.2 Ciliophora

The Ciliates (=ciliata, Ciliophora) are unicellular, with cells that can be large (up to 2 mm in length). More than 10,000 species have been described. The oldest fossils date from 580 Ma in the Doushantuo formation in China. At the genetic level, a striking feature is that the codons that code usually for "stop" in the "universal" genetic code can code for amino acids; in some species, UAA and UAG code for glutamine; in others, UGA codes for cysteine or tryptophan. These are derived, not ancestral, characters (Kim et al. 2005).

Ciliates are named after their coating of hundreds or thousands of cilia (relatively short undulipodiums) arranged in longitudinal rows called kineties. Cilia play a role in the movement and capture food. In some taxa, thick round bundles of cilia, called cirri, act like "legs" and enable the organism to "walk" over a surface (Fig. 7.30). Alveoli are present under the plasmalemma (cytoplasmic membrane) and are packed against it to form a pellicle maintaining the cell shape. Trichocysts (extrusomes) are membrane-bound structures which can discharge their content outside the cell for the purpose of attack or defense. Unlike most eukaryotes, ciliates have two different sorts of nuclei, generally a small micronucleus and a large macronucleus. However, there may be several micronuclei or macronuclei (Gülkiz Şenler and Yildiz 2003; Hansen and Fenchel 2006). The micronucleus is the true nucleus; it is diploid, with chromosomes, serves as the germ line nucleus but does not express its genes and it is involved in sexual reproduction. In contrast, the macronucleus is polyploid (many copies of the chromosomes), undergoes direct division without mitosis, and controls the non-reproductive cell functions, such as metabolism. Many species have a "cell mouth" (cytostome), a part of the cell specialized for phagocytosis, with a cytopharynx through which food is ingested and included within a digestive vacuole (Fig. 7.30). Contractile vacuoles collect water and expel it from the cell to maintain osmotic pressure. Mitochondria have tubular cristae, a shared character in ciliates; however, some ciliates have no mitochondria, but double-membraned hydrogenosomes (probably derived from mitochondria), which produce adenosine triphosphate (ATP) fermentatively (Dyall et al. 2004).

Ciliates reproduce asexually by binary fission. During fission, the micronucleus undergoes mitosis and the macronucleus elongates and splits in half; the cell then divides into two, and each new cell has a copy of the micronucleus and the macronucleus. When conditions are unfavorable, some species produce resistant cysts. The sexual reproduction is a kind of cystogamy (=conjugation). Mating cells meet and form a bridge between them; the macronuclei disappear while the diploid micronuclei undergo meiosis producing four haploid micronuclei; three of these micronuclei disintegrate, while the fourth undergoes mitosis, producing two micronuclei, one of them being smaller than the other; the two cells exchange over the bridge the smaller micronuclei, which play the role of male gametes; in each cell, the small and large micronuclei fuse. Subsequently, a macronucleus is formed from micronuclei through a complex process (Fig. 7.31). The life cycle can be considered as monogenetic,

7 Taxonomy and Phylogeny of Unicellular Eukaryotes

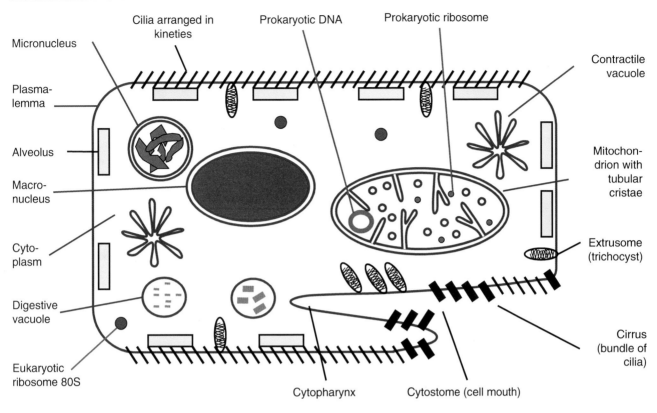

Fig. 7.30 *Theoretical scheme of a cell of Ciliophora (ciliate)*. This scheme combines features that do not correspond to any particular species. The kinetosomes at the base of the cilia are not represented

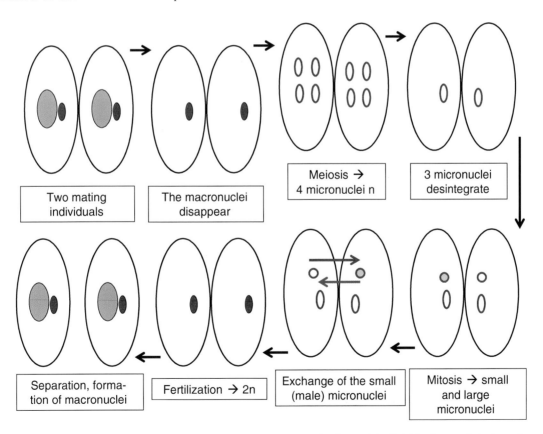

Fig. 7.31 *Sexual reproduction in ciliates*. *Orange*: macronuclei (polyploid). *Red*: micronuclei (diploid). *Open red circles* and *ovals*: haploid micronuclei. The "male" haploid micronucleus is filled in *light blue*

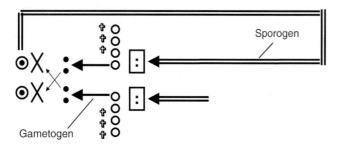

Fig. 7.32 *The life cycle of a ciliate species.* Though the life cycle can be considered as monogenetic, with a single diploid generation, it has been regarded here as digenetic, with a sporogen (diploid genetation) and a vestigial gametogen (haploid generation). For the meaning of the symbols used, see Fig. 7.8

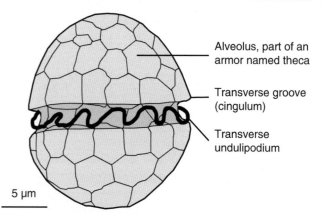

Fig. 7.33 *General appearance of a cell of Dinobionta.* The transverse groove (cingulum), the transverse undulipodium, and the alveoli of the theca in *Tovellia coronata*. The second undulipodium and its longitudinal groove (sulcus) are located on the other side of the cell and are therefore not visible here (Drawn from a photo of Lindberg et al. (2005))

with a single diploid generation; however, the final division of the nucleus originating from meiosis, corresponding to a spore, can be regarded as the remnant of the haploid generation (gametogen). In this case, the biocycle could be interpreted as digenetic (Fig. 7.32).

Ciliates are generally heterotrophic organisms. They are predators of bacteria, unicellular eukaryotes (including other ciliates) and larvae of some "invertebrates."[16] *Ichthyophthirius multifilis* is a parasite of freshwater teleosts, which is the main pathogen worldwide (Jousson et al. 2007). Ciliates can also use dead organic matter. Some species of ciliates host in their cytoplasm photosynthetic organisms (probably mutualistic symbionts) such as *Cyanobacteria*, Chlorobionta, and Cryptophyta. In some cases, ciliates recover chloroplasts (kleptoplasty) and possibly the nucleus (karyoklepty) of their prey (*cf.* Sect. 5.4.4). All intermediates between strict heterotrophy and mandatory mixotrophy exist (Smith and Hansen 2007).

Continental habitats (fresh and brackish water, soil) harbor more ciliate species than marine habitats. In the marine environment, ciliates are important not only through the blooms they determine (*e.g., Myrionecta rubra*) but also as a link in the food web between the bacterial level and the micro-metazoans (zooplankton) level.

7.8.3 Dinobionta

The Dinobionta,[17] sensu stricto or *sensu lato*, are also named Dinophyceae, Dinophyta, Dinozoa, Dinoflagellates, Dinoflagellida, and Peridinians. The above names are either synonyms, or nested sets, or correspond to taxa whose limits vary according to the authors. Here, the term "Dinobionta" is used *sensu lato*. Most Dinobionta are unicellular. However, some colonial species, or intermediate between the colonial and the multicellular state, have been described, for example *Haplozoon axiothellae* (Fig. 5.6e; Leander et al. 2002). The Dinobionta have long been considered as intermediate between prokaryotes and eukaryotes, due to a number of special characteristics of their nucleus (Herzog et al. 1984); these characters are actually not ancestral but derived characters. The age of Dinobionta is controversial. For some authors, on the basis of fossils that may belong to the ancestors of Dinobionta, although there is no indication they had photosynthesis, they were already present 540–420 Ma ago (Lecointre and Le Guyader 2006). For others, the acquisition of photosynthesis by secondary or tertiary endosymbiosis is much more recent: 225–250 Ma ago (Fig. 5.15; Kokinos et al. 1998; Falkowski et al. 2004). The Dinobionta seem to have dominated the marine plankton until about 65 Ma, when they were partially supplanted by diatoms (Falkowski et al. 2004). More than 2,000 living species have been described so far.

The alveoli contain cellulose, a polysaccharide consisting of linear chains of thousands of β-1,4 linked glucose units, with hydrogene bonds between chains, holding the chains firmly together, side by side. In armoured Dinobionta, overlapping alveoli create a sort of armor called the theca (Fig. 7.33).

The number of alveoli and their arrangement with each other varies considerably from one species to another and provides many taxonomic criteria. In addition, resistant stages (cysts) have an external cell wall consisting of dinosporine. Dinosporine is a polymer poorly understood from a chemical point of view, non-biodegradable and virtually indestructible, so that the cyst wall is preserved in sediments, not only for decades but also at a geological timescale (Kokinos et al. 1998; Versteegh et al. 2004). The

---

[16] The customary notion of 'invertebrates' does not correspond to a monophyletic taxon (modern meaning), but encompasses a paraphyletic group of taxa.

[17] Dinobionta: from the Ancient Greek '*dinô*', meaning spinning top, vortex. This name refers to the fact that cells frequently turn on themselves like tops. A similar but different Greek root, '*deinos*', meaning terrible, awesome, is the origin of the name of the dinosaurs (=terrible lizards).

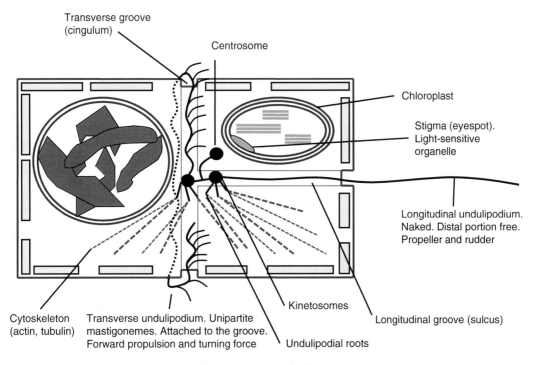

Fig. 7.34 Theoretical scheme of the kinetic apparatus of a photosynthetic Dinobionta

kinetic apparatus of Dinobionta is in a unique form in eukaryotes; it generally has two undulipodiums inserted into grooves. The transverse undulipodium beats in a groove called the cingulum; it is attached to the groove at several points, so that it can only undulate; it has unipartite[18] mastigonemes and produces forward propulsion and also a turning force. The longitudinal undulipodium lies in a groove named sulcus, its distal portion projecting freely behind the cell; it is generally naked (*i.e.* without mastigonemes) and plays a role of propeller and rudder (Fig. 7.34).

The nucleus presents unique characters in eukaryotes. It is very large, occupying up to half the volume of the cell (Fig. 7.35). The reason is that many genes are present in several hundred to several thousand copies. The chromosomes are attached to the nuclear membrane, lack histones, and remain condensed throughout interphase rather than just during mitosis. This sort of nucleus, called a dinokaryon, was once considered to be intermediate between the nucleoid region of prokaryotes and the true nucleus of eukaryotes; this is now considered a derived trait rather than an ancestral one. A second nucleus, a relic of tertiary endosymbioses, may be present.

In some Dinobionta, the chloroplasts are derived from a secondary endosymbiosis with a Rhodobionta. In this case, it contains chlorophyll *a* and chlorophyll *c2*; an apocarotenoid pigment, peridinin, is very abundant and forms a complex with chlorophyll. The chloroplast is bound by three (sometimes four) membranes and thylakoids are in groups of three (lamellae). However, the origin of photosynthesis in Dinobionta is complex, chloroplasts being incorporated *via* a variety of endosymbiotic events (Fig. 5.15). In addition, the chloroplast was secondarily lost in many taxa, in some cases eventually re-incorporated *via* a variety of tertiary endosymbioses (Falkowski et al. 2004). The Dinobionta involved in these endosymbioses exhibit the characteristics of the taxon which is at the origin of their chloroplasts: chlorophyll *b*, *c1*, fucoxanthin, and even phycobilin (Fig. 7.35). The main polysaccharide derived from photosynthesis is the dinamylon, a starch close to that of Viridiplantae but with a more branched amylopectin (a glucose polymer with α 1–4 and α 1–6 bonds) (Seo and Fritz 2002); it is stored within the cytoplasm. Extrusomes (trichocysts) may be present; they can discharge their content outside the cell for the purpose of attack or defense. Mitochondria have tubular cristae, an ancestral characteristic in Chromalveolata.

The Dinobionta multiply asexually by binary fission. When conditions are unfavorable briefly, temporary cysts are formed. When unfavorable conditions are lasting, male and female gametes (n) are formed and fuse. The zygote (2n) becomes a resistant cyst that can survive at least a decade. When conditions become favorable, it germinates, undergoes meiosis, and gives haploid individuals (Fig. 7.36; Genovesi-Giunti 2006). Fertilization is a planogamy and the life cycle is monogenetic haplophasic (Fig. 7.9b). In some taxa, *e.g. Noctiluca*, the life cycle is monogenetic diplophasic (Fig. 7.9c).

---

[18] The unipartite mastigoneme of Dinobionta corresponds to the terminal part of the tripartite mastigoneme of Chromobionta (*cf.* Sect. 7.9.3).

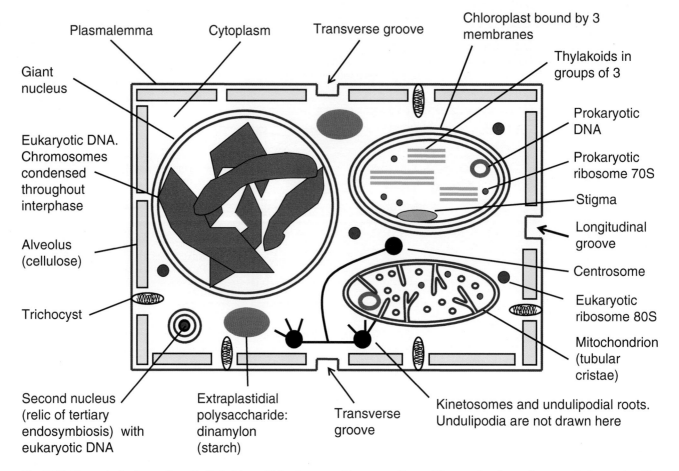

Fig. 7.35 *Theoretical scheme of a cell of Dinobionta*. This scheme combines organelles and features that do not fit any particular species. The undulipodiums are not represented

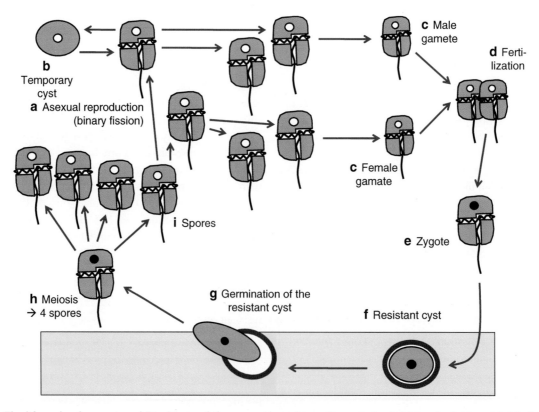

Fig. 7.36 *The life cycle of a species of Dinobionta of the genus Alexandrium.* Open circles (*white*) = haploid nuclei (n). *Solide circles* (*black*) = diploid nuclei (2n). The cycle takes place in the water column, with the exception of resistant cysts which are found within the sediment

Most Dinobionta species thrive in marine and freshwater plankton. About half of the species are photosynthetic. A number of photosynthetic species live in mutualistic symbiosis with a variety of hosts: *e.g.* metazoans (sponges, jellyfish, anemones, mollusks, scleractinian corals, gorgonians), ciliates, radiolarians, and foraminifera. *Symbiodinum*, in particular, is associated with scleractinian corals; it provides them with most of the products of photosynthesis and they in return receive waste (nitrogen, phosphorus); this mutualism, which keeps the nutrients in a closed circuit, is the key to the extraordinary richness of coral reef ecosystems in very **oligotrophic*** waters, *i.e.* very poor in nutrients. Although photosynthetic, many species are mixotrophic. This is the case of *Karlodinium armiger*; it has a chloroplast, but only in the presence of light, growth is low; it is capable of preying on a wide variety of unicellular species; it extracts the contents of prey through a feeding tube, but may also ingest whole prey cells; in the dark, predation is not enough to ensure the growth and survival; thus, *K. armiger* is an omnivorous obligate mixotrophic species (Berge et al. 2008). Finally, many species of Dinobionta are obligate heterotrophs. *Haplozoon axiothella* lives in the intestine of an annelid, *Axiothella rubrocincta* (Fig. 5.6e); *Noctiluca* spp. are predators that engulf their food, *e.g.* diatoms, other Dinobionta, metazoan eggs, and bacteria, by phagocytosis; *Pfiesteria* spp. are parasites of teleosts which are claimed to be responsible for large fish kills (Vogelbein et al. 2002); *Dissodinium pseudolunula* parasites the eggs of planktonic copepods (Drebes 1981).

A number of species of Dinobionta may proliferate to the point where they generate what is known as "red tides" (more correctly named "discolored waters"). The cell density can reach 20 million per liter. The color of the water can become red, but sometimes brown, yellow, *etc*. Many of these species produce toxins, so that red tides are often associated with mass mortalities of marine and lagoon animals (*e.g.*, fish) or problems related to human health. Generally speaking, blooms of planktonic unicellular organisms that cause negative impacts to other organisms *via* production of natural toxins, mechanical damage to other organisms, or by other means, are named "Harmful Algal Blooms" (HABs). Although HABs can be caused by taxa other than Dinobiontes, the latter are responsible for 75 % of HABs (Genovesi-Giunti 2006). In tropical and subtropical waters, shallow benthic Dinobionta belonging to a complex of cryptic species (*e.g., Gambierdiscus toxicus, G. caribaeus,* and *G. carolinianus*; Litaker et al. 2009) produce toxins (*e.g.,* maitotoxin) transmitted through the food chain with magnification *via* bioaccumulation; the consumption of predator teleosts, near the top of the food chain, causes ciguatera poisoning; hallmark symptoms of ciguatera in humans include gastrointestinal and neurological effects; severe case of ciguatera can result in long-term disability and even death.

7.8.4 Apicomplexa

The Apicomplexa[19] (also referred to as Apicomplexia and Sporozoa) are unicellular intracellular parasites of metazoans. There are about 5,000 known species, but this is almost certainly a gross underestimate of the actual number.

The cell of Apicomplexa is characterized by an apical complex (a structure involved in penetrating the host's cell) which is constituted in particular by a conoid (a set of spirally arranged microtubules) and secretory bodies (rhoptry and micronemes) which produce enzymes allowing penetration into the host cell. The name of the taxon is derived from this apical complex. The centriole, where present, has an unconventional structure; it consists of a central single microtubule surrounded by nine singlet microtubules, a deviation from the classical structure in eukaryotes, characterized by nine sets of microtubules triplets. The size of the genome is very small (3,800 genes in *Cryptosporidium parvum*, 5,300 in *Plasmodium falciparum*); it is among the smallest in eukaryotes, which is a consequence of parasitism (Abrahamsen et al. 2004). Flattened alveoli are present below the plasmalemma and mitochondria have tubular cristae, which are shared characters in Alveolata.

Most Apicomplexa have an apicoplast, an organelle surrounded by 3–4 membranes which is the non-photosynthetic relict of a former chloroplast (McFadden and Waller 1997; McFadden et al. 2001). The ancestors of the present Apicomplexa were therefore photosynthetic organisms. Photosynthesis originated from a secondary endosymbiosis with a Rhodobionta (Archaeplastida) (Bhattacharya et al. 2003). The discovery of *Chromera velia* and *Vitrella brassicaformis* (Chromerida; Fig. 7.28), sort of 'living fossils' close to the supposed ancestor of the Apicomplexa, which have chloroplasts, confirms the ancestry of photosynthesis in Apicomplexa (Keeling 2008; Moore et al. 2008; Oborník et al. 2012).

The life cycle is often extremely complex and involves two successive hosts: an intermediate host and a definitive host, in which sexual reproduction occurs. In the case of *Plasmodium falciparum*, a species causing malaria in humans, the intermediate host is the man and the definitive host is a mosquito of the genus *Anopheles* (Fig. 7.37). The fertilization is of the fucogamy type and the life cycle can be interpreted as digenetic monophasic, the single phase being a haplophase (Fig. 7.38); however, it can also be interpreted as showing an alternation between a haplophase and diplophase (Lecointre and Le Guyader 2006). In *Toxoplasma gondii*, that causes the disease toxoplasmosis, rodents are the intermediate hosts while felids, *e.g.* domestic cats, are the definitive hosts; it can reproduce sexually only within the intestines of members of the felid family. *T. gondii* appears to manipulate the behavior of rodents in ways that would increase their predation by cats

---

[19] The name Apicomplexa is derived from le Latin *apex* (top) and *complexus* (infolds).

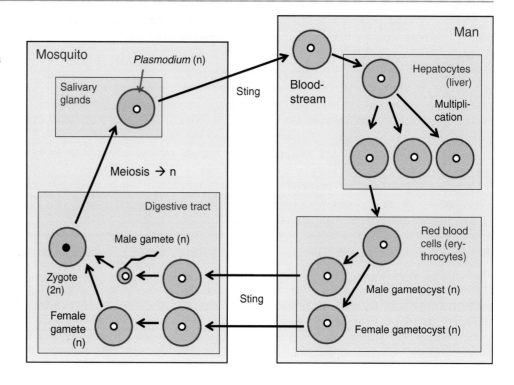

Fig. 7.37 *The life cycle of Plasmodium falciparum (Apicomplexa). This representation of the life cycle is extremely simplified. Haploid nuclei are represented by open circles, the diploid nuclei by a solid circle*

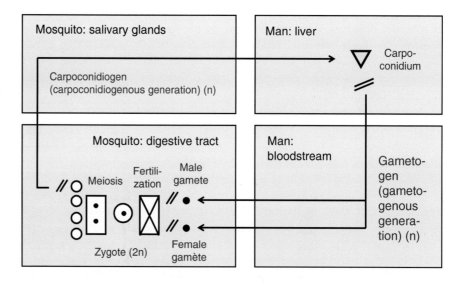

Fig. 7.38 *A possible interpretation of the life cycle of Plasmodium falciparum (Apicomplexa). This interpretation is based on the representation, extremely simplified, in Fig. 7.37. For the meaning of symbols, see Fig. 7.8*

and transmission to them, an adaptation that helps complete the life cycle of the parasite. It diffuses into the rodent's brain chemicals that make it more active, less fearful of predators, and that give it somehow "risk appetite". *T. gondii* is capable of infecting humans, that play the role of the intermediate host. Cats rarely eat humans, so there should be little advantage for the parasite to manipulate human, which definitely constitutes a dead end. Still, it influences human behavior, as if it were that of a mouse and it was necessary to rush into the cat's claws: "risk-taking" and machismo increase significantly in human individuals infected with *T. gondii*; the frequency of traffic accidents for example is 2.7 times higher in infected individuals compared to healthy ones (Webster 2001; Lafferty 2006).

7.9 Kingdom Stramenopiles

7.9.1 General Remarks

The kingdom Stramenopiles[20] (=Straminopiles, Stramenopila, Heterokonta) brings together taxa that were customarily classified within animals (Actinophryda, opalines), fungi (customary meaning; *e.g.* Oobionta, Labyrinthulobionta) and "algae" (Chromobionta). The bicosoecids have the

---

[20] Stramenopiles: from the Latin *stramen* (straw) and *pilus* (hair), in reference to the undulipodiums covered by mastigonemes.

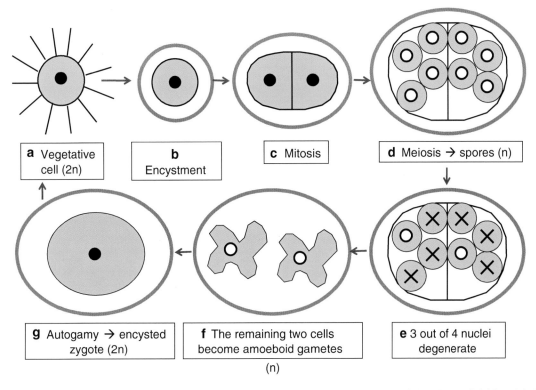

Fig. 7.39 *False sexual reproduction (autogamy) in an Actinophryda. Open circles* = haploid nuclei, *solid circles* = diploid nuclei. The encysted zygote will germinate when environmental conditions become favorable

particularity of being claimed by traditional botanists, under the name Bicosoecophyceae,[21] while they are predators basically devoid of photosynthesis (Fig. 7.1). The term "stramenopiles" was proposed by Patterson (1989) to account for all taxa, unicellular or not, with tripartite tubular mastigonemes on their undulipodiums. Tripartite mastigonemes, which may be taken as the defining characteristic of stramenopiles, are composed of a short basal segment, a long tubular shaft, and two or three thin terminal filaments.

Stramenopiles are a major lineage of eukaryotes currently containing more than 100,000 species (van den Hoek et al. 1998; Lecointre and Le Guyader 2006). By the number of species, stramenopiles (mainly Chromobionta) are one of the "big three" taxa of eukaryotes, after opisthokonts (mainly Metazoa and Fungi, modern meaning) and Archaeplastida (mainly embryophyta).

Actinophryda[22] (=actinophryida, actinophryids, actinophrids) were for a long time placed within Heliozoa, "sun animacules," with *e.g.* Radiolaria (today: Rhizaria) and Centrohelida (today: *insertae sedis*; maybe Archaepastida),

due to their radiating long pseudopodia (axopods). Subsequently, genetic analyses showed that Heliozoa were actually a polyphyletic group and that actinophryids were emerging within the stramenopiles (Fig. 7.1; Nicolaev et al. 2004). Actinophryda are unicellular and spherical in shape, 50–1,000 μm in diameter, without any cell wall or test and with long raylike pseudopodia, supported internally by a microtubule (axopods), radiating from the cell body, as in Centrohelida (*cf.* Sect. 7.5.2). Asexual reproduction takes place by binary fission. Under unfavorable conditions, false sexual reproduction (autogamy or self-fertilization) occurs; the organism forms a multi-walled cyst; within the cyst, two gametes are produced, which then fuse together (Fig. 7.39). The Actinophryda thrive mainly in the freshwater plankton, especially lakes and rivers; a few species can be found in soil and marine habitats. They passively feed on some unicellular species (*e.g.*, small Ciliophora and Dinobionta) which adhere to the axopods.

Bicosoecida (=Bicosoecea, Bicosoecophyceae, Bicosoecids) are unicellular heterotrophic organisms. The cell bears two unequal undulipodiums, is generally less than 10 μm in diameter and may or may not have a lorica, a covering that does not completely enclose the cell. It may be attached to the substrate by the "posterior" undulipodium or swim freely. There is a groove on the side of the cell, through which small prey (*e.g.*, bacteria) may be ingested. *Cafeteria roenbergensis* is a voracious predator very

---

[21] The suffix 'phyceae' is used to designate taxa (taxonomic rank: class) belonging to 'algae' (traditional sense). 'Phyceae' comes from the ancient Greek 'phykos' which means 'seaweed'.

[22] Actinophryda: from the ancient Greek *aktina* (ray) and *ophrys* (eyebrow).

common in marine habitats. Other species thrive in freshwater environments. Bicosoecida reproduce only through asexual binary division.

Opalinida[23] are parasites or, more probably, commensals in the digestive tract of metazoans, mainly frogs and toads. They are large (500 μm or more) and flattened. The cells have multiple nuclei (two to several hundred). The undulipodiums are numerous and short, which makes Opalinida resemble ciliates. Nutrition is saprotroph.

Labyrinthulobionta[24] (=labyrinthulomycetes, Labyrinthulomycota, Labyrinthulea, slime nets) belong to the polyphyletic group formerly called "fungi" (customary meaning). They produce within their host a network of tubes, which serve as tracks for the cells to glide along. They live in marine or estuarine environment. They are saprotrophic or parasites. The saprotrophic species play an important role in the degradation of dead leaves of Magnoliophyta, especially in the mangroves. Heterotrophic pico-stramenopiles (1 μm in diameter), close to Labyrinthulobionta, are very abundant within the mucus of *Fungia* (corals, metazoans); their role (mutualistic symbionts?) is not known (Kramarsky-Winter et al. 2006).

7.9.2 Oobionta

Oobionta[25] (=Oomycetes, Oomycota) are part of the polyphyletic set previously called "fungi" (customary meaning). The fusion of the Oobionta with the Chromobionta and the split from Fungi (modern meaning), on the basis of cytological and biochemical characteristics, was proposed early (Mereschkowsky 1910; Cavalier-Smith 1981), well before molecular phylogenies confirmed their position within the kingdom Stramenopiles. Present-day Oobionta are all heterotrophic, but the sequencing of the genome of *Phytophthora* showed that their ancestors were photosynthetic; they had acquired photosynthesis through secondary endosymbiosis with a Rhodobionta; they have in fact hundreds of genes possibly inherited from Rhodobionta and *Cyanobacteria* (Tyler et al. 2006). We know about 500 species of Oobionta, mainly belonging to Saprolegniales and Perososporales (Molds 2009).

Oobionta are characterized by their coenocytic-type vegetative apparatus with numerous scattered nuclei in a single cytoplasm. Their cell wall is mainly composed of cellulose (β 1,4 polymer of glucose, with hydrogen bonds between molecules). The main reserve polysaccharide is chrysolaminarin (β 1,3 polymer of glucose with a number of β 1,6 bonds). Unlike other eukaryotes, Oobionta do not synthesize sterols and must obtain them from their environment; the loss of this metabolism is secondary (Tyler et al. 2006). Their mitochondria have tubular cristae, a characteristic common to all Stramenopiles and even to Chromalveolata (Beakes 1989). The nuclear membrane does not disappear during mitosis (designated "closed mitosis"). Reproductive cells are mobile due to two unequal undulipodiums, one being covered by tripartite mastigonemes, while the other is smooth and usually longer than the first (Barr and Désaulniers 1989).

Asexual reproduction is often the only one known. It occurs by means of conidia equipped with two undulipodiums. In *Saprolegnia*, primary conidia with terminal undulipodiums encyst (conidial cysts) and germinate into secondary conidia that yield again multinucleate vegetative stage cells (Fig. 7.40a, e). The life cycle is monogenetic; the single, diploid (diplophase) generation can be interpreted as a sporogen or as a gametogen; meiosis occurs in male and female sporocysts, spores functioning as gametes; gametes are not spread and male gametes are not materialized; the male nucleus reaches the female nucleus through a siphon (siphonogamy). The zygote that has two apical undulipodiums, encysts (carpoconidian cyst) and then germinates, yielding a carpoconidium with two lateral undulipodiums that in turn yield the multinucleated vegetative stage (Fig. 7.40 a and f–h).

Oobionta are always heterotrophic, sometimes saprotrophic but mostly parasites. Peronosporales generally live in open habitats; they are in particular parasites of nematodes, of vertebrates (Metazoa), and Magnoliophyta (Archaeplastida). *Phytophthora infestans*, a parasite of potato, is famous for causing the Irish potato famine in the nineteenth century, which resulted in more than one million deaths and forced more than two million Irish to migrate (Gray 1995). *Plasmopara viticola*, the downy mildew of vine, is an introduced species from North America that destroyed European vineyards between 1878 and 1883, until a treatment with copper sulfate was discovered (Chadefaud 1978). *Pythium insidiosum* can be transmitted to dogs and horses (that are not its normal hosts); these diseases are called pythioses. Finally, a number of Peronosporales (*Pythium, Maullinia*, etc.) parasitize marine Rhodobionta and Chromobionta. The Saprolegniales are aquatic. They parasitize in particular teleosts, nematodes, rotifers, and crustaceans (Metazoa); they are thus a problem in fish farms. *Ectrogella eurychasmoides* is a parasite of

---

[23] The name of Opalinida is derived from their opalescent appearance when illuminated with sunlight.

[24] The name of Labyrinthulobionta is derived from *Labyrinthula*, a diminutive of the Latin *Labyrinthus* (labyrinth) and the ancient Greek *biont* (living thing).

[25] Oobionta: From the Greek "oon", egg. The name refers to the female gametes, that are rounded, large, and non motile, that ancient writers named 'eggs'.

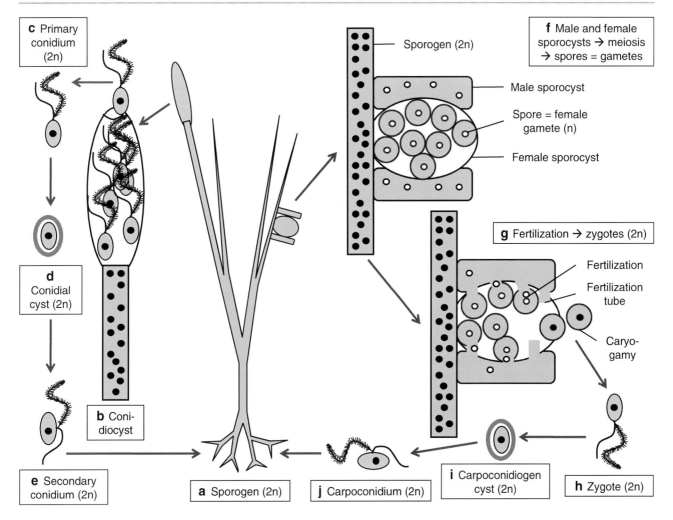

Fig. 7.40 *The life cycle of an Oobionta of the genus Saprolegnia.* The cells from meiosis (spores) functioning as gametes, the sporogen and the sporocysts might as well be named gametogen and gametocysts, respectively. *White circles* = haploid nuclei, *black circles* = diploid nuclei (According to Chadefaud (1978), modified, redrawn and reinterpreted)

diatoms of the genus *Licmophora* (Feldmann and Feldmann 1955). *Eurychasma* spp. are widespread parasites of marine Phaeophyceae (Chromobionta).

7.9.3 Chromobionta

The Chromobionta[26] (=Chromobiota, Chromophyta, Ochrophyta[27]) are a very large taxon within stramenopiles, both by the number of species and the phylogenetic diversity (Fig. 7.41). These include in particular organisms, mostly photosynthetic, that tradition has called "brown algae" (Phaeophyceae), "golden algae" (Chrysophyceae and Synurophyceae) and "yellow algae" (Xanthophyceae). Most taxa of Chromobionta are unicellular and/or colonial (Table 7.5). Several taxa encompass very small-sized unicellular organisms (<3 μm); eukaryotic organisms whose size is similar or identical to that of prokaryotes are designated under the general term of pico-eukaryotes[28] (Box 5.2; *cf.* Sect. 5.1.2); this is the case of Bolidophyceae (*e.g., Bolidomonas*), Eustigmatophyceae (*e.g., Nannochloropsis*), Pelagophyceae

---

[26] The name of Chromobionta is derived from the ancient Greek *khrôma* (color) and *biont* (living thing). The name refers to the fact that the green color of chlorophyll, when present, is usually masked by the abundance of pigments in a different color (brown, yellow, gold, *etc.*).

[27] Some authors (*e.g.* Andersen 2004) refer to the Chromobionta as 'Heterokonta', which causes confusion. For most authors, Heterokonta are what we refer to here as the stramenopiles.

[28] The term 'pico-eukaryote' refers to the small size of these organisms; '*pico*' is a Spanish expression ('y pico') meaning 'and some'. It should be noted that the size of these organisms (up to 3 μm) is much greater than the picometer.

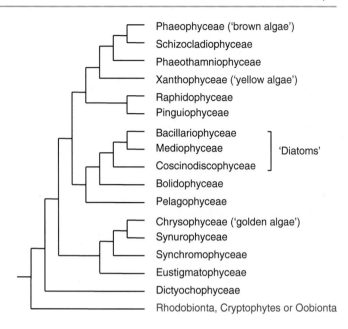

Fig. 7.41 *Simplified phylogenetic tree of Chromobionta (Stramenopiles)*. This tree is a synthesis of the data published by Kawai et al. (2003), Sims et al. (2006) and Horn et al. (2007), based on rbcL and SSU rDNA genes. The length of the branches is not proportional to the distances. Branch length is not proportional to distance. In *red*, outgroups used by the authors

Table 7.5 Structure of the vegetative apparatus in the classes of Chromobionta. + = present, − = absent. Taxa are listed in the same order as in the phylogenetic tree of Fig. 7.41. A cormus is a vegetative apparatus consisting of specialized tissues organized into specialized organs, unlike the thallus

Class	Unicellular	Colonial	Filamentous (thallus)	Two- and three-dimensional tissues (thallus)	Three-dimensional tissues (cormus)
Phaeophyceae[a]	−	−	+	+	+
Schizocladiophyceae	−	−	+	−	−
Phaeothamniophyceae	−	−	+	−	−
Xanthophyceae[b]	+	−	+	−	−
Raphidophyceae	+	−	−	−	−
Pinguiophyceae	+	−	−	−	−
Bacillariophyceae[c]	+	+	−	−	−
Bolidophyceae	+	−	−	−	−
Pelagophyceae	+	−	−	−	−
Chrysophyceae	+	+	+	−	−
Synurophyceae	+	+	−	−	−
Synchromophyceae	+	+	−	−	−
Eustigmatophyceae	+	−	−	−	−
Dictyochophyceae	+	−	−	−	−

[a]Phaeophyceae = Fucophyceae
[b]Xanthophyceae = Tribophyceae
[c]*Sensu lato*, i.e. Coscinodiscophyceae, Mediophyceae (= Fragilariophyceae) and Bacillariophyceae (*sensu stricto*). Bacillariophyceae *sensu lato* have been named by some authors 'Diatomophyceae'

(*e.g., Aureococcus, Aureoumbra, Pelagomonas*), Chrysophyceae (*e.g., Picophagus*), and Pinguiophyceae (*e.g., Pingiococcus*) (Fuller et al. 2006).

Diatoms (Fig. 7.41), which are all the most important taxon of unicellular Chromobionta, both by the number of species and by the role they play in the global primary production, are relatively recent at the scale of geological times. In fact, the oldest fossils are dated 180, 120 Ma only for centric diatoms (Coscinodiscophyceae), and it was during the last 100 Ma that diatoms have gradually emerged as the dominant taxon of photosynthetic plankton (Falkowski et al. 2004; Medlin and Kaczmarska 2004; Raven and Waite 2004; Sinninghe-Damsté et al. 2004).

Given the extraordinary diversity of Chromobionta (Andersen 2004), it is not possible to detail here all the structures, or any combination of characters, which can be observed. It is in some way an "average cell" that is described here. The cell wall is mainly composed of alginic acid or silica. Alginic acid is an uronic acid formed by the polymerization of D-mannuronic acid and L-guluronic acid, with β-1.4 bonds. In diatoms, the cell wall, mainly made of silica, is composed of two valves connected by one or two cingula (girdles, sorts of belts) with ligulae (sorts of belt buckles) (Fig. 7.42). Siliceous or calcareous scales may be present outside the cell wall (Synurophyceae, Chrysophyceae). The kinetic apparatus has two anterior undulipodiums. The first

Fig. 7.42 Scheme of a diatom of the genus *Cyclotella*. Drawn from a photo of Prasad et al. (1990)

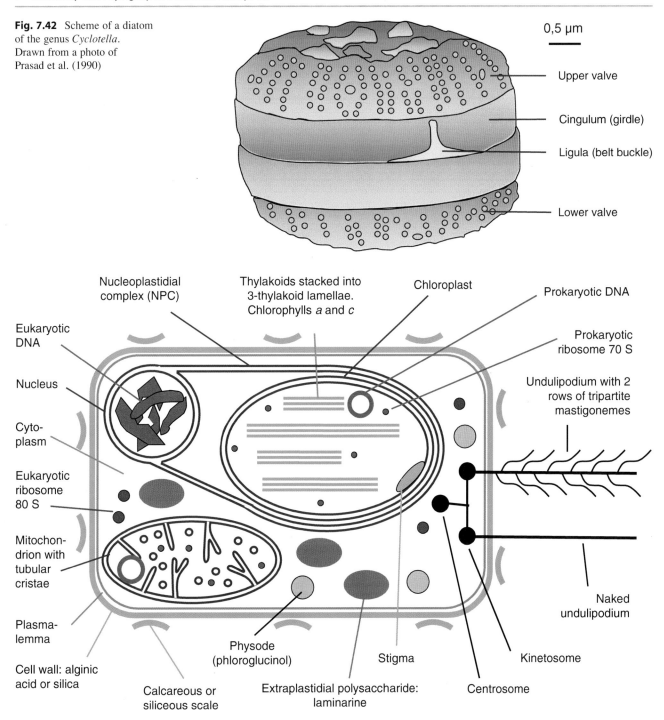

Fig. 7.43 *Theoretical scheme of a cell of Chromobionta*. This scheme combines organelles and features that do not fit any particular species. The vacuole, which can occupy a large portion of the cell, cytoskeleton, undulipodial roots and, within the chloroplast, the girdle lamella, have not been shown

has an axoneme covered with tripartite mastigonemes arranged in two rows. A tripartite mastigoneme comprises: (i) a basal portion in the form of a carrot, 0.1–0.3 μm long, (ii) a cylindrical hollow intermediate section, 0.7–2.0 μm long and 13–20 nm wide, and (iii) a terminal section consisting of 1–3 very thin hairs. The movements of the axoneme produce a forward displacement of the cell, while the movements of mastigonemes produce backward movement. The second undulipodium[29] is smooth, *i.e.* without mastigonemes; it plays the role of a rudder (Fig. 7.43).

---

[29] This undulipodium is often referred to in the literature as the 'posterior' undulipodium, which is incorrect. Rather, as in all bikonts, both undulipodiums are anterior.

The chloroplast, when present, is surrounded by a four-membrane envelope. The double outer envelope, a remnant of the secondary endosymbiosis, merged with the nuclear envelope, forming the nucleoplastidial complex (NPC), a structure characteristic of Chromobionta and Haptobionta (Andersen 2004). The thylakoids are stacked into three-thylakoid lamellae; there is a lamella, termed the girdle lamella, also composed of three thylakoids, that runs around the whole periphery of the chloroplast, parallel to and just beneath the chloroplast envelope. In addition to the chlorophyll *a*, Chromobionta have chlorophyll *c* (except for Eustigmatophyceae): *c1 + c2* (Bacillariophyceae, Chrysophyceae, Dictyochophyceae, Pelagophyceae, Phaeophyceae, Pinguiophyceae, Raphidophyceae, and Xanthophyceae), *c2 + c3* (Bacillariophyceae and Bolidophyceae), *c1* (Synurophyceae), and *c2* (Synchromophyceae) (Jeffrey 1989; Andersen 2004; Horn et al. 2007). Carotenoid pigments (carotenes, xanthophylls) are very abundant, so that their own color usually masks that of chlorophyll; four dozen or so carotenoid pigments, including fucoxanthin, have been reported (Bjornland and Liaaen-Jensen 1989). The polysaccharides arising from photosynthesis are chrysolaminarin (a polymer of glucose, with β-1,3 bonds and some β-1,6 bonds) or laminarin (a chrysolaminarin whose some chains can end with a mannitol molecule) (de Reviers 2003); chrysolaminarin and laminarin are stored within the cytoplasm. Mannitol, a hexaalcool, is an important storage compound, mainly in Phaeophyceae. The main sterol is fucosterol. Polymers of phloroglucinol (1,3,5-tri-hydroxy-benzene) are phenolic compounds restricted to the Phaeophyceae and a few other taxa; they are stored in organelles named physodes. Physodes and phloroglucinols play a structural role in the formation of the cell wall; in addition, they have a role in antiherbivory defense, photoprotection against UV, and heavy metal sequestration (Pellegrini 1974; McInnes et al. 1984; Schoenwaelder 2002). Mitochondria have tubular cristae (Fig. 7.43).

Asexual reproduction is by binary fission or by conidia. In diatoms (Bacillariophyceae, Mediophyceae, and Coscinodiscophyceae), which are enclosed within a cell wall made of silica (two valves), one of the valves (external) is larger than the other (internal); during binary fission, each daughter cell receives one of the parent cell's two valves; this valve is used by the daughter cell as the larger valve (external) within which a small new valve (internal) is constructed. This form of division results in wide diversity in cell size and an average cell size reduction over time (Fig. 7.44). In *Synchroma grande* (Synchromophyceae), whose cell is enclosed in a shell, termed the lorica, one of the daughter

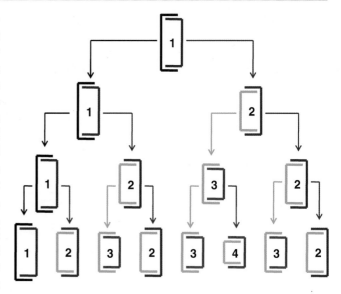

Fig. 7.44 *Asexual reproduction in diatoms (Chromobionta)*. Asexual reproduction results in wide diversity in cell size. *1, 2, 3,* and *4* = size classes of individuals from asexual reproduction (From Feldmann (1978), redrawn)

cells retains the lorica and the other comes out and becomes a "migrant amoeba" (Horn et al. 2007).

Sexual reproduction is known only in some classes of Chromobionta. The fertilization belongs to the planogamy type (Chrysophyceae, Xanthophyceae, Phaeophyceae), cystogamy (Bacillariophyceae), fucogamy (Phaeophyceae), or to the oogamy type (Coscinodiscophyceae, Phaeophyceae, Xanthophyceae). The life cycle is often monogenetic, with only a haploid generation (Fig. 7.9b), *e.g.* in *Dinobryum cylindricum* (Chrysophyceae) and in Xanthophyceae. In Phaeophyceae ("brown algae"), the life cycle is generally digenetic, with alternation between a haploid gametogen and a diploid sporogen (Fig. 7.9a). As in Viridiplantae (Archaeplastida), the evolutionary trend toward a reduction of the gametogenic generation is obvious in Phaeophyceae (*cf.* Sect. 7.5.5); the gametogen is reduced to a few cells (*e.g.*, in *Laminaria*) and may even disappear (*e.g.*, in *Fucus*); the life cycle is then monogenetic, with only a diploid generation, in contrast with monogenetic life cycles observed in possibly ancestral species (see above) (Fig. 7.9c). The monogenetic life cycle is present in diatoms; sexual reproduction also enables them to reverse the decrease in size due to the asexual reproduction (Fig. 7.45).

Chromobionta are basically photosynthetic organisms. At least in some diatoms and Phaeophyceae, photosynthesis belongs to the C4 type: the first organic compound within which is incorporated the carbon atom proceeding from CO_2 (mineral carbon) consists of four atoms of carbon. This type of photosynthesis is particularly efficient under intermittent

Fig. 7.45 *Sexual reproduction in diatoms (Chromobionta). Open circles* = haploid nuclei, *solid circles* = diploid nuclei, *red crosses* = degenerating nuclei

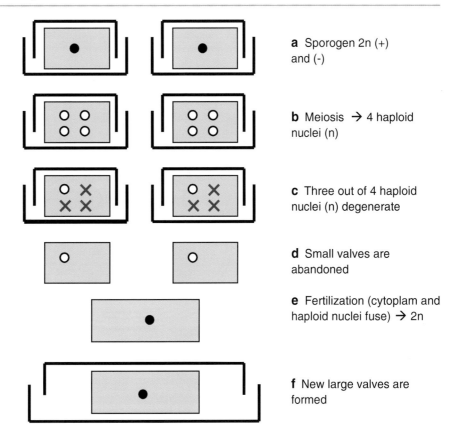

a Sporogen 2n (+) and (−)

b Meiosis → 4 haploid nuclei (n)

c Three out of 4 haploid nuclei (n) degenerate

d Small valves are abandoned

e Fertilization (cytoplam and haploid nuclei fuse) → 2n

f New large valves are formed

illumination. However, many species are mixotrophic: while possessing photosynthesis, they are saprotrophic and/or predatory. For example, *Dinobryon* (Chrysophyceae) can ingest three bacteria every 5 min, this prey representing ~50 % of its carbon supply (Bird and Kalff 1986). *Synchroma grande* (Synchromophyceae) captures, with its pseudopodia, bacteria and diatoms (Horn et al. 2007). A strain of the diatom *Melosira nummuloides* (Coscinodiscophyceae) can grow without light using various amino acids, which explains its success in polluted waters (McLean et al. 1981). Finally, some species have lost photosynthesis and are therefore obligate heterotrophs. This is the case for example of the genus *Spumella* (Chrysophyceae) and of *Heterosigma nitzschia* (Bacillariophyceae). *Bolidomonas mediterraneus* (Bolidophyceae) and *Picophagus flagellatus* (Chrysophyceae) are active predators and grazers, the latter including in its diet mainly *Prochlorococcus*, a cyanobacterium very abundant in the marine plankton.

Chromobionta are thriving in all continental and marine habitats, more especially in aquatic ones, both pelagic and benthic. Diatoms account for over 40 % of the net primary production of the world ocean (Falkowski et al. 2004; Sinninghe-Damsté et al. 2004). Some taxa of Chrysophyceae and diatoms live on moist soils and mosses. For example, a diatom, *Orthoseira gremmenii*, lives on the mosses of freshwater seeps at Gough Island, in the South Atlantic Ocean (Van de Vijver and Kopalova 2008).

7.10 Haptobionta

Haptobionta[30] (=Haptophytes, Haptophyta, Haptophyceae, Coccolithophorida, coccolithophores, coccolithophorids[31]) are unicellular organisms. About 500 species have been described, which are divided into two classes, Pavlovophyceae and Prymnesiophyceae. The position of Haptobionta in the eukaryotic tree (Fig. 7.1) is still very uncertain (Bhattacharya et al. 2003; Andersen 2004; Nakayama et al. 2005; Cuvelier et al. 2008). The oldest fossils date from 300 Ma, but it was only 180 Ma ago that the abundance of Haptobionta became conspicuous (Lecointre and Le Guyader 2006).

Haptobionta do not have a cell wall (de Reviers 2003). Below the plasmalemma (cytoplasmic membrane), a peripheral envelope of endoplasmic reticulum (Peripheral Endoplasmic Reticulum, PER) runs around the cell. This envelope is characteristic of the taxon. Above the plasmalemma, in Prymnesiophyceae, scales (generally calcareous)

---

[30] Haptobionta, from the ancient Greek *hapsis* (binding) and *biont* (living thing). The name refers to the possible role of the haptonema.

[31] In fact, coccolithophorids (=coccolithophores, Coccolithophorida) only correspond to a class of Haptobionta, namely the Prymnesiophyceae (de Reviers 2003; Andersen 2004), although sometimes incorrectly used with a wider meaning.

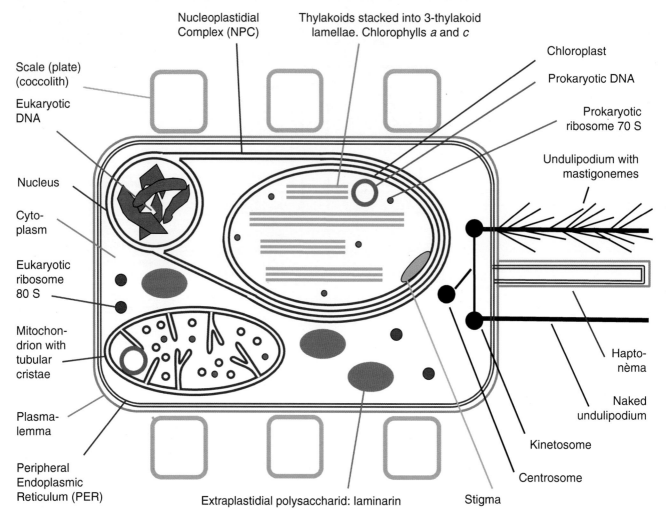

Fig. 7.46 *Theoretical scheme of a cell of Haptobionta*. This scheme combines organelles and features that do not fit any particular species or generation. The vacuole, cytoskeleton, and undulipodial roots are not shown. The scales are only shown in one part of the cell to allow room for the legends

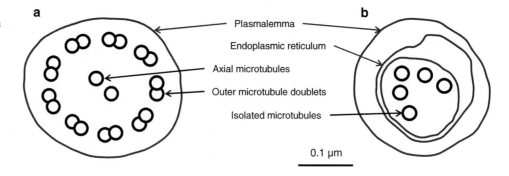

Fig. 7.47 *The haptonema of Haptobionta*. Cross sections of an undulipodium (**a**) and of a haptonema (**b**) of *Chrysoculter rhomboideus* (Haptobionta) (Drawn from photos of Nakayama et al. 2005)

are present. These scales, sometimes large and with a shape that is extremely diverse, are often called "coccoliths." Within chloroplasts, when present, thylakoids are stacked into three-thylakoid lamellae; in contrast with Chromobionta, there is no girdle lamella. The chloroplast is surrounded by a four-membrane envelope; the double outer envelope, a remnant of the secondary endosymbiosis, merges with the nuclear envelope, forming the nucleoplastidial complex (NPC), a structure characteristic of Chromobionta, Haptobionta, and Cryptophytes (Andersen 2004). In addition to chlorophyll *a*, chlorophylls *c1* and *c3* are present (Andersen 2004). The polysaccharide arising

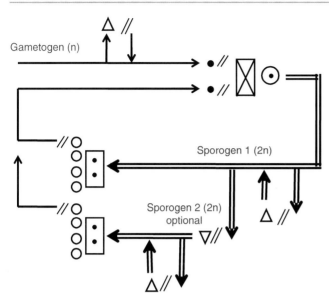

Fig. 7.48 *Life cycle of Emiliania huxleyi (Haptobionta)*. For the meaning of the used symbols, see Fig. 7.8 (From reinterpreted data of Green et al. (1996), original drawing)

from photosynthesis is a glucose polymer close to the laminarin, with short chains, β-1,3 bonds, and a few β-1,6 bonds; it is stored within the cytoplasm (de Reviers 2003). Mitochondria have tubular cristae (Fig. 7.46).

The kinetic apparatus consists of two anterior undulipodiums and a haptonema, a peg-like organelle attached near the undulipodiums and unique to the taxon. The first undulipodium can be covered with non-tubular and very thin mastigonemes, arranged around the axoneme. The second is generally naked (Leadbeater 1989) (Fig. 7.46). The structure of the haptonema is very different from that of undulipodiums: in cross section, instead of the classical undulipodial "9 + 2" structure, it has 5–8 irregularly arranged microtubules (Fig. 7.47). The role of the haptonema would be obstacle detection, attachment to the substrate, food (including prey) capture, and transport, in the process of phagocytosis, to the posterior part of the cell, where food is ingested (de Reviers 2003; Andersen 2004).

In species in which sexual reproduction is known (some Prymnesiophyceae), the life cycle includes several generations. In *Emiliania huxleyi*, the stage without undulipodiums, provided with calcareous scales, which determines planktonic blooms, is the sporogen (diplophase). It can produce, by meiosis, spores that give rise to a gametogen (haplophase) with undulipodiums and organic scales. The gametogen produces gametes which fuse (fertilization) and restore the sporogen. The spore can also give rise to a third generation, optional, with neither scales nor undulipodiums (Fig. 7.48). The basic life cycle is therefore digenetic (optionally trigenetic), with alternation of a diplophase and a haplophase.

Most Haptobionta are photosynthetic. However, a number of species are mixotrophs. For example, species of the genus *Chrysochromulina*, while photosynthetic, capture food particles through their haptoneme, which wraps around the cell to bring them to the rear end of the cell, where they are ingested (Andersen 2004). Finally, *Balaniger balticus* is and obligate heterotroph (de Reviers 2003).

Haptobionta are mainly present in the marine plankton. For example, species of the genus *Phaeocystis* are entering the food chain through euphausiids (Crustacea), euphausiids which are themselves eaten by birds and marine mammals. Haptobionta, e.g. *Emiliania huxleyi*, cause spectacular planktonic blooms; these blooms constitute a sink for CO_2 since, after the death of cells, calcareous scales fall into the sediment. This causes the formation of limestone, particularly Cretaceous chalk. Some species causing these blooms are toxic. *Prymnesium parvum*, a component of plankton in many parts of the world, produces toxins (allelopathy) that have hemolytic, ichthyotoxic, and cytotoxic effects; cyanobacteria (and other bacteria) and Dinobionta are simply inhibited, while diatoms and ciliates (potential predators) are eliminated; the planktonic community is therefore controlled by this species (Fistarol et al. 2003). This is the giant virus EHV-86 that causes the sudden disappearance of plankton blooms of *Emiliania huxleyi* (Monier et al. 2009).

Some Haptobionta (*Phaeocystis antarctica, Emiliania huxleyi*) synthesize DMSP (dimethyl sulfonium propionate) whose degradation by bacteria can produce DMS (dimethyl sulfide), which, when released into the atmosphere, serves as a nucleus to the condensation of cloud. DMS fluxes of marine origin, mainly due to Haptobionta, are estimated at 15–30 Tg/a (Valiela 1991; Tréguer 2002; Todd et al. 2007).

Telonemia (=telonemids) is an *incertae sedis* taxon. Considering the tripartite mastigonemes and mitochondria with tubular cristae, it is close to the stramenopiles. Considering the alveolae, it is close to Alveolata. Finally, some phylogenies based on genes place Telonemia near Cryptophyta (Shalchian-Tabrizi et al. 2006; Burki et al. 2012). Telonemia are unicellular, lack chloroplasts, and are therefore heterotrophic. Although only two species have been described formally (*Telonema antarctica* and *T. subtile*), DNA sequences collected from seawater suggest there are many more species which have not yet been described (Shalchian-Tabrizi et al. 2006). They live in the marine plankton; they have also been found in freshwater.

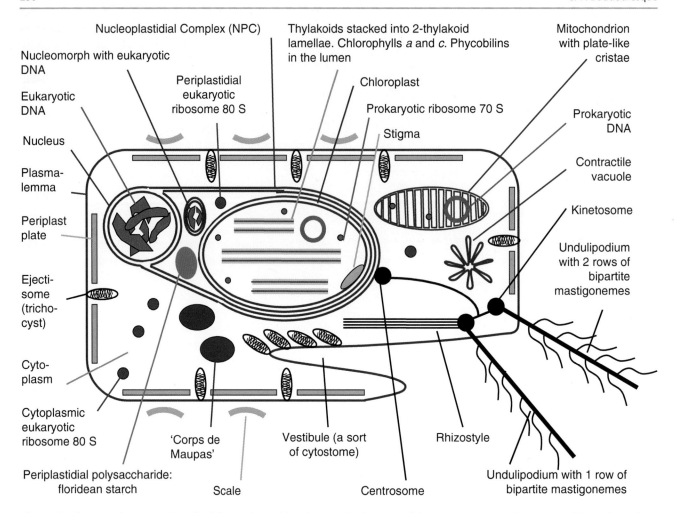

Fig. 7.49 *Theoretical scheme of a cell of Cryptophyta*. This scheme combines organelles and features that do not fit any particular species. The vacuole, cytoskeleton, and undulipodial roots are not shown. Scales and periplast plates are shown in only part of the cell periphery, to make room for the legends

7.11 Cryptobionta

Here, we call Cryptobionta[32] the group consisting of Cryptophyta, Katablepharida, and possibly Biliphyta (Fig. 7.1). According to Andersen (2004), they are the sister group of the Chromalveolata (*i.e.* Alveolata, Stramenopiles, and Haptobionta), while Burki et al. (2012) consider that they are the sister group of Archaeplastida (Fig. 7.1). In fact, the position of Cryptobionta in the eukaryotic tree remains unclear (*incertae sedis*).

Cryptophyta (=Cryptista, Cryptophyceae, Cryptomonadea, cryptomonads) are unicellular organisms. They encompass about 200 species (Lecointre and Le Guyader 2006).

A major originality of Cryptophyta, which acquired photosynthesis through a secondary endosymbiosis with Rhodobionta, is having retained a considerable number of remnants of this event: not only the chloroplast itself, but also a second two-membrane envelope around the chloroplast, the nucleomorph (the rest of the Rhodobionta nucleus), mitochondria with plate-like flat cristae (a characteristic feature in Archaeplastida) and even photosynthetic pigments (phycobilins) and polysaccharides arising from photosynthesis (floridean starch; *cf*. Sect. 7.5.3) (Fig. 7.49).

Cryptophyta have no true cell wall. Outside the plasmalemma (cytoplasmic membrane), there may be scales. Under the plasmalemma, there is a continuous layer of plates, perhaps composed of proteins, called periplast. The chloroplast is surrounded by a four-membrane envelope; the double outer envelope, a remnant of the secondary endosymbiosis, merges with the nuclear envelope, forming the nucleoplastidial complex (NPC). Within the periplastidial space (the space located between the two 2-membrane envelopes of the chloroplast), a nucleomorph, 80S

---

[32] Cryptobionta, from the ancient Greek *kruptos* (hidden) and *biont* (living thing). This name is not based on a general characteristic of the taxon, but the name of one of the genera belonging to it, *Cryptomonas*.

eukaryotic ribosomes, and floridean starch are present. The nucleomorph is the remnant of the nucleus of the Rhodobionta cell that is at the origin of the chloroplast, *via* secondary endosymbiosis. In contrast with the other taxa (Chromalveolata) which acquired photosynthesis from endosymbiosis with a Rhodobionta, the Rhodobionta nucleus has been preserved, but has lost most of its genes (Bhattacharya et al. 2003; Tanifuji et al. 2011). The thylakoids contain chlorophylls *a* and *c2* (Jeffrey 1989). In addition, they have preserved phycobilins (phycocyanin or phycoerythrin); however, unlike cyanobacteria, Glaucocystobionta and Rhodobionta, phycobilins are not localized within phycobilisomes, granules located on the thylakoid (Figs. 7.13 and 7.15), but within the thylakoid lumen (Fig. 7.49). The alloxanthin is a xanthophyll specific to Cryptophyta; its presence in the Dinobionta *Dinophysis norvegica* is due to a tertiary endosymbiosis with a Cryptophyta (Meyer-Harms and Pollehne 1998). The thylakoids are stacked into two-thylakoid lamellae, an uncommon feature in eukaryotes, which Cryptophyta only share with Chlorarachnobionta (*cf.* Sect. 7.6.3).

Cryptophyta have two undulipodiums, of identical length or not, inserted into a vestibular groove (a sort of cytostome) or in the vicinity of this groove. These undulipodiums are covered with bipartite mastigonemes: a cylindrical hollow section and a terminal very thin hair; compared to the tripartite mastigonemes of Chromobionta, the basal section is missing. One of the undulipodiums has two rows of mastigonemes, the other a single one (Fig. 7.49; Leadbeater 1989). The undulipodial root system comprises a rhizostyle which originates near the kinetosome, is longitudinally elongated, and has 6–10 microtubules (de Reviers 2003). The trichocysts (ejectisomes) are of a particular type, different from those of Alveolata and Stramenopiles. The "corps de Maupas" is a large vesicular structure, specific to Cryptophyta, whose main function could be that of disposing of unwanted cytoplasmic structures by digestion.

In Cryptophyta, asexual reproduction is by binary fission. The sexual reproduction is unknown in most species. It is known in *Proteomonas sulcata*, whose life cycle is digenetic and diphasic (haplo-diplophasic), with slightly heteromorphic generations (de Reviers 2003).

Most species of Cryptophyta are photosynthetic, but some have secondarily lost photosynthesis; some of these have retained a remnant chloroplast (*Chilomonas*), others not (*Goniomonas*). Cryptophyta live in both marine and freshwater plankton, more rarely in the intersticial water of soils and beach sand. In the marine ultra-phytoplankton (<5 μm), Cryptophyta are one of the three dominant taxa (with Haptobionta and Chrysophyceae) (McDonald et al. 2007).

Several unicellular taxa seem to be related to Cryptophyta, on the basis of molecular phylogenies. Katablepharida are the sister-group of Cryptophyta (Burki et al. 2012). They live in marine and freshwater habitats; they lack chloroplasts and are therefore heterotrophic, though some species are kleptoplastic (Okamoto and Inouye 2005). The Picobiliphyta (=Biliphyta) are a recently discovered taxon, present and sometimes abundant in the nanoplankton (2–6 μm in length) of all the seas of the world, while still poorly known. As Cryptophyta, they have phycobilins and a nucleomorph (Not et al. 2007; Cuvelier et al. 2008). Picobiliphyta have been considered as autotrophic organisms; there is, however a strong question mark over this hypothesis: they are more likely heterotrophic than autotrophic (Kim et al. 2011a).

7.12 Kingdom Discicristates

7.12.1 General Remarks

The Discicristates[33] are often met with Excavates within a single kingdom, which seems to be monophyletic (Simpson 2003; Hampl et al. 2009). Discicristates are characterized, from a cytological point of view, by their mitochondria whose cristae are flattened at the ends, like ping-pong rackets. Discicritates encompass in particular Euglenobionta (Euglenoidea, Kinetoplastida, and Pseudociliata), Percolobionta, and Acrasiobionta (=Heterolobosa).

Percolobionta (=Percolozoa) are unicellular organisms which are either in the form of amoeba (*e.g.*, *Vahlkampfia* and *Pseudovahlkampfia*), or in the form of a cell with undulipodiums (*e.g.*, *Percolomonas*, *Lyromonas*, and *Psalteriomonas*), but can also alternate amoeboid forms (when the environment is nutritionally rich) and cells with undulipodiums (when fast moving is a priority). The "brain-eating amoeba" *Naegleria fowleri*, a Percolobionta, infects humans by entering the body through the nose, which occurs when people go swimming in warm freshwater places.

Acrasiobionta (=Acrasiomycota, Acrasiomycetes, Acrasidae) are sometimes included within Percolobionta, or alternatively placed in their vicinity (Heterolobosa). They occur only in the amoeboid stage.

---

[33] Discicritates, from the Latin *discus* (disk) and *crista* (crest). This name refers to the mitochondrial cristae which resemble ping-pong rackets.

Fig. 7.50 *The strips of the pellicle, in Euglenoidea.* In this figure, these strips were isolated from the rest of the cell. Their helical arrangement, from the anterior to the posterior pole of the cell, is here partly disorganized, which is an artifact. Scale bar = 5 µm

7.12.2 Euglenoidea

The Euglenoidea (=Euglenophyta, Euglenophyceae, euglenids) are unicellular, more rarely colonial (*Colacium*), organisms. They comprise about 1,500 species. Some Euglenoidea acquired the photosynthesis by means of a secondary endosymbiosis with a Chlorobionta (Bhattacharya et al. 2003). Photosynthesis has been preserved in many species (*e.g., Euglena* spp.), while secondarily lost in others (*e.g., Astasia* and *Hyalophacus*). Some other Euglenoidea (*e.g., Rhabdomonas* and *Distigma*) may never have acquired photosynthesis (de Reviers 2003).

Euglenoidea have no true cell wall. Under the plasmalemma, proteinaceous strips helically arranged constitute the pellicle, a characteristic feature of the taxon that gives the cell its shape (Figs. 7.50 and 7.51). The pellicle is supported by dorsal and ventral microtubules. In many Euglenoidea, the pellicle strips can slide past one another, causing an inching motion called metaboly; otherwise, the cells move using the undulipodium.

In photosynthetic species, the chloroplast is surrounded by a three-membrane envelope. The thylakoids are stacked into three-thylakoid lamellae and contain chlorophylls a and b. The stigma, when present, is located outside the chloroplast, a rare feature in photosynthetic eukaryotes. The polysaccharide arising from photosynthesis, called paramylon, is a glucose polymer with β-1,3 bonds and many β-1,6 bonds giving rise to ramifications; it is stored within the cytoplasm. Trichocysts (extrusomes) may be present, beneath the plasmalemma (Fig. 7.51).

There are generally two undulipodiums, inserted in an anterior depression, the cytostome; one of them is reduced to its base and merges with the other. Mastigonemes are present, inserted in a single spiraled row (Fig. 7.51).

Asexual reproduction is by binary fission. When conditions become unfavorable, some species produce resistant cysts. Sexual reproduction is unknown, except perhaps in the genus *Scytomonas* (de Reviers 2003).

Euglenoidea mainly live in freshwater habitats. A few species can be found in marine habitats and in terrestrial soils. Even when they have chloroplasts, euglenoids use, optionally or mandatorily, organic matter; this explains why they usually thrive in habitats rich in organic matter. Many species are obligate heterotrophs, which are saprotrophic, parasites or predators of bacteria, and unicellular eukaryotes (de Reviers 2003; Lecointre and Le Guyader 2006).

7.12.3 Kinetoplastida

Kinetoplastida (=Kinetoplastea, kinetoplastids) are unicellular organisms. About 400 species have been described. The major distinguishing feature is the presence of a kinetoplast, a DNA-containing organelle located within the anterior part of the single giant mitochondrion, close to the kinetosomes of the undulipodiums. The kinetoplast, after which the taxon is named, contains many copies of the mitochondrial genes.

Most Kinetoplastida have two anterior undulipodiums, sometimes reduced to a single one. In *Trypanosoma*, the single long undulipodium is longitudinally attached along the cell body; the undulipodium function is twofold, locomotion *via* oscillations along the attached undulipodium and cell body, and attachment to the host gut. An anterior depression (cytostome) is generally present, although reduced or absent in *Trypanosoma*.

The life cycle of Kinetoplastida is often complex, with multiple hosts and various morphological stages. Most Kinetoplastida are parasites, such as *Leishmania* and *Trypanosoma*. *Trypanosoma brucei* causes African tripanosomiasis, also known as sleeping sickness in humans; the insect vector is the tsetse fly *Glossina*. One hundred or so

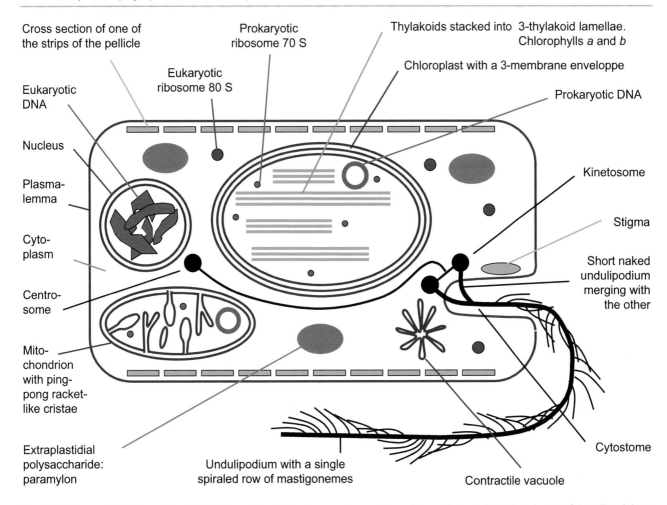

Fig. 7.51 *Theoretical scheme of a cell of Euglenoidea.* This scheme combines organelles and features that do not fit any particular species. The vacuole, cytoskeleton, and undulipodial roots have not been shown. Strips of the pellicle are shown in only part of the cell periphery, to make room for the legends

species of the genus *Leishmania* are responsible for the disease leishmaniasis; they are spread through sandflies of the genus *Phlebotomus* in the Old World, and of the genus *Lutzomyia* in America; their primary hosts are vertebrates (*e.g.*, dogs, rodents, and humans). Bodonidae (*Bodo*) live in freshwater and feed on organic matter or are predators of bacteria. Finally, many taxa of Kinetoplastida were discovered in hydrothermal vents of the Mid-Atlantic Ridge (Lopez-Garcia et al. 2003).

7.13 Kingdom Excavates

7.13.1 General Remarks

For many authors, Excavates (*sensu lato*) include Discicristates (*e.g.*, Simpson 2003; Hampl et al. 2009). Here they are reduced to Parabasalia, Diplomonadida, Oxymonadida, Retortamonadida, and related taxa (Fig. 7.1). The three latter taxa are generally met within the phylum Metamonada. The Excavates include about 650 species, all unicellular (Lecointre and Le Guyader 2006). They are unusual eukaryotes in lacking mitochondria. For this reason, together with an artifact in phylogenetic and clustering analyses ("long branch attraction"), they were originally considered as "primitive" eukaryotes, diverging from the others before the endosymbiosis at the origin of mitochondria; however, as in Microsporidia (Opisthokonta), they are now known to have lost mitochondria secondarily, while retaining both organelles and nuclear genes derived from them. Mitochondrial relics include hydrogenosomes, which produce hydrogen, and mitosomes (Germot et al. 1997; Simpson 2003).

In Oxymonadida, Retortamonadida, and some Diplomonadida, there is a ventral groove that plays a role in feeding: the beating of the undulipodiums pushes food particles into the groove. The presence of the ventral groove is almost a characteristic feature in Excavates, although it may be slightly marked or absent. Its presence in Percolobionta (Discicristates) indicates either an ancestral character to Excavates and Discicristates, or a convergent evolution (homoplasy) (Simpson 2003).

The association of the nucleus (or of each nucleus, when there are several) with a group of undulipodiums is considered by some authors as evidence of the endosymbiotic origin of the kinetic apparatus in eukaryotes (Margulis and Dolan 1997).

7.13.2 Parabasalia

The Parabasalia (=Parabasala, Parabasalidea, parabasalids) encompass the trichomonads and a number of taxa formerly grouped as the "hypermastigids." The kinetosomes are linked to parabasal fibers that attach to prominent Golgi complexes, constituting the "parabasal apparatus" which is distinctive to Parabasalia. The undulipodiums are grouped by four ("kinetid"). There may be a single kinetid in Trichomonads (*Trichomonas* and *Tritrichomonas*), but these can be quite numerous in "hypermastigids" (up to 100,000 undulipodiums). When there are many nuclei, each kinetid is associated with a nucleus (*e.g.*, in *Calonympha*).

Sexual reproduction exists, or has existed until recently, in Parabasalia, but it is poorly understood. In *Trichomonas vaginalis* (trichomonads), the meiosis-specific genes are present, although meiosis was not actually observed. In "hypermastigids," cell divisions reminiscent of meiosis, but with a single division (unlike most other eukaryotes, in which there are normally two), were observed (Dacks and Roger 1999; Malik et al. 2008).

Parabasalia are obligate heterotrophic organisms. Trichomonads live freely or are parasites. *Trichomonas vaginalis* is a parasite of the human urogenital tract and is the main agent of sexually transmitted disease (STD) on a global scale, with about 175 million new cases each year (Malik et al 2008). "Hypermastigids" are mutualistic symbionts in the gut of termites and cockroaches. *Calonympha grassii* lives in the gut of the termite *Cryptotermes brevis* and participates in the digestion of cellulose and hemicellulose, due to endosymbiotic bacteria. *Hoplonympha* sp., which lives in the gut of the termites *Hodothermopsis sjoestedti* and of cockroaches, is associated with ectosymbiotic bacteria (*Bacteroidales*) (Noda et al. 2006).

7.13.3 Diplomonadida

In Diplomonads, there are usually two diploid nuclei per cell, both being transcriptionally active. They have identical copies of the genes. Each nucleus is associated with a set of microtubules and four undulipodiums; this structure is called karyomastigont. The *Giardia lamblia* genome is relatively small (only 11.7 Mbp), with many genes acquired by horizontal gene transfer (HGT) from *Bacteria* and *Archaea*. The Golgi apparatus is absent (Lecointre and Le Guyader 2006; Morrison et al. 2007).

The life cycle involves alternation between cells with undulipodium ("trophozoites"), actively swimming, and cysts that ensure dissemination and infection. Although sexual reproduction has not been observed directly, the presence in *Giardia* of genes specifically involved in meiosis and the observation of genetic recombination are evidence of its presence (Logsdon 2007; Malik et al. 2008).

Diplomonads live in habitats poor in oxygen and rich in organic matter, especially in the digestive tract of vertebrates. *Giardia lamblia* is a parasite of the intestine of man, in whom it causes giardiasis, a disease characterized in particular by diarrhea (Simpson 2003; Lecointre and Le Guyader 2006).

7.14 Kingdom Opisthokonta

7.14.1 General Remarks

The Opisthokonta[34] (=Opisthochonta, opistokonts) are by far, with about 1,350,000 species, the largest taxon within the eukaryotes. They include Metazoa (=animals in the modern sense), choanoflagellates, Fungi (modern meaning; only a part of what was called fungi, customary meaning; Fig. 5.27), Microsporidia (a "protozoan" taxon that early molecular phylogenies considered as close to ancestral eukaryotes) and a number of poorly known taxa such as nucleariid amoebas and Mesomycetozoa (see below).

The grouping of Metazoa and Fungi (modern meaning) within the same kingdom, Opisthokonta, was unexpected, on the basis of available cytological and biochemical data; this grouping was therefore one of the surprises resulting from the early molecular phylogenies (Cavalier-Smith 1987b). From a genetic point of view, Opisthokonta clearly appear as a monophyletic and robust taxon (Steenkamp et al. 2006). For example, the gene tyrosyl-tRNA synthetase, of Archean origin, clearly separates Opisthokonta from all other eukaryotes (Huang et al. 2005). From a biochemical point of view, the presence of chitin as a structural macromolecule is an ancestral characteristic of Opisthokonta, although it may be present in other eukaryotic taxa (*e.g.*, in Rhodobionta). Considering the cytological characters,

---

[34] Opisthokonta: from the ancient Greek *opistho* (hind) and *kontos* (pole, in reference to the undulipodium).

Opisthokonta have mitochondria with plate-like flattened cristae, as in Archaeplastida and Cryptophytes. Finally, these are the only eukaryotes to have a single undulipodium, located at the rear of the cell and acting as a propellant (Fig. 7.10; *cf.* Sect. 7.4).

In the simplified phylogenetic tree presented here (Fig. 7.1), four higher taxa of Opisthokonta are shown: metazoans, choanoflagellates, Fungi (modern meaning; including Microsporidia), and Mesomycetozoa. This simplified tree does little to account for the complexity of the deep roots of the phylogenetic tree, still largely uncertain, and the diversity of amoeboid taxa located close to the roots of Opisthokonta. For example, *Nuclearia* would fall toward the root of the Fungi (modern meaning), *Amoebidium corallochytrium* toward the root of choanoflagellates, and *Ministeria* toward the root of Metazoa (Steenkamp et al. 2006). *Ministeria* and *Nuclearia* are amoebae equipped with thin pseudopodia (filopodia), the latter superficially resembling Centrohelida (*cf.* Sect. 7.5.2) and Actinophryda (*cf.* Sect. 7.9.1).

7.14.2 Microsporidia

Microsporidia are considered either as a sister group of the Fungi (Liu et al. 2006), or a taxon of Fungi (Hibbett et al. 2007), or placed at the base of the Fungi together with the "Zygomycota" to which they could belong (Dyer 2008; Lee et al. 2008), or the sister group of Chytridiomycota, or within the Chytridiomycota to which they may belong. The choice to treat them in a separate section of this chapter is therefore more pedagogic than scientific as it allows description of their highly derived and unique characters. About 1,500 species have been described and their true number could exceed one million (Larsson 2009).

Microsporidia are characterized by very small genomes, even smaller than bacterial genomes. Mitochondria are absent, which is not an ancestral state, as hydrogenosomes or mitosomes, considered as remnants of mitochondria, are present (Germot et al. 1997; Williams et al. 2002). Cytoplasmic ribosomes are of the 70S type and not of the 80S as generally observed in other eukaryotes. This could also be a derived, and not ancestral, character as mitochondria and their 70S ribosomes are absent (Cavalier-Smith 2002). The cells are not mobile and undulipodiums have never been observed. All these derived characters could be explained by the fact that microsporidia are all obligate intracellular parasites.

A cell wall is only present at the cyst stage. It is made of two layers, the endospore composed of chitin and the exospore. Within the cyst, a polar tube coiled in a helix, derived from the Golgi apparatus, is fixed to an anchoring disk at the anterior part of the cell, next to a lamellar structure called the polaroplast (Fig. 7.52; Dyer 2008; Larsson 2009). A vacuole is located at the posterior part of the cell. The polar tube serves to inject the cyst content into the host cell (Fig. 7.53). Injection results from the sudden swelling of the vacuole and is completed in less than 2 s.

Once the cyst content is within the host cell, it expands as a multinucleated cytoplasm or coenocyte, which at the end of the process occupies almost all of the host cell volume and starts producing cysts also named "spores" in the literature (Fig. 7.53). The life cycle of Microsporidia does not include an obvious sexual stage. However, sexuality could be present, or has been present in the past, as suggested from the occurrence of genes linked to sexual processes in Microsporidia's genomes (Dyer 2008; Lee et al. 2008; Larsson 2009).

Microsporidia are thus obligate intracellular parasites mostly of teleost fishes and of arthropods. *Antonospora locustae* is a grasshopper pathogen commercialized for the biological control of locusts. *Nosema bombycis* is responsible for a silk worm disease. *Nosema ceranae* is a benign pathogen of the Japanese bee *Apis cerana* which turned out to be a severe, even deadly, pathogen for the European bee *A. melifera* upon its introduction in the far East. Following re-importation of contaminated bee colonies from Asia to Europe and North America, this pathogen is now responsible for disease outbreaks in bees in these geographical areas (Chauzat et al. 2007). Finally, some Microsporidia parasite mammals as exemplified by *Enterocytozoon bieneusi* which causes diarrhoea in humans (Dyer 2008).

7.14.3 Fungi (Modern Meaning)

Fungi[35] in the current meaning of this taxonomic term form a monophyletic group from which have been successively excluded in the last decade's organisms belonging to the bacteria (actinobacteria and planctomycetes), Rhizaria (Phytomyxea; *cf.* Sect. 7.6.4), Alveolata (Ellobiopsidae; *cf.* Sect. 7.8.1) and Stramenopiles (Oobionta and Labyrinthulobionta; *cf.* Sects. 7.9.1 and 7.9.2). With more than 100,000 described species (Hawksworth 2001; Lecointre and Le Guyader 2006), the Fungi include species as diverse as lichen (generally belonging to Ascomycota) and mushroom-forming species (generally belonging to Basidiomycota, such as saffron milk cap, button mushroom, cepe, and death cap). The true number of species in the

---

[35] Fungi is the plural of the Latin word '*Fungus*', which means 'mushroom'. By convention, the term 'Fungi' is used here even when used in the singular.

Fig. 7.52 (a) *Transmission electron microscopy of a cyst of Episeptum inversum (Microsporidia)*. *A* anchor disk, *EN* endospore (cell wall), *EX* exospores (cell wall), *MB* membranes, *N* nucleus, *P* polar bag, *PA* anterior polaroplast, *PF* polar filament (polar tube); it is helically wound and cut many times, *PL* plasmalemma, *PP* posterior polaroplast, *R* ribosomes, *RU* rough endoplasmic reticulum, *S* septum in the exospores, *V* posterior vacuole, * outer layer of the endospore. Scale bar = 0.2 µm. (b) *Detail of the rough endoplasmic reticulum*. Scale bar = 0.1 µm (Courtesy of Ronny Larsson 2009)

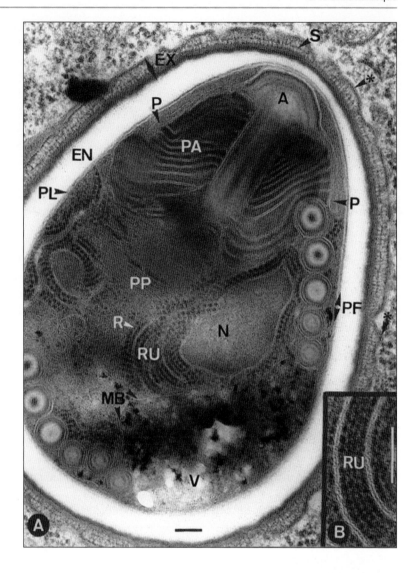

Fig. 7.53 *The life cycle of a Microsporidia*. Nuclei are drawn in *red* color

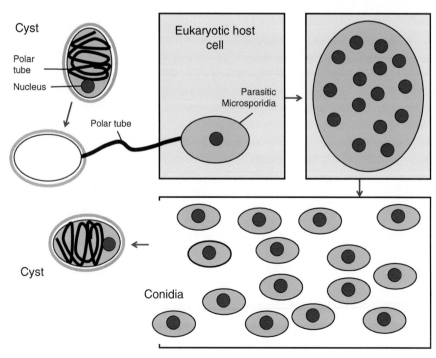

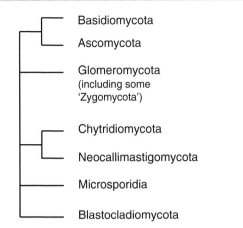

Fig. 7.54 *Simplified phylogenetic tree of the phylum Fungi (modern meaning)* (From Hibbett et al. (2007), modified and redrawn)

Fungi could actually exceed 1,500,000 (Bruns 2006; Mueller and Schmit 2007).

A majority of the Fungi are multicellular. Some species are however unicellular such as *Nuclearia simplex* which branches at the base of the fungal tree of life (Steenkamp et al. 2006). Other unicellular species, named yeasts, as illustrated by *Saccharomyces cerevisiae*, derive from multicellular taxa (Raven et al. 2000). Some fungal species are also characterized by a coenocytic vegetative stage (Chytridiomycota). Inclusion of the Fungi in the field of microbiology, though most of them are multicellular and macroscopic, results from the fact that they share with bacteria numerous functional roles in the degradation of organic matter, soil biology, or in many diseases.

The vegetative stage of multicellular fungal species is constituted of branched filaments. These filaments are called by traditional botanists as "hyphae." They can agglomerate to form a sort of tissue called plectenchyme. This is the case in lichens, stable mutualistic associations between a fungal species (belonging to the Ascomycota with few cases in the Basidiomycota) and a photosynthetic partner. This latter belongs most of the time to the Chlorobionta in the classes Trebouxiophyceae (*Trebouxia*), Trentepohliophyceae (*Trentepohlia*), or Chlorophyceae (*Coccomyxa*), but also to the *Cyanobacteria*. Mushroom **sporocarps***, picked and eaten by fungal collectors, are also made of agglomerated filaments forming a plectenchyme.

Chytridiomycota (chytrids) may represent one of the most ancestral fungal taxon (Fig. 7.54). Chytrids still display an undulipodium which has been lost in all other fungal groups. Both Chytridiomycota *sensu lato* and "Zygomycota" represent paraphyletic groups (Bruns 2006; Hibbett et al. 2007). As for the Ascomycota and Basidiomycota, they present numerous derived characters.

Cellulose is absent (or present in marginal amounts) in the cell wall whose main constituents are chitin (a polymer of *N*-acetyl D-glucosamine linked in β1-4), chitosan (a polymer of D-glucosamine linked in β1-4) (Robert and Catesson 1990) and β1-3 glucans (polymers of glucose linked in β1-3). Cells communicate with each other by septal pores. In "higher" Basidiomycota, these pores look like the synapses of Rhodobionta but their structure is different; they are called dolipores. Except in the chytrids, the kinetic apparatus is reduced to a centrosome (Bornens and Azimzadeh 2007). The loss of the undulipodium and of the kinetosome is a secondary event which occurred at least four times independently (Bruns 2006; James et al. 2006). The Golgi apparatus is also absent. Absence of chloroplasts is ancestral. The Fungi, as opposed to many higher eukaryotic taxa, had never been photosynthetic. The main form of cellular carbon storage is glycogen. Ergosterol is a fungal-specific sterol which is used as a specific chemical marker of their presence in specific habitats (Pasanen et al. 1999). Finally, within the eukaryotes, Fungi are characterized by a specific lysine biosynthetic pathway, the alpha-aminoadipic acid pathway.

In Fungi, modes of sexual reproduction are extraordinarily diverse and it is beyond the scope of this chapter to detail them all. In taxa which have lost the kinetic apparatus, such as the Ascomycota and the Basidiomycota, the ancestral life cycle could be trigenetic triphasic (Fig. 7.9e). This life cycle is similar to the cycle found in Rhodobionta (*cf.* Sect. 7.5.4). As in Rhodobionta, fertilization is of the trichogamy type. In contrast to Rhodobionta, in these Fungi, fusion of the cytoplasms (plasmogamy) is not immediately followed by nuclear fusion (karyogamy). Karyogamy is indeed delayed and takes place just before meiosis. As a result, a diploid phase sensu stricto does not occur; it is replaced by a micthaploid and a dikaryotic phases which alternate with a haploid phase (Fig. 7.9e). In many Basidiomycota this life cycle has been simplified, fertilization is a somatogamy which does not occur between two differentiated gametes but instead between two undifferentiated somatic cells (Fig. 7.55). In this case the resulting dikaryotic filaments ("mycelium") usually represent the permanent stage of the Fungi in nature and it is this mycelium which differentiates sporocarps (improperly called fruit bodies), prominent macroscopic structures in which meiosis takes place and the subsequent formation of spores which are disseminated (Fig. 7.56). Finally, especially among Ascomycota, sexual reproduction seems to have been lost in many species.

Fungi are always heterotrophs. Phagotrophy is absent (James et al. 2006). This absence is a derived character if we consider that phagotrophy could be an ancestral state in the eukaryotes (Cavalier-Smith 1987a, 2002). To obtain the organic matter necessary to their metabolism, Fungi can sometimes be commensal, more frequently saprotrophic, and many of them are mutualistic or parasitic of photosynthetic or non-photosynthetic organisms (James et al. 2006).

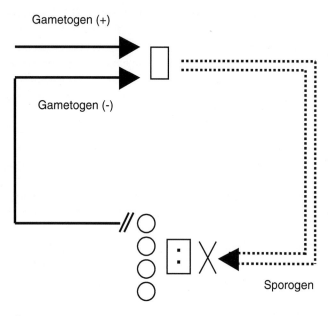

Fig. 7.55 *The biocycle of Basidiomycota of the "mushroom cap" (sporocarp) type*. There are no real gametes or zygote. The fusion of haploid somatic cells (n) directly gives rise to the dikaryotic sporogen (n + n). Compared to other Basidiomycota and some Ascomycota (Fig. 7.9e), the carpoconidiogen generation disappeared. For the meaning of symbols, *cf.* Fig. 7.8

Although not phagotrophic, about 200 fungal species (from the "Zygomycota," Ascomycota and Basidiomycota) prey on Metazoa, in general on nematodes. Nematodes are trapped in constricting rings formed by filaments ("hyphae"). Filaments subsequently penetrate the nematode body and digest it from the inside. Predatory fungi play an important role in soils. Such Fungi were already present in the Cretacean, more than 100 Ma ago (Raven et al. 2000; Schmidt et al. 2007).

The lichen mutualistic association could be more ancient as it has seemingly been observed in samples from the 600 Ma-old Chinese deposit of Doushantuo (Yuan et al. 2005). Lichens play an important role in extreme environment and in the initial stages of successions on land. The mycorrhizal mutualism is also very ancient; at least 400 Ma-old (Selosse and Le Tacon 1998). It associates the roots of most Embryophyta (land plants) with filaments from Basidiomycota, Ascomycota, or Glomeromycota (Bruns 2006; James et al. 2006). This association is called "mycorrhiza." Within mycorrhizas, the Fungi provides its partner plant with the following:

1. Mineral nutrients, such as for example nitrogen and phosphorus: This could seem surprising as plant roots possess the (theoretical) capacity to acquire these nutrients by themselves. Compared to plant roots, Fungi display a better scavenging activity toward many soil nutrients and continue to assimilate them when their concentrations become extremely low. Furthermore,

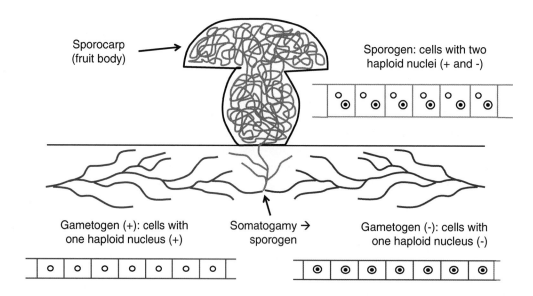

Fig. 7.56 *The two generations (gametogen and sporogen) of a Basidiomycota with sporocarp*. The filaments of cells were enlarged to show the nuclei. The colors, *red* for the gametogen (+), *blue* for the gametogen (−) and *brown* for the sporogen, do not correspond to reality but only to facilitate visualization

Fig. 7.57 *Mycorrhizal mutualism*. Exchanges and interactions between the Fungi and the Embryophyta (green plants: Viridiplantae)

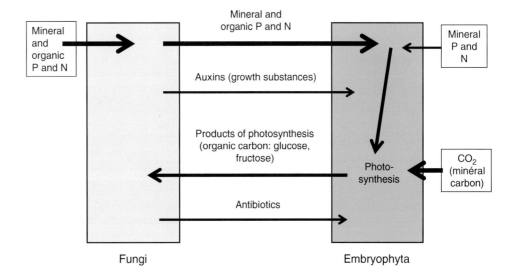

Fungi are capable of using nutrients in organic forms (*e.g.*, amino acids, nucleotides) and of solubilizing nutrients present in insoluble forms. Finally, fungal filaments connected to the roots explore a larger volume of soil compared to the root system alone. As a consequence, plants in undisturbed ecosystems are highly dependent upon their fungal partners for their supply of N and P.

2. Auxins and other growth regulators:
3. By either antibiosis or by efficiently competing with the soil microflora, mycorrhizal fungi protect their host plants against soil borne pathogens.
4. Finally, especially Basidiomycota species protect calcifuge Embryophyta species from an excessive influx of calcium.

In return, mycorrhizal Fungi receive simple sugars, directly derived from the plant photosynthesis. This can represent between 5 and 30 % of the net primary production of the host Embryophyta (Fig. 7.57; Selosse and Le Tacon 1998).

Fungi are also implicated in many other mutualistic associations with, for example, insects (termites, ants).

Saprotrophic Fungi play a major role at the biosphere scale, in terrestrial as well as in marine environments, by degrading organic matter produced by primary producers. As such, they represent an essential link in the global carbon cycle. Individual saprotrophic Fungi can degrade either one or several of the plant cell wall components such as cellulose, hemicelluloses, and also lignin. Numerous Fungi, as exemplified by the brewing yeast *Saccharomyces cerevisiae*, also participate in fermentation (degradation of simple sugars into ethanol and CO_2) and as such play an important economic role in the food industry (production of bread, wine, beer, *etc.*).

Fungal parasites are numerous in all habitats, either terrestrial or aquatic. Plant pathogens have always been a major concern in agriculture and all major plant crop species are affected; they are responsible for several billion dollars of losses every year. More recently, major outbreaks of infections produced by fungal animal pathogens have been reported. The yeast *Candida albicans* (Ascomycota), a common member of healthy human mucosal surface microflora, is also one of the most frequent human fungal pathogens responsible for digestive, urinary, and skin infections. The chytrid *Batrachochytrium dendrobatidis* is a fatal skin pathogen for amphibians which is responsible for global population decline and also species extinctions in this group of vertebrate (Berger et al. 2005). It could have been disseminated worldwide by the export of one of its hosts, the South African toad *Xenopus laevis*, that has been used for several decades in pregnancy tests. Chytrids are also pathogens of plankton cells such as diatoms (Stramenopiles). As such they control algal blooms and it has been suggested that many of the plankton "flagellates" reported in oceanic and lacustrian studies may actually be chytrid reproductive cells (Lefèvre et al. 2007).

7.15 Kingdom Amoebobionta

7.15.1 General Remarks

The members of the kingdom Amoebobionta (or Amoebozoa) owe their name to the fact that in at least one stage of their life cycle, amoeboid cells are present. It is noted however that amoeboid taxa are present in all eukaryotic kingdoms, and it is therefore not a characteristic of the Amoebobionta alone. About 750 species of Amoebobionta have been described (Lecointre and Le Guyader 2006).

The Amoebobionta constitute a monophyletic group of organisms that are closer to Opisthokonta than to Archaeplastida and other eukaryotes (Baldauf and Doolittle 1997; Baldauf et al. 2000; Nicolaev et al. 2004; Eichinger et al. 2005). Amoebobionta and Opisthokonta are included in the group of Unikonta (*cf.* Sect. 7.4), characterized by a single undulipodium (as in Opisthokonta) but located on the front of the cell (as in Bikonta). However, this undulipodium may be secondarily split, in Mycetobionta. Within Amoebobionta, we can distinguish Archamoebae, Mycetobionta, and Lobosa (or Lobosea); the latter seems to be paraphyletic (Fig. 7.58; Lecointre and Le Guyader 2006; Smirnov et al. 2009).

The cell wall, when present, consists of cellulose. Mitochondria, when present, have flattened cristae, as in Opisthokonta.

7.15.2 Archamoeba

The Archamoebae are always unicellular cells with amoeboid shape. Mitochondria have often been lost secondarily. However, mitosomes in *Entamoeba histolytica* are remnants of mitochondria (Embley and Martin 2006). This species has lost many genes but has also gained others by HGT (horizontal gene transfer) from bacteria (Loftus et al. 2005).

Entamoeba histolytica is a parasite of the digestive tract of humans; it causes diarrhea that can be lethal. *Polymyxa palustris* is common on the floor of mangroves and plays an important role in the passage of bacteria to micropredators and from there to the rest of the food web (Liao et al. 2009).

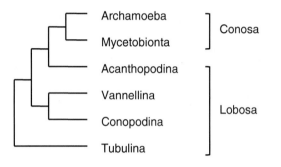

Fig. 7.58 *Phylogenetic tree of Amoebobionta*. Simplified and redesigned from Smirnov et al. (2009)

7.15.3 Mycetobionta

The Mycetobionta (=Mycetozoa) have long been placed in the polyphyletic group of "fungi" (customary meaning), specifically in the Myxomycetes which are also polyphyletic (Chadefaud 1960).

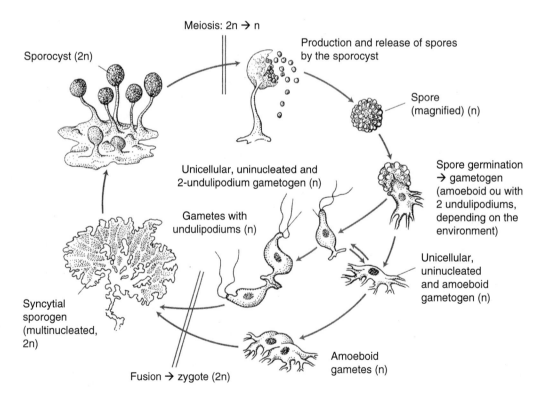

Fig. 7.59 *The life cycle of Fuligo septica (Mycetobionta, Amoebobionta)* (According to Lecointre and Le Guyader (2006) with modified legends. Courtesy of Editions Bélin, France)

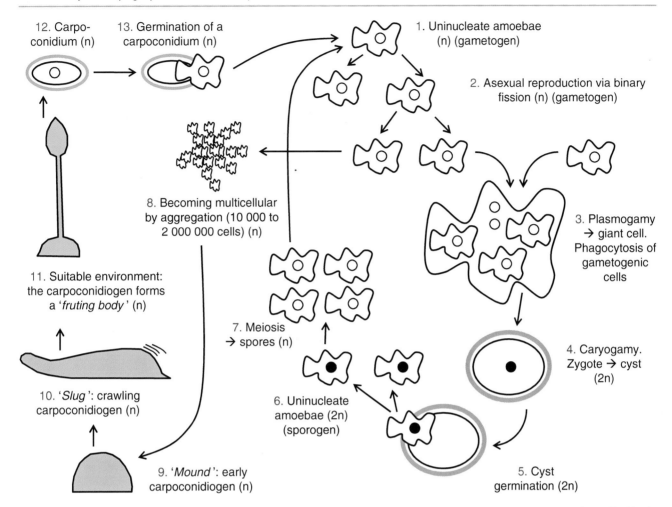

Fig. 7.60 *The life cycle of Dictyostelium discoideum (Mycetobionta, Amoebobionta).* The scale is changed in figures *7.1–7 + 12–13* (the amoeboid cell measuring 10–20 μm), with figure 8 (smaller) and figures *9–11* (represented structures measuring a few mm to cm)

A characteristic of Mycetobionta is the alternation between one or two generations of uninucleated unicellular populations (amoeboid cells or with undulipodiums) and one generation of multinucleated (coenocytic) or multicellular populations. From this point of view, Mycetobionta provide an extraordinary model for the study of the emergence of multicellularity in eukaryotes. The absence of undulipodiums in *Dictyostelium* is a derivative character. The cell wall, when it exists, is made of cellulose (Eichinger et al. 2005).

In *Fuligo septica* ("tin flower"), the life cycle has two generations (digenetic) and two phases (diphasic: haploid and diploid phases). The haploid gametogens are unicellular and uninucleated; depending on environmental conditions where they live, they have the form of amoeboid cells or of cells with two undulipodia (Fig. 7.59). The diploid sporogens are present in the form of a multinucleated coenocyte on which sporocysts form, the site of meiosis, which gives haploid spores (Lecointre and Le Guyader 2006).

In *Dictyostelium discoideum*, the life cycle is more complex. Haploid and uninucleated amoeboid cells constitute the gametogen. In the dark and when the humidity is high, two cells of the opposite sex (+ and −) behave as gametes and merge together (plasmogamy) to give a so-called "giant" cell. This giant cell ingests amoeboid cells, keeps for some time their nucleus, and may appear as multinucleated. Then the two gamete nuclei merge to form a diploid zygote protected by a triple envelope. Zygote germination releases a diploid amoeboid and uninucleated cell (sporogen), the site of meiosis that forms new gametogens. This sporogen – gametogen alternation constitutes the sexual part of the life cycle (Fig. 7.60). When environmental conditions become unfavorable (*e.g.*, nutrient limitation), the asexual part (=carpoconidiogen) of the life cycle takes place and it is this part of the life cycle for which *Dictyostelium* owes its celebrity. For this reason, this taxon and neighboring taxa are called "social amoebae." Under the stimulating effect of a chemical produced by one or more cells, between 10,000 and 2,000,000 cells group together to form a cluster of non-amoeboid cells (mound stage); this multicellular individual has specialized cells and cell regions, and takes the form of a slug (slug stage), able to orient itself relative to light and temperature, crawling on the substrate in search of a favorable zone for fixation. The reorganization of cells and cell regions gives rise to a carpoconidiocyst (=sporocarp, "fruiting body") in which form carpoconidia ("spores") surrounded by an envelope, which are disseminated and can remain dormant for several years, waiting for favorable conditions. Carpoconidiocysts and carpoconidia can also be interpreted as conidiocysts and conidia. During germination, carpoconidia give rise to haploid, amoeboid, and uninucleated cells (gametogen) (Fig. 7.60; Macinnes and Francis 1974; Erdos et al. 1975; Eichinger et al. 2005; Schaap et al. 2006; Gaudet et al. 2008).

In general, Mycetobionta have a preference for continental habitats rich in decaying organic matter. They are heterotrophic and feed by phagocytosis of bacteria, Fungi, and organic particles (Eichinger et al. 2005; Gaudet et al. 2008).

7.16 Conclusion

The current classification of eukaryotes, as it emerges from recent cytological and biochemical approaches and more recently from molecular phylogeny studies, is very different from the traditional classification, more or less derived from the ideas of Linnaeus in the eighteenth century, and those of his successors. However, polyphyletic groups (plants, protozoa, fungi, algae, *etc.*) continue to be used by many authors. These polyphyletic sets have no common character that could define them. "Algae," for example, scattered in seven of the ten taxa of higher order (kingdoms) distinguished here (Fig. 7.1) have biochemical, cytological, and biological characteristics of each of these kingdoms; they cannot either be defined by their morphology as some are unicellular, colonial, coenocytic or multicellular, nor by their ecology, since they are present in all continental and aquatic habitats. We cannot even define them by a set of characters, of which none would be characteristic, but which

Table 7.6 *Some chemical, cytological and biological "markers" of the taxa of eukaryotes described in this chapter*. The discriminant value of markers is not homogeneous. Some are probably characteristics of a taxon. Others are more or less characteristic of a taxon, but exceptions are known, although they have not been reported here. Finally, some markers are shared by several taxa. The marker list is not exhaustive

Taxa	Chemical 'markers'	Cytological 'markers'	Biological 'markers'
Archaeplastida		Mitochondria with plate-like flattened cristae	
Centrohelida		Microtubules radiating from the centroplast	
		Absence of undulipodium	
Glaucocystobionta	Chlorophyll *a* only	Cyanobacterian cell wall maintained between the two chloroplast membranes	
	Phycobilins		
	True starch		
	Cell wall: cellulose	Thylakoids unstacked, with phycobilisomes	
		Phycobilisomes	
Rhodobionta	Chlorophyll *a* only	Thylakoids unstacked, with phycobilisomes	Life cycle trigenetic
	Phycobilins		Trichogamy
	Floridean starch	Pit connections between adjacent cells	Mainly marine
	Cell wall : polymers of ester sulfated galactose (agar-agar, carrageenan and porphyran)	Absence of kinetic apparatus	Well adapted to dim light
	Glycerol combined with sugars → heterosides		
	Low lipid content		
	Bromine metabolism		
	Sterol: cholesterol		
Viridiplantae	Chlorophylls *a* and *b* (*c*)	Thylakoids stacked	Types of fertilization varied: planogamy, cystogamy, siphonogamy
		Long and short thylakoids	
	True starch	Amyloplasts	
	Cell wall: cellulose, lignin, pectin, etc.	Polysaccharides (starch): intraplastidial	
	Sterol: β sitosterol (in Embryophyta)	2–30 undulipodiums naked and identical	
		Absence of a dynein arm on one of the microtubule doublets of the undulipodiums	

(continued)

Table 7.6 (continued)

Taxa	Chemical 'markers'	Cytological 'markers'	Biological 'markers'
Rhizaria			
Radiolaria		Internal siliceous skeleton	
		Pseudopodia (axopods and filopods)	
Chlorarachniobionta	Chlorophylls *a* and *b*	Meroplasmodium (anastomosed pseudopodia between neiboring individuals)	
	Paramylon	Nucleomorph	
		Undulipodium wound helically around the cell	
Phytomyxea		Coenocyte	
		Cruciform nuclear division	
		2 undulipodiums naked of unequal length	
Foraminifera		Test with chambers	Fertilization: planogamy
		Pseudopodia (filopodes)	
		2–4 undulipodiums naked, of inequal length, ending in a single mastigoneme	Life cycle: alternation between uninucleated gametogen and multinucleated sporogen
Alveolata		Mitochondria with tubular cristae	
		Alveoli under the plasmalemma	
Ciliophora		Undulipodiums very numerous (cilia) and/or bundles of cilia (cirri)	Fertilization: cystogamy
		Micronucleus and macronucleus	Codons 'stop' ('universal' code) code for glutamine, cysteine or triptophan
		Extrusomes (trichocysts)	
Dinobionta	Alveoli: cellulose	Chloroplast with 3–4 membranes	Fertilization: planogamy
	Dinamylon	2 undulipodiums inserted in perpendicular grooves	
	Wall of the cysts: dinosporine		Life cycle: monogenetic
	Nucleus lacking histones	Unipartite mastigonemes on the transverse undulipodium	Chromosomes condensed throughout interphase
		Giant nucleus	
		Extrusomes (trichocysts)	
Apicomplexa		Apicoplast (relict of a former chloroplast)	Fertilization: fucogamy
		Apical complex	Life cycle: digenetic with alternation of 2 hosts
		Centriole: 9 singlet microtubules	
Stramenopiles		Mitochondria with tubular cristae	
		Tripartite mastigonemes on one of the undulipodiums, the other naked	
Oobionta	Chrysolaminarin	Vegetative apparatus : coenocyte	Fertilization : siphonogamy
	Cell wall: cellulose		
	Sterols: absent		Life cycle: monogenetic
Chromobionta	Chlorophylls *a* and *c*	Chloroplast: 4-membrane envelope. 2-membrane outer envelope merged with the nuclear envelope → nucleoplastidial complex (NPC)	Fertilization: planogamy, cystogamy, fucogamy, oogamy
	Laminarin ou chrysolaminarin, mannitol		
	Cell wall: alginic acid, silica	Thylakoids stacked intro 3-thylakoid lamellae	Life cycle : monogenetic or digenetic
	Sterol: fucosterol	Physodes	
Haptobionta	Chlorophylls *a* and *c*	Chloroplast: 4-membrane envelope. 2-membrane outer envelope merged with the nuclear envelope → nucleoplastidial complex (NPC)	Life cycle: digenetic (optionally trigenetic)
	Laminarin		Major role in DMSP production
		Thylakoids stacked intro 3-thylakoid lamellae	
		Peripheral envelope of endoplasmic reticulum (PER)	
		Mitochondria with tubular cristae	
		Haptonema	
		Scales, sometimes giant and calcareous, on the cell	

(continued)

Table 7.6 (continued)

Taxa	Chemical 'markers'	Cytological 'markers'	Biological 'markers'
Cryptobionta	Chlorophylls *a* and *c*	Chloroplast: 4-membrane envelope. 2-membrane outer envelope merged with the nuclear envelope → nucleoplastidial complex (NPC)	
Cryptophytes	Phycobilins within thylakoids		
	Xanthophyll: alloxanthin		
	Floridean starch	Thylakoids stacked intro 2-thylakoid lamellae	
		Nucleomorph	
		Mitochondria with plate-like flattened cristae	
Discicristates		Mitochondria with 'ping-pong racket' cristae	
Euglenoidea	Chlorophylls *a* and *b*	Chloroplast: 3-membrane envelope	
	Paramylon	Thylakoids stacked intro 3-thylakoid lamellae	
		Extraplastidial stigma	
		Pellicle (proteinaceous strips helically arranged, under the plasmalemma)	
		Mastigonemes : a single spiraled row	
Kinetoplastida		Single giant mitochondrion	
		Kinetoplast within the mitochondrion	
Excavates		Absence of mitochondria (secondary loss)	
		Mitosomes or hydrogenosomes	
		Ventral groove (feeding role)	
		Nucleus associated with a group of undulipodiums	
Parabasalia		Parabasal apparatus	Meiosis with a single division?
		Undulipodiums grouped by 4 ('kinetid')	
Diplomonadida		2 diploid nuclei per cell	
		Golgi apparatus: absent	
Opisthokonta	Chitin	Mitochondria with flattened cristae	
		Single undulipodium, at the rear of the cell	
Microsporidia (Fungi)		Absence of mitochondria (secondary loss)	
		Absence of kinetic apparatus	
		Cytoplasmic ribosomes 70S	
		Polar tube	
Other fungi (modern meaning)	Sterol: ergosterol	Absence of kinetic apparatus (except in Chytridiomycota)	(in Asco-, Basidio- and Glomeromycota)
		Dolipores between cells	Fertilization: trichogamy or somatogamy
		Absence of Golgi apparatus	Life cycle (ancestral) trigenetic
			Presence of a dikaryotic phase
Amoebobionta	Cellulose	Single anterior undulipodium	
Archamoeba		Absence of mitochondria (secondary loss)	
Mycetobionta			Alternation between an unicellular generation and a coenocytic or multicellular generation

in combination would be unique. The same goes for "protozoa", "fungi" (customary meaning), and "plants".

Extracting information from old, or even recent literature, to replace the current framework is therefore a difficult task. The fact that specialists in each discipline (botanists, zoologists), in each sub-discipline (mycologists, phycologists, *etc.*) of certain families, sometimes of a single genus (*e.g., Euglena* and *Dictyostelium*), have forged their own jargon, makes it difficult to detect possible homologies. This is further complicated by the fact that, from one discipline to another, the same term (spore, endospore, *etc.*) is used for non-homologous structures. Finally, the information available for a taxon, for example the nature of synthesized sterol or the form of mitochondrial cristae, may not be available for another taxon.

It is impossible to summarize the extraordinary diversity of unicellular eukaryotes in ~70 pages. Hence this presentation is necessarily highly simplified. On one hand, many taxa of the sub-kingdom level have not been described here (*e.g.,* Haplosporidia, Labyrinthulobionta, Jakobida,

Retortamonadida, choanoflagellates, *etc.*). On the other hand, for the taxa considered, the description is deliberately simplified and does not mention the many special cases and exceptions. For example, when it is stated that the life cycle of Rhodobionta is trigenetic (*cf.* Sect. 7.5.4 and Table 7.6), it is because this is probably the ancestral cycle and because it is widespread; the description of life cycles derived by modification or simplification from the trigeneric cycle would be beyond the scope of this book. Similarly, in Mycetobionta, the diversity and complexity of life cycles is not limited to the two selected examples, *Dictyostelium* and *Fuligo*.

In Table 7.6, an attempt is made to present in a synthetic manner a number of chemical, cytological, and biological characteristics that are unique to a taxon, or that are shared by only a few taxa, so that they can serve as "markers" of these taxa.

References

Abrahamsen MS, Templeton TJ, Enomoto S, Abrahante JE, Zhu G, Lancto CA et al (2004) Complete genome sequence of the Apicomplexan, *Cryptosporidium parvum*. Science 304:441–445

Andersen RA (2004) Biology and systematics of heterokont and haptophyte algae. Am J Bot 91:1508–1522

Andersen RA (2005) Algae and the vitamin mosaic. Nature 438:33–34

Baker BJ, Tyson GW, Goosherst L, Banfield JF (2009) Insights into the diversity of eukaryotes in acid mine drainage biofilm communities. Appl Environ Microbiol 75(7):2192–2199

Baldan B, Andolfo P, Navazio L, Tolomio C, Mariani P (2001) Cellulose in algal cell wall: an '*in situ*' localization. Eur J Histochem 45:51–56

Baldauf SL (2003) The deep roots of eucaryotes. Science 300:1703–1706

Baldauf SL (2008) An overview of the phylogeny and diversity of eukaryotes. J Syst Evol 46:263–273

Baldauf SL, Doolittle WF (1997) Origin and evolution of the slime molds (Mycetozoa). Proc Natl Acad Sci U S A 94:12007–12012

Baldauf SL, Roger AJ, Wenk-Siefret I, Doolittle WF (2000) A kingdom-level phylogeny of eukaryotes based on combined protein data. Science 290:972–977

Barr DJS, Désaulniers NL (1989) The flagellar apparatus of the oomycetes and hyphochytriomycetes. In: Green JC, Leadbeater BSC, Diver WL (eds) The chromophyte algae. Problems and perspectives. Clarendon, London, pp 343–355

Beakes GW (1989) Oomycete fungi: their phylogeny and relationship to chromophyte algae. In: Green JC, Leadbeater BSC, Diver WL (eds) The chromophyte algae. Problems and perspectives. Clarendon, London, pp 325–342

Berge T, Hansen PJ, Moestrup Ø (2008) Feeding mechanism, prey specificity and growth in light and dark of the plastidic dinoflagellate *Karlodinium armiger*. Aquat Microb Ecol 50:279–288

Berger L, Hyatt AD, Speare R, Longcore JE (2005) Life cycle stages of the amphibian chytrid *Batrachochytrium dendrobatidis*. Dis Aqua Org 68:51–63

Bert JJ, Dauguet JC, Maume D, Bert M (1991) Recherche des acides gras et des stérols chez deux Rhodophycées: *Calliblepharis jubata* et *Solieria chordalis* (Gigartinales). Cryptogamie Algol 12:157–162

Bhattacharya D, Yoon HS, Hackett JD (2003) Photosynthetic eukaryotes unite: endosymbiosis connects the dots. Bioessays 25:50–60

Bird DF, Kalff J (1986) Bacterial grazing by planktonic lake algae. Science 231:493–494

Bittner L et al (2013) Diversity patterns of uncultured Haptophytes unravelled by pyrosequencing in Naples Bay. Mol Ecol 22:87–101

Bjornland T, Liaaen-Jensen S (1989) Distribution patterns of carotenoids in relation to chromophyte phylogeny and systematics. In: Green JC, Leadbeater BSC, Diver WL (eds) The chromophyte algae. Problems and perspectives. Clarendon, London, pp 37–61

Bodyl A, Stiller JW, Mackiewicz P (2009) Chromalveolate plastids: direct descent or multiple endosymbioses? Trends Ecol Evol 24:119–121

Boerjan W, Ralph J, Baucher M (2003) Lignin biosynthesis. Annu Rev Plant Biol 54:519–546

Bornens M, Azimzadeh J (2007) Origin and evolution of the centrosome. In: Jékeli G (ed) Origins and evolution of eukaryotic endomembranes and cytoskeleton. Landes Bioscience, Austin, pp 119–129

Bouchet P (2000) L'insaisissable inventaire des espèces. La Recherche 333:40–45

Boudouresque CF (2011) Taxonomie et phylogénie des Eucaryotes unicellulaires. In: Bertrand JC, Caumette P, Lebaron P, Matheron R, Normand P (eds) Ecologie microbienne. Microbiologie des milieux naturels et anthropisés. Presses Universitaires de Pau et des Pays de l'Adour (PUPPA), Pau, pp 203–260

Boudouresque CF, Ruitton S, Verlaque M (2006) Anthropogenic impacts on marine vegetation in the Mediterranean. In: Proceedings of the second Mediterranean symposium on marine vegetation, Athens, 12–13 Dec 2003. Regional Activity Centre for Specially Protected Areas Publ, Tunis, pp 34–54

Braselton JP (1995) Current status of the plasmodiophorids. Crit Rev Microbiol 21:263–275

Braselton JP (2009) Plasmodiophorid home page. http://oak.cats.ohiou.edu/~braselto/plasmos

Brodie J, Maggs CA, John DM (eds) (2007) Green seaweeds of Britain and Ireland. British Phycological Society, UK

Brown MW, Kolisko M, Silberman M, Roger AJ (2012) Aggregative multicellularity evolved independently in the eukaryotic supergroup Rhizaria. Curr Biol 22(12):1123–1127

Bruns T (2006) A kingdom revised. Nature 443:758–760

Bulman SR, Kühn SF, Marshall JW, Schnepf E (2001) A phylogenetic analysis of the SSU rRNA from members of the Plasmodiophorida and Phagomyxida. Protist 152:43–51

Burki F, Okamoto N, Pombert JF, Keeling PJ (2012) The evolutionary history of haptophytes and cryptophytes: phylogenomic evidence for separate origins. Proc Roy Soc B 279:2246–2254

Butterfield NJ (2000) *Bangiomorpha pubescens* n. gen., n. sp.: implications for the evolution of sex, multicellularity, and the Mesoproterozoic/Neoproterozoic radiation of eukaryotes. Paleobiology 26:386–404

Calderon-Saenz E, Schnetter R (1989) Morphology, biology and systematics of *Cryptochlora perforans* (Chlorarachniophyta), a phagotrophic marine alga. Plant Syst Evol 163:165–176

Cavalier-Smith T (1981) Eukaryote kingdoms: seven or nine? BioSystems 14:461–481

Cavalier-Smith T (1987a) Eukaryote cell evolution. In: Greuter W, Zimmer B (eds) Proceedings of the XIV international botanical congress, Berlin, 24 Jul–1 Aug 1987. Koeltz, Koenigstein, pp 203–223

Cavalier-Smith T (1987b) The origin of fungi and pseudofungi. In: Rayner ADM (ed) Evolutionary biology of fungi. Cambridge University Press, Cambridge, pp 339–353

Cavalier-Smith T (2002) The phagotrophic origin of eukaryotes and phylogenetic classification of protozoa. Int J Syst Evol Microbiol 52:295–354

Cavalier-Smith T, Chao EE (2002) Molecular phylogeny of centrohelid Heliozoa, a novel lineage of bikont eukaryotes that arose by ciliary loss. J Mol Evol 56:387–396

Cavalier-Smith T, Chao EEY (2003) Phylogeny and classification of phylum Cercozoa (Protozoa). Protist 154:341–358

Cavalier-Smith T, von der Heyden S (2007) Molecular phylogeny, scale evolution and taxonomy of centrohelid Heliozoa. Mol Phylogenet Evol 44:1186–1203

Chadefaud M (1960) Les végétaux non vasculaires (cryptogamie). In: Chadefaud M, Emberger L (eds) Traité de botanique systématique, Tome 1. Masson & Cie, Paris, pp i–xv + 1–1018

Chadefaud M (1978) Les champignons. In: Des Abbayes H, Chadefaud M, Feldmann J, De Ferré Y, Gaussen H, Grassé PP, Prévot AR (eds) Précis de botanique, Tome 1. Végétaux inférieurs. Deuxième édition. Masson & Cie, Paris, pp 321–518

Chauzat MP, Higes M, Martin-Hernandez R, Meana A, Cougoule N, Faucon JP (2007) Presence of Nosema ceranae in French honey bee colonies. J Apicul Res 45:127–128

Courties C, Perasso R, Chétiennot-Dinet MJ, Gouy M, Guillou L, Trousselier M (1998) Phylogenetic analysis and genome size of Ostreococcus tauri (Chlorophyta, Prasinophyceae). J Phycol 34:844–849

Croft MT, Lawrence AD, Raux-Deery E, Warren MJ, Smith AG (2005) Algae acquire vitamin B12 through a symbiotic relationship with bacteria. Nature 438:90–93

Cuvelier ML et al (2008) Widespread distribution of a unique marine protistan lineage. Environ Microbiol 10:1621–1634

Dacks J, Roger AJ (1999) The first sexual lineage and the relevance of facultative sex. J Mol Evol 48:779–783

de Reviers B (2003) Biologie et phylogénie des algues, Tome 2. Belin, Paris

Drebes G (1981) Possible resting spores of Dissodinium pseudolunula (Dinophyta) and their relation to other taxa. Br Phycol Bull 16:207–215

Dyall SD, Brown MT, Johnson PJ (2004) Ancient invasions: from endosymbionts to organelles. Science 304:253–257

Dyer PS (2008) Evolutionary biology: microsporidia sex – a missing link to Fungi. Curr Biol 18:R1012–R1014

Eichinger L et al (2005) The genome of the social amoeba Dictyostelium discoideum. Nature 435:43–57

Embley TM, Martin W (2006) Eukaryotic evolution, changes and challenges. Nature 440:623–630

Erdos GW, Raper KB, Vogen LK (1975) Sexuality in the cellular slime mold Dictyostelium giganteum. Proc Natl Acad Sci U S A 72:970–973

Evert RF, Eichhorn SE (2013) Raven biology of plants, 8th edn. W.F. Freeman Publishers

Falkowski PG, Katz ME, Knoll AH, Quigg A, Raven JA, Schofield O, Taylor FJR (2004) The evolution of modern eukaryotic phytoplankton. Science 305:354–360

Feldmann J, Feldmann G (1945) Sur le metabolism du glycérol chez les Rhodophycées. C R Acad Sci 220:467–469

Feldmann J, Feldmann G (1955) Observations sur quelques Phycomycètes marins nouveaux ou peu connus. Rev Mycol 20:231–251

Fistarol GO, Legrand C, Granéli E (2003) Allelopathic effect of Prymnesium parvum on a natural plankton community. Mar Ecol Prog Ser 255:115–125

Foster RA, Carpenter EJ, Gergman B (2006) Unicellular cyanobionts in open ocean dinoflagellates, radiolarians and tintinnids: ultrastructural characterization and immuno-localization of phycoerythrin and nitrogenase. J Phycol 42:458–463

Fuller NJ, Campbell C, Allen DJ, Pitt FD, Zwirglmaier K, Le Gall F, Vaulot D, Scanlan DJ (2006) Analysis of photosynthetic picoeukaryote diversity at open ocean sites in the Arabian Sea using a PCR biased towards marine algal plastids. Aquat Microb Ecol 43:79–93

Gaudet P, Williams JG, Fey P, Chisholm RL (2008) An anatomy ontology to represent biological knowledge in Dictyostelium discoideum. BMC Genom 9:1–12

Gazzaniga M (2009) Aquariofilia e microscopia ottica. http://www.acquariofiliaemicroscopia.it

Genovesi-Giunti B (2006) Initiation, maintien et récurrence des efflorescences toxiques d'Alexandrium catenella (Dinophyceae) dans une lagune méditerranéenne (Thau, France): rôle du kyste dormant. Thèse Doct, Univ Montpellier II

Germot A, Philippe H, Le Guyader H (1997) Evidence for loss of mitochondria in microsporidia from a mitochondrial- type HSP70 in Nosema locustae. Mol Biochem Parasitol 87:159–168

Gray P (1995) L'Irlande au temps de la grande famine. Gallimard, Paris

Gray AP, Lucas IAN, Seed R, Richardson CA (1999) Mytilus edulis chilensis infested with Coccomyxa parasitica (Chlorococcales, Coccomyxaceae). J Mollus Stud 65:289–294

Green JC, Course PA, Tarran GA (1996) The life cycle of Emiliana huxleyi: a brief review and a study of relative ploidy levels analysed by flow cytometry. J Mar Syst 9:33–44

Gretz MR, Sommerfeld MR, Aronson JM (1982) Cell wall composition of the generic phase of Bangia atropurpurea (Rhodophyta). Bot Mar 25:529–535

Gretz MR, Aronson JM, Sommerfeld MR (1984) Taxonomic significance of cellulosic cell walls in the Bangiales (Rhodophyta). Phytochemistry 23:2513–2514

Groisillier A, Massana R, Valentin K, Vaulot D, Guillou L (2006) Genetic diversity and habitats of two enigmatic alveolate lineages. Aquat Microb Ecol 42:277–291

Grzymski J, Schofield OM, Falkowski PG, Bernhard JM (2002) The function of plastids in the deep-sea benthic foraminifer, Nonionella stella. Limnol Oceanogr 47:1569–1580

Gülkiz Şenler N, Yildiz I (2003) Infraciliature and other morphological characteristics of Enchelyodon longikineta n.sp. (Ciliophora, Haptoria). Eur J Protistol 39:267–274

Hampl V, Hug L, Leigh JW, Dacks JB, Lang BF, Simpson AGB (2009) Phylogenetic analyses support the monophyly of Excavata and resolve relationships among eukaryotic "supergroups". Proc Natl Acad Sci U S A 106:3859–3864

Hansen PJ, Fenchel T (2006) The bloom-forming ciliate Mesodinium rubrum harbours a single permanent endosymbiont. Mar Biol Res 2:169–177

Hawksworth DL (2001) The magnitude of fungal diversity: the 1.5 million species estimate revisited. Mycol Res 105:1422–1432

Herzog M, Von Boletzky S, Soyer MO (1984) Ultrastructural and biochemical nuclear aspects of Eukaryote classification: independent evolution of the dinoflagellates as a sister group of the actual Eukaryotes? Orig Life 13:205–215

Hibbett DS, Binder M, Bischoff JF, Blackwell M, Cannon PF et al (2007) A higher-level phylogenetic classification of the Fungi. Mycol Res 111:509–547

Horn S, Ehlers K, Fritzsch G, Gil-Rodriguez MC, Wilhelm C, Schnetter R (2007) Synchroma grande sp. nov. (Synchromophyceae class. nov., Heterokontophyta): an amoeboid marine alga with unique plastid complexes. Protist 158:277–293

Huang J, Xu Y, Gogarten JP (2005) The presence of a holoarchaeal type tyrosyl-tRNA synthetase marks the opisthokonts as monophyletic. Mol Biol Evol 22:2142–2146

Inouye I (1993) Flagella and flagellar apparatuses of algae. In: Berner T (ed) Ultrastructure of microalgae. CRC Press, Boca Raton, pp 99–133

James TY, Kauff F, Schoch CL, Matheny PB, Hoffstetter V, Cox CJ et al (2006) Reconstructing the early evolution of Fungi using a six-gene phylogeny. Nature 443:818–822

Jeffrey SW (1989) Chlorophyll c pigments and their distribution in the chromophyte algae. In: Green JC, Leadbeater BSC, Diver WL (eds) The chromophyte algae. Problems and perspectives. Clarendon, London, pp 13–36

Jousson O, Di Bello D, Donadio E, Felicioli A, Pretti C (2007) Differential expression of cysteine proteases in developmental stages of the parasitic ciliate Ichtyophthirius multifilis. FEMS Microbiol Lett 269:77–84

Kawai H, Maeba S, Sasaki H, Okuda K, Henry EC (2003) *Schizocladia ischiensis*: a new filamentous marine chromophyte belonging to a new class, Schizocladiophyceae. Protist 154:211–228

Keeling PJ (2008) Bridge over troublesome plastids. Nature 451:896–897

Kim OTP, Yura K, Go N, Harumoto T (2005) Newly sequenced eRF1s from ciliates: the diversity of stop codon usage and the molecular surfaces that are important for stop codon interactions. Gene 346:277–286

Kim E, Harrison JW, Sudek S, Jones MDM, Wilcox HM, Richards TA, Worden AZ, Archibald JM (2011a) Newly identified and diverse plastid-bearing branch on the eukaryotic tree of life. Proc Natl Acad Sci 108(4):1496–1500

Kim E, Harrison JW, Sudek S, Jones MD, Wilcox HM, Richards TA, Worden AZ, Archibald JM (2011b) Newly identified and diverse plastid-bearing branch on the eukaryotic tree of life. Proc Natl Acad Sci USA 108:1496–1500

Kling SA, Boltovskoy D (2002) What are radiolarians? http://www.radiolarian.org

Kokinos JP, Eglinton TI, Goni MA, Boon JA, Martoglio PA, Anderson DM (1998) Characterization of a highly resistant biomacromolecular material in the cell wall of a marine dinoflagellate cyst. Org Geochem 28:265–288

Kramarsky-Winter E, Harel M, Siboni N, Ben Dov E, Brickner I, Loya Y, Kushmaro A (2006) Identification of a protist-coral association and its possible ecological role. Mar Ecol Prog Ser 317:67–73

Kroken S, Taylor JW (2000) Phylogenetic species, reproductive mode, and specificity of the green alga *Trebouxia* forming lichens with the fungal genus *Lethraria*. Bryologist 103:645–660

Lafferty KD (2006) Can the common brain parasite, *Toxoplasma gondii*, influence human culture? Proc Roy Soc Lond B 273:2749–2755

Lara E, Moreira D, Vereshchaka A, López-García P (2009) Pan-oceanic distribution of new highly diverse clades of deep-se diplonemids. Environ Microbiol 11(1):47–65

Larsson R (2009) Cytology and taxonomy of the microsporidia. http://www.cob.lu.se/microsporidia

Lass-Flörl C, Mayr A (2007) Human prototothecosis. Clin Microbiol Rev 20(230):242

Leadbeater BSC (1989) The phylogenetic significance of flagellar hairs in the Chromophyta. In: Green JC, Leadbeater BSC, Diver WL (eds) The chromophyte algae. Problems and perspectives. Clarendon, London, pp 145–165

Leander BS, Saldarriaga JF, Keeling PJ (2002) Surface morphology of the marine parasite *Haplozoon axiothellae* Siebert (Dinoflagellata). Eur J Protistol 38:287–297

Le Calvez J (1953) Ordre des Foraminifères. In: Grassé PP (ed) Traité de zoologie. Anatomie, systématique, biologie, Tome I. Protozoaires: Rhizopodes, Actinopodes, Sporozoaires, Cnidosporidies. Fascicule II. Masson & Cie, Paris, pp 149–265

Lecointre G, Le Guyader H (2006) Classification phylogénétique du vivant, 3rd edn. Belin, Paris

Lecroq B, Lejzerowicz F, Bachar D, Christen R, Esling P, Baerlocher L, Østeras M, Farinelli L, Pawlowski J (2011) Ultra-deep sequencing of foraminiferal microbarcodes unveils hidden richness of early monothalamous lineages in deep-sea sediments. Proc Natl Acad Sci USA 108:13 177–13 182

Lee SC, Corradi N, Byrnes EJ III, Torres-Martinez S, Dietrich FS, Keeling PJ, Heitman J (2008) Microsporidia evolved from ancestral sexual Fungi. Curr Biol 18:1675–1679

Lefèvre E, Bardot C, Noël C, Carrias JF, Viscogliosi E, Amblard C, Sime-Ngando T (2007) Unveiling fungal zooflagellates as members of freshwater pico-eucaryotes: evidence from a molecular diversity study in a deep meromictic lake. Environ Microbiol 9:61–71

Lejzerowicz F, Pawlowski J, Fraissinet-Tachet L, Marmeisse R (2010) Molecular evidence for widespread occurrence of Foraminifera in soils. Environ Microbiol 12(9):2518–2526

Lejzerowicz F, Esling P, Majewski W, Szczucinski W, Decelle J, Obadia C, Martinez Arbizu P, Pawlowski J (2013) Ancient DNA complements microfossil record in deep-sea subsurface sediments. Biol Lett (in press)

Leliaert F, Smith DR, Moreau H, Herron MD, Verbruggen H, Delwiche CF, De Clerck O (2012) Phylogeny and molecular evolution of the green algae. Crit Rev Plant Sci 31:1–46

Lemieux C, Otis C, Turmel M (2000) Ancestral chloroplast genome in *Mesostigma viride* reveals an early branch of green plant evolution. Nature 403:649–652

Liao QY, Li J, Zhang JH, Li M, Lu Y, Xu RL (2009) An ecological analysis of soil sarcodina at Dongzhaigang mangrove in Hainan Island, China. Eur J Soil Biol 45:214–219

Lindberg K, Moestrup O, Daugbjerg N (2005) Studies on wolozynskioid dinoflagellates I:*Wolozynskia coronate* re-examined using light and electron microscopy and partial LSU rDNA sequences, with description of *Tovellia* gen. nov. and *Jadwigia* gen. nov. (Tovelliaceae fam. nov.). Phycologia 44:416–440

Litaker RW, Vandersea RW, Faust MA, Kibler SR, Chinain M, Holmes MJ, Holland WC, Tester PA (2009) Taxonomy of *Gambierdiscus* including four new species, *Gambierdiscus caribaeus, Gambierdiscus carolinianus, Gambierdiscus carpenteri* and *Gambierdiscus ruetzleri* (Gonyaulacales, Dinophyceae). Phycologia 48(5):344–390

Littler MM, Littler DS, Blair SM, Norris JN (1985) Deepest known plant life discovered on an uncharted seamount. Science 227:57–59

Liu YJ, Hodson MC, Hall BD (2006) Loss of the flagellum happened only once in the fungal lineage: phylogenetic structure of kingdom Fungi inferred from RNA polymerase II subunit genes. BMC Evol Biol 6:1–13

Loftus B, Anderson I, Davies RU, Alsmark CM, Samuelson J, Amedeo P et al (2005) The genome of the protist parasite *Entamoeba histolytica*. Nature 433:865–868

Logares R, Audic S, Santini S, Pernice MC, de Vargas C, Massana R (2012) Diversity patterns and activity of uncultured marine heterotrophic flagellates unveiled with pyrosequencing. ISME J 6:1823–1833

Logsdon JM Jr (2007) Evolutionary genetics: sex happens in *Giardia*. Curr Biol 18:R66–R68

López-García P, Philippe H, Gail F, Moreira D (2003) Autochthonous eukaryotic diversity in hydrothermal sediment and experimental microcolonizers at the Mid-Atlantic Ridge. Proc Natl Acad Sci U S A 100:697–702

Macinnes MA, Francis D (1974) Meiosis in *Dictyostelium mucoroides*. Nature 251:321–324

Maldonado M, López-Acosta M, Sitjà C, Aguilar R, García S, Vacelet J (2013) A giant foraminifer that converges to the feeding strategy of carnivorous sponges: *Spiculosiphon oceana* sp. nov. (Foraminifera, Astrorhizida). Zootaxa 3669(4):571–584

Malik SB, Pightling AW, Stefaniak LM, Schurko AM, Logsdon JM Jr (2008) An expanded inventory of conserved meiotic genes provides evidence for sex in *Trichomonas vaginalis*. PLoS One 3:1–13

Margulis L, Dolan MF (1997) Swimming against the current. In: Margulis L, Sagan D (eds) Slanted truths. Copernicus Publ, New York, pp 47–58

Martone PT, Estevez JM, Lu F, Ruel K, Denny MW, Somerville C, Ralph J (2009) Discovery of lignin in seaweed reveals convergent evolution of cell-wall architecture. Curr Biol 19:169–175

Massana R, Terrado R, Forn I, Lovejoy C, Pedrós-Alió C (2006) Distribution and abundance of uncultured heterotrophic flagellates in the world oceans. Environ Microbiol 8:1515–1522

May R (1997) L'inventaire des espèces vivantes. In: *L'évolution*. Dossier Hors-série *Pour la Science,* pp 40–47

McDonald SM, Sarno D, Scanlan DJ, Zingone A (2007) Genetic diversity of eukaryotic ultraphytoplankton in the Gulf of Naples during an annual cycle. Aquat Microb Ecol 50:75–89

McFadden GI, Waller RF (1997) Plastids in parasites of humans. Bioessays 19:1033–1040

McFadden GI, Waller RF, Ralph SA, Foth B, Tonkin C, Su V et al (2001) The relict plastid of malaria parasites. Phycologia 40(4 suppl):16–17

McInnes AG, Ragan MA, Smith DG, Walter JA (1984) High-molecular-weight phloroglucinol based tannins from brown algae: structural variants. Hydrobiologia 116–117:597–602

McLean RO, Corrigan J, Webster J (1981) Heterotrophic nutrition in *Melosira nummuloides*, a possible role in affecting distribution in the Clyde Estuary. Br Phycol J 16:95–106

Medlin LK, Kaczmarska I (2004) Evolution of the diatoms: V. Morphological and cytological support for the major clades and a taxonomic revision. Phycologia 43:245–270

Mereschkowsky C (1910) Theorie der zwei Plasmaarten als Grundlage der Symbiogenesis, einer neuen Lehre von der Entstehung der Organismen. Biol Zentralbl 30:278–367

Meyer-Harms B, Pollehne F (1998) Alloxanthin in *Dinophysis norvegica* (Dinophysiales, Dinophyceae) from the Baltic Sea. J Phycol 34:280–285

Miller JJ, Delwiche CF, Coats DW (2012) Ultrastructure of *Amoebophrya* sp. and its change during the course of infection. Protist 163:720–745

Molds W (2009) Introduction to Oomycota. http://www.ucmp.berkeley.edu/chromista/oomycota.html

Monier A, Pagarete A, De Vargas C, Allen MJ, Read B, Claverie JM, Ogata H (2009) Horizontal gene transfer of an entire metabolic pathway between a eukaryotic alga and its DNA virus. Genome Res 19:1441–1449

Moore RB, Obornik M, Janouškovec J, Chrudimsky T, Vancová M, Green DH et al (2008) A photosynthetic alveolate closely related to Apicomplexan parasites. Nature 451:959–963

Mora C, Tittensor DP, Adl S, Simpson AGB, Worm B (2011) How many species are there on earth and in the ocean? PLoS Biol 9(8):1–8

Moreira D, Von Der Heyden S, Bass D, López-García P, Chao E, Cavalier-Smith T (2007) Global eukaryote phylogeny: combined small- and large subunit ribosomal DNA trees support monophyly of Rhizaria, Retaria and Excavata. Mol Phylogen Evol 55:255–266

Morrison HG, McArthur AG, Gillin FD, Aley SB, Adam RD, Olsen GJ et al (2007) Genomic minimalism in the early diverging intestinal parasite *Giardia lamblia*. Science 317:1921–1926

Motte J (1971) Le biocycle. Introduction à l'étude des grands groupes végétaux. Opuscula Botanica, Montpellier, 10:1–253 + 23 plates

Mueller GM, Schmit JP (2007) Fungal biodiversity: what do we know? What can we predict? Biodivers Conserv 16:1–5

Müller DG, Maier I, Gassman G (1985) Survey on sexual pheromone specificity in Laminariales (Phaeophyceae). Phycologia 24:475–477

Nakayama T, Yoshida M, Noel MH, Kawashi M, Inouye I (2005) Ultrastructure and phylogenetic position of *Chrysoculter rhomboideus* gen. et sp. nov. (Prymnesiophyceae), a new flagellate haptophyte from Japanese coastal waters. Phycologia 44:369–383

Neustupa J, Němková Y, Veselá J, Steinová J, Škaloud P (2013) *Parachloroidium* gen. nov. (Trebouxiophyceae, Chlorophyta), a novel genus of coccoid green algae from subaerial corticolous biofilms. Phycologia 52(5):411–421

Nicolaev SI, Berney C, Fahrni JF, Bolivar I, Polet S, Mylnikov AP, Aleshin VV, Petrov NB, Pawlowski J (2004) The twilight of Heliozoa and rise of Rhizaria, an emerging supergroup of amoeboid eukaryotes. Proc Natl Acad Sci U S A 101(21):8066–8071

Noda S et al (2006) Identification and characterization of ectosymbionts of distinct lineages in *Bacteroidales* attached to flagellated protists in the gut of termites and a wood-feeding cockroach. Environ Microbiol 8:11–20

Not F et al (2007) Picobiliphytes: a marine picoplanctonic algal group with unknown affinities to other Eukaryotes. Science 315:253–255

Nozaki H, Maruyama S, Matsuzaki M, Nakada T, Kato S, Misawa K (2009) Phylogenetic position of Glaucophyta, green plants (Archaeplastida) and Haptophyta (Chromalveolata) as deduced from slowly evolving nuclear genes. Mol Phylogenet Evol 55:872–880

Obornik M, Modrý D, Lukeš M, Cernotíková-Stříbrná E, Cihlář J, Tesařová M, Kotabová E, Vancová M, Prášil O, Lukeš J (2012) Morphology, ultrastructure and life cycle of *Vitrella brassicaformis* n. sp., n. gen., a novel chromerid from the Great Barrier Reef. Protist 163(2):306–323

Okamoto N, Inouye I (2005) The Katablepharids are a distant sister group of the Cryptophyta: a proposal for Katablepharidophyta divisio nova/Katablepharida phylum novum based on SSU rDNA and beta-tubulin phylogeny. Protist 156:163–179

Ota S, Ueda K, Ishida KI (2007) *Norrisiella sphaerica* gen. et sp. nov., a new coccoid Chlorarachniophyte from Baja California, Mexico. J Plant Res 120:661–670

Ota S, Kudo A, Ishida KI (2011) *Gymnochlora dimorpha* sp. nov., a new chlorarachnophyte with unique daughter cell behaviour. Phycologia 50(3):317–326

Palmer JD, Soltis DE, Chase MW (2004) The plant tree of life: an overview and some points of view. Am J Bot 91:1437–1445

Pasanen AL, Yli-Pietila K, Pasanen P, Kalliokoski P, Tarhanen J (1999) Ergosterol content in various fungal species and biocontaminated building materials. Appl Environ Microbiol 65:138–142

Paterson HL, Pesant S, Clode P, Knott B, Waite AM (2007) Systematics of a rare radiolarian – *Coelodiceras spinosum* Haecker (Sarcodina: Actinopoda: Phaeodaria: Coelodendridae). Deep Sea Res II 54:1094–1102

Patterson DJ (1989) Stramenopiles: chromphytes from a protistean perspective. In: Green JC, Leadbeater BSC, Diver WL (eds) The chromophyte algae. Problems and perspectives. Clarendon, Lonon, pp 357–379

Pawlowski J et al (2012) CBOL Protist Working Group: barcoding eukaryotic richness beyond the animal, plant, and fungal kingdoms. PLoS Biol 10:e1001419

Pedrós-Alió C (2003) Diversity of microorganisms. In: Vilà M, Rodà F, Ros J (eds) Seminar on biodiversity and biological conservation. Institut d'Estudis Catalans Publ, Barcelona, pp 339–353

Pellegrini L (1974) Origine et modifications ultrastructurales du matériel osmiophile contenu dans les physodes et dans certains corps iridescents des cellules végétatives apicales chez *Cystoseira stricta* Sauvageau (Phéophycée, Fucale). C R Acad Sci 279:903–906

Peters AF, Müller DG (1985) On the sexual reproduction of *Dictyosiphon foeniculaceus* (Phaeophyceae, Dictyosiphonales). Helgol Meeresunters 39:441–447

Pitombo LF, Teixeira VL, Kelecom A (1989) Feromônios sexuais de algas pardas. Uma visão quimiosistemática. Insula, Brazil 19(suppl):229–248

Prasad AHSK, Nienow JA, Livingstone RJ (1990) The genus *Cyclotella* (Bacillariophyta) in Choctawhatchee Bay, Florida, with special reference to *C. striata* and *C. choctawhatcheeana* sp. nov. Phycologia 29:418–436

Puglisi E, Nicelli M, Capri E, Trevisan M, Del Re AAM (2003) Cholesterol, b-sitosterol, ergosterol, and coprostanol in agricultural soils. J Environ Qual 32:466–471

Raven PH, Evert RF, Eichhorn SE, Bouharmont J (2000) Biologie végétale. De Boeck publication, p 968

Raven JA, Waite AM (2004) The evolution of silicification in diatoms: inescapable sinking and sinking as escape? New Phytol 162:45–61

Robert D, Catesson AM (1990) Biologie végétale, tome 2. Caractéristiques et stratégie évolutive des plantes. Organisation végétative. Doin publication, Paris, pp viii + 256

Rodriguez F, Derelle E, Guillou L, Le Gall F, Vaulot D, Moreau H (2005) Ecotype diversity in the marine picoeukaryote *Ostreococcus* (Chlorophyta, Prasinophyceae). Environ Microbiol 7:853–859

Rokas A (2006) Genomics and the tree of life. Science 313:1897–1898

Sakaguchi M, Inagaki Y, Hashimoto T (2007) Centrohelida is still searching for a phylogenetic home: analyses of seven *Raphidiophrys contractilis* genes. Gene 405(1–2):47–54

Schaap P et al (2006) Molecular phylogeny and evolution of morphology in the social amoebas. Science 314:661–663

Schmidt AR, Dörfelt H, Perrichot V (2007) Carnivorous Fungi from Cretaceous amber. Science 318:1743

Schnepf E, Kühn SF, Bulman S (2000) *Phagomyxa bellerocheae* sp. nov. and *Phagomyxa odontellae* sp. nov., Plasmodiophoromycetes feeding on marine diatoms. Helgol Mar Res 54:237–241

Schoenwaelder MEA (2002) The occurrence and cellular significance of physodes in brown algae. Phycologia 41:125–139

Scott JL, Baca B, Ott FD, West JA (2006) Light and electron microscopic observations on *Erythrolobus coxiae* gen. et sp. nov. (Porphyridiophyceae, Rhodophyta) from Texas U.S.A. Algae 21:407–416

Selosse MA, Le Tacon F (1998) The land flora: a phototroph- fungus partnership? Trends Ecol Evol 13:15–20

Seo KS, Fritz L (2002) Diel changes in pyrenoid and starch reserves in dinoflagellates. Phycologia 41:22–28

Shalchian-Tabrizi K et al (2006) Telonemia, a new protest phylum with affinity to chromist lineages. Proc Roy Soc B 273:1833–1842

Silberman JD, Collins AG, Gershwin LA, Johnson PJ, Roger AJ (2004) Ellobiopsids of the genus *Thalassomyces* are Alveolates. J Eukaryot Microbiol 51:246–252

Simpson AGB (2003) Cytoskeletal organization, phylogenetic affinities and systematic in the contentious taxon Excavata (Eukaryota). Int J Syst Evol Microbiol 53:1759–1777

Sims PA, Mann DG, Medlin LK (2006) Evolution of the diatoms: insights from fossil, biological and molecular data. Phycologia 45:361–402

Sinninghe-Damsté JS et al (2004) The rise of rhizosolenid diatoms. Science 304:584–587

Six C, Worden AZ, Rodríguez F, Moreau H, Partensky F (2005) New insights into the nature and phylogeny of Prasinophyte antenna proteins: *Ostreococcus tauri*, a case study. Mol Biol Evol 22:2217–2230

Šlapeta J, López-García P, Moreira D (2005a) Global dispersal and ancient cryptic species in the smallest marine eukaryotes. Mol Biol Evol 23:23–29

Šlapeta J, Moreira D, López-García P (2005b) The extent of protist diversity: insights from molecular ecology of freshwater eukaryotes. Proc Roy Soc B 272:2073–2081

Smirnov A, Berney C, Nikolaev S, Pochon X, Pawlowski J (2009) Molecular phylogeny of amoeboid protists. www.biani.unige.ch/msg/Amoeboids/Amoebozoa/html

Smith M, Hansen J (2007) Interaction between *Mesodinium rubrum* and its prey: importance of prey concentration, irradiance and pH. Mar Ecol Prog Ser 338:61–70

Steenkamp ET, Wright J, Baldauf SL (2006) The protistan origin of animal and fungi. Mol Biol Evol 23:93–106

Stevenson RN, South GR (1974) *Coccomyxa parasitica* sp. nov. (Coccomyxaceae, Chlorococcales), a parasite of giant scallops in Newfounland. Br Phycol Bull 9:319–329

Stevenson RN, South GR (1976) Observations on phagocytosis of *Coccomyxa parasitica* (Coccomyxaceae: Chlorococcales) in *Placopecten magellanicus*. J Invertebr Pathol 25:307–311

Swanberg NR, Anderson OR (1981) *Collozoum caudatum* sp. nov.: a giant colonial radiolarian from equatorial and Gulf Stream waters. Deep Sea Res 28A(9):1033–1047

Takahashi F, Okabe Y, Nakada T, Sekimoto H, Ito M, Kataoka H, Nozaki H (2007) Origins of the secondary plastids of Euglenophyta and Chlorarachniophyta as revealed by an analysis of the plastid-targeting, nuclear-encoded gene *psbO*. J Phycol 43:1302–1309

Tanifuji G, Onodera NT, Wheeler TJ, Dlutek M, Donaher N, Archibald JM (2011) Complete nucleomorph genome sequence of the nonphotosynthetic alga *Cryptomonas paramecium* reveals a core nucleomorph gene set. Genome Biol Evol 3:44–54

Thüs H, Muggia L, Pérez-Ortega S, Favero-Longo SE, Joneson S, O'Brien H, Nelsen MP, Duque-Thüs R, Grube M, Friedl T, Brodie J, Andrew CJ, Lücking R, Lutzoni F, Gueidan C (2011) Revisiting photobiont diversity inthe lichen family Verrucariaceae (Ascomycota). Eur J Phycol 46(4):399–415

Todd JD, Rogers R, Li YG, Wexler M, Bond PL, Sun L, Curson ARJ, Malin G, Steinke M, Johnston AWB (2007) Structural and regulatory genes required to make the gas dimethyl sulfide in bacteria. Science 315:666–669

Todo Y, Kitazato H, Hashimoto J, Gooday AJ (2005) Simple foraminifera flourish at the ocean deepest point. Science 307:689

Tregouboff G (1953a) Classe des Radiolaires. In: Grassé PP (ed) Traité de zoologie. Anatomie, systématique, biologie, Tome I. Protozoaires: Rhizopodes, Actinopodes, Sporozoaires, Cnidosporidies. Fasicule II. Masson & Cie, Paris, pp 321–436

Tregouboff G (1953b) Classe des Héliozoaires. In: Grassé PP (ed) Traité de zoologie. Anatomie, systématique, biologie, Tome I. Protozoaires: Rhizopodes, Actinopodes, Sporozoaires, Cnidosporidies. Fasicule II. Masson & Cie, Paris, pp 437–489

Tréguer P (2002) Les algues et le souffle d'Eole. La Recherche 355:52–53

Tyler BM, Tripathy S, Zhang X, Dehal P, Jiang RHY, Aerts A et al (2006) *Phytophtora* genome sequences uncover evolutionary origins and mechanisms of pathogenesis. Science 313:1261–1266

Valiela I (1991) Ecology of coastal ecosystems. In: Barnes RSK, Mann KH (eds) Fundamentals of aquatic ecology. Blackwell Scientific Publications, Oxford, pp 57–76

van den Hoek C, Mann DG, Jahns HM (1998) Algae. An introduction to phycology. Cambridge University Press, Cambridge

Van De Vijver B, Kopalova K (2008) *Orthoseira gremmenii* sp. nov., a new aerophylic diatom from Gouth Island (southern Atlantic Ocean). Cryptogamie Algol 29(2):105–118

Verbruggen H, Maggs CA, Saunders GW, Le Gall L, Yoon HS, De Clerck O (2010) Data mining approach identifies research priorities and data requirements for resolving the red algal tree of life. BMC Evol Biol 10(16):1–15

Versteegh GJM, Blokker P, Wood GD, Collinson ME, Damsté JSS, de Leeuw JW (2004) An example of oxidative polymerization of unsaturated fatty acids as a preservation pathway for dinoflagellate organic matter. Org Geochem 35:1129–1139

Viprey M, Guillou L, Ferréol M, Vaulot D (2008) Wide genetic diversity of picoplanktonic green algae (Chloroplastida) in the Mediterranean Sea uncovered by a phylumbiased PCR approach. Environ Microbiol 10:1804–1822

Vogelbein WK, Lovko VJ, Shields JD, Reece KS, Mason PL, Haas LW, Walker CC (2002) *Pfiesteria shumwayae* kills fish by micropredation not by exotoxin secretion. Nature 418:967–970

Webster JP (2001) Rats, cats, people and parasites: the impact of latent toxoplasmosis on behaviour. Microbes Infect 3:1037–1045

Wellman CH, Osterloff PL, Mohiuddin U (2003) Fragments of the earliest land plants. Nature 425:282–285

Williams BAP, Hirt RP, Lucocq JM, Embley TM (2002) A mitochondrial remnant in the microsporidian *Trachipleistophora hominis*. Nature 418:865–869

Yabuki A, Inagaki Y, Ishida KI (2010) *Palpitomonas bilix* gen. et sp. nov.: a novel deep-branching heterotroph possibly related to Archaeplastida or Hacrobia. Protist 161:523–538

Yoon HS, Ciniglia C, Wu M, Comeron JM, Pinto G, Pollio A, Bhattacharya D (2006a) Establishment of endolithic populations of extremophilic Cyanidiales (Rhodophyta). BMC Evol Biol 6:78

Yoon HS, Müller KM, Sheath RG, Ott FD, Bhattacharya D (2006b) Defining the major lineages of red algae (Rhodophyta). J Phycol 42:482–492

Yuan X, Xiao S, Taylor TN (2005) Lichen-like symbiosis 600 million years ago. Science 308:1017–1020

Zhao S, Burki F, Bråte J, Keeling PJ, Klaveness D, Salchian-Tabrizi K (2012) *Collodictyon* – an ancient lineage in the tree of Eukaryotes. Mol Biol Evol 29(6):1557–1568

Part III

Microbial Habitats: Diversity, Adaptation and Interactions

Biodiversity and Microbial Ecosystems Functioning

Philippe Normand, Robert Duran, Xavier Le Roux, Cindy Morris, and Jean-Christophe Poggiale

Abstract

All ecosystems are composed of multiple species performing numerous functions. This plurality identified as biodiversity has become a research topic of general importance for understanding how ecosystems function. The word "biodiversity" has subsequently received different interpretations we aim to describe. Microbial systems can be used to illustrate these definitions and to clarify paradigms that have emerged in general ecology. Microbial biodiversity can be characterized by the kind of biodiversity (taxa or functional groups present), the representativeness of samples, and the culturability of samples or taxa, in terms of genetic, functional, or physiological characteristics. Biodiversity is a major driver of ecosystem functioning, and this relation is described in the case of several biotopes. Several mathematical approaches have been used to quantify microbial diversity, and the various indices are described and discussed.

Keywords

Alpha diversity • Beta diversity • Biodiversity • Canonical correspondence analysis • Chimeras • Cluster analysis • Community structure • Cultural approach • Denaturing gradient gel electrophoresis (DGGE) • Diversity indices • DNA arrays • DNA reassociation • Fingerprints • Functional groups • Identification • Linkage disequilibrium analysis • Lipids • Metabolic capacities of communities • Metabolites • Microbial communities • Multivariate analysis • Nei index • Nonparametric test • OTU (operational taxonomic unit) • Parametric test • Phylotype • Pigment • Principal component analysis • Rarefaction curve • Ribosomal intergenic spacer analysis (RISA) • Richness • Saturation analysis • Shannon alpha diversity index • Simpson index • Single-strand conformational polymorphism (SSCP) • Spearman correlation tests • Species richness index • Temperature gradient gel electrophoresis (TGGE) • Terminal-restriction fragment length polymorphism (T-RFLP) • Variables

---

* Chapter Coordinator

P. Normand* (✉) • X. Le Roux
Microbial Ecology Center, UMR CNRS 5557 / USC INRA 1364,
Université Lyon 1, 69622 Villeurbanne, France
e-mail: philippe.normand@univ-lyon1.fr

R. Duran
Institut des Sciences Analytiques et de Physico-chimie pour l'Environnement et les Matériaux (IPREM), UMR CNRS 5254, Université de Pau et des Pays de l'Adour, B.P. 1155, 64013 Pau Cedex, France

C. Morris
Unité de Recherches de Pathologie Végétale, INRA, Domaine St-Maurice, BP 94, 84143 Montfavet Cedex, France

J.-C. Poggiale
Institut Méditerranéen d'Océanologie (MIO), UM 110, CNRS 7294 IRD 235, Université de Toulon, Aix-Marseille Université, Campus de Luminy, 13288 Marseille Cedex 9, France

8.1 Introduction

The existence of a plurality of species in ecosystems, identified as biodiversity, has become a research topic of general interest important nationally and internationally since Wilson published his groundbreaking eponymous book "Biodiversity" (1988). The word "biodiversity" has subsequently received different interpretations, e.g., Gast and collaborators (1991) defined it as the variety, distribution, and structure of plant and animal communities, including all age groups, arranged in space and time. In addition, Wilcox (1984) added the concept of genetic diversity.

Takacs (1996) illustrated the rise of the concept by the fact that in 1988, "biodiversity" was not used as keyword in biological abstracts, and the expression "biological diversity" appeared in them only once. In 1993, 5 years later, "biodiversity" had 72 occurrences, and "biological diversity" 19. Fifteen years later, it would be difficult to quantify the use of these terms in daily use by scientists, politicians, journalists, and the general public. A Google search (www.google.com) in June 2013 showed more than 53 million websites where the word is used.

This word is now used very broadly, by environmentalists who want to designate the whole of the organisms of a biotope or the functions they perform, by evolutionists who thus designate all members of a phylogenetic clade with their variations, by molecular biologists who study the variations of an enzyme to better understand the motifs that are important for a biological function or the structure of a protein or metabolite, and by population geneticists who want to quantify the rate of genetic exchange or the range of morphological features of a taxon.

Regarding microbial biodiversity, one faces several problems when characterizing it:
- The question of what kind of biodiversity should be targeted: global diversity of taxa (bacteria, fungi present, etc.), diversity within functional groups, diversity of functional groups, etc.
- The problems of representativity of the samples to use. It has thus been estimated that the Earth is home to about 10^{30} prokaryotes (Whitman et al. 1998). One gram of soil may contain 10^4 or even 10^5 different taxa.
- The difficulty to cultivate most microorganisms, which limits our ability to understand their diversity using this approach.

These problems overlap with issues that arise about approaches and criteria for characterization of microbial activities carried out by consortia or other complex sets of complementary microorganisms. Despite these complexities and difficulties, studies of microbial biodiversity are increasing rapidly for several reasons that this chapter will seek to explain.

8.2 The Pertinence of the Paradigms of Biodiversity and Ecosystem Functioning for Microbial Ecology

Biological variability between populations and communities of organisms (macroorganisms and microorganisms) is the result of selection pressures imposed by the physical and biological conditions of the environment coupled with mutations or recombinations of the genome of these organisms. Many of the processes that create biological variety occur on timescales or under environmental conditions that are difficult to reproduce in the laboratory. Biodiversity of populations or communities is an archive of the adaptive history of these organisms. It reflects complex phenomena creating diversity by spontaneous point mutations, duplications or gene losses, lateral transfers, and selection by the environment. A major reason to study biodiversity is to open these archives to understand the factors controlling the ecology and evolution of organisms and how this biodiversity is related to the way ecosystems function. This information is essential for managing the biodiversity of organisms on our planet, to limit the spread of pests, to better exploit organisms that have industrial or medical applications, or to preserve the organisms involved in important ecosystem processes.

It is clear that natural populations of organisms do not grow infinitely either in terms of the number of individuals or in the abundance of different taxa (species, races, biotypes, etc.). The twentieth century saw intense debates about the processes governing growth and diversification of populations and species. These discussions helped to establish ecology as a discipline with a fluctuating set of paradigms on control processes. These paradigms have been developed based almost exclusively on works on macroorganisms, and few were assessed for their relevance to microorganisms. A challenge to the discipline of microbial ecology is to evaluate the relevance of these paradigms for communities of microorganisms.

Among the processes that regulate population growth and the composition, structure, and functioning of communities, the concepts of niche, food webs, succession, and the link between biodiversity and ecological stability of a community or ecosystem functioning as a whole are particularly pertinent to microbial ecology. The exploitation of resources in a habitat is considered one of the fundamental factors determining abundance and diversity. "Niche" defines resource fields in terms of nutrients and space in particular, in which a species exists. One of the main paradigms of ecology is that two species cannot occupy the same niche. Coexistence implies that each organism has a different niche, even if these niches may overlap. For many microbial experimental systems, it has been shown that the coexistence of different species or biotypes of microorganisms increases when the environment is structured (as summarized by Jessup et al. 2004). Spatial structure essentially leads to an increase in the number of niches, especially in terms of microenvironments. Trophic relationships between organisms are also considered important for the overall structure of a community. Primary producers may limit the abundance of heterotrophic ones, and the complexity of the network can be a buffer against the collapse of a network if a primary producer were to disappear. Using macrocosms, laboratory microbiologists have shown that the length of the food chain depends on the productivity of primary producers (Jessup et al. 2004). The term "succession" describes the changing composition of organisms during the maturation of a community. Initial colonizers of habitats are named "pioneer species." Their presence alters the habitat and thereby creates new opportunities for others. The succession process in the establishment of forests has been well described, with herbs, willow (Salix), and nitrogen-fixing species such as alder (*Alnus*) and coniferous species (*Pinus*, *Abies*, *Juniperus*, etc.) colonizing coastal sand dunes, which are then replaced by deciduous species (oak, maple, etc.) as organic matter and nitrogen accumulate in the soil and as the growth of large and long-lived species gradually limits the entry of light into the canopy (Lichter 1998). Microorganisms play a central role in the colonization process given the scarcity of nitrogen in most pioneer-stage terrestrial ecosystems. Thus, during glaciers retreat, the moraine is first colonized by species (*Dryas*, *Alnus*) that fix dinitrogen in symbiosis with the actinobacterium *Frankia* and then give way to climax species such as spruce or poplar (Lawrence et al. 1967). Comparable phenomena are described following volcanic eruptions (Vitousek et al. 1987), landslides, or forest fires (Yelenik et al. 2013).

Biodiversity of a community is considered a potentially important element for the functioning of an ecosystem in its entirety. However, there have been many debates about the general nature of such causation: Does the ecosystem and its associated processes determine the level of biodiversity, or are the functions of the ecosystem the result of the level of biodiversity (Naeem 2002)? Whatever the cause or effect, a paradigm is well established that the functioning of communities and ecosystems and the stability of operation are strongly related to biodiversity, low-diversity systems are the most vulnerable, and those with high diversity are the most robust and resilient. Details of the intense debate on the relationship between diversity and stability were summarized by McCann (2000). An example of the role of diversity in community stability is given by Mitchell and collaborators (2002) who have highlighted the role of the level of diversity of grassland plant communities for resistance to fungal diseases. Symptom severity (% leaf area infected) in monocultures was three times larger than that observed in diversified prairies. Plant diversity appears to play a role in limiting the spread of pathogens between individuals of susceptible species. In plots with intermediate plant diversity, community composition plays a major role and compensates for the absence or presence of a particular species with high or low susceptibility that can have a major influence on the overall reaction of the community. Regarding microorganisms, microcosms were used to assess the relationship between diversity and stability. These microcosms allow good control of physicochemical factors. Using mixtures of algae, bacteria, protists, and small metazoans, several authors have shown that the microcosms with the highest diversity were the most stable in terms of CO_2 fluxes, for example, and were more resistant to invasions. The use of microcosms to test such hypotheses in microbial ecology has been reviewed by Jessup and collaborators (2004).

There are many other hypotheses about the establishment and structuring of communities. A major contribution to ecological concepts as applied to microbiology was provided by Kinkel and collaborators (Andrews et al. 1987; Kinkel et al. 1987a, b, 1989) in their test of the theory of Island Biogeography. The theory of Island Biogeography proposed by MacArthur and Wilson (1963) postulates that the number of species found on an island (the number at equilibrium) is determined by two factors: the distance to the mainland and the size of the island. These two factors affect the rate of **extinction*** on the islands and the level of **immigration***. For microbiology, the concept of island is of course relative to the scale of microorganisms, and it will therefore be applied to fertile habitats separated by stretches of low fertility. By sterilizing leaves in an apple orchard, Kinkel et al. (1987a, b, 1989) created new islands for colonization. They were also able to estimate the rate of immigration, emigration, multiplication, and mortality of leaf colonizers such as *Aureobasidium* spp. and the plant pathogen *Venturia inaequalis* and demonstrate that the pattern of colonization corresponded to theoretical predictions.

Other aspects of ecological theory have been studied in microbiology, e.g., eutrophication and algal blooms clearly contributed to our understanding of the concept of biological carrying capacity of the habitat. Obviously, in plant ecology, for example, the important parameter for explaining ecosystem productivity in large-scale experiments such as those conducted in Jena (Roscher et al. 2007) is not so much the number of plant species but rather the functional groups defined a posteriori (dinitrogen fixers, small herbs, etc.) to which they belong (Le Roux et al. 2013). A comparable question for microorganisms would address functional groups such as photosynthetic organisms, dinitrogen fixers, cellulolytic ones, antibiotic producers, as well as others that will emerge as a result of research on microbial ecology such as those described here. Even within a given microbial functional group, microbial functional traits are better predictors of functional complementarity and ecosystem functioning than taxonomic diversity (Salles et al. 2012).

Clearly, microbial ecological research can help to test hypotheses developed in general ecological theory. A major and current motivation for the study of microbial diversity is to understand the role that biodiversity plays in the functioning of microbial communities to better manage, protect, or handle them. Some examples of goals of microbial biodiversity studies are to:

(i) Identify the role of community complexity in the degradation and chemical modifications of chemicals in biogeochemical cycles of elements in general
(ii) Describe the biogeography of microorganisms to highlight the limits to their dissemination and the importance of their adaptation to habitats
(iii) Discover organisms playing key roles in environmental processes or with a strong potential for industrial utility
(iv) Assess the impact of human activities on microbial biodiversity, especially to avoid the destruction of useful diversity (i.e., beneficial to the environment) or to manage negative organisms (limit the emergence of harmful organisms: pathogens, hypercompetitive organisms, etc.)

Despite the relevance of these objectives to contemporary issues of environmental protection, most studies in microbial diversity have long referred to the simple description of biodiversity and have spotlighted species or variants that were previously unknown (Morris et al. 2002). The biological richness revealed in these studies has occasionally found practical applications in the development of diagnostic tools (e.g., medical or plant health assessment) and in industrial processes or biotechnology. The thermostable DNA polymerase from *Thermus aquaticus* used in PCR reactions, discovered during a study of microbial diversity in extreme environments, is the archetypal example of such an application. Several DNA polymerases were known before, but it was the discovery of a thermostable enzyme which has made chain reaction possible and opened the market for PCR, which is now an industrial product that represents a 300 million dollar market.

8.3 Mathematical Approaches and Tools for the Study of Microbial Biodiversity

A large proportion of studies on microbial biodiversity have been mainly descriptive. In reading the literature, a novice student could easily get the impression that the motto "look and see" is the accepted approach to the study of microbial diversity. This trend was revealed by a systematic study of the literature on microbial biodiversity for the period 1977–2002 (Morris et al. 2002). A major challenge for research on microbial biodiversity that has been partially addressed in recent years is the conceptualization and implementation of experimental design.

Despite recent calls to do descriptive, hypotheses-free ecological studies, we feel that the implementation of appropriate experimental designs for the study of microbial biodiversity involves the application of the scientific method common to all research in the natural sciences. This implies:

(i) Formulating hypotheses relevant to the objectives
(ii) Developing sampling procedures to distinguish experimental error from effects due to the environmental processes studied
(iii) Demonstrating the repeatability of the results obtained
(iv) The use of nonsubjective (statistical) tests of hypotheses

The nature of microbial communities makes sampling especially critical for studies of biodiversity. In addition, advances in molecular biology have rendered a wide variety of techniques for characterizing microorganisms and microbial communities more available to researchers, thereby facilitating the elucidation of subtle differences between members of the same species. All this leads to a paradox faced by many researchers in the field of microbial biodiversity: the relatively small number of samples that can be characterized reveals a level of diversity that effectively and severely limits the power of statistical analyses. Some of the steps for defining experimental design for the study of microbial biodiversity that can help resolve this paradox will be presented in details later in this chapter.

Formulating clear objectives and identifying the underlying assumptions are one of the first steps in any research project. The hypotheses to be tested will be crucial to determine the choice of methods, the nature of the experimental setup, and the type of statistical analyses to be used. However, microbial biodiversity studies often face trade-offs between the optimal size of samples and the limits of time and labor necessary for characterizing the samples.

Therefore, formulating hypotheses in studies of microbial biodiversity is by necessity, more dramatically so than in other fields of biology, a dialectical process between ideal objectives and strong experimental and analytical constraints.

8.3.1 Approaches Based on Diversity Indices

To formulate hypotheses about the structure and composition of microbial communities, it is useful to examine the statistical tests available. A variety of parametric and nonparametric statistical tests have been developed and used (Morris et al. 2002). These tests are based on calculations of indices (single variables summarizing information on the original multivariable data) that integrate different diversity parameters. Unfortunately, for "cultural" reasons, microbiologists do not always have sufficient expertise in statistics to know all the available tests and their limitations.

There are two major differences between parametric and nonparametric analyses, which can have a significant impact on the formulation of hypotheses and the way the study is conducted. Firstly, parametric tests generally require diversity measures for variables with normal distribution. Several diversity indices, including the Shannon, are inherently randomly distributed (i.e., they are random variables). In the absence of data on this point, a preliminary measure should be used to verify this. Given the sampling and technical constraints for characterizing microbial diversity, it is not surprising that this step is generally neglected if it cannot be done by mathematical simulations. In contrast, nonparametric tests can be used with data that are not randomly distributed. In addition, nonparametric tests can be used with a large number of quantitative as well as qualitative biodiversity measures, while parametric tests require more formal measures of diversity. Many examples of how statistical tests were coupled to different indices for quantification of diversity have been described by Morris and collaborators (2002).

8.3.2 Multivariate Approaches

The accumulation of microbial diversity data and information on related environmental parameters not only permit to better define patterns of diversity but also to better understand what are the spatial and temporal parameters that explain these patterns.

These complex microbial ecology issues often lead to data sets that are very difficult to exploit by conventional statistical tests. These difficulties can be solved by using multivariate statistical analyses. These methods of analysis can be exploratory or explanatory (Table 8.1).

Table 8.1 The main methods of multivariate statistical analysis

Multivariate statistical analysis	
Exploratory	Explanatory
Principal component analysis (PCA)	Redundancy analysis (RDA)
Correspondence analysis (CA)	Canonical correspondence analysis (CCA)
Principal coordinate analysis (PCoA)	Linear discriminant analysis (LDA)
Nonmetric multidimensional scaling (NMDS)	Rank-order similarity correspondence (ROSC)
Cluster analysis	

These tools, developed for the study of patterns of diversity of higher organisms (plants or animals), may be applied in microbial ecology. Although these methods of analysis are well described in the literature, they are rarely used in microbial ecology and often only for exploratory approaches. Indeed, in literature searches to find examples of the use of multivariate analyses, microbial ecology ranks third after studies of plants and fish (Ramette and Tiedje 2007). The complex data sets are explored primarily through principal component analysis and cluster analysis. Techniques such as canonical correspondence analysis (CCA) or Spearman correlation tests of similarity ranks are very rarely used. These methods of analysis should, however, be indispensable tools in microbial ecology because they allow the identification of trends and the major parameters influencing this distribution. However, it is important to remember that if the multivariate statistical analyses applied to data sets acquired in in situ situations may suggest causes or factors, researchers should, after a round of multivariate analysis, formulate hypotheses and then test them. All these multivariate methods and their applications and limitations are well described by Ramette and Tiedje (2007).

Principal component analyses (PCA), the correspondence analysis (CA), and multidimensional analysis are useful for comparing communities for which data on the presence and relative abundance of many species (or OTUs) are obtained. The processing of data obtained by cultivation and molecular approaches by these methods will bring together communities with similar structures. Structurally remote communities are scattered across graphs generated by computational tools. These tools can also highlight the important and characteristic bacterial populations within the community. These data analyses can explain the origin of the observed variability between populations.

Canonical correspondence analyses (CCA), multiple regression tests, or Spearman rank correlation of similarity are used to establish correlations between community structure and environmental parameters or functions. These statistical analyses are used to understand the structure/function relationships (Patra et al. 2007), a central question in microbial ecology. Populations highly correlated with measured parameters will also be spotlighted as a means to identify potential indicator species. The importance of these species must be tested to confirm their "indicator" status. For example, the processing of data by CCA has highlighted the interactions between various abiotic parameters and dynamics of bacterioplankton in the North Sea and the identification of bacterial phylotypes responding specifically to certain environmental factors (Sapp et al. 2007). Other examples of the use of CCA to assess the impact of abiotic factors or pollutants on the structure of microbial mats are given in this chapter.

8.4 Variables and Methods for Studying Microbial Diversity

Different levels of diversity are studied in microbial ecology to answer different questions. A first level of questioning is that of bacterial species, the only taxonomic level for which an operational definition exists, based on DNA hybridization or 16S sequences comparison (*cf.* Chap. 6). At a finer level, approaches targeting DNA can also reach subspecies levels.

There are two types of biodiversity used in the literature (Whittaker 1972). Alpha diversity is defined as the small-scale diversity observed within a community. Beta diversity is defined as the large-scale diversity observed between communities. This distinction is rarely used in microbial ecology, because the definition of "local" microbial communities is not easy in many environments like soil (Box 8.1) that are highly heterogeneous even at the microscale so that environmental samples (e.g., a soil core) are often already a mixture of local communities. In addition, numerous horizontal gene transfers can jeopardize the definition of species identity and associated functional traits as classically accepted in ecology of higher organisms. The

---

Box 8.1: Example of a Study of Biodiversity and Soil Management

Xavier Le Roux

The following example illustrates the interaction between statistical tests, indices, and assumptions. Management strategies are important for soil fertility and for the abundance of plant pathogens and of microorganisms that contribute to the degradation and recycling of compounds. It may therefore be interesting to evaluate the impact of a fallow period, for example, on the soil microbial community. For a period of fallow, a field is colonized by a wide variety of weeds rather than a single species as in classic monoculture. Therefore, several hypotheses can be posed such as:
1. A fallow season will increase the overall diversity of microorganisms in the soil.
2. A fallow season will reduce the frequency or abundance of aggressive strains of a plant pathogen specific to a particular host (e.g., *Fusarium oxysporum*).

In the case of the first hypothesis, it is necessary to express the overall microbial diversity as an index incorporating the relative abundance of different taxonomic groups present in the soil. The Shannon index, for example, could be used for this purpose. This would involve sampling and characterizing by the most appropriate method (see below) under field conditions and comparing microbial communities before and after the fallow. To calculate the index, many strains or clones must be characterized before and after the fallow period. These observations must then be integrated into a single index value for each situation. However, parametric tests require replicated observations for each treatment (e.g., *t*-test or analysis of variance) in order to assess the variance. This will clearly lead to a significant amount of work. Some researchers have used numerical simulations to estimate the variance of diversity indices in order to compare the indices without replicate measurements. It should be kept in mind that the simulation can only be based on the observed variability in the sample and cannot be based on "real" variability in situ, especially if the real variability is greater than that observed in the sample. An alternative approach might be to compare populations using nonparametric techniques based on gross multinomial values. For populations with bi- or trinomial distributions, χ^2 tests could be used to compare populations. However, if the aim is to compare global biodiversity, it is likely that the samples will reveal many taxonomic groups and require the comparison of multinomial distributions. At present, few statistical techniques are available to compare multinomial complexes.

The second hypothesis mentioned above is more tractable than the first because the target of the study of biodiversity is more precisely defined than the first hypothesis. For this study, the abundance of *Fusarium oxysporum* in soil will be evaluated and strains tested for their degree of pathogenicity on different hosts. To statistically compare the abundance of the *Fusarium*

(continued)

Box 8.1 (continued)

strains (e.g., number of fungal propagules/g soil), replicate measurements are required for each test parameter for each different soil treatment. To compare the frequency of highly aggressive strains for different treatments, several options are available. For each sample replicate of each treatment, the proportion of strains of a given level of aggressiveness is calculated and the differences can be evaluated with parametric tests. An alternative option would be to characterize a random sample of strains for 2 or 3 levels of aggressiveness (low vs. strong, or no, intermediate, high) and then compare the bi- or trinomial frequencies obtained with nonparametric tests. It would be necessary to estimate the effect of sample size on the power of statistical tests for these two options in order to assess which analysis strategy is the most efficient in terms of time and labor involved. For a basic statistical approach such as an analysis of variance and a χ^2, such estimates are easily obtained with statistical tools.

---

distinction between alpha and beta diversity is thus not used hereafter.

It is important to take into account the sensitivity thresholds related to the techniques used to characterize microbial diversity. Techniques related to PCR can in theory provide a signal, e.g., a band detectable on a gel, from a targeted sequence in the reaction volume if it yields at least one billion fragments after 30 amplification cycles. However, several factors diminish the amplification yield below the theoretical level. The first is the resistance of the polymerase used. The latter is a thermostable enzyme but it must undergo cycles of heating at the limits of its resistance, and part of the molecules is thus lost during each cycle. The second is the presence of polymerase inhibitors: like any enzyme, Taq polymerase functions optimally in the environment proposed by the supplier, but factors such as the presence of soil humic acids (Watson and Blackwell 2000) or damaged nucleotides (Jenkins et al. 2000) are known to decrease the efficiency of amplification. Finally, although there are thousands of bacterial species per gram of soil (Torsvik et al. 1990) or other complex environments, the literature does not indicate the presence of more than a few tens of taxa in most collections of amplicons. Indeed, in a given reaction mixture, some DNA sequences with low GC are likely to be more rapidly amplified and will thus eventually monopolize the Taq polymerase. Overall, it is estimated that below 1 % of initial abundance, a taxon cannot be detected even if the experience of intensive sequencing for complex environments has never been attempted convincingly. This is the objective of ongoing projects (Vogel et al. 2009).

Beyond this intrinsic limit, a wide range of methods exist today to characterize microbial diversity, each method apprehending a different aspect of diversity.

8.4.1 Approaches Targeting Phenotype Diversity

8.4.1.1 Cultural Approaches

For a long time, diversity studies were performed using cultivation techniques, which implies the isolation and cultivation of microorganisms. Counting techniques, such as the most probable number (MPN), or the determination of usage patterns of growth substrates such as done with BIOLOG (Bossio and Scow 1998), have been developed to quantify microorganisms and estimate their metabolic profiles. Although these methods have provided the fundamental basis for the study of microbial communities in their environment (Garland and Mills 1991), they have some limitations, the main ones being:

(i) Representativeness: Media and culture conditions impose constraints such that only a small fraction of microorganisms from a given environment can be cultured. According to some estimates, the cultivable diversity represents only 0.1–1 % of the actual bacterial diversity (Amann et al. 1995).

(ii) Identification: These methods require to further characterize the bacteria by cultivating them to better know their taxonomic affiliation.

In particular, it is recognized that the general use of rich media can strongly bias the type of bacterial strains able to grow as compared to the taxa most common in natural, often oligotrophic environments. The use of a range of selective media may allow better access to a range of microbial diversity, but the multiplication of media may prove to be time-consuming and tedious. The development of new selective media and of automatic approaches using robots should partly alleviate these limitations and allow the isolation and characterization of new bacterial species in the future, which would help quantifying the actual level of microbial diversity in situ.

8.4.1.2 Approaches Targeting the Diversity of Metabolites

Lipids

Different lipid classes can be targeted to assess microbial diversity. The most widely used are:

(i) The phospholipid fatty acids (PLFA) characterized by chemical analysis and/or molecular isotopy (Frostegard et al. 2011). Several of these biomolecules, present only in viable cells, can be used as markers to identify major taxonomic and functional groups and even some bacterial taxonomic groups (*cf.* Sect. 17.6.1). They have been used for the study of soil microbial communities (Ritchie et al. 2000), deep-sea ecosystems (Fang et al. 2003; Schrenk et al. 2003), marine ecosystems contaminated by petroleum hydrocarbons (Mazzella et al. 2005), or different stages of composting (Eiland et al. 2001).

(ii) Hopanes (pentacyclic triterpenoids to 30 carbon atoms), whose complex chemical structure has to withstand the processes of diagenesis and **catagenesis***, are biomarkers that are used to identify the location and condition of synthesis of petrol oil and its level of maturity. The hopanes and steranes are also used in environmental geochemistry. In effect, each oil provenance has a specific chemical composition, a true "molecular fingerprint" that can pinpoint the source of an oil spill (*cf.* Chap. 16).

(iii) Glycerol ethers that are specific to *Archaea* and which have demonstrated the involvement of these microorganisms in the anaerobic biodegradation of methane (*cf.* Box 14.9).

(iv) Quinones which have been used to characterize the evolution of microbial communities (bacteria and fungi) during the thermophilic phase during a composting process (Tang et al. 2007).

Pigments

Some microorganisms contain pigments such as chlorophylls, bacteriochlorophylls, carotenoids, and phycobiliproteins, with characteristics allowing their use as taxonomic markers:

(i) They are present in photosynthetic microorganisms that can be distinguished from other members of the microbial community.

(ii) Certain pigments are limited to certain microbial classes or genera, allowing characterization of the taxonomic composition of photosynthetic microbial communities at the class or genus levels.

(iii) They are highly colored, and considering the chlorophylls and phycobiliproteins, they are fluorescent at wavelengths in the visible spectrum, allowing their detection at very low sensitivity levels.

(iv) They are labile and rapidly degraded after cell death, allowing to target mainly living cells.

The main drawbacks to the use of pigments as taxonomic markers are:

(i) Since they are labile, they must be protected from light, oxygen, acidity, and alkali during sampling and analysis. In solution they form isomers spontaneously.

(ii) Their distribution is complex, some pigments being species specific and others covering many different microbial classes.

(iii) Their expression is variable, and the pigment level per cell varies depending on environmental parameters such as irradiation or nutrients.

Despite these drawbacks, pigment analysis is a method of choice to monitor the abundance and composition of photosynthetic microorganisms. It has been used to analyze the diversity of picophytoplankton in the Pacific Ocean by semi-automatic methods such as flow cytometry (Blanchot and Rodier 1996) of cyanobacteria in microbial mats by confocal microscopy (Fourcans et al. 2004) or of microbial mats after extraction and HPLC analysis (Nubel et al. 1999). Another interesting example is the HPLC analysis of pigments from microbial mats growing in arctic ecosystems (Vincent et al. 2004), which highlighted the presence of cyanobacteria in cryo-ecosystems.

8.4.1.3 Analysis of Metabolic Capacities of Communities

BIOLOG plates that permit to establish patterns of individual carbon substrate utilization were originally designed for the identification of microbial isolates. Ease of use and the possibility afforded to obtain rapid results resulted in their utilization for the analysis of metabolic capacities of microbial communities for the first time by Garland and Mills (1991), despite the often-heard criticism of lack of reproducibility: the latter would be due to the nonlinearity of substrate utilization over time and the difficulties of standardizing the inoculum (Haack et al. 1995). In particular, the inoculum density and incubation time are critical, and the interpretation of results has to be done with caution. Moreover, BIOLOG analysis is a method based on cells expressing metabolic activities and each metabolic profile observed therefore reflects only partly the nature and structure of the targeted microbial community derived from an environmental sample.

BIOLOG-ECO plates were developed to include 31 substrates among the most used in many natural environments. These 31 carbon sources are repeated three times, thus providing data replication. Metabolic fingerprints of microbial communities can then be processed by multivariate statistical analyses.

Gamo and Shoji (1999) have developed a method using BIOLOG-MPN plates to enumerate bacteria by the MPN method on a range of substrates. This method has been used to characterize the temporal changes of functional diversity for bacterioplankton communities (Matsui et al. 2001). BIOLOG plates have been widely used in soil microbiology (Garland and Mills 1991). They helped to highlight the resistance of bacterial communities of forest soils during application of glyphosate at concentrations used commercially, whereas stronger effects can be observed when higher concentrations are used (Ratcliff et al. 2006). Similarly, the effect of agricultural practices on soil bacterial communities has been demonstrated with BIOLOG plates, which has been confirmed by the T-RFLP fingerprint method (Widmer et al. 2006).

8.4.2 Approaches Targeting Genotype Diversity

8.4.2.1 Approaches Targeting Nucleic Acids (DNA)

DNA Reassociation

For this technique, DNA is extracted from a given biotope, denatured, and allowed to renature. Originally it was used to determine the size of a genome: At a given set of concentration and temperature, the rate of reassociation of denatured DNA strands is a function of the complexity of the DNA. The DNA of a single taxon will then reassociate very quickly, whereas DNA retrieved from a biotope containing thousands of taxa will take much more time to reassociate. Using this technology, Torsvik et al. (1990) estimated that agricultural soils typically contain thousands of taxa per gram of soil, making it the most diverse habitat known (2002). Gans and collaborators (2005) refined the approach and obtained even higher values (around 10^6 bacterial taxa per 10 g of soil). This technology is applicable to any habitat. However, its application is complex and it gives no indication of the taxa present, which explains why it has been rarely used.

All the approaches described below target nucleic acids and are based on PCR using DNA primers, which are short sequences that anneal to conserved regions of genes. The concept of "universal primers" is often used, but this concept is to be taken with caution because the primers are designed based on known taxa, for example, the symbiotic bacteria *Nanoarchaeum* could not be amplified using these primers, and it had to be characterized by Southern/cloning/sequencing of its 16S rRNA gene. This permitted to show that there were two mismatches in one of the so-called universal primers, which prevented polymerization (Huber et al. 2002). Conversely, adequately targeting a narrow taxonomic group, e.g., Actinobacteria (Stach et al. 2003), poses less problems.

Linkage Disequilibrium Analysis Through Polymorphism Association (on Isolates)

Two populations of microorganisms can be compared by the tools of population genetics in order to quantify the frequency of genetic exchange among populations. Bacteria belonging to the same species may exchange (by conjugation or otherwise) genes, which will cause genetic mixing. The frequency of these exchanges is quantified by the study of linkage disequilibrium between two markers. This can be extended to quantify the genetic diversity among several isolates.

Denaturing Gradient Gel Electrophoresis (DGGE) and Temperature Gradient Gel Electrophoresis (TGGE)

These techniques were developed for the study of human polymorphism (Sheffield et al. 1989) to distinguish very similar DNA or RNA sequences because they still have different denaturation parameters that will result in differences in gel migration. Indeed, the concentration of denaturant (DGGE) or the temperature (TGGE) at which a double-stranded sequence is separated into two single-stranded chains varies between very similar sequences. It is also known that a sequence of double-stranded nucleic acid being more compact will migrate much faster than a single-stranded sequence in an acrylamide gel. In these approaches, a GC-rich clamp (GC-clamp) keeps the two strands together in order to have a higher sensitivity of the method. This technique was initially applied to microbial ecology by Muyzer and Smalla (1998) in a series of works on different biotopes that began in 1993. This revealed the existence of hundreds of microbial taxa in the environments studied (Muyzer et al. 1993). In January 2013, a Web of Science search with DGGE or TGGE as keywords permitted to retrieve 7,753 references, most of them corresponding to microbial ecology papers. DGGE has been increasingly used since the early 1990s, though no such increased use is observed after 2010 (Fig. 8.1).

Single-Strand Conformational Polymorphism (SSCP)

While the principle of DGGE is based on the complementarity between the two strands of DNA, the one of SSCP is based on sequence complementarity within strands. SSCP differs from DGGE in that DNA is denatured and then allowed to renature before (not during) gel migration. The denatured DNA sequences, even closely related ones, have very different migration rates. This technique was used, for instance, to characterize the diversity of the archaeal community in 44 anaerobic digesters (Leclerc et al. 2004). A Web of Science search in January 2013 with SSCP as keyword permitted to retrieve 2,081 references. This method has been increasingly used between 1990 and 1998, whereas a progressive decrease of its use is observed during the 1998–2012 period (Fig. 8.1).

Ribosomal Intergenic Spacer Analysis (RISA)

Gene order is highly conserved in ribosomal operons with a primary transcript comprising the 16S (*rrs*) at the 5′ end, the 23S (*rrl*) in the middle, and the 5S (*rrf*) at the 3′ end in all microbial species studied so far (Normand et al. 1996). The only known exception is *Wolbachia* where *rrs* genes are located away from the *rrf* and *rrl* genes (Bensaadi-Merchermek et al. 1995). By contrast, what varies greatly from one bacterial species to another is the length of the intergene between *rrs* and *rrl* genes, which led Fisher and Triplett (1999) to use a pair of primers to amplify this region,

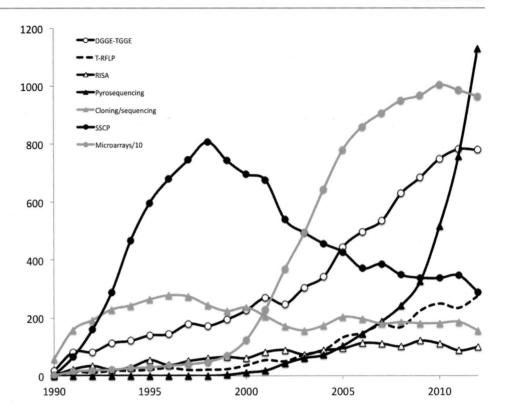

Fig. 8.1 Temporal variations of the numbers of papers published each year which use particular methods to target genotypic diversity. The search was done on the "Web of Science" database for the 1990–2012 period. All numbers correspond to absolute values, except those for microarrays that were divided per 10. The total number of papers is 7,753, 2,081, 1,683, 10,255, 3,614, 4,728, and 95,838 for (DGGE and TGGE), T-RFLP, RISA, SSCP, pyrosequencing, cloning/sequencing, and microarrays, respectively

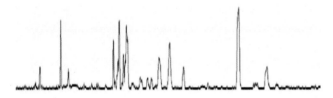

Fig. 8.2 T-RFLP profile of the bacterial community from a microbial mat. The DNA of all of the microorganisms present is extracted, purified, and amplified with primers targeting a conserved gene (16S rRNA), one of which is fluorescent allowing its detection. This composite amplicon is then digested with a restriction enzyme which cuts at different sites in each amplicon. These gene fragments having a variable length can be separated by electrophoresis, detected by a fluorescence reader, and each peak represents a different sequence whose amplitude is a function of its abundance in the community

with a primer at the 3′ end of the *rrs* gene and the other at the 5′ end of *rrl*. The approach gives a fingerprint that allows us to characterize and compare microbial communities (Fig. 8.2). This technique was applied later (Ranjard et al. 2000) to the study of soils. A Web of Science search with RISA as keyword in January 2013 permitted to retrieve 1,683 references and demonstrated that the use of RISA remained low during the 2006–2012 period (Fig. 8.1). This lower use of RISA is probably related to the fact that the number of intergene sequences available in databases is much lower (8,788 entries) than that for the 16S (724,000) and 23S (21,427). In actinobacteria, a hypervariable region at the 5′ end of the 23S gene can even be used to make high-resolution phylogeny (Honerlage et al. 1994) or fine identification of strains (Roller et al. 1992).

Terminal-Restriction Fragment Length Polymorphism (T-RFLP)

For this technique DNA is extracted from different communities to be compared, and a conserved target region such as 16S rRNA gene is amplified with at least one fluorescent primer. The amplified DNA fragments are then purified and digested with a restriction enzyme, and the collection of fragments obtained is then separated by capillary electrophoresis with an automatic sequencer. Each restriction fragment, or T-RF, corresponds to a sequence differing from the others by its size, i.e., the distance between the fluorescent primer and the nearest restriction site from the fluorescent primer. For instance, this technique was used to assess the bacterial community dynamics in the hyporheic zone of an intermittent stream (Febria et al. 2012). A Web of Science search with T-RFLP as keyword in December 2012 permitted to retrieve 2,081 references. An increased use of the method is observed between the 2002–2012 period, though its use remains rather low (Fig. 8.1).

Cloning/Sequencing to Analyze the Diversity of Ribosomal Genes or That of Their Complementary DNA (cDNA)

After PCR (or RT-PCR), amplified fragments are cloned in *E. coli* using a TA plasmid (to accommodate the A base added by *Taq* polymerases at the 3′ end of the amplified fragment). The clones are grouped according to their RFLP pattern (restriction fragment length polymorphism), and the fragments are then sequenced. This approach provides a molecular inventory of microbial populations. It was used in particular to characterize the composition of bacterial communities of sediments contaminated by hydrocarbons (Bordenave et al. 2007) and extreme environments such as acid mine drainage (Bruneel et al. 2006). Banks are analyzed and compared by statistical tools as LIBSHUFF, and saturation curves are used to assess the depth of diversity assessment (see below). A Web of Science search with cloning/sequencing as keyword done in December 2012 permitted to retrieve 4,728 references. However, the use of this method slightly decreased during the 1996–2012 period (Fig. 8.1).

8.4.2.2 Data Analysis for Comparison of Community Structure and Diversity Between Samples

Whatever the method used and the targeted level (i.e., metabolic diversity, genetic diversity of phylogenetic or functional markers, using DNA or RNA), adequate approaches and statistical tests must be used to compare diversity or community structure between samples (Box 8.2).

Box 8.2: Testing Hypotheses About Biodiversity

Cindy Morris

Objective: Determine the impact of a genetically modified bacterium improving plant growth (PGPR) on the community of bacteria associated with plant roots.

Hypothesis: A genetically modified strain of PGPR *Pseudomonas fluorescens* introduced by seed coating has a significant effect on the structure (i.e., the richness) of the bacterial community associated with cucumber roots compared to the wild strain (Mahaffee and Kloepper 1997).

Testing the hypothesis: The diversity of culturable bacteria in the rhizosphere and roots was quantified by two indices – richness (total number of bacterial genera) and the Hill N_1 and N_2 indices:

$N_1 = e^{H'}$ where $H' = \Sigma\text{-}[(n_i/n)\ln(n_i/n)] = $ and $N_2 = 1/\lambda$ where $\lambda = \text{-}\Sigma[n_i(n_i\text{-}n)]/[n(n\text{-}1)]$

In both cases, n is the total number of individuals and n_i is the number of individuals in the ith class (genus level in this study). Significant differences between populations for plots inoculated with the two types of bacteria are determined by an analysis of variance.

Experimental design: The wild and genetically modified strains of *P. fluorescens* were used as inoculum in field plots. Six replicate plots for each bacterial treatment were established in a random pattern. The same field experiment was conducted for 2 years. For each replicate plot, the size of the bacterial populations in the rhizosphere and roots was determined by cultural methods. At each sampling time (7, 14, 28, 42, and 70 days after planting), 35 bacterial strains were collected at random from the populations in the rhizosphere and 25 among the endophytic populations. Among samples harvested at the same date, the strains collected from the same dilution allow comparisons and pooling. Strains were identified to genus level by fatty acid methyl ester profiling.

Calculation of diversity indices and statistical tests: Diversity indices were calculated for each sampling date. In the case of rhizosphere bacteria, 210 strains in total were collected at each date for all six plots combined. For endophytic populations, a total of 150 strains were collected at each date. So at each sampling date, the authors had two independent measures of diversity for each diversity index – a value for each year in which the experiment was conducted. This led to an analysis of variance to compare differences between treatments but with only one degree of freedom.

Note: For the same work load, an alternative experimental design would have led to two independent tests of the hypothesis. Instead of six repetitions in each field, an alternative plan could have been two independent fields, each containing three replicates. Indices would then have been calculated based on half the number of strains at each site, and statistical comparisons could be made each year. Nevertheless, this study helps to illustrate how an intense sampling effort to estimate diversity may have a very low statistical power. Even though the sample size was at each date between 150 and 210 strains, the true statistical power is only equal to the number of repeated measurements of the random variable – the diversity indices themselves.

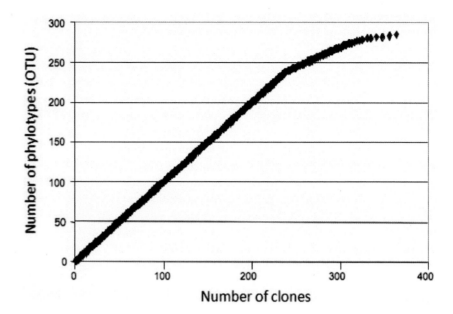

Fig. 8.3 Rarefaction analysis of a bank of 16S rRNA genes from a microbial mat. In the example shown, 364 clones were analyzed revealing a richness of 285 OTUs that is based on the Chao1 estimator (1125 OTUs) with a recovery of 25.33 %

Comparison of Phenotypic or Genetic Structures Among Samples Based on Fingerprints

Several methods like BIOLOG, DGGE, TGGE, SSCP, RISA, and T-RFLP yield fingerprints, i.e., patterns of metabolic levels, peaks, or bands which are characteristics of the community structure, though a given signal (e.g., a given band) cannot be attributed to a single known taxon. To compare these fingerprints, a range of multivariate methods can be applied. Though ANOVA can be used, NMDS (nonmetric multidimensional scaling) based on a Bray Curtis-like coefficient is particularly suitable to analyze matrices including a lot of zeros and to compare samples according to the (dis)similarity of the corresponding fingerprints. To analyze genetic fingerprints, the intensity and relative position of each DNA band/peak for each sample are digitally analyzed using an adequate software. Rank similarity matrices are computed for each community and used to construct "maps" highlighting the similarity/dissimilarity of genetic structures among samples (Kruskal and Wish 1978). Two- or three-dimensional maps can be chosen so that the stress factor (i.e., distortion factor between actual similarity rankings and the corresponding distance rankings in the map) is sufficiently low. Then, analysis of similarities (ANOSIM) can be performed to test for possible treatment effects on the structure of each bacterial community, one-way ANOSIM being performed to compare the genetic structures of samples pairs. ANOSIM results in the computation of p values (level of significance) and R statistics values (degree of discrimination between treatments: values around 0–1 for no discrimination and perfect discrimination, respectively). This approach was used to analyze RISA profiles, T-RFLP profiles, and DGGE fingerprints targeting four different communities for grassland soils by Patra and collaborators (2006).

Comparison of Community Diversity and Composition Between Samples Based on Sequence Banks

Methods like cloning/sequencing or pyrosequencing provide a list of sequences for each sample, which can be used to characterize the community diversity and composition (e.g., richness, evenness, major versus minor taxa). Typically, three steps of data analysis are used:

(i) *Chimeras*: The use of PCR on DNAs extracted from complex environments can cause the formation of chimeras, or hybrid DNA sequences where one fragment is used as primer on a different sequence. The program CHECK_CHIMERA of the "Ribosomal Database Project" (RDP) detects this kind of event by phylogenetic analysis of sequences by sliding window. If a chimera has been formed, the two ends of an amplicon will have different nearest phyletic neighbors. This type of event can of course occur naturally due to conjugation events followed by recombination, though very rarely.

(ii) *Rarefaction curves*: This technique (Colwell et al. 2004) allows us to estimate that the number of clones in a given bank is sufficient to yield a correct estimate of the sequence diversity of a sample (http://purl.oclc.org/estimates). The principle is to subsample a set of data and determine whether each new subgroup provides new sequences. The result is a curve that gradually becomes flat, the asymptote representing the maximum possible diversity (Fig. 8.3). Extrapolation of the curves can also be used to better estimate microbial richness, i.e., the likely number of sequences in each sample, by using a Chao1 estimate, for instance.

(iii) *Testing for differences between community composition*: LIBSHUFF ("LIBrary SHUFFling") (Singleton et al. 2001) can be used to determine if two clone libraries are significantly different. A first analysis will determine if the bank is representative of the diversity present in the sample. Then in a second step, the analysis compares two banks using the recovery value (coverage). The analysis can be done online (http://libshuff.Mib.Uga.Edu/doc.Html). In general, researchers conduct analysis whereas two sequences are different when they have less than 97 % similarity, which is consistent with the criteria for distinguishing the two species.

DNA Arrays

Sorting of sequences in a mixture of nucleic acids is most often done by capillary electrophoresis which yields a pattern of presence/absence of data for all lengths of fragments. An alternative is cloning/sequencing that is very expensive and low speed but yields phylogenetic information for each fragment analyzed. Before the development of pyrosequencing, a compromise had emerged between the two types of approaches with DNA microarrays that are medium to high throughput, are relatively inexpensive, and provide phylogenetic information and semiquantitative abundance data. The chips are still not widely used: indeed, whereas the per-use price is low, the necessary equipment is expensive and available in only a few laboratories. For this approach, the amplicons of a marker gene such as the 16S rRNA gene are amplified by incorporating a fluorochrome and hybridized on a glass slide with hundreds or thousands of oligonucleotides designed after the study of known sequences to represent a large number of individual bacterial taxonomic groups. The detection is performed by laser excitation and fluorescence reading for each of the plots, and quantification is made by comparison with a standardized scale. However, these data should be viewed as semiquantitative. This method was used to compare rhizobacterial community composition in soils suppressive or conducive to tobacco black root rot disease (Kyselkova et al. 2009). A Web of Science search with DNA arrays or microarrays as keywords until December 2012 permitted to retrieve 95,838 references, though microbial ecology represents only a small part of these papers. A strong increase in the use of this method was observed during the 1998–2005 period, with a maximum observed in 2010. However, a decrease in the use of microarrays is observed since 2010 (Fig. 8.1).

Pyrosequencing: An Increasingly Used Method to Quantify Microbial Diversity

The pyrosequencing technique is detailed in Sect. 18.1.2. Shortly, it consists in amplifying a large number of targeted sequences from genomic DNA and sequencing them, which allows a direct access to sequence identity and relative abundance. This approach has been developed during the last years, and it will greatly reduce the price of sequences and make the acquisition of environmental sequences easier and therefore more prevalent in microbial ecology projects. In particular, the approach does not require any step of preselection of sequences by restriction analysis. Pyrosequencing has been used to quantify microbial diversity in a range of environments such as plant rhizospheres (Lundberg et al. 2012), the human intestinal tract (Tasse et al. 2010), or marine biotopes (Bittner et al. 2013). A Web of Science search with pyrosequencing as keyword until December 2012 permitted to retrieve 3,614 references. An exponential increase in the use of this method was observed over the 2000–2012 period, and in 2012 this method became the most widely used after microarrays (Fig. 8.1).

Approaches Targeting RNA

DNA is a stable molecule selected over evolution to be the long-term memory of cell lineages. However, the stability of the DNA molecule is such that DNA of inactive cells present in a given biotope will be quantified as well as the DNA of active cells. Moreover, the DNA of dead cells remains detectable for years after cell death (Willerslev et al. 2004) until lesions (alkylation, oxidation) accumulate, making PCR less effective. However, there are now techniques that are used to separate the DNA of viable cells from DNA that is not associated with viable cells, prior to performing PCR reactions (Nocker et al. 2007). If the aim is the characterization of active cells/communities in a given biotope, it is possible to extract RNA from a given sample and study it. However, RNA is a more unstable molecule that is recycled rapidly by the cell so that the cell's instructions are modulated according to the changing environment. This implies one needs to be particularly cautious when studying environmental samples, for instance, by immediately freezing in liquid nitrogen the soil, sediment, or water samples in the field just after harvesting from their natural habitats.

Functional Genes

In theory, any gene can be used to study, by the PCR-based methods described above, the diversity of microorganisms performing a function in a given environment. However, various factors explain that some genes are more or less interesting for this kind of approach. The first factor is the sequence conservation required for primers to hybridize and therefore amplify all genes associated with a given function and only them. However, it is generally difficult to be certain that all proteins performing a function are homologous: for instance, the second step of denitrification ($NO_2^- \Rightarrow > NO$) is performed by two proteins, namely, the Cu-NIR enzyme with a copper cofactor and the NIR-cd1 enzyme with a heme cytochrome cofactor. These two proteins have no detectable sequence similarity and the two should thus be targeted in parallel (Coyne et al. 1989 ; Philippot et al. 2007). In other

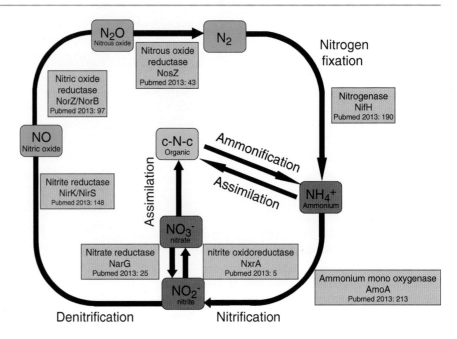

Fig. 8.4 Main functional genes used in microbial ecology to characterize the diversity (and sometimes abundance) of microorganisms involved in the nitrogen cycle. The figures indicate the number of sequences available for each functional gene in the NCBI database in May 2013 (Courtesy of Laurent Philippot)

cases, the sequence divergence of genes linked to a function can vary too much between phylogenetic groups for a single pair of primers to amplify all sequences corresponding to this gene. For instance, the second step of nitrification ($NO_2^- > NO_3^-$) is performed by different groups of nitrite-oxidizing bacteria which harbor the gene encoding nitrite oxidoreductase NxrA. However, groups such as *Nitrobacter* versus *Nitrospira* have sequences that are too different for this gene to be amplified by the same primers (Wertz et al. 2008). The second factor is, on the contrary, that it is also necessary that there is enough variation for the targeted sequence in different taxa to be distinguished. Finally, the concept of conservation of function between homologous genes, which is the foundation on which homology is defined, is sometimes difficult to assess. For example, in the case of alternative iron or vanadium nitrogenases, there is clearly a common origin, the function is retained, but with a non-negligible modification, i.e., the change of metal in the cofactor (Normand and Bousquet 1989). Despite these difficulties, many genes other than *rrs* are studied to characterize the diversity of microbial communities. This is particularly important when microbial functions rather than taxonomic diversity or population dynamics per se are studied. Indeed, the link between phylogenetic markers and functional ones is sometimes rather weak (Salles et al. 2012).

Nitrogen Cycle

The nitrogen cycle is probably the one where knowledge of the sequences of key genes of the various functions involved is the most advanced (Fig. 8.4). Given the importance of the nitrogen cycle for ecosystem functioning and dynamics, this explains the importance of studies of the diversity of microbial functional groups involved in this cycle. In particular,

many studies focused on the diversity of genes for nitrate reductase (*nar* and *nap*), nitrite reductase (*nirK* and *nirS*), and N_2O reductase (*nosZ*) involved in denitrification (Philippot 2002), whereas there have been few studies focused on NO reductase (*nor*). Similarly, many studies have characterized the diversity of nitrifiers targeting the *amoA* gene of nitrate-oxidizing bacteria and archaea (Kowalchuk et al. 1997) and more recently the gene *nxrA* of nitrite-oxidizing bacteria (Attard et al. 2010). In addition, the *nifH* gene involved in the synthesis of nitrogenase (the enzyme involved in the binding of dinitrogen), a function often studied in agronomy and microbial ecology due to the fact that nitrogen is often limiting for primary productivity, is the most conserved gene. In addition, this gene has a phylogeny overall congruent to that of the gene coding for 16S rRNA. It has been targeted by Zehr and McReynolds (1989) to study marine habitats and by Poly and collaborators (2001) to study soil environments.

Beyond DNA-based approaches targeting these functional genes, other approaches can also study their diversity, for example, by targeting mRNAs for better focus on active microorganisms (Zani et al. 2000) or by using 15 N-enriched DNA to target specifically diazotrophs (Buckley et al. 2007).

Phototrophy (*rbc*)

The *rbc* gene (ribulose bisphosphate carboxylase) codes for RuBisCO, the enzyme responsible for carbon fixation in photosynthetic organisms and in most autotrophic microorganisms. This enzyme also present in photosynthetic plants is qualified as the most abundant protein on earth. This gene has a complex phylogeny linked to specialization, duplications, and gene transfers (Delwiche and Palmer 1996; Watson and Tabita 1997). Nevertheless, this gene was used to study the composition of phytoplankton in lakes (Paul et al. 1990), microbial mats of volcanic deposits (Nanba et al. 2004), soils (Selesi et al. 2005), and even communities in subglacial Lake Vostok (Lavire et al. 2006).

Anoxygenic phototrophic bacteria play a prominent role in the colonization of ecosystems. They belong to different phylogenetic groups, and a way to analyze them is based on the analysis of the gene encoding a protein PufM which is the reaction center of photosynthesis enzyme (Achenbach et al. 2001).

Oxidation of Methane (*pmo*)

Methane-oxidizing bacteria oxidize the methane produced by methanogens or released from the deep earth. The gene responsible for this function, *pmo* (particulate methane monooxygenase), were used in PCR-sequencing approaches (Cebron et al. 2007). In addition, the survey technique with stable isotopes (stable isotope probing or SIP) was also used to identify the 16S bacteria involved in this function (Radajewski et al. 2000; Cebron et al. 2007).

Sulfur Cycle (*dsrAB* and *aprA*)

Bacteria of the sulfur cycle (reducing or oxidizing sulfur) are important in habitats such as bacterial mats, symbiotic invertebrate tissues, sediments, and black smokers from oceanic ridges. Several functional genes have been used to monitor sulfur-metabolizing bacterial communities, such as *dsrAB* genes (disulfite reductases) and *aprA* (coding the dissimilatory adenosine-5′-phosphosulfate (APS) reductase) (Meyer and Kuever 2007). In addition, it has been shown that *dsrAB* genes can confirm the phylogenetic links revealed by analysis of 16S rRNA genes (Wagner et al. 1998).

Other Genes Linked to a Function

Other functions can be listed as the assimilation of sulfur (sulfur esterase), phosphate uptake (phosphatase), the synthesis of protective pigments, synthesis of protective squalene lipids (squalene hopene cyclase), the synthesis of secondary metabolites (NRPS, PKS), and the catabolism of xenobiotics (diox), which have permitted or may permit diversity studies.

Metagenomic approaches (Sect. 18.3.3) are also increasingly yielding a mass of data to analyze. For instance, work has been done on riboswitches (*cf.* Box 9.1) in microbial communities (Kazanov et al. 2007).

8.4.3 How to Measure This Biodiversity?

Quantifying the biodiversity of an ecological community is usually done by computing a number which increases with biodiversity that is called a "biodiversity index." The value of this number may not have any meaning in itself, but a comparison of the number of different communities must let us know which harbors the largest diversity. Whatever the method used, different indices can be calculated to analyze the microbial diversity of the samples studied. As a first step, some indices are presented to illustrate the process of computation, and they are discussed in the context of microbial ecology. There are many indices of biodiversity in the literature, and it is not easy to make a wise choice in a particular context. That is why, in a second step, an experiment-based approach is presented. The idea consists in choosing an index of biodiversity which takes into account the specificity of the method used to acquire the data set.

Diversity indices are of two types: primary indices and subindices (or composite). The primary indices are direct measures of population parameters and do not require calculations confusing the number of species (richness) with the frequency of each species and their identity. The number of species or OTUs (Box 8.3) in a community and the frequency of a given OTU in a community are examples of primary indices. Secondary or composite indices are most often used in studies of biodiversity indices such as Shannon, Simpson, and Nei described below. It is important to distinguish between primary and secondary indices because they have an impact on the experimental devices that more or less facilitate statistical analyses. In general, it is often easier to perform parametric analyses (e.g., analysis of variance) on the primary indices than on subindices, which require some refinements. Trends in statistical processing of microbial biodiversity data demonstrate this situation (Morris et al. 2002). For a given workload (e.g.,

---

Box 8.3: OTU (Operational Taxonomical Unit)

Philippe Normand

Bacterial taxonomy includes different levels from phylum to subspecies, including order, family, genus, and species, as discussed in Chap. 1. The issue of diversity is studied at each of these levels, as long as individuals can be identified in a particular group. In the microbial world, it is often difficult to agree on the belonging of isolates to a species. It is for this reason that different authors have developed the concept of OTU used in different softwares for individuals or groups resulting from different aggregation procedures (Sneath 2005).

The definition of an OTU would be "a group of phylogenetically related organisms used in a study without specifying its taxonomic rank."

The concept of OTU has already led to a second one, that of molecular operational taxonomic units or MOTU (Floyd et al. 2002), to designate the OTUs obtained based on molecular approaches.

number of samples), it is often easier to have access to measures of variance of primary indices than those of secondary indices, thus allowing parametric analyses. In contrast, indices such as Shannon and Simpson are random variables that are quite suitable for parametric analyses. However, the need for experimental measurements of variance can be important (Box 8.1).

8.4.3.1 Species Richness Index

S is the number of species. This index whose meaning is a priori obvious and is widely used in animal and plant biology stumbles on the problem of the species concept, in the case of microorganisms, a group where sexuality is rare and atypical (*cf.* Box 6.1). There are many bacterial species, but the approaches are difficult to link with species sensu stricto. For these reasons, this descriptor is rarely used in microbial ecology. This problem of the taxonomic level to use is constant in the field of microbial ecology. In the remainder of the text, the word "species" may be replaced by different taxonomic levels, thus the widespread use of the term OTU, and notice that and the factor to consider is thus the resolution power of available techniques. In a hypolithic desert ecosystem, where water is the limiting factor for life, a correlation has been shown between species diversity (DGGE with 16S primers, with a 99 % identity threshold) and the availability of water, but not with temperature or rainfall (Pointing et al. 2007).

8.4.3.2 Shannon Alpha Diversity Index (*H'*)

This index is computed as

$$H' = -\sum_{i=1}^{S} p(i)\ln(p(i)) \qquad (8.1)$$

where i varies between 1 and S (=number of species) and $p(i)$ is the proportion of individuals belonging to the ith species or taxon. In a sample, the true value of $p(i)$ is unknown but is estimated by $n(i)/S$.

The value of H' is generally between 1.5 and 3.5 and rarely exceeds 4.5. This number is 0 if only one species is present (low biodiversity) and takes the maximum value $\ln(S)$ when S species are present and are evenly distributed: $p(i) = 1/S$.

The Shannon index, sometimes mistakenly called Shannon-Weiner or Shannon-Weaver, originally developed for use in the field of information theory (Shannon 1948), is often used in microbial ecology by changing "species" by taxon, and using a level of resolution in accordance with the tool. The major problem of this descriptor is that it does not take into account the identity of individuals. If in a sample 50 % of *Pseudomonas aeruginosa* and 50 % *Pseudomonas syringae* are present while in another sample 50 % of *Pseudomonas fluorescens* and 50 % of *Rhizobium meliloti* are present, both will yield the same Shannon index for composition while environmental functions are quite different. A Google search for "Shannon diversity index" yields 972,000 responses and 632 responses are provided by PubMed. This index has been used, for example, to quantify the impact of the retention time on the diversity of bacterial communities in a digester (Saikaly et al. 2005).

8.4.3.3 Simpson Index (*D*)

The Simpson index is computed as

$$D = \frac{1}{\sum_{i=1}^{S} p(i)^2} \quad \text{or} \quad D = 1 - \sum_{i=1}^{S} p(i)^2 \qquad (8.2)$$

where $p(i)$ is again the proportion of individuals belonging to the ith species and S is again the total number of species in the community. This descriptor, similar to the Shannon index, is a simple mathematical measure to quantify the diversity of species in a community. The meaning of this formula is precised below. If the number of individuals is large enough (which is the case in microbial ecology), the probability of randomly selecting an individual of species i is $p(i)$ and the probability of randomly selecting two individuals of the same species i is approximately $p(i)^2$. It follows that the probability, when S couples of individuals are randomly chosen, to have at least one set of two individuals of the same species is equal to the sum of the $p(i)^2$. In addition, if only one species is present (low diversity), this probability is 1, it decreases when the number of species increases and tends to 0 when all species are well distributed (high diversity). In order to have an index that increases with biodiversity to take the inverse of the sum of squares or to consider the probability of the opposite event, the second formulation is the probability that, choosing S pairs of individuals at random, all are made up of individuals of different species.

8.4.3.4 Nei Index

The index of Nei (1973) is more sophisticated because the measure makes the difference between a set comprising very different organisms and a set containing the same number of relatively close organisms. It is more limited in its applications because it considers only the level of a given population. It is also the most used in the study of population genetics. It indicates the average level of heterozygosity of populations and is therefore unsuitable for bacteria and archaea with only one chromosome, whereas it can be used for eukaryotic microorganisms. The way to calculate it is

$$H_S = \frac{1}{k}\sum_{i=1}^{S} H_{S_i} = \frac{1}{k}\sum_{i=1}^{S}\left(1 - q_i^2 - (1 - q_i)^2\right) \qquad (8.3)$$

where k is the total number of loci studied, $H_{S_i} = 1 - q_i^2 - (1 - q_i)^2$, and q_i is the frequency of one of the two alleles at the ith diallelic locus.

The Shannon, Simpson, Nei, and other indices permit to summarize quantitative parameters describing community structure. As described above for the Shannon index, they all have the same weakness, in that they do not take into account the identity of the organisms in a community.

Because diversity indices compile information obtained from a set of several species in a single value for each observation before comparison, it is not surprising that complex changes in patterns of diversity are not well converted into changes in these indices. For example, Widmer and collaborators (2006) have not observed significant changes in soil bacterial communities subjected to various treatments using diversity indices, while molecular analyses did show significant changes in bacterial fingerprint, as well as in the structure of communities, through multivariate analyses. Two strategies have been proposed to ensure consistent results between diversity indices and multivariate analysis techniques: the strategy "CA-species-richness" suitable for data where rare species are important and the strategy "nonsymmetric CA–Simpson" which is more appropriate for data where a species dominates (Pelissier et al. 2003).

Interestingly, indices described above can be integrated into a more general family that will allow to propose a theoretical framework for estimating biodiversity. The methods used to get the data on which the indices are calculated may be taken into account in the choice of these indices. This would enhance their relevance as it will be discussed briefly. Here is thus an example describing how to determine theoretically the choice of an index. The example is based on a counting method for studying biodiversity, based on the reassociation kinetics of single-stranded DNA. This example is taken from the article by Haegeman et al. (2008). All indices described in this section have been proposed in contexts other than microbial ecology. However, as already mentioned, the methods of observation and study of microbial diversity induce bias in the estimation of biodiversity through the previous indices. To summarize, each index corresponds to a few facets of biodiversity (specific, structural, functional, etc.), and its expression involves the number of individuals of a species, which cannot be achieved in practice by methods that skew their estimates. For example, the index of specific richness does not distinguish the relative abundance of species and includes all in the same way. But in microbial ecology, there is a large number of rare species. Simpson or Shannon indices seem therefore better suited to this context, in that they attribute gradually the relative importance of the different species according to their abundance. The few indices mentioned above does not constitute an exhaustive list and can be imbedded into the general family of diversity indices of Rényi (1961), defined by

$$R_\alpha = \frac{1}{1-\alpha} \ln\left(\sum_{s=1}^{S} p_s^\alpha\right) \qquad (8.4)$$

where α is a positive number and where R_1 must be understood as the limit of R_α as α tends toward 1. This results in the following relationships between Rényi indices and the indices presented previously (specific richness S, Shannon index H', and Simpson index D):

$$R_0 = \ln(S)$$
$$R_1 = H'$$
$$R_2 = \ln(D)$$

This approach (now α can be any positive number, which defines the family of indices) allows us to determine a priori which index should be used when a method is chosen. The principle is to try to get a relationship between the experimental methodology and the biodiversity indices considered. This process is easier if the choice of indices is large, which is the case with infinite number of indices provided by the family of Rényi.

Making assumptions about the distribution of specific abundances, we can more easily treat the case of rare species. Indeed, a rare species may be poorly sampled, and estimated abundances by molecular methods usually add uncertainty about the value of this estimate. However, estimates of diversity using indices depend heavily (sometimes several orders of magnitude) on the assumptions about the distribution of specific abundances.

Regarding the usual methods of microbial enumeration, they are based on the analysis of 16S rRNA and require amplification methods or cloning. The RNA strands are then amplified by PCR and analyzed by "fingerprint"-type methods, for instance. Other types of counting methods are also available, such as methods of reassociation kinetics of single-stranded DNA.

A simple model, based on the mass action law, to represent the reassociation kinetics is:

$$\frac{dC}{dt} = kC_s^2 \qquad (8.5)$$

where C_s is the concentration of single-stranded DNA corresponding to species s and k is a nonspecific parameter characterizing the reactivity of the DNA strands. The solution of this equation is

$$C_s(t) = \frac{C_s(0)}{1+kC_s(0)t} \qquad (8.6)$$

Consider the total concentration of the single strands of DNA:

$$C(t) = \sum_{s=1}^{S} C_s(t) = \sum_{s=1}^{S} \frac{C_s(0)}{1+kC_s(0)t} \qquad (8.7)$$

The relative abundance p_s is given by $p_s = C_s(0)/C(0)$, making it possible to rewrite the concentration of the single strands as follows:

$$C(t) = \sum_{s=1}^{S} \frac{p_s C(0)}{1 + k p_s C(0) t} \qquad (8.8)$$

Finally, a time rescaling is performed by considering $\tau = kC(0)t$, which allows to express the ratio of the concentration of single strands reassociated at any moment:

$$c(\tau) = \frac{c(\tau)}{c(0)} = \sum_{s=1}^{S} \frac{p_s}{1 + p_s \tau} \qquad (8.9)$$

This shows what the theoretical relationship is between the kinetics of pairing up and diversity index and allows us, using this relationship, to obtain a good estimate of microbial diversity. Denoted by τ_β is the time required for a fixed proportion β of the single strands initially present to be reassociated. Since the expression of $c(\tau)$ decreases with time τ, there is a unique moment τ_β for which we have $c(\tau_\beta) = \beta$. It is difficult to determine τ_β in the general case, but in the particular case where all species are equally distributed ($p_s = 1/S$), the expression of $c(\tau_\beta)$ simplifies as follows:

$$c(\tau) = \frac{1}{1 + \frac{\tau}{S}} \qquad (8.10)$$

From this formula, it can be deduced that $c(\tau_\beta) = \beta$ is equivalent to $\tau_\beta = S(1/\beta - 1)$. Moreover, the Rényi index, when species are equally distributed, is

$$R_\alpha = \frac{1}{1-\alpha} \ln\left(\sum_{s=1}^{S} \left(\frac{1}{S}\right)^\alpha\right) = \frac{1}{1-\alpha} \ln(S^{1-\alpha}) = \ln(S) \qquad (8.11)$$

Thus, in this particular case, there is a relationship between the time of reassociation τ_β and the Rényi index: R_α. Using a simulation model for the distributions of specific abundances, Haegeman et al. (2008) showed that the previous relationship still exists for more realistic specific abundance distributions, similar to what can be observed in microbial ecology. The authors show the following very interesting result: There is a strong relationship between ln (τ_β) and R_α for couples (α, β) satisfying a particular relationship. This relationship combines, for example, $\alpha = 0.5$ and $\beta = 0$ or $\alpha = 1$ and $\beta = 0.5$ or also $\alpha = 2$ and β close to 1. This means that the estimation of biodiversity by the method of pairing up is chosen, first, by fixing a value for β and, if this value is, for instance, 50 % of reassociation, then by using the index corresponding to Rényi $\alpha = 1$ that is the Shannon index. On the contrary, if a value of β close to 100 % of reassociation is set, then the Simpson index (e.g., Rényi with $\alpha = 2$) must be chosen for measuring biodiversity.

8.5 Procedures for the Study of Relations Between Microbial Biodiversity-Ecosystem Function

An important question for microbial ecologists is to assess the role of biodiversity in operations of microbial ecosystems. Most of the studies on biodiversity-function relationships have focused on macroorganisms and have shown that high levels of biodiversity can increase the performance of ecosystems (Loreau et al. 2001). The presence of a large diversity of species that respond differently to perturbations can moreover stabilize the functioning of ecosystems after disturbances. Very few similar studies have been conducted on microorganisms (Le Roux et al. 2006). This may seem surprising, since they play key roles in the functioning of ecosystems. Beyond the methodological difficulties that make very difficult quantitative and qualitative assessments of microbial diversity, a major reason that so few studies have been adequately conducted to analyze the relationships between diversity functioning is the immense diversity of these communities. It is estimated that there are tens or hundreds of thousands of bacterial OTUs per gram of soil (Gans et al. 2005; Torsvik et al. 2002) to be compared with approximately 10,000 bird species and 300,000 plant species on earth (Villenave et al. 2011). This has led some authors to consider that the level of functional redundancy (that is to say, the extent to which species are interchangeable in terms of the functions they provide) was probably very high in the case of microbial communities. In fact, the diversity of microbial communities is a tremendous reservoir of genes and functions, which may be interesting in response to extreme or new conditions, making the notion of functional redundancy relative. A good example of this is the de novo formation in soil bacteria of a gene permitting the degradation of lindane, a small molecule resulting from human activities; this gene was likely created over several decades by combining genetic material from different microbial soil species capable of metabolizing compounds structurally related to lindane (Boubakri et al. 2006).

In this context, different approaches have been used to analyze the relationship between diversity and functioning in microbial communities (Le Roux et al. 2006). Much work simply sought to analyze the (inverse) correlations between diversity and functioning of microbial communities by comparing different experimental situations. Typically, the question is: are changes in microbial activity levels, observed between treatments or in response to a disturbance related to concomitant changes in diversity of these microbial communities (Cavigelli and Robertson 2000)? Very often, the results show that changes in activity of a microbial community are not or poorly related to changes in composition/diversity of this community. Thus, the operation of methanogenic bioreactors is maintained over time despite significant temporal variations in the composition of the microbial community in the bioreactor (Fernandez et al. 1999). Similarly, the impact of different types of organic and mineral fertilization on the activity of the denitrifying community cannot be systematically linked to simultaneous changes in the diversity of the community (Enwall et al. 2005). However, some studies have reported significant correlations between microbial diversity and functioning. For instance, intake of mercury can induce a reduction in the diversity of the bacterial community of soils concomitant with a reduction in respiration and in resistance of this function against disruption by heat (Muller et al. 2002).

It has recently been shown that changes in the activity of microbial communities in situ are more correlated with some aspects of the diversity of these communities (such as abundance of some dominant species) than to their diversity taken very broadly (Patra et al. 2006). These authors showed that changes in the activity of the soil nitrifying, denitrifying or N-fixing communities observed between differently managed grassland ecosystems are weakly correlated with changes in overall diversity of this community but strongly correlated to changes in relative abundance of some major populations of nitrifiers, denitrifiers, or N-fixers, respectively. Moreover, the response of the activity of the nitrifying community of grassland soil after a nitrogen input strongly depended on the type of dominant nitrifying populations, only some of them being able to ensure a high level of nitrification in the presence of high levels of urea (Webster et al. 2005). However, these studies do not really target the causal relationships between microbial diversity and ecosystem functioning. Indeed, the impact of diversity changes on ecosystem functioning can be confused with the impact of changes in environmental factors, because some of these factors covary with the microbial diversity in situ. In addition, the abundance of microorganisms can also vary between the situations studied, which can largely explain changes in the level of activity of these communities rather than community diversity (Attard et al. 2011).

Another way to assess the importance of the diversity of microbial communities for their functioning in situ, while avoiding some of the biases presented above, is to "displace" the microbial communities from one environment to another, or to compare the functioning of microbial communities retrieved from different environments displaced into a unique environment. It was found that after transplantation of soil cores between forest and grassland sites, the denitrification activity varied while the diversity of denitrifiers did not change (Boyle et al. 2006). This type of study allows one in some cases to show that the functioning of microbial communities is more linked with environmental conditions than to their diversity.

In all the field studies described above, the experimenter does not choose the diversity of the communities to be studied and can only use the natural microbial communities with their preexisting diversity. However, experimental approaches permit to manipulate microbial diversity under conditions as controlled as possible and with equal population levels are necessary to analyze the causal relationships (or lack thereof) between diversity and functioning of microbial communities.

An approach to study more thoroughly the relationship between microbial diversity and ecosystem functioning is species removal. This approach consists in reducing, by following a predefined scenario, the diversity of microbial communities present in situ. The possible changes induced by this diversity erosion on ecosystem functioning are then characterized. In a study on soil microbial communities (Griffiths et al. 2001), it was shown that the decomposition of organic matter (mostly done by soil microorganisms) was not affected by erosion of microbial diversity. It has even been shown that the functioning and the resistance and resilience capabilities are largely insensitive to a marked reduction of diversity even for microbial communities with reduced diversity and providing specialized functions, such as the denitrifying and nitrifying communities (Wertz et al. 2007). In this last study, it was estimated that more than 99 % of the soil bacterial taxa had been removed without affecting the functioning of microbial communities.

Another approach to study the diversity-functioning relationships is to proceed by assembling species. Synthetic microbial communities with different levels of diversity are made in order to study their functioning. The approach allows one to analyze the interaction mechanisms and the complementarity between species that explain the observed biodiversity-functioning relationships. Such studies have been conducted on different fungal microorganisms or bacteria (Bell et al. 2005; van der Heijden et al. 1998). Some of these studies have shown a positive relationship between level of functioning and richness of the communities studied. However, the maximum number of species used in these approaches never did exceed a few tens, which is not commensurate with the richness of natural microbial communities. Using this approach, Salles and collaborators (2009) demonstrated the paramount importance of functional diversity among bacteria for diversity-functioning relationships and proposed a simple index, called community niche, to predict the functioning of bacterial assemblages.

A more systematic application of these approaches, for the most part recent ones, will teach us a lot about the importance of the diversity of microbial communities in the functioning of ecosystems and the underlying mechanisms. More generally, the microorganisms will become fabulous models to test ecological theories. Their short generation time implies that the responses of microbial communities functioning in the environment include both functional mechanisms, demographic and evolutionary aspects. In this context, the analysis of relationships between diversity and functioning requires to combine the approaches of functional ecology and evolutionary ecology.

8.5.1 Soils

Issues that arise with respect to the soil relate to plant pathogens, to the capacity to metabolize agrochemical products, better use of microbial communities for a sustainable agriculture and forestry, or effect of transgenic plants on microbial communities and on various biogeochemical cycles. It was found in recent years that soils and sediments were the biotopes with the richest microbial diversity (Torsvik et al. 2002). In these heterogeneous biotopes, the most adapted taxa can multiply without eliminating the less adapted, so they can remain at very low levels in some microhabitats waiting for more appropriate conditions, i.e., until the arrival of the symbiotic plant or recalcitrant substrate. The DNA of dead microorganisms can also be maintained on charged soil components such as clay for years (Picard et al. 1992) until a competent bacterium integrates it in its genome and modifies its metabolic capabilities.

Inventory by PCR sequencing of bacteria in a typical soil shows an abundance of Proteobacteria (alpha, gamma), of actinobacteria, and of acidobacteria (Herrera et al. 2007), the latter being essentially a taxonomic group of bacteria very rarely grown in pure culture and thus poorly characterized. This distribution changes depending on soil conditions and the addition of compounds such as nickel in soils around mining sites in New Caledonia decreases the relative proportion of alphaproteobacteria and increases that of betaproteobacteria (Hery et al. 2005).

8.5.2 Microbial Mats

Microbial mats, laminated structures of microorganisms that thrive in coastal areas, are particularly vulnerable to pollution by hydrocarbons. By their extreme metabolic wealth, these structures are considered model bacterial ecosystems to study the degradation of hydrocarbons. To estimate the impact of pollutants on the structure of microbial mats, mats from Camargue and from the Etang de Berre were studied. The Camargue mat, considered unpolluted, is located in a pre-concentration basin of seawater in the salt marshes of Salins-de-Giraud. The Etang de Berre mat is located close to an outlet from a treatment plant from an oil refinery and is therefore considered to be highly contaminated.

The vertical structure of microbial mats from Camargue was defined by combining molecular approaches, analysis of lipids, and confocal microscopy (Fourcans et al. 2004). Then the dynamics (Fourcans et al. 2006) and the phototrophic communities of sulfate-reducing bacteria (SRB) were highlighted during a circadian cycle. Population dynamics of anoxygenic purple bacteria (APB) were followed by T-RFLP targeting the *pufM* genes coding for a protein of the photosynthesis reaction center. Despite the identification of possible horizontal transfer of this gene, it remains, on the whole, a good genetic marker for monitoring APB (Achenbach et al. 2001). CCA on the T-RFLP data showed that the distribution of anoxygenic phototrophic bacteria (APB) was directly influenced by the abundance of cyanobacteria and *Microcoleus chthonoplastes* and *Halomicronema excentricum* and indirectly through their metabolism (Fig. 8.5). The main parameters influencing the vertical migration of APB are pH and oxygen concentrations, suggesting that the mechanisms of air and energy taxis control the positioning of these bacteria in the depth of the mat (Fourcans et al. 2006).

SRB analysis was performed by targeting specific sequences of the 16S rRNA genes. Since the SRB are divided into five classes, pairs of primers were designed to specifically target each of these classes (Daly et al. 2000). The SRB can be targeted by specific functional genes such as those involved in sulfate reduction. The *dsr*AB genes encoding subunits of the disulfite reductase are most commonly used because they allow on the one hand to follow the same phylogenetic relationships as the 16S rRNA and on the other hand many sequences are available in databases (Wagner et al. 1998). Oxygen and sulfur are the main parameters influencing the behavior of SRB. Molecular analyses of diversity helped highlight new populations of SRB tolerant to high concentrations of oxygen. The study of such bacterial strains will help to understand the mechanisms of resistance to oxygen.

These mats (Fig. 8.6) were maintained in laboratory microcosms to assess the impact of oil on bacterial communities and in turn their effect on degradation (Bordenave et al. 2007) as they are on the front line of marine pollution. The most important observation from this study was the resilience of the bacterial community following contamination. The bacterial community structure changed after more than a month. However, the analysis of the active community (rRNA analysis) showed that the change occurred upon addition of the pollutant.

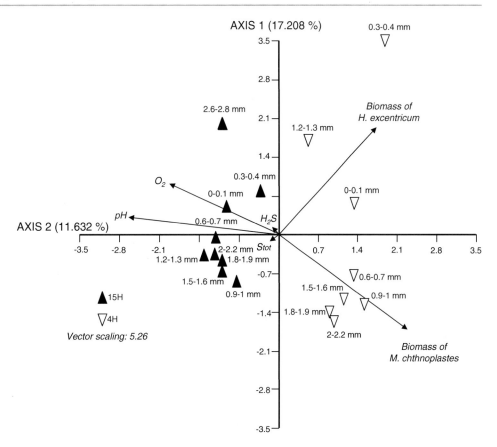

Fig. 8.5 Canonical correspondence analysis (CCA): comparing anoxygenic communities of phototrophic bacteria (APB) for each microbial mats depth during the daytime at 15 h (*closed triangles*) to those during the night at 4 h (*open triangles*), and the following environmental parameters: *Microcoleus chthonoplastes* biomass, pH, and concentration of O_2, H_2S, and Stot (total sulfur) (Courtesy of Robert Duran)

Gammaproteobacteria dominate initially the community and then *Alphaproteobacteria* and *Bacillaceae* take over. These developments confirm that the bacterial communities adapt to the presence of the pollutant, and then the microbial mat regains its original structure 1 year after infection. Functional studies have been undertaken to highlight the mechanisms of adaptation to the presence of oil.

Mats of the Etang de Berre exposed to chronic contamination were characterized by molecular and cultural approaches. Analyses by DGGE and RISA showed that the two sites exposed to different levels of contamination possessed different bacterial communities but similar diversity indices (Villanueva et al. 2007). Comparing the community composition showed that *Gamma-* and *Alphaproteobacteria* were abundant in both sites, while phylotypes associated with *Deltaproteobacteria* and group WS3 were detected in the most polluted site. Consortia capable of degrading hydrocarbons were obtained from enrichment media. The identification of bacterial species present in these consortia showed that the species involved in oil degradation belonged to the order *Rhodobacterales*. These observations suggest that pollution does not affect the richness of microbial mats, and it seems that the populations involved in the degradation represent only a minor fraction of the bacterial community.

8.5.3 Aquatic Biodiversity, Water Blooms

Issues that arise with respect to aquatic pathogens involve animals and man, pollution, currents, and especially algal blooms. Rivers carry bacterial communities that vary in size and composition depending on inocula from sewage treatment plants, fecal contamination (septic tanks, breeding animals), runoff from surfaces, etc. Bacterial communities also fluctuate depending on irradiation, temperature, nutrients, and predation by phage or eukaryotic (Jardillier et al. 2005). These fluctuations are especially studied to identify the arrival of fecal coliforms associated to agricultural and water treatment plants (Lemarchand and Lebaron 2003). A body of work based on a comparison of molecular and cultural approaches reveals that the physiological state of bacteria in the community plays a central role in the transmission of diseases to humans because of the presence of viable but non-culturable strains (*cf.* Box 15.4), it thus seems important to use more than one type of approach.

The other major problem with bacterial diversity in aquatic biotopes is understanding the factors controlling algal blooms. These sudden increases in the proportion of cyanobacteria are not only spectacular and threatening to fish, there is also a public health risk due to the presence of toxin-producing species that are taken up the food chain and in particular in

Fig. 8.6 Microbial mats at the Polynesian atoll, called Kopara. As the majority of microbial mats, these mats dominated by cyanobacteria (green layer on the surface), have a layered structure in which the sulfur photosynthetic bacteria and nonsulfur red bacteria (pink layer) are organized according to gradients of light, oxygen, and sulfur. The low iron content of these mats is not sufficient to trap the sulfide produced by sulfate-reducing bacteria. Thus, the absence of precipitated iron sulfide allows to observe the different microbial pigments such as the carotenoids (gelatinous layer) (Courtesy of Pierre Caumette)

shellfish. Several studies have sought to predict factors associated with blooms in general and toxin production in particular. It is clear from this work that the main factor limiting the production of water blooms is the amount of phosphorus, which is in contradiction with the general principle that nitrogen is the factor most often limiting primary production, e.g., in soil, in marine systems, and freshwater. The debate between proponents of phosphorus and nitrogen as responsible for blooms is not only a scientific discussion but also an intense political debate with active lobbyings, that of farmers that use nitrate and the producers of phosphorus-based detergent powder, who both wish to be granted a responsibility for blooms as low as possible (Barroin 2003). It is clear from these discussions that the addition of nitrogen does indeed increase primary production but also the addition of phosphate, particularly with regard to nitrogen-fixing bacteria like many cyanobacteria.

However, the problem is more complex in terms of toxin production in the cyanobacterium *Planktothrix* because toxin-producing clones coexist alongside nonproducing clones, and one of the factors explaining the proportion seems to be cell density (Yepremian et al. 2007). It has also been shown that clones of *Microcystis* toxin producers were disadvantaged under conditions of higher cell density (Briand et al. 2008).

8.5.4 Sea

Regarding the oceans, many studies have focused on the natural functioning of the carbon and nitrogen cycles and changes induced by anthropogenic influences and also molecular approaches for the identification of microbes present in the water column (benthic) and in the sediment (pelagos). Multiphasic approaches coupling markers (16S rRNA and *nifH*) and activity measurements (nitrogenase) in a tropical lagoon have demonstrated daily fluctuations and the presence of a high proportion of unknown taxa (Bauer et al. 2008). The distribution of OTUs followed by RISA and activity measurements in the basin of San Pedro in Southern California seems uncorrelated with variations related to depth (Hewson et al. 2007). Work on molecular identification of picoplankton cells (that range from 0.2 to 3 μm) have demonstrated a large diversity compared to that previously known but with OTUs having a worldwide distribution (Moreira and Lopez-Garcia 2002 ; Vaulot et al. 2008). Finally, work on ocean metagenomic approaches has been initiated. This work, focused on DNA sequencing of microbial communities in several sites of the ocean (Venter et al. 2004), has permitted to identify many unknown sequences in particular and also permitted to see that among photosynthetic bacteria those belonging to the genera *Prochlorococcus* and *Synechococcus* were clearly dominant.

Regarding sediments, they have been described as the type of habitat with the highest diversity (Torsvik et al. 2002) with 5×10^{10} genome equivalents per gram. This is an often anoxic biotope, receiving a large variety of molecules, some of which are resistant and therefore accumulate. Moreover, in that biotope are found burrowing animals that therefore mix and homogenize the sediment ("turbators"), contributing to the activities of associated microflora and to microbial diversity. There are also many uncultivated endolithic taxa including several chemolithotrophic bacteria capable of oxidizing Fe (II) (Edwards et al. 2003).

Finally, a relatively recent development concerns study of viruses in the oceans. These small organisms are naturally present in all habitats but are more difficult to analyze than from soil, sediment, or the digestive tract. They are now considered to be responsible for controlling the numbers of microorganisms and also the composition of the microbial community with a hypothesis called "phage kills the winner" (Hoffmann et al. 2007), which would favor scarce taxa. This hypothesis is strongly debated (Bouvier and del Giorgio

2007), and a lot of work in the coming years can be expected to support the different hypotheses. Eukaryotic viruses are also abundant in this habitat (Monier et al. 2008), and metagenomic approaches thus now include also phages (Williamson et al. 2008).

8.5.5 Digestive Tract and Microbial Communities

The questions that arise regarding the digestive tract relate to pathogens, to competition and gas production, and to the effect of food on the comparison between populations and between age groups.

Microbial ecology of the human gut has been studied extensively especially regarding some diseases, especially gastric ulcers and Crohn's disease.

Foods sometimes contain large amounts of microorganisms, especially fermented or stale foods. In addition, birth and contact with the mother and the outside world are important bacterial inocula to colonize the digestive tract. Molecular inventories of bacteria in the stomach have demonstrated the presence of 128 phylotypes, belonging mainly to *Proteobacteria*, *Firmicutes*, *Actinobacteria*, *Bacteroidetes*, and *Fusobacteria* (Bik et al. 2006). Most of these bacteria do not persist in inhospitable habitats such as the stomach with a pH of 3. However, *Helicobacter pylori* has the ability, based on the presence of urease to convert urea present in the stomach to CO_2 and ammonium, which will then neutralize the acid pH of the stomach. This capability has the consequence that this bacterium is present in 80 % of human stomachs and causing in many of them ulcers and tumors. Warren and Marshall have also won the Nobel Prize for Medicine in 2005 for their demonstration of the role of *Helicobacter pylori* in this disease (Forbes et al. 1994) and the ensuing possibility of treating ulcers with antibiotics or with bismuth. In fact, the bacteria is transmitted within the family, probably swallowed by the child from the mother's saliva, and the bacteria is then implanted in the gastric mucosa. This bacterium mutates quickly so that typing it allows to distinguish human population, based on samples taken from patients with ulcers, which allowed Wirth and collaborators (2004) to show by analysis of samples taken from human populations that Ladakh Muslims and Buddhists had very different *Helicobacter pylori* populations. In fact, this pathogen to commensal ratio behaves as a very sensitive marker of human populations.

The normal human gut has a much larger and more diverse microbial community, including enterobacteria, *Bifidobacterium*, and *Bacteroidetes*. The composition of the community varies along the digestive tract; it also fluctuates throughout human life (Hayashi et al. 2003), in particular as a function of the diet (Hayashi et al. 2002) and of course according to the physiological state of the host. Many pathogens are also known and discussed in Chap. 11. If the intestinal microbial community is roughly comparable from one individual to another within a human population, methanogens on the contrary are only present in about half of the individuals, which means that half the human population produces methane, while the other half produces CO_2.

Microorganisms in the gut play several important roles in the absorption of nutrients, energy metabolism, and defense against microbial pathogens (Braun and Wei 2007). There are several inflammatory bowel diseases (IBD), the best known being Crohn's disease and ulcerative colitis. The etiology of Crohn's disease is unknown. It results in irritation and inflammation of all or part of the gastrointestinal tract, which leads to digestive difficulties. Several pathogens have been tentatively associated to it such as *Mycobacterium* spp. or *Fusobacterium* or an imbalance in the composition of the intestinal community with a decrease in the proportion of *Firmicutes* and *Bacteroidetes* (Frank et al. 2007), but no consensus could be reached. However, antibiotic treatment resulting in lower counts in some members of the communities of the intestine was correlated with a decrease in symptoms of colitis (Nomura et al. 2005). It is now recognized that the normal gut community acts as a barrier against pathogens, preventing them to establish themselves, to multiply, and cause various diseases.

Other animals have intestinal flora that are very different from those of man. Ruminants in particular consume large amounts of cellulose, a polymer that mammals cannot metabolize in the absence of cellulase. However, some cellulolytic bacteria can grow in the digestive tract, either in the rumen or in the single mammalian stomach; these bacteria are only in the cecum and the glucose produced can only be absorbed through coprophagia, a practice found in some animals such as horses, rabbits, mice, rats, and insects. In the rumen, cellulolytic bacteria are found such as *Fibrobacter succinogenes*, *Ruminococcus albus*, and *Ruminococcus flavefaciens* (Mosoni et al. 2007). These two genera have contrasting physiological characteristics regarding the production of methane. If several foods did not affect the microbial balance, the addition of live yeast has allowed to change it in a favorable direction (increase in the proportion of *Ruminococci*), which had the effect of reducing the amount of methane produced. The synthesis of amino acids (Torrallardona et al. 1996), vitamins (Coburn et al. 1989), and dinitrogen fixation (Bergersen and Hipsley 1970; Potrikus and Breznak 1977) or the synthesis of a sleep-inducing factor (Brown et al. 1988) have all been proposed as related to intestinal bacteria.

Another aspect that will continue to expand in the coming years is mycotoxins humans and animals consume in their normal diet and their effects on health, starting with the carcinogenic aflatoxin produced by the wheat-colonizing *Aspergillus flavus*, the psychotropic ergotamine produced by the rye ergot pathogen *Claviceps purpurea*, and the immunotoxic patulin produced by several apple-colonizing molds, such as *Aspergillus* and *Penicillium*. The list of such toxins (ochratoxin, citrinin, fumonisins, trichothecenes, zearalenone, beauvericin, enniatins, butenolide, equisetin, fusarins) is thus likely to expand in the coming years along with studies on their effects on human health (Juan-García et al. 2013).

It is likely that work on nutrition will be increased in the coming years to identify risk factors for various diseases. We can already see the work on the link between nutrition, intestinal metabolites, colorectal cancer, and microbial communities (O'Keefe et al. 2007) or between microflora and irritable bowel syndrome (Lee and Pimentel 2006). We should also see work on other aspects such as the link between microbial communities and cardiovascular diseases, immune diseases, and neurological diseases.

8.5.6 Skin and Other Body Parts

The digestive tract is the part of the body with the highest density of microbial cells, as compared to the skin and mucous membranes such as the urogenital or the nasal cavity that also host microbial communities both pathogenic and nonpathogenic. A study by pyrosequencing targeting the 16S rRNA gene has permitted to identify on the hands' skin over 150 phylotypes per individual out of a total of 4,742 phylotypes in the sample with strong variations between individuals and between the two hands and as explanatory "factors of diversity, gender, laterality, and time since last hand washing" (Fierer et al. 2008). In the case of superficial and deeper skin of the elbow, one study showed the presence of bacteria belonging to six divisions, with a dominance of Proteobacteria and a structure similar to that of mouse skin (Grice et al. 2008). Another study on the skin of the arms instead showed the presence of Actinobacteria, Firmicutes, and Proteobacteria as dominant taxa (Gao et al. 2007). Another area studied in humans for its bad smell is the armpit where long-chain fatty acids secreted by the skin are converted into smelly volatile compounds. One study showed the presence of four major bacterial groups (staphylococci, aerobic coryneform, micrococci, and propionibacteria) and yeasts of the genus *Malassezia* with a correlation between odor intensity and presence of aerobic coryneform (Taylor et al. 2003).

The healthy vagina has been studied in comparison with that of patients infected with HIV or by bacterial vaginosis by pyrosequencing of 16S rRNA. This study showed a greater bacterial diversity in patients with bacterial vaginosis, which was not the case of HIV-infected women. Healthy women had a predominance of *Lactobacillus* spp. (96 %) with taxa such as *Prevotella*, *Megasphaera*, *Gardnerella*, *Coriobacterineae*, *Lachnospira*, *Sneathia*, *Propionibacterineae*, *Citrobacter*, and *Anaerococcus* present below 1 %. The most surprising is the existence of a greater diversity in women with vaginitis and HIV (Spear et al. 2008).

Milk and colostrum have also been studied, showing the presence of dominant bacteria such as *Staphylococcus epidermidis* and *Enterococcus faecalis*, followed by *Streptococcus mitis*, *Propionibacterium acnes*, and *Staphylococcus lugdunensis* without pathogenic bacteria (Martin et al. 2007; Jimenez et al. 2008). Similar work on cow's milk was also conducted.

8.5.7 Depollution

Anaerobic digesters are attractive biotechnological tools to reduce pollution caused by organic waste. To successfully carry out such processes, the biological anaerobic ecosystem must be fully characterized. The microorganisms involved in this process are generally divided into four trophic groups. The first three steps (hydrolysis, acidogenesis, and acetogenesis) are performed by *Bacteria*, while the last step (methanogenesis) is performed by *Archaea*. Acidogenic and methanogenic groups have nutritional needs, physiological pH optima, and growth requirements that are completely different. Digesters combine conventional acidogenic bacteria and archaeal methanogens in a single reactor imposing uniform conditions for both groups. To optimize the process, it makes sense to separate the two steps in two reactors. Digesters have been developed where two-phase acidification and methanogenesis take place in two reactors connected in series. Monitoring of bacterial communities by SSCP showed that bacterial communities are different between the two reactors. During the implementation process, the bacterial communities were significantly altered, and a significant reduction in bacterial diversity was observed in both reactors (Khelifi et al. 2009). It seems that both reactors are two different ecosystems with one population being common to both reactors. The challenge is to maintain control of these communities when these digesters are used to treat other types of pollutants such as polycyclic aromatic hydrocarbons (Bernal-Martinez et al. 2007). A metagenomics approach has demonstrated the presence in an anaerobic digester of bacteria belonging to an abundant taxon unknown so far, named WWE3 (Waste Water of Evry 3) (Guermazi et al. 2008).

8.5.8 Deep Earth

Analysis of aquifers and deep environments requires the development of appropriate sampling methods. Contamination of the samples, where diversity is low, can be a major problem. Nevertheless, an adapted sampling method based on T-RFLP analysis of bacterial communities has been developed (Basso et al. 2005). Bacterial communities were characterized by molecular inventories and cultivation methods. This approach has allowed access to a new source of diversity, including bacterial strains trapped in geological layers, permitting to address issues of paleomicrobiology and the origins of life on earth. Energy sources and carbon are limited in this environment, apart from hydrothermal vents, where hydrogen flow from the thermolysis of water allows chemolithotrophic bacteria (hydrogen, reduced iron) to thrive (Kimura et al. 2005). Sources of carbon and nitrogen are rare but nevertheless constitute hot spots where many taxa are active and have been thriving probably for considerable periods of time (Parkes et al. 2005).

8.5.9 Agricultural and Forestry Systems

The effects of agricultural and forestry practices on the functioning and diversity of soil microbial communities have been quantified in many studies. Some studies have characterized the impact of agricultural practices and land uses on key microbial functional groups such as denitrifiers or nitrifiers and the possible feedbacks for agroecosystem functioning and services (Attard et al. 2011; Philippot et al. 2007). Similarly, forestry practices and tree species have been demonstrated to influence taxonomic and functional diversity of soil bacteria and fungi (Buee et al. 2009).

8.5.10 Polluted Environments: The Case of Acid Mine Drainages

Acid mine drainages (AMD) are considered extreme environments because on the one hand they have low pH and on the other they have high concentrations of toxic compounds. These environments that are hostile to life are also sometimes considered good models to address the problem of the origin of life on earth. The composition of bacterial communities in the AMD of a former lead mine was followed in relation to seasonal variations and with changes in the concentration of dissolved oxygen in the water. These acidic waters (pH ~ 2) are rich in arsenic (300 mg/L), iron (20,000 mg/L), and sulfate (7,500 mg/L). The main bacteria isolated from this environment belonged to the genera *Acidithiobacillus* and *Thiomonas*, genera frequently encountered in these environments (Bruneel et al. 2006). T-RFLP analyses showed a low diversity. The molecular inventory has highlighted the role of sulfate-reducing bacteria (SRB) in this environment (Bruneel et al. 2006). Efforts are underway to isolate SRB strains that play a role in the transformation of arsenic. More recent metagenomic studies allowed to identify putative bacteria involved in the AMD functioning (Bertin et al. 2011), and the challenge is now to isolate such bacteria and characterize their role.

8.6 Conclusion

Biodiversity is increasingly seen as a resource to be preserved, and microbial biodiversity is no exception to this rule. Microbial biodiversity has given us antibiotics, beer, cheese, and bioactive molecules of many kinds, and it also allows wastewater treatment and the growth of plants in soil (Box 8.4). It is moreover an overall indicator of the status of certain biotopes impacted by anthropogenic pressures, biotopes that must be preserved so that they can fulfill various functions and deliver services, on which our well-being depends. This is why it is important to quantify and compare the different disturbances that biotopes can withstand before reaching tipping points.

The approaches used to study microbial biodiversity should also make significant progress both technically due to the tools currently developed, e.g., for high-throughput sequencing, and analytically because of new websites which democratize mathematical and bioinformatic approaches.

Microbial biodiversity should continue to be addressed by scientists for basic reasons but also due to societal issues linked to pollution, sustainable agriculture and forestry, health and productivity of oceans and freshwaters, or global change. Fundamental work should further seek to establish the causal link between microbial diversity and ecosystem functioning. In this context, the capacity to acquire and analyze very large data sets is a major challenge, but sequencing of nucleic acids increases continuously with associated bioinformatic approaches, rapidly offering new perspectives. These approaches should help identify taxa present in a given biotope but also to understand which genes are expressed in a given situation. A sticking point is our ability to identify the function(s) of genes; this bottleneck will certainly not be broken for years, which is the time necessary to do solid work on microbial physiology. More generally, the functional traits concept used in ecology has been increasingly used in

> **Box 8.4: Industrial Exploitation of Microbial Biodiversity**
>
> Robert Duran
>
> The food industry is a major user of microbial biodiversity. The French cheese industry, for example, produces hundreds of soft or semihard cheese, using several species of lactic acid bacteria (*Lactococcus*, *Streptococcus*, *Lactobacillus*, *Leuconostoc*) that contribute to the acidification of milk and also participate in sensory characteristics through the synthesis of lytic enzymes and propionic bacteria (*Propionibacterium*) that produce a variety of compounds, in particular carbon dioxide that in turn produces holes. Fungi (yeasts of the genera *Kluyveromyces*, *Debaryomyces*, and *Saccharomyces* and molds of the genus *Penicillium*) are used for the production of aromas; *Geotrichum candidum* changes the appearance of goat cheese imparting a "toad skin," *P. camembertii* contributes to the formation of a layer of felt on Camembert, and *P. roqueforti* causes a marbling of blue cheese. Hundreds of strains have been selected for each type of cheese over the centuries (www.inra.fr/layout/set/print/la_science_et_vous/apprendre_experimenter/monde_microbien/menons_l_enquete/la_biodiversite_des_microorganismes_des_produits_laitiers) (Box Fig. 8.1).
>
>
>
> **Box Fig. 8.1** Photograph of a cheese plate illustrating the variety of cheeses, diversity that is related to the microorganisms selected by man over the centuries (Photograph: Robert Duran)
>
> **Example of Green Chemistry**
>
> Acrylamide is used mainly to produce polyacrylamide that is a water-soluble thickener useful for treatment of wastewaters, gel electrophoresis, papermaking, ore processing, oil production, and the manufacture of fabrics. Production of acrylamide from acrylonitrile is made using bacteria that have nitrile hydratase. It is one of the first economic successes in microbial biotechnology in the field of chemistry.
>
> In 2001 the global demand for acrylamide was estimated at more than 200,000 tons/year, ¾ of which being produced by a microbiological process. In the 1990s, the Japanese company Nitto developed a microbiological process involving the strain *Pseudomonas chlororaphis* B23 (Nishiyama et al. 1991). This process is capable of producing 10,000 tons of acrylamide per year. Currently, a strain of *Rhodococcus rhodochrous*, a third-generation biocatalyst, is used by SNF Floerger located in Saint Etienne (France). This company produces about 100,000 tons/year by a microbiological method with different production units installed worldwide. Molecular tools targeting the reaction site of nitrile hydratase were developed to explore the presence of this gene in different ecosystems (Precigou et al. 2001).

microbial ecology during the last years, which may provide further insights in the causal relationships between microbial biodiversity and ecosystem functioning. Another sticking point is our ability to link the diversity of metabolites and the diversity and activities of microorganisms in a range of ecosystems. Integrative microbiology made on model microorganisms and collaboration between microbial ecologists and scientists working on natural substances should help defining these links. Ecological approaches have immense prospects for current ecotoxicological issues; they now largely guide studies on the impact of pollutants in the environment on environmental and human health and ecosystem biodiversity and functioning. The ability of researchers to assess to what extent the diversity of microbial communities are important for ecosystem functioning and services, and the resilience capacity of these facing global change factors is a great challenge for which social and political expectations are high.

References

Achenbach LA, Carey J, Madigan MT (2001) Photosynthetic and phylogenetic primers for detection of anoxygenic phototrophs in natural environments. Appl Environ Microbiol 67:2922–2926

Amann RI, Ludwig W, Schleifer KH (1995) Phylogenetic identification and in situ detection of individual microbial cells without cultivation. Microbiol Rev 59:143–169

Andrews JH, Kinkel LL, Berbee FM, Nordheim EV (1987) Fungi, leaves, and the theory of island biogeography. Microb Ecol 14:277–290

Attard E, Poly F, Commeaux C, Laurent F, Terada A, Smets BF, Recous S, Le Roux XL (2010) Shifts between *Nitrospira*- and *Nitrobacter*-like nitrite oxidizers underlie the response of soil potential nitrite oxidation to changes in tillage practices. Environ Microbiol 12:315–326

Attard E, Recous S, Chabbi A, De Berranger C, Guillaumaud DN, Labreuche J, Philippot L, Schmid B, Le Roux X (2011) Soil environmental conditions rather than denitrifier abundance and diversity drive potential denitrification after changes in land-uses. Global Change Biol 17:1975–1989

Barroin G (2003) Gestion des risques. Santé et environnement : le cas des nitrates, Phosphore, azote et prolifération des végétaux aquatiques. Courr Environ 48:13–26

Basso O, Lascourreges J-F, Jarry M, Magot M (2005) The effect of cleaning and disinfecting the sampling well on the microbial communities of deep subsurface water samples. Environ Microbiol 7:13–21

Bauer K, Diez B, Lugomela C, Seppala S, Borg AJ, Bergman B (2008) Variability in benthic diazotrophy and cyanobacterial diversity in a tropical intertidal lagoon. FEMS Microbiol Ecol 63:205–221

Bell T, Newman JA, Silverman BW, Turner SL, Lilley AK (2005) The contribution of species richness and composition to bacterial services. Nature 436:1157–1160

Bensaadi-Merchermek N, Salvado JC, Cagnon C, Karama S, Mouchès C (1995) Characterization of the unlinked 16s rDNA and 23S-5s rRNA operon of *Wolbachia pipientis*, a prokaryotic parasite of insect gonads. Gene 165:81–86

Bergersen FJ, Hipsley EH (1970) The presence of N2-fixing bacteria in the intestines of man and animals. J Gen Microbiol 60:61–65

Bernal-Martinez A, Carrère H, Patureau D, Delgenès JP (2007) Ozone pre-treatment as improver of PAH removal during anaerobic digestion of urban sludge. Chemosphere 68:1013–1019

Bertin PN et al (2011) Metabolic diversity among main microorganisms inside an arsenic-rich ecosystem revealed by meta- and proteo-genomics. ISME J 5:1735–1747

Bik EM, Eckburg PB, Gill SR, Nelson KE, Purdom EA, Francois F, Perez-Perez G, Blaser MJ, Relman DA (2006) Molecular analysis of the bacterial microbiota in the human stomach. Proc Natl Acad Sci U S A 103:732–737

Bittner L et al (2013) Diversity patterns of uncultured Haptophytes unravelled by pyrosequencing in Naples Bay. Mol Ecol 22:87–101

Blanchot J, Rodier M (1996) Picophytoplankton abundance and biomass in the western tropical Pacific Ocean during the 1992 El Nino year: Results from flow cytometry. Deep Sea Res Part I: Oceanogr Res Pap 43:877–895

Bordenave S, Goni-Urriza MS, Caumette P, Duran R (2007) Effects of heavy fuel oil on the bacterial community structure of a pristine microbial mat. Appl Environ Microbiol 73:6089–6097

Bossio DA, Scow KM (1998) Impacts of carbon and flooding on soil microbial communities: phospholipid fatty acid profiles and substrate utilization patterns. Microb Ecol 35:265–278

Boubakri H, Beuf M, Simonet P, Vogel TM (2006) Development of metagenomic DNA shuffling for the construction of a xenobiotic gene. Gene 375:87–94

Bouvier T, del Giorgio PA (2007) Key role of selective viral-induced mortality in determining marine bacterial community composition. Environ Microbiol 9:287–297

Boyle SA, Rich JJ, Bottomley PJ, Cromack K, Myrold DD (2006) Reciprocal transfer effects on denitrifying community composition and activity at forest and meadow sites in the Cascade Mountains of Oregon. Soil Biol Biochem 38:870–878

Braun J, Wei B (2007) Body traffic: ecology, genetics, and immunity in inflammatory bowel disease. Annu Rev Pathol 2:401–429

Briand E, Escoffier N, Straub C, Sabart M, Quiblier C, Humbert JF (2008) Spatiotemporal changes in the genetic diversity of a bloom-forming *Microcystis aeruginosa* (cyanobacteria) population. ISME J 3:419–429

Brown R, Price RJ, King MG, Husband AJ (1988) Autochthonous intestinal bacteria and coprophagy: a possible contribution to the ontogeny and rhythmicity of slow wave sleep in mammals. Med Hypotheses 26:171–175

Bruneel O, Duran R, Casiot C, Elbaz-Poulichet F, Personne JC (2006) Diversity of microorganisms in Fe-As-rich acid mine drainage waters of Carnoules, France. Appl Environ Microbiol 72:551–556

Buckley DH, Huangyutitham V, Hsu SF, Nelson TA (2007) Stable isotope probing with 15N2 reveals novel noncultivated diazotrophs in soil. Appl Environ Microbiol 73:3196–3204

Buee M, Reich M, Murat C, Morin E, Nilsson RH, Uroz S, Martin F (2009) 454 Pyrosequencing analyses of forest soils reveal an unexpectedly high fungal diversity. New Phytol 184:449–456

Cavigelli MA, Robertson GP (2000) The functional significance of denitrifier community composition in a terrestrial ecosystem. Ecology 81:1402–1414

Cebron A, Bodrossy L, Stralis-Pavese N, Singer AC, Thompson IP, Prosser JI, Murrell JC (2007) Nutrient amendments in soil DNA stable isotope probing experiments reduce the observed methanotroph diversity. Appl Environ Microbiol 73:798–807

Coburn SP, Mahuren JD, Wostmann BS, Snyder DL, Townsend DW (1989) Role of intestinal microflora in the metabolism of vitamin B-6 and 4'-deoxypyridoxine examined using germfree guinea pigs and rats. J Nutr 119:181–188

Colwell RK, Mao CX, Chang J (2004) Interpolating, extrapolating, and comparing incidence-based species accumulation curves. Ecology 85:2717–2727

Coyne MS, Arunakumari A, Averill BA, Tiedje JM (1989) Immunological identification and distribution of dissimilatory heme cd1 and nonheme copper nitrite reductases in denitrifying bacteria. Appl Environ Microbiol 55:2924–2931

Daly K, Sharp RJ, McCarthy AJ (2000) Development of oligonucleotide probes and PCR primers for detecting phylogenetic subgroups of sulfate-reducing bacteria. Microbiology 146:1693–1705

Delwiche CF, Palmer JD (1996) Rampant horizontal transfer and duplication of rubisco genes in eubacteria and plastids. Mol Biol Evol 13:873–882

Edwards KJ, Bach W, Rogers DR (2003) Geomicrobiology of the ocean crust: a role for chemoautotrophic Fe-bacteria. Biol Bull 204:180–185

Eiland F, Klamer M, Lind AM, Leth M, Baath E (2001) Influence of initial c/n ratio on chemical and microbial composition during long term composting of straw. Microb Ecol 41:272–280

Enwall K, Philippot L, Hallin S (2005) Activity and composition of the denitrifying bacterial community respond differently to long-term fertilization. Appl Environ Microbiol 71:8335–8343

Fang J, Chan O, Kato C, Sato T, Peeples T, Niggemeyer K (2003) Phospholipid FA of piezophilic bacteria from the deep sea. Lipids 38:885–887

Febria CM, Beddoes P, Fulthorpe RR, Williams DD (2012) Bacterial community dynamics in the hyporheic zone of an intermittent stream. ISME J 6:1078–1088

Fernandez A, Huang S, Seston S, Xing J, Hickey R, Criddle C, Tiedje J (1999) How stable is stable? Function versus community composition. Appl Environ Microbiol 65:3697–3704

Fierer N, Hamady M, Lauber CL, Knight R (2008) The influence of sex, handedness, and washing on the diversity of hand surface bacteria. Proc Natl Acad Sci U S A 105:17994–17999

Fisher MM, Triplett EW (1999) Automated approach for ribosomal intergenic spacer analysis of microbial diversity and its application to freshwater bacterial communities. Appl Environ Microbiol 65:4630–4636

Floyd R, Eyualem A, Papert A, Blaxter ML (2002) Molecular barcodes for soil nematode identification. Mol Ecol 11:839–850

Forbes GM, Glaser ME, Cullen DJ, Warren JR, Christiansen KJ, Marshall BJ, Collins BJ (1994) Duodenal ulcer treated with *Helicobacter pylori* eradication: seven-year follow-up. Lancet 343:258–260

Fourcans A et al (2004) Characterization of functional bacterial groups in a hypersaline microbial mat community (Salins-de-Giraud, Camargue, France). FEMS Microbiol Ecol 51:55–70

Fourcans A, Solé A, Diestra E, Ranchou-Peyruse A, Esteve I, Caumette P, Duran R (2006) Vertical migration of phototrophic bacterial populations in a hypersaline microbial mat from Salins-de-Giraud (Camargue, France). FEMS Microbiol Ecol 57:367–377

Frank DN, St Amand AL, Feldman RA, Boedeker EC, Harpaz N, Pace NR (2007) Molecular-phylogenetic characterization of microbial community imbalances in human inflammatory bowel diseases. Proc Natl Acad Sci U S A 104:13780–13785

Frostegard A, Tunlid A, Baath E (2011) Use and misuse of PLFA measurements in soils. Soil Biol Biochem 43:1621–1625

Gamo M, Shoji T (1999) A method of profiling microbial communities based on a most-probable-number assay that uses BIOLOG plates and multiple sole carbon sources. Appl Environ Microbiol 65:4419–4424

Gans J, Wolinsky M, Dunbar J (2005) Computational improvements reveal great bacterial diversity and high metal toxicity in soil. Science 309:1387–1390

Gao Z, Tseng CH, Pei Z, Blaser MJ (2007) Molecular analysis of human forearm superficial skin bacterial biota. Proc Natl Acad Sci U S A 104:2927–2932

Garland JL, Mills AL (1991) Classification and characterization of heterotrophic microbial communities on the basis of patterns of community level sole-carbon-source utilization. Appl Environ Microbiol 57:2351–2359

Gast WRJ, Scott DW, Schmitt C et al (1991) Blue Mountains forest health report—new perspectives in forest health. U. S. Department of Agriculture, Forest Service, Pacific Northwest Region, Portland, p 182

Grice EA et al (2008) A diversity profile of the human skin microbiota. Genome Res 18:1043–1050

Griffiths BS et al (2001) An examination of the biodiversity-ecosystem function relationship in arable soil microbial communities. Soil Biol Biochem 33:1713–1722

Guermazi S et al (2008) Discovery and characterization of a new bacterial candidate division by an anaerobic sludge digester metagenomic approach. Environ Microbiol 10:2111–2123

Haack SK, Garchow H, Klug MJ, Forney LJ (1995) Analysis of factors affecting the accuracy, reproducibility, and interpretation of microbial community carbon source utilization patterns. Appl Environ Microbiol 61:1458–1468

Haegeman B, Vanpeteghem D, Godon J-J, Hamelin J (2008) DNA reassociation kinetics and diversity indices: richness is not rich enough. Oikos 117:177–181

Hayashi H, Sakamoto M, Benno Y (2002) Fecal microbial diversity in a strict vegetarian as determined by molecular analysis and cultivation. Microbiol Immunol 46:819–831

Hayashi H, Sakamoto M, Kitahara M, Benno Y (2003) Molecular analysis of fecal microbiota in elderly individuals using 16S rDNA library and T-RFLP. Microbiol Immunol 47:557–570

Herrera A, Héry M, Stach JEM, Jaffré T, Normand P, Navarro E (2007) Species richness and phylogenetic diversity comparisons of soil microbial communities affected by nickel-mining and revegetation efforts in New Caledonia. Eur J Soil Biol 43:130–139

Hery M, Herrera A, Vogel TM, Normand P, Navarro E (2005) Effect of carbon and nitrogen input on the bacterial community structure of Neocaledonian nickel mine spoils. FEMS Microbiol Ecol 51:333–340

Hewson I, Jacobson Meyers ME, Fuhrman JA (2007) Diversity and biogeography of bacterial assemblages in surface sediments across the San Pedro Basin, Southern California Borderlands. Environ Microbiol 9:923–933

Hoffmann KH, Rodriguez-Brito B, Breitbart M, Bangor D, Angly F, Felts B, Nulton J, Rohwer F, Salamon P (2007) Power law rank-abundance models for marine phage communities. FEMS Microbiol Lett 273:224–228

Honerlage W, Hahn D, Zepp K, Zeyer J, Normand P (1994) A hypervariable region provides a discriminative target for specific characterization of uncultured and cultured *Frankia*. Syst Appl Microbiol 17:433–443

Huber H, Hohn MJ, Rachel R, Fuchs T, Wimmer VC, Stetter KO (2002) A new phylum of Archaea represented by a nanosized hyperthermophilic symbiont. Nature 417:63–67

Jardillier L, Boucher D, Personnic S, Jacquet S, Thenot A, Sargos D, Amblard C, Debroas D (2005) Relative importance of nutrients and mortality factors on prokaryotic community composition in two lakes of different trophic status: microcosm experiments. FEMS Microbiol Ecol 53:429–443

Jenkins GJ, Burlinson B, Parry JM (2000) The polymerase inhibition assay: a methodology for the identification of DNA-damaging agents. Mol Carcinog 27:289–297

Jessup CM, Kassen R, Forde SE, Kerr B, Buckling A, Rainey PB, Bohannan BJ (2004) Big questions, small worlds: microbial model systems in ecology. Trends Ecol Evol 19:189–197

Jimenez E, Delgado S, Fernandez L, Garcia N, Albujar M, Gomez A, Rodriguez JM (2008) Assessment of the bacterial diversity of human colostrum and screening of staphylococcal and enterococcal populations for potential virulence factors. Res Microbiol 159:595–601

Juan-García A, Manyes L, Ruiz MJ, Font G (2013) Applications of flow cytometry to toxicological mycotoxin effects in cultured mammalian cells: a review. Food Chem Toxicol 56:40–59

Kazanov MD, Vitreschak AG, Gelfand MS (2007) Abundance and functional diversity of riboswitches in microbial communities. BMC Genomics 8:347

Khelifi E, Bouallagui H, Touhami Y, Godon JJ, Hamdi M (2009) Bacterial monitoring by molecular tools of a continuous stirred tank reactor treating textile wastewater. Bioresour Technol 100:629–633

Kimura H, Sugihara M, Yamamoto H, Patel BK, Kato K, Hanada S (2005) Microbial community in a geothermal aquifer associated with the subsurface of the Great Artesian Basin, Australia. Extremophiles 9:407–414

Kinkel LL, Andrews JH, Berbee FM, Nordheim EV (1987a) Leaves as islands for microbes. Oecologia 71:405–408

Kinkel LL, Andrews JH, Nordheim EV (1987b) Microbial introduction to apple leaves: influences of altered immigration on fungal community dynamics. Microb Ecol 18:161–173

Kinkel LL, Andrews JH, Nordheim EV (1989) Fungal immigration dynamics and community development on apple leaves. Microb Ecol 18:45–58

Kowalchuk GA, Stephen JR, De Boer W, Prosser JI, Embley TM, Woldendorp JW (1997) Analysis of ammonia-oxidizing bacteria of the beta subdivision of the class Proteobacteria in coastal sand dunes by denaturing gradient gel electrophoresis and sequencing of PCR-amplified 16S ribosomal DNA fragments. Appl Environ Microbiol 63:1489–1497

Kruskal JB, Wish M (1978) Multidimensional scaling. Sage University Series, Beverly Hills

Kyselkova M, Kopecký J, Frapolli M, Défago G, Ságová-Marecková M, Grundmann GL, Moënne-Loccoz Y (2009) Comparison of rhizobacterial community composition in soil suppressive or conducive to tobacco black root rot disease. ISME J 3:1127–1138

Lavire C, Normand P, Alekhina I, Bulat S, Prieur D, Birrien J-L, Fournier P, Hänni C, Petit J-R (2006) Presence of *Hydrogenophilus thermoluteolus* DNA in accretion ice in the subglacial Lake Vostok, Antarctica, assessed using *rrs*, *cbb* and *hox*. Environ Microbiol 8:2106–2114

Lawrence D, Schoenike R, Quispel A, Bond G (1967) The role of *Dryas drummondii* in vegetation development following ice recession at Glacier Bay, Alaska, with special reference to its nitrogen fixation by root nodules. J Ecol 55:793–813

Le Roux X, Philippot L, Degrange V, Poly F, Wertz S (2006) Relations entre biodiversité et fonctionnement chez les communautés microbiennes. Biofutur 268:50–53

Le Roux X et al (2013) Soil environmental conditions and microbial build-up mediate the effect of plant diversity on soil nitrifying and denitrifying enzyme activities in temperate grasslands. PLoS One 8: e61069. doi:10.1371/journal.pone.0061069

Leclerc M, Delgenes JP, Godon JJ (2004) Diversity of the archaeal community in 44 anaerobic digesters as determined by single strand conformation polymorphism analysis and 16S rDNA sequencing. Environ Microbiol 6:809–819

Lee HR, Pimentel M (2006) Bacteria and irritable bowel syndrome: the evidence for small intestinal bacterial overgrowth. Curr Gastroenterol Rep 8:305–311

Lemarchand K, Lebaron P (2003) Occurrence of *Salmonella* spp and *Cryptosporidium* spp in a French coastal watershed: relationship with fecal indicators. FEMS Microbiol Lett 218:203–209

Lichter J (1998) Primary succession and forest development on coastal Lake Michigan dunes. Ecol Monogr 68:487–510

Loreau M et al (2001) Biodiversity and ecosystem functioning: current knowledge and future challenges. Science 294:804–808

Lundberg DS et al (2012) Defining the core *Arabidopsis thaliana* root microbiome. Nature 488:86–90

MacArthur RH, Wilson EO (1963) An equilibrium theory of insular zoogeography. Evolution 17:373–387

Mahaffee WF, Kloepper JW (1997) Bacterial communities of the rhizosphere and endorhiza associated with fieldgrown cucumber plants inoculated with a plant growth-promoting rhizobacterium or its genetically modified derivative. Can J Microbiol 43:344–353

Martin R, Heilig HG, Zoetendal EG, Jimenez E, Fernandez L, Smidt H, Rodriguez JM (2007) Cultivation-independent assessment of the bacterial diversity of breast milk among healthy women. Res Microbiol 158:31–37

Matsui K, Jun MS, Ueki M, Kawabata Z (2001) Functional succession of bacterioplankton on the basis of carbon source utilization ability by BIOLOG plates. Ecol Res 16:905–912

Mazzella N, Molinet J, Syakti AD, Barriol A, Dodi A, Bertrand J-C, Doumenq P (2005) Effects of pure *n*-alkanes and crude oil on bacterial phospholipid classes and molecular species determined by electrospray ionization mass spectrometry. J Chromatogr B 822:40–53

McCann KS (2000) The diversity-stability debate. Nature 405:228–233

Meyer B, Kuever J (2007) Molecular analysis of the diversity of sulfate-reducing and sulfur-oxidizing prokaryotes in the environment, using *aprA* as functional marker gene. Appl Environ Microbiol 73:7664–7679

Mitchell CE, Tillman D, Groth JV (2002) Effects of grassland species diversity, abundance and composition on foliar fungal disease. Ecology 83:1713–1726

Monier A, Claverie JM, Ogata H (2008) Taxonomic distribution of large DNA viruses in the sea. Genome Biol 9:R106

Moreira D, Lopez-Garcia P (2002) The molecular ecology of microbial eukaryotes unveils a hidden world. Trends Microbiol 10:31–38

Morris CE et al (2002) Microbial biodiversity: approaches to experimental design and hypothesis testing in primary scientific literature from 1975 to 1999. Microbiol Mol Biol Rev 66:592–616

Mosoni P, Chaucheyras-Durand F, Bera-Maillet C, Forano E (2007) Quantification by real-time PCR of cellulolytic bacteria in the rumen of sheep after supplementation of a forage diet with readily fermentable carbohydrates: effect of a yeast additive. J Appl Microbiol 103:2676–2685

Muller AK, Westergaard K, Christensen S, Sorensen SJ (2002) The diversity and function of soil microbial communities exposed to different disturbances. Microb Ecol 44:49–58

Muyzer G, Smalla K (1998) Application of denaturing gradient gel electrophoresis (DGGE) and temperature gradient gel electrophoresis (TGGE) in microbial ecology. Antonie Van Leeuwenhoek 73:127–141

Muyzer G, de Waal EC, Uitterlinden AG (1993) Profiling of complex microbial populations by denaturing gradient gel electrophoresis analysis of polymerase chain reaction-amplified genes coding for 16S rRNA. Appl Environ Microbiol 59:695–700

Naeem S (2002) Ecosystem consequences of biodiversity loss : the evolution of a paradigm. Ecology 83:1537–1552

Nanba K, King GM, Dunfield K (2004) Analysis of facultative lithotroph distribution and diversity on volcanic deposits by use of the large subunit of ribulose 1,5-bisphosphate carboxylase/oxygenase. Appl Environ Microbiol 70:2245–2253

Nei M (1973) Analysis of gene diversity in subdivided populations. Proc Natl Acad Sci U S A 70:3321–3323

Nishiyama M, Horinouchi S, Kobayashi M, Nagasawa T, Yamada H, Beppu T (1991) Cloning and characterization of genes responsible for metabolism of nitrile compounds from *Pseudomonas chlororaphis* B23. J Bacteriol 173:2465–2472

Nocker A, Sossa-Fernandez P, Burr MD, Camper AK (2007) Use of propidium monoazide for live/dead distinction in microbial ecology. Appl Environ Microbiol 73:5111–5117

Nomura T et al (2005) Mucosa-associated bacteria in ulcerative colitis before and after antibiotic combination therapy. Aliment Pharmacol Ther 21:1017–1027

Normand P, Bousquet J (1989) Phylogeny of nitrogenase sequences in *Frankia* and other nitrogen-fixing microorganisms. J Mol Evol 29:436–447

Normand P, Ponsonnet C, Nesme X, Neyra M, Simonet P (1996) ITS analysis of prokaryotes. In: Akkermans ADL, van Elsas JD, De Bruijn FJ (eds) Molecular microbial ecology manual. Kluwer Academic Publishers, Dordrecht, pp 1–12

Nubel U, Garcia-Pichel F, Kuhl M, Muyzer G (1999) Quantifying microbial diversity: morphotypes, 16S rRNA genes, and carotenoids of oxygenic phototrophs in microbial mats. Appl Environ Microbiol 65:422–430

O'Keefe SJ, Chung D, Mahmoud N, Sepulveda AR, Manafe M, Arch J, Adada H, van der Merwe T (2007) Why do African Americans get more colon cancer than Native Africans? J Nutr 137:175S–182S

Parkes RJ et al (2005) Deep sub-seafloor prokaryotes stimulated at interfaces over geological time. Nature 436:390–394

Patra AK et al (2006) Effects of management regime and plant species on the enzyme activity and genetic structure of N-fixing, denitrifying and nitrifying bacterial communities in grassland soils. Environ Microbiol 8:1005–1016

Patra AK, Le Roux X, Abbadie L, Clays-Josserand A, Poly F, Loiseau P, Louault F (2007) Effect of microbial activity and nitrogen mineralization on free-living nitrogen fixation in permanent grassland soils. J Agron Crop Sci 193:153–156

Paul JH, Cazares L, Thurmond J (1990) Amplification of the *rbcL* gene from dissolved and particulate DNA from aquatic environments. Appl Environ Microbiol 56:1963–1966

Pelissier R, Couteron P, Dray S, Sabatier D (2003) Consistency between ordination techniques and diversity measurements: two strategies for species occurrence data. Ecology 84:242–251

Philippot L (2002) Denitrifying genes in bacterial and archaeal genomes. Biochim Biophys Acta 1577:355–376

Philippot L, Hallin S, Schloter M (2007) Ecology of denitrifying prokaryotes in agricultural soil. Adv Agron 96:249–305

Picard C, Ponsonnet C, Paget E, Nesme X, Simonet P (1992) Detection and enumeration of bacteria in soil by direct DNA extraction and polymerase chain reaction. Appl Environ Microbiol 58:2717–2722

Pointing SB, Warren-Rhodes KA, Lacap DC, Rhodes KL, McKay CP (2007) Hypolithic community shifts occur as a result of liquid water availability along environmental gradients in China's hot and cold hyperarid deserts. Environ Microbiol 9:414–424

Poly F, Ranjard L, Nazaret S, Gourbiere F, Monrozier LJ (2001) Comparison of *nifH* gene pools in soils and soil microenvironments with contrasting properties. Appl Environ Microbiol 67:2255–2262

Potrikus CJ, Breznak JA (1977) Nitrogen-fixing *Enterobacter agglomerans* isolated from guts of wood-eating termites. Appl Environ Microbiol 33:392–399

Precigou S, Goulas P, Duran R (2001) Rapid and specific identification of nitrile hydratase (NHase) encoding genes in soil samples by polymerase chain reaction. FEMS Microbiol Lett 204:155–161

Radajewski S, Ineson P, Parekh NR, Murrell JC (2000) Stable-isotope probing as a tool in microbial ecology. Nature 403:646–649

Ramette A, Tiedje JM (2007) Biogeography: an emerging cornerstone for understanding prokaryotic diversity, ecology, and evolution. Microb Ecol 53:197–207

Ranjard L, Brothier E, Nazaret S (2000) Sequencing bands of ribosomal intergenic spacer analysis fingerprints for characterization and microscale distribution of soil bacterium populations responding to mercury spiking. Appl Environ Microbiol 66:5334–5339

Ratcliff AW, Busse MD, Shestak CJ (2006) Changes in microbial community structure following herbicide (glyphosate) additions to forest soils. Appl Soil Ecol 34:114–124

Rényi A (1961) On measures of entropy and information. In: Neyman GE (ed) Proceedings of the fourth Berkeley symposium on mathematical statistics and probability. University of California Press, Berkeley, pp 547–561

Ritchie NJ, Schutter ME, Dick RP, Myrold DD (2000) Use of length heterogeneity PCR and fatty acid methyl ester profiles to characterize microbial communities in soil. Appl Environ Microbiol 66:1668–1675

Roller C, Ludwig W, Schleifer KH (1992) Gram-positive bacteria with a high DNA G + C content are characterized by a common insertion within their 23S rRNA genes. J Gen Microbiol 138:1167–1175

Roscher C, Schumacher J, Weisser WW, Schmid B, Schulze ED (2007) Detecting the role of individual species for overyielding in experimental grassland communities composed of potentially dominant species. Oecologia 154:535–549

Saikaly PE, Stroot PG, Oerther DB (2005) Use of 16S rRNA gene terminal restriction fragment analysis to assess the impact of solids retention time on the bacterial diversity of activated sludge. Appl Environ Microbiol 71:5814–5822

Salles JF, Poly F, Schmid B, Le Roux X (2009) Community niche predicts the functioning of denitrifying bacterial assemblages. Ecology 90:3324–3332

Salles JF, Le Roux X, Poly F (2012) Relating phylogenetic and functional diversity among denitrifiers and quantifying their capacity to predict community functioning. Front Microbiol 3:0209

Sapp M, Wichels A, Gerdts G (2007) Impacts of cultivation of marine diatoms on the associated bacterial community. Appl Environ Microbiol 73:3117–3120

Schrenk MO, Kelley DS, Delaney JR, Baross JA (2003) Incidence and diversity of microorganisms within the walls of an active deep-sea sulfide chimney. Appl Environ Microbiol 69:3580–3592

Selesi D, Schmid M, Hartmann A (2005) Diversity of green-like and red-like ribulose-1,5-bisphosphate carboxylase/oxygenase large-subunit genes (*cbbL*) in differently managed agricultural soils. Appl Environ Microbiol 71:175–184

Shannon CE (1948) A mathematical theory of communication. Bell Syst Tech J 27:379–423

Sheffield VC, Cox DR, Lerman LS, Myers RM (1989) Attachment of a 40-base-pair G + C-rich sequence (GC-clamp) to genomic DNA fragments by the polymerase chain reaction results in improved detection of single-base changes. Proc Natl Acad Sci U S A 86:232–236

Singleton DR, Furlong MA, Rathbun SL, Whitman WB (2001) Quantitative comparisons of 16S rRNA gene sequence libraries from environmental samples. Appl Environ Microbiol 67:4374–4376

Sneath PHA (2005) Numerical taxonomy. In: Garrity GM (ed) Systematic bacteriology. Springer, East Lansing, pp 39–42

Spear GT, Sikaroodi M, Zariffard MR, Landay AL, French AL, Gillevet PM (2008) Comparison of the diversity of the vaginal microbiota in HIV-infected and HIV-uninfected women with or without bacterial vaginosis. J Infect Dis 198:1131–1140

Stach JE, Maldonado LA, Ward AC, Goodfellow M, Bull AT (2003) New primers for the class Actinobacteria: application to marine and terrestrial environments. Environ Microbiol 5:828–841

Takacs D (1996) The idea of biodiversity: philosophies of paradise. The Johns Hopkins University Press, Baltimore, 393 p

Tang JC, Shibata A, Zhou Q, Katayama A (2007) Effect of temperature on reaction rate and microbial community in composting of cattle manure with rice straw. J Biosci Bioeng 104:321–328

Tasse L et al (2010) Functional metagenomics to mine the human gut microbiome for dietary fiber catabolic enzymes. Genome Res 20:1605–1612

Taylor D, Daulby A, Grimshaw S, James G, Mercer J, Vaziri S (2003) Characterization of the microflora of the human axilla. Int J Cosmet Sci 25:137–145

Torrallardona D, Harris CI, Fuller MF (1996) Microbial amino acid synthesis and utilization in rats: the role of coprophagy. Br J Nutr 76:701–709

Torsvik V, Salte K, Sorheim R, Goksoyr J (1990) Comparison of phenotypic diversity and DNA heterogeneity in a population of soil bacteria. Appl Environ Microbiol 56:776–781

Torsvik V, Ovreas L, Thingstad TF (2002) Prokaryotic diversity-magnitude, dynamics, and controlling factors. Science 296:1064–1066

van der Heijden MGA, Klinomoros JN, Ursic M, Moutoglis P, Streitwolf-Engel R, Boller T, Wiemken A, Sanders IR (1998) Mycorrhizal fungal diversity determines plant biodiversity, ecosystem variability and productivity. Nature 396:69–71

Vaulot D, Eikrem W, Viprey M, Moreau H (2008) The diversity of small eukaryotic phytoplankton (<or =3 microm) in marine ecosystems. FEMS Microbiol Rev 32:795–820

Venter JC et al (2004) Environmental genome shotgun sequencing of the Sargasso Sea. Science 304:66–74

Villanueva L, Navarrete A, Urmeneta J, White DC, Guerrero R (2007) Analysis of diurnal and vertical microbial diversity of a hypersaline microbial mat. Arch Microbiol 188:137–146

Villenave C, Saj S, Attard E, Klumpp K, Le Roux X (2011) Grassland management history affects the response of the nematode community to changes in above-ground grazing regime. Nematology 13:995–1008

Vincent WF, Mueller DR, Bonilla S (2004) Ecosystems on ice: the microbial ecology of Markham Ice Shelf in the high Arctic. Cryobiology 48:103–112

Vitousek P, Walker L, Whiteaker L, Mueller-Dombois D, Matson P (1987) Biological invasion by *Myrica faya* alters ecosystem development in Hawaii. Science 238:802–804

Vogel TM et al (2009) TerraGenome: a consortium for the sequencing of a soil metagenome. Nat Rev Microbiol 7:252. doi:10.1038/nrmicro2119

Wagner M, Roger A, Flax JL, Brusseau GA, Stahl DA (1998) Phylogeny of dissimilatory reductase supports an early origin of sulphate respiration. J Bacteriol 180:2975–2982

Watson R, Blackwell B (2000) Purification and characterization of a common soil component which inhibits the polymerase chain reaction. Can J Microbiol 46:633–642

Watson GM, Tabita FR (1997) Microbial ribulose 1,5-bisphosphate carboxylase/oxygenase: a molecule for phylogenetic and enzymological investigation. FEMS Microbiol Lett 146:13–22

Webster G, Embley TM, Freitag TE, Smith Z, Prosser JI (2005) Links between ammonia oxidizer species composition, functional diversity and nitrification kinetics in grassland soils. Environ Microbiol 7:676–684

Wertz S, Degrange V, Prosser JI, Poly F, Commeaux C, Guillaumaud N, Le Roux X (2007) Decline of soil microbial diversity does not influence the resistance and resilience of key soil microbial functional groups following a model disturbance. Environ Microbiol 9:2211–2219

Wertz S, Poly F, Le Roux X, Degrange V (2008) Development and application of a PCR-denaturing gradient gel electrophoresis tool to study the diversity of *Nitrobacter*-like *nxrA* sequences in soil. FEMS Microbiol Ecol 63:261–271

Whitman WB, Coleman DC, Wiebe WJ (1998) Prokaryotes: the unseen majority. Proc Natl Acad Sci U S A 95:6578–6583

Whittaker RH (1972) Evolution and measurement of species diversity. Taxon 21:213–251

Widmer F, Rasche F, Hartmann M, Fliessbach A (2006) Community structures and substrate utilization of bacteria in soils from organic and conventional farming systems of the DOK long-term field experiment. Appl Soil Ecol 33:294–307

Wilcox BA (1984) In situ conservation of genetic resources: determinants of minimum area requirements. In: McNeely JA, Miller KR (eds) National parks, conservation and development: the role of protected areas in sustaining society. Smithsonian Institution Press, Washington, DC, pp 639–647

Willerslev E, Hansen AJ, Rønn R, Brand TB, Wiuf C, Barnes I, Gilichinsky DA, Mitchell D, Cooper A (2004) Long-term persistence of bacterial DNA. Curr Biol 14:R9–R10

Williamson SJ et al (2008) The Sorcerer II Global Ocean Sampling Expedition: metagenomic characterization of viruses within aquatic microbial samples. PLoS One 3:e1456

Wilson EO (1988) Biodiversity. National Academy Press, Washington, DC

Wirth T, Wang X, Linz B, Novick RP, Lum JK, Blaser M, Morelli G, Falush D, Achtman M (2004) Distinguishing human ethnic groups by means of sequences from *Helicobacter pylori*: lessons from Ladakh. Proc Natl Acad Sci U S A 101:4746–4751

Yelenik S, Perakis S, Hibbs D (2013) Regional constraints to biological nitrogen fixation in post-fire forest communities. Ecology 94:739–750

Yepremian C, Gugger MF, Briand E, Catherine A, Berger C, Quiblier C, Bernard C (2007) Microcystin ecotypes in a perennial *Planktothrix agardhii* bloom. Water Res 41:4446–4456

Zani S, Mellon MT, Collier JL, Zehr JP (2000) Expression of *nifH* genes in natural microbial assemblages in Lake George, New York, detected by reverse transcriptase PCR. Appl Environ Microbiol 66:3119–3124

Zehr JP, McReynolds LA (1989) Use of degenerate oligonucleotides for amplification of the *nifH* gene from the marine cyanobacterium *Trichodesmium thiebautii*. Appl Environ Microbiol 55:2522–2526

Adaptations of Prokaryotes to Their Biotopes and to Physicochemical Conditions in Natural or Anthropized Environments

Philippe Normand, Pierre Caumette, Philippe Goulas, Petar Pujic, and Florence Wisniewski-Dyé

Abstract

Microorganisms live in a constantly changing environment and must modify their physiology and morphology to cope with these changes. The main systems for molecular adaptation to modifications of environmental conditions and the behavioral responses of prokaryotes in various habitats, excluding extreme habitats, are discussed. The main regulation systems that are described are transcription, signal transduction, and protein modifications. Three specialized systems are also presented in details: quorum sensing, phase variation, and antibiosis. Quorum sensing allows bacteria to trigger some responses when their density is high enough to permit the function to be successful. Phase variation is an adaptive process by which a bacterial subpopulation undergoes frequent, usually reversible phenotypic changes resulting from genetic or epigenetic alterations, allowing rapid modification of the cells physiology. Antibiosis is the ability to synthesize molecules that will impact other taxa and eventually provide a selective advantage to which some microbes respond by resisting to these molecules.

Finally are described the physiological responses to various environmental parameters such as temperature, oxidants, salinity, acidity, pressure, desiccation, and how this translates in different biotopes such as soil, water bodies, sediments, biofilms, mats, air, and manmade biotopes.

Keywords

Adaptation • Adaptability • Antibiosis • Fitness • Glycosylation • Homeostasis • Metabolism • Morphology • Physiology • Protein modification • Quorum sensing • Phase variation • Signal transduction • Transcription

---

* Chapter Coordinators

P. Normand* (✉) • P. Pujic • F. Wisniewski-Dyé
Microbial Ecology Center, UMR CNRS 5557 / USC INRA 1364,
Université Lyon 1, 69622 Villeurbanne, France
e-mail: philippe.normand@univ-lyon1.fr

P. Caumette* (✉) • P. Goulas
Institut des Sciences Analytiques et de Physico-chimie pour l'Environnement et les Matériaux (IPREM), UMR CNRS 5254, Université de Pau et des Pays de l'Adour, B.P. 1155, 64013 Pau Cedex, France
e-mail: pierre.caumette@univ-pau.fr; philippe.goulas@univ-pau.fr

9.1 Introduction

Adaptation*, from the Latin prefix "ad," indicating the purpose of the action, and "Aptus" capable, is a word that designates all morphological and physiological characteristics deriving from selection that enables organisms to grow and multiply in a given biotope. This definition implies that any physiological characteristic is an adaptation, but this chapter will be restricted to adaptations that allow prokaryotic microorganisms to change their shape and physiological and enzymatic setup, beyond the basic metabolism. Adaptability is the ability to adapt, and **fitness***, which derives from the Old English "fitte" meaning able, suitable, is a term now used in many languages in evolutionary biology to indicate the ability of an organism to reproduce and transmit its genes. By their small size relative to their volume and metabolism associated with their envelopes, prokaryotic cells are in constant interaction with their environment. They are constantly forced to adapt to changing conditions of environmental parameters. These conditions can vary over very short distances and very short time scales. Prokaryotic cells must be able to respond to changes in micro-spatial and temporal gradients in a timely manner. To follow the evolution of various environmental parameters, such as osmotic pressure, ionic strength, pH, temperature, concentrations of nutrients, or toxic substances, is a necessity for survival. Among microorganisms, prokaryotes are those who are most capable of wide adaptability and resistance to changing conditions of the environment. Prokaryotes are practically the only organisms able to resist or even thrive over the extreme conditions of life that exist on our planet (*cf.* Chap. 10). For this purpose, prokaryotic cells have developed adaptive systems that induce morphological changes (such as variations in size, changes in envelopes features, productions of protective features such as capsules or **spores***, etc.) or changes in the metabolism and physiology of the cell. These systems are based either on signal transduction proteins that are able to perceive changes in environmental parameters and to transmit information to modify other proteins by different molecular mechanisms, resulting in metabolic or behavioral adjustments, or on the regulation of transcription at the genetic level and modifications in the protein expression pattern. In a complex microbial community, other responses may occur linked to the "quorum sensing" (*cf.* Sect. 9.3) or the phenomena of competition and antibiosis (*cf.* Sect. 9.4).

This chapter presents the main systems for molecular adaptation to environmental conditions (*cf.* Sects. 9.2, 9.3, 9.4, and 9.5) and the behavioral responses of prokaryotes in various habitats, including biofilms (*cf.* Sects. 9.6, 9.7, and 9.8). The answers to the most extreme conditions encountered in archaea but also in some bacteria are presented in Chap. 10.

9.2 Main Regulation Systems

One of the major modes of adaptation is the ability to respond to changing conditions of the physicochemical parameters of the biotope; this ability is called regulation, and it is done at different levels, that of **transcription***, of **translation***, and at the posttranslational level.

9.2.1 Regulation of Transcription

In eukaryotes, the availability of large regions for expression of the genome depends on the structure of chromatin, which can be altered due to the attachment of specific proteins, the histones, which is controlled by DNA methylation, by noncoding RNAs (ncRNA), or by proteins binding to DNA. These adaptation strategies are also used by bacteria and archaea: the histones have been described in *Archaea* (Reeve et al. 2004); ncRNA exists in bacteria (Box 9.1) as well as proteins that bind to DNA.

Box 9.1: Small RNAs for Important Functions
Wafa Achouak

Bacteria use different mechanisms to respond to environmental stresses. Small noncoding RNA regulators (ncRNAs) are integrated into regulatory networks and are involved in regulating the stress response, iron homeostasis, virulence, quorum sensing, sporulation, and many other functions. The ncRNAs are found in eukaryotes, bacteria, and archaea.

History and Prediction Tools
The ncRNAs have long escaped biochemical and genetic studies because they do not code for proteins, and they are not subject to nonsense mutations. Since they are often encoded in intergenic regions, they also escape transcriptomic analyses using microarrays containing only those genes that code for proteins. The first ncRNAs were discovered four decades ago, and for a long time, only a dozen were known in *Escherichia coli*. Approximately 80 have currently been identified (Wassarman et al. 2001), and approximately 20 have been described in other bacteria (*Bacillus subtilis*, *Vibrio cholerae*, *Pseudomonas aeruginosa*, *Staphylococcus aureus*, and *Listeria monocytogenes*). They have a size ranging from 50 to 500 bases, are located in intergenic regions, but can also be encoded by the antisense strand of genes encoding proteins.

Tools for Predicting ncRNA
- Bioinformatics: The algorithms developed for the prediction of ncRNA target noncoding regions and

---
W. Achouak
UMR 7265 CNRS-CEA-Aix Marseille University,
Institute of Environmental Biology and Biotechnology
IBEB/DSV/CEA, CEA Cadarache,
13108 Saint-Paul-lez-Durance, France

(continued)

Box 9.1 (continued)

are based on homology with known ncRNA. These prediction tools cannot detect antisense ncRNAs encoded by ncRNAs genes and proteins that are not conserved, which genomes are not yet sequenced. However, hundreds of ncRNAs have been predicted and are waiting to be validated experimentally (Altuvia 2007).
- Direct sequencing of ncRNAs: the RNAs are separated by electrophoresis after metabolic labeling or end-labeling.
- Microarray-based genome oligonucleotide. Several experimental conditions have to be tested for targeting ncRNAs that are expressed in response to certain stimuli.
- RNomics: RNAs of a size between 20 and 500 bases are retro-transcribed, isolated by denaturing PAGE, and cloned. Reverse transcription is still not effective on highly structured RNA or modified; thus, only transcribed and abundant RNAs will be identified (Vogel et al. 2003).
- Genomic SELEX ("systematic evolution of ligands by exponential enrichment"): This is an experimental approach for identifying protein-binding ncRNAs, independently of the conditions of their expression. The ncRNA-protein complex may be essential for the regulatory function of ncRNAs that act as antisense regulators that interact with the protein Hfq (Wassarman et al. 2001). Enrichment steps are based on a ncRNAs size selection and affinity for protein ligands (Lorenz et al. 2006).

Functional Approach of RNomics

The function of ncRNAs identified by different techniques mentioned above could be elucidated by different approaches (Hüttenhofer and Vogel 2006): since most functional ncRNAs are part of ribonucleoprotein complex (RNP), the protein component of ncRNAs can be sought and identified by affinity trapping technique by labeling the ncRNA. Some ncRNAs act as antisense mRNA; the search for complementary ncRNAs by bioinformatics tools should identify the targets of these ncRNAs. Overexpression or construction of mutants that do not express the ncRNA is expected to highlight the phenotypes associated with them.

Mode of Action of ncRNA

The ncRNAs interact with other RNA molecules either to stabilize them and translate them or to destabilize them and lead to their degradation (Fig. 9.15). NcRNA-mRNA interactions are reinforced by Hfq-like proteins. The ncRNA can also interact directly with proteins to sequester them (Box Fig. 9.1).

Base pairing with the transcript (RNA) target:
- The matching is perfect if the ncRNAs are encoded (in cis) by the complementary strand of the target RNA. This mainly concerns the plasmid ncRNAs that act as antitoxins to toxins: it is the case of the *hok/sok* ("host killing/suppressor of killer") locus – the ncRNA *sok* and the *hok* mRNA form a duplex that is rapidly degraded by RNase III (Gerdes et al. 1992). If the bacterium loses the plasmid, the level of *sok* decreases faster than that of *hok* which is translated into the Hok protein, leading to cell death (Hayes 2003).
- Pairing is imperfect if the target RNA and ncRNAs are encoded (in trans) by genes located in different parts of chromosome as in the case of ncRNA RyhB or SgrS which respectively control the level of iron and glucose-phosphate in the cells of *E. coli*. A ncRNA can target multiple mRNAs; this is the case for RyhB, which regulates the level of iron-containing proteins expression in cases of iron deficiency (Massé and Gottesman 2002).

Modification of the Activity of Proteins

6S RNA from *E. coli* is one of the first ncRNA detected and identified (Lee et al. 1978). Its expression level increases when the medium is depleted and bacteria enter the stationary phase. 6S RNA binds to RNA polymerase (RNA polymerase-σ70) by mimicking the DNA, thus preventing gene transcription that is dependent upon the 70s factor (Lee et al. 1978). 6S RNA is essential for the transition from exponential phase to stationary phase and for long-term survival of bacteria.

Modulation of Translation

Most ncRNAs modulate the stability of transcripts and regulate translation by masking the ribosome binding site on the mRNA. Some ncRNAs form a ribonucleoprotein complex with the protein Hfq and RNaseE to inhibit translation of target mRNA and to accelerate its degradation (Morita et al. 2005). This is the mode of action of RhyB and SgrS; a family of homologous proteins to CsrA ("carbon storage regulator A") in *E. coli* and to RsmA ("repressor of stationary phase of metabolites") in species of *Pseudomonas* is involved in the regulation of carbon metabolism, virulence genes, mobility, and many other functions. These proteins bind to the 5′ end of

(continued)

Box 9.1 (continued)

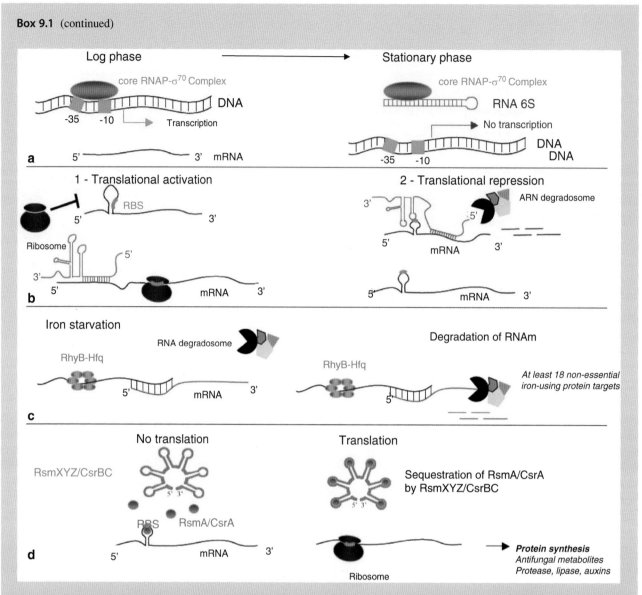

Box Fig. 9.1 Mechanisms of action of noncoding regulatory RNAs (ncRNAs). (**a**) RNA polymerase complex is sequestered by the 16S RNA during stationary phase to prevent the transcription of genes whose expression is under the control of alternative sigma70 factor. (**b-1**) Activation of the translation of genes through association ncRNA-mRNA to unlock access to RBS ribosome. (**b-2**) Repression of translation by blocking access to RBS, the hybrid ncRNA-mRNA is degraded. (**c**) Repression of translation by the ncRNA-Hfq protein complex that binds to mRNA of target genes to induce their degradation. (**d**) Activation of the translation of target genes by sequestration of proteins that inhibit transcription by binding of the mRNA to the ncRNA

mRNA target, at the Shine Dalgarno sequence, thus preventing the initiation of translation (Fig. 9.16). The ncRNAs CsrB/C and RsmY/Z, respectively, sequester CsrA and RsmA and remove repression. The CsrA/RsmA and CsrB-C/RsmY-Z systems are under the control of the Bary/UvrY (Suzuki et al. 2006) and GacS/GacA (Lapouge et al. 2008) two-component regulation systems, respectively, and constitute a signal transduction pathway that regulates the social behavior of gammaproteobacteria.

Conclusion

The world of ncRNA is vast with a wide variety of structures and functions, and their role in the adaptation of bacteria to the environment remains to be explored. It would be interesting to study their expression in situ and to determine their importance in the physiological adaptation of bacteria to environmental fluctuations, extreme environments, and also in different types of interaction with hosts. The development of bioinformatics tools for predicting ncRNA should help identify a large number of ncRNA from metagenomes being sequenced.

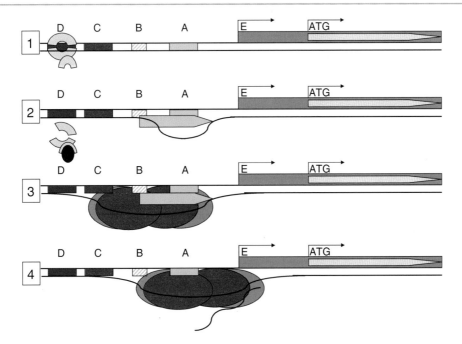

Fig. 9.1 Structure of a prokaryotic promoter. Schematic diagram showing the function of a regulator, illustrated here is a transcriptional repressor. "A" indicates the "Pribnow box" at coordinates −10, with consensus sequence TATAAT (in *green*) which will bind to the sigma factor (70) of RNA polymerase (*green arrow*); in "B" is the "−35" consensus sequence TTGACA that will also bind the sigma factor. The upstream activator sequence (UAS, "upstream activator sequence") allows the attachment of the alpha subunit (in *red*) in "C." "D" is the site of attachment to the transcription factor (a repressor is shown here in *blue*). In a situation where the metabolite specifically recognized by it is absent, the transcriptional repressor is bound to DNA and prevents the insertion of the RNA polymerase (*1*). When a specific metabolite (*black oval*) appears in the cell, the transcription factor interacts with it and its conformation is changed; it then releases the DNA (*2*), which allows the attachment of the sigma factor (70) of RNA polymerase (*3*). Then the alpha and beta units join the sigma factor (*3*) and initiate transcription (the *dark gray rectangle* represents the message) from "E" (*4*) of a messenger RNA that will be later translated on ribosomes (Lewin 1981)

In the past three billion years of evolution, microorganisms have been selected so as to synthesize messenger RNAs only to fulfill their immediate physiological requirements because the synthesis of messengers is expensive in energy. This level of regulation, called "transcriptional regulation," is more economical in metabolites than the others described below, but also slower because once suppression of messengers synthesis is lifted, cells proceed to synthesize mRNAs, translate them into proteins, and then possibly synthesize metabolic effectors before the cells are adapted to the changed environment. The level of transcriptional regulation is nevertheless very useful and widespread as evidenced by the recent knowledge derived from many genomics programs.

The synthesis of a messenger RNA depends on RNA polymerase, a protein that binds to a locally distorted region of DNA and from there initiates the synthesis of a messenger RNA that will later be translated into a protein (Fig. 9.1). The **RNA polymerase*** consists of several subunits, one of which, detachable from the rest and called sigma factor, serves to recognize a particular region upstream of genes called "promoter" (Fig. 9.1). In the absence of sigma factor, RNA polymerase is still able to polymerize mRNA but with little specificity, which is a waste of resources for the cell.

Bacterial cells contain three RNA polymerases. Each of these enzymes targets the transcription of a type of RNA: RNA polymerase I transcribes ribosomal RNA (rRNA), RNA polymerase II transcribes messenger RNAs (mRNAs), and finally the RNA polymerase III transcribes transfer RNA and small RNA (tRNA). In addition, there are several genes encoding sigma subunits because some microorganisms realize transcription of many genes through the synthesis of special sigma factors. *Streptomyces coelicolor* has 65 genes encoding sigma factors of which 45 belong to the extracellular category (ECF "extracellular family") dedicated to non-cytoplasmic functions such as synthesis of aerial hyphae, response to the disulfide stress, or wall homeostasis (Bentley et al. 2002).

Conversely, there are proteins called **antisigma***, often present in the same operon as the sigma factor itself and whose function is to inhibit the interaction of the sigma factor with DNA promoters; there are even anti-antisigma. The rationale for regulating with such anti-sigma factors is the need to finely tune sigma factor activity due to the heavy energy investment that follows. For example, in *Bacillus subtilis*, the transcription of genes associated with sporulation is controlled by a sigmaF factor, but the ability of this factor to attach to DNA is inhibited by another factor called SpoIIAC that attaches to sigmaF factor; this factor is called anti-sigma and interacts with a third factor, SpoIIAA, which is called anti-anti-sigma, which then triggers sporulation (Duncan and Losick 1993).

Gene transcription by RNA polymerase is sometimes modified by the presence of proteins that attach to DNA; such proteins are called "**transcriptional regulators***." The attachment of the regulator to DNA is not irreversible, because this protein oscillates between two states, attached or not attached, according to various physicochemical factors. When the regulator is in a free, non-attached state, it cannot bind to DNA, and synthesis of messengers can then take place (Fig. 9.1).

Some transcriptional regulators are **repressors***; they are attached to the DNA as a base situation. When the metabolite or metal for which the repressor has a high affinity is present in the cell, it attaches to the repressor, which then has its conformation changed, reducing its affinity for the DNA sequence, and it is then released from DNA. Other transcriptional regulators are **activators***; they change their conformation as repressors do following environmental stimuli such as the presence of a metabolite or a metal. They attach to the DNA near the **promoters*** of the genes under their control and then activate their expression. Finally, some regulators can operate either as an activator or as a repressor.

There are about a dozen families of transcriptional regulators identified in bacteria; among these are the families AraC (arabinose), CRP (catabolite repressor protein), LacI (lactose), LysR (lysine biosynthesis), and MerR (metabolism of mercury) families, whose structures and mechanisms have been best characterized (Ma et al. 2004).

For example, the regulator Fur is a small protein of 17kD with high affinity for iron. Its name means "<u>f</u>erric ion <u>u</u>ptake <u>r</u>egulator" because the function that was attributed originally to this protein was to regulate the intracellular level of Fe (II), which is a cofactor for many enzymes such as nitrogenase, superoxide dismutase, or uptake hydrogenase. Since free iron can generate oxidative stress through the Fenton effect, it is necessary to closely regulate its level in the cell. In case of low intracellular iron, there is little risk to the cell, and the Fur protein remains attached to the upstream sequences of genes involved in iron metabolism and oxidative stress. When the intracellular iron level rises, this iron is recognized by the Fur protein that has a high affinity for it, and Fur then sees its conformation changed; it then releases the DNA sequences to which it was attached (Andrews et al. 2003). Many proteins exist that are homologous to Fur; these proteins are derived from duplications that occurred before the emergence of lineages such as that of the *Actinobacteria* (Santos et al. 2009).

The transcriptional regulators bind to DNA through areas of attachment to the DNA of different types; the best known of these is the "helix-turn-helix" domain which has two sites of interaction with the major DNA groove

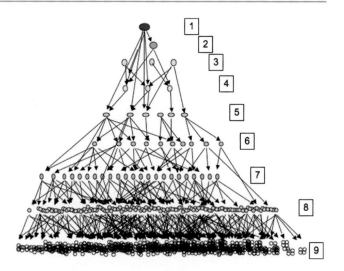

Fig. 9.2 Hierarchical structure of the network that controls transcription in *Escherichia coli*. This network includes 1,278 factors (*circles*) and 2,274 interactions (*lines* connecting the circles). There are nine levels in the regulatory network, the highest ones controlling those below. The first level at the top of the pyramid is CRP; the second is RpoS; the third includes CspE, Ihf, and PhoB; the fourth includes CytR, SoxR, and DnaA; and the fifth includes RpoE, HNS, and RpoN. The self-regulatory loops are not represented in this network (Modified from Ma et al. 2004)

("helices") and a hinge region ("turn"). Other known domains are called "zinc finger domain," "winged helix," or "leucine zipper."

Some genes are under the control of several of these transcription factors, which makes sense since they are involved in the response to several physicochemical factors. For example, the gene *kdg*N is responsible for the synthesis of a carrier for the entry of galacturonate in the phytopathogenic bacterium *Dickeya dadantii* formerly known as *Erwinia chrysanthemi* (Condemine and Ghazi 2007). No less than five regulators modulate the transcription of this gene, in particular KdgR (galacturonate), PecS (pectinase), OmpR (outer membrane protein), HNS (histone nucleoid structuring), and CRP (catabolite repressor protein). It is thus estimated that complex regulatory networks relate the genes and operons, thus constituting **regulons***. The different regulons are related to each other; thus, ultimately all genes of the genome interact with others in the global network of transcriptional regulation as shown in Fig. 9.2.

The genome sequencing programs have revealed a strong correlation between genome size and the number of transcriptional regulators (Fig. 9.3). A strict symbiont such as *Wolbachia*, whose genome has undergone a sharp reduction, is thus dependent on its host not only for food but also for its regulation as it has only six transcriptional regulators (Wu et al. 2004). Conversely, *Frankia* bacteria that live in soil and symbiotically in root nodules of the plant and have

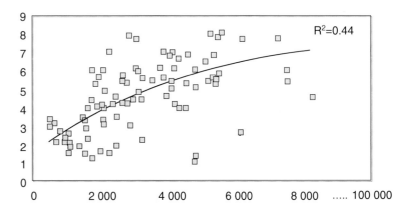

Fig. 9.3 Correlation between genome size and number of genes of category K (transcription). The relationship between the number of genes in the genome (*x*-axis) and the % of the genome (*y*-axis) for 99 sequenced genomes based (Modified from Konstantinidis and Tiedje 2004)

genomes of more than 9 Mb have 150 identified transcriptional regulators (Normand et al. 2007).

It is also estimated that seven transcriptional regulators control 50 % of regulated genes in *E. coli*, while 60 transcriptional regulators control only one promoter (Martinez-Antonio and Collado-Vides 2003).

9.2.2 Signal Transduction Systems

Bacteria have developed systems for the **transduction of signals***, which associate proteins able to perceive changes in environmental parameters and transmit these informations to other proteins through different molecular mechanisms that will result in adaptation of the metabolism, the physiology, or the behavior of the bacteria.

Depending on the signal transduction mechanisms, several major types of **signaling systems*** can be distinguished in bacteria (Galperin 2005):

(i) Systems with histidine protein kinases (HK and MCP)
(ii) Systems with serine/threonine/tyrosine protein kinases (STYK)
(iii) Systems with an intracellular secondary messenger such as di-cGMP (GGDEF/EAL/HD-GYP) and cAMP (AC)

Generally the number of signaling proteins is correlated with genome size; however, it tends to increase with the complexity of the lifestyle of the organism (Table 9.1).

The detailed structural study of a number of these signal transduction systems coupled with comparative genomic studies made possible by the rapid increase in the number of sequenced genomes has revealed the modular structure of bacterial signal transduction proteins (Fig. 9.4). This modular organization resulting from the combination of distinct protein domains has led to the generation of a wide variety of signaling proteins by multiple combinations of a limited number of domains (Table 9.2) (Galperin and Gomelsky 2005).

9.2.2.1 Histidine Protein Kinase Systems

Histidine protein kinase systems are the signal transduction systems most common in the bacterial world. They consist of two proteins, a histidine kinase protein (HK) and a response regulator (RR), that interact through the transfer of a phosphoryl group from the HK protein onto the RR protein (Fig. 9.5). Phosphorylation of the response regulator leads to its activation and to generation of a signal along the pathway. According to the method of perception of the signal, two types of histidine protein kinase systems can be distinguished:

(i) The conventional two-component systems in which the HK perceives the signal.
(ii) The chemotaxis signaling systems. The perception of the signal is carried by transmembrane receptors distinct from the HK, the MCP proteins.

9.2.2.2 The Two-Component Systems (TCS)

With the exception of mycoplasmas, most bacteria and many archaea have several **two-component systems*** (Stock et al. 2000). *Escherichia coli* has 30 HKs, *Bacillus subtilis* 37, the cyanobacterium *Nostoc* sp. PCC7220 134, and an archaea *Methanospirillum hungatei* 76 (Table 9.3). The TCS are involved in many processes such as the regulation of carbon and nitrogen metabolism, phosphate uptake, growth under aerobic or anaerobic conditions, osmoregulation, sporulation, and biofilm formation (Table 9.3).

TCS are much less common in eukaryotes where cascades involving phosphorylation of Tyr and Ser/Thr protein kinases are predominant. Nevertheless, such systems have been identified in yeasts such as *Saccharomyces cerevisiae* and *Schizosaccharomyces pombe*; fungi such as *Candida albicans*, *Neurospora crassa*, and *Aspergillus nidulans*, where they are involved in osmoregulation and development; but also in plants such as *Arabidopsis thaliana* in the genome of which 11 HKs have been identified. HKs are homodimers (Fig. 9.6) consisting of:

Table 9.1 Distribution of different signaling systems in the prokaryotic world

	Total number of proteins	HK	MCP	STYK	GGDEF/EAL/HD-GYP	AC	RR	Total
Bacteria								
Actinobacteria								
Frankia alni	6,787	45	–	59	9	3	49	179
Mycobacterium tuberculosis	3,927	13	–	13	2	16	13	57
Streptomyces coelicolor	8,154	95	–	37	8	1	84	225
Cyanobacteria								
Nostoc sp. PCC7120	6,033	134	3	52	17	6	96	286
Synechocystis sp. PCC 6,803	3,567	46	4	12	28	3	47	128
Alphaproteobacteria								
Agrobacterium tumefaciens	5,402	53	20	2	29	3	58	162
Bradyrhizobium japonicum	8,317	92	35	4	42	35	96	298
Caulobacter crescentus	3,737	62	18	2	14	3	45	142
Rhodopseudomonas palustris	4,813	66	29	5	42	7	54	197
Rickettsia prowazekii	835	4	–	1	2	–	4	10
Wolbachia (Drosophila endosymbiont)	1,195	2	–	1	1	–	2	5
Betaproteobacteria								
Bordetella pertussis	3,436	18	5	3	9	–	29	57
Dechloromonas aromatica	4,171	103	25	9	75	3	98	292
Gammaproteobacteria								
Alcanivorax borkumensis	2,755	30	1	6	13	2	31	81
Escherichia coli K12	4,242	30	5	12	29	1	32	99
Pseudomonas aeruginosa	5,567	63	26	8	41	2	73	206
Salmonella enterica	4,758	30	6	3	21	1	32	93
Vibrio cholerae	3,835	43	45	1	62	1	49	196
Deltaproteobacteria								
Desulfovibrio vulgaris	3,531	64	28	1	41	1	76	200
Geobacter sulfurreducens	3,446	92	33	1	39	–	91	240
Epsilonproteobacteria								
Wolinella succinogenes	2,044	39	31	2	26	1	48	138
Firmicutes								
Bacillus subtilis	4,105	36	10	4	6	–	35	91
Clostridium acetobutylicum	3,848	37	38	5	20	1	43	144
Mycoplasma pneumoniae	689	–	–	1	–	–	–	1
Staphylococcus aureus	2,624	17	–	2	1	–	17	37
Spirochaetes								
Treponema denticola	2,767	7	20	–	13	9	11	59
Archaea								
Sulfolobus acidocaldarius	2,223	–	–	10	–	1	–	11
Haloarcula marismortui	4,240	59	17	8	–	1	25	109
Methanospirillum hungatei	3,139	76	27	9	–	1	61	169

Modified after Galperin (2005) http://www.ncbi.nlm.nih.gov/Complete_Genomes/SignalCensus.html
HK histidine protein kinase, *MCP* methyl-accepting chemotaxis proteins, *STYK* Ser/Thr/Tyr kinase, *GGDEF*, *EAL*, *HD-GYP* proteins involved in the metabolism of di-cGMP, *AC* adenylate cyclase involved in the metabolism of cAMP, *RR* response regulator

(i) An N-terminal domain of signal perception (sensor) that can be periplasmic or bound to the cytoplasmic membrane (Table 9.3., Mascher et al. 2006)

(ii) A conserved C-terminal cytoplasmic region that includes a dimerization domain with two α helices (HisKA/DHp) containing the phosphorylation site, which is a highly conserved histidine residue and the catalytic domain that has a kinase activity (HATPase_c)

The periplasmic domains of membrane HK (EnvZ) are poorly conserved which could reflect specific sensor types for given stimuli. Conversely, the cytoplasmic sensor HK that can be anchored to the membrane (ArcB) or soluble (KinA) often has a conserved GAF or PAS domain, a domain sensitive to changes in redox potential, in O_2 concentration, or in light intensity, and varies also as a function of the prosthetic group, heme, or FAD chromophore linked to the PAS domain.

Fig. 9.4 Modular structure of bacterial signaling proteins. The names of the different domains are given in Table 9.2. Transmembrane helices. Domain names from databases SMART (http://smart.embl.de) and Pfam (http://pfam.sanger.ac.uk) (Modified and redrawn from SMART). Drawing: M.-J. Bodiou

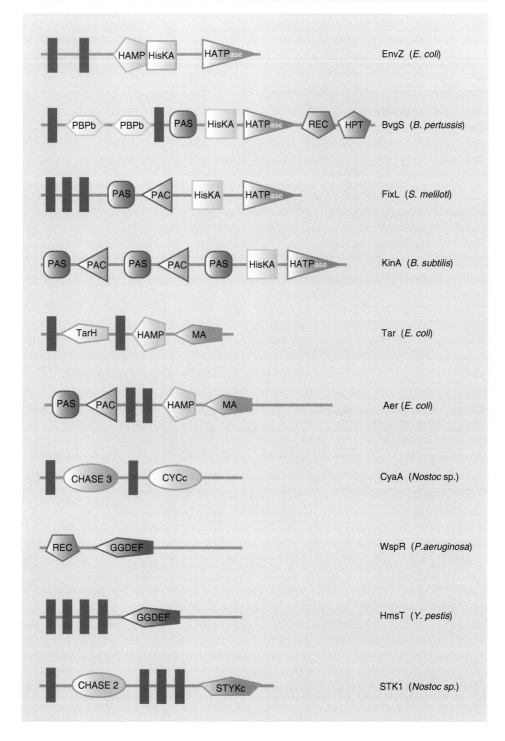

The perception of the signal by the sensor leads to changes in the intramolecular conformation of HK, which leads in turn to activation of the kinase activity and results in ATP-dependent autophosphorylation of the highly conserved His residue dimerization domain. The autophosphorylation of the dimer is a bimolecular reaction in which one monomer catalyzes the phosphorylation of the conserved His residue of the other monomer. Most periplasmic domain HKs have a HAMP domain (histidine kinase, adenylyl cyclase, methyl-binding proteins, and phosphatase domain) essential for signal transduction between the sensor and the catalytic domain (Inouye 2006).

The response regulator RR has a conserved domain with phosphotransferase activity (REC) that uses phosphorylated histidine of the HK as donor group to autophosphorylate a highly conserved Asp residue and thus regulates the activity

Table 9.2 Main bacterial protein signal transduction domains

Domains	Origin of the name	Localization	Size (aa)	Characteristics
Perception				
CACHE	Ca^{2+} **c**hannels and **Che**motaxis receptors	Periplasmic	~80	Fixation of small molecules
CHASE	**C**yclase/**H**is kinase **A**ssociated **S**ensing **E**xtracellular	Periplasmic	150–300	Fixation of amino acids and peptides
GAF	c**G**MP phosphodiesterase, **A**denyl cyclase, **F**hlA	Cytoplasmic	~150	Fixation of AMPc and GMPc
MASE	**M**embrane-**a**ssociated **s**ensor domain	Transmembrane	~280	Unknown stimuli
PAS	Identified in **P**ER, **A**RNT, and **S**IM proteins	Cytoplasmic	~110	Variation of the redox potential, oxygen, light (may bind heme, FAD, cinnamic acid, ATP), often associated with an extension on the C-terminal end of the PAC domain
PBPb	**P**eriplasmic solute-**b**inding **p**roteins, **b**acterial	Periplasmic	~220	Fixation of amino acids, opines
TarH	**Tar** **h**omologous protein	Periplasmic	~150	Ligand-binding domain of the chemotaxis receptor (MCP)
Transduction				
CYCc	Adenylyl, guanylyl **cyc**lase **c**atalytic domain	Cytoplasmic	~190	Adenylate cyclase
EAL	Conserved amino acids	Cytoplasmic	~250	c-di-GMP phosphodiesterase
GGDEF	Conserved amino acids	Cytoplasmic	~180	Diguanylate cyclase
HAMP	**H**is kinases, **a**denylyl cyclases, **m**ethyl-binding proteins, **p**hosphatases domain	Cytoplasmic	~50	Junction domain involved in signal transduction
HATPase_c	**H**is kinase like **ATPase** **c**atalytic domain	Cytoplasmic	~140	Catalytic domain of histidine kinase proteins
HD-GYP	Conserved amino acids	Cytoplasmic	~170	c-di-GMP phosphohydrolase
HisKA/DHp	**His** **k**inase **A** domain/**d**imerization, **H**is **p**hosphotransfer	Cytoplasmic	~80	Dimerization and phosphoacceptor domain (His) of histidine kinase proteins
HPt	**H**is **p**hospho**t**ransfer domain	Cytoplasmic	~100	Phosphotransfer domain having an active His
MA	**M**ethyl-**a**ccepting chemotaxis domain	Cytoplasmic	~260	Signal transduction domain of chemotaxis receptors (MCP)
PP2C	**P**rotein **p**hosphatase **2C**	Cytoplasmic	~250	Protein phosphatase
REC	**Rec**eiver domain, CheY homologous	Cytoplasmic	~100	Domain having phosphotransferase activity, contains a phosphoacceptor site (Asp)
STYKc	**S/T/Y** **k**inases, **c**atalytic domain	Cytoplasmic	~250	Serine, threonine, tyrosine protein kinases
Response				
wHTH	**w**inged **h**elix-**t**urn-**h**elix	Cytoplasmic	~240	DNA-binding domain
HTH	**H**elix-**t**urn-**h**elix	Cytoplasmic	~240	DNA-binding domain
FIS	**F**actor **i**nversion **s**timulation	Cytoplasmic	~170	DNA-binding domain
AAA+	**A**TPases **a**ssociated with a variety of cellular **a**ctivities	Cytoplasmic	~220	Domain interacting with σ^{54}, having ATPase activity
ANTAR	**A**miR **N**asR **t**ranscription **a**nti-termination **r**egulator	Cytoplasmic	~180	RNA-binding domain

A protein domain is a sequence of 50–300 amino acid residues with its own three-dimensional structure, which is associated with a biological function, such as the specific attachment of a ligand or a catalytic activity. The areas of signaling proteins can be grouped into three groups: domains involved in the perception of signal, domains involved in transduction of the signal, and domains involved in response to the signal. Domain names from data banks (SMART http://smart.embl.de) or Pfam (http://pfam.sanger.ac.uk)

of an effector domain that is generally a DNA-binding domain. In a phosphorylated form, the response regulator (RR-P) will activate or repress target genes and trigger the adaptive response of the cell.

In the case of the EnvZ/OmpR system of *E. coli* (Fig. 9.7), an increase in osmotic pressure will result in an increased rate of autophosphorylation of EnvZ that in turn will phosphorylate OmpR. OmpR-P represses transcription of *ompF*, the gene that encodes a porin (protein that allows the passage of metabolites across the outer membrane of Proteobacteria) with a wide pore and inversely will activate transcription of *ompC* that encodes a porin with a smaller pore size.

Besides, the classical two-component systems are systems that have phosphorelay domains with supplementary Asp (REC) and His (HPT) residues (Fig. 9.7). These areas can be present in the HK; these HKs will then be referred to as hybrid as in the ArcB/ArcA system, which regulate the facultative anaerobic metabolism of *E. coli*, or as independent proteins in the signaling system of *Bacillus subtilis* controlling sporulation (KinA/Spo0F/Spo0B/Spo0A). Phosphorelay systems likely allow to multiply the regulatory sites, thus providing more flexibility in the signaling pathway. In eukaryotes, hybrid histidine kinases constitute the majority of HK.

Dephosphorylation of the RR that allows the system to return to the initial state or to modulate the intensity of the response is provided by an autophosphatase RR activity. The half-life of phosphorylated RRs ranges from a few seconds to several hours. In many cases, the dephosphorylation of the RR is accelerated by external phosphatases or by a protein phosphatase activity of the HK.

The analysis of sequenced bacterial genomes has allowed to classify RRs depending on the nature of their effector domain (Galperin 2006). Transcriptional regulators represent about 66 % of all RRs (Table 9.4). Some RRs have an effector domain with enzymatic activity, diguanylate cyclase (GGDEF), phosphodiesterase (EAL), or protein phosphatase (PP2C), for example, and are integrated into complex signaling networks. 14 % of RRs are made up of an isolated receiver domain (REC) and can be part of a phosphorelay (Spo0F) or a molecular component of a system involved in chemotaxis as in the case of *E. coli* CheY.

9.2.2.3 The Chemotaxis Signaling System

Mobile bacteria can move toward attractive molecules (sugars, amino acids) and inversely move away from repellent compounds. This phenomenon is called chemotaxis. The movement of flagellated bacteria such as *Escherichia coli* is characterized by a series of straight runs punctuated by rapid tumblings, corresponding to counterclockwise (CCW) and clockwise (CW) rotations, respectively, of the flagellar motor (Fig. 9.8). The probability that an *E. coli* cell stops its run and tumbles depends on its immediate chemical environment compared to that encountered a few seconds before. The tendency to tumble is enhanced when the

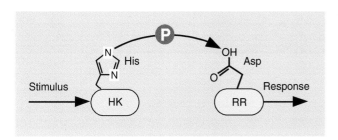

Fig. 9.5 Schematic representation of histidine protein kinase systems. A phosphoryl group is transferred to a conserved His residue of the protein histidine kinase (HK) to a conserved Asp residue of the RR (response regulator) protein, which activates it and initiates the response (Modified and redrawn from Gao et al. 2007). Drawing: M.-J. Bodiou

Table 9.3 Examples of bacterial two-component systems

TCS	Environmental signal	Localization of the HK sensor	Adaptive response	Source
EnvZ/OmpR	Osmotic pressure	Periplasmic	Regulation of expression of two porins OmpF and OmpC	*Escherichia coli*
PhoR/PhoB	Phosphate	Periplasmic	Assimilation of phosphate	*Escherichia coli*
PhoQ/PhoP	Antimicrobial peptides	Periplasmic	Virulence and resistance genes	*Salmonella typhimurium*
NarX/NarL	Nitrate, nitrite	Periplasmic	Nitrogen metabolism	*Escherichia coli*
CitA/CitB	Citrate	Periplasmic	Anaerobic degradation of citrate	*Escherichia coli*
DcuS/DcuR	Fumarate	Periplasmic	Anaerobic degradation of fumarate	*Escherichia coli*
BvgS/BvgA	Temperature, magnesium sulfate	Periplasmic	Virulence and resistance genes	*Bordetella pertussis*
LytR/LytS	Stress	Periplasmic	Hydrolysis of peptidoglycan	*Staphylococcus aureus*
ComP/ComA	Quorum sensing	Membrane	Competence	*Bacillus subtilis*
ArcB/ArcA	Redox state of the quinones pool	Cytoplasmic (HK anchored to membrane)	Anaerobic respiration	*Escherichia coli*
FixL/FixJ	Oxygen	Cytoplasmic (HK anchored to membrane)	Nitrogen fixation	*Rhizobium meliloti*
NtrB/NtrC	Ammonium	Cytoplasmic (soluble HK)	Assimilation of nitrogen	*Escherichia coli*
KinA/Spo0A	Energetic state, ATP	Cytoplasmic (soluble HK)	Sporulation	*Bacillus subtilis*

TCS are listed in the Kyoto Encyclopedia of Genes and Genomes, KEGG PATHWAY database (http://www.genome.jp/kegg/pathway/ko/ko02020.html). Localization of sensor in Mascher et al. 2006

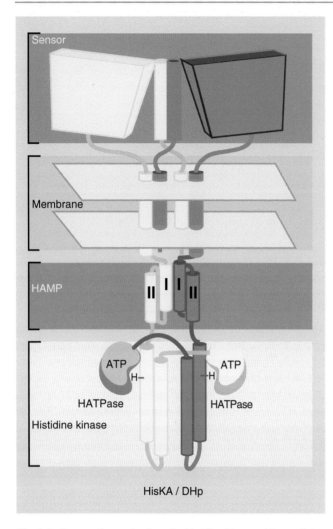

Fig. 9.6 Structural organization of a histidine kinase with a periplasmic sensor. The dimerization domains (HisKA/DHp) form a bundle with four α helices and have a conserved His residue at the site of phosphorylation. ATP binds to the catalytic domains (HATPase_c) (Modified and redrawn from Khorchid and Ikura (2006) and Inouye (2006)). Drawing: M.-J. Bodiou

bacterium perceives conditions that may be harmful, conversely the number of flips decreases, and the bacterium stays its course when chemical conditions improve, allowing the bacterium to climb down gradients of repellents concentration and climb up gradients of attractants.

The molecular mechanism of chemotaxis is complex and depends on the existence of transmembrane receptors called MCPs ("methyl-accepting chemotaxis proteins"), a specific histidine protein kinase (CheA), two response regulators (CheY and CheB, an RR with a methylesterase activity), and a constitutive methyltransferase (CheR), which allows methylation from S-adenosylmethionine of glutamate residues of the MCP receptor (Wadhams and Armitage 2004).

Binding of an attractive or repulsive molecule to the receptor will change the autophosphorylation activity of the normal histidine protein kinase CheA, coupled to the receiver via an adapter protein CheW. The formation of CheA-P allows the phosphorylation of the response regulator CheY. It is the phosphorylated form of the response regulator CheY-P that induces a change in the direction of rotation of the flagellum by binding to the flagellar motor.

In *E. coli* (Fig. 9.9), a decrease in attractant concentration in the medium reduces the number of attractant molecules bound on the MCP receiver, which in turn stimulates the activity of autophosphorylation of CheA and therefore the amount of intracellular CheY-P with, as a consequence, an increase in the frequency of tumbles. One phosphatase, CheZ, increases the speed of spontaneous dephosphorylation of CheY-P and ensures a quick stop of the signal.

CheA-P also phosphorylates CheB, but at a lower speed, resulting in increased methylesterase activity and thus demethylation of the MCP receiver. When demethylated, the MCP receiver has a lower capacity to induce phosphorylation of CheA for the same concentration of attractant. Thus, the rate of autophosphorylation of CheA and the frequency of tumbles return to the level preceding the stimulation. The system is then adapted and able to receive any further increase or decrease in the number of bound ligands.

Conversely, an increase in the concentration of attractant inhibits the autophosphorylation of CheA, which will decrease the concentration of CheY-P and, ultimately, allow runs to occur for longer periods in a straight line. Phosphorylation of CheB and thus its methylesterase activity will also be reduced, which allows the constitutive CheR methyltransferase to increase the rate of methylation of MCPs. When highly methylated, for a given concentration of attractant, MCP receptors will have a higher capacity to stimulate autophosphorylation of CheA, bringing the system to the level of activity preceding the stimulation despite the continued presence of attractant, leading to a normal frequency of tumbles.

MCPs are transmembrane homodimers with a periplasmic effector-binding domain, between two transmembrane helices, a HAMP domain for signal transduction at a highly conserved cytoplasmic signaling domain (AD) (Fig. 9.10). MA has sites for methylation/demethylation that are glutamate residues that allow cells to adapt the response of receptors to a change in concentration of the chemoattractant over time, as well as a binding site for the adapter protein CheW and the histidine kinase CheA. This site is also involved in the association of clusters of MCP receptors.

CheA is a homodimer that has a structure different from HKs encountered in conventional two-component systems (Fig. 9.10). Each monomer is indeed made up of five domains: P1 is a Hpt domain carrying the active His, P2 a binding domain for CheY and CheB, P3 a dimerization domain, P4 is a HATPase catalytic domain, and P5 is a domain coupling CheA to CheW and to the MCP receiver. Phosphorylation of the His residue of the P1 domain of one of the monomers is done by the P4 catalytic domain of the

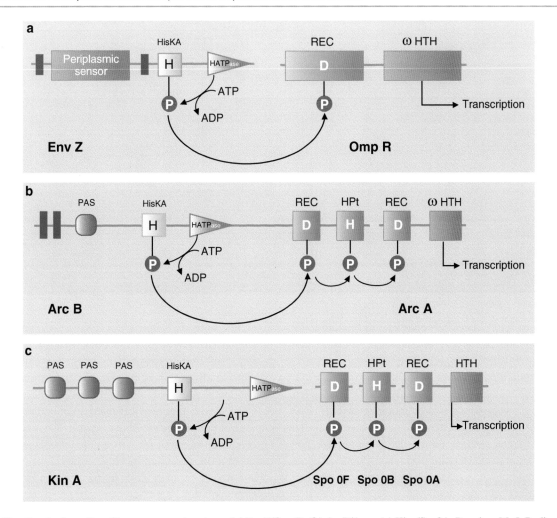

Fig. 9.7 Functional schematics of two-component systems. (**a**) EnvZ/OmpR; (**b**) ArcB/Arca; (**c**) Kina/Spo0A. Drawing: M.-J. Bodiou

Table 9.4 Distribution, structure, and function of the main bacterial response regulators

Function of the effector domain		Examples of structure		Representatives
Binding to DNA	66 %			
		REC-wHTH	29.7 %	OmpR /PhoB
		REC-HTH	16.5 %	NarL/FixJ
		REC-AAA$^+$-FIS	9.7 %	NtrC
		REC-LytTR	2.9 %	LytR
Binding to RNA	1 %			
		REC-ANTAR	0.9 %	AmiR/NasR
Enzymatic activity	12 %			
Methyltransférase		REC-CheB	2.3 %	CheB
Diguanylate cyclase		REC-GGDEF	3.4 %	
di-cGMP phosphodiesterase		REC-EAL	1.8 %	
di-cGMP phosphodiesterase		REC-HD-GYP	1.6 %	
Proteine phosphatase		REC-PP2C	1 %	
Histidine kinase		REC-HisKA-HATPase	1.9 %	
Isolated receiver domain	14 %			
Component of a phosphorelay		REC		Spo0F
Component of chemotaxis		REC		CheY
Others	7 %			

The definition of domains is given in Table 9.2 (from Gao et al. 2007); http://www.ncbi.nlm.nih.gov/Complete_Genomes/SignalCensus.html

Fig. 9.8 Chemotaxis. (**a**) Flagellar motility: straight run followed by a tumble and another rectilinear run in another direction; (**b**) movement of a bacterium in a gradient of attractant with longer strokes in the direction of attractive conditions (Modified and redrawn from Webre et al. 2003). Drawing: M.-J. Bodiou

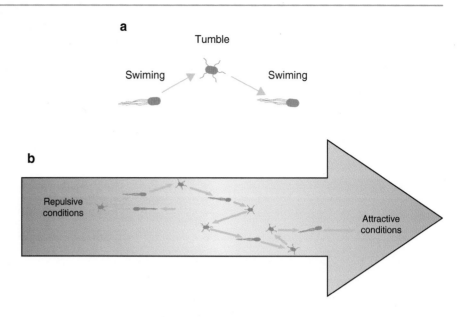

other monomer. CheY and CheB compete for binding to P2. CheA transfers phosphoryl groups faster to CheY than to CheB, which allows triggering of the response before adaptation of the MCP receiver.

Chemotactic molecules bind to the periplasmic domain of MCPs at the interface between two monomers. Many MCPs are capable of binding two different ligands, such as the Tar receptor of *E. coli*, which binds directly aspartate and, through a periplasmic binding protein (PBP), maltose. MCPs are predominantly localized to one pole of the cell, where they form clusters involving several thousands of receptors (Fig. 9.11). *E. coli* has five MCPs, Tar, Tsr (serine), Trg (ribose, galactose), Tap (dipeptides), and Aer, a particular MCP involved in aerotactism. Aer lacks a periplasmic domain but has at its end N-terminal a cytoplasmic PAS domain that complexes FAD whose oxidation state is related to changes in oxygen concentration of the medium (Fig. 9.4).

Tar and Tsr are present at about 3,000 copies per cell, while Trg, Tap, and Aer at a few hundred. Although the precise molecular architecture of these clusters is not established, it is clear that interactions between cytoplasmic domains of receptors are crucial for the formation of clusters and their operation. This cluster organization involving different MCP receptors creates a network capable of amplifying allosteric stimuli of different chemoeffectors before transmission of the signal to other proteins of the signaling system (Kentner and Sourjik 2006).

9.2.2.4 Ser/Thr/Tyr Kinases (STYKs)

Protein kinase systems play a key role in the signaling mechanisms in eukaryotic cells. The tyrosine protein kinase receptors are an important class of membrane receptors, and serine/threonine protein kinases are involved in the regulation of many metabolic pathways. Systematic analysis of sequenced bacterial genomes has permitted to reveal the presence of numerous proteins with Ser/Thr/Tyr kinase domains in all bacterial phyla (Krupa and Srinivasan 2005).

The fact that many of these proteins have transmembrane helices suggests that many of them are membrane receptors capable of binding extracellular ligands. Thus, STK PknB of *Mycobacterium bovis*, for which counterparts are present in different groups of Gram-positive bacteria, has an extracellular PASTA domain able to fix the D-alanyl-D-alanine dipeptide of the peptidoglycan and the β-lactam ring of penicillins. Stimulatory molecules as well as the cellular targets of bacterial STYKs are poorly known. The membrane STK, AfsK of *Streptomyces coelicolor* phosphorylates a global regulator of the secondary metabolism, AfsR. Another membrane STK of *Mycobacterium tuberculosis*, PknH phosphorylates EmbR, a transcription regulator of the gene-coding arabinosyltransferase, an enzyme involved in the synthesis of arabinogalactan, a key molecule of the cell wall of mycobacteria. Various studies have shown the involvement of STYKs in the formation of fruiting bodies in *Myxococcus xanthus* and the development of aerial hyphae in *Streptomyces coelicolor*.

9.2.2.5 Second Messenger Systems
Systems with di-cGMP

Di-cGMP is a secondary messenger found in most bacteria. It regulates many cellular processes including motility, biofilm formation, and expression of virulence genes (Tamayo et al. 2007). Di-cGMP was first identified as an allosteric activator of cellulose synthase in *Gluconacetobacter xylinus*. The diguanylate cyclase (DGC) and phosphodiesterase A (PDEA), enzymes catalyzing the synthesis and the degradation of di-cGMP, respectively, were purified

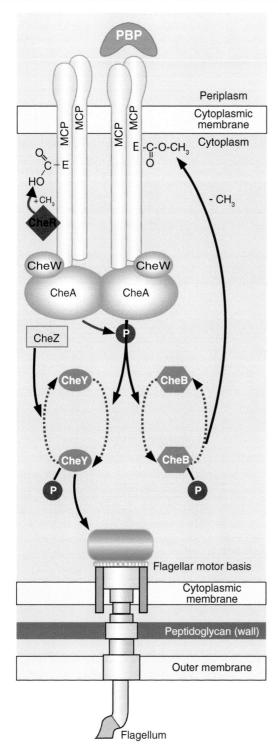

Fig. 9.9 Schematic diagram of the signaling system of chemotaxis in *E. coli*. CheA a homodimer interacts through CheW with two MCP dimer receptors. Chemotactic molecules can bind directly to the receiver or through a periplasmic binding protein (PBP). A decrease in attractant concentration induces trans-autophosphorylation of the dimer CheA, which phosphorylates the response regulator CheY. CheY-P binds to the flagellar motor and induces a tumble. CheA also phosphorylates the methylesterase CheB. CheB-P is in competition with a constitutive methyltransferase CheR to control the degree of methylation of specific glutamate residues of the MCP receptors. Dephosphorylation of CheY-P is accelerated by the phosphatase CheZ (Modified and redrawn from Wadhams and Armitage 2004). Drawing: M.-J. Bodiou

from this organism and used in reverse genetics to identify the corresponding genes. Analysis of these genes has highlighted two conserved domains, called GGDEF and EAL (named based on the conserved amino acid residues), that were found in other bacterial proteins. Phylogenetic analysis of genes encoding GGDEF domains or EAL shows they are widely distributed in bacteria but absent from archaea and eukaryotes, suggesting that the di-cGMP is limited to the bacterial world. The biochemical study of many GGDEF domain proteins has shown that this domain catalyzed the synthesis of di-cGMP from two GTPs (Fig. 9.12) (Ryjenkov et al. 2005).

The di-cGMP is hydrolyzed to 5′-pGpG by the EAL domain of PDEA. Recently, another protein domain, HD-GYP, whose phylogenetic distribution is similar to that of GGDEF and EAL domains, capable of directly hydrolyzing di-cGMP into two GMP, was described. The reasons for the presence of two systems of hydrolysis of di-cGMP in the same genome are still unknown. Most proteins containing GGDEF and EAL domains have a modular structure combining to these two areas, other conserved domains involved in both perception (MASE, CACHE, PAS, GAF) and in transduction (REC) and signal response (HTH). The GGDEF and EAL domains are often found in tandem in the same protein, but usually only one enzyme activity is expressed in vivo by the native protein (Fig. 9.13). The structural diversity of proteins with GGDEF and EAL domains contributes to a fine-level regulation of the concentration of di-cGMP in the cell. Bacteria with a large number of GGDEF and EAL domain proteins have in common the ability to survive in a wide variety of environmental media. For example, *Pseudomonas aeruginosa* encodes 17 GGDEF domain proteins, 5 with an EAL domain, 16 with a GGDEF-EAL domain, and 3 with an HD-GYP domain; *Vibrio cholerae* in turn encodes 31 GGDEF domain proteins, 22 with an EAL domain, 10 with a GGDEF-EAL domain, and 9 with an HD-GYP-domain.

Various works on the physiological role of di-cGMP have made it clear that the di-cGMP activates biofilm formation and inhibits cell motility, thus regulating the transition between planktonic and sessile states of bacterial cells.

The *fimX* mutation of *P. aeruginosa*, a gene that encodes a PDEA (Fig. 9.13), leads to an increase in the intracellular concentration of di-cGMP and an absence of type IV pili on the cell surface with a consequent loss of twitching motility (Kazmierczak et al. 2006). Also in *P. aeruginosa*, constitutive activation of WspR (Fig. 9.13), the response regulator with a DGC activity of the MCP receptor chemosensor Wsp, causes motility inhibition and activation of biofilm formation (Hickman et al. 2005). The decrease in flagellar motility has been demonstrated in *Salmonella enterica* serovar *typhimurium* and *Vibrio cholerae*. In *Salmonella*, ectopic expression of AdrA, a DGC (Fig. 9.13) leads to an inhibition of motility, the opposite effect being observed with YhjH, a PDEA (Simm et al. 2004). Similar experiments

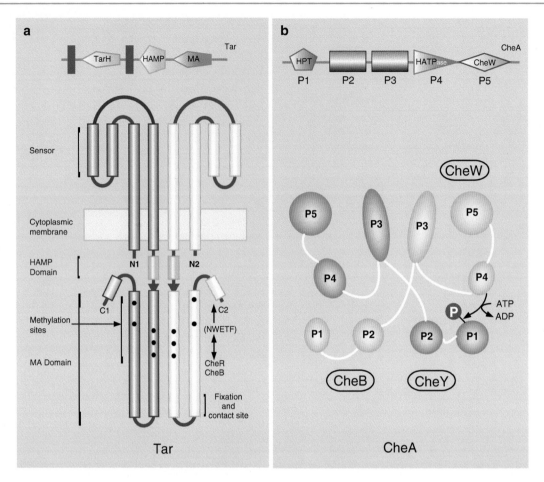

Fig. 9.10 Structural organization of MCP receptors and histidine protein kinase CheA. (**a**) Modular structure and schematic representation of the Tar homodimer of *E. coli*. NWETF is a pentapeptide at the C-terminus of Tar which binds CheR and CheB; sites of methylation/demethylation, MA is the cytoplasmic domain that has binding sites for CheA and CheW and contact sites for the formation of trimers (Modified and redrawn from Parkinson et al. 2005). (**b**) Modular structure and schematic representation of the homodimer of HK CheA (P1: HPT field carrying the His active P2 binding domain of CheY and CheB, P3 dimerization domain, P4: HATPase catalytic domain, P5: CheW binding domain) (Modified and redrawn from Kentner and Sourjik 2006). Drawing: M.-J. Bodiou

were performed in *V. cholerae*. Overexpression of a DGC, VCA0956, abolished swimming, while expression of a PDEA, VieA (Fig. 9.13) greatly increased it (Tischler and Camilli 2005). Transcriptome analysis of *V. cholerae* following an increase in the concentration of intracellular di-cGMP clearly shows a repression of genes involved in the biosynthesis of flagella, motility, and chemotaxis (Beyhan et al. 2006).

Another process regulated by the di-cGMP is the production of exopolysaccharides, especially those involved in biofilm formation. In *V. cholerae*, the enzymes of the biosynthetic exopolysaccharide pathway (VPS) are encoded by two operons under the control of two transcriptional activators, VpsR and VpsT. A mutation in the gene for PDEA, VieA is sufficient to cause a significant increase in *vps* genes transcription and biofilm formation (Tischler and Camilli 2005). Work on the El Tor biotype, which is responsible for several recent cholera outbreaks, shows that overexpression of the DGC, VCA0956, increased biofilm formation and, conversely, overexpression of the PDEA, VieA prevented biofilm development. The regulation of the biosynthesis of VPS by di-cGMP is achieved through transcriptional activation of *vpsR* and *vpsT* (Beyhan et al. 2006).

P. aeruginosa also uses di-cGMP to regulate biofilm formation, which in this organism involves multiple aspects such as the synthesis of exopolysaccharides, chemotaxis, quorum sensing (*cf.* Sect. 9.3), and twitching mobility (*cf.* Sect. 9.7.2). An analysis of the different phenotypes associated with mutations in the 39 genes encoding proteins with GGDEF and EAL domains shows a strong correlation between a high intracellular concentration of di-cGMP and a high production of biofilm (Kulasakara et al. 2006).

In *Yersinia pestis*, a DGC, HmsT (Fig. 9.13) and a PDEA, HmsP control biofilm formation (Kirillina et al. 2004), while in *Salmonella typhimurium*, Adra and YhjH control the biosynthesis of cellulose (Simm et al. 2004).

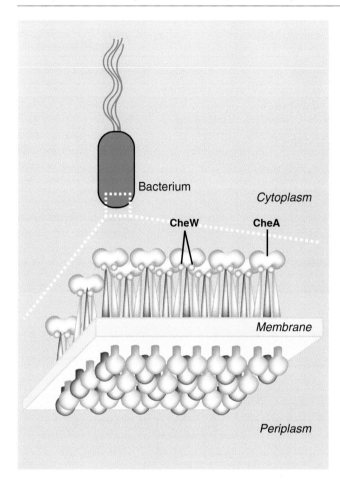

Fig. 9.11 Organization of MCP receptors in clusters at one pole of the cell. In *E. coli* the different proteins could be involved in MCP trimers of dimers and interact with CheA and CheW (Modified and redrawn from Wadhams and Armitage 2004). Drawing: M.-J. Bodiou

Several studies on pathogenic bacteria also show a regulation of the expression of virulence genes by di-cGMP (Kulasakara et al. 2006). This is the case for *V. cholerae* for which it was shown that the di-cGMP inhibited virulence genes. The *vieSAB* operon encodes a two-component system VieS/VieA. The response regulator VieA has three conserved domains (REC-EAL-HTH) and is essential for the expression of *toxT*, which encodes a transcriptional activator of virulence genes encoding cholera toxin and associated pili. The prejudicial effect of a mutation in VieA on the expression of virulence genes is attributed to the loss of the PDEA activity of VieA and an increase in the intracellular concentration of di-cGMP in the mutant. These results are confirmed by inhibiting the production of cholera toxin in vitro by overexpression of a DGC (Tischler and Camilli 2005). The proposed model for activation of virulence genes involves activation by an unknown factor of the histidine kinase VieS that will phosphorylate the REC domain of VieA. VieA-P will autoactivate transcription of the *vieSAB* operon with as a consequence a significant increase in the PDEA VieA which will reduce the intracellular concentration of di-cGMP and allow the expression of virulence factors (Fig. 9.14).

A number of targets of c-di-GMP have been identified in recent years highlighting the great diversity of mechanisms of action of c-di-GMP that may act at the functional, translational, or transcriptional level. Among the receptors c-di-GMP are (Krasteva et al. 2012):

- PilZ domain proteins such as the YcgR protein in *E. coli* which functions, after binding to c-di-GMP, as a molecular flagellar motor brake or BscA and Alg44, which are the subunits of the cellulose synthase of *G. xylinus* and

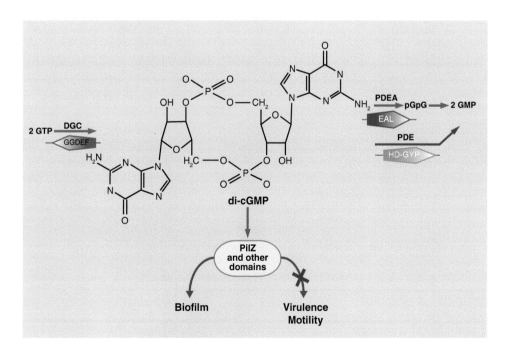

Fig. 9.12 Metabolism of di-cGMP. *DGC* diguanylate cyclase, *PDE(A)* phosphodiesterase (A), *PilZ* a domain binding cGMP-di (Modified and redrawn from Tamayo et al. 2007). Drawing: M.-J. Bodiou

Fig. 9.13 Structural organization of protein domains GGDEF and EAL. Domain names from databases SMART (http://smart.embl.de) or Pfam (http://pfam.sanger.ac.uk) (Modified and redrawn from SMART, Pfam, Christen 2007). Drawing: M.-J. Bodiou

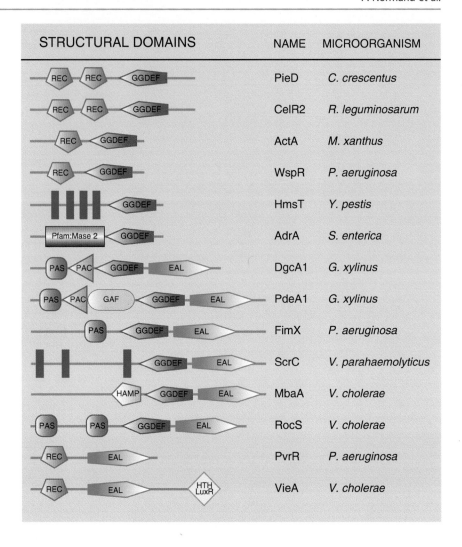

alginate synthetase of *P. aeruginosa* enzyme complexes, respectively, responsible for the synthesis of exopolysaccharides (Amikam and Galperin 2006).
- GGDEF domain proteins or degenerate EAL that have lost their enzymatic activity, but not their ability to bind c-di-GMP, such as LapD *P. fluorescens*, which acts via a protease LapG on maintaining a LapA adhesin on the outer membrane protein essential for biofilm formation
- Riboswitches that are untranslated sequences of messenger RNA which control the expression of downstream genes in response to changes in concentration of a specific ligand, here the c-di-GMP
- Transcription factors as FleQ in *P. aeruginosa* and VpsT in *V. cholerae* or Clp in *X. campestris* that control the expression of genes for the biosynthesis of exopolysaccharide or virulence

cAMP-Based Systems

The cAMP synthesized by adenylate cyclase is known to activate the CAP ("catabolite activator protein," also known as CRP, "cyclic AMP receptor protein") transcription factor that regulates various catabolic operons. Cellular signaling proteins involving on the one hand a membrane domain such as MASE or CHASE as well as PAS or GAF and on the other hand a domain having cytoplasmic adenylate cyclase activity (ACyc) were identified in many genomes, particularly those of cyanobacteria, mycobacteria, and alphaproteobacteria. The molecular mechanisms involved remain largely unknown; however, it has been shown that cyanobacterial adenylate cyclase whose activity is modulated by light regulates the concentration of intracellular cAMP and motility of cells (Ohmori and Okamoto 2004). In *P. aeruginosa*, an adenylate cyclase with a sensory membrane domain plays an important role in pathogenicity by regulating the expression of a type III secretion system (Lory et al. 2004).

An Uridylation System

After carbon, the metabolite probably the most limiting and therefore the most important for the prokaryotic cell is ammonium, an essential constituent of proteins, nucleic acids, and many other components of the cell. Cells that

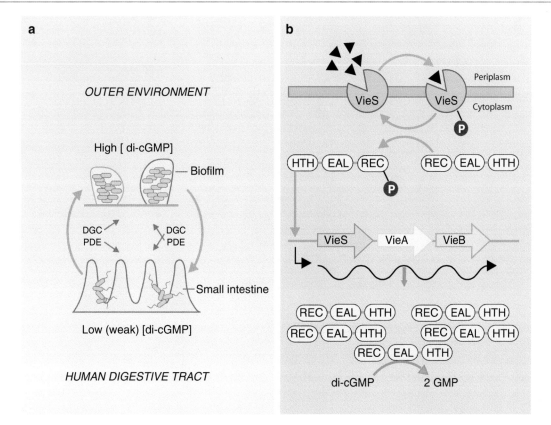

Fig. 9.14 Role of di-cGMP in the human gut colonization by *Vibrio cholerae*. (**a**) Transition between the biofilm state in environmental reservoirs and dispersion of cells in the human gut. (**b**) Role of two-component system VieS/VieA in regulating the amount of intracellular di-cGMP. The activation of the HK VieS leads to phosphorylation of the REC domain of VieA which, through its HTH domain, activates transcription of the operon *vie*SAB resulting in an increase in phosphodiesterase (EAL domain) of VieA and therefore a decrease in the cellular concentration of di-cGMP (Modified and redrawn by Tamayo et al. 2007). *DGC* diguanylate cyclase, *PDE* phosphodiesterase

use other forms of nitrogen (nitrate, nitrite, nitrous) convert them first into ammonium, which is costly in energy terms. Ammonium is then converted into amino acids primarily by two enzymes, glutamine synthetase (GS) and glutamine: 2-oxoglutarate aminotransferase (GOGAT), which produce glutamate and glutamine. Ammonium deficiency in the bacterial cell leads to a rapid, proportionate, and multicomponent response, in order to maintain a level of $NH4^+$ compatible with the maintenance of cellular functions. The perception of abundance is made through a protein called GlnB or PII nitrogen regulator. The protein is covalently modified by uridylic tyrosyl residues located on the outer part of the protein (Fig. 9.15).

The GlnB protein perceives the amount of intracellular alpha-ketoglutarate, which is an intermediary metabolite of the Krebs cycle and also a nitrogen-free acceptor used in the assimilation of ammonium by the proteins glutamine synthetase (GS) and GOGAT. The protein GlnB also senses the amount of intracellular glutamine, an amino acid rich in nitrogen, and most importantly is also the reaction product of GS after transfer of ammonium to glutamate.

The first role of the regulatory protein GlnB is to control the level of GS adenylation of the protein, a double hexamer, which can contain from 0 to 12 adenylate residues and thus be more or less active in the transformation of glutamate and ammonium into glutamine. Other roles of GlnB are to control the level of transcription of several genes and the level of activation of several proteins (PipX, NAGK kinase, AmtB, AmtR, Drat, DraG, NifL, NifA, Atase, TnrA, etc.) all involved in the acquisition of ammonium directly or indirectly.

9.2.3 Translational Regulation

Transcription is the step most often used to regulate gene expression in bacteria; however, the response includes the conversion of DNA into RNA and then translation into protein which are both time consuming. Some functions that require a faster response are under the control of an RNA transcript that is not translated before the environmental stimulus. This is the case of cold-shock proteins

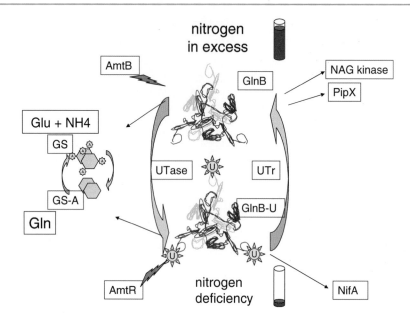

Fig. 9.15 Regulation of GlnB through uridylation linked to the nitrogen status of the cell. The cell is subjected to periods of excess nitrogen figured above by a red cursor almost full and periods of nitrogen deficiency figured at the bottom by an almost empty cursor. The protein PII trimer or GlnB exists in two forms, uridylated (*low*) or not (*top*) on a tyrosine residue of the outer loop. The transition between these two forms is performed by a protein called uridylyltransferase (*UT*) that adds UTP residues to GlnB and thus activates it as well or, conversely in the presence of glutamine residues, removes UMP and inactivate GlnB. The activated form of GlnB, GlnB-U (*bottom*), inhibits the transcription of repressor AmtR (and activate transport of ammonium) and activates regulator NifA (nitrogen fixation). The inactivated form of GlnB modulates activity of other proteins (NAG kinase, PipX, DRAT) and especially the adenyl GS which can then assimilate ammonium. The double hexamer of the protein glutamine synthetase (*GS*) alternates between an adenylated state (*bottom*) and a non-adenylated one (*top*), allowing the GS to perform ammonification of glutamate to glutamine (Redrawn and modified from Leigh and Dodsworth 2007)

(CSP "cold-shock proteins") found in archaea and bacteria. These proteins have a high affinity for RNA.

The gene domain involved in the attachment to the RNA is now called the CSP domain; it has been found in many proteins involved in different functions: adaptation to low temperatures, cell growth, nutritional stress, and stationary phase (Graumann and Marahiel 1998).

9.2.4 Protein Glycosylation

Protein glycosylation in eukaryotes is mainly a means of cellular addressing of proteins to the cell membrane, for it to be incorporated into it or to be secreted. The presence of glycosylated motifs is also a way to modulate the biochemical properties and activities of many proteins and is dependent on posttranslational modifications. If many cases of glycosylation exist in archaea, it is only recently that some cases have been described in bacteria, especially in relation to virulence (Abu-Qarn et al. 2008).

Glycosylation enzyme binds specific saccharides to residues, thus altering their structure and function. These saccharides are attached to different parts of a few amino acids, mainly asparagine (*N*-glycosylation of the amide function), serine, threonine (*O*-glycosylation of the hydroxyl-oxygen), etc.

Several aspects of the physiology of proteins are influenced by glycosylation, in particular adhesion, recognition, stability, and folding. They are also important mediators in different processes of signaling and targeting.

In the metazoan pathogen, *Campylobacter jejuni*, a heptasaccharide (Table 9.5) is assembled through the activity of five glycosyltransferases starting with a membrane lipid carrier intermediate ("lipid carrier protein"), galactose derivatives, and *N*-acetyl-glucosamine before it passes through the membrane and is transferred as a whole to an asparagine residue of a protein outside the membrane. The mutants for this biosynthetic pathway have an attenuated phenotype (Abu-Qarn et al. 2008).

In the *Proteobacteria Neisseria* spp. and *Pseudomonas aeruginosa*, pili contain *O*-glycosylated proteins. In *Neisseria* spp., a trisaccharide composed of galactose and of the unusual 4-diacetamido-2,4,6-trideoxyhexose residue is attached to a conserved serine of the pilin. In *Pseudomonas* and *Helicobacter*, an unusual residue is also present on the pilin, pseudaminic acid.

Given these saccharides appear to be important for pilus assembly and virulence; they constitute targets for new drugs searches.

Table 9.5 Structure of glycans bound to proteins in some bacteria and archeae

Bacteria
N-glycosylated
Campylobacter jejuni
GalNAc-a1,4-GalNAc-a1,4-(Glc-b1,3)-GalNAc-a1,4-GalNAc-a1,4-GalNAc-a1,3-BacAc2-[Asn]
O-glycosylated
Neisseria gonorrhoeae
Gal-b1,4-Gal-a1,3-DATH-[Ser]
Pseudomonas aeruginosa
5N(3-OH)But7NFmPse-a2,4-Xyl-b1-3FucNAc-b-[Ser]
Archaea
N-glycosylated
Halobacterium salinarum
[GalNAc-3-1-(3fGal)GalA-4-1-(6GalA3OCH3)GlcNAc-4-](10–15)-1-GalNAc-[Asn]
(OSO3)GlcA-[b1,4-GlcA(OSO3)]2-b1,4-Glc-[Asn]
Methanococcus voltae
ManNAcA6Thr-b1,4-GlcNAc3NAcA-b1,3-GlcNAc-[Asn]

Modified from Abu-Qarn et al. (2008)
The amino acid to which the oligosaccharide is attached is framed on the right
Abbreviations used are *BacA2* di-N-acetylbacillosamine, *DATH* 2,4-diacetamido-2,4,6-trideoxyhexose, *FmPse* formyl-pseudaminic acid, *FucNAc* N-acetylfucosamine, *Gal* galactose, *GalA* galacturonic acid, *GalNAc* N-acetylgalactosamine, *Glc* glucose, *GlcA* glucuronic acid, *GlcNAc* N-acetylglucosamine, *GlcNAc3NAcA* 2,3-diacetamido-2,3-dideoxy-glucuronic acid, *ManNAcA* N-acetylmannuronic acid, *Xyl* xylose

9.3 Quorum Sensing*, a System for Perception of Cell Density

9.3.1 Discovery of Quorum Sensing

In the prokaryotic world where each individual cell reproduces by binary fission and is constantly struggling for access to nutritional resources, recognition and cooperation between cells seemed highly unlikely. Early studies undermining the paradigm of bacterial unicellular life and evoking the existence of bacterial pheromones have concerned the formation of fruiting bodies in *Myxococcus xanthus*, the production of streptomycin in *Streptomyces griseus*, the induction of competence in *Streptococcus pneumoniae*, and control of bioluminescence in *Vibrio fischeri*. In *V. fischeri*, a bioluminescent marine bacterium living in symbiosis with the squid *Euprymna scolopes*, bioluminescence was observed only at high cell density, an extracellular compound accumulating during growth could induce the phenomenon of bioluminescence in a low cell density culture. This autoinducer compound was purified and its structure elucidated; it is *N*-(3-oxohexanoyl) homoserine lactone (3-oxo-C6-HSL) (Eberhard et al. 1981) (Fig. 9.16).

It was not until 1992 that production of 3-oxo-C6-HSL was detected in other bacterial species, particularly in *Erwinia carotovora* now called *Pectobacterium carotovorum*, where the signal molecule regulates the biosynthesis of carbapenem antibiotic, of the β-lactams family (Bainton et al. 1992). It was then that the term "quorum sensing" (QS) appeared to describe the phenomenon that allows bacteria to assess their population density via the production of small sensor molecule ("sensing") and initiate a coordinated response when a certain cell density is reached ("quorum") (Fuqua et al. 1994). Detection of *N*-acyl-homoserine lactones (AHLs) was made possible through the development of biological systems based on the induction of a reporter gene or production of a pigment in response to the addition of exogenous AHL (Bainton et al. 1992). Several AHLs, which differ in acyl chain length (from 4 to 18 carbons), and the substitution on the 3rd carbon (H, O, or OH; Fig. 9.16) have been identified in various Gram-negative bacteria belonging to subdivisions α, β, and γ of *Proteobacteria* (Williams et al. 2007), and recently the production of AHLs was discovered in a cyanobacterium (Sharif et al. 2008). Most AHL-producing bacteria produce a set of AHLs in various proportions (Lithgow et al. 2000).

QS regulation based on AHLs is used to control various phenotypes (Williams et al. 2007) such as production of virulence factors (plant pathogen *Erwinia*, opportunistic pathogens such as *Pseudomonas aeruginosa* or *Burkholderia pseudomallei*), production of pigment or antibiotics (*Chromobacterium violaceum*, *Erwinia carotovora*, *Pseudomonas aureofaciens*), conjugative transfer of plasmids (*Agrobacterium tumefaciens*, *Rhizobium*), mobility through swimming or swarming (*Serratia marcescens*, *Yersinia enterocolitica*), development of

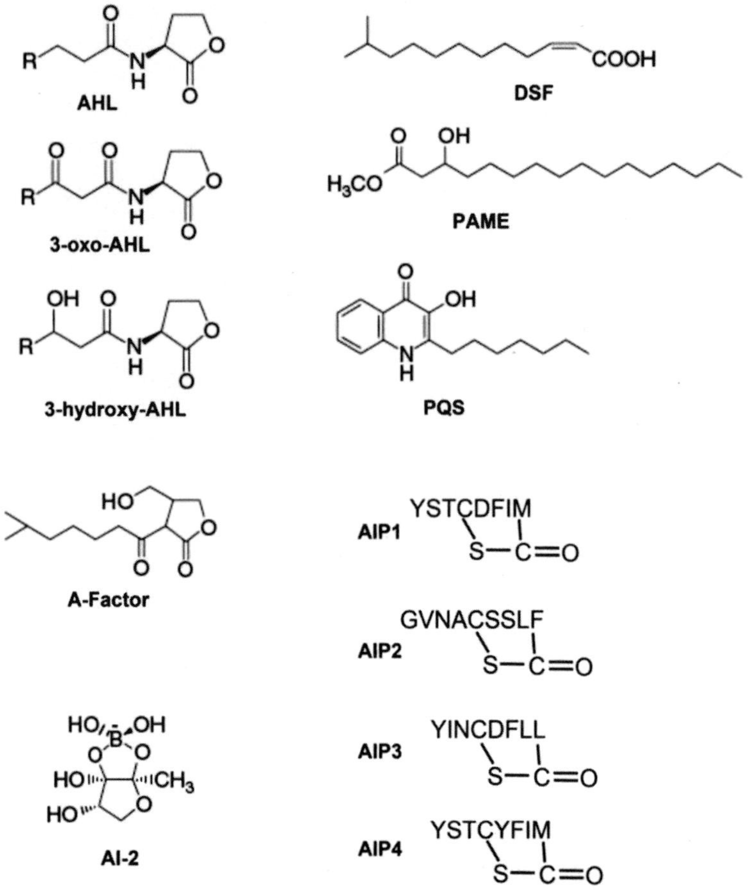

Fig. 9.16 Structures of signaling molecules involved in quorum sensing. *AHL* N-acyl-homoserine lactone, *3-oxo-AHL* N-(3-oxoacyl) homoserine lactone, *3-hydroxy-AHL* N-(3-hydroxyacyl) homoserine lactone, with a hydrocarbon chain with 1–15 C, which may also contain one or more unsaturated bonds, *factor A* 2-isocapryloyl-3-hydroxy-methyl-γ-butyrolactone, *AI-2* autoinducer-2, ester form of boric acid furanosyl, *PQS* "*Pseudomonas* quinolone Signal" 2-heptyl-3-hydroxy-4-quinolone, *DSF* "diffusible factor"-methyl-dodecenoic, *PAME* methyl ester hydroxypalmitic acid, *AIP1* to *AIP4 Staphylococcus aureus* "autoinducer peptide" (Adapted and modified from Williams et al. 2007)

biofilm (*P. aeruginosa*, *S. marcescens*), nodulation (*Sinorhizobium meliloti*, *Rhizobium etli*), etc. The phenotypes regulated by AHLs are often crucial for interaction with a eukaryotic host, and production of virulence factors in the plant pathogen *Erwinia* allows coordinated action of bacterial cells to evade the defense reactions of the host.

9.3.2 Genes Involved in Communication by Acyl Homoserine Lactones (AHLS) and Regulatory Cascade

AHLs synthesis is based on the presence of one or more bacterial genes encoding AHLs synthases. Of the three families of AHLs synthases, the most common gathers LuxI-like proteins (the first member of this family has been identified in *V. fischeri*) (Fuqua et al. 2001), the second family includes LuxM-type proteins identified in several species of *Vibrio*, and the third is represented by the enzyme HdtS identified in *P. fluorescens* (Laue et al. 2000). LuxI-type proteins synthesize AHLs by catalyzing the amide bond between the fatty acid chain carried by a carrier protein of an acyl group (ACP) and the amino group of S-adenosyl methionine. Lactonization of the molecule then takes place with the release of 5′-methylthioadenosine. AHLs diffuse through cell wall and accumulate in the extracellular medium. When a threshold concentration is reached, AHLs bind to a receptor that is a transcriptional regulator of the LuxR family; the LuxR/AHL complex binds upstream target genes at sequences designated "*lux*" boxes and activates (or in some cases represses) target genes, such as the *lux* operon responsible for bioluminescence in *V. fischeri*. In many cases, the *luxI* gene is one of the targets of the LuxR/AHL complex, leading to a positive feedback loop of regulation.

Many bacteria possess multiple LuxR/LuxI/AHL modules that are often interconnected. Most *Rhizobium* strains studied harbor several LuxR/LuxI couples (up to four in *R. leguminosarum* bv. *viciae*) under the control of the CinR/CinI system (Wisniewski-Dyé and Downie 2002). Several LuxR homologs that are not related to a *luxI*-type gene ("orphans" LuxR) are also involved in these regulatory cascades. Except the symbiotic plasmid transfer, the functions regulated by QS are not common among the strains studied and the majority of QS mutants retain their ability to nodulate. The functions controlled by QS (swarming motility, EPS production, growth inhibition, nodulation efficiency) are involved in adaptation to the rhizosphere and in optimizing the interaction with the plant. In *P. aeruginosa*, the *las* system (*lasI*/*lasR*) exerts a transcriptional control of the *rhl* system (*rhlI*/*rhlR*) (Latifi et al. 1996). Transcriptomic studies have revealed that the *las* and *rhl* systems regulate over 300 genes scattered throughout the genome (Schuster et al. 2003). *P. aeruginosa* also has two "orphan" LuxR, QscR and VqsR, which interact with the *las* and *rhl* systems. QscR represses *lasI* at low cell density, whereas VqsR positively regulates expression of virulence factors via LasI (Chugani et al. 2001; Juhas et al. 2005). Quorum sensing regulation is often integrated into other regulatory networks of the bacterial cell; hence, AHL production depends not only on cell density but is also modulated by physiological parameters that control bacterial growth and is therefore related to changes in the extracellular medium. Two-component systems can thus exert a regulation on *luxI*/*luxR* genes, as the GacS/GacA system in various *Pseudomonas* (Lapouge et al. 2008) or the PprB/PprA system in *P. aeruginosa* (Dong et al. 2005). In *P. aeruginosa*, regulation of the *las* and *rhl* systems involves a large number of regulators in addition to the two-component systems, such as Vfr, a homolog of Crp ("catabolite repressor protein," a receptor binding cyclic AMP), the stationary phase-specific sigma factor RpoS, the alternative sigma factor RpoN, the stress response protein RelA, the transcriptional regulators RsaL and MvaT connected to growth phase, the ANR regulator involved in the control of anaerobic respiration, and VqsM, a global regulator of the AraC family (Williams et al. 2007). Expression of genes involved in the production of AHLs can also be controlled at a posttranscriptional level; the most studied system is the system of secondary metabolite Rsm repressors ("repressor of secondary metabolites") present in *Erwinia carotovora* and in various *Pseudomonas* (Lapouge et al. 2008).

Since a given AHL molecule is not specific to a bacterial population, it is likely that AHLs produced by a bacterial population are seen and recognized by another population, resulting in cross talk. For example, AHLs produced by *P. aeruginosa* can activate production of virulence factors in *B. cepacia* in vitro but also in vivo under conditions of mixed biofilm on murine models (McKenney et al. 1995). An interspecies communication via AHLs has also been demonstrated in the wheat rhizosphere (Pierson et al. 1998). Some bacteria are unable to produce but may perceive them via an orphan LuxR-type regulator and regulate target genes; as such, SdiA of *Escherichia coli* senses exogenous AHLs and regulates genes involved in acidity tolerance while saving the cost related to AHL production (Van Houdt et al. 2006). Furthermore, *E. coli* might sequester AHLs produced by other cohabiting bacteria and therefore interfere with the process of AHLs accumulation.

Recently, some bacteria interacting with plants were shown to possess orphans LuxR that instead of responding to AHLs are able to respond to plant compounds (Subramoni et al. 2011).

9.3.3 Interference with Communication by Bacterial AHLs

Temperature and pH are the main abiotic parameters affecting AHLs half-life. Indeed, alkaline pH (>8) and high temperature conditions favor hydrolysis of the AHLs lactone ring, yielding a *N*-acylhomoserine derivative that is inactive as a signaling molecule (Yates et al. 2002). In the environment, particularly in soil, AHLs producing bacteria interact with other organisms capable of degrading these signaling molecules, an interference phenomenon designated by the term "quorum quenching." Three families of enzymes with different enzymatic activities have been identified so far: AHL-lactonases belonging to the AiiA family, acylases/amidohydrolases that are homologous to AiiD, and oxidoreductases (Uroz et al. 2009). Lactonases have initially been identified in several species of *Bacillus* and then detected in other Gram-positive strains isolated from soil but also in several species of Gram-negative bacteria (*Klebsiella pneumoniae*, *Agrobacterium tumefaciens*). Bacterial degradation of AHLs could confer several advantages such as the use of AHLs as a nutrient source, inhibition of QS regulated functions in other bacteria, and resistance toward the antibiotic activity displayed by certain AHLs (Leadbetter and Greenberg 2000). Degradation of AHLs by soil bacteria can be exploited agronomically to protect crops, notably potato plants from the worldwide pathogen *Pectobacterium* (Cirou et al. 2011).

Some AHL-producing bacteria have the ability to degrade AHLs, a property that allows them to finely tune AHL production (Sio et al. 2006). In animals, enzymes belonging to the paraoxonases family, which have no known counterparts in bacteria, degrade AHLs in the same way that lactonases do. Moreover some plants have the ability to degrade AHLs, but the mechanisms involved remain to be identified (Götz et al. 2007).

Interferences are not only due to inactivation of AHLs; different organisms produce compounds able to "mimic" AHLs. Cyclic dipeptides (diketopiperazines, DKP) isolated from culture supernatant of different bacteria (including *P. aeruginosa*) are able to activate biological systems used for AHLs detection; yet the role of these dipeptides in signaling remains to be demonstrated (Degrassi et al. 2002). The benthic alga, *Delisea pulchra*, produces halogenated furanones that interfere with QS regulation by binding to the LuxR-type receptor and inducing its proteolysis (Manefield et al. 2002). Production of furanones is thought to limit bacterial colonization and prevent expression of virulence phenotypes controlled by AHLs. The unicellular alga *Chlamydomonas reinhardtii* produces a dozen compounds that stimulate functions regulated by QS (Teplitski et al. 2004). Plants, especially legumes like peas, soybean, and *Medicago*, exude different compounds mimicking AHLs that stimulate or inhibit phenotypes regulated by QS (Keshavan et al. 2005). Although most of these compounds have not been characterized, their discovery suggests that interactions established between plants and pathogenic symbiotic or saprophytic bacteria can be "manipulated" by plants.

9.3.4 Acyl Homoserines Lactones (AHLs), Signals Only for Bacteria?

Several studies have revealed that eukaryotic organisms might respond to AHLs. The first demonstration of a eukaryotic detection of AHLs comes from the stimulation of the production of interleukin-8 by epithelial cells exposed to the 3-oxo-C12-HSL produced by *P. aeruginosa* (DiMango et al. 1995). Subsequently, immunomodulatory effects of AHLs have been shown to play a direct role in pathogenicity (Smith et al. 2002; Khajanchi et al. 2011).

In the marine environment, AHLs released by bacterial biofilms can attract zoospores of the marine Viridiplantae *Ulva* (Joint et al. 2002). When zoospores detect the AHL, their rate of swimming is reduced and they aggregate near the AHL source; AHLs would act as indicators for the attachment of zoospores and would influence the biogeography of the seaweed.

In the rhizosphere, the model legume *Medicago truncatula* responds to nanomolar or micromolar concentrations of two different long acyl chain AHLs and shows significant changes in the accumulation of more than 150 proteins (Mathesius 2003); these proteins are involved in functions related to defense, stress, transcriptional regulation, and hormone response. In addition, AHLs induce changes in the exudation of compounds that mimic AHLs (Mathesius et al. 2003). Plants may use the information encoded by AHLs as a reliable means of detection, in order to activate defense responses before the bacterial infection occurs.

Inoculation of *Serratia liquefaciens* onto tomato plants infected by the leaf pathogen *Alternaria alternata* increases systemic resistance, a phenotype that is not observed when a strain of *Serratia* defective in the synthesis of AHLs is used (Schuhegger et al. 2006); in addition, C6-HSL induces a systemic accumulation of salicylic acid and expression of defense genes of the ethylene pathway in tomato. The contact of *Arabidopsis thaliana* roots with C6-HSL induces changes in gene expression in roots and also in aerial parts, modifies the auxin/cytokinin ratio, and increases root elongation, whereas long-chain AHLs reinforce plant resistance (von Rad et al. 2008; Schenk et al. 2012); such modifications could be an integral part of the phytobeneficial effect of some rhizosphere AHL-producing bacteria.

9.3.5 Other Bacterial Signal Molecules

N-acyl homoserine lactones are not the only class of signaling molecules used by prokaryotes. In Gram-negative bacteria, fatty acids (*Xanthomonas campestris*, *Stenotrophomonas maltophilia*), esters of fatty acids (*Ralstonia solanacearum*), and quinolones (*P. aeruginosa*, *Burkholderia* spp.) also have a role as signal molecules (Fig. 9.16) (Vial et al. 2008). In *P. aeruginosa*, PQS signals are involved in the hierarchical regulatory cascade leading to the synthesis of AHLs. During nutritional deficiency, *Myxococcus xanthus* secretes an unknown compound, which causes cells to clump together and form a fruiting body; within this globular aggregate which can consist of more than 100,000 bacteria, some cells then turn into myxospores.

With the exception of *Streptomyces* that produce γ-butyrolactones such as A-factor (Fig. 9.16, a compound that controls sporulation and antibiotic production), *Firmicutes* generally use autoinducer peptides (AIPs) (Fig. 9.16). These peptides (from 5 to 34 amino acids) are produced in the cytoplasm and are then modified (some are cyclized) after translation and exported via a specific transporter. Above a threshold concentration, AIPs specifically interact with a membrane histidine kinase; this interaction stimulates the kinase activity, leading to phosphorylation of a response regulator which can then bind to DNA and control the transcription of target genes. The most studied phenotypes regulated by AIPs are competence and sporulation in *Bacillus subtilis*, competence in *Streptococcus pneumoniae*, expression of virulence factors and biofilm formation in *Staphylococcus aureus*, and virulence in *Enterococcus faecalis* (Lyon and Novick 2004). Differences in the primary structure of AIPs result in a high degree of specificity. In the human pathogen *S. aureus*, each of the four AIPs is specifically recognized by a receptor (Lyon and Novick 2004).

The use of peptides does not seem to be restricted to *Firmicutes* since a linear pentapeptide involved in cell death has recently been identified in *E. coli* (Kolodkin-Gal et al. 2007).

Although no universal bacterial QS system has been discovered, many Gram negative and Gram positive produce autoinducer-2 (AI-2) (Fig. 9.16, a collective term for a group of interconvertible furanones). The *luxS* gene required for

AI-2 production is indeed present in the genome of more than 60 species (*Proteobacteria*, *Spirochaetes*, *Firmicutes*, etc.), suggesting that AI-2 could be part of an interspecies language (Xavier and Bassler 2003). The LuxS/AI-2 system was analyzed in *Vibrio*, including *V. harveyi* and *V. cholerae*, where it helps regulate bioluminescence and the production of virulence factors. However, the precise role of AI-2 in other species remains controversial, partly because LuxS also plays a metabolic role in the activated methyl cycle (Williams et al. 2007).

Given the large number of extracellular bacterial metabolites and the low proportion of bacteria that are culturable in the laboratory, it is likely that the diversity of QS molecules known to date represents only the "tip of the iceberg."

The term "quorum sensing" does not describe satisfactorily all the situations in which bacteria use diffusible compounds. In the case of AHLs, the quorum size is not defined and may vary depending on the spatial distribution of cells, rates of synthesis, and loss (by degradation or diffusion). Hence, the notion of "efficiency sensing" was proposed (Hense et al. 2007): autoinducers produced in the environment would have a role of proxy in assessing the effectiveness of producing extracellular costly effectors (such as exoenzymes).

In pathogenic bacteria, bacterial communication provides a new target for novel antibacterials; while conventional antibiotics target bacterial growth and lead to the emergence of resistant bacteria, using molecules that block the QS could precisely target pathogenesis (Bjarnsholt and Givskov 2007). Thus, many studies currently aim at identifying natural QS inhibitors, especially from plants, or at developing synthetic QS inhibitors (Chung et al. 2011).

9.4 Phase Variation*

9.4.1 Phase Variation and Antigenic Variation

One of the clearest manifestations of the phase variation phenomenon is the appearance of a minority of colonies or sectors of colonies having a different appearance on agar. Phase variation or phenotypic conversion leads to the coexistence within the same population of different cell types (wild type and variants), some of which may have a significant selective advantage under some physicochemical conditions. Through phase variation, the expression of a given phenotype is either turned "ON" or turned "OFF"; these events are generally reversible ("ON" ↔ "OFF") but can be irreversible ("ON" → "OFF" or "OFF" → "ON") and result from genetic or epigenetic alterations at specific loci (van der Woude and Baumler 2004; Wisniewski-Dyé and Vial 2008).

In contrast to spontaneous mutations which occur at a frequency of about 10^{-8} to 10^{-6} mutations per growing cell per generation, phase variation occurs at a frequency higher than 10^{-5} events per cell per generation and always affects the same phenotype(s). Phase variation has been described for many different bacterial genera belonging to diverse taxonomic groups and displaying various lifestyles (pathogens, saprophytes, symbionts). It contributes to regulate many phenotypes: synthesis of pili, flagella, surface lipoproteins, secondary metabolites, etc.

Antigenic variation refers to the expression of a number of alternative forms of a cell surface antigen (lipoproteins, pili, etc.); this generates within a clonal population, individual cells that are antigenically distinct, allowing bacterial pathogens to escape the host immune system. At the molecular level, certain mechanisms of antigenic variation have common features with mechanisms of phase variation.

9.4.2 Phase Variation Through Modification of the Genome

9.4.2.1 Gene Conversion*

This gene conversion mechanism, extensively documented for antigenic variation, implies a recombination event between a silent copy of a gene and another copy that is expressed, leading to a new chimeric gene. When several copies of the silent gene are present, many chimeric sequences can be theoretically generated, which allows the expression of various forms of a given antigen. There is no common mechanism for gene conversion, but in some cases, proteins of the homologous recombination pathway are involved. One of the best documented cases of gene conversion concerns type IV pili in the human pathogen *Neisseria gonorrhoeae*; this phenomenon generates multiple antigenic forms of pilin (Howell-Adams and Seifert 2000).

9.4.2.2 Site-Specific Inversion

This process involves specific enzymes (recombinases) and requires short homologous regions. These recombinases recognize inverted repeat sequences (IR) located on either side of the element to be inverted. Most inversion systems are independent of RecA activity. However, antigenic variation of surface layer proteins (SLP) in *Campylobacter fetus* via an inversion event requires the action of RecA.

Campylobacter fetus, an opportunistic parasite of humans and animals that interferes with reproductive functions, possesses eight SLP gene cassettes, clustered on the chromosome, encoding proteins of 97–149 kDa. All these

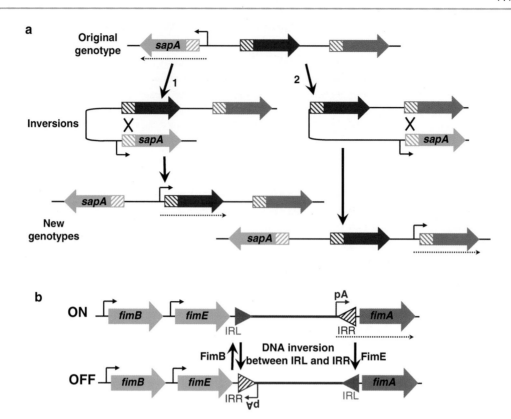

Fig. 9.17 Phase variation and antigenic variation by site-specific inversion. (**a**) Antigenic variation of surface layer proteins (SLP) of *Campylobacter fetus*. The 5′ conserved region and variable regions of SLP cassettes are represented respectively by small *stripped box* and *thick arrows*. Only one SLP gene, *sapA* (*green*), has a functional promoter (*over bent arrow*). DNA inversion takes place between two oppositely oriented cassettes following DNA exchange within the 5′ conserved region. Inversion of DNA containing the *sapA* promoter allows expression of alternative SLP cassettes (mRNAs depicted by *thin dashed arrows*). Two inversion events (1 and 2) and the resulting new genotypes are represented. For clarity, only three cassettes SLP are represented (Adapted from Dworkin and Blaser 1997). (**b**) Phase variation of type I pili in *Escherichia coli*. The relative positions of promoters (*over bent arrows*), genes (*thick colored arrows*), and inverted repeats IRR and IRL (*triangles*) are shown. The DNA element which undergoes inversion is represented by a *blue line*. IRR and IRL are inverted repeat sequences located within the binding sites for the recombinases FimB and FimE (Modified and redrawn from Wisniewski-Dyé and Vial 2008)

cassettes share a 600-pb homology beginning 74 pb before the start codon, followed by divergent sequences for the rest of the gene. Only one copy (*sapA*) has a functional promoter. SLP variation occurs through the inversion of a DNA fragment containing the *sapA* promoter (Fig. 9.17a); this inversion event moves the *sapA* promoter upstream one of the SLP cassette, allowing the exclusive expression of this cassette. This inversion may concern only the promoter or one or more SLP cassettes (Fig. 9.17a) (Dworkin and Blaser 1997). SLP surface proteins undergo significant selective pressure and the ability to synthesize different SLPs allows *C. fetus* to escape the host immune system. Moreover, virulence varies according to the SLP expressed, the strongest virulence being observed with a strain expressing the 97-kDa SLP. Other SLP would be produced in order to survive in hostile environments or colonize specialized microenvironments inside the host.

Type I pili, the most common adhesins of *E. coli*, are important for colonization and attachment of the bacteria to eukaryotic cells. *E. coli* can generate, at high frequency, cells lacking pili, via inversion of a DNA fragment located in the *fim* operon encoding type I pili (Fig. 9.17b) (Freitag et al. 1985). The *fimA* gene, encoding the major subunit of type I pili, can either be transcribed ("ON" position) or silent ("OFF" position). Upstream of the *fimA* gene lies an invertible element (314 bp) containing an essential promoter for transcription of *fimA* (Fig. 9.17b). Inversion of this element changes the orientation of the *fimA* promoter and abolishes *fimA* transcription ("OFF" position). This inversion requires two site-specific recombinases: FimB and FimE. FimB permits inversion in both directions, while FimE only mediates the transition from the "ON" to the "OFF" position (Klemm 1986).

Recently, phenotype switching between normal and small colonies of *Staphylococcus aureus* was correlated to a reversible large-scale chromosome inversion; this inversion switches "ON" or "OFF" bacterial phenotypes, including colony morphology, antibiotic susceptibility, hemolytic activity, and expression of dozens of genes (Cui et al. 2012).

9.4.2.3 Insertion-Excision

Phase variation by insertion-excision involves a mobile genetic element, such as an insertion sequence (IS), and is reversible when excision of the element is perfect.

Many examples involving IS have been described, such as surface properties in *Shigella flexneri* (Mills et al. 1992) and biofilm formation in *Staphylococcus aureus* (Kiem et al. 2004).

In *Legionella pneumophila*, the causative pathogen of legionellosis, an avirulent variant form devoid of LPS and flagella, can be isolated at high frequency. During the transition from wild type to variant, a 30-kb region is excised from the chromosome and replicates in the variant like a high-copy number plasmid (Luneberg et al. 2001). During reversion, this region is inserted back into the chromosome at the initial location. Variants, although less competitive for infection of the host, could be adapted to the aquatic environment where *L. pneumophila* usually lives (Luneberg et al. 2001).

9.4.2.4 Duplication

Pseudomonas tolaasii, the causal agent of brown blotch disease of mushroom *Agaricus bisporus*, can generate in vitro nonpathogenic, non-mucoid variants with enhanced mobility. The *pheN* regulatory gene, whose deduced product displays homology to both sensor and regulator domains of the conserved family of two-component systems, is duplicated during the phenotypic conversion (Han et al. 1997). A duplication of 661 bp within the sensor domain of *pheN* leads to the formation of two truncated ORFs. Reversion of the variant form to the wild type occurs via the precise deletion of the 661-bp duplicated region. Wild type and variants are adapted to different environmental niches; the wild type would penetrate and proliferate in fungal tissues, inducing their degradation, whereas the variant type would be more adapted to telluric life.

In *Streptococcus pneumoniae*, a human pathogen responsible for otitis, pneumonia, and meningitis, the appearance of small colonies lacking capsule is due to a duplication (11–239 bp) in the first gene of the biosynthetic pathway of capsules (Waite et al. 2001). Reversion to capsulated cells can occur via a precise excision of the duplication. The ability to regulate the synthesis of capsule is advantageous when invading eukaryotic cells; indeed, adherence and invasion would be 200-fold less efficient for a capsulated strain than for a noncapsulated strain, but after cellular invasion, the capsule prevents *S. pneumoniae* from being eliminated by phagocytosis.

9.4.2.5 Deletion

Phenotypic conversion by deletion is an irreversible phenomenon concerning regions of varying sizes (from a few bp to several hundred kb). In *Yersinia pestis*, the causal agent of bubonic plague, avirulent nonpigmented mutants are frequently observed, and these phenotypes are due to the deletion of a 102-kb region flanked by a repeated element (Fetherston et al. 1992). In the plant growth-promoting bacteria *Azospirillum lipoferum*, the emergence of variants affected in the assimilation of carbohydrates and mobility is correlated with the loss of a 750-kb plasmid (Vial et al. 2006).

9.4.2.6 Slipped-Strand Mispairing

Phase variation may be due to frequent and reversible changes in the length of short DNA repeats, generally composed of stretches of polypurines and/or polypyrimidines. Loss or gain of repeat units implies a mechanism of slipped-mispairing strain (SSM), a RecA-dependent process occurring during chromosomal replication, DNA repair, and recombination processes requiring DNA synthesis. When this event occurs in the coding region of a gene, changes in the number of repeat units result in frameshift mutations, thereby switching the expression of the encoded protein "ON" or "OFF." Gain or loss of repeat units can also affect transcription initiation by altering the relative positioning of the RNA polymerase-binding sites at the promoter or the termination site.

Opacity proteins of *Neisseria gonorrhoeae* and *Neisseria meningitidis* allowing bacterial adhesion and invasion of host tissues undergo antigenic variation through gene conversion and phase variation by SSM. Expression of other surface components (capsule, outer membrane proteins) can also be regulated by SSM (van der Woude and Baumler 2004; Wisniewski-Dyé and Vial 2008).

Synthesis of pili in *Haemophilus influenzae* is regulated at the transcriptional level by SSM of two divergently oriented genes, *hifA* and *hifB*, involved in the biosynthesis of pili. Their overlapping promoter region contains repetitive TA units; variation in the number of TA units changes the spacing between the −35 and −10 sequences and results in the control of transcription initiation (van Ham et al. 1993).

9.4.2.7 Multiple Events Affecting a Single Gene

Ralstonia solanacearum, the causal agent of bacterial wilting of over 200 plants, can undergo a phenotypic conversion in vitro and in planta: the variants are avirulent or less virulent. Different types of mutations in the *phcA* gene, encoding a regulator, have been characterized: deletions (in the reading frame or in the promoter region), duplications, IS insertions, or base substitutions (Brumbley et al. 1993). Revertants recovering a functional *phcA* gene can be isolated in the presence of the plant for variants having a duplication of 64 bp or IS inserted into *phcA* (Poussier et al. 2003). The variant form, with increased mobility, would be more adapted to survive in soil and to colonize new ecological niches.

In rhizospheric *Pseudomonas* exhibiting antagonistic activity toward fungal phytopathogens, the two-component system GacA/GacS is involved in regulating the production of secondary metabolites and the synthesis of exoenzymes (chitinase, lipase, protease). Variants devoided of biocontrol activity have mutations in *gacA* and *gacS*: point mutation, insertion, and deletion. In some cases, revertants can be observed (van den Broek et al. 2005). Strains mutated in *gacS* display a shorter lag phase compared to the wild type. In an environment like the rhizosphere, variants would be more competitive and could adapt more easily to the heterogeneous and challenging rhizosphere ecosystem (van den Broek et al. 2005).

9.4.3 Phase Change in Epigenetics Control

In contrast to the genetic mechanisms described above, **epigenetic regulation*** of phase variation occurs without modification of the DNA sequence. This involves differentially methylated sequences in regulatory regions of genes or operons that undergo phase variation operons. The state of methylation of these sequences affects fixation of a transcriptional regulator.

This epigenetic mechanism of phase variation was first elucidated for the expression of *pap* ("pyelonephritis-associated") pili among uropathogenic *Escherichia coli*. In the "OFF" state, the global regulator Lrp is bound to sites in the vicinity of the *pap* operon promoter, preventing transcription. Transition to the "ON" state requires methylation of a GATC sequence present in the binding site; methylation prevents binding of Lrp and allows the *pap* operon to be expressed (Hernday et al. 2003).

Synthesis of *pef* pili in *Salmonella typhimurium* is also regulated by methylation of specific sites in the promoter region of the *pef* genes (Nicholson and Low 2000).

Historically, phase variation has been investigated mainly in bacterial pathogens of humans and animals. However, recent studies have shown that phase variation also occurs in bacteria interacting with plants. When phase variation involves proteins of the cell surface, it allows the bacterium to evade the immune system. But it would be simplistic to consider phase variation as a strategy intended solely to evade the immune system. Phase variation can also modulate virulence; indeed, variants of *R. solanacearum* or *L. pneumophila* are less virulent than the wild type (Poussier et al. 2003). For these variants with reduced virulence, phase variation could allow to save energy, preventing the cell to synthesize some factors dispensable under certain conditions. In other cases, the appearance of variants could allow these cells to colonize new environments.

9.5 Antibiosis and Antibiotic Resistance*

The diversity of microorganisms has been estimated to range between 10^3 and 10^6 taxa/g soil (Gans et al. 2005; Schloss and Handelsman 2006). Among these, environmental microorganisms compete for energy sources and nutrients. Some organisms live symbiotically or coexist with other species of microorganisms, many of which produce antibiotics (Donadio et al. 2007). These are toxic molecules, bactericidal or bacteriostatic, which limit the existence or the number of other microorganisms in their neighborhood. The number of different antibiotics from natural sources is estimated at about 10^4 (Challis and Hopwood 2003). However, only a small part or a few dozen of these molecules are used for human and animal health including antibiotics of synthetic origin, for example, as sulfonamides, oxazolidinones, and quinolone (Table 9.6). Others are not used for various reasons such as toxicity to the host or else too short a lifetime in the body, problems that can be circumvented by combinatorial chemistry (Fig. 9.18).

9.5.1 Biosynthesis of Antibiotics

The biosynthesis of polyketide antibiotics, aminoglycosides, and peptides typically involves the enzymes of the family polyketide synthase (PKS), glycosyltransferases (GT), and nonribosomal peptide synthetases (NRPS). In bacteria, three types of polyketide synthases (PKS I, II, III) have been described: the PKS type I multifunctional enzymes are organized into modules where each module has a specific activity for the assembly of specific substrates to synthesize ultimately the antibiotic. For example, for the synthesis of erythromycin (Long et al. 2002), a molecule of propionate and mevalonate is recognized by the adenylation domain (AT), activated by binding to coenzyme A (CoA), and each molecule is transferred to an transporter of acyl moieties or acyl carrier protein (ACP). The condensation of the two substrates that is accompanied by the release of a molecule of CO_2 is catalyzed by another domain called keto-synthase (KS). These activities are common to all modules of type I PKS. The product of these reactions gives a keto acyl group which may be reduced to an alcohol by a keto reductase domain (KR), then dehydrated by a dehydratase domain (DH), and finally reduced by an enoyl reductase domain (ER), if they are present in the same module. Finally the thioesterase domain (TE) catalyzes the final reaction by cutting the thio-ester bond and thus releasing a free molecule of lactone (Fig. 9.19). Among type II PKS enzyme, activities are carried by different proteins that can iteratively catalyze the biosynthesis of polyketides (Hopwood and Sherman 1990). The PKS type III or chalcone synthases lack an

Table 9.6 Types and mode of action of the major types of antibiotics

Target	Antibiotic	Mode of action	Resistance mechanism	Biosynthesis genes
The cell wall				
Transpeptidase transglycosylase	β-lactams: penicillin	Transpeptidase inhibitor	β-lactamase: transpeptidase mutants	PS
Lipid II peptidoglycan precursor	Vancomycin	Interaction with D-Ala-DAla	Structural reprogramming of D-Ala-D-Ser	PS, glycosylase
	Lantibiotics: nisin, ramoplanin	Pore formation, lipid II-dependent	Protease specific	*nis*B, C, T, P
Arabinosyl transferase	Ethambutol	Arabinan biosynthesis inhibitor	Mutations in arabinosyl transferase genes	Synthetic
Ribosomes				
	Macrolides: erythromycin, azithromycin	Protein biosynthesis	rRNA methylation, efflux pump	PKS, glycosylase, synthetic
Peptidyl transferase	Tetracyclines	Protein biosynthesis	Efflux pump	PKS, Me transferase
	Aminoglycosides: kanamycin spectinomycin	Protein biosynthesis, translation of RNA messages	Acetylation, phosphorylation, adenylation	Glycosyltransferases
	Oxazolidinones: linezolid	Proteins biosynthesis	Not known	Synthetic
Enzymes involved in replication, repair or biosynthesis of DNA				
Gyrases	Fluoroquinolones: ciprofloxacin	DNA replication and repair	mutations in the *gyr* gene	Synthetic
	Sulfonamides: sulfamethoxazole trimethoprim	Folate biosynthesis	Mutations in the synthase gene	Synthetic
Lipids lipids biosynthesis	Isoniazid	Enoyl-ACP reductase	Mutations in genes involved in activation of isoniazid (katG) and dehydrogenases	Synthetic
	Thiolactomycin	Keto-acid synthase (KAS)	Mutations in the *kas* gene	PS, PKS
	3-decynoyl-N	3- hydroxyacyl-ACP dehydratase	Not described	Synthetic

ACP domain as PKS type I and II (Shen 2003) but incorporate the substrates previously activated as acyl-CoA derivatives (Funa et al. 1999).

The modular NRPS are enzymes similar to type I PKS except that they incorporate amino acids into the structure of the antibiotic. An amino acid is recognized by the adenylation domain (A), activated by ATP, and then forms an aminoacyl-AMP after transfer to the peptidyl carrier protein (PCP) also known as thiolation domain (T). This domain that contains the T phosphopantetheine acts as a cofactor essential for its enzymatic activity (Stack et al. 2007). The peptide bond formation between the two amino acids is catalyzed by the condensation domain (C). NRPS modules may have other domains with enzymatic activities leading to epimerization of amino acids (domain E) converting the amino acid "L" shape D or cyclization (Cy domain) or the oxidation of certain acids (Ox domain). Vancomycin is an antibiotic synthesized in part by the NRPS (Fig. 9.20).

Both types of NRPS and PKS enzymes, united in a molecule (hybrid molecule), may participate in the biosynthesis of an antibiotic. The same appears to be the case in the biosynthesis of rapamycin. Other enzymes may also contribute to the modification of the antibiotic molecule such as acyl transferases or glycosyltransferases that add sugars or acyl groups. This is the case in the biosynthesis of vancomycin. These changes alter the solubility of antibiotics or their half-life or even their activity, so many studies have focused on this theme. PKS and PS genes are widely distributed in the genomes of *Actinobacteria*, *Bacillus*, and *Pseudomonas* (Donadio et al. 2007). These genes are the largest known in bacteria; the genome of *Frankia alni*, for example, has a gene encoding a PKS 19299nt coding an enzyme of a molecular weight larger than 660 kDa. That of *Streptomyces coelicolor* contains a gene encoding a 22392nt NRPS of about 800 kDa, or 25 times the average size of bacterial genes. This enormous size is related to the need for a long series of reactions involving highly reactive intermediates without setting these intermediaries free in the cytoplasm where they could block normal cellular functions.

The phenylpropanoid pathway, which assembles units of isopentenyl diphosphate into long linear chains partially

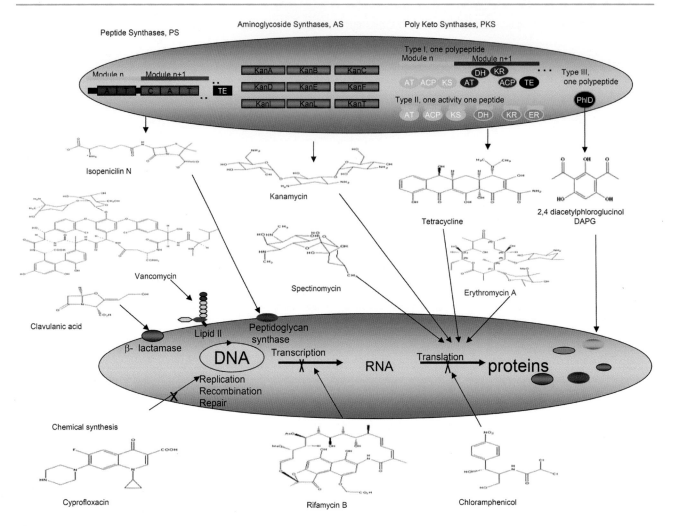

Fig. 9.18 Schematic drawing of a microbial cell and major antibiotics as well as their site of action. The microbial cell is shown with the main proteins involved in the synthesis of three antibiotics, left to right with the peptide synthases involved in the synthesis of N isopenicillin, the aminoglycoside synthases involved in the synthesis of kanamycin, and polyketide synthases involved in the synthesis of tetracycline. The molecular structure of these and several other antibiotics is shown, as well as their site of action in the cell. These sites are indicated in a clockwise direction starting from translation, which is the target of many types of antibiotics (spectinomycin, kanamycin, tetracycline, erythromycin), transcription (rifamycin), replication, recombination and DNA repair (ciprofloxacin), peptidoglycan synthesis (isopenicillin), elimination of penicillin (clavulanic acid), and lipopolysaccharide synthesis (vancomycin)

cyclized, is known in plants. It is involved in the synthesis of antibiotic molecules such as flavonoids, furanocoumarins, aucubin, or cinnamate. These molecules are not generally regarded as antibiotics and few examples are known in microorganisms. Aminoglycosides such as kanamycin, neomycin, and gentamicin are antibiotics containing aminocyclitol and amino hexoses. Biosynthetic genes of kanamycin and gentamicin have been described in *Streptomyces kanamyceticus* (Kharel et al. 2004).

The biosynthesis of peptidoglycan is the target of antibiotics β-lactam such as penicillin, amoxicillin, carbenicillin, and cephalosporin. The antibiotic molecule inactivates one of the two active sites of the enzyme known as penicillin-binding protein (PBP). The protein possesses transpeptidase and glycosyltransferase activities.

The antibiotics mentioned form a covalent bond with the —OH group of serine in the active site of transpeptidase, after which it cannot be active anymore. The polymers of peptidoglycan are no longer connected by peptide bonds between them, and the structure of peptidoglycan is weakened, which, coupled to the osmotic pressure, can cause lysis of the cell. The cell can also inactivate antibiotics if it is able to synthesize a β-lactamase enzyme that can hydrolyze the antibiotics into a nonactive form. Also some microorganisms such as streptomycetes produce inhibitors of β-lactamase as clavulanic acid and react synergistically with other β-lactams (Challis and Hopwood 2003).

Vancomycin is a glycopeptide that inhibits transpeptidation of peptidoglycan covering two D-alanine (D-Ala) into the ends of the peptide monomers of peptidoglycan,

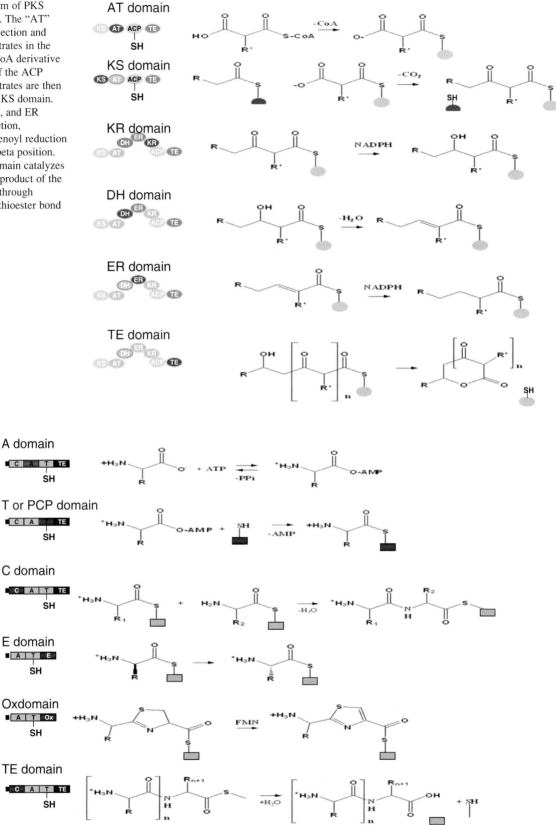

Fig. 9.19 Diagram of PKS enzymes domains. The "AT" domain allows selection and activation of substrates in the form of an acyl-CoA derivative and the transfer of the ACP domain. The substrates are then condensed on the KS domain. Domains KR, DH, and ER catalyze the reduction, dehydration, and enoyl reduction on carbon in the beta position. Finally the TE domain catalyzes the release of the product of the enzyme complex through hydrolysis of the thioester bond

Fig. 9.20 Diagram of NRPS enzymes domains. The "A" domain allows activation by ATP and therefore the adenylation of the starting amino acid and then a thiolation domain called "T" or "PCP" that attaches the amino acid to phosphopantetheine. Then a "C" domain allows the condensation of two activated amino acids. Then an "E" domain allows epimerization to change amino acids to "D" forms. Then an "Ox" domain allows the oxidation of certain amino acids. Finally, a "TE" domain releases the molecule formed by thioesterization

known as lipid II. One of the two D-alanine may be replaced by a D-lactate (D-Lac), or D-serine (D-Ser) structures and D-Ala-D-Lac-Ala-D or D-Ser, and thus become resistant to vancomycin (Healy et al. 2000). The lipid II, which is the precursor of peptidoglycan, also interacts with antibiotics, proteins that form pores in the membrane (Breukink and de Kruijff 2006). Multicellular organisms such as insects, amphibians, mammals, and plants produce oligopeptides that have antimicrobial activity (Zasloff 2002). The peptide antibiotics destroy or weaken the microbial membrane, leading to a loss of metabolites induced by pore formation, resulting in cell death.

Ribosomes are the target of macrolide antibiotics types, oxazolidinones (target 50S ribosomal subunit), tetracyclines, and aminoglycosides (target 30S subunit). The activity of protein biosynthesis is inactivated because the active site of transpeptidation and more precisely the molecule rRNA of the large subunit of ribosomes is inactivated in the presence of antibiotic molecules.

Replication enzymes and DNA repair are targets of fluoroquinolones, synthetic antibiotics that inhibit the activity of topoisomerases or gyrases (Drlica and Malik 2003). The double-stranded DNA once broken cannot be repaired by inactivated topoisomerase. The accumulation of breaks in the DNA causes cell death.

Several antibiotics of natural origin, produced by actinobacteria, interact with DNA. Antibiotics such as enediyne produce direct covalent bonds with the DNA molecule (Liu et al. 2002), which prevents DNA replication, the termination of transcription, and cell death.

It has been shown that antibiotics such as quinolones (norfloxacin), the β-lactams (ampicillin), and aminoglycosides (kanamycin) are bactericidal agents that cause the formation of free radicals (hydroxyl) via the Fenton reaction, leading eventually to cell death. In the same experiment, it was shown that bacteriostatic drugs do not stimulate the production of free radicals (Kohanski et al. 2007). The sulfonamides are derivatives of para-aminobenzoic acid, which block the biosynthesis of tetrahydrofolate, and while trimethoprim is an analog of a part of folic acid dihydropterin, thereby inhibiting the enzyme dihydrofolate reductase, these two molecules are therefore involved in the synthesis of dTMP nucleosides and UTP, causing a depletion of the cell.

9.5.2 Types of Antibiotic Resistance

Development in an organism of a specific resistance to an antibiotic can be achieved by accumulation of spontaneous mutations or be induced by the stress response (SOS), followed by positive selection. If the SOS system is inactivated, the evolution of antibiotic resistance is slowed (Cirz et al. 2005).

The presence of resistance genes in the genomes of microorganisms has been described repeatedly. Antibiotic resistance through methylation of the active site of the ribosome, the export of antibiotics from the cells by efflux pumps or modification of antibiotics is well described. These involve acetylation, phosphorylation, or adenylylation of antibiotic molecules by specific enzymes such as the kanamycin modifying enzymes.

Each microorganism producing a given antibiotic either has no target for that antibiotic or also has resistance genes to that same antibiotic, in order not to commit suicide. The set of determinants of the genome that may provide resistance against antibiotics is defined as the resistome (D'Costa et al. 2006) These resistance genes encode specific or nonspecific transporters, modifying enzymes or enzyme that cause degradation of the antibiotic or modification of the target molecule. These genes are located in the chromosome and often mobile elements such as plasmids, transposons, or insertion sequences (IS) which enable their mobilization to other microorganisms. It has been shown that bacteria resistant to natural and synthetic antibiotics are able to use antibiotics as sole carbon source (Dantas et al. 2008). The horizontal transfer of resistance genes and degradation of antibiotics to the pathogens of plants and animals represent a potential danger to the health of living organisms. The discovery of new antibiotics from natural or synthetic is therefore necessary to prevent the spread of pathogenic microorganisms.

9.6 Physiological Responses to Abiotic Stress*

9.6.1 Temperature Stress

Temperature is one of the most important factors that influence the growth and survival of microorganisms. When the temperature increases, the speed of chemical and enzymatic reactions increases until a maximum temperature beyond which cellular components (enzymes, nucleic acids, proteins, etc.) are sensitive and denatured (Fig. 9.21a). This leads in turn to inactivation of cellular functions, a growth arrest, or cell death. Before cell death, there may be repair and growth recovery if the temperature drops again.

At low temperatures, enzyme activities and bacterial growth can continue more slowly. When the temperature decreases, cell functions continue so long as the cytoplasmic membrane remains fluid. When the temperature is too low, there is a tendency to gel, thus solidifying which causes blockage of growth by stopping cell transport, respiration, and energy production. However, at low temperatures the prokaryotic cells in general are not destroyed, only resting as dormant cells. This allows them

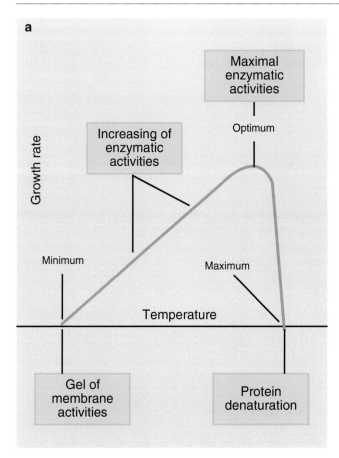

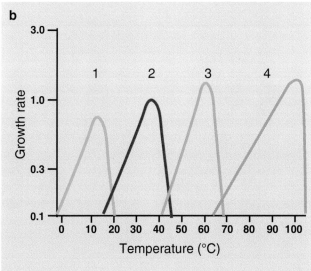

Fig. 9.21 (a) Effect of temperature on cellular activities. (b) Definition of microorganisms based on the temperature: *1* psychrophilic, *2* mesophilic, *3* thermophilic, *4* extreme thermophilic (Modified and redrawn after Madigan et al. 2010). Drawing: M.-J. Bodiou

to be maintained in international collections (*cf.* Sect. 6.2). When the temperature increases again, cellular functions are reactivated.

The majority of microorganisms living at temperatures average between 15 and 45 °C with optimum growth between 25 and 37 °C are considered mesophilic (Fig. 9.21b). Microorganisms growing at low temperature are called psychrophiles, and those living at high temperatures are called thermophiles even extreme thermophiles (Fig. 9.21b).

9.6.1.1 High Temperatures

After Brock (1997), "bacteria can develop at any temperature where water remains liquid." The discovery of hydrothermal vents at the bottom of oceans with temperatures of around 250 °C raised the question of the limits of life and enabled the discovery of many hyperthermophilic archaea; however, it was not possible to obtain bacteria able to grow at these extreme temperatures (*cf.* Sect. 10.3). Any organism able to develop at temperatures above 45–50 °C is now described as thermophilic. Maximum tolerated temperature varies between groups of microorganisms (Table 9.7) with the following rules:

(i) Prokaryotes accept higher temperatures than eukaryotes.
(ii) Non-photosynthetic thermophiles may develop a higher temperatures than photosynthetic thermophiles.
(iii) Above 70 °C, there are only prokaryotes able to grow.
(iv) Above 100 °C, there are only archaea able to grow.

Prokaryotes can be classified into three groups of thermophiles:

(i) The facultative thermophiles with a maximum temperature at 50–60 ° C; these are bacteria or archaea living at room temperature (mesophilic), but tolerating high temperatures not exceeding 60 °C. However, these prokaryotes are generally used in the food industry (*Bacillus*, *Lactobacillus*, etc.).
(ii) The strict thermophiles that do not grow below 40 °C and have an optimum at 60–65 °C.
(iii) The extreme thermophiles whose minimum temperature for growth is 55–65 °C and have optima between 80 and 110 to 115 °C. The maximum temperature is in general superior to 110 °C. They are for the most part archaea but some bacteria are also hyperthermophiles (*Thermotoga*, *cf.* Sect. 10.2). Most hyperthermophiles are also acidophilic and are developing at pH between 2 and 5.

The first two groups corresponding to moderate thermophiles are found in hot springs where they can form microbial mats, in food, warm soils, and fermentation products (compost, manure, etc.). These bacteria are present in various metabolic groups, aerobic or anaerobic, either respiratory, fermentative, or photosynthetic (*cf.* Chap. 3). They are able to maintain themselves at low temperature and develop in a hot environment when favorable.

Among eukaryotic microorganisms, enzymes are not thermally stable and are denatured beyond 65–70 °C. Moreover, internal membranes (nuclear, mitochondrial, and chloroplast) are very sensitive to heat and are destroyed when the temperature exceeds a few degrees the temperature for maximum growth.

Table 9.7 Temperature limits for the growth of thermophilic organisms present in different groups of living organisms

Groups of organisms		Maximum tolerable temperature (°C)
Animals (metazoans)	Fish	38
	Insects	45–50
	Crustaceans	49–50
Plants	Magnoliophytes	45
	Bryophytes	50
Eukaryotic microorganisms	Non-photosynthetic unicellular	55–60
	Photosynthetic microorganisms	60
	Fungi	60–65
Bacteria	Cyanobacteria	70–75
	Anoxygenic phototrophic bacteria	75
	chemoorganotrophic bacteria	75
	Chemolithotrophic bacteria	80
Archaea	Methanogenic archaea	105
	Hyperthermophilic archaea	115–120
	Halophilic archaea	70

In thermophilic prokaryotes, enzymes are thermostable and protected from denaturation. Thus, the metabolism remains stable at high temperatures (*cf.* Sect. 10.2). Stability seems to be maintained by supplementary hydrogen bonds and bridges between polar groups. Changes in the amino acid composition also affect the thermostability. Ribosomes are also more resistant to heat: in the moderately thermophilic bacterium *Bacillus stearothermophilus*, ribosomes resist to 77–82 °C, while in *E. coli*, they resist to 66 C only. Similarly, nucleic acids contain a greater percentage of G + C, which presents a stabilizing effect for the double strands. Finally, the membranes are rich in saturated fatty acids, or polycyclic as hopanoids, to maintain their stability at high temperatures. In contrast, phototrophic bacteria are much more sensitive because of the thermolability of their photosynthetic apparatus rich in unsaturated lipids. In hyperthermophilic archaea, many changes occur at several specific molecules, such as unique membrane lipids, transfer RNA, nucleic acids, and certain specific enzymes.

The answers to the increase in temperature result in an increase in K^+ present in the form of 2,3-diphosphoglycerate potassium, by the change in the percentage of polyamines with more penta-amines and in some cases by the supercoiling of DNA.

Finally, it has been shown that thermophiles in general had developed a specific usage of the code for certain amino acids such as arginine and leucine (Singer and Hickey 2003); a modified proportion of amino acids glutamic acid, lysine, and arginine (Tekaia et al. 2002); generally an increase of hydrophobic amino acids in theoretical proteomes (Lieph et al. 2006); and finally a change in the fluidity of the membrane by the synthesis of hopanoid lipids (Hermans et al. 1991).

9.6.1.2 Low Temperatures

Bacteria that inhabit cold environments will be different depending to their adaptations to low temperatures either permanently or periodically. We therefore distinguish psychrophilic microorganisms (living in permanently cold environments) and psychrotrophic microorganisms. The psychrotrophic microorganisms are generally mesophilic bacteria that are able to continue to grow slowly even at low temperatures, less than 5–7 °C. For example, mesophilic *Bacillus* (*B. subtilis* and *B. megaterium*), natural inhabitants of soils, can develop at temperatures below 5 °C. The pathogen *Yersinia pestis* is able to live from −2 °C to +40 °C. These psychrotrophic bacteria, also called facultative psychrophiles, are more widespread than the strict psychrophiles. In fact, the latter are inhibited when temperature rises by a few degrees, thus leaving space for psychrotrophs who have a great adaptability to the variations of temperature, and thus can colonize a majority of cold environments either permanently or periodically. These bacteria can be harmful in certain preserved foods chilled or frozen badly as *Listeria monocytogenes* which can grow at 4 °C in meat and cheese, called for this reason the bacteria of the refrigerator. There are also eukaryotic microorganisms, such as yeasts and filamentous fungi, that live in psychrotrophic environments (food, snow, etc.) and contaminate it.

Strict psychrophilic microorganisms are essentially bacteria, some archaea (*cf.* Sect. 10.2), and micro-eukaryotic (micro-algae in the snow) restricted to permanently cold environments, due to their high thermolability when the temperature rises a few degrees. Indeed, these microorganisms are inhibited and even killed when they are "warmed up" for a brief moment at room temperature.

Therefore, it was very difficult to isolate them from permanently cold environments. Isolation itself is not a difficult process, but it requires some precautions: the cold chain must not be interrupted, which means that all sampling and laboratory operations (spreading, culturing, isolation, etc.) must be realized at temperatures below 10 °C. Similarly, all the laboratory equipment (pipettes, culture media, flasks, etc.) must be kept permanently cold, before and during use. Psychrophilic bacteria have temperature optima between 8 and 15 °C and are generally inhibited and killed over 15–20 °C. Their marked sensitivity at room temperature (quick death) was such that many scientists have long sought to isolate them without success, and thus, these prokaryotes have been little studied. During handling, psychrotrophs were selected during the warming period and were considered as more abundant. But in permanently cold environments, strict psychrophiles are likely abundant. Today, they are increasingly isolated thanks to the more stringent technical control of the temperature. Microeukaryotes that continuously live in cold environments are qualified of cryophilic microorganisms.

The minimum temperature of psychrophilic bacteria development is difficult to identify because it is often in the freezing zone. Formation of colonies has often been observed at −11 °C on agar surfaces. In the cell, biosynthesis stops at −30 °C and the limit of biochemical reactions (tested in liquid with glycerol as antifreeze added) was found to be at −140 °C. Glycerol penetrates into cells and protects them by preventing the freezing of water molecules.

Growth maximum occurs around 6–10 °C, sometimes 15 °C (Fig. 9.21b). One to 2 °C above it, the syntheses of macromolecules (DNA, RNA, proteins) stop, and 5 °C above, destruction of cellular material occurs with a leak of macromolecules through membranes such as amino acids and enzymes, resulting in cell death.

In psychrophilic bacteria, different abnormally thermolabile enzymes have been identified. These enzymes which exhibit psychrophiles traits have optimum of activity at 15–20 °C and are inhibited and denatured beyond 25 °C. However, if there are some unusually labile enzymes in a psychrophilic bacterium, other enzymes are normally resistant to heat. In addition, some enzymes are thermolabile in a psychrophile and not thermolabile in another. Common enzymes presenting psychrophilic characters are malate dehydrogenase, lactate dehydrogenase, succinate dehydrogenase, hexokinase, aldolase, and phosphoglucose isomerase.

These enzymes are adapted to low temperatures due to their structure being different from their mesophilic homologs: they contain more polar amino acids and fewer and fewer weak interactions between their different domains, thus promoting flexibility at low temperature. Their secondary structure comprises a higher proportion of helices and less sheets, thus contributing to a reduced rigidity at low temperature.

Psychrophiles also possess membranes containing a high proportion of unsaturated fatty acids with several double bonds (up to 4 or 5) for a better functionality and fluidity of the cytoplasmic membrane at low temperature, whereas in the mesophilic bacteria, the membrane which contains more saturated fatty acids becomes waxy, rigid, and nonfunctional at low temperature. The high proportion of unsaturated fatty acids in psychrophilic prokaryotes causes a fragility of the membrane at ambient temperature and thus favors leakage of macromolecules through the membrane and cell death. An inability to acylate tRNA at room temperature was also observed in psychrophilic prokaryotes, whereas acylation occurs at low temperature. All these peculiarities allow adaptation to low temperatures, but result in high brittleness at room temperature.

9.6.2 Oxidative Stress

Molecular oxygen, O_2, possesses two electrons in its outer orbitals. It presents a high redox potential and is a powerful oxidizing agent. The two electrons are separated on two outer orbitals, and thus oxygen is not directly reactive and needs to be activated. The inactive state is called triplet oxygen. The high-energy form is called singlet oxygen. This is the most reactive and more toxic with a super-oxidant power; it is also the most widespread in the living world, produced in the cells of aerobic organisms. This form is also chemically produced and present in the air, especially in mists. In cells, singlet oxygen is produced by different enzymes (myeloperoxidases, photosynthetic systems) and during photooxidation reactions involving photosensitive molecules (pigment molecules); its superactive state can be unwanted and can be destroyed by oxidation vital compounds in cells. Organisms that are in contact with this compound possess molecules, such as carotenoids, which can deactivate this compound by transforming it in triplet oxygen.

During aerobic respiration, triplet oxygen is reduced to H_2O:

$$O_2 + 4e^- + 4H^+ \rightarrow 2H_2O$$

This reduction which requires four electrons occurs in four stages and produces four toxic forms of oxygen:
1. $O_2 + e^- \rightarrow O_2^-$ (superoxide ion)$\rightarrow HO_2^\bullet$ (peroxide radical)
2. $O_2^- + e^- + 2H^+ \rightarrow H_2O_2$ (hydrogen peroxide)
3. $H_2O_2 + e^- + H^+ \rightarrow H_2O + OH^\bullet$ (hydroxyl radical)
4. $OH^\bullet + H^+ + e^- \rightarrow H_2O$

The superoxide ion and radical peroxide are formed in small quantities during respiration and are also produced by photooxidation in the atmosphere. In cells, they are produced by the action of flavoproteins, Fe/S proteins, and quinones during the electron transfer in the respiratory chain. These compounds are the most stable of all the toxic forms of oxygen and thus their time life is longer. They can move from one cell to another, resulting in the oxidative destruction of lipids. They are responsible for the sensitivity to oxygen of strict anaerobic bacteria and archaea.

Hydrogen peroxide is stable too. It is formed by respiration at flavoproteins among aerobic organisms and microorganisms.

The hydroxyl radical is the most reactive and least stable of all forms of oxygen. It is the most oxidizing agent that can oxidize and destroy all cell compounds (macromolecules). It is formed during aerobic respiration, but also by ionizing radiation (destroying effect of radiation by destructive superoxydation).

Eukaryotes have also developed a response when they get in contact with the bacteria that involve reactive species of oxygen. They enzymes (NADPH oxidase, polyamine oxidase, peroxidases) can produce either peroxide or superoxide ions in what is called the oxidative burst (Bolwell 1996). This is why pathogenic or symbiotic microorganisms must be able to manage such a rapid increase in oxidative ion concentration.

9.6.2.1 Proteins Induced by Oxidative Stress

At least 30 proteins are induced during a hydrogen peroxide stress in response to increasing H_2O_2. The stimulus for induction is still unknown. In *E. coli*, 8 proteins are necessary and are regulated by the regulon *oxyR*. The most important of these are catalases and peroxidases which can destroy the molecule of H_2O_2 according to the following equations:

Catalase : $H_2O_2 + H_2O_2 \rightarrow 2\,H_2O + O_2$
NADH-dependent peroxidase : $H_2O_2 + NADH + H^+ \rightarrow 2\,H_2O + NAD^+$

The *oxyR* gene encodes a protein which acts at *oxyR* transcription interacting with RNA polymerase for transcription of defense enzymes. When the level of superoxide anion (O_2^-) is high, bacteria respond by using a different stimulon. More than 30 proteins are also produced through the action of gene regulators *soxR* and *soxS*. At least six proteins are known and among them the superoxide dismutase which converts superoxide to ion peroxide:

$$O_2^- + O_2^- + 2\,H^+ \rightarrow H_2O_2 + O_2$$

The production of hydrogen peroxide leads then to the induction of enzymes of the peroxide regulon. The mechanism that allows the activation of the transcription of genes involved in the synthesis of two proteins SoxR and SoxS is little known. These proteins possess four cysteines in the carboxyl terminal region, that probably bind to metal. The state of the redox metal could act as a signal for transcription.

The presence of hydrogen peroxide in the cytoplasm is already a significant cellular stress, but the presence of iron (II) will exacerbate the situation by generating hydroxyl ions in the so-called Fenton reaction which includes two distinct reactions:

$$Fe^{2+} + H_2O_2 \rightarrow Fe^{3+} + OH^\bullet + OH^-$$
$$Fe^{3+} + H_2O_2 \rightarrow Fe^{2+} + O \cdot OH + H^+$$

Iron is therefore a catalyst, and the produced highly reactive ions will oxidize a wide range of compounds including lipids, DNA, and proteins. In several bacteria, especially *Actinobacteria*, the regulator Fur is involved in the response to oxidative stress, by the detection of ferric ions, potential generators of Fenton reaction.

9.6.2.2 Responses to Oxidative Stress

During oxidative stress, the cellular responses and defenses are of two kinds: preventive defenses and remedial defenses. The preventive defenses can destroy the toxic oxygen species and result in the direct production of detoxification enzymes: catalase, peroxidase, and superoxide dismutase according to the toxic form involved. The remedial defenses help to repair the damages caused by reactive forms of oxygen such as by inducing the production of glutathione reductase that lowers glutathione oxide and reduce intracellular redox potential. There is also a significant production of glucose 6-phosphate dehydrogenase which allows the synthesis of NADPH needed for glutathione reductase and other reductases used to repair different enzymes damaged by oxidation. If nucleic acids are also damaged, there may be a stimulation of the production of endonuclease IV to repair the DNA oxide, for example. Further damage can occur on oxidized membranes which require also the presence of reductases for repair.

In eukaryotes adapted to living conditions in the presence of oxygen, different enzymatic equipments for preventive and remedial defenses are present.

In prokaryotes (bacteria and archaea), the responses to oxidative stress will vary according to the enzyme setup present. Thus, four different behaviors in the presence of oxygen may exist (Fig. 9.22):

(i) The bacterium is strictly **aerobic***: it realizes aerobic respiration (*cf*. Sect. 3.3.2) and therefore possesses the systems of detoxification of toxic forms of oxygen. This bacterium cannot live without oxygen.
(ii) The bacterium is facultatively anaerobic: it lives by aerobic respiration and therefore possesses the systems necessary to cope with oxidative stress, but it can also live in the absence of oxygen by fermentation or anaerobic respiration. Enzymes involved in fermentation or anaerobic respirations are derepressed and/or induced and active in the absence of oxygen. The bacterium can also be air tolerant, that is to say, it can live in

9 Adaptations of Prokaryotes to Their Biotopes and to Physicochemical Conditions...

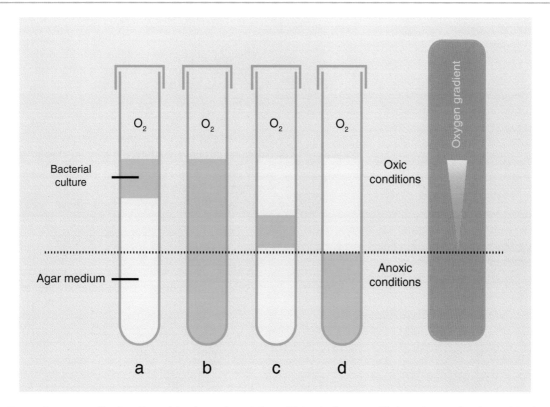

Fig. 9.22 Types of oxygen utilization. Bacterial cultures (*green*) in tubes containing nutrient agar whose surface is in contact with oxygen. Depending on the case, the bacterial culture was developed in a descending gradient of oxygen from the surface to the bottom of the tube where there is almost no oxygen. (**a**) strictly aerobic bacterium; (**b**) facultative anaerobic bacterium or air tolerant; (**c**) microaerophilic bacterium growing in an intermediate zone where partial oxygen concentration is low; (**d**) strictly anaerobic bacteria developing at the bottom of the tube in the absence of oxygen. Drawing: M.-J. Bodiou

the presence of oxygen by realizing fermentation and thus not using oxygen for respiration. This is the case of lactic acid bacteria (*Lactobacillus*) that do not have a respiratory transport system and can only ferment (lactic acid fermentation), but possess the detoxification systems for the reactive forms of oxygen and therefore can live in the presence of oxygen without using it.

(iii) The bacterium is **microaerophilic***, that is to say, it can live only in an environment containing low concentrations of oxygen. It realizes aerobic respiration, but its detoxification system is not efficient enough to resist the normal atmospheric pressure of oxygen.

(iv) The bacterium is strictly **anaerobic***: it realizes the anaerobic respirations or fermentation. It cannot live in the presence of oxygen because it does not possess the necessary enzymes and regulation systems to respond to oxidative stress. However, some strictly anaerobic bacteria can tolerate traces of oxygen, others do not.

9.6.3 Chemical Stress

9.6.3.1 Responses to Salinity

Bacterial growth is influenced by the presence or absence of salt (NaCl), which exerts an osmotic pressure, making homeostasis difficult. Some bacteria and archaea are not able to develop in the presence of NaCl or at very low concentrations of NaCl. They are sensitive to and inhibited by salt and are called non-halotolerant or halophobic prokaryotes. It is generally the case of prokaryotes living in freshwater or in some soils. Most bacteria and archaea are capable of living in the presence of NaCl concentrations higher or lower. There are salt-tolerant microorganisms that can develop in the absence of salt and can tolerate the presence of NaCl. Depending on the tolerated concentration, these bacteria and archaea are called slightly, moderately, or extremely salt tolerant. Among them, some bacteria and archaea cannot live in the absence of NaCl. They are called halophiles. They therefore require the presence of NaCl in their environment and are called slightly, moderately, or extremely halophilic depending on their NaCl needs (Fig. 9.23) (*cf.* Sect. 10.4).

Influence of Inorganic Ions on Growth

We must distinguish the needs and requirements in salts and in NaCl. A bacterium is called halophilic when it needs NaCl for growth. According to the NaCl concentration tolerated and needed, the bacterium is known as slightly, moderately, or extremely halophilic. However, at maximum concentration of NaCl, the bacterium cannot develop in the absence of other salts. Growth rate may be affected by monovalent (K^+) or divalent (Mg^{++}, Ca^{++}) ions. These ions are required

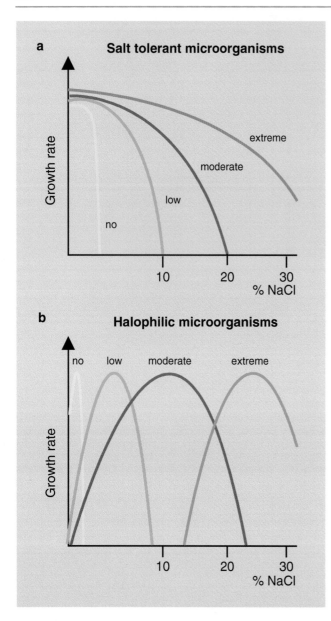

Fig. 9.23 Bacterial growth according to the concentration of NaCl. (**a**) Salt-tolerant bacteria and archaea depending on the concentration of tolerated NaCl: Low salt tolerant, 0–8 % NaCl; moderately salt tolerant, 0–20 % NaCl; and extremely salt tolerant, 0–30 % NaCl. (**b**) Halophilic bacteria and archaea for which are presented the lowest, highest, and optimum concentrations of NaCl in which they can develop: Lowly halophilic: 1–3 – 8 % NaCl; moderately halophilic: 2–3 – 10–12 – 20–25 % NaCl; and extremely halophilic: 12–15 – 25–30 – 30 % NaCl. Drawing: M.-J. Bodiou

and must be supplied to the prokaryotes, at concentrations suitable for their development. The minimum concentration of each salt required for maximum growth rate must in general be defined (Fig. 9.24a).

Metabolism-Dependent Sodium
It has been shown that in halophilic bacteria, protons (H^+) are exchanged with sodium ions (Na^+) using a Na^+/H^+ anti-port system. This system is necessary to reduce the rate of intracellular sodium and allows at the same time the production of a gradient that generates a sodium membrane potential, a "sodium motive force" similar to the proton motive force (*cf.* Sect. 3.3.1), and results in the return of sodium ions into the cell where they are continuously exchanged with protons. This Na^+ motive force allows activation of symports causing penetration of substrates accompanied by sodium and activation of flagella by the flow of sodium. It has even been shown that sodium ions can also be excreted through the activity of the respiratory chain in the same way as protons, thus increasing sodium motive force (Fig. 9.24b).

Resistance to Salt and Osmoregulation
- Accumulation of inorganic compounds in organic divalent ions (Mg^{++}, Ca^{++}) act at lower concentrations than monovalent ions (K^+) and are therefore more effective. They have a role in maintaining the integrity of the cell envelope forming bridges between the hydrophilic poles of phospholipids or between the carbonyl groups of membrane proteins.

  In extremely halophilic archaea (*Halobacterium*, etc.), there are Na^+/K^+ anti-ports that are very active and can replace the sodium with potassium, which in turn accumulates in the cytoplasm. This accumulation of potassium compensates the osmotic pressure due to the extreme salinity. The cell wall is stabilized by the Na^+ ions that bind to negatively charged amino acids (aspartate, glutamate) glycoproteins. Thus, the cell is surrounded by sodium ions binding to the outside of the wall. When the sodium concentration decreases, the wall breaks because of the negative charge of the proteins, and the cell is lysed. The cytoplasmic proteins are also very acidic and Na^+ and K^+ are required to maintain the internal pH. Similarly, K^+ ions stabilize the ribosomes.

- Accumulation of organic compounds (*cf.* Sect. 10.4.2)
  In halophilic bacteria, osmotic pressure compensation is effected mainly through the accumulation of organic compounds in the cytoplasm. These osmoregulatory compounds are accumulated and do not interfere with the cellular metabolism; they are called "compatible solutes," that is to say that their accumulation is neutral and compatible with cell life. They are generally sugars or disaccharides especially betaines which are organic compounds derived from amino acids (Fig. 9.25). The most frequently observed compound is glycine betaine. This compound is present in many halophilic or salt-tolerant bacteria and in many organisms or animals where it plays an osmoregulatory role. Glycine betaine has been detected and analyzed in aerobic heterotrophic bacteria submitted to salt stress (*Pseudomonas*, *Escherichia coli*) (Galinski and Trüper 1982; Le Rudulier and Bouillard 1983; Imhoff 1988; Mouné et al. 1999).

Fig. 9.24 Effect of cation concentrations on the bacterial cell. (**a**) Influence of cations on bacterial growth. (**b**) Process of membrane transport in halophilic bacteria showing evidence of the use of sodium motive force. *1* symport substrate-sodium, *2* Na$^+$/H$^+$, *3* excretion of protons and sodium ions by the respiratory chain, *4* activation of flagella basal body by the flow of Na$^+$, *5* anti-port Na$^+$/K$^+$. Drawing: M.-J. Bodiou

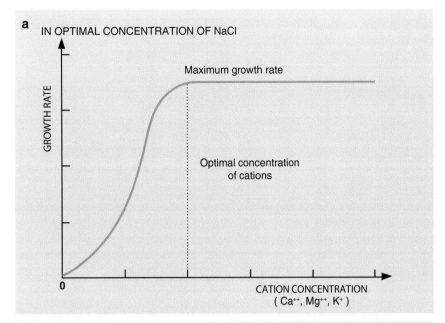

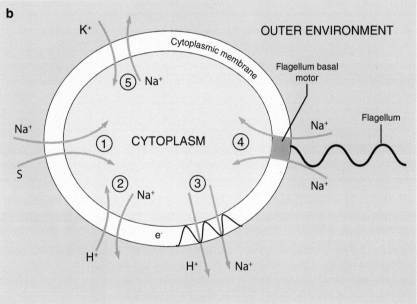

It has also been evidenced in the halophilic sulfate-reducing bacteria (Welsh et al. 1996) and in halophilic fermentative bacteria. Most of these bacteria accumulate glycine betaine by active absorption, but some are capable of synthesizing such as phototrophic bacteria and sulfate-reducing bacteria. Aerobic heterotrophic bacteria and fermentative bacteria can realize a partial synthesis from a precursor of the biosynthetic pathway (proline) they must obtain outside in their environment. In general, it is rare that the glycine betaine is the only compatible solute present in a bacterium subjected to salt stress. It is often accompanied by sugars (sucrose, trehalose), polyalcohols (sorbitol, glycerol), or other amino acids and their derivatives (glutamate, ectoine). Glycine betaine is largely present in hypersaline environments. It can also serve as a substrate for some bacteria. Thus, a halophilic fermentative bacterium has been recently isolated that was capable of using glycine betaine as a substrate for energy, reducing it to trimethylamine via a Stickland reaction (Mouné et al. 1999).

Halophilic Bacteria and Archaea and Their Habitats (*cf.* Sect. 10.4.1)

Salty aquatic environments represent the largest part of the aquatic environment with oceans. Oceans have a salinity level between 35 and 37 ppm for the more salty biotopes (Mediterranean Sea). These are environments where the pH is neutral to slightly alkaline (7.5–8). Halophilic prokaryotes

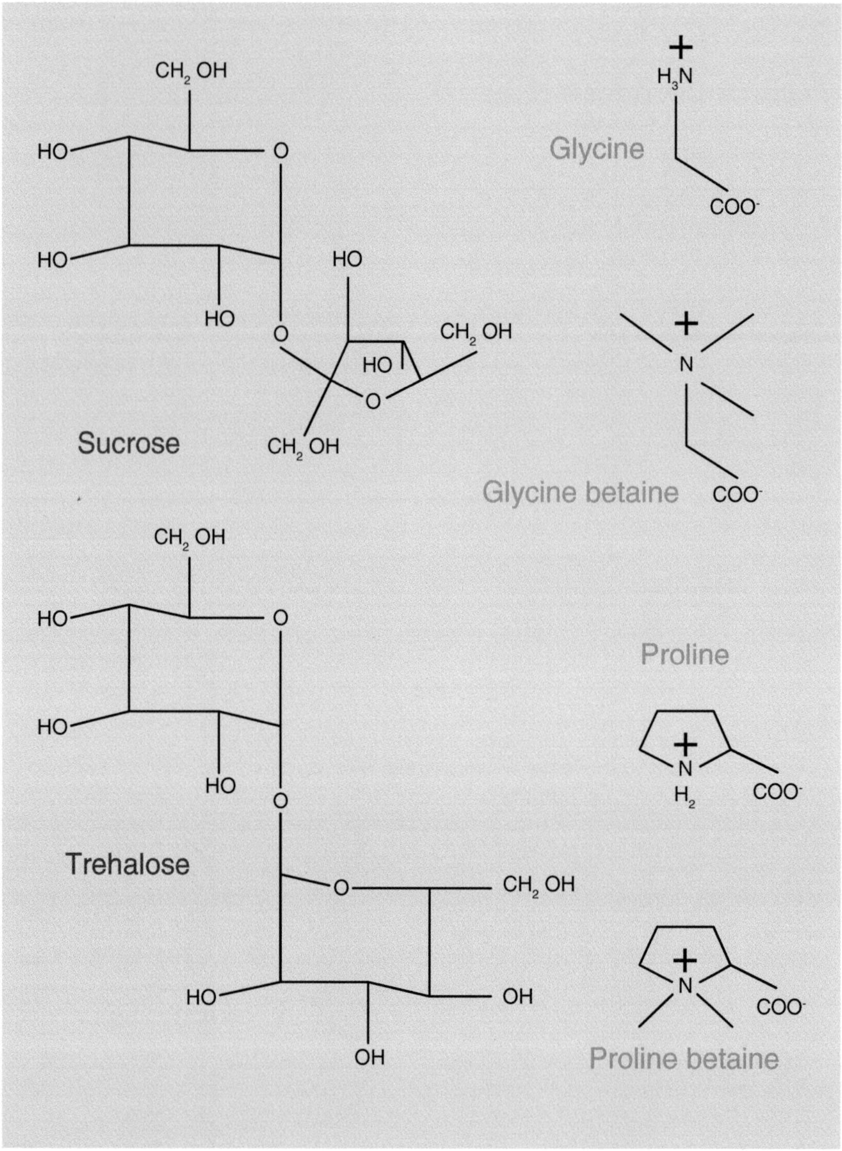

Fig. 9.25 Main osmoregulatory organic compounds. These compounds are compatible solutes that accumulate in the cytoplasm of halophilic bacteria to compensate for the osmotic pressure. Drawing: M.-J. Bodiou

that are living in these environments are part of the domain bacteria and are qualified of slightly halophilic also called marine bacteria. There are also many salt-tolerant bacteria derived from continental areas and discharged into the marine environment where they may develop because of their relative salt tolerance. These latter cannot be regarded as true marine bacteria in the strict sense of the term. Marine bacteria are found in almost all bacterial groups with the exception of certain groups typically associated with soils. Thus, marine bacteria are present in most families where they constitute differentiated genera or differentiated species. From a metabolic point of view, there exists chemoorganotrophic chemolithotrophic and phototrophic marine bacteria, realizing aerobic or anaerobic respirations, fermentations, and oxygenic or anoxygenic photosynthesis. The presence of archaea was also demonstrated in the oceanic waters (Karner et al. 2001; Giovannoni and Sting 2005).

- Hypersaline environments
  Seawater, which is most of the water on Earth, contains dissolved mineral salts in concentrations that are remarkably constant; the main ones are chlorine (18.98 %), sodium (10.556 %), sulfate (2.649 %), magnesium (1.272 %), calcium (0.4 %), potassium (0.38 %), and carbonate (0.14 %). Hypersaline environments are those which exhibit concentrations of mineral salts superior to those of seawater. This definition is not specific and does not take into account the types of salts and their proportions. However, the majority of hypersaline environments originate from salts of sea water, either directly by evaporation and concentration or indirectly by the dissolution of fossil deposits (evaporites).

Therefore, their mineral salts will be in similar proportions and percentages to those contained in the sea water, at least until the precipitation thresholds are not reached. Hypersaline waters obtained by partial evaporation of sea water are called "thalassohaline," while those obtained by the dissolution of salty fossil deposits are called "athalassohaline." These latter may have different proportions of salts well different from those of seawater according to the nature of the deposits. Some contain a high percentage of sodium carbonate and are very alkaline (alkaline lakes).

Hypersaline environments have been formed over a long period in the history of the Earth. Some authors have shown that the salinity of seawater has been constant for at least 200 million years, due to a subtle balance between the dissolution of igneous rocks, the contribution of elements in the Earth's crust by oceanic ridges, and chemical reactions in water.

When seawater is concentrated by evaporation, all salts present an increase of their concentrations in the same proportions until reaching their respective precipitation thresholds. Carbonates precipitate as calcium carbonate when the salinity reaches 6 %. Then, sulfate precipitate and form deposits of gypsum (calcium sulfate) when the salinity exceeds 10 %. Above 25 % sodium chloride begins to precipitate and form halite precipitates up to 35 % where it is fully precipitated (10 times the concentration of sea water). The waters are then enriched in magnesium and potassium salts which precipitate at salinities 20 times superior to those of the sea water.

Completely evaporated sea water is the origin of deposits of salt formations which constitute through time evaporites. These last are considered as hypersaline fossil environments and are present on all continents. Their dissolution in water can be at the origin of athalassohaline environments. All these hypersaline environments formed during the evolution of the Earth represent reservoirs for an evolution over a long-term of halophilic life.

- The halophilic bacteria and archaea

Regarding extreme halophilic prokaryotes belonging to the domain *Bacteria*, most were isolated from anoxic hypersaline environments. They are included in two families: *Haloanaerobiaceae* (fermentative bacteria) and *Ectothiorhodospiraceae* with the genus *Halorhodospira* (phototrophic purple bacteria). These bacteria are isolated from environments with salinities of 20–25 %. Some are halophilic and alkaliphilic, and are isolated from alkaline lakes with pH superior to pH 9. Extreme halophilic archaea (also called halobacteria or haloarchaea) are found in hypersaline habitats such as salt marshes, salt lakes and the Dead Sea, soda lakes, and salty foods (fish, meat).

These microorganisms require NaCl concentrations of at least 1.5 M (9 %) to develop. They live well in environments with NaCl concentrations between 15 and 30 %. Some are able to multiply until the saturation limit of NaCl concentration (35 %). Extreme halophilic archaea isolated from hypersaline thalassohaline environments are neutrophiles (*Halobacterium*); those isolated from athalassohaline environments can be, depending on the environment, neutrophilic or alkalophilic (*Natronobacter*, *Natronococcus*, etc.).

In halophilic bacteria, some "marine" genes that encode the active transport and biosynthesis of betaines and the active transport of potassium have been identified. Osmoregulation processes are controlled by the osmotic pressure that acts as a signal for the synthesis of transport systems and porins via the system *omp*. The accumulation of potassium is regulated by the kdp operon; accumulation of proline as precursor of glycine betaines in halophilic chemoorganotrophic bacteria is controlled by *proU*.

9.6.3.2 Responses to pH

Prokaryotes are present in environments with pH extremes, either very acidic (pH 1) or very alkaline (pH 11), but the majority of them live a pH around neutrality even moderately acidic or alkaline, within a range from pH 5 to pH 8. Among bacteria and archaea, it is possible to distinguish **acidophiles***, **neutrophiles***, and **alkaliphiles*** by their ability to grow at pH acidic, neutral, or alkaline. For each species, we can define a minimum growth pH, an optimum pH, and a maximum pH beyond which growth is no longer possible. Acidophilic prokaryotes can be divided into moderately or strictly acidophilic as the optimum pH is below pH 5.5, and extremely acidophiles with pH lower than pH 3. In general, pH for neutrophilic prokaryotes ranges from pH 5.5 to pH 8.5, with an optimum between pH 6.8 and 7.2. Alkaliphilic prokaryotes can be either tolerant alkaliphilic (pH optimum at pH 7 and trends has higher values), moderately alkaliphilic when their pH range for growth is above pH 7 and below 10, and extreme alkalophilic when growth pH is between pH 9 and pH 12 or more (Fig. 9.26).

When an organism develops, it modifies rapidly the pH of its environment through its metabolism by absorbing or excreting substrates or products more or less acidic or alkaline: excretions of protons (H^+), excretion of acidic or alkaline products, absorption of CO_2, and uses of acidic or alkaline substrates. It will therefore adapt to variations in pH, but these must remain within the range of tolerable pH so that it can continue to develop.

On our planet, there are many acidic environments such as some hot springs, volcanic soils, some lakes, hydrothermal vents at the bottom of oceans, drainage of mining

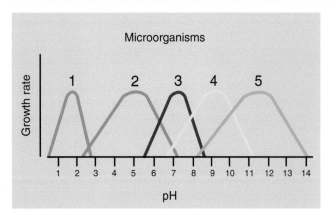

Fig. 9.26 Rate of growth of microorganisms according to pH. Scale ranging from extremely acidophilic (optimum pH inferior to 3) to extremely alkaliphilic (optimum pH superior to 9). *1* Extremely acidophilic, *2* moderately acidophilic, *3* neutrophilic, *4* moderately alkaliphilic, *5* extremely alkaliphilic Drawing: M.-J. Bodiou

operations, etc. Some are alkaline environments such as soda lakes (salt lakes containing sodium carbonate at saturation level), some soils, etc.

Ecology and Diversity of Prokaryotes Living in Extreme pH

For now, there are few described bacteria that can live at pH extremes. The majority of prokaryotes living where pH is extreme are alkaliphilic or hyperacidophilic archaea. Acidophilic archaea are generally hyperthermophilic and are described as thermo-acidophilic and are isolated for the most part from hydrothermal vents. They can tolerate pH ranging from pH 1 to pH 3–4. Alkaliphilic archaea are also extreme halophiles, also called natrono-archaea isolated from soda lakes and developing in habitats at pH 11–12.

Bacteria (domain *Bacteria*) living at pH extremes are found in living organisms and in biological media or coexist with archaea in extreme environments such as acidic drainage or soda lakes. For example, bacteria living in the digestive tract must pass from a very acidic pH (pH 1.5 in the stomach) to a very alkaline pH (pH 10 at the beginning of the intestine). Some optional anaerobic bacteria (lactobacilli, *Sarcina ventriculi*) are tolerant to pH ranges from pH 1 to pH 9.5. In the gastrointestinal tract, anaerobic bacteria produce acids, as fermentation products, and thereby lower the pH leading to a variability of bacterial community: at pH 7 *Clostridium* is dominating the community, whereas at pH 5 it is *Bifidobacterium*. In carbohydrate fermenters (silage, compost), the production of lactic acid by fermentation lowers the pH to pH 3.5–5, thus promoting the development of moderately or strictly acidophilic bacteria (*Lactobacillus*, *Pediococcus*, *Leuconostoc*). In contrast, in the fermenters of proteins, the release of NH3 leads to an increase in pH (pH 8–9), promoting the development of enterobacteria and *Clostridium*. In mining operation drainages, *Acidithiobacillus* proliferates, and their sulfur and iron oxidizing metabolism, producer of sulfuric acid, maintains a very acid pH in their environment around pH 1.5–2.5 which allows to solubilize iron ore (pyrite), freeing the iron which will then be oxidized to generate energy.

Acidophilic Bacteria and Archaea (*cf.* Sect. 10.6)

Obligatory and extreme acidophilic prokaryotes cannot develop at neutral pH. They are essentially thiobacilli (*Thiobacillus*, *Acidithiobacillus*) and hyperthermophilic archaea (*Acidianus*, *Thermoplasma*, *Sulfolobus*, etc.). They use reduced sulfur or iron compounds as electron donors in their chemolithotrophic metabolism, or organic compounds. The mechanisms of adaptation to extreme pH in acidophilic prokaryotes do not consume energy. For example, *Acidithiobacillus*, who lives in an environment at pH 2, maintains its intracellular pH at pH 6.5 based on the simple and strict mechanisms that control the membrane flow of carbon and proton. The membranes are modified and membrane transport proteins are synthesized. Cell division is slower and produces greater amounts of cytoplasmic membrane with a higher surface/cell volume ratio. The membrane surface allows a greater flux of protons and increased compensatory activation of potassium via the Kdp pumping system, allowing an exchange of H^+ and K^+.

Alkaliphilic Bacteria and Archaea (*cf.* Sect. 10.7)

There are many alkali tolerant bacteria, but few are strict to extreme alkaliphiles. The extreme alkaliphiles are mainly halophilic archaea.

Among bacteria, some NH_3-using nitrifying bacteria are moderately alkaliphilic, as are some aerobic (some *Bacillus*) or anaerobic (some *Clostridium*) chemoorganotrophs. In the marine environment, there are many alkali tolerant and some strictly alkaliphiles (*Vibrio*). Some alkaliphilic bacteria are also phototrophic, such as cyanobacteria (*Spirulina*, *Synechococcus*) or purple bacteria (*Halorhodospira*) isolated from soda lakes. There are also some alkaliphiles living in alkaline soils (some *Actinomyces*, and *Corynebacterium*).

Alkaliphilic archaea have been isolated from soda lakes containing natron (*Natronococcus*, *Natronobacter*, etc.). They are also extremely halophilic and maintain their energy metabolism using a sodium flow like all halophilic bacteria. Some methanogenic archaea are also halophilic and alkaliphilic (*Methanosalsum*, *Methanobacterium*).

Strict alkaliphiles are dependent upon sodium as halophiles. In alkaliphiles as well as in halophiles, a Na^+/H^+ anti-port has been observed as well as symports, allowing intake of substrates in exchange for sodium. Sodium motive force created by the delocalization of sodium during the transfer of electrons in the respiratory chain

Fig. 9.27 Mechanism of neutralization of gastric acidity by urea synthesis in *Helicobacter pylori*. The bacterium *Helicobacter pylori* (*upper drawing*) has a urease (*yellow*), which converts urea present in the stomach to carbamate which is transformed spontaneously into carbonic acid, and ammonium (*red*). This ammonium reacts then with hydrochloric acid (*blue*) generated by the wall of the stomach and thus neutralized, thereby increasing the pH locally. This creates regions of the stomach where the pH is higher, thus favoring the bacterium to multiply and causing finally stomach ulcers (*Bottom drawing*) which can lead, in some cases, to stomach cancer

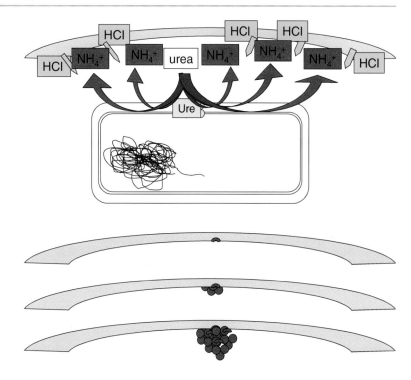

allows activation of the basal body of the flagella through the flow of sodium and the generation of ATP by sodium motive force. The internal pH of alkaliphilic prokaryotes is generally maintained at two pH units below the pH optimum in their environment. The Na^+/H^+ anti-port allows an internal accumulation of H^+ and a decreased pH.

Alkaline enzymes have been identified in alkaliphilic bacteria. Alkaline proteases and amylases have been isolated and characterized from alkalophilic *Bacillus*. These enzymes have optimum activity at pH 10 to 10.5 and still present 50 % of their maximum of activity at pH 9 or 11 (Hamamoto et al. 1994).

Maintenance of Homeostasis*

The bacteria and archaea regulate their internal pH through the accumulation of weak acids (salicylic acid) or weak bases (methylamines). Amino acids can also play the role of cytoplasmic buffer (anions or cations). There may also be an induced production of acids or bases. For example, at alkaline pH, malic acid is synthesized from glucose, while at acidic pH, it is converted into pyruvate by malate decarboxylase. Finally, active transports of H^+ or OH^- are regulated by the internal pH: H^+ is replaced by K^+ or Na^+, and OH^- is replaced by Cl^-. In alkaliphilic bacteria, regulation of genes for alkaliphilicity was demonstrated based on the presence of two DNA fragments, pALK11 and pALK 2, that are present in alkaliphilic *Bacillus* and absent in non alkaliphilic *Bacillus* and can confer alkaliphilicity when they are transferred via the help of a plasmid.

An original mechanism of pH homeostasis is that elaborated by the bacterium *Helicobacter pylori* which lives in the stomach of mammals where hydrochloric acid is produced during the digestion of food. The bacterium is exposed transiently to very low pH as low as 2. *H. pylori* then synthesizes a membrane urease which hydrolyzes urea according to the reaction below, thus producing ammonium which neutralizes the HCl from the wall of the stomach (Fig. 9.27) and therefore maintains a pH compatible with local cell physiology.

$$\underset{(urea)}{H_2N\text{-}CO\text{-}NH_2} + H_2O \underset{(urease)}{\rightarrow} NH_3 + \underset{(carbamate)}{H_2N\text{-}CO\text{-}OH}$$

$$\underset{(carbamate)}{H_2N\text{-}CO\text{-}OH} + H_2O \underset{(spontaneous)}{\rightarrow} NH_3 + \underset{(carbonic\ acid)}{H_2CO_3}$$

$$2NH_3 + 2H_2O \underset{(spontaneous)}{\rightarrow} 2NH_4^+ + 2OH^-$$

A similar mechanism was found in the gammaproteobacteria *Edwardsiella ictaluri*, a pathogen of catfish, or in *Ictalurus punctatus* and *Yersinia enterocolitica* that are pathogens of the gastrointestinal tract (De Koning-Ward and Robins-Browne 1995).

9.6.4 Adaptation to Pressure

At the surface of the earth, the mean pressure is 1 bar or 1,000 hectoPascals (hPa) or 1 atmosphere (atm). Deep environments are habitats with high pressures: the deep seas (77 % of marine waters are environments with pressures higher than 300,000 hPa), the deep layers of the earth (at 1,000 m deep below the earth's subsurface, the pressure is 300,000–600,000 hPa), and deep lakes (with a pressure increase of 1,000 hPa or 1 atm for every 10 m depth) (*cf*. Sect. 10.5).

9.6.4.1 Piezo-Tolerant and Piezophilic Prokaryotes

The majority of piezo-tolerant and piezophilic or barophilic microorganisms were isolated from oceanic depths. They are confronted with three unusual parameters: low temperatures (psychrophilic microorganisms), high pressures (piezophilic or barophilic microorganisms), and low nutrients (oligotrophic microorganisms). In the marine environment such as in lakes, pressure increases by approximately 1,000 hPa or 1 atm every 10 m depth. The majority of piezo-tolerant or piezophilic prokaryotes isolated from environments between 4,000 and 5,000 m deep have to withstand pressures of 400–500 atm. Piezo-tolerant prokaryotes are those that are able to live at normal atmospheric pressure and can tolerate high pressures ranging up to 500 atm for some. They predominate in deep environments down to 5,000 m depth. Below, in environments between 5,000 and 6,000 m depth, moderately piezophilic prokaryotes are found mainly those with an optimum growth occurring at pressure of around 400 atm. In deeper environments (10,000 m or more), extreme piezophiles are found that require high pressures for growth of up to 800 or 1,000 atm, such as the bacterium *Moritella* that may develop at pressures greater than 400 atm with an optimum at 700–800 atm. These extreme piezophilic bacteria are capable of supporting decompression and resist for short periods at normal atmospheric pressure, but it is necessary to reapply pressure quickly to maintain their cellular integrity, which is destroyed during prolonged exposure to 1 atm.

The majority of piezophiles are psychrophilic and live optimally at temperatures of 2–4 °C.

However, around deep hydrothermal vents, thermophilic and piezophilic bacteria and archaea have been isolated. These microorganisms in most cases use chemoorganotrophic aerobic or anaerobic respiration (nitrate, iron, or thiosulfate), or are capable of fermentation.

9.6.4.2 Adaptation of Microorganisms to Pressure

Enzymes of piezophilic bacteria have modified their folding in order to bind to their substrates. Similarly enzymes associated with membrane transport are actively synthesized when pressure is higher. Microorganisms grown under pressure accumulate unsaturated fatty acids in their membrane; this adaptive response has the effect of allowing a better fluidity of the membrane and an increased transport of substances at high pressures. In piezophiles, growth at high pressure is possible through the synthesis of a specific protein of the outer membrane OmpH ("outer membrane protein H"), which is part of the family of porins. This specific protein that forms channels allows diffusion of small molecules necessary to cell metabolism through the outer membrane. It replaces the usual porins that are not functional at high pressure. Therefore, when the cell is subjected to high pressure (300 atm), this new protein is synthesized and allows the transport of substances efficiently.

9.6.5 Adaptation to Desiccation (*cf.* Box 9.2)

Liquid water is the main factor affecting life. Thus, life is possible at high temperature if liquid water is still present, e.g., hydrothermal vents at more than 100 °C, or at lower temperature. Thus, scientists working in the field of exobiology have tried to detect the presence of water on other planets, Mars in particular; this would be considered as a strong indication for the possibility of life (*cf*. Sect. 10.8).

The lack of water in general occurs in a gradual way, the first consequence for prokaryotic cells being an increase in osmotic pressure. To cope, the cells have implemented response systems via mechanisms of homeostasis of osmotic pressure discussed above, followed by the appearance of reactive species of oxygen causing oxidative stress also discussed above.

Finally, in the situation of extreme lack of water, microorganisms have adapted mainly through the synthesis of structures such as dormancy spores or cysts, a phenomenon triggered by a nutritional deficiency. The spores are cells with thickened walls that contain little free water and in which respiratory activity is low (*Actinobacteria*) or absent (*Firmicutes*). When they return to open water, the spores germinate and transform into "vegetative" cells which metabolism is back to the level anterior to the transformation in spore. Spores can be classified as "endospores" in *Firmicutes* when they are formed inside a mother cell or as "*exospores*" formed by septation at one end of a sporophore as in *Actinobacteria*. In cyanobacteria and many taxa of *Proteobacteria*, there are forms of survival called cysts, which are cells with thickened walls, tolerant to desiccation. This term should not be confused with the term heterocyst that refers to non-photosynthetic cells of cyanobacteria (heterocysts) specialized in nitrogen fixation that also have a thickened wall to prevent the diffusion of oxygen. The term cyst refers to a sack containing cells that are generally dormant, with a less intimate contact between the wall of the cyst and the cells as in the case of spores; nevertheless the two terms are considered by some as synonymous mainly differentiated by habit. In the case of eukaryotes, the distinction between the two terms is clearer with the term spore linked to a sexual differentiation while cysts are dormant vegetative structures.

Box 9.2: Microbial Life in the Deserts
Thierry Heulin

Introduction
The desert is a paradigm of an extreme environment for life, to the extent that water limits the growth and development of living organisms during much of the day. In hot deserts, this lack of water is compounded by excessive heat and light during the day.

The main features of a hot desert are:
- Scarce and irregular precipitation [average less than 200 mm per year (Miller 1961)]
- Morning dew forming the only source of water for living species
- Evaporation exceeding precipitation and large temperature range between day and night
- Poor soil organic matter (sandy or rocky)
- Flora and fauna scarce, consisting of a few adapted species

Some Characteristics of the Sahara
If the cold deserts of the polar regions are true deserts because of a lack of available water due to freezing, hot deserts are the most extensive and mainly located near the tropics of Cancer and Capricorn (about 50 million km^2 on the surface of the Earth, one-third of land area). In Africa, the Kalahari and Namib deserts are located in the south and the Sahara to the north. The Sahara is by far the largest hot desert on the face of the earth (9,000,000 km^2) extending from Mauritania to Egypt, extending through semidesert or desert northeastward, to the Gobi Desert. The sand is of sedimentary origin and comes from the destruction of other rocks mainly through erosion.

Microbial Communities in Hot Deserts
A desert like the Sahara is far from "sterile," as shown by several studies of microbial ecology published recently.

Some Results from the Project "Treasures of the Sahara"
On the site of Merzouga (Morocco), the sand is of the homometric type, mainly composed of quartz with a mean grain diameter of 300 μm. One gram of sand contains about 20,000 grains of sand. On the surface of each grain, an average of 10 bacteria was detected by fluorescence (approximately 200,000 bacteria per g of sand), with approximately 14 % of bacteria that could be cultured by analyzing the sand grain by grain (130 grains analyzed) (Gommeaux et al. 2005). The overall analysis of the diversity of culturable bacteria in the form of colonies growing on nutrient media, or by cloning-sequencing of 16S rDNA gene, shows that the following are dominant bacterial groups: *Firmicutes*, *Actinobacteria*, and *Proteobacteria* with other groups *Flexibacter-Bacteroides-Cytophaga* (FBC), GNS bacteria (green non-sulfur), *Acidobacteria*, and *Planctomycetes* at much lower frequencies (Gommeaux et al. 2005). Importance of diversity was estimated between 400 and 1,400 equivalent species, indicating that in this extreme environment, the total number of bacteria (2×10^5 per gram of soil, direct observation) was much lower than in a cultivated soil of temperate areas (10^8 bacteria per gram of soil), while diversity is roughly comparable (Torsvik et al. 2002). On the site of Tataouine (Tunisia), conclusions about the wide variety of bacterial species are identical. These species belong mostly to the *Firmicutes*, *Actinobacteria*, *Proteobacteria*, and CFB. On these samples, the presence of non-thermophilic archaea has also been demonstrated (Chanal et al. 2006). A search for bacteria tolerant to ionizing radiation was undertaken among these desiccation-tolerant bacteria. A sample of Tataouine sand was subjected to a gamma irradiation of 15,000 Gy, which revealed the presence of spore-forming bacteria (*Bacillus*), of *Proteobacteria* (*Chelatococcus*), and, as expected, of *Deinococcus* (de Groot et al. 2005, 2009; Chanal et al. 2006). Some similarities between the two sites are worth highlighting:
- About 70–80 % of 16S rDNA sequences revealed by the molecular approach (diversity of culturable bacteria and non-culturable) do not match any bacterial species described to date.
- Bacterial diversity is dominated by bacteria belonging to phyla and genera with known mechanisms of desiccation tolerance, *Firmicutes* and *Actinobacteria* (sporulation) and *Deinococcus* (DNA repair), but also soma bacteria in the phylum Proteobacteria that we do not know the mechanisms of adaptation to desiccation, with the exception of encystment. A new bacterial genus, *Ramlibacter* isolated from the Tataouine sand, has this property.

T. Heulin
UMR 7265 CNRS-CEA-Aix Marseille University,
Institute of Environmental Biology and Biotechnology
IBEB/DSV/CEA, CEA Cadarache,
13108 Saint-Paul-lez-Durance, France

(continued)

Box 9.2 (continued)

The Ramlibacter Bacterium and Its Cell Cycle

In 1931, a meteorite was found near the town of Tataouine, in southern Tunisia. The largest fragments of the meteorite were sent the next day to the National Museum of Natural History in Paris (Lacroix 1931). New samples of small fragments of the meteorite were recovered in 1994. Analysis by scanning electron microscopy revealed areas of dissolution on the surface of these fragments, associated with the precipitation of calcite crystals, although such areas were absent from the fragments preserved in the Museum (Barrat et al. 1998; Benzerara et al. 2003). A search for microbes responsible for this alteration permitted to isolate a candidate bacterium (strain TTB310) that was identified as belonging to a new genus and therefore a new bacterial species: *Ramlibacter tataouinensis* (Raml-sand in Arabic; a bacterium from the Tataouine sand) (Heulin et al. 2003). We have shown that this bacterium was actually able to colonize and alter fragments of pyroxene rock (meteoritic silicate mineral component) and to induce biomineralization under controlled conditions (Benzerara et al. 2004). The originality of this bacterium is that it has a complex cell cycle involving two very different cell types: (1) a spherical form with a diameter of 0.8 μm with bacterial cyst properties and (2) a mobile rod-shaped form (0.2 μm in diameter and 2–3 μm in length) permitting dissemination (Gommeaux et al. 2005). The cell cycle consists of two phases that can explain its mode of reproduction and dissemination: (1) a multiplication phase during which a cyst divides to yield two cysts and (2) a dissemination phase during which a cyst differentiates into rods, which can divide and move in search of more favorable conditions, followed by redifferentiation of rods into cysts. The cysts can then divide and form a new satellite colony. The cell cycle is an adaptation to extreme conditions of hot deserts. The main factor limiting growth and reproduction of living beings in such a medium is water. For this bacterium, the form sensitive to desiccation (rods) permits dissemination, while the form tolerant to desiccation (cysts) is able to divide, which optimizes the narrow "window" of availability of water (dew mainly late at night during winter). Its genome has recently been studied (De Luca et al. 2011).

Rhizobia and the Desert

Many studies mention the presence of rhizobia in desert soils, and, in particular, strains can nodulate acacias (Zerhari et al. 2000). One adaptive response often noted in these bacteria is the biosynthesis of osmoprotecting molecules. Looking for the dominant bacteria that produce exopolysaccharides (EPS) in many desert soils in Algeria, Kaci et al. (2005) have highlighted the presence of *Rhizobium* producing heteroglycans. These studies suggest that EPS production is stimulated in *Rhizobium* by the rhizosphere of plants other than legumes.

Cyanobacteria and the Desert

A study of cyanobacterial diversity in "crusts" developing at the surface of a Utah desert has highlighted the existence of a cluster named "*Xeronema*" consisting of thin-walled cyanobacteria, close to *Phormidium* (Garcia-Pichel et al. 2001). This diversity of cyanobacteria was found in the meteorite Tataouine, with genera *Oscillatoria, Anabaena, Nostoc,* and *Symploca* (Benzerara et al. 2006).

Conclusion

Far from being barren, hot deserts, both low in nutrients (carbon, nitrogen) and subjected to various stresses (water, UV, temperature), harbor a great diversity of microorganisms. This diversity reveals only very few bacterial species indicating an adaptation of thermophilic microorganisms to water stress, but little or no adaptation to temperature (though very high in this ecosystem). Water being available, and thus growth of microorganisms being only possible during the few cool hours of the day (dew, late at night during the winter). Bacteria in fact, have not developed mechanisms of heat tolerance: when it is hot, bacteria "sleep." Within this diversity, families and genera of bacteria have developed mechanisms of desiccation tolerance: sporulation (firmicutes endospores and actinobacteria sporophores), DNA repair (*Rubrobacter, Deinococcus*), encystment (*Ramlibacter, Azotobacter*), or the production of exopolysaccharides and osmoprotectants (*Rhizobium*).

In contrast, these studies show that:

- Bacterial genera described for their desiccation tolerance (*e.g., Arthrobacter, Cystobacter*), but for which involved mechanisms remain unknown
- Bacterial genera already known (most genera belonging to *Proteobacteria*: e.g., *Chelatococcus, Pseudomonas, Agrobacterium*) but with so far unknown tolerance to water stress
- A very large number of bacteria not belonging to any species described so far

(continued)

> Box 9.2 (continued)
>
> The analysis of the diversity of microorganisms in hot deserts makes us think it represents the true treasure of the desert: both for coping with water stress it imposes of all life forms and also for new molecules of medical interest, food and agriculture still hidden in it.

9.7 Adaptation to Biotopes

9.7.1 Soil

Soil is the substrate (together with sediments) that supports the greatest microbial diversity with 20,000 different microbial genome equivalents per gram (Torsvik et al. 2002). This means that soil is a permissive biotope for a large number of microbial taxa and that it contains a varied array of trophic resources. It is however a selective biotope because of hydric alternances and mainly because it is heavily populated by diverse microflora with which must deal any newcomer. In this biotope are particularly present filamentous bacteria (*Actinobacteria*), gliding bacteria, spore-forming bacteria (*Firmicutes*, *Actinobacteria*), and bacteria producing antibiotics (*Actinobacteria*).

Soil is considered as a homogeneous habitat, but there are hundreds of soils that vary for a number of parameters including the concentration of organic matter, sand, acidity, concentration of nutrients, trace elements, and pollutants. However, most soils experience alternating daily or seasonal water regime. These fluctuations mean that microorganisms spend part of their lives with an abundance of water and therefore hypoxia or at the other extreme with limiting water and thus dehydration.

Hypoxia is a physicochemical condition to which microorganisms adapt via the synthesis of electron transport alternatives that use nitrate or sulfate as the final acceptor. Such systems are widespread in soil microorganisms, such as nitrite reductase (nirS) and nitrate reductase (NarGHIJ), especially in rhizosphere microorganisms that have access to a large amount of carbon substrates compared to the electron acceptors available, oxygen, nitrate, and nitrite. The heterogeneous structure of the soil with numerous discontinuities also seems to have selected a particular morphology, filamentous morphology, to move through the soil to areas more favorable for growth. This filamentous structure is of course also found in cyanobacteria in the ocean or animal or plant pathogens, but it is particularly prevalent among actinobacteria, oomycetes, and fungi (basidiomycetes, ascomycetes). Hyphae allow not only to grow faster (in terms of tip speed) but also to overcome the discontinuities in the soil.

9.7.2 Water and Sediment

9.7.2.1 Adaptation to Aquatic Environment
The Prokaryotic Form
In water, prokaryotes must swim or float. Cocci are generally not motile with rare exceptions; they are less numerous than rods (straight, curved, or spiral) that are most numerous in the aquatic environment. These may represent up to 80–90 % of the prokaryotic community of aquatic environments. These forms are generally motile in the presence of flagella or gas vacuoles (*cf.* Sect. 3.1); it is the cell forms best adapted to aquatic life.

Movement
In prokaryotes that move through water, the movement is due to the presence of flagella or vacuoles (vesicles) of gas.
- Prokaryotes motile with flagella

  These are mostly rods, straight, curved, or spiral. According to the positioning or flagella, these microorganisms are called monotrichous (single polar flagellum), lophotrichous (a tuft of polar flagella), amphitrichous (two tufts of flagella, one at both ends), or peritrichous (flagella distributed over the entire cell) (*cf.* Sect. 3.1.1). Flagella grow at their tips: flagellin molecules synthesized in the cytoplasm passing through the hollow filament of the flagellum and assembled to each other at the end of the flagellum, thereby forming a "rigid" structure consisting of subunits of the same type. In *E. coli*, flagella are two to three times as long as the cell, approximately 5–7 µm. They are constantly built. Flagella are activated by a basal body consisting of various proteins inserted in the cell envelope and extending through a rigid hook which constitutes the base of the flagellum. The basal body is like a motor that drives the hook and so all coiled flagellum. The energy required comes from the proton motive force (*cf.* Sect. 3.1.1; Figs. 3.6b and 3.4). The fastest movement recorded is 200 rotations per second. Spirilla are among the faster bacteria.

  The energy required to activate the basal body of the flagellum is about 1,200 protons per rotation of the flagellum (Khan 1992). The rotary movement is known only to the flagellum prokaryotes. Because of this, movement may be bidirectional. The change of direction is instantaneous. Periods of activation of the flagellum are generally one second, interspersed with stop periods

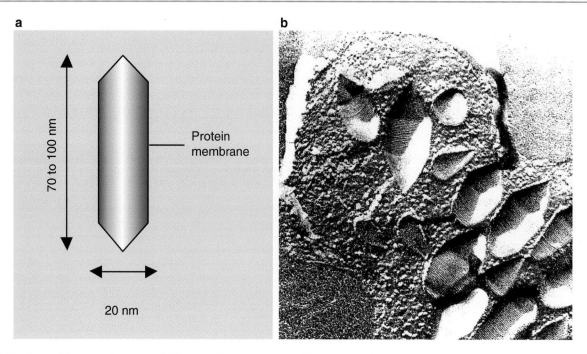

Fig. 9.28 Gas vesicles in prokaryotes. (**a**) Diagram of a gas vesicle. (**b**) Electron microscopy microphotograph of vacuolated bacteria (cell of *Prosthecomicrobium pneumaticum*) containing gas vesicles (Photography of Branton and Walsby in Walsby (1978, 1994); copyright: courtesy of SGM "Society for General Microbiology"). Drawing: M.-J. Bodiou

of about 1 millisecond during which the bacteria tumble on itself before going in another direction; the displacement due to the activation of flagellum is controlled by the chemotaxis signaling systems (*cf.* Sect. 9.2.2). In liquid medium, the speed of movement by the flagella varies depending on cells: *Bacillus megaterium* (large rod) moves from 1.5 to 2 mm/min, whereas *Vibrio cholerae* (curved rod) can move at the speed of 12 mm/min which is approximately 3,000 times the length of the bacterial body. The generally accepted average is 2–4 mm/min.

- Bacteria possessing gas vacuoles
  One adaptation to aquatic life is the synthesis of flotation vesicle, through a set of proteins called "gas vesicles proteins." These vesicles are small organelles found in different aquatic bacteria such as cyanobacteria (*Nostoc muscorum*, *Anabaena flos-aquae*, etc.), heterotrophic bacteria [(*Prosthecomicrobium pneumaticum*, *Aquabacter*, *Ancylobacter aquaticus* (Konopka et al. 1977)], phototrophic purple and green bacteria (*Thiocapsa*, *Thiodictyon*, *Pelodictyon*, etc.), Firmicutes (*Bacillus megaterium*), and also archaea (*Halobacterium halobium*, *Haloferax* sp.) (Halladay et al. 1992) or even soil actinobacteria (van Keulen et al. 2005), probably so that they can remain at the top of the water column. The assembly of vesicles in the cell leads to the formation of gas vacuoles visible under an optical microscope as clear spaces in the cell. They provide buoyancy by decreasing cell density. The vesicles that form these vacuoles are hollow cylinders tapered at their ends, of about 75 nm in diameter and 150–300 nm in length (Fig. 9.28). They comprise a rigid wall protein consisting of hydrophobic protein bands arranged in crystalline lines forming the cylindrical tube and the end cones and a second protein consisting of a repeated amino acid sequence, which adheres to the outside of the vesicle and stabilizes the structure (Walsby 1994). The vesicle is permeable to gases and creates an inner void space that can be filled by gas. They are rigid and cannot be inflated; the cavity is created by the molecular array and fills immediately by the diffusion of gas from the external environment when it is soluble in water (typically N_2). The water is kept out of the vesicle by the surface tension of the hydrophobic outer side of the vesicle protein. The pressure is maintained by the rigid wall and does not compress the gas inside. The gas vesicle is capable of withstanding pressures of several bars; at a critical pressure, determined by the mechanical properties of the proteins and the diameter of the cylindrical structure, the vesicle collapses permanently and cannot return to its initial shape and be refilled with gas. Thus, new vesicles need to be produced when pressure decreases.

  There is a balance between the gas dissolved in the liquid and the gas in the vesicle. The amount of gas which enters the vesicle depends on the volume of the vesicle and the pressure of the gas dissolved in the liquid outside. The gases diffuse freely through the pores in the wall protein.

The gas vesicles confer of buoyancy to prokaryotic cells and are involved in the regulation of buoyancy. This regulation allows microorganisms to position themselves at an optimal depth in the water column or to make vertical migrations based on gradients favorable to their development. The production of gas vesicles is dependent on the light in phototrophic organisms, nutrients and oxygen gradients, temperature, salinity, or pH in chemolithotrophic or chemoorganotrophic vacuolated prokaryotes living in aquatic environments.

9.7.2.2 Adaptation to Supports

In aquatic environments, the majority of prokaryotic microorganisms are living on fixed supports, like particles suspended in the water, borders and banks, sediments and dumped materials, etc. Prokaryotes can colonize these substrates by forming biofilms (*cf.* Sect. 9.7.3) or move at their surface by gliding. Microorganisms that move by gliding are usually filamentous; this is the group of gliding filamentous bacteria. The gliding velocity is much lower than that obtained when moving based on flagellar movement. These bacteria move slowly and the speed of movement depends on the length of the filament. However, many nonfilamentous bacteria can also move by gliding on supports (McBride 2001). Gliding mechanisms are complex.

Chemoorganotroph unicellular filamentous prokaryotes such as *Cytophaga*, some *Flavobacterium*, or some *Myxococcus* may have different sliding modes. For example, *Flavobacterium* (formerly *Cytophaga*) *johnsoniae* colonizes the surface of plant debris on the soil or sediment; it has particulate structures (proteins) inserted into envelopes. A series of proteins anchored in the cytoplasmic membrane is in contact with a second set of proteins inserted into the outer membrane (Fig. 9.29a). The cytoplasmic membrane proteins are activated through the proton motive force and induce activation of proteins of the outer membrane which are set in motion and which, positioned and moving on the support, move the cell in the opposite direction by gliding (Fig. 9.29b). *Myxococcus*, a gliding unicellular bacterium, moves to the surface of the supports (particulate organic debris) by secreting a detergent that reduces the surface tension and allows the bacteria to glide. These bacteria are well adapted to their organic materials that biodegrade slowly by excreting cellulases, chitinases, and other extracellular enzymes capable of hydrolyzing the solid polymers.

Other unicellular prokaryotes and some multicellular filamentous prokaryotes move on their support by means of the activation of pili that vibrate and advance the filament. These pili called type IV pili allow cell adhesion to the supports and induce advance of the cell by retracting rhythmically. This type of motility is called motility by pulling or jerking ("twitching motility"). Many nonflagellated but also flagellated bacteria can move on supports such as biological surfaces (cells, tissues), organic (biofilms, etc.), or inert and inorganic (sediment, metals, concrete, etc.), in aquatic sediments or biofilms by using this type of motility with type IV pili in polar position (Mattick 2002). The type IV pili have a diameter of 5–7 nm and a length of 1 to several microns. They are activated by a protein basal body and are made up of a protein (pilin). They act by retracting and extending rhythmically through a "back-and-forth" movement that allows the cells to move on the supports and also aggregating and moving out of the bacterial colonies. This mode of movement is mistakenly called "gliding motion," but should instead be called "moving by jerking or by rhythmic contractions" (Mattick 2002).

Multicellular filamentous prokaryotes such as *Beggiatoa*, *Thiothrix*, or *Leucothrix* as well as filamentous cyanobacteria excrete exopolysaccharides (EPS) from pores on their outer membrane at the junction of septa between cells. These filaments form exopolymer mucoid layers around the filament. This layer, adhering to the support, also allows gliding displacement of these filamentous bacteria by movement due to the excretion of filaments of exopolysaccharides (McBride 2001). These bacteria agglomerating together form mucoid masses that adhere to and colonize surfaces. For example, sulfur-oxidizing filamentous bacteria such as *Beggiatoa* or *Thiothrix* form complex clumps that adhere to the walls of sulfur thermal water pools (*cf.* Sect. 14.4.4, Fig. 14.38). Branched filamentous actinobacteria (*Streptomyces*) are well suited to supports (organic particles, plant debris) on which they grow without moving, but by developing a filamentous network that little by little colonizes all the support.

9.7.3 Adhesion to Surfaces, Biofilms, and Microbial Mats

9.7.3.1 Mechanisms of Adhesion

Most prokaryotes live by adhering to surfaces in the natural environment. Cells can join individually or form mono- or multispecies biofilms, more or less thick (Fig. 9.30). Many bacteria in aquatic environments adhere to surfaces in response to stress (survival method), especially Gram-negative bacteria. Some bacteria (*Caulobacter*, *Myxococcus*) include a phase of adhesion in their cell cycle. Others need to bind to solid particles for feeding, such as bacteria that degrade cellulose plant debris, bacteria that degrade lipid droplets in an aquatic environment, etc.

Ionic Interactions

Prokaryotes can adhere through ionic interactions (electric charges). Cell surface is negatively charged, which allows binding to positively charged surfaces.

Fig. 9.29 Diagram of the movement mechanism on a support of *Flavobacterium johnsoniae*. (**a**) Two protein structures are inserted over the entire periphery of the cell, a protein in the cell membrane and a second protein in extension in the outer membrane. (**b**) The cytoplasmic protein is activated by the proton gradient, vibrates, and causes the external protein which oscillates by pressing on the support in the same direction. The movement created causes movement of the cell in the opposite direction (Modified and redrawn from McBride 2001). Drawing: M.-J. Bodiou

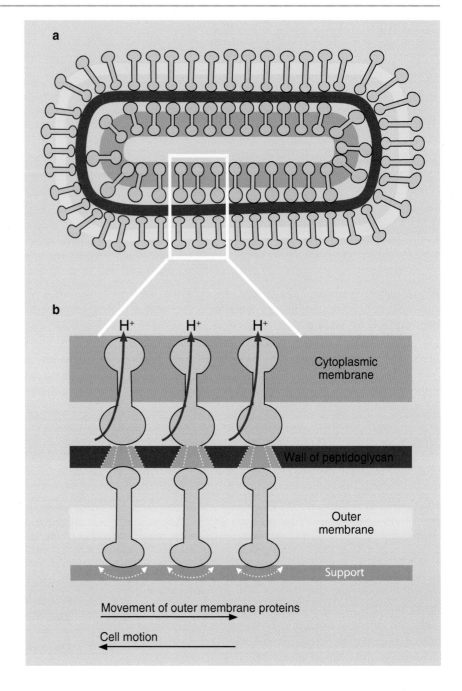

Hydrophobic Interactions

Many surfaces are hydrophobic in nature: oil drops in water, cell membranes, waxy surfaces, etc. Many prokaryotes have hydrophobic surfaces that allow them to adhere directly. However, microbes covered with exopolysaccharides are hydrophilic. They can produce fimbriae or pili which are hydrophobic to help them fasten onto surfaces.

Presence of Lectins

Lectins are proteins linked to carbohydrates produced by prokaryotes and eukaryotes. They are located at the end of pili in prokaryotes and allow adhesion with receptor sites specific to each lectin type. They are specific to certain surfaces or membranes or heterotrophic or photosynthetic prokaryotes. For instance, enteropathogenic *Escherichia coli* have lectins which enable them to adhere specifically to cells of the gut, whereas uropathogenic *Escherichia coli* adhere to the cells of the bladder.

Other Adhesion Mechanisms

There are other proteins of mammalian cells such as fibronectin or vitronectins which are receptors for different pathogenic bacteria.

Flagella of certain bacteria (*Vibrio*) allow their adherence to epithelial cells of the intestine wall. Similarly, lipoteichoic acids of Gram-positive bacteria (streptococci, staphylococci) also help adherence to epithelial cells or skin.

Fig. 9.30 Prokaryotic biofilm formation. Free cells in the water attach themselves to surfaces and initiate biofilm formation by the production of exopolymers (exopolysaccharides, EPS) that adhere to the substrate. When the biofilm grows and matures, flow channels of the surrounding water appear in the matrix of EPS. This creates physical and chemical gradients which determine oxic areas on the surface and anoxic zones in deeper layers. When the biofilm is highly developed, fragments break off and are carried by the current to downstream places they can colonize. Drawing: M.-J. Bodiou

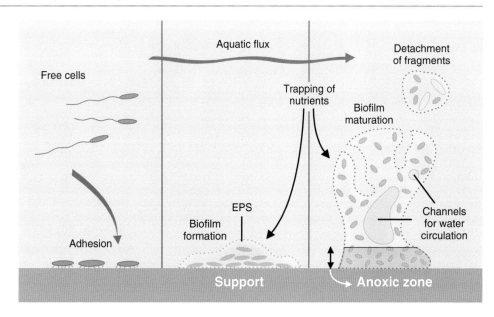

Biofilm Formation*

In aquatic environments, when microorganisms adhere to surfaces, they can grow by forming biofilms. These structures consist of a matrix of exopolymers excreted by bacteria, forming a mucoid mass spread over the surface and in which the bacteria are glued. They develop by constituting a biomass in this biofilm, being protected from the environment. Biofilms trap nutrients dissolved in the water but also contaminants (metals, organic pollutants) which are concentrated in the matrix of exopolymers. Nutrients allow the growth of bacteria enmeshed in biofilms. These bacteria form a community that can be mono- or multispecies.

The formation of a biofilm depends primarily on the accession of free microorganisms to a surface and production of exopolysaccharides (EPS). These prokaryotes are considered pioneer microorganisms. They allow the installation of biofilm on a surface (Fig. 9.30). Communication between cells is important and necessary for the development of biofilm. Upon attachment of cells to a surface, intercellular communication signals are used to enable the production of EPS which will form the matrix. Then in the biofilm, cells communicate via signals, depending on the number of cells and on the cell density ("quorum sensing," *cf.* Sect. 9.3). Multiplying bacterial cells allows the development and maturation of the biofilm. Other bacterial cells may be attracted by chemotaxis and live and multiply in the biofilm; these are secondary colonizers who take advantage of the matrix structure of the biofilm, the physicochemical conditions and nutrient sources prepared by primary colonizers, and the cells pioneers.

When the biofilm is well developed, channel flows of water appear deep in the biofilm (maturation period, Fig. 9.30). These channels allow water to flow into the surrounding biofilm and provide nutrients and oxygen to the bacteria that thrive in the sticky EPS matrix. However, in the deeper parts, anoxic conditions settle and lead to the development of anaerobic metabolism (anaerobic respiration or fermentation). For example, a biofilm is the seat of various metabolisms coexisting within the matrix based on environmental conditions and physicochemical gradients that move from the surface to the deepest area of the biofilm even if it is only a few millimeters.

When the biofilm is highly developed, cells can break off and migrate to colonize other areas, and portions of the biofilm can detach and be driven by erosion into the liquid phase.

Many types of prokaryotes can constitute biofilms. It is mainly the case of heterotrophic bacteria, aerobic chemoorganotrophic, which can be opportunistic pathogens (*Pseudomonas*, *Flavobacterium*, enterobacteria, etc.), fermentative bacteria, or microbes that respire anaerobically (sulfate-reducing bacteria). There are also photosynthetic bacteria which often constitute specific biofilms (microbial mats; see below, Fig. 9.31). Prokaryotes in general form biofilms under conditions of nutritional stress. Indeed, the formation of a biofilm allows prokaryotic cells to protect themselves from adverse environmental physicochemical conditions and benefit from favorable nutritional conditions either by trapping nutrients in the matrix from the surrounding liquid flow or by the use of the organic carrier (cell surface, plant debris, etc.). Finally, the formation of a biofilm allows cells to be concentrated together in close association, thus promoting cellular communication ("quorum sensing") which often allows the expression of metabolic genes not expressed in general. The microbial biofilms occur on all types of media. In living organisms, bacteria can colonize different surfaces (mucosa, organs,

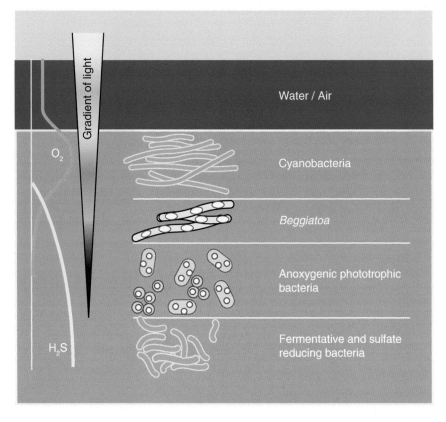

Fig. 9.31 Example of cyanobacterial microbial mats in the coastal marine environment. Different bacterial groups coexist in superimposed layers based on vertical gradients of oxygen, sulfide, and light. Drawing: M.-J. Bodiou

teeth). One common example is the formation of dental plaque, which begins with the formation of a biofilm by pioneer bacteria or early colonizers (*Streptococcus oralis*, *Stomatococcus*) that adhere to the teeth and produce the biofilm matrix. Then, other bacteria (staphylococci, *Enterobacteriaceae*, fermentative bacteria) grow on the first deposit (secondary colonizers) and cause, through their development, the formation of anoxic zones in the deepest part in contact with the tooth where fermentation produces organic acids and releases sulfides that are responsible for the attack of the tooth. Biofilms are also often installed on inert surfaces. In tubes where liquids circulate, many types of biofilms are common. *Pseudomonas aeruginosa* is responsible for the formation of catheters biofilms in hospital settings, making it the top infectious agent responsible for nosocomial infections. Heterotrophic bacteria are responsible for aerobic and anaerobic biofilms in water pipes, gas, and oil and are often the cause of bacterial corrosion problems in pipelines (sulfate-reducing bacteria, Box 14.12). Microbial biofilms also form on submerged structures (metal or concrete) in ports, boats, and marine platforms and are responsible for corrosion.

Microbial Mats

Microbial mats are particular biofilms. They thrive on immersed inert surfaces (lake sediments and marine rocks seashores or rivers, coral reef, etc.) and on coastal rocks wetted by spray. In the area illuminated by sunlight (photic zone), most microbial mats are formed by photosynthetic filamentous bacteria (cyanobacteria), which excrete the EPS matrix onto which other organisms agglomerate. These microbial mats, composed of cyanobacteria as early colonizers and subsequent settlers, are considered as representing the oldest life on the planet. The dating of fossil forms (stromatolites) is estimated at 3.5 billion years (*cf.* Sect. 4.2). Actual microbial mats of cyanobacteria can also be carbonated and "fossilized." These mats are mainly encountered in tropical atolls. They are called recent stromatolites or microbialites.

In deeper areas of the ocean, where no light reaches, non-photosynthetic microbial mats may develop, around hydrothermal vents or on the deep sediments. They are usually formed by filamentous gliding bacteria such as *Beggiatoa*, *Thiomargarita*, or *Thiothrix* as early colonizers together with other chemoorganotrophic bacteria producing mucoid EPS matrix.

In coastal sediments of shallow waters, cyanobacterial microbial mats may colonize very large areas in deltas, coastal sandy areas, etc. Prokaryotes are not the only inhabitants of the microbial mats; different micro-eukaryotes are also present such as micro-photosynthetic eukaryotes (diatoms) that colonize the surface of the mat and eukaryotic micro-ciliates or flagellates living immerged in the matrix of the mat. These microbial mats are typically layered structures in thin layers ranging from one to a few mm thick. The total thickness of microbial mats is variable depending on the situation: for some mats, the total thickness is a few mm or cm, and others that grow on the coral reefs in tropical lagoons and called "Koparas" in local language can reach thicknesses of several tens of cm up to 40 cm thicker. In these microbial mats, cyanobacteria form the basic structure of the belt in producing the EPS matrix. They are only active in the surface of the mat where they receive enough light to grow and produce oxygen through their oxygenic photosynthesis (*cf.* Sect. 3.3.4). At a few mm below the anoxic obscure zone, the structure of the mat is due to dead cyanobacterial filaments stuck in the mucoid matrix. Indeed, these microbial mats have developed over the years and even centuries with the accretion of a new layer of cyanobacteria in surface every year. In coastal marine microbial mats, gradients of oxygen, sulfide, and light determine the distribution of strata with metabolically different bacterial groups (Fig. 9.31; *cf.* Sect. 14.4.4 and Fig. 14.44). At the surface of the mat, the metabolically active cyanobacteria produce excess oxygen forming a gradient with reduced oxygen in deeper layers, generally a few mm below. In deeper anoxic layers, the concentration of organic material from dead cyanobacteria promotes the development of fermentative and sulfate-reducing bacteria that are stimulated by high concentrations of seawater sulfate. In the intermediate zone, the chemolithotrophic aerobic sulfur-oxidizing bacteria (*Beggiatoa*, *Thiobacilli*, etc.) develop at the interface between the upper part containing oxygen and the deeper part with sulfide. If the light reaching the anoxic zone is sufficient in quantity and quality, the sulfur-oxidizing purple bacteria can grow and form a purple layer (Figs. 9.31 and 14.44). Thus, these mats are vertical models of cooperation based on trade between bacteria of the biogeochemical sulfur cycle (*cf.* Sect. 14.4.4).

Microbial mats and biofilms are well-structured systems where different prokaryotic and eukaryotic microorganisms coexist on the basis of trophic interactions. Man has exploited these capabilities in purification processes such as the formation of biofilms in trickling filters or biodiscs used for the biological purification of wastewater treatment plants. The ability of biofilms and microbial mats to trap specific molecules including organic pollutants and heavy metals can also be exploited to improve water decontamination processes.

9.7.4 Air

Air contains many microorganisms from soil, water, animals, or plants in response to gusts, especially in situations of partial canopy as in the steppes and deserts. Microorganisms which are carried by the wind and suspended in the air can travel great distances, thousands of kilometers away from some dust sources (Sahara Desert, Gobi Desert) to sites where precipitations will bring them to the ground (Central Europe, the Pacific Islands). These microorganisms are more easily carried away as they are smaller, the diameter of a bacterial cell varying by several orders of magnitude. The smaller cells are the spores that are favored here and will therefore be carried farther. A second factor that will affect the ability for traveling of microorganisms is their resistance to ionizing radiation to which they are subjected in the atmosphere. Again microorganisms capable of synthesizing spores can survive the journey as well as *Deinococcus* and *Geodermatophilus* found in soils after gamma irradiation (Rainey et al. 2005). In addition, microorganisms must withstand solar radiation including UV which causes cellular damage by photooxidation. Microorganisms that produce carotenoid pigments can withstand the effects of photooxidation due to radiation by diverting a portion of the solar energy and acting as photocatalysts, thus protecting the cell. In fact, in prokaryotic communities suspended in the air, the percentage of pigmented cells (carotenoids) is generally very high. Finally, the microorganisms must complete the trip and so get down to the earth. This will happen when they join the regions where the temperature and the concentration of water vapor will condense them into droplets or snowflakes. Moreover, some microorganisms (*Pseudomonas*) contain proteins that facilitate this condensation and therefore the process whereby they can be brought to the ground (Morris et al. 2008). Sampling microorganisms in precipitation, glaciers, or in the air has begun to be published (Xiang et al. 2005).

9.8 Adaptation to Manmade Environments (*cf.* Chap. 16)

Man has profoundly changed the different biotopes in the world especially since the beginning of the industrial age. This effect called "human impact" was made in different ways, for example, the synthesis of new molecules called "xenobiotics" in agriculture (herbicides, fungicides, etc.) or industry (PCBs, PAHs), or by intensive use of natural molecules like metals in agriculture (copper) or in industry (mercury, nickel), or organic compounds such as aromatics. All these compounds are found in soils, sediments, or water bodies, where they alter microbial activities and end up in the food chain, causing many problems that are beyond the scope of this text. The use of reactive molecules such as copper in agriculture dates back more than a century, especially in the Bordeaux mixture (copper sulfate neutralized with lime) against various fungal pathogens of grapevine as downy mildew. As this compound is allowed at levels of up to 6 kg Cu/ha/year by "organic" agriculture charters, it should continue to be used in large quantities in the coming years despite the toxicity observed for microorganisms, for ground animals (snails), and even for grapes and yeasts that are responsible for the aroma of wine. This metal is therefore found in the soil where it accumulates over the years; it changes the composition of microbial communities and has a cumulative effect on their activity (Ranjard et al. 2006; Anderson et al.

2009). Bacteria have developed resistance to many toxic metals. Resistant bacteria were detected and isolated mainly in natural metallic sites, for example, copper in the mining area of Katanga in Congo (Monchy et al. 2006) and arsenic in India or Portugal (Muller et al. 2006). Among the mechanisms of resistance described, a few should be mentioned: specific metal carrier or broad spectrum, oxidases or reductases that transform metal into a volatile compound as in the case of mercury (Barkay and Wagner-Dobler 2005) or in a less toxic compounds as in the case of arsenic (Cervantes et al. 1994), or even the formation of structures able to immobilize toxic compound such as in the case of the uranium fixed by extracellular melanin (Turick et al. 2008).

The official dumps or wild bioreactors are very efficient for the selection of new strains able to degrade xenobiotic compounds, which were still unknown 60 years ago. Thus, lindane (cyclohexane hexachloride) is an insecticide widely used around the world for agriculture, forestry, or even in shampoo for the elimination of lice. Bacteria belonging to the genera *Pseudomonas* or *Sphingomonas* (Suar et al. 2004; Manickam et al. 2008) have been described as capable of degrading this new compound. The mechanism by which this biodegradation pathway seems to have been obtained is the result of the mixing of gene domains, yielding new enzymes with novel functionalities. Several genes are grouped and carried by plasmids, which leads to their dispersion by conjugation into vulnerable ecosystems (*cf*. Sect. 11.2.2). The addition of synthetic pesticides to the environment is thus a powerful catalyst for the selection of recombinant bacteria (Davison 1999).

Anthropisation also affects habitats beyond the introduction of new chemical compounds. The intensive use of fertilizers, as well as urbanization with little or untreated wastes discharged, results in the addition of considerable amounts of nitrogen and phosphorus to rivers and the ocean. These additions cause eutrophication with algal blooms, some of which are toxic to humans.

A current problem concerns radioactive waste. Since the discovery of radioactivity by Marie Curie in 1898, many applications have been developed as weapons (bomb uranium or plutonium, H-bomb, depleted uranium munitions), power plants, accelerators, and medical care that all create wastes. These wastes are dumped in soils, sediments, and water courses where they create difficult conditions for microorganisms due to their chemical toxicity described above or due to the fact that they emit radiation and cause damage mainly to nucleic acids. The influence of radiation on microbial ecosystems has been well explored, in particular to identify taxa that are most resistant to gamma irradiation. Soils exposed to high doses of gamma radiation permit to select bacteria belonging mainly to the well-known genus *Deinococcus*, but also *Geodermatophilus*, etc. (Rainey et al. 2005). *Deinococcus* seems to have been selected in environments exposed to drought causing a strong oxidative stress at the origin of damages to DNA (Fredrickson et al. 2008). Gamma radiation causes direct damage to DNA by breaking the phosphodiester bridge, while UV damages DNA by fusion of neighboring thymidine bases, thus blocking replication fork of DNA and indirectly generating reactive species of oxygen which also attack the DNA. Different mechanisms of DNA repair have been described (Cox 1998) in bacteria and in archaea (Grogan 2004). *Deinococcus* combines the traditional mechanisms and the existence of multiple copies of the chromosome per cell, copies that complement each other in order to recreate an intact chromosome. The mode of resistance of other taxa remains unknown for now (Minton 1994).

9.9 Conclusion

Adaptation of microorganisms to different biotopes constitutes one of the most fascinating research domains of modern microbiology. Extreme environments have of course provided numerous examples of unusual metabolisms (*cf*. Chap. 10), and many more discoveries from genomes studies are expected. Microorganisms from complex biotopes such as soil, sediments, or the digestive tract should also yield their share of new metabolites and new regulation modes. All these approaches should also provide us with an overview of the adaptability of the microbial world, yielding new metabolites and new ways to fight pathogens.

References

Abu-Qarn M, Eichler J, Sharon N (2008) Not just for Eukarya anymore: protein glycosylation in *Bacteria* and *Archaea*. Curr Opin Struct Biol 18:544–550

Altuvia S (2007) Identification of bacterial small non-coding RNAs: experimental approaches. Curr Opin Microbiol 10:257–261

Amikam D, Galperin MY (2006) PilZ domain is part of the bacterial c-di-GMP binding protein. Bioinformatics 22:3–6

Anderson JAH, Hooper MJ, Zak JC, Cox SB (2009) Molecular and functional assessment of bacterial community convergence in metal-amended soils. Microb Ecol 58:10–22

Andrews SC, Robinson AK, Rodriguez-Quinones F (2003) Bacterial iron homeostasis. FEMS Microbiol Rev 27:215–237

Bainton NJ, Stead P, Chhabra SR, Bycroft BW, Salmond GP, Stewart GS, Williams P (1992) *N*-(3-oxohexanoyl)-L-homoserine lactone regulates carbapenem antibiotic production in *Erwinia carotovora*. Biochem J 288:997–1004

Barkay T, Wagner-Dobler I (2005) Microbial transformations of mercury: potentials, challenges, and achievements in controlling mercury toxicity in the environment. Adv Appl Microbiol 57:1–52

Barrat JA, Gillet P, Lecuyer C, Sheppard HM, Lesourd M (1998) Formation of carbonates in the meteorite Tatahouine. Science 280:412–414

Bentley SD et al (2002) Complete genome sequence of the model actinomycete *Streptomyces coelicolor* A3(2). Nature 417:141–147

Benzerara K, Menguy N, Guyot F, Dominici C, Gillet P (2003) Nanobacteria-like calcite single crystals at the surfaces of the meteorite Tataouine. Proc Natl Acad Sci U S A 100:7438–7442

Benzerara K, Menguy N, Guyot F, Skouri F, De Luca G, Barakat M, Heulin T (2004) Biologically controlled precipitation of calcium phosphate by Ramlibacter tataouinensis. Earth Planet Sci Lett 228:439–449

Benzerara K et al (2006) Nanoscale detection of organic signatures in carbonate microbialites. Proc Natl Acad Sci U S A 103:9440–9445

Beyhan S, Tischler AD, Camilli A, Yildiz FH (2006) Transcriptome and phenotypic responses of *Vibrio cholerae* to increased cyclic di-GMP level. J Bacteriol 188:3600–3613

Bjarnsholt T, Givskov M (2007) Quorum-sensing blockade as a strategy for enhancing host defences against bacterial pathogens. Philos Trans R Soc Lond B Biol Sci 362:1213–1222

Bolwell GP (1996) The origin of the oxidative burst in plants. Biochem Soc Trans 24:438–442

Breukink E, de Kruijff B (2006) Lipid II as a target for antibiotics. Nat Rev Drug Discov 5:321–332

Brock T (1997) The value of basic research: discovery of *Thermus aquaticus* and other extreme thermophiles. Genetics 146:1207–1210

Brumbley SM, Carney BF, Denny TP (1993) Phenotype conversion in *Pseudomonas solanacearum* due to spontaneous inactivation of PhcA, a putative LysR transcriptional regulator. J Bacteriol 175:5477–5487

Cervantes C, Ji G, Ramirez JL, Silver S (1994) Resistance to arsenic compounds in microorganisms. FEMS Microbiol Rev 15:355–367

Challis GL, Hopwood DA (2003) Synergy and contingency as driving forces for the evolution of multiple secondary metabolite production by *Streptomyces* species. Proc Natl Acad Sci U S A 100(Suppl 2):14555–14561

Chanal A et al (2006) The desert of Tataouine: an extreme environment hosts a wide diversity that of microorganisms and bacteria radiotolerant. About Microbiol 8:514–525

Christen B (2007) Principles of C-di-GMP signaling. PhD thesis, Faculty of Sciences, University of Basel, Basel

Chugani SA, Whiteley M, Lee KM, D'Argenio D, Manoil C, Greenberg EP (2001) QscR, a modulator of quorum-sensing signal synthesis and virulence in *Pseudomonas aeruginosa*. Proc Natl Acad Sci U S A 98:2752–2757

Chung J, Goo E, Yu S, Choi O, Lee J, Kim J, Kim H, Igarashi J, Suga H, Moon JS, Hwang I, Rhee S (2011) Small-molecule inhibitor binding to an N-acyl-homoserine lactone synthase. Proc Natl Acad Sci U S A 108:12089–12094

Cirou A, Mondy S, An S, Charrier A, Sarrazin A, Thoison O, DuBow M, Faure D (2011) Efficient biostimulation of native and introduced quorum-quenching Rhodococcus erythropolis populations is revealed by a combination of analytical chemistry, microbiology, and pyrosequencing. Appl Environ Microbiol 78:481–492

Cirz RT, Chin JK, Andes DR, de Crecy-Lagard V, Craig WA, Romesberg FE (2005) Inhibition of mutation and combating the evolution of antibiotic resistance. PLoS Biol 3:e176

Condemine G, Ghazi A (2007) Differential regulation of two oligogalacturonate outer membrane channels, KdgN and KdgM, of *Dickeya dadantii* (*Erwinia chrysanthemi*). J Bacteriol 189:5955–5962

Cox MM (1998) A broadening view of recombinational DNA repair in bacteria. Genes Cells 3:65–78

Cui L, Neoh HM, Iwamoto A, Hiramatsu K (2012) Coordinated phenotype switching with large-scale chromosome flip-flop inversion observed in bacteria. Proc Natl Acad Sci U S A 109:E1647–E1656

D'Costa VM, McGrann KM, Hughes DW, Wright GD (2006) Sampling the antibiotic resistome. Science 311:374–377

Dantas G, Sommer MO, Oluwasegun RD, Church GM (2008) Bacteria subsisting on antibiotics. Science 320:100–103

Davison J (1999) Genetic exchange between bacteria in the environment. Plasmid 42:73–91

de Groot A, Chapon V, Servant P, Christen R, Saux F, Sommer S, Heulin T (2005) Deinococcus deserti sp. November, a gamma-radiation-tolerant bacterium isolated from the Sahara Desert. Int J Syst Evol Microbiol 55:2441–2446

de Groot A et al (2009) Genomics and proteomics of Alliance to unravel the specificities of Sahara bacterium *Deinococcus deserti*. PLoS Genet DOI: 10. 1371/

De Koning-Ward TF, Robins-Browne RM (1995) Contribution of urease to acid tolerance in *Yersinia enterocolitica*. Infect Immun 63:3790–3795

De Luca G, Barakat M, Ortet P, Jourlin-Castelli C, Ansaldi M, Py B, Fichant G, Coutinho PM, Voulhoux R, Bastien O, Maréchal E, Henrissat B, Quentin Y, Noirot P, Filloux A, Méjean V, DuBow M, Barras F, Barbe V, Weissenbach J, Mihalcescu I, Verméglio A, Achouak W, Heulin T (2011) The cyst-dividing bacterium *Ramlibacter tataouinensis* genome reveals a well-stocked toolbox for adaptation to a desert environment. PLoS One 6(9):e23784

Degrassi G, Aguilar C, Bosco M, Zahariev S, Pongor S, Venturi V (2002) Plant growth-promoting *Pseudomonas putida* WCS358 produces and secretes four cyclic dipeptides: cross-talk with quorum sensing bacterial sensors. Curr Microbiol 45:250–254

DiMango E, Zar HJ, Bryan R, Prince A (1995) Diverse *Pseudomonas aeruginosa* gene products stimulate respiratory epithelial cells to produce interleukin-8. J Clin Invest 96:2204–2210

Donadio S, Monciardini P, Sosio M (2007) Polyketide synthases and nonribosomal peptide synthetases: the emerging view from bacterial genomics. Nat Prod Rep 24:1073–1109

Dong YH, Zhang XF, Soo HM, Greenberg EP, Zhang LH (2005) The two-component response regulator PprB modulates quorum-sensing signal production and global gene expression in *Pseudomonas aeruginosa*. Mol Microbiol 56:1287–1301

Drlica K, Malik M (2003) Fluoroquinolones: action and resistance. Curr Top Med Chem 3:249–282

Duncan L, Losick R (1993) SpoIIAB is an anti-sigma factor that binds to and inhibits transcription by regulatory protein sigma F from *Bacillus subtilis*. Proc Natl Acad Sci U S A 90:2325–2329

Dworkin J, Blaser MJ (1997) Nested DNA inversion as a paradigm of programmed gene rearrangement. Proc Natl Acad Sci U S A 94:985–990

Eberhard A, Burlingame AL, Eberhard C, Kenyon GL, Nealson KH, Oppenheimer NJ (1981) Structural identification of autoinducer of *Photobacterium fischeri* luciferase. Biochemistry 20:2444–2449

Fetherston JD, Schuetze P, Perry RD (1992) Loss of the pigmentation phenotype in *Yersinia pestis* is due to the spontaneous deletion of 102 kb of chromosomal DNA which is flanked by a repetitive element. Mol Microbiol 6:2693–2704

Fredrickson JK et al (2008) Protein oxidation: key to bacterial desiccation resistance? ISME J 2:393–403

Freitag CS, Abraham JM, Clements JR, Eisenstein BI (1985) Genetic analysis of the phase variation control of expression of type 1 fimbriae in *Escherichia coli*. J Bacteriol 162:668–675

Funa N, Ohnishi Y, Fujii I, Shibuya M, Ebizuka Y, Horinouchi S (1999) A new pathway for polyketide synthesis in microorganisms. Nature 400:897–899

Fuqua WC, Winans SC, Greenberg EP (1994) Quorum sensing in bacteria: the LuxR-LuxI family of cell density – responsive transcriptional regulators. J Bacteriol 176:269–275

Fuqua C, Parsek MR, Greenberg EP (2001) Regulation of gene expression by cell-to-cell communication: acylhomoserine lactone quorum sensing. Annu Rev Genet 35:439–468

Galinski EA, Trüper HG (1982) Betaine, a compatible solute in the extremely halophilic phototrophic bacterium, *Ectothiorhodospira halochloris*. FEMS Microbiol Lett 13:357–361

Galperin MY (2005) A census of membrane-bound and intracellular signal transduction proteins in bacteria: bacterial IQ, extroverts and introverts. BMC Microbiol 5:35

Galperin MY (2006) Structural classification of bacterial response regulators: diversity of output domains and domain combinations. J Bacteriol 188:4169–4182

Galperin MY, Gomelsky M (2005) Bacterial signal transduction modules: from genomics to biology. ASM News 71:326–333

Gans J, Wolinsky M, Dunbar J (2005) Computational improvements reveal great bacterial diversity and high metal toxicity in soil. Science 309:1387–1390

Gao M, Mack TR, Stock AM (2007) Bacterial response regulators: versatile regulatory strategies from common domains. Trends Biochem Sci 32:225–234

Garcia-Pichel F, Lopez-Cortes A, Nubel U (2001) Phylogenetic and morphological diversity of cyanobacteria in soil desert crusts from the Colorado Plateau. Appl Environ Microbiol 67:1902–1910

Gerdes K, Nielsen A, Thorsted P, Wagner EG (1992) Mechanism of killer gene activation. Antisense RNA-dependent RNase III cleavage Ensures rapid turnover of the stable, hok, RSNB effector messenger RNAs and NADP. J Mol Biol 226:637–649

Giovannoni SJ, Sting V (2005) Molecular diversity and ecology of microbial plankton. Nature 437:343–348

Gommeaux M, Barakat M, Lesourd M, Thiery J, Heulin T (2005) A morphological transition in the pleomorphic bacterium *Ramlibacter tataouinensis* TTB310. Res Microbiol 156:1026–1030

Götz C et al (2007) Uptake, degradation and chiral discrimination of N-acyl-D/L-homoserine lactones by barley (*Hordeum vulgare*) and yam bean (*Pachyrhizus erosus*) plants. Anal Bioanal Chem 389:1447–1457

Graumann PL, Marahiel MA (1998) A superfamily of proteins that contain the cold-shock domain. Trends Biochem Sci 23:286–290

Grogan DW (2004) Stability and repair of DNA in hyperthermophilic Archaea. Curr Issues Mol Biol 6:137–144

Halladay JT, Ng WL, DasSarma S (1992) Genetic transformation of a halophilic archaebacterium with a gas vesicle gene cluster restores its ability to float. Gene 119:131–136

Hamamoto T, Hashimoto M, Hino M, Kitada M, Seto Y, Kudo T, Horikoshi K (1994) Characterization of a gene responsible for the Na+/H+antiporter system of alkalophilic *Bacillus* species strain C-125. Mol Microbiol 14:939–946

Han B, Pain A, Johnstone K (1997) Spontaneous duplication of a 661 bp element within a two-component sensor regulator gene causes phenotypic switching in colonies of *Pseudomonas tolaasii*, cause of brown blotch disease of mushrooms. Mol Microbiol 25:211–218

Hayes F (2003) Toxins-antitoxins: plasmid maintenance, programmed cell death, and cell cycle arrest. Science 301:1496–1499

Healy VL, Lessard IA, Roper DI, Knox JR, Walsh CT (2000) Vancomycin resistance in enterococci: reprogramming of the D-ala-D-Ala ligases in bacterial peptidoglycan biosynthesis. Chem Biol 7:R109–R119

Hense BA, Kuttler C, Muller J, Rothballer M, Hartmann A, Kreft JU (2007) Does efficiency sensing unify diffusion and quorum sensing? Nat Rev Microbiol 5:230–239

Hermans MA, Neuss B, Sahm H (1991) Content and composition of hopanoids in *Zymomonas mobilis* under various growth conditions. J Bacteriol 173:5592–5595

Hernday AD, Braaten BA, Low DA (2003) The mechanism by which DNA adenine methylase and PapI activate the pap epigenetic switch. Mol Cell 12:947–957

Heulin T, Barakat M, Christen R, Lesourd M, Sutra L, De Luca G, Achouak W (2003) *Ramlibacter tataouinensis* gen. November, sp. November, and *Ramlibacter henchirensis* sp. November, cyst-producing bacteria isolated from soil in Tunisia subdesert. Int J Syst Evol Microbiol 53:589–594

Hickman JW, Tifrea DF, Harwood CS (2005) A chemosensory system that regulates biofilm formation through modulation of cyclic diguanylate levels. Proc Natl Acad Sci U S A 102:14422–14427

Hopwood DA, Sherman DH (1990) Molecular genetics of polyketides and its comparison to fatty acid biosynthesis. Annu Rev Genet 24:37–66

Howell-Adams B, Seifert HS (2000) Molecular models accounting for the gene conversion reactions mediating gonococcal pilin antigenic variation. Mol Microbiol 37:1146–1158

Hüttenhofer A, Vogel J (2006) Experimental approaches to identify non-coding RNAs. Nucleic Acids Res 34:635–646

Imhoff J (1988) Halophilic phototrophic bacteria. In: Rodriguez-Valera F (ed) Halophilic bacteria. CRC Press, Boca Raton, pp 85–108

Inouye M (2006) Signaling by transmembrane proteins shifts gears. Cell 126:829–831

Joint I, Tait K, Callow ME, Callow JA, Milton D, Williams P, Camara M (2002) Cell-to-cell communication across the prokaryote-eukaryote boundary. Science 298:1207

Juhas M, Eberl L, Tummler B (2005) Quorum sensing: the power of cooperation in the world of *Pseudomonas*. Environ Microbiol 7:459–471

Kaci Y, Heyraud A, Barakat M, Heulin T (2005) Isolation and identification of EPS-producing *Rhizobium* year strain from arid soil (Algeria): characterization of EPS icts and the effect of inoculation on wheat rhizosphere soil structure. Res Microbiol 156:522–531

Karner MB, Delong EF, Karl DM (2001) Archeal dominance in the mesopelagic zone of the Pacific Ocean. Nature 409:507–510

Kazmierczak BI, Lebron MB, Murray TS (2006) Analysis of FimX, a phosphodiesterase that governs twitching motility in *Pseudomonas aeruginosa*. Mol Microbiol 60:1026–1043

Kentner D, Sourjik V (2006) Spatial organization of the bacterial chemotaxis system. Curr Opin Microbiol 9:619–624

Keshavan ND, Chowdhary PK, Haines DC, Gonzalez JE (2005) L-Canavanine made by *Medicago sativa* interferes with quorum sensing in *Sinorhizobium meliloti*. J Bacteriol 187:8427–8436

Khajanchi BK, Kirtley ML, Brackman SM, Chopra AK (2011) Immunomodulatory and protective roles of quorum-sensing signaling molecules N-acyl homoserine lactones during infection of mice with Aeromonas hydrophila. Infect Immun 79:2646–2657

Khan S (1992) Motility. In: Lederberg J (ed) Encyclopedia of microbiology. Academic, New York, pp 193–202

Kharel MK, Subba B, Basnet DB, Woo JS, Lee HC, Liou K, Sohng JK (2004) A gene cluster for biosynthesis of kanamycin from *Streptomyces kanamyceticus*: comparison with gentamicin biosynthetic gene cluster. Arch Biochem Biophys 429:204–214

Khorchid A, Ikura M (2006) Bacterial histidine kinase as signal sensor and transducer. Int J Biochem Cell Biol 38:307–312

Kiem S et al (2004) Phase variation of biofilm formation in *Staphylococcus aureus* by IS 256 insertion and its impact on the capacity adhering to polyurethane surface. J Korean Med Sci 19:779–782

Kirillina O, Fetherston JD, Bobrov AG, Abney J, Perry RD (2004) HmsP, a putative phosphodiesterase, and HmsT, a putative diguanylate cyclase, control Hms-dependent biofilm formation in *Yersinia pestis*. Mol Microbiol 54:75–88

Klemm P (1986) Two regulatory fim genes, fimB and fimE, control the phase variation of type 1 fimbriae in *Escherichia coli*. EMBO J 5:1389–1393

Kohanski MA, Dwyer DJ, Hayete B, Lawrence CA, Collins JJ (2007) A common mechanism of cellular death induced by bactericidal antibiotics. Cell 130:797–810

Kolodkin-Gal I, Hazan R, Gaathon A, Carmeli S, Engelberg-Kulka H (2007) A linear pentapeptide is a quorum-sensing factor required for *mazEF*-mediated cell death in *Escherichia coli*. Science 318:652–655

Konopka AE, Lara JC, Staley JT (1977) Isolation and characterization of gas vesicles from *Microcyclus aquaticus*. Arch Microbiol 112:133–140

Konstantinidis KT, Tiedje JM (2004) Trends between gene content and genome size in prokaryotic species with larger genomes. Proc Natl Acad Sci U S A 101:3160–3165

Krasteva PV, Giglio KM, Sondermann H (2012) Sensing the messenger: The diverse ways that bacteria signal trough c-diGMP. Protein Sci 21:929–948

Krupa A, Srinivasan N (2005) Diversity in domain architectures of Ser/Thr kinases and their homologues in prokaryotes. BMC Genomics 6:129

Kulasakara H et al (2006) Analysis of *Pseudomonas aeruginosa* diguanylate cyclases and phosphodiesterases reveals a role for bis-(3′-5′)-cyclic-GMP in virulence. Proc Natl Acad Sci U S A 103:2839–2844

Lacroix A (1931) On the recent fall (June 27, 1931) of a meteorite asiderite in the extreme south of Tunisia. C R Acad Sci Paris 193:305–309

Lapouge K, Schubert M, Allain FH, Haas D (2008) Gac/Rsm signal transduction pathway of gamma-proteobacteria: from RNA recognition to regulation of social behaviour. Mol Microbiol 67:241–253

Latifi A, Foglino M, Tanaka K, Williams P, Lazdunski A (1996) A hierarchical quorum-sensing cascade in *Pseudomonas aeruginosa* links the transcriptional activators LasR and RhIR (VsmR) to expression of the stationary-phase sigma factor RpoS. Mol Microbiol 21:1137–1146

Laue BE et al (2000) The biocontrol strain *Pseudomonas fluorescens* F113 produces the *Rhizobium* small bacteriocin, N-(3-hydroxy-7-cis-tetradecenoyl) homoserine lactone, via HdtS, a putative novel N-acylhomoserine lactone synthase. Microbiology 146:2469–2480

Le Rudulier D, Bouillard L (1983) Glycine betaine, an osmotic effector in *Klebsiella pneumoniae* and other members of the Enterobacteriaceae. Appl Environ Microbiol 46:152–159

Leadbetter JR, Greenberg EP (2000) Metabolism of acyl-homoserine lactone quorum-sensing signals by *Variovorax paradoxus*. J Bacteriol 182:6921–6926

Lee SY, Bailey SC, Apirion D (1978) Small stable RNAs from *Escherichia coli*: evidence for the existence of new molecules and for a new ribonucleoprotein particle containing 6S RNA. J Bacteriol 133:1015–1023

Leigh JA, Dodsworth JA (2007) Nitrogen regulation in bacteria and archaea. Annu Rev Microbiol 61:349–377

Lewin B (1981) Gene expression. Wiley, New York

Lieph R, Veloso FA, Holmes DS (2006) Thermophiles like hot T. Trends Microbiol 14:423–426

Lithgow JK, Wilkinson A, Hardman A, Rodelas B, Wisniewski-Dyé F, Williams P, Downie JA (2000) The regulatory locus *cinRI* in *Rhizobium leguminosarum* controls a network of quorum-sensing loci. Mol Microbiol 37:81–97

Liu W, Christenson SD, Standage S, Shen B (2002) Biosynthesis of the enediyne antitumor antibiotic C-1027. Science 297:1170–1173

Long PF et al (2002) Engineering specificity of starter unit selection by the erythromycin-producing polyketide synthase. Mol Microbiol 43:1215–1225

Lorenz C, von Pelchrzim F, Schroeder R (2006) Evolution of Genomic systematic ligands by exponential enrichment (Genomic SELEX) for the identification of protein-binding RNAs independent of Their Expression Levels. Nat Protoc 1:2204–2212

Lory S, Wolfgang M, Lee V, Smith R (2004) The multi-talented bacterial adenylate cyclases. Int J Med Microbiol 293:479–482

Luneberg E et al (2001) Chromosomal insertion and excision of a 30 kb unstable genetic element is responsible for phase variation of lipopolysaccharide and other virulence determinants in *Legionella pneumophila*. Mol Microbiol 39:1259–1271

Lyon GJ, Novick RP (2004) Peptide signaling in *Staphylococcus aureus* and other Gram-positive bacteria. Peptides 25:1389–1403

Ma HW, Kumar B, Ditges U, Gunzer F, Buer J, Zeng AP (2004) An extended transcriptional regulatory network of *Escherichia coli* and analysis of its hierarchical structure and network motifs. Nucleic Acids Res 32:6643–6649

Madigan M, Martinko JM, Stahl D, Clark DP (2010) Brock: biology of microorganisms, 13th edn. Pearson Benjamin-Cummings, San Francisco

Manefield M, Rasmussen TB, Henzter M, Andersen JB, Steinberg P, Kjelleberg S, Givskov M (2002) Halogenated furanones inhibit quorum sensing through accelerated LuxR turnover. Microbiology 148:1119–1127

Manickam N, Reddy MK, Saini HS, Shanker R (2008) Isolation of hexachlorocyclohexane-degrading *Sphingomonas* sp. by dehalogenase assay and characterization of genes involved in gamma-HCH degradation. J Appl Microbiol 104:952–960

Martinez-Antonio A, Collado-Vides J (2003) Identifying global regulators in transcriptional regulatory networks in bacteria. Curr Opin Microbiol 6:482–489

Mascher T, Helmann JD, Unden G (2006) Stimulus perception in bacterial signal-transducing histidine kinases. Microbiol Mol Biol Rev 70:910–938

Massé E, Gottesman S (2002) A small RNA regulates the expression of genes involved in iron metabolism in *Escherichia coli*. Proc Natl Acad Sci U S A 99:4620–4625

Mathesius U (2003) Conservation and divergence of signalling pathways between roots and soil microbes – the *Rhizobium*-legume symbiosis compared to the development of lateral roots, mycorrhizal interactions and nematode-induced galls. Plant Soil 255:105–119

Mathesius U, Mulders S, Gao M, Teplitski M, Caetano-Anolles G, Rolfe BG, Bauer WD (2003) Extensive and specific responses of a eukaryote to bacterial quorumsensing signals. Proc Natl Acad Sci U S A 100:1444–1449

Mattick JS (2002) Type IV pili and twitching motility. Annu Rev Microbiol 56:289–314

McBride MJ (2001) Bacterial gliding motility: multiple mechanisms for cell movement over surfaces. Annu Rev Microbiol 55:49–75

McKenney D, Brown KE, Allison DG (1995) Influence of *Pseudomonas aeruginosa* exoproducts on virulence factor production in *Burkholderia cepacia*: evidence of interspecies communication. J Bacteriol 177:6989–6992

Miller AA (1961) Climatology. Methuen, London

Mills JA, Venkatesan MM, Baron LS, Buysse JM (1992) Spontaneous insertion of an IS1-like element into the virF gene is responsible for avirulence in opaque colonial variants of *Shigella flexneri* 2a. Infect Immun 60:175–182

Minton KW (1994) DNA repair in the extremely radioresistant bacterium *Deinococcus radiodurans*. Mol Microbiol 13:9–15

Monchy S et al (2006) Transcriptomic and proteomic analyses of the pMOL30-encoded copper resistance in *Cupriavidus metallidurans* strain CH34. Microbiology 152:1765–1776

Morita T, Maki K, Aiba H (2005) RNase E-based ribonucleoprotein complexes: mechanical mRNA destabilization basis of noncoding RNAs mediated by bacterial. Genes Dev 19:2176–2186

Morris CE et al (2008) The life history of the plant pathogen *Pseudomonas syringae* is linked to the water cycle. ISME J 2:321–334

Mouné S, Manac'h N, Hirschler A, Caumette P, Willison JC, Matheron R (1999) *Haloanaerobacter salinarius* sp. nov., a novel halophilic fermentative bacterium that reduces glycine-betaine to trimethylamine with hydrogen or serine as electron donors; emendation of the genus Haloanaerobacter. Int J Syst Bacteriol 49:103–112

Muller D et al (2006) *Herminiimonas arsenicoxydans* sp. nov., a metalloresistant bacterium. Int J Syst Evol Microbiol 56:1765–1769

Nicholson B, Low D (2000) DNA methylation-dependent regulation of pef expression in *Salmonella typhimurium*. Mol Microbiol 35:728–742

Normand P et al (2007) Genome characteristics of facultatively symbiotic *Frankia* sp. strains reflect host range and host plant biogeography. Genome Res 17:7–15

Ohmori M, Okamoto S (2004) Photoresponsive cAMP signal transduction in cyanobacteria. Photochem Photobiol Sci 3:503–511

Parkinson JS, Ames P, Studdert CA (2005) Collaborative signalling by bacterial chemoreceptors. Curr Opin Microbiol 8:116–121

Pierson EA, Wood DW, Cannon JG, Blachere FM, Pierson LS (1998) Interpopulation signaling via N-acylhomoserine lactones among bacteria in the wheat rhizosphere. Mol Plant Microbe Interact 11:1078–1084

Poussier S, Thoquet P, Trigalet-Demery D, Barthet S, Meyer D, Arlat M, Trigalet A (2003) Host plant-dependent phenotypic reversion of *Ralstonia solanacearum* from non-pathogenic to pathogenic forms via alterations in the phcA gene. Mol Microbiol 49:991–1003

Rainey FA et al (2005) Extensive diversity of ionizing-radiation-resistant bacteria recovered from Sonoran Desert soil and description of nine new species of the genus *Deinococcus* obtained from a single soil sample. Appl Environ Microbiol 71:5225–5235

Ranjard L, Lignier L, Chaussod R (2006) Cumulative effects of short-term polymetal contamination on soil bacterial community structure. Appl Environ Microbiol 72:1684–1687

Reeve JN, Bailey KA, Li WT, Marc F, Sandman K, Soares DJ (2004) Archaeal histones: structures, stability and DNA binding. Biochem Soc Trans 32:227–230

Ryjenkov DA, Tarutina M, Moskvin OV, Gomelsky M (2005) Cyclic diguanylate is a ubiquitous signaling molecule in bacteria: insights into biochemistry of the GGDEF protein domain. J Bacteriol 187:1792–1798

Santos CL, Tavares F, Thioulouse J, Normand P (2009) A phylogenomic analysis of bacterial helix-turn-helix transcription factors. FEMS Microbiol Rev 33:411–429

Schenk ST, Stein E, Kogel KH, Schikora A (2012) Arabidopsis growth and defense are modulated by bacterial quorum sensing molecules. Plant Sig Beh 7:178–181

Schloss PD, Handelsman J (2006) Toward a census of bacteria in soil. PLoS Comput Biol 2:e92

Schuhegger R et al (2006) Induction of systemic resistance in tomato by N-acyl-L-homoserine lactone-producing rhizosphere bacteria. Plant Cell Environ 29:909–918

Schuster M, Lostroh CP, Ogi T, Greenberg EP (2003) Identification, timing, and signal specificity of *Pseudomonas aeruginosa* quorum-controlled genes: a transcriptome analysis. J Bacteriol 185:2066–2079

Sharif DI, Gallon J, Smith CJ, Dudley E (2008) Quorum sensing in Cyanobacteria: N-octanoyl-homoserine lactone release and response, by the epilithic colonial cyanobacterium *Gloeothece* PCC6909. ISME J 2:1171–1182

Shen B (2003) Polyketide biosynthesis beyond the type I, II and III polyketide synthase paradigms. Curr Opin Chem Biol 7:285–295

Simm R, Morr M, Kader A, Nimtz M, Romling U (2004) GGDEF and EAL domains inversely regulate cyclic 360 di-GMP levels and transition from sessility to motility. Mol Microbiol 53:1123–1134

Singer GA, Hickey DA (2003) Thermophilic prokaryotes have characteristic patterns of codon usage, amino acid composition and nucleotide content. Gene 317:39–47

Sio CF et al (2006) Quorum quenching by an N-acylhomoserine lactone acylase from *Pseudomonas aeruginosa* PAO1. Infect Immun 74:1673–1682

Smith RS, Harris SG, Phipps R, Iglewski B (2002) The *Pseudomonas aeruginosa* quorum-sensing molecule N-(3-oxododecanoyl) homoserine lactone contributes to virulence and induces inflammation *in vivo*. J Bacteriol 184:1132–1139

Stack D, Neville C, Doyle S (2007) Nonribosomal peptide synthesis in *Aspergillus fumigatus* and other fungi. Microbiology 153:1297–1306

Stock AM, Robinson VL, Goudreau PN (2000) Two-component signal transduction. Annu Rev Biochem 69:183–215

Suar M, van der Meer JR, Lawlor K, Holliger C, Lal R (2004) Dynamics of multiple lin gene expression in *Sphingomonas paucimobilis* B90A in response to different hexachlorocyclohexane isomers. Appl Environ Microbiol 70:6650–6656

Subramoni S, Gonzalez JF, Johnson A, Péchy-Tarr M, Rochat L, Paulsen I, Loper JE, Keel C, Venturi V (2011) Bacterial subfamily of LuxR regulators that respond to plant compounds. Appl Environ Microbiol 77:4579–4588

Suzuki K, Babitzke P, Kushner SR, Romeo T (2006) Identification of a novel regulatory protein (CSRD) That the targets Global Regulatory RNAs CsrB and CSRC for degradation by RNase E. Genes Dev 20:2605–2617

Tamayo R, Pratt JT, Camilli A (2007) Roles of cyclic diguanylate in the regulation of bacterial pathogenesis. Annu Rev Microbiol 61:131–148

Tekaia F, Yeramian E, Dujon B (2002) Amino acid composition of genomes, lifestyles of organisms, and evolutionary trends: a global picture with correspondence analysis. Gene 297:51–60

Teplitski M et al (2004) *Chlamydomonas reinhardtii* secretes compounds that mimic bacterial signals and interfere with quorum sensing regulation in bacteria. Plant Physiol 134:137–146

Tischler AD, Camilli A (2005) Cyclic diguanylate regulates *Vibrio cholerae* virulence gene expression. Infect Immun 73:5873–5882

Torsvik V, Ovreas L, Thingstad TF (2002) Prokaryotic diversity-magnitude, dynamics, and controlling factors. Science 296:1064–1066

Turick CE, Knox AS, Leverette CL, Kritzas YG (2008) In situ uranium stabilization by microbial metabolites. J Environ Radioactiv 99:890–899

Uroz S, Dessaux Y, Oger P (2009) Quorum sensing and quorum quenching: the yin and yang of bacterial communication. Chembiochem 10:205–216

van den Broek D, Chin AWTF, Bloemberg GV, Lugtenberg BJ (2005) Molecular nature of spontaneous modifications in gacS which cause colony phase variation in *Pseudomonas* sp. strain PCL1171. J Bacteriol 187:593–600

van der Woude MW, Baumler AJ (2004) Phase and antigenic variation in bacteria. Clin Microbiol Rev 17:581–611

van Ham SM, van Alphen L, Mooi FR, van Putten JP (1993) Phase variation of *H. influenzae* fimbriae: transcriptional control of two divergent genes through a variable combined promoter region. Cell 73:1187–1196

Van Houdt R, Aertsen A, Moons P, Vanoirbeek K, Michiels CW (2006) N-acyl-L-homoserine lactone signal interception by *Escherichia coli*. FEMS Microbiol Lett 256:83–89

van Keulen G, Hopwood DA, Dijkhuizen L, Sawers RG (2005) Gas vesicles in actinomycetes: old buoys in novel habitats? Trends Microbiol 13:350–354

Vial L et al (2006) Phase variation and genomic architecture changes in *Azospirillum*. J Bacteriol 188:5364–5373

Vial L, Lepine F, Milot S, Groleau MC, Dekimpe V, Woods DE, Deziel E (2008) *Burkholderia pseudomallei*, *B. thailandensis*, and *B. ambifaria* produce 4-hydroxy- 2-alkylquinoline analogues with a methyl group at the 3 position that is required for quorum-sensing regulation. J Bacteriol 190:5339–5352

Vogel J, Bartels V, Tang TH, Churakov G, Slagter-Jager JG, Huttenhofer A, Wagner EG (2003) RNomics in Escherichia coli detects new sRNA species and data and identify parallel transcriptional output in bacteria. Nucleic Acids Res 31:6435–6443

von Rad U et al (2008) Response of *Arabidopsis thaliana* to N-hexanoyl-DL-homoserine-lactone, a bacterial quorum sensing molecule produced in the rhizosphere. Planta 229:73–85

Wadhams GH, Armitage JP (2004) Making sense of it all: bacterial chemotaxis. Nat Rev Mol Cell Biol 5:1024–1037

Waite RD, Struthers JK, Dowson CG (2001) Spontaneous sequence duplication within an open reading frame of the pneumococcal type 3 capsule locus causes high-frequency phase variation. Mol Microbiol 42:1223–1232

Walsby AE (1978) The gas vesicles of aquatic prokaryotes. In: Relations between structure and function in the prokaryotic cells. Symposium of Society for General Microbiology, vol. 28. Cambridge/New York: Published for the Society for General Microbiology, Cambridge University Press, pp 327–358

Walsby AE (1994) Gas vesicles. Microbiol Rev 58:94–144

Wassarman KM, Repoila F, Rosenow C, Storz G, Gottesman S (2001) Identification of novel small RNAs using comparative genomics and microarrays. Genes Dev 15:1637–1651

Webre DJ, Wolanin PM, Stock JB (2003) Bacterial chemotaxis. Curr Biol 13:1247–1249

Welsh DT, Bourgues S, Herbert RA, De Wit R (1996) Seasonal variation in nitrogen fixation (acetylene reduction) and sulphate reduction rates in the rhizosphere of *Zostera noltii*: nitrogen fixation by sulphate reducing bacteria. Mar Biol 125:619–628

Williams P, Winzer K, Chan WC, Camara M (2007) Look who's talking: communication and quorum sensing in the bacterial world. Philos Trans R Soc Lond B Biol Sci 362:1119–1134

Wisniewski-Dyé F, Downie JA (2002) Quorum-sensing in *Rhizobium*. Antonie Van Leeuwenhoek 81:397–407

Wisniewski-Dyé F, Vial L (2008) Phase and antigenic variation mediated by genome modifications. Antonie Van Leeuwenhoek 94:493–515

Wu M et al (2004) Phylogenomics of the reproductive parasite *Wolbachia pipientis* wMel: a streamlined genome overrun by mobile genetic elements. PLoS Biol 2:e69

Xavier KB, Bassler BL (2003) LuxS quorum sensing: more than just a numbers game. Curr Opin Microbiol 6:191–197

Xiang S, Yao T, An L, Xu B, Wang J (2005) 16S rRNA sequences and differences in bacteria isolated from the Muztag Ata glacier at increasing depths. Appl Environ Microbiol 71:4619–4627

Yates EA et al (2002) *N*-acylhomoserine lactones undergo lactonolysis in a pH-, temperature-, and acyl chain length-dependent manner during growth of *Yersinia pseudotuberculosis* and *Pseudomonas aeruginosa*. Infect Immun 70:5635–5646

Zasloff M (2002) Antimicrobial peptides in health and disease. N Engl J Med 347:1199–1200

Zerhari K, Aurag J, Khbaya B, Kharchaf D, Filali-Maltouf A (2000) Phenotypic characteristics of rhizobia isolates nodulating acacia species in the arid and Saharan regions of Morocco. Lett Appl Microbiol 30:351–357

The Extreme Conditions of Life on the Planet and Exobiology

10

Jean-Luc Cayol, Bernard Ollivier, Didier Alazard, Ricardo Amils, Anne Godfroy, Florence Piette, and Daniel Prieur

Abstract

Extreme physicochemical conditions (low and high temperatures, high salinity, low and high pH, high hydrostatic pressure, etc.) existing on Earth are compatible with the occurrence of microbial life. The diversity and metabolic features of microbial trophic groups inhabiting extreme environments (cold, hot, saline, acidic, alkaline, and deep marine) are described. They include hydrothermal vents, acid springs, hypersaline and/or alkaline lakes, permafrost, and deep-sea environments, etc.

To live or survive under such drastic conditions, prokaryotes (*Bacteria* or *Archaea*) have developed a variety of physiological and metabolic strategies allowing them to adapt to in situ extreme conditions. Many of these extremophiles are recognized to be of industrial interest or to be potential candidates for future biotechnological applications.

Clearly, the discovery of extremophiles living in terrestrial, subterrestrial, and deep marine environments has changed our perception of microbial life. One or a combination of extreme physicochemical conditions that they have to face may have prevailed in the primitive atmosphere and favored early extremophilic life not only on Earth but perhaps also on other planets.

Keywords

Acidophiles • Alkaliphiles • Exobiology • Extremophiles • Halophiles • Piezophiles • Psychrophiles • Thermophiles/hyperthermophiles

* Chapter Coordinators

J.-L. Cayol* (✉) • B. Ollivier* (✉) • D. Alazard
Institut Méditerranéen d'Océanologie (MIO), UM 110, CNRS 7294, IRD 235, Université de Toulon, Aix-Marseille Université, Campus de Luminy, 13288 Marseille Cedex 9, France
e-mail: jean-luc.cayol@univ-amu.fr; bernard.ollivier@univ-amu.fr

R. Amils
Centro de Biología Molecular Severo Ochoa (CSIC-UAM), Universidad Autónoma de Madrid, Cantoblanco, 28049 Madrid, Spain

Centro de Astrobiología (CSIC-INTA), 28850 Torrejón de Ardoz, Madrid, Spain

A. Godfroy • D. Prieur
Laboratoire de Microbiologie des Environnements Extrêmes, UMR 6197 (UMR CNRS/IFREMER/Université de Bretagne Occidentale), BP70 29280 Plouzané, France

F. Piette
Laboratoire de Biochimie, Université de Liège, 4000 Liège, Belgique

10.1 Introduction

To our knowledge, Earth is the only inhabitable and inhabited planet in the solar system and even the universe. All its inhabitants, from the simplest to the most complex, from the smallest to the biggest are under the control of a variety of physicochemical parameters within their environments. According to time and space, the reached values by these parameters including temperature, salinity, pH, or pressure vary considerably. Combinations of minimal and maximal values reached by these parameters within a particular geographical area define biotopes, or the whole set of environmental conditions suitable for a given species.

For most of the major parameters listed above, specific limits beyond which no organism can totally accomplish its life cycle have been determined. Any environment wherein one or several parameters show permanently values close to these lower or upper limits for life is named an extreme environment: pH lower than 3, temperature above 80 °C, hydrostatic pressure above 20 MPa, etc. Deep-sea hydrothermal vents; cold deep-sea; acidic pools; salt marshes; and permafrost are examples of extreme environments.

Regarding organisms that thrive in such environments, one can distinguish several physiological categories.

When certain organisms are exposed to a given parameter with value incompatible with their life cycle requirements, they can face this aggression. For instance, some bacteria form spores, which are resistant to heat, and can remain under this almost inert stage for very long periods (a lifetime of 250 million years have been reported) (Vreeland et al. 2000).

Other organisms have developed diverse and complex cellular and molecular mechanisms that allow them to resist, for instance, to elevated concentrations of toxic heavy metals or to repair their DNA damaged by ionizing radiations. For both categories, extreme conditions are not required for the accomplishment of the life cycle. But some organisms have "learned" during evolution how to protect themselves or to repair damages.

A third category of microorganisms can accomplish their life cycles only if the environmental conditions are extreme. This is the case of a hyperthermophile (*cf.* Sect. 10.3) which cannot grow at temperature below 65 °C, a temperature which is lethal for psychrophiles (*cf.* Sect. 10.2). A piezophilic microorganism (*cf.* Sect. 10.5) will stop its growth when exposed to atmospheric pressure. All of them are called extremophiles. All these organisms really deserve the name of extremophile. Despite some of them belonging to the *Eukarya* domain, mostly belong to the *Bacteria* and *Archaea* domains. In all cases, their study is in relation with microbial ecology, and this is why they have been highlighted in this book.

10.2 Psychrophilic Bacteria

10.2.1 Presentation

Psychrophilic bacteria are microorganisms adapted to cold (literally "cold loving" *cf.* Sect. 9.6.1). They represent a considerable biomass since cold environments are the most important part on Earth: indeed they comprise not only polar regions and Arctic and Antarctic regions (Fig. 10.1) but also high mountain areas, glaciers, permafrost zones, and oceans below 1,000 m where temperature is below 5 °C regardless of the latitude. The temperature range in which these organisms grow is very wide: it varies in the free water of oceans from −1.9 °C, corresponding to an average freezing temperature of seawater, to 5 °C; temperatures for which bacterial activity were detected extend to −20 °C and perhaps even less in sea ice. Cell concentrations can rise to 10^5 bacteria mL^{-1}. Temperature limits for bacterial life

Fig. 10.1 A typical cold Antarctic ecosystem (Photography: courtesy from Jean-Claude Marx)

which still seems possible within the permafrost are probably even lower than this value because it has been shown that even at temperatures as low as 170 °K (−60 °C), liquid water could still exist under saline solutions. This also brings us the temperatures encountered on Mars (an average of 225 °K or −5 °C) and makes plausible the hypothesis that anaerobic microorganisms were able to survive until now on this planet in ice (Price 2007). The importance of a detailed study of these organisms is therefore significant not only because of their abundance but also because they form the basis of a plausible alternative theory which proposes that the first representative of life on Earth was a psychrophilic microorganism (Last Universal Common Ancestor -LUCA) although a mesophilic or thermophilic origin of life would be the most commonly accepted (*cf.* Chap. 4).

10.2.2 Terminology

In the scientific literature regarding cold-adapted organisms, various terms constituting many attempts to classify these particular microorganisms in subdivisions are encountered. Oddly these subdivisions are defined in terms of their apparent optimum temperature for growth and maximum temperature for cell division. Thus, according to the most common classification (D'Amico et al. 2006), psychrophilic

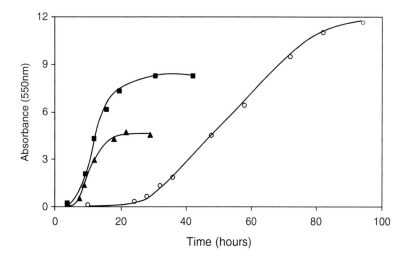

Fig. 10.2 Growth of *Pseudoalteromonas haloplanktis* at 4 °C (*circles*), 18 °C (*squares*), and 25 °C (*triangles*) (A_{550} absorbance at 550 nm)

microorganisms have an optimum growth ≤ 15 °C and a maximum temperature for growth ≤ 20 °C. They would differ from psychrotolerant microorganisms with optimum and maximum temperatures being ≤ 20 °C and ≤ 35 °C, respectively. The psychrotrophic term also occurs and corresponds more or less to the psychrotolerant term; these terms are unnecessary and can be confusing. Indeed culture experiments indicate for these microorganisms that:

1. There is a continuum in the adaptation to cold.
2. Any classification is restrictive.
3. The psychrophilic term is the only one correct as it unambiguously indicates that the bacterium is adapted to cold.
4. The psychrotolerant term is based on the misconception that bacteria grow particularly well in what is called the optimal growth temperature obtained from the growth rate ($\times h^{-1}$) or doubling time ($\times h$) of the microorganism measured according to the temperature of cultivation. In fact, this temperature is a critical temperature resulting from the crossing of two curves: one is the exponential increase in the growth rate as a function of temperature kinetic effect corresponding to the Arrhenius law and the second decreasing corresponds to the deleterious effect of temperature on cellular components such as proteins (*cf.* Sect. 9.6.1). The real optimum temperature for growth is often very different from those defined above; it is probably close to the temperature of the environment for endemic species. This can easily be demonstrated by considering the cell density obtained at the end of the exponential growth phase. It is observed in *Pseudoalteromonas haloplanktis* (Fig. 10.2) that at the apparent optimum temperature, the cell density is much lower than that obtained at lower temperatures. The cells are then in a state of metabolic distress. The psychrotolerant term is particularly inappropriate for many species belonging to this category which have doubling times shorter than psychrophilic species originating from the same environment. It is therefore better to use this term to mesophilic species which can survive in cold environments. The psychrotrophic term, introduced in 1960, is often used in the food industry to identify microorganisms capable of degrading frozen foods; it is not more appropriate because of its inadequacy with any bacterial characteristic. The stenopsychrophilic (psychrophilic) and eurypsychrophilic (psychrotolerant) terms have also been proposed to designate organisms, respectively, capable of dividing in narrow or wide areas of low temperatures, respectively (D'Amico et al. 2006). In this chapter, the psychrophilic term will be the only one used to refer to any psychrophilic bacterium living permanently in low-temperature environments and therefore ecologically well adapted to this type of habitat.

10.2.3 Environnements and Biodiversity

The inventory will be limited to surface waters and deep waters, sea ices, permafrost areas, snow, and glaciers as well as subglacial lakes (Mikucki et al. 2011).

Evaluation of bacterial populations remains a difficult problem due to the fact that only 1–10 % of the microbial species are considered as cultivable (Nogi 2011). Independent cultures techniques have been developed. They are based either on the determination of the gene sequence of the small subunit ribosomal 16S RNA, either on in situ hybridization with appropriate probes (*cf.* Sect. 17.3.5). Even more recent, metagenomic techniques have opened up enormous opportunities in this area as they consider to determine from total DNA recovered from original samples and cloned in suitable vectors, the sequence of all genomes present (*cf.* Chap. 18). The magnitude of the task is enormous, since a thousand of bacterial genomes may be present in the natural sample collected bringing the number of pair basis to about $3-5.10^9$. Large sequencing centers have already harnessed to this type of business and the results are promising.

10.2.3.1 Arctic and Antarctic Oceans

In these environments, below 60° south latitude, the temperature of the water varies between −1.9 and 5 °C. The selection pressure is high due to the relative stability of these temperatures as compared to the same latitude terrestrial environments. In the Arctic Ocean, below the permanent ice, densities of 10^2 mL^{-1} of heterotrophic bacteria have been reported. Populations do not appear very different when comparing the Arctic to the Antarctic; so at the two poles, *Alpha-* and *Gammaproteobacteria* and *Bacteroidetes* dominate with 97 % of the phylotypes identified being common to Arctic and Antarctic. In Antarctica a large fluctuation of marine biomass depending on the season, which reflects changes in temperature and light, is observed. Large populations of *Flavobacteria* whose abundance is connected with chlorophyll and nutrients are found on surface. *Archaea* are also abundant and represent about 30 % of the biomass of prokaryotes. The most common types of bacteria found in Arctic and Antarctic seawaters are *Alteromonas Colwellia, Glaciecola, Pseudoalteromonas, Shewanella,* and *Polaribacter*.

10.2.3.2 Deep Waters

Sixty percent of the Earth's surface is covered by seas deeper than 1,000 m. The hydrostatic pressure of these environments is more than 110 MPa and temperatures often below 4 °C (except in hydrothermal vents where temperature can reach 370 °C). Microorganisms capable of withstanding these pressures and temperatures are called psychropiezophiles. The first bacterium of this type was discovered in 1979. It is a *Spirillum*-like bacterium (strain CNPT-3) that grows well at 50 MPa and is unable to grow at atmospheric pressure. Numerous psychropiezophilic bacteria have been isolated and characterized. It turns out that they all belong to the group of *Gammaproteobacteria* according to the classification based on the sequences of 16S and 5S rRNA genes. These bacteria are divided into five predominant genera: *Photobacterium, Colwellia, Moritella, Shewanella,* and *Psychromonas*. Regardless of the type to which these bacteria belong, their membranes have a high rate of unsaturated fatty acids to maintain fluidity at low temperature and high pressure as well. Depending on genus to which bacteria belong, they also produce long-chain polyunsaturated fatty acids including eicosapentaenoic acid and/or docosahexaenoic acid (*cf.* Sect. 17.7.8). The proportion of these polyunsaturated fatty acids in the membrane varies between 50 and 70 %. However, production of these fatty acids does not seem to be required for growth of psychropiezophiles. As an example, *Psychromonas profunda* strain SS9 only produces monounsaturated fatty acids.

10.2.3.3 Sea Ice

Sea ice represents a particular habitat (Fig. 10.1) because of temperature possibly dropping to −35 °C and its semisolid state. Its formation causes the expulsion of the dissolved salts of the solid matrix but also the formation of a labyrinth with liquid veins varying with temperature. Salinity can reach a value of 14.5 % at −10 °C compared to a value of 3.4 % observed at −1.9 °C. A very large concentration of dissolved organic matter (DOM) is recorded, which transforms these veins in habitats where many microorganisms develop. The composition of these veins is not constant, resulting in a desalination process increasing with the age of ice. Temperature gradients also exist within it, from the surface to the ice-water interface, which makes this habitat a particularly heterogeneous environment. The total areas of these liquid streams are of the magnitude order of 0.6–4.0 m^2 kg^{-1} of ice. Heterotrophic bacteria are the main groups of prokaryotes detected. They are psychrotrophic but also halotolerant; cyanobacteria are also present. Unlike free waters, archaea are poorly represented with about 0–3 % of total cells. The presence of exopolysaccharides produced by microorganisms may act as cryoprotectants but also as nutrient traps (see below) thus favoring the colonization of habitats. More than 100 strains belonging to five phylogenetic groups: *Alpha-* and *Gammaproteobacteria, Bacillus-Clostridium* group, *Actinobacteria,* and *Bacteroidetes* were highlighted in the sea ice of the Spitzberg; *Gammaproteobacteria* dominate the ecosystem. All strains are psychrophilic (broadly defined) with a maximum production of extracellular enzymes between 4 and 10 °C. Active bacteria up to −20 °C have been identified in the sea ice of the Arctic. In Antarctica, 100 distinct phylotypes were identified with a distribution similar to that encountered in the Arctic ice. They include *Alpha-* and *Gammaproteobacteria, Bacteroidetes,* Gram-positive bacteria, and members of the orders *Chlamydiales* and *Verrucomicrobiales*.

10.2.3.4 Snow and Glaciers

Snow and glaciers are interrelated ecosystems as ice-glacier formed from snow by natural compression. The majority of glaciers is found in Greenland and Antarctica and contain about 80 % of the freshwater on the planet; their thickness varies from a few tens of meters to 4 km. Although the low salinity of freshwater is not conducive to the formation of liquid jets, they do exist and their diameters vary from 1 μm at about −50 °C to 10 μm at 2 °C (Price 2007). Other liquid films form on the surface of insoluble mineral grains and generate liquid inclusions hosting an abundant bacterial life.

The number of cells is on average between 10^2 and 10^4 mL^{-1} ice, but in the basal region in contact with the ground highly charged in minerals, cell densities can reach 10^9 mL^{-1}. It was also demonstrated that the supply of organic carbon in liquid veins would maintain a population of 10–100 cells.mL^{-1} for 400,000 years with possible metabolic flux without apparent limit temperature (Price 2007). Conservation of species would be therefore possible in a very old ice. In this respect, surviving species were isolated from a Tibetan glacier from samples of 750,000 years old ice. Analysis of populations is very difficult due to the low average density (10^2–10^4 cells mL^{-1}), which indeed complicates the DNA extraction. In addition, cultivation methods are performed in poor culture media with incubation time of several months at low temperatures. It appears that due to the narrowness of the liquid jets and the change in diameter of them as a function of temperature, a selective distribution of microorganisms is observed depending on the cell size. Cells whose size is greater than 2 μm are fixed in ice crystals and those whose size is less than 1 μm are free in the liquid streams. The dominant species belong to *Proteobacteria* followed by *Bacteroidetes*. *Thermus*, *Bacteroides*, *Eubacterium*, and *Clostridium* genera have also been highlighted. Gram-positive species dominate in this type ecosystems and this is certainly related to the ability of some of these bacteria to form spores.

10.2.3.5 Permafrost

Permafrost areas occupy more than 20 % of the land surface and are defined as soils whose temperature remains at or below 0 °C for at least two consecutive years. Despite regarded as a reservoir of ancient microbial life, recent experiments have provided evidence of an active microbial life at temperatures between 0 and −20 °C in these environments.

At this place, the ice-water interfaces are of primary importance and the "eutectophiles" term was proposed to describe psychrophilic microorganisms existing at these critical interfaces (D'Amico et al. 2006). Oxidative activities of organic compounds and $^{14}CO_2$ fixation have been demonstrated in Alaska within permafrost areas with temperatures up to −40 °C. Special techniques to enrich microorganisms in solid phase from permafrost samples have been described; they mainly use microcrystalline cellulose, mineral-based nutrients, and ethanol as substrates.

Measurements of CO_2 production to estimate the growth at different temperatures were made.

Enrichments in the liquid phase conducted to the identification of bacteria at temperatures ranging from −1 °C (*Polaromonas*) to −17 °C (*Pseudomonas* and *Arthrobacter*) with close relationships to microorganisms isolated from glaciers or polar sea ice. The solid-phase enrichment enabled the preferential isolation of eukaryotes such as unicellular yeasts and mycelial ascomycetes. To emphasize, once again, the inconsistency of the distinction between psychrophilic and psychrotolerant ascomycetes strains still show growth at −17 °C! To fix ideas, over temperatures between −18 and −20 °C, microorganisms retain their characteristics; they develop exponentially, consume ethanol, and convert it into CO_2 by half while the other half is used for biosynthesis. The doubling time of less efficient microorganisms is about 14–35 days at −8 °C. Below −20 °C, stress is greater with a transitory activation phase that probably prepares microorganisms for prolonged survival (dormancy), thanks to the biosynthesis of intra- and extracellular components. Bacterial population densities of 10^2–10^8 viable cells per gram of 1–3-million-year-old permafrost have been observed in the Arctic and Antarctic, and the relationship between aerobic and anaerobic bacteria varies with the geological history of soils. Studies performed in Siberian samples with a temperature of −10 °C showed that the proportion of unfrozen water was ± 3 % and that permafrost contained 10^8 cells g^{-1}. The incorporation of lipid ^{14}C-acetate was used to measure doubling times at different temperatures. They were of 8 days at −10 °C and 200 days at −20 °C. This suggests that functional ecosystems exist in the permafrost and that melting might cause a very important greenhouse effect since at the planetary level, methane represents 30 % of the total carbon content in soils. Moreover, these ecosystems are of primary interest for scientists because of their similarity with glacial environments that have been discovered on Mars.

Regarding the existing biodiversity, recent studies showed that within permafrost samples of the Canadian Arctic (80 °N), collected at 9 m depth and at temperature about −15 °C, the majority of microorganisms were psychrotolerant according to the definition of Morita (1975); 50 % of them were halotolerant. Cultivated bacteria belonged to the phyla *Firmicutes*, *Actinobacteria*, and *Proteobacteria*, this latter phylum being less represented in Siberian soils. A total of 42 bacterial and 11 archaeal phylotypes were identified.

10.2.3.6 Antarctic Subglacial Lakes

In the 1950s, geothermal flow modeling has predicted that beyond a certain thickness of ice, the interface between the ice and the underlying rock could be in a liquid state. In 1973, radio soundings demonstrated the presence of 17 subglacial lakes in eastern Antarctic and nearly 150 of these lakes in Antarctica were detected later on using different techniques. The most famous of these lakes is Lake Vostok which name originates from the Russian Antarctic station below which this lake was discovered. Its surface is 14,000 km^2 with depth ranging from 500 m north to 800 m south. The thickness of ice above is about 4 km and because of the difference in ice thickness between north and south, ± 300 m, the northern part is warmer (+0.3 °C). This creates a flow of water vapor from north to south which refreezes at the basal ice of south, forming what is called ice accretion between −3,540 and −3,743 m. The formation of this lake dated from the Miocene period with age of the water being estimated between 1 and 15 million years. The drilling of ice lake was undertaken in 1989 by an international team of researchers from Russia, United Kingdom, France, and

Belgium and stopped in 1998 at a depth of 3,623 m within ice accretion, at about 120 m from the surface of the water. In 2006, additional drilling was performed by a Russian team up to 3,651 m. Ice samples collected at different depths between 3,540 and 3,623 m in this lake have predicted that the organic carbon concentrations would range between 86 and 160 μM, cell densities between 150 and 460 cells.mL^{-1}, and solutes between 1.5 and 34 mM. The concentration of dissolved O_2 was estimated to be 50 times higher than at the equilibrium water/air surface due to the pressure and the high concentration of gas present in ice in the form of clathrate (complex gas hydrates). Samples from melted ice accretion were analyzed and the microorganisms were identified after isolation by sequencing of their 16S rRNA gene. Representatives of *Proteobacteria* (subdivisions alpha, beta, gamma, and epsilon), *Firmicutes*, *Actinobacteria*, and *Bacteroidetes* were recovered from these samples. However, to confirm the indigenous nature of the microorganisms detected, further investigations of the underlying water and direct sampling will need the implementation of a considerable logistics because of the risk of external contamination of the huge groundwater covering the Antarctic by a layer of water 1 m thick. In addition, it appears that these habitats which have been regarded for long time as closed are in fact connected by extensive networks of underground channels that could be sources of contaminations. The rapid transfer of water masses from one place to another by underground roads would lead to a substantial redistribution of solutes and microorganisms.

10.2.4 Molecular Adaptations

At low temperatures, bacteria have to face several constraints such as (1) a decrease in membrane fluidity, (2) a decrease in the rate of enzymatic reactions, (3) the stabilization of secondary structures of nucleic acids resulting in a decrease in the efficiency of transcription and translation, and (4) the slowing down of the process of protein folding. In response to these constraints, microorganisms may adapt as described below (*cf*. Sect. 9.6.1).

10.2.4.1 Membranes

Low temperatures induce a decrease in membrane fluidity. They favor synthesis of unsaturated fatty acids, polyunsaturated fatty acids, and branched fatty acids. These changes aim to introduce a steric constraint and reduce the number of interactions within the membrane, thus increasing its fluidity (D'Amico et al. 2006). A decrease in the length of the hydrocarbon chains is also observed.

For anaerobic bacteria, the fatty acid synthase is able to synthesize saturated and unsaturated fatty acids. This is due to an elongation system resulting in an intermediate that can be dehydrated to synthesize unsaturated fatty acids. In aerobic bacteria, the fatty acid synthase does not produce unsaturated fatty acids. They therefore have an additional enzyme: a desaturase able to desaturate fatty acids. This enzyme, with other proteins and cofactors, form complexes in the membranes and act as a small respiratory chain by transferring electrons from the fatty acids to an electron acceptor (e.g., dioxygen). When the temperature increases, there is an induction of a new enzyme (desaturase) in aerobic bacteria, whereas in anaerobic the existing enzymes switch their function.

There are two types of branched fatty acids, the iso-branched and the anteiso branches ones, resulting from the use of a derivative of leucine and isoleucine as the first molecule in the synthesis of the fatty acid. Temperature controls acyl-CoA: transacylase responsible for the conversion of amino acid derivatives (keto acids) in ACP (acyl carrier protein)-thioester. These molecules are then used by the fatty acid synthase to produce branched fatty acids. Some bacteria show a higher number of short-chain fatty acids. This is the case for *Micrococcus cryophilus*, a psychrophilic bacterium which has a ratio of oleic acid/palmitoleic acid four times higher at 25 °C than at 4 °C. The transition from oleic acid (18 carbons) to palmitoleic acid (16 carbons) and vice versa is catalyzed by an elongase bonded to the membrane. In other bacteria, there is synthesis of new short-chain fatty acids using a fatty acid synthase.

10.2.4.2 Enzymes

Most enzymatic reactions obey the Arrhenius law, a law which reflects the variation of reaction rate as a function of temperature: $k = A \exp^{(-Ea/RT)}$ where k is the rate constant, A is a pre-exponential factor, Ea is the activation energy, R is the perfect gas constant, and T the absolute temperature in Kelvin. According to this law, a decrease in temperature induces an exponential decrease of reaction rate. Enzymes of psychrophilic microorganisms have adapted to maintain an appropriate reaction rate at low temperatures. Many psychrophilic enzymes have been studied and the three-dimensional structure of more than 20 of these enzymes is currently known (Feller 2013). These studies show that:

1. Specific activity of psychrophilic enzymes is higher at low and mesothermic temperatures than their mesophilic counterparts.
2. The apparent optimal temperature of activity of psychrophilic enzymes is significantly shifted to lower temperatures.
3. Specific activity of psychrophilic enzymes is generally lower than their mesophilic homologues. This suggests that adaptation to cold is not quite complete (D'Amico et al. 2006) (Fig. 10.3).

The adaptation of these enzymes to their environment is explained by the relation "activity-stability-flexibility." This suggests that psychrophilic enzymes increase their overall

Fig. 10.3 Comparison of the effect of temperature on the activity of two homologous enzymes. The *curves* represent the activity in function of temperature for a psychrophilic alpha-amylase from *Pseudoalteromonas haloplanktis* (*dark squares*) and a mesophilic alpha-amylase from *Bacillus amyloliquefasciens* (*empty circles*) (Modified and redrawn from D'Amico et al. 2006)

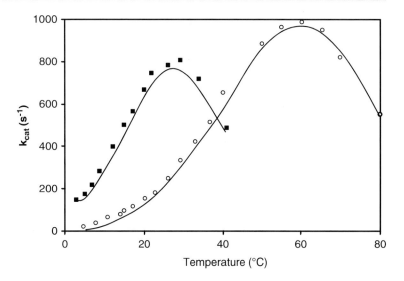

flexibility or that of the active site to compensate for the cold-induced stiffening. Thus, they lose their stability but become more active (D'Amico et al. 2006). If the thermodynamic parameters of these enzymes are compared with those of their mesophilic counterparts, it appears that the former have a high catalytic constant (k_{cat}) at low temperature, thanks to a decrease in the enthalpy activation (k_{cat} is the constant rate designating the number of molecules of substrates converted into product per molecule of enzyme and per unit of time). Many psychrophilic enzymes also have a high Michaelis constant (k_M) (representing the affinity of the enzyme for the substrate) in relation to the significant flexibility of these proteins (D'Amico et al. 2006).

10.2.4.3 Other Informations
Genome Sequencing
Several genomes of psychrophilic bacteria and archaea, including methanogenes, have been completely sequenced (Feller 2013). They include *Desulfotalea psychrophila*, *Colwellia psychrerythraea*, and *Pseudoalteromonas haloplanktis* genomes. In these genomes, several genes of cold-shock proteins (CSPs) together with genes involved in fatty acid desaturation have been identified. Among them, a lipid desaturase and two groups of genes probably involved in the regulation of membrane fluidity were found in *Pseudoalteromonas haloplanktis* (D'Amico et al. 2006).

In *Colwellia psychrerythraea*, a beta-ketoacyl-CoA synthase and a fatty acid, cis/trans isomerase, possibly involved in the increase in membrane fluidity were also identified. In addition to the adaptation to cold, psychrophilic bacteria must protect themselves from the increased oxygen concentration. Indeed, the solubility of gases in water increases with decreasing temperature. *P. haloplanktis* does not possess the molybdenum-dependent metabolic pathway that produce ROS (reactive oxygen species), known as toxic compounds for cells. The lipid desaturases protect bacteria against oxygen and synthesize polyunsaturated fatty acids thus rendering the membrane more flexible. In *C. psychrerythraea* and *D. psychrophila*, genes encoding catalase and superoxide dismutase increasing their antioxidant capacity were found. In terms of amino acid content of proteins, *P. haloplanktis* showed a higher content in asparagine than mesophilic and thermophilic bacteria. This residue is heat labile and is therefore subject to deamination at high temperature. A study comparing psychrophilic and thermophilic archaea showed that psychrophilic bacteria contain leucine and rarely glutamine and threonine (D'Amico et al. 2006). In general, all studies do not go in the same way, and it is therefore not possible to draw definitive conclusions about the preferential use of certain amino acids for protein constitution in psychrophilic bacteria.

Exopolysaccharides
Polysaccharides are polymers consisting of chains of monosaccharide or disaccharide units. Among bacteria, polysaccharides are present either (1) in the cell wall, where they constitute the main part of the lipopolysaccharides (LPS); (2) outside of the cell but associated to it (capsular exopolysaccharide), or (3) released into the culture medium as exopolysaccharides (EPS). Polysaccharide production in psychrophilic bacteria seems to be produced mainly by bacteria belonging to the genera *Alteromonas*, *Pseudoalteromonas*, *Shewanella*, and *Vibrio*. Studies conducted in the Arctic and Antarctic on microorganisms living in the ice suggested that the EPS produced by phytoplankton and bacteria significantly contribute to the presence of organic carbon in the ice-water interface (D'Amico et al. 2006). The EPS are used for attachment of bacteria on supports possibly leading to (1) the formation of biofilms, the uptake and concentration of nutrients, (2) the bacterial defense toward the external environment, or (3) the retention of water to avoid desiccation. In strain CAM025, a

psychrotrophic bacterium of the genus *Pseudoalteromonas*, there is an increase of the production of high-molecular-weight polyanionic EPS at temperatures below the optimum temperature for growth of this bacterium. The EPS may have a role in cryoprotection in cold environments, such as liquid veins of sea ice. The EPS produced by strain CAM025 at low temperatures contain more uronic acid (negatively charged) than those produced at a higher temperature. Due to their polyanionic property, the EPS could complex metals such as iron which is poorly available in Antarctica (D'Amico et al. 2006). In *Lactobacillus sakei* strain 0–1, the low temperatures and the use of glucose as carbon source also increased EPS production.

Nucleating Proteins and Antifreeze Proteins

The formation of intra- or extracellular ice is harmful to the cells. Indeed, ice crystals can lyse the cells mechanically. Furthermore, the formation of ice results in the concentration of solutes causing osmotic shock and cell dehydration. Some bacteria are able to withstand freezing, thanks to several types of proteins and glycoproteins that they possess. Nucleating proteins allow the formation of ice crystals at a temperature above the freezing temperature of water. They are secreted outside the bacteria.

Therefore, when the temperature decreases, small crystals of ice are formed outside the cell which conduct to a gradual dehydration by concentrating the intracellular fluid. Consequently the freezing point also decreases. Beside nucleation proteins, there are intra- or extracellular antifreeze proteins which inhibit ice crystal formation by a complementary binding process to them, thus preventing the enlargement of existing crystals by water. In these conditions, the antifreeze proteins are 500 times more effective to decrease the freezing point of water as compared to NaCl at the same concentration. Antifreeze proteins were mainly known in fish but they have been discovered in a bacterium originating from Antarctic, *Marinomonas primoryensis* (D'Amico et al. 2006). The rhizobacterium *Pseudomonas putida* strain GR 12-2 secretes an antifreeze protein having a nucleating activity. This is the first report of an antifreeze protein having these two opposing activities. This phenomenon still remains to be elucidated (D'Amico et al. 2006).

Cold-Shock Proteins

When bacteria are suddenly transferred to a colder temperature, they generally enter in a lag phase with no more cell division. During this period, the synthesis of most proteins is inhibited. However, some proteins called "cold-shock proteins" or Csp are synthesized and help to reduce the harmful effects due to the drop in temperature.

The phenomenon of cold shock has been widely studied in *E. coli* where the most abundant Csp is CspA which

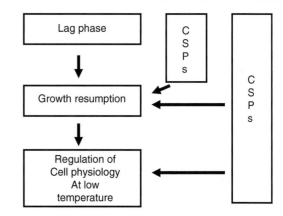

Fig. 10.4 Schematic representation of the hypothetical interactions between cold-induced proteins and acclimation to cold by psychrophilic bacteria (Modified and redrawn from Hébraud and Potier 2000)

would act as a chaperone destabilizing RNA secondary structures thus improving the efficiency of translation. Other proteins are involved in protein folding, regulation of membrane fluidity, transcription, and translation (D'Amico et al. 2006). In psychrophilic bacteria, the effects of a cold shock are different from those found in mesophilic bacteria. In addition to Csp produced during the lag phase, there is synthesis of "cold-acclimation proteins" or Cap, not only during the lag phase following the drop in temperature but also later on, when cell division starts again (Fig. 10.4).

The psychrophilic bacterium, *Arthrobacter globiformis* strain SI55, possessing Cap would play the role of proteases removing denatured proteins (Hébraud and Potier 2000). In another psychrophilic bacterium, *Pseudomonas fragi* having proteins homologous to CspA were identified among the Cap. They may regulate the synthesis of other proteins (Hébraud and Potier 2000).

10.2.5 Biotechnological Applications

Despite the extreme cold conditions encountered in some terrestrial environments and the slowdown in biochemical reactions together with increasing of water viscosity in such extreme conditions, psychrophilic bacteria demonstrate an astonishing adaptation to colonize these habitats. Strategies used by these microorganisms to face cold are quite fascinating. Future studies on the cellular mechanisms allowing cell survival at low temperatures will most probably provide a better understanding on the relationship "structure-function" involved in these mechanisms. These studies will provide the opportunity to identify future applications in biotechnology as it is already the case for the use of psychrophilic enzymes in detergents, food processing, decontamination of polluted environments, etc. (Feller 2013).

10.3 Thermophilic and Hyperthermophilic Prokaryotes

10.3.1 Presentation

According to their behavior regarding temperature, microorganisms are classified into three groups: psychrophiles (*cf.* Sect. 10.2.2), mesophiles, and thermophiles. Several decades ago, a microorganism was considered as a thermophile when its optimal temperature for growth was above 45–50 °C (Fig. 10.5). In the 1970s, the discovery of microorganisms able to grow at temperature largely above this limit resulted to define again and to extend the concept of thermophily. Indeed, today the term "thermophily" is used for all microorganisms having an optimal temperature for growth above 60 °C and the term "hyperthermophily" to prokaryote having their optimal temperature for growth above 80 °C. Some microorganisms, sometimes named extreme thermophiles, have growth temperature above 105 °C (*cf.* Sect. 9.6.1).

10.3.2 Thermophilic Microorganism Habitats

Natural geothermal habitats are widely widespread on Earth and are mainly associated with tectonically active areas. Also, some hot habitats result from human activities.

10.3.2.1 Terrestrial Geothermal Ecosystems

Terrestrial geothermal ecosystems are usually associated with volcanic activity. In general, run off waters penetrate in the deep where they are heated. When the temperature of the percolating fluid become high enough, the resulting pressure lead it to the surface where it emerges to form hot springs or geysers. During its circulation through the Earth's crust, it loads in gas and mineral elements. Thermal water composition will therefore depend on the nature of the rocks and the temperature and therefore the type of volcanic activity that they are associated with. In sites where the volcanic activity is very high, the heating source (magmatic chamber) is located between 2 and 5 km depth; the water temperature is from 150 to 350 °C depending on the depth. Water emerges at the surface as steam creating fumaroles, enriched in volcanic gases (mainly N_2 and CO_2, but also H_2, H_2S, CH_4 CO, and NH_3). These springs are generally acidic due to hydrogen sulfide (H_2S) oxidation into elemental sulfur then into sulfuric acid. This oxidation process is due to the presence of dioxygen into the surface soil layers that chemically reacts with sulfides but also to biological activity. Another type of acidic thermal spring is acidic mud pools that result from the alteration of surrounding rocks by acidified waters and heated by the fumaroles. Neutral thermal springs are generally located at the periphery of active areas; they occur only when high water flow is present at low depth. Such systems are frequently unstable and the upper oxidized soil layer has from 1 to 2 cm thickness, the lower layers being at neutral pH.

Alkaline hot springs are located outside active volcanic fields. When it reaches the surface, water has temperature under 150 °C; it is heated by deep circulating lava and is highly mineralized (silicates), enriched in CO_2, and depleted in H_2S. In upper layers, surface oxidation has no effect on pH that stays near neutrality, but on the other hand, CO_2 release and silicates precipitation lead to a pH increase and stabilization between 9 and 10. These various geological systems are present in tectonically actives areas; some of them have been particularity studied by microbiologists. These include Iceland springs (this country is crossed by the sole emerged part of the mid Atlantic ridge), the Azores, Napoli region and Aeolian Island (both in Italy), and the famous hot springs of Yellowstone National Park in the United States (Fig. 10.6a, b).

10.3.2.2 Deep Oceanic Hydrothermal Vents

Hydrothermalism is an indirect consequence of the extension and accretion of tectonics plates. On mid-oceanic ridges, those are accretion areas; magma (1,200 °C) can ascend and form magmatic chambers at a few kilometers depth. While cooling, newly formed oceanic crust retracts and so generates anfractuosities where cold seawater (2 °C) seeps down at several kilometers depth where it heats up close to the magma. Due to high temperature and pressure, solubility power is enhanced and the low-density heated fluid rise up back to the oceanic floor, washing the surrounding rock. It becomes acidic and enriched in metallic elements. Fluid composition depends on the temperature and the nature of the passed through rocks. Hydrothermal fluid is

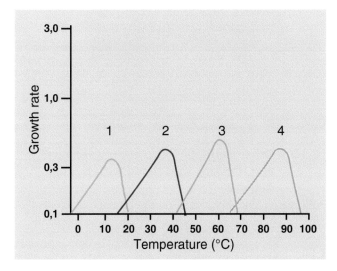

Fig. 10.5 Classification of prokaryotes according to their growth temperature: *1*, psychrophiles; *2*, mesophiles; *3*, thermophiles; *4*, hyperthermophiles (Modified and redrawn from Madigan and Martinko (2007). Drawing: M.-J. Bodiou)

Fig. 10.6 Hot habitats. (**a**) Terrestrial hot spring, Furnas, Sao Miguel, Azores. (**b**) Geyser, geysir, Iceland. (**c**) Active hydrothermal chimney, Rainbow site, (Mid-Atlantic Ridge, 2,700 m depth). (**d**) Hydrothermal chimney sampling by the remote operated vehicle VICTOR 6000 on TAG site, 3,500 m depth) (Photography 10.6a: courtesy Joël Quérellou, Ifremer, Brest; Photography 10.6b: courtesy J. Briffotaux, Ifremer, Brest; Photographies 10.6c, d: courtesy Ifremer, Oceanographic cruise EXOMAR)

then emitted on the seafloor at focused points forming hydrothermal vents. Hydrothermal fluid highly differs from seawater (cold, slightly alkaline, and oxygenated) by their temperature that can reach 400 °C and by their pH generally acidic, being anoxic, and having high concentrations in dissolved gasses (H_2S, CH_4, CO, CO_2, H_2) and metals (Mn^{2+}, Fe^{2+}, Si^+, Zn^{2+}). Due to these physicochemical characteristics, the contact between hydrothermal fluid and seawater results in the precipitation of minerals and the creation of metal-rich mineral structures, so called black smokers. Black smokers are formed when the hydrothermal fluid is not diluted by seawater while so-called white smokers or diffuser result from seawater diluted hydrothermal fluids. They are the main observed hydrothermal structures, but all intermediates can exist. Hydrothermal fluid seepage does not always result in the formation of mineral structures; indeed within "*Pillow lava*," structures with low-temperature fluid (6–23 °C) can be emitted while another particular hydrothermal seepage is the one observed in the Guaymas Basin (Gulf of California) where superheated fluids are emitted through a thick organic-rich sediment layer. In these ecosystems, chemolithotrophic microorganisms (free or living in symbiosis) are the first level of the food web and allow the development of high-density animal communities at the vicinity of the fluid emissions despite conditions that look unfriendly for life.

10.3.2.3 Coastal Marine Hydrothermal Vents

Coastal marine hydrothermal vents are associated with terrestrial volcanic activity as, for example, in Iceland, Azores, and Italy (Vulcano, Aeolian Islands). Like in deep-sea sites, they are characterized by small mineral structures emitting hydrogen sulfide-rich fluids but also by percolated sediments. Their temperatures are generally close to 100 °C.

10.3.2.4 Other Habitats: Petroleum Reservoirs and Habitats Associated with Human Activity

Both marine and continental oil reservoirs are habitats that can be colonized by thermophilic or hyperthermophilic prokaryotes. This is particularly true for oil reservoirs with depths below the surface ranging from 1.5 to 4 km. Such reservoirs exhibit not only high temperatures (between 60 and 130 °C), but also high lithostatic pressure. The pH of the oil-field waters is generally close to neutrality. These waters

Table 10.1 Energy-yielding reactions in thermophilic and hyperthermophilic *Archaea* and *Bacteria*

	Energy-yielding reaction	Metabolic type	Examples
Chemoorganotrophs	Organic compounds + $S°$ → $H_2S + CO_2$	Anaerobic respiration	*Thermoproteus, Thermoplasma, Thermococcus, Desulfurococcus, Thermofilum, Pyrococcus*
	Organic compounds + SO_4^{2-} → $H_2S + CO_2$	Anaerobic respiration	*Archaeoglobus, Thermodesufobacterium, Thermodesulfatator*
	Organic compounds + O_2 → $H_2O + CO_2$	Aerobic respiration	*Sulfolobus, Aeropyrum, Marinithermus, Oceanithermus, Vulcanithermus*
	Organic compounds → $CO_2 + H_2$ + fatty acids	Fermentation	*Staphylothermus, Pyrodictium, Pyrococcus, Thermococcus, Thermotoga, Thermosipho, Marinitoga, Fervidobacterium, Petrotoga, Caminicella, Tepidibacter*
	Organic compounds + Fe^{3+} → $CO_2 + Fe^{2+}$	Anaerobic respiration	*Pyrodictium, Deferribacter*
	Pyruvate → $CO_2 + H_2$ + acetate	Fermentation	*Pyrococcus*
Chemolithotrophs	$H_2 + S° → H_2S$	Anaerobic respiration	*Acidianus, Pyrodictium, Thermoproteus, Stygiolobus, Ignicoccus, Desulfurobacterium, Caminibacter*
	$H_2 + NO_3^- → NO_2^- + H_2O$ (NO_2^- is reduced into N_2 by some species)	Anaerobic respiration	*Pyrobaculum*
	$4 H_2 + NO_3^- + H^+ → NH_4^+ + 2 H_2O + OH^-$	Anaerobic respiration	*Pyrolobus, Caminibacter*
	$H_2 + 2 Fe^{3+} → 2 Fe^{2+} + 2 H^+$	Anaerobic respiration	*Pyrobaculum, Pyrodictium, Archaeoglobus, Deferribacter, Thermotoga*
	$2 H_2 + O_2 → 2 H_2O$	Anaerobic respiration	*Acidianus, Sulfolobus, Pyrobaculum, Thermocrinis, Aquifex, Persephonella, Balnearium*
	$2 S° + 3 O_2 + 2 H_2O → 2 H_2SO_4$	Aerobic respiration	*Sulfolobus, Acidianus, Thermocrinis, Aquifex, Persephonella*
	$2 FeS_2 + 7 O_2 + 2 H_2O → 2 FeSO_4 + 2 H_2SO_4$	Aerobic respiration	*Sulfolobus, Acidianus, Metallosphaera*
	$2 FeCO_3 + NO_3^- + 6 H_2O → 2 Fe(OH)_3 + NO_2^- + 2 HCO_3^- + 2 H^+ + H_2O$	Anaerobic respiration	*Ferroglobus*
	$4 H_2 + SO_4^{2-} + 2 H^+ → 4 H_2O + H_2S$	Anaerobic respiration	*Archaeoglobus, Thermodesufobacterium, Thermodesulfatator*
	$4 H_2 + CO_2 → CH_4 + 2 H_2O$	Anaerobic respiration	*Methanopyrus, Methanocaldococcus, Methanothermus*

may contain various concentrations in sulfur compounds (e.g., sulfate), metals, and dissolved gases such as H_2, H_2S, CO_2, and CH_4 as well as organic molecules (C_2 to C_7 compounds) such as acetate which can accumulate up to 20 mM in some oil reservoirs. The salinity varies from one reservoir to another and most likely in a same reservoir. In some cases (hypersaline oil-field waters), salinity saturation may be observed.

Also there are some hot ecosystems that result from human activities. Coal-refuse pile contains coal fragments, pyrite (FeS), and organic compounds extracted from coals; these mining wastes can be heated by spontaneous combustion that generate hot environment favorable for thermophilic prokaryotes. Some industrials activities are artificial heat sources that can be also habitats for thermophilic microorganisms: we can cite sugar refineries, paper mills, power plants, hot water pipes, and domestic or industrial boilers.

10.3.3 Metabolic and Phylogenetic Overview of Thermophilic and Hyperthermophilic Prokaryotes

The discovery of hyperthermophilic prokaryotes is contemporaneous of the discovery of the *Archaea* domain, thanks to the pioneer work of Dr Karl Woese (Woese et al. 1990).

Thermophilic and hyperthermophilic prokaryotes that were isolated from both continental and marine geothermal areas belong to both the *Bacteria* and *Archaea* domains (Stetter 2011). They exhibit highly diverse physiological and metabolic characteristics and are involved in all biogeochemical cycles (Table 10.1). Regarding their phylogenetic position, based on their 16S rRNA gene sequences, we will present here an overview of this diversity with some focus on some selected genera presented in more detail (Madigan and Martinko 2007).

10.3.3.1 Bacteria Domain

In the domain *Bacteria*, thermophilic representatives belong to several lineages with some of them having both mesophilic and thermophilic species. Two lineages located at the root of the bacterial phylogenetic tree are constituted by only thermophilic and hyperthermophilic species.

Within the *Cyanobacteria*, some thermophilic species were described: the most thermophiles are members of the genus *Chloroflexus* and *Synechococcus*. They are moderate thermophilic phototrophs and were retrieved from terrestrial thermal systems.

Within Gram-positive bacteria, numbers of thermophilic species belonging to the *Firmicutes* were described; they are species belonging to the genera *Thermoanaerobacterium*, *Carboxydibrachium*, *Caminicella*, and *Tepidibacter* and are sporulating anaerobes fermentative bacteria.

Also species from the *Thermus-Deinococcus* group belonging to the genera *Thermus*, *Marinithermus*, *Rhabdothermus*, *Meiothermus*, *Vulcanithermus*, and *Oceanithermus* were isolated from both continental or marine hot ecosystems. They are nonsporulating aerobic (while some of them can grow under anaerobic conditions) chemoorganotrophs.

More recently, thermophilic species belonging to the Epsilon division of the *Proteobacteria* were isolated from marine hydrothermal ecosystems; they belong to the genera *Caminibacter*, *Nautilia*, and *Lebetimonas* and are autotrophic anaerobic sulfur reducers. Other bacterial lineages as *Alpha-* and *Deltaproteobacteria*, *Bacteroidetes*, and *Deferribacteres* have some thermophilic members.

Three bacterial lineages located at the root of the bacterial phylogenetic tree (*Aquificales*, *Desulfurobacteriales*, and *Thermodesulfobacteria*) are constituted by one or more genera including only thermophiles and hyperthermophiles. Species of the genera *Thermodesulfobacterium* and *Thermodesulfatator* are thermophilic sulfate reducers (optimal temperature around 70 °C) while the newly described species *Thermosulfurimonas dismutans* mediate elemental sulfur disproportionation. They are autotrophs and form a separate lineage between *Aquificales* and *Thermotogales*. Microorganisms from the genera *Aquifex*, *Persephonella* and *Desulfurobacterium*, *Balnearium*, and *Thermovibrio* are chemolithotrophs. Some species from the *Aquificales* use H_2 as electron donor (hydrogenotrophic species) but can also use $S°$ or S_2O_3 and O_2 and NO_3 as electron acceptors; they can grow at temperatures up to 95 °C and tolerate only low dioxygen concentration; they are (with some archaeal species) the rare known aerobic (more specifically microaerophilic) hyperthermophiles.

Thermocrinis ruber is an organism closely related to *Aquifex* that grows in the outflow from some hot spring from Yellowstone National Park. It forms pink streamers consisting of a filamentous form of the cells attached to siliceous sinter. This hyperthermophilic organism (optimal growth at 80 °C) is chemolithotroph being able to oxidize H_2, $S°$, or thiosulfate with O_2 as terminal electron acceptor.

The *Thermotogales* that also deeply branched in the bacterial phylogenetic tree include more than 20 species isolated from a great diversity of hot ecosystems. First described as being composed only by thermophilic species, this order was recently shown to include mesophilic species of the genera *Mesotoga* (Ben Hania et al. 2011). Some *Thermotoga* and related species are thermophilic fermentative chemoorganotrophic bacteria; their optimal temperature for growth is between 60 and 80 °C. They were isolated from terrestrial hot spring (*Thermotoga*, *Fervidobacterium*) as well as from either coastal or deep marine hot springs (*Thermotoga*, *Thermosipho*, *Marinitoga*) but also from petroleum reservoirs (*Geotoga*, *Petrotoga*). They have in common the unique characteristic of having an extern sheet-like envelope called the "Toga" (Fig. 6.10e).

10.3.3.2 Archaea Domain

Numerous thermophilic and hyperthermophilic species belong to the two main branches of the archaeal domain: the *Euryarchaeota* and the *Crenarchaeota* (Fig. 10.7).

These two lineages also include nonthermophilic species as extremes halophiles, mesophilic methanogens (*Euryarchaeota*), and species mainly uncultivated today belonging to the *Crenarchaeota* and the newly described domain *Thaumarchaeota* (Brochier-Armanet et al. 2008).

Within the *Crenarchaeota*, hyperthermophilic species cluster at the basis of the tree and gather a majority of organisms that are mainly chemolithotrophs, autotrophs, and some chemoorganotrophs. *Euryarchaeota* gather hyperthermophiles as diverse as the methane-producing methanoarchaea and chemoorganotrophic species.

10.3.3.3 Hyperthermophilic Euryarchaeota

Most of hyperthermophilic methanoarchaea (methanogenic species) belong to the order *Methanococcales* which includes four hydrogenotrophic genera *Methanococcus*, *Methanocaldococcus*, *Methanothermococcus*, and *Methanotorris*. These species were isolated from both submarine and terrestrial hydrothermal systems. Species belonging to the *Methanosarcinales*, which are acetoclastic species, were also described. While not being isolated yet, thermophilic anaerobic methane oxidation (AOM) mediated by ANME *Methanoarchaea* was reported from hot environment (Biddle et al. 2012).

Methanopyrus kandleri has a specific phylogenetic position far from other methanoarchaea and is a hydrogenotrophic methanogen isolated from hydrothermal chimneys and sediments. This species exhibits rapid growth rate at 100 °C and is the most thermophilic described methanogen (up to 110 °C) so far.

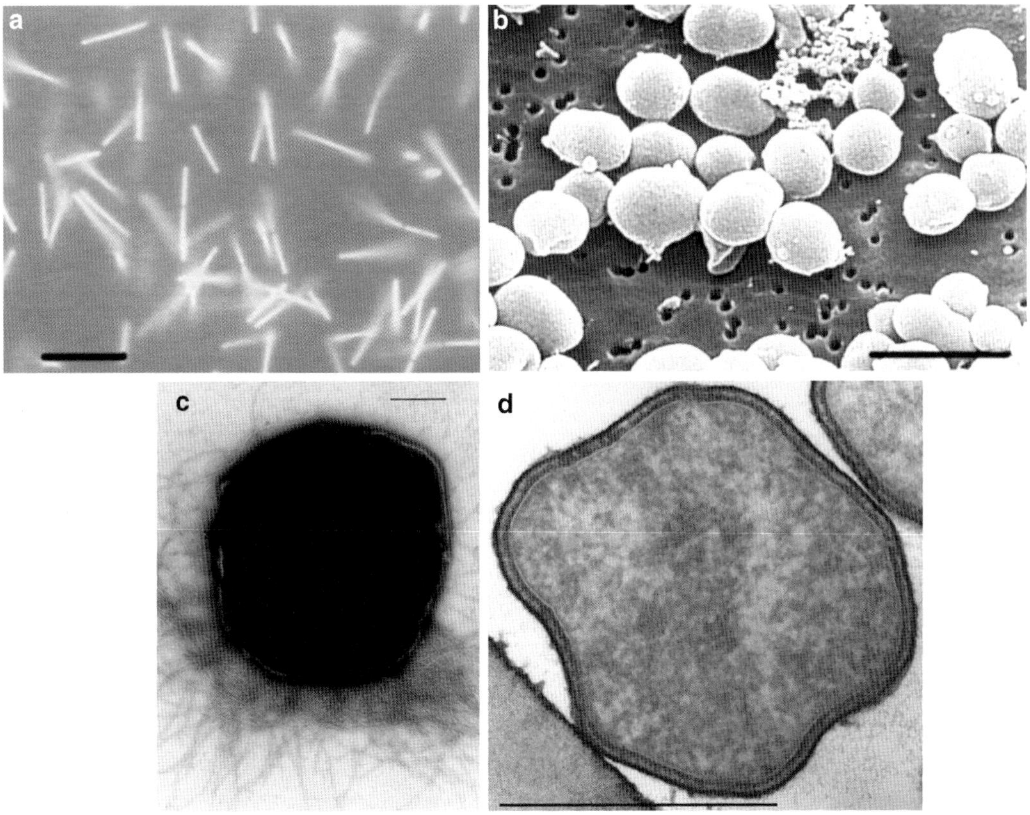

Fig. 10.7 Microphotographs of some hyperthermophilic Archaea. (**a**) *Methanopyrus kandleri*, (optimal temperature for growth 98 °C), observed by epifluorescence microscopy. (**b**) *Thermococcus fumicolans* (optimal temperature for growth 85 °C), scanning electron micrograph. (**c**) *Methanocaldococcus indicus* (optimal temperature for growth 85 °C), negative-staining electron micrograph. (**d**) *Pyrolobus fumarii*, the most thermophilic known organism (optimal temperature for growth 106 °C), transmission electron micrograph of an ultrathin section (Photographies 10.7a, c: courtesy Stéphane L'Haridon, IUEM, Brest; Photography 10.7b: courtesy Anne Godfroy Ifremer, Brest; Photography 10.7d: courtesy Reinhard Rachel and Karl O. Stetter, University of Regensburg, Germany). Bars represent: **a**, 10 μm; **b**, 2 μm; **c**, 0.5 μm; **d**, 1 μm

Thermoplasmatales include one of the most acidophilic known thermophiles. Species of the genus *Thermoplasma* are lacking cell wall organisms, chemoorganotrophs, and facultative anaerobes that use S° as terminal electron acceptor. They colonize acidic hot habitats as coal-refuse pile and terrestrial hot springs. Species of the genus *Picrophilus* that were isolated from acidic solfatara are certainly the most acidophilic species described being able to grow at extremely low pH (<1). Recently the first member of the group DHVE2 (deep-sea hydrothermal vent *Euryarchaeota*) that was only known from environmental sequences was isolated and named "*Aciduliprofundum boonei*." It is the first thermoacidophilic species isolated from marine environments (Flores et al. 2012).

Species from the order *Thermococcales* belongs to three genera: *Thermococcus*, *Pyrococcus*, and *Palaeococcus*. They are all hyperthermophilic organisms that were isolated from deep hydrothermal and swallow hot springs but also from oil reservoir waters. They were first described as forming a homogenous metabolic group being chemoorganotrophs growing quickly on complex proteinaceous substrates or sugar and reducing elemental sulfur into hydrogen sulfide. Recently it was shown that some species from the genus *Thermococcus* are able to grow on carbon monoxide and that some species are able to perform iron reduction. They are (with the *Thermotogales*, domain *Bacteria*) within the most extensively studied hyperthermophilic species in terms of their physiology and metabolism, and many whole genome sequences are now available. Some species within this order are piezophiles (*cf.* Sect. 10.5).

Within the order *Archaeoglobales*, species of the *Archaeoglobus* are the sole sulfate reducers within the Archaea while *Geoglobus* and *Ferroglobus* are organisms involved in iron reduction (*cf.* Sect. 14.5).

10.3.3.4 Hyperthermophilic Crenarchaeota

Within the *Crenarchaeota*, *Thermoproteales* and *Sulfolobales* seem to be specific of terrestrial hot habitats while organisms from the order *Desulfurococcales* seem to colonize marine volcanic habitats.

Within the *Sulfolobales*, species of the genera *Sulfolobus* and *Acidianus* were isolated from various terrestrial volcanic areas (solfataras and hot springs). They are acidophiles, facultative chemolithotrophs *Sulfolobus* species oxidizing sulfur compounds (H_2S or $S°$) into sulfuric acid while *Acidianus* species are facultative anaerobes that can oxidize $S°$ in H_2SO_4 or reduce it in H_2S.

The order *Desulfurococcales* gathers around ten genera with highly diverse phenotypic characteristics. Indeed, it includes aerobic chemoorganotrophs (*Aeropyrum, Sulfurococcus*), anaerobic ones (*Hyperthermus, Desulfurococcus, Staphylothermus, Pyrodictium* ...), as well as strict chemolithotrophs (*Ignicoccus, Pyrolobus*), which are frequently sulfur reducers. One of the major characteristic of this order is that it includes some of the most thermophilic known species. Species from the genera *Pyrodictium* and *Pyrolobus* are able to grow at temperature above 100 °C. *Pyrolobus fumarii* isolated from a hydrothermal chimney of the Mid-Atlantic Ridge holds the record for growth temperature (113 °C) within the prokaryotes.

Desulfurococcus, Staphylothermus, and *Hyperthermus* are hyperthermophilic chemoorganotrophic sulfur metabolizers, and *Aeropyrum* is one of the rare aerobic hyperthermophiles. Species of the genus *Ignicoccus* were isolated from both swallow and deep hydrothermal vents. They are chemolithoautrophic organisms that use elemental sulfur as terminal electron acceptor. Some *Ignicoccus* species live in association with a small prokaryote, *Nanoarchaeum equitans*. This species form a unique lineage at the archaeal phylogenetic tree. It is one of the smallest described prokaryote due to either its cell volume or genome size. The absence in its genome of most of the genes encoding metabolic function made it being totally dependent from its host, *Ignicoccus* (*cf.* Sect. 6.6.1).

The number of hyperthermophilic prokaryotes is considerably increasing in the last 3 years (from a few tens in the 1970s to several hundreds in 2013). Like in many other ecosystems, cultivated species represent a small fraction of the microbial diversity of hot ecosystems and molecular surveys based on 16S rRNA allowed to point out an astonishing diversity. Thus, numerous lineages (both *Bacteria* and *Archaea*) in which there is no cultivated representative were detected in many hot environments.

Within the *Archaea*, many uncultivated lineages within the *Euryarchaeota* were evidenced as well as the lineage *Korarchaeota* that gathers uncultured organisms that are supposed to be thermophiles due to the position of this lineage at the basis of the phylogenetic tree (*cf.* Sect. 6.6.1).

The development of techniques based on the search of gene encoding metabolic functions as well as metagenomic and metatranscriptomic approaches would lead to enlarge our vision of this diversity. It clearly appears that only the combination of both molecular and innovative cultural techniques would allow to characterize thermophilic microbial communities of hot ecosystems on Earth.

10.3.4 Life at High Temperature

Thermophilic and hyperthermophilic prokaryotes live at temperatures much higher than those tolerated by most other prokaryotes. Indeed, their macromolecules (proteins, lipids, and DNA) have specificities that make them good candidates for biotechnological applications.

10.3.4.1 Protein Stability

Thermophilic proteins have similar amino acids composition to that of their mesophilic counterparts. Nevertheless, thermostable proteins exhibit specific characteristics that are (1) the presence of highly hydrophobic centers that avoid the protein to unfold, (2) a diminution of the loop size at the surface of the protein, and (3) the presence of numerous ionic interactions at the surface that also contribute to maintain proteins together (Luo and Robb 2011). It was demonstrated that it is more the protein folding that contributes to its resistance to temperature. Thereby small changes in the amino acids sequence will result in significant changes in the protein folding and so its behavior at high temperature. Also, the presence of chaperons proteins (heat-shock proteins) helps to the folding on partially unfolded proteins.

10.3.4.2 DNA Stability

Several mechanisms that contribute to DNA molecule stability were identified in hyperthermophilic prokaryotes. On one hand, the presence of compatible solutes as high intracellular concentration of cyclic potassium 2,3-diphosphoglycerate in *Methanopyrus* will contribute to chemical lesion as depurination. On the other hand, hyperthermophiles possess a reverse DNA gyrase which produces positive supercoiling (while DNA gyrase from mesophiles produces negative supercoiling) of DNA that will ensure better stability to high temperatures. Other DNA-linked proteins were identified, sometimes similar to histones from the eukaryotes (histone-like proteins). They contribute to the compaction of the DNA molecule and so to its stability.

10.3.4.3 Membrane Lipids

Membrane lipids of hyperthermophilic *Archaea* are composed of biphytanyl tetraethers. They are resistant to temperature due to the presence of a covalent link between the phytanyl units that allows the formation of a one-layer membrane by opposition to the classic bilayer phospholipids membrane (*cf.* Sect. 4.1.5). This structure shored up by covalent link is more resistant to temperature which tends to separate the phospholipids bilayer. This type of lipids was also found in some thermophilic members of the *Bacteria* domain.

10.3.5 Specific Constraints for the Cultivation of Thermophilic Prokaryotes

The cultivation of thermophilic microorganisms needs the adaptation of classic cultural techniques. As an example, working at high temperature above 60 °C prevents to use agar for solid media. Isolation on Petri dishes or using roll tubes is generally performed using gellan gum, a thermostable polymer that resists to temperature above 80 °C (Erauso et al. 1995). Numerous hyperthermophilic microorganisms are anaerobes, and so cultivation techniques require strict precautions with respect to dioxygen by using anaerobic chambers for example (*cf.* Sect. 17.8.2). At high temperature, solubility of gases decreases and so their availability for microorganisms is lower (Henry law). It is therefore necessary to incubate aerobes under agitation and under pressure (0.1 or 0.2 MPa above atmospheric pressure) for the cultivation of chemolithotrophic autotrophs such as methanogens that utilize H_2 and CO_2 for methane production.

Not only for the interest in evolution process and ecology, hyperthermophilic prokaryotes were very early identified as a potential sources of thermostable biomolecules (*cf.* Sect. 10.3.5). Thus, the study of their physiological and metabolic characteristics appeared necessary to concretize their biotechnological potential. One of the main limits was the difficulty to obtain large amount of biomass. The cultivation of anaerobic marine hyperthermophilic microorganisms as sulfur reducers belonging to the order *Thermococcales* raises a number of technical problems linked to the use of seawater-based culture medium and the production of hydrogen sulfide (as a result of sulfur reduction). Combined with high temperature these conditions promote corrosion. One of the key factors for the study of these hyperthermophiles was the implementation of cultures in specially designed bioreactors that had allowed the study of their physiology and metabolism and the production of enough quantities biomass (Godfroy et al. 2006).

10.3.6 Biotechnological Applications

As previously mentioned, the interest for hyperthermophiles, which resulted in the isolation, during the 30 past years, of numerous new species, was highly boosted by the search for biotechnological applications and more specifically in the biocatalyst area: enzymes from hyperthermophiles are in most cases more resistant than their mesophilic homologues. We will present here only a few examples.

The more known is the use for PCR (*polymerase chain reaction*) of the Taq polymerase from *Thermus aquaticus*. Many other polymerases from hyperthermophilic prokaryotes were described and are marketed: Vent pol™ from *Pyrococcus furiosus* that was isolated from a swallow vent (Vulcano, Italy) and the Deep Vent pol™ and the polymerase Isis™ from *Pyrococcus* GBD and *Pyrococcus abyssi*, respectively, that were isolated from deep-sea hydrothermal vents.

Other applications deal with hydrolase that can be used in the food industry for starch treatment, for example, or biomass transformation for hydrogen or biogas production (Basen et al. 2012).

10.3.7 Hyperthermophiles and Evolution

Due to their position at the basis of the phylogenetic tree, hyperthermophiles whatever there are *Archaea* or *Bacteria* could be the sign of a slow evolution process, probably due to the habitats in which extremophilic prokaryotes thrive with hard physical and chemical constraints. These ecosystems as they exist today existed on the early Earth and could be those wherein life could have appeared. Thereby extreme microorganisms (both *Archaea* and *Bacteria*) are perhaps the closest relatives of a primitive life form. It is also interesting to notice that hydrogen metabolism, widely widespread in thermophilic microorganisms, could be one ancient metabolism adapted to primitive life conditions (*cf.* Sect. 4.1).

10.4 Halophilic and Extreme Halophilic Microorganisms

10.4.1 Presentation

Halophiles can be found in all three domains of life. The term "halophilic" is used to qualify microorganisms that require the presence of salts (usually sodium chloride) for their growth. A distinction must be made between halophiles and microorganisms that can tolerate salt even if present in large quantities. They are called as salt-tolerant microorganisms.

In 1962, Helge Larsen defined three categories of halophilic microorganisms based on the salt concentration that allows optimal growth (Fig. 10.8). The slightly halophiles have optimum growth between 2 and 5 % NaCl, while moderate halophiles organisms exhibit optimum growth between 5 and 20 % (*cf.* Sect. 9.6.3).

The term "extreme halophiles" is applied to microorganisms which have better growth at concentrations above 20 % salts. These microorganisms can generally grow in a saturated salt environment (about 35 %), although some species grow very slowly at this salinity. By definition, extreme halophiles require at least 9 % NaCl for growth.

10.4.2 Habitats of Halophilic Microorganisms

Salty habitats are common throughout the world (especially marine ones), but hypersaline environments are relatively infrequent, most of them being located in the hot and dry areas on the planet (Fig. 10.9), but some are also found in temperate and even polar regions. Microbiology of hypersaline

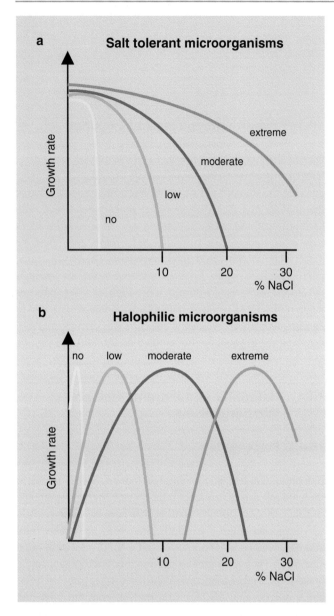

Fig. 10.9 Examples of a hypersaline ecosystem: Retba Lake (or Pink Lake in Senegal) (Photography: courtesy of Pierre Roger)

Fig. 10.8 Classification of prokaryotic microorganisms according to their behavior toward NaCl. According to Madigan and Martinko (2007). Drawing: M.-J. Bodiou

environments will be mainly discussed in this chapter. Regarding salt lakes, their ionic composition may significantly vary depending on (1) the surrounding topography, (2) geology (soil leaching), and even (3) the weather. The hypersaline environments are defined as environments that have concentrations of salts greater than that of seawater. This definition is not specific and does consider neither the salt types nor their proportions. They are found in hypersaline seas, salt evaporation pools, salt marshes, but also in subterranean salt deposits, dry soils, or salted meats (Oren 2011).

10.4.2.1 Ecosystems

Seawater, which represents the greater part of the water on Earth, contains minerals in remarkably constant concentration. The main ones are chloride (18.98 %), sodium (10.56 %), sulfate (2.65 %), magnesium (1.27 %), calcium (0.4 %), potassium (0.38 %), and carbonate (0.14 %).

The majority of hypersaline environments originate from salts seawater, either directly by evaporation and concentration or indirectly by the dissolution of evaporitic deposits found in the form of fossil deposits. The minerals are in similar proportions to those contained in seawater, at least until the precipitation thresholds are not reached.

Hypersaline waters formed by partial evaporation of seawater are called "thalassohalines" while those obtained by the dissolution of fossil salt deposits are called "athalassohalines." These can have proportions in salts quite different from those of seawater depending on the nature of the deposits. Some contain a high percentage of sodium carbonate and are very alkaline. These hypersaline systems are of continental origin, since they are generated by dissolution of salts contained in rocks or crossed geological layers after the action of rain or runoff waters. Then, these salts are concentrated when waters accumulate in impermeable basins. In this case, each salt lake is unique due to its peculiar chemical composition. These lakes are classified into two categories according to the predominant anion present. They include lakes containing sulfate and those containing carbonate like the alkaline lakes in East Africa. (Oren 2011).

While in sulfate-rich environments, ubiquitous microorganisms can be found. There is often a restricted and specific biodiversity in the carbonate-rich ones.

Salt marshes may also contain extreme halophilic prokaryotes. They are small and closed seawater pools, which concentrate salts after evaporation.

Since the 1980s, hypersaline and anaerobic deep-sea trenches were discovered in the Eastern Mediterranean Sea (Box 10.1).

Some soils containing high salt concentrations are also considered as hypersaline.

Box 10.1: Deep Hypersaline Anoxic Basins in Eastern Mediterranean Sea

Danielle Marty

In the early 1980s, the first deep depressions filled with NaCl-saturated anoxic brines, known under the English acronym "DHABs" for *"deep hypersaline anoxic basins,"* were discovered in the Eastern Mediterranean Sea. These anoxic basins were usually named after the names of the oceanographic vessel that first explored them.

"Tyro" was the first basin discovered in 1983, in the Strabo Trench, and within the year after "Bannock" (1984), on the southwestern slopes of the Mediterranean Ridge. Ten years or so after, three more basins were discovered on the eastern part of the Mediterranean Ridge, "l'Atalante" (1993), "Urania"(1993), and "Discovery" (1993–1994). In 2008, a novel basin was discovered on the western part of the Mediterranean Ridge and named "Thetis."

Some similar deeps were previously observed in smaller oceanic basins: in the Red Sea, where, among some 25 deeps found since the late 1960s; "Atlantis II Deep" discovered in 1968 is the largest (52 km^2); and in the Northern Gulf of Mexico, where "Orca Basin," discovered in 1975, covers an area of 400 km^2.

All these basins have a common feature, the presence of evaporitic deposits in the subsurface sediments at shallow depths; these evaporites are Jurassic in age (Mesozoic Era) in the Gulf of Mexico and Messinian in age (Late Miocene in Cenozoic Era) in the Red Sea and Mediterranean Sea (Cita 2006).

Mediterranean DHABs are extreme deep-sea biotopes, characterized by extremely high salinity and corresponding density, elevated hydrostatic pressure, absence of light and oxygen, and a sharp chemocline between seawater and brines, some meters in thickness. These unique physicochemical characteristics demonstrate that DHABs were physically isolated from other habitats of the planet during thousands of years.

The Detection of DHABs
All DHABs have a similar sonar signature, with a well-defined seismic reflection at the interface between normal seawater and high saline brines.

D. Marty
Institut Méditerranéen d'Océanologie (MIO), UM 110, CNRS 7294 IRD 235, Université de Toulon, Aix-Marseille Université, Campus de Luminy, 13288 Marseille Cedex 9, France

Once the presence of brine accumulation recognized by sound reflector, it is necessary to sample and verify the presence of brines, to approve the hypersaline deep.

Hydrological and Chemical Characteristics of Various DHABs (Box Table 10.1)

The Formation of Mediterranean DHABs
About 5.96–5.33 million years ago, during the Messinian (Late Miocene), tectonic processes of convergence between the African and Eurasian plates caused the occlusion of the ancient Tethys Sea and interruption of water inflow from the Atlantic Ocean. The nearly complete desiccation of the Mediterranean, called "the Messinian salinity crisis," came along with a decrease in Mediterranean sea level of about 1.5 km and massive evaporite deposits on the seafloor: these salt deposits could be as thick as some hundred meters, one thousand meters, and even several thousand meters. These gigantic evaporite deposits underlying all the Mediterranean Basin, with more than 1 million km^3 salts covering more than 2 million km^2, are widely regarded as one of the biggest evaporitic episodes, if not the biggest that our planet knew. About 5.3 million years ago, at the Messinian/Pliocene boundary, the opening of the Gibraltar Strait allowed drastic refilling of the Mediterranean by Atlantic waters intrusion, dissolution of the Messinian evaporites, and formation of brine lakes at the bottom of the deepest depressions.

Location and Description of the DHABs of the Mediterranean Sea
Mediterranean DHABs represent unique, extreme, and largely unexplored habitats, lying at more than 3,500 m below sea level, containing very stable brines entrapped into basins, deeper than the surrounding seafloor: the brine columns exhibit thickness ranging from 80 to 500 m and are homogeneous or stratified into different layers (Box Fig. 10.1). Each of the six brine lakes, physically isolated from each other (Box Fig. 10.2), presents environmental peculiarities, suggesting that the brines are derived from the dissolution of different levels of the Messinian evaporitic succession; evaporites are essentially halite for Bannock, Tyro, Urania, and L'Atalante and bischofite for Discovery.

Tyro. The smallest basin: diameter c.a. 4 km, funnel shaped, with a rounded rim; it lies east of Kretheus and west of Poseidon; both basins are not more anoxic and brine filled since about 3,000 years. Tyro shows

(continued)

Box 10.1 (continued)

Box Table 10.1 Hydrological characteristics (maximal depth, salinity, pH, density) and chemical compositions (sodium, chlorine, magnesium, methane, hydrogen sulfide, and sulfate concentrations) of different DHABs

Area	Depth m	S ‰ psu	pH	Na$^+$ mM	Cl$^-$ mM	Mg mM	Density 10^3/kg/m^3	CH$_4$ µM	H$_2$S mM	SO$_4^{2-}$ mM
"Normal" seawater		26–38	8.2	528	616	60	1.03	2–6 nM	0	31.8
Tyro	3,437	321	6.6	5,300	5,350	71	1.21		2.1	52
Bannock	3,790	321	6.5	4,235	5,360	650	1.21	450	3.0	137
Libeccio Basin										
Urania	3,570	211	6.8	3,503	3,729	316	1.13	5,560	16	107
L'Atalante	3,522	320		4,674	5,289	410	1.23	520	2.9	397
Discovery	3,533	>120	4.5	68	9,491	4,995	1.32	50	0.7	96
Thetis	3,258	348		4,760	5,300	604	1.22		2.1	265
Orca – Gulf of Mexico	2,400	250	7.0	4,155	4,136	42.4	1.19	899	0.003	38
Atlantis II – Red Sea	2,192	310	5.2	4,880	5,110	32.8	–	6.3	–	26

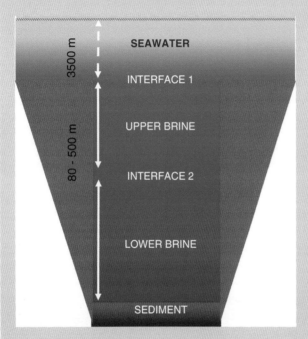

Box Fig. 10.1 Schematic depth structure of a DHAB containing two brine layers

homogeneous brine throughout the anoxic layer, which thickness is limited, less than 60 m.

Bannock. A complex and irregular depression about 15 km in diameter; with an area of 46.5 km^2 occupied by 4.1 km^3 of salt brine, it is the largest Mediterranean anoxic basin. Within this area, nine satellite basins have been identified and named after various local winds: Libeccio, Sirocco, Ponante, Tramontane, Levante, etc. (Box Fig. 10.2). The Libeccio sub-basin, the deepest and the most studied, is referred as the Bannock Basin: brines, constituted by seawater concentrated 8–10 times, are enriched in potassium, brome, and magnesium and saturated in gypsum and dolomite, with formation of gypsum crystals within sediments. Brines are stratified and consist of two separate layers with different physical and chemical properties: the upper layer is 150 m thick and the lower layer is 300 m thick. Bannock brines would have formed 150,000 years ago.

Urania. A horseshoe-shaped depression about 6 km width, filled with brines which thickness ranges from 80 to 200 m, and stratified in two separated layers, and the presence in the western arm, of a deep hole fed with very fluid black mud (so called the "mud pit") at up to 65 °C. Urania brines are strongly H$_2$S enriched, up to 10 mM, and so exhibiting the highest concentration of sulfide among the Earth aquatic environments; in addition, brines contain 10–100 times more dissolved methane than other DHABs, as well as other hydrocarbons, such as ethane and propane (Borin et al. 2008; Sass et al. 2001).

L'Atalante. The brine layer is homogeneous in chemical composition; however, thermal stratification is observed, with three layers: seawater 13.84 °C, the top 16-m thick layer 13.82 °C, the middle 30-m thick layer 13.91 °C, and the bottom 40-m thick layer 14.06 °C. Brines exhibited the highest sulfate concentrations (397 mM) measured among Mediterranean DHABs and are also strongly enriched in potassium (up to 390 mM), suggesting a derivation from the potassium chloride, which is a characteristic of the uppermost levels of the evaporitic suite.

Discovery. An elongated basin in arc of a circle, with a surface area of about 7.5 km^2 and a volume of nearly 0.2 km^3, contains a single homogeneous brine

(continued)

Box 10.1 (continued)

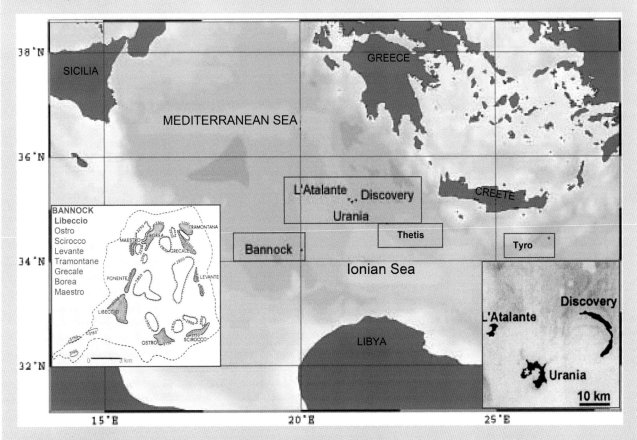

Box Fig. 10.2 Location of the six DHABs in the Eastern Mediterranean Sea

layer. Among Mediterranean DHABs, Discovery is unique, as almost exclusively filled with a solution of magnesium chloride that exhibits the highest density ever found in aquatic environments. Numerous nodules of magnesium and carbonate are present in the sediment, as well as hydrocarbon seeps.

Thetis. An elongated narrow depression of elliptical shape, with a surface area of about 11 km^2 and a total volume of approximately 0.7 km^3, filled by a brine 157 m thick, almost nine times more saline than seawater, one of the highest salinity reported for DHABs. Hydrochemistry of Thetis brines seems quite similar to L'Atalante and Bannock.

The Life in DHABs

Although deep hypersaline basins in the Mediterranean Sea are extreme habitats, these basins are not biogeochemical dead ends but support prokaryotic activities, including Discovery, in which chloride concentrations close to the saturation (Mg ~5 M) are considered anathema to life (van der Wielen et al. 2005).

Microbial activities (heterotrophy, sulfate reduction, and methanogenesis) were evidenced in brines and sediments of various basins, indicating the presence of bacterial populations adapted to these extreme biotopes. These results were corroborated with studies by epifluorescence microscopy, fluorescent in situ hybridization (FISH), and rRNA gene sequencing (Daffonchio et al. 2006). These studies demonstrated the presence of bacterial and archaeal communities, with a higher diversity within the *Bacteria*. Some microorganisms are similar to those previously described, such as *Proteobacteria* and *Sphingobacteria* within *Bacteria* and *Halobacteria* within *Archaea*; other microorganisms correspond to a new division or a new order of yet-to-be-cultivated *Archaea* (van der Wielen et al. 2005; La Cono et al. 2011).

10.4.2.2 The Physical Chemistry of Lakes

The Great Salt Lake, located in Utah, USA, is an example of hypersaline environment at neutral pH corresponding in ionic composition to that of concentrated seawater. Sodium is the dominant cation, while chlorine is the dominant anion. Significant concentrations of sulfate are also present. In contrast, in the Dead Sea, another hypersaline basin located between Israel and Jordan, having a neutral pH, the sodium concentration is relatively low, while that of magnesium is high. Both ecosystems have salt concentrations that can reach saturation. Regarding the water chemistry of alkaline lakes, it is similar to that of Great Salt Lake. However, because of the presence of large amounts of carbonates in the surrounding rocks, pH is relatively high. Waters with pH ranging from 10 to 12 are common in these extreme environments. Ca^{2+} and Mg^{2+} ions are absent from these lakes as they precipitate at alkaline pH under high carbonate concentrations. Several lakes in the Great Basin of the western United States, whose salinity varies from 9 to 10 %, are alkaline with pH values measured between 9 and 10.

When seawater salts are concentrated by evaporation, carbonate precipitates into calcium carbonate when salinity reaches 6 %. Thereafter, sulfate precipitates forming gypsum deposits (calcium sulfate) when salt concentration exceeds 10 %. Beyond 25 %, sodium chloride starts to precipitate as halite and fully precipitates at 34 % corresponding approximately to ten times the salt concentration in seawater. Waters are subsequently enriched in magnesium and potassium which precipitate at salinities 20 times higher than that of seawater.

Evaporated seawater is at the origin of salt deposit formations which with time lead to the formation of evaporites. The latter are considered as fossil hypersaline environments which are present on all continents. Athalassohaline environments may result from their dissolution by seawater. These extreme saline environments become appropriate sites for a progressive adaptation of microorganisms to high salt conditions.

The temperature of the inland lakes can widely vary throughout the year. It can range from 30 °C in winter to over 48 °C in summer in the region of the Great Salt Lake. A large majority of the halophilic microflora inhabiting this lake is heterotrophic, aerobic, mesophilic, or thermotolerant with an optimum growth temperature between 40 and 55 °C. Thermotolerance was observed in particular among the halophilic phototrophic anaerobic bacteria belonging to the genus *Ectothiorhodospira*. Their optimal growth temperature is between 45 and 50 °C. In contrast, some halophilic bacteria isolated from hypersaline lakes in Antarctica are psychrophilic, but this is an exception. Within the domain *Bacteria*, thermotolerance of microorganisms may result from the presence in their cytoplasm of solutes such as betaine that protects enzymes from heat. Although being mesophilic, halophilic eukaryotic organisms of the genus *Dunaliella* withstand high temperatures. This resistance to high temperature is linked to the presence of glycerol, which allows intracellular enzyme stabilization.

The hypersaline systems are anoxic environments below the surface of the water, because of high temperatures and salinities that limit dioxygen dissolution. This creates favorable conditions for the development of anaerobes, although aerobes are widely present. Some microorganisms pertaining to the domain *Archaea* possess gas vacuoles enabling them to float, thus staying in direct contact with dioxygen.

Finally, the different chemical composition of hypersaline habitats selected a large variety of halophilic microorganisms. Some are endemic while others are found in various habitats around the world. Despite the extreme conditions existing in hypersaline environments, primary production is important notably via autotrophy. Organic matter resulting from this primary production by aerobic or anaerobic phototrophic organisms allows the development of alternative heterotrophic extremophilic microbial populations.

10.4.3 Origin of Organic Matter

Hypersaline ecosystems generally have limited life forms. Indeed, the largest salt concentration tolerated by a vertebrate organism is 10 % (*Tilapia* spp.). Above this limit within *Eukarya*, only crustaceans (*Artemia salina*) and algae (*Dunaliella salina*) were identified. At higher concentrations of salt, aerobic and anaerobic heterotrophic prokaryotes, together with anaerobic phototrophic bacteria and methanoarchaea, are known to inhabit these environments. Plants and algae that develop along the banks of hypersaline lakes can be a significant source of organic matter after immersion of shoreline by water.

Invertebrates and microorganisms within the domains *Archaea* or *Bacteria* deliver a significant source of organic matter in these environments. They include sugars, lipids, proteins, and even more complex carbon and energy sources such as chitin, which are important constituents of crustaceans. A significant proportion of mineral and organic matter can be brought by birds when transiting by these areas.

Intracellular compounds of low molecular weight allowing most of these microorganisms to face osmotic pressure are preferential substrates for halophilic/hyperhalophilic prokaryotes.

Methylated compounds resulting from the degradation of glycine betaine in particular are used as energy sources by methanoarchaea (*cf.* Sect. 14.2.7).

It therefore appears that a broad range of potential substrates is available in hypersaline ecosystems, and despite existing unfavorable physicochemical conditions, they allow growth of adapted microorganisms via autotrophic, heterotrophic, or the phototrophic ways both under aerobic and anaerobic conditions.

10.4.3.1 Degradation of Organic Matter in Hypersaline Environments

In most habitats, the macromolecules are decomposed by hydrolytic and fermentative bacteria into simpler products such as organic acids, fatty acids, alcohols, dihydrogen, and carbon dioxide. The dihydrogen can then be used as an electron donor for different types of prokaryotes including homoacetogenic bacteria, sulfate-reducing bacteria, or methanoarchaea leading to an optimized mineralization of organic matter via acetate which can be completely converted to methane or sulfide and carbon dioxide. Increase in salinity leads to dihydrogen and various volatile fatty acids accumulation in sediments. These results suggest that oxidation of organic matter is incomplete in high saline environments in contrast to other ecosystems such as marine environments or digesters. Beyond salinity higher than 15 %, mineralization of organic compounds through methanogenesis is limited because of the absence or low activity of methanoarchaea regarding dihydrogen and acetate utilization. In contrast, hydrogenotrophic sulfate-reducing activities still occur but decrease while increasing salinity up to saturation. It is noteworthy that analysis of the Great Salt Lake sediments demonstrated that up to 200 mM dihydrogen accumulated in deep layers.

These hydrogenotrophic activities, reduced in case of sulfate-reducing bacteria or absent in methanoarchaea, lead to an accumulation of fermentation products that are ultimately preserved in sediments and may result at geological scales in oil formation.

It was evidenced that substrates such as methylamines were used only by methanoarchaea and not by sulfate-reducing bacteria, known as methanogenic competitors for dihydrogen in most of the habitats of the planet. This gas and formate are preferably used by sulfate-reducing bacteria when sulfate is not limiting in the environment, and this is particularly the case in hypersaline environments.

Bacterial sulfate reduction is therefore an important process in the mineralization of organic matter in these salted and anoxic environments with a peculiar emphasis with regard to dihydrogen oxidation. However, beside dihydrogen, the oxidation of volatile fatty acids such as acetate and butyrate by sulfate-reducing bacteria has not been established at salinity that exceeds 19 %. Consequently, there is an incomplete anaerobic oxidation of organic matter in hypersaline environments. This oxidation may lead to acetate accumulation in such environments, reaching up to several hundred millimolars.

10.4.3.2 The Compatible Solutes

A difference of solute concentration on both sides of a cell membrane causes a force called "osmotic pressure." This pressure has been estimated at about 0.1 bar in freshwater and to more than 100 bars in hypersaline environments. In order to maintain their viability, cells must always adjust the intracytoplasmic osmotic pressure to that of its immediate environment. For this purpose, halophilic microorganisms have developed two strategies to prevent osmotic shocks:
1. The transport of ions across the membrane by means of ion pumps (salt-in strategy).
2. The synthesis or accumulation of organic molecules. Many halophilic or halotolerant prokaryotic species accumulate organic osmolytes by synthesizing or transporting them from the surrounding environment (compatible-solute strategy).

In both cases, these mechanisms are of a high energy cost to the cell that usually prefers transport to biosynthesis when osmolytes are already present in the environment. The organic molecules involved are generally of low molecular weight. They are polar, highly soluble, neutral at physiological pH, and unable to cross the cell membrane without the intervention of specific transporters. They include amino acids and derivatives, betaines, and other polyols which do not interfere with the cell mechanisms. The term "compatible solutes" was introduced to define these protective molecules (*cf.* Sect. 9.6.3, Fig. 10.9).

Accumulation of ions was evidenced in aerobic, extremely halophilic *Archaea* of the family *Halobacteriaceae* but also in anaerobic bacteria. This is the case of K^+ which replaces advantageously Na^+ in the cytoplasm as the latter is toxic for cells. Cl^- is the dominant anion in the cytoplasm.

10.4.4 Moderate to Extremely Halophilic Bacteria

Anaerobic halophilic microorganisms belong to the order of *Halanaerobiales* (Ollivier et al. 1994) which includes two families *Halanaerobiaceae* and *Halobacteroidaceae*.

All species of these families ferment carbohydrates with the exception of a homoacetogenic bacterium (*Acetohalobium arabaticum*) that reduces CO_2 by producing acetate and grows on betaine and trimethylamine. More recently, *Fuchsiella alkaliacetigena*, the first alkaliphilic anaerobic hydrogenotrophic homoacetogenic bacterium, was isolated from sediments of the soda-depositing soda lake in Russia. *Halocella cellulolytica* is the only species capable of cellulolysis among all described species. Species belonging to the genera *Orenia* and *Sporohalobacter* are the only ones able to sporulate. This phenotypic feature differentiates them from all other species.

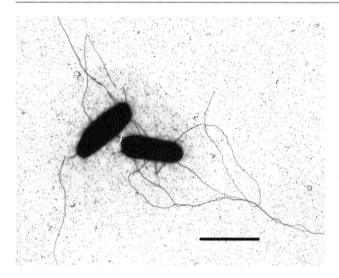

Fig. 10.10 Example of one halophilic anaerobic bacterium: *Halanaerobium lacusrosei* (Photography: Jean-Luc Cayol). The bar represents 2 μm

These species can be considered as moderate halophilic microorganisms because their optimum growth is between 3 and 15 % NaCl. *Halanaerobium lacusrosei* is the only extreme halophilic anaerobic organism (Fig. 10.10), its optimum growth being at about 20 %.

All described species belonging to *Halanaerobiales* are mesophilic organisms. In contrast, *Halothermothrix orenii* (Fig. 6.15l) is the only thermophilic organism described to date; a second fermentative halophilic and thermophilic microorganism was isolated from a salt marsh but it was found to be closely phylogenetically related to members of the order *Clostridiales*. In addition to their ability to colonize terrestrial saline environments, these halophilic anaerobic microorganisms may also occur below the surface and particularly in oil wells. Surprisingly, the microorganisms isolated from these reservoirs belonged only to the genus *Halanaerobium* (*H. kuschneri*, *H. acetethylicum*, *H. salsuginis*, *H. congolense*). It is only recently that a new species (*Halanaerocella petrolearia*), belonging to the family of *Halobacteroidaceae*, was isolated from a deep subsurface hypersaline oil reservoir (Gales et al. 2011).

Many photosynthetic microorganisms have been described including cyanobacteria but also anoxygenic photosynthetic bacteria represented by the genera *Halorhodospira*, *Halochromatium*, and *Ectothiorhodospira*, etc.

10.4.5 Extreme Halophilic Archaea

Extreme halophilic archaea are a diverse group of microorganisms that live in environments with high salt concentrations, the surfaces of salted fish or meat.

Members of the aerobic halophilic archaea are classified in the family *Halobacteriaceae* that require 2 M of NaCl as a minimum for growth. They are heterotrophic organisms.

Various criteria, including sequence analysis of genes encoding 16S rRNA, were used to define the different genera within the extreme aerobic halophiles (Litchfield 2004; Oren 2006). They are frequently called haloarchaea. The genus *Halobacterium*, first described in this group, remains the most studied of the representatives of these haloarchaea. *Natronobacterium*, *Natronomonas* genera and relatives were different from other extreme halophilic by growing optimally at alkaline pH (9–11). These archaeons have also optimal growth at very low concentrations of magnesium.

Extreme halophilic divide by binary fission and do not produce any form of resistance like spores. Most haloarchaea are not motile, but few species may float due to their ability to produce gas vesicles. The structure of the genome of *Halobacterium* and *Halococcus* species are unusual because of the presence of large plasmids which may represent 25–30 % of the total DNA and whose G + C content (57–60 mol%) is significantly different from that of chromosomal DNA (66–68 mol%). Plasmids of extreme halophilic archaea are the largest known so far (DasSarma et al. 2008). Haloarchaea use amino acids and organic acids as energy sources and require growth factors (mainly vitamins) for optimal growth. Some species of the genus *Halobacterium* oxidize sugars, but this ability is relatively rare. Electron transport chains include cytochromes of a, b, and c types which are present in *Halobacterium*. Energy is produced during the aerobic phase, by means of a proton-motive force from chemo-osmotic membrane events. For some species, anaerobic growth has been demonstrated (e.g., sugar fermentation, nitrate, or fumarate reduction).

Anaerobic extreme halophilic archaea also include members of the genera *Methanohalophilus* and *Methanohalobium* of the family *Methanosarcinaceae*. They are essentially methylotrophic using methanol and methylamines as energy sources.

10.4.6 The Halophilic Microorganisms Within the *Eukarya*

Although prokaryotic organisms are dominant in hypersaline ecosystems, eukaryotic microorganisms may inhabit such ecosystems. A variety of obligate and facultative halophytic plants can grow in saline soils. Other invertebrates are either adapted or surviving at high salt concentrations. They comprise algae, diatoms, protozoa, fungi, and even rotifers, tubellarian worms, copepods, or ostracods. Studies carried out in Slovenia have uncovered a variety of fungi pertaining in particular to the genus *Cladosporium* while those conducted in the Dead Sea identified a large number of filamentous fungi belonging to *Zygomycetes* and *Ascomycetes* (Buchalo et al. 2000). Algae of the genus *Dunaliella* were often encountered in the Dead Sea but also in artificial ecosystems such as salt marshes.

10.4.7 Biotechnological Applications

Moderate halophilic microorganisms can be the source of very promising potential technological applications, because some of them produce molecules that are of industrial interest (polymers or compatible solutes, bioemulsifiers, etc.). Many studies on hydrolases produced by halophiles have been reported because of their stability in solutions containing organic solvents (Moreno et al. 2013).

Most of these organisms require few nutrients to grow. They may also use a fairly wide range of organic compounds as carbon and energy sources. These organisms growing at high salt concentrations significantly reduce the risk of contamination in industrial processes.

The biotechnological applications where moderate/extreme halophiles are involved include the fermentation of soy sauce by halotolerant/moderately halophilic bacteria and archaea which is used in the preparation in a Vietnamese fish dipping sauce called "nuoc mam." Moderate halophilic bacteria accumulate large amounts of molecules of low molecular weight in order to face the osmotic pressure. The compatible solutes glycine, betaine, and ectoine can be used as protective agents against salinity stress, heat denaturation, freezing, or drying (Shivanand and Mugeraya 2011). They also have a high interest in the pharmaceutical and cosmetics industries. Recent advances in fermentation processes and genetics of halophilic prokaryotes have rendered possible the overproduction of these compatible solutes (Vargas et al. 2004).

It is also possible to perform gene transfer from moderately halophiles to give to plants, such as wheat, rice, and barley, the capacity to adapt salted environmental conditions and drought. Indigenous halophilic microorganisms to oilfield waters could be extremely valuable for enhancing petroleum recovery by producing polysaccharides having bio-emulsifying properties. These compounds have the properties to resist to heat denaturation. They have higher viscosity values at high temperatures as compared to polymers currently commercialized such as xanthan gum.

For example, *Halomonas eurihalina* can produce large amounts of a polyanionic extracellular polysaccharide (EPS V2-7) with emulsifying properties that can be used in food and pharmaceutical industries.

In recent years, there is an increase in the amount of saline waters rejected by chemical and pharmaceutical industries. Conventional microbiological treatments are not applicable to these salted wastes, but the use of halotolerant/halophilic microorganisms should be taken into consideration for treating them either aerobically or anaerobically. As an example, the treatment of high saline wastewater was successfully for the removal of phenolic or aromatic compounds and highly toxic molecules such as organophosphates (Oren et al. 1992).

Many other potential applications for halophilic microorganisms can be considered for biotechnology (Kivisto and Karp 2011), among which are: the recovery of phosphate and the search for new bioactive molecules such as antibiotics, surfactants, or new enzymes (restriction endonucleases).

Regarding the food industry, carotenes can be extracted from halophilic microorganisms to be used as food additives or as food-coloring agents.

A fascinating application has been considered for a protein originating from the cytoplasmic membrane of *Halobacterium halophilum* (bacteriorhodopsin). This protein acts as a proton pump to establish a gradient between both sides of the membrane (*cf.* Sect. 3.3.4). An electronic microchip composed of a thin layer of bacteriorhodopsin would be able to store more informations than a silicon chip, thus processing the information faster. These microchips will be probably helpful in high performance-computer production and even mass-producing vision systems for robotics.

Considered for long as environments hostile to all forms of life, it is only recently that hypersaline ecosystems have been intensively studied. Microorganisms that inhabit such ecosystems have a great diversity in terms of phylogenetic characteristics, shapes, metabolisms, and also strategies to develop in these hypersaline environments. They have promising biotechnological potentialities that can be used in different industrial sectors (e.g., pharmacy, oil production). Further investigations are however still needed to improve our knowledge on the existing microbial diversity and associated metabolisms in these extreme environments.

10.5 Piezophilic Microorganisms

10.5.1 Presentation

Hydrostatic pressure is a physical parameter, typical for aquatic environments. Hydrostatic pressure is the ratio of a force applied onto an area. A force of 1 N onto a surface of $1\ m^2$ corresponds to a pressure of 1 Pascal (Pa). Other units are sometimes used as shown below but the legal unit is Pascal:

$$1\ Pa = 1.10^{-5} bar = 9.8692.10^{-4} atm = 1.4503.10^{-4} PSI$$
$$= 7.5003.10^{-3} Torr$$

To calculate the hydrostatic pressure applied by a liquid onto a given area, gravity acceleration, volumic mass of the liquid, and depth must be taken into account. More simply, for water (freshwaters or oceans), hydrostatic pressure increases by 10 MPa per kilometer. For instance, at a depth of 3 km in the ocean, hydrostatic pressure is 30 MPa, 300 atm or 300 bars.

The choice of the oceans to explain the different pressure units is justified, since the study of this environment gave birth to the discovery and study of piezophilic microorganisms, sometimes called piezomicrobiology. Since about 2000, the word piezophile (from the Greek "πίεση" which likes pressure) progressively replaced the word barophile (from the Greek "βάροςτΘ", which likes weight) that was used before (cf. Sect. 17.2.3).

Seventy percent of planet Earth is covered by seas and oceans, with an average depth of 3,800 m. However, the maximum depth recorded in the Marianna Trench is 10,790 m, which corresponds to an hydrostatic pressure of 110 MPa. For English-speaking authors, the term "deep sea" corresponds to depth below 1,000 m. Such a depth exists for 88 % of the oceanic area, corresponding to 75 % of the total volume of the oceans, and 62 % of the volume of the biosphere. At the time of these estimations (about 1980), the deep sea was one of the more important biotopes of the planet influenced by hydrostatic pressure. Since that time, subsurface habitats, exposed to elevated pressures and inhabited by microorganisms, have been discovered: deep aquifers, oil reservoirs, or deep-sea sediments. As we will see, most of the Earth's biosphere is inhabited by piezophiles (cf. Sect. 9.6.4).

10.5.2 Deep-Sea Piezopsychrophiles: A Brief History

In addition to depth and darkness, a main feature of deep-sea habitats is a cold temperature, about 2 °C, which is suitable for psychrophiles (cf. Sect. 10.1).

The occurrence of bacteria living in oceanic waters and sediments at a depth of 5,000 m has been shown by the French microbiologist A. Certes, who studied samples collected by the ships "Le Travailleur" and "Le Talisman" (1882–1883). After this pioneer work, and 10 years later that of the US scientist B. Fisher, deep-sea microbiology remained sleepy until the major works of the US microbiologists Claude Zobell and Richard Morita, who can be considered as the "fathers" of marine microbiology (cf. Sect. 2.5).

Because of total darkness, there is no photosynthesis in the deep sea, and consequently, the deep sea is an oligotrophic environment. One of the first topics addressed by deep-sea microbiologists was the degradation of organic matter under deep-sea conditions: low temperature and elevated hydrostatic pressure. Many experiments under atmospheric and hydrostatic pressure were carried out using surface and deep-sea bacterial strains, deep-sea samples decompressed or not. Most experiments showed that heterotrophic activity and aerobic respiration of organic molecules were considerably slowed down by these deep-sea conditions. Probably, very few free-living bacteria adapted to the deep-sea environments existed (Jannasch and Wirsen 1973). The study of some enzymatic activities such as chitinases under elevated hydrostatic pressure also gave similar results: activities were higher under atmospheric pressure. But the microorganisms used for these various experiments, although collected in the deep sea, were mostly originated from the upper layers of the oceans. Further, the analysis of deep-sea amphipods (crustaceans) or holothurians and fishes showed the existence of true deep-sea piezophiles, particularly in their digestive tracts.

10.5.2.1 Some Piezopsychrophilic Bacteria

The first piezopsychrophilic bacteria were isolated and described by A. Yayanos from the Scripps Institution for Oceanography (California, USA). Strain CNPT3, belonging to the genus *Spirillum,* was isolated from a dead amphipod, captured in a trap at a depth of 5,600 m. This organism isolated at 2 °C and 57 MPa grows optimally in between 2 and 4 °C, with a doubling time in between 4 and 13 h. At the same temperature, but under atmospheric pressure, the doubling time increases up to 3–4 days. Later, still from a dead amphipod trapped at a depth of 10,476 m, strain MT41 was isolated. This organism grows optimally at 2 °C and 69 MPa, with a doubling time of 25 h. No growth was observed for pressures below 38 MPa, and the strains lost their ability to form colonies after a few hours of exposure at atmospheric pressure. Later on, a variety of aerobic heterotrophic piezopsychrophilic organisms were isolated often from deep-sea animals, mostly by US and Japanese scientists. Thanks to molecular phylogeny techniques, these strains were described as novel species within the genera *Photobacterium* or *Shewanella* or even novel genera such as *Colwellia* or *Moritella*.

From the study of the growth of several piezophilic strains, Yayanos showed several unique features for these organisms. Particularly, he showed that piezophily was a common feature of cold oceanic waters, at depths below 2,000 m, that seems to be a threshold for piezophily. He also noted that the response to hydrostatic pressure varied with the capture depth of the strain concerned. But in all cases, the optimal pressure for growth of a given strain was always lower than the pressure existing at the place where the original sample was collected.

Although he produced a major contribution to deep-sea microbiology, Yayanos himself noted that many questions remained unanswered. Particularly, all these piezophilic bacteria had been isolated from environments (deep-sea animal digestive tracts) enriched in organic matter, while the deep sea is globally oligotrophic (Yayanos 1986).

Also, total mineralization of organic matter requires several categories of microorganisms (oligoheterotrophs, chemolithoautotrophs); none have been detected by the cultivation techniques used. Discovery of such organisms would allow to claim that piezophily is a unique feature for a majority of deep-sea microorganisms. But according to the slow growth of deep-sea heterotrophic piezopsychrophiles, one can expect weeks or even months for the growth of deep-sea chemolithoautotrophs, which makes their studies very difficult.

10.5.3 Taxonomy and Phylogeny of Piezopsychrophiles

Most of piezopsychrophilic bacteria that have been isolated were also studied for their taxonomy and phylogeny. All belong to the gamma subdivision within the *Proteobacteria*. No correlation was observed between the phylogeny of these organisms and depth, location, or original habitat. While many other extremophiles (halophiles, hyperthermophiles) often belong to distinct phylogenetic lineages, piezopsychrophilic organisms studied up to now are distributed within phylogenetic groups together with piezosensitive species. Adaptations to elevated hydrostatic pressures probably appeared rather recently during a speciation process. The recent utilization of molecular techniques indeed revealed more diversity than the cultivation methods (alpha and delta subdivisions within *Proteobacteria*, *Actinobacteria*, *Planctomycetales*, etc.), but no specific piezophilic lineage was reported so far.

10.5.4 Adaptations to Elevated Hydrostatic Pressure

First studies about adaptation to elevated pressures concerned the fatty acid composition of cellular membranes for the strain CNPT3. When this strain is grown under elevated pressures, $C_{14:1}$, $C_{16:0}$, and $C_{14:0}$ fatty acids decreased, but $C_{16:1}$ and $C_{18:1}$ fatty acids increased (Simonato et al. 2006). More globally, the ratio of unsaturated fatty acids versus saturated fatty acids shifts from 1.9 to 3, when hydrostatic pressure increased from 0.1 to 68 MPa. A similar increase of unsaturated fatty acids was also observed for *Vibrio marinus*, when temperature for incubation decreased from 25 to 15 °C. Similar experiments for strains isolated from depths in a range 1,200 to 10,476 m also showed an increase of concentration of unsaturated fatty acids, particularly for those with carbon chains up to 22 carbons. Such synthesis of long-chain unsaturated fatty acids, which are not very usual for prokaryotes, could contribute to maintain the fluidity of the cellular membrane.

In the case of the strain *Photobacterium* SS9, molecular adaptations to hydrostatic pressures have been shown that again took place in the outer cellular membrane. When this strain originated from a depth of 2,500 m is cultivated under hydrostatic pressure, several outer membrane proteins are repressed while some novel proteins are expressed. Among them the 37-KDa protein OmpH (outer membrane protein high pressure) is a nonspecific porin, the transcription level of which increases under hydrostatic pressure (Campanaro et al. 2005). Other environmental factors, such as cellular density or carbon availability, may also influence the expression of *Omph* gene: the same promotor can be activated either by a high cellular density at 0.1 MPa or a low cellular density at 27 MPa. Other adaptive mechanisms controlled by genes organized in operons and regulated by hydrostatic pressure may exist, but most of adaptations to hydrostatic pressure reported for piezopsychrophiles concern the cellular membrane and transmembrane transports. The modification of membrane fatty acids or associated proteins aims at maintaining functionality of exchanges between the cell and its environment (*cf*. Sect. 9.6.4).

10.5.5 A Particular Case: Deep-Sea Hydrothermal Vents

Deep-sea hydrothermal vents were discovered in 1977, at the East Pacific Rise. They are mostly located along oceanic ridges at depths from 800 to 4,000 m. Steep gradients exist particularly for temperature. Within a few decimeters, temperature shifts from 2 °C (deep-sea water) to 350 °C or more for the hydrothermal fluid (water remains liquid at such elevated temperatures because of high pressure). These very hot temperatures incited microbiologists to search for hyperthermophilic microorganisms, and, therefore, many novel species and genera, particularly in the *Archaea* domain, have been described (*cf*. Sect. 10.3).

10.5.5.1 Responses to Hydrostatic Pressure

The majority of hyperthermophilic microorganisms isolated from deep-sea hydrothermal vents were cultivated under a pressure of 0.2 MPa, in order to avoid boiling of culture media. However, despite their deep-sea origins, very few were then exposed to elevated pressures to study their responses to high pressure. However, some strains appeared piezosensitive, other piezotolerant, and some others piezophilic. For heterotrophic anaerobic piezophiles, an increase of the growth rate and an increase of the maximum temperature for growth (in average 1–4 °C) were observed as a response to increasing pressures. Piezophily was also observed for hyperthermophilic methanogenic *Archaea*, but for the more elevated temperatures, growth and methane

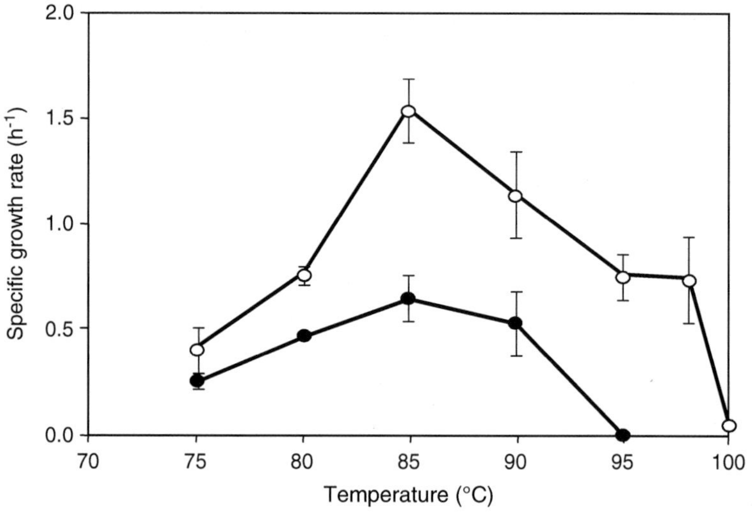

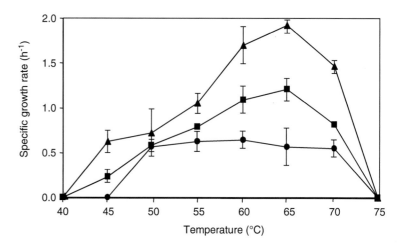

Fig. 10.11 Effects of temperature and hydrostatic pressure on the growth rate of *Thermococcus barophilus*. *Open circles*, growth at 0.3 Mpa; *black circles*, growth at 40 Mpa (Courtesy of General Society of Microbiology)

Fig. 10.12 Effects of temperature and hydrostatic pressure on the growth rate of *Marinotoga piezophila*. *Black circles*, growth at 0.3 Mpa, after previous growth under hydrostatic pressure; *black squares*, growth at 0.3 Mpa after previous cultures under atmospheric pressure; *black triangles*, growth at 40 Mpa (Courtesy of General Society of Microbiology)

production were uncoupled: pressure increased maximal temperature for methanogenesis only.

Several microbiologists carried out enrichment cultures from deep-sea hydrothermal vent samples under elevated temperatures and pressures, using specific pressurized bioreactors. This method allowed to isolate *Thermococcus barophilus*, an heterotrophic anaerobic hyperthermophilic Archaea. For this organism which also grows under atmospheric pressure, growth rate and maximal temperature for growth increased under pressure (Fig. 10.11). When grown at atmospheric pressure, this organism expressed a stress protein that confirmed its piezophilic feature. Similarly, *Marinotoga piezophila*, isolated from another deep-sea hydrothermal vent sample using the same procedure, grew much better under hydrostatic pressure, up to 40 MPa (Fig. 10.12). Under these conditions, cells of *Marinitoga* are coccoid shaped, while for lower pressures and particularly at atmospheric pressure, cells did not separate and formed filaments (Fig. 10.13). Interestingly, similar observations were reported for *Escherichia coli*, but for increasing pressures. For these two organisms, cell division is affected by pressure, but in opposite directions.

During experiments on deep-sea hyperthermophiles cultivated under elevated pressures, an interesting point was reported by several microbiologists. For piezophilic species, hydrostatic pressure allowing maximum growth is always higher than pressure existing at the place of sample collection. This is the opposite of what was reported for deep-sea psychrophiles (*cf.* Sect. 10.2). These observations lead John Baross and Jody Deming to ask a question: do hyperthermophilic microorganisms isolated from black smoker walls live permanently there? May they come from reservoirs located beneath the oceanic floor and therefore exposed to higher pressures? (Deming and Baross 1993). Several observations following deep-sea volcanic eruptions could support this hypothesis. But, also, several deep environments, all exposed to high pressures, were observed these last years.

Fig. 10.13 Microphotographs under phase-contrast microscopy of *Marinotoga piezophila* after growth at 40 Ppa (**a**), 10 Mpa (**b**) and 0.3 Mpa (**c**) (Photograph: courtesy of Daniel Prieur)

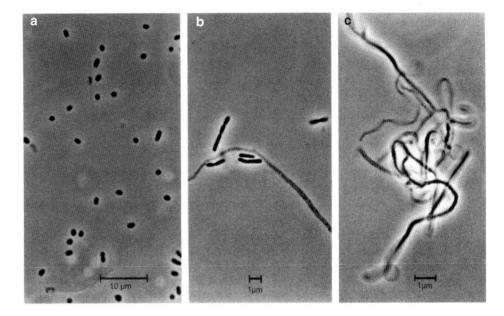

Two interesting studies, carried out independently, were published on this topic. The first study reported the discovery of an obligate piezophilic hyperthermophile isolated from a deep-sea hydrothermal vent sample, collected at a depth of 4,200 m: *Pyrococcus yayanosii*. This organism grows in a temperature ranges from 85 to 105 °C. The optimum pressure for growth is 52 MPa at 98 °C, but no growth was observed for pressure below 20 MPa. At 105 °C, optimum pressure is extended to 60 MPa. Growth was still detectable at 130 MPa and 110 °C, and *P. abyssi* cells could survive at 125 °C and 145 MPa for at least 48 h. The second study concerned analysis of marine sediments collected at 1,600 m beneath the seafloor, itself at a depth of 3.5 km. In these old sediments (110 million years) and most probably hot (100 °C), novel molecular signatures of *Pyrococcus* and *Thermococcus* (hyperthermophilic *Archaea*) were detected.

Since these independent studies showed that obligate piezophilic hyperthermophiles and deep sediment hyperthermophiles (certainly piezophilic) exist, the hypothesis of the deep, thermophilic, and piezophilic biosphere is enhanced.

10.5.6 Other Piezophiles to Be Discovered

Aquifers, oil reservoirs, and deep sediment layers are environments exposed to high pressures. Although the occurrence of microorganisms living in these ecosystems has been reported and studied, very few data about pressure are available. Nevertheless, according to estimations by Whittman, the majority of Prokaryotes would live in subsurface environments, particularly beneath the seafloor (Whitman et al. 1998). But if the effect of pressure remains a question, the most important questions concern the energy and carbon sources available. Do these nutrients come from the surface? Are these deep subsurface microorganisms dependant, although not directly, from photosynthesis and solar energy? Are they definitely buried in deep sediment layers or transported by deep circulating fluids? Are they active or under a dormant stage? What is their doubling time (certain estimations suggest thousands of years)? Do they use dihydrogen and carbon dioxide produced during abiotic reactions as energy and carbon sources? If the answer to the last question is yes, these organisms would be totally independent from solar energy, like the first organisms living about 4 billion years ago. Certainly, the study of these recently discovered subsurface environments and their inhabitants, all obligate piezophiles, would contribute to raise novel hypothesis about the origin of life on Earth.

10.6 Acidophilic Microorganisms

10.6.1 Presentation

Organisms that thrive under extreme conditions have recently attracted considerable attention due to their peculiar physiology, ecology, as well as their biotechnological potential (Rawlings 2002). Acidic environments are especially interesting because, in general, the low pH of the habitat is the consequence of microbial metabolism (Ehrlich 2002) and not to condition imposed by the system, as is the case for other extreme environments (temperature, ionic strength, high pH, radiation, pressure, etc.). The case of the animal stomach was discussed in Chap. 9 (*cf.* Sect. 9.6.3).

Acidic, metal-rich environments have two major origins. The first one is associated with volcanic activities. The acidity in these locations may derive from the microbial oxidation of elemental sulfur:

$$S^\circ + 1.5O_2 + H_2O \rightarrow SO_4^{2-} + 2H^+$$

which is produced as a result of the condensation reaction between oxidized and reduced sulfur-containing gases:

$$2H_2S + SO_2 \rightarrow 3S^\circ + 2H_2O$$

In these environments, temperature gradients are easily formed. These sites may therefore be colonized by a variety of acidophilic microorganisms, with different optimal temperatures. Acidic, metal-rich environments can also be found associated to mining activities. Metal and coal mining operations expose sulfidic minerals to the combined action of water and oxygen, which facilitate microbial attack, producing the so-called acid mine drainage (AMD), a serious environmental problem. The most abundant sulfidic mineral, pyrite, is of particular interest in this context.

The biooxidation of pyrite has been studied in detail (*cf.* Sects. 14.5.2 and 16.9.2). The process occurs in several steps, with the overall reaction being:

$$FeS_2 + 3.5H_2O + 3.75O_2 \rightarrow Fe(OH)_3 + 2SO_4^{2-} + 4H^+$$

These environments vary greatly in their physicochemical characteristics and also in their microbial ecology. High temperature may also occur as a consequence of biological activity, facilitating colonization by thermophilic acidophiles.

The mechanism by which microbes obtain energy by oxidizing sulfidic minerals, a process of biotechnological interest known as bioleaching, has remained controversial for many years (Ehrlich 2002). The recent demonstration that ferric iron, present in the cell envelopes and in the extracellular polysaccharides of leaching microorganisms is responsible for the electron transfer from the sulfidic minerals to the electron transport chain, has clarified this issue, with important fundamental and applied consequences (Sand et al. 2001).

The differences observed during bioleaching of diverse metallic sulfides (MS) depend on the type of the chemical oxidation promoted by ferric iron, which is determined by the crystallographic characteristics of the mineral. Under normal conditions of pressure and temperature, pyrite and two other sulfides, tungstenite and molybdenite, can only be oxidized by ferric iron through the so-called thiosulfate mechanism with sulfate as the main product:

$$FeS_2 + 6Fe^{3+} + 3H_2O \rightarrow S_2O_3^{2-} + 7Fe^{2+} + 6H^+$$

$$S_2O_3^{2-} + 8Fe^{3+} + 5H_2O \rightarrow 2SO_4^{2-} + 8Fe^{2+} + 10H^+$$

Most other sulfides (e.g., galena, chalcopyrite, sphalerite, etc.) are susceptible to proton attack as well as to ferric iron oxidation through a different mechanism, the so-called polysulfide mechanism:

$$8MS + 8Fe^{3+} + 8H^+ \rightarrow 8M^{2+} + 4H_2S_n + 8Fe^{2+} (n \geq 2)$$

$$4H_2S_n + 8Fe^{3+} \rightarrow S_8^\circ + 8Fe^{2+} + 8H^+$$

In this case, elemental sulfur is produced which can be further oxidized by sulfur-oxidizing bacteria to sulfuric acid.

The reduced iron (Fe^{2+}) produced in all these reactions is reoxidized by iron-oxidizing microorganisms to ferric iron (Fe^{3+}):

$$4Fe^{2+} + O_2 + 4H^+ \rightarrow 4Fe^{3+} + 2H_2O$$

The main role of acidophilic chemolithotrophic microorganisms is to maintain a high concentration of the chemical oxidant, ferric iron, in solution.

The acidophilic strict chemolithotroph, *Acidithiobacillus ferrooxidans* (formerly *Thiobacillus ferrooxidans*), was first isolated from an acidic pond in a coal mine, more than 50 years ago (Colmer et al. 1950). Although *A. ferrooxidans* can obtain energy oxidizing both reduced sulfur and ferrous iron, much more attention has been paid to the sulfur oxidation reaction due to bioenergetic considerations (Ehrlich 2002). The discovery that some strict chemolithotrophs (*Leptospirillum ferrooxidans*) could grow using ferrous iron as its only source of energy and that this microorganism is mainly responsible for metal bioleaching and AMD generation has completely changed this perspective (Rawlings 2002). Furthermore, it is now well established that iron can be oxidized anaerobically, coupled to anoxygenic photosynthesis or to anaerobic respiration using nitrate as an electron acceptor (Widdel et al. 1993; Benz et al. 1998).

Most of the characterized strict acidophilic microorganisms have been isolated from volcanic areas or AMD from mining activities (Hallberg and Johnson 2001). Río Tinto (Iberian Pyritic Belt, SW Spain) (Fig. 10.14) is an unusual ecosystem due to its acidity (mean pH 2.3, buffered by Fe^{3+} iron), size (100 km), high concentration of heavy metals (Fe, As, Cu, Zn, Cr...), and an unexpected level of microbial diversity (Amaral-Zettler et al. 2003). Recently, it has been proved that the extreme conditions of the Tinto system are much older than the oldest mining activities known in the area, strongly suggesting that they are natural and not the consequence of industrial contamination (Fernández-Remolar et al. 2003, 2005). Due to its size and accessibility, Río Tinto is considered a convenient model system for acidic environments.

10.6.2 Diversity of Extreme Acidic Environments

Acidic environments are poorly characterized environments due to the physiological peculiarities of the microorganisms associated to them. Furthermore, strict acidophilic chemolithotrophs are, in general, not easy to grow, especially in solid media (Hallberg and Johnson 2001).

Prokaryotic microorganisms that are metabolically active in extreme acidic environments are distributed in the domains *Bacteria* and *Archaea* and can be classified according to their energy and carbon sources (*cf.* Sect. 9.6.3).

A variety of chemoautolithotrophic microorganisms capable of oxidizing iron and sulfur-containing minerals have been isolated from acidic environments. The mineral-oxidizing bacteria found in these natural environments are ubiquitous and the most commonly encountered have been characterized as *Acidithiobacillus ferrooxidans*, *Leptospirillum* spp., *Acidithiobacillus thiooxidans*, *Acidithiobacillus caldus* and *Sulfobacillus* spp. in *Bacteria* Domain, and members of the archaeal *Ferroplasmaceae* family.

In acidic environments there is also a variety of acidophilic heterotrophs such as those belonging to the genera *Acidiphilium* or facultative heterotrophs such as *Acidimicrobium* spp. and *Ferromicrobium* spp. Acidophiles can also be categorized using other physiological criteria, such as temperature (mesophiles, moderate thermophiles, thermophiles, and hyperthermophiles), optimal pH, or on the basis of their carbon source (autotrophs, heterotrophs, and mixotrophs). In general, the most extremely thermophilic acidophiles correspond to the domain *Archaea*.

10.6.2.1 Bacteria Domain

The majority of the best characterized acidophilic microorganisms correspond to the bacterial domain. Acidophiles are widely spread out on the bacterial phylogenetic tree (Fig. 10.15). Most of them are Gram-negative (all the species of the *Acidithiobacillus* genus, *Leptospirillum*

Fig. 10.14 View of Peña de Hierro, in the head section of Río Tinto in Spain (Photography: Ricardo Amils)

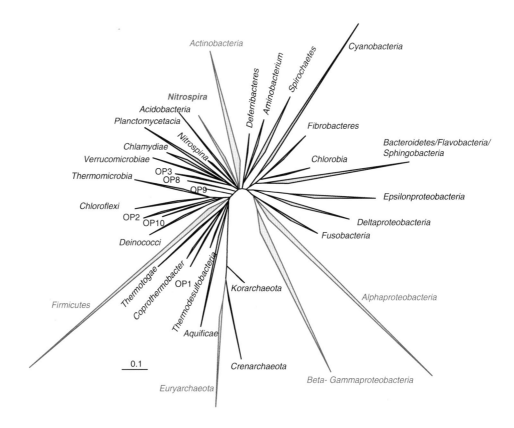

Fig. 10.15 Phylogenetic relationship (in *red*) of the acidophilic microorganisms detected in the Tinto Basin (Modified from Aguilera et al. 2010)

Table 10.2 Comparison of phenotypic properties of acidophilic iron-oxidizing bacteria

Characteristics	*Leptospirillum ferrooxidans*	*Acidithiobacillus ferrooxidans*	*Ferrimicrobium acidophilus*	*Sulfobacillus acidophilus*	*Acidimicrobium ferrooxidans*
Cell morphology	Vibrios 1 µm, spirilla	Rods 1–2 µm	Rods 1–3 µm	Rods 3–5 µm	Rods 1–1.5 µm, filaments
Optimal pH	1.5–2.0	2.5	2.0–2.5	2.0	2.0
Mol %G + C	51–56	58–59	52–55	55–57	67–68
S^0 oxidation	−	+	−	+	−
Iron reduction	−	+	+	+	+
Endospores	−	−	−	+	−
Motility	+	+	+	+	+
Utilization of yeast extract	−	−	+	+	+
Growth at 50 °C	−	−	−	+	−

spp., and *Acidiphilium* spp.), but recently several Gram-positive microorganisms corresponding to different genera have been isolated from acidic environments and biohydrometallurgical operations (*Sulfobacillus* spp., *Acidimicrobium* spp., and *Ferrimicrobium* spp.). Table 10.2 compares the main characteristics of representative chemolithotrophic acidophilic bacteria.

Regarding *Acidithiobacillus*, *A. ferrooxidans* was the first strict chemolithotrophic acidophilic microorganism isolated from acidic mining waters. This Gram-negative bacterium is a 0.5 × 1.0 µm motile rod, using CO_2 as a carbon source and obtaining its energy from the oxidation of ferrous iron, elemental sulfur, and reduced sulfur compounds. *A. ferrooxidans* was believed for many years to be the dominant bacterium responsible for metal sulfides solubilization. *A. ferrooxidans* is generally assumed to be an obligate aerobic organism. However, under anaerobic conditions, ferric iron can replace dioxygen as an electron acceptor for the oxidation of elemental sulfur (Brock and Gustafson 1976). At pH 2, the free energy of the reaction is negative ($\Delta G^{\circ\prime} = -314$ kJ.mol^{-1}), so it can be used for energy transduction reactions.

$$S^0 + 6Fe^{3+} + 4H_2O \rightarrow H_2SO_4 + 6Fe^{2+} + 6H^+$$

Acidithiobacillus thiooxidans is a phylogenetically close relative of *A. ferrooxidans*. As *A. ferrooxidans*, *A. thiooxidans* is an autotrophic Gram-negative sulfur oxidizer, but the main difference is its inability to oxidize iron (Kelly and Wood 2000).

Acidithiobacillus caldus is a Gram-negative moderately thermophilic acidophile found in mining environments such as coal spoil heaps or industrial bioreactors, where the oxidative dissolution of sulfide-containing minerals occurs. This bacterium obtains its carbon by reductive fixation of atmospheric CO_2. *A. caldus* can oxidize a wide range of reduced sulfur compounds but, like *A. thiooxidans*, is incapable of oxidizing ferrous iron or pyrite. The main product of the oxidation of reduced sulfur compounds is H_2SO_4, so this bacterium is able to grow at rather low pH.

Other acidophilic microorganisms are grouped within the genus *Leptospirillum* and are also Gram-negative autotrophic acidic bacteria (optimal pH 1.5–1.8) that can only grow using reduced iron as a source of energy. As a result, *Leptospirillum* spp. have a high affinity for ferrous iron, and, unlike *A. ferrooxidans*, its ability to oxidize ferrous iron is not inhibited by ferric iron. It is aerobic and extremely tolerant to low pHs. Based on phylogeny studies, three species have been defined: *L. ferrooxidans* (optimal growth temperature from 26 to 30 °C), *L. ferriphilum* (optimal temperature between 30 and 40 °C), and *L. ferrodiazotrophum*. Interestingly enough, *L. ferrooxidans* is the dominant species in a natural ecosystem like Río Tinto, while *L. ferriphilum* dominates in other acidic environments, like Iron Mountain and many industrial bioreactors.

Species of genus *Acidiphilium* are acidophilic heterotrophic Gram-negative bacteria, belonging to the *Alphaproteobacteria* phylum, which normally appears to be associated with strict chemolithoautotrophic bacteria like *Acidithiobacillus* or *Leptospirillum* spp. These microorganisms are able to respire reduced carbon compounds using ferric iron as an electron acceptor. Some strains can reduce ferric iron in the presence of dioxygen. They are used in the isolation of strict chemolithotrophs to eliminate organic acids that inhibit their growth.

Among low G + C Gram-positive bacteria, *Sulfobacillus* spp. are acidophilic moderate thermophilic, endospore-forming bacteria. They have been isolated from geothermal environments and biohydrometallurgical processes (*cf.* Sect. 16.9). *Sulfobacillus* spp. grow autotrophically between 40 and 60 °C, using ferrous iron, sulfur, and sulfide minerals as energy sources. The members of this genus are able to grow mixotrophically in media with ferrous iron and yeast extract or heterotrophically using glucose as carbon and energy source.

Regarding *A. ferrooxidans*, it is a moderately Gram-positive thermophilic bacterium that grows autotrophically oxidizing iron in aerobic conditions or reducing it in anaerobic conditions. Different strains have been isolated with different carbon requirements.

Within acidophilic Gram-positive bacteria, *Desulfosporosinus* spp. are able to use sulfate as an electron acceptor for anaerobic respiration (commonly named sulfate-reducing bacteria (SRB)). The presence of SRB in acidic environments has been clearly established (González-Toril et al. 2003, 2005) where they are likely to use hydrogen and glycerol as energy sources. It is the case of the first acidophilic sulfate-reducing bacterium which has been recently isolated and named *Desulfosporosinus acidophilus* (Alazard et al. 2010); metabolism of these microorganisms is very appropriate to develop methodologies for metal removal from industrial contaminated waters.

10.6.2.2 Archaea Domain

Acidophilic aerobic or facultative anaerobic archaea that colonize biotopes such as AMD and solfatara fields, where iron and sulfur are in reduced forms, are represented by two different phylogenetic groups, the orders *Sulfolobales* and *Thermoplasmatales*. These groups differ in their phenotypic properties. Some representatives of the *Sulfolobales*, e.g., *Acidianus brierleyi* or members of the genus *Metallosphaera*, obtain energy by oxidizing sulfur, sulfidic minerals, and reduced iron. Other species of the order *Sulfolobales*, e.g., *Sulfolobus acidocaldarius* and *S. solfataricus*, utilize only sulfur and reduced sulfur compounds. *Sulfolobus metallicus* exploits sulfidic ores and elemental sulfur as energy sources. Although other members of the *Sulfolobales* are able to grow chemolithoautotrophically, *S. metallicus* and *Acidianus ambivalens* are the only obligate chemolithoautotrophs known. In contrast, in the order *Thermoplasmatales*, only species of the genus *Ferroplasma*, *F. acidiphilum* and *F. acidarmanus*, are acidophilic chemolithotrophs capable to oxidize iron, although they require traces of yeast extract for autotrophic growth. *F. acidarmanus*, able to grow at pH near zero, has been isolated from biofilms of pyritic sediments in the Iron Mountain mine in California (Edwards et al. 2000). *Ferroplasma*-like microorganisms are detected recurrently in different commercial bioleaching plants, indicating that they must play an important role in this type of artificial ecosystem.

The rest of the genus *Thermoplasma* and *Picrophilus* corresponds to heterotrophic archaea. In the Tinto ecosystem, only members of the *Thermoplasmatales* order, *Thermoplasma acidophilum*, and *Ferroplasma acidiphilum* have been identified.

10.6.2.3 Eukarya Domain

An important level of eukaryotic diversity has been reported associated to acidic environments, mainly AMD (Amaral-Zettler et al. 2003). Unfortunately there is scarce information about the eukaryotes that are able to deal with a proton gradient of five orders of magnitude through their membranes, most of them without any extracellular structure that could help to control the acidity of the habitat in which they develop. In addition, they are exposed, so they must be resistant, to extremely high concentrations of toxic heavy metals, normally present in acidic environments. In the Tinto Basin, members of the phylum *Chlorophyta* (*Chlamydomonas*, *Chlorella*, and *Euglena*) are the most frequent species followed by two filamentous algae belonging to the genera *Klebsormidium* and *Zygnemopsis*. The most acidic part of the river is inhabited by eukaryotic community dominated by two species of the genera *Dunaliella* and *Cyanidium*, well known for their acidity and heavy-metal tolerance. Among the eukaryotic decomposers, fungi are very abundant and exhibit great diversity, including yeast and filamentous forms; the mixotrophic community is dominated by cercomonads and stramenopiles. The protistan consumer community is characterized by two species of ciliates, tentatively assigned to the genera *Oxytrichia* and *Euplotes*. Amoebas related to the genera *Valkampfia* and *Noegleria* can be found in the most acidic part of the river, and one species of heliozoan belonging to the genus *Actinophyris* seems to be the most characteristic predator of the benthic food chain of the river.

10.6.3 Microbial Ecology of an Extreme Acidic Environment: Río Tinto

The combined use of conventional (enrichment cultures, isolation, and phenotypic characterization) and molecular ecology methods (DGGE, FISH, and cloning) has led to identify the most representative microorganisms of the Río Tinto Basin (González-Toril et al. 2003). Eighty percent of the diversity of the water column corresponds to microorganisms belonging to two species, *Leptospirillum ferrooxidans*, *Acidithiobacillus ferrooxidans*, and several species of the genus *Acidiphilium*, all conspicuous members of the iron cycle.

All *L. ferrooxidans* isolates from Río Tinto are aerobic iron oxidizers. *A. ferrooxidans* can oxidize iron aerobically and reduce it anaerobically. All *Acidiphilium* isolates can oxidize reduced organic compounds using ferric iron as an electron acceptor (anaerobic respiration) and some isolates can do so in the presence of oxygen. Although other iron-oxidizing (*Ferroplasma* spp., *Thermoplasma acidophilum*, *Acidimicrobium* spp., and *Ferrimicrobium* spp.) or iron-reducing (*Acidimicrobium* spp. and *Ferrimicrobium* spp.) microorganisms have been detected in the Tinto ecosystem, their low numbers suggest that they play a minor role in the operation of the iron cycle, at least in the water column.

Concerning the sulfur cycle, only *A. ferrooxidans* is found in significant numbers in the water column of the Tinto Basin. This bacterium can oxidize ferrous iron and reduced sulfur compounds. The oxidation of reduced sulfur compounds can be carried out aerobically and anaerobically.

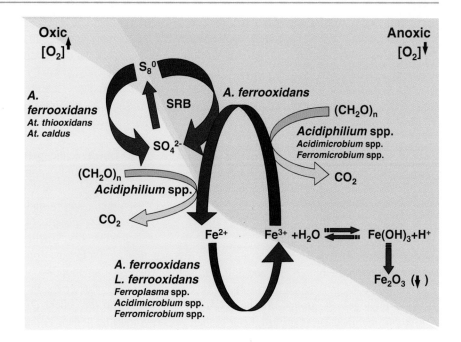

Fig. 10.16 Geomicrobiological model of Río Tinto. The role of the different isolated and identified microorganisms is shown associated to the iron and sulfur cycles operating in the Tinto Basin

Some sulfate-reducing activity has been detected associated to sediments in certain parts of the river.

Considering the geomicrobiological characteristics of the Tinto ecosystem, we postulate that the river is predominantly under the control of iron (González-Toril et al. 2003). Iron is the main product of bioleaching of pyrite and other iron-bearing sulfidic mineral, like chalcopyrite, both of which are present in high concentrations in the Iberian Pyritic Belt. The activity of iron-oxidizing prokaryotes is responsible for both the solubilization of sulfidic minerals and the correspondent high concentration of ferric iron, sulfate, and protons in the water column. Ferric iron is a strong oxidant able to oxidize exposed sulfidic minerals.

Sulfur-oxidizing microorganisms are responsible for the generation of sulfuric acid from the oxidation of the elemental sulfur produced by the polysulfide mechanism attack of acid soluble sulfides. Figure 10.16 shows a geomicrobiological model of the different acidophilic microorganisms interacting with the iron and sulfur cycles in the Tinto ecosystem (*cf.* Sect. 14.5.2).

Iron has different properties of ecological interest, which gives to the Tinto ecosystem an interesting perspective. Iron is not only a source of energy for iron-oxidizing prokaryotes but can also be used as an electron acceptor for anaerobic respiration. In addition, ferric iron is responsible for the constant pH of the ecosystem.

$$Fe^{3+} + 3H_2O \leftrightarrow Fe(OH)_3 + 3H^+$$

The dilution of the river by rain and/or neutral tributaries promotes the precipitation of ferric hydroxide and the generation of protons, which maintain the acidity of the river. Although this reaction is reversible, the dilution factor is stronger than evaporation, thus an important part of the iron remains precipitated along the river, giving rise to iron bioformations. Accordingly, the concentration of soluble iron decreases gradually from the origin to the mouth of the river. Furthermore, acidic ferric iron solutions readily absorb harmful UV radiation, protecting the organisms growing in its waters. Figure 10.16 shows the geomicrobiological model of the Tinto ecosystem in which the role of the different components in the iron and sulfur cycles is displayed.

This iron-controlled scenario seems reasonable for the chemolithotrophic prokaryotes found in the Tinto ecosystem and other acidic environments. However, given that eukaryotic diversity in these environments is much greater than prokaryotic diversity and that most of the primary production of the system derives from the activity of photosynthetic protists, what is the advantage, if any, for eukaryotes to develop in an extreme acidic environment? A possible answer to this question may be linked to the limited availability of iron in a neutral pH world.

Although iron is an extremely important element for life, it is a limiting factor for growth at neutral pH. Organisms have developed very specific and elaborate mechanisms to trap iron anywhere they can find it. Why is this so, if iron is one of the most abundant elements on Earth? In an oxidized atmosphere at neutral pH, soluble iron is rapidly oxidized into insoluble compounds, which are incorporated into anaerobic sediments where sulfate-reducing bacteria (SRB) may further transform them into pyrite, an even less reactive iron mineral. The geological recycling of these sediments and the geomicrobiology associated to the iron cycle are the only ways to reintroduce this critical element into the biosphere. A possible advantage for the eukaryotes thriving in the extreme conditions of the Tinto Basin is an unlimited iron supply provided by the chemolithotrophs growing on the rich iron sulfides of the Iberian Pyritic Belt.

Has this iron-limiting scenario been a constant in the history of life on Earth? According to different authors, the Archaean oceans held important concentrations of soluble iron with extremely low concentration of oxygen. In these conditions, it is reasonable to postulate the appearance of microorganisms able to obtain energy oxidizing or reducing iron, in other words, maintaining operative an anaerobic iron cycle. It has been suggested that the sulfur cycle was not fully operative at that time. Although microbial sulfur oxidation was probably active, different authors consider that sulfate-reducing activity was not operative until the Proterozoic, which gave rise to a massive sequestering of heavy metals and a correspondent severe nitrogen crisis with important evolutionary consequences.

As can be seen the history of iron in the biosphere is still an open question. There is no doubt about the importance of microorganisms in iron mobilization. At present, it is not clear whether the Tinto Basin and the related ecosystems are a relic of the period in the history of life on our planet when iron was freely available or simple a recent adaptation to an extreme environment product of chemolithotrophy and able to maintain enough ferric iron in solution to generate a constant acidic pH environment. Concerning adaptation, preliminary physiological studies of acidophilic photosynthetic protists suggest that they can adapt rather easily to extreme acidic environments. If complex cellular systems can do it, it is reasonable to assume that less complex systems, like prokaryotes, should not have any problem. As mentioned, the level of eukaryotic diversity in the Tinto ecosystem seems to be much higher than the prokaryotic one. This apparent contradiction could result from the limitations of our preliminary knowledge of the system. It could well be that prokaryotic diversity is much higher in the unexplored anaerobic sediments or in the subsurface of the Iberian Pyritic Belt. In any case, it should be kept in mind that we are dealing with extant ecosystems. Conveniently addressed questions should facilitate a deeper characterization of these systems, which could help to clarify its origin and the role of the different components of the habitat.

In conclusion, the exploration of different extreme acidic environments has been able to answer some basic questions concerning:

1. The microbial generation of the extreme conditions of the habitats
2. The isolation and characterization of the iron- and sulfur-metabolizing microorganisms responsible for the active iron and sulfur cycles fully operative in these systems

3. The identification of an unexpected level of eukaryotic diversity
4. The characterization of iron bioformations

Understanding these ecosystems is necessary not only to increase our knowledge of unusual habitats with implications in different areas – geomicrobiology, paleontology, evolution, astrobiology, etc. – but to improve the performance of biotechnology processes, such as biomining, and to search for new applications for the microorganisms thriving at extreme low pH.

10.7 Alkaliphilic Microorganisms

10.7.1 Presentation

Microorganisms whose development is favored at high pH values can be divided into two categories: (*1*) alkaline-tolerant microorganisms able to grow at a pH greater than or equal to 9, with optimal at neutral pH, and (*2*) alkaliphilic microorganisms with optimal growth at pH above 9 (*cf.* Sect. 9.6.3). Alkaliphiles can be further subdivided into facultative alkaliphiles and obligate alkaliphiles. The former group of bacteria grows at a pH greater than 9 but also in the neutral pH range, while the latter group of bacteria cannot grow at pH below 8.5–9.

Some alkaliphiles are also halophiles (haloalkaliphiles); they are able to grow in the presence of NaCl concentration up to saturation. Alkaliphilic microorganisms are ubiquitous; unlike other extremophiles found mainly in the corresponding extreme environment, alkaliphiles can be found not only in high pH environments but also in neutral pH environments. Alkaliphiles are of considerable interest regarding the production of enzymes.

There are also big challenges in discovering novel microorganisms and novel mechanisms that they use to adapt to these harsh conditions of life.

10.7.2 Alkaline Environments

The alkalinity of the natural alkaline environments results from the presence of high levels of sodium carbonate. Globally there are two types of alkaline environments differing by their salinity (NaCl): the alkaline saline environments and the weakly salted alkaline environments. The formers are represented by the soda lakes; they correspond to the most stable natural alkaline environments on Earth. They are found in closed ponds situated in the dry zones where high rates of evaporation contribute to concentrate salts and carbonates. Their salinity can reach 35 % in NaCl and pH values range between 8 and 12. They are characterized by

Table 10.3 Worldwide distribution of natural soda environments (soda lakes and deserts)

Africa	
Botswana	Sua salt pan
Libya	Lake Fezzan
Egypt	Wadi El Natrun
Ethiopia	Lake Abjata, Lake Aranguadi, Lake Kilotes, Lake Shala, Lake Chilu, Lake Herlate, Lake Metahara
Sudan	Dariba lakes
Kenya	Lake Bogoria, Lake Nakuru, Lake Elmenteita, Lake Magadi, Lake Simbi, Lake Sonachi
Ouganda	Lake Katwe, Lake Mahega, Lake Kikorongo, Lake Nyamunuka, Lake Munyanayange, Lake Murumuli, Lake Nunyampaka
Tanzania	Lake Natron, Lake Embagi, Lake Magad, Lake Manyara, Lake Balangida, Basotu, Lake Crate, Lake Kusare, Lake Tulusia, El Kekhooito, Lake Momela, Lake Lekandiro, Lake Reshitani, Lake Lgarya, Lake Ndutu, Lake Ruckwa North
Tchad	Lake Bodu, Lake Rombou, Lake Djikare, Lake Momboio, Lake Yoan
North America	
Canada	Manito
United States	Alkali Valley, Albert Lake, Soap Lake, Big Soda Lake, Owens Lake, Mono Lake, Deep Springs, Rhodes Marsh, Soap Lake, Summer Lake, Surprise Lake, Pyramid Lake, Walker Lake
Central America	
Mexico	Lake Texcoco
South America	
Venezuela	Langunilla Valley
Chili	Antofagasta
Asia	
Central Asia	Tuva lakes
Turquie	Lake Van
India	Lake Looner, Lake Sambhar, Lake Lonar, Chilika lagoon, Pangong Lake
China	Dali Lake, Qinghai Hu, Sui-Yian, Heilungkiang, Horsemeno Lake, Kirin, Jehol, Chahar, Shansi, Shensi, Kansu, Meizhou bay, Tengchong, Xiarinaoer Lake, Yuncheng Lake, Yunnan
Japan	Toyotomi, Iwayama bay (Palau)
Korea	Keusman
Tibet	Lake Soda
Australia	Lake Corangamite, Lake Red Rock, Lake Waerowrap, Lake Chidnup
Europe	
Hungary	Lake Fehér, Kelemen-szék, Kiskunsag National Park
Slovenia	Pecena Slatina
Russia	Kulunda Steppe, Lakes Tanatar, Karakul, Araxes lowlands, Chita, Barnaul, Slavgorod
Greenland	Ikka fjord
Ekho Lake (Antarctica), Sargasoo Sea	

very high concentrations of sodium carbonate (Na_2CO_3) and very low level in ions calcium (Ca^{2+}) and magnesium (Mg^{2+}). Soda lakes and soda deserts are widely distributed throughout the world (Table 10.3). Soda lakes can be found in Asia (Turkey, India, China), in Europe (Hungary, ex-Yugoslavia), and in countries of the ex-USSR. The best studied soda lakes are the lakes of the Rift Valley Eastern African. Their formation results from the combination of geological, geographical, and climatic factors: leaching of the surrounding rocks rich in carbonates by ground and runoff waters, accumulation in a depression, and high rate of evaporation. Generation of alkalinity is the result of a displacement of the balance $CO_2/ HCO_3^-/ CO_3^{2-}$ toward a dominance of CO_3^{2-} due to the absence of Ca^{2+}. When they are present, Ca^{2+} ions decrease the alkalinity and cause the precipitation of carbonates in insoluble calcite ($CaCO_3$) as in the case of brine waters derived from seawater. These waters generally have high concentrations of calcium ions and have a neutral pH value, even when being highly concentrated because the calcium ion concentration always exceeds that of carbonate. The weakly salted alkaline environments rarely occur.

Thermal springs with a high calcium content are another type of highly alkaline natural environment found in different regions: California, Oman, Cyprus, the former Yugoslavia, and Jordan. The chemical composition of such waters is determined by CO_2-mediated weathering of the calcium and magnesium minerals such as olivine and pyroxene, which result in the release of OH⁻, and subsequent generation of alkaline conditions. OH⁻ together with Mg^+ ions precipitates to serpentine and lime $(Ca(OH)_2)$. The pH of such waters can reach values higher than 11. Some preliminary studies of these sources report low bacterial populations whose composition is similar to that of soils and less alkaline waters (Grant and Sorokin 2011).

Termites harbor a variety of microorganisms in different compartments of their gut. The first compartment of the termite gut has an elevated pH, between 10 and 12, and is very rich in K^+ ions. However, most of alkaliphilic bacteria isolated from the gut of termites are alkalitolerant. Microorganisms developing in alkaline pH are also frequently isolated from soils. Soil microbial processes, especially ammonification, can cause localized soil alkalinization. It is recognized that the metabolism of some alkaliphilic bacteria can locally modify the pH, thus creating alkaline microniches. In other cases, it is the metabolic activity of other microorganisms which raises the pH sufficiently high to allow the development of alkaliphiles. Most of isolated alkaliphilic bacteria from such non-alkaline environments are Gram-positive, spore-forming bacteria belonging to the genus *Bacillus*. Nonspore-forming alkalitolerant bacteria belong to genera *Aeromonas*, *Corynebacterium*, *Micrococcus*, *Paracoccus*, *Pseudomonas*, and *Streptomyces*.

Finally, there are anthropogenic alkaline environments. They are associated with the use of alkaline products (e.g., sodium hydroxyde) in various industrial processes (electroplating, food and paper industries) or derived from the production of lime or from calcium hydroxide (cement, blast furnace). The few studies conducted in microbiology indicate that these environments are inhabited by alkalitolerant and alkaliphilic bacteria similar to those isolated from non-alkaline environments.

10.7.3 Microbial Diversity of Soda Lakes

Despite the extreme conditions existing in such environments, the different trophic groups composing the microflora are similar to those composing the microbial communities already reported in traditional environments, except that all are alkalitolerant or alkaliphilic (Grant 1992). Table 10.4 shows all the bacterial genera isolated from alkaline environments. The most studied haloalkaliphilic communities were primary producing communities (cyanobacteria and purple anoxygenic bacteria) and anaerobic fermentative bacteria. The fermentative bacteria are mainly represented by Gram-positive bacteria with low G + C%.

Recently, the bacteria responsible for the oxidation of reduced inorganic compounds such as methane, dihydrogen, sulfide, and ammonia produced in haloalkaline anaerobic sediments of soda lakes have been identified. However,

Table 10.4 Genera comprising alkalitolerant and alkaliphilic microorganisms

Domain Bacteria
Phylum Cyanobacteria
Cyanobacteria
    Arthrospira, Spirulina, Anabaenopsis, Cyanospira
Phylum Actinobacteria
 Actinobacteria
  Actinomycetales
   Cellulomonadaceae
    Cellulomonas
   Dietziaceae
    Dietzia
   Micrococcaceae
    Arthrobacter, Nesterenkonia
   Nitriliruptoraceae
    Nitriliruptor
   Bogoriellaceae
    Bogoriella, Georgenia
Phylum Bacteroidetes
 Bacteroidia
  Bacteroidales
   Marinilabiliaceae
    Alkaliflexus (O_2 tolerance)
 Cytophagia
  Cytophagales
   Cytophagaceae
    Rhodonellum, Litoribacter
   Cyclobacteriaceae
    Indibacter, Mongoliicoccus, Nitritalea
Phylum Firmicutes (Gram-positive bacteria)
Bacilli
 Bacillales
  Bacillaceae
   Anoxybacillus, Amphibacillus, Alkalibacillus, Bacillus
   Caldalkalibacillus, Exiguobacterium, Oceanobacillus
   Paenibacillus, Natronobacillus
  Staphylococcaceae
   Salinicoccus
 Lactobacillales
  Carnobacteriaceae
   Alkalibacterium, Carnobacterium, Marinilactibacillus
Clostridia
 Clostridiales
  Clostridiaceae
   Clostridium (cluster XI), *Alkaliphilus*, Anaerovirgula
   Anoxynatronum, Natronincola, Tindallia
  Eubacteriaceae
   Alkalibacter
  Peptococcaceae
   Desulfotomaculum
  Syntrophomonadaceae
   Anaerobranca, Dethiobacter
  Heliobacteriaceae
   Heliorestis

(continued)

Table 10.4 (continued)

Halanaerobiales	*Oceanospirillaceae*
Halobacteroidaceae	*Marinospirillum*
Fuchsiella, Halonatronum, Natroniella	*Pseudomonadales*
Natranaerobiales	*Pseudomonadaceae*
Natranaerobiaceae	*Pseudomonas*
Natranaerobius	*Thiotrichales*
Phylum Proteobacteria	*Piscirickettsiaceae*
Alphaproteobacteria	*Methylophaga, Thioalkalimicrobium*
Rhizobiales	*Vibrionales*
Bradyrhizobiaceae	*Vibrionaceae*
Nitrobacter	*Salinivibrio*
Rhodobacterales	Gammaproteobacteria non classified
Rhodobacteraceae	*Alkalimonas*
Maritimibacter; Rhodobaca, Roseinatronobacter, Roseibaca	**Phylum Spirochaetes**
	Spirochaetes
Betaproteobacteria	*Spirochaetales*
Nitrosomonadales	*Spirochaetaceae*
Nitrosomonadaceae	*Spirochaeta*
Nitrosomonas	***Domain Archaea***
Deltaproteobacteria	**Phylum Euryarchaeota**
Desulfobacterales	Halobacteria
Desulfobacteraceae	*Halobacteriales*
Desulfonatronobacter	*Halobacteriaceae*
Desulfobulbaceae	*Natrialba, Natrinema, Natronoarchaeum, Natronobacterium*
Desulfurivibrio, Desulfopila	
Desulfovibrionales	*Natronomonas, Natronorubrum, Natronolimnobius*
Desulfohalobiaceae	*Natronococcus, Halorubrobacterium (Halorubrum)*
Desulfonatronospira, Desulfonatronovibrio	Methanobacteria
Desulfonatronaceae	*Methanobacteriales*
Desulfonatronum	*Methanobacteriaceae*
Desulfuromonadales	*Methanobacterium*
Geobacteraceae	Methanomicrobia
Geoalkalibacter	*Methanosarcinales*
Gammaproteobacteria	*Methanosarcinaceae*
Alteromonadales	*Methanosalsum, Methanohalophilus*
Alteromonadaceae	
Nitrincola, Salinimonas	
Moritellaceae	
Paramoritella	
Chromatiales	
Chromatiaceae	
Thioalkalicoccus, Marichromatium	
Ectothiorhodospiraceae	
Alkalilimnicola, Alkalispirillum, Ectothiorhodospira, Halorhodospira, Thiorhodospira, Thiohalospira, Thioalkalivibrio	
Thioalkalispiraceae	
Thioalkalispira	
Methylococcales	
Methylococcaceae	
Methylomicrobium (Methylohalobius)	
Oceanospirillales	
Halomonadaceae	
Halomonas	

some metabolic pathways which are possible in terms of thermodynamics have not yet been identified within haloalkaliphiles. It is noteworthy that microorganisms which may oxidize acetate in soda lakes were not identified so far, nor those who could benefit from the oxidation of NO during the course of heterotrophic denitrification.

10.7.3.1 Phototrophic Primary Producers

A characteristic feature of the majority of soda lakes is their color. According to the chemistry of their water, soda lakes can be green, pink, red, or orange because of their permanent or seasonal abundance of microorganisms. This reflects an extremely high primary carbon production more than 10 g C m^{-2} d^{-1}. Indeed, soda lakes are mainly located in areas of the world that benefit of high light intensities and where the temperature is high (30–35 °C). Phosphate availability and virtually unlimited availability to CO_2 in the carbonate-rich waters also promote photosynthesis. The main contributors to primary production of organic matter in the soda lakes are prokaryotic microorganisms: cyanobacteria and anaerobic phototrophic bacteria also called purple bacteria.

Filamentous cyanobacteria (*Cyanospira*) and unicellular organisms (*Chroococcus* and *Pleurocapsa*) have been described. In some soda lakes, algal mats result mainly from development of filamentous cyanobacteria of the genera *Spirulina*, *Anabaenopsis*, and *Arthrospira*. Cyanobacteria also contribute in primary production by their ability to atmospheric nitrogen fixation.

Weakly alkaline lakes are generally dominated by large clusters of cyanobacteria while hyper-alkaline lakes such as Lake Magadi contain both cyanobacteria and anoxygenic phototrophic bacteria belonging to the genera *Ectothiorhodospira* and *Halorhodospira*. Primary productivity of these haloalkaliphilic phototrophic bacteria is probably at the origin of all biological processes in soda lakes. In the mid-1990s, new anoxygenic phototrophic bacteria have been isolated from soda lakes. These bacteria include not only purple bacteria but also heliobacteria (*Clostridia*). The latter differ from haloalkaliphilic species such as *Halorhodospira*, both in terms of phylogeny and physiology, by requiring in particular low concentrations of NaCl for growth. These isolates are obligate alkaliphilic. Many of these isolates belong to the genera *Rhodobaca*, *Thiorhodospira* and *Thioalkalicoccus* (all purple bacteria), and *Heliorestis* (*Heliobacteria*). Members of the genus *Rhodobaca* present a particular interest because:

1. They are devoid of light collecting device type II antena, considered as an original property in non-sulfur purple bacteria.
2. These phototrophic microorganisms synthesize several original carotenoids responsible for their yellow color in cultures.

Highly salted soda lakes (from 20 % NaCl to NaCl saturation) have a different primary microflora. Cyanobacteria are missing and it is *Ectothiorhodospira* spp., which are responsible for primary production. The dominant microorganisms are haloalkaliphilic *Archaea*, with population reaching up to 10^7–10^8 archaeons per mL. Haloalkaliphilic archaea are classified in two genera: *Natronobacterium* and *Natronococcus*, containing 5 and 4 species, respectively.

10.7.3.2 Alkaliphilic Aerobic Microflora

Although the alkaline environments are usually shallow and limited in dioxygen (eutrophic), they have very dense populations of aerobic non-phototrophic organotrophic bacteria that use the products of photosynthesis as well as the products of anaerobic degradation. Relatively few isolates were identified and deposited in public collections of microorganisms. Regarding the Gram-negative bacteria, they belong to the *Gammaproteobacteria* and include many proteolytic microorganisms related to the genus *Halomonas*. Proteolytic Gram-negative isolates were also affiliated to genera *Pseudomonas* and *Stenotrophomonas*, while other isolates were phylogenetically related to typical aquatic bacteria such as *Aeromonas*, *Vibrio*, and *Alteromonas*. Recent studies on bacterial communities of low saline lakes of Siberia and lakes of the Rift Valley in Kenya have revealed the presence of aerobic chemolithotrophic sulfur-oxidizing bacteria of the genera *Thioalkalimicrobium*, *Thioalkalivibrio*, and *Thioalkalispira* belonging to the *Gammaproteobacteria*. More recently, a new aerobic heterotrophic bacterium, *Alkalilimnicola halodurans* of the family *Ectorhodospirillaceae*, was isolated from sediments of Lake Natron. Furthermore, several alkaliphilic methanotrophic bacteria of the family *Methylococcaceae* were isolated from alkaline moderately saline lakes of Central Asia and Kenya. They include *Methylomicrobium alcaliphilum*, *Methylomicrobium buryatense*, and *Methylomicrobium kenyense*. At high pH values (pH 10–10.5), *M. kenyense* was shown to oxidize ammonia to nitrite and organic sulfur compounds.

Chemolithotrophic nitrite-oxidizing bacteria (*Nitrobacter alkalicus*) belonging to the *Alphaproteobacteria* were isolated from soda lakes. These nitrifying bacteria play an important role in the nitrogen cycle by converting inorganic nitrogen into nitrate.

Aerobic Gram-positive bacteria of the high and low G + C% phyla have been described. Bacteria with low G + C% mainly belong to the genus *Bacillus* while those with high G + C% were affiliated to *Dietza*, *Arthrobacter*, and *Terrabacter* genera. Several *Bacillus* species have recently been reclassified into new genera phylogenetically related to the genus *Bacillus*. They include *Anoxybacillus*, *Natronobacillus*, and *Oceanobacillus* genera. Members of genus *Natronobacillus* are capable of dinitrogen fixation.

Alkalispirillum and *Alkalilimnicola* genera were originally described as aerobic non-photosynthetic heterotrophic microorganisms with remarkable tolerance to alkalinity. Although the possibility of anaerobic growth with nitrate was briefly mentioned for *Alkalilimnicola*, the denitrification potential has not been studied in detail, nor the capacity to grow lithoautotrophically. A study of the arsenic cycle in Mono Lake in California revealed a new chemolithoautotrophic metabolism with arsenic as the sole electron donor and nitrate as electron acceptor. During this process, arsenite is oxidized to arsenate with the simultaneous reduction of nitrate to nitrite. The bacterium responsible for this reaction is strain MLHE-1. It is a new member of the group *Alkalispirillum-Alkalilimnicola*. In addition, this bacterium is able to grow lithoautotrophically with sulfide and hydrogen as electron donors and nitrate as electron acceptor. It also grows heterotrophically on acetate under aerobic and denitrifying conditions. This is a clear demonstration of a remarkable metabolic versatility possibly existing in the group of haloalkaliphilic *Gammaproteobacteria* which was not reported so far.

Screening of many alkaline lakes with high salinity in different parts of the world indicated that haloalkaliphilic archaea belonging to the family *Halobacteriaceae*, also known under the generic name of "halobacteria" or "haloarchaea," were always present in such environments. These microorganisms are the most halophiles that we know. They are part of the dominant microflora when salinity conditions reach saturation. They are responsible for the red color of the water due to their carotenoid content. The soda lakes are inhabited by members of the genera *Natronobacterium*, *Natronococcus*, *Natronomonas*, *Natrialba*, *Natrorubrum*, and *Halorubrum* (Wiegel 2011).

10.7.3.3 Alkaliphilic Anaerobic Microflora

Compared to aerobic environments, haloalkaline anaerobic environments have been poorly studied. A biological process that predominates in soda lakes is sulfate reduction. This process is responsible in not just the final stage of degradation of organic matter but also, in part, the conditions that generate alkalinity as a result of the conversion of sulfate to sulfide. The first alkaliphilic sulfate-reducing bacterium *Desulfonatronovibrio hydrogenovorans*, a member of the *Deltaproteobacteria*, was isolated from the soda lake Magadi in Kenya. This species was also found in the soda lakes of the region of Tuva in Central Asia, which prompted some authors to suggest that this species might play a universal role as final hydrogen acceptor in alkaliphilic anaerobic communities producing sulfide. Another sulfate-reducer member of the *Deltaproteobacteria*, *Desulfonatronum lacustre*, was isolated from an oligotrophic alkaline lake, the lake Khadyn (Tuva). Both genera comprise mesophilic nonsporulated Gram-negative species, which require the presence of carbonate, and are unable to grow at pH values below 8. A spore-forming alkaliphilic and moderately thermophilic sulfate reducer, *Desulfotomaculum alkaliphilum*, was isolated from pig slurry at neutral pH, indicating that the adaptation of microorganisms to alkaline environments does not result from important genetic changes.

There is no doubt that methanogenesis occurs in soda lakes. To date, the methanoarchaea isolated from soda lakes are predominantly methylotrophic microorganisms using methylated compounds such as methanol or methylamines. These compounds are abundant in soda lakes. They probably originated from the anaerobic degradation of cyanobacterial mats and/or from compatible solutes (e.g., betaine and ectoine) transported or synthesized by alkaliphilic organotrophs. Methylotrophic methanoarchaea isolated from several soda lakes with salt concentrations close to saturation belong to the genera *Methanosalsus* and *Methanohalophilus*, family *Methanosarcinaceae*, phylum *Euryarchaeota*. Hydrogenotrophic methanogens within genus *Methanobacterium* were also identified in several soda lakes. However, these microorganisms appear to be alkalitolerant rather than true alkaliphiles. Generally, studies on alkaliphilic methanogens reveal that methanogenesis from H_2/CO_2 increases when salinity decreases, while under high salinity conditions, methanogenesis from methylamines predominates.

Acetogenesis (*Natroniella acetigena*) and nitrate reduction (*Halomonas campisalis*, *Thioalkalivibrio nitratireducens*) are metabolic activities already detected in alkaline environments. *Natroniella acetigena* is a haloalkaliphilic bacterium, which belongs to *Halobacteroidaceae* (order *Halanaerobiales*). *Halonatronum saccharophilum*, a chemoorganotrophic moderately haloalkaliphilic member of this order, was isolated from sediments of the lake Magadi, and more recently, the first alkaliphilic anaerobic hydrogenotrophic homoacetogenic bacterium (*Fuchsiella alkaliacetigena*) was isolated from sediments of the soda-depositing soda lake Tanatar III (Altay, Russia).

Microorganisms fermenting aminoacids were isolated in soda lakes. Both species *Natronincola histidinovorans* and *Tindallia magadiensis* were also isolated from lake Magadi. They are closely phylogenetically related to the group XI of the *Clostridium* taxon.

10.7.4 Homeostasis of the pH

Alkaliphilic bacteria have to face a bioenergetical problem which arised from the necessity of maintaining a cytoplasmic pH that is much lower than optimal external pH for growth. These bacteria have developed during the course of evolution adaptation mechanisms to prevent alkalinization of their cytoplasm. Regulatory mechanisms of ion transport exist in the cell membrane. In normal conditions, these mechanisms maintain the cytoplasmic pH in a narrow range around neutrality, even when the extracellular pH varies significantly. They are known under the term "pH homeostasis" (Krulwich et al. 2011).

Alkaliphilic microorganisms, which grow abundantly at a pH higher than 9, require the presence of Na^+ ions in the surrounding environment for the effective transport of solutes across the cytoplasmic membrane. Studies on alkaliphilic *Bacillus* showed that a Na^+ gradient provided the energy necessary to transport and motility by the creation of a proton-motive force engendering ATP synthesis. According to the chemiosmotic hypothesis, proton-motive force in the cell is generated by the electron transport chain or by the excretion of protons originated from ATP metabolism by the ATP synthase. The protons are then reintroduced into cells by co-transportation with other substrates. In Na^+-dependent transport systems, protons are exchanged with Na^+ by an H^+/Na^+ antiporter, generating a sodium-motive force which transports the substrates with Na^+ ions into the cells. The specific requirement of alkaliphilic bacteria for Na^+ is dependant of an efficient pathway of Na^+ ions reintroduction into the cell, thus allowing the antiporter to operate. At least two mechanisms are important, the Na^+/solute symporters and the sodium channels. The precise functioning of these processes remains an interesting aspect of the work to be conducted with alkaliphilic bacteria (*cf.* Sect. 9.6.3).

10.7.5 Biotechnological Applications of Alkaliphiles

Microbial communities of natural environments such as alkaline soda lakes have for a long time attracted the attention of industry because of the many possible uses of their enzymes or metabolites in industrial processes (Horikoshi 2011). Alkaliphiles have a great commercial interest in the production of enzymes such as alkaline proteases commonly used in biological washing powders and in the leather industry. Starch industry uses alkaline cyclodextrin glycotransferases in order to obtain cyclodextrins from starch for use in the cosmetics, pharmacology, chemistry, and food industry. Alkaline amylases and xylanases have also many applications in the paper industry. Studies of alkaline enzymes mainly focused on the most commonly observed microorganisms in natural environments for which isolation is easy. So, many alkaliphilic *Bacillus* species have been isolated through the work of Horikoshi (2011).

The biotechnological potential of some archaea in particular was also evaluated as they contain lipids with a stereochemical configuration very different from that of bacteria. The archaeal haloalkaliphilic *Natronococcus* strain Ah-36 synthesizes an extracellular amylase which allows the production of maltotriose. The gene for this enzyme was cloned and expressed in *Haloferax volcanii*. The majority of haloalkaliphiles contain pigments involved in the translocation of ions across the cell membrane. Bacteriorhodopsin, as a light-driven proton pump, and the halorhodopsin, as a pump of chlorides for entering the cell, open large biotechnological perspectives. They may be used in diverse applications such as holography and information storage. The production of *Spirulina* biomass as a source of proteins is still used in Africa and in other parts of the world. The European Space Agency is considering its use for organic matter recycling in long-duration space flights.

Despite harsh settled conditions existing (extreme pH values, most often coupled to high salinity), in soda lakes, these environments are inhabited by a wide range of microorganisms that include the major phylogenetic and trophic groups of bacteria and archaea including fermentative, nitrate- and sulfate-reducing bacteria, methanoarchaea, and haloarchaea. The functioning of basic carbon, nitrogen, and sulfur cycles is particularly active in soda lakes. Although alkaline environments have been intensively studied during the last years, microbial diversity and community structure that develop and the role played by some alkaliphiles are not fully understood. Subsequently our knowledge of the distribution and taxonomic diversity of such extremophiles will undoubtedly improve, thanks to the help of (1) microbiologists with isolation and characterization of novel microorganisms and (2) genome sequencing programs and total DNA sequencing directly from alkaliphilic environment.

10.8 Conclusion

10.8.1 Extremophiles and Exobiology

The term "exobiology" (now astrobiology for English-speaking authors) was first used by the Nobel Prize winner Joshua Lederberf to name the science devoted with life, its origin, evolution, and distribution in the universe. Such a topic probably exists (although unnamed) since the beginning of humanity. This topic is very complex because those who study it (human beings) have no reference on life for comparison, with the exception of their own references as inhabitants of Earth.

There is no doubt that planet Earth, formed by accretion more than 4.5 billion years ago, was more hostile at its origin, than today. Particularly, temperature was too much more elevated to keep water liquid. Now, liquid water is obligatory required for life (such as we know). 100–200 million years later, liquid water existed on Earth, temperature was still rather high, volcanism was very active, and huge meteorites frequently hit the surface of Earth (the impact of an object with about the size of Mars was at the origin of the Moon) up to 3.8 billion years ago.

One billion years after accretion, it seems that life had arose, as suggested by analysis of ancients rocks where fossils assigned to ancient life forms, presumably anaerobes, were reported. About one billion years later, dioxygen is present in the atmosphere; one billion years more and the first Eukaryotes appeared (*cf.* Chap. 4).

The discovery of extremophiles and particularly the hyperthermophiles has deeply changed our vision of life on Earth. Life is not only installed in environments suitable for human beings, animals, and plants: life abounds (essentially via microscopic and prokaryotic forms) in places of Earth where humans cannot venture without a protection or using complex technologies.

Since physicochemical conditions were most probably extreme on the primitive Earth, one can imagine that actual extremophiles are the progeny of the first inhabitants of Earth. This hypothesis is particularly attractive, looking at the universal phylogenetic tree of life based on 16S rRNA genes. For that tree, the most ancient branches for *Bacteria* and *Archaea* correspond to hyperthermophilic organisms. However, this hypothesis received several criticisms, but one must say that the question of temperature (high, moderate, low) that allowed the origin of life is still unanswered.

In addition to the extreme values of physiochemical parameters to which extremophiles are exposed nowadays, the metabolism of extremophiles has a great interest. Some of them utilize organic carbon as carbon and energy sources through various processes of fermentation. But many extremophiles are lithotrophic (inorganic energy source), anaerobic (their terminal electron acceptor is not molecular oxygen), and autotrophic, using carbon dioxide as sole carbon source. Actually, many extremophiles and mostly hyperthermophiles thrive in environments where nutritional conditions may resemble to those expected on primitive Earth. Some extremophiles that live in subsurface environments seem to be independent from solar energy and compounds produced by photosynthesis: it is what occurred more than 2.5 billion years ago. In this respect, the study of actual extremophiles may considerably help us to understand our far origins. But, for sure, the whole planet and its biological diversity are not totally explored yet.

Several centuries ago, the Italian Dominican monk Giordano Bruno was burned after he claimed that many planets moved around their sun, such as our 7 (right number at that time) planets around the sun, and many organisms lived there. While in 2010 NASA's Phoenix probe explored Mars, the European ExoMars probe is preparing to drill the red planet, and a sophisticated satellite should go to the moons of Jupiter in less than 10 years to finally prove the existence of an ocean of liquid water beneath the ice crust of Europa.

Clearly, there was liquid water on Mars when life appeared on Earth. What happened then on the red planet? Did life start? Did life still exist, perhaps in the subsurface? Or, are there some fossils or signatures somewhere? In a further step, after automatic space vessels that carried out in situ analysis, one may expect that other vessels will collect and bring to Earth Martian samples, for analysis after severe quarantine. Later on, manned vehicles will land on Mars. Dreams?

Europa, a natural satellite of Jupiter, is an interesting target since liquid water is suspected beneath its thick ice cover, probably because a tide effect from Jupiter. At the bottom of this 100-km deep ocean, a silicate core should exist. Consequently, water and rocks are in contact and a hydrothermal circulation could exist. In the case of Europa, exobiologists will wait for a long time to get answers. After one or several orbiter missions, a lander mission is required, then a drilling mission (at least a few meters, but the ice cover to drill to reach the ocean is several kilometers thick). And a journey to Jupiter takes about 5–7 years!

Hopefully, on Earth, analogues to Europa ocean exist: the subglacial lakes on Antarctica, such as lake Vostok. Deep drilling through Vostok ice revealed molecular signatures of microorganisms, some being thermophiles that could indicate a hydrothermal activity at depth.

If we move out of the solar system, more than 100 extrasolar planets (moving around remote stars) have been discovered. Some recently observed have masses equivalent to several Earths. In the next future, planets with the size of Earth, located in the habitable zone from their star (liquid water, acceptable temperatures, etc.), are expected.

For all these targets, exobiologists, planetologists, and astrophysicists search for traces, marks, and signatures for life. The outstanding properties and capabilities of terrestrial extremophiles have pushed further the physiochemical, metabolic, and geographical limits for life. Consequently, plausible habitats are more and more numerous. Technologies required for exploration are more and more powerful. Will the twenty-first century bring any answer?

References

Aguilera A, Gonzalez-Toril E, Souza-Egipsy V, Amaral-Zettler L, Zettler E, Amils R (2010) Phototrophic biofilms from Rio Tinto, an extreme acidic environments, the prokaryotic component. In: Seckbach J, Oren A (eds) Microbial mats. Modern and ancient microorganisms in stratified systems. Springer, Heidelberg, pp 471–481

Alazard D, Joseph M, Battaglia-Brunet F, Cayol JL, Ollivier B (2010) *Desulfosporosinus acidiphilus* sp. nov.: a moderately acidophilic sulfate-reducing bacterium isolated from acid mining drainage sediments. New taxa: Firmicutes (Class *Clostridia*, Order *Clostridiales*, Family *Peptococcaceae*). Extremophiles 14:305–312

Amaral-Zettler LA, Gómez F, Zettler E, Keenan BG, Amils R, Sogin ML (2003) Eukaryotic diversity in Spain's River of Fire. Nature 417:137

Basen M, Sun J, Adams M (2012) Engineering a hyperthermophilic archaeon for temperature-dependent product formation. mBio 3: e00053-12. doi:10.1128/mBio.00053-12

Ben Hania W, Ghodbane R, Postec A, Brochier-Armanet C, Hamdi M, Fardeau M, Ollivier B (2011) Cultivation of the first mesophilic representative ("mesotoga") within the order *Thermotogales*. Syst Appl Microbiol 34:581–585

Benz M, Brune A, Schink B (1998) Anaerobic and aerobic oxidation of ferrous iron at neutral pH by chemoheterotrophic nitrate-reducing bacteria. Arch Microbiol 169:159–165

Biddle J, Cardman Z, Mendlovitz H, Albert D, Lloyd K, Boetius A, Teske A (2012) Anaerobic oxidation of methane at different temperature regimes in Guaymas Basin hydrothermal sediments. ISME J 6:1018–1031

Borin S et al (2008) Sulfur cycling and methanogenesis primarily drives microbial colonization of the highly sulfidic Urania deep hypersaline basin in the Mediterranean Sea. Proc Natl Acad Sci USA 106:9151–9156

Brochier-Armanet C, Boussau B, Gribaldo S, Forterre P (2008) Mesophilic *Crenarchaeota*: proposal for a third archaeal phylum, the *Thaumarchaeota*. Nat Rev Microbiol 6:245–252

Brock TD, Gustafson J (1976) Ferric iron reduction by sulfur- and iron-oxidizing bacteria. Appl Environ Microbiol 32:567–571

Buchalo AS, Nevo E, Wasser SP, Volz PA (2000) Newly discovered halophilic fungi in the Dead Sea (Israel). In: Seckback J (ed) Journey to diverse microbial worlds. Adaptation to exotic environments. Kluver Academic, Dordrecht, pp 241–252

Campanaro S, Vezzi A, Vitulo N, Lauro FM, D'Angelo M, Simonato F, Cestaro A, Malacrida G, Bertoloni G, Valle G, Bartlett DH (2005) Laterally transferred elements and high pressure adaptation in *Photobacterium profundum* strains. BMC Genomics 6:122

Cita MB (2006) Exhumation of Messinian evaporites in the deep-sea and creation of deep anoxic brine-filled collapsed basins. Sediment Geol 188–189:357–378

Colmer AR, Temple KL, Hinkle HE (1950) An iron-oxidizing bacterium from the acid drainage of some bituminous coal mines. J Bacteriol 59:317–328

D'Amico S, Collins T, Marx JC, Feller G, Gerday C (2006) Psychrophilic microorganisms: challenges for life. EMBO Rep 7:385–389

Daffonchio D et al (2006) Stratified prokaryote network in the oxic–anoxic transition of a deep-sea halocline. Nature 440:203–207

DasSarma S, Capes M, DasSarma P (2008) Haloarchaeal megaplasmids. In: Schwartz E (ed) Megaplasmids. Springer, Berlin/Heidelberg, pp 3–30

Deming JW, Baross JA (1993) Deep-sea smokers: windows to a subsurface biosphere. Geochim Cosmochim Acta 57:3219–3230

Edwards KJ, Bond PL, Gihrin TM, Banfield JF (2000) An archaeal iron-oxidizing extreme acidophile important in acidic mine drainage. Science 287:1796–1798

Ehrlich HL (2002) Geomicrobiology, 4th edn. Marcel Dekker, New York

Erauso G, Prieur D, Godfroy A, Raguenes G (1995) Plate cultivation techniques for strictly anaerobic, thermophilic, sulfur-metabolizing archae. In: Robb FT, Place AR (eds) Archaea: a laboratory manual. Thermophiles. Cold Spring Harbor Laboratory Press, Cold Spring Harbor, pp 25–29

Feller G (2013) Psychrophilic enzymes: from folding to function and biotechnology. Scientifica. Article ID 512840, 28 pages. doi:10.1155/2013/512840

Fernández-Remolar D, Rodríguez N, Gómez F, Amils R (2003) The geological record of an acidic environment driven by iron hydrochemistry: the Tinto River system. J Geophys Res 108:6. doi:10.1029/2002JE001918

Fernández-Remolar D, Morris RV, Gruener JE, Amils R, Knoll AH (2005) The Río Tinto basin, Spain: mineralogy, sedimentary geobiology and implications for interpretation of outcrop rocks at Meridiani Planum, Mars. Earth Planet Sci Lett 240:149–167

Flores GE, Wagner I, Liu Y, Reysenbach A-L (2012) Distribution, abundance, and diversity patterns of the thermoacidophilic "Deep-sea Hydrothermal Vent Euryarchaeota 2" (DHVE2). Front Microbiol 3:47

Gales G, Chehider N, Joulian C, Battaglia-Brunet F, Cayol J-L, Postec A, Borgomano J, Neria-Gonzalez I, Lomans BP, Ollivier B, Alazard D (2011) Characterization of *Halanaerocella petrolearia* gen. nov., sp. nov., a new anaerobic moderately halophilic fermentative bacterium isolated from a deep subsurface hypersaline oil reservoir. Extremophiles 15:565–571

Godfroy A, Postec A, Raven N (2006) Growth of hyperthermophilic microorganisms for physiological and nutritional studies. In: Rainey FA, Oren A (eds) Methods in microbiology, extremophiles. Academic, Oxford, pp 93–108

González-Toril E, Llobet-Brossa E, Casamayor EO, Amann R, Amils R (2003) Microbial ecology of an extreme acidic environment, the Tinto River. Appl Environ Microbiol 69:4853–4865

Gonzalez-Torril E, Garcia-Moyano A, Amils R (2005) Phylogeny of prokaryotic microorganisms from the Tinto River. In: Harrison STL, Rawlings DE, Pedersen J (eds) IBS-2005. Compress, Cape Town, pp 737–749

Grant WD (1992) Alkaline environments. In: Lederberg J (ed) Encyclopedia of microbiology, vol 2. Academic, New York, pp 73–84

Grant WD, Sorokin DY (2011) Distribution and diversity of soda lakes alkaliphiles. In: Horikoshi K, Antranikian G, Bull AT, Robb FT, Stetter KO (eds) Extremophiles handbook, vol 1. Springer, Tokyo/Dordrecht/Heidelberg/London/New York, pp 27–54

Hallberg KB, Johnson DB (2001) Biodiversity of acidophilic prokaryotes. Adv Appl Microbiol 49:37–84

Hébraud M, Potier P (2000) Cold acclimation and cold shock response in psychrotrophic bacteria. In: Inouye M, Yamanaka K (eds) Cold shock response and adaptation. Horizon Scientific Press, Norfolk, pp 41–60

Horikoshi H (2011) Enzymes isolated from Alkaliphiles. In: Horikoshi K, Antranikian G, Bull AT, Robb FT, Stetter KO (eds) Extremophiles handbook, vol 1. Springer, Tokyo/Dordrecht/Heidelberg/London/New York, pp 163–181

Jannasch HW, Wirsen CO (1973) Deep-sea microorganisms: in situ response to nutrient enrichment. Science 180:641–643

Kelly DP, Wood AP (2000) Reclassification of some species of *Thiobacillus* to the newly designated genera *Acidithiobacillus* gen. nov., *Halothiobacillus* gen. nov. and *Thermithiobacillus* gen. nov. Int J Syst Evol Microbiol 50:511–516

Kivisto AT, Karp MT (2011) Halophilic anaerobic fermentative bacteria. J Biotechnol 152:114–124

Krulwich TA, Liu J, Morino M, Fujisawa M, Ito M, Hicks DB (2011) Adaptative mechanisms of extreme alkaliphiles. In: Horikoshi K, Antranikian G, Bull AT, Robb FT, Stetter KO (eds) Extremophiles handbook, vol 1. Springer, Tokyo/Dordrecht/Heidelberg/London/New York, pp 119–139

La Cono V et al (2011) Unveiling microbial life in new deep-sea hypersaline Lake Thetis. Part I: prokaryotes and environmental settings. Environ Microbiol 13:2250–2289

Larsen H (1962) Halophilism. In: Gunsalus IC, Stanier RY (eds) The bacteria, vol 4. Academic, New York, pp 297–342

Litchfield CD (2004) Microbial molecular and physiological diversity in hypersaline environments. In: Ventosa A (ed) Halophilic microorganisms. Springer, Berlin/Heidelberg, pp 49–58

Luo H, Robb FT (2011) Thermophilic protein folding systems. In: Horikoshi K, Antranikian G, Bull AT, Robb FT, Stetter KO (eds) Extremophiles handbook, vol 1. Springer, Tokyo/Dordrecht/Heidelberg/London/New York, pp 583–599

Madigan M, Martinko J (2007) Brock biologie des micro-organismes. Pearson Education France, Paris

Mikucki JA, Han SK, Lanoil BD (2011) Ecology of psychrophiles: subglacial and permafrost environments. In: Horikoshi K, Antranikian G, Bull AT, Robb FT, Stetter KO (eds) Extremophiles handbook, vol 1. Springer, Tokyo/Dordrecht/Heidelberg/London/New York, pp 755–775

Moreno ML, Perez D, Garcia MT, Mellado E (2013) Halophilic bacteria as a source of novel hydrolytic enzymes. Life 3:38–51

Morita RY (1975) Psychrophilic bacteria. Bacteriol Rev 39:144–167

Nogi Y (2011) Taxonomy of psychrophiles. In: Horikoshi K, Antranikian G, Bull AT, Robb FT, Stetter KO (eds) Extremophiles handbook, vol 2. Springer, Tokyo/Dordrecht/Heidelberg/London/New York, pp 777–792

Ollivier B, Caumette P, Garcia JL, Mah RA (1994) Anaerobic bacteria from hypersaline environments. Microbiol Rev 58:27–38

Oren A (2006) Molecular ecology of extremely halophilic *Archaea* and *Bacteria*. FEMS Microbiol Ecol 39:1–7

Oren A (2011) Ecology of halophiles. In: Horikoshi K, Antranikian G, Bull AT, Robb FT, Stetter KO (eds) Extremophiles handbook, vol 1. Springer, Tokyo/Dordrecht/Heidelberg/London/New York, pp 343–361

Oren A, Gurevich P, Azachi M, Henis Y (1992) Microbial degradation of pollutants at high salt concentrations. Biodegradation 3:387–398

Price PB (2007) Microbial life in glacial ice and implications for a cold origin of life. FEMS Microbiol Ecol 59:217–231

Rawlings DE (2002) Heavy metal mining using microbes. Annu Rev Microbiol 56:65–91

Sand W, Gehrke T, Jozsa PG, Schippers A (2001) (Bio)chemistry of bacterial leaching-direct vs. indirect bioleaching. Hydrometall 59:159–175

Sass AM, Sass H, Coolen MJL, Cypionka H, Overmann J (2001) Microbial communities in the chemocline of hypersaline deep-sea basin (Urania Basin, Mediterranean Sea). Appl Environ Microbiol 67:5392–5402

Shivanand P, Mugeraya G (2011) Halophilic bacteria and their compatible solutes – osmoregulation and potential applications. Curr Microbiol 100:1516–1521

Simonato F, Campanaro S, Lauro FM, Vezzi A, D'Angelo M, Vitulo N, Valle G, Bartlett DH (2006) Piezophilic adaptation: a genomic point of view. J Biotechnol 126:11–25

Stetter KO (2011) History of discovery of hyperthermophiles. In: Horikoshi K, Antranikian G, Bull AT, Robb FT, Stetter KO (eds) Extremophiles handbook, vol 1. Springer, Tokyo/Dordrecht/Heidelberg/London/New York, pp 403–426

Van Der Wielen PWJJ et al (2005) The enigma of prokaryotic life in deep hypersaline anoxic basins. Science 307:121–123

Vargas C, Calderon MI, Capote N, Carrasco R, Garcia R, Moron MJ, Ventosa A, Nieto JJ (2004) Genetics of osmoadaptation by accumulation of compatible solutes in the moderate halophile *Chromohalobacter salexigens*: its potential in agriculture under osmotic stress conditions. In: Ventosa A (ed) Halophilic microorganisms. Springer, Berlin/Heidelberg, pp 135–153

Vreeland RH, Rosenzweig WD, Powers DW (2000) Isolation of a 250 million-year-old halotolerant bacterium from a primary salt crystal. Nature 407:897–900

Whitman WB, Coleman DC, Wiebe WJ (1998) Prokaryotes: the unseen majority. Proc Natl Acad Sci U S A 95:6578–6583

Widdel F, Schnell S, Heising S, Ehrenreich A, Assmus B, Schink B (1993) Ferrous iron oxidation by anoxygenic phototrophic bacteria. Nature 362:834–836

Wiegel J (2011) Anaerobic alkaliphiles and alkaliphilic poly-extremophiles. In: Horikoshi K, Antranikian G, Bull AT, Robb FT, Stetter KO (eds) Extremophiles handbook, vol 1. Springer, Tokyo/Dordrecht/Heidelberg/London/New York, pp 81–97

Woese CR, Kandler O, Wheelis ML (1990) Towards a natural system of organisms: proposal for the domains *Archaea*, *Bacteria*, and *Eucarya*. Proc Natl Acad Sci U S A 87:4576–4579

Yayanos AA (1986) Evolutional and ecological implications of the properties of deep-sea barophilic bacteria. Proc Natl Acad Sci U S A 83:9542–9546

Microorganisms and Biotic Interactions

Yvan Moënne-Loccoz, Patrick Mavingui, Claude Combes,
Philippe Normand, and Christian Steinberg

Abstract

Most ecosystems are populated by a large number of diversified microorganisms, which interact with one another and form complex interaction networks. In addition, some of these microorganisms may colonize the surface or internal parts of plants and animals, thereby providing an additional level of interaction complexity. These microbial relations range from intraspecific to interspecific interactions, and from simple short-term interactions to intricate long-term ones. They have played a key role in the formation of plant and animal kingdoms, often resulting in coevolution; they control the size, activity level, and diversity patterns of microbial communities. Therefore, they modulate trophic networks and biogeochemical cycles, regulate ecosystem productivity, and determine the ecology and health of plant and animal partners. A better understanding of these interactions is needed to develop microbe-based ecological engineering strategies for environmental sustainability and conservation, to improve environment-friendly approaches for feed and food production, and to address health challenges posed by infectious diseases. The main types of biotic interactions are presented: interactions between microorganisms, interactions between microorganisms and plants, and interactions between microorganisms and animals.

Keywords

Agronomy • Antagonism • Biogeochemical cycles • Commensalism • Competition • Ecosystem functioning • Infectious diseases • Parasitism • Predation • Symbiosis • Trophic networks

---

* Chapter Coordinators

Y. Moënne-Loccoz* • P. Mavingui* (✉) • P. Normand
Microbial Ecology Center, UMR CNRS 5557 / USC INRA 1364,
Université Lyon 1, 69622 Villeurbanne, France
e-mail: yvan.moenne-loccoz@univ-lyon1.fr; patrick.mavingui@univ-lyon1.fr; philippe.normand@univ-lyon1.fr

C. Combes
Member of the Académie Française des Sciences (French Academy of Sciences), 16, rue du Vallon, 66000 Perpignan, Paris, France

C. Steinberg
Pôle des Interactions Plante-Microorganismes,
INRA UMR 1347 Agroécologie, AgroSup-INRA-Université de Bourgogne, BP 86510, 21065 Dijon Cedex, France
e-mail: christian.steinberg@dijon.inra.fr

11.1 Main Types of Interactions

11.1.1 Interaction: A Key Aspect of Living

Since the end of the era known as prebiotic era, which happened about 3.8 billion years ago, life has existed on earth in the form of cells, more or less complex in structure and operation, but still built on the same basic pattern (membrane + cytoplasm + nucleic acids). Conversely, if the pattern has remained fundamentally the same throughout evolution, diversification of genetic information – for all kinds of processes and not only by point mutations – has given birth to a biosphere composed of myriads of different organisms, from bacteria to the more organized multicellular eukaryotes.

A striking feature of the biosphere, repeatedly stressed among others by Stephen Jay Gould (Gould 1989), is the appearance of new body plans, which has not consistently resulted in the elimination of body plans that had formed previously. Thus, even though individuals have constantly disappeared and the species succeeded each other, the main types of living beings continue to exist to this day, whether they appeared 3 billion years ago or recently. The most striking example is provided by prokaryotes, which have remained ubiquitous and diversified in all areas of the world, despite the emergence of all kinds of unicellular or multicellular organisms. The evolution of life has led to the emergence of multiple interactions between all levels of organization, the first species to appear also taking advantage of the emergence of new ones, as shown, for example, by the colonization of eukaryotes by prokaryotes.

Interactions are so much the rule that we may even wonder if, in a given ecosystem – inasmuch as there is unity of place and time – there are living species not interacting directly or indirectly with other species. Interaction thus appears as one of the fundamental features of life, alongside metabolism and reproduction. We should not be surprised that interactions take many forms and those involving microorganisms are both diverse and particularly important for the exchange of matter and energy.

A fundamental character of interactions of living beings is that any individual can interact with any other individual. For example, viruses interact with prokaryotes, unicellular eukaryotes, or humans; prokaryotes interact with multicellular eukaryotes; individuals of a given species interact with others of the same species; etc. Even individual cells of the same multicellular organism interact with other cells of the same organism, through continuous exchange of electrical (nerve impulses) or chemical (hormones) signals between cells that are genetically identical but express different genes.

Curiously, this universality of interactions is a strong argument in favor of the unity of life. This concept of unity, of which Geoffroy Saint-Hilaire was the defender in the middle of the nineteenth century, is evidenced, for example, by the construction of proteins from a list determined once and for all of 20 amino acids, by the similarities of metabolic pathways and biophysical processes, and of course by the universal genetic code. The ability of a transposon to insert foreign DNA, the ability for bacteria to invade animal tissues, the ability of *Toxoplasma* (protozoan) to manipulate the behavior of a rodent, and the possibility of limitless pairwise interactions are all

Table 11.1 Different types of interactions and their effects on partners

	Effect on partner #1	Effect on partner #2
Mutualism (symbiosis)	+	+
Commensalism	+	0
Parasitism, predation	+	−
Neutralism	0	0

0 means absence of effect, + means a positive effect, − means a negative effect

further evidence that life probably arose only once, or at least only the descendants of a particular life form survived (*cf.* Sect. 2.1). For example, any pathogen and its host share a common ancestor. The result is that signals can be exchanged between organisms whose common ancestor existed billions of years ago, and this might explain why a pathogen can use and manipulate to his advantage the biochemical processes of its host. What has changed throughout evolution is the complexity of organisms, but not the basis for their functioning or the nature of the molecular signaling pathways: this could only facilitate the establishment of multiple interactions (Table 11.1).

11.1.2 Conflictual Interactions

2.7 billion years ago, cyanobacteria developed molecular tools allowing photosynthesis, a feature that allows us to divide most living beings into two major groups, photosynthetic organisms such as algae and green plants that get their energy from the sun and those who, directly or indirectly, derive their energy from the former. Food chains and webs in ecosystems are essentially based on these two types of organisms, typically referred to as primary producers (comprising also chemoautotrophic bacteria such as nitrifiers) and consumers, respectively, as well as on microbial decomposers. These food webs involve several types of conflictual interactions.

Conflictual interactions within ecosystems are of multiple types. They can be classified according to trophic level, distinguishing, e.g., primary consumers (such as herbivores), secondary consumers (e.g., carnivores that eat herbivores), etc. We can distinguish the interactions between organisms from different trophic levels from interactions between individuals at the same trophic level, which may lead to intra- or interspecific competition. **Competition*** is an interaction in which partners use the same resource, whether nutrients, water, or even space.

Intraspecific competition can be particularly intense, when the partners involved have needs that are very close and use comparable if not identical means for the resource acquisition. Competition can also occur between individuals belonging to different trophic levels, and even different kingdoms, such as the competition for nitrate between plant roots and microorganisms in soil.

Competition is sometimes assisted by the production of toxic compounds in the environment, such as bacteriocins or antibiotics. It corresponds to **interference competition***, which relies on chemical warfare mechanisms, to be distinguished from competition by exploitation. The distinction between the two types of competition is sometimes difficult to establish. Another difficulty lies in the distinction between interference competition and another conflictual interaction, **amensalism***. Amensalism (synonym of antagonism) is an interaction that has a negative effect on one partner but no effect on the other. Amensalism assumes that the two species do not compete significantly with each other. Amensalism is based on a physical or chemical modification of the environment and, in the latter case, often involves the release of toxic compounds. Thus, the production of antibiotics (i.e., antibiosis) may correspond to interference competition or antagonism, depending on the nature of the partner affected and on the energy invested in the synthesis of these secondary metabolites. Allelopathy, i.e., the synthesis by a higher organism of compounds bioactive on another higher organism, is conceptually close to antibiosis and deserves to be mentioned in this chapter because microorganisms can metabolize some of these compounds and thus interfere with the interaction.

Finally, there are two types of conflictual interactions for which the interaction is negative for one partner but beneficial for the other, i.e., predation and parasitism. **Predation*** is of short duration (a cat that eats a mouse or a protozoan that feeds on a bacterium) and usually leads to destruction of the genetic information of the prey (in this case the mouse or the bacterium). **Parasitism*** is sustained in time (the parasite in its host) and the genetic informations of the partners remain in intimate contact over time; here we recognize the traditional distinction between predator–prey and parasite–host relationships. The predator is free, while the parasite is physically associated with its host, at least for part of its life cycle. It should be noted that the importance of signals is very different in the two types of interactions. When a cat chases a mouse, signals are visual, auditory, or olfactory, and they last only the duration of the hunt. When a virus or bacterium settles in a host, the exchange of molecular signals can persist for months or years, for example, involving the immune system of the host and processes by which pathogens circumvent these defenses. One of the most amazing aspects of these sustainable interactions is manipulation: *Toxoplasma* are parasitic protozoa whose complex life cycle requires them, when they are in a rodent, to pass into a cat to perform their sexual reproduction; *Toxoplasma* settles in the nerve cells of the rodent and induces in them a suicidal attraction to cats, as well as altered social behavior. Although we do not know the precise mechanisms involved, it is clear that this manipulation of rodent behavior must rely on biochemical compounds produced by *Toxoplasma*. The more research is done on it, the more manipulation appears to be widespread – the result of natural selection – in pathogen–host interactions.

11.1.3 Beneficial Interactions

Some interactions, unlike the previous ones in which conflict predominates, provide benefits to the different partners. Among beneficial interactions, **commensalism*** is the only one for which the positive effects are exerted on only one of the two partners. This interaction is therefore the counterpart of amensalism. **Cooperation*** is a mutually beneficial, facultative relationship, this facultative aspect distinguishing it from symbiosis.

In the case of **mutualism*** or symbiosis (Box 11.1), both species benefit from the interaction. It is interesting to note that, as in conflictual interactions, genetic distance is not a barrier to partnership: one can observe mutualistic interactions between plants and animals, bacteria and vertebrates, and fungi and plants; a complete list would be endless. Mutualistic interactions have played a very important role in evolution (Fig. 11.1). Mitochondria and chloroplasts – and certainly other structures – of eukaryotic cells are nothing more than ancient bacteria. In fact, the modern eukaryotic cell, the cornerstone of all multicellular organisms, can be seen as a collection of bacteria. Such interactions are also reversible: for example, the agent of malaria, the *Plasmodium* protozoan, has lost its chloroplasts (which its ancestors had, as indicated by recently discovered remnants of these chloroplasts), the same way tapeworms lost their gut.

Box 11.1: Story of a Word, Symbiosis

Yvan Moënne-Loccoz

Symbiosis, from the Greek "syn" (with) and "biosis" (life), illustrates the ongoing interaction (obligatory) of two or more organisms, their lack of autonomy. It has been defined, at least initially, as the intimate and sustained interaction of two organisms belonging to different species. Currently, the symbiosis is generally regarded, especially in Microbial Ecology, as a mutually beneficial obligatory interaction.

The paternity of the term has been attributed to Heinrich Anton de Bary (de Bary 1879), a German botanist (1831–1888) who described in his book *Die Erscheinung der Symbiose* (*The Phenomenon of Symbiosis*) life in association of different organisms, definition thus including the parasitism of fungi, on which de Bary worked all his life. This term has quickly gained popularity and has been applied in 1885 to mycorrhizae and in 1889 to nodules induced by *Rhizobium* on peas by the German botanist Albert Bernhard Frank (1885, 1889) who would prove to be an effective proselytizer of the word. Root nodules form following sophisticated molecular dialogues and are inhabited by nitrogen-fixing Actinobacteria (*Frankia*; Box Fig. 11.1) or Proteobacteria (*Rhizobium* and many other taxa), depending on the type of plant partner.

The term has been modified by different prefixes or adjectives. We thus speak of endosymbiosis and ectosymbiosis whether the host is penetrated or not, of obligatory or facultative symbiosis, depending on whether or not the partners can live without the other. In the latter case, which corresponds to cooperation, partners are sometimes termed aposymbiotic, asymbiotic, or pre- or post-symbiotic. The word protosymbiosis is also used, when the relationship is with little mutual benefit as in the case of yogurt where *Lactobacillus bulgaricus* and *Streptococcus thermophilus* coexist without strong metabolic complementarity. The term associative symbiosis is also used as a synonym for cooperation, especially in the case of saprophytic microorganisms in the rhizosphere that enhance plant growth. Parasymbiosis is a symbiosis where an additional organism secondarily joins an already formed symbiosis.

The meaning of symbiosis has evolved over time, and today the term is rarely used anymore in biology to talk of parasitism. On the contrary, some confusion has emerged about the distinction between mutualism (sustainable, mutually beneficial interaction) and symbiosis. According to Odum (1971), true symbiosis (obligatory) and mutualism are equivalent, which also corresponds to the definition given in the manual of Microbial Ecology of Atlas and Bartha (1981). However, others, such as Barbault (1997), distinguish mutualism (not obligatory) from symbiosis (mandatory).

A Google search permitted to identify in June 2014 more than 10 million sites with "symbiosis" in the fields of biology, management, politics, and culture.

Box Fig. 11.1 Nitrogen-fixing symbiosis between the nodulating actinobacterium *Frankia* and the actinorhizal plant alder. On the left, a black alder growing along the bank of river Rhône in Lyon; on the right, a longitudinal section of an alder nodule stained with toluidine blue showing the central stele and, on both sides, the enlarged cortical cells (*deep blue*) filled with nitrogen-fixing *Frankia* cells (Photographs: Ph. Normand).

It should be added that the distinction between parasitism and mutualism is not always easy to make. In some cases it is difficult to measure mutual benefits; in other cases we can question the "honesty" of the interaction. To cite just one example, many organisms living in deep ocean hydrothermal ecosystems harbor sulfate-reducing bacteria that provide the bulk of their energy resources. This interaction can be seen as a simple exchange of protection against food, but if we take into account that the host regularly consumes part of the bacterial population, one can also see true parasitism. The length of the interaction is not an absolute guarantee of peace between the partners. For instance, "mitochondrial bacteria," which in sexually reproducing organisms are usually transmitted to zygotes by females, "endeavor" to skew the sex ratio at the expense of males. This is a well-known strategy in the interaction between *Wolbachia* bacteria and their arthropod hosts (*cf.* Sect. 11.4.3).

The mutualistic interaction reaches perhaps its perfection in multicellular organisms. While in a population of unicellular organisms, competition for resources is the rule, it is logically excluded between members of multicellular organism that constitute a clone: there is a real change in target

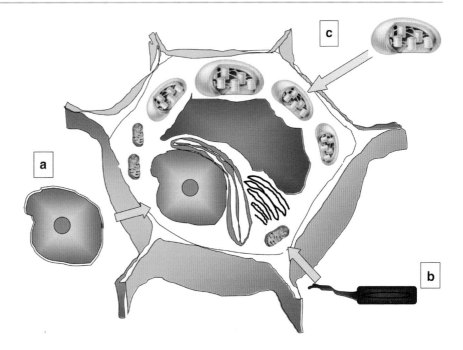

Fig. 11.1 Plant cell comprising a nucleus (*gray*) originating from an ancestor such as *Thermoplasma* (**a**), mitochondria (*red*) originating from an ancestor such as an alphaproteobacterium (**b**), chloroplasts (*green*) originating from an ancestor such as cyanobacteria (**c**) (Redrawn from John H. Miller (jhmiller@uh.edu))

selection; it does not oppose cellular individuals but rather groups of cells. Of course, this totally peaceful interaction is made possible by the perfect genetic similarity of the cells constituting the group. Evidence of the importance of this similarity is provided by instances where mutations lead to the emergence of rival groups within the same population.

11.1.4 Dynamics of Interactions

Interactions, whether involving predator–prey relationships in a forest or more complex relationships between bacterial populations in the soil, can only be understood in a perspective that is both dynamic and Darwinian. Interaction dynamics are important to consider because the terms of the interactions change over time, even if sometimes they oscillate momentarily around an equilibrium position. A Darwinian perspective is also needed because biotic interactions are the result of selective pressures that organisms exert on each other, even if other factors such as genetic drift can also affect the reproductive success of individuals.

When interactions are conflictual, they usually give rise to an arms race. Arms races represent one of the fundamental drivers of evolution, because selective pressures that organisms exert on each other remain the *sine qua non* of evolutionary change. In a 1973 paper, the American evolutionist Leigh Van Valen gave for this driver of evolution an explanation known as the Red Queen hypothesis, which has become a classic (Fig. 11.2). He suggested that changes constantly generate new adaptations to meet adaptations in the competing species; if species share, even partially, the same spatial or energy resources, and if one of them increases its fitness, the other species have then to compensate for this adaptive advantage or else will disappear. Species that adapt will in turn modify the fitness of the others; and so forth, says Van Valen, life is in a self-sustaining perpetual movement.

Van Valen's hypothesis leads to the concept that humans and bacteria, although differing profoundly in their complexity, may be equally adapted to their environment. Indeed, microorganisms can thrive in environments where vertebrate cannot survive and adapt quickly to drastic changes in their environment.

In mutualistic interactions, one can have the impression that the arms race concept does not apply. However, as we have seen, even an interaction as ancient as that of the eukaryotic cell with mitochondria is not without discordant notes. If we add that, throughout the history of the association, the majority of genes in mitochondria (it is the same for chloroplasts) was transferred to the nucleus, this interaction – capital for evolution – can be seen as a trusteeship followed by genetic looting.

11.1.5 Weapons Specific to Microorganisms

In Red Queen-type clashes, infectious disease agents have their own weapons. One of them, particularly formidable, is their very short generation time, inherited from the early ages of life, which can lead to high production of individuals with mutated genes. In contrast, hosts of microbial parasites must compose with one major constraint that evolution has imposed onto multicellular organisms, i.e., their long generation time. Indeed, the advent of multicellularity was accompanied by increased body size, which resulted also, inevitably, in longer generation time. While the time between two divisions, thus opportunities for mutations and selection, is sometimes counted in minutes in bacteria, many multicellular organisms only acquire the ability to reproduce, and thus to mutate, after several years (somatic

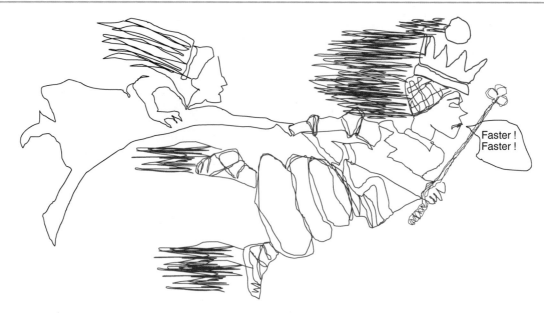

Fig. 11.2 Alice and the Red Queen, illustration by John Tenniel for "Through the Looking-Glass" by Lewis Carroll. "... Alice never could quite make out, in thinking it over afterward, how it was that they began: all she remembers is, that they were running hand in hand, and the Queen went so fast that it was all she could do to keep up with her: and still the Queen kept crying 'Faster! Faster!' but Alice felt she could not go faster, though she had no breath left to say so. The most curious part of the thing was, that the trees and the other things round them never changed their places at all: however fast they went, they never seemed to pass anything. 'I wonder if all the things move along with us?'". (http://www.victorianweb.org/art/illustration/tenniel/lookingglass/2.4.html)

mutations that arise during development and postembryonic life are not transmitted and are thus useless for evolution). This issue is discussed by Ochman and Wilson (1987). They argue that replication is not the only time when mutations occur; instead these may occur at any time over the lifespan of organisms. They conclude that mutations generally take place at constant rates, independently of generation time, which of course is not always true, as, for instance, stressful situations lead to the emergence of hypermutator clones (Foster 2007).

In addition to their short generation time, another advantage enjoyed by infectious agents is their capacity to change their genetic information. Mutation rates in microorganisms can be high. In addition, viruses and bacteria are capable of altering their genomes by exchanging and incorporating foreign nucleic acid sequences, while the nuclear membrane has made such exchanges rarer in eukaryotes; evolutionists even think that one of the reasons for genetic recombination in meiosis is to maintain gene flow.

Living organisms have developed many types of interaction during evolution; the interactions of microorganisms with each other, with plants, and finally with animals will be detailed in the following text.

11.2 Interactions Between Microorganisms

Most biotopes contain many microbial taxa that coexist and interact in many ways, of which few details are known or even considered. These interactions are in most cases not specialized, based mainly on trophic aspects because resources are always limiting, but in some cases there are instances of mutualistic or parasitic relations.

11.2.1 Conflictual Interactions

Martin (2002) has classified into a few categories the known strategies of predation or parasitism: pack predation, epibiotic attachment, direct cytoplasmic invasion, and periplasmic invasion. These strategies constitute a continuum toward more and more specialized forms.

11.2.1.1 Parasitism

The best-known example of conflictual interaction involving two bacterial taxa concerns *Bdellovibrio bacteriovorus* (Fig. 11.3). This interaction is often described as predation, but corresponds rather to parasitism because it is a lasting interaction, with several cell doublings of *Bdellovibrio* while inside the periplasm of the target cell. In fact, this bacterium should more accurately be called parasitoid since it leads to the death of the host. The small deltaproteobacterium *B. bacteriovorus* consumes motile cells belonging to many Gram-negative taxa. It starts by entering the periplasm and sealing the pore entrance and then replicates in the periplasm, yielding daughter cells without flagella that will invade all the cell space. Thereafter, *Bdellovibrio* hydrolyzes cell constituents, forms filamentous cells that become septate, lyses the host cell, and releases flagellated offsprings.

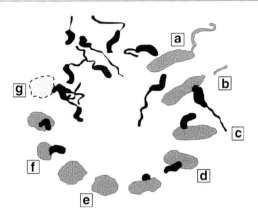

Fig. 11.3 Clockwise starting at the top, two cells of *Bdellovibrio* (in *blue*) approaching their bacterial prey (in *gray*) (**a**), one *Bdellovibrio* cell attaches to the prey (**b**) and loses its flagellum (**c**), penetrates the periplasm (**d**), multiplies in the cell (**e**) before the daughter cells emerge (**f**), leaving a lysed prey (**g**). (Redrawn from Stephan Schuster, Tübingen, Germany (scs@bx.psu.edu))

Fig. 11.4 Paris mushrooms (*Agaricus bisporus*) attacked by *Pseudomonas tolaasi* causing bacterial brown rot (bacterial blotch disease) (Photo Ph. Normand)

This way of life and the lack of attack on mammalian cells have even led some to consider the use of *Bdellovibrio* for the treatment of infections in humans (Stolp and Starr 1963). The genus was divided to yield *Bacteriovorax* to accommodate marine strains physiologically distinct and thereafter *Peredibacter starrii*, but all these genera are phylogenetically very close to each other.

In addition, there are many bacteriophages that attack bacteria and act as parasites. Finally, the mushroom *Agaricus bisporus* is parasitized by *Pseudomonas tolaasi*, forming brown spots on carpophores, i.e., brown blotch disease (Fig. 11.4), and by various other parasitic bacteria and fungi.

11.2.1.2 Predation

The main microbial predators are protozoa, which regulate bacterial populations in various ecosystems. In soil, however, many bacterial taxa belonging to Actinobacteria and Proteobacteria were described as predators of other bacteria and fungi (Zeph and Casida 1986), but few studies have been carried out subsequently, except on the actinobacterium species *Agromyces ramosus* and the proteobacterial genera *Ensifer* (reclassified recently in genus *Sinorhizobium*) and *Pseudomonas*.

Soils and many other biotopes contain bacterivorous organisms belonging to different taxa. This is the case of *Caenorhabditis* nematodes and *Brachionus* rotifers.

Another type of predation, quite dramatic, involves fungi such as *Hirsutella minnesotensis* or *Arthrobotrys robusta* that catch and metabolize nematodes. Nematodes are captured using a constrictor ring or adhesive structures such as nets or buttons (http://www.edslides.nematologists.com/fungal.html), after which lytic enzymes attack the nematode (Fig. 11.5). This interaction is used to fight against nematode pests in mushroom farms, intestinal nematodes parasiting sheep (Waller and Larsen 1993), or even phytoparasitic nematodes in soil.

11.2.1.3 Antibiosis

Antibiosis is generally negative for some of the taxa sharing the same habitat as the antibiotic producer. Therefore, it is often a case of interference competition, i.e., an interaction between two species in which one of the two inhibits the development of the other and thus gains greater access to food resources in the biotope. This strategy is widespread in prokaryotes and eukaryotes and has been extensively studied since Fleming (Fleming 1922) (*cf.* Sect. 9.5), mainly because of its developments in public health. Compounds involved belong to several chemical classes ranging from simple molecules such as aminoglycosides to complex compounds such as macrolides or polypeptides, targeting several cellular functions such as protein synthesis (kanamycin) or RNA synthesis (rifampin).

Taxa known to produce antibiotics are bacteria especially soil actinobacteria and fungi, and the types of compounds produced and their mechanisms of action are described in Sect. 9.5. Many discussions have been held to determine if antibiosis was positive for the organism that synthesizes antibiotics, if not we should speak of antagonism. It is difficult to determine the cost of antibiotic synthesis, which includes the genetic burden of maintaining dozens of

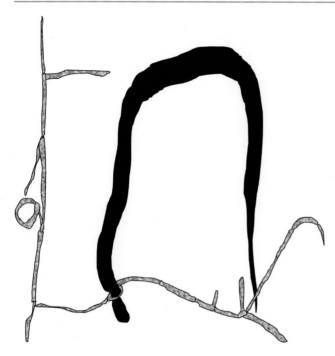

Fig. 11.5 Nematode (*in purple*) captured using a constrictor ring by the fungus *Arthrobotrys dactyloides* (*in gray*). Another ring at the left is ready to catch another nematode (Redrawn from B.A. Jaffee (plpnemweb.ucdavis.edu))

biosynthetic genes, resistance genes, and genes necessary for their transport out of the cell. The presence of an antibiotic is not necessarily detrimental for a given taxon, either because it has acquired genes permitting its degradation, and thus to feed on it, or a mutation has occurred in the genes whose product is the antibiotic target (Birge and Kurland 1969). This kind of phenomenon could in principle occur with any antibiotic.

There has been a debate on the relationship between antibiotics and bacteriocins. In contrast to antibiotics, bacteriocins affect bacteria closely related to the producing organism, often belonging to the same species. They are often of a proteinaceous nature. The mode of action is varied and includes pore formation in the membrane inducing leakage of cellular constituents, DNA or RNA degradation in the target strain, and inhibition of the production of murein in cell walls. They have been described in various bacteria especially *Escherichia coli, Vibrio, Lactococcus,* and *Pediococcus* and have a high potential in biotechnology, for example, in yogurt manufacturing. The distinction between antibiotics and bacteriocins is not always clear, because some bacteriocins have targets in distant taxa, such as *Lactobacillus* spp. against *E. coli* or *Listeria* (Millette et al. 2007), and also because antibiotics and bacteriocins are less and less defined by their modes of action.

Antibiotics have been given several definitions, initially by Waksman (1969) who used this term for any natural compound active on other organisms, to distinguish from synthetic compounds such as sulfonamides. Given the progress of synthetic chemistry that helped change natural compounds to modify their properties, they are now designated as synthetic–natural compounds. The word antibiotic now refers to compounds, natural or otherwise, with antimicrobial activity and that are used in human and veterinary medicine or in biotechnology.

Genes for antibiotic synthesis, carried on plasmids, are often exchanged by conjugation or transformation. They are especially abundant in soil organisms with large genomes such as *Streptomyces* (Bentley et al. 2002), particularly in telomeric regions subject to high recombination rates. Bacterial isolates are generally resistant rather than sensitive to antibiotics produced by isolates from the same soil (Davelos et al. 2004), which probably comes from the phenomenon of coadaptation.

It has long been argued that the synthesis of antibiotics belongs to the same type of evolutionary event as the emergence of wings, called exaptation (spandrel) by Gould and Vrba (1982) to indicate that their primary function could be quite different from the final one. The genes encoding the synthesis of antibiotics belong to secondary metabolism; that is to say, they are expressed mainly at the end of the exponential phase and later, and furthermore they are not always essential. They would therefore be genes resulting from random recombination as ORFans, initially without detectable function but that have gained one due to reactivity toward various cell constituents, and their selective advantage would have resulted in their fixation in the genome and their subsequent dispersion by lateral transfer (Daubin and Ochman 2004).

11.2.1.4 Competition

Competition* is an interaction defined as a simultaneous demand by two or more organisms for a limited environmental resource, such as a nutrient, water, living space, or light. This is of course the rule in the microbial world where many habitats include several taxa with closely related metabolic capabilities. For example, the addition to the soil of a complex carbon source like deciduous tree litter stimulates many fungi and bacteria capable of metabolizing polymers such as cellulose, hemicelluloses, and pectin. Microbial populations that grow in the soil thus vary according to the carbon sources added (Hery et al. 2005). A wastewater treatment plant is another biotope (anthropized), in which is found a rich mixture of carbon molecules inducing fluctuations of the microbial community present (Snaidr et al. 1997).

11.2.2 Beneficial Interactions

When comparing various beneficial interactions between microorganisms, it appears that the terms cooperation, mutualism, and syntrophy have been used for conceptually similar phenomena. The term cooperation is sometimes used to designate situations encompassing the other two. Mutualism is little used to describe microbe–microbe relations, but rather is reserved for relationships between microbes and higher organisms. The term **syntrophy*** would cover just cases where microorganisms have complementary metabolisms and are in mutualistic situations, each partner providing substrate(s) to the other(s). At one extreme of the spectrum, there are situations of loose or non-exclusive relationships, based on trophic exchanges thus designated syntrophic; at the other extreme are situations of mutualism or symbiosis involving potentially more than trophic aspects of detoxification and entailing toxin synthesis, signaling, etc. It has been argued that the origin of eukaryotic cells was a syntrophy that has evolved into mutualism and finally into an obligatory interaction (López-García and Moreira 1999).

11.2.2.1 Cometabolism

Cometabolism is a prominent case of beneficial interaction taking place between different microorganisms. It corresponds to cooperation, syntrophy, or even symbiosis when metabolisms are different and complementary and when the interaction is sustained. Microorganisms that live in complex habitats such as soils, sediments, or digestive tracts are typically in contact with complex trophic resources they are unable to catabolize alone. For a single organism, cometabolism is the transformation of a compound that cannot serve as sole carbon and energy source (i.e., a non-growth substrate), which is made possible by parallel degradation of a growth substrate. We also speak of cometabolism when several microorganisms must cooperate as a consortium to synthesize all the enzymes necessary for a catabolic pathway. This is the case of soil isolates individually unable to metabolize polycyclic aromatic compounds, but which are able to do so when they are grown in a consortium (Bouchez et al. 1999). It is likely that this consortium includes strains able to compensate for the inhibition caused by a metabolic intermediate by degrading it or when the product of a strain is used as a substrate by a different strain.

11.2.2.2 Mutualism

The deltaproteobacterium *Syntrophus aciditrophicus* can metabolize various saturated or unsaturated fatty acids, hexanoate and butyrate esters, or benzoate when in coculture with microorganisms capable of metabolizing hydrogen or formate (Mouttaki et al. 2007). Anaerobic degradation of saturated fatty acids and aromatic acids in the absence of a terminal electron acceptor necessitates the presence of an organism capable of maintaining a hydrogen partial pressure low enough so that these reactions are thermodynamically possible. We therefore find it associated with hydrogenotrophic microorganisms such as *Desulfovibrio* or the archeon *Methanospirillum hungatei* (*cf.* Chap. 3 and Sect. 14.2.5). Mutualism is also found during methanogenesis which is a complex process described in detail in Chaps. 3 and 14. It involves many taxa that perform various steps in the complex pathway that transforms organic compounds into acetate and eventually into methane (*cf.* Sects. 3.3 and 14.2.5).

Sometimes the metabolic basis of the association is unknown, as in the case of the two archaea *Nanoarchaeum equitans* (Nanoarchaeota; *cf.* Chap. 6) and *Ignicoccus hospitalis* (Crenarchaeota) present in hydrothermal environments, which are in close interaction in a relationship described as symbiosis or parasitism (Jahn et al. 2008). These two organisms are closely nested into one another and cannot be cultured separately. Another interesting case concerns *Symbiobacterium thermophilum* (Firmicute) (Watsuji et al. 2006), which cannot be cultivated in the absence of a *Bacillus* sp. Both thermophilic bacteria are found in composts. *Bacillus* sp. provides *S. thermophilum* with CO_2 from its respiration, which allows *S. thermophilum* to compensate for the absence of a carbonic anhydrase, an enzyme that allows different processes such as photosynthesis, respiration, pH homeostasis, and ion transport. It also catabolizes indolic compounds, which are self-inhibitors to *S. thermophilum*. The advantage for the *Bacillus* sp. is not clear; it may simply be a situation of commensalism or catabolism of recalcitrant compounds found in the compost.

Mutualism is not restricted to catabolic functions. Sometimes, diseases of plants and animals are caused by consortia or teams consisting of very different partners. This is the case of seedling blight of rice, caused by the fungus *Rhizopus microsporus*, which contains in its cytoplasm endosymbiotic *Burkholderia* bacteria necessary for production of the virulence factor rhizotoxin and plant disease (Partida-Martinez and Hertweck 2005). The presence of *Burkholderia* within fungal tissues has been reported in many fungi.

Biofilms (*cf.* Sect. 9.7.3) found in soils, sediments, higher organisms, and man-made environments are complex assemblies that can include many taxa, some more active than others for the synthesis of exopolysaccharides. These polymers, which trap heavy metals or antibiotics, allow the development of a three-dimensional structure and protect all cells present, those synthesizing the polymers as well as the others. Microbial mats (*cf.* Sect. 9.7.3) include many microbial taxa with complementary physiological properties such as sulfate reduction and sulfate oxidation, photosynthesis, and heterotrophy. Despite the synthesis in these mats of compounds with antibiotic activity (Socha et al. 2007), the taxa present complement and protect each other during the development of microbial mats (Fourcans et al. 2006) and are therefore in situations of mutualism.

11.2.2.3 Cooperation

Trophic relations such as the ones above would correspond to cooperation if facultative. One of the main cases of cooperation between microorganisms is quorum sensing. Bacteria have mechanisms that allow them to control certain physiological functions (such as conjugation) according to cell density, which is useful to ensure success. This quorum sensing mechanism is based on synthesis and perception of signals such as *N*-acyl homoserine lactone (described in details in Sect. 9.3) and is sometimes disturbed by higher plants or algae (Teplitski et al. 2000; Bauer and Robinson 2002). Thus, it has been shown that gamma-amino butyric acid (GABA) (synthesized by plants upon wounding or infection by *Agrobacterium*) induces the catabolism of *N*-acyl homoserine lactone and thus may modulate quorum sensing (Chevrot et al. 2006). Quorum sensing may also be affected by bacteria that synthesize lactonases, especially *Bacillus thuringiensis* and *Agrobacterium tumefaciens*, a mechanism that is being evaluated for biological control of various infectious agents.

11.2.2.4 Commensalism

It is difficult to identify situations of true commensalism between microorganisms, i.e., where a taxon derives benefits from an interaction but the other derives none. This could be the case of bacteria involved in nitrification, a process that occurs in two steps. Chemolithotrophic ammonia-oxidizing archaea and bacteria convert ammonium to nitrite, while nitrite-oxidizing bacteria *Nitrobacter* and *Nitrospira* convert nitrite into nitrate. Nitrite oxidizers obviously depend on the provision of nitrite by ammonia oxidizers, while the benefit for the latter is less obvious. However, nitrite is toxic for many taxa, including certain ammonia oxidizers, which would then benefit from nitrite removal. The interaction involved would rather be a cooperation in this case.

In many habitats, an element is limiting for bacterial growth and taxa with access to this element will be at the base of a food chain. This is the case of cyanobacteria capable of fixing carbon dioxide through photosynthesis, which are trophic partners of marine bacteria particularly those of the microbial loop (*cf.* Chap. 13). This is also the case for nitrogen-fixing bacteria in the rhizosphere and soils.

There are other examples of commensalism, in particular, *E. coli* that is optionally aerobic, consumes oxygen, and renders the intestinal tract anaerobic and therefore suitable for *Bacteroides*, a strict anaerobe. Similarly, bacteria involved in milk souring have a fermentative metabolism and release acidic compounds that acidify the medium and provide favorable growth conditions for acid-tolerant lactic acid bacteria.

Plants and microorganisms also synthesize and secrete compounds that are toxic to many microorganisms. In the same vein, many toxic compounds from human activity are found in the environment, where they selectively inhibit microbial taxa. Various microorganisms may degrade them, as illustrated with olive wastes metabolized by composting (Zenjari et al. 2006) or pentachlorophenol used as a wood treatment that is degraded by *Sphingobium* (Dams et al. 2007). These biodegrading microorganisms enable or enhance the growth of others.

11.2.2.5 Horizontal Gene Transfer

Horizontal gene transfer (*cf.* Chap. 12) may be mutually beneficial and would fall within the framework of cooperation but is not easy to qualify in terms of ecological interaction. The best-known case is that of antibiotic treatment in a medium comprising more than one microbial taxon, where initially most bacteria are sensitive to a compound to which they have never been in contact. Mutants resistant to this antibiotic will proliferate. Finally, the DNA fragment conferring resistance to the antibiotic may be transferred to sensitive cells of the same taxon or belonging to a remote taxon. One of the first described cases of transfer of resistance genes took place in Birmingham, England, in 1960. A care unit for burn victims found waves of nosocomial infections starting initially with *Klebsiella aerogenes* resistant to carbenicillin and carrying a resistance plasmid (RK2), followed by a second wave of *Pseudomonas aeruginosa* also carrying a closely related resistance plasmid. It has been shown that this phenomenon involved conjugative plasmid transfer (Ingram et al. 1973). Such cases of transfer of antibiotic resistance genes are common nowadays; they threaten our ability to curb infections and are one of the public health problems of greatest concern (Fig. 11.6).

Another case of gene transfer following a massive chemical selection pressure was found regarding mercury, a metal present in the bedrock and abundant in many natural or man-made habitats such as hydroelectric reservoirs or gold panning workshops. It was shown that *mer* genes carried by conjugative plasmids or transposons and conferring the ability to volatilize mercury were transferred between phylogenetically distant bacteria by conjugation (Mindlin et al. 2002). A similar phenomenon has also been observed with man-made compounds such as atrazine, currently the most widely used herbicide in the USA (*cf.* Sect. 16.7.2). When the herbicide is added to a field and ends up in the soil, it induces massive transfer to the soil microbiota of a plasmid carrying genes *atz* and *trz* for atrazine degradation (Devers et al. 2005). *atz* and *trz* are often on plasmids, but also on the chromosome near insertion sequences, suggesting that transposition plays an important role in the dispersion of this metabolic competence (*cf.* Sect. 16.7.2).

Maintaining large plasmids is not neutral in terms of fitness if they do not carry essential genes, which is reflected in the existence of strains having lost their plasmids in the soil, as is the case for many bacteria in particular

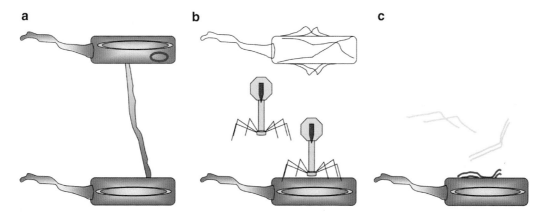

Fig. 11.6 (**a**) Bacterial conjugation between a donor cell (*upper*) and a recipient cell (*below*) by means of a conjugative pilus (*medium*) within which DNA (*red*) is transferred is transferred and will remain independent (plasmid) or integrate by recombination into the chromosome of the recipient cell. (**b**) Transduction between lysed bacterial cell (*above*), which has released a phage (*medium*) containing a bacterial gene (*red*) that will be integrated into the genome of the infected cell by the phage (*low*). (**c**) Transformation of a bacterial cell (*bottom*) by the naked DNA (*top*); bacteria that can be transformed with naked DNA are called competent. Two other processes of horizontal gene transfer have been proposed, i.e., via intercellular nanotubes and phage-like gene transfer agents (GTA) (Redrawn from Popa and Dagan 2011)

Agrobacterium (Krimi et al. 2002). It was shown that the presence of a host plant may increase the rate of conjugative transfer, as in *Sinorhizobium loti*, and give rise to many strains able to nodulate (Sullivan et al. 1995).

11.3 Interactions Between Plants and Microorganisms

11.3.1 Introduction

Plants play a major ecological role as the main primary producers of organic matter in terrestrial ecosystems. Part of this organic material is available for plant-associated microorganisms. The latter are present within the plant, at the surface of the plant (roots or shoots), or in the immediate vicinity of the roots, i.e., in rhizosphere (the soil under the direct influence of living roots) (Fig. 11.7). Indeed, plants release a fraction of photosynthates in the form of exudates or more generally (in the case of roots) rhizodeposits (Nguyen 2003). The availability of these organic nutrients stimulates plant-associated microorganisms, resulting in increased population size and physiological activity (Garbeva et al. 2004; Bais et al. 2006). In aquatic ecosystems, algae and other photosynthetic eukaryotes play a similar role in terms of primary production, but the relationship between algae and microorganisms is poorly documented, and these interactions are not considered in this chapter.

In return, the microorganisms associated with the plant will have a significant impact on the nutrition, development, growth, and health of the plant partner. Some of these microorganisms especially pathogens (parasites) have a negative impact on the plant. Conversely, others will have direct phytobeneficial effects, noticeably those in symbiosis (Mathesius 2003; Brundrett 2009) or cooperation with the plant (Dobbelaere et al. 2003; Mantelin et al. 2006). Indirect phytobeneficial effects are also possible, in particular, for microorganisms competing with or antagonistic toward phytoparasites (Alabouvette et al. 2006; Raaijmakers et al. 2009). Therefore, microorganisms associated with plants have considerable practical significance, which explains why historically they have often been studied to address concerns in the fields of agronomy and phytopathology. In this context, newer issues pertaining to food quality, environmental health, and global climate change are receiving increased research attention, along with long-standing issues about plant nutrition and health and soil fertility (van Elsas et al. 2007).

11.3.2 Location and Population Levels of Microorganisms Associated with Plants

Microorganisms associated with plants are mainly bacteria and fungi, as well as protozoa to a lesser extent. Depending on their location on/in plants, these microorganisms will be exposed to contrasting environmental conditions. A difference can be made between the root system and aerial parts. Microorganisms at the surface of aerial parts are confronted to fluctuating microclimatic situations, and they may be exposed to stressful conditions in particular with regard to desiccation and ultraviolet radiations (Kinkel et al. 2000). In addition, organic exudates are released in smaller quantities by leaves than by roots (Lucas and Sarniguet 1998; Leveau

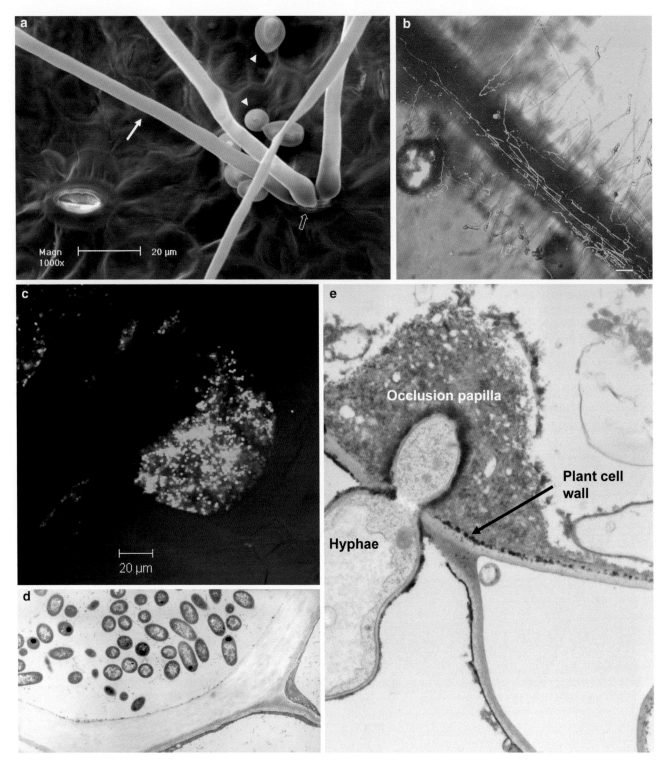

Fig. 11.7 Microorganisms associated with plants (Photos by electron microscopy). (**a**) The downy mildew agent *Plasmopara viticola* on the surface of a grapevine leaf. The oomycete developed within the leaf tissues and initiated its sporulation phase by generating sporangiophores (*solid arrow*) emerging from stomata (*open arrow*). These sporangiophores produce many sporangia (*arrowheads*), which contribute to parasite spread and buildup of a secondary inoculum source (Photo S. Trouvelot, UMR INRA 1347 Agroécologie, Dijon, France). (**b**) Nonpathogenic *Fusarium oxysporum* strain Fo47 in the root hair zone of a tomato root in a cultivated soil previously inoculated with the fungus. The strain was labeled by transformation, using a gene encoding a green fluorescent protein (GFP), and the sample was observed by confocal laser microscopy. Scale bar = 100 μm (Photo C. Humbert, UMR INRA 1347 Agroécologie, Dijon, France). (**c**) Microcolonies formed by the PGPR bacterium *Azospirillum brasilense* Sp245 on wheat root. The strain was labeled using the *egfp* gene, and the sample was observed by confocal laser microscopy (Photo C. Prigent-Combaret, UMR CNRS 5557 Microbial Ecology, Villeurbanne, France). (**d**) Pathogenic bacterium *Xanthomonas campestris* pv. *manihotis* in a cassava stem (×22,000) (Photo B. Boher, UR IRD 075 Resistance des Plantes, Montpellier, France). (**e**) Pathogenic fungus *Fusarium oxysporum* f. sp. *lycopersici* in tomato root tissues. The growth of the fungus is inhibited by the defense reactions of the plant. This involves parietal deposits of callose and phenolics, which form an occlusion papilla. Scale bar = 1 μm (Photo C. Olivain, UMR INRA 1347 Agroécologie, Dijon, France)

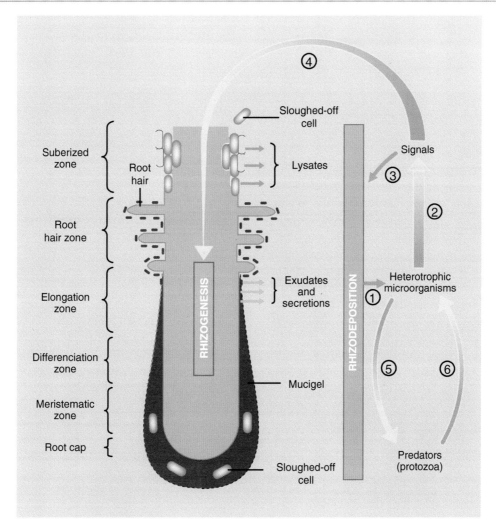

Fig. 11.8 Rhizodeposition and role of multitrophic interactions in the rhizosphere. Rhizodeposits correspond to exudates/diffusates, secretions, lysates, and the fraction of the mucigel that is of plant origin. They stimulate the growth of heterotrophic microorganisms in the rhizosphere (*1*) and their production of molecular signals (*2*), such as indole-3-acetic acid (IAA) and 2,4-diacetylphloroglucinol. These microbial metabolites stimulate rhizodeposition (*3*), as well as root system branching and root growth (*4*), and the latter effects on rhizogenesis stimulate rhizodeposition even further. The proliferation of rhizosphere microorganisms also stimulates predators, including protozoa (*5*). Selective predation and ammonia excretion by protozoa favor some of the rhizosphere microorganisms (*6*), including IAA producers and nitrifying bacteria, further stimulating rhizogenesis and rhizodeposition and so on (Inspired from Bonkowski 2004. Drawing: M.-J. Bodiou)

and Lindow 2001). Microorganisms colonizing aboveground plant surfaces are called epiphytes. Microbial numbers on leaves depend on the plant species, with densities in the order of 10^{5-7} bacteria per cm^2 (i.e., 10^{5-8} per g fresh weight) and 10^4 fungal propagules per cm^2 (Kinkel et al. 2000; Lindow and Brandl 2003). The stems can also be colonized by microorganisms, but at lower levels. It is especially documented for nitrogen-fixing bacteria, such as photosynthetic *Bradyrhizobium* that induce stem nodules on *Aeschynomene* (Wong et al. 1994) and *Frankia* on *Casuarina equisetifolia* (Prin et al. 1991).

In contrast, microorganisms present on the surface of roots (i.e., at the rhizoplane) and in rhizosphere soil have access to significant amounts of organic rhizodeposits (Nguyen 2003) (Fig. 11.8), and their habitat is buffered (Lucas and Sarniguet 1998) by the presence of the soil (porous solid phase) and the root mucigel (a gelatinous layer on the root surface that is composed of plant and microbial polysaccharides). They can reach high numbers, in the order of 10^{8-9} bacteria per g and 100–150 cm fungal hyphae per cm^2 root surface (Cavagnaro et al. 2005). These numbers are modulated by the amount of available organic substrates, which are mainly released by growing roots during the vegetative growth of the plant (Nguyen 2003) and by deleterious interactions (including competition between microorganisms and predation by protozoa). Microbial populations well adapted to the rhizosphere produce compounds (such as phytohormones, nitric oxide, or

2,4-diacetylphloroglucinol) that stimulate root growth and/or exudation, thereby amplifying the rhizosphere effect (Bonkowski 2004) (Fig. 11.8).

There are also differences according to the microbial habitat within the phytosphere, in particular at the scale of the root system. The quantitative importance of root systems varies depending on pedoclimatic and plant community properties, with root densities up to 100–200 cm roots per cm^3 of soil and roots occupying typically 1–5 % of the soil volume. However, most exchanges between the plant and the soil, both in terms of nutrients/water acquisition and rhizodeposition, occur at root tips (Cardon and Whitbeck 2007). Rhizosphere microbial populations are thus higher in the root elongation and root hair zones compared with the root cap (Cardon and Whitbeck 2007). Cell differentiation is accompanied by a decrease in the permeability of plant cell walls, and therefore older root parts are rather involved in sap transport. In the case of aerial parts, microbial numbers depend on leaf position and age, and they are often higher on the lower side of leaves (Kinkel et al. 2000). Whether on roots or leaves, microbial growth is stimulated following lysis of plant cells resulting from attacks by phytoparasites.

Microorganisms on plant surfaces are often found as microcolonies (sometimes with a cell density of over 10^4 bacteria per microcolony) and biofilms (Wimpenny and Colasanti 1997; Ramey et al. 2004). This may concern up to 80 % of the bacterial populations on the surface of leaves (Morris and Monier 2003). Biofilms occur particularly at the junction of epidermal cells, at wounds, and (for aerial parts) at stomata, hydathodes, and along veins (Beattie and Lindow 1999). Biofilm distribution becomes more erratic on older leaves (Leben 1988) and root tissues (Lübeck et al. 2000). Biofilms, which include the mucigel matrix in the case of roots, can be several tens of micrometers thick (Fig. 11.7c). Their buildup reflects the availability of nutrients (Leveau and Lindow 2001), and by comparison with other types of environments, their development level is intermediate (phyllosphere) to high (rhizosphere) (Wimpenny and Colasanti 1997).

Finally, a difference can be made between the plant surface and the interior of the plant, which can also be colonized by microorganisms (Gamboa et al. 2003). Microorganisms present in the plant are found in the cortex, more rarely in the stele (vascular cylinder). They often have intercellular localization, but some (often parasites or symbionts) may be within plant cells (Fig. 11.7). In the case of fungi, the mycelium is either entirely in the plant or only partly within the plant, as in the case of ectomycorrhizal fungi (Brundrett 2009). It should be noted that certain mycorrhizal fungi harbor bacteria, forming multitrophic associations (Bertaux et al. 2003). Microorganisms present in the plant but that do not lead to physiological dysfunctions (as pathogens do) or host differentiation (as bacterial symbionts nodulating roots do) are termed endophytes. The plant tissues provide particular environmental conditions, inasmuch as microorganisms have direct access to available substrates in the plant, while being directly exposed to defense mechanisms of the host. The microbial numbers in the plant are generally low (10^{2-5} bacteria per g), except for some endophytic bacteria (up to 10^8 per g) and especially symbionts and parasites, when they are engaged in a successful interaction with the host (McInroy and Kloepper 1995; Gamboa et al. 2003).

11.3.3 Sources of Microorganisms Associated with Plants

Microorganisms interacting with plants have various provenances. First, many originate from soil. They can find themselves in contact with plants if they are present in an area of the soil that is penetrated by a root. However, many terrestrial microorganisms have the ability to actively come into contact with seeds or roots, through a process of chemotaxis (for bacteria) or chemotropism (for fungi) triggered by the perception of plant exudates (Bais et al. 2006). Terrestrial microorganisms in contact with plants will mostly be associated to the root system, but some may colonize plant shoots or fruits. This situation is mainly documented for pathogens, including fungi such as *Fusarium* spp. (Deuteromycetes) that produce mycotoxins (Leplat et al. 2013).

Second, certain microorganisms that interact with plants come in contact with the plant through precipitation and irrigation water, the fall of atmospheric dust, or wind (Agrios 1997; Morris et al. 2010). Precipitations also lead to runoff events along the aerial parts of the plants and the soil surface. This runoff will lead to a redistribution of some phyllospheric microorganisms. The presence of microorganisms in the atmosphere can be explained by the formation of aerosols, i.e., microdroplets, which originate from liquid phases that are turbulent or that face convection, e.g., by wind. This presence can also be explained by the transportation, caused by wind or air currents, of microorganisms or microparticles colonized by microorganisms, including for resting stages such as many fungal spores (Savage et al. 2012).

Third, animals can act as carriers of microorganisms. This will particularly involve phytophagous arthropods (especially insects) and herbivorous mammals, which will inoculate the plant with their mouthparts, as well as parasitic animals through their stylet or spear (nematodes) or rostrum (biting insects) (Villate et al. 2012). However, the mere contact between the animal (or its feces) and the plant can also allow plant contamination by microorganisms from the surface of the animal (or its digestive system), whether microarthropods or livestock.

Fourth, importing seeds, seedlings, and plants from remote areas brings microorganisms that interact with plants. Unintentional deliveries can also occur through the use of agricultural tools or machinery (farming practices), while some phytobeneficial microorganisms are deliberately introduced to inoculate crops (Agrios 1997; Dobbelaere et al. 2001; Alabouvette et al. 2006).

Fifth, the plant itself can be a source of microorganisms. Indeed, the microorganisms present on living plants or plant remnants (litter, crop residues) are a source of inoculum for plants that will later develop in the same place. For example, mycotoxin-producing *Fusarium* can survive in overwintered corn stalk residues and may colonize wheat in the following growing season (Leplat et al. 2013). In addition, certain microorganisms are present in the seed, whether as internal or external contamination. These microorganisms are in a favorable position to colonize plant seedlings emerging from these seed. Finally, endophytic microorganisms can be found in plants of the next generation if these plants are propagated vegetatively, as with sugarcane cuttings.

In the end, microorganisms associated with plants have contrasting sources, and they have undergone dissemination phenomena at different spatial scales. In terms of biogeography, even though many of them have an endemic distribution, some microorganisms appear to be transported over long distances (Ramette and Tiedje 2007). This is particularly the case of microorganisms associated with cultivated plants, which colonize plant shoots and/or are easily spread by wind, and some of them are plant pathogens.

11.3.4 Diversity and Activity of Microorganisms Associated with Plants

11.3.4.1 Ecological Factors

Due to high microbial numbers and the importance of microbial interactions in the rhizosphere, the ecological specificities of this microbial habitat are summarized below. In comparison with non-rhizosphere soil, the rhizosphere has specific physical and (bio) chemical features. From a physical point of view, the root surface is not homogeneous in terms of topography and rhizodeposit flux, noticeably with the occurrence of a groove and the preferential release of exudates at the junction of epidermal cells (Beattie and Lindow 1999). The level of hydrophobicity of the root surface (and thus its contact properties) may vary depending on the plant considered. Root growth results in the creation of a root-based soil macroporosity, which leads to compaction of the rhizospheric soil. It is also in the rhizosphere that diffusion of soluble rhizodeposits occurs, resulting in stimulation of microbial populations. Insoluble rhizodeposits render the mucigel gelatinous, which facilitates adhesion between root and soil particles. The mucigel also affects water movement in the rhizosphere, thereby modulating the availability of water to roots.

From a chemical point of view, the properties of the rhizosphere are largely related to the presence of rhizodeposits (formerly referred to as root exudates). Depending on the mode of release of rhizodeposits in the soil (Fig. 11.8), we distinguish exudates (in the modern sense of the term) or diffusates (soluble or gaseous low-molecular-weight compounds released by passive transport), secretions (compounds actively transported out of the root cells), cell lysates (compounds released during the lysis of epidermal cells, cortical cells, or senescent sloughed-off cells detached from the root cap), and the mucigel (plant polysaccharides, to which microbial exopolysaccharides can be added) (Nguyen 2003). Quantitatively, the rhizodeposits represent 15–20 % of the net organic carbon from photosynthesis, i.e., about half the organic carbon translocated to roots (Nguyen 2003). The organic rhizodeposits have a high functional and chemical diversity. They include the following:

1. Simple substrates, namely, a very large number of amino acids, organic acids, sugars, fatty acids, sterols, nucleic acid derivatives, etc.
2. Insoluble polymers (cellulose and other carbohydrates)
3. Vitamins and other growth factors (biotin, inositol, thiamine, etc.)
4. Phytohormones (auxins, cytokinins, etc.)
5. Enzymatic proteins (amylases, phosphatases, proteases, etc.)
6. Toxins (e.g., calystegines) and defense compounds (such as phytoalexins and glucosinolates)
7. Signals acting on microorganisms as chemoattractants (sugars, organic acids, etc.) and/or transcription inducers (e.g., flavonoids), etc. (Bais et al. 2006).

The rhizosphere exhibits other chemical particularities, which are also related to root functioning. Plant nutrition requires selective uptake of mineral nutrients by roots, which leads to accumulation and sometimes also precipitation (e.g., calcium carbonate or iron oxide) of excess constituents and especially depletion for metabolized nutrients (to a larger extent for soluble anions such as nitrate than for poorly soluble anions such as phosphate or for cations) in the rhizosphere. Cations and anions uptake involves parallel excretion of proton or hydroxide anion, respectively, which can lead to local pH change in the rhizosphere (Hinsinger et al. 2003). Finally, root (and microbial) respiration may lead to a decrease in oxygen concentration, with a parallel increase in carbon dioxide content, in particular, when soil is very wet. However, these processes can be offset by the ease of gas exchange in the root macroporosity. This is particularly the case in soils containing swelling clays, which retract during drying. It can be noted that certain plants that thrive in submerged conditions (e.g., rice) can release oxygen through its roots (Liesack et al. 2000). Depending on these factors, the availability of oxygen in the root zone will be lower (most often) or higher than in non-rhizosphere soil.

11.3.4.2 Microbial Diversity

To colonize plants, a microorganism must not only use the trophic resources they provide (in a context of competition with the rest of the microbial community) but also tolerate abiotic stress conditions (including desiccation and UV in the phyllosphere) as well as toxins and other host defense compounds. The best adapted microorganisms will be thus favored over the others. This selection process is amplified by the molecular dialogues with the plant, including microbial chemotaxis/chemotropism phenomena (*cf.* Sect. 9.2) and the ability of certain plants to interfere with bacterial quorum sensing (Bauer and Robinson 2002). Finally, all terrestrial microorganisms are not affected the same way by the antimicrobial compounds produced by the other microorganisms associated with plant (Gilbert et al. 1993; Raaijmakers et al. 2009) and by predation by protozoa (Bonkowski 2004) and nematodes (Jousset et al. 2009). In total, the trophic stimulation of microbial growth by the plant is therefore selective. This usually results in a lower microbial diversity when comparing bulk soil with rhizosphere soil or the surface with the inside of the plant. However, the species diversity of plant-associated microorganisms is important, and many bacterial and fungal taxa have been identified on leaves or roots (Lindow and Brandl 2003; Morris and Monier 2003; Garbeva et al. 2004).

Selection of microorganisms by plants can be observed at different taxonomic levels, from the family to the strain (Mavingui et al. 1992; Lindow and Brandl 2003; Garbeva et al. 2004; Cardon and Whitbeck 2007). At the lowest taxonomic levels, especially the intraspecies level, horizontal gene transfer can influence microbial selection. Indeed, plants favor horizontal gene transfer events, including the exchange of conjugative plasmids between bacteria (Bjorklof et al. 2000). Many abiotic and biotic factors modulate microbial selection by the plant, mainly through an effect on plant photosynthetic activity (and hence on exudation). This can be reflected by the occurrence of microbial successions in the rhizosphere, which can be related to host phenology (Garbeva et al. 2004; Mougel et al. 2006). Plant selection processes are also scale dependent. Since both the soil habitat and plant organization/physiology display spatiotemporal heterogeneity, different parts of a same individual plant may display different microbial communities, e.g., from one root microsite to another (Garbeva et al. 2004).

In the rhizosphere, the reduction of microbial diversity resulting from plant selection may be partly compensated by other processes that enhance genetic variability of microorganisms. First, it is believed that certain constituents released by the plant, particularly the seed, may have mutagenic effects (Miché et al. 2003). Second, certain bacteria display phenotypic variation (also termed phase variation), a process often reversible that generates at high frequency several different cell types (e.g., with different mobility patterns) from a single strain (Vial et al. 2006). Phenotypic variation facilitates root colonization of symbiotic and pathogenic bacteria that have this property (Achouak et al. 2004).

11.3.4.3 Microbial Activities

For microorganisms adapted to plants, trophic stimulation by exudates leads to an increase in the rates of cell physiological activities, the synthesis of many metabolites (including microbial exopolysaccharides of root mucigel), the rate of proliferation, and (for bacteria with a viable nonculturable state) the culturability level of cells (Bjorklof et al. 2000; Lübeck et al. 2000; Bais et al. 2006; Troxler et al. 2012). This trophic stimulation results in an increase in overall physiological activity of the microbial community (van Elsas et al. 2007), for example, in terms of carbon dioxide production (Griffiths et al. 2004). When considering a particular microbial function, the plant may have an impact on the size of the corresponding functional group (by a factor of 100 or more for rhizosphere ammonifiers) and/or the implementation level of the function (Cardon and Whitbeck 2007; Patra et al. 2007). This impact varies according to the stage of development of the plant, the age of the root or leaf, environmental factors influencing photosynthesis (which also impacts on microbial numbers and diversity), and the microbial function considered.

Generally, microbial transformations involving heterotrophic/saprophytic microorganisms are strongly stimulated by plants (Cardon and Whitbeck 2007). This is the case for many biodegradation functions for which the substrate comes exclusively/predominantly from plant exudates (Griffiths et al. 2004), as in the case of calystegines (Guntli et al. 1999) and opines, or both from exudates and soil organic matter (organic nitrogen compounds, cellulose, etc.). Substrates not of plant origin such as exogenous aromatic compounds degraded by cometabolism are transformed faster in the rhizosphere (Cardon and Whitbeck 2007). Plants also favor nitrogen fixation (whose energy cost is high) by nonsymbiotic aerobic bacteria (Patra et al. 2007). Conversely, transformations carried out by autotrophic microorganisms (e.g., nitrification) are little affected by the presence of the plant (Patra et al. 2006). Finally, there are transformations that are either stimulated or inhibited by the plant depending on the availability of oxygen and overall redox conditions in the rhizosphere. This is mainly the case of fermentations and anaerobic respirations, e.g., denitrification (Patra et al. 2006), which are favored by the presence of organic rhizodeposits but also require specific redox conditions partly determined by soil structure and soil

Fig. 11.9 Symptoms of plant disease. (**a**) Bacterial wilt of tobacco plant caused by the bacterium *Ralstonia solanacearum* (*left*). The bacterium colonizes spiral protoxylem vessels, which leads to vessel clogging and stops sap flow (Photo R. Pépin, UMR CNRS 5557 Microbial Ecology, Villeurbanne, France). (**b**) Leaf coalescent necrosis (*arrows*) on beet due to the epiphytic fungus *Cercospora beticola* (Deuteromycete) (Photo C. Steinberg)

moisture (Liesack et al. 2000). Redox conditions also modulate aerobic transformations including nitrification (Bohrerova et al. 2004).

Besides microbial transformations, most functions related to biotic interactions are also stimulated in the presence of plants. Most of them involve the plant directly, whether they are favorable (cooperation and symbiosis), unfavorable (competition, antagonism, and parasitism), or without effect on the plant (commensalism). Interactions not directly involving the plant can also be stimulated. These are positive interactions such as gene transfer (Bjorklof et al. 2000) or quorum sensing (a mechanism involving signals such as *N*-acyl homoserine lactones and used to perceive microenvironmental changes; Hense et al. 2007) and negative interactions such as predation of bacteria by protozoa (Bonkowski 2004). It also applies to interactions between microorganisms and soil fauna, for instance, the antagonism of certain bacteria toward plant-parasitic nematodes (Dabiré et al. 2005). Overall, the stimulation of biotic interactions by the plant often has a positive impact on the latter, except when phytoparasites are favored.

11.3.5 Biotic Interactions of Microorganisms Associated with Plants

11.3.5.1 Overview

The plant is exposed to high numbers of microorganisms, but most of them probably have little effects on the plant if any. Phytobeneficial microorganisms are much less prevalent comparatively. Cooperating microorganisms predominate over symbiotic ones on plant surfaces or in the rhizosphere, but this is less significant or even the opposite within the plant. Plant-parasitic microorganisms are a minority, except sometimes in infected plants, but their impact on the latter can be high. It is important to note that there is a continuum of plant–microbe interactions, ranging from mutualism to parasitism, and that the implementation of many plant–microbe interactions is highly dependent on plant genotype, interactions with other members of the microbial community, and environmental conditions.

11.3.5.2 Parasitism
Plant Parasites

Most parasitic diseases of plants (Fig. 11.9) are caused by fungi or oomycetes with the remainder due to bacteria or viruses (Agrios 1997) (Box 11.2). Phytopathogenic fungi and oomycetes are often facultative parasites called necrotrophic, whose saprophytic and parasitic phases alternate. They are responsible for penetrating injuries and necroses on a broad host range. Others are obligate parasites (biotrophic) developing only in living host plant. They require specialized bodies intrusion (**appressorium*** and **haustorium***), their host range is narrow, and often they cause different types of symptoms. The majority of phytopathogenic fungi and oomycetes affect the aerial parts of plants, causing diseases such as downy mildew (caused by oomycetes *Phytophthora parasitica* on tomato or *Plasmopara viticola* on vines), powdery mildew (*Erysiphe graminis* on herbaceous or *Uncinula necator* on vines), corn smut (*Ustilago maydis* on corn), rust (*Puccinia* spp. on cereals and *Uromyces* spp. on Fabaceae), or rot (*Botrytis cinerea* on grapevine) (Agrios 1997; Heitman 2011). In some cases, the disease symptoms are subtle, like ergot (*Claviceps purpurea*) or head blight of cereals (*Fusarium* spp.) because the phytoparasite produces mycotoxins such as lysergic acid [*C. purpurea* (Lorenz et al. 2007)]; trichothecenes, fumonisins, and zearalenone [*Fusarium* spp. (Osborne and Stein 2007)]; and ochratoxins and aflatoxins [(*Penicillium* spp. and *Aspergillus* spp. (Berthiller et al. 2013)]. These mycotoxigenic fungi are present all around the world and cause great losses to the world's agriculture. The disease caused by these molds reduces seed vigor, crop yield, and grain quality, making them poisonous to animals and humans. Mycotoxins are produced in the cereals, fruits, as well as in vegetables, in the field, and during transportation, storage, and processing. The problem of mycotoxins fluctuates from year to year, due to changes in the environmental conditions favorable for the production of mycotoxins and the development of the producing fungi. The Food and Agriculture Organization (FAO) of the United Nations and the European Union have set limits for the most important mycotoxins in different crops, to avoid the adverse effects toward animals and humans.

Box 11.2: Impact of Parasitic Microorganisms on Crop Plants

Christian Steinberg

Diseases caused by parasitic microorganisms on crops can result in food shortages and famines, and historically some of them have had a major impact on human populations (Rosenzweig et al. 2000). Late blight of potato, caused by the oomycete *Phytophthora infestans*, was a major factor in the great famine in Ireland (more than one million deaths out of a population of eight million inhabitants) and Irish emigration to the USA (two million emigrants) in the years 1845–1849. In more recent times (1943), the fungus *Helminthosporium oryzae* (Deuteromycete) resulted in the death of more than two million people in India and Bangladesh due to malnutrition caused by the destruction of rice crops.

The importance of crop losses due to diseases varies according to the type of pathogen, pathogen inoculum size, the development stage, and the genotype of the plant, as well as soil properties, climatic factors, and agronomic conditions (Agrios 1997). For the present time, the estimation of Oerke et al. (1995) based on eight major crops is 42 % of the yield (before harvest) is lost due to biotic factors across the globe (31 % for North America). Plant-parasitic microorganisms are responsible for a third of these losses, the rest being caused by insects and weeds. Loss due to diseases ranges from 28 % (for corn, soybean, and tobacco) to 40 % (potato) of all losses of biotic origin. Each year, yield loss due to diseases (before harvest) exceeds $15 billion for potato, $20 billion for wheat, and $50 billion for rice for the whole world (2008 estimate).

Currently, the extent of these damages increases mainly due to the development of trade in seeds and agricultural products and changes in farming methods, such as agriculture intensification and a switch from crop rotation to crop monoculture (Rosenzweig et al. 2000). The ongoing climate change will likely have major repercussions on crops and diseases affecting them.

Soil-borne fungi and oomycetes that infect the seed or roots cause important diseases. Damping-off is generally caused by oomycetes such as *Aphanomyces cochlioides* on sugar beet and more particularly *Pythium* spp. on most crops, but fungi such as *Rhizoctonia solani* can also cause damage in vegetables or forest nurseries. Conversely, necroses and root and crown rots are mainly due to fungi (*Thielaviopsis basicola, R. solani*), although oomycetes such as *A. euteiches* are a real obstacle to grow legumes in temperate countries (Persson et al. 1999). White mold of many vegetable crops is caused by *Sclerotinia sclerotiorum* and leads to general wilting symptoms associated with water-soaked lesions on stems, while sclerotia are formed inside stems. Other fungi such as *Fusarium oxysporum* and *Verticillium dahlia* also cause partial or general wilting, but they are tracheomycoses because the fungi are able to penetrate the roots to rapidly reach and colonize the vascular tissues (Olivain et al. 2006).

Many foliar bacteria are phytopathogenic, such as *Erwinia amylovora* (fire blight pathogen of pear and apple), *Xanthomonas campestris* (which causes stalk lesions and necrosis), or *Pseudomonas syringae* (causing cankers in many fruit trees or brown necrotic spots on the glumes of wheat) (Agrios 1997). Part of *P. syringae* epiphytes produce IceC membrane protein, which has an ice nucleation activity and promotes frost damage on strawberry (Lindow and Brandl 2003). Some are soil-borne phytopathogenic bacteria and are responsible for vascular disease (*Ralstonia solanacearum*), soft rot (*Pectobacterium carotovorum*), or gall (*Agrobacterium tumefaciens, Streptomyces scabies*). Certain *Agrobacterium tumefaciens* carry the Ti plasmid that confers them the ability to transfer genes to the plant (Krimi et al. 2002; Tzfira and Citovsky 2006).

Phytopathogenic bacteria are often facultative parasites, and an inoculum threshold is generally required for infection (10^4 cells of *P. syringae* per g of bean leaf). Some protozoa including *Phytomonas* are phytopathogens. These trypanosomes are responsible for heart-rot disease of coconut palms, and they can be transmitted back and forth from host plants to insects. Finally, some viruses such as poxvirus cause significant damages, but most of the time, they need vectors to infect their host plants. For instance, parasitic nematodes (e.g., *Xiphinema* spp.) serve as vectors of viruses such as the grapevine fanleaf virus (GFLV). Moreover, the wound caused by stylet penetration of the parasitic nematodes may also facilitate the plant infection by fungal (*Verticillium*), oomycete (*Phytophthora*), or bacterial (*Clavibacter michiganense*) pathogens. Besides nematodes, the main virus vectors are Hemiptera insects, but some viruses can also be transmitted to the plant by fungi (*Olpidium* spp.) or protists (*Polymyxa betae*) or between plants through parasitic plants such as dodder (Agrios 1997).

Infection Processes

Infection of the plant is done in several steps. The contact of the plant with spores (constituting the primary inoculum) of fungal pathogens involves, as presented above, several environmental factors including wind, water, insects, and microfauna that vectorize the pathogens, but also the biochemical and mechanical infectious potential of these pathogens. Production of fungal mucilage consisting of polysaccharides and/or glycoproteins ensures the adhesion of spores and germ tubes to the plant tissues. Enzymatic hydrolysis of the cuticle surface (*U. necator* on grapevine) or production of small proteins such as hydrophobins Mgp1 (*Magnaporthe grisea* on rice) enhances the adhesion of the propagules to the plant (Ebbole 2007). In most cases, there is no specific site of infection on the surface of the plant. Fungi penetrate the aerial part of the plant by piercing the cuticle, growing through the anthesis, exploiting wounds (opportunistic fungi), or growing between the stomata guard cells. The fungal germ tube generally forms an appressorium that is simply a swelling of the tip of the hypha (*B. cinerea*) or results from cell differentiation (case of rusts) to puncture through plant cell wall using high physical pressure, cell wall degrading enzymes, or both. In the case of pathogenic bacteria too, penetration occurs most often through natural openings, such as nectaries of flowers (*E. amylovora*), stomata or hydathodes (*X. campestris*), or wounds (*P. syringae*). Their lytic enzymes play an important role in virulence (Kazemi-Pour et al. 2004). Quorum sensing is sometimes involved in the infectious process (Smadja et al. 2004). Different types of trophic relationships can then be established. Some fungi such as *Venturia inaequalis*, responsible for apple scab, establish under the leaf cuticle and use epidermal walls as trophic base thanks to pectinases they produce. Others, such as powdery mildews, exploit the cytoplasm of epidermal cells thanks to a sucker they develop from the surface of the leaf. In rust or mildew agents, colonization of a first cell is followed by the development of secondary hyphae that gradually colonize other cells. Similar strategies are observed among soil-borne fungi and oomycetes when penetrating the plant root and crown. For instance, no specific site of infection was found for *F. oxysporum* attacking tomato roots (Olivain et al. 2006), and a set of extracellular enzymes is produced by the pathogens to degrade cellulose, hemicelluloses, and pectins of the plant epidermis and parenchyma leading to necroses, generalized rot, or tracheomycoses.

Interplay with Plant Defense Systems

As we saw previously, plants interact with many microorganisms, some of which being neutral or beneficial, while others are deleterious. It is important for plants to recognize and discriminate between them and to be able to respond accordingly. Conversely, it is important for beneficial and pathogenic microbes to modulate the host immune system to establish an intimate relationship or to prevent defense reactions. The plant, for instance, can respond to the presence of pathogens by the production of volatile compounds that will either promote or block microbial development, as for *B. cinerea* on strawberry (Abanda-Nkpwatt et al. 2006). It now appears that plants rely on the innate immunity of each cell and on systemic signals emanating from the infection site. This immune system allows them to recognize and respond specifically to invading pathogens (Jones and Dangl 2006).

Upon infection, compounds such as lipopolysaccharides, flagellin, peptidoglycans, and chitin are released from the pathogens. These microbial elicitors called microbe-associated molecular patterns (MAMPs), also referred to as pathogen-associated molecular patterns (PAMPs), are recognized at the surface of the host cell by receptor proteins called pattern recognition receptors (PRRs). MAMPs are basic components of all classes of pathogens. They may be constituents of the wall of the pathogen (exogenous elicitors) that are recognized by receptors. They may also correspond to the constituents of the plant cell wall that are released after degradation by microbial enzymes and recognized by receptors. A number of these common MAMPs and their role in the elicitation process have been recently described (Pel and Pieterse 2013). Recognition by the plant of these non-self compounds is a first step toward an effective immune response. PRRs generally consist of an extracellular leucine-rich repeat (LRR) domain and an intracellular kinase domain. Stimulation of PRRs leads to PAMP-triggered immunity (PTI), which provides a first line of defense against most nonspecific pathogens (Jones and Dangl 2006).

The second line of defense involves recognition by intracellular receptors of pathogen virulence molecules called effectors. The pathogens deliver effectors into the host cell by type III secretion (for bacteria) or using haustoria or other intracellular structures by an unknown mechanism (for fungi

Table 11.2 Nature of the interaction (compatible or incompatible) between pathogen and host plant in relation to the gene-for-gene concept (Based on Flor 1956)

Pathogen genotype[b]	Host plant genotype[a]	
	R1, r2	r1, R2
Avr1, avr2	Incompatible	Compatible
avr1, Avr2	Compatible	Incompatible

Recognition of the microbial elicitor (encoded by the avirulence gene *avr*) by the plant receptor (encoded by the resistance gene *R*) gives an incompatible interaction leading to the absence of infection (avirulence), while the absence of one or the other leads to a compatible interaction and the infection of the host plant, which reacts most often by a hypersensitive reaction

[a]Functional resistance allele *R*, non-functional resistance allele *r*
[b]Functional avirulence allele *Avr*, non-functional avirulence allele *avr*

and oomycetes) (Dodds and Rathjen 2010). These intracellular effectors often act to suppress PTI. However, many effectors are recognized by intracellular nucleotide-binding (NB)-LRR receptors; these sentry proteins are present at various levels in the plant (wall, cytoplasm, nucleus, cell membrane), and they allow a rapid response upon pathogen intrusion by inducing effector-triggered immunity (ETI). The gene-for-gene interaction takes place in that context (Table 11.2). The pathogen avirulence genes encode the elicitor. Host plant resistance genes corresponding to avirulence gene in the pathogen encode plant receptors recognizing this elicitor. This recognition leads to cascades of defense reactions involving essential metabolic pathways such as the salicylic acid and jasmonic acid pathways, as well as a massive influx of calcium ions into the cell together with potassium and chlorine efflux. Then, alarm signals are transmitted to the interior of the cell inducing the production of oxidizing radicals capable of inhibiting pathogen development. They lead to the activation of defense genes, synthesis of phytoalexins (plant antibiotics), and defense proteins called PR proteins (pathogenesis-related proteins), whose spectrum of activity is broad or narrow (van Loon et al. 2006). Some correspond to chitinases (PR3) and glucanases (PR2), whereas others are poorly known (PR17). To reduce the spread of the pathogen, the plant cell wall thickens (Fig. 11.7e) or, in the case of the hypersensitive response (HR), the infected cell transmits alarm signals to the neighboring cells that then destroy them (**apoptosis***). It has to be noted that NB-LRR-mediated disease resistance is effective against pathogens that grow only (obligate biotrophs) or during part of the life cycle (hemibiotrophs) on living host tissue, but not against pathogens that kill host tissues during colonization (necrotrophs) (Jones and Dangl 2006).

According to Dodds and Rathjen (2010), ETI and PTI reveal different coevolution dynamics between the plant and the pathogen, as, in contrast to PAMPs, effectors are variable and unessential. Similarly, the diversity of ETI receptors and pathogen effectors both within and between species is frequent, whereas PRR functions are widely conserved across families. Usually, PTI and ETI provide similar responses, but ETI is qualitatively stronger and faster and often it is responsible for the hypersensitive response (HR). PTI is generally effective against nonspecific pathogens, whereas ETI is active against specific pathogens. However, these relationships are not exclusive and depend on the elicitor molecules present in each infection (Dodds and Rathjen 2010).

Jones and Dangl (2006) proposed a simple zigzag scheme including four steps to illustrate the functioning of the plant immune system: First, PRRs recognize PAMPs, which results in PTI and may stop further pathogenic invasion. Then, successful pathogens produce effectors that contribute to pathogen virulence. They negatively interfere with PTI. In a third step, a given effector is "specifically recognized" by one of the NB-LRR proteins, which leads to ETI. ETI is an accelerated and amplified PTI response, resulting in disease resistance and, usually, a hypersensitive cell death response at the infection site. At last, on one side, natural selection drives pathogens to avoid ETI by diversifying the recognized effector gene or by acquiring additional effectors that suppress ETI. On the other side, natural selection favors new plant NB-LRR alleles that can recognize one of the newly acquired effectors, resulting again in ETI. Therefore, the interplay between microbial pathogens and plants presents several similarities with the case of animal diseases (*cf.* Sect. 11.4).

Host Range and Pathogen Diversity

Some pathogens such as *Phoma betae* are restricted to a plant species (sugar beet), while others like *Rhizoctonia solani* have a wide range of host plants belonging to different families. Plant-pathogenic fungi are drawing specific attention not only because of their negative impact on plant growth, but also because studying the interactions between the fungi and their host plant has revealed the potential of fungi to adapt to changing external constraints such as the ones we saw before. For instance, the *Fusarium* genus includes many soil-borne species that can infect roots and crowns as well as shoots, thanks to their ability to adapt infection mechanisms to the host plant, the plant phenology, and the plant's organs. In *F. oxysporum*, the large intraspecific diversity allows the fungus to infest a broad spectrum of host plants through very narrow fungus–host plant interactions. Indeed, more than 80 formae speciales have been described so far, each gathering one or more *F. oxysporum* populations that are able to infest one and only one plant species. Diversity studies revealed that different and probably independent genetic events within the species have led to the "speciation" of *F. oxysporum* into different formae speciales (Rep et al. 2005). In some cases, we can assume that the fungus–host plant interaction was such that the plant set up defense reactions that were circumvented by the fungus. This process

> **Box 11.3: Resistance and Host Specificity in Plant–Pathogen Relationships**
>
> Christian Steinberg
>
> In fungi and bacteria, isolates of the same species causing the same symptoms on a particular plant genotype, whether a cultivated (cultivar) or wild variety, are a race.
>
> The concepts of race and variety result from a pathogen–plant coevolution. Indeed, the emergence of a resistance gene in the host plant gives rise to a new variety, which will no longer be infested by the original pathogen (race 0), as is the initial variety. The appearance by mutation of a gene for virulence in the pathogen, allowing it to circumvent this resistance, gives birth to race 1, capable of infecting both varieties, while race 0 is limited to the first variety, and so on.
>
> The nature of the resistance of the host plant can be divided into two types. In horizontal resistance, many genes are involved (polygenic resistance). Vertical resistance in a single gene gives the plant (cultivar or variety) total resistance to a very specific biotype (race) of the parasite, but the variety concerned is sensitive to other biotypes of the pathogen. Any biotype able to overcome a vertical resistance factor is said to be virulent.

led to the definition of resistant plant genotypes and adapted pathogenic fungal races (Takken and Rep, 2010) (*cf.* Box 11.3). The gene-for-gene relationship is well known and exists in many host–pathogen interactions, but in *F. oxysporum*, such interactions occur within a formae specialis (Rep et al. 2005). In rare cases, the formae speciales are quite homogenous. Finally, the same *F. oxysporum* species also includes nonpathogenic strains (Edel et al. 2001).

Phytopathogen Control

Control of microbial pathogens of crops relies on a combination of various methods. Fungicides, which represent about 20 % of the pesticides used in agriculture, are ineffective against soil-borne bio-aggressors, and for human health and environmental quality, their use must be reduced. Prophylaxis should be preferred. This involves the regular cleaning of farm machinery and irrigation systems, but also the removal of dead plant matter from fields as well as the management (burial) of crop residues to minimize potential reservoirs for the plant pathogens. Similarly, many pathogens can live on and in seeds. Therefore, seed companies must ensure the quality of plant seeds. Innovative farming practices should be promoted. This includes new rotation schemes (longer and more diverse than the present ones), the use of intermediate crops having sanitizing impact, and appropriate tillage systems to preserve both soil structure and biological activity (multitrophic interactions), so that pathogen populations decline between host plantings. These practices should be combined with the use of resistant cultivars when possible as well as biological control agents if any in given conditions. Based on what was shown previously, these control methods (when used) can only limit the extent of damage (Box 11.2).

11.3.5.3 Symbiosis

Certain bacteria and fungi form a mutualistic symbiosis with plants (Cardon and Whitbeck 2007). Those between bacteria and plants consist mainly of nitrogen-fixing symbioses involving the floating fern *Azolla* (with the cyanobacterium *Anabaena*), phylogenetically diverse plants called actinorhizal (with the actinobacterium *Frankia*), and Fabaceae (with selected alphaproteobacteria, such as *Rhizobium*, and selected betaproteobacteria) (Mathesius 2003; Huguet et al. 2005). The *Rhizobium*–Fabaceae dialogue involves molecular signals produced by the plant (including flavonoids) and leads to bacterial synthesis of a lipo-chito-oligosaccharide termed Nod factor, which is a signal inducing root nodulation (Mathesius 2003) (Fig. 11.10a). Host specificity depends largely on the chemical specificity of these two types of signals. In the case of *Frankia* and of photosynthetic *Bradyrhizobium*, the bacterial signal involved is different, and indeed their genomes are devoid of canonical *nod* genes (Giraud et al. 2007; Normand et al. 2007). The Fabaceae nodule is formed by proliferation of cells from the cortex and the pericycle. The bacteria usually enter the plant through root hairs, which are deformed during the interaction, and find themselves in an infection thread of plant origin. This thread enables proliferation and delivery of bacteria to plant cells in the nodule. Once endocytosed by nodule plant cells, the bacteria differentiate into bacteroids but remain separated from the cytoplasm of the plant cell by the peribacteroid membrane. In *Medicago truncatula*, this transformation into bacteroids is triggered by nodule-specific cysteine-rich peptides resembling antimicrobials from the plant innate immune response (Van de Velde et al. 2010). The peribacteroid membrane and the bacteroid(s) it contains constitute the symbiosome. The nitrogen fixed by the bacteroids will serve as nitrogen source for the plant, which in return provides a source of carbon and energy (25–35 ATP necessary for each N_2 reduced) in the form of dicarboxylic acids (mainly malate). The nitrogenase is sensitive to oxygen but is protected by low oxygen diffusion within

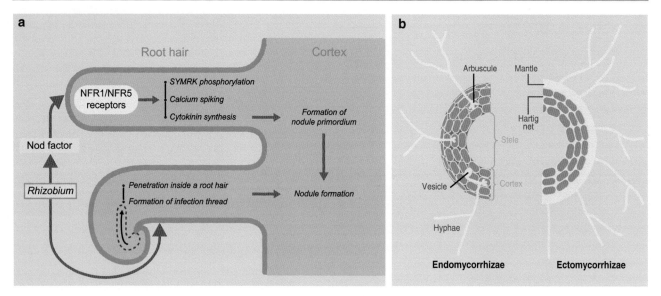

Fig. 11.10 Symbiotic interactions between microorganisms and plants. (**a**) Root nodulation process. Nod factors released by rhizobia are perceived by receptors of the epidermal cells of the root, which alters calcium flux and activates a signaling pathway leading to the production of cytokinins. The local increase in cytokinin concentration activates division of cortical cells and induces the formation of the nodule primordium. *Rhizobium* infects and invades the nodule via the infection thread, initiated in the root hair (From Oldroyd 2007). (**b**) Structural comparison of endomycorrhizae and ectomycorrhizae. In the case of endomycorrhizal fungi, the fungal filaments colonize the intercellular spaces of the cortex without reaching the central cylinder. Most endomycorrhizae are characterized by the differentiation of arbuscules (intracellular sites of exchange with the plant partner) and vesicles (intra or intercellular storage structures). In the case of ectomycorrhizae, the surface of the root tips is covered by a mantle made up of fungal mycelium that also colonizes the intercellular spaces of the cortex (forming the Hartig net) (Drawing: M.-J. Bodiou)

the nodule. The oxygen supply to the bacteroid is provided by leghemoglobin, a carrier of both bacterial (heme) and plant origin (globin). The leghemoglobin allows a high flux of oxygen without the latter reaching a concentration deleterious for the nitrogenase. The nitrogen fixed by the bacteroid is transferred to the plant in the form of ammonia and more rarely alanine. It will be transported in the phloem sap as amides (for temperate Fabaceae, with indeterminate nodules) or ureides (for tropical Fabaceae, with determinate nodules). In the bacterial partner, key genes (over fifty) involved in symbiosis are generally clustered as islands of genes, often plasmid-borne. They may be transferred horizontally, which probably explains their presence in both alphaproteobacteria and betaproteobacteria (Moulin et al. 2001).

The mycorrhizal symbioses implicate plant roots (Fig. 11.10b) (Mathesius 2003; Bailly et al. 2007). The fungus provides the plant with minerals taken from the soil (phosphorus and to a lesser extent nitrogen), whereas the plant provides carbon substrates derived from photosynthesis (Brundrett 2009). The vast majority of plants are mycorrhized in nature, the few plant species not forming mycorrhizae belong mainly to the family Chenopodiaceae (beets) and Brassicaceae (rapeseed, *Arabidopsis*). There are several types of mycorrhizae, particularly ectomycorrhizae, arbuscular endomycorrhizae, and ectendomycorrhizae (Mukerji et al. 2000; Brundrett 2009).

Ectomycorrhizal fungi (Ascomycetes and Basidiomycetes) develop extensively at root tips, forming fungal sheaths over short roots, which is very characteristic and easily visible to the naked eye (Brundrett 2009). Less than 5 % of plant species are concerned, among which mainly forest species growing in temperate areas. The mycelium grows between root cortical cells but does not penetrate living cells, thus forming an intercellular network called the Hartig net, which is involved in nutrient exchanges between the two partners. The functioning of ectomycorrhizae is under the metabolic and genetic control of both partners (Bailly et al. 2007).

Arbuscular endomycorrhizal fungi (Glomeromycota division) do not develop a mantle around the root (Brundrett 2009). The endomycorrhizae are widespread and affect about 90 % of plant species. They are found mainly in herbaceous plants and in some woody species. The fungus penetrates through plant cell walls and develops arbuscules and vesicles in cortical cells. Ectendomycorrhizae are intermediate between ectomycorrhizae and endomycorrhizae; an external mantle may be produced, but the fungus penetrates into root cells, as coils (arbutoïd mycorrhizae; among Ericaceae) or very short hyphae (monotropoid mycorrhizae; among Pyrolaceae). The establishment of mycorrhizal symbiosis shows common features with that of bacterial nitrogen-fixing symbioses (Mathesius 2003). Before contact, the mycorrhizal fungus stimulates the production of fine roots to increase the contact sites, secreting hormones including the auxin indole-3-acetic acid (IAA). After contact, the fungus enters the root and must overcome defense mechanisms. It switches from a saprophytic to a biotrophic phase, which recalls the case of biotrophic parasites.

Signaling mechanisms between partners in the establishment of a nitrogen-fixing symbiosis or a mycorrhizal symbiosis involve common symbiosis genes (*SYM*) in the plant, including *SYMRK/NORK/DMI2* and *DMI3* that code for receptor-like leucine-rich kinases, *CASTOR/DMI1* and *POLLUX* encoding ion channels ensuring transmembrane calcium flux, and (in the case of nitrogen-fixing symbiosis in Fabaceae) *ENOD40* (early nodulin 40) (Oldroyd and Downie 2006; Reinhardt 2007; Gherbi et al. 2008). The structure of the Myc factors, i.e., sulfated and non-sulfated lipo-chito-oligosaccharides consisting of four substituted *N*-acetylglucosamines, is very close to that of the Nod factors, which reinforces the view that the two types of symbioses are functionally very close (Maillet et al. 2011). In terms of evolution, the bacterial symbioses (relatively young; about 65 million years) and the ectomycorrhizal symbiosis (also recent; about 180 million years) probably recruited *SYM* genes already involved in the arbuscular mycorrhizal symbiosis, which is older (probably 400 million years) (Simon et al. 1993).

11.3.5.4 Cooperation

Several bacteria and fungi actively cooperate (syn. associative symbiosis) with the plant. In the case of bacteria, this ability is mainly found in **plant growth-promoting rhizobacteria*** (PGPR) (Fig. 11.11a). PGPR are documented mainly in Proteobacteria and Firmicutes and to a lesser extent in Actinobacteria. Growth stimulation usually results from a combination of direct and indirect phytobeneficial effects (Dobbelaere et al. 2001). Direct phytobeneficial effects may entail improved mineral nutrition of the plant, e.g., via free-living nitrogen fixation or phosphate solubilization (Dobbelaere et al. 2003) and/or improved water uptake through rhizosphere soil structuration by bacterial exopolysaccharides (Amellal et al. 1998) or aquaporin stimulation (Groppa et al. 2012). Interference with plant hormonal metabolism may also be involved, via bacterial production of phytohormones (auxins and cytokinins) and/or bacterial deamination of 1-aminocyclopropane-1-carboxylate (ACC), the ethylene precursor in the plant (Dobbelaere et al. 2003). The auxin indole-3-acetic produced by PGPR helps getting around plant defense mechanisms, thereby facilitating PGPR colonization of the plant (Remans et al. 2006). Finally, phytobeneficial effects also include triggering of systemic resistance in the plant, in particular, induced systemic resistance (ISR) pathways relying on jasmonate and ethylene (van Loon et al. 2006).

Indirect phytobeneficial effects of PGPR entail inhibition of parasitic bacteria, oomycetes, fungi, nematodes, and even parasitic plants (e.g., *Striga*), mainly through competition or antagonism (Chapon et al. 2002; Raaijmakers et al. 2009). Competition can take place for macronutrients (such as organic carbon), micronutrients (such as soluble ferric iron via high-affinity siderophores), and/or infection sites. Antagonism (syn. amensalism) may involve extracellular lytic enzymes (e.g., cellulases, chitinases, proteases), which act on the cell wall of pathogenic microorganisms, their virulence factors (such as fusaric acid from *Fusarium oxysporum*, degraded by some *Burkholderia*), or their intercellular signals (such as *N*-acyl homoserine lactone in *Pectobacterium carotovorum*) (Raaijmakers et al. 2009). Antagonism may also involve type III secretion system effectors (*cf.* Sect. 11.4.3) with the ability to reduce the virulence of certain pathogens (Rezzonico et al. 2005). Finally, antagonism can rely on production of antimicrobial secondary metabolites (antibiosis), such as 2,4-diacetylphloroglucinol, and a single antagonistic strain often produces several of these antimicrobial metabolites (Raaijmakers et al. 2009). Some pathogens can defend themselves by suppressing the production of these metabolites (case of 2,4-diacetylphloroglucinol, via fusaric acid from *Fusarium*) or the expression of genes involved in root colonization by the antagonistic bacterium (Fedi et al. 1997; Raaijmakers et al. 2009).

The metabolites released by PGPR are important for their interactions with the plant and plant pathogens, and many of them play multiple roles. Indeed, 2,4-diacetylphloroglucinol (enabling antagonism toward phytoparasites) and some siderophores such as pyoverdine (involved in iron competition) can also induce plant resistance (Raaijmakers et al. 2009). Phenylacetic acid, synthesized by *Azospirillum brasilense* Sp245 and others, is both an auxinic phytohormone and an antimicrobial compound. Therefore, certain PGPR are multifunctional in that they can act both on plant (via induction of resistance and ACC deamination) and phytopathogens (antagonism or competition).

Fungi in cooperation with the plant remain poorly documented compared with PGPR. Their direct phytobeneficial effects include solubilization of mineral nutrients (*Trichoderma* and *Gliocladium*) and the induction of systemic resistance in plants (*Trichoderma*, *Gliocladium*, and nonpathogenic *F. oxysporum*) (Harman et al. 2004). However, indirect phytobeneficial mechanisms are more important. Hyperparasitism toward different phytoparasitic oomycetes and fungi is one of the modes of action of *Trichoderma*. The fungus coils around its target, e.g., *R. solani*, and secretes lytic enzymes (chitinases and cellulases) that alter phytoparasite cell wall. The degradation

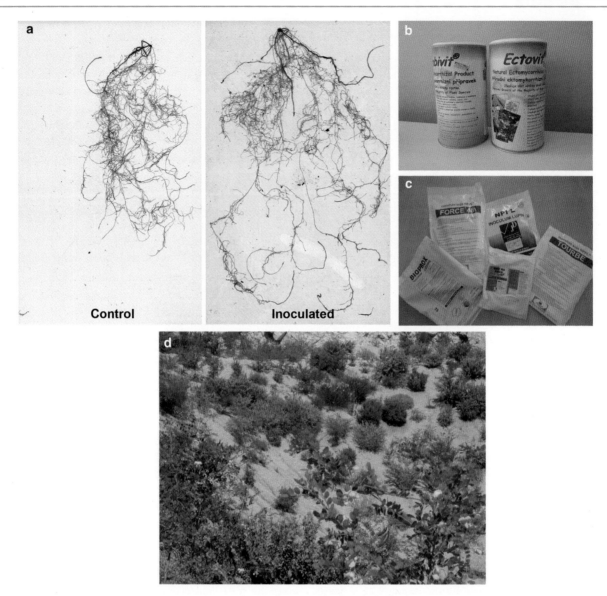

Fig. 11.11 Biotechnological use of beneficial plant–microorganism interactions. (**a**) Effect of a PGPR bacterium on the development of the root system of wheat seedlings (Photo C. Prigent-Combaret, UMR CNRS 5557 Microbial Ecology, Villeurbanne, France). (**b**) Commercial inoculum preparations based on endo- or ectomycorrhizal fungi (Photo Y. Moënne-Loccoz). (**c**) Commercial inoculum preparations based on nitrogen-fixing bacteria for Fabaceae crops (Photo C. Steinberg). (**d**) Rehabilitation of bare slopes in an abandoned limestone quarry, based on revegetation using the Fabaceae *Medicago arborea*, *Coronilla glauca* (yellow flowers in the foreground), and *Dorycnium hirsutum* associated with symbiotic nitrogen-fixing bacteria. The photo is taken after 3 years of plant growth (Photo J.C. Cleyet-Marel, UMR 113 IRD/CIRAD/SupAgro/UM2 Laboratoire des Symbioses Tropicales et Méditerranéennes, Montpellier, France)

products will allow chemotropism toward the phytoparasite. Physical contact between fungi will trigger external colonization and then penetration of the phytoparasite by *Trichoderma*. *Trichoderma* and *Gliocladium* may also antagonize phytoparasites through the production of antimicrobial secondary compounds (Harman et al. 2004). Finally, competition for trophic resources can be a major mode of action, as in nonpathogenic *F. oxysporum* (Alabouvette et al. 2006). Overall, as for PGPR, phytobeneficial fungi often display a combination of different modes of action.

11.3.6 Significance for the Plant

Plant–microbe interactions have a major impact on plant functioning and plant community ecology. At the scale of a given plant, negative effects of parasitic microorganisms are usually easy to identify, based on observation of disease symptoms. Many other plant-associated microorganisms have positive effects on plant growth or development, but these effects are often more difficult to visualize. Phytobeneficial microorganisms promote plant development primarily by altering plant hormonal balance, while their impact on plant growth may also involve trophic microbial effects. For example, the potential for symbiotic nitrogen fixation is typically in the order of 60–200 kg N/ha/year, for which the plant will invest 3–25 % of net photosynthates, versus only 5–25 kg N/ha/year for free nitrogen fixation. Nevertheless, the symbiotic nitrogen fixation potential is often poorly exploited in modern agriculture, which relies on massive use of chemical nitrogen fertilizers. Mycorrhization improves plant nutrition in phosphorus and nitrogen, and to a lesser extent in potassium and iron, especially in low-fertility soils (Brundrett 2009). Phosphorus can be recovered by mycorrhizal hyphae at more than 7 cm from the root. Mycorrhizal fungi also improve plant resistance to biotic stress (disease, herbivory; Bennett and Bever 2007) and abiotic stress (drought, metal pollution; Auge 2004).

Regarding plant health, the role of plant-protecting bacteria and fungi is important in the rhizosphere. In soils that are suppressive to disease, their interactions are sufficient to limit disease severity despite the presence of the pathogen, plant susceptibility to disease, and environmental conditions favorable for infection (Alabouvette et al. 1996; Garbeva et al. 2004). These disease-suppressive soils are documented for root diseases caused by fungi and to a lesser extent by bacteria or nematodes. In some pathosystems, suppressiveness gradually develops during crop monoculture in soils that were initially conducive to the disease. This is the case of wheat take-all caused by *Gaeumannomyces tritici*; take-all severity increases over the early years of wheat monoculture, reaches a peak (usually in years 4–8), and decreases thereafter to minor levels in the following years (Lucas and Sarniguet 1998; Raaijmakers and Weller 1998). In other pathosystems, disease suppressiveness is a natural property of the soil and does not require monoculture, as for soils suppressive to Fusarium wilt (caused by *F. oxysporum*) (Alabouvette et al. 1996) or to Thielaviopsis black root rot (caused by *Thielaviopsis basicola*) (Kyselková et al. 2009). Soil suppressiveness is often attributed to *Pseudomonas* PGPR subpopulations producing 2,4-diacetylphloroglucinol, although other phytobeneficial microorganisms might also play an important role (Kyselková et al. 2009).

Plant–microbe interactions have a significant impact on natural plant communities (Cardon and Whitbeck 2007). Nitrogen-fixing and mycorrhizal symbioses may be important to promote the establishment of pioneer plant groups. This is the case for plant recolonization of ground surfaces following deglaciation (at the end of ice ages or with current global warming), volcanic eruptions, or forest fires (van der Maarel 2005). Pioneer plant species will modify soil microbial community composition/functioning which, in turn, may then influence plant succession and the composition of intermediate groups, both by favoring certain plant species and counterselecting others. Finally, when the climax stage is reached, pathogens adapted to the main plant species may decrease their dominance or eliminate enough individuals to allow re-installation of pioneer species (within forest clearings) and reinitiate plant succession (van der Maarel 2005). In the particular case of invasive plant species, the soil microbial community may facilitate their establishment at the expense of native plants. Plant–microbe interactions are also important for the functioning of plant communities, particularly in terms of recycling of plant litter by microbial decomposers and nitrogen fixation (van der Maarel 2005). Organic matter transformations contribute to pedogenesis and to soil functioning, and as such they impact both on plant growing conditions and plant community successions.

11.3.7 Biotechnological Uses of Plant–Microbe Interactions

Certain microorganisms are used as inoculum in agriculture (Fig. 11.11b), for biofertilization, phytostimulation, or biocontrol purposes (Dobbelaere et al. 2001; Alabouvette et al. 2006). The goal is to improve crop yield (productivity) and/or reduce chemical inputs (environmental quality). Biofertilizers aim at improving plant mineral nutrition, e. g., symbiotic nitrogen-fixing bacteria (*Rhizobium* and other genera) and to a lesser extent mycorrhizal fungi. For PGPR having mainly hormonal effects on the plant, the term phytostimulation is used instead of biofertilization. The main phytostimulators used belong to the genus *Azospirillum*, especially for cereals (0.5–1 million ha of corn inoculated each year). *Azospirillum* inoculation resulted in increased cereal yield (by 10–30 %) in two-thirds of the cases (Dobbelaere et al. 2001). Biological control is most often implemented in commercial greenhouse, using bacteria or fungal inoculants. The objective is to protect crops against phytopathogens (often fungi or oomycetes, sometimes bacteria), parasitic nematodes or plants, or weeds. The microorganisms used correspond to Firmicutes (*Bacillus*), Proteobacteria (*Pseudomonas*), and Actinobacteria (*Streptomyces*) for bacteria as well as Deuteromycetes (*Trichoderma, Gliocladium, Coniothyrium minitans*) for fungi (Gilbert et al. 1993; Harman et al. 2004; Rezzonico et al. 2005). Inoculum formulation is a key issue determining inoculant performance, and spore-forming microorganisms have better storage capacities. The inoculum is often applied to the seed before sowing or in the furrow, more rarely on the soil surface or on the plant during growth. Current regulations on the use of inoculants in agriculture vary greatly from one country to the next (regarding the need for toxicological assessment, proof of effectiveness, guarantee for inoculum level in the product), which can make it expensive in certain countries to develop new microbial products, especially for biological control purposes.

In a broader sense, the biotechnological use of plant–microbe interactions in agriculture also includes the use of plants benefiting from symbiotic nitrogen fixation (Moulin et al. 2001). This is particularly the case of Fabaceae, as forage or cash crops, traditionally used to improve soil fertility, particularly in crop rotations. Some of the Fabaceae are used as green manure, i.e., once grown they are incorporated into soil rather than being harvested. In agroforestry, Fabaceae shrub species and actinorhizal plants (e.g., *Casuarina*) can be used for alley cropping with herbaceous companion crops. Two other types of nitrogen-fixing symbioses, which do not lead to the formation of nodules, are also important from a biotechnological point of view. One involves aquatic ferns and cyanobacteria, such as the *Azolla–Anabaena* symbiosis, which can provide in the order of 30–60 kg N/ha/year for submerged rice fertilization (Roger et al. 1993). The other is formed by sugarcane and endophytic Proteobacteria such as *Gluconacetobacter diazotrophicus*. These nitrogen-fixing symbioses may provide up to 100–150 kg N/ha/year, making sugarcane particularly attractive for biofuel production. This crop contributes to atmospheric flux of the greenhouse gas N_2O, but much less than other crops (Crutzen et al. 2007). Finally, crop plants are often involved in cooperative interactions with PGPR and/or endophytic fungi. This capacity is of agronomic interest, but it varies depending on plant genotype (Picard and Bosco 2006), and so far it has been largely ignored in crop breeding schemes.

In nonagricultural soils, plant–microbe interactions are of interest for the establishment of windbreaks and landscaping of roadside and other anthropogenic sites. Nitrogen-fixing symbioses based on Fabaceae or actinorhizal shrubs are often used in this context, all the more as these species also benefit from mycorrhizal symbioses. These symbioses are particularly useful for revegetation of land low in organic matter (abandoned quarries or eroded slopes; Fig. 11.11c), degraded, highly acidic, and/or rich in heavy metals (mining areas, etc.) (Roy et al. 2007). In the latter case, revegetation (*cf.* Sect. 16.2.1) can stabilize the contaminated soil (phytostabilization), and interaction with PGPR displaying ACC deaminase activity and/or with mycorrhizal fungi is important for improving plant tolerance to the metals (Belimov et al. 2005).

In the case of contaminated soils, interactions between plants and microorganisms are also of interest for two phytoremediation techniques, i.e., phytoextraction and rhizoremediation (Cardon and Whitbeck 2007; Roy et al. 2007). Regarding metal pollution, some of the rhizosphere bacteria and fungi can solubilize metals such as chromium, lead, and arsenic (Khan 2005), while promoting their absorption by the roots (phytoextraction). For some organic pollutants, rhizodeposit availability can promote their biodegradation by the rhizosphere microbial community, whose action may be supplemented by enzymes released by the root. Organic pollutants that can be treated through such rhizoremediation include pesticides, simple hydrocarbons, polycyclic aromatic hydrocarbons (PAHs), and polychlorinated biphenyls (PCBs) (Rentz et al. 2005). Microbial biodegradation by rhizospheric microorganisms also helps protect the plant from phytotoxic effects of organic pollutants, at the same time promoting site revegetation.

Plant–microbe interactions also have an interest for purification of effluents. This is the case of rhizoremediation systems, including those based on reeds or bamboo, used to decompose organics in certain water treatment operations (*cf.* Chap. 16). However, the role of plant–microbe interactions in these systems is poorly documented. A simplified system based on unicellular microalgae (*Chlorella*) and bacteria (*Azospirillum*) co-immobilized in alginate beads was also developed for tertiary treatment of wastewater, allowing removal of nitrogen (ammonium mainly) and phosphorus (de-Bashan et al. 2004).

Finally, the ability of *Agrobacterium* to transfer genes into higher plants is one of the means used to obtain transgenic plants (Tzfira and Citovsky 2006). This ability relies on the pTi plasmid, which allows the transfer of a particular DNA segment (the T-DNA) located between 25-bp flanking repeats and that includes genes involved in the synthesis of phytohormones (auxins and cytokinins) and opines. The T-DNA is transferred in single-stranded form, combined with VirD2 and VirE2 proteins, and is randomly integrated into the nuclear genome of dicotyledonous plants. To obtain transgenic plants, the genes of interest replace genes between the flanking sequences. This technology has been enhanced to allow genetic modification of monocots, a transfer that does not occur under natural conditions.

11.3.8 Epilogue

Interactions between microorganisms and plants have played a fundamental role in the evolution of plants, allowing the emergence of the first plant cells (thanks to cyanobacteria) and, later, land colonization by plants thanks noticeably to the endomycorrhizal symbiosis (Simon et al. 1993). Plant–microbe interactions are very diverse, ranging from facultative relationships with minor consequences to obligatory partnership with major evolutionary consequences. The microbial community contributes significantly to the functioning and ecology of the plant partner and to soil quality. Indeed, microorganisms modulate the growth, nutrition, and health of plants. Conversely, plants grow in biotopes where thousands of microbial taxa lead an existence whose focal point is the plant as a source of exudates or litter or as a partner to attack or collaborate with. It is therefore inevitable that plants have evolved and will continue to do so according to the microorganisms interacting with them. Mankind must improve plant growth to meet the global challenge of feeding billions of people in a changing environment and to do so must better integrate the constraints and benefits of plant–microbe interactions. In the field of agronomy, this concerns in particular the principles of crop breeding schemes, cultivation techniques, and the development of ecological engineering approaches in cropping systems.

11.4 Interactions Between Microorganisms and Animals

11.4.1 Introduction

Man and animals, whether vertebrate or invertebrate, live in environments (air, water, soil) constantly populated by microorganisms consisting mainly of bacteria, fungi, protozoa, and viruses. Such cohabitation usually leads to more or less intimate associations between microorganisms and animals. Spatial proximity, contact time, and the degree of dependence between the interacting partners influence the type and evolution of these associations. Microorganisms use animals as habitats in which they live and from which they retrieved nutrients necessary for their multiplication. They colonize the outlying areas, such as the surface of the skin or mucous membranes, and can penetrate the tissues and organs of host animals. Some microorganisms can colonize the cytoplasm of germ cells, oocytes, and sperm and are thus subject to possible vertical transmission (from parents to offsprings), which can allow a greater ability to spread among host population. The presence and activities of microorganisms will affect development, reproduction, and survival of host animals. The nature of the effects caused by microorganisms and the animal responses to these infections are very diverse and are part of a continuum from parasitism to mutualism (Buchner 1965). Whatever the outcome of the interaction (neutral, beneficial, or deleterious), animals respond to infection by implementing defense systems, either by inducing immune responses or setting up physical barriers to prevent or limit the spread of microorganisms considered as potentially infectious agents. In parallel, microorganisms develop strategies to circumvent the defenses of animals by manipulation host traits or molecular mimicry. It is the trade-off between cost and benefits, that is to say, the balance of the association in terms of adaptive value, which defines the type of interaction (Moran 2007). Knowledge and understanding of the interactions between microorganisms and animals represent major challenges in various areas of biology. In the fundamental domain, microorganisms and their animal hosts entail a great diversity of species potentially giving rise to a variety of interactions. These interactions are of great interest in cell biology, molecular biology, evolutionary biology, and ecology. In human and animal health, the interactions between microorganisms and their vertebrate or invertebrate hosts are important because they are the cause of many diseases involving emerging human and animal pathogens.

11.4.2 Diversity, Distribution, and Abundance of Microorganisms Associated with Animals

Since the first microscopic observations of microbes by Dutchman Antonie van Leeuwenhoek, followed by successful cultivation of some of them by Louis Pasteur and other scientists, and, more recently, the development of molecular methods for the detection and identification without prior culturing (culture-independent methods), the list of microbes documented in animals is increasing. It remains that, for most microbes, animals are hostile habitats due to the physico-chemical conditions for growth and survival. One can mention, in healthy individuals, the often constant temperature inside the body but variable at the surface exposed to environmental stress (radiation), the pH that can be either neutral or acidic in the stomach, the osmotic pressure (variation of ion concentration), the partial pressure of oxygen (aerobic and anaerobic), the presence of various microbial inhibitors, and limited energy resources both in quality and quantity.

Compared to the great diversity of microbial phyla (DeLong and Pace 2001) present in areas where animals live, only certain groups or certain strains or variants of a group have the capacity for colonization (adhesion, secretion systems, lytic enzymes, metabolism) and are able to adapt and associate transiently or permanently with animals.

11.4.2.1 Microbial Diversity

Bacteria are the group most frequently encountered both at the surface and in the internal parts of animal bodies (*cf.* Sect. 8.1.5). Bacteria associated with humans belong to the mesophiles and have their growth optimum around 37 °C. The culturable bacteria from the skin mostly belong to two groups (Noble 1993). One is the Gram-positive cocci family *Micrococcaceae*, with a dominance of staphylococci, and the other consists of corynebacteria, in particular, bacteria belonging to genus *Corynebacterium*, which are found on all parts of the human body surface. Bacteria belonging to genera *Propionibacterium* and *Brevibacillus* occupy more limited areas with high production of lipids such as the sebaceous glands, the face, and the scalp. Microbial flora of the skin of pets (cats, dogs) and farm animals (pigs, cows) is also dominated by staphylococci and some fungi (Nagase et al. 2002).

Review of various studies on the microbiota (cultivable or unculturable microbes associated to an individual) of human or other vertebrate species confirmed previous dominance of bacteria, but it has also expanded the repertoire of microbial taxa (fungi, protozoa, viruses) residing in external

Fig. 11.12 (**a**) Distribution of bacterial phyla in healthy individuals. The area of each part is an average number of distinct phylotypes based on *rrs* gene sequences (16S rRNA). The average number of phylotypes per individual is given in brackets (Adapted from Dethlefsen et al. 2007). (**b**) Richness in taxonomic units of bacteria in the human gut. Estimated percentage of *rrs* sequence identity to the genus level (95 % identity), species (98 % identity), and strain (unique sequence) (Adapted from Bäckhed et al. 2005)

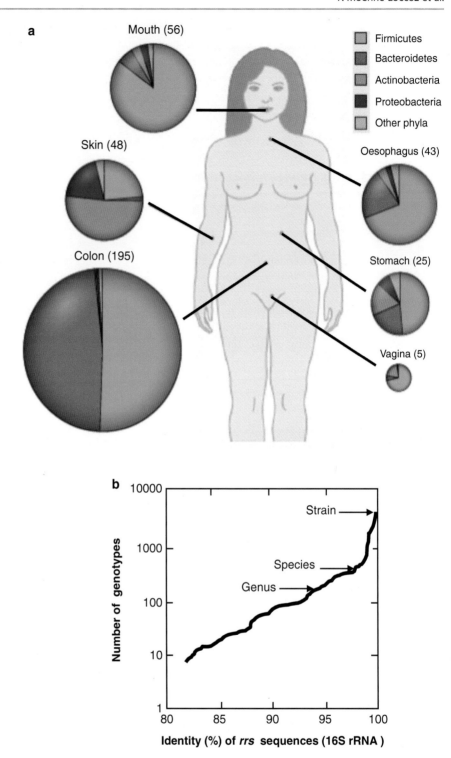

parts or internal organs (Nardon and Charles 2001; Dethlefsen et al. 2007). Extraction of DNA from human tissues and large-scale sequencing of the gene encoding 16S ribosomal RNA have revealed bacterial phyla of the skin, mouth, intestines, and genitals (Fig. 11.12). Among more than 50 bacterial phyla described, only four phyla are commonly found in humans: Actinobacteria, Bacteroidetes, Firmicutes, and Proteobacteria (Dethlefsen et al. 2007). These four phyla coexist with nine other secondary phyla (*Chlamydiaceae, Cyanobacteria, Deferribacteria, Deinococcus-Thermus, Fusobacteria, Spirochaeta, Verrucomicrobiota*, and two new phyla TM7 and SR1) in humans as well as in other vertebrate hosts (Ley et al. 2006). The low diversity in bacterial phyla (13 of 50) contrasts with a wide variety of genera, species, and strains (Fig. 11.12b) identified in both humans and animals (Bäckhed et al.

2005; Ley et al. 2006). Commensal bacteria, symbiotic bacteria (*Bacteroides*, *Lactococcus*), or pathogens such as enteropathogenic *Escherichia coli*, *Pseudomonas aeruginosa*, *Streptococcus*, or *Chlamydia* are those most often found (Bäckhed et al. 2005). Diversity and structure of the microbiota reflect contrasting types of evolutionary selection pressures and the nature of interacting organisms (Ley et al. 2006).

Invertebrates (arthropods and nematodes) consist of phyla including a large number of species, 1 million species described for 4–5 million estimated for the single group of insects (Novotny et al. 2002). Invertebrates harbor diverse microbial communities, some of which contribute to their development and sometimes play an important role in their ability to adapt to extreme environments (Buchner 1965; Nardon and Charles 2001). Most wood-feeding termites depend on microbial partners (bacteria, fungi, protozoan flagellates) to degrade cellulose and lignified compounds in their intestinal tract (Brune 2003). More than 700 species of bacteria belonging to different phyla (Spirochetes, Bacteroidetes, Proteobacteria, Actinobacteria, Mycoplasma) have been described in the gut of termites *Reticulitermes speratus* and *Macrotermes gilvus*. This diversity can vary from one termite type to another (Hongoh et al. 2006; Fall et al. 2007). *Drosophila*, the fruit fly very familiar to geneticists, hosts Proteobacteria (*E. coli*, *Pseudomonas*, *Sphingomonas*), *Bacteroidetes*, and *Mycoplasma* (Mateos et al. 2006). Besides the parasites (*Plasmodium*) or viruses (dengue or Chikungunya) they transmit, mosquitoes belonging to genera *Anopheles* and *Aedes* harbor many acetic acid bacteria including *Asaia*, *Acetobacter*, *Gluconobacter*, and *Sphingomonas* (Favia et al. 2007) or even genera *Aeromonas*, *Acidovorax*, *Bacillus*, *Paenibacillus*, *Pseudomonas*, and *Stenotrophomonas* (Lindh et al. 2005).

Although little studied, worms also host microbes, especially bacteria. For example, filarial nematodes such as *Onchocerca volvulus* (etiologic agent of onchocerciasis or river blindness) and *Brugia malayi* (etiologic agent of lymphatic filariasis such as elephantiasis) shelter in ovarian tissues the bacteria of genus *Wolbachia* that synthesize polysaccharides in infected animals, which constitutes an inflammatory factor aggravating disease (Saint André et al. 2002; Taylor 2003). As it will be described later on, the entomophagous nematodes (feeding on arthropods) of the genera *Steinernema* and *Heterorhabditis* host enterobacteria of the genera *Xenorhabdus* and *Photorhabdus*, respectively, with which they establish symbiotic associations (Forst et al. 1997).

11.4.2.2 Abundance and Location of Microbes

Microbial populations vary in size according to their intracellular or extracellular localization, the surface or inner organ colonized, and depending on the host considered. These numbers can reach relatively high values in certain organs of human and animals. In healthy adult men, densities are 10^{6-7} bacteria per cm^2 of skin (Roth and James 1988), 10^{11-12} bacteria per ml in the colon (Whitman et al. 1998), and 10^{14} bacteria in the entire intestinal tract. The microbes of the human body account for 10 times the total number of somatic and germ cells. The bacterial densities can greatly increase in cases of bacteremia or sepsis.

Cytological observations showed that in some insects, bacteria can also be localized in specialized cells called bacteriocytes (or bacteriomes) and in oocytes, which promotes vertical transmission from parents to offspring (Baumann et al. 2000). This is the case of bacteria *Buchnera* and *Rickettsia* in aphids or *Wolbachia* and *Spiroplasma* in many Diptera and Hymenoptera (Fig. 11.13). The whitefly *Bemisia tabaci*, the vector of viral pathogens of many plants, can host in a single cell many different bacterial genera, including *Arsenophonus*, *Cardinium*, *Hamiltonella*, *Portiera*, *Rickettsia*, and *Wolbachia* (Gottlieb et al. 2006). The bacteriocytes of the beetle *Sitophilus oryzae* (Fig. 11.13), a pest of stored grain, contain an average of 10^3 bacteria per cell or 10^6 bacteria per insect (Heddi et al. 1999). Apart from these few fragmentary data on insect models, investigations on the associated microbiota remain very limited despite their important role in the adaptation and development of invertebrates. A large effort is under way to study the microbes of insects since many of the latter are devastating pests of crops or vectors of pathogens that cause serious diseases in plants, animals, and humans.

11.4.3 Types of Interactions Between Microorganisms and Animals/Humans

Microbes that establish interactions with animals mostly have environmental origins (air, water, soil, food), but some of them move from one animal to another. An exemplary case of inter-animal microbial traffic is the transmission of the microbiota from mother to newborn. The vast majority of microorganisms are harmless and some can even be used as nutrients for animals. Fungivorous nematodes eat fungi, while bacterivorous ones feed on bacteria, up to 10^6 bacterial cells per day (Blanc et al. 2006). In soils, this predation may significantly alter the composition and structure of microbial communities and therefore the biological functioning of soil, as microorganisms play an important role in various processes such as decomposition of organic matter, solubilization of phosphates, nitrogen-fixing symbiosis, or denitrification. Other microorganisms have spectacular effects on their host, such as light organ differentiation in the squid *Euprymna scolopes* or manipulation of invertebrate reproduction by the bacterium *Wolbachia*.

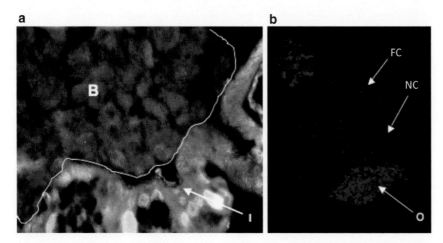

Fig. 11.13 Location of symbiotic bacteria in the tissues of insects by fluorescence in situ hybridization. (**a**) Bacteriome delimited by the white line containing (B) bacteria-filled bacteriocytes SOPE (*red*) *Sitophilus*. The intestine (I) is located in the lower level (Photo A. Heddi, Insects and Interactions Functional Biology, INSA, Villeurbanne, France). (**b**) Egg chamber of the mosquito *Aedes albopictus* with *Wolbachia* (*red*) in the follicular cells (FC), nurse cells (NC), and the oocyte (O) (Photo P. Mavingui and D. Voronin, UMR CNRS 5557 Microbial Ecology, Villeurbanne, France). Bacteria are detected using specific oligonucleotidic probes coupled to the rhodamine. The cell nuclei are stained (*blue*) with DAPI

The type and evolution of relationships between microorganisms and animals, especially along the parasitism-mutualism continuum, depend on the mode of transmission of the microbe (Edwald 1987). Horizontal transmission, by contact between infected and uninfected individuals, tends to favor the evolution to parasitism with intermediate levels of virulence resulting from a trade-off between maximum exploitation of the host and the probability of transmission of the microbe. In contrast, vertical transmission from parents to offspring promotes a reduction in virulence and evolution toward mutualism, fitness depending then on host reproduction. Examples of interactions sustainable or not between microbes and animals are presented hereafter, as well as the mechanisms involved and consequences on the evolution and adaptation of interacting entities.

11.4.3.1 Parasitic Interactions

Microorganisms are the cause of most diseases of man and animals. From the 1,500 microbial pathogens identified in man, over half would be zoonotic (i.e., of animal origin), and many of them are transmitted by hematophagous arthropods (Woolhouse et al. 2005). Nosological entities (microorganisms responsible for a disease) are involved in zoonoses or anthropozoonoses (animal diseases transmissible to humans), in epizooties (animal epidemics), and in human pathologies. These diseases include prion disease (bovine spongiform encephalopathy or BSE), bacterioses (tuberculosis, leprosy, pneumonia), fungal infections (ringworm, candida), parasitoses (malaria, filariasis, sleeping sickness), and various virus diseases (influenza, acquired immunodeficiency syndrome or AIDS, severe acute respiratory syndrome, or SARS, Ebola). Apart from the viruses, in terms of prevalence and incidence, bacteria represent the largest group involved in human and animal diseases; fungal diseases are a minority (*cf.* Chap. 15).

Microbes and Infectious Diseases

Since their appearance on earth about 3 billion years ago (*cf.* Chap. 4), microbes have caused major epidemics and pandemics in humans and animals. Examples of infectious diseases include cholera since antiquity, the plague in the fourteenth century, or tuberculosis that decimated human populations in the nineteenth century and that is unfortunately reemerging in recent years. The Spanish flu of the early twentieth century caused many damages estimated at more than 40 million people. Since its discovery in 1981, AIDS has become pandemic and continues to cause victims. Animals are not spared with outbreaks of diseases such as BSE (or mad cow disease), which is a form that causes the Creutzfeldt–Jakob disease to humans. Since the 1960s, the foot and mouth disease and swine fever have killed and led to the slaughter of millions of animals. Recently, the outbreak of avian influenza due to the highly pathogenic avian influenza virus H5N1 has killed or led to the culling of millions of poultry.

Many parasitoses are among the most devastating infectious diseases in humans and animals, of which vector-borne diseases occupy an important part. One of them is the Chagas disease caused by *Trypanosoma cruzi* transmitted by blood-sucking bugs. Trypanosomiasis is endemic in tropical South and Central America, infecting 300,000 people each year and causing 13,000 deaths. Schistosomiasis is itself considered the second most important parasitic infection after malaria. This disease is caused by flatworms mainly *Schistosoma haematobium*, *S. japonicum*, and *S. mansoni* and is contracted in water infested with larvae that develop in freshwater snails. Schistosomiasis is endemic in 76 countries mainly located in Africa, South America, Caribbean islands, eastern Mediterranean, and Southeast Asia. More than 600 million people are at risk of infection and 200 million are infected with schistosomiasis, with 20 million cases of severe illness.

Malaria also deserves to be cited as an example since it is the first global pandemic parasitosis. In fact, 107 countries are at risk of malaria, representing three billion people (46 % of the estimated world population of 6.5 billion people). An estimated 350–500 million clinical malaria episodes annually cause the death of 1.5–2.5 million people, including many children (Murray et al. 2012). *Plasmodium*, the causative agent of malaria, is transmitted by mosquitoes of the genus *Anopheles* that suffer only marginally from the parasitic infection. A negative effect of the parasite on fertility of the host mosquito has been identified (Ahmed and Hurd 2006). In humans, *Plasmodium* infects red blood and liver cells during its development cycle, which is one of the most complex cycles in the parasitic world (Box 11.4, Fig. 11.14).

Emergence and Reemergence of Diseases

In the context of epidemics, many new diseases are termed emerging diseases. An emerging disease means a disease whose incidence increases significantly in a given population in a given region, compared to the usual situation of this disease. An exemplary emerging disease is the toxic shock syndrome caused by diffusion into the body of toxins produced by the bacterium *Staphylococcus aureus*, more commonly known as golden staph. Toxic shock syndrome is increasing especially among women as a result of using tampons (during the menstrual cycle), which increases the risk of infection. Avian influenza, as indicated above with highly pathogenic H5N1 virus, decimates sensitive wild birds and poultry and is emerging in different regions of the world. Under certain conditions, the H5N1 virus can cross the species barrier and cause fatal infections in humans, especially in Southeast Asia where the proximity between human population and livestock is the major risk of contamination.

Reemerging diseases are those that reappear after a silent period, short or long, and often in a different form, sometimes more severe. The AIDS pandemic causes multiple microbial coinfections involved in the reemergence of diseases such as tuberculosis. Epidemics of chikungunya or dengue hemorrhagic fevers, human diseases due to viral agents transmitted by blood-sucking mosquitoes, reemerge after long silent periods (7–30 years) in the regions of Africa and Southeast Asia. In 2005, a new strain of highly pathogenic Chikungunya virus carrying a mutation at position 221 (alanine substitution by valine) of the envelope protein E1 was involved in severe and debilitating symptoms as well as in human deaths in the islands of the Indian Ocean (Schuffenecker et al. 2006).

Factors involved in disease emergence and reemergence can vary considerably and are generally poorly known, but there are many risk factors worsening epidemiological changes. These factors include general environmental perturbations (global warming) that will interfere with the circulation pattern of infectious agents and their vectors, urbanization and increased human populations that lead to human impact on the environment, aging of human populations that become weakly immunocompetent, and new land uses and practices. Manufacture of animal meal from the carcass of dead animals, potentially contaminated with prion, and its use in animal feed are one of the causes of the emergence and reemergence of mad cow disease. As well as viruses, prion is an entity that is not independent of the host cell for its multiplication. It should be noted that the prion protein (PRoteinaceous Infectious particle ONly) is present naturally in a nonpathogenic form in mammals (including humans), where it is involved in the development of the nervous system of the embryo. It would also play a protective role against oxidative stress and programmed cell death. It is the mutated form of the prion protein that becomes pathogenic. With the ability to multiply exponentially and self aggregate, the mutated prion destroys neuronal cells and causes deposits in the brain.

International trade can lead to the spread of pathogens, as suggested for H5N1 avian influenza outbreaks which reflect the poultry trade on the Trans-Siberian route (Gautier-Clerc et al. 2007). Other causes are related to the intrinsic adaptive capacity of infectious agents and their ability to acquire genetic information by horizontal transfer (*cf.* Chap. 12). These genetic events may lead to the selection of resistance to biocides (antibiotics and other antimicrobials), which poses serious problems in human and veterinary medicine (Box 11.5).

The Conflictual Nature of Animal–Microbe Interactions

Any pathology arises from uncontrolled conflictual interactions between potentially infectious agents and their hosts. The example of commensal bacteria in animals including humans is striking. These bacteria are called normal flora, confined to specific areas and live in balance with the immune system of healthy individuals. Upon environmental (rain, cold, heat) or immune (injury, poisoning) disturbance, the balance of the interaction can be broken, and the resulting imbalance can lead to the expression of a more or less severe disease in the healthy individual carrier. The infectious agent escapes the control of the host, colonizes unusual tissues and reaches excessive numbers, produces toxins, and can eventually kill its host.

Box 11.4: Cycle of *Plasmodium*, the Etiologic Agent of Malaria

Patrick Mavingui

Injection of *Plasmodium* sporozoites infectious forms is performed by infected females *Anopheles* mosquito as they bite. The sporozoites migrate to the liver via the blood or lymphatic circulation, invade hepatocytes, and differentiate into schizonts that release merozoites into the blood (Fig. 11.14a). In some species, *Plasmodium ovale* and *P. vivax*, a stage known as cryptozoic remains hidden in the liver before waking up several months or years later to restart the cycle. Merozoites infect red blood cells where they differentiate as characteristic rings called amoeboid or trophozoites. The active parasites (merozoites) and erythrocytes pass from the blood to invade new erythrocytes. The sexual stages are then produced and can again be drawn by a female *Anopheles* mosquito following a bite of the infected individual.

In mosquitoes (Fig. 11.14b), the replication cycle takes place in the gut. Gametocytes differentiate into gametes, fuse to generate entire zygote ookinetes which differentiate by passing the barrier peritoneal epithelial cells, change in oocyst which releases a large number of sporozoites. Infective sporozoites migrate through the hemolymph of the mosquito where they are ready to be injected into the blood of the next host and the cycle restart.

Malaria fevers are due to activities of parasites in the blood, which lyse erythroid cells and release toxic substances.

Corresponding to the complexity of the infectious cycle of *Plasmodium*, the triggered immune response is equally complex: activation of macrophages and NK cells (natural killer) that recruit and activate other immune cells such as neutrophils (Baratin et al. 2005). The complexity of the life cycle of these parasites and the high variability of associated antigens makes it difficult to develop a vaccine.

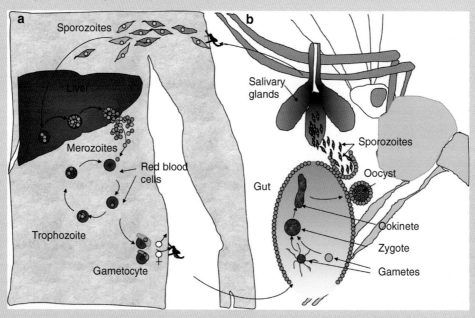

Fig. 11.14 Cycle of *Plasmodium*. (**a**) In human; (**b**) in the mosquito. The female mosquito bites humans and injects *Plasmodium* under the form of sporozoites. They move to the liver, where they develop through several stages into merozoites that invade and multiply in red blood cells, via the trophozoite form. Up to 10 % of red blood cells can thus be infected. The clinical symptoms of malaria, including fever and chills, anemia, and cerebral malaria are associated with infected red blood cells, and most current drugs target this stage of the life cycle. The merozoites of a subpopulation of infected red blood cells develop into gametocytes. Following a sting on an infected individual, the mosquito sucks gametocytes (gametocytes) contained in the blood, which differentiate into male and female gametes. In the gut of the mosquito, the gametes fuse to form a zygote. The zygote develops into ookinetes that pass through the intestinal wall and forms an oocyst filled with sporozoites. When the oocyst bursts, it releases sporozoites that will migrate to the salivary glands (Adapted from Wirth 2002)

Box 11.5: Antibiotic Resistance and Infectious Diseases

Patrick Mavingui

An antibiotic is a chemical substance, natural or synthetic, capable of compromising the growth of bacteria (antibacterial antibiotic) or fungi (antifungal antibiotic). Antibiotics act on specific targets by inhibiting or disrupting biosynthetic pathways essential for the development of microbes. Different microbial taxa such as Actinobacteria and some eukaryotic microorganisms are capable of producing antibiotics (*cf.* Chap. 9). Following major use in human and veterinary therapeutics to fight against microbial infections, the emergence of antibiotic resistance of pathogenic microorganisms, usually sensitive, has become a major public health problem.

Several factors are at the origin of the current resistance. First, they include genetic factors intrinsic to the microbes. Random mutations at the DNA level that alter the targets of antibiotics (mutational resistance) are infrequent and represent only 10–20 % of the resistance encountered in hospitals. The nonspecific mechanisms such as efflux pumps or the production of matrix polysaccharides are also involved in tolerance to antibiotics. Second, there is resistance acquired by horizontal transfer of plasmids carrying the genes for antibiotic resistance (hydrolytic or inactivating enzymes). They are the most numerous and correspond to 80–90 % of resistant isolates in the clinic. This is the case of *Staphylococcus aureus* MRSA (methicillin resistant) and XDR *Mycobacterium tuberculosis* (ultra-resistant TB). Extrinsic predisposing factors are those that facilitate the passage and movement of resistant microbes found in the environment (air, water, soil), plants, and animals. The massive use of antibiotics in different sectors of human activity (human and veterinary medicine, feed, food, etc.) is to be related to the proliferation of resistance, as evidenced by the high frequency of resistant strains in hospitals where antibiotics are commonly used.

The resurgence of multiple resistances in microbes requires the constant development of new antimicrobial molecules but also the application of more appropriate preventive strategies. The use of antibiotics should be carefully considered and rigorously applied.

Box 11.6: AIDS and Its Opportunistic Infectious Process

Patrick Mavingui

The Syndrome of Acquired Immune Deficiency or AIDS is caused by infection by the HIV (Human Immunodeficiency Virus) retrovirus identified by the team of Luc Montagnier (Barré-Sinoussi et al. 1983), which earned him and Françoise Barré-Sinoussi the Nobel Prize for Medicine in 2008. HIV is transmitted through body fluids (blood, breast milk, vaginal secretions, semen). Since its discovery in 1981, AIDS has reached pandemic levels because it has killed more than 25 million people, and 40 million people live with HIV in almost all regions of the world (UNAIDS, January 2006 http://www.unaids.org). In the absence of a vaccine, the disease progression can be delayed for several years by the administration of an antiretroviral triple therapy and improved hygiene of people with HIV. Uncontrolled, the disease leads to a weakening of the immune system and opens the door to opportunistic infections responsible for diseases such as pneumonia and Kaposi's sarcoma that kill infected patients. Many fungi are among the opportunistic pathogens and include the genera *Candida* which causes candidiasis; *Cryptococcus* responsible for cryptococcosis; *Histoplasma*, the causative agent of histoplasmosis; or *Aspergillus* causing aspergillosis. New fungal pathogens are increasingly identified. For some species, the pathogenesis mechanisms are still unknown and cause problems. There are also a large number of bacteria involved in these opportunistic infections (*cf.* Chap. 15).

A dramatic example is that of immunocompromised patients (AIDS (Box 11.6), cystic fibrosis, transplantation, etc.) who are victims of nosocomial infections (from the Greek "nosokomeone", which means hospital and describes what is contracted in hospital) by opportunistic microbial taxa (bacteria, fungi, parasites). In these individuals whose immune system is affected, opportunistic enterobacteria, usually nonpathogenic, such as *Pseudomonas aeruginosa*, *Klebsiella pneumoniae*, or the *Burkholderia cepacia* complex (Vandamme et al. 2007) colonize organs such as the lungs, where they multiply to very high densities, produce biofilms that clog arteries, and cause tissue necrosis. Biofilms consist of microbial communities of cells attached to a surface (cell matrix) and generally embedded in a viscous substance rich in polysaccharides. Biofilms protect bacteria against phagocytosis by immune cells and prevent the penetration of biocides used to treat patients, at the same time promoting systemic infections and chronic diseases. Some microbes, such as atypical mycobacteria involved in opportunistic lung infections, spread and survive in macrophages, which are immune cells of the host paradoxically specialized in the fight against infectious agents (Sundaramurthy and Pieters 2007).

Another amazing case of weakening of the innate immunity system with a disease outcome in insects lies in the criminal alliance between the entomophagous nematodes *Steinernema* and *Heterorhabditis* and the entomopathogenic bacteria *Xenorhabdus* and *Photorhabdus*, respectively (Goodrich-Blair and Clarke 2007). *Xenorhabdus* and *Photorhabdus* bacteria live symbiotically in intestinal vesicles of the nematodes. When the larvae of nematodes, living freely in the soil, encounter and infect an insect host, they metamorphose into the adult stage, which weakens the immune system of the insect. Bacteria are then released into the blood of the immunocompromised insect and act as opportunistic agents. They multiply and release through their secretion systems toxins that cause septicemia and toxemia, causing the death of the insect. Necrotic tissues in turn are eaten by the entomophagous nematode. Inoculation of the asymbiotic nematode or the bacterium alone does not lead to death of the insect whose immune system can then effectively eliminate the intruders.

In contrary to opportunistic pathogens that exploit immune vulnerability of hosts to cause damage, certain groups of bacteria such as enteropathogens (*Campylobacter, E. coli, Legionella, Listeria, Salmonella, Yersinia*) responsible for diarrheal crises that affect hundreds of millions of people around the world (of which millions will die) are natural pathogens with invasive capacity and production of toxins, which give them a virulent character (Cossart and Sansonetti 2004).

11.4.3.2 Virulence Factors of Pathogens

One of the first virulence factors shared by pathogens of animals such as enteropathogenic bacteria is the ability to overcome the physical barrier (skin and mucosa) through an intimate attachment via structures such as adhesins (Ofek et al. 2003).

Adhesins are the first virulence factors of pathogenic bacteria involved in the infection process (Mainil 2013). They interact with the components, called receptors, present on the surface of eukaryotic cells or on the surface of the extracellular matrix leading to the attachment of interacting cells. Dozens of bacterial adhesins are known and three types of adhesin–receptor interactions with a certain degree of specificity have been described. Bacterial lectins are fimbriae-like structures or have fibrillar proteic nature, which are classified according to their hemagglutination properties, and carbohydrate receptors are components of glycoproteins or glycolipids of cell membranes of the host animal. The second type includes proteins of the cell wall or membrane associated or not to bacterial lipopolysaccharides and peptidoglycans and protein receptors of cytoplasmic membranes of the host cells. Finally, hydrophobins, present on the surface of both bacteria (adhesins) and eukaryotic cells (receptors), are mostly composed of lipid or hydrophobic domains of proteins.

The adhesion of bacteria to the surface of host cells is followed by injection of invasive effector molecules via secretion systems. These secretion systems constitute the second group of virulence factors, which play a role as important as adhesins in the establishment of interactions between microorganisms and hosts. Seven secretion systems, designated type I to type VII (Fig. 11.15) (Abdallah et al. 2007; Filloux et al. 2008), have been described and characterized mainly in pathogenic bacteria and to a lesser extent in the mutualists (Thanassi and Hultgren 2000). The type I secretion system (or ABC transporters) is involved in the transport of a wide variety of substrates (proteins, sugars, lipids) and many processes of bacterial life *per se*. This is by far the most represented secretion system found in many bacteria. Type II (or typical sec-dependent secretion) and type V (or autotransporters) systems are mainly used for the secretion of enzymes and/or other proteins (Sandkvist 2001). The type III (T3SS) and type IV (T4SS) secretion systems have evolved for interaction with other organisms (Cascales and Christie 2003). The T3SS develops a path allowing the bacterium to inject cytotoxic or cyto-destabilizing factors directly into the cytoplasm of the target cell. The T4SS, meanwhile, allows the translocation of nucleoprotein complexes to hosts, either directly inside the target cell or via the extracellular medium (Llosa and O'Callaghan 2004). Mammal cells such as human cells but also plant cells are potential targets, providing an agronomic and medical interest to the study of these systems. Pili formed by the T3SS and T4SS are also involved in the attachment process of bacteria to target cells. The relationships between bacteria and host cells (animal or plant) via the T3SS and T4SS often have a pathogenic outcome and sometimes a mutualistic character (Dale et al. 2002).

Like the bacterial symbiotic islands, genetic determinants of virulence factors of pathogenic bacteria can be located in a region of the genome designated pathogenicity island (or PAI) (Hacker and Kaper 2000). Discovered for the first time in pathogenic *E. coli,* genome analysis showed the presence of PAIs in many other bacterial pathogens of animals. With a

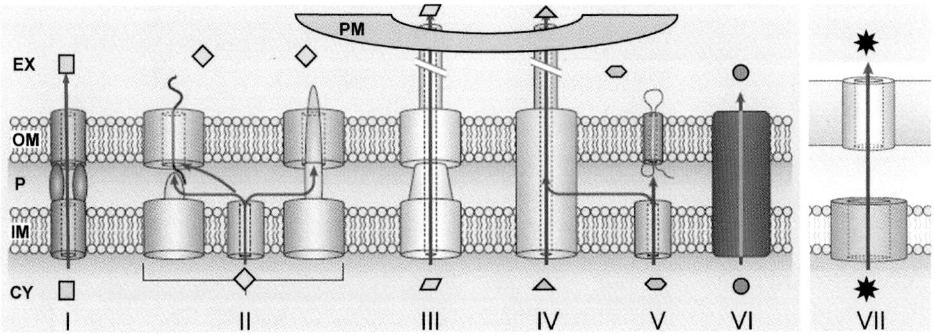

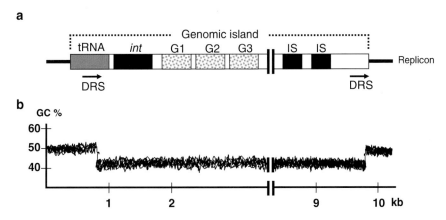

Fig. 11.15 The bacterial secretion systems. In Gram-negative bacteria, the six secretion systems described (named types I to VI) are anchored into internal (IM) and outer (OM) membranes and periplasm (P). Usually, types I, III, IV, and VI carry effectors from the cytoplasm (CY) to the cell surface or to the external environment (EX) in a single step. Types III and IV produce a syringe which will cross the plasma membrane (PM) of the host cell, thereby discharging directly effectors into the cytosol. Types II and V carry the effector molecules in two steps, passing through the periplasm (P) via Sec or Tat. The newly described type VII in Gram-positive bacteria would carry effectors in a single step. *Arrows* indicate the route followed by transported effectors (Adapted from Abdallah et al. 2007 and Filloux et al. 2008)

Fig. 11.16 Schematic representation of a genomic island carried by a bacterial replicon. (**a**) The genetic map of the DNA fragment transferred contains a tRNA and two direct repeat sequences (DRS) at the ends. (**b**) The GC content of the island is generally different from that of the entire replicon. Int, integrase; G1 to G3, genes encoding specific functions; IS, insertion sequence; kb, kilobase (Adapted from Hacker and Carniel 2001)

size varying from 10 to 200 kilobases (Fig. 11.16), the PAIs carry genes that encode virulence proteins (adhesins, protein secretion machinery, toxins, invasins, etc.). There is typically a transfer RNA (tRNA) that serves as the site of insertion of PAIs in a genome, sequence repeats generated during the events of horizontal transfer, or different genetic elements (integrases, transposases, insertion sequences) known to provide mobility features of gene modules (Hacker and Carniel 2001). PAIs can move from one region of the bacterial genome to another or be transferred from one strain to another by horizontal transfer. Some PAIs and other virulence factors are in fact carried by mobile elements such as plasmids of bacteria *Shigella* and *Yersinia* (Parsot and Sansonetti 1999) or bacteriophages of enteric bacteria and *Vibrio cholerae* (Karolis et al. 1999). Environmental factors (temperature, pH, osmolarity, oxygen partial pressure, etc.) affect positively (activation) or negatively (inhibition) the expression of virulence genes and thus the pathogenicity of invasive bacteria (Altier 2005).

The case of enteropathogenic bacteria is exemplary; its toxins are injected into target cells through T3SS systems and/or T4SS inducing perturbations of cellular processes and suppression of their defenses. For example, the IpaB shiga toxin of *Shigella* and its *Salmonella* counterpart SipB induce apoptosis (a form of programmed cell death; *cf*. Sect. 10.4.2) in macrophages involved in immune defense (Hersh et al. 1999). Through bypassing the host defenses, enteric bacteria cause septicemia and toxemia resulting in inflammation and diarrhea that allow microbial growth and ecological success.

In addition to toxins excreted by classical secretion systems, metabolic enzymes are increasingly found on the surface of the cell walls of pathogenic microorganisms, without the mode of secretion clearly established. Recent studies indicate the involvement of some excreted glycolytic enzymes in microbial virulence. These enzymes include glyceraldehyde-3-phosphate dehydrogenase (GAPGH), enolase, aldolase, and pyruvate kinase in staphylococci, mycobacteria or the fungus *Candida albicans* (Pancholi and Chhatwal 2003).

11.4.3.3 Mutualistic Interactions

Relationships between animals and their "normal" microbial communities may evolve toward mutualism, with sometimes spectacular effects on the host biology. These mutualistic associations are maintained stably and sustainably by fitness gains of partners because of their interdependence, as already observed and noted by Charles Darwin (1859). Genetic studies suggest population coevolution cases for mutualistic associations which are often highly specific (Futayama 1986). The phenomenon of coevolution refers to transformations that occur during the evolution of species and the resulting reciprocal influences of interacting partners. Mainly studied in bipartite relations, coevolution may involve mutualistic associations as well as multi-parasitic ones. They may include, among others, transfer mechanisms and interspecies genomic coadaptations, but also qualitative and quantitative controls of microbial communities by the host. Mutualistic interactions between microorganisms and invertebrates are the most studied at the molecular and evolutionary levels, especially insect–bacteria endosymbioses.

Insect Nutritional Symbioses

The example of aphids (Homoptera, Insecta) and their symbiotic bacteria is one of the best described. These insect pests attack almost all plant species feeding off their phloem rich in carbohydrates but low in amino acids. This imbalance is compensated by a mutualistic association with symbiotic bacteria *Buchnera*, which are housed in specific organs of the host, bacteriomes, or bacteriocytes where they play nutritional functions (Douglas 1998). *Buchnera* genomes contain multiple copies of genes that encode the biosynthesis of essential amino acids such as tryptophan and appropriate gene regulators (Shigenobu et al. 2000). The intracellular localization of the bacteria facilitates the transfer of microbial compounds produced to the host. In return, the host provides the bacteria with energy for growth in an environment free of competitors.

Like most obligate intracellular symbionts, unable to grow outside the host cell, the genome of *Buchnera* has undergone a significant size reduction by eliminating a large number of genes (for membranes, metabolism, etc.), whose functions are performed by the aphid. It is possible to identify genes lost in *Buchnera* by comparative analysis with the genome of its closest relative, the bacterial saprophyte *E. coli* (Moran and Mira 2001). From an evolutionary point of view, the mutual dependence between aphids and *Buchnera* dates back nearly 200 million years and the congruence of

Fig. 11.17 Low- and high-magnification views of the giant worm *Riftia pachyptila* (Copyright: Ifremer-Nautile/Campagne Mescal 2010)

phylogenies of the partners reflects a coevolutionary process supported by vertical transmission of *Buchnera* across generations of aphids (van Ham et al. 2003). In addition to *Buchnera*, aphids are hosts of many bacteria termed secondary symbionts. The role of these clandestine passengers is now better understood. Some of these secondary symbionts are involved in tolerance to heat stress, and resistance to parasitoids, which are insects that develop as aliens at the expense of other insects they parasitize, which content they consume and kill (Oliver et al. 2003; Dunbar et al. 2007; Moran et al. 2008).

A spectacular multipartner symbiotic interaction in the world of insects is that between social ants of the genus *Atta* and some fungi (Bacci et al. 1995). Unlike humans and herbivorous animals that depend on endosymbiotic microbial partners to digest cellulolysic compounds of the plants they eat, ants have adopted an ectosymbiotic strategy that is surprising but effective. They grow in their nest cellulolytic fungi of the genera *Leucoagaricus* and *Lepiota* that they feed with plant material brought from the outside, often also causing considerable damage in the tropics. Fungi decompose the plant material, use it as nutrient, and develop numerous hyphae that can grow to represent large biomass. In return, ants, exclusively mycophagous, feed the fungal hyphae. Even more remarkable, these ants establish a symbiotic association with bacteria of the genus *Burkholderia* that produce an antibiotic involved in the fight against other fungi that contaminate nests (Santos et al. 2004).

Symbiosis with Marine Animals

The discovery of the giant worm, *Riftia pachyptila* (Fig. 11.17), present in hydrothermal vents has permitted to study an equally singular nutritional symbiosis (Jones 1981; Cavanaugh et al. 1981). This annelid devoid of a digestive tract has a specialized organ, the trophosome, which hosts symbiotic sulfur-oxidizing bacteria belonging to the Proteobacteria (Distel et al. 1988). To feed,

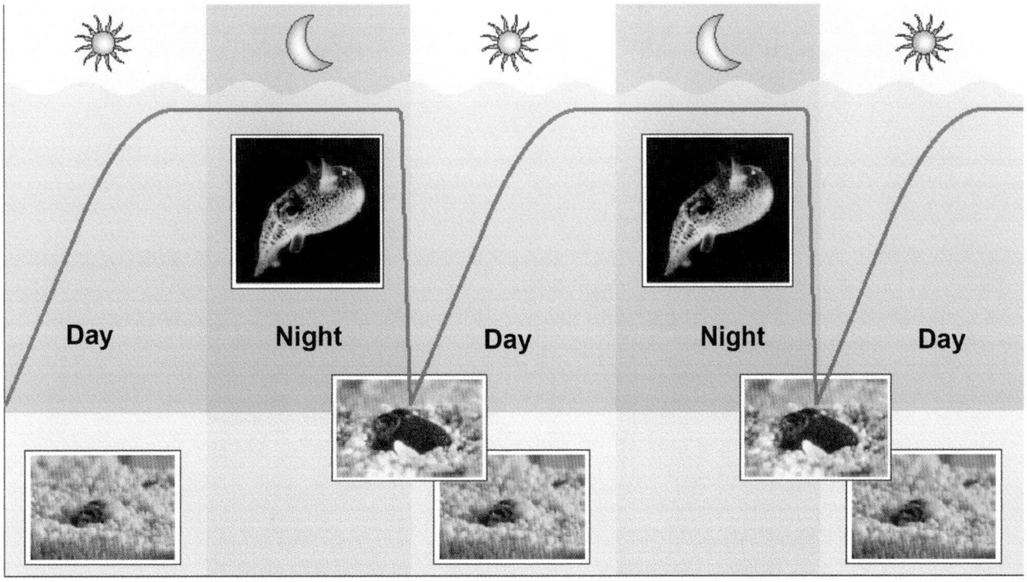

Fig. 11.18 Symbiosis between the squid *Euprymna scolopes* and *Vibrio fischeri* and the phenomenon of chemiluminescence. The *E. scolopes* squid hosts in its light organ the domesticated bacterium *V. fischeri* whose density increases over the day to reach a maximum after dark (blue curve), thereby producing a strong luminescence. This behavior allows the squid to hunt at night while escaping predators by camouflage. At daybreak, almost all (~90 %) bacteria are expelled into the environment; *E. scolopes* buries itself in the sand. During the day, the remaining bacterial population (~10 %) in the light organ is multiplied in order to reach again the maximum density required to produce luminescence at night (Adapted from Nyholm and McFall-Ngai 2004)

R. pachyptila absorbs through its gill filaments oxygen and hydrogen sulfide that are transported via the blood to the trophosome, where bacteria perform chemosynthesis, generating organic matter necessary for the growth of the host (*cf.* Chap. 10).

Another exemplary mutual interaction of the marine world is observed in the association between the squid *Euprymna scolopes* and the alphaproteobacterium *Vibrio fischeri* (Nyholm and McFall-Ngai 2004). This squid that lives mainly on the seabed is exposed to predation when swimming back to the surface at night. To escape predators, *E. scolopes* emits a ventral light that serves as camouflage concealing its silhouette against the light of the stars. The emission of light by the squid results from the phenomenon of bioluminescence produced by *V. fischeri* (formerly *Photobacterium fischeri*) located in a specialized light organ (photophore) of the host (Fig. 11.18). Although the initiation of morphogenesis of the photophore occurs without the intervention of the bacterium, the latter is involved in postembryonic maturation among others acting on actin filaments of the cytoskeleton (Montgomery and McFall-Ngai 1994; Kimbell and McFall-Ngai 2004). In the photophore, *V. fischeri* can reach high densities in the order of 10^{10} cells per ml, and the light is emitted at a certain bacterial density (or quorum perceived by homoserine lactone-like molecules; *cf.* Chap. 9). At dusk, *E. scolopes* expels almost all bacteria and quorum (optimal number) required for light emission is reached again at night.

Many other bacteria are found associated with marine animals, particularly in oceanic ridge ecosystems, whose functions are gradually beginning to be elucidated. This is the case of sulfur-oxidizing and methanotrophic gammaproteobacteria in clams of Mytilidae group *Bathymodiolus*. These symbiotic bacteria can reach 10^{10} to 10^{11} cells per g of gill tissue (Yamamoto et al. 2002) where they use sulfide, thiosulfate, and methane as energy sources (Nelson and Fisher 1995). These symbiotic chemosynthetic systems are a significant component of the carbon cycle in the ocean, and organic matter resulting from autotrophy, whether thiotrophy and/or methanotrophy (*cf.* Chap. 14), is a major source of carbon for animals that populate these particular environments.

Microbe-Vertebrate Mutualism

In vertebrates, the mutualistic associations between the indigenous microbiota of the intestine and the animals are clearly established in the carbon metabolism of ruminants and monogastric coprophages. The microbiota of the intestinal tract is structured in trophic network that plays an essential role in the degradation of polysaccharides such as cellulose and lignin of plant material ingested by the hosts (Zhanq et al. 2007). From this mutualistic association, ruminants and coprophages draw energy as carbohydrates, while microbes are protected in a germ-free biotope and are supplied with nutrients.

The mutualistic nature of the human intestinal microbiota has often been deduced from the data obtained in animals.

Indeed, mammalian cells are also unable to digest most polysaccharides of plant origin except starch, which therefore depends on the human microbiota. Active anaerobic intestinal bacteria in the gut, anoxic, are similar between humans and animals. Recent work confirms the existence of mutualistic bacteria in the human gut. Indeed, the genomes of the obligate anaerobic bacteria *Bacteroides*, the dominant human gut bacterial genus, confirm the presence of genetic determinants and proteins involved in the degradation of polysaccharide compounds (Xu et al. 2003). In addition, inoculation to laboratory mice devoid of microbiota and maintained under axenic conditions, of strain *Bacteroides thetaiotaomicron* VPI-5482 isolated from the human intestine, leads to a pleiotropic effect: activation of carbon metabolism and storage of lipid compounds (Bäckhed et al. 2004), stimulation of cells producing antimicrobial compounds, and microvilli of the intestinal wall (Stappenbeck et al. 2002; Hooper et al. 2003). Another effect called **probiotic*** was supported by some data. The concept of probiotics has been proposed by the Russian Nobel Prize winner Elie Metchnikoff. It is defined as any living microorganism which, when ingested at a certain amount, exerts beneficial effects beyond basic nutritional functions. The probiotic microorganisms most frequently cited are lactobacilli and bifidobacteria, and the reported effects range from changing the intestinal ecology to the stimulation of the immune system or the decrease of the risk of cancer (Ouwehand et al. 2002). The probiotic issue has generated many controversies.

The prevalence and persistence across generations of microbes that play an important role in nutrition, in immunity, and in development are clear proof of the mutualistic nature of animal-microbe associations.

11.4.3.4 Commensalism, Parasitism, and Other Interactions with Intermediate Phenotypes

Overview

A large number of microorganisms that live on the surface and in the internal organs of animals benefit from this habitat without producing observable effects in their hosts. These are microorganisms known as commensals, among which are mainly bacteria such as *Pseudomonas, Acinetobacter,* and *Bacillus* in insects or *E. coli, Staphylococcus, Corynebacterium,* and *Lactobacilli* in mammals. Some of these bacterial taxa are also, as noted above, the pathogenic or mutualistic flora. Commensal microbiota that colonizes the surface of a given organ can prevent through a physical barrier effect the colonization of other bacteria, including pathogens. In contrast, the commensal microflora may represent a hazard to animal health as some groups carry genes for antibiotic resistance that may be acquired by pathogens by transformation, conjugation, or transduction. Transfer of antibiotic resistance genes between bacteria of animal and human origin has been identified in the gastrointestinal tract of animals (Moubareck et al. 2003). It has been seen that many commensal microbes colonize the skin or integument of animals. Some of them are involved in the production of volatile and odorous molecules that may have a direct or indirect impact on health. This is particularly the case of compounds of microbial origin that act as attractants toward mosquitoes (Brady et al. 1997) whose bites cause viral or parasitic infections.

Genomics can allow comparative research on the differences and similarities at the genetic and functional relationship between commensal microbes, parasites, and mutualistic animals. The distinction between these interactions is sometimes difficult to establish. Whether pathogenic or mutualistic, commensal or parasitic, microorganisms share common genetic determinants and adaptations that allow them to colonize many environments (Hentschel et al. 2000; Goebel and Gross 2001). The similarity of the processes involved in the expression of a bipartite or multipartite symbiotic interaction can lead to intermediary phenotypes that are not always obvious to position in traditional categories (pathogenesis, parasitism, cooperation, mutualism, etc.).

Wolbachia and Phenotypic Pleiotropy

The case of bacteria of genus *Wolbachia* is exemplary because they establish associations with their hosts yielding multiple phenotypes (**pleiotropy***) which lie in the parasitism–mutualism continuum. These alphaproteobacteria of the order *Rickettsiales* infect filarial nematodes of the family *Onchocercidae* and many arthropods such as mites, crustaceans, and insects (Werren et al. 1995; Bandi et al. 2001). *Wolbachia* is mutualistic of filarial nematodes that include human pathogens such as *Onchocerca volvulus* involved in human onchocerciais or river blindness or *Wuchereria bancrofti* or *Brugia malayi* both responsible for elephantiasis. Bacteria are located in the hypodermal cells of the lateral chords of larvae but also in the ovaries in adult females and are thus transmitted transovarially to offsprings (Taylor et al. 1999; Fisher et al. 2011). Polysaccharides of *Wolbachia* represent an aggravating factor in inflammatory processes associated with filariasis (Saint André et al. 2002). The elimination of *Wolbachia* by antibiotics (rifampin, tetracycline) from infected nematodes leads to inhibition of embryogenesis and larval development (Bandi et al. 2001). Therapeutic trials combining antifilarial drugs (albendazole and ivermectin) and antibiotics (tetracyclines) have been conducted with some success against infection with *Wuchereria bancrofti* (Turner et al. 2006; Johnston and Taylor 2007).

In arthropods, *Wolbachia* induces multiple effects to the hosts. A pathogenic effect has been shown in drosophila; *Wolbachia* strain *w*MelPop is able to grow at high densities in nervous tissue and causes fatal bacteremia (McGraw et al. 2002). However, in most cases, *Wolbachia* manipulates the reproduction of their arthropod hosts. Four effects of reproductive manipulation are known: feminization of genetic males, male-killing by larval development arrest, parthenogenesis in haplodiploid Hymenoptera, and cytoplasmic incompatibility, which is a non-fertile cross between infected males with *Wolbachia* and uninfected female or carrying a different strain of *Wolbachia* (Werren et al. 1999). *Wolbachia* is thus qualified as a parasite of the reproduction of its host. These reproductive manipulations have resulted in biased sex ratios disadvantaging males, thus generating more females that transmit *Wolbachia* transovarially to their descendants. This explains the high prevalence of *Wolbachia* in natural populations of arthropods, estimated at up to 76 % in insects (Jeyaprakash and Hoy 2000; Hilgenboecker et al. 2008). A phenotype of mutual dependence was found in some arthropods where *Wolbachia* is required for normal reproduction or development. In the hymenopteran *Asobara tabida*, *Wolbachia* has become necessary for oogenesis in females (Dedeine et al. 2001). The elimination of *Wolbachia* by antibiotics generates asymbiotic individuals completely devoid of oocytes, and ovarian tissues showed early manifestations of apoptosis (Pannebakker et al. 2007). The bedbug, *Cimex lectularius*, hosts *Wolbachia* in a bacteriome and the two partners seem to establish a nutritional mutualistic symbiosis through the provisioning of vitamins (Hosokawa et al. 2010). Remarkably, *Wolbachia* is also able to protect drosophila against viral pathogens (Hedges et al. 2008), and in some cases this bacterium inhibits the transmission of pathogens by the mosquitoes (Moreira et al. 2009; Mousson et al. 2012).

There is a unique situation where the same bacterium induces various effects on the hosts, differing in nature and intensity, either on their physiology or reproduction, and is ranged in a continuum from parasitism to mutualism. These associations thus raise many questions in evolutionary biology (evolution of mutualism, coevolution, role in the speciation of hosts), epidemiology (diffusion process), cell biology (cellular targets, signaling), and molecular biology (structure, functioning, and evolution of the genome). The mechanisms underlying these interactions are unknown. Effectors secreted via the *Wolbachia* T4SS (Rancès et al. 2008) have been proposed. Many genes encoding proteins with ankyrin domains known to be involved in protein–protein interactions are potential candidates present in complete genomes of *Wolbachia* (Wu et al. 2004; Foster et al. 2005; Klasson et al. 2008). Transfers of *Wolbachia* genes to insects and from the insect to the bacteria have been reported following the sequencing of genomes (Hotopp et al. 2007; Klasson et al. 2009).

11.4.4 Defense and Counteroffensive in Microbe–Animal Interactions

Animals interacting with microbes have developed a set of mechanisms and weapons to contain or eliminate potentially infectious agents. These mechanisms include phagocytosis by immune cells such as macrophages, secretion of toxic compounds, or inhibitors such as cytokines and reactive oxygen species. Discovered in drosophila and in many other arthropods and vertebrates, the production of antimicrobial peptides is part of the humoral response against microbial infection. The synthesis of antimicrobial peptides is regulated by molecules of the nuclear factors kappa B (NF-kappaB) family, such as DIF and Relish in *Drosophila melanogaster*. Nuclear factor DIF is mainly activated in response to fungal infection or by Gram-positive bacteria, whereas Relish plays a role during infection by Gram-negative bacteria (Ferrandon et al. 2007).

Programmed cell death (PCD) is a manifestation of defense commonly observed in response to infection by microbes, either pathogenic or mutualistic (Williams 1994; Vavre et al. 2008). In contrast, many microorganisms have developed mechanisms to circumvent the host defenses by detoxification, immunosuppression, or molecular mimicry for their survival and proliferation (Silver et al. 2007).

11.4.4.1 PCD as a Means of Defense of Animals

PCD is an integral part of the development program of higher organisms. **Apoptosis*** is the best known and described PCD in insects, nematodes, and mammals (Jacobson et al. 1997). It takes place during development and is involved in the elimination of damaged cells and in maintaining the homeostasis of immune cells (Vaux et al. 1994). Apoptosis involves the caspase enzymes that are activated by many stimuli. Two apoptotic pathways are known: the extrinsic pathway is triggered by TNFR family receptors (Tumor Necrosis Factor Receptor) and the intrinsic pathway that is activated by intracellular signals such as oxidative stress and involves the disruption of mitochondria (Fig. 11.19). From a phenotypic point of view, apoptotic cells have their nuclear DNA fragmented.

In addition to its role in development, apoptosis is involved in the defense against microbes and participates in the elimination of infected cells. Its induction is initiated after contact and perception by host cell receptors of microbial components or structures designated MAMPs (microbe-associated molecular patterns) or PAMPs (pathogen-associated molecular patterns). These components that are specific to microorganisms are composed of lipopolysaccharide (LPS), peptidoglycan (PGN), and lipoteichoic acids of the cell walls. Host cell receptors that are on the front line for detecting MAMPs and PAMPS are designated PRRs (pattern recognition receptors), the most reknown being the Toll-like receptors or TLRs. Discovered first in the innate antibacterial response in drosophila (Hoffmann 2003), TLR homologues were subsequently identified in other animals including mammals (Medzhitov et al. 1997). TLRs are also involved in the induction of apoptosis in

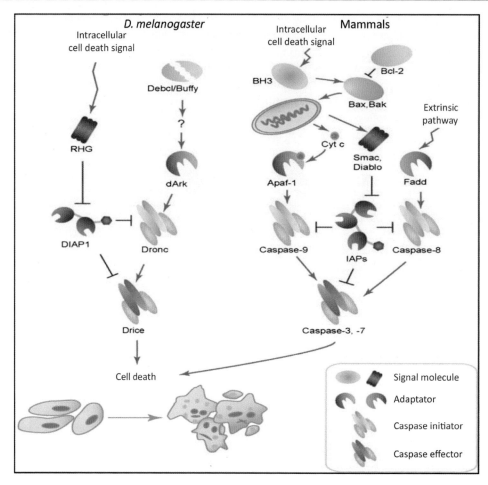

Fig. 11.19 Comparison of apoptotic pathways between drosophila and mammals. In *D. melanogaster,* dArk adapter (protein homologous to Apaf-1 in mammals) activates the initiator caspase Dronc. Proteins having pro- and antiapoptotic domains Debcl and Buffy, belonging to the Bcl-2 family, may regulate the activation through dArk, but this step has not been demonstrated. The apoptosis inhibitor protein, DIAP-1, negatively regulates activity of Dronc and of the caspase effector Drice. The RGH family proteins (such as Rpr, Hid and Grim) activate apoptosis by inhibiting the action of DIAP-1 on caspases. In mammals, the intrinsic pathway is characterized by the activation of caspase 9 (functional homologue of Dronc) by Apaf-1. But the activity of Apaf-1 depends on the Bcl-2 family proteins: Bax and Bak have a proapoptotic activity, while Bcl-2 inhibits this activity to prevent apoptosis. Intracellular death signals activate BH3 domain proteins, thereby promoting activity of Bax and Bak. The latter form pores in the mitochondrial membrane resulting in the release of proteins that either promote the activation of the Smac/Diablo complex, preventing the action of the inhibitor of apoptosis proteins (IAPs), or activate Apaf-1 (role of cytochrome C). In the extrinsic pathway, the cell death receptor is activated by its specific ligand which causes the recruitment of the FADD adapter. FADD can then activate caspase 8, an initiator caspase. Finally, the two routes meet in the activation of effector caspases (such as caspases 3 and 7, Drice homologues) that lead to cell death (Adapted from Hay and Guo 2006)

response to microbial infection (Salaun et al. 2007). For example, the perception of bacterial LPS by TLR4 on endothelial cells leads to the induction of apoptosis (Bannerman and Goldblum 1997; Aliprantis et al. 1999). In mammalian cells such as mice, apoptosis mediated by TLR4 confers resistance to infection due to the perception of the pneumococcal virulence factor, pneumolysin (Srivastava et al. 2005). To eliminate or stop the progression of the infection with *Shigella* or *Mycobacterium*, infected macrophages can undergo apoptosis resulting in concomitant death of these bacterial pathogens (Zychlinsky and Sansonetti 1997).

In insects, the involvement of apoptosis in the defense mechanism was shown for the first time in response to infection of the butterfly *Spodoptera frugiperda* by baculovirus (Clarke and Clem 2003). Phenomena associated with apoptotic caspase activity were observed in mosquitoes infected with *Plasmodium* (Hurd et al. 2006).

11.4.4.2 Modulation of Apoptosis by Microorganisms

Microbial pathogens have developed various mechanisms to inhibit apoptosis in host cells to facilitate their replication and persistence. Microbes can also induce apoptosis in immune cells to destroy these cells thereby facilitating their escape. The ability to circumvent host defenses by modulating apoptosis is found in various microbial taxa,

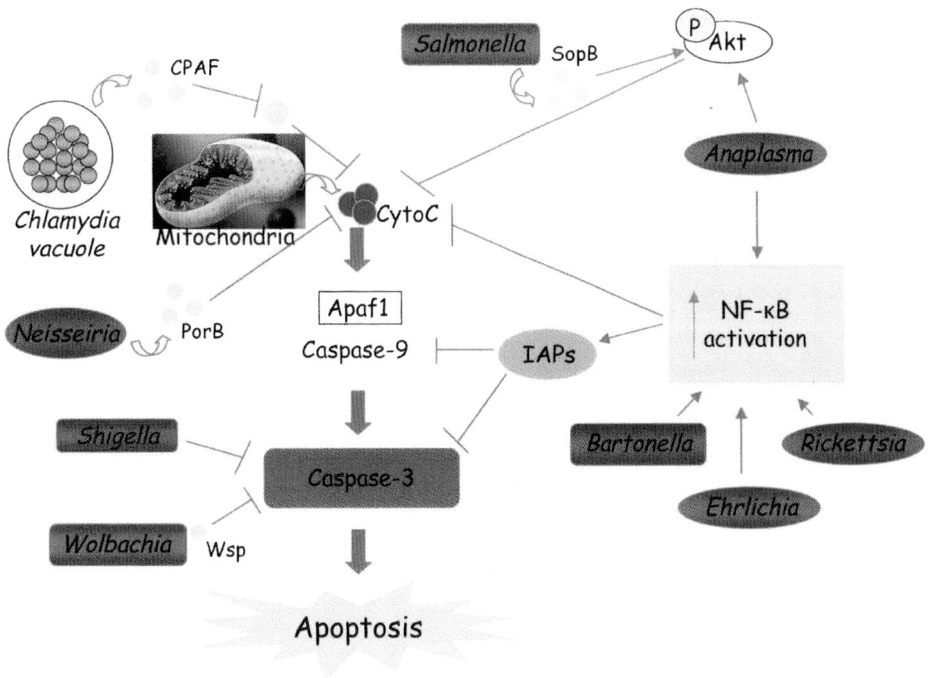

Fig. 11.20 Different bacterial mechanisms involved in the inhibition of apoptosis. *Chlamydia* sp. secretes the CPAF (Chlamydial Protease/proteasome-like Activity Factor) that has proteasomal activity and degrades a proapoptotic protein, thereby preventing the output of cytochrome C in mitochondria. *Neisseria* sp. secretes the PorB protein which inhibits the release of cytochrome C. *Salmonella enterica* SopB secretes the protein that activates the phosphatidylinositol 3-kinase pathway-Akt (PI3K/Akt) inhibiting the release of cytochrome C. *Anaplasma phagocytophilum* activates the PI3K-Akt and NF-kappaB activation allows inhibitor of apoptosis proteins (IAPs). *Bartonella* sp., *Ehrlichia chaffeensis*, and *Rickettsia rickettsii* activate NF-kappaB. *Shigella flexneri* inhibits caspase activation. *Wolbachia* may inhibit the activation of caspase-3 using proteins such as WSP or certain proteins with ankyrin domains. *Red lines*: inhibition of apoptosis. *Green arrows*: activation of apoptosis inhibitory pathways (Adapted from Faherty and Maurelli 2008)

especially in intracellular microbes (Roulston et al. 1999; Gao and Abu Kwaid 2000).

Inhibition of Apoptosis by Microorganisms

Commensal and pathogenic bacteria can inactivate apoptosis by modulating the extrinsic and intrinsic pathways (Fig. 11.20). Facultative intracellular bacterium *Mycobacterium tuberculosis*, the causative agent of chronic tuberculosis, can prevent apoptosis in macrophages where it multiplies and persists. To this end, *Mycobacterium* increases the production of TNF receptors which in turn activate the antiapoptotic pathway NF-kappaB (Balcewicz-Sablinska et al. 1998). Another facultative intracellular bacterium, *Bartonella henselae*, involved in bacillary peliosis, prevents endothelial cell apoptosis by inhibiting caspase activation and DNA fragmentation (Kirby and Nekorchuk 2002). Obligate intracellular bacteria of the genus *Chlamydia*, involved in various human infectious diseases, can protect infected cells against apoptosis during the early phase of infection by blocking the proapoptotic caspases or by preventing the release of cytochrome c from mitochondria (Fan et al. 1998). Similarly, strict intracellular pathogens belonging to the order *Rickettsiales* are able to inhibit apoptosis. This is the case of *Rickettsia rickettsii*, the causative agent of Rocky Mountain spotted fever, which inhibits apoptosis of infected cells by activation of NF-kappaB (Clifton et al. 1998), while *Anaplasma phagocytophilum*, responsible of human anaplasmosis, prevents apoptosis of infected neutrophils through the transcriptional control of the antiapoptotic BFL1 and inhibition of the activation of apoptosis mediated by mitochondrial caspase 3 (Ge et al. 2005). Recalling that, as indicated above, the invasive success of the symbiotic bacteria *Wolbachia* would involve inhibition of apoptosis among other traits.

Finally, protozoan parasites such as *Leishmania*, *Plasmodium*, *Toxoplasma*, and *Trypanosoma* have antiapoptotic capabilities that allow them to invade mammalian cells by modulating different known control points of apoptosis, including Bcl-2 pathways and NF-kappaB (Heussler et al. 2001).

Activation of Apoptosis by Microbes

Activation of apoptosis in host cells often allows microbes to destroy immune cells or escape from infected target cells (Weinrauch and Zychlinsky 1999). Again caspase-mitochondria-dependent pathways are concerned. Upon infection of macrophages by mycobacteria, apoptosis is induced after attachment to TLR2 and activation of proapoptotic pathways and caspase-1 TNF-alpha (Rojas et al. 1999). Exemplary cases of apoptosis induced by microbes involved pathogenic *Shigella* and intracellular *Salmonella*. To escape from macrophages, these bacteria excrete effectors via a secretion system directly into the cytoplasm of the host cell (Mills et al. 1997). Injected toxins activate caspase-1 and induce apoptosis, thus facilitating systemic infection as demonstrated in mice (Monack et al. 1998).

Functional Duality: Inhibition and Activation of Apoptosis

Many microbial pathogens can use pro- and antiapoptotic activities to promote their replication and diffusion in the host cells. This seemingly paradoxical behavior is found in many viruses and some bacteria such as *Chlamydia* and mycobacteria (Miyairi and Byrne 2006). It is suggested that the antiapoptotic activity allows replication and generation of enough infectious entities in primo-infected host cells. Then, apoptosis is activated in the latter stages of infection to facilitate the spread of microbes to other surrounding host cells. Cell tropism has been observed in *Chlamydia* that has a functional duality: inhibition of apoptosis occurs primarily in phagocytic cells in which infectious agents multiply, while apoptosis is activated in immune response cells such as T cells (Miyairi and Byrne 2006). The complex interactions that occur during the modulation of apoptosis often involve the expression of many genes. However, in some cases the product of a single gene of microbial origin may be at the origin of apoptosis modulation. First shown in baculovirus (Clarke and Clem 2003), the involvement of a single gene in the control of apoptosis has been found in many other pathogens. For example, in *Photorhabdus*, the *mcf* gene alone is sufficient to induce apoptosis in insect cells (Daborn et al. 2002). In a similar way, the *nuoG* virulence gene in *Mycobacterium tuberculosis* is sufficient to inhibit apoptosis of host cells (Velmurugan et al. 2007).

11.4.5 Applications in the Interactions Between Microorganisms and Animals

11.4.5.1 Microbes and Nutrition

Microbes play an important role in human and animal nutrition. Following the ingestion of food by animals, the natural process of digestion through the action of enzymes leads to the conversion of macromolecules into simple compounds (sugars, amino acids) that can be assimilated and that are involved in the production of the energy needed for growth and development of organisms. However, we have seen that through digestive enzymes eukaryotic cells are unable to degrade certain macromolecules of plant origin, such as cellulose or chitin. This function is then performed by microbial communities (bacteria, fungi, protozoa) of the digestive tract.

The products of metabolism of the microbiota also contribute to animal nutrition, such as amino acids and various vitamins (B, C), whose deficiencies are the cause of many diseases such as anemia or beriberi. Food industries use microorganisms to produce these essential compounds that can be incorporated as dietary supplements.

Finally, microbial fermentation has been used since antiquity in the manufacture of dairy products (milk, yogurt, cheese), but also in the production of alcoholic beverages such as beer and wine, which excessive consumption may in turn adversely affect human health.

11.4.5.2 Microbes and Health

In the field of human and animal health, antibiotics of bacterial origin are the major molecules used in treatments against nosocomial and community-acquired infections. The group of Actinobacteria, in particular, the genus *Streptomyces*, is the major producer of antibiotics (glycosides, penicillins, tetracyclines, *cf.* Chap. 9), antifungals (nystatin, amphotericin B), and immunosuppressants such as cyclosporine used during organ transplantations. Some anticancer compounds are also produced by microbes. Other microbial molecules used for public health purposes include dextrans, a substitute for blood plasma, numerous steroids for hormone treatments. Microbes themselves, once inactivated by heat or ionizing radiation, can be used in immunization.

Many infectious diseases are transmitted by blood-sucking arthropods such as mosquitoes, ticks, and lice. The absence of a vaccine against many of these diseases renders indispensable the control of arthropod vectors. Spores of the bacterium *Bacillus thuringiensis* have been used since the 1950s to fight against a large number of crop pests and vectors of animal and human pathogens. However, vectors increasingly develop resistance toward the products used, and, on the one hand, the production of new chemical compounds is hindered by the financial costs for research and development and, on the other hand, by the increasing risk linked to the toxicity of these products, both for human and animal populations but also for the environment. In this context, the exploitation of microbial communities as biopesticides for vector control, in addition to genetic and chemical agents, is a strategy globally promoted worldwide. Knowledge of interaction mechanisms and effector molecules of pathogens can lead to the development of antagonistic molecules that block infection or spreading of infectious agents. It should be recalled that the commensal microflora that colonizes the surface of a given organ such as the skin can prevent by simple physical barrier the colonization by pathogens.

Finally, it should be noted, in addition to many cancer-causing chemicals, that microorganisms are also responsible, at least for the initial mechanisms, of cancer diseases. For example, the bacterium *Helicobacter pylori*, which infects about two-thirds of the world population, plays a major role in the development of gastric cancer. Human papilloma virus (HPV) is itself responsible for almost all cervical cancers, while Kaposi's sarcoma is linked to infection with the virus of the herpes family (HHV8, Human Herpes Virus 8). Progress in the identification and characterization of these microbial pathogens as well as knowledge of infectious processes and physiopathology can exploit the microbes own weapons to develop effective therapies against these diseases that too often prove fatal.

11.4.6 Epilogue

Interactions between microorganisms and animals are part of a continuum from parasitism to mutualism. They generate associated entities whose properties often exceed the sum of preexisting structures in the interacting partners. The complexity of associations is a major driver of the evolution and adaptation of species. A better understanding of the interactions between microorganisms and animals in the environment where they operate provides knowledge with multiple outcomes in all areas. We can mention basic biological functions of genes and their products, as well as regulatory networks at the molecular and cellular levels. We can also quote medicine in the development of diagnostics and the development of new therapies against infectious diseases including recurrent reemerging ones or biotechnology on the production of antimicrobial molecules or interest in food and livestock production. The advent of DNA sequencing of the complete genomes of microorganisms and their animal hosts, including the human genome, provides new insights for understanding the interactions between microbes and animals, which should be optimized to the benefit of humans and its environment.

11.5 Conclusion

Biotic interactions, whether transient or sustainable and intraspecific or interspecific, have played and continue to play a major role in the development of microorganisms and macroorganisms with which they are associated, substantially altering the structure and the level of organization. They were decisive in the evolution of organisms, as exemplified by the endosymbiosis as the origin of the eukaryotic cell structures, and have contributed to the adaptation and biodiversity of interacting organisms. Microorganisms interact with each other but also with mobile genetic elements and macroorganisms occupying the same ecosystems. In addition, biotic interactions involve a large variety of entities, unicellular and multicellular, and concern different levels of complexity, starting from the molecule and the gene through the cell, tissues, and organs to the shape of organisms.

Microorganisms stand prominently in the associations between living entities as they are involved in all types of known biotic interactions. The complexity of these interactions and the importance of multiple interactions make them difficult to understand, and many of them are difficult to describe with the scientific terminology currently available. Moreover, the meaning of concepts has changed significantly over time, including improved knowledge, and the meaning of terms describing biotic interactions can vary significantly from one discipline to another.

The importance of biotic interactions implies that microorganisms must first be defined by the number of interactions in which they participate (and thus their social life) and microbial communities (and possibly associated with eukaryotic hosts) in terms of networks of interactions. In food webs, microorganisms provide essential metabolic functions, from primary production to decomposition, and they are essential to the mobilization of flow of matter and energy in the environment. Interactions between microorganisms and multicellular eukaryotes also have a major impact on the health of plants, animals, and human in ecosystems and on ecosystem health itself.

References

Abanda-Nkpwatt D, Krimmb U, Allison Coiner H, Schreiber L, Schwab W (2006) Plant volatiles can minimize the growth suppression of epiphytic bacteria by the phytopathogenic fungus *Botrytis cinerea* in co-culture experiments. Environ Exp Bot 56:108–119

Abdallah MA, van Pittius NC, Champion PAD, Cox J, Luirink J, Vandenbroucke-Grauls CMJE, Appelmelk BJ, Bitter W (2007) Type VII secretion-mycobacteria show the way. Nat Rev 5:883–891

Achouak W, Conrod S, Cohen V, Heulin T (2004) Phenotypic variation of *Pseudomonas brassicacearum* as a plant root-colonization strategy. Mol Plant Microbe Interact 17:872–879

Agrios GN (1997) Plant pathology, 4th edn. Academic, San Diego

Ahmed AM, Hurd H (2006) Immune stimulation and malaria infection impose reproductive costs in *Anopheles gambiae* via follicular apoptosis. Microbes Infect 8:308–315

Alabouvette C, Höper C, Lemanceau P, Steinberg C (1996) Soil suppressiveness to diseases induced by soilborne plant pathogens. In: Stotsky G, Bollag JM (eds) Soil biochemistry. Marcel Dekker Inc., New York, pp 371–413

Alabouvette C, Olivain C, Steinberg C (2006) Biological control of plant pathogens: the European situation. Eur J Plant Pathol 114:329–341

Aliprantis AO, Yang RB, Mark MR, Suggett S, Devaux B, Radolf JD, Klimpel GR, Godowski P, Zychlinsky A (1999) Cell activation and apoptosis by bacterial lipoproteins through toll-like receptor-2. Science 285:736–739

Altier G (2005) Genetic and environmental control of *Salmonella* invasion. J Microbiol 43:82–92

Amellal N, Burtin G, Bartoli F, Heulin T (1998) Colonization of wheat roots by an exopolysaccharide-producing *Pantoea agglomerans* strain and its effect on rhizosphere soil aggregation. Appl Environ Microbiol 64:3740–3747

Atlas RM, Bartha R (1981) Microbial ecology: fundamentals and applications. Addison-Wesley, Reading

Auge RM (2004) Arbuscular mycorrhizae and soil/plant water relations. Can J Soil Sci 84:373–381

Bacci M Jr, Anversa MM, Pagnocca FC (1995) Cellulose degradation by *Leucocoprinus gongylophorus*, the fungus cultured by the leaf-cutting ant *Atta sexdens rubropilosa*. Anton Leeuw Int J G 67:385–386

Bäckhed F, Ding H, Wang T, Hooper LV, Koh GY, Nagy A, Semenkovich CF, Gordon JI (2004) The gut microbiota as an environmental factor that regulates fat storage. Proc Natl Acad Sci U S A 101:15718–15723

Bäckhed F, Ley RE, Sonnenburg JL, Peterson DA, Gordon JI (2005) Host-bacterial mutualism in the human intestine. Science 307:1915–1920

Bailly J, Debaud JC, Verner MC, Plassard C, Chalot M, Marmeisse R, Fraissinet-Tachet L (2007) How does a symbiotic fungus modulate expression of its hostplant nitrite reductase? New Phytol 175:155–165

Bais HP, Weir TL, Perry LG, Gilroy S, Vivanco JM (2006) The role of root exudates in rhizosphere interactions with plants and other organisms. Annu Rev Plant Biol 57:233–266

Balcewicz-Sablinska MK, Keane J, Kornfeld H, Remold HG (1998) Pathogenic *Mycobacterium tuberculosis* evades apoptosis of host macrophages by release of TNFR2, resulting in inactivation of TNF-alpha. J Immunol 161:2636–2641

Bandi C, Trees AJ, Brattig NW (2001) *Wolbachia* in filarial nematodes: evolutionary aspects and implications for pathogenesis and treatment of filarial diseases. Vet Parasitol 98:215–238

Bannerman DD, Goldblum SE (1997) Endotoxin induces endothelial barrier dysfunction through protein tyrosine phosphorylation. Am J Physiol 273:217–226

Baratin M et al (2005) Natural killer cell and macrophage cooperation in MyD88-dependent innate responses to *Plasmodium falciparum*. Proc Natl Acad Sci U S A 102:14747–14752

Barbault R (1997) Ecologie Générale. Structure et fonctionnement de la biosphère, 4th edn. Masson, Paris, 281 p

Barré-Sinoussi F et al (1983) Isolation of a T-lymphotropic retrovirus from a patient at risk for Acquired Immune Deficiency Syndrome (AIDS). Science 4599:868–871

Bauer WD, Robinson JB (2002) Disruption of bacterial quorum sensing by other organisms. Curr Opin Biotechnol 13:234–237

Baumann P, Moran NA, Beaumann L (2000) Bacteriocyte- associated endosymbionts of insects. In: Dworkin M (ed) The prokaryotes. Springer, NewYork, pp 1–67

Beattie GA, Lindow SE (1999) Bacterial colonization of leaves: a spectrum of strategies. Phytopathology 89:353–359

Belimov AA, Hontzeas N, Safronova VI, Demchinskaya SV, Piluzza G, Bullitta S, Glick BR (2005) Cadmium-tolerant plant growth-promoting bacteria associated with the roots of Indian mustard (*Brassica juncea* L. Czern.). Soil Biol Biochem 37:241–250

Bennett AE, Bever JD (2007) Mycorrhizal species differentially alter plant growth and response to herbivory. Ecology 88:210–218

Bentley SD et al (2002) Complete genome sequence of the model actinomycete *Streptomyces coelicolor* A3(2). Nature 417:141–147

Bertaux J, Schmid M, Prevost-Boure NC, Churin JL, Hartmann A, Garbaye J, Frey-Klett P (2003) In situ identification of intracellular bacteria related to *Paenibacillus* spp. in the mycelium of the ectomycorrhizal fungus *Laccaria bicolor* S238N. Appl Environ Microbiol 69:4243–4248

Berthiller F et al (2013) Masked mycotoxins: a review. Mol Nutr Food Res 57:165–186

Birge EA, Kurland CG (1969) Altered ribosomal protein in streptomycin-dependent *Escherichia coli*. Science 166:1282–1284

Bjorklof K, Nurmiaho-Lassila EL, Klinger N, Haahtela K, Romantschuk M (2000) Colonization strategies and conjugal gene transfer of inoculated *Pseudomonas syringae* on the leaf surface. J Appl Microbiol 89:423–432

Blanc C, Sy M, Djigal D, Brauman A, Normand P, Villenave C (2006) Nutrition on bacteria by bacterial-feeding nematodes and consequences on the structure of soul bacterial community. Eur J Soil Biol 42:S70–S78

Bohrerova Z, Stralkova R, Podesvova J, Bohrer G, Pokorny E (2004) The relationship between redox potential and nitrification under different sequences of crop rotations. Soil Till Res 77:25–33

Bonkowski M (2004) Protozoa and plant growth: the microbial loop in soil revisited. New Phytol 162:617–631

Bouchez M, Blanchet D, Bardin V, Haeseler F, Vandecasteele JP (1999) Efficiency of defined strains and of soil consortia in the biodegradation of polycyclic aromatic hydrocarbon (PAH) mixtures. Biodegradation 10:429–435

Brady J, Costantini C, Sagnon N, Gibson G, Coluzzi M (1997) The role of body odours in relative attractiveness of different men to malarial vectors in Burkina Faso. Ann Trop Med Parasitol 91:S121–S122

Brundrett MC (2009) Mycorrhizal associations and other means of nutrition of vascular plants: understanding the global diversity of host plants by resolving conflicting information and developing reliable means of diagnosis. Plant Soil 320:37–77

Brune A (2003) Symbionts aiding digestion. In: Cardé RT, Resh VH (eds) Encyclopedia of insects. Academic, New York, pp 1102–1107

Buchner P (1965) Endosymbiosis of animals with plant microorganisms. Wiley Interscience, New York

Cardon ZG, Whitbeck JL (2007) The rhizosphere: an ecological perspective. Elsevier, Burlington

Cascales E, Christie PJ (2003) The versatile bacterial type IV secretion systems. Nat Rev Microbiol 1:137–150

Cavagnaro TR, Smith FA, Smith SE, Jakobsen I (2005) Functional diversity in arbuscular mycorrhizas: exploitation of soil patches with different phosphate enrichment differs among fungal species. Plant Cell Environ 28:642–650

Cavanaugh CM, Gardiner SL, Jones ML, Jannasch HW, Waterbury JB (1981) Prokaryotic cells in the hydrothermal vent tube worm *Riftia pachyptilia* Jones: possible chemoautotrophic symbionts. Science 213:340–342

Chapon A, Guillerm A-Y, Delalande L, Lebreton L, Sarniguet A (2002) Dominant colonisation of wheat roots by *Pseudomonas fluorescens* Pf29A and selection of the indigenous microflora in the presence of the take-all fungus. Eur J Plant Pathol 108:449–459

Chevrot R, Rosen R, Haudecoeur E, Cirou A, Shelp BJ, Ron E, Faure D (2006) GABA controls the level of quorum-sensing signal in *Agrobacterium tumefaciens*. Proc Natl Acad Sci U S A 103:7460–7464

Clarke TE, Clem RJ (2003) Insect defenses against virus infection: the role of apoptosis. Int Rev Immunol 22:401–424

Clifton DR, Goss RA, Sahni SK, van Antwerp D, Baggs RB, Marder VJ, Silverman DJ, Sporn LA (1998) NF-kappaB-dependent inhibition of apoptosis is essential for host cell survival during *Rickettsia rickettsii* infection. Proc Natl Acad Sci U S A 95:4646–4651

Cossart P, Sansonetti PJ (2004) Bacterial invasion: the paradigms of enteroinvasive pathogens. Science 304:242–248

Crutzen PJ, Mosier AR, Smith KA, Winiwarter W (2007) N_2O release from agro-biofuel production negates global warming reduction by replacing fossil fuels. Atmos Chem Phys Discuss 7:11191–11205

Dabiré KR, Ndiaye S, Chotte J-L, Fould S, Diop MT, Mateille T (2005) Influence of irrigation on the distribution and control of the nematode *Meloidogyne javanica* by the biocontrol bacterium *Pasteuria penetrans* in the field. Biol Fertil Soils 41:205–211

Daborn PJ, Waterfield N, Silva CP, Au CP, Sharma S, Ffrench-Constant RH (2002) A single *Photorhabdus* gene, *makes caterpillars floppy* (*mcf*), allows *Escherichia coli* to persist within and kill insects. Proc Natl Acad Sci U S A 99:10742–10747

Dale C, Plague GR, Wang B, Ochman H, Moran NA (2002) Type III secretion system and the evolution of mutualistic endosymbiosis. Proc Natl Acad Sci U S A 99:12397–12402

Dams RI, Paton GI, Killham K (2007) Rhizoremediation of pentachlorophenol by *Sphingobium chlorophenolicum* ATCC 39723. Chemosphere 68:864–870

Darwin C (1859) On the origin of species by means of natural selection, or the preservation of favoured races in the struggle for life. John Murray, London

Daubin V, Ochman H (2004) Bacterial genomes as new gene homes: the genealogy of ORFans in *E. coli*. Genome Res 14:1036–1042

Davelos AL, Kinkel LL, Samac DA (2004) Spatial variation in frequency and intensity of antibiotic interactions among *Streptomycetes* from prairie soil. Appl Environ Microbiol 70:1051–1058

de Bary HA (1879) Die Erscheinung der Symbiose. KJ Trübner, Strasbourg

de-Bashan L, Hernandez J-P, Morey T, Bashan Y (2004) Microalgae growth-promoting bacteria as "helpers" for microalgae: a novel approach for removing ammonium and phosphorus from municipal wastewater. Water Res 38:466–474

Dedeine F, Vavre F, Fleury F, Loppin B, Hochberg ME, Bouletreau M (2001) Removing symbiotic *Wolbachia* bacteria specifically inhibits oogenesis in parasitic wasps. Proc Natl Acad Sci U S A 98:6247–6252

Delong EF, Pace NR (2001) Environmental diversity of bacteria and archae. Syst Biol 50:470–478

Dethlefsen L, McFall-Ngai M, Relman DA (2007) An ecological and evolutionary perspective on human-microbe mutualism and disease. Nature 449:811–818

Devers M, Henry S, Hartmann A, Martin-Laurent F (2005) Horizontal gene transfer of atrazine-degrading genes (*atz*) from *Agrobacterium tumefaciens* St96-4 pADP1::Tn5 to bacteria of maize-cultivated soil. Pest Manag Sci 61:870–880

Distel DL, Lane DJ, Olsen GJ, Giovannoni SJ, Pace B, Pace NR, Stahl DA, Felbeck H (1988) Sulfur-oxidizing bacterial endosymbionts: analysis of phylogeny and specificity by 16S rRNA sequences. J Bacteriol 170:2506–2510

Dobbelaere S et al (2001) Responses of agronomically important crops to inoculation with *Azospirillum*. Aust J Plant Physiol 28:871–879

Dobbelaere S, Vanderleyden J, Okon Y (2003) Plant growth-promoting effects of diazotrophs in the rhizosphere. Crit Rev Plant Sci 22:107–149

Dodds PN, Rathjen JP (2010) Plant immunity: towards an integrated view of plant-pathogen interactions. Nat Rev Genet 11:539–548

Douglas AE (1998) Nutritional interactions in insect-microbial symbioses: aphids and their symbiotic bacteria Buchnera. Annu Rev Entomol 43:17–37

Dunbar HE, Wilson AC, Ferguson NR, Moran NA (2007) Aphid thermal tolerance is governed by a point mutation in bacterial symbionts. PLoS Biol 5:e96

Ebbole DJ (2007) Magnaporthe as a model for understanding host-pathogen interactions. Annu Rev Phytopathol 45:437–456

Edel V, Steinberg C, Gautheron N, Recorbet G, Alabouvette C (2001) Genetic diversity of *Fusarium oxysporum* populations isolated from different soils in France. FEMS Microbiol Ecol 36:61–71

Edwald PW (1987) Transmission modes and evolution of the parasitism-mutualism continuum. Ann N Y Acad Sci 503:295–306

Fall S, Hamelin J, Ndiaye F, Assigbetse K, Aragno M, Chotte JL, Brauman A (2007) Differences between bacterial communities in the gut of a soil-feeding termite (*Cubitermes niokoloensis*) and its mounds. Appl Environ Microbiol 73:5199–5208

Fan T, Lu H, Hu H, Shi L, McClarty GA, Nance DM, Greenberg AH, Zhong G (1998) Inhibition of apoptosis in *Chlamydia*-infected cells: blockade of mitochondrial cytochrome c release and caspase activation. J Exp Med 187:487–496

Favia G et al (2007) Bacteria of the genus *Asaia* stably associate with *Anopheles stephensi*, an Asian malarial mosquito vector. Proc Natl Acad Sci U S A 104:9047–9051

Faherty CS, Maurelli AT (2008) Staying alive: bacterial inhibition of apoptosis during infection. Trends Microbiol 16:173–180

Fedi S, Tola E, Moënne-Loccoz Y, Dowling DN, Smith LM, O'Gara F (1997) Evidence for signaling between the phytopathogenic fungus *Pythium ultimum* and *Pseudomonas fluorescens* F113: *P. ultimum* represses the expression of genes in *P. fluorescens* F113, resulting in altered ecological fitness. Appl Environ Microbiol 63:4261–4266

Ferrandon D, Imler JL, Hetru C, Hoffmann JA (2007) The *Drosophila* systemic immune response: sensing and signaling during bacterial and fungal infections. Nat Rev Immunol 7:862–874

Filloux A, Hachani A, Bleves S (2008) The bacterial type VI secretion machine: yet another player for protein transport across membranes. Microbiology 154:1570–1583

Fisher K, Beatty WL, Jiang D, Weil GJ, Fisher PU (2011) Tissue and stage-specific distribution of *Wolbachia* in *Brugia malayi*. PLoS Negl Trop Dis 5:e1174

Fleming A (1922) On a remarkable bacteriolytic element found in tissues and secretions. Proc R Soc Ser B 93:306–317

Flor JM (1956) The complementary genetic systems in flax and flax rust. Adv Genet 8:29–54

Forst S, Dowds B, Boemare N, Stackebrandt E (1997) *Xenorhabdus* and *Photorhabdus* spp.: bugs that kill bugs. Annu Rev Microbiol 51:47–72

Foster PL (2007) Stress-induced mutagenesis in bacteria. Crit Rev Biochem Mol Biol 42:373–397

Foster J et al (2005) The *Wolbachia* genome of *Brugia malayi*: endosymbiont evolution within a human pathogenic nematode. PLoS Biol 3:599–614

Fourcans A, Sole A, Diestra E, Ranchou-Peyruse A, Esteve I, Caumette P, Duran R (2006) Vertical migration of phototrophic bacterial populations in a hypersaline microbial mat from Salins-de-Giraud (Camargue, France). FEMS Microbiol Ecol 57:367–377

Frank AB (1885) Uber die auf Wurzelsymbiose beruhende Ernahrung gewisser Baume durch unteriridische Pilze. Ber Deut Bot Ges 3:128–145

Frank AB (1889) Uber die Pilzsymbiose der Leguminosen. Ber Deut Bot Ges 7:332–346

Futayama D (1986) The evolution of interactions among species. In: Davis A, Vesely J (eds) Evolutionary biology. Sinauer Associates, Inc., Sunderland, pp 482–504

Gamboa MA, Laureano S, Bayman P (2003) Measuring diversity of endophytic fungi in leaf fragments: does size matter? Mycopathologia 156:41–45

Gao LY, Abu Kwaid Y (2000) The modulation of host cell apoptosis by intracellular bacterial pathogens. Trends Microbiol 87:306–313

Garbeva P, van Veen JA, van Elsas JD (2004) Microbial diversity in soil: selection of microbial populations by plant and soil type and implications for disease suppressiveness. Annu Rev Phytopathol 42:243–270

Gautier-Clerc M, Lebarbenchon C, Thomas F (2007) Recent expansion of highly pathogenic avian influenza H5N1: a critical review. Ibis 149:202–214

Ge Y, Yoshiie K, Kuribayashi F, Lin M, Rikihisa Y (2005) *Anaplasma phagocytophilum* inhibits human neutrophil apoptosis via upregulation of *bfl-1*, maintenance of mitochondrial membrane potential and prevention of caspase 3 activation. Cell Microbiol 7:29–38

Gherbi H et al (2008) SymRK defines a common genetic basis for plant root endosymbioses with arbuscular mycorrhiza fungi, rhizobia, and *Frankia* bacteria. Proc Natl Acad Sci U S A 105:4928–4932

Gilbert GS, Parke JL, Clayton MK, Handelsman J (1993) Effects of an introduced bacterium on bacterial communities on roots. Ecology 74:840–854

Giraud E et al (2007) Legumes symbioses: absence of *nod* genes in photosynthetic bradyrhizobia. Science 316:1307–1312

Goebel W, Gross R (2001) Intracellular survival strategies of mutualistic and parasitic prokaryotes. Trends Microbiol 9:267–273

Goodrich-Blair H, Clarke DJ (2007) Mutualism and pathogenesis in *Xenorhabdus* and *Photorhabdus*: two roads to the same destination. Mol Microbiol 64:260–268

Gottlieb Y et al (2006) Identification and localization of a *Rickettsia* sp. in *Bemisia tabaci* (Homoptera: Aleyrodidae). Appl Environ Microbiol 72:3646–3652

Gould SJ (1989) Wonderful life – the burgess shale and the nature of history. W.W. Norton & Company, New York

Gould SJ, Vrba ES (1982) Exaptation – a missing term in the science of form. Paleobiology 8:4–15

Griffiths RI, Manefield M, Ostle N, McNamara N, O'Donnell AG, Bailey MJ, Whiteley AS (2004) $^{13}CO_2$ pulse labelling of plants in tandem with stable isotope probing: methodological considerations for examining microbial function in the rhizosphere. J Microbiol Methods 58:119–129

Groppa MD, Benavides MP, Zawoznik MS (2012) Root hydraulic conductance, aquaporins and plant growth promoting microorganisms: a revision. Appl Soil Ecol 61:247–254

Guntli D, Burgos S, Moënne-Loccoz Y, Défago G (1999) Calystegine degradation capacities of microbial rhizosphere communities of *Zea mays* (calystegine-negative) and *Calystegia sepium* (calystegine positive). FEMS Microbiol Ecol 28:75–84

Hacker J, Carniel E (2001) Ecological fitness, genomic islands and bacterial pathogenicity. EMBO Rep 2:376–381

Hacker J, Kaper JB (2000) Pathogenicity islands and the evolution of microbes. Annu Rev Microbiol 54:641–679

Harman GE, Howell CR, Viterbo A, Chet I, Lorito M (2004) *Trichoderma* species – opportunistic, avirulent plant symbionts. Nat Rev Microbiol 2:43–56

Hay BA, Guo M (2006) Caspase-dependent cell death in *Drosophila*. Ann Rev Cell Dev Biol 22:623–650

Heddi A, Grenier AM, Khatchadourian C, Charles H, Nardon P (1999) Four intracellular genomes direct weevil biology: nuclear, mitochondrial, principal endosymbiont, and *Wolbachia*. Proc Natl Acad Sci U S A 96:6814–6819

Hedges LM, Brownlie JC, O'Neill SL, Johnson KN (2008) *Wolbachia* and virus protection in insects. Science 322:702

Heitman J (2011) Microbial pathogens in the fungal kingdom. Fungal Biol Rev 25:48–60

Hense BA, Kuttler C, Muller J, Rothballer M, Hartmann A, Kreft JU (2007) Does efficiency sensing unify diffusion and quorum sensing? Nat Rev Microbiol 5:230–239

Hentschel U, Steinert M, Hacker J (2000) Common molecular mechanisms of symbiosis and pathogenesis. Trends Microbiol 8:226–231

Hersh D, Monack DM, Smith MR, Ghori N, Falkow S, Zychlinsky A (1999) The *Salmonella* invasin SipB induces macrophage apoptosis by binding to caspase-1. Proc Natl Acad Sci U S A 96:2396–2401

Hery M, Herrera A, Vogel TM, Normand P, Navarro E (2005) Effect of carbon and nitrogen input on the bacterial community structure of Neocaledonian nickel mine spoils. FEMS Microbiol Ecol 51:333–340

Heussler VT, Küenzi P, Rottenberg S (2001) Inhibition of apoptosis by intracellular protozoan parasites. Int J Parasitol 31:1166–1176

Hilgenboecker K, Hammerstein P, Schlattmann P, Telschow A, Werren JH (2008) How many species are infected with *Wolbachia*? A statistical analysis of current data. FEMS Microbiol Lett 281:215–220

Hinsinger P, Plassard C, Tang C, Jaillard B (2003) Origins of root-mediated pH changes in the rhizosphere and their responses to environmental constraints: a review. Plant Soil 248:43–59

Hoffmann J (2003) The immune response of *Drosophila*. Nature 426:33–38

Hosokawa T, Koga R, Kikuchi Y, Meng XY, Fukatsu T (2010) *Wolbachia* as a bacteriocyte-associated nutritional mutualist. Proc Natl Acad Sci U S A 107:769–774

Hongoh Y, Ekpornprasit L, Inoue T, Moriya S, Trakulnaleamsai S, Ohkuma M, Noparatnaraporn N, Kudo T (2006) Intracolony variation of bacterial gut microbiota among castes and ages in the fungus-growing termite *Macrotermes gilvus*. Mol Ecol 15:505–516

Hooper LV, Stappenbeck TS, Hong CV, Gordon JI (2003) Angiogenins: a new class of microbicidal proteins involved in innate immunity. Nat Immunol 4:269–273

Hotopp JC, Clark ME, Oliveira DC (2007) Widespread lateral gene transfer from intracellular bacteria to multicellular eukaryotes. Science 317:1753–1756

Huguet V, Gouy M, Normand P, Zimpfer JF, Fernandez MP (2005) Molecular phylogeny of Myricaceae: a reexamination of host-symbiont specificity. Mol Phylogenet Evol 34:557–568

Hurd H, Grant KM, Arambage SC (2006) Apoptosis-like death as a feature of malaria infection in mosquitoes. Parasitology 132(Suppl): S33–S47

Ingram LC, Richmond MH, Sykes RB (1973) Molecular characterization of the R factors implicated in the carbenicillin resistance of a sequence of *Pseudomonas aeruginosa* strains isolated from burns. Antimicrob Agents Chemother 3:279–288

Jacobson MD, Weil M, Raff MC (1997) Programmed cell death in animal development. Cell 88:347–354

Jahn U, Gallenberger M, Junglas B, Eisenreich W, Stetter KO, Rachel R, Huber H (2008) *Nanoarchaeum equitans* and *Ignicoccus hospitalis*: new insights into a unique, intimate association of two Archaea. J Bacteriol 190:1743–1750

Jeyaprakash A, Hoy MA (2000) Long PCR improves *Wolbachia* DNA amplification: *wsp* sequences found in 76 % of sixty-three arthropod species. Insect Mol Biol 9:393–405

Johnston KL, Taylor MJ (2007) *Wolbachia* in filarial parasites: targets for filarial infection and disease control. Curr Infect Dis Rep 9:55–59

Jones ML (1981) *Riftia pachyptila* Jones: observations on the vestimentiferan worm from the Galapagos rift. Science 213:333–336

Jones JDG, Dangl JL (2006) The plant immune system. Nat Rev 444:323–329

Jousset A, Rochat L, Péchy-Tarr M, Keel C, Scheu S, Bonkowski M (2009) Predators promote defence of rhizosphere bacterial populations by selective feeding on non-toxic cheaters. ISME J 3:666–674

Karolis DKR, Somara S, Maneval DR Jr, Johnson JA, Kaper JB (1999) A bacteriophage encoding a pathogenicity island, a type-IV pilus and a phage receptor in cholera bacteria. Nature 399:375–379

Kazemi-Pour N, Condemine G, Hugouvieux-Cotte-Pattat N (2004) The secretome of the plant pathogenic bacterium *Erwinia chrysanthemi*. Proteomics 4:3177–3186

Khan AG (2005) Role of soil microbes in the rhizospheres of plants growing on trace metal contaminated soils in phytoremediation. J Trace Elem Med Biol 18:355–364

Kimbell JR, McFall-Ngai M (2004) Symbiont-induced changes in host actin during the onset of beneficial animal-bacterial association. Appl Environ Microbiol 70:1434–1441

Kinkel LL, Wilson M, Lindow SE (2000) Plant species and plant incubation conditions influence variability in epiphytic bacterial population size. Microb Ecol 39:1–11

Kirby JE, Nekorchuk DM (2002) *Bartonella*-associated endothelial proliferation depends on inhibition of apoptosis. Proc Natl Acad Sci U S A 99:4656–4666

Klasson L et al (2008) Genome evolution of *Wolbachia* strain wPip from *Culex pipientis* group. Mol Biol Evol 25:1877–1887

Klasson L, Kambris Z, Cook PE, Walker T, Sinkins SP (2009) Horizontal gene transfer between *Wolbachia* and the mosquito *Aedes aegypti*. BMC Genomics 10:33

Krimi Z, Petit A, Mougel C, Dessaux Y, Nesme X (2002) Seasonal fluctuations and long-term persistence of pathogenic populations of *Agrobacterium* spp. in soils. Appl Environ Microbiol 68:3358–3365

Kyselková M, Kopecký J, Frapolli M, Défago G, Ságová-Marečková M, Grundmann GL, Moënne-Loccoz Y (2009) Comparison of rhizobacterial community composition in soil suppressive or conducive to tobacco black root rot disease. ISME J 3:1127–1138

Leben C (1988) Relative humidity and the survival of epiphytic bacteria with buds and leaves of cucumber plants. Phytopathology 78:179–185

Leplat J, Friberg H, Abid M, Steinberg C (2013) Survival of *Fusarium graminearum*, the causal agent of Fusarium head blight. A review. Agron Sustain Dev 33:97–111

Leveau JH, Lindow SE (2001) Appetite of an epiphyte: quantitative monitoring of bacterial sugar consumption in the phyllosphere. Proc Natl Acad Sci U S A 98:3446–3453

Ley RE, Peterson DA, Gordon JI (2006) Ecological and evolutionary forces shaping microbial diversity in the human intestine. Cell 124:837–848

Liesack W, Schnell S, Revsbech NP (2000) Microbiology of flooded rice paddies. FEMS Microbiol Rev 24:625–645

Lindh JM, Terenius O, Faye I (2005) 16S rRNA gene-based identification of midgut bacteria from fieldcaught *Anopheles gambiae* sensu lato and *A. funestus* mosquitoes reveals new species related to known insect symbionts. Appl Environ Microbiol 71:7217–7223

Lindow SE, Brandl MT (2003) Microbiology of the phyllosphere. Appl Environ Microbiol 69:1875–1883

Llosa M, O'Callaghan D (2004) Euroconference on the biology of type IV secretion processes: bacterial gates into the outer world. Mol Microbiol 53:1–8

López-García P, Moreira D (1999) Metabolic symbiosis at the origin of eukaryotes. Trends Biochem Sci 24:88–93

Lorenz N, Wilson EV, Machado C, Schardl CL, Tudzynski P (2007) Comparison of ergot alkaloid biosynthesis gene clusters in *Claviceps* species indicates loss of late pathway steps in evolution of *C. fusiformis*. Appl Environ Microbiol 73:7185–7191

Lübeck PS, Hansen M, Sørensen J (2000) Simultaneous detection of the establishment of seed-inoculated *Pseudomonas fluorescens* strain DR54 and native soil bacteria on sugar beet root surfaces using fluorescence antibody and in situ hybridization techniques. FEMS Microbiol Ecol 33:11–19

Lucas P, Sarniguet A (1998) Biological control of soilborne pathogens with resident versus introduced antagonists: should diverging approaches become strategic convergence? In: Barbosa P (ed) Conservation biological control. Academic, New York, pp 351–370

Maillet F et al (2011) Fungal lipochitooligosaccharide symbiotic signals in arbuscular mycorrhiza. Nature 469:58–63

Mainil J (2013) *Escherichia coli* virulence factors. Vet Immunol Immunopathol 152:2–12

Mantelin S, Desbrosses G, Larcher M, Tranbarger TJ, Cleyet-Marel J-C, Touraine B (2006) Nitrate-dependent control of root architecture and N nutrition are altered by a plant growth-promoting *Phyllobacterium* sp. Planta 223:591–603

Martin MO (2002) Predatory prokaryotes: an emerging research opportunity. J Mol Microbiol Biotechnol 4:467–477

Mateos M, Castrezana SJ, Nankivell BJ, Estes AM, Markow TA, Moran NA (2006) Heritable endosymbionts of *Drosophila*. Genetics 174:363–376

Mathesius U (2003) Conservation and divergence of signalling pathways between roots and soil microbes – the *Rhizobium*- legume symbiosis compared to the development of lateral roots, mycorrhizal interactions and nematode-induced galls. Plant Soil 255:105–119

Mavingui P, Laguerre G, Berge O, Heulin T (1992) Genetic and phenotypic diversity of *Bacillus polymyxa* in soil and in the wheat rhizosphere. Appl Environ Microbiol 58:1894–1903

McGraw EA, Merritt DJ, Droller JN, O'Neill SL (2002) *Wolbachia* density and virulence attenuation after transfer into a novel host. Proc Natl Acad Sci U S A 99:2918–2923

McInroy JA, Kloepper JW (1995) Population dynamics of endophytic bacteria in field-grown sweet corn and cotton. Can J Microbiol 41:895–901

Medzhitov R, Preston-Hulburt P, Janeway CA Jr (1997) A human homologue of the *Drosophila* Toll protein signals activation of adaptive immunity. Nature 388:394–397

Miché L, Belkin S, Rozen R, Balandreau J (2003) Rice seedling whole exudates and extracted alkylresorcinols induce stress-response in *Escherichia coli* biosensors. Environ Microbiol 5:403–411

Millette M, Luquet FM, Lacroix M (2007) In vitro growth control of selected pathogens by *Lactobacillus acidophilus*- and *Lactobacillus casei*-fermented milk. Lett Appl Microbiol 44:314–319

Mills SD, Boland A, Sory MP, van der Smissen P, Kerbourch C, Finlay BB, Cornelis GR (1997) *Yersinia enterocolitica* induces apoptosis in macrophages by a process requiring functional type III secretion and translocation mechanisms and involving YopP, presumably acting as an effector protein. Proc Natl Acad Sci U S A 94:12638–12643

Mindlin SZ, Bass IA, Bogdanova ES, Gorlenko ZM, Kaliaeva ES, Petrova MA, Nikiforov VG (2002) Horizontal transfer of mercury resistance genes in natural bacterial populations. Mol Biol (Mosk) 36:216–227

Miyairi I, Byrne GI (2006) *Chlamydia* and programmed cell death. Curr Opin Microbiol 9:102–108

Monack DM, Mecsas J, Bouley D, Falkow S (1998) *Yersinia*-induced apoptosis in vivo aids in the establishment of a systemic infection of mice. J Exp Med 188:2127–2137

Montgomery MK, McFall-Ngai MJ (1994) Bacterial symbionts induce host organ morphogenesis during early postembryonic development of the squid *Euprymna scolopes*. Development 120:1719–1729

Moran NA (2007) Symbiosis as an adaptive process and source of phenotypic complexity. Proc Natl Acad Sci U S A 104(Suppl 1):8627–8633

Moran NA, Mira A (2001) The process of genome shrinkage in the obligate symbiont *Buchnera aphidicola*. Genome Biol 2:54.1–54.12

Moran NA, McCutcheon JP, Nakabachi A (2008) Genomics and evolution of heritable bacterial symbionts. Annu Rev Genet 42:165–190

Moreira LA et al (2009) A *Wolbachia* in *Aedes aegypti* limits infection with dengue, Chikungunya, and *Plasmodium*. Cell 139:1268–1278

Morris CE, Monier JM (2003) The ecological significance of biofilm formation by plant-associated bacteria. Annu Rev Phytopathol 41:429–453

Morris CE, Sands DC, Vanneste JL, Montarry J, Oakley B, Guilbaud C, Glaux C (2010) Inferring the evolutionary history of the plant pathogen *Pseudomonas syringae* from its biogeography in headwaters of rivers in North America, Europe, and New Zealand. mBio 1:3

Moubareck C, Bourgeois N, Courvalins P, Doucet-Populaire F (2003) Multiple antibiotic resistance gene transfer from animal to human enterococci in the digestive tract of gnotobiotic mice. Antimicrob Agents Chemother 47:2993–2996

Mougel C, Offre P, Ranjard L, Corberand T, Gamalero E, Robin C, Lemanceau P (2006) Dynamic of the genetic structure of bacterial and fungal communities at different developmental stages of *Medicago truncatula* Gaertn. cv. Jemalong line J5. New Phytol 170:165–175

Moulin L, Munive A, Dreyfus B, Boivin-Masson C (2001) Nodulation of legumes by members of the β-subclass of Proteobacteria. Nature 411:948–950

Mousson L, Zouache K, Arias-Goeta C, Raquin V, Mavingui P, Failloux AB (2012) The native *Wolbachia* symbionts limit transmission of dengue virus in *Aedes albopictus*. PLoS Negl Trop Dis 6: e1989

Mouttaki H, Nanny MA, McInerney MJ (2007) Cyclohexane carboxylate and benzoate formation from crotonate in *Syntrophus aciditrophicus*. Appl Environ Microbiol 73:930–938

Mukerji KG, Chamola BP, Singh J (2000) Mycorrhizal biology. Kluwer Academic Publishers, New York

Murray CJL et al (2012) Global malaria mortality between 1980 and 2010: a systematic analysis. Lancet 379:413–431

Nagase N, Sasaki A, Yamashita K, Shimizu A, Wakita Y, Kitai S, Kawano J (2002) Isolation and species distribution of staphylococci from animal and human skin. J Vet Med Sci 63:245–250

Nardon P, Charles P (2001) Morphological aspects of symbiosis. In: Seckbach J (ed) Symbiosis. Kluwer Academic Publishers, Dordrecht

Nelson DC, Fisher CR (1995) Chemoautotrophic and methanotrophic endosymbiontic bacteria at deep-sea vents and seeps. In: Karl DM (ed) Microbiology of deep-sea hydrothermal vents. CRC Press, Boca Raton, pp 125–167

Nguyen C (2003) Rhizodeposition of organic C by plants: mechanisms and controls. Agronomie 23:375–396

Noble WC (1993) The skin microflora and microbial skin disease. Cambridge University Press, Cambridge

Normand P et al (2007) Genome characteristics of facultatively symbiotic *Frankia* sp. strains reflect host range and host plant biogeography. Genome Res 17:7–15

Novotny V, Basset Y, Miller SE, Weiblen GD, Bremer B, Ciezk L, Drozd P (2002) Low host specificity herbivor insect in tropical forest. Nature 416:841–844

Nyholm SV, McFall-Ngai MJ (2004) The winnowing: establishing the squid-vibrio symbiosis. Nat Rev Microbiol 2:632–642

Ochman H, Wilson AC (1987) Evolution in bacteria: evidence for a universal substitution rate in cellular genomes. J Mol Evol 26:74–86

Odum EP (1971) Fundamentals of ecology. Saunders, Philadelphia

Oerke EC, Dehne HW, Schohnbeck F, Weber A (1995) Crop production and crop protection: estimated losses in major food and cash crops. Elsevier, Amsterdam/New York

Ofek I, Hasty DL, Doyle RJ (2003) Bacterial adhesion to animal cells and tissues. ASM Press, Washington, DC, 416 p

Oldroyd GE (2007) Nodules and hormones. Science 315:52–53

Oldroyd GE, Downie JA (2006) Nuclear calcium changes at the core of symbiosis signalling. Curr Opin Plant Biol 9:351–357

Olivain C, Humbert C, Nahalkova J, Fatehi J, L'Haridon F, Alabouvette C (2006) Colonization of tomato root by pathogenic and nonpathogenic *Fusarium oxysporum* strains inoculated together and separately into the soil. Appl Environ Microbiol 72:1523–1531

Oliver KM, Russell JA, Moran NA, Hunter MS (2003) Facultative bacterial symbionts in aphids confer resistance to parasitic wasps. Proc Natl Acad Sci U S A 100:1803–1807

Osborne LE, Stein JM (2007) Epidemiology of Fusarium head blight on small-grain cereals. Int J Food Microbiol 119:103–108

Ouwehand AC, Salminen S, Isolauri E (2002) Probiotics: an overview of beneficial effects. Anton Leeuw Int J G 82:279–289

Pancholi V, Chhatwal GS (2003) Housekeeping enzymes as virulence factors for pathogens. Int J Med Microbiol 293:391–401

Pannebakker B, Loppin B, Elemans CP, Humblot L, Vavre F (2007) Parasitic inhibition of cell death facilitates symbiosis. Proc Natl Acad Sci U S A 104:213–215

Parsot C, Sansonetti PJ (1999) The virulence plasmid of Shigellae: an archipelago of pathogenicity island? In: Kaper JB, Hacker J (eds) Pathogenicity islands and other mobile virulence elements. ASM Press, Washington, DC, pp 151–165

Partida-Martinez LP, Hertweck C (2005) Pathogenic fungus harbours endosymbiotic bacteria for toxin production. Nature 437:884–888

Patra AK et al (2006) Effects of management regime and plant species on the enzyme activity and genetic structure of N-fixing, denitrifying and nitrifying bacterial communities in grassland soils. Environ Microbiol 8:1005–1016

Patra AK, Le Roux X, Abbadie L, Clays-Josserand A, Poly F, Loiseau P, Louault F (2007) Effect of microbial activity and nitrogen mineralization on free-living nitrogen fixation in permanent grassland soils. J Agron Crop Sci 193:153–156

Pel MJC, Pieterse CMJ (2013) Microbial recognition and evasion of host immunity. J Exp Bot 64:1237–1248

Persson L, Larsson-Wikstrom M, Gerhardson B (1999) Assessment of soil suppressiveness to Aphanomyces root rot of pea. Plant Dis 83:1108–1112

Picard C, Bosco M (2006) Heterozygosis drives maize hybrids to select elite 2,4-diacetylphloroglucinol-producing *Pseudomonas* strains among resident soil populations. FEMS Microbiol Ecol 58:193–204

Popa O, Dagan T (2011) Trends and barriers to lateral gene transfer in prokaryotes. Curr Opin Microbiol 14:615–623

Prin Y, Duhoux E, Diem H, Roederer Y, Dommergues Y (1991) Aerial nodules in *Casuarina cunninghamiana*. Appl Environ Microbiol 57:871–874

Raaijmakers JM, Weller DM (1998) Natural plant protection by 2,4-diacetylphloroglucinol-producing *Pseudomonas* spp. in take-all decline soils. Mol Plant Microbe Interact 11:144–152

Raaijmakers JM, Paulitz TC, Alabouvette C, Steinberg C, Moënne-Loccoz Y (2009) The rhizosphere: a playground and battlefield for soilborne pathogens and beneficial microorganisms. Plant Soil 321:341–361

Ramette A, Tiedje JM (2007) Biogeography: an emerging cornerstone for understanding prokaryotic diversity, ecology, and evolution. Microb Ecol 53:197–207

Ramey BE, Koutsoudis M, von Bodman SB, Fuqua C (2004) Biofilm formation in plant-microbe associations. Curr Opin Microbiol 7:602–609

Rancès E, Voronin D, Tran-Van V, Mavingui P (2008) Genetic and functional characterization of the type IV secretion system in *Wolbachia*. J Bacteriol 190:5020–5030

Reinhardt D (2007) Programming good relations – development of the arbuscular mycorrhizal symbiosis. Curr Opin Plant Biol 10:98–105

Remans R, Spaepen S, Vanderleyden J (2006) Auxin signaling in plant defense. Science 313:171

Rentz JA, Alvarez PJJ, Schnoor JL (2005) Benzo[*a*]pyrene cometabolism in the presence of plant root extracts and exudates: implications for phytoremediation. Environ Pollut 136:477–484

Rep M, Meijer M, Houterman PM, van der Does HC, Cornelissen BJC (2005) *Fusarium oxysporum* evades I-3-mediated resistance without altering the matching avirulence gene. Mol Plant Microbe Interact 18:15–23

Rezzonico F, Binder C, Défago G, Moënne-Loccoz Y (2005) The type III secretion system of biocontrol *Pseudomonas fluorescens* KD targets the phytopathogenic Chromista *Pythium ultimum* and promotes cucumber protection. Mol Plant Microbe Interact 18:991–1001

Roger PA, Zimmerman WJ, Lumpkin TA (1993) Microbiological management of wetland rice fields. In: Metting B (ed) Soil microbial ecology: applications in agricultural and environmental management. Marcel Dekker Inc., New York, pp 417–455

Rojas M, Olivier M, Gros P, Barrera LF, García LF (1999) TNF-alpha and IL-10 modulate the induction of apoptosis by virulent *Mycobacterium tuberculosis* in murine macrophages. J Immunol 162:6122–6131

Rosenzweig C, Iglesias A, Yang XB, Epstein PR, Chivian E (2000) Climate change and US agriculture: the impacts of warming and extreme weather events on productivity, plant diseases, and pests. Center for Health and the Global Environment, Boston. http://www.med.harvard.edu/chge

Roth RR, James WD (1988) Microbial ecology of the skin. Annu Rev Microbiol 42:441–464

Roulston A, Marcellus RC, Branton PE (1999) Viruses and apoptosis. Annu Rev Microbiol 53:577–628

Roy S, Khasa DP, Greer CW (2007) Combining alders, frankiae, and mycorrhizae for the revegetation and remediation of contaminated ecosystems. Can J Bot 85:237–251

Saint André AV et al (2002) The role of endosymbiotic *Wolbachia* bacteria in the pathogenesis of river blindness. Science 295:1892–1895

Salaun B, Romero P, Lebecque S (2007) Toll-like receptors' two-edged sword: when immunity meets apoptosis. Eur J Immunol 37:3311–3318

Sandkvist M (2001) Biology of type II secretion. Mol Microbiol 40:271–283

Santos AV, Dillon RJ, Dillon VM, Reynolds SE, Samuels RL (2004) Occurrence of the antibiotic producing bacterium *Burkholderia* sp. in colonies of the leaf-cutting ant *Atta sexdens rubropilosa*. FEMS Microbiol Lett 15:319–323

Savage D, Barbetti MJ, MacLeod WJ, Salam MU, Renton M (2012) Seasonal and diurnal patterns of spore release can significantly affect the proportion of spores expected to undergo long-distance dispersal. Microb Ecol 63:578–585

Schuffenecker I et al (2006) Genome microevolution of chikungunya viruses causing the Indian Ocean outbreak. PLoS Med 3:e263

Shigenobu S, Watanabe H, Hattori M, Sakaki Y, Ishikawa H (2000) Genome sequence of the endocellular bacterial symbiont of aphids *Buchnera* sp. APS Nat 407:81–86

Silver AC, Kikuchi Y, Fadl AA, Sha J, Chopra AK, Graf J (2007) Interaction between innate immune cells and a bacterial type III secretion system in mutualistic and pathogenic associations. Proc Natl Acad Sci U S A 104:9481–9486

Simon L, Bousquet J, Levesque RC, Lalonde M (1993) Origin and diversification of endomycorrhizal fungi and coincidence with vascular land plant. Nature 363:67–69

Smadja B, Latour X, Faure D, Chevalier S, Dessaux Y, Orange N (2004) Involvement of N-acylhomoserine lactones throughout plant infection by *Erwinia carotovora* subsp. *atroseptica* (*Pectobacterium atrosepticum*). Mol Plant Microbe Interact 17:1269–1278

Snaidr J, Amann R, Huber I, Ludwig W, Schleifer KH (1997) Phylogenetic analysis and in situ identification of bacteria in activated sludge. Appl Environ Microbiol 63:2884–2896

Socha A, Long R, Rowley D (2007) Bacillamides from a hypersaline microbial mat bacterium. J Nat Prod 70:1793–1795

Srivastava A et al (2005) The apoptotic response to pneumolysin is Toll-like receptor 4 dependent and protects against pneumococcal disease. Infect Immun 73:6479–6487

Stappenbeck TS, Hooper LV, Gordon JI (2002) Developmental regulation of intestinal angiogenesis by indigenous microbes via Paneth cells. Proc Natl Acad Sci U S A 99:15451–15455

Stolp H, Starr MP (1963) *Bdellovibrio bacteriovorus* gen. et sp. n., a predatory, ectoparasitic, and bacteriolytic microorganism. Anton Leeuw Int J G 29:217–248

Sullivan JT, Patrick HN, Lowther WL, Scott DB, Ronson CW (1995) Nodulating strains of *Rhizobium loti* arise through chromosomal symbiotic gene transfer in the environment. Proc Natl Acad Sci U S A 92:8985–8989

Sundaramurthy V, Pieters J (2007) Interactions of pathogenic mycobacteria with host macrophages. Microbes Infect 9:1671–1679

Taylor MJ (2003) *Wolbachia* in the inflammatory pathogenesis in human filariasis. Ann N Y Acad Sci 990:444–449

Taylor MJ, Bilo K, Cross HF, Archer JP, Underwood AP (1999) 16S rDNA phylogeny and ultrastructural characterization of *Wolbachia* intracellular bacteria of the filarial nematode *Brugia malayi, B. pahangi*, and *Wuchereria bancrofti*. Exp Parasitol 91:356–361

Teplitski M, Robinson JB, Bauer WD (2000) Plants secrete substances that mimic bacterial N-acyl homoserine lactone signal activities and affect population density-dependent behaviors in associated bacteria. Mol Plant Microbe Interact 13:637–648

Thanassi DG, Hultgren SJ (2000) Multiple pathways allow protein secretion across the bacterial outer membrane. Curr Opin Cell Biol 12:420–430

Troxler J, Svercel M, Natsch A, Zala M, Keel C, Moënne-Loccoz Y, Défago G (2012) Persistence of a biocontrol *Pseudomonas* inoculant as high populations of culturable and non-culturable cells in 200-cm-deep soil profiles. Soil Biol Biochem 44:122–129

Turner JD, Mand S, Debrah AY, Muehlfeld J, Pfarr K, McGarry HF, Adjei O, Taylor MJ, Hoerauf A (2006) A randomized, double-blind clinical trial of a 3-week course of doxycycline plus albendazole and ivermectin for the treatment of *Wuchereria bancrofti* infection. Clin Infect Dis 42:1081–1089

Tzfira Y, Citovsky V (2006) *Agrobacterium*-mediated genetic transformation of plants: biology and biotechnology. Curr Opin Biotechnol 17:147–154

Van de Velde W et al (2010) Plant peptides govern terminal differentiation of bacteria in symbiosis. Science 327:1122–1126

van der Maarel E (2005) Vegetation ecology. Blackwell Science, Oxford

van Elsas JD, Jansson JK, Trevors JT (2007) Modern soil microbiology. CRC Press, Boca Raton

van Ham RC, Kamerbeek J, Palacios C (2003) Reductive genome evolution in *Buchnera aphidicola*. Proc Natl Acad Sci U S A 100:581–586

van Loon LC, Rep M, Pieterse CMJ (2006) Significance of inducible defense-related proteins in infected plants. Annu Rev Phytopathol 44:135–162

Van Valen L (1973) A new evolutionary law. Evol Theor 1:1–30

Vandamme P, Govan J, LiPuma J (2007) Diversity and role of *Burkholderia* spp. In: Coenye T, Vandamme P (eds) *Burkholderia*: molecular microbiology and genomics. Horizon Bioscience, Wymondham, pp 1–28

Vavre F, Kremer N, Pannebakker BA, Mavingui P (2008) Is symbiosis influenced by pleiotropic role of programmed cell death in immunity and development? In: Bourtzis K, Miller TA (eds) Insect symbiosis, vol 3. CRC Press/Taylor and Francis Group, Boca Raton, pp 57–75

Vaux DL, Haecker G, Strasser A (1994) An evolutionary perspective on apoptosis. Cell 76:777–779

Velmurugan K et al (2007) *Mycobacterium tuberculosis nuoG* is a virulence gene that inhibits apoptosis of infected host cells. PLoS Pathol 3:e110

Vial L, Lavire C, Mavingui P, Blaha D, Haurat J, Moënne-Loccoz Y, Bally R, Wisniewski-Dyé F (2006) Phase variation and genomic architecture changes in *Azospirillum*. J Bacteriol 188:5364–5373

Villate L, Morin E, Demangeat G, Van Helden M, Esmenjaud D (2012) Control of *Xiphinema index* populations by fallow plants under greenhouse and field conditions. Phytopathology 102:627–634

Waksman SA (1969) Successes and failures in the search for antibiotics. Adv Appl Microbiol 11:1–16

Waller PJ, Larsen M (1993) The role of nematophagous fungi in the biological control of nematode parasites of livestock. Int J Parasitol 23:539–546

Watsuji TO, Kato T, Ueda K, Beppu T (2006) CO_2 supply induces the growth of *Symbiobacterium thermophilum*, a syntrophic bacterium. Biosci Biotechnol Biochem 70:753–756

Weinrauch Y, Zychlinsky A (1999) The induction of apoptosis by bacterial pathogens. Annu Rev Microbiol 53:155–187

Werren JH, Windsor D, Gao L (1995) Distribution of *Wolbachia* among neotropical arthropods. Proc R Soc Lond B 262:197–204

Werren JH, Zhang W, Guo LR, Stouthamer R, Breeuwer JA, Hurst GD (1999) *Wolbachia pipientis*: microbial manipulator of arthropod reproduction. Annu Rev Microbiol 53:71–102

Whitman WB, Coleman DC, Wiebe WJ (1998) Prokaryotes: the unseen majority. Proc Natl Acad Sci U S A 95:6578–6583

Williams GT (1994) PCD: a fundamental protective response to pathogens. Trends Microbiol 12:463–464

Wimpenny JWT, Colasanti R (1997) A unifying hypothesis for the structure of microbial biofilms based on cellular automaton models. FEMS Microbiol Ecol 22:1–16

Wirth DF (2002) The parasite genome: biological revelations. Nature 419:495–496

Wong FY, Stackebrandt E, Ladha JK, Fleischman DE, Date RA, Fuerst JA (1994) Phylogenetic analysis of *Bradyrhizobium japonicum* and photosynthetic stem-nodulating bacteria from *Aeschynomene* species grown in separated geographical regions. Appl Environ Microbiol 60:940–946

Woolhouse ME, Haydon DT, Antia R (2005) Emerging pathogens: the epidemiology and evolution of species jumps. Trends Ecol Evol 20:238–244

Wu M et al (2004) Phylogenomics of the reproductive parasite *Wolbachia pipientis* wMel: a streamlined genome overrun by mobile genetic elements. PLoS Biol 2:327–341

Xu J, Bjursell MK, Himrod J, Deng S, Carmichael LK, Chianq HC, Hooper LV, Gordon JI (2003) A genomic view of the human-*Bacteroides thetaiotaomicron* symbiosis. Science 299:2074–2076

Yamamoto H, Fujikura K, Hiraishi A, Kato K, Mayi Y (2002) Phylogenetic characterization and biomass estimation of bacterial endosymbionts associated with invertebrates dwelling in chemosynthetic communities of hydrothermal vents and cold seeps. Mar Ecol Prog Ser 245:61–67

Zenjari B, El Hajjouji H, Ait Baddi G, Bailly JR, Revel JC, Nejmeddine A, Hafidi M (2006) Eliminating toxic compounds by composting olive mill wastewater-straw mixtures. J Hazard Mater 138:433–437

Zeph LR, Casida LE (1986) Gram-negative versus Gram-positive (Actinomycete) nonobligate bacterial predators of bacteria in soil. Appl Environ Microbiol 52:819–823

Zhanq Y, Gao W, Meng Q (2007) Fermentation of plant cell walls by ruminal bacteria, protozoa and fungi and their interaction with fibre particle. Arch Anim Nutr 61:114–125

Zychlinsky A, Sansonetti PJ (1997) Perspectives series: host/pathogen interactions. Apoptosis in bacterial pathogenesis. J Clin Invest 100:493–495

Horizontal Gene Transfer in Microbial Ecosystems

Céline Brochier-Armanet and David Moreira

Abstract

Microorganisms live in fluctuating environments. At the microscopic scale, their habitats are contrasted and have highly variable physical, chemical, and biological parameters. The versatility of microbial ecosystems implies that microorganisms must constantly adapt to these. Adaptation can result from spontaneous point mutations of the genetic material that creates genetic diversity at the population level. However, this mechanism by itself is not capable of explaining the extraordinary adaptive capacity of microorganisms. Research performed in the first part of the twentieth century has revealed that this capacity results very often from the ability to import genetic material from other microorganisms. This phenomenon is called horizontal (or lateral) gene transfer (HGT). The mechanisms of HGTs and their evolutionary consequences are discussed.

Keywords

Adaptation • *Archaea* • *Bacteria* • Bioinformatics • Conjugation • Eukaryotes • Evolution • Genome evolution • Phylogeny • Transformation • Transduction

12.1 Introduction

Microorganisms live in fluctuating environments. At the microscopic scale, their habitats are contrasted and highly variable in their physical, chemical, and biological conditions. The versatility of microbial ecosystems implies that microorganisms must constantly adapt to them. Adaptation can result from spontaneous point mutations of the genetic material that create genetic diversity at the population level. However, this mechanism by itself is not capable of explaining the extraordinary adaptive capacity of microorganisms. Research performed in the first part of the twentieth century has revealed that this capacity results very often from the ability to import genetic material from other microorganisms. This phenomenon is called horizontal (or lateral) gene transfer (HGT). The mechanisms of HGTs and their evolutionary consequences will be discussed in this chapter.

12.2 Mechanisms of Horizontal Gene Transfer

12.2.1 DNA Acquisition Through Horizontal Gene Transfer

12.2.1.1 Transformation

Transformation is the mechanism by which naked DNA present in the environment penetrates into cells, which are called receptor cells (for additional information, see Chen and Dubnau 2004; Thomas and Nielsen 2005). This capacity is very important for microorganisms because DNA can be a source of nutrients (i.e., phosphorus, carbon, nitrogen) for

* Chapter Coordinator

C. Brochier-Armanet* (✉)
Laboratoire de Biométrie et Biologie Évolutive, UMR CNRS 5558, Université Claude Bernard Lyon 1, 69622 Villeurbanne Cedex, France
e-mail: celine.brochier-armanet@univ-lyon1.fr

D. Moreira
Unité d'Écologie, Systématique et Évolution, UMR CNRS 8079, Université Paris-Sud, 91405 Orsay Cedex, France

cells. In addition, the acquired genetic material (1) can be used as template for DNA repair if the foreign DNA originated from a representative of the same species and/or (2) can carry genes (or alleles) conferring a selective advantage (e.g., resistance to antibiotics, new metabolic capabilities, virulence factors, etc.) to the recipient cell. The DNA involved in transformation, called transforming DNA, may have various origins: lysed or decomposed cells, dislocated viral particles, living cells secreting DNA, etc. While the transforming DNA is usually of chromosomal origin, the transfer of plasmids is also possible but involves additional steps (see below). The rate of transformation in natural populations is very difficult to measure, because it is likely linked to the persistence of DNA in the environment. The half-life of DNA depends on several factors such as temperature, pH, presence of enzymes degrading DNA, etc. It can vary from a few minutes to several weeks depending on the environments.

Transformation was discovered by Griffith (1928; Downie 1972), who studied in the 1920s the virulence of *Streptococcus pneumoniae* (*Firmicutes*) in mice. He showed that the injection of *S. pneumoniae* S1, a virulent strain carrying capsular antigens of type S1, caused the death of animals. In contrast, mice survived when the cells were previously killed by heat. In parallel, Griffith realized another series of experiments using *S. pneumoniae* R2, a non-virulent strain derived from the virulent strain S2, which had lost its capsular antigens. He showed that the simultaneous injection of killed S1 cells together with living R2 cells caused the death of animals. The examination of dead mice revealed living *S. pneumoniae* S1 cells. Based on these results, Griffith hypothesized that the R2 cells have acquired S1 capsular antigens from the dead S1 cells. This acquisition would have restored the virulence of the R2 strain. The principle of transformation has been correctly formulated by Avery and colleagues in the 1940s (Avery et al. 1944; Downie 1972).

Transformation has been put in evidence in prokaryotes belonging to very different bacterial phyla: *Cyanobacteria*, *Firmicutes*, *Deinococcus-Thermus*, *Chlorobi*, etc., and in *Archaea* (Lorenz and Wackernagel 1994). It requires that the recipient cells reach a genetically programmed stage of development called **competence***, which involves 20–50 proteins. To a few exceptions (e.g., *Neisseria gonorrhoeae*, *Betaproteobacteria*), natural competence is a transient state depending on environmental conditions (e.g., nutrient depletion, high cell density, etc.). Entry and exit into competence, as well as the proportion of cells entering into competence within a population, are influenced by environmental and genetic factors, which vary from one species (or strain or isolate) to another (Maughan and Redfield 2009). For instance, in the case of *Bacillus subtilis* (*Firmicutes*), 20 % of cell population on average enters into competence for a period lasting several hours; in the case of *Streptococcus*, the whole population is capable of entering into competence but remains in this state only for a few minutes. This explains why it is so difficult to determine whether a species, strain, or isolate is capable of natural transformation or not. Nevertheless, current data suggest that only 1 % of bacterial described species are capable of natural transformation. It is important to note that while many microbial species (e.g., *Escherichia coli* strains, *Gammaproteobacteria*) are not naturally transformable, artificial methods of transformation based on the alteration of the permeability of membranes (e.g., electroporation) have been developed and are routinely used in the laboratory. It has been proposed that electrical storms could create comparable physicochemical conditions in soils and result in transformation by electroporation of cells present in soil (Ceremonie et al. 2004).

The second step of transformation involves the non-covalent binding of the DNA present in the environment to DNA-binding proteins (e.g., ComAE proteins in *Bacillus subtilis*, *Firmicutes*) located at the surface of the cell (Fig. 12.1a). DNA must be double-stranded and present a free end. In Gram-negative bacteria, the outer membrane is an additional obstacle to the entry of transforming DNA into the cells. Secretins (e.g., PilQ in *N. gonorrhoeae*, a betaproteobacterium) form a channel and allow the transforming DNA to cross the outer membrane. Other mechanisms involve small membrane vesicles, called transformasomes, protruding on the external surface of the cell. The transforming DNA is captured by these vesicles and transported into the periplasmic space. Some microorganisms are able to filter selectively the incoming DNA during its translocation. This selective step involves the recognition of specific 10 base pair sequences called uptake signal sequences (USS), such as 5'-GCCGTCTGAA-3' in *N. gonorrhoeae* or 5'-AAGTGCGGT-3' in *Haemophilus influenzae*, a gammaproteobacterium (Danner et al. 1980; Goodman and Scocca 1988). USS are highly repeated in the genomes of the species carrying them, whereas according to their relative large size, the expected occurrence of a given USS in any genome is low. For instance, 1,900 copies are present in *N. gonorrhoeae* and 1,471 in *H. influenzae* (Smith et al. 1995). Their overrepresentation in these genomes suggests a positive selection pressure to maintain a high number. It has been shown that the recipient cells translocate preferentially transforming DNA carrying USS similar to their own USS. This suggests that USS recognition systems exist. Even if they have not been identified yet, in *N. gonorrhoeae* their activity appears to depend on the pilus-coding genes (Chen and Dubnau 2004). It is important to note that the existence of USS seems to be an exception rather than a rule because many microorganisms are able to translocate efficiently transforming DNA regardless of its sequence. The entry rate of foreign DNA into cells is modulated by the presence of endonucleases (e.g., NucA in *B. subtilis*) which create double-strand breaks in the transforming DNA (third step, Fig. 12.1b). This DNA fragmentation step limits the effectiveness of the transfer of plasmids, which must be reconstituted at the end of the transformation process to be successful. The fourth step of transformation consists in the translocation of the foreign DNA through the plasma membrane (Fig. 12.1b). Some data suggest that this process can be very fast: 100 bases per second in *S. pneumoniae* (*Firmicutes*), 60 bases per second in *Acinetobacter baylyi* (*Gammaproteobacteria*). It usually involves proteins which are homologous to components of **type IV pili*** and of type II secretion systems (Chen and Dubnau 2004). However, components of other systems can also be used, such as the type IV secretion system in *Helicobacter pylori*. This step also requires a permease (e.g., ComCE in *B. subtilis*) and a DNA-binding protein (e.g., ComFA in *B. subtilis*). During its translocation across the cytoplasmic membrane, the double-stranded DNA is usually converted into single-stranded DNA (Fig. 12.1b). Thus, only one of the two strands penetrates in the cytoplasm, while the second strand is degraded into nucleotides that are released to the extracellular medium (in Gram-positive bacteria) or to the periplasm (in Gram-negative bacteria). The fifth and final step of transformation consists in the integration of the transforming DNA into the chromosome of the recipient (Fig. 12.1c, d) or the reconstitution of the plasmid if the transforming DNA was plasmid.

12.2.1.2 Conjugation

In contrast with transformation and transduction, conjugation requires cell-to-cell contact or the establishment of a bridge between the donor and the recipient cells. Accordingly, this mechanism is often compared to a form of sexuality in prokaryotes. Physical contact between bacteria was known for microbiologists for a long time, but it is only in the 1940s that the principle of DNA transfer via conjugation was put in evidence (Lederberg 1947; Tatum and Lederberg 1947). At that time, the microbiologists Lederberg and Tatum tried to understand the emergence of new phenotypes from mixed cultures of $B^-M^-P^+T^+$ and $B^+M^+P^-T^-$ *E. coli* K-12 strains. These two strains lacked the capacity to synthesize some coenzymes and essential amino acids: (B) biotin, (M) methionine, (P) phenylalanine, and (T) threonine. Due to their various auxotrophies, none of these strains should have been able of growth on minimal medium. However, the coculture of these two strains on minimal medium led to the development of colonies, indicating the appearance of $B^+M^+P^+T^+$ strains. Lederberg and Tatum ruled out the possibility of reversions because they considered that the simultaneous reversion of different characters was highly

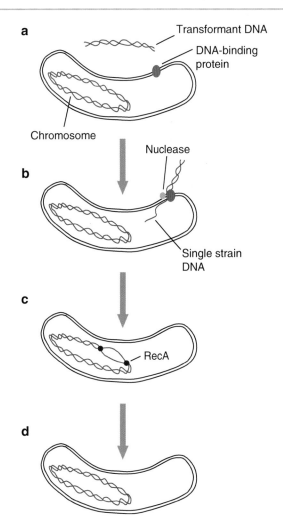

Fig. 12.1 The transformation. (**a**) A double-stranded DNA fragment from the environment touches a competent cell. It is recognized by a DNA-binding membrane protein. (**b**) A nuclease degrades one of the two DNA strands, while the other strand is transported inside the cell. (**c**) The DNA may be integrated into the host genome by various mechanisms, such as the homologous recombination that involves the RecA protein. (**c**) The transforming DNA is stably integrated into the genome of the recipient cell (**d**)

unlikely, and instead, they proposed that character recombination events occurred between the two parental strains. They also proposed that this phenomenon should be different from that described by Griffith few years ago, because the reproduction of Griffith's experiments on *E. coli* K-12 did not lead to the appearance of $B^+M^+P^+T^+$ strains (remind that *E. coli* is not naturally transformable, see above) (Tatum and Lederberg 1947). Few years later, Davis showed that the physical contact between the parental $B^-M^-P^+T^+$ and $B^+M^+P^-T^-$ strains was required for the emergence of $B^+M^+P^+T^+$ strains. Indeed, $B^+M^+P^+T^+$ colonies did not appear when the two parental strains were isolated by a barrier allowing the transit of small molecules but not the passage of cells (Davis 1950).

An important difference between natural transformation and conjugation is that the latter allows transferring plasmids. Plasmids are double-stranded DNA autonomous genetic elements able of self-replication, meaning that they do not need to be integrated into the chromosome to be replicated in cells. Most plasmids are small and circular. They are found in *Bacteria*, but also in *Archaea* and *Eucarya*. Besides replication, plasmids carry genes involved in various processes such as antibiotic resistance, detoxification, heavy metal resistance, etc., that may help organisms carrying plasmids to survive in changing environments. Among plasmids, conjugative plasmids encode the genes required for conjugation (Thomas and Nielsen 2005). In contrast, non-conjugative plasmids are not able to initiate the conjugation process, even if they can be transferred with the assistance of conjugative plasmids. In fact, the transfer of chromosomes is also possible via conjugation, but it involves their mobilization by a conjugative plasmid called episome. Several mechanisms of conjugation have been described, among which the best known occurred in Gram-negative bacteria. It allows the transfer of large plasmids and requires the formation of sexual **pilus*** by the type IV secretion system (Cascales and Christie 2003) which allows the donor cell to come into contact with the recipient cell (Fig. 12.2a). The contraction of the pilus brings the two cells in close contact (Fig. 12.2b). The contact of the cell membranes leads to the formation of a pore through which the conjugative plasmid will be transferred (Fig. 12.2c). The transfer is initiated by the cleavage of one of the two DNA strands of the plasmid with an enzyme encoded by the plasmid itself. This protein has also a helicase activity ensuring the separation of the two DNA strands of the plasmid. One of them is transferred to the recipient cell through the channel (Fig. 12.2c). A DNA polymerase synthesizes the complementary strand during the entry of the DNA strand in the cell through a process called rolling circle replication (Fig. 12.2d). A similar process allows synthesizing the complementary DNA strand of the single DNA strand that remains into the donor cell (Fig. 12.2e). At the end of the replicative process, both cells harbor an identical copy of the original plasmid (Fig. 12.2f). Because conjugation involves genes that are carried by the plasmid, their acquisition allows the recipient cell to be able to transfer the plasmid in turn. The conversion of recipient cells into donor cells allows the efficient dissemination of conjugative plasmids in microbial populations. However, in the absence of a positive selection pressure for their maintenance, plasmids can disappear from the population relatively quickly and especially if they belong to the same incompatibility group (Novick 1987).

As mentioned above, conjugation may promote the horizontal transfer of chromosomes provided that they are mobilized by conjugative plasmids called episomes. In

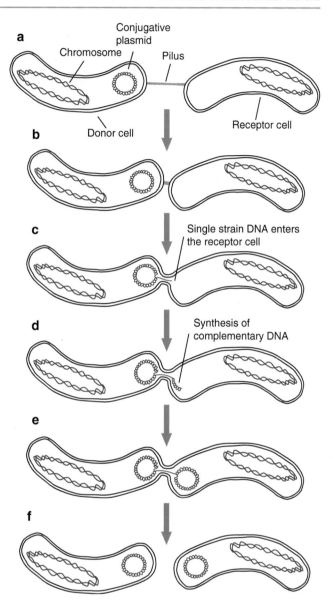

Fig. 12.2 The conjugation. (**a**) A contact between a donor cell containing a conjugative plasmid and a recipient cell is established by means of a pilus. (**b**) The contraction of the pilus brings the two cells into physical contact. (**c**) The contact between the membranes of the donor and the recipient cells induces the formation of a channel. At the same time, one of the two DNA strands of the plasmid is cut and transferred to the recipient cell. (**d**) During the transfer, a replication process called rolling circle replication allows synthesizing the complementary strand of the transferred single-strand DNA into the recipient cell. (**e**) The same mechanism is responsible for the synthesis of the complementary strand of the single-strand DNA plasmid that remains in the donor cell. The conjugation process continues until the complete achievement of the transfer of the plasmid from the donor to the recipient cell. (**f**) At the end of the process, each cell harbors a copy of the plasmid (**f**)

principle, the transfer of complete circular chromosomes is possible, but this phenomenon is unlikely in natural conditions because it takes time (more than one hour), and in most cases the physical contact between the donor and the recipient cells is broken before the achievement of the process. Episomes carry specific sequences called insertion sequences. If a chromosome possesses similar insertion sequences, homologous recombination events can occur between the chromosome and the episome, leading to the insertion of the plasmid into the chromosome. Once the episome is integrated, the entire chromosome becomes mobilizable and thus transferrable via the conjugative process. Cells with a conjugative plasmid integrated into their chromosomes are called HFR for high-frequency recombination because they are able to transfer part of their chromosomes to recipient cells by conjugation. HFR bacteria have played an important role in the history of microbiology by allowing establishing the first physical maps of bacterial chromosomes. In fact, it is possible to determine the relative position of genes in chromosomes by measuring the frequency of their cotransfer. For example, if *E. coli* K-12 markers B^+ and M^+ are systematically cotransferred during conjugation, this indicates that the corresponding loci are close on the chromosome. Conversely, if two loci are distant on the chromosome, their cotransfer during conjugation will be less frequent as the probability that the two cells are separated before the transfer of the second marker is high. Precise chromosomal maps have been established by measuring the time required for the transfer of each genetic marker. Loci located near to the episome insertion site penetrate first into the recipient cell and are thus transferred faster than loci located at the other end. The insertion of an episome into a chromosome is reversible as the plasmid can be excised by reverse recombination events involving the same insertion sequences. Because plasmids and chromosomes generally possess multiple copies of these insertion sequences, excision events can occur in such a way that chromosome fragments carrying chromosomal genes can be excised with the episome. If the resulting episome is involved in new conjugation events, the genes of chromosomal origin can be transferred to the recipient cell.

At the time of its discovery, conjugation was considered a mechanism of horizontal gene transfer occurring among cells belonging to the same species. This idea was questioned after the discovery of similar conjugative plasmids in cells belonging to very different taxonomic groups. Such taxonomic distribution suggested that conjugation could occur among distant microbial species. This hypothesis has been confirmed by laboratory experiments which have also showed that conjugation can occur among organisms from different domains (e.g., from bacterial to yeast cells, namely, between bacteria and eukaryotes) (Heinemann and Sprague 1989).

12.2.1.3 Transduction

This mechanism consists in transferring DNA between bacterial cells via bacteriophages. Bacteriophages were discovered in 1915 (Twort 1915). At that time, Twort did not make the link between these agents and the viruses infecting the cells of plants and animals that had been described a few years earlier (Duckworth 1976). This link was made few years later by Felix d'Herelle, who isolated agents capable of destroying the bacteria responsible of dysentery. He proposed to call bacteriophages the viruses having the capacity of killing bacteria (d'Herelle 1917). These viruses would have been now called "lytic phages" because they are able to induce the immediate lysis of the target cells through a lytic cycle. They are opposed to the "temperate phages," which are able to integrate at a specific site of the genome of the target cells, where they can remain silent for many generations. Temperate phages are able to induce a lysogenic cycle in response to an environmental stimulus or stress. In that case, the genome of the phage is excised from the bacterial chromosome and the lytic cycle is initiated. This leads to the lysis of the cell and the release of new phage particles.

The involvement of bacteriophages in HGT was demonstrated later by Zinder during his 2-year stay in the laboratory headed by Lederberg (Zinder 1992). His work aimed at extending the work of Lederberg on bacterial conjugation. To do so, he worked on *Salmonella typhimurium* strains (Zinder and Lederberg 1952) and observed that the crossing of strain LT-22 (auxotroph for phenylalanine and tryptophan, due to a double mutation in the pathway of aromatic amino acid synthesis) and strain LT-2 (auxotroph for methionine and histidine) produced recombinant phenotypes at higher frequencies than in the case of *E. coli* (10^{-5} compared to $10^{-6}/10^{-7}$). Rapidly, significant differences became apparent between the studied phenomenon and conjugation. First, the physical contact between the two partners was not required. Second, self-crosses between pairs of *S. typhimurium* mutants belonging to the same strain but carrying complementary mutations failed, whereas all the *E. coli* K-12 strains were interfertile. In addition, irrespective of the considered markers, the recombination rates were always constant (10^{-5}). Furthermore, in contrast to transformation, the filter-sterilized supernatants from the separate cultures were without effect, whereas the supernatant of a coculture could convert LT-22 but not LT-2 mutants. Finally, the recombinant colonies presented a genomic background similar to that of strain LT-22, suggesting that strain LT-22 was the recipient. Shortly after, Zinder and Lederberg put in evidence that the strain LT-22 carries a lysogenic phage (PLT-22 or P22) that can infect and lyse strain LT-2. During this process, genome fragments of LT-2 can be incorporated to the genomes of newly produced phages and transferred to LT-22 cells, restoring the aromatic amino acid biosynthesis pathway of this strain (Zinder 1992). This new mechanism of HGT was named by its discoverers "transduction."

Research works have revealed two types of transduction mechanisms: the generalized transduction and the specialized transduction (sometimes qualified as restricted transduction) (Thomas and Nielsen 2005). The generalized transduction involves virulent phages (lytic) and allows transferring any region of a bacterial genome. By contrast, phages involved in specialized transduction are called temperate. This mechanism allows transferring specific regions of the bacterial chromosome. The generalized transduction starts with infection of a bacterium by a virulent phage that injects its DNA into the cell cytoplasm (Fig. 12.3a). The viral DNA is replicated by the replication machinery of the host cell (Fig. 12.3b), whereas the genome of the cell is degraded (Fig. 12.3c). At the same time, viral capsids are synthesized. The packaging of the newly synthesized genomes allows forming new viral particles (Fig. 12.3d). At this stage, portions of the genome of the infected cell can be packaged into capsids instead of viral genomes (Fig. 12.3d). The resulting particles are called transducing particles. The incorporation of some bacterial genetic material occurs in ~1 % of the new virions. During the lysis of the infected cells, transducing particles are released into the environment like other viral particles. They can bind to (Fig. 12.3e) and infect other cells (Fig. 12.3f). However, because transducing particles do not contain viral DNA, they are not replicated into the host cell and do not cause its lysis. In some cases, the DNA of the transducing particle is integrated through recombination processes into the chromosome of the host and can thus be transmitted to its offspring (Fig. 12.3g). If no recombination occurs, the genes carried by the DNA of the transducing particle can be expressed for a while, but the corresponding genomic fragment disappears rapidly because it will be transmitted to only one of the daughter cells during the division of the recipient cell. Therefore, it cannot be fixed in the cell population. This phenomenon is called abortive transduction.

The specialized transduction involves temperate phages that are able to integrate at a specific location of the cell chromosome. Integrated temperate phages are called prophages. They lose their replicative autonomy. They are replicated at the same time that the chromosome of the host and are transmitted to the two daughter cells during cell division. The integration sites are very specific, meaning that a given bacteriophage will always be integrated at the same location in the genome of its host. For instance, in E. coli, the integration site of the λ phage is located close to an operon involved in galactose degradation. A cell carrying an integrated phage is called a lysogen, and the replication cycle of the phage is called lysogenic. The viral genes can be expressed by the cell and impact its phenotype (this is called lysogenic conversion). For example, some phages encode genes repressing the expression of their own genes. The main consequence is that host cells become immunized against infection by phages of the same type.

The integration of prophages is not irreversible. They can be excised from the chromosome of the host spontaneously, but their excision can also be induced by environmental or physiological stimuli (e.g., exposure to UV or X-rays, stress of the host cell). The excision of the prophage initiates a lytic cycle, leading to the formation of virus particles that are released during the cell lysis. In some cases, the excision of the prophage can be abnormal, leading to the formation of a hybrid DNA molecule made of viral and host DNA. In the case of the λ phage of E. coli, a piece of the galactose operon can be integrated to the phage genome. In return, a part of the prophage genome remains integrated into the host chromosome. The viral particles carrying pieces of the host genome can serve as shuttle for HGT during the next cycle of infection. Indeed, because they contain a part of viral DNA, these phages are able to replicate in the newly infected cells, increasing the probability of inserting the genes they carry on the bacterial chromosome. The effectiveness of specialized transduction is thus much higher than that of generalized transduction, even if only a subset of genes can be transferred by this mechanism.

Viruses are extremely abundant in many natural environments. For instance, the amount of viral particles in marine environments can exceed by several orders of magnitude than that of cells (Suttle 2007). The main consequence is that marine organisms, and especially microorganisms, are constantly exposed to viral infections. Viruses are thus major players of the regulation of bacterial populations and of HGT.

12.2.2 Fate of the Transferred DNA

Following its entry into the recipient cell (Fig. 12.4a), the "foreign" (or exogenous) DNA is the target of a series of processes that determine its fate, namely, its maintenance in the recipient cell and the transmission to its offspring or its fast disappearance. First, the foreign DNA can be directly degraded (Fig. 12.4b) or be cut into pieces by the endonucleases present in the cytoplasm of the recipient cell (Fig. 12.4c). This cleavage step involves generally a system of modification/restriction (M/R) such as the *hsd* system of *E. coli*. Several M/R systems have been described in prokaryotes, but all involve a methylase and an endonuclease, also called restriction enzyme (for a detailed review of these systems, see Murray 2000). The functioning of these systems is simple. The methylase recognizes and methylates specific bases of the DNA. This methylation process protects the DNA molecule from the action of the endonuclease. The M/R systems are highly variable from one species to another and sometimes even from one strain to another. The main consequence is that the methylation process allows very specific DNA labeling and therefore an accurate distinction between

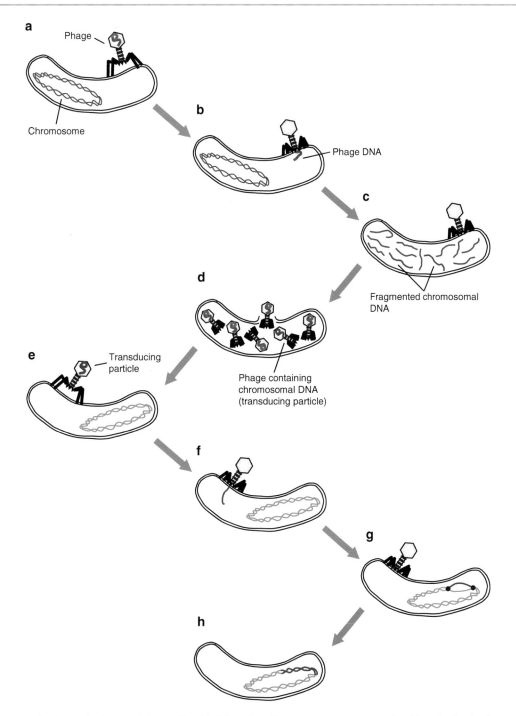

Fig. 12.3 The generalized transduction. (**a**) A bacterial cell is infected by a phage. (**b**) The phage injects its DNA into the cell. (**c**) The viral DNA is replicated many times by the replication machinery of the bacterium. The chromosome of the bacterium is fragmented and degraded. (**d**) Viral capsids are synthesized. The new copies of viral DNA are encapsulated to form new phages. This process ends with the lysis of the infected cell. At that step, fragments of the bacterial chromosome can be encapsulated by mistake, leading to the formation of transducing particles. (**e-f**) Transducing particles can infect new bacterial cells, leading to the injection of bacterial chromosomal DNA fragments in the recipient cell. (**g**) The DNA carried by transducing particles will not be replicated and may recombine with the chromosome of the newly infected cell. (**h**) Stable integration of the DNA of the transducing particle into the genome of the recipient

native and exogenous DNA. Indeed, most of the time the exogenous and the native DNA harbor different methylation patterns. Therefore, the exogenous DNA is not protected against the action of the endonucleases encoded by the host and is rapidly degraded (Fig. 12.4d). Overall, the M/R systems limit the frequency of HGT ensuring the cohesion of the genomes of recipient cells. In addition, it increases the resistance of the cells to phages. The importance of M/R

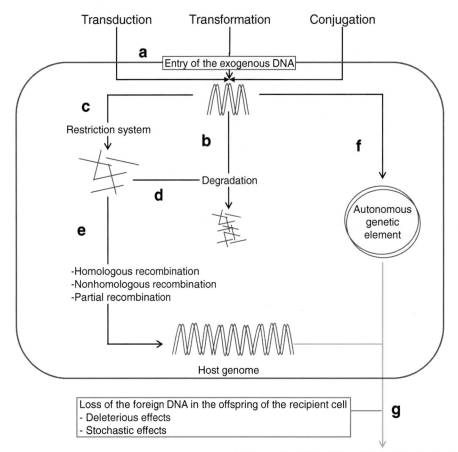

Fig. 12.4 Fate of the transferred DNA. (**a**) Entry of the exogenous DNA into the recipient cell by one of the three HGT mechanisms. (**b**, **d**) Degradation of the exogenous DNA by the restriction system of the host. (**c**) Fragmentation of the exogenous DNA by the restriction system of the host. (**e**) Stable integration of the exogenous DNA fragments into the host genome. (**f**) Maintenance of the exogenous DNA as an autonomous replicating element. (**g**) The transmission and fixation of the exogenous DNA in microbial population depends on many factors

systems is underlined by the fact that most prokaryotic genomes encode them. Moreover, many genomes encode several different M/R systems. For example, the genome of *H. pylori* contains eleven M/R systems, representing ~1 % of its genes. Despite being highly efficient, M/R systems are not infallible. For instance, if the exogenous DNA harbors methylation patterns similar to those of the recipient cell, it will not be degraded by the restriction system. When the foreign DNA is not methylated, it could be methylated by the methylases before being degraded by the endonucleases. If the exogenous DNA is single-stranded (as in the case of transformation), the synthesis of the complementary strand by the replication machinery of the recipient cell may allow it to escape to the M/R system of the host. Phages and plasmids have developed strategies to escape to the M/R systems. These include (1) the presence of specific modifications in their DNA that prevent the action of endonucleases, (2) the underrepresentation of sequences recognized by endonucleases, and (3) the presence of genes encoding proteins that interfere with the M/R systems.

The endonucleases of the M/S systems cut the exogenous double-stranded DNA. The ends of DNA resulting from the action of these endonucleases are then degraded by exonucleases. Eventually, these two processes lead to a significant reduction of the size of the exogenous DNA, preventing its integration in the genome of the host. Alternative systems of exogenous DNA recognition exist but will not be described here. Despite their effectiveness, it can happen that exogenous DNA fragments escape the defensive M/R systems of the cells.

The integration of exogenous DNA into the chromosome of the recipient cell involves recombination (Fig. 12.4e). Three mechanisms of recombination have been described: homologous, nonhomologous, and homology-facilitated illegitimate recombination. Homologous recombination is the most frequent mechanism. It occurs between regions (20–200 DNA base pairs) presenting a high similarity. These regions initiate a pairing between the two DNA molecules, leading to an exchange of strands. This mechanism involves proteins such as the recombinase RecA. This protein and its protein partners are usually involved in DNA repair. However, due to its activity, RecA also allows the integration of exogenous DNA. Proteins, such as MutS and MutL in *E. coli*, have antagonistic activities to RecA. The MutS/MutL/MutH proteins detect and correct mismatches in double-stranded DNA during replication, ensuring the fidelity of this process. The function of the MutS/MutL/MutH system is very important because the inactivation of *mutS* is responsible for the emergence of strains with very high mutation rates (Taddei et al. 1996). The relative level of expression of RecA compared to MutS/MutL has important consequences for the cell: a higher level of expression of RecA can increase the ability of the cell to initiate homologous recombination and thus to integrate foreign DNA. Stressful situations, through the modulation of the protein levels, can change the capacity of microorganisms to detect and repair point mutations and thus to maintain the integrity of their genetic material. At the population level, this phenomenon can lead to the emergence of individuals carrying beneficial adaptations. Then, HGT may facilitate the dissemination of these adaptations to a larger number of individuals.

The rate of homologous recombination varies from one organism to another. For instance, it has been shown in vitro that, under optimal conditions, it varies from 0.1 % in *A. baylyi* to 25–50 % in *B. subtilis* and *S. pneumoniae*. Because it requires a high level of sequence similarity between chromosomal and recombinant DNA, in most cases homologous recombination allows acquiring new alleles but not new genes. In some cases however, homologous recombination may lead to a gain of genetic material and thus to an increase of the size of the recipient cell genome. This may happen when the flanking regions of the exogenous DNA fragment are very similar to the sequence of the genome of the recipient cell. In this case, homologous recombination is initiated through the intermediary of the flanking regions leading to the integration of exogenous DNA into the genome of the recipient.

DNA recombination can occur even if chromosomal and exogenous DNA are not sufficiently similar to induce homologous recombination. This involves another process called nonhomologous (or illegitimate) recombination. Several genes acting in this mechanism have been identified in *E. coli*. It involves double-strand breaks and reassembly via the ends of DNA strands (Ikeda et al. 2004). In principle, illegitimate recombination can lead to the integration of exogenous DNA at any location of the recipient cell genome, except when the illegitimate recombination occurs via the activity of site-specific recombinases. Nevertheless, this process promotes the integration of new DNA fragments and thus leads to an increase of the genome size of the recipient. Laboratory experiments have shown that the frequency of illegitimate recombination is very low in the case of linear DNA and slightly higher in the case of circular DNA. The low frequency of illegitimate recombination events could result from selection pressures ensuring the conservation of the integrity of the genome of the recipient cell.

The third mechanism of recombination is called homology-facilitated illegitimate recombination (HFIR). It requires the presence of a single small region of high similarity between exogenous and chromosomal DNA (de Vries and Wackernagel 2002). Once recombination is initiated in the region of high similarity, it continues in the adjacent regions which present little or no similarity at all. As in the case of illegitimate recombination, HFIR leads to a net gain of DNA and thus can contribute to increase the size of the recipient genomes. In the case of plasmids (or circular DNA), this mechanism can result in their complete integration into the chromosome of the recipient cell. HFIR can lead to the integration of fragments of relatively large size (>1,000 bp). Experiments have shown that the effectiveness of HFIR is relatively low. However, it can significantly increase when more than one area of strong similarity are shared by the exogenous and host DNA (for additional details on HFIR, see Thomas and Nielsen 2005).

Irrespective of the mechanism of recombination involved, the integration of exogenous DNA into the genome of the host (Fig. 12.4e) or its maintenance as an autonomous genetic element (Fig. 12.4f) do not mean that the HGT has been successful. Indeed, the long-term persistence of the exogenous DNA in populations depends on several factors (Fig. 12.4g), such as possible deleterious effects resulting from its integration. Indeed, depending on its location, the integration of the exogenous DNA can lead to the disruption of important genes (or groups of genes or operons) and thus cause deleterious effects on the recipient cell and its offspring. In this case, due to natural selection, the transferred DNA will disappear rapidly at the same time as the cells that carry them. **Stochastic phenomena*** can also lead to the disappearance of the exogenous DNA more or less rapidly even if it carries alleles or genes conferring an advantage to the receiving cell. Finally, natural selection determines the long-term persistence of the exogenous DNA in populations.

In conclusion, to be successful, HGT must go throughout each of these steps, and therefore, only a minority succeeds. This means that there is a large difference between the number of HGT events measured in terms of amount of exogenous DNA entering into cells and the number of HGT events that can be detected in present-day microbial genomes. These two concepts are often wrongly confused.

12.3 Detection

The detection and quantification of HGT has been a hotly debated issue in the scientific community for many years. Indeed, at the end of the 1990s, the analysis of the first

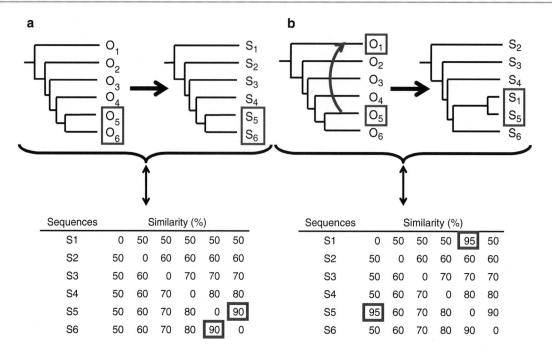

Fig. 12.5 HGT detection based on sequence similarity. The matrices show the percentage of similar residues observed between each pair of sequences. Phylogenies with leaves annotated O_1 to O_6 represent the phylogeny of organisms, while those with leaves annotated S_1 to S_6 represent phylogenies of the genes carried by organisms O_1 to O_6. (**a**) The high similarity observed between S_5 and S_6 sequences reflects the close relationship between O_5 and O_6. (**b**) The strong similarity observed between S_1 and S_5 sequences is unexpected because the O_1 and O_5 organisms are not closely related. This suggests that an HGT occurred between ancestors of O_5 and O_1 (*curved arrow*)

complete genome sequences revealed many cases of HGTs and that HGT rates might have been likely underestimated. This issue is now a major challenge in the genomic and post-genomic eras, as exemplified by the multiplication of the studies aiming at estimating the frequency of HGT in genomes and microbial populations. In fact, despite the apparent simplicity of the question (how many genes in a genome have been acquired through HGT?), providing an answer has turned out to be very difficult. Two main approaches have been developed to address this question. Methods of the first category are called extrinsic because they are based on the comparison of sequences from different genomes. They are presented in Sects. 12.3.1, 12.3.2, and 12.3.3. By contrast, methods belonging to the second category rely on the detection of irregularities within genome sequences. They are called intrinsic and are presented in Sect. 12.3.4.

12.3.1 Methods Based on Sequence Similarity

12.3.1.1 Principle

The more intuitive approach to detect HGT is based on the comparison of homologous sequences from different genomes. The rationale is that the more similar are two homologous sequences, the more related they are. Indeed, homologous sequences carried by two closely related organisms are expected to be more similar than sequences carried by more distantly related organisms (S_5 and S_6 on Fig. 12.5a). By contrast, if a genome contains a sequence that is more similar to the sequence from a distant relative than to the sequence from a closer relative, one can hypothesize that an HGT occurred between ancestors of the two distant organisms (S_1 and S_5 on Fig. 12.5b). The implementation of this approach consists in comparing the sequence of a given genome with sequences contained in databases. Most often, BLAST-based procedures are used. The comparison can be done at the nucleotide level if DNA sequences are directly compared or at the protein level if coding sequences are considered. In any case, the detection of highly similar sequences in distantly related organisms can highlight HGT cases.

12.3.1.2 Biases

The major limitation of HGT detection approaches based on sequence similarity is linked to the confusion between sequence similarity and relationship. Indeed, these two notions are equivalent if and only if the considered sequences evolve at the same rate (Fig. 12.6a), namely, if they verify the molecular clock assumption (Box 5.3). However, this condition is rarely verified (Koski and Golding 2001), as each sequence usually presents its own evolutionary rate. The main consequence is that in most cases the similarity observed among sequences does not reflect their relationships, and the sequences harboring the

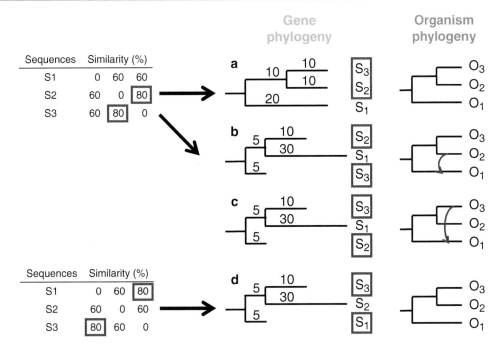

Fig. 12.6 Wrong inference of HGT due to sequence evolutionary rate differences. The matrices show the percentage of similar residues between each pair of sequences. Phylogenies with leaves annotated O_1 to O_3 represent the phylogeny of organisms, while those with leaves annotated S_1 to S_3 represent phylogenies of genes carried by organisms O_1 to O_3. Numbers at the branch represent the evolutionary distance between the sequences. The evolutionary distance is equal to 1 minus the observed similarity between sequences. The highest similarity observed between S_2 and S_3 sequences can reflect the close evolutionary relationship between O_2 and O_3 (**a**). In this case a close relationship between S_2 and S_3 is expected in the corresponding phylogeny. However, such a sequence similarity pattern is also compatible with an HGT from an ancestor of O_2 (**b**) or O_3 (**c**) to an ancestor of O_1, coupled with a high evolutionary rate of S_1. According to sequence similarity only, it is not possible to distinguish among these three scenarios. To do so, it is necessary to reconstruct the phylogeny of the sequences. Finally, the higher evolutionary distance observed between S_2 and S_3 than between S_1 and S_3, whereas O_3 is more closely related to O_2 than to O_1, can be wrongly interpreted as a the consequence of an HGT event, while it reflects only the higher evolutionary rate of S_2 (**d**)

highest similarity may not be the most related. Due to evolutionary rate variations, a sequence may present a greater evolutionary distance with its closest relatives than with more distant relatives, without the need to invoke any HGT event (Fig. 12.6d). Conversely, HGT may be not detected if the acquired sequence evolved faster than the other sequences (Fig. 12.6b, c). This point is important because evolutionary rate accelerations are frequently observed in sequences acquired by HGT after their integration in the recipient genome.

Biases related to differences in rates of evolution between sequences are aggravated by the fact that some taxonomic groups of organisms are underrepresented in sequence databases (Fig. 12.7). Indeed, all taxonomic groups are not of equal interest for the scientific community, leading to an overrepresentation of organisms presenting medical, industrial, or agronomic interests, easy to isolate or cultivate, abundant in the environment, etc. Such biases are exemplified by the heterogeneous number of complete genome sequences available for each phylum (Table 12.1).

Taxonomic sampling biases can lead to strong overestimation of HGT rates. Finally, sequence-based detection of HGT does not allow inferring the direction of the transfer. For instance, according to the case presented on Fig. 12.5b, it is not possible to say if the high sequence similarity observed between S_1 and S_5 is the consequence of an HGT from the ancestor of O_1 to an ancestor of O_5 or vice versa. Answering this question would require the reconstruction of the phylogeny of S_1, S_2, S_3, S_4, S_5, and S_6.

Because of multiple biases and weaknesses, the methods based on sequence similarity (including BLAST-based approaches) are too rough and not adapted to the accurate detection of HGT (Zhaxybayeva 2009).

12.3.2 Methods Based on Taxonomic Distribution

12.3.2.1 Principle

To overcome the biases of methods based on pair-wise sequence similarity, alternative approaches that take into account the taxonomic distribution of sequences have been proposed. These approaches are widely used to identify HGT in closely related genomes. The strategy consists in

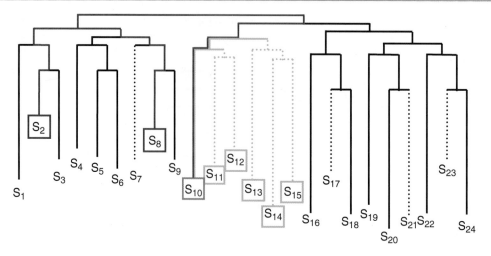

Fig. 12.7 Wrong inference of HGT due to poor or biased taxonomic sampling. Based on sequences available in public databases (branches in *solid lines*), S_{10} the sequence under study (in *blue*) displays the highest similarity with sequences S_2 and S_8 (in *red*), although it is more closely related to sequences S_{16}, S_{18}, S_{19}, S_{20}, S_{22}, and S_{24}. In fact, due to evolving rate differences, S_{10} appears as more similar to S_2 and S_8 that evolve slowly than to its closest relatives, which display higher evolutionary rates. Accordingly, based on sequence similarity, an HGT from lineages S_2 or S_8 to S_{10} will be postulated. This conclusion is however erroneous and could have been avoided if the phylogeny of the sequences had been rebuilt and/or if the taxonomic diversity of the studied groups had been better represented, that is to say, if sequences in green and/or S_{17} and S_{23} had been available

Table 12.1 Number of fully sequenced genomes deposited in the public database GenBank

Phylum/class	Nb of genomes	Phylum/class	Nb of genomes	Phylum/class	Nb of genomes
<u>Crenarchaeota</u>	53	Chlamydiae	113	Gammaproteobacteria	557
<u>Euryarchaeota</u>	101	Chlorobi	11	Gemmatimonadetes	1
<u>Korarchaeota</u>	1	Chloroflexi	20	Ignavibacteriae	2
<u>Nanoarchaeota</u>	1	Chrysiogenetes	1	Lentisphaerae	0
<u>Thaumarchaeota</u>	4	Cyanobacteria	74	Nitrospirae	4
Acidobacteria	8	Deferribacteres	4	Planctomycetes	6
Actinobacteria	274	Deinococcus-Thermus	20	Spirochaetes	58
Alphaproteobacteria	246	Delta/epsilon-proteobacteria	158	Synergistetes	5
Aquificae	12	Dictyoglomi	2	Tenericutes	77
Armatimonadetes	1	Elusimicrobia	1	Thermodesulfobacteria	3
Bacteroidetes	95	Fibrobacteres	2	Thermotogae	16
Betaproteobacteria	144	Firmicutes	581	Verrucomicrobia	5
Caldiserica	1	Fusobacteria	9		

Source: NCBI, 20 October 2013
Archaeal phyla are underlined

identifying core genes, i.e., the families of homologous genes (also called gene repertories) present in the genomes under study (Fig. 12.8). Together, these genes form the core genome. In contrast the pangenome is composed of all genes (i.e., core genes and other genes) present in the genome under study (*cf.* Sect. 12.4.1.2). Core genes are supposed to have been inherited from the last common ancestor of the organisms and preserved during the evolution of the group (Fig. 12.8a). In contrast, families represented in only one (or a few) genome are interpreted as the result of recent acquisitions via HGT of genes that were absent in the last common ancestor of the group under study (Fig. 12.8b). However, HGT is not the only process that can lead to such gene distribution patterns. They can correspond to genes of recent origin, namely, which emerged in a specific genome (or lineage). In this case, no homologue of the studied gene is expected to be present in other genomes (or lineages). By contrast, if homologues of the gene are detected in genomes (or lineages) other than those under study, the hypothesis of an acquisition through HGT would be the most likely (Fig. 12.8b).

12.3.2.2 Limits

Although its principle is very simple and intuitive, this approach may lead to erroneous results. First, the fact that a gene is common to all members of a group does not

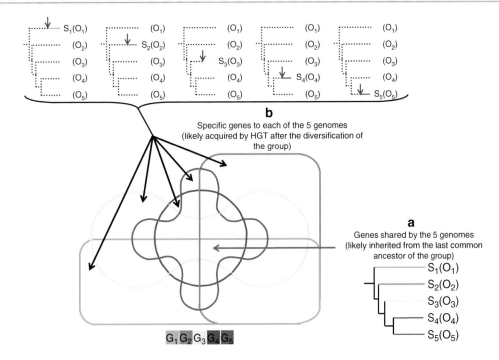

Fig. 12.8 Comparison of five closely related genomes. The Venn diagram shows the gene shared among the five organisms under study (O_1 to O_5). The overlapping areas correspond to families of homologous genes represented in more than one genome. (**a**) The central area of the Venn diagram indicated by a *gray arrow* corresponds to the gene families represented in all genomes. These genes are supposed to have been inherited from the last common ancestor of the five organisms. (**b**) The nonoverlapping areas of the Venn diagram (indicated by *black arrows*) correspond to genes that are specific to a given genome. Most of the time, these genes are interpreted as resulting from HGTs that occurred during the diversification of the group (*red arrows*)

guarantee that it has not been affected by HGT. Indeed, a gene inherited from the last common ancestor of a given lineage may have been replaced in one or several of its descendants (Fig. 12.9a). This phenomenon is called homologous gene replacement, because the native gene is lost after the acquisition by HGT of a homologous gene from a distant relative. Furthermore, the presence of a gene in a single genome does not imply that it has been systematically acquired by HGT. Depending on the branching position of the considered lineage, gene losses and HGT can provide similar gene taxonomic distributions (Fig. 12.9b, c). Most of the time, it is difficult to discriminate between hypotheses involving some HGTs and those involving some losses because these two kinds of events are relatively frequent in microorganisms (especially in prokaryotes).

Maximum parsimony and maximum likelihood models trying to integrate gene loss, gene birth, and HGT events have been proposed to improve the estimation of HGT rates. The main difficulties encountered by these models are that the frequency of each type of events is not known a priori and must be estimated. To do so, complex statistical models have been developed (Szollosi et al. 2012).

Finally, another difficulty is linked to the fact that the absence of a gene in a given genome is difficult to infer with confidence. Indeed, due to fast evolutionary rates, some genes cannot be affiliated to a given gene family (Fig. 12.9d), although they are actually present, leading to an overestimation of HGT (Zhaxybayeva et al. 2007).

12.3.3 Methods Based on Phylogenetic Inference

12.3.3.1 Principle

Detection of HGT with phylogenetic methods aims at comparing the evolutionary history of genes and that of the organisms carrying them. Indeed, at each generation, the genomic DNA is transmitted vertically (from parents to descendants) during the cell division process. This implies that the evolutionary histories and, therefore, the phylogenies of genes are expected to be similar to that of organisms carrying these genes (Fig. 12.10a). If the phylogeny of a gene under study differs from that of organisms, this may indicate that HGTs have occurred during the evolutionary history of the gene (Fig. 12.10b). Detection of HGT based on phylogenetic methods is more efficient than that based on sequence similarity or taxonomic distribution. It is less susceptible to be misled by differences of evolutionary rates, taxonomic sampling biases, and/or gene losses. Moreover, in contrast with methods based on sequence similarity,

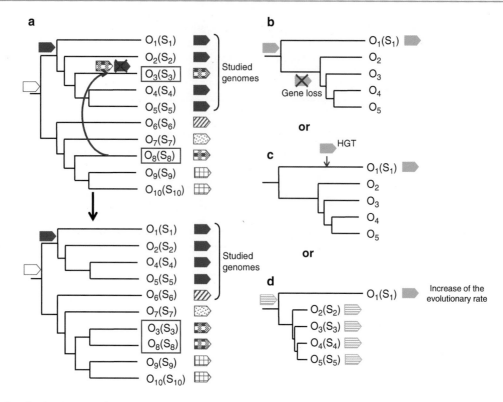

Fig. 12.9 Limits of HGT detection based on the taxonomic distribution of genes. Genomes are designated by O and sequences studied by S. (**a**) Homologous gene replacement. Sequence S_3 of O_3 was acquired by HGT from an ancestor of O_8. The native copy which was present in O_3 was subsequently lost. Based on taxonomic distribution, this HGT is undetectable, whereas it would have been revealed by a phylogenetic analysis. The presence of a sequence in one genome (or in a few genomes) may result to (**b**) gene losses in other genomes. (**c**) Gene acquisition by HGT. (**d**) The non-detection of the homologues present in other genomes due to high evolutionary rates

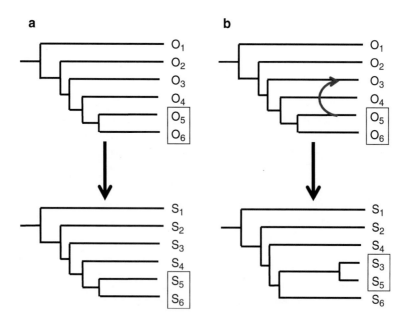

Fig. 12.10 Detection HGTs by molecular phylogeny. The organisms under study are designated as O_1 to O_6 and sequences from these organisms as S_1 to S_6. (**a**) The phylogenies of organisms (*top*) and that of a gene (*bottom*) are consistent, suggesting that this gene has been vertically transmitted all along the cell lineages that lead to organisms O_1 to O_6. (**b**) The phylogenies of organisms (*top*) differ from that of the gene (*bottom*). S_3 is more related to S_5, while O_5 is more closely related to O_6 than to O_3. This strongly suggests that an HGT occurred from an ancestor of O_5 to an ancestor of O_3

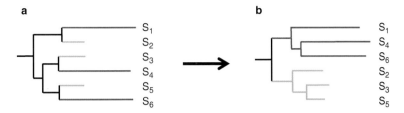

Fig. 12.11 Long branch attraction artifact. (**a**) Phylogenetic relationships among 6 sequences (S_1 to S_6). S_1, S_4, and S_6 (in *red*) evolve much faster than S_2, S_3, and S_5 (in *green*). (**b**) Due to extreme differences in evolutionary rates, phylogenetic reconstructions may be affected by a tree reconstruction artifact called the long branch attraction artifact, which tends to group the fastest and the slowest evolving sequences in two separate parts of the trees

phylogenetic approaches allow determining the direction of transfer or, in other words, distinguishing between the donor and the recipient lineages.

Applying phylogenetic approaches, it is possible to identify the genes (or groups of genes) presenting similar phylogenetic histories. Among these sets of genes, one corresponds to genes that have been vertically transmitted during the evolution of the organisms under study, while others represent groups of genes with atypical stories which may result from HGT. The study of these sets of genes can provide a lot of information for understanding the evolution of genomes and thus of organisms.

12.3.3.2 Limits

Although regarded as one of the most efficient methods to detect HGT, the implementation of phylogenetic approaches is actually complex. Indeed, besides the phylogenetic inference itself, these approaches require the identification of the homologues of the gene under study, the construction of a multiple alignment, and the trimming of the resulting alignment in order to remove from the analysis the regions where the alignment is ambiguous. Although these steps can be automated, phylogenetic reconstruction is time consuming, especially when the number of sequences to analyze is high. In addition, applying these approaches necessitates the knowledge of the phylogenetic relationships among the organisms studied. However, most of the time, this phylogeny of organisms is not known a priori. In that case, a preliminary step in the analysis consists in establishing the phylogeny of organisms. Genes that are most frequently used for this purpose are those encoding the RNA component of the small subunit of the ribosome, ribosomal proteins, or housekeeping genes (i.e., genes expressed in a constitutive manner and/or whose product is essential to the cell) (*cf.* Sect. 6.4). Finally, accurate identification of HGT requires that the phylogeny of the gene under study is well resolved (i.e., statically significant), that is to say, the branching pattern of sequences must be unambiguous (i.e., supported by significant bootstrap values or posterior probabilities). If this condition is not met, it is not possible to determine whether the phylogeny of the gene is consistent or inconsistent with the phylogeny of organisms. Last but not least, the phylogenies of genes shall not be affected by tree reconstruction artifacts, which make the topology of the reconstructed tree different from the evolutionary history of sequences. Tree reconstruction artifacts may have several origins. For instance, among them, the long branch attraction artifact is caused when the sequences under study evolve at very different evolutionary rates. As a result, the sequences displaying the longest branches (i.e., the fast-evolving ones) will be grouped together in the resulting trees, irrespectively of their true relationships. Accordingly, the slowly evolving sequences will be grouped together in another part of the tree (Fig. 12.11). Parsimony and distance methods are more sensitive to long branch attraction artifact than maximum likelihood and Bayesian approaches. Other artifacts exist, for instance, those resulting from compositional biases that lead to the artifactual grouping of sequences sharing similar nucleotide (or amino acid) compositions in the reconstructed trees. It is important to keep in mind that biases affecting phylogenetic methods for HGT detection affect also and, more severely, those based on sequence similarity or taxonomic distribution.

Another difficulty relies on the interpretation of the results. Inconsistencies among gene and organism phylogenies should not be systematically interpreted as the consequence of HGTs. For example, hidden paralogy, the duplication of a gene followed by differential gene losses, can be a source of inconsistency (Fig. 12.12). The differential fixation of alleles in populations may also be responsible for incongruence between the phylogenies of genes and the phylogenies of organisms (Fig. 12.13). Indeed, in populations, different alleles of a gene coexist. Following a speciation event (the process by which a population is split into two separate lineages, leading to the formation of new species), the new lineages harbor alleles similar to that of the ancestral population (Fig. 12.13a). Over time, in each new lineage some alleles disappear whereas others remain. This phenomenon is called incomplete allele sorting. The lost and kept alleles may be different from a lineage to another (Fig. 12.13a). Since the phylogenies of genes are in fact those of alleles, their evolutionary history may differ from that of organisms (Fig. 12.13b).

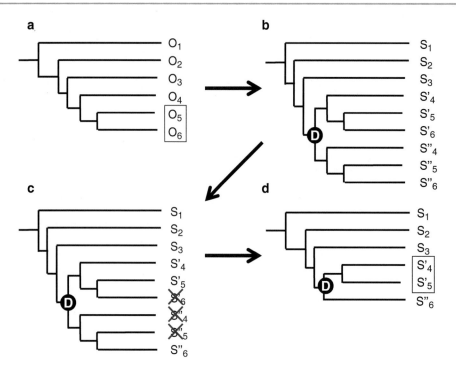

Fig. 12.12 Inconsistency between gene and organism phylogenies resulting from hidden paralogy. (**a**) Phylogeny of organisms (O_1 to O_6). (**b**) A gene duplication event (indicated by the *black circle* D) occurred in a common ancestor shared by O_4, O_5, and O_6. The two resulting paralogous copies are transmitted to the offspring of this ancestor. (**c**) During the diversification of lineages leading to O_4, O_5, and O_6 independent gene losses of one or the other paralogue occurred (*red crosses*). (**d**) While O_5 is more closely related to O_6 than to O_4, S_5' and S_4' will be grouped together in the phylogeny because they descend from the same ancestral paralogue. The unexpected close relationship between S_5' and S_4' can be wrongly interpreted as the consequence of an HGT that occurred between an ancestor of O_4 and O_5, whereas it results from differential losses of paralogous sequences

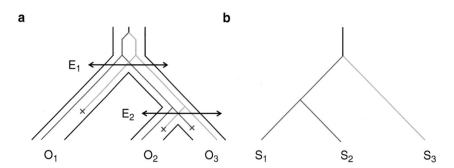

Fig. 12.13 Inconsistency between gene and organism phylogenies resulting from incomplete allele sorting. (**a**) According to organism phylogeny shown in *black*, O_2 and O_3 are closely related. In the ancestral population, two alleles (in *pink* and *green*) of the same gene coexisted. These two alleles were present in the three populations resulting from the two speciation events E_1 and E_2 but were independently lost during the evolution of these populations (*red crosses*). This phenomenon is called incomplete allele sorting. (**b**) Because gene phylogeny reflects in fact the evolutionary history of alleles, it will group S_1 and S_2. This relationship contradicts the phylogeny of organism and may be wrongly interpreted as the result of an HGT

Finally, phylogenetic methods are unable to detect HGTs that occurred between sister-lineages, because the corresponding gene phylogenies will be similar to that of organisms. For instance, if the HGT shown on Fig. 12.10 had occurred between ancestors of the organisms O_5 and O_6, then the sequence S_5 would emerge as the sister group of S_6, as it would have been expected in the absence of HGT.

In summary, although the methods of phylogenetic reconstruction are more adapted to HGT detection than those based on sequence similarity, they are more complex to

12.3.4 Methods Based on Compositional Biases

12.3.4.1 Principle

These methods rely on the fact that genomes are more than a collection of genes. They are dynamic entities whose sequence evolves constantly through the mutational process. Depending on the constraints acting on genomes, some types of mutations are fixed more or less frequently. The main consequence of this phenomenon is that each genome harbors relatively homogeneous characteristics. In contrast, pressures acting on genomes vary from one organism to another, and thus genomes from different organisms usually present different features, which may constitute signatures.

Genomic signatures can be of different types. The most obvious is the oligonucleotide composition of genomic sequences (Burge et al. 1992; Gautier 2000). It has been noticed very early that the frequencies of oligonucleotides are homogeneous within a genome but variable from one genome to another. Mononucleotides composition (i.e., the frequency of bases A, C, G, and T) is the easiest to measure and is usually expressed in terms of G + C content. In present-day genomes, the G + C content varies from 16 to 75 % (cf. Chap. 6). Di-, tri-, and tetra-nucleotides (i.e., words of size two, three, or four) are also frequently used to characterize genomes. Intuitively, it is obvious that the frequencies of the different types of oligonucleotides are correlated. For instance, the G + C content will influence the frequency of dinucleotides. Accordingly, under the assumption that nucleotides are randomly distributed in genomes, a depletion of A- and T-containing dinucleotides (i.e., AA, TT, AT, and TA) and, conversely, an excess of G- and C-containing dinucleotides (i.e., CC, CG, GG, and GG) is expected in G + C rich genomes. However, genome analyses reveal differences between expected and observed oligonucleotide frequencies. For instance, while the genome of *B. subtilis* contains 43.5 % of G + C, the dinucleotides AA, GC, and TT are more frequent than expected by chance, and the dinucleotides TA, GT, and AC are underrepresented (Rocha et al. 1998). The over- and underrepresentation of oligonucleotides vary among genomes. This suggests that nucleotides are not randomly distributed within genomes. This also supports that the frequencies of oligonucleotides of length equal or greater than 2 can be used for the fine and reliable detection of atypical genomic regions.

A second type of signature is based on the genetic code usage. Indeed, because of the degeneration of the genetic code, most amino acids can be coded by different codons (Fig. 12.14). This does not mean that when an amino acid is coded by different codons, each codon is equally used. Instead, it is frequent that one or two codons are used preferentially to others. For instance, among the two codons encoding glutamine, CAG is two times more represented than CAA in *E. coli* APEC O1 (*Gammaproteobacteria*) genes (Fig. 12.14). Several hypotheses have been proposed to explain codon usage biases, such as the G + C content of genomes or the copy number of genes encoding each tRNA. Regardless of its origin, it has been noticed very early that the codon usage varies from one organism to another. The comparison of *E. coli* APEC O1 and *Staphylococcus aureus* Mu50 (*Firmicutes*) revealed that these bacteria do not use each codon at the same frequency. For instance, while CAA (Glu) is seven times more frequently used than CAG in *S. aureus*, *E. coli* APEC O1 uses preferentially CAG (Fig. 12.14). The most obvious example regards the codons coding for leucine: CTG is used preferentially by *E. coli* APEC O1 while TTA is majority in *S. aureus* Mu50 (Fig. 12.14). The difference of codon usage in these two bacteria may reflect the fact that the genomes of *Firmicutes* are A and T rich compared to that of *E. coli* APEC O1 (the genomic G + C of these two bacteria is 32 % and 50.5 %, respectively). Therefore, it is not surprising that the A + T rich codons are more frequently represented in the genome of *S. aureus* Mu50.

Finally, the amino acid composition of proteins is also distinctive among prokaryotes. The frequency of each amino acid varies from a given proteome to another. For instance, UGG (the only codon for tryptophan) is almost two times more abundant in *E. coli* APEC O1 than in *S. aureus* Mu50 (Fig. 12.14). This indicates that the proteins of the gammaproteobacterium are enriched with tryptophan compared to that of the firmicute. Conversely, proteins of *S. aureus* Mu50 are richer in lysine (codons AAA and AAG) than those of *E. coli* APEC O1. The mechanisms at the origin of amino acid composition variations are not fully understood. It has been hypothesized that these variations reflect, at least partially, environmental selective constraints. For example, it has been shown that the proteomes of hyperthermophilic organisms are enriched in isoleucine, valine, tyrosine, tryptophan, arginine, leucine, and glutamate (Zeldovich et al. 2007). Conversely, they tend to be depleted in amino acids such as asparagine. Similarly, it has been shown that cytoplasmic proteins from halophiles are enriched in acidic amino acids and depleted in basic amino acids (Kennedy et al. 2001).

Nucleotide composition, codon usage, and amino acid composition biases are well documented (Gautier 2000). They can also be used for detecting HGT events based on the fact that if the donor and the recipient organisms belong to different species, their composition biases may be different. At the time of transfer, the transferred DNA harbors genomic signatures which are specific of the donor genome whereas the receptor cell will present different genomic

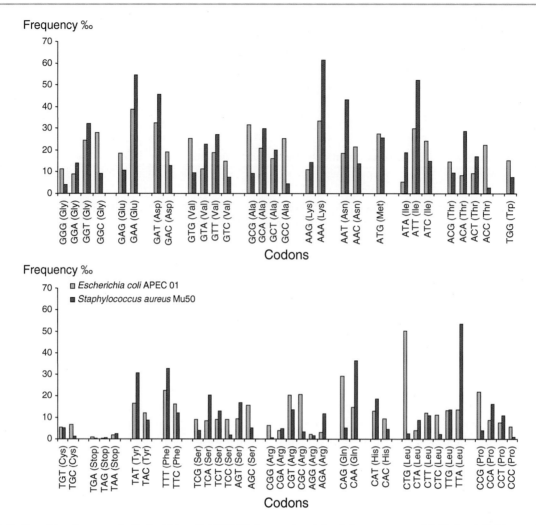

Fig. 12.14 Preferential codon usage in *Escherichia coli* APEC O1 (in *green*) and *Staphylococcus aureus* Mu50 (in *pink*). The graphs show the frequency (‰) of use of the 64 codons. The data have been extracted from the database "codon usage database" (http://www.kazusa.or.jp/codon/)

features. Thus, a way to detect HGT events consists in searching for genomic regions presenting unusual features compared to the background of the genome (Rocha et al. 1998). In contrast to methods based on phylogeny, taxonomic distribution, or sequence similarity, the detection of HGT based on compositional signatures does not require external data as it is based only on the analysis of the genome sequence of the organism under study. This is one of the great advantages of these approaches.

12.3.4.2 Limits

Detection of HGT based on oligonucleotide compositional signatures is easy to apply. The main problem of these approaches is linked to the fact that even if genomes harbor global specific features, local variations exist and can be wrongly interpreted as the consequence of HGT. For instance, it has been shown that the G + C content is greater in coding regions than in intergenic regions (Rocha et al. 1998). This is at the origin of local deviations of the oligonucleotide frequencies observed at the genome scale. Furthermore, HGT detection based on variation of the global codon usage and amino acid composition can be applied only to genomic coding regions. In addition, the codon usage and amino acid composition vary according to the type of genes. For instance, highly expressed genes (e.g., ribosomal proteins) present amino acid compositions and codon usage different from those of other proteins (Gouy and Gautier 1982) (Fig. 12.15). Importantly, atypical nucleotide and amino acid content and codon usage variation allow detecting especially the recent HGT events. Indeed, over time, the transferred DNA will acquire the characteristic of the recipient genome, because it becomes subject to the same constraints and mutational process. As a consequence, the foreign DNA will gradually acquire the features of the host genome. This process is called amelioration (Lawrence and Ochman 1997). At the end, the transferred DNA fragment will become indistinguishable

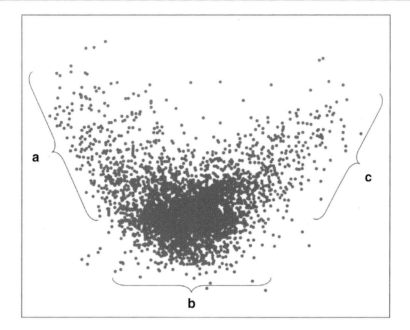

Fig. 12.15 Biased codon usage in *Escherichia coli*. The codon usage of each gene is calculated. These data are gathered into a matrix which is then subjected to a principal component analysis allowing visualizing each gene in a two-dimensional space (instead of the original 64 dimensions). The resulting graph shows three distinct sets of genes: (**a**) highly expressed genes (including ribosomal proteins). These genes are known to have a codon usage different from that of other genes due to their high expression. (**b**) Moderately expressed genes, gathering the majority of genes. (**c**) Genes having very atypical codon usage. This biased codon usage has been interpreted as an indication that the corresponding genes were acquired by HGT from donors having preferential very different codon usage from that of *E. coli*. Applying this method allows rapid identification of potential candidates for HGT in given organisms

from the genomic DNA of the host, meaning that the detection of HGT based on genomic features becomes ineffective. Studies conducted on *E. coli* suggest that only the most recent HGTs (<10 million years) can be detected efficiently by these approaches, even if it has been proposed that a few older HGT events (up to 100 million years) could still be detected (Lawrence and Ochman 1997).

12.3.5 Quantification

As detailed in the previous section, many methods of HGT detection have been implemented. They have been used extensively to quantify this phenomenon in prokaryotes. For instance, applying methods based on the taxonomic distribution (*cf.* Sect. 12.3.2) to the comparison of genomes of five *Salmonella* strains (*Gammaproteobacteria*): *S. enterica* subsp. *typhimurium* (LT2 strains and S21), *S. enterica* subsp. *enterica* serovar Typhi (M229), *S. enterica* subsp. *enterica* serovar Muenchen (S71), and *S. enterica* subsp. *enterica* serovar Typhimurium (M321), suggests that 20 % of the genome of LT2 might be of exogenous origin (Hayashi et al. 2001). These results are based on subtractive hybridization experiments among the chromosomal DNA of LT2 and the chromosomal DNA of each of the four other *Salmonella*. The total size of the non-hybridized fragments in LT2 varies from ~100 kb 1,286 kb (Lan and Reeves 1996). A more precise analysis of M321 non-hybridized fragments indicates that they do not represent divergent (i.e., fast evolving) regions, but that they correspond to DNA regions actually missing in LT2. Among the 39 non-hybridized fragments that have been sequenced, 6 contain genes with known functions and 16 have a G + C content below 45 % (while the average G + C content of *Salmonella* is around 52 %), suggesting an exogenous origin (Lan and Reeves 1996).

The case of *Salmonella* is not exceptional. The comparison of closely related genomes reveals in some cases significant gene content differences. For instance, the genome of *E. coli* strain O157:H7 contains 859 additional Kb compared to the K-12 strain MG1655 (i.e., 5.5 Mb compared to 4.6 Mb, respectively) (Hayashi et al. 2001). A detailed analysis showed that the core genome of these two strains represents 4.1 Mb. The 1.4 Mb present in O157:H7 but absent in K-12 represent 1,632 coding genes (including 131 genes involved in virulence) and 20 tRNA genes. These differences in gene content were interpreted as the consequence of recent gene acquisitions through HGT. Strengthening this hypothesis, analyses revealed that most of this DNA is of phage origin (Hayashi et al. 2001).

A similar study was conducted in three *Frankia* strains presenting less than 2.2 % of 16S rRNA divergence

(Normand et al. 2007). This genus of N_2-fixing plant symbionts belongs to the *Actinobacteria*. The genome size of the three strains varies considerably: 5.43 Mb (*Frankia* sp. HFPCcI3), 7.50 Mb (*Frankia alni* ACN14a), and 9.04 Mb (*Frankia* sp. EAN1pec). Comparative genomic studies revealed a core genome composed of 35–50 % of the genes present in these strains, whereas 578, 1,563, and 1,289 genes were specific to each strain.

Most studies based on gene taxonomic distribution provide similar results. Under the assumption that specific genes result from recent HGT, estimated HGT rates are high. Extrapolating these estimated rates to larger evolutionary scales would lead to the conclusion that the amount of genes acquired by a lineage during its evolution is several orders of magnitude greater than the complete genome size, meaning that all genes in a given genome have been exchanged several times during evolution. Sequence similarity analyses (*cf.* Sect. 12.3.1) performed in the 1990s suggest also high HGT rates among prokaryotes, especially among organisms living in similar ecological niches, such as hot ecosystems. Accordingly, HGTs are considered as very abundant and even to represent the major evolutionary driving force in prokaryotes (Lan and Reeves 2000). Although more recent studies suggest that the HGT rates estimated by these methods are likely overestimated, they remain relatively high.

Phylogenetic methods (*cf.* Sect. 12.3.3) are also frequently used to estimate HGT rates. For instance, the phylogenetic analysis of 1,128 single copy genes present in at least nine of eleven cyanobacterial genomes (*Anabaena* sp. PCC7120, *Trichodesmium erythraeum* IMS101, *Synechocystis* sp. PCC6803, *Prochlorococcus marinus* CCMP1375, *P. marinus* MED4, *P. marinus* MIT9313, *Synechococcus* WH8102, *Thermosynechococcus elongatus* BP-1, *Gloeobacter violaceus* PCC7421, *Nostoc punctiforme* ATCC29133, and *Crocosphaera watsonii* WH8501) identified 685 phylogenies inconsistent with the phylogeny of the 16S rRNA gene (Zhaxybayeva et al. 2006). Assuming that those inconsistencies result from HGT leads to the conclusion that ~61 % of the studied genes have been affected by HGT during the evolutionary history of these nine cyanobacteria. This result is somewhat surprising because the analyzed genes are conserved and well distributed in these cyanobacteria (i.e., present in one copy in at least 9 of the 11 genomes), which suggests that they are ancient and may have been inherited from the last common ancestor of the eleven cyanobacteria. Besides multiple independent acquisitions from non-cyanobacterial donors, these HGTs may result from HGT among cyanobacteria and/or homologous gene replacement from non-cyanobacterial donors (Figs. 12.9 and 12.10). To go further, the authors analyzed the 700 genes for which homologues are found in archaeal and bacterial lineages (other than *Cyanobacteria*) in public databases. In 160 cases (23 %), the gene phylogenies did not recover the monophyly of cyanobacteria, suggesting multiple HGTs from cyanobacteria to other prokaryotes or vice versa. This result illustrates one of the main weaknesses of HGT detection based on taxonomic distribution. Indeed, these approaches allow detecting HGT resulting in the introduction of new genes in genomes, that is to say, genes with no homologues in the recipient genome. In contrast, they are inefficient to detect HGT leading to the replacement of a preexisting gene. Nevertheless, the analysis of cyanobacterial genomes reinforces the idea that HGTs are frequent in prokaryotes.

Approaches based on atypical oligonucleotide and amino acid composition and codon usage suggest also high rates of HGT in prokaryotes. Among them, the most famous was performed on *E. coli* K-12 MG1655 by Lawrence and Ochman at the end of the 1990s. It shows that ~18 % of the genome of *E. coli* K-12 (i.e., 755 genes on 4,288) results from 234 HGT events since its divergence from the *Salmonella* lineage (Lawrence and Ochman 1998). Because of the amelioration process (*cf.* Sect. 12.3.4.2), these estimates are viewed as an underestimation of the real number of HGT events. The average age of the detected HGTs was estimated to be 14.4 Myr, leading to a rate of transfer of 16 kb/Myr/lineage. Given that the divergence of the *Salmonella* and *Escherichia* genera is thought to have occurred 100 million years ago, the *E. coli* K-12 chromosome would have acquired 1.6 Mb since this speciation event (Lawrence and Ochman 1998). Extrapolating this rate over larger evolutionary scales implies that HGT have likely affected each gene present in the genome of *E. coli*. This suggests that prokaryotic chromosomes are fluidic entities where the continual introduction of new genes through HGT is balanced by gene losses preventing the endless increase of prokaryotic genome size.

HGT rate estimations vary considerably from one lineage to another. For instance, the study of oligonucleotide composition of 116 genomes reveals HGT rates ranging from 0.6 % in the case of *Buchnera* (a gammaproteobacterial intracellular symbiont of aphids) to 25.2 % in the case of the archaeon *Methanosarcina acetivorans* C2A (Nakamura et al. 2004), with an average of 14 % over all the studied genomes. The low HGT rate in *Buchnera* can be explained by a relative isolation due to their very particular lifestyle (intracellular symbionts) compared to other bacteria. In the case of *M. acetivorans*, a large fraction of the detected HGT involved bacterial donors. However, other studies suggest lower rates of HGT. For instance, a phylogenetic analysis of 205 single copy genes present in 13 gammaproteobacteria (*Buchnera aphidicola*, *E. coli*, *H. influenzae*, *Pseudomonas aeruginosa*, *Pasteurella multocida*, *S. enterica* subsp. *Enterica* serovar Typhimurium, *Vibrio cholerae*,

Wigglesworthia brevipalpis, *Xanthomonas axonopodis*, *X. campestris*, *Xylella fastidiosa*, *Yersinia pestis*, and *Y. pestis* KIM CO_92) shows only two cases of inconsistency between gene and organism phylogenies (Lerat et al. 2003). This suggests that in these gammaproteobacteria, homologous gene replacement of ancestral genes is rare, despite their very different lifestyles. Analyses of nucleotide compositions and codon usage patterns revealed a very low HGT rate in the spirochete *Borrelia burgdorferi* (the agent of Lyme disease), the alphaproteobacterium *Rickettsia prowazekii* (the etiologic agent of epidemic typhus), and the firmicute *Mycoplasma genitalium* (responsible for infections of the genital and respiratory tracts) (Ochman et al. 2000). Finally, a phylogenetic analysis of 297 genes present in 40 prokaryotes only shows 33 HGT cases (11 %) (Ge et al. 2005).

Studies aiming at quantifying HGT rates in prokaryotes are numerous but failed to reach a consensus. Indeed, the main conclusion of these studies is that the HGT rates vary from one study to another. This suggests that HGT rates vary among lineages, maybe due to different capacities to catch exogenous DNA. For instance, some species are not naturally transformable. HGT rate variations may also reflect variations in the effectiveness of the restriction and recombination systems. External factors, such as the persistence of DNA in the environment, microbial diversity of the ecosystems, and selective pressures, may also affect HGT rates. More surprisingly, HGT rate estimations vary according to the methods used to detect the HGT events. For example, a detailed analysis of the *E. coli* and *S. enterica* genomes reveals that a large number of *E. coli* genes with normal nucleotide composition have no apparent orthologues in *S. typhi*, and conversely, many *E. coli* genes of atypical composition have orthologues in *S. typhi*. The former group of genes would have been detected as the result of HGT by methods based on taxonomic distribution but not by methods based on nucleotide composition. In contrast, the latter group of genes would have been detected as the result of HGT by methods based on nucleotide composition but not by methods based on taxonomic distribution. Finally, phylogenetic methods show that a number of putative HGT events detected by their atypical base composition and biased codon usage are in fact native genes, and conversely, some genes previously classified as native appeared to be the result of HGT (Koski et al. 2001). This illustrates that HGT rate estimations are impacted by the methods used irrespectively of the organisms under study. It was proposed that the use of different methods of HGT detection could help to overcome these methodological biases. Indeed, if the approaches used to detect HGT are reliable, we expect that HGT events detected by different approaches represent real HGT, whereas those detected by only one method represent false positives. However, most published studies are based on the use of a single method.

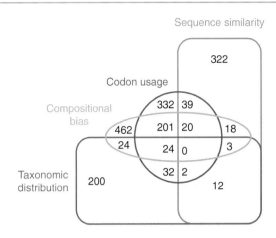

Fig. 12.16 Comparison of four methods of HGT detection. The Venn diagram shows the number of genes detected as resulting from HGTs in *E. coli* K-12: nucleotide composition (*green*), codon usage (*blue*), sequence similarity (*orange*), and taxonomic distribution (*pink*) (Data extracted from Ragan 2001)

These observations have led some researchers to compare the reliability and effectiveness of different approaches for HGT detection. To do so, they have measured the rate of false positives (i.e., genes detected as the result of HGT whereas they are in fact native) and false negatives (i.e., genes detected as native whereas they have been transferred). Among them, the study performed by Ragan in 2001 is particularly interesting (Ragan 2001). Using the *E. coli* K-12 as model organism, he applied four methods of HTG detection:

1. Atypical nucleotide composition
2. Atypical codon usage
3. Sequence similarity
4. Taxonomic distribution

Without surprise, the number of HGTs detected by these methods is highly variable (752, 650, 416, and 297 genes, respectively). More surprisingly, the four sets of genes detected as HGTs show little overlap (Fig. 12.16). Among them, methods based on atypical nucleotide composition and codon usage provide the most similar results (though not identical). One possible explanation is that these methods are ineffective and provide random results. However, this does not seem to be the case as the set of HGT detected by these methods is much more disjointed than expected by chance. For instance, 41 on the 752 genes showing atypical nucleotide composition are also detected by methods based on sequence similarity, whereas under the assumption that both methods provide random results, 71 genes should have been detected simultaneously by both methods (Ragan 2001). It was thus proposed that the great differences observed reflect the fact that each method has its own specificities, that is to say, all methods are not equally efficient to detect all types of HGT. Some are more effective for detecting:

1. Recent or ancient HGT
2. HGT occurring between close or distant lineages
3. HGT among genomes with different or similar genomic features
4. HGT leading to the replacement of preexisting genes or the acquisition of new genes, etc.

12.4 A Major Evolutionary Mechanism?

HGT is considered a major evolutionary mechanism by many evolutionists and microbiologists. Indeed, although point mutations are the only mechanism creating genetic diversity, HGT is a powerful and efficient vector for gene and allele dissemination in populations and ecosystems.

At the population level, the probability of fixation of a new gene (or allele) is very low. This probability depends on the selective advantage it confers. In a population of N haploid cells, the probability of fixation of a neutral allele (i.e., one that does not confer any particular selective advantage or disadvantage) is equal to $1/N$. This process would require in average $2N$ generations. If the allele is not neutral, that is to say, it is associated to a coefficient of selection s, its fixation will require $(2/s)\ln(N)$ generations. This number of generation decreases as the selective advantage brought by the allele increases. For instance, in a population of 10^6 haploid individuals having a generation time of 24 h, the probability of fixation of a new neutral allele is equal to 10^{-6} and will require 2×10^6 days (~5,480 years) in average. If the new allele confers a selective advantage of 1 % ($s = 0.01$), it will be fixed in $(2/0.01)\ln(10^6)$ generations, that is, 2,763 days (~7.6 years). This example shows that even in the case of a haploid population of relatively small size, the time required to fix a gene or allele under positive selection can be relatively high.

HGT may accelerate the rate of diffusion and thus of fixation of genes and alleles in microbial populations. Beyond intrapopulation diffusion, conjugation and natural transformation allows the transfer of genetic material among species living in similar environments. As a consequence, if a gene conferring a strong selective advantage appears in a given ecosystem, it may be transferred to other organisms living in the same ecosystem. Finally, transduction allows HGT across organisms and species also living in distant ecosystems.

12.4.1 HGT in Prokaryotes

12.4.1.1 HGT and Adaptation

Irrespective of its frequency, HGT is considered as an important way for microorganisms to adapt to their habitats, allowing for instance the acquisition of resistance to toxic compounds or phages or the acquisition of metabolic capabilities allowing the use of new resources and the conquest of new ecological niches. The scientific literature on the adaptive role of HGT in prokaryotes is extremely rich. Therefore, the few examples that will be developed here should not be considered as exhaustive.

From a historical perspective, the adaptive role of HGT was uncovered in the 1950s following the worldwide emergence of antibiotic resistant strains (Barlow 2009) (cf. Sect. 9.5). From a biological point of view, antibiotic resistance can have different origins: alterations or changes in the regulation of enzymes targeted by the antibiotics; changes in the permeability of the membrane to antibiotics, either through a decrease in influx or an increase in efflux capacities; activation of secondary metabolic pathways bypassing the enzymes targeted by the antibiotics; presence of competitors presenting a stronger affinity to the antibiotic compared to its target, etc. The literature on this subject is very abundant and a few examples will be detailed here. The interested reader may refer to Lambert (2005).

Irrespective of their origin, it has been clearly shown that HGTs have played a major role in the dissemination of antibiotic resistance among microorganisms. Although mechanisms involved in HGT vary from one species to another (e.g., conjugation in enterococci, natural transformation in streptococci, transduction or conjugation in staphylococci) (Barlow 2009), the dissemination of antibiotic resistance is rapid even among different strains and species and over large geographical areas. For instance, recent surveys suggest that less than 3 years are required between the time of emergence of resistance factors in bacterial populations and their worldwide dissemination (for additional details, see references in Barlow 2009). Such rapidity is linked to the fact that antibiotics induce very strong selection pressures on microorganisms. As a consequence, carrying resistance genes provides a strong advantage that favors the fixation of the corresponding genes in microbial populations through both vertical and horizontal gene transfer. Therefore, the appearance of resistance factors in a very small number of individuals may be a sufficient starting point for their dissemination over large scales through HGT.

Recombination events, which are common in prokaryotic genomes, are another factor explaining the rapidity of dissemination of antibiotic resistance. Recombination may approach in the same genome region genes carrying various resistances, leading to the formation of groups (or cassettes) of resistance genes. For instance, such a cassette has been characterized in a bird pathogenic strain of E. coli (Gilmour et al. 2004), which contains resistance factors to the following:

- Various aminoglycosides (e.g., amikacin, gentamicin, kanamycin, neomycin, netilmicin, paromomycin, streptomycin, tobramycin, etc.)
- Sulbactam a β-lactamase inhibitor

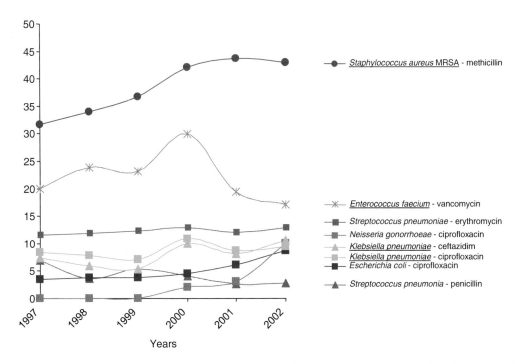

Fig. 12.17 Evolution of antibiotic resistant bacterial strains. These data result from blood analyses performed by the UK Health Protection Agency in England and Wales (Data from Livermore 2004). *Underlined* bacterial species names correspond to hospital samples

- Tetracycline, an antibiotic which blocks translation by inhibiting the attachment of charged aminoacyl-tRNAs to the A site on the ribosome
- Quaternary ammonium compounds that destroy cell membranes and proteins
- Silver ions which are toxic to cells
- Tellurite

Phylogenetic analyses show that resistance cassettes are heteroclite assemblies of genes of very diverse origins which have been brought together on the genome by successive recombination events. Successive HGT through different hosts favors the emergence of multiresistance cassettes because at each step, the acquired cassette may be enriched by new resistance factors which are present in the genome of the recipient cell. Resistance cassettes are key players in the dissemination process of resistances through HGT, because a single HGT event allows the recipient cell acquiring a large set of resistance genes.

The emergence of antibiotic resistances raises serious public health concerns. First worldwide antibiotic resistances were reported few years after the massive clinical use of penicillin and sulfonamide against microbial infections during the first half of the twentieth century (Livermore 2004). In Europe, the rapid expansion of antibiotic resistances has been taken seriously, by adopting, at the end of the 1960s, the first measures aiming at restricting the use of antibiotics, especially in the food industry. Microbial antibiotic resistance was overcome in the 1970s and 1980s by the development of second-generation antimicrobial drugs such as cephalosporins and fluoroquinolones (Livermore 2004). However, the 1990s witnessed the emergence of the first Gram-negative strains resistant to this new generation of antibiotics and soon after the emergence of the first multiresistant Gram-positive strains (i.e., staphylococci, pneumococci, enterococci, etc.) (Bhavnani et al. 2005).

In order to slow down the spread of antibiotic resistance, at the end of the 1990s many European countries adopted measures aiming at restricting the prescription of antibiotics. However, the effectiveness of these measures remains controversial (Livermore 2004). For instance, in the UK, they seem to have been effective in the case of penicillin with a decline of resistant pneumococci (from 6.9 to 2.8 %) following a diminution of 23.4 % of antibiotic prescription (Fig. 12.17). In contrast, no significant diminution of erythromycin resistance was observed in pneumococci over the same period. Similarly, despite drastic efforts aiming at limiting their prescription outside the hospitals, the emergence of strains resistant to fluoroquinolones (such as ciprofloxacin) increases rapidly, especially in *E. coli* and *N. gonorrhoeae* species. In hospitals, the continuous use of antibiotics exerts a strong selective pressure on microbial pathogens that favors the emergence and the dissemination of antibiotic resistant

Fig. 12.18 Organization of the genomic island conferring the ability to respire nitrates in *Thermus thermophilus* (Modified and redrawn from Ramirez-Arcos et al. 1998). *Red*, genes encoding the NADH dehydrogenase type II; *pink*, genes encoding the nitrate reductase; *green*, genes regulating the expression of these two sets of genes; *black*, genes coding for highly conserved proteins of unknown function; *gray*, probably inactive Fe^{2+} transporter; *orange*, cryptic origin of replication. Region in *blue* may correspond to regulatory regions

strains, as exemplified by the case of methicillin and ciprofloxacin resistances in *S. aureus* and *Klebsiella pneumoniae*, respectively. The only significant decrease of antibiotic resistance in hospitals was observed in the case of vancomycin-resistant *Enterococcus faecalis* strains (Fig. 12.17). The main conclusion of these studies is that, to a few exceptions, there is no clear relation between antibiotic restriction policies and the decrease of antibiotic resistance (Livermore 2004). Nevertheless, they may have contributed to slow down the progression of antibiotic resistance.

Health studies report the emergence of new resistance factors in some pathogens. This is the case, for example, of IMP-1, a metallo-beta-lactamase (i.e., a β-lactamase involved in resistance to β-lactams) that is beginning to emerge in *Pseudomonas* and *Acinetobacter* in different places (Livermore and Woodford 2000). Although carrying IMP-1 does not provide a full resistance, this report anticipates the emergence of fully resistant strains (Livermore 2004). The emergence of resistance has had a considerable impact on public health policies. Indeed, if pharmaceutical industries have invested considerable amount of money for the development of new antibiotics in the 1980s and 1990s, an opposite trend is now observed. Instead, alternative approaches are under study, such as the use of biological agents (phages) to control microbial populations. This opens the door for the design of evolutionary systems, in which the biological agents and the pathogens could coevolve, meaning that the emergence of new microbial mutants may be counteracted by compensatory mutants occurring in the phages, ensuring the long-term efficiency of the system.

The role of HGT in antibiotic resistance spreading illustrates its importance in microbial populations. Genomic studies show that HGT is also a key factor for the dissemination of many resistances, such as those to heavy metals or radionuclides (i.e., atoms with unstable nuclei that may undergo radioactive decay, resulting in the production of ionizing radiations that may damage cells). Scientific surveys show that microorganisms are able to survive and thrive in heavily contaminated sites. Genomic and metagenomic studies of microbial populations living in these environments have revealed the presence of gene cassettes allowing transforming and precipitating heavy metals and radionuclides, reducing significantly the toxicity of these compounds. As in the case of antibiotics, phylogenetic analysis indicates that HGT has played a major role in the formation and dissemination of these gene cassettes across diverse taxonomic groups.

Beyond the dissemination of resistance factors allowing microorganisms to adapt to environmental changes, HGT is regarded as an important process for the acquisition of new capabilities. This may lead to radical changes in lifestyle and allow conquering new ecological niches. This is for instance the case of the ability to grow in anaerobic conditions in *Thermus thermophilus*. The first isolates of this species were strictly aerobic and used O_2 as sole terminal electron acceptor. Later, facultative anaerobic strains able to use alternative electron acceptors have been described. This is the case of *T. thermophilus* HB8, which can use both nitrate and O_2 as terminal electron acceptors, whereas strain B27 is strictly aerobic. Interestingly, HB8 strains are able to transfer by conjugation their capacity to reduce nitrate to B27 strains. Detailed analyses have revealed that nitrate respiration, and thus the capacity to grow anaerobically, is linked to the presence of a 30-kb DNA fragment (Fig. 12.18) (Ramirez-Arcos et al. 1998). This DNA fragment carries two operons: one corresponds to the *nar* operon that codes for a membrane nitrate reductase and a nitrate transporter, whereas the second corresponds to the *nrc* operon which codes for a type II NADH dehydrogenase and a Fe^{2+} carrier. Both operons are separated by regulatory elements (regAB genes). A cryptic origin of replication (ori) is located in the vicinity of these genes. Thus, this genomic island carries genes coding for the terminal enzyme of a respiratory chain and elements allowing regulating its expression. Its acquisition by HGT would thus enable strict aerobic strains to acquire the capacity to respire nitrate and thus to colonize anoxic environments rich in nitrate. Scientific literature is full of such examples and there is no doubt about the importance of HGT for the adaptation of microorganisms to new ecological niches and the use of new resources. An additional example concerns the proteorhodopsins and bacteriorhodopsins, which have been extensively transferred between bacteria and archaea and among bacteria. It is important to keep in mind that the physical clustering of functionally linked genes on genomes facilitates the dissemination of new capabilities among microorganisms by reducing the number of HGTs required for the acquisition of a new function.

Table 12.2 Fraction of genes in complete genomes of five bacteria more similar to archaeal than to bacterial genes

Bacterial genome (phylum)	Number of genes displaying higher similarity with archaeal than with bacterial homologues
Aquifex aeolicus (*Aquificae*)	246/1,529 (16.2 %)
Bacillus subtilis (*Firmicutes*)	207/4,177 (5 %)
Synechocystis sp. (*Cyanobacteria*)	126/3,172 (4 %)
Borrelia burgdorferi (*Spirochaetes*)	45/851 (3.6 %)
Escherichia coli (*Gammaproteobacteria*)	99/4,149 (2.3 %)

Data extracted from Aravind et al. (1998)

Beyond the acquisition of new abilities, massive HGT may drive more radical changes of lifestyle. A seminal study was performed on *Aquifex aeolicus*, a hyperthermophilic bacterium belonging to the *Aquificae*. The analysis of its 1,529 genes revealed that 16 % of them (i.e., 246 genes) were more similar to genes present in hyperthermophilic archaea than in bacteria, whereas genomes of mesophilic species show significant lower proportions (<5 %) (Table 12.2) (Aravind et al. 1998). A similar trend was observed in *Thermotoga maritima*, a hyperthermophilic bacterium belonging to the *Thermotogae*, containing 24 % genes (450 on 1,877) displaying higher similarity to archaeal than to bacterial genes (Nelson et al. 1999).

These two studies concluded that the higher proportion of genes more similar to archaeal than to bacterial homologues observed in hyperthermophilic bacteria results from HGT (red arrows, Fig. 12.19a). Supporting this hypothesis, 81 of

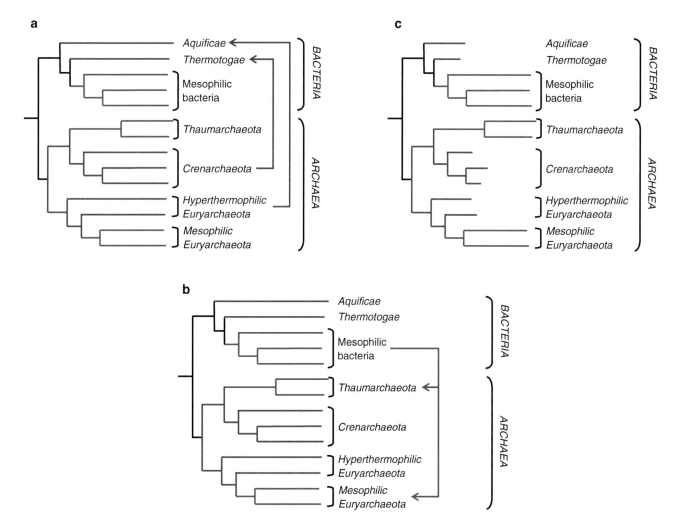

Fig. 12.19 Scenarios that could explain the strong similarity observed between a large number of bacterial and archaeal genes. (**a**) Massive HGTs occurred from hyperthermophilic archaea to hyperthermophilic bacteria. (**b**) Massive HGTs occurred from mesophilic bacteria to mesophilic archaea. (**c**) Protein sequences of hyperthermophiles evolve slower than proteins from mesophilic organisms

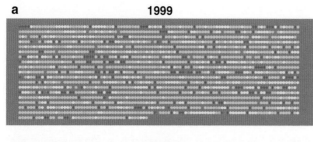

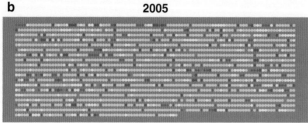

Fig. 12.20 Gene conservation maps (taxmaps) of the genome of the bacterium *Thermotoga maritima* determined in 1999 (**a**) and 2005 (**b**). Each *dot* represents a gene from the genome of *T. maritima*. The colors indicate the taxonomic origin of the homologue found in public databases displaying the highest similarity with each *T. maritima* gene: archaea (*yellow*), bacteria (*blue*), eukaryote (*pink*), or virus (*gray*). Protein sequences for which no homologues are known are shown in *dark green*. Crosses correspond to noncoding RNA genes (tRNA, rRNA, etc.)

the 450 genes supposed to have been acquired by HGT from archaeal donors in *T. maritima* map in 15 genomic regions with a gene order similar to that observed in archaea, suggesting that they result from cotransfer events. Both observations led to the assumption that acquisition of genes of archaeal origin occurred rapidly in *T. maritima* through a relatively limited number of HGTs (Nelson et al. 1999).

However, the hypothesis of massive HGT from archaeal to bacterial hyperthermophiles has been criticized. Indeed, in these first studies, the detection of HGT was based on the sequence similarity approach, one of the less reliable approaches for HGT detection (*cf*. Sect. 12.3.1). Overestimation of HGT rates is thus likely because this approach is very sensitive to differences in evolutionary rates and taxonomic sampling biases. As a matter of fact, recent studies have shown that genes of hyperthermophilic prokaryotes evolve significantly slowly than those of mesophilic organisms (Fig. 12.19c). Moreover, at the time of the analysis of the *T. maritima* and *A. aeolicus* genomes, hyperthermophilic prokaryotes were underrepresented in public sequence databases. Both biases may have led to an overestimation of HGT rates in hyperthermophilic bacteria. Strengthening this hypothesis, a more recent analysis reveals that the proportion of *T. maritima* genes more similar to archaeal than to bacterial homologues drops from 24 % to 14 % (Fig. 12.20). Moreover, the combination of sequence similarity and taxonomic distribution approaches suggests that only 11.3 % of *T. maritima* genes have been acquired by HGT (Podell and Gaasterland 2007). These values have been confirmed by phylogenetic analyses (Zhaxybayeva et al. 2009).

Even if recent analyses have led to a significant reduction of the HGT rates observed between archaeal and bacterial hyperthermophiles, they remain sufficiently high as to have played an important role in the evolution and adaptation of bacteria to very hot environments (*cf*. Sects. 9.6.1 and 10.3). Indeed, the acquisition of proteins already adapted to hot environments from hyperthermophilic donors could have helped the colonization of these ecosystems by mesophilic or thermophilic bacteria (Aravind et al. 1998). This hypothesis is supported by phylogenetic analyses of a few key genes that show clear cases of HGT from archaeal to bacterial hyperthermophiles. Among them, the most studied gene codes for the reverse gyrase, a protein present without exception in all hyperthermophilic organisms and in some thermophilic organisms (Brochier-Armanet and Forterre 2007). This enzyme is composed of two protein domains (an N-terminal helicase domain and a C-terminal topoisomerase domain) and induces positive DNA supercoiling in vitro, although its role in vivo remains poorly understood. The systematic occurrence of this enzyme in hyperthermophiles suggests an essential role of reverse gyrase for life at high temperature (Forterre 2002). The phylogeny of this enzyme shows many discrepancies with the phylogeny of organisms (Forterre et al. 2000). In particular, bacterial sequences do not form a monophyletic group but instead appear intermixed with their archaeal counterparts (Fig. 12.21), suggesting that the gene coding for the reverse gyrase was acquired several times by bacteria from archaea and spread among hyperthermophilic bacterial lineages by HGT (Brochier-Armanet and Forterre 2007). This observation together with the relatively high level of HGT from archaeal to bacterial hyperthermophiles is consistent with the hypothesis that hyperthermophilic bacteria derive from mesophilic ancestors which adapt to very hot environments via the acquisition of archaeal genes already adapted to these particular environments. Contradicting this hypothesis, recent analyses have shown that genes of archaeal origin present in *T. maritima* are also present in some mesophilic bacteria (Nesbo et al. 2001).

If secondary adaptation to hyperthermophilic lifestyle in bacteria may have been driven by HGT from archaea, the opposite hypothesis, namely, the secondary adaptation of archaea to mesophilic lifestyles from hyperthermophilic ancestors, has also been proposed. The analysis of genomic sequence of mesophilic archaea has highlighted many genes similar to genes present in bacterial in archaeal mesophiles. For instance, 1,043 of the 3,371 genes (~31 %) present in *Methanosarcina mazei* (a methanogenic archaeon found in semiaquatic environments such as sewage receptacles and anoxygenic, moist soils) are more similar to bacterial than to non-methanosarcinales archaeal genes (Deppenmeier et al.

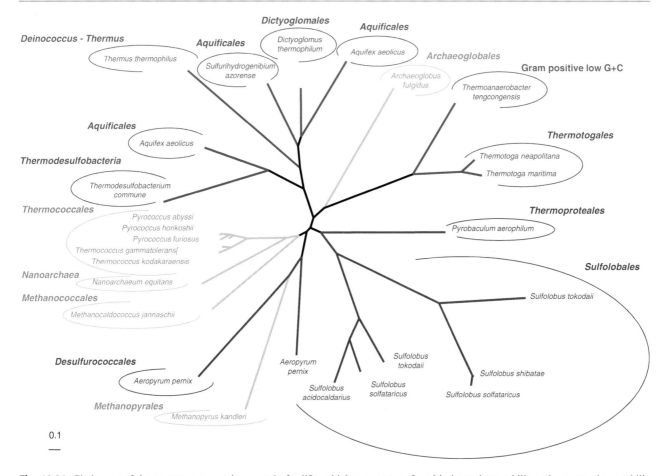

Fig. 12.21 Phylogeny of the reverse gyrase, a key protein for life at high temperature found in hyperthermophilic and extreme thermophilic prokaryotes. Bacterial sequences are shown in *blue*, archaeal sequences in *red* (*Crenarcheota*) and in *gray* (*Euryarchaeota*).

Table 12.3 Number (and fraction) of genes showing higher similarity to bacterial than to archaeal homologues

Archaeal genome	Number of genes displaying higher similarities with bacterial than with archaeal homologues
Methanosarcina mazei	1,043/3,371 (31 %)
Halobacterium sp.	664/2,822 (24 %)
Pyrococcus abyssi	417/1,919 (22 %)
Archaeoglobus fulgidus	529/2,530 (21 %)
Methanothermobacter thermautotrophicus	372/1,918 (19 %)
Methanocaldococcus jannaschii	269/1,738 (15 %)
Sulfolobus solfataricus	358/2,977 (12 %)
Pyrobaculum aerophilum	167/1,567 (11 %)

Data extracted from Deppenmeier et al. (2002)
Hyperthermophilic archaea are in *bold*

2002). While this proportion is significantly lower in hyperthermophilic or thermophilic archaea, similar trends are observed in other mesophilic archaea (*Halobacterium* sp., Table 12.3).

As in the case of hyperthermophilic bacteria, the higher fraction of genes more similar to bacterial than to archaeal homologues was interpreted as the result of HGT, but in the opposite direction, from mesophilic bacteria to mesophilic archaea (Fig. 12.19b). Strengthening this hypothesis, among the 1,043 genes displaying high similarity with genes found in mesophilic bacteria, 544 have no homologues in archaeal lineages other than Methanosarcinales, suggesting secondary acquisitions by HGT from bacteria (Deppenmeier et al. 2002). A more detailed analysis shows that most often the donors are strict (e.g., *Clostridium*, *Desulfitobacterium hafniense*) or facultative (e.g., Enterobacteria) anaerobes living in environments similar to those occupied by *M. mazei*. Another study shows that the majority of the 41 genes encoded on a large genomic fragment (33 kbp) from a noncultivated thaumarchaeon belonging to Group I.1A had been acquired by HGT from mesophilic bacterial and archaeal donors (Lopez-Garcia et al. 2004). High frequency of HGT in Thaumarchaeota has been confirmed by the large-scale genomic analyses, which shows that members of this phylum share an unusually high fraction of genes with mesophilic euryarchaeota and mesophilic bacteria,

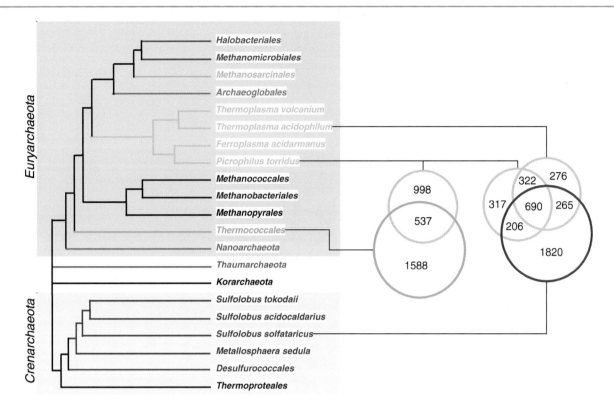

Fig. 12.22 Number of homologous genes shared among the genomes of four archaea: *Picrophilus torridus* and *Thermoplasma acidophilum* (two aerobic thermoacidophilic euryarchaeota belonging to the *Thermoplasmatales* order), *Sulfolobus solfataricus* (an aerobic thermoacidophilic crenarchaeon belonging to the *Sulfolobales* order), and *Pyrococcus furiosus*, an anaerobic hyperthermophilic euryarchaeon belonging to *Thermococcales* (Data extracted from Futterer et al. 2004)

suggesting high rates of HGT from the two latter to the former (Fig. 12.19b) (Hallam et al. 2006; Makarova et al. 2007; Brochier-Armanet et al. 2011). As in the case of hyperthermophilic bacteria, it was proposed that massive HGT of genes from mesophilic prokaryotes could have help hyperthermophilic archaea to adapt to colder environments.

The hypothesis that massive HGT allows prokaryotes to colonize new ecological niches is not limited to hot/cold environments. Similar scenarios have been proposed in the case of thermoacidophiles (Ruepp et al. 2000). For instance, *Thermoplasmatales* (thermoacidophilic heterotrophic aerobic euryarchaeota) share many genes with *Sulfolobales* (thermoacidophilic aerobic crenarchaeota) (*Futterer* et al. 2004). More precisely, the two members of *Thermoplasmatales Picrophilus torridus* and *Thermoplasma acidophilum* share 65 % of their proteomes and, unexpectedly, share approximately the same fraction of genes (58 % and 61 %) with *Sulfolobus solfataricus*, a very distant relative, whereas only 35 % of their genes are present in *Pyrococcus furiosus*, a closer relative belonging to the *Thermococcales* (*cf.* Sect. 6.6.1) (Fig. 12.22). This strongly suggests that massive HGT occurred between the two thermoacidophilic lineages because they occupy similar ecological niches. These HGTs may have also contributed to their adaptation to this extremophilic lifestyle (Ruepp et al. 2000).

While adaptation to hyperthermophilic, mesophilic, and thermoacidophilic lifestyles illustrates the important role of HGT for adaptation of microorganisms to new ecological niches, this is not a rule. This is exemplified by *Salinibacter ruber*, a heterotrophic hyperhalophilic bacteroidetes that adapted secondarily to hyperhalophilic environments (Mongodin et al. 2005). This bacterium lives in the same ecosystems and presents similar characteristics to archaeal hyperhalophiles, including a high cytoplasmic concentration of potassium ions, a proteome enriched in acidic amino acids and depleted in basic amino acids, requirement of high NaCl concentration for growth (>2 M), presence of many carotenoids producing red pigmentation, etc. In addition, it uses the same range of organic compounds and sources of energy as hyperhalophilic archaea. The numerous similarities observed between *S. ruber* and hyperhalophilic archaea could suggest that many HGTs have occurred between these lineages and may have helped the former to secondarily adapt to hyperhalophilic environments. This is not the case. In fact, the genome analysis of *S. ruber* shows very few cases of HGT (<100) between *S. ruber* and hyperhalophilic archaea (Fig. 12.23a) (Mongodin et al. 2005). The analysis of the *S. ruber* proteome also reveals HGT cases from non-hyperhalophilic donors (Fig. 12.23a), including the HGT of key genes for adaptation to

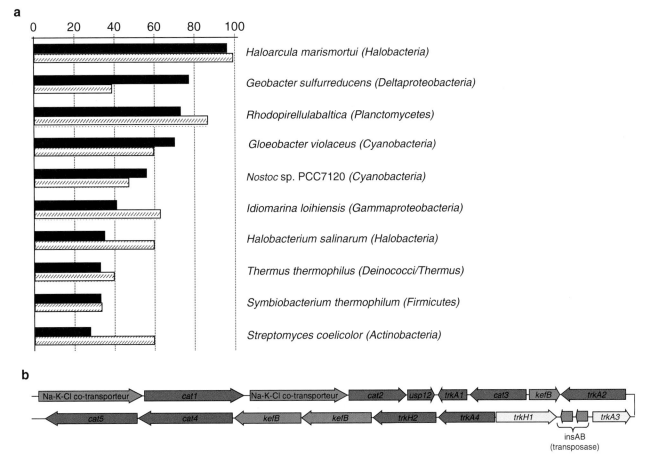

Fig. 12.23 Horizontal gene transfer in the bacterium *Salinibacter ruber* (*Bacteroidetes*). (**a**) Histogram showing the taxonomic distribution of organisms that have contributed to gene acquisition through HGT in *S. ruber*. The taxonomic origin of the genes has been determined by sequence similarity (*black bars*) and phylogenetic analyses (*hatched bars*) of genes present in the genome of this *S. ruber*. (**b**) Structure of a genomic island involved in adaptation to halophilic environment in *S. ruber*. The colors indicate the taxonomic origin of genes: *Halobacteria* (*purple*), *Cyanobacteria* (*blue*), methanogenic archaea (*orange*), and *Firmicutes* (*yellow*) (Modified and redrawn from Mongodin et al. 2005)

hyperhalophilic environments. This is for instance the case of proteins responsible for the red pigmentation in *S. ruber* that are clearly of bacterial origin and thus have not been acquired from archaeal hyperhalophiles. This is also the case of a genomic island that contains 19 genes (Fig. 12.23b), some of them being important for life in hyperhalophilic environments, such as transport systems for K^+ (trk genes), cationic amino amines, etc. (Mongodin et al. 2005). The analysis of this island reveals that it results from the assembly of genes with very different origins, including non-hyperhalophilic archaea and bacteria (Fig. 12.23b). These results were unexpected and suggest that adaptation to new ecological niches can result from different processes: adaptive evolution of native proteins, punctual HGT from organisms already present in the new environment or from other environments, etc., and does not necessary requires massive HGT.

The extent of HGT among prokaryotes sharing similar ecological niches is still highly debated: does it just reflect genetic exchanges among organisms living in the same environments? Or does it have significantly contributed to the colonization of new ecological niches? Providing an answer to this question would require a better knowledge of the biology of these microorganisms together with a better characterization of their genomes.

12.4.1.2 HGT, Evolution and the Species Concept

Thanks to the improvement of sequencing technologies, the past few years have witnessed a burst of comparative genomics studies. Among them, many aim at identifying genomic determinants responsible for phenotypic variation among organisms belonging to the same species or closely related species. For instance, why some strains of bacteria, such as *E. coli* K12 MG1655, are commensal whereas others members of the same species, such as *E. coli* IAI39 or S88, are pathogenic? Similarly, why *E. coli* S88 causes neonatal meningitis while UTI89 is uropathogenic? A tempting hypothesis is that differences exist at the genome level among commensal and

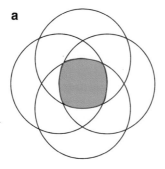
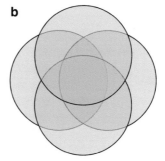

Fig. 12.24 Core genome and pan genome. (**a**) The core genome (in *orange*) corresponds to the set of homologous genes shared by all genomes under study. The areas in white represent the variable shell, i.e. genes that are present in only one or a subset of the genomes understudy. (**b**) The pan genome (in *blue*) corresponds to all genes present in the genomes under study

pathogenic *E. coli*. Sequencing of genomes from related strains provides the material to address this type of question. For instance, a study aiming at determining how the genetic variability drives pathogenesis in *Streptococcus agalactiae* (*Firmicutes*) has been performed in 2005 (Tettelin et al. 2005). This pathogenic bacterium is the major cause of bacterial septicemia of newborns. To do so, complete genome sequences of eight *S. agalactiae* strains belonging to five of the nine described serotypes have been obtained. The genomes are very similar in size (from 2.18 to 2.24 Mb) and code for 2,034–2,481 genes. They display high sequence similarity (from 85 to 95 % depending on the considered strains). The core genome of these eight strains is composed of 1,806 genes (~80 %), whereas the number of unique genes in a genome varies from 13 to 61 depending on the considered strain. This shows that although a majority of genes is common to all *S. agalactiae* genomes, the genomic variability among strains is not negligible, underlying rapid genomic evolutionary processes (including gene acquisitions and losses) even at very small evolutionary scales. The comparison of the gene order in the *S. agalactiae* genomes reveals a conserved common structure with large regions of synteny interrupted by 69 genomic islands. The syntenic regions gather mainly core genes (i.e., shared by all genomes), and the order of genes within these regions is highly conserved. The most parsimonious hypothesis to explain these results is that these syntenic regions have been inherited by the eight strains from their common ancestor. Their high conservation suggests that they may be functionally important. In contrast, the genomic islands contained genes that are specific to or shared by a few strains. These genomic islands result thus of secondary acquisitions and losses of genetic material that occurred during the diversification of the *S. agalactiae* strains. Interestingly, although the gene content of the islands differs from one genome to another, they are located at similar positions, suggesting that genomes contain specific regions in which the newly acquired DNA is inserted preferentially and thus that selective pressures maintaining the overall structure of genomes exist. Many comparative genomic studies, including a large-scale survey of *E. coli* genomes (Touchon et al. 2009), provide similar conclusions. This has led to the emergence of the core and pan genome concepts. The former gathers genes having homologues in all representatives of a given taxon (in orange on Fig. 12.24). They are supposed to have been inherited from the last common ancestor of the taxon. They are usually considered as essential for the taxon because they have been conserved all along the diversification of the lineage, despite HGT, gene duplications, and gene losses, which make the gene content of genomes to evolve continuously. The core genome is opposed to the variable shell, which is defined as the set of genes that do not belong to the core genome, namely, genes present in one or in a few representatives of the taxon (in white on Fig. 12.24a). These genes have two origins. They may derive from ancient genes that were present in the last common ancestor of the taxon and secondarily lost in some representatives of the lineage and, alternatively, they may be of more recent origin, meaning that they have been secondarily acquired by some representatives of the taxon during its diversification, through either HGT or genetic innovation. Irrespective of their origin, these genes are often regarded as nonessential for the taxon, because they are missing in some representatives of the taxon. However, it is important to remind that the notion of gene essentiality or dispensability should be considered carefully because it varies from one organism to another and from one environment to another. For instance, the genes coding for the oxygen reductase that are part of the core genome of *Thermus thermophilus* are essential in strains B27 and B8 under aerobic condition, whereas they are dispensable in strain B8 if nitrates are available to allow anaerobic respiration. Conversely, when nitrates are available in anaerobic conditions, genes coding for the nitrate reductase that are part of the variable shell become essential in strain B8. Together, the core genome plus the variable shell forms the pan genome (Fig. 12.24b), which represents all the genes present in a given taxon.

It is worth to note that the core genome and pan genome notions refer to a given taxon. There are thus as many core and pan genomes as taxonomic levels. Accordingly, some *E. coli* K-12 genes may be part of the core genome of *E. coli* and the *Enterobacteriaceae* and be part of the pan

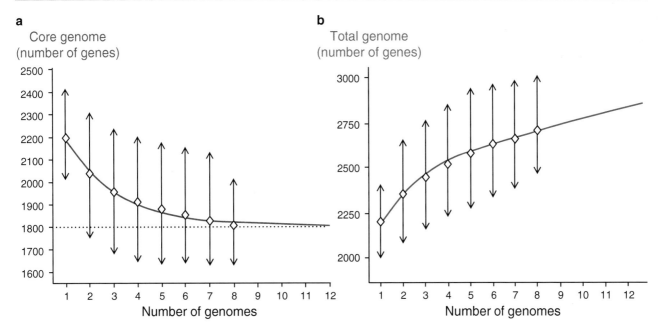

Fig. 12.25 Estimation of the size of the core genome (**a**) and pan genome (**b**) of *Streptococcus agalactiae* (*Firmicutes*) (data from Tettelin et al. 2005). Each simulation consists in generating many times 1–8 artificial genomes of similar size to those of the eight available genomes of *Streptococcus agalactiae* (x-axis). Based on these artificial genomes, the size of the corresponding core genome and pan genome is determined (y-axis). The *red curve* shows that the size of the core genome decreases as the number of considered genomes increases (from 2,200 to 1,800 genes). However, a plateau is reached, indicating that when the number of considered genomes is sufficient, the addition of new genomes has less and less impact on the size of the core genome and, therefore, that the core was correctly delimited. In contrast, the size of the pan genome continues to increase as the number of genomes considered increases (*blue curve*)

genome of Enterobacteriales. The size of the core genome decreases with time. Accordingly, the size of the core genome at lower taxonomic levels (e.g., strains and species) is expected to be greater than at higher taxonomic levels. For instance, the core genome of *Rickettsia* is predicted to contain more genes than the core genome of *Alphaproteobacteria* as many genes present in *Rickettsia* may be absent in representatives of other alphaproteobacterial genera (e.g., *Sphingopyxis*, *Azorhizobium*, etc.). Conversely, the size of the pan genome increases accordingly to the taxonomic level. The core genome is an essential notion in comparative genomics as many microbiologists think that core genomes define the "essence" of taxa and give them a biological sense. This is particularly important in prokaryotes because most taxonomic levels (including species) are defined based on arbitrary measures (*cf.* Chap. 6, Box 6.1).

Even if the notion of core and pan genomes is easy to understand, their characterization is more difficult. In particular, this asks the question of how many genomes should be sequenced to accurately delineate these two sets of genes. For instance, how many and which strains should be sequenced to determine the core and the pan genome of a species? In theory, all representatives of the species should be taken into consideration. In practice, such a strategy is very often not feasible. However, it is possible to estimate the size of the core and the pan genomes of a given taxon using bioinformatic simulations. The robustness and the accuracy of such estimations increases as the number of genomes compared increases. For instance, if a single genome is considered to represent a given taxon, it will correspond to the core genome (Fig. 12.25), although this is likely a very bad estimation of the core genome. If a second genome is included in the analysis, it will uncover genes belonging to the variable shell (i.e., genes present in one or the other genome, but not in both). The more genomes are added, the more precise the distinction between core genome and variable shell will be. However, at some point, the inclusion of new genomes will have little influence on the estimated size of the core and pan genomes, because more and more genes carried by these new genomes will already be present in the previously considered genomes and have thus already been taken in consideration. This prediction is illustrated by bioinformatic simulations using the eight complete genomes of *S. agalactiae* (Fig. 12.25a), which show that the inclusion of seven to eight genomes is sufficient to delimitate the core genome of this bacterial species. According to these simulations, the core genome of *S. agalactiae* is estimated to contain ~1,800 genes (Fig. 12.25b). In contrast, the pan genome is only partially characterized because each new genome carries a number of specific genes. Precise estimations of the pan genome would require the sequencing of many more strains. This illustrates again the great genetic diversity of prokaryotes even at the species level. It is important to remind that two representatives of the same species with similar lifestyle and occupying the same ecological niche may present important genomic variations. This is for instance the case of *Bradyrhizobium*. These symbionts of legumes have very similar lifestyles, but they use a different set of genes to establish their symbiotic interaction with their host.

12.4.2 HGT in Eukaryotes

Our knowledge about the impact of HGT on the genome content and the evolution of eukaryotes remains very limited in comparison with that of prokaryotes. In fact, despite the relatively large number of complete eukaryotic genome sequences available since several years ago, very few studies have addressed this question in a systematic way. Thus, the data about HGT in eukaryotes are often anecdotal and most likely do not reflect the actual importance of this phenomenon in eukaryotes. Several reasons can explain this delay in the investigation of HGT in eukaryotes. The first one is historical since, for long time, the scientific community was dominated by the idea that eukaryotes are naturally reluctant to HGT due to the presence of the nuclear membrane. This membrane, an essential character of all eukaryotic cells, represents an additional barrier that exogenous DNA has to overcome to have the opportunity to integrate in the cell genome. The absence of such a membrane in prokaryotes was considered as a reason to explain the facility with which these microorganisms are able to acquire exogenous DNA. The second reason is linked to the classical view of the major evolutionary mechanisms acting on eukaryotes. For a long time, eukaryotic evolution has been studied using multicellular species with sexual reproduction (most often animals). In these species, an HGT can be successful only if it occurs in the germ cells, which is a very improbable event. Thus, even the most fervent advocates of a major role of HGT in evolution conclude that this phenomenon is certainly marginal in eukaryotes (Bapteste et al. 2009). Hence, the evolution of eukaryotes and prokaryotes would follow different ways because of their different cellular structures. However, this vision does not take into account that all eukaryotes are subjected (at present and all along their history) to a type of event extremely rare in prokaryotes but which represents a very rich source of HGT: symbiosis and, more particularly, endosymbiosis.

12.4.2.1 Symbioses and Acquisition of Bacterial Genes

Symbiotic associations are very frequent in eukaryotes and have been described in all eukaryotic lineages. The eukaryotic cell itself is the product of a very ancient endosymbiosis. In fact, it is now widely accepted that all contemporary eukaryotes have, or have had in their past, mitochondria. These organelles derive from the endosymbiosis of an alphaproteobacterium within a proto-eukaryotic host (*cf.* Chap. 4, Sect. 4.3.1) (Roger 1999; Philippe et al. 2000). That endosymbiosis had very important consequences for eukaryotes, in particular by allowing them to respire oxygen and, thus, to conquer numerous ecological niches. Without aerobic respiration, it seems likely that multicellular eukaryotes had never evolved. A second endosymbiosis has also played a very important role in the evolutionary history of eukaryotes. It concerns the acquisition by the ancestor of the supergroup *Plantae* (red algae, green algae and plants, and glaucophytes; *cf.* Chap. 7) of a photosynthetic plastid by the endosymbiosis of a cyanobacterium. Plastids derived from this event are called primary plastids. It is thanks to them that eukaryotes have acquired the capacity to carry out oxygenic photosynthesis. Other eukaryotes also have this capacity (e.g., *Euglenozoa*, *Cryptophyta*, *Haptophyta*, etc.). These are eukaryotic lineages that have acquired plastids through secondary endosymbioses involving green or red algae (these plastids are called secondary plastids).

A characteristic common to both the mitochondrial and plastidial endosymbionts is that their genomes are extremely reduced in comparison to those of their bacterial counterparts (*Alphaproteobacteria* and *Cyanobacteria*). In general, mitochondrial and plastidial genomes only have a few dozens of genes, whereas alphaproteobacterial and cyanobacterial species have several thousands. The first eukaryotic genome sequences have shown that this genome reduction had two origins: the clearest one is that many genes originally present in the ancestral mitochondrial and plastidial genomes were lost because they become nonessential in the new environment provided by the eukaryotic host cytoplasm. In addition, many other genes were not lost but transferred into the host genome. These genes are expressed by the host and their products are carried to the plastids and mitochondria allowing the functioning and maintenance of these organelles. This implies that, during eukaryotic evolution, complex targeting and transport systems have emerged (Cavalier-Smith 1999). HGT events linked to these endosymbioses are called endosymbiotic gene transfers (EGTs). These EGTs provide selective advantages to the host because it takes the control of the endosymbiont (of its cell cycle, apoptosis, metabolic activity, etc.).

Although this is an important question, few studies have tried to quantify in an accurate way the number of genes in the eukaryotic genomes acquired by EGT from mitochondria and plastids. This seems simple because it just requires identifying the evolutionary origin of each gene in a eukaryotic genome, but very often it is very difficult or even impossible to identify precisely the bacterial donors. That is, it can be very difficult to differentiate between classical HGT and EGT. Consequently, numbers can be overestimated. Nevertheless, available data suggest that they can represent a very significant proportion of the eukaryotic genomes (Doolittle et al. 2003). For example, the analysis of the genome of the yeast *Saccharomyces cerevisiae* has shown that ~75 % of genes of bacterial origin are likely of mitochondrial origin (Esser et al. 2004). In the case of the chloroplast endosymbiosis, a large quantity of EGTs has been detected in the host genomes. For example, more than 15 % of nuclear genes in *Arabidopsis thaliana* are close to cyanobacterial homologues and could thus have been acquired from the plastid endosymbiont (Martin et al. 2002).

Even if all known mitochondrial and plastidial genomes only contain a small number of genes, their reduction by EGT appears to be an ongoing process. The identification of multiple recent and independent EGTs of the plastidial gene *infA* towards the nuclear genome in different lineages of angiosperms offers a good example (Millen et al. 2001). Mitochondrial genomes also illustrate this phenomenon, as shown by the study of 40 mitochondrial genes in 280 angiosperm species, which reveals 16 cases of parallel EGT during the evolution of these organisms (Adams et al. 2002). This contrasts with the extreme homogeneity of the animal mitochondrial genomes, which suggests that very few EGTs have occurred in this lineage since their first diversification. Thus, EGTs occur at different rates in different eukaryotic phyla. Similarly, certain mitochondrial or plastidial genes can be transferred in certain lineages but not in others.

Mitochondria and plastids are not the only actors in EGT events since this phenomenon is general in all endosymbioses. In fact, the study of the nuclear genomes of diverse eukaryotes has revealed numerous gene acquisitions from other types of endosymbionts. For example, fragments of the genome of the alphaproteobacterium *Wolbachia*, an endosymbiont of nematodes and arthropods, have been detected in the nuclear genomes of some of its hosts (Hotopp et al. 2007). An even more surprising example comes from the discovery of an almost complete copy of the *Wolbachia* chromosome inserted in the genome of a beetle (*Callosobruchus chinensis*) hosting this bacterium. This insertion is probably very recent since it is not found in another beetle species closely related to *C. chinensis*. The possibility that this is an artifact or a nonfunctional sequence can be discarded since it has been shown that at least 2 % of the *Wolbachia* genes inserted in the beetle genome are transcribed (Hotopp et al. 2007; Nikoh et al. 2008). These results show not only that, in contrast with what is commonly assumed, there is no strong barrier that impedes the incorporation of exogenous genes into the eukaryotic genomes but also that this is possible even in animals, where, as stated above, only transfers occurring in the germ line can be fixed.

These examples illustrate that during the evolution of the endosymbiotic relationships, transfers of genes from the endosymbiont towards the host nuclear genome occur. Nevertheless, if in the case of mitochondria and plastids the selective advantage of such transfers is clear, for many other genes acquired from other endosymbionts a possible selective advantage remains to be identified. Even the hypothesis that they are kept in the nuclear genome just in a transient way cannot be discarded.

EGT is a major source of gene acquisition by eukaryotes and plays an important role in the evolution of these organisms. However, eukaryotes can also acquire exogenous DNA by other mechanisms. Many eukaryotic phyla have phagotrophic species, namely, they feed on other cells (prokaryotic and/or eukaryotic). This mode of nutrition may offer to these species the opportunity to acquire DNA from their preys. This idea is summarized in an axiom, derived from the German proverb *"Man ist was man isst,"* reformulated by Doolittle in 1998 as "You are what you eat" (Doolittle 1998). The implicit hypothesis is that a larger number of exogenous genes has to be found in phagotrophic eukaryotes than in non-phagotrophic ones. Moreover, it is expected that these genes should have very diverse origins in the nonspecialized phagotrophs, namely, those with a large spectrum of preys. Even if few studies have analyzed this hypothesis, some elements appear to support it. It has been shown that typical phagotrophic eukaryotes, such as the amoebas, seem to have acquired a variety of genes from bacterial preys. This is the case for *Dictyostelium discoideum*, which possesses at least 18 bacterial genes in its genome. Some of them assure essential functions, as the one coding for a glycoprotein involved in cell aggregation (Eichinger et al. 2005). The number of genes of bacterial origin detected in this amoeba is certainly small, but some have also been detected in other amoebas very distantly related to *D. discoideum*, such as species of the genus *Naegleria*. This suggests independent HGT events and points to an important role of those proteins in these amoebas. Higher HGT rates have been found in another group of phagotrophic eukaryotes, the ciliates, especially in anaerobic species that thrive in the rumen of ruminants. These ciliates share a relatively large number of genes (148) with prokaryotes. Those genes are absent in other sequenced eukaryotic genomes, suggesting that they are specific acquisitions (Ricard et al. 2006). Some of these genes have allowed these ciliates carrying out essential function for their life in the rumen, for example, the capacity to degrade complex carbohydrates, which are very abundant in the rumen due to the herbivorous nutrition of ruminants.

These studies appear to confirm the existence of a correlation between phagotrophy and the acquisition of genes by HGT, but other cases have also been characterized in non-phagotrophic species, such as chlorophytes and fungi. For example, the analysis of the complete genome sequence of the unicellular green alga *Ostreococcus tauri* has revealed many cases of HGT, likely from bacterial donors. Surprisingly, almost all those genes are concentrated in a single chromosome of this alga and have no homologues in other related algal species, suggesting that they were acquired recently (Derelle et al. 2006). This is in sharp contrast with the case of another green alga, *Chlamydomonas reinhardtii*, for which no HGT at all has been detected (Keeling and Palmer 2008). In fungi (osmotrophic eukaryotes), a certain number of genes acquired by HGT has been identified. For instance, 13 genes are of bacterial origin in the yeast *S. cerevisiae*. These genes represent a tiny proportion of the yeast genome, but their functions are very important, including the capacity to grow in anaerobic conditions and the synthesis of some vitamins (Hall et al. 2005). This reminds the case of ciliates and fungi inhabiting the rumen, which have also acquired bacterial genes that allow them adapting to this anaerobic habitat, such as the glycosyl hydrolases involved in the degradation of cellulosic polymers (Garcia-Vallve et al. 2000). All these examples show that phagotrophy is not indispensable to acquire genes by HGT in eukaryotes.

An interesting trend shown by these analyses is that HGT from bacterial donors has played an important role in the adaptation of very diverse eukaryotes to anaerobic life. The example of rumen ciliates and fungi has already been commented, but many other cases are known. For instance, in trypanosomatids, parasites belonging to the phylum Euglenozoa, several dozens of genes of bacterial origin have been characterized (Berriman et al. 2005). This is also the case for *Cryptosporidium*, a parasite belonging to the phylum Alveolata (Huang et al. 2004), and for several parasitic diplomonad species such as *Giardia lamblia* and *Spironucleus salmonicida* (Andersson et al. 2007). In all these protists, genes acquired from bacteria are involved in functions linked to their anaerobic lifestyle. These genes have certainly played a crucial role in the adaptation of these eukaryotes to oxygen-depleted environments, either permanently in Diplomonadida or transiently in the case of the complex life cycles of trypanosomatids and *Cryptosporidium*, which alternate aerobic and anaerobic phases (Andersson et al. 2007). It is important to note that these changes in lifestyle have happened several times independently during the evolutionary history of eukaryotes. It is paradoxical to notice that if these eukaryotes have adapted to aerobic life thanks to bacteria (through the mitochondrial endosymbiosis), bacteria have also allowed them to readapt to anaerobic niches as donors of genes in HGT events.

12.4.2.2 HGT Among Eukaryotes

In the previous section, we have shown that eukaryotes, even those belonging to sexual multicellular lineages, have experienced HGT and incorporated genes of bacterial origin into their genomes, but are there also HGT events among eukaryotic species? The dominant view in the scientific community is that this type of HGT is rare. However, this may rather reflect the lack of data necessary to address this question and the persistence of strong preconceptions. Methodological difficulties explain in part the scarcity of studies focused on this question. In fact, it is relatively easy to detect genes of bacterial origin in a eukaryotic genome, but it is much more difficult to detect a gene acquired from another eukaryotic lineage, since this requires a much more precise phylogenetic signal. Despite these technical limitations, some cases of HGT between eukaryotic lineages have been described. Many of them concern EGTs found in photosynthetic eukaryotes with secondary plastids, such as Cryptophyta, Haptophyta, Heterokonta, Alveolata, and Chlorarachniophyta. As in the case of mitochondria and primary plastids (those directly derived from the endosymbiosis of a cyanobacterium, see above), secondary plastids have very reduced genomes, which in part reflects the massive transfer of genes by EGT from the endosymbiont to the host nucleus (Timmis et al. 2004). The study of the genome sequences of some secondary photosynthetic eukaryotes, such as chlorarachniophytes and diatoms, has shown that the nuclear genomes of these organisms have acquired a relatively large number of genes from their secondary plastids (Archibald et al. 2003; Curtis et al. 2012).

In contrast with the HGT cases of genes of bacterial origin, which very often have provided new functions, several HGTs among eukaryotes concern genes with functions that already existed in the receptor cell. Thus, these can be considered as homologous replacements (Keeling and Palmer 2008). Those HGTs can even affect essential genes, such as the alpha-tubulin of the jakobid *Andalucia*, which has been replaced by a homologue from a diplomonad species (Simpson et al. 2008). Nevertheless, the acquisition of new functions by eukaryote-to-eukaryote HGT has also been described, especially in fungi. As in the case of the bacterial genes acquired by fungi, those of eukaryotic origin have also provided important adaptive functions. This is the case, for example, of the cell recognition systems involved in sexual reproduction (Inderbitzin et al. 2005), of the systems of absorption of molecules from the environment (Slot and Hibbett 2007), or of virulence factors (Friesen et al. 2006). This latter example is remarkable because it concerns the very recent acquisition of a toxin-coding gene by a non-pathogenic fungus that has led to the emergence of a new disease in the 1940s. Cases of HGT in the opposite direction, namely, from fungi to other unrelated eukaryotes, are also known. For example, the analysis of the complete genome sequences of oomycete species (filamentous parasites belonging to the phylum Heterokonta) has shown the presence of 11 genes acquired by HGT from filamentous fungi (Richards et al. 2006). Most of these genes have functions involved in osmotrophy, which suggests that this nutritional character may have evolved secondarily in oomycetes by convergence with fungi thanks to the transfer of those genes. These examples show the importance of HGT in the adaptation of eukaryotes to new ecological niches.

12.5 Conclusion

Horizontal gene transfer and the mechanisms associated have been discovered during the first half of the twentieth century. Despite intensive studies, the detailed molecular mechanisms have not been completely understood. This is nevertheless a fundamental question in biology as HGT plays a major role in the adaptation of microorganisms to their environment. However, microbiologists have considered for a long time that this phenomenon was restricted to genes involved in the rapid adaptation of prokaryotes to sudden changes in their ecosystems. The arrival of the genome era has revealed that the quantitative and qualitative significance of HGT had been underestimated. In fact, HGTs get rid of the geographical barriers allowing the exchange of genetic material among individuals living in different ecosystems. Similarly, through the acquisition of genes from donors belonging to different species, HGT has compelled to reexamine the barriers between species, a central concept in biology. It has also been demonstrated that HGT affects all living beings, including eukaryotes.

Today, most evolutionists consider that HGT has contributed (and still contributes) to build up the prokaryotic genomes. The most fervent advocates propose that HGT is so dominant that concepts as notion of species or of evolution have to be completely revised. However, there is a debate on this topic in the scientific community that is still alive and far from settled, so that a cautious approach to these questions remains necessary. For example, despite the increasing availability of genome data and the improvement of detection methods, the accurate quantification of HGT remains very difficult to achieve. There are multiple methodological problems, and, up to now, any approach has not been able to provide a satisfying solution. In addition, even if HGT is frequent, it is important to remind that genes also experience a vertical evolution. For example, the *E. coli* strains K12 and O157:H7 have about 200 different genome fragments (~30 % of their genomes), which most likely correspond to ~200 events of insertion or deletion linked to HGT. Knowing that the two strains diverged ~10 million years ago, this corresponds to an HGT event every $2 \times 10^7/200 = 10^5$ years. Assuming 100 generations per year for this bacterial species, the number of generations that have existed since the divergence of the two strains is of $10^7 \times 100 = 10^9$. Thus, vertical inheritance is much more frequent than horizontal transfer. This implies that even if HGT is common at the scale of complete genomes, it has to be cautiously considered by taking into account that it has occurred over very long time spans.

The development, commercialization, and massive utilization of a large number of new antibiotics, fungicides, and biotechnological molecules have induced a strong selection pressure on microorganisms. These can use HGT as a way to adapt as they did before the arrival of these human-induced stresses. Large-scale analyses of genomes and metagenomes will allow in forthcoming years acquiring a better knowledge on the processes and the prevalence of HGT in the environment.

References

Adams KL, Qiu YL, Stoutemyer M, Palmer JD (2002) Punctuated evolution of mitochondrial gene content: high and variable rates of mitochondrial gene loss and transfer to the nucleus during angiosperm evolution. Proc Natl Acad Sci U S A 99:9905–9912

Andersson JO et al (2007) A genomic survey of the fish parasite *Spironucleus salmonicida* indicates genomic plasticity among diplomonads and significant lateral gene transfer in eukaryote genome evolution. BMC Genomics 8:51

Aravind L, Tatusov RL, Wolf YI, Walker DR, Koonin EV (1998) Evidence for massive gene exchange between archaeal and bacterial hyperthermophiles. Trends Genet 14:442–444

Archibald JM, Rogers MB, Toop M, Ishida K, Keeling PJ (2003) Lateral gene transfer and the evolution of plastid-targeted proteins in the secondary plastid-containing alga *Bigelowiella natans*. Proc Natl Acad Sci U S A 100:7678–7683

Avery OT, MacLeod CM, McCarty M (1944) Studies on the chemical nature of the substance inducing transformation of pneumococcal types. Induction of transformation by a deoxyribonucleic acid fraction isolated from *Pneumococcus* type III. J Exp Med 79:137–158

Bapteste E et al (2009) Prokaryotic evolution and the tree of life are two different things. Biol Direct 4:34

Barlow M (2009) What antimicrobial resistance has taught us about horizontal gene transfer. Methods Mol Biol 532:397–411

Berriman M et al (2005) The genome of the African trypanosome *Trypanosoma brucei*. Science 309:416–422

Bhavnani SM, Hammel JP, Jones RN, Ambrose PG (2005) Relationship between increased levofloxacin use and decreased susceptibility of *Streptococcus pneumoniae* in the United States. Diagn Microbiol Infect Dis 51:31–37

Brochier-Armanet C, Forterre P (2007) Widespread distribution of archaeal reverse gyrase in thermophilic bacteria suggests a complex history of vertical inheritance and lateral gene transfers. Archaea 2:83–93

Brochier-Armanet C, Deschamps P, López-García P, Zivanovic Y, Rodríguez-Valera F, Moreira D (2011) Complete-fosmid and fosmid-end sequences reveal frequent horizontal gene transfers in marine uncultured planktonic archaea. ISME J 5:1291–1302

Burge C, Campbell AM, Karlin S (1992) Over- and underrepresentation of short oligonucleotides in DNA sequences. Proc Natl Acad Sci U S A 89:1358–1362

Cascales E, Christie PJ (2003) The versatile bacterial type IV secretion systems. Nat Rev Microbiol 1:137–149

Cavalier-Smith T (1999) Principles of protein and lipid targeting in secondary symbiogenesis: euglenoid, dinoflagellate, and sporozoan plastid origins and the eukaryote family tree. J Eukaryot Microbiol 46:347–366

Ceremonie H, Buret F, Simonet P, Vogel TM (2004) Isolation of lightning-competent soil bacteria. Appl Environ Microbiol 70:6342–6346. 509

Chen I, Dubnau D (2004) DNA uptake during bacterial transformation. Nat Rev Microbiol 2:241–249

Curtis BA et al (2012) Algal genomes reveal evolutionary mosaicism and the fate of nucleomorphs. Nature 492:59–65

d'Herelle F (1917) The nature of bacteriophage. Br Med 2:289–293

Danner DB, Deich RA, Sisco KL, Smith HO (1980) An eleven-base-pair sequence determines the specificity of DNA uptake in *Haemophilus* transformation. Gene 11:311–318

Davis BD (1950) Nonfiltrability of the agents of genetic recombination in *Escherichia coli*. J Bacteriol 60:507–508

de Vries J, Wackernagel W (2002) Integration of foreign DNA during natural transformation of *Acinetobacter* sp. by homology-facilitated illegitimate recombination. Proc Natl Acad Sci U S A 99:2094–2099

Deppenmeier U et al (2002) The genome of *Methanosarcina mazei*: evidence for lateral gene transfer between bacteria and archaea. J Mol Microbiol Biotechnol 4:453–461

Derelle E et al (2006) Genome analysis of the smallest free-living eukaryote *Ostreococcus tauri* unveils many unique features. Proc Natl Acad Sci U S A 103:11647–11652

Doolittle WF (1998) You are what you eat: a gene transfer ratchet could account for bacterial genes in eukaryotic nuclear genomes. Trends Genet 14:307–311

Doolittle WF, Boucher Y, Nesbo CL, Douady CJ, Andersson JO, Roger AJ (2003) How big is the iceberg of which organellar genes in nuclear genomes are but the tip? Philos Trans R Soc Lond B Biol Sci 358:39–57; discussion 57–38

Downie AW (1972) Pneumococcal transformation – a backward view. Fourth Griffith Memorial Lecture. J Gen Microbiol 73:1–11

Duckworth DH (1976) Who discovered bacteriophage? Bacteriol Rev 40:793–802

Eichinger L et al (2005) The genome of the social amoeba *Dictyostelium discoideum*. Nature 435:43–57

Esser C et al (2004) A genome phylogeny for mitochondria among alpha-proteobacteria and a predominantly eubacterial ancestry of yeast nuclear genes. Mol Biol Evol 21:1643–1660

Forterre P (2002) A hot story from comparative genomics: reverse gyrase is the only hyperthermophile-specific protein. Trends Genet 18:236–237

Forterre P, Bouthier De La Tour C, Philippe H, Duguet M (2000) Reverse gyrase from hyperthermophiles: probable transfer of a thermoadaptation trait from *Archaea to Bacteria*. Trends Genet 16:152–154

Friesen TL et al (2006) Emergence of a new disease as a result of interspecific virulence gene transfer. Nat Genet 38:953–956

Futterer O et al (2004) Genome sequence of *Picrophilus torridus* and its implications for life around pH 0. Proc Natl Acad Sci U S A 101:9091–9096

Garcia-Vallve S, Romeu A, Palau J (2000) Horizontal gene transfer of glycosyl hydrolases of the rumen fungi. Mol Biol Evol 17:352–361

Gautier C (2000) Compositional bias in DNA. Curr Opin Genet Dev 10:656–661

Ge F, Wang LS, Kim J (2005) The cobweb of life revealed by genome-scale estimates of horizontal gene transfer. PLoS Biol 3:e316

Gilmour MW, Thomson NR, Sanders M, Parkhill J, Taylor DE (2004) The complete nucleotide sequence of the resistance plasmid R478: defining the backbone components of incompatibility group H conjugative plasmids through comparative genomics. Plasmid 52:182–202

Goodman SD, Scocca JJ (1988) Identification and arrangement of the DNA sequence recognized in specific transformation of *Neisseria gonorrhoeae*. Proc Natl Acad Sci U S A 85:6982–6986

Gouy M, Gautier C (1982) Codon usage in bacteria: correlation with gene expressivity. Nucleic Acids Res 10:7055–7074

Griffith MB (1928) The significance of pneumococcal types. J Hygiene 27:113–159

Hall C, Brachat S, Dietrich FS (2005) Contribution of horizontal gene transfer to the evolution of *Saccharomyces cerevisiae*. Eukaryot Cell 4:1102–1115

Hallam SJ et al (2006) Genomic analysis of the uncultivated marine crenarchaeote *Cenarchaeum symbiosum*. Proc Natl Acad Sci U S A 103:18296–18301

Hayashi T et al (2001) Complete genome sequence of enterohemorrhagic *Escherichia coli* O157:H7 and genomic comparison with a laboratory strain K-12. DNA Res 8:11–22

Heinemann JA, Sprague GF Jr (1989) Bacterial conjugative plasmids mobilize DNA transfer between bacteria and yeast. Nature 340:205–209

Hotopp JC et al (2007) Widespread lateral gene transfer from intracellular bacteria to multicellular eukaryotes. Science 317:1753–1756

Huang J, Mullapudi N, Lancto CA, Scott M, Abrahamsen MS, Kissinger JC (2004) Phylogenomic evidence supports past endosymbiosis, intracellular and horizontal gene transfer in *Cryptosporidium parvum*. Genome Biol 5:R88

Ikeda H, Shiraishi K, Ogata Y (2004) Illegitimate recombination mediated by double-strand break and end-joining in *Escherichia coli*. Adv Biophys 38:3–20

Inderbitzin P, Harkness J, Turgeon BG, Berbee ML (2005) Lateral transfer of mating system in *Stemphylium*. Proc Natl Acad Sci U S A 102:11390–11395

Keeling PJ, Palmer JD (2008) Horizontal gene transfer in eukaryotic evolution. Nat Rev Genet 9:605–618

Kennedy SP, Ng WV, Salzberg SL, Hood L, Das-Sarma S (2001) Understanding the adaptation of *Halobacterium* species NRC-1 to its extreme environment through computational analysis of its genome sequence. Genome Res 11:1641–1650

Koski LB, Golding GB (2001) The closest BLAST hit is often not the nearest neighbor. J Mol Evol 52:540–542

Koski LB, Morton RA, Golding GB (2001) Codon bias and base composition are poor indicators of horizontally transferred genes. Mol Biol Evol 18:404–412

Lambert PA (2005) Bacterial resistance to antibiotics: modified target sites. Adv Drug Deliv Rev 57:1471–1485

Lan R, Reeves PR (1996) Gene transfer is a major factor in bacterial evolution. Mol Biol Evol 13:47–55

Lan R, Reeves PR (2000) Intraspecies variation in bacterial genomes: the need for a species genome concept. Trends Microbiol 8:396–401

Lawrence JG, Ochman H (1997) Amelioration of bacterial genomes: rates of change and exchange. J Mol Evol 44:383–397

Lawrence JG, Ochman H (1998) Molecular archaeology of the *Escherichia coli* genome. Proc Natl Acad Sci U S A 95:9413–9417

Lederberg J (1947) Gene recombination and linked segregations in *Escherichia Coli*. Genetics 32:505–525

Lerat E, Daubin V, Moran NA (2003) From gene trees to organismal phylogeny in prokaryotes: the case of the gamma-Proteobacteria. PLoS Biol 1:E19

Livermore D (2004) Can better prescribing turn the tide of resistance? Nat Rev Microbiol 2:73–78

Livermore DM, Woodford N (2000) Carbapenemases: a problem in waiting? Curr Opin Microbiol 3:489–495

Lopez-Garcia P, Brochier C, Moreira D, Rodriguez-Valera F (2004) Comparative analysis of a genome fragment of an uncultivated mesopelagic crenarchaeote reveals multiple horizontal gene transfers. Environ Microbiol 6:19–34

Lorenz MG, Wackernagel W (1994) Bacterial gene transfer by natural genetic transformation in the environment. Microbiol Rev 58:563–602

Makarova KS, Sorokin AV, Novichkov PS, Wolf YI, Koonin EV (2007) Clusters of orthologous genes for 41 archaeal genomes and implications for evolutionary genomics of archaea. Biol Direct 2:33

Martin W et al (2002) Evolutionary analysis of *Arabidopsis*, cyanobacterial, and chloroplast genomes reveals plastid phylogeny and thousands of cyanobacterial genes in the nucleus. Proc Natl Acad Sci U S A 99:12246–12251

Maughan H, Redfield RJ (2009) Extensive variation in natural competence in *Haemophilus Influenzae*. Evolution 63:1852–1866

Millen RS et al (2001) Many parallel losses of infA from chloroplast DNA during angiosperm evolution with multiple independent transfers to the nucleus. Plant Cell 13:645–658

Mongodin EF et al (2005) The genome of *Salinibacter ruber*: convergence and gene exchange among hyperhalophilic *Bacteria* and *Archaea*. Proc Natl Acad Sci U S A 102:18147–18152

Murray NE (2000) Type I restriction systems: sophisticated molecular machines (a legacy of Bertani and Weigle). Microbiol Mol Biol Rev 64:412–434

Nakamura Y, Itoh T, Matsuda H, Gojobori T (2004) Biased biological functions of horizontally transferred genes in prokaryotic genomes. Nat Genet 36:760–766

Nelson KE et al (1999) Evidence for lateral gene transfer between *Archaea* and *Bacteria* from genome sequence of *Thermotoga maritima*. Nature 399:323–329

Nesbo CL, L'Haridon S, Stetter KO, Doolittle WF (2001) Phylogenetic analyses of two "archaeal" genes in *Thermotoga maritima* reveal multiple transfers between archaea and bacteria. Mol Biol Evol 18:362–375

Nikoh N, Tanaka K, Shibata F, Kondo N, Hizume M, Shimada M, Fukatsu T (2008) *Wolbachia* genome integrated in an insect chromosome: evolution and fate of laterally transferred endosymbiont genes. Genome Res 18:272–280

Normand P et al (2007) Genome characteristics of facultatively symbiotic *Frankia* sp. strains reflect host range and host plant biogeography. Genome Res 17:7–15

Novick RP (1987) Plasmid incompatibility. Microbiol Rev 51:381–395

Ochman H, Lawrence JG, Groisman EA (2000) Lateral gene transfer and the nature of bacterial innovation. Nature 405:299–304

Philippe H, Germot A, Moreira D (2000) The new phylogeny of eukaryotes. Curr Opin Genet Dev 10:596–601

Podell S, Gaasterland T (2007) DarkHorse: a method for genome-wide prediction of horizontal gene transfer. Genome Biol 8:R16

Ragan MA (2001) On surrogate methods for detecting lateral gene transfer. FEMS Microbiol Lett 201:187–191

Ramirez-Arcos S, Fernandez-Herrero LA, Marin I, Berenguer J (1998) Anaerobic growth, a property horizontally transferred by an Hfr-like mechanism among extreme thermophiles. J Bacteriol 180:3137–3143

Ricard G et al (2006) Horizontal gene transfer from *Bacteria* to rumen Ciliates indicates adaptation to their anaerobic, carbohydrates-rich environment. BMC Genomics 7:22

Richards TA, Dacks JB, Jenkinson JM, Thornton CR, Talbot NJ (2006) Evolution of filamentous plant pathogens: gene exchange across eukaryotic kingdoms. Curr Biol 16:1857–1864

Rocha EP, Viari A, Danchin A (1998) Oligonucleotide bias in *Bacillus subtilis*: general trends and taxonomic comparisons. Nucleic Acids Res 26:2971–2980

Roger AJ (1999) Reconstructing Early events in eukaryotic evolution. Am Nat 154:S146–S163

Ruepp A et al (2000) The genome sequence of the thermoacidophilic scavenger *Thermoplasma acidophilum*. Nature 407:508–513

Simpson AG, Perley TA, Lara E (2008) Lateral transfer of the gene for a widely used marker, alpha-tubulin, indicated by a multi-protein study of the phylogenetic position of *Andalucia* (Excavata). Mol Phylogenet Evol 47:366–377

Slot JC, Hibbett DS (2007) Horizontal transfer of a nitrate assimilation gene cluster and ecological transitions in fungi: a phylogenetic study. PLoS One 2:e1097

Smith HO, Tomb JF, Dougherty BA, Fleischmann RD, Venter JC (1995) Frequency and distribution of DNA uptake signal sequences in the *Haemophilus influenzae* Rd genome. Science 269:538–540

Suttle CA (2007) Marine viruses – major players in the global ecosystem. Nat Rev Microbiol 5:801–812

Szollosi GJ, Boussau B, Abby SS, Tannier E, Daubin V (2012) Phylogenetic modeling of lateral gene transfer reconstructs the pattern and relative timing of speciations. Proc Natl Acad Sci U S A 109:17513–17518

Taddei F, Matic I, Radman M (1996) Du nouveau sur l'origine des especes. La recherche 291:52–59

Tatum EL, Lederberg J (1947) Gene recombination in the bacterium *Escherichia coli*. J Bacteriol 53:673–684

Tettelin H et al (2005) Genome analysis of multiple pathogenic isolates of *Streptococcus agalactiae*: implications for the microbial "pangenome". Proc Natl Acad Sci U S A 102:13950–13955

Thomas CM, Nielsen KM (2005) Mechanisms of, and barriers to, horizontal gene transfer between bacteria. Nat Rev Microbiol 3:711–721

Timmis JN, Ayliffe MA, Huang CY, Martin W (2004) Endosymbiotic gene transfer: organelle genomes forge eukaryotic chromosomes. Nat Rev Genet 5:123–135

Touchon M et al (2009) Organised genome dynamics in the Escherichia coli species results in highly diverse adaptive paths. PLoS Genet 5:e1000344

Twort FW (1915) An investigation on the nature of ultra-microscopic viruses. Lancet 2:1241–1243

Zeldovich KB, Berezovsky IN, Shakhnovich EI (2007) Protein and DNA sequence determinants of thermophilic adaptation. PLoS Comput Biol 3:e5

Zhaxybayeva O (2009) Detection and quantitative assessment of horizontal gene transfer. Methods Mol Biol 532:195–213

Zhaxybayeva O, Gogarten JP, Charlebois RL, Doolittle WF, Papke RT (2006) Phylogenetic analyses of cyanobacterial genomes: quantification of horizontal gene transfer events. Genome Res 16:1099–1108

Zhaxybayeva O, Nesbo CL, Doolittle WF (2007) Systematic overestimation of gene gain through false diagnosis of gene absence. Genome Biol 8:402

Zhaxybayeva O et al (2009) On the chimeric nature, thermophilic origin, and phylogenetic placement of the *Thermotogales*. Proc Natl Acad Sci U S A 106:5865–5870

Zinder ND (1992) Forty years ago: the discovery of bacterial transduction. Genetics 132:291–294

Zinder ND, Lederberg J (1952) Genetic exchange in *Salmonella*. J Bacteriol 64:679–699. 512 512

Part IV

Role and Functioning of Microbial Ecosystems

Microbial Food Webs in Aquatic and Terrestrial Ecosystems

Behzad Mostajir, Christian Amblard, Evelyne Buffan-Dubau, Rutger De Wit, Robert Lensi, and Télesphore Sime-Ngando

Abstract

In microbial food webs, different types of interactions occur between microorganisms themselves and with meio- and macroorganisms. After an historical and general introduction, the biological components of the microbial food webs in the pelagic and benthic marine and lake ecosystems, as well as in the terrestrial ecosystems, are presented. The functioning of the microbial food webs in different ecosystems is illustrated and explained, including the trophic pathways and transfer of matter from microbial food webs toward meio- and macroorganisms of the superior trophic levels, the nutrient recycling in the aquatic environments, and the decomposition of organic matter in soils. Finally, the factors regulating microbial food webs, primarily "top-down" and "bottom-up" controls, are described with a special focus on the role of viruses in the aquatic microbial food webs.

Keywords

Biodiversity • Biogeochemical cycles • Ecological interactions • Microbial food webs • Microbial loop

* Chapter Coordinators

B. Mostajir* (✉) • R. De Wit
Écologie des systèmes marins côtiers (ECOSYM, UMR5119),
Universités Montpellier 2 et 1, CNRS-Ifremer-IRD, 34095 Montpellier Cedex 05, France
e-mail: Behzad.Mostajir@univ-montp2.fr;
rutger.de-wit@univ-montp2.fr

C. Amblard* (✉) • T. Sime-Ngando
Laboratoire Microorganismes: Génome et Environnement (LMGE),
UMR CNRS 6023, Université Blaise Pascal, Clermont Université, B.P. 80026, 63171 Aubère Cedex, France
e-mail: christian.amblard@univ-bpclermont.fr;
telesphore.sime-ngando@univ-bpclermont.fr

E. Buffan-Dubau
Laboratoire d'Écologie Fonctionnelle et Environnement (ECOLAB),
UMR CNRS 5245, Université Paul Sabatier, 31062 Toulouse Cedex 9, France
e-mail: evelyne.buffan-dubau@univ-tlse3.fr

R. Lensi
Centre d'Écologie Fonctionnelle et Évolutive (CEFE), Département d'Écologie Fonctionnelle, UMR 5175, 34293 Montpellier Cedex 5, France

13.1 Introduction

13.1.1 Pelagic Ocean and Lake Ecosystems

The quantitative and functional importance of microorganisms in aquatic food web was considered seriously only since around 40 years (Pomeroy 1974). This is in spite of the fact that Lohmann, at the beginning of the 1900s (quoted in Laval-Peuto et al. 1986), already recognized the importance of the microzooplankton. In the same way, Vernadskii (1926, quoted in Pomeroy 1991) referred frequently to microorganisms as an important component of the general functioning of the oceans, making reference, especially, to protozoan and to their roles in the flows of matter and energy and in the biogeochemical cycles. In fact, before the discovery of the functional role of microorganisms, the interactions between the organisms, in particular heterotrophic ones, were considered in a simplistic way. The structure of the food web for the aquatic systems was so considered as being established by three levels: the production of organic matter (OM) by the phytoplankton, its

Table 13.1 Size-classes classification of virus and planktonic microorganisms

Category	Size classes	Virus and microorganisms
Femtoplankton	0.02–0.2 μm	Virus (mostly), archaea, and small bacteria
Picoplankton	0.2–2 μm	Larger virus, archaea and bacteria (mostly), cyanobacteria, small auto- and heterotrophic eukaryotes
Nanoplankton	2–20 μm	Hetero-, auto-, and mixotrophic flagellates, small ciliates, small naked amoeba, zoospores, and filaments of fungi
Microplankton	20–200 μm	Ciliates (mostly), amoeba (mostly), other Sarcodina (Foraminifera, heliozoaires), dinoflagellates, pigmented unicellular eukaryotes (Desmidiae, diatoms, etc.)
Mesoplankton	0.2–20 mm	Filamentous cyanobacteria

consumption (grazing) by the zooplankton (metazoan), and the predation of this one by fishes. As, in this simple and linear trophic chain, the grazing was mainly as a central role, it had so been identified in the aquatic system, "grazer food chain," and the trophic chain based on the detritus "detritus food chain" (Wetzel et al. 1972).

This simplified vision of the aquatic food web resulted from methods of sampling, preservation, and especially observation, which were adapted well to the large-sized planktonic organisms but were not appropriate for the smallest microorganisms. For practical reasons, the planktonic organisms were separated in several size classes (Sieburth et al. 1978): femto (0.02–0.2 μm), pico (0.2–2 μm), nano (2–20 μm), micro (20–200 μm), meso (0.2–20 mm), macro (2–20 cm), and mega (20–200 cm) (Table 13.1).

The conceptual simplistic model of the food web was thus progressively modified. First of all, the observations which come from numerous oceanic and lake sites showed that the pico- and nanoplankton (<20 μm) were responsible for the largest part of the aerobic respiration in the pelagic community (Pomeroy 1974). Afterward, the observation and the quantification of the auto- and heterotrophic picoplankton in the aquatic ecosystems became possible by using the epifluorescence microscopy (Hobbie et al. 1977) and, additionally, by the electronic microscopy (cf. Sect. 17.2.1). The phototrophic picoplankton has proved of a major importance, in particular in the oligotrophic environments where more than 50 % of the total primary production is due to organisms < 3 μm (Li et al. 1983). These discoveries concerning the major role of the pico- and the nanoplankton in the functioning of the aquatic ecosystems ended in the development of a new conception of the aquatic microbial ecology, one of the fields of study of which is microbial food web.

The investigations led in the lake ecosystems during the last three decades allowed in the same way to show that the flows of matter and energy are not organized only according to the linear classic trophic pathway based on the photosynthetic assimilation (phytoplankton → zooplankton → fishes), but also borrow the microbial food web formed by the very small-sized microorganisms (pico- and nanoplankton; Carpenter 1988) to form a real food web.

The structure of the microbial food web depends, on one hand, on the type of environments and, on the other hand, on forcing factors exerting on these environments (cf. Sect. 13.4). In aquatic environment, pelagic and benthic domains are mostly contrasted and this as well in the oceans, that in lagoons or in lakes. The pelagic ecosystems are situated between the atmosphere and the benthic zone. The coastal marine ecosystems are also the interface between the continent and the oceanic zone. Therefore, the coastal pelagic ecosystems are the buffer zones, on the vertical plan between the air and the benthic zone and/or on the horizontal plan, between the continent and the oceanic zone. The pelagic ecosystem is, consequently, very often under the influence of the process taking place in the adjacent ecosystems (light, rain, wind, what comes from watershed, resuspension of sediments or upwelling, etc.).

13.1.2 Benthic Ecosystems

The benthos is defined as the community living at the bottom of the aquatic systems. The sediment systems represent a particularly important physical support for many benthic organisms. It comprises two phases, i.e., solid represented by the mineral and organic particles and liquid as the water occupying the interstitial spaces between these particles. By comparisons, soils comprise three different physical phases (solid, liquid, and gas). The benthic organisms can live at the surface and in the interior of the sediments. The microbial communities that develop on the surfaces of benthic macrophytes (periphyton) and hard bottoms also belong to the benthos.

A large part of the benthic microbial prokaryotes and eukaryotes adhere to the solid surface and are really sessile, while others are mobile and move by gliding over surfaces or by swimming in the interstitial water space. The densities of prokaryotes living in the first mm of the sediment exceed their densities in the overlying water column often by a factor of 10^3 to 10^4. Hence, the total quantity of prokaryotes living in the top cm of the sediments is equivalent to that occurring in a water column of tens of meters height. The sediment thus represents a biogeochemical reactor, which is particularly involved in the mineralization of the organic matter (OM – cf. Sect. 14.2.4). In deep systems (the ocean offshore and larger deep lakes), the OM input into the sediment originates mainly from the water column and arrives by sedimentation. In shallower aquatic ecosystems, this OM can also be produced on site by benthic macrophytes (aquatic Magnoliophyta and macroalgae), benthic microalgae, and cyanobacteria. The benthos of lakes and coastal systems also receive OM input from their watersheds through advective transport. The benthic systems covered by an aquatic vegetation of macrophytes bear strong resemblances with the terrestrial systems.

Many benthic prokaryote species and microalgae release large quantities of extracellular polymeric substances (EPS). The polymers of the EPS play several important roles for the benthic microorganisms including (1) their adhesion to the surfaces, (2) their mobility by gliding, and (3) as an important stock of energy that can be used later under starvation conditions.

EPS is generally very rich in carbon (C) and poor in nitrogen (N) and phosphorus (P). Hence, the excretion of EPS can, in some cases, also be the result of unbalanced growth of benthic microorganisms under N and/or P limiting conditions.

The EPS thus represents a stock of OM that represents a source of food for any different benthic organisms, and, therefore, EPS represents an important compartment that should be considered for benthic microbial food webs. EPS is particularly important in biofilms; here the polymers form an organic matrix in which the microorganisms are embedded. Photosynthetic biofilms are a special case where microalgae and cyanobacteria are the primary producers.

13.1.3 Terrestrial Ecosystems

As described for the aquatic ecosystems, the soils of terrestrial systems also deliver important ecosystem services to humans. Soils occur at the interface between the atmosphere, the lithosphere, the hydrosphere, and the biosphere. The soil ecosystems contribute to the major biogeochemical cycles, including the water cycle, and the major element (C, N, P, and others) cycles. The soils are the support for agriculture, forestry, and pastoral exploitation. These ecosystems contribute to the regulation of the climate and to detoxification as well to erosion control (Lavelle et al. 2006). For the soil constituents, four different organization levels can be identified:
1. The aggregate (ranging from a couple of mm to several dm in size), which represents a coherent group of elementary soil particles
2. The assemblage (an association of aggregates)
3. The horizon (ranging from a couple of cm to several m), soil layers morphologically homogeneous
4. Pedological cover: three-dimensional all horizons

Because aggregation of soil particles (1) and assemblage of soil aggregates (2) have such a profound effect on soil structure, it is often observed that the topsoil layers show a complex three-dimensional distribution of the main environmental variables (dioxygen partial pressure, OM contents, redox potential). This is in contrast to most sediments, where a one-dimensional depth stratification of these variables overwhelms their heterogeneity in the horizontal plane. However, sediments that are submitted to important bioturbation are an exception in this respect.

The presence of a gas phase in soils obviously results in important differences in oxygenation among (1) the zone in between the aggregates and (2) the zone located in the interior of the aggregates and (3) oxic-anoxic microgradients across the aggregates (Sextone et al. 1985; Renault and Stengel 1994). The way of how the different particles and other constituents are ordered within and at the surface of a soil aggregate is always very complex. However, the resulting three-dimensional heterogeneity of environmental conditions and the occurrence of micro-gradients vary a lot among different soils. These three-dimensional structures are very well developed in clearly fragmented structures (soil layers where aggregates are visible to the naked eye) and become progressively less important in structures built of particles (where soil particles remain dissociated from each other) and continuous structures (where no visible aggregates are present).

The biological functioning of the soils encompasses the functioning of all the organisms, which, in interaction with the chemical and physical components of the soil, contribute to nutrient recycling and water flows. These functions are realized by organisms that range in size from microorganisms (*Archaea*, *Bacteria*, unicellular heterotrophic eukaryotes) to invertebrates. Moreover, strong interactions exist between these organisms, which contribute to the quality of the above-mentioned ecosystem services. Despite this obvious fact, the concept of a food web has not shaped the studies of the soil to a comparable extent as it has done for aquatic ecosystems. One of the main reasons appears to be that the function of primary productivity in terrestrial ecosystems is mainly realized by higher plants and only to a very minor extent by microorganisms. Hence, the functioning of the biological component of belowground terrestrial systems is mainly dependent on the vegetation. In the first place, the plants deliver almost all the energy to this belowground system. In addition, through their root systems, the plants exert an important control of the biophysicochemical interactions in the soil and of the structuring of the soils and can, therefore, be considered as important **soil engineers***.

Another difficulty for developing food web studies in soils is that it represents an environment comprising three phases (water, solid, and gas). In contrast to aquatic environments, the proportion of these three phases can rapidly change as a function of the climatic conditions. This implies that the organisms in the soil need a large degree of adaptability and adapt quickly. As an example that will be treated in more detail later, protozoa in soils need a water film particularly to exert their predatory activity. These protozoa become inactive and form cysts when the water content drops below a certain level (Bardgett 2005a). Finally, as in many other ecosystems, the large majority of the microorganisms in the soil cannot be cultured using the currently available techniques. Admittedly, the recent development of molecular techniques (which are independent of culturing) allows exploring the reel diversity of the microorganisms in the environment (Bardgett 2005b; *cf.* Sect. 17.6). Nonetheless, the efficient extraction of DNA from soils remains an important methodological problem to overcome.

13.2 Biological Components in the Microbial Food Webs

13.2.1 Microorganisms in Pelagic Ocean and Lake Ecosystems

Progress in techniques for identifying, counting, and measuring metabolic activity, in particular using epifluorescence microscopy and molecular biology, has allowed a glimpse of the extraordinary diversity of microorganisms, the extent of their living conditions, and their abundances hitherto largely underestimated, especially in aquatic environments. In addition, the sensible improvement of various chromatographic methods allowed to describe the biochemical composition of communities and address the qualitatively transfer of matter in the food webs.

In pelagic marine and lake ecosystems, planktonic lifestyle (floating) is best adapted. Prokaryotic or eukaryotic planktonic microorganisms may be heterotrophs, **mixotrophs***, or autotrophs. These microorganisms provide almost all of the biomass production and ensure a large part of the pelagic ecosystem functions. More generally, pelagic marine and lake ecosystems are the only ecosystems that the biomass is mainly microbial.

Differentiation by size classes of microorganisms, as mentioned in the introduction, is a practical way to study and is also linked to the functional role of these microorganisms in pelagic environments. The major components in the pelagic microbial food webs (Table 13.1) are presented according to the classification of Sieburth et al. (1978): femtoplankton (0.02–0.2 μm), picoplankton (0.2–2 μm), nanoplankton (2–20 μm), microplankton (20–200 μm), and mesoplankton (0.2–20 mm). The main methods of counting and estimating biomass of pelagic microorganisms are flow cytometry and epifluorescence microscopy for pico- and nanoplankton and for microplankton is reversed optical microscopy. The high-performance liquid chromatography (HPLC) is also used to estimate phytoplankton biomass from the quantification of pigments.

13.2.1.1 Femtoplankton

In addition to very small archaea and bacteria, the femtoplankton is mainly composed of virioplankton or virus-like particles (VLP) which are ubiquitous and the most abundant biological entities in aquatic environments with concentrations ranging from 10^4 to 10^8 mL^{-1} (Sime-Ngando 1997; Sime-Ngando et al. 2003; *cf.* Sect. 13.4.1).

13.2.1.2 Picoplankton

Picoplankton is composed of prokaryotic and eukaryotic organisms: archaea, heterotrophic bacteria, picophytoplankton (cyanobacteria and eukaryotic autotrophs), and heterotrophic picoflagellates. More specifically, the bacterioplankton is represented by two main domains: *Archaea* and *Bacteria*. *Archaea* are not confined environments called "extreme" and can represent up to 30 % of prokaryotes counted in oxygenated euphotic zones of lake (Jardillier et al. 2005) and marine ecosystems (Delong 1992).

The abundance of heterotrophic bacteria in pelagic marine and lake ecosystems varies from 10^5 to 10^7 mL^{-1} in very oligotrophic (poor in nutrients) and more eutrophic (rich in nutrients) ecosystems, respectively. Generally, a weak seasonal variability of the heterotrophic bacteria abundance is observed with, however, a significantly higher biomass during spring development of phytoplankton. The abundance of cyanobacteria in coastal marine environments is highly variable but is generally about 10^4 mL^{-1} (7–13 × 10^4 mL^{-1} in Northwestern Mediterranean Sea and can reach 17 × 10^4 mL^{-1} in Mediterranean lagoons). The abundance of eukaryotic picoplankton as *Ostreococcus tauri* is approximately 10^3 mL^{-1}, particularly in coastal marine environments (lagoon).

In lake environments, the abundance of autotrophic picoplankton (picocyanobacteria and eukaryotic autotrophic picoplankton) fluctuates between 10^3 and 10^6 ml^{-1} in oligotrophic and eutrophic lakes, respectively. However, autotrophic picoplankton is a major part (20–50 %) of picoplankton biomass due to a mean cell biovolume higher than heterotrophic bacteria.

The density of autotrophic eukaryotic picoplankton is generally highest during spring because of their well-adapted pigments to low irradiance of this period, while picocyanobacteria develop preferentially in summer when temperature rises.

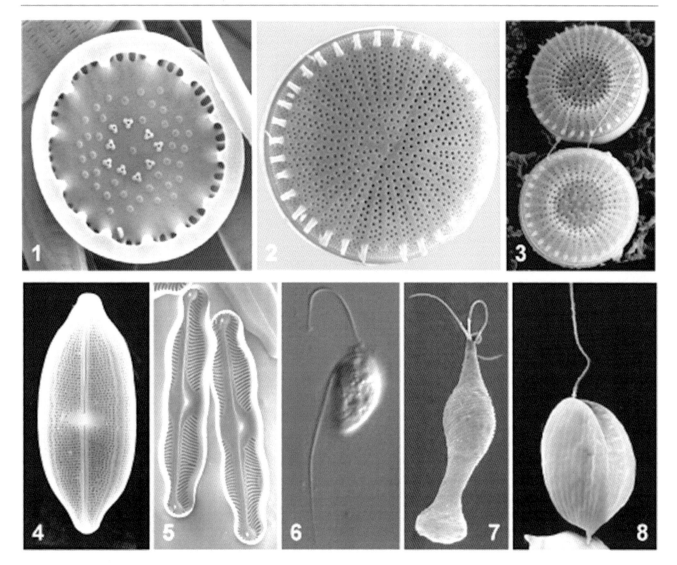

Fig. 13.1 Some examples of nano- and microplanktonic eukaryotic microorganisms in lake ecosystems. (*1*) *Cyclotella comta* (size between 8 and 40 μm), (*2*) *Stephanodiscus astrea* (10–50 μm), (*3*) *Cyclostephanos dubius* (7–20 μm), (*4*) *Neidium sp.* (30–45 μm), (*5*) *Pinnularia mesolepta* (30–60 μm), (*6*) *Bodo caudatus* (10–20 μm), (*7*) *Distigma proteus* (40–80 μm), (*8*) *Phacus sp.* (45–90 μm) (Photos: LMGE UMR CNRS 6023 – Christian Amblard and Télesphore Sime-Ngando)

The main predators of picoplankton in marine and lake ecosystems are the heterotrophic pico- and nanoflagellates which their abundance is about 1,000 cells mL^{-1}.

13.2.1.3 Nanoplankton

Nanoplankton consists of phytoplankton cells and protozooplankton (Fig. 13.1). The nanophytoplankton together with the picophytoplankton represents important components of the microbial food webs in marine and lake ecosystems. The abundance of nanophytoplankton is generally an order of magnitude less than that of picophytoplankton. The main nanoheterotrophic component is protozooplankton consisting essentially of flagellates and ciliates and, incidentally, of amoebae. Heterotrophic nanoflagellates form an abundant ubiquitous community in marine and lake ecosystems. They are considered the main predators of picoplankton and are thus a key component of microbial food web. This group is particularly heterogeneous both morphologically and physiologically. They are characterized by the presence of one or more flagella involving to cells movement and food capture. These protists flagellates can be autotroph, mixotroph, or heterotroph.

Heterotrophic nanoflagellates feed by **phagotrophy*** or **osmotrophy***. Their abundance in the marine ecosystems is less than 1,000 cells mL^{-1}. In the lake ecosystems, they are more abundant in eutrophic (10^5 mL^{-1}) than in oligotrophic (10^2 mL^{-1}) waters with a size varying generally between 2 and 30 μm, but in temperate lakes most species have a size less than 5 μm. The ciliated protozooplankton has long been ignored in pelagic environments because of its extreme sensitivity to methods of sampling and fixing commonly used. With the improvement of these methods, over 7,000 species of ciliates, free or parasites, have been described in aquatic ecosystems. This is a group characterized by the presence of a ciliate cortex in the form of simple cilia, or ciliated structures which are much more complex. Most ciliates also have a nuclear dimorphism because they have a macronucleus and usually several micronuclei.

Some groups of ciliates, such as Oligotrichia, Scuticociliates, and Haptoria, are particularly well represented in the lakes. Their size ranges from 5–10 to over 200 μm. Their nutritional mode is very diverse being mixotroph, bacterivore, algaevore, detritivore, or even omnivore. Their abundance in marine and lake ecosystems is an average of 1 cell mL^{-1}. Temperature, available light, and nutrient resources induce a heterogeneous vertical distribution of ciliates through the water column. Their abundance also varies according to nutrient concentrations.

13.2.1.4 Micro- and Mesoplankton

The microphytoplankton contributes significantly to primary production in pelagic environments. Generally, the biomass is estimated by inverted microscopy cell counts or by measuring chlorophyll (Chl) or other pigment biomarkers by HPLC (*cf.* Chap. 17). It is very difficult to determine an average value for the phytoplankton biomass in the marine environment, while Chl a concentrations fluctuated between 0.1 and 70 μg Chl a mL^{-1} in coastal lagoons. The phytoplankton biomass in the lake environments ranges from <1 to > 25 mg m^{-3} according to nutrient concentrations.

13.2.1.5 Microscopic Fungi

Recent environmental 18S-rDNA surveys of microbial eukaryotes have unveiled major infecting agents in pelagic systems, consisting primarily of the fungal order of Chytridiales (chytrids) (Lefèvre et al. 2007). Chytrids are considered the earlier branch of the Eumycetes and produce motile, flagellated zoospores, characterized by a small size (2–6 μm) and a single, posterior flagellum. The existence of these dispersal propagules includes chytrids within the so-called group of zoosporic fungi, which are particularly adapted to the plankton lifestyle where they infect a wide variety of hosts, including fishes, eggs, zooplankton, algae, and other aquatic fungi but primarily freshwater phytoplankton (Sime-Ngando 2012). In Lake Pavin (France), chytrid zoospores can contribute up to 40 % of the total diversity and 60 % of the total abundance of small-sized heterotrophic flagellates in the plankton. These flagellates are exclusively considered as grazers of prokaryotes. Ecological implications are huge because parasitic chytrids can also kill their hosts, but release substrates for microbial processes, and provide nutrient-rich particles as zoospores (Kagami et al. 2007) and short fragments (from the fragmentation of filamentous inedible hosts) for the grazer food chain. Furthermore, based on the observation that phytoplankton chytridiomycosis preferentially impacts the larger size species, blooms of such species (e.g., filamentous cyanobacteria) may not totally represent trophic bottlenecks. In the case of chytrid epidemics, this can also represent an important driving factor in phytoplankton seasonal successions and maturation, in addition to the sole seasonal forcing (Sime-Ngando 2012).

It is thus becoming evident that non-bacterivorous heterotrophic microbial eukaryotes such as chytrids, but also the members of the genus *Perkinsus* or alveolates, which are parasites or saprotrophs, represent an important but as yet overlooked ecological driving force in aquatic food web dynamics. In addition to being able to resist adverse conditions and use different sources of carbon and nutrients, these parasites can indeed affect the plankton food web functions and ecosystem properties and topology, such as stability and trophic transfer efficiency (Grami et al. 2011). We are perhaps approaching a new paradigm-shift point in the development of aquatic microbial ecology (Sime-Ngando and Niquil 2011).

13.2.2 Microorganisms of Benthic Ocean and Lake Ecosystems

Benthic consumers of microorganisms are unicellular and metazoan organisms. Unicellular consumers, such as heterotrophic ciliates, are important potential consumers of benthic microalgae and bacteria. Their density can reach up to 5,000 and 11,000 $ind.cm^3$ in marine and freshwater sediments, respectively. Other groups such as amoebozoan and benthic foraminifera are also present. Microbial biofilms constitute one of their typical microhabitats (Giere 2009).

Most of metazoan consumers of microorganisms belong to meiofauna (benthic invertebrates which pass through a 1 mm mesh and are retained on a 40 μm mesh sieve). They are abundant and diversified in superficial sediments (on average 10^6 $ind.m^2$). The patchy spatial distribution of meiofauna is associated with that of microorganisms in superficial sediments. These invertebrates are particularly abundant in phototrophic microbial aggregates which serve them both as refuge against flow (Majdi et al. 2012a) and as food source (Majdi et al. 2012b). Moreover, the patchy distribution of meiofauna and benthic microorganisms is associated in superficial sediments (Giere 2009). Meiofauna are generally dominated, in terms of density and biomass, by free-living nematodes, harpacticoid copepods, and rotifers.

Among macrobenthic invertebrates (larger than meiofaunal invertebrates), consumers of microorganisms mainly belong to gastropods, insect larvae (e.g.,

Chironomidae, Ephemeroptera), and oligochaetes (e.g., Lawrence et al. 2002; Majdi et al. 2012b).

In sediments, the extraction techniques are focused on some groups of microorganisms. However, to date, none of these techniques allow to obtain all resuspended benthic microorganisms neither to totally separate them from sediment and detrital particles. Thus, the efficiency of the extraction techniques are variable according to the studied groups, and consequently, samples of benthic microorganisms collected using resuspension techniques are not totally representative of the diversity of benthic microbial communities.

Moreover, it can be noticed that classification of benthic microorganisms according to their size is rarely used in contrast to pelagic systems. Determination of benthic microorganisms is based on morphology, cell functioning, or phylogenetic studies using microscopic techniques, biochemistry, or DNA fingerprints.

13.2.2.1 Microalgae and Cyanobacteria

Among microorganisms, microalgae (eukaryotic cell) and cyanobacteria (prokaryotic cell) are the primary producers. Filamentous pluricellular forms of cyanobacteria and diatoms are commonly present in benthic systems. Photoautotrophic microorganisms are able to perform oxygenic photosynthesis. They can be studied using either direct microscopic observation since they have very characteristic cell morphologies or indirect methods based on HPLC analyses of signature photosynthetic pigments (e.g., Buffan-Dubau and Carman 2000). Dioxygen production can be quantified using microsensors and CO_2 uptake using ^{14}C or ^{13}C isotopic tracers (cf. Chap. 17).

Among microalgae, diatoms and particularly pennate diatoms with a raphe are common in superficial sediments and biofilms. Pennate diatoms and most of filamentous cyanobacteria are able to migrate by sliding through the benthic community. These organisms produce important amounts of extracellular polymeric substances (EPS). These EPS may be present under different forms:
1. Characteristic sheath-like structures
2. Structured mucus masses enveloping the cells
3. Mucus masses spread in the environment

Except diatoms, other microalgal taxa belonging to the green microalgal group are commonly observed in biofilms and in superficial sediments.

13.2.2.2 Phototrophic Anoxygenic Bacteria

Anoxygenic photrophic bacteria represent another group of phototrophic microorganisms in benthic systems. Because these organisms have very characteristic pigments (bacteriochlorophylls and specific carotenoids), it is possible to reveal their presence by HPLC pigment analyses when they are abundant. Instead of water, these organisms use another electron donor for their photosynthesis as H_2S, $S_2O_3^{2-}$, or small organic molecules; they use CO_2 or dissolved small organic molecules as the source of carbon. While these organisms convert also light energy into biomass, they differ from the classical oxygenic primary producers by the fact that they use electron donors from biological origin which have a higher energy content than the water used by oxygenic phototrophs. Therefore, these organisms have sometimes been described as para-primary producers. Hence, the energy for the fixation of the CO_2 is partially delivered by the electron donor and partially by the photons.

Many species of anoxygenic phototrophic bacteria have a very characteristic morphology that can be recognized by optical microscopy. Examples include the *Chromatiaceae* (which store sulfur intracellularly) and the green filamentous phototrophic bacteria. The latter may, however, be confounded with the very thin filamentous cyanobacteria (e.g., *Leptolyngbya*). In contrast, the morphology of other anoxygenic phototrophic bacteria including the green sulfur bacteria (*Chlorobiaceae*), *Ectothiorhodospiraceae,* and purple nonsulfur bacteria are easily confounded with the chemoheterotrophic and chemolithoautotrophic bacteria.

13.2.2.3 Chemolithotrophic Bacteria

The chemolithoautotrophic bacteria comprise different functional groups that all fix CO_2 and are distinguished according to different electron donors, i.e., H_2, H_2S and other reduced inorganic sulfur compounds, NH_4^+, and NO_2^-. Only a very small number of sulfur-oxidizing bacteria have a morphology that is easily recognizable under the light microscope (e. g., the members of the *Beggiatoa* and *Thioploca* genus). This functional group also comprises some giant bacteria, which are visible to the naked eye as some members of *Beggiatoa*, *Thioploca*, and *Thiomargarita* (Schulz and Jorgensen 2001; cf. Chap. 5).

13.2.2.4 Heterotrophic Bacteria and Archaea

The chemoorganoheterotrophic prokaryotes (hereafter referred to as heterotrophs) use organic compounds as the electron donor and as the carbon source. They comprise a large number of functional groups, which can be defined according to the electron acceptor used for energy generation (O_2, NO_3^-, organic compounds, metal oxides [Mn^{4+}, Fe^{3+}], SO_4^{2-}, and CO_2; cf. Chap. 3). In general, microscopic observation does not allow determining to which functional group they belong.

Within the sediment, the counting of prokaryotes may be difficult and problematic because of the large quantities of detritus. Nevertheless, their quantification has improved a lot since the introduction of specific fluorochromes (DAPI) and the use of epifluorescent microscopy (cf. Chap. 17). The density of prokaryotes in the sediment varies in the surface layers between 10^8 and 10^{10} cells mL^{-1}. Prokaryotes that still occur in oceanic sediment columns below 500 m depth

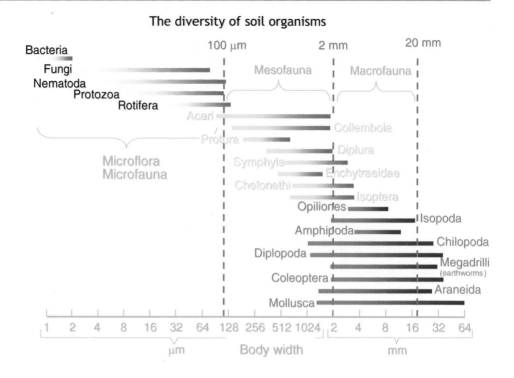

Fig. 13.2 Classification of soil organisms as a function of their body size (Modified and redrawn according to Swift et al. 1979)

densities of 10^5 to 10^6 have been reported for these deep sediment layers. With epifluorescence microscope techniques it is equally possible to observe and count virus-like particles and densities of about 10^{11} have frequently been reported for sediments. Surprisingly, however, the number of prokaryotes showing viral infections appears often to be very low in sediments, and the role of viruses in sediments still remains enigmatic (Filippini et al. 2006).

As it is the case for many other ecosystems, the use of the concept of functional groups is particularly useful for the analyses of the biogeochemistry of the benthic communities. However, functional groups do often not correspond to phylogenetic groups. Therefore, the analysis of 16S rRNA gene clone libraries is often not sufficient to infer the importance of biogeochemical reactions. Rather, the amplification and sequencing of functional groups is a better method to establish a link between biogeochemical activity and the phylogeny of the main responsible prokaryote actors.

The use of the concept of functional groups can also be problematic for food web studies, as the main predators select their prey according cell size and microhabitat, rather than according to their biogeochemical functions. Moreover, the concept of functional groups may also be problematic at the level of the species and individual cells, because several prokaryotes are generalists and highly versatile in their metabolism and could, therefore, belong to several different functional groups at the same time. This is particularly the case for the mixotrophs that perform autotrophy and heterotrophy simultaneously.

During longtime, the methanogenic archaea were considered as the only archaea in most sediments where they proliferate in the anoxic part. Other archaea were considered as characteristic for extreme environments as hypersaline and high-temperature environments. The sampling and amplification of 16r RNA genes has shown, however, that archaea are much more widespread and more diverse in sediments, where they also may include the ammonium-oxidizing archaea.

13.2.3 Microorganisms of Terrestrial Ecosystems

Studies on terrestrial environments mainly concern organisms living on the soil (animals and higher plants), and knowledge of soil biodiversity and its functional role remains insufficient (Mittelbach et al. 2001). Numerous microorganisms and telluric animals constitute the main soil food webs (see summary in Fig. 13.2). The most abundant and diverse members are microorganisms: archaea, bacteria, protozoa, and fungi. Other groups also occur in variable densities, depending on the type of soil: microfauna (<0.1 mm; i.e., micro-arthropods and enchytraeids) and macrofauna (>2 mm; i.e., earthworms, termites, and diplopods).

Microorganisms form the main food of many microfauna (de Ruiter et al. 1995). However, certain authors regard

Table 13.2 Abundance and diversity of prokaryotes in different soils (according to Torsvik et al. 2002)

Habitat	Abundance (cells mL^{-1})	Diversity (equivalent to genome)
Forest soils	4.8×10^9	6,000
Grassland soils	1.8×10^7	3,500–8,800
Cultivated soils	2.1×10^{10}	140–350

many soil organisms as omnivorous and capable of feeding from different trophic levels.

Since the models of carbon, energy, and nutrient flows throughout the food web of the soil are dominated by trophic links between bacteria, fungi, protozoa, and nematodes (de Ruiter et al. 1997), the analysis will be limited to these for the rest of this section. They may be classed into primary and secondary consumers.

13.2.3.1 Primary Consumers

Primary consumers play an essential role in the degradation of complex organic substances. The first role of fungi (oobiontes and fungi) is the decomposition of OM, due to their capacity to produce a large range of extracellular enzymes. In any ecosystem, the fungal biomass may greatly exceed that of other biotic components (prokaryotes, vegetables, animals). For example, in a prairie system, fungal hyphae may reach a dry matter biomass of 250 kg per ha in the superficial soil layer (5 cm). It may be ten times greater in acid forest soils. Besides their role in degrading OM, fungi may be pathogenic for plants, promoting soil structure and forming an abundant resource for the fauna that consume microorganisms. Here, it is important to mention mycorrhizal fungi that are capable of forming a mutualistic association with plant roots: the fungi facilitate the plant's access to often scarce elements (N and P), and in return the plant brings the fungi carbon-based products from their roots and reduces competition (*cf.* Chap. 11).

The great abundance of bacteria in soil has been long known. Molecular methods have enabled us to identify the immense, and unexpected, diversity of telluric bacteria (Table 13.2). Soil DNA analysis can be carried out to quantify this great diversity. Torsvik et al. (2002) estimated that 5 cm^3 of soil could contain up to 10,000 types of genetically distinct prokaryotes. According to Curtis et al. (2002), there are about 4×10^6 prokaryotic taxa in a ton of soil. Bacteria, like fungi, also play a major role in the degradation of OM. Like their taxonomic diversity, their metabolic diversity is considerable (*cf.* Chap. 6).

13.2.3.2 Secondary Consumers

Protozoa and nematodes are the most abundant of secondary consumers in soil. Unicellular protozoa can be divided into three principal groups: amoebae, flagellates, and ciliates. A film of water is necessary for their locomotion in soil, affecting their ability to feed. Their activity is thus linked to water-filled pore space. Where water is insufficient, their activity ceases, but they are able to form resistant cysts. The number of unicellular organisms varies greatly depending on soil type and climate. However, very often densities ranging from 10^4 to 10^6 per g. are found in the superficial layer of soil, but their density does not exceed 100 cells per g. of soil in deeper levels where the organic content and the number of prey become greatly limiting. There is also much diversity: it is estimated that more than 1,000 species of ciliates exist in the soils of the Earth (Foissner 1997a) and about 150 were found on one site (Foissner 1997b). Most protozoa are bacterivorous, but some eat fungi and saprophytes.

Nematodes are the most abundant and diverse multicellular organisms in soil. They can reach densities of 10^7 per m^2 (Yeates et al. 1997). One study of forest soil in Cameroon evidenced more than 400 species. They may also colonize environments that are, a priori, hostile. For example, 27 species were found in Scotland on a site of sandy dunes with no vegetation. As with protozoa, a film of water is necessary for their activity (especially nutrition and reproduction), and they have the ability to resist unfavorable conditions by greatly reducing their metabolism. Nematodes can be classified on the morphology of their mouthparts whose structures reflect their modes of nutrition. The most common species eat plants, bacteria, fungi, and other nematodes; some are omnivorous.

13.3 Functioning of Microbial Food Webs

13.3.1 The Functioning of Microbial Food Webs in Oceanic Pelagic Ecosystems

In oceanic pelagic ecosystems, two types of food web can be distinguished: herbivorous or classic food web (Cushing 1989) and the microbial food web (Azam et al. 1983).

13.3.1.1 Herbivorous Food Web

The herbivorous food web (Table 13.3) is dominated by microphytoplanktonic communities (notably diatoms) assuring the essential of productivity of some regions and in some periods (resurgence, spring bloom, etc.). These phytoplanktonic communities assimilate preferentially the nitrate (coming from rivers, rain, or the upwelling deep waters rich in

Table 13.3 General characteristics of two types of food web in marine ecosystem

Food webs	Dominant organisms size	Trophic status	Type of primary production	Dominant nutrient (N)	Productivity of system	Exportation or sedimentation
Herbivorous	<200 μm	Eutroph	New	NO_3^-	High	High
Microbial	<20 μm	Oligotroph	Regenerated	NH_4^+	Low	Low

nitrate) and are consequently involved in the new production.

The classic food web (herbivorous), by its important productivity, feeds the large size of metazooplankton and then fishes. The settling rate of microphytoplankton and associated products (fecal pellets produce by herbivores) being high, this type of food web is responsible of strong exportation of OM outside the euphotic zone.

13.3.1.2 Microbial Food Web

The microbial food web, including the auto- and heterotrophic microorganisms, is dominated by pico- and nanoplankton communities (cyanobacteria, heterotrophic bacteria, and flagellates) and is characteristic of oligotrophic zones (Table 13.3). The primary production is rather based on regenerated form of nitrogen (NH_4^+) and so considered as regenerated production.

The classical concept, which has been developed in offshore oceanography, in which the new production is based on utilization of NO_3^- and the regenerated production on NH_4^+ does not necessarily apply in certain environments such as shallow coastal and lagoon areas or most lakes. This is due to the fact that much of NH_4^+ is not directly associated with the mineralization of organic nitrogen but is produced in the sediment from NO_3^- by dissimilatory reduction of nitrate to ammonium (DRNA, *cf.* Sect. 14.3.3) and then be released into the shallow water column. The sedimentation rate of particulate organic matter (POM) in oligotrophic areas is very low; the cells do not therefore contribute significantly to export out of the euphotic zone, and the OM is recycled on site rather by protozooplankton (see flagellates and ciliates) that provides the essential link between microorganisms and macroorganisms.

13.3.1.3 Microbial Loop

The complexity of pelagic food webs, with all existing interactions between the various components, is shown in Fig. 13.3. Dissolved organic matter (DOM) and POM play an important role in the microbial food web. They provide food support for the development of microbial food web and derived, in part, the phytoplankton exudation, zooplankton grazing, and bacteriovory, as well as viral lysis.

Presently, there is a debate in the scientific community regarding the ecological phytoplankton carbon dependency of heterotrophic bacteria, postulating that bacteria can use other carbon sources than those freshly exudated by phytoplankton (Fouilland and Mostajir 2010, 2011), although there is no unanimity of this new ecological concept (Moran and Alonso-Saez 2011). The DOM is so a necessary substrate for bacterial production. In the microbial food web, heterotrophic bacteria may compete with phytoplankton depending on the concentration of dissolved inorganic matter (DIM) notably the dissolved inorganic nitrogen (DIN) (Fouilland et al. 2007). When the DIN concentration is low, heterotrophic bacteria are winner in competition with phytoplankton for ammonium. In this case, system drives to the dominance of the microbial loop. The microbial loop is defined within the microbial food web as the community of heterotrophic microorganisms: archaea, bacteria, flagellates, and ciliates (Azam et al. 1983). Predation of prokaryotic prey by heterotrophic flagellates and ciliates, as well as flagellates by the ciliates, produces OM and allows the recycling of various nutrients (ammonium, phosphorus, etc.) that are mobilized for bacterial growth.

13.3.1.4 Multivorous Food Webs

There is nevertheless an intermediate food web between herbivorous and microbial food webs, namely, "multivorous" food web (Legendre and Rassoulzadegan 1995). In multivorous food web, grazing by metazooplankton (e.g., copepods) and by protozooplankton (e.g., flagellates and ciliates) have both a significant role.

13.3.1.5 Trophic Pathways and Transfer of Matter from the Microbial System Toward Macroorganisms

The flow of matter resulting from the activities of autotrophic and heterotrophic microorganisms of different food webs is channeled to higher trophic levels by two distinct trophic pathways (Fig. 13.3). The autotrophic pathway is based on the activity of phototrophic microorganisms of herbivorous and multivorous food webs, while the heterotrophic pathway relies on the activity of heterotrophic microorganisms in the microbial loop and the microbial food web, specially ciliates and flagellates.

Therefore, protozooplankton is a key link in the food webs functioning by transferring matter and energy from lower trophic levels (picoplankton: heterotrophic bacteria, cyanobacteria, and eukaryotic picophytoplankton) to metazooplankton, corals, oysters, and fish larvae. In pelagic marine ecosystems, the relative importance of matter flow (e.g., carbon), channeled by heterotrophic pathway, is

Fig. 13.3 Complexity of pelagic food webs including microbial loop (in *blue*) and microbial (in *red*), multivorous (in *green*), and herbivorous food webs (in *light green*). *DIM* and *DOM* dissolved inorganic and organic matter, respectively; POM, particulate organic matter; *blue arrows*, predation; *mauve arrows*, DIM and DOM uptakes; *black arrows*, MID and MOD releases; *orange arrows*, return of matter (e.g., fecal pellets) from higher trophic levels supplying microorganisms. For supplementary information about aerobic anoxygenic phototrophic bacteria, see Kolber et al. (2000)

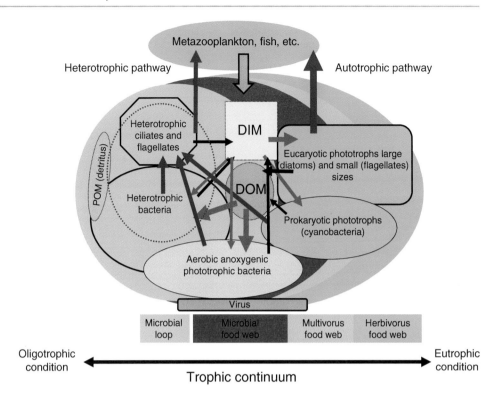

generally lower than that transferred by the autotrophic pathway. It is obvious that a part of the matter and energy required for microorganisms' metabolic activities is provided by the return flow of matter from the higher trophic level activities, such as fecal pellets.

13.3.2 Functioning of Microbial Food Webs in Lake Pelagic Ecosystems

Research into lake ecosystems has shown that matter and energy flows are organized not only according to the classical linear food chain, based on photosynthetic assimilation (Phytoplankton → Zooplankton → Fish), but also involve a microbial loop chain formed of small heterotrophic organisms (pico- and nanoplankton) that constitute a true food web (Fig. 13.4; Amblard et al. 1998).

13.3.2.1 Transfer of Matter and Energy Toward Superior Trophic Levels

It is generally accepted that a highly significant proportion (10–50 %) of phytoplankton primary production is not consumed directly by metazooplankton but is excreted into the environment and passes through the microbial loop.

DOC realized by phytoplankton, usually of low molecular weight, is rapidly used by heterotrophic bacteria which act in transferring energy and matter in aquatic ecosystems by mineralization and production of biomass. Heterotrophic flagellated protists are generally regarded as the principal predators of picoplankton cells and especially of bacteria.

Mixotrophic flagellates, ciliates, cladocera, and, to a lesser degree, rotifers also prey on picoplankton communities. Experimental work carried out under controlled conditions has revealed that macrozooplankton, apart from cladocera, are not capable of consuming picoplankton cells effectively, because they are too small with a relatively low density. Predatory activity, which may be selective depending on the size, mobility, and metabolic activity of the predator cells, alters the structure of bacterial communities and promotes large and/or filamentous strains.

The predation of phagotrophic protists guarantees the transformation of picoplankton production to larger-sized particles that are accessible to metazooplankton. The microbial loop transforms the DOM issuing from phytoplankton into particulate matter while moving picoplankton production, via phagotrophic protists, to higher trophic levels due to the predation of metazooplankton on phagotrophic protists.

While discussing the microbial loop in aquatic food webs, an important question concerns the relative importance of the carbon flows due to the protist predation on picoplankton and of losses due to respiration related to the multiplication of trophic levels between primary producers and metazooplankton. This debate, known as "link or sink" (i.e., trophic link or carbon sink), was very lively in the 1980s–1990s (Sherr and Sherr 1988). The functional importance of the microbial food web, in comparison with the herbivorous food web (classic), varies according to the

Fig. 13.4 Schematic representation of the microbial food web in pelagic lake ecosystems. Variations of the abundance of the different planktonic microbial communities according to the trophic level of lakes (Modified from Amblard et al. 1998)

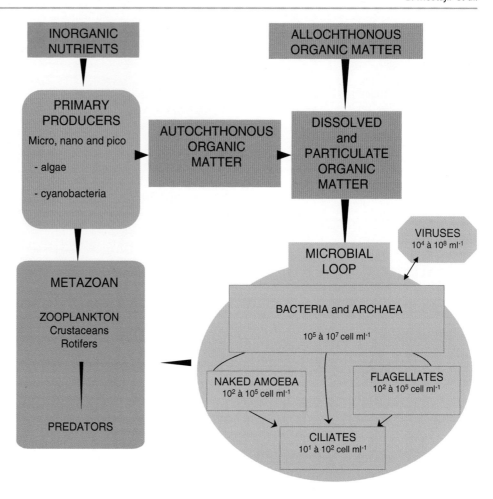

trophic level of different environments, the seasons, and the depth and structure of populations (Amblard et al. 1995; Carrias et al. 1998). For example, it is known that lacustrine cladocera, particularly *Daphnia,* can directly consume bacterial communities with great efficacy, making them highly competitive with protozooplankton. In this case, there is a direct trophic link between picoplankton and metazooplankton communities.

Moreover, because of their biochemical composition, microorganisms are the source of essential elements (fatty acid, sterols, amino acids, etc.) vital for the development of their predators. Any deficiency of these essential elements reduces the growth and fertility of animal predators. In this context, heterotrophic protists of the microbial loop improve the nutritional quality of autotrophic picoplankton which they ingest, thanks to their capacity to convert fatty acids by trophic upgrading, and this improves the nutritive quality of biomass throughout the food chain (Bec et al. 2006).

More generally, this trophic upgrading seems to play a major role in the functioning of lake ecosystems.

13.3.2.2 Nutrient Recycling

Inorganic nutrients, particularly N and P, are essential for phytoplankton growth. Although they may be airborne (atmospheric transport, precipitations), allochtonic inorganic nutrients issue mainly from the earth, directly via rivers or by seepage into a river's catchment area. The resuspension of sediments may also constitute an important source of inorganic nutrients.

Heterotrophic microorganisms are also highly involved in the biogeochemical cycles of N and P, some acting as assimilators and others as regenerators. Long seen as the only mineralizer of OM in stagnant aquatic ecosystems, bacteria are considered of capital importance for phytoplankton growth, especially when minerals are limited. However, bacteria can incorporate more inorganic nutrients than they produce. It has been demonstrated that bacteria, because of their high surface/volume ratio, are more competitive than phytoplankton in assimilating inorganic nutrients, particularly at low concentrations. Thus, bacteria constitute a pool of P and N that cannot be used by strictly autotrophic phytoplankton.

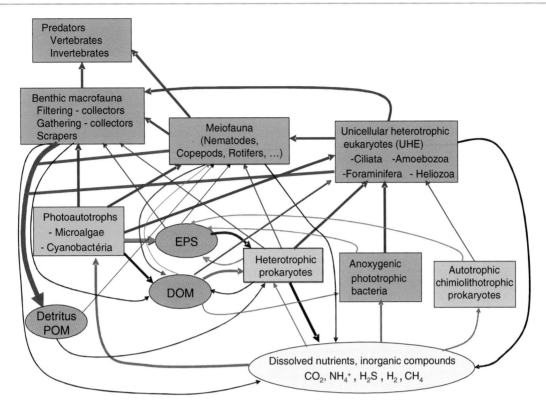

Fig. 13.5 Trophic interactions identified in benthic communities. Microalgae and cyanobacteria represent the oxygenic photoautotrophic primary producers. Other autotrophic prokaryotes (chemolithoautotrophic prokaryotes and anoxygenic photoautotrophic bacteria) also fix CO_2 by using the electron donors (NH_4^+, CH_4, H_2S, H_2) that are produced in the anoxic zone. Organic matter issued from macrophytes enters the microbial food web mainly as detritus. Trophic transfers and material recycling are indicated by *arrows* (*dark blue arrows*, predation; *rose arrows*, uptake of dissolved inorganic nutrients and of dissolved organic matter (DOM); *black arrows*, excretion of dissolved inorganic nutrients and of DOM; *brown arrows*, return flow from higher trophic levels to the microorganisms through the detritus compartment; *green arrows*, excretion of EPS; *purple arrows*, hydrolyses of particulate organic matter and uptake of organic matter)

Certain experiments (Dolan 1997) have showed that heterotrophic flagellates could play a major role in the mineralization of OM. Indeed, by excreting a part of the assimilated nutrients, principally in the form of ammonia for N and soluble phosphates in the case of P, heterotrophic flagellates contribute to phytoplanktonic growth in two ways:

1. By their bacterivorous activity, limiting bacterial populations and hence reducing their competition with phytoplankton
2. By producing inorganic nutrients stimulating phytoplanktonic growth

13.3.3 Functioning of Microbial Food Webs in Oceanic and Lake Benthic Ecosystems

Many trophic relationships have been identified in benthic systems implying that in these systems the food web structure is more complex than the simple classical food web (benthic photoautotrophs → invertebrate herbivores → invertebrate and vertebrate predators). A large number of potential interactions have also been identified among the benthic microorganisms, which strongly suggests that a microbial food web also exists in benthic systems (Fig. 13.5), even though it is still difficult to quantify these trophic transfers.

The high density of the microorganisms and the relatively short distances that separate them in space favor the trophic transfer of dissolved compounds based on molecular diffusion and osmotrophy. Molecular diffusion is an efficient transport mechanism on the microscale, while it becomes a constraint for larger distances. Therefore, osmotrophic organisms, which take up dissolved organic compounds as their energy resource, are generally small with some exceptions of chemolithotrophic bacteria from different sediments (Schulz and Jorgensen 2001). Thus, *Thioploca* is capable by moving to create a convective current, which enhances the mass transfer of solutes to the cells. Other giant bacteria, as, e.g., certain species of *Beggiatoa*, and particularly *Thiomargarita namibiensis*, are capable to stock large quantities of sulfur and nitrate, which they use as electron donor and electron acceptor, respectively. This allows them to proliferate in environments that are characterized by large spatial variability and/or temporal fluctuations in the concentrations of their electron donors and acceptors (Schulz and Jorgensen 2001).

In non-permeable sediments as muds, the movement of the interstitial water is extremely slow and the solute transport mainly depends on molecular diffusion. The characteristic diffusion time for O_2 to travel 1 μm by diffusion is equal to 0.3 ms, while the time to travel 1 cm is equal to 7 h. The microhabitat becomes anoxic when the diffusive delivery is lower than the local demand. When comparing soils and sediments, it is important to consider that the diffusion coefficient of O_2 in air exceeds its value in water by a factor of 10^4. Hence, the diffusive transport of O_2 is much more efficient in soils that are not saturated with water than in sediments and in the waterlogged soils of wetlands. As a result, in many benthic systems, anoxic zones occur in close proximity to oxic microhabitats, which implies that functional groups of prokaryotes tend to stratify along redox gradient (*cf.* Sect. 14.2.4). The heterotrophic archaea and bacteria that live in the anoxic parts of the sediments function in concert based on their trophic interactions that are relatively well known. Hence, the methanogenic archaea and the sulfate-reducing bacteria are only capable to catabolize dihydrogen gas and a small number of low-weight compounds and thus cannot mineralize all the OM in the anoxic sediment. These prokaryotes are organized in consortia together with fermenting and acetogenic bacteria, which produce the intermediate metabolites (low-weight fatty acids, acetic acid, and dihydrogen) that are used by the sulfate-reducing bacteria and the methanogenic archaea (*cf.* Sect. 14.2.4). The products of the anaerobic metabolism, i.e., NH_4^+, Mn^{2+}, Fe^{2+}, H_2S, and CH_4, can diffuse into the oxic habitats of the sediment or directly into the water column, where they can be used as electron donors for chemolithoautotrophic prokaryotes.

The direct strong coupling between aerobic and anaerobic microbial processes is, therefore, another characteristic feature of benthic systems. The coupling between nitrification and denitrification represents a nice example of such a strong coupling. Osmotrophy is also particularly widespread among the unicellular eukaryotes thriving in sediments where they may benefit from the excretion by and lysis of other microorganisms as well as from their hydrolytic activities which results in hydrolysis of extracellular polymers.

Macrofauna (i.e., macrobenthic invertebrates) are divided into different functional feeding groups according to their trophic behavior (Tachet et al. 2010). Hence, macrofauna can be classified (Fig. 13.5) according to their trophic behavior (functional feeding groups). "Scrapers" collect microbial patches by removing them from their supports before ingestion (e.g., gastropods). "Gathering collectors" (e.g., chironomid larvae) ingest superficial particles and their associated microbiota. Thus, these two groups potentially feed on both microalgae, cyanobacteria, and other bacteria as well as their associated EPS. The benthic "filtering collectors" (e.g., Simuliidae larvae) feed by filtering the overlaying water. Thus, these invertebrates feed on plankton and constitute a trophic link between benthic and pelagic communities. However, benthic microorganisms that are resuspended in the overlaying water are also used as food sources by the benthic "filtering collectors." Meiofauna actively move through sediments and biofilms and may consume microalgae such as diatoms, bacteria, and EPS (Mathieu et al. 2007; Buffan-Dubau and Carman 2000; Majdi et al. 2012b). Nematodes, for example, are divided into four feeding groups based on morphological characteristics (Moens et al. 2006). "Deposit feeders or swallowers" feed on bacteria and unicellular eukaryotes that are swallowed whole. Epistrate feeders (tear-and-swallow feeders having no teeth) feed on bacteria, unicellular eukaryotes, diatoms, and other microalgae. "Chewers" (having a small tooth) feed on unicellular eukaryotes and smaller meiofauna. "Suction feeders" (voluminous buccal cavity with teeth and denticles) are potentially omnivorous using algae, fungi, plants, and animals. Heterotrophic unicellular eukaryotes used microalgae, cyanobacteria and other bacteria, and DOM as trophic sources by phagocytosis and osmotrophy, respectively. Most of benthic ciliates are bacterivores; however, they are also detritivores and often even selective algivorous grazers feeding on diatoms of selected size (Giere 2009).

Faunal activity in sediments is not only linked to their trophic role since benthic invertebrates profoundly alter the distribution of sediment particles, solutes, and microbial communities, through sediment reworking and irrigation. They particularly burrow galleries and actively ventilate them, contributing to enhance sediment irrigation. Thus, faunal activity directly affects the decomposition and remineralization of organic matter in sediments (e.g., Mathieu et al. 2007; Pischedda et al. 2012). These whole activities are described using the term "bioturbation." The burrowing activity of benthic macrofauna induces a very complex spatial structure of sediments with relatively welled oxygenated micro-horizons such as the worm burrows and anoxic zones (Pischedda et al. 2012). Some meiofaunal invertebrates (e.g., nematodes) have also an intensive burrowing activity modifying the oxygen turnover in biofilms (Riemann and Helmke 2002; Mathieu et al. 2007).

Within permeable sediments, which are composed of sand, gravels, and pebbles, advection currents, which percolate through these sediments by hydrodynamic forces, can facilitate the transport of dissolved O_2 and other solutes. Detritus plays a key role in benthic ecosystems. Sediments covered by aquatic vegetation can be compared to terrestrial soil systems. The larger herbivores can be represented by Anatidae. Nevertheless, a large part of the macrophyte production enters the food web as detritus. Benthic macrofauna comprise shredders that only consume this matter partially and cut the larger detritus into smaller pieces that can be colonized by heterotrophic microorganisms. These microorganisms are subsequently consumed by small heterotrophic eukaryotes and invertebrates.

So far, little is known about the quantitative importance of predation on prokaryotes and trophic transfer in sediments. One of the reasons why the microbial food web has been less studied in sediment systems compared to the water column is that some of the classical techniques like dilution and retrieval of specific size classes by filtration cannot be used in sediments without complete destruction of the spatial structure of the sediments. Novel techniques based on the use of stable isotopes as tracers are promising. Carbon dioxide and specific organic compounds can be enriched in ^{13}C and added to the benthic system as substrates for oxygenic phototrophs and heterotrophic bacteria, respectively. Their fate in the food web can be followed by studying the $^{13}C/^{12}C$ ratio in the different trophic compartments. After a certain incubation time, specimens of the main macrofauna and meiofauna are collected and analyzed for their $\partial^{13}C$ signatures by mass spectrometry. In contrast, it is very difficult or often impossible to physically separate sufficient microalgae and cyanobacteria from the heterotrophic bacteria. The $\partial^{13}C$ signatures for these functional groups can be evaluated by using group specific biomarkers and measuring the $\partial^{13}C$ signatures for these compounds by using LC-MS or GC-MS. The addition of $NaH^{13}CO_3$ to a diatom biofilm in intertidal sediments has allowed to detect a progressive increase of the ^{13}C tracer in the diatoms during the first 4 h and a measurable trophic transfer of ^{13}C to the nematodes from t = 1 h to 3 days of incubation. Simultaneously, some fatty acids that are specific to heterotrophic bacteria also became enriched in ^{13}C; their labeling peaked after 1 day of incubation (Middelburg et al. 2000). This shows that there is a quick trophic transfer of OM from diatoms to heterotrophic bacteria, probably through the EPS compartment (Fig. 13.5). In another experiment, heterotrophic bacteria have been labeled by adding ^{13}C glucose to the same type of biofilms. This way it has been estimated that the heterotrophic bacteria constitute 8 % and 11 % of the diet of meiofauna and macrofauna, respectively.

13.3.4 Functioning of Microbial Food Webs in Terrestrial Ecosystems

The major actors constituting the terrestrial food webs have been presented in the Sect. 13.2.3. Here, we focus on the functional role of these actors and their interactions. It is now widely accepted that the structure of the soil food webs has a direct impact on its functioning in general (Beare et al. 1995). The biotic components contribute to functional soil processes via direct and indirect mechanisms. Direct mechanisms are the result of metabolic degradation of complex molecules and release of energy, C, N, and minerals through respiration, excretion, and chemotrophy. Indirect mechanisms include many effects of feedback and agencies are classified as ecosystem engineers (Anderson 2000).

13.3.4.1 Decomposition of Organic Matter in Soils

The function of decomposing of OM is obviously one of the major roles of the food webs in soil ecosystems. The decomposition of OM produces CO_2 but also soluble compounds such nutrients for plants. In natural unfertilized soils, almost all the nitrogen needs and part of the phosphorus needs come from the decomposition of OM. Even in **amended systems***, the role of the decomposition of OM remains central in maintaining an acceptable biological production in soils.

In uncultivated areas, 80–90 % of the primary production returns to the soil as litter and dead roots. Heterotrophic fungi and prokaryotes are the main actors in decomposing these plant incomes. Through fragmentation and ingestion, the detritivorous fauna essentially play a facilitating role for the subsequent microbial activity (mainly through the increase of the colonization surfaces). The incredible range of metabolic capabilities of microorganisms allows them to be able to substantially degrade almost all the molecules present in plants. In many ecosystems, fungi are the most abundant "primary" decomposers. Compared to bacteria, they are better adapted to start the decomposition of insoluble plant materials, due to their hyphae which can penetrate and proliferate inside plant cells. Like bacteria, they exhibit a wide range of metabolic capabilities that, in particular, allow them to degrade recalcitrant components of plant cells, such as lignin. In addition, being able to mobilize nutrients across their hyphae networks gives to fungi the opportunity to exploit new substrates, even when the local levels of P and N are low. These characteristics of active use of substrates by fungi contrast with the immobility or passive mobility (e.g., the movements of water in the soil) of bacteria. As a result, the latter are probably more frequently inactive due to lack of resources in their immediate environment.

13.3.4.2 Role of Selective Predation

Fungi and heterotrophic prokaryotes are the main actors of the MO transformation in soils, particularly, for its degradation by mineralization. However, the organisms belonging to higher trophic levels strongly affect the activity of these microorganisms (Bardgett 2005a).

Selective feeding of microorganisms can modify the abundance, structure, and activity of fungal and bacterial communities. Although a number of studies report the generalist and/or opportunistic feeding characteristics of secondary consumers, other studies mention a real "selectivity" of the preys by their predators. Some nematodes, for example, have a buccal morphology adapted to feed on either bacteria (bacterial feeders) or fungi (hyphal feeders) (Yeates et al. 1993). It is also known that some heterotrophic unicellular eukaryotes are able to detect chemical signals to differentiate between bacterial and microalgal potential preys (Verity 1991). This selective consumption of preys can have an important impact on ecosystem functioning. In forest systems, for example, the selective feeding of the fungus *Marasmius androsaceus* by *Onychiurus latus* (Collembola) induces a decrease in abundance of the fungus and allows an increase in density of another fungus species *Mycena galopus*. The latter is not used as food source by this collenbolan. This change in composition of the fungal community induces a decrease of the degradation rate of conifer litters resulting from a difference in efficiency between *M. androsaceus* and *M. galopus*, the latter being less efficient for litter degradation.

Predation of a microorganism species by a consumer may also stimulate the prey activity without any influence of the competition processes. Hedlund et al. (1991) have shown that the specific amylase activity of the fungi *Mortierella isabellina* is higher when this fungus is consumed by a collembolan than when it is protected against predation.

Finally, among the effects of larger organisms on the communities which are involved in litter degradation as fungi and heterotrophic prokaryotes, the ingestion of these microorganisms and their exposure to the digestive activity of the consumers must be considered. On this subject, literature data are contrasted. Brown (1995) has shown that bacterial community is more abundant after its passage through the gut of earthworms although Moody et al. (1996) have observed a decrease in spore germination of fungi following their ingestion by an earthworm.

13.3.4.3 Role of Other Biotic and Abiotic Interactions

Some interactions are more based on physical modifications of the habitat than on trophic relationships. They have a drastic importance in soil functioning. This is the case for organisms which ingest phytal detritus and produce fecal pellets having smaller sizes than the ingested material. The production of these biological aggregates fundamentally modifies physical characteristics of soils (e.g., high compaction, improved ability to retain water). This modified litter becomes a very attractive medium for bacteria resulting in a considerable increase of the bacteria/fungi ratio. For example, a multiplication by 100 of bacteria and only by 3 for fungi has been shown in fecal pellets of arthropods, when compared with the initial material. It is likely that the compaction of the material is the main factor involved in this differential effect. This effect can be considered as a mutualistic interaction since (1) the stimulation of the microbiota in fecal pellets goes with an increase of the OM mineralization and (2) several ingestions of the modified pellets by the fauna contribute to improve its diet. This system functions like an "external rumen" and becomes very important when the fauna feed on a litter which is still not modified by microbial communities and has a low nutrient content.

Besides the mutualistic interaction described above, the engineer organisms of ecosystems and engineer organisms of soils induce physical modifications of the environment and create new habitats for microorganisms and microfauna of soils. This is the case for earthworms, termites, and ants. As an example, many authors have shown that microbiota was stimulated in structures created by earthworms both through an improvement of oxygenation and hydric conditions of soils and by favoring the accessibility of resources (e.g., litter burying) (Brown 1995). It should be noted that in some cases, this stimulating effect is only a transient effect and that, at long term, these structures can, in contrast, reduce the degradation of OM.

The input of OM via the phytal litter constitutes a fundament of the functioning of these webs including a very high diversity of telluric organisms which are mainly heterotrophs. This is close to the "detritus food chain" of aquatic ecosystems, mentioned in the introduction. Under a considered climactic and edaphic context, both the amount and the quality of litters strongly control the composition of food webs. Studies that focus on understanding how this structure is regulated in the soil by the environment were rare a few years ago (Wardle 2002), but their numbers are currently increasing. This is particularly due to the new methodological developments such as the use of the natural variations of abundance of C and N stable isotopes ($^{13}C/^{12}C$ and $^{15}N/^{14}N$) (Schmidt et al. 2004).

13.4 Factors Controlling Microbial Food Webs

13.4.1 Factors Controlling Pelagic Ocean and Lake Microbial Food Webs

In pelagic environments, different food web structures can coexist. The predominance of a food web relative to another one is controlled by environmental forcing factors such as the nature and concentration of nutrients, temperature (responsible for stratification or vertical mixing), and pollution. Thus, the difference in structure of the pelagic food webs between the offshore and coastal marine environments is linked to different types of environmental forcing factors exerted in these ecosystems.

The coastal marine environment is regularly enriched with nutrients that come from rivers and the rain. This leads to the predominance of herbivorous food web in coastal zone (with high phytoplankton biomass) relative to oceanic area which is characterized generally by a microbial food web. Environmental forcing factors do change the structure of the food web for the benefit of microorganisms that best adapt to the new environment. This temporal change is close to the concept of ecological succession. The succession of food webs may lead, reversibly, from the microbial loop to microbial, multivorous, and herbivorous food webs.

The notion of "equilibrium change and status reversibility" between different food webs means that a situation can switch from one to the other as a result of environmental forcing factors. The consequences of the dominance of a food web relative to another one are very important because they lead to the ecosystem completely different in terms of production, regeneration, material flow, grazing, sequestration, or export of matter (fisheries can be considered as an exportation of biomass). Herbivorous food web and the microbial loop would be transitory and unstable systems, while multivorous and microbial food webs are much more stable and sustainable (Legendre and Rassoulzadegan 1995).

Apart from climate forcing factors (wind, temperature, light, etc.), two main groups of factors controlling pelagic microbial food webs are usually considered. The "bottom-up" factors consist of resources and affect more the growth rate of organisms than their biomass. The "top-down" factors are represented by all the predation pressure that leads to biomass removal.

By extension, all factors affecting the biomass, such as viral lysis or parasitism, are considered as top-down factors, although their effects on the structure and functioning of the food webs are fundamentally different. According to McQueen et al. (1989), the bottom-up and top-down factors are not exclusive and the relative intensity of each type of control varies in space and time. Predation on picoplankton communities and nutrient recycling within the food webs and the microbial loop are discussed in Sects. 13.3.1 and 13.3.2; only control via viral lysis is discussed below.

13.4.1.1 Viruses in Aquatic Microbial Food Webs
Abundance and Diversity
Since the discovery two to three decades ago that viruses of microbes are abundant in marine ecosystems (Bergh et al. 1989), aquatic viral ecology has grown increasingly to reach the status of a full scientific discipline in environmental sciences. A dedicated ISVM society, the *International Society for Viruses of Microorganisms* (http://www.isvm.org/), was recently launched. Viruses are omnipresent components of the microbial food web dynamics in a great variety of marine (Fuhrman 1999; Suttle 2005, 2007) and freshwater ecosystems (Sime-Ngando and Colombet 2009). They often represent the most abundant biological entity in pelagic and benthic habitats where abundances generally fluctuate between 10^4 and 10^8 viruses mL^{-1}, that is, 10–1,000-fold higher than the abundances of prokaryotes. The highest abundances (10^8 to 10^{10} viruses mL^{-1}) were reported in the sediments of Lake Gilbert, Canada (Maranger and Bird 1996). Values generally increase with the increasing productivity of aquatic ecosystems and, as a consequence, decrease from freshwater to marine ecosystems, from costal to oceanic zones and from the surface to the bottom of the euphotic layer. The abundance of viruses in individual aquatic systems appears to be independent of salinity but is related to the biomass of primary and secondary producers, as well as to the seasonal forcing (Weinbauer 2004; Mann 2003).

The most recent 9th report of the International Committee on Taxonomy of Viruses (ICTV, http://www.ictvonline.org/index.asp?bhcp = 1) includes 6 orders, 87 families, 19 subfamilies, 349 genera, and 2,284 viral types or "species." This is mostly based on isolated viral hosts in laboratory cultures which, in the case of environmental samples, may not exceed 1 % of the total prokaryotes. This implies that the diversity of environmental viruses is huge, although the bulk of the estimated 10^{31} viruses in the biosphere remains unknown. We now consider that environmental viruses represent the greatest reservoir of noncharacterized genetic resources on earth (Suttle 2007). In aquatic samples, viral phenotypes are limited, mainly including tailed or untailed particles with capsid heads characteristic of bacteriophages (Fig. 13.6). Tailed phages belong to the order *Caudovirales* (Ackermann 2003), all of which are double-stranded DNA viruses that generally represent about 10–50 % of the total abundance of viruses in aquatic systems. Within *Caudovirales*, three families emerge as quantitatively dominant: *Siphoviridae* with long noncontractile tails (e.g., phage lambda), *Podoviridae* with a short noncontractile tail (e.g., phage T7), and *Myoviridae* with contractile tails of variable length (e.g., phage T4). Phenotypic traits and viral morphs in aquatic viruses are cryptic of the selective pressures faced by these communities and provide insight into host range, viral replication, and function (Suttle 2005). For instance, myoviruses are mostly lytic with a large spectrum of sensitive hosts, which is a competitive advantage that can be assimilated to **r-strategist*** species thriving with high proliferation rates in fluctuating environments. In contrast, podoviruses are more highly specific to their hosts, with siphoviruses being intermediate between myo- and podoviruses. In addition, several siphoviruses can encode their genome into their hosts for several generations (lysogeny), which can be rather assimilated to **K-strategist*** species' characteristics of stable environments. Combined with the capacity of viruses to potentially face almost all types of environments and the related interfaces, the ability of viruses

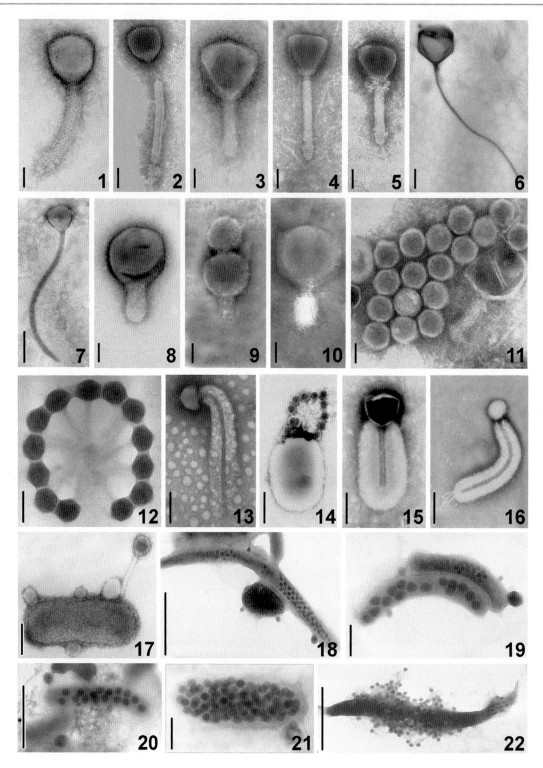

Fig. 13.6 Micrographs of viruses (phages) in aquatic ecosystems:- phenotypic diversity (micrographs *1–16*) and visibly infective viruses in prokaryotic cells (*17–22*). The three main families typical of the plankton are myoviruses (*1–5*), siphoviruses (*6–7*), and podoviruses (*8–10*). Note the presence of untailed phages (*11*) which can be very abundant in the plankton and of atypical morphologies (*12–16*) sampled in particular environments, herein in the permanently anoxic monimolimnion of Lake Pavin, France (Colombet 2008). Scale bars increase in length as follows: 50 nm (*1–6, 8–11*), 100 nm (*7, 12–17, 19, 21*), and 200 nm (*18, 20, 22*) (Micrographs are courtesy of Jonathan Colombet)

to develop along the r-K-selection continuum, i.e., from typical r (e.g., prokaryotes) to typical K (e.g., vertebrates) strategist organisms, may help to explain their ubiquity, hence the notion of virosphere (i.e., viral biosphere).

From a genomic point of view, whole genome comparisons have shown that there are conserved genes shared among all members within certain viral taxonomic groups (Breitbart and Rohwer 2005). These conserved genes can be targeted using PCR-amplification and sequencing for diversity studies of groups of cultured and environmental viruses. Examples of such genes are structural proteins such as *gp20* which codes for the capsid formation in T4 phage-like viruses, DNA polymerases for T7-like podophages, or the RNA-dependent RNA polymerase fragment, which has been used to identify novel groups of marine viruses. For the whole environmental communities, molecular fingerprinting approaches that separate polymerase chain reaction PCR-generated DNA products, such as denaturing gradient gel electrophoresis (DGGE) and pulse-field gel electrophoresis (PFGE), have been widely used but with limited results, restricted to double-stranded DNA viruses. With this approach, the genome size of aquatic viruses fluctuates from 10 to about 900 kb, with mean ranges of 10–650 kb in marine and freshwater pelagic systems. Finally, metagenomics has revolutionized microbiology by paving the way for a culture-independent assessment and exploitation of microbial and viral communities present in complex environments (Edwards and Rohwer 2005). Metagenomic analyses of 184 viral assemblages collected over a decade and representing 68 sites in four major oceanic regions showed that most of the viral DNA and protein sequences were not similar to those in the current databases (Angly et al. 2006). Global diversity was very high, presumably several hundred thousand "species" and regional richness varied on a north–south latitudinal gradient. However, most viral "species" were found to be widespread, supporting the idea that viruses are widely dispersed and that local environmental conditions enrich for certain viral types through selective pressure.

Overall, genomic approaches suggest that environmental viral diversity is high and essentially uncharacterized. They also suggest novel patterns of evolution and are increasingly changing the existing ideas on the composition of the virus world, while revealing novel groups of viruses and virus-like agents. The gene composition of aquatic DNA viromes (i.e., metagenomes of viruses) is dramatically different from that of known bacteriophages.

Roles of Viruses in Aquatic Systems

Environmental viruses exhibit various lifestyles that intimately depend on the deep-cellular mechanisms and are ultimately replicated by all members of the three domains of cellular life (*Bacteria*, *Eukarya*, *Archaea*), as well as by some giant viruses of eukaryotic cells. Most of the free-occurring viruses in aquatic systems are infectious and this infectivity can last for a long time, several tens of years for benthic viruses, for example (Suttle 2005). This establishes lytic viruses as major players in key ecological processes such as microbial mortality and the related biogeochemical cycling of nutrients and OM (Fuhrman 1999). It is now well accepted that the energy requirement for the lytic lifestyle of viruses is one of the main causes of microbial mortality in aquatic systems. Based on the direct observation of infected cells (Fig. 13.6), viral-mediated mortality averages 10–50 % of the daily production of heterotrophic prokaryotes and approximately equals the bacterivory from grazers, primarily in both fresh and marine pelagic waters. In the benthos, viral abundances are largely higher compared to the water column, but, paradoxically, few visible infected cells were reported in the sediments (Filippini et al. 2006). Viral lysis of cells has the strong feedback effect of preventing species dominance and enhanced species cohabitation within microbial communities, i.e., the so-called *phage kills the winner* hypothesis (Thingstad 2000). By this way, viruses also play major role in governing microbial diversity and structuring microbial food webs through complex mechanisms that are summarized in Fig. 13.7 (Weinbauer and Rassoulzadegan 2004).

It was estimated that the absolute abundance of oceanic viruses results in about 10^{29} infection events per day, causing the release of 10^8–10^9 tons of carbon per day from the living biological pool (Suttle 2005, 2007). By exploding microbial cells, lytic viruses are strong catalyzers of the transformation of living organisms to detrital and dissolved phases available to noninfected microbes (Fig. 13.7). This biogeochemical reaction increases the retention time of OM and its respiration in the water column and weakens the trophic efficiency of the food web but also provides nutrients (e.g., directly or indirectly from mineralization and photodegradation of DOM) to primary producers. For example, it has been shown that the iron content in the viral lysis products could fulfill the metabolic requirements of marine phytoplankton. Primary production in the size fraction 2–200 nm can be depleted by about half due to an increase in viral abundance as low as 20 %. Similarly, a modeling exercise suggested that viral lysis of 50 % of bacterial production could increase bacterial respiration by 27 % while decreasing the grazing efforts from protists and metazooplankton by 37 and 7 %, respectively. When adding 7 % of viral-mediated loss of phytoplankton and 3 % grazing of viruses by phagotrophic flagellates, bacterial respiration increases to 33 % (Fuhrman 1999).

One of the key explanations for the omnipresence of viruses in natural ecosystems is undoubtedly through the existence of several lifestyles, of which two major pathways, namely, lysis and lysogeny, are prevalent in aquatic systems.

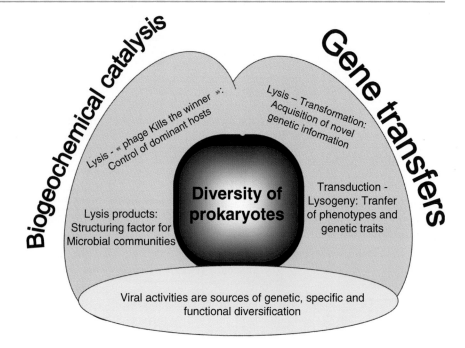

Fig. 13.7 Potential roles of viruses in the dynamics of microbial diversity and the functioning of aquatic ecosystems. This figure is centered on the diversity of prokaryotes which provide the greatest host reservoirs for aquatic viruses. The interactions between these two communities are best studied in aquatic viral ecology (Modified from Weinbauer and Rassoulzadegan 2004)

Lytic infections are by far the best studied of the virus-host interactions. Lysogenic activity has been less studied in aquatic environments where the temperate phage can alternatively integrate into the host genome as prophage which is replicated for several generations. The lysogenic conversion of competent prokaryotes allows the dissemination and maintenance of viral populations primarily in harsh conditions while promoting horizontal transfers of viral and host genes between microorganisms through the process of transduction. This conversion can thus influence microbial speciation, providing novel phenotypic, genotypic, and metabolic traits to the host cell, such as resistance to superinfection or the capacity to produce toxins. A spectacular case for such a phage conversion is the finding that cholera infection is due to a lysogenic strain of the *Vibrio cholerae* bacterium.

Large-scale metagenomics has shown that viruses contain diverse genes of interest, including virulence genes such as the cholera toxin, respiration, nucleic acid, carbohydrate, and protein metabolism genes, as well as genes involved in vitamin and cofactor synthesis, in stress response, and in motility and chemotaxis, which are more common in viromes (metagenomics of viruses) than in their corresponding microbiomes (metagenomics of microbes) (Edwards and Rohwer 2005). Microbes that take up these genes increase their competitive ability and extend their ecological niches. More interestingly, virally encoded host genes also include crucial photosynthetic genetic elements present in cyanophage genomes, which can be used to maintain the targeted function in dead hosts and accomplish the lytic cycle and can be transferred between hosts as well. About 10 % of total global photosynthesis could be carried out as a result of phage genes originally from phages (Lindell et al. 2005). Given the prevalence of phage-encoded biological functions and the occurrence of recombination between phage and host genes, phage populations are thus expected to serve as gene reservoirs that contribute to niche partitioning of microbial species in aquatic ecosystems.

Overall, our conceptual understanding of the function and regulation of aquatic ecosystems, from microbial to global biogeochemical processes, has changed with the study of viruses. Aquatic viral ecology, although is a relatively recent discipline, is a source of novel knowledge related to the biodiversity of living things, the functioning of ecosystems, and the evolution of the cellular world (Forterre 2007). For example, it is becoming clear that most of the environmental viruses are not microbial pathogens but mutualistic cell partners that provide helper functions (Rohwer and Thurber 2009; Roossinck 2011). The discovery of giant viruses which encode trademark cellular functions (La Scola et al. 2003) is weakening the gap between inert and living things.

13.4.2 Main Factors Controlling Benthic Marine and Lake Microbial Food Webs

Grazing effects of microbial communities by benthic macrofauna ("top down") are well known for biofilms. These effects are mainly direct and negative reducing both diversity and biomass of microalgae (e.g., Liess and Hillebrand 2004). In contrast, for meiofauna, bioturbation effects are stronger than grazing effects, enhancing the photosynthetic activity of photoautotrophic microorganisms in biofilms (Mathieu et al. 2007; Majdi et al. 2012b).

It should be noted that the behavior of predators is also an important factor to consider when assessing the control systems of benthic trophic webs. So, feeding of gastropods and insect larvae (e.g., Trichoptera: Psychomyiidae) in marine and freshwater systems is linked to their movements and life cycle, respectively. Their feeding activity on biofilms often induces drastic effects (e.g., Lawrence et al. 2002; Majdi et al. 2012a) and, for a matter of fact, is called "trophic raid." After a trophic raid, a new colonization cycle can start allowing the growth of a new biofilm when the macrofaunal grazing pressure is low.

Predation pressure exerted by macrobenthic invertebrates often avoids an accumulation of detritic OM as observed in microbial mats. Microbial mats mainly occur in an area where predation pressure is low, i.e., in extreme environments (hyperhaline waters, high and low temperatures, high hydrodynamics). Moreover, it must be noticed that during the Precambrian, i.e., before eukaryotic metazoan occurrence, cyanobacterial mats were largely widespread (*cf.* Chap. 4).

Production of benthic photosynthetic microorganisms in intertidal superficial sediments is mainly submitted to "bottom-up" control processes. This means that their production is mainly limited by nutrient availability and light. However, studies have shown that biomass and production of benthic microorganisms are tightly linked to the rate of the grazing and bioturbation activities exerted by meiofaunal consumers (Buffan-Dubau and Carman 2000; Mathieu et al. 2007, 2012b). Grazing pressure is not the main control process; however, production of benthic microphytes is indirectly enhanced by meiofaunal activity modifying "bottom-up" control potential effect (i.e., enhance light penetration and nutrient regeneration) in coastal superficial sediments and in biofilms. The sediment reworking resulting from meiofaunal activity can also increase the porosity and the solute transport rates in sediments and microbial aggregates, improving availability of dissolved nutrients for photoautotrophic microorganisms (Aller and Aller 1992; Mathieu et al. 2007).

The feeding activity of meiofauna induces indirect effects. The most significant are described for nematode activity. They produce copious mucous secretions that in some cases attract bacteria, stimulate their growth, and modify the bacterial community composition and diversity in river superficial sediments (Moens et al. 2006).

13.4.3 Factors Controlling Terrestrial Microbial Food Webs

Given the fact that almost all of the primary production of terrestrial ecosystems come from higher plants, it is obviously necessary to consider plants as central to the regulation of soil organisms and their interactions, at least as important as environmental factors and disturbances.

13.4.3.1 Regulation by Environmental Factors and Perturbations

As in aquatic environments, the microbial food web of soil depends on both "top-down" and "bottom-up" controlling factors. It is generally accepted that bacteria are primarily regulated by predation, while fungi are mainly regulated by the resources. However, there are many cases of "top-down" control of fungi. For example, the presence of protozoa and nematodes in the rhizosphere of trees reduces the colonization of tree roots by mycorrhizal fungi (Wardle 2002). Diversity within a given group in the soil depends on many factors such as resource availability, temperature, pH, or water status, as well as disturbances such as setting culture or rapid and extreme climate events.

Some studies show that an increase in resources is accompanied by a stimulation of microbial functional diversity of soil (Degens and Harris 2000). This observation goes against the theory developed for the plants that the increase in resources should lead to a reduction in biodiversity. In the study by Degens and Harris (2000), diversity is measured in vitro by the ability of microorganisms in the soil to use substrates made, and it is likely that the observed results reflect further metabolic abilities of bacteria relative to fungi. The bacteria are indeed able to grow rapidly on the substrates and are, therefore, less regulated by the competition that groups with slower growth like fungi. Yet, it is generally recognized that the diversity of soil fauna is maintained even when conditions become more favorable.

Yeates and Bongers (1999) found that the diversity of nematodes (assuming that this group is poorly regulated by competition) gradually decreases with an increase in the concentration of heavy metals and is not reduced by dietary calcium ground. More generally, the diversity of many groups in the soil is more important on a humus of *Fagus sylvatica* where relatively favorable conditions prevail for the activity of organisms (pH: 4.3–6.8, high cation exchange capacity) relative to the humus in the same vegetation where the conditions are much less favorable (pH: 3–4, with low cation exchange capacity). Most studies tend to show that the diversity of decomposers does not decrease with the improvement of environmental conditions. These organisms do not seem, from this point of view, to follow the models most frequently observed in plants. In contrast, and as noted earlier in this paragraph, certain groups such as fungi can be controlled by the competition. Thereby, Egerton-Warburton and Allen (2000) show that the addition of N reduces the diversity of fungi. This result suggests that the increased resources promote **competitive exclusion***.

13.4.3.2 Regulation by Biotic Factors

Studies on the biotic regulation of soil organisms and their activities are scarce, with the notable exception of the interactions between plants and microorganisms. The presence of a microfauna that consumes fungi generally, but not always, influences the fungal diversity. This can be explained by the combined effect of the importance of the selective nature of consumers (see above) and of the role of competition in structuring fungal communities (McLean and Parkinson 2000).

Besides, organisms considered as engineers such as earthworms can have an impact on the diversity – especially that of fungi – when modifying the soil structure, leading to the buildup of new habitats. McLean and Parkinson (2000) have found, based on experiments in microcosms, that the worm *Dendrobaena octaedra* could alter the structure of the fungal community in the soil and, in some conditions, increase the fungal diversity by decreasing the intensity of competition. In other experiences, these same authors show that the presence of *D. octaedra* can, however, reduce the fungal diversity by promoting a fast-growing species which becomes dominant.

Numerous studies have shown that plant species can force diversity, abundance, and activity of microorganisms in soils. For example, Badejo and Tian (1999) have shown in monoculture plots in Nigeria, Africa, that the diversity of some groups of microfauna was greater under the plant species *Leucaena leucocephala* than under three other plant species investigated. Similarly, it has been shown that the species richness of mycorrhizal fungi on the roots of plants is very different for five different plant species on the same site.

Finally, the role of the composition of the vegetation on the diversity of soil organisms was also shown for various stages of plant successions which are dominated by different functional types of plants (Wardle 2002).

13.5 Conclusion

The microbial food web (MFW) organisms, including viruses, archaea, bacteria, fungi, phytoplankton, benthic algae, and protozoa, play an essential role in the functioning of aquatic (pelagic and benthic) and terrestrial ecosystems. However, there are fundamental differences between the different ecosystems in the functional role of microorganisms. The functioning of pelagic ecosystems is largely depending on the advection, vertical mixing, and stratification of the water column and the associated effects. Planktonic mode life is best suited to this dynamic environment. Therefore, pelagic aquatic ecosystems are particularly original because microbial communities provide almost all primary production, constitute the essential of the biomass present, and form thereby the basis of all aquatic food webs.

In connection with the physical nature of the sediments and soils, the functioning of benthic and terrestrial MFW is significantly different. In terrestrial ecosystems, unlike aquatic pelagic ecosystems and even if there are the same groups of microorganisms in water with a diversity at least as important, the functional basis of these ecosystems is formed by higher plants as the main primary producers. Moreover, the plants exercise, through their root system, an important control on the soil structuring and on the biophysicochemical interactions. In this context, microorganisms play, however, an important role in terrestrial biogeochemical cycles and they act as "bioreactors," transforming and recycling the primary organic matter that comes from plants. On this last point, sediments in some cases share similarities with the soil when the bulk of primary production provides by phanerogam. The similarities between sediments and pelagic systems can be also observed when the essential of organic matter in the water column comes from the resuspension of autotrophic microorganisms living in the sediment surface or even resuspension of settled matter on the sediment. Because of the biophysical characteristics of soil at macro-, meso-, and microscales, terrestrial MFW are usually apprehended in substantially different ways, and the concept of food web has much less impregnated studies on soil ecosystems than those in aquatic ecosystems. The aggregation process, favoring a three-dimensional structure of environmental factors, leading to a three-dimensional distribution of microorganisms. As in the sediment, this structuring results in a partial dissociation of the types of microorganisms in space. In addition, the presence of gaseous dioxygen in the soil, associated with aggregation mechanisms, has the effect of promoting microscale coexistence of aerobic and anoxic zones. These microgradients of oxygenation guide the "colonization" of aggregates by microorganisms according to their strict aerobic, aerobic-anaerobic, or strict anaerobic characteristics. Such oxygenation gradients exist also in the immediate vicinity of the roots and affect the distribution of microorganisms according to the availability of electron acceptors that can be used in the respiration process. It should be noted, however, that the pelagic aquatic environments are not either strictly homogeneous, particularly because of microaggregates that can lead to the presence of anoxic zones in microscale within an overall aerobic water column. More generally, the structure of MFW in aquatic (pelagic and benthic) and terrestrial environments reflects the adaptation of microorganisms against physical, chemical, and biological environmental forcing factors. Understanding the mechanisms responsible for these adaptations to changing environmental conditions is a major challenge for the coming years.

References

Ackermann H-W (2003) Bacteriophage observations and evolution. Res Microbiol 154:245–251

Aller RC, Aller JY (1992) Meiofauna and solute transport in marine muds. Limnol Oceanogr 37:1018–1033

Amblard C, Carrias JF, Bourdier G, Maurin N (1995) The microbial loop in a humic lake: seasonal and vertical variations in the structure of the different communities. Hydrobiologia 300(301):71–84

Amblard C, Boisson JC, Bourdier G, Fontvieille D, Gayte X, Sime-Ngando T (1998) Ecologie microbienne en milieu aquatique: des virus aux protozoaires. Rev Sc. Eau, N° Spécial: Les Sciences de l'Eau: Bilan et perspectives, 145–162

Anderson JM (2000) Food web functioning and ecosystem processes. In: Colemab DC, Hendrix PF (eds) Invertebrates as webmasters in ecosystems. CABI Publishing, Wallunford, pp 3–24

Angly FE et al (2006) The marine viromes of four oceanic regions. PLoS Biol 4:e368

Azam F, Fenchel T, Field JG, Gray JS, Meyer-Reil LA, Thingstad F (1983) The ecological role of water-column microbes in the sea. Mar Ecol Prog Ser 10:257–263

Badejo MA, Tian G (1999) Abundance of soil mites under four agroforestry tree species with contrasting litter quality. Biol Fertil Soils 30:107–112

Bardgett RD (2005a) Organism interactions and soil processes. In: Crawley MJ, Little C, Southwood TRE, Ulfstrand S (eds) The biology of soil, a community and ecosystem approach. Oxford University Press, Oxford, UK, pp 57–85

Bardgett RD (2005b) The diversity of life in soil. In: Crawley MJ, Little C, Southwood TRE, Ulfstrand S (eds) The biology of soil, a community and ecosystem approach. Oxford University Press, Oxford, UK, pp 24–56

Beare MH, Coleman DC, Crossley DA, Hendrix PF, Odum EP (1995) A hierarchical approach to evaluating the significance of soil biodiversity to biogeochemical cycling. Plant and Soil 170:5–22

Bec A, Martin-Creuzburg D, von Elert E (2006) Trophic upgrading of autotrophic picoplankton by the heterotrophic nanoflagellate *Paraphysomonas* sp. Limnol Oceanogr 51:1699–1707

Bergh O, Børsheim KY, Bratbak G, Heldal M (1989) High abundance of viruses found in aquatic environments. Nature 340:467–468

Breitbart M, Rohwer F (2005) Here a virus, there a virus, everywhere the same virus? Trends Microbiol 13:278–284

Brown GG (1995) How do earthworms affect microfloral and faunal community diversity. Plant and Soil 170:209–231

Buffan-Dubau E, Carman KR (2000) Diel feeding behavior of meiofauna and their relationships with microalgal resources. Limnol Oceanogr 45:381–395

Carpenter SR (1988) Complex interactions in lake communities. Springer, New York

Carrias JF, Amblard C, Bourdier G (1998) Seasonal dynamics and vertical distribution of planktonic ciliates and their relationship to microbial food resources in the oligomesotrophic lake Pavin. Arch Hydrobiol 143:227–255

Colombet J (2008) Importance de la variabilité verticale dans un lac méromictique profond: diversité et activité lysogène des communautés virales. Thèse de Doctorat, Université Blaise Pascal, 204 p

Curtis TP, Sloan WT, Scannel JW (2002) Estimating prokaryotic diversity and its limits. Proc Natl Acad Sci U S A 99:10494–10499

Cushing DH (1989) A difference in structure between ecosystems in strongly stratified waters and in those that are only weakly stratified. J Plank Res 11:1–13

De Ruiter PC, Neutel AN, Moore JC (1995) Energetics, patterns of interactions strengths and stability in real ecosystems. Science 269:1257–1260

De Ruiter PC, Neutel AN, Moore JC (1997) Soil foodweb interactions and modelling. In: Benckiser G (ed) Fauna in soil ecosystems. Dekker, New York, pp 363–386

Degens B, Harris JA (2000) Decreases in organic C reserves in soil can reduce the catabolic diversity of soil microbial communities. Soil Biol Biochem 32:189–196

Delong EF (1992) Archaea in coastal marine environments. Proc Natl Acad Sci U S A 89:5685–5689

Dolan JR (1997) Phosphorus and ammonia excretion by planktonic protists. Mar Geol 139:109–122

Edwards RA, Rohwer F (2005) Viral metagenomics. Nat Rev Microbiol 3:504–510

Egerton-Warburton L, Allen EB (2000) Shifts in arbuscular mycorrhizal communities along an anthropogenic nitrogen deposition gradient. Ecol Appl 10:484–496

Filippini M, Buesing N, Bettarel Y, Sime-Ngando T, Gessner MO (2006) Infection paradox: high abundance but low impact of freshwater benthic viruses. Appl Environ Microbiol 72:4893–4898

Foissner W (1997a) Global soil ciliate (Protozoa: Ciliophora) diversity: a probability-based approach using large sample collections from Africa, Australia and Antarctica. Biodivers Conserv 5:1627–1638

Foissner W (1997b) Soil ciliates (Protozoa: Ciliophora) from evergreen rain forest of Australia, South America and Costa Rica: diversity and description of new species. Biol Fertil Soils 25:317–339

Forterre P (2007) Microbes de l'enfer. Edition Belin, Paris

Fouilland E, Mostajir B (2010) Revisited phytoplanktonic carbon dependency of heterotrophic bacteria in freshwater, transitional, coastal and oceanic waters. FEMS Microbiol Ecol 73:419–429

Fouilland E, Mostajir B (2011) Complementary support for the new ecological concept of "bacterial independence on contemporary phytoplankton production" in oceanic Waters. FEMS Microbiol Ecol 78:206–209

Fouilland E, Gosselin M, Rivkin RB, Vasseur C, Mostajir B (2007) Nitrogen uptake by heterotrophic bacteria and phytoplankton in Arctic surface waters. J Plankton Res 29:369–376

Fuhrman JA (1999) Marine viruses and their biogeochemical and ecological effects. Nature 399:541–548

Giere O (2009) Meiobenthology: the microscopic motile fauna of aquatic sediments. Springer, Berlin/Heidelberg

Grami B, Rasconi S, Niquil N, Jobard M, Saint-Béat B, Sime-Ngando T (2011) Functional effects of parasites on food web properties during the spring diatom bloom in Lake Pavin: a linear inverse modeling analysis. PLoS One 6:e23273

Hedlund K, Boddy L, Preston CM (1991) Mycelial response of the soil fungus, *Mortierella Isabellina*, to grazing by *Onychiurus armatus* (collembolan). Soil Biol Biochem 23:361–366

Hobbie JE, Daley RJ, Jasper S (1977) Use of Nuclepore filters for counting bacteria by fluorescence microscopy. Appl Environ Microbiol 33:1225–1228

Jardillier L, Boucher D, Personnic S, Jacquet S, Thenot A, Sargos D, Amblard C, Debroas D (2005) Relative importance of nutrients and mortality factors on prokaryotic community composition in two lakes of different trophic status: microcosm experiments. FEMS Microbiol Ecol 53:429–443

Kagami M, De Bruin A, Ibelings BW, Van Donk E (2007) Parasitic chytrids: their effects on phytoplankton communities and food-web dynamics. Hydrobiologia 578:113–129

Kolber ZS, Van Dover CL, Niederman RA, Falkowski PG (2000) Bacterial photosynthesis in surface waters of the open ocean. Nature 407:177–179

La Scola B, Audic S, Robert C, Jungang L, de Lamballerie X, Drancourt M, Birtles R, Claverie JM, Raoult D (2003) A giant virus in amoebae. Science 299:2033

Laval-Peuto M, Heinbokel JF, Anderson R, Rassoulzadegan F, Sherr BF (1986) Role of micro- and nanozooplankton in marine food webs. Insect Sci Appl 7:387–395

Lavelle P, Decaëns T, Aubert M, Barot S, Blouin M, Bureau F, Margerie P, Mora P, Rossi JP (2006) Soil invertebrates and ecosystem services. Eur J Soil Biol 42:3–15

Lawrence JR, Scharf B, Packroff G, Neu TR (2002) Microscale evaluation of the effects of grazing by invertebrates with contrasting feeding modes on river biofilm architecture and composition. Microb Ecol 43:199–207

Lefèvre E, Bardot C, Noël C, Carrias J-F, Viscogliosi E, Amblard C, Sime-Ngando T (2007) Unveiling fungal zooflagellates as members of freshwater picoeukaryotes: evidence from a molecular diversity study in a deep meromictic lake. Environ Microbiol 9:61–71

Legendre L, Rassoulzadegan F (1995) Plankton and nutrient dynamics in marine waters. Ophelia 41:153–172

Li KW, Subba Rao DV, Harrison GW, Smith CJ, Cullen JJ, Irwin B, Platt T (1983) Autotrophic picoplankton in the tropical ocean. Science 219:292–295

Liess A, Hillebrand H (2004) Direct and indirect effects in herbivore–periphyton interactions. Arch Hydrobiol 159:433–453

Lindell D, Jaffe JD, Johnson ZI, Church GM, Chisholm SW (2005) Photosynthetic genes in marine viruses yield proteins during host infection. Nature 438:86–89

Majdi N, Mialet B, Boyer S, Tackx M, Leflaive J, Boulêtreau S, Ten-Hage L, Julien F, Fernandez R, Buffan-Dubau E (2012a) The relationship between epilithic biofilm stability and its associated meiofauna under two patterns of flood disturbance. Fresh Sci 31:38–50

Majdi N, Tackx M, Buffan-Dubau E (2012b) Trophic positioning and microphytobenthic carbon uptake of biofilm-dwelling meiofauna in a temperate river. Fresh Biol 57:1180–1190

Mann NH (2003) Phages of the marine cyanobacterial phytoplankton. FEMS Microbiol Rev 27:17–34

Maranger R, Bird DE (1996) High concentrations of viruses in the sediments of Lac Gilbert, Quebec. Microb Ecol 31:141–151

Mathieu M, Leflaive J, Ten-Hage L, de Wit R, Buffan-Dubau E (2007) Free-living nematodes affect oxygen turn-over of artificial diatom biofilms. Aquat Microb Ecol 49:281–291

McLean MA, Parkinson D (2000) Field evidence of the effects of the epigenic earthworm Dendrobaena octaedra on the microfungal community in pine forest floor. Soil Biol Biochem 32:351–360

McQueen DJ, Johannes MRS, Post JR, Steward TJ, Lean DRS (1989) Bottom-up and top-down impacts on freshwater pelagic community structure. Ecol Monogr 59:289–310

Middelburg JJ, Barranguet C, Boschker HTS, Herman PMJ, Moens T, Heip CHR (2000) The fate of intertidal microphytobenthos carbon: an in situ ^{13}C labelling study. Limnol Oceanogr 45:1224–1234

Mittelbach GG, Steiner CF, Scheiner SM, Gross KL, Reynolds HL, Waide RB (2001) What is the observed relationship between species richness and productivity? Ecology 82:2381–2396

Moens T, Traunspurger W, Bergtold M (2006) Feeding ecology of free-living benthic nematodes. In: Abebe E, Traunspurger W, Andrassy I (eds) Freshwater nematodes: ecology and taxonomy. CABI Publishing, Wallingford, pp 105–131

Moody SA, Piearce TG, Dighton J (1996) Fates of some fungal spores associated with wheat straw decomposition on passage through the guts of Lumbricus terrestris and Aporrectodea longa. Soil Biol Biochem 28:533–537

Moran XA, Alonso-Saez L (2011) Independence of bacteria on phytoplankton? Insufficient support for Fouilland and Mostajir's (2010) suggested new concept. FEMS Microbiol Ecol 78:203–205

Pischedda L, Cuny P, Esteves JL, Poggiale J-C, Gilbert F (2012) Spatial oxygen heterogeneity in a Hediste diversicolor irrigated burrow. Hydrobiologia 680:109–124

Pomeroy LR (1974) The ocean's food web, a changing paradigm. BioSci 24:499–504

Pomeroy LR (1991) Status and future needs in protozoan ecology. In: Reid PC, Turley CM, Burkill PH (eds) Protozoa and their role in marine processes. Nato Asi series. Springer, New York, pp 475–492

Renault P, Stengel P (1994) Modelling oxygen diffusion in aggregated soils: anaerobiosis inside the aggregates. Soil Sci Soc Am J 58:1017–1023

Riemann F, Helmke E (2002) Symbiotic relations of sediment agglutinating nematodes and bacteria in detrital habitats: the enzyme-sharing concept. PSZN I Mar Ecol 23:93–113

Rohwer F, Thurber RV (2009) Viruses manipulate the marine environment. Nature 459:207–212

Roossinck MJ (2011) The good viruses: viral mutualistic symbioses. Nat Rev Microbiol 9:99–108

Schmidt O, Curry JP, Dyckmans J, Rota E, Scrimgeour CM (2004) Dual stable isotope analysis (delta C-13 and delta N-15) of soil invertebrates and their food source. Pedobiologia 48:171–180

Schulz HN, Jorgensen BB (2001) Big bacteria. Annu Rev Microbiol 55:105–137

Sextone AJ, Revbesh NP, Parkin TB, Tiedje JM (1985) Direct measurement of oxygen profiles and denitrification rates in soil aggregates. Soil Sci Soc Am J 49:645–651

Sherr EB, Sherr BF (1988) Role of microbes in pelagic food webs: a revised concept. Limnol Oceanogr 33:1225–1227

Sieburth JMCN, Smetacek V, Lenz J (1978) Pelagic ecosystem structure: heterotrophic compartments of the plankton and their relationship to plankton size fractions. Limnol Oceanogr 23:1256–1263

Sime-Ngando T (1997) Importance des virus dans la structure et le fonctionnement des réseaux trophiques microbiens aquatiques. Ann Biol 36:181–210

Sime-Ngando T (2012) Phytoplankton chytridiomycosis: fungal parasites of phytoplankton and their imprints on the food web dynamics. In: Grossart HP, Reimann L, Tang KW (eds) Molecular and functional ecology of aquatic microbial symbionts, vol 3. Frontiers Microbiology, p 361

Sime-Ngando T, Colombet J (2009) Viruses and prophages in aquatic ecosystems. Can J Microbiol 55:95–109

Sime-Ngando T, Niquil N (2011) Disregarded microbial diversity and ecological potentials in aquatic systems. Development in hydrobiology, vol 216. Springer, Dordrecht

Sime-Ngando T, Bettarel Y, Chartogne C, Sean K (2003) The imprint of wild viruses on freshwater microbial ecology. Recent Res Dev Microbiol 7:481–497

Suttle CA (2005) Viruses in the sea. Nature 437:356–361

Suttle CA (2007) Marine viruses – major players in the global ecosystem. Nat Rev 5:801–812

Swift MJ, Heal OW, Anderson JM (1979) Decomposition in terrestrial ecosystems. University of California Press, Berkeley

Tachet H, Richoux P, Bournaud M, Usseglio-Polatera P (2010) Invertébrés d'eau douce: systématique, biology, écologie. CNRS Éditions, Paris

Thingstad TF (2000) Element of a theory for the mechanisms controlling abundance, diversity, and biogeochemical role of lytic bacteria viruses in aquatic systems. Limnol Oceanogr 45:1320–1328

Torsvik V, Ovreas L, Thingstad TF (2002) Prokaryotic diversity: magnitude, dynamics and controlling factors. Science 296:1064–1066

Verity PG (1991) Feeding in planktonic protozoans – evidence for non-random acquisition of prey. J Protozool 38:69–76

Wardle DA (2002) Linking the aboveground and belowground components. Princeton University Press, Princeton

Weinbauer MG (2004) Ecology of prokaryotic viruses. FEMS Microbiol Rev 28:127–181

Weinbauer MG, Rassoulzadegan F (2004) Are viruses driving microbial diversification and diversity? Environ Microbiol 6:1–11

Wetzel RG, Rich PH, Miller MC, Allen HL (1972) Metabolism of dissolved and particulate detrital carbon in a temperate hard-water lake. In: Melchiorri-Santolini U, Hopton JW (eds) Detritus and its role in aquatic ecosystems, vol 29. Memoire Inst Ital Idrobiol, pp 185–243

Yeates GW, Bongers T (1999) Nematode biodiversity in agroecosystems. Agric Ecosyst Environ 74:113–135

Yeates GW, Bongers T, De Groede RGM, Freckman DW, Georgieva SS (1993) Feeding habits in soil nematode families and genera – an outline for ecologists. J Nematol 25:315–331

Yeates GW, Bardgett RD, Cook R, Hobbs PJ, Bowling PJ, Potter JF (1997) Faunal and microbial diversity in three Welsh grassland soils under conventional and organic managements regimes. J Appl Ecol 34:453–471

Biogeochemical Cycles

14

Jean-Claude Bertrand, Patricia Bonin, Pierre Caumette, Jean-Pierre Gattuso, Gérald Grégori, Rémy Guyoneaud, Xavier Le Roux, Robert Matheron, and Franck Poly

Abstract

All living organisms contribute to the biogeochemical cycles, but microorganisms, due to their high abundance, their tremendous metabolic capacities and adaptation potential, play a key role in the functioning and the evolution of biogeochemical cycles. Consequently, they are keyplayers in adaptation, resistance and resilience of ecosystems. The role of microorganisms in the main biogeochemical cycles (carbon, nitrogen, sulfur, phosphorus, silicon, metals), in soils, freshwater and marine ecosystems is presented. Microbial processes involved in the turnover of biogeochemical cycles are discussed from gene to ecosystem (natural and anthropogenic ecosystems), at global, regional and local scales, as well as in targeted microenvironments (such as particles or microniches). The biodiversity of microorganisms is highlighted and their metabolic pathways on which are based exchanges and biotransformations of organic and mineral components within ecosystems are described in details. The impacts of human activities on the microbial actors and processes of biogeochemical cycles, and the cascading ecological effects (greenhouse gas emissions, acid rains, dystrophic crises, etc.), are also discussed.

Keywords

Anoxic zones • Biogeochemical cycles • Carbon cycle • Ecosystem functioning • Iron cycle • Lake ecosystems • Manganese cycle • Marine ecosystems • Mercury cycle • Microbial functions • Nitrogen cycle • Oxic zones • Phosphate cycle • Silicon cycle • Soils • Sulfur cycle

---

* Chapter Coordinator

J.-C. Bertrand* (✉) • P. Bonin • G. Grégori
Institut Méditerranéen d'Océanologie (MIO), UM 110, CNRS 7294
IRD 235, Université de Toulon, Aix-Marseille Université,
Campus de Luminy, 13288 Marseille Cedex 9, France
e-mail: jean-claude.bertrand@univ-amu.fr

P. Caumette • R. Guyoneaud
Institut des Sciences Analytiques et de Physico-chimie pour l'Environnement et les Matériaux (IPREM), UMR CNRS 5254, Université de Pau et des Pays de l'Adour, B.P. 1155, 64013 Pau Cedex, France

J.-P. Gattuso
Observatoire Océanologique, Laboratoire d' Océanographie, CNRS-UPMC, BP 28, 06234 Villefranche-sur-Mer Cedex, France

X. Le Roux • F. Poly
Microbial Ecology Center, UMR CNRS 5557 / USC INRA 1364, Université Lyon 1, 69622 Villeurbanne, France

R. Matheron
Institut Méditerranéen de Biodiversité et d'Ecologie marine et continentale (IMBE), UMR-CNRS-IRD 7263, Aix-Marseille Université, 13397 Marseille Cedex 20, France

14.1 Introduction

In order to grow, to reproduce and to maintain their structural and functional integrity, living organisms select a limited number of chemical compounds in their surrounding environment to build organic biomolecules that form the basic units of cells (Fig. 14.1). When dying, the cells release their organic constituents back to the environment as simple minerals, primarily as the result of the activity of microorganisms. Any biological element thus undergoes a continuous cycle, called "biogeochemical cycle", in which it passes alternately from a mineral, non-living status to a status of living matter. The oxidation state of organic elements such as carbon, nitrogen, or sulfur forms can also be modified without being incorporated into living organisms because these elements can be used as electron donors or acceptors. This is valid for almost all natural elements susceptible to serve as electron donors or acceptors (*e.g.*, arsenic), or to be modified as a consequence of a biological activity (Box 14.15). Different biotic, chemical abiotic (autooxidation, photooxidation, etc.), and physical processes (dissolution, precipitation, volatilization, etc.) ensure the transformation of biological components and their movement between the different compartments of the biosphere, i.e., lithosphere (emerged continents and sediments), hydrosphere, and atmosphere.

Cycles of matter are accompanied by a unidirectional flow of energy because, unlike elements, energy cannot be regenerated. Energy is inexhaustible but the amount of chemicals available on Earth is constant, only fueled by contributions from meteorites and micrometeorites. For life to be maintained on Earth, the recycling of chemical elements is thus essential.

The elemental composition of living matter is very different from the composition of the three compartments of the biosphere (Fig. 14.1). Among 92 natural elements, only about 30, considered as biogenic elements, are critical to living organisms. The four most abundant elements are oxygen, carbon, hydrogen, and nitrogen which constitute about 97–99 % of the mass of most cells. Other elements (iron, sulfur, phosphorus, manganese, magnesium, calcium, copper, zinc, molybdenum, etc.) represent only a very small fraction of living matter, nervertheless they are essential for life.

14.1.1 A Biogeochemical Cycle Includes Three Key Steps (Figs. 14.1 and 14.2)

1. Mineral elements (MI) are fixed by autotrophic organisms, qualified as primary or paraprimary producers. These organisms use light energy (oxygenic and anoxygenic photosynthesis) or chemical energy (aerobic and anaerobic chemolithotrophy) (Fig. 14.2, *fluxes 1 and 2*) to fix CO_2 and build their organic biomolecules (*cf.* Sect. 3.2).

   In terrestrial environments, primary producers are mainly vascular plants; the contribution of unicellular

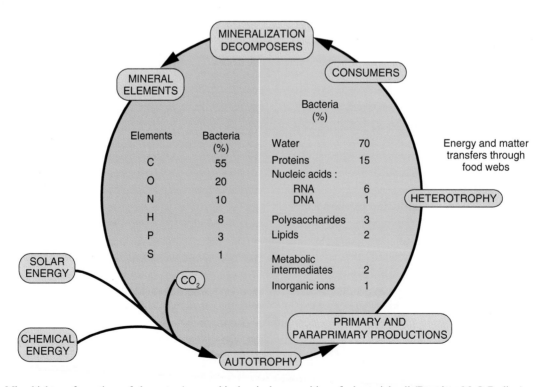

Fig. 14.1 Microbial transformations of elements. Average biochemical composition of a bacterial cell (Drawing: M.-J. Bodiou)

14 Biogeochemical Cycles

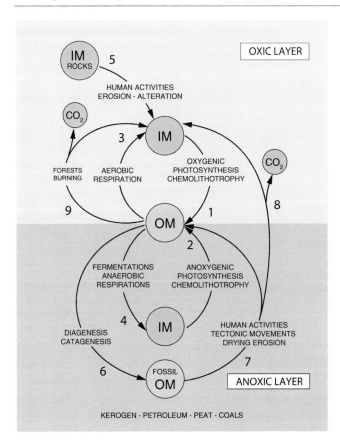

Fig. 14.2 Transformation of organic and mineral matter in ecosystems. *IM* Inorganic Matter, *OM* Organic Matter, *1,2* Organic matter production from mineral compounds in oxic (*1*) and anoxic (*2*) conditions, *3,4* Organic matter mineralization in oxic (*3*) and anoxic (*4*) conditions, *5* Rock degradation and dissolution of mineral elements, *6* Storage and trapping of organic matter in soils and anoxic sediments (fossilization), *7,8* Release of organic matter from fossil stocks in anoxic (*7*) and oxic (*8*) conditions, *9* Mineralization of organic matter by burning. The study of the degradation of organic matter can be company at different levels; it can be analyzed from a global approach (content of carbon, nitrogen, etc..) through the analysis of a class of compounds (proteins, carbohydrates, lipids, humic acids, etc.) till a precise analysis of the compounds at the molecular level (fatty acids, sterols, quinones, hopanes, saturated hydrocarbons, and aromatic pesticides, etc.). The terms "oxic" and "anoxic" characterize environments containing or not containing free oxygen (dioxygen). Aerobic microorganisms are living in oxic environment and anaerobic microorganisms in anoxic environments (Drawing: M.-J. Bodiou)

algal in soils being of minor importance. This contrasts with aquatic environments where phytoplanktonic microorganisms play the major role; large marine algae (i.e., seaweed) only develop on coastal areas.

Chemolithotrophic and photolithotrophic bacteria are mainly autotrophic (*cf.* Sect. 3.3.2). When biomass is produced from carbon and electron donors known from a metabolism pathway, the production is then qualified as paraprimary production. This is the case of photosynthetic purple and green bacteria that consume sulfide produced by sulfate-reducing bacteria. Paraprimary production also applies to the amount of biomass produced by methanogenic archaea that consume hydrogen produced by fermentative bacteria. The term of primary production when applied to bacteria (prokaryotes) relates only to microorganisms that typically use light as energy source and electron donors of geochemical origin (H_2, H_2S, NH_3, Fe^{2+}). This is for example the case of purple or green bacteria that perform anoxigenic photosynthesis by oxidizing sulfide from ground via volcanism;

2. The organic matter synthesized by autotrophs flows through food webs (Fig. 14.1) and is used by heterotrophic consumers that release carbon dioxide gas and produce organic waste or minerals;

3. At the death of organisms, organic matter (MO) is transformed into mineral form (MI) by the action of aerobic (aerobic respiration; Fig. 14.2, *pathway 3*) and anaerobic (fermentation and anaerobic respiration; Fig. 14.2, *pathway 4*) heterotrophic microorganisms known as the main decomposers (Fig. 14.1), because they recycle elements from all trophic levels of the food chains. It is estimated that 90 % of the CO_2 produced in the biosphere results from the activity of bacteria and fungi. The microbial decomposers can, in turn, serve as food for higher trophic organisms (feeding of bacteria, worms, filter feeders).The mineral elements released by decomposers will be reused later. They can also come from abiotic sources such as erosion of rocks or human exploitation of mineral deposits (Fig. 14.2, *pathway 5*).

In the absence of oxygen, a small proportion of organic matter can escape biodegradation processes and accumulate in the form of peat, with thicknesses which can reach several hundred meters. These deposits could end as coals over time scale of several hundred millions of years. Another form of accumulation of organic matter recalcitrant to biodegradation is the appearance of buried **kerogens*** which can, ultimately, lead to the formation of petroleum or natural gases (such as methane) under the combined effects enhanced temperature and pressure (processes of diagenesis and catagenesis). Such organic matter deposits, sequestered in the sediments and isolated from contemporaneous biogeochemical cycles (Fig. 14.2, *pathway 6*), can be injected back to the surface by tectonic movements, or by an erosion following drying, and by human activity as well (Fig. 14.2, *pathway 7*). The human utilization of fossil fuels (Fig. 14.2, *pathway 8*), coupled to forest fires (Fig. 14.2, *pathway 9*), will be accompanied by a significant release of CO_2 into the atmosphere, primarily as the so called "black carbon".

The knowledge of biodegradation of organic compounds requires both knowledges of their synthesis as well as their degradation pathways.

For mineral forms, transformation during the biogeochemical cycle can be performed with or without change of valence. The phosphorus atom, for example, does not change valence unlike carbon, oxygen, nitrogen, sulfur, iron or manganese.

Moreover, the use of an element will not always depend on its concentration in the environment but its chemical status. For example, molecular nitrogen which is the most abundant gas in the atmosphere (80 %) is not used by plants which are able to assimilate nitrogen only in the form of nitrate or ammonia.

14.1.2 The Qualitative and Quantitative Evolution of Biological Elements Depends on the Spatial and Temporal Scales of Interest

At the spatial scale, the evolution of an element can be studied from a global point of view (at global, regional, or local scales) to that of a targeted microenvironment (of a particle or a microniche).

The global cycle of an element can be expressed as the speed of movement of an element from a compartment to another, each of which is considered a "black box" or a reservoir wherein the elements are stored for a period of time (residence time). The term reservoir can be applied to different compartments of the biosphere (e.g., hydrosphere, lithosphere, atmosphere) or to the different trophic levels of a food chain (primary producers → primary consumers → secondary consumers → tertiary consumers → decomposers). The speed of transfer from one compartment to another, so-called "flow", provides quantitative information on the intensity of a biogeochemical cycle.

In contrast, the study of biogeochemical cycles may be addressed at a micro-scale. For example, particles sedimenting in the water column can form anaerobic microhabitats within an environment that is globally well oxygenated (Fig. 14.3).

Aerobic bacteria will develop on the surface of the sinking particles, primarily nitrifying bacteria which extensively use the available oxygen at the periphery of the particles. Below this aerobic zone, the oxygen concentration will quickly decrease, setting the conditions for certain steps of the denitrification process ($NO_3^- \rightarrow NO_2^- \rightarrow N_2O$) (cf. Sect. 14.3.5). Anoxic conditions will then be established in the center area of the particles and favor the development of fermentative denitrifying ($NO_3^- \rightarrow N_2$) and methanogenic bacteria, recruited primarily from the digestive tract of zooplankton (Karl and Tilbrook 1994). In the absence of oxygen, a detrital anaerobic food chain becomes prevalent. Different aerobic and anaerobic bacterial communities can thus coexist in a microenvironment.

Particles as aggregates are sites of intense microbial activity often called "hotspots". Such heterogeneous distribution of microbial activity which is associated with a heterogeneous distribution of the organic matter is found in sediments and soils.

In addition, some elements are produced, used, and regenerated on a limited geographical area. This may be the case of nitrogen in soils. In contrast, for other elements, the production site may be quite distant from the site where the elements are used. This is the case for gas products (oxygen, molecular nitrogen, nitrogen oxides, carbon dioxide, etc.) which can be transported via atmospheric circulation (winds, vertical circulation) around the Earth, far away from their source of production. An autotrophic organism can use the CO_2 released into the atmosphere by another organism living in a far different geographical province.

The processes to which an element is subjected during its biogeochemical cycle can occur in time scales that can be very long, concerning events that took place in geologic times (millions of years in the cases of oil or coal formation), or extremely brief (the duration of an enzymatic reaction). It is therefore important to know the lifetime of an element which is defined by the time elapsed between its fixation in the organic material by autotrophic organisms, and its regeneration in the form of minerals. Thus, in the ocean, the carbon fixed in bacterial biomass is recycled in a few hours, while the carbon of macroorganisms such as fishes and mammals is sequestered for several years.

14.1.3 Role of Microorganisms in Biogeochemical Cycles

Because of their high numbers everywhere on Earth, coupled to tremendous metabolic capacities and the related adaptation potentials, microorganisms will have a key role in the functioning and transformation of biogeochemical cycles and, consequently, of ecosystems (Madsen 2011).

In the biosphere, microorganisms include producers, consumers, and decomposers. During three billion years, all stages of biogeochemical cycles were performed only by microorganisms. Currently, they are involved in almost all steps of the cycles and they are the sole organisms capable of performing some of these steps.

For example, processes such as sulfate-reduction, denitrification, nitrogen fixation, and the production and consumption of methane are specifically related to the activity of microorganisms. In their absence, all essential elements for life would remain trapped in organic molecules of cadavers and wastes. The maintenance of life for plants, animals and

Fig. 14.3 Representation of an organic particle in oxic aquatic environment, showing associated bacterial metabolisms: aerobic metabolism in the oxic micro upper-layer and anaerobic metabolism in the inner anoxic zone. *DN* Denitrification, *OM* Organic Matter (Drawing: M.-J. Bodiou)

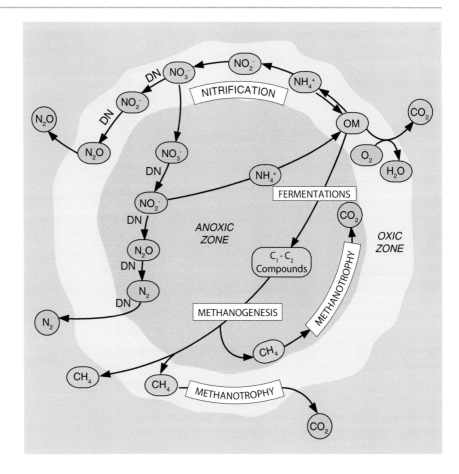

humans is totally dependant upon microbial activity. This chapter addresses the biogeochemical cycles of carbon, nitrogen, sulfur, phosphorus, iron, manganese, and in small boxes of silica and mercury, with a particular focus on the role of involved microorganisms.

14.2 The Carbon Cycle

14.2.1 Introduction

On Earth, carbon undergoes a set of transformations characterized by redox processes from the most oxidized (CO_2, +IV oxidation state) to the most reduced form (the Methane CH_4, −IV oxidation state). Carbon in its neutral state, that is to say, pure (oxidation state 0), is in the form of graphite or diamond. The carbon associated with organic oxygen, hydrogen, nitrogen, sulfur, and phosphorus, can be found in oxidation forms ranging from 0 (glucose, acetate) to −IV (methane) depending on the compound. However, the oxidation state for mineral carbon is always between 0 and +IV; examples: CO_2, $CaCO_3$ (+IV), CO (+II), etc. The whole oxidation–reduction process of the carbon is the carbon cycle (Fig. 14.4).

The carbon cycle is expressed through fluxes between its major reservoirs: atmosphere, hydrosphere (mainly oceans), mainland biosphere (including soils), and lithosphere (fossil carbon) (Prentice et al. 2001; Houghton et al. 2004; IPCC 2007). Despite its very low concentration in the crust (about 0.27 %), carbon is present in all living organisms and it is at the basis of organic chemistry. Associated with oxygen, it forms CO_2, the main carbon source allowing the growth of autotrophic organisms and in particular plants. Combined with hydrogen, it forms hydrocarbons. Associated with oxygen and hydrogen, it gives rise to carbohydrates and lipids. Finally, with nitrogen and sulfur, carbon forms amino acids which are the basic units of proteins, and the addition of phosphorus allows the formation of nucleotides, the basic units of nucleic acids that encode the genetic information.

The importance of carbon in the cycling of matter and in processes that serve as support for human activities can be illustrated by the photosynthetic fixation of CO_2, the transfer of organic matter in food webs (which is a great provider of services and resources to Mankind), the storage of detritus, and the formation of fossil carbon (Figs. 14.1 and 14.2). The increasingly intensive use of fossil carbon reservoir as a source of energy for human activities causes a major

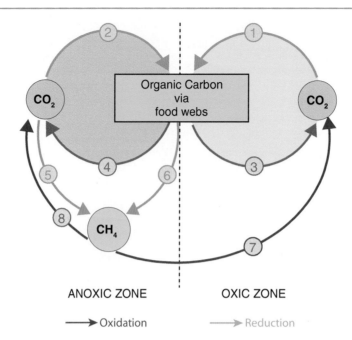

Fig. 14.4 Role of microorganisms in the carbon cycle. *1* Organic carbon production (biomass) from CO_2 reduction (autotrophy) Under oxic conditions: chemolithotrophic bacteria and archaea, photosynthetic micro-eukaryotes, cyanobacteria, *2* Organic carbon production from CO_2 reduction (autotrophy) under anoxic conditions: chemolithotropic bacteria and archaea, anoxygenic photolithotrophic bacteria, *3,4* Use of organic carbon as energy source and oxidation to CO_2 (mineralization) under oxic conditions (*3*, aerobic respiration) and anoxic conditions (*4*, anaerobic respiration), *5* Methane production (methanogenesis) by methanogenic archaea from CO_2 reduction (CO_2 respiration), *6* Methane production by methanogenic archaea from reduction of C1 organic compouds (methanol, formic acid, etc.), C_2 (acetate) or C3 (trimethylamine), *7,8* methane oxidation to CO_2 (methanotrophy) by methylotrophic bacteria under oxic conditions (*7*) or by methanotrophic archaea under anoxic conditions (*8*); see Sect. 14.2.5 and Box 14.10; *Blue arrows* represent reduction processes; *red arrows* are oxidation processes (Drawing: M.-J. Bodiou)

disruption in the global carbon cycle, with major consequences such as enhanced greenhouse effect, and a large-scale change in the climate system of the Earth and in the global ocean chemistry.

14.2.2 The Global Carbon Cycle

14.2.2.1 The Major Reservoirs

The carbon cycle of the 1990s is shown in Fig. 14.5. The CO_2 content of the atmosphere (0.0368 % or 368 ppm) is very low compared with that of nitrogen (78 %) and oxygen (21 %). The size of this reservoir is about 762 Pg C (1 Pg = 1 petagram carbon = 10^{15} g). The atmosphere also contains methane (CH_4: ~1.7 ppm) and carbon monoxide (CO: 0.1 ppm) as trace gases. Carbon is present in both organic and inorganic forms in continental and oceanic reservoirs. Oceanic inorganic carbon is involved in the carbon cycle at a short time scale (years to a few hundred years), whereas the continental inorganic carbon is involved at a longer time scale (millions of years).

The continental reservoir of biomass is about 2,261 Pg C, which is divided into 550 Pg C for vegetation and 1,711 Pg C for soils and detritus. The terrestrial vegetation thus contains less carbon than the atmosphere (550 versus 762 Pg C), while the first meter of soil contains 2–3 times more carbon (1,500–2,000 Pg C) than the atmosphere. Half of these two sub-reservoirs of organic carbon is in dense forests then, by decreasing order of importance, in open forests, grasslands, tundra, wetlands, and agricultural ecosystems as well. Almost all terrestrial living carbon is in plants; animals (including the Human species) only represent 0.1 % of this carbon.

The oceans contain 37,000 Pg of inorganic carbon and 1,000 Pg of organic carbon, i.e., 50 times more than the atmosphere and 70 times more than the terrestrial vegetation. The vast majority of carbon is found in the mesopelagic and bathypelagic ocean. The sediment contains 6,000 Pg C, with a very slow renewal ("turnover") time.

The chemistry of inorganic carbon in seawater is more complex than that of oxygen or that of CO_2 in the

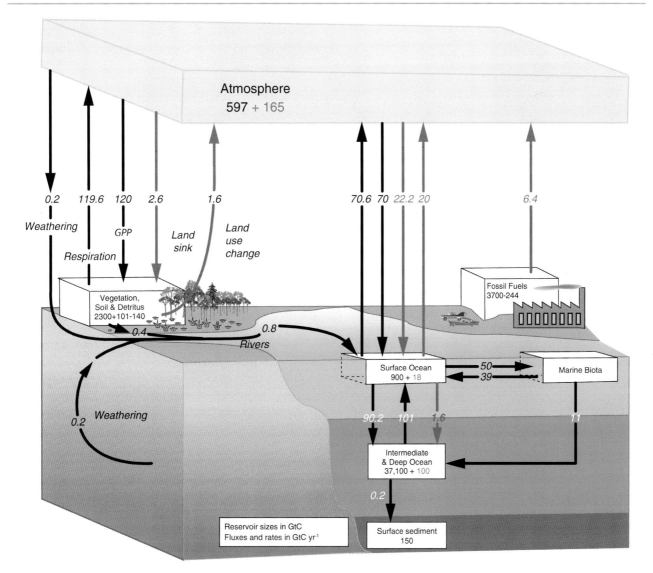

Fig. 14.5 The global carbon cycle in the 1990s. The tanks are in *bold* (Pg C) and flows in *italics* (Pg C year^{-1}). Values before human disturbance are shown in *black* while those resulting from human activities are in *red*; respiration concerns only aerobic respiration. 1 Pg (1 petagramme) = 10^{15} g; *PMG* Primary Modified Gross, *GPP* Gross Primary Production, *GtC* Gigatons carbon (and redesigned by IPCC 2007 Production, courtesy of Cambridge University Press) (Drawing: M.-J. Bodiou)

atmosphere. The reason is that the inorganic carbon is present in various forms, in equilibrium with each other:

$$CO_2 + H_2O \Leftrightarrow HCO_3^- + H^+ \Leftrightarrow CO_3^{2-} + 2H^+$$

Because of this equilibrium, the concentration of inorganic carbon in seawater is much larger than that of other gases. For example, the ratio of oxygen between the atmosphere and ocean is 99:1, while it is 1.5:98.5 for inorganic carbon. CO_2 gas is less than 1 % of the inorganic carbon in seawater, the rest being in the form of bicarbonate and carbonate.

Out of 1,000 Pg of organic carbon (Fig. 14.5), living marine organisms represent only 3 Pg, against 550 Pg C for organisms living on land. The carbon turnover is indeed very fast in the oceans and its accumulation in living organisms is low, compared with terrestrial ecosystems where carbon accumulates in long-lived organisms (trees). The distribution of organic carbon between living and dead matter is also very different: about 1:3 for continents and 1:300 in the oceans.

Fossil fuels represent an important carbon reservoir formed over long periods of time (see next section). Gas, oil, and coal are the residues of organic matter accumulated during geological time scale which escaped oxidation and have been transformed into a fossil form and buried in sediments (Fig. 14.2). The energy stored in this fossil

Fig. 14.6 The carbon cycle in the long term (Modified and redrawn from Berner 2004). *Org C* Organic Carbon, *CaCO₃* Calcium Carbonate, *HCO₃* bicarbonate (Drawing: M.-J. Bodiou)

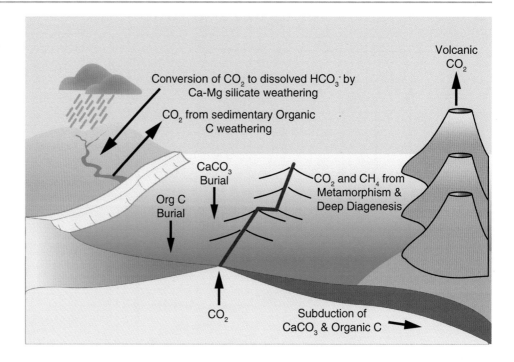

material is released during combustion. The reservoir of fossil carbon is the second in size after the deep ocean.

14.2.2.2 Long Term Carbon Cycle

The long-term carbon cycle is distinguished from the short-term one (discussed in the next paragraph) by the carbon transfer between rocks and other reservoirs (Fig. 14.6). The long-term cycle has played a major role in the control of atmospheric CO_2 on geological time scales because rocks contain much more carbon than the oceans, atmosphere, biosphere, and soils combined. This resulted in changes in the partial pressure of CO_2 (pCO_2) which are much larger than those generated in the short-term carbon cycle.

The four main reactions of the long-term cycle are:

1. The aerial dissolution of silicate rocks ($2CO_2 + 3H_2O + CaSiO_3 \rightarrow Ca^{2+} + 2\ HCO_3^- + H_4SiO_4$). The dissolved species are transported to the ocean where calcium and bicarbonate precipitate in the form of calcium carbonate ($Ca^{2+} + 2HCO_3^- \rightarrow CaCO_3 + CO_2 + H_2O$), while silicic acid precipitates as biogenic silica ($H_4SiO_4 \rightarrow SiO_2 + 2\ H_2O$). These two minerals are stored in sediments. The final result of these reactions ($CO_2 + CaSiO_3 \rightarrow CaCO_3 + SiO_2$) illustrates the transfer of CO_2 from the atmosphere to the rocks through dissolution and sedimentation.
2. The decarbonation of silicate rocks by volcanism, metamorphism, and diagenesis ($CaCO_3 + SiO_2 \rightarrow CO_2 + CaSiO_3$) that ensure the return of CO_2 back to the atmosphere.
3. The net storage of organic carbon in sediments ($CO_2 + H_2O \rightarrow CH_2O + O_2$) which is at least 1,000 times smaller than the fluxes generated by photosynthesis and respiration.
4. The "georespiration" ($CH_2O + O_2 \rightarrow CO_2 + H_2O$), i.e., the oxidation of fossil organic matter trapped in rocks by oxidative dissolution of continental sedimentary rocks or by microbial or thermal decomposition (including the special case of the anthropogenic combustion of fossil carbon).

14.2.2.3 Short Term Carbon Cycle

The short-term cycle is characterized by very intense **gross annual fluxes*** between the carbon reservoirs and **sub-reservoirs***, the **net fluxes*** being extremely low. CO_2 is chemically stable and its residence time in the atmosphere is about 4 years before it enters the ocean or is incorporated into the continental biomass. In general, the processes involved take place on time scales shorter than a century. Three reactions need to be considered:

1. Photosynthesis which uses light to synthesize organic matter by fixing carbon into carbohydrates (CH_2O).
2. Respiration which oxidizes organic matter to CO_2.
3. Some fermentations which can produce CO_2 (*cf.* Sect. 3.3.3).

The short-term carbon cycle is mainly based on photosynthesis and respiration.

Land-Atmosphere Fluxes

Photosynthesis fixes atmospheric CO_2 in the plant biomass as according to the following very simplified reaction:

$$CO_2 + H_2O \rightarrow CH_2O + O_2$$

This CO_2 reduction in the form of organic matter uses photons provided by the sun. The quality of organic matter is very diverse, but it has the major characteristic of storing reduced carbon and energy. Respiration is the reverse process: i.e., the oxidation of organic matter which converts the energy stored in the chemical links between organic molecules into energy which can be used by living cells. The CO_2 fluxes derived from continental photosynthesis (or **Gross Primary Production***, GPP) and respiration are 120 and 119.6 Pg C year^{-1}, respectively (Fig. 14.5). Half of GPP is respired by plants themselves (**autotrophic respiration***), leaving a **Net Primary Production*** (NPP) of about 60 Pg C year^{-1}. A large fraction of the NPP is used by heterotrophic respiration (*Bacteria*, *Archaea*, Fungi, and primary consumers), leaving a **Net Ecosystem Production*** (NEP) of the order of 10 Pg C year^{-1}. Finally, the NEP is affected by other carbon losses (fire, harvest, erosion, export of particulate, and dissolved carbon into the ocean), leaving the net biome production (NBP) which represents the net amount of carbon accumulated in the continental biosphere. The NBP is subject to important inter-annual variations (0.2 and 1.4 Pg C year^{-1} in the years 1980 and 1990, respectively). The continental biosphere is thus currently a source of carbon sink, despite large releases of CO_2 from deforestation and, more generally, from changes in land use.

Ocean–Atmosphere Fluxes

Three **pumps*** allow the ocean to store carbon:

1. The chemical pump is the property, described above, of the seawater carbonate system to absorb inorganic carbon in the form CO_2 and store it as bicarbonate and carbonate;
2. The solubility pump is based on the fact that the CO_2 solubility increases with decreasing temperature, so that cold and CO_2-rich waters at high latitudes sink and contribute to the deep circulation, allowing the ocean surface to continuously absorb atmospheric CO_2;

The biological pump* also transfers carbon in the deep ocean. Only part of the primary production of phytoplankton is respired in the surface waters, the remainder is exported in the deep ocean in the form of **Particulate Organic Matter (POM)*** and **Dissolved Organic Matter (DOM)***. This exported fraction is either respired in the deep waters and transported back to surface water by oceanic circulation sooner or later, or is stored in the sediments. The combined action of the solubility and biological pumps is to maintain a concentration of Dissolved Inorganic Carbon (DIC) about 10 % lower in surface than in deep water.

In addition to these three pumps, the so-called carbonate pump produces the opposite effect since it is a source of CO_2 during the precipitation calcium carbonate ($CaCO_3$):

$$Ca^{2+} + 2HCO_3^- \rightarrow CaCO_3 + CO_2 + H_2O$$

In the short term, the ocean–atmosphere fluxes are of the order of 90 year Pg C year^{-1} in both directions (Fig. 14.5). They depend on the gradient of CO_2 between the ocean and atmosphere. The direction and intensity of the air-sea fluxes vary considerably depending on temperature, wind forcing, oceanic circulation, and the strength of the biological pump. In contrast to warm waters, cold waters tend to absorb CO_2. The most important control is exercised, however, by the vertical movements of the water masses, primarily upwellings. The gross primary production is of the same order of magnitude in the ocean and on land (103 *vs.* 120 Pg C year^{-1}). Only 11 Pg C year^{-1} escape respiration in the surface layer and is exported in the deep ocean where the bulk is oxidized. Indeed, only 0.01 Pg C year^{-1} is stored in the sediments for long-life time carbon cycling.

Continent-Ocean Fluxes

Rivers transfer directly, that is without passing through the atmosphere, a small amount of inorganic carbon from the continents to the ocean. This contribution, about 0.4 Pg C year^{-1}, is roughly offset by the storage of calcium carbonate in sediment (0.2 Pg C year^{-1}) and by a CO_2 emission to atmosphere (0.1 Pg C year^{-1}). In addition to inorganic carbon, rivers also provide organic carbon up to 0.3–0.5 Pg C year^{-1}.

14.2.2.4 Anthropogenic Changes in the Carbon Cycle and the Feedbacks on the Climate

Emissions from the Fossil Carbon Combustion

The natural fluxes mentioned in the previous section do not alter the total amount of carbon flowing in the short-term carbon cycle. They only cause a redistribution of carbon between its reservoirs. In contrast, the oxidation of fossil fuels represents a net gain for the short-term carbon cycle. Until the 1800s, the contribution of the fossil reservoir to this cycle was marginal. This is no longer the case since the industrial revolution as the CO_2 emissions produced by human activities have increased exponentially (Fig. 14.7a) to reach 6.4 Pg C year^{-1} between 1990 and 2000, and 7.2 Pg C year^{-1} between 2000 and 2005. Although these emissions are responsible for most of the increase in atmospheric carbon (Fig. 14.5), they are small compared with the natural fluxes from photosynthesis and respiration (~120 Pg C year^{-1}) or from the exchanges between ocean and atmosphere (~90 Pg C year^{-1}). Emissions related to the changes in land use are about 1.6 Pg C year^{-1} for the period 1990–2000. The cumulative amount of this transfer of carbon from the long- to the short-term cycle is about 245 Pg C since 1750. CO_2 emissions from human activities can accumulate in the atmosphere, or be absorbed by the terrestrial biosphere or by the ocean.

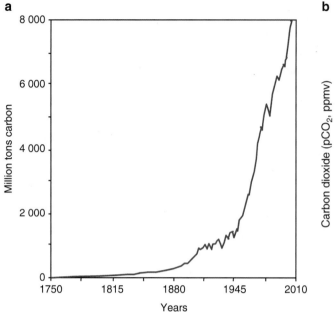
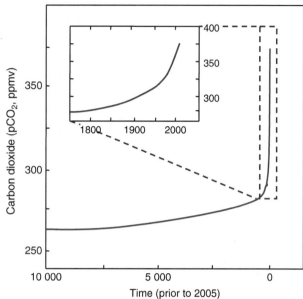

Fig. 14.7 Evolution of CO$_2$ in the atmosphere of the Earth. (**a**) annual carbon emissions from the burning of fossil energies (coal, oil, and gas). CO$_2$ production during the production of cement is also included, though it is not technically a fossil fuel use, but a thermal decomposition of limestone (Modified and redrawn after Marland et al. 2003); (**b**) atmospheric CO$_2$ partial pressure for 10,000 years. The magnification represents an enlargement of the 1751–2005 period (Modified and redrawn after IPCC 2007)

Increase in Atmospheric CO$_2$

The atmospheric CO$_2$ concentration has increased from about 280 ppm during the pre-industrial era to 379 ppm in 2005 (Fig. 14.7b). This value is the highest recorded since 800,000 years. Although the annual increase is variable, the value for the last 10 years is greater than for the average value for the period 1960–2005 (that is 1.9 against 1.4 ppm year^{-1}). Although there is no simple evidence of a causal relationship between carbon emissions and the concentration of atmospheric CO$_2$, at least four arguments seem compelling:

1. The CO$_2$ emissions from human activities alone can explain the increase in atmospheric CO$_2$: Fossil fuel use and land use changes have produced 244 and 140 Pg C, respectively, while the increase in the atmospheric reservoir was 165 Pg C, representing 40 % of the total emission;
2. The pCO$_2$ varied by ±10 ppm during the millennium prior to the industrial revolution, and then increased synchronously with the use of fossil fuels, by about 100 ppm since 1750;
3. Despite a rapid atmospheric transport of CO$_2$, the anthropogenic signal is evident, pCO$_2$ being higher in the most industrialized areas, that is mainly in the mid latitudes of the northern hemisphere;
4. the distribution of **biogeochemical tracers*** is consistent with a fossil origin of the carbon recently accumulated in the atmosphere. For example, the ^{14}C/^{12}C ratio of atmospheric CO$_2$ has decreased since 200 years due to CO$_2$ emissions from fossil carbon known to have a low ^{14}C/^{12}C ratio.

The atmospheric concentration of methane has increased from 715 ppb (parts per billion by volume) during the pre-industrial era to 1,774 ppb in 2005, a value that is greater than the natural variability during the last 650,000 years (320–790 ppb).

Vegetation–Atmosphere Fluxes

The net flux of CO$_2$ between the atmosphere, vegetation, and land was originally estimated as the difference between the carbon emitted by human activities during the period 1850–2000 (244 Pg C) and the sum of the carbon accumulated in the atmosphere (165 Pg C) and oceans (118 Pg C). This difference is a net source of 39 Pg C (Fig. 14.5).

The impact of changes in land use is variable. Few human activities increase carbon storage but they are less important than those which decrease it (oxidation of the organic matter contained in plants and soils). The general trend over the last 300 years is a decrease in the surface area of forests and an increase in the surface area of agricultural land. The resulting net balance over the period 1850–2000 is a carbon source of 140 Pg C, which is much greater than the amount of carbon accumulated in the continental biosphere (40 Pg C). This difference can be explained by a terrestrial reservoir of carbon in the order of 100 Pg C linked to the stimulation of primary production by environmental changes and/or the underestimation of carbon storage by human practices. A projection of the future continental carbon reservoir is difficult and is far beyond the scope of this chapter. Only a few major points are given below:

1. CO_2 increase. While there is no doubt that elevated CO_2 stimulates photosynthesis of many plants, long-term enrichment experiments conducted at the community level do not always increase CO_2 uptake and when the uptake is detected, it does not necessarily lead to an accumulation of carbon;
2. Nitrogen availability. Photosynthesis is often limited by nitrogen, the availability of which is increased by human activities, especially by the use of fertilizers. It should in principle lead to an increase in biomass production, but again, the long-term effect on plant communities is uncertain. Nitrogen enrichment indeed results in an increase or a decrease in growth rates depending on the case study and, at times, can increase the residence time of carbon in soils;
3. Atmosphere chemistry. Plants are very sensitive to the concentration of atmospheric ozone that reduces the efficiency of photosynthesis. Soil acidification resulting from the input of nitrate and sulfate by rain decreases the concentration of ions essential to plant metabolism (Ca^{2+}, Mg^{2+}, K^+). This can cause a decrease in primary production and therefore in the carbon storage capacity of many ecosystems;
4. Climatic factors. Climatic variables (temperature, humidity, light) influence carbon storage in the terrestrial biosphere via their impact on plant photosynthesis, respiration, and growth, and on the oxidation of organic matter in soils as well;
5. Plant regrowth. The continental carbon reservoir also includes carbon storage due to the regrowth of plants in ecosystems which is disturbed by human activities such as fire prevention, reforestation of agricultural lands, etc. This component seems to have been largely underestimated.

Modeling the numerous impacts and feedbacks detailed above is complicated by the lack of information on their interactions: For example, the impact of a parameter can be strengthened or inhibited by the impact of another parameter. Finally, most perturbation experiments have been performed over short periods of time (a few weeks or months, rarely a few years), which does not take into account the biological processes of acclimatization and adaptation.

Oceanic Absorption of CO_2

The future evolution of the oceanic carbon sink depends, as the land sink, on several parameters and feedback:

1. Buffering effect and the carbonate chemistry. The fact that less than 1 % of the dissolved inorganic carbon (DIC) is present in the form of CO_2 in the ocean implies that the DIC increases ten times less than CO_2. This explains why the ability of the global ocean to absorb CO_2 is lower than might be expected from its considerable size. Thus, the increase in atmospheric CO_2 of about 30 % since 1850 has resulted in an increase of only 3 % of the DIC in surface waters. Hence, the oceanic reservoir will absorb less and less atmospheric CO_2 over decades. Finally, as CO_2 is weak acid, its increase will also cause the seawater acidification and a decrease in the concentration of carbonate ions:

$$CO_2 + CO_3^{2-} + H_2O \rightarrow 2HCO_3^-$$

pH has decreased by about 0.1 units in the surface ocean since 1850 and could decrease by 0.2–0.4 units by 2100. Ocean acidification can be measured as shown in time series even though they are short (10–20 years);
2. Impact of elevated CO_2. Some primary producers are not limited by the availability of inorganic carbon because they can use HCO_3^- ions which are much more abundant than CO_2 in seawater. However, calcification may decrease due to the lower availability of CO_3^{2-} ions. The net balance of the impact of the biological response to CO_2 fluxes is uncertain;
3. Temperature. The solubility of CO_2 in seawater decreases as the temperature increases. This positive feedback reduces the capacity of the oceans to absorb CO_2;
4. Vertical mixing. Surface warming will increase the difference in the density of seawater between surface and deep layers, thereby increasing the stratification of the ocean and decreasing deep water formation. This can lower the efficiency of the solubility pump;
5. Increase in the rate of CO_2 emissions. If the CO_2 emissions are rising faster than the potential for absorption by the ocean, more CO_2 will remain in the atmosphere;
6. Nutrients. The strength of the biological pump depends on the availability of nutrients which may be affected by the reduction vertical mixing and by some microbial activities such the fixation of atmospheric nitrogen (diazotrophy) and denitrification (release of dinitrogen).

Given the complexity of the mechanisms mentioned above, and their interactions, the prediction of the future evolution of the oceanic carbon cycle is very uncertain.

After the above presentation of the global cycle of carbon, the functioning of this biogeochemical cycle will be described in the oceanic water column, marine sediment, soils, and lake ecosystems.

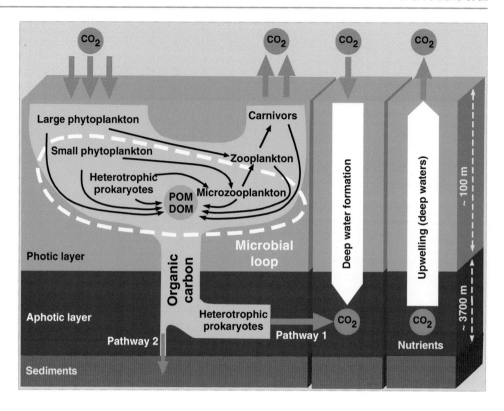

Fig. 14.8 Conceptual diagram greatly simplified of the marine carbon cycle in the water column. *Pathway 1* represents a full remineralization of organic matter in the water column; *Pathway 2* shows schematically an incomplete remineralization of organic matte leading to its burial in sediments (Modified and redrawn from Chisholm 2000)

14.2.3 Carbon Cycle in the Oceanic Water Column

The carbon cycle in the coastal and open ocean water column is dominated by two subsets each involving a couple of antagonistic processes: photosynthesis/respiration (for organic carbon) and precipitation/dissolution of calcium carbonate (for inorganic carbon). These processes are coupled because they involve in some cases the same types of microorganisms. The cycle of organic carbon is by far the most studied (Fig. 14.8). In the illuminated oceanic surface layer (or euphotic zone), the autotroph biological activity assimilates different forms of inorganic carbon, primarily through photosynthesis. Carbon assimilation, known as primary production, tends to decrease the ocean pCO_2 and enhance the role of the Ocean as a CO_2 sink.

This biological uptake of inorganic carbon is mainly carried out by small size phytoplankton (of the order of μm) that will be described later, and whose activity is mainly controlled by light and nutrient concentrations. The organic matter synthesized in the photic layer can be:

1. Partially mineralized as CO_2 and nutrients within the food web;
2. Exported to the deep ocean by **advection*** of water masses, by sedimenting particles (organic detritus, faeces), or through transfer to higher trophic levels (zooplankton, fish, etc.).

During its sinking down to the deep ocean, the surface organic carbon keeps being oxidized by living organisms: it is either partially degraded, fueling the pool of dissolved organic matter, or is mineralized (*cf.* Sect. 3.3.5). Thus, in open ocean, a tiny (less than 1 %) fraction of organic matter synthesized in surface waters reaches the water-sediment interface (*cf.* Sect. 14.2.3). This degradation process maintains an increasing gradient of dissolved inorganic carbon and nutrients from surface to bottom waters.

During the transfer from surface to sediment, most of the dissolved organic material, mainly carbohydrates and amino acids, is subjected to physicochemical processes (condensation, polymerization, photo-oxidation, etc.) that make it recalcitrant to biodegradation (this fraction is thus described as "refractory"). Thus, only a small fraction, known as "labile" dissolved organic matter, is readily usable by microorganisms.

The carbon cycle is not isolated, but has important coupling with the cycles of nitrogen, phosphorus, and minor nutrients that exhibit very low concentrations in the ocean, such as silica (Box 14.1), iron, zinc, etc.

Box 14.1: The Biogeochemical Cycle of Silicon

Bernard Quéguiner

Silicon alone constitutes 27 % of the Earth's lithosphere. It is also an important component of the biogenic material that accumulates in coastal and deep-sea sediments. The role of this element in the marine biogeochemical cycles is directly correlated to its necessary requirement by diatoms. These organisms are a major component of marine phytoplankton, ensuring alone more than 40 % of the gross primary production of carbon on a global scale. However, this global contribution reflects highly contrasting situations since diatoms provide an average of 35 % of the primary production of the poorest oligotrophic areas, against nearly 75 % of highly fertilized ecosystems (coastal ecosystems, upwellings, etc.). Individual situations are even more contrasted as recent studies have shown that the contribution of diatoms is almost zero in the ultra-oligotrophic South Pacific area as opposed to situations of spring blooms in temperate systems where they represent almost all of the autochthonous production of organic carbon. One of the main characteristics of diatoms is the presence of a siliceous wall (the frustule) surrounding the cell. This wall shows significant morphological differentiation and is composed of biogenic silica (or opal), whose general formula is (SiO_2 nH_2O). These unicellular algae are dependent for their development of silicon and DNA replication, a prelude to cell division, is impossible in the absence of this element (Box Fig. 14.1).

Silica plays an important role in natural systems. From the first eukaryotes to the most advanced higher plants, it indeed provides support, protection, or building structures. In unicellular microorganisms, plausible function of the biogenic silica is to ensure the robustness of the living cell, thereby ensuring some protection against predators. In the marine environment, silicification especially concerns frustule of diatoms and siliceous endoskeleton of silicoflagellates and radiolarians.

B. Quéguiner
Institut Méditerranéen d'Océanologie (MIO), UM 110, CNRS 7294 IRD 235, Université de Toulon, Aix-Marseille Université, Campus de Luminy, 13288 Marseille Cedex 9, France

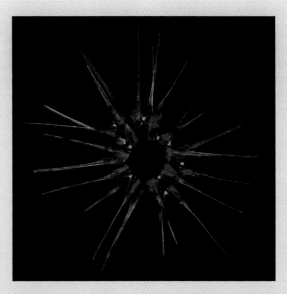

Box Fig. 14.1 Marine Diatom (*Asterionella glacialis*) observed under epifluorescence light. Biogenic silica forming the frustule appears *green* after fixation of a specific marker (PDMPO). The *red* fluorescence is that of chlorophyll a (Photo courtesy of LOPB – UMR6535, K. Leblanc)

In marine ecosystems silicon influence phytoplankton cycles on at least two levels:

1. The dissolved silicon (silicic acid, H_4SiO_4) controls the composition of phytoplankton assemblages when its availability with respect to nitrogen and phosphorus is low. Silicon has a quite different behaviour of the other two major nutrients since recycling is slow, at least in the spring period, being essentially dependent on the dissolution of diatom frustules, a physico-chemical process tightly controlled by temperature. Under these conditions the biogenic silica formed during the spring bloom leaves the surface layer and dissolves very slowly in deep water (or sediment); nitrogen and phosphorus, in turn, are regenerated quickly (at time scales often less than a day and by processes of excretion by heterotrophic organisms) and can be used again by phytoplankton (regenerated production). The surface layer therefore "loses" its silicon gradually during the season and acts as a silicon pump. Only the return of favourable hydrodynamic conditions

(continued)

Box 14.1 (continued)

Box Fig. 14.2 Operation of the silicon pump in deep ocean environments (Modified and redrawn from Del Amo et al. 1997)

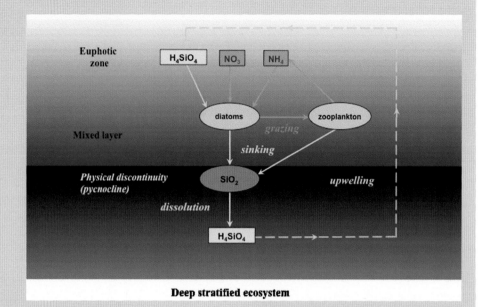

(increased vertical mixing) may allow replenishment of silicic acid in surface waters during winter (Box Fig. 14.2);

2. In disturbed coastal ecosystems, the increase in nitrogen and/or phosphorus inputs result in a decrease of Si/N and Si/P because silicon is hardly rejected by the human activities. These conditions can lead to changes in the factors controlling diatoms (from a limitation by N or P to a limitation by Si) and even alter the essential nature of phytoplankton communities (from a community dominated by diatoms to a community dominated by non-siliceous flagellates some of which, such as dinoflagellates, can be toxic). However, in shallow ecosystems, trapping of biogenic silica in the surficial sediments (either directly or through the activity of benthic filter feeders) allows a continuous replenishment of silicic acid during the productive period, while limiting the adverse effects of flagellates developments. Thus, the coastal ecosystem is a silicon pump, which, unlike the offshore ecosystem, retains silicon instead of eliminating it (Box Fig. 14.3).

At the global scale, the annual net input of silicic acid to the ocean was estimated at 6.1 ± 2.0 teramoles (10^{12} mol). The major source is the major rivers alone accounting for nearly 80 % of the total input. The inputs are balanced by the annual burial in sediment that is estimated to be 7.1 ± 1.8 teramoles. Gross production of biogenic silica surface water is 240 ± 40 teramoles per year, so that the preservation rate of silicon in sediments (opal sedimentary accumulation vs. gross production) is particularly high (averaging 3 %) compared with that of other major elements. In the World Ocean the residence time of silicon (defined as the ratio between the total amount of dissolved silicon and total inputs) is 16 000 years. At this time scale it is clear that the recent perturbations in riverine silicon (reduction of inputs related to river eutrophication and/or resulting from the construction of large dams) have not yet generally impacted on the global biogeochemical cycle. Biological residence time (ratio between the total amount of dissolved silicon and gross output) is, meanwhile, in the order of 400 years. Thus each silicon atom finally buried in the abyssal sediments have previously completed nearly 40 successive cycles of uptake by siliceous organisms and dissolution in marine waters (Box Fig. 14.4).

(continued)

Box 14.1 (continued)

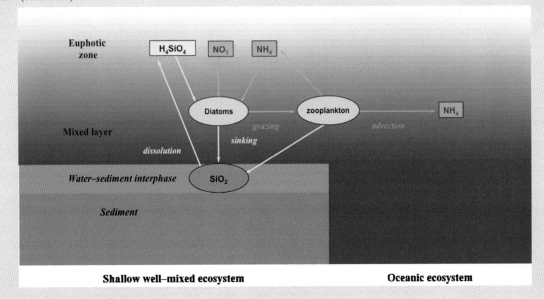

Box Fig. 14.3 Operation of the silicon pump in coastal ecosystems (Modified and redrawn from Del Amo et al. 1997)

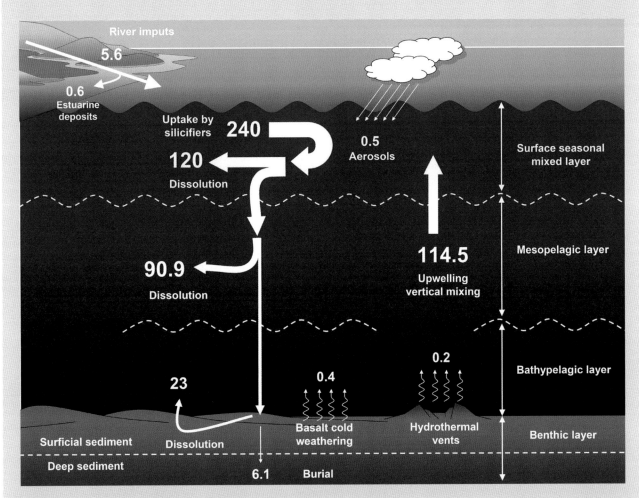

Box Fig. 14.4 Global biogeochemical cycle of silicon in the oceans. Units are teramoles of Si.year^{-1} at the planet scale; the inputs to the marine environment came from riverine inputs, partial dissolution of atmospheric dust, the basalts of the ocean floor and direct discharges from deep hydrothermal vents. These inputs are balanced by burial in ocean sediments and continental margins (Modified and redrawn from Tréguer et al. 1995)

14.2.3.1 The Importance of Planktonic Microorganisms

Primary production from the phytoplankton photosynthetic activity can be considered as a major biological process compared with the primary production from benthic organisms that only thrive on the littoral fringe representing less than 3 % of the World Ocean surface.

In the neritic zone (including both littoral fringe and continental shelf) that is shallow (less 200 m depth), organic matter mineralization is mainly due to the activity of benthic microorganisms. In the open ocean, about 90 % of the World Ocean surface, mineralization occurs primarily in the water column where planktonic microorganisms play a major role.

Since the years 1970–1990, significant technological advances (epifluorescence microscopy, flow cytometry, etc.) (*cf.* Chap. 17) lead to an increasing understanding of the structure and functioning of planktonic microbial communities and their importance in the carbon cycle. Microorganisms (autotrophic, heterotrophic, and **mixotrophic***) are the major actors in the production and mineralization of organic matter and represent the bulk of oceanic biomass. For instance, picoplankton (Table 14.1) alone contributes 50–90 % of the total phytoplankton biomass in sub-tropical and tropical oligotrophic zones.

Within the picoplankton size-fraction, two marine cyanobacteria belonging to the genera *Synechococcus* (Waterbury et al. 1979) and *Prochlorococcus* (Chisholm et al. 1988) were found in high abundance (Fig. 14.9), and playing a crucial role in the oceanic primary production. Indeed, they can make more than half of the total biomass and primary production.

The major role of microorganisms in the production and mineralization of organic matter lead Azam and collaborators (1983) to propose the "microbial loop" concept (*cf.* Chap. 13). This concept originally consisted of only three compartments: non-living organic matter, bacteria consuming this organic matter, and unicellular eukaryotes feeding on bacteria. It was progressively made more complex in order to take into account all trophic relationships between auto- and heterotrophic microorganisms (Fig. 14.8).

In the marine environment, there is an inverse relationship between the size of organisms and their abundance. The unicellular microorganisms with size in the range 0.5–100 µm are present in concentrations from 10^2 to 10^6 cells.cm^{-3}, making more than 50 % of the planet total biomass (Table 14.1).

14.2.3.2 Primary Production and Carbon Export: Role of Phytoplankton

The Major Primary Producers

Phytoplankton consists of photosynthetic organisms of polyphyletic origin, typically consisting of microscopic and unicellular floating cells. Although they only represent about 2 % of the Earth photosynthetic biomass, they contribute up to 45 % to the annual Earth photosynthesis. The number of phytoplankton species is ten times higher in the continental domain that in the marine realm (about 275,000 against 25,000), but the phylogenetic diversity of marine phytoplankton is much more greater (eight clades against one).

Three phyla can be distinguished. The first one is that of cyanobacteria, class of bacteria which includes all oxygenic phototrophic prokaryotes. The biomass of these microorganisms whose estimated number amounts to 10^{24} cells is the most important in the World Ocean. Cyanobacteria played a major role in the genesis of atmospheric oxygen (Sect. 4.4.2) and, more generally, in the evolution of the carbon and nitrogen biogeochemical cycles. All other planktonic photosynthetic organisms are either aerobic anoxygenic phototrophic bacteria, or eukaryotes which are divided between "green" and "red" phyla. "The green phylum," from which the higher plants originate, contains chlorophyll *b* as accessory pigment and played an important role in biogeochemical cycles until the appearance of the "red phylum", likely about 250 million years ago. The latter phylum includes taxa that nowadays play a major role in the carbon cycle, such as diatoms, dinoflagellates, haptophytes, and chrysophytes (*cf.* Sect. 5.4.3).

Net Primary Production

Three major mechanisms of photosynthetic production take place in the water column (Fig. 14.10):
1. Oxygenic photosynthesis (OP) that is the major process;
2. Anoxygenic anaerobic photosynthesis (AnAnP) that is limited to coastal and estuarine areas;
3. Anoxygenic aerobic photosynthesis (AnAP).

Two other mechanisms may also occur: one based on rhodopsin (*cf.* Sect. 3.3.4), and the other on a phytochrome (PC).

There is currently no quantitative assessment of the relative importance of these five metabolisms, but the OP is the major metabolism.

Table 14.1 Distribution of planktonic microorganisms (<200 µm) according to size classes

Terminology	Size (µm)	Fresh Weight (g cell^{-1})	Concentration (cell mL^{-3})	Biomass (g C mL^{-3})	Examples
*Femto*plankton	0.02–0.2	10^{-15}	10^7–10^8	10^{-8}–10^{-7}	Virus
*Pico*plankton	0.2–2.0	10^{-12}	10^5–10^6	10^{-7}–10^{-6}	Bacteria, Archaea, Algae
*Nano*plankton	2.0–20	10^{-9}	10^2–10^3	10^{-7}–10^{-6}	Algae
*Micro*plankton	20–200	10^{-6}	0.2–1	10^{-7}–10^{-6}	Ciliates

According to Sieburth et al. (1978)
The average concentrations and biomasses (fresh weight) have been estimated from the literature

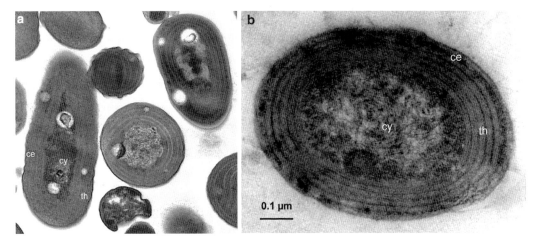

Fig. 14.9 Oceanic cyanobacteria. (**a**) *Synechococcus* observed by transmission electron microscopy (Source: Wikipedia; authors: Wenche and Eikrem Jahn Throndsen); (**b**) *Prochlorococcus* observed by transmission electron microscopy, *ec* cell envelope, *cy* cytoplasm, *th* thylakoids (Source: Wikipedia, author: William K. Li and Frederick Partensky)

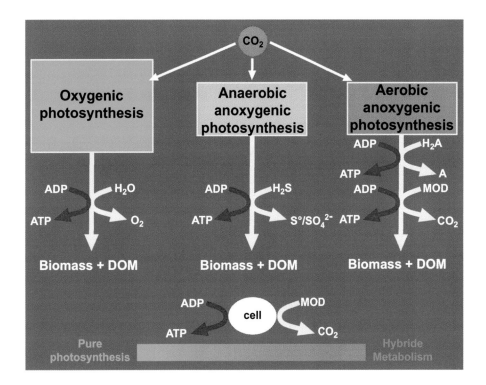

Fig. 14.10 Relationship between illumination, dissolved organic matter (DOM), and the biological production processes. Schematic representation of the three main mechanisms of photosynthetic production that exist in the water column. Although there is no quantitative assessment of the importance of these processes in marine environment, oxygenic photosynthesis (water photolysis with released dioxygen) is known as the dominant process. H_2A electron donor (Modified and redrawn from Karl 2002)

The main environmental parameters that control photosynthesis in the ocean are light and nutrient concentration.

Control by Light

Light is absorbed and scattered by seawater and also by dissolved and particulate matter (including phytoplankton itself). The penetrating light decreases exponentially with depth. The layer where light is sufficient to enable a positive net primary production is called euphotic zone (Fig. 14.11). Its lower limit is called compensation depth because the specific production balances the specific respiration. The thickness of the euphotic layer varies from a few meters to over 100 m depth, depending on seawater turbidity. The lower limit of the euphotic layer is often defined as the depth at which light intensity is 1 % of the surface irradiance, but photosynthesis remains possible at lower light intensity. This is particularly the case in oceanic regions with very low nutrient concentrations (Oligotrophic areas) such as, for

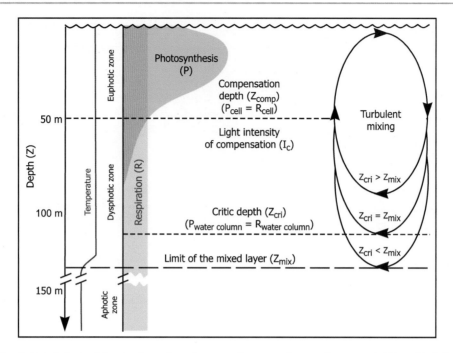

Fig. 14.11 Interaction light-turbulent mixing in the water column. Turbulent mixing generated by the action of wind creates a layer of homogeneous mixture to a certain depth. When this layer is shallower than the photic zone, phytoplankton is maintained under conditions of favorable lighting (except phenomenon of photo-inhibition) which allows daily positive net photosynthesis. If the mixed layer is deeper than the euphotic zone, the phytoplankton lies some time in an area where too little photons no longer allow growth. The critical depth is defined as the maximum depth of the mixed layer at which net growth is possible. The net primary production is positive when the mixture layer is less deep than the critical depth and the nutrients are not limiting. $Zcri$ critical depth, $Zcomp$ depth compensation, $Zmix$ depth mixture (Modified and redrawn from Parsons et al. 1984)

example, in the Pacific Ocean where the deepest euphotic layer depth ever measured was recorded (about 170 m; Morel et al. 2007).

The quantity of light received by phytoplankton depends not only on light penetration into the water column, but also on vertical mixing (Fig. 14.11). The wind-generated turbulent-mixing diminishes with depth. An homogeneous mixed layer is thus formed down to a depth where turbulence becomes too weak to homogenize further down the water column. This vertical mixing therefore exposes phytoplankton to a variable light regime and the received quantity of photons decreases with increasing mixed-layer depth. The net photosynthesis then depends on the interaction between the euphotic zone and the mixed layer. When the mixed layer is shallower than the euphotic zone, phytoplankton thrives under favorable light conditions (excluding the possibility of photo-inhibition when the light intensity is too high) enabling a positive daily net photosynthesis. However, when the mixed layer is deeper than the euphotic zone, phytoplankton thrives in a depth where the number of photons received is too low to enable a net growth. The maximum depth of the mixed layer where a net growth is possible is called the critical depth. When nutrients are not limiting, NPP is positive when the mixed layer is shallower than the critical depth. This interaction between light and the vertical mixing has a strong latitudinal variability. NPP varies greatly depending on season in temperate zones, because light is limiting in winter when vertical mixing is maximal and temperature cold. These conditions lead to a low primary production.

In contrast, in the tropical and sub-tropical zones where the major oceanic gyres are located, light intensity varies much less over the year and vertical mixing is episodic and essentially associated with storm and cyclone events. As a consequence, the primary production does not exhibit large seasonal variations.

Control by Nutrient Concentration

In 1934, A. C. Redfield noted that the composition of particulate organic matter was very similar to the deep ocean chemical composition, with C:N:P atomic ratios of 106:16:1, respectively (Redfield 1934). He concluded that these deep ocean atomic ratios resulted from the remineralization of particulate organic matter exported from the surface ocean. Remineralization is a process which is largely aerobic, implying that the above atomic ratios may be coupled to the oxygen consumption. The oxidation of organic matter can be represented as follows:

$$(CH_2O)_{106}(NH_3)_{16}H_3PO_4 + 138O_2$$
$$\Leftrightarrow 106CO_2 + 122H_2O + 16NO_3^- + PO_4^{3-} + 19H^+$$

The above Redfield ratios are obviously mean values and they have been shown to largely vary. For instance, in the **deep ocean***, the C:N:P ratios were in fact of the order of 14:7:1, respectively. However, the concept of Redfield ratios continues to be very used in biological oceanography because it remains a powerful concept to understand the processes that limit primary production.

Among the six major constitutive elements (C, H, N, O, P, and S) that are essential to phytoplankton activity and biomass synthesis, nitrogen and phosphorus (Box 14.2) are often limiting because of their uptake by microorganisms. In contrast, they are abundant in deep waters where most of the organic matter exported from the euphotic layer is mineralized. With the exception of coastal areas where nutrient can be supplied by continents, vertical mixing, atmospheric transport, and upwellings are the main mechanisms that supply nutrients to the euphotic zone, thereby canceling out the nutrient limitation of primary production. These physical mechanisms do not take place in summer when the high surface temperature combined with moderate winds, generates a strong thermal stratification of the water column that prevents vertical mixing. Under these conditions primary production is strongly limited by nutrients despite high levels of incident light.

Other minerals may also limit primary production. This is the case of silicon (Box 14.1), a frustule constituent of diatoms whose contribution to primary production and carbon export toward the deep ocean is very important (Box 14.1). Iron is weakly abundant in oceans where its importance was only recognized in the 1980s. It is present as a co-factor of enzymes responsible for the reduction of nitrates and fixation of atmospheric dinitrogen (*cf.* Sect. 14.3). Iron sources are numerous (atmospheric dust, deep waters, and continental rocks) but large oceanic zones away from continents (equatorial zones, sub-arctic Pacific, and Southern Ocean) exhibit iron concentrations which are limiting for primary production. This explains the paradox of the so-called high nutrients but low chlorophyll (HNLC) areas where the phytoplankton biomass is low despite high nitrate, phosphate, and silicon concentrations.

Fate of the Net Primary Production

Regarding the fate of net primary production (NPP) it is necessary to distinguish three scenarios (Fig. 14.12):

1. All NPP is consumed by heterotrophs in the euphotic zone. Primary production and respiration balance each other and then there is no organic matter available for export to the deep ocean. This situation, that is largely under the control of the microbial network is the most common in the open ocean;
2. The total respiration exceeds NPP. In contrast, there is necessarily an input of allochthonous material to sustain heterotrophic respiration. The system is therefore open. The microbial assemblage will control the export of organic matter to the deep ocean;
3. A portion of photo-autotrophs and/or of heterotrophs escapes remineralization and is exported. The vertical flux of the sinking particulate and dissolved organic matter corresponds to the **exported primary production***, most of which is remineralized in the water column before reaching the sediments. Except in upwelling areas, the products of this remineralization cannot return back to the surface when the water column depth becomes greater than the ventilation depth (i.e., permanent **pycnocline***). CO_2 and nutrients accumulate in the deep ocean, creating a strong vertical gradient. This is one of the mechanisms of the biological pump, the other being the long-term storage of carbon in sediments.

Ammonium and other reduced forms of nitrogen (urea and amino acids) are excreted when the NPP is remineralized in the euphotic layer.

However, remineralization that takes place in the deep ocean produces nitrate that can be brought back to the euphotic zone by vertical mixing and upwelling. There are therefore two dinitrogen sources for phytoplankton: one autochthonous source (i.e., ammonium) which leads to **regenerated primary production***, and one allochthonous source (nitrate and atmospheric dinitrogen fixed by diazotrophy, *cf.* Sect. 14.3.2), which leads to **new primary production***.

The **f ratio*** represents the fraction of total primary production due to the regenerated production. It should be noted that a complication was not sufficiently evaluated at the definition of these terms: nitrification (*cf.* Sect. 14.3.4) can occur in the photic layer, particularly in coastal waters, and this can lead to the presence of autochthonous nitrate. Another term often used is the **e ratio*** that is the fraction of total primary production that is exported.

Under steady-state conditions, new primary production, exported primary production and the net **community production*** should be in balance. However, *in situ* measurements only rarely express a steady state of the productive oceanic system.

Global Production

Remote sensing is the only tool that gives access to global primary production, i.e., on a planetary scale. Satellites provide ocean color images from which are derived parameters used by numerical models to estimate net primary production:

1. The amount of phytoplanktonic chlorophyll;
2. The photosynthetically active radiation (wavelength range of 400–700 nm) reaching the ocean surface;
3. Temperature.

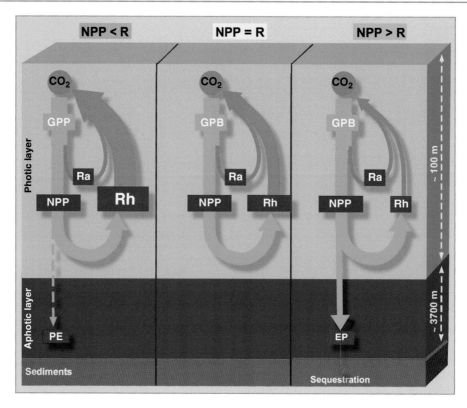

Fig. 14.12 Fate of Gross Primary Production (GPP). (**a**) Location to which the overall respiration (*R*) in the photic zone (total of respirations of autotrophic organisms (*Ra*) and heterotrophic (*Rh*), is greater than the net primary production (*NPP*). Allochtonous organic matter supports respiration; the system is called open. The microbial assembly will determine the export of organic matter (*PE*) to the deep ocean; b: Situation in which the entire NPP is consumed by heterotrophic in the euphotic layer as primary production and respiration offset each other, there is then more organic matter available for export to the deep ocean. All production is recycled. This situation, which is largely controlled by the microbial network, is widespread in the open ocean; c: Situation in which the primary production is greater than the overall respiration. Part of the organic matter produced by photoautotrophic and/or heterotrophic escapes to remineralization in the euphotic layer and is exported to the deep ocean. Source of organic matter may be autochthonous and/or allochthonous. The vertical downward flow corresponds to particulate and dissolved organic material in the exported production (*EP*). A small part (about 1 %) is sequestered in sediments

Table 14.2 Seasonal and annual net primary production (Pg C year^{-1}, According to Field et al. 1998). Biogeographic zones are distinguished by their chlorophyll a concentration <0.1 mg m^{-3} (oligotrophic), 0.1–1 mg m^{-3} (mesotrophic) and >1 mg m^{-3} (eutrophic). Macrophytes (obviously benthic) are shown to provide an overall estimate of NPP

	Ocean	Continent
Seasonal averages		
April–June	10.9	
July–September	13.0	
October–December	12.3	
January–March	11.3	
Annual averages by biogeographic zone		
Oligotrophic	11.0	
Mesotrophic	27.4	
Eutrophic	9.1	
Macrophytes	1.0	
Total	48.5	56.4

Model outputs are generally compared with *in situ* determination of production made by using the ^{14}C method, whose limits will be discussed later (*cf.* Sect. 17.4.2).

The various estimates of global oceanic NPP vary from 45 to 57 Pg C year^{-1}, about half of the world NPP (Table 14.2). It is important to note that these estimates do not take into account the highly productive coastal waters. This is partially explained by the inability to infer chlorophyll concentrations from ocean color data from these regions.

With a biomass less than 1 Pg C, phytoplankton represents less than 1 % of the total biomass of the Earth primary producers but is responsible for about 45 % of the global NPP. The turnover of marine plant organic matter is of the order of 1 week. The contrast with the continental domain is important because the biomass of terrestrial plants (which includes wood) is of the order of 650 Pg C for a NPP of the order of 56.4 Pg C year^{-1}, implying a turnover of approximately 10 years.

Respiration

"Respiration" is a generic term that groups together very different biochemical mechanisms (*cf.* Sect. 3.2.2). For example, aquatic photolithotrophs exhibit at least six different metabolic pathways that consume dioxygen: Mehler reaction, RuBisCo, glycolate oxidase, chlororespiration, cytochrome oxidase and other oxidases (del Giorgio and Williams 2005).

The main goal of cell respiration is to produce the energy required by cell needs. To survive, cell needs a continuous supply of energy to achieve the various cell works (mechanical work for moving; chemical work in the biosynthesis of biological macromolecules, osmotic work during active transports of molecules across membranes). In heterotrophic organisms, these energy requirements are covered by catabolism, that is to say all reactions that break down and transform complex organic compounds (carbohydrates, lipids, and proteins) into smaller and simpler molecules. The biological oxidation metabolisms lead to the formation of reduced compounds (NADH and NADPH) that fuel electron transport systems, resulting in the reduction of a terminal electron acceptor which is usually supplied by diffusion from the external environment (*cf.* Sect. 3.3.1). Fifteen electron acceptors were described in the ocean, but oxygen is the main one in the water column where it is almost always the terminal acceptor (*cf.* Sect. 3.3.2). Electron transfer is coupled to proton translocation through bioenergetic membranes, namely the mitochondrial inner membrane in eukaryotic cells, and the plasmic membrane in prokaryotic cells. The resulting difference in the electrochemical potential of the proton activates the phosphorylation of adenosine diphosphate (ADP) into adenosine triphosphate (ATP) via the activity of ATP synthase. These mechanisms of energy conversion are described by the chemio-osmotic theory of Mitchell (1961). ATP is the main energy vector for living cells and its hydrolysis into ADP and inorganic phosphate (Pi) during subsequent couplings enables the recovery of the reaction free energy to perform most of the cellular functions.

Respiration is the most important sink of organic matter on the planet. In addition, its reagents (O_2, NO_3^-, SO_4^{2-}) and end-products (*e.g.*, CO_2, CH_4, N_2, H_2S) play a major role in biogeochemical cycles. It is therefore surprising that marine respiration has received little attention compared with the reverse process of primary production. Though respiration would be impossible without primary production, coupling between both processes is far from being as tight as thought until very recently. Production and respiration can be decoupled with respect to time (during the transfer of organic matter into the trophic network) and space, because a large fraction of respiration takes place away from the surface where organic matter is produced.

Respiration at the Ecosystem Scale

Respiration in the main aquatic ecosystems was reviewed by Williams and del Giorgio (2005). Values related to the marine field are reported in Table 14.3. It should be noted that most of these estimates were made from often limited data sets and biased both geographically and seasonally. For

Table 14.3 Marine pelagic respiration (Pg C year^{-1}; from del Giorgio and Williams, 2005)

	Respiration
Coastal Ocean	14.4
Open Ocean	
Epipelagic zone (0–150 m)	108–144
Mesopelagic zone (150–1,000 m)	16.8
Bathypelagic zone (>1,000 m)	2
Total Open Ocean	126.8–162.8
Total	141.2–177.2

instance, most of the respiration rate values of the water column were determined in spring and summer, mainly in the North Atlantic Ocean.

Respiration per volume unit expressed in terms of carbon (CO_2 production), varies by four orders of magnitude between surface and deep ocean. The mean value is 2.9 mmol C m^{-3} d^{-1} in the open-ocean epipelagic-zone, and 7.4 mmol C m^{-3} d^{-1} in coastal ocean. Given the difference in water volume between these two areas, respiration in coastal ocean is nearly ten times lower than it is in the open-ocean epipelagic-zone. In the **mesopelagic*** waters, the average value of the few available direct measurements of respiration (Fig. 14.13) are 100–200 times lower than those in the epipelagic zone (0.014 against 2.9 mmol C m^{-3} d^{-1}). In this zone, total respiration is close to that of the coastal ocean. There is no direct measurement of respiration in the bathypelagic area and the available estimates are based on the balance of dissolved oxygen and dissolved organic carbon. This zone undoubtedly exhibits the lowest volumetric respiration rates, estimated at about 0.5×10^{-3} mmol C m^{-3} d^{-1}, which is 4 orders of magnitude lower than in the surface waters. However, because of its important volume, the bathypelagic zone (Fig. 14.13) shows a total respiration which is only two orders of magnitude lower than in the surface ocean (Table 14.3).

Pelagic respiration in **oligotrophic systems*** is generally due to micro-organisms smaller than 2 μm in size, primarily prokaryotes (*Bacteria* and *Archaea*) that thus control most of the remineralization of dissolved (DOM) and particulate (POM) organic matter, both autochthonous and allochthonous. Bacterial production will thus reincorporate dissolved organic carbon into higher trophic levels through predation (bacterivory) by microzooplanktonic organisms (unicellular eukaryotes), that are not able to feed directly on dissolved materials. The contribution of organisms with larger sizes tends to increase with productivity (**mesotrophic*** and **eutrophic*** systems).

Balance Between Primary Production and Respiration

In a given geographical area, occasional measurements of primary production and respiration rates are rarely equivalent. For example, during a phytoplankton bloom, the daytime primary production in surface waters is largely higher

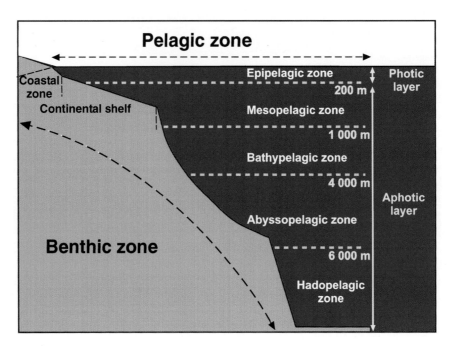

Fig. 14.13 Large vertical subdivisions of the ocean environment. The pelagic realm is divided in substantially subdomains in relation to the depth:
The epipelagic zone extends from the surface to about 200 m depth. It corresponds to the photic layer, where the illumination is sufficient for photosynthesis;
The mesopelagic zone extends between 200 and 1,000 m depth. There is more than enough light for photosynthesis. It is from this area that begins the aphotic zone.
The bathypelagic zone extends between 1,000 and 4,000 m depth.
The abyssopelagic zone extends betwen 4,000 m down to the oceanic crust (down to 6,000 m).
The hadopelagic zone beyond to 6,000 m.

than respiration. Conversely, in the aphotic layer where primary production does not occur, respiration is largely prevailing. However, the ocean is a perennial and spatially constraint thermodynamic system, considered in equilibrium at a given temporal scale (typically 1 year). In terms of carbon budget, production and respiration should be in balance. But this is not the case; the comparison of NPP and respiration in the open ocean reveals a major problem. Indeed NPP, which is of the order of 47.5 Pg C year^{-1} (Table 14.2), cannot sustain a respiration level of the order of 108–144 Pg C year^{-1} (Table 14.3). Allochthonous inputs of organic carbon from continents (rivers), coastal zones (soil washing), and atmosphere (rain, wind) are insufficient to explain the imbalance between autochthonous production (NPP) and community respiration. Consequently, one must admit that NPP is underestimated and/or respiration is overestimated. This discrepancy may be explained either by an insufficient number of measurements of both activities, or by the methods involved in their evaluation.

Regarding aerobic respiration, current measurement methods are not sufficiently sensitive to accurately determined *in situ* activity (O_2 uptake or CO_2 release), the intensity of which is often too low due to low cell concentrations. All commonly used methods require an incubation phase, such as the conventional Winkler method for determining dioxygen concentration. These incubation-based techniques assume that communities remain stable and respiration constant during incubation. In addition, the water column integrated respiratory activity is deduced from a limited number of small volumes of samples isolated from their environment. We must therefore take into account the bias generated by sampling (representativeness, environment modification, etc.), and by the constraints linked to *ex situ* incubations.

It is thus likely that methodological limitations in respiration rate estimations may be responsible for significant errors. It seems however that the main uncertainties are related to primary-production determinations:

1. Photosynthesis is usually estimated by incorporation of ^{14}C, a radioactive carbon isotope (*cf.* Sect. 17.4.2). Depending on the sampling conditions (light conditions), this approach provides an estimation of carbon assimilation in between gross primary production and NPP. This intermediate rate depends on incubation duration and conditions (temperature, light). When incubation lasts 24 h, the method then estimates NPP. For a few hour incubations, the value obtained corresponds to gross primary production. The maximum correction thus corresponds to the respiration of autotrophic organisms and amounts to approximately 40 %. This could raise the open ocean NPP from 47.5 to 66.5 Pg C year^{-1};

2. The ^{14}C method only estimates the production of particulate organic matter and, in most cases, the production of dissolved organic matter is not accounted for. This can represent an underestimation of NPP of about 15 %, which could raise the open ocean NPP from 66.5 to 76.4 Pg C year^{-1};
3. The models used to generate global maps of primary production are inefficient in low latitudes. For example, these models yield NPP values of 4.2–8.3 mol C m^{-2} years^{-1} in the subtropical gyre of the North Atlantic Ocean, which are 2–4 times lower than values derived from *in situ* measurements (15–20 mol C m^{-2} years^{-1}). Given that such warm oligotrophic waters contribute 30 % of the global ocean surface, NPP could be underestimated by 25 %, leading to a corrected open-ocean NPP up to 95.5 Pg C year^{-1}.

Although there are other sources of underestimation that are difficult to address here, it seems clear, from the above arguments, that taking into account the limitations in NPP estimations could considerably reduce the current huge difference between NPP and respiration in the open ocean.

14.2.3.3 Calcium Carbonate Cycle

Precipitation and dissolution of calcium carbonate are integral parts of the oceanic carbon cycle because they consume and produce bicarbonate and CO_2, respectively:

$$Ca^{2+} + 2HCO_3^- \Leftrightarrow CaCO_3 + CO_2 + H_2O$$

This calcium carbonate cycle is much less known than the organic carbon cycle, mainly due the small number of available data. Uncertainties are thus considerable. The main fluxes are presented in Table 14.4. The calcification estimates in the euphotic zone range from 0.5 to 1.6 Pg $CaCO_3$ years^{-1}. Unlike primary production, most of the produced $CaCO_3$ (0.4–1.8 $CaCO_3$ Pg year^{-1}) is exported below the mixed layer. A very significant part of exported $CaCO_3$ is dissolved in the water column between 200 and 1,500 m depth. This dissolution was long unsuspected because the related waters are already supersaturated with $CaCO_3$ and, from a thermodynamic point of view, $CaCO_3$ dissolution should not occur. It seems that biological activity is responsible for the dissolution. For example, bacterial respiration can increase the CO_2 partial pressure while reducing the pH (and thus the water $CaCO_3$ saturation) in aggregates of organic matter. A large fraction of $CaCO_3$ that escapes dissolution in the water column is dissolved in sediments (0.4 Pg C year^{-1}) and only 0.1 Pg C, corresponding to 6–20 % of $CaCO_3$ produced in surface waters, is preserved.

Beyond its role in CO_2 fluxes at the air-sea interface, calcification is also involved in the biological pump. Indeed, the skeletons of calcareous (coccolithophorids, *cf.* Sect. 7.10) and siliceous (diatoms) organisms are excellent ballasts for organic matter. They favor the carbon export to the deep ocean (Box 14.2).

Table 14.4 Principal elements of the calcium carbonate cycle (Pg $CaCO_3$ year^{-1}; from Berelson et al. 2007)

Flux	Estimation
Production (calcification) in the euphotic layer	0.5–1.6
Exportation	0.4–1.8
Dissolution between 150 and 1,500 m	1
Exportation at 2,000 m	0.6
Dissolution in sédiments (>2,000 m)	0.4
Sedimentary long-term storage	0.1

Box 14.2: The Oceanic Phosphate Cycle

Thierry Moutin and France Van Wambeke

Phosphorus is an essential component of living organisms, generally, representing around 1 % of the dry matter weight. Due to its low availability, compared to organism requirement, it was quickly recognized as one of the main factors controlling organism growth and biomass. In the ocean, it is considered as one of the elements restricting the transfer of carbon to the deep ocean via biological processes (biological pump). On a decadal time scale, any change in its availability at the ocean's surface could influence the concentration of atmospheric carbon dioxide which recent increase is considered the main cause of global warming.

A little history: Phosphorus is the 11th most abundant element in the Earth's crust. It was discovered in 1669 by the German alchemist Hennig Brand who noted that this white solid glowed in the dark and ignited spontaneously in the air. Its name means literally "light carrier" from the Greek phos 'light' and phorus 'carrier'. This spontaneous oxidation has the result that phosphorus exists only in the form of phosphates in the ocean[1]. Highlighting the importance of phosphate as nutrient was conducted by Justus Liebig in 1840. He demonstrated that plants absorb the chemical elements they need in the form of simple compounds, and that it is possible to directly provide

---
[1] The chemical element phosphorus (^{31}P) exists in waters only in forms of phosphates (oxidation degree V), organic or mineral, particulate or dissolved, non-reducible under natural conditions. It is then preferable to talk about the phosphate cycle instead of the phosphorus cycle.

T. Moutin (✉) • F. Van Wambeke
Institut Méditerranéen d'Océanologie (MIO), UM 110, CNRS 7294 IRD 235, Université de Toulon, Aix-Marseille Université, Campus de Luminy, 13288 Marseille Cedex 9, France

(continued)

Box 14.2 (continued)

these nutrients for the plants in the form of mineral salts (fertilizers). It was only in the twentieth century that the role of phosphate in the transfer of energy has been highlighted with the discovery of adenosine triphosphate (ATP), which has revolutionized the understanding of the cellular energy metabolism of all living organisms. The identification of the molecular structure of DNA by Watson and Crick in 1953 formally established the importance of phosphate in the growth and development of all organisms. During the past decades, much research in the field of biochemistry has continued to demonstrate the importance of phosphate in a wide variety of metabolic processes (Benitez-Nelson 2000). Since the first work of Redfield (1934) to the recent synthesis of Sarmiento and Gruber (2006), a major role has been assigned to phosphate in the control of the oceanic carbon cycle.

Measurable fractions: Phosphate exists in the ocean under various forms, mineral and organic, dissolved and particulate, that would be difficult, if not impossible, to quantify individually. The classically considered fractions are dissolved inorganic (mineral) phosphate (PO_4), Dissolved Organic Phosphate (DOP), and Particulate Organic Phosphate (POP). Dissolved mineral phosphate essentially corresponds to the orthophosphoric acid which is essentially in the form HPO_4^{2-} in seawater. With an average concentration of 2.3 µM, dissolved mineral phosphate is by far the largest reservoir of phosphate in the ocean. The concentration is minimal close to the surface as a result of biological activity and can reach really low values, of the order of the nanomole per liter in the highly stratified oceanic areas. The concentration increases in depth at the level of the phosphacline and then reach a maximum value near 1,000 m depth. The concentration is then usually constant and high toward the bottom. Particulate organic phosphate (POP) represents essentially living organisms mainly in the surface ocean (phospholipids and nucleic acids). A detrital part (from biological degradation processes, lysis, and cell death) difficult to measure must nevertheless be considered in this pool. The concentration varies generally between 0.01 and 0.30 µM even if exceptionally higher values may have been observed during phytoplankton blooms. POP decreases with depth reaching hardly measurable values below 1,000 m.

Dissolved organic phosphate (DOP) corresponds to the set of molecules originating from mineralization processes that would be difficult, or even impossible, to quantify individually. The DOP concentration varies from 0.2 to 1.7 µM in surface waters. In depth, the concentration decreased to very low values near the bottom.

The oceanic cycle of phosphate (Box Fig. 14.5) is presented together with the processes at the origin of the main fluxes. Inputs into the ocean are mainly coming from rivers through erosion of soils. Atmospheric depositions are generally low but may be locally significant. Loss, at the scale of the entire ocean, corresponds to sediment burial. Inputs of phosphate in the ocean's surface (using the simple view of an ocean with 2 reservoirs, surface and depth, separated by the thermocline) come mainly from the deep ocean (upwelling, convection, diffusion) and the losses are dominated by sedimentation of biogenic particles. Although other chemical processes such as adsorption of phosphate on atmospheric dust or precipitation of calcium phosphate under specific conditions should be reported, the fluxes in the biogeochemical cycle of phosphate are essentially constrained by biological processes (assimilation, excretion, regeneration, mineralization) that mainly take place in the ocean's surface. Recent discoveries have critically questioned the understanding of the functional roles of microorganisms in the oceanic phosphate cycle (Thingstad et al. 1997). Heterotrophic prokaryotes (Eubacteria and Archaea), originally considered as simple organic matter remineralizers (from POP to PO_4), finally use mineral phosphate. Their growths can therefore be limited by its availability, like phytoplankton. The resulting competition appears as a key process structuring the trophic chain that directly influences the flux of material exported from the euphotic to the deep ocean layer (the biological pump, Moutin 2000 and references included, Moutin et al. 2012). Furthermore, it was recently shown that some autotrophic organisms, originally considered to be only consumers of mineral phosphate, actively participate in the mineralization of dissolved organic phosphate through ectoenzyme activities[2] (Box Fig. 14.6). Particularly

---

[2] Enzyme excreted by the cell and remaining attached to the outer membrane (or in the periplasmic space of gram-negative bacteria). Prokaryotic heterotrophic microorganisms, prokaryotic autotrophic (cyanobacteria) and single-celled eukaryotes (flagellates, diatoms) have for some species the genetic ability to produce ectoenzymes such as the alkaline phosphatase or the phosphonatase. The production of these enzymes is usually induced by mineral phosphate deficiency, and enables hydrolyze the phospho-ester (C-O-P) or the phosphonate (C–P) bonds of organic molecules of the DOP. The mineral phosphate released can then be incorporated by microorganisms.

(continued)

Box 14.2 (continued)

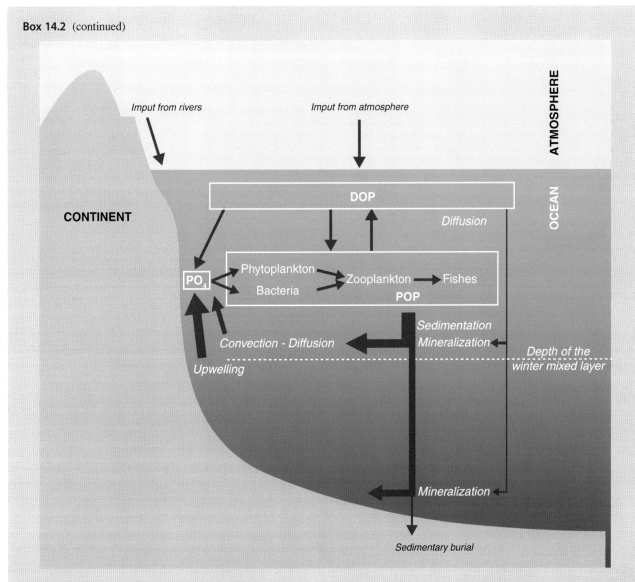

Box Fig. 14.5 Schematic representation of the main processes and flow in the oceanic phosphate cycle. *PO4* dissolved inorganic phosphate, *DOP* dissolved organic phosphate, *POP* particulate organic phosphate

important efforts are currently made to characterize the role of microorganisms in the phosphate cycle at the level of functional groups (Dyhrman et al. 2007; van Wambeke et al. 2008; Van Mooy et al. 2009; Duhamel et al. 2011).

Phosphate availability and climate change: The increase of atmospheric carbon dioxide concentration due to the burning of fossil fuels is the main cause of global warming. It depends on the fraction that the ocean can absorb. If it is commonly accepted that the oceanic penetration of anthropogenic CO_2 depends so far mainly on chemical and physical processes, it seems that the future penetration will strongly depend on biological processes mainly controlled by nutrient availability in the ocean's surface (Box Fig. 14.7, from Sarmiento and Gruber 2006). It seems that the expected climate change would lead to a decrease in the surface ocean phosphate availability (Karl 2007; Moutin et al. 2008) that could significantly curb the

(continued)

Box 14.2 (continued)

penetration of anthropogenic carbon dioxide in the ocean at the decadal time scale, and thus increase its accumulation in the atmosphere and global warming. We must nevertheless remain very careful on these predictions because many biological processes are still poorly known. To improve predictions, it will be necessary to obtain a better understanding of the processes controlling all important steps of the biological pump, i.e., the surface production, mineralization, and export. It is particularly important to be able to quantify the fraction of organic phosphate which is exported outside the annually winter mixed layer and whose mineralization will not enrich in phosphate the surface layer at the decadal time scale.

Box Fig. 14.6 Image representing a culture of heterotrophic bacteria (*Alteromonas infernus*). *Blue* microorganisms (DAPI staining) and *yellow* active sites of alkaline phosphatase (marking ELF) (Photo: France van Wambeke)

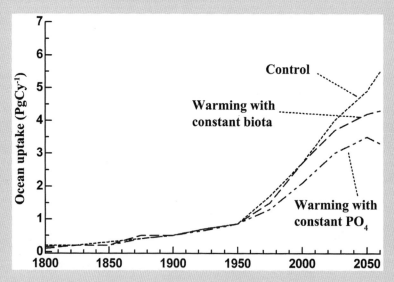

Box Fig. 14.7 Penetration rate of anthropogenic carbon dioxide in the ocean as a function of time. Simulations were obtained considering a coupled ocean-atmosphere model with a pre-industrial climate (*dotted line curve end*) and a climate changed by global warming according to the usual scenario with increased concentration of carbon dioxide from the atmosphere. *Wide dotted line curve*: the biological pump is considered constant; *Dotted line curve* with alternating and wide: the biological pump is supposed to adapt in order to maintain the phosphate concentration at the surface constant (Modified and redrawn from Sarmiento and Gruber 2006)

14.2.4 Carbon Cycle in Marine Sediments

14.2.4.1 Sinking Organic Carbon at Water-Sediment Interface

Marine sediments are the largest organic carbon reservoir on Earth: 80 % being stored in coastal sediments (that represent only 10 % of the ocean surface), 17 % in continental slope sediments, and 1.5 % in open ocean sediments. The quantity and chemical composition of organic matter that reaches the water-sediment interface depends on its location (continental inputs in coastal areas *vs* atmospheric inputs in offshore areas), on primary production in the euphotic zone, on water column depth, on local hydrodynamism and on the sinking rate of organic matter.

On average, the organic matter yielded by primary production that reaches sediment surface amounts up to 25–50 % in coastal areas, but does not exceed 1 % in abyssal zones. It is estimated that only 1–0.1 % of primary production is sequestered in marine sediments. This heterogeneous organic material free of nitrogen and phosphorus, is subject to the action of microorganisms as it reaches the water-sediment interface.

14.2.4.2 Sediment Ecosystem

Sediment ecosystem is very complex. It consists of:
1. A mineral phase that will favor prokaryote and organic substrate fixation;
2. An aqueous phase that plays an essential role in exchanges of dissolved elements (e.g., dioxygen, nutrients) between sediment and overlying waters;
3. An organic phase constituted of dissolved compounds in interstitial waters and as particles of various sizes (up to 90 % of total organic matter), offering substrates and binding sites for prokaryotes;
4. Living organisms: animals (zooplankton) and primary producers (phytobenthos) whose distribution within sediments depends on physical and chemical gradients, e.g., organic matter and nutrient concentrations, partial pressure of dioxygen and other dissolved gases (CO_2, N_2, CH_4), oxido-reduction potential, pH, temperature, salinity, light.

Zoobenthos in sediments can be divided up in several size classes: mega- and macrofauna encompassing organisms larger than 1 mm, and meiofauna which comprise organisms smaller than 1 mm and that are retained on a sieve of 63–40 μm mesh size. Microorganisms are collected on filters of lower porosity.

When the amount of light energy reaching the sediment surface is sufficient, phototrophic organisms constituting the phytobenthos can develop, frequently forming microbial mats.

The number of individuals that constitute macrobenthic, meiobenthic, and bacterial communities considerably varies, depending on sediment structure, organic matter inputs, and depth. Indeed, organic matter is not evenly distributed in sediment, in particular under the action of bioturbation (Box 14.3). This distribution results in "*hot spot*" formations, zones where biological activities are more intense in connection with organic matter accumulation.

Macrofauna is mainly located in the first 20 cm, but some organisms thrive more deeply, such as the fish *Boleophthalmus boddarti* whose burrow may extend up to 2 m. Most of meiofauna organisms are mainly distributed in the first 2 cm depth.

The fate of organic matter in sediments is largely under the control of prokaryotes (Jørgensen 2000). Counting them by epifluorescence microscopy in the upper 2 cm of sediment, yields concentrations between 10^8 and 10^{11} cells per gram of dry weight sediment in areas rich in organic matter.

Even if prokaryotes play a major role in organic carbon mineralization in sediments, the action of other organisms (unicellular eukaryotes, meiofauna, macrofauna) must also be taken into account (Box 14.3 and *cf.* Sects. 13.3.3 and 13.4.2).

Box 14.3: Macrobenthic Functional Groups Responsible of Sediment Reworking

Georges Stora

Macrobenthic organisms can be classified into five functional groups:

The biodiffusors (Box Fig. 14.8a). They move sediment particles in a random manner over short distances. The movement of organisms leads, more or less rapidly, to a homogeneous distribution of the particles in the sedimentary column.

The downward conveyors (Box Fig. 14.8b). They are head-up vertical oriented species causing an "active" transport of sediment through their gut from the sediment water interface to their egestion depth.

The upward conveyors (Box Fig. 14.8c). They are head-down vertically oriented species that remove sediment at depth in the substratum and expel it as faeces at the sediment surface. They cause an "active" transport of sediment from the bottom up through their gut and a "passive" transport all around them from the interface to the bottom of their feeding zone, due to sediment discharge at the sediment-water interface

---

G. Stora
Institut Méditerranéen d'Océanologie (MIO), UM 110, CNRS 7294 IRD 235, Université de Toulon, Aix-Marseille Université, Campus de Luminy, 13288 Marseille Cedex 9, France

(continued)

Box 14.3 (continued)

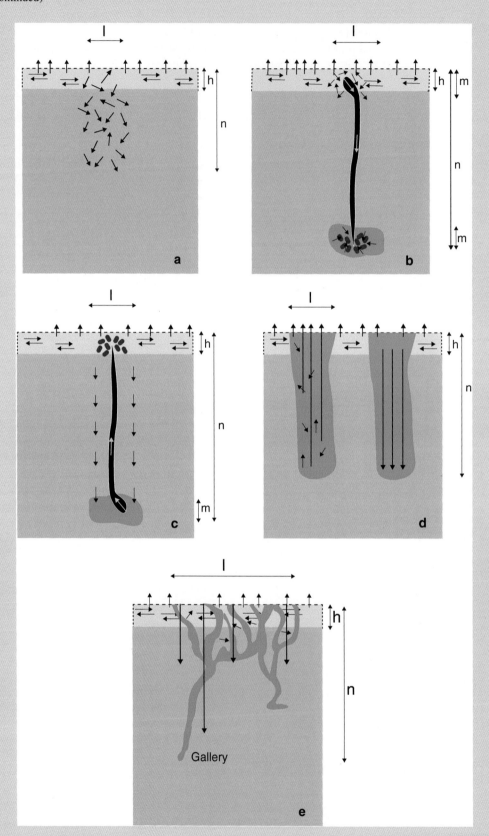

Box Fig. 14.8 (f) Resin cast of galleries of *Nereis diversicolor*, a polychaete belonging to the group of galleries diffusers (Photograph: Georges Stora))

(continued)

Box 14.3 (continued)

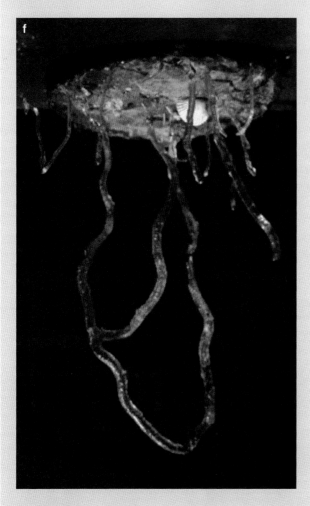

Box Fig. 14.8 (continued)

The listed parameters are: (*l*), the diameter of the bioturbated zone; (*h*), the height of the physical mixing due to local currents; (*n*), the depth of the bioturbated zone by the organisms, (*m*), the height of the ingestion/egestion zone. Arrows represent movement of sediment particles.

14.2.4.3 Organic Matter Degradation in Sediment

According to oxido-reduction (OR) potential values, three zones are distinguished (Fig. 14.14): an oxic zone in which OR potentials are greater than +100 mV, where aerobic, facultative anaerobic, and microaerophilic organisms live. Below, there are two dioxygen-depleted sediment layers: one suboxic layer where OR potential is in the range −100 to +100 mV, and an anoxic zone where OR values are lower than −100 mV. Both areas are typical environments for the expression of anaerobic metabolisms. Their distribution will depend on the OR potential of the electron donor-acceptor couple that is considered (Lovley 1991; Jørgensen 2000).

Under non-limiting conditions of substrates and terminal electron acceptors, the metabolism vertical distribution in sediments is typically stratified, on the basis of their thermodynamic efficiency (Fig. 14.14). The metabolism gradient from surface to deeper sediments comprises successively: aerobic respiration, denitrification, respirations that use manganese and iron as terminal electron acceptors, sulfate reduction, and methanogenesis (*cf*. Sect. 3.3.2). This vertical distribution is theoretical because there is no clear boundary between the expression levels of the different metabolisms, but there are instead levels separated by gradual transitions. Thus, in the same sediment layer, different respiratory activities can be detected, especially in coastal areas receiving large organic matter inputs.

Organic Matter Degradation in the Presence of Dioxygen

The organic matter degradation by microorganisms mainly depends on the energy efficiency of their metabolism, when the electron donor (growth substrates) and the terminal electron acceptors are abundant enough in the environment. Thermodynamically, aerobic respiration is the most efficient metabolism; when dioxygen is present, it is preferably used as the terminal electron acceptor by heterotrophic aerobic prokaryotes.

Some of these latter will firstly attack macromolecules (cellulose, chitin, proteins, lipids, etc.) by means of hydrolytic enzymes (cellulases, chitinases, proteases, lipases, etc.). These enzymes are usually located outside the cytoplasmic membrane or in the periplasmic space and are rarely free or adsorbed on particles. In the first case, the hydrolytic enzymes are labeled "ecto-enzymes" and, in the second case, "extracellular" enzymes. Heterotrophic prokaryotes will then mineralize molecules produced by enzymatic hydrolysis.

and to subsidence of the sediment column into the ingestion cavity.

The regenerators (Box Fig. 14.8d). These are organisms that dig burrows or galleries, with wide openings to the sediment-water interface, which creates initially a mass expulsion of deep sediment to the surface. In a second time after their abandonment by the organisms, these structures act as sediment traps, particles coming from the surface through gravity and that settle to the bottom of the burrows.

The gallery-diffusors (Box Fig. 14.8e). This group corresponds to species that dig gallery networks in sediments, involving a biodiffusive tracer distribution in the upper layer of the sediments intensively burrowed by organisms and to an advective transport of matter from the surface to the deep part of the tubes due to the dragging of particles down to the tube bottoms by animal movements.

Fig. 14.14 Pathways of organic matter transformation in aquatic sediments. $\Delta G^{0'}$ values are expressed in KJ.mole^{-1} of oxidized acetate (Thauer et al. 1977, 1077). For fermentations, $\Delta G^{0'}$ values range from +70 to −260 KJ.mole^{-1} according to the fermented substrate. Interspecies hydrogen transfers are included. Alcohols and fatty acids are also utilized by denitrifiers (**a**), Mn (**b**), iron (**c**) and sulfate reducers (**d**). Sulfate reducers (**d**) and methanogens (**e**) use also hydrogen and formate (Drawing: M.-J. Bodiou)

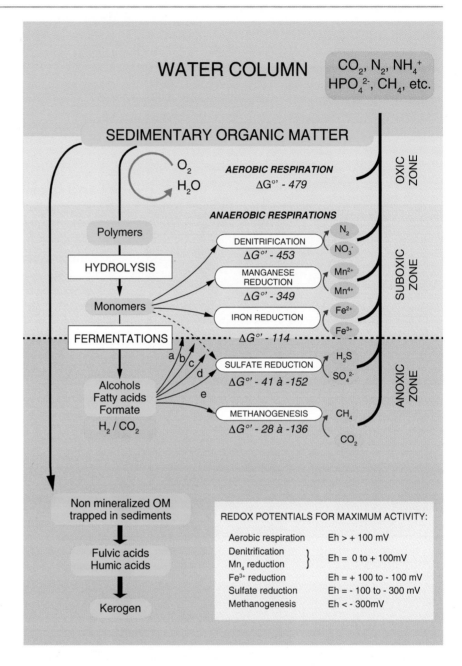

In organic-matter rich-sediments (coastal zones), the activity of aerobic heterotrophic prokaryotes is intense and the dioxygen provided by the overlying waters is rapidly depleted, before full oxidation of all organic carbon. Thus, below a thin (1–10 mm) and oxygenated sediment layer, anoxic conditions quickly take place. The situation is different in the open ocean deep sediments that are weakly supplied with organic matter and where the oxygenated layer thickness can exceed 1 m.

Organic Matter Degradation in the Absence of Dioxygen

The amount of retrieved energy depends on the used substrate (*cf.* Sect. 3.3.2) and it will be especially higher as the OR potential difference between electron donor and acceptor is larger.

Depending on the nature of the electron acceptors, the following metabolisms will take place in sequence: denitrification, manganese reduction and iron reduction in the suboxic zone, and sulfate reduction and methanogenesis in the anoxic zone (Fig. 14.14).

In both suboxic and anoxic conditions, prokaryotes are practically the only organisms capable to develop. Organic matter mineralization is achieved by several prokaryotic communities with different and complementary metabolic activities. A first step is accomplished by fermentative prokaryotes that are the only microorganisms able to degrade complex polymers non-mineralized in the oxic layer, by help of their hydolytic enzymes (Fig. 14.14). They thus produce monomers (sugars, amino acids, long chain fatty acids, etc.) in the interstitial sediment waters, at concentrations 100–1,000 times higher than those found in the overlying waters. Part of these monomers is transported by diffusion into oxygenated zones where they are degraded by aerobic heterotrophic prokaryotes. The monomers remaining in the sediment are used by different prokaryotic communities. Some are growth substrates for bacteria and archaea, denitrifying, oxidizing iron and manganese, and/or reducing sulfate (*cf*. Sects. 14.3, 14.4, 14.5). All monomers are assimilated by fermentative microorganisms, primarily those responsible for the complex polymer hydrolysis. These fermentation products are: lactate, ethanol, volatile fatty acids (such as acetate, propionate, butyrate), fumarate, hydrogen, and carbon dioxide. These compounds are growth substrates for the majority of anaerobic prokaryotic communities.

Role of Fermentative Microorganisms

Based on energy efficiency, fermentative prokaryotes are disadvantaged compared with prokaryotes that perform aerobic and anaerobic respiration (*cf*. Sect. 3.3.3). However, in anoxic ecosystems (marine and lake sediments, soils, digestive tracts of many animals and humans, anaerobic digesters), fermentative microorganisms play an essential role. Indeed they are the only ones able to use macromolecules (enzymatic hydrolysis) that other microorganisms cannot harness. They generate substrates that can fuel different respiratory metabolic pathways. Fermentative prokaryotes are the first trophic link in the anaerobic detrital food chain and, in their absence, all mineralization processes would be impossible.

In Marine Systems, Sulfate Reduction Is the Dominant Metabolism

In coastal sediments, 25–50 % of organic carbon is mineralized by prokaryotic sulfate reduction, sulfate being generally present at non-limiting concentrations, in contrast with nitrate that seems to play a minor role as an oxidant. The denitrification contribution to degradation processes is limited, only 3 % of organic carbon being oxidized by denitrifying prokaryotes (Jørgensen 1983). Similarly, the methanogen activity is also low. Because of their lower energetic metabolism, methanogens are outcompeted by sulfate-reducers for the use of acetate, formate, and hydrogen (Fig. 14.14). However, the methanogen activity in the anaerobic mineralization of organic matter is not negligible. This stems from their ability to use substrates such as methanol (source of 10 % methane) or trimethylamine, that are not or poorly accessible to sulfate-reducers (*cf*. Sect. 14.2.6).

Hydrogen Production and Consumption

During organic matter degradation in anoxic sediments, hydrogen is produced by different fermentation processes involving monomers, products of macromolecule hydrolysis by fermentative prokaryotes (Fig. 14.14). Hydrogen is also produced by microorganisms that transform their fermentative substrates (ethanol, propionate, butyrate, fatty acids) into acetate, CO_2, and H_2. The energetic balance of these reactions that produce hydrogen, under standard conditions (*cf*. Sect. 3.3.3), is endergonic. For efficiency, these reations must be coupled with reactions consuming hydrogen. The hydrogen consumption considerably decreases the hydrogen partial pressure, thus enabling the reaction producing hydrogen to become exergonic. For example, under standard conditions (H_2 partial pressure of 1 atm or 101.3 kPa), the free energy ($G^{\circ\prime}$) variation in the ethanol oxidation into acetate is +9.7 kJ:

$$CH_3CH_2OH + H_2O \rightarrow CH_3COO^- + H^+ + 2H_2$$

This reaction becomes exergonic (-44 kJ) at 10^{-4} atm H_2. This transfer of hydrogen is called "interspecific hydrogen transfer" between organisms that establish a syntrophic relationship (*cf*. Sect. 3.3.3): substrates are degraded by the combined action of prokaryotes that produce hydrogen (hydrogenogenic) and others who consume hydrogen (hydrogenotrophs). In marine sediments, the majority of hydrogenotrophs are sulfate-reducers, the contribution of methanogenes to hydrogen removal being much more limited.

Bioturbation and Its Consequences

Bioturbation is defined as all activities generated by benthic organisms that have direct or indirect effects on biotic and abiotic mechanisms and processes that occur in sediment (Aller 1994). Bioturbation exerts a strong influence on the sedimentary ecosystem functioning, in particular on the microorganism living conditions, by altering the sediment physical, chemical, and biological properties.

In the sediment layer subject to bioturbation, generally between 0 and 20 cm thickness, 75 % of the sinking organic carbon settled on the sediment surface, are degraded. The bioturbation influence on the sedimentary organic matter was mainly studied in relation to macrobenthic organisms pooled in functional groups. There are bioturbation functional groups, each one containing different species that generate equivalent sedimentary stirring through their ethology (food, motion, burrow construction, etc.).

From examining the different functional group activity that is presented in Box 14.3, it is possible to figure out the bioturbation consequences on the distribution, structure, and activity of microbial communities. Macrobenthic organisms will :
1. Bring about a redistribution of organic and mineral materials, especially of electron acceptors (dioxygen, nitrate, manganese, iron, etc.; Satoh et al. 2007), resulting in modifications of the microbial-metabolism distribution and of their stratification below the water-sediment interface (Box 14.3). For example, bioturbation will induce deep inputs of dioxygen in sediment (down to 10 cm), enabling proliferation of aerobic, facultative aerobic, and microaerophilic prokaryotes in typically anoxic sediment spots. This sediment oxygenation is generated by an

intense water circulation inside sediment tunnels built by the so-called "gallery-diffusing" organisms, such as polychaete worms that can pump up to 6.6 ml of water per minute through its gallery. It was also demonstrated that microbial communities present in these galleries were different from those outside (Satoh et al. 2007);
2. Draw down organic matter (downward conveyors, regenerators, and gallery diffusors) or bring back to the oxygenated surface sediment, materials buried in an anoxic zone (upward conveyor and regenerators). One of the consequences of these organic material movements will be the appearance of oxygenated microniches in typically anoxic sediment, and conversely. These microenvironments correspond to areas of high microbial activity ("*hot spot*");
3. Create alternating oxic and anoxic conditions that amplify organic matter degradation (Aller 1994);
4. Regulate the microbial biomass. It was shown that 50 % of microbial production is consumed by benthic invertebrates, encompassing macrofauna (2/3) and meiofauna (the remaining);
5. Change the organic matter composition during gut transit and provide more labile organic materials to microorganisms; any organic compound in the sediment will pass, at least once, through the digestive tract of benthic animals;
6. Change the composition of microbial populations fixed on ingested particles that pass through the digestive tract (Grossi et al. 2006);
7. Favor exchanges of dissolved organic and inorganic materials at the water-sediment interface.

Organic Matter Sequestration in the Sediment

A small fraction of organic carbon is not degraded in sediment and will be lost for the carbon cycle; sediment then acts as a carbon "sink." This loss which average 0.1 % could reach 4 % in strictly anoxic sediments. The sediment-trapped organic matter will lead to the formation of **fulvic and humic acids***, and later, kerogen. In addition, the presence of biomacromolecules that are particularly resistant to chemical and microbial transformation, such as those found in some higher plants (cutans and suberans), chlorobiontes (algaenans), dinobiontes (dinosporine) and cyanobacteria (bacterans), suggests that the organic matter amount sequestered in sediment is underestimated.

14.2.5 The Carbon Cycle in Soils

The soil, whose key characteristics have been described in Sect. 13.1.3, plays a key role in the carbon cycle at all scales on Earth (*cf.* Sect. 14.2.2). Organic matter in soils represents a stock of about 1,500 Pg of carbon (PgC), and a further amount of 500 PgC is contained in soil microbial biomass.

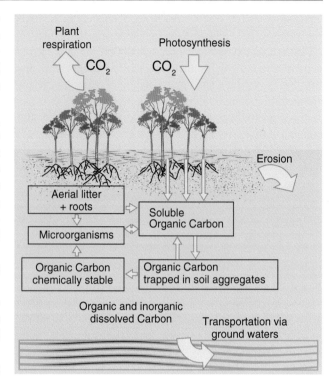

Fig. 14.15 Representation of the carbon cycle in the soil seen in the context of the carbon balance of an ecosystem (Drawing: M.-J. Bodiou)

These stocks are particularly high when compared with that (762 PgC) contained in the atmosphere (Fig. 14.5). This implies that any alteration of carbon cycle or stocks in soils can have a major impact at a global scale. Soil carbon cycling is very important for the net carbon balance of terrestrial ecosystems where **soil respiration*** consumes a large fraction of the net ecosystem primary production (Fig. 14.15). The carbon input into the soil is mainly derived from vegetation (via root exudates and root or air litter), representing about 200 gC m^{-2}year^{-1} for a tundra ecosystem, and up to 4,500 gC m^{-2}year^{-1} for a sugar cane field. **Fresh organic matter*** is then transformed, primarily by the microbial activity, following three different pathways. It can be:

1. Decomposed and the associated carbon released in the atmosphere as CO_2 and, to a lesser extent, as methane;
2. Assimilated and converted into biomass;
3. Converted into **humic substances***.

14.2.5.1 The Organic Matter in Soils

Primary production in terrestrial systems is essentially due to plants and not micro-organisms (unlike in many aquatic environments for example). This explains why organic matter inputs to soils are largely dominated by plant macromolecules: cellulose (polymer of glucose) being the most important (35–50 % of the plant biomass) (Lynd et al. 2002), followed by lignin (polymer composed in varying proportions of coniferyl, *p*-coumaryl and sinapic alcohols). Although the inputs of carbon to soils largely correspond to

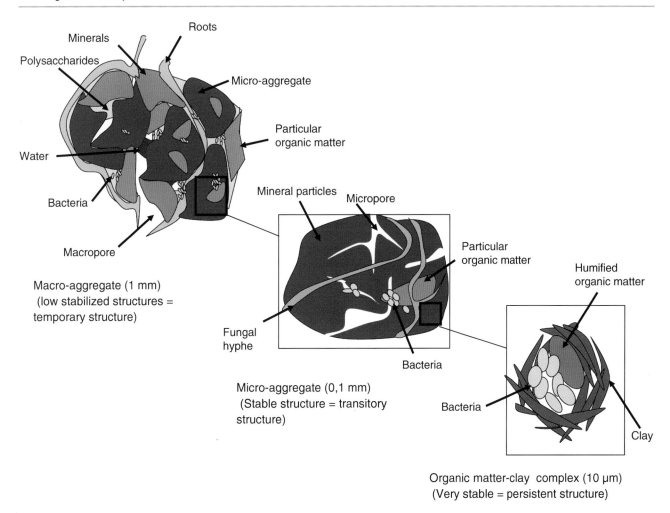

Fig. 14.16 Location of particulate and humified organic matter and in the soil matrix (Modified and redrawn from Chenu, unpublished). The residence time of the aggregates in soil and the associated organic matter increases as the size decreases

root exudates and root- and aboveground-litter (Fig. 14.15), microorganisms have a fundamental role in the carbon cycle in soil, since they are strongly involved in the transformation of organic carbon compounds. In particular, they are known to assimilate root exudates very quickly (Hutsch et al. 2002). In addition, most of the roots are tightly linked to mycorrhizal fungi, so that the carbon assimilated by plants goes through these mycorrhizal fungi.

The organic matter content is critical for biological activities and is often used as an indicator of soil quality (De Bona et al. 2008), both for agricultural purposes (such as fertility) and for environmental issues as such carbon sequestration in soils, which can influence carbon concentration in the atmosphere and therefore the greenhouse effect. The abundance, diversity, and activity of soil fauna and microorganisms are generally closely related to organic matter concentration and quality. In addition, organic matter and the related biological activities have a major influence on the physical and chemical properties of soils: water retention capacity, aerobic conditions, aggregation, and stability of soil structure, all increase with its organic matter content. The latter also affects the dynamics and availability of the main nutrients and key toxic elements (pesticides, heavy metals, etc.) (Farenhorst 2006; van Herwijnen et al. 2007).

However, as in sediments, organic carbon compounds are very numerous in soils and they have very different reactivity and turnover characteristics (Fig. 14.16). While the residence time of organic material in a soil as a whole is typically 20–50 years, its various chemical fractions may have residence times ranging from days to thousands of years (Fig. 14.16). Different approaches are possible to characterize the status of soil carbon, from a characterization of the total amount to the characterization of large compartments with contrasted residence times, such as microbial biomass, litter, **fresh organic matter***, and humified organic matter, up to the characterization of particular compounds. These characterizations are based either on the molecular size or on the biochemical properties, when targeting certain families of key molecules (lignin, sugar, etc.).

The chemical heterogeneity of soil organic matter partly reflects its spatial location: The characteristics of the fresh organic matter brought to the soil as litter and other debris differ from those of the more labile organic matter associated with macro-aggregates, and from those of the stable organic matter associated with micro-aggregates or clay-organic matter complexes (Fig. 14.16).

This biochemical and spatial heterogeneity of soil organic matter has led very often to drastically simplify the characterization of its properties in the analysis of soil carbon cycle.

14.2.5.2 The Soil Biota

Numerous organisms, with size ranging from few centimeters to a micrometer, are involved in soil carbon cycle. These ecosystems contain macrofauna (earthworms, termites, ants, etc.), mesofauna (insects, mites, microarthropods, etc.), microfauna (nematodes, etc.), and eukaryotic and prokaryotic microorganisms which are abundant and extremely diverse. All these organisms coexist and can interact. Their actions are complementary or redundant (i.e., functional redundancy) and ultimately contribute to the main functions in soils, including the recycling of organic material whose degradation can lead to its mineralization and to the release of essential inorganic compounds (nitrogen, phosphorus, etc.) for plant growth.

The smallest organisms are also the most abundant and the most diversified ones. In particular, the diversity of soil micro-organisms is huge, the estimated number of bacterial taxa ranging from 2000 to several millions per gram of soil (Gans et al. 2005; Schloss and Handelsman 2006; Roesch et al. 2007).

With prokaryotes, fungi, symbiotic or not, have a particularly important role in carbon cycle in soils compared with aquatic environments (cf. Sect. 13.3.4). Indeed, saprophytic *Fungi* play a key role in the decomposition of organic matter (Koide et al. 2005); Mycorrhizal *Fungi* play the role of interface between the plant roots and soil in many ecosystems. Few studies have so far focused on soil *Archaea* that generally are considered to play a minor role, excepted in carbon-rich and anoxic soils such as rice crop fields where methanogens (Goevert and Conrad 2009) and ammonia-oxidizing chemolithotrophs (Schauss et al. 2009) are numerous.

Within a soil, the activity of the diverse organisms varies in time and space. Indeed, because the availability of carbon depends on plants, all physiological changes in plant activity alter the amount and the nature of exudates and litters and ultimately micro-organisms over time. Moreover, carbon inputs and transformations are not uniform throughout the soil, but instead are localized in locations of intense activity known as "*hot spots*" (Box 14.4), in particular the rhizosphere (area under the influence of roots), the surroundings of organic detritus, or the physical structures produced by soil fauna (earthworm or termite burrows, etc.). As in the sediments, microbial activity is often higher in these "*hot spots*" (Ruiz-Rueda et al. 2009; Hernesmaa et al. 2005), due to favorable conditions and increased availability of organic compounds. "*Hot spots*" generally are associated with intense biochemical transformations, and are considered as "oases" that are important for the functioning of soil ecosystems as a whole, seen as a very heterogeneous environment. Given the size and the generation time of microorganisms, microbial communities react quickly to the favorable conditions provided by "*hot spots*", even at small spatial scale (<1 mm). This leads to strong variations in the activity of microorganisms that can also be accompanied by changes in the composition of soil microbial communities over time and space (Hernesmaa et al. 2005; Weisskopf et al. 2005).

If fungi and bacteria are the main actors in the chemical alteration of organic material, the macrofauna (earthworms, termites, ants, snails, crustaceans, etc.) is also significantly involved in several degradation processes of organic matter (OM), such as:

1. Fragmentation of OM that increases the surface offered for colonization by micro-organisms;
2. Redistribution of OM by soil bioturbation;
3. Oxygenation and, more generally, promotion of gas and water movements through the creation of a network of burrows within the soil;
4. Changes in the physical properties of soils. For example, a compact and impermeable soil may become soft and porous under the action of burrowing macrofauna;
5. Neutralization of the soil pH, for example, after the transit of organic material in the digestive tract of earthworms;
6. Grazing pressure and selective grazing that can shape the microbial community structure and influence microbial turnover (cf. Sect. 13.3.4).

Some data highlight the importance of bioturbation by macrofauna. A soil may contain 100–500 earthworms per m^2, and up to 2,000 individuals per m^2 in some soils (Lavelle and Spain 2001). Given that a worm ingests three times its own mass of soil per day, for a soil containing 100 worms per m^2, the latter can incorporate 5 t of dry leaves per year and 50 t (dry mass) of soil (Frontier and Pichod-Viale 1998).

Box 14.4: Soil Engineers and Greenhouse Gas Emissions

Alain Brauman and Eric Blanchart

Is Soil a Desert or an Oasis for Microorganisms?

One gram of soil encompasses more than one billion of microorganisms and shelter between 3,000 and 10,000 different species; despite this putative richness, soil from a bacterial point of view, is more a desert than an oasis. In fact, microorganisms occupy only a tiny portion of the soil: less than 6 % of the pore volume. The limited accessibility to organic resources (mineral material represents up to 95 % of soil) involves that only a small portion of the microbial biomass is active, and most of the microorganisms are dormant. Therefore, the microbial functions involved in the biogeochemical cycles are mostly performed in oasis (or microbial patches) generally associated with a higher availability in carbon substrates. These oases are multiple (organic fractions, aggregates, litter, soil pores). However, the formation, distribution, and maintenance of these microbial habitats are mainly under the dependence of soil organisms like soil macrofauna.

Soil Engineers and Soil's Microbial Compartment (Lavelle 2002)

Soil invertebrates play a key role in soil functioning: (1) they decompose the litter which is further incorporated in the soil; (2) thanks to their building activities (galleries, nest etc.), they deeply influence the soil structure and its aggregation pattern; (3) they influence the diversity and activities of the soil microbiota; (4) they contribute to plant protection against diseases and pathogens. These organisms, like earthworm, ants, and termites (in tropical soils) regulate the microbial activity in a non-trophic manner by modifying the physical environment and the availability of nutrients for microorganisms. These regulation properties of the soil microbial compartment are at the origins of the soil engineer's concept (Lavelle et al. 1997). The soil volumes affected by the engineer's activity constitute some major soil functional domains such as the drilosphere for earthworms and the termitosphere for termites. These functional domains are in constant interactions. The formation of galleries and cast (dejections) by earthworms influences the root dynamics (rhizosphere) and modify the litter distribution (residusphere) in the soil. In these functional domains, the microbial activity is not obligatory stimulated, but regulated in time and space. In fresh earthworms casts (=dejections), microbial activities are deeply stimulated, resulting in a strong mineralization of the organic material, which leads to mineral nitrogen liberalization while within the "old" and dry earthworm casts, the microbial activity is reduced resulting in a weak mineralization. Soil engineers are affecting all microbial parameters, as density, diversity, and/or the genetic structure. However, the intensity of this modification varies depending on the organism and the microbial function studied. To illustrate this point, we will now discuss the regulation by earthworms and termites of an important ecosystemic microbial function: greenhouse gas emission (methane and nitrous oxide).

Earthworms and N_2O (Drake and Horn 2007; Majeed et al. 2013)

Earthworms are present in the majority of soils where they represent the principal soil fauna in terms of biomass. They process between 500 and 1,000 t of soil per hectare and per year. Thanks to this huge activity, the major part of the surface layer of the soil is transiting at least once in their digestive tube. During this intestinal transit, the soil microbial community will find more appropriate conditions for its activity (neutral pH, rich substrates and labile form of mucus, weak pO_2 etc.). This gut environment will particularly stimulate the facultative anaerobic bacterial communities like the denitrifiers, rather than the strict anaerobic one such as methanogens. This explains the relative important impact of earthworms on the biogenic N_2O emissions rate. In the laboratory, this emission rate varies according to species, but is between 5 and 225 ng $N_2O\ h^{-1}\ g^{-1}$ (fresh weight of earthworms). This leads to a potential annual estimate of 3×10^8 kg $N_2O^{-1}\ year^{-1}$ (1 % of the total N_2O annual emissions). Surface casting is also one of the major aspects of earthworm activities with a production rate varying from few tens to several hundreds of tons dry soil. Within the fresh casts, the microbial community behave mostly the same than in the gut (stimulation of the denitrifying activity without selection of the microbial community). However, in the older and dryer casts, the scenario becomes more complex: diminution of the microbial biomass but increase of the metabolic activities. However, whatever the location (gut or cast), the intensity of these N_2O emission rates vary according to the different feeding behaviour (epigeic, endogeic, and anecic) of worms. Globally, epigeic earthworms emitted much more N_2O than endogeic species.

(continued)

A. Brauman (✉) • E. Blanchart
IRD, UMR 210 ECO&SOLS, Montpellier, France

Box 14.4 (continued)

Box Fig. 14.9 Conceptual diagram of the in situ conditions and methane emissions in a humivore termite (Modified and redrawn from Brune, 1998). *P1* first part of posterior gut, *P3* Panse, *P5* Rectum

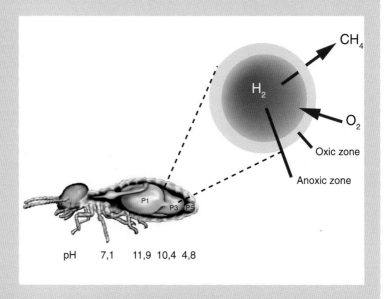

Termites and Methane (Brauman et al. 2008)

If the earthworms constitute the principal biomass in temperate soils, they are in competition, in the tropical soils, with termites whose density can reach 10,000 ind.m^2. Lavelle (2002) indicated that termite nest recover almost 10 % of African soils and constitute important microbial oasis. The relationship termites-microorganisms is complex and highly dependent of the trophic affiliation (wood feeders, fungus growing termites, soil feeders). However, among these different feeding habits, only the soil feeding termites (50 % of the 2,600 species described) live and feed on the soil; thus, they are considered, through their foraging activities, as true soil engineers comparable to earthworms. If the environmental conditions of the digestive tube are particularly favorable to microbial communities, on the other hand, the highly alkaline pH (up to 11, Box Fig. 14.9) and the anoxic environment (Eh −300 to −400 mv) of their gut make this environment highly selective which favors the development of specific communities as methanogenic archaea. Their gut density (up to 10^8–10^{10} organisms mL^{-1}) explains why termites and their associated biogenic structures (such as termite nests), represent, with the ruminants, one of the main biological CH_4 producer. They emit between 5 and 40 % of CH_4 annual emissions rate from natural origin (IPCC 1995). This imprecision originated from methodological problems and calculation methods link to CH_4 emission measurement. However, if we take into account the CH_4 oxidation originated from the soil or termite nest, their contribution to the annual CH_4 budget is reduced and comprised between 0.4 and 1.8 %. The specificity of the physico-chemical termite gut environment and to a lesser extent of the termite nest, lead to a highly phylogenetic specificity of the microorganisms associated with termite environment. Their gut ecosystems harbor more than 300 different phenotypes, most of them being specific of this environment and constitute a monophyletic group, the "Termite group phylum."

Therefore, despite their apparent functional behavior, these two soil engineers, earthworms and soil feeding termites, harbor different relationship with their gut microorganisms; mutualistic for earthworms, more symbiotic or specific for termites. At the soil scale, they both deeply modify the physico-chemical environment for the microorganisms and constitute therefore a major regulation factor of greenhouse gas emission.

14.2.5.3 Degradation of Organic Matter in Soils (the Plant – Microorganisms – Soil System)

Main Metabolic Pathways of the Carbon Cycle in Soils

Carbon cycling in soils occurs across a mosaic of oxic and more or less anoxic habitats whose distribution varies in space (*e.g.*, oxygen gradients observed between the surface and the interior of soil aggregates). In addition, some habitats are subject to important temporal fluctuations. The soil macroporosity may experience an oscillation of oxic/anoxic conditions in relation to major flood events, rises of groundwater, or more generally precipitation regime. Soil is more heterogeneous than sediment, and several types of metabolisms may thus complement each other over time or across microhabitats with different levels of oxygenation.

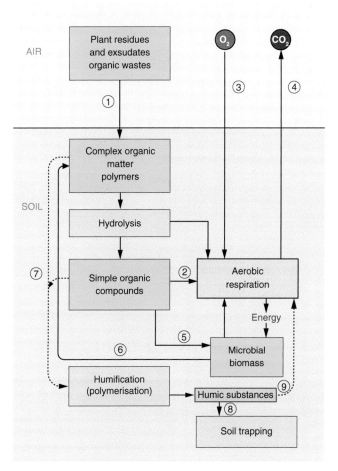

Fig. 14.17 Schematic representation of the decomposition of the organic matter in the soil by aerobic respiration. *1* Physical-Chemical degradation and polymerization of organic compounds in ground litter, *2* Energy Catabolism: aerobic respiration of prokaryotic and eukaryotic micoorganisms, organic compound consumers, *3* Consumption of oxygen by prokaryotic and eukaryotic microorganisms performing aerobic breathing, *4* Release of CO_2, the ultimate product of the mineralization of organic matter, in aerobic breathing, *5* Anabolic biomass production by prokaryotic microorganisms and eukaryotes that multiply in the soil, *6* Microbial biomass formed contributes partly to the weld material organic soil, *7* A portion of the soil organic matter is directly polymerized and contributes to the formation of humus (humic and fulvic acids), *8* humic organic matter, not or poorly biodegradable, is stored in the soil, *9* Part of the humic polymerized organic matter can be degraded by specialized microorganisms (Drawing: M.-J. Bodiou)

Chemo-organotrophic aerobic respiration in which oxygen is the electron acceptor is a major source of carbon (CO_2) loss from soils (Fig. 14.17, *pathway 4*). When the oxygen concentration becomes limiting, other respiratory pathways are possible using other electron acceptors such as nitrate or sulfate, which can be used by microorganisms to recover energy. However, the energetic yields are increasingly low for the following electron acceptors: $O_2 > NO_3^- > Mn^{4+} > Fe^{3+} > SO_4^{2-} > CO_2$ (Fig. 14.14, and Sect. 3.3.2). When the oxygen concentration becomes limiting, nitrate (derived from nitrification or fertilization) may serve as an electron acceptor by denitrifiers (Fig. 14.14), though the decomposition of OM in anoxic environments is slower.

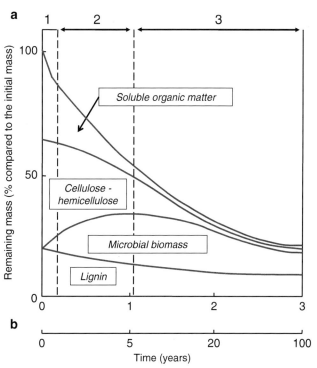

Fig. 14.18 Schematic representation of the three stages of litter decomposition traditionally identified in soil presenting the main chemical components. (**a**) Hot environments (tropical) (**b**) cold environments (Arctic) (Modified and redrawn from Chapin 2002)

In the complete absence of oxygen and when nitrate is deficient, other microorganisms become competitive and decomposition switches to fermentative pathways (Fig. 14.14, and *cf.* Sect. 3.3.3), although these pathways are often fairly marginal in soils. The products of fermentation (acetate, other organic acids, hydrogen) may be used by sulfate-reducers or methanogens that transfer electrons to sulfate or CO_2 to produce sulfides or methane. In soils, and unlike in the marine environment, the low availability of sulfate explains that methane production by methanogenic archaea can become quantitatively important. Methane produced in anoxic habitats of soils can be oxidized to CO_2 by methanotrophic bacteria when exposed to oxic habitats (*cf.* Sect. 14.2.6).

Degradation of Organic Matter in Soils

Fresh plant organic matter (root- and aboveground-litters) input to soil is fragmented by invertebrates that live in the soil or on its surface, and then is rapidly colonized by microorganisms. A several-phase-process of organic matter degradation is then observed (Fig. 14.18) (Chapin 2002).

During the first phase, the more labile substances with low molecular weights are used and rapidly mineralized by a diverse and abundant microbial community that declines when these substrates have been exhausted.

During the second phase, a more specialized community uses hemicelluloses and cellulose which are the main constituents of plant cell walls (10–30 % and 35–50 %, respectively). This implies in particular cellulolytic bacteria and fungi that usually have several endo- or exo-cellulases, cellobiohydrolases and β-glucosidases acting in synergy as extracellular enzymes. These enzymes ensure the breakdown of cellulose chains, the release of cellobiose and, ultimately, of molecules of glucose. Some organisms such as fungi have a broad spectrum of enzymes while others are only able to produce few enzymes but take advantage of microbial consortia to gain their energy from the decomposition of cellulose. Similarly, the decomposition of hemicelluloses involves the action of extracellular enzymes, but these are more diverse. Indeed, hemicelluloses are more complex molecules compared with cellulose, including different hexoses, pentoses, and uronic acids. The release of simple and soluble sugars from the degradation of hemicelluloses and cellulose allows the development of a microbial community that differs from that typical of Phase I and is subjected to enhanced competition among microorganisms. The end-products of the degradation of cellulose act as inhibitors of the activity of cellulases. By consuming these products, non-cellulolytic microorganisms ensure the completion of cellulose mineralization by cellulolytic organisms. Other important polymers degraded during the decomposition of soil organic matter are starch (linear or branched polymers of glucose molecules degraded by extracellular amylases produced by bacteria and fungi), chitin (linear polymer of N-acetylglucosamine from fungal cell walls and exoskeletons of arthropods) and peptidoglycans (polymers of N-acetylglucosamine and N-acetylmuramic acid from bacterial cell walls).

Some compounds such as lignin (which represents 5–30 % of the plant tissues), waxes, resins etc. are heterogeneous and recalcitrant to decomposition. Their decomposition occurs during a third phase and is typically slow and often incomplete even under oxic conditions. The decomposition of these recalcitrant compounds requires the stimulating presence of simple sugars acting as "starters" for the decomposition process, i.e., the so-called priming effect (Fontaine et al. 2003). This is a key mechanism for understanding and predicting phenomena of carbon storage/release from soil (Box 14.5). Decomposition of the most recalcitrant compounds is essentially done by fungi (white rot) in the groups of basidiomycetes and ascomycetes. From an evolutionary point of view, the biochemical heterogeneity of recalcitrant compounds probably has limited the synthesis of specific enzymes involved in their biodegradation, which involves the action of enzymes with broad-spectrum activity, primarily peroxidases and laccases.

The activity of these extracellular enzymes can be heavily modified by their attachment/adsorption to the surface of roots or micro-organisms which often favor their activity, or to the surface of mineral particles such as clays which often restrict their activity. The quality of the fresh organic matter entering the soil ecosystem is often a major factor controlling the rate of the decomposition of recalcitrant compounds, other factors being especially climatic ones (temperature and humidity).

In conclusion, fungi play an essential role in the above various steps of the decomposition of litter. Fungal successions are generally observed during the 3 phases of decomposition (Fig. 14.19), thereby emphasizing the more or less specialization of fungal species involved in the decomposition process (Koide et al. 2005). It is currently possible to identify fungal and prokaryotic taxa involved in the different degradation phases in soil or compost samples, by molecular biology approaches using DNA extracted from environmental samples (Danon et al. 2008).

Box 14.5: Stimulation of Decomposition of Native Soil Organic Matter by the Supply of Fresh Organic Matter: The Priming Effect

Franck Poly and Xavier Le Roux

A positive relationship may be expected between the inputs of fresh organic matter (OM) to the soil and the carbon (C) storage by the soil. However, this relationship is rarely observed in field conditions; fresh OM and soil C content are generally not correlated or negatively correlated. A lab experiment demonstrated that in a nutrient-poor soil, a fresh OM input (^{13}C labeled cellulose) can lead to a reduction of soil C stock (Fontaine et al. 2004).

In this experiment, an exogenous carbon (495 mg C-cellulose.kg^{-1} soil) labeled with ^{13}C was added to a soil with low or high nutrient availability. In the low nutrient soil (LNS), a large part of organic C (exogenous ^{13}C + endogenous ^{12}C) was mineralized and lost as CO_2 (365 + 140 mg C.kg^{-1}) (Box Table 14.1). In contrast, in the high nutrient soil (HNS), the total C loss by respiration was only 318 + 72 mg C.kg^{-1}. Thus, 110

Box Table 14.1 Soil carbon balance in poor and rich nutrient soils after amendment of ^{13}C-cellulose

	Low nutrient soil	High nutrient soil
^{13}C-cellulose supplied to soil	495	495
^{13}C-CO_2 released	−365 ± 21	−318 ± 3
^{13}C remaining in soil	110 ± 11	140 ± 4
^{12}C-CO_2 released	**−140 ± 3**	**−72 ± 15**
Soil C balance after organic matter input	−30 ± 11	+68 ± 19

Means ± standard errors are expressed in mg C kg^{-1} soil
The priming effect is highlighted in bold

(continued)

Box 14.5 (continued)

and 140 mg C.kg^{-1} of fresh organic C input remained in low and high nutrient soils, respectively.

In particular, the input of exogenous C stimulated heterotrophic microorganisms in both soils leading to the decomposition of endogenous stable OM: It is the so-called priming effect. This effect was higher in the LNS than in the HNS, resulting in a negative balance of total C in nutrient-poor soil. In LNS, after the fresh C input, the total soil C content decreased, whereas in HNS the balance was positive and the soil further stored C.

Two main mechanisms were proposed by Fontaine and collaborators (2003) to explain this stimulation of stable and old OM decomposition following the fresh organic matter input. Two kinds of microbial decomposers in soil are involved: r-strategists using efficiently fresh OM and K-strategists degrading preferentially stable OM. Following mechanism 1, enzymes produced by r-strategists to degrading fresh OM could also digest old and stable OM. According to mechanism 2, abundances and enzymatic pools of K strategists can be stimulated by catabolites coming from fresh OM degradation by r-strategists, which can increase degradation of stable OM (Box Fig. 14.10)

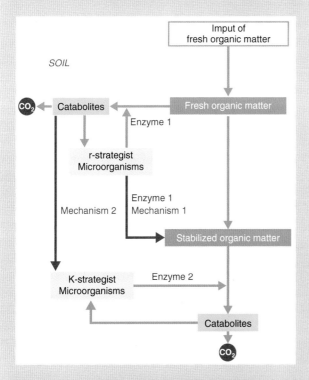

Box Fig. 14.10 The priming effect mechanism in the soil (Modified and redrawn from Fontaine et al. 2003)

Assimilation of Organic Carbon into the Microbial Biomass

While some carbon of organic substrates used for redox reactions is ultimately lost as CO_2, another part is assimilated and used for the growth of micro-organisms (Fig. 14.17, *pathway 5*). The efficiency of substrate utilization, calculated as the ratio between the quantity of carbon assimilated to the total amount of carbon used, is higher for simple sugars compared with polymers or to more recalcitrant compounds such as lignin. This efficiency is generally substantial for fungi (35–55 %), low for aerobic bacteria (<10 %), and even lower for anaerobic bacteria (<5 %).

Nitrogen and carbon are essential components of biomass and their assimilation is essential for microbial growth (Fig. 14.1), with a mean stoichiometric ratio of 1 nitrogen for 10 carbon atoms. Thus, for an utilization efficiency of 50 %, the consumption of a substrate containing 20 atoms of carbon (of which 10 will be oxidized as CO_2 and 10 assimilated) requires one atom of nitrogen. Therefore, if the C:N ratio of the substrate is less than 20, extra nitrogen from the external medium is needed (nitrogen immobilization by micro-organisms). In contrast, if the C:N ratio of the substrate is comprised between 12 and 15, the microbial community will release nitrogen in the surrounding medium (net mineralization). The quality of organic matter based on C:N ratio thus ultimately determines both the rate of its degradation and the intensity of the competition between heterotrophic microorganisms and plants for nutrients: The two communities are typically in competition for C:N ratios between 15 and 20, while heterotrophic microorganisms supply plants with nutrients for substrate C:N ratios less than 15.

Humus Formation

Humus results from the transformation of organic matter under the influence of biotic and abiotic processes. It is a complex material, more heterogeneous than lignin, and chemically poorly defined. The fraction of organic material that is not mineralized or assimilated (consisting mainly of tannins, waxes, suberin, polyphenolic residues of lignin, etc.) constitutes the phenolic precursors of humus. These precursors include reactive centers from which binding products from microbial metabolism or bodies (protein residues, nucleic acids) condense spontaneously or under the action of enzymes to gradually become humus (Figs. 14.17, 14.20). This process involves several steps, including the formation of highly reactive compounds called quinones under the action of polyphenoloxidase and peroxidases (enzymes synthesized by fungi to degrade lignin).

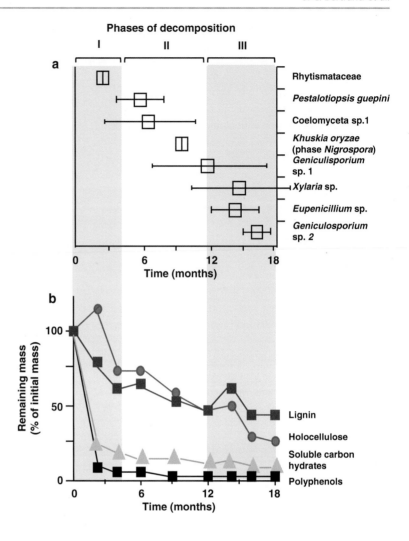

Fig. 14.19 Successions of microorganisms often observed during the key stages of the carbon cycle in soils. The process of litter decomposition can be divided into three phases. (**a**) Species of Fungi involved during the decomposition of the litter (here leaves of Camellia japonica) (**b**) The mass loss over time for soluble sugars (*green triangles*), polyphenols (*black squares*), holocellulose (*blue circles*) and lignin (*red squares*) (Modified and redrawn from Koide et al. 2005)

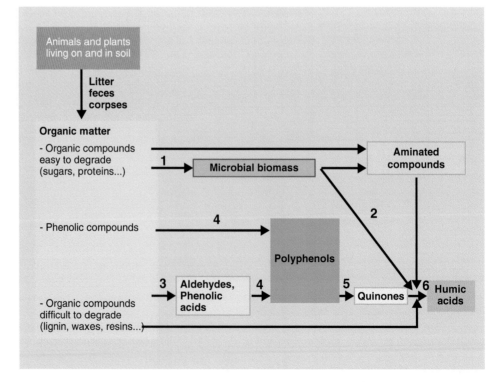

Fig. 14.20 Main routes of humus formation (Modified and redrawn from Chapin 2002). *1* Microbial Assimilation, *2* Release and alteration of microbial phenolic polymers, *3* selective preservation of recalcitrant compounds, *4* Production of polyphenols, *5* Oxidation of polyphenols, *6* Abiotic condensation/polymerization (Drawing: M.-J. Bodiou)

During these various reactions and transformations, the molecular weight of humic compounds continuously increases due to the enrichment by compounds from microbial activity: bacterial polysaccharides, fungal pigments, etc. Humic substances obtained this way have a molecular weight which may range from 700 to 300,000 Da, and are often classified into three main types: fulvic acids (rich in lateral chains and relatively soluble in water), humic acids (rich in aromatic nuclei and poor in lateral chains, relatively poorly soluble), and humins (moderately rich in lateral chains, slightly soluble). These compounds have a "sponge" structure and offer a large surface for exchanges, with hydrophilic and hydrophobic zones. They are often associated with inorganic particles, forming three-dimensional structures which are poorly accessible to any enzymatic attack. Their properties are very important to explain the physico-chemical characteristics of soils, such as cation exchange capacity, adsorption of solutes, etc. Note that direct humification (i.e., without any role of biological activity) may occur in areas where pedoclimatic conditions hinder the development of microorganisms. In this case, humic compounds result from slow oxidations and physico-chemical condensation, and have a lower molecular weight.

The turnover time of humus is typically 2–5 % per year and varies depending on soil and climatic conditions. For comparison, about 50 % of the lignin supplied in a soil typically disappears within a year (Maier et al. 2000). When protected in microsites which are hardly accessible to microorganisms or when the availability of labile energetic substrates is low, humic components are protected from biodegradation and contribute to carbon sequestration in soils. Conversely, if accessible to microorganisms and if the availability labile energetic substrates is sufficient, a "priming effect" may occur (Box 14.4) (Fontaine et al. 2004): Humus is then a source of carbon, nitrogen (humus is rich in nitrogen), and energy for microbial populations (secondary mineralization). This explains why soil cultivation from forest or grassland greatly accelerates the mineralization of organic matter already present through a priming effect-like process: without organic matter amendment, the quantity of humus decreases after land conversion to cropping systems, slowly but steadily.

14.2.6 Carbon Cycle in Lakes

The physical, chemical, and biological processes that govern the major biogeochemical cycles in general, and that of carbon in particular, are globally similar in marine and lacustrian ecosystems (Box 14.6). This is particularly the case for pelagic areas, highly original and characterized by microbial communities ensuring almost all of the primary

---

Box 14.6: The Carbon Cycle in Freshwater Ecosystems

Gerard Fonty

Lakes are closed and spatially well-delimited ecosystems and represent stagnant water surfaces that occupy natural or arranged depressions on the earth. Although it is possible to describe a general dynamics of carbon cycle in these aquatic biotopes, each lake possesses its own characteristics which can only be understood in regards to its morphometrical, physico-chemical, and biological features. These characteristics depend on the geographical localization of the lake and on its mode of alimentation. Lakes are very diverse by their size, shape, depth, and also genesis.

Morphometrical features of lakes and stratification of physico-chemical parameters. Morphometrical characteristics condition the physical, chemical, and biological functioning of the lakes. Depth of light penetration in the water column, wind action, temperature, and oxygen concentration of water are major factors determining the ecological function of lakes. The vertical stratification of light determines a euphotic upper layer in photosynthesis takes place, and an aphotic zone where the intensity of light is too weak to authorize this process. Due to the variation of water density with temperature, the water column stratifies vertically in the absence of turbulences according to a vertical gradient of temperature. This stratification depends on the lake depth. In summer time, in deep lakes, three zones can be distinguished: the upper warm zone called epilimnion, an intermediate compartment where temperature rapidly decreases with depth named metamolimnion, and finally a deep zone where temperature remains low, the hypolimnion. In deep lakes, hypolimnion temperature is near to 4 °C. The sharp variation of temperature in hypolimnion determines a thermocline that plays a fundamental role in the functioning of the lakes because it constitutes a barrier for the transfers between the hypo- and the epilimnion. The vertical thermocline position changes with the seasons according to the warming and the water turbulence in the epilimnion. Seasonal variations of water density lead to a more or less pronounced mixing of the water column. According to the frequency of water mixing lakes are classified into: (1) meromictic lakes (one annual period of destratification); (2) dimictic lakes (water

---

G. Fonty (✉)
Laboratoire des Microorganismes: Génome et Environnement (LMGE), UMR CNRS 6023, Université Blaise Pascal, Clermont Université, B.P. 80026, 63171 Aubère Cedex, France

(continued)

Box 14.6 (continued)

mixing in spring and autumn); (3) polymictic lakes (several successive annual periods of mixing and stratification); (4) meromictic lakes in which the upper layers of water are never mixed with the deep layers.

Lakes also present a stratification of oxygen concentrations that is the consequence of a dynamic balance of gas exchanges between atmosphere, photosynthesis, and respiration of organisms. The stratification can be temporary like in dimictic lakes where the deep water layer is anoxic in summer time or permanent as in meromictic lakes where the hypolimnion is always anoxic. The mixing of water is a major event in the life of lakes because at the stratification of abiotic factors due to stratification matches a stratification of biological processes involved in biogeochemical cycles.

Major biotic components of freshwater lake ecosystems. The organisms are distributed in three main zones. The littoral zone is colonized by the primary producers that include submerged or floating high-sized plants (the macrophytes: hydrophytes and helophytes) and various types of aquatic metazoa. The macrophytes belong to Streptobiontes and Chlorobiontes (*cf.* Sect. 7.5.5); they can be attached to different types of surfaces (periphyton). The pelagic zone harbors the plankton and the nekton. As in oceans, the plankton comprises the zooplankton and the phytoplankton. The phytoplankton is constituted by oxygenic photosynthetic prokaryotes (cyanobacteria), by photosynthetic eukaryotes (diatoms, Chlorophyceae) and by non-photosynthetic eukaryotes such as Chytridiomyctes (Fungi). The zooplankton is composed of eukaryotes belonging to a large diversity of taxonomic units (ciliates, rotifers, crustaceans such as cladocerans and copepods). It also includes insect larvas. The nekton includes animals that are able to freely swim (fishes, amphibians). The benthic zone comprises organisms living on or nearby the bottom of the lake (insect larvas, molluscs, worms, crustaceans). According to the trophic level (mineral and intact organic matter concentrations), different types of lakes are distinguished: the oligotrophic lakes that are poor in biogenic mineral materials and in plankton, the eutrophic lakes that are rich in nutrients and the mesotrophic lakes that represent an intermediate between oligotrophic and eutrophic lakes.

Forms of Carbon in Lakes
- *Inorganic carbon* is the result of both physical (exchanges of CO_2 with atmosphere and dissolution from chalky rocks) and biochemical processes (primary production and respiration). The atmospheric CO_2, very soluble in water, is hydrated to form carbonic acid (H_2CO_3) that is then dissociated in bicarbonate (HCO_3^-) and carbonate (CO_3^-) according to the pH. In most of the freshwater ecosystems the pH varies from 6 to 9, so the bicarbonate is usually dominant. The carbonic acid attacks the superficial rocks, and transforms the calcium carbonate ($CaCO_3$) in calcium bicarbonate [$Ca(CO_3)_2$] that is more soluble.
- *Organic carbon* is located in the dissolved organic matter (DOM) and in the particulate organic matter (POM) and constitutes the dissolved organic carbon (DOC) and the particulate organic carbon (POC), respectively. In lacustrine systems, the DOC storage is 5–10 times higher than that of POC and remains more stable with time. The POM and DOM origin can be autochthonous or allochthonous. The autochthonous POM comprises the living organisms mentioned above and dead organic matter. The living fraction usually represents only a small part of total organic carbon. DOM storage is supplied by the compounds excreted by organisms and by the intermediary products of microbial degradation. In non-anthropic lakes, the allochthonous organic matter is brought by interstitial and run-off water. When the basin side is wooded and is of small surface, leaves are one of the main source of this type of organic matter. In most of cases, this matter arrives in the dissolved form, but the allochthonous POC can be dominant in strongly eroded basins. The allochthonous organic matter is essentially composed of recalcitrant compounds (lignin). DOM and POM are constituted of complex but non-well characterized molecules.

Food webs. In pelagic zones of lakes, carbon and energy circulates within a complex web of interactions between the organisms and their biotopes, and between organisms themselves. This web is summarized in Box Fig. 14.11: The organic carbon (primary production) is synthetized from mineral carbon (CO_2) by photosynthesis performed by macrophytes, phytoplankton and at a lesser extent by photo-and chemosynthetic bacteria. Some heterotrophic bacterial species are also

(continued)

Box 14.6 (continued)

Box Fig. 14.11 The microbial loop in lake ecosystems (Modified and redrawn from Amblard et al. 1998)

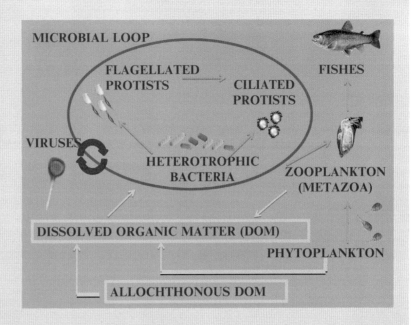

able to incorporate CO_2 by anaplerotic reactions. Then, the phytoplankton is ingested by herbivorous zooplankton (predation) which itself consumed by planktonophagous fishes. Finally, these planktonophagous fishes are prey for carnivorous fish species. Metabolites released by living organisms and those resulting in the *post mortem* degradation of phytoplankton produces DOM that initiates the functioning of a trophic network named "microbial loop" (Amblard et al. 1998).

Then, these products are degraded and metabolized by heterotroph prokaryotes that are abundant in pelagic zones of lakes and by pico-cyanobacteria and pico-eukaryotes. Then, these microorganisms are ingested by the zooplankton; their major predators are the unicellular mixotrophic- and heterotrophic eukaryotes (ciliates, Dinobiontes, "amibs," etc.). Other unicellular eukaryotes serve also as prey to Ciliates which are then consumed by the metazoan of the zooplankton. The detritic allochthonous OM, particular or dissolved, is also degraded by the microorganisms. This transfer of matter and energy accumulated in prokaryotes, via the bacterivorous organisms, ensures a better recycling of the organic matter.

The microbial processes of the organic matter degradation involve the same steps and the same mechanisms than those described in oceans (Fig. 14.12). In oxygenated compartments, the OM is mineralized by aerobic microorganisms to form H_2O and CO_2. Anaerobic mineralization is the fact, through successive steps, of specific functional microbial groups organized in trophic chains. Except in the first step, mineral degradation processes are redox reactions in which the oxidation of OM is coupled to the reduction of various electron acceptors. In limnic ecosystems, usually poor in sulfates, the end-products of anaerobic OM degradation are CO_2 and CH_4 (and sometimes reduced forms of iron and manganese). The metallic components that are usually soluble accumulate in the upper part of the water column, and the gases (CO_2 and CH_4) can escape to atmosphere. During this diffusion, the CO_2 is partially used for photosynthesis in the euphotic zone and CH_4 is partly oxidized by aerobic methanotrophic bacteria (methylotrophs). In some lakes, such as the lake Pavin (Central Massif, France), the unique French meromictic lake, a part of CH_4 is oxidized in the anaerobic compartment (Lehours et al. 2010) *via* a process not yet elucidated but probably different from that observed in marine sediments. It is possible that this mechanism involved, in a syntrophic association, methanogenic archaea performing reverse methanogenesis and nitrate (NO_3^-)- or iron (Fe^{3+})-reducing microorganisms. Methane oxidation is a major process in carbon cycle and is fundamentally important in regards to environment because it avoids the emission in atmosphere of a greenhouse gas that possesses a radiative forcing (greenhouse effect) 20 times more powerful than that of CO_2.

production that constitutes the bulk of the biomass, i.e., the basis of all aquatic food webs. However, there are substantial differences between lacustrian and marine ecosystems, related primarily to:

1. Element distribution and concentration: concentrations of sodium chloride and sulfates are high in marine environments and low in lacustrian ones.
2. The involved organisms. Indeed, microbial communities are depending on the nature of electron acceptors available in the environment. Hence, under anoxic conditions, the high sulfate concentration in marine environment favors the development of sulfate-reducing bacteria, whereas in lacustrian sediments, methanogenic *Archaea* predominate.
3. Environment forcing variables (hydrology, climatology, geology, etc.) that differently vary in space and time in both ecosystem types. In particular, the ratio (R) between the food zone and the water body surface is much higher in lakes (R is generally comprised between 10 and 20) than in marine environment ($R = 0.32$). As a result, the importance of sediments and allochthonous organic matter is much greater in lacustrian than marine environments. Likewise, continental climate conditions have a larger impact on lacustrian ecosystems.

14.2.7 Methane Production and Consumption

14.2.7.1 What Is Methane?

Methane is an odorless gas whose chemical formula is CH_4. This is a very simple hydrocarbon molecule that represents the most reduced form of carbon. It was discovered as an inflammable gas by the Italian physicist Alessandro Volta in 1776. It is responsible for the phenomena of will-o'-the-wisp and firedamp explosion. CH_4 played a major role in the carbon cycle throughout the Earth history and still is a major component in all anoxic ecosystems: Part of the carbon fixed by primary producers is regenerated as methane by methanogenic *Archaea*. Moreover, as a greenhouse gas, CH_4 is involved in global warming.

14.2.7.2 Methane Sources

It is very difficult to know the exact amount of CH_4 released into the atmosphere; proposed values are orders of magnitude. There is likely that between 500 (Fung et al. 1991) and 600 million tons of CH_4 (Lelieveld et al. 1998) are released each year in the atmosphere.

The sources are both biological (natural) and abiotic (human activities). The values for these CH_4 sources may vary with their authors; those considered by Kvenvolden and Rogers (2005) are listed below:

1. The main biological sources (methanogenic habitats) expressed in Tg.year^{-1} (Tg = 10^{12} g) are from ruminants such as cows, sheep (115), termites (20), rice fields (110), natural wetlands (125), waste discharge areas (55), oceans (10), and freshwater (5). Plants that cover a large part of the Earth, would also represent a significant source of CH_4, but the opinions differ on the amount of CH_4 emitted by plants.
2. The main abiotic sources expressed in Tg.year^{-1} are coal mines (40), losses of natural gas (70), combustion of biomass (40), and methane hydrates (10) (Box 14.7).

It appears, however, that many sources of fossil methane from volcanoes, geothermal areas, and ocean ridges have been underestimated at 45–48 Tg.year^{-1} (Kvenvolden and Rogers 2005).

Box 14.7: Methane Hydrates

Danielle Marty

Definition and Structure (Buffet 2000; Beauchamp 2004)

"Methane hydrates" are the most abundant naturally-occurring "clathrates". "Clathrates" (from the Greek word *Khlatron*, barrier, lattice) are crystalline compounds in which small guest atoms or molecules are physically sequestered within cage-like structures formed by a host molecule or a large complex network of host molecules. These "inclusion compounds" are called "clathrate hydrates" when they contain water molecules, and "gas hydrates" when the enclosed molecules are natural gases. Within natural gas hydrates, methane is the most common encountered guest molecule; however, other gases, such as carbon dioxide or hydrogen sulfide, and less frequently, hydrocarbons, may also be present, they stabilize the aqueous structure, and form gas hydrates.

About 90 % of natural gas hydrates contain methane; these methane hydrates compose a unique class of solid ice-like substances in which a rigid cage structure of hydrogen-bonded water molecules enclose methane molecules, without chemical bonding between the gas molecules and the aqueous structure. The presence of a sufficient number of gas molecules trapped within the network of host water molecules acts to stabilize thermodynamically the crystalline structure through physical bonding; in addition, very

D. Marty (✉)
Institut Méditerranéen d'Océanologie (MIO), UM 110, CNRS 7294 IRD 235, Université de Toulon, Aix-Marseille Université, Campus de Luminy, 13288 Marseille Cedex 9, France

(continued)

Box 14.7 (continued)

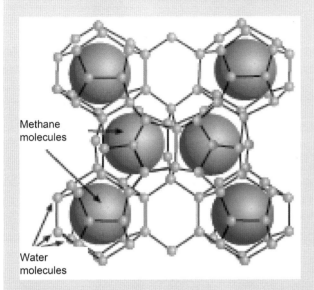

Box Fig. 14.12 Structure of a CH_4 hydrate (www.eawag.ch/medien/pub/.aenews/news_58)

Box Fig. 14.13 The depressurization of CH_4 hydrate leads to dissociation and release of CH_4 gas that can ignite spontaneously

specific conditions of pressure and temperature are required, any shift leading to the destabilization and dissociation of the clathrate (Box Fig. 14.12).

History

An intriguing chlorine hydrate, a crystalline structure in which two components are associated without chemical bonding, was synthetized in 1810 by Sir Humphrey Davy. However, this unusual compound was not expected to be encountered in the natural world, and remained a laboratory curiosity for more than a century, until the dangerousness of natural gas hydrates was evidenced. In the 1930s, hydrates were observed to form spontaneously within natural gas pipelines located in cold regions (Alaska, Siberia, etc.). These deposits were a nuisance for pipeline transportation, due to their tendency to plug the pipelines, leading to explosions and destruction of the equipment. In the 1940s, during submarine drillings for offshore petroleum and gas exploration, the penetration in a clathrate zone did result in hydrate destabilization, massive gas release, and possibly the loss of the well.

In the late 1960s, "solid natural gas" or "methane hydrate" was observed as a naturally-occurring constituent of subsurface sediments in the Siberian Messoyakha gas field; afterward, methane hydrates were evidenced in sub-permafrost sediments in Alaska. Consequently, in the 1970s, scientists began to speculate on the extensive presence of methane hydrate, not only in the permafrost, but also under deep oceans, two geologic configurations where thermobaric conditions favorable for the development and stabilization of gas-hydrate are combined. It was only in 1974 that the first nodules of marine gas hydrates were recovered from the sub-seafloor of the Black Sea (Box Fig. 14.13).

Formation and Stability of Methane Hydrates

A limited range of temperatures and pressures controls formation and stability of gas hydrates. For example, for a pressure corresponding to a water column of 600 m, and a temperature up to 7 °C, water and methane combine to form a stable hydrate; nonetheless, a temperature increase of less than 1 °C could cause the dissociation of the two compounds, which return to their own state, liquid and gas. Methane hydrates are found in two regions:
1. In the continental regions, where the temperatures are low enough to allow the formation of

(continued)

Box 14.7 (continued)

Box Fig. 14.14 Methane bubbles produced by dissociation of hydrate mound on Gulf of Mexico seafloor (www.netl.doe.gov/scng/hydrate/pdf)

permafrost; pressure is low, but temperature very cold, far below 0 °C (−10 to −20 °C). Gas hydrate deposits are generally found a few hundred meters below the ground in circumpolar permafrost regions (Canada, Siberia, or Alaska); average thickness of the deposits is about 400 m, but sometimes extends over more than one kilometer.

2. In marine sediments along continental slopes and rises; gas hydrates are distributed through seafloor sediments around all continent. Gas hydrate deposits occur where the bottom water temperatures are cold (between 0 and 10 °C) and the pressures high (>3 MPa). Marine gas hydrate are typically buried within sediment at water depths below 150–200 m at high latitudes, and more than 500 m at low and middle latitudes; the average thickness of marine hydrate deposits is about 0.5 km (Box Fig. 14.14).

Detection of Methane Hydrates

The presence of gas-hydrate deposits is commonly inferred from the observation of a seismic reflection called "Bottom-Simulating-Reflector" or BSR, because it mimics the seafloor. On the one hand, different sonic velocities induce different seismic reflections, and on the other hand, hydrate, being a solid substance, alters the velocity at which seismic energy is transmitted. Consequently, in the presence of natural submarine gas hydrates, BSR marks an interface between the high velocity hydrate-bearing sediment and, immediately below, the low-velocity free-gas-bearing sediment, that mimics the seafloor "ordinary sediment".

Occurrence of Methane Hydrates

Commonly, natural methane hydrates are dispersed in the sediments, occurring as small crystals, nodules, veins, and fracture fillings, as are much of the oil, coal, and conventional gas; and only occasionally, clathrates occur in massive layers of pure hydrate. Because of this dispersal, and the intrinsic difficulty of the measurements, assessment of gas hydrate reservoir is a difficult task, and estimates were and remain highly speculative. Since the 1970s and the first global estimates of CH_4 in gas hydrates of the order of 3.10^{18} m^3, the range decreased by roughly one order of magnitude per decade, until 10^{15}–10^{16} m^3, while remaining considerable. The global amount of carbon bound in gas hydrates is estimated to about 11,000 G tons for oceanic reservoir and 400 G tons for permafrost reservoir, exceeding the reserves of conventional oilfields, coal, and natural gas on Earth.

Natural Gas Hydrates as Potential Energy Source for the Future

Gas hydrates, with their huge amount of methane locked up, have attracted considerable interest as a potentially vast energy reservoir. However, recovery of methane from hydrates will pose enormous technical challenges, and as gas hydrates mainly occur thinly dispersed in the sediments, extraction will not come cheap, economically, energetically, or environmentally. To date, has been only an example of successful gas-hydrate exploitation, the Messoyakha gas field in Western Siberia, and it is reassessed, because hydrate destabilization and methane release may be secondary, as a response to production of free gas from deeper conventional gas zones.

Gas Hydrates and the Global Climate Change (Macdonald 1990)

As CH_4 is a greenhouse gas about 20 times more potent than CO_2, any massive and uncontrollable emission of CH_4 through the destabilization of gas hydrates deposits could have an impact on CH_4 atmospheric concentration and greatly multiply the greenhouse effect.

14.2.7.3 Consequences of Methane Emissions: Greenhouse Effect

CH_4 is a major greenhouse gas that plays a significant role in climate change. Indeed, for the same mass as CO_2, CH_4 absorbs 24–30 times more energy; 1 kg CH_4 heats the planet 24–30 times more than 1 kg CO_2. The CH_4 contribution to the greenhouse effect is at about 12 %.

Through the analysis of gases trapped in tiny gas bubbles preserved in ice cores from Greenland and Antarctica, it was observed that an increase in the amounts of CO_2 and CH_4 released into the atmosphere since the beginning of the preindustrial era. This increase continues at about the same rate as the intensification of human activities: development of natural gas exploitation, proliferation of waste discharges and garbage, increase of flooded agricultural areas (especially rice fields), increase in livestock.

The atmosphere CH_4 content that was estimated between 0.5 and 0.7 ppmv (parts per million by volume) 15,000 years ago, was at 715 ppmv before the industrial era and steadily increased since then: 1,720 ppmv in 1990, 1,745 ppmv in 2001, and 1,770 ppmv in 2005.

14.2.7.4 Methanogenesis

Microorganisms : Methanogenic *Archaea*

The CH_4 biological production is due to the activity of strictly anaerobic *Archaea*: methanoarchaea. Methanoarchaea are all included in the phylum *Euryarchaeota* that is described in Sect. 6.6.1. The main CH_4 production pathways are presented in Sect. 3.3.2 and in the book of Vandecasteele (2008).

All methanogenic microorganisms belong to the phylum *Euryarchaeota* that includes four classes and six orders (*cf.* Sect. 6.6.1). It is only recently that *Methanomassiliicoccus luminyensis* has been isolated (Dridi et al 2012). The latter belongs to a seventh order which is still under consideration.

Most of methanogens grow at pH 6–8. They are able to grow in a wide range of temperature (4–110 °C), with psychrophilic, mesophilic, and thermophilic species. For example, during a field campaign carried out by the submersible Alvin of the coast of Mexico, one hyperthermophilic species of the genus *Methanopyrus*, whose optimum growth temperature is 98 °C, was isolated at 2,000 m depth. Halophilic and hyperhalophilic species were also described.

Methanogens Habitats

Methanogens live in all wet ecosystems, highly reductive (redox potential < −300 mV), organic matter rich, and completely free of oxygen. They are very sensitive to oxygen; low dioxygen partial pressures (5–10 ppm) cause lethal and irreversible dissociation of some enzymatic complexes (*e.g.*, F_{420}-hydrogenase) involved in methanogenesis. Thus, their isolation, study and conservation require the implementation of highly stringent deoxygenation techniques: anaerobic jars, Hungate tubes, anaerobic chambers maintained reduced atmosphere (5 % H_2 in the presence of palladium, *cf.* Sect. 17.8.2). They are also inhibited by some pollutants (carbon tetrachloride, chloroform), light and 2-bromoethanesulfonic acid (structural analogue of coenzyme M).

Fully anoxic environments, favorable to methanogens development, are abundant and very diverse:

1. Sediments: wetlands, ponds, lakes, oceans, rice fields, flooded anoxic zones.
2. Rumen of ruminants (cows, sheep, deer, elk, camel).
3. Cecum of many herbivores (horse, rabbit, etc.).
4. Human intestines (about one third of healthy adult humans excrete methane), most mammals, some insects (digestive tract of insects), etc.
5. Hydrothermal vents along ocean ridges.
6. Hot springs (*e.g.*, those of Yellowstone National Park).

Processes of industrial interest, using bioreactors, rely on the activity of methanogenic *Archaea* (methanation, *cf.* Sect. 16.5.4); the same happens for the anaerobic treatment of wastewaters and related sludge, and for the treatment of solid wastes in waste discharges as well.

Role of Methanogenesis During Anaerobic Biodegradation of Organic Matter

The first step of organic material degradation starts with hydrolysis of complex polymers leading to the formation of monomers which, by fermentations, provide substrates to be used by methanogenic *Archaea* (H_2, formate, acetate, alcohols, methylated compounds) in the absence of sulfate. Indeed, it is known that in the presence of sulfate (with the exception of methylated compounds), sulfate-reducing bacteria may outcompete methanogens. The heterotrophic methanogenic archaea develop most often in close association with other organisms that supply hydrogen. The consumption of hydrogen, and to a lesser extent of acetate by methanogenic archaea, is a crucial step in the complete oxidation of organic matter under anaerobic conditions (*cf.* Sect. 3.3.3).

The Different Methane Production Pathways

There are three pathways for CH_4 formation (Fig. 14.21):

1. Through anaerobic respiration: most methanogens use CO_2 as terminal electron acceptor (*e.g.*, *Methanobacterium*, *Methanococcus*). Electron donors are hydrogen or formate. Changes in free energy, $\Delta G^{\circ\prime}$, expressed in kJ.mol^{-1} CH_4, are given in brackets:

$$CO_2 + 4H_2 \rightarrow CH_4 + 2H_2O \quad (-131)$$

Some methanoarchaea are able to oxidize hydrogen using methanol as terminal electron acceptor (*e.g.*, *Methanosphaera*, *Methanomassiliicoccus*):

$$CH_3OH + H_2 \rightarrow CH_4 + H_2O \quad (-113)$$

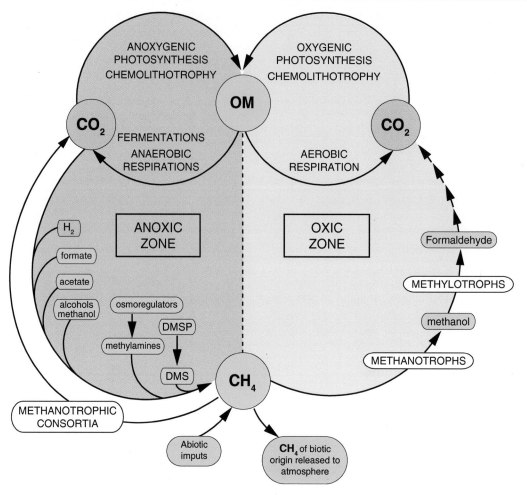

Fig. 14.21 Production and transformation of methane in the carbon cycle. Methane production from CO_2 (CO_2 respiration) or simple organic compounds by methanogenic archaea; methane oxidation to CO_2 by methanotrophic and methylotrophic bacteria under oxic conditions, or by a bacterial consortium under anoxic conditions (Drawing: M.-J. Bodiou)

2. By a process of pseudofermentation (disproportionation) from compounds containing one atom of carbon (C_1 compounds):

   Methanol (*e.g., Methanosarcina, Methanolobus*):

   $$4CH_3OH \rightarrow 3CH_4 + CO_2 + 2H_2O \; (-130)$$

   Formate (*e.g., Methanobacterium, Methanococcus*):

   $$4HCOO^- + 2H^+ \rightarrow CH_4 + CO_2 + 2HCO_3^- \; (-119.5)$$

   Trimethyamine (*e.g., Methanosarcina, Methanolobus*):

   $$4\,(CH_3)_3NH^+ + 6H_2O \rightarrow 9CH_4 + 3CO_2 + 4NH_4^+ (-76)$$

   Dimethylsulfide (*e.g., Methanomethylovorans*):

   $$(CH_3)S + H_2O \rightarrow 1.5CH_4 + 0.5CO_2 + H_2S \; (-73.8)$$

3. From acetate: acetoclastic reaction (disproportionation) (*e.g., Methanosarcina, Methanosaeta*):

   $$CH_3COO^- + H_2O \rightarrow CH_4 + HCO_3^- \; (-36)$$

Given the energetic efficiency of reactions described earlier, methanogenesis will be efficiently expressed only if electron acceptors such as dioxygen, nitrate, iron, manganese, and sulfate, are absent or present in limiting concentrations, or if the corresponding metabolisms are disadvantaged by physico-chemical conditions of the concerned ecological niches.

Biological Methane Production Depending on Biotope

The CH_4 formation pathway and its intensity vary as a function of biotope. It is possible to group the CH_4-producing ecosystems into three categories (Garcia et al. 2000):

1. Marine and lacustrian sediments, wetlands, ponds, rice fields, sludge digesters where organic matter is fully degraded. In the wastewater treatment in bioreactors

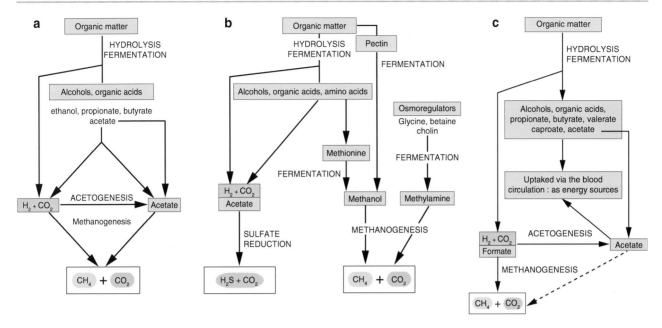

Fig. 14.22 Pathways of methane formation in environments. (**a**) In bioreactor or in lacustrine ecosystems; (**b**) In marine ecosystems; c: In the digestive tract or in the rumen (Drawing: M.-J. Bodiou)

(Fig. 14.22a), the anaerobic biodegradation is ensured by cooperation between different microbial communities (*cf.* Sect. 16.4.3). Macromolecules are hydrolyzed into simple sugars, amino acids, peptides, fatty acids, and alcohols that are further fermented or oxidized to produce, depending on the case, acetate, CO_2 and hydrogen. Methanoarchaea consume hydrogen whose partial pressure changes from 1 to 10^{-3}–10^{-4} atm, thereby favoring the organic-matter oxidation process. This is an example of interspecies hydrogen transfer between synthrophic organisms (obligatory producers of H_2) and methanoarchaea as H_2 consumers; 70 % CH_4 derive from CO_2 and 30 % from acetate. The produced CH_4 can be collected, burned, or used as fuel for heating and power facilities. If the methanogenesis processes are not normally expressed, there is H_2 overproduction and accumulation of volatile fatty acids, that jeopardize the process functioning; monitoring hydrogen concentration is thus a good way to control the process functioning. Anaerobic organic-matter mineralization in limnic ecosystems is roughly comparable with that described in a sludge digester where methanogenesis is very active. In contrast, methanogenesis is of minor importance in sulfate rich biotopes, such as marine sediments where, despite everything, hydrogenotrophic methanoarchaea are still frequently isolated.

In anoxic marine ecosystems (Fig. 14.22b), hydrogen is consumed by sulfate-reducing bacteria which, thermodynamically, are particularly effective competitors of methanogens for hydrogen and acetate utilization. In such habitats, the major precursors of CH_4 are methylated substrates such as methanol and methylamines. It is noted that, to date, no sulfate-reducing bacteria able to use methylamines have been described. It follows from this observation that these compounds are essential substrates for methanogenic archaea in marine sediments and hypersaline habitats (Ollivier et al. 1994).

2. Rumen and the intestinal tract of most living beings (animals, men) (Box 14.8) where the organic matter degradation processes are incomplete and where the derived intermediate products (such as volatile fatty acids) are absorbed by the bloodstream and contribute to nutritional feeding. The rumen (Fig. 14.22c), "true fermentation chamber," is a very reductive environment (OR potential varies between −250 and −400 mV). This ecosystem hosts a highly complex microflora, organized as a food chain, consisting of prokaryotes (about $10^{10}.cm^{-3}$), unicellular heterotrophic eukaryotes ($10^6.cm^{-3}$), cellulolytic anaerobic Fungi. This microflora ensures the degradation of plant materials (cellulose, hemicellulose, pectin) consumed by animals that otherwise are not able to digest them. Enzymatic hydrolysis of these polymers by microorganisms releases oligosaccharides that, through fermentation, lead to the formation of volatile fatty acids (propionate, butyrate, caproate, valerate), hydrogen and CO_2. Much of the acids is absorbed by the animal bloodstream and represents an important energy source. The methanogenesis which is essentially based on hydrogen (80 %) is expressed by the transfer of hydrogen between two microbial communities,

hydrogen-producing and hydrogenotrophic prokaryotes (mostly methanoarchaea). These organisms lower the partial pressure of hydrogen produced by hydrogen-producing bacteria and favor essential processes such as cellulose degradation. By reducing the volume of gas produced during digestion, methanogenesis also has a beneficial effect on the animal. The CH_4 origin and production in the human digestive tract is presented in Box 14.8.

3. In the absence of organic matter (e.g., in hot springs), methanogenesis is exclusively fueled by hydrogen originating from geochemical processes.

Box 14.8: Methanogenic Archaea of the Human Intestinal Microbiota

Joël Doré

The human intestinal tract is a mesophilic, physico-chemically stable biotope, receiving complex dietary polymers and endogenous mucins that supply carbon and energy for microbial fermentations (Box Fig. 14.15). As for any complex anaerobic ecosystem on the planet, hydrogen is one of the dominant fermentation gases with an output of approximately 300 mL.g^{-1} fermented substrate (Macfarlane and Gibson 1994).

Box Fig. 14.15 Schematic representation of the intestinal lumen illustrating the biodiversity of the microbiota of the digestive tract (Photo courtesy of Biocodex)

J. Doré (✉)
Unité Micalis (INRA, UMR 1319), Paris, France

This sets ideal conditions for the establishment of a hydrogenotrophic and methanogenic microbiota, although a fraction of the hydrogen produced by fermentative procaryotes is excreted as breath-gas and flatus (Christl et al. 1992). Methanogenic archaea, responsible for this metabolism are nonetheless not always found in the human colon, and only a fraction of the population (30–50 %) harbors them. Long before methanogenic archaea could be cultured in the laboratory, the ability to detect methane in breath-gas had led to the distinction of methane excretors and non-methane excretors within the human population, based on a threshold of 1 ppm corresponding to the partial pressure of methane in the air we breathe (Bond et al. 1971). The factors that determine the status of humans toward methane-excretion remain unknown to date but the analysis of breath-gas did suggest an impact of familial factors.

The enumeration of methanogenic archaea (Miller and Wolin 1986) was made possible with the use of culture conditions under 2 atm of pressurized hydrogen and CO_2. This allowed to confirm the direct relationship between the presence of dominant methanogens ($>10^8$.g^{-1} faeces) and the methano-excretor status. Once established, neither population levels of methanogenic Archaea nor the status of methane excretion appear influenced by the diet (Miller and Wolin 1986). Yet among the three recently described enterotypes of the human population (Arumugam et al. 2011), methanogenic archaea were observed associated with only one ecological arrangement also dominated by *Firmicutes* that gather major hydrogen producers of the gut ecosystem.

Isolates from two species of methanogenic Archaea have been obtained from human intestinal contents. The predominant species in methane-excretors is *Methanobrevibacter smithii* (Miller and Wolin 1986; Eckburg et al. 2005), that converts H_2 and CO_2 to methane. The other species, *Methanosphaera stadtmaniae,* is present at lower population levels and uses H_2 to reduce methanol to methane. These two species are obligate hydrogenotrophs that use as substrates products of prokaryotic fermentations. As for the rumen, acetoclastic methanogenesis appears too slow and energetically unfavorable to allow the corresponding archaea to establish in dominance in the human colon (Miller and Wolin 1986). The genome of *Methanobrevibacter smithii* was recently sequenced (Samuel et al. 2007). This did highlight potential

(continued)

Box 14.8 (continued)

adaptations for persistence in the intestine such as the production of surface glycans analogous to mucus structures, adhesin-like proteins and an enzymatic repertoire to efficiently use nitrogen sources and products of saccharolytic prokaryotes.

Although they only come into the ecological framework at the end of the trophic chain, hydrogenotrophic microorganisms play an essential role in favoring a more complete oxidation of substrates and thereby maximizing the total ATP harvest of the microbiota (Wolin 1974). Hydrogen is nonetheless always detectable in the colon (Bond et al. 1971) and concentrations of short chain fatty acids are not significantly different whatever the status of the individual. This could come from the co-existence of different metabolisms in the proximal colon (acetogenesis) and the distal colon (methanogenesis) conditioned by substrate availability but also by differences in pH and osmolarity. In turn, being methano-excretor or not, and hence the enterotype of an individual could have an incidence on the prevalence of different cellulolytic species present in the human digestive tract.

The production of methane by methanogenic archaea is only one of the three major pathways of H_2 metabolism in the colon. In non-methane-excretors, alternative hydrogenotrophic pathways were evidenced among which reductive acetogenesis and sulfate reduction (Bernalier et al. 1996).

Finally, no causal link was ever established between the presence of dominant methanogenic archaea and intestinal pathologies. Cancer patients repeatedly showed a higher prevalence of methano-excretors, but these proportions return to normal after surgical resection of tumors.

Methane as Energy Source

Fermentation processes of organic matter from animal or plant (from food industries, domestic wastes, urban wastewater) origin, yields a mixture of combustible gases, known as "biogas."

14.2.7.5 Photochemical Methane Destruction

Much of CH_4 emissions in the atmosphere is photo-oxidized (Vandecasteele 2008), mainly in the troposphere (first 10–12 km altitude) where CH_4 is oxidized by hydroxyl radicals (420 ± 80 Mt year^{-1}) into carbon monoxide, ozone, and water vapor. These reactions correspond to an average lifetime of CH_4 at 8–12 years. In the stratosphere (12–15 km altitude), CH_4 consumption amounts to 10 ± 5 Mt year^{-1}.

14.2.7.6 Methane Removal by Methanotrophic Bacteria in the Presence of Oxygen

Most of the produced CH_4 is not released into the atmosphere but is oxidized near the production areas. For example, for CH_4 produced in rice fields, the percentage of gas released into the atmosphere with respect to the overall production would not exceed 17 %. Similarly, CH_4 emissions from marine sediments are very low compared with the total production (anaerobic and aerobic biodegradation in sediment and water column, respectively). Biological CH_4 removal is carried out by methanotrophic bacteria for which CH_4 is a carbon and energy source (Hanson and Hanson 1996). Methanotrophs, that also use methanol, are called methylotrophs, i.e., organisms that use methylated compounds.

Methanotrophs are Gram-negative and strict aerobes, omnipresent wherever there is a source of CH_4 and dioxygen. They grow on the surface of soil and sediment, in water, around roots of plants in the rhizosphere. Some of these microorganisms form exospores that are resistant to heat and desiccation (*Methylosinus*, *Methylocystis*), while others form cysts which are resistant to desiccation (*Methylomonas*, *Methylobacter*).

CH_4 biodegradation in the presence of dioxygen results from four successive oxidations:

$$CH_4 \rightarrow \underset{(\text{Methanol})}{CH_3OH} \rightarrow \underset{(\text{formaldehyde})}{HCHO} \rightarrow \underset{(\text{formate})}{HCOOH}$$

The first step is the action of a monooxygenase that converts CH_4 to methanol; this enzyme may be linked to the cell membrane or soluble in the cytosol. The lack of specificity of the soluble methane-oxidase enables it to oxidize more than 150 different compounds: saturated or unsaturated hydrocarbons (ethylene, propylene), aromatics (naphthalene), chloroform, dichloroethane, tri- and dichloroethylene, etc. These oxidative capacities, especially xenobiotic oxidation, have been used in the decontamination of polluted biotopes, e.g., soils contaminated with organochlorines (*cf.* Sect. 16.7.3).

Other steps of methanol oxidation supply electrons to an electron transport chain for ATP synthesis. The assimilation of compounds with a single carbon atom (C1 compound) occurs at the level of formaldehyde through two distinct pathways.

There are three types of methanotrophic bacteria that essentially differ by their cell membrane organization, their mode of using formaldehyde, their taxonomic position (Chistoserdova et al. 2005). In type I methanotrophs (*Methylococcus*, *Methylobacter*, *Methylomonas*), membrane system consists of vesicles with a disc shape distributed throughout the cell. Formaldehyde is assimilated by a pathway known as ribulose monophosphate (*cf.* Sect. 3.4.2). Taxonomically, type I methanotrophs have been affiliated to the group of gammaproteobacteria, based on the analysis of the gene encoding RNA16S. Among type II methanotrophs (*Methylosinus*, *Methylocystis*), the inner membranes are arranged at the cell periphery, and formaldehyde is incorporated into an amino acid, serine (*cf.* Sect. 3.4.2). They often contain a soluble monooxygenase, in addition to a particulate monooxygenase, and belong to the group of alphaproteobacteria. The third type of methanotrophs belongs to the genus *Methylococcus* (deltaproteobacteria) and combines the features of both former types: disc-shaped flattened vesicles and ribulose monophosphate pathway, but elements of the serine cycle and a soluble monooxygenase are also present.

14.2.7.7 Anaerobic Degradation of Methane

CH_4 can also be oxidized anaerobically by anaerobic prokaryotes (Box 14.9).

Box 14.9: Anaerobic Oxidation of Methane (AOM)

Vincent Grossi

It has long been considered that the degradation of methane in natural environments could only be driven by aerobic microbial processes. During the last decades, however, diverse studies have demonstrated that methane can be oxidized in the absence of oxygen and that anaerobic oxidation of methane (AOM) essentially controls the biogeochemical cycle of this potent greenhouse gas.

Following different geochemical observations indicating the removal of methane within marine anoxic sediments and seawater (e.g., Barnes and Goldberg 1976), field and laboratory experiments led to the hypothesis that, in the marine environment, AOM is carried out by a consortium of methanogenic archaea and sulfate-reducing bacteria (SRB), despite the low thermodynamic energy yield from this reaction (Hoeler et al. 1994). In this model of obligate syntrophic interaction, SRB use the hydrogen produced during the oxidation of methane by methanogenic Archaea, which perform the reverse reaction of the reduction of CO_2 to methane (referred to as "reverse methanogenesis"). Under such conditions, the final reaction becomes thermodynamically favorable (Box Scheme 14.1).

The principle of "reverse methanogenesis" has been extensively discussed and generally adopted, but the complete mechanism of sulfate-dependent AOM remains incompletely understood. A key issue is the nature of the intermediate produced by the methanotrophic archaea and scavenged by the sulfate-reducing bacterial partner. Various alternatives have been proposed, often with typical methanogenic growth substrates serving as intermediates (such as H_2, formate, acetate, or methanol) that could eventually feed several SRB populations (see example in Box Scheme 14.2 showing the use of acetate and hydrogen by two different populations of SRB; Hoeler et al. 1994; Valentine and Reeburgh 2000). Other studies have suggested that the SRB partner uses (in addition to H_2) environmental CO_2 as a carbon source, or methyl sulfides produced by the archaea (e.g., CH_3SH) as electron donor (Moran et al. 2008).

Confirmation of the existence of anaerobic methanotrophs (ANME) at AOM sites came from the measurement of stable carbon isotopes in lipid biomarkers specific to archaea (see examples of structures in Box Scheme 14.3a), which appeared strongly depleted in ^{13}C compared with the source methane ($-130‰ < \delta^{13}C$ archaeal lipids $< -50‰$), indicating that methane was the carbon source for this archaeal community (Hinrichs et al. 1999; Michaelis et al. 2002). Phylogenetic (16S rRNA) analysis of the same samples further revealed the presence of novel archaeal phylotypes likely linked to methane oxidation, whereas FISH microscopy and ion microprobe mass spectrometry were elegantly used to confirm the occurrence of consortia of ANME and SRB mediating the AOM process (Boetius et al. 2000; Orphan et al. 2001; Strous and Jetten 2004).

The stable carbon isotope composition of other lipid biomarkers proposed to derive from the SRB partners involved in AOM (see examples of structures in Box Scheme 14.3b) further supports a syntrophy between ANME and SRB. These lipids are usually less depleted ($-90‰ < \delta^{13}C < -40‰$) than the archaeal lipids. Moreover, differences in biomarker profiles from one AOM site to the other typify distinct anaerobic microbial consortia involved in AOM (Hinrichs et al. 2000; Michaelis et al. 2002; Elvert et al. 2003; Blumenberg et al. 2004). In such settings, the diversity of ^{13}C-depleted biomarkers of prokaryotic origin are indicative of methane carbon transferring via ANME to several SRB populations (see example in Box Scheme 14.2) and then, eventually, to other microbial populations (Pancost and Sinninghe Damsté 2003).

Investigations based on 16S rRNA or *mcrA* (the key gene of methanotrophic and methanogenic archaea) gene phylogeny have shown that three distinct clusters

---

V. Grossi (✉)
Laboratoire de Géologie de Lyon: Terre, Planètes,
Environnement, UMR CNRS 5276, Université Claude Bernard
Lyon 1, 69622 Villeurbanne Cedex, France

(continued)

Box 14.9 (continued)

Reactions		Organisms	$\Delta G^{0'}$ (kJ)
$CH_4 + 2 H_2O \longrightarrow CO_2 + 4 H_2$		Archaea	+ 131
$SO_4^{2-} + 4 H_2 + H^+ \longrightarrow HS^- + 4 H_2O$		SRB	− 156
$\Sigma: CH_4 + SO_4^{2-} \longrightarrow HCO_3^- + HS^- + H_2O$		Consortium Archaea-SRB	− 25

Box Scheme 14.1 AOM coupled to sulfate-reduction

Reactions	Organisms	$\Delta G^{0'}$ (kJ)
$2 CH_4 + 2 H_2O \longrightarrow CH_3COOH + 4 H_2$	Archaea	+ 166
$SO_4^{2-} + 4 H_2 + H^+ \longrightarrow HS^- + 4 H_2O$	SRB 1	− 152
$CH_3COOH + SO_4^{2-} \longrightarrow 2 HCO_3^- + H^+ + HS^-$	SRB 2	− 47
$\Sigma: CH_4 + SO_4^{2-} \longrightarrow HCO_3^- + HS^- + H_2O$	Consortium Archaea-SRB	− 33

Box Scheme 14.2 Alternative model for AOM coupled to sulfate-reduction

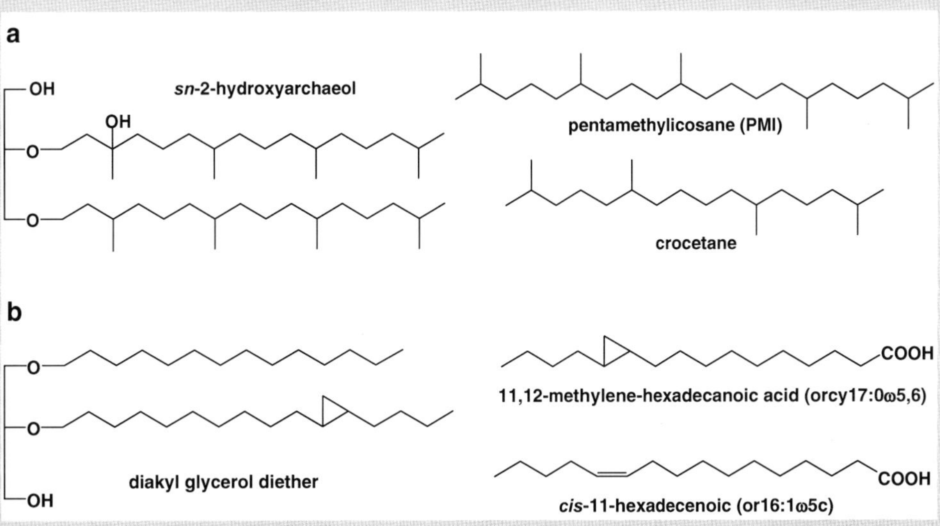

Box Scheme 14.3 (a) Examples of lipid structures produced by ANME. (b) Examples of lipid structures produced by the BSR partners of ANME

of Euryarchaeota, namely ANME-1, ANME-2, and ANME-3, can mediate AOM in the marine environment (Reeburgh 2007; Knittel and Boetius 2009 and references therein). These groups are not monophyletic and several sub-groups of ANME-2 with low gene sequence similarity have also been described (Martinez et al. 2006). The ANME-2 and ANME-1 groups are the most frequently encountered, and are distantly or closely related to the orders *Methanosarcinales* and *Methanomicrobiales*. The ANME-3 group appears mainly related to *Methanococcoides* spp. (Knittel et al. 2005). The SRB partners of ANME are mostly affiliated with the δ-Proteobacteria subdivision and are close relatives of *Desulfosarcina/Desulfococcus* species (the so-called DSS cluster; Boetius et al. 2000).

A strong research effort during the last decade produced significant improvements in our understanding of the AOM process, but further highlighted variations in the distribution, diversity, physiology, and morphology of anaerobic methanotrophic consortia (Strous and Jetten 2004; Knittel and Boetius 2009 and references therein). The long-term *in-vitro* incubation of marine sediments showing AOM activity led to the partial enrichment of methanotrophic consortia (Nauhaus et al. 2007) but no defined co-culture of ANME-SRB is available to date. Alternatively, the coupling of methane oxidation to electron acceptors energetically more favorable than sulfates was demonstrated. AOM coupled to the reduction of iron- or manganese-oxides, presumably mediated by a microbial consortium as for sulfate-dependent

(continued)

Box 14.9 (continued)

Box Scheme 14.4 AOM coupled to reduction of sulfate to zero-valent sulfur and associated disulfide disproportionation

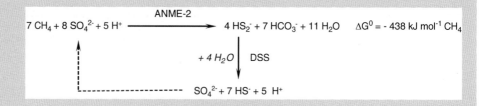

Box Scheme 14.5 AOM by nitrite reduction and dismutation (the "intra-aerobic" pathway)

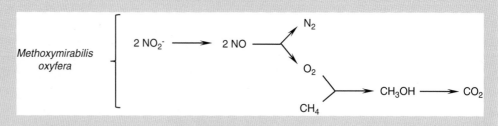

Box Scheme 14.6 AOM coupled to nitrate reduction

$$CH_4 + 4\,NO_3^- \xrightarrow[\text{Methanoperedens nitroreducens}]{\text{ANME-2d}} CO_2 + 4\,NO_2^- + 2\,H_2O \quad \Delta G^0 = -503 \text{ kJ mol}^{-1}\,CH_4$$

AOM (cf. Box Schemes 14.1 and 14.2), has been reported (Beal et al. 2009), but the organisms responsible for this process and the metabolic intermediates involved have yet to be determined.

The long-term enrichment of cultures mediating AOM and the combination of a large array of geochemical, molecular biological, and microbiological techniques led, however, to the recent discovery of three alternative mechanisms of AOM in which a single microorganism, either a bacteria or an archaea, can mediate both the oxidative and reductive processes of AOM using previously undescribed metabolic pathways. After eight years of culturing microorganisms enriched from sea-floor sediments, it was demonstrated that AOM coupled to sulfate-reduction might not be an obligate syntrophic process (as shown in Box Schemes 14.1 and 14.2) but may be carried out by archaea of the ANME-2 clade alone (Milucka et al. 2012). These microorganisms employ an unusual strategy to concomitantly mediate both AOM and dissimilatory sulfate-reduction, by reducing sulfate to zero-valent sulfur (S^0) which is likely released as disulfide (HS_2^-, Box Scheme 14.4). By doing so, a bacterial partner is not required for the shuttling of electrons or metabolites, and most of the energy derived from both reactions benefits the archaea. Additionally, the disulfide produced by the archaea can be transformed (i.e., disproportionated) into hydrogen sulfide (HS^-) and sulfate by associated bacteria from the DSS cluster (Box Scheme 14.4; Milucka et al. 2012), although this association of ANME-2 and bacteria is not an obligate cooperative metabolism, but rather an association of convenience.

Bioreactors capable of AOM coupled to denitrification have been obtained from freshwater sediments (Raghoebarsing et al 2006; Ettwig et al. 2008) and resulted in enrichment of the unusual bacterium "Candidatus Methoxymirabilis oxyfera" which couples AOM to the reduction of nitrite (NO_2^-) without involving archaea. M. oxyfera is able to convert nitrite to nitric oxide (NO) and to dismutate NO into nitrogen (N_2) and oxygen (O_2), which is then used to oxidize methane (Box Scheme 14.5; Ettwig et al. 2010). Thus, despite its anaerobic lifestyle, this bacterium is thought to employ an intra-aerobic pathway for methane oxidation in anoxic environments.

In parallel with the enrichment of M. oxyfera, bioreactors coupling AOM to denitrification also resulted in the enrichment of a novel ANME lineage, ANME-2d (Raghoebarsing et al 2006; Hu et al. 2009). It was recently demonstrated that a single ANME-2d population (Candidatus "Methanoperedens nitroreducens") can independently couple AOM to nitrate reduction to nitrite (Box Scheme 14.6; Haroon et al. 2013). Interestingly, the nitrite produced by ANME-2d is reduced to dinitrogen gas (N_2) by an associated anaerobic ammonium-oxidizing bacterium (anammox from the genus Kuenenia), which outcompetes "M. oxyfera" in the presence of ammonium (Haroon et al. 2013).

Due to their recent discovery, the environmental significance of AOM coupled to electron acceptors energetically more favorable than sulfates (i.e., nitrate, nitrite, or metal oxides), and the environmental contribution of microorganisms performing AOM alone, remain relatively unexplored and need to be further documented. Our understanding of AOM functioning keeps evolving and progressively reveals a strong diversity of the modes of AOM existing in nature. Ongoing research efforts will undoubtedly yield new exciting discoveries concerning this process.

14.3 Nitrogen Cycle

14.3.1 Introduction

Nitrogen is a chemical element with symbol N and atomic number 7. Nitrogen is present in natural state in the form of two stable isotopes (^{14}N, ^{15}N). The standard ($^{15}N/^{14}N$)$_{atm}$ for the atmosphere is at 3.613×10^{-3}.

Nitrogen is one of the major nutrients used by plants and animals. Combined nitrogen refers to all forms of nitrogen other than dinitrogen (N_2). Organic nitrogen (N_{org}) is nitrogen which is bound to carbon to form complex molecules. In living beings, organic nitrogen is present mainly in the form of proteins (nitrogen constitutes 18 % of proteins), nucleic acids (DNA, RNA), in certain polysaccharides such as chitin (shell of arthropods) or peptidoglycan (membrane wall of bacteria). The nitrogen content in plants can be particularly low, less than 0.5 % of the total mass, while the content can reach 15 % in bacteria. Living organisms obtain nitrogen from their surrounding environment as dissolved chemicals or by predation. Organic nitrogen returns back to the environment in the form of organic matter (excretion – urine, feces, mucus – or dead organisms). The atmosphere consists of 78 % of dinitrogen. Nitrogen means "lifeless," its name comes from its low chemical reactivity. Indeed, very few organisms are able to directly use dinitrogen as a nitrogen source, despite its wide availability.

Nitrogen is present in various reservoirs on Earth (Table 14.5, Fig. 14.23). Inputs and outputs from the different reservoirs are due to physical, chemical, and biological reactions. The nitrogen cycle is complex and consists of multiple redox reactions in which nitrogen passes through many states of oxidation, varying from –III for ammonium to +V for nitrate (Fig. 14.24).

The inter-conversion between the different nitrogen forms represents the biogeochemical cycle of nitrogen which is mainly made of biological processes in which microorganisms play a predominant role.

The nitrogen biogeochemical cycle comprises a series of redox reactions for transforming nitrogen compounds. This complex cycle is shown in Fig. 14.25.

The nitrogen cycle is of particular interest since it will regulate the amount of nitrogen available for the food chain. Nitrogen is assimilated as ammonium (−III) or nitrate (+V) (Fig. 14.25, *pathway 1a and 1b*); ammonium is derived from the mineralization of organic matter or from the reduction of nitrate. Dinitrogen is relatively inert and may be directly used only by some microorganisms in a process called nitrogen fixation (Fig. 14.25, *pathway 3*) that converts inorganic nitrogen into biologically available substrates: ammonium (NH_4^+) and its conjugate acid, ammonia (NH_3). This is the main mechanism for the introduction of nitrogen into the biosphere. Ammonium and ammonia can also be converted to nitrate (NO_3^-) and nitrite (NO_2^-) during the nitrification (Fig. 14.25, *pathway 4*), an aerobic process which is performed by specialized microorganisms. During denitrification (Fig. 14.25, *pathway 5*), the nitrate is transformed into gaseous compounds: nitric oxide (NO), nitrous oxide (N_2O), and finally dinitrogen (N_2) which is quickly released back to the atmosphere. It is important to note that different processes can lead to the same product (Herbert 1999).

Table 14.5 Different pools of nitrogen (from Galloway et al. 2003)

Réservoirs	Nitrogen Tg N year^{-1}	Percentage (%)
Atmosphèr N_2	3,950,000,000	79.5
Sediment	999,600,000	20.1
Ocean N_2	20,000,000	0.4
Ocean NO_3^-	570,000	0.0
Organic soils	190,000	0.0
Terrestrial biotope	10,000	0.0
Marine biotope	500	0.0

14.3.2 Organic Nitrogen Assimilation of Nitrogen Compounds

Microorganisms preferentially uptake ammonium while plants and some microorganisms preferentially assimilate nitrate. Animals use nitrogen only in its organic form (amino acid).

14.3.2.1 Ammonium Assimilation

NH_4^+, whatever its origin, will be taken in charge within the cell by an enzymatic system, GS-GOGAT, consisting of two enzymes. The first enzyme (GS, glutamine synthetase) adds a group NH_3 to a molecule of glutamate (*cf.* Sect. 3.4.4). The resulting glutamine transfers one of its nitrogenous groups to α-ketoglutarate under the action of the second enzyme GOGAT (glutamine α-ketoglutarate aminotransferase). This transaminase allows the transfer of a group ammonium to another molecule of α-ketoglutarate to form two molecules of glutamate (Fig. 14.26). Glutamate serves as a template for the synthesis of most amino acids or for the reactions of transamination with keto acids. Unlike animals and plants (which in general can only synthesize about 10 amino acids known as essentials), a number of microorganisms are capable of synthesizing the 20 amino acids.

14.3.2.2 Nitrate Assimilation

Assimilatory nitrate reduction is an energy-consuming reaction which results in the formation of ammonium necessary for cellular biosynthesis. This reduction is usually a preferential pathway in higher plants, but occurs only in the absence of ammonium in microorganisms. The involved

Fig. 14.23 Global nitrogen cycle and fluxes between different compartments of the Earth (Modified from Planète Terre, Bourque et Dansereau, Laval University Québec)

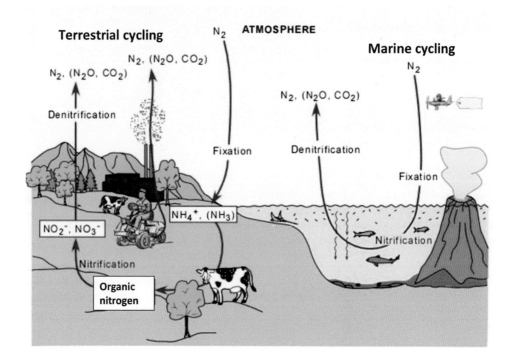

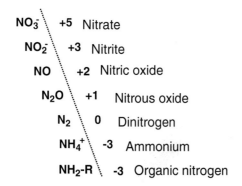

Fig. 14.24 Various oxidation states of nitrogen

nitrate reductase is a soluble enzyme known as Type B-nitrate reductase (assimilatory nitrate reductase, NAS Table 14.6), which is inhibited by chlorate and the presence ammonium.

14.3.2.3 Nitrogen (N_2) Fixation

The fixation of dinitrogen (N_2), which relies on several mechanisms, is done by free-living micro-organisms for by microorganisms living in symbiosis with certain plants (*cf.* Sect. 3.4.4).

The N_2 fixation process was discovered in 1880. Organisms capable of N_2 fixation, called diazotrophs, are all prokaryotes (*Bacteria* and *Archaea*) and are phylogenetically very diverse (Figs. 14.27a, b). Indeed, diazotrophs may belong to groups as diverse as proteobacteria, Gram-positive bacteria, cyanobacteria, or *Archaea*. Cyanobacteria have long been considered the main N_2 fixers in the open ocean and in many marine ecosystems. However, the importance of other diazotrophic groups such as gamma and deltaproteobacteria has been demonstrated. These microorganisms are diverse in terms of phylogeny but also of metabolism. Indeed, multiple sources of carbon and energy can be used by diazotrophs, which can be autotrophic or heterotrophic, phototrophic (anoxygenic photosynthetic cyanobacteria or bacteria), or chemotrophic.

The high diversity of these organisms allows them to colonize various environments such as the open ocean, the deep ocean, microbial mats, soils, digestive tracks of some invertebrates, or to live in symbiosis with some legumes. Bacteria in the family Rhizobiaceae, primarily those in the genus *Rhizobium*, can infect the roots of legumes resulting in the formation of nodules (*cf.* Sect. 11.3). The nodules of the host plant offer a micro-habitat that is exceptionally favorable to the bacteria, by providing carbon substrates from photosynthesis. In the nodules, bacteria fix dinitrogen and provide the host plant with labile nitrogen in the form of ammonia (NH_3). This beneficial association between legume and bacteria, known as mutualistic symbiosis, allows the fixation of atmospheric nitrogen by symbiotic plants, estimated between 300 and 400 kg of dinitrogen per hectare and per year.

Despite their high diversity, all diazotrophs use a similar enzymatic complex to fix dinitrogen, consisting of two cytoplasmic metallo-enzymes: dinitrogenase and dinitrogenase

Fig. 14.25 Nitrogen cycle. *1a* ammonium assimilation, *1b* nitrate assimilation, *2* Ammonification: mineralization, *3* fixation, *4* nitrification, *5* denitrification, *6* DRNA dissimilative reduction of nitrate to ammonium, *7* anammox: anaerobic ammonium oxidation (Drawing: M.-J. Bodiou)

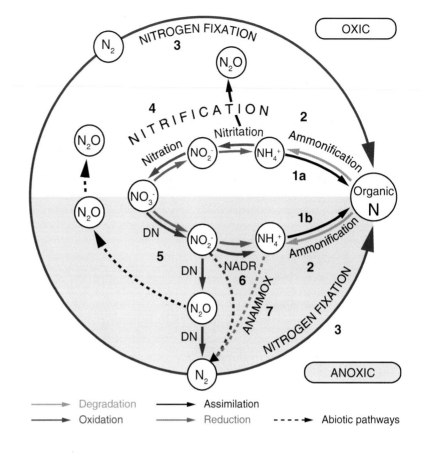

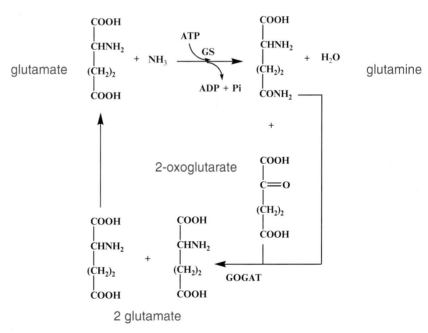

Fig. 14.26 Ammonium assimilation with GS-GOGAT System

reductase (*cf.* Chapter 3.4.4, Figure 3.42). The specificity of these two enzymes is weak. Indeed, the dinitrogenase is capable of catalyzing many reactions other than the fixation of dinitrogen, such as the reduction of various compounds: cyanide, cyclopropene, acetylene, azide, methyl isocyanide, diazirine, and nitrous oxide. The ability to reduce acetylene down to ethylene is commonly used to quantify the activity of nitrogen-fixing microorganisms in natural samples. This inexpensive method, easy to implement, is an alternative to the isotopic method.

Table 14.6 Different processes leading to the reduction of nitrate and the involved enzymes

	Denitrification	DNRA	Nitrate assimilation
$NO_3 \rightarrow NO_2$	NAR or/and NAP	NAR or/and NAP	NAS
$NO_2 \rightarrow NH_4$		NAD(P)H NiR or Formate NIR	
$NO_2 \rightarrow NO$	Cu-NIR or cd1-NIR		
$NO \rightarrow N_2O$	NOR (Nitric oxide reductase)		
$N_2O \rightarrow N_2$	N2OR (nitrous oxide reductase)		

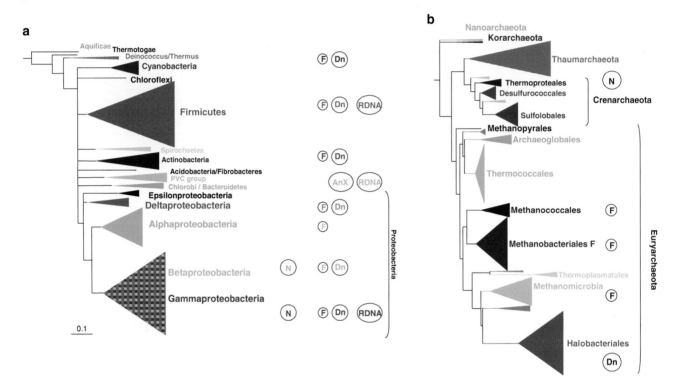

Fig. 14.27 Phylogenetic diversity of bacteria (**a**) and archea (**b**) involved in the nitrogen cycle. *Dn* denitrification, *N* nitrification, *F* N-fixation, *DRNA* dissimilatory nitrate reduction to ammonium, *AnX* Able to perform the anammox reaction

The functioning of these enzymes consumes high quantities of energy: 8 electrons and 16 ATP are necessary for the fixation of one molecule of dinitrogen:

$$N_2 + 8H^+ + 8e^- + 16ATP \rightarrow 2NH_3 + H_2 + 16ADP + 16Pi$$

Under natural conditions, between 20 and 30 ATP would be used. This system, energetically costly, works only in the absence of other nitrogen sources in the medium. Therefore, dinitrogenase complex is highly dependent on environmental conditions such as the presence of other sources of available nitrogen and oxygen. The presence of other nitrogen sources which can be more easily assimilated (nitrate, ammonium) generally inhibits the N_2 fixation. Similarly, oxygen (or more likely an oxidized product which is more reactive such as superoxide ions or hydroxyl radicals) would oxidize dinitrogenase reductase, causing a conformational change in this protein and its inactivation. However, many diazotrophs are aerobic and have had to develop various strategies to prevent the irreversible inactivation of enzymes (Fay 1992):

1. The first strategy consists of increasing the respiratory activity depending on the partial pressure of oxygen present in the cell, thereby maintaining an active dinitrogenase complex.
2. Another type of protection has been discovered in *Klebsiella pneumoniae* and *Rhodopseudomonas*, known as conformation protection or "switch-off-switch-on" phenomenon, consisting of a reversible inactivation of the enzyme when the partial pressure of oxygen becomes too high. This conformation change involves a protein (FeS) which binds to dinitrogenase in the presence of oxygen. This mechanism does not allow fixation but avoids the irreversible inactivation of dinitrogenase.
3. This weakly specific enzyme, capable of forming hydrogen, is at the origin of another protection mechanism involving a unidirectional hydrogenase, that allows O_2 elimination (Knallgas reaction: $O_2 + 2H_2 \rightarrow H_2O$).

Fig. 14.28 Heterocystous midsole with microorganisms involved in nitrogen fixation. *Anabaena* with polar heterocyst (B. Becker); Trichomes belonging to Nostocales: heterocysts (H in *red*) and Akinetes (A. Couté, MNHN).

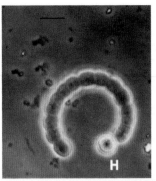

Fig. 14.29 Examples of deamination reactions (**a**) and deamidation (**b**)

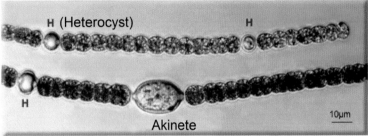

All these mechanisms protect the dinitrogenase complex from the oxygen present in the medium.

In addition, some phototrophic organisms produce oxygen through photosynthesis. This is the case of many cyanobacteria which use different protection mechanisms:

1. Photosynthesis and N_2 fixation can be uncoupled during the daily cycle, with fixation occurring at night. In this case, ATP required for fixation is provided by respiration and not through photosynthesis. This process has been demonstrated in *Gloeothece* sp, *Synecchococcus*, *Microcoleus chthonoplastes*, and *Oscillatoria* under natural conditions.
2. Another mechanism exists in some filamentous cyanobacteria that have specialized cell structures for N_2 fixation, the heterocysts, which promote a spatial uncoupling of the two types of metabolism (Fig. 14.28). The formation of heterocysts requires a significant change in vegetative cells. The cell wall is modified and narrow junctions appear between heterocysts and vegetative cells.

14.3.3 Processes Leading to the Production of Ammonium

14.3.3.1 Mineralization, Ammonification

Production of ammonium from nitrogenous organic matter is called ammonification. Depending on the complexity of the organic matter, ammonification can correspond either to a simple reaction of deamination or deamidation (Fig. 14.29) or to a series of complex metabolic reactions comprising several steps where polymers are hydrolyzed into simple soluble compounds (dissolved organic nitrogen currently known as DON) such as amines, amino acids, urea, and pyrimidines (see Sect. 14.2).

In the absence of specific method to quantify the process of ammonification, most authors quantify this process as follows:
1. *In situ* via incubation of natural samples and measurements of the kinetic of ammonium production;
2. In laboratory experiments via measurements of the kinetics of ammonium production or using techniques of isotopic dilution (Box 14.10).

> **Box 14.10: Methods for Determining the Denitrifying Activity: Approaches for a Difficult Problem**
> Patricia Bonin
>
> Denitrification, the reduction of nitrate to nitrite and gaseous products (NO, N_2O, and N_2) is an important microbial process regulating both primary production through the availability of nitrates, the water quality and chemistry of the atmosphere, both locally and globally. Unfortunately, quantifying this process on a natural sample from the soil or aquatic environments is very complex. This is mainly due to the fact that the substrate denitrification can be used in many biological processes and the final product (N_2) is the major component of the atmosphere.
>
> The main approaches developed to quantify denitrification are presented and discussed in this part.
>
> Methods involving the quantification of the consumption of nitrate (substrate for denitrification) are unreliable because depending of the ecosystems denitrifying activity may be either (1) overestimated, nitrates are used by processes other than denitrification such as dissimilatory reduction to ammonium or (2) underestimated since the production of nitrate through the nitrification-denitrification coupling is not taken into account or assimilation The calculations of nitrate fluxes from nitrate profiles determined in the pore water lead to the same problem. Moreover, in most cases only the passive diffusion of nitrate is taken into account and not the diffusion resulting from sediment reworking.
>
> Other methods are based on the quantification of the products of denitrification, N_2O or N_2, have been developed. Direct measurement of di-nitrogen production is difficult because nitrogen (molar mass 14 g/mol) is the main component of air (78.09 %). In some studies, the authors incubated their samples under atmosphere without N_2, by bubbling vials with an inert gas such as argon or helium, to directly quantify the production of N_2. But this operation is difficult because N_2 contaminations are difficult to control. Other authors quantify the evolution of N_2/Ar ratio mass spectrometry. Another type of approach involves the use of a stable isotope of nitrogen, the heavy nitrogen (molar mass of 15 g/mol). This isotope, accounting for only 0.36 % of the total nitrogen is in low quantities in the natural environment. Different isotopic methods have been developed. These labelling methods are more accurate and sensitive than previous methods, but are more expensive and difficult to implement. Different approaches based on the use of ^{15}N were used: methods of isotopic fractionation, isotope dilution or direct measurement of denitrification by mass spectrometry assay of 28/29/30 N_2 formed from $^{15}NH_4$ or $^{15}NO_3$ added.
>
> **Acétylène Inhibition Technique (AIT)**
>
> The development of inhibition method developed for the study of terrestrial and aquatic ecosystems has stimulated research on nitrogen cycle in the late 1970s. Acetylene amendment allows inhibition of the last step of denitrification, that is the reduction of nitrous oxide by di-nitrogen, and allows assimilating to the speed of accumulation of nitrous oxide to denitrifying activity (Yoshinari and Knowles 1976). Quantification of nitrous oxide concentration in the samples is facilitated by gas chromatography (electron capture detector). This method is inexpensive, easy to implement and not too time consuming (Box Fig. 14.16).
>
> However, the limitations of this method, did not take long to be demonstrated: (1) incomplete inhibition of nitrous oxide reduction in the presence of low nitrate concentrations and/or high organic matter

(continued)

Box 14.10 (continued)

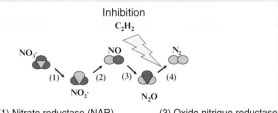

(1) Nitrate reductase (NAR)
(2) Nitrite reductase (NIR)
(3) Oxide nitrique reductase
(4) Nitrous oxide reductase

Box Fig. 14.16 Inhibition of the last stage of denitrification with acetylene

content, (2) bacterial degradation acetylene, (3) interactions between acetylene and sulfide leading to inhibition of blocking, (4) inhibition of nitrification, that is a significant source of nitrate, leads to underestimating the denitrifying activity. In addition, the kinetics of N_2O production must be followed in very short time, few hours.

However AIT has been and is still used in many studies. This technique also permits to determine the potential activity, i.e., the activities determined in the optimum conditions of expression in the absence of limitation by the electron donor (carbon source) or electron acceptor (nitrate). The addition of chloramphenicol is used to inhibit the synthesis of enzymes involved in this process during the incubation, provides access to the enzymatic activity (DEA, activity of enzymes present in situ at the sampling time). The use of chloramphenicol has been widely discussed as some authors believe that it could inhibit bacterial activity, while others feel it would only inhibit protein synthesis de novo during the incubation period. This approach is widely used to compare study sites or the effect of different treatments (change in temperature, salinity, or contribution of a xenobiotic) or to determine the nature of the factors that control the expression of this activity or as an indirect measure of the functional microbial diversity.

This approach has led to wonder about the type of reaction systems used to perform slurry systems, sediment suspension in sealed vials or direct incubation corers sample. Each approach presents advantages and limitations: "slurry," requires suspension of sediment in bottles sealed anaerobic, the breakdown of the structure of the matrix of soil or sediment and thus natural gradients but we can expect under these conditions good homogeneity of acetylene added and no loss of nitrous oxide formed as the gas being injected directly into the chromatograph. In contrast, direct incubation of the sample in the sampling core maintains the structure of the sample of soil or sediment but homogeneous distribution of acetylene or substrates injected into the sampler is difficult to control and sub-sampling of the gases produced during the incubation is very delicate.

Isotopic Labelling Approaches

The addition of $^{15}NO_3$ in soil samples directly to quantify denitrification was used for the first time by Hauck and Bouldin (1961). These authors demonstrated that the ratio of the two species $^{29}N_2$ and $^{30}N_2$ produced depend on the added $^{14/15}NO_3$, showing that isotopic matching $^{14}N^{14}N$, $^{15}N^{15}N$, or $^{14}N^{15}N$ followed a binomial distribution. More recently (Nielsen 1992) applied this principle to quantify the denitrifying activity in sediment samples.

This method has the advantage to discriminate the production of dinitrogen coming from nitrified nitrate to that from nitrate of the water column.

It was sensitive and robust method used in many studies. However, it is based on various assumptions that it is vital to take into account: (1) the theoretical distribution of the two isotopes must be uniform and the percentage should be kept constant during the incubation; (2) the denitrification from $^{14}NO_3$ (D14) must be independent of the amount of added $^{15}NO_3$; (3) the denitrification from $^{15}NO_3$ (D15) must be proportional to the amount of added $^{15}NO_3$ and $^{14}NO_3$ produced via nitrification from $^{14}NH_4$ (Box Fig. 14.17).

Therefore, it is essential to achieve kinetic incubation in the presence of different concentrations of added $15NO_3$ to determine the incubation conditions for which these assumptions are met. For details of the calculations, see the work of Nielsen (Nielsen 1992). In recent years, some authors have attempted to improve the IPM to simultaneously quantify denitrification and nitrogen fixation (An et al. 2001) or to include from incomplete denitrification stopping stage N_2O (Master et al. 2005).

More recently the process of anammox – acronym for anaerobic ammonium oxidation – was highlighted in different ecosystems and can reach up to 67 % of the

(continued)

Box 14.10 (continued)

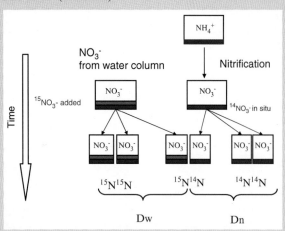

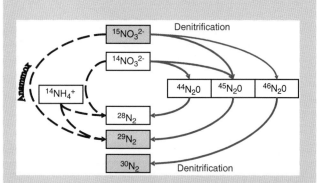

Box Fig. 14.17 Determination of denitrifying activity and discrimination between Dn and Dw by isotope tracing. *Red* fraction of the stable isotope (15N), *blue* 14 N, *Dw* Denitrification of nitrate from the water column, quantified from 15NO3-Added; Dn denitrification of nitrate from the nitrification traced from 14N

total flow of N_2 (Thamdrup and Dalsgaard 2002). This process involves one molecule of ammonium and nitrite to oxidize ammonium directly in di-nitrogen, ammonium being used as an electron donor and nitrite electron acceptor. As a result, the production of N_2 measured by IPM is not exclusively produced by denitrification but may result from both denitrification and anammox that occur simultaneously in the sediment.

The expression of the anammox violates the basic assumptions of the calculations of IPM since anammox can only lead to the production of $^{29}N_2$ without the production of $^{28/30}N_2$. When both processes take place simultaneously, binomial distribution is not respected and calculations originally proposed by Nielsen cannot be applied. A modification of the original method has recently been proposed to quantify the proportion of N_2 from anammox in the total N_2 fluxes. If anammox occurs, different alternative methods to quantify denitrification and anammox have been proposed and are discussed (Thamdrup and Dalsgaard 2002; Risgaard-Petersen 2003; Trimmer et al 2006. Minjeaud et al. 2008). It appears that: (1) change in stocks $^{14}NO_3$ via nitrification during the incubation period has no effect on the denitrifying activity measured, (2) it is vital to determine the range of added $^{15}NO_3$ concentrations for which the activity is independent of the $^{15}NO_3$ concentration (3) IPM can be used only after verifying the absence of anammox and (4) modification of the original

Box Fig. 14.18 Diagram showing all the reactions involved in the formation of different isotopes to study the nitrogen cycle

method of Nielsen allows to calculate the total production of N_2 and the ratio Ra (anammox/production of N_2) and thus simultaneously quantify both these processes. One advantage of this method lies also in the fact that other processes associated with the nitrogen cycle such as ammonification, nitrification, or nitrate (or dissimilatory nitrate reduction to ammonium, DNRA see Chapter 14) can be determined simultaneously in the same system, by isotopic dilution of the initially added $^{15}NO_3$ with $^{14}NO_3$ produced from $^{14}NH_4$ in the case of nitrification, by the production of $^{15}NH_4$ in the case of DNRA (Box Fig. 14.18).

14.3.3.2 Dissimilatory Reduction of Nitrate to Ammonium (DRNA) or Nitrate Ammonification

Although denitrification is often considered as the sole process of dissimilatory reduction of nitrate (using nitrate as an electron acceptor), the dissimilatory nitrate reduction to ammonium (DRNA, Fig. 14.25, *pathway 6*), also called ammonification of nitrate, can be the dominant process in some ecosystems. This process presents a major ecological interest because, unlike denitrification which corresponds to a net loss of nitrogen in the ecosystem, DRNA ended by ammonium production which is then biologically available for the food web. However, the production of ammonium via DRNA is lower compared with ammonification.

Dissimilatory reduction differs from the assimilatory reduction by two important criteria: its inhibition by oxygen and its non-inhibition by excess of ammonium. There are two types of DRNA depending on microorganisms:

1. A respiratory reduction where energy used for growth is produced by oxidative phosphorylation during electrons transfer by the cytochromes chain.

2. A fermentative reduction where the energy production is derived from the substrate phosphorylation and is not directly coupled to nitrite reduction. In strict anaerobic microorganisms, the DRNA seems to be essentially fermentative whereas in facultative aerobes, some of the steps are respiratory and others fermentative.

In all facultative anaerobic prokaryotes that reduce nitrate, the first step of nitrate reduction is not different from that encountered in denitrifying bacteria. The dissimilatory nitrate reductase (Type A, Nar) of a non-denitrifying strain of *Escherichia coli* has been studied in detail and its properties are common to those of nitrate reductase type A of denitrifying bacteria (*cf.* Sect. 14.3.5). In strict anaerobic fermentative prokaryotes, nitrate would, in this case, allow the reoxidation of NADH, H^+ produced during glycolysis.

The reduction of nitrite to ammonia comprises a single step, although the involved enzymes are diverse. Two main types of nitrite reductases were identified: a cytoplasmic nitrite reductase working with NADH, H^+ (fermentative pathway), and a nitrite reductase-formate dehydrogenase (respiratory pathway).

14.3.4 Processes Leading to the Production of Nitrate: Nitrification

Nitrification is the unique process that leads to the production of organic nitrate in the environment. This reaction is performed in two steps: the oxidation of ammonium to nitrite (or ammonium oxidation, AO) and the oxidation of nitrite to nitrate (nitrite oxidation, NO) (*cf.* Sect. 3.3.2, Fig. 3.18a, b):

$$2NH_4^+ + 3O_2 \rightarrow 2NO_2^- + 4H^+ + 2H_2O$$
$$NO_2^- + \tfrac{1}{2} O_2 \rightarrow NO_3^-$$

During nitrification, ammonia or nitrite are used as reducing power in the respiratory chain. These oxidation reactions lead to the establishment of a gradient of protons which enables the formation of ATP used to synthesize organic matter via CO_2 fixation (Calvin cycle), oxygen being the final electron acceptor. The energetic efficiency is very low.

Each step of the nitrification reaction is carried out by a different bacterial community. The first step of ammonium oxidation is performed by ammonium oxidizing bacteria or archaea (AOB, AOA). In the second reaction, the oxidation of nitrite, is performed by nitrite oxidizing bacteria (NOB).

Most of the AOB and NOB are chemiolitho-autotrophic bacteria, meaning that they use energy obtained by ammonium oxidation and fix CO_2 to synthesize their biomass. Chemolitho-autotrophic nitrification dominates in most ecosystems. However, some organisms can perform heterotrophic nitrification. In forest ecosystems for example, the contribution of heterotrophic nitrification, which is a poorly understood mechanism, may constitute a significant part of nitrification. Such data is lacking for marine ecosystems. Compared to autotrophic nitrification, heterotrophic nitrification would increase with increasing C:N ratio, i.e., when microorganisms are in competition for nitrogen.

Concerning AO, the electrons from the oxidation of a molecule of ammonium are transferred to the respiratory chain to form ATP, generally during aerobic respiration. Most of these micro-organisms are microaerophilic. AOB generally are strict aerobic *Beta* or *Gammaproteobacteria*. Most of the known AOB belong to the phylum of *Betaproteobacteria*: *Nitrosospira* and *Nitrosomonas*. Some marine AOB belong to the phylum of *Gammaproteobacteria*, particularly to the genus *Nitrosococcus*. A limited number of genera are capable of oxidizing ammonium: *Nitrosomonas, Nitrosococcus, Nitrosospira, Nitrosovibrio*, and *Nitrosolobus*. These bacteria are found in many environments, such as soils, marine sediments, and lakes. In addition, a nitrifying *Archaea*, "*Candidatus* Nitrosopumilis maritimus" belonging to Crenarchaeota, has been isolated. In different marine ecosystems, nitrifying *Archaea* seems more abundant than nitrifying bacteria (Nicol and Schleper 2006).

Nitrite oxidizing nitrite bacteria (NOB) are also chemiolitho-autotrophs, using the oxidation of nitrite to nitrate to produce energy, oxygen being the final electron acceptor. These bacteria belong to phylogenetic groups which are relatively close to each other, affiliated to *Alpha* and *Gammaproteobacteria*, in addition to the genus *Nitrospira*. Four known genera oxidize nitrite: *Nitrobacter, Nitrospina, Nitrococcus*, and *Nitrospira* (Fig. 14.27).

Two enzymes are involved in the oxidation of ammonium to nitrite: ammonia monooxygenase (i.e., the membrane-bound enzyme AMO encoded by three genes: *amo*A, *amo*B, and *amo*C, which are grouped under the same operon), and hydroxylamine oxidoreductase (the periplasmic enzyme HAO, encoded by the genes *hao*, *hcy*, and *cyc* which also encode for the associated cytochrome *c554*). AMO can oxidize ammonium into hydroxylamine, which is then oxidized to HAO (Fig. 14.30). The oxidation of nitrite to nitrate is performed by a nitrite oxidoreductase bounded to the cell membrane (Fig. 14.30).

In some *Nitrosomonas* species, the presence of nitrite (NIR) and nitric oxide reductase (NOR), enzymes of the denitrification, is surprising. It has been shown that under low oxygen conditions, these enzymes are expressed and allow to accumulate nitrous oxide. However, the exact conditions of this expression are not clear at all. The inactivation of gene *nirK* in *N. europea* reduces the tolerance to nitrite in aerobic cultures, indicating its potential role in nitrite detoxification. Moreover, *Nitrosomonas eutropha*

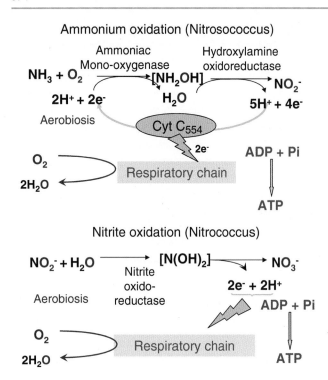

Fig. 14.30 Different steps of ammonium oxidation and nitrite oxidation by nitrifiers

can oxidize ammonium anaerobically using nitrite or tetraoxide nitrogen (N_2O_4) as electrons acceptor; denitrification in these oxygen-limiting conditions thus also allows energy production in both nitrating and denitrifying bacteria (Schmidt et al. 2004). The importance of this process remains poorly known neither at the ecosystem level nor at the physiology level for growth and survival of specific nitrifying organisms. This needs to be further investigated, considering the ecological role of nitrous oxide as a greenhouse gas.

Nitification is often attributed to chemiolitho-autotrophic microorganisms in the genera *Nitrosomonas* or *Nitrosobacter*. However, although there is little information on this process, heterotrophic nitrification performed by microorganisms (Fungi and prokaryotes) is known for long time (Pedersen et al. 2002) but often been considered as a minor process. Studies on a limited number of bacteria show that heterotrophic nitrifying bacteria use similar mechanisms to those of autotrophic bacteria. In contrast, heterotrophic nitrification by Fungi uses different mechanisms. Targeted studies on the competition between *Paracoccus denitrificans* (nitrifying heterotroph, sometimes also called *Thiospora pantotropha*) and *Nitrosomonas*, conducted in varying oxygen conditions and different C:N ratios, show that under limiting oxygenation conditions with high C:N ratios (10), heterotrophic nitrification becomes the dominant process and can contribute up to 60 % of the total nitrification (Jetten et al. 1997).

Many environmental factors influence the reactions involved in nitrification. The presence of oxygen is essential for the development of these microorganisms that carry out aerobic respiration. The oxidation of nitrite is reduced in the presence of light, but the mechanisms involved remain poorly understood. pH (optimum between 7.5 and 8) and temperature are also important factors because they can affect the affinity of the enzyme for its substrate. Salinity seems to have a very marked effect on the structure of the nitrifying community.

Assimilation of organic carbon has been demonstrated in different AOB, but does not seem to confer an advantage on the growth of these microorganisms. The AOB could play an important role in the degradation of organic pollutants in oligotrophic environments. The low specificity of ammonium monooxygenase would allow oxidation of compounds such as methanol, bromoethane, cyclohexane, benzene, propylene, or phenol. In contrast, AOB can grow under mixotrophic conditions. This could explain the abundance and the ubiquity of these microorganisms, and the absence of nitrite accumulation of in most ecosystems.

14.3.5 Production of Dinitrogen

For a long time, denitrification has been considered as the only process leading to the production of dinitrogen and responsible of the net loss of nitrogen from the ecosystems. Another process producing dinitrogen called anammox (acronym for "*an*aerobic *amm*onium *ox*idation "(Fig. 14.25)) was discovered in wastewater treatment systems. This process allows the anaerobic conversion of ammonium and nitrite into molecular nitrogen.

14.3.5.1 Denitrification

Denitrification is the anaerobic process of microbial respiration, alternative to the respiration of oxygen, where nitrate instead of oxygen is used as the terminal acceptor of electrons in the anaerobic mineralization of organic matter. Denitrification corresponds to a dissimilatory reduction of nitrate, as opposed to the assimilation of nitrogenous compounds which leads to the synthesis of organic matter.

Nitrous oxide and dinitrogen are the end products of denitrification whereas ammonium is the end product of the dissimilatory nitrate reduction to ammonium (DNRA). The reduction of nitrate to nitrite is common to both metabolisms, but in the case of DNRA, nitrite formed is then reduced to ammonium which is not a gas. However, some denitrifying bacteria can only achieve some steps: reduction of nitrate to nitrite (nitrate reduction) and reduction of nitrous oxide to dinitrogen. In these cases, denitrification is known as incomplete denitrification.

In an ecological point of view, denitrification is an important process that allows biological elimination of nitrogen. Indeed, 15–70 % of the ammonium derived from the mineralization of organic matter could be eliminated through the coupling nitrification/denitrification.

In denitrification, the transformation of nitrate to dinitrogen is carried out by four steps during which the initial nitrate (NO_3^-) is successively reduced to nitrite (NO_2^-), nitric oxide (NO), nitrous oxide (N_2O), up to the production of molecular nitrogen (N_2). Each step of denitrification is catalyzed by a distinct enzymatic system.

$$NO_3^- \rightarrow NO_2^- \rightarrow NO \rightarrow N_2O \rightarrow N_2$$

Nitrate reductase, Nitrite reductase, Nitric oxyde reductase, Nitrous oxyde reductase

Denitrification is an anaerobic respiration process where nitrate is used as the terminal acceptor of electrons. The energy efficiency of denitrification is lower compared with aerobic respiration. Denitrification yields 24 molecules of ATP for one molecule of reduced nitrate while aerobic respiration yield to 32 molecules of ATP for one molecule of oxygen reduced. Facultative anaerobic denitrifying bacteria use preferentially aerobic respiration in the presence of oxygen, but when oxygen is absent or present in very low concentrations, NO_3 is used as alternative electron acceptor to oxygen respiration. Despite the initial controversy, the phenomenon of aerobic denitrification is now clearly established. *Paracoccus denitrificans* was the first isolated bacteria which can denitrify in the presence of oxygen (*cf.* Sect. 3.3.2, Fig. 3.21). Since then, other strains with the same property were isolated (*Microvirgula aerodenitrificans* and *Thauera mechernichensis*). *P. denitrificans* and *M. aerodenitrificans* can even denitrify in the presence of saturating concentrations of oxygen (Zumft 1997).

Denitrification is mainly observed in prokaryotes, although some Fungi (*Fusarium oxysporum* or *Cylindrocarpon tonkinense*) are able to denitrify. *F. oxysporum* and *C. tonkinense* are typically denitrifying organisms, even if *F. oxysporum* exhibits original features. In this fungus, denitrification indeed requires a basic concentration of oxygen (Laughlin and Stevens 2002).

From a physiological point of view, the majority of denitrifying prokaryotes are facultative aerobic, heterotrophic, and chemo-organotrophic microorganisms. Within prokaryotes, the ability to denitrify exists in diverse phylogenetic groups, with wide ranges of physiological and metabolic characteristics (Fig. 14.27). Phylogenetically, the ability to denitrify is found in *Archaea* and in both Gram negative and Gram positive bacteria. Isolated denitrifying bacteria mainly belong to the phyla of *Alpha-* and *Betaproteobacteria*, while no isolate was found in enterobacteria. Very few *Archaea* able to denitrify have been isolated, and the majority belong to the extreme halophilic genera of *Haloferax*, *Haloarcula*, and *Halobacterium*. So far, only one strain of denitrifying hyperthermophile has been isolated: *Pyrobaculum aerophilum*. This biodiversity explains the presence of denitrifying activity in many ecosystems, including extreme environments. Some microbial denitrifiers are also able to fix molecular nitrogen or to perform nitrification (*Paracoccus denitrificans* or *Nitrosomonas* sp). The large phylogenetic heterogeneity within denitrifiers makes their study difficult on the sole basis of the gene encoding 16S rRNA, which led to the investigation of novel specific markers targeting functional genes.

The dissimilatory reduction of nitrate to nitrite is a common enzymatic step to two respiratory processes: denitrification and dissimilatory reduction of nitrate to ammonium (DNRA). There are two types of dissimilatory nitrate reductases which have a different location in the cell: one is periplasmic (Nap) and the other is membraneous (Nar). The structure and biochemical properties of these enzymes are fairly well known and conserved among different microorganisms. The majority of biochemical and genetic studies have been conducted on *E. coli* where two isoenzymes were differentiated according to their regulation mode. One contributes to 90 % of the total activity and is induced under anaerobic conditions and in the presence of nitrate (Nar), whereas the other is expressed constitutively (Narz) and independently of the partial pressures of oxygen. It seems that this second membraneous nitrate reductase plays a role during the transition from aerobic to anaerobic metabolism (Table 14.6).

The genes *narG*, *narH*, *narI*, and *narJ* that encode for the subunits α, β, γ, and δ of Nar A, respectively, are organized in operon. The periplasmic nitrate reductase (Nap) was also isolated and their biochemical and genetic properties characterized in some denitrifiers. Currently, the role of this enzyme is poorly understood and different physiological functions have been proposed. Periplasmic nitrate reductase may help to maintain the redox balance in cells when adapting to some environmental changes, including the transition between aerobic and anaerobic conditions. Because Nap is not sensitive to oxygen, some denitrifiers can realize aerobic denitrification by coupling this enzyme to nitrite reductase or to nitric oxide reductase. This is particularly favorable for microorganisms that thrive in natural environments subject to changing conditions in oxygen availability.

The nitrite formed this way is released into the cytoplasm and then transferred into the periplasmic space. Nitrates and nitrites should thus be transported through the cytoplasmic membrane. The mode of this transport remains unclear. One believes that these carriers are nitrate/nitrite antiport transporters or nitrate/proton symport transporters, although they have not yet been discovered.

The nitrite transferred into the periplasmic space can get in contact with nitrite reductase, a periplasmic enzyme which catalyzes the reduction of nitrite to nitric oxide. There are two types of nitrite reductases which are very different in their structure, but although have a similar function: One is made of cytochrome cd_1 (Nir-cd_1) and the other of copper (Nir-Cu). They are encoded by the genes *nir*S and *nir*K, respectively. Until now, these two types of nitrite reductases were never found simultaneously in the same organism.

Among the enzymes of denitrification, nitric oxide reductase was the one discovered more recently. Nitric oxide reductase forms a complex membraneous cytochrome bc with two monomers.

The conversion of nitrous oxide (N_2O) to dinitrogen (N_2) is the final step of denitrification (*cf*. Sect. 3.3.2, Fig. 3.21). The enzyme that catalyses this reduction, the nitrous oxide reductase, is a generally soluble homodimer, located in the periplasm. Mechanisms and factors that inactivate nitrous oxide reductase are still unknown, but oxygen seems to play an active role in this phenomenon. Moreover, this enzyme is inhibited by acetylene (C_2H_2), thus leading to an accumulation of nitrous oxide. This acetylene inhibition technique is used to measure denitrifying activity in pure cultures and natural samples (Box 14.10).

In Gram-positive bacteria which have no periplasm, denitrification enzymes are associated with the membrane. However, the structure of these enzymes is very close to that of periplasmic enzymes in Gram-negative bacteria.

Denitrification rates are influenced by many factors such as oxygen, nitrate and organic matter concentrations, pH, temperature, and the presence of macrofauna. These factors interact with each other and make the control of denitrification complex in spatial and temporal scales. In 1982, Tiedje and colleagues (Tiedje et al. 1982) presented a conceptual model of control based on three factors with the following hierarchy: oxygen > nitrate > organic matter. The availability of nitrate and of labile carbon control denitrification rates and the importance of the ratio denitrification: DNRA. In the presence of high nitrate concentrations, denitrifying community is favored. Dissolved oxygen concentration controls denitrification either directly (enzyme inhibition) or indirectly *via* nitrification (nitrate production). In oxic area, denitrifying activity can exist in "hot spots" of denitrification defined as anoxic micro-niches and characterized by a strong coupling between nitrification and denitrification. Such anoxic micro-niches in globally oxygenated ecosystems (Fig. 14.3), colonized by denitrifying microorganisms were also reported in Mediterranean waters rich in suspended particles. Denitrification rates measured by the acetylene inhibition technique (AIT) remain low compared with those observed in sediments. Apart from some special cases, denitrification enzymes are induced under anaerobic conditions and inhibited in the presence of oxygen. However, depending on bacterial species, these enzymes do not all have the same sensitivity to oxygen. In *Pseudomonas nautica*, Bonin and Gilewicz (1992) showed that after two cycles of alternating aerobic and anaerobic conditions, the

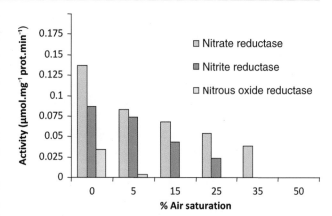

Fig. 14.31 Effect of dissolved oxygen concentration on nitrate, nitrite, and nitrous oxide reduction on *P. nautica* IP 617

rate of reduction of nitrate remained unchanged, that of nitrite decreased, and that of nitrous oxide was irreversibly inhibited (Fig. 14.31).

14.3.5.2 Anammox

Anammox is a biological process where energy is produced, in which nitrite and ammonium are converted directly into dinitrogen, ammonium being the electron donor and nitrite the electron acceptor (Mulder et al. 1995) (Fig. 14.27, *pathway 7*):

$$NH_4^+ + NO_2^- \rightarrow N_2 + 2H_2O \ \left(\Delta G° = -357 \text{ kJ mol}^{-1}\right)$$

Until 2008, no microorganism capable of performing this process has been obtained in pure culture. In contrast, microorganisms capable of anammox activity were observed in enrichment cultures, with up to 99.7 % enrichment of a targeted strain, following culture optimization. This strain was identified using molecular approaches without the conventional step of isolation; i.e., a chemolitho-autotrophic microorganism with an unusual morphology (Fig. 14.32), phylogenetically affiliated to *Planctomycetes*. This is surprising because all known Planctomycetes hitherto were aerobic heterotrophic bacteria. Planctomycetes belong to a distinct phylum (*Planctomycetes*), the order of *Planctomycetales*, and the family of *Planctomycetaceae*. The presence of *Planctomycetes* is not limited to wastewater treatment systems and has also been reported in many ecosystems, including marine and freshwaters. In 2008, bacteria capable of performing anammox are affiliated to five "*Candidatus*," genera: "Brocadia," "Kuenenia," "Scalindua," "Anammoxoglobus," and "Jettenia."

Bacteria that perform the anammox activity are anaerobic chemolitho-autotrophs and exhibit very long generation times (up to 3 weeks). All Planctomycetes able to achieve anammox possess an organelle, the anammoxosome. This cytoplasmic compartment is delimited by a membrane and

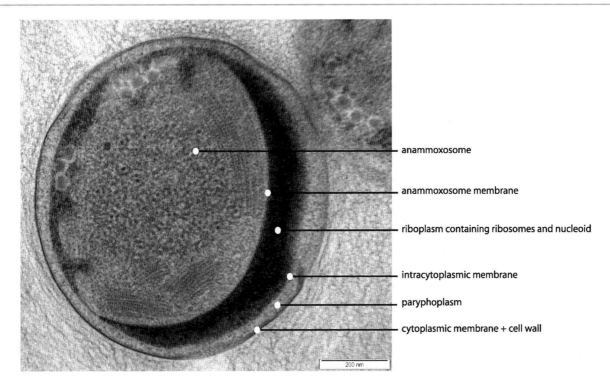

Fig. 14.32 Cellular organization of the strain "*Candidatus* Kunenia stuttgartiensis" containing anammoxosome (Gift of Laura Niftrick, Ultrecht University)

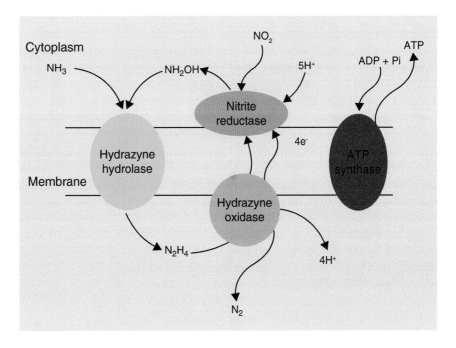

Fig. 14.33 Proposed pathway for the formation of di nitrogen from nitrite and ammonium by anammox (N_2H_4 Hydrazine, NH_2OH hydroxylamide) (Drawing: M.-J. Bodiou)

occupies 30–60 % of the cell volume. The membrane of anammoxosome contains a particular lipid: the ladderane (Damste et al. 2002). The singular properties of this lipid, which is found only in anammoxosomes, are considered the main driver of the anammox reaction.

During the anammox reaction, the terminal acceptor of electrons, i.e., nitrite, is reduced to hydroxylamine by a membrane-associated enzyme, a cytochrome *c* nitrite reductase. Nitric oxide but not hydroxylamine is the reaction intermediate. Hydroxylamine or nitric oxide is associated with ammonium to form hydrazine in the anammoxosome (Fig. 14.33).

A particular interest is given to hydrazine as the intermediate in this reaction, because this compound is commonly known as a poison for most living organisms. Oxidation of hydrazine takes place in the anammoxosome and forms dinitrogen. The enzyme performing this reaction can either be the hydroxylamine oxidoreductase (HAO) or the hydrazine dehydrogenase. This latter reaction provides four electrons used for the reduction of nitrite to hydroxylamine. All these reactions allow the transfer of 4 protons through the membrane of anammoxosome and, therefore, the promotion of a proton gradient for the synthesis of ATP molecules, ATP synthases being present in the membrane of anammoxosomes (Van Niftrik et al. 2004).

The anammox process is expressed only in anaerobic conditions and is inhibited in the presence of oxygen, but this inhibition is reversible. Organic matter also has an effect on the production of dinitrogen via anammox, by increasing the rate of mineralization and therefore the production of ammonium. The activity of anammox has been demonstrated in the wastewater treatment systems rich in organic matter, which can reach up to 60 % of the total production of N_2 in some ecosystems.

Temperature also has an effect on the anammox activity. Indeed, *in situ* studies have shown that the production of molecular nitrogen by anammox is maximum around 15 °C and becomes zero around 37 °C. At a temperature of 6 °C, 80 % of dinitrogen is still produced by anammox whereas at 37 °C, denitrification becomes dominant.

14.3.6 Nitrous Oxide Production

Nitrous oxide is a very stable gas, with a lifetime of about 120 years in the troposphere. It is a greenhouse gas which, similar to methane (CH_4) or carbon dioxide (CO_2), contributes to the increase in temperature at the surface of the Earth via absorption of infrared radiations. In the troposphere, this gas has a concentration of 310 ppb. Because of its long lifetime, its concentration, and its high radioactive absorption coefficient, the contribution of one molecule of N_2O to the increasing greenhouse effect would be 180–300 times greater than that of one molecule of CO_2. Since 1980, the concentration of N_2O in the troposphere is continuously increasing by 0.2–0.3 % per year, which yield an overall increase of 15 % between the pre-industrial era and the present time. A stabilization of the current rate requires a reduction in anthropogenic emissions of more than 50 %.

Beyond 0.3 ppm, N_2O can flow into the stratosphere where it undergoes the following photochemical reaction: $N_2O + O \rightarrow 2NO$. N_2O is thus a natural source of nitric oxide, which then catalyzes the destruction of the ozone layer as follows:

$$NO + O_3 \rightarrow NO_2^- + O_2 \text{ and } NO_2^- + O \rightarrow NO + O_2$$

Modeling studies predict that a severe increase in the denitrification rates in the Arctic could increase by more than 30 % of the ozone hole over the boreal pole.

Table 14.7 Natural sources of nitrous oxide

Sources	Emission of N_2O (Tg.N.an^{-1})	
	45–20°N	World
Aquatic		
Ocean	0.58	3.5
Limnic environments	0.16	0.65
Rivers	0.67	0.22
Estuarine	0.14	1.05
	1.56	5.42
Terrestrial		
Abiotiques	0.03	0.06
Biofuel	0.11	0.16
Energy/transportation	0.19	0.46
Industry	0.33	0.68
Biotiques	0.56	
Cultivated soil	0.43	0.96
Animal excression	0.08	1.02
Combustion organique	1.19	0.54
Natural soil	2.26	6.6
Total sources Terrestres	2.6	9.12

From IPCC (1998)

Production of nitrous oxide has so far been studied mainly in soils where 70 % of nitrous oxide originates from microbial activity (Table 14.7).

N_2O production was reported as a phenomenon accompanying different processes involved in the nitrogen cycle: denitrification, DNRA, or nitrification.

During denitrification, nitrous oxide production is due to the inhibition of the end reaction (reduction of nitrous oxide to dinitrogen) by oxygen, and/or to the loss of the ability to reduce nitrous oxide down to N_2. However, the released nitrous oxide associated with nitrification is expressed in limiting oxygen conditions, the metabolism of ammonium oxidizing bacteria being modified. Nitric oxide and nitrous oxide are produced by oxidation of ammonium using nitrite as electrons acceptor through a process called nitrifying denitrification, known as OLAND "Oxygen Limited Autotrophic Nitrifying Denitrification").

A third possible source of nitrous oxide is DNRA, but its physiological role remains unknown.

14.3.7 Nitrogen Cycle in Natural Ecosystems

Physical and chemical environmental changes, primarily in organic matter (eutrophization level), nitrogen compounds, oxygen, pH, temperature, or relations with macro-organisms, will influence the eco-physiology of microorganisms involved in the nitrogen cycle and the functioning of natural ecosystems. Most of these factors interact, which makes complex the control of nitrogen cycle in space and time. The distribution of nitrogen compounds in different oxidation states and the great biodiversity of microorganisms involved explain why the nitrogen cycle is expressed in most ecosystems.

14.3.7.1 Nitrogen Cycle and Agriculture

Naturally present in the soil, nitrates play a central role in the nitrogen cycle. Nitrates are natural and soluble compounds. They diffuse in soils and groundwaters and are discharged in running waters, but are also provided from nitrogen fertilizers. Nitrates are one of the main causes of the deterioration of the quality of natural waters. European Framework Directive on water (2000/60/EC) is a directive of the European Parliament which establishes a framework for a comprehensive management of European waters. This is the major European policy concerning the overall protection of natural fresh-, brackish, salt-, ground-, and coastal water resources. State members should encourage consultation and active participation of all parties involved in the implementation of this directive, including the development of management plans. The European Directive on "nitrate" is the main regulatory instrument in the fight against pollution related to agricultural nitrogen enrichments in Europe. It concerns nitrogen from different sources: chemical fertilizers, breeding, and food wastes. The nitrogen of urea was used and is still used as a natural fertilizer. However, today, the agricultural fertilizers are mainly produced industrially from chemical synthesis. Nitrogen is one of the important nutrients for agricultural crop growth and yields. Indeed, an optimum amount of nitrogen in soil enhances the absorption of other nutrients and promotes the development of plants and roots. In contrast, an excess of nitrogen can lead to yield losses, favor diseases and infestations by insects, and increase pollution of rivers and groundwaters. It is therefore important to optimize the nitrogen fertilization in order to obtain expected levels of agricultural yields while avoiding these disadvantages.

Mineralization of organic compounds presents a set of fundamental processes in the functioning of agroecosystems. Microorganisms in soils produce extracellular enzymes that transform polymers into soluble and easily assimilable compounds such as amino acids. This soluble fraction represents an important source of nitrogen even in cultivated soils receiving large inputs of inorganic nitrogen. The nitrates from anthropogenic (fertilization) or natural (produced by nitrification) sources will be: absorbed by plants, immobilized by the microflora, leached in surface or deep waters, or denitrified and therefore eliminated from the ecosystem.

The only biological process allowing the retention of nitrogen in soils is the biological fixation. Some fixing micro-organisms are saprotrophic, others live in association with the roots of plants. This is the case of bacteria belonging to the genus *Rhizobium* that live in association with the roots of *Fabaceae*. These bacteria cause the formation of nodules on the roots that are sites of intense symbiotic activity wherein the plant provides carbon sources and energy from photosynthesis and, in turn, benefits from amino acids which are produced by the dinitrogen-fixing bacteria. Other microorganisms are involved in the fixation of dinitrogen in association with other types of plants. Bacteria of the genus *Frankia* belonging to the class of *Actinobacteria*, fix dinitrogen in symbiosis with a wide range of dicotyledonous plants called actinorhizal, including 24 genera in 8 families such as Betulaceae (alder), Myricaceae (*Myrica* or wood feels good), Casuarinaceae (Casuarinas). These plants, with their symbiotic bacteria are collectively responsible for approximately 15 % of nitrogen inputs from biological fixation on the Earth. They are found in different ecosystems where nitrogen is limiting, e.g., aged ice moraines, volcanic ash, mine backfills, or burning.

14.3.7.2 Nitrogen Cycle in Forest Ecosystems

Nitrogen cycle in most forest ecosystems can be characterized by a relative balance between inputs (precipitation, biological fixation) and outputs of nitrogen (nitrification/denitrification, volatilization, and leaching).

14.3.7.3 Nitrogen Cycle in Wetlands

In the form of nitrate or ammonia, nitrogen is highly soluble in water and mobile in ecosystems. Runoff and rains carry nitrogen compounds into lakes and oceans. Wetland or riparian zone is the zone that surrounds the riverbed between the low water level and high water. These areas have specific characteristics due to the mixing of aquatic and terrestrial factors that influence their structure and functioning. They play an important role in the maintenance of the water quality in agricultural and urban areas, by acting as significant sinking sites for diffuse pollution, nutrients (especially nitrogen), and micropollutants (pesticides).

The flows of particulate and dissolved matter within the alluvial system are multiple and complex: surface and subsurface cross flows from the adjacent hillsides, surface (rivers, streams) and groundwater longitudinal flows, and vertical flows (water exchanges along the continuum groundwater–soil–atmosphere). They are affected by physico-chemical and biological processes, some of which are specific to these areas or are expressed there with particular intensity (dilution, sedimentation, adsorption, absorption by plants, or denitrification). The transfer of nitrogen in river slopes occurs mainly in its dissolved oxidized form (nitrate) which is hardly retained by soils. Wetlands in valley floors, located between farmlands and running waters, are recognized to be true "nitrate filters" against diffuse inputs from groundwater or surface inputs from watersheds. The purifier potential of these areas can often be expressed only partially because it is attenuated by the presence of natural (deep circulation or runoff, macropores) or artificial (ditches, drains) short circuits.

This role of filter is due to two biological processes favored by the wetland ecology: plant assimilation which corresponds to more or less long-term storage of nitrogen, and microbiological denitrification which allows elimination of nitrates (Forshay and Stanley 2005).

Potentially, these processes are active throughout the wetland, although nitrogen recycling of the groundwater is carried out in a few meters distance from its entrance in wetlands. The boundary between agricultural parcels and riparian areas appears to be the key point (active area) of the buffering capacity against diffuse nitrogen pollution. However, this recycling activity is heterogeneous in time and space. This process of nitrogen elimination is difficult to measure *in situ*. Little assessments of emitted gases are available today. Most studies estimate that the denitrification potential can reach up to 40 kg nitrogen ha^{-1} day^{-1}. Many factors influence the denitrifying activity in the scale of prokaryotes (temperature, oxygen, mineralizable carbon, nitrate), soil (waterlogging of soils, organic material), the targeted site (topography, vegetation), and wetland (geomorphology, hydrology, uses). The effectiveness of this phenomenon depends on two factors: the immersion time (few days are necessary to remove oxygen from ground) and the connectivity of water flows (which allow incomes of nitrate and carbon). Denitrification is at its maximum in winter (high water period) if temperatures are not too low (weak activity of microorganisms).

14.3.7.4 Nitrogen Cycle in Marine Ecosystems

In the sea, as on the land, dissolved nitrogen is absorbed by photosynthetic organisms and in the food chain. However, important inputs of nitrogen in marine systems can cause **eutrophication***. This phenomenon seems to become more frequent, taking different forms in the enclosed seas and in gulfs or bays where water is not renewed: Blooms of planktonic or photosynthetic multicellular organisms occur there, causing anoxic conditions. The risk is highest in areas where agricultural nutrient inputs from estuaries or rains (dissolved nitrogen) are excessive. The only sustainable solution is to limit upstream agricultural nitrogen losses. The phenomenon of eutrophication may occur in some muddy estuaries, at certain times of the year, for example, as a result of riverine inputs in autumn or after heavy rains with flooding. Waterproofing and agriculture have exacerbated these phenomena. These episodes of eutrophication have limited duration, often less than 15 days during which intense denitrifying activity allows the removal of nitrogen in excess. Because of their high productivity (>30 t ha^{-1} $year^{-1}$), estuaries are recognized as carbon sinks. When they are eutrophic or dystrophic, CO_2, N_2O (via nitrification and denitrification) and CH_4 can be emitted in very large quantities from sediments. These emissions could correspond to 8 % of total emissions of greenhouse gases in Europe.

Competition between denitrification and DNRA presents a major ecological interest as denitrification led to the loss of nitrogen from the ecosystem while DNRA converts nitrate to ammonium which is available to the food chain. The competition between these two processes seems to be essentially controlled by the amounts of electron donors and acceptors (nitrate), DNRA being favored when carbon is limiting while denitrification becomes the dominant process when nitrate is limiting (Bonin et al. 1998). A study was conducted in coastal marine sediments collected in Mediterranean Sea, study in which different sampling stations were selected for their contrasting physico-chemical characteristics: sediments from muddy areas rich in organic material (2.5–7.3 %), sandy stations where organic matter content were lower (0.3–0.8 %). Nitrate concentrations were comparable regardless of the stations. For the study period, microbial activities are generally higher in muddy and organic matter rich stations than in sandy sites. The seasonal effect is indisputable: DNRA activity is maximum in autumn when C:N ratios are high, due to the death and sedimentation of a phytoplankton bloom; maximum denitrifying activities are observed in winter when the nitrate concentration is highest (period of heavy rains). In the samples which were very rich in organic matter (15 %), up to 98 % of nitrate can be reduced to ammonium and thus remain available to the food chain. This process is therefore optimally expressed in marine sediments, but has been little studied in soils. Similarly, the anammox process, initially reported in wastewater treatment ponds, has also been observed in aquatic ecosystems. This process which is in direct competition with denitrification, contributes significantly to the production of molecular nitrogen in some aquatic ecosystems. In sediments where the supply of organic matter is low, more than 67 % of dinitrogen would be formed due to anammox (against 33 % for denitrification) (Thamdrup and Dalsgaard 2002). Nevertheless, this percentage becomes negligible (2 %) in eutrophic coastal areas, where the presence of significant quantities of organic matter promotes dissimilatory reduction processes (denitrification and/or DNRA).

In the euphotic zone of coastal marine ecosystems, magnoliophytes may affect the nitrogen cycling and in particular denitrification. Indeed, oxygen from photosynthesis which passes from leaves to roots/rhizomes is released in certain zones of the sediment, promoting nitrification and denitrification in the adjacent anoxic areas. Depending on the species, the percentage of oxygen emitted by the roots compared with the total percentage emitted by the plant can be variable. It is at 100 % for *Lobelia* sp. and at 1–4 % for *Zostera* sp. Plants can also stimulate denitrifying activity by increasing the amount of labile organic matter in the sediments. This phenomenon occurs by excretion of dissolved organic compounds through the roots of plants. However, the presence of benthic algae covering the sediment surface may also cause a decrease in denitrification when the concentrations of oxygen released are too important.

Sediment bioturbation processes generated by benthic macrofauna (Box 14.3) also affects the nitrogen cycling in marine sediments. Irrigation generated by the activity of burrowing macroorganisms enables the transport of nitrate and oxygen from the water column to the depths of the sediment. This input causes two direct and antagonistic effects on denitrification: a positive effect due to an increase of the amount of nitrate and/or a negative effect due to oxygen inhibition (Gilbert et al. 1995). However, the presence of dissolved oxygen and the release of ammonium through excretions of these animals promote nitrification in areas near burrows, and thus indirectly stimulate denitrification in sediments. Another aspect concerns the production of fecal pellets by suspension and deposit feeders. These fecal pellets are sites of intense microbial activity whose high oxygen demand causes the formation of anoxic microniches. These latter facilitates contact and exchanges between aerobic (nitrifying) and anaerobic (denitrifying) microorganisms.

14.4 Sulfur Cycle

14.4.1 Introduction

Similar to carbon, nitrogen, oxygen, phosphorus and selenium, sulfur, atomic number 16, is classified among the "non-metallic" elements. Sulfur is present in its natural state as four stable isotopes (^{32}S, ^{33}S, ^{34}S and ^{36}S). ^{32}S (95 % of total) and ^{34}S (4 %) are the most abundant while ^{35}S is a synthetic radioactive isotope. Sulfur exists in many oxidation states (Fig. 14.34) which vary from +VI for sulfate (SO_4^{2-}) to −II for sulfide. The term "sulfides" includes several chemical forms such as the sulfide ion (S^{2-}), the sulfhydryl ion (HS^-) and hydrogen sulfide or sulfhydric acid (H_2S). In nature, the most abundant forms are: sulfate, elemental sulfur ($S°$), as well as sulfides.

Sulfur is an abundant element and exists under different solid, liquid, soluble, or gaseous forms. Unlike nitrogen, sulfur is very scarce in the atmosphere where it is present in both oxidized (SO_2) and reduced (H_2S, organo-sulfur compounds such as dimethyl sulfide or DMS) forms. Sulfur compounds in the atmosphere (Fig. 14.35) have a natural or anthropogenic (combustion of fossil materials) origin. In the atmosphere, sulfur is oxidized to sulfur dioxide, and to sulfate, which can cause acid rain phenomena due to the formation of sulfuric acid (H_2SO_4). In addition, sulfate is an extremely hygroscopic compound that promotes cloud formation and thus participates in the reverberation of a fraction of solar radiations.

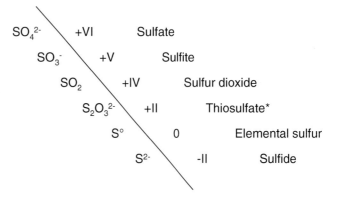

Fig. 14.34 Main sulfur species and their oxydation state. *Concerning thiosulfate, the indicated value corresponds to the mean oxidation state. In fact, one sulfur atom is reduced and the other one is oxidized

Rocks and sediments are the main natural reservoirs of sulfur. In these reservoirs, sulfur can be found in its oxidized form in gypsum for example (SO_4Ca), or in its reduced form in pyrite (FeS_2), cinnabar (HgS), galena (PbS) or other metallic compounds. Native sulfur (the main form is called orthorhombic sulfur, S_8), visibly corresponds to yellow crystals. Another sulfur reservoir, the most important for the biosphere, are oceans which contain large amounts of assimilable sulfur in the form of soluble sulfate. The poorer sulfur reservoir is living matter which contains only 1 % of sulfur (dry weight). In organic matter, sulfur is present in its reduced form in sulfur amino acids (methionine, cysteine, homocysteine), or in its oxidized form in molecules derived from amino acids (taurine, chondroitin sulfate). Sulfur plays a very important role in the assembly and structure of proteins. It is also present in the iron-sulfur complexes of metalloproteins, in certain vitamins such as thiamine (vitamin B1, cofactor of dehydrogenases) or biotin (vitamin H or B7, cofactor of carboxylases) or other cofactors (lipoic acid, coenzyme M). Sulfur is also included in the composition of dimethylsulfoniopropionate (DMSP), an osmoregulatory compound (*cf.* Sect. 9.6.3) present in marine phytoplankton, from which dimethylsulfide (DMS) is emitted in the atmosphere (Fig. 14.35).

Sulfur flows between its different reservoirs depending on complex physical, chemical, and biological processes. In addition, in Nature, sulfur undergoes a series of chemical or biological transformations occurring in oxic or anoxic conditions. All of these redox reactions constitute the biogeochemical cycle of sulfur in which biological processes, primarily those from prokaryotes, are predominant (Fig. 14.36). This cycle includes the sulfur assimilation (Fig. 14.36, *pathway 1*) by living organisms, either directly as sulfide or indirectly as sulfate. In the latter case, an assimilatory sulfate-reduction step is indeed necessary, in order to incorporate sulfur into organic matter. The mineralization of organic matter associated with the death of living

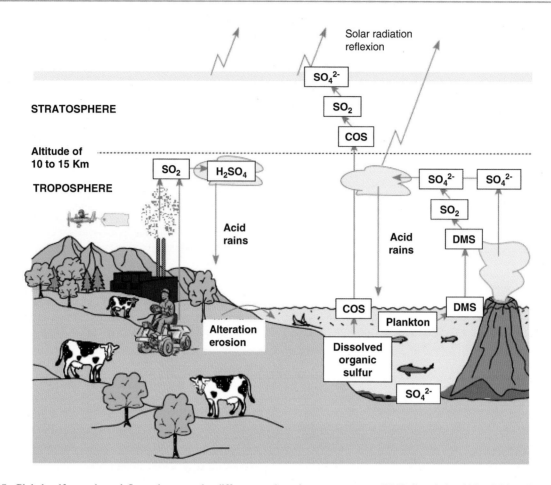

Fig. 14.35 Global sulfur cycle and fluxes between the different earth main compartments. *DMS* dimethyl sulfide, *COS* carbon oxysulfide (Redrawn from Pierre Andre Bourque and Pauline Dansereau, University Laval, Québec, Canada)

organisms results in the release of inorganic sulfur, mainly in the form of sulfide, a phenomenon called sulfhydrization (Fig. 14.36, *pathway 2*). Another process contributing to the production of reduced sulfur is grouped under the terms of dissimilatory sulfate-reduction or sulfate respiration (Fig. 14.36, *pathway 3*). During this process, all the oxidized forms of sulfur such as sulfate (SO_4^{2-}), thiosulfate ($S_2O_3^{2-}$), sulfite (SO_3^{2-}), or elemental sulfur (S^0) may be used as electron acceptors by prokaryotes (bacteria and *Archaea*) and reduced to sulfide (S^{2-}). This activity most often occurs in reduced and anoxic environments. The sulfur oxidation processes (Fig. 14.36, *pathway 4*) will use reduced or partially oxidized sulfur compounds as electron donors. Sulfur oxidation involves photolithotrophic anoxygenic metabolisms in anoxic conditions, and chemolithotrophic metabolisms in oxic and anoxic conditions.

Although sulfate, sulfur, and sulfide are the three major reservoirs of sulfur, intermediate compounds such as thiosulfate ($S_2O_3^{2-}$), have a very important role in the functioning of the biogeochemical cycle of sulfur. The thiosufate used as electron donor or acceptor generates, in the sulfur cycle, a phenomenon called the "thiosulfate shunt" (Jørgensen 1990). This compound can actually either be oxidized to sulfate (SO_4^{2-}), dismutated to sulfate (SO_4^{2-}) and sulfide (HS^-) (disproportionation of thiosulfate), or be reduced to sulfide.

14.4.2 The Production of Sulfide

14.4.2.1 Organic Sulfur: Biosynthesis and Degradation

The organic matter is a reservoir of sulfur and thus a form of immobilization of this element. Organic forms of sulfur can originate either from the assimilatory reduction by plants or micro-organisms, or from the direct assimilation of sulfide leading to the incorporation of sulfur in cellular material (amino acid biosynthesis, osmoregulators, coenzymes).

Under the action of animal decomposers and of aerobic or anaerobic microorganisms, organic sulfur in death organisms is mineralized and released into the environment as inorganic sulfur, mainly in the form of sulfide. Sulfur can also be transformed in many ways in oxic or anoxic conditions, resulting in the production of volatile organic

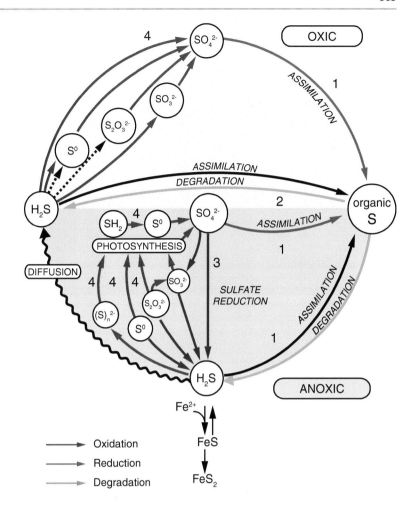

Fig. 14.36 Biogeochemical sulfur cycle under oxic and anoxic conditions, showing the main biological transformations (Drawing: M.-J. Bodiou)

compounds (dimethyl sulfide or DMS, carbon oxysulfide or COS) found in the atmosphere (Lomans et al. 2002).

Sulfides and polysulfides (S_x^{2-}) may react with methylated groups or long-chain aliphatic or aromatic organic compounds, to produce organic sulfur compounds during biotic and abiotic reactions. Biodegradation of amino acids and dimethylsulfoniopropionate (DMSP) contributes to the formation of other organic sulfur compounds such as methanethiol (MSH: CH_3SH), dimethyl sulfide (DMS: $[CH_3]_2S$), dimethyl disulfide (DMDS: $[CH_3]_2S_2$), or 3-mercaptopropionate (3-MPA: $HSCH_2CH_2COOH$). These volatile organic sulfur compounds exhibit small molecular weight and can be used as electron donors in anaerobic conditions. Some anoxygenic phototrophic bacteria can use the DMS as electron donor when fixing CO_2, and the produced dimethyl sulfoxide (DMSO) can be used in respiratory processes, as electron acceptor, by phototrophic bacteria (also called photosynthetic bacteria) in the absence of light energy and by facultative aerobic microorganisms. Methanogens are also able to consume DMS and MSH. Under aerobic conditions, sulfur-oxidizing chemolithotrophic bacteria and the micro-organisms of the genus *Hyphomicrobium* are also able to achieve this degradation (Lomans et al. 2002).

Sulfur is thus subjected to many changes between its organic and mineral forms. Degradation of sulfur organic compounds releases simple organic compounds (formaldehyde, formate, acetate, propionate) and sulfide. Production of sulfides through this step of the sulfur cycle represents a small fraction (1–5 %) of sulfides produced in marine ecosystems where dissimilatory reduction of sulfate is predominant (Jørgensen 1977), but represents the largest source of reduced sulfur compounds in sulfate poor freshwater environments.

14.4.2.2 Dissimilatory Sulfate Reduction

One of the main mechanisms for the production of reduced mineral sulfur compounds is often grouped under the general term of dissimilatory sulfate-reduction or dissimilatory reduction of sulfates, a specific metabolism of bacteria and *Archaea*. This term, widely used, is not appropriate because it includes organisms that use sulfate but also other oxidized sulfur compounds (thiosulfate, sulfite, elemental sulfur) as terminal electron acceptors. This anaerobic respiration metabolism, commonly called "sulfate respiration" (*cf.* Sect. 3.3.2, Figs. 3.22, 3.23; Table 3.9), largely contributes to the degradation of organic compounds in salted anoxic

Fig. 14.37 Photographs showing microbial sulfureta in shallow coastal lagoons. (**a**) Development of chemolithotrophic sulfur-oxidizing bacteria (*white patches*) at the surface of an anoxic sediment rich in black iron sulfide; (**b**) development of anoxygenic phototrophic bacteria on the leaves of decaying macroalgae (Photographs from R. Guyoneaud and Pierre Caumette)

environments rich in sulfur such as marine water or salt marshes (Jørgensen and Fenchel 1974). One of the main groups of microorganisms involved in this process is represented by sulfate-reducing bacteria (SRB), but in addition few *Archaea* are also involved in this activity. The general term of sulfate-reducers will be used hereafter in this chapter.

Fermentation and anaerobic respiration processes (denitrification, sulfate-reduction and methanogenesis), which are essential in the mineralization of organic matter, produce large amounts of CO_2 as well as reduced inorganic compounds which can be either trapped in anoxic zones (FeS, FeS_2), recycled (H_2, CH_4), or released in surface waters and atmosphere (CH_4, N_2, H_2S). Among these metabolisms, the dissimilatory sulfate reduction is largely favored in sulfate-rich marine environments where some authors consider that 50 % of the organic matter is mineralized this way (Jørgensen 1977). In some eutrophic areas, sulfate-reduction activity is so intense that the environment is saturated with sulfur. In such ecosystems called sulfuretum, the sulfur cycle is particularly active (Fig. 14.37).

During their activity measured by incorporation of $^{35}SO_4^{2-}$ in sediments (Box 14.11), sulfate-reducers produce large quantities of sulfides which will either react and precipitate with metals and stored as metallic sulfide precipitates (mainly pyrite) in the sediments, or reoxidized via chemical or biological processes (Fig. 14.36). Overall, the activity of sulfate-reducers can have beneficial impacts on the ecosystem because in addition to the mineralization of organic matter, it contributes to the precipitation of potentially toxic metals. However, it is shown that sulfate-reducers have many adverse effects on the environment. Production of sulfide is responsible for bad smells and possibly animal or human poisoning in natural environments or in sewers. The sulfides are indeed extremely toxic or even lethal at low doses, and act by forming complexes with metalloproteins (respiratory obstruction, hepatic intoxication). Sulfate-reducers are also known for their role in bacterial corrosion phenomena (Box 14.12), and in the phenomenon of sulfide accumulation ("Souring") existing in oil wells. Finally, some sulfate-reducing bacteria are involved in the production of methylmercury, a neurotoxic compound implicated in the Minamata disease (Box 14.14).

Physiologically, sulfate-reducers are a very diverse group of microorganisms; they use many organic or inorganic compounds, both as electron donors and carbon sources (Pfennig 1989). Organic substrates most commonly used are low molecular weight compounds from the fermentation of organic matter. Simple organic acids (acetate, lactate, propionate, or butyrate) are excellent electron donors which are oxidized by these microorganisms either completely to CO_2 or incompletely, the final product in the latter case generally being acetate. Other substrates can be used by sulfate-reducing bacteria, including long chain fatty acids (up to C_{20}), some dicarboxylic acids (succinate, malate, fumarate), alcohols (ethanol, butanol, propanol, glycerol), some amino acids (glycine, alanine, serine) or sugars such as glucose and fructose (e.g., *Desulfotomaculum*). Some strains metabolize certain aliphatic hydrocarbons such as hexadecane, or aromatic hydrocarbons such as toluene or naphthalene (*cf.* Sect. 16.8.2). Sulfate-reducers can also use molecular hydrogen (H_2) as an electron donor, in this case the source of carbon can be either an organic compound or CO_2. Others have the opportunity to grow by fermenting substrates such as

Box 14.11: Measurements of Sulfate-Reducing Activity

The measurement of the sulfate-reducing activity in sediment is performed by determining the reduced forms of sulfur (elemental sulfur, iron monosulfide, pyrite, etc.) produced after the addition of radioactive sulfate ($^{35}SO_4^{2-}$). Most often, this technique is applied to sediment cores by the method of injection into strata. The method involves, after incubation, a sediment distillation to recover the reduced forms of sulfur and trap the insoluble form of zinc sulfide (ZnS). After these steps of injection/incubation and distillation, the method is based in part on the analysis by spectrophotometry of sulfate concentration (Tabatabai 1974) and sulfide concentration (Cline 1969) in the different fractions obtained after distillation, and second, on the measurement of the radioactivity emitted by the fraction containing the sulfide. This method was originally described by Jørgensen and Fenchel (1974). The authors used a sediment distillation under acid conditions which permitted the recovery as a fraction of all the reduced forms of sulfur, essentially free sulfide and acid-volatile sulfide (FeS). The method has subsequently been modified to retrieve all elements of radioactive sulfur from sulfate (Fossing and Jørgensen 1989). This new technique of distillation, commonly called single step method involves a distillation under acidic and reducing conditions with the presence of chromium (CrIII). It allows to take into account a greater number of reduced forms of sulfur, including some products of sulfide reoxidation such as elemental sulfur and sulfide trapped in the form of pyrite (FeS_2), which were not taken into account in the original method. The latter method has subsequently been the subject of many optimizations according to study sites (marine or freshwater, temperate, hot or cold) by considering the injected dose of radioactive sulfate, the distillation process or the time incubation. The method presented here is an example used for brackish and temperate sediments. The sample on which the measurement is made is removed by means of a core barrel provided with apertures (Box Fig. 14.19) sealed with silicone to allow the injection of radioactive sulfate (concentration of 1–10 $\mu Ci.mL^{-1}$) on the entire height of the sediment

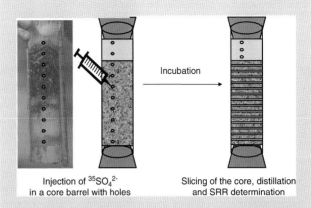

Injection of $^{35}SO_4^{2-}$ in a core barrel with holes

Slicing of the core, distillation and SRR determination

Box Fig. 14.19 Schematic drawing of the cores used for radioactive sulfate injection. The cores are fitted with injection holes every 2 cm along the cores; each hole is closed with silicone. Injection is done through every hole

sample. The injected volume is between 2 and 10 µL. After incubation (30 min to 6 h, depending on the type of sample), the sediment core was sliced between 1 and 2 cm thick fractions stored then in zinc acetate 2 %.

The activity of sulfate-reducing (Sulfate reduction rate, SRR) is calculated using the formula

$$SRR = \frac{[SO_4^{2-}] \cdot A \cdot 1.06}{(A + a) \cdot h} \text{ (in mmoles sulfate.cm}^{-3}.h^{-1})$$

$[SO_4^{2-}]$ = concentration of sulfate (mmoles.cm^{-3}) a = radioactivity of the fraction containing sulfide (Zn^{35}S) A = radioactivity of the fraction containing sulfate ($^{35}SO_4^{2-}$) 1.06 = conversion factor, taking into account isotope discrimination related to the use of ^{35}S compared with ^{32}S. h = incubation time (h^{-1})

fumarate or malate. In most natural environments, sulfate-reducers preferably use dihydrogen and acetate.

The electron acceptor most commonly used by sulfate-reducers is the sulfate. Other sulfur compounds such as sulfite and thiosulfate can serve as electron acceptors, as well as elemental sulfur for a few particular strains belonging to the genera *Desulfovibrio* and *Desulfomicrobium*.

Box 14.12: Bacteria and Corrosion of Steel

Michel Magot

Bacterial corrosion, biocorrosion, MIC (microbially Influenced – or Induced – Corrosion), biological corrosion, corrosion by microorganisms: it seems that there is not much consensus to designate the phenomenon than there is to interpret its mechanisms. But whatever you call it, it is undoubtly a harmful consequence of bacterial proliferation against which many industries would like to guard, as it is considered that its economic weight just in France in this early twenty-first century is about the billion € per year (Féron et al. 2002). Under these various names, one usually refers to a set of special cases of biodeterioration of materials (Lemaitre et al. 1994), that of the steel corrosion caused by bacterial growth, and more specifically by that of sulfate-reducing bacteria. The corrosion of a metal is a phenomenon of electrochemical nature. In contact with an electrolyte, the metal atoms can spontaneously move as cations in the aqueous solution (anodic reaction), and its electrons are used to reduce the elementary particles of the aqueous phase (cathodic reaction). These two half-reactions are associated with electric currents, namely anodic and cathodic currents. The net result is the oxidation of the metal, the most commonly observed form being the rust that occurs under aerobic conditions, when the metal electrons are transferred to oxygen. Under anaerobic conditions, the oxidation reaction of the metal results from electron transfer to the protons of the electrolyte, and leads to the formation of hydrogen. In all industries using steel equipments in contact with water, whether aerated or not, most cases of bacterial corrosion occur under anaerobic conditions. Indeed, within a biofilm in which aerobic bacteria grow by consuming oxygen, micro-niches favorable for the development of anaerobic bacteria are created, including sulfate-reducing bacteria (SRBs) that are most frequently implicated in cases of bacterial corrosion listed in the scientific literature.

By the way, their role in this process had been demonstrated very early in the history of microbiology, since this was in 1934 that Von Wolzogen Kur and Van der Vlugt proposed a mechanism of bacterial corrosion called "the cathodic depolarization theory". Briefly, this theory states that SRBs accelerate the spontaneous reaction of steel corrosion in contact with water (two half-reactions mentioned above) by consuming its products: metal ions by precipitation as sulfide minerals, and cathodic hydrogen used as an energy source. This theory has been, and is still, extremely popular within the community of microbiologists more than 70 years after its statement. Actually, some experimental approaches tended to confirm it, and discoveries made after its statement, e.g., the hydrogenases of SRBs, brought new arguments to the concept of cathodic depolarization. Electrochemists marked nevertheless less effusive, until they show that advances in the field of thermodynamics made after the 1930s were not compatible with this theory (Crolet 1992). It thus may still take many years before the traditional citation of the cathodic depolarization theory gives way to other interpretations of the mechanism of corrosion by SRB. Nevertheless, other tracks have been or are being explored, among which is the role of sulfides, resulting from the bacterial reduction of sulfate or thiosulfate. Their extreme corrosivity, especially in acidic medium, is well documented in the corrosion literature. Precisely, the local acidification of the steel surface under a biofilm resulting from SRB metabolic activity is a mechanism that has been hypothesized, and then demonstrated experimentally (Daumas et al. 1993). The putative role of phosphorus compounds was also suggested.

The direct or indirect role of bacterial sulfide production was reinforced by the discovery that other sulfidogenic bacteria could cause unusually fast steel corrosion in very special circumstances. By analogy, the acronym TRB, thiosulfate-reducing bacteria, was proposed to designate this group of strictly anaerobic bacteria that reduce thiosulfate but not sulfate, to sulfide (see Sect. 14.4). Although thiosulfate-reduction does not seem to produce energy for these fermentative anaerobic bacteria, the amount of sulfides that may be produced in culture could be 2–10 times greater than sulfide production by SRB. In vitro electrochemical experiments in anaerobic fermenters showed that *Dethiosulfovibrio peptidovorans* can corrode steel coupons at a rate in the order of 1 cm per year in the presence of thiosulfate, while SRB, tested under the same conditions corrode at 5–10 times slower rates. It seems that the most spectacular industrial cases of

M. Magot (✉)
Institut des Sciences Analytiques et de Physico-chimie pour l'Environnement et les Matériaux (IPREM), UMR CNRS 5254, Université de Pau et des Pays de l'Adour, B.P. 1155, 64013 Pau Cedex, France

(continued)

Box 14.12 (continued)

Box Fig. 14.20 Crater corrosion of a steel pipeline approximately 1 cm thick. (Photograph: courtesy of the Company Total)

Box Fig. 14.21 Corrosion of a bottom plate of a storage tank of crude oil. (Photography: courtesy of the Company Total)

bacterial corrosion are sometimes associated with the particular action of these bacteria, especially when the presence of thiosulfate in the waters may be suspected (Box Fig. 14.20).

Mechanisms of bacterial corrosion are still poorly understood, and are likely more complex than the interpretations that have been given, since the SRBs and TRBs grow within multispecies microbial communities, wherein each microbial population contributes to the development and activity of the biofilm. So there is currently no preventive or curative treatment that is truly specific (and reliable!) of bacterial corrosion. It is therefore considered by the industry first by monitoring the presence of SRBs and sometimes TRBs in facilities, and second by treating industrial water with unspecific, wide spectrum antibacterial agents (biocides). Detection is mainly done by the method of most probable number method in liquid culture media. So-called SRNB and TRB "Test-kits" are commercially available, and some are proving surprisingly much more sensitive and specific than the culture media of BSR traditionally used in research laboratories. Other methods have been developed, patented, and sometimes marketed as the detection of hydrogenase or APS-reductase enzymes, but they are either poorly insensitive or nonspecific, and always tedious to implement. Considering techniques based on gene amplification, they are still considered maladjusted to monitoring constraints in industrial environments and have not been commercially developed yet.

Treatment of industrial equipments are designed to limit the development of biofilms on one hand, and to deal more specifically with SRB on the other. At eliminate biofilms, mechanical treatments are sometimes applied when possible, as the scraping (pigging) of pipelines in the oil industry. Chemical treatments involve the injection of biocides, a term which includes a very wide range of more or less active commercial products. Some are particularly effective in limiting the development of SRBs as glutaraldehyde, or broad spectrum like THPS (TetrakisHydroxymethylPhosphoniumSulfate) (Box Fig. 14.21).

Sulfate-reducing bacteria have many other metabolic alternatives. Electron acceptors are not only limited to sulfur compounds: nitrate, fumarate, and certain metals can play this role. Finally, some studies have challenged the question of dioxygen toxicity for sulfate-reducers, some of the strains of which are not only able to tolerate low dioxygen tensions, but are also capable of reducing dioxygen to H_2O (*cf.* Sect. 3.3.2). The ecological significance of this process and its possible contribution in terms of energy production remain to be elucidated (Cypionka 2000).

Finally, sulfate-reducing bacteria are involved in the dismutation of thiosulfate, sulfite, or elemental sulfur, as discovered in *Desulfovibrio sulfodismutans* (Bak and Cypionka 1987). Dismutation is an inorganic fermentation during which sulfite and thiosulfate are used both as electron donors and acceptors, thus contributing to the formation of sulfide (reduction of the electron acceptor) and sulfate (oxidation of the electron donor) (Fig. 14.36). Thus, sulfate-reducers exhibit a great number of metabolic potentials allowing them to thrive in highly diverse environmental situations, and play a major role in the transformation of organic and inorganic compounds.

The phylogenetic dispersion of sulfate-reducers makes difficult the study of this community based exclusively on the 16S rRNA gene (*cf.* Sect. 6.6.2), which led to the investigation of novel molecular markers among specific sulfate-reducing enzymes. Among these enzymes, adenosine phosphosulfate reductase (APS reductase) and sulfite reductase catalyze the reduction of APS to AMP and sulfite, and the dissimilatory reduction of sulfite to sulfide, respectively (*cf.* Sect. 3.3.2). The genes encoding these enzymes, including dsrAB genes, are becoming increasingly used for the study of the communities of sulfate-reducers in natural environments (Wagner et al. 2005).

14.4.2.3 Other Sulfide-Producing Metabolisms

Micro-organisms which are unable to use sulfate as an acceptor of electrons participate in the production of reduced sulfur compounds by dissimilatory reduction of elemental sulfur. Sulfur reducers constitute a physiologically disparate group of microorganisms, comprising both *Bacteria* (genera *Desulfuromonas*, *Desulfurella*) and *Archaea* (*Crenarchaeota* and *Euryarchaeota*). Some of these microorganisms completely oxidize simple organic substrates (acetate, lactate, pyruvate, succinate, ethanol). Those belonging to the genus *Desulfurella* are strict sulfur reducers, while in the genus *Desulfuromonas*, fumarate or malate may be used as electron acceptors by certain strains (Widdel and Pfennig 1992). Some fermentative sulfur reducers are not involved in respiration but rather use sulfur as an electron trap to increase energy production and synthesize additional ATP through substrate phosphorylation.

Sulfur reducers are abundant in inland and oceanic geothermal environments, rich in elemental sulfur. Indeed, many sulfo-reducing species are extreme (especially among *Archaea*) or moderate (bacteria) thermophiles, although mesophilic or even psychrophilic sulfo-reducing species exist. They are also abundant in marine or brackish waters where sulfides produced by sulfate-reducers are chemically or biologically oxidized, contributing to the formation of elemental sulfur used by sulfo-reducing bacteria (Fig. 14.36). These bacteria can live in mutualistic association with phototrophic green bacteria, based on the oxidative reduction of sulfur.

Finally, some microorganisms, such as phototrophic purple sulfur bacteria, chemolithotrophic *Beggiatoa* and certain cyanobacteria, have the possibility of exploiting the sulfo-reduction as a mechanism to conserve energy, the sulfur being utilized as the final electron acceptor during oxidation of cellular polysaccharide reserves in the dark.

Another group of organisms, the thiosulfate-reducers, are specialized in the use of thiosulfate as electron acceptor.

These micro-organisms have a fermentative metabolism. In thiosulfate-reducers, and in fermentative microorganisms which reduce sulfur, the sulfur compound react by trapping electrons and contributes to the reduction of high dihydrogen concentrations generated by fermentative metabolism. During the fermentation of substrates, the reduction of sulfur compounds increases the oxidation level of fermentative products, thereby enhancing the production of energy and the growth rates of microorganisms. This is not a respiratory process but the use of thiosulfate as an electron trap by fermentative microorganisms (which is similar in some sulfur reducers), involves strict anaerobic thermophiles, primarily those from the genera *Thermotoga*, *Thermosipho* and *Fervidobacterium*. The thiosulfate-reduction by these non-sulfate-reducers has been identified in soil and subsoil ecosystems. They seem to have a corrosion power far superior to that of sulfate- or sulfur reducers (Box 14.12).

14.4.3 The Production of Sulfate

The oxidation of sulfides or of other sulfur-containing compounds of higher oxidation state depends on both chemical and biological processes. In the presence of dioxygen, sulfide is chemically oxidized to elementary sulfur or thiosulfate. Similarly, sulfite, which is a highly reactive compound is rapidly oxidized to sulfate in the presence of molecular oxygen. In anoxic environments, sulfide can be trapped in the sediments in the presence of ferrous iron to form iron sulfide (FeS) and pyrite (FeS_2), while ferric iron reacts with sulfide to form elementary sulfur. The biological oxidation of sulfur compounds involves many bacterial groups with diverse metabolisms: chemolithotrophic sulfo-oxidizing prokaryotes (CSOP), anoxygenic phototrophic bacteria (APB), sulfate-reducing bacteria, and cyanobacteria. While the two first groups are typical sulfur-oxidizing bacteria (*i.e.* the sulfur compounds are used as electron donors), the contribution of microorganisms in the two other groups to the oxidation of sulfur compounds are occasional or circumstantial. Hence, sulfate-reducing bacteria capable of dismutation contribute to the production of both reduced and oxidized sulfur compounds. In the presence of sulfide which is an inhibitor of the photosystem II, some cyanobacteria can also carry out an anoxygenic photosynthesis identical to that of APB by photometabolizing sulfide (electron donor) through the photosystem I (Cohen et al. 1975).

14.4.3.1 Chemotrophic Production of Sulfate: Colorless Sulfo-Oxidizing Microorganisms

Chemolithotrophic sulfo-oxidizing prokaryotes (CSOP) are also called colorless sulfo-oxidizing microorganisms, by opposition to phototrophic bacteria in which the presence of photosynthetic pigments provides a color to cellular suspensions and, in some cases, to the whole natural communities (e.g., bacterial mats, red tide phenomena). CSOP are highly diverse, comprising phylogenetically distant taxa (*Archaea* and *Bacteria*) (*cf.* Chapter 6.6). Depending on the species, they are able to oxidize various forms of reduced sulfur compounds, such as sulfides, elemental sulfur, sulfite, thiosulfate, or tetrathionate. The end product of this oxidation is sulfate, together with intermediate compounds, primarily elemental sulfur which can be stored as intra- or extracellular sulfur globules. It is possible to distinguish four groups of CSOP depending on their carbon and energy sources (Robertson and Kuenen 2006).

The first group, known as obligatory chemolithotrophs, consists of specialized microorganisms in which the electron donor is a mineral sulfur compound, the carbon being obtained by CO_2 fixation via the Calvin cycle. However, most of these microorganisms can utilize exogenous carbon sources and ferment their cytoplasmic glycogen reserves when environmental conditions are unfavorable. Yet, in this group of organisms that includes bacteria (*Thiomicrospira*, *Thiobacillus*, *Acidithiobacillus*) and archaea (*Sulfolobus*), these heterotrophic metabolisms are secondary and mainly represent an adaptation to adverse environmental conditions. The second group comprises facultative chemolithotrophs which exhibit the same possibilities compared with the first group, but can also grow as heterotrophs or mixotrophs. The third group, the chemolithoheterotrophs, is represented by few *Thiobacillus* and *Beggiatoa* who are unable to fix mineral carbon but use reduced sulfur compounds as electron donors and energy source; they thus absolutely require an organic carbon source for growth. Finally, the **chemoorganoheterotrophs*** (*Beggiatoa* sp., *Macromonas*, *Thiobacterium* and *Thiothrix*) constitute the fourth group. They can oxidize sulfur compounds without apparent energetic benefit.

Dioxygen is the prime electron acceptor for CSOP, but depending on the species, the need and tolerance to dioxygen vary greatly. Many CSOP have metabolic potentials to grow or survive in anoxic conditions; the electron acceptor being nitrate (reduced to nitrite or dinitrogen) or nitrite (with release of dinitrogen). Under these conditions, some chemolithoautotrophic CSOP (*Thiobacillus denitrificans*, *Sulfurimonas denitrificans*) continue to grow by using reduced sulfur compounds as electron donors. Others lose their ability to use sulfur compounds and then grow as denitrifying chemoorganoheterotrophs.

CSOP live mainly at the interface between oxic (containing dioxygen) and anoxic (containing sulfides) environments. Their development is especially visible in the water-sediment interfaces in both continental and oceanic sulfur sources. The development of some filamentous forms (genera *Beggiatoa*, *Thiotrix*) can be visualized in the form of white filaments (Fig. 14.38). Their role in the detoxification of sulfides is especially important in the absence of light and thus of anoxygenic photosynthesis. In some ecosystems, such as the deep oceanic (sulfurous hot springs) or coastal (sulfide-rich sediments) environments, CSOP live in mutualistic symbiosis with molluscs or worms; these latter provide a favorable environment (habitat, substrates) to CSOP which, in turn, eliminate toxic sulfides and produce biomass and energy used by their hosts.

In some cases, aerobic sulfur oxidation activity has adverse effects on poorly buffered ecosystems, primarily via production of large quantities of sulfuric acid (H_2SO_4) which promotes corrosion of certain materials (concrete, stone, metals). This property, coupled with the ability of certain CSOP to oxidize ferrous iron, is used to improve the recovery of ore deposits (bioleaching) or the bioremediation of soils contaminated with metals (*cf.* Sect. 16.9).

14.4.3.2 Phototrophic Sulfate Production: Photosynthetic Sulfo-Oxidizing Microorganisms

Anoxygenic phototrophic bacteria (APB) have only one photosystem and use compounds which are more reduced than water as electron donors during their photosynthesis. In this case, the electron donors are hydrogen sulfide or other reduced sulfur compounds, dihydrogen, or organic compounds of low molecular weight. This photosynthesis, which does not lead to the production of dioxygen, is called anoxygenic photosynthesis (*cf.* Sect. 3.3.4). Therefore, it represents an essential step in matter cycle under anoxic conditions, wherein the couple SO_4^{2-}/H_2S can be compared with the couple O_2/H_2O under aerobic conditions. Anoxygenic phototrophic bacteria are considered as paraprimary producers because they use compounds derived from the degradation of organic material and from anaerobic respirations (Pfennig 1975).

From a metabolic and ecological point of view, APB can be separated into two main groups: phototrophic sulfur bacteria that mainly use sulfur compounds as electron donors (sulfide, polysulfides, elemental sulfur, thiosulfate, sulfite), and phototrophic nonsulfur bacteria which preferentially photo-oxidize organic compounds (Overmann and Garcia-Pichel 2006). When electron donors are reduced sulfur compounds, their oxidation during photosynthesis leads to the formation of elemental sulfur and sulfate. From a phenotypic point of view, anoxygenic phototrophic bacteria were classified into two major groups according to their photosynthetic pigments: purple and green phototrophic bacteria. The differentiation of these two groups is based, on one hand, on the ultrastructure of the photosynthetic apparatus where the reaction centers are located and, on the other hand, on the nature of the pigments, bacteriochlorophylls and carotenoids which determine the natural colors of these microorganisms (Fig. 14.39; Table 14.8).

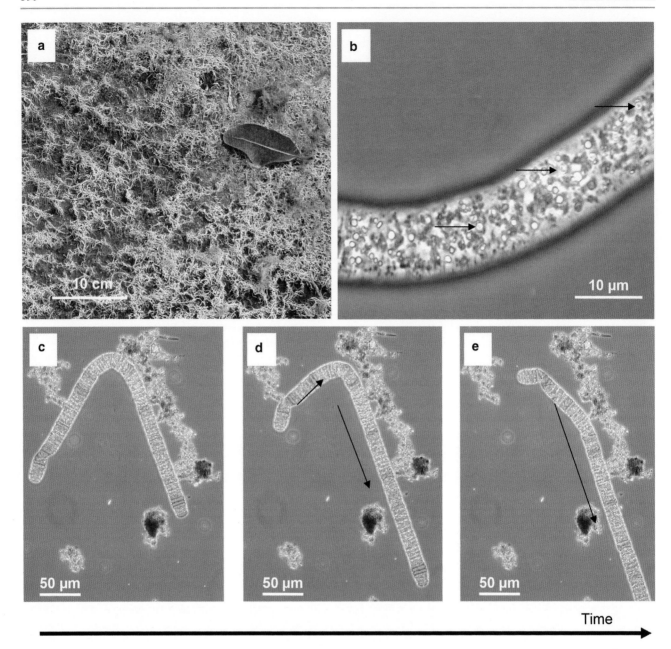

Fig. 14.38 Aerobic chemolithotrophic sulfur-oxidizing bacteria from a sulfuretum at the surface of coastal sediments. (**a**) Development of white filaments at the sediment surface; (**b**) microphotograph of a *Beggiatoa* filament showing the intracellular sulfur globules (*arrows*); (**c, d** and **e**) time series microphotographs showing the gliding movement of the filaments on a microscopic slide (Photographs from Rémy Guyoneaud)

From a phylogenetic point of view (*cf.* Sect. 6.6.2), anoxygenic phototrophic bacteria are distributed among four lineages of the domain *Bacteria*:
The purple bacteria (sulfur and nonsulfur) belong to the *Alpha-*, *Beta-*, and *Gammaproteobacteria*;
The green sulfur bacteria which constitute an independent phylum (*Chlorobi*);
The filamentous green bacteria are found in the *Chloroflexi*;
The gram positive *Heliobacteria* in the phylum *Firmicutes*.

Within each lineage, phototrophic species are grouped with nonphotosynthetic members, except for green sulfur bacteria which form a phylogenetically coherent group.

Because of their phylogenetic heterogeneity, the use of molecular techniques for the *in situ* study of these microorganisms is most often based on the study of functional genes, such as *puf* (involved in the photosystem of purple bacteria) and *fmo* (involved in the photosystem of green sulfur bacteria) genes. However, approaches based on

Fig. 14.39 Pure cultures of anoxygenic phototrophic bacteria showing their natural colour. (**a**, **b**, and **c**) purple sulfur and non-sulfur phototrophs; (**d**) Green sulfur bacteria containing chlorobactene and bacteriochlorophyll c as main pigments; (**e**) Green sulfur bacteria containing isorenieratene and bacteriochlorophyll e as main pigments (Photographs from Pierre Caumette)

Table 14.8 Biosynthetic pathways and main carotenoids found in anoxygenic phototrophic bacteria

Main biosynthetic pathways	Carotenoids	Color of cell suspensions	
Normal spirilloxanthin	Lycopene, Rhodopin, Spirilloxanthin	Pink, red to brown-red	Purple bacteria
Rhodopinal	Lycopene, lycopenal, lycopenol, rhodopine, rhodipinal, rhodopinol	Violet	
Alternative spirilloxanthin	spheroidene, hydroxyspheroidene, spheroidenone, hydroxyspheroidenone	Yellow to brown	
Okenone	Okenone	Purple-red	
Chlorobactene	γ-carotene, chlorobactene, hydroxychlorobactene	Green	Green bacteria
Isorenieratene	β-carotene, isorenieratene	Brown	

the analysis of the gene encoding rRNA 16S are possible for *Chromatiaceae* and green sulfur bacteria.

Phototrophic Purple Bacteria

Phototrophic purple (or red) bacteria are distributed among *Proteobacteria:* three families of sulfur bacteria within the *Gammaproteobacteria* (*Chromatiaceae*, *Ectothiorhodospiraceae*, *Halorhodospiraceae*), and a set of physiologically close species (phototrophic purple nonsulfur bacteria) within the alpha- (*Rhodospirillum*, *Rhodobacter*, etc.) and beta- (*Rhodoferax*, *Rubrivivax*, etc.) proteobacteria. In these phototrophic microorganisms, reaction centers are located in cytoplasmic invaginations forming cellular vesicles, plates, or tubules (*cf.* Sect. 3.3.4) that contain photosynthetic pigments: carotenoids and bacteriochlorophylls *a* or *b* (Overmann and Garcia-Pichel 2006).

The family *Chromatiaceae* (Table 14.9) comprises photolithotrophic microorganisms that use reduced sulfur compounds as electron donors under anoxic conditions. They accumulate elemental sulfur as intracellular globules (Fig. 14.40). In these microorganisms, CO_2 fixation is achieved through the reductive cycle of pentoses-phosphates (Calvin-Benson Cycle). Two major physiological groups can be differentiated among *Chromatiaceae*. The first group comprises strict anaerobic microorganisms and obligatory photoautotrophs (*Chromatium okenii*, *Thiospirillum jenense*, and *Thiococcus pfennigii*). These bacteria photo-oxidize sulfide or sulfur and can only photo-assimilate few simple organic compounds (acetate, pyruvate) in the presence of CO_2 and sulfide. They are unable to grow on thiosulfate and dihydrogen; they also do not grow under photoheterotrophic or chemotrophic conditions as well. The second group includes species which are more versatile, capable to photo-assimilate many organic compounds in the presence (mixotrophy) or in the absence of reduced sulfur compounds. These microorganisms are less sensitive to dioxygen can be grown in chemolithotrophic or in chemoorganotrophic conditions, in the presence of low dioxygen tensions. The genera *Thiocapsa*, *Amoebobacter*, and *Allochromatium* are the main representatives of this group.

The families *Ectothiorhodospiraceae* and *Halorhodospiraceae* (Table 14.9) include anaerobic photolithoautotrophic microorganisms, and form extracellular sulfur globules during the oxidation of reduced sulfur compounds

Table 14.9 Main anoxygenic phototrophic bacterial groups and their characteristics

Phototrophic microorganisms	Ultrastructural localisation of the pigments[a]	Photosynthetic pigments	Election donors[b]	Carbon sources
Purple nonsulfur bacteria	IMS (V. L, P. T)	Bchl. a/b carotenoids	Organic ($H_2S/S°/S_2O_3^{2-}/H_2$)	Organic, CO_2
Chromatiaceae	IMS (V)[c]	Bchl. a/b carotenoids	$H_2S/S°/S_2O_3^{2-}/H_2$ (organic)	CO_2, Organic
Ectothiorhodospiraceae	IMS (P)	Bchl. a/b carotenoids	$H_2S/S°/S_2O_3^{2-}/H_2$ (organic)	CO_2, Organic
*Aerobic anoxygenic phototrophs[d]	Cytoplasmic membrane	Bchl. a/b carotenoids	Organic	Organic
Heliobacteriaceae	Cytoplasm	Bchl. g carotenoids (neurosporene)	Organic	Organic
Green nonsulfur bacteria (multicellular, filamentous)	Chlorosomes	Bchl. c/d carotenoids	Organic ($H_2S/S°/H_2$)	Organic, CO_2
Green sulfur bacteria	Chlorosomes	Bchl. c/d/e carotenoids	$H_2S/S°/H_2$	CO_2

[a]IMS, internal membrane system due to membrane invaginations organized as vesicles (V), Lamella (L), lamella organized in plates (P) or tubules (T)
[b]Main and accessory (in *brackets*) electron donors
[c]Some strains within this group contain lamellar stacks as internal membrane system
[d]Aerobic anoxygenic phototrophs are heterotophs and photosynthesis contributes only partially to their energetic metabolism

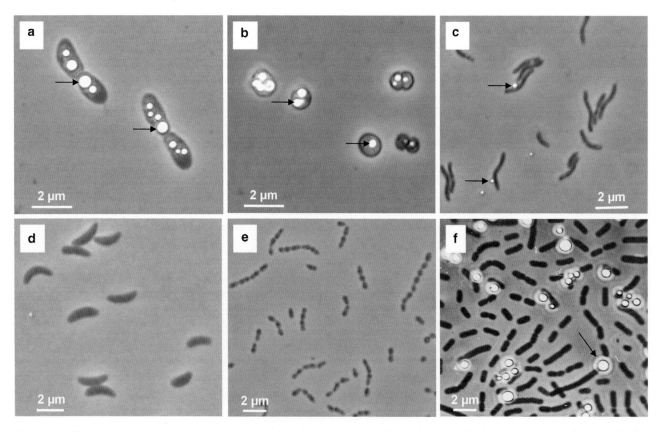

Fig. 14.40 Phase contrast microphotographs of anoxygenic phototrophic bacteria. (**a, b**) *Chromatiaceae* with intracellular sulfur globules (*arrows*); (**c**) *Ectothiorhodospiraceae* with extracellular sulfur globules (*arrows*); (**d, e**) purple nonsulfur bacteria; (**f**) green sulfur bacteria with extracellular sufur globules (*arrows*) (Photographs from Rémy Guyoneaud and Pierre Caumette)

(Fig. 14.40). Similar to *Chromatiaceae*, some species are able to photo-assimilate simple organic compounds or to grow in micro-oxic conditions or in the darkness. The two families group weak, moderate, and extreme halophilic species. Some of these species are able to grow at very high pH. For example, the species *Halorhodopsira abdelmalekii*, isolated in sodium-carbonate-rich lakes (*i.e.* saturated with Na_2CO_3) in Egypt, grows at pH 11 and in salinity conditions up to 250 **psu***. Their reaction centers are located in specific structures consisting of stacks of lamellae, due to cytoplasmic invaginations which are fixed to the cytoplasmic membrane by a single attachment point, thereby forming almost a differentiated organelle (*cf.* Chap. 3.4; Fig. 3.27).

Phototrophic purple nonsulfur bacteria (Table 14.9) represent a more heterogeneous group. This diversity is reflected in their morphology (Fig. 14.40), their modes of cell division (binary or by budding), the types of inner membrane structures containing various carotenoids, and in their high metabolic diversity. Under anaerobic conditions and in the presence of light, these organisms preferentially use organic compounds as electron donors (photoorganotrophy), but are capable of using dihydrogen or sulfur compounds (photolithotrophy). In this latter case, the oxidation of sulfides stops most often at the stage of elemental sulfur, deposited extra-cellularly. Only a few species belonging to genera *Rhodobacter*, *Rhodovulum*, *Rhodobium*, and *Rhodopseudomonas* oxidize sulfur compounds to sulfate. The majority of these microorganisms is tolerant to dioxygen and can perform aerobic respiration. The respiratory capabilities of these bacteria can also be expressed under anoxic conditions, with electron acceptors such as nitrate, dimethyl sulfoxide, or trimethylamine oxide. These microorganisms are also capable of fermentation.

Phototrophic Green Bacteria
They include phototrophic green sulfur bacteria (phylum *Chlorobi*) and multicellular filamentous green bacteria (Phylum *Chloroflexi*). These microorganisms exhibit bacteriochlorophyll *c*, *d*, or *e*, and carotenoids in the biosynthetic series of chlorobactene or isorenieratene (Tables 14.8 and 14.9). These pigments are contained in chlorosomes, *i.e.* vesicles which are attached to the inner cytoplamic membrane and act as powerful light energy sensors. Small amounts of bacteriochlorophyll *a* are also present in these bacteria, mainly localized in the reaction centers, at the attachment point of chlorosomes to the cytoplasmic membrane.

Green sulfur bacteria (Table 14.9) are strict anaerobes and photolithoautotrophs. They fix CO_2 via the reverse-cycle of tricarboxylic acids. Few simple organic compounds may be photo-assimilated in the presence of CO_2 and sulfide, the latter electron donor being oxidized to sulfate during photosynthesis. Sulfur globules (intermediate compounds of this oxidation) are deposited outside the cells (Fig. 14.40). Only few strains use dihydrogen or thiosulfate as electron donors. Based on their carotenoid pigments (Table 14.8), it is possible to distinguish between brown and green bacteria.

The brown ones contain bacteriochlorophyll *e* and carotenoid from the biosynthetic series of isorenieratene, while the green ones exhibit bacteriochlorophyll *c* and/or *d* and carotenoids of the chlorobactene series.

Multicellular filamentous green bacteria (Table 14.9) are represented by four genera, each comprising one or two species. Bacterial cells form flexible and motile filaments which allow individual cells to move by gliding. These facultative aerobic microorganisms preferably use organic compounds via phototrophic or chemotrophic metabolism, reduced sulfur compounds being used very little. In the latter case, the oxidation stops at the stage of elemental sulfur which is deposited extracellularly, *i.e.* outside the filaments. Photosynthetic pigments are bacteriochlorophylls *c* of *d*, while the carotenoids are beta and gamma carotenes (β- and γ-carotenes).

14.4.4 Sulfur Cycle in Natural Ecosystems

Changes in physical and chemical environments, including organic matter (eutrophication level), concentration of sulfur compounds, light, and oxygen, will influence the sulfur cycle and the dominance of targeted microorganisms, depending of their ecophysiological potentials. The sulfur cycle is strongly linked to the carbon cycle: microorganisms that produce reduced compounds (sulfate-reducers in particular) are involved in the degradation of organic matter and are known as chemoorganotrophs and heterotrophs, while those oxidizing reduced sulfur compounds are often known as para-primary chemoautotrophs or phototrophs. The concentration of oxidized sulfur compounds will eventually be a limiting factor for the activity of sulfate-reducing microorganisms. The concentration of the resulting reduced sulfur compounds will determine the oxidation pathways of these compounds by anoxygenic phototrophic and chemolithotrophic sulfur oxidizing microorganisms. The affinity of these microorganisms to sulfide, and their ability to oxidize completely into sulfate or incompletely into elemental sulfur, are parameters that may constitute a decisive advantage for some species. In these microorganisms, light (phototrophs) or dioxygen (chemolithotrophs) will also represent key factors for the functioning of the sulfur cycle.

14.4.4.1 Stratified Lakes
Many aquatic ecosystems are stratified by the presence of a **chemocline***. Unlike coastal lagoons and some athalossohaline environments whose stratification is the result of a **halocline***, freshwater lakes often present a temporary summer stratification (holomictic lakes), due to the warming of surface waters by solar radiation. The resulting establishment of a temperature gradient in the water column is known as

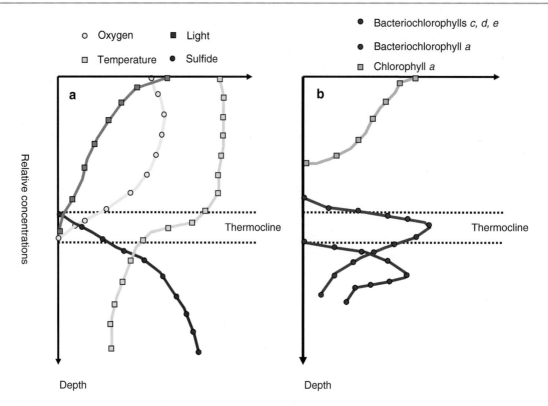

Fig. 14.41 Distribution of the main physico-chemical parameters (**a**) and biological parameters (**b**) as a function of depth in a stratified lake

thermocline*, *i.e.* the water layer where the temperature decreases rapidly. This thermocline separates a surface oxic and hot layer (**epilimnion***), from a deep anoxic layer (**hypolimnion***) where water is colder and hence denser (Fig. 14.41).

Due to the resulting density gradient created, this physical separation makes temporarily impossible the mixing of these two layers. Stratified lakes can be characterized by a unique mixing event in autumn (monomictic lakes), but more generally, in the temperate regions of the globe, by two mixing events, one in spring and one in fall (dimictic lakes). In stratified aquatic ecosystems, the anoxic hypolimnion constitutes a closed environment. This hypolimnion nevertheless receives organic matter from the primary production (algae, cyanobacteria) which occurs in the epilimnion. The degradation of this organic matter under anoxic conditions is performed by various groups of anaerobic microorganisms, including sulfate-reducers that produce sulfides which diffuse up to the transition zone between the two water layers (thermocline and chemocline). The sulfur cycle is very active in this zone.

If the light energy is available for anoxygenic phototrophic bacteria (APB), these microorganisms develop in the upper part of the hypolimnion and in the chemocline and constitute a mixed and stratified community of phototrophic bacteria that color the water in red, green, or brown, depending on the type of the dominant microorganisms (Parkin and Brock 1981). The maintenance of these bacteria in the zone where the growth conditions are optimal (light, sulfide) is either due to their motility or to the presence of gas vesicles. A stratification within the phototrophic community is often observed. Phototrophic purple sulfur bacteria grow above a proliferation of green or brown sulfur bacteria (Fig. 14.41). This stratification depends on the gradients of sulfide and dioxygen, as well as on the intensity and the nature of the light radiation. In deep lakes (chemocline between 10 and 25 m), low light intensity and the nature of light radiation select phototrophic green sulfur bacteria, whose peripheric photosynthetic apparatus (chlorosomes, *cf.* Sect. 3.3.4) enhances the light capture at low radiation. This is characteristic of depth layers where only the light radiations included between 450 and 550 nm (blue-green) are present. Under these conditions, phototrophic green sulfur bacteria will be more competitive because their carotenoid (isorenieratene and β-isorenieratene) and bacteriochlorophyll *e* pigments are particularly effective in capturing these wavelengths (Overmann et al. 1992).

In the oxygenated waters immediately above the proliferation of phototrophic bacteria, chemolithotrophic sulfur-oxidizing prokaryotes (CSOP) also oxidize the sulfide which escaped the phototrophic community underneath without being oxidized. PSB and CSOP communities re-oxidize sulfide from sulfate-reduction, and regenerate sulfate for sulfate-reducers. The stratification of all these microorganisms depends on the gradients of sulfide (electron donors for APB and CSOP), dioxygen (from the oxic epilimnion), and light (intensity and type of light radiation). These communities are maintained as long as the water column remains stratified.

Sometimes, stratification also occurs in marine water systems with the occurrence of a minimum oxygen zone in the water column. It has been demonstrated that a cryptic sulfur cycle is extremely active in these zones where sulfate-reducing organisms take advantage of low oxygen concentrations to develop.

14.4.4.2 Lagoons and Other Coastal Environments

Located at the sea-continent interface, coastal environments receive freshwater inflows (rivers, rain, and run-off waters), as well as inputs of marine waters. The former are most often rich in both organic matter (urban or industrial effluents) and nutrients from the leaching of agricultural soils. This explains the high productivity of these environments. The aerobic mineralization of both allochthonous and autochthonous organic matter creates a high biological oxygen demand (BOD), promoting the establishment of anoxic zones in sediments or deep waters of these lagoons. The presence of sulfate from seawater promotes the development of sulfate-reducers and the accumulation of toxic sulfides in the sediments and possibly in the upper waters (Caumette 1986).

It is important to distinguish shallow lagoons with homogeneous water column, from deep lagoons or estuaries where the water column can undergo stratification, most often due to vertical gradients in salinity. These gradients result from the superposition of a low-salinity surface layer over a high-salinity and denser deep layer of waters. For deep lagoons, we must also distinguish holomictic lagoons in which there is mixing of the water column (one or more times per year), from meromictic lagoons with a permanent stratification. In these environments, processes related to the sulfur cycle are similar to those described for stratified lakes.

In shallow lagoons where light penetrates to the sediment surface, the interface between the oxic and anoxic zones is localized, mostly often, in the few first millimeters of the water column. Sulfate-reducing activity is the most important metabolism in these shallow sediments (input of organic matter and of sulfate from the overlying water), where phototrophic sulfur-oxidizing (need of light) and/or chemolithotrophic sulfur-oxidizing (need of dioxygen) microorganisms develop in the water-sediment interface. However, the low depth of the water column causes high instability of these environments which are subjected to strong and rapid variations in the sulfide and dioxygen gradients.

Thus, during warm and calm periods, anoxia can develop in the whole water column (Fig. 14.42). This promotes a shift in the oxic–anoxic boundary up to the air–water interface. The entire water column is then invaded by sulfide derived from the sulfate reduction. These phenomena are called **dystrophic crises*** (Caumette 1986). Phototrophic purple sulfur bacteria subsequently grow and heavily colonize the water column, causing red colored water and oxidizing the accumulated sulfide to sulfate (Fig. 14.42). Indeed, high light intensity reaching the water column promotes the development of phototrophic purple sulfur bacteria that outcompete the green sulfur bacteria. These red-water phenomena are temporary and last until the disappearance of sulfide in the water column. Biotopes other than shallow lagoons, such as seawater pools, swamps, and sheltered parts of bay and estuaries, are susceptible to the occurrence of these red water phenomena.

The microorganisms responsible for these red waters are often very versatile in terms of their metabolic capacity, and are adapted to enhanced environmental fluctuations (De Wit and van Gemerden 1990). Thus, among these organisms, are found phototrophic purple sulfur bacteria of the genus *Thiocapsa*, which can tolerate high concentrations of sulfide (2–8 mM), store elemental sulfur in the form of intracellular sulfur, and tolerate dioxygen. Their ability to store sulfur is an advantage for these microorganisms which can use it in the absence of light, and in the presence of dioxygen by respiration. In addition, in the dark and under anaerobic conditions, this sulfur may be used as an electron acceptor during endogenous fermentation of polysacharidic reserves. Dioxygen inhibits the biosynthesis of photosynthetic pigments (bacteriochlorophylls) in phototrophic bacteria and thus their growth. However, the microorganisms in the genus *Thiocapsa* tolerate the presence of dioxygen and, under these conditions, can grow via chemoautotrophic or mixotrophic metabolisms, or by oxidizing sulfide, thiosulfate or sulfur (intra- or extra-cellular). However, these organisms are not competitive with colorless chemolithotrophic sulfur-oxidizing bacteria whose generation time is shorter. During bacterium-forced red tide phenomena, phototrophic bacteria dominate, because of the absence dioxygen.

During dystrophic crises that lead to the appearance of red bacterial water, the development of anoxygenic phototrophic bacteria ensures a resiliency toward the initial environmental conditions. Indeed, these phenomena are only the visible part of a set of processes involved in the mineralization of excess organic materials in highly productive environments. They also highlight the role of anoxygenic phototrophic bacteria in the re-oxidation of toxic sulfides from an intense sulfate-reducing activity. All of these processes are highlighted in Winograsky columns (Box 14.13). The activity of these microorganisms results in an important biomass production which is at the basis of the establishment of a new food chain, at the end of the dystrophic crisis. The biomass production by anoxygenic phototrophic bacteria is considered a para-primary production. In the case of a dystrophic crisis or not, this biomass production can serve, in addition or not to primary production, as the first level of a food chain which develops in parallel to the classical trophic chain (Fig. 14.43).

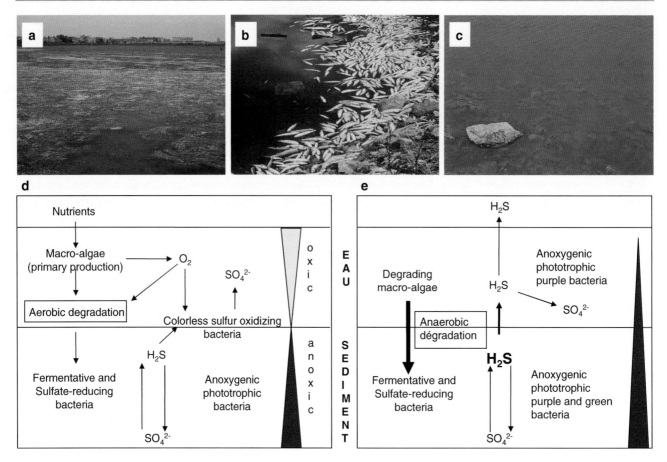

Fig. 14.42 Development of a dystrophic crisis in a shallow coastal lagoon. (**a**) Massive development of macroalgae in the eutrophic lagoon; (**b**) impact of the toxic sulfide released throught anaerobic sulfate-reduction; (**c**) development of anoxygenic phototrophic bacteria (Purple sulfur bacteria) causing red waters; (**d** and **e**) processes leading to the dystrophic crisis and red waters due to the excess of organic material (algae) in eutrophic lagoons (Photographs from Pierre Caumette)

Box 14.13: Winogradsky Columns

Developed by Sergei Winogradsky (1856–1953), there are nearly a century, Winogradsky columns constitute a model for studying processes related to the sulfur cycle. The experimental apparatus consisted of enrichment cultures performed in glass cylinders filled with a mixture of river water, plant fragments (rhizomes of *Butomus*), and gypsum ($CaSO_4$), inoculated with sediments and then sealed (allowing the installation of anoxic conditions) and exposed to light from a north window (no direct sunlight). After consumption of the remaining oxygen, the anaerobic degradation of organic matter by bacteria (mainly fermentative) produces acids and alcohols that serve as carbon source for sulfate-reducers. Their activity is revealed by the formation of black iron sulfide precipitates. After several weeks of incubation, blooms of red and green photosynthetic bacteria that oxidize sulfide appear on the inner surface of the glass cylinder at the light side. There is thus a balance between the activity of photosynthetic sulfur bacteria and sulfate-reducing bacteria, the sulfur being from the reduced form to the oxidized form and vice versa. These closed microcosms can be maintained indefinitely, the only exogenous energy consisting of light. Winogradsky original columns are still functional today and visible at the Pasteur Institute (Paris). Many variants of this device can be imagined. In particular, the producing of columns – constituted

(continued)

Box 14.13 (continued)

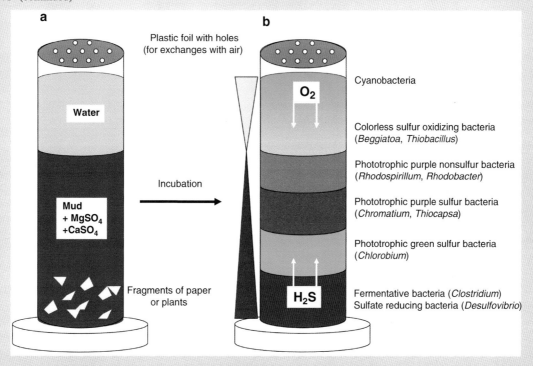

Box Fig. 14.22 Winogradsky columns open to air when settled (**a**) and after one to some weeks of incubation (**b**)

by a third of anoxic sediment covered by two thirds of water of the same environment – not hermetically closed (cylinder cover so as to limit the evaporation of the water, but not the entry of air) allows the creation of a gradient of oxygen in the column (Box Fig. 14.22a). After several weeks of incubation, a stratification of various microorganisms is observed (Box Fig. 14.22b). The production of sulfide by the sulfate-reducing bacteria causes blackening of the sediment (iron sulfide) and partly diffuse up into the water column. Proliferation of phototrophic microorganisms red and/or green (phototrophic sulfur bacteria) is visible in the sediment against the glass wall, light side. A red color is also visible in deep water, due to the development of phototrophic purple sulfur or non-sulfur bacteria. In the upper oxic part of the column, the development of cyanobacteria and photosynthetic eukaryotic microorganisms is indicated by a green color. Between the two colored layers, there is the developing of aerobic sulfur-oxidizing bacteria, generally forming a white water.

14.4.4.3 Photosynthetic Microbial Mats

Microbial mats are widely scattered across the globe and can be considered as actual stromatolites, the oldest of which dating from 3.5 billion years (*cf.* Sect. 4.2.2). The ubiquity and the presence of these ecosystems for several millions of years exhibit their capability to adapt to different environments, including hostile and fluctuating conditions. They can indeed be observed in a wide range of environments, including coastal, hypersaline, polar lake and soil, hot spring, or hypersaline lake environments. These organo-sedimentary structures have a role in stabilizing coastal areas subjected to erosion by waves, tides, and winds.

Microbial mats are perfectly organized ecosystems in which the interactions between different functional groups are very strong (*cf.* Sect. 9.7.3). They constitute an excellent example of microbial consortium which combines different populations and is organized in the form of horizontal strata, due to physicochemical gradients of light, dioxygen, and sulfur (Fig. 14.44). In microbial mats, microorganisms of simple structure (compared with multicellular organisms) are organized in a form of **ectosymbiosis*** made by genetically

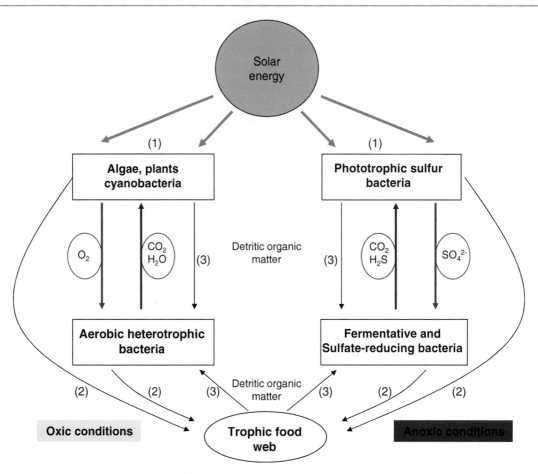

Fig. 14.43 Comparison of the primary production (oxygenic photosynthesis due to algae, macrophytes, and cyanobacteria) and para-primary production (anoxygenic phototrophic bacteria) in aquatic ecosystems and their contribution to food webs. *1* photosynthetic production, *2* organic matter incorporation in food webs, *3* organic matter degradation

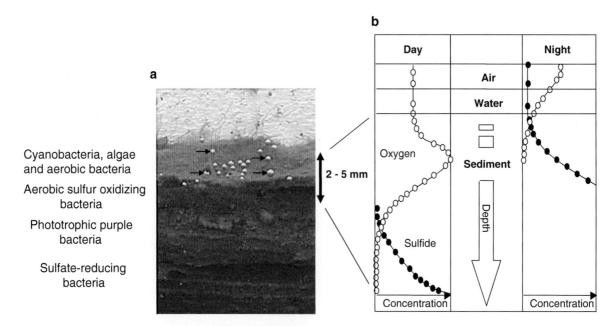

Fig. 14.44 Microbial mats. (**a**) Photograph showing spatial distribution of the main metabolic groups and the presence of oxygen bubbles at the surface (*arrows*); (**b**) distribution of oxygen and sulfide during the day and the night in a microbial mat (Modified from van Gemerden 1993) (Photograph from Rémy Guyoneaud)

distinct populations with a wide variety of metabolic functions, but which are coordinated within a small space.

The association of various bacterial metabolisms present in microbial mats induces the formation of chemical microgradients (Des Marais 2003). For example, in the presence of light, oxygenic photosynthesis produces dioxygen at the surface (up to 0.5–1 mm), thereby establishing a gradient of dioxygen from the surface to the bottom of the mat. In contrast, a gradient of sulfide appears due to the sulfate-reducing activity in the deeper zones of the mat (1–3 mm). This sulfide concentration decreases in the upper layers, primarily because of the activity of aerobic and anaerobic sulfo-oxidizing bacteria. In addition, a pH gradient is established from the surface to the bottom of the mat, created by the high photosynthetic activity of cyanobacteria coupled with the low buffering capacity of microbial mats. The absorption of solar radiations by photosynthetic organisms and by sediments may also establish a light gradient in microbial mats. The formation of these gradients relies on the metabolic richness of the system and the proximity of different micro-environments created by the superposition of gradients. The unstable equilibrium of this system can be observed when changing the gradient of light that affects the activity of photosynthetic microorganisms. In fact, physico-chemical gradients within microbial mats fluctuate strongly with the circadian cycle (Fig. 14.44), typically with the spatial rearrangements of the associated microbial communities (van Gemerden 1993). Thus, during a circadian cycle, vertical shifts in anoxygenic phototrophic bacteria, denitrifying bacteria, sulfate-reducing microorganisms, and chemolithotrophic sulfur-oxidizing bacteria, could be observed within microbial mats.

The establishment of physical and chemical gradients at the micrometer scale in microbial mats allows the coexistence of various metabolisms (Fig. 14.44). Cyanobacteria are the pioneer builders of microbial mats due to their ability to produce exopolymeric matrices. Cyanobacteria are also considered the main producers of dioxygen on earth. Both filamentous and unicellular forms group complex microorganisms, with huge diversity (*cf.* Sect. 6.6.2). These primary producers are typically oxygenic photosynthetic organisms, producing organic matter and dioxygen from solar energy used to fix CO_2. In addition, many cyanobacteria have the ability to fix dinitrogen. These characteristics explain why they can colonize surface habitats that are originally poor in organic matter and in nutrients. This colonization enriched the surrounding medium in organic matter, in dioxygen, and in nitrogenous compounds, allowing the development of other microorganisms. Among these microorganisms, aerobic heterotrophic bacteria will use the organic matter as carbon and energy source, by performing an aerobic respiration based on the dioxygen released by cyanobacteria. Facultative anaerobic heterotrophs are also able to develop in such ecosystems. This is particularly the case of denitrifying bacteria and other nitrate respiring bacteria that will benefit from ammonium oxidizing activities by nitrifying bacteria or by those microorganisms that are able to oxidize ammonium under anaerobic conditions (*cf.* Sect. 14.3.5). Fermentative bacteria, *i.e.* strict or facultative anaerobes, also have an important role in microbial mats, by participating in the degradation of organic matter and by providing organic substrates (organic acids, alcohols) or minerals (dihydrogen) for other anaerobic microorganisms. Among them, the activity of sulfate-reducing microorganisms is often very important in these ecosystems. From their activity, the generated sulfide will be oxidized by anoxygenic phototrophic sulfur bacteria and chemolithotrophic sulfur-oxidizing bacteria. In these shallow ecosystems, high wavelengths (near-infrared) penetrate the most deeply, and among anoxygenic photosynthetic bacteria, those containing bacteriochlorophyll *a* and *b* are favored (phototrophic purple bacteria). Other microorganisms such as methanogenic archaea and iron-reducing or iron-oxidizing bacteria are also present. Finally, eukaryotic microorganisms, photosynthetic or not, are also present in these ecosystems and have a role in the regulation of prokaryotic populations, primarily as predators. Thus, a photosynthetic microbial mat groups, on a very narrow spatial scale (a few mm deep), all functions that sustain a proper functioning of biogeochemical cycles, particularly those of carbon, nitrogen, sulfur, and iron.

14.4.4.4 Deep Ocean Hydrothermal Vents

These ecosystems, located in hot mid-ocean ridges (contraction areas with faults), are life oases characterized by an abundant fauna thriving in inhospitable, cold, and dark deep seafloors. In these environments, seawater enters faults of the Earth's crust and, subjected to high temperatures and pressures, becomes a hot, acidic, and reduced fluid that is enriched in metal elements, in sulfide and in dihydrogen, before being injected back to the oceanic waters as hydrothermal vents.

Hydrothermal vents (*cf.* Sect. 10.3.1) are resurgences of fluid whose characteristics may vary depending on environmental conditions. Some of these hydrothermal vents, called black or white smokers, consist of hot (200–400 °C) and anoxic fluid which is ejected into cold (4 °C) and oxygenated deep oceanic waters. These hydrothermal vents are enriched in H_2S whose concentration varies depending on the content of metallic elements (formation of iron metal sulfides, manganese, zinc, copper), differentiating black from white smokers. These sources are also enriched in oxidized (CO, CO_2) and reduced (CH_4) carbon, and in calcium which precipitates in the form of $CaSO_4$ in contact with oceanic waters. This $CaSO_4$ is at the origin of the chimney wall formation of smokers, which then are enriched in metal sulfides and in iron and manganese oxides.

Ecosystems that develop in deep hydrothermal vent receive no solar radiation and biomass production is due to chemosynthetic processes (Jannasch and Mottl 1985). Autotrophic microorganisms that grow in these ecosystems used reduced sulfur compounds (sulfur-oxidizing chemolithoautotrophs) and other reduced mineral compounds (such as NH_3, H_2, Fe^{2+}, or Mn^{2+}) as energy sources. Methylotrophic microorganisms also colonize these environments. Many hyperthermophilic and thermophilic microorganisms grow directly on the walls of smokers in the form of microbial mats, or in the water around the plume of the hydrothermal source. Chemosynthetic microorganisms are at the origin of a food chain involving planktonic or benthic animals, filter-feeding organisms and grazers, including worms, crustaceans, molluscs, and fishes. The distribution of animal populations is established according to the distance to the smoker, the different species being arranged in concentric rings around the source-ejection depending on the prevailing extreme conditions (temperature, pH, concentrations of sulfide, and metal).

In addition, in these ecosystems, symbioses between chemolithotrophic sulfur-oxidizing microorganisms and metazoans (worms, bivalves) have been described (*cf.* Sect. 10.3.1). For example, red giant tubeworms (*Riftia pachyptila*), which lack digestive tube, feed through the symbiotic activity of sulfur-oxidizing microorganisms (endobionts) housed in a particular organ, the trophosome. The substrates required for the growth of microorganisms (H_2S, CO_2, and O_2) are provided by the circulatory system of the worm, and part of the organic matter synthesized by bacteria is used by worms for their diet. Other symbioses involving exobiontes, often located in the gills of aquatic organisms in bacteriocytes, have also been described. Such symbioses also exist in coastal marine sediments between bivalves (clams) and chemosynthetic sulfur-oxidizing bacteria.

Because of the total absence of solar light, deep hydrothermal ecosystems therefore seem to exclusively depend on biomass production by chemosynthetic processes. However, it was discovered that these areas also contain phototrophic green sulfur bacteria (*Prosthecochloris* sp.) which use geothermal radiations (near-infrared radiations between 750 and 1,050 nm) as energy source. The question of their exact role still remains open (Beatty et al. 2005).

14.5 Cycles of Metals and Metalloids

14.5.1 Introduction

Many microorganisms are involved in the biogeochemical cycles of metals and metalloids, primarily by modifying their oxidation level and/or their chemical forms. Among these elements, some are essential to life (iron, manganese, cobalt) while others are toxic at low doses (mercury, cadmium, arsenic). Their biological transformations are typically related either to energetic metabolisms (donor or electron acceptor) or to resistance and detoxification mechanisms. They can also be due to incidental or indirect mechanisms (biomethylation, indirect reduction). Metals and metalloids are also subject to non-enzymatic redox reactions and it is often difficult to distinguish between changes of biotic origin and those generated by strictly physico-chemical processes. The relative importance of these biotic versus abiotic transformations depends on environmental conditions, such as the presence or absence of oxygen, and the prevailing pH (*cf.* Chap. 16.9).

This chapter will focus on the biogeochemical cycles of iron and manganese which are among the most abundant metals on the planet, and whose cycles are closely related. The standard potential of the couple Mn(IV)/Mn(II) is at +770 mV, and that of the couple Fe(III)/Fe(II) at +380 mV. The oxidized forms of iron Fe(III) and manganese Mn(IV) are excellent respiratory electron acceptors. Thus, the degradation of many organic compounds is coupled to the dissimilatory reduction of Fe(III) and Mn(IV), which play an important role in the anaerobic degradation of organic matter (*cf.* Sect. 3.3.2). Microorganisms involved in the dissimilatory reduction of these two metals are frequently the same. For example, *Shewanella putrefaciens* produces the energy necessary to its growth by coupling the oxidation of formate into CO_2 to the reduction of iron or manganese:

$$\text{Formate}^- + 2\,\text{Fe(III)} + H_2O \rightarrow HCO_3^- + 2\,\text{Fe(II)} + 2H^+$$

$$\text{Formate}^- + 2\,\text{Mn(IV)} + H_2O \rightarrow HCO_3^- + 2\,\text{Mn(II)} + 2H^+$$

In the presence of these electron acceptors, lactate and pyruvate are also oxidized but not completely mineralized (acetate production). In part against the dissimilatory reduction, many bacterial species derive their energy from the oxidation of iron and manganese. These organisms can be either aerobic (*Thiobacillus* spp. for example) or denitrifying anaerobic chemolithotrophs, or anoxygenic phototrophs (*Rhodovulum robiginosum* and *Rhodovulum iodosum*).

Because of their abundance, iron and manganese have been the most extensively studied metals, but the spectrum of metals used by microorganisms is much larger. Microorganisms are involved in the biogeochemical cycles of many metals and metalloids, including reduction processes and, more rarely, oxidation activities. Some microorganisms are capable of growth by using reduced forms of arsenic and selenium as electron donors. The case of arsenic to which certain microorganisms oxidize arsenite (As(III)) to arsenate (As(V)) is the best documented (Oremland and Stolz 2003) and particularly interesting. Indeed, in the case of arsenic, the reduced form, *i.e.* arsenite, is more toxic because its availability is higher compared with the oxidized form. Some microorganisms that oxidize As(III) use this ability as a resistance mechanism to detoxify As(III) to As (V), which is less toxic and less prone to enter the cell. Other organisms, chemolithotrophs, are capable of coupling the oxidation of As (III) (electron donor) to the reduction of oxygen or nitrates involved as terminal electrons acceptors.

Among the many metals which can be reduced by microorganisms, uranium U(VI) can be reduced to uranium U(IV) or chromium Cr(VI) to chromium Cr(III), but arsenic (As), selenium (Se), vanadium (V), molybdenum (Mo), copper (Cu), gold (Au), mercury (Hg; Box 14.14), or silver (Ag)

Box 14.14: Iron in Ocean

Stephane Blain

The Iron Cycle in the Ocean

Concentrations of dissolved iron in the modern ocean are extremely low (often less than 1 nanomole. L^{-1}). This is partly due to the low solubility of this element in seawater, which is an oxygenated lightly alkaline environment (pH = 8). However, over geological time this has not always been the case. Notably, iron concentrations were much higher during the Archean (3.8–2.5 Ma) and Proterozoic (2.5–0.5 Ma), periods when oxygen was absent or found at very low levels. The first living organisms appeared in the oceans during these periods have therefore widely incorporated iron into their metabolisms. Indeed, iron has two oxidation states, Fe (II) and Fe (III), which makes it particularly useful for electron transport within the cells. Iron has become an essential element in many biochemical functions, such as respiration, photosynthesis, dinitrogen fixation, and nitrate reduction.

The discovery of low levels of iron in the ocean is relatively recent. Indeed, measuring iron concentrations in the ocean requires so-called ultra-clean techniques first used in oceanography in the 80s (Bruland et al. 1979). They are based on non-metallic sampling equipments and they follow strict protocols for storage and analysis of the samples. The low iron supply to the ocean compared with the physiological demand of the marine microorganisms results in a limitation of photosynthesis and growth of phytoplankton in very large ocean areas (e.g., Southern Ocean, Equatorial sub-Arctic Pacific). This has profound climatic implications because the process called biological pump of CO_2, which includes photosynthesis and transport of organic carbon in the deep layers of the ocean and ultimately to the sediments, is the main oceanic mechanism which removes CO_2 from the atmosphere (Sarmiento and Gruber 2006).

A major breakthrough in the demonstration of iron limitation of biological production in the ocean has clearly resulted from the realization of a dozen of artificial fertilization experiments on a small scale (approximately 100 km^2) between 1993 and 2009. During these experiments, the addition of iron (several tons) stimulates organic production in the surface layer (Boyd et al. 2007). This results in an increase in the concentration of chlorophyll and a decrease of the partial pressure of CO_2 in water. Fertilization alters the structure of the phytoplankton community in promoting growth diatoms. In some experiments, the effect on the second trophic level (zooplankton) was also observed. These experiments did not, however, clearly demonstrated that the carbon was then transferred into the deep ocean. But, this was demonstrated in the study of a region naturally fertilized by iron: the Kerguelen Plateau in the Southern Ocean (Blain et al 2007). This region has high chlorophyll concentrations, while the surrounding waters are much poor (Box Fig. 14.23). In this study, it was demonstrated that the increase in biological activity was caused by an enhanced iron supply, and that this led to an increase in the deep export of carbon.

Natural fertilization of surface waters of the Southern Ocean therefore enhanced the biological pump of CO_2.

In the present ocean, the main external source of iron is the deposition of dust from natural (deserts) or anthropogenic (industry, transport) origins. Two types of depositions are distinguished: the wet deposition (rain) and the dry deposition (Jickells et al. 2005). The impact of these deposits on the biological

S. Blain (✉)
Laboratoire d'Océanographie Microbienne, LOMIC UMR CNRS-UPMC, Observatoire Océanologique de Banyuls, 66650 Banyuls-sur-Mer, France

(continued)

Box 14.14 (continued)

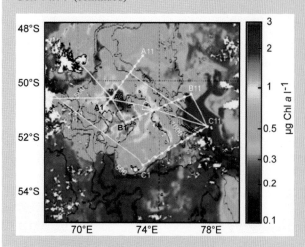

Box Fig. 14.23 Satellite image showing high concentrations of chlorophyll due to the natural iron fertilization of the surface water located above the Kerguelen Plateau in the Southern Ocean

activity is still uncertain, since these events are sporadic and relatively difficult to study with conventional oceanographic campaigns. However, there are some clues suggesting that dust inputs in oligotrophic regions of the ocean (e.g., Mediterranean, subtropical gyres) could benefit diazotrophic organisms. Indeed, the contribution of atmospheric iron would favor N_2 fixation metabolism that has a high iron requirement, and thus provide a competitive advantage to these organisms in a medium poor in nitrogen (Mills et al. 2005).

Major Unknowns of Iron Cycle, Speciation, and Bioavailability

Iron is present in very many physical–chemical forms in the ocean. It could be present as particles, colloids, or dissolved species. Dissolved iron is mainly in its oxidized form (Fe III) complexed by organic molecules. Microorganisms have therefore developed appropriate strategies to acquire iron from of these different forms (Box Fig. 14.24).

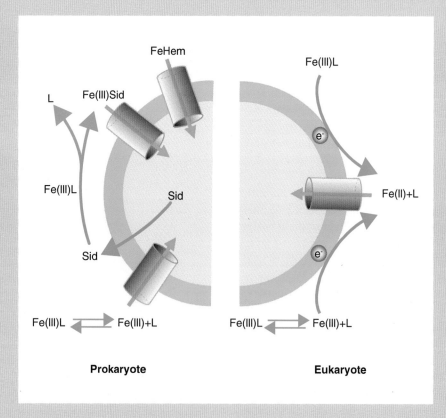

Box Fig. 14.24 Schematic drawing of a few examples of iron acquisition mechanisms by marine microorganisms. Abbreviations: *Sid* siderophore, *Hem* Hemin, *L* organic ligand. The cylinders are membrane proteins for the transport of the various forms of iron through the membrane. The *circles* represent reductases that reduce Fe(III) to Fe(II)

(continued)

Box 14.14 (continued)

All of these mechanisms are not yet known but some pathways have been described (Morel et al. 2008). When iron is limiting, a number of bacteria excrete molecules, siderophores (*cf.* Sect. 14.5.2) that forms strong complex with iron. That allows them to extract iron from other chemical species present in the environment. The iron-siderophore complex is then taken up by bacteria (Butler 1998). Eukaryotes seem to have adopted a different strategy, which consists in reducing Fe (III) at the surface of the cell to form iron (II). This facilitates the dissociation of the organic complexes and allows the transport iron into the cell. Other mechanisms are likely to be discovered in the future, such as direct transport of other iron complex like heme-type. Detailed analysis of the growing number of genomes available for marine microorganisms should allow rapid progress in this direction.

The Iron Cycle and Climate Geoengineering

The major role of iron in controlling the biological pump of CO_2 have inspired geo engeneering proposals describing large scale iron fertilization of the ocean to mitigate the atmospheric CO_2 increase. A very large majority the scientific community does not support this idea. Indeed, the efficiency of a such manipulation of the ocean is not firmly proven. The amount of carbon that would be eventually stored would be very difficult to assess. In addition, the side effects of such large scale manipulations of a global ecosystem are not known and could be dangerous (Strong et al. 2009).

can also be reduced. These reduction processes can be related to energetic metabolisms (dissimilatory reduction), or can be the result of an indirect reduction (reduction by reduced iron or sulfide). The reduction of these elements is interesting in the case of treatment processes of contaminated sites, because the reduced forms are often less toxic and/or less soluble (unlike iron and manganese).

14.5.2 The Cycle of Iron

14.5.2.1 Iron in the Environment

Iron is quantitatively the fourth constitutive element of the earth's crust. It is abundant in the continents and rare in the oceans where iron resources are often a limiting factor for primary production (Box 14.15).

Iron is present in the environment as oxides, carbonates, sulfides, and hydroxides. It exists in different oxidation states, ranging from 0 to +VI (ferrate), but only two of these states are involved in biological pathways: Fe^{2+}(II) which is the reduced ferrous form, and Fe^{3+}(III) which is the oxidized ferric form. Iron is an essential element in some enzymes: cytochromes, catalases, peroxidases, nitrite reductases, and iron-sulfur metalloprotein complexes. It is therefore an essential element in respiratory and photosynthetic processes.

The greater part of iron found in the environment is insoluble, the solubility of ferric ion being very low at physiological pH. The solubility of Fe(II) form is substantially greater, but in anoxic conditions. However, under these conditions, iron is usually precipitated as black iron sulfide (FeS) or even as pyrite (FeS_2), the last one forming a crystalline structure, highly insoluble and very difficult to oxidize (Fig. 14.45, *pathways 1 and 3*).

The oxidation of pyrite requires the sum of chemical and biological reactions involving two electron acceptors: dioxygen and ferric ions. In the presence of sulfide (S^{2-}) and when anaerobic conditions prevail, ferric iron (Fe^{3+}) is chemically (*i.e.* by abiotic processes) reduced to sulfide (FeS) and ferrous iron (Fe^{2+}) (Fig. 14.45, *pathways 1 and 2*). During this reaction, a small portion of sulfide (S^{2-}) is oxidized to sulfur ($S°$) which participates in the formation of pyrite by association with sulfide iron (Fig. 14.45, *pathway 3*). Sulfide iron can be chemically oxidized, in the presence of oxygen, to ferric iron by releasing sulfur and thiosulfate ($S_2O_3^{2-}$) (Fig. 14.45, *pathway 4*).

14.5.2.2 Iron Assimilation by Microorganisms

The inventory of biological functions of iron shows that this element is essential to life. In the cell, the iron must be in the Fe^{2+} state. Due to its low solubility, living organisms have developed systems to capture and incorporate iron (Fig. 14.45, *pathway 5*). Indeed, many microorganisms produce a wide variety of small organic molecules, siderophores, capable of fixing Fe^{3+} with a very high efficiency, ensuring its transport within the cell where it is reduced and incorporated into the organic material.

Some aquatic bacteria, described as magnetotactic microorganisms, such as *Aquaspirillum magnetotacticum*, *Magnetotacticum bavaricum*, and *Magnetospirillum magnetotacticum*, can transform iron extracellularly to crystal of magnetite (Fe_3O_4) and construct "intracellular magnetic compasses" called magnetosomes (Bazylinski and Frankel 2004), which are included in cytoplasmic vesicles (Figure 14.46). Some magnetotactic bacteria do not synthesize magnetite but greigite (Fe_3S_4) (Farina et al. 1990). With this structure, these motile bacteria can move to areas rich in organic matter (water-sediment interface), by following the Earth's magnetic field. High concentrations of magnetotactic bacteria were detected in anoxic marine sediments at 3,000 m depth (Petermann and Bleil 1993).

Box 14.15: Biogeochemical Mercury Cycle

Rémy Guyoneaud

Mercury, whose chemical symbol Hg comes from the Latin name Hydrargyrum (liquid silver) has different chemical forms in the environment. Main species are elemental mercury (Hg^0, volatile), bivalent inorganic mercury (Hg^{2+} or Hg(II)), mostly combined with organic matter, chloride, hydroxydes or sulfide (cinabar, HgS), monomethylmercury (CH_3Hg^+) and dimethylmercury (organic volatile species, $(CH_3)_2Hg$). Mercury toxicity, is due to its capacity to combine with thiol groups in proteins. In contrast to biogeochemical cycles of carbon, nitrogen, or sulfur, toxic mercury in not an essential element for life. Micro-organisms participate in the mercury biogeochemical cycle but it is not used through energetic metabolisms (electron donor or acceptor) nor assimilation/degradation mechanisms. Biological transformations of mercury are linked to resistance mechanisms (transformation of the toxic forms to the less toxic volatile elemental mercury) and to mercury methylation mechanisms that enhance mercury accumulation in food webs (Box Fig. 14.25).

The reduction of inorganic mercury to volatile Hg^0 is a prokaryotic resistance mechanism whose genetic determinism is well known (Barkay et al. 2003). Proteins involved in mercury reduction are encoded by genes organizsed in the inducible mer operon, present in many Bacteria. In the absence of mercury (or at low concentrations) MerR transcriptional regulatory protein is linked to the mer promoter (PmerT) and represses the genes expression. In the presence of inorganic mercury or methylmercury, the modification of the complex MerR-operator-promoter allows the operon transcription and thus the synthesis of transport proteins, catalytic enzymes, and regulation protein (MerD) involved in the repression when mercury concentration decreases. Mercury reduction is not linked to an energetic metabolism (energy conservation). After its transport inside the cell, inorganic mercury (Hg(II)) is reduced to volatile elemental mercury (Hg(0)) by a mercury reductase (MerA). The mer operon may comprise one particular gene (merB) encoding for an organomercurial lyase involved in methylmercury detoxication. The presence of this gene and its related enzyme allows a broad spectrum resistance (inorganic mercury and methylmercury). The methylmercury is first demethylated to CH_4 and inorganic mercury and the later is afterward reduced to elemental mercury through MerA reductase. This

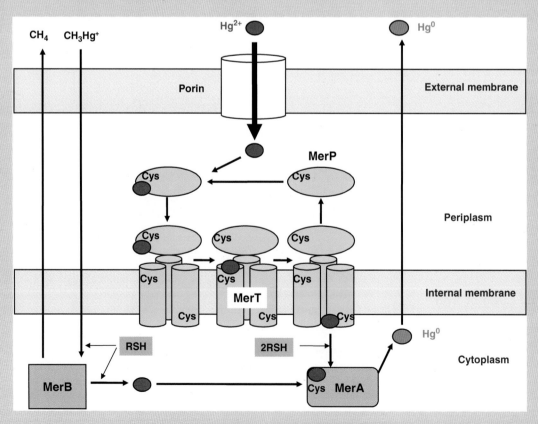

Box Fig. 14.25 Resistance mechanisms to inorganic mercury and methylmercury through proteins encoded on Tn501 transposon (*merT, merP, merA*) as well as organomercurial lyase (*MerB*). Mercury binds to proteins through cysteine residues, cytoplasmic transport occurs through small peptidic thiols (*RSH*) (Redrawn from Barkay et al. 2003)

(continued)

Box 14.15 (continued)

demethylation is called reductive demethylation. An oxidative demethylation, producing CO_2 and inorganic mercury, called oxidative demethylation, has been demonstrated under anoxic conditions (sulfate reducing or methanogenic conditions).

In ecosystems numerous abiotic processes (photochemical or chemical) and biotic processes are involved in methylmercury or dimethylmercury production. The methylmercury accumulates much more easily in food webs, due to its lipophilic characteristics. Its half life in biota is much longer than the one of inorganic mercury and it accumulates in the central nervous system. Thus, activities that contribute to methylmercury production are of interest for human health. Indeed, methylmercury production (due to acethaldehyde production) and its accumulation in food webs was implicated in the Minameta disease in Japan (1950–1960). Methylmercury, may also be produced through biological activities. Some are indirect mechanisms (non-enzymatic) due to biogenic molecules such as iodomethane, methylated metals, and other metabolites. Mercury methylation through enzymatic processes mainly occur under anoxic conditions and it has been demonstrated that sulfate-reducing bacteria are among the main mercury methylators (Compeau and Bartha 1985) even if methylmercury production capacities may vary a lot in between strains/populations (Ranchou-Peyruse et al. 2009). Some other microorganisms such as some iron reducers (*Geobacter* spp.) may also producemethylmercury (Fleming et al. 2006). Finally, the genetic determinisms of methylmercury production has been elucidated (Parks et al. 2013) and these genes have been found to be distributed in many other organisms such as methanogens.

Mercury biomethylation increases mercury bioaccumulation in organisms and mercury bioamplification through food webs to higher trophic levels (carnivorous fishes, humans) where its concentration may be 10^7 times the concentration in the water. Since mercury anthropic emissions have increased three to four times since the industrial era (fossil energy burning, mining activities, other industrial activities), the global biogeochemical mercury cycle has been modified in terms of fluxes. In addition, in some ecosystems such as the Amazonian region, goldmining activities has implications for mercury contamination and methylmercury bioamplification in food webs. Methylmercury contamination and poisoning is of public health concern (Box Fig. 14.26).

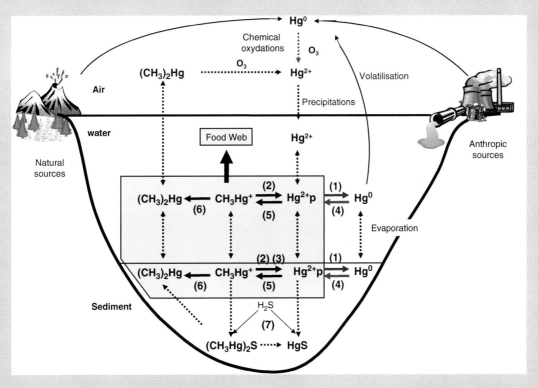

Box Fig. 14.26 Biogeochemical mercury cycle showing transformation and transfers between different ecosystem compartments. Transformations that may be due to direct biological processes are indicated with *plain arrows*. *1* inorganic mercury reduction through mercury reductase (*MerA*) or small metabolites originated from primary production, *2* reductive demethylation, *3* oxydative demethylation, *4* elemental mercury oxidation (catalase), *5* biological methylation in sediment or particulate material (anoxic conditions), *6* dimethylmercury production, *7* reactions with sulfide producing cinabar (HgS), dimethylmercury sulfide. $Hg^{2+}p$, particulate inorganic mercury bond to organic or inorganic particles

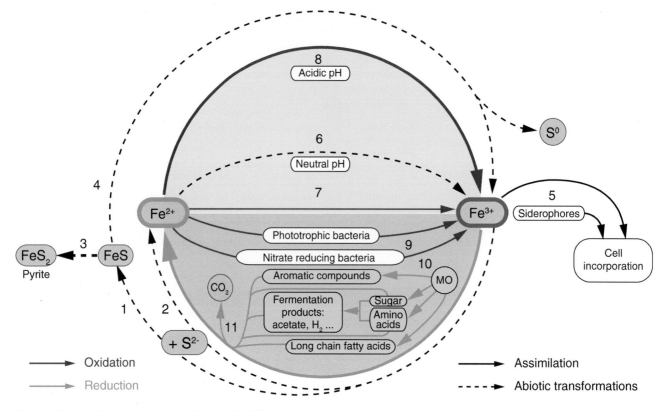

Fig. 14.45 Iron biogeochemical cycle showing the different biological pathways (Drawing: M.-J. Bodiou)

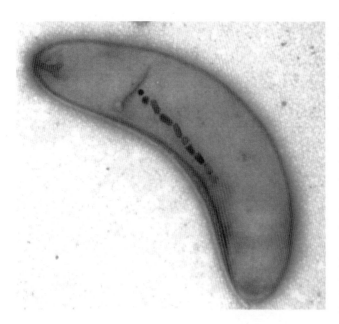

Fig. 14.46 Magnetotactic bacteria (*Vibrio* sp.) The magnetosomes form chains inside the cell (1.5 × 0.4 μm) (Courtesy from Long-Fei Wu)

14.5.2.3 Iron Oxidation

The oxidation of Fe(II) to Fe(III) iron by chemical or microbial processes depends on pH and oxygen concentration. In oxic conditions, the oxidation pathways of iron will be a function of pH. At neutral pH, ferrous iron is spontaneously oxidized (by abiotic pathway) to ferric iron (Fig. 14.45, *pathway 6*). However, at oxic–anoxic interfaces, it can also be oxidized by bacteria such as *Gallionella ferruginea*, *Leptothrix ochracea*, or *Sphaerotilus natans* (Fig. 14.45, *pathway 7*). The biological oxidation of iron is the most important at acidic pH (less than 3) where the Fe^{2+} form is stable (Fig. 14.45, *pathway 8*). This oxidation is carried out by autotrophic acidophilic micro-organisms such as *Acidithiobacillus ferrooxidans* or *Leptospirillum ferrooxidans*, which fix CO_2 via the Calvin cycle and use iron as electrons donor and oxygen as terminal electrons acceptor. The respiratory chain involved in this mechanism is described in Chap. 3 (Fig. 3.18d). These bacteria are commonly encountered in polluted acidic environments and their metabolism is used in the bioleaching process (*cf.* Sect. 16.9.2). They must oxidize a large amount of iron to grow. Indeed, the energy released during the oxidation of ferrous iron (Fe^{2+}) to ferric iron (Fe^{3+}) is low, because only one electron is transferred in this reaction:

$$2\,Fe^{2+} + 2H^+ + \tfrac{1}{2}O_2 \rightarrow 2\,Fe^{3+} + H_2O \left(\Delta G^{0'} = -65.8\,\text{kJ}\right)$$

The oxidation of iron in such environments causes significant deposits of iron oxides which may be visualized

Fig. 14.47 Ferric iron accumulation in an acidic stream (Carnoules, France) (Photograph Rémy Guyoneaud)

as rusty-colored deposits at the surface of sediments (Fig. 14.47).

The oxidation processes of iron are at the origin of specific geological formations, visible as brown oxidized iron strata in ancient sedimentary rocks (*i.e.* BIFs for "Banded Iron Formation"). These accumulations of oxidized iron consist of hematite (Fe_2O_3) and magnetite (Fe_3O_4). The majority of these structures, dated between 2.5 and 1.8 billion years, likely result from iron oxidation by oxygen produced by cyanobacteria and contain 20 times the amount of oxygen in the atmosphere of the present planet. Other hypotheses suggest that the formation of these geological layers is due to the anaerobic oxidation of iron either by chemosynthesis or by anoxygenic photosynthesis.

The oxidation of Fe(II) does not happen only in oxic environments, it also occurs in anoxic conditions at neutral pH. Indeed, in addition to aerobic acidophilic and neutrophilic chemolithotrophs, anaerobic oxidation of iron is also performed by bacteria which use nitrate as electrons acceptor (Straub et al. 1996, 2004) (Fig. 14.45, *pathway 10*). The description of this process in many natural ecosystems suggests that it plays a significant role in the coupling of nitrogen and iron cycles in anoxic sediments. Moreover, some anoxygenic photosynthetic microorganisms oxidize iron to fix CO_2 by using solar energy (Ehrenreich and Widdel 1994; Hegler et al. 2008) (Fig. 14.45, *pathway 9*). Phylogenetically diverse, they are represented by purple sulfur (*Thiodictyon* sp. strain F4), purple non-sulfur (*Rhodovulum iodosum* and *R. robiginosum*) and green sulfur (*Chlorobium ferrooxidans*) bacteria.

14.5.2.4 Iron Reduction

Iron reduction is carried out via both chemical and biological processes. The latter case refers to the so-called anaerobic respiration where iron is the terminal electrons acceptor. This type of respiration is common in anoxic sediments, soils, marshes, bogs, and has also been highlighted in deep aquifers or fossil waters, oil reservoirs, continental hot springs, and marine hydrothermal vents.

Because the oxidized forms of iron (and manganese) are highly insoluble, they behave differently compared with other potential soluble electron acceptors (oxygen, nitrate, sulfate, or carbon dioxide), which diffuse within the cell. To reduce these compounds, organisms will develop two strategies (Lovley et al. 2004): either establishing a direct contact with the electron acceptor (e.g. *Geobacter metallireducens*) or produce "shuttle carriers" that transfer electrons from the surface membrane to iron (and manganese) oxides (*Shewanella* spp, *Geothrix fermentans*).

The phylogenetic diversity of microorganisms capable of dissimilatory reduction of iron is high, with representatives in both *Bacteria* and *Archaea* domains. The most studied micro-organisms belong to the genera *Shewanella*, *Geobacter*, and *Geospirillum*. Hyperthermophiles reduce iron, such as *Thermotoga maritima* and *Thermodesulfobacterium commune* (in the domain *Bacteria*), and *Pyrobacterium islandicum* and *Ferroglobus placidus* (in the domain *Archaea*) (Lovley et al. 2004). Other electron acceptors can be used by microorganisms which carried out the dissimilatory reduction of iron: oxygen, other metals (manganese as already mentioned, uranium, cobalt, chromium, gold), extracellular quinones (humic substances are the most abundant source of extracellular quinones), some sulfur compounds (S^0), nitrate, fumarate (Lovley et al. 2004).

Concerning the electron donors, the most frequently used are organic acids (Fig. 14.45, *pathway 11*). If hydrogen is an electron donor for several species of *Geobacter* and *Shewanella*, acetate, which is the most important organic electron donor in many environments, is the most frequently used. The total oxidation of this molecule by *Geobacter metallireducens* is as follows:

$$\text{Acetate}^- + 8\,Fe^{3+} + 4\,H_2O \rightarrow 2\,HCO_3^- + 8\,Fe^{2+} + 9H^+ \left(\Delta G^{0'} = -233\,\text{kJ}\right)$$

This bacterium is also able to develop on a wide spectrum of compounds, by coupling their oxidation to iron reduction: benzaldehyde, benzoate, benzyl alcohol, butanol, butyrate, ethanol, *p*-hydroxybenzaldehyde, *p*-hydroxybenzoate, *p*-hydrybenzyl alcool, *p*-cresol, isobutyrate, isovalerate, phenol, propanol, propionate, pyruvate, toluene, valerate (Lovley 2006). Some examples (Lovley 1991):

$$benzoate^- + 30\ Fe(III) + 19\ H_2O$$
$$\rightarrow 7\ HCO_3^- + 30\ Fe(II) + 36\ H^+$$

$$toluene + 36\ Fe(III) + 21\ H_2O$$
$$\rightarrow 7\ HCO_3^- + 36\ Fe(II) + 43\ H^+$$

$$phenol + 28\ Fe(III) + 17\ H_2O$$
$$\rightarrow 6\ HCO_3^- + 28\ Fe(II) + 34\ H^+$$

$$p\text{-cresol} + 34\ Fe(III) + 20\ H_2O$$
$$\rightarrow 7\ HCO_3^- + 34\ Fe(II) + 41\ H^+$$

Metabolism of sugars has also been demonstrated. It produces enough energy to allow growth; oxidation may be total (Küsel et al. 1999) or incomplete (Coates et al. 1998):

$$C_6H_{12}O_6 + 2\ H_2O + 8\ Fe(III)$$
$$\rightarrow 2\ CH_3COOH + 2\ CO_2 + 8\ Fe(II) + 8\ H^+$$

$$C_6H_{12}O_6 + 24\ Fe(III) + 24\ OH^-$$
$$\rightarrow 6\ CO_2 + 24\ Fe(II) + 18\ H_2O$$

The metabolic capacities of microbial iron-reducers have many environmental applications. The *Geobacteraceae* can play an important role in the rehabilitation of deep anoxic environments contaminated with aromatic hydrocarbons (benzene, toluene, ethylbenzene, *o*-xylene, *p*-cresol, phenol). Their action in the cleanup of groundwaters or sediments contaminated by hydrocarbons was observed and provides opportunities for the development of remediation techniques against contaminated sites (Jahn et al. 2005) (*cf.* Sect. 16.8.3).

14.5.3 The Cycle of Manganese

14.5.3.1 Manganese in the Environment

The abundance of manganese in the earth's crust was estimated to be 0.1 %. Manganese can exist in different oxidation states ranging from 0 to +VII (permanganate). In the nature, only the states +II, +III, and +IV are commonly found. The oxidation state of manganese is influenced by chemical and physical characteristics of their environment. In reduced sediments and at low pH, the form Mn(II), equivalent to the soluble manganese ion Mn^{2+}, is the most abundant. The appearance of the insoluble (in water) oxidized form, Mn(IV) that corresponds to the manganese dioxide (MnO_2), is favored at high pH and in oxidized environments.

However, even if environmental conditions favor changes in the oxidation state, these chemical conversions are slow. The oxidation and reduction of manganese in many biotopes are mainly related to bacterial metabolism.

In some marine and freshwater habitats, precipitation of manganese forms the so-called polymetallic nodules which also contain other elements. Chemical analysis of these nodules shows that they mainly consist of manganese (20–30 %) and iron (6–15 %). Other metallic elements such as nickel (1.34 %), copper (1.25 %), cobalt (0.25 %), titanium (0.60 %), and aluminum (2.90 %) are also present in these nodules. They also contain sodium, magnesium, silica, oxygen, and hydrogen. The problem of the genesis of these nodules is far from being resolved. Four origins are possible:

1. "Hydrogenated", slow precipitation of metallic elements from seawater which leads to concretions with equivalent concentrations of Fe and Mn and relatively high Ni, Cu, and Co contents;
2. "Hydrothermal", giving concretions which are generally rich in iron and poor in Mn, Ni, Cu, and Co;
3. "Diagenetic," known from the remobilization of manganese in the sediments and its precipitation at the water-sediment interface, giving nodules which are rich in Mn and poor in Fe, Ni, Cu, and Co;
4. "Halmyrolitic," where the source of metallic compounds is the attack of basaltic debris by seawater

These theories, termed "mineral theories," contrast with the so-called "biologic theories" (Graham and Cooper 1959) where detritus from organisms are responsible for the enrichments in Cu and Ni.

14.5.3.2 Biological Functions

Manganese is an essential trace element in several biological systems. It is essential in the diet of micro-organisms, plants, and animals. It is necessary in the oxygenic photosynthesis where it is involved in the functioning of photosystem II. It is also a cofactor of various enzymes such as pyruvate kinase which is involved in glycolysis. For some bacteria, the Mn (II) can be used as an energy source and the forms Mn(III) and Mn(IV) as terminal electron acceptors. The importance of microorganisms that are capable to oxidize and reduce the manganese is demonstrated by their ubiquity in terrestrial, freshwater, and marine environments. After oxygen, manganese oxides are among the most effective oxidizing agents encountered in the environment.

14.5.3.3 Oxidation by Microorganisms

In the nature, Mn oxidation takes place in two ways: either by spontaneous reaction which is enhanced in alkaline conditions or by enzymatic reactions (Fig. 14.48). The

Fig. 14.48 Manganese biological transformations through reduction and oxidation processes. Some implicated microorganisms are listed

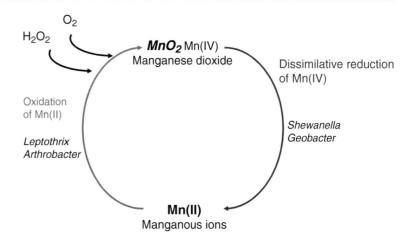

oxidation of manganese involves the transfer of two electrons (Mn(II) → Mn(IV)), and comprises an intermediate step involving a single electron Mn(III). As in the case of iron(II), Mn(II) is soluble but its oxidation product Mn(IV) is very insoluble. The resulting oxides can sometimes form concretions around sediment grains, pebbles, dead, or biological structures (mollusc shell, fragment of coral, or other debris) or dark layers in sedimentary rocks.

However, unlike iron(II), Mn(II) is stable in oxic conditions at neutral pH. Chemical oxidation indeed becomes significant only when pH is equal or higher than 8. Because Mn is less prone to chemical oxidation, Mn(II) will be present at a concentration greater than that of iron in aerobic zones (marine and fresh waters) as well as in anoxic soils.

Microorganisms capable of oxidizing manganese have been isolated in many biotopes (marine and fresh waters, soils, sediments, manganese nodules, hydrothermal vents). They are particularly prevalent in oxic–anoxic interface areas and in hydrothermal systems which are a major source of dissolved Mn(II) in the oceans. Mn(II) is reoxidized by many bacteria, present in several phylogenetic lineages (*Firmicutes*, *Proteobacteria*) (Tebo et al. 2005). Most studied species are:

Pseudomonas putida strain GB-1 and MnB1;

Leptothrix discophora strain SS-1 (characterized by a double precipitation of iron and manganese oxides in the sheet which surrounds the cells, yielding dark brown or black colors);

Gallionella spp.;

Bacillus sp. strain SG-1; it is important to note that some marine *Bacillus* produce spores that oxidize Mn(II) and thereby are recovered with a precipitate of manganese dioxide.

Some Fungi can also oxidize manganese (Thompson et al. 2005). The main related enzymatic mechanisms involve a copper oxidase. Its existence is demonstrated by the inhibition of its oxidizing activity by azide (inhibitor of copper-protein), and by a stimulation in the presence of copper (Francis et al. 2001).

Based on physiological criteria, it is possible to classify microorganisms that oxidize manganese into three major groups (Ehrlich 2002).

Some (group I) oxidize soluble species of Mn(II) using oxygen as terminal acceptor of electrons; this oxidation can be coupled or not to the synthesis of ATP. The overall reaction is:

$$Mn^{2+} + 0.5\, O_2 + H_2O \rightarrow MnO_2 + 2H^+$$

Organisms belonging to group II oxidize Mn(II) providing that it is bound to a solid extracellular substrate Mn(IV) oxide, ferromanganese, montmorillonite, kaolinite). These organisms also use oxygen as terminal electrons acceptor. For example, the reaction catalyzed by these bacteria when manganese is bound to a hydrated oxidized form such as $MnO_2 \cdot H_2O$ (H_2MnO_3) can be summarized as follows:

$$Mn^{2+} + H_2MnO_3 \rightarrow MnMnO_3 + 2\,H^+$$

$$MnMnO_3 + 0.5\, O_2 + 2\, H_2O \rightarrow 2\, H_2MnO_3$$

Microorganisms of group III oxidize dissolved Mn(II) with H_2O_2 using a catalase as enzyme. The reaction can be summarized as follows:

$$Mn^{2+} + H_2O_2 \rightarrow MnO_2 + 2\,H^+$$

14.5.3.4 Reduction by Microorganisms

As in the case of oxidation, the reduction of manganese can be performed enzymatically or not. The reduction of Mn(III) and Mn(IV) forms, which is particularly important in anoxic environments, generally leads to the solubilization of manganese.

The most common biological pathway is the bacterial reduction (Fig. 14.48) which corresponds to a respiration where the oxidized manganese, which serves as a terminal electron acceptor, is reduced to Mn(II). This is a dissimilatory reduction as well as that of iron, nitrate, or sulfur. Some Fungi can also reduce manganese.

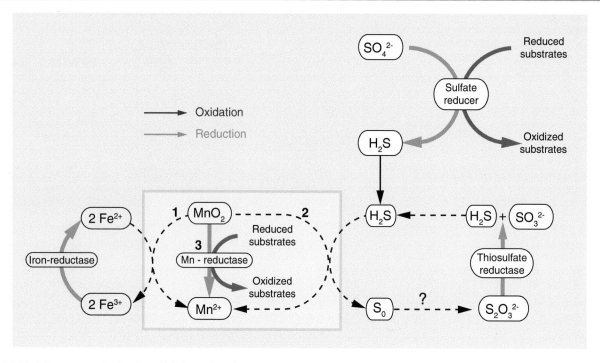

Fig. 14.49 Manganese reduction through indirect (reactions 1 and 2) or direct (reaction 3) biological pathways. Indirect reduction of manganese oxide (MnO_2) is due to Fe^{2+} originated from iron reduction (reaction 1) or to sulfide (reaction 2). Sulfide may be produced through thiosulfate reduction or dissimilative sulfate and sulfur reduction. Direct manganese reduction is due to microorganisms harvesting a manganese reductase (Redrawn from Nealson and Myers 1992) (Drawing: M.-J. Bodiou)

Some bacteria can reduce manganese aerobically or anaerobically. Others perform reduction only in anoxic conditions where electron donors vary depending of species. These species include *Geobacter metallireducens* which can use butyrate, propionate, lactate, succinate, acetate, and other compounds which are fully oxidized to CO_2. *Shewanella putrefaciens* uses pyruvate and lactate as electron donors but oxidizes them up to only acetate; formate and H_2 are also electron donors. During their development on solid nutrient medium, colored in black by the addition of manganese oxide (MnO_2), bacteria that reduce manganese can be identified by the formation of a clear halo around the colonies, which is due to the dissolution of MnO_2.

In the environment, after oxygen and nitrate depletion, the oxidation of organic matter coupled to the reduction of Mn(IV) oxides to return back to the reduced and soluble form Mn(II) is carried out according to the following primary reaction:

$$246\ MnO_2 + MO + 364\ CO_2 + 104\ H_2O \rightarrow 470\ HCO_3^- + 8\ N_2 + 236\ Mn^{2+} + HPO_4^{2-}$$

(MO = Organic Matter of theoretical composition $C_{106}H_{263}O_{110}N_{16}P$)

Manganese oxides can also undergo spontaneously abiotic reduction in the presence of Fe(II) and sulfides (Fig. 14.49). This reduction is indirectly facilitated by the activity of certain iron(II) and sulfide (S^{2-}) producing microorganisms. For example, in *Shewanella putrefaciens*, Fe(III) and thiosulfate are used as electron acceptors to indirectly reduce Mn(III) from the action of a Fe-reductase and a thiosulfate reductase which, respectively, transform Fe(III) to Fe(II) and thiosulfate ($S_2O_3^{2-}$) to hydrogen sulfide (H_2S) and sulfite (SO_3^{2-}) (Nealson and Myers 1992). By producing sulfides, sulfate-reducing bacteria also contribute to this indirect reduction (Fig. 14.49). Another example of indirect chemical reduction is that achieved by organic acids produced by fermentative bacteria or by certain cyanobacteria. *Escherichia coli* produces formic acid from glucose which reduces MnO_2:

$$MnO_2 + HCOO^- + 3H^+ \rightarrow Mn^{2+} + CO_2 + 2\ H_2O$$

14.6 Conclusion

Each element has a specific biogeochemical cycle, but cycles of all elements are tightly coupled and interdependent. The growth of living beings is only possible when all required biogenic elements are simultaneously available in well-defined concentrations. The synthesis of cellular materials associates carbon, as cell carbon and/or energy source, to the incorporation of a wide variety of mineral elements (oxygen, hydrogen, nitrogen, phosphorus, sulfur, iron, etc.). Some of them also act as electron acceptors or donors during redox reactions that are at the root of the various biogeochemical cycles. Organic matter mineralization returns mineral elements back to the natural environment under many chemical forms, enabling their recycling.

In this chapter, the cycles of iron, manganese, nitrogen, and sulfur were described as if they operated independently. But all these cycles are intimately linked by oxido-reduction reactions, even if all the involved actors are not yet known. For example, in some cases, there is an anaerobic oxidation of sulfur or ammonium in the presence of manganese (IV) or iron oxide as electrons acceptor. In some ecosystems, CH_4 can be oxidized in the anoxic zone by one or several syntrophic mechanisms that associate methanogenic archaea with other bacterial functional groups such as sulfate-reducing bacteria or denitrifying bacteria.

14.6.1 The Critical Role of Microorganisms

All living beings participate into the functioning of biogeochemical cycles as known these days. However, higher organisms are not essential to these cycle functioning. Prokaryotic microorganisms express all the functions that are necessary to biogeochemical cycles, enabling them to ensure all biotransformations of chemical elements. They have acquired and achieve these functions over geological times. For 3 billion years, life was strictly prokaryotic and the functioning of biogeochemical cycles was totally ensured by prokaryotic microorganisms (*cf.* Chap. 4). From their biogeochemical functions, prokaryotes have created environmental conditions in terrestrial habitats in which eukaryotic life forms could appear, grow, and diversify.

14.6.2 Microorganism Plasticity

An important property of prokaryotes is their metabolic plasticity that enables them to use different sources of carbon, energy, and electron acceptors, thereby contributing directly to the functioning of several biogeochemical cycles. For example:
1. autotrophic and chemolithotrophic *Thiobacillus denitrificans* uses sulfide as electron donor and nitrate as electron acceptor to produce its energy, and CO_2 as carbon source. These bacteria are thus directly involved in the biogeochemical cycles of sulfur, nitrogen and carbon;
2. several bacteria within the genus *Shewanella* can live under chemolithotrophic conditions using hydrogen, or under chemo-organotrophic conditions with simple organic compounds. Under both conditions, they grow either aerobically with oxygen as electron acceptor, or anaerobically with various electron acceptors such as iron, manganese, vanadium, arsenic, fumarate, the TMAO (trimethylamine oxide) and DMSO (dimethylsulfoxide). These very bacteria are also involved in the biogeochemical cycles of many pollutants (aromatic compounds, metals) and contribute to contaminated ecosystem bioremediation;
3. some photosynthetic bacteria are able to grow under very different environmental conditions. For example, the sulfur-oxidizing *Thiocapsa roseopersicina* uses light as energy source under anaerobic conditions and oxide sulfur compounds (sulfides, thiosulfate). In the dark and in the presence of dioxygen, this bacterium is able to achieve aerobic respiration using the same sulfur compounds or simple organic compounds (pyruvate, lactate, acetate, etc.) as energy sources. In addition, this species can use CO_2 (autotrophy) or the same simple organic compounds (heterotrophy) as carbon sources.

The prokaryote plasticity is confirmed by findings related to the strain "*Candidatus* Desulforudis audaxviator." This bacterium alone makes 99.9 % of the microbial community observed in a fault of the earth'crust located at 2.8 km depth (Chivan et al. 2008). The metagenomic investigation established that this organism has all the metabolic capacities necessary for an autonomous life and was perfectly adapted to its environment. Thus, this anaerobic, thermophilic, sporulated, and mobile bacterium uses sulfate as the main electron acceptor, contains dehydrogenases, and makes all its cell syntheses with organic or mineral carbon as carbon source and ammoniacal nitrogen or dinitrogen as nitrogen source. The absence of a complete protection system with respect to dioxygen suggests an ancient isolation from the Earth surface.

All functions performed by microorganisms (functional diversity) are variously distributed among the different known microorganisms (taxonomic diversity). Hence, if some steps in biogeochemical cycles are carried out by very diverse microorganisms present in almost all phyla (denitrifying, organic carbon users), others are conducted by more specialized groups (CO_2 users, nitrogen fixers) even limited to some phyla (nitrifiers, methanogens, methanotrophs, sulfate-reducers, iron-, and manganese bacteria). It must be underlined that prokaryotes are the greatest source of biodiversity still unknown on Earth. Regularly new microorganisms are discovered. This knowledge leads to complete the array of microbial actors involved in some functions, for example, the 2005 discovery of archaea able to oxidize ammonium.

The metabolic plasticity of prokaryotes is probably far from being completely understood, since new functions and physiological characteristics are part of the most recent discoveries in the functioning of biogeochemical cycles and, more generally, in microbial ecology. By the intrinsic characteristics of microorganisms (metabolic plasticity, short generation time, exchange of genetic materials), new metabolic functions appear under the influence of human activities on relatively short time scales (few decades), such as the metabolism of xenobiotics. This metabolic plasticity confers to prokaryotes this large ecological valence.

The discovery of new metabolic pathways broadens the concepts related to biogeochemical cycle functioning, and many microbial functions were recently discovered. For example, the anaerobic methane or ammonium oxidation, the thiosulfate disproportionation, the hydrocarbon anaerobic mineralization, or the iron oxidation by phototrophic anoxygenic bacteria, are known prokaryotic activities that remain to be explored. Similarly, some well-characterized functional groups have shown new metabolic capabilities; for example, some sulfate-reducing bacteria, identified as anaerobic microorganisms, are able to process sulfate-reduction and grow under oxic conditions. Similarly, some nitrifying bacteria are able to achieve denitrification. These findings tend to demonstrate that our knowledge of the role and capabilities of microorganisms in biogeochemical cycles, even if already large, is not exhaustive and new discoveries are still expected in that domain.

14.6.3 Evolutive Cycles

For millions of years, the global cycles of elements were stable and only environmental changes, often global changes, affected them. In contrast, since two centuries and especially during the past 50 years, demographic pressure and human activity development have trigered important imbalances in biogeochemical cycles. Hence, disturbances in carbon and nitrogen cycles led to the acceleration of greenhouse effects from emissions of CO_2, CH_4, and nitrogen oxides resulting from excessive use of fossil fuels, waste disposal, and intense fertilisation of agricultural soils. Similarly, disturbances in the sulfur cycle appear in some environments to be subject to anthropogenic sulfur-oxide emissions that generate acid rains.

References

Aller RC (1994) Bioturbation and remineralisation of sedimentary organic matter: effects of redox oscillation. Chem Geol 114:331–345

Amblard C, Boisson J-C, Fontvielle D, Gayte X, Sime-Ngando T (1998) Microbial ecology in aquatic systems: a review from viruses to protozoa. Rev Sci Eau N° Special: 145–162

An SM, Gardner WS, Kana T (2001) Simultaneous measurement of denitrification and nitrogen fixation using isotope pairing with membrane inlet mass spectrometry analysis. Appl Environ Microbiol 67:1171–1178

Arumugam M et al (2011) Enterotypes of the human gut microbiome. Nature 473:174–180

Azam F, Fenchel T, Gray JG, Meyer-Reil LA, Thingstad TF (1983) The ecological role of water-column microbes in the sea. Mar Ecol Prog Ser 10:257–263

Bak F, Cypionka H (1987) A novel type of energy metabolism involving fermentation of inorganic sulphur compounds. Nature 326:891–892

Barkay T, Miller SM, Summers AO (2003) Bacterial mercury resistance from atoms to ecosystems. FEMS Microbiol Rev 27:355–384

Barnes RO, Goldberg ED (1976) Methane production and consumption in anoxic marine sediments. Geology 4:297–300

Bazylinski DA, Frankel RB (2004) Magnetosome formation in prokaryotes. Nat Rev 2:217–230

Beal EJ, House CH, Orphan VJ (2009) Manganese- and iron-dependent marine methane oxidation. Science 325:184–187

Beatty JT et al (2005) An obligately photosynthetic bacterial anaerobe from a deep-sea hydrothermal vent. Proc Natl Acad Sci USA 102:9306–9310

Beauchamp B (2004) Natural gas hydrates: myths, facts and issues. C R Geosci 336:751–765

Benitez-Nelson CR (2000) The biogeochemical cycling of phosphorus in marine systems. Earth Sci Rev 51:109–135

Berelson WM, Balch WM, Najjar R, Feely RA, Sabine C, Lee K (2007) Relating estimates of CaCO3 production, export, and dissolution in the water column to measurements of CaCO3 rain into sediment traps and dissolution on the sea floor: a revised global carbonate budget. Glob Biogeochem Cycles 21, GB1024

Bernalier A, Rochet V, Leclerc M, Doré J, Pochart P (1996) Diversity of H_2/CO_2-utilizing acetogenic bacteria from feces of non-methane-producing humans. Curr Microbiol 33:94–99

Berner RA (2004) The Phanerozoic carbon cycle: CO_2 and O_2. Oxford University Press, Oxford

Blain S et al (2007) Effect of natural iron fertilization on carbon sequestration in the Southern Ocean. Nature 446:1070–1075

Blumenberg M, Seifert R, Reitner J, Pape T, Michaelis W (2004) Membrane lipid patterns typify distinct anaerobic methanotrophic consortia. Proc Natl Acad Sci USA 101:11111–11116

Boetius A et al (2000) A marine microbial consortium apparently mediating anaerobic oxidation of methane. Nature 407:623–626

Bond JH, Engel RR, Levitt MD (1971) Factors influencing pulmonary methane excretion in man. An indirect method of studying the in situ metabolism of the methane-producing colonic bacteria. J Exp Med 133:572–588

Bonin P, Gilewicz M (1992) A direct demonstration of "co-respiration" of oxygen and nitrogen oxides by *Pseudomonas nautica*. FEMS Microbiol Lett 80:183–188

Bonin P, Omnes P, Chamalet A (1998) Simultaneous occurence of denitrification and nitrate ammonification in sediments of French Mediterranean Coast. Hydrobiologia 389:169–182

Boyd PW et al (2007) Synthesis and future directions. Science 315:612–617

Brauman A, Fonty G, Roger P (2008) La Méthanisation. Éditions Tech & Doc-Lavoisier, Paris

Bruland K et al (1979) Sampling and analytical methods for the determination of copper, cadmium, zinc and nickel at the nanogram per liter level in sea water. Anal Chim Acta 105:223–245

Brune A (1998) Termite guts: the world's smallest bioreactors. Trends Biotechnol 16:16–21

Buffet BA (2000) Clathrates hydrates. Annu Rev Earth Planet Sci 28:477–507

Butler A (1998) Acquisition and utilization of transition metals ions by marine organisms. Science 281:207–210

Caumette P (1986) Phototrophic sulfur bacteria and sulfatereducing bacteria causing red waters in a shallow brackish coastal lagoon (Prevost Lagoon, France). FEMS Microbiol Ecol 38:113–124

Chapin FS (2002) Principles of terrestrial ecosystem ecology. Springer, New York

Chisholm SW (2000) Oceanography – stirring times in the Southern Ocean. Nature 407:685–687

Chisholm SW, Olson RJ, Zettler ER, Goericke R, Waterbury JB, Welschmeyer NA (1988) A novel free living prochlorophyte abundant in the oceanic euphotic zone. Nature 334:340–343

Chistoserdova L, Vorholt JA, Lidstrom ME (2005) A genomic view of methane oxidation by aerobic bacteria and anaerobic archaea. Genome Biol 6:208

Chivan D et al (2008) Environmental genomics reveals a single-species ecosystem deep within earth. Science 322:275–278

Christl SU, Murgatroyd PR, Gibson GR, Cummings JH (1992) Production, metabolism and excretion of H2 in the large intestine. Gastroenterology 102:1269–1277

Cline JD (1969) Spectrophotometric determination of hydrogen sulfide in natural waters. Limnol Oceanogr 14:454–458

Coates JD, Councell T, Ellis DJ, Lovley DR (1998) Carbohydrate oxidation coupled to Fe (III) reduction, a novel form of anaerobic metabolism. Anaerobe 4:277–282

Cohen Y, Jørgensen BB, Padan E, Shilo M (1975) Sulfide-dependent anoxygenic photosynthesis in the cyanobacterium *Oscillatoria limnetica*. Nature 257:486–492

Compeau GC, Bartha R (1985) Sulfate-reducing bacteria: principal methylators of mercury in anoxic estuarine sediment. Appl Environ Microbiol 50:498–502

Crolet JL (1992) From biology and corrosion to biocorrosion. Oceanol Acta 15:87–94

Cypionka H (2000) Oxygen respiration by Desulfovibrio species. Annu Rev Microbiol 54:827–848

Damste JSS et al (2002) Linearly concatenated cyclobutane lipids form a dense bacterial membrane. Nature 419:708–712

Danon M, Franke-Whittle IH, Insam H, Chen Y, Hadar Y (2008) Molecular analysis of bacterial community succession during prolonged compost curing. FEMS Microbiol Ecol 65:133–144

Daumas S, Magot M, Crolet JL (1993) Measurement of the net production of acidity by a sulphate-reducing bacterium: experimental checking of theoretical models of microbially influenced corrosion. Res Microbiol 144:327–332

De Bona FD, Bayer C, Dieckow J, Bergamaschi H (2008) Soil quality assessed by carbon management index in a subtropical acrisol subjected to tillage systems and irrigation. Aust J Soil Res 46:469–475

de Wit R, van Gemerden H (1990) Growth and metabolism of the purple sulfur bacterium *Thiocapsa roseopersicina* under combined light/dark and oxic/anoxic regimens. Arch Microbiol 154:459–464

Del Amo Y, Quéguiner B, Tréguer P, Breton H, Lampert L (1997) Impacts of high-nitrate freshwater inputs on macrotidal ecosystems. II. Specific role of the silicic acid pump in the year-round dominance of diatoms in the Bay of Brest (France). Mar Ecol Prog Ser 161:225–237

del Giorgio PA, Williams PJB (2005) The global significance of respiration in aquatic ecosystems: from single cells to the biosphere. In: del Giorgio PA, Williams PJB (eds) Respiration in aquatic ecosystems. Oxford University Press, Oxford, pp 267–303

Des Marais DJ (2003) Biogeochemistry of hypersaline microbial mats illustrates the dynamics of modern microbial ecosystems and the early evolution of the biosphere. Biol Bull 204:160–167

Drake HL, Horn MA (2007) As the worm turns: the earthworm gut as a transient habitat for soil microbial biomes. Annu Rev Microbiol 61:169–189

Dridi B, Fardeau ML, Ollivier B, Raoult D, Drancourt M (2012) *Methanomassiliicoccus luminyensis*, gen. nov., sp. nov., a methanogenic archaeon isolated from human faeces. Int J Syst Evol Microbiol 62:1902–1907

Duhamel S, Björkman KM, van Wambeke F, Moutin T, Karl DM (2011) Characterization of alkaline phosphatase activity in the North and South pacific subtropical gyres: implications for phosphorus cycling. Limnol Oceanogr 56:1244–1254

Dyhrman ST, Ammermann JW, Van Mooy BA (2007) Microbes and marine phosphorus cycle. Oceanography 20:110–116

Eckburg PB, Bik EM, Bernstein CN, Purdom E, Dethlefsen L, Sargent M, Gill SR, Nelson KE, Relman DA (2005) Diversity of the human intestinal microbial flora. Science 308:1635–1638

Ehrenreich A, Widdel F (1994) Anaerobic oxidation of ferrous iron by purple bacteria, a new type of phototrophic metabolism. App Environ Microbiol 60:4517–4526

Ehrlich HL (2002) Geomicrobiology. Marcel Dekker, New York

Elvert M, Boetius A, Knittel K, Jørgensen BB (2003) Characterization of specific membrane fatty acids as chemotaxonomic markers for sulphate-reducing bacteria involved in anaerobic oxidation of methane. Geomicrobiol J 20:403–419

Ettwig KF et al (2008) Denitrifying bacteria anaerobically oxidize methane in the absence of *Archaea*. Environ Microbiol 10:3164–3173

Ettwig KF et al (2010) Nitrite-driven anaerobic methane oxidation by oxygenic bacteria. Nature 464:543–548

Farenhorst A (2006) Importance of soil organic matter fractions in soil-landscape and regional assessments of pesticide sorption and leaching in soil. Soil Sci 70:1005–1012

Farina M, Esquivel DMS, Lins de Bzarros HGP (1990) Magnetic iron-sulfur crystals from a magnetotactil microorganism. Nature 343:256–258

Fay P (1992) Oxygen relations of nitrogen-fixation in cyanobacteria. Microbiol Rev 56:340–373

Féron D, Compère C, Dupont I, Magot M (2002) Biodétérioration des matériaux métalliques ou biocorrosion. In Béranger G, et Mazille H (eds) Corrosion des métaux et alliages. Lavoisier, Paris, pp 385–405

Field CB, Behrenfeld MJ, Randerson JT, Falkowski P (1998) Primary production of the biosphere: integrating terrestrial and oceanic components. Science 281:237–240

Fleming EJ, Mack EE, Green PG, Nelson DC (2006) Mercury methylation from unexpected sources: molybdate-inhibited freshwater sediment and iron-reducing bacterium. Appl Environ Microbiol 64:457–464

Fontaine S, Mariotti A, Abbadie L (2003) The priming effect of organic matter: a question of microbial competition? Soil Biol Biochem 35:837–843

Fontaine S, Bardoux G, Abbadie L, Mariotti A (2004) Carbon input to soil may decrease soil carbon content. Ecol Lett 7:314–320

Forshay KJ, Stanley EH (2005) Rapid nitrate loss and denitrification in a temperate river floodplain. Biogeochemistry 75:43–64

Fossing H, Jørgensen BB (1989) Measurement of bacterial sulfate reduction in sediments: evaluation of single step chromium reduction method. Biogeochemistry 8:205–222

Francis CA, Co EM, Tebo BM (2001) Enzymaticmanganese(II) oxidation by a marine alpha-proteobacterium. Appl Environ Microbiol 67:4024–4029

Frontier S, et Pichot-Viale D (1998) Ecosystèmes: Structure, fonctionnement, évolution, 2e édition. Dunod, Paris

Fung I, John J, Lerner J, Matthews E, Prather M, Steele LP, Fraser PJ (1991) Three-dimensional model synthesis of the global methane cycle. J Phys Res 96(D7):13033–13065

Galloway JN, Aber JD, Erisman JW, Seitzinger SP, Howarth RW, Cowling EB, Cosby BJ (2003) The nitrogen cascade. Bioscience 53:341–356

Gans J, Wolinsky M, Dunbar J (2005) Computational improvements reveal great bacterial diversity and high metal toxicity in soil. Science 309:1387–1390

Garcia J-L, Patel BKC, Ollivier B (2000) Taxonomic, phylogenetic, and ecological diversity of methanogenic Archaea. Anaerobe 6:205–226

Gilbert F, Bonin P, Stora G (1995) Effect of bioturbation on denitrification in a marine sediment from the West Mediterranean littoral. Hydrobiologia 304:49–58

Goevert D, Conrad R (2009) Effect of substrate concentration on carbon isotope fractionation during acetoclastic methanogenesis by *Methanosarcina barkeri* and *M. acetivorans* and in rice field soil. Appl Environ Microbiol 75:2605–2612

Graham JW, Cooper SC (1959) Biological origin of manganese-rich deposits of the sea floor. Nature 183:1050–1051

Grossi V, Cuny P, Caradec S, Nerini D, Pancost R, Gilbert F (2006) Impact of feeding by *Arenicola marina* (L.) and ageing of faecal material on fatty acid distribution and bacterial community structure in marine sediments: an expérimental approach. J Exp Mar Biol Ecol 336:54–64

Hanson RS, Hanson TE (1996) Methanotrophic bacteria. Microbiol Rev 60:439–471

Haroon MF et al (2013) Anaerobic oxidation of methane coupled to nitrate reduction in a novel archaeal lineage. Nature 500:567–570

Hauck RD, Bouldin DR (1961) Distribution of isotopic nitrogen gas during denitrification. Nature 191:871–872

Hegler F, Popsth NR, Jiang J, Kappler A (2008) Physiology of phototrophic iron (II)-oxidizing bacteria: implications for modern and ancient environments. FEMS Microbiol Ecol 66:250–260

Herbert RA (1999) Nitrogen cycling in coastal marine ecosystems. FEMS Microbiol Rev 23:563–590

Hernesmaa A, Bjorklof K, Kiikkila O, Fritze H, Haahtela K, Romantschuk M (2005) Structure and function of microbial communities in the rhizosphere of Scots pine after tree-felling. Soil Biol Biochem 37:777–785

Hinrichs K-U, Hayes JM, Sylva SP, Brewer PG, De Long EF (1999) Methane-consuming archaebacteria in marine sediments. Nature 398:802–805

Hinrichs K-U, Summons RE, Orphan V, Sylva SP, Jayes JM (2000) Molecular and isotopic analysis of anaerobic methane-oxidizing communities in marine sediments. Org Geochem 31:1685–1701

Hoeler TM, Alperin MJ, Albert DB, Martens CS (1994) Field and laboratory studies of methane oxidation in an anoxic marine sediment: evidence for a methanogen-sulfate reducer consortium. Glob Biogeochem Cycle 8:451–463

Houghton JT, Ding Y, Griggs DJ, Noguer M, van der Linden PJ, Dai X, Maskell K, Johnson CA (2004) Climate change 2001: the scientific basis. Contribution of working group I to the third assessment report of the Intergovernmental Panel on Climate Change. Cambridge University Press, Cambridge

Hu S, Zeng RJ, Burow LC, Lant P, Keller J, Yuan Z (2009) Enrichment of denitrifying anaerobic methane oxidizing microorganisms. Environ Microbiol Rep 1:377–384

Hutsch BW, Augustin J, Merbach W (2002) Plant rhizodeposition – an important source for carbon turnover in soils. J Plant Nut Soil Sci 165:397–407

IPCC (1995) Greenhouse gas inventory reference manual. IPCC Guidelines for National Greenhouse Gas Inventories. United Nations Environment Programme (UNEP), the Organisation for Economic Co-operation and Development (OECD), the International Energy Agency (IEA) and the Intergovernmental Panel on Climate Change (IPCC), Bracknell

IPCC (2007) Climate change 2007: the physical science basis. Contribution of working group I to the fourth assessment report IPCC. In: Solomon S, Qin D, Manning M, Chen Z, Marquis M, Averyt KB, Tignor M, Mille HL (eds). Cambridge University Press, Cambridge/New York

Jahn MK, Haderlein SB, Meckenstock RU (2005) Anaerobic degradation of benzene, toluene, ethylbenzene, and o-xylene in sediment-free iron-reducing enrichment cultures. Appl Environ Microbiol 71:3355–3358

Jannasch HW, Mottl MJ (1985) Geomicrobiology of deep-sea hydrothermal vents. Science 229:717–725

Jetten MS, Logemann S, Muyzer G, Robertson LA, de Vries S, van Loosdrecht MC, Kuenen JG (1997) Novel principles in the microbial conversion of nitrogen compounds. Antonie Van Leeuwenhoek 71:75–93

Jickells T et al (2005) Global iron connections between desert dust, ocean oiogeochemistry, and climate knowledge. Science 303:67–71

Jørgensen BB (1977) The sulfur cycle of a coastal marine sediment. Limnol Oceanogr 22:817–832

Jørgensen BB (1983) The microbial sulphur cycle. In: Krumbein WE (ed) Microbial geochemistry. Blackwell Scientific Publication, Oxford, pp 91–124

Jørgensen BB (1990) A thiosulfate shunt in the sulfur cycle of marine sediments. Science 249:152–154

Jørgensen BB (2000) Bacteria and marine biogeochemistry. In: Schulz HD, Zabel M (eds) Marine geochemistry. Springer, Berlin/Heidelberg/New York, pp 173–207

Jørgensen BB, Fenchel T (1974) The sulfur cycle of a marine sediment model system. Mar Biol 24:189–201

Karl DM (2002) Microbiological oceanography – hidden in a sea of microbes. Nature 415:590–591

Karl DM (2007) The marine phosphorus cycle. In: Hurst CJ et al (eds) Manual of environmental microbiology, 3rd edn. American Society of Microbiology, Washington, DC, pp 523–539

Karl DM, Tilbrook BD (1994) Production and transport of methane in oceanic particulate organic matter. Nature 368:732–734

Knittel K, Boetius A (2009) Anaerobic oxidation of methane: progress with an unknown process. Annu Rev Microbiol 63:311–334

Knittel K, Lösekann T, Boetius A, Kort R, Amann R (2005) Diversity and distribution of methanotrophic Archaea at cold seeps. Appl Environ Microbiol 71:467–479

Koide K, Osono T, Takeda H (2005) Fungal succession and decomposition of *Camellia japonica* leaf litter. Ecol Res 20:599–609

Küsel K, Dorsch T, Acker G, Stackebrandt E (1999) Microbial reduction of Fe(III) in acidic sediments: isolation of *Acidiphilium cryptum* JF-5 capable of coupling the réduction of Fe(III) to the oxidation of glucose. Appl Environ Microbiol 65:3633–3640

Kvenvolden KA, Rogers B (2005) Gaia's breathglobal methane exhallations. Mar Petrol Geol 22:579–590

Laughlin RJ, Stevens RJ (2002) Evidence for fungal dominance of denitrification and codenitrification in a grassland soil. Soil Sci 66:1540–1548

Lavelle P (2002) Functional domains in soils. Ecol Res 17:441–450

Lavelle P, Spain AV (2001) Soil ecology. Kluwer Academic, Dordrecht/Boston/London

Lavelle P, Bignell D, Lepage M, Wolters V, Roger P, Ineson P, Heal OW, Dhillion S (1997) Soil function in a changing world: the role of invertebrate ecosystem engineers. Eur J Soil Biol 33:159–193

Lehours A-C, Carrias J-F, Amblard C, Sime-Ngando T, Fonty G (2010) les bactéries du Lac Pavin. Pour la Science 387:30–35

Lelieveld J, Crutzen PJ, Dentener FJ (1998) Changing concentration, lifetime and climate forcing of atmospheric methane. Tellus 50B:128–150

Lemaitre C, Pébère N, Festy D (1994) Biodétérioration des matériaux. EDP Sciences, Les Ulis

Lomans BP, van der Drift C, Pol A, Op den Camp HJM (2002) Microbial cycling of volatile organic sulfur compounds. Cell Mol Life Sci 59:575–588

Lovley DR (1991) Dissimilatory Fe(III) and Mn (IV) reduction. Microbiol Rev 55:259–287

Lovley DR (2006) Dissimilatory Fe(III)- and Mn(IV)- reducing prokaryotes. In: Dworkin M, Falkow S, Rosenberg E, Schleifer KH, Stackebrandt E (eds) The prokaryotes. Springer, New York, pp 635–658

Lovley DR, Holmes DE, Nevin KP (2004) Dissimilatory Fe(III) and Mn(Iv) reduction. Adv Microbiol Physiol 49:219–286

Lynd LR, Weimer PJ, van Zyl WH, Pretorius IS (2002) Microbial cellulose utilization: fundamentals and biotechnology. Microbiol Mol Biol Rev 66:506–577

Macdonald GJ (1990) Role of methane clathrates in past and future climates. Clim Change 16:247–281

Macfarlane GT, Gibson GR (1994) Metabolic activities of the normal colonic flora. In: Gibson SAW (ed) Human health. The contribution of microorganisms. Springer, London, pp 17–52

Madsen EL (2011) Microorganisms and their roles in fundamental biogeochemical cycles. Curr Opin Biotechnol 22:456–464

Maier RM, Pepper IL, Gerba CP (2000) Environmental microbiology. Academic, San Diego

Majeed MZ et al (2013) Emissions of nitrous oxide from casts of tropical earthworms belonging to different ecological categories. Pedobiologia 56:49–58

Marland G, Boden TA, Andres RJ (2003) Global, regional, and national CO_2 emissions. In: Trends: a compendium of data on global change. Carbon Dioxide Information Analysis Center, Oak Ridge National Laboratory, U.S. Department of Energy, Oak Ridge

Martinez RJ, Mills HJ, Story S, Sobecky PA (2006) Prokaryotic diversity and metabolically active microbial populations in sediments from an active mud volcano in the Gulf of Mexico. Environ Microbiol 8:1783–1796

Master Y, Shavit U, Shaviv A (2005) Modified isotope pairing technique to study N transformations in polluted aquatic systems: theory. Environ Sci Technol 39:1749–1756

Michaelis W et al (2002) Microbial reefs in the Black Sea fueled by anaerobic oxidation of methane. Science 297:1013–1015

Miller TL, Wolin MJ (1986) Methanogens in human and animal intestinal tracts. Syst Appl Microbiol 7:223–229

Mills MMC, Ridame MT, Davey J, La Roche R, Geider J (2005) Iron and phophorus co-limit nitrogen fixation in the eastern tropical North Atlantic. Nature 435:292–294

Milucka J et al (2012) Zero-valent sulphur is a key intermediate in marine methane oxidation. Nature 491:541–546

Minjeaud L, Bonin PC, Michotey VD (2008) Nitrogen fluxes from marine sediments: quantification of the associated co-occurring bacterial processes. Biogeochemistry 90:141–157

Mitchell P (1961) Coupling of phosphorylation to électron and hydrogen transfer by chemi-osmotic type of mechanism. Nature 191:144–148

Moran JJ, Beal EJ, Vrentas JM, Orphan VJ, Freeman KH, House CH (2008) Methyl sulfides as intermediates in the anaerobic oxidation of methane. Environ Microbiol 10:162–173

Morel A, Gentili B, Claustre H, Babin M, Bricaud A, Ras J, Tieche F (2007) Optical properties of the "clearest" natural waters. Limnol Oceanogr 52:217–229

Morel FMM, Kustka AB, Shaked Y (2008) The role of unchelated Fe in the iron nutrition of phytoplankton. Limnol Oceanogr 53:400–404

Moutin T (2000) Cycle biogéochimique du phosphate: rôle dans le contrôle de la production planctonique et conséquences sur l'exportation de carbone de la couche éclairée vers l'océan profond. Océanis 36:643–660

Moutin T, Karl DM, Duhamel S, Rimmelin P, Raimbault P, van Mooy BAS, Claustre H (2008) Phosphate availability and the ultimate control of new nitrogen input by nitrogen fixation in the tropical Pacific Ocean. Biogeosciences 5:95–109

Moutin T, van Wambeke F, Prieur L (2012) Introduction to the biogeochemistry from the oligotrophic to the ultraoligotrophic mediterranean (BOUM) experiment. Biogeosciences 9:3817–3825

Mulder A, Vandegraaf AA, Robertson LA, Kuenen JG (1995) Anaerobic ammonium oxidation discovered in a denitrifying fluidized-bed reactor. FEMS Microbiol Ecol 16:177–183

Nauhaus K, Albrecht M, Elvert M, Boetius A, Widdel F (2007) *In vitro* cell growth of marine archaeal-bacterial consortia during anaerobic oxidation of methane with sulphate. Environ Microbiol 9:187–196

Nealson KH, Myers CR (1992) Microbial reduction of manganese and iron: new approches to carbon cycling. Appl Environ Microbiol 58:439–443

Nicol GW, Schleper C (2006) Ammonia-oxidising Crenarchaeota: important players in the nitrogen cycle? Trends Microbiol 14:207–212

Nielsen LP (1992) Denitrification in sediment determined from nitrogen isotope pairing. FEMS Microbiol Ecol 86:357–362

Ollivier B, Caumette P, Garcia JL, Mah RA (1994) Anaerobic bacteria from hypersaline environments. Microbiol Rev 58:27–38

Oremland RS, Stolz JF (2003) The ecology of arsenic. Science 300:939–944

Orphan VJ, House CH, Hinrichs K-U, McKeegan KD, DeLong EF (2001) Methane-consuming archaea revealed by directly coupled isotopic and phylogenetic analysis. Science 293:484–487

Overmann J, Garcia-Pichel F (2006) The phototrophic way of life. In: Dworkin M, Falkow S, Rosenberg E, Schleifer KH, Stackebrandt E (eds) The prokaryotes. Springer, New York, pp 32–85

Overmann J, Cypionka H, Pfennig N (1992) An extremely low-light-adapted phototrophic sulfur bacterium from the Black Sea. Limnol Oceanogr 37:150–155

Pancost RD, Sinninghe Damsté JS (2003) Carbon isotopic compositions of prokaryotic lipids as tracers of carbon cycling in diverse settings. Chem Geol 195:29–58

Parkin TB, Brock TD (1981) The role of phototrophic bacteria in the sulfur cycle of a meromictic lake. Limnol Oceanogr 26:880–890

Parks JM et al (2013) The genetic basis for bacterial mercury methylation. Science 339:1332–1335

Parsons TR, Takahashi M, Hargrave BT (1984) Biological oceanographic processes. Pergamon Press, Oxford

Pedersen H, Dunkin KA, Firestone MK (2002) The relative importance of autotrophic and heterotrophic nitrification in a conifer forest soil as measured by 15N tracer and pool dilution techniques. Biogeochemistry 44:135–150

Petermann H, Bleil U (1993) Detection of live magnetotactile bacteria in South Atlantic deep-see sediments. Earth Planet Sci Lett 117:223–228

Pfennig N (1975) The phototrophic bacteria and their role in sulfur cycle. Plant and Soil 43:1–16

Pfennig N (1989) Metabolic diversity among the dissimilatory sulfate-reducing bacteria – Albert Jan Kluyver memorial lecture. Ant Leeuw Int J G 56:127–138

Prentice IC et al (2001) The carbon cycle and atmospheric carbon dioxide. In: Houghton J, Ding Y, Griggs DJ, Noguer N, van der Linden PJ, Dai X (eds) Climate change 2001: the scientific basis. Cambridge University Press, Cambridge, pp 183–239

Raghoebarsing AA et al (2006) A microbial consortium couples anaerobic methane oxidation to denitrification. Nature 440:918–921

Ranchou-Peyruse M, Monperrus M, Bridou R, Duran R, Amouroux D, Salvado JC, Guyoneaud R (2009) Overview of mercury methylation capacities among anaerobic bacteria including representatives of the sulphate-reducers : implications for environmental studies. Geomicrobiol J 26:1–8

Redfield AC (1934) On the proportions of organic derivatives in sea water and their relation to the composition of plankton. In: Daniel RJ (ed) James Johnstone memorial volume. University Press of Liverpool, Liverpool, pp 177–192

Reeburgh WS (2007) Oceanic methane biogeochemistry. Chem Rev 107:486–513

Risgaard-Petersen N (2003) Coupled nitrification-denitrification in autotrophic and heterotrophic estuarine sediments: on the influence of benthic microalgae. Limnol Oceanogr 48:93–105

Robertson LA, Kuenen JG (2006) The colorless sulfur bacteria. In: Dworkin M, Falkow S, Rosenberg E, Schleifer KH, Stackebrandt E (eds) The prokaryotes. Springer, New York, pp 985–1011

Roesch LF et al (2007) Pyrosequencing enumerates and contrasts soil microbial diversity. ISME J 1:283–290

Ruiz-Rueda O, Hallin S, Baneras L (2009) Structure and function of denitrifying and nitrifying bacterial communities in relation to the plant species in a constructed wetland. FEMS Microbiol Ecol 67:308–319

Samuel BS, Hansen EE, Manchester JK, Coutinho PM, Henrissat B, Fulton R, Latreille P, Kim K, Wilson RK, Gordon JI (2007) Genomic and metabolic adaptations of *Methanobrevibacter smithii* to the human gut. Proc Natl Acad Sci USA 104:10643–10648

Sarmiento JL, Gruber N (2006) Ocean biogeochemical dynamics. Princeton University Press, Princeton

Satoh H, Nakamura Y, Okabe S (2007) Influences of infaunal burrows on the community structure and activity of ammonia-oxidizing bacteria in intertidal sediments. Appl Environ Microbiol 73:1341–1348

Schauss K et al (2009) Dynamics and functional relevance of ammonia-oxidizing Archaea in two agricultural soils. Environ Microbiol 11:446–456

Schloss PD, Handelsman J (2006) Toward a census of bacteria in soil. Plos Comput Biol 2:786–793

Schmidt I, van Spanning RJ, Jetten MS (2004) Denitrification and ammonia oxidation by *Nitrosomonas europaea* wild-type, and NirK- and NorB-deficient mutants. Microbiology 150:4107–4114

Sieburth JM, Smetacek V, Lenz J (1978) Pelagic ecosystem structure: heterotrophic compartments of the plankton and their relationship to plankton size fractions. Limnol Oceanogr 23(6):1256–1263

Straub KL, Benz B, Schink B, Widdel F (1996) Anaerobic, nitrate-dependent microbial oxidation of ferrous iron. Appl Environ Microbiol 62:1458–1460

Straub KL, Schönhuber WA, Buchholz-Cleven BEE, Schink B (2004) Diversity of ferrous iron-oxidizing, nitrate-reducing bacteria and their involvement in oxygen-independent iron cycling. Geomicrobiol J 21:371–378

Strong A, Chisholm S, Miller C, Cullen J (2009) Iron fertilization: time to move on. Nature 461:347–348

Strous M, Jetten SM (2004) Anaerobic oxidation of methane and ammonium. Annu Rev Microbiol 58:99–117

Tabatabai MA (1974) Determination of sulfate in water samples. Sulphur Inst J 10:11–14

Tebo BM, Johson HA, McCarthy JK, Templeton AS (2005) Geomicrobiology of manganese (II) oxidation. Trends Microbiol 13:421–428

Thamdrup B, Dalsgaard T (2002) Production of N_2 through anaerobic ammonium oxidation coupled to nitrate reduction in marine sediments. Appl Environ Microbiol 68:1312–1318

Thauer RK, Jungermann K, Deker K (1977) Energy conservation in chemotrophic anaerobic bacteria. Microbiol Mol Biol Rev 41:100–180

Thingstad TF, Hagstrom A, Rassoulzadegan F (1997) Accumulation of degradable DOC in surface waters: is it caused by a malfunctioning microbial loop? Limnol Oceanogr 42:398–404

Thompson IA, Huber DM, Guest CA, Schulze DG (2005) Fungal manganese oxidation in a reduced soil. Environ Microbiol 7:1480–1487

Tiedje JM, Sextone AJ, Myrold DD, Robinson JA (1982) Denitrification, ecological niches competition and survival. Ant Leeuw Int J G 48:569–583

Tréguer P, Nelson DM, Van Bennekom AJ, Demaster DJ, Quéguiner B, Leynaert A (1995) The silica budget of the World Ocean: a re-estimate. Science 268:375–379

Trimmer M, Risgaard-Petersen N, Nicholls J-C, Engstrom P (2006) Direct measurement of anaerobic ammonium oxidation (anammox) and denitrification in intact sediment cores. Mar Ecol Prog Ser 326:37–47

Valentine DL, Reeburgh WS (2000) New perspectives on anaerobic methane oxidation. Environ Microbiol 2:477–484

van Gemerden H (1993) Microbial mats, a joint venture. Mar Geol 113:3–25

van Herwijnen R, Laverye T, Poole J, Hodson ME, Hutchings TR (2007) The effect of organic materials on the mobility and toxicity of metals in contaminated soils. Appl Geochem 22:2422–2434

van Mooy BAS et al (2009) Phytoplankton in the ocean use non-phosphorus lipids in response to phosphorus scarcity. Nature 458:69–72

van Niftrik LA, Fuerst JA, Damste JSS, Kuenen JG, Jetten MSM, Strous M (2004) The anammoxosome: an intracytoplasmic compartment in anammox bacteria. FEMS Microbiol Lett 233:7–13

van Wambeke F, Nedoma J, Duhamel S, Lebaron P (2008) Alkaline phosphatase activity of marine bacteria studied with ELF97 phosphate: success and limits in P-limited Mediterranean Sea. Aquat Microb Ecol 52:245–251

Vandecasteele J-P (2008) Petroleum microbiology, vols 1 and 2. Editions Technip, Paris

Von Wolzogen Kur CAH, van der Vlugt IS (1934) De graphiteering van gietijzer ais electrobiochemisch proces in anaerobe gronden. Water 18:147–165

Wagner M, Loy A, Klein M, Lee N, Ramsing NB, Stahl DA, Friedrich MW (2005) Functional marker genes for identification of sulfate-reducing prokaryotes. Methods Enzymol 397:469–489

Waterbury JB, Watson SW, Guillard RRL, Brand LE (1979) Widespread occurrence of a unicellular marine planktonic cyanobacterium. Nature 277:293–294

Weisskopf L, Fromin N, Tomasi N, Aragno M, Martinoia E (2005) Secretion activity of white lupin's cluster roots influences bacterial abundance, function and community structure. Plant and Soil 268:181–194

Widdel F, Pfennig N (1992) The genus *Desulfuromonas* and other Gram negative sulfur-reducing bacteria. In: Balows A, Truper HG, Dworkin M, Harder W, Schleifer KH (eds) The prokaryotes. Springer, New York, pp 3379–3389

Williams PJB, del Giorgio PA (2005) Respiration in aquatic ecosystems: history and background. In: del Giorgio PA, Williams PJB (eds) Respiration in aquatic ecosystems. Oxford University Press, Oxford, pp 1–17

Wolin MJ (1974) Metabolic interactions among intestinal microorganisms. Am J Clin Nutr 27:1320–1328

Yoshinari T, Knowles R (1976) Acetylene inhibition of nitrous oxide reduction by denitrifying bacteria. Biochem Biophys Res Commun 69:705–711

Zumft WG (1997) Cell biology and molecular basis of denitrification. Microbiol Mol Biol Rev 61:533–589

Environmental and Human Pathogenic Microorganisms

15

Philippe Lebaron, Benoit Cournoyer, Karine Lemarchand, Sylvie Nazaret, and Pierre Servais

Abstract

As the study of interactions between pathogenic microorganisms and their environment is part of microbial ecology, this chapter reviews the different types of human pathogens found in the environment, the different types of fecal indicators used in water quality monitoring, the biotic and abiotic factors affecting the survival and the infectivity of pathogenic microorganisms during their transportation in the environment, and the methods presently available to detect rare microorganisms in environmental samples. This chapter exclusively focuses on human pathogens.

Keywords

Pathogens • Wastewater treatment plant • Dissemination • Antibiotic resistance • Environmental reservoirs • Toxins • Sanitary microbiology • Biological pollution

Abbreviations

AIDS	Acquired immunodeficiency syndrome
BOD	Biological oxygen demand
CSO	Combined sewer overflow
HAV	Hepatitis A virus
HEV	Hepatitis E virus
GI	Genomic islands
ID50	The dose of an infectious organism required to produce infection in 50 % of the experimental subjects
Ig	Immunoglobulin
IS	Insertion sequence
LPS	Lipopolysaccharide
MID	Minimal infectious dose
OECD	Organization for Economic Co-operation and Development
PAI	Pathogenicity islands
SS	Suspended solids
IPCC	Intergovernmental Panel on Climate Change
VNC	Viable but non-culturable
WWTP	Wastewater treatment plant

---

* Chapter Coordinator

P. Lebaron* (✉)
Observatoire Océanologique de Banyuls, Laboratoire de Biodiversité et Biotechnologie Microbiennes (LBBM), Sorbonne Universités, UPMC Univ Paris 06, USR CNRS 3579, 66650 Banyuls-sur-Mer, France
e-mail: lebaron@obs-banyuls.fr

K. Lemarchand
Laboratoire de Bactériologie marine et Écotoxicologie microbienne, Institut des sciences de la mer de Rimouski, Rimouski, QC G5M1E1, Canada

B. Cournoyer • S. Nazaret
Microbial Ecology Center, UMR CNRS 5557 / USC INRA 1364, Université Lyon 1, 69622 Villeurbanne, France

P. Servais
Laboratoire d'Écologie des Systèmes Aquatiques (ESA), Université Libre de Bruxelles, Campus de la Plaine, 1050 Bruxelles, Belgium

15.1 Introduction

A large majority of microorganisms are completely harmless and play essential functions in ecosystems as well as in human life sustaining. Nevertheless, some microorganisms, called **pathogens***, can cause diseases of varying severity in humans and animals. Human pathogens are relatively diverse, belonging to viruses, bacteria, fungi, protozoa, and algae. They may be present in many different environments such as water, soil, gastrointestinal tract, and air. The contamination of water by pathogenic microorganisms arguably represents one of the most significant risks for human health on a global scale, and there have been numerous cases of poisoning and disease outbreaks throughout history resulting directly of poorly treated or untreated water. Water-related diseases are a major cause of morbidity and mortality worldwide, and contamination of drinking water remains a significant problem, even in OECD countries. Improving the quality of drinking water is a major public health concern for the twenty-first century. Moreover, monitoring of drinking water quality is part of a broad control of the microbiological quality of water systems, including groundwater and surface water (freshwaters and marine waters), which are increasingly used as drinking water supply. If the health problems associated with direct consumption of poor water quality are by far the most important concerns, it is also important to consider the quality of surface waters used for irrigation of agricultural lands, aquaculture, and recreation.

The availability of reliable, easy to use, and effective methods is crucial to ensure the microbiological safety of water. Whatever the method used, cultivation or direct observation, the great variety of pathogenic microorganisms does not allow the search for all pathogens and requires the use of "pollution indicators" such as fecal bacteria (e.g., *Escherichia coli* and enterococci). Moreover, culture-based methods are time-consuming, and when they are used to enumerate indicators or pathogenic bacteria, the time required to obtain a diagnostic is often too long to protect people from contamination. As an example, at least 4–5 days are required to diagnose the presence of *Salmonella* spp. or *Legionella* spp. in water samples by culture processes. The use of the fecal bacteria indicators has been very helpful in reducing cases of cholera and typhoid and to protect against pathogenic bacteria such as *Salmonella* and *Shigella*. Nevertheless, it is now clearly demonstrated that these indicators have only a limited predictive efficiency when the contamination is due to viruses or protozoa whose behavior in the environment is often very different from the behavior of bacteria in water systems. As a consequence, an important challenge in modern water microbial quality monitoring is the rapid, specific, and sensitive detection of reliable microbial indicators and specific waterborne pathogens.

As the study of interactions between pathogenic microorganisms and their environment is part of microbial ecology, this chapter reviews the different types of human pathogens found in the environment, the different types of fecal indicators used in water quality monitoring, the biotic and abiotic factors affecting the survival and the infectivity of pathogenic microorganisms during their transportation in the environment, and the methods presently available to detect rare microorganisms in environmental samples. This chapter exclusively focuses on human pathogens.

15.2 Different Types of Pathogens

Pathogens in the environment are very diverse: viruses, bacteria, fungi, protozoa, and cyanobacteria. The main pathogens of water systems and the diseases they cause are shown in Table 15.1. We recommend consulting the English version of the French Institute of Health Surveillance website (http://www.invs.sante.fr/en) as an example for information regarding the mandatory reporting of certain diseases. This site provides a wealth of important information about the epidemiological surveillance of infections. The reader should also refer to the following paper on emerging issues in water and infectious diseases: http://www.who.int/water_sanitation_health/emerging/emerging.pdf.

15.2.1 Viruses

A virus is a biological entity that requires a host cell and its constituents to multiply. It consists of nucleic acid (DNA or RNA) surrounded by a protein coat called **capsid***. Viruses are obligate **parasites*** of living cells, either prokaryotic or eukaryotic. They are able to quickly adapt to changes in their host and to select processes to bypass the defenses of the host. Frequent mutations make defenses against viruses very difficult to implement. They do not require nutrients to survive, and they can potentially persist in the environment for long periods of time (weeks to months). Their size is very small (mostly between 10 and 200 nm). If the virus can infect bacteria, animals, plants, and humans, we will only focus on human viruses. There are a wide variety of viruses in the environment that are likely to cause disease in humans, some of which may have a virus of animal origin such as avian influenza.

Modern means of communication allow epidemics to propagate at a speed hitherto unparalleled. New viruses and bacteria are identified daily, and at the same time ancient and almost forgotten diseases such as cholera, avian flu, or others are back to the actuality. Scientists believe that the twenty-first century will be that of viruses, and many researchers around the world are trying to develop methods for the detection, prediction, and monitoring to address this growing concern.

There are a variety of viruses that cause disease in humans with more than 150 types of viruses that infect the intestinal

Table 15.1 Major human pathogenic and toxin-producing microorganisms detected in aquatic environments

Group	Pathogen	Disease or symptoms	Source	Survival potential in the environment	Incubation time
Virus	Enterovirus (polio, echo, Coxsackie)	Meningitis, paralysis, myocarditis, fever, respiratory disease, etc.	Human	Unknown	2–15 days
	Hepatitis A and E virus	Hepatitis	Human	Unknown	30–40 days
	Norwalk virus	Diarrhea	Human, pig?	Unknown	24–48 h
	Astrovirus	Diarrhea	Human	Unknown	1–2 days
	Calicivirus	Diarrhea	Human	Unknown	1–3 days
	Adenovirus	Diarrhea, eye infections, respiratory diseases	Human	Resistant	8–10 days
	Reovirus	Respiratory diseases, intestinal pains	Human	Unknown	24–72 h
Bacteria	Salmonella	Typhoid, severe diarrhea	Human, domestic and wild animals, poultry. Healthy carriers	Few days	6–72 h; 7–28 days for S. typhi; 1–10 days for S. paratyphi
	Shigella	Diarrhea	Human	2–3 days	1–7 days
	Campylobacter	Diarrhea	Birds and animals	2–5 days	1–10 days
	Vibrio cholerae	Diarrhea	Human, water, environment	Weeks	12 h–5 days
	Yersinia enterocolitica	Diarrhea	Animals	Weeks	2–7 days
	Escherichia coli (pathogenic strains)	Diarrhea, cystitis, hemorrhagic colitis	Human, animals	Few days in water, longer in soil	2–8 days
	Legionella	Pneumonia, respiratory infections	Water	Resistant	2–10 days
Protozoa	Naegleria	Meningoencephalitis		Months (cysts)	3–7 days
	Entamoeba histolytica	Amoebic dysentery	Human	Months (cysts)	Variable from days to weeks
	Giardia lamblia	Diarrhea	Human, domestic and wild animals	Months (cysts)	3–25 days
	Cryptosporidium parvum	Diarrhea	Human, domestic and wild animals	Months (cysts)	1–12 days
Cyanobacteria	Microcystis	Diarrhea	Water		
	Anabaena	Diarrhea, carcinogens	Water		
	Aphanizomenon	Diarrhea, carcinogens	Water		

tract of man. The virus that can infect and multiply in the intestines are **enteric*** viruses. Some enteric viruses can also replicate in other organs such as the liver, heart, eyes, etc. Enteric viruses often have a high host specificity. During infection, many virus particles are produced and released by feces (10^8–10^{12} particles per gram) and then transported by the sewer. These sewer networks provide a reservoir of viruses in the environment. Data on the behavior of viruses in the environment are very scarce, particularly regarding their ability to survive in water, soil, and sediments.

15.2.1.1 Enteroviruses

These viruses were the first viruses isolated from sewage and water systems and therefore the most studied. The most common are the poliovirus (3 types), *coxsackievirus* (30 types), and echovirus (34 types). They can cause a variety of human diseases, some of which can be very dangerous but most are often benign. *Coxsackieviruses* can cause severe diseases (meningitis, paralysis, heart disease). For the latter, the ID_{50} (infectious dose which means the dose of microorganisms required to cause infection in 50 % of the experimental animals) is low (15–20 infectious units). This dose is often not known for other viruses.

15.2.1.2 The Hepatitis A Virus (HAV) and Hepatitis E (HEV)

The HAV and HEV are transported by water or food contaminated with feces. Both forms of hepatitis are very common in developing countries. The hepatitis A virus is ubiquitous, and its **prevalence*** varies among countries. It is very resistant in the environment and is often found in sewage, seafood, bathing water, or irrigation water. This virus is often related to enteroviruses although it does not belong to this family. Today there are fewer than 10 cases per year and per 100,000 inhabitants, and the mortality is very low (less than 0.02 %). More than 98 % of the population produces antibodies against HAV. HAV is highly resistant to heat. HEV causes hepatitis E and it was discovered in 1990. It is, as HAV, a non-enveloped virus with a positive single-stranded RNA. It is excreted in the feces over a period of more than 6 weeks after infection, thereby facilitating its spread in the environment and its transmission. Contamination is foodborne and occurs mainly by drinking contaminated water. Its incidence in adults is higher than that of HAV and is particularly harmful to pregnant women. Mortality is higher than for HAV and can reach 2–3 %.

15.2.1.3 Rotavirus

Rotavirus is the main cause of infantile gastroenteritis (95 % of children are infected). This non-enveloped virus with double-stranded RNA is responsible for severe gastroenteritis in children from 4 months to 3 years. Transmission is from person to person by fecal-oral route, through contaminated food and water.

15.2.1.4 Norwalk Virus

Norwalk virus is non-enveloped and has a single-stranded RNA. Like other viruses, the main reservoirs are feces. Contamination is by fecal-oral route but also by drinking and bathing contaminated water. It is widespread and causes diarrhea and vomiting in infected patients. The infectious dose appears to be very low.

Virus concentration in surface and drinking waters is often very low due to the high dilution rate after contamination with feces. Their detection therefore often requires a concentration step from tens of liters of water.

15.2.2 Bacteria

Bacteria are prokaryotic microorganisms (without a differentiated nucleus) surrounded by a membrane and a cell wall. Their size varies between 0.1 and 10 μm, and they may have different morphologies. Two types of morphologies dominate, coccus (spherical) and rod (cylindrical shape, sometimes curved), and the cells can remain together in groups or clusters. Depending on the nature of the cell wall structure, we differentiate between Gram-negative (Gram-) and Gram-positive (Gram+) bacteria according to the response of bacterial cells to a special staining procedure, the Gram stain. Some bacterial species can produce a resistance form (the endospore) when the environment is unfavorable, and endospores are able to remain dormant for many years and, then, convert back to a vegetative cell when the environment is favorable. The bacteria that multiply in the digestive tract of humans or warm-blooded animals are called enteric bacteria (sometimes simply called fecal bacteria).

The existence of pathogenic enteric bacteria has been known for over a century. The main bacteria involved belong to the genera *Salmonella*, *Shigella*, *Campylobacter*, *Yersinia*, *Escherichia coli*, and *Vibrio*. Other pathogens such as *Legionella* and some species of the genus *Vibrio* are not enteric bacteria but pathogenic bacteria whose natural habitat is the environment, especially aquatic systems. **Opportunistic pathogenic*** bacteria may be of fecal origin or not. They belong to a variety of types of bacteria such as *Pseudomonas*, *Aeromonas*, *Klebsiella*, *Flavobacterium*, *Enterobacter*, *Citrobacter*, *Serratia*, *Acinetobacter*, *Proteus*, *Providencia* and *Mycobacterium,* and *Nocardia*. Infections by opportunistic pathogens mainly concern newborns, the elderly, **immunocompromised persons***, and persons hospitalized. Several organizations such as nosocomial infection prevention centers and foundations that assist individuals with cystic fibrosis organize or promote the structuring of monitoring networks for opportunistic infections.

Below is a brief description of some pathogenic bacteria, but we recommend consulting the following website for more comprehensive information regarding waterborne pathogens: http://www.hc-sc.gc.ca/ewh-semt/pubs/water-eau/pathogens-pathogenes/index-eng.php.

Salmonella is a very important genus that brings together more than 2,400 different serotypes. All serotypes are pathogenic for humans. They belong to the family *Enterobacteriaceae* and are responsible for salmonellosis. The disease results in acute gastroenteritis caused by ingestion of a dose that varies depending on the person, from 10 to 10^7 bacterial cells. There are healthy carriers of *Salmonella*. Transmission is by fecal-oral route, food consumption directly or indirectly (seafood) contaminated, or by direct contact with infected animals. *Salmonella paratyphi* and *Salmonella typhi* are only found in humans. *S. paratyphi* is responsible for gastroenteritis and paratyphoid fever. The infectious dose (ID_{50}) is approximately 1,000 cells, and contamination occurs through ingestion of contaminated food or water after incubation for 6–72 h. The ID_{50} is 100,000 for *S. typhi* infection and leads to typhoid fever after 7–28 days of incubation. Transmission is similar to *S. paratyphi*, but the incubation is from 1 to 10 days.

Cattle and poultry farms are *Salmonella* reservoirs in the environment, and they contribute together with the wastewater discharges to the contamination of aquatic ecosystems. Contamination of oysters in shellfish farming areas is often linked to *Salmonella* inputs via the discharge of surface waters.

Shigella also belongs to the family of *Enterobacteriaceae* and is responsible for shigellosis. This disease, also called "bacillary dysentery," results in pain in the colon and in the small intestine, with severe diarrhea, fever, and nausea. The disease affects humans and primates and requires the ingestion of 10 and 200 bacterial cells to be triggered after incubation for 1–7 days. Contamination occurs directly or indirectly by fecal-oral route or by contaminated water (drinking water and recreational water). There are healthy carriers of the germ.

Campylobacter organisms are microaerophilic microorganisms with a spirillum shape. *Campylobacter* bacteria are the number one cause of food-related gastrointestinal illness in the United States. The main reservoirs are animals (pigs, sheep, cats, dogs, rodents) and birds (wild and farmed). Humans usually become infected after eating poorly prepared meat, especially undercooked chicken, through ingestion of contaminated water or by contact with the animal. Ingestion of approximately 500 cells and an incubation of 2–5 days are enough to cause diarrhea, fever, and abdominal cramps. This germ is responsible for 5–14 % of diarrheal diseases in the world.

Yersinia *enterocolitica* and *Yersinia pseudotuberculosis* are two pathogenic species of the *Yersinia* genus. Some strains produce a heat-labile enterotoxin. The main reservoirs are animals (pigs, rodents, rabbits, sheep, cattle, etc.). *Yersinia* infection is a yersiniosis, an enterocolitis or pseudotuberculosis. The main symptoms are bloody diarrhea, fever, and abdominal pain. ID_{50} is on the order of 10^6 bacterial cells and incubation of 2–7 days. Transmission is by contact from person to person by fecal-oral route or by ingestion of contaminated food or water. The main reservoir is the animal.

Escherichia ***coli*** belongs to the family of *Enterobacteriaceae*. Its main habitat is the gastrointestinal tract of humans and warm-blooded animals, and most are harmless to humans. However, some strains are pathogenic and responsible for gastroenteritis. In the environment it is mainly enterohemorrhagic *E. coli* (EHEC), enteroinvasive *E. coli* (EIEC), enterotoxigenic *E. coli* (ETEC), and enteropathogenic *E. coli* (EPEC). EHEC is responsible for food poisoning, but the waterborne transmission is recognized today. Infections can result from ingesting contaminated drinking water or by the contact with contaminated water during recreational activities. Serotype 0157: H7 is the most common, and ingestion of 10–100 cells is enough to trigger **colitis***. The incubation period is 2–8 days. EPEC strain is rare, and it is necessary to ingest a large amount of cells (more than 10^8) to trigger an intestinal infection and acute diarrhea. Contamination pathways are similar to those of EHEC, but waterborne infections are rare due to the high ID_{50}. EIEC strain has characteristics of infection rather similar to those of the EHEC strain. It triggers bacillary dysentery and infects only humans. The ETEC strain is mainly found in hot and humid countries and is responsible for the traveler's diarrhea. It produces heat-stable **enterotoxins*** and others that are thermolabile. Many strains of *E. coli* are uropathogenic (UPEC) and responsible for **cystitis***, sometimes associated with swimming in contaminated water.

The genus *Vibrio* includes a large number of species, but only a few species can infect humans. The most common example is *Vibrio cholerae* which is responsible for cholera in humans only. It is necessary to ingest 10^6–10^{11} cells to

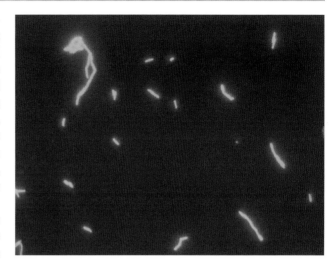

Fig. 15.1 Epifluorescence microscopy photography of *Legionella pneumophila* after cells labeling by a fluorescent antibody labeled with FITC (magnification × 1,000) (Photography by Philippe Catala, Laboratory Arago (Banyuls-sur-Mer) library)

trigger the disease after an incubation period of a few hours to 5 days. Contamination occurs by ingestion of food or water contaminated with feces. The disease can be cured easily, but it must be treated quickly to minimize loss of fluid and minerals. The bacterium survives very long in water and soil. It grows well in waters whose temperature is above 15 °C, alkaline (pH >8), and salty wetlands. *Vibrio* also grows well in **brackish*** water rather warm, but also survives for years in deep waters. It has been observed in Bangladesh that the resurgence of endemic cholera is correlated with the seasonal increase in the temperature of surface waters. **Serogroups*** O1 and O139 are responsible for the disease, and serogroup O1 has spread all over the planet who actually knows the seventh cholera pandemic. In 2003, 111,575 cases and 1,894 deaths were reported to the WHO by 45 countries, and among those, African countries were the most affected. However, this number is underestimated since, for example, between 100,000 and 600,000 cases per year are reported in Bangladesh. Serogroup 0139 has emerged in 1992 and is now spreading in Asia. It could be the origin of the eighth cholera pandemics. There are other pathogenic *Vibrio* species: *V. alginolyticus*, *V. parahaemolyticus*, *V. vulnificus*, and *V. mimicus*. All these species are responsible for gastroenteritis characterized by watery diarrhea and abdominal cramps. They can also infect open wounds. The ID_{50} is the order of 10^6 cells, and the incubation period ranges from 4 to 96 h. Contamination is through raw or undercooked seafood or by direct contact of a wound with contaminated water.

Legionella pneumophila is responsible for Legionnaire's disease (Fig. 15.1). The genus contains 42 species representing 64 different serogroups. The ID_{50} for

L. pneumophila is still unknown. The disease was discovered in 1976 during the annual meeting of veterans of the American Legion in Philadelphia (USA). Thirty people died of the disease, and since the reporting is mandatory, there are more and more reported cases (2,000 to 3,000 cases in France in 1995 and from 8,000 to 18,000 cases in the United States each year). The main reservoir is freshwater, including water relatively warm (35–45 °C) and stagnant encountered in the natural environment, sanitation, and air conditioning. Pathogen transmission is by aerosolization from these reservoirs. The incubation period is 2–10 days depending on the dose ingested. The pathogen can survive long enough in the water, but its life cycle is still poorly understood. Interactions with some aquatic protozoa are particularly important. This organism is able to grow and protect in the interior of some **amoebae*** and stay well protected inside the **cysts***. *Legionella* are highly resistant to conventional disinfection treatments such as chlorination (Box 15.1).

> **Box 15.1: Classification of Infectious Agents in Risk Groups According to the World Health Organization (WHO)**
>
> Philippe Lebaron
>
> In many countries, infectious agents are categorized in risk groups based on their relative individual, collectivity, and environmental risks. The "American Biological Safety Association" (ASBA) presents a well-documented website on risk groups (http://www.absa.org/riskgroups/index.html). Depending on the country and/or organization, this classification system might take different factors into consideration, such as type of infectious agent (virus, bacteria, fungi, protozoa), pathogenicity of the organism, mode of transmission and host range, and availability of effective preventive measures (e.g., vaccines) or effective treatment (e.g., antibiotics). Despite these regional differences, there is a worldwide agreement on the four risk group classification:
> - *Risk Group 1* (no or low individual and community risk): A microorganism that is unlikely to cause human disease or animal disease.
> - *Risk Group 2* (moderate individual risk, low community risk): A pathogen that can cause human or animal disease but is unlikely to be a serious hazard to laboratory workers, the community, livestock, or the environment. Laboratory exposures may cause serious infection, but effective treatment and preventive measures are available and the risk of spread of infection is limited.
> - *Risk Group 3* (high individual risk, low community risk): A pathogen that usually causes serious human or animal disease but does not ordinarily spread from one infected individual to another. Effective treatment and preventive measures are available.
> - *Risk Group 4* (high individual and community risk): A pathogen that usually causes serious human or animal disease and that can be readily transmitted from one individual to another, directly or indirectly. Effective treatment and preventive measures are not usually available.

15.2.3 Particular Case of Toxic Cyanobacteria (Box 15.2)

Cyanobacteria (formerly blue-green algae) contain pigments such as chlorophylls, carotenoids, and phycobilins, giving them a blue-green color characteristic of the group (*cf.* Sect. 6.2.2; Figs. 6.12 and 6.13). Their size is highly variable (1–100 µm), some cells being filamentous or colonial. During blooms cyanobacteria can give a very characteristic color to surface waters where they grow ranging from green to red. Only some species produce toxins harmful to humans and animals. Cyanobacteria can withstand very specific living conditions like those encountered in highly saline lakes or hot springs (Box 15.2).

> **Box 15.2: Cyanobacteria, Environmental Pathogenic Agents?**
>
> Jean-François Humbert
>
> Cyanobacteria have played a major role in the evolution of the earth's ecosystem by allowing, through photosynthesis, the enrichment of our planet's atmosphere with oxygen. Currently, these microorganisms contribute to a large fraction of the primary production in the ocean and are fundamental actors in the marine cycles of carbon and nitrogen. In continental aquatic ecosystems, particularly in lentic ecosystems, cyanobacteria are involved as well
>
> ---
> J.-F. Humbert
> INRA, UMR BIOEMCO, ENS, Paris, France
>
> (continued)

Box 15.2 (continued)

Box Fig. 15.1 Proliferation of cyanobacteria found in a river in Australia (Photograph J.-F. Humbert, INRA Thonon)

as eukaryotic microalgae to primary production, and therefore both belong to the first trophic level of the food chain. Under certain environmental conditions, these microorganisms may experience phases of proliferation (Box Fig. 15.1), also called efflorescence or "bloom," during which considerable biomass (up to 400 μg/L chlorophyll a) can be achieved.

Even if these blooms occur in mesotrophic systems, they are very common in eutrophic systems (phosphorus being the key element) (Downing et al. 2001), especially in summer when weather conditions favor a stratification of the water column of lakes.

Depending on the species involved, cyanobacterial biomass can be distributed throughout the water column (e.g., *Planktothrix agardhii*) or only in the upper layer (e.g., *Microcystis aeruginosa*). *Planktothrix rubescens* is a species able to grow for several months in the metalimnion of some alpine lakes, at 15 m depth, thanks to special adaptive traits (red pigmentation, growth at low temperature). Cyanobacterial blooms occur most often under conditions favoring competition between autotrophic microorganisms (no limiting nutrient factors, physical stability of the environment). The competitive success of cyanobacteria in these conditions can be explained by specific characteristics such as the presence, in some species, of gas vesicles that allow them to move vertically in the water column and occupy the surface water bodies or differentiated cells for fixing atmospheric nitrogen (heterocysts) or to withstand adverse environmental conditions (akinetes) (Box Fig. 15.2). In addition, the cellular organization of some species in filaments or colonies and their ability to synthesize toxins can limit predation of cyanobacteria by zooplankton.

Works has been devoted to study the changes in diversity and genetic composition of populations of cyanobacteria during their phases of proliferation. Among them it has been shown for *Microcystis aeruginosa* that during the development of a proliferation of this species in a reservoir of the Loire River, the dominant genotypes in the population at the beginning of the proliferation process were gradually replaced by a minor genotype at the first stage of the bloom (Briand et al. 2009). In addition, another study on the same species in several aquatic ecosystems in France revealed that the selection of these dominant genotypes during blooms was performed at the level of each ecosystem, even when they are geographically close to each other and if they exchange at some periods of the year. Finally, on a broader scale, no biogeographical differentiation was observed between populations of *M. aeruginosa* from several continents.

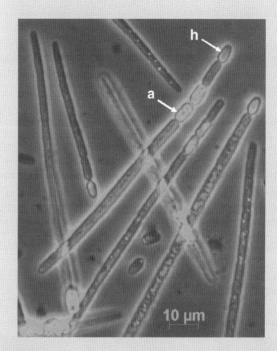

Box Fig. 15.2 *Cylindrospermopsis raciborskii* filaments, *a* akinete, *h* heterocyst (Photograph A. Caruana, INRA Thonon)

This result suggests that the high cellular abundances of the populations and migratory flows prevent the establishment of such a differentiation. By comparison, a well-marked biogeographical structure was demonstrated, using the same genetic markers for another cyanobacterium found in continental aquatic ecosystems: *Cylindrospermopsis raciborskii*. This differentiation could be explained by the recent extension of the surface of distribution of this species from tropical areas to northern latitudes, probably in connection with global warming. The sequencing of the genomes of several genera of freshwater cyanobacteria should allow us to better understand the

(continued)

Box 15.2 (continued)

evolutionary and ecological strategies of these microorganisms. Thus, the genome of *M. aeruginosa* is characterized by a very high plasticity, which would give this species the ability to colonize and proliferate in very diverse environments (Frangeul et al. 2008).

The consequences of these blooms affect both the functioning of the ecosystems in which they occur and their uses. The high biomass generated during these proliferation events are associated to a decrease in diversity in the phytoplankton community (one or two species becoming dominant), mainly due to strong competition for access to light. This decrease in diversity may affect the entire food webs because colonial and filamentous cyanobacteria are poorly consumed by grazers. In addition, the large amounts of organic matter generated during these blooms are nutritional resource for heterotrophic bacteria, which results in an increase in respiratory activity and therefore a decrease in oxygen concentrations in the water column up to anoxia. The consequences of this anoxia, which moves progressively from the bottom to the upper layer of the water body, are twofold. On the one hand, it can promote the release of phosphorus trapped in sediments and make it available to cyanobacteria, and, on the other hand, it can cause mass mortality of fishes. Regarding the use of water, cyanobacterial blooms disrupt recreational activities in water bodies, drinking water production, and fishing. Recent guidelines in many countries have also set thresholds in cyanobacteria cell and microcystin concentrations beyond which bathing should be suspended or the production of drinking water should be stopped.

This regulation on cyanobacteria is justified by the fact that many types of cyanobacteria can produce toxins harmful to human and animal health. These toxins are of different chemical natures, and they have different targets. The most frequently encountered are hepatotoxins, especially microcystins (Box Fig. 15.3a) and nodularins, which are small cyclic peptides, as well as cylindrospermopsin (Box Fig. 15.3b) and its analogues which are alkaloids. They are followed by neurotoxins, such as anatoxins (Box Fig. 15.3c), saxitoxin and its derivatives, and β-N-methylamino-L-alanine (BMAA), which is a non-proteinogenic amino acid. The last category of cyanotoxins is dermatotoxins/irritant toxins, like the lyngbyatoxin-a (Box Fig. 15.3d), the aplysiatoxin, or the debromoaplysiatoxins which are alkaloids or lipopolysaccharides constituting the wall of cyanobacteria and other Gram-negative bacteria.

These toxins have been involved, on all continents, in many cases of intoxication of animals and humans. However, if there are numerous cases of animal's death due to cyanotoxins, fortunately few cases of human deaths due to cyanotoxins have been reported because of water treatment processes, which allow removing a very large fraction of these toxins in drinking water treatment plants. Many questions remain about these toxins, and, among them, two are currently attracting a lot of work.

The first of these issues relates to the observation that, when a proliferation of cyanobacteria occurs, it is very difficult to predict what kind of toxin will be produced and how much it will be. For microcystins, it has been shown, for example, that all species of cyanobacteria were not able to synthesize them, some of them having lost the genes necessary for the biosynthesis of these molecules during their evolution. In the same way, for species known to be able to produce microcystins, it has been shown that their populations contain a mix of cells having the genes for the synthesis of this toxin and of cells that have lost these genes or have nonfunctional genes. In some ecosystems, it has been found, for two very frequent bloom-forming species, *Planktothrix agardhii* and *M. aeruginosa*, that the proportions of microcystin-producing cells could vary significantly during the course of the bloom (between 20 and 80 %). Finally, by using an experimental approach based on co-cultures of microcystin-producing and non microcystin-producing strains, it has been shown that under limiting conditions for the growth of cyanobacteria, the fitness of toxic strains was better than that of the nontoxic ones, and, generally, the inverse was observed in non-limiting conditions for growth (Box Fig. 15.4). All these results suggest that the cost/benefit ratio to produce microcystins varies with environmental conditions encountered by the cyanobacteria.

These results lead us to the second question, which concerns the identification of the functional role of cyanobacterial toxins, knowing that this issue is common to almost all secondary metabolites produced by bacteria. For microcystins, some authors have suggested that these molecules could:

1. Play a major role in defense against predators, zooplankton, or have allelopathic functions
2. Constitute carbon or nitrogen reserves
3. Be involved in the iron intracellular metabolism
4. Be signal molecules involved in cell communication (quorum sensing)

None of these assumptions is presently validated. Very recent works, based on a comparative transcriptomic approach between a microcystin-producing

(continued)

Box 15.2 (continued)

Box Fig. 15.3 Chemical structures of microcystins (**a**); cylindrospermopsin (**b**); anatoxin-a (**c**); lyngbyatoxin-a (**d**)

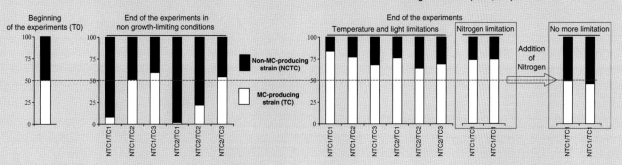

Box Fig. 15.4 Co-culture experiments among several strains of *Planktothrix agardhii* producing or nonproducing microcystins under different environmental conditions (From Briand et al. 2008)

(continued)

> **Box 15.2** (continued)
>
> strain (*M. aeruginosa* PCC7806) and its nontoxic mutant, seem to suggest that microcystins may have pleiotropic actions in the cellular metabolism.
>
> To conclude, freshwater cyanobacteria are the focus of many studies aiming to better understand why they proliferate in these ecosystems, what are the health risks associated with these blooms, and what is the functional role of cyanotoxins. The interest for these microorganisms for the scientific community, but also for managers of aquatic systems as well as for health authorities, is increasing, in conjunction with the increase in the number of ecosystems displaying cyanobacterial blooms in the last 30 years. This increase was linked to local and global changes and, in particular, to the pollution of many freshwater ecosystems by nutrients, especially phosphorus. In addition, very recent works also suggested that more global pressures, such as global warming, could contribute to the development of cyanobacterial blooms in freshwater ecosystems (Paerl and Huisman 2009).

Cyanobacteria produce toxins that attack various organs such as the liver (hepatotoxins) and nervous system (neurotoxins) and less dangerous toxins that can affect the skin or cause gastroenteritis. The toxins are usually released at the time of death and cell **lysis***. Neurotoxins are alkaloids, and some of them are closely related to the paralytic toxin (saxitoxin). Hepatotoxins or microcystins are heptapeptides with more than 50 types that have been described. The best known is microcystin-LR which can retain its activity for several weeks after release into the environment (Box 15.2).

There are about forty species of cyanobacteria producing various toxins. The neurotoxins are produced by the genera *Anabaena*, *Aphanizomenon*, and *Oscillatoria*. Concerning hepatotoxins, they are also synthesized by these three genera plus *Microcystis*, *Nodularia*, and *Nostoc*. The species most often involved in the formation of algal blooms in freshwater are *Anabaena flosaquae*, *Aphanizomenon flosaquae*, *Microcystis aeruginosa*, and *Planktothrix agardhii* (Box 15.2).

Microcystis aeruginosa is a potentially toxic cyanobacteria present in freshwater (lakes, ponds, rivers) where it can develop important blooms. This organism synthesizes hepatotoxins, microcystins, which can lead to fish (threat to aquaculture) and birds' death. These toxins also pose a risk to human health because they cause liver disorders and skin problems. In Brazil in 1996, renal dialysis done with badly purified water caused 60 deaths due to these toxins. The prevalence of *Microcystis aeruginosa* in France is not well known in the absence of specific environmental studies, but problems associated with the proliferation of these cyanobacteria have been reported in many countries (UK, Finland, Australia, USA, China, etc.).

15.2.4 Fungi

Fungi are classified in their own kingdom (*cf.* Sects. 5.5.3 and 7.14.3). They are heterotrophic eukaryotic organisms that acquire their nutrients by absorption. They secrete powerful enzymes that digest their food externally. They have body structures and modes of reproduction unlike those of any other organism. Fungi can be unicellular or multicellular. A typical fungus consists of thin filaments called **hyphae***. Hyphae can branch repeatedly forming a feeding network called **mycelium***. Most fungi, including the molds and mushrooms, are multicellular but yeasts are unicellular. The fungal cells are composed of a core in which we find the chromosomes. Fungi are divided roughly into two main groups, the "macro-fungi" (or macrofungi) and "micro-fungi" (or microfungi). A hundred species of microfungi can cause infections in humans, infections called mycoses. These species are differentiated into primary pathogens, that is to say, can infect healthy individuals, or opportunistic agents infecting individuals that can be vulnerable for various reasons including a weakening of their immune system. These agents are classified according to their morphology and some molecular markers. The main genera that have pathogenic species are *Aspergillus*, *Candida*, *Epidermophyton*, *Microsporum*, *Trichophyton* (fungi ascomycetes), and *Cryptococcus* (fungi basidiomycetes).

Fungal infections (mycoses) can be divided according to the severity of symptoms into two broad categories, superficial and systemic infections. Superficial infections affecting the skin, the hair, and the nails are due to dermatophytes. Systemic fungal infections occur when fungi get into the bloodstream and colonize organs including the lungs and even the central nervous system. These infections are often associated with immunodeficiency. The incidence of fungal infections is increasing due to the increase in diabetes, AIDS, and treatments requiring the use of immunosuppressive drugs. Aspergillosis, candidiasis, and cryptococcosis are among the main human mycoses, and microfungi responsible for these infections belong, respectively, to the genera *Aspergillus*, *Candida*, and *Cryptococcus*.

Aspergillus fumigatus is a filamentous fungus and is the principal agent of aspergillosis. It produces cleistothecium, closed ascocarps with asci scattered rather than gathered in a hymenium. This fungus is found in the environment and plays a role in the recycling of organic matter. Its spores are in frequent contact with humans and may migrate into the respiratory tract to the pulmonary alveoli, where they can cause severe disease in immunocompromised patients. The bronchopulmonary aspergillosis can progress to abscess formation. Invasive aspergillosis is characterized by blood spread, and the outcome is often fatal.

Candida are non-encapsulated white yeasts that multiply by budding. *Candida albicans* is the most frequently encountered species in pathologies, but other species such as *Candida (Torulopsis) glabrata*, *C. krusei*, and *C. tropicalis* can also infect humans. *Candida albicans* is a commensal species naturally present in the gut, mouth, and vagina. This species may be involved in severe systemic infections that can be fatal, but it is mainly involved in infection of the mucous membranes and skin.

Cryptococcus neoformans is a yeast involved in cases of cryptococcosis observed in individuals with acquired immunodeficiency syndrome (AIDS). These yeasts are of varying shapes and reproduce by budding. They have the distinction of having a polysaccharide capsule around their wall. They are found in soil and pigeon droppings. These yeasts can enter the human body through the pulmonary tree and spread throughout the body to the central nervous system where they can cause subacute meningoencephalitis.

Dermatophytes, mainly the genera *Epidermophyton*, *Trichophyton*, and *Microsporum*, are fungi involved in skin infection and its appendages. These mycoses can be associated with peeling or hair loss. They can cause circular or annular lesions. In nutrient broth, these fungi can produce a mycelium, conidiophores, and conidia. The genus *Microsporum* forms both macroconidia (large asexual reproductive structures, fusiform with thick cell walls) and microconidia (smaller asexual reproductive structures, pyriform to clavate and smooth walled) on short conidiophores. The genus *Trichophyton* forms macroconidia that are hyaline, multiseptate, variable in form, fusiform, and spindle shaped to obovate with roughness cell walls. Its microconidia are round or pyriform. The *Epidermophyton* produce macroconidia but do not seem to produce microconidia. Several species are anthropophilic dermatophytes, and transmission to humans generally occurs in wetlands or by contact with contaminated clothing.

15.2.5 Unicellular Eukaryotes

Even if some fungi are unicellular (see above), unicellular eukaryotes that are discussed in this section belong to four kingdoms: the discicristates (*Heterolobosea: Naegleria*), excavates (*Fornicata: Giardia*), Amoebozoans (*Entamoeba*), and alveolates (ciliates: *Balantidium; Apicomplexa: Cryptosporidium*). Their size varies between 10 and 300 μm. Most of these microorganisms can transform into cysts when environmental conditions are unfavorable. The encystment is particularly common among parasitic forms and corresponds to a state of dormancy. Cysts provide protection and allow the transfer of an individual to another by passing through an unfavorable environment, as, for example, the aquatic environment. There are many cysts in river sediments. The very thick envelope of cyst gives them excellent resistance to disinfectants.

Giardia is a *Fornicata* (excavate) whose size varies between 9 and 21 μm long and 2–4 μm wide. It has two nuclei at the front end and five undulipodiums (flagella) allowing rapid movement. The encysted form develops in the colon. The reservoir and host of this pathogen are humans, wild (bears, beavers), and domestic (cats and dogs) animals. The ID_{50} for humans is less than 10 ingested cysts, and incubation can be long (3–25 days). The disease is called giardiasis or lambliasis for the species *Giardia lamblia*. It results in severe diarrhea and abdominal cramps, malaise, and weight loss. Recovery is often spontaneous and occurs within a few weeks in healthy individuals. Transmission is by fecal-oral ingestion of food or water contaminated with human or animal feces. Survival in the environment is very good, and cysts can remain infectious for several months. This pathogen is very common and poses serious problems of contamination in drinking water when there is no filtration step in drinking water treatment because it is very resistant to chlorine.

Cryptosporidium is an apicomplexan (alveolate) whose size varies between 6 and 8 μm long and 2–4 μm wide. They are unique in their method of invading host cells, especially by the presence of the apical complex and apicoplast. The sexual and asexual cycle is performed within a single host. Sporozoites, trophozoites, and merozoites remain attached to the intestinal epithelial cells. The mature oocyte contains four sporozoites. The symptoms of cryptosporidiosis are similar to those of giardiasis with watery diarrhea, abdominal cramps, and headaches. Immunocompromised people such as those with AIDS or cancer are particularly vulnerable to these parasites but especially *Cryptosporidium*. For these people, several deaths were encountered. *ID50* is about 130 organisms, and incubation is between 1 and 12 days. Contamination is by fecal-oral route, by ingesting contaminated food or water. Reservoirs are man for the species *C. parvum* and animals (birds, fish, reptiles, mammals).

Entamoeba histolytica is a pathogen whose reservoirs are humans. This obligate parasite of the digestive tract of humans can be in the form of trophozoite (12–50 μm in diameter) or cyst (a form of resistance from 10 to 15 μm in diameter). The ingestion of 1–10 cysts is enough to cause amoebiasis or amoebic dysentery in humans and primates. Ingestion is through water or food contaminated with feces. Cysts can persist for several weeks in the environment.

Naegleria fowleri is a discicristate pathogen. It is a naturally occurring amoeba in freshwater and soil. Trophozoites are between 8 and 20 μm in diameter and produce lobopods roughly rounded. Cysts are spherical and their size ranged from 8 to 12 μm in diameter. They are coated with a single

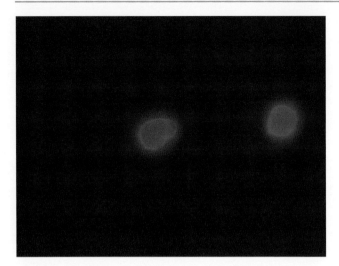

Fig. 15.2 Epifluorescence microscopy photography of *Naegleria fowleri* (*Heterolobosea* discicristates) after cells labeling by a fluorescent antibody labeled with CY3 (magnification × 1,000) (Photography by Philippe Catala, Laboratory Arago (Banyuls-sur-Mer) library)

wall (Fig. 15.2). Humans are the exclusive hosts of the amoeba. After an incubation period of 3–7 days, the infected person develops a naegleriasis or amoebic meningoencephalitis, which is often fatal. The ID_{50} is unknown but certainly very small (a few cysts). Infection is transmitted nasally when swimming in contaminated water (pool, pond, lake or spring, spa). This pathogen grows preferentially in water.

15.3 What Is a Pathogen?

A pathogen is a parasitic organism that selected mechanisms to promote directly or indirectly the colonization of a host organism and enable its growth inside this host. This colonization gradually generates a dysfunction of some organs or weakens certain barriers, which leads to infection of the host that can range from a simple skin infection to a septic shock leading to death.

We can distinguish two types of pathogens, the primary and opportunistic ones:

1. Primary pathogens can be responsible for diseases in "healthy" subjects (e.g., typhoid fever, cholera, salmonellosis, etc.), despite the various host barriers which can be triggered to resist invasion. However, there might be some situations where the presence of the pathogen (e.g., *Salmonella typhi*, *Vibrio cholerae*, etc.) will not cause the infection of the host organism. This corresponds to the definition of a host being a healthy carrier. This person may disseminate a pathogen without being affected. The reasons for this tolerance are not fully understood and may be related, among other things, to certain genetic characters of the individual and its diet.

2. Opportunistic pathogens differ, by definition, from the primary ones because of their main involvement in the colonization of weakened individuals. These hosts may be compromised for a variety of reasons:
    - Weakening of the immune system
    - Injury
    - Organ dysfunction
    - Malnutrition
    - Invasive surgery
    - Antibiotic treatment

These opportunistic infections are responsible for many deaths each year in most countries. They represent a growing concern for our societies. These etiological agents are highly competitive in the environment and highly transmissible and show great adaptability including the acquisition of antibiotic resistance genes by horizontal gene transfer events.

In general, because of the many parameters that can influence the behavior of a pathogen, we speak about their pathogenic potentialities. These represent all the biological properties of a microorganism which contribute to the development of an infection in a specified host. These potentials govern the type and severity of an infection. Pathogenic potentialities may vary for a same host-pathogen model because of the multiple factors that can affect the outcome of an infection, which are driven by the host heredity but also the genetic lineage of a pathogen. Pathogenic potentialities reflect the evolutionary history of the infectious agent and are the result of a series of selective rounds leading to clonal lineages that can be epidemic. In fact, a pathogenic species can be defined as a taxonomic unit grouping multiple strains and complexes of strains (also termed clonal complexes) showing relatively high genetic conservation (e.g., genomes showing more than 70 % of DNA reassociation values), similar phenotypes, and biochemical properties including the ability to infect a host. However, the severity of infections between strains for a given species can be variable and indicative of differences in pathogenic potentialities.

These potentialities are qualitative estimates that can be completed by the notion of virulence (or pathogenicity) which brings quantitative elements into the understanding of a disease. Virulence properties can be divided into two groups: (1) those involved in colonization and growth in the host and (2) those involved in toxigenic activities (ability to produce toxic substances). For a given outcome, there can be different etiological species or strains which are involved, and these can show major differences in their virulence profiles. Virulence also describes the degree or intensity of a colonization and infection. This can be estimated by the number of deaths (mortality) and/or the number of cases of a particular infection (morbidity). Virulence can be measured by two approaches: the lethal dose 50 (LD50) and the infectious dose 50 (ID50), sometimes called the minimal infectious dose (MID). LD50 is the number of microorganisms that cause death in 50 % of the inoculated animals in a defined time interval. ID50 is the number of pathogens required to cause disease in 50 % of animals after inoculation and incubation for a well-defined time period. ID50 for humans is less than 1,000 for *Salmonella enterica* Typhi but is between 1 and 10 for *Mycobacterium tuberculosis* or *Francisella tularensis* and 1 for the amoeba *Naegleria fowleri*. The ID50 seems higher for opportunistic pathogens, although the measurement is difficult for these pathogens because of the specificities of opportunistic infections.

15.3.1 Entry Routes

15.3.1.1 Skin and Conjunctival Routes

The skin barrier is a good protection, but a simple wound can lead to an infection. Skin and conjunctival infections (e.g., *P. aeruginosa* keratitis) are often benign but can become severe if the wounds are deeply infected and if primary pathogens such as *Staphylococcus aureus* or opportunist ones such as *Nocardia* are involved. *S. aureus* can be found in recreative waters while being shed by healthy swimmers. *Nocardia* (part of the Actinobacteria) are of environmental origin (Fig. 15.3) and can be found in soils and waters.

15.3.1.2 Respiratory Route

In this case, inhalation of contaminated aerosols can occur naturally or through artificial mechanical ventilators, implying invasive or noninvasive procedures. Contamination of the inhaled aerosols can be due to infected individuals or environmental sources such as showers, air conditioners, and cooling towers (legionellosis). Certain agricultural practices such as spreading of composts and manure or a close proximity with wastewater treatment plants or combined sewer overflows can lead to human infections through inhalation of contaminated aerosolized particles.

15.3.1.3 Oral Route

This is the most important route of entry of microorganisms into the human body. Most food products harbor a microflora, and ingestion of food contaminated by infectious agents can rapidly lead to an infection. Children and the elderly but also immunocompromised individuals are the most sensitive to such exposures, e.g., the recent *E. coli* O104:H4 outbreaks.

15.3.2 Colonization Processes

15.3.2.1 Adhesion Properties

Most microbial infections imply colonization through the main entry points of the human body and affect or involve the mucous membranes of the respiratory, intestinal, or urinary tracts and most particularly interactions with the epithelial cells. The pathogen must be able to recognize and adhere to molecules at the surface of host cells which are often proteins. These molecules generally serve specific function in the host organism. For example, the AIDS virus (or HIV) can adhere to T cells that have **CD4 receptors*** on their surface and whose major function is the immune system (white blood cells). After infection, this role in host defense is inhibited and the host becomes vulnerable to other pathogens. Furthermore, some pathogenic bacteria can use pili (fimbriae) to adhere to mucosal cells. Some will develop biofilms after an intense production of exopolysaccharides.

15.3.2.2 Growth Within Host Tissues

Some microorganisms are able to produce a number of extracellular enzymes or toxins among which we can mention the hyaluronidases (which facilitate the spread of bacteria in tissues), fibrinolysins (to dissolve fibrin clots produced by the host), proteases (collagenase, elastase), nucleases and pilases (increases invasiveness), hemolysins (which attack the red blood cell membranes), and leukocidins (which can lyse white blood cells and reduce host resistance).

15.3.2.3 The Ability to Enter and Survive Within Eukaryotic Cells

Some pathogens can enter and survive within non-phagocytic eukaryotic cells such as epithelial cells of mucosal surfaces or endothelial cells of blood vessels.

15.3.2.4 The Ability to Escape the Immune Defenses of the Host

Some microorganisms, including bacteria, have evolved mechanisms to circumvent host defenses. The most important ones relate to the production of antiphagocytic factors such as **capsules*** and surface proteins but also the ability to survive and reproduce within macrophages (internalized *Legionella pneumophila* being able to withstand acidity, proteases, and free radicals from the phagolysosome), resistance to serum bactericidal activity, the development of **siderophores*** to chelate iron from the host proteins, the development of IgA proteases, the variation of surface antigens, and molecular mimicry.

15.3.3 Toxigenic Activities

Beyond extracellular toxic factors that promote invasiveness, bacteria can produce toxins that are responsible for symptoms or even some syndromes associated with infection. Toxins act as poisons which are lethal to the host and are sometimes used to penetrate tissues. For example, *Clostridium perfringens* can produce a collagenase which degrades collagen tissue and allows penetration by the infectious agent inside the tissue. There are two types of toxins: (1) endotoxins which are

Fig. 15.3 Different entry routes of human contamination by pathogenic microorganisms (Drawing: m-j Bodiou)

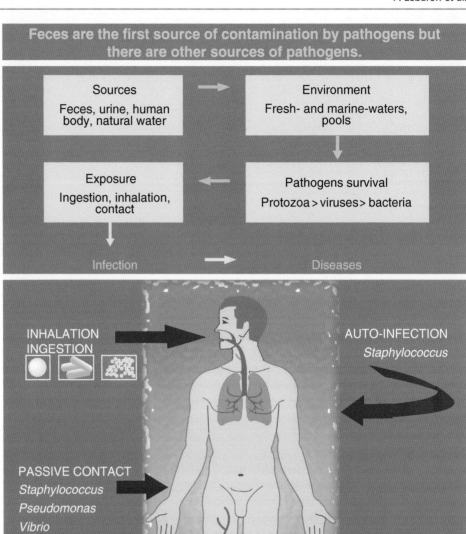

released after cell lysis and (2) exotoxins which are released more readily in the environment by growing bacteria (Table 15.2).

15.3.3.1 Exotoxins

Exotoxins are generally more virulent than endotoxins. They generally have a domain A (activity) and a domain B (binding domain). The B domain is responsible for the introduction of the A domain into the cytoplasm of the eukaryotic cell through the membrane. The organization of these domains varies from one toxin to another (diphtheria toxin, cholera, and anthrax ones). Nevertheless, other exotoxins termed cytotoxins do not include a B domain as they penetrate into the cell by a secretion system (e.g., type III secretion system or TTSS). Examples of diseases due to the action of exotoxins are listed in Table 15.3.

Table 15.2 Characteristics of exotoxins and endotoxins

Exotoxins	Endotoxins
Excreted by the Gram-positive and Gram-negative bacteria	Produced during cell lysis by Gram-negative bacteria
Proteins which are generally thermolabile	Lipopolysaccharide associated with the outer membrane of the cell wall. They are thermostable
Highly antigenic	Weakly antigenic
Easily transformed into toxoids	No transformation into toxoids
Often specific mode of action (cytotoxins, enterotoxins, neurotoxins)	Diffuse mode of action (fever, diarrhea, vomiting, hypotension)
Does not usually produce fever	Generally produce fever
Highly toxic	Low toxicity

Table 15.3 Examples of diseases caused by the action of exotoxins

Bacteria	Toxins	Disease
Clostridium tetani	Neurotoxin	Tetanus
Clostridium botulinum	Neurotoxin	Botulism
Clostridium perfringens	Lecithinase, hemolysins, collagenases, proteases, etc.	Gas gangrene
	Enterotoxin	Food poisoning
Corynebacterium diphtheriae	Inhibition of protein synthesis	Diphtheria
Staphylococcus aureus	Enterotoxins	Food poisoning
	TSS toxins	Toxic shock syndrome
	Exfoliative toxin (EF)	
Streptococcus pyogenes	Erythrogenic toxin	Scarlet
	SPE toxin	Toxic shock syndrome
Vibrio cholerae	Enterotoxin (activation of the AMP cyclase)	Cholera
Escherichia coli	Enterotoxins	Gastroenteritis
Pseudomonas aeruginosa	Exotoxin A and cytotoxins	Otitis, keratitis, opportunistic infections

15.3.3.2 Endotoxins

Endotoxins are part of the membrane of Gram-negative bacteria (i.e., the lipopolysaccharides or LPS) (*cf*. Sect. 3.1). The toxin is composed of three parts: an O-specific chain, a "core," and the lipid A. The O-specific chain is highly variable and composed of saccharides. It carries the antigenic determinants responsible of their specificity. The "core" is much less variable in its composition and is made of saccharides. The lipid A consists of a framework of glucosamine with aliphatic chains and is responsible for the toxicity of the molecule. The saccharide constituents of lipid A ensure its solubility. Unlike exotoxins, endotoxins are heat resistant and of low toxicity and remain attached to the cells. They generally cause a high fever, diarrhea and vomiting, hypotension, and coagulation disorders. They can be lethal. This is the case of *Legionella pneumophila* LPS, a pathogenic species commonly found in aquatic systems, including hot water systems, and is responsible for legionellosis.

Virulence can vary under the influence of these factors; it may increase during an outbreak or decrease under the effect of physicochemical agents or by subculturing on minimal culture media.

15.3.4 Intercellular Communication and Virulence

Contrary to what was long thought, several bacteria have a collective behavior and are able to sense chemicals produce by neighboring cells and migrate toward more favorable environments. This intercellular communication named "quorum sensing" is based on a density-dependent chemical factor which could also be involved in controlling the expression of virulence factors.

The chemical "language" used by bacteria is variable. The best known system is the one of Proteobacteria, which can secrete small "signal" molecules called "autoinducers." Through these molecules, in a concentration-dependent manner, these bacteria can change their behavior. When the bacterial density is high, concentration of these molecules will become high. These high concentrations will favor interactions with other cellular components and induce or repress a number of target genes that will change the metabolic activities of these bacteria. Among Proteobacteria, these autoinducers are part of the N-acylhomoserine lactones (AHL). At high concentrations, these AHL will bind and

Table 15.4 Examples of quorum sensing in Gram-negative bacteria

Bacteria	Main autoinducer	Regulatory proteins	Phenotype
Vibrio fischeri	3-Oxo-C_6-HSL	LuxI/LuxR	Bioluminescence
Pseudomonas aeruginosa	3-Oxo-C_{12}-HSL	LasI/LasR	Extracellular enzymes, biofilm formation
	C_4-HSL	RhlI/RhlR	Extracellular enzymes, secondary metabolites
Agrobacterium tumefaciens	3-Oxo-C_8-HSL	TraI/TraR	Ti plasmid conjugation
Burkholderia cepacia	C_8-HSL	CepI/CepR	Siderophores, proteases
Aeromonas hydrophila	C_4-HSL	AhyI/AhyR	Exoproteases
Aeromonas salmonicida	C_4-HSL	AsaI/AsaR	Extracellular exoproteases

HSL homoserine lactone

activate a transcriptional activator protein (named R) that can modify the expression of a set of target genes (Table 15.4). In Gram-positive bacteria, chemically different autoinducers are produced, but their structure remains poorly characterized. These play similar functions, that is to say, being involved in conjugation, competence toward DNA or virulence, as demonstrated in *S. aureus*.

15.3.5 Pathogenicity Islands (PAI) and Other Mobile Genomic Islands (GI)

The term "pathogenicity island" (PAI) was introduced by Knapp et al. (1986) to describe the genetic elements conferring high virulence among uropathogenic *Escherichia coli* strains. It was subsequently used to describe all mobile genomic islands involved in virulence among pathogenic bacteria. Nowadays, PAI are a part of the vast group of mobile genomic islands (GI) being found in bacteria. Generally speaking, GI have a variable length (5–10 kb to over 100 kb) with a specific organization often showing a G + C % distinct from the core genome. GI show short repeated sequences at their ends (*attL* and *attR*). These are involved in the GI DNA integration process. GI harbor genes involved in various functions including a tRNA gene in 75 % of the cases and, to a lesser extent, a gene encoding an integrase (Fig. 15.4). PAI harbor genes involved in the synthesis and secretion of toxins, genes involved in iron acquisition (siderophore synthesis), the synthesis of **adhesins*** such as fimbriae, etc. PAI may also contain repeated elements such as IS (insertion sequences encoding a transposase) and **transposons*** (similar to IS but also encoding additional genes such as those involved in antibiotic and heavy metal resistances). PAI can have a free form, that is to say, be found as an autonomous excised form. Some genetic convergences between recognition sites (*att*) of integrases can induce phenomena of PAI cross excisions, a phenomenon that could explain the loss of integrase genes on some PAI (Fig. 15.4).

15.3.5.1 Evidence of PAI Transfers Among Bacteria

The observation of distinct G + C% in the composition of PAI and core genome DNA and the presence of plasmid-borne genes or phage ones on PAI suggest acquisitions by horizontal transfer events involving conjugative plasmids, integrating conjugative elements (ICE), transposons, or **phages***. These observations were confirmed by a demonstration of transfer events of such elements by transduction or conjugation between phylogenetically closely related species or strains of a same species. To illustrate, we can cite the case of a *Vibrio cholerae* PAI named VPI which is involved in the synthesis of a TCP pilus (toxin co-regulated pilus). This pilus acts as an anchor of the filamentous bacteriophage (CTXφ) encoding the cholera toxin (Karaolis et al. 1999). Analysis of culture supernatant of a strain of *V. cholerae* encoding this PAI showed the presence of a filamentous phage (VPIφ) harboring a full VPI element. This phage was shown to infect nonpathogenic strains of this species. Interestingly, a protein encoded by the VPI was found to be a component of the phage structural proteins while also having a role in the synthesis of the TCP pilus. These findings suggest a coevolution between bacteria and bacteriophage, with the expression of proteins performing functions in both organisms. PAI transfers involving phage particles were also observed in *S. aureus*. The SaPI1 PAI of *S. aureus* encodes a toxin and showed phage-related DNA sequences favoring its excision and transfer by φ13 and 80α phages into other staphylococci (Lindsay et al. 1998). Plasmids have also been shown to have contributed at the dissemination of PAI. We can cite the case of the PAPI-1 (*Pseudomonas aeruginosa* pathogenicity island-1) which was shown to be derived from pKLC102. This plasmid is 103.5 kb long and was detected in free and integrated forms among *P. aeruginosa* clone C. PAPI-1 was shown to be transferable by conjugation between *P. aeruginosa* strains

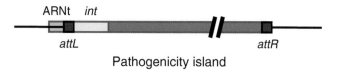

Fig. 15.4 Illustration of a pathogenicity island integrated into a bacterial genome. *Int* integrase gene, *attL* and *attR* short repeated sequences generated by the integration process, *tRNA* (in *green*) gene encoding a tRNA which is a frequent but not exclusive target of integrases. A DNA segment that can carry functional genes such as those involved in virulence is shown in *blue*. This segment is highly variable in length

(Qiu et al. 2006). PAPI-1 was classified in the group of mobile elements termed "ICE" that encodes integration and conjugative properties (Burrus and Waldor 2004).

Phylogenetic analyses suggest PAI distribution to be mainly genus or species specific. However, a transfer of the SaPI1 PAI by a bacteriophage was demonstrated between *S. aureus* and *Listeria monocytogenes* (Chen and Novick 2009). This suggests that the environmental cycle of these pathogenic bacteria may represent an opportunity to acquire PAI from a wide range of eubacteria. Thousands of species of bacteria can be found per gram of soil or milliliter of water and be reservoirs of PAI. Nevertheless, many PAI do not harbor elements that could trigger their transfer. The absence of a transfer function in some PAI may be an indication of an ancient acquisition. These PAI might have been subjected to genetic purges and selective rearrangements leading to a counterselection of genes having low benefits. Indeed, it seems that mobility could be a counterselected trait under some circumstances if other functions encoded by the PAI confer selective advantages under a greater genomic stability. An example of such a situation could be the one of the ExoU island of *P. aeruginosa* (Kulasekara et al. 2006). The evolutionary history of this PAI suggests a transposition event leading to the presence of the *exoU* gene onto an ICE. This ICE would have been acquired by *P. aeruginosa* and been integrated into a tRNA gene. Subsequently, this ICE would have been affected by multiple rearrangements leading to a loss of mobility. Many other PAI were found to match this kind of configuration. We can cite HP1 of *Yersinia pestis*; PAI I_{536}, II_{536}, III_{536}, and IV_{536} of uropathogenic *E. coli* (UPEC); LIPI 2 of *Listeria ivanovii*; etc. (Hacker and Kaper 2000).

Nowadays, the GI concept was enriched by extending its range of functional benefits to other contexts than virulence. This extension was performed on the basis of observations made on full genome DNA sequences of non-pathogenic bacteria. These genomes were found to harbor genomic islands with configurations similar to those of PAI but without the presence of virulence genes. These GI are sometimes referred to as ecological islands (EI) because they were found to play part in the environmental degradation of phenols or the assimilation of sugars. These can also be termed symbiotic islands when they contribute to the establishment of symbiotic associations with other organisms such as plants. Some of these "ecological" GI can be found among pathogens and could play part in their environmental cycle.

15.3.6 Antibiotic Use and the Fight Against Bacterial Infectious Disease

The discovery of antibiotics and their introduction in clinical practices was a therapeutic revolution and contributed to lower mortality due to infectious diseases (diphtheria, plague, typhoid, tuberculosis, leprosy, meningitis, syphilis, etc.). Penicillin and streptomycin used in the 1940s followed by the golden era of discovery of novel antibiotics classes between the 1950s and 1970s had raised many hopes in the fight against infectious diseases. However during these last 60 years, we observed an increase of the number and diversity of the species developing antibiotic resistance. The emergence of bacterial resistances is preferentially made in a context of disadvantaged social circles, the poverty favoring self-medication, the use of out-of-date antibiotics, and consequently an often inappropriate consumption. In parallel the intensive use of antibiotics in the hospitals of the western countries, but also in the context of practices without any therapeutic aims (i.e., prophylactic purposes and stimulation of animal growth), associated with an increase of immunocompromised individuals participates in the emergence and spread of antibiotic resistance.

15.3.6.1 Emergence and Spread of Antibiotic Resistance Among Bacterial Pathogens

Adaptation and Spread Mechanisms

Antibiotics are categorized as bacteriostatic (i.e., growth inhibition) or bactericidal (i.e., cell death) and show a narrow or wide range of activity toward bacterial species. Basic mechanisms of antibiotic action against bacterial cells are inhibition of cell wall synthesis (penicillin, cephalosporin, or vancomycin), of protein synthesis (chloramphenicol, streptomycin, tetracycline), of nucleic acid synthesis (quinolones), or of metabolic processes (sulfonamides). Bacterial species differ in terms of their sensitivity toward antibiotics, some exhibiting a high level of intrinsic resistance. Enhanced resistance toward an antibiotic or resistance toward new antibiotic can also be acquired through gene mutation or gain of new genes. Resistance mechanisms and genetic determinants have been widely investigated, and many of them have been identified (Levy and Marshall 2004). There are several molecular mechanisms that include modifying or protecting the target, reducing the membrane permeability, inactivating the antibiotic through enzyme activity, or actively exporting an antibiotic from the bacterial cell (efflux pump). These mechanisms are more or less specific of an antibiotic and can act simultaneously. Dissemination of resistance is mediated by clonal spread of a particular resistant strain and/or by spreading of resistance genes through horizontal gene transfer. The former is the main explanation for the first resistance described within bacterial pathogenic species. The latter was further involved in the spread of resistance between taxonomically distantly related species (Martinez et al. 2009). The location of resistance genes on plasmids or transposons contributes to intercellular transfer. These genes are also often associated to **integrons***, which capture genes by site-specific recombination, allowing intracellular mobility then favoring multiresistance acquisition.

Antibiotics: A Selective Pressure at the Hospital

It is now accepted that excessive and improper antibiotic use in hospitals has played a major role in the emergence and rapid spread of antibiotic-resistant bacteria in humans. Overuse/misuse of antibiotics was the major factor responsible for the development of nosocomial infections (hospital-acquired infections). Penicillin-resistant bacteria and methicillin-resistant *Staphylococcus aureus* (MRSA) arose within a few years of penicillin and methicillin introduction in the 1940s and 1959, respectively (Taubes 2008). MRSA are now common in clinical care settings but also appeared a decade ago as community-acquired infections. Similarly, in the 1980s the physicians responded to enterococci resistance by using vancomycin to treat enterococci infections. Vancomycin-resistant *Enterococcus* was first reported in 1986, and the *van*A gene soon spread throughout bacterial species including the MRSA.

Within Gram-negative bacteria, *Enterobacteriaceae* highly contribute to nosocomial and community-acquired infections. Penicillin and its derivates were effective treatments against these bacterial infections until the emergence of β-lactamase-producing strains. To counteract the effect of these β-lactamases, novel classes of β-lactams were then developed such as cephalosporins, carbapenems, or monobactams. But new β-lactamases emerged rapidly, and more than 300 enzymes are presently described (Kong et al. 2010). Moreover, the worldwide spread of extended-spectrum β-lactamases, such as SHV, TEM, and OXA types, which conferred resistance to most β-lactams, as well as carbapenemases, such as VIM and IMP types, currently poses a major public health problem. These genes are not only present among *Enterobacteriaceae* but are also widespread among *Pseudomonas aeruginosa* or *Acinetobacter baumannii*.

Opportunistic pathogens (i.e., ubiquitous bacteria infecting immuno-debilitated patients such as young children or elderly people, cystic fibrosis individuals, severely burn, or under chemotherapy patients) are becoming important new emergent infectious agents. Among the most frequently observed, *Pseudomonas aeruginosa* is characterized by a high intrinsic resistance to antibiotics, thanks to a low permeability of its outer membrane, constitutive expression of numerous efflux pumps, and production of various enzymes enabling antibiotic inactivation (i.e., cephalosporinases). It was estimated that 1.8 % of the genome of this species contributes to its intrinsic resistance (Fajardo and Martínez 2008). This species is also characterized by its important ability to acquire new resistance determinants through horizontal gene transfer (Mesaros et al. 2007). *P. aeruginosa* as well as other opportunistic species (i.e., *Acinetobacter baumannii* and *Stenotrophomonas maltophilia*) is developing resistance toward all available antibiotics (Taubes 2008).

Antibiotic Use Outside the Hospital

Similarly the use of antibiotics to treat animal diseases and/or promote animal growth in livestock production also contributed to the development of drug-resistant infections among animals. Resistant bacteria from animal food may be passed through the food chain to humans resulting in resistant infections. A recent French survey of antibiotic resistance among *Escherichia coli*, the most common pathogen isolated from cattle, pigs, goats, or horses, evidenced a high level of resistance toward the oldest and most frequently used antibiotics, amoxicillin and tetracyclines. It also showed numerous resistances toward beta-lactams including the third generation of cephalosporins (cefotaxime, ceftazidime) and the presence of *bla*CTX-M genes similar to those initially detected among *E. coli* strains isolated from human. Furthermore, the presence of these genes on plasmids that also carry resistance genes for other antibiotics (florfenicol) suggests that the selective pressure associated to one antibiotic can select for diverse unrelated resistance mechanisms. In Europe, the use of the glycopeptide antibiotic avoparcin as a growth promoter for pig, and thus its use at subtherapeutic levels, also led to an increase of vancomycin-resistant enterococcus prevalence in animal but also in human populations. Despite the ban of avoparcin as a feed additive in some countries more than 10 years ago, a recent study proved that acquired vancomycin-resistant gene *van*A can still be detected in food-producing animals raising the question of gene persistence in the absence of the selective agent and its potential reemergence and risk for human health.

Outside hospital and animal farms, the selective pressure of antibiotics can also occur in various natural or human-made environmental compartments and then impact and modify the evolution rate of antibiotic resistance within indigenous microbial communities. High concentrations of antibiotics are likely to occur in hospital effluents, sewage treatment facilities, streams with large amounts of agricultural runoff from livestock areas, or aquaculture units. Most of them are not eliminated during wastewater treatments due to some limits of present sewage treatment plants. Manure and other animal farm effluents also stand for an important source of veterinary products, including antibiotics, in the environment. As many antibiotics are poorly adsorbed in the guts, about 20–90 % of the administered antibiotics are excreted rapidly after the treatment (Kummerer 2003). Antibiotics such as tetracycline and tylosin have been detected in various animal wastes, i.e., cattle or poultry manure and pig slurry. Values as high as 0.50 mg.kg^{-1} and 35.5 mg.kg^{-1} for oxytetracycline and sulfachloropyridazine, respectively, in poultry manures were reported. In chicken manure maximum concentrations of enrofloxacin up to 2.8 mg.kg^{-1} were measured. Application of antibiotic-polluted manure in agricultural fields would then lead to the introduction of various antibiotics in soils. It has been documented for sulfamethazine, tetracycline family, as well as tylosin (Schwaiger et al. 2009). Maximum concentrations of various antibiotics in soils fertilized with animal manure were 0.3 mg.kg^{-1}, 0.015 mg.kg^{-1}, and 0.37 mg.kg^{-1} for tetracyclines, sulfonamides, and fluoroquinolones, respectively. Despite antibiotic detection in the amendments, their occurrences and concentrations in soil might strongly differ due to differences in stability. Such differences can exist between compounds with structural similarities. For instance, whereas oxytetracycline was frequently detected at high levels in soil samples, chlortetracycline has been rarely detected (Karci and Balcioğlu 2009).

Whether environmental concentrations of antibiotics are sufficient to exert a selective pressure in the environment and present a risk for human health in terms of resistance, development is under investigation. Tello et al. (2012) evaluated that for clinically relevant bacteria, measured environmental concentrations in some environments, i.e., river sediments, swine feces lagoons, liquid manure, and farmed soil, are enough to inhibit wild-type populations in up to 60 %, 92 %, 100 %, and 30 % of bacterial genera, respectively. The authors concluded that concentrations are high enough to exert a selective pressure on clinically relevant bacteria that may lead to an increase in the prevalence of resistance. Similarly antibiotic-resistant bacteria (ARB) present in these amendments can be added to soils and favor antibiotic resistance gene (ARG) spread. The frequency of bacteria carrying antimicrobial resistance genes seems to be especially high for pigs as compared to cattle or sheep which correlates with the amounts of antibiotics used in the husbandry of these animal species (Schwaiger et al. 2009). Evidence of increase ARG genes in soils over time has been demonstrated by Knapp et al. (2010) in the Netherlands during their survey of five long-term soil series that spanned 1940–2008. Data showed that ARG from all classes of antibiotics tested (tetracyclines, vancomycin, erythromycin, ampicillin, methicillin, and beta-lactams) has significantly increased since 1940, but especially within the tetracyclines. A study from Heuer and Smalla (2007) showed that the manure added to sulfadiazine increased levels of antibiotic resistance genes and mobile genetic elements carrying these genes in manured soil over at least 2 months. A review from Heuer et al. (2011) on ARG spread due to manure application on agricultural fields concluded on the very likely significant effect of the agricultural antibiotic use on the spread of resistance among human beings because of the high exposure rate to environmental contamination with transferable antibiotic resistance genes.

Impact of Nonantibiotic Chemicals in the Selection of Antibiotic Resistance

An increasing number of studies has evidenced the role of additional chemicals acting as selective pressure for antibiotic-resistant bacteria in the environment. Some biocides, detergents, and organic solvents can select for mutants exhibiting increase expression of multidrug-resistant determinants. The most documented ecological studies are those focusing on the role of metal as a selective agent in the proliferation of antibiotic resistance. Early studies on Gram-negative fecal bacteria of primates exposed to mercury (Hg) consecutively to the use of dental amalgam demonstrated that mercury-resistant strains are more likely to be multiresistant to antibiotics than mercury-sensitive strains are (Ready et al. 2007). Similarly positive correlation between the presence of metals in ecosystems and the resistance toward both metal and antibiotics in water (Stepanauskas et al. 2006), sediments, or soils has been reported. Huysman et al. (1994) observed higher frequencies of co-resistance to antibiotics and metal among copper-resistant bacteria when compared to copper-sensitive isolates from fields amended with copper-contaminated pig manure. Similarly Berg et al. (2005b) compared bacterial communities from copper polluted and nonpolluted soils and showed an enrichment in copper-resistant populations, an enrichment in antibiotic-resistant populations and a higher prevalence of multiresistance (resistance to more than three antibiotics) in the polluted soil. These authors (Berg et al. 2010) also evidenced that soil Cu exposure provides a sufficient selection pressure to co-select for resistance to vancomycin.

Several mechanisms underlie the co-selection process: co-resistance (different genetic determinants are present on the same genetic element), cross-resistance (the same determinant responsible for resistance to antibiotic and metal), or co-regulation (metal or antibiotic act as inducer of a common regulator system responsible of various metal and antibiotic-resistant determinant expression). These three mechanisms were shown to be responsible for antibiotic resistance within the opportunistic bacterial pathogen, *Pseudomonas aeruginosa*:

1. Plasmids and/or transposons frequently harbor genes encoding metal and metalloid resistance as well as genes encoding antibiotic resistance.
2. Presence of an efflux pump enabling resistance to vanadium and increase resistance to ticarcillin, netilmicin, and clavulanic acid (Aendekerk et al. 2002).
3. Exposure to zinc select for strains resistant to metal (zinc, cadmium, and cobalt) and imipenem (Perron et al. 2004). The latter resistance mechanism involves a two-component system which activates expression of a metal-controlling efflux pump and repressed protein membrane expression.

15.3.6.2 Origin and Evolution of Antibiotic Resistance: The Resistome Concept

Environment and especially soils are known to harbor antibiotic-producing organisms, mainly fungi and actinomycetes, from which over 80 % of antibiotics in clinical use originated. There is now growing evidence that not only these organisms have biosynthesis genetic pathways but also have associated resistance mechanisms enabling their self-protection. Kanamycin and neomycin, two antibiotics synthesized by bacteria belonging to the genus *Streptomyces*, can be degraded by these bacteria, thanks to acetylation and phosphorylation mechanisms. These mechanisms have been identified among clinical pathogens resistant to these antibiotics. An exhaustive inventory of resistance susceptibilities among actinomycete isolates from different soils revealed widespread resistance including multiresistance up to seven or eight antibiotics. Then, ecological and evolutionary studies led to the conclusion that antibiotic resistances are widespread in environment, emerged long before our use of antibiotics, and coevolved with biosynthesis genes (D'Costa et al. 2011). Phylogenetic studies on resistance genes also suggest a physiological role other than that of the resistance to antibiotics. For instance, beta-lactamases would have evolved from transpeptidases involved in the synthesis of the wall, and enzymes modifying aminoglycosides would have evolved from kinases and acyltransferases. Considering the efflux pumps responsible from multidrug resistance among some opportunistic pathogens (*P. aeruginosa*, *B. cepacia*, *S. maltophilia*), it has been understood that they also have physiological roles (Martinez et al. 2009). They participate in the processes of detoxification of intracellular metabolites, confer resistance to host-defense molecules, and contribute to bacterial colonization and persistence in the host. They are also involved in cell-to-cell communication, and antibiotics can play a role of signal molecules. Finally natural and semisynthetic antibiotics can be metabolized by various bacterial species, which were found closely related to pathogenic ones as *Serratia marcescens* and the *Burkholderia cepacia* complex. These species also showed the ability to resist to high level of antibiotics and exhibited a multidrug resistance phenotype. We are now aware that the environment, especially the soil compartment (as well as aquatic sediments), is a reservoir of an extent diversity of indigenous resistant bacteria and of large panel of genes that can potentially be involved in antibiotic resistance (resistome concept: addition of genes involved in the antibiotic-resistant phenotypes *plus* cryptic genes able to confer resistance but without the expressed phenotype *plus* precursors of resistant determinants; Wright 2007).

15.4 Fecal Indicator Bacteria

The search for pathogenic microorganisms in natural environments is a difficult, long, and tedious task. These microorganisms are very diverse, and their detection often requires very specific, sophisticated, and expensive methods that are sometimes difficult to implement for routine analysis, including in accredited laboratories. As a consequence, indicator organisms such as fecal indicator bacteria (FIB) are generally used to evaluate the level of microbial contamination by fecal material in different environments (water, soil, etc.). The abundance of these FIB in environmental samples is supposed to be related with the presence of pathogenic microorganisms from fecal origin. Indeed, even if some pathogens have a primary habitat other than feces, the majority of human pathogens encountered in the environment are from enteric origin and released in the environment by feces of humans and warm-blooded animals. Therefore, the detection of FIB provides an indication of the sanitary risk associated with the various water utilizations. The criteria governing the choice of a good fecal indicator of fecal contamination are presented in Table 15.5. Even if it is very difficult to find an indicator that meets all these criteria, several organisms meet most of them and are now used in the regulations on environmental microbiological quality. The best example of application of indicators of fecal contamination is certainly the monitoring of recreational water quality (Box 15.3).

15.4.1 Total Coliforms, Thermotolerant Coliforms (TTC), and *Escherichia coli*

The coliform group includes different bacteria whose natural habitat is most often the digestive tract of warm-blooded animals (homeotherms). This group is composed of

Table 15.5 Characteristics of an ideal indicator of fecal contamination (Rose et al. 2004)

Properties	Features an ideal indicator of water contamination
Pathogenicity	Nonpathogen
Occurrence	Present when pathogens are present; absent in the absence of fecal contamination
Survival	Kinetics similar to the survival of pathogens
Reproduction	No multiplication in natural waters
Inactivation	Inactivated in a similar way than pathogens by the water treatment lines (including disinfection)
Source	The only source in natural waters is fecal contamination
Detection methods	Detection methods rapid, cheap, and easy to perform

Box 15.3: Recreational, or Bathing, Water Quality: World Health Organization (WHO)

Philippe Lebaron

Bathing is a very popular hobby worldwide. Yet, there may also be adverse health effects associated with recreational use if the water is polluted or unsafe. Maintaining safe recreational waters requires a concerted effort from all of its stakeholders. From government at all levels, to local businesses and industry, to beach managers, community members, and recreational water users, all have a role to play in helping keep our beaches clean and our swimming waters safe. The most effective way to ensure that our waters remain safe for use is to become aware of the types of hazards (microbiological, chemical, and physical) that can impact a bathing area. The WHO produces international guidelines on recreational water quality (Box Fig. 15.5).

The first edition of the WHO Guidelines for Safe Recreational Water Environments can be downloaded on the WHO website at the following URL: http://www.who.int/water_sanitation_health/bathing/en/.

Other informations on the regulations that apply to recreational waters can be found at the following site: http://water.epa.gov/lawsregs/lawsguidance/beachrules/act.cfm.

Box Fig. 15.5 Summer beach on the French Mediterranean Coast (Photograph: Philippe Lebaron; photograph collection of Laboratoire Arago, Banyuls)

Gram-negative rods belonging to the *Enterobacteriaceae* family. These microorganisms are not sporulated oxidase negative, aerobic, or facultative anaerobes and able to grow in the presence of bile salts or surfactants having the same properties and fermenting lactose with the production of acid, gas, or aldehydes in 24–48 h at 37 °C. Habitat specificity of total coliforms is weak, and some coliforms are present in the environment without any source of fecal contamination. The interpretation of their presence in a given environment is often very difficult, and their use as FIB is becoming less frequent. At the opposite, thermotolerant coliforms (TTC), also called fecal coliforms, are natural inhabitants of the digestive tract of humans and warm-blooded animals; they are therefore a good indicator of the contamination of a natural environment by feces. They are able to ferment lactose at a temperature of 44.5 °C. Their biochemical and physiological properties allow identifying them among very complex environmental microflora by the use of specific and very selective culture media. The major representative of the thermotolerant coliform group is *Escherichia coli*. This species represents between 70 and 95 % of TTC and is highly specific from feces. *Escherichia coli* is now the most often used FIB in the context of microbiological analysis of water quality (Edberg et al. 2000; Fewtrell and Bartram 2001).

15.4.2 Intestinal Enterococci

The group of intestinal enterococci is a very heterogeneous bacterial group belonging to the *Streptococcaceae* family and including the Lancefield group D streptococci. These microorganisms are characterized by a coccoid form, 0.5 to 1 μm in diameter, and are Gram-positive aerobic or facultative anaerobic bacteria, catalase negative with the group D antigen, forming chains of cells during cell divisions. This group includes enterococci (e.g., *E. faecalis, E. faecium, E. durans, E. avium*) and non-enterococci (e.g., *S. bovis, S. equinus*). Such as TTC, enterococci are abundant in human and animal feces. In the natural environment, enterococci have a better resistance to different environmental stresses than non-enterococci.

The enumeration of both intestinal enterococci and TTC (or *Escherichia coli*) is therefore a good indicator of aquatic contamination by human or animal feces. As the survival time in the environment is generally greater for enterococci than for TTC, a higher proportion of intestinal enterococci in environmental samples may be interpreted as a contamination of animal origin if the pollution is recent or as a former contamination. In practice, this distinction is often difficult because of the complexity of events and potential sources of pollution. In general, the specific detection of pathogenic microorganisms is only performed in environments where the concentration of FIB is higher than the guideline values.

Since the beginning of their routine use in legislation purposes, TTC, *Escherichia coli*, and intestinal enterococci were found to be effective indicators to assess the bacteriological quality of drinking and bathing waters. However, these indicators are weak indicators of the potential presence

Table 15.6 Characteristics of alternative fecal indicators in terms of prediction of fecal pollution and associated pathogens

Criteria for fecal indicators	Alternative fecal indicators					
	Bacteroides spp.	*Bifidobacterium* spp.	*Clostridium perfringens*	F-specific RNA coliphage	*B. fragilis* phage	Fecal sterols
Presence in the feces and environmental waters	Significant portion of fecal bacteria in warm-blooded animals; host-specific distribution; wide geographical distribution of human-specific genetic markers	Significant portion of fecal bacteria in warm-blooded animals; host-specific distribution; similar frequency and distribution of species of wastewater and human fecal isolates	Entirely of fecal origin from warm-blooded animals; present in wastewater and sediment; species distribution varies among animal species; may indicate remote or old fecal pollution	Present in human and animal feces as well as in wastewater; well correlated with the source of fecal pollution	Mainly found in human feces; could be used as a specific index of human fecal pollution; isolated from sewage and fecally contaminated water	Considerable amount in animal feces; indicator of recent contamination in water and of old or remote contamination in sediment; suitable for temperate and tropical regions
Ability to multiply and survive in the water environment	Short survival time and inability to replicate outside the host; effect of temperature and predation on survival rates; long persistence of DNA markers in water	Short survival time outside the host; unable to replicate in oxygenated nonhost environment; low frequency of detection in warm waters; inhibition of the detection due to background bacteria	Extreme stability in the environment (more stable than pathogens in environmental waters); no effect of temperature and predation on survival	Possibility to replicate in sewage; sensitive to high temperatures, solar radiation in conditions of high salinity; similar persistence with enteric viruses especially in freshwater	No replication in aquatic environments; high persistence in the environment; always more abundant than human enteric viruses	Aerobic degradation in the water column; incorporation in sediments under anoxic conditions
Correlation with pathogens or waterborne disease	N.D.	N.D.	Good correlation with some pathogens; might be an indication of the effectiveness of inactivation and removal of viruses and cysts; high correlation with diarrhea disease levels	Valuable model for viral contamination and inactivation	Relate to the level of human enteric viral contamination	N.D.
Applied methodology	Increasing use of direct molecular methods to identify the origin of fecal pollution using host-specific genetic markers	Use of direct molecular methods to identify the origin of fecal pollution using host-specific genetic markers	Indirect culture-based methods	Indirect (culture) and direct (molecular biology) methods; expensive and labor intensive due to their low abundance in feces and environment	Complex methodology requiring the analysis of large sample volumes; indirect (culture) and direct (molecular biology) methods	Simple and sensitive analytical method using gas chromatography-mass spectrometry (GC-MS)

Modified from Savichtcheva and Okabe (2006)

of viruses or pathogenic protozoa from fecal origin. In fact, due to different physiological traits such as virus reproduction modes and the ability of protozoa to develop resistance forms (cysts), viruses and protozoa are more resistant than FIB to disinfection treatments that are used in drinking water and wastewater treatment plants. As a consequence, other indicators are being sought that could complement usual FIB. Looking for some bacteriophages from specific fecal bacteria such as coliphages is currently performed, but, up to now, no indicator has been officially adopted. Enumeration of *Clostridium perfringens* has also been proposed because this fecal microorganism has the capacity to sporulate and, thus, may be representative of the behavior of resistance forms of some pathogenic protozoa. However, detection of these alternative indicators is more complex than the enumeration of classical FIB. The characteristics of alternative indicators of fecal contamination that are sometimes used in addition to conventional indicators are summarized in Table 15.6.

15.5 Soil and Dissemination of Pathogenic Agents

Soil environments are extremely diverse both considering their microbial composition and the extent of their chemical, physical, and structural characteristics. A current global estimate suggests that one gram of soil contains as many as 10^{10}–10^{11} bacteria and as many as 10^4–10^6 bacterial species. The distribution of these species depends on abiotic soil characteristics, on their ability to grow under anaerobic or aerobic conditions, as well as on their interactions with higher organisms including plants and micro(macro)-fauna.

Soil can act as a reservoir of pathogenic agents either because they are indigenous soil inhabitants or as transient visitors introduced through anthropogenic activities. It is now recognized or hypothesized that soil constitutes the natural habitat of several bacterial pathogens belonging to genus *Clostridium* (*C. perfringens*, *C. tetani*, *C. botulinum*), *Bacillus* (*B. anthracis*), *Nocardia*, *Mycobacterium* (*M. tuberculosis*), and *Burkholderia* (*B. pseudomallei*). Phylogenetically related species (i.e., *B. cereus*, *M. bovis*) to these pathogens have been isolated from soils (Kuske et al. 2006). Researches on *C. difficile*, a pathogen responsible for nosocomial infections, evidence its occurrence in 21–37 % of the soil samples tested (Simango 2006). Intrinsic properties as spore-forming ability could favor their long-term survival. The occurrence of *B. pseudomallei*, the agent responsible for melioidosis, a potentially fatal septicemic infectious disease among humans and an endemic disease in Southeast Asia (particularly Thailand) and Northern Australia, is intensively studied in these countries. Palasatien and collaborators (2008) investigated the presence of *B. pseudomallei* in relation to the physicochemical properties of soil from the Khon Kaen Province, northeast Thailand. The authors evidenced that the organism was found mainly at a depth of 30 cm and was significantly associated with some soil physicochemical parameters, including a pH of 5.0–6.0, a moisture content >10 %, and a higher chemical oxygen demand and total nitrogen than those reported for negative sites. They concluded on an uneven distribution of *B. pseudomallei* in this area, with the pH of the soil being the major determinant for the presence of the organism. Analysis of 104 collected soil samples in northern Australia has revealed a significant association between *B. pseudomallei*-positive sites and the presence of animals at these locations and also with moist, reddish brown to reddish gray soils (Kaestli et al. 2007).

Among opportunistic pathogens, many of them are abundant populations in plant rhizosphere. Various species involved in infections among cystic fibrosis individuals (i.e., *Burkholderia cenocepacia*, *B. vietnamiensis*, *Stenotrophomonas maltophilia*, *Achromobacter xylosoxidans*, and various members of the *Ralstonia* genera) (Fig. 15.5) are frequently isolated from agricultural soil samples and roots from wheat, rice, or maize and further studied as potential agent for the biological control of plant pathogens (Berg et al. 2005a). Similarly many strains of *P. aeruginosa*, *B. cepacia*, or *A. xylosoxidans* have been isolated from soils contaminated with pesticides (Vedler et al. 2004). The active role of these pathogenic species as well as some species belonging to the actinomycete group (*Nocardia asteroides*, *Dietzia cinnamea*) has also been reported for the degradation of hydrocarbons in polluted soils (Von der Weid et al. 2007). These species have several common properties as an important metabolic versatility; intrinsic capabilities to colonize different niches, i.e., denitrification ability of *P. aeruginosa* favoring its growth under anaerobic conditions in the mucus of cystic fibrosis lung (Philippot 2005); and virulence properties shared by clinical and environmental strains (Alonso et al. 1999). In parallel many of these bacteria are able to produce antimicrobial compounds

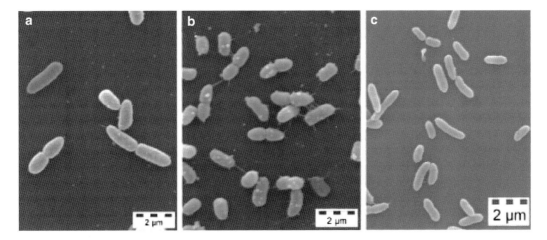

Fig. 15.5 Scanning electron microscopy photographies of pathogenic bacteria found in human and in the environment: (**a**) clone of *Pseudomonas aeruginosa* isolated from a waste stabilization pond (France), (**b**) clone of *Burkholderia cenocepacia* responsible for nosocomial outbreak in ICU (France), (**c**) clone of *Stenotrophomonas maltophilia* isolated from a soil under barley cultivation (France)

(pyocyanin synthesis by *P. aeruginosa*, production of the antifungal pyrrolnitrin, and the alkyl-quinolone-like compounds by *B. cepacia*) that could facilitate the colonization of various environments and enhance competitivity with indigenous microbial populations (Norman et al. 2004).

Pathogens can also enter soil environments due to human activities. Agricultural practices such as organic amendment (cattle manure, pig slurry, sewage sludge), animal grazing, or irrigation with microbiologically contaminated wastewater are potential sources of pathogens. The presence of foodborne pathogens (*Listeria monocytogenes*, *Salmonella* spp.), enteric bacteria, and opportunistic pathogens (*P. aeruginosa*) in organic amendments has been reported (Lavenir et al. 2007). The application of 1–2 t of sewage sludge on French soils would led to the introduction of 10^6–10^8 *L. monocytogenes* cells in soil per ha and per year if wastes were not composted before land application (Garrec et al. 2003). Numbers of 10^2–10^5 cells of *E. coli* O157:H7 per gram of bovine feces are mentioned. Transfer of pathogens from contaminated soil to rivers through leaching and runoff water has been shown (Ogden et al. 2001) such as transfer from contaminated irrigation water to soil and crops (Solomon et al. 2002). In Australia, a study from Vally et al. (2004) suggested that the water used to irrigate a soccer field was the source of an outbreak due to *Aeromonas hydrophila* among several soccer players.

Various studies aiming at evaluating the persistence of introduced pathogens have reported long-term survival in soil up to 1 year for *Salmonella enterica*, *S. typhimurium*, and *E. coli* O157:H7 whether experiments were conducted at the field (manure amendment) or the laboratory level (inoculated soil microcosm) (You et al. 2006). On the opposite some studies revealed that *E. coli* O157:H7 was not detectable 7 days (Ogden et al. 2001) or a few weeks (Gagliardi and Karns 2002) after soil addition. Among the various abiotic factors influencing pathogen survival temperature is the most frequently studied one. Data showed *E. coli* O157:H7 survival being longer at high temperature (15–21 °C) than at low temperature (5 °C) (Mukherjee et al. 2006). However, it has to be noticed that the impact of temperature, pH, or clay content on pathogen survival strongly varied from one pathogen to another (*E. coli* O157:H7, *Campylobacter*, *S. typhimurium*, *Yersinia enterocolitica*) (Guan and Holley 2003; Kidd et al. 2007). Furthermore, several factors and their combined effects were not yet investigated and should be taken into account to better identify the environmental conditions favoring pathogen persistence, growth, and dispersal in soil.

Soils as any environment can also offer conditions allowing a nonpathogenic bacterium to evolve toward a pathogenic one. This is not yet investigated, but we must be aware that new selective pressures occurred in various ecosystems as a consequence of increasing worldwide population and intensive industrialization processes. Climate change, intensive farming, and changes in agricultural practices (see below paragraph on global change) introduce new constraints on biota and impact ecological balances. Microorganisms due to their short generation time can be quickly affected by these changes. New genetic variants are probably emerging within the microbial life, but we are not yet able to predict their future impact on humans.

15.6 Sources of Pathogens Present in Aquatic Environments

Even if some pathogens of concern (bacteria of the genus *Legionella*, *Pseudomonas aeruginosa*, *Aeromonas*, *Burkholderia pseudomallei*, and cyanobacteria) are natural inhabitants of aquatic systems, most microbial waterborne pathogens originate in the intestinal tract of humans and animals and enter the aquatic environment via fecal contamination. This section devoted to the sources of waterborne pathogens will be focused only on these enteric pathogens, for which aquatic natural systems are not a primary environment.

Among the sources of contamination, point sources are usually distinguished from nonpoint (diffuse) sources. Major point sources include discharges of raw urban wastewater, domestic wastewater treatment plant (WWTP) effluents, industrial wastewater effluents, and urban runoff waters when they are collected by a separate sewer. In addition to contribution from point sources, the natural environment also receives fecal microorganisms by diffuse sources of contamination, which are more difficult to locate and quantify. No clear definition of the term "diffuse sources" exists in the literature. In this chapter, we consider it in its broadest sense, i.e., including both diffuse contamination from animal origin (wild animals and livestock) reaching the aquatic environment via runoff and soil leaching and diffuse contamination from human origin (leakage from sanitation systems such as septic tanks, leaks in networks sanitation defaulters, etc.) (Fig. 15.6). The relative importance of these various sources is highly variable both in space and time; it depends of course on the aquatic environment considered but also on hydrological and meteorological conditions. If point sources from WWTPs can be easily studied independently of other sources of contamination (and are also the subject of most studies on the presence of enteric microorganisms in natural environments), the flux of pathogenic microorganisms brought by nonpoint sources, which can be dominant in rural watersheds, is much more difficult to evaluate.

15.6.1 Point Sources

Domestic wastewater contains very high concentrations of microorganisms of fecal origin. For example, *E. coli*

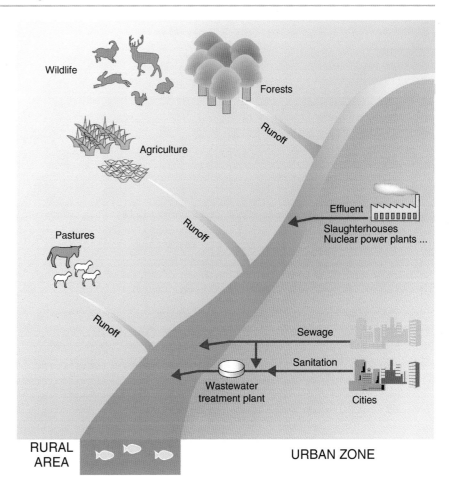

Fig. 15.6 Different sources of pathogenic microorganisms potentially present in a drainage network (Drawing: m-j Bodiou)

concentrations in raw sewage arriving in WWTPs are usually around 10^7/100 mL (Servais et al. 2007). While there are still 20 or 30 years, the wastewaters were often discharged without treatment into the aquatic environment, most of the urban domestic wastes are today, in industrialized countries, treated in wastewater treatment plants (WWTPs).

15.6.1.1 Wastewater Treatment Plants Effluents

The goal of a WWTP is primarily to remove suspended solids, organic matter (especially its biodegradable fraction), nitrogen, and phosphorus. Although wastewaters carry out many microorganisms, including some pathogens, very few WWTPs are equipped with treatments specifically designed to eliminate microorganisms. Presently, there is no European standard on the bacteriological quality of treated wastewater discharged into the natural environment.

Usually urban WWTPs include a pretreatment step (screening, grit removal, oil removal), a primary settling treatment which aims to retain suspended solids (SS) responsible for the turbidity of the water. This is often followed by secondary treatment, i.e., biological treatment involving bacteria that will oxidize the biodegradable organic matter (measured by biological oxygen demand, BOD). Generally, the biological treatment used is the "activated sludge." In more and more WWTPs, secondary treatment is now followed by a biological treatment of nitrogen; it may seek to convert ammonium to nitrate (nitrification treatment) or to remove nitrogen (nitrification treatment followed by denitrification). A biological or physicochemical treatment stage devoted to remove phosphates can also be inserted in the treatment line. According to European legislation (Directive on Urban Waste Water, DERU), the treatment of nitrogen and phosphorus is now compulsory for all WWTPs treating wastewater of more than 10,000 inhabitants. Studies on the removal of fecal bacteria in WWTPs usually show for plants using settling followed by activated sludge, a 2-Log units reduction of FIB present in the raw effluent corresponding to a removal of 99 % (Servais et al., 2007) (Fig. 15.7). The Log removal is calculated as the difference between the Log of the FIB concentration before treatment and the Log of the concentration after treatment. This elimination begins in the primary treatment in which fecal bacteria attached to suspended solids are removed with these particles by settling. It continues during the biological treatment where they can be consumed by protozoa. In WWTPs where a tertiary treatment of nitrogen or phosphorus is added, one can observe an improvement in the removal of fecal bacteria (up to 3-Log removal). It is the

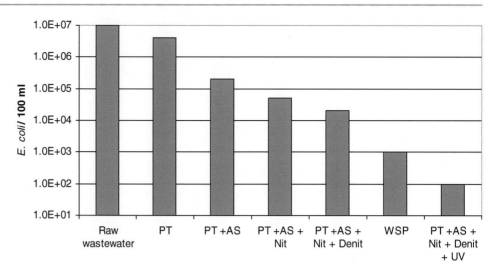

Fig. 15.7 Average *E. coli* concentrations in wastewaters at different stages of treatment in WWTPs (average values for several French WWTPs, based on data from George et al. 2002; Servais et al. 2007 and Ouattara et al. 2011). *PT* primary treatment, *AS* activated sludge, *Nit* nitrification, *Denit* denitrification, *WSP* waste stabilization ponds, *UV* disinfection by UV irradiation

case when the retention time of the water in the activated sludge is greatly increased to allow nitrification and denitrification but also when biofilters are used for this (Fig. 15.7). It appears from this that the treated domestic wastewater still contains significant amounts of fecal bacteria (*E. coli* concentration output from WWTPs in the range 10^4–10^6/100 mL).

Besides intensive wastewater treatments described above and used in most urban areas, there is a wide range of extensive treatments that can be used in rural areas, such as waste stabilization ponds. During this treatment, the water flows over long periods in successive shallow depth basins where natural processes allow for water purification (sedimentation of suspended solids, BOD biodegradation by bacteria, consumption of mineralized nutrients (N, P) by phytoplankton growth, grazing of phytoplankton by zooplankton). Even if this type of treatment is often limited to rural areas due to the large surface required, it may be as effective as conventional intensive treatments on parameters such as SS, BOD, nitrogen, and phosphorus. The pond can also efficiently remove pathogens and indicators. Studies have shown that this removal increased when the retention time in the ponds increased. Sunshine is also an important depuration factor since a part of the enteric microorganism load is destroyed by UV-B and UV-A solar radiations (Davies-Colley et al. 1997, 1999).

15.6.1.2 Wastewater Disinfection

As previously mentioned, most WWTPs are not designed to achieve a specific goal in terms of microbiological quality of the treated effluent. However, some WWTPs, located in sensible areas (bathing, shellfish harvesting, and drinking water sources), possess a specific disinfection treatment stage aiming to eliminate pathogens as well as indicators. Options for disinfection of domestic wastewater mainly include chlorination, ozonation, and UV irradiation. Chlorination of effluents is now discouraged due to the formation of chlorination by-products that could be carcinogenic and mutagenic, due to a reaction between chlorine and organic molecules contained in wastewater. Ozonation has the advantage to be very fast acting and effective against bacteria and viruses and to present a low propensity to generate undesirable by-products. Nevertheless, this method is highly expensive in terms of investment and operation. For these reasons, UV irradiation is the predominant option in Europe for wastewater disinfection. The main advantages of this technology are the absence of formation of undesirable by-products and the ease and safety of operation compared to chemical methods. The effectiveness of UV irradiation depends on the type of radiation (monochromatic at a wavelength of 254 nm or polychromatic), the diffusion of radiation through the effluent (mainly depending on the SS content), and the type of microorganisms considered (Hijnen et al. 2006). Thus, a monochromatic radiation of 40 mJ/cm^2 (as conventionally applied in WWTPS including such disinfection stage) allows around a 4-Log units removal of culturable *E. coli*. However, studies have shown that some fecal bacteria, which lose their ability to multiply immediately after UV irradiation, could recover after several hours of incubation in the light, a phenomenon known as the photoreactivation. Moreover, it is known that UV irradiation at a given dose resulted in a greater reduction in the abundance of FIB (*E. coli*, intestinal enterococci) than that of some pathogens (viruses, protozoan parasites, etc.). Because of these observations (photoreactivation and modification of ratio FIB/pathogen by UV irradiation), there is still a debate regarding the merits of the widespread use of this technique for wastewater disinfection.

15.6.1.3 Combined Sewer Overflows (CSO)

The input to the aquatic natural environments of pathogenic microorganisms from domestic wastewater is not restricted to discharges of WWTPs' effluents. In fact, even today in industrialized countries, some wastewaters are not yet processed in WWTPs leading to the release of untreated sewages directly into the receiving environment. This is especially true during rainy or storm conditions. In cities equipped by a combined sewer network, urban runoff waters significantly increase the flow entering the WWTP during rain events. It may become unable to handle the entire flow of wastewater and untreated or partially treated (bypass certain stages of purification treatment) waters that are released into the environment. The content of microorganisms and indicators of CSO is generally lower than those of raw sewage entering the WWTP during dry weather due to the dilution of domestic waters by runoff waters less rich in microorganisms (Marsalek and Rochfort 2004). However, the sudden and important flow increase occurring during storm events can cause resuspension of fecally contaminated sediments accumulated in the pipes of the sewer during dry weather periods. The type of situation encountered depends on network characteristics (size, slope, etc.) and those of the rainy event (intensity, duration, etc.). In the case of urban areas with separate sewer networks, runoff waters are commonly discharged directly into the receiving water without any treatment. In both scenarios, the intake flow of pathogenic microorganisms is increasing in rainy conditions. These observations explain why the most critical situations of microbiological environmental water quality are monitored just after storm or rain periods (e.g., degraded microbiological quality of bathing waters after summer storms).

15.6.2 Diffuse Sources

The inputs of microorganisms from fecal origin via nonpoint sources are relatively difficult to quantify. The concentrations of FIB has been determined in small streams upstream of any source of domestic contamination in order to get a rough estimate of the levels of contamination due to runoff waters and soil leaching. These studies have shown that the measured concentrations were very different depending on the landuse of the watershed. Streams whose catchment area is mainly covered with pastures presented contamination levels one order of magnitude higher than the streams whose catchment area is covered mainly by forest or crops. For example, in a study conducted in the Seine river watershed (France), average *E. coli* concentrations around 400/100 mL were found in small rivers flowing in pastured areas, while they were around 40/100 mL in streams flowing in forest and culture areas (Garcia Armisen and Servais 2007). This clearly highlights the importance of manure grazing livestock as a source of fecal contamination of surface waters in agricultural areas.

It was also shown that the contribution of fecal bacteria in grazing areas depended on the intensity of precipitation. In rainy periods, the flux exported to aquatic systems is increasing due to the greater contribution of runoff waters that are richer in fecal microorganisms than baseflow fed by water that seeped through the ground. Indeed, the percolation of water through the soil leads to a natural removal of microbial contamination. Indeed, the technique to infiltrate water through soil horizons is also used in some countries for the production of drinking water.

15.7 Fate and Transport of Enteric Microorganisms in the Environment

The fate of enteric microorganisms in the environment is very difficult to assess because current methods (*cf.* Sect. 14.10) make it difficult to follow them from their natural reservoir to another point. In addition, their fate is variable depending on the nature and intensity of environmental pressures experienced throughout their route to the point of contact with humans. Environmental factors that may influence the fate of a pathogen into the environment are multiple (Fig. 15.8). They contribute to the self-purification of the natural environment, and they are sometimes used to purify water, as in the case of waste stabilization ponds.

The transport of enteric microorganisms in aquatic environments is controlled by the hydrodynamics of the system in which they are released and in some cases by settling. Concerning the fate of fecal microorganisms, it is quite usual to distinguish two types of controlling factors: the biotic and the abiotic factors.

15.7.1 The Hydrodynamical Factors

Enteric microorganisms released in rivers (e.g., by the effluent of a WWTP) are diluted by the river water, and their transport is mainly controlled by the river flow. The impact of dilution, diffusion, and transport is much more complex in coastal areas. It is thus sometimes necessary to conduct tracing operations to locate the point of discharge of sewage and understand their fate in the ecosystem. Figure 15.9 shows aerial tracing with **rhodamine*** in a coastal area. Rhodamine is a red dye. It is injected in high concentration at a given point (here it was mixed in the effluent of a WWTP) and then emerges a few minutes later at sea outfall (end of the pipe often equipped with a vertical diffuser) and diffuses into coastal waters. Measurements of the color intensity over time and at various points are used to determine the diffusion parameters and thus to better understand the impact of wastewater in the coastal environment.

Sedimentation is an important factor contributing to the removal of some pathogens from the water column of aquatic environments and to the accumulation of these microorganisms in the sediments. Protozoa but also viruses and pathogenic bacteria associated with particles can settle in areas with little turbulence. They accumulate on the bottom where they constitute a secondary reservoir which can be a source of water contamination in some of the

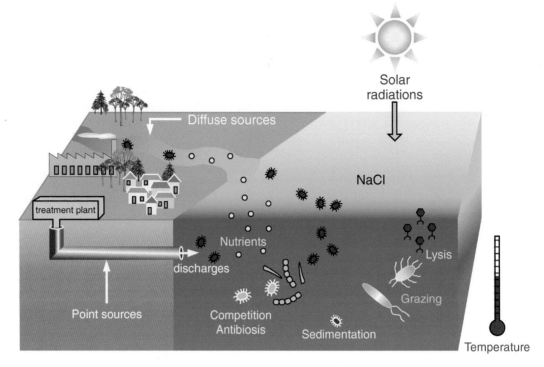

Fig. 15.8 Direct and diffuse sources of pathogenic microorganisms to coastal areas and environmental factors that change their physiological state and survival capability in the environment (Drawing: m-j Bodiou)

Fig. 15.9 Tracer experiment with rhodamine in a coastal area designed to determine the impact of the release of a wastewater treatment plant in the coastal environment. Rhodamine was injected in high concentration at the outlet of the WWTP. It diffuses in the sea from the end of the output (pipe), and the red plume is followed by aerial photo over time. At the same time, samples are collected at various locations and times to determine the dye concentration

hydrological conditions (rapid increase of the discharge in rivers, tidal impact in coastal zones, etc.). In some cases, accumulation in sediments led to the death of pathogens, especially strict aerobic species that are dependent on oxygen. However, sediments are also a favored place for survival; indeed, microorganisms find in the sediments conditions that fully or partly protect them against solar radiation and predators and that are rich in nutrients.

15.7.2 The Abiotic Factors

15.7.2.1 Temperature
Temperature plays a very important role, and it has often been suggested that low temperatures favor the survival of pathogens in reducing their energy costs (slowed metabolic activity). However, freezing helps maintain viruses but has a lethal effect on most bacteria and protozoa or alters their physiological state. The optimum temperature for growth of pathogens whose primary habitat is the digestive tract of humans and warm-blooded animals is close to 37 °C. Their ability to grow in the natural environment is greatly reduced or grow with the exception of some ecosystems where more appropriate conditions may be encountered.

However, human activities can occasionally and locally change the temperature of aquatic ecosystems. It can be due to the release of hot waters from cooling systems of nuclear power plants which can promote the growth of some bacterial pathogens and also the development of thermophilic amoeba such as *Naegleria fowleri* (Pelandakis and Pernin 2002; Behets et al. 2007). In addition, men have built artificial ecosystems where conditions favoring the development of certain human pathogenic species are met. Artificial hydrosystems such as hot water distribution systems, air conditioning circuits, spas, cooling towers, etc., have become preferred habitat for Legionella (Fields et al. 2002). In several countries in Europe and Asia, it has been revealed that the abundance and dispersion of *Legionella pneumophila* follow a seasonal dynamic with peaks during the warmer months of the year (Wery et al. 2008; Lin et al. 2009).

15.7.2.2 Light
Natural light plays a very important role in the self-purification of the environment. If the very bactericidal (maximum absorption at 256 nm DNA) ultraviolet radiation UV-C (wavelength between 100 and 290 nm) do not reach the earth's surface and are used only in the treatment processes for disinfection of drinking water or wastewater, UV-B (wavelength between 290 and 320 nm) and UV-A (wavelength between 320 and 400 nm) radiations have a bactericidal effect on many pathogens causing direct or indirect cellular damages (Davies-Colley et al. 1997). Cellular damages linked to UV-B and UV-A are rather indirect damages to DNA by photooxidation and production of reactive oxygen species that can damage DNA. Thus, solar radiation plays a very important role in the mechanisms of self-purification of the natural environment. Obviously, this effect is much more important in the shallow ecosystems with clear water (containing low SS concentration). This self-purification process is used in waste stabilization ponds.

15.7.2.3 Nutrients
Most heterotrophic pathogens require high concentrations of organic nutrients to multiply; they are copiotrophic microorganisms, while autochthonous microorganisms are **oligotrophic*** organisms adapted to poor environments. This nutritional requirement makes them less competitive with regard to the native oligotrophic microflora of natural environments. However, some habitats such as sediments of fluvial and coastal aquatic systems may have favorable conditions in terms of energy, especially at low temperature when metabolic activity is reduced.

15.7.2.4 Salinity
For a long time, it was thought that the salinity of the seawater was sufficient to remove most pathogens. However, it is now demonstrated that many pathogens are very tolerant to salinity. In fact, the salinity of seawater is much closer to the salinity encountered in the gastrointestinal tract than to the salinity of freshwater, which often have a hypo-osmotic effect on pathogenic microorganisms. Some protozoa such as *Cryptosporidium* survive very well in coastal waters and can contaminate shellfish like oysters that are most often eaten without cooking. Most bacteria of enteric origin are halotolerant and can survive in brackish and marine waters.

15.7.3 The Biotic Factors

Various biotic factors may play a role in the demographic behavior of microorganisms in the environment. Most of these factors are based on the interactions within microorganisms or between microorganisms and higher organisms. These are essentially predation, parasitism, and competition. Pathogenic microorganisms are subject to predation by protozoan and viral lysis as it is the case for all autochthonous bacteria.

Predation by protozoa (nano- and microzooplankton) is recognized as the leading cause of mortality of autochthonous bacteria in natural aquatic environments. Heterotrophic nanoflagellates (size between 2 and 20 μm) are the main grazers of bacteria followed by ciliates. These organisms also consume bacteria of enteric origin released into the water, and many studies have shown that predation by protozoa was the first biotic factor responsible for the decrease in the abundance of fecal bacteria in aquatic environments. Predation is particularly important when the

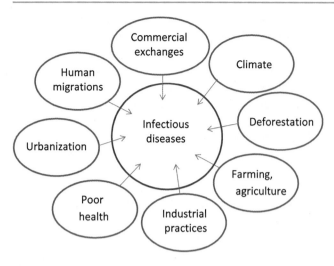

Fig. 15.10 The key factors of global change in relation with infectious diseases

concentrations of pathogens are high but also when the abundance of protozoa is important. The latter varies from one environment to another; in a given environment, it generally increases with increasing temperature. As the rate of ingestion of bacteria by protozoa also increases with temperature, this implies a larger grazing in summer than in winter. This may partly explain the better survival of pathogenic microorganisms at low temperatures reported by some authors.

Viral lysis is also concentration dependent and is exerted especially when a pathogenic species becomes abundant. It may be important in the case of blooms of cyanobacteria. The exact effect of viral lysis in the natural environment is difficult to estimate.

Competition for food resources is also important, especially in environments where this resource is limited and where native species are much more competitive.

15.7.4 Global Change Effects

Global changes relate to a variety of factors including climate (temperature, precipitation, wind strength) and socioeconomic developments taking place at the world scale (Fig. 15.10). These factors affect or may affect the incidence of infectious diseases. The socioeconomic factors include a set of practices and processes, causing a significant proportion of human population exposure to infectious agents. We can include agricultural and industrial practices, urban development, hygiene practices, poverty, patterns of waste management, population displacements, and commercial exchanges including animal transfers worldwide, etc. These factors have in many cases a local origin a priori less global than climate change. However, these local changes occur across the globe resulting in a global change

such as chemical changes in the atmosphere leading to an increase in the concentration of CO_2, methane, and gas from industrial origin such as chlorofluorocarbons.

Urbanization and the increase in human population at the world level represent major challenges for our societies and will directly affect key parameters of human exposure to infectious agents. On several continents, predictions suggest an increase from 50 to 100 % of the urban population giving an estimated five billion people located in urban centers in 2030. Cities will be centers of attraction for most people and will lead to an increase in waste and contacts with pathogen sources. Poor countries with deficient treatment system or ways of recycling wastes will be strongly affected by overcrowding. The increase in human population also induces increases in demand for food resources and leads to changes in agricultural practices (intensive livestock, crops) leading to major changes in landscapes and ecosystems. Two consequences of these agricultural practices are deforestation and the introduction of irrigation system on a large scale. Irrigation can lead to major changes in aquatic ecosystems and reduce water resources available for human consumption. Deforestation results in the fragmentation of habitats for many biological species and therefore in a change in the probability of contact between the host, the vector, and the pathogen. Deforestation is at the origin of the emergence of viral hemorrhagic fevers in several countries in South America. Changes in biodiversity with disappearance of predators and increase and diversification of reservoir hosts can also promote the spread of pathogens. This is particularly well illustrated by the higher incidence of Lyme disease after one cycle of deforestation/reforestation in 1976 in the United States. This disease is caused by the pathogenic bacteria *Borrelia burgdorferi* whose main vector is the *Ixodes* sp. that can be found in deer and rodents. The cycle of deforestation/reforestation has resulted in a loss of predators (wolves, coyotes) of deer and rodents, leading to an increase in their populations and, consequently, increasing the density of ticks. The expansion of peri-urban areas near these reforested forests subsequently increased human exposure to ticks carrying *B. burgdorferi*. Another effect of deforestation is linked to its contribution to global warming.

Another important aspect related to the increase of the human population is increasing movements of people in the world and the globalization of trade. Today, more than one million people are engaged on international travels every day, and more than one million individuals perform weekly trips between developed and developing countries (Garrett 1996). These exchanges and travel increase the risk of exposure of individuals to certain pathogens and promote the transfer of unwanted species (mosquitoes, rodents, ticks, etc.) between countries. Transfers of animal species are also causing changes in the exposure scenarios of human populations to infectious agents.

15.7.4.1 Special Case of Global Climate Change

Today, the seasonal variations in climate that affect the survival or spatiotemporal dynamics of pathogens are combined to the phenomenon of global climate change. The Intergovernmental Panel on Climate Change (IPCC http://www.ipcc.ch/) gathered data sets from 1850 to the present concerning changes in air temperature across the world and has demonstrated in his 2007 report a significant increase of 0.76 °C (+/−0.19 °C) of the earth's temperature, with an acceleration of the increase since the 1980s (Delecluse 2008). This increase appears to induce a series of chain reactions that may change humidity rates, precipitation levels, exposure to sunlight, etc. The consequences of climate change can be multiple, including increased sea levels, a reduction in snowy areas, an increase in forest fires, but also changes in the balance between living organisms and changes in the distribution of some species. Climate change could induce many species extinctions, and today we can observe significant effects on amphibians. These changes will also affect pathogens and vectors at the origin or source of transmission to humans.

There are only few solid demonstrations of an effect of climate change on infectious diseases. However, from 460 to 377 BC, Hippocrates described the climate as a determinant of many diseases. The most frequently cited case of impact that these changes could have on human diseases concerns the transmission of pathogens via mosquitoes and cholera. For example, epidemiological investigations on cases of encephalitis caused by a flavivirus transmitted by ticks suggest a correlation between changes in spatial distribution of these vectors, global warming for some Nordic countries such as Sweden, and the impact of this disease (Lindgren and Gustafson 2001). As another example, monitoring programs of cholera in Bangladesh suggest a replacement of the traditional strain by *Vibrio cholerae* by strain El Tor. This replacement would be due to a change in the weather by changing rainfall during the monsoon and, therefore, affecting salinity (Koelle et al. 2005). El Tor strain would be more competitive with low salinity, thereby increasing the exposure of local populations to this strain.

In addition, it should be noted that the cumulative data on the impact of cyclical climate phenomena such as ENSO (El Niño Southern Oscillation – abnormally high ocean temperatures in the eastern part of the South Pacific Ocean causing heavy rainfall) on infectious diseases suggest highly significant effects of global climate change on the incidence of infectious diseases. To illustrate, a positive correlation between ENSO and the annual incidence of visceral leishmaniasis has been observed in Brazil (Ready 2008). In addition, the cholera epidemic affecting South America in 1991 would also be linked to this phenomenon (Salazar-Lindo et al. 2008).

Global changes can thus significantly alter the ecology of infectious agents, and current predictions likely underestimate their consequences.

15.8 Viable but Non-culturable (VNC) Bacteria

In 1982, Xu and coworkers described viable bacteria in a state of survival or dormancy making them non-culturable by classical culture-based methods. These non-culturable bacterial forms were obtained by exposing *Escherichia coli* and *Vibrio cholera* cells resuspended in sterile seawater, at a low temperature without nutrient supplements. Numbering these viable but non-culturable bacteria by direct epifluorescence microscopy as those which are able to elongate in the presence of yeast extract and nalidixic acid (Kogure et al. 1979) showed counts significantly higher than those obtained from plate count agar or most probable number approaches. These bacteria were able to persist for several days or weeks in seawater while they were unable to form colonies on agar plates (Box 15.4). This finding sparked a major interest in the international community. Indeed, the health significance of these viable but non-culturable (VNC) forms became of concern, especially in the case of pathogenic bacteria that could have been maintained in a viable state without being detectable so far by the classical methodologies implemented in health risk assessment.

> **Box 15.4: The Viable but Non-culturable (VNC) State in Bacteria**
>
> Philippe Lebaron
>
> The proportion of culturable bacteria in the environment is usually very low and often less than 1 %. This percentage is often expressed in reference to the total number of bacterial cells that can be enumerated on the basis of direct quantification by epifluorescence microscopy or flow cytometry (*cf*. Chap. 17), after labeling of nucleic acids by a fluorescent probe (Box Fig. 15.6). Three main reasons can explain this "non-culturable state":
>
> – Many bacteria have specific needs, which are frequently not satisfied by the formulation of the available commercial culture media, to complete their metabolism and as a consequence to grow. Moreover, in the environment, bacteria live in multispecies communities including other bacteria, macroorganisms, and plants and are often organized in complex structure such as biofilms. This multispecies organization allows many interactions between the different members of the community that are, sometimes, required for growth of a specific member.
>
> (continued)

Box 15.4 (continued)

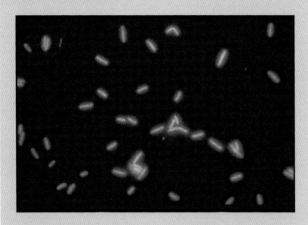

Box Fig. 15.6 Epifluorescence microscopy photography of active but non-culturable cells of *Escherichia coli*. Fluorescence of cells is induced by the hydrolysis of a fluorogenic substrate specific for esterase activity (Photograph: Philippe Catala, photograph collection of Laboratoire Arago, Banyuls)

– Some bacteria of the community may be moribund or dead. These cells can persist in the environment and be detected by direct observation by flow cytometry until their effective lysis.
– Some other bacterial cells, which are normally able to form colony on an agar culture medium, may have undergone various stress factors (temperature, salinity, ultraviolet radiation, nutrient deficiency, etc.), and although they conserve a cellular activity, they can be temporarily unable to grow on this medium. This physiological state often called "viable but non-culturable state (VNC)" or "active but non-culturable state (ANC)" is extensively studied. The viability is partly based on the ability of an organism to grow, but this concept is difficult to apply when considering a single cell. The term "active but non-culturable state" is more suitable since it relies on the detection of cellular functions (*cf*. Chap. 17) and not to the replication ability of the cells.

All known pathogenic bacteria present in the environment can be cultivated on synthetic culture media. For these cells, the ANC state may exist due to the multiple stress factors encountered by these bacteria in the environment, but their sanitary impact remains to be demonstrated. In addition, as pathogens are often in low numbers in natural microbiota, it is difficult to accurately detect and quantify their ANC forms in the environment even if a combination of physiological sensors and taxonomic probes can theoretically be used to achieve this goal. New cytometry techniques such as solid-phase cytometry could offer new perspectives in this field.

Since the work of Xu et al. (1982), VNC forms have been described in many bacterial groups including several pathogenic species, e.g., *Legionella pneumophila*, *Francisella tularensis*, *Enterococcus faecium*, *Helicobacter pylori*, *Listeria monocytogenes*, *Burkholderia pseudomallei*, *Burkholderia* of the *cepacia* complex, and many others (for a review see Oliver 2010; Colwell and Grimes 2000). The VNC state has now been induced by a variety of factors such as temperature, nutrient deficiency, hydrostatic pressure, pH, or salinity changes. These factors induce a series of events resulting in morphological changes (usually a reduction in cell size) and changes in the concentration of major polymers (proteins, membrane lipids, nucleic acids). They lead to a progressive loss of the cell ability at multiplying on agar plates or in broths normally used as standard growth media for these bacteria. The nature of the changes to be considered in defining the VNC status, however, remains debated. Operational definitions have been introduced to clarify the situation such as the terms bacteria "immediately cultivated," "ultimately cultivated," or "active but non-culturable" highlighting that procedures used for growing bacteria on agar media could change the proportions of estimated VNC forms. VNC bacteria are also sometimes defined as "dormant" or "inactive." This ambiguity is based on the fact that the criteria to define the state of bacterial viability remain unclear, and therefore it remains hard to declare a cell as being dead, dying, or alive. Some authors consider that the VNC forms shall evolve to cell death, while others believe that the non-culturable state is a part of the life cycle of the cell, while also reflecting a survival strategy when environmental conditions are unfavorable. This latter vision of the VNC state would need to be built on the existence of reversible bacterial processes regulated by environmental factors to be identified.

Work has been performed to verify the existence of processes inducing a transition to the VNC form in *E. coli*. A significant finding was the observation of an increase in oxidized proteins in VNC forms of *E. coli* associated with an induction of the defense processes against reactive oxygen species (ROS) (Dukan and

Nyström 1998). In addition, increases in the synthesis of heat shock resistance proteins, glutathione reductase, superoxide dismutase A, etc., have been described in *E. coli* stationary phase indicating that these cells (a mixture of culturable bacteria and VNC) increase their resistance toward denaturation and protein aging. These observations led Desnues et al. (2003) to develop centrifugation techniques for separating the VNC and culturable forms and approaches for in situ detection of oxidized proteins in an *E. coli* cell. These technological developments allowed showing that VNC *E. coli* cells can lose their culturability due to a sharp deterioration of their protein content and not because of an induction of an adaptive process changing their physiological state to dormant or non-culturable one. The return of a VNC cell to a culturable state would therefore be related to processes allowing some kind of protection of the cell content such as an induction of activities favoring a repair of damages caused by a period of stasis (period of deceleration of bacterial growth that could go down to a complete stop and cell death). The results obtained with VNC *E. coli* must now be verified with other bacterial species to conclude on the general meaning to be given to the VNC status among bacteria. However, the existence of VNC cells among bacterial pathogens, whether damaged or dormant, remains an undeniable reality, and the possibility of their return to a physiological state representing a health hazard remains a public health concern. Return of VNC to culturable forms was also obtained after passage in a higher organism such as amoebas for the particular case of *L. pneumophila*. VNC cells of *Vibrio* spp. were found to have kept their virulence. By extrapolation, man could thus be infected by VNC bacteria. Man could be acting as a favorable host for the return of VNC into culturable cells.

On an operational point of view, it is difficult to detect in routine a cellular activity that may provide evidence of viability of pathogenic VNC bacteria and establish a relationship with any potential colonization of man. The difficulty lies, among other things, on the fact that this activity should be determined without a culturing step and at the scale of a single cell. Fluorescence techniques in combination with currently available probes targeting cell function have limited resolution, either because the targeted activity is low or because the detection tools are not sensitive enough. Moreover, there is no single function to inform on cell viability. It is now recognized that several tests are necessary to establish a satisfactory diagnosis of the physiological state of a cell because it goes along a continuum of physiological conditions from active and dividing ones to inactive and irreversible ones, i.e., cell death.

15.9 Detection Methods

The primary goal of monitoring water quality is to highlight the presence or absence of pathogens and avoid human exposure. Monitoring waters for the many different human pathogens by using conventional methods such as the most probable number approach based on bacterial broths, the membrane filtration, and the culture-based methods are not convenient enough. These methods are time-consuming, are laborious, and do not allow the detection of all the cells present in a sample (VNC cells). Therefore, many alternative detection methods, including direct methods and indirect methods, of specific pathogens have now been developed (Fig. 15.11).

Detection of pathogens via culture-based methods requires an enrichment of a target population over other environmental microorganisms, followed by pathogens' isolation and identification (Fig. 15.12). Culture-based methods are indirect methods since they require a culture step before the diagnostic. Even if these methods are still in use in many monitoring agencies, their major limitations when addressing the detection of pathogens in environmental samples are that they (1) are time-consuming, (2) might not grow the targeted microorganisms because of environmental factors leading to VNC (*cf.* Sect. 15.8), and (3) will not detect many enteric viruses and protozoa (*cf.* Sect. 15.3).

In the last two decades, scientists have developed direct methods to rapidly detect the presence of fecal indicators and fecal pathogens in food, soil, and water (*Thompson* et al. 2010). These methods are much faster than culture-based methods and are able to detect specific microorganisms, in only few hours, with a high degree of sensitivity and specificity without the need for complex cultivation. These methods include cell-based and nucleic acid-based methods.

15.9.1 Cell-Based Methods

These methods allow an accurate numbering of microorganisms and involve detection tools such as epifluorescence microscopy (EFM), flow cytometry (FC), and solid-phase cytometry (SPC). Unfortunately, EFM and FC are not suitable for the detection of low cell numbers (as it is the case of pathogens in natural environments) due to their intrinsic detection limits (*cf.* Sect. 17.3). SPC, with a limit of detection of one target cell in a filterable sample, enables the application of cell-based methods for the detection of pathogenic microorganisms in the environment.

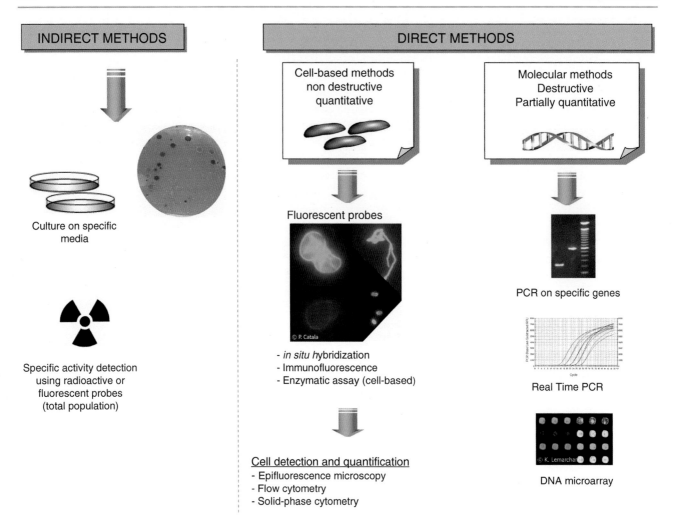

Fig. 15.11 Direct (destructive and nondestructive) and indirect methods used for the detection and quantification of pathogenic microorganisms

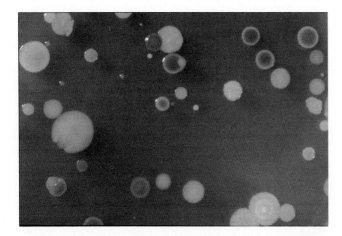

Fig. 15.12 Bacterial colonies formed on a solid medium culture presenting very different morphological and tinctorial characteristics. Each colony is the result of the multiplication of one bacterial cell (Photography by Philippe Lebaron, Laboratory Arago (Banyuls-sur-Mer) library)

15.9.1.1 Fluorescent In Situ Hybridization (FISH)

FISH enables detection of specific nucleic acid sequences inside intact cells. Probes, oligonucleotides, or peptide nucleic acids (PNA), complementary to specific sequences in the cellular DNA or RNA molecules, are combined with fluorochromes to allow the detection of the hybridized cells (*cf.* Sect. 17.7). The main DNA targets are genes (*rrs* and *rrl*) encoding ribosomal RNA (16S rRNA and 23S rRNA). Although this technique is quite specific and sensitive, it has limitations when applied to the detection of physiologically stressed cells or less active ones. This limit can be overcome by coupling FISH and direct viable count (DVC) (Box 15.5).

15.9.1.2 Immunofluorescence (IF)

Immunofluorescence is based on the formation of a specific antibody-antigen complex. This method allows the detection, identification, and enumeration of targeted cells in a complex matrix. Direct detection requires the use of a

Box 15.5: DVC-FISH: In Situ Hybridization Coupled with a Viability Test for the Enumeration of Viable *E. coli*

Philippe Lebaron

Fluorescent in situ hybridization (FISH, *cf.* Sect. 15.9.1) can be coupled with a viability test to specifically detect specific viable bacteria by microscopy or cytometry. The direct viable count (DVC) test combines the culture of the bacteria in a rich medium and the use of antibiotics that inhibit the formation of the septum and, as a result, the cell division. After incubation, the viable cells appear as elongated cells, whereas nonviable cells remain at their initial size. The viable cell elongation is concomitant with an increase of the RNA content of the cells that could be easily detected by FISH after staining with a specific fluorescent molecular probe, without any modification of the total number of bacteria detected. This combined method has doubled benefits since it allows an easy enumeration of active cells (showing a metabolic activity) of a specific species in complex communities (Garcia Armisen and Servais 2004). This method was successfully used for the enumeration of viable *E. coli* cells in different types of water (Box Fig. 15.7). Unlike methods based on cultivation, which are limited to the detection of culturable cells, this technique allows the enumeration of all viable *E. coli* cells in an environmental sample. Nevertheless, this method is relatively hard to implement and requires highly qualified people in molecular biology and microscopy.

Box Fig. 15.7 Epifluorescence microscopy photography of *E. coli* cells in a water sample labeled with an in situ fluorescent molecular probe (Photograph: Philippe Catala, photograph collection of Laboratoire Arago, Banyuls)

fluorescent primary antibody, while the indirect detection involves binding of a non-fluorescent primary antibody followed by the addition of a fluorescent secondary antibody directed against the primary one.

The properties of the antigen-antibody complexes can also be used to specifically concentrate the targeted cells by immunocapture. One of the most commonly immunocapture method used in microbiology is **ELISA*** (enzyme-linked immunosorbent assay allowing the quantification of an antibody-antigen reaction from an enzymatic assay), but its quantitative limits do not allow a reliable quantification of rare microorganisms. A second immunocapture technique is the immuno-magnetic separation (IMS). This method uses magnetic beads coated with monoclonal or polyclonal antibodies. The antibody-antigen reaction allows efficient separation of the cells from complex matrices. After this separation, the target microorganisms can be identified by various techniques such as culture-based or nucleic acid-based methods.

The major limitations of IF and IMS are linked to the lack of specificity of the antibodies used in the natural environment and to the absence of physiological information on the detected cells detected.

15.9.1.3 Enzymatic Methods

The detection of bacterial species in environmental samples by enzymatic test relies on the detection of a specific enzymatic activity. Such methods were developed and tested to estimate *Escherichia coli* counts in wastewaters and surface waters. These assays are based on the detection of β-glucuronidase (GUD) activities that can cleave 4-methylumbelliferyl β-D-glucuronide (MUG) and release 4-methylumbelliferone (MUF). When exposed to UV light, MU exhibits a bluish fluorescence that is easily detected. Over 95 % of the *E. coli* strains can produce GUD, including anaerogenic (non-gas-producing) ones. Exceptions are the enterohemorrhagic *E. coli* (EHEC) strains of serotype O157:H7, which are consistently GUD negative.

The measurement of the GUD activity can be achieved at a global scale (Box 15.6) and a cellular one. As the fluorescence intensity is directly correlated to the number of microorganisms in environmental samples, global scale measurements are used to describe the intensity of the β-D-glucuronidase activity of a water sample and thus, indirectly, quantify *E. coli* cells. At the cellular level, these methods can allow a more accurate detection and quantification of *E. coli* cells expressing β-D-glucuronidase through an SPC fluorescence detection.

The estimation of β-D-glucuronidase activities is now widely used to monitor *E. coli* among environmental samples, and several commercial kits making use of the MUG substrate are now available (Box 15.7).

Box 15.6: The β-D-Glucuronidase Assay: A Tool for Real-Time Management of *E. coli* Concentration in Water

Philippe Lebaron

Measuring the activity of the β-D-glucuronidase of *E. coli* while avoiding the use of culture media has been proposed as an alternative to estimate *E. coli* concentrations in water quality monitoring. It involves filtering a sample of water, incubating the filter in the presence of the fluorogenic substrate 4-methylumbelliferyl β-D-glucuronide (MUG) and measuring, under standardized conditions, of the increase of fluorescence over time due to MUF production which results from the hydrolysis of the substrate by the β-D-glucuronidase of *E. coli*. The appearance of fluorescence is proportional to the amount of enzyme present in the sample and thus to the number of *E. coli* in the sample. This protocol allows measuring the activity of glucuronidase activity in less than 1 h. A regression line linking enzyme activity to the number of *E. coli* allows converting the results expressed in terms of enzymatic activity in abundance of *E. coli*. Speed, simplicity of the experimental protocol, and low cost compared to the reference method (microplate method) allow to propose this approach as a fast method for real-time control of the microbiological quality of freshwater (Servais et al. 2005) and marine ones (Lebaron et al. 2005) (Box Fig. 15.8).

This method is presently used for the real-time monitoring of the microbial quality of many French beach bathing waters. The main drawbacks of this method are the need of a calibration with culturable bacteria and the presence of some false positives in the environment (non-*E. coli* cells showing a ß-glucuronidase activity).

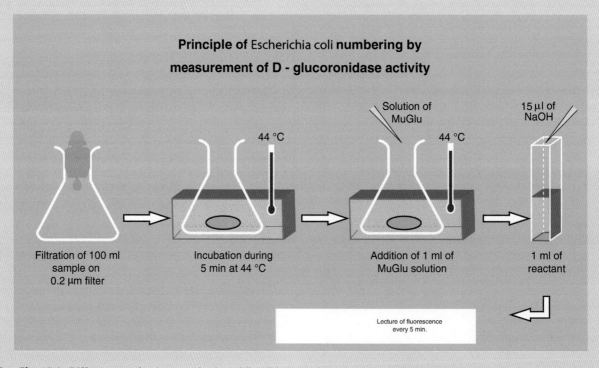

Box Fig. 15.8 Different steps for the quantification of *E. coli* in water through an enzymatic approach

> **Box 15.7: The MUG Microplate Method**
>
> Philippe Lebaron
>
> **Determination of *Escherichia coli* Most Probable Number (MPN) by the Microplate Method**
>
> MPN consists at inoculating tenfold serial dilution in tubes (tube assay) or wells (microplate assay) containing a specific culture medium and to observe the growth of the bacteria in each tube or well. Then, statistics are used to calculate the MPN on the basis of the proportion of positive tubes (where bacterial growth appeared) for each dilution (Box Fig. 15.9).
>
> A miniaturized method, based on this principle, is commercially available for the enumeration of *E. coli* in water. This method is now standardized (ISO 9803-3) and commonly used for routine water microbiological quality monitoring in France. Briefly, a 96-well microplate containing a fluorogenic substrate that can be hydrolyzed by the β-D-glucuronidase of *E. coli* cells, the 4-methylumbelliferyl-beta-D-glucuronide (MUG), is inoculated with 200 µl of tenfold dilution series of water sample to be tested. After 36 h incubation at 44 °C, positive wells (fluorescent) are recorded to determine the *E. coli* MPN in the sample.
>
>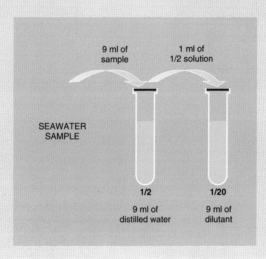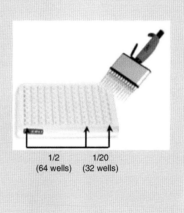
>
> **Box Fig. 15.9** Inoculation protocol of a 96-well microplate for *E. coli* numbering

15.9.2 Nucleic Acid-Based Methods

Nucleic acid-based methods have been developed to detect bacterial fecal indicators as well as pathogens without growing the targeted microorganisms. These are based on the identification of specific genes or specific nucleic acid sequences. Environmental samples often do not contain enough cells of the targeted pathogenic microorganisms to produce a direct detection. This limit can be overcome by the use of the polymerase chain reaction (PCR) to exponentially amplify a specific target sequence and, as consequence, increase its relative concentration, following extraction of total DNA from the sample. This gene amplification significantly increases the probability of detecting the presence of a relatively low number of target microorganisms and to confirm the presence or absence of specific pathogens in natural samples.

Identification systems making use of nucleic acid-based methods involve predominantly molecular hybridization strategies targeting DNA sequences through nucleic acid probings. The probes can be single-stranded (oligonucleotide probes) or denatured double-stranded DNA probes. Hybridizations can occur between the DNA probe and its targeted genomic DNA (DNA-DNA hybridization) or between the DNA probe and rRNA or tRNA sequences (DNA-RNA hybridization); these latter probings make possible assessment regarding both the physiological state and classification of the detected microorganisms. Nucleic acid-based methods are increasingly used in microbial ecology since they are much faster than culture-based ones and are able to detect targeted microorganisms with a high degree of sensitivity and specificity.

Even if one the major drawbacks of these methods, which is the lack of quantification of cells initially present in the sample, was overcome by the rise of quantitative PCR process, the precise numbering of pathogens present in low concentrations remains difficult by this approach (if not impossible). Another limitation of these DNA tests is the lack of indication of the physiological state of the detected cells in the original sample.

15.9.2.1 Polymerase Chain Reaction (PCR) Applications

PCR amplification (*cf.* Sect. 17.7) is frequently used for the detection and identification of pathogenic microorganisms (e.g., *Escherichia coli*, Salmonella, *Pseudomonas aeruginosa*) in food, feces, soils, sediments, and waters. Different DNA regions or genes can be chosen as targets for molecular identification. For detection of specific bacteria, the most frequently used target is the 16S rDNA gene. For pathogenic organisms, genes encoding virulence determinants (e.g., LT, SLT I, and SLT II in pathogenic *E. coli* strains) are frequently used and often allow a very specific and sensitive identification of the bacterial species. By targeting virulence genes for PCR amplification, pathogenic strains can be discriminated from nonpathogenic ones among the analyzed samples.

Despite the fact that PCR is now recognized as a highly specific method, its application to the detection of enteric organisms in natural waters can present some limitations. DNA is stable enough to be amplified many days after its cellular release. Therefore, a classic PCR will not differentiate DNA targets, and these targets could be coming from growing, VNC, or "dead" cells. This drawback can be overcome by targeting messenger or ribosomal RNA (mRNA, rRNA) sequences instead of DNA sequences. Since RNA molecules degrade at a faster rate among damaged or killed bacteria than DNA and can be present at thousands of copies per cell, they can act as indicators of the physiological state of the detected cells. Presently, RT-PCR (reverse transcriptase PCR) allows the detection of enteric RNA viruses in sewage and environmental samples. Despite their potential advantage, mRNA-based approaches have proved difficult to operate in environmental sample monitoring. This is mainly because mRNAs are produced in much less numbers than rRNAs. In addition, mRNA molecules are quite unstable and phylogenetically less informative.

A quantitative PCR method including an inhibition of free DNA from dead cells was developed. This procedure involves the use of ethidium monoazide (EMA) or propidium monoazide (PMA) molecules that bind covalently to free DNA after a photoactivation. These molecules cannot enter viable cells. After EMA treatment of a sample, DNA extraction of the viable cells can be performed. These extracts will contain EMA-treated and EMA-free DNA. Only EMA-free DNA will be amplifiable by PCR. This very promising approach allows the quantification of culturable and VNC cells.

15.10 Conclusions

The study of human pathogens represents a vast field of research in microbial ecology. Changes in environmental conditions at local or regional levels (whether natural or anthropogenic) as well as global changes such as climate change are likely to affect the biogeography of infectious diseases. Research aimed at linking infectious diseases and environmental changes is still underdeveloped, but this is a growing field of investigation to better predict health risks and to anticipate environmental actions. The methods used for the detection of pathogens are also constantly changing and improving in relation to technological advances, particularly in the field of instrumentation and molecular biology. The challenge for the future is to increase the sensitivity of detection methods and the time that is necessary to provide results, which should tend toward real time.

In the major issues underlying the future research, the problem of the emergence or reemergence is important and should take into account the global scale effects of climate and environmental changes on the distribution, dissemination, and survival of pathogens in the environment. These problems also raise the question of increased virulence in connection with the modification of the adaptability of microorganisms to environmental stresses, sometimes stimulating horizontal genes transfers. Finally, the question of the survival and viability of pathogens in environmental reservoirs (niches) remains an important point for potential reemergence of pathogens maintained in dormancy.

References

Aendekerk S, Ghysels B, Cornelis P, Baysse C (2002) Characterization of a new efflux pump, MexGHI-OpmD, from *Pseudomonas aeruginosa* that confers resistance to vanadium. Microbiology 148:2371–2381

Alonso A, Rojo F, Martinez JL (1999) Environmental and clinical isolates of *Pseudomonas aeruginosa* show pathogenic and biodegradative properties irrespective of their origin. Environ Microbiol 1:421–430

Behets J, Declerck P, Delaedt Y, Verelst L, Ollevier F (2007) Survey for the presence of specific free-living amoebae in cooling waters from Belgian power plants. Parasitol Res 100:1249–1256

Berg G, Eberl L, Hartmann A (2005a) The rhizosphere as a reservoir for opportunistic human pathogenic bacteria. Environ Microbiol 7:1673–1685

Berg J, Tom-Petersen A, Nybroe O (2005b) Copper amendment of agricultural soil selects for bacterial antibiotic resistance in the field. Lett Appl Microbiol 40:146–151

Berg J, Thorsen MK, Holm PE, Jensen J, Nybroe O, Brandt KK (2010) Cu exposure under field conditions coselects for antibiotic resistance as determined by a novel cultivation-independent bacterial community tolerance assay. Environ Sci Technol 44:8724–8728

Briand E, Yépremian C, Humbert JF, Quiblier C (2008) Comparative studies on the fitness of microcystin-producing and non-producing *Planktothrix agardhii* strains cultivated under different environmental conditions. Environ Microbiol 10:3337–3348

Briand E, Escoffier N, Straub C, Sabart M, Quiblier C, Humbert JF (2009) Spatiotemporal changes in the genetic diversity of a bloom-forming *Microcystis aeruginosa* (cyanobacteria) population. ISME J 3:419–429

Burrus V, Waldor MK (2004) Shaping bacterial genomes with integrative and conjugative elements. Res Microbiol 155:376–386

Chen J, Novick RP (2009) Phage-mediated intergeneric transfer of toxin genes. Science 323:139–141

Colwell R, Grimes DJ (2000) Nonculturable microorganisms in the environment. ASM Press, Washington, DC

Davies-Colley RJ, Donnison AM, Speed DJ (1997) Sunlight wavelengths inactivating faecal indicator microorganisms in waste stabilization ponds. Water Sci Technol 35:219–225

Davies-Colley RJ, Donnison AM, Speed DJ, Ross CM, Nagels JW (1999) Inactivation of faecal indicator microorganisms in waste stabilisation ponds: interactions of environmental factors with sunlight. Water Res 53:1220–1230

D'Costa VM et al (2011) Antibiotic resistance is ancient. Nature 477:457–461

Delecluse P (2008) The origin of climate changes. Rev Sci Tech Off Int Epiz 27:309–317

Desnues B, Cuny C, Grégori G, Dukan S, Aguilaniu H, Nyström T (2003) Differential oxidative damage and expression of stress defense regulons in culturable and non-culturable *Escherichia coli* cells. EMBO Rep 4(4):400–404

Downing JA, Watson SB, McCauley E (2001) Predicting Cyanobacteria dominance in lakes. Can J Fish Aquat Sci 58:1905–1908

Dukan S, Nyström T (1998) Bacterial senescence: stasis results in increased and differential oxidation of cytoplasmic proteins leading to developmental induction of the heat shock regulon. Genes Dev 12:3431–3441

Edberg SC, Le Clerc H, Robertson J (2000) *Escherichia coli*: the best biological drinking water indicator for public health protection. J Appl Microbiol 88:1068–1168

Fajardo A, Martínez JL (2008) Antibiotic as signals that trigger specific bacterial responses. Curr Opin Microbiol 11:161–167

Fewtrell L, Bartram J (2001) Water quality: guidelines, standards and health. World Health Organization water series. IWA Publishing, London

Fields BS, Benson RF, Besser RE (2002) Legionella and Legionnaires' disease: 25 years of investigation. Clin Microbiol Rev 15:506–526

Frangeul L et al (2008) Highly plastic genome of *Microcystis aeruginosa* PCC7806, a ubiquitous toxic freshwater cyanobacterium. BMC Genomics 9:274

Gagliardi JV, Karns JS (2002) Persistence of *Escherichia coli* O157:H7 in soil and on plant roots. Environ Microbiol 4:89–96

Garcia Armisen T, Servais P (2004) Enumeration of viable *E. coli* in rivers and wastewaters by fluorescent in situ hybridization. J Microbiol Methods 58:269–279

Garcia Armisen T, Servais P (2007) Respective contributions of point and non point sources of *E. coli* and Enterococci in a large urbanised watershed (the Seine river, France). J Environ Manage 82(4):512–518

Garrec N, Picard-Bonnaud F, Pourcher AM (2003) Occurrence of *Listeria* sp. and *L monocytogenes* in sewage sludge used for land application: effect of dewatering, liming and storage in tank on survival of *Listeria* species. FEMS Immunol Med Microbiol 35:275–283

Garrett L (1996) The return of infectious disease. Foreign Aff 75:66–79

George I, Crop P, Servais P (2002) Fecal coliforms removal in wastewater treatment plants studied by plate counts and enzymatic methods. Water Res 36:2607–2617

Guan TY, Holley RA (2003) Pathogen survival in swine manure environments and transmission of human enteric illness–a review. J Environ Qual 32:383–392

Hacker J, Kaper JB (2000) Pathogenicity islands and the evolution of microbes. Annu Rev Microbiol 54:641–679

Heuer H, Smalla K (2007) Manure and sulfadiazine synergistically increased bacterial antibiotic resistance in soil over at least two months. Environ Microbiol 9:657–666

Heuer H, Schmitt H, Smalla K (2011) Antibiotic resistance gene spread due to manure application on agricultural fields. Curr Opin Microbiol 14:236–243

Hijnen WAM, Beerendonck EF, Medema GJ (2006) Inactivation credit of UV radiation for viruses, bacteria and protozoan (oo)cysts in water: a review. Water Res 40:3–22

Huysman F, Verstraete W, Brookes PC (1994) Effect of manuring practices and increased copper concentrations on soil microbial populations. Soil Biol Biochem 26:103–110

Kaestli M et al (2007) Sensitive and specific molecular detection of *Burkholderia pseudomallei*, the causative agent of melioidosis, in the soil of Tropical Northern Australia. Appl Environ Microbiol 73:6891–6897

Karaolis DK, Somara S, Maneval DR Jr, Johnson JA, Kaper JB (1999) A bacteriophage encoding a pathogenicity island, a type-IV pilus and a phage receptor in cholera bacteria. Nature 399:375–379

Karci A, Balcioğlu IA (2009) Investigation of the tetracycline, sulfonamide, and fluoroquinolone antimicrobial compounds in animal manure and agricultural soils in Turkey. Sci Total Environ 407:4652–4664

Kidd SE, Chow Y, Mak S, Bach PJ, Chen H, Hingston AO, Kronstad JW, Bartlett KH (2007) Characterization of environmental sources of the human and animal pathogen, *Cryptococcus gattii*, in British Columbia, Canada, and Pacific Northwest USA. Appl Environ Microbiol 73:1433–1443

Knapp S, Hacker J, Jarchau T, Goebel W (1986) Large, unstable inserts in the chromosome affect virulence properties of uropathogenic *Escherichia coli* O6 strain 536. J Bacteriol 168:22–30

Knapp CW, Zhang W, Sturm BS, Graham DW (2010) Differential fate of erythromycin and beta-lactam resistance genes from swine lagoon waste under different aquatic conditions. Environ Pollut 158:1506–1512

Koelle K, Pascual M, Yunus M (2005) Pathogen adaptation to seasonal forcing and climate change. Proc R Soc B 272:971–977

Kogure K, Simidu U, Taga N (1979) A tentative direct microscopic method for counting living marine bacteria. Can J Microbiol 25:415–420

Kong KF, Schneper L, Mathee K (2010) Beta-lactam antibiotics: from antibiosis to resistance and bacteriology. APMIS 118:1–36

Kulasekara BR, Kulasekara HD, Wolfgang MC, Stevens L, Frank DW, Lory S (2006) Acquisition and evolution of the exoU locus in Pseudomonas aeruginosa. J Bacteriol 188:4037–4050

Kummerer J (2003) The significance of antibiotics in the environment. J Antimicrob Chemother 1:5–7

Kuske CR, Barns SM, Grow CC, Merrill L, Dunbar J (2006) Environmental survey for four pathogenic bacteria and closely related species using phylogenetic and functional genes. J Forensic Sci 51:548–558

Lavenir R, Jocktane D, Laurent F, Nazaret S, Cournoyer B (2007) Improved reliability of *Pseudomonas aeruginosa* PCR detection by the use of the species-specific *ecfX* gene target. J Microbiol Methods 70:20–29

Lebaron P, Henry A, Lepeuple A-S, Pena G, Servais P (2005) An operational method for real-time monitoring of *E. coli* in bathing waters. Mar Polluut Bull 50:652–659

Levy SB, Marshall B (2004) Antibacterial resistance worldwide: causes, challenges and responses. Nat Med 10:S122–S129

Lin H, Xu B, Chen Y, Wang W (2009) Legionella pollution in cooling tower water of air-conditioning systems in Shanghai, China. J Appl Microbiol 106:606–612

Lindgren E, Gustafson R (2001) Tick-borne encephalitis in Sweden and climate change. Lancet 358:16–18

Lindsay JA, Ruzin A, Ross HF, Kurepina N, Novick RP (1998) The gene for toxic shock toxin is carried by a family of mobile pathogenicity islands in *Staphylococcus aureus*. Mol Microbiol 29:527–543

Marsalek J, Rochfort Q (2004) Urban wet-weather flows: sources of fecal contamination impacting on recreational waters and threatening drinking-water sources. J Toxicol Environ Health Part A Curr Issues 67(20–22):1765–1777

Martinez JL, Fajardo A, Garmendia L, Hernandez A, Linares JF, Martínez-Solano L, Sánchez MB (2009) A global view of antibiotic resistance. FEMS Microbiol Rev 33:44–65

Mesaros N et al (2007) *Pseudomonas aeruginosa*: resistance and therapeutic options at the turn of the new millennium. Clin Microbiol Infect 13:560–578

Mukherjee A, Cho S, Scheftel J, Jawahir S, Smith K, Diez-Gonzalez F (2006) Soil survival of *Escherichia coli* O157: H7 acquired by a child from garden soil recently fertilized with cattle manure. J Appl Microbiol 101:429–436

Norman RS, Moeller P, Mc Donald TJ, Morris PJ (2004) Effect of pyocyanin on a crude-oil-degrading microbial community. Appl Environ Microbiol 70:4004–4011

Ogden LD, Fenlon DR, Vinten AJ, Lewis D (2001) The fate of *Escherichia coli* O157 in soil and its potential to contaminate drinking water. Int J Food Microbiol 66:111–117

Oliver JD (2010) Recent findings on the viable but nonculturable sate in pathogenic bacteria. FEMS Microbiol Rev 34:415–425

Ouattara KN, Passerat J, Servais P (2011) Faecal contamination of water and sediment in the rivers of the Scheldt drainage network. Environ Monit Assess 183:243–257

Paerl HW, Huisman J (2009) Climate change: a catalyst for global expansion of harmful cyanobacterial blooms. Environ Microbiol 1:27–37

Palasatien S, Lertsirivorakul R, Royros P, Wongratanacheevin S, Sermswan RW (2008) Soil physicochemical properties related to the presence of *Burkholderia pseudomallei*. Trans R Soc Trop Med Hyg 1:S5–S9

Pelandakis M, Pernin P (2002) Use of multiplex PCR and PCR restriction enzyme analysis for detection and exploration of the variability in the free-living amoeba *Naegleria* in the environment. Appl Environ Microbiol 68:2061–2065

Perron K, Caille O, Rossier C, Van Delden C, Dumas JL, Köhler T (2004) CzcR-CzcS, a two-component system involved in heavy metal and carbapenem resistance in *Pseudomonas aeruginosa*. J Biol Chem 279:8761–8768

Philippot L (2005) Denitrification in pathogenic bacteria: for better or worst ? Trends Microbiol 13:191–192

Qiu X, Gurkar AU, Lory S (2006) Interstrain transfer of the large pathogenicity island (PAPI-1) of *Pseudomonas aeruginosa*. Proc Natl Acad Sci U S A 103:19830–19835

Ready PD (2008) Leishmaniasis emergence and climate change. Rev Sci Tech Off Int Epiz 27:399–412

Ready D, Pratten J, Mordan N, Watts E, Wilson M (2007) The effect of amalgam exposure on mercury- and antibiotic-resistant bacteria. Int J Antimicrob Agents 30:34–39

Rose JB, Farrah SR, Harwood VJ, Levine AD, Lukasik J, Menendez P, Scott T (2004) Reduction of pathogens, indicators bacteria and alternative indicators by wastewater treatment and reclamation processes. WERF final report. IWA Publishing, London

Salazar-Lindo E, Seas C, Gutierrez D (2008) ENSO and cholera in South America: what can we learn about it from the 1991 cholera outbreak? Int J Environ Health 2:30–36

Savichtcheva O, Okabe S (2006) Alternative indicators of fecal pollution: relations with pathogens and conventional indicators, current methodologies for direct pathogen monitoring and future application perspectives. Water Res 40:2463–2476

Schwaiger K, Harms K, Hölzel C, Meyer K, Karl M, Bauer J (2009) Tetracycline in liquid manure selects for co-occurrence of the resistance genes tet(M) and tet(L) in *Enterococcus faecalis*. Vet Microbiol 139:386–392

Servais P, Garcia Armisen T, Lepeuple AS, Lebaron P (2005) An early warning method to detect fecal contamination of river waters. Ann Microbiol 55:67–72

Servais P, Garcia-Armisen T, George I, Billen G (2007) Fecal bacteria in the rivers of the Seine drainage network: source, fate and modeling. Sci Tot Environ 375:152–167

Simango C (2006) Prevalence of *Clostridium difficile* in the environment in a rural community in Zimbabwe. Trans R Soc Trop Med Hyg 100:1146–1150

Solomon EB, Yaron S, Matthews KR (2002) Transmission of *Escherichia coli* O157:H7 from contaminated manure and irrigation water to lettuce plant tissue and its subsequent internalization. Appl Environ Microbiol 68:397–400

Stepanauskas R, Glenn TC, Jagoe CH, Tuckfield RC, Lindell AH, King CJ, McArthur JV (2006) Coselection for microbial resistance to metals and antibiotics in freshwater microcosms. Environ Microbiol 8:1510–1514

Taubes G (2008) The bacteria fight back. Science 321:356–361

Tello A, Austin B, Telfer TC (2012) Selective pressure of antibiotic pollution on bacteria of importance to public health. Environ Health Perspect 120:1100–1106

Thompson JR, Marcelino LA, Polz MF (2010) Diversity, sources, and detection of human bacterial pathogens in the marine environment. In: Belkin S, Colwell RR (eds) Oceans and health, pathogens in the marine environment. Springer, New York, pp 29–68

Vally H, Whittle A, Cameron S, Dowse GK, Watson T (2004) Outbreak of *Aeromonas hydrophila* wound infections associated with mud football. Clin Infect Dis 38:1084–1089

Vedler E, Vahter M, Heinaru A (2004) The completely sequenced plasmid pEST4011 contains a novel IncP1 backbone and a catabolic transposon harboring tfd genes for 2,4-dichlorophenoxyacetic acid degradation. J Bacteriol 186:7161–7174

Von der Weid I, Marques JM, Cunha CD, Lippi RK, Dos Santos SC, Rosado AS, Lins U, Seldin L (2007) Identification and biodegradation potential of a novel strain of Dietzia cinnamea isolated from a petroleum-contaminated tropical soil. Syst Appl Microbiol 4:331–339

Wery N, Bru-Adan V, Minervini C, Delgénes JP, Garrelly L, Godon JJ (2008) Dynamics of *Legionella* spp. and bacterial populations during the proliferation of *L. pneumophila* in a cooling tower facility. Appl Environ Microbiol 74:3030–3037

Wright GD (2007) The antibiotic resistome: the nexus of chemical and genetic diversity. Nat Rev Microbiol 5:175–186

Xu HS, Roberts NC, Adams LB, West PA, Siebeling RJ, Huq A, Huq MI, Rahman R, Colwell RR (1982) An indirect fluorescent antibody staining procedure for detection of *Vibrio cholerae* serovar 01 cells in aquatic environmental samples. J Microbiol Methods 2:221–231

You Y, Rankin SC, Aceto HW, Benson CE, Toth JD, Dou Z (2006) Survival of *Salmonella enterica* serovar Newport in manure and manure-amended soils. Appl Environ Microbiol 72:5777–5783

Applied Microbial Ecology and Bioremediation

Microorganisms as Major Actors of Pollution Elimination in the Environment

Jean-Claude Bertrand, Pierre Doumenq, Rémy Guyoneaud, Benoit Marrot, Fabrice Martin-Laurent, Robert Matheron, Philippe Moulin, and Guy Soulas

Abstract

The large diversity of metabolic capacities and the high genetic plasticity of microorganisms allow them to degrade virtually all organic compounds of natural or anthropogenic (xenobiotics) origin including those that are sources of environmental pollution. Thus microorganisms are major actors to eliminate or alleviate pollutions in the environment. The natural attenuation processes due to microbial activities (biodegradation and/or biotransformation) as well as the possibilities of using microorganisms in preventive treatments and bioremediation – biostimulation, bioaugmentation, rhizostimulation, bioleaching, and bioimmobilization – are presented. The main methods for microbial treatment of pollution, the chemical structure and the origin of the major pollutants, as well as the mechanisms of degradation by microorganisms – on the basis of physiological, biochemical, and genetic approaches – are described. Examples of treatments are presented for urban wastewater (activated sludge, lagoons, and planted beds), solid wastes (aerobic treatment or composting, anaerobic treatment and methanization, discharges), gaseous effluents, pesticides, polychlorobiphenyls, and finally hydrocarbons and petroleum products in the marine environment.

Keywords

Bioaugmentation • Biodegradation • Bioimmobilization • Bioleaching • Bioremediation • Biostimulation • Hydrocarbons • Pesticides • Petroleum • Pollutants • Polychlorobiphenyls • Rhizostimulation • Solid wastes treatments • Wastewater treatments • Xenobiotics

---

* Chapter Coordinator

J.-C. Bertrand* (✉)
Institut Méditerranéen d'Océanologie (MIO), UM 110, CNRS 7294 IRD 235, Université de Toulon, Aix-Marseille Université, Campus de Luminy, 13288 Marseille Cedex 9, France
e-mail: jean-claude.bertrand@univ-amu.fr

P. Doumenq
Laboratoire de Chimie de l'Environnement, Aix-Marseille Université, FRE CNRS LCE 3416, Batiment Villemin, Europôle Environnement, 13545 Aix-en-Provence Cedex, France

R. Guyoneaud
Institut des Sciences Analytiques et de Physico-chimie pour l'Environnement et les Matériaux (IPREM), UMR CNRS 5254, Université de Pau et des Pays de l'Adour, B.P. 1155, 64013 Pau Cedex, France

B. Marrot • P. Moulin
Laboratoire de Mécanique, Modélisation et Procédés Propres, Aix-Marseille Université, M2P2-UMR 6181, Europole de l'Arbois, BP 80, 13545 Aix en Provence Cedex, France

F. Martin-Laurent
Laboratoire Microbiologie du Sol et de l'Environnement, INRA, UMR 1347 Agroécologie, 17 rue Sully, BP 86510, 21065 Dijon Cedex, France

R. Matheron
Institut Méditerranéen de Biodiversité et d'Ecologie marine et continentale (IMBE), UMR-CNRS-IRD 7263, Aix-Marseille Université, 13397 Marseille Cedex 20, France

G. Soulas
Œnologie INRA, 33405 Talence Cedex, France

16.1 Introduction

Industrial development and increasing population are the cause of environmental pollution that occurs globally and affects all terrestrial and aquatic ecosystems. The deleterious effects of pollution on human health, the functioning of biogeochemical cycles, erosion of biodiversity and more broadly on the entire biosphere, have become a major concern for modern societies. Awareness of the seriousness of this situation has led to the implementation of actions against environmental pollution at local, regional, national and international levels.

Microorganisms are key players in the depollution of the environment. Indeed, because of the huge diversity of their metabolic capabilities (*cf.* Chap. 3), expressed whatever environmental conditions are, including the most extreme conditions (*cf.* Chap. 10), microorganisms can process almost all natural pollutants when sufficient time is allowed. They are also able, in most cases, to deal with artificial substances, **xenobiotics*** that never existed in nature. The mechanisms involved are numerous and diverse: **biodegradation***, **biotransformation***, **cometabolism***, **syntrophy***, **bioaccumulation***, bioimmobilization, and bioleaching. The adaptive capacities of microbial communities to pollutants are explained by their ability to express enzymatic tools via genetic changes such as mutations, endogenous rearrangements or acquisition of exogenous DNA fragments, etc. These changes include the horizontal gene transfers (HGTs) which play an important role in enabling the spread of catabolic genes often located on conjugative and integrative plasmids, integrative transposons, or conjugative integrative elements. These elements can exist in the form of plasmids (linear or circular) or are integrated at dedicated chromosomal sites of the "host organism."

Thus, research in microbial ecology has important practical implications for environmental depollution. In elucidating the mechanisms by which microorganisms attack pollutants, this research use the decontaminating power of microorganisms in the development of many biological processes that lead to the elimination or mitigation of pollution.

Treatments by microbes depend on the pollutant and the biotope (water, soil, sediment, air) and can be broadly grouped into two domains of application. The first domain is the preventive treatment of pollution for which the organic (domestic or industrial **sewage***, wastes, etc.) or mineral pollutant is either totally or partially eliminated or transformed into a nontoxic product (compost or biogas) having a market value or is reduced to an acceptable level before discharge into the environment. The second domain of application is the use of microorganisms to rehabilitate a site contaminated with one or more pollutants. During the implementation of this type of biotechnology known as bioremediation, the pollutant of organic or inorganic origin can be treated in situ (without excavation) or ex situ (after excavation) (Iwamoto and Nasu 2001; Pandey et al. 2009). In the latter case, the treatment of the pollutant can be done on site or off-site after transport to a depollution center.

In addition to these two applications of microbial activities, microorganisms can be used in the preservation of food, management of agricultural soil, and the fight against biodeterioration (metals, paints, concrete, etc.) or in the control and prevention of pollution through qualitative and quantitative highlighting of microbial pathogens in the environment (*cf.* Sect. 15.6).

In all cases, the development of a microbial-based depollution technique cannot be considered without thorough knowledge of:

(i) The chemical structure of the pollution source and the products formed during its degradation, which most often require the implementation of sophisticated analytical techniques.
(ii) Its origin and its concentration in the environment; waste.
(iii) Its behavior, i.e., of its biogeochemical cycle.
(iv) The degradation mechanisms of pollutant by microorganisms which can only be elucidated by experiments conducted in the laboratory that should simulate, as realistically as possible, the environmental conditions in which the microorganisms will be subjected to in the environment; this step involves studies on the physiology, biochemistry, and genetics of the involved microorganisms.
(v) The effects of the implemented treatment on the flora, fauna, and humans.

In this chapter, the following are described:

(i) The chemical structure and the origin of the major pollutants
(ii) The natural biogeochemical cycles (i.e., without human action) of the major pollutants, emphasizing the role of microorganisms in their natural elimination, which include all physicochemical and biological processes to which pollutants are subjected in the natural environment
(iii) The major microbiological treatments that may be applied both for prevention and rehabilitation of a polluted environment
(iv) Examples of the treatment of major sources of pollution using microbial-based techniques that enhance **natural attenuation***.

It is impossible to present the natural attenuation and the treatment of all types of pollution; the chosen examples are representative of major pollutions affecting the environment, and for a given type of pollution, only few examples of treatments described in the scientific literature will be presented.

16.2 Preventive and Bioremediation Treatments

The aim is to exploit the degradation capabilities of microorganisms by implementing ways that will increase and accelerate the natural process of biodegradation. For example, wastewater treatment plants reproduce, under controlled conditions, processes that occur in the environment. The composting process is based on the acceleration of the humification of organic residues into natural **humic substances*** as it occurs in soils.

The acceleration and amplification of the biodegradation process are obtained by inputs of nutrients, electron acceptors (oxygen, nitrate, sulfate, etc.), and/or microorganisms (pure strains, **consortium***, **communities, genetically modified organisms, GMO***) whose metabolic capabilities are specifically adapted to the biodegradation of the treated pollutant.

16.2.1 Microorganisms as Bioremediation Agents

Overall, there are five possibilities of using microorganisms in preventive treatments and bioremediation: biostimulation, bioaugmentation, rhizostimulation, bioleaching, and bioimmobilization.

16.2.1.1 Biostimulation
Biostimulation is based on the stimulation of the activity of natural communities on the polluted site by adding the following:
  (i) Nutrients (e.g., nitrogen, phosphorus) (Prince 1993)
  (ii) Electron acceptors (oxygen for aerobic biodegradation, nitrate or sulfate for anaerobic biodegradation) (Cunningham et al. 2001)
  (iii) **Surfactants*** of chemical or biological origin which will increase the bioavailability of the more hydrophobic pollutants (hydrocarbons, PCBs, etc.) (Iwamoto and Nasu 2001; Ron and Rosenberg 2002; Lawniczak et al. 2013).

Ideally, the goal is to suppress the action of any limiting factor for biodegradation, allowing a maximal activity of microorganisms.

16.2.1.2 Bioaugmentation
Bioaugmentation is based on the input of microorganisms capable of degrading the pollutant in the polluted area (Timmis and Pieper 1999; Sayler and Ripp 2000; Gentry et al. 2004; El Fantroussi and Agathos 2005; Thompson et al. 2005). The strains used are indigenous strains which are isolated from the polluted site and selected in laboratory conditions according to the pollutant; after growth (usually in a bioreactor), microorganisms are then reinjected on site in the polluted area.

It is also possible to make use of genetically modified microorganisms (GMMs). A.M. Chakrabarty was the first to file a patent in 1971 for a GMM capable of degrading alkanes and aromatic hydrocarbons after transfer of plasmids of three different strains in a strain of *Pseudomonas*. Potentially, GMMs can optimize the biodegradation of a pollutant: by increasing degradation rates, by creating new biodegradation pathways, and by broadening the range of biodegraded substrates. However, the introduction of GMMs in a natural environment is subject to stringent laws, which severely limits their use. In general, the greatest difficulty in using bioaugmentation techniques is the competition between introduced and indigenous microorganisms.

16.2.1.3 Rhizostimulation
Many plants grow in the presence of a pollutant that can extract, accumulate, transform, or eliminate (Pilon-Smits 2005). In the case of rhizodegradation, the pollutant is degraded by enzymes released by the plant roots or by phytostimulation of microorganisms associated with the rhizosphere (*cf.* Sect. 11.3.6) (Gerhardt et al. 2009). Hydrocarbons (e.g., PAHs) and PCBs are organic compounds that are degraded in the rhizosphere. This treatment, which lasts for several years, is a low-cost treatment because the energy source is provided by the sun that ensures photosynthetic processes. It also offers a landscape which is esthetically acceptable.

16.2.1.4 Bioleaching
Bioleaching corresponds to solubilization by microorganisms and to leaching by entrainment in the aqueous phase of pollutants (ores, heavy metals, phosphorus, etc.) fixed or trapped in a solid matrix (*cf.* Sect. 16.9.2).

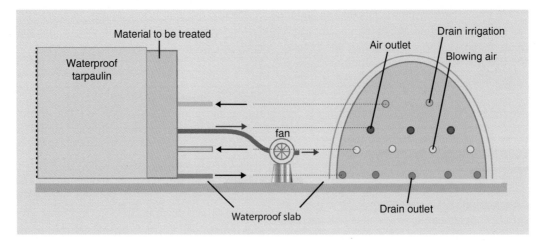

Fig. 16.1 Schematic drawing of artificially ventilated biopile (air insufflation). The extraction of the air (*red arrows*) is ensured by a fan; before release into the atmosphere, the air can be treated (treatment with biofilter). Irrigation is carried out by a network of drains. Water is collected by a network of drainage instead of on a waterproof slab. The biopile is often covered by a plastic membrane or by a layer of old compost and "mature" (see Sect. 16.5.3) (Modified and redesigned from Bocard 2007). Drawing: M.-J. Bodiou

16.2.1.5 Bioimmobilization

Bioimmobilization is the ability of certain microorganisms (prokaryotes and eukaryotes, primarily fungi) to immobilize some inorganic (and radioactive mining waste) or organic pollutants (hydrocarbons in the matrix of a microbial mat) by precipitating or by trapping them in a mineral or organic matrix.

16.2.2 The Main Methods for Microbial Treatment of Pollution

16.2.2.1 Bioreactors

Biodegradation of a pollutant can be performed in a bioreactor operated in batch or continuous mode (*cf.* Sects. 17.8.3 and 19.3.4), the microorganisms being free or immobilized on a support. This technique allows a good control of the treatment (control of oxygen concentration, temperature, nutrient intake, pH, etc.) and a maximal accessibility to pollutants; it is particularly well suited for use in bioaugmentation techniques. The design and dimensions of the reactor are functions of the type of culture (aerobic or anaerobic), the sample volume to be treated, and the nature of the pollutant. For decontamination of solid samples (soil, sediment), these may be in the form of more or less liquid sludge held in suspension in an aqueous phase ("bioslurry"), with a percentage of the solid fraction generally between 10 and 50 % (weight). It is possible to include the biological processes in wastewater treatment plants, including membrane bioreactors, in this treatment category.

16.2.2.2 Treatment in Ponds and Wastewater Plants

For example, aqueous **effluents*** produced by an oil refinery are initially treated in an aerated pond where biodegradation takes place; then, they pass through a settling basin and the sludge formed is treated in another basin where the biodegradation of residual contaminants continues. If the wastewaters obtained meet the standard requirements, they are rejected; otherwise, they undergo a new treatment cycle.

Another example is the biological ponds of wastewater plants that allow the reduction of the concentration of organic matter, but also a decrease in the concentrations of nitrates and phosphates (*cf.* Sect. 16.4.3); indeed, treatment in ponds can also use techniques to control the most important parameters such as oxygen or organic matter load (*cf.* Sect. 16.4.3).

16.2.2.3 Biopiles, Windrows, and Landfarming

In a biopile, the sample, after excavation, is scraped up in a waterproof slab (Fig. 16.1). A biopile can be 3–4 m height with a volume of several hundred m^3 (Vandecasteele 2008) and is equipped to enable the injection of air, water, nutrients (bioaugmentation), and possibly microorganisms (biostimulation).

In the case of windrow treatment, the biopile height does not exceed 1–1.5 m; the windrow is moved or turned on regularly. Aeration is provided either by natural or passive movement of air or through pipes buried in the windrow with or without a ventilation system. The air is supplied by a fan that can work an even larger heap. If a ventilation system is set up, the pile is not moved after the start of the operation.

Deposition over large areas ("landfarming") consists of spreading out the material (e.g., soil polluted by oil products) with a thickness of a few ten centimeters. An alternative is to use wetlands.

These techniques are easy to implement, do not require heavy infrastructure, and are of low cost. However, operations lasted a long time (year or decade) and are likely to produce odors and visual nuisances. In addition, to avoid any risk of infiltration, a waterproof cover must be in place before the start of the treatment.

16.2.2.4 Bioventing and Air Sparging

The method consists of injecting flowing air through the contaminated area. This artificial aeration is essential in the case of a bioremediation by composting (*cf.* Sect. 16.5.3). Applied to the in situ treatment of soil and groundwater, air injection, whose the flow rate must be sufficient to enable effective biodegradation, can be made in the following way:

(i) Through strainer wells penetrating unsaturated soil (Fig. 16.2). The air flows into the polluted area and the resulting volatile organic compounds are recovered by extraction wells placed at the periphery of the treatment area. The treatment of gaseous effluents is carried out in the surface by biofilters fixed on the outputs of the wells (*cf.* Sect. 16.6). This method of ventilation (venting) also brings oxygen to the microorganisms in the soil, stimulating the biodegradation of pollutants; the term of bioventing is used to describe this type of treatment. In fact, venting and bioventing are simultaneous. The concomitant injection of (bio)surfactants can also be implemented to promote the release of hydrophobic compounds strongly bound to the soil matrix (e.g., non-volatile hydrocarbons) and the accessibility of contaminants to microorganisms. The effectiveness of bioventing depends on the permeability of the soil.

(ii) By the establishment of wells penetrating the polluted groundwater. In this case, the air injection takes place under the contaminated area, and this sparging process called biosparging ("air sparging" Fig. 16.2) promotes the release of the most volatile pollutants present in the saturated zone, but also those in the unsaturated soil located above. The dissolution of oxygen in the aqueous phase increases the degradation of pollutants by microorganisms; this degradation continues in the unsaturated soil. In such a process, compounds such as benzene, toluene, ethylbenzene, and xylene (called BTEX mixture) are removed by being driven into the soil where biodegradation continues.

"Bioventing" and "biosparging" treatments are in situ treatments, without excavation, and may take several months or even several years.

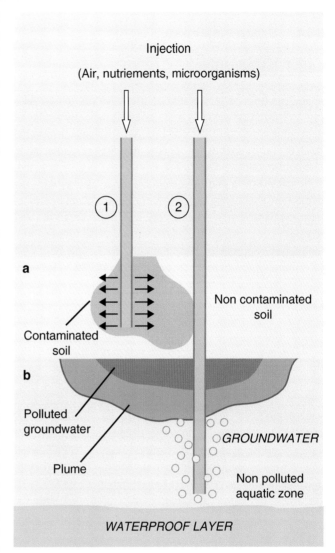

Fig. 16.2 Schematic drawing of a processing device by bioventilation (*1*) and sparging (*2*). Drawing: M.-J. Bodiou

16.3 Main Pollution Types

European legislation defines pollution as "the introduction of substances or energy into the environment, resulting in deleterious effects of such a nature as to endanger human health, harm living resources and ecosystems, and impair or interfere with amenities and other legitimate uses of the environment".

Pollution types can be characterized according to their origin: physical pollution (radioactivity, heat, noise, etc.), chemical pollution (detergents, pesticides, hydrocarbons, endocrine disruptors, trace metals, etc.), biological pollution (microbial contamination of water, soil air, toxins, etc.), and aesthetic nuisances (landscape degradation etc.). Pollution is also characterized by their "target": air pollution, water, soil, visual pollution, light pollution, and genetic pollution.

16.4 Urban Wastewater

16.4.1 Introduction

Water can be used very often because it gathers an outstanding collection of physical and chemical properties. Out of any domestic activity, it can become a solvent, a thermal fluid, or simply an easy- and handle-liquid. Sewage discharge has considerably changed in quantity and quality together with urbanization and industrialization, as well as with the evolution of consumption patterns. Environmental protection and particularly water supply have become a vital and strategic issue, which has passed from a quantitative to a qualitative approach. Requirements for implementation have been added to reach this challenge. Urban wastewater (UWW) was discharged in fresh and oceanic waters without prior treatment. The self-cleaning power of the natural environment – particularly the activity of microorganisms – was in most cases enough to eliminate the pollutants that they contained. The significant increase in population involved the discharge of ever larger important quantities of pollutants in the receiving environments. The self-cleaning power has rapidly proved inefficient, and the collection and treatment of wastewater have become necessary to cope with the increasing discharges. Simple domestic discharge includes more complex substances (pesticides pharmaceuticals, PCB, etc.) and sewerage collected industrial, commercial, and craft discharges with very different characteristics. As the need for water increases, it is important to consider both the quantity available and also its quality, which is often ignored. It is therefore vital to treat wastewater. Several treatment processes already exist, but the **activated sludge*** biological treatment is the most widely used; lagoon treatment is another process to treat wastewater (Box 16.1). Before specifying the treatment to implement, it is first of all necessary to know the quantity and the quality of pollution in the water as well as the characteristics of the site where the wastewater treatment plant will operate. For example, lagoon treatment requires very large areas.

16.4.2 Origin and Content of Wastewater

Usually there are two main categories of wastewater (also called effluents):
(i) Urban wastewater (or domestic) resulting from household or commercial activities together with rainwater (stormwater). It contains grease, soap, and detergents; suspended and organic or mineral dissolved solids; and a very large diversity of microorganisms. Household wastewater consists of **grey waters*** and waters from toilets, baths, and sinks.

Box 16.1: Extensive Water Treatments in Lagoons and Planted Beds

Pierre Caumette

Wastewater treatment has to be objective to reach water quality criteria in terms of suspended solids, organic matter, and nutrients before its release into the environment. Many systems use biological activities, which, in wastewater treatment plants, are optimized to achieve maximum efficiency. These biological activities may also be used in extensive treatment processes such as lagoon treatment plants or macrophytic treatment plants. These treatments are less efficient in terms of performance (degradation rate, residence time) but not in the quality of the treated water. In both cases, the role of microorganisms is essential (Box Figs. 16.1 and 16.2).

The lagoon treatment process is a natural alternative to other treatment systems for the purification of urban sewage and industrial effluents (e.g., food, paper). The establishment of efficient lagoon treatment systems dates from the 1970s, and among the most successful, those from the Ecosite of Meze (Box Fig. 16.1) and the station of Rochefort (both in France) are noteworthy. In lagoon treatment systems, raw sewage will stay in several basins (usually three), shallow (maximum 1.5 m) and of great area. The biological activities of aerobic and anaerobic degradation, as well as primary production and grazing by zooplankton, allow purification in the successive lagoons. The residence time in this case is very long (50–80 days), which makes these systems having large buffer capacities with respect to changes in load and effluent flow.

As in the large majority of treatment systems, the first steps include a screening for the removal of large size solids, a grit chamber to eliminate sand and other solid deposits, and oil/water separation that removes fat that may disrupt the treatment system. The pre-treated effluent is then discharged into the first and largest basin, which provides aerobic (in the water) and anaerobic (sludge and sediment) bacterial degradation of most of the organic load. The main anaerobic metabolisms involved are fermentations, nitrates respiration (including denitrification), and

---

P. Caumette
Institut des Sciences Analytiques et de Physico-chimie pour l'Environnement et les Matériaux (IPREM), UMR CNRS 5254, Université de Pau et des Pays de l'Adour, B.P. 1155, 64013 Pau Cedex, France

(continued)

Box 16.1 (continued)

Box Fig. 16.1 Aerial view of the lagoon treatment plant of Meze (Hérault, France) showing successive ponds used for wastewater purification. Pools *1*, *2*, and *3* correspond to those set up in 1980 during the installation of the station and using a classical lagoon treatment process (see text). Pools *4* (two) correspond to tertiary treatments installed in 1996. The pools *5*, installed in 1997, permit methanogenesis. If the lagoon treatment contains brackish water (coastal environments), sulfate reducing activity can cause problems related to the production of smelly and toxic sulfide. These mineralization activities produce large quantities of mineral salts (NH_4^+, NO_2^-, PO_4^{3-}) and gas (N_2, CH_4, CO_2, H_2S). The second, shallower pond, allows primary production by microalgae that will use the nutrients originating from organic matter degradation in the first pond. Due to the presence of oxygen arising by oxygenic photosynthesis, degradation proceeds through aerobic metabolism. The third pond allows the development of primarily herbivorous zooplankton (protozoans and metazoans), which will reduce the microalgae load.

There are variations to this lagoon treatment process:

(i) Aerated lagoons replace oxygen production by microalgae by forced ventilation. Only two ponds are then required (degradation and sedimentation), thus reducing the surface. This system generates costs due to forced aeration.

pretreatment (flocculation of raw material) under anaerobic conditions (deep ponds) with, in addition, aeration of the surface waters to reduce odors. Finally, in 2002, four high load material ponds (pools *6*) have been installed to allow a more efficient aerobic treatment (Photograph courtesy from CCNPT, Lagunage de Mèze)

(ii) Anaerobic lagoons require deeper ponds (more than 4 m) and high organic load for the establishment of anoxic conditions. For proper operation, the temperature should be high (at least 25 °C).

Epuration yields through lagoon treatment are good, and most important, it allows excellent treatment of pathogenic microorganisms due to residence times that are long enough to permit competition between microorganisms (indigenous microorganisms out-competing nonnative microorganisms). Pathogen elimination is also due to the action of light radiation (UV). This explains why lagoon treatment systems are sometimes used as tertiary treatment after activated sludge or extended aeration. This process has very low operational costs and excellent integration into the landscape. The investment depends on the size of the basins, which, in turn depend on the volumes of effluent to be treated. Thus, for big cities, the area required (10–20 m^2 per capita equivalent) is often a limiting factor.

In lagoon treatment processes, primary production is driven by microalgae. It can also be driven by

(continued)

Box 16.1 (continued)

Box Fig. 16.2 Photograph of a macrophytic plant (reeds) used for the treatment of wastewaters of a small village (Photograph: P. Caumette)

macrophytes, but the use of the latter takes place in the context of planted beds. These processes use macrophyte plants (reeds, iris) and rhizosphere microorganisms for treatment (Box Fig. 16.2). The roots of plants provide oxygen and simple organic molecules that promote bacterial aerobic growth. Depending on the distance to the roots of the plants, degradation will be aerobic or anaerobic. Planted beds are often coupled with sand filters (vertical or horizontal) for individual effluent treatments. The planted beds, in addition to good overall degradation, accumulate large quantities of metals and thus ensure their effective decontamination. Obviously, the biomass has no value and in this case should be incinerated. The metals contained in the ash can be recovered by bioleaching (*cf.* Sect. 16.9.2).

(ii) Industrial wastewater discharged by factories. Unlike domestic wastewater, they are characterized by their wide diversity according to the use of water during the industrial process. Pollution varies according to the industrial sector even for each company. On the whole, there are four categories:
- *Waters from cooling systems*, which are important and generally not very polluted because they are not in contact with the manufactured products. They can therefore be recycled.
- *Wash waters*, which are full of raw materials (hydrocarbons and machine oils, detergents, bactericidal compounds, etc.). The production and degree pollution may therefore vary.
- *Process waters*, whose chemical composition and discharged quantity vary significantly from an industry to an other. The polluting discharges come from the contact of water with solids, liquids, or gases.
- *Service wastewater*, which corresponds to the domestic waters of the factory combined with the kitchen discharges.

The diversity of pollutants in industrial discharges makes difficult and even impossible the installation of a treatment of these waters that could be extended. The treatment is adapted to the nature of the pollution. Therefore, in this chapter, the study will be limited to the operation of domestic wastewater treatment plants. It has to be pointed out that water resulting from treatment plants is not drinkable. There are two types of works: the first one makes water drinkable before distributing it and the second one cleans up wastewater before discharging it in natural environment (the price of water includes both treatments). For example, in 2008, wastewater of 95 % of the French population was cleaned with a wastewater treatment plant (WWTP; more than 17,000 WWTP exist in France) or with **an autonomous sewerage system***.

UWW is characterized by different parameters, the most important of which are suspended solids (VSS) and the

weight in oxidizable matters, mainly organic. Oxidizable matters in waters are characterised by two quantitative indices:
- Chemical Oxygen Demand (COD), which corresponds to the amount of oxygen necessary to the chemical oxidation of the totality of the reduced organic and mineral compounds
- Biochemical Oxygen Demand (BOD_5), which corresponds to the amount of oxygen consumed by organisms in 5 days to oxidize the reduced organic and mineral compounds

Other parameters are taken into account: nitrogenous matters expressed in total nitrogen (TN) corresponding to the sum of organic and ammonium nitrogen ($N-NH_4^+$) and phosphorus matters expressed in total phosphorus (TP) and/or in phosphorus ($P-PO_4^{3-}$) (Table 16.1). It is noteworthy that 95 % of urban wastewater consist of biodegradable matter. The remaining part discharged in environment is then submitted to the action of natural biodegradation processes.

16.4.3 Wastewater Treatment

Because of the discharge diversity, treatments (and pretreatments) are carried out through physicochemical and biological processes. A microbial ecologist who deals with wastewater cannot ignore the physicochemical treatments to which the wastewater is submitted before the microbiological treatment. The different treatment processes depend on the origin of the pollution, on its chemical composition (organic or mineral), and on its soluble colloidal or particulate form (Fig. 16.3). Various processes have been implemented and take into account the UWW volume to treat, the economical constraints linked to the functioning of the WWTP, and the quality of water required which depends on the discharge area. There is an important literature about wastewater treatment. Only a few articles have been quoted in this chapter (Henze et al. 2002; Tchobanoglous et al. 2003; Vesilind 2003; van Haandel and van der Lubbe 2007; Degrémont 2008).

16.4.3.1 Pretreatments

Pretreatments enable the treatment plant to operate properly, but they generate some operational constraints linked to the refusal recovery, sanitation, and maintenance. The WWTP are usually equipped with the following preliminary treatments:
- Screening, Sieving: separation by size
- Grit removal, Degreasing, Settling: separation by density or density ratio to water

These pretreatments enable the elimination of the coarse matters, which are liable to damage mechanical parts and/or hamper the efficiency of the subsequent steps such as settling or aeration of the biological tank.

From Coarse to Fine Screening

The purpose of screening when water passes between the grids or through the holes is to keep any matter bigger than the cross section (Fig. 16.4). So, it protects mechanical equipments and prevents pipe clogging. The separation limit is based on the more or less fine meshes of the grids, which can be straight or curved:
(i) Coarse screening or prescreening for the mechanical retention of the heaviest waste (papers, textile fibers, plastics, pieces of wood, etc.): width > 40 mm
(ii) Average screening for the mechanical screening of all the little compounds (plastics, etc.): width > 10 mm
(iii) Fine screening: width < 10 mm

Table 16.1 Average concentration values of pollutants in urban wastewater

Parameter	Average value
pH	7.8
Conductivity	1,100 $\mu S \cdot cm^{-1}$
Temperature	12 à 20 °C
DCO	700–750 $mg \cdot L^{-1}$
DBO_5	300 $mg \cdot L^{-1}$
MES	250 $mg \cdot L^{-1}$
NK	75–80 $mg \cdot L^{-1}$
$N-NH_4^+$	60 $mg \cdot L^{-1}$
PT	15–20 $mg \cdot L^{-1}$
$P-PO_4^{3-}$	13–18 $mg \cdot L^{-1}$

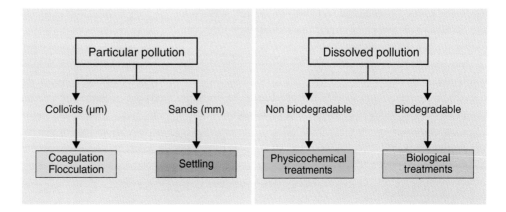

Fig. 16.3 General outline of the various methods used in a wastewater treatment plant

Fig. 16.4 Example of screening. (**a**) Grid screening; (**b**): grid and comb for elimination of accumulated pollution on the grid (Photographs courtesy of Société des Eaux de Marseille)

Periodically, a rake or comb moves up along the grid to get rid of the waste stuck and then spills it into a receptacle placed behind the grid. It is then compacted, put in dumpsters, and conveyed in a final waste storage center (*cf.* Sect. 16.5.5), and it will not be treated by sludge process.

After grit removal, water must be raised from a low level (wastewater gravity collection in the WWTP low point) to a high level (+10 m) to allow a gravity flow in the WWTP (run-of-river). No pumping equipment will be used in the WWTP. This raise must not chop oil slicks; hence, Archimedean screws are used (Fig. 16.5).

Grit Removal Degreasing

Water will be submitted to a degritting and a degreasing phase. The purpose is to eliminate the heaviest Volatile Suspended Solids (VSS) by settling and grease residues by flotation. To recover grease residues and sand, a surface and bottom scraper placed on a bridge crane pushes the compounds towards the end. Air is transferred in an air diffuser to accelerate the elimination of greases (Fig. 16.6). Thus, the created bubbles ($d = 1$ mm) will bring greases to the surface. The more important the incoming flow is, the higher the rate of climb will be.

The recovering of sand in wastewater avoids the following:
(i) Operating problems of installations linked to the sedimentation of sands in aeration basins, in clarifiers, and in digesters;
(ii) Problems of excessive wears of stirrers as well as pipe clogging or intermediate structures.

Fig. 16.5 Screw lift (treatment plant of the town of Vitrolles, France) (Photo: courtesy of Société des Eaux de Marseille)

Sand settling is a grain settling with a constant fall velocity (the size, form, and density of the solid particle do not vary). Some fall velocities are given in Table 16.2. After this step, a fine screening is proposed (1 cm) to eliminate the little VSS and particularly packing of fibers. Sand and fibers are evacuated towards final waste storage centers. At this processing stage, the fine VSS are still present and leave

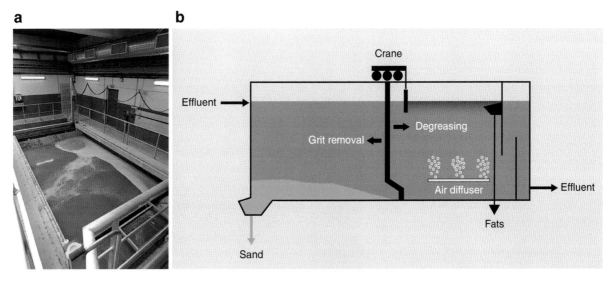

Fig. 16.6 Basin grit and oil removal with crane to scrape the surface and eliminate oils and deposited sands. (**a**) Photograph of a disposal (Photo: courtesy of Société des Eaux de Marseille). (**b**) Longitudinal section of a basin

Table 16.2 Examples settling velocity

Diameter (mm)	Type	v (Stocks) (mm·s^{-1})	Settling time for 1 m
10	Gravel	1,000	1 s
1	Sand	100	10 s
0.1	Sand	8	2 min
0.01	Limon	0.154	2 h
0.001	Bacteria	0.00154	5 j

them settling by gravity would be too long compared with the flow coming into the plant (Table 16.2). This quantity of VSS represents about 10 % of the incoming VSS. To increase the settling velocity, VSS will be artificially thickened (coagulation–flocculation) because these fine particles can be assimilated to resin grains exchanging generally negatively charged ions.

16.4.3.2 Primary Treatments or Physicochemical Treatments

Coagulation–Flocculation

The particles involved by these treatments are colloids. Their origin varies: soil erosion, dissolution of mineral substances, decomposition of organic matters, etc. These particles have a very small diameter (less than 1 μm) and are particularly responsible for the color and the turbidity of the surface water. The decantation velocities of these particles are very slow: coagulation–flocculation is therefore necessary. The main purpose of coagulation is the destabilization of the suspended particles, to make easy their agglomeration. Flocculation facilitates contact between destabilized particles. They begin to agglomerate to form a floc that will be eliminated by settling. However, as colloids are negatively charged, they repel one another and make any agglomeration impossible. If repulsions are eliminated, particles will agglomerate, grow, and settle quickly. The addition of high oxidation state cations such as Fe^{3+} and Al^{3+} avoids the repulsion of electric charges. So, in the WWTP, charges will be brought by a small quantity of charged coagulants ($FeCl_3$, $Al_2(SO_4)_3$). During this process, phosphates precipitate in the form of insoluble salts.

Settling

Settling is the step after coagulation–flocculation. It is carried out in various types of settlers: horizontal, vertical, or lamellar settlers (Degrémont 2008).

Since inlet of the plant, the water has generally been submitted to these pretreatments in confined premises and at low pressure to prevent the diffusion of unpleasant odors in the atmosphere. These odors will generally be treated with a chlorinated solution and soda. Flocculation, coagulation, and settling are also part of the elimination process of microorganisms, particularly of enteric origin, among which some of them are pathogenic.

Fig. 16.7 Basin for biological treatment of wastewater with surface aerator (Photo: courtesy of Société des Eaux de Marseille)

16.4.3.3 Secondary Treatments

Secondary treatments have biological treatments with or without oxygen and a settling step in the activated sludge process.

Aerobic Biological Treatments

Treatment of urban and industrial effluents mainly requires a biological process either under aerobic or anoxic, even anaerobic conditions (terminology used in process engineering will be specified later). Processes vary according to the type of culture either free or fixed. In the first case, microorganisms are maintained in suspension in the liquid treated (activated sludge). In the second case, bacterial cultures are fixed on a solid support (pozzolan, gravel, Raschig rings) where they develop to form a biofilm, which will eliminate pollution. In processes of fixed cultures, microorganisms constitute **bacteria beds*** or develop on disks or other granular beds. These processes are not detailed here because they are less and less attractive due to their excessive dimensioning to obtain a suitable purification rate.

The activated sludge process consists in a suspended biomass aeration tank where air is injected in order to provide oxygen. This is the most widespread process in the world to treat wastewater (Fig. 16.7). This results from the composition of urban wastewater constituted at 95 % of biodegradable matters.

The purpose of wastewater treatment plants is to degrade a maximum of organic matters, the remaining part is submitted to the action of the natural processes of biodegradation. For example, in France, this process represents about 60 % of the wastewater treatment plants and 80 % of the treatment capacity.

In the aeration tank, pollutants in waters are next to living microorganisms which can metabolize them partly or totally according to their metabolic capacity.

Free microorganisms are very few, most of them are agglomerated in a mucilage composed of exopolymers they synthesize. They form biomasses or bacterial flocs. *Zoogloea ramigera*, one of the microorganisms in the flocs, is associated to a great number of other bacterial species. To avoid the settling of flocs, activated sludge must be stirred. By malfunctions of the plant, an excessive amount of filamentous bacteria (*Thiothrix*, *Leucothrix*, actinobacteria, etc.) usually present in activated sludge may hamper the floc settling in the secondary settler; this dysfunction is called bulking sludge.

Eukaryotic and heterotrophic microorganisms (flagellates, ciliates free or fixed on the flocs, amoeba, etc.) and metazoa (rotifers, nematodes, etc.) living in the activated sludge feed on free bacteria, the biggest also catch small eukaryotic microorganisms; some (amoeba, nematodes) can ingest the floc biomass. Sludge eukaryotes take part to the elimination of pathogens and facilitate the development of flocs, whose exopolymers bring a better protection of prokaryotes against environmental factors.

Based on the observation of this microfauna, which is very sensitive to the variations of in situ conditions, it can be verified if the ecosystem works well and how are its perturbations. For example, an important quantity of rotifers shows a good stability of the purifying biomass and good water purification; in contrast, a lot of small flagellate or amoeba communities are the sign of recent sludge, sudden high load, a bad aeration, and a poor quality of purified water (Degrémont 2008).

Fig. 16.8 Design parameters of the activated sludge process. X: biomass concentration at the outlet of the aeration basin; X_0: initial biomass concentration; X_1: biomass concentration entering in the aeration basin and taking into account the recycled biomass concentration (X_r); X_s: outgoing biomass concentration; S: substrate concentration (pollution); Q: daily outflow ; Q_p: daily outflow of evacuation of produced excess sludge; Q_r: recycled flow, V: volume of the reactor. Drawing: M.-J. Bodiou

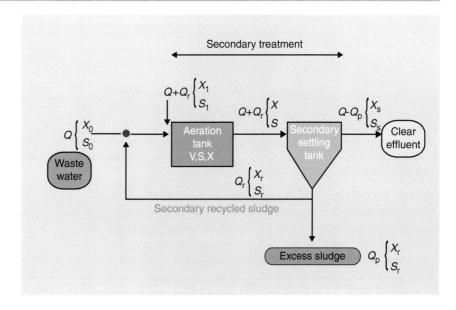

Aerobic degradations are obtained by an adapted biomass in constant and sufficient concentration in the aerated tank. The biomass adsorbs and eliminates a more or less important part of pollution according to the detention time in the tank and to the organic load received by the treatment plant. The dissolved pollution is partly transformed into sludge by bacterial assimilation. Another part of the biodegradable organic matter is mineralized and releases as carbon dioxide, water, sulfate ions, ammonium ions, etc. In the same tank, in presence of oxygen, ammonium ions resulting from degradation of organic matter as well as those brought by the effluent are oxidized to nitrate ions by two communities of aerobic, chemolithotrophic, and autotrophic microorganisms. Ammonium oxidizing prokaryotes achieve the conversion from NH_4^+ to NO_2^- and nitrite oxidizing prokaryotes achieve the conversion from NO_2^- to NO_3^-. Most of the wastewater treatment plants eliminate properly carbon compounds, whereas nitrogen and phosphorus compounds are processed only in the most performing plants.

Two major parameters must be determined in the activated sludge process; dimensioning and aeration of the tanks:

- *Dimensioning of tanks*

  For an optimum functioning, the structure will depend on dimensioning parameters, which are primarily mass and volume load, **hydraulic detention time***, and sludge detention time called "sludge age" (Fig. 16.8).

  Load represents the quantity of pollution measured in COD or in BOD_5 supplied everyday to the treatment plant and in relation either to the volume of the aeration tank (volume load/flow) or to the quantity of measured biomass: Volatile Suspended Solids (VSS) or mass load (ml), which represents the biomass active part.

$$\text{Mass load } (C_m) = \frac{\text{kg of biodegradable injected per day (BOD}_5)}{\text{quantity of biomass in the system (VSS)}}$$

Mass load is expressed in kg of BOD_5 to be treated per kg of VSS per day.

$$\text{Volume load } (C_v) = \frac{\text{kg of biodegradable injected per day (BOD}_5)}{\text{reactor volume (m}^3)}$$

The volume load is expressed in kg of BOD_5 to process/ m^3 of reactor/day. This notion of mass load is important because it influences:

(i) The purification efficiency of an activated sludge: The low mass load corresponds to high purification rates and high mass load to lower rates.

(ii) The excessive production of biological sludge: For a low load, endogenous respiration is more important than for a high load because of the substrate restriction.

(iii) The requirements for oxygen in relation to the eliminated pollution: The importance of endogenous respiration with a weak load leads to consumption of oxygen in relation to eliminated pollution more than those obtained with high load.

Low mass load also increases the age of sludge, which means an increase in a chemolithotrophic and autotrophic community and the possibility to obtain better stabilized sludge and better flocs. The advantages of a low mass load are obvious but have a major drawback: the tank requires a larger volume.

Values of C_m and C_v define the type of load of the wastewater treatment plants (Table 16.3). A facility with a low volume tank receiving a lot of pollutants is a high load plant. WWTP with activated sludge usually work with low load. The device performances (assimilation velocity of biomass) depend on the dimensioning and functioning parameters previously set (C_m, aeration and quantity of VSS in the reactor) and on their stability. An "optimal" dimensioning cannot in any case only rely on the value of

Table 16.3 Loads according to their order of magnitude

Type of load	F/M ($kg_{BOD5} \cdot d^{-1} \cdot kg_{MLVSS}^{-1}$)	L_v ($kg_{BOD5} \cdot d^{-1} \cdot m^{-3}$)
High load	0.4 to 1	1.5
Medium load	0.15–0.35	0.5–1.5
Low load	0.1	0.3
Extended aeration	0.07	0.25

the mass or volume load. The issue is far more complex since the "sludge age" is also a decisive parameter to maintain purification capacities. This is particularly true of nitrifying autotrophic bacteria, which require a "sludge age" greater than heterotrophic bacteria to develop. "Sludge age" is directly linked to the sludge production resulting from the applied mass load. For important "sludge ages," the substrate–microorganisms (C_m) ratio reaches a maximum value, as well as the biomass concentration; the growth velocity is in balance with mortality. This limit in mass load C_m, due to a more important "sludge age" (>50 days), magnifies the endogenous respiration phenomenon, which generates a decrease of the production of sludge. However, the protracted "sludge age" generates detrimental effects on the sludge biological activity, which tends to become mineralized.

- *Aeration of activated sludge*
  The biological aeration tank is the core of the wastewater treatment plant since the suspended aerobic microbial culture degrades the pollutant load of wastewater in presence of dissolved oxygen. This oxygen is brought by an aeration device whose functioning amounts to 50–70 % of the total energetic cost of the wastewater plant, which corresponds to one-third of the total operational cost approximately. With such costs, optimizing oxygen transfer or the energetic control of wastewater treatment processes has become a major challenge. The efficiency of aerobic bacteria in the wastewater treatment strongly depends on the oxygen supply and consequently on its transfer from the gaseous phase towards the liquid phase, which must not be a restrictive step of the process operation.

  Whatever the medium may be studied, an increase in the viscosity of the liquid phase or of the suspension involves a decrease of the oxygenation capacity of the medium; thus, it generates degradation in the performances of the aerobic biological systems.

Biological Nitrogen and Phosphorus Removal

The nitrates constituted during the aerobic process in the activated sludge tank are removed during denitrification by heterotrophic and chemolithoautrophic autotrophic bacteria responsible for anammox (*cf.* Sect. 14.3.3) by recycling waters of the aeration tank in an anoxic tank (no free oxygen but presence of bonded oxygen, particularly under the form of NO_3^-) upstream of the aerobic reactor (Fig. 16.9a).

The organic or mineral molecules in wastewater will bring by oxidation the energy necessary to the reduction of nitrates to nitrogen. In this type of process, the reduction of nitrates can be incomplete and stopped to the stage N_2O (greenhouse gas) which is released to the atmosphere. In these conditions, a tertiary treatment is necessary. As far as phosphorus removal is concerned, the phosphate removal of the effluent is achieved in an anaerobic tank (absence of free and bonded oxygen, in particular under the form of NO_3^-) upstream of the anoxic tank and immediately after the primary treatment (Fig. 16.9b).

The operators of wastewater treatment used to describe the different tanks of a WWTP as the expression of aerated tank (aerobic) to indicate oxygenated environments (presence of free oxygen) and the expression of anoxic tank to indicate environments without free oxygen but containing combined oxygen (particularly under the form of nitrate). The expression of anaerobic tank is only for environments without free or bonded oxygen (i.e., nitrate).

For biologists, these expressions have different meanings. Indeed, oxic (or aerated) and anoxic, respectively, characterize biotops in which oxygen is present or absent; bonded oxygen is not taken into account in this definition. Besides, aerobic and anaerobic, respectively, characterize organisms (or biological processes), which require the presence or absence of molecular oxygen to live.

The principle of phosphorus removal is the following one: in anaerobic conditions (in anaerobic tanks), various bacterial communities, probably most of them belonging to *Betaproteobacteria*, use their polyphosphates to synthetize from acetate or other volatile organic acids intracellular polyhydroxyalkanoates (PHA), releasing phosphate in the tank. These bacteria under aerobic conditions reconstitute stocks of cellular polyphosphates more important than the previous ones. They will be eliminated with the cells during the secondary settling (Fig. 16.10) (Seviour et al. 2003).

Frequently in the WWTP, the aerobic activated sludge treatment is combined in the same tank to anoxic treatments (denitrification) and anaerobic treatments (phosphorus removal). The activated sludge tank is compartmentalized. Each compartment is intended for a specific treatment. Such devices can be observed, for example, at the WWTP of Gréasque (France), where the central annular zone of the tank is only for the anaerobic process and the peripheral zone for aerobic and anoxic processes (Fig. 16.11). Another example is represented by the WWTP of Skaekinge (Denmark), whose tank is divided into three annular compartments. The central compartment is devoted to phosphorus removal, the most peripheral one to aerobic process,

Fig. 16.9 Removal of nitrates and phosphates by microbiological processes. (**a**) Removing of nitrates. The basin is called anoxic because it does not contain free dioxygen. The dioxygen may be present in the combined form of nitrate. (**b**) Removing of nitrates and phosphates. The diagram shows the method of the STEP Phoredox (Modified and redrawn from Degrémont 2008). There are several other processes which differ by the number of reactors and their respective positions as well as the mode of effluent recirculation and secondary sludge (Degrémont 2008). Drawing: M.-J. Bodiou

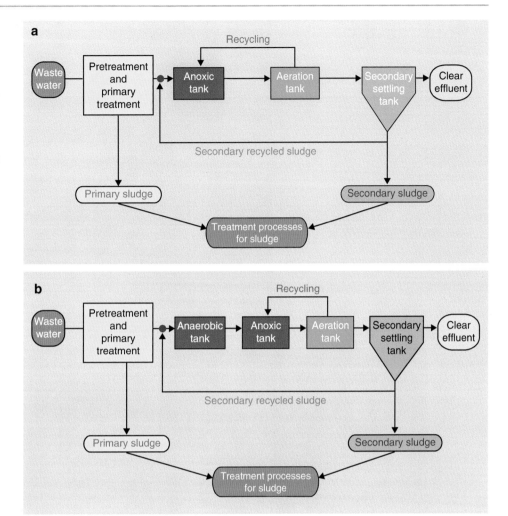

Fig. 16.10 Mechanism of bacterial phosphorus removal. *P* phosphate, *PHA* polyhydroxy alkanoate. In anaerobic conditions, the synthesis of PHA transforms cellular storages of polyphosphate; these last are reconstituted massively in aerobiosis, via an aerobic metabolism. Drawing: M.-J. Bodiou

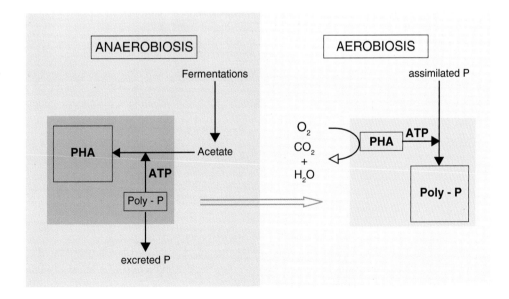

and in the intermediate one to denitrification (Degrémont 2008). Alternation in the aerobic tank of aerated and non-aerated periods (sequential aeration) enables the combination of aerobic treatment and denitrification in the same tank. Wastewater professionals use the expression "biological treatement by referring to the whole processes that they define as aerobic, anoxic, and anaerobic wastewater are submitted to.

Fig. 16.11 Aerobic treatment tank. *1* Anaerobic zone, *2* aerobic and anoxic zone

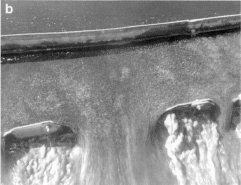

Fig. 16.12 Secondary clarifier. Overflow of the decanter: (**a**) effective settling (the effluent is clear), (**b**): insufficient settling, the effluent still contains suspended materials (presence of activated sludge) (Photographs Benoit Marrot and Philippe Moulin, courtesy of Société des Eaux de Marseille)

Secondary Settling in Aerobic Activated Sludge Treatments

Another stage to separate microorganisms and the treated liquid is necessary to prevent any discharge of suspended solids in the natural environment. Flocs are therefore separated from the treated water by a secondary settling (Fig. 16.12).

A part of the settled sludge is recycled in order to enable the reseeding of the biological aerated tank (aeration tank) and thus to maintain a biomass concentration enough to purify the effluent (Fig. 16.8). The other part, namely, the excessive sludge, is periodically extracted to be treated in composting centers, for example.

In practice, activated sludge process remains limited by a lack of flexibility towards the variations of the composition of the pollutant loads and the biomass concentration; it is expressed by significant secondary settling problems if the settling is more than 4 g·L^{-1}. Besides, because of the increase in the production capacity of the industrial sites and of stricter discharge standards, biological wastewater treatment plants are generally undersized and cannot meet the specification required by the legislator. These drawbacks have been a tremendous challenge for the scientific community and the wastewater professionals. This challenge has been solved by the coupling of biological processes (treatment in bioreactor) and physical processes (processes with porous membranes) with biomass recycling which led to the membrane bioreactor (MBR).

Anaerobic Biological Treatments

The purification of effluents can also be completely carried out in anaerobic conditions (in the biological sense) in anaerobic reactors. During the degradation, the organic matter is transformed into gas (carbon dioxide, methane). Compared with aerobic purification, anaerobic purification is slower and generates a less important biomass production, since the involved microorganisms have a metabolism

Fig. 16.13 Place of membrane bioreactors in the two disposal of wastewater treatment. The physicochemical processes correspond to the primary treatment; in biological processes, primary treatment is supplemented with secondary treatment. Drawing: M.-J. Bodiou

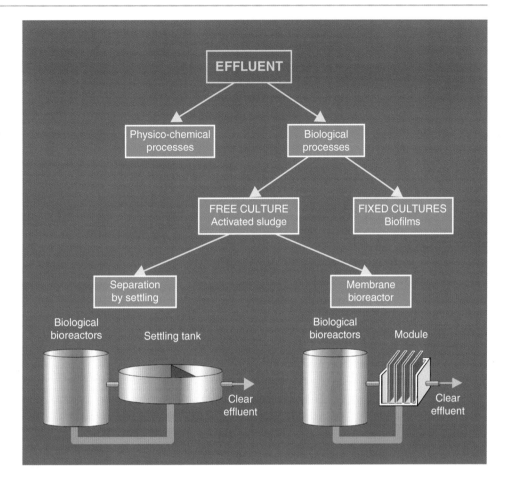

which releases very little energy (*cf.* Sects. 3.3.2 and 3.3.3). This treatment mode is generally used to purify agricultural or industrial wastewater and to stabilize primary sludge (sludge methanization).

16.4.3.4 Tertiary Treatments: Nitrogen and Phosphorus Removal by Other Processes; Complementary Refining Treatments

If after secondary treatments nitrogen and phosphorus concentrations do not meet standards specified by legislation, tertiary physicochemical treatments are implemented. Nitrogen removal under the form of nitrogen can be obtained after alkalinization and chlorination, whereas phosphates can be removed by precipitation with salts of iron, aluminum, or calcium. Other more specific tertiary treatments (non-exhaustive list) contribute to refine the quality of the discharged water:

(i) Activated carbon filtration to remove organic matters, which are resistant to biological treatments
(ii) Reduction of the load of pathogen agents by coagulation–flocculation
(iii) Ultrafiltration, nanofiltration, and reverse osmosis

16.4.3.5 An Innovative Process: Membrane BioReactors (MBR)

In the MBR system, the bioreactor and the membranes, tubular or plane, have each a specific function (Fig. 16.13):

(i) The organic degradation of the dissolved organic pollution is achieved in the bioreactor with adapted bacteria.
(ii) The separation of particulate pollution is obtained via membranes, which constitute a physical barrier for suspended solids (VSS) that cannot get through.

Membrane bioreactors are innovative systems. Their development has been undeniable since the beginning of the twenty-first century (Stephensons 2000; van der Roest et al. 2002; Cornel and Krause 2008). The related scientific studies have been developing exponentially, and consequently this treatment process has been implemented more and more worldwide. The expansion of membrane bioreactors is due mostly to limitations of the secondary settler in the classical treatment and to the evolution of effluents to be treated. Because of effluents are more or less polluted, containing various and new types of pollutants such as xenobiotics, with varying loads, it is necessary to apply adaptable, flexible, and highly efficient purification processes in order to meet the more and more strict

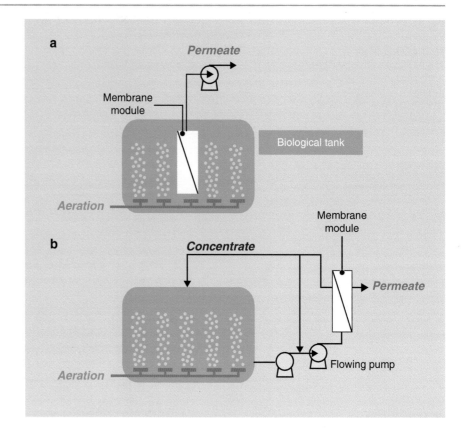

Fig. 16.14 Drawings illustrating the devices with immersed (**a**) and external (**b**) membranes in a treatment with membrane bioreactors. Drawing: M.-J. Bodiou

discharge standards. The two sets of membrane reactors (submerged or external membranes) meet these requirements for urban as well as industrial effluents (Figs. 16.14 and 16.15a, b).

Unlike the conventional biological systems, MBR enables the purification improvement with a higher biomass concentration in the biological tanks (about 10–15 g·L^{-1} wet weight). Besides, they produce less sludge, are more compact, and provide a total retention of suspended solids. A further development of MBR mainly depends on the control of the phenomena which limit the performances of the process. In the first place, the membrane clogging is impossible to eliminate totally; therefore, it should be controlled with one of the following solutions:

(i) Preventive: air is blown at the surface of the membrane to avoid clogging.
(ii) Curative: optimization of cycles and wash solutions.

The oxygen supply to microorganisms in biological tanks is among the limiting phenomena. However, the clear advantages of membrane bioreactors largely compensate for these restricting phenomena. Nowadays, MBR are being optimized, and their future is very promising in a wide range of application fields, such as Food and Drink Industries, petrochemical industries, landfills, or urban wastewater.

It is still possible to argue about the more important cost of the MBR system compared with the conventional one.

16.5 Solid Waste

16.5.1 Definition of a Waste

The definition of a waste varies according to countries. For United Nation statistical division, "Wastes are materials that are not prime products (products produced for the market) for which the initial user has no further use in terms of his/her own purposes of production, transformation or consumption, and of which he/she wants to dispose. Wastes may be generated during the extraction of raw materials, the processing of raw materials into intermediate and final products, the consumption of final products, and other "human activities". The European Union defines waste as "an object the holder discards, intends to discard or is required to discard".

16.5.2 Classification of Waste

The distinction is made between liquid waste (such as manure (slurry), sewage sludge, etc.) and solid waste consisting of household waste (domestic waste, municipal waste, or urban waste), industrial waste, agricultural waste, hospital waste, radioactive waste, etc. Waste is not only

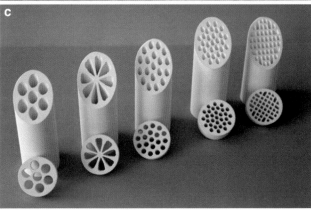

Fig. 16.15 Example of a pilot plant with flat membranes. (**a**) Flat membranes. (**b**) Reactor. (**c**) Tubular ceramic membranes (Photographs: courtesy of the Company Orelis Environment)

made up of organic matter; manure, for example, contains nitrates, toxic garbages, etc.

World production of waste is about 4 billion tons per year and only 1 billion is currently valued. The two major degradation pathways and biological recovery of waste, composting and methanogenesis, mainly provided by microorganisms, tend to grow at the expense of physicochemical treatment and disposal. Generally, methanogenesis is called "solid" when the waste to be treated has a dry matter content higher than 15–20 %. Biological processes involved in the degradation of organic waste are similar to processes that are expressed in the natural environment during the degradation of organic matter; however they are enhanced and controlled.

In this manual, the treatment of radioactive, industrial, and hospital wastes will not be discussed. Readers can refer to books published by Williams (2005); Woodard and Curran, Inc (2006); and Degrémont (2008).

16.5.3 Aerobic Treatment or Composting

16.5.3.1 Definition

Composting is a biological process in which organic matter is degraded by aerobic microbial communities whose composition varies according to the different stages of transformation of organic matter and its maturity. All organic wastes are biodegradable and compostable. This process therefore applies to a wide spectrum of organic materials: green waste (garden waste from the public and private gardens), household organic wastes (a preliminary waste sorting step mandatory to eliminate nonbiodegradable components such as glass, metals, etc.), sludge from wastewater treatment plants which, because of their high moisture content, are mixed with a structuring compound (chips, sawdust wood, straw, bark, etc.) before being composted, agro-food waste (waste from the sugar beet industry), animal manure (compost is produced from cattle, sheep, pigs, and poultry manure), etc.

The benefits of composting are numerous:
(i) The main benefit is the value of the resulting product, called compost, which is sold and used as an organic amendment resulting in the improvement of soil fertility.
(ii) Composting will reduce the volume of the treated organic biomass (about 50 %) by concentrating minerals; this mass loss results from the **mineralization*** of organic matter (release of CO_2 and water).
(iii) The substantial rise in temperature during the composting (which can reach 75 °C) causes the destruction of most pathogenic microorganisms.

16.5.3.2 The Different Steps

The mechanisms involved in the composting process are numerous and complex. They can be theoretically divided into four phases (Fig. 16.16):

(i) A mesophilic stage in which the easily degradable organic substances (sugars, amino acids) are mineralized by mesophilic microorganisms whose optimum growth temperature ranges between from 20 to 45 °C. The activity of mesophiles markedly increases the temperature within the compost, resulting in the inhibition of their metabolism.

(ii) The second phase, referred to as the thermophilic one, corresponds to the development of heat-resistant microorganisms whose activity continues the biodegradation of organic matter. The increase in temperature can exceed 70 °C, leading to the elimination of most pathogens and to a drastic loss of water of the compost with enhanced water evaporation. To overcome this loss of water, a watering of compost is often done, the optimum moisture for composting usually being between 50 and 80 % of the total raw biomass.

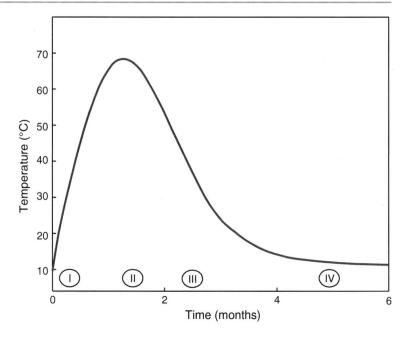

Fig. 16.16 Evolution of temperature during the various stages of composting

(iii) A cooling phase follows the thermophilic phase. If the first two phases correspond to an intensive degradation of organic matter, the cooling phase corresponds to a slowing in microbial activity with the decrease in the amount of readily biodegradable organic matter; mesophilic microorganisms again colonize the compost.

(iv) A final phase, known as the maturation phase, corresponds to the formation of humic substances during which the compost color becomes dark brown to black and the compost is then considered as "mature." However, the maturity of compost is difficult to define. Overall, two approaches can be used to evaluate this maturity. The first considers the stability of the composition of the organic material obtained after the composting where microbial biodegradation is extremely low: the rate and amplitude of organic matter stabilization depending on the type of waste and the composting method. The second approach considers the effects of compost which should not harm the plants; for example, an "immature" compost can lead to deficiencies in nitrogen or have phytotoxic effects.

To take into account the previously defined two approaches to assess the maturation of compost, the following parameters are measured:
- The consumption of oxygen measured by respirometric tests. This is based on the respiratory activity of endogenous communities in the compost.
- The carbon–nitrogen ratio which decreases during composting. The value of this ratio to characterize a mature compost varies according to the authors: it must be less than 25 for some, for others ratios less than 20 or even than 15 are preferable.
- The distribution of mineral forms of nitrogen: a mature compost is poor in ammonium and rich in nitrate.
- The ratio of the humic to the fulvic fraction which increases during composting.
- The phytotoxicity, for example, by studying the germination of cress.

An important point must be raised: to describe the first two phases of the degradation of organic matter, the term "fermentation" (anaerobic microbial process) is often used by the specialists of composting. It is obvious that this terminology is inappropriate: composting is an aerobic process.

16.5.3.3 Microorganisms

As previously mentioned, the composition of the microbial communities varies depending on the evolution phase of the compost and its chemical composition. In the composting processes, prokaryotes are the most important microorganisms in both quantitative and qualitative terms because of the diversity of their metabolic capabilities and their ability to live in very wide ranges of temperature and pH.

The microorganisms found in compost are composed of bacteria and fungi whose concentrations and compositions will vary depending on the materials treated during the composting. Examples of these variables are proposed by Maier and collaborators (2000). When starting the process, with temperature below 40 °C, in one gram of dry compost, the concentrations of mesophilic and thermophilic bacteria are at 10^8 and 10^4, respectively; that of mesophilic fungi is at 10^6 and that of thermophilic fungi at 10^3. When the temperature increases between 40 and 70 °C, thermophilic bacteria become dominant (10^9) and mesophiles decrease (10^6). In particular, thermophilic actinobacteria that also develop during this phase (10^8) are capable of degrading cellulose and solubilizing lignin and tolerate higher temperatures than fungi. They are therefore important agents of lignocellulose degradation during the thermophilic phase. Concentration of mesophilic fungi decreases (10^3) and that of the thermophiles increases (10^7). Thermophilic fungi presenting ligninocellulolytic capacity have been isolated from various composts (Tuomela et al. 2000). During the cooling phase, mesophilic bacteria reappear (10^{11}), the concentration of thermophiles decreases but remains high (10^7).

Regarding the thermophilic microorganisms in the cattle and poultry compost manure, 34 thermophilic bacteria grouped into five genera were isolated based on phylogenetic analysis of 16S rRNA: *Ureibacillus*, *Bacillus*, *Geobacillus*, *Brevibacillus*, and *Paenibacillus* (Wang et al. 2007). Thermophilic actinobacteria belonging to the genera *Nocardia*, *Streptomyces*, *Thermoactinomyces*, and *Micromonospora* were also isolated (Tuomela et al. 2000)

Works by Ryckeboer and collaborators (2003) illustrate the variations in the concentrations of different populations in compost. These authors report the almost complete disappearance of fungi and streptomycetes when the temperature reaches 76 °C and their reappearance during the cooling phase.

16.5.3.4 The Different Composting Methods

Composting processes are numerous and vary according to the type of waste to be treated and their volume. However, for all type of treatments, i.e., implemented by an individual (home composting) or at an industrial level, the processes that occur within compost remain the same. A major concern for the success of composting is to maintain aerobic conditions throughout the compost, heterogeneous environment in which gradients of temperature and humidity establish, with the formation of numerous anaerobic microniches, particularly during the two phases of intensive degradation when the consumption of oxygen by microorganisms is particularly high. The challenge is that the oxygen content in the pores of the compost should not be less than 5 % of the total volume of the treated organic material, in order to prevent the occurrence of complete anaerobic conditions. Two types of processes capable of ensuring good oxygenation of the compost can be implemented:

(i) Windrow composting: compost is thus arranged in rows to facilitate engine handling during the necessary regular shuffling. Another technique is to blow air at the base of compost heaps using perforated pipes embedded in the swath. It is also possible to combine shuffling and ventilation.
(ii) Composting in sealed rooms with ventilation by mechanical turning. A first device is to contain the compost in lockers (walls with usually a roof) or between walls that form long and narrow corridors. Compost can also be treated in silos or in bioreactors (Sharma et al. 1997).

In both types of processes, there is a biostimulation of the activity of aerobic microorganisms by oxygen supplying. Microorganisms can be added during the processing (biostimulation).

16.5.4 Anaerobic Treatments and Methanization

16.5.4.1 Definition
Unlike composting, biological processes involved are expressed under anaerobic conditions and result in the formation of biogas, especially methane (*cf.* Sect. 14.2.6).

The objectives of this biotechnology are to reduce the amount of organic biomass waste and to retrieve and use the biogas produced during the treatment. A remediation technique is coupled with a technique for production of renewable energy. Compared to aerobic degradation of waste, the anaerobic degradation has three advantages:

(i) In terms of energy, it is self-sufficient and even can produce extra energy, because the biogas can be converted into heat, electricity, or fuel for vehicles (cars, buses) (e.g., the city of Lille (France)).
(ii) Because the degradation is done in sealed reactors, odors are greatly reduced.
(iii) The surface used is limited.

The main disadvantage of this treatment is its cost which is higher than that of composting.

16.5.4.2 Principle of Methane Fermentation and Microorganisms Involved in the Process

Mechanisms and microorganisms involved in methanization, which is a natural anaerobic process, have been described previously (*cf.* Sect. 14.2.6).

Anaerobic digestion involves four main steps (Figs. 14.14 and 16.18):
(i) Hydrolysis of macromolecules
(ii) Formation of volatile fatty acids
(iii) Formation of acetic acid and production of hydrogen and CO_2
(iv) Production of biogas

The degraded organic matter is largely identical to that treated by composting, essentially consisting of urban sewage, food waste, the organic fraction of domestic waste, and livestock effluents. Waste is treated in a digester (whose volume can vary from a few hundred to several thousand m^3), which can operate in continuous, semicontinuous, or discontinuous mode.

Depending on the temperature, three modes of production are defined, called psychrophilic (15–25 °C), mesophilic (25–45 °C), and thermophilic (55–65 °C). Microorganisms can be free or attached (biofilm formation). The process results in the formation of biogas, and the degradation yields are variable depending on the type of organic matter treated and of the technology implemented. The average amount of biogas obtained varies with the treated waste; for example, a ton of household organic waste yields 80–150 m^3 of biogas. In addition, depending on the nature of the starting biomass, it is possible to obtain a product usable as fertilizer or organic amendment (e.g., green waste).

Biogas composition depends on the nature of the organic material used, but the average composition is as follows: CH_4 (50–90 %), CO_2 (10–40 %), H_2 (1–3 %), N_2 (0, 5–2 %), H_2S (0.1–0.5 %), CO (0.0–0.1 %) (La Farge 1995). This is an energy form which is as noble as natural gas. The presence of H_2S and CO_2 in biogas makes it corrosive and requires, for installations, the use of appropriate materials or the implementation of technologies that ensure the purification of biogas; the elimination of H_2S is carried out by physicochemical or biological treatments (elimination by sulfur-oxidizing bacteria; cf. Sect. 14.4) which transform H_2S into elemental sulfur. It should be noted the presence of H_2S in fossil gas, for example, the H_2S content, is 15 % in Lacq gas (France) and some deposits contain 50 %. This similarity in composition may be explained by the fact that fossil gas and contemporary gas are obtained from similar fermentation processes.

Biogas is a renewable and inexhaustible energy source, and the possibility of using biogas as a renewable energy source is now proved; facilities exist around the world, ranging from domestic digesters to industrial units of several thousand cubic meters. Given the increasing production of waste, "industrial development potential which is susceptible to be generated by large-scale production of biogas is of the same order of magnitude as that derived from the exploitation of natural gas" (La Farge 1995).

16.5.5 Discharges

There are different types of waste including the household waste and similar (HWS) produced by households: for example, daily, a Frenchman produces about 1 kg of domestic waste. For the protection of the environment, discharges and installations for elimination of waste become Ultimate Waste Storage Facilities (UWSF or landfills). In the field of HWS, the ultimate waste is defined as the "non-recoverable fraction of waste." In 2004 in France, over 47 % of the HWS were sent to landfills. This technique is also widely used in European countries, such as Germany, Spain, and the United Kingdom. A UWSF is constituted by a set of bins, hydraulically independent, themselves composed of boxes, in which the wastes are stored. Each compartment is isolated by an impervious membrane. The operating time of a site is about 20 years. The operation of these landfills may be considered as a bio-physicochemical reactor resulting in reactions and complex changes that lead to chemical, physical, and biological processing of wastes. Water promotes the conversion of wastes and therefore the evolution of landfills. All of these phenomena lead to the generation of two pollutants flow:

(i) A gaseous effluent, the biogas, produced by natural fermentation of the organic matter contained in waste.
(ii) An aqueous effluent, the leachate, which results from the percolation, through the solid waste, of the water contained in the waste or added. This water promotes biodegradation of fermentable organic matter and then produces leachates that concentrate organic substances and/or minerals from waste or by-products of waste degradation.

Understanding the genesis of leachates not only requires knowledge of the nature of buried waste and the operating mode of the landfills (waste height, operated, compaction, etc.) but also the study of interactions between water and waste. The mechanisms of the genesis of leachate are very complex: they are of biological and physicochemical nature (Fig. 16.17).

Microbial processes play an important role. Indeed, buried waste is used as a substrate for heterotrophic organisms (prokaryotes, yeasts, fungi) responsible for the degradation of the organic fraction. The effects of this activity may be multiple, direct, or indirect and trigger secondary physicochemical phenomena. This then results in a change in environmental conditions (pH, temperature, redox potential) that affect microbial mechanisms. In the presence and/or absence of oxygen, by the action of extracellular enzymes secreted by microorganisms, the organic waste content (proteins, carbohydrates, lipids, hydrocarbons, phenols, etc.) is degraded in intermediary metabolites (polypeptides, amino acids, amines, fatty acids, aldehydes, quinones, diacids, mono- and disaccharides, etc.). The metabolites formed are then degraded by aerobic and anaerobic microorganisms:

(i) The aerobic degradation is possible as long as the density of the waste allows the diffusion of oxygen. It leads, in particular, the formation of the final metabolites (H_2O, CO_2, HCO_3^-, CO_3^{2-}, NO_3^-, PO_4^{3-}, SO_4^{2-}). The term "aerobic fermentation" is commonly used in the treatment of discharges; formally, this term is inappropriate, and the processes involved are occurring in the presence of oxygen. The early stages of aerobic degradation of discharges treatment are identical to the early stages of composting, including a rise in temperature that can reach 60 °C within the waste mass.

Fig. 16.17 Schematic drawing of the transformation of the mineral matter in a waste storage facility. Biological processes are illustrated in Fig. 16.18

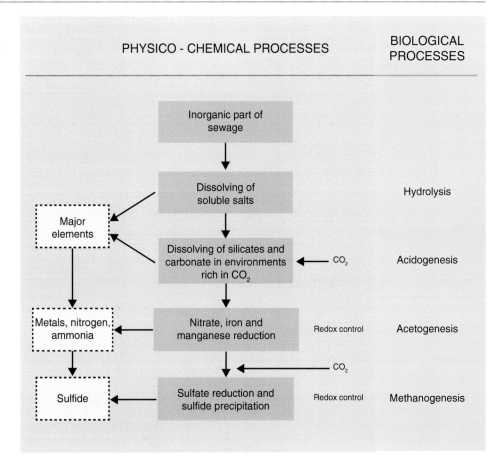

(ii) The anaerobic degradation of organic matter is the dominant mechanism in landfills. Under anaerobic conditions, the metabolites resulting from hydrolysis are converted into intermediate metabolites consisting mainly of volatile fatty acids (VFA), monocarboxylic linear saturated short chain (acetic, propionic, butyric acids, etc.), and end gaseous metabolites, responsible for the offshoot of biogas in landfills (CO_2, CH_4, H_2S). Figure 16.18 is a synthetic representation of the biological degradation of organic matter under anaerobic conditions.

16.6 The Gaseous Effluents

16.6.1 Sources

Increase in energy consumption, development of industries, increased road and air traffic, garbage incineration, and increasing number of sewage treatment plants are sources of air pollution, especially in urban areas. Air pollutants can be gaseous, organic volatile, or particulate. Gaseous pollutants are mainly the sulfur compounds (SO_3, H_2S), carbon oxides (CO, CO_2), nitrogen oxides (NOx), ammonia (NH_3), ozone, acid aerosols (Cl^-, F^-), and cyanides. The main volatile organic compounds (VOCs) are:

(i) *n*-Alkanes (methane, ethane, pentane, heptane, octane, etc.)
(ii) Aromatic hydrocarbons (benzene, toluene, xylene, benzopyrene, etc.), olefins (ethylene, butadiene)
(iii) Halogenated hydrocarbons
(iv) Oxygenate compounds (alcohols, aldehydes, ketones)
(v) Sulfur compounds (mercaptans, dimethylsulfide, etc.)

16.6.2 Treatments

Treatment of waste gases can be classified according to the type of process: regenerative or destructive. All of these processes, as well as going technological variants, are detailed in Fig. 16.19. Each process will be more efficient according to the specific gas to be processed (flow ranges, concentrations, and nature of chemical pollutants) and the peculiarities of the production site (reusability of pollutants, availability of refrigerant vapor or water, etc.). The final choice will be made primarily on physicochemical considerations, since they will determine the technical feasibility of a particular process. Then, it will be based on economic criteria like the annual capital and operating costs of industrial equipment. To facilitate the process selection, many authors have evaluated and represented areas of

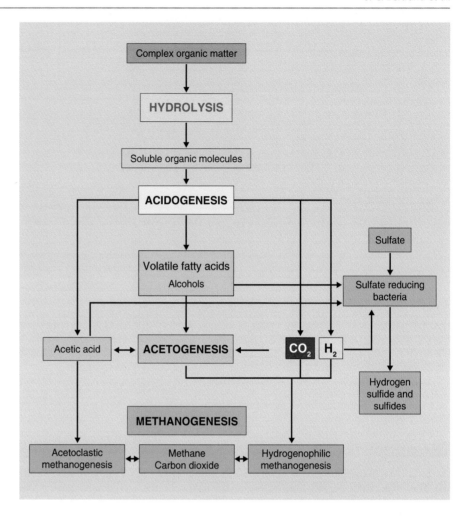

Fig. 16.18 Schematic drawing of biodegradation of organic matter in a sanitary landfill. Drawing: M.-J. Bodiou

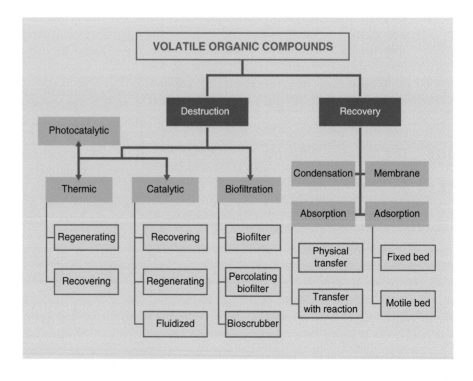

Fig. 16.19 Simplified organization chart of different methods for the treatment of volatile organic compounds (VOCs). Drawing: M.-J. Bodiou

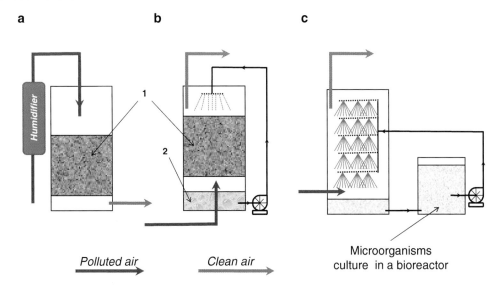

Fig. 16.20 Biological treatments (cross section) (Adapted from Vandecasteele 2008). (**a**) Conventional biofilter. (**b**) Bioreactor or biotrickling filtration systems. (**c**) Bioscrubber. *1*: Biofilm on solid porous media, *2*: aqueous phase with nutrients

economic competitiveness of different treatment methods based on a diagram flow/concentration. It should be noted that the proposed schemes keep a general appearance, and the limits of economic competitiveness data are dependent on the nature of the processed air.

16.6.2.1 The Recuperative Treatment

The condensation is achieved by reaching the **dew point*** of the gas by reducing the temperature or by increasing the pressure. For economic and technical reasons, the organic compounds to be treated must have a dew point temperature above 40 °C.

Absorption consists to transfer the organic compounds of the air to a liquid phase. The effectiveness of treatment depends mainly on the affinity of the pollutant with the liquid phase. From the technical point of view, the contact of the two phases is performed by dispersing the liquid phase in the gas phase using columns (filled or with beds) or spray towers. This method has the advantage of being well known.

Adsorption is a process of transferring the pollutant from the gas phase to the surface of a solid. Conventionally, the process is done in a fixed bed and operated alternately with two adsorbent beds in a regeneration system. This regeneration is usually done at warm temperature. Gas permeation is also conceivable for transferring the pollutant from a gas phase to another concentrated gas phase. This technology has emerged through the use of semipermeable membranes. According to the separation factor of the membrane, a permeate gas more or less concentrated is obtained downstream of the membrane simultaneously depleting the retentate.

16.6.2.2 Destructive Treatments

Incineration Treatments

It is commonly found in the industry and led to the destruction of molecules forming CO_2 and water if combustion is complete. Two types of incineration can be distinguished:

(i) At hot temperature from 600 to 850 °C depending on the compounds.
(ii) Catalytic which involves oxidation catalysts and lower temperatures ($370 < T < 480$ °C). These incinerators are known as regenerative or recuperative according to the mode of recovered energy obtained during combustion. This process of easy implementation is suitable for a wide range of flow and enhances the recovered energy. However, it is not suitable when the gas to be treated contains dust (dirt), heteroatoms (chlorine atoms lead to the formation of hydrochloric acid), or poisons. Systems for pretreatment or posttreatment of gases are possible (filters, washing).

Biological Treatments

These treatments use the ability of microorganisms (bacteria, yeasts, etc.) to degrade organic compounds used as metabolic substrate. The kinetics of degradation are generally slow and therefore require large filter surfaces. However, significant flows at low concentrations and at room temperature can be treated by this process using either biofilters, bioreactors (also called trickling biofilters) or bioscrubbers (Vandecasteele 2008)

Whatever the biological process implemented, the treatment is carried out in two stages:

- A step of absorption of the gaseous compounds in a liquid phase or a biofilm.
- A step of aerobic biodegradation of pollutants in the solution or in the biofilm. These pollutants are used as carbon and energy source by microorganisms.

There are three types of biological processes:

(i) Conventional biofilters: Biomass is immobilized on a solid support containing nutrients (compost, peat) to which are often associated with "structuring" compounds (bark or wood chips) (Fig. 16.20a). The incoming air which contains the pollutant is humidified in order to ensure the transfer of the pollutant through the biofilter and the optimal development of the microorganisms. To avoid compaction, the height of the biofilter generally does not exceed 1 m. This device is used when the ground surface is not limited. In this case, the filter is flat and allows the processing of large amounts of low gas pollutant concentration (treatment of odors from sewage treatment plants, water stations, or composting facilities). For example, a 2,000 m^2 biofilter can handle about 160,000 m^3/h of air; in some facilities, gas flow rates can reach 360,000 m^3/h. Biofilters can also be used to treat either VOCs from contaminated sites (soil, groundwater) where a bioremediation is operated or industrial gases containing, for example, BTEX (benzene, toluene, ethylbenzene, xylenes). The main limitation of this treatment is the inability to control the pH.

(ii) Bioreactors: In a bioreactor, wherein the pH and the temperature are regulated, the microorganisms constituting the biofilm are immobilized on a column (Fig. 16.20b) consisting of an inorganic carrier with high packing vacuum level (e.g., Raschig rings, ceramic, etc.). The polluted air passes through the filter bed where biodegradation occurs. The liquid phase, which is recycled, flows against the flow and allows a nutrient input. Aeration is also operated.

(iii) Bioscrubbers: These devices have two compartments (Fig. 16.20c). In the first compartment, the contaminated gas is brought into contact with the liquid phase (introduced by spraying). This phase is then pumped into the second chamber (reactor) where the biodegradation of dissolved pollutants take place by suspended biomass (activated sludge). Bioscrubbers that do not contain a solid phase are effective only for soluble chemical compounds in the water.

16.7 Xenobiotics

16.7.1 Presentation

The term xenobiotic (in ancient greek *xenos*, stranger; *bios*, life) includes all synthetic chemicals that do not occur naturally and are produced by human activity. In terms of environment hazard, polychlorinated biphenyls (PCBs), chlorophenols, pesticides, chlorinated solvents, chlorinated aliphatic hydrocarbons, plastics, ammunition, etc. are the most problematic. PCBs and pesticides are discussed in this chapter. These chemicals have the greatest impact on ecosystems resulting from an extensive use and their inherent toxicity. Additional information regarding other xenobiotics and their degradation can be found in the reference books listed at the end of the chapter (Leisinger 1983; Connel 1997).

Since a century microorganisms have to face with these new molecules that do not occur previously in the geological eras. They have shown an extraordinary potential to attack most of these compounds, even if biodegradation may be extremely slow and take several years. Often biodegradation cannot be performed by a single strain and involves microbial communities. Frequently cometabolism is the sole possible route of degradation. This is the case of trinitrotoluene which is degraded in the presence of starch as a source of carbon and energy and numerous halogenated aliphatic compounds (dichloromethane and trichloroethene) which are degraded in the presence of methane.

16.7.2 The Adaptation of Microorganisms to Xenobiotic Compounds

The adaptation of microorganisms to the degradation of xenobiotic compounds is based on a wide variety of enzymes naturally distributed in microbial communities with a high genome plasticity favorable to the appearance of novel gene combinations or sequences suitable for the evolution of new biochemical capacities. Considering the time scale required for de novo appearance of a new metabolic capacity, it is likely that, in the most general case, the transformation of xenobiotic compounds can be initiated through the "recruitment" of enzymes assigned to the degradation of compounds that are sources of nutrition for microorganisms (Leslie-Grady 1985). The ability of an enzyme to be recruited depends on its substrate specificity. These occasional transformations have been collectively called "gratuitous" (Knackmuss 1981) or "fortuitous" (Slater and Bull 1982) metabolism.

Xenobiotic compounds bearing a common structural basis with natural substrates may be misleading for the enzyme which can accept them as new substrates. This is indeed not sufficient. The presence of functional groups or halogens may generate **steric*** and **mesomeric effects*** that can affect the fixation of the substrate and/or the reaction mechanisms at the active site of the enzyme. For instance, dioxygenases involved in the cleavage of aromatic rings of the soil organic matter cannot accept as substrates the corresponding chlorinated compounds. They are degraded by specific dioxygenases (Dagley 1978).

Enzyme recruitment is a key process to explain the main features that control the degradation of xenobiotic compounds in natural environments. Diverse situations have been observed. They are not mutually exclusive. Sometimes, degradation is brought about by enzymes belonging to different functional classes or pathways producing different reaction intermediates. For these chemicals, competitive degradation may occur with parallel degradation pathways running simultaneously. For instance, triazine herbicides can be degraded either slowly via successive *N*-dealkylations and accumulation of stable metabolites or much more rapidly via dechlorination and breaking of the side chains to form cyanuric acid which is then further mineralized. Soil pH regulates the balance between these two metabolic pathways (Houot et al. 2000). In some cases the enzymatic machinery can be operated via microbial consortia, each member carrying out a few metabolic steps forward. For a number of chemicals, we could expect the preexistence of a potential for degradation distributed in different microbial hosts, immediately efficient, or after structural changes of recruited enzymes. It has long been realized that the intrinsic biodegradation of a xenobiotic compound was better evaluated in environments showing high specific microbial richness such as soils or sewage sludge. More than an increased metabolic efficiency of microbial communities compared to pure strains, it is now established that they also represent a framework where different microorganisms harboring gene exchange may contribute to the emergence of new gene combinations making possible emergence of novel biochemical capacities. The possibility that extrachromosomal elements can cause dispersion in the soil microflora of characters they code for was suggested as soon as the early 1970s (Waid 1972). "Horizontal gene transfer" involves all the mechanisms of transformation, conjugation, and transduction. The evolutionary advantages of gene transfer may be extended inside the cell by different mechanisms of transposition that affect gene order and may contribute to the emergence of new operons. The dispersion of a newly acquired metabolic "competence" in a degrading metapopulation has been regarded as a process of stabilization (Bull 1980). Some have even seen in such a dispersion process the opportunity for genes to survive according to the evolutionary theory of "selfish gene" (Reanney 1976).

16.7.3 Pesticides

Since its settlement and development of agriculture and livestock, man had to protect his products and crops against various pests and diseases. To control them, he naturally turned to chemical substances of natural origin of which he had previously recognized the biocidal character. Thus, since antiquity, civilizations around the Mediterranean Sea have used as insecticides extracts of pyrethrum flower heads or tobacco leaves. From the fifth century, the Chinese used preparations of arsenic for the same purpose. The Egyptians used petroleum oil to control the growth of phytopathogenic fungi. In 1878 the "Bordeaux mixture," a solution of copper sulfate supplemented with lime, was prepared by the chemist Ulysse Gayon (1845–1929). With the botanist Alexis Millardet (1838–1902), they developed a means of protecting vineyards against mildew. It has been widely used to fight against diseases related to fungal attack.

Many synthetic pesticides appear early in the twentieth century. This is the case, for example, of dinitro-ortho-cresol (DNOC), which was synthesized in the late 1920s and patented in France by Truffaut and Pastac in 1932 for weeding grain. In 1935, the pesticide phenothiazine was commercialized by DuPont factory. In 1938, the Swiss chemist Paul Müller (1899–1965) discovered the insecticidal properties of DDT further marketed in 1946. In 1948, he received the Nobel Prize in Physiology and Medicine. DDT has been widely used by the Allies during the Second World War. The discovery just after the Second World War of the first selective herbicides, the hormone herbicides, contributed to extensive development of chemical control. In 1944, Mitchell and Hamner of the USDA mentioned the possibility of using synthetic growth regulators, including 2,4-D, as selective herbicides. Since that time and during the last 60 years, industrial countries have developed farming practices largely based on an extensive use of pesticides that have contributed to improving the quality and yield of crops with increasingly detrimental consequences for the environment. With rare exceptions, surface and deep-water reserves contain residues of pesticides that pose a potential danger to human health. Fortunately, these residues mainly pass through the soil with the purifying action of abundant and active microorganisms that help to reduce their concentration. Advances in knowledge of the physiological and biochemical aspects of microbial degradation of pesticides are needed both to define environmental friendly molecules and to develop strategies for remediation of contaminated sites.

16.7.3.1 Pesticides in the Environment: A Few Figures

The sale and use of synthetic organic molecules that help fight against pests was regulated by a European Directive (Directive 91/414/EC), which has established a procedure for assessing the risk to the consumer, the user, and the environment. This Directive was recently repealed by the regulation (EC) N° 1107/2009 adopted with the new Directive 2009/128/EC "establishing a framework for Community action to achieve the sustainable use of pesticides." The need to register all of the molecules within the new EU law has contributed to a significant reduction of authorized pesticides. For instance, in 2013, a total of 1,297 active substances were listed in the EU Pesticide database. Only 428 are approved in Europe and 317 in France, the largest market of pesticides in Europe (http://e-phy.agriculture.gouv.fr/). Since 1999, the use of pesticides has decreased from 89,100 tons of synthetic products to 48.800 tons in 2011 (www.uipp.org/). However, this trend should be taken with caution. Indeed, the ban on old molecules applied at high doses and the placing on the market of new active products applied at very low doses have probably contributed as much as various measures implemented to reduce pesticide usage.

Whatever the conditions and characteristics of the environment, it is estimated that less than 5 % of applied pesticides reach their target, the excess being dispersed in different environmental compartments (air, soil, water, sediment, etc.) and in foodstuffs. They may involve immediate or long-term risks and hazards for humans, animals, and other environmental non-target organisms (NTO). The data on soil pollution are difficult to collect and still relatively scarce to have real significance. The compilation of data for aquatic environments gives a more reliable view of the state of pollution of aquatic bodies. The results of a survey conducted in France between 2007 and 2009 were published in 2011 by the French Ministry in charge of the environment. Data reported the overall presence of pesticides in surface and ground waters on the entire territory. Pesticides were detected at 91 % of the 2,889 monitoring points in rivers with 21 % showing a total pesticide concentration greater than 0.5 µg/L. The presence of pesticides was quantified in 70 % of the 2,321 monitoring points in groundwaters. A little more than 27 % of groundwater points did not meet water quality requirements in at least 1 year between 2007 and 2009. The pesticides most often encountered in surface waters were essentially herbicides: glyphosate and AMPA, its major degraded compound, and deethylatrazine, a metabolite of atrazine banned 10 years ago. Atrazine and its main dealkylated and hydroxylated metabolites are most commonly found in groundwaters.

16.7.3.2 Pesticide Biodegradation
The Early Studies of Biodegradation: The Hormone Herbicides

The studies of Mitchell and Hamner on synthetic hormone herbicides were considered as "the greatest advance in the history of weeding" (Brian 1976). It is based on two important observations: the selective biological activity on mono- and dicotyledonous plants and the persistent action of synthetic compounds compared to the short-lived natural auxin, β-indole acetic acid. However, even with synthetic compounds, it became obvious that the herbicidal activity decreased over time, especially when the conditions of moisture, temperature, and organic content of the soil were more favorable to microbial activity (Brown and Mitchell 1948). It also appeared that the exponential decrease of the herbicidal activity following a lag phase could be related to the development of a specific degrading microorganisms (Audus 1949). The definite conclusion of the involvement of the microflora in the process of inactivation was supported by an extension of the herbicidal activity after sterilization of the soil (Audus 1951) and by the isolation from soil of microbial species capable of utilizing 2,4-D as the sole source of carbon and energy. In 1950, Audus isolated a microbial strain, *Bacterium globiforme*, capable to degrade 2,4-D from a soil previously perfused with a solution of this herbicide. Since then, many other microbial strains were isolated from agricultural soils or even pristine soil with no previous 2,4-D application. In general, bacteria belonging to different genera or species were isolated rather than fungi. Some of them show limited metabolic capabilities that do not allow them to mineralize 2,4-D.

Principal Biochemical Pathways Involved in Pesticide Degradation

All the chemicals which are currently authorized for crop protection all over the world show a large variety of chemical structures. An illustration is given on Table 16.4 for herbicides which are most commonly used. The degradation of these chemicals is based on biochemical transformations controlled by a broad spectrum of enzymes. A detailed review of different degradation pathways of pesticides is presented in a database at the University of Minnesota (http://umbbd.ethz.ch/index.html). This database has been used to develop a computational metabolic pathway prediction system to derive plausible metabolic pathway of chemicals for which biodegradation data is currently lacking (Ellis and Wackett 2012). A limited presentation of the major degradation pathways and associated enzymes will be given here. More exhaustive presentations can be found in the literature (Van Eerd et al. 2003; Scott et al. 2008).

Pesticide degradation involves two stages. A first series of transformations such as oxidations, reductions, and hydrolysis contribute to modify the initial structure, and biological properties, of the parent molecule. Metabolites are produced which are generally more soluble and less toxic with some exceptions such as chlorophenols and chloroanilines. These early changes have the effect of reducing the initial xenobiotic character of the chemicals which, in the most favorable case, can integrate the central metabolic pathways for further mineralization and synthesis of microbial biomass. When this metabolic process cannot go to its completion, the accumulation of transformation products may result. In the soil, most of these metabolites do not remain as free solutes. Rather they bind to organic constituents of the matrix to form high molecular compounds, the so-called bound residues.

(i) Oxidation Reactions (Fig. 16.21)

Oxidoreductases are a large group of enzymes that catalyze the transfer of electrons from one electron donor (the reducing agent) to an electron acceptor, NAD $(P)^+$, or a cytochrome for dehydrogenases and reductases, molecular oxygen for oxidases, quinone, etc. Some require an additional cofactor. Degradation of glyphosate into aminomethylphosphonic acid (AMPA) and glyoxylate is mediated by a dehydrogenase operating with a flavoprotein as a cofactor. An enzyme of the

16 Applied Microbial Ecology and Bioremediation

Table 16.4 Main chemical families of herbicides and examples of widely used compounds

Amide	*Metolachlor*	
Anilide		
Arylalanine		
Sulfonanilide		
Sulfonamide		
Thioamide	*Chlorthiamid*	
Aromatic acid		
Benzoic acid	*2,3,6-TBA*	
Picolinic acid	*Picloram*	
Benzoyl cyclohexanedione	*Sulcotrione*	
Benzothiazole	*Methabenzthiazuron*	
Carbamate		
Dinitroaniline	*Trifluralin*	
Dinitrophenol	*DNOC*	
Diphenyl ether		
Halogenated aliphatic acid	*TCA*	
Dithiocarbamate		
Imidazolinone		
Nitrile	*Ioxynil*	
Organophosphonates	*Glyphosate*	
Oxadiazolone		
Phenoxyalcanoic acid	*2,4-D*	
Pyrazole		
Pyridazine		
Pyridazinone		
Pyridine	*Triclopyr*	
Pyrimidinediamine		
Quaternary ammonium	*Diquat*	

(continued)

Table 16.4 (continued)

Thiocarbamate		
Triazine		
Chlorotriazine	*Atrazine*	
Methoxytriazine		
Methylthiotriazine		
Triazinone		
Triazole	*Amitrole*	
Triazolone		
Uracil		
Urea		
Phenylurea	*Isoproturon*	
Sulfonylurea		
Triazinylsulfonylurea	*Chlorsulfuron*	

same class brings about N-demethylation of isoproturon with an unspecified acceptor. N-dealkylation of atrazine in soils is mediated via a monooxygenase, a **cytochrome P450***, in the presence of molecular oxygen. This enzyme brings about the incorporation of one atom of oxygen on the substrate, the other atom reacts with the electrons a cofactor (often NAD(P)H) to form a water molecule. Some monooxygenases and dioxygenases are involved in the degradation of aromatic rings, which are often the basic structure of xenobiotic chemicals. They are central in the enzyme machinery of soil microorganisms as they are key players in the degradation of natural polymers that make up the soil organic matter. A specific subclass, the hydroxylases, contributes to the formation of dihydrodiols and/or catechols preparing the opening of aromatic rings by another group of dioxygenases. They catalyze the incorporation of one or both atoms of molecular oxygen in the presence of NAD(P)H as a donor. Complete mineralization of aromatic compounds may also depend on intramolecular rearrangements. The transformation of geometric isomers is mediated by isomerases (Fig. 16.22). Nitroreductases are also oxidoreductases which catalyze the reduction of NO_2 groups present in some cycles such as parathion and trifluralin in the presence of NAD (P) H. These amino derivatives are probably involved in the formation of bound residues. Microorganisms also produce a wide range of other oxidative enzymes, peroxidase, laccase, and tyrosinase that catalyze very specific transformations producing minor structural changes. These transformations are collectively called "**detoxification reactions***" as they generally contribute to a decrease in the biological activity. Soil fungi are mainly responsible for these transformations. Some of the products of these transformations, e.g., quinone derivatives, have the ability to react spontaneously to form polymers which are included in the pool of non-extractable bound residues. With 3,4-dichloroaniline however (a metabolite of phenyl carbamates and phenylurea), one can observe the formation of tetrachloroazobenzenes with carcinogenic activity. Other oxidative transformations by conjugation such as methylation and acetylation have been reported. They are still very poorly documented.

(ii) Hydrolysis Reactions (Fig. 16.23)

Hydrolases are another class of enzymes that frequently bring about degradation of pesticides. Ester, amide, or carbon-halogen bonds are broken by hydrolases. They generally operate in the absence of redox cofactors.

Fig. 16.21 Examples of transformations brought about by oxydoreductases. (**a**) Acting on the CH–CH bond of donors. (**b**) Acting on the CH–NH bond of donors. (**c**) Acting on a donor with incorporation of molecular oxygen. (**d**) Acting with a reduced cofactor with incorporation of molecular oxygen. (**e**) Other oxydoreductases

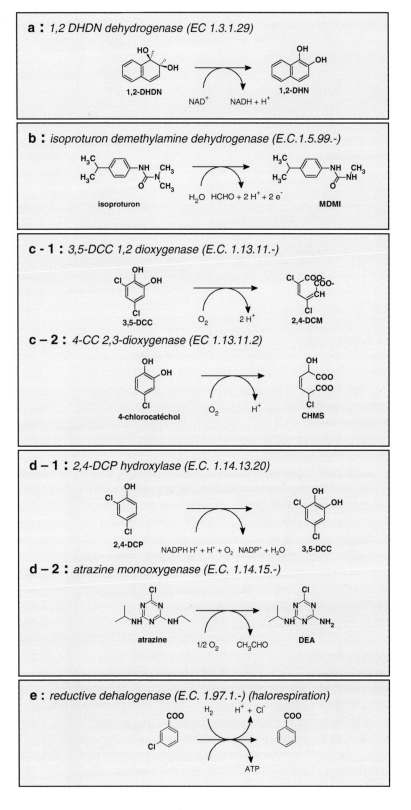

Phosphotriesterases are involved in the transformation of organophosphorus insecticides: parathion is converted either into paraoxon by hydrolysis of the P–S bond or into paranitrophenol and diethylthiophosphoric acid by hydrolysis of the O–P bond. There are several examples of hydrolases that target the C–N bond. A microbial nitrilase converts bromoxynil into 3,5-dibromo-4-hydroxybenzoate. Amido and chlorohydrolases play a

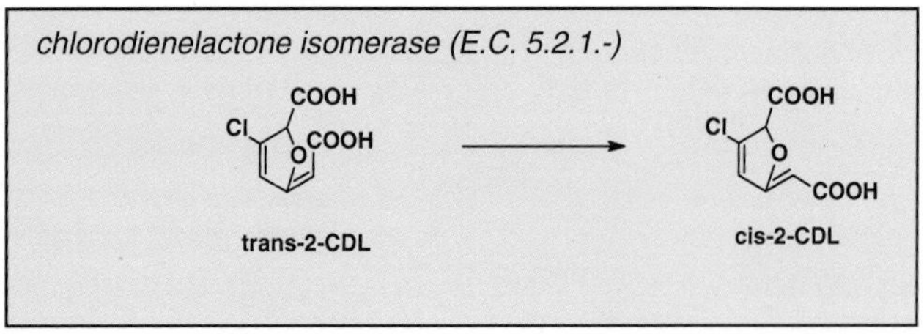

Fig. 16.22 Example of transformation brought about by an isomerase

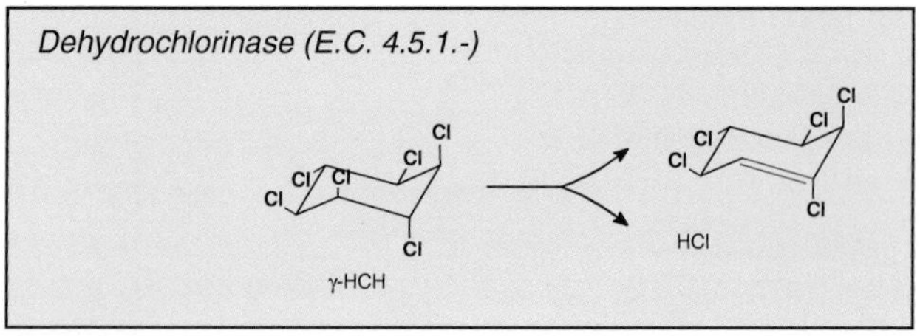

Fig. 16.24 Example of transformation brought about by a lyase (acting on the C–Cl bond)

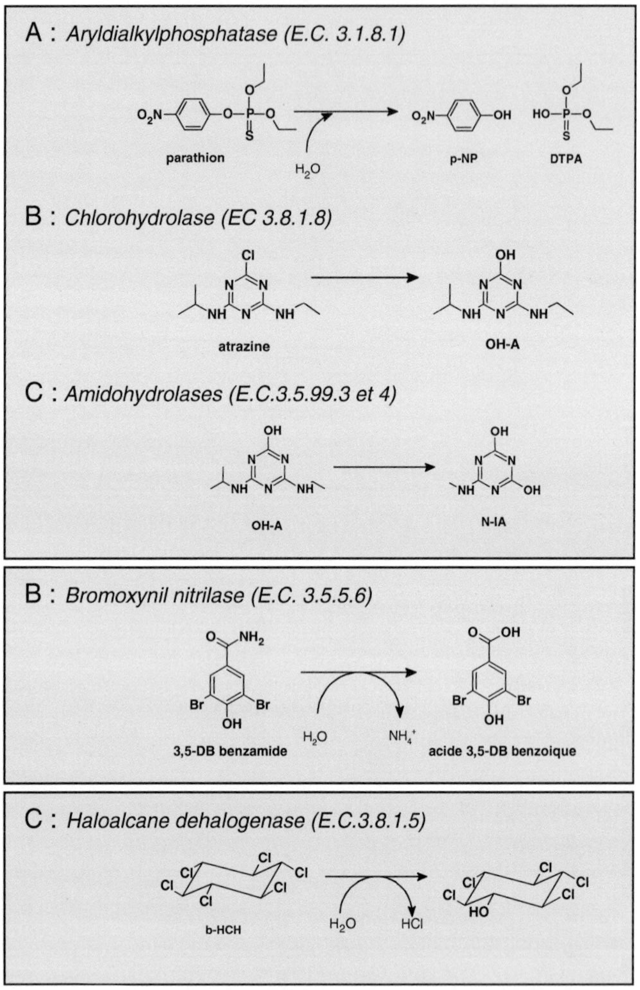

Fig. 16.23 Examples of transformations brought about by hydrolases. (**a**) Acting on the ester bond. (**b**) Acting on the C–N bond (not the peptide bond). (**c**) Acting on the C–Cl bond

hydroxyatrazine. This intermediate product is itself further transformed into cyanuric acid by two amidohydrolases which cut the C–N bonds of both side chains. Another amidohydrolase transforms the cyanuric acid intermediate into biuret that is readily degraded and mineralized.

(iii) Dehalogenation Reactions (Fig. 16.24)

Dehalogenations are central in the metabolism of chemical compounds where one or several hydrogen atoms of the core structure are replaced by halogen atoms which give them their xenobiotic character and contribute to their chemical stabilization. Several types of enzymes are involved in these transformations. The haloalkane dehalogenases and the 4-chlorobenzoate dehalogenase are hydrolases which bring about substitution of the Cl atom by an OH group coming from the water molecule. This is the case with the β isomer of HCH. The γ isomer, lindane, and DDT are degraded by a dehydrochlorinase, a lyase which releases a molecule of HCl, with both atoms being derived from the pesticide molecule. Many bacteria have been isolated which are capable of reductive dehalogenation of aliphatic (tetrachloroethylene) and aromatic compounds (3-chlorobenzoate). The energy released in the process of dehalogenation is coupled to ATP synthesis. This type of transformation under the special control of oxidoreductases, the reductive dehalogenases, is called "halorespiration." Chlorinated substrates are used as terminal acceptor of electrons supplied by the hydrogen or another compound such as formate.

Which Organisms Are Involved?

As a general rule, as demonstrated for hormone herbicides, the soil microorganisms are mainly responsible for the breakdown of xenobiotic compounds. This potential results from a large diversity of metabolic capacities and from a high plasticity of microbial genome that allows gene sequence rearrangements that are the basis for developing new metabolic pathways.

key role in the degradation of chlorinated triazines in soils where they have contributed to the emergence of the phenomenon of accelerated degradation. A chlorohydrolase is first responsible for the formation of

The bacteria and fungi seem to play different roles. Fungi are often responsible for detoxification reactions (dehalogenation, hydroxylation, etc.) associated with small changes in the molecular structure that often result in solubility increase and a decrease in biological activity. By comparison, bacteria appear to be involved in transformations associated in complex metabolic schemes catalyzed by more specific enzymes. Products may enter central biochemical pathways to generate energy and new biomass. For the last 10 years, the potential for biotransformation and biodegradation of *Actinomycetes* was found very important (De Schrijver and De Mot 1999). Members of this group of Gram-positive bacteria, sometimes associated in consortia, are capable of degrading products as diverse as organochlorines, s-triazines, carbamates, organophosphates and organophosphonates, acetanilides, sulfonylureas etc. The molecular mechanisms of these transformations are less well known than those that contribute to the degradation of pesticides by Gram-negative bacteria.

Physiological Bases of Pesticide Degradation

Biodegradation can be studied by using pure strains or microbial consortia:

(i) Degradation by pure cultures: metabolism and cometabolism

It has been possible to isolate from soil a large corpus of microbial strains that show very diverse catabolic capacities according to different modes of physiological functioning.

- *Degradation of pesticides as the sole source of carbon and energy: mineralization*

  When a single microbial strain can mobilize a number of enzymes sufficient to use a "foreign" substrate as sole source of carbon and energy, degradation is called "metabolism" and results in the mineralization of the chemical. This mode of degradation is functionally very similar to those which occur for the consumption of "natural" nutrient sources. It concerns only a limited number of chemicals. In the soil environment, metabolism may contribute to a stepwise increase of the microbial community as applications of the chemical are repeated. This microbial enrichment is the basis of the phenomenon of "accelerated degradation," also called "adaptation" or "acclimatization." It may be detrimental for chemicals which are directly applied to the soil as they show a progressive loss of efficacy. This phenomenon has been observed in the early 1950s with 2,4-D (Audus 1951). Since, it seems to occur on a larger scale and for a growing number of chemicals. Most recent examples of this behavior are shown by the insecticide carbofuran and the herbicides atrazine and isoproturon.

- *Degradation of pesticides in the presence of other sources of carbon and energy*

  For a large majority of chemicals, it was not possible to isolate microbial strains capable of using them as sole sources of carbon and energy. However, it is not uncommon to isolate degrading strains with the addition in the culture medium of a second substrate serving as a source of carbon or nitrogen and energy. This procedure reveals the ability of a number of microorganisms to transform a pesticide molecule without being able to drive the metabolic process at a stage where energy can be derived. These "failures" may have several causes: lack of or low catalytic activity of the enzymes, independent enzyme induction leading to the accumulation of toxic intermediates, etc. All transformations that are not directly connected to cell growth and require the presence of a co-substrate to progress at a significant rate were collectively called "cometabolism" (Horvath 1972). In the soil environment, fresh or humidified organic matter is the main source of co-substrates. The concept of cometabolism is a generalization proposed by Jensen (1963) of the "co-oxidation" process previously developed by Leadbetter and Foster (1959) who observed the oxidation of certain hydrocarbons in the presence of methane without growth of the responsible strain *Pseudomonas methanica*. Cometabolism includes transformations such as dehalogenations that are not oxidation strictly speaking. Originally presented as a new microbial physiological trait, the use of the term cometabolism was criticized by Hurlbert and Krawieck (1977) who saw no specific biochemical characteristics but rather the outcome of dysfunctional enzyme. Nevertheless, the term was maintained to describe transformations with particular kinetics, the paradigmatic first-order law, and/or ecotoxicological consequences when accumulation of hazardous end products occurs.

(ii) Degradation by complex microbial communities. Primary and ultimate degradation

In 1965, Alexander concluded that degradation studies conducted with mixed cultures and microbial communities from natural ecosystems were more appropriate for assessing the inherent biodegradability of xenobiotic compounds. However, even in the presence of complex microbial communities, some pesticides are relatively persistent and/or are not transformed enough to be further converted by catabolic reactions to provide the energy needed by the cells. "Primary degradations" characterize incomplete transformations which result in the production of dead end products. In contrast, "ultimate degradations" are those which drive a complete set of transformations enabling mineralization of the pesticide molecule. These two categories of degradations are also not mutually exclusive and may coexist in the same environment. At the community level, primary and ultimate degradations are the functional equivalents of metabolism and cometabolism described for pure strains. The ecological role of complex microbial communities was later reconsidered by Bull (1980) who recognized that pure culture studies had benefits for the elucidation of the metabolic functioning of microorganisms and transformation pathways. However, they had also many limitations for extrapolation to natural environments that allow enzyme complementarity and redundancy required to sustain a collaborative process of degradation. The presence, within a microbial community of isofunctional enzymes, is likely to ensure better enzyme efficacy especially under varying environmental conditions. This is also the case for consortia that drive alternative routes of transformation. At last, the diversification of microbial hosts resulting from horizontal gene transfer contributes to stabilize the genetic determinants of degradation and to maintain the overall degrading capacity. Nevertheless, at the present state of technical advances, the isolation of degrading strains or microbial communities requires the use of liquid culture media. This is a major drawback because it focuses on culturable microorganisms which are likely to have limited contribution to the degradation of persistent compounds with ecological significance. It may also explain some of the pitfalls and failures encountered in bioremediation experiments as will be discussed in the following.

Genetic Basis of Pesticide Biodegradation

The knowledge about genetic processes responsible for the capacity of soil microorganisms to degrade pesticides mainly results from advanced studies carried out on accelerated biodegradation of pesticides. This phenomenon that relies on the growth and development of the size of degrading-microbial community diminished pesticide persistence in the environment (for reviews, see Topp et al. 2004; Arbeli and Fuentes 2007). It is well known that accelerated biodegradation of pesticides can last for several crop rotations. In several cases, accelerated biodegradation was shown to diminish the efficacy of pesticide treatment: this phenomenon is particularly well described for the insecticide carbofuran (Felsot et al. 1982). The stability of pesticide-degrading capabilities observed in situ is in contradiction with the frequently reported instability of degrading capabilities of pure microbial strains maintained under laboratory conditions (Changey et al. 2011). Indeed, it has been observed by different researchers that pesticide-degrading potential can easily be lost under laboratory conditions by simply sub-cultivating degrading strains on agar medium without the selection pressure exerted by the pesticide. Recently, Changey et al. (2011) suggested that in absence of the selection pressure, pesticide-degrading-genetic potential represents a genetic burden favoring the selection of strains having selectively lost this burden by different genetic mechanisms. Keeping in mind this instability observed on pure strains kept under laboratory conditions, several hypotheses have been drawn to tentatively explain the stability of pesticide-degrading ability of soil microbial community observed in situ. Among these hypotheses, survival of pesticide-degrading populations and maintenance of the integrity of their pesticide-degrading-genetic potential might result from the presence in the soil of natural compounds sharing enough structural homologies with the pesticide to maintain the selective advantage favorable for their survival. This hypothesis was verified for 2,4-D biodegradation for which it was found that both 4-chlorocatechol (intermediary metabolite of 2,4-D) and homogentisic acid (2,5-dihydroxyphenylacetic acid, intermediary metabolite of the biosynthesis of aromatic amino acid) are able to induce the expression of 2,4-D biodegradation pathway in *Arthrobacter* sp. (Sandmann and Loos 1988). Similarly, for atrazine biodegradation, Udikovic-Kolic and collaborators (2012) suggested that the prevalence of *trzN* (responsible for the dechloration of atrazine) might results from its ecological advantages: It is faster (Vmax 31 mmol min^{-1} mg^{-1} of protein), it has a higher affinity for the substrate (Km 25 µM), and it has a higher range of substrates including natural ones. One could also hypothesize that under oligotrophic conditions found in the soil, induction and repression mechanisms do not act with the same intensity than under **pleiotropic*** conditions. It is now well admitted that a lot of the genes coding for pesticide catabolic enzymes are plasmid-borne. A lot of these plasmids are conjugative and are part of the so-called mobilome which can be transferred from a donor to a receptor population by horizontal gene transfer (HGT) mediated by bacterial conjugation (Frost et al. 2005). The first demonstration of the plasmidic location of pesticide-degrading genes was done at the end of the 1970s on *Alcaligenes paradoxus* able to degrade 2,4-D, thanks to degrading genes located on the pJP1 plasmid (Pemberton and Fisher 1977). Thereafter, several other plasmids harboring 2,4-D-degrading genes were isolated among which pJP4, isolated from *Alcaligenes eutrophus* JMP 134 (Don and Pemberton 1985) later on renamed *Ralstonia eutropha* and then *Cupriavidus necator* JMP 134, has particularly been studied. All the genes coding for the catabolic enzymes, transporter, and regulators have been identified on pJP 4. In addition to these plasmids, several others have been identified. Based on sequence similarities observed between them, it was suggested that they may derive from a common ancestor (van der Meer et al. 1992). This hypothesis was later reinforced by the observation of independent recruitment of diverse gene cassettes during the assembly of 2,4-D-degrading plasmids in different bacterial strains (Vallaeys et al. 1999). It was also shown that genetic recombination mediated by transposable elements, insertion sequences (ISs) being their simplest representative, played an important role in the construction of pesticide-degrading pathways. These transposable elements play the role of shuttle for degrading genes favoring the construction of catabolic operon (Box 16.2).

However, whatever the transfer mechanism involved, successful transfer of pesticide-degrading genes between different populations of soil-degrading community does not necessarily mean that transferred genes are expressed in the host. Indeed if chromosomal-borne regulatory elements are not transferred to the new host, the expression of pesticide-degrading genes might be compromised. In addition, following HGT structural modifications of catabolic plasmid to adapt to genetic environment of the host might also alter the expression of pesticide-degrading genes. In that way, several observations suggest that catabolic-plasmid dispersion among different populations of the bacterial community is accompanied by its structural and genetic diversification (van der Meer et al. 1992). It has been demonstrated that different catabolic plasmids, in particular those involved in the degradation of halogenated aromatic compounds, are sharing important sequence similarities. This observation reveals the existence of conserved sequences, notably those involved in the degradation. Recently, fine analyses of data issued from large-scale sequencing program of the full genome of numerous microorganisms allowed the identification of those conserved sequences which are often part of ICE elements referring to group of mobile DNA elements, which are known as conjugative transposons or plasmids or genomic islands (van der Meer and Sentchilo 2003). These components have particularly well been observed for *clc* element coding for chlorobenzoate and chlorocatechol degradation. Indeed, Springael et al. (2001) showed that *Tn*4371-like mobile elements are involved in the mobility of *clc* genes. Later on, Gaillard and collaborators (2006) showed that the *clc* element responsible for 3- and 4-chlorocatechol degradation was part of a mobile genomic island in *Pseudomonas* sp. strain B13 which was not only showing catabolic properties but also exhibiting pathogenic characters. These observations are again in favor of an ancestral origin of the different catabolic plasmids that evolved following their transfer to different hosts to adapt to their lifestyle. These evolutions are often mediated by mobile genetic elements that are providing

---

Box 16.2: Pesticide Biodegradation: Genetic Basis. The Examples of Hormone Herbicides and *s*-Triazines

Guy Soulas* and Fabrice Martin-Laurent

2,4-D and Hormone Herbicides

The genes *tfdA*, *B*, *C*, *D*, *E*, and *F* encode the transformation of 2,4-D into chloromaleylacetate further transformed into maleylacetate by a chromosome encoded maleylacetate reductase. They were first identified on a plasmid pJP4 of a 2,4-D-degrading strain *Cupriavidus necator* JMP 134 (formerly *Alcaligenes eutrophus* and *Ralstonia eutropha*) (Don and Pemberton 1985). pJP4 is a member of the IncP1 family of plasmids which may be viewed as among the most potent vehicles for the spread and accumulation of multiantibiotic resistance and many kinds of accessory genes between a broad range of host bacteria. DNA signature analysis demonstrated that recombination events have allowed them to rapidly adapt to changing bacterial hosts and environmental conditions (Norberg et al. 2011; Sen et al. 2011). The first step in the degradation of 2,4-D is performed by an α-ketoglutarate-dependent 2,4-D dioxygenase, encoded by the *tfdA* and *tfdA*-like genes. The analyses of the sequences in isolated strains have defined the existence of four groups of *tfdA* genes: classes I, II, and III in β- and g-*Proteobacteria* (Mcgowan et al. 1998) and *tfdA*a genes in some a-*Proteobacteria* (Itoh et al. 2002). The continuous exposure of soil microcosms to different hormone herbicides produced an impressive variety of new *tfdA* gene-related sequences (Gazitúa et al. 2010). The abundance and diversity of these sequences in soils may be explained by the role of α-ketoglutarate dioxygenases in catabolism of natural, more common, compounds (Hogan et al. 1997). The product of the *tfdB* genes performs the crucial step responsible for the initial hydroxylation of the benzene ring. It catalyzes the FAD-dependent oxidative hydroxylation of 2,4-DCP and its homologues, in the presence of O_2 and NADPH/NADH as an electron donor. Many chlorophenol hydroxylases have been isolated. Most of them display high activity towards 2,4-DCP and 4-chloro-2-methylphenol. A novel *tfdB* gene, designated as *tfdB*-JLU, was identified (Lu et al. 2011). The gene product shows only moderate sequence identity (48 % identity) with other known TfdB proteins. TfdB-JLU exhibited monooxygenase activity towards many chlorinated phenols with the highest activity towards

(continued)

Box 16.2 (continued)

Box Fig. 16.3 Degradation of the 2,4-D herbicide and organization of *tfd* genes on plasmid pJP4 (Modified and redrawn after Laemmli et al. 2000)

3-chlorophenol. The four genes *tfdC, D, E,* and *F* encode, respectively, the dichlorocatechol-1,2-dioxygenase, the chloromuconate cycloisomerase, the chlorodienelactone isomerase, and the chlorodienelactone hydrolase. In the reference plasmid pJP4, these genes are organized in an operon. The expression of genes *tfdA* and *tfdCDEF* operon is modulated by a transcriptional regulator of the LysR family produced by the regulatory genes *tfdR* (Harker et al. 1989) and *tfdS* which are inverted repeat sequences that are likely to play a similar role. Three other genes, *tfdT*, *tfdBII*, and *tfdK*, have been located upstream of the *tfdC-F* cluster. The protein TfdT has 49 % homology with the product of *tfdR* (and *tfdS*) but is inactive due to the presence of an insertion sequence (ISJP4) in *tfdT*. The gene *tfdK* is responsible for producing a carrier involved in the facilitated transport of 2,4-D (Leveau et al. 1998). The plasmid pJP4 carries a second group of genes *tfdDIICIIEIIFII* phylogenetically different from those of the *tfdC-F* cluster. It is preceded by a sequence highly homologous to the regulator genes *tfdS* and *tfdR* (Box Fig. 16.3) (Laemmli et al. 2000). The genes of the *tfdDII-FII* operon, which was suspected to have widespread occurrence in soil bacteria (Kim et al. 2013), differ from those of the *tfdC-D* operon by the reverse order of the genes C and D; a higher percentage of G + C%, 66 % against 58 %; by an identity in the amino-acid sequence ranging from 16 to 60 %; and with a lower catalytic activity. These characteristics argue in favor of a different phylogenetic evolution by duplication divergence.

The gene clusters of the oxidative degradation pathway of hormone herbicides located in different bacteria or plasmids may be organized as described for pJP4. For instance, plasmids pDB1 (*Variovorax* sp. DB1) and p712 (*Pseudomonas pickettii* 712) carry *tfd* genes sharing significant similarity in gene sequences and cluster organization compared to the *tfdR*, *tfdBII*, *tfdCII*, *tfdDII*, *tfdEII*, *tfdFII*, *tfdK*, and *tfdA* genes of plasmid pJP4 (Kim et al. 2013). However, other gene assemblages have been described. Plasmids pAKD25 and pAKD26 show *tfd* gene organization and sequence identity closely related to each other and to those found on the plasmids pEST4011 (*Pseudomonas putida* PaW85) and pIJB1 (*Burkholderia cepacia* strain 2a) and are less similar to the *tfdI* and *tfdII* clusters of the catabolic plasmid pJP4. Both plasmids lack *tfdF*, and on pAKD26, a segment containing *tfdD* has been deleted (Sen et al. 2011). Considering the contribution of plasmids to horizontal gene transfer and prokaryotic evolution, discovering the biological mechanisms by which plasmids evolve is important to better understand the principles underlying the emergence and spread of catabolic capacities in bacterial communities. Plasmids of the incP-1 group carry a wide spectrum of genes coding for the degradation of halogenated aromatic chemicals. They have two distinct regions, the backbone region that contains

(continued)

Box 16.2 (continued)

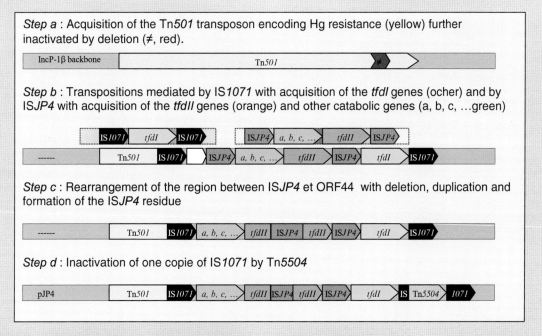

Box Fig. 16.4 Construction of pJP4 (Modified and redrawn after Trefault et al. 2004)

genes involved in the control of plasmid replication, transfer, and maintenance and a region with accessory genes that confer beneficial traits to the bacterial host such as the ability to metabolize xenobiotic compounds. Accessory genes are often associated with mobile genetic elements such as transposons that facilitate their incorporation in or expelling from the backbone. Plasmids with similar backbones can have different accessory genes (Schlüter et al. 2007). To illustrate these genetic events, a pattern of evolution of the plasmid pJP4 has been proposed (Trefault et al. 2004) (Box Fig. 16.4). It would have been constructed from an ancestral plasmid backbone IncP-1 β further modified by successive transposition events. The first, controlled by the transposon Tn501, resulted in the acquisition of the *mer* genes encoding mercury resistance. Two other transposition events respectively controlled by the transposon IS1071 for the *tfdC-F* operon and ISJP4 for the *tfdDII-FII* operon have imported genes encoding the oxidative degradation pathway of hormone herbicides. Deletions occurred later that have contributed to the inactivation of these transposons. These structural changes seem to optimize the expression of catabolic genes. In particular, both *tfd* and *tfdII* operons seem essential to ensure degradation of 2,4-D, as the *tfd* operon alone only controls degradation of 3-chlorobenzoate.

s-Triazines

s-Triazines family comprises several herbicides including atrazine which was world widely used to control weeds development in corn cropping for several decades. Over this period, atrazine has been long recognized as a moderately persistent herbicide, with half-life spans ranging from a few days to a few months. Until the early 1990s, the only known degradation pathway relied on a series of N-desalkylations leading to the formation of dealkylated metabolites deisopropylatrazine (2-chloro-4-ethylamino-6-amino-s-triazine, CEAT) and deethylatrazine (2-chloro-4-amino-6-isopropylamino-s-triazine, CIAT) (Mulbry 1994; Shao et al. 1995). Several chlorohydrolases and aminohydrolases are then taking part to the transformation of dealkylated metabolites generating a series of intermediary metabolites such as melamine, ammeline, ammelide, and cyanuric acid. These metabolites are produced in low amounts and strongly adsorbed to soil components. This degradation pathway contributed only slightly to open the s-triazine ring and mainly resulted in the accumulation of CEAT and CIAT in the soil. The genetic background of this degradation pathway was well studied thanks to the discovery, on one hand, of *Rhodococcus corallinus* NRRL B-15444R that dechlorinates and deaminates the

(continued)

Box 16.2 (continued)

dealkylated *s*-triazine compounds CEAT and CIAT but not atrazine and of TrzA, a *s*-triazine hydrolase (Cook and Hutter 1984), on the other hand, of *Rhodococcus* sp. strain TE1 that can degrade atrazine efficiently to produce the dealkylated metabolites deisopropylatrazine (CEAT and CIAT) due to AtrA, an enzyme able to catalyze *N*-dealkylation reactions. Based on these reports, it was suggested that in soil, atrazine was dealkylated under the action of AtrA and then dealkylated metabolites were dechlorinated by TrzA. It was recently shown as well that TrzA was able to transform melamine to ammeline like TriA (Dodge et al. 2012). It is noteworthy that cloning and expression of *trzA* in *Rhodococcus* sp. strain TE1 successfully led to the dechlorination of the dealkylated metabolites of atrazine (Shao et al. 1995). The sequence of the gene *trzA* is known, but until now the sequence of *atrA* remained unknown. This gene was found to be located on a 77 kb plasmid in *Rhodococcus* sp. strain TE1 which is able to degrade atrazine and the thiocarbamate herbicide EPTC (*S*-ethyl dipropylcarbamothioate). It was hypothesized that AtrA could be a P450 monooxygenase. This is also the case for *thcB* (coding for a P450 monooxygenase), *thcC*, and *thcD* genes coding for EPTC-degrading enzymes in *Rhodococcus* sp. NI86/21 which was also shown to degrade atrazine (Nagy et al. 1995). Similarly, other *trz* genes coding for enzymes of *s*-triazines catabolism have been identified in *Pseudomonas* sp. NRRL B-12 227, *Pseudomonas* sp. NRRL B-12 228, and *Klebsiella pneumoniae*.

In the early 1990s, following the observation of several cases of accelerated degradation of atrazine in soils repeatedly cropped with corn and consequently frequently exposed to this herbicide, several bacterial consortia and even pure bacterial strains able to mineralize this herbicide were isolated (for review see Udikovic-Kolic et al. 2012). Among these different atrazine-degrading isolates, *Pseudomonas* sp. ADP (Mandelbaum et al. 1995) has been particularly well studied. In the presence of citrate, this microbial strain is capable of using atrazine as the sole source of nitrogen, carbon of atrazine being almost entirely mineralized. All the genes coding for the enzymes catalyzing the mineralization of atrazine have been identified (Martinez et al. 2001). These are the *atz* genes coding for a chlorohydrolase (AtzA) transforming atrazine to hydroxyatrazine, a first amidohydrolase (AtzB) transforming hydroxyatrazine to *N*-isopropylammelide and a second one (AtzC) producing cyanuric acid starting from *N*-isopropylammelide. *s*-Triazine ring of cyanuric acid is then degraded by three other enzymes encoded by *atzDEF* genes organized as an operon on pADP1 (Garcia-Gonzalez et al. 2005). Comparison of the conserved N-terminal sequences rich in histidine residues of proteins AtzA, B, and C region suggests that these three genes are derived from a common ancestor. It is noteworthy that the amino-acid sequence of AtzA shared only 41 % of similarity with that of TrzA. On the contrary, the very high similarity (99 %) found between these genes with *atz* genes found in various atrazine-degrading strains suggests their recent divergence. More recently, the *trzN* gene coding for a chlorohydrolase has been shown in *Nocardioides* sp. C190 (Topp et al. 2000). Although *trzN* is contributing to the transformation of atrazine in hydroxyatrazine, it is significantly differing from *atzA* and *trzA*, which suggest a different origin. This gene has since been isolated in many bacterial strains and detected in many agricultural soils cropped with corn and regularly treated with atrazine (Devers et al. 2007b).

In *Pseudomonas* sp. ADP, *atzABCDEF* genes are located on pADP1, a 96 kb conjugative plasmid, harboring IncP-1β incompatibility group and most likely derived from pR751 plasmid. Horizontal gene transfer (HGT) of pADP1 was studied in vitro, and thereafter, this phenomenon was observed among indigenous microflora of a soil cropped with corn and treated with atrazine (Devers et al. 2005). In addition, like for phytohormones of synthesis, HGT of catabolic plasmid harboring *atz* and *trz* genes was suggested following the observation of these genes on plasmid differing in size detected in various atrazine-bacterial isolates (Devers et al. 2007a). The observation of IS surrounding *atz* genes on these catabolic plasmids further indicates that genetic rearrangement mediated by IS might occur. This last hypothesis was otherwise confirmed by the full sequencing of pADP1 showing that *atzA*, *B*, and *C* genes are dispersed on the pADP1 plasmid, but each one is surrounded by IS*1071* and IS*801* sequences. The involvement of these IS in the transposition of *atzAB* gene cassette by homologous recombination from pADP1 to the bacterial chromosome was shown for *Variovorax* sp. MD1 and MD2 strains (Devers et al. 2007b). Elsewhere, the *atzA* gene was already observed on the chromosome of *Arthrobacter* sp.

(continued)

Box 16.2 (continued)

Box Fig. 16.5 Regulation of expression of the operation *atzDEF* by *atzR*. The regulating dimeric structure formed by AtzR and NtrC is hypothetical

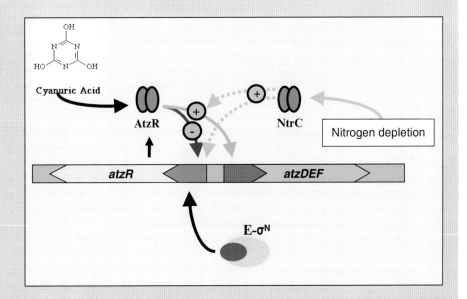

strain MCM B436. In addition, for *Pseudomonas* sp. ADP, the involvement of IS in the duplication of the gene *atzB* was shown to improve the rate of mineralization of atrazine giving a gain of fitness to the newly evolved population under atrazine selection pressure (Devers et al. 2008). Conversely, under cyanuric acid selection pressure, homologous recombination mediated by IS*Pps1* led to the selective loss of a 47-kb fragment containing the *atzABC* genes (Changey et al. 2011). In addition, the full sequencing of the pTC1 plasmid of *Arthrobacter aurescens* TC1 showed the existence of six copies of the gene *trzN* which are most likely responsible for the high atrazine-degrading efficiency of this strain. Beyond these transposition and recombination events, the recent discovery of the *trzN* gene in phage DNA suggests that transduction might as well play a role in the dissemination of atrazine-degrading genes.

Up to know, not much is known regarding the regulation of atrazine-degrading pathway. Thereby, the *atzA, B,* and *C* genes are basally expressed in *Pseudomonas* sp. ADP (Devers et al. 2004). On the contrary, the hypothesis of the transcriptional regulation of the *atzDEF* operon (Martinez et al. 2001) was validated by the identification of the gene *atzR*, coding for a LysR-type transcription factor. *atzR* was shown to regulate the expression of *atzDEF* operon in response to nitrogen limitation thereby confirming that at least for atrazine-mineralizing strain, atrazine constitutes a nitrogen source (Garcia-Gonzalez et al. 2005). Recently, it was shown that higher amount of *atzD* mRNA were detected in soil treated with atrazine than in control soil suggesting that this regulation is effective at the bacterial community level under field conditions (Monard et al. 2013) (Box Fig. 16.5).

genetic plasticity giving genetic support with new properties. They indicate that following pesticide exposure, soil microbial communities are able to rapidly adapt to pesticide biodegradation by different mechanisms allowing the recruitment, the establishment, and the dispersion of genes coding enymes responsible for the pesticide-degrading pathway among the community. From this point of view, one could consider that the selection of efficient pesticide-degrading communities is the result of an intense genetic patchwork leading to the emergence of new catabolic pathways. Recent observation done on emerging contaminants such as antibiotics suggests that similar processes are ongoing among soil microflora for other types of xenobiotics (Topp et al. 2013).

Genes coding for chlorocatechol dioxygenases are of prime interest to understand how widespread catabolic pathways have emerged from the β-ketoadiapate pathway, the central bacterial metabolic pathway for the aerobic degradation of aromatic compounds (Harwood and Parales 1996). Numerous lignin monomers, aromatic hydrocarbons, and amino aromatics are degraded along protocatechuate and catechol branches of the β-ketoadiapate pathway. This metabolism is relying on several classes of dioxygenases comprising those involved in the degradation of soil organic matter (El Azhari et al. 2010). Interestingly, although chlorocatechol dioxygenases from the ortho-modified pathway (i.e., type II dioxygenase) part of the *clc* element are belonging to this

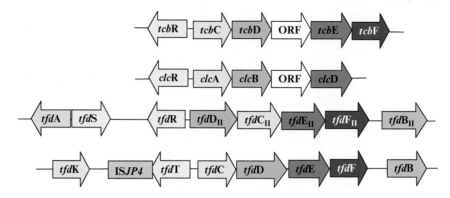

Fig. 16.25 Comparison of the gene organization of different operons involved in the modified ortho-pathway for intradiol clivage of chlorinated aromatic rings. The *tcb* genes are carried on the pP51 plasmid of *Pseudomonas* sp. P51. The *clc* genes are carried on the pAC27 plasmid of *Pseudomonas putida*. The *tfd* genes are carried on the pJP4 plasmid of *Cupriavidus necator* JMP 134 (isofunctional genes have the same color) (Modified and redrawn from van der Meer 1997)

class of enzyme, they are highly specific for the halogenated compounds. It has been shown that chlorocatechol dioxygenases of the modified ortho-pathway (i.e., *ccd* or *clc* genes for the degradation of chlorocatechol) are sharing relatively high level of similarities of amino-acid sequences (i.e., up to 22 % of similarity) with catechol dioxygenases of the ortho-pathway (i.e., β-ketoadipate pathway) such as *catA* which are widely dispersed within soil microbiota (El Azhari et al. 2010). The most well-known type II dioxygenases of the modified ortho-pathway are those involved in the degradation of 2,4-D, 3- and 4-chlorbenzoate and 1,2,4-trichlorobenze. These enzymes are sharing between 50 and 60 % of similarities at the amino-acid levels. Corresponding genes are located on different plasmids: *clcA* on pAC27 (*Pseudomonas putida*), *tbcB* on pP51 (*Pseudomonas* sp. strain 51), and *tfdC* on pJP4 (*Cupriavidus necator* JMP 134), pEML159 (*Alcaligenes* sp.), pRC10 (*Flavobacterium* sp.), pEST4011 (*Pseudomonas putida* PaW85), or pMAB1 (*Burkholderia cepacia* CSV90) (Fig. 16.25). These three genes are coding for enzymes sharing high similarities but having different substrate specificity: ClcA preferentially converts 3-chlorocatechol, TcbC 3,4-dichlorocatechol, and TfdC 3,5-dichlorcatechol into an intermediary metabolite formed during 2,4-D degradation.

16.7.3.3 Pesticide Bioremediation
The Methods for Soil Bioremediation

Keeping in mind that soil microorganisms are able to adapt to pesticide biodegradation by different means, different techniques of bioremediation have been proposed to use this property to clean contaminated soil from pesticides as an alternative to physicochemical treatment of the pollution. It is noteworthy that the use of microbial inoculum for soil bioremediation was suggested at the beginning of the 1950s (Audus 1951). Since then several attempts have been made to develop a wide range of bioremediation technologies taking advantages of the capabilities of microorganisms in the removal of pesticides. These technologies appeared to be promising, relatively efficient, and cost-effective (for review, see Megharaj et al. 2011). It is noteworthy that International Standard Organization (ISO) is currently studying a new work item proposal entitled "Procedure for site specific ecological risk assessment of soil contamination (TRIAD approach), ISO/TC 190/SC 7 N 286" which gives a guideline to estimate the ecological risk of a contaminated site and gives indications to stakeholders to handle related risks. Several classes of bioremediation mediated by soil microorganisms could be considered for *in situ* and *ex situ* treatments including (**i**) natural attenuation, (**ii**) biostimulation, and (**iii**) bioaugmentation. The choice of the technology to be applied to a contaminated site depends on a careful analysis of the cause of the pollution. This step consists in performing the environmental site assessment which requires a strategy defining the type of sampling and of chemical analyses to be done. As a part of this environmental evaluation, the capability of the soil to recover from the pollution can usually be classified in four principal categories:

(i) Microbial transformation is occurring at the contaminated site.
(ii) The chemical form of the pesticide makes it unavailable to microorganisms susceptible to degrade it.
(iii) The pesticide-degrading populations present at the contaminated site are not active enough
(iv) The pesticide-degrading populations are not abundant enough to be successful.

At these different causes, different technical solutions can be proposed. When microbial transformation is occurring at the contaminated site at a satisfactory rate (case (**i**)), the bioattenuation (natural attenuation) responsible for the pollutants' transformation to less harmful or immobilized forms can be applied. Monitoring of bioattenuation efficacy is usually done to give information to stakeholder in charge of the management of the site. This natural way is usually appropriate for contamination on large pieces of land without risks for animal and human populations. It is well accepted as methods for treating fuel components (e.g., BTEX) but not for many classes (Atteia and Guillot 2007).

When degrading microorganisms are present but the pesticide poorly available (case (**ii**)), the solution to be applied consists in adding different agents like synthetic surfactants to improve the availability of the compounds like it is done for petroleum compounds. This can also be done by inoculating biosurfactant-producing microbial inoculums which are producing anionic or nonionic surfactant like rhamnolipid which were found to enhance the solubility of PAHs and their subsequent biodegradation (Pei et al. 2010). Recently, dual application of biosurfactant-producing and pollutant-degrading microbes has been suggested presenting the advantages of a continuous supply of biodegradable surfactant and the ability to degrade pollutants (Megharaj et al. 2011).

When degrading microorganisms are present but not enough active (case (**iii**)) to clean the contaminated site, microbial turnover of pesticide degradation by indigenous microbial populations can be promoted by supplying with carbon, nutrients such as N and P, and oxygen or by modifying soil pH or redox potential. Electrobioremediation a hybrid technology combining bioremediation and electrokinetics was proposed to treat soil contaminated with hydrocarbons and metal (Chilingar et al. 1997). It consists in applying to polluted site of a current between electrodes and uses of microbiological phenomenon for pollutant degradation. All these techniques consisting in modifying the proxies of the contaminated environment to favor the activity of pesticide-degrading populations are grouped under the generic term of biostimulation.

When pesticide-degrading populations are present but in too less abundance (case (**iv**)) or even absent from the contaminated site, bioaugmentation consisting in inoculating soil with an appropriate pesticide-degrading population can be applied. It is noteworthy that the concept of "soil activation" which is based on the cultivation of a degrading-microbial biomass starting from a composite soil sample collected at the contaminated site and its subsequent use as an inoculum for the same soil was proposed (Otte et al. 1994). One of the most frequent problems encountered with bioaugmentation is the poor competitiveness of the inoculum to the indigenous members of natural communities and also the poor adaptability of these inoculums to physicochemical environments, thereby impairing the efficiency of bioaugmentation treatment.

Finally, it is noteworthy that phytoremediation techniques relying on the use of different plant species to clean polluted soils can be used as well. Phytoremediation was mainly used to remove heavy metals from soil, but nowadays it is also used to treat soil polluted with organic pollutants such as pesticides. In general, plants can promote dissipation of organic pollutants by immobilization (phytostabilization), removal (phytoextraction), degradation (phytodegradation), and promotion of microbial degradation (rhizodegradation). As an example, the natural resistance of corn plants to herbicide atrazine results from hydroxylation by benzoxazinones, conjugation reaction catalyzed by glutathione-*s*-transferase, and dealkylation under the control of P450 monooxygenases. It seems that to some extent, these different processes are shared by different plant lines, with, for each of them, differences in the resistance level. Thereby for the Vetiver grass (*Chrysopogon zizanioides* Nash), the resistance to atrazine seems to mainly result from its conjugation to the glutathione (Marcacci et al. 2005). This physiological property of the Vetiver grass is one of the traits explaining its large ecological tolerance. This physiological feature is of interest to use Vetiver grass in buffer zone (i.e., grass buffer strip naturally tolerant to different herbicides) to protect water resources, notably under Mediterranean and subtropical climates. On the contrary, atrazine hydroxylation and dealkylation are the two dominating reactions in poplar trees (Burken and Schnoor 1997; Chang et al. 2005). Recently, several studies were conducted with the aim of constructing genetically modified plants (GMO) in which microbial degrading genes are introduced. With this aim, Wang et al. (2005) showed that it was possible to transform different models of plants (*Medicago sativa*, *Nicotiana tabacum,* and *Arabidopsis thaliana*) with the *atzA* gene of *Pseudomonas* sp. ADP coding for the first enzyme of the atrazine-degrading pathway, catalyzing its transformation to hydroxyatrazine. In addition, T-DNA mutant of *Arabidopsis thaliana* was shown to be able to phytoaccumulate ^{14}C-atrazine in presence of sucrose offering potential for phytoremediation (Sulmon et al. 2007). On the reverse way several studies showed that plant can be used to establish microbe-assisted phytoremediation naturally promoting the microbial biodegradation of pesticide via root exudation (Piutti et al. 2002).

Whatever, these different bioremediation techniques can be applied *in situ* directly on the contaminated site or *ex situ*, after soil excavation and delocalized treatment in specifically built areas or in pilot reactor to protect surrounding environment from contamination and to protect workers from pollutant exposure. *In situ* approaches (bioattenuation, biostimulation or bioaugmentation) present the advantage of being cost-effective, but usually they are poorly efficient on a short term and required long-term treatment to reach acceptable results. In addition, the efficiency of *in situ* treatment is spatially variable mainly in reason of natural variability of soil physicochemical properties. On the contrary, *ex situ* treatments of contaminated soils in biopiles or bioreactors are expensive but more effective. This gain in efficacy is most likely due to soil manipulation which leads to the homogenization of the soil matrix. In addition, soil manipulation offers the possibility to thoroughly mix it and to apply the different bioremediation treatments (biostimulation, bioaugmentation, etc.) in a control manner. Indeed, *ex situ* treatments allow to control the principal parameters that are regulating the activity of pesticide-degrading community such as degree of aeration, water saturation, and nutrient supply. *Ex situ* treatment is applied to treat punctual pollutions of soil due to accidental cause or to treat high-level spots of contamination in a contaminated site. Recently, it was suggested that following *ex situ* treatments, treated soil can be used to construct a new soil profile at the polluted site, named Technosol, made of treated soil, recycled waste, and industrial by-products (Sere et al. 2008; WRB IWG 2006). Mainly applied to treat PAHs and heavy metal contamination, it was shown that the abundance and the activity of N-cycling microbial guilds in Technosols were in the same range as in the other terrestrial ecosystems suggesting that construction of Technosols is a promising technology for the restoration of degraded lands and recycling of industrial wastes (Hafeez et al. 2012). It has to be noticed that during the past two decades, on-farm bioremediation devices designed to effluents of pesticides have been developed and are now on the market being recognized by the authorities as one of the solution to treat pesticides effluents (for review, see De Wilde et al. 2007).

Although bioremediation strategies are considered as potentially simple and economically and environmentally friendly, their effectiveness must be monitored. Monitoring should be part of the bioremediation process (Megharaj et al. 2011). Usually it includes: (**i**) chemical analyses of the soil under treatment to monitor pollutant dissipation and search for possible metabolites formed during the process and (**ii**) ecotoxicological analyses to test the impact of the treatment on soil living organisms. Recently, several studies also proposed to monitor the abundance and activity of pollutant-degrading populations which are actively involved in the bioremediation process but also to estimate the impact of the bioremediation treatment on the overall soil biodiversity and, more particularly, on microbial diversity driving important soil ecosystemic services (Petric et al. 2011).

A Study Case of Soil Bioremediation

Since its discovery in the early 1950s, the herbicide atrazine belonging to the *s*-triazine family has largely been used in Americas and in European countries to control the development of broadleaf weeds in corn crop. Almost ten years ago, the use of atrazine was ban by the European Commission (2004/248/EC) mainly in reason of the detection of atrazine and of its main metabolites in water resources found in higher amount than the authorized concentration (i.e., 0.1 $\mu g \cdot L^{-1}$, according to water framework directive 91/414/EC). It is noteworthy that atrazine still remains in use in Americas where the US Environmental Protection Agency (EPA) approved its continued use and in many other countries worldwide. In France, atrazine was ban by the French government in year 2003 after the observation of widespread water pollution. Indeed, the French Agency for the Environment (IFEN) reported the detection of atrazine and its dealkylated metabolites in most of surface and groundwater analyzed even 3 years after its ban (Annual report of IFEN, 2006). Recent report on surface water quality reports that almost ten years after atrazine ban, this herbicide remains detected in 12 % of surface water analyzed being the fifth pesticide the most frequently detected. More disturbing desethyl-atrazine, one metabolite of the atrazine, is found in 33 % of water resources in France, being the second metabolite the most frequently found after AMPA, resulting from partial glyphosate transformation (Report 2013 with data obtained in 2011, French Ministry of the Environment). Also, next to bioremediation studies carried out on sites contaminated with petroleum residues, atrazine is likely the pesticide for which many bioremediation techniques have been proposed and tested. In most of the cases, these reports proposed prototype for atrazine bioremediation tested under laboratory conditions using both biostimulation and bioaugmentation approaches.

- Biostimulation approach for atrazine bioremediation

  It has been shown that the occurrence of the phenomenon of accelerated degradation for atrazine occurred only under certain conditions of soil pH, the lower limit value falling to 6.5 (Houot et al. 2000). In acidic soils, it has been shown that liming allows stimulating atrazine-degrading-microbial community which cannot be expressed at low pH in the unmodified soil. In several soils, it has been shown that liming alone is not enhancing atrazine degradation and must be associated with the inoculation of an atrazine-degrading population to be successful (Rousseaux et al. 2003). Demonstration that environmental conditions rather than the presence of microbial agents may be detrimental to the degradation of atrazine was provided by another study reporting that application of dairy effluent stimulated the activity of atrazine-degrading bacterial community native soil of a maize field (Topp et al. 1996). From this point of view, it has been shown that not all the organic substrates are promoting atrazine-degrading activity in soil (Barriuso et al. 1997). Several of them like straw, manure, and compost produce a two-step response on atrazine-degrading-microbial community: one could observe in the early stages, immediately following the addition of organic matter, an initial phase of stimulation of atrazine degradation followed by a second stage during which atrazine degradation is then slowed down, most likely reflecting stabilizing residues of atrazine on the organic matter during humidification (Barriuso et al. 1997). Later on several reports further confirmed that addition of simple or complex carbohydrates, thus creating a nitrogen-limited environment, stimulates the degradation of atrazine being viewed as a nitrogen source for degrading microorganisms (Udikovic-Kolic et al. 2012). Massive organic amendments often have the effect of stimulating the fraction of the most opportunistic high growth microorganisms (r-strategy). More limited, but more frequent, carbon inputs would avoid this nutritional discrimination. This carbon delivery can be achieved by plants which are known to release up to 20 % of their photosynthates as root exudates in their rhizosphere. This allows microorganisms to grow in the rhizosphere where their abundance can reach up to an order of magnitude higher than that it is in the non-rhizosphere soil. From these considerations originate the idea of developing microbe-assisted phytostimulation (Anderson et al. 1994), in which plant is not playing part of pesticide biodegradation but acts as a source of carbon that can be used by rhizospheric microorganisms as co-substrates to degrade pesticides (Alvey and Crowley 1996; Piutti et al. 2002). Thus, a significant increase in the degradation of atrazine applied at concentrations far exceeding the recommendations dose, in the presence of other herbicides such as metolachlor and trifluralin, was obtained by cropping contaminated soil with *Kochia scoparia* (L) Roth, an annual plant commonly found on pesticide site production and tolerant to different herbicides (Perkovich 1996). Moreover, *K. scoparia* has also been successfully used to promote microbial degradation of other pesticides such as metolachlor and lindane (Singh 2003). Other plants like *Pennisetum clandestinum* or *Zea maize* have also been shown to enhance the activity of atrazine-degrading-microbial populations in their rhizosphere (Piutti et al. 2002; Singh et al. 2004). Conversely, atrazine-degrading-microbial activity was not improved in the rhizosphere of several grasses such as *Sorghum vulgare*, *Lolium perenne*, *Festuca arundinacea*, and *Agropyron desertorum* (Anderson et al. 1995). The impression left by the literature on the subject is that of a certain disparity of situations from which it seems difficult to draw general conclusions and guidance on the involved processes.

- Bioaugmentation approach for atrazine bioremediation

Those that have been first implemented in the mid 1990s are based on the use of pure strains or microbial consortia able to degrade atrazine. Most of the bioaugmentation studies carried out in microcosms incubated under laboratory conditions have concluded to an immediate increase of degrading capacity of atrazine-contaminated soil in response to inoculation of degrading inoculums (Rousseaux et al. 2003). Only few bioaugmentation studies have been done at field scale. Indeed atrazine-contaminated soil from the Purdue Agricultural Experiment Station (Indiana, USA) was treated *ex situ* by bioaugmentation leading to a 20-fold increase of initial atrazine-degradation rate (Grigg et al. 1997). In another study, bioaugmentation carried out in simulated wetland sediment was shown to be successful (Runes et al. 2001). These studies also showed the major limitations inherent to the use of living microorganisms for bioremediation which are related to the characteristics of applied atrazine-degrading-microbial strains, the physicochemical and microbiological characteristics of receptor environment, and the composition, preparation, storage, and application of the inoculums.

- Inoculum characteristics

A first distinction is clear: it is that which differentiates bacterial and fungal inoculums. The last being *a priori* good candidates for bioaugmentation application, especially because these microorganisms can be produced in large amount by solid-state fermentation using cheap plant by-products (Durand 2003). Fungi species are known to produce a wide range of enzymes such as peroxidases, laccases, and phenoloxidases. These enzymes presenting strong oxidizing capabilities and having a wide substrate range are involved in the conversion of lignin and of its derivatives but also of various compounds including several pesticides. Thus, several studies have shown that, in a liquid medium, *Phanerochaete chrysosporium* was able to transform atrazine to its *N*-dealkylated derivative (Mougin et al. 1997). In addition, it was shown that inoculation of atrazine-contaminated soil with *Phanerochaete chrysosporium* contributes to atrazine dissipation but did not have any effect on *s*-triazine ring (Hickey et al. 1994). Indeed, inoculation of *Phanerochaete chrysosporium* to soil microcosms led to the production of *N*-dealkylated derivatives and to the accumulation of bound residues (Entry et al. 1996). If *Phanerochaete chrysosporium* inoculation leads effectively to atrazine dissipation from contaminated soil, accumulation of *N*-dealkylated metabolites in treated soils is not to recommend.

From this point of view, bacteria whose biochemical activity tends to orientate the flow of carbon and of electrons to biosynthetic and energetic metabolisms are more environmentally friendly. Indeed, a wide range of pure bacterial strains able to mineralize atrazine (i.e., transformation of atrazine to simple compounds such as CO_2 and NH_4) are now available (for review, see Udikovic-Kolic et al. 2012). It has to be underlined that this has not always been the case since *Pseudomonas* sp. ADP, the first known strain to mineralize atrazine, was

Fig. 16.26 Atrazine mineralization and dissipation in soil after inoculation with two degrading-microbial strains (Modified and redrawn from Topp 2001)

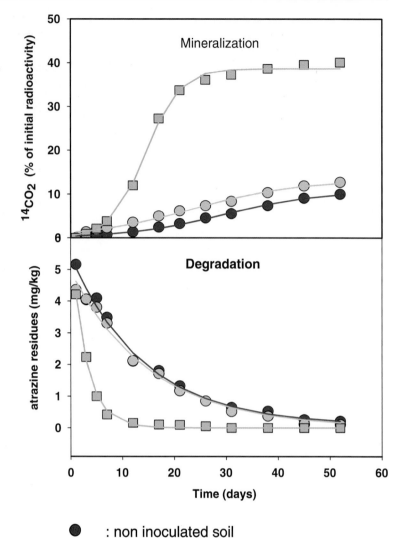

● : non inoculated soil

○ : soil inoculated with *Pseudomonas* sp ADP

■ : soil inoculated with *Nocardioides* sp C190

isolated by Mandelbaum et al. (1995) almost 50 years after its discovery and that atrazine from this point of view represents an exception since for the vast majority of xenobiotics, no mineralizing-bacterial strains are known. It turns out that isolation of pesticide-degrading strains, usually by enrichment cultures with or without co-substrate carried out under laboratory conditions, is an essential prerequisite. Their biotechnological use for land reclamation is not always successful, especially over time, including after large inoculations. The most cited reason is the difficulty for the introduced strain to find an ecological niche allowing its permanent in situ settlement. This may appear as a consequence of the Pareto law (Dejonghe et al. 2001) showing that 20 % of microorganisms in an ecosystem control 80 % of the energy flow. One could hypothesize that only physiological capacities consistent with those of the dominant microbial populations can ensure the survival of the introduced pesticide-degrading strain. From this point of view, bioaugmentation experiment carried out with *Pseudomonas* sp. ADP is exemplary. Indeed, despite its known ability to mineralize atrazine in liquid culture, its inoculation must be at two orders of magnitude higher than that of *Pseudoaminobacter* sp. and *Nocardioides* sp., two other degrading strains known, to produce similar effect, was not yielding in extensive atrazine mineralization. This poor bioremediation performance was attributed to the impossibility of *Pseudomonas* sp. ADP to use aminoalkylated carbon side chains branched on *s*-triazine ring and the necessity to find a carbon source for its survival (Fig. 16.26) (Topp 2001).

For the same reason, it is likely that strains with restricted degradation abilities, performing only a limited number of reactions of the degradation pathway, are not good candidates to be used as inoculums for bioaugmentation. Their inability to carry out the metabolic process completion does not allow them to convert their metabolic peculiarity in a selective advantage favoring their implanting within indigenous soil microbial community. However, several reports have shown that these degrading strains can express their catabolic potential in complex microbial consortia in which each microorganism brings its metabolic competence to reconstruct a coherent set of steps leading to the mineralization of the pesticide (Smith et al. 2005; Kolic et al. 2007). For atrazine biodegradation, the literature gives a number of examples of such microbial associations where no individual species has a complete set of catabolic gene, but together they have the full degrading pathway (Udikovic-Kolic et al. 2012). These associations have various taxonomic compositions (Gram + and Gram – negative bacteria) and have also various atrazine-degrading-genetic combinations (*atzABC*, *trzNatzBC*, *trzNatzBCtrzD*, etc.) (Martin-Laurent et al. 2006). To some extent, pesticide-degrading-microbial consortia might be considered more resilient than pure degrading strain, as they are hosting various metabolic pathways that may constitute a selective advantage favoring their survival in oligotrophic conditions found in the soil environment (Soulas 2003). However, despite their ecological interest, until now the recourse to the use of degrading-microbial consortia for the bioaugmentation is compromised by two yet unresolved questions. The first concerns a practical issue on how to produce pesticide-degrading consortia inoculum having stable taxonomic- and degrading-genetic compositions for ensuring the maintenance of its degrading ability. Indeed, the transience of pesticide-degrading ability observed in pure strains portends major difficulties when the degrading ability will have to be held in several strains simultaneously. The second concerns their stability in the natural environment where the selection pressure that allowed their isolation and their maintenance under laboratory conditions may decrease and where ambient nutrient conditions may drastically change with consequences on the cohesion of the whole. This is an important point knowing that the loss of a single member of the consortium may be sufficient to stop all the degrading activity of the functional unit.

However, it has to be noticed that beside technical limitations to use living microorganisms for bioaugmentation to bioremediate pesticide-contaminated soils, there are some limitations due to the regulations in particular in Europe. Indeed, prior to its commercial use, any microorganism inoculum has to get an official agreement delivered by EU authorities after the constitution of a homologation dossier made of several elements including a deep toxicological analysis of the impact of the inoculum on animal and human health. Several studies report the use of dead microorganisms or cell-free enzymes to avoid this strict regulation. As an example, field-scale atrazine bioremediation has been achieved using killed and stabilized transgenic *E. coli* (Strong et al. 2000). More recently, the use of cell-free TrzN (as a DNA-free cell lysate) to treat a 1.5-ML dam reservoir contaminated with atrazine was shown to be successful (Scott et al. 2010). Interestingly, the encapsulation of AtzA and of nonviable atrazine degraders was found to be successful to treat atrazine-contaminated water under laboratory conditions (Reátegui et al. 2012).

- What solutions for the future?

To improve the efficiency of soil bioremediation and remove obstacles sometimes found when trying to apply this technology, solutions have been proposed. The most immediate is that of improving the microbial-assisted phytoremediation by associating bioaugmentation to rhizostimulation. The beneficial effects of this double treatment on atrazine biodegradation were demonstrated long ago (Alvey and Crowley 1996). These authors showed that maize seedlings increased atrazine mineralization catalyzed by a composite microbial inoculated to the soil. A better survival of the composite inoculum was reported in the maize rhizospheric soil as compared to the bare soil (Alvey and Crowley 1996). This first observation was later on confirmed by several other studies confirming that the activity of degrading-microbial communities was stimulated in the maize rhizosphere (Piutti et al. 2002). This process combining bioaugmentation to rhizostimulation where plant rhizosphere is used both as vector exploration of the soil environment and as source of carbon and electrons has not yet been applied for field-scale bioremediation of atrazine-polluted sites. However, its feasibility was shown in a recent study reporting the stimulation of the degradation of α-, β-, γ-, and δ-HCH in the rhizosphere of maize inoculated with four *Sphingomonas* sp. strains previously isolated from maize rhizospheric soil (Böltner et al. 2008).

Molecular ecology is offering some interesting perspectives to the issue of the efficiency, the survival, and/or the maintenance of the pesticide-degrading populations inoculated to the contaminated site. On atrazine, the possibility of introducing degrading genes of different origins in a microbial host was illustrated by the transfer in *Rhodococcus* sp. TE1, naturally possessing the *atrA* gene encoding enzyme catalyzing *N*-dealkylations, the *trzA* gene from *Rhodococcus corallinus* encoding enzyme dechlorinating dealkylated derivatives of atrazine (Shao et al. 1995).

They showed that both genes were expressed in the transformant strain resulting in the transformation of atrazine to cyanuric acid which is easily degraded by the soil microflora. Taking in consideration that rhodococci are ubiquitous and that members of this genus have already been successfully used in industry, this makes them ideal candidates for manipulation of pesticide-degrading-genetic potential to exploit bioremediation technologies (Shao et al. 1995). The isolation of a wide diversity of atrazine-degrading strains having different atrazine-degrading gene modules constitutes a stock of atrazine-degrading-genetic modules that could be used to "design on request" recombinant microbial strains able to cope with different *s*-triazine type of pollution. One could even think that application of new genetic technologies such as directed mutagenesis or DNA shuffling could allow extending the range of substrate that an enzyme can catalyze (Scott et al. 2009). Experiments of directed evolution applying "DNA shuffling" have indeed shown that they could generate functional diversity (Raillard et al. 2001) (Box 16.3). Thus, starting from two genes, *atzA* and *triA*, encoding hydrolases responsible for the dechlorination of atrazine and the deamination of aminoatrazine, respectively, it was possible to produce up to 1,600 recombinant sequences. Of these, some encode enzymes whose catalytic constants are 150-times higher than those of the two original enzymes. Others were capable of hydrolyzing a range of *s*-triazines which were not recognized as substrates by the two original enzymes (Raillard et al. 2001).

For reasons already given, the introduced strain, from natural origin or genetically modified, may have difficulties to survive in the soil environment at the population level consistent with desired degrading activity. In order to avoid this problem, another solution has recently been considered. It comprises of introducing a mobile genetic element carrying the catabolic genes using a microorganism vector. This strategy has proven its feasibility through the example of 2,4-D degradation (Dejonghe et al. 2000). A significant increase of 2,4-D degradation in particular in the B horizon (Fig. 16.27) where it was very low has actually been observed following the transfer pEMT1 and pJP4 plasmids harboring *tfd* genes from a donor strain, *Pseudomonas putida*, to indigenous soil recipient bacterial strains belonging to *Burkholderia*, *Ralstonia*, and *Pseudomonas* genus, the *Betaproteobacteria* being large majority (Dejonghe et al. 2000).

Also under greenhouse conditions, this strategy has shown its potential in the case of soil treated with atrazine and inoculated with a donor strain *Agrobacterium tumefaciens* sp. 96-4 having the plasmid pADP1:: Tn5 harboring *atzABCDEF* genes, which showed better ability to degrade

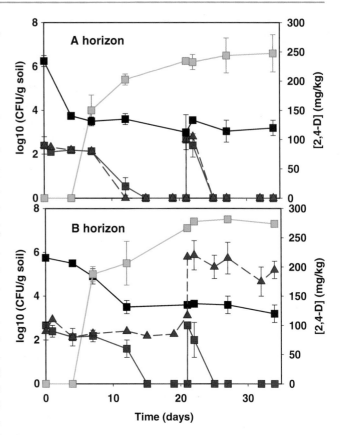

Fig. 16.27 Transfer of the plasmid pJP4 in two soil horizons. *Black*: Survival of donor cells. *Green*: Transconjugant formation. *Red*: Dissipation of 2,4-D after two applications at a 21-day interval in the inoculated soil (*continuous line*) and the non-inoculated soil (*dashed line*). (Plasmide pJP4 is labeled with a mini-Tn5 KmlacZ transposon) (Modified and redrawn from Dejonghe et al. 2000)

atrazine and from which transconjugants belonging to *Variovorax* strains have been isolated (Devers et al. 2005). Among these recombinants, genetic rearrangement mediated by IS was observed leading to the transfer of the *atzAB* cassette from the plasmid pADP1:: Tn5 to the bacterial chromosome (Devers et al. 2007b). Given the high transfer frequency commonly observed in these studies, this "gene-augmentation" approach has a chance to succeed only if the transconjugants, initially present in a very few number, have a selective advantage to allow their rapid proliferation. With this regard, the pesticide which represents a source of carbon and of energy for degrading microorganisms is undoubtedly exerting a favorable selection pressure enabling rapid growth of the transconjugants. The question remaining is: does the transconjugant retained catabolic-mobile-genetic-element become useless when the contaminant has disappeared from the environment? As already discussed in the context of accelerated biodegradation, currently one cannot give a satisfactory answer. Recent observation of selective loss of atrazine-degrading gene in *Pseudomonas* sp. ADP in response to a

Box 16.3: Involvement of Horizontal Gene Transfer in Chlorinated Compounds Biological Degradation: Example of the Biological Degradation of Lindane

Sibel Berger, Maude M. David, Timothy M. Vogel, and Pascal Simonet

Utilization and Toxicity of Lindane

Lindane (gamma-hexachlorocyclohexane) is a wide activity spectrum pesticide against plant eating, soil living insects, and man and animal parasites. It was extensively used in forest industry, agriculture, and in treatments against parasites. Because of a low solubility in water (7 mg·L^{-1}), lindane can persist in soil for months (half-life varying from 30 to 300 days), leading to its accumulation in numerous environmental compartments. In addition, the Cancer International Research Center classifies lindane as a potential carcinogenic agent for humans (class 2B, IARC, 1991). Effects on the nervous, respiratory, digestive, and endocrine (reproduction) systems were reported leading to its ban in most western countries. However, lindane is still used in numerous developing countries because of its low cost.

Biological Degradation of Lindane

Only one degradation pathway of lindane under aerobic conditions was identified in bacteria that belong to only one genus, *Sphingobium*. The complete lindane degradation pathway in *Sphingobium japonicum* involves successively LinA to LinJ enzymes, the regulator LinR, and a putative ABC-type transporter, LinKLMN (Box Fig. 16.6) (Nagata et al. 2007). The corresponding coding genes (*lin* genes) are well conserved among all *Sphingobium* isolates that degrade lindane whatever the country the strains were isolated (Lal et al. 2006). Under laboratory conditions, the degradation phenotype can be detected by a halo surrounding colonies developing on solidified rich media supplemented with lindane. The degradation phenotype is very instable with, for instance, in *Sphingobium francense* sp.+ (Thomas et al. 1996), a very high mutation frequency leading to the loss of the degradation capacity in up to 4 % of cells (Cérémonie et al. 2006). Molecular characterization of mutants revealed significant genomic rearrangements including deletions or insertion sequence (IS) movements affecting *lin* genes and particularly the dehydrochlorinase encoding *linA* gene, a key enzyme in lindane degradation pathway. This gene was found to be affected by mutations occurring at high frequency in addition to a strong plasticity of the *Sphingobium francense* genome involving mobile genetic elements.

Putative Origin of Xenobiotic Compound Degradation Genes

Studies on xenobiotic compound-contaminated ecosystems, including lindane, confirm that these compounds are highly recalcitrant to biological degradation because of the lack of degrading bacteria in the soil. Conventional or DNA-based microbiological studies systematically lead to negative results regarding detection of degrading bacteria or their genes in most soils. However, some degradation-adapted bacteria were successfully isolated on sites polluted for years by the xenobiotic compounds. The hypothesis of HGT and genetic recombination involving the soil bacterial community DNA pool were proposed to explain the assembly of new catabolic genes in some bacterial cells. The experimental demonstration that *linA* and other new catabolic genes result from the natural and recent shuffling of DNA fragments initially spread in various bacteria would confirm the major adaptive role of HGT. The assembly of such new genes would indicate the extremely high frequency of transfer of DNA fragments occurring at random among soil bacteria necessary to provide enough different genetic combinations including the one corresponding to a catabolic gene susceptible to increase the fitness of its bacterial host in presence of the xenobiotic compound.

In Silico and Experimental Approach

The molecular origin of the new degrading genes can be deduced from their similarity level with proteins in other catabolic pathways. For instance, *linD*, *linE*, and *linF* genes share some similarity with, respectively, *pcpC*, *pcpA*, and *pcpE* genes involved in pentachlorophenol degradation in *Sphingobium chlorophenolicum*, suggesting that they could have been acquired from this host. In addition, presence of mobile genetic elements (IS) in the vicinity of the *lin* genes and their localization on plasmids suggest the role of HGT events for spreading them among soil bacteria including *Sphingobium francense* (Cérémonie et al. 2006; Nagata et al. 2006). However, the

---

S. Berger (✉) • M.M. David • T.M. Vogel • P. Simonet
Environ. Microb. Genom. Group, École Centrale de Lyon, UMR CNRS 5005, Laboratoire Ampère, Écully, France

(continued)

Box 16.3 (continued)

Box Fig. 16.6 Scheme of the degradation pathways of lindane in *Sphingobium japonicum* UT26. Enzymes: LinA: Dehydrochlorinase; LinB: halidohydrolase; LinC and LinX: Dehydrogenases; LinD: Dechlorinase; LinE: Dioxygenase; LinF: Maleylacetate reductase; in GHI: succinyl-CoA transferase; LinJ: b-ketoadipyl-CoA thiolase. The enzymes of the "track upstream" ("upstream pathway") are expressed and constitutively active. 2,5-DCHQ (8) then initiates the downstream pathway by activation of the LinR regulator that induces the expression of genes *linD*, *linE*, and *LinR* (Nagata et al. 2007) (Copyright: Springer)

dehydrochlorinase encoding gene *linA* whose products are involved in the initial lindane dechlorination step does not exhibit any sequence similarity with genes in databases (Imai et al. 1991) in spite of the use of very sensitive bioinformatics tools including PSI-BLAST (*cf.* Sect. 17.7.5). These data can be seen as the first indication of an unknown molecular origin of the *linA* gene that led to the hypothesis of a molecular shuffling to explain emergence of this gene in bacteria isolated from different polluted sites. However, validation of this hypothesis requires that bacterial populations in nonpolluted soils do not contain a functional *linA* gene. It also requires presence in one or several bacteria hosts of the various DNA fragments whose shuffling could create this gene. Experimental confirmation of this hypothesis led to define several investigation steps that were applied to *linA* from *S. francense* as models (Boubakri et al. 2006). *Step 1*: Bioinformatics analysis of the *linA* gene to define its structure. *Step 2*: Analysis of not polluted soil metagenomes to detect presence of *linA* gene fragments and concomitantly to confirm absence of a complete and functional *linA* gene. *Step 3*: Experimental demonstration that a functional *linA* gene can

(continued)

Box 16.3 (continued)

Box Fig. 16.7 Cutting the *linA* gene using the SAME bioinformatics software. The different patterns are represented as rectangles. The Ar-A6, promoter (*P*), and Shine–Dalgarno sequence (*SD*) are indicated (Boubakri et al. 2006) (Copyright: Elsevier)

be assembled from the shuffling of the soil metagenome DNA.

Step 1

The "MEME" ("Multiple Expectation maximization for Motif Elicitation") bioinformatics program was used in order to detect sequence similarities after protein or nucleic motives were established in the complete sequence for detecting structural similarity among proteins. Applied to the *linA* gene, "MEME" detected several distinct domains or blocks suggesting a mosaic-type structure (Boubakri et al. 2006; Box Fig. 16.7) similar to the structure in other catabolic genes including those involved in the degradation of 3-chlorobenzene (van der Meer 1997) and other dehalogenase encoding genes. For instance, the fusion of a DNA fragment from an halohydrin dehalogenase encoding gene *hheB* with a segment from an haloalkane dehalogenase almost identical to DhaA would be responsible of the emergence of the new gene *dhaA* in *Mycobacterium* (Poelarends et al. 1999).

Step 2

The metagenomic DNA extracted from a nonpolluted soil was used as template for PCR amplification of the various *linA* gene motives leading to a pool of PCR products that were used as template for an in vitro shuffling step. This resulted in the production of several hybrid molecular combinations from which the putative capacity of resulting in a functional gene was tested by transforming a bacterial host and searching for a degradation halo. More than 33 % of the sequence of a functional *linA* gene was assembled with DNA originating from the metagenome confirming presence in the genome of soil bacteria of a DNA pool mobilizable for generating new genetic determinants that increased the fitness of their hosts under the elective pressure of the pollutant.

Step 3

The in vitro shuffling process leads to the assembly of a *linA* gene from metagenomic DNA filling successfully the central part of the gene and its 3′ end. However, although informative on the potential of HGT and recombination for generating genetic diversity, these in vitro experiments are not sufficient to assess that these mechanisms were those actually involved in the emergence of a functional *linA* gene in *Sphingobium francense* sp.+. Under natural conditions, several barriers limit the extent of HGT, including the critical step of recombination-mediated integration of the incoming DNA into the host genome. Only DNA shuffling experiments under in vivo conditions could definitely validate the proposed hypothesis, and this objective is addressed by testing the capacity of totally or partly *linA* deleted *E. coli* or *Sphingobium* strains to restore a functional gene when transformed with nonpolluted soil metagenomic DNA.

Conclusion

Sphingobium isolates have rapidly acquired the potential to use lindane as a carbon source, but the final genetic construction is far from being optimized as indicated by a strong phenotypic and genetic instability and a mosaic structure of the degradation pathway key gene. The "fine tuning" of the degradation molecular machinery will take place much more slowly by point mutations and endogenous rearrangements from this preliminary molecular sketch. These minor and repeated genetic changes will improve little by little the genetic stability of the construction, the rearrangements leading to assemble genes in operons contributing to increase the expression level of the various genes involved in the degradation process.

Definitely, the experimental confirmation that the *linA* gene detected in *Sphingobium francense* sp. + resulted from a shuffling of genetic information initially spread among different soil bacteria would bring new clues for demonstrating the extent of HGT in the environment and its critical role in adaptation and genome evolution of bacteria.

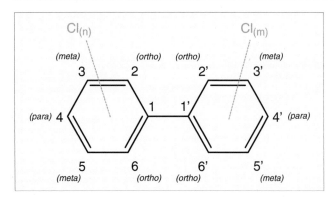

Fig. 16.28 Chemical structure of polychlorinated biphenyls (PCBs)

depletion of the selection pressure is in favor of the hypothesis of the loss or at least of the dilution of the genetic potential within the pesticide-degrading community (Changey et al. 2011), suggesting that pesticide-degrading ability remains a versatile function. Finally, among the techniques associated to the bioaugmentation, there is one relying on the use of killed *E. coli* recombinant strains that have been genetically modified to overexpress the *atzA* gene (Strong et al. 2000). It is rather, in this case, an enzyme treatment, the microbial strain having no other role to ensure than the encapsulation of enzymes.

16.7.4 Polychlorobiphenyls (PCBs)

16.7.4.1 Synthesis, Chemical Structure, and Use

Polychlorobiphenyls (PCBs) are xenobiotics industrially produced by nonselective chlorination of biphenyl. They are made up with two phenyl cycles, each one substituted by 1–5 chlorine atoms; their general formula is as follows: $C_{12}H_{10-(n+m)}Cl_{(n+m)}$ (Fig. 16.28). The PCBs success that leads to intensive use is explained in particular by their synthesis straightforwardness and the great industrial interest for their physicochemical properties. The obtained products that are of great thermal stability possess large dielectric properties and are hardly ignitable. They were universally used in capacitors and current transformers as insulating fluid and coolant, as hydraulic fluid, and coolant for heat exchangers. Moreover, they were widely used in different industrial compositions such as cutting oil (metal work) and, to a lesser extent, as additive in the composition of insecticides, bactericides, plasticizer compounds, adhesives, inks, bleach, etc.

Theoretically, 209 congeners can be synthesized and are distinguished by the number and position of chlorine atoms on the biphenyl molecule. A specific nomenclature from CB1 to CB209 was proposed by Ballschmister and Zell (1980). The number and position of chlorine atoms on the molecule determine both their physicochemical properties and their eco-toxicity. Commercial products, characterized by a mean chlorine content between 30 and 70 %, are complex mixtures of tens of different congeners and are identified by different names depending on the producing country: Aroclor (USA), Clophen (Germany), Kanechlor (Japon), Fenclor (Italy), Pyralène (France), and Soval (Russia). A product is usually identified by a number corresponding to its chlorination level. For instance, in the USA, Aroclor is identified by a 4 digit number. The first two correspond to the number of carbon atoms in the parent molecule (biphenyl), whereas the last two express the average weight (% w/w) chlorine content. Thus, Aroclor 1242 contains 12 carbon atoms and 42 % chlorine. The 7 congeners that are usually investigated in natural environments are listed in Table 16.5.

16.7.4.2 Presence in the Environment

The industrial PCB production started about in 1930; but considering the product toxicity and their impact on the environment their production, use and importation are currently forbidden: since 1970 in Sweden, 1972 in Japan, 1977 in the USA, 1987 in France, etc. The 2001 Stockholm international agreement completely prohibited their production. Before this ban, the ecosystem contamination was accidental, occurring during transportation, generated by leaking installations, exploding current transformers, etc. Wastes containing these xenobiotics were often buried thus leading to an important contamination of soils, groundwater, and sediments. Nowadays, PCBs are found in the air, water, sediment, soil, and food (particularly shellfish and fish); contamination continues, particularly through contaminated soil washing, prompting pollution of rivers followed by their outflow into the sea where PCBs are concentrated, more particularly in estuarine areas. Diluted and carried away by the large ocean circulation, they are still detected in sites very distant from their emission point and are now present in all ecosystems, from the Arctic to the Antarctic. Their very strong chemical stability has two main consequences:

(i) They are very hardly degraded and thus will remain in the environment over decades, particularly in soils and sediments where they are strongly adsorbed on organic matter. For instance, Magar et al. (2005), under reducing conditions (lacustrine sediments), observed the resistance to biological dechlorination of chlorine atoms in *ortho* position and the attack of chlorine atoms in *para* and *meta* position leading to the accumulation of less chlorinated *ortho*-congener. The dechlorination rate measured in the environment showed that, in average, 16 years were necessary to eliminate *para*- and *meta*-chlorines. In contrast, *ortho*-chlorine were retained for 50 years.

(ii) They will pile up in biological materials with the highest lipid content due to their hydrophobic feature, characterized by high octanol/water partition coefficient

Table 16.5 The European 7 indicators PCBs (iPCBs[a]); nomenclature and structure

Specific nomenclature	Chlorine Atom number	IUPAC nomenclature	Chemical structure
CB 28	3	2,4,4′-trichlorobiphenyl	
CB 52	4	2,2′,5,5′-tetrachlorobiphenyl	
CB 101	5	2,2′,4,5,5′-pentachlorobiphenyl	
CB 118	5	2,3′,4,4′,5-pentachlorobiphenyl	
CB 138	6	2,2′,3,4,4′,5′-hexachlorobiphenyl	
CB 153	6	2,2′,4,4′,5,5′-hexachlorobiphenyl	
CB 180	7	2,2′,3,4,4′,5,5′-heptachlorobiphenyl	

[a]In Europe, a set of 7 congeners, called "indicator" *PCBs* (iPCBs), are currently used rather than Aroclor ® or similar *PCB* technical mixtures

(log Ko/w), from 4.5 for monochlorobiphenyls up to over 8 for polychlorinated congeners. The ensuing bioaccumulation leads to an occasionally very important bioamplification. For instance, the progression of PCB concentration in a trophic chain of the North America Great Lakes is significant: water (0.005 ppb) → phytoplankton (0.0025 ppm) → zooplankton (0.123 ppm) → rainbow smelt (1.04 ppm) → made up trout (4.83 ppm) → silver seagull eggs (124 ppm), let a bioconcentration factor (BCF) close to 25106 (Ramade 2002).

16.7.4.3 Microbial Biodegradation

Microbial biodegradation proceeds according to two distinct and complementary ways (Fig. 16.29): the reductive anaerobic way and the oxidative aerobic way (Abramowicz 1990; Mhiri and Tandeau de Marsac 1997; Borja et al. 2005; Pieper 2005; Field and Sierra-Alvarez 2008). Biodegradation essentially depends on the following:

(i) Environmental factors such as temperature, pH, environment oxygenation, salinity, and presence of toxic compounds. Indeed, contaminated biotopes often contain other xenobiotics such as metals that may have a toxic effect on microorganisms able to degrade PCBs. Some PCB catabolic products may also have an inhibitory effect if they are not eliminated. It is the case of chlorobenzoic acids, chloroacetophenones, and compounds produced by chlorocatechol cleavage (Mhiri and Tandeau de Marsac 1997).

(ii) Congener solubility: the most water-soluble compounds are more easily attained by microorganisms than those with a low water solubility.

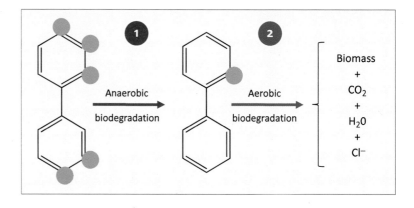

Fig. 16.29 Two-step combined anaerobic (*1*) and aerobic (*2*) process to biodegrade PCBs (Adapted from Abramowicz 1990)

But, if PCBs are lipid soluble, they are in contrast very little water soluble. For a given PCB, its solubility is inversely proportional to its chlorination level. It is indeed in the range 6.36–29.15 µM for monochlorobiphenyls and $2.4\ 10^{-6}$ µM for decachlorobiphenyl. This partially explains the biodegradation resistance of highly chlorinated congeners (bioavailability) (Hurst et al. 2002). The solubility importance is confirmed by the biodegradation stimulation in the presence of biosurfactants that increase the PCB bioavailability (Ohtsubo et al. 2004).

(iii) Congener structure, in particular, the number and position of chlorine atoms.
(iv) PCB concentration: generally, when the pollutant concentration is too small, the enzymes involved in their biodegradation are not induced and the growth of relevant microorganisms is not favored. In contrast, a too large concentration will inhibit growth.

The anaerobic biodegradation leads to the elimination of chlorine atoms of highly chlorinated PCBs that are likely mineralized under aerobic conditions where only weakly chlorinated congeners are attacked. Indeed, there are substrates for aerobic bacteria that are able to open the related benzene cycle. This situation means that:

(i) Generally, a single bacterial strain is unable to mineralize polychlorinated congeners.
(ii) Anaerobic processes primarily occur in the degradation of industrially made compounds that usually contain 5–7 chlorine atoms. These biodegradation processes will thus be presented before those occurring in the presence of dioxygen.

Anaerobic Biodegradation

Anaerobic biodegradation implies a reductive dechlorination during which the halogenated compound serves as electron acceptor and is replaced by hydrogen:

$$RCl + 2e + H^+ \rightarrow RH + Cl^-$$

During this reaction, the number of Cl atoms decreases while the biphenyl skeleton remains unchanged. Stains able to perform dechlorination under anaerobiosis were, in a first time, identified but not isolated (Cutter et al. 2001). Their isolation was undertaken later on. This is particularly the case for the following strains:

(i) The strain *Dehalococcoides ethenogenes* that removes chlorine atoms in position *para* and *meta* on 2,3,4,5,6-PCB (Fig. 16.30)
(ii) The strain DF-1 ultramicrobacterium that, using hydrogen and formate as electron donor, turns 2,3,4,5-PCB into 2,3,5-PCB and ensures a significant dechlorination of Aroclor 1260 present in a contaminated soil. This strain can only grow when cultured with a strain of *Desulfovibrio* sp. (May et al. 2008)
(iii) The strain *Dehalococcoides* CBDB1 that performs an important dechlorination of Aroclor 1248 and 1260. After a 4 month experiment, a 60 % decrease of the 16 major Aroclor-1260-congeners was observed (Adrian et al. 2009).

A large amount of evidence was collected regarding PCB anaerobic elimination in the environment (sediments, anoxic soils). One of the first evidences was provided by investigations carried out on anaerobic aquatic sediments collected in the Hudson River (Massachusetts, USA), contaminated by PCBs. It was demonstrated that the most chlorinated congeners were preferentially biodegraded while the percentage of the less chlorinated congeners (mono- and dichlorobiphenyl) was simultaneously increasing (Brown et al. 1987); these initial observations are in agreement with a preferential dechlorination of chlorine atoms in *meta* and *para* positions. Thereafter, a large number of studies conducted both in situ in a great number of polluted sites (soils, sediments) and in vitro (in microcosms and bioreactors) confirmed the works of Brown and collaborators.

Nies and Vogel (1990) showed that addition of glucose, acetone, and methanol to Hudson River sediments contaminated by Aroclor 1242 resulted in dechlorination in *meta* and *para* positions of highly chlorinated congeners. As an example, after 22 weeks, penta- and hexachlorobiphenyls and tetrachlorobiphenyls were about 80 and 70 % biodegraded, respectively, after glucose and acetone addition. Simultaneously, the accumulation of mono- and

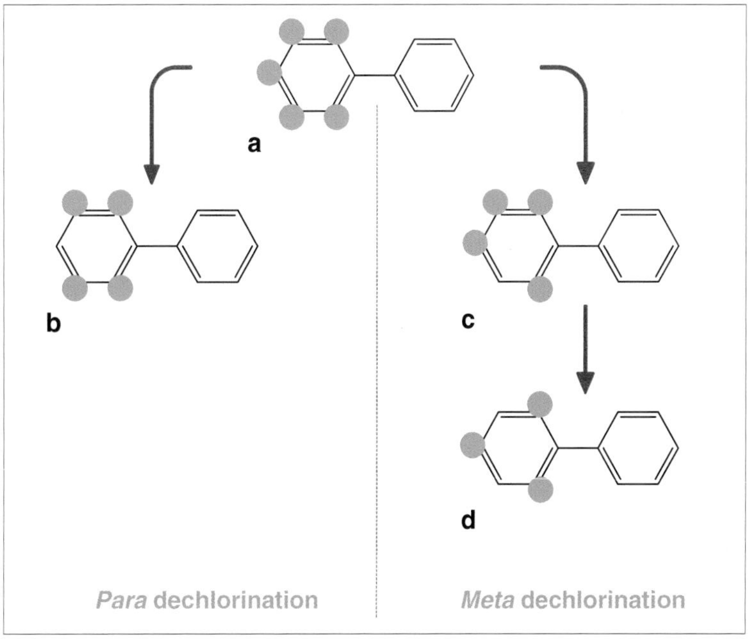

Fig. 16.30 Biodegradation pathway of (**a**) 2,3,4,5,6-pentachlorobiphenyl; (**b**) 2,3,5,6-tetrachlorobiphenyl; (**c**) 2,3,4,6-tetrachlorobiphenyl; (**d**) 2,4,6-trichlorobiphenyl (reductive dechlorination) by *Dehalococcoides ethenogenes* (According to Fennell et al. 2004)

dichlorobiphenyls was observed. However, in cultures without enrichment, biodegradation could not be observed. Finally, anaerobiosis in all cultures was demonstrated by an intense methanogenic activity.

A similar approach was conducted on PCB contaminated sediments from 3 different ecosystems: a river (Hudson River), a lake (Silver Lake), and an estuary (New Bedford Harbor) (Alder et al. 1993). The 3 sediment types were supplemented with Aroclor 1242 and a carbon source in the form of a short-chain free fatty acid (SCFA) mixture, consisting of acetate, propionate, butyrate, and hexanoic acid. In flasks rigorously maintained in anaerobiosis, PCB evolution was monitored under methanogenesis and sulfato-reduction conditions (by adding carbonate and sulfate in excess, respectively). In the 3 tested sediments, the authors could provide evidence of dechlorination processes. The fastest were observed with the Hudson River sediments supplemented with volatile free fatty acids or short-chain fatty acids (SCFA), with preferential elimination of chlorine atoms in *meta* and *para* positions (65 % were eliminated after 2 month incubation), leading to the accumulation of mono-, di-, and tri-*ortho*-substituted PCB congeners. In contrast, in the absence of SCFA supplementation, the same dechlorination level was only achieved after 11 months.

From the whole set of studies dedicated to PCB anaerobic biodegradation that was most often observed under methanogenesis conditions, it turns out that the same trends were observed whatever the considered sites and PCBs: *para*- and *meta*-disubstituted congeners are more frequently degraded than *ortho*-ones. Dechlorination in *ortho* position is weak. Dechlorinations usually produce mono- and dichlorinated congeners. Some works demonstrated the existence of slow anaerobic pathways that may lead to the production of trace biphenyl (Natarajan et al. 1996) (Fig. 16.31). Moreover, microorganisms present in a given site have specific capacities to degrade some PCBs (Quensen et al. 1990).

Aerobic Biodegradation

Weakly chlorinated congeners produced by dechlorination of highly chlorinated congeners are substrates for Gram-negative aerobic bacteria (genus *Pseudomonas*, *Alcaligenes*, *Achromobacter*, *Comamonas*, *Sphingomonas*, *Ralstonia*, *Acinetobacter*, *Burkholderia*) and Gram-positive bacteria (genus *Rhodococcus*, *Corynebacterium*, *Bacillus*) (Pieper 2005; Field and Sierra-Alvarez 2008).

PCBs are also metabolized by different strains of white-root fungi. The trend reflects the general observation that aerobic PCB biodegradability decreases with increased chlorine content (Field and Sierra-Alvarez 2008). Biodegradation of Aroclors 1242, 1254, and 1260 (42, 54, and 60 % chlorine, respectively) by *Phanerochaete chrysosporium* decreases their concentration by 60.9, 30.5, and 17.6 %, respectively (Yadav et al. 1995).

Many experiments demonstrated aerobic biodegradation of PCBs. They were conducted with microorganisms cultured in liquid medium, sediment, or soil kept under aerobic conditions in microcosms in the presence of cells immobilized on materials at the surface of which they formed biofilms. For instance, PCB biodegradation was achieved by bacterial biofilms in packed-bed batch bioreactors (Fava et al. 1996), with a bacterial co-culture

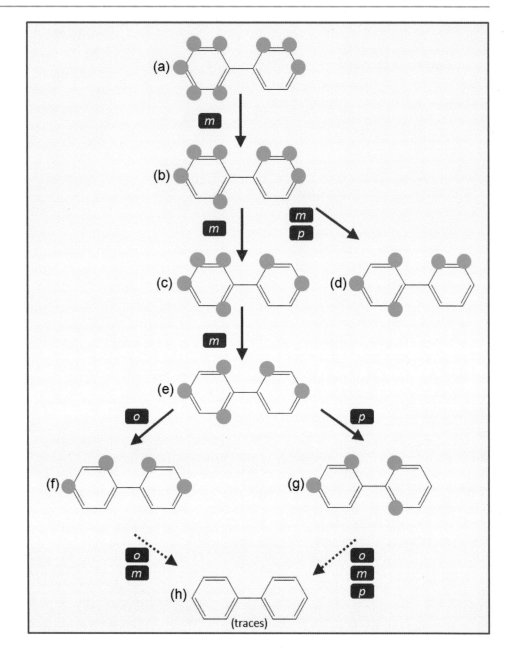

Fig. 16.31 Examples of anaerobic pathways of reductive dechlorination observed in sediment microcosms and anaerobic enrichment cultures (according Field and Sierra-Alvarez 2008). *Dotted arrows* correspond to slow reactions; *o*: ortho-dechlorination; *m*: meta-dechlorination; *p*: para-dechlorination.
(**a**) 2,2′,3,3′,4,4′,5,6-octachlorobiphenyl.
(**b**) 2,2′,3,3′,4,4′,6-heptachlorobiphenyl.
(**c**) 2,2′,3,4,4′,6-hexachlorobiphenyl.
(**d**) 2,2′,3′,4,6-pentachlorobiphenyl.
(**e**) 2,2′,4,4′,6-pentachlorobiphenyl.
(**f**) 2,2′,4,4′-tetrachlorobiphenyl.
(**g**) 2,2′,4,6′-tetrachlorobiphenyl.
(**h**) Biphenyl

(called ECO3) immobilized on different supports. The co-culture was made with 3 strains:

(i) *Alcaligenes* sp. that uses and totally dechlorinates 3-, or 4-chlorobenzoic, and 3,4-dichlorobenzoic acids
(ii) *Pseudomonas* sp. capable of using and completely dechloronating 2-chlorobenzoic and 2,5-dichlorobenzoic acids
(iii) Another strain of *Pseudomonas* sp. that can use biphenyl or 4-chlorobiphenyl as sole carbon source and cometabolize several weakly chlorinated PCBs in the presence of biphenyl

This co-culture was immobilized on three different supports: cubic polyurethane foam (side: 0.8 mm), glass beads (diameter: 0.8 mm), and silica beads. Biodegradation was more important with immobilized than with free cells. The tested congeners (2,5-, 3,5-, 2,3′, 2,4′-, 3,3′-, and 4,4′-dichlorobiphenyls) as well as Aroclor 1221 were intensively degraded (48–99 %). The biodegradation intensity varied as a function of the PCB nature and the material. For Aroclor 1221, the dechlorination percentage was 94.4, 79.4, and 49.2 % with polyurethane foam, glass, and silica beads, respectively. These results prove that it is possible to use such a consortium, immobilized, in a reactor with fixed bed, for the treatment of wastewaters contaminated by PCBs. Biodegradation under aerobic conditions was also investigated upon addition of oxygen, molecules increasing the PCB availability, or co-substrates such as biphenyl.

For instance, Borja et al. (2006) studied the biodegradation of highly chlorinated PCBs (Aroclor 1260) by a mixed culture isolated from a contaminated soil after biphenyl enrichment of the soil. The consortium was immobilized on cement particles in a three-phase fluidized-bed reactor running in a discontinuous mode where it was progressively adapted to PCB by alternated supplies of biphenyl and PCB. The formed biofilm degraded with a high-efficiency PCBs added to the reactor: 80 % biodegradation after 1 day and 91 % after 5 days; the important decline of Aroclor the first day is accounted for by the pollutant adsorption on the biofilm followed by a progressive PCB degradation.

Biphenyl was used as co-substrate to stimulate PCB biodegradation in soils and sediments. However, as biphenyl is itself a pollutant, alternative solutions were proposed, based in particular on the use of natural compounds such as carvone extracted from mint (*Mentha spicata*). This non-toxic ketone (terpenoid) induces in *Arthrobacter* sp. strain B1B, Aroclor 1242 cometabolism and is used at a 50 mg dm^{-3} concentration that is too low for the bacterium growth in the absence of an additional input (fructose). Aroclor 1242 cometabolism in the presence of carvone was demonstrated in microcosms by a significant disappearance of dominant congeners, formation of split products from phenylhexadienoate, and production of various chlorobenzoates that are intermediate compounds of PCB degradation (Gilbert and Crowley 1997). This important finding suggests that it would be possible to use plants secreting monoterpene inside the rhizosphere to achieve in situ bioremediation of contaminated soils. PCB biodegradation in chronically polluted soils (about 350 mg kg^{-1} of soil, mainly Aroclor 1242 and 1254) was also investigated in the presence (and absence) of cyclodextrines that increase the availability of water-insoluble molecules (Fava et al. 1998); the PCBs' low solubility was mentioned here above. The soil that contained microorganisms able to grow on cyclodextrins as the sole source of carbon and energy was supplemented with biphenyl (4 g kg^{-1} of soil) and inorganic nutrients (0.3 NH_4NO_3, g kg^{-1} of soil, and 0.04 KH_2PO_4, g kg^{-1} of soil), N and P concentrations being very low in the soil. The presence of cyclodextrins effectively increased the biodegradation level of some congeners.

Results above presented concern studies conducted in microcosms on small volumes of soil or sediment. An experiment was conducted in situ to evaluate the aerobic biodegradation of a preexistent PCB contamination in Hudson River sediments till then submitted to a natural anaerobic biodegradation (Harkness et al. 1993). It was performed in steel cylinders of 1.8 m diameter and 5 m high, inserted in the initially anoxic sediment. Supplementing dioxygen, inorganic nutrients and biphenyl increased PCB biodegradation (37–55 % of disappearance), the latter being evidenced by chlorobenzoate production occurring during PCB biodegradation.

Attempts of bioamplification were also undertaken in order to increase the PCB aerobic biodegradation. They essentially focus on improving the efficiency of strains able to degrade PCBs and amenable to inoculation in polluted sediments and soils. Several strategies were developed:
(i) Addition of one (several) strain(s) able to degrade PCBs in a polluted site (seeding) knowing that a given strain can only attack some congeners.
(ii) Grouping together in a single microorganism the genes involved in the different degradation steps. For instance, grouping in the same organism genes ensuring under aerobiosis the degradation of PCBs and that of chlorobenzoic acid.
(iii) Using biodegradation capacities of a microorganism by amplifying its activity through biosurfactant implementation: coupling biostimulation and bioamplification. Under these conditions, by coupling *Arthrobacter* sp. B1B and *Ralstonia eutropha* H850 addition to that of a surfactant (sorbitol trioleate), the PCB elimination in a sediment polluted by Aroclor 1242 reached about 60 %.

One the major problems faced by microorganisms used in bioamplification experiments is their elimination by microorganisms already present in the environment. In order to overcome this competition, one solution is to have one microorganism able to use for its growth a substrate not taken up by endogenous communities that consume biphenyl. To reach this objective, the genes *bhpABCD* (Fig. 16.32), present in the strain *Pseudomonas* sp. ENV307, were inserted via a plasmid into the strain *Pseudomonas paucimobilis* 1IGP4 that uses as a growth substrate a nonionic surfactant, Igepal CO-720, water-soluble, and nontoxic unlike biphenyl (Lajoie et al. 1993). A soil contaminated by Aroclor 1242 and seeded with the 1IGP4 strain (about 4.10^6 cells·g^{-1} of soil) was added with Igepal CO-720 (1 % w/w). This treatment lead to biodegradation of many congeners: 60 % biodegradation of 2′,3,4- and 2,5,2′,6′ congeners, after 24 days.

Several conclusions can be drawn from these experiments on PCB aerobic biodegradation (Furukawa 2000; Field and Sierra-Alvarez 2008):
(i) Monochlorinated compounds are growth substrates for several bacteria species. Generally, biodegradation of PCBs with more than one chlorine atom proceeds through cometabolism in the presence of biphenyl which is then used as a carbon and energy source and as inducing biodegradation enzymes. The aerobic biodegradation of polychlorinated congeners in the absence of biphenyl is nevertheless possible. Thus, *Pseudomonas aeruginosa*, strain TMU56, isolated from a soil contaminated for 35 years by a current transformer fluid, degrades in the absence of biphenyl 2,4-diCB; 2,2′,5,5′-tetraCB; and 2,2′,4,4′,5,5′-hexaCB as well as several Aroclor 1442 polychlorinated congeners (Hatamian-Zarmi et al. 2009).

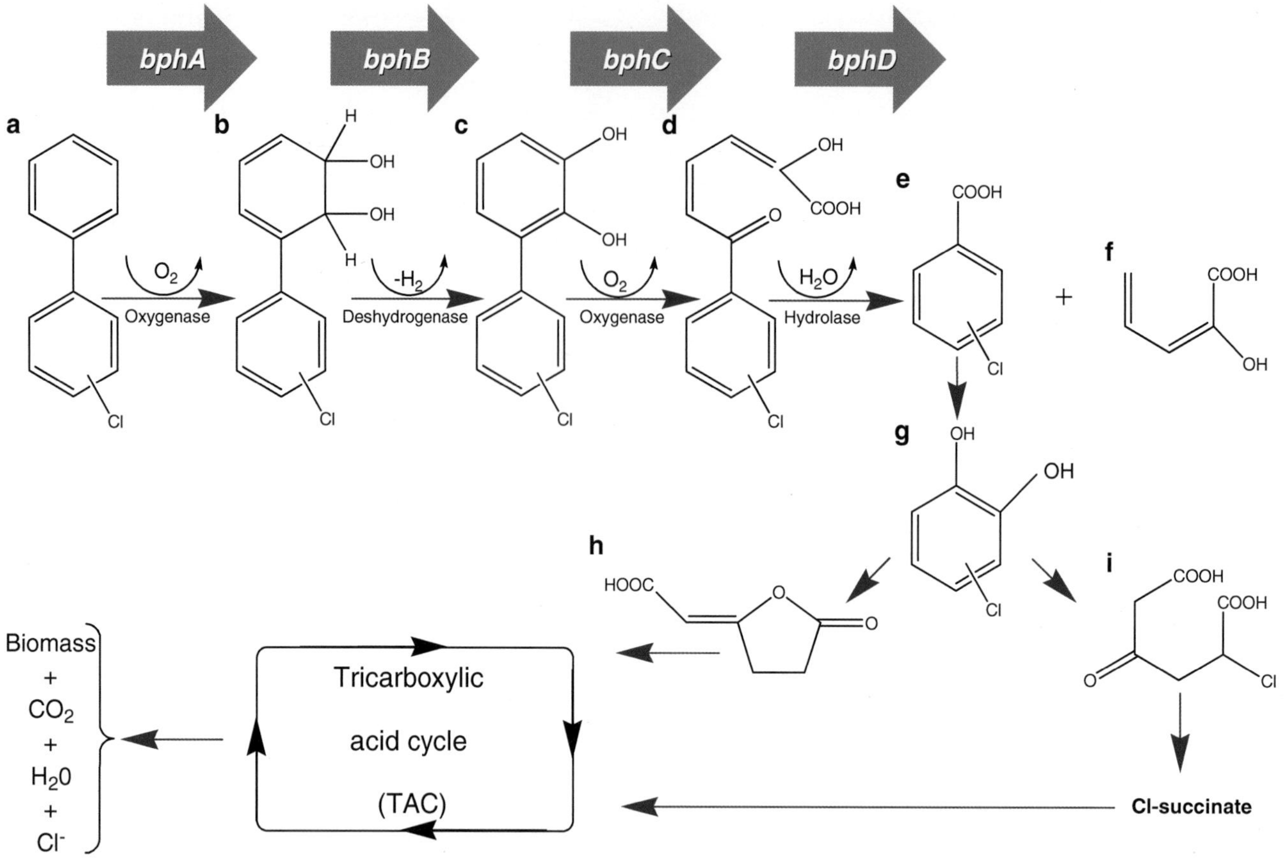

Fig. 16.32 Major steps of aerobic biodegradation pathway (Adapted from Mhiri and Tandeau de Marsac 1997; Borja et al. 2005; Field and Sierra-Alvarez 2008). Letters between brackets correspond to metabolites produced during the biodegradation. Inside the *arrows* are indicated *bph* genes that code enzymes involved in the transformation of PCBs in chlorobenzoic acid. (**a**) PCB; (**b**): *cis*-dihydrodiol-PCB; (**c**): 2,3-dihydroxy-PCB; (**d**): 2-hydroxy-6-oxo-6-chlorophenyl-hexa-2,4-dienoic acid; (**e**): chlorobenzoic acid; (**f**): 2-hydroxy-penta-2,4-dienoic acid; (**g**): chlorocatechol; (**h**): *cis*-dienelactone; (**i**): 5-chloro-3-oxo-adipate

(ii) Biodegradation levels decrease with the number of chlorine atoms of the congeners.

(iii) Two chlorine atoms in *ortho* position on the same aromatic ring (2,6) or on different rings (2,2′), are biodegradation resistant.

(iv) If chlorine atoms are present on both rings, that with the lower number of chlorine atoms will be first hydroxylated.

(v) Congeners chlorine substituted on the same aromatic ring are more rapidly metabolized than those with the same chlorine atom number but distributed on both aromatic rings. Thus, 3,4-CB will be more easily metabolized than 3′,4-CB.

(vi) PCBs containing chlorine atoms on 2 and 3 positions (2,2′3,3′-tetrachlorobyphenyl), 2,2′,3,5′-tetrachlorobiphenyl and 2,2′,3′,4,5-pentachlorobiphenyl, are more rapidly biodegraded than tetra- and pentachlorinated congeners.

(vii) An aromatic ring cleavage preferentially takes place on an unsubstituted ring.

A series of enzymatic reactions takes place during biodegradation of weakly chlorinated congeners (Fig. 16.32). The first one corresponds to the introduction of 2 oxygen atoms in positions −2 and −3 of the non-chlorinated PCB ring or of the less chlorinated (**a**), thanks to a dioxygenase to form *cis*-dihydrodiol derivatives (**b**). The later compounds are dehydrogenated into 2,3-dihydroxybiphenyl upon the action of a dehydrogenase (**c**) and then split in position 1,2 by a 2,3-dihydroxybiphenyl dioxygenase to end in forming a "metasplit" product (**d**) that is hydrolyzed to give chlorobenzoic acid (**e**) and a compound with 5 carbon atoms (**f**) that can be further degraded in pyruvic acid and acetaldehyde. The chlorobenzoic acid is in its turn attacked by a 1,2 dioxygenase and then a dehydrogenase to produce chlorocatechol (**g**). There are different ways to attack chlorocatechol, leading to its mineralization, producing in particular as intermediate compounds (**h**) and (**i**).

The attack in the presence of dioxygen implies the involvement of two gene clusters. The first one is responsible for transforming congeners into chlorobenzoic acid and the second ensures the degradation of the chlorobenzoic acid.

The Coupling of Anaerobic and Aerobic Treatments Is Necessary to Mineralize Polychlorinated Compounds

It comes out from the above results that PCB mineralization can only be achieved by the joint and complementary action of anaerobic and aerobic communities (Fig. 16.29). In this last paragraph will be presented experiment examples where was applied on sediments or soils an anaerobic treatment that ensures dechlorination followed by an aerobic treatment that leads to PCB mineralization. The efficiency of such a biotreatment was demonstrated in laboratory:

(i) With contaminated sediments from the Hudson River (USA). In a first step, the sediment was put under anaerobic conditions in sealed flasks, during 20 weeks, in the presence of methanol: tri-, tetra-, penta-, and hexachlorobiphenyls were reduced and the increase of mono- and dichlorobiphenyls was observed. In a second step, the anaerobic cultures were flushed out with dioxygen and inoculated with an aerobic bacterium isolated from the sediment. After 96 h incubation, most of the mono- and about 25 % of dichlorobiphenyls were degraded (Borja et al. 2005).

(ii) On a soil containing Aroclor 1260 (Master et al. 2002) firstly submitted during 4 months to an anaerobic treatment (initial PCB concentration 59 mg kg^{-1} of soil). This treatment produced dechlorination of the congener majority (mean chlorine atom number shifting from 6.2 to 5.2) with in particular formation of 2,2′,4,4′-tetrachlorobiphenyl and 2,2′,4,6′-tetrachlorobiphenyl. In contrast, the overall PCB amount did not decrease. Then, the anaerobic degradation products were biodegraded by an aerobic treatment under the action of the strain *Burkholderia* sp. LB400. After 28 days of aerobic treatment, the PCB concentration was reduced to about 28 mg kg^{-1} of soil which corresponded to 68 % disappearance.

(iii) With sediment from the Red Cedar River (Michigan, USA) contaminated by Aroclor 1242 (70 mg PCB kg^{-1} of soil) (Rodrigues et al. 2006). One-year incubation under anaerobiosis with microorganisms present in the sediment leads to the elimination of about 32 % chlorine atoms of Aroclor 1242 (14 % of Aroclor in weight), essentially at the level of *meta* substitutions. After this incubation period under anaerobiosis, the sediment was placed under aerobic conditions and two genetically built strains were inoculated: (**1**) *Burkholderia xenovorans* LB400 containing the biphenyl degradation pathway and genetically modified to insert the ohb operon enabling growth on *ortho*-chlorobenzoate and thus fully mineralize *ortho*-chlorobiphenyl and (**2**) *Rhodococcus* sp. strain RHA1 modified to insert the fcb operon enabling growth on *para*-chlorobenzoate. Both strains grew on PCBs (growth estimated by counting with epifluorescence microscopy and by real time PCR) and remained genetically stable during incubation. After 30 days incubation, 57 % of the remaining PCBs were degraded, whereas only 4 % were eliminated in the absence of both strains, thus demonstrating that their joint presence was necessary for the degradation of Aroclor anaerobic dechlorination products. At the end of the experiment, the content was about 25 mg kg^{-1} of soil.

(iv) In a biofilm developed on a granular support made of anaerobic sludge particles encapsulated by a polymer and placed in a bioreactor to monitor the fate of Aroclor 1242 (Tartakovsky et al. 2001). In the thickness of the biofilm built by prokaryotic populations present in the mud, a dioxygen gradient took place (oxic zone at surface, anoxic one at depth); this procedure enables the simultaneous establishment of aerobic and anaerobic conditions. The analyses showed a PCB disappearance of about 80 %. This suggests that an entire PCB mineralization can be achieved in a system where aerobic and anaerobic conditions are simultaneously established.

16.7.4.4 Bioremediation of PCB Contaminated Sites

The most used and efficient process is burning, but it is costly and produces dangerous compounds such as polychlorodibenzofuranes and polychlorodibenzodioxines formed under incomplete PCB burning. Under these conditions, researches were considered on alternative treatments involving microorganisms with PCB degradation capacities (Mhiri and Tandeau de Marsac 1997; Ohtsubo et al. 2004). However, bioremediation is generally at a development stage. Proofs of its efficiency and cost-effectiveness at large scale are limited. Nevertheless, experiments conducted by Harkness et al. (1993) demonstrated the feasibility of contaminated sediment bioremediation under in situ conditions, and promising results were also obtained with contaminated soils treated by composting. As an example, soil contaminated by PCB whose composition was close to that of Aroclor 1248 was excavated and mixed with plant wastes to form 19 m^3 biopiles with a volume to surface ratio of 1.5–1.9 (m^2/m^3) (Michel et al. 2001). In bio-hummocks containing 80 % plant wastes, temperature rose to 60 °C after 10 days. Under these aerobic conditions (monthly reversal of the bio-hummocks), 40 % PCBs were eliminated after 370-day treatment (<1 % evaporation). The most easily degraded congeners contained 1–3 chlorine atoms. However, beyond 4 chlorine atoms, no significant alteration was observed, which is in agreement with the aerobic biodegradation processes previously described.

16.8 Hydrocarbons and Petroleum Products

16.8.1 Hydrocarbons

Each year, huge amounts (about $3 \cdot 10^9$ tons) of hydrocarbons, mainly from petroleum, are consumed. Because crude oil is a mixture of hydrocarbons, the industrial core refining process starts with a simple distillation. This basic operation consists to separate the crude oil into lighter, middle, and heavy fractions. These latter heavy products are recovered, sometimes at temperatures over 500 °C.

The simplest refineries stop at this point. Other refineries reprocess the heavier fractions into lighter products to maximize the output of the most desirable products, conducting to changes in the molecular structure of the input with various chemical reactions, some of them in the presence of a catalyst and some with thermal reactions (cracking or reforming processes).

A catalytic cracker, for instance, uses the gasoil (heavy distillate) output from crude distillation as its feedstock and produces additional finished distillates (heating oil and diesel) and gasoline. A reforming unit produces higher-octane components for gasoline from lower-octane feedstock that was recovered in the distillation process. Sulfur removal is accomplished in a hydrotreater. Some of the obtained products are then recombined in order to correspond to technical or safety guidelines.

From lighter to heaviest, petroleum fractions correspond to liquid petroleum gases (LPG) (mainly propane and butane), naphtha or ligroin (distillate intermediate further processed to make gasoline), gasoline, "white spirit" (containing less than 5 % benzene), lamp oil, kerosene, gas oil or diesel distillate, lubricating oil, heavy fuel oil, and residuals (coke, asphalt, tar, waxes, etc.).

Some chemical bases (olefins, aromatics, etc.) are also produced and used as the initial reactants for the petrochemical industry, particularly to produce polymers.

As a consequence of the hydrocarbon exploitation and intensive use, almost all ecosystems are polluted. Microorganisms are major players in their natural elimination and their metabolic capacities are also made use of for the remediation of polluted sites.

16.8.1.1 Different Kinds of Hydrocarbons

Hydrocarbons are, strictly speaking, composed exclusively of carbon and hydrogen and can be classified into saturated, unsaturated, and aromatic compounds (Fig. 16.33).

Saturated Hydrocarbons
They are divided into three structural groups: linear alkanes, branched alkanes, and cycloalkanes. Normal *n*-alkanes have a straight carbon chain: their molecular formula is C_nH_{2n+2} and their carbon atom number ranges between 5 and 60. Branched alkanes have a skeleton with one or several offshoots, usually a methyl group ($-CH_3$); pristane (2,6,10,14-tetramethylpentadecane) and phytane (2,6,10,14-tetramethylhexadecane) (acyclic isoprenoids) are the most abundant branched alkanes and are products of the chlorophyll side chain. Cycloalkanes (also called alicyclic hydrocarbons or naphthenes) have a saturated ring and are essentially made of cyclopentane or cyclohexane derivatives. Polycyclic derivatives (bi-, tri-, tetra-, pentacyclic) are also present: drimanes, tricyclicterpane (or sterane), and hopanes.

Unsaturated Hydrocarbons
Unsaturated hydrocarbons are hydrocarbons that have double or triple covalent bonds between adjacent carbon atoms. Those with at least one double bond are called alkenes and those with at least one triple bond are called alkynes. For instance, phytadienes are found in zooplankton, lamellibranch mollusks, and fishes. Cyanobacteria contain linear alkenes (*n*-alkenes, C_{19}–C_{29}).

Aromatic Hydrocarbons
They have 1 aromatic ring (mono-aromatic compounds) or several aromatic rings (polyaromatic compounds or PAH). Benzene and alkylbenzene(toluene, ethylbenzene, and the 3 isomers *ortho*-, *meta*-, and *para*-xylene that make the BTEX group) contain only one benzene ring with methyl- and ethyl-substituents. Naphthalene, phenanthrene, and benzo(a)pyrene contain 2, 3, and 5 rings, respectively. Some aromatic structures (or polyaromatics) are associated to saturated rings (rings with 5 or 6 carbons). Such compounds (e.g., steroids), are qualified as naphtenoaromatics.

16.8.1.2 Hydrocarbon Origin
Hydrocarbons found in the environment may have a natural or anthropogenic origin (Fig. 16.34).

Hydrocarbons from Natural Origin
Natural sources are divided in four categories:

(i) Compounds synthesized or produced by living organisms and released in the environment. They are ubiquitous compounds whose chemical structure (alkanes, alkenes, isoprenoids, etc.) often varies with the organisms that synthesized them. For instance, regarding n-alkanes, compounds with a long chain and an odd carbon atom number (n-C_{23} to n-C_{33}) are typical of higher plants (Magnoliophytes). The analysis of n-alkanes isolated from marine phytoplankton usually points out the dominance of n-C_{15}, n-C_{17}, and n-C_{21} compounds. Benthic and pelagic macro-algae essentially synthesize n-14 to n-32 hydrocarbons with an observed maximum for n-C_{15} or n-C_{17}. Hydrocarbons with odd carbon atom chains, n-C_{21} to n-C_{29}, are dominant with some benthic macro-algae. The simplest alkane, methane, is synthesized by

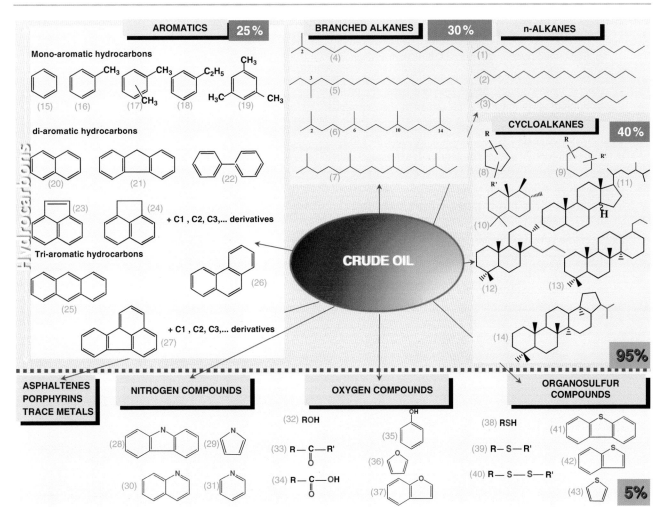

Fig. 16.33 Chemical composition of a crude oil. Displayed values correspond to the composition (w/w%) of the crude oil "arabian light" (Persian gulf). n-Alkanes: *1*: n-octadecane; *2*: n-heptadecane; *3*: n-hexadecane. Branched alkanes: *4*: isooctadecane; *5*: anteisooctadecane; *6*: 2,6,10,14-tetramethylpentadecane (pristane); *7*: 2,6,10,14-tetramethylhexadecane (phytane). Cycloalkanes (naphtenes): *8*: cyclopentane and alkylated derivatives; *9*: cyclohexane and alkylated derivatives; *10*: drimanes (bicyclics); *11*: steranes; *12*: tricyclic terpanes; *13*: tetracyclic terpanes; *14*: pentacyclic terpanes (hopanes). Aromatics: *15*: benzene; *16*: toluene; *17*: xylenes; *18*: ethylbenzene; *19*: 1,3,5-trimethylbenzène; *20*: naphthalene; *21*: fluorene; *22*: biphenyl; *23*: acenaphthylene; *24*: acenaphtene; *25*: anthracene; *26*: phenanthrene; *27*: fluoranthene. Nitrogen compounds: *28*: carbazole; *29*: pyrrole; *30*: quinoline; *31*: pyridine. Oxygen compounds: *32*: Alcohols; *33*: ketones; *34*: acids; *35*: phenols; *36*: furanes; *37*: benzofuranes. Organosulfur compounds: *38*: thiols; *39*: sulfides; *40*: disulfides; *41*: dibenzothiophene; *42*: benzothiophene; *43*: thiophene. C_1-, C_2-, and C_3 derivatives correspond to alkylated congeners with 1 (methyl), 2 (dimethyl or ethyl), or 3 (trimethyl, methyl, and ethyl, propyl or isopropyl) substituents, respectively

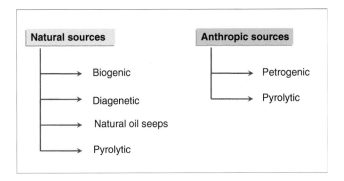

Fig. 16.34 Hydrocarbon sources in the environment

methanogenic archae (*cf.* Sect. 14.2.6). Branched alkanes (iso and anteiso) with short carbon chain length and with odd carbon chain length predominance (C_{15} and C_{17}) are considered as bacterial **biomarkers***. Pristane and phytane are widely distributed; they were detected in photosynthetic microorganisms, heterotrophic bacteria, some zooplankton species, and macrophytes. The unsaturated compounds, alkenes, are synthesized by a number of marine organisms and higher plants. These compounds are highly unstable and quickly transformed by diagenetic processes, which explains their absence in fossil resources. As to aromatic hydrocarbons found in small amounts in some living organisms (bacteria, algae, plants), it was not clearly demonstrated whether these compounds were really biosynthesized or their presence resulted from bioaccumulation.

(ii) Diagenetic hydrocarbons. Some hydrocarbons are produced through early diagenetic processes of the organic matter (time scale from a few days to tens of years). As an example, in sediments, phytol that is a degradation product of the chlorophyll *a* phytyl moiety leads to pristane and phytane under oxic and anoxic conditions, respectively.

(iii) Hydrocarbons from natural marine seepages. Petroleum hydrocarbons are released through Earth crust faults, whether continental or oceanic. One of the most important seepage is that of Santa Barbara (California) daily releasing about 80 tons of methane and 35 tons of non-methanic hydrocarbons (Hornafius et al. 1999). Most methane is oxidized by the water column microorganisms; according to Mau and collaborators (2007), only 1–10 % are finally released into the atmosphere.

(iv) Pyrolytic hydrocarbons. They are released into the atmosphere in large amount during forest fires (Kim et al. 2003). The main hydrocarbon families above mentioned are present. For instance, the chemical composition of pine wood smoke is as follows: terpenoids and derivatives (39 %), PAH and alkyl-PAH (17 %), alkenes (15 %), alkanes and steroids (8 % each), alkanoic acid (6 %), monosaccharides (4 %), and phenols (3 %) (Simoneit et al. 2000). During fires, essentially light hydrocarbons are released in the atmosphere where they can be carried away over long distances. Thus, HAPs produced by forest fires occurring in Indonesia were detected more than 1,000 km away (Okuda et al. 2002).

Hydrocarbon from Anthropogenic Origin

The use of petroleum products by mankind causes a very large environment contamination and two kinds of sources contribute to this contamination. The first kind of sources is represented by chronic inputs with particularly rejections in the city or industrial networks, inputs linked to exploiting and producing petroleum and to transportations. A more detailed description of inputs into the ocean will be provided further. The second kind of sources concerns accidental inputs that may happen at each production and exploitation stage and represent an important contamination risk. The most spectacular are oil-well-head accidents, mishaps linked to storage, pipeline breakdown, and shipwrecks of oil tankers and gas carriers. Anthropogenic inputs are from pyrolytic and petroleum origin:

(i) Pyrolytic origin. This source is supplied by burning fossilized (hydrocarbon, coal) or recent (wood) organic matter used for house or city heating; industrial processes (steel factories) have a strong impact on the amount and structure of released compounds, specially HAPs. These residues are released into the environment through

Table 16.6 Solubility in water[a] of n-alkanes and aromatics ($mg \cdot L^{-1}$) at 25 °C (Mackay et al. 1992)

n-alkanes	Solubility ($mg \cdot L^{-1}$)	Aromatics	Solubility ($mg \cdot L^{-1}$)
n-C_5	40	Benzene	1,700
n-C_6	10	Toluene	530
n-C_7	3	Ethylbenzene	170
n-C_8	1	p-Xylene	150
n-C_{12}	0.01	Naphthalene	30
n-C_{30}	0.002	1-methyl naphthalene	28
		1,3-dimethyl naphthalene	8
		1,3,6-trimethyl naphthalene	2
		Fluorene	2
		Phenanthrene	1
		Anthracene	0.075
		Chrysene	0.002

[a]In seawater, the displayed solubilities are about 25 % lower

the atmosphere (directly or by soil washing) and by transportations. The combustion temperature and the organic matter nature determine the kind of released hydrocarbon.

(ii) Petroleum origin. This origin corresponds to all pollutions induced by the mankind use of hydrocarbons except their burning.

16.8.1.3 Biodegradation by Microorganisms

Hydrocarbons are biodegraded by many microorganisms called hydrocarbonoclastics (Timmis 2010). These microorganisms carry the necessary enzymes for the first transformation steps of these compounds, which corresponds to the metabolism peripheral or high pathways. Hydrocarbons are then transformed into key intermediate metabolites: alcohol for alkanes, catechol or benzyl-succinate with aromatic hydrocarbons, etc. These metabolites will be further involved in metabolic pathways present in many microorganisms and finally reach the central or low metabolic pathways such as β-oxidation and the tricarboxylic acid cycle (Krebs cycle). The initial reactions of hydrocarbon transformation involve a large diversity of hydrocarbonoclastic bacteria-specific enzymes. These enzymes are coded by different genes whose localization (plasmidic, chromosomic, associated to mobile elements like "**cassettes***"), organization, and regulation also appear diversified. It is worth noting that some non-catabolic genes that are involved in a decisive way in hydrocarbon biodegradation can be associated and expressed together with catabolic genes (Samanta et al. 2002).

Hydrocarbon Transfer to Cell Surface

Hydrocarbon represents hydrophobic molecules very weakly water soluble (Table 16.6). Consequently, microorganisms have developed mechanisms to transfer substrates to the cell in order to assimilate and then degrade them (Husain et al. 1997; Bonin and Bertrand 1998; Abbasnezhad et al. 2011) (Fig. 16.35; Box 16.4).

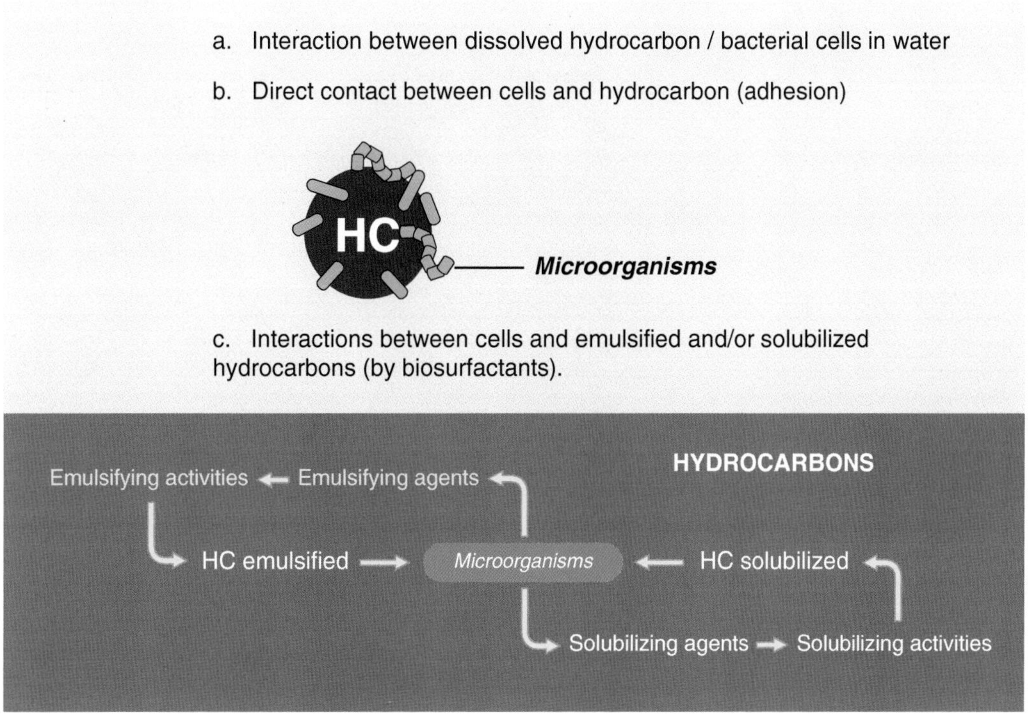

a. Interaction between dissolved hydrocarbon / bacterial cells in water
b. Direct contact between cells and hydrocarbon (adhesion)
c. Interactions between cells and emulsified and/or solubilized hydrocarbons (by biosurfactants).

Fig. 16.35 Modes of transfer of hydrocarbons to the bacterial cell surface. *HC* hydrocarbons. Drawing: M.-J. Bodiou

---

Box 16.4: Hydrocarbon Degradation in Marine Environment by Biofilm Formation at the Water-Hydrocarbon Interfaces

Regis Grimaud

In the marine environment, hydrocarbons resulting from accidental or chronic pollution are partly eliminated by biodegradation. The search for bacterial strains responsible for this biodegradation led to the discovery of the group of marine hydrocarbonoclastic bacteria. These bacteria, present in low numbers in uncontaminated waters, proliferate and become dominant when oil is introduced in seawater. Marine hydrocarbonoclastic bacteria belong, among others, to the genera *Alcanivorax, Marinobacter, Thalassolituus,* and *Cycloclasticus Oleispira* (Yakimov et al. 2007).

These bacteria have the remarkable ability to adapt their physiology to the use of hydrocarbons as carbon and energy sources. Cellular functions involved in the assimilation of hydrocarbons are diverse and remain for most of them poorly understood. The recent availability of the genome sequence of some marine hydrocarbonoclastic bacteria facilitates the study of the biodegradation of hydrocarbons by allowing to find its genetic determinants, using genetic, genomics, and proteomics approaches (Schneiker et al. 2006) (Grimaud 2010a). *Marinobacter hydrocarbonoclasticus* SP17 is one model chosen for the study of hydrocarbons degradation in the marine environment. Bacteria of the genus *Marinobacter* are found in most seas and oceans. They are able to colonize very different ecosystems, including extreme environments with high salt concentration (up to 2.5 M NaCl) or high pH. The strain *M. hydrocarbonoclasticus* SP17 was isolated in 1992 from sediments of the Gulf of Fos sur Mer (France), near an oil terminal (Gauthier et al. 1992). This strain has a pronounced hydrocarbonoclastic activity and is able to use linear alkanes of 8–28 carbon atoms as the sole carbon and energy source. The degradation pathway is initiated by oxidation of the alkane into a primary alcohol, followed by successive oxidations into aldehyde and fatty acid which then enters into the β-oxidation pathway. During growth on linear alkanes, *M. hydrocarbonoclasticus* SP17 forms a biofilm at the alkane–water interface. Biofilms were also observed at the interfaces between the aqueous phase and other substrates nearly insoluble in water, such as long-chain alkanes, fatty alcohols, and triglycerides (Box Fig. 16.8) (Klein et al. 2008).

R. Grimaud (✉)
Environ. Microb. Genom. Group, Ecole Centrale de Lyon, UMR CNRS, 5005, Laboratoire Ampére, Écully, France

(continued)

Box 16.4 (continued)

Box Fig. 16.8 Image with confocal laser scanning microscopy of a biofilm of *M. hydrocarbonoclasticus* SP17 covering a hexadecane droplet. *Green*, polysaccharide matrix of the biofilm revealed by fluorescent lectin. The hexadecane is visualized by the Red Nile fluorochrome (*orange*) at the non-coated areas by the biofilm of the droplet (Photo courtesy of Pierre-Joseph Vaysse and Thomas R. Neu, Helmholtz Center for Environmental Research – UFZ, Magdeburg, Germany)

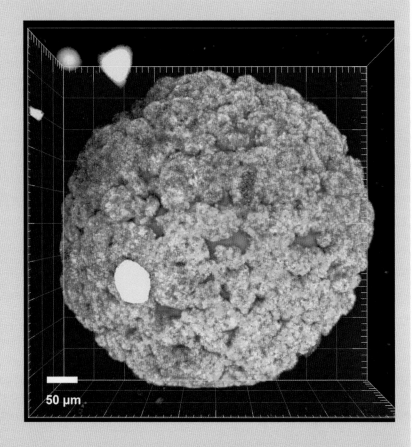

Biofilm formation at hydrocarbon–water interfaces has been observed in many strains or consortia degrading aliphatic or aromatic hydrocarbons. These biofilms develop on interfaces qualified of nutritive since they play both the role of substrate and substratum. This feature distinguishes them from conventional biofilms growing on inert supports, such as minerals, metals or plastics. The bacteria responsible for the formation of this type of biofilm form preferentially biofilms on metabolizable compounds poorly soluble in water suggesting that the determinism of these biofilms is regulated according to the solubility of the substrate (Johnsen and Karlson 2004; Grimaud 2010b).

Hydrocarbons in water form a two-phase systems, comprising an organic phase consisting of solid or liquid hydrocarbon and the aqueous phase in which a small fraction of the hydrocarbon dissolves. At the interface between the two phases, there is a layer of stagnant water called the boundary layer. The hydrocarbons dissolved in the water at the interface and then diffuse through the boundary layer to reach the free aqueous phase. This creates a concentration gradient of hydrocarbons in the boundary layer. Planktonic bacteria floating in the aqueous phase can only assimilate the hydrocarbon fraction dissolved in water. The degradation rate and consequently the growth rate of the bacteria depend on the rate of mass transfer of the hydrocarbon from the organic phase to the cells in aqueous phase. The mass transfer process includes the dissolution of the molecules in the water and their diffusion through the boundary layer. Planktonic cells can increase the rate of mass transfer of hydrocarbons by secreting biosurfactants (Box Fig. 16.9a). In the case of bacteria forming biofilms, it has often been observed that they do not release biosurfactants in the aqueous phase. In addition, the growth rate of cell within the biofilm is much higher than the growth rate suggested by the rate of mass transfer of the hydrocarbon in the absence of bacteria. In light of these observations, it is clear that the biofilm lifestyle favors the mass transfer of hydrocarbons and thus stimulates the growth of hydrocarbons. Biofilm formation at hydrocarbon–water interfaces is an adaptive process, allowing bacteria to assimilate substrates weakly soluble in water (Johnsen et al. 2005).

(continued)

Box 16.4 (continued)

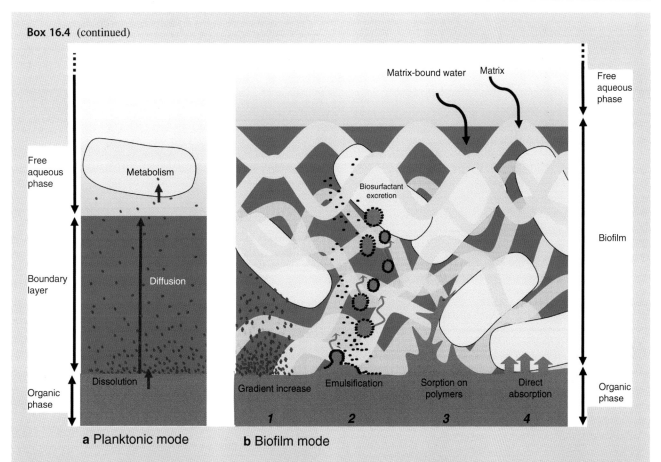

Box Fig. 16.9 Theoretical mass transfer mechanisms of hydrocarbons. (**a**) Planktonic mode. Hydrocarbons are solubilized at the interface, diffused through the boundary layer to reach the free aqueous phase where they are assimilated. Bacteria can stimulate these processes by producing biosurfactants. (**b**) In biofilm. Mode: (*1*) Solubilization and establishment of a diffusion gradient of hydrocarbon concentration in the aqueous phase immobilized in the matrix. The consumption of oil by the cells of the biofilm increases the concentration gradient which stimulates their solubilization and diffusion. (*2*) Biofilm cells produce biosurfactants which accumulate in the matrix until reaching a concentration above their Critical Micelle Concentration (CMC). Then, there is formation of micelles which carry the hydrocarbons to cells. (*3*) Hydrocarbon are adsorbed onto the polymer matrix. Cells use them directly or after desorption. (*4*) The cells absorb the hydrocarbons directly at the interface

Studies conducted on biofilms growing on the surface of solid hydrocarbons allowed to elucidate a mechanism of mass transfer facilitation. The rate of diffusion of hydrocarbons through the boundary layer is controlled by the steepness of the concentration gradient created in this layer. Bacteria bound to the interface consume hydrocarbons within the boundary layer, thus contributing to increase the gradient steepness and consequently stimulate the diffusion of the hydrocarbons (Box Fig. 16.9b) (Bouchez et al. 1997; Wick et al. 2001, 2002). However, the mechanisms of assimilation of hydrocarbons within biofilms are most likely multiple and remain, for most of them, poorly understood. Nevertheless, on the basis of biofilms characteristics, it is possible to envisage some hypothetical mechanisms (Box Fig. 16.9b). The hydrocarbons may be transported inside the cell without being dissolved in the aqueous phase. In other words, the cells absorb the oil directly at the interface. Biofilm formation would allow cells to contact the interface. At present, it has never been demonstrated that a bacterial cell is able to assimilate a compound that is not in aqueous solution. However, this hypothesis cannot be excluded. Another possible mechanism could involve the extracellular matrix which is a ubiquitous component of biofilms. Extracellular matrices of biofilms are complex environments, highly hydrated, consisting of biopolymers such as polysaccharides, proteins, and nucleic acids. They could facilitate the transfer of hydrocarbons to the cells by

(continued)

> **Box 16.4** (continued)
>
> optimizing mechanisms used in planktonic growth mode, such as the secretion of biosurfactants. If it has been noticed that strains growing as a biofilm on the hydrocarbon–water interface do not release bioemulsifiers or biosurfactants in the environment, it is quite conceivable that biosurfactants would be produced within the biofilm and could accumulate in the matrix without diffusing into the surrounding medium. Thus biosurfactants could easily reach a concentration above their CMC (Critical Micelle Concentration) which would allow the formation of micelles, ensuring a fast diffusion of hydrocarbons within the biofilm. Polymers of the biofilm matrix could also serve as adsorbents for hydrocarbons making them possibly more accessible to the cells.
>
> The biofilm mode of growth is characterized by a profound change in the physiology and in the behavior of cells, requiring regulation of the expression of hundreds of genes. The metabolic activities of cells with diffusion processes generate microchemical gradients which in turn affect cell physiology (Stewart and Franklin 2008). As a result, biofilms are highly heterogeneous structures, containing cells with different phenotypes, metabolism, and physiological states. Given such a functional and structural remodeling, it is not difficult to imagine that the biofilm lifestyle offers the possibility of several mechanisms to increase the mass transfer of hydrocarbons.

In addition to the spontaneous solubilization in the aqueous phase (Table 16.6), transfer may take place through a cell-substrate contact that depends on the cell intrinsic adherence capacities. Another transfer mode that is the most widespread brings into play production of tensioactive molecules, synthesized by microorganisms, called biosurfactants, that emulsify hydrocarbons or increase their solubilization and, consequently, favor their assimilation. The chemical nature of biosurfactants is highly variable and depends on the organism that synthesize them (phospholipids, glycolipids, lipopolysaccharides, lipoproteins, etc.) (Rosenberg and Ron 2000).

Aerobic Biodegradation

Microorganisms that biodegrade hydrocarbons under aerobiosis are ubiquitous (van Hamme et al. 2003; Prince 2010; Prince et al. 2010) and essentially belong to *Proteobacteria*: *Gammaproteobacteria* (genus *Acinetobacter*, *Alcanivorax*, *Pseudomonas*, *Marinobacter*, *Mycobacterium*, *Nocardia*, *Rhodococcus*, *Streptomyces*, etc.), *Betaproteobacteria* (genus *Achromobacter*, *Alcaligenes*, *Burkholderia*, etc.), *Alphaproteobacteria* (genus *Sphingomonas*); *Firmicutes* (genus *Bacillus*, *Geobacillus*), as well as some eukaryotic microorganisms (genus *Candida*, *Hansenula*, *Saccharomyces*, *Aspergillus*, *Fusarium*, etc.) are also able to biodegrade hydrocarbons. Biodegradation by cyanobacteria was not established. However, within some microbial mats and biofilms, these organisms seem to act indirectly by producing dioxygen, which would favor the activity of hydrocarbonoclastic microorganisms. A number of review were dedicated to biodegradation mechanisms leading to hydrocarbon mineralization (van Hamme et al. 2003; Berthe-Corti and Höpner 2005; Vandecasteele 2008). These mechanisms will not be described into details, only hydrocarbon primary oxidation will be described as it is the essential step of catabolic pathways for these compounds (Fig. 16.36).

From a chemical point of view, alkanes are relatively inert molecules. To make possible their degradation, they must be activated before undergoing other metabolic stages. Under aerobic conditions, in most cases, n-alkanes are oxidized by monooxygenases/hydroxylases to form the corresponding primary alcohol; thus an oxygen atom is inserted into the molecule. The substrate attack is usually terminal, but may take place in biterminal or subterminal position (Fig. 16.36a–c). Depending on the alkane carbon chain length, different enzymatic systems will be necessary to insert dioxygen within the substrate and initiate biodegradation. For instance, for alkanes with chain length over 4 carbon atoms, different enzyme families are involved, such as cytochrome P450-like enzymes (e.g., bacterial enzymes CYP153) and membranous hydroxylases (pAHs). This last type of enzyme corresponds to monooxygenases with nonhemic iron with dinuclear active site of AlkB type, enabling a number of actinomycetal proteobacteria to use C_5–C_{16} n-alkanes. These enzymes work within a complex containing two electron transporters, one rubredroxin and one rubredroxin reductase enabling electron transfer from NADH to the active site of the alkane hydroxylase (Fig. 16.37). After the initial oxidation of the n-alkane, the corresponding alcohol is then oxidized by an alcohol dehydrogenase to produce aldehyde and finally the corresponding fatty acid. The latter serves as substrate to acyl-CoA synthetase and the formed acyl-CoA is then transformed by β-oxidation.

In the case of aromatic compounds, two key steps take place in the aerobic biodegradation of mono- or polycyclic aromatic hydrocarbons. In the first step, stable benzene rings are activated by incorporation of one or two oxygen atoms upon the action of monooxygenases or dioxygenases, respectively (Fig. 16.36d, e). For instance, in the case of naphthalene, this reaction involves the naphthalene-dioxygenase (NDO) system that usually comprises one NADH oxidoreductase, ferredoxin, and one oxygenase containing the catalytic site. This system

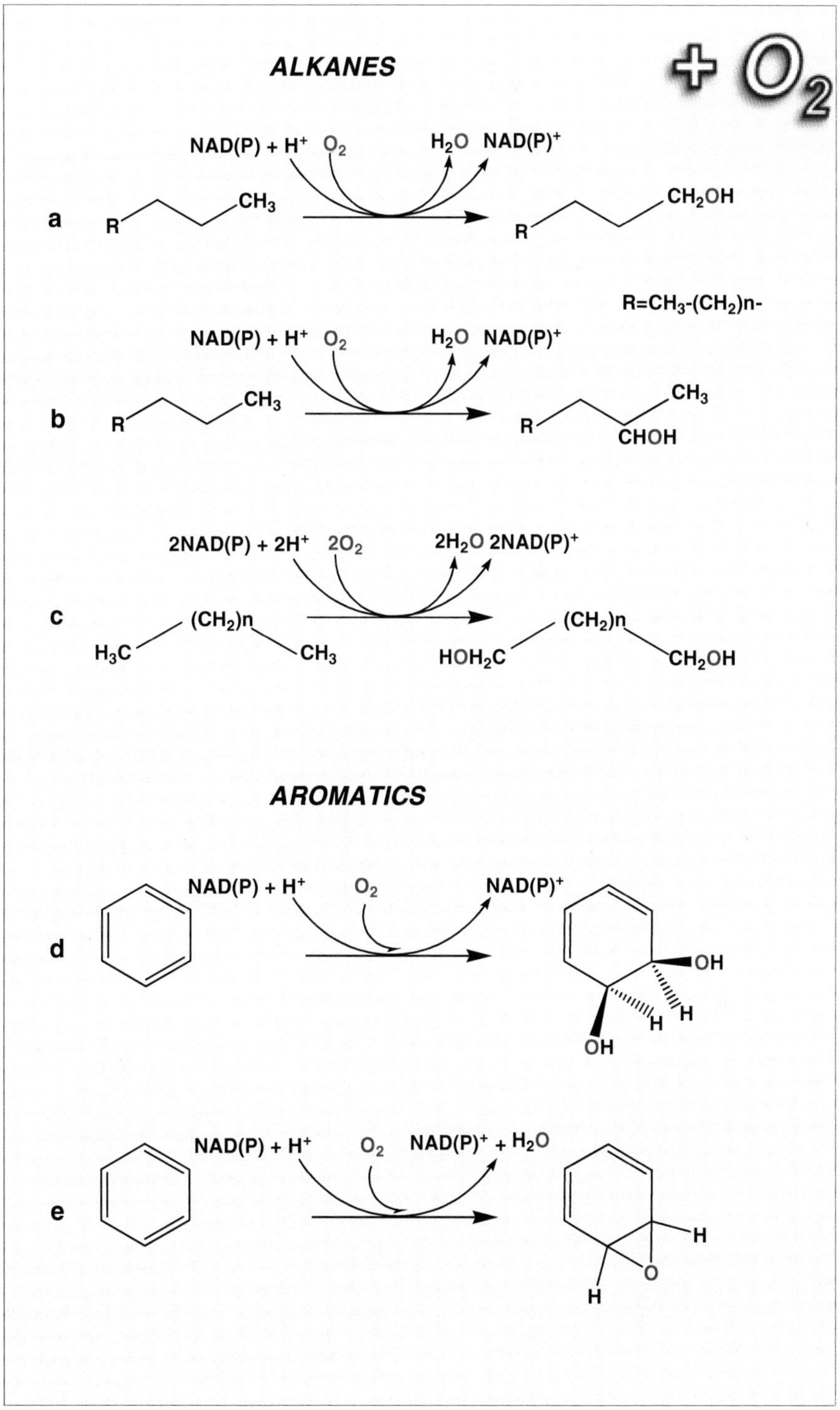

Fig. 16.36 First oxidation steps in aerobic biodegradation of saturated (**a** to **c**) and aromatic hydrocarbons (**d** and **e**) (According to Berthe-Corti and Höpner 2005)

enables electron transfer to the nonheme iron of the NDO catalytic site. The second step, corresponding to the ring split, is achieved by an extradiol- or intradiol-dioxygenase-type enzyme. This set of enzymatic reactions, whose complexity depends on the number of hydrocarbon aromatic rings, produces a limited number of metabolites (e.g., catecol, protocatechuic acid, gentisic acid, hydroquinone) that will be later on transformed into intermediaries of the tricarboxylic acid cycle (TCA). These transformations imply, among others, the pathways of a-ketoadipate, phenylacetate, 2-hydroxypentadienoate, or else gentisate. Some steps of these metabolic pathways are common and catalyzed by similar enzymes. However, some genetic variability may exist through the different prokaryotic species (isoenzymes, gene regulation, and organization). The genes involved in the biodegradation of n-alkanes and aromatic compounds are described in the Box 16.5. The different hydrocarbon classes are not biodegraded with

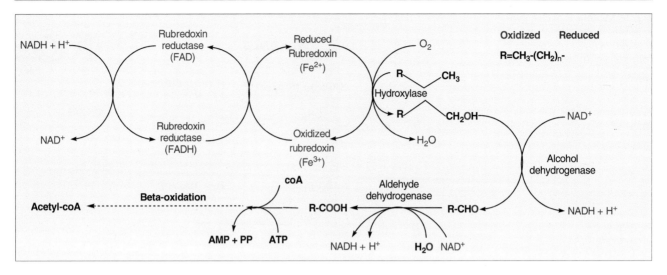

Fig. 16.37 Enzymes involved in aerobic biodegradation of n-alkanes

---

Box 16.5: Aerobic Hydrocarbon Biodegradation: Bacterial Genes and Enzymes

Philippe Cuny and Cécile Militon

The genes involved in the enzymatic transformations of *n*-alkanes have been particularly well studied in *Pseudomonas putida* GPo1 (formerly known as *P. oleovorans*) capable of using the *n*-alkanes in C_5 to C_{14} (Rojo 2009). This strain carries the OCT-plasmid containing the operon *alkBFGHJKL* with the catabolic genes as well as the operon *alkST* encoding a rubredoxin reductase (AlkT) and the positive regulator (AlkS) of the *alkBFGHJKL* operon induced in the presence of *n*-alkanes (Box Fig. 16.10a). These two operons are located end to end on the plasmid and are separated by a 9.7 kb fragment encoding in particular a methyl acceptor transduction protein (AlkN), possibly involved in chemotaxis. AlkL protein seems to play an important role in hydrophobic substrate mass transfer over the outer membrane, enabling efficient AlkB hydroxylase catalysis (Julsing et al. 2012) (Box Fig. 16.10b). It is worth noting that the G + C content of *alk* genes and insertion sequences in *P. putida* GPo1 is lower than that of the rest of the chromosome and the OCT-plasmid. This observation, also remarked in other strains with these genes, suggests that they are part of an integrated mobile element (Schneiker et al.

P. Cuny (✉) • C. Militon
Institut Méditerranéen d'Océanologie (MIO), UM 110, CNRS 7294 IRD 235, Université de Toulon, Aix-Marseille Université, Campus de Luminy, 13288 Marseille Cedex 9, France

2006). In addition, there is no correlation between the phylogenetic affiliation of the strains and the diversity of *alkB* genes, supporting the hypothesis of a horizontal transfer of these genes.

It is not uncommon for strains to have multiple genes encoding for different hydroxylases. Mention may be made of the case of the marine bacterium *Alcanivorax borkumensis* strain which has 2 hydroxylases (AlkB1 and AlkB2). It is a cosmopolitan strain highly specialized in the use of hydrocarbons as it grows almost exclusively on alkanes (linear, branched, and isoprenoid) and alkylated compounds. This strain, which is considered a major player in the process of bioremediation of marine ecosystems contaminated by hydrocarbons, also has three cytochrome P450 genes that are expressed preferentially in the presence of isoprenoid alkanes (Schneiker et al. 2006).

Before we look at the genes involved in the breakdown of aromatic hydrocarbons, it should also be pointed out that in these compounds, in some strains (e.g., *Acinetobacter* sp. *strain M1*) activation of the alkane is carried out by a dioxygenase leading directly to the formation of the corresponding aldehyde (with no alcohol formation) via *n*-alkyl hydroperoxides.

Aromatic hydrocarbon biodegradation involves a large number of peripheral catabolic pathways related to the variety of aromatic compounds present in the environment (e.g., *Rhodococcus* sp. RHA1 could produce enzymes involved in at least 26 different peripheral catabolic pathways). However, similar paths can be found in phylogenetically diverse species even if

(continued)

Box 16.5 (continued)

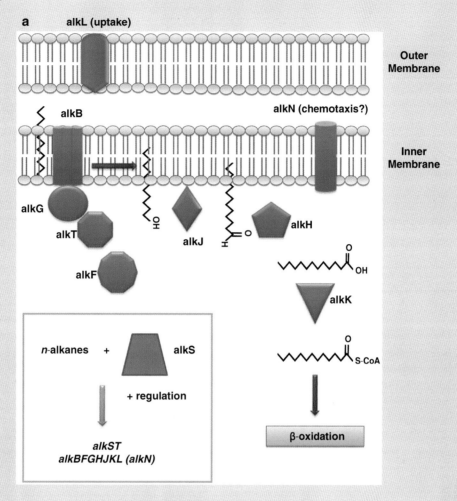

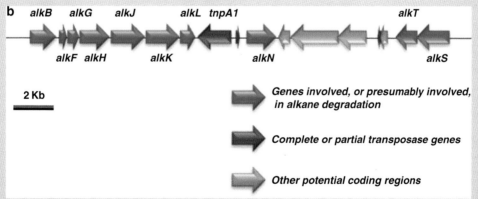

Box Fig. 16.10 Metabolic pathway of n-alkane degradation in *Pseudomonas putida* GPo1. (**a**) Role and cellular localization of Alk enzymes and regulation of the alk genes: AlkB (alkane hydroxylase), AlkF and AlkG (rubredoxins), AlkH (aldehyde dehydrogenase), AlkJ (alcohol dehydrogenase), AlkK (acyl-CoA synthetase), AlkL (outer-membrane protein functioning as an alkane uptake facilitator), AlkN (methyl-accepting transducer protein, putative chemotactic transducer for alkanes), AlkT (rubredoxin reductase), and AlkS (positive regulator of the expression of the alkBFGHJKL operon) (Modified from van Hamme et al. 2003). (**b**) Genomic map showing the *alk* genes organization (Modified from van Beilen et al. 2001)

(continued)

Box 16.5 (continued)

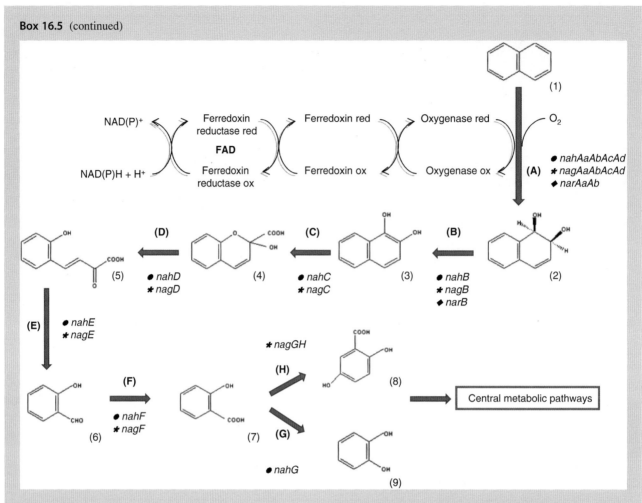

Box Fig. 16.11 Upper catabolic pathways and genes involved in naphthalene biodegradation in *Pseudomonas putida* G7 (l), *Ralstonia* sp. U2 («), and *Rhodococcus* sp. NCIMB12038 (t). The compounds are naphthalene (*1*), *cis*-naphthalene dihydrodiol (*2*), 1,2-dihydroxynaphthalene (*3*), 2-hydroxy-2H-chromene-2-carboxylic acid 1-hydroxy-2-naphthoic acid (*4*), *trans-o*-hydroxybenzylidene-pyruvic acid (*5*), salicylaldehyde (*6*), salicylic acid (*7*), gentisic acid (*8*), and catechol (*9*). The enzymes involved in each reaction step are naphthalene dioxygenase (*A*), *cis*-naphthalene dihydrodiol dehydrogenase (*B*), 1,2-dihydroxynaphthalene dioxygenase (*C*), 2-hydroxy-2H-chromene-2-carboxylate isomerase (*D*), *trans-o*-hydroxybenzylidenepyruvate hydratase-aldolase (*E*), salicylaldehyde dehydrogenase (*F*), salicylate hydroxylase (*G*), and salicylate 5-hydroxylase (*H*) (Modified from Habe and Omori 2003)

there may be some genetic variability (e.g., isoenzyme, regulation, organization of genes). To illustrate this phenomenon, the biodegradation of naphthalene by *Pseudomonas putida* G7 can be taken as an example. In this strain, the catabolic genes involved in the biodegradation of naphthalene (*nah*) are organized in two operons including the genes encoding for the enzymes involved in the conversion of naphthalene to salicylic acid (peripheral aromatic pathways) (Box Fig. 16.11) and in the conversion of salicylic acid to tricarboxylic acid cycle intermediates (central aromatic pathways), respectively. Within the genus *Pseudomonas*, the nucleotide sequences of catabolic genes (e.g., *nah* genes in *P. putida* G7, *ndo* in *P. putida* NCIB9816, *dox* in *Pseudomonas* sp. C18 and *pah* in *P. aeruginosa* PAK1) are presenting more than 90 % of identity and similar spatial arrangement (Box Fig. 16.12a) (van Beilen et al. 2001). Thus, in these strains, the genes encoding the upper pathway enzymes are organized according to the following scheme: naphthalene-dioxygenase system (NDO) composed of *nahAa* (ferredoxin reductase), *nahAb* (ferredoxin), *nahAc* (α-subunit of naphthalene dioxygenase), *nahAd* (β-subunit of the naphthalene dioxygenase), *nahB* (naphthalene-*cis*-dihydrodiol dehydrogenase),

(continued)

Box 16.5 (continued)

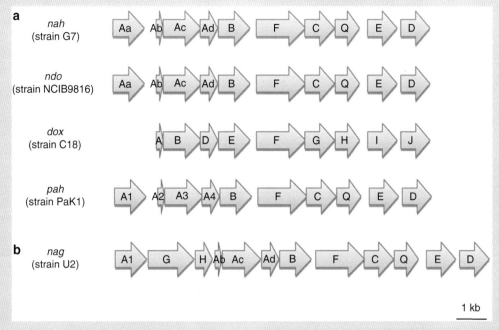

Box Fig. 16.12 Gene clusters coding for enzymes involved in naphthalene upper catabolic pathway in *Pseudomonas putida* G7, *P. putida* NCIB9816, *Pseudomonas* sp. strain C18, *P. aeruginosa* PaK1 (**a**), and *Ralstonia* sp. U2 (**b**). The arrows represent the size, location, and direction of transcription of the ORF (Open Reading Frame) (Modified from Peng et al. 2008)

nahF (salicylaldehyde dehydrogenase), *nahC* (1,2-dihydroxynaphthalene dioxygenase), *nahQ* (unknown gene), *nahE* (*trans-o*-hydroxybenzylidene-pyruvate hydratase-aldolase), and *nahD* (2-hydroxy-2H-chromene-2-carboxylate isomerase) (Habe and Omori 2003).

These similarities can also be found in other bacterial genera. In *Ralstonia* sp. U2, the operon genes encoding the upper pathway enzymes (*nag*) have the same organization as the *nah* genes except for the presence of two additional genes *nagG* and *nagH* encoding two subunits of the salicylate-5-hydroxylase involved in the conversion of naphthalene to gentisate (Box Fig. 16.12b) (Peng et al. 2008). On the contrary, in the *Rhodococcus* sp. *strain* NCIMB 12038, the naphthalene-dioxygenase system (NDO) is formed only by the genes *narAa*, *narAb*, and *narB* encoding, respectively, for the subunits α and β of the NDO and for the *cis*-naphthalene dihydrodiol dehydrogenase. In this strain, the genes encoding for the ferredoxin and reductase components of the NDO do not appear to be localized in the *nar* operon.

With the explosion of genomic data, microbial metabolism knowledge involved in the breakdown of hydrocarbons have been largely enriched in recent years. However, the field of investigation is still large especially for anaerobic mechanisms. Thanks to recent advances in sequencing technologies, we can access to the genomes of most cultivated microorganisms and thus to their metabolic potential. The coupling of these techniques to the analysis of transcriptomic (e.g., RNAseq) and proteomics (e.g., LC/MS/MS) will undoubtedly enable clearer identification of the metabolic pathways and regulations implemented by microorganisms in response to the presence of organic contaminants in their environments.

the same efficiency (Fig. 16.38). Biodegradation starts by n-alkane attack, followed successively by that of branched alkanes, low molecular weight aromatic compounds, acyclic alkanes, high molecular weight aromatic compounds, and cycloalkanes (steranes, hopanes) (van Hamme et al. 2003).

Finally, hydrocarbonoclastic prokaryotes generally have a specific degradation capacity: they biodegrade either n-alkanes or aromatic compounds. However, some biodegrade both saturated and aromatic compounds (White et al. 1997; Tapilatu et al. 2010).

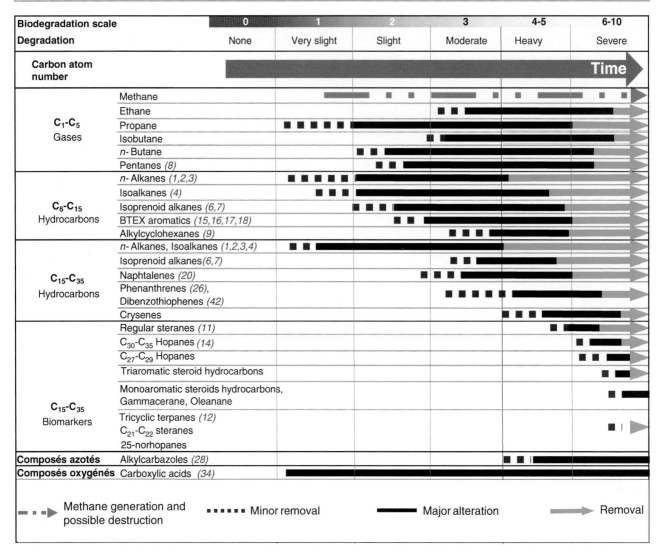

Fig. 16.38 Schematic diagram of physical and chemical changes occurring during crude oil and natural gas biodegradation. Figures in *red italics* refer to structure numbering of Fig. 16.33 caption (Modified and redrawn from Head et al. 2003)

Anaerobic Biodegradation

Anaerobic hydrocarbon biodegradation that was controversial for a long while is now well established (Widdel and Rabus 2001; Rabus 2005; Carmona et al. 2009). A number of hydrocarbons, essentially aromatics, but also alkanes and alkenes, are degraded by bacteria mainly belonging to proteobacteria that use nitrate, sulfate, and even iron as terminal electron acceptors (Widdel and Rabus 2001; van Hamme et al. 2003; Heider 2007; Kunapuli et al. 2008; Callaghan et al. 2009). A hyperthermophile sulfate-reducing *Archaea*, *Archaeoglobus fulgidus*, oxidize *n*-alk-1-enes in the presence of thiosulfate as a terminal electron acceptor (Khelifi et al. 2010). Other strains are fermenting these substrates (*Pelobacter acetylenicus* ferment acetylene) (Seitz et al. 1990); one photosynthetic strain (*Blastochloris sulfoviridis*, strain TOP1) assimilates toluene by using light as energy source (Zengler et al. 1999). The used aromatic compounds make a long list: benzene, alkylbenzenes (toluene, xylene, ethylbenzene, propylbenzene, ethyltoluene, cymene), naphthalene, 2-methylnaphthalene, etc. The alkane biodegradation was described for long carbon chains, between C_3 and C_{20}, whereas that of alkenes occurs between C_7 and C_{23} (Grossi et al. 2008). Alike for aerobic biodegradation, hydrocarbon primary oxidation is the critical step (Fig. 16.39). As to *n*-alkanes, the initial biodegradation step takes place either by fumarate addition or by carboxylation. For aromatic compounds, the initial step is carried out either by fumarate addition, hydroxylation or carboxylation (Fuchs et al. 2011).

16.8.2 Petroleum Products

Petroleum hydrocarbons are currently the most important primary energy source to be commercialized: they will remain so over several decades. The increasing demand of emergent

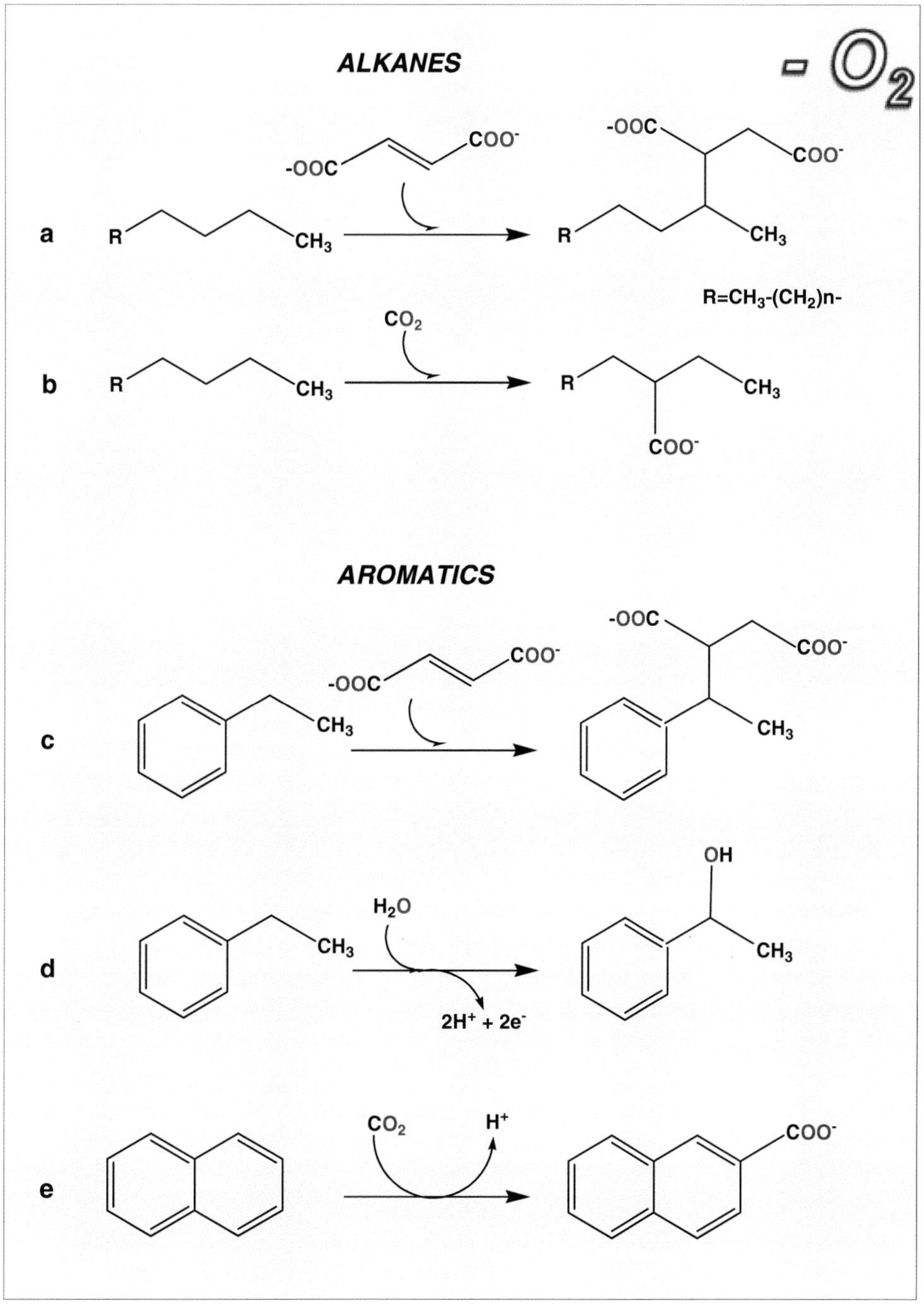

Fig. 16.39 Anaerobic degradation of saturated (**a** and **b**) and aromatic (**c** to **e**) hydrocarbons; first oxidation steps (Modified and redrawn from Widdel and Rabus (2001))

countries, associated to the petroleum cost increase, will lead to an increased exploration and drilling exploitation in areas free of any human activity (Siberia, Arctic, etc.). The price increase of the petroleum barrel will make profitable the exploitation of new resources in spite of complex and dangerous drilling conditions (deep environments, reservoirs under high pressure or high temperature). Prospecting in extreme environments will unavoidably increase the number of polluted sites. Consequently, petroleum hydrocarbons are and will remain the source of pollutions that had, have, and will have detrimental consequences on human health, air and water quality, the ecosystem functioning, etc.

16.8.2.1 The Two Petroleum Biogeochemical Cycles

The petroleum origin and fate imply a long biochemical cycle, corresponding to their geological formation and a short cycle related to their fate in the contemporary environment (Fig. 16.40).

16.8.2.2 Petroleum Chemical Composition

The average elementary composition of crude oil is as follows (American Petroleum Institute, 20038002-05-9): carbon (84 %), hydrogen (14 %), sulfur (1–3 %), nitrogen (1 %), and others (0.1 %). More than 20 000 different

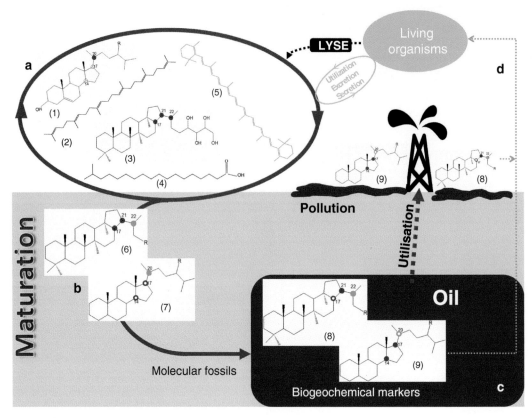

Fig. 16.40 The two biogeochemical cycles of the petroleum. (**a**) Molecules used as substrate, secreted, excreted by living organisms, or issued from cell lysis. (*1*) Sterols (Eukarya): stereochemistry 14α (H), 17α (H), 20S, (*2*) Squalene, (*3*) Bacteriohopanetetrol (Procarya) stereochemistry: 17β (H), 21β (H), 22R, (*4*) branched fatty acid, (*5*) β-carotene. (**b**, **c**) Maturation of organic matter in sediments: this continuum of processes is divided in three successive steps called diagenesis, catagenesis, and metagenesis (for petroleum only the first two steps are achieved). Example of the formation of steranes and hopanes from sterols and bacteriohopanetetrol, respectively. During these processes, molecules first undergo thermal defunctionalization and saturation, but their stereochemistry is preserved as the starting molecule (**b** biological configuration). (*6*) 17β(H) 21β(H), 22R – hopane (biological configuration) (*7*) 14α (H), 17α (H), 20R – sterane (biological configuration). Then, during the formation of petroleum, there is change from biological (B) to geological (C) configuration which corresponds to isomerization phenomena (cyclanic: red stereocenters; optical: blue stereocenters). (*8*) 17α(H), 21β(H), 22R-hopane; (*9*) 14β(H), 17β(H), 20S-sterane. (**d**) Anthropic use of fossil hydrocarbons and release in the environment due to extraction, transportation, etc. The *blue dashed line* illustrates the concept of "biogeochemical marker" (see text): with fossil molecules from (*8*) and (*9*) families, it is possible to characterize a crude oil ("molecular fingerprinting" see text). For "oil and gas" exploration and prospection aims, it is possible to evaluate oil maturity in the reservoir rock or to monitor the fate of an oil spill in the recent environment. Due to their stability, these molecules can be used as endogenous or "internal conserved standards." In some cases, due to the stability of the carbon skeleton, it is possible to identify the fossil organisms at the origin of this molecule. Stereochemistry: Cyclanic isomerization (positions 14 and 17 for steranes and positions 17 et 21 for hopanes) ○ α hydrogen atom (bond directed into page); ● β hydrogen atom (bond directed out of page). Optical isomerization (position 20 for steranes and 22 for hopanes); ○ S enantiomer; ● R enantiomer

structures were described for the molecules involved in crude oil composition. Hydrocarbons make the majority and represent about 95 % of petroleum (Fig. 16.33). The other petroleum components are asphaltenes, resins, sulfur, oxygen, nitrogen compounds, and metals. The latter are only present as traces: the most abundant are vanadium, nickel, iron, sodium, copper, and uranium. Asphaltenes and resins represent 5–20 % of crude oil. They are compounds with a complex structure built with condensed aromatic cycles of naphtenoaromatics, branching, and heteroatoms (O, N, and S) (Fig. 16.41). It is worth noting that asphaltenes are not soluble in organic solvents (pentane, hexane) and this solubility criterium defines them as a class.

The chemical composition of a given petroleum depends on its geographical area (Table 16.7), the physicochemical conditions that prevailed when it was formed (diagenesis, catagenesis) and on the living organisms from where it stems. Thus, each petroleum has a specific chemical composition. The notion of "molecular fingerprint" derived from studies in petroleum geochemistry enables to identify the petroleum origin when a pollution occurs. The molecular

Fig. 16.41 2D structural model (conceptual) of an asphaltene molecule. ● = $-(CH_2)_n-$ or $-(CH_2)_n-CH_3$. Aromatic rings are figured with an *orange circle*, *colored letters* represent major heteroatoms (O,N,S)

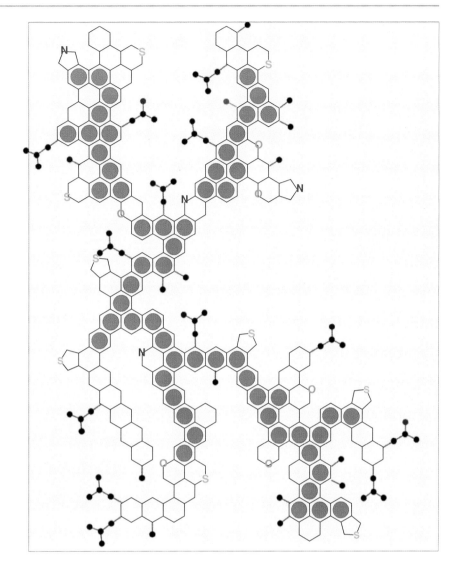

Table 16.7 Characteristics of different crude oils as a function of their geographical origin

Origin	Total saturated hydrocarbons %	Aromatics %	Polars %	Asphaltens %
Venezuela	16	43	29	12
Saudi Arabia (Safaniya)	28	41	19	12
Saudi Arabia (Arabian Light)	43	38	15	4
North Sea (Ninian)	53–45	33–29	13–26	1–0
North Sea (Alwyn Se-Ellon)	63	28	9	0
Algeria (Hassi Messaoud)	64	27,5	8	0,5

Adapted from Vandecasteele (2008)

families recalcitrant to biodegradation processes that enable such identifications are typically pentacyclic terpanes (hopanes), regular and reordered steranes, etc.

16.8.2.3 Origin of Petroleum Products in the Marine Environment

The following part in this chapter will be dedicated to the origin and fate of petroleum products in the marine environment: it is the environment that is quantitatively submitted to the largest hydrocarbon pollutions. The assessment of the different hydrocarbon sources released into the sea is, at the Earth scale, difficult to establish. The analysis of estimations provided by different expert groups shows that the huge and spectacular oil input generated by ships running aground is less important than chronic inputs linked to transportation, refining operations, city, and industrial wastes (Table 16.8). If hydrocarbon inputs into the sea remain important, the overall amount of released hydrocarbons diminished the past 10 years; the "National Research Council" estimates that the average total amount of petroleum released in 1985 to the sea was 9×10^6 tons. The same organism gives an estimate of 1.3×10^6 tons for the period 1990–1999 (National Research Council 2002).

Table 16.8 Worldwide average, annual releases (1990–1999) of petroleum by source (in thousands of tons) in marine environment

Source	($\times 10^3$ tons)
Natural seeps	600
Extraction of petroleum	38
Platforms	0.86
Atmospheric deposition	1.3
Produced waters	36
Transportation of petroleum	150
Pipeline spills	12
Tank vessel spills	100
Operational discharges (cargo washings)	36
Coastal facility spills	4.9
Atmospheric deposition	0.4
Consumption of petroleum	480
Land based (river and runoff)	140
Spills (non-tank vessels)	7.1
Operational discharges (vessels $\geq$100 GT)	270
Atmospheric deposition	52
Aircraft fuel	7.5
Total	1,300

Adapted from National Research Council, USA, 2002

16.8.2.4 Natural Elimination of Petroleum Hydrocarbons in the Marine Environment

As soon as released in the environment, the petroleum is submitted to a natural elimination upon the joint action of physical, chemical, and biological processes (Fig. 16.42). The relative importance of these different processes depends on the input chemical composition, the concerned environments (pelagic, coastal, or lagoon area), and their geographical location (Leahy and Colwell 1990; Harayama et al. 1999; Peters et al. 2005).

Physicochemical Processes

In a first time, after spread and dispersal of the oil spill, the lightest compounds evaporate, prompting atmospheric pollution. Compounds whose boiling point is < 150 °C (from C_1 to C_{10}) are eliminated within 2 h after their input into the sea and those whose boiling point is < 250 °C (from C_{10} to C_{15}) are lost in one day. In the Amoco Cadiz case, in 3 days, almost 40 % of the oil mass was transferred to the atmosphere. Though petroleum is globally weakly soluble in seawater, a non-negligible fraction that may represent 9–15 % is dissolved and leave the oil spill (Fig. 16.42). In parallel, petroleum is submitted to the action of photooxidation processes; there are oxygenation reactions that occur under the combine action of dioxygen and sun radiations with wavelength > 290 nm. These reactions are grouped in two categories: direct reactions where light is directly absorbed by hydrocarbons (aromatics) induced reactions where molecules qualified as photosensitizing (chlorophyll, pheopigments, humic and fulvic acids, some quinones like anthraquinone) absorb light and transfer the received energy to hydrocarbons. Photooxidation essentially affects aromatic compounds; among aliphatic compounds, the branched ones are more easily oxidized than *n*-alkanes. Photooxidation products (hydroperoxides, phenols, carboxylic acids, ketones, aldehydes, etc.) are found in the polar fraction of petroleum. The light action, very important at surface, is very quickly attenuated with depth and is usually limited to the upper 25 m of the water column. It is worth noting that photoproducts, though often more toxic than the initial hydrocarbons, are more soluble and thus more easily eliminated by bacteria (Ni'Matuzahro et al. 1999). Another petroleum fraction is also emulsified, producing emulsions of the types water in oil and oil in water. The former, called "chocolate mousse", appear as a sticky mass, highly viscous, and more or less dark brown: they contain 50–80 % water and are very difficult to mechanically collect, treat, or burn. They increase the initial petroleum volume (three- to fivefold), whereas its density becomes higher than that of seawater which prompts its sedimentation. Viscosity also increases, inducing the formation of a semisolid material, precursor of the tarry residues, and/or to "tarballs" found on beaches. This kind of emulsion reduces the petroleum product spreading, their evaporation, their oxidation by photooxidation, the size of the hydrocarbon exposed surface, and, consequently, their biodegradation. Under the influence of turbulence, oil-in-water emulsions are formed, resulting from the oil spill fragmentation leading to build-up of a few μm to 100 μm microdroplets that favor biodegradation processes. This process is enhanced by the presence of emulsifying factors (biosurfactants). Petroleum degradation by evaporation, dissolution, emulsification, photooxidation, and biodegradation contributes to increasing its density and also causes its sedimentation that can be accelerated by petroleum (hydrophobic) interaction with water column suspended particles. Thus, about 20–30 % petroleum may reach the water-sediment interface.

Biodegradation Processes

Under in vitro conditions, the degradation of the different hydrocarbon classes involved in the petroleum composition is relatively well known for most of the compounds (Atlas 1981; Bertrand et al. 1983; Leahy and Colwell 1990; van Hamme et al. 2003; Peters et al. 2005; Timmis 2010). However, the degradation of several polyaromatic compounds and the polar fraction is still incompletely cleared up, particularly that of asphaltenes and resins; however, it was shown that asphaltenes could be biotransformed (Rontani et al. 1985) and even biodegraded (Pineda-Flores et al. 2004). In situ, within the water column and sediments, the degradation by microorganisms is a major process in the hydrocarbon elimination. However, taking into account the chemical petroleum complexity, its biodegradation can only be achieved by bacterial communities with complementary metabolic capacities and whose activity will depend on the following:

(i) The hydrocarbon chemical structure and concentration.
It comes out from all in situ studies, particularly

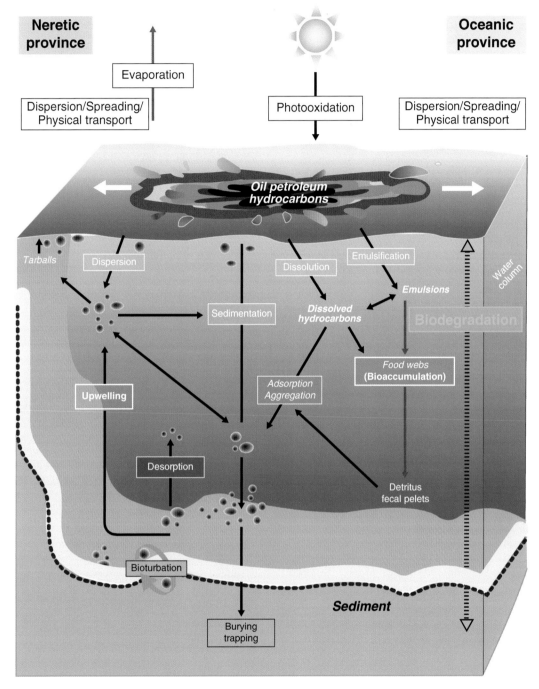

Fig. 16.42 Fate of crude oil in the marine environment. Drawing: M.-J. Bodiou

those addressing oil slicks, that most of petroleum compounds are degraded except asphaltenes. In addition, some molecules are very resistant (recalcitrant) to biodegradation, and their degradation only takes place at the very end of hydrocarbon degradation (Head et al. 2003). These molecules called "biomarkers" are used as "internal conserved biodegradation standards" to determine, in a relative way, the hydrocarbon degradation level of a crude oil by measuring the disappearance of less recalcitrant compound families (Fig. 16.38) (Prince et al. 1994; Le Dréau et al. 1997). In a first time, isoprenoid alkanes (pristane and phytane) were used as biomarkers and the time course of the ratios C_{17}/pristane and C_{18}/phytane was considered as a good biodegradation indicator. However, it was later shown that isoprenoids were degraded in situ at the early stage of hydrocarbon degradation (1–6 months) (Le Dréau et al. 1997).

Thus, tri-, tetra-, and pentacyclic (hopanes) terpanes, and particularly 17α(H)21β(H) hopane, that are degraded at the ultimate stage of petroleum degradation (10–15 years after their input into the environment) are selected as endogenous standards.

(ii) Petroleum chemical composition (Head et al. 2003).
(iii) Petroleum physical state (*cf.* emulsion water-in-oil and oil-in-water).
(iv) Nutrient concentration: The petroleum input in the ecosystem corresponds to an important carbon supply that generates a major imbalance of C/N and C/P ratios.
(v) Dioxygen concentration: The aerobic biodegradation is usually more rapid than the anaerobic one.
(vi) Several physical factors such as hydrodynamism, pH, temperature, salinity. For example, *Bacteria* (Margesin and Schinner 2001) and *Archaea* (Aiken et al.; Rudolph et al. 2001; Khelifi et al. 2010) oxidize hydrocarbons at low and elevated temperatures. In the same way, *Bacteria* (Bertrand et al. 1993; Gauthier et al. 1992; Fernandez-Linarez et al. 1995; Margesin and Schinner 2001) and *Archaea* (Bertrand et al. 1990; Al-Mailem et al. 2010) degrade hydrocarbons at high salt concentrations.
(vii) The presence of emulsifying compounds that may be chemical (dispersal agents) or biological (biosurfactants).

In the water column, biodegradation is essentially aerobic. In the sediments, it is aerobic in the superficial layer (less than 2–3 mm in coastal areas) and anaerobic in deeper layers where it is extremely slowed down. For instance, in an in situ experiment conducted in coastal sediments at 20 m depth, Miralles and collaborators (2007) showed, 16 months after implementation in the sediment of Arabian Light crude oil, an important biodegradation of *n*-alkanes (80–90 %) and isoprenoids (55–60 %) under oxic conditions. Under anoxic conditions, biodegradation is markedly lower, for instance, that of eicosane (saturated hydrocarbon C_{20}) is 90 % in oxic zone and 17 % in sediment anoxic layer. It is important to note that in vitro aerobic biodegradation of the same hydrocarbon by hydrocarbonoclastic single strains or communities isolated from the same site is substantially faster: eicosane is 90 % biodegraded within 24 or 48 h (time scale is no more month but hour).

To explain hydrocarbon evolution within sediments, the action of microorganisms (bioturbation) must be also taken into account (Grossi et al. 2002; *cf.* Sect. 14.2.4). They exert two antagonistic effects: the first one is burying in the sediment anoxic layers which may lead to hydrocarbon sequestration in a zone where biodegradation becomes impossible and the second one is resuspension in the water column and sediment oxygenation (tunnel formation) that increases the hydrocarbon availability and favors aerobic biodegradation processes.

16.8.3 Bioremediation of Contaminated Sites

16.8.3.1 Bioremediation of Marine Ecosystems; Physical, Chemical, and Biological Treatments

Means to fight against accidental and huge oil spills can be grouped into three categories (Fig. 16.43): mechanical means (pollution confinement by setting up barriers and oil recovery (Fig. 16.44), chemical means (in situ burning; washing contaminated substrates with pressurized water, cold or warm; use of dispersal agents; etc.), and biological means (Atlas 1995; Swannell et al. 1996; Bocard 2007; Singh et al. 2012; Dash et al. 2013). Treatments susceptible to be implemented depend on the nature of petroleum products discharged at sea. For instance, in the case of *Erika* shipwreck in December 1999 that occurred in the Atlantic off Belle-Ile, the spilled oil, 19000 tons of heavy fuel oil "N° 2", was not crude oil but a refining product, highly viscous and rich in aromatic compounds, resins, and asphaltenes (saturated: 25 %; aromatics: 55 %; resins: 13 %; asphaltenes: 7 %) (Vandecasteele 2008). This product was poorly biodegradable; biodegradation undertaken in laboratory under optimal conditions was less than 10–15 %. Under such conditions, the only solution to decontaminate the affected areas was the implementation of mechanical and physicochemical means (washing rocks, oil recovery with shovels and buckets, etc.). A similar situation took place with the pollution generated by the *Prestige* shipwreck (2002).

In different circumstances, implementation of microbiological fighting means can be considered. This means are regrouped under the term bioremediation; they rely on metabolic capacities of hydrocarbonoclastic microorganisms. Biostimulation and bioincrease strategies that were developed for coastal area bioremediation are very difficult to implement in open sea where it is impossible to keep nutrients in contact with the oil spill and where hydrocarbonoclastic bacteria concentrations are extremely low. Furthermore, up to now, the efficiency of these treatments in open sea was never demonstrated. In contrast, promising results were obtained on the coastline through biostimulation or by combining bioaugmentation and biostimulation.

Bioaugmentation (or Bioincrease)

Microorganisms, isolated from the environment and/or modified by genetic engineering, must satisfy criteria already mentioned:

(i) To have a large metabolic activity spectrum enabling the degradation of the different components present in the petroleum product.
(ii) To be adapted to the geographical area affected by the pollution. Hydrocarbonoclastic communities that are

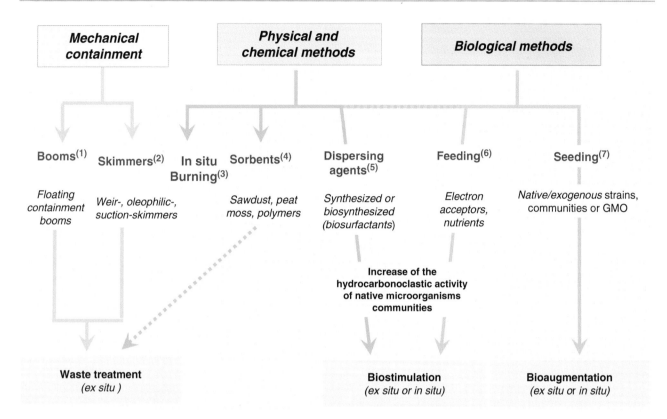

Fig. 16.43 Oil spill control methods in marine environment. (*1*) Used to control the spread of oil and to reduce the possibility of polluting shorelines. (*2*) Device for recovering spilled oil from water's surface. (*3*) Because it releases pollutants into air, it depends on numerous factors to comply with safety rules for human health. (*4*) Insoluble materials (floating or not), ideally both oleophilic (oil attracting) and hydrophobic (water repellent) are used to recover spread liquids through the mechanism of absorption and/or adsorption. (*5*) Dispersing agents, also called dispersants, are chemicals that contain (bio) surfactants or compounds which act to break liquid oil into small droplets allowing their dispersion into the water column. (*6*) Feeding or biostimulation is the addition of nutrients (P,N), electron acceptors (NO^{3-}, SO_4^{2-}, O_2, etc.), to stimulate the growth of existing native oil-degrading communities, and/or indirectly, by the addition of (bio) surfactants as they enhance the oil bioavailability. (*7*) Seeding or bioaugmentation is the introduction of hydrocarbonoclastic microorganisms in the existing native communities to speed up biodegradation

Fig. 16.44 Examples of mechanical techniques to fight oil spills. (**a**) Floating containment booms (Marseille's Bay, June 1998). (**b**) Helicopter spraying of dispersing agents (Courtesy of BMP Marseille and SOS Dépollution)

Table 16.9 Formulation of two additives designed for bioaugmentation: EAP 22 (Inipol®) and Customblem®

Compounds	Formula	Function
Inipol EAP 22	N-P-K ratio %: 7.4:0.7:0	Oleophilic fertilizer
Oleic acid	$CH_3(CH_2)_7CH=CH(CH_2)_7COOH$	Primary ingredient: oleophilic Easily degradable carbon source
Tri(laureth-4)-phosphate	$[C_{12}H_{25}(OC_2H_4)_3O]_3PO$ About 10 % of final product by weight	Phosphorus source and surfactant
2-Butoxyethanol	$HO-C_2H_4-O-C_4H_9$[a]	Co-surfactant and emulsion stabilizer
Urea	$NH_2-CO-NH_2$	Nitrogen source
Water	H_2O	Solvent
Customblen	N-P-K ratio %: 28.0:3.5:0	Fertilizer
Ammonium nitrate	NH_4NO_3	Nitrogen source
Calcium phosphate	$Ca_3(PO_4)_2$	Phosphorus source
Ammonium phosphate	$(NH_4)_3PO_4$	Nitrogen and phosphorus source

[a]This substance is potentially hazardous for environment and/or harmful for health

active in warm or temperate waters are different from those active in cold waters.

(iii) To not produce toxic metabolites
(iv) To not be pathogens
(v) To not be eliminated by endogenous communities during the period of bioremediation.

Conversely, it is necessary that the massive input of exogenous microorganisms does not affect on the long term the equilibrium of the communities initially present in the ecosystem.

The treatment efficiency by bioincrease was demonstrated in laboratory; the situation is completely different for experiments conducted in natural environment where very few studies were undertaken. None rigorously demonstrated positive effects of such a treatment.

Attempts of in situ seeding were undertaken in Japan with a mixture of cultured bacteria called "TerraZyme," after the Nakhodia tanker beaching on the Japan coasts (1997). The only published data related to this operation are visual observations or analyses of digitized images that only showed the cleaning of polluted surfaces. A bioincrease action was also undertaken in the USA after the *Mega Borg* beaching on Texan coasts (June 1990), where a microorganism mixture (called "Alpha BiopSea") was applied on contaminated surfaces. In the latter case, independent observers reported that the contaminated zone look only changed after an emulsification assigned to the implementation of bacterial cultures. Finally, chemical analyses of treated samples never rigorously demonstrated the efficiency of this seeding. Indeed, whatever the sample, the calculated ratio n-C_{18}/phytane revealed no hydrocarbon selective alteration which would have proved a biodegradation beginning (Means 1991).

Biostimulation

Hydrocarbonoclastic microorganisms are ubiquitous in the environment. This property is built on in biostimulation techniques in order to stimulate hydrocarbon biodegradation capacities of microorganisms already present in the site submitted to pollution. Indeed, it was observed that after a massive petroleum input, the biomass of hydrocarbonoclastic microbial communities was extensively increased. To be efficient, an additive must display four major properties:

(i) To keep ratios C/N/P compatible with hydrocarbon biodegradation. The petroleum input on the site corresponds to a massive carbon input. To achieve the largest biodegradation, it is necessary that the applied product contains nitrogen and phosphorus amounts enabling the establishment of ratios carbon-nitrogen-phosphorus supporting the optimal microorganism growth; ratios 100:10:1 are usually considered as ideal.
(ii) To be oleophilic microemulsion so that the added nutrients stay in contact with petroleum. Indeed, water-soluble products, easily washed away by waves and sea sprays or lixiviated by rain falls, are very quickly ineffective and require almost daily implementations.
(iii) To have surfactant properties that increase the hydrocarbon bioavailability which favors biodegradation processes.
(iv) To be nontoxic.

Such a treatment was applied after the *Exxon Valdez* beaching. Two products, called Inipol EAP22 and Customblen, were used over large scale (Table 16.9). Inipol EAP22 is a product that stands as an oleophilic microemulsion containing urea (nitrogen source), lauryl phosphate (biosurfactant and phosphate source), 2-butoxy ethanol (co-surfactant stabilizing the emulsion), and oleic acid (source of easily assimilated carbon) and whose aim was to increase the hydrocarbonoclastic biomass. The Custoblen™ stands as pellets containing nutrients (ammonium nitrate, calcium phosphate, and ammonium phosphate) in a polymerized vegetal oil; the drawback of this fertilizer is its quick dissolving and, consequently, its move away from the polluted areas (particularly under the effect of tides), unlike Inipol EAP22 that, because of its oleophilic properties, better stay in contact with petroleum. Photographical surveys, comparing EAP22-treated and EAP22-untreated beach sections, showed a very important petroleum disappearance on contaminated sites. But the efficiency of this treatment was objected by some observers for whom the Inipol EAP22 action was limited to a "washing" of contaminated beaches (in line with the surfactant properties of the additive), the microorganisms having used not petroleum but oleic acid as carbon and energy source. However, the monitoring of petroleum biodegradation in different plots by using 17α (H)21β (H) hopane (Fig. 16.33) as internal conserved standard showed that the biodegradation rates were two to seven times higher than that in untreated plots, according to the treated zones (Pritchard et al. 1992; Bragg et al. 1994; Prince et al. 1994).

16.8.3.2 Bioremediation of Soils and Groundwaters

Petroleum product pollutions have a number of causes: emission from the production sites, often linked to the plant dilapidation, during transportation from the production site to refineries (oil pipeline network, rail and road transport), and contamination of oil processing and storage places (storage in refineries, petrol stations, storage tanks of individuals). All these pollution can be chronic or accidental. For instance, during the Gulf war, 400 km^2 soil were contaminated by fallouts from 800 set ablaze oil wells, and 50 millions m^3 crude oil were discharged by burning well eruptions. Once in contact with soil, hydrocarbon will penetrate more or less quickly and deeply depending on the soil structure, its texture, its clay and organic matter composition, hydrological conditions, and pollutant nature (**log kow***, hydrophobicity, etc.); in some cases, hydrocarbons may reach the groundwater level.

In soils, hydrocarbons are submitted to a set of physicochemical (dissolving, evaporation, adsorption, dispersal, etc.) and biological (biodegradation) (Vandecasteele 2008) processes, similar to those taking place in marine ecosystems.

Though a natural hydrocarbon disappearance occurs in these biotopes, treatments based on microorganism biodegradation capacities through microbiological means (biostimulation, bioincrease, rhizostimulation (Box 16.6)) were developed in addition to physicochemical treatments (burning, thermal desorption), to comply with increasingly constraining regulations.

Treatments implemented in situ and ex situ will be separately considered:

(i) In situ treatments are implemented when polluted areas represent large surfaces and/or are deeply located; under these conditions, digging cannot be considered. In situ biostimulation of hydrocarbonoclastic community activity can be achieved through bioventilation or biobubbling: injection of air for aerobic biodegradation and electron acceptors other than dioxygen if the site is under anoxic conditions (supply of nitrate, sulfate, iron). For instance, nitrate (15–125 mg dm^{-3}) and sulfate (15–100 mg dm^{-3}) were injected over a 15-month period in groundwaters contaminated by BTEX (Cunningham et al. 2001). The set of compounds was degraded at the end of the treatment, particularly xylenes whose biodegradation seems to be linked to the use of sulfate by the sulfate-reducing community. Bioincrease can also be considered. Introduction of a genetically modified microorganism was undertaken to treat a soil artificially contaminated with HAPs (naphthalene, anthracene, phenanthrene). The strain *Pseudomonas fluorescens* HK44 that is able to degrade these hydrocarbons survived 22 months after inoculation. The efficiency of this bioincrease treatment in terms of remediation was difficult to establish because of the heterogeneous contaminant distribution in the soil, but their concentration significantly decreased within zones occupied by the microorganisms (Ripp et al. 2000).

(ii) Ex situ treatments can be applied when the polluted zone is limited and shallow (a few meters); in such a case, soil is transferred out and treated. This operational mode enables a better control of bioincrease and/or biostimulation treatments that will be applied. A first solution involves one fermenter (of variable size, with a volume up to 1,000 m^3) to treat the collected material introduced as sludge; it is thus kept in suspension in the aqueous phase. This technique enables the establishment of biodegradation optimal conditions and makes easier supplementation, if necessary, of nutrients (biostimulation) and microorganisms adapted to the biodegradation of the related pollution source (bioincrease). This technology allows the treatment of highly contaminated soils (up to 250 g kg^{-1}), but the equipment and energetic costs induced by ventilation limit its development.

A second approach consists of treating polluted soils in biopiles (*cf.* Sect. 16.2.2). To stimulate biodegradation, soil can be enriched with nutrients (nitrogen, phosphorus). Structuring agents, in particular organic debris (straw, wood chips, etc.), can also be added. Stimulation of aerobic biodegradation is insured by two ways:

(i) Through regular reversal of the contaminated soil with mechanical machines that homogenize and ventilate the biopile (dynamic biopile)
(ii) By insufling or pumping out air through a tubing network spread inside the structure

The biopile treatment proved to be efficient to treat sludges from urban STEPs contaminated by PAH. Oleszczuk (2007) treated STEP sludges from different origins, faintly contaminated by one PAH mixture with 3, 4, 5, and 6 aromatic rings. Sludges were piled up and their ventilation was insured by insufling air and by mechanical reversal (every 15

Box 16.6: Phytoremediation of Polycyclic Aromatic Hydrocarbons (PAHs): From Laboratory to Field Trials

Corinne Leyval

PAH Origins and Properties

PAHs are major pollutants of the environment. They are formed by incomplete combustion of organic matter and can be found, for example, during forest fires. However, the high PAH concentrations in soils are mainly due to industrial activities (coking plant, gas plant, and wood treatment). PAHs are a group of compounds with 2–6 aromatic rings. PAH properties depend on their molecular weight, but all of them are characterized by a low solubility and high hydrophobicity; they are highly toxic, especially the ones with high molecular weight, that are mutagenic and carcinogenic. In contaminated soils, PAHs can reach high concentrations up to one to tens of gram per kilo (Leyval 2005).

PAH Biodegradation

Contrary to metallic pollutants that cannot be eliminated and can only be transferred from a compartment to another, PAHs can be transformed and biodegraded. The first bacteria able to degrade PAHs were described fifty years ago. However, if the ability to degrade low molecular weight PAHs, from 2 to 4 cycles, has been described for a large number of bacterial species, the biodegradation of 5 to 6 cycles PAHs is far less documented and seems to imply cometabolism processes. The metabolic pathways and the genes involved in the biodegradation are well described in the literature. The quantification of genes and their expression in situ is currently developing using quantitative PCR or microarrays to estimate the potential of soils to biodegrade these pollutants. Some remediation techniques for PAH-contaminated soils use microbial degradation activity, either stimulated by the addition of nutrients or oxygen (biostimulation) or improved by addition of microorganisms (bioaugmentation). Such techniques, for example, biopiles, landfarming, and bioreactors, are efficient and can reduce the concentration of PAHs from thousands to hundreds of milligrams per kg within a few months. However, the low molecular weight PAHs are mostly degraded.

C. Leyval (✉)
Laboratoire Interdisciplinaire des Environnements Continentaux (LIEC, UMR 7360), Université de Lorraine, CNRS, Vandoeuvre-les-Nancy, France

Biodegradation of PAHs in Plant Rhizosphere

The rhizosphere is the zone of soil influenced by the presence of plants, especially root exudates that promote microbial activity. It is a zone of intense microbial activity, where microorganism density and activity is higher than in bare soil and microbial diversity is also different. It has been shown that the presence of plants increase PAH degradation in plant rhizosphere. Such activity has been mainly attributed to the stimulation of microbial degradation activity, since the plants take up little or no PAHs (Binet et al. 2000). This is called rhizodegradation.

However, the results published in the literature show variable results depending on the experiments: in some of them, the biodegradation is significantly increased in presence of plants, and in others there is no or little effect of plants. Such differences in the results can be due to the use of different plants, different PAHs, different experimental conditions in terms of duration, and nutrient conditions. Many studies were performed with soil and simplified substrates such as sand artificially spiked with one or many PAHs. In order to better understand the dissipation of PAHs in the rhizosphere, experiments using columns or compartment devices were performed. They clearly showed a gradient in PAH concentration that decreases when the distance to roots increases. Such decrease in PAH concentration was associated with changes in microbial community structure. The biodegradation also differed with the type of PAH, for example, phenanthrene, a 3-ring PAH, is more rapidly degraded than 5- or 6-ring PAHs, such as benzo(a) pyrene. The artificial spiking of soils or substrates with PAHs allows to have non-contaminated controls; however, the bioavailability of pollutants is much higher than in historically contaminated industrial soils, where the contamination is often complex. The so-called aging phenomena have a strong influence on PAH degradation in soil and in rhizosphere.

PAH-degradation phenomena in the rhizosphere are complex, since microorganism and plant are involved. In laboratory conditions, it was shown that PAH biodegradation was increased in presence of plants inoculated with PAH-degrading bacteria such as *Pseudomonas* or *Mycobacterium* species (Ho et al. 2007). The selection of plants for phytoremediation can therefore be focused on their ability to select or promote the bacterial community harboring PAH-degradation genes. The release of root exudates could favor cometabolism processes and promote the

(continued)

Box 16.6 (continued)

Box Fig. 16.13 Phytoremediation assays in lysimeter plots (Homecourt) (Photograph: Corinne Leyval)

biodegradation of high molecular weight compounds. However, such root exudates represent organic carbon sources that may be easier to degrade than the PAH and could reduce pollutant biodegradation rate.

Which Plant for the Phytoremediation of PAHs

Plants like fescue, ryegrass and other grasses, legumes such as alfalfa, clover, or trees such as willow and birch have been tested, and their ability to increase PAH degradation in soil was shown. Among a large number of plants tested, crushed roots of celery, carrots, and potatoes showed interesting pyrene degradation rates that were attributed to their content in linoleic acid, promoting PAH degradation (Yi and Crowley 2007). The association of plant roots with a symbiotic fungus also increased PAH degradation in comparison to a non-mycorrhizal plant (Joner and Leyval 2003).

Bioavailability of PAHs in the Rhizosphere

The dissipation (decrease of extractible concentration) of PAHs also depends on their bioavailability. Due to their high hydrophobicity, they tend to adsorb to soil organic matter, which limits their accessibility and ability to be biodegraded. However, the production of molecules with tensioactive properties can modify the availability of PAHs and their biodegradation. The production of linoleic acid and saponins by some plants, of linoleic acid, or rhamnolipids by some rhizosphere bacteria or fungi could increase PAH availability (Yi and Crowley 2007). The concentration of PAHs in plant rhizosphere is therefore the result of degradation processes, but also of mobilization processes, which can explain some of the contrasted results.

PAH Phytoremediation of PAH *In Situ*: Which Perspectives

A large number of studies on PAH phytoremediation were performed in laboratory conditions, in growth chamber, pots, and columns. In situ studies are less numerous and not always convincing. PAH rhizodegradation has been studied in situ, but real application of this remediation technique is still limited. As an example, in Nord Pas de Calais region, a phytoremediation treatment using poplar was set up after a biological treatment in a pile (Info Chimie Magazine, Mars/Avril 2005, N°462). It was a complementary treatment associating phytostabilization and landscaping to pollution remediation.

A field phytoremediation trial set up in 2005 with alfalfa plots (www.gisfi.fr) did not show significant differences in PAH concentration between planted and unplanted plots (weed cleared manually) (Box Fig. 16.13). However, in such study, the volume of leaching water was much lower in the planted plots, so

(continued)

> **Box 16.6** (continued)
>
> the presence of plants reduced the risk of pollutant dispersal. In a seven-year study using poplar, a reduction of PAH concentration was observed after a few years but was limited to 3-ring PAHs and was limited within the time, which was attributed to the low dissolution of PAHs from the solid phase (Widdowson et al. 2005). The challenge of such phytoremediation technique is high, since it has a low cost and could be well adapted to the treatment of large contaminated areas, which are not compatible with many other remediation techniques. However, to control the feasibility and potential of PAH rhizodegradation, it is still necessary to further study and understand the processes involved and the interactions of plant–microorganisms–pollutant. Many plants showed a potential for phytoremediation, but the parameters such as nutrient conditions, soil conditions, and PAH availability should be taken into account.

days). During the 76-day treatment, the different steps specific to composting succeeded each other: mesophilic (30–40 °C) and thermophilic (40–53 °C) phases, cooling, and maturation. The decrease in PAH concentration that varied according to STEP origin was in the range 15.8–57.9 %. Experiments of this kind, conducted by other authors with other STEP sludges, lead to similar results and sometimes better (70 % decrease). The PAH elimination is explained by their biodegradation achieved by microorganisms, but a fraction was also immobilized or sequestered within compost.

The treatment can also be implemented according to the controlled spreading method (*landfarming*; *cf*. Sect. 16.2.2). In that case, the polluted soil (particularly by gasoil, fuel oil, oil), eventually enriched with nutrients and microorganisms, is laid down in thin layers (10 to a few tens cm) over large surfaces and regularly mechanically reversed.

This process enables the treatment of large soil volumes (several thousands m^3) at relatively low coast. In 1987, 1.22×10^6 tons hydrocarbon per year were treated by *landfarming* (Amaral 1987).

16.9 Metals and Metalloids

16.9.1 Introduction

Metals constitute the majority of elements in the periodic table of elements. With the exception of hydrogen, they include all the elements on the left of the metalloids (to the left of a line from boron to polonium). Thus, apart from noble gases, halogens, nonmetal, and metalloid elements, all the other elements are metals. Because of this diversity, metals are classified into different groups such as posttransition metals (e.g., aluminum and lead), transition metals (the most numerous including iron, manganese, silver, cobalt), or alkaline earth metals (magnesium, calcium). Metals are also characterized by their chemical properties, including their ability to be oxidized (electron loss) in the form of cationic ions. In addition to their classification is the name of "heavy metals" which remain very empirical and variable depending on the authors. There is often a confusion between the terms "heavy metal", "toxic metals," or even "metallic trace elements." For example, arsenic, classified as a metalloid, is often included among the heavy metals because of its toxicity.

Metals and metalloids are present in soils, sediments, and at lower concentrations in water. They naturally originate from the weathering of rocks, but higher concentrations may arose from anthropogenic origins such as mining, waste storage and incineration, or industrial activities (smelting plants). In the case of mining, microorganisms are used for the recovery of metals of interest through a process called bioleaching and bio-oxidation. These processes tend to increase the levels of available metals in the ecosystem, but can also be used for the removal of metals from soils or sediments.

The majority of treatment techniques for polluted sites focused on the removal of organic pollutants. Treatment of wastes or sites containing metals is considered only recently (Summers 1992). Unlike for organic compounds, it is not possible to speak of metal degradation but only of modification of their chemical form and/or their precipitation. Toxicity, availability, and transport of metals in the environment depend on their speciation, i.e., the chemical form in which they are found. The microorganisms may be involved in the speciation of metals and thus affect their fate in an ecosystem. This change in chemical form may be related to energetic metabolisms where metals are used as electron donors or acceptors (*cf*. Chap. 14). The processes of metal reduction most often lead to a decrease of their solubility and toxicity and thus can be applied in the treatment of polluted sites. The elimination of metals from ecosystems or wastes can also make use of long-term trapping or immobilization processes. In this case, both physicochemical and biological processes are used: adsorption, immobilization with chelating agents, indirect precipitation, and accumulation. Finally, some resistance mechanisms to metals can also participate to their removal from a matrix.

16.9.2 Use of Microorganisms in Mining Activities: Bioleaching and Bio-oxidation

Some microorganisms increase the solubility of metals through oxidation processes and/or acidification. This is

Table 16.10 Correspondence between some of the well-known sulfide ores and their common name

Name	Formula
Argentite	Ag_2S
Arsenopyrite	$FeAsS$
Calcocyte	Cu_2S
Chalcopyrite	$CuFeS$
Covellite	CuS
Galena	PbS
Pyrite	FeS_2
Millerite	NiS
Orpiment	As_2S_3
Sphalerite	ZnS

particularly the case of microorganisms involved in the process of bacterial bioleaching.

Metals are found in nature in the form of ores and therefore trapped as oxides (bauxite), silicates (mica, feldspar), carbonates (malachite, dolomite), or sulfides (galena, chalcopyrite, pyrite, cinnabar). Many metals of interest are extracted in the latter form (Table 16.10), and the metal production from sulfide ores can involve bacterial activity. These processes use the capacity of microorganisms to oxidize sulfides and metals, especially iron. If the metabolisms put at stake in both cases are the same, in terms of processes, bioleaching and bio-oxidation have two distinct objectives:

(i) The bioleaching aims, from a sulfide ore, to leach the metal of interest (copper or zinc) which is at the end of the process present in the liquid phase, whereas the solid residue is a waste.

(ii) The bio-oxidation aims to remove metal sulfides from an ore rich in gold or silver, for example. In this case, the metal of interest (gold or silver) is in the solid phase and the leachate is a waste.

In the case of mining, copper, for example, is found in the form (among others) of chalcocite (Cu_2S). It must be extracted from the ore and then reduced in the form of zerovalent metal. Biological metabolisms that will allow the separation of copper sulfide by oxidation of the latter exist (Fig. 16.45, reactions 1 and 2). However, this process is much slower than the chemical process (Fig. 16.45, reaction 3) involving an oxidation by ferric iron (Fe^{3+} or FeIII). This led to the production of copper in the soluble ionic form that can be reduced to metal copper (Fig. 16.45, reaction 4). Reactions 3 and 4 in Fig. 16.45 produce large amounts of reduced iron (Fe^{2+}), a by-product of the oxidation–reduction reactions that can be recycled in order to maintain the sulfide oxidation process. This upgrading of ferrous iron (Fe^{2+}) is realized via chemolithotrophic prokaryotic microorganisms achieving the oxidation of iron (Fig. 16.45, reaction 5). These microorganisms must be acidophiles since they must grow under the highly acidic pH of the leachates that are rich in sulfuric acid originating from the oxidation of sulfides (Fig. 16.45, reaction 3). The acid leachate containing oxygen and oxidized iron produced by these microorganisms will be spread again on the ore to maintain the chemical oxidation of sulfides and the release of copper. The best known microorganism, commonly cited for its involvement in this process, is *Acidithiobacillus ferrooxidans* that can oxidize both iron and sulfide. Other microorganisms are involved directly or indirectly in bioleaching processes such as *Leptospirillum ferrooxidans* or the thermophilic *Acidimicrobium ferrooxidans*. The bioleaching process is accelerated by microorganisms oxidizing elemental sulfur, a by-product of the reaction between the sulfide ore (MS) and ferric iron.

$$MS + 2Fe^{3+} \rightarrow M^{2+} + S° + 2Fe^{2+}$$

The oxidation of sulfur by specialized sulfur-oxidizing microorganisms (*Acidithiobacillus thiooxidans*) prevents the accumulation of sulfur that can decrease the oxidation of ore and produces sulfuric acid, which contributes to improve the bioleaching efficiency. These bioleaching processes can be controlled during the leaching of ores (in this case they occur in oxidation lagoons designed for this purpose), but they also take place naturally when sulfide ores (e.g., mining wastes) are exposed to air and rainfall. In all cases, the mining production tends to produce acidic effluents, often characterized by their color (Fig. 16.46) due to their richness in metals, and among them iron. A large part of the metals that are not of interest are not recovered and contaminate mines catchments. One of the best known examples is the Rio Tinto in Spain (Leblanc et al. 2000) which is the repository of metal contamination related to copper mining.

Acid mine drainage waters represent a risk for water resources downstream of the mining areas. Their impact in terms of metal contamination can be felt until their sea outlet (Elbaz-Poulichet et al. 2001). Some of the treatments sometimes implemented aim to prevent natural bioleaching of mining wastes in the presence of oxic water. Wastes (mine tailings) can be covered in order to isolate them from rainfall and water runoff. Another solution is to submerse them in a water column under anoxic conditions. It is also possible to treat these wastes through long-term and costly processes. These treatments can include chemical neutralization/precipitation/flocculation (called active methods). Passive methods, often preferred, are based on the following:

(i) Bacterial reduction and precipitation of metals in wetlands
(ii) Bio-oxidation and precipitation of the obtained leachate (Roanne et al. 1996; *cf.* Sect. 16.9.5)

Despite their extensive use in mining, bioleaching processes have been little used in bioremediation of soils and sediments. Some successes were obtained for agricultural use, after decontamination, of wastewaters, and sludges form wastewater treatment plants contaminated with metals (Couillard and Zhu 1992). Some studies have also been

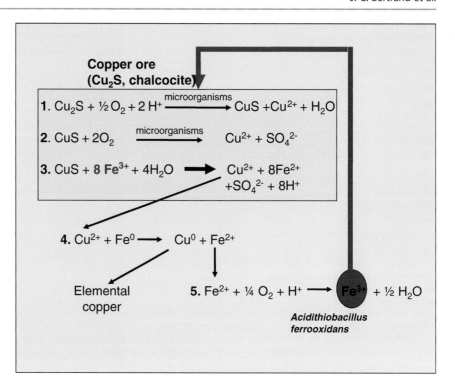

Fig. 16.45 Schematic diagram of bacterial bioleaching for mining. The numbers in the diagram correspond to the reactions cited in the text

Fig. 16.46 Photographs of river water contaminated by acid mine drainage (Religious river in Carnoules, France). (**a**) The color is due to the presence of large amounts of iron; (**b**) the acid drainage waters are also characterized by high concentrations of arsenic

carried out for the recovery of uranium from contaminated soils by nuclear wastes. However, in this case, the process of bacterial metal reduction seems more effective (Lloyd et al. 2003).

16.9.3 Microbial Bioreduction of Metals

16.9.3.1 Involved Mechanisms

Microorganisms may be involved in metal availability by modifying their speciation through oxidation and reduction processes (Barton et al. 2003). Some microorganisms will reduce metals and contribute to decrease their solubility and therefore their availability and consequently their toxicity (Fig. 16.47). The reduction can be direct (e.g., metal used as electron acceptor) or indirect (*cf.* Sect. 14.5). A large number of metals can be directly reduced by microorganisms (respiratory metabolism) or indirectly reduced (Lloyd et al. 2003). These metals include those arising from the nuclear industry (*cf.* Sect. 14.5.1) and metalloids (arsenic, selenium, and tellurium). This reduction can be done under oxic or anoxic conditions. In the latter case, the role of sulfate-reducing microorganisms and bacteria of the genus *Geobacter* are well documented. Finally, some metals can be reduced through detoxification mechanisms as it is the case for mercury (Box 14.15).

Fig. 16.47 Diagram showing the different possibilities for microorganisms to be involved in speciation, sequestration precipitation, or accumulation of metals and metalloids

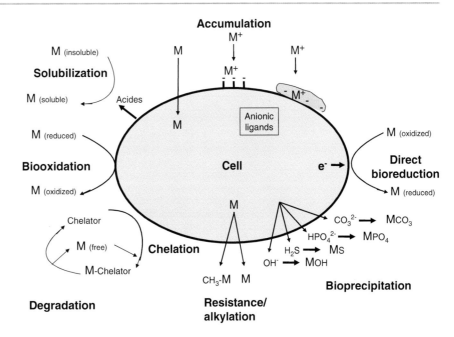

16.9.3.2 Environmental Applications

Contamination by chromium which is widely used in the automotive, metallurgic, mining, or tannery industries may reach high levels in some ecosystems. The most oxidized form, Cr(VI) or chromate, is highly soluble and highly toxic (mutagenic compound), while trivalent chromium (Cr(III), reduced form) is very insoluble and 1,000 times less toxic (Lofroth and Ames 1978). Processes for the treatment of wastes contaminated by chromium combine chemical reduction of Cr(VI) to Cr(III) and precipitation of the reduced chromium by increasing the pH (usually by the addition of calcium carbonate). Biological reduction of chromium is also possible as an alternative treatment. Indeed, the ability to use chromium as electron acceptor is present in many strict or facultative anaerobic prokaryotes belonging to the genera *Bacillus*, *Pseudomonas*, *Desulfovibrio*, *Aeromonas*, or some enterobacteria (Wang 2000). Depending on the organisms involved, this reduction can be achieved under oxic or anoxic conditions. To be effective, the bacterial biomass must be high, and thus the treatments need high amounts of electron donors, most of the time organic ones. This can be an advantage since chromium reduction can be coupled to the degradation of organic molecules such as phenols commonly found in the chromium-rich industrial effluents. The main limitation of the use of methods involving bacterial reduction of chromium is related to its inhibition by the presence of other metals such as zinc or copper which may be present in the chromium-contaminated effluent. Nevertheless, some successful biological treatments were obtained on the basis of biofilms (fixed cultures) grown in bioreactors (Wang 2000).

Microbial bioreduction of uranium (Lovley et al. 1991) is especially designed for the treatment of wastes from the nuclear industry. Some microorganisms can develop using uranium in its oxidized form (U(VI), UO_2^{2+} soluble) and reduce it to U (IV) insoluble (UO_2). This is the case of *Geobacter metallireducens* in the presence of acetate and *Shewanella putrefaciens* in the presence of hydrogen (H_2), which are capable of growth in the presence of uranium concentrations up to 8 mM.

$$\text{Acetate} + 4\,U(VI) \rightarrow 2HCO_3 + 4\,U(IV) + 9H^+$$

The addition of acetate in situ stimulates the activity of bacteria belonging to the genus *Geobacter* which indicates a significant reduction of the oxidized form of uranium in 50 days (70 % decrease of the amount initially present). These studies have demonstrated the possibility of immobilization treatments for uranium that could be present in an aquifer (Anderson et al. 2003). The ability to reduce uranium in insoluble UO_2 demonstrated in sulfate-reducing microorganisms has been used to concentrate uranium in radioactive wastes mixed with other chemicals. Some other studies demonstrated the efficiency of processes for soils contaminated with oxidized uranium. This treatment includes a first chemical step (uranium extraction with bicarbonate) and a second step involving the reduction of the extracted uranium through sulfate-reducing bacteria activity (Phillips et al. 1995).

Finally, numerous investigations focused on the speciation of metalloids by microorganisms. Their role in the speciation of tellurium (Te) and selenium (Se) is studied because they are part of the group VI of the periodic table of the elements, as well as sulfur and polonium. Selenium can be incorporated into organic matter (as sulfur) in the form of selenium amino acids. Thus, selenium is an essential trace element that undergoes a complex biogeochemical cycle. Tellurium is an

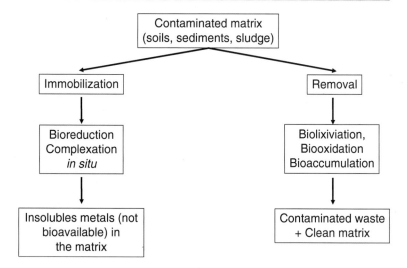

Fig. 16.48 Comparison of strategies for bioremediation of metals and metalloids from polluted sites

extremely rare element, well studied because of its use in the industry due to its semiconductor properties. It is also chemically similar to polonium (the first radioactive element discovered by Marie and Pierre Curie), which is a product of the desintegration of the radioactive uranium (U238). Another metalloid, arsenic, is studied because of its toxicity and its frequent occurrence in groundwaters, potential sources of drinking water. Leaching of rocks rich in arsenic (arsenopyrite), its use as an herbicide (lead arsenate), and in wood treatments also contribute to increase its concentration in water in the form of arsenite (AsIII) or arsenate (ASV). Unlike most metals, the reduced form is the most toxic. Organic arsenic (methyl forms) is much less toxic and it is in this form that the biological removal is the most effective. Numerous microorganisms are involved in the mobilization, reduction, oxidation, methylation, and precipitation of arsenic (Oremland and Stolz 2003).

The main risk associated with treatment of metals by in situ microbial reduction processes concerns the possible long-term changes under environmental conditions that may influence their speciation. A metal present in an ecosystem in its reduced form (less toxic most of the times) may be remobilized (reoxidized) by both physicochemical and biological processes and end up in a toxic and mobile form. It is therefore necessary in the context of an in situ treatment to monitor metal speciation to ensure that their chemical form does not change with time (Fig. 16.48).

16.9.4 Bioaccumulation and Production of Metal Chelating Agents

16.9.4.1 Involved Mechanisms

The toxicity of metals depends on their bioavailability, that is, to say, their soluble fraction and accessibility. It depends on their speciation (see above), on the physicochemical conditions (including pH), but also on the fraction immobilized by adsorption on organic, inorganic, or colloidal particles. Prokaryotic or eukaryotic cells are not inert particles but are instead very active surfaces. This activity is linked to membranes, peptidoglycan, and other cell walls as well as extracellular polymers they can produce (sheaths, capsules, S-layer, etc.). These molecules, through their carboxyl, phosphate, sulfonate, or hydroxyl groups, are all potential sites for attraction of cations (Fig. 16.47). These mechanisms represent extracellular accumulation of metals and lead to prevent their penetration into the cell. These mechanisms will promote metal complexation and thus their inactivation as potential toxic compounds.

Microorganisms do not act only as passive samplers for metals and metalloids but are also able to adapt to improve their trapping. Some microbial exudates may play an important role as ligands in metal-trapping phenomena. They act either as resistance processes or as chelating agents necessary for the cell, as is the case of siderophores known for their ability to capture the iron when present as trace metal.

Other organisms will promote the penetration of metals in the cell in the view of their intracellular sequestration which involves active, passive, or facilitated transport systems. Once inside the cell, sequestration can be achieved through metal-thionein (Robinson and Tuovinen 1984) and the metal will be stored inside the cell in the form of granules. This ability to accumulate metals directly promotes their bioconcentration and can be operated for recovery, reprocessing, or disposal (Fig. 16.47).

16.9.4.2 Environmental Applications

Bioaccumulation (passive and active extracellular accumulation, intracellular accumulation) can be used in bioremediation of polluted sites (Gadd 1990). These treatments are very attractive as compared to chemical treatments, often

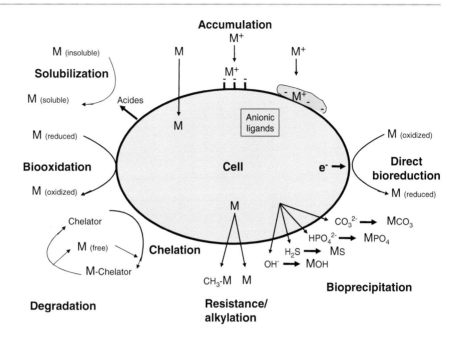

Fig. 16.47 Diagram showing the different possibilities for microorganisms to be involved in speciation, sequestration precipitation, or accumulation of metals and metalloids

16.9.3.2 Environmental Applications

Contamination by chromium which is widely used in the automotive, metallurgic, mining, or tannery industries may reach high levels in some ecosystems. The most oxidized form, Cr(VI) or chromate, is highly soluble and highly toxic (mutagenic compound), while trivalent chromium (Cr(III), reduced form) is very insoluble and 1,000 times less toxic (Lofroth and Ames 1978). Processes for the treatment of wastes contaminated by chromium combine chemical reduction of Cr(VI) to Cr(III) and precipitation of the reduced chromium by increasing the pH (usually by the addition of calcium carbonate). Biological reduction of chromium is also possible as an alternative treatment. Indeed, the ability to use chromium as electron acceptor is present in many strict or facultative anaerobic prokaryotes belonging to the genera *Bacillus*, *Pseudomonas*, *Desulfovibrio*, *Aeromonas*, or some enterobacteria (Wang 2000). Depending on the organisms involved, this reduction can be achieved under oxic or anoxic conditions. To be effective, the bacterial biomass must be high, and thus the treatments need high amounts of electron donors, most of the time organic ones. This can be an advantage since chromium reduction can be coupled to the degradation of organic molecules such as phenols commonly found in the chromium-rich industrial effluents. The main limitation of the use of methods involving bacterial reduction of chromium is related to its inhibition by the presence of other metals such as zinc or copper which may be present in the chromium-contaminated effluent. Nevertheless, some successful biological treatments were obtained on the basis of biofilms (fixed cultures) grown in bioreactors (Wang 2000).

Microbial bioreduction of uranium (Lovley et al. 1991) is especially designed for the treatment of wastes from the nuclear industry. Some microorganisms can develop using uranium in its oxidized form (U(VI), UO_2^{2+} soluble) and reduce it to U (IV) insoluble (UO_2). This is the case of *Geobacter metallireducens* in the presence of acetate and *Shewanella putrefaciens* in the presence of hydrogen (H_2), which are capable of growth in the presence of uranium concentrations up to 8 mM.

$$\text{Acetate} + 4\,U(VI) \rightarrow 2HCO_3 + 4\,U(IV) + 9H^+$$

The addition of acetate in situ stimulates the activity of bacteria belonging to the genus *Geobacter* which indicates a significant reduction of the oxidized form of uranium in 50 days (70 % decrease of the amount initially present). These studies have demonstrated the possibility of immobilization treatments for uranium that could be present in an aquifer (Anderson et al. 2003). The ability to reduce uranium in insoluble UO_2 demonstrated in sulfate-reducing microorganisms has been used to concentrate uranium in radioactive wastes mixed with other chemicals. Some other studies demonstrated the efficiency of processes for soils contaminated with oxidized uranium. This treatment includes a first chemical step (uranium extraction with bicarbonate) and a second step involving the reduction of the extracted uranium through sulfate-reducing bacteria activity (Phillips et al. 1995).

Finally, numerous investigations focused on the speciation of metalloids by microorganisms. Their role in the speciation of tellurium (Te) and selenium (Se) is studied because they are part of the group VI of the periodic table of the elements, as well as sulfur and polonium. Selenium can be incorporated into organic matter (as sulfur) in the form of selenium amino acids. Thus, selenium is an essential trace element that undergoes a complex biogeochemical cycle. Tellurium is an

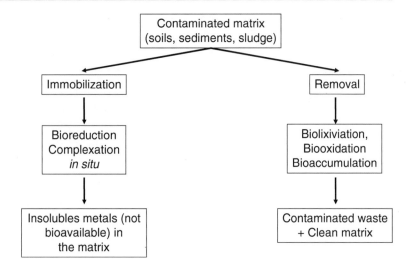

Fig. 16.48 Comparison of strategies for bioremediation of metals and metalloids from polluted sites

extremely rare element, well studied because of its use in the industry due to its semiconductor properties. It is also chemically similar to polonium (the first radioactive element discovered by Marie and Pierre Curie), which is a product of the desintegration of the radioactive uranium (U238). Another metalloid, arsenic, is studied because of its toxicity and its frequent occurrence in groundwaters, potential sources of drinking water. Leaching of rocks rich in arsenic (arsenopyrite), its use as an herbicide (lead arsenate), and in wood treatments also contribute to increase its concentration in water in the form of arsenite (AsIII) or arsenate (ASV). Unlike most metals, the reduced form is the most toxic. Organic arsenic (methyl forms) is much less toxic and it is in this form that the biological removal is the most effective. Numerous microorganisms are involved in the mobilization, reduction, oxidation, methylation, and precipitation of arsenic (Oremland and Stolz 2003).

The main risk associated with treatment of metals by in situ microbial reduction processes concerns the possible long-term changes under environmental conditions that may influence their speciation. A metal present in an ecosystem in its reduced form (less toxic most of the times) may be remobilized (reoxidized) by both physicochemical and biological processes and end up in a toxic and mobile form. It is therefore necessary in the context of an in situ treatment to monitor metal speciation to ensure that their chemical form does not change with time (Fig. 16.48).

16.9.4 Bioaccumulation and Production of Metal Chelating Agents

16.9.4.1 Involved Mechanisms

The toxicity of metals depends on their bioavailability, that is, to say, their soluble fraction and accessibility. It depends on their speciation (see above), on the physicochemical conditions (including pH), but also on the fraction immobilized by adsorption on organic, inorganic, or colloidal particles. Prokaryotic or eukaryotic cells are not inert particles but are instead very active surfaces. This activity is linked to membranes, peptidoglycan, and other cell walls as well as extracellular polymers they can produce (sheaths, capsules, S-layer, etc.). These molecules, through their carboxyl, phosphate, sulfonate, or hydroxyl groups, are all potential sites for attraction of cations (Fig. 16.47). These mechanisms represent extracellular accumulation of metals and lead to prevent their penetration into the cell. These mechanisms will promote metal complexation and thus their inactivation as potential toxic compounds.

Microorganisms do not act only as passive samplers for metals and metalloids but are also able to adapt to improve their trapping. Some microbial exudates may play an important role as ligands in metal-trapping phenomena. They act either as resistance processes or as chelating agents necessary for the cell, as is the case of siderophores known for their ability to capture the iron when present as trace metal.

Other organisms will promote the penetration of metals in the cell in the view of their intracellular sequestration which involves active, passive, or facilitated transport systems. Once inside the cell, sequestration can be achieved through metal-thionein (Robinson and Tuovinen 1984) and the metal will be stored inside the cell in the form of granules. This ability to accumulate metals directly promotes their bioconcentration and can be operated for recovery, reprocessing, or disposal (Fig. 16.47).

16.9.4.2 Environmental Applications

Bioaccumulation (passive and active extracellular accumulation, intracellular accumulation) can be used in bioremediation of polluted sites (Gadd 1990). These treatments are very attractive as compared to chemical treatments, often costly. Indeed, large amounts of bacterial, fungal, or algal biomass originating from waste treatment plants, agriculture, or food processing can be made available for such treatments. Most often they complete physicochemical treatment (flocculation) to remove the remaining contamination. The economic value of these techniques has also been shown as part of the decontamination of radionuclides polluted sites (Kalin et al. 2005). Today they tend to be developed and improved through genetic modification of bacterial strains to increase their ability to complex metals of interest. The improvement also concerned processes, particularly with the development of continuous bioreactors.

In addition to bioremediation, the biorecovery uses the capabilities of microorganisms to extract toxic metals or metals of economic interest. One example is gold, which is accumulated and reduced on biological membranes allowing its recovery in the form of usable metal (Au0).

16.9.5 Microbial Bioprecipitation of Metals

16.9.5.1 Involved Mechanisms

The precipitation of metals leads to ore production from soluble ionic forms. The precipitation produces an inert form of the metal (Fig. 16.47) obtained by the precipitation of metal by products of bacterial metabolism (sulfides, hydroxides, carbonates, phosphates, silica, etc.). Bioprecipitation has been studied for many elements:

(i) Many studies concern iron (formation of iron hydroxides) because of its quantitative importance (fourth most abundant element in the earth's crust) and qualitative interest (many metabolisms are based on iron) as well as, in a lesser extent, because of its involvement in the formation of manganese oxides.

(ii) Other studies focus on the role of microorganisms in the formation of clays and in precipitation processes of carbonates, phosphates, and sulfate (for more details, see Konhauser 2007). In the case of carbonates, calcite formation can be achieved via cyanobacteria (see Microbialites) through the adsorption of calcium at the surface of the cells (bioadsorption) and then precipitation of the carbonates with calcium. Microorganisms are also involved in the formation of dolomite (CaMg$(CO_3)_2$), siderite ($FeCO_3$), and rhodochrosite ($MnCO_3$).

(iii) Finally, the most successful processes in terms of bioremediation techniques involve the formation of sulfidic ores that can be of abiotic origin (including hydrothermal) or biogenic (Ehrlich 1995). In the latter case, the process involves sulfate-reducing microorganisms. These chemoorganotrophic or chemolithotrophic microorganisms share the ability to use oxidized sulfur compounds (sulfate, sulfur, sulfite, thiosulfate) as terminal electron acceptor in a conservative energy metabolism. The result of their metabolism is the production of sulfide (S^{2-}), which precipitates many metals, including divalent metals (cadmium, cobalt, copper, iron, lead, manganese, mercury, nickel, zinc).

16.9.5.2 Environmental Applications

The efficiency of bioprecipitation has been demonstrated by White and Gadd (1996) on the removal of metals from acidic leachate. In this context, the removal of metals could reach 95 %. There is a commercial application of this method of treatment (THIOPAQ technology, Paques, The Netherlands). Such processes have demonstrated their efficiency in terms of treatment on several sites (Budelco BV, Netherlands and Kennecott Utah, USA) for the removal of different metals such as zinc, aluminum, copper, iron, and manganese. The treatment process is carried out in two phases. The first allows for the production of sulfide by sulfate-reducing microorganisms in the presence of hydrogen as an electron donor. Hydrogen sulfide is used in the precipitation of metals that can be recovered and the sulfides produced in excess can be subsequently reoxidized by chemolithotrophic aerobic microorganisms that will produce elemental sulfur. The use of such methods has also been proposed for the treatment of groundwater rich in arsenic (Bangladesh).

16.9.6 Other Microbial Activities Related to Metal Contaminants

The metals can be highly toxic to living organisms, especially those who have no known biological function, such as mercury, lead, cadmium, or tin. Microorganisms have developed different strategies to withstand the presence of these metals in the environment:

(i) Mechanisms preventing the penetration in the cell
(ii) Mechanisms for their excretion (transport) out of the cell

The active mechanisms of excretion involve sequestration phenomena, active transport, and speciation. Some genes encoding systems for active transport of toxic metals are known (Silver and Phung 2005) as it is the case of cadmium (*cad*), copper (*cop*), lead (*pbr*), or arsenic (*ars*). Other resistance systems, specific or not, involve methylation processes that produce volatile forms of tin, arsenic, selenium, lead, etc. (Thayer and Brinkman 1982). These volatile forms must be considered as losses from the ecosystem under consideration and therefore induce a reduction in the toxicity of these metals.

These resistance systems have been described in natural environment and are rarely used as bioremediation techniques. Similarly, the microbial reduction of mercury through the Mer detoxification system (Box 14.15) implies a loss of mercury by volatilization. The use of this metabolic capacity (reduction of inorganic mercury to volatile elemental mercury) has been proposed for contaminated sites, wastes or wastewaters contaminated with mercury (Saouter et al. 1994). In this context, the volatile mercury product can be recovered by chemical trapping.

16.10 Conclusion

The immense diversity of metabolic capabilities of microorganisms (*cf.* Sect. 3.3) allows them to degrade virtually all organic compounds of natural or anthropogenic (xenobiotics) origins, including those that are a source of environmental pollution. These processing capacities of molecules rely on the complementary metabolisms between different populations within a community subject to a pollutant (*cf.* Sect. 16.8). These capabilities are also due to the high genetic plasticity characteristic prokaryotic that results in genetic transfer and recombination or even in the "creation" of new genes (e.g., lindane and atrazine). All these properties explain the importance of the role of microorganisms in the natural attenuation or the natural elimination of pollutants in the environment. The observation of these natural attenuation phenomena has led to numerous studies conducted using cultures in flasks and fermenters, in microcosms containing low volume samples which have yield a sophisticated knowledge of degradation pathways and parameters that regulate them. The efficiency of microbiological treatments is first confirmed by tests carried out in the field (experimental sites, greenhouses) that resulted in treatments at the industrial scale of soils, sediments, aquifers (groundwater), sand beaches. The success of these technologies is confirmed by the number of patents in this area and by the proliferation of commercial companies that offer and commercialize microbiological treatments, thus proving that these treatments are economically feasible. Like any emerging technology, bioremediation develops in a scientific context that is still far from being fully mastered.

Questions that remain unanswered are essentially twofold. One concerns the bioaugmentation in natural ecosystems and its possible consequences (**i**) on the indigenous microbial community and (**ii**) on the risks of releasing nonindigenous and/or genetically modified microorganisms into the environment. Strict regulation is now in place to control this type of treatment. The introduction of isolated and/or genetically modified microorganisms in environments where they were not present naturally raises the question of their future and their effects on the evolution of these ecosystems. The few available observations tend to reinforce the idea of the existence of a direct link between the decline of a given function and the survival of microbial operators. The mobility and transience of the genetic elements involved do not allow making it a general rule. Under favorable conditions of saprophytism, some strains could loss plasmids that are becoming useless. Thus, as generations increase, some catabolic properties involved in the biodegradation of pollutants tend to disappear, even though the supporting microorganisms continue to proliferate. These microbial invasions, some of which are visible in aquatic ecosystems, remain cryptic in the soil and are difficult to characterize except by their potential impact on the functionality of the ecosystem.

The other limitation is operational and concerns the transition from the laboratory to the natural environment. A perfect knowledge of the site, its physical, chemical, and biological characteristics, is the only likely way to identify limiting factors. Thus, treatments that are mature and have resulted in large-scale applications – sewage treatment, waste treatment – are implemented in closed/sealed systems (sewage plants, lagoons, ponds, fermentors) that allow a better control of the treatment functioning.

Despite these restrictions, microbiologically based pollution treatments have a good future and their application is undeniable in terms of sustainable development.

References

Abbasnezhad H, Gray M, Foght JM (2011) Influence of adhesion on aerobic biodegradation and bioremediation of liquid hydrocarbons. Appl Microbiol Biotechnol 92:653–675

Abramowicz DA (1990) Aerobic and anaerobic biodegradation of PCBs: a review. Crit Rev Biotechnol 10:241–251

Adrian L, Dudková V, Demnerová K, Bedard DL (2009) "*Dehalococcoides*" sp. Strain CBDB1 extensively dechlorinates the commercial polychlorinated biphenyl mixture Arochlor 1260. Appl Environ Microbiol 75:4516–4524

Alder AC, Häggblom MM, Oppenheimer SR, Young LY (1993) Reductive dechlorination of polychlorinated biphenyls in anaerobic sediments. Environ Sci Technol 27:530–538

Alexander M (1965) Biodegradation: problems of molecular recalcitrance and microbial infallibility. Adv Appl Microbiol 7:35–80

Al-Mailem DM, Sorkhoh NA, Al-Awadhi H, Eliyas M, Radwan SS (2010) Biodegradation of crude oil and pure hydrocarbons by extrême halophilic archaea from hypersalin coasts of the Arabian Gulf. Extremophiles 14:321–328

Alvey S, Crowley DE (1996) Survival and activity of an atrazine-mineralizing bacterial consortium in rhizosphere soil. Environ Sci Technol 30:1596–1603

Amaral S (1987) Landfarming of oily wastes: design and operation. Water Sci Technol 19:75–86

Anderson TA, Kruger EL, Coats JR (1994) Enhanced degradation of a mixture of three herbicides in the rhizosphere of a herbicide-tolerant plant. Chemosphere 28:1551–1557

Anderson TA, White DC, Walton BT (1995) Degradation of hazardous organic compounds by rhizosphere microbial communities. In: Signh VP (ed) Biotransformations: microbial degradation of health risk compounds, vol V. Elsevier Science, BV, Amsterdam, pp 205–225

Anderson RT et al (2003) Stimulating *in situ* activity of *Geobacter* species to remove uranium from groundwater of a uranium-contaminated aquifer. Appl Environ Microbiol 69:5884–5891

Arbeli Z, Fuentes CL (2007) Accelerated biodegradation of pesticide: an overview of the phenomenon, its bias and possible solutions; and a discussion on the tropical dimension. Crop Prot 26:1733–1746

Atlas RM (1981) Microbial degradation of petroleum hydrocarbon: an environmental perspective. Microbiol Rev 45:180–209

Atlas RM (1995) Bioremediation of petroleum pollutants. Int Biodeterior Biodegrad 35:317–327

Atteia O, Guillot C (2007) Factors controlling BTEX and chlorinated solvents plume length under natural attenuation conditions. J Contam Hydrol 90:81–104

Audus LJ (1949) The biological detoxification of 2,4-dichlorophenoxyacetic acid in soil. Plant Soil 2:31–36

Audus LJ (1951) The biological detoxification of hormone herbicides in soil. Plant Soil 3:170–192

Ballschmister K, Zell M (1980) Analysis of PCB by glass capillary gas chromatography. Fresen J Anal Chem 302:20–31

Barriuso E, Houot S, Serra-Wittling C (1997) Influence of compost addition to soil on the behaviour of herbicides. Pestic Sci 49:65–75

Barton LL, Plunkett RM, Thomson BM (2003) Reduction of metals and nonessential elements by anaerobes. In: Ljungdahl LG, Adams MW, Barton LL, Ferry JG, Johnson MK (eds) Biochemistry and physiology of anaerobic bacteria. Springer, New York, pp 220–234

Berthe-Corti L, Höpner T (2005) Geobiological aspects of coastal oil pollution. Paleogeogr Paleoclimatol Paleoecol 219:171–189

Bertrand J-C, Rambeloarisoa E, Rontani J-F, Guisti G, Mattei G (1983) Microbial degradation of crude oil in sea water in continuous culture. Biotechnol Lett 5:567–572

Bertrand J-C, Al Mallah M, Acquaviva M, Mille G (1990) Biodegradation of hydrocarbons by an extremely halophilic archaeabacterium. Lett Appl Microbiol 11:260–263

Bertrand J-C, Bianchi M, Almallah M, Acquaviva M, Mille G (1993) Hydrocarbon biodegradation and hydrocarbonoclastic bacterial communities composition grown in seawater as a function of sodium chloride concentration. J Exp Mar Biol Ecol 168:125–138

Binet P, Portal JM, Leyval C (2000) Fate of polycyclic aromatic hydrocarbons in the rhizosphere and mycorrhizosphere of ryegrass. Plant Soil 227:207–213

Bocard C (2007) Marine oil spills and soils contaminated by hydrocarbons. Editions Technip, Paris

Böltner D, Godoy P, Muñoz-Rojas J, Duque E, Moreno-Morillas S, Sánchez L, Ramos J-L (2008) Rhizoremediation of lindane by root-colonizing *Sphingomonas*. Microb Biotechnol 1:87–93

Bonin P, Bertrand J-C (1998) Involvement of a bioemulsifier in heptadecane uptake in *Pseudomonas aeruginosa*. Chemosphere 38:1157–1164

Borja J, Taleon DM, Auresenia J, Gallardo S (2005) Polychlorinated biphenyls and their biodegradation. Process Biochem 40:1999–2013

Borja JQ, Auresenia JL, Gallardo SM (2006) Biodegradation of polychlorinated biphenyls using biofilm grown with biphenyl as carbon source in fluidized bed reactor. Chemosphere 64:555–559

Boubakri H, Beuf M, Simonet P, Vogel TM (2006) Development of metagenomic DNA shuffling for the construction of a xenobiotic gene. Gene 375:87–94

Bouchez M, Blanchet D, Vandecasteele J-P (1997) An interfacial uptake mechanism for the degradation of pyrene by a *Rhodococcus* strain. Microbiology (UK) 143:1087–1093

Bragg JR, Prince RC, Harner EJ, Atlas RM (1994) Effectiveness of bioremediation for *Exxon Valdez* oil spill. Nature 368:413–418

Brian RC (1976) The history and classification of herbicides. In: Audus LJ (ed) Herbicides. Physiology, biochemistry, ecology. Academic, New York, pp 1–54

Brown JW, Mitchell JW (1948) Inactivation of 2,4-dichloropenoxyacetic acid in soil as affected by soil moisture, temperature, the addition of manure and autoclavong. Bot Gaz 109:314–323

Brown JF Jr, Bedard DL, Brennan MJ, Carnahan JC, Feng H, Wagner RE (1987) Polychlorinated biphenyl dechlorination in aquatic sediments. Science 236:709–712

Bull AT (1980) Biodegradation: some attitudes and strategies of microorganisms and microbiologists. In: Ellwood DC, Hedger JN, Latham MJ, Lynch JM, Slater JH (eds) Contemporary microbial ecology. Academic, London/New York/Toronto/Sydney/San Francisco, pp 107–136

Burken JG, Schnoor JL (1997) Uptake and metabolism of atrazine by poplar trees. Environ Sci Technol 31:1399–1406

Callaghan AV, Tierney M, Phelps CD, Young LY (2009) Anaerobic biodegradation on *n*-hexadecane by a nitrate-reducing consortium. Appl Environ Microbiol 75:1339–1344

Carmona M, Zamarro MT, Blázquez B, Durante-Ro-dríguez G, Juárez JF, Valderrama JA, Barragán MJL, Garcia JL, Díaz E (2009) Anaerobic catabolism of aromatic compounds: a genetic and genomic view. Microbiol Mol Biol Rev 73:71–133

Cérémonie H, Boubakri H, Mavingu IP, Simonet P, Vogel TM (2006) Plasmid-encoded gamma-hexachlorocyclohexane degradation genes and insertion sequences in *Sphingobium francense* (ex-*Sphingomonas paucimobilis* Sp+). FEMS Microbiol Lett 257:243–252

Chang SW, Lee SJ, Je CH (2005) Phytoremediation of atrazine by poplar trees: toxicity, uptake, and transformation. J Environ Sci Health B 40:801–811

Changey F, Devers-Lamrani M, Rouard N, Martin-Laurent F (2011) *In vitro* evolution of an atrazine-degrading population under cyanuric acid selection pressure: evidence for the selective loss of a 47 kb region on the plasmid ADP1 containing the *atzA*, *B* and *C* genes. Gene 490:18–25

Chilingar GV, Loo WW, Khilyuk LF, Katz SA (1997) Electrobioremediation of soils contaminated with hydrocarbons and metals: progress report. Energy Source 19:129–146

Connel DW (1997) Basic concepts of environmental chemistry. CRC Press LLC., Boca Raton

Cook AM, Hütter R (1984) Deethylsimazine: bacterial dechlorination, deamination and complete degradation. J Agric Food Chem 32:581–585

Cornel P, Krause S (2008) Membrane bioreactors for wastewater treatment. In: Li NN, Fane AG, Ho WAG, Mat-suura T (eds) Advanced membrane technology and applications. Wiley, Holoken, pp 217–238

Couillard D, Zhu S (1992) Bacterial leaching of heavy metals from sewage sludge for agricultural application. Water Air Soil Poll 63:67–80

Cunningham JA, Rahme H, Hopkins GD, Lebron C, Reinhard M (2001) Enhanced in situ bioremediation of BTEX-contaminated groundwater by combined injection of nitrate and sulfate. Environ Sci Technol 35:1663–1670

Cutter LA, Watts JEM, Sowers KR, May HD (2001) Identification of a microorganism that links its growth to the reductive dechlorination of a 2,3,5,6-chlorobiphenyl. Environ Microbiol 3:699–709

Dagley S (1978) Determinants of biodegradability. Q Rev Biophys 11:577–602

Dash HR, Mangwani N, Chakraborty J, Kumari S, Das S (2013) Marine bacteria : potential candidates for enhanced bioremediation. Appl Microbiol Biotechnol 97:561–571

De Schrijver A, De Mot R (1999) Degradation of pesticides by actinomycetes. Crit Rev Microbiol 25:85–119

De Wilde T, Spanoghe P, Debear C, Ryckeboer J, Springael D, Jaeken P (2007) Overview of on-farm bioremediation systems to reduce the occurrence of point source contamination. Pest Manag Sci 63:111–128

Degrémont (2008) Mémento techniques de l'eau. 10e édition, tomes 1 et 2. Lavoisier, Cachan

Dejonghe W, Goris J, El Fantroussi S, Hofte M, De Vos P, Verstraete W, Top EM (2000) Effect of dissemination of 2,4-dichlorophenoxyacetic acid (2,4-D) degradation plasmids on 2,4-D degradation and on bacterial community structure in two different soil horizons. Appl Environ Microbiol 66:3297–3304

Dejonghe W, Boon N, Seghers D, Top EM, Verstraete W (2001) Bioaugmentation of soils by increasing microbial richness: missing links. Environ Microbiol 3:649–657

Devers M, Soulas G, Martin-Laurent F (2004) Real-time reverse transcription PCR analysis of expression of atrazine catabolism genes in two bacterial strains isolated from soil. J Microbiol Methods 56:3–15

Devers M, Henry S, Hartmann A, Martin-Laurent F (2005) Horizontal gene transfer of atrazine-degrading genes atz from *Agrobacterium tumefaciens* St96-4 pADP1::Tn5 to bacteria of maize-cultivated soil. Pest Manag Sci 61:870–880

Devers M, Rouard N, Martin-Laurent F (2007a) Genetic rearrangement of the atzAB atrazine-degrading gene cassette from pADP1::Tn5 to the chromosome of *Variovorax* sp. MD1 and MD2. Gene 392:1–6

Devers M, El Azhari N, Udikovic- Kolic N, Martin-Laurent F (2007b) Detection and organization of atrazine-degrading genetic potential of seventeen bacterial isolates belonging to divergent taxa indicate a recent common origin of their catabolic functions. FEMS Microbiol Lett 273:78–86

Devers M, Rouard N, Martin-Laurent F (2008) Fitness drift of an atrazine-degrading population under atrazine selection pressure. Environ Microbiol 10:676–684

Dodge AG, Wackett LP, Sadowsky MJ (2012) Plasmid localization and organization of melamin degradation genes in *Rhodococcus* sp. strain Mel. Appl Environ Microbiol 78:1397–1403

Don RH, Pemberton JM (1985) Genetic and physical map of the 2,4-Dichlorophenoxyacetic acid-degradative plasmid pJP4. J Bacteriol 161:466–468

Durand A (2003) Bioreactor designs for solid-state fermentation. Biochem Eng J 13:113–125

Ehrlich HL (1995) Biogenesis and biodegradation of sulfide minerals on the earth's surface. In: Ehrlich HL (ed) Geomicrobiology. Dekker, New York, pp 578–614

El Azhari N, Devers-Lamrani M, Chatagnier G, Rouard N, Martin-Laurent F (2010) Molecular analysis of the catechol-degrading bacterial community in a coal wasteland heavily contaminated with PAHs. J Hazard Mater 177:593–601

El Fantroussi S, Agathos SN (2005) Is bioaugmentation a feasible strategy for pollutant removal and site remediation? Curr Opin Microbiol 8:268–275

Elbaz-Poulichet F, Morley NH, Beckers JM, Nomerange P (2001) Dissolved metals fluxes through the Strait of Gibraltar – the influence of the Tinto and Odiel Rivers (SW Spain). Mar Chem 3–4:193–213

Ellis LB, Wackett LP (2012) Use of the University of Minnesota biocatalysis/Biodegradation Database for study of microbial degradation. Microb Inform Exp 2:1

Entry JA, Donnelly PK, Emmingham WH (1996) Mineralization of atrazine and 2,4-D in soils inoculated with *Phanerochaete chrysosporium* and *Trappea darkeri*. Appl Soil Ecol 3:85–90

Fava F, Di Gioia D, Marchettei L, Quattroni G (1996) Aerobic dechlorination of low-chlorinated biphenyls by bacterial biofilms in packed-bed batch bioreactors. Appl Microbiol Biotechnol 45:562–568

Fava F, Di Gioia D, Marchetti L (1998) Cyclodextrin effects on the *ex situ* bioremediation of a chronically polychlorobiphenyl-contaminated soil. Biotechnol Bioeng 58:345–355

Felsot SA, Wilson JG, Kuhlamn DE, Steey KL (1982) Rapid dissipation of carbofuran as a limiting factor in corn root-worm control in ¢ elds with histories of continuous furadan use. J Econ Entomol 75:1098–1103

Fennell DE, Nijenhuis I, Wilson SF, Zinder SH, Häggblom MM (2004) *Dehalococcoides ethenogenes* strain 195 reductively dechlorinates diverse chlorinated aromatic pollutants. Environ Sci Technol 38:2075–2081

Fernandez-Linares L, Acquaviva M, Bertrand J-C, Gauthier M (1995) Effect of sodium chloride concentration on growth and degradation of eicosane by the marine halotolerant bacterium *Marinobacter hydrocarbonoclasticus*. Syst Appl Microbiol 19:113–121

Field JA, Sierra-Alvarez R (2008) Microbial transformation and degradation of polychlorinated biphenyls. Environ Pollut 155:1–12

Frost LS, Leplae R, Summers AO, Toussaint A (2005) Mobile genetic elements: the agents of open source evolution. Nat Rev Microbiol 3:722–732

Fuchs G, Boll M, Heider J (2011) Microbial dégradation of aromatic compounds – from one strategy to four. Nat Rev Microbiol 9:803–816

Furukawa K (2000) Biochemical and genetic bases of microbial degradation of polychlorinated biphenyls (PCBs). J Gen Appl Microbiol 46:283–296

Gadd GM (1990) Heavy metal accumulation by bacteria and other microorganisms. Experientia 46:834–840

Gaillard M, Vallaeys T, Vorhöleter FJ, Minoia M, Werlen C, Sentchilo V, Pülher A, van der Meer JR (2006) The *clc* element of *Pseudomonas* sp. strain B13, a genomic island with various catabolic properties. J Bacteriol 188:1999–2013

Garcia-Gonzalez V, Govantes F, Porrua O, Santero E (2005) Regulation of the Pseudomonas sp. strain ADP cyanuric acid degradation operon. J Bacteriol 187:155–167

Gauthier M, Lafay B, Christen R, Fernandez L, Acquaviva M, Bonin P, Bertrand JC (1992) *Marinobacter hydrocarbonoclasticus* gen. nov., sp. nov., a new, extremely halotolerant, hydrocarbon-degrading marine bacterium. Int J Syst Bacteriol 42:568–576

Gazitúa MC, Slater AW, Melo F, González B (2010) Novel a-ketoglutarate dioxygenase *tfdA*-related genes are found in soil DNA after exposure to phenoxyalkanoic herbicides. Environ Microbiol 12:2411–2425

Gentry TJ, Rensing C, Pepper IL (2004) New approaches for bioaugmentation as a remediation technology. Crit Rev Environ Sci Technol 34:447–494

Gerhardt KE, Huang X-D, Glick BR, Greenberg BM (2009) Phytoremediation and rhizoremediation of organic soil contaminants: potential and challenges. Plant Sci 176:20–30

Gilbert EC, Crowley DE (1997) Plant compounds that induce polychlorinated biphenyl biodegradation by *Arthrobacter* sp. Strain B1B. Appl Environ Microbiol 63:1933–1938

Grigg BC, Assaf NA, Turco RF (1997) Removal of atrazine contamination in soil and liquid systems using bioaugmentation. Pestic Sci 50:211–220

Grimaud R (2010a) Marinobacter. In: Timmis KN, Mc Genity TJ, Merr JR, Lorenzo V (eds) Handbook of hydrocarbon and lipid microbiology. Springer, Berlin/Heidelberg, pp 1289–1296

Grimaud R (2010b) Biofilm development at interfaces between hydrophobic organic compounds and water. In: Timmis KN, Mc Genity TJ, Merr JR, Lorenzo V (eds) Handbook of hydrocarbon and lipid microbiology. Springer, Berlin/Heidelberg, pp 1491–1499

Grossi V, Massias D, Stora G, Bertrand J-C (2002) Burial, exportation and degradation of acyclic petroleum hydrocarbons following a simulated oil spill in bioturbed Mediterranean coastal sediments. Chemosphere 48:947–954

Grossi V, Cravo-Laureau C, Guyoneaud R, Ranchou- Peyruse A, Hirschler-Réa A (2008) Metabolism of *n*-alkanes by anaerobic bacteria: a summary. Org Geochem 39:1197–1203

Habe H, Omori T (2003) Genetics of polycyclic aromatic hydrocarbon metabolism in diverse aerobic bacteria. Biosci Biotechnol Biochem 67:225–243

Hafeez F, Spor A, Breuil MC, Schwartz C, Martin-Laurent F, Philippot L (2012) Distribution of bacteria and nitrogen-cycling microbial communities along constructed Technosol depth-profiles. J Hazard Mater 231:88–97

Harayama S, Kishira H, Kasai Y, Shutsubo K (1999) Petroleum biodegradation in marine environments. J Mol Microbiol Biotechnol 1:63–70

Harker AR, Olsen RH, Seidler RJ (1989) Phenoxyacetic acid degradation by the 2,4-dichlorophenoxyacetic acid (TFD) pathway of plasmid pJP4: mapping and characterization of the tfd regulatory gene, tfd R. J Bacteriol 171:314–320

Harkness MR et al (1993) In situ stimulation of aerobic PCB biodegradation in Hudson river sediments. Science 259:503–507

Harwood CS, Parales RE (1996) The β-ketoadipate pathway and the biology of self-identity. Annu Rev Microbiol 50:553–590

Hatamian-Zarmi A, Shojaosadati A, Vasheghani-Farahani E, Hosseinkhani S, Emamzadeh A (2009) Extensive biodegradation of highly chlorinated biphenyl and Aroclor 1242 by Pseudomonas aeruginosa TMU56 isolated from contaminated soils. Int Biodeterior Biodegrad 63:788–794

Head IM, Jones DM, Larter ST (2003) Biological activity in the deep subsurface and the origin of heavy oil. Nature 426:344–352

Heider J (2007) Adding handles to unhandy substrates: anaerobic hydrocarbon activation mechanisms. Curr Opin Chem Biol 11:188–194

Henze M, Harremoës P, La Cour Jansen J, Arvin E (2002) Wastewater treatment: biological and chemical processes. Springer, Berlin

Hickey WJ, Fuster DJ, Lamar RT (1994) Transformation of atrazine in soil by *Phanerochaete chrysosporium*. Soil Biol Biochem 26:1665–1671

Ho C, Applegate B, Banks MK (2007) Impact of microbial/plant interactions on the transformation of polycyclic aromatic hydrocarbons in the rhizosphere of *Festuca arundinacea*. Int J Phytoremediation 9:107–114

Hogan DA, Buckley DH, Nakatsu CH, Schmidt TM, Hausinger RP (1997) Distribution of the *tfdA* gene in soil bacteria that do not degrade 2,4- dichlorophenoxyacetic acid (2,4-D). Microb Ecol 34:90–96

Hornafius TS, Quigley DC, Luyendik BNP (1999) The world's most spectacular marine hydrocarbon seeps (Coil Oil Point, Santa Barbara Channel, California): quantification and emission. J Geophys Res 104:20703–20711

Horvath RS (1972) Microbial co-metabolism and the degradation of organic compounds in nature. Bacteriol Rev 36:146–155

Houot S, Topp E, Yassir A, Soulas G (2000) Dependence of accelerated degradation of atrazine on soil pH in French and Canadian soils. Soil Biol Biochem 32:615–625

Hurlbert MH, Krawieck S (1977) Cometabolism: a critique. J Theor Biol 69:287–291

Hurst CJ, Crawford RL, Knudsen GR, McInerney MJ, Stetzenbach LD (2002) Manual of environmental microbiology, 2nd edn. ASM Press, Washington, DC

Husain DR, Goutx M, Bezac C, Gilewicz M, Bertrand J-C (1997) Morphological adaptation of *Pseudomonas nautica* strain 617 to growth on eicosane and modes of eicosane uptake. Lett Appl Microbiol 24:55–58

Imai R, Nagata Y, Fukuda M, Takagi M, Yano K (1991) Molecular cloning of a *Pseudomonas paucimobilis* gene encoding a 17-kilodalton polypeptide that eliminates HCl molecules from g-hexachlorocyclohexane. J Bacteriol 173:6811–6819

Itoh K, Kanda R, Sumita Y, Kim H, Kamagata Y, Suyama K, Yamamoto H, Hausinger RP, Tiedje JM (2002) tfdA -like genes in 2,4-dichlorophenoxyacetic acid-degrading bacteria belonging to the *Bradyrhizobium-Agromonas-Nitrobacter-Afipia* cluster in a-proteobacteria. Appl Environ Microbiol 68:3449–3454

Iwamoto T, Nasu M (2001) Current bioremediation practice and perspective. J Biosci Bioeng 92:1–8

Jensen HL (1963) Carbon nutrition of some microorganisms decomposing halogen-substituted aliphatic acids. Acta Agr Scand 13:404–412

Johnsen AR, Karlson U (2004) Evaluation of bacterial strategies to promote the bioavailability of polycyclic aromatic hydrocarbons. Appl Microbiol Biotechnol 63:452–459

Johnsen AR, Wick LY, Harms H (2005) Principles of microbial PAH-degradation in soil. Environ Pollut 133:71–84

Joner EJ, Leyval C (2003) Phytoremediation of organic pollutants using mycorrhizal plants; a new aspect of rhizosphere interactions. Agronomie 23:495–502

Julsing MK, Schrewe M, Cornelissen S, Hermann I, Schmid A, Bühler B (2012) Outer membrane protein AlkL boosts biocatalytic oxyfunctionalization of hydrophobic substrates in *Escherichia coli*. Appl Environ Microbiol 78:5724–5733

Kalin N, Wheeler WN, Meinrath G (2005) The removal of uranium form mining waste water using algal/microbial biomass. J Environ Radioact 78:151–177

Khelifi N, Grossi V, Hamdi M, Dolla A, Tholozan J-L, Ollivier B, Hirschler-Réa A (2010) Anaerobic oxidation of fatty acids and alkenes by the hyperthermophilic sulfate-reducing archaeon *Archaeoglobus fulgidus*. Appl Environ Microbiol 76:3057–3060

Kim E-J, Oh JE, Chang Y-S (2003) Effects of forest fire on the level and distribution of PCDD/Fs and PAHs in soil. Sci Total Environ 311:177–189

Kim DU, Kim MS, Lim JS, Ka JO (2013) Widespread occurrence of the *tfd*-II genes in soil bacteria revealed by nucleotide sequence analysis of 2,4-dichlorophenoxyacetic acid degradative plasmids pDB1 and p712. Plasmid 69:243–248

Klein B, Grossi V, Bouriat P, Goulas P, Grimaud R (2008) Cytoplasmic wax ester accumulation during biofilm-driven substrate assimilation at the alkane-water interface by *Marinobacter hydrocarbonoclasticus* SP17. Res Microbiol 159:137–144

Knackmus HJ (1981) Degradation of halogenated and sulfonated hydrocarbons. In: Leisinger T, Cook AM, Müller R, Nüsch J (eds) Microbial degradation of xenobiotic and recalcitrant compounds. Academic, London, pp 189–212

Kolic NU, Hrsak D, Begonja Kolar A, Petric I, Stipicevic S, Soulas G, Martin-Laurent F (2007) Combined metabolic activity within an atrazine-mineralizing community enriched from agrochemical factory soil. Int Biodeter Biodegr 60:299–307

Konhauser K (2007) Biomineralisation. In: Kohnhauser K (ed) Introduction to geomicrobiology. Blackwell, Oxford, pp 139–191

Kunapuli U, Griehler C, Beller HR, Meckenstock RU (2008) Identification of intermediates formed during anaerobic benzene degradation by iron-reducing enrichment culture. Environ Microbiol 10:1703–1712

La Farge B (1995) Le biogaz. Masson, Paris/Milan/Barcelona

Laemmli CM, Leveau JHJ, Zehnder AJB, van der Meer JR (2000) Characterization of a second *tfd* gene cluster for chlorophenol and chlorocatechol metabolism on plasmid pJP4 in *Ralstonia eutropha* JMP134 (pJP4). J Bacteriol 82:165–4172

Lajoie CA, Zylstra CJ, DeFlaun MF, Strom PF (1993) Development of field application vectors for remediation of soils contaminated with polychlorinated biphenyls. Appl Environ Microbiol 59:1735–1741

Lal R, Dogra C, Malhotra S, Sharma P, Pal R (2006) Diversity, distribution and divergence of *lin* genes in hexachlorocyclohexane-degrading sphingomonads. Trends Biotechnol 24:121–130

Lawniczak L, Marecik R, Chrzanowski L (2013) Contributions of biosurfactants to natural or induced bioremediation. Appl Microbiol Biotechnol 97:2327–2339

Le Dréau Y, Gilbert F, Doumenq P, Bertrand J-C, Mille G (1997) The use of hopanes to track *in situ* variations in petroleum composition in surface sediments. Chemosphere 34:1663–1672

Leadbetter ER, Foster JW (1959) Oxidation products formed from gaseous alkanes by the bacterium *Pseudomonas methanica*. Arch Biochem Biophys 82:491–492

Leahy JG, Colwell RR (1990) Microbial degradation of hydrocarbons in the environment. Microbiol Rev 54:305–315

Leblanc M, Morales JA, Borrego J, Elbaz-Poulichet F (2000) 4,500-year-old mining pollution in southwestern Spain: long-term implications for modern mining pollution. Econ Geol 95:655–662

Leisinger T (1983) General aspects. Microorganisms and xenobiotic compounds. Cell Mol Life Sci 39:1183–1191

Leslie-Grady CP Jr (1985) Biodegradation: its measurement and microbiological basis. Biotechnol Bioeng 27:660–674

Leveau JHJ, Zehnder AJB, van Der Meer JR (1998) The *tfdK* gene product facilitates uptake of 2,4-dichlorophenoxyacetate by *Ralstonia eutropha* JMP134 (pJP4). J Bacteriol 180:2237–2243

Leyval C (2005) Pollutions organiques agricoles, urbaines ou industrielles: cas des hydrocarbures aromatiques polycycliques. In: Girard M-C, Walter C, Remy JC, Berthelin J, Morel JL (eds) Sols et environnement. Dunod, Paris

Lloyd JR, Lovley DR, Macaskie LE (2003) Biotechnological application of metal-reducing microorganisms. Adv Appl Microbiol 53:85–128

Lofroth G, Ames BN (1978) Mutagenicity of inorganic compounds in salmonella typhimurium: arsenic, chromium and selenium. Mutat Res 53:65–66

Lovley DR, Phillips EJP, Gorby YA, Landa E (1991) Microbial reduction of uranium. Nature 350:413–416

Lu Y, Yu Y, Zhou R, Sun W, Dai C, Wan P, Zhang L, Hao D, Ren H (2011) Cloning and characterisation of a novel 2,4-dichlorophenol hydroxylase from a metagenomic library derived from polychlorinated biphenyl-contaminated soil. Biotechnol Lett 33:1159–1167

Mackay D, Shiu WY, Ma KC (1992) Illustrated handbook of physical-chemical properties and environmental fate for organic chemicals, vol 1, Monoaromatic hydrocarbons, chlorobenzenes and PCBs. Lewis Publishers, Ann Arbor

Magar VS, Brenner RC, Johnson GW, Quensen JF III (2005) Long-term recovery of PCB-contaminated sediments at the lake Hartwell superfund site: PCB dechlorination 2. Rates and extend. Environ Sci Technol 39:3548–3554

Maier RM, Pepper IL, Gerba CP (2000) Environmental microbiology. Academic, San Diego/San Francisco/New York/Boston/London/Sydney/Tokyo

Mandelbaum RT, Allan DL, Wackett LP (1995) Isolation and characterization of a *Pseudomonas* sp. that mineralizes the s-triazine herbicide atrazine. Appl Environ Microbiol 61:1451–1457

Marcacci S, Raveton M, Ravanel P, Schwitzguébel J (2005) The possible role of hydroxylation in the detoxification of atrazine in mature vetiver (*Chrysopogon zizanioides* Nash) grown in hydroponics. Z Naturforsch C 60:427–434

Margesin R, Schinner F (2001) Biodegradation and bioremediation of hydrocarbons in extreme environments. Appl Microbiol Biotechnol 56:650–663

Martinez B, Tomkins J, Wackett LP, Wing R, Sadowsky MJ (2001) Complete nucleotide sequence and organization of the atrazine catabolic plasmid pADP-1 from *Pseudomonas* sp. Strain ADP. J Bacteriol 183:5684–5697

Martin-Laurent F, Barret B, Wagschal I, Piutti S, Devers M, Soulas G, Philippot L (2006) Impact of the maize rhizosphere on the genetic structure, the diversity and the atrazine-degrading gene composition of cultivable atrazine-degrading communities. Plant Soil 282:99–115

Master ER, Lai VWM, Kuipers B, Cullen WR, Mohn WW (2002) Sequential anaerobic-aerobic treatment of soil contaminated with weathered Aroclor 1260. Environ Sci Technol 36:100–103

Mau S, Valentine DL, Clark JF, Reed J, Camilli R, Washburn L (2007) Dissolved methane distribution and air-sea flux in the plume of a massive seep field, Coal Oil Point, California. Geophys Res Lett 34:1–5

May HD, Miller GS, Kjellerup BV, Sowers KR (2008) Dehalorespiration with polychlorinated biphenyls by an anaerobic ultramicrobacterium. Appl Environ Microbiol 74:2089–2094

McGowan C, Fulthorpe R, Wright A, Tiedje JM (1998) Evidence for interspecies gene transfer in the evolution of 2,4-dichlorophenxyacetic acid degraders. Appl Environ Microbiol 64:4089–4092

Means AJ (1991) Observation of an oil spill bioremediation activity in Galveston Bay, Texas. US Department of Commerce, National Oceanic and Atmospheric Administration, National Ocean Service, Seattle

Megharaj M, Ramakrishnan B, Venkateswarlu K, Sethunathan N, Naidu R (2011) Bioremediation approaches for organic pollutants: a critical perspective. Environ Int 37:1362–1375

Mhiri C, Tandeau de Marsac N (1997) Réhabilitation par les microorganismes de sites contenant du pyralène : problématique et perspectives d'étude. Bull Inst Pasteur 95:3–28

Michel FC Jr, Quensen J, Reddy CA (2001) Bioremediation of a PCB-contaminated soil *via* composting. Compost Sci Util 9:274–284

Miralles G, Grossi V, Acquaviva M, Duran R, Bertrand J-C, Cuny P (2007) Alkane degradation and dynamics of phylogenetic subgroups of sulfate-reducing bacteria in an anoxic coastal marine sediment artificially contaminated with oil. Chemosphere 68:1327–1334

Mitchell JW, Hamner CL (1944) Polyethyleneglycol as carriers for growth regulating substances. Bot Gaz 105:474–483

Monard C, Martin-Laurent F, Lima O, Devers-Lamrani M, Binet F (2013) Estimating the biodegradation of pesticide in soils by monitoring pesticide-degrading gene expression. Biodegradation 24:203–213

Mougin C, Laugero C, Asther M, Chaplain V (1997) Biotransformation of s-triazine herbicides and related degradation products in liquid cultures by the white rot fungus Phanerochaete chrysosporium. Pestic Sci 49:169–177

Mulbry WW (1994) Purification and characterization of an inducible s-triazine hydrolase from *Rhodococcus corallinus* NRRL B-15444R. Appl Environ Microbiol 60:613–618

Nagata Y, Kamakura M, Endo R, Miyazaki R, Ohtsubo Y, Tsuda M (2006) Distribution of γ-hexachlorocyclohexane-degrading genes on three replicons in *Sphingobium japonicum* UT26. FEMS Microbiol Lett 256:112–118

Nagata Y, Endo R, Ito M, Ohtsubo Y, Tsuda M (2007) Aerobic degradation of lindane (gamma-hexachlorocyclohexane) in bacteria and its biochemical and molecular basis. Appl Microbiol Biotechnol 76(4):741–752

Nagy I, Schoofs G, Compernolle F (1995) Degradation of the thiocarbamate herbicide EPTC (S-ethyl dipropylcarbamothioate) and biosafening by *Rhodococcus* sp. strain NI86/21 involve and inducible cytochrome P450 system and aldehyde deshydrogenase. J Bacteriol 177:676–687

Natarajan MR, Wu WM, Nye J, Wang H, Bhatnagar L, Jain MK (1996) Dechlorination of polychlorinated biphenyl congeners by an anaerobic microbial consortium. Appl Microbiol Biotechnol 46:673–677

National Research Council (1985) Oil in the sea: inputs, fates and effects. National Academy of Sciences, Washington, DC

National Research Council (2002) Oil in the sea III: inputs, fates and effects. National Academy of Sciences, Washington, DC

Ni'Matuzahro N, Gilewicz M, Guliano M, Bertrand J-C (1999) *In vitro* study of interaction between photo-oxidation and biodegradation of n-methylphenanthrene by *Sphyngomonas* sp. 2MPII. Chemosphere 38:2501–2507

Nies L, Vogel TM (1990) Effects of organic substrates on dechlorination of Aroclor 1242 in anaerobic sediments. Appl Environ Microbiol 56:2612–2617

Norberg P, Bergström M, Jethava V, Dubhashi D, Hermansson M (2011) The IncP-1 plasmid backbone adapts to different host bacterial species and evolves through homologous recombination. Nat Commun 2:268. doi:10.1038/ncomms1267

Ohtsubo Y, Kudo T, Tsuda M, Nagata Y (2004) Strategies for bioremediation of polychlorinated biphenyls. Appl Microbiol Biotechnol 65:250–258

Okuda T, Kumata H, Zakaria MF, Naroaka H, Ishiwatari R, Takada H (2002) Source identification of Malaysian atmospheric polycyclic aromatic hydrocarbons nearby forest fire using molecular and isotopic compositions. Atmos Environ 36:611–618

Oleszczuk P (2007) Changes of polycyclic aromatic hydrocarbons during composting of sewage sludges with chosen physicochemical properties and PAHs content. Chemosphere 67:582–591

Oremland RS, Stolz JF (2003) The ecology of arsenic. Science 300:939–944

Otte MP, Gagnon J, Comeau Y, Matte N, Greer CW, Samson R (1994) Activation of an indigenous microbial consortium for bioaugmentation of pentachlorophenol-creosote contaminated soils. Appl Microbiol Biotechnol 40:926–932

Pandey J, Chauhan A, Jain RK (2009) Integrative approaches for assessing the ecological sustainability on *in situ* bioremediation. FEMS Microbiol Rev 33:324–375

Pei XH, Zhan XH, Wang SM, Lin YS, Zhou LX (2010) Effects of a biosurfactant and a synthetic surfactant on phenanthrene degradation by a *Sphingomonas* strain. Pedosphere 20:771–779

Pemberton JM, Fisher PR (1977) 2,4-D plasmids and persistence. Nature 268:732–733

Peng RH, Xiong AS, Xue Y, Fu XY, Gao F, Zhao W, Tian YS, Yao QH (2008) Microbial biodegradation of polyaromatic hydrocarbons. FEMS Microbiol Rev 32:927–955

Perkovich BS (1996) Enhanced mineralization of [^{14}C]atrazine in *Kochia scoparia* rhizospheric soil from a pesticide-contaminated site. Pestic Sci 46:391–396

Peters KE, Walters CC, Moldowan JM (2005) The biomarker guide, vols 1 and 2. Cambridge University Press, Cambridge

Petric I, Bru D, Udikovic-Kolic N, Hrsak D, Philippot L, Martin-Laurent F (2011) Evidence for shifts in the structure and abundance of microbial community in a long-term PCB-contaminated soil under bioremediation. J Hazard Mater 195:254–260

Phillips EJP, Landa E, Lovley DR (1995) Remediation of uranium contaminated soils with bicarbonate extraction and microbial U (VI) reduction. J Ind Microbiol 14:203–207

Pieper DH (2005) Aerobic degradation of polychlorinated biphenyls. Appl Microbiol Biotechnol 67:170–191

Pilon-Smits E (2005) Phytoremediation. Annu Rev Plant Biol 56:15–39

Pineda-Flores G, Boll-Argüello G, Lira-Galeana C, Mesta-Howard AM (2004) A microbial consortium isolated from a crude oil sample that uses asphaltenes as carbon and energy source. Biodegradation 15:145–151

Piutti S, Marchand AL, Lagacherie B, Martin-Laurent F, Soulas G (2002) Effect of successive cropping cycles with different plants and repeated herbicide applications on the degradation of the diclofop-methyl, bentazon, diuron, isoproturon and pendimethalin in soil. Pest Manag Sci 58:303–312

Poelarends GJ, van Hylckama Vlieg JE, Marchesi JR, Freitas Dos Santos LM, Janssen DB (1999) Degradation of 1,2-dibromoethane by *Mycobacterium* sp. strain GP1. J Bacteriol 181(7):2050–2058

Prince RC (1993) Petroleum spill bioremediation in marine environments. Crit Rev Microbiol 19:217–242

Prince RC (2010) Eucaryotic hydrocarbon degraders. In: Timmis KN, Mc Genity TJ, Merr JR, Lorenzo V (eds) Handbook of hydrocarbon and lipid microbiology. Springer, Berlin/Heidelberg, pp 2065–2078

Prince RC, Elmendorf DL, Lute JR, Hsu CS, Halth CE, Senlus JD, Dechert GJ, Douglas GS, Butler EL (1994) 17. alpha. (H)-21. beta. (H)-hopane as a con- served internal marker for estimating the biodegradation of crude oil. Environ Sci Technol 28:142–145

Prince RC, Gramain A, McGenity TJ (2010) Procaryotic hydrocarbon degraders. In: Timmis KN, Mc Genity TJ, Merr JR, Lorenzo V (eds) Handbook of hydrocarbon and lipid microbiology. Springer, Berlin/Heidelberg, pp 1671–1692

Pritchard PH, Mueller JG, Rogers JC, Kremer FV, Glaser JA (1992) Oil spill bioremediation : experiences, lessons and results from the *Exxon Valdez* oil spill in Alaska. Biodegradation 3:315–335

Quensen JF III, Boyd SA, Tiedje JM (1990) Dechlorination of four commercial polychlorinated biphenyl mixtures (Aroclors) by anaerobic microorganisms from sediments. Appl Environ Microbiol 56:2360–2369

Rabus R (2005) Biodegradation of hydrocarbon under anoxic conditions. In: Ollivier B, Magot M (eds) Petroleum microbiology. ASM Press, Washington, DC, pp 277–299

Raillard S et al (2001) Novel enzyme activities and functional plasticity revealed by recombining highly homologous enzymes. Chem Biol 8:891–898

Ramade F (2002) Dictionnaire encyclopédique de l'écologie et des sciences de l'environnement, 2eth edn. Dunod, Paris

Reanney DC (1976) Extrachromosomal elements as possible agents of adaptation and development. Bacteriol Rev 40:552–590

Reátegui E, Reynolds E, Kasinkas L, Aggarwal A, Sadowsky M, Aksan A, Wackett L (2012) Silica gel-encapsulated AtzA biocatalyst for atrazine biodegradation. Appl Microbiol Biotechnol 96:231–240

Ripp S, Nivens DE, Ahn Y, Werner C, Jarrell J IV, Easter JP, Cox CD, Burlage RS, Sayler GS (2000) Controlled field release of a bioluminescent genetically engineered microorganism for bioremediation process monitoring and control. Environ Sci Technol 34:846–853

Roanne TM, Pepper IL, Miller RM (1996) Microbial remediation of metals. In: Crawford RL, Crawford DL (eds) Bioremediation, principles and application. Cambridge University Press, Cambridge, pp 312–340

Robinson JB, Tuovinen OH (1984) Mechanisms of microbial resistance and detoxification of mercury and organomercury compounds: physiological, biochemical, and genetic analyses. Microbiol Rev 48:95–124

Rodrigues JLM, Kachel CA, Aiello MR, Quensen JF, Maltseva OV, Tsoi TV, Tiedje JM (2006) Degradation of Aroclor 1242 dechlorination products in sediments by *Burkholderia xenovorans* LB400 (*ohb*) and *Rhodococcus* sp. Strain RHA1 (*fcb*). Appl Environ Microbiol 72:2476–2482

Rojo F (2009) Degradation of alkanes by bacteria. Environ Microbiol 11(10):2477–2490

Ron EZ, Rosenberg E (2002) Biosurfactants and oil bioremediation. Curr Opin Biotechnol 13:249–252

Rontani JF, Bosser-Jpulak F, Rambeloarisoa E, Bertrand J-C, Giusti G (1985) Analytical study of asthart crude oil: asphaltenes biodegradation. Chemosphere 14:1413–1422

Rosenberg E, Ron EZ (2000) Biosurfactant. In: Dworkin M, Falkow S, Rosenberg E, Schleifer K-H, Stackebrandt E (eds) The prokaryotes. Springer, New York, pp 834–849

Rousseaux S, Hartmann A, Lagacherie B, Piutti S, Andreux F, Soulas G (2003) Inoculation of an atrazine-degrading strain, *Chelatobacter heintzii* Cit1, in four different soils: effects of different inoculum density. Chemosphere 51:569–576

Rudolph C, Wanner G, Huber R (2001) Natural communities of novel *Archaea* and *Bacteria* growing in cold sulfurous springs with a string-of-pearls-like morphology. Appl Environ Microbiol 67:2336–2344

Runes HB, Jenkins JJ, Bottomley PJ (2001) Atrazine degradation by bioaugmented sediment from constructed wetlands. Appl Microbiol Biotechnol 57:427–432

Ryckeboer J, Mergaert J, Coosemans J, Deprins K, Swings J (2003) Microbiological aspects of biowaste during composting in a monitored compost bin. J Appl Microbiol 94:127–137

Samanta SK, Singh OV, Jain RK (2002) Polycyclic aromatic hydrocarbons: environmental pollution and bioremediation. Trends Biotechnol 20:243–248

Sandmann ERIC, Loos MA (1988) Aromatic metabolism by 2,4-G degrading *Arthrobacter* sp. Can J Microbiol 34:125–130

Saouter E, Turner R, Barkay T (1994) Microbial reduction of ionic mercury for the removal of mercury from contaminated environments. Ann N Y Acad Sci 721:423–427

Sayler GS, Ripp S (2000) Field applications of genetically engineered microorganims for bioremediation processes. Curr Opin Biotechnol 11:286–289

Schlüter A, Szczepanowski R, Pühler A, Top EM (2007) Genomics of IncP-1antibiotic resistance plasmids isolated from wastewater treatment plants provides evidence for a widely accessible drug resistance gene pool. FEMS Microbiol Rev 31:449–477

Schneiker S et al (2006) Genome sequence of the ubiquitous hydrocarbon-degrading marine bacterium Alcanivorax borkumensis. Nat Biotechnol 24:997–1004

Scott C et al (2008) The enzymatic basis for pesticide bioremediation. Indian J Microbiol 48:65–79

Scott C, Jackson CJ, Coppin CW, Mourant RG, Hilton ME, Sutherland TD, Russell RJ, Oakeshott JG (2009) Catalytic improvement end evolution of atrazine chlorohydrolase. Appl Environ Microbiol 75:2184–2191

Scott C, Lewis SE, Milla R, Taylor MC, Rodgers AJW, Dumsday G, Brodie JE, Oakeshott JG, Russell RJ (2010) A free-enzyme catalyst for the bioremediation of environmental atrazine contamination. J Environ Manage 91:2075–2078

Seitz H-J, Siñeriz F, Schink B, Conrad R (1990) Hydrogen, production during fermentation of acetoin and acetylene by *Pelobacter acetylenicus*. FEMS Microbiol Lett 71:83–88

Sen D, Van der Auwera GA, Rogers LM, Thomas CM, Brown CJ, Top EM (2011) Broad-host-range plasmids from agricultural soils have IncP-1 backbones with diverse accessory genes. Appl Environ Microbiol 77:7975–7983

Sere G, Schwartz C, Ouvrard S, Sauvage C, Renat JC, Morel JL (2008) Soil construction: a step for ecological reclamation of derelict lands. J Soils Sediments 8:130–136

Seviour RJ, Mino T, Onuki M (2003) The microbiology of biological phosphorus removal in activated sludge systems. FEMS Microbiol Rev 27:99–127

Shao ZQ, Seffens W, Mulbry W, Behki RM (1995) Cloning and expression of the s-triazine hydrolase gene (*trzA*) from *Rhodococcus corallinus* and development of *Rhodococcus* recombinant strains capable of delkylating and dechlorinating the herbicide atrazine. J Bacteriol 177:5748–5755

Sharma VK, Canditelli M, Fortuna F, Cornacchia G (1997) Processing of urban and agro-industrial residues by aerobic composting: review. Energy Convers Manage 38:453–478

Silver S, Phung LT (2005) A bacterial view of the periodic table: genes and proteins for toxic inorganic ions. J Ind Microbiol Biotechnol 32:587–605

Simoneit BRT, Rogge WF, Lang Q, Jaffé R (2000) Molecular characterization of smoke from campfire burning of pine wood (*Pinus elliottii*). Chemosphere: Global Chang Sci 2:107–122

Singh N (2003) Enhanced degradation of hexachlorocyclohexane isomers in rhizosphere soil of *Kochia* sp. Bull Environ Contam Toxicol 70:0775–0782

Singh N, Megharaj M, Kookana RS, Naidu R, Sethunathan N (2004) Atrazine and simazine degradation in *Pennisetum* rhizosphere. Chemosphere 56:257–263

Singh A, Kumar V, Srivastava J (2012) Bioremediation: for petroleum cleanup: bioremediation: a promising tool for petroleum contamination. Lambert Academic Publishing, Sarrebruck

Slater JH, Bull AT (1982) Environmental microbiology biodegradation. Philos Trans R Soc Lond B Biol Sci 297:575–597

Smith D, Alvey S, Crowley DE (2005) Cooperative catabolic pathways within an atrazine-degrading enrichment culture isolated from soil. FEMS Microbiol Ecol 53:265–273

Soulas G (2003) Pesticide degradation in soils. In: Encyclopedia of environmental microbiology. Wiley, Oxford, pp 2385–2402

Springael D, Ryngaert A, Merlin C, Toussaint A, Mergeay M (2001) Occurrence of TN4371-related mobile elements and sequences in chlorobiphenyl-degrading bacteria. Appl Environ Microbiol 67:42–50

Stephensons T (2000) Membrane bioreactors for wastewater treatment. IWA Publishing, London

Stewart PS, Franklin MJ (2008) Physiological heterogeneity in biofilms. Nat Rev Microbiol 6:199–210

Strong LC, McTavish H, Sadowsky MJ, Wackett LP (2000) Field-scale remediation of atrazine-contaminated soil using recombinant *Escherichia coli* expressing atrazine chlorohydrolase. Environ Microbiol 2:91–98

Sulmon C, Gouesbet G, Binet F, Martin-Laurent F, El Amrani A, Couée I (2007) Soluble sugar amendment of plants enhances phytoaccumulation of organic contaminants and phytoremediation of contaminated soil. Environ Pollut 145:507–515

Summers AO (1992) The hard stuff: metal in bioremediation. Curr Opin Biotechnol 3:271–276

Swannell RPJ, Lee K, McDonagh M (1996) Field evaluations of marine oil spill bioremediation. Microbial Rev 60:342–365

Tapilatu HY, Grossi V, Acquaviva M, Militon C, Bertrand J-C, Cuny P (2010) Isolation of hydrocarbon-degrading extremely halophilic archaea from an uncontaminated hypersaline pond (Camargue, France). Extremophiles 14:225–231

Tartakovsky B, Michotte A, Cadieux J-CA, Lau PCK, Hawari J, Guiot SR (2001) Degradation of Aroclor 1242 in a single-stage coupled anaerobic/aerobic bioreactor. Water Res 35:4323–4330

Tchobanoglous G, Burton FL, Stensel HD (2003) Wastewater engineering: treatment and re-use. McGraw-Hall Companies, New York

Thayer JS, Brinkman FE (1982) The biological methylation of metals and metalloids. Adv Organomet Chem 20:313–357

Thomas JC, Berger F, Jacquier M, Bernillon D, Baud-Grasset F, Truffaut N, Normand P, Vogel TM, Simonet P (1996) Isolation and characterization of a novel γ-hexachlorocyclohexane-degrading bacterium. J Bacteriol 178:6049–6055

Thompson IP, van der Gast CJ, Ciric L, Singer AC (2005) Bioaugmentation for bioremediation: the challenge of strain selection. Environ Microbiol 7:909–915

Timmis KN (2010) Handbook of hydrocarbon and lipid microbiology. Springer, Berlin

Timmis KN, Pieper DH (1999) Bacteria designed for bioremediation. Trends Biotechnol 17:201–204

Topp E (2001) A comparison of three atrazine-degrading bacteria for soil bioremediation. Biol Fert Soils 33:529–534

Topp E, Tessier L, Gregorich EG (1996) Dairy manure incorporation stimulates rapid atrazine mineralisation in an agricultural soil. Can J Soil Sci 76:403–409

Topp E, Mulbry WM, Zhu H, Nour SM, Cuppels D (2000) Characterization of S-Triazine herbicide metabolism by a *Nocardioides* sp. isolated from agricultural soils. Appl Environ Microbiol 66:3134–3141

Topp E, Martin-Laurent F, Hartmann A, Soulas G (2004) Bioremediation of atrazine-contaminated soil. In: Gan JJ, Zhu PC, Aust SD, Lemley AT (eds) Pesticide, decontamination and detoxification. American Chemical Society, Washington, DC, pp 141–154

Topp E, Chapman R, Devers-Lamrani M, Hartmann A, Martin-Laurent F, Marti R, Sabourin L, Scott A, Sumaraha M (2013) Accelerated biodegradation of veterinary antibiotics in agricultural soil following long-term exposure, and isolation of a sulfonamide-degrading microbacterium. J Environ Qual 42:173–178

Trefault N, de la Iglesia R, Molina AM, Manzano M, Ledger T, Pérez-Pantoja D, Sánchez MA, Stuardo M, González B (2004) Genetic organization of the catabolic plasmid pJP4 from *Ralstonia eutropha* JMP134 (pJP4) reveals mechanisms of adaptation to chloroaromatic pollutants and evolution of specialized chloroaromatic degradation pathways. Environ Microbiol 6:655–668

Tuomela M, Vikman M, Hatakka A, Itävaara M (2000) Biodegradation of lignin in a compost environment: a review. Bioresour Technol 72:169–183

Udikovic-Kolic N, Scott C, Martin-Laurent F (2012) Evolution of atrazine-degrading capabilities in the environment. Appl Microbiol Biotechnol 96:1175–1189

Vallaeys T, Courde L, Mc Gown C, Wright AD, Fulthorpe RR (1999) Phylogenetic analyses indicate independent recruitment of diverse gene cassettes during assemblage of the 2,4-D catabolic pathway. FEMS Microbiol Ecol 28:373–382

van Beilen JB, Panke S, Lucchini S, Franchini AG, Röthlisberger M, Witholt B (2001) Analysis of Pseudomonas putida alkane degradation gene clusters and flanking insertion sequences: evolution and regulation of the alk genes. Microbiology 147:1621–1630

van der Meer JR (1997) Evolution of novel metabolic pathways for the degradation of chloroaromatic compounds. Antonie Van Leeuwenhoek 71:159–178

Van der Meer JR, Sentchilo V (2003) Genomic islands and the evolution of catabolic pathways in bacteria. Curr Opin Biotechnol 14:248–254

van der Meer JR, de Vos WM, Harayama S, der Zehn- AJB (1992) Molecular mechanisms of genetic adaptation to xenobiotic compounds. Microbiol Rev 56:677–694

van der Roest HF, Lawrence DP, van Bentem AGN (2002) Membrane bioreactors for municipal wastewater treatment. IWA Publishing, London

van Eerd LL, Hoagland RE, Zablotowicz RM, Hall JC (2003) Pesticide metabolism in plants and microorganisms. Weed Sci 51:472–495

van Haandel A, van der Lubbe J (2007) Handbook biological wastewater treatment. Quist Publishing, Leidschendam

van Hamme JD, Singh A, Ward OP (2003) Recent advances in petroleum microbiology. Microbiol Mol Biol Rev 67:503–549

Vandecasteele J-P (2008) Petroleum microbiology. Tomes 1 et 2. Editions Technip, Paris

Vesilind PA (2003) Wastewater treatment plant design. IWA Publishing, London

Waid JS (1972) The possible importance of transfer factors in the bacterial degradation of herbicides in natural ecosystems. Res Rev 44:65–71

Wang YT (2000) Microbial reduction of chromate. In: Lovley DR (ed) Environmental microbe-metal interactions. ASM Press, Washington, DC, pp 225–235

Wang L, Samac DA, Shapir N, Wackett LP, Vance CP, Olszewski NE, Sadowsky MJ (2005) Biodegradation of atrazine in transgenic plants expressing a modified bacterial atrazine chlorohydrolase (*atzA*) gene. Plant Biotechnol J 3:475–486

Wang C-M, Shyu C-L, Ho S-P, Chiou S-H (2007) Species diversity and substrate utilisation patterns of thermophilic bacterial communities in hot aerobic poultry and cattle manure composts. Microbiol Ecol 54:1–9

White C, Gadd GM (1996) Mixed sulphate-reducing cultures for the bioprecipitation of toxic metals: factorial and response analysis of the effects of dilution rate, sulphate and substrate concentration. Biodegradation 14:139–151

White LG, Bourbonnière L, Greer CW (1997) Biodegradation of petroleum hydrocarbons by psychrotrophic *Pseudomonas* strains possessing both alkane (*alk*) and naphthalene (*nah*) catabolic pathways. Appl Environ Microbiol 63:3719–3723

Wick LY, Colangelo T, Harms H (2001) Kinetics of mass transfer-limited bacterial growth on solid PAHs. Environ Sci Technol 35:354–361

Wick LY, De Munain AR, Springael D, Harms H (2002) Responses of *Mycobacterium* sp. LB501T to the low bioavailability of solid anthracene. Appl Microbiol Biotechnol 58:378–385

Widdel F, Rabus R (2001) Anaerobic biodegradation of saturated and aromatic hydrocarbons. Curr Opin Biotechnol 12:259–276

Widdowson MA, Shearer S, Andersen RG, Novak JT (2005) Remediation of polycyclic aromatic hydrocarbon compounds in groundwater using poplar trees. Environ Sci Technol 39:1598–1605

Williams PT (2005) Waste treatment and disposal. Wiley, Chichester

Woodard and Curran, Inc, (2006) Industrial waste treatment handbook. Elsevier, Burlington

WRB IWG (2006) World reference base for soil resource, 2nd ed. World Resources Report 103, FAO, Rome

Yadav JS, Quensen JF III, Tiedje JM, Reddy CA (1995) Degradation of polychlorinated biphenyl mixtures (Aroclors 1242, 1254, and 1260) by the white rot fungus *Phanerochaete chrysosporium* as evidenced by congener-specific analysis. Appl Environ Microbiol 61:2560–2565

Yakimov MM, Timmis KN, Golyshin PN (2007) Obligate oil-degrading marine bacteria. Curr Opin Biotechnol 18:257–266

Yi H, Crowley DE (2007) Biostimulation of PAH degradation with plants containing high concentrations of linoleic acid. Environ Sci Technol 41:4382–4388

Zengler K, Heider J, Roselló-Mora R, Widdel F (1999) Phototrophic utilisation of toluene under anoxic conditions by a new strain of *Blastochloris sulfoviridis*. Arch Microbiol 172:204–212

Part V

Tools of Microbial Ecology

Methods for Studying Microorganisms in the Environment

17

Fabien Joux, Jean-Claude Bertrand, Rutger De Wit, Vincent Grossi, Laurent Intertaglia, Philippe Lebaron, Valérie Michotey, Philippe Normand, Pierre Peyret, Patrick Raimbault, Christian Tamburini, and Laurent Urios

Abstract

The main methods for the study of microorganisms in the environment (water, soil, sediment, biofilms), the different techniques of sampling for measuring biomass, the activities, and the diversity of the microorganisms are presented. To respond to these various issues, techniques as varied as those of flow cytometry, molecular biology, biochemistry, molecular isotopic tools, or electrochemistry are implemented. These different techniques are described with their advantages and disadvantages for different types of biotopes. The question of the isolation, culture, and conservation of microorganisms from the environment are also addressed. Without being exhaustive, this chapter emphasizes the importance of using appropriate and efficient methodological tools to properly explore the still mysterious compartment of microorganisms in the environment.

Keywords

Bacterial isolation • Biomarkers • Cultural techniques • Cytometry • DNA microarray • Microbial activities • Microelectrodes • Molecular fingerprints • PCR • Phospholipid fatty acid analyses • Pigment analyses • Sampling techniques

---

* Chapter Coordinator

F. Joux* (✉)
Laboratoire d'Océanographie Microbienne (LOMIC), UMR 7621
CNRS-UPMC, Observatoire Océanologique de Banyuls,
66650 Banyuls-sur-Mer, France
e-mail: fabien.joux@obs-banyuls.fr

J.-C. Bertrand • V. Michotey • P. Raimbault • C. Tamburini
Institut Méditerranéen d'Océanologie (MIO), UM 110, CNRS 7294
IRD 235, Université de Toulon, Aix-Marseille Université,
Campus de Luminy, 13288 Marseille Cedex 9, France

R. De Wit
Écologie des systèmes marins côtiers (ECOSYM, UMR5119),
Universités Montpellier 2 et 1, CNRS-Ifremer-IRD, 34095 Montpellier Cedex 05, France

V. Grossi
Laboratoire de Géologie de Lyon: Terre, Planètes, Environnement,
UMR CNRS 5276, Université Claude Bernard Lyon 1, 69622 Villeurbanne Cedex, France

L. Intertaglia
Observatoire Océanologique de Banyuls sur Mer UPMC/CNRS – UMS 2348, Plateforme BIO2MAR, 66650 Banyuls-sur-Mer, France

P. Lebaron
Observatoire Océanologique de Banyuls, Laboratoire de Biodiversité et Biotechnologie Microbiennes (LBBM), Sorbonne Universités, UPMC Univ Paris 06, USR CNRS 3579, 66650 Banyuls-sur-Mer, France

P. Normand
Microbial Ecology Center, UMR CNRS 5557 / USC INRA 1364,
Université Lyon 1, 69622 Villeurbanne, France

P. Peyret
EA CIDAM 4678, CBRV, Université d'Auvergne, 63001 Clermont-Ferrand, France

L. Urios
Institut des Sciences Analytiques et de Physico-chimie pour l'Environnement et les Matériaux (IPREM), UMR CNRS 5254, Université de Pau et des Pays de l'Adour, B.P. 1155, 64013 Pau Cedex, France

17.1 Introduction

This chapter presents the main methods for the study of microorganisms in the environment. This study passes first through an adapted sampling for the different environments explored. The various components of the microbial community can then be characterized from the point of view of their biomass, their activity, and their diversity. To respond to these various issues, techniques as varied as those of flow cytometry, molecular biology, biochemistry, molecular isotopic tools, or electrochemistry are implemented. These different methods are described with their advantages and disadvantages for different types of biotopes (water, soil, sediment, biofilms). A final important point discussed in this chapter concerns the isolation of microorganisms from the environment and their culture in the laboratory. Without being exhaustive, this chapter emphasizes the importance of using appropriate and efficient methodological tools to properly explore the still mysterious compartment of microorganisms in the environment.

17.2 Sampling Techniques in Microbial Ecology

17.2.1 Soils Sampling Techniques

Sampling of soil for microbial ecology studies can be achieved with a shovel, but more usually using a hand auger (diameter between 1 and 10 cm) or vehicle-mounted hydraulic auger, which allows to collect at a constant depth. The depth is important because it influences many parameters (oxygen, temperature, water content, concentration of organic matter, etc.). The depth studied is variable in function of the questions. It can reach several meters (Box 17.1), but more often it is the layer between 0 and 30 cm that is studied. Techniques described above (use of a shovel or an auger) concern masses of soil from few hundred grams to several kilograms.

A major element to be taken into account during the sampling is the quantity and the quality of the organic materials that will in part determine the functioning of heterotrophic microflora representing the majority of the soil microorganisms. For this reason, many studies take into account the organization of the soil horizon. The superficial horizons (0–30 cm) are the most studied, because the most intense microbial activity is observed in these horizons. For the same reason, the rhizospheric soil (under the influence of the roots) is often the object of particular attention. Sampling rhizospheric soil is done by shaking the roots to remove not adhering soil, the soil still adhering constituting rhizospheric soil.

Box 17.1: Sampling the Deep Biosphere

Michel Magot

The results of a study published in 1998 sounded like a thunderclap in the sky of microbial ecology. The authors suggested that the amount of biomass hidden in the deep Earth subsurface, down to about 4,000 m deep, would be equivalent or higher to that of all living organisms on its surface (Whitman et al. 1998)! The total population of prokaryotes, only denizens of the deep, was estimated between 25 and 250 $\times$ 10^{28} cells, while about 15 years ago most microbiologists believed that the Earth subsurface was sterile beyond a few tens meters deep. Why did it take more than a century of research in microbiology for the estimation of such a great unknown biodiversity? The main reason undoubtedly lies in the difficulty of obtaining uncontaminated geological samples or deep fluids. Three main ways to collect microbiologically representative deep subsurface samples will be briefly described below.

Collecting Fluid Samples from Pre-existing Wells

This is a priori the simplest and cheapest method. This is what the American microbiologist Edson Bastin did during the 1920s to study sulfate-reducing bacteria from American oilfield waters, suggesting for the first time that bacteria could live in the deep subsurface. Whether collecting petroleum fluids or water samples from deep aquifers, the technique is the same: in principle, it just needs a production or monitoring well drilled in the geological formation of interest to collect liquid samples by following some basic rules to maintain aseptic conditions and avoid contact with oxygen. Studies based on such samples, either by conventional microbiology or molecular methods, show that things are really not that easy! Multiple sources of contamination can lead to erroneous conclusions. The two main ones result from (i) the drilling fluids used during the implementation of the well, which will be discussed below, and (ii) the colonization of the entire inner part of the well (the tubing) by a biofilm, which most often has nothing to do with the indigenous subsurface microorganisms (Magot 2005). A solution is nevertheless available to collect representative subsurface

---

M. Magot
Institut des Sciences Analytiques et de Physico-chimie pour l'Environnement et les Matériaux (IPREM), UMR CNRS 5254, Université de Pau et des Pays de l'Adour, B.P. 1155, 64013 Pau Cedex, France

(continued)

Box 17.1 (continued)

samples from wellheads: it necessitates the sterilization of the tubing throughout its length, i.e., sometimes a few thousand meters! The operation is obviously extremely cumbersome and costly, since it uses mechanical and chemical treatments involving specific equipments and specialized teams and lasts several weeks. But it is efficient, and the relevance of this approach was demonstrated (Basso et al. 2005).

Deep Drillings Dedicated to Microbiological Studies

Specific drilling operations taking into account all the constraints of microbiological sampling is the dream of any scientist interested in the deep biosphere. This has been done, albeit with some compromises, in the context of two major research programs ("Subsurface Science Microbiology" of the US Department of Energy – DOE – and Integrated Ocean Drilling Program – IODP) and some more specific operations. In the DOE program for which drilling down to a few hundred meters deep has been made, specific drilling tools were designed and tested. Samples (cores) could be packaged in sterile materials, protecting them from contamination before being brought to the surface (Box Fig. 17.1). More generally, precautions were taken to minimize or accurately assess the contamination introduced by drilling fluids. Actually, beyond several tens of meters deep, using more or less viscous fluid is needed to lubricate and cool the drill bit and bring the cuttings up to the surface. These fluids may contaminate the porous geological material. As they are virtually impossible to sterilize, it is generally preferable to estimate the level of sample contamination by the use of tracers. For this, either chemical agents such as boron or physical techniques such as fluorescent latex microspheres having the size of bacteria can be used. Quantification of these tracers in samples later allows to estimate the level of potential contamination. Finally, the estimation of contamination ideally also needs the microbiological study of all tools or drilling fluids used: comparison of bacterial communities of drilling fluid samples with that of collected cores provides informations on exogenous contamination. Core sampling for microbiological studies also imposes harsh technical and logistical constraints: collected cores must be rapidly transferred in a sterile anaerobic device, e.g., anaerobic glove box, and then subsampled or crushed under sterile anaerobic conditions. The implementation of such research programs requires considerable funding that can be obtained only through large national and international collaborative projects.

Access to Deep Geological Formations Through Excavations

The last and simplest way to access a deep geological formation is simply to go down there with all the needed equipment! Mines are an obvious means of access to deeper geological layers, and highly relevant studies were, for instance, conducted in the South African gold mines that reach depths of about 5,000 m (Onstott et al. 1997). Underground laboratories represent other privileged sites. Some were developed for the study of underground storage of nuclear waste in several countries, one of the aims being to study the impact of microorganisms on the storage containers. Several sites exist in North America or Europe, in various geological formations such as granites (Aspö in Sweden) (Pedersen 2001) or clays (e.g., Bure in France, Mont Terri in Switzerland, etc.). In all these cases, and whatever the depth, access to the geological site under study is direct, and drilling small diameter, several meters deep cores in the wall of the excavation is just needed to collect samples. Precautions to prevent contamination must obviously be taken, but the difficulties and constraints are incommensurable with deep wells mentioned above.

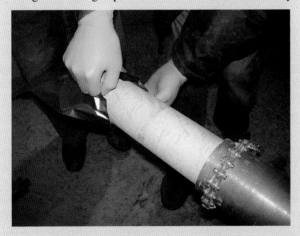

Box Fig. 17.1 Extraction of a core from a geological clay formation (Photograph: Laurent Urios)

(continued)

> **Box 17.1** (continued)
>
> The extension of research on the deep biosphere through these techniques will allow to describe the immense biodiversity of the subsurface of our planet and try to understand how bacterial communities have persisted in such extremely nutrient-limited environment, isolated from any contribution from the surface for tens of thousands, even hundreds of millions of years. The adventure is just beginning.

In all studies, it is important to adapt the size and the number of samples in order to consider the spatial heterogeneity existing in the soil (due both to the soil itself and the presence of vegetation cover), so that this heterogeneity does not hide the studied effect or get confused with it. To obtain a representative sample of the study site, it is better to collect several soil cores, group them, mix them, and sift them (with a 2 mm mesh sieve) to create a homogeneous sample before starting the analysis. To avoid the contamination between different sites, it is necessary between each site to clean the auger by ridding it of a maximum of soil and possibly to rinse it with ethanol (70 %) and then to dry it with a clean paper.

It is sometimes necessary to look the soil organization at a microscale, the scale of the soil aggregate. In this case, soil aggregates (for the desired size, often by a few millimeters) (Grundmann and Debouzie 2000) are sorted from the soil core, under a binocular loupe with sterile material before analysis. Some approaches are intended to dissociate the microorganisms located in or outside the aggregates (and therefore more or less exposed to mobile water with charged ions and soluble elements circulating in the ground). In this case, methods with moderate soil washing can be used to recover the microorganisms localized on the surface of aggregates (Ranjard et al. 1997), the microorganisms remaining with the aggregates being predominantly those located inside.

17.2.2 Sampling Techniques in Aquatic Environment

Although the development of automatic sensors now permits the in situ acquisition of many hydrological parameters related to the physiology of living organisms, water sampling still represents a primary and obligatory step in the collection of large amounts of data from the marine environment and especially for the collection of microorganisms. Collection can be considered the process of obtaining an aliquot of the studied aquatic environment. Sampling consists of retaining, preserving, and storing a portion of this collected water for analytical purposes. It should be noted that sampling is only relevant if the water collection is conducted correctly and is representative of the studied environment. In some cases, water collection and sampling are merged, especially when contamination absolutely has to be prevented. Collection and pretreatment of water samples and sediment in the field present certain difficulties in comparison to laboratory work. There are also difficulties associated with working on a vessel, where the working conditions are rarely comparable to those in laboratories. This chapter documents methods and useful equipment for collecting marine samples and processing them for various downstream applications, including analyses of nutrients and gases, biomass concentration, cell abundance and experimentation, both offshore rather than at the coast and in shallow lagoons.

17.2.2.1 Equipment for Water Collection

Collection of Surface Water by Hand

This procedure is only applicable to collection of surface water from the water's edge (e.g., at a beach or harbor deck) or from a small boat at a short distance from the coast. This type of sampling does not require specific equipment. A surface sample can be collected using a simple bucket, assuming that this simple device is compatible with the subsequent analyses. Nevertheless, if possible, it is better to collect water samples directly in the flasks that will be used to store the samples, to reduce contamination due to intermediate handling. In all cases, the hands of the sampler have to be protected by polyethylene gloves, and sampling is performed by immersing the bucket or bottle under the surface, as far as possible from the boat or deck. A "fishing rod" system can be helpful in collecting water at a distance from a boat or deck.

Sampling with Hydrological Bottles

This technique is used to collect water at depths below the surface as well as at the surface when hand sampling is impossible or inappropriate. First, the boat must be equipped with a winch system to unwind a cable on which hydrological bottles are attached. On small boats, hand winches are suitable for this purpose. In all cases, it is necessary to ensure that the chemical quality of the cable is suitable for the chemical analysis and experiments to be conducted on the samples. In addition, it is essential to estimate the depth of sampling accurately. One suitable approach is to equip the winch with a counting pulley; another is to mark some lengths along the cable. However, the best approach is to install a pressure sensor or gauge at the end of the cable.

Nansen The first hydrological bottle was developed in 1910 by the oceanographer Fridtjof. This bottle was entirely made of brass and remained the typical oceanographic instrument until 1966, when Shale Niskin designed a new bottle from a PVC

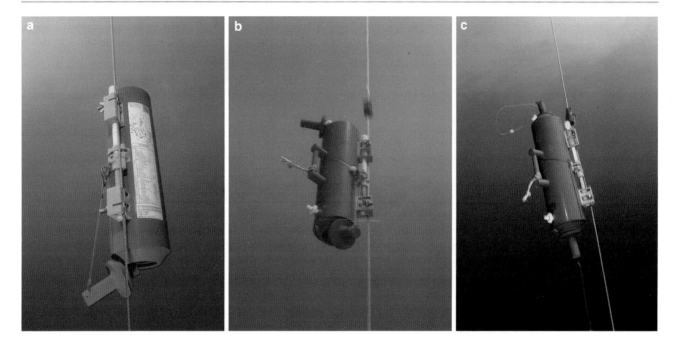

Fig. 17.1 A Niskin bottle at sea. (**a**) Bottle positioned at the sampling depth. Valves are open. (**b**) The "messenger" triggers the closure. (**c**) The valves are closed. The water contained in the cylinder is isolated from the outside. The bottle can then be brought on board (Photographs: Courtesy of Centre d'Océanologie de Marseille)

(polyvinyl chloride) material that was cheaper and easier to use. The Niskin bottle is currently the most commonly used device for all hydrology work. The simplest bottle is a PVC cylinder that can be hermetically sealed at both ends by removable valve systems and that have systems for attaching them to a cable (Fig. 17.1). Each hydrological bottle is equipped with an air inlet screw in the upper part and a draw valve at the bottom that permits sampling of the collected water. These two parts are always closed during water collection at depths below the surface. The bottle is opened before it is hooked on the cable and positioned under water at the required depth. Once this depth is reached, the bottle is closed with the help of a stainless steel block, called a messenger, launched along the cable from the surface (Fig. 17.2). Automatic closure can also be achieved with either a pressure sensor or by remote control using an electrical signal sent from the surface. Niskin bottles are available in various volumes from 1 to 100 l.

However, the closure mechanism (elastic, spring) is often located inside the bottle, i.e., in contact with the water sample. Some analyses require equipment made of materials that are even more "clean," i.e., totally inert and easily decontaminable, such as glass, polycarbonate, or Teflon-coated plastic. In addition, to avoid contamination, some bottles are equipped with Teflon-coated tensioners. In some extreme cases, it is preferable to use bottles with external springs to avoid contact with the sample water. Most of the time, a hydrological bottle is opened when it is attached to the cable. However, this technique may cause some contamination during the period of exposure to the air prior to immersion or during the passage through the film surface of seawater, which is known to be enriched with contaminants (Box 17.2). To avoid this type of problem, especially whenever uncontaminated samples need to be obtained, for instance, for chemical analysis of trace metals in seawater, the use of specific hydrological bottles that are equipped with external elastic for closure and that can be closed when immersed (Go-Flo bottle; Fig. 17.3) is recommended. The bottle is then opened only under the surface of the water (usually at a depth of approximately 10 m), just before the vertical cast. In this case, it is imperative that the cable is stainless steel or Kevlar.

Go-Flo bottles are tubular and are available in various sizes (5, 10, 30 l, and so on). The top and bottom of each tube are equipped with stopper balls, which must be rotated 90° to close them. The bottle goes into the water closed to avoid any surface-level contaminants or pollutants entering it. The bottle is lowered into the sea on a steel cable. As the bottle is lowered past the surface, the increased water pressure causes the pressure release valve to pop in and the balls at the top and the bottom of the bottle to roll open by 90°. As a result, the bottle is neither contaminated on deck nor as it is lowered into the water by the uppermost layer of seawater at the surface that is contaminated by interaction with the air.

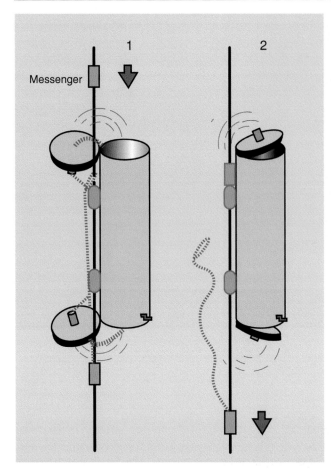

Fig. 17.2 How does a Niskin bottle function? The bottle is lowered to the selected depth with valves opened. A messenger made of steel is launched from the surface and glides along the cable. When this messenger hits the bottle support, it releases the two spring-loaded valves. Water is then trapped in the bottle and isolated from the water outside the bottle (Drawing: M.-J. Bodiou)

Box 17.2: Sampling the Surface Microlayer of Aquatic Ecosystems

Philippe Lebaron

The surface microlayer (SML) of aquatic ecosystems has been defined as the top 1–1,000 μm of the water surface, i.e., the interfacial region where many important bio-physicochemical processes and exchange of gases are taking place. SML plays an important role in photochemical and biologically mediated transformations of organic matter at sea surface. Depending on the type of transformations occurring in these layers, they may have important consequences in the transfer of pollutants to marine food webs.

Therefore, the SML have a crucial role in the marine environmental protection and global changes (Liss and Duce 1997). SML of natural waters have generally been considered to be enriched relative to underlying water, with various chemical and microbiological components. There are more than 20 published techniques to collect the SML, but very few are suitable for simultaneous sampling of chemical and biological parameters (Agogué et al. 2004) (Box Fig. 17.2).

Membranes

To collect bacteria living in the first 10–20 μm of the SML, hydrophobic membranes are very efficient since they can collect viable bacteria by electrostatic forces. The hydrophilic filters are much less efficient to collect these bacteria. Membranes are placed on the surface of the water using a clamp. The hydrophilic Teflon membrane is often used for the isolation of bacteria because the adsorption of bacteria on the surface is lower and it provides higher rates of recovery.

The Glass Plate

Glass plates used to collect a slightly thicker layer (approximately 200 μm) are often preferred for the analysis of organic pollutants and heavy metals. Before using the glass plate (~0.25 × 0.35 m and 4 mm thick), it is cleaned thoroughly. The plate is introduced vertically in the water and is then removed vertically from the water very slowly (0.1 m·s^{-1}) and wrung between two Teflon plates. The film adhering to the sides of the plate is then transferred to a clean glass bottle. Sampling is very long (about 1 l collected in 45 min).

The Rotating Drum

The SML collector consists of a rotating cylinder of glass or stainless steel roller coated with a very resistant ceramic material. A large Teflon blade is pressed tightly to the surface of the cylinder to remove continuously the film and adhering water (around 60–100 μm) which are collected in a clean glass bottle. During the collection operation, the apparatus is pushed ahead of the boat at low speed by the means of an electric motor. The speed has to be adapted to the currents, wind speed, etc. This equipment is suitable for relatively calm weather but cannot be used effectively in rough water.

(continued)

Box 17.2 (continued)

Box Fig. 17.2 Techniques for sampling the surface microlayer of aquatic ecosystems. (**a**) Membranes; (**b**) rotating drum; (**c**) glass plate; (**d**) metal screen (Photographs: Philippe Lebaron)

The Metal and Nylon Screens

A rectangular (more or less 0.6 m × 0.75 m) stainless steel (or aluminum when applicable) framed screen is lowered vertically through the water and then oriented horizontally and raised through the SML. An alternative consists in lowering merely the sampler to touch the surface. This second procedure may be of great interest when the concentration of particles is important in surface waters. In all cases, efficiency is dependent on the mesh size and open space, and it explains why different thicknesses are reported in the literature. This technique is nonselective because it does not depend upon adsorption. Metal can be replaced by nylon when working on the chemistry of the SML.

Sampling Strategies

The time required to collect the water from the SML varies a lot depending on the type of sampler. Therefore, if we assume that the SML is patchy, important variations may exist at both spatial and temporal scales. This is probably the most difficult challenge in analyzing this biotope. One way to integrate this natural variability is to collect the SML during a few hours at the same station and to homogenize the sample. The glass plate technique generally requires the largest time of sampling (a few hours to collect a few liters). However, because all techniques did not require the same time to collect a given volume, it is important to sample at different intervals of time in order to allow further comparisons. The time intervals should be defined at regular intervals during the largest time of sampling (generally the glass plate).

Once the required depth has been reached, a "messenger" weight is slid along the cable to the bottle and drops onto the closing mechanism. The bottle, now hermetically sealed, is then brought back up to the surface.

Originally, to obtain multiple water samples during the same vertical cast, several hydrological bottles had to be suspended one after the other on the cable at certain distances from each other. A messenger was attached to each bottle (except the deepest one) before lowering. Once all the bottles were on the line, the line was lowered to the desired depth and 2 min were allowed to pass. A messenger was sent down and a few minutes were allowed to pass to ensure that all the bottles closed. The line was reeled in and the bottles were unloaded as they came up. Now, water samples are collected more rapidly with the help of a "rosette" system on which a large number of hydrological bottles (12–36) are attached to a circular steel frame (Fig. 17.4). The rosette of hydrological bottles is typically

Fig. 17.3 Go-Flo on the line ready to be dropped into the water. The Go-Flo has the advantage of passing through the sea surface microlayer closed, and thereby avoids its inner surface being coating with a surface film rich in contaminants (Photograph: Patrick Raimbault)

connected to sensors attached to the center of the frame that provide real-time information on the hydrologic characteristics of the water column (e.g., temperature, salinity, oxygen, pH, fluorescence, and optical properties). This information is typically used to select the water collection depths precisely. Closure of the bottles is activated by the operator using an electrical impulse, usually as the rosette is being lifted. Water samples are obtained during the upcast, and the closure of the bottles is tripped while the package is stopped or still moving slowly.

The collected water is then brought on board very quickly and can be immediately used for chemical analysis or for experiments with microorganisms (Fig. 17.5). Nevertheless, the rapid rise from deep waters to the surface may cause large variations of some physical parameters such as pressure. Thus, for physiological studies of microorganisms that are adapted to grow under very high pressures (such as bacteria), microbiologists have developed bottles able to maintain the hydrostatic pressure during water collection and during sampling on board (*cf.* Sect. 17.3).

It is a great challenge to sample seawater across interfaces, such as the halocline or the redoxcline, to investigate trace metal distribution. With the use of 5- to 10-l sampling bottles mounted on a wire or a CTD rosette, it is possible to obtain a maximum vertical resolution of 5 m. For the detection of small vertical structures in the vertical distribution of trace metals across the redoxcline, a CTD-Bottle-Rosette is not sufficient. Therefore, a PUMP–CTD System has been developed that permits water sampling at high resolution (1 m maximum) along a vertical profile.

 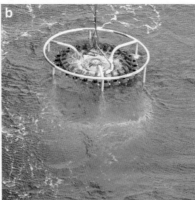 

Fig. 17.4 Sampling rosette with 24 Niskin bottles. (**a**, **b**) Niskin bottles equipped with draw pipes, ready for sampling. (**c**) Rosette on the deck (Photographs: Patrick Raimbault)

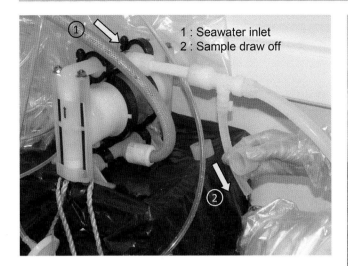

Fig. 17.5 Teflon pump used for "ultra-clean" sampling (Photograph: Courtesy of Sophie Bonnet)

Sampling with a Pump

Some analyses or experiments dealing with trace elements (e.g., some metals or organic compounds) require water collection and sampling without any contact with the air. In these cases, a pumping system is better than hydrological bottles (Fig. 17.6). The upstream end of the pump is located below the surface while the downstream end is located in a clean room to which seawater is brought by pumping. Thus, the water sample is never in contact with the atmospheric air of the ship. Large quantities of water can be collected in a very short time (10 l in a few minutes), depending on the pump flow and the pressure drop along the pipes. In all cases, the objective is to not contaminate the water sample during the collection step at sea or the sampling operation on board. Thus, the whole system (pump and pipes) is entirely Teflon coated and the environment is protected from external contamination. This type of sampling requires more equipment and heavier installation onboard but allows the collection of large volumes of water and, above all, better protection against contamination. Unfortunately, the maximum depth of sampling reached is clearly dependent on the depression that can be created on the surface and cannot be deeper than 140 m. For deeper clean samples, only the use of Go-Flo bottles attached to a cable of inert material is possible.

17.2.2.2 How to Collect Particulate Matter?

To collect and often to concentrate suspended matter from seawater, the most commonly used technique is filtration, which permits the dissolved component to be separated easily from the particulate fraction. The choice of filtration membrane or filter depends on many criteria, such as the size of the particles to be collected, the volume to be filtered, and the analytical methods that will subsequently be used. Glass fiber filters, which have high filtration capacities, are commonly

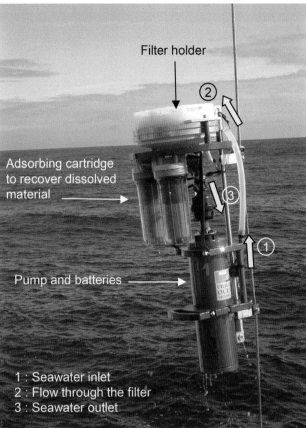

Fig. 17.6 In situ pump on a hydrological cable (Photograph: Patrick Raimbault)

used for analysis of many parameters in the marine environment (e.g., phytoplanktonic pigments and the chemical composition of suspended matter). However, because their nominal porosity is on the order of 0.7 μm, microbiologists often prefer to use membranes that have better retention efficiency, with porosities on the order of 0.2 μm. These membranes are made of many materials (e.g., cellulose, cellulose esters, polycarbonate, Teflon, and metal). The choice of membrane filters is therefore very varied and can satisfy the majority of the purposes of analyses and experiments concerning particulate matter and organisms collected by filtration.

In the ocean, particles are primarily of biological origin (e.g., phytoplankton, zooplankton, bacteria, and fecal pellets) and a small fraction is of nonbiological origin (e.g., dust, aerosols, suspended sediment, and continental inputs). Biological activity is concentrated in the surface layer, in approximately the first 100 m, where the light intensity is sufficient to permit algae to achieve photosynthesis. Below this photic zone, the particulate matter content decreases exponentially, and as a result, the water volumes needed to collect sufficient quantities of suspended particles for experimentation increase considerably. Hydrological bottles typically have volumes of some liters that are often insufficient to be able to measure the chemical composition of deep particles. Only in situ pumps (ISP) are able to collect large amounts of particulate matter at depths below the water surface (Fig. 17.6). These autonomous pumps, powered by batteries, are suspended along a cable. Once at the selected depth, the pump is started and filters several hundred to more than 1,000 l of water in a few hours. On the same cable, between 5 and 12 pumps can be deployed simultaneously at different depths. In situ pumps can be submerged into ocean water for up to several hours at a time. All of the water that the pump pulls in is passed through filters, which catch the very fine particles. Several filters can be stacked to achieve a size-fractionated filtration. When the pumps are brought up to the surface after several hours, the material captured on the filters can be analyzed.

17.2.2.3 How to Collect Living Organisms in the Water Column?

The simplest and most commonly used device for collecting living organisms (plankton) floating and drifting in the water column is the plankton net. The amount of water filtered is large and the gear is suitable for both qualitative and quantitative studies. The plankton nets are of various sizes and types. The sampling success will largely depend on the selection of suitable gear, the mesh size of the netting material, the time of collection, the water depth of the study area, and the sampling strategy. With minor variations, the plankton net is conical in shape and consists of a ring (either rigid or flexible and either round or square), a filtering cone made of nylon, and a collecting bucket for collection of organisms (Fig. 17.7). The collecting bucket can be detached and easily transported to a laboratory. The different nets can be broadly placed two categories: open nets used mainly for horizontal and oblique hauls and closed nets with messengers for collecting vertical samples from desired depths. Horizontal collections are mostly carried out at the surface and subsurface layers. In oblique hauls, the net is usually towed above the bottom. The disadvantage of this method is that the sampling depth may not be accurately known. A vertical haul is made to sample the water column. The net is lowered to the desired depth and hauled slowly upward. The zooplankton sample collected is from the water column traversed by the net.

The standard zooplankton net has a mesh size of 250 µm, a 25-cm frame diameter and 50-cm long net. A 250-µm net will retain small animals such as crustaceans while allowing most algae and protozoa to pass through. For unicellular microorganisms (mainly phytoplankton), it is necessary to use a very fine mesh, on the order of 64 µm. For quantitative plankton sampling, it is imperative to know the actual amount of water that passed through the net. For this purpose, an instrument called a flow meter is installed at the mouth of the net.

Fig. 17.7 A zooplankton triple net is brought up on board (Photograph: Courtesy of Nicole Garcia)

17.2.2.4 Collection of Particles

A significant fraction of organic matter produced in the oceanic surface layer has negative buoyancy and will tend to settle to the bottom. This sedimentation exports a large amount of organic matter and chemical elements that constitute an uninterrupted flow often called "marine snow." It is a significant means of exporting energy from the light-rich photic zone to the aphotic zone below. Although most organic components of marine snow are consumed by microorganisms in the first 1,000 m of sedimentation, a fraction reaches the ocean floor and sustains the development of the deep benthic ecosystem. The small percentage of material not consumed in shallower waters becomes incorporated into the ocean floor, where it is further decomposed through biological activity.

The quantification of this vertical flux of matter and microorganisms, including dead or dying animals and plants (plankton), fecal matter, sand, soot, and other inorganic dusts, may be performed with specific devices called sediment traps. A sediment trap normally consists of an upward-facing funnel that directs sinking particulate materials toward a mechanism for collection and preservation. There are many types of sediment traps, ranging from detachable cylindrical sampling tubes mounted with lead weights at the bottom to sophisticated systems for sequential collection over long periods of time, both in the water column and at the bottom of the ocean (Fig. 17.8). Typically, traps operate over an extended period of time (weeks to months), and their collection mechanisms may consist of a series of sampling flasks that are cycled through to allow the trap to record the changes in sinking flux with time. A trap is often moored at a specific depth in the water column (usually below the euphotic zone or mixed layer) in a particular location. However, in some cases, a floating trap, also called a Lagrangian trap, which drifts with the surrounding ocean current, can be used. In any case, the

 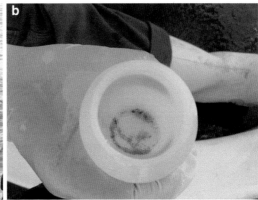

Fig. 17.8 Launching a model PPS5 sequential sediment trap, consisting of a huge cone with an opening of a square meter (**a**) and a revolving wheel with sampling flasks in which particles are collected during sequential periods (**b**) (Photographs: Courtesy of Nicole Garcia)

construction of the sediment trap array ensures a permanently vertical position of the sampling tubes during deployment.

17.2.2.5 Sediment Sampling

The main problems posed by the study of marine sediment are associated with its heterogeneity, both horizontal and vertical. In addition, for many workers in the field of oceanography, it is necessary to recover unmixed continuous sediment samples, including sediment/water interfaces, that represent sediment fractions and surface deposits that are as little disturbed as possible. For large samples, bins or grabs that can recover up to 1 m^3 of sediment are frequently used to study meio- and macrobenthic communities, for instance. A grab is used to obtain sediment samples from the seafloor. The grab is lowered to the seabed on a steel cable with its "jaws" open. As soon as the jaws touch the bottom, the valve that holds them open is released. As the grab is pulled back up, the jaws close, scooping up sand and sediment from the seabed. However, the use of bins or grabs is not recommended for work on physiology at the water–sediment interface, as these types of equipment permanently mix and disturb sedimentary layers. To avoid mixing and disturbing the materials sampled, it is preferable to use a coring technique.

In areas directly accessible by foot (e.g., in a low-tide zone) or by diving, sampling of small volumes can easily be performed "by hand" using simple tools such as hand corer sediment samplers. In shallow waters, a sampler can be pushed into sediment using the handles on the head. If the water depth permits, extension handles of 4.5 m or 6 m can be used to sample from boats or docks. In deeper water, the sampler can be dropped by attaching a line to the clevis located on the head assembly between the handles. A simple valve allows water to flow through the sampler during descent and close tightly upon removal, minimizing sample loss. A corer takes a 50-mm-diameter sample and measures 50 cm long. A box corer is a marine geological sampling tool for use in soft sediments in lakes or oceans. It is deployed from a research vessel with a deep sea wire and is suitable for sampling at any water depth. It is designed to minimize disturbance of the sediment surface by bow wave effects, which is important for quantitative investigations of the benthos micro- to macrofauna, geochemical processes, sampling of bottom water, and sedimentology.

To obtain more accurate and less disturbed samples, it is preferable to use coring "tubes" recommended primarily for sampling the water–sediment interface for biological and physiological studies. Coring tubes can obtain long cores but have small cross sections. A coring tube is a long tube, typically made of metal, with a sealing system at its base. The coring tube is driven into the sediment to collect a truly undisturbed sediment sample from the seabed, including the sediment–water interface and overlying supernatant water. Many models have been developed that differ primarily in their modes of penetration (e.g., gravity, sinking, and vibration), shutter modes, and base types (e.g., diaphragms and pneumatic valves).

Recent developments in coring technology have enabled the use of an assembly of multiple corers for collecting several cores simultaneously in a small area (Fig. 17.9). The Multicorer is designed to obtain multiple samples of sediment from the seabed at great depths. The Multicorer is used for sampling in chemical, geochemical, and biological applications.

The Multicorer is used to obtain undisturbed sediment samples from the surface of the seabed. It consists of a system to which a series of tubes, measuring approximately 4 cm in diameter, are attached. Above the system, a weight is mounted, and this falls onto the assembly system when the Multicorer touches the sediment. The falling weight drives the tubes into the seabed so that when they are raised again, each of them contains a drilling core with sediment from the seafloor.

17.2.3 Deep-Seawater Sampling

While the dark ocean, below 200 m depth, represents the largest habitat, the largest reservoir of organic carbon, and the largest pool of microbes in the biosphere (Whitman et al. 1998), this

Fig. 17.9 Octopus-type multitube corer is brought up on board (a). This is a sampler that permits simultaneous collection of 8 small sediment cores (less than 1 m in length) to study biogeochemical processes at the water-sediment interface (b) (Photographs: Courtesy of Nicole Garcia)

realm has been much less studied than euphotic ocean certainly because of difficulty in sampling (particularly maintaining high pressure conditions) and the time and expenses involved.

17.2.3.1 From Adapted-to-Pressure Microorganisms Discovery to Physiological and Molecular Adaptation Face to High-Pressure Condition

The Challenger Expedition (1873–1876) is commonly credited as the historical beginning of deep-sea biology. The finding of live specimens at great depths obliterated the azoic-zone theory, below 600-m depth, that had been suggested in the 1840s by Edward Forbes. A very good summary of the controversial concept of the azoic zone is provided in Jannasch and Taylor (1984). In 1884, Certes (1884) examined sediment and water collected from depths to 5,000 m and cultured bacteria from almost every sample. In 1904, Portier used a sealed and autoclaved glass-tube device as a bacteriological sampler and reported counts of colonies from various depths and location (Jannasch and Wirsen 1984). Research into the effects of high pressure on the physiology of deep-sea bacteria was developed during the last century. A synthesis of that work can be found in ZoBell (1970), pioneer biologist in the study of the effects of hydrostatic pressure on microbial activities (ZoBell and Johnson 1949). ZoBell and Johnson (1949) began studies of the effect of hydrostatic pressure on microbial activity using pure cultures. "Barophilic" was the first term used to define optimal growth at a pressure higher than 0.1 MPa, or for a requirement for increased pressure for growth (ZoBell and Johnson 1949), but was subsequently replaced by Yayanos (1995), who suggested "piezophilic" (from the Greek "piezo," meaning pressure). Current terminology (reviewed by Fang et al. 2010 and Kato 2011) defines pressure-adapted microorganisms either as piezotolerant (similar growth rate at atmospheric pressure and high pressure), piezophilic (more rapid growth at high pressure than atmospheric pressure), or hyperpiezophilic (growth only at high pressure), with pressure maxima increasing in rank order (highest for hyperpiezophiles). Organisms that grow best at atmospheric pressure, with little to no growth at increased pressure, are termed piezosensitive.

Pressure-adapted microorganisms have been isolated from many deep-sea sites by researchers around the world. Isolates include representatives of the Archaea (both Euryarchaea and Crenarchaea kingdoms) mainly from deep-sea hydrothermal vents, and Bacteria from cold, deep-sea habitats. Most of the bacterial piezophiles have been identified as belonging to the genera *Carnobacterium*, *Colwellia*, *Desulfovibrio*, *Marinitoga*, *Moritella*, *Photobacterium*, *Psychromonas*, and *Shewanella* (reviewed by Bartlett et al. 2007). The membrane properties of piezophiles have been described and other characteristics of piezophiles, including motility, nutrient transport, and DNA replication and translation under elevated hydrostatic pressure, have been explored (Lauro et al. 2008). Protein structural adaptation to high pressure has also been

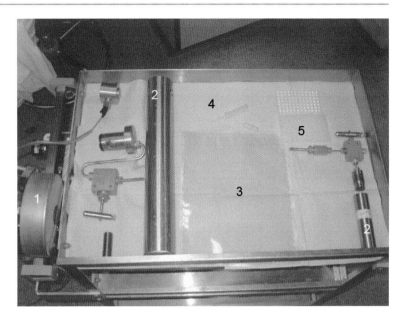

Fig. 17.10 Hyperbaric apparatus classically used to perform high-pressure cultivation. (*1*) Hand-operated high-pressure pump; (*2*) high-pressure vessels. Various systems used to hold culture incubated within high-pressure vessel: (*3*) sterile bags, (*4*) transfer pipettes, (*5*) multi-well plates (Photograph: Courtesy of Douglas H. Bartlett)

described in comparative studies of piezophilic and piezosensitive microorganisms (Kato et al. 2008). Recent studies have also highlighted that hydrostatic pressure influences not only the unsaturation ratio of membrane fatty acids but also the unsaturation ratio of intracellular wax esters (storage lipids) which accumulated in the cells of a piezotolerant hydrocarbonoclastic bacterium under the form of individual lipid bodies (Grossi et al. 2010). Finally, piezophilic bioluminescent bacteria shall produce more light under high pressure than at atmospheric pressure conditions conferring them an ecological advantage (Martini et al. 2013).

To obtain such results, experiments were performed using diverse "classical" high-pressure systems. Figure 17.10 presents a picture of apparatus presently used, for example, in the Douglas H. Bartlett Lab (Scripps Institution of Oceanography, USA) to perform high-pressure cultures and study pressure effect on the physiological and metabolisms of microbial strains.

17.2.3.2 Microbial Activities Measured Under In Situ Pressure Conditions

Although the deep ocean supports a diversity of prokaryotes with functional attributes interpreted as adaptation to a pressurized environment (Lauro and Bartlett 2007; Nagata et al. 2010), the contribution of the natural microbial assemblages to the carbon cycle of the biosphere remains poorly understood. Recent reviews (Arístegui et al. 2009; Nagata et al. 2010; Robinson et al. 2010) strongly suggest reconsidering the role of microorganisms in mineralizing organic matter in the deep pelagic ocean. However, majority of estimates of prokaryotic activities have been made after decompression and at atmospheric pressure conditions. However, (i) results from experiments mimicking pressure changes experienced by particle-associated prokaryotes during their descent through the water column show that rates of degradation of organic matter (OM) by surface-originating microorganisms decrease with sinking and (ii) analysis of a large data set shows that, under hydrologic-stratified conditions, deep-sea pelagic communities are adapted to in situ conditions of high pressure, low temperature, and low OM (Tamburini et al. 2013). Measurements made using decompressed samples and atmospheric pressure thus underestimate in situ activity (Tamburini et al. 2013). To obtain such results, specific pressure-retaining samplers have been developed, which are deployed using research vessels, using manned-submersible (e.g., Shinkai 6500, JAMSTEC, Japan; Nautile, IFREMER, France; DSV Alvin, WHOI, USA) or remotely operated vehicle (ROVs).

Initial estimates of deep-sea microbial activity under elevated pressure were based on the unintentional experiment involving the "sandwich in the lunchbox" from the sunken research submarine Alvin, "incubated" in situ more than 10 months at 1,540 m depth in the Atlantic Ocean (Jannasch et al. 1971). According to Jannasch et al. (1971), the crew's lunch was recovered and "from general appearance, taste, smell, consistency, and preliminary biological and biochemical assays, [...] was strikingly well preserved." Based on subsequent studies carried out employing in situ conditions of high pressure and low temperature, the Jannasch team concluded that deep-sea microorganisms were relatively inactive under in situ pressure and not adapted to high pressure and low temperature. However, Jannasch and Taylor (1984) offered the caveat that the type of substrate influenced the results and concluded, from laboratory experiments, that "barophilic growth characteristics have been unequivocally demonstrated." These early observations of deep-sea microbial activity were accompanied by the development of pressure-retaining water samplers, with the conclusion from results of experiments employing these samplers that "elevated pressure decreases rates of growth and metabolism of natural microbial populations collected from surface waters as well as from the deep sea" (Jannasch and Wirsen 1973). Contrary to this early conclusion, virtually all subsequently collected data from the water column under in situ conditions have shown that the situation is the reverse, namely, those microorganisms autochthonous to depth are adapted to both the high pressure and low temperature of their environment (Tamburini et al. 2013).

17.2.3.3 High-Pressure Retaining Deep-Seawater Samplers

A limited number of high-pressure vessels have been constructed during the past 50 years to measure microbial activity in the cold deep ocean and evaluate the effects of hydrostatic pressure, as well as decompression, on deep-sea microbial activity. Sterilizable pressure-retaining samplers for retrieving and subsampling undecompressed deep-seawater samples have been developed independently by three laboratories including Jannasch/Wirsen at the Woods Hole Oceanographic Institution (USA), Colwell/Tabor/Deming at the University of Maryland (USA), and Bianchi at the Aix-Marseille University (Marseille, France) (Jannasch and Wirsen 1973; Jannasch et al. 1973; Tabor and Colwell 1976; Jannasch and Wirsen 1977; Deming et al. 1980; Bianchi and Garcin 1993; Bianchi et al. 1999; Tholosan et al. 1999; Tamburini et al. 2003). Extensive sampling equipment for cold deep-sea high-pressure work has also been developed by Horikoshi and his team (Jamstec, Japan) exclusively devoted to recovering new piezophilic microorganisms and to study the effect of pressure on those isolates, as described in the Extremophiles Handbook (Horikoshi 2011). At least two other groups are developing pressure-retaining samplers, the Royal Netherlands Institute for Sea Research (NIOZ) and the National University of Ireland (Galway), but the designs or initial results have not yet been published.

Relatively few interactions occurred between these laboratories; then only the high-pressure serial sampler (HPSS), always used today, is briefly presented in Fig. 17.11. More details can be found in Bianchi et al. (1999) and in Tamburini et al. (2003). The HPSS is based on a commercially available multi-sampling device that includes a CTD (Sea-Bird Carousel) and equipped with 500-ml high-pressure bottles (HPBs) fitted on polypropylene boards, totally adaptable to Sea-Bird Rosette 12 or 24 bottles (Fig. 17.11a). The HPSS allows collection of several ambient-pressure water samples at different depths during the same hydrocast down to a depth of 3,500 m.

HPBs are 500-ml APX4 stainless steel cylinders (75-mm OD, 58-mm ID, and 505-mm total length) with a 4-mm thick polyetheretherketone (PEEK®) coating. The PEEK® floating piston (56-mm total length) is fitted with two O-rings. The screw-top endcap is covered with a sheet of PEEK to avoid contact between the sample and the stainless steel. Viton® O-rings are used to ensure that the system is pressure-tight; Viton® O-rings are chosen instead of nitrile O-rings to eliminate possible carbon contamination of the sample. The screwed bottom endcap is connected, via a stainless steel tube, to the piloted pressure generator. HPBs are autoclaved to ensure sterility of sampling. Functioning of the HPBs is depicted in Fig. 17.11b; when the filling valve is opened at depth by the operator via the deck unit connected to the electric wire, the natural hydrostatic pressure moves the floating piston downward and seawater enters the upper chambers of two HPBs. The distilled water flushes from the lower chambers to the exhaust tanks through a nozzle that acts as a hydraulic brake avoiding decompression during filling. Samples are collected into two identical high-pressure bottles filled at the same time through the same two-way valve (Fig. 17.11b). One of these bottles retains the in situ pressure, while the other which lacks a check valve is progressively decompressed when the sampler is climbed back up to the surface. The decompressed bottle, like a classical retrieving bottle (e.g., Niskin bottle), is used in comparative fashion to estimate the effect of decompression on measurements of deep-sea microbial activity. Decompressed and high-pressure (ambient) samples are treated in exactly the same way. High-pressure sampling, subsampling, and transfer are performed using hydrostatic pressure (instead of gas pressure) thanks to a piloted pressure generator (described in Tamburini et al. 2009) in order to diminish risk during manipulation. In contrast to classical high-pressure pumps based on alternating movement of a small piston associated with inlet and exhaust valves, the piloted pressure generator is based on a step motor-driven syringe. A step motor with a high starting torque is connected to a high-precision worm screw operating the high-pressure syringe. Adjustment of the movement of the syringe is controlled by a digital computer fitted with a high-power processor. Subsampling is performed by the use of counterpressure (a maximum of 0.05 MPa) from the bottom of the HP bottle; opening the sampling valve allows the counterpressure to move the floating piston upward. The primary sample within the HPB is maintained under in situ pressure while the secondary sample (decompressed) is analyzed directly or else fixed (by adding a preservative solution like formaldehyde), stored, and analyzed later.

To evaluate the state of the field of piezomicrobiology, Tamburini et al. (2013) have compiled data from published studies of deep samples where prokaryotic activities were measured under conditions of in situ pressure and the results compared with those obtained using incubation at atmospheric pressure after decompression. The pressure effect (Pe) has been calculated from this compiled data set ($n = 52$ pairs of samples maintained under in situ pressure condition versus decompressed and incubated at atmospheric pressure condition). Pe is defined as the ratio between activity obtained under HP and that obtained under DEC conditions (Pe = HP/DEC), where a ratio >1 indicates piezophily (adaptation to high pressure) and a ratio <1, piezosensitivity. Calculation of Pe values has proven to be a useful diagnostic tool for evaluating the effect of decompression

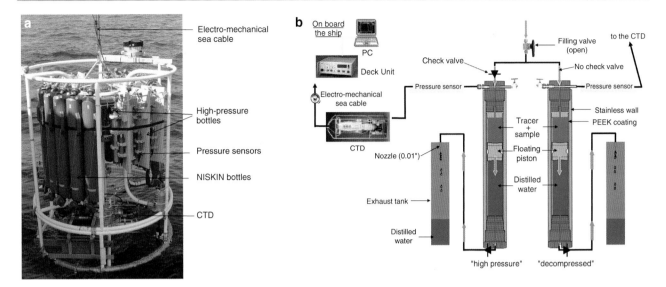

Fig. 17.11 High-pressure serial sampler. (**a**) Photograph of the high-pressure serial sampler (HPSS) on board the R/V Urania (Italy) during the CIESM-Sub cruise in the Tyrrhenian Sea. Six high-pressure bottles (HPBs) were mounted with 12 Niskin bottles on a Sea-Bird Carousel equipped with a CTD (Conductivity – Temperature – Density). (**b**) Diagrammatic representation of the filling of HPBs. For more details see Bianchi et al. (1999) and Tamburini et al. (2003) and see http://www.com.univ-mrs.fr/~tamburini (Photograph: Christian Tamburini)

on metabolic rate in deep-sea samples. A Pe ratio >1 indicates the deep-sea prokaryotic assemblage is adapted to predominantly the in situ pressure, and prokaryotic activity will be underestimated if the sample is decompressed and incubated at atmospheric pressure. On the other hand, if the Pe < 1, inhibition by high pressure is indicated and metabolic activity will be overestimated if the sample is decompressed (Tamburini et al. 2013). From deep-seawater collected during stratified conditions ($n = 120$), the mean Pe ($n = 120$) was 4.01 (median 2.11), with 50 % of values between 1.50 and 2.82 and 90 % between 1.12 and 8.17. During stratified conditions, the prokaryotic assemblage was adapted to high pressure (Wilcoxon rank test Pe >1, $p < 2.2 \times 10^{-16}$) and the metabolic rate has to be determined under in situ pressure conditions to avoid underestimating activity (Tamburini et al. 2013).

Hyperbaric System to Simulate Particles Sinking Throughout the Water Column

Biogenic aggregates (>500 μm in diameter), including marine snow and fast-sinking fecal pellets of large migrating macrozooplankton, constitute the majority of vertical particle flux to the deep ocean (Fowler and Knauer 1986; Bochdansky et al. 2010). Enzymatic dissolution and mineralization of particulate organic matter (POM) by attached prokaryotes during descent can provide important carbon sources for free-living prokaryotes, thereby playing important biogeochemical roles in mesopelagic and bathypelagic carbon cycling (Cho and Azam 1988; Smith et al. 1992; Turley and Mackie 1994, 1995). Attached prokaryotes, however, tend to comprise a small fraction (5 %) of the total prokaryotic biomass (Cho and Azam 1988), reaching somewhat higher proportions (10–34 %) only when the concentration of aggregates is high (Turley and Mackie 1995). The extent to which sinking particles contribute to microbial community structure in the deep sea remains an open question.

To better understand the metabolic capacity of prokaryotes of shallow-water origin, which are carried below the euphotic zone on sinking particles, to degrade organic matter in the deep sea, different approaches proved informative. For instance, Turley (1993) applied increasing pressure to collections of sinking particles, obtained by trapping for 48 h at 200 m depth and containing microbial assemblages. These samples were placed in sealed bags incubated in pressure vessels at 5 °C. Pressures of 0.1, 10, 20, 30, and 43 MPa were applied in step function (within 30 min, then maintained constant for 4 h) to simulate pressure at the deep water–sediment interface. In the early work on heterotrophic microbial activity associated with particulate matter in the deep sea, comparative responses to moderate (surface water) versus extreme (abyssal) temperatures and pressures were used to diagnose prokaryotic origin (Deming 1985). Samples of sinking particulates, fecal pellets, and deposited sediments were collected in bottom-moored sediment traps and boxcores at station depths of 1,850, 4,120, and 4,715 m in the North Atlantic and incubated for 2–7 days under both surface water and simulated deep-sea conditions (the latter in sterile syringes in pressure vessels at 3 °C).

To simulate more accurately the increase in pressure (and decrease in temperature) prokaryotes associated with particles experience in sinking to depth, Tamburini et al. (2009) created a PArticulate Sinking Simulator (PASS) system (Fig. 17.12). Same high-pressure bottles (HPBs), developed for the HPSS, were used to incubate samples while pressure was increased continuously (linearly) by means of a piloted pressure generator. The HPBs were rotated

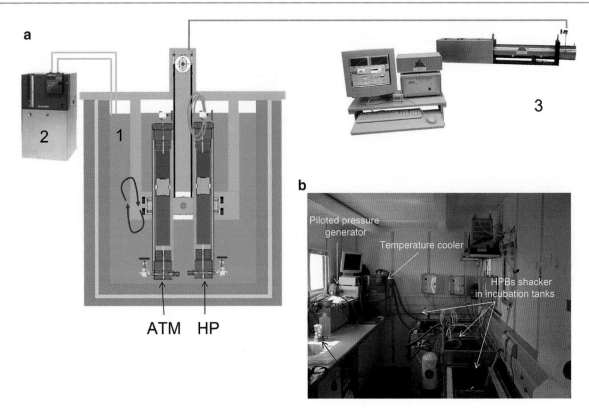

Fig. 17.12 PArticles Sinking Simulator (PASS). (**a**) Diagrammatic representation of the PASS composed of (*1*) several water baths where high-pressure bottles are rotated to maintain particles in suspension, (*2*) a cooler to control the temperature of water baths, and (*3*) a programmable computer-driven: piloted pressure generator. (**b**) Photograph of the PASS on board the R/V Endeavor (USA) during the MedFlux cruise where 3 HPBs were maintained at atmospheric pressure (ATM), while 3 others were connected to the piloted pressure generator, which continuously increased the hydrostatic pressure to simulate a fall through the water column (HP). See details in Tamburini et al. (2009) (Photographs – **a3**: Metro-Mesures Sarl, France; **b**: Christian Tamburini)

(semi-revolution) to maintain particles in suspension during incubation in water baths reproducing temperature changes with depth. The PASS system can be used in the laboratory or at sea, depending on samples being analyzed and objectives of the study. Tamburini et al. (2006, 2009) focused on prokaryotic processes and particle degradation in the mesopelagic zone, at the time just after particles exit the euphotic zone and before they arrive on the deep seafloor, employing a realistic settling velocity.

In summary, the effect of pressure on surface-derived bacteria attached to sinking organic matter is that their contribution to decomposition and dissolution of organic matter decreases with depth. This reinforces the conclusion that rapidly settling particles are less degraded during passage through the mesopelagic water column and, therefore, this phenomenon results in a labile food supply for bathypelagic and epibenthic communities (Honjo et al. 1982; Turley 1993; Wakeham and Lee 1993; Goutx et al. 2007). It also fits the results of in situ experimentation (Witte et al. 2003) and the calculation of recently proposed models (Rowe and Deming 2011) that show effective competition between metazoa and microorganisms for resources reaching the deep seafloor from the sea surface.

17.2.3.4 Concluding Remarks

Microbial communities found in the deep ocean comprise those microorganisms autochthonous to the deep sea and adapted by some degree to in situ temperature and pressure of the deep-sea environment and allochthonous microorganisms transported from the sea surface via sedimenting particles, deep migrating zooplankton, or other mechanisms.

Microbial metabolic rates are best measured under in situ conditions, which in the case of deep-sea microbial populations include high pressure, low temperature, and appropriate concentration (usually low) of ambient nutrient. Metabolic activity of an allochthonous community decreases with depth, limiting its capacity to degrade organic matter sinking through the water column (Turley 1993; Turley et al. 1995; Tamburini et al. 2006). Such microbial communities may be inactive (not dead) under conditions of low temperature and elevated pressure of the deep sea, but they can become dominant, i.e., more numerous and metabolically active when incubated under atmospheric pressure. Thus, community activity measured at atmospheric versus deep-sea pressure can reflect an entirely different mixture of community components.

17.3 Cytometry Techniques

17.3.1 Historical

Cytometry include a set of techniques to characterize and to measure the physical properties of individual cells and cellular components. The best known of these techniques and the oldest is the microscope. It appeared in the seventeenth century and quickly gave birth to an extraordinary period of discoveries in biology, in particular under the leadership of two Dutch, Antonie van Leeuwenhoek (1632–1723) and Jan Swammerdam (1637–1680), and one Italian, Marcello Malpighi (1628–1694).

Progress in the understanding of the fundamental laws of optics and in the construction of instruments in the fields of astronomy and navigation then led to a considerable evolution of microscopes in the eighteenth century. The increasing interest in biology and the discoveries done in the field of medicine in the nineteenth century have stimulated the need for observation and for the development of more effective instruments. The microscope allowed Pasteur to undermine the theory of spontaneous generation and Koch to discover several pathogens. In the twentieth century, advances in electronics, informatics, and optics have enabled the design of new generations of photonic and electronic microscopes with previously unmatched performance. Meanwhile, progress in the field of chemistry and methods for fluorescence labeling of cellular constituents have allowed the development of new applications of cytometry techniques in biology.

The more and more important need to automate the counting of cells, especially in the medical field for the analysis of blood cells, has led to the emergence of a new technique based on flow cytometry analysis of cells in liquid vein: flow cytometry. Designed by Moldavan in the early twentieth century, it was mainly developed in the 1970s by researchers at Los Alamos and Stanford in the USA who linked methods of measuring individual volume or fluorescence of cells driven by a stream with methods for electrostatic cell sorting in vital conditions.

This technique helped to make extraordinary discoveries in the field of biological oceanography by observing phytoplankton cells on the basis of their natural pigments. Thus, Sallie Chisholm and Robert Olson teams made the discovery of two new cyanobacteria in the years 1979–1986, *Synechococcus* and *Prochlorococcus* genera, and were discovered by this technique, and their quantification allowed to reveal the essential role played by these cyanobacteria in the functioning of the global ocean and CO_2 fixation.

Flow cytometry was quickly used in biological oceanography by specialists of the phytoplankton under the leadership of Daniel Vaulot in Roscoff (Marine Biological Station, France) who participated in the work carried out in the USA. Later with the development of fluorescent cellular markers that could be applied to non-photosynthetic organisms, this technique has been developed and used in microbial ecology, and in the 1990s, it has become a routine technique in many laboratories for the enumeration and characterization of the physiological state of bacterial cells. All these instrumental developments combined with advances in parallel in the field of molecular biology have revolutionized microbiology, in general, and particularly environmental microbiology.

17.3.2 The Different Cytometry Techniques

This chapter does not aim to describe the different microscopy techniques since there are books and websites dedicated to these techniques. Some of them are listed below to which the reader should refer for further analysis and more technical details on instruments. Our goal is to provide the reader with a list of techniques that are available today to present their principles and main applications in the field of microbial ecology.

17.3.2.1 Microscopy Techniques
Optical Microscopes
This is the oldest technology used by Renaissance scientists, but they are still present in laboratories. Light microscopes can be bright-field, dark-field, polarization, fluorescence, phase contrast, interference, and confocal scanning laser.

Direct-Light Microscopes
Visible light is passed through the specimen and then through glass lenses. One of the limitations of the bright-field microscopy is insufficient contrast. The lenses refract the light in such a way that the image of a specimen is magnified as it is projected into the eye. The observed structures can be naturally colored, but dyes can also be used to stain cells and increase their contrast so that they can be more easily seen in the bright-field microscope. Depending on the intensity of the colored parts, the light will differently be absorbed and these parts will appear more or less dark. The coloration is due either to a dye that binds preferentially to a particular molecule or family of molecules or to a dark precipitate. This precipitate generally results from the action of an enzyme and occurs at the place where the protein is located, which allows to observe the distribution of the enzyme in the biological structure.

In these techniques, the light rays coming from the condenser can penetrate into the lens (bright-field) or indirectly through a set of mirrors (dark-field). The black-field is used for the observation of specimen whose structures have significant variations in refractive index and, because of the lack of contrast, are only slightly or not visible in bright-field microscopy (bacteria, nucleus, vacuoles, flagellar, skeletons of diatoms, etc.). The dark-field microscopy is very suitable for fresh observations, and its resolution is quite high. It is also an excellent way to observe the motility of microorganisms, as bundles of flagella are often resolvable in bright-field or phase contrast microscopes.

In a bright-field microscope, it is possible to insert a polarizer between the light source and the specimen and an analyzer. Polarizer and analyzer are polarizing filters. In this case, different areas of the preparation deviate differently the polarization plane. This microscope called "polarizing" will highlight the fibrous structures, lamellar, granular crystals and examination is often fresh.

Phase Contrast Microscopy

This is one of the few techniques that can be used to observe living cells. Phase contrast microscopy is based on the principle that cells differ in refractive index from their surroundings and hence bend some of the light rays that pass through them. Light passing through a specimen of refractive index different from that of the surrounding medium is retarded. This effect is amplified by a special ring in the objective lens of a phase contrast microscope, leading to the formation of a dark image on a light background. The phase contrast microscope is widely used in research applications because it can be used to observe wet-mount (living) preparations. It is thus widely used in cellular engineering.

Differential Interference Contrast Microscope

Differential interference contrast is a form of light microscope that employs a polarizer to produce polarized light. It uses either unpolarized light or polarized light for the observation of transparent or reflective objects. The most interesting ones use polarized light and produce images whose relief can be surprising and astonishing. The study of the light interference leading to the formation of images in different interference microscopes is beyond the scope of this chapter. It is widely used to perform micromanipulation of living cells but also to observe surfaces. Its use in microbial ecology is still quite limited.

Fluorescence Microscopy

It is widely used in microbial ecology. It exploits the ability of certain molecules such as fluorophores or fluorochromes to emit fluorescent light upon absorption of photons, which means after excitation by a light source. A fluorophore can be excited only by radiation that it can absorb. The absorption spectra are band spectra that include therefore the excitation spectra. Obviously, the specimen to study must be illuminated by a spectral interval that belongs to the excitation spectrum of the fluorophore. Generally, a spectral interval centered on the maximum wavelength of the excitation spectrum of the fluorophore is used. It is therefore important to ensure consistency between the excitation source whose spectral bands vary qualitatively and quantitatively from one source to another and the excitation spectrum of the fluorophore.

Light sources can be lamps, lasers, and more and more often diodes. Most lasers have a major and narrow band and must correspond to a spectral region of strong absorption to excite the fluorophore. Conversely, lamps (i.e., mercury vapor lamp) emit a spectrum with many bands of varying intensity, which is reflected on the spectrum with peaks more or less important. We choose a fluorophore which absorbs strongly in one of these spectral ranges that are produced by the light source. The fluorescence intensity is important when working with very small particles such as viruses or bacteria, and it can be increased by using light sources with a powerful light flow.

If the fluorophore is a molecule naturally present in the cell of interest, such as a chlorophyll pigment, the source will be selected based on the excitation spectrum of the pigment. Conversely, if the fluorescence is not constitutive and is based on the staining of a cellular component with a specific fluorochrome, in this case one may choose the fluorophore depending on the light source of the cytometer (microscope or flow cytometer). For a given cellular target, such as nucleic acids, there is a wide variety of fluorochromes with very varied peak wavelengths of excitation and emission (see website of Molecular Probes®).

In all cases the wavelength of the fluorescence emitted upon excitation is higher (lower energy) than the wavelength used to excite the molecule. This results from a loss of energy between excitation and emission phases. The selection of spectral bands is done with optical filters whose characteristics are very different. A band-pass filter is a device that passes frequencies within a certain range and rejects (attenuates) frequencies outside that range. These filters are mainly used as excitation filters. Other filters, high-pass or low-pass, can pass the wavelengths above or below a given value and are used for transmission, that is to say that the fluorescence light will be visible to the eye (for details, see the many websites that specialize in cytometry instruments). A dichroic mirror is used to reflect the light rays in order to direct them to the specimen for the excitation phase and the eyepiece for the transmission phase. The excitation and emission filters and the dichroic mirror are often grouped together in a removable block, and the microscope may be equipped with a set of different blocks in order to combine several fluorochromes within the same preparation. It is possible to use multiple fluorochromes that are excited simultaneously by a single light source by exploiting one or more spectral bands and emit with different wavelengths. This produces a multicolor staining. For example, it is common to combine a fluorochrome associated with a physiological function with a second fluorochrome associated with a taxonomic probe.

The Confocal Scanning Laser Microscopy (CSLM)

CSLM is a computerized microscope that couples a laser light source to a light microscope. A thin laser beam (argon ion) is bounced off a mirror that directs the beam through a scanning device. Then the laser is directed through a pinhole that precisely adjusts the plane of focus of the beam to a given vertical layer within the specimen. By precisely illuminating only a single plane of the specimen, illumination intensity drops off rapidly above and below the plane of focus, and because of this, stray light from other plane of focus are minimized. Therefore, in a relatively thick specimen such as a microbial film, not only the cells on the surface can be visible as is the case with a conventional light microscope but cells from the different layer scan can also be observed by adjusting the plan of focus of the laser beam. The images are of exceptional clarity. The image processing software associated with these microscopes allows us to reconstruct a complete image from images of different planes and reconstruct the three-dimensional arrangement of the various components of the preparation. In microbial ecology, this technique is mainly used to study microbial assemblages after staining with fluorescent probes including oligonucleotide probes to recognize specific cells or physiological probes to target specific functions. An important field of application is the study of biofilms and more generally interactions between organisms and the environment. The resolution is limited to that of a light microscope.

Electronic Microscopes

The main limitation of optical microscopy is its resolution because the smallest details that can be observed in theory correspond to the half wavelength of the light source. By optical microscopy, using the shortest possible wavelengths (UV radiation), the limit is around 0.25 μm which means that details of the cellular contents are beyond its capabilities. Electron microscopes are widely used for studying the detailed structure of cells. Since the photon does not go further, the idea was to use another elementary particle, the electron. The wavelength associated with the electron is indeed much lower than that of ultraviolet photon and the final resolution is much higher and in the nanometer range. Electron microscopy essentially allows to observe structures and sometimes macromolecules, such as proteins bound to DNA. Sample preparation is much more complex than light microscopy.

The Transmission Electron Microscope (TEM)

By far the most effective, this technique is similar to the direct light microscopy. The electron beam is produced by an electron gun, commonly fitted with a tungsten filament cathode as the electron source. The electron beam is accelerated by an anode typically at +100 keV (40–400 keV) with respect to the cathode, focused by electrostatic and electromagnetic lenses, and transmitted through the specimen that is in part transparent to electrons and in part scatters them out of the beam. When it emerges from the specimen, the electron beam carries information about the structure of the specimen that is magnified by the objective lens system of the microscope. The magnification is much higher than optical microscopy. This technique requires

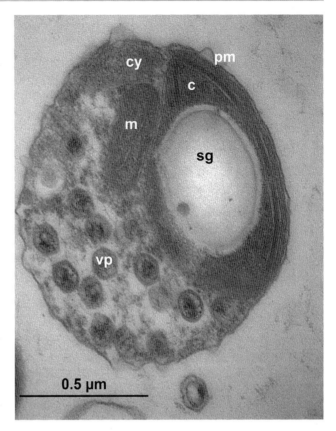

Fig. 17.13 Observation by transmission electron microscopy of *Ostreococcus tauri* infected by the double-stranded DNA virus OtV5. *cy* cytoplasm, *pm* plasma membrane, *m* mitochondria, *c* chloroplast, *sg* starch granule, *vp* viral particle (Photographs: Courtesy of Marie-Line Escande and Nigel Grimsley)

ultrathin sections with a thickness of about 50 nm or less made after hardening of the sample by inclusion. A staining step is sometimes required depending on the contrast of the preparation. In microbial ecology, this technique is used to describe the cellular structure of microorganisms, to observe viral particles inside the host cell (Fig. 17.13), to observe the organization of the nucleus in eukaryotic cells, etc.

The Scanning Electron Microscope (SEM)

Unlike the TEM, where electrons of the high-voltage beam carry the image of the specimen, the electron beam of the Scanning Electron Microscope (SEM) does not at any time carry a complete image of the specimen. This technique provides images absolutely spectacular in pseudo 3D. When the electron beam bombards the preparation, a part of the electrons pass through the preparation while others are backscattered (secondary electrons) and used to construct the image using an electron collector. The result is a representation of the surface of the studied object (cell, particle, etc.). The biological sample must be metallic because electrons scattered by the metal are those who are collected to produce an image. A large range of magnifications can be

Fig. 17.14 FACSAria™ flow cytometer (Becton Dickinson) equipped with a broadband cell sorter. *1* Optical bench; *2* cell sorting chamber; *3* screen control (Photograph: Courtesy of Jennifer Guarini)

obtained with the SEM. In microbial ecology, this technique is used to describe surfaces, cell morphology, and more generally the interaction between the cell and its environment. A new generation is derived from SEM microscope: it is the environmental microscope in which a partial vacuum is produced at the place of the sample in order to preserve its "environment". In this case the metallization is no longer necessary.

17.3.2.2 Flow Cytometry (FCM)

In cell biology, FCM is a laser-based, biophysical technology employed in cell counting, cell sorting, biomarker detection, and protein engineering by suspending cells in a stream of fluid and passing them by an electronic detection apparatus. A beam of light (usually laser light) of a single wavelength is directed onto a hydrodynamically focused stream of liquid. A number of detectors are aimed at the point where the stream passes through the light beam: one in line with the light beam (Forward Scatter or FSC) and several perpendicular to it (Side Scatter or SSC) and one or more fluorescence detectors. Each suspended particle from 0.2 to 150 μm passing through the beam scatters the ray, and fluorescent chemicals found in the particle or attached to the particle may be excited into emitting light at a longer wavelength than the light source. This combination of scattered and fluorescent light is picked up by the detectors, and, by analyzing fluctuations in brightness at each detector (one for each fluorescent emission peak), it is then possible to derive various types of information about the physical and chemical structure of each individual particle.

It differs from microscopy because cells are no longer fixed on a support but driven by a liquid stream, hence the term "flow." For a more technical description of the instrument, the reader is referred to one of the many websites dedicated to flow cytometry platforms.

The measured signals are essentially:
(i) Physical signals which correspond to the properties of light scattering related to the dimensions of the particle, its internal structure, or form
(ii) Optical signals which correspond to the constitutive fluorescence properties for cells that are naturally fluorescent (photosynthetic cells) or whose fluorescence is induced by a specific staining on the basis of one or more probes

Each cell is analyzed individually on the basis of several parameters. Multiparametric data are acquired over thousands of events per second, and it is thus possible to obtain statistics on very large populations. These data are shown as monoparametric histograms or cytograms combining two parameters.

Flow cytometry is a very common technique in microbial ecology not only for the enumeration of bacteria, pico- and micro-phytoplankton, and more recently microzooplankton but also for viruses (even if more difficult). FCM allows the study and quantification of the properties of cells. This technique has participated to the development of microbial ecology by providing informations on the abundance of microorganisms in a variety of environments and not only to discover many new groups (*Synechococcus*, *Prochlorococcus*, *Ostreococcus*, etc.) but also to reveal their importance in different oceanic regions and in the element cycles in the sea. Flow cytometers are increasingly onboard oceanographic vessels and provide near real-time data that are essential to improve sampling strategies. Cell sorters are flow cytometers that can sort cells based on their optical properties (Fig. 17.14). It is possible to sort cells on the basis of one or more cellular parameters in order to complete their analysis by other analytical techniques. The sort function can be very useful in microbial ecology to better understand the functional role of targeted cell populations sorted on the basis of taxonomic or physiological properties.

Fig. 17.15 Solid-phase cytometer ChemScan RDI ® (Chemunex). *1* solid-phase cytometer; *2* screen control to locate the cells detected on the membrane; *3* epifluorescence microscope for eye validation (Photograph: Courtesy of Philippe Catala)

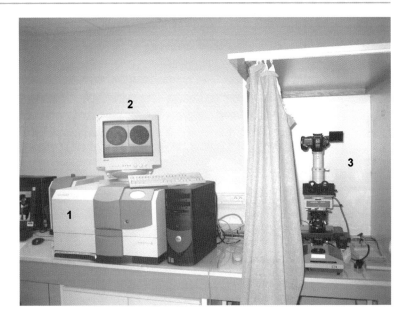

17.3.2.3 Solid-Phase Cytometry (SPC)

SPC is not very common but it is a very powerful technique for studying rare events (Fig. 17.15). This technique is at the interface between microscopy and flow cytometry. In this case, the cells are fixed and deposited on a support as it is the case for microscopy. The light source is a laser which is directed toward a set of oscillating mirrors which can scan the entire surface of the support within a few minutes. Each particle is detected and fluorescent signals are collected and processed similarly to the processing implemented in a flow cytometer. The collected data are analyzed and discriminants (which can be changed depending on the application) can eliminate noncellular and/or nonspecific signals. SPC is connected to a fluorescence microscope whose stage is driven and controlled by the solid-phase cytometer. It is thus possible to have an eye control and to validate by microscopy all fluorescent events that were detected.

The solid-phase cytometer has been proposed as a tool to enable the very rapid detection of rare events in many different products and as an alternative to traditional plate counts. The advantage of this technique is that it allows the detection of rare events, which is not possible in flow cytometry or microscopy. It is therefore particularly suitable for pathogenic microorganisms when searching for a single cell within a sample.

17.3.3 Applications Related to Counting

The enumeration of microorganisms by cytometry techniques is most often done by fluorescence microscopy, flow cytometry, and more rarely by solid-phase cytometry. In microbial ecology, cell counting is important for different applications and in particular to assess the biomass of targeted microorganisms, including bacteria.

In fluorescence microscopy, the organisms are often concentrated on a membrane filter. When this membrane is then observed through the eyepieces, only a small fraction can be analyzed because the surface of a microscopic field (SMF – the surface of the membrane visible through an eyepiece) represents only a small fraction of the total area of the filter. For the enumeration of bacteria, we often use membrane filters (generally 25 mm diameter and 0.2 μm porosity) to concentrate the cells. The surface of the filter on which the bacteria are spread (UFS – usable surface of the filter) represents only a part of the total surface of the filter. SMF represents between 1/5,000 and 1/10,000 of UFS. This means that it should be necessary to observe more than 5,000 microscopic fields for counting the bacteria on the total surface of the filter. In practice, this is impossible and the operator randomly selects and counts 30 microscopic fields (SMF) to integrate the spatial heterogeneity in the distribution of bacteria. Then he reports the average number of bacteria per microscopic field at USF. This approach is tedious, time consuming, and often dependent on the experimenter. It is often difficult to count more than a few hundred microorganisms. With this approach being used, it is necessary to concentrate at least 5,000 bacteria on the filter in order to detect at least one cell per SMF and to quantify, which prohibits quantifying populations poorly represented in the studied communities (Lemarchand et al. 2001).

In flow cytometry, it is possible to count a large number of cells in a very short time. The flux used for microorganisms whose with a size close to or less than 1 μm is often close to 40 $\mu l \cdot min^{-1}$. It means that it is possible to analyze 0.2 ml in 5 min or 1 ml in 25 min. Moreover, it is necessary to detect at least a thousand cells to be able to analyze the cytograms. It is therefore necessary that microorganisms are present at a sufficient concentration to be analyzed.

In the aquatic environment, where there is usually between 100,000 and 500,000 bacterial cells per milliliter of water or thousands of cyanobacteria, it is possible to count thousands of cells in a few minutes when the objective is to

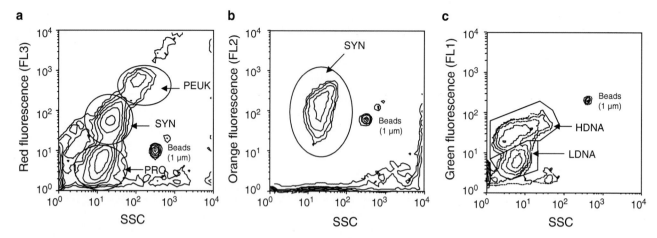

Fig. 17.16 Example of flow cytometric cytograms that displays the different groups that can be found in a marine microbial community. (**a**) Different groups of picophytoplankton are visible on a cytogram combining the relative red fluorescence (an index of the chlorophyll content) versus side scatter (SSC, an index of the cell size). *PRO Prochlorococcus*, *SYN Synechococcus*, *PEUK* photosynthetic picoeukaryotes; (**b**) *Synechococcus* are discriminated and counted on a cytogram combining the relative orange fluorescence (FL2 – an index of the phycoerythrin content) and the side scatter ssc; (**c**) the heterotrophic bacteria are detected after labeling with SYBR Green I on a cytogram combining ssc and green fluorescence (FL1). Two clusters of bacteria with different DNA content (FL1 values) and cell size (ssc values) are shown on a contour plot of relative green fluorescence (an index of the DNA content) and SSC. It is possible to discriminate two groups called HNA ("high nucleic acid content") cells and LNA ("low nucleic acid content") (Lebaron et al. 2001). Fluorescent beads of 1 μm are added to the samples during analysis, to relativize the cellular characteristics of microbial populations compared to beads (Joux et al. 2005) (Copyright: Courtesy of Life and Environment edition)

enumerate total bacteria or cyanobacteria (Fig. 17.16). Conversely, it is difficult to detect and enumerate populations poorly represented (such as a specific taxonomic or functional groups), and it is difficult to concentrate the sample without introducing bias in the quantification.

With solid-phase cytometry (SPC), it is possible to count rare cells after concentration on a membrane. This technique is very complementary to the FCM and microscopic tool for quantification. We can quantify a single cell of a pathogenic bacteria in a given volume of sample (e.g., 100 ml of water are filtered through a membrane filter). The only limit is the high sensitivity of this instrument to the fluorescence background of the filter or introduced by the presence of organic and/or inorganic particles. Many applications are currently under development. Figure 17.17 presents the limits of the different techniques in quantitative terms and an application to the quantification of bacteria (Lemarchand et al. 2001).

17.3.4 Activity Measurement at the Cellular Level

17.3.4.1 Use of Fluorescent Probes

There are many molecular probes that can be used to analyze the physiological state of the cells (Fig. 17.18). These probes conjugated to a fluorophore often have different targets (Joux and Lebaron 2000), and they can learn about the physiological state of individual cells within a population. They may be combined with each other or with a taxonomic probe provided that the appropriate excitation wavelength of the different probes are provided by the light source and that the emission wavelengths of the fluorophores can be easily distinguished. Fluorescent probes specific to nucleic acids are often used in microbial ecology for cell counting but also to analyze the nucleic acid content of individual cells. Within bacterial communities in aquatic environments, it is often possible to distinguish populations having different nucleic acids content. This discrimination is now often used in microbial ecology because the cells that have a high nucleic acid content are considered to be more active than those with a lower content of nucleic acids (Lebaron et al. 2001).

17.3.4.2 Microautoradiography

It is possible to characterize the metabolic activity of a cell by the use of radioactive substrates of commercial origin or produced by a biological organism (amino acids, carbohydrates, lipids, etc). The method consists in providing a given substrate to bacterial communities for a period of time depending on the biological activity. Cells that metabolize the substrate become radioactive, and this activity can be revealed by a photographic emulsion and the precipitation of silver salts around each active cell. The observation is made by light microscopy. Nevertheless, it is possible to combine this approach with a fluorescent nucleic acid probe (see below) and the method is called Microfish.

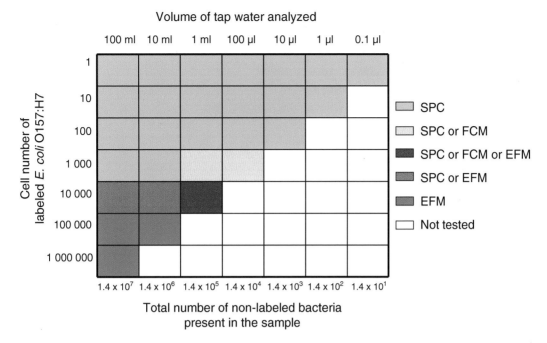

Fig. 17.17 Domains of application of epifluorescence microscopy (*EFM*), flow cytometry (*FCM*), and solid-phase cytometry (*SPC*) depending on both the absolute number of labeled *Escherichia coli* cells and the volume of tap water analyzed. For each volume, the number of non-labeled cells present in the tap water is indicated. Each color represents the domain of application of the different instruments (Modified and redrawn from Lemarchand et al. (2001))

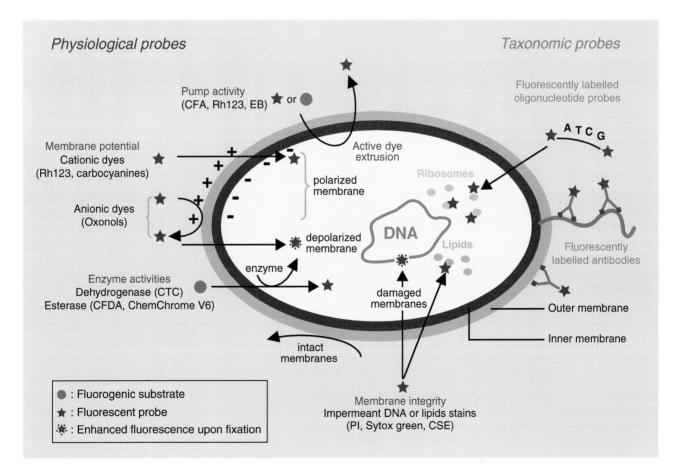

Fig. 17.18 Different targets used to analyze the physiological state and identity of individual cells using fluorescent probes (Modified and redrawn from Joux and Lebaron (2000). Drawing: M.-J. Bodiou)

17.3.5 A Taxonomic Approach at the Cellular Level

17.3.5.1 Immunofluorescence

The immunoassay techniques are widely used in microbiology as a diagnostic tool. They can be used to search for a pathogen, its identification, and counting (especially in controlling the quality of the water). These techniques are based on the use of specific antibodies to pathogens. There are monoclonal antibodies (recognizes only one type of epitope) which are very specific and reproducible and polyclonal antibodies (mixture of antibodies recognizing different epitopes on the same antigen) that are less specific. Among the various immunoassays, immunofluorescence consists of chemically modifying an antibody by adding a fluorochrome without altering the specificity of the antibody (Fig. 17.18). The targeted microorganisms are then detected using an instrument that can detect fluorescent cells (epifluorescence microscopy, flow cytometry, or solid-phase cytometry). These techniques are mainly used in the field of food microbiology, because antibodies are commercially available for most pathogenic microorganisms. For taxonomic purposes, these probes are increasingly substituted by DNA probes whose specificity is often better and also because these probes can be defined for all microorganisms, without the need for prior cultivation.

17.3.5.2 In Situ Hybridization with Fluorescent Probes

In situ hybridization with fluorescent probes ("fluorescent in situ hybridization" FISH) is a cytogenetic approach based on the principle of complementary between nucleic acid chains and allows the specific pairing of a labeled oligonucleotide probe to an RNA or DNA sequence, ribosomes, or messengers. So it is an approach conceptually close to the gene amplification by PCR followed by sequencing, with the advantage of a possible eye control by microscopy; but since it is a short sequence, the disadvantage is that the resolution is lower.

This technique was originally developed for the detection of mutations in eukaryotic genomes after spreading of the chromosome on a glass slide. Subsequently adapted to the bacterial cells, the technique allows to observe cells without culturing.

Bacterial cells are surrounded by a lipid membrane and a cellular wall consisting of sugar (peptidoglycan) and proteins that can be made porous to allow the passage of small oligonucleotide sequences but not the ribosomes, much bigger, which therefore remain inside cells. Ribosomes are made of RNA sequences. Their number varies from 20,000 per active cell in E. coli to 0 in starved cells. The effectiveness of the FISH technique is proportional to the number of ribosomes and ultimately to the physiological state of the cell.

It is also possible to detect specific mRNA by in situ hybridization (Pernthaler and Amann 2004). In the latter case, the situation will depend on the level of expression of the selected gene, considering that some genes do not have enough expression to be detected by FISH. The probes are oligonucleotide sequences of about twenty bases. It is necessary that they have a short size to allow their penetration into the cells and also to have a melting temperature close to the room temperature to protect the cell structures. The sequences of the target group are aligned and analyzed, for example, using the ARB software (Ludwig et al. 2004), which allows to design an oligonucleotide sequence whose specificity can be checked in the database. When possible, mismatches are located at the center to increase the instability of hybrid and thus to reduce false positives. Cells undergo different treatments before hybridization, starting with diethylpyrocarbonate to inactivate intracellular RNases and then with lysozyme (which catabolizes the peptidoglycan) and proteinase K to permeabilize the cellular wall, then with paraformaldehyde to preserve the subcellular structures. The cells are incubated in a hybridization solution containing the labeled oligonucleotides for a few hours and then washed to remove probes that are unbound to their specific target. A second nonspecific dye (i.e., DAPI) is used to stain all the cells. Finally, the cells are observed under a common fluorescence microscope.

It is possible to use two primers in the same experiment to determine, for example, the proportion of cells belonging to two nested taxa, which is the case of cells belonging to the *Enterobacteriaceae* and to the genus *Erwinia*. In this case, two probes are used with different fluorochromes. It is also possible to use two primers, one targeting ribosomes and the other a functional gene in the same cell to link taxonomic and functional informations. Finally, it is possible to have two primers targeting two different taxa, such as one for *E. coli* and one for *Pseudomonas fluorescens* when the objective is to quantify their co-occurrence in a complex environment. One of the most limiting problems in the use of FISH is the electrostatic repulsion due to the negative charge of the phosphodiester groups of the DNA that make some ribosome regions inaccessible to the classic FISH method. It is then possible to use, instead of DNA, APN molecules (or acid peptido-nucleic) where sugar phosphates are replaced by neutral links based on polyamide and following the Watson–Crick matching rules, but which hybridize more rapidly and more efficiently (Stender et al. 1999). Finally, it is possible to use molecular beacons which operate on the basis of FRET (fluorescence resonance energy transfer) and are already used in quantitative PCR to reduce the background noise (Lenaerts et al. 2007) due to nonspecific hybridization.

17.4 Measurements of Microbial Biomass and Activities

17.4.1 Measurements of Biomass

17.4.1.1 Biochemical Methods

Biochemical methods for quantifying biomass are based on the measurement of characteristic cellular compounds or intracellular metabolites. To be meaningful, the selected compounds should comply with the following criteria:

(i) The selected compounds should be specific, i.e., restricted to a genus, a functional group, etc.
(ii) The selected compounds should disappear rapidly upon cellular death and, therefore, should be absent in the surrounding environment
(iii) The selected compounds should be present in the cells at a fixed concentration, which level should be independent of both the physiological status of the cells and of the substrates it uses as a source of carbon and energy.

Lipid Analyses

The biochemical methods are mainly based on the analyses of membrane lipids and lipid components of bacterial cell wall. These particularly include the **phospholipid fatty acids** (**PLFA**)*. The lipids are extracted using the method of Bligh and Dyer (1959) and the extract is eluted and fractioned using a siliceous column to separate them from the apolar compounds as the aliphatic hydrocarbons (Syakti et al. 2006). The PLFA are methylated to produce fatty acid methyl esters. The position and the geometry of the double bonds can be determined by forming methyldisulfide-type derivatives (Nichols et al. 1986), and the position of the methyl groups can be determined after their derivatization into n-acyl pyrroline type compounds according the method of Andersson and Holman (1974). PLFA are separated, identified, and quantified by using gas chromatography coupled to mass spectrometry (GC-MS) (Syakti et al. 2006). The analyses of the PLFA allow estimating the viable bacterial biomass. Phospholipids are present in all bacterial membranes, are characterized by a high turnover rate, and are rapidly hydrolyzed into diglycerides by the phospholipases after cellular death. Hence, the estimation of the bacterial biomass is based on the PLFA weight using a conversion factor that has been determined for *E. coli*. In this species, PLFA represent 100 $\mu mol \cdot g^{-1}$ dry weight and 1 g of cells is equivalent to $4.3 \pm 1.2 \cdot 10^{12}$ cells (Balkwill et al. 1988). By knowing the concentration of PLFA in a natural sample, we can thus estimate the viable biomass concentration. This method has been applied successfully in different environments, although it may present specific problems, which are related to:

(i) Differences in PLFA concentrations between species
(ii) Variability of PLFA due to variations of cell size
(iii) Variability of PLFA contents related to different environmental conditions, as, e.g., temperature, growth substrate
(iv) Variability of PLFA contents related to the physiological status of the microorganisms

Despite these limitations, this method is one of the most efficient methods currently used for estimating the bacterial biomass. In certain cases, this method yields comparable results as those obtained by other approaches as, e.g., the measurement of the uraminic acid and the beta hydroxyl groups, which are characteristics for the lipopolysaccharides (Balkwill et al. 1988).

The Other Biochemical Methods

The PLFA are not the only cellular constituents that can be used for estimating the bacterial biomass. Muramic acid is specific for the bacteria and thus a good biomarker for estimating bacterial biomass. However, the cellular contents are highly variable and particularly divergent between the two main groups of Gram-positive and Gram-negative bacteria. ATP can potentially also be used, although its conversion to bacterial biomass is very difficult as its cellular levels are very variable among species and dependent on the physiological status of the cells.

17.4.1.2 Biovolume and Conversion Factors

Estimation of microbial biomass carbon is possible by the use of conversion factors indicating the average carbon per cell content and the result of cell counts obtained by microscopy or flow cytometry (*cf*. Sect. 17.2.3). These conversion factors are logically linked to the size of the cells, as is necessary in the case of the bacteria to determine their average biovolume by epifluorescence microscopy after cell staining with DAPI, for example. The most commonly used formula for this determination is as follows:

$$\text{Biovolume } (\mu m^3) = \pi/4 \times D^2 \times (L - D/3)$$

with L: length of the cell (μm)
D: width of the cell (μm)

Relationships between biovolume and carbon biomass have been established by different authors, such as Norland (1993):

$$\text{pg C cell}^{-1} = 0.12 \times \left(\mu m^3 \text{cell}^{-1}\right)^{0.7}$$

In the marine environment, carbon biomass of bacteria varies from 12 to 30 fg of carbon per cell depending on the particular environment (Fukuda et al. 1998). Similarly, relationships between biovolume and carbon content exist for other microorganisms (Davidson et al. 2002a).

Concerning viruses, a factor of 0.055 fg of carbon by virus was estimated (Steward et al. 2007).

17.4.1.3 Quantification of Phytoplanktonic Biomass

The cycle of organic matter in the marine environment mainly results from the activity of extremely varied and numerous organisms, among which we can distinguish phytoplankton and algae as primary producers, zooplankton as secondary producers, and predatory chain and heterotrophic bacteria that remineralize organic matter. In general, the organisms in these groups are difficult to isolate and separate from other to observe and measure their abundance. The exception is phytoplankton, which has specialized chloroplasts containing pigments that allow it to capture the light energy necessary for photosynthesis. Although a wide variety of pigments exist, chlorophyll a (Chl a) is the main pigment in aerobic photosynthetic organisms (excluding cyanobacteria). The cell content of Chl a is 1–2 % by dry weight. Following collection of phytoplankton by filtration, it is quite easy to quantify this pigment by extraction using a nonpolar solvent. The method is simple: phytoplankton is captured on filters capable of retaining all cells (see Sect. 17.1.2) and chlorophyll is extracted with a nonpolar solvent, with or without mechanical action.

Measurement of extracted chlorophyll is based on two spectroscopic characteristics:

(i) Its ability to absorb light at a well-defined wavelength (663 nm), which enables quantification by spectrophotometry
(ii) Its ability, after excitation at a wavelength of 430 nm, (blue) to rapidly return a portion of the energy in the form of red light at a wavelength of 669 nm, according to the fluorescence process

The spectrophotometric method is the oldest method of measuring extracted chlorophyll (Richard and Thompson 1952), but has been replaced by the fluorimetric method (Yentsch and Menzel 1963) which is much more sensitive. However, it should be noted that spectrophotometric measurement is intrinsically calibrated because the absorption coefficients of the pigments are known precisely, while the fluorimetric method requires regular calibration of the fluorometer.

Many variants of the procedure have been developed, differing in terms of the quality of the solvent (e.g., acetone, ethanol, and methanol), the grinding process, and the duration of extraction. The literature is abundant in this area (cf. Jeffrey et al. 2005), and due to the interest in studying the productivity of aquatic environments, successive working groups have been formed since 1964, with the support of UNESCO (Scor-UNESCO 1966), to identify the best analytical procedures. A methodological synthesis was conducted under the auspices of the International Council of the Exploration of the Sea (Aminot and Rey 2002). Although these fast and simple methods are still widely used for the quantification of phytoplankton biomass, more efficient techniques now make it possible to study the entire pigment spectrum. Chief among these more efficient techniques are high-performance liquid chromatography (HPLC, Jeffrey et al. 1997) and spectrofluorimetry (Neveux and Lantoine 1993) (cf. Sect. 17.6.6).

Unfortunately the conversion of Chl a concentration in terms of total phytoplankton biomass is difficult and is still highly criticized because the chlorophyll content of cells varies greatly, depending on many factors. This is especially true for the Chl a carbon ratio, which changes over the life of the cell and depends on the available nutrients and light conditions. It is also well known that the chlorophyll content of phytoplankton cells tends to increase in low intensity light, whereas, in contrast, a nutrient deficiency may cause a destruction of chlorophyll (chlorosis). Chlorophyll: nitrogen and chlorophyll: phosphorus ratios seem to be more consistent (Strickland 1960), and in optimal growing conditions, 1 mg of Chl a is assumed to correspond to 14 mg of nitrogen.

The in vivo fluorescence technique (Lohrenzen 1966) permits the direct determination of chlorophyll in sea samples without the filtration step. However, the large variability often observed in the in vivo fluorescence/Chl ratio greatly limits the value of this parameter as a measure of biomass.

On the other hand, the determination of the optical properties of phytoplankton using satellite imagery has made possible significant advances in the study of the spatial and temporal variability of phytoplankton populations (Fig. 17.19). Indeed, by absorbing light, especially in the blue part of the visible spectrum, the Chl a selectively modifies photon flux that passes through the photic zone of the ocean. This absorption changes the spectrum of the sunlight reflected from the ocean (reflectance). Because the color in the visible light region (wavelengths of 400–700 nm) in most of the world's oceans varies with the concentration of chlorophyll and other plant pigments present in the water, the more phytoplankton that is present, the greater the concentration of plant pigments and the greener the water. This property has been exploited by space agencies that have launched the so-called "ocean color" satellites to map chlorophyll content at the ocean surface (Gordon et al. 1988). The first demonstration was made during the testing of the Coastal Zone Color Scanner (CZCS), a sensor launched by NASA in 1978. CZCS was able to provide measurements of ocean color over large geographic areas in short periods of time in a way that was not previously possible with other measurement techniques, such as from surface ships, buoys, and aircraft. These measurements allowed oceanographers to infer the global distribution of the standing stock of

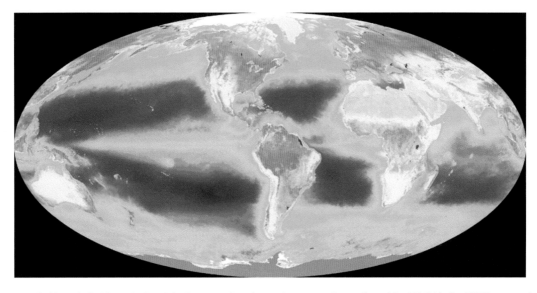

Fig. 17.19 Map of chlorophyll (phytoplankton) in the ocean based on color sensor data collected by NASA's SeaWiFS sensor, for the period September 1997–August 2000 (© SeaWiFS Project. NASA/Goddard Space Flight Center. ORBIMAGE – Authorization Gene Feldman)

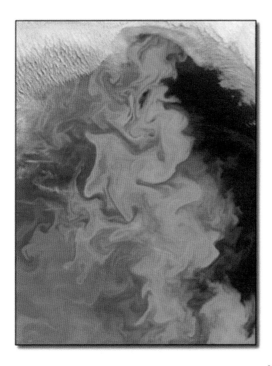

Fig. 17.20 Significant development, over 150,000 km^2, of coccolithophores in the surface waters of the Barents Sea, detected by the MODIS sensor (August 1, 2007) (Authorization Gene Feldman)

phytoplankton for the first time. Collection of ocean color data has become very common with the launch of many other sensors: OCTS (NASDA) and POLDER (CNES), launched in 1996, which worked for only 8 months; SeaWiFS, MODIS-T, and MODIS-A (NASA), launched in 1997, 1999, and 2002, respectively; and MERIS (from the European Space Agency), launched in 2002. In addition, permanent improvements in both the quality and sensitivity of the sensors and algorithms make it possible now to extract data on other biogeochemical components. Information on three groups of phytoplankton (micro-, nano-, and picoplankton), which differ in the proportions of their accessory pigments (i.e., other than Chl *a*) can be obtained (Uitz et al. 2006). Similarly, under certain conditions, populations of coccolithophorids (Fig. 17.20) and *Phaeocystis*, which have a very strong impact on the color of ocean water through the limestone pieces or the mucus they produce, can be identified. Recently developed algorithms make it possible to quantify the amount of organic carbon present from chlorophyll estimation (Stramski et al. 2008). Chl *a* being the main compound involved in the photosynthetic process, it is therefore reasonable to presume that primary production is directly related to the concentration of this pigment (Ryther and Yentsch 1957). The phenomenon is complex, but bio-optical models are constantly being improved, and data on in situ chlorophyll, such as those obtained by "ocean color" satellites, are now commonly used to estimate global primary production.

17.4.2 Measurements of Primary Production

Primary production, i.e., the amount of organic matter produced each day by autotrophic microorganisms from atmospheric or aquatic carbon dioxide, is a fundamental parameter in the cycling of matter in aquatic environments. Almost all life on earth is directly or indirectly reliant on primary production. The organisms responsible for primary production are known as primary producers or autotrophs and form the base of the food chain. Phytoplankton are the organisms (primary producers or autotrophs) responsible for primary production in oceanic areas and form the base of the food chain. Primary production is distinguished as either net or gross, the former accounting for losses to processes such as cellular respiration and the latter not accounting for these losses. Indeed, research on oceanic primary production has increased dramatically since the development of the carbon-14 (^{14}C) tracer method by Steemann Nielsen (1951). ^{14}C is a radioactive isotope that is easy to use because it has little energy and has a very long half-life or period (5,730 years). The experimental procedure is simple. A small known amount (a few ml) of ^{14}C as sodium bicarbonate is introduced into a water sample collected in a transparent flask. After placing the sample under conditions as close as possible to the sample's initial environmental conditions for a period of time, called the incubation period, the particulate material is collected on a filter, and its radioactivity is measured using a scintillation counter. Primary production, or the quantity of carbon assimilated during the incubation period (PP), can then be calculated as follows:

$$PP = \left(^{14}C_{phyto}/^{14}C_{int}\right) \times C_t$$

where

$^{14}C_{phyto}$ = amount of ^{14}C measured in the particulate fraction collected on the filter,
$^{14}C_{int}$ = quantity of ^{14}C initially added to the sample,
C_t = Inorganic carbon concentration in the sample.

The main variations of the method are:
(i) The incubation period (from a few minutes to 24 h)
(ii) The volume of the sample (a few ml to some liters)
(iii) The material of the incubation flasks (glass, polycarbonate, quartz)
(iv) The incubation conditions

Incubation can be carried out in incubators under natural or artificial light. It is imperative to strive to maintain a temperature equivalent to the initial temperature of the sample. The method of incubation that best reflects the initial conditions of the sample is the so-called in situ incubation method, in which the sample is returned to its original depth immediately after the addition of tracers. To carry out incubations at different depths, a mooring line is prepared (Fig. 17.21) with mooring points for positioning the incubation flasks at different depths. It should be noted that some bottles that are opaque to light (called dark bottles) are used to account for the biological processes that occur in darkness (dark carbon assimilation).

Since the pioneering work of Dugdale and Goering (1967), the isotopic tracer technique has been generalized to detection of nitrogen compounds through the use of the stable isotope ^{15}N. Quantifying the rate of uptake of nitrogen by microorganisms is necessary because the mineral nitrogen is often present at very low concentrations in surface waters and appears to be a factor that limits the growth of primary producers. In contrast, the carbon dioxide substrate for photosynthesis is always present in large quantities in natural waters. The incubation technique for quantification of nitrogen fluxes using ^{15}N tracer is identical to that of ^{14}C. A sample is enriched with an inorganic or organic nitrogen compound artificially labeled with nitrogen-15, and then the isotopic enrichment of the particulate fraction recovered on a filter after the incubation period is measured with a mass spectrometer to quantify the relative proportion of the isotopic ^{15}N/^{14}N ratio. It would be noted that dual tracer measurements using the stable isotopes ^{13}C and ^{15}N can be used to simultaneously estimate the uptake rates of dissolved inorganic carbon and nitrogen. The inorganic nitrogen available for microorganism growth in aquatic environments is present in several forms (mainly nitrate and ammonium), which allows this tracer method to differentiate between two types of production supported by substrates that differ in their origin (Dugdale and Goering 1967):

Fig. 17.21 In situ incubation flasks on a mooring line ready for use (Photographs: Courtesy of Joséphine Ras)

(i) The regenerated production is the carbon assimilation rate, essentially supported by the assimilation of nitrogen substrates from recycling bacterial and zooplankton excretion (ammonium, nitrite, urea).
(ii) The new production is the carbon assimilation based on the assimilation of nitrogen inputs from mineral reserves outside the area of primary production. These are mainly nitrate, which is the main reservoir in deep water, and molecular nitrogen from the atmosphere.

Recent developments in the measurement of the enrichment of dissolved inorganic and organic fractions have made it possible to complete the study of the nitrogen cycle by estimating the rate of excretion of dissolved organic nitrogen (Bronk and Glibert 1991; Slawyk and Raimbault 1995) and regeneration rates, based on the technique known as isotope (ammonium nitrate, urea) dilution.

17.4.3 Measurements of Heterotrophic and Chemoautotrophic Bacterial Production

Heterotrophic bacterial production is classically measured by radioisotopic techniques using thymidine or leucine (^{3}H or ^{14}C) incorporation into DNA and protein, respectively (Furhman and Azam 1980; Kirchman et al. 1985). These techniques are particularly suitable for samples of aquatic environments, but they can also be used for sediments and soils. Due to their high sensitivity, these measurements can be performed in low productive environments (e.g., deep seawater). These techniques also offer the advantage of limited sample handling (the measurement can be carried out on the raw sample without prefiltering) and a short incubation time (from 0.5 to 4 h depending on the activity of bacteria). Samples (1–10 ml) are incubated with a saturating concentration of radioactive tracer (to be determined beforehand by the means of a saturation curve with a range of concentrations of radioisotopes) in the dark at in situ temperature and stopped by addition of trichloroacetic acid (TCA). Bacteria are then recovered by filtration or centrifugation and cleared of radiolabeled molecules not incorporated by different stages of washing with cold TCA (Smith and Azam 1992). Radioactivity incorporated by bacteria is measured with a liquid scintillation counter, and the results are expressed in rate of thymidine or leucine incorporated per volume unit of sample and per unit of time. It is also possible to express these results in bacterial carbon production (e.g., $\mu gC \cdot l^{-1} \cdot d^{-1}$) using experimental conversion factors which consist to follow in parallel the rate of tracers incorporation and the rate at which cell numbers increased for natural samples after grazers have been removed by filtration or dilution. Theoretical factor can be also used to make this conversion. Unfortunately both approaches introduce uncertainty in the results. The common use of incubation in the dark was also criticized, as sunlight can stimulate or inhibit bacterial activity for various reasons (Gasol et al. 2008). An alternative measurement of bacterial production without radioactive materials has been proposed: it determines the incorporation of bromodeoxyuridine (BrdU) into DNA by means of immunological detection (Nelson and Carlson 2005).

The prokaryotic chemoautotroph activity can be measured by dark [^{14}C]bicarbonate assimilation (Herndl et al. 2005). After incubation (24–72 h), the prokaryotes are recovered by filtration (0.2 μm). After rinsing, the filters are exposed to concentrated hydrochloric acid fumes and then the radioactivity of each filter is counted with a liquid scintillation counter.

17.4.4 The Measurement of Bacterial Respiration

17.4.4.1 Measurement of Respiration in the Aquatic Environment

Heterotrophic bacteria are involved for a variable and sometimes important part (e.g., oligotrophic ocean) to the planktonic community respiration. To determine more specifically the respiration related to the heterotrophic bacteria, it is necessary to eliminate the rest of the microorganisms by filtration (0.8 or 1 μm). This step can introduce bias because:

(i) The possible size range overlaps between heterotrophic bacteria and cyanobacteria and photosynthetic protists.
(ii) The possibility that heterotrophic nanoflagellates (2–3 μm) sneak through pores of size smaller than their own size.
(iii) The particles-attached bacteria, which are generally the most active bacteria, are retained on the filter. Respiration can be measured by different ways.

The chemical method, known as Winkler method, is still the reference method for the determination of dissolved O_2. This technique consists to distribute the filtrate in glass bottles with calibrated volume (60–200 ml) taking care to not introduce air bubbles. A part of the bottles are fixed at the beginning of the incubation and the other part at the end of the incubation. The variation of the amount of O_2 dissolved measured during the period of incubation allows to estimate bacterial respiration. The time incubation needed to measure a significant difference of O_2 depends on the activity of the bacteria in the sample. In oligotrophic systems, it is often necessary to incubate 24 h or more in order to measure a significant O_2 variation in the bottles. These long incubations can introduce bias due to bottle effect (i.e., change in bacterial activity and composition due to confinement).

The principle of dissolved O_2 determination involves different chemical reactions: a solution of manganese, potassium iodate, and a strong base is added to the sample to attach the O_2 to manganese. After acidification, the precipitate is dissolved and the iodine ions are oxidized in iodine. The iodine released is then determined by thiosulfate. The equivalence point for the determination of iodine by thiosulfate is determined through an electrochemical (potentiometry) or photometric approach with an accuracy of the order of 2 $\mu g \cdot O_2 \cdot l^{-1}$ (Carrignan et al. 1998). A spectrophotometric approach has also proposed to directly measure the iodine released after acidification by absorbance at 288 or 430 nm (Roland et al. 1999). This method provides a speed of analysis far greater than the potentiometric approach (about 2 min per analysis) but requires a calibration curve to transform the absorbance values measured in quantity of O_2.

The bacterial respiration can be also measured with chemical and optical microsensors (*cf.* Sect. 17.5). These measurements can be achieved continuously or semicontinuously in closed bottles or in dedicated microchambers (Briand et al. 2004). The chemical and optical microsensors technology is changing rapidly with sensibilities that become more and more important. However, the number of samples that could be monitored at the same time with chemical and optical microsensors remains lower compared to the chemical approach.

The amount of O_2 respired ($\mu M \cdot l^{-1} \cdot h^{-1}$) can be subsequently transformed into the amount of released CO_2 using a respiratory quotient (RQ = CO_2/O_2). RQ is generally estimated between 0.85 and 1, although lower values were also measured (Robinson and Williams 1999).

When Bacterial Production (BP) Bacterial Respiration (BR) are measured on the same sample and expressed in the same unit ($\mu g\ C \cdot l^{-1} \cdot h^{-1}$), it is possible to calculate the bacterial growth efficiency (BGE), i.e., the effectiveness of the heterotrophic bacteria to transform organic carbon assimilated into biomass, as well as the bacterial carbon demand (BCD), i.e., the total quantity of organic carbon assimilated by heterotrophic bacteria (del Giorgio and Cole 1998):

$$BGE = BP/(BP + BR)$$

$$BCD\ (\mu g\ C \cdot l^{-1} \cdot h^{-1}) = BP + BR$$

The calculation of BGE and BCD is complicated because BP and BR are measured at different time scales: while the extent of the BP is virtually instantaneous, BR requires sometimes more than 24 h of incubation.

17.4.4.2 Measurements of Respiration in Soils and Sediments

The bacterial respiration in soils and sediments is more complicated to measure than in planktonic environments due to the difficulty to dissociate the bacterial part from the metabolic activities of the other heterotrophic and autotrophic components. For this reason we generally measure in these environments the global flux of CO_2 or O_2. The soil respiration is an important component of the ecosystem carbon balance, with a strong contribution related to microorganisms. Respiration of the soil can be measured by recording the variations of CO_2 concentrations by means of a respiration chamber with an infrared gas analyzer (Fig. 17.22). By measuring accumulation of soil CO_2 productivity released from the soil surface, chambers are unable to provide information about soil profiles. Moreover, these

Fig. 17.22 Soil respiration measurement in the field with the use of a respiration chamber SRC-1 and a probe CIRAS-2 (PP Systems, International, Inc., Amesbury, USA) (Photograph: Courtesy of M.L. Doyle)

measurements include some methodological biases that are largely related to the very low but sufficient air pressure changes induced by the rooms that can modify the CO_2 flux coming out of the ground (Davidson et al. 2002b). The soil respiration can be also measured by analyzing regularly the gradient of CO_2 into the soil using infrared sensors placed at different depths. Transmitters placed on the ground receive signals from the probes and send them to a datalogger (Tang et al. 2003).

Measurements of respiration and photosynthesis in interstitial waters in microbial mats and coastal sediment can be performed using chemical microprobes (*cf.* Sect. 17.5.2). Benthic metabolism can be also measured on the intertidal areas using a benthic chamber coupled with a CO_2 infrared analyzer (Migné et al. 2002). The use of dark incubation allows measuring respiration, while light incubation allows measuring the net community production.

17.4.5 The Measurement of Bacterial Enzymatic Activities

17.4.5.1 Profile of Enzyme Activities

The potential for degradation of different carbon substrates by bacterial communities can be measured using the Biolog EcoPlate™ system (Biolog Inc., CA, USA). Each 96-well microplate contains 31 carbon substrates (amino acids, sugars, carboxylic acids, phenolic compounds, polymers) in triplicate, as well as controls (wells without carbon sources). Additionally, each well contains a colorless tetrazolium dye. Environmental samples are inoculated in the microplate directly (water samples) or after suspension (e.g., soil, sediment, sludge) (Zak et al. 1994; Montserrat-Sala

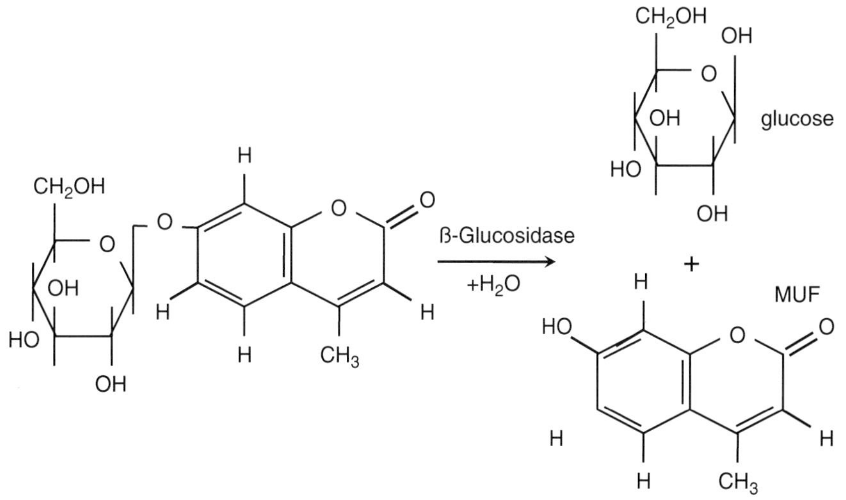

Fig. 17.23 Hydrolysis reaction of a fluorogenic substrate. In the presence of ß-glucosidase, the MUF-ß-D-glucopyranoside releases glucose and the MUF (fluorescent molecule)

et al. 2005). The response of the bacterial community is monitored regularly during incubation over a period of 2–10 days. The use of the substrates is indicated by the development of a purple coloration due to respiratory activity of cells that reduces the tetrazolium salt. Color development is measured spectrophotometrically (optical density) at 590 nm with a microplate reader. Various indices of functional diversity can be calculated from the metabolic profiles obtained (Preston-Mafham et al. 2002).

17.4.5.2 Measurement of Bacterial Exoenzymatic Activity

In aquatic environments, organic matter comes mostly in the form of macromolecules that do cannot be directly assimilated by bacteria. Exoenzymatic hydrolysis of the polymeric material into low molecular weight organic matter (<600 Da) is required before its use by the bacteria. Hoppe (1983) showed that the bacteria contributed to the bulk of extracellular enzyme activities (EEA) (glucosidases, proteases). Bacterial EEA is measured with the use fluorogenic substrates composed of a fluorescent molecule (methylumbelliferone (MUF) or the 4-methylcoumarinyl-7-amide (MCA)), linked to one or several natural molecules (e. g., amino acids, glucose). This complex is devoid of fluorescence. After enzymatic hydrolysis (breakdown of the specific binding), the chromophore (MUF or MCA) is released and the fluorescence appears (Fig. 17.23). During incubation (from 1 to 3 h), the emitted fluorescence is quantified by a spectrofluorometer, and the EEA is expressed in units of fluorescence per unit time. The EEA is then converted to moles of substrate hydrolyzed per unit time and per unit volume through a calibration curve. In general, EEA follow a Michaelis–Menten kinetic (Hoppe 1991). The use of a range of different substrate concentrations (at least 5) allows to determine the enzymatic parameters of this kinetic: the constant K_m (which indicates the affinity of the enzyme for the substrate) and the maximum rate of hydrolysis (V_m). The measure with a single concentration of substrate can also be done by choosing a saturating concentration to determine the V_m, or a very low concentration of substrate to determine the hydrolysis rate near the in situ rate.

A wide range of fluorogenic substrate analogues is available commercially to measure the EEA of different enzymes: aminopeptidase, glucosidase, phosphatase, etc. This technique can be applied for freshwater and marine waters, but also with sediment samples (Talbot and Bianchi 1995).

17.4.6 Determination of the Bacterial Mortality Mediated by Heterotrophic Protozoa and Viruses

17.4.6.1 Rates of Predation by Heterotrophic Protozoa

The measurements of the rate of protistan grazing may be classified into two types: those that use tracers to track bacteria in predator organisms and those that manipulate the communities in order to alter the rate of encounter between predators and bacteria (Strom 2000). These different approaches may yield results contrasted in the estimation of the rate of predation of bacteria, with often lower values for measurements using tracers compared to methods using manipulations of the whole community (Vaqué et al. 1994).

Methods Using Tracers

Fluorescent labeled bacteria are added to the sample. After incubation, the fluorescent bacteria present in the digestive vacuoles of the bacterivores can be visualized by epifluorescence microscopy. The decrease of free labeled bacteria can be also determined by epifluorescence microscopy or flow cytometry. The main limitation of this approach is the use of fluorescent bacteria derived from a culture which are not necessarily representative of the bacteria from the natural community. It is also possible to label radioactively the bacteria naturally present in the sample studied (e.g., ^{3}H thymidine incorporation). The amount of ingested bacteria is measured after a short incubation period (<2 h), the protozoa are separated from bacteria by filtration (~2 μm), and the number of radioactive bacteria ingested is counted by a liquid scintillation counter. This approach also has drawbacks such as the release of radioactivity during digestion of bacteria or the imprecision of the size fractionation. Finally, it is possible to use some of *E. coli* minicells radioactively marked as tracers. These minicells are approaching the size of bacteria from the environment and have no possibility of multiplication. The decrease in radioactivity is followed within the bacterial compartment during incubation.

Methods Using the Sample Manipulation

Predation rates can be estimated from samples diluted to varying degrees with natural water filtered through 0.2 μm to remove all particles. The bacterial growth is followed during 12–48 h incubation in order to deduce the rate of predation. The idea is that dilutions will proportionally reduce the chances of encounter between bacteria and their predators, when the rate of bacterial growth remains unchanged. The latter assumption is certainly not respected knowing that bacterial growth is largely dependent on other organisms and, in particular, on phytoplankton.

Predation rate can be estimated by comparing the bacterial growth of a gross sample with a sample for which heterotrophic protozoa were eliminated by filtration. Despite its simplicity, this approach raises the problem of the choice of the membrane porosity to make this separation knowing that an overlap of size may exist between the bacteria and protozoa. Furthermore, the disadvantages described for the dilution method are found here (disruption of interactions between bacteria and phytoplankton). Finally, predation rate can be measured by comparing the bacterial response between a raw sample and samples treated with inhibitors of prokaryotic cells (penicillin, streptomycin) or eukaryotic (cycloheximide, colchicine). The use of these inhibitors can be problematic, because their actions can be insufficient (e.g., the activity of predation is removed incompletely by eukaryotic inhibitor) or nonselective (e.g., prokaryotic inhibitors also affect the rate of predation of protozoa).

17.4.6.2 Viral Production and Mortality Induced by the Virus

Different approaches have been proposed to measure the viral production and the mortality induced by the virus, but none of them has risen to the rank of standard method (Furhman 2000; Weinbauer 2004). The transmission electron microscopy allows the detection of bacteria infected by viruses when they arrive at the final stage of infection (i.e., with a sufficient number of viruses within the cell). The total infection rate can be calculated from this count using a simple model and taking into account the number of bacteria infected but not visible by microscopy.

Viral production is function of the number of cells infected by viruses and the burst size (i.e., the number of phages produced per infected bacterium). Viral production can be used to calculate the number of host cells lysed. Viral production can be calculated by measuring the decrease of the number of viruses in a sample where the viral production is stopped by the addition of cyanide, but where the elimination of the virus continues. Viral production can be also measured by adding viruses labeled with SYBR Green I to the sample. During incubation, the number of labeled viruses and the total number of viruses are followed. Viral production adds viruses unlabeled in the sample, thus reducing the proportion of labeled virus originally added. The elimination of viruses is supposed to be equivalent between labeled and unlabeled viruses. From these measurements, it is possible to simultaneously determine the rates of production and elimination of viruses.

The rate of DNA and RNA viral synthesis can be measured via the incorporation of thymidine-^{3}H or orthophosphate-^{32}P. After incubation with radiotracers, viruses are separated from the bacteria by filtration to measure the radioactivity in the viral fraction. The amount of radioactivity is converted into viral abundance using conversion factors which, unfortunately, vary greatly from one study to another.

Dilution technique similar to that used for the measurement of predation by protozoa has been proposed to measure bacterial mortality induced by viruses. In this case, water sample is diluted in different proportions with ultrafiltrated water free of viruses to reduce the chances of viral infection.

The proportion of inducible lysogens can be determined with specific treatments (e.g., addition of Mitomycin C, exposure to UV-C radiation) to induce a lytic cycle for the lysogenic bacteriophages. These treatments result in prophage induction by direct DNA damage. Induction was detected when we observed simultaneously a significantly lower bacterial abundance and higher viral abundance in the treatments relative to control.

17.5 Chemical and Optical Microsensors

17.5.1 Measuring Principles of Chemical Microsensors

Microelectrodes or electrochemical microsensors have been used in animal physiology since the 1950s and only been introduced in microbial ecology since the end of the 1970s.

Their first application in microbial ecology was for the measurement of the concentration profiles of O_2 and S^{2-}, as well as of the pH, in the interstitial pore waters of microbial mats and coastal sediments. Due to impressive methodological developments, presently a large number of compounds can be measured with microelectrodes. The range of compounds has been further enlarged by the application of immobilized enzymes and cells in the electrochemical microelectrodes design, which are known as biosensors. Optical microsensors have been developed and introduced in microbial ecology since the mid-1980s. These comprise two classes, i.e.,

(i) Microsensors designed for measuring different variables that characterize optical properties as, e.g., the flux of photons in a microhabitat.
(ii) Microsensors designed for measuring the concentration of chemical compounds. The latter, also known under the name of optodes, use an indicator pigment that is sensitive to the concentration of the targeted compound.

The use of microsensors allows us to describe the chemistry at the local scale of the microhabitat of the microorganisms with a minimal disturbance. In sediments, biofilms, and microbial mats, it is particularly important to measure the local chemistry at a submillimetric resolution. For example, the measurement of O_2 at high spatial resolution is important for a precise delimitation of the oxic and anoxic zones in microbial habitats, and thus to know where aerobic, microaerophilic, and anaerobic microorganisms may proliferate. In addition to this fine scale description of microhabitats, the use of chemical microsensors also allows us to study the metabolic activities of the microbes and to quantify the reduction and consumption of some key solutes. Hence, for O_2 it is thus possible to study the processes of aerobic respiration and oxygenic photosynthesis at a high spatial resolution within biofilms.

17.5.1.1 How Do Microelectrodes Function?
Potentiometric Microelectrodes

The surface of the working electrode functions as an ion-selective membrane and a voltage is generated by charge separation. Three different types of membrane are used:

(i) A specific type of glass (e.g., pH-sensitive glass for pH electrodes)
(ii) A precipitate of an oxide or a sulfide coating the metal surface of the working electrode (e.g., a precipitate of Ag_2S on the surface of silver allows to measure S^2)
(iii) A liquid ion exchange (LIX) membrane. This technique is very attractive, because the specificity of the electrode response is determined by the incorporation of a synthetic molecule which is called the ionophore. Most ionophores are highly specific for their target compounds.

Nowadays, many different ionophores have become available on the market and novel ionophores are continuously being developed (Table 17.1). The liquid membrane separates the environment from the liquid electrolyte in the interior of the working electrode. In many cases PVC is incorporated in the liquid membrane to stiffen it, which is particularly important for applications in sediments and other environments where mechanical forces may operate on the electrode. After equilibration the potential across the membrane of the working electrode related to the charge separation is described according the Nernst equation:

$$\Delta E = \frac{RT}{zF} \ln\left(\frac{a_e}{a_i}\right)$$

where R is the ideal gas constant, T the absolute temperature, z is the ion charge, F the Faraday constant, and a_e and a_i represent the activities of this ion in the environment and in the interior of the working electrode. The potential of the working electrode is measured with respect to a reference electrode (Ag/AgCl or calomel electrode). The activity of the ion in the interior of the working electrode (a_i) is constant, while the activity in the environment is directly proportional to its concentration ($a_e = f[\text{ion}]$). The potential between working and reference electrode is thus described accordingly:

$$\Delta E = E_0 + \frac{RT}{zF} \ln(a_e) \quad \text{or} \quad \Delta E = E_0' + \left(\frac{k}{z}\right) \log[\text{ion}]$$

where k/z corresponds to the slope factor, which ideally is equal to 59.2 mV at 25 °C for monovalent ions and 29.6 mV for bivalent ions ($z = 2$). However, ideal conditions are hardly never achieved and, in practice, potentiometric microelectrodes showing values of k between 56 and 59 mV are often used. The measured potential is thus proportional to the logarithm of the ion concentration. Detection limits of potentiometric microelectrodes are often between 10^{-5} and 10^{-6} M depending on the targeted ion.

The advantages of the potentiometric electrodes are related to the large spectrum of ions that can be analyzed by this technique (Table 17.1) and their ease of use. The disadvantages are related to their often slow response times, interfering compounds for specific cases, and the nonlinearity of the response as the signal depends on the logarithm of the concentration of the ion. Hence, a geometric series of ion concentrations needs to be prepared for the target ion to obtain a good quality calibration.

Polarographic Microelectrodes

A voltage is imposed on the working electrode with respect to the reference electrode. The target compound that is analyzed using a polarographic method undergoes an oxidation or reduction at the surface of the working electrode. This process generates a current between the working and

Table 17.1 Chemical microsensors used in microbial ecology

Solute (target)	Measurement principle[a]	Interference/difficulty	Application domain[b]
O_2	Polarography	H_2S interferes during first use, afterward signal stabilizes	FrW, EM, Bf
O_2	Optode		FrW, EM, Bf
H_2S	Polarography	Good response if pH <7.5	FrW, EM, Bf
N_2O	Polarography	O_2, but a combined O_2/N_2O electrode avoids this problem	FrW, MaW, Bf
S_2^-	P, membrane = Ag_2S	O_2	FrW, MaW, Bf
pH	P, glass membrane	Very fragile	FrW, MaW, Bf
pH	P, membrane LIX		FrW, MaW, Bf
NO_3^-	P, membrane LIX	Cl^-	FrW, Bf(Fr)
NO_2^-	P, membrane LIX	Cl^-	FrW, Bf(Fr)
NH_4^+	P, membrane LIX	Na^+, K^+	FrW, Bf(Fr)
Ca^{2+}	P, membrane LIX	Bivalent cations	FrW, MaW, Bf, calcification studies (corals, foraminifera)
CO_3^{2-}	P, membrane LIX	SO_4^{2-}	FrW, MaW, Bf, calcification studies (corals, foraminifera)
CO_2	Biosensor, diffusion reaction chamber with pH electrode and enzyme		FrW, MaW, Bf
Sum of $NO_3^- + NO_2^-$	Bio sensor, diffusion reaction chamber with denitrifying bacteria and N_2O microelectrode		FrW, MaW, Bf
CH_4	Bio sensor, diffusion reaction chamber with methanotrophic bacteria and O_2 microelectrode		FrW, MaW
$S_2O_3^{2-}$ and CH_3SH	Bio sensor, diffusion reaction chamber with sulfur-oxidizing bacteria and O_2 microelectrode		MaW
Mn^{2+}	Voltammetry	Fe^{2+}	MaW
Fe^{2+}	Voltammetry	Mn^{2+}	MaW

[a]P potentiometry
[b]*FrW* sediments in fresh water systems, *MaW* sediments in marine or brackish waters, *Bf* biofilm, *Bf(Fr)* biofilm only in fresh water systems

the reference electrodes that is directly proportional to the concentration of the compound in the environment of the working electrode. The voltage used will determine which compounds will be subjected to an oxidoreduction process. The environment between the working and the reference electrode needs to be electrically conductive; hence, spatially separated electrodes can be used in a salty environment. However, the use of combined polarographic electrodes (Fig. 17.24) is preferred. A combined polarographic microelectrode houses the working electrode and the reference electrode within the same micropipette that is filled with a conducting electrolyte (e.g., 3 M KCl). In many cases, the specificity of the combined electrode can be improved by using of selective membranes as, e.g., for O_2, H_2S, N_2O, and Cl_2 microelectrodes (Fig. 17.24). The polarographic combined O_2 microelectrode has been the most used microsensor in microbial ecology. In this case, the gold-plated working electrode is charged −0.8 V with respect to the reference electrode. O_2 diffuses through a silicone membrane, which is only permeable for small uncharged molecules as dissolved gases and reaches the surface of the working electrode, where it is reduced to OH^-. A small silicone membrane is also used for the H_2S microelectrode. Inside the micropipette, H_2S is oxidized into S at the surface of the working electrode (charged +85 mV with respect to the reference electrode) and the electrons are transferred through intermediate compounds as the ferri-/ferrocyanide redox couple. The signal of this combined electrode is directly proportional to the concentration of the non-dissociated H_2S. Therefore, measures of the pH at the same spot are needed to calculate the sum of the total dissolved sulfides (sum of H_2S, HS^-, and S^{2-}).

The response time of the polarographic combined microelectrodes can be extremely rapid (less than 0.2 s). However, this response time depends on their design, particularly the thickness of the membrane, the geometry of the sensing tip, and the insertion of the working electrode in the micropipette. The extremely low response times are a great advantage for studies of metabolic rates, particularly when transition states are analyzed. The perfect linear relation between the signal of the electrode (often in pA = 10^{-12} A) and the concentration of the target compound greatly facilitates their calibration. The disadvantage is related to the consumption of the compound by the electrode, which results in an artifact. However, for most microelectrodes this consumption is so low and the measuring can be neglected.

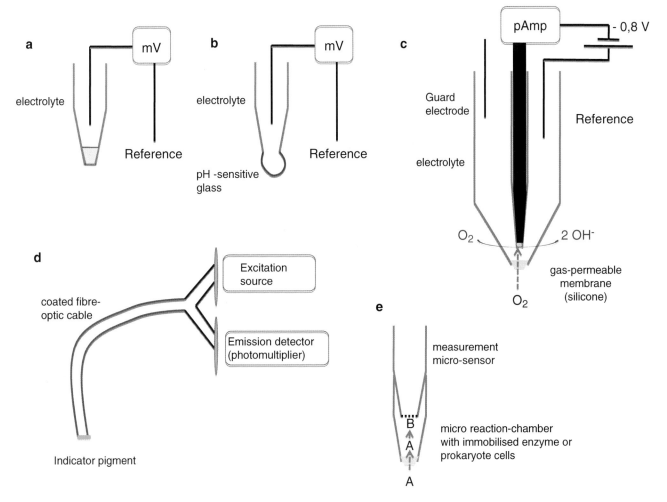

Fig. 17.24 Designs of different microsensors used in microbial ecology. (**a**) Couple of potentiometric microelectrodes: working electrode with a liquid ion exchange (LIX) membrane; (**b**) couple of potentiometric microelectrodes for pH measurements, the working electrode has a sensing tip made of pH-sensitive glass; (**c**) combined polarographic O_2 microelectrode. Both the working electrode and the reference electrode are located in the same micropipette filled with an electrolyte. This micropipette also contains a guard electrode which improves the performance of this sensor; (**d**) micro-optode; (**e**) miniaturized biosensor; the compound A passes through a membrane and is converted into compound B catalyzed by an enzyme or a prokaryotic cell

Voltammetry with Microelectrodes

A microelectrode measuring system based on voltammetry has been developed by geochemists, which, so far, has rarely been used in microbial ecology. This system allows the simultaneous measurement of a whole suite of compounds and ions. A gold microelectrode plated with mercury (Hg) is polarized with respect to a reference electrode and the targeted compounds or ions are oxidized or reduced as in polarography. However, this system applies rapidly changing voltages according a programmed cycle. For example, in the specific case of linear sweep voltammetry, the voltage is varied with time according to a linear function (e.g., from + 0.1 V to −2 V) and the instantaneous current is scanned and plotted as a function of the voltage applied. The relationship between the voltage applied and the current measured often shows stepwise increases with several plateaus, and the differences in current between plateaus allow, for example, to calculate the concentrations of Mn^{2+}, H_2O_2, and O_2. The advantage of this system is its capacity to provide measurements of dissolved Mn^{2+} and Fe^{2+}, which are important products of anaerobic metabolism and can be used as electron donors for chemolithotrophy in the presence of O_2. However, the signal of Mn^{2+} is masked by the signal of Fe^{2+} when the ratio of Fe^{2+}/Mn^{2+} >20. It appears particularly interesting to measure both reduced metals in the presence of O_2. Nevertheless, the measurement of O_2 by this voltammetric system has a lower spatial resolution and is less precise than can be achieved with the polarographic oxygen microelectrodes described above (Fig. 17.24c).

Micro-optodes

Optodes are based on using an indicator fluorochrome pigment coated on the surface at one end of a fiber optic cable. The fluorescence signal of this pigment depends on the concentration of the target compound. Based on this principle, micro-optodes have been developed for measuring O_2, Li^+, pH, and temperature. The other end of the fiber optic cable is separated into two different light-transmitting channels. The first channel is coupled to a monochromatic light source, which light is used for excitation of the fluorochrome. The other channel is coupled to a detector which allows to measure the intensity of the signal and sometimes even the kinetics if the intensity changes. The intensity of the fluorescence signal is not linearly proportional to the concentration, but rather described according the equation of Stern–Volmer for ideal behavior. However, the immobilization of the pigment indicator induces a deviation from ideal behavior, and for the specific case of the O_2 optode, the relation can be described according to the following adaptation of the Stern–Volmer equation:

$$\frac{I_c}{I_0} = \left(\frac{1-\alpha}{1+K_{sv}\cdot C}\right) + \alpha$$

where C is the concentration of O_2 and I_C and I_0 represent the intensity of the fluorescence signal in the presence and absence, respectively, of O_2. K_{SV} is a constant that is characteristic for the pigment indicator and α represents a so-called non-quenchable fraction, which is often about 0.1.

The advantages of the optodes are their ease of use, their high stability, and the absence of autoconsumption. Some disadvantages are related to a more complex calibration procedure and a slower response time than observed for polarographic oxygen microelectrodes.

Miniaturized Biosensors

In a miniaturized biosensor, a very small-sized reaction chamber is combined with a potentiometric or polarographic microelectrode or with an optode (Fig. 17.24). An enzyme or prokaryotic cells catalyze the reaction, and the product formed is detected by the appropriate microsensor. For the CO_2 microsensor, a silicon membrane is used to separate the micro-chamber from the exterior environment. The CO_2 gas diffuses through this membrane into the chamber where it is hydrated by the carbonic anhydrase enzyme ($CO_2 + H_2O \rightarrow H_2CO_3$). The dissociation of this weak acid induces a change of pH that is monitored in the micro-chamber with a pH microelectrode. At steady state when the system has equilibrated with its environment, the pH measured inside is related to the CO_2 concentration outside according to a hyperbolic relationship. This biosensor has a detection limit of about 10^{-5} M and a response time of about 10 s. The biosensor for measuring NO_3^- and NO_2^- represents another example. This microsensor has been developed because the LIX nitrate microelectrode cannot be used in marine environments as Cl^- interferes with the measurements. NO_3^- and NO_2^- diffuse across a membrane into the micro reaction chamber where these compounds are converted into N_2O by a specific culture of denitrifying bacteria. The N_2O produced diffuses through a membrane to a polarographic N_2O microelectrode, and the induced current is proportional to the sum of NO_3^- and NO_2^- concentrations in the direct environment of this biosensor.

17.5.2 Applications of Microsensors for Chemical Compounds

17.5.2.1 Applications in Homogeneous Liquids

The microsensors for chemical compounds can be used in extremely small volumes (e.g., pH or O_2 concentrations in a water droplet or in microtiterplate wells). The polarographic oxygen microelectrodes are equally used in micro-respiration systems in volumes of 500 µl (e.g., measurement of the respiration of a small aquatic animal).

The polarographic O_2 microelectrodes have been used for the measurement of oxygen production by photosynthesis and consumption by aerobic respiration in samples of larger volumes, because of the advantages of microelectrodes with respect to macro-electrodes. Hence, microelectrodes for O_2 have very fast response times and very low autoconsumption of O_2. This has allowed monitoring aerobic respiration at high time resolution in marine waters following a transition from light to dark. During the light phase, net oxygen production is calculated from the net increase of O_2 with time. During darkness, the O_2 concentration decreases because of aerobic respiration. Nevertheless, the rate of respiration is often very high directly after the transition from light to darkness and decreases thereafter. This important observation shows that we have to reject a classical hypothesis according which the respiration during light and dark periods are equal. Rather, during light there is enhanced respiration, which can be explained by the production of organic excretion products by the photosynthesizers and their subsequent consumption by chemoorganoheterotrophs. During the first minutes after the shift from light to darkness, these prokaryotes still continue to benefit from these excretion products, and this effect ebbs away later (Pringault et al. 2007).

17.5.2.2 Microsensor Measurements at Interfaces and Within Sediments and Biofilms

Chemical microsensors are used to measure solute gradients within sediments and biofilms and across the interface of these systems with their overlying water. Therefore, the microsensor is introduced in these systems with a micromanipulator in a specific direction (normally following the vertical axe for sediments and benthic biofilms). The microelectrode moves in this direction and is stopped at regular space intervals to take the measurements; the sensor signal is recorded after stabilization of the signal, which depends on the response time of the microsensor (*cf.* Sect. 17.5.1). The spatial resolution is about twice the dimension of the sensing tip. Hence, for an O_2 microelectrode with a sensing tip of 40–50 µm, measurements can be taken about every 100 µm, while these intervals can be reduced to 20 µm for an electrode with a sensing tip of 5–10 µm.

The shape of the solute gradients is determined by the spatial separation of the sources and sinks for these solutes and by the mass transfer rate between sources and sinks. For example, for an aerobic heterotrophic prokaryote living in the sediment, the overlying water represents the source of O_2, while the aerobic respiration processes by himself and other prokaryotes represent a sink for oxygen. Nevertheless, during daytime, cyanobacteria or microalgae living close to the sediment surface may represent another source of O_2. The solute could be transported from the source to the sink by water currents that may percolate through permeable sediments or through the so-called water channels in some three-dimensionally structured biofilms. However, in many cases interstitial water does not move, as it is the case for non-permeable sediments and dense biofilms. In addition, most sediments and biofilms are covered by a fine water layer of 200–500 µm thickness that does not move and which is called the Diffusion Boundary Layer (DBL) or non-stirred layer. In quiet water, the mass transfer of the solutes occurs through molecular diffusion, which is based on the random movement of these molecules in water. The mass transfer for molecular diffusion can be described by the diffusion laws of Fick. Hence, using Fick's laws, microbial ecologists have developed methods to infer the metabolic rates of microbial populations in solute gradients. This approach represents a particularly interesting extension for the application of microsensor studies in microbial ecology. Nonetheless, before applying Fick's diffusion laws, the microbial ecologists need to check that no water movements occur in their sediment or biofilm samples.

The first law of Fick is used to calculate the diffusive flux, which corresponds to the mass transfer rate per unit surface of the concerned compound. Thus, the following equation describes the molecular diffusion in water for a one-dimensional system:

$$J(x) = -D_0 \bullet \frac{\delta C(x)}{\delta x}$$

where $J(x)$ represents the diffusive flux along the axe, $C(x)$ is the concentration of the solute at place x, and D_0 is a proportionality constant known as the diffusion coefficient. According to this equation, the diffusive flux is directly proportional to the slope $\delta C(x)/\delta x$. This equation needs to be adapted for sediments, where it takes the following expression:

$$J(x) = -\sigma \bullet D_s \bullet \frac{\delta C(x)}{\delta x} \quad \text{with} \quad D_s = \frac{D_0}{\theta^2}$$

where σ represents the porosity of the sediment (the fraction of the sediment volume occupied by interstitial water, value ranging from 0 to 1). D_s is the sediment specific diffusion coefficient, which corresponds to the ratio of D_0/tortuosity (the tortuosity is represented by the symbol θ^2, which corrects for the fact that in the sediment the shortest diffusion distance in the interstitial water space of sediments is longer than the rectilinear distance as diffusion has to get round the sediment particles). In practice, the porosity (σ) can be easily determined (loss of water of a known sediment volume upon complete drying); in contrast, it is often very difficult to determine experimentally the tortuosity (θ^2), thus also the D_s. Therefore, the values used for D_s are often based on approximations. The second law of Fick is used to calculate the rate of the metabolic rates for microbial populations living in diffusion gradients both in sediments and in biofilms. Hence, the second law of Fick for a one-dimensional diffusion system adapted to sediments and biofilms takes the following expression:

$$\frac{\delta C(x,t)}{\delta t} = \sigma \bullet D_s \bullet \frac{\delta^2 C(x,t)}{\delta x^2} + P(x,t) - K(x,t)$$

where the metabolic rates are represented by $P(x,t)$, the metabolic production rate of the solute and $K(x,t)$, its metabolic consumption rate. For O_2, the terms $P(x,t)$ and $K(x,t)$ represent its oxygenic photosynthetic production rate and its respiration rate, respectively. The term $C(x,t)$ represents the concentration of the solute located at x in space and at time t.

Hence,

$$\frac{\delta C(x,t)}{\delta t}$$

represents the change in time of the concentration of x and the term

$$\sigma \bullet D_s \bullet \frac{\delta^2 C(x,t)}{\delta x^2}$$

describes the mass transfer by molecular diffusion.

Generally, while a biofilm or a sediment is experimentally exposed to constant environmental conditions, the gradients tend to become stable in time and reflect steady-state conditions. Under steady-state conditions, i.e.,

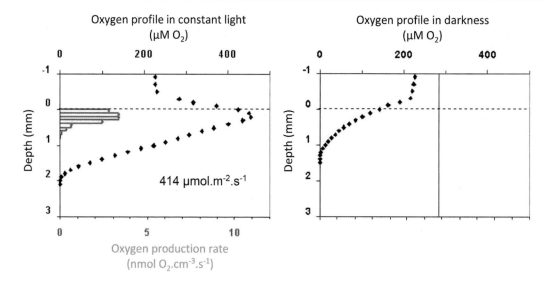

Fig. 17.25 Profiles of oxygen (O_2) concentrations and gross oxygenic photosynthesis rates measured in a biofilm sampled from the mudflats in the Kaw estuary in French Guiana

$$\frac{\delta C(x,t)}{\delta t} = 0$$

After rearranging this equation, one obtains:

$$P(x) - K(x) = -\sigma \bullet D_s \bullet \frac{\delta^2 C(x)}{\delta x^2}$$

Accordingly, the net result of the metabolic rates ($P(x) - K(x)$) is directly proportional to the second derivative off the solute concentration $C(x)$ versus x. However, this approach requests that the porosity (σ) is constant at all depth layers in the sediment profile. A mathematical approach for calculating these rates has been proposed by Berg et al. (1998). Another application of Fick's second one-dimensional diffusion law is used to calculate the gross photosynthetic oxygen production rate using the light–dark shift method. Accordingly, the biofilm is exposed to constant light conditions, and the experimenter waits until steady-state conditions are established.

To realize a measurement of gross photosynthesis at place x, a particularly fast-responding O_2 microelectrode is positioned at this spot. Subsequently, a shift is realized from light to darkness. During the light phase the steady-state conditions are checked, which means that:

$$\frac{\delta C(x,t)}{\delta t} = \sigma \bullet D_s \bullet \frac{\delta^2 C(x,t)}{\delta x^2} + P(x,t) - R(x,t) = 0$$

where $C(x,t)$ represents the O_2 concentration at time t and at the spot x, $P(x,t)$ represents the rate of gross photosynthesis and $R(x,t)$ the respiration rate. During the dark phase, a decrease of O_2 concentration with time is observed according to the following equation:

$$\frac{\delta C(x,t)}{\delta t} = \sigma \bullet D_s \bullet \frac{\delta^2 C(x,t)}{\delta x^2} - R(x,t)$$

because $P(x,t) = 0$ in darkness.

By assuming that the respiration rate remains constant for a couple of seconds after the transition from light to dark, we can deduce that:

$$P(x,t)(\text{light phase}) = -\frac{\delta C(x,t)}{\delta t}(\text{dark phase})$$

This experimental measurement is repeated at different depth horizons in the biofilm to obtain a vertical distribution of gross photosynthesis rates in the biofilm. Nevertheless, for each measurement the steady-state conditions need to be checked for the light phase, before imposing the light–dark transition.

Figure 17.25 represents an example measured in cyanobacterial biofilm on mudflats in French Guiana for the profiles of gross photosynthesis and O_2 concentrations. During darkness, O_2 decreased with depth and the mud was completely anoxic below 1.2 mm depth. During the illumination phase using an artificial source illuminating with 414 µmol photons·m^{-2}·s^{-1}, O_2 accumulated in the surface layer of the mud showing a maximum at 0.3 mm depth. The green bars represent the gross photosynthetic rates determined according the light–dark shift technique.

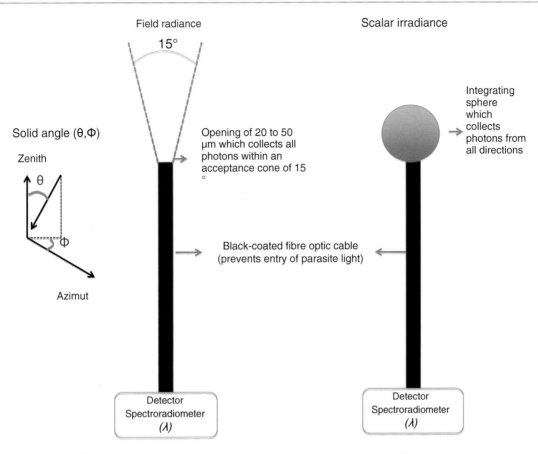

Fig. 17.26 Designs for two different optical microsensors used for measuring the flux of photons in different spots of photosynthetic biofilms and sediments

17.5.3 Optical Microsensors

Different optical microsensors have been developed using coated fiber optic cables (Fig. 17.26). One end corresponds to the sensing part where the light conditions are probed, while the other end is connected to a spectroradiometer used as the detector. The black coating prevents that parasite light enters the detector.

The simplest form of an optical microsensor corresponds to a fiber optic cable of 20–50 µm diameter. This sensor has an acceptance angle of about 30° and is used to measure the photons according the direction of their flux. The direction in a three-dimensional space is described by the solid angle sr (θ,Φ). The variable that is measured with this microsensor is the field radiance which describes the flux of photons traveling through the probed spot in the experimentally fixed direction and is expressed per surface and solid angle, hence in mol photons·m^{-2}·s^{-1}·sr^{-1}. By repeating measurements in different directions, one can obtain a detailed picture of the intensity and directionality of the light fluxes at different spots in photosynthetic biofilms. This field radiance microsensor has also been used to detect the localization of photosynthetic bacteria in biofilms by using their spectral signatures.

Another optical microsensor consists of such a fiber optic cable with an integrating sphere. This sensor allows measuring the total integrated flux of photons passing through the measurement spot independent of their direction. The measured variable corresponds to the scalar irradiance expressed in mol photons·m^{-2}·s^{-1}. The scalar irradiance determines the light energy available for photosynthesis at the measurement spot. The use of a spectroradiometer allows to quantify both variables according the wavelength (λ).

17.6 Stable Isotopes and Lipid Biomarkers

17.6.1 Concepts and Definitions

Organic biomarkers are compounds that have a biological specificity, in the sense that they are synthesized by a limited number of (micro)organisms or classes of (micro)organisms. In this section, the biomarker concept refers to the lipid components of prokaryotes (i.e., lipid biomarkers) such as phospholipid fatty acids (PLFAs), hopanoids, or some lipids specific to Archaea (Box 14.9).

The combined study of prokaryotic lipids and their natural stable isotopic composition (e.g., $^{13}C/^{12}C$; D/H), the so-called compound specific isotope analysis (CSIA), is often used to characterize the carbon cycle and associated biogeochemical processes in recent or ancient ecosystems. CSIA is also useful for linking the structure of communities (phylogeny) with the functions of uncultivable microorganisms. This approach relies on the analysis of the natural stable isotopic composition (usually $^{13}C/^{12}C$) of individual biomarkers (Boxes 4.1 and 14.9) or of their isotopic composition following uptake of a labeled substrate (enriched in stable isotopes).[1] We present below a few examples of both approaches.

17.6.2 Natural Stable Isotopic Composition of Lipid Biomarkers

The parameters controlling the stable carbon isotopic composition of prokaryotic lipids are varied and sometimes difficult to understand. In particular, they include the origin of the carbon substrate, the mechanism by which this carbon is assimilated, the biosynthetic pathways through which lipids are formed, and environmental and physiological conditions. Although the diversity and the variability of these factors complicate the interpretation of lipid $\delta^{13}C$ values (cf. Box 4.1 for δ notation), such data still can provide valuable information about the biology and the chemistry of microorganisms and about the ecosystems in which they thrive.

17.6.2.1 Origin of the Carbon Assimilated by (Micro)organisms

The isotopic composition of heterotrophic bacterial populations is generally quite similar to that of their nutritional carbon source (in principle, "you are what you eat") and may (inter alia) be used to characterize carbon cycling (e.g., food chains) in continental or marine sedimentary ecosystems (Boschker and Middelburg 2002; Pancost and Sinninghe Damsté 2003). For example, the analysis of biomarker ^{13}C composition (such as in PLFAs) in some coastal environments has shown that organic matter derived from aquatic higher plants does not always contribute to bacterial growth, which can be supported through other carbon sources such as phytoplankton (Canuel et al. 1997; Boschker et al. 1999).

In studying the carbon cycle using CSIA (organic compounds being considered as representatives of the bacterial biomass), it is important to take into account the isotopic variability that exists between individual compounds, which arises from fractionation that occurs during biosynthesis. Lipids from heterotrophic organisms are generally depleted in ^{13}C by 3–6 ‰ compared to bulk biomass and assimilated carbon, but different isotopic fractionations (in the range of +4 to −9 ‰) have been observed in some heterotrophic organisms metabolizing different substrates. Thus, establishing a link between the isotopic composition of biomarkers and a specific carbon growth substrate requires that the isotopic fractionation occurring during biosynthesis is known as precisely as possible and is relatively constant. Correction factors are often obtained by using appropriate control experiments (Boschker et al. 1999).

17.6.2.2 Identification of (Micro)organisms Involved in Biogeochemical Processes

Some prokaryotes use a carbon source with a very specific isotopic signature that is thereafter recorded in their biomarkers (following an eventual additional fractionation related to metabolism). These specific biomarker isotopic signatures can thus be used to identify parent microbial populations in the environment. This is especially the case for microorganism involved in the methane cycle since methane is commonly strongly depleted in ^{13}C (due to fractionation occurring during its biological or thermal production[2]). For example, highly depleted $d^{13}C$ values (which can be lower than −110 ‰) of certain lipids, such as the hopanes derived from aerobic bacteria, may serve as indicators of aerobic oxidation of methane (i.e., by aerobic methanotrophs). However, synthesis of hopanes by some strict anaerobes warrants caution in this interpretation (Birgel and Peckmann 2008). Similarly, depleted ^{13}C signatures in specific archaeal lipids (e.g., isoprenoid hydrocarbons and glycerol ethers with isoprenoid alkyl chains) may be used as markers of methanogenesis (Freeman et al. 1994). Like methanotrophs, autotrophic and methylotrophic methanogenic communities can exhibit biomass strongly depleted in ^{13}C relative to the carbon growth substrate. It should be noted that the parameters controlling the isotopic composition of methanogens are still poorly constrained and that ^{13}C depletion in their biomass and lipids may not occur systematically (Pancost and Sinninghe Damsté 2003).

The strongly depleted carbon isotopic composition of glycerol ether lipids specifically synthesized by Archaea has also provided the first irrefutable evidence of the involvement of these organisms in the anaerobic oxidation

---

[1] The use of substrates enriched in radioisotopes (such as ^{14}C) is less often considered in molecular studies and is not considered in this chapter.

[2] However, $\delta^{13}C$ values of biogenic and thermogenic methane can exhibit a large range of variability.

of methane (AOM) in habitats where the process was previously inferred (Hinrichs et al. 1999; Thiel et al. 1999). Since then, many additional isotopic studies coupled with microscopic observations and phylogenetic analyses have helped to refine our knowledge of AOM and to specify that it often involves a syntrophic association between anaerobic methanotrophic archaea (ANME) and sulfate-reducing bacteria (AOM consortia; Box 14.9). Our understanding of AOM, as well as demonstrations of its occurrence in different ecosystems and of its direct or indirect involvement in biogeochemical processes (e.g., precipitation of carbonates or iron sulfide nodules), is steadily increasing and relies largely on the analysis of lipid biomarker ^{13}C composition.

17.6.2.3 Mechanism of Carbon Assimilation in (Photo)autotrophic (Micro)organisms

The differences in isotopic composition between fixed CO_2 and the organic carbon synthesized by (photo)autotrophic bacteria may be characteristic of the modes of carbon fixation and assimilation.

Many autotrophic organisms synthesize biomass using the Calvin cycle, in which the enzyme Rubisco catalyzes the incorporation of $^{12}CO_2$ preferentially to $^{13}CO_2$. The cellular material produced by these organisms is consequently depleted in ^{13}C (or isotopically lighter) by ca. 20–25 ‰ relative to CO_2. Further isotopic fractionations also occur during the biosynthesis of specific cellular components. This produces distinct isotopic compositions for different compound classes (sugars, lipids, etc.), but also for individual compounds within the same class (Pancost and Sinninghe Damsté 2003). Thus, for the Calvin cycle, lipids with linear carbon chains (also called acetogenic lipids), such as PLFAs, are generally depleted in ^{13}C by ca. 4 ‰ relative to biomass, while isoprenoid lipids are slightly less depleted (thus appearing slightly ^{13}C-enriched relative to acetogenic lipids).

The isotopic relationships between the inorganic carbon source, biomass, and lipids vary differently in prokaryotes using assimilation pathways other than the Calvin cycle. Studies with pure cultures indicate that the reverse citric acid cycle existing notably in green sulfur bacteria, or the hydroxypropionate cycle used by the phototrophic bacterium *Chloroflexus* and some hyperthermophilic Archaea, generate smaller isotopic fractionations during biomass formation than the Calvin cycle (van de Meer et al. 1998, 2001). In addition, acetogenic lipids in (micro)organisms using the reverse citric acid cycle are enriched in ^{13}C relative to biomass and to isoprenoid lipids (van der Meer et al. 1998). For the hydroxypropionate cycle, linear lipids are slightly depleted (by ca. 1–2 ‰) relative to biomass but are enriched relative to isoprenoid lipids (van der Meer et al. 2001).

17.6.3 Isotopic Labeling

Isotopic labeling is based on the (partial) consumption of a substrate artificially enriched in stable isotopes (e.g., ^{13}C, ^{2}H, or D) by (micro)organisms growing in laboratory micro- or mesocosms, or in the environment. In microbial ecology, this approach is often called the "SIP method" (for stable isotope probing, see Box 17.3).

Box 17.3: "Who Does What?" Isotope Probing

Pierre Peyret

One of the biggest challenges that microbiologists face is to identify which microorganisms are carrying out a specific set of metabolic processes in the natural environment. New approaches of isotope labeling (Box Fig. 17.3) allow a better ecophysiological understanding of microbial communities (Neufeld et al. 2007b). SIP (stable isotope probing) was first applied in the analysis of phospholipid fatty acids (PFLA) that can be extracted from microorganisms and analyzed by isotope-ratio mass spectrometry (IRMS). Although PFLA analysis offers great sensitivity, the use of labeled nucleic acids as biomarkers has the potential to identify a wider range of bacteria with a greater degree of confidence. DNA-based SIP (DNA-SIP) is increasingly being used in attempts to link the identity of microorganisms to their functions (Dumont and Murrell 2005). This approach has been used to characterize bacteria metabolizing C1 compounds such as methane, methanol, and methyl halides in various environments, as well as multicarbon compounds. The incorporation of a high proportion of ^{13}C into DNA greatly enhances the density of labeled DNA compared with unlabeled (^{12}C) DNA. The DNA was isolated and subjected to caesium chloride (CsCl) buoyant density-gradient centrifugation with ethidium bromide. The heavy ^{13}C-DNA can be purified away from the light ^{12}C-DNA by needle collection and used as a template in PCR, with general primer sets that amplify rRNA genes. It is possible to target "functional" genes as it was demonstrated for methanotroph bacteria (Cebron et al. 2007). FISH-microautoradiography (FISH-MAR) and the isotope array both use radioactive tracers to monitor the incorporation of substrate (Dumont and Murrell 2005). The isotope array involves incubating an environmental sample with a ^{14}C-labeled substrate, after which the RNA is extracted from the sample, labeled

(continued)

Box 17.3 (continued)

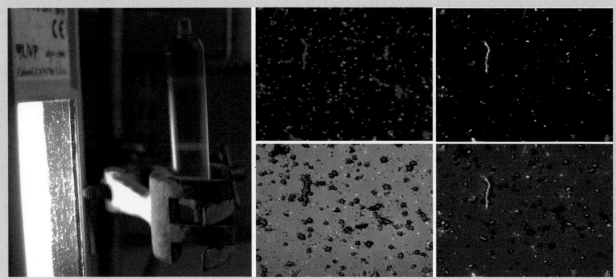

Box Fig. 17.3 The SIP technique. (**a**) Stable isotope probing (SIP). The heavy ^{13}C-DNA can be purified away from the light ^{12}C-DNA to caesium chloride (CsCl) buoyant density-gradient centrifugation; fluorescent staining and micro-autoradiography. (**b**) DAPI staining of prokaryotic cells from a lacustrine ecosystem; (**c**) fluorescent in situ hybridization (FISH) using EUB338 probe; (**d**) DAPI staining and autoradiography after ^{3}H-Thymidine labeling; (**e**) FISH using eub338 probe and autoradiography after ^{3}H-Thymidine labeling (Photographs **b–e**: Courtesy of Delphine Boucher)

with a fluorophore and analyzed with an oligonucleotide array that targets 16S rRNA. The array is then scanned for fluorescence and incorporation of radioactive isotope to determine which community members have metabolized the substrate. Alternatively, secondary ion mass spectrometry (SIMS) can be combined with in situ hybridization to reveal the relationship between phylogeny and naturally occurring variation in stable isotope ratios, indicative of particular metabolic processes such as anaerobic methane oxidation (Orphan et al. 2001). SIMS imaging and Raman microspectroscopy are suitable for detecting and quantifying stable isotope labeling of single microbial cells in complex microbial communities and can be combined with in situ hybridization to identify the active cells (Wagner 2009).

Depending on the process investigated (autotrophy/heterotrophy, metabolic pathways, etc.), the labeled substrate may be inorganic (e.g., $^{13}CO_2$, $NaH^{13}CO_3$) or organic (e.g., $^{13}CH_3COOH$, $^{13}CH_4$, labeled pollutants, or planktonic cells). Following incubation, the cellular components (including lipids) of bacteria that have metabolized the substrate are enriched in the isotope being considered.

17.6.3.1 Deciphering Active Populations in Biogeochemical Processes

The use of labeled substrates together with biomarker investigation provides the ability to identify the part of the prokaryotic community involved in a biogeochemical process. To do so, it is necessary to compare the distribution of labeled biomarkers with known lipid compositions of (micro)organisms. For this type of study, the most commonly used biomarkers are phospholipid fatty acids (i.e., PLFAs-SIP method). In addition to identification of active organisms, degradation rates and growth yields can sometimes be estimated since lipid biosynthesis is closely linked to the growth of (micro)organisms.

Sulfate-reducing bacteria metabolizing acetate, one of the main degradation products of organic matter in anoxic environments, were studied by incubating uniformly labeled acetate ($^{13}CH_3^{13}COOH$) in different sediments (Boschker et al. 1998; Boschker and Middelburg 2002). The study of PLFAs showed that the labeled carbon predominantly occurred in compounds with even-numbered carbon chains (i.e., 16:1ω7, 16:1ω5, 16:0, and 18:1ω7) and only to a limited extent in fatty acids typical of Gram-negative sulfate-reducing bacteria (e.g., i17:1 and 10Me-16:0). In these experiments, the strong resemblance of labeled PLFAs profiles to those of the Gram-positive sulfate-reducers *Desulfotomaculum acetoxidans* and *Desulfofrigus* spp.

suggests that these genera are involved in acetate mineralization. The same kind of approach with labeled propionate ($^{13}CH_3CH_2COOH$) showed that this other ubiquitous intermediate may be mineralized without acetate production by populations of sulfate-reducing bacteria that are distinct from those involved in the oxidation of acetate (Boschker and Middelburg 2002).

The PLFAs-SIP method is often used to highlight the activity of a population composed of a small number of cells or with a low growth rate. A typical example is the consumption of atmospheric methane by soil bacterial communities. Ambient methane concentrations are generally low and soil methanotrophic populations are sparse, making measurement of methane oxidation rates difficult, as well as identification of the populations involved in the process. By continuously supplying portions of soils with small amounts of $^{13}CH_4$, several studies (Neufeld et al. 2007a) have demonstrated the variable activity of two different methanotrophic bacterial populations (type I and/or type II) depending on ambient methane concentration (both populations having distinct PLFA profiles).

Use of the PLFAs-SIP method is not limited to natural substrates, and it can also be employed to characterize the populations involved in degradation of xenobiotics or of toxic substances such as toluene (Hanson et al. 1999) or phenanthrene (Johnsen et al. 2002).

17.6.3.2 Primary Production and Food Chains

Labeling of bicarbonate (i.e., $NaH^{13}CO_3$) coupled with PLFA analysis can be used to distinguish the primary producers (bacterial communities vs. phytoplankton) in aquatic ecosystems and to trace carbon transfer between autotrophic and heterotrophic populations (Boschker and Middelburg 2002). For example, microcosm incubations made with sediments from a brackish estuary have shown that, in this ecosystem, primary production may involve different organisms at different times of day. Carbon is essentially fixed by phytoplankton [characterized by polyunsaturated PLFAs (PUFAs), such as 18:3ω3 for green algae and 20:5ω3 for diatoms] under illuminated conditions, while most of the primary production occurring in the dark is due to chemoautotrophic bacteria (which do produce PUFAs). During an in situ study, monitoring of labeled carbon in the different compartments of a benthic ecosystem suggested that heterotrophic bacterial communities metabolized extracellular polymers originally formed by the phytoplanktonic community initially fixing carbon. The role of heterotrophic bacteria in food chains can also be directly inferred using labeled organic substrates (e.g., planktonic cells enriched in ^{13}C and/or ^{15}N). The transfer of carbon to higher trophic levels (meiofauna, macrofauna) can also be estimated in this manner.

17.6.3.3 Biodegradation Pathways

In addition to characterizing the microbial populations metabolizing specific organic substrates, labeled molecules can further be used to unambiguously determine their biodegradation (or biotransformation) pathways. This is particularly useful for monitoring the fate of pollutants in the environment.

For example, incubation of anaerobic bacteria (pure strains, populations, or communities) using labeled hydrocarbons as the sole source of carbon and energy has helped to elucidate some mechanisms involved in the anaerobic oxidation of aliphatic (Grossi et al. 2008) and aromatic (Foght 2008) compounds, which have long been considered to be refractory in the absence of oxygen. These studies also identified specific metabolites arising from the anaerobic oxidation of non-methane hydrocarbons (i.e., specific degradation intermediates which are not produced from other compounds or by abiotic processes). Some of these metabolites can then be used as indicators of anaerobic hydrocarbon degradation activity during in vitro or in situ investigations (e.g., Gieg and Suflita 2002; Young and Phelps 2005).

17.6.3.4 Concluding Remarks

Despite the clear promise of combined studies of prokaryotic lipids and of their natural stable isotopic composition, it is also worth keeping in mind the limitations of this approach.

The interpretation of CSIA data inherently depends on knowing the biomarker composition of existing prokaryotic populations, whose taxonomic identification may remain hypothetical and whose involvement in biogeochemical processes may be difficult to define precisely. The identification of novel molecular proxies specific to (micro)organisms and/or certain biochemical processes remains an ongoing objective. The diversity and the variability of factors controlling the natural ^{13}C composition of prokaryotic lipids also require that we improve our understanding of natural isotopic fractionation, notably by performing additional studies based on isolated organisms.

An advantage of the PLFAs-SIP method is based on the fact that all biomarkers of an active (micro)organism can be labeled. In this case, biomarker specificity is a priori less essential than for studies based on the natural stable isotopic composition of individual compounds. However, even it is occasionally possible to identify uncultivable (micro)organisms by comparing their PLFA profiles with those of isolated and identified species, biomarkers generally provide limited information on the phylogeny of most uncultivable (micro)organisms. For such purposes, the DNA-SIP technique is preferred (Box 17.3). Moreover, methods of isotopic labeling are limited to the study of living (micro)organisms and/or of actual biogeochemical processes, whereas the natural stable isotopic composition of biomarkers can also be used to study geological samples (dating up to several million years old). However, in this latter case, the geological and thermal history of the samples may complicate the interpretation of biomarker $\delta^{13}C$ values, and thus multidisciplinary (biogeochemistry, microbiology, sedimentology, etc.) or "multi-proxy" (organic /inorganic) approaches may be preferred.

Unlike studies based on the natural abundance of stable isotopes, isotopic labeling does not systematically require the use of an isotope-ratio mass spectrometer, since the identification of labeled lipids may be performed using a conventional mass spectrometer (GC-MS). The lower isotopic sensitivity of a GC-MS relative to a GC-IRMS may however require the use of extensive of labeling. It then becomes necessary to consider the commercial availability and the cost of the labeled substrate. If the substrate is not readily available and/or its price is too high, it can be synthesized in the laboratory using cheaper and commercially available labeled precursors.

17.7 Techniques for Microbial Diversity Studies

17.7.1 Nucleic Acids Extraction from Environmental Samples

Organisms contain DNA and RNA, DNA being the cell's long-term biological memory with a lifetime that is significantly longer than that of RNA. This molecule, which is the cell's short-term memory, is a fragile molecule that is regularly recycled to ensure responsiveness to biochemical and physical changes in the biotope. Extraction techniques targeting the DNA are much simpler and are based either on a direct extraction of bulk DNA by physical processes of sonication, thermal shocks, and purification by reverse phase chromatography on Elutip columns (Picard et al. 1992) or when a prior separation of cells in the inorganic matrix by Nycodenz gradient centrifugation, the latter approach is particularly appropriate when the matrix contains a lot of humic acids that are inhibitors of the polymerases used for RNA retrotranscription or to amplify DNA by PCR (Berry et al. 2003). Approaches targeting RNA have a supplementary step included for inactivation of RNases that uses guanidinium isothiocyanate, which allows to recover up to 26 % of the RNA initially present (Ogram et al. 1995). In many ecosystems, part of the microorganisms present die, and their nucleic acids are no longer repaired by enzymes normally present and thus accumulate damages due to ionizing radiation or related to the presence of reactive compounds in the environment. It is thus known that as DNA ages, for instance, after a few hundreds or thousands of years, it will become more and more difficult to analyze (Mitchell et al. 2005). The released nucleic acids can bind to the inorganic matrix, such as clay layers of the soil and keep well for several years (Frostegård et al. 1999). Nucleic acids have a life expectancy that varies depending on the types of environments, especially the presence of a matrix, of nucleases, of other microorganisms, of ionizing radiations. Reagent kits comprising a step of chemical lysis and column chromatography are now offered by various companies such as Mo-Bio ™ (www.mobio.com) ™ Bio101 (www.bio101.com) or Soil Master ™ kit (www.epibio.com) for this kind of approach, allowing for greater reproducibility.

17.7.2 The Different PCR Techniques

17.7.2.1 The Regular PCR

Polymerase chain reaction (PCR) method relies on the properties of DNA polymerase to extend from $5'$ to $3'$ end direction, an incomplete strand of a partially double-stranded DNA using the other strand as template (Saiki et al. 1988). The double-stranded zone corresponds to the sequence on which a small fragment of about 20 nucleotides of DNA called primer can bind. This primer serves as a starting point for DNA synthesis. Using primers, it is possible to copy the two strands of a DNA fragment, with a primer binding to the sense strand (for the synthesis of the antisense strand) and the other primer binding to the antisense strand (for the synthesis and the sense strand) (Fig. 17.27). The DNA polymerase used in PCR is thermostable, i.e., it is resistant to high temperatures (around 95 °C). To enable the copy of these fragments numerous times, it is necessary that DNA be single-stranded. In consequence, after their synthesis, double-stranded fragments need to be denatured by heating (denaturation step). Then, in order to generate partially double-stranded DNA, it is necessary to favor the annealing of primers to their complementary sequence, and the temperature is rapidly reduced to reach favorable temperature (hybridization step). Hybridization temperature depends on the length of the primers and their base composition (roughly the number of hydrogen bonds linking the two strands, that is, 3 and 2 between C-G and A-T, respectively). An empirical formula for determining the temperature is to count 4 °C for each G or C and 2 °C for each A or T of the primers; otherwise a number of software calculate a more accurate value, but this latter will depend on the amount of mono- and divalent ions in the PCR buffer. During the PCR reaction, when the temperature drops, the primers being more numerous than the DNA fragments, their attachment to the DNA strands is favored over the re-pairing of the two complementary DNA strands. Then, the temperature increases again to reach the optimum temperature of DNA polymerase (around 72 °C), and complementary strand synthesis occur by primer extension (elongation step)

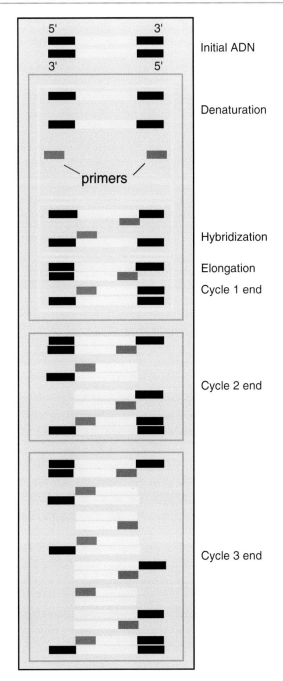

Fig. 17.27 Amplification by PCR of a DNA fragment. Each cycle contain a denaturation, hybridization and an elongation step (Drawing: M.-J. Bodiou)

(Fig. 17.27). The duration of each cycle is variable, generally about 20–30 s for the denaturation and hybridization. Regarding elongation duration, it depends on the length of the fragment to be synthesized and the synthesis rate of the chosen polymerase. In general, it is about 1,500 bases synthesized by minute. This cycle of denaturation, annealing, elongation is repeated 20–40 times and corresponds to three steps PCR. Some enzymes present an elongation activity at lower temperature (68 °C), and two steps PCR can be performed in some case with only denaturation and annealing–elongation steps. In PCR, since complementary strand of target strand is synthesized at each cycle, the amount of DNA doubles each round of PCR during an exponential phase. After this phase, one or more components become limiting and a plateau is observed. Although the initial DNA fragment can be very long, the vast majority of amplified ones would have a size corresponding to the distance between the two primers. It is possible to visualize the amplified fragment by electrophoresis and check its size and the absence of nonspecific fragments. For each PCR, it is necessary to include a tube corresponding to the positive control (containing a tube having the DNA fragment to be amplified) and a tube corresponding to the negative control (with ultrapure water instead of DNA to verify the absence of any contamination). The choice of primers is essential for specific and efficient amplification. If the sequence of the gene is known, it is possible to use softwares designed to provide compatible primers with close hybridization temperatures, without sequences complementary to each other and separated with a close distant location (the more a fragment will be long, the lower the PCR efficiency will be unless using a specific polymerase). It is also necessary to verify the specificity of the primers by comparing their sequence in gene banks (using, e.g., Blastn software on the NCBI) (see Sect. 17.6.5). If the sequence to be amplified is not known for the target organism, one possibility is to compare all of the known gene sequences for different organisms (e.g., for global sequence alignment with the Clustal software) and then select conserved areas as potential sequences of PCR primers. The primers can be degenerated, i.e., it is possible to synthesize primer having a nucleotide variation at a given position. To increase their specificity and avoid mismatch (hybridization of noncomplementary bases), primers containing specific nucleotides (LNAs or locked Nucleic Acids) can be used. In these nucleic acids, furanose is chemically blocked, which increases its affinity for the complementary nucleotide (Koshkin et al. 1998; Vester and Wengel 2004). Only the DNA fragments can be amplified by PCR.

17.7.2.2 RT-PCR

PCR method is rapid, efficient, and frequently used. However, DNA can come from dead or living organisms and does not inform about the expression level of a given gene. In contrast, due to the instability of RNA, its detection indicates a recent synthesis from living organism; furthermore its quantity is an indication of the level of expression of the corresponding gene. However, PCR cannot be applied on RNA directly. To amplify a fragment of RNA, it is first necessary to synthesize the complementary sequence of RNA into DNA (cDNA) with a reverse transcriptase. This enzyme is used in nature by retroviruses and mobile element called retransposon to convert their genome constituted of RNA into DNA before their insertion into DNA genome of their host. The starting point of cDNA synthesis is a partly doubled stranded nucleic acid composed of RNA for one strand and a primer for the other strand. Primer must harbor complementary sequence to RNA strand. It must also bind far from the 5′ end of the RNA since the cDNA is synthetized by 3′ end extension of the primer in direction of the 5′ end of the RNA. Different types of primer can be used for cDNA synthesis:

(i) For studying eukaryotic mRNA, as they comport at their 3′ end a poly A tail (about 200 nucleotides long), a poly T primer can be used.
(ii) To study a specific ARN, the reverse primer of PCR can be used for reverse transcription step.
(iii) To analyze the whole transcriptome of a sample, the use of random hexanucleotide can be useful. One should keep in mind that mRNA constitutes only about 5 % of total RNA. Removal of rRNA before reverse transcription might be necessary. After cDNA synthesis, PCR can be applied on the product of the reverse transcription step; depending on supplier, different types of polymerase can be found and RT-PCR can be performed by one (same mixture for reverse transcription and PCR) or two steps (reverse transcription and PCR being two independent reactions). In order to validate result of RT-PCR, control checking contamination of RNA extract by DNA should be included: The reverse transcription step must include a tube containing all constituents of the assay except reverse transcriptase. After PCR step, negative amplification should be observed for this tube.

17.7.2.3 Quantitative PCR: Real-Time PCR

Different methods have been proposed to use PCR for gene quantification. The most popular is call quantitative PCR (qPCR) or real-time PCR (be aware that RT-PCR term that can lead confusion with reverse transcription). As seen previously, in the exponential phase of PCR, the number of amplified fragments doubles theoretically after each PCR cycle. After N cycles, the theoretical amplification yield is 2^n. If the number of genes initially present in the PCR tube was X_0, after N cycles, the number of X_n gene will be theoretically: $X_n = X_0 \times 2^n$. Since X_n is measurable after staining of DNA, it is theoretically possible to calculate X_0. If this gene is characteristic of a type of organism, and the number of genes per organism is known, the concentration of this organism in the considered sample can be assessed. Experimental data have shown, however, that the number of gene does not double after each PCR cycle and different inhibitors could alter the activity of the thermostable polymerase (humic acids, phenolic compounds, etc.). The yield of the PCR can be determined from the relation between X_n and X_0 using a calibration curve constructed with different concentrations of solutions of known gene fragment (e.g., cloned on a plasmid).

The principle is to use a heat resistant dye (e.g., SYBR Green), which fluoresces only when fixed on double-stranded DNA. Fluorescence is proportional to the amount of DNA, and it is measured in each tube at the end of each elongation step (Fig. 17.28). Quantification is usually made by amplifying relatively small fragments (<500 bp, preferably around 200 bp). This technique requires relatively sophisticated equipment coupled with computer analysis of data. Peculiar PCR tubes or plates should be used and be particularly transparent. The experimenter determines a fluorescence threshold that must be in the exponential part of the curve of DNA amplification. In this part of the curve, amplification follows equation:

$$X_n = X_0 R^n$$

with R, the yield of PCR,
n the number of cycles,
X_0 initial amount of the gene, and
X_n the amount after n cycles.

The software calculates then the number of cycles necessary to reach the fluorescence threshold. This number of cycle is determined for each point of the standard curve. A relation between the base 10 logarithm of the initial concentration of fragments and the number of cycles required to reach the fixed fluorescence can be assessed (Fig. 17.29). This relationship is a straight line whose slope is $10^{-1/R}$. The determination of these parameters is then used to calculate X_0 for unknown samples.

17.7.3 The Molecular Fingerprints

Molecular fingerprinting techniques include different ways to quickly view and analyze diversity between gene fragments amplified by PCR. The most common are the RISA, RFLP (T-RFLP), DGGE, and SSCP.

17.7.3.1 RISA (Ribosomal Intergenic Sequence Amplification)

The method is based on amplification of intergenic region between two genes, in most cases those encoding the 16S and 23S rRNA (for prokaryotes) or the 18S and 28S rRNA (for eukaryotes). Ribosomal genes form an operon (adjacent genes and transcribed at the same time). The beginning and end of these genes are conserved allowing the fixing of PCR primers. Sometimes the intergenic region containing genes coding for transfer RNA and/or noncoding DNA of variable length. After PCR reaction, the amplified fragments size is analyzed on gel electrophoresis (agarose or acrylamide) or capillary. On a single genome, there are

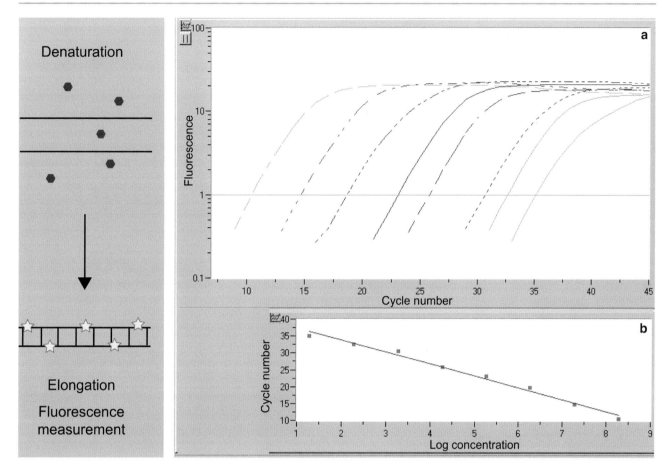

Fig. 17.28 Real-time PCR. Fluorescence of SYBR Green depending on its fixation on DNA (*left part*); example of result allowing to construct a standard curve for real-time PCR quantification (*right part*); (**a**) fluorescence in each PCR tube during each PCR cycle; (**b**) relation between cycle number necessary to reach a given fluorescence and the base 10 logarithm of the target gene initial concentration

several ribosomal operons genes (between 1 and 15 for prokaryotic organisms and hundreds for eukaryotes) and difference between copies may be observed. In consequence, for a given organism, it is possible to obtain several intergenic fragments with different lengths. As mutations accumulate much faster in noncoding regions of the genome, this method can differentiate phylogenetically related organisms; however, as the separation is done only on the criterion of the fragment size, its resolution will depend on that of the electrophoresis method (better to lower: capillary, acrylamide, agarose). In addition, phylogenetically different organisms may yield RISA intergenic fragments of the same size. To overcome this drawback, some authors incubate restriction enzymes with PCR fragments, to increase the resolution of the method. A phylogenetic sequences analysis can be performed on fragment from RISA using preferentially sequences corresponding to the ends of the ribosomal genes since the number of intergenic sequences is still quite limited in gene banks. The automated method is called ARISA (Automated Ribosomal Intergenic Sequence Amplification).

17.7.3.2 RFLP (Restriction Fragment Length Polymorphism)

This method permit to analyze sequence diversity of PCR fragments after their hydrolysis with restriction enzymes and determination of the size of the hydrolyzed products by electrophoresis (Fig. 17.29). The chosen enzymes generally recognize a sequence of four nucleotides and on a statistic point of view are enzymes that cut frequently DNA. The resolution of the technique will depend on the enzyme used and for a given study, different tests must initially be performed. The multitude of bands can increase the difficulty of analysis and variations of this technique have been proposed. To limit the number of bands after electrophoresis, one of the fragments is analyzed by the method called T-RFLP (Terminal Restriction Fragment Length Polymorphism). For this, one of the terminal moiety of the PCR fragment is detected by fluorescence because during PCR, one of the two primers used included a fluorochrome group. If the RFLP and T-RFLP are applicable to any gene, the ARDRA analysis method (ARDRA amplified rDNA restriction analysis) focuses on the diversity of ribosomal genes

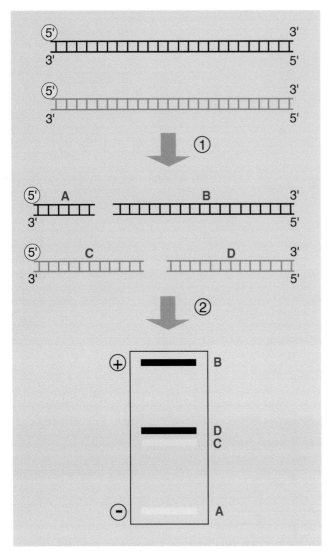

Fig. 17.29 RFLP and T-RFLP techniques. PCR amplified fragments are hydrolyzed with a restriction enzyme (*1*). The length of generated fragments (*A, B, C, D*) is visualized by electrophoresis (*2*). For RFLP, all fragment are stained, whereas for T-RFLP, only the terminal fragments (A and C) are visualized by fluorescence since during PCR, one of the two primers harbored a fluorochrome (*yellow spot*) (Drawing: M.-J. Bodiou)

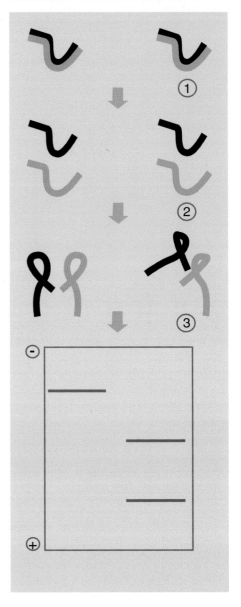

Fig. 17.30 SSCP. DNA strands are denatured by heating then cooled suddenly (*1*). Thanks to dilution, each single strand hybridized on itself (*2*). For a given strand, one or different three-dimensional foldings can occur. The different conformations can be visualized by electrophoresis (*3*) (Drawing: M.-J. Bodiou)

only. The primers and enzymes are standardized allowing the identification of organisms thanks to a data bank containing size of the fragments generated corresponding to different reference bacterial strains.

17.7.3.3 SSCP (Single Strand Conformational Polymorphism)

The sequence diversity of the DNA fragments is analyzed by electrophoresis. To generate different electrophoretic profiles from fragments of the same size but of more or less variable sequence, the fragments are firstly denatured in dilution, after temperature drops, the strands do not hybridize with their complementary strands, but intramolecular reannealing occurs, between small complementary regions (Fig. 17.30). This will generate molecules with varying spatial structures which migrate differently in electrophoresis. The migration is performed on an acrylamide gel or in capillary at low temperature.

17.7.3.4 DGGE (Denaturing Gradient Gel Electrophoresis) and TGGE (Thermal Gradient Gel Electrophoresis)

These are electrophoretic techniques that allow the gradual denaturation of PCR fragments during migration. This denaturation is due either to a gradient of concentration of formamide and urea in the gel (DGGE) or to a temporal increase

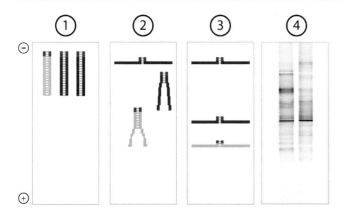

Fig. 17.31 DGGE. The denaturation of double strand DNA is gradual during electrophoretic migration due to the presence of gradient of denaturing compounds. Double-stranded DNA fragment (*1*), two of them are partially denatured (*2*), all of them are denatured except on one end due to the presence of a GC tail (*red*) (*3*). Migration is slowed down with the increase of partial denaturation. Example of a DGGE result (*4*)

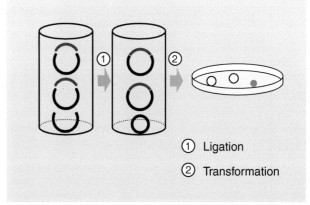

Fig. 17.32 Cloning and transformation of *E. coli* cells. Cloning vector (*black*) and fragment to be inserted (*red*) are ligated together with a DNA ligase (*1*). Cloning vector might close on itself. Ligation mix is introduced in *E. coli* cells (*2*); if the fragment has been inserted in the beta galactosidase gene, after growth on agar plate, it is possible to differentiate a colony harboring a vector with a inserted fragment (*white colony*) from those containing the vector closed on itself (*blue colony*). Since the vector harbors also an antibiotic resistance gene, only vector harboring cells are able to growth on antibiotic containing agar medium (Drawing: M.-J. Bodiou)

of temperature during migration (TGGE). Partial denaturation of DNA fragments slows their migration because they become spherically larger and progress with more difficulty in the meshes of the gel (Fig. 17.31). To increase the resolution of the migration, a modified primer is used during the PCR. One of the two PCR primers comprises an additional part of forty pairs composed only of C and G (GC tail). This additional sequence is never denatured during migration and the two strands of the same fragment are always matched by that part. This lead to a merely stops of the fragment migration when it is denatured except by the GC tail. According to their sequence, the different fragments will stop migrating sooner or later and will be spatially separated in the gel. Resolution of this technique is a sequence variation for about 500 bp fragment.

17.7.4 Cloning, Sequencing

Cloning involves integrating a DNA fragment into a cloning vector i.e., on an extra chromosomal element like plasmid, cosmid or BAC (Bacterial Artificial Chromosome), for easier work. The principle is to linearize the cloning vector by using a restriction enzyme, to mix together the cloning vector, the fragment to integrate with a DNA ligase, an enzyme which will generate a covalent bond between the adjoining linear DNA strands (Fig. 17.32). For successful cloning, the molarity of the cloning vector and the fragment must be close, the ends of the linearized cloning vector and the fragment must be compatible in term of the form (blunt end with two strands ending at same level, or with a cohesive end with protruding edges strand relative to the other, and in the case of protruding edges, complementary between them). In order to generate compatible ends, it is possible to linearize the vector and generate the fragment to be cloned with the same restriction enzyme or enzymes giving end of the same type (this information is usually provided in catalogs of restriction enzyme suppliers). Particular attention must be paid to the ligation of PCR fragments. Indeed, according to the polymerase, the edges of the fragments are either blunt or have a protruding A $3'$. In the first case, a linear vector with blunt ends should be used while in the second case, ligation should be performed with a cloning vector with a protruding $5'$ T (some providers offer this type of plasmid that are already linearized). Another alternative is to make blunt end the $3'$ A protruding PCR fragment. If the starting material is a mixture of different DNA fragments (e.g., fragments from the amplification of ribosomal genes of a community), cloning allow to separate them after transformation in *E. coli* cells, since each clone harbors a plasmid carrying a single type of fragment. Transformation of *E. coli*, may be carried out either by heat shock of $CaCl_2$ treated cells, or by electroporation (electric shock treatment of cells in a solution without ions). If the sequence of the plasmid is characterized, it will be possible to sequence the cloned fragment with a primer binding to the plasmid by the Sanger method (see sequencing technique). The sequencing technique has been proposed in 1975 by Sanger method and allows the determination of the order of succession of different nucleotides comprising the DNA (Fig. 18.1) (Sanger et al. 1977). It is based on the synthesis by DNA polymerase of a complementary strand from an oligonucleotide primer. This synthesis is performed by the polymerization of deoxynucleoside triphosphates (reaction between the hydroxyl at the $3'$ nucleotide of the $n-1$ and the $5'$ phosphate of the nucleotide n). If the reaction is made in four tubes and if each of the

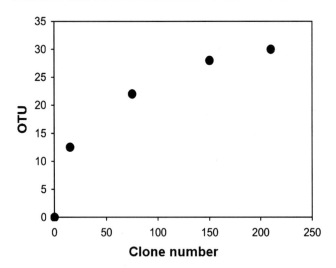

Fig. 17.33 Example of relation between OTU number and number of sequenced clones

tubes a modified nucleotide (one of the four dideoxynucleotide triphosphates, devoid of hydroxyl 3′) is also introduced it will lead to stop the polymerization reaction. In each tube, fragments whose synthesis was stopped randomly have different sizes and all end with the same nucleotide. In the initial method, the various DNA fragments were radiolabeled, and were subjected to electrophoresis in a polyacrylamide gel. Negatively charged fragments migrated the quicker their size is small. On photographic film, it will be possible to determine the sequence by comparing the size of the different fragments and knowing the di-deoxynucleotide terminal 3′ of each fragment. This initial method as evolved now and different sequencing methods are available (see Chap. 18).

Initially cloning sequencing was used in diversity studies. To compare the diversity between samples one must be sure to have exhausted their diversity. In consequence the number of sequenced clones should be important. In such study, it is necessary to analyze the number of different sequences (OTU, Operational Taxonomic Unit) based on the number of clones sequenced. This curve has an asymptote of scarcity which is the number of different sequences in the sample (Fig. 17.33). The traditional approach is less and less used to the benefit of new sequencing techniques (new generation sequencing, NGS) (cf. Sect. 18.1.2), because they can increase drastically the number of processed sequences with lower costs. Number of techniques of sequencing and softwares for sequences processing are growing. As an example, a study of human intestinal microbiomes has analyzed nearly two million 16S rRNA sequences (Turnbaugh et al. 2009).

17.7.5 Bioinformatics Analysis of Sequences

Microorganisms all contain nucleic acids, be they genes or intergenic sequences and these sequences are transmitted vertically with occasional mutation events that make them gradually more and more different in various lineages. This mechanism of mainly vertical transmission should not obscure the fact that some genes are transmitted laterally; however, several genes are considered as molecular markers of the organism as a whole, first and foremost the ribosomal genes that have been considered by Woese et al. (1990) as molecular clock. The 16S gene in particular is now present in databases as more than 4,000,000 entries and is now used as a first approach for the identification of a new organism; it has actually become the gold standard of bacterial taxonomy.

The 16S rRNA gene is obtained by amplification using "universal" (cf. Sect. 8.4.2) primers targeting highly conserved sites at both ends of the gene, for instance, FGPS4-281bis (59-ATGGAGAAGTCTT-GATCCTGGCTCA-39) and FGPS1509′-153 (59-AAGGAGGGGATCCAGCCGCA-39) (Normand 1995) or a partial sequence of about 500nt with other primers (com1 and com2; Lane et al. 1985). The sequence can be obtained by sequencing with the amplification primers or following cloning into *E. coli* if there are copies with slight variations as in *Thermomonospora* (cf. Sect. 6.1) that make direct sequencing impossible. Clones or amplicons can be sequenced by any of several private companies such as MWG (http://www.eurofins.com/en.aspx) or Genomex (http://www.genomex.com/) for a few euros per read.

The analysis of a given DNA sequence is done by comparing it to a set of other sequences using a computerized approach called "Basic Local Alignment Search Tool" (BLAST). This approach is based on a bioinformatics algorithm, where speed is privileged to the expense of accuracy. Speed is essential to compare a new sequence to huge databases containing 130 billion nucleotides in 110 million sequences and still growing at an exponential rate. It was thus not surprising that the paper by Altschul et al. (1990) describing BLAST has been the most cited in the years 1990, in January 2012, the total number of citations had reached 31,530. The BLAST approach begins by eliminating low complexity regions such as repeats, the sequence is then cut into short overlapping sub-sequences, of a length varying depending on the version (16–64 for Blastn as implemented at the NCBI), then these sub-sequences or "words" are compared to the database, receiving a positive score (1–5) for identity or negative (-1 to -4) in other cases. "Words" with a low score are eliminated, while the others are retained in the search tree. Neighborhoods upstream and downstream of the "words" with high scores are then examined for identities. These identities are then quantified and a second score calculated by taking into account gaps. These are then called "High Scoring Pairs" (HSP). The last step is the calculation of BLAST significance score for each HSP, which can be expressed as the probability that a given score is reached by chance, depending on the length of the sequence analyzed and the size of the database. Blast

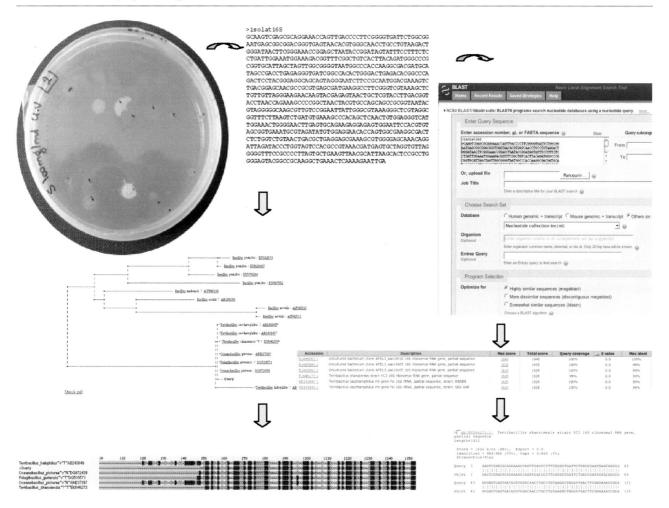

Fig. 17.34 Identification of a bacterial isolate from the 16SrRNA sequence. Soil bacterial isolates were grown in a Petri dish (Photo: Courtesy of P. Pujic), their DNA obtained following cell lysis was amplified with two universal 16S primers (Normand 1995), the sequence obtained was treated either by BLASTN in a general database, the NCBI (*right*), or in a dedicated database, the BIBI database (*left*). In both cases, an identification to the genus *Terribacillus* was obtained

analysis thus yields three values, the % similarity, the score and the likelihood the sequence was obtained by chance or E-value. If the sequence to be analyzed is a protein sequence, the steps are roughly similar with simply a more detailed way to quantify the similarities between amino acids (tables WFP). Different variants of BLAST exist, for instance, BLASTP to compare protein sequences to a protein database, TBLASTN to compare one or several protein sequences to a nucleic acid database that will be translated into protein sequences in the six reading frames, BLASTX to compare a nucleotide sequence translated in the six reading frames against a database of protein sequences, or finally TBLASTX to compare nucleic acid sequence(s) translated into the six reading frames with a nucleic database also translated into proteins.

There are several databases used for this type of analysis. The database of the National Center for Biotechnology Information (NCBI, http://www.ncbi.nlm.nih.gov/blast/) is the best known. It is a database in which new sequences are deposited daily and shared with two other databases that operate in a coordinated manner, the DNA Data Bank of Japan (DDBJ) and the European Molecular Biology Laboratory (EMBL). The three databases have developed a collaboration 19 years ago, called the International Nucleotide Sequence Database Collaboration (http://insdc.org/) to ensure a regular exchange of data and their backups. These sites are sometimes victims of their popularity and can be difficult to access at certain times of the day. This problem is circumvented by the creation of many sites that download data regularly and that allow analyses, such as the PRABI site (www.prabi.fr/) in France. An example of bacterial identification using BLASTN is shown in Fig. 17.34.

Large numbers of input sequences can be treated in a single step via the command-line "megablast," which is much faster than running BLAST several times. Many input sequences are concatenated to form a large sequence before searching the BLAST database, then treated to obtain individual alignments and statistical values.

Other possibilities exist for comparing a given sequence to a database. The best known and most widely used by microbial ecologists is the Ribosomal Database Project (RDP, http://rdp.cme.msu.edu/) where the new sequences in FASTA format are compared to a sequence alignment of 16S gene from bacteria and archaea with a Bayesian approach simplified for increased speed based on a reference hierarchy (Wang et al. 2007). The 16S sequences of type strains in the database are divided into "words" of eight nucleotides whose frequency is calculated. When a sequence is submitted for analysis, the joint probability of finding every "word" is calculated for each genus of the base. Subsets of "words" having a high probability are then used to recalculate the probability a hundred times. Each type strain sequence in the database then receives a number that is the sum of the probabilities of co-occurrence of the presence of "words." For larger taxonomic entities, identification is made by summing the probabilities for each sequence. Other databases have been developed for those who want to work on genes other than 16S. It is the case of BIBI (http://umr5558-sudstr1.Univ-lyon1.fr/lebibi/lebibi.cgi) that allows identifying bacteria using the sequence of one of the following genes *gyrB*, *recA*, *sodA*, *rpoB*, *tmRNA*, *tuf*, *groES*, *groEL*, *dnaK*, *dnaJ*, *fusA* (bacteria), *groel2-hsp65*, and *beta-lactamase*. The result is given in the form of a phylogenetic tree with 30 leaves with the positioning of the unknown sequence and a sequence alignment (Devulder et al. 2003). It is possible and even necessary to automate some of these steps to identify, for instance, 1,000 16S sequences from a metagenomic sequencing project. The sequences must then be formated (FASTA format, the first comment line starting with a ">," the sequence is on the following line) and use the "batch" option in the megablast version.

It is generally considered that an identity of >99 % with the entire 16S sequence of a bacterial sequence present in the base is sufficient to consider assigning it to a given species. This is true in general, but some distinct species have the same sequence, conversely some strains belonging to a given species have different 16S sequences. In general, the sequence of a gene cannot as such enable the identification of a species, but it must be complemented by the analysis of other genes ideally covering the whole genome or by other tests. Other servers do this kind of analysis locally, for example, MultHoSeqI (http://pbil.univ-lyon1.fr/software/HoSeqI/), which also implements a detection tool to detect the presence of chimeras (Arigon et al. 2008).

Databases should continue to expand in the years ahead with the exploration of complex environments using metagenomics approaches and characterization of complete genomes of many organisms at the genomic level. The sequencing capabilities will also increase with new sequencing technologies such as high-throughput pyrosequencing or the Solexa and Solid technologies (Chap. 18). New developments in computer science are also underway to expedite data processing and allow analysis of large data sets.

17.7.6 DNA Microarrays

DNA microarrays (DNA chips, microchips, biochips, gene chips) technology is a powerful, high-throughput experimental system that allows the simultaneous analysis of thousands to hundreds of thousands of genes at the same time. Originally developed in 1995 for monitoring whole-genome gene expressions (Schena et al. 1995), microarrays have been used for the first time in microbial ecology in 1997 (Guschin et al. 1997). Nine probes targeting rRNA 16S genes have been used to identify key genera of nitrifying bacteria. The application of microarray technology for microbial ecology is a rapidly developing approach (Dugat-Bony et al. 2012a). After a short description of the principle of the DNA microarrays approach, various platforms and applications in microbial ecology will be described.

17.7.6.1 DNA Microarrays Principle

DNA microarrays technology is based on nucleic acids hybridizations (Fig. 17.35). However, contrary to Northern-blot or Southern-blot, probes are attached to the solid surface and targets are labeled (Ehrenreich 2006). Under conditions suitable for hybridization, the probes on the chip are exposed to a solution containing a complex sample of fluorescent-labeled targets. DNA macroarrays (dot blots on nitrocellulose or nylon membranes) have the disadvantage of moderate throughput and uncontrolled binding of probes. Planar glass microarrays have become the most widely used type of array. DNA microarrays are solid surfaces to which arrays of specific DNA fragments of various lengths have been attached (ex situ) or in situ synthesized (photolithography technology of Affymetrix or the ink-jet technology of Agilent) at discrete locations. Oligonucleotide arrays are becoming the most widely used type of arrays with the exponential growth in available complete genome sequences, metagenomic data sets, and the low cost of DNA synthesis. Furthermore, with the advancement of microarray technology (in situ synthesis), high-density oligonucleotide microarrays can hold billions of probes on a single microscopic glass slide with multiplexing capacities. These molecular tools can be easily synthesized on-demand, in small batches, and at low cost. This flexibility combined to rapid data acquisition, management, and interpretation allows oligonucleotide microarrays to continue to challenge next-generation sequencing on various applications. However, a very detailed attention to the design of probes is needed for the development of accurate tool (Rimour et al. 2005; Militon et al. 2007). The cross-hybridization is the major point that limits the determination of specific probes.

DNA Microarray Platforms
Owing to advances in microarray fabrication technology, many choices of DNA microarray platforms and physical

(several millions probes). Affymetrix GeneChip® array uses photolithographic method and phosphoramidite chemistry for in situ synthesis of high-density 25-mer oligonucleotide probes (Fig. 17.37a). Adapting technologies used in the semiconductor industry, manufacturing begins with quartz wafer as solid surface (Dalma-Weiszhausz et al. 2006). Nimblegen activity is stopped so we will not describe this technology close to Affymetrix development. A different method for in situ synthesis of oligonucleotide probes (60-mer) using ink-jet technology (Wolber et al. 2006) is proposed by Agilent (Fig. 17.37b).

Probes Design and Targets Preparation

PCR products and cDNA have been first used as probes in transcriptomic assays. For comparative genomics or identification of close genomes in environmental samples, gDNA could also be used. As previously indicated due to ease of synthesis and quality control, oligonucleotides are largely used to date as probe improving specificity detection (Kreil et al. 2006). However, sensitivity detection could decrease with reduced length and needs new design strategy such as GoArrays algorithm (Rimour et al. 2005). One of the major drawbacks of DNA microarrays lied on a sequence a priori with constraints to survey only genes with available sequences in public databases. New probe design strategies like PhylArray (Militon et al. 2007) and KASpOD (Parisot et al. 2012) for phylogenetic microarrays and MetabolicDesign (Terrat et al. 2010) and HiSpOD (Dugat-Bony et al. 2011) for functional microarrays can get around the limitation of sequence availability and make possible the detection of uncharacterized sequences (Dugat-Bony et al. 2012a). Gene capture has been recently applied to environmental samples (Fig. 17.38) with such explorative probes (Denonfoux et al. 2013).

Many different fluorescent dyes and other labeling agents have been described in the literature to label the targets but the cyanine dyes Cy-3 and Cy-5 are most commonly used, offering strong fluorescence, similar chemical properties, well-separated fluorescence spectra, and little adherence to chip surface. Hybridization of DNA microarrays is done by placing labeled, denatured target on a slide. After washing for non-targets elimination, the DNA microarrays is scan to detect fluorescence revealing specific interaction between the probe and the target. The scanners mostly use lasers for exciting the surface of the hybridized microarray. The fluorescence emitted from the dyes linked to the targets is collected and quantified by photomultiplier tubes or charge-coupled device (CCD) cameras. To quantify the fluorescence of the features via image analysis, pixels have to be assigned either to a spot or the background. PCR amplification of the marker gene(s) is applied to improve the detection sensitivity and to limit nonspecific hybridization. Naturally amplified RNA molecules (rRNAs) offer a potential for PCR-free, direct detection,

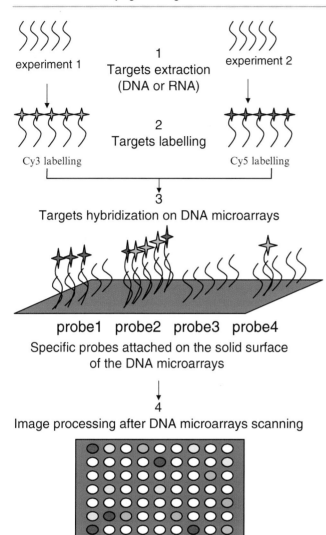

Fig. 17.35 Schematic representation of the different steps involved in the DNA microarrays approach (Drawing: Martine Chomard)

formats are available (Dharmadi and Gonzalez 2004). Based on the arrayed material, currently there are two different microarray platforms (ex situ and in situ).

Ex Situ DNA Microarrays

Through surface derivatization, all kinds of nucleic acids (PCR products, cDNA, gDNA, and oligonucleotides) can be arrayed on solid surface like glass slides using robotic pin spotting or ink-jet printing (Fig. 17.36). One disadvantage of ex situ DNA microarrays is that each probe must be synthesized, purified, and stored prior to microarray fabrication (Schena et al. 1998). Such situation could be expensive for high-density DNA microarrays. Furthermore, probes quality is difficult to validate.

In Situ DNA Microarrays

In situ probe synthesis (oligonucleotide) allows a very flexible DNA microarray fabrication in a very high density

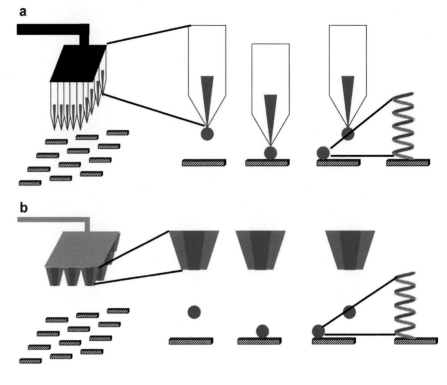

Fig. 17.36 Schematic representation of the probes spotting on the solid surface of the DNA microarrays. (**a**) Contact spotting. (**b**) Ink-jet spotting (Drawing: Martine Chomard)

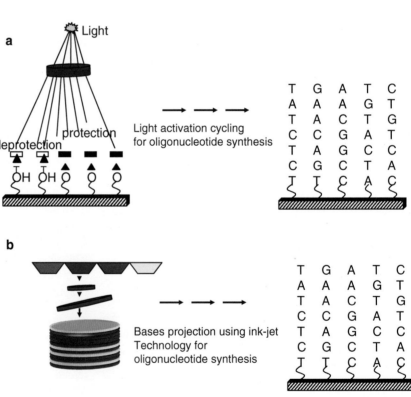

Fig. 17.37 In situ synthesis of the oligonucleotide probes. (**a**) Photolithographic method used by Affymetrix. (**b**) Ink-jet technology used by Agilent (Drawing: Martine Chomard)

thus avoiding the inherent bias in consensus PCR (von Wintzingerode et al. 1997).

Hybridization and DNA Microarrays Analysis

Some factors affecting duplex formation on DNA microarrays include: probe density, microarray surface composition and the stabilities of probe-target duplexes, intra- and inter-molecular self-structures and secondary structures (Pozhitkov et al. 2006). Microarray hybridization has conventionally been conducted in a manual manner by placing the fluorescently labeled probe onto the array under a slide cover-slip and incubating in a humidified chamber

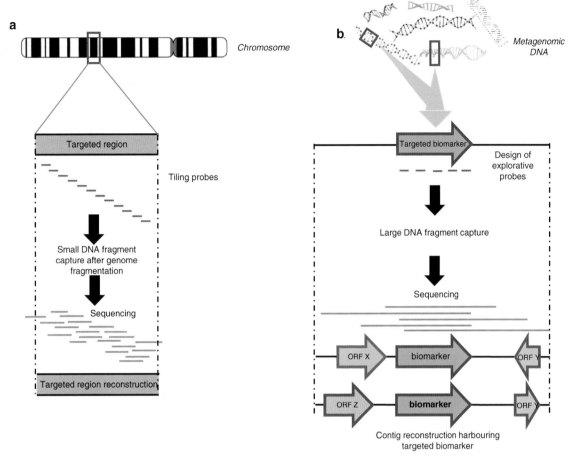

Fig. 17.38 Schematic representation of the gene capture technique. (**a**) Classical approach used for large genomic region re-sequencing. (**b**) Innovative approach for metagenomics targeting (Drawing: Jérémie Denonfoux)

overnight. Although conventional hybridizations are used due to their low cost and ease of implementation, the hybridization reaction is solely dependent on diffusion as the means of probe dissemination, limiting the ability of the probe to react with the entire bound target on the array. Hybridization performance could be improved by automated hybridization (Peeva et al. 2008). Scanning a microarray is a fairly simple task to execute, but it involves selection of a variety of parameters that can have profound effects on the resulting data (Timlin 2006). Image processing through several steps including quality controls allows efficient representation to facilitate interpretations (Ehrenreich 2006).

17.7.6.2 DNA Microarrays for Microbial Ecology

DNA microarrays have been primarily developed and used for gene expression profiling of pure cultures of individual organisms, but major advances have been made in their application to environmental samples (Gentry et al. 2006). Different categories of DNA microarrays have been applied for comparative genomics, transcriptomics assays, phylogenetic identification and functional characterization to precisely describe microbial communities (structure and function) and their dynamics but also to discriminate strains (Fig. 17.39).

Genome-Based DNA Microarrays

Community genome arrays (CGAs) contain the whole genomic DNA of cultured organisms and can describe a community based on its relationship to these cultivated organisms (Wu et al. 2004). Metagenomic arrays (MGA) are a potentially powerful technique because, unlike the other arrays, they contain probes produced directly from environmental DNA itself and can be applied with no prior knowledge of the community (Sebat et al. 2003). Whole-genome open reading frame (ORF) arrays (WGA) contain probes for all of the ORFs in one or multiple genomes. Dong et al. (2001) used a WGA containing 96 % of the annotated ORFs in *E. coli* K-12 to comparatively interrogate the genome of the closely related (97 % based on 16S rRNA gene) *Klebsiella pneumonia* 342, which is a maize endophyte (Dong et al. 2001). More recently, a "pangenome" probe set provides coverage of core *Dehalococcoides* genes as well as

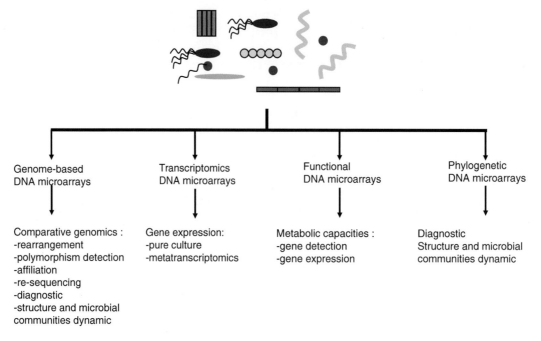

Fig. 17.39 DNA microarrays for microbial ecology (Drawing: Martine Chomard)

strain-specific genes while optimizing the potential for hybridization to closely related, previously unknown *Dehalococcoides* strains (Hug et al. 2011).

Transcriptomics DNA Microarrays

DNA microarray studies are usually carried out as a comparison of two samples to identify differentially expressed genes (Schena et al. 1998). The application of the microarrays to analyze global gene expression of the microbial community in response to oxidative stress has been conducted with success (Scholten et al. 2007). DNA microarray studies of biofilm formation have also addressed numerous questions such as what genes are required for biofilm formation, what environmental signals regulate biofilm formation, how different are biofilm cells, and is biofilm formation a developmental process (Lazazzera 2005). However, it is difficult to evaluate the spatial genes expression in such structures. Probably, biofilm microdissection will help to solve such limits.

Functional DNA Microarrays

Functional gene arrays (FGAs) are designed for key functional genes that encode for proteins involved in various metabolic processes. Currently, the most comprehensive tool developed are the GeoChip 3.0 with ~28,000 probes covering approximately 57,000 gene variants from 292 functional gene families involved in carbon, nitrogen, phosphorus, and sulfur cycles, energy metabolism, antibiotic resistance, metal resistance, and organic contaminant degradation (He et al. 2010). Recently, an efficient functional microarray probe design algorithm, called HiSpOD (High Specific Oligo Design), was proposed to detect unknown genes (Dugat-Bony et al. 2012a). A microarray focusing on the genes involved in chloroethene solvent biodegradation was developed as a model system and enabled the identification of active cooperation between *Sulfurospirillum* and *Dehalococcoides* populations in the decontamination of a polluted groundwater (Dugat-Bony et al. 2012b). Another software program called Metabolic Design ensures in silico reconstruction of metabolic pathways and the generation of efficient explorative probes through a simple convenient graphical interface (Terrat et al. 2010).

Phylogenetic DNA Microarrays

Phylogenetic oligonucleotide arrays (POAs) are designed based on a conserved marker such as the 16S ribosomal RNA (rRNA) gene, which is used to compare the relatedness of communities in different environments. The most comprehensive POA developed so far are the high-density PhyloChip, with nearly 500,000 oligonucleotide probes to almost 9,000 operational taxonomic (Brodie et al. 2006). Currently, very few software dedicated to POAs that allows the design of explorative probes have been developed. The PhylArray program relies on group-specific alignments before the probe design step to identify conserved probe-length regions (Militon et al. 2007). KASpOD is a web service dedicated to the design of signature sequences using a *k*-mer–based algorithm. Such highly specific and explorative oligonucleotides are then suitable for various goals, including Phylogenetic Oligonucleotide Arrays (Parisot et al. 2012).

17.7.7 Pigment Analyses

The main biological functions of pigments in microorganisms are related to light harvesting and processing as well as for photo-protection. Photoprotecting pigment absorbs and neutralizes the photons that would be potentially damaging for the maintenance of the cellular structures. The analyses of pigments in microbial communities can thus tell us something about the ecological importance of both functions. The photosynthetic pigments can also be used as taxonomic biomarkers allowing the estimation of the quantitative importance of anoxygenic phototrophic bacteria, cyanobacteria and different classes of phototrophic eukaryotes in microbial communities. Pigment analyses of pure cultures are also used for ecophysiological and taxonomic studies. Hence, a detailed description of pigment composition is necessary for describing novel species of photosynthetic bacteria and phototrophic eukaryotes. However, ecophysiological studies have shown that the specific contents of pigments may change with environmental conditions as well as the relative ratio between the different pigments in a single species.

The color of the pigments is determined by its absorption spectrum, which is a graphical representation of its absorption of photons as a function of their wavelength (λ). The photosynthetic pigments have absorption maxima limited to the visible wavelength range (400–700 nm), with the exception of the Bacteriochlorophylls (BChl), which also show maxima in the ultraviolet (BChl a) and in the near infrared (BChl a, BChl b, BChl c, BChl d, BChl e, and BChl g). A first approach for analyzing photosynthetic pigments of a phototrophic microorganism in liquid culture is to measure an in vivo absorption spectrum using a spectrophotometer equipped with an integrating sphere. When a cuvette with a culture of phototrophic microorganisms is placed in the spectrophotometer, the photon flux density decreases along the optical path both because of light absorption by the pigments as well as due to diffraction, because of the optical behavior at the interfaces between the cells and their liquid environment. The integrating sphere allows to recover all the photons that have been deviated from their optical path by diffraction and thus to obtain good measurement of the photon absorption alone according their wavelengths (λ). Within the cells, the absorption spectra of the pigments can, however, be modified by biochemical and biophysical interactions and this is particularly the case for the chlorophylls. Therefore, the in vivo absorption spectra of living photosynthetic microorganisms is particularly relevant for the study of biophysical features and often less useful for calculating the specific contents of the different pigments in the cell. The latter can be achieved by extracting the pigments in solution.

There is not a single liquid that allows extracting all known photosynthetic pigments from the phototrophs. Hence, the choice of the extraction solvent determines which pigments are targeted for the analyses. An organic solvent as methanol or acetone is typically used to extract lipophilic pigments as chlorophylls and carotenoids. The hydrophilic pigments, as, e.g., the phycobiliproteins and mycosporine-like amino acids of cyanobacteria, are extracted using a buffered aqueous solution. In order to improve the extraction efficiency different treatments can be used, as, e.g., a freezing thawing cycle or using a French press. The pigment abstracts are normally centrifuged or filtered to obtain a solution that is free of cellular debris to prevent interference of turbidity during spectrophotometric analyses. When the pigment composition of the sample is well known, it is possible measure the concentrations by using multi wavelength (multi λ) spectrometry. Quantification is based on the Lambert-Beer law, using the following equation for a single pigment solution:

$$I_{(\lambda,x)} = I_{(\lambda,0)}\varepsilon^{-KC}$$

where $I_{(\lambda,0)}$ is the incident flux of photon of wavelength λ, $I_{(\lambda,x)}$ is the flux of photons of wavelength λ leaving the cuvette, x is the optical path length in the cuvette, K is the absorption coefficient of the dissolved pigment at wavelength λ, and C is the concentration of the pigment in solution. The absorbance is defined by the following equation:

$$A_{(\lambda,x)} = -\log\frac{I_{(\lambda,x)}}{I_{(\lambda,0)}} = \varepsilon_\lambda \bullet C \bullet x$$

where $A_{(\lambda, x)}$ represents the absorbance at wavelength λ (dimensionless), ε_λ the molar extinction coefficient (en l. mol^{-1}.cm^{-1}). Note the difference in the base of the logarithm; hence $\varepsilon_\lambda = K/2.3$. After rearrangement, the concentration is directly proportional to the Absorbance according the following equation:

$$C = \frac{A_{(\lambda,x)}}{\varepsilon_\lambda \bullet x}$$

This law is additive and can be adapted to calculate the concentrations of several pigments by using a multi wavelength approach (the number of different wavelengths should at least be equal to the number of pigments in solution). However, for pigment mixtures it is often preferable to separate the pigments in order to achieve a better quantification.

Liquid chromatography (LC) allows to separate pigments. LC is often used at high pressure and then referred to as High Pressure Liquid Chromatography (HPLC) where pressures range from 50 to 200 bars. HPLC is most commonly used to separate lipophilic pigments in organic solvents, although methods for separating hydrophilic pigments have also been developed. The pigment molecules are separated in a column filled with a stationary phase. A solvent flux, or mobile phase, elutes through the column and the separation is based on the fact that each pigment equilibrates in a different way between the stationary and mobile phases. The molecules that have a high affinity for the stationary phase are retained on the column for a long time, while the molecules with a lower affinity elute faster from the column. Therefore, retention time (R_t) has been defined as the time between injection on the column and the time it leaves the column and enters the detector.

In most currently used HPLC protocols, the pigments are mainly separated according their hydrophobicity, while the molecule weight and its stereochemistry interfere less strongly. For historical reasons, the term normal phase is used for protocols where the most hydrophobic compounds have the shortest retention times and reversed phase is used for the contrary. Nowadays, most HPLC pigment protocols are based on reversed phase chromatography. An injector is located upstream the column and the mobile phase is delivered by pumps or a solvent delivery system. Isocratic conditions imply that the composition of the mobile phase remains constant. Nevertheless, many solvent delivery systems allow to change the composition of the mobile phase according a programmed solvent gradient. Thus, under reversed phase conditions, the degree of hydrophobicity of the mobile phase is increased during the chromatography in order to optimize pigment separation.

The outflow of the column is connected with a detector. A diode array spectrophotometer is often used for pigment analyses as it allows the instantaneous measurement of a full absorption spectrum. This way a three-dimensional data matrix is generated as the Absorbance (A) is expressed as a function of R_t and λ. As in classical spectrometry the Lambert-Beer law is applicable and there is thus a direct proportionality between the response $A(R_t,\lambda)$. The software often allows the operator to visualize his data as three-dimensional graphs or to choose two-dimensional representations for selected chromatograms (λ fixed) or absorption spectra (R_t fixed).

Other type of detectors can equally be used for the detection of photosynthetic pigments, which can often be coupled in series with a diode array detector. A fluorimetric detector may be particularly interesting for the detection and quantification of chlorophylls as the fluorescence signal is more sensitive than absorption, which thus allows to lower their detection limits. However, the non-linear response of the fluorescence signal is a drawback for quantifications and requests more elaborate calibration. Coupling of HPLC with mass spectrometry, known as LC-MS (Liquid Chromatography Mass Spectrometry) has been developed for pigment analyses. Carotenoids do poorly fragment, and the main information obtained thus concerns the total molecular weight of these compounds. In contrast, the chlorophylls fragment very well and LC-MS thus allows to deduce the following information:

(i) The molecular weight of the chlorophyll.
(ii) The molecular weight of the esterified alcohol.
(iii) The molecular weight of the macrocycle of the chlorophyll molecule.
(iv) The presence of substitutions on the macrocycle and their molecular weight. This type of information allows to determine the exact structure of different allomers of BChl *c* and BChl *d* and can be very useful for discovering novel pigments.

17.7.8 The Analyses of Phospholipid Fatty Acids

The analyses of **phospholipid fatty acids** (**PLFA**)* provides a quick way to study and compare the biodiversity of microbial communities and to reveal the impact of changing environmental conditions on these communities (variations of temperatures, oxygen concentrations, impact of pollutants as, e.g., hydrocarbons and heavy metals). Some PLFA can be used as **biomarkers*** of certain functional groups and of certain genera within a functional group (Spring et al. 2000) (Table 17.2). For example, some genera of sulfate-reducing bacteria are characterized by specific AFLP; i.e., *Desulfobulbus* spp. contain the 15:1ω6 and 17:1ω6 fatty acids, *Desulfovibrio* spp. contain the i17:1ω7c fatty acid, while *Desulfobacter* spp. contain the 10me16:0 and the Cy17:0 fatty acids.

While other characteristic PLFA have been detected in many different microorganisms, their contents are often highest in bacterial species. This is the case for the highly branched iso and anteiso 15:0 and 17:0. The iso and anteiso fatty acids constitute 75 % of the total lipids of *Micrococcus agilis*. For the species *Micrococcus halobius* the iso and anteiso fatty acids with aliphatic chain comprising from 14 to 17 atoms of carbon (14:0 to 17:0) represent almost the total amount of fatty acids, the branched 17:0 fatty acid alone representing already 45 %. Cyclopropane acids are major PLFA of numerous Gram-positive bacteria and *Desulfobacteria* spp. (Dowling et al. 1986). Other PLFA, e.g., palmitic acid (16:0) and linoleic acid (18:2ω6), are widely distributed among living organisms and can thus not be used to infer taxonomic affiliation or physiological status in microbial communities.

The following other type of information can be obtained from PLFA analyses:

(i) The isomerization of the monounsaturated fatty acids, i.e., the conversion of the cis isomer into the trans isomer, can be used as an indicator for stress for bacteria that have been exposed to toxic organic compounds (phenol, toluene) (Heipieper et al. 1995).

Table 17.2 Compilation of the major microbial lipid biomarkers and their occurrence (modified after Spring et al. 2000)

Compound	Structure	Distribution
Saturated FA	16:0	*Bacteria, Eukarya*
	12:0	*Eukarya*
Polyunsaturated FA	18:2ω6c	Fungi, algae, protozoa, Cyanobacteria
	18:3ω6c	Microeukaryotes, Fungi, marine algae
	18:3, 20:3, 20:4	Mycorrhizae
	20:5ω3	*Psychrophilic Shewanella* spp.
	20:3ω6, 20:4ω6, 20:5ω3	Microeukaryotes, diatoms
	Polyunsaturated > C_{20}	*Eukarya*
	22:6	Dinophyceae
	22:6ω3	*Colwellia psychrerythraea*
Monounsaturated FA	16:1ω7c, 16:1ω7t, 16:1ω5c, 17:1ω6, 17:1ω9, 18:1ω9c, 18:1ω7c, 18:1ω7t, cy17:0, cy19:0	Gram-negative bacteria
	16:1ω7, 18:1ω7 predominant	Thio-oxidizing bacteria
	16:1ω7, 18:1ω7, 18:1ω9	*Eukarya*
	16:1ω8c, 16:1ω6c, 16:1ω5t, 16:1ω5c predominant	Type I methanotrophic bacteria (e.g., *Methylomonas, Methylococcus*)
	16:1ω5c	Mycorrhizae
	16:1ω13t	Photosystem I
	15:1ω6, 17:1ω6	*Desulfobulbus* spp.
	18:1ω8c predominant	Type II methanotrophic bacteria (*e.g., Methylosinus, Methylocystis*)
Terminally branched FA	i14:0, i15:0, a15:0, i16:0, i17:0, a17:0	*Arthrobacter* spp. and other Gram-positive bacteria, Gram-negative sulfate-reducing bacteria (*e.g., Desulfovibrio* spp.) *Cytophaga, Flavobacterium*
Monounsaturated branched FA	10me16:0	*Desulfobacter* spp.
	10me16:0, 10me18:0	Actinomycetes
Branched monoenoic FA	i17:1ω7c	*Desulfovibrio* spp. *Syntrophobacter wolinii, Syntrophobacter pfennigii*
Hydroxylated FA	Most common 3-OH	Gram-negative bacteria
Mycolic acids (complex long-chain (C60-C90)	β–OH, α–branched	*Mycobacterium, Nocardia*
Isoprenoid ether-linked glycerol lipids	Caldarchaeol	Hyperthermophilic Archaea, some methanogens,
	Archaeol	Methanogens, extreme haliphilic Archaea,
	α-, β-hydroxyarchaeol	Methanosarcinales
Sterols	Ergosterol	Fungi (*cf.* Chapter 7)
	Sitosterol	Higher plants (*cf.* Chapter 7)
	Cholesterol	Animals
Hopanoids	Bacteriohopanetetrol, aminobacteriohopanetriol	*Bacteria* (*cf.* Sec.4.2.5)
Benzoquinones	Ubiquinones, coenzyme Q	Aerobic Gram-negative bacteria
Naphtoquinones	Menaquinones, dimethylmenaquinone	Aerobic Gram-positive bacteria, anaerobic Gram-negative bacteria extreme halophiles

Fatty acids are linear or branched alkyl chains with a carboxylic acid group on one terminal carbon. Chain-length may contain even or odd number of carbon atoms. Their nomenclature is based on the following criteria:
The fatty acid nomenclature generally follows that recommended by the IUPAC-IUB. A saturated hexadecanoic acid is designated as 16:0, with the first number representing the number of carbon atoms in the acyl group and the second number representing the number of double bonds present. A monounsaturated hexadecanoic acid, such as hexadec-5-enoic acid, is designated 16:1ω5.

(ii) The ratio of vaccenic acid (18:1ω7) to oleic acid (18:1ω9) is very high for bacteria (25) and much lower in diatoms (1) and green algae (0.2).

(iii) An index for hydrocarbonoclastic activity has been proposed by Aries and his collaborators (2001). This index allows inferring the proportion of bacterial growth sustained by use of hydrocarbons as a substrate compared to the proportion of growth sustained by use of hydrophilic growth substrates as, e.g., acetate. This index is calculated by summing up (i) the saturated linear fatty acids with an odd number of C atoms, (ii) the saturated and monounsaturated branched fatty acids, and (iii) the other monounsaturated fatty acids with an odd number of C atoms. This sum is divided by the total amount of monounsaturated fatty acids with an even number of C atoms. Values ranging between 0.8 and 1.3 are characteristic for cultures growing on petrol, while the value is systematically lower than 0.1 for cultures growing on acetate.

The analyses of PLFA allows to follow the temporal dynamics of microbial communities. For example, the chronology of the different phases that characterize compost formation have been studied by following the PLFA. Hence, it has been observed that initial communities are dominated by Gram-positive bacteria, and that this community shifts to increasing dominance of Gram-negative bacteria, actinobacteria, and fungi (Klamer and Bääth 1998).

The study of PLFA contributes very useful information about microbial communities, albeit it must be used with caution. While, the specific biomarkers for microorganisms have been discovered in axenic culture studies of these species under laboratory conditions, which may be very different from natural conditions in the environment. Moreover, the qualitative and quantitative screening for lipid biomarkers among microorganisms, so far, has been limited to a restricted number of species, which is probably insufficient. Novel biomarkers will certainly be discovered in the future as the number of studied strains increases. Therefore, interpretations will need to be updated and revisioned. The polyunsaturated fatty acids provide a good example of such a shift in interpretation. Long time it was considered that bacteria do not contain these polyunsaturated fatty acids, while they have been recently discovered in bacteria living under very high pressure in deep ocean trenches (extremely piezophilic bacteria) (Fang et al. 2002).

Two rather novel approaches in the study of lipids appear particularly promising:
(i) Study of the intact phospholipids. For example, Mazella and his collaborators (2005) have shown that the phospholipid composition changes in the presence of hydrocarbons.
(ii) Analyses of quinones. An example is provided by Tang and his collaborators (2004), who showed how microbial communities changed during the thermophilic phase during compost formation.

17.8 Methods of Isolation, Culture, and Conservation

17.8.1 Cultures in Aerobiosis

Cultivating microorganisms means to put them in conditions favorable enough to allow their development. These conditions include the definition of physicochemical and metabolic parameters (temperature, pH, salinity, oxygen): the cells must have access to an energy source and nutrients. During the culturing of a sample, microorganisms are placed in a new environment. Whether carried out on a solid or liquid medium, the choice of the medium is essential (Fig. 17.40).

For over a century, the list of culture media, presented as more or less selective and more or less "rich" has increased. Developed to be specific or nonspecific according to the data at the time of their formulation, often in a medical or health context, they are still widely used, sometimes simply out of habit. Knowledge and subjects have evolved and thus it is now necessary to reassess the characteristics of these media before considering their use.

Like any environment, any culture medium is necessarily selective because not all microorganisms could develop on it. Similarly, no medium can be considered as specific due to highly variable capabilities of the microorganisms. Consequently, the notion of specificity or "universality" should be considered carefully, especially according to the aim of the study for which the medium has to be used.

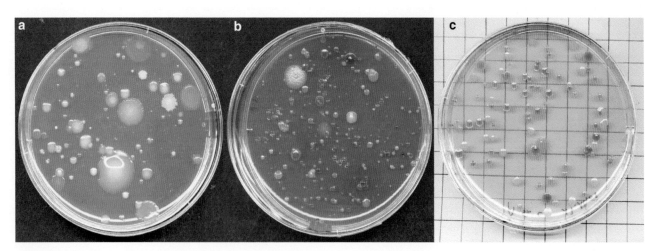

Fig. 17.40 Bacterial colonies obtained after spreading seawater sample on different agar plates culture media and an incubation of few days. Bacterial colonies obtained from a coastal lagoon after 2 weeks of growth on R2A medium (**a**) and on Marine Agar 2216 medium (**b**).The grid on photo (**c**) represents 1 cm^2 (Photographs: L. Intertaglia)

These "classical" culture media may be of defined composition, when all the components are known, or undefined when they contain substances of poorly precise composition (e.g., yeast extract). They must contain at least an energy source and nutrients providing all the necessary elements for growth. In the case of undefined media (e.g., containing cell homogenates), many elements present in trace amounts are present. In the case of defined media, trace elements and micronutrients have to be added (metals, vitamins, growth factors).

Defined media have the advantage of being perfectly controlled because their composition can be adapted and optimized, a modulation of the proportions of each ingredient being theoretically possible. Mathematical tools exist to facilitate what may become an extremely long and heavy work. The optimization is obviously more difficult for undefined media because of the partial ignorance of the compositions of their components.

Once the medium has been chosen, the physicochemical parameters need also to be determined. Incubating temperature, salinity of the medium and pH have to be defined. It seems reasonable to mimic the conditions of the original environment. Depending on growing conditions, microorganisms will have a tolerance range, with an optimum value, for each of these parameters. This range varies for each microorganism and may be as wide as very reduced. To maximize the chances of success, the cultures conditions should approach as close as possible environmental conditions of the biotope of origin.

In the context of a study of the most exhaustive cultivable biodiversity, it is necessary to choose the best culture conditions. This option requires beforehand the knowledge of the main characteristics of the environment of origin to reproduce them instead of selecting one or more media simply because they are usually used for years or even decades.

To limit the selectivity of the medium chosen, it will be rather preferable to work with a set of media. Choices are guided by the objective of the work, therefore by the control the known conditions and not by a historical practice. Moreover, it is also possible to define culture conditions (medium and parameters) to no longer seek to increase the cultivable biodiversity present in a sample, but to target a fraction of this diversity. For example, the search of microorganisms having abilities to resist to compounds, to degrade, transform, or use certain molecules, may be oriented by the choice of cultural constraints imposed voluntarily.

Despite all the precautions taken to reach the balance between the objective of the work, the environment of origin and culture conditions, a bias of selectivity will remain. Technical choices cannot be unlimited (combinations of several culture media and different temperatures, pH, etc.). The strategy for culturing in itself is a factor of selection (solid, liquid, batch). The current development of other practices (alternative techniques, continuous culture) shows how culturing, despite more than a century of microbiology, is still a field of exploration.

17.8.2 Dioxygen Requirements and Cultures Under Anaerobic Conditions

In the environment, concentrations of dioxygene are highly variable. There are a range of intermediates among the microorganisms that grow in the presence of oxygen and those growing in the total absence of dioxygen (see Sect. 3.3). Thus, it is necessary to distinguish between microorganisms (Fig. 17.41):

 (i) Obligate aerobes that require dioxygen to grow; their respiration is aerobic.
 (ii) Microaerophiles, which cannot develop at dioxygen concentration equivalent to that of atmospheric level (20 %), but which still need dioxygen concentration between 2 and 10 %; their respiration is also aerobic.
(iii) The facultative anaerobes, which are able to live either in the presence or in the absence of dioxygen; for example, denitrifying bacteria grow in the presence of dioxygen, but can grow in the absence of this acceptor of electrons if nitrate is available.
(iv) Anaerobic air tolerant that tolerate dioxygen and grow in its presence; it is the case of some fermentative bacteria.
 (v) Obligate anaerobes which are inhibited or killed by dioxygen; these organisms obtain energy by fermentation or anaerobic respiration (Sects 3.3.2 and 3.3.3).

Isolation and growth of aerobic microorganisms take place in the presence of air and sometimes it is necessary to ensure maximum growth to aerate the culture medium by stirring or insufflation of sterile air. For handling anaerobic microorganisms, various techniques for eliminating all traces of dioxygen are implemented (Fig. 17.42). Anaerobiosis can be obtained by:

 (i) The elimination of air which is replaced by nitrogen usually CO_2 enriched for the growth of many anaerobic microorganisms. The operation is performed in either a sealed chamber (Fig. 17.42a), either in tubes, and particularly tubes called "Hungate tubes" which are closed by a stopper "butyl rubber" for withdrawals and transfers (Fig. 17.42b). These tubes are used for enumeration of anaerobic microorganisms by the most probable number (MPN) technique and for the measurement of their activity.
 (ii) The use of a culture medium containing a reducing agent (thioglycolate, cysteine, etc.).
(iii) The removal of dioxygen by catalysis, in an "anaerobic" jar which contains hydrogen and CO_2; in the presence of palladium as catalyst, hydrogen reacts with the

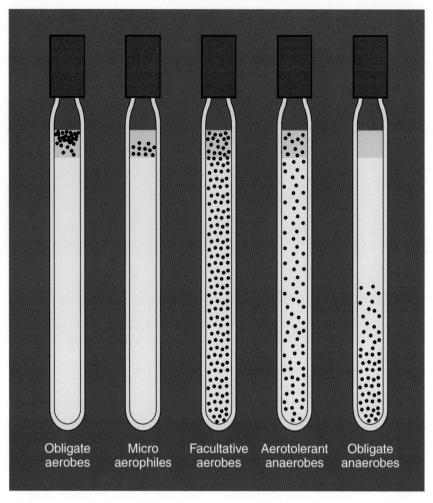

Fig. 17.41 Development of microorganisms as a function of oxygen concentration. Test tubes of small diameter (about 5 mm) are filled to three quarters of their height by a culture medium containing a reducing agent (thioglycolate) and a small amount of agar (7 gl^{-1}, called "deep agar media"), which is supplemented with resazurin indicator of oxidation-reduction; this indicator, colorless in reducing conditions, becomes pink in the presence of traces of dioxygen. The tubes completely devoid of dioxygen after sterilization (autoclaving for 20 min at 120 °C), are immediately immersed in cold water. After cooling, the dioxygen in the air dissolves at the agar surface defining an oxic zone revealed by the pink color of resazurin. Below this zone, the dioxygen concentration decreases, the bottom of the tube is completely devoid of dioxygen. After inoculation of the tubes over the entire culture medium with a platinum wire (seeding "pitting"), the microorganisms will develop according to the presence of dioxygen in the tubes. Colonies are represented by *black dots*, the pink area corresponds to pink resazurin in the presence of dioxygen (Drawing: M.-J. Bodiou)

oxygen present in the jar which is thus removed (Fig. 17.42c).

(iv) The use of an "anaerobic chamber" (or anoxic glove bag) (Fig. 17.43) containing an atmosphere completely anoxic, which allows all the techniques used for aerobic bacteria (usually the air is replaced by nitrogen). A sas attached to the anaerobic chamber, wherein anaerobic conditions are established by replacement air by dinitrogen, allows the equipment input and output though this sas. This is the most effective technique, which requires a significant financial investment, but it is required in laboratories fully specialized in the study of anaerobic microorganisms.

17.8.3 Continuous Cultures

Conventional culture methods allow to cultivate a part, often estimated very low, of the total population of microorganisms in a sample. This problem is due to the artificial culture conditions in vitro which can be very different from those of the environmental context of origin. In addition, except the incubation temperature, the other parameters are defined at the beginning of the experimentation and are not further controlled. Thus, the development of certain microorganisms will be accompanied by a change in the composition of the medium due to the consumption of components and the production of compounds. This

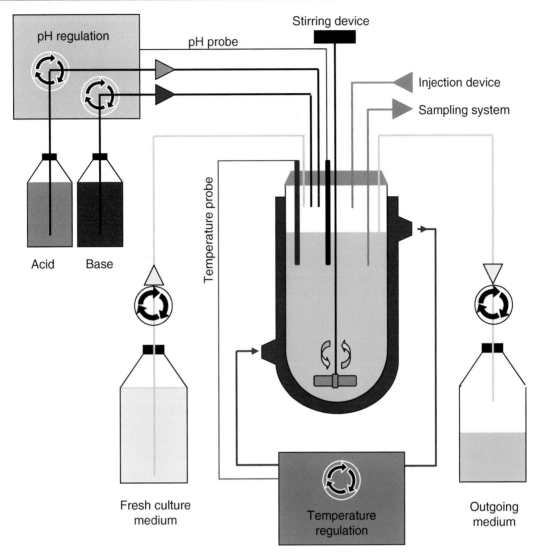

Fig. 17.44 Flow diagram of a bioreactor for continuous culture

(i) The term "isolation" means the complete loss of all interactions with the natural environment and the whole bacterial community (metabolic consortia, Quorum sensing).
(ii) The use of inadequate media that could be too rich (nutrient) and in any case that could not completely mimic the environment targeted.
(iii) The toxicity of some common products like the phenol traces in agar. So, it is a real challenge to overcome these difficulties, that is why alternative techniques have been developed during last decades (Alain and Querellou 2009; Pham and Kim 2012). These methods, sometimes cumbersome to implement and/or expensive, have enabled isolating strains of ecological (widespread) and/or biotechnological interest (Joint et al. 2010).

17.8.5.1 Micro-manipulation

It was in the late 1960s that these isolation methods of a single cell were born (Johnstone 1969). The first system consisted in the aspiration of a cell through a microcapillary. This technique is suitable for the isolation of large cells (eukaryotes), but for small cells (bacteria, spores), a highly accurate technique was developed: the optical tweezers. This method selects a single cell using an infrared laser under a light microscope coupled to a motorized stage. The cell of interest is then moved with the same laser through a capillary glass and then transferred into a nutrient medium (Ashkin et al. 1987 and Ericsson et al. 2000). The major advantage of this technique is the accuracy in which a single cell can be selected from a complex natural sample. The system is very expensive; the choice of the targeted cell is arbitrary and does not guarantee the future culturability.

17.8.5.2 Micro-encapsulation Coupled to Cell Sorting (GMDs)

In 2002, Zengler and colleagues (2002) described an original approach to isolate bacterial strains. The first step consists in the concentration of a natural sample and to mix it with a pre-heated (40 °C) agarose emulsion. After cooling the mixture, the emulsion will generate statistically much agarose micro-droplets (gel microdroplets or GMDs) than cells. Afterward, the GMDs that contain one cell (microcolonies) are selected under light microscope. The second step is the growth of these microcolonies, their selection, and their retrieval. Each GMD is deposited in selected chromatography columns added with continuous nutrient flow. Then, each GMD is deposited on microplate wells containing rich nutrient medium. To confirm the encapsulated cell growth, the GMDs were finally sorted by flow cytometry and double checked by microscopy. Here, the major advantage is the encapsulation a single cell that will grow slowly in a nutrient flow with free cells in the sample. Furthermore, this technique is applicable to high-speed (microplate). This system is expensive, hard to implement, and does not guarantee long-term culturability of the selected cells.

17.8.5.3 Dilution to Extinction

Dilution to extinction culture has emerged in the mid-1990 (Button et al. 1993) and developed in 2000s (Connon and Giovannoni 2002; Stingl et al. 2007) It consists in doing serial dilution on microplate until reaching 1 cell by well at the end using natural sampling water. It is also possible to distribute directly a small number of cells (1–5 for example) in each well. The main benefit of this technique is to allow a slow and gradual adaptation (time incubation for several weeks) of the bacterial cells in conditions that mimics the natural environment studied. In addition, the very low number of cells eliminates opportunistic bacteria that overgrow and inhibit the growth of slow growers of interest. Even if time-consuming, this technique has proved its evidence and continues to be improved. Indeed, many previously uncultured bacteria were isolated for the first time like those that dominate marine ecosystems such as SAR11 or OMG gammaproteobacteria (Rappé et al. 2002; Cho and Giovannoni 2004; Stingl et al. 2007) or some rare bacteria from the rumen (Kenters et al. 2011).

17.8.5.4 In Situ Colonizers and Traps

The principle of this method is to use organic or inorganic supports directly incubated in the natural environment. This method allows the colonization of microorganisms in their natural conditions. Many types of colonizers were described in the literature such as the diffusion chambers (Kaeberlein et al. 2002; Gavrish et al. 2008) used on marine sediment or soil, polyurethane foams (Yasumoto-Hirose et al. 2006) on marine samples, or more recently on steel metal minitraps use for the isolation of previously uncultured oral bacteria (Sizova et al. 2012).

17.8.5.5 Modifications of Growth Media and Conditions

Some modifications done on the sample and/or directly to the culture media can increase the culturability (Nyonyo et al. 2012):

(i) Supplementation: It is the addition of non-traditional nutrient sources such as the culture supernatant of a species which stimulates the growth of another one (Tanaka et al. 2004), the use of cell signaling molecules such as cAMP or acyl homoserine lactone (Bruns et al. 2002) that can play on quorum sensing or growth inhibitors (antibiotics).

(ii) Filtration: the sample is filtered prior on 0.22 µm polycarbonate filters in order to eliminate most of the bacterial cells and work on the filtrate fraction to recover Ultramicrobacteria or previously dormant cells (Hahn et al. 2003).

(iii) An alternative to conventional gelling agents: classic agar is potentially toxic (phenol) and its physicochemical parameters limited. That is why alternative techniques have been developed, such as the use of floating filters (De Bruyn et al. 1990) or the use of other gelling agents such as agarose, gellan gum also called Gelrite or Phytagel (Nyonyo et al. 2012). This gellan gum was first used for the cultivation of thermophilic and acidophilic given its thermal stability and tolerance to pH variations. More recently, this gelling agent was used with success in the isolation of new bacterial diversity (Tamaki et al. 2005).

17.8.6 Management of Culture Collections

17.8.6.1 Culture Collections of Microorganisms

Isolation of microorganisms in the environment has many interests such as: accessibility to the genotype of a strain, the use of strain as experimental model, the production of molecules with high biotechnological potential, and/or the development of scientific collaborations. That's why more and more laboratories around the world develop culture collections despite the requirement of significant human and material resources. In mid-2013, the World Federation for Culture collections or WFCC (http://wdcm.nig.ac.jp/wfcc/) referenced 645 culture collections in the world from 70 different countries with 5,248 people that are fully dedicated to it (Fig. 17.45). Among the 2,244,376 "microbials" stored in these collections, 977,858 are bacterial strains, 633,901 are fungi, and the rest is shared by virus and cell lines in the same proportion.

Fig. 17.45 Distribution of the 645 culture collections referenced by the WFCC (May 2013)

Culture collections should ensure the long-term preservation of isolates, their viability, purity and access to the strains. The storage of the microorganisms is a crucial prerequisite for all the culture collections. The organisms should be long-term preserved in a revival conditions as long as possible. To do this, the metabolic activities have to be blocked in order to reduce the risk of cell damages. The two commonly used methods are freeze-drying and cryopreservation.

Freeze-Drying

This method was born in the 1950s. This technique consists in (Bimet 2007) dehydrating the cells at low temperature under vacuum from an added (e.g., skim milk, sucrose) culture. The freeze-dried powder can be stored at room temperature or at 4–8 °C (Heckly 1978). This method is performed in two steps: freezing and desiccation. For freezing, two methods are used: immersion in a dry ice-alcohol mixture (−78 °C) or centrifugation-freezing to −7 °C. For drying, three factors are important: the void, a very low temperature, and the apparatus for retaining water sublimation. The success by freeze-drying depends on the following parameters:
 (i) The number of cells
 (ii) The strain
 (iii) The cell size and its complexity
 (iv) Resuspending the medium
 (v) Maintaining the vacuum over time

This technique allows a long-term preservation (decades) due to the dehydrated conditions of the cells and an easy storage. Nevertheless, this technique requires specific equipment (freeze dryer, specific glassware) and cannot guarantee full success. It should be noted that very often, freeze-dried bacteria lose their plasmids during the process.

Cryopreservation

This method consists in keeping the cells alive at very low temperatures with the addition of cryoprotectants, compounds that limit adverse effects of freezing. In theory, the more temperature is lower, the better the storage is. But biological structures can be very disturbed by:
 (i) Mechanic breaks
 (ii) Membrane topography changes
 (iii) Water crystallization (and biochemical changes)
 (iv) Mechanical injuries (crystals)
 (v) The increasing of electrolyte concentration

The choice of the cryoprotectant is crucial for cryopreservation success. About 50 different kind of molecules were tested on cultures (Hubálek 2002) as:
 (i) Sulfoxides (e.g., DMSO)
 (ii) Alcohols (e.g., methanol, glycerol)
 (iii) Proteins (e.g., BSA)
 (iv) Polysaccharides (e.g., trehalose)
 (v) Complex compounds (e.g., yeast extract)

Their actions on the cells are multiple. Highly hydrophilic, they involve interactions with the water molecules and thus protect proteins. The permeable cryoprotectants (glycerol and DMSO) limit hyper concentrations of salts and prevent the formation of large ice crystals.

They lower the freezing point of water and biological fluids. The most effective cryoprotectants are: the dimethylsulfoxide (DMSO) and the glycerol. Temperatures are commonly used around −80 °C, but some may go down −130 °C or −196 °C (liquid nitrogen). Maximum survival is observed in a so-called "transition area" during which the formation of intracellular ice hyper concentration and saline are attenuated. This means that freezing must be slow (1 °C /min) and thawing needs to be fast.

As the freeze-drying, cryopreservation also depends on many parameters:
- Cell wall (Gram + > Gram −)
- Cell size and shape
- Growth phase (stationary)
- Incubation temperature
- Culture medium composition
- pH
- Osmolarity
- Cell water content
- Membrane lipid content
- Composition of cryoprotectant
- Cooling rate
- Storage temperature
- Storage time
- Thawing rate
- Revival medium

Cryopreservation can be done easily and quickly on many samples and allows long-term preservation (decades). Unfortunately, the success of this method is also highly variable given the many parameters that influence it. However, the freezing/thawing successive steps could be lethal for cells.

17.9 Conclusion

As shown in this chapter, the range of approaches developed in microbial ecology is extremely broad. Technological developments in this area evolve very rapidly and allow to describe more finely and precisely the structure and the activities of the communities of microorganisms in their biotopes. It is undeniable that molecular techniques have revolutionized microbial ecology and that they are now an integral part of research and teaching in this discipline. The recent application of high-throughput molecular biology methods to natural microbial communities is profoundly changing our view on the microbial world. By combining these new technologies with ecosystem and biogeochemical measurements, it becomes possible to identify more precisely environmental controls on microbial processes and the specific roles of microbes in biogeochemical cycles.

References

Agogué H et al (2004) Comparison of samplers for the biological characterization of the air-seawater interface. Limnol Oceanogr Methods 2:213–225

Alain K, Querellou J (2009) Cultivating the uncultured: limits, advances and future challenges. Extremophiles 13:583–594

Altschul SF, Gish W, Miller W, Myers EW, Lipman DJ (1990) Basic local alignment search tool. J Mol Biol 215:403–410

Aminot A, Rey F (2002) Chlorophyll *a* determination by spectrometric methods. ICES Techn Mat Environ Sci 30:18p

Andersson BA, Holman RT (1974) Pyrrolidides for mass spectrometric determination of the position of the double bond in monounsaturated fatty acids. Lipids 9:185–190

Aries E, Doumenq P, Artaud J, Acquaviva M, Bertrand J-C (2001) Effects of petroleum hydrocarbons on the phospholipid fatty acid composition of a consortium composed of marine hydrocarbon-degrading bacteria. Org Geochem 32:891–903

Arigon AM, Perriere G, Gouy M (2008) Automatic identification of large collections of protein-coding or rRNA sequences. Biochimie 90:609–614

Arístegui JG, Josep M, Duarte CM, Herndl G (2009) Microbial oceanography of the dark ocean's pelagic realm. Limnol Oceanogr 54:1501–1529

Ashkin A, Dziedzic JM, Yamane Y (1987) Optical trapping and manipulation of single cells using infrared laser beams. Nature 330:769–771

Balkwill DL, Leach FR, Wilson JT, McNabb JF, White DC (1988) Equivalence of microbial biomass measures based on membrane lipid and cell wall components, adenosine triphosphate, and direct counts in subsurface aquifer sediments. Microbiol Ecol 16:73–84

Bartlett DH, Lauro FM, Eloe EA (2007) Microbial adaptation to high pressure. In: Gerday C, Glandsdorf N (eds) Physiology and biochemistry of extremophiles. American Society for Microbiology Press, Washington, DC, pp 333–348

Basso O, Lascourrèges JF, Jarry M, Magot M (2005) The effect of cleaning and disinfecting the sampling well on the microbial communities of deep subsurface water samples. Environ Microbiol 7:13–21

Berg P, Risgaard-Petersen N, Rysgaard S (1998) Interpretation of measured concentration profiles in sediment pore water. Limnol Oceanogr 43:1500–1510

Berry AE, Chiocchini C, Selby T, Sosio M, Wellington M (2003) Isolation of high molecular weight DNA from soil for cloning into BAC vectors. FEMS Microbiol Lett 223:15–20

Bianchi A, Garcin J (1993) In stratified waters the metabolic rate of deep-sea bacteria decreases with decompression. Deep-Sea Res I 40:1703–1710

Bianchi A, Garcin J, Tholosan O (1999) A high-pressure serial sampler to measure microbial activity in the deep sea. Deep-Sea Res 46:2129–2142, Part 1

Bimet F (2007) Conservation des bactéries. Actualités permanentes en bactériologie clinique. Editions ESKA, Paris

Birgel D, Peckmann J (2008) Aerobic methanotrophy at ancient marine methane seeps: a synthesis. Org Geochem 39:1659–1667

Bligh EG, Dyer WJ (1959) A rapid method of total lipid extraction and purification. Can J Biochem Physiol 35:911–917

Bochdansky AB, van Aken HM, Herndl GJ (2010) Role of macroscopic particles in deep-sea oxygen consumption. Proc Natl Acad Sci U S A 107:8287–8291

Boschker HTS, Middelburg JJ (2002) Stable isotopes and biomarkers in microbial ecology. FEMS Microbiol Ecol 40:85–95

Boschker HTS, Nold SC, Wellsburry P, Bos D, de Graaf W, Pel R, Parkes RJ, Cappenberg TE (1998) Direct linking of microbial populations to specific biogeochemical processes by ^{13}C-labelling of biomarkers. Nature 392:801–805

Boschker HTS, de Brouwer JFC, Cappenberg TE (1999) The contribution of macrophyte-derived organic matter to microbial biomass in salt-marsh sediments: stable carbon isotope analysis of microbial biomarkers. Limnol Oceanogr 44:309–319

Briand E, Pringault O, Jacquet S, Torréton J-P (2004) The use of oxygen microprobes to measure bacterial respiration for determining bacterioplankton growth efficiency. Limnol Oceanogr Methods 2:406–416

Brodie EL et al (2006) Application of a high-density oligonucleotide microarray approach to study bacterial population dynamics during uranium reduction and reoxidation. Appl Environ Microbiol 72:6288–6298

Bronk DA, Glibert P (1991) A ^{15}N tracer method for the measurement of dissolved organic nitrogen release by phytoplankton. Mar Ecol Prog Ser 77:171–182

Bruns A, Cypionka H, Overmann J (2002) Cyclic AMP and acyl homoserine lactones increase the cultivation efficiency of heterotrophic bacteria from the central Baltic Sea. Appl Environ Microbiol 68:3978–3987

Button DK, Schut F, Quang P, Martin R, Robertson BR (1993) Viability and isolation of marine bacteria by dilution culture: theory, procedures, and initial results. Appl Environ Microbiol 59:881–891

Canuel EA, Freeman KH, Wakeham SG (1997) Isotopic compositions of lipid biomarker compounds in estuarine plants and surface sediments. Limnol Oceanogr 42:1570–1583

Carrignan R, Blais A-M, Vis C (1998) Measurement of primary production and community respiration in oligotrophic lakes using Winkler method. Can J Fish Aquat Sci 55:1078–1084

Cebron A, Bodrossy L, Stralis-Pavese N, Singer AC, Thompson IP, Prosser JI, Murrell JC (2007) Nutrient amendments in soil DNA stable isotope probing experiments reduce observed methanotroph diversity. Appl Environ Microbiol 73:798–807, Epub 2006 Nov 22

Certes A (1884) Sur la culture, à l'abri des germes atmosphériques, des eaux et des sédiments rapportés par les expéditions du Travailleur et du Talisman. C R Acad Sci Paris 98:690–693

Cho BC, Azam F (1988) Major role of bacteria in biogeochemical fluxes in the ocean's interior. Nature 332:441–443

Cho J-C, Giovannoni SJ (2004) Cultivation and growth characteristic of a diverse group of oligotrophic marine gammaproteobacteria. Appl Environ Microbiol 70:432–440

Connon SA, Giovannoni SJ (2002) High-throughput methods for culturating microorganisms in very-low-nutrient media yield diverse new marine isolates. Appl Environ Microbiol 68:3878–3885

Dalma-Weiszhausz DD, Warrington J, Tanimoto EY, Miyada CG (2006) The affymetrix GeneChip platform: an overview. Methods Enzymol 410:3–28

Davidson EA, Savage K, Verchot LV, Navarro R (2002a) Minimizing artifacts and biases in chamber-based measurements of soil respiration. Agric For Meteorol 113:21–37

Davidson K, Roberts EC, Gilpin AC (2002b) The relationship between carbon and biovolume in marine microbial mesocosm under different nutrient regimes. Eur J Phycol 37:501–507

De Bruyn JC, Boogerd FC, Bos P, Gijs Kuenen J (1990) Floating filters, a novel technique for isolation and enumeration of fastidious, acidophilic, iron-oxidizing, autotrophic bacteria. Appl Environ Microbiol 56:2891–2894

Del Giorgio P, Cole J-J (1998) Bacterial growth efficiency in natural aquatic systems. Annu Rev Ecol Syst 29:503–541

Deming JW (1985) Bacterial growth in deep-sea sediment trap and boxcore samples. Mar Ecol Prog Ser 25:305–312

Deming JW, Tabor PS, Colwell RR (1980) Deep ocean microbiology. In: Diemer F, Vernberg J, Mirkes D (eds) Advanced concepts in Ocean Measurements for Marine Biology. University of South Carolina Press, Columbia, pp 285–305

Denonfoux J, Parisot N, Dugat-Bony E, Biderre-Petit C, Boucher D, Morgavi DP et al (2013) Gene capture coupled to high-throughput sequencing as a strategy for targeted metagenome exploration. DNA Res 20:185–196

Devulder G, Perriere G, Baty F, Flandrois JP (2003) BIBI, a Bioinformatics Bacterial Identification Tool. J Clin Microbiol 41:1785–1787

Dharmadi Y, Gonzalez R (2004) DNA microarrays: experimental issues, data analysis, and application to bacterial systems. Biotechnol Prog 20:1309–1324

Dong Y, Glasner JD, Blattner FR, Triplett EW (2001) Genomic interspecies microarray hybridization: rapid discovery of three thousand genes in the maize endophyte, *Klebsiella pneumoniae* 342, by microarray hybridization with *Escherichia coli* K-12 open reading frames. Appl Environ Microbiol 67:1911–1921

Dowling NJE, Widdel F, White DC (1986) Phospholipid ester-linked fatty acid biomarkers of acetate-oxidizing sulphate-reducers and other sulfide-forming bacteria. J Gen Microbiol 132:1815–1825

Dugat-Bony E, Missaoui M, Peyretaillade E, Biderre-Petit C, Bouzid O, Gouinaud C et al (2011) HiSpOD: probe design for functional DNA microarrays. Bioinformatics 27:641–648

Dugat-Bony E, Biderre-Petit C, Jaziri F, David MM, Denonfoux J, Lyon DY et al (2012a) In situ TCE degradation mediated by complex dehalorespiring communities during biostimulation processes. Microb Biotechnol 5:642–653

Dugat-Bony E, Peyretaillade E, Parisot N, Biderre-Petit C, Jaziri F, Hill D et al (2012b) Detecting unknown sequences with DNA microarrays: explorative probe design strategies. Environ Microbiol 14:356–371

Dugdale RC, Goering JJ (1967) Uptake of new and regenerated forms of nitrogen in primary productivity. Limnol Oceanogr 12:196–206

Dumont MG, Murrell JC (2005) Stable isotope probing – linking microbial identity to function. Nat Rev Microbiol 3:499–504

Ehrenreich A (2006) DNA microarray technology for the microbiologist: an overview. Appl Microbiol Biotechnol 73:255–273

Ericsson M, Hanstorp D, Hagberg P, Enger J, Nyström T (2000) Sorting out bacterial viability with optical tweezers. J Bacteriol 182:5551–5555

Fang J, Barcelona MJ, Abrajano T, Nogi Y, Kato C (2002) Isotopic composition of fatty acids of extremely piezophilic bacteria from the Mariane Trench at 11,000 m. Mar Chem 80:1–9

Fang J, Zhang L, Bazylinski DA (2010) Deep-sea piezosphere and piezophiles: geomicrobiology and biogeochemistry. Trends Microbiol 18:413–422

Foght J (2008) Anaerobic biodegradation of aromatic hydrocarbons: Pathways and prospects. J Mol Microbiol Biotechnol 15:93–120

Fowler SW, Knauer GA (1986) Role of large particles in the transport of elements and organic compounds through the oceanic water column. Prog Oceanogr 16:147–194

Freeman KH, Wakeham SG, Hayes JM (1994) Predictive isotopic biogeochemistry: hydrocarbons from anoxic marine basins. Org Geochem 21:629–644

Frostegård A et al (1999) Quantification of bias related to the extraction of DNA directly from soils. Appl Environ Microbiol 65:5409–5420

Fukuda R, Ogawa H, Nagata T, Koike I (1998) Direct determination of carbon and nitrogen contents of natural bacterial assemblages in marine environments. Appl Environ Microbiol 64:3352–3358

Furhman JA (2000) Impact of viruses on bacterial processes. In: Kirchman DL (ed) Microbial of the Oceans. Wiley-Liss, New York, pp 351–386

Furhman JA, Azam F (1980) Bacterioplankton secondary production estimates for coastal waters of British Columbia, Antarctica and California. Appl Environ Microbiol 39:1085–1095

Gasol JM et al (2008) Towards a better understanding of microbial carbon flux in the sea. Aquat Microb Ecol 53:21–38

Gavrish E, Bollmann A, Epstein S, Lewis K (2008) A trap for in situ cultivation of filamentous actinobacteria. J Microbiol Methods 72:257–262

Gentry TJ, Wickham GS, Schadt CW, He Z, Zhou J (2006) Microarray applications in microbial ecology research. Microb Ecol 52:159–175

Gieg LM, Suflita JM (2002) Detection of anaerobic metabolites of saturated and aromatic hydrocarbons in petroleum-contaminated aquifers. Environ Sci Technol 36:3755–3762

Godfroy A, Raven ND, Sharp RJ (2000) Physiology and continuous culture of the hyperthermophilic deep-sea vent archaeon *Pyrococcus abyssi* ST549. FEMS Microbiol Lett 186:127–132

Gordon HR, Brown OB, Evans RH, Brown JW, Smith KS, Baker KS, Clark DK (1988) A semi analytical radiance model of ocean color. J Geophys Res 93:10909–10924

Goutx M, Wakeham SG, Lee C, Duflos M, Guigue C, Liu Z, Moriceau B, Sempéré R, Tedetti M, Xue J (2007) Composition and degradation of sinking particles with different settling velocities. Limnol Oceanogr 52:1645–1664

Grossi V, Cravo-Laureau C, Guyoneaud R, Ranchou-Peyruse A, Hirschler-Réa A (2008) Metabolism of *n*alkanes and *n*-alkenes by anaerobic bacteria: a summary. Org Geochem 39:1197–1203

Grossi V, Yakimov MM, Al Ali B, Tapilatu Y, Cuny P, Goutx M, La Cono V, Giuliano L, Tamburini C (2010) Hydrostatic pressure affects membrane and storage lipid compositions of the piezotolerant hydrocarbon-degrading *Marinobacter hydrocarbonoclasticus* strain #5. Environ Microbiol 12:2020–2033

Grundmann GL, Debouzie D (2000) Geostatistical analysis of the distribution of NH4+ and NO2- oxidizing bacteria and serotypes at the millimeter scale along a soil transect. FEMS Microbiol Ecol 34:57–62

Guschin DY, Mobarry BK, Proudnikov D, Stahl DA, Rittmann BE, Mirzabekov AD (1997) Oligonucleotide microchips as genosensors for determinative and environmental studies in microbiology. Appl Environ Microbiol 63:2397–2402

Hahn MW, Lünsdorf H, Wu Q, Schauer M, Hölfe MG, Boenigk J, Stadler P (2003) Isolation of novel ultramicrobacteria classified as Actinobacteria from five freshwater habitats in Europe and Asia. Appl Environ Microbiol 69:1442–1451

Hanson J, Macalday JL, Harris D, Scow KM (1999) Linking toluene degradation with specific microbial populations in soil. Appl Environ Microbiol 65:5403–5408

He Z, Deng Y, Van Nostrand JD, Tu Q, Xu M, Hemme CL et al (2010) GeoChip 3.0 as a high-throughput tool for analyzing microbial community composition, structure and functional activity. ISME J 4:1167–1179

Heckly RJ (1978) Bacterial culture preservation methods. Adv Appl Microbiol 24:1–53

Heipieper HJ, Loffeld B, Keweloh H, de Bont JAM (1995) The *cis/trans* isomerisation of unsaturated fatty acids in *Pseudomonas putida* S12: an indicator for environmental stress due to organic compounds. Chemosphere 30:1041–1051

Herndl GJ, Reinthaler T, Teira E, van Aken H, Veth C, Pernthaler A, Pernthaler J (2005) Contribution of *Archaea* to total prokaryotic production in the deep Atlantic Ocean. Appl Environ Microbiol 71:2303–2309

Hinrichs K-U, Hayes JM, Sylva SP, Brewer PG, DeLong EF (1999) Methane-consuming archaea molecular- isotopic and phylogenetic evidence. Nature 398:802–805

Honjo S, Manganini SJ, Cole JJ (1982) Sedimentation of biogenic matter in the deep ocean. Deep-Sea Res 29:609–625

Hoppe H-G (1983) Significance of exoenzymatic activities in the ecology of brackish water: measurements by means of methylumbelliferyl-substrates. Mar Ecol Prog Ser 11:299–308

Hoppe H-G (1991) Microbial extracellular enzyme activity: a new key parameter in aquatic ecology. In: Chrøst RJ (ed) Microbial enzymes in aquatic environments. Springer, New York, pp 60–83

Horikoshi K (ed) (2011) Extremophiles handbook, vol 1. Springer, Tokyo

Hubálek Z (2002) Protectants used in the cryopreservation of microorganisms. Cryobiology 46:205–229

Hug LA, Salehi M, Nuin P, Tillier ER, Edwards EA (2011) Design and verification of a pangenome microarray oligonucleotide probe set for Dehalococcoides spp. Appl Environ Microbiol 77:5361–5369

Jannasch HW, Taylor CD (1984) Deep-sea microbiology. Annu Rev Microbiol 38:487–514

Jannasch HW, Wirsen CO (1973) Deep-sea microorganisms: in situ response to nutrient enrichment. Science 180:641–643

Jannasch HW, Wirsen CO (1977) Retrieval of concentrated and undecompressed microbial populations from the deep sea. Appl Environ Microbiol 33:642–646

Jannasch HW, Wirsen CO (1984) Variability of pressure adaptation in deep sea bacteria. Arch Microbiol 139:281–288

Jannasch HW, Eimhjellen K, Wirsen CO, Farmanfarmaian A (1971) Microbial degradation of organic matter in the deep sea. Science 171:672–675

Jannasch HW, Wirsen CO, Winget CL (1973) A bacteriological pressure-retaining deep-sea sampler and culture vessel. Deep-Sea Res 20:661–664

Jeffrey SW, Mantoura F, Wright SW (1997) Phytoplankton pigments in oceanography: guidelines to modern methods. In: Jeffrey SW, Montana F, Wright SW (eds) Mongraphs on Oceanographic Methodology. UNESCO Pub, Paris

Jeffrey SW, Mantoura F, Wright SW (2005) Phytoplankton pigments in oceanography: guidelines to modern methods. In: Jeffrey SW, Montana F, Wright SW (eds) Mongraphs on Oceanographic Methodology, 2nd edn. UNESCO Pub, Paris

Johnsen AR, Winding A, Karlson U, Roslev P (2002) Linking of microorganisms to phenanthrene metabolism in soil by analysis of 13C-labeled cell lipids. Appl Environ Microbiol 68:6106–6113

Johnstone KI (1969) The isolation and cultivation of single organisms. In: Norris JR, Ribbons DW (eds) Methods in Microbiology, vol 1. Academic, New York, pp 455–471

Joint I, Mühling M, Querellou J (2010) Culturing marine bacteria – an essential prerequisite for biodiscovery. Microb Biotechnol 3:564–575

Joux F, Lebaron P (2000) Use of fluorescent probes to assess physiological functions of bacteria at the single cell level. Microbes Infect 2:1523–1535

Joux F, Servais P, Naudin J-J, Lebaron P, Oriol L, Courties C (2005) Distribution of picophytoplankton and bacterioplankton along a river plume gradient in the Mediterranean Sea. Vie Milieu 55:197–208

Kaeberlein T, Lewis K, Epstein SS (2002) Isolating "uncultivable" microorganisms in pure culture in a simulated natural environment. Science 296:1127–1129

Kato C (2011) Distribution of Piezophiles. In: Horikoshi K (ed) Extremophiles handbook. Springer, Tokyo, pp 643–655

Kato C et al (2008) Protein adaptation to high-pressure environments. In: Thomas T, Siddiqui KS (eds) Protein adaptation in extremophiles, Molecular anatomy and physiology of proteins series. Nova Science Publisher, Hauppauge, pp 167–191

Kenters N, Henderson G, Jeyanathan J, Kittelmann S, Janssen PH (2011) Isolation of previously uncultured rumen bacteria by dilution to extinction using a new liquid culture medium. J Microbiol Methods 84:52–60

Kirchman DL, K'ness E, Hodson R (1985) Leucine incorporation and its potential as a measure of protein synthesis by bacteria in natural aquatic systems. Appl Environ Microbiol 49:599–607

Klamer M, Bääth E (1998) Microbial community dynamics during composting of straw material studied using phospholipid fatty acid analysis. FEMS Microbiol Ecol 27:9–20

Kreil DP, Russel RR, Russel S (2006) Microarray oligonucleotide probes. Methods Enzymol 410:73–98

Koshkin AA, Nielsen P, Meldgaard M, Rajwanshi VK, Singh SK, Wengel J (1998) LNA (locked nucleic acid): an RNA mimic forming exceedingly stable LNA:LNA duplexes. J Am Chem Soc 120:13252–13253

Lane DJ, Pace B, Olsen GJ, Stahl DA, Sogin ML, Pace NR (1985) Rapid determination of 16S ribosomal RNA sequences for phylogenetic analyses. Proc Natl Acad Sci U S A 82:6955–6959

Lauro F, Bartlett D (2007) Prokaryotic lifestyles in deep sea habitats. Extremophiles 12:15–25

Lauro FM, Tran K, Vezzi A, Vitulo N, Valle G, Bartlett DH (2008) Large-scale transposon mutagenesis of *Photobacterium profundum* SS9 reveals new genetic loci important for growth at low temperature and high pressure. J Bacteriol 90:1699–1709

Lazazzera BA (2005) Lessons from DNA microarray analysis: the gene expression profile of biofilms. Curr Opin Microbiol 8:222–227

Lebaron P, Servais P, Agogué H, Courties C, Joux F (2001) Does the nucleic acid content of individual bacterial cells allow us to discriminate between active cells and inactive cells in aquatic systems? Appl Environ Microbiol 67:1775–1782

Lemarchand K, Parthuisot N, Catala P, Lebaron P (2001) Comparative assessment of epifluorescence microscopy, flow cytometry and solid-phase cytometry used in the enumeration of specific bacteria in water. Aquat Microb Ecol 25:301–309

Lenaerts J, Lappin-Scott HM, Porter J (2007) Improved fluorescent in situ hybridization method for detection of bacteria from activated sludge and river water by using DNA molecular beacons and flow cytometry. Appl Environ Microbiol 73:2020–2023

Liss PS, Duce RA (1997) The sea surface and global change. Cambridge University Press, Cambridge, UK

Lohrenzen CJ (1966) A method for the continuous measurement of in vivo chlorophyll concentration. Deep-Sea Res 13:223–227

Ludwig W et al (2004) ARB: a software environment for sequence data. Nucleic Acids Res 32:1363–1371

Magot M (2005) Indigenous microbial communities in oil fields. In: Ollivier B, Magot M (eds) Petroleum Microbiology. ASM Press, Washington, DC, pp 21–33

Martini S, Al Ali B, Garel M, Nerini D, Grossi V, Casalot L, Cuny P, Tamburini C (2013) Effects of hydrostatic pressure on growth and luminescence of a moderately-piezophilic luminous bacteria *Photobacterium phosphoreum* ANT-2200. PLoS One 8:e66580

Mazzella N, Molinet J, Syakti AD, Barriol A, Dodi A, Bertand J-C, Doumenq P (2005) Effects of *n*-alkanes and crude oil on bacterial phospholipid classes and molecular species determined by electrospray ionization mass spectrometry. J Chromatogr B 822:40–53

Migné A, Davoult D, Spilmon N, Menu D, Boucher G, Gattuso J-P, Rybarczyk H (2002) A closed-chamber CO2-flux method for estimating intertidal primary production and respiration under emersed conditions. Mar Biol 140:865–869

Militon C et al (2007) Phyl Array: phylogenetic probe design algorithm for microarray. Bioinformatics 23:2550–2557

Mitchell D, Willerslev E, Hansen A (2005) Damage and repair of ancient DNA. Mutat Res 571:265–276

Montserrat-Sala M, Arin L, Balagué V, Felipe J, Guadayol Ò, Vaqué D (2005) Functional diversity of bacterioplankton assemblages in western Antarctic seawaters during late spring. Mar Ecol Prog Ser 292:13–21

Nagata T et al (2010) Emerging concepts on microbial processes in the bathypelagic ocean – ecology, biogeochemistry and genomics. Deep Sea Res II 57:1519–1536

Nelson CE, Carlson CA (2005) A nonradioactive assay of bacterial productivity optimized for oligotrophic pelagic environments. Limnol Oceanogr Methods 3:211–220

Neufeld JD, Dumont MG, Vohra J, Murrell JC (2007a) Methodological considerations for the application of stable isotope probing in microbial ecology. Microb Ecol 53:435–442

Neufeld JD, Wagner M, Murrell JC (2007b) Who eats what, where and when? Isotope-labelling experiments are coming of age. ISME J 1:103–110

Neveux J, Lantoine F (1993) Spectrofluorometric assay for chlorophylls and phaeopigments using the least squares approximation technique. Deep-Sea Res 40:1747–1765

Nichols PD, Guckert JB, White DC (1986) Determination of monounsaturated fatty acids double bond position and geometry for microbial monocultures and complex consortia by capillary GC-MS of their dimethyl disulphide adducts. J Microbiol Methods 5:49–55

Norland S (1993) The relationship between biomass and volume of bacteria. In: Kemp PF, Sherr BF, Sherr EB, Cole JJ (eds) Handbook of methods of aquatic ecology. Lewis Publishers, Boca Raton, pp 303–307

Normand P (1995) Utilisation des séquences 16S pour le positionnement phylétique d'un organisme inconnu. Oceanis 21:31–56

Nyonyo T, Shinkai T, Tajima A, Mitsumori M (2012) Effect of media composition, including gelling agents, on isolation of previously uncultured rumen bacteria. Lett Appl Microbiol 56:63–70

Ogram A, Sun W, Brockman FJ, Fredrickson JK (1995) Isolation and characterization of RNA from low-biomass deep-subsurface sediments. Appl Environ Microbiol 61:763–768

Onstott TC et al (1997) The deep gold mines of South Africa: Windows into the subsurface biosphere. Proc SPIE Int Soc Opt Eng 3111:344–357

Orphan VJ, House CH, Hinrichs KU, McKeegan KD, DeLong EF (2001) Methane-consuming archaea revealed by directly coupled isotopic and phylogenetic analysis. Science 293:484–487

Pancost RD, Sinninghe Damsté JS (2003) Carbon isotopic compositions of prokaryotic lipids as tracers of carbon cycling in diverse settings. Chem Geol 195:29–58

Parisot N, Denonfoux J, Dugat-Bony E, Peyret P, Peyretaillade E (2012) KASpOD–a web service for highly specific and explorative oligonucleotide design. Bioinformatics 28:3161–3162

Pedersen K (2001) Subterranean microorganisms and radioactive waste disposal in Sweden. Eng Geol 59:163–176

Peeva VK, Lynch JL, Desilva CJ, Swanson NR (2008) Evaluation of automated and conventional microarray hybridization – a question of data quality and best practice? Biotechnol Appl Biochem 50:181–190

Pernthaler A, Amann R (2004) Simultaneous fluorescence in situ hybridization of mRNA and rRNA in environmental bacteria. Appl Environ Microbiol 70:5426–5433

Pham VHT, Kim J (2012) Cultivation of unculturable soil bacteria. Trends Biotechnol 30:475–484

Picard C, Ponsonnet C, Paget E, Nesme X, Simonet P (1992) Detection and enumeration of bacteria in soil by direct DNA extraction and polymerase chain reaction. Appl Environ Microbiol 58:2717–2722

Postec A, Le Breton C, Fardeau ML, Lesongeur F, Pignet P, Querellou J, Ollivier B, Godfroy A (2005a) *Marinitoga hydrogenitolerans* sp. nov., a novel member of the order Thermotogales isolated from a black smoker chimney on the Mid-Atlantic Ridge. Int J Syst Evol Microbiol 55:1217–1221

Postec A, Pignet P, Cueff-Gauchard V, Schmitt A, Querellou J, Godfroy A (2005b) Optimisation of growth conditions for continuous culture of the hyperthermophilic archaeon *Thermococcus hydrothermalis* and development of sulphur-free defined and minimal media. Res Microbiol 156:82–87

Postec A, Urios L, Lesongeur F, Ollivier B, Querellou J, Godfroy A (2005c) Continuous enrichment culture and molecular monitoring to investigate the microbial diversity of thermophiles inhabiting deep-sea hydrothermal ecosystems. Curr Microbiol 50:138–144

Postec A, Lesongeur F, Pignet P, Ollivier B, Querellou J, Godfroy A (2007) Continuous enrichment cultures: insights into prokaryotic diversity and metabolic interactions in deep-sea vent chimneys. Extremophiles 11:747–757

Pozhitkov A, Noble PA, Domazet-Loso T, Nolte AW, Sonnenberg R, Staehler P et al (2006) Tests of rRNA hybridization to microarrays suggest that hybridization characteristics of oligonucleotide probes for species discrimination cannot be predicted. Nucleic Acids Res 34:e66

Preston-Mafham J, Boddy L, Randerson PF (2002) Analysis of microbial community functional diversity using sole-carbon-source utilisation profiles: a critique. FEMS Microbiol Ecol 42:1–14, 869

Pringault O, Tassas V, Rochelle-Newall E (2007) Consequences of light respiration on the determination of production in pelagic systems. Biogeosciences 4:105–114

Ranjard L, Richaume A, Jocteur-Monrozier L, Nazaret S (1997) Response of soil bacteria to Hg(II) in relation to soil characteristics and cell location. FEMS Microbiol Ecol 24:321–331

Rappé MS, Giovannoni SJ (2003) The uncultured microbial majority. Annu Rev Microbiol 57:369–394

Rappé S, Connon SA, Vergin KL, Giovannoni SJ (2002) Cultivation of the ubiquitous SAR11 marine bacterioplankton clade. Nature 418:630–633

Raven N, Ladwa N, Cossar D, Sharp R (1992) Continuous culture of the hyperthermophilic archaeum *Pyrococcus furiosus*. Appl Microbiol Biotechnol 38:263–267

Richard FA, Thompson TG (1952) The estimation and characterization of plankton by pigments analyses. II. A spectrophotometric method for the estimation of plankton pigments. J Mar Res 21:155–172

Rimour S, Hill D, Militon C, Peyret P (2005) Go Arrays: highly dynamic and efficient microarray probe design. Bioinformatics 21:1094–1103

Robinson C, Williams PJ leB (1999) Plankton net community production and dark respiration in the Arabian Sea during September 1994. Deep-Sea Res part II 46:745–765

Robinson C et al (2010) Mesopelagic ecology and biogeochemistry – a synthesis. Deep-Sea Res II 57:1504–1518

Roland F, Caraco NF, Cole JJ, del Giorgio P (1999) Rapid and precise determination of dissolved oxygen by spectrophotometry: evaluation of interference from color and turbidity. Limnol Oceanogr 44:1148–1154

Rowe GT, Deming JW (2011) An alternative view of the role of heterotrophic microbes in the cycling of organic matter in deep-sea sediments. Mar Biol Res 7:629–636

Ryther JH, Yentsch CS (1957) The estimation of phytoplankton, production in the ocean from chlorophyll and light data. Limnol Oceanogr 2:281–286

Saiki RK et al (1988) Primer-directed enzymatic amplification of DNA with a thermostable DNA polymerase. Science 239:487–491

Sanger F, Nicklen S, Coulson AR (1977) DNA sequencing with chain-terminating inhibitors. Proc Natl Acad Sci U S A 74:5463–5467

Schena M, Shalon D, Davis RW, Brown PO (1995) Quantitative monitoring of gene expression patterns with a complementary DNA microarray. Science 270:467–470

Schena M, Heller RA, Theriault TP, Konrad K, Lachenmeier E, Davis RW (1998) Microarrays: biotechnology's discovery platform for functional genomics. Trends Biotechnol 16:301–306

Scholten JC, Culley DE, Nie L, Munn KJ, Chow L, Brockman FJ, Zhang W (2007) Development and assessment of whole-genome oligonucleotide microarrays to analyze an anaerobic microbial community and its responses to oxidative stress. Biochem Biophys Res Commun 358:571–577

Sebat JL, Colwell FS, Crawford RL (2003) Metagenomic profiling: microarray analysis of an environmental genomic library. Appl Environ Microbiol 69:4927–4934

Sizova MV, Hohmann T, Hazen A, Paster BJ, Halem SR, Murphy CM, Panikov NS, Epsteina SS (2012) New approaches for isolation of previously uncultivated oral bacteria. Appl Environ Microbiol 78:194–203

Slawyk G, Raimbault P (1995) A simple procedure for the simultaneous recovery of dissolved inorganic and organic nitrogen in 15N-tracer experiments on oceanic waters improving the mass balance. Mar Ecol Prog Ser 124:289–299

Smith DC, Azam F (1992) A simple, economical method for measuring bacterial protein synthesis rates in seawater using 3H-leucine. Mar Microb Food Webs 6:107–114

Smith DC, Simon M, Alldredge AL, Azam F (1992) Intense hydrolytic enzyme activity on marine aggregates and implications for rapid particle dissolution. Nature 359:139–142

Spring S, Schulze R, Overmann J, Schleifer K-H (2000) Identification and characterization of ecologically significant prokaryotes in the sediment of freshwater lakes: molecular and cultivation studies. FEMS Microbiol Rev 24:573–590

Stanley JT, Konopka A (1985) Measurement of in situ activities of nonphotosynthetic microorganisms in aquatic and terrestrial habitats. Annu Rev Microbiol 39:321–346

Steemann Nielsen E (1951) Measurement of the production of organic matter in the Sea. Nature 167:684

Stender H, Lund K, Petersen KH, Rasmussen OF, Hongmanee P, Miorner H, Godtfredsen SE (1999) Fluorescence In situ hybridization assay using peptide nucleic acid probes for differentiation between tuberculous and nontuberculous mycobacterium species in smears of mycobacterium cultures. J Clin Microbiol 37:2760–2765

Steward GF, Fandino LB, Hollibaugh JT, Whitledge TE, Azam F (2007) Microbial biomass and viral infections of heterotrophic prokaryotes in the sub-surface layer of the central Arctic Ocean. Deep-Sea Res I 54:1744–1757

Stingl U, Tripp HJ, Giovannoni SJ (2007) Improvements of high-throughput culturing yielded novel SAR11 strains and other abundant marine bacteria from the Oregon coast and the Bermuda Atlantic Times Series study site. Int Soc Microbiol Ecol 1:361–371

Stramski D et al (2008) Relationships between the surface concentration of particulate organic carbon and optical properties in the eastern South Pacific and eastern Atlantic oceans. Biogeosciences 5:171–201

Strickland JDH (1960) Measuring the production of marine phytoplankton. Bull Fish Res Bd Can 122:1–172

Strom SL (2000) Bacterivory: interactions between bacteria and their grazers. In: Kirchman DL (ed) Microbial of the oceans. Wiley-Liss, New York, pp 351–386

Syakti AD, Mazzella N, Torre F, Acquaviva M, Gilewicz M, Guiliano M, Bertrand J-C, Doumenq P (2006) Influence of growth phase on the phospholipidic fatty acid composition of two marine bacterial strains in pure and mixed cultures. Res Microbiol 157:479–486

Tabor P, Colwell RR (1976) Initial investigation with a deep ocean in situ sampler. In: Proceedings of the MTS/IEEE OCEANS'76. IEEE, Washington, DC, pp 13D-11–13D-14

Talbot V, Bianchi M (1995) Utilisation d'un substrat modèle fluorogène pour mesurer l'activité enzymatique extracellulaire (AEE) bactérienne. Oceanis 21:247–260

Tamaki H, Sekiguchi Y, Hanada S, Nakamura K, Nomura N, Matsumura M, Kamagata Y (2005) Comparative analysis of bacterial diversity in freshwater sediment of a shallow eutrophic lake by molecular and improved cultivation-based techniques. Appl Environ Microbiol 71:2162–2169

Tamburini C, Garcin J, Bianchi A (2003) Role of deep-sea bacteria in organic matter mineralization and adaptation to hydrostatic pressure conditions in the NW Mediterranean Sea. Aquat Microb Ecol 32:209–218

Tamburini C, Garcin J, Grégori G, Leblanc K, Rimmelin P, Kirchman DL (2006) Pressure effects on surface Mediterranean prokaryotes and biogenic silica dissolution during a diatom sinking experiment. Aquat Microb Ecol 43:267–276

Tamburini C et al (2009) Effects of hydrostatic pressure on microbial alteration of sinking fecal pellets. Deep Sea Res II 56:1533–1546

Tamburini C, Boutrif M, Garel M, Colwell RR, Deming JW (2013) Prokaryotic responses to hydrostatic pressure in the ocean – a review. Environ Microbiol 15:1262–1274

Tanaka Y, Hanada S, Manome A, Tsuchida T, Kurane R, Nakamura K, Kamagata Y (2004) *Catellibacterium nectariphilum* gen. nov., sp. nov., which requires a diffusible compound from a strain related to the genus *Sphingomonas* for vigorous growth. Int J Syst Evol Microbiol 54:955–959

Tang JW, Baldocchi DD, Qi Y, Xu LK (2003) Assessing soil CO2 efflux using continuous measurements of CO2 profiles in soils with small solid-state sensors. Agric For Meteorol 118:207–220

Tang J-C, Kanamori T, Inoue Y, Yasuta T, Yoshida S, Katayama A (2004) Changes in the microbial community structure during thermophilic composting of manure as detected by the quinone profile method. Process Biochem 39:1999–2006

Terrat S, Peyretaillade E, Goncalves O, Dugat-Bony E, Gravelat F, Mone A et al (2010) Detecting variants with Metabolic Design, a new software tool to design probes for explorative functional DNA microarray development. BMC Bioinformatics 11:478

Thiel V, Peckmann J, Seifert R, Wehrung P, Reitner J, Michaelis W (1999) Highly isotopically depleted isoprenoids: Molecular markers for ancient methane venting. Geochim Cosmochim Acta 63:3959–3966

Tholosan O, Garcin J, Bianchi A (1999) Effects of hydrostatic pressure on microbial activity through a 2000 m deep water column in the NW Mediterranean Sea. Mar Ecol Prog Ser 183:49–57

Timlin JA (2006) Scanning microarrays: current methods and future directions. Methods Enzymol 411:79–98

Turley CM (1993) The effect of pressure on leucine and thymidine incorporation by free-living bacteria and by bacteria attached to sinking oceanic particles. Deep-Sea Res I 40:2193–2206

Turley CM, Mackie PJ (1994) Biogeochemical significance of attached and free-living bacteria and the flux of particles in the NE Atlantic Ocean. Mar Ecol Prog Ser 115:191–203

Turley CM, Mackie PJ (1995) Bacterial and cyanobacterial flux to the deep NE Atlantic on sedimenting particles. Deep-Sea Res I 42:1453–1474

Turley CM, Lochte K, Lampitt RS (1995) Transformation of biogenic particles during sedimentation in the northeastern Atlantic. Philos Trans R Soc Lond B Biol Sci 348:179–189

Turnbaugh PJ et al (2009) A core gut microbiome in obese and lean twins. Nature 457:480–484

Uitz J, Claustre H, Morel A, Hooker S (2006) Vertical distribution of phytoplankton communities in open ocean: an assessment based on surface chlorophyll. J Geophys Res 111:C08005. doi:10.1029/2005JC003207

UNESCO (1966) Determination of photosynthetic pigments in seawater. UNESCO Monographs on oceanograph methodology. UNESCO, Paris

Van der Meer MTJ, Schouten S, Sinninghe Damsté JS (1998) The effect of the reversed tricarboxylic acid cycle on the 13C contents of bacterial lipids. Org Geochem 28:527–533

Van der Meer MTJ et al (2001) Biosynthetic controls on the 13C contents of organic components in the photoautotrophic bacterium *Chloroflexus aurantiacus*. J Biol Chem 276:10971–10976

Vaqué D, Gasol JM, Marasse C (1994) Grazing rates on bacteria: The significance of methodology and ecological factors. Mar Ecol Prog Ser 109:263–274

Vester B, Wengel J (2004) LNA (locked nucleic acid): high-affinity targeting of complementary RNA and DNA. Biochemistry 43:13233–13241

Von Wintzingerode F, Gobel UB, Stackebrandt E (1997) Determination of microbial diversity in environmental samples: pitfalls of PCR-based rRNA analysis. FEMS Microbiol Rev 21:213–229

Wagner M (2009) Single-cell ecophysiology of microbes as revealed by Raman microspectroscopy or secondary ion mass spectrometry imaging. Annu Rev Microbiol 63:411–429

Wakeham SG, Lee C (1993) Production, transport, and alteration of particulate organic matter in the marine water column. In: Engel M, Macko S (eds) Organic geochemistry. Plenum Press, New York, pp 145–169

Wang Q, Garrity GM, Tiedje JM, Cole JR (2007) Naive Bayesian classifier for rapid assignment of rRNA sequences into the new bacterial taxonomy. Appl Environ Microbiol 73:5261–5267

Weinbauer MG (2004) Ecology of prokaryotic viruses. FEMS Microbiol Rev 28:127–181

Whitman WB, Coleman DC, Wiebe WJ (1998) Prokaryotes: the unseen majority. Proc Natl Acad Sci U S A 95:6578–6583

Witte U, Wenzhofer F, Sommer S, Boetius A, Heinz P, Aberle N, Sand M, Cremer A, Abraham WR, Jorgensen BB, Pfannkuche O (2003) In situ experimental evidence of the fate of a phytodetritus pulse at the abyssal sea floor. Nature 424:763–766

Woese CR, Kandler O, Wheelis ML (1990) Towards a natural system of organisms: proposal for the domains *Archaea*, *Bacteria*, and *Eucarya*. Proc Natl Acad Sci U S A 87:4576–4579

Wolber PK, Collins PJ, Lucas AB, De Witte A, Shannon KW (2006) The Agilent in situ-synthesized microarray platform. Methods Enzymol 410:28–57

Wu L, Thompson DK, Liu X, Fields MW, Bagwell CE, Tiedje JM, Zhou J (2004) Development and evaluation of microarray-based whole-genome hybridization for detection of microorganisms within the context of environmental applications. Environ Sci Technol 38:6775–6782

Yasumoto-Hirose M, Nishijima M, Ngirchechol MK, Kanoh K, Shizuri Y, Miki W (2006) Isolation of marine bacteria by in situ culture on media-supplemented polyurethane foam. Mar Biotechnol 8:227–237

Yayanos AA (1995) Microbiology to 10,500 meters in the deep-sea. Annu Rev Microbiol 49:777–805

Yentsch CS, Menzel DW (1963) A method for the determination of phytoplankton chlorophyll and phaeophytin by fluorescence. Deep Sea Res 10:221–231

Young LY, Phelps CD (2005) Metabolic biomarker for monitoring in situ anaerobic hydrocarbon degradation. Environ Health Perspect 113:62–67

Zak J, Willig M, Moorhead D, Wildman H (1994) Functional diversity of microbial communities : a quantitative approach. Soil Biol Biochem 26:1101–1108

Zengler K, Toledo G, Rappé M, Elkins J, Mathur EJ, Short JM, Keller M (2002) Cultivating the uncultured. Proc Natl Acad Sci U S A 99:15681–15686

ZoBell CE (1970) Pressure effects on morphology and life processes of bacteria. In: Zimmerman HM (ed) High pressure effects on cellular processes. Academic, New York, pp 85–130

ZoBell CE, Johnson FH (1949) The influence of hydrostatic pressure on the growth and viability of terrestrial and marine bacteria. J Bacteriol 57:179–189

Contributions of Descriptive and Functional Genomics to Microbial Ecology

18

Philippe N. Bertin, Valérie Michotey, and Philippe Normand

Abstract

Originally, "genomics" was used only to describe a scientific discipline which consisted in mapping, sequencing, and analyzing genomes. Nowadays, this term is widely used by a growing number of people in a broader sense to describe global techniques for studying genomes including from a functional point of view. These include the analysis of messenger RNAs (transcriptomics), protein contents (proteomics), and metabolites (metabolomics). At a higher level of complexity, it also describes the so-called "meta" approaches that allow to investigate the ecology of microbial communities, including uncultured microorganisms. Based on the use of recent technological developments, the numerous examples provide an integrated view of how microorganisms adapt to particular ecological niches and participate in the dynamics of ecosystems.

Keywords

Bacterial artificial chromosome (BAC) • Cloning • Cosmid • Cultured and uncultured strains • DNA chips • Genomics • High-performance liquid chromatography or high-pressure liquid chromatography (HPLC) • High-resolution liquid chromatography (HRLC) • Isoelectric focusing (IEF) • Matrix-assisted laser desorption ionization-time of flight (MALDI-TOF) • Metabolomics • Metagenomics • Plasmid • Proteomics • Pyrosequencing • Sequencing • Shotgun • Sodium dodecyl sulfate-polyacrylamide gel electrophoresis (SDS-PAGE) • Synteny • Transcriptomics

---

* Chapter Coordinator

P.N. Bertin* (✉)
Génétique Moléculaire, Génomique, Microbiologie (GMGM), UMR 7156, Université de Strasbourg, 67084 Strasbourg Cedex, France
e-mail: philippe.bertin@unistra.fr

V. Michotey
Institut Méditerranéen d'Océanologie (MIO), UM 110, CNRS 7294 IRD 235, Université de Toulon, Aix-Marseille Université, Campus de Luminy, 13288 Marseille Cedex 9, France

P. Normand
Microbial Ecology Center, UMR CNRS 5557 / USC INRA 1364, Université Lyon 1, 69622 Villeurbanne, France

The word **genomics*** is currently a term widely used by the scientific community and the general public. Originally its sense was more restrictive and defined a specific discipline which consisted in mapping, sequencing, and analyzing **genomes***. A growing number of people currently use genomics in a broader signification to denote the techniques for studying the genomes also from a functional point of view. This functional analysis includes gene expression (mRNA analysis, transcriptomics), protein content analysis (proteomics), and analysis of metabolism through quantification of metabolites (metabolomics). In addition a number of new "omics" terms have emerged. For example, the discipline that uses genomic methods to analyze the ecology of natural communities is called ecological genomics, genomics community, or environmental genomics. The suffix "ome" is becoming more widely associated to different objects to designate the global study of secreted proteins, small molecules, effectors, lipids, amino acids, etc. In this chapter, the term genomics is used in its original meaning.

18.1 Descriptive Genomics

Genomics is the biological study of organism through the analysis of all information contained in its genome.

This concept emerged at the end of World War II (Bertin et al. 2008). It grew out the US Atomic Energy Commission which later became the State Department of Energy or DOE. After the explosion of the thermonuclear bombs of Hiroshima and Nagasaki, DOE found necessary to analyze the important genetic alterations due to radiations. Knowledge of genetic information of an organism became, in consequence, indispensable. The achievement of this ambitious project required to overcome several conceptual and technical concepts.

18.1.1 The Structure of DNA and Sequencing of DNA

The structure of the DNA molecule of heredity was characterized in 1953 (Watson and Crick), allowing theoretically to study mutations caused by exposure to radiations. However, a delay of more than 20 years was necessary to develop a method able to determine the sequence of the different nucleotides on DNA molecule (Fig. 18.1)

(Sanger et al. 1977). It is based on the synthesis by DNA polymerase of a complementary strand from elongation of an oligonucleotide primer. This synthesis is carried out by the polymerization of deoxynucleotide triphosphates (reaction between the hydroxyl at the 3′ nucleotide of the n−1 and the 5′ phosphate of the nucleotide n). If the copy is made in four tubes and if each tube contains a modified nucleotide (one of four dideoxynucleotide triphosphates, lacking the hydroxyl at 3′), incorporation will result in stopping the polymerization reaction. In each tube, fragments which synthesis was stopped randomly will have different sizes and will end all at the same modified nucleotide. Radioactively labeled, the various fragments will be subjected to an electric field in a polyacrylamide gel (electrophoresis). The speed of migration of these negatively charged fragments is inversely correlated with their size. After analysis of photographic film, the sequence can be obtained taking into account the size of the different fragments and the nature of their 3′ terminal dideoxynucleotide.

This technology was time and labor consuming, thus largely insufficient to decrypt a large genome. Two major technological innovations emerged in roughly 10 years, partly induced by the Human Genome Project (HGP) which gathers an international consortium of various public institutions (http://www.ornl.gov/sci/techresources/Human_Genome/project/about.shtml) (Fig. 18.2). The first was the replacement of radioactivity by fluorescent markers that allowed the simultaneous migration of reactions involving the four bases A, T, C, and G on the same track of a gel with automatic detection by a reading laser. The second was the

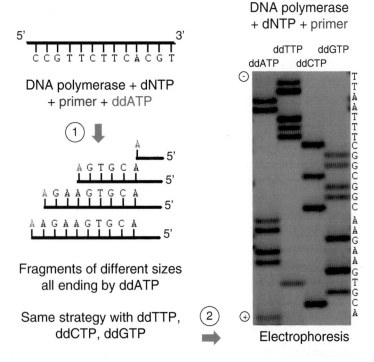

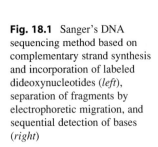

Fig. 18.1 Sanger's DNA sequencing method based on complementary strand synthesis and incorporation of labeled dideoxynucleotides (*left*), separation of fragments by electrophoretic migration, and sequential detection of bases (*right*)

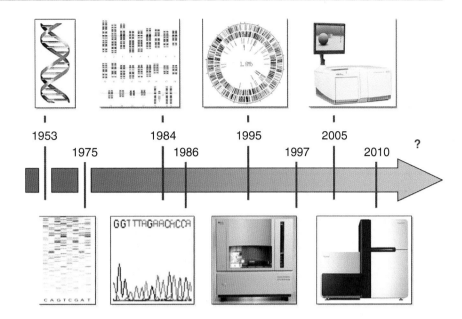

Fig. 18.2 Some steps that have marked the development of sequencing programs, from the initial project to recent technological developments. *1953*: Discovery of the structure of DNA, *1975*: the sequencing method of Sanger, *1984*: emergence of the Human Genome Project, *1986*: sequencing using fluorescent markers, *1995*: first microbial genomes, *1997*: development of capillary sequencers, *2005*: commercialization of the GS20 pyrosequencer, *2010*: commercialization of the HiSeq 2000 sequencer

appearance in the laboratories of capillary sequencers (the migration is performed in long miniature hair-size gel that can be used several times) to analyze large number (96) samples in a single operation. With this, the sequencing capabilities were multiplied ten-fold between 1995 and 1997 and again ten-fold between 1997 and 1999. For comparison, projects such as the sequencing of reference prokaryotic microorganisms such as *Bacillus subtilis* or the eukaryote *Saccharomyces cerevisiae* required the collaboration of many laboratories and lasted almost 10 years. Such projects now require not more than 1 year, much of the raw data being acquired in a few days. The increase in sequencing capacity of specialized centers coupled with a significant cost reduction after 2005 has led to the publication of the complete genome sequence of many organisms.

18.1.2 The Sequencing of Genomes

Bacterial genome size is variable. Generally, it ranges from 600 kb in *Buchnera*, an insect symbiont belonging to the gammaproteobacteria (Gil et al. 2002) to more than 12,000 kb (or 12 Mb) for the myxobacteria *Sorangium cellulosum*, a soil saprophyte (Pradella et al. 2002). However, in the race to discover extreme sizes, mini-genomes were described such as that of *Carsonella ruddii*, the endosymbiont of the psyllid insects that feeds on the sap of plants and that contains only 160 kb (Nakabachi et al. 2006). This bacterium with very low G + C % genomes and high density of genes, often overlapping and many of them truncated, seems to be undergoing transformation into organelles.

A sequencing reaction by the Sanger method permits to read about 700 nucleotides. To determine the entire sequence of a genome, one thus needs to establish a strategy for adding sequences end to end. Two approaches have been classically considered to sequence a genome:

1. "Stepwise" sequencing where the successive position on the genome is conserved. This was done by cloning large ordered fragments.
2. **"Shotgun"*** sequencing where random genomic fragments are sequenced and are ordered afterward. This was done by assembling sequence data from multiple clones to determine overlaps and establish a contiguous sequence. The second strategy was initially considered as risky because of possible ambiguities due to repeated sequences that may complicate the re-sorting of partial sequences, is now largely used because of its simplicity and its speed relative to "stepwise" sequencing previously favored.

The "shotgun" sequencing genome strategy usually implies to construct three gene banks in *Escherichia coli* whose size of cloned fragments and cloning vectors differed: 3 kb fragments in a high-copy number cloning vector, 10 kb fragments into an intermediate copy number vector and fragments >100 kb in a vector with low-copy number such as BAC (bacterial artificial chromosome). The use of vectors with low-copy numbers permits to overcome the problems of toxicity of certain genes. The differences in length make it possible to order the sequence on the chromosome. Once the banks are constructed, the first wave of sequencing will allow to evaluate their quality according to the rate of contamination with sequences from other organisms mainly *E. coli*. The complete sequence of a genome is the result of tens of thousands of readings (or "runs"), sometimes conflicting, to be assembled, i.e., the primary sequences will have to be reordered, overlapping areas removed, and each region read several times to ensure errors are eliminated through consensus building. This work is done by computers and different statistical softwares. Missing zones ("gaps") are filled by priming the library clones of 10 kb when there is such a clone or by subcloning from the bank made from the BAC. Smoothing will then be made to improve the quality of areas that do not fulfill international standard quality called "Bermuda Standards" which are <1 error per 10,000 bases DNA, corresponding to 99.99 % accuracy (Schmutz et al. 2004). This high degree of certainty is obtained by sequencing several times any given area of a genome (about 15 times).

The 2000s have seen the development of new methods that offer the chance of further increasing the rate of sequencing centers. Several have been largely adopted by the scientific community. The first one is called **pyrosequencing*** or also commonly "454 technique" (http://www.454.com/) (Ronaghi 2001) (Fig. 18.3). It is based on the binding onto beads of nebulized DNA fragments to which adapters are previously added. One adapter allows the binding of fragment onto the beads. The fragment bound is then replicated by PCR. The other adapter located at the other extremity of the fragment is used as binding zone for the sequencing primer. The four nucleotides are added one by one and induce elongation only when the fragment contains the complementary nucleotide. In this condition, the liberation of pyrophosphate in the presence of luciferase and sulfurylase will cause an emission of light which can be detected by a CCD camera. Pyrosequencers offer huge yields (0.5 Gb per cycle of 8 h) without problems linked to high GC% or cloning difficulties and allow reading of fragments of approximately 500 nucleotides (nt); only regions containing more than eight identical nucleotides remain ambiguous. The second next-generation sequencing technique was developed by Solexa/Illumina and is based on a reversible termination of chains of fragments bound on a microarray. Fluorescent nucleotides are added and camera detects the color corresponding to the incorporated dideoxynucleotide. This method offers yields of several Gb for readings of about 200 nucleotides and is very useful for several applications, particularly when genome information is available. Others have emerged later, in particular the technique of single-molecule sequencing developed by Pacific Biosciences (Metzker 2010), but this technique still remains to be adopted on a large scale by sequencing centers due to its high error rate. Each technique has different characteristics such as total number of runs and their length and degree of accuracy of the sequences, details that should be taken into account in the choice of the technique in a specific study. The rapid evolution in this area illustrates the importance of sequencing centers in the implementation of genomics projects.

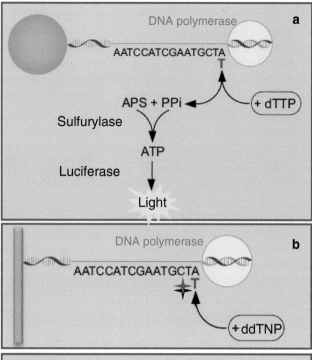

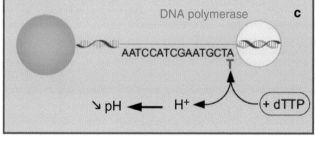

Fig. 18.3 Schematic representation of high-throughput sequencing methods: (**a**) 454 pyrosequencing: DNA fragment is fixed on a bead. Nucleotides are added one by one. Insertion of a nucleotide releases pyrophosphate (*PPi*). Sulfurylase synthesizes ATP from adenosine 5′ phosphosulfate (*APS*) and PPi. Luciferase emits light in the presence of ATP. If the added nucleotide is not incorporated, no light is emitted. (**b**) Illumina (Solexa) sequencing: DNA fragment is fixed on a plate. Fluorescent dideoxynucleotides are all added together (*ddNTP*); each of them contains a different fluorochrome. Only one nucleotide can be incorporated at a time, leading to a color-specific fluorescence. (**c**) Ion Torrent semiconductor sequencing: DNA fragment is fixed on a bead. Nucleotides are added one by one. Insertion of a nucleotide releases a proton that induces a decrease of pH. If the added nucleotide is not incorporated, no pH change is recorded

18.1.3 Annotation and International Databases

Nucleotide sequences are then analyzed to identify genes and other genetic elements present in the genomes (repeats, **operons***, regulatory areas, the presence of **phages***, **transposons***, etc.). The tools used are multiple and rapidly changing; they can identify automatically the elements that can then be annotated by experts of a function or a metabolic

pathway using an annotation platform such as the MAGE platform (Vallenet et al. 2006). For each putative gene, the expert annotation will consist in assigning it a name and possibly a function based on:
1. Similarities between genes from different organisms.
2. The order of distribution on the genome (by comparing several genomes to identify **synteny***, i.e., groups of neighboring genes because the genes of a metabolic pathway are often clustered in one area).
3. The experimental knowledge obtained in other organisms. The result of all these steps will be a primary genomic sequence of DNA, with a description of thousands of genes that will be stored in database such as EMBL, freely available to anyone on the Internet and especially to the biologists' community.

18.1.4 The Comparison of Genomes

One consequence of the continuous growing number of genomic sequences is that comparison of genomes is becoming systematic. Thus, many, if not most, sequencing projects are now made with multiple genomes to identify common/peculiar presence or absence of gene. The explanation of gene repartition will favor the hypothesis of a single event from several independent ones (loss or addition). For example, if a gene is present in all genomes but one, it can be assumed that it has disappeared from the latter genome (single event). On the contrary, if a gene is present in a single genome, it can be considered that it was obtained through lateral transfer (single event). In addition, a gene may also have been duplicated or even created from assembly of different sequences in a given microbe. It is then possible to identify categories of genes for each of these three situations and analyze this distribution in the light of the knowledge of the ecology and the physiology of the microorganism.

By comparing two genomes and in parallel the physiological properties of both corresponding strains, it is also possible to identify relatives of the additional genes and to assign a function or a participation in a function to uncharacterized genes. An example could be the determination of pathogenic genes after comparison of pathogenic and nonpathogenic strains of the same species. By comforting many genomes, it may also be possible to identify genes common to all strains, the core genome, and subsequently all the genes absolutely necessary for the life of a prokaryotic cell. Genome analysis can also provide insight on the phenomena of **coevolution*** between host and symbiont or the transformation steps in a symbiont organelle. This approach was used in *Frankia* to correlate with specific aspects of the ecology and evolution of host plants (Normand et al. 2007) while the genome organization of small-sized *Carsonella*

ruddii seems to be an example illustrating the latter phenomenon of becoming a symbiont organelle (Nakabachi et al. 2006).

It is also possible for a set of genes to identify those that are subjected to positive (diversifying) or negative (homogenizing) selection, genes for which selection takes a different amount of mutations in the protein chain compared to what is expected depending on the model of neutral evolution (Kimura 1968). Such a situation was shown for the human papillomavirus (Chen et al. 2005) or the agent of late blight *Phytophthora infestans* (Liu et al. 2005). A conceptually similar approach is to measure the linkage disequilibrium of genes (Bailly et al. 2006).

Synteny or conservation of gene order also provides information on the selection pressures that were experienced during the evolution from the moment when two genomes diverged. It is known that genes with similar functions are often co-located on the genome so as to have similar expression level (polycistronic messenger RNA), in the same location of the cell and be cotransferred by conjugation. That is why the co-localization of genes on a genome is an indication that their functions are similar, an indication even stronger when co-localization is phylogenetically widespread. Synteny is quantifiable, for example, by the number of homologous gene pairs with less than five nonhomologous genes. It is known that the rate of synteny is generally a function of phylogenetic distance between two strains except for strains that suffered massive loss of genes (minimizing selection) and have become symbionts and or vice versa for taxa inhabiting soil or other environments where permissive genomes tend to accumulate genes (permissive selection).

Having multiple genomes available can also help identify and quantify the laterally transferred genes, i.e., acquired by **transformation***, **conjugation***, or **transduction***. These genes are recognizable because their phylogeny is different from that of the 16S RNA of the organism (Daubin and Ochman 2004a). The genome comparison also identifies the ORFans, genes present in a particular organism with no known counterpart (Daubin and Ochman 2004b).

Finally, several comparative analyzes are now possible such as codon usage (frequency of a codon to encode a particular amino acid), the use of amino acids, the G + C %, related to the physiology, habitat, or ecology. It is of course possible to represent these results graphically in particular to identify regions bearing specific genes of a bacterial species or focusing hot spot of genomic rearrangements.

18.1.5 Results

If the official birth of genomics took place in 1995 with the publication of the first complete genome of

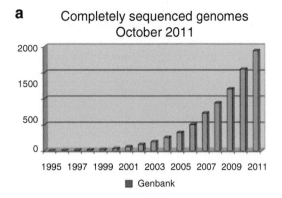
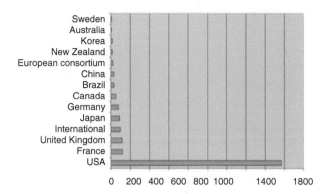
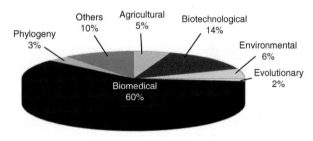
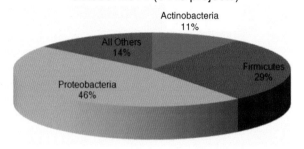

Fig. 18.4 Number of complete genomes sequenced over time (**a**), by countries (**b**), by the type of project (**c**), and according to the taxonomic groups (**d**)

an free-living microorganism, *Haemophilus influenzae* (Fleischmann et al. 1995), it was preceded by a long gestation comprising numerous discussions on the usefulness of spending very large budgets for a few model genomes, *B. subtilis*, *E. coli*, *Arabidopsis thaliana*, and of course humans. For example, using conventional methods, sequencing a genome of 5 Mb as that of *E. coli* costs around 200,000 euros, including salaries. The publication in 2001 (Lander et al.) of the human genome permitted to soften the original controversy. The opponents to the project had argued that to devote as many resources and investment to the purchase of sequencers, the development of appropriate technologies and the training of technicians for the sequencing of several model organisms would automatically get them to spend much less on the thousands of other organisms living in the real world. However, things have followed their own logic, and the number of complete genomes published today is over 15,000 (www.genomesonline.org/), including 300 for archaea and 1,000 for eukaryotes, showing an almost exponential growth (Fig. 18.4). Studied genomes belong mostly to Proteobacteria, the phylum that comprises the majority of bacteria described in taxonomy, pathology, or biotechnology, followed by the Firmicutes and Actinobacteria. This has led in 2010 to the establishment of the project GEBA (Genomic Bacteria and Archaea Encyclopedia gold) which aims to sequence the genomes of many bacteria and archaea from all branches of the tree of life.

These sequenced genomes are primarily determined in the USA and Europe and are studied by a few institutes, with different capacity such as the JGI and JCVI (USA) that weigh for 19 % and 13 %, respectively, of the global effort, or the Sanger Institute (UK) and the Genoscope (Fr), counting for only a few %.

18.2 Functional Genomics

Genome sequencing has made a complete inventory of the repertoire of genes in living organisms. However, little information can be deducted about their regulation under environmental conditions. Therefore, various approaches have been developed in recent years for addressing the complex issue of adaptive responses in microorganisms.

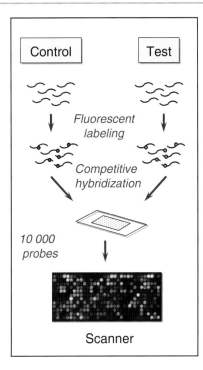

Fig. 18.5 Principle of the analysis of differential gene expression by the microarrays method (DNA chips)

18.2.1 Transcriptomics*

This approach involves analyzing the genes expressed (**transcripts***) under given conditions. Knowledge of the genomes of many organisms now allows to draw the expression profiles in their entirety (transcriptome). Thus, instead of studying the individual expression of a gene or even a group of genes, it becomes possible to perform a differential analysis to assess the overall effect of multiple stresses or various growth conditions on microbial physiology. For this, macroarrays (nylon membranes commonly used in hybridization experiments) were initially used. Currently, the technique relies on the use of microarrays (or DNA chips), glass slides chemically treated to permit the fixation of DNA fragments (Bertin et al. 2008). These materials are deposited on duplicate probes corresponding to each of the genes of the studied organism and usually consist of short oligonucleotides (~25 nucleotides or 25-mers) that can be synthesized directly on the support or longer (~70-mers).

In each case, the total RNAs extracted from the cells provide the target for the synthesis of complementary DNAs using a reverse transcriptase. Nucleotides carrying a fluorescent group are incorporated during this step for signal detection. The use of two different fluorochromes, usually Cy3 and Cy5 in chip technology on glass slides, allows a competitive hybridization (Fig. 18.5). Using an appropriate fluorescence scanner and dedicated softwares (GenePix, e. g.), the intensity of the transmitted signal at each deposit is analyzed, and a comparative analysis of this signal between the two conditions tested is performed. Statistical methods can then identify genes whose expression is preferentially induced or repressed in one or the other condition (Nadon and Shoemaker 2002). As an example, with transcriptomics, gene expression in a biofilm was compared to that of planktonic cells. The results showed that many genes expressed during the biofilm formation were not specific to **biofilms*** because individually, they could also be expressed in other circumstances. On the contrary, it appeared that their coordinated expression induced biofilm formation (De-Vriendt et al. 2004). In 2010, it was shown that the bacterial response to arsenic took place in two phases. Initially, the bacterium activates defense mechanisms, particularly those involved in resistance and resting on the synthesis of efflux pumps. In a second step, metabolic transformation of the element is initiated, especially its oxidation leading to its detoxification (Cleiss-Arnold et al. 2010).

Arrays have some drawbacks, mainly the short dynamic range and high background, so alternatives have been investigated. The latest approach is called RNAseq and consists in extracting all RNAs, converting them to cDNA, and using high-throughput sequencing technology to generate millions of reads that are then ascribed to specific genes (Croucher and Thomson 2010). However, rRNAs are far more abundant than mRNAs that are specific to metabolic activities. The huge amount of rRNAs may mask the metabolic signal and should be removed. Prokaryotes do not present polyA-mRNA, in contrast to eukaryotes. In consequence, polyA binding cannot be used to purify mRNA from abundant ribosomal RNAs, and alternative purification step such as rRNA removal by hybridization with specific oligonucleotides bound to beads has been applied. However, the analysis of the data obtained by RNAseq represents a more consequent task (Coppée 2008; Wang et al. 2009).

18.2.2 Proteomics*

This method consists in the global analysis of the full set of proteins synthesized by an organism under given conditions. This set is called the proteome. Proteins are generally the effectors of the biological response that an organism sets up to face changes in environmental conditions, which is not always the case of mRNAs. Indeed, the modulation of metabolic activities does not always depend on a variation in the expression level of the corresponding genes but may result from posttranslational modification or proteolysis. Under certain conditions, transcriptome analysis may not be the appropriate approach when we consider the genetic determinants enabling organisms to adapt to such environmental stresses or to grow under sometimes extreme conditions (Arsène-Ploetze et al. 2012).

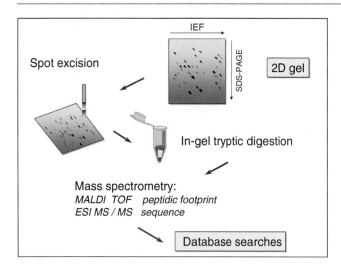

Fig. 18.6 The various stages of proteomic analysis, from the separation of proteins by two-dimensional gel electrophoresis to compare the peptide profiles obtained with the data contained in the biological databases

Proteome analysis is usually based on protein profiling by two-dimensional electrophoresis, i.e., on the separation of proteins according to two physicochemical criteria. By this method, proteins are separated initially according to their charge (**IEF***) within a pH gradient. The proteins have a charge that depends on the pH of the solution in which they are located. In a pH gradient, and subjected to an electric field, each protein migrates toward a specific area where its charge annuls. When this zone is reached, migration is stopped (migration in the first dimension). Strips containing the separated proteins are then deposited on a polyacrylamide gel (**SDS-PAGE***) and further separated according to their molecular mass (Fig. 18.6). The principle is to use a detergent (SDS) that binds to proteins and makes them all negatively charged. The proteins subjected to an electric field then migrate into the mesh of gel as a function of their mass (migration in the second dimension). The proteins separated in this way are then visualized using an organic dye (Coomassie Blue), by reduction of metal salts (silver nitrate) or by fluorescent labeling (DIGE). Protein separation by two-dimensional electrophoresis has been used for many years, but protein identification was difficult and required large amounts of protein because it was based on methods such as the Edmann's chemical degradation. With the development of methods of functional genomics related to genome sequencing, it became possible to characterize polypeptides by mass spectrometry, which allowed considerable progress in their identification, this technique requires indeed only very small amounts (Rabilloud 2002). The proteins of interest are excised from the gel and digested enzymatically (e.g., using trypsin which cleaves specifically the polypeptide chain). All the peptides thus obtained are then analyzed by mass spectrometry, which allows to define the molecular weight, at least for a fraction of them (MALDI-TOF). The peptide fingerprint obtained is then compared with protein sequences contained in databases. To do this, algorithms (e.g., ProFound, MS-Fit, or Mascot) generate theoretical mass profiles for all of these sequences and compare them with peptide fingerprints obtained experimentally by mass spectrometry. The likelihood that the protein of interest corresponds to a particular sequence is then assessed. If the results are not satisfactory or if the genome of the organism studied is not yet available (Carapito et al. 2006), tandem mass spectrometry can be a useful alternative (ESI-MS/MS). In this technique, the mass of fragments of several specific peptides is determined by mass spectrometry, and comparative analysis of these masses by inference allows to define short sequences of amino acids. These are then used to identify homologous proteins present in databases using sequence comparison (Aebersold and Mann 2003).

The proteomic analysis of two-dimensional electrophoresis gels has several known limitations, such as the lack of visualization of all proteins whatever the detection method used or the difficulty of dealing with membrane proteins. Despite this, the global or partial (cytoplasmic fraction, membrane or extracellular) proteomic maps of several organisms have been established. Such approaches have been used foremost for human pathogenic bacteria such as *Mycobacterium tuberculosis*, the agent of tuberculosis, *Helicobacter pylori*, responsible for gastric ulcer, or *Clostridium perfringens*, the cause of gas gangrene (Jungblut 2001; Arsène-Ploetze et al. 2012). It has been extended subsequently to environmentally significant microbes, for example, to study the toxic metal resistance in *Cupriavidus metallidurans* or *Herminiimonas arsenicoxydans* (Noel-Georis et al. 2004; Weiss et al. 2009), adaptation to cold in the archaeal *Methanococcoides burtonii* (Saunders et al. 2005), or repair mechanisms in bacteria radioresistant *Deinococcus* (Liedert et al. 2010).

18.2.3 Metabolomics

One of the last-born in the family of "omics" approaches is metabolomics or the analysis of metabolome, which aims "ultimately" to identify all metabolites present in a culture. A bacterial cell contains thousands of genes, many of which encode proteins with enzymatic activity, and should therefore perform hundreds of metabolic reactions. Several research groups have developed metabolomics approaches based on analytical chemistry techniques such as nuclear magnetic resonance (NMR), Fourier transform mass spectrometry (FT-MS), or high-performance liquid chromatography (HPLC). It is estimated that for humans, the 30,000 genes code for 3,000 proteins and produce about 1,400 metabolites (Wishart 2005), and we now know how to analyze a few hundred at a time. It is estimated that microorganisms also contain hundreds of different types of metabolites (proteins, lipids, saccharides, aromatic compounds, etc.). Their characterization improves our understanding of the activities of microorganisms in their environment. In 2002, the Nobel Prize in chemistry was awarded to John B. Fenn and Koichi Tanaka for mass spectrometry (MS) and Kurt Wüthrich for nuclear magnetic resonance (NMR) for significant advances in analytical techniques applied to biology. The development of these innovative methods to study complex biological molecules allows the elucidation of their planar and three-dimensional structures and the molecular mechanisms governing their interactions. Nowadays, one can identify and quantify tens of metabolites by high-field NMR (300 MHz), and new devices with higher power (900 MHz) that should improve the number of analyzable metabolites are under development. In combination with the approaches of descriptive and functional genomics, it will be possible, for example, to make the connection in model microorganisms between a mutation and entire biosynthetic pathways, the so-called systems biology or integrative biology. An approach applicable to model organisms can be generally extended to environmental organisms after a few years, and we expect to see works performed on integrative environmental bacterial communities soon. Results based on the use of these techniques have thus suggested a major role in the protist *Euglena mutabilis* in the functioning of the microbial community on a site contaminated with arsenic (Halter et al. 2012).

18.3 The Contribution of "Omics" Sciences in Microbial Ecology

18.3.1 Knowledge of Cultivated and Uncultivated Strains

Initially most of genomics studies have been focused on organisms pathogenic to man, whether bacteria or parasites or to higher eukaryotes. In addition to medical consideration, this approach allows to reinforce the basic understanding of physiological processes common between several organisms including human.

Whether pathogenic microorganisms explored during the first programs on microbial genomics and natural isolates examined latter, the result of genome sequencing is still interesting since it enables to examine the body as a whole. In recent years, renewed interest has been focused on environment microorganisms. The goal is to use the available methodologies to identify specific properties that allow these organisms to grow in various environments, including extreme ones. In most cases, the data obtained induce a major improvement of existing knowledge on the physiology of organisms. The results are too numerous to be listed here (http://www.genomesonline.org), but they include the nitrogen metabolism in *Nitrosomonas europaea* (Chain et al. 2003) or that of arsenic in *H. arsenicoxydans* (Muller et al. 2007).

Multiple studies have been conducted to study the adaptive mechanisms adopted by microorganisms grown in pure culture and subjected to variations in environmental conditions. Resistance to thermal shock, anaerobic growth, oxidative stress, or the effect of various deficiencies have been studied under laboratory conditions, especially in reference organisms such as *E. coli* or *B. subtilis* (Rhodius and LaRossa 2003; Hecker and Volker 2004) or on strains recently isolated from the environment or presenting original features such as *Shewanella oneidensis*. The genus *Shewanella* is composed of a large number of species and belongs to the gammaproteobacteria. These bacteria are widespread in aquatic environments (fresh or salt water) and are abundant in environments with redox interfaces where nutrients are in significant quantities. The *Shewanella* species have considerable respiratory versatility and are believed to be of significant importance in the cycling of carbon in the sediments. Indeed, they can "breathe" (use as electron acceptor) besides oxygen various organic or inorganic substrates such as fumarate, nitrate, nitrite, thiosulfate, elemental sulfur, trimethylamine N-oxide (TMAO), dimethyl sulfoxide (DMSO), and anthraquinone 2,6 disulfonate (AQSD) and soluble or insoluble various metals such as chromium, cobalt, iron, manganese, technetium, uranium, and vanadium (Beliaev et al. 2005). They also have a very large number of methyl-accepting chemotaxis protein (MCP), which are involved in **chemotaxis*** and probably allow these bacteria to move along numerous gradients. The genomes of more than 24 strains of *Shewanella* have been sequenced, and the genome of one of them, *S. oneidensis*, has been annotated. Proteomics and transcriptomics studies have been performed upon different conditions including respiration (aerobic, anaerobic, different electron acceptors), stress (low and high temperature, radiation, UV, NaCl), carbon sources, or biofilm formation. The transcriptomic study in the presence of different electron acceptors has shown that reductive dissimilation of metals or nitrate is accompanied by the induction of proteins involved in detoxification and stress resistance. The relatively low efficiency of metal reduction and the induction of stress proteins show that these electron acceptors are probably secondary for this microorganism. Yet, by this way *S. oneidensis* can adapt to the toxic accumulation of reduced metals that induce oxidative stress and whose concentration may reach consistent values in flooded soils or sediments. The ability of *S. oneidensis* to move in the gradients in order to locate in an area that is thermodynamically favorable, while minimizing oxidative damage, is surely a key to its ecological success.

Another advance in the catabolism of chitin has also been achieved through studies of *S. oneidensis* (Yang et al. 2006). In nature, chitin is an organic polymer, the second in biomass abundance after cellulose. It is composed of GlcNAc (N-acetyl glucosamine) and is a source of carbon and energy for a significant number of organisms. Chitin is present in the wall of fungi, in cuticles, and in exoskeletons of worms, mollusks, and arthropods. Most *Shewanella* can use chitin as a carbon source. However, the annotation of the genome revealed that some genes described in the catabolism of chitin in *E. coli* were absent in *Shewanella*. Comparison of *Shewanella* genomes with those of various Proteobacteria has permitted to reconstruct the chitin catabolic pathway, showing to be a variant of the system described in *E. coli*.

Among the recently published data on natural isolates, several have provided some answers of particular interest, highlighting the novelty of the methods used. For example, the sequencing of the unicellular alga *Cyanidioschyzon merolae* living in hot springs rich in sulfates has revealed a compact genome whose components provide information about the genes essential to the development of photosynthetic activity, including higher plants (Matsuzaki et al. 2004). Similarly, *Idiomarina loihiensis*, a bacterium isolated from cracks on the ocean floor, presents many distinct adaptive features compared to other gammaproteobacteria. Thus, the existence of many peptidases and transport systems of amino acids and peptides suggests that these compounds exist in the environment as particles and that they are the basis of carbon and energy metabolisms, at the expense of the metabolism of carbohydrates (Hou et al. 2004). This type of metabolism may challenge the respiration measurements made with sugars labeled with carbon 14 (with measurement of radioactive CO_2), if these substrates are not used. Finally, the analysis of the whole genome of psychrophilic organisms such as *Pseudoalteromonas haloplanktis* (Médigue et al. 2005) allows us to better understand the strategies of resistance to cold, a condition commonly encountered in our biosphere. Indeed ocean waters that constitute the majority of our planet have temperatures of about 2–4 °C below 100 m depth. Added to this are the polar ice caps, glaciers, or tundra. The results obtained from this psychrophilic bacterium showed that growth at low temperatures is not subservient to the presence of a single group of genes. Instead, these processes are based on various metabolic activities, such as the presence of many enzymes to support the oxidative stress induced by to the increased solubility of oxygen. Moreover, the comparison of the proteome of psychrophilic bacteria with those of mesophilic or thermophilic organisms indicates a preferential utilization of certain amino acids in proteins, particularly asparagine, because of its thermal labile nature.

The comparative analysis of transcriptomes of the piezophilic (or barophilic) bacteria *Photobacterium profundum* obtained under low and high pressures and the sequence of its genome indicate that this organism has developed mechanisms to adapt to these disturbances, not only through normal variations of stress factors abundance but also by profound changes in the metabolism of the carbon sources used (Vezzi et al. 2005). A differential analysis carried out in *Frankia alni* of free-living cells or in symbiosis with their host allowed the detailed study of the processes involved in nitrogen fixation in this bacterium (Alloisio et al. 2010). In addition, an integrated analysis was performed on two strains of *Burkholderia pseudomallei*, an opportunistic bacterium pathogenic to humans and animals. The comparison of genomes but also transcriptomes and proteomes allowed to highlight the presence or absence in these organisms of many genomic regions acquired by horizontal transfer, which contribute to the phenotypic diversity of this pathogen (Ou et al. 2005). Finally, a comparative analysis using microarrays on *Thiomonas* bacterial strains stressed the importance of horizontal transfer in the acquisition of new functions in response to environmental pressure (Arsène-Ploetze et al. 2010).

Many microorganisms, although known for a long time, resist to conventional cultivation such as *Mycobacterium leprae*, the causative agent of leprosy that still needs to be inoculated in the footpads of the armadillo to produce biomass. For such organisms, molecular biology could circumvent the problems of cultivation and help to improve knowledge about them. Thus, many sequencing programs focused initially on pathogenic microorganisms in the hope of elucidating the mechanisms of pathogenesis, enabling the development of new drugs, and discovering of new therapeutic targets (Schmid 2004; Scarselli et al. 2005). In this context, the medical sciences have been a key driver and have helped indirectly the development of genomics in ecology. Culture difficulties concern various pathogens and symbionts of plants and animals; they may be linked to their genome undergoing degeneration or very slow growing organisms living in extreme habitats. The reasons for these difficulties in culture of microorganisms vary and are not discussed here, but the genomic approach may permit to overcome these problems and generate lots of data on corresponding organisms. Thus, we have seen the publication of many genomes of uncultivated pathogenic microorganisms as *Mycobacterium leprae* (Cole et al. 2001) or *Tropheryma whipplei* (Bentley et al. 2003), the symbiotic *Buchnera aphidicola* (van Ham et al. 2003) or that have remarkable characteristics such as *Kuenenia stuttgartiensis* (Strous et al. 2006).

18.3.2 Communities Analysis and Identification of Metabolic Interactions

Our interest in studying microorganisms relies not only on the fact that they have evolved over three billion years and have succeeded to colonize practically all ecological niches, including extreme environments in terms of salinity, acidity, pressure, or temperature but also on their metabolic activities that control the biogeochemical cycles and affect soil productivity and water quality. The advent of genomics has yield the emergence of a comprehensive approach, focusing not on a particular microorganism but on an entire microbial community. Bacterial communities are complex assemblies, with fluctuating variables, and their study is emblematic of microbial ecology. The questions range from "who is this" to "what are the metabolic pathways involved in specific biogeochemical transformation." In recent years, their study has been addressed by genomic approaches that circumvent the problems of cultivation. The complexity of communities varies from 2 to 3 taxa when the energy source is poor as in mineral media to thousands of taxa when many carbon molecules exist as in the soil or in sediments. In this case, the term "metagenome" is used to describe the genomic approach used. Currently, hundreds of metagenomic analyzes are underway, too numerous to be listed here, covering a number of diverse ecological niches, from the human intestinal microbiome to the coral reef (http://www.genomesonline.org/).

As examples, habitats such as mine tailings (Tyson et al. 2004; Bertin et al. 2011), the surface layer of the ocean (Venter et al. 2004), or the skin of a marine worm (Woyke et al. 2006) have been subject to massive sequencing, and complete or nearly complete genomes were obtained. The metagenome of an acid mine tailing (Tyson et al. 2004) yielded the genomes of *Leptospirillum* and *Ferroplasma,* two chemolithotrophic bacteria that usually require complex cultivation techniques (Matsumoto et al. 2000). The metagenome of the Sargasso Sea has meanwhile shown the presence of hundreds of thousands of unknown genes and helped to uncover a new type of phototrophy in the oceans (Beja et al. 2000). Genes never found before in the seas have been identified, and the corresponding proteins were named proteorhodopsins because they are similar to bacteriorhodopsins, molecules capable of generating a chemoosmotic potential across membranes in response to light in extremely halophilic Archaea. The unusual fact is that the proteorhodopsin gene (PR) present in the metagenome of the Sargasso Sea was associated with bacterial ribosomal gene. Consecutively, uncultivated strains harboring PR have been identified by molecular methods as gamma proteobacteria belonging to the SAR 86 group that is commonly found in marine environments. Then proteorhodopsin gene could be expressed in *E. coli* and has been shown to be functional (Walter et al. 2007). Subsequently, PRs were found in most seas and oceans of the world and have been associated with very diverse groups (gamma- and deltaproteobacteria or Thermoplasmatales belonging to Euryarchaeota). This large phylogenetic distribution range and diversity of location on the genome are strong evidence for horizontal transfer of this gene between bacterioplankton and Euryarchaeota. However, and surprisingly, horizontal gene transfer appears to be limited to the photic zone because, although there are Thermoplasmatales throughout the water column, those containing PR are localized in the euphotic layer. The first isolate harboring PR is *Pelagibacter ubique* (Giovannoni et al. 2005), and the physiologic role of this pigment seems to be strain specific. According to data from genomics, it appears that this strain is heterotrophic. Although no difference of growth between dark and light was observed for *Pelagibacter ubique*, proteorhodopsin could be involved in some other heterotrophic strains in the enhancement of survival upon starvation conditions of electron donor (Johnson et al. 2010; Gomez-Consarnau et al. 2010). Therefore, the capture of energy by the proteorhodopsin does not seem to be linked to CO_2 fixation as in the case of phototrophy involving chlorophyll. These data indicate that there is alternative way to transform sunlight energy into chemical energy, different from photosynthesis and chlorophyll. This process is not marginal, since according to the average size of a prokaryotic genome, the genomic data indicate that approximately 13 % of prokaryotes present in oceans have the gene coding for proteorhodopsin.

The genomic data were also used to study the physiology of surprising organisms. The marine worm *Olavius algarvensis* has neither mouth nor intestine or anus. How this worm feeds was an enigma and one of the hypotheses dealing with the implication of the flora of the skin was explored. Four genomes including two complete of bacteria were obtained. Based on genome analysis, it seems that there is interdependence between the worm and bacteria but also between bacteria. It appears that one of the bacteria uses sulfur compounds in sediments and rejects sulfate that is used by a second one. Both bacteria also possess enzymes allowing them to use the worm wastes and to provide it with amino acids. In exchange the worm, by moving in marine sediments, provides them accessibility to sulfur (Woyke et al. 2006).

More recently, metagenomic sequences from an arsenic-rich environment have led to de novo reconstruction of seven genomes corresponding to dominant organisms, including uncultured bacteria belonging to a new phylum and maybe involved in the recycling of organic matter (Bertin et al. 2011). Similarly, a new class of archaea has been identified in hypersaline environments (Narasingarao et al. 2012).

The methods of functional genomics that are the differential analysis of transcriptome and proteome are also of considerable interest in the study of interactions between organisms by their ability to reveal all the mechanisms induced in partners. Such approaches have been developed, for example, in *Sinorhizobium meliloti* to understand the physiological changes resulting from growth within plant nodules (Bestel-Corre et al. 2004; Pühler et al. 2004) or synergistic/antagonistic interactions between fungi and rhizobacteria (Moretti et al. 2010). However, whether human pathogens or symbiotic organisms to plants, the focus was so far on the host response rather than on the physiology of the microorganism itself. Thus, microarrays have been used to study the inflammatory response of different types of cell lines in contact with organisms such as bacteria, viruses, or protozoa (Jenner and Young 2005). Similarly, several proteomic or transcriptomic approaches have enabled to investigate the adaptive response of plants interacting with mycorrhizal rhizobacteria (Kim et al. 2004; Hocher et al. 2011).

Ultimately, such works should not only enable a better understanding of ecosystem functioning and interactions among organisms (Faust and Raes 2012) but also to identify the microorganisms responsible for diffuse pathologies, genes for the biosynthesis of secondary metabolites, or new functions usable in biotechnology.

18.3.3 Identify New Organizations, Especially with Original Physiological Characteristics

Diversity data obtained, for example, by using molecular methods suggest that in some ecosystems, nearly 99 % of prokaryotes are not yet cultivated. Development of genomes analysis present in an environment (metagenomes) has opened the way for the exploration of organisms still inaccessible by culture under laboratory conditions, with a goal to discover new genes of interest (Streit and Schmitz 2004; Bertin et al. 2008). The available scientific data from many genomic approaches should permit an optimal utilization, in areas as diverse as industry, agriculture, or medicine, of the properties of microorganisms by promoting the beneficial aspects to the detriment of unwanted effects.

The introduction of molecular methods has transformed the vision established for a number of years on the role of a particular group of microorganisms in the ecosystem functioning. The sequencing of several metagenomes showed that, for example, the presence of Archaea was not only restricted to extreme environments but that these microorganisms were rather widely distributed in many marine and terrestrial ecosystems. However, if molecular methods are useful to identify genes, a better understanding of organisms still requires physiological studies on cultivated species. Nevertheless, genomics can help in identifying missing genes or on the contrary present metabolic pathways that will enable to cultivate and isolate the strain of interest over the others present in the ecosystem. For example, if the strain to isolate is autotrophic, it probably cannot grow in the presence of organic carbon that would moreover favor fast growing heterotrophic prokaryotes. The culture media must therefore exclude these compounds. If the original strain is the only one with the genes for nitrogen fixation, then the culture media must be free of mineral and organic nitrogen to support its selection. This strategy was used to isolate a strain of *Leptospirillum ferrodiazotrophum*, shown from metagenomic study to be the only strain capable of nitrogen fixation in an acid mine tailing (Tyson et al. 2005). Another example is the discovery of nitrifying archaea (Schleper et al. 2005). Nitrification is a microbial process that allows the conversion of ammonium to nitrite (ammonium-oxidizing bacteria, AOB) then nitrate (nitrite-oxidizing bacteria, NOB). From an ecological point of view, this step of the nitrogen cycle provides a link between nitrogen input into the ecosystem (nitrogen fixation producing ammonia from N_2) to output from the ecosystem by production of N_2 (denitrification and anammox producing N_2 is from nitrate or nitrite for denitrification or from ammonium or nitrite for anammox). Studies on the metagenomes of the Sargasso Sea have revealed the coexistence on the same DNA strand of an archaeal ribosomal gene and of a gene coding for ammonia monooxygenase, a key enzyme in nitrification. This discovery has revolutionized our view of nitrifying bacteria (AOB) that were deemed to be a group phylogenetically little diverse and restricted to the beta- and gammaproteobacteria. However, this function was hypothetical and had to be demonstrated. A year later, the first archaeal nitrifier was isolated, and its physiological study has shown that the function was actually expressed (Könneke et al. 2005). Without the metagenomic approach, these microorganisms would have remained hidden for a long time because the oligonucleotides used in PCR to amplify a fragment of the gene coding for ammonium monooxygenase betaproteobacteria do not hybridize to the gene of Archaea making amplification impossible. Nitrifying archaea have been found not only in the oceans but also in soil and in sewage treatment plants.

Since, diversity data obtained, for example, by using 16S RNA analyzes indicate that only a small fraction of organisms in a given environment is cultivable, a new aspect including a global proteomic analysis centered on the whole community emergence in several studies. Metaproteome was assigned to this approach, and the first proteomic analysis of this type was applied to activated sludge treatment plant (Wilmes and Bond 2004). This study allowed to characterize some major proteins belonging to a non-cultivable *Rhodocyclus* type, accumulating polyphosphate and allowing the removal of phosphates by cell sedimentation in wastewater treatment plants. Ultimately, this should improve phosphate removal responsible for eutrophication of aquatic environments. More recently, the analysis of a metaproteomics microbial community of mine drainage has helped to highlight the proteins involved in biomineralization of iron and arsenic, leading to natural attenuation of the load in these elements (Bertin et al. 2011). This type of approach presents a tremendous interest in understanding the functioning of ecosystems, and, despite the technical difficulties inherent in the methods used, it seems to have a promising future. It is the same for metatranscriptomic analysis whose objective is to develop an inventory of functions expressed by all the organisms making up a microbial community. Such an approach has been recently used to characterize a microbial community involved in the production of biogas. The results have identified the main pathways of methanogenesis while stressing the important role exerted by Archaea (Zakrzewski et al. 2012). Similarly, the establishment of expression profiles of the symbiotic bivalve *Solemya velum* has identified various pathways involved in sulfur metabolism. The data obtained emphasize the importance of thioautotrophy in symbiosis located in marine reductive environment (Stewart et al. 2011).

The interest in environmental microorganisms was anterior to the development of microbial genomics. Indeed, from there, enzymes have been retrieved several times, especially from extremophile organisms. It includes the most famous enzyme of thermophile organisms, DNA polymerase (Taq polymerase) that has revolutionized the techniques of molecular biology, and, from a more practical point of view, some medical and sanitary diagnostics. Other thermophile enzymes are also used by industry because of their stability and/or their high specific activity enabling them to operate in industrial processes of degradation or synthesis (Demirjian et al. 2001): such is the case for many proteases, lipases, or glycosidases, whose properties have often been optimized in the laboratory by various methods such as mutagenesis or directed molecular evolution. Several other enzymes having high potential for industrial or pharmaceutical applications were also identified especially esterases and oxidoreductases as well as proteins involved in the synthesis of vitamins and antibiotic molecules (Lorenz and Eck 2005). However, other environments may also harbor interesting metabolism, and new hydrolases with activities of hemicellulose degradation were also identified from a metagenome of compost (Dougherty et al. 2012). Moreover, from the sequencing data from many (meta)genomes, it has been possible to predict the existence of enzymes with potential biotechnological interest (Yamada et al. 2012).

Besides the implementation of new processes in bioremediation, these studies also led to develop new biosensors that can detect the presence of chemicals in the environment (Parales and Ditty 2005; Zylstra and Kukor 2005; Eberly and Ely 2008).

18.4 Conclusion and Outlook

Genome sequencing has, for the first time, made available the knowledge of all the elements involved in the development and functioning of a cell. Just as molecular biology many years ago, genomics thus represents a true paradigm shift. For a long time indeed, biology has focused on the description of separate elements and their classification. Thus, a cell has often been regarded as a collection of objects. Genomics can now turn to living organisms from a global perspective by considering them as a set of elements acting within a complex network of interactions. This is fundamental in understanding the biological mechanisms since no living organism can be reduced to a single gene or even any family of genes expressed at one time or another cell cycle (Bertin et al. 2008). There is currently an increasingly important and diverse genome sequencing effort, and, if the tendency continues, all taxa described in Bergey's Manual, the reference publication in bacterial taxonomy, would have their genome sequenced. The number of taxa is currently around 5,000 (Holt et al. 1994) and corresponds roughly to the number of over 5,500 high-quality sequences of 16S referenced (https://www.msu.edu/~garrity/taxoweb/datasets.html). This number is probably underestimated due to the large number of non-cultivated taxa (e.g., symbionts of insects) and to the ongoing exploration of new habitats (extreme environments).

The first genome sequencing projects have shown, however, that even in reference organisms such as *E. coli* or *B. subtilis* (Blattner et al. 1997; Kunst et al. 1997), 25–40 % of the coding sequences have unknown function and are not related to any gene of known function. This observation was verified later, and the role of genes of unknown function in newly sequenced genomes is sometimes close to 50 %. This observation alone justifies the various programs of functional genomics undertaken, at least if the results are not limited to this simple analysis, but can instead lead to the discovery of new functions and regulatory mechanisms that are associated. Whole sections of the physiology of organisms thus remain undiscovered. Increasing knowledge of the complete genome sequence of many organisms has induced profound changes in the methods of investigation, whether the characterization of genes whose function could not be identified by conventional methods or the study of stress response because of its multifactorial nature. Thus, we have seen the increasing use of global analysis methods, also called "extensive analysis." Indeed, instead of studying genes, proteins, or metabolic products individually, profiles of organisms can now be apprehended in their entirety. In particular, the differential analysis under various conditions of growth of total proteins (proteome), transcripts (transcriptome), or metabolites (metabolome) to quantify simultaneously the expression of genes or accumulation of the corresponding products in an organism (Kahn 1995; Velculescu et al. 1997; Tweeddale et al. 1998).

Whether to screen the metagenomic DNA libraries based on enzymatic activity or sequence similarity, these methods have their limitations. In particular, the expression in a host can lead to absence of activity by the lack of an appropriate secretion system, by misfolding proteins, or by the presence of different regulatory signals. In addition, a search based on molecular biology methods, whether PCR or microarrays, leads to identify only sequences homologous to those already described, since it relies on sequence conservation. Metagenomic approaches based on a systematic sequencing should therefore evolve quickly to become an everyday tool. An overview of reconstructed metagenomes of all extreme environments (oceanic ridges, deep earth, irradiated media) or complex (human gut, sewage sludge, microbial mats) is important to obtain and first results now begin to emerge. These methods becoming more widespread, they should be coupled with measures of activity, assays of metabolites or analyzes of metatranscriptome and metaproteome. It will be important, however, to focus on real biological questions to be sure to get an appropriate response.

In parallel, biomathematic approaches should also progress greatly and lead to better understand the genesis of new forms of genes, such as ORFans (Siew and Fischer 2003), gene duplications, lateral transfer (Daubin et al. 2003), or how the domains constituting the genes are organized (Bru et al. 2005; Mulder et al. 2005). This should enable us to reconstruct the evolutionary history of genomes and more precisely understand how microorganisms adapt optimally to each of their ecological niche.

The availability to the scientific community of all these data and their comparison offers an unprecedented opportunity to study how various components of a living organism but also the organisms themselves work together and meet environmental stresses. A better understanding of the elements involved, their spatial and temporal distribution in the cell, the metabolic pathways to which they belong, and their interactions, should provide an integrated picture of the biological processes studied and if necessary enable optimal use of the properties of microorganisms by promoting the beneficial aspects to the detriment of unwanted ones.

Sites Presently Available for the Study of Genomes

www.genomesonline.org/ Site that lists genomes of microorganisms (archaea or bacteria), eukaryotes, and metagenomes, published or underway. In June 2014, it lists more than 6,000 completed isolate genomes.

www.sanger.ac.uk/Projects/Microbes/ Site of the third genomic center, specialized in pathogens. Some calculations can be done there (BlastN, BlastP, FTPdownload).

www.genoscope.cns.fr/ Site of the French sequencing center. More than 100 genomes listed.

www.jgi.doe.gov Site of the Joint Genome Institute, largest sequencing center for microorganisms' genomes with 19 % of the world total.

www.venterinstitute.org/ Site of the J. Craig Venter Institute, a private, nonprofit institution of which one of the research themes is the genomics of environmental microorganisms.

References

Aebersold R, Mann M (2003) Mass spectrometry-based proteomics. Nature 422:198–207

Alloisio N et al (2010) The *Frankia alni* symbiotic transcriptome. Mol Plant Microbe Interact 23:593–607

Arsène-Ploetze F et al (2010) Structure, function and evolution of *Thiomonas* spp. inferred from genome sequencing and comparative genomic analysis. PLoS Genet 6:e1000859

Arsène-Ploetze F, Carapito C, Plewniak F, Bertin PN (2012) Proteomics as a tool for the characterization of microbial isolates and complex communities. In: Heazlewood J, Petzold CJ (eds) Proteomic applications in biology. InTech, Croatia, pp 69–92

Bailly X, Olivieri I, De Mita S, Cleyet-Marel JC, Bena G (2006) Recombination and selection shape the molecular diversity pattern of nitrogen-fixing *Sinorhizobium* sp. associated to *Medicago*. Mol Ecol 15:2719–2734

Beja O et al (2000) Bacterial rhodopsin: evidence for a new type of phototrophy in the sea. Science 5486:1902–1906

Beliaev AS et al (2005) Global transcriptome analysis of *Shewanella oneidensis* MR-1 exposed to different terminal electron acceptors. J Bacteriol 20:7138–7145

Bentley SD et al (2003) Sequencing and analysis of the genome of the Whipple's disease bacterium *Tropheryma whipplei*. Lancet 361:637–644

Bertin PN, Médigue C, Normand P (2008) Advances in environmental genomics: towards an integrated view of micro-organisms and ecosystems. Microbiology 154:347–359

Bertin PN et al (2011) Metabolic diversity among main microorganisms inside an arsenic-rich ecosystem revealed by meta- and proteo-genomics. ISME J 5:1735–1747

Bestel-Corre G, Dumas-Gaudot E, Gianinazzi S (2004) Proteomics as a tool to monitor plant-microbe endosymbiosis in the rhizosphere. Mycorrhiza 14:1–10

Blattner FR et al (1997) The complete genome sequence of *Escherichia coli* K-12. Science 5:1453–1474

Bru C, Courcelle E, Carrere S, Beausse Y, Dalmar S, Kahn D (2005) The ProDom database of protein domain families: more emphasis on 3D. Nucleic Acids Res 3(Database issue):D212–D215

Carapito C et al (2006) Identification of genes and proteins involved in the pleiotropic response to arsenic stress in *Caenibacter arsenoxydans*, a metalloresistant beta-proteobacterium with an unsequenced genome. Biochimie 88:595–606

Chain P et al (2003) Complete genome sequence of the ammonia-oxidizing bacterium and obligate chemolithoautotroph *Nitrosomonas europaea*. J Bacteriol 185:2759–2773

Chen Z, Terai M, Fu L, Herrero R, DeSalle R, Burk RD (2005) Diversifying selection in human papillomavirus type 16 lineages based on complete genome analyses. J Virol 79:7014–7023

Cleiss-Arnold J et al (2010) Temporal transcriptomic response during arsenic stress in *Herminiimonas arsenicoxydans*. BMC Genomics 11:709

Cole ST et al (2001) Massive gene decay in the leprosy bacillus. Nature 409:1007–1011

Coppée JY (2008) Do DNA microarrays have their future behind them? Microbes Infect 10:1067–1071

Croucher NJ, Thomson NR (2010) Studying bacterial transcriptomes using RNA-seq. Curr Opin Microbiol 13:619–624. PMCID: PMC3025319

Daubin V, Ochman H (2004a) Recognizing lateral gene transfer by quartet mapping. Mol Biol Evol 21:48–51

Daubin V, Ochman H (2004b) Bacterial genomes as new gene homes: the genealogy of ORFans in *E. coli*. Genome Res 14:1036–1042

Daubin V, Lerat E, Perriere G (2003) The source of laterally transferred genes in bacterial genomes. Genome Biol 21:48–51

Demirjian DC, Moris-Varas F, Cassidy CS (2001) Enzymes from extremophiles. Curr Opin Chem Biol 5:144–151

De-Vriendt K, Sandra K, Desmet T, Nerinckx W, Van Beeumen J, Devreese B (2004) Evaluation of automated nano-electrospray mass spectrometry in the determination of non-covalent protein-ligand complexes. Rapid Commun Mass Spectrom 18:3061–3067

Dougherty MJ, D'haeseleer P, Simmons BA, Adams PD, Hadi MZ (2012) Glycoside hydrolases from a targeted compost metagenome, activity-screening and functional characterization. BMC Biotechnol 12:38

Eberly JO, Ely RL (2008) Thermotolerant hydrogenases: biological diversity, properties, and biotechnological applications. Crit Rev Microbiol 34:117–130

Faust K, Raes J (2012) Microbial interactions: from networks to models. Nat Rev Microbiol 10:538–550

Fleischmann RD et al (1995) Whole-genome random sequencing and assembly of *Haemophilus influenzae*. Science 269:496–512

Gil R, Sabater-Munoz B, Latorre A, Silva FJ, Moya A (2002) Extreme genome reduction in *Buchnera* spp.: toward the minimal genome needed for symbiotic life. Proc Natl Acad Sci U S A 99:4454–4458

Giovannoni SJ et al (2005) Proteorhodopsin in the ubiquitous marine bacterium SAR11. Nature 438:82–85

Gomez-Consarnau L et al (2010) Proteorhodopsin phototrophy promotes survival of marine bacteria during starvation. PLoS Biol 8:e1000358

Halter D et al (2012) In situ proteo-metabolomics revealed metabolite secretion by the acid mine drainage bioindicator, *Euglena mutabilis*. ISME J 6:1391–1402

Hecker M, Volker U (2004) Towards a comprehensive understanding of *Bacillus subtilis* cell physiology by physiological proteomics. Proteomics 4:3727–3750

Hocher V et al (2011) Transcriptomics of actinorhizal symbioses reveals homologs of the whole common symbiotic signaling cascade. Plant Physiol 156:700–711

Holt JG, Krieg NR, Sneath PH, Staley JT, Williams ST (eds) (1994) Bergey's manual of determinative bacteriology. Williams & Wilkins, Baltimore

Hou S et al (2004) Genome sequence of the deep-sea gamma-proteobacterium *Idiomarina loihiensis* reveals amino acid fermentation as a source of carbon and energy. Proc Natl Acad Sci U S A 101:18036–18041

Jenner RG, Young RA (2005) Insights into host responses against pathogens from transcriptional profiling. Nat Rev Microbiol 3:281–294

Johnson E, Baron D, Naranjo B, Bond D, Schmidt-Dannert C, Gralnick J (2010) Enhancement of survival and electricity production in an engineered bacterium by light-driven proton pumping. Appl Environ Microbiol 76:4123–4129

Jungblut PR (2001) Proteome analysis of bacterial pathogens. Microbes Infect 3:831–840

Kahn P (1995) From genome to proteome: looking at a cell's proteins. Science 270:369–370

Kim ST et al (2004) Proteomic analysis of pathogen-responsive proteins from rice leaves induced by rice blast fungus, *Magnaporthe grisea*. Proteomics 4:3569–3578

Kimura M (1968) Evolutionary rate at the molecular level. Nature 217:624–626

Könneke M, Bernhard AE, de la Torre JR, Walker CB, Waterbury JB, Stahl DA (2005) Isolation of an autotrophic ammonia-oxidizing marine archaeon. Nature 7058:543–546

Kunst F et al (1997) The complete genome sequence of the gram-positive bacterium *Bacillus subtilis*. Nature 20:249–256

Lander ES et al (2001) Initial sequencing and analysis of the human genome. Nature 409:860–921

Liedert C et al (2010) Two-dimensional proteome reference map for the radiation-resistant bacterium *Deinococcus geothermalis*. Proteomics 10:555–563

Liu Z et al (2005) Patterns of diversifying selection in the phytotoxin-like scr74 gene family of *Phytophthora infestans*. Mol Biol Evol 22:659–672

Lorenz P, Eck J (2005) Metagenomics and industrial applications. Nat Rev Microbiol 3:510–516

Matsumoto N, Yoshinaga H, Ohmura N, Ando A, Saiki H (2000) High density cultivation of two strains of iron-oxidizing bacteria through reduction of ferric iron by intermittent electrolysis. Biotechnol Bioeng 70:464–466

Matsuzaki M et al (2004) Genome sequence of the ultrasmall unicellular red alga *Cyanidioschyzon merolae* 10D. Nature 428:653–657

Médigue C et al (2005) Coping with cold: the genome of the versatile marine Antarctica bacterium *Pseudoalteromonas haloplanktis* TAC125. 1. Genome Res 15:1325–1335

Metzker ML (2010) Sequencing technologies – the next generation. Nat Rev Genet 11:31–46

Moretti M et al (2010) A proteomics approach to study synergistic and antagonistic interactions of the fungal-bacterial consortium *Fusarium oxysporum* wild-type MSA 35. Proteomics 10:3292–3320

Mulder NJ et al (2005) InterPro, progress and status in 2005. Nucleic Acids Res 33(Database issue):D201–D205

Muller D et al (2007) A tale of two oxydation states: bacterial colonization of arsenic-rich environments. PLoS Genet 3:e53

Nadon R, Shoemaker J (2002) Statistical issues with microarrays: processing and analysis. Trends Genet 18:265–271

Nakabachi A, Yamashita A, Toh H, Ishikawa H, Dunbar HE, Moran NA, Hattori M (2006) The 160-kilobase genome of the bacterial endosymbiont *Carsonella*. Science 314:267

Narasingarao P et al (2012) De novo metagenomic assembly reveals abundant novel major lineage of Archaea in hypersaline microbial communities. ISME J 6:81–93

Noel-Georis I et al (2004) Global analysis of the *Ralstonia metallidurans* proteome: prelude for the large-scale study of heavy metal response. Proteomics 4:151–179

Normand P et al (2007) Genome characteristics of facultatively symbiotic *Frankia* sp. strains reflect host range and host plant biogeography. Genome Res 17:7–15

Ou K et al (2005) Integrative genomic, transcriptional, and proteomic diversity in natural isolates of the human pathogen *Burkholderia pseudomallei*. J Bacteriol 187:4276–4285

Parales RE, Ditty JL (2005) Laboratory evolution of catabolic enzymes and pathways. Curr Opin Biotechnol 16:315–325

Pradella S, Hans A, Sproer C, Reichenbach H, Gerth K, Beyer S (2002) Characterisation, genome size and genetic manipulation of the myxobacterium *Sorangium cellulosum* So ce56. Arch Microbiol 178:484–492

Pühler A, Ariat M, Becker A, Göttfert M, Morrissey JP, O'Gara F (2004) What can bacterial genome research teach us about bacteria-plant interaction? Curr Opin Plant Biol 7:137–147

Rabilloud T (2002) Two-dimensional gel electrophoresis in proteomics: old, old fashioned, but it still climbs up the mountains. Proteomics 2:3–10

Rhodius VA, LaRossa RA (2003) Uses and pitfalls of microarrays for studying transcriptional regulation. Curr Opin Microbiol 6:114–119

Ronaghi M (2001) Pyrosequencing sheds light on DNA sequencing. Genome Res 11:3–11

Sanger F, Nicklen S, Coulson AR (1977) DNA sequencing with chain-terminating inhibitors. Proc Natl Acad Sci U S A 74:5463–5467

Saunders NF, Goodchild A, Raftery M, Guilhaus M, Curmi PM, Cavicchioli R (2005) Predicted roles for hypothetical proteins in the low-temperature expressed proteome of the Antarctic archaeon *Methanococcoides burtonii*. J Proteome Res 4:464–472

Scarselli M, Giuliani MM, Adu-Bobie J, Pizza M, Rappuoli R (2005) The impact of genomics on vaccine design. Trends Biotechnol 23:84–91

Schleper C, Jurgens G, Jonuscheit M (2005) Genomic studies of uncultivated archaea. Nat Rev Microbiol 3:479–488

Schmid MB (2004) Seeing is believing: the impact of structural genomics on antimicrobial drug discovery. Nat Rev Microbiol 2:739–746

Schmutz J et al (2004) Quality assessment of the human genome sequence. Nature 429:365–368

Siew N, Fischer D (2003) Twenty thousand ORFan microbial protein families for the biologist? Structure 11:7–9

Stewart FJ, Dmytrenko O, DeLong EF, Cavanaugh CM (2011) Metatranscriptomic analysis of sulfur oxidation genes in the endosymbiont of *Solemya velum*. Front Microbiol 2:1–10

Streit WR, Schmitz RA (2004) Metagenomics – the key to the uncultured microbes. Curr Opin Microbiol 7:492–498

Strous M et al (2006) Deciphering the evolution and metabolism of an anammox bacterium from a community genome. Nature 440:790–794

Tweeddale H, Notley-McRobb L, Ferenci T (1998) Effect of slow growth on metabolism of *Escherichia coli*, as revealed by global metabolite pool ("metabolome") analysis. J Bacteriol 180:5109–5116

Tyson GW et al (2004) Community structure and metabolism through reconstruction of microbial genomes from the environment. Nature 428:37–43

Tyson GW, Lo I, Baker BJ, Allen EE, Hugenholtz P, Banfield JF (2005) Genome-directed isolation of the key nitrogen fixer *Leptospirillum ferrodiazotrophum* sp. nov. from an acidophilic microbial community. Appl Environ Microbiol 71:6319–6324

Vallenet D et al (2006) MAGE: a microbial genome annotation system supported by synteny results. Nucleic Acids Res 34:53–65

van Ham RC et al (2003) Reductive genome evolution in *Buchnera aphidicola*. Proc Natl Acad Sci U S A 100:581–586

Velculescu VE et al (1997) Characterization of the yeast transcriptome. Cell 24:243–251

Venter JC et al (2004) Environmental genome shotgun sequencing of the Sargasso Sea. Science 304:66–74

Vezzi A et al (2005) Life at depth: *Photobacterium profundum* genome sequence and expression analysis. Science 307:1459–1461

Walter JM, Greenfield D, Bustamante C, Liphardt J (2007) Light-powering *Escherichia coli* with proteorhodopsin. Proc Natl Acad Sci U S A 104:2408–2412

Wang Z, Gerstein M, Snyder M (2009) RNA-Seq: a revolutionary tool for transcriptomics. Nat Rev Genet 10:57–63

Watson JD, Crick FH (1953) Molecular structure of nucleic acids; a structure for deoxyribose nucleic acid. Nature 17:737–738

Weiss S et al (2009) Enhanced structural and functional genome elucidation of the arsenite-oxidizing strain *Herminiimonas arsenicoxydans* by proteomics data. Biochimie 91:192–203

Wilmes P, Bond PL (2004) The application of two-dimensional polyacrylamide gel electrophoresis and downstream analyses to a mixed community of prokaryotic microorganisms. Environ Microbiol 6:911–920

Wishart DS (2005) Metabolomics: the principles and potential applications to transplantation. Am J Transplant 5:2814–2820

Woyke T et al (2006) Symbiosis insights through metagenomic analysis of a microbial consortium. Nature 7114:950–955

Yamada T et al (2012) Prediction and identification of sequences coding for orphan enzymes using genomic and metagenomic neighbours. Mol Syst Biol 8:581

Yang C et al (2006) Comparative genomics and experimental characterization of N-acetylglucosamine utilization pathway of *Shewanella oneidensis*. J Biol Chem 40:29872–29885

Zakrzewski M et al (2012) Profiling of the metabolically active community from a production-scale biogas plant by means of high-throughput metatranscriptome sequencing. J Biotechnol 158:248–258

Zylstra GJ, Kukor JJ (2005) What is environmental biotechnology. Curr Opin Biotechnol 16:243–245

Modeling in Microbial Ecology

Jean-Christophe Poggiale, Philippe Dantigny, Rutger De Wit, and Christian Steinberg

Abstract

The bases and the principles of modeling in microbial community ecology and biogeochemistry are presented and discussed. Several examples are given. Among them, the fermentation process is largely developed, thus demonstrating how the model allows determining the microbial population growth rate, the death rate, and the maintenance rate. More generally, these models have been used to increase the development of bioenergetic formulations which are presently used in biogeochemical models (Monod, Droop, DEB models). Different types of interactions (competition, predation, and virus–bacteria) are also developed. For each topic, a complete view of the models used in the literature cannot be presented. Consequently, the focus has been done on the demonstration how to build a model instead of providing a long list of existing models. Some recent results in sediment biogeochemistry are provided to illustrate the application of such models.

Keywords

Biofilm models • Biotic interactions • Chemostat • Fermenter models • Metabolic models • Population dynamics

* Chapter Coordinator

J.-C. Poggiale* (✉)
Institut Méditerranéen d'Océanologie (MIO), UM 110, CNRS 7294 IRD 235, Université de Toulon, Aix-Marseille Université, Campus de Luminy, 13288 Marseille Cedex 9, France
e-mail: jean-christophe.poggiale@univ-amu.fr

P. Dantigny
Laboratoire des procédés alimentaires et microbiologiques, UMR PAM, AgroSup Dijon et Université de Bourgogne, 21000 Dijon, France

R. De Wit
Écologie des systèmes marins côtiers (ECOSYM, UMR5119), Universités Montpellier 2 et 1, CNRS-Ifremer-IRD, 34095 Montpellier Cedex 05, France

C. Steinberg
Pôle des Interactions Plante-Microorganismes, INRA UMR 1347 Agroécologie, AgroSup-INRA-Université de Bourgogne, BP 86510, 21065 Dijon Cedex, France

19.1 Introduction

Ecological studies do not focus on enzymatic reactions, cells, organs, neither on individuals but on populations or communities interacting in dynamics ecosystems. It is then a field which needs the collaboration between various other natural disciplines (systematic, genetic, physiology, microbiology) or exact ones (chemistry, physics, mathematics). Furthermore, it covers a wide range of spatial and temporal scales, which makes their integration difficult.

Microbial ecology is actually a part of general ecology. It is however a relatively young discipline, and it still has to discover its own set of methods for investigations, tools, and concepts. In the case of microbial ecology, the reasons are probably related first to the accessibility of microorganisms (cultivable or not), the scale changes, whether they are quantitative (from 1 to 10^{10} bacteria/g or mL of substrate and from 50 cm to 250 m mycelial hyphae/g substrate) or spatial (from microaggregate to plot or from biofilm to water flow and from in situ microbial activity of a given population to the biogeochemical cycles), and second to their rate of evolution (generation time, mutation frequency, horizontal gene flow, metabolic versatility). In addition, technical constraints (how to observe, characterize, identify, quantify microbial communities, their activities, and functions in the environment?) also made an obstacle to the conceptualization of ecosystem functioning. Molecular tools helped to overcome a number of these constraints. They now allow to monitor the fate of a population in a complex environment and to assess changes in community structure of bacterial, fungal, and protozoan in response to environmental perturbations.

Whatever the methods to monitor and ultimately control the fate of microbial populations in the environment (soil, lakes, rivers or oceans, air, food), it is necessary to understand how the regulation of populations occurs, that is to say to know and to characterize the mechanisms involved in the dynamics of microbial populations. The environment is complex, and it is not easy to consider separately each abiotic factor or each of the mechanisms of regulation (of biotic origin) of microbial populations. Therefore, the contrasting situations (the experimenter varies the environment) are first experimental and involve the use of simplified systems (chemostat, disinfected water or soil) and/or the introduction of a microbial population model, identifiable within the indigenous microflora (antibiotic resistance, serological identification, molecular monitoring, etc.). Changes in population densities are then quantified and correlated with the factor studied. Mathematical models are proposed to describe the dynamics and test the weight of the factor assumed. The interest in modeling is multiple, but above all, it is a tool to simplify a complex system to understand how it works and to estimate nonmeasurable values. This is why microbial ecology needs mathematical modeling.

Mathematical modeling in ecology has its origins in the mid-nineteenth century including the Verhulst model (now called logistic equation) describing the kinetics of logistic growth with a phase of exponential growth at low density and self-limitation of growth at high density. The use of mathematical formulas has always been motivated by quantifying process. Mathematical modeling in biology has grown considerably in the twentieth century and became popular especially after the development of computers. There are actually many modeling softwares in ecology to make system simulations representative of more or less complex ecosystems. This chapter lays the foundation for a good use of such softwares with a specific focus on mathematical modeling in microbial ecology. We will first describe some goals fairly briefly.

The first objective of modeling in ecology is forecasting. In this context, the model is a mathematical object (formula) or a computing tool (software) in which we introduce the known relationships between some quantities, some of which can be measured. Then the model is used to predict the value of quantities of interest, including those to which access is difficult by measurement. It is of course necessary that the model has been validated in operating conditions that will be encountered in the context of its use. The validation of current ecological models is usually obtained in very restrictive management, and use of such models for predictive purposes may be ineffective. More of this will be discussed in this chapter, certain phenomena are inherently unpredictable, and it is then necessary to take specific steps.

Other objectives of mathematical modeling in ecology seem a priori less challenging in practice but are actually more rewarding in terms of acquisition of knowledge and for the understanding of ecosystem functioning. First, the realization of a model requires a leveling of knowledge and effort of synthesis and concision which may in some cases facilitate the integration of knowledge. Once the model is carried out in well-defined conditions, it allows to test different scenarios and understand the role of the assumptions that underlie them. As surprising as it may seem at first, a model that does not work can be more interesting than a model that works. Indeed, if a model based on carefully selected assumptions does not work, that means that our understanding of the system is inadequate and requires deeper reflection, thus leading to a better knowledge of the system and possibly implementation of new experiments. Finally, a model can also be used to estimate quantities that are difficult or impossible to obtain by measurement or observation.

In summary, mathematical modeling facilitates the rigorous synthesis of processes and their relationships, allows to test hypotheses and scenarios to deepen our knowledge on the subject of study, and in some cases enables to predict the evolution of a system or estimate quantities difficult to measure.

Mathematical modeling is to write mathematical equations to represent natural phenomena. When the model is written, two main approaches can be used to run the model; these two approaches are generally conducted in parallel. The first approach is to try to understand how, based on the values of the model parameters, the solutions will behave: it allows anticipating the variation over time of the state variables. This approach is rather technical and will be little discussed in this chapter. The second approach is to implement the model on a computer to perform numerical simulations. This second approach is sometimes called numerical modeling. It has the advantage of being performed in the case of complex models and it is also in this situation it is the most interesting. However, it raises a technical problem; it is important to keep in mind when implementing it. A computer is a tool that does not have the ability to calculate derived function accurately. Like most models that involve derivatives, it is necessary to provide approximate expressions of these derivatives; we construct

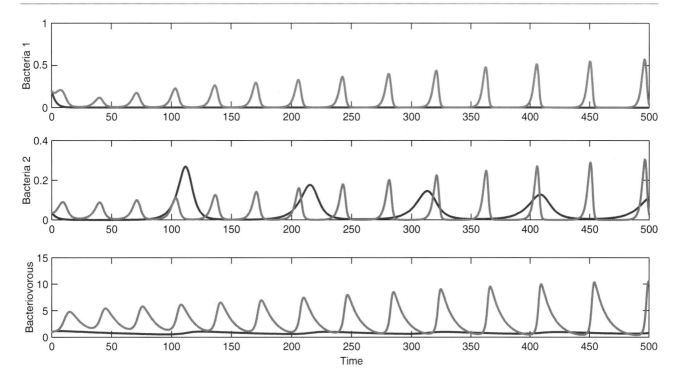

Fig. 19.1 This figure represents the numerical solutions of a model of interactions between two species of bacteria and a predator. The solutions are obtained with two different numerical methods. According to the choice of the numerical methods, the ecological interpretations are quite different. This highlights the fundamental role of the numerical simulations tools

what is called a numerical scheme. However, these approximations can lead to substantial errors between the result of numerical simulation and those that the simulation would have given if no approximation had been made; they are called numerical errors. We illustrate this problem with an example on the interaction between two populations of bacteria grown in competition for a substrate and a population of bacterivores (Fig. 19.1). The corresponding model is described in details at the end of the chapter. In this example, depending on the numerical method used, the ecological interpretation varies. Indeed, one of the simulations led to the extinction of bacterial populations, while the other one shows their coexistence (as shown by further analysis, it is the correct result for this example). Yet this is the same model, and the difference between the simulations results from an artifact of the method used to integrate the model with the computer. The choice of this method is not only a technical detail, but it can also be very important in terms of the ecological interpretation. To facilitate the work of the modeler who does not want to invest in this difficult task and however wants to perform numerical simulations, many modeling softwares are currently on the market. They provide an interface that allows the user to enter mathematical expressions of the model, and the software already contains numerical schemes. Care should be taken however because even very simple models are not suitable for standard numerical schemes that are found on the software.

What makes the problem more difficult is that any program or software toward which calculation is subjected makes the calculation, even if the method is not suitable. It is therefore preferable to have a good understanding of what the model can give or define a battery of tests to verify that the simulation results are consistent with what they have to give.

Finally, if we accept the fact that an ecological model is difficult to validate, we must admit the corollary: implementation of a model is never a finished work in itself, each realization bringing new issues and reformulation of the model. This remark allows us to introduce the approach of modeling environment in general and the necessary interaction between modeling and observation. In a somewhat schematic way, this approach can be described as a series of steps (see diagram in Fig. 19.2):

(i) The starting point is to formulate a specific question about a system under study. This a priori trivial step is not at all trivial because it will determine the next steps. The question should be simple and precise and require a simple answer. If the question is too complex or poorly formulated, it must be broken down into several points that meet the following steps separately.

(ii) One must then determine the parameters that characterize the system.

(iii) Experiments or observations about the system studied are made to acquire data for these variables.

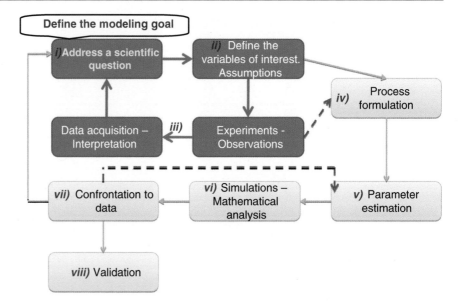

Fig. 19.2 This schematic drawing describes how the modeling approach (in *green*) can be inserted into a general scientific activity (in *blue*). The starting point lies in the definition of set of scientific questions. Details of this procedure are provided in the text

(iv) Prior knowledge (when available) on these variables or those acquired through experimentation can establish relationships between them and build a first model (formalized process). As the reader will realize through the examples in this chapter, an effort of simplification is necessary and beneficial during this stage.

(v) The relationships between the variables usually involve parameters whose values are not always known. However, the comparison between the outputs of models and observations is designed to estimate the unknown parameters.

(vi) This step provides a calibrated model, that is to say the equations and parameter values they contain are known.

(vii) To validate the model, it is necessary to confront them with data that were not used to estimate the values of the parameters.

(viii) A method for evaluating the quality of the model should then be defined. A current evaluation mode is the distance between model outputs and observations. The smaller the distance, the better the model is able to reproduce the data. If this step does not work, then we must return to the beginning of the process, by defining new questions whose answer can explain what did not work and produce a model better suited to the original question.

The remainder of this chapter is organized as follows. In the second section, the principles for building a model in ecology are presented. The third section is devoted to the modeling of microbial dynamics in culture. This section adopts the point of view of process engineering. In the fourth section, we present the modeling of biotic interactions and show through examples how these models can advance knowledge in microbial ecology. The fifth section is the problem of scaling and outlines the potential of modeling it. The sixth section illustrates the representation of space in the models in ecology. The seventh section presents the modeling of biofilms. Finally, the eighth section briefly introduces the modeling in biogeochemistry. The chapter ends with a concluding section.

19.2 Principles of Model Building in Ecology

The construction of a model is based on a number of basic principles valid in all areas as we have seen above. By cons, some principles are specific to the exact sciences (physics, chemistry), and others are specific to natural sciences such as ecology. For example, modeling in biology and ecology is not based on any general law to obtain a set of equations. It is for this reason that a rigorous approach to the construction of equations is required and the principles described below can therefore be very useful. They will be included in the next section in more specific contexts, in particular the section on methods. Two main families of models can be defined in terms of the representation of time: discrete-time models and continuous-time models. In the first case, the system is described at times 0, Δt, $2\Delta t$, etc., and nothing is expressed during the interval between two instants. In the second case, the system is described at each instant; time is a continuum. The first situation can be applied, for example, to a population in which all individuals are duplicated simultaneously, at times more or less accurate. By cons, in a population with a large number of individuals whose duplications are not synchronized, we can assume that in any small time interval duplication can occur, and in this case the second approach is preferable. This dichotomy is formal, but it facilitates the presentation of the construction of models, and we use it in this section.

19.2.1 Discrete-Time Models

A natural system can be studied through the measurement of different sizes. Each of these variables, which characterizes the system and evolves over time, is called *state variable*. It is denoted by X_t where t is the time when we consider this variable state X and $X_{t+\Delta t}$ is the value of the quantity at the moment $t + \Delta t$. In very general terms, we can formalize the dynamics of the state variable X as follows:

$$X_{t+\Delta t} = X_t + (\text{source} - \text{sink})\Delta t \quad (19.1)$$

where the term "source" (respectively "sink") represents the contribution per time unit to the growth (respectively decrease) of the variable X_t. For example, in the framework of population dynamics, we can define the density X_t of a population at time t and write

$$X_{t+\Delta t} = X_t + (\text{birth} - \text{death})\Delta t \quad (19.2)$$

It is possible to further specify the previous example by adding assumptions about the process of birth and death. For example, assume that the birth rate and the death rate are constant. The birth rate is the number of individuals produced by present individual per time unit; the mortality rate is the proportion of individuals who die per time unit. Let b and m be the birth rate and mortality rate, respectively. According to the previous definitions, the model reads

$$X_{t+\Delta t} = X_t + (bX_t - mX_t)\Delta t \quad (19.3)$$

The numbers b and m are called parameters; they are assumed to be constant here. In this example, the mathematical expression of the model can be simplified as follows:

$$X_{t+\Delta t} = (1 + (b - m)\Delta t)X_t \quad (19.4)$$

Setting $r = 1 + (b - m)\Delta t$ allows writing the previous expression as

$$X_{t+\Delta t} = rX_t \quad (19.5)$$

The parameter r is the *population growth rate*. A first result from this is that if $r > 1$, then the population grows between t and $t + \Delta t$, while it declines if $r < 1$ (Fig. 19.3a as an illustration).

The interpretation of this result is straightforward from the definition of r. If $r > 1$, then it means that birth is larger than death; as a consequence, the population density increases. Another information can be extracted from model (19.5), by noting that $X_{t+\Delta t} = rX_t$ implies $X_k = r^{k\Delta t}X_0$. This last formula allows to determine the population density at time $k\Delta t$ directly from the initial density (obtained at $t = 0$) and the parameter r. Figure 19.3 illustrates the dynamics obtained from this model, depending on the relative value of r with respect to 1.

From a general point of view, using Eq. (19.1), it is possible to express the dynamics of a state variable X_t by means of the following equation:

$$X_{t+\Delta t} = F(X_t) \quad (19.6)$$

where F is a function defined from $\mathbb{R}$ to $\mathbb{R}$ where $\mathbb{R}$ denotes the set of real numbers. It is assumed from now that this

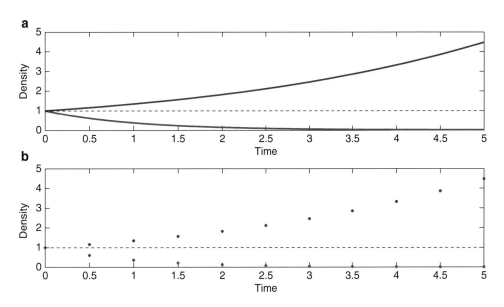

Fig. 19.3 Simulation of the linear model: the time discrete case, Eq. (19.5), is plotted on graph (**a**) with $r = 0.6$ in *red* and $r = 1.16$ in *blue*; the time-continuous case, Eq. (19.10), is plotted on graph (**b**) with $r = -1$ in *red* and $r = 0.3$ in *blue*

function is differentiable. In the example provided by Eq. (19.5), the function F is defined by $F(X) = rX$.

19.2.2 Continuous-Time Models

Equation (19.1) (or 19.6) allows seeing the system only at time $k\Delta t$ where k is an integer. In order to get some information at other times, it is possible to reduce the time step Δt and even to make it tend to zero. Model (19.1) allows thus to write the variation of the state variable per unit of time:

$$\frac{X_{t+\Delta t} - X_t}{\Delta t} = \text{source} - \text{sink} \quad (19.7)$$

The left-hand term is the variation rate. Its limit when Δt tends to zero is what is called the time derivative of X which is denoted by

$$\frac{dX}{dt} = \lim_{\Delta t \to 0} \frac{X_{t+\Delta t} - X_t}{\Delta t} \quad (19.8)$$

The general equation of a model describing a state variable dynamics in continuous time reads then

$$\frac{dX}{dt} = \text{source} - \text{sink} = F(X) \quad (19.9)$$

where F is a function which is assumed to be differentiable. Consider again the example of an isolated population with constant birth and death rates. The time-continuous model reads

$$\frac{dX}{dt} = bX - mX = rX \quad (19.10)$$

where now $r = b - m$. If the birth rate is larger than the death rate, then $r > 0$, and since X, as population density, is a positive number, it follows that the time derivative of X is positive; thus, this density increases. In the same way, if the birth rate is lower than the death rate, the population density declines. Eq. (19.10) can be written as

$$\frac{dX}{X} = r\,dt \quad (19.11)$$

By integrating the right- and the left-hand sides separately, it follows that $\log(X) = rt + C_1$ where C_1 is a constant. It allows to get the population density explicitly:

$$X(t) = X(0)\exp(rt) \quad (19.12)$$

Once again, the solution provides an explicit relation between the population density and the population growth rate. The linear population growth model in continuous time (19.10) has a dynamics similar as its discrete counterpart (19.5), as it is illustrated in Fig. 19.3b.

19.3 Microbial Kinetics in a Culture

In this section, models used in process engineering for instance are presented. These models form the basis of more general ones often used in microbial ecology. Moreover, this section presents the vocabulary and the modeling approach associated to experiments in laboratories.

19.3.1 The Process

This section focuses on processes operating in liquid medium. Unlike solid media, it will be assumed that the treated liquid media are well mixed, so that they are homogeneous: at any point in a liquid medium, the concentration of the components is the same. We will also assume that all organisms are in the liquid medium, which could be not true if the experiment is not conducted properly. In fact, some microorganisms, in amounts which can be neglected, are in the headspace. This explains why it is necessary to condense and then filter the gas out of a fermenter especially when pathogenic microorganisms are grown. Moreover, the liquid medium may go up the edges of a fermenter by capillarity and then be colonized by microorganisms. It is interesting to note that in this case the microorganisms form a film along the wall of the fermenter.

In a perfectly mixed liquid medium, microorganisms are found in planktonic state; the cells are individualized. Thus, this chapter mainly concerns unicellular microorganisms such as bacteria (e.g., *Clostridium pasteurianum*) and yeast (e.g., *Saccharomyces cerevisiae*). Metabolic products of the activity of the microbial population introduced into the fermenter under anaerobic conditions have an industrial or medical interest. Fatty acids, alcohols, gas, or other products can be used either directly (e.g., ethanol) or in chemical processes for the production of biopolymers. Note that the filamentous fungi (e.g., *Fusarium graminearum* producing myco-protein of food interest) can grow in liquid media, but these microorganisms form pellets when growing and environment cannot be considered homogeneous at the end. Similarly, animal cells will form biofilms on the surface of microcarriers. Within the non-planktonic systems, there are concentration gradients which in fact are not homogeneous.

19.3.2 Different Fermentation Modes

According to the values of rate of input and output flows, three main modes of fermentation can be defined (Table 19.1).

The fermentation in batch mode is widely used. The culture medium is sterilized directly in the fermenter, in

19 Modeling in Microbial Ecology

Table 19.1 Different fermentation modes

Fermentation mode	Input rate Q_e	Output rate Q_s	Volume
Batch (not continuous)	0	0	Constant
Fed-batch (semicontinuous)	Different from 0	0	Variable
Chemostat (continuous)	Different from 0	Q_e	Constant

situ, by steam injection; then after cooling, fermentation begins by adding sterile inoculum. All components necessary for growth are present from the start of fermentation, which takes place until exhaustion of the substrate or to the maximum product yield. In general, this type of fermentation lasts from a few hours to a week.

In the case of a semicontinuous fermentation, the procedure starts the same way. However, the originality of this method of fermentation is to add the substrate by varying the input rate Q_e, as the substrate is consumed by microorganisms. It is possible to adjust the input rate so that the substrate concentration in the fermenter (S) remains constant over time. Since the input rate is not zero and that the output flow rate is zero, the volume increases in the fermenter. Therefore, it must be retained in fed-batch that it is essential to think in mass (g) instead of concentration (g.L^{-1}); otherwise, errors occur.

Strictly speaking, the chemostat is a continuous mode of fermentation to which the inflow is constant. This type of fermentation is potentially very productive because, theoretically, fermentation never stops, which prevents the frequent and typical sterilization and cleaning steps of other modes of fermentation. In practice, this type of fermentation is used industrially for the production of molecules due to the high risk of contamination. In the laboratory, the chemostat is used to determine the characteristic parameters of cellular metabolism as yields or maintenance coefficient (see Sect. 19.3.5). Indeed, in a chemostat at steady state (when all state variables are constant), metabolism depends only on the dilution rate defined by

$$D = \frac{Q_e}{V} \quad (19.13)$$

It is possible to consider further that the chemostat with recycling is close to the effluent treatment, with the difference that, in function of time, the flow of effluent to be treated may vary frequently and strongly. Therefore, it is unrealistic to think that in these transient conditions, a steady state (where all state variables of the process are constant) can exist. However, as mentioned in previous chapters, the natural microbial ecosystems that are installed in sewage basins are stable enough not to fear contamination that occurs when essentially pure culture is used.

The chemostat is a continuous culture for which the input rate is generally kept constant. It should be noted that there are other types of continuous fermentations where the inflow varies so as to maintain the pH constant (pH-stat fermenter) or the biomass concentration constant (turbidostat).

Finally, there is another type of fermentation, mostly used for animal cells, called cascade gradostat, which can be related to successive batch system.

19.3.3 Process Modeling

19.3.3.1 State Variables

As in the previous models, the set of state variables allows to completely describe the process at each time t. A complete modeling process can be made by considering these variables. The first step is to define the process. This approach is characteristic of what is known as process engineering. The process consists in the liquid medium in the fermenter or other container. There are four variables that usually describe the process as shown schematically in Fig. 19.4.

19.3.3.2 Mass Balance Equation

The basis of any deterministic model is the correct expression of the balance equation. Eq. (19.14) is a very general equation and is applied for all budgets, for either material or enthalpy. Depending on the budget, the units will differ. Note that all of these units are additive and of course as concentration units (g.L^{-1}) are excluded. It must therefore be emphasized that

- For mass balance, the unit for biomass, substrate, and product is g.h^{-1}.
- For volume balance, the unit is L.h^{-1}:

$$\text{(Rate of variation)} = \begin{pmatrix} \text{inflow} - \text{ouflow} \\ + \text{generated flow} - \text{consumed flow} \end{pmatrix} \quad (19.14)$$

Rate of Variation: It is defined by the variation of a quantity per time unit. The quantity is one of the state variables of the process. Thus, considering Fig. 19.4, the only variables that can be written are X, S, P, or V. If this last term is negative, the state variable decreases with time. Otherwise, it increases.

Inflow: It represents the flow of the quantity entering in the fermenter per time unit by means of the input flow Q_e. Of course this flow tends to increase the size of the corresponding process and is assigned a positive sign. Note that if there are multiple input flows, the input will consist of all terms of different input rates.

Fig. 19.4 Schematic drawing of a fermenter biomass concentration: $X(\text{g.L}^{-1})$, limiting substrate concentration: $S(\text{g.L}^{-1})$, product concentration: $P(\text{g.L}^{-1})$, fermenter volume: $V(\text{L})$

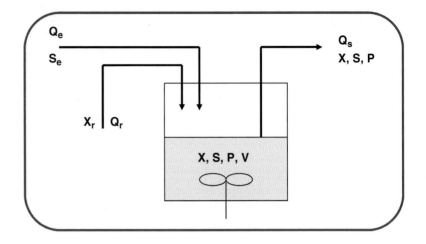

- biomass concentration: $X\ (\text{g.L}^{-1})$
- limiting substrate concentration: $S\ (\text{g.L}^{-1})$
- product concentration: $P\ (\text{g.L}^{-1})$
- fermenter volume: $V\ (\text{L})$

Outflow: It represents the flow of the quantity leaving the fermenter per time unit by means of the output flow Q_s. This flow tends to reduce the size of the process and is assigned a negative sign.

Generated Flow: It is what tends to increase the size of the process but cannot fit through the inflow. More precisely, this term refers to the production of biomass or of products inside the fermenter. It is assigned a positive sign.

Consumed Flow: It is what tends to reduce the size of the process but cannot fit through the outflow. For instance, this term could include the consumption of substrate inside the fermenter. It is assigned a negative sign.

The general way to proceed to write a mass balance equation is based on this equation:

$$\frac{d(\cdots)}{dt} = Q_e \cdot (\cdots) - Q_s \cdot (\cdots) + \cdots - \cdots \quad (19.15)$$

Before applying this general framework to the different state variables of the process, it is advised to refer to the diagram of the process, in Fig. 19.4. The common mistake when writing different equations is to mix sizes, namely, biomass, substrate, and product. Consider first the biomass alone.

19.3.3.3 Biomass Balance (g.h^{-1})

In the general case, where recycling takes place, the biomass balance is

$$\frac{d(X \cdot V)}{dt} = Q_r \cdot X_r - Q_s \cdot X + r_x''' \cdot V - k_d \cdot X \cdot V \quad (19.16)$$

Units are very useful to define the new quantities in Eq. (19.16). r_X''' is the rate at which biomass is produced, expressed in g.L^{-1}.h^{-1}. In other words, it is the amount of biomass produced in grams per liter of fermenter and per hour. In process engineering, the exponent ($'''$) means a unit per volume, and the exponent ($''$) means a unit per surface. In all cases, this production rate can be relied to the specific growth rate μ. The index X means that the rate is associated to biomass:

$$r_X''' = \mu X \quad (19.17)$$

The calculation of the specific growth rate, more simply known as growth rate, is the basis of microbial kinetics. We omit that it is specific, i.e., it is related to the amount of biomass. It will be seen later that only the specific rates allow characterizing the metabolism. Actually, μ represents the quantity in grams of biomass produced per 1 g of biomass for 1 h. This explains why μ can remain constant, while the biomass concentration increases. However, Eq. 19.17 shows that r_X''' is not constant and depends on X. It is important to keep this point in mind when integrating certain mathematical expressions in the section on yield calculations.

We will see in the subsection devoted to secondary models that as long as substrate is present, growth occurs. In this case, the cells divide. It is thus logical that the generated flow is nonzero. However, when there is no substrate, there is no growth (see Monod relationship, below), and one can observe a decrease in the biomass concentration. Therefore, it is necessary to introduce a consumed flow $k_d \cdot X \cdot V$ where k_d (h^{-1}), like μ, represents the constant cell death rate:

$$r_{\text{death}}''' = k_d X \quad (19.18)$$

Note that in the case where there is growth, so substrate, the consumed flow is null, $k_d = 0$. However, the absence of substrate causes growth stop, $\mu = 0$, and after a stationary phase, cell death is observed.

19.3.3.4 Substrate Balance (g.h^{-1})

$$\frac{d(S \cdot V)}{dt} = Q_e \cdot S_e + Q_r S_r - Q_S \cdot S - r_s''' \cdot V \quad (19.19)$$

This very general balance illustrates that it is possible to get an input rate of substrate related to the inflow Q_e and to the recycling flow Q_r. The substrate concentration in the reservoir is denoted S_e (g.L^{-1}). In the case of a settling pond for wastewater treatment, it can be considered that recycled sludge contains microorganisms in high concentration and, therefore, that the substrate is exhausted $S_r = 0$. However, in some industrial applications constituted by an integrated fermentation ultrafiltration system, the substrate may be present in the material which is recycled.

19.3.3.5 Product Balance (g.h^{-1})

$$\frac{d(P \cdot V)}{dt} = -Q_S \cdot P + r_P''' \cdot V - k_{\text{dissoc}} \cdot P \cdot V \quad (19.20)$$

The goal is to synthesize the product in the fermenter, so there is no inflow. As for the coefficient of cell death, a coefficient of dissociation k_{dissoc} (h^{-1}) appears as a factor in the sink flow. Note also that the consumed flow is proportional to the concentration P.

19.3.3.6 Volume Balance (L.h^{-1})

$$\frac{d(V)}{dt} = Q_e + Q_r - Q_S \quad (19.21)$$

19.3.4 Primary Models

These models describe the dynamics of the state variables from the basic equation (19.14) described above, according to the different operating modes. In this section, in order to introduce the reader in the use of equations, we will make three simplifying assumptions, and we will indicate in which cases they can be considered reasonable:

- The specific rate of growth, $\mu = C^{te}$.
- The death constant is zero, $k_d = 0$.
- The maintenance coefficient is zero, $m = 0$.

When the concentration of limiting substrate S is greater than a value that can be set, in this example around 1 g.L^{-1}, μ is constant even when S varies. We will verify this hypothesis in the paragraph devoted to secondary models and, particularly, to the Monod relation. Also, when S is greater than the arbitrary value 1 g.L^{-1}, microbial growth occurs, and as we have previously pointed out, there is no cell death. The coefficient of maintenance, $m(g(S) \cdot g(X)^{-1} \cdot h^{-1})$, is a new parameter that we need to introduce. Due to its unit, m can be simply defined as the amount of substrate in grams required for maintenance (i.e., covering the energy needs of the cell) of one gram of biomass for 1 h. We will discuss further the significance of the coefficient of maintenance in the section devoted to the metabolism. From now, note that one can neglect the maintenance during the exponential phase of growth. However, maintenance becomes significant when the substrate concentration is low or zero, i.e., during the stationary phase. A nonzero coefficient is responsible for maintaining cell death of certain cells during the period of decline due to the lack of substrate.

19.3.4.1 Batch

The aim of this section is to show how it is straightforward to determine the expressions of the specific growth rate μ and of the generation time t_g (h) from the mass balance equation. When there is no recycling and no cell death, the total biomass equation reads

$$\frac{d(X \cdot V)}{dt} = r_X''' \cdot V \quad (19.22)$$

Replacing r'''_x by $\mu \cdot X$ leads to

$$\frac{d(X \cdot V)}{dt} = \mu \cdot X \cdot V \quad (19.23)$$

In batch, the volume V is constant; thus, the previous expression can be simplified as

$$\frac{d(X)}{dt} = \mu \cdot X \quad (19.24)$$

Separating the variable X on the left and t on the right leads to

$$\frac{d(X)}{X} = \mu \cdot dt \quad (19.25)$$

Integration of both terms provides

$$\int_{X_0}^{x} \frac{d(X)}{X} = \int_0^t \mu \cdot dt \quad (19.26)$$

Note that the relation between the integration bounds is important. It is recommended when the fermentation is a sequence of several phases (latency, exponential, stationary, decline), in a batch for instance, to start each phase by resetting the time to 0 at each phase start and then to recalculate the new values of X_0. On the other hand, to find X, it is convenient to take X as the upper bound in the integration (Eq. 19.26).

The assumption $\mu = C^{te}$ is now essential, since it allows to extract it from the integration to get

$$\text{Ln}\frac{X}{X_0} = \mu \cdot t \quad (19.27)$$

This shows that growth is exponential; these results are similar to those obtained in Sect. 19.2:

$$X(t) = X_0 \exp(\mu \cdot t) \quad (19.28)$$

Furthermore, the relation (19.27) allows determining the relation between μ and the generation time. For $t = t_g$, the initial biomass is doubled: $X(t_g) = 2X_0$. By replacing, it reads

$$\text{Ln}(2) = \mu \cdot t_g \quad (19.29)$$

19.3.4.2 Fed-Batch

The biomass balance in fed-batch is the same as that in batch:

$$\frac{d(X \cdot V)}{dt} = r'''_X \cdot V = \mu \cdot X \cdot V \quad (19.30)$$

On the other hand, unlike in the batch situation, the volume is not constant. It can be set $M = X \cdot V$ to perform a change of variables. The new bounds of integration are $M_0 = X_0 \cdot V_0$ at $t = 0$:

$$\int_{M_0}^{M} \frac{d(M)}{M} = \int_0^t \mu \cdot dt \quad (19.31)$$

After integration, it becomes

$$\text{Ln}\left(\frac{M}{M_0}\right) = \text{Ln}\left(\frac{X \cdot V}{X_0 \cdot V_0}\right) = \mu \cdot t \quad (19.32)$$

It is noteworthy that the reasoning to calculate μ is the same as the batch, but for the fed-batch, the volume varies. Thus, the observation unit for biomass is the gram. Also, fed-batch quantity in grams of biomass varies exponentially as follows:

$$X \cdot V = X_0 \cdot V_0 \cdot \exp(\mu t) \quad (19.33)$$

19.3.4.3 Chemostat

$$\frac{d(X \cdot V)}{dt} = -Q_s \cdot X + r'''_x \cdot V \quad (19.34)$$

Recall that in a chemostat, $Q_e = Q_s$, and replacing r'''_x by $\mu \cdot X$,

$$\frac{d(X \cdot V)}{dt} = -Q_e \cdot X + \mu \cdot X \cdot V \quad (19.35)$$

Dividing by the constant volume, the dilution rate appears (Eq. 19.14):

$$\frac{d(X)}{dt} = -D \cdot X + \mu \cdot X = (\mu - D) \cdot X \quad (19.36)$$

By again separating the variables,

$$\frac{d(X)}{X} = (\mu - D) \cdot dt \quad (19.37)$$

This expression is thus very close to what was obtained in the batch. The dilution rate is subtracted to the growth rate μ. The dilution rate term comes from the dilution of the biomass concentration X in the fermenter.

If the dilution rate is abruptly changed at $t = 0$, a transient regime is observed during which the biomass concentration varies. The integration of the previous formula explains how X varies in time:

$$\text{Ln}\left(\frac{X}{X_0}\right) = (\mu - D) \cdot t \quad (19.38)$$

19 Modeling in Microbial Ecology

with $X = X_0$ at $t = 0$. On the other hand, if the dilution rate remains constant during a long time (from 20 to 50 times the generation time), an equilibrium takes place, for which $X = C^{te}$. When the stationary state is reached, the previous equation leads to

$$\mu = D \qquad (19.39)$$

Strictly speaking, it should be noted that if the maximum dilution rate is given by the maximum flow rate of the pump, the maximum rate of growth of a microorganism is called μ_{max}. Of course, if the dilution rate imposed is higher than maximal growth rate μ_{max} of the microorganism, the biomass concentration at steady state is null.

On the other hand, at the stationary state, X does not vary (as a function of t), but is a function of D. Thus, for each dilution rate, a biomass concentration is obtained. When the dilution rate increases, the biomass concentration tends to decrease. The critical value of the dilution rate for which the biomass concentration decreases until zero is called the washout dilution rate, denoted by D^*, and the relation $D^* = \mu_{max}$ is obtained.

19.3.4.4 Chemostat with Recycling

A method capable of increasing dilution rate without causing the biomass washout in the chemostat consists in recycling a portion of the cells at the output, by a separation method (ultrafiltration module for the production of cells or molecule decanter for wastewater treatment), as shown in Fig. 19.5.

The separation step is an integral part of the system. On the one hand, it allows increasing the concentration of biomass in the fermenter by recycling a fraction of the biomass, and on the other hand, it allows to separate the components of the fermentation medium. Let's call ω the recycling rate and Q_r the recycling flow rate; then the relation becomes

$$\omega = \frac{Q_r}{Q_e} \qquad (19.40)$$

From Fig. 19.5, a budget of biomass in the fermenter, assumed to be a chemostat at steady state, reads

$$\frac{d(X \cdot V)}{dt} = Q_r \cdot X_X - Q \cdot X + r'''_x \cdot V \qquad (19.41)$$

Let Q_e be the inflow:

$$\frac{d(X \cdot V)}{dt} = \omega \cdot Q_e \cdot X_X - Q \cdot X + \mu \cdot X \cdot V \qquad (19.42)$$

The fermenter volume is constant if

$$Q = Q_e + Q_r = (1 + \omega) \cdot Q_e \qquad (19.43)$$

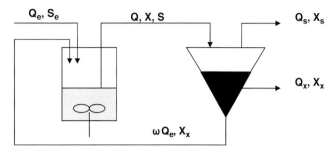

- Biomass concentration: X (g.L^{-1})
- Limiting substrate concentration: S (g.L^{-1})
- Product concentration: P (g.L^{-1})
- Fermenter volume: V (L)
- Recycling rate: ω

Fig. 19.5 Schematic drawing of a chemostat with recycling. Biomass concentration: X(g.L^{-1}), limiting substrate concentration: S(g.L^{-1}), product concentration: P(g.L^{-1}), fermenter volume: V (L), recycling rate: ω

By substituting this equality in Eq. (19.41), it reads

$$\frac{d(X \cdot V)}{dt} = \omega \cdot Q_e \cdot X_X - (1+\omega) \cdot Q_e \cdot X + \mu \cdot X \cdot V \qquad (19.44)$$

At stationary state, X is constant and thus $\frac{d(X \cdot V)}{dt} = 0$
It follows

$$\omega \cdot Q_e \cdot X_X = (1+\omega) \cdot Q_e \cdot X - \mu \cdot X \cdot V \qquad (19.45)$$

Dividing all the terms by V, the dilution rate is governed by

$$\omega \cdot D \cdot X_X = (1+\omega) \cdot D \cdot X - \mu \cdot X \qquad (19.46)$$

Finally, after some simple algebra, the relation between D and ω reads

$$\mu = D \left[1 + \omega \left(1 - \frac{X_X}{X} \right) \right] \qquad (19.47)$$

Eq. 19.47 shows that it is possible to impose a dilution rate D larger than μ, if the recycling rate, ω, is not null and if the decanter functions normally (i.e., $X_x/X > 1$). Note that the term in parentheses in the right side of equation (19.47) is negative and thus that the term in brackets is lower than 1.

It is also possible to realize a biomass budget in the chemostat:

$$QX = Q_S X_S + Q_X X_X + \omega Q_e X_X \qquad (19.48)$$

In practice, Q_S represents the flow of purified water released. It is possible to consider, if the decanter worked well, that X_S is negligible.

If the chemostat volume is constant, the volume budget reads

$$Q = Q_S + Q_X + \omega Q_e \qquad (19.49)$$

The next formula expresses the flows Q and Q_X, as functions of Q_e and Q_S:

$$\begin{aligned}(1+\omega)Q_e X &= (Q - Q_S - \omega Q_e)X_X + \omega Q_e X_X \\ &= ((1+\omega)Q_e - Q_S - \omega Q_e)X_X \\ &\quad + \omega Q_e X_X \end{aligned} \qquad (19.50)$$

It allows calculating the ratio:

$$\frac{X_X}{X} = \frac{1 + \omega - \frac{Q_S X_S}{Q_e X}}{1 + \omega - \frac{Q_S}{Q_e}} \qquad (19.51)$$

Replacing Eq. 19.51 in Eq. 19.47 provides the specific growth rate expression:

$$\mu = D\left(1 + \omega\left(1 - \frac{1+\omega}{1+\omega - \frac{Q_S}{Q_e}}\right)\right) \qquad (19.52)$$

This relation allows calculating the ratio μ/D as a function of the recycling rate ω and of the ratio Q_S/Q_e (Fig. 19.6). Of course, the aim is to get a ratio Q_S/Q_e close to 1 in order to release in the natural environment the most treated water as possible. The remaining fraction is composed of mud which has to be regularly extracted from the decanter.

19.3.5 Metabolism

In the previous chapter, different biomass balances were made for the main types of fermentation. The relative role of biomass balance, substrates, and products in the metabolism of microorganisms is addressed. Figure 19.7 shows how the speed of biomass production is connected with, on the one hand, the rate of consumption of substrate and, on the other hand, with the rate of metabolite production.

From the biomass budget, the expression of the rate of production of biomass can be determined according to the fermentation mode. Then, the expressions of the rates r'''_S and r'''_X will be deducted, respectively, from the substrate and product budgets. In this way, we can calculate the value of the parameters of the two equalities shown in Fig. 19.6.

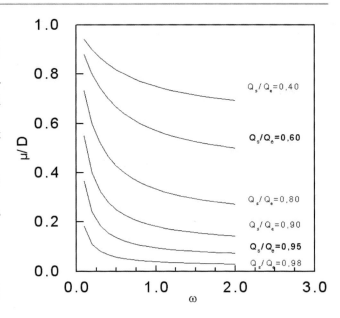

Fig. 19.6 Relation between μ/D and recycling rate ω in the case of wastewater treatment decanter functioning as a chemostat with recycling for different values of the ratio: Q_S/Q_e

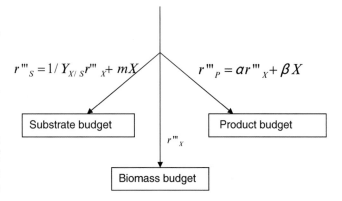

Fig. 19.7 Relations between biomass production and metabolite production rates on the one hand and biomass production and substrate consumption rates on the other hand

19.3.5.1 Cellular Yield and Maintenance Coefficient

The cellular yield $Y_{X/S}$ is defined as the ratio between the produced biomass in grams (g) of dry matter and the consumed substrate in grams also:

$$Y_{X/S} = \text{biomass production (g)/substrate consumption (g)} \qquad (19.53)$$

The cellular yield, between 0 and 1, can easily be determined in batch because in this case the maintenance mX can be neglected. It follows thus from Fig. 19.7 that

$$r'''_S = \frac{1}{Y_{X/S}} r'''_X \qquad (19.54)$$

The substrate consumption rate, from the substrate budget, reads

$$\frac{d(SV)}{dt} = -r'''_S V \quad (19.55)$$

Simplifying by the volume (which is constant) leads to

$$r'''_S = -\frac{d(S)}{dt} \quad (19.56)$$

Since, on the other hand,

$$r'''_X = \frac{dX}{dt} \quad (19.57)$$

it follows that the cellular yields satisfies

$$Y_{X/S} = -\frac{dX}{dS} \quad (19.58)$$

Again, separating the variables X and S provides

$$\int_{S_0}^{S} Y_{X/S} dS = -\int_{X_0}^{X} dX \quad (19.59)$$

And the cellular yields then read

$$Y_{X/S} = \frac{X - X_0}{S_0 - S} \quad (19.60)$$

The maintenance coefficient is not determined in batch conditions, but in a chemostat, at the steady state, for different values of the dilution rate:

$$r'''_S = \frac{1}{Y_{X/S}} r'''_X + mX \quad (19.61)$$

Using the expression $r'''_X = \mu X$ and dividing the different terms by X lead to

$$\frac{r'''_S}{X} = \frac{1}{Y_{X/S}} \mu + m \quad (19.62)$$

19.3.5.2 Specific Rates

To determine the metabolic parameters, $Y_{X/S}$ and m, it is sufficient to draw the specific rates of substrate consumption, $v_s = r'''_s/X$ as a function of the specific growth rate (see Fig. 19.8). This specific rate, v_S, represents the amount of substrate (g) consumed per one unit of biomass (g), per unit of time (h). The substrate consumption rate is proportional to biomass concentration. However, the specific consumption rate of substrate is constant if the metabolism is constant, in

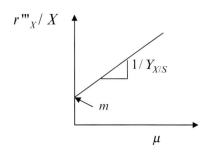

Fig. 19.8 Graphical determination of the cell yields $Y_{x/s}$ and of the maintenance coefficient m at steady state

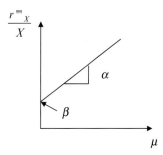

Fig. 19.9 Graphical determination of the coefficients α and β at stationary state

particular if $\mu = C^{te}$. It is the reason why the specific rates must be calculated to characterize the metabolism.

Here is the example of a metabolite production P:

$$r'''_P = \alpha r'''_X + \beta X \quad (19.63)$$

Two types of metabolites are distinguished: the primary metabolites, for which the production is associated to the catabolic pathways (degradation of substrate with energy production), and the secondary metabolites, for which the production is associated to anabolic pathways (production of molecules from primary metabolites with the use of energy). The primary metabolite production is associated to growth. These metabolites are not produced when the stationary phase of growth starts in batch, $\beta = 0$. It follows that

$$\alpha = \frac{r'''_P}{r'''_X} \quad (19.64)$$

In the general case, the production of secondary metabolites is initiated during the growth phase and continues during the stationary phase in batch. To determine the parameters α and β (which characterize the metabolism) graphically, the specific rates are again useful (see Fig. 19.9):

$$\frac{r'''_P}{X} = \alpha \mu + \beta \quad (19.65)$$

In conclusion, to determine the parameters of metabolism, it is necessary that the specific growth rate μ remains constant. If a batch-type fermentation is carried out, the cell yield can be calculated, but not the maintenance coefficient. Indeed, there will be a single value $\mu = \mu_{max}$, because the substrate is not limited except in a very small period at the end of the exponential phase. To get several values of μ constant, it is necessary that the substrate or another environmental factor be limiting. Relations between μ and limiting or inhibiting factors are given by the secondary models.

19.3.6 Secondary Models

The goal of secondary models is to determine the relationships between the growth rate μ and the environmental factors.

19.3.6.1 Limiting Substrate

A famous relation to explicit the limitation by a substrate is due to Monod:

$$\mu = \mu_{max} \frac{S}{k_S + S} \quad (19.66)$$

k_S (g.L^{-1}) represents the substrate concentration for which the growth rate is half of its maximal value $\mu = \mu_{max}/2$. In practice, this value is about one-tenth of a gram. During the most part of the batch, k_S is negligible with respect to the substrate concentration S, which explains why $\mu = \mu_{max}$. On the other hand, in chemostat, the dilution rates are lower than the washout rate $D^* = \mu_{max}$. The Monod law shows that in chemostat, the substrate concentration in the fermenter is very low (of the order of k_S).

19.3.6.2 Inhibiting Substrate

In some cases, the substrate concentration can inhibit the growth. From the Monod law, many other laws can be suggested; here is an example, the so-called Andrews law:

$$\mu = \mu_{max} \frac{S}{k_S + S + \frac{S^2}{k_i}} \quad (19.67)$$

k_i (g.L^{-1}) is an inhibition constant. Note that if the substrate concentration is low, the term S^2/k_i can be neglected. The Monod law is then obtained again. However, when the substrate concentration is high, then k_S can be neglected, leading to $\mu = \mu_{max} \frac{S}{S + \frac{S^2}{k_i}}$.

This expression can be simplified as follows:

$$\mu = \mu_{max} \frac{1}{1 + \frac{S}{k_i}} \quad (19.68)$$

19.3.6.3 Inhibition by Product

The denominator characterizes an inhibition factor which can also be used to represent a product like

$$\mu = \mu_{max} \frac{1}{1 + \frac{P}{k_P}} \quad (19.69)$$

It is possible to associate the limitation to the substrate and the inhibition to a product according to a relation of a multiplicative type, which implicitly assumes that the different factors do not interact:

$$\mu = \mu_{max} \frac{S}{k_S + S} \frac{1}{1 + \frac{P}{k_P}} \quad (19.70)$$

19.3.6.4 Quota Models

Quota models involve another state variable, called *quota*, which represents the amount of limiting factor inside each cell after ingestion. This concept has been developed by Droop (1968). It allows explaining the growth of microbial populations not on the external substrate concentration, but as a function of the absorbed substrate. In other terms, this concept of cell quota considers that the growth rate μ is a function of the nutrient inside the cell.

There is, therefore, no direct relationship between the growth rate and the extracellular substrate concentration. The model implies that the cells (phytoplankton or bacterial) can store a nutrient (taken from the environment or culture medium) and then the rate of growth will depend on the amount of stored nutrients. The consumption rate of the nutrient concentration depends on the element outside the cell and also, in some cases, from that stored element. Thus, the cell can store this when it is available outside and continue to grow and divide using its reserves when it becomes limited or nonexistent in the environment, until its own reserves are exhausted. This approach is generally accepted for the phytoplankton (Smith 1997) but not used for bacteria (van den Berg et al. 1998) (see Fig. 19.10). In the case of primary production in aquatic environments, this approach allows, for example, to reproduce the ability of phytoplankton populations to develop when two limiting factors are not present at the same time. Indeed, considering the case of a phytoplankton population in the mixed layer, it will have access to the necessary light for photosynthesis when it is near the surface and will access the nutrients when it is deep. Monod-type models, where the population growth is directly related to the presence of the limiting factor in the environment, do not allow the representation of these phenomena.

Here, the Droop model is presented as an example. In his original work (Droop 1968), the author had pointed out a relationship between the specific growth rate μ of a

Fig. 19.10 This figure illustrates that a population can still grow when the limiting resource is exhausted in the environment. The *top panel* (**a**) represents the growth of a population of *Escherichia coli* limited by potassium. The original data (circles and diamonds) are from Mulder (1988). The curves have been obtained with the Droop model. The *bottom panel* (**b**) represents the growth of a population of *Chlamydomonas reinhardtii* limited by nitrite, from Cunningham and Maas (1978). In both cases, the Monod model cannot explain the growth of population with the decay of resource simultaneously

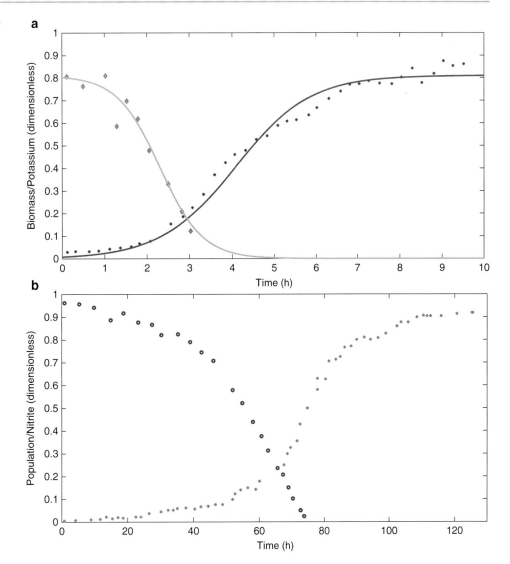

phytoplankton population limited by vitamin B12 and the internal quota, denoted by Q. This relation has the following formulation:

$$\mu = \mu_{max}\left(1 - \frac{Q_{min}}{Q}\right) \quad (19.71)$$

where Q_{min} is the minimal quota in the sense that it is the minimum amount of quota to get the population to grow. If the quota is lower that this value, the population will decay. In a constant volume, the complete model reads

$$\begin{cases} \dfrac{dS}{dt} = D(S_e - S) - \dfrac{V_{max}S}{k_S + S}N \\ \dfrac{dQ}{dt} = \dfrac{V_{max}S}{k_S + S} - \mu_{max}(Q - Q_{min}) \\ \dfrac{dN}{dt} = \mu_{max}\left(1 - \dfrac{Q_{min}}{Q}\right)N \end{cases} \quad (19.72)$$

where V_{max} is the maximal absorption rate for one cell and N is the number of cells. The biomass can be calculated with $X = QN$. The second equation of the previous model is obtained by mass conservation. Indeed, in order to get the idea, consider that the biomass is in nitrogen, S is a mass in nitrogen, Q is a mass in nitrogen per cell, and thus $X = QN$ is the nitrogen biomass. The mass conservation reads $S + QN = $ Constant. Admitting equations 1 and 3 in the model (19.72), equation 2 follows. Note that lots of biogeochemical models are based on this model for the primary production.

19.3.6.5 Environmental Factors

Environmental factors like substrates, inhibiting products, are generally not in optimal conditions. It is useful to take this fact into account in the secondary models. Many secondary models describe the influence of factors like temperature, water activity (mainly in solid media), or pH. A few of them are detailed here. The choice is focused on those which are based on the cardinal values (Rosso et al. 1993).

The global structure of the model, which takes account for the combined effect of the factors, is founded on the γ-concept suggested by Zwietering et al. (1992):

$$\frac{\mu}{\mu_{opt}} = \gamma(T)\gamma(a_w)\gamma(pH) \qquad (19.73)$$

μ_{opt} represents the optimal growth rate for a microbial population, when all the environmental factors, like temperature T, water activity a_w, and the pH, are optimal.

If, in the culture medium, there is neither limitation nor inhibition, then $\mu_{opt} = \mu_{max}$. Some examples of gamma-functions, normalized between 0 and 1, are presented hereafter:

$$\gamma(T) = \frac{(T - T_{max})(T - T_{min})^2}{(T_{opt} - T)\left[(T_{opt} - T_{min})(T - T_{opt}) - (T_{opt} - T_{max})(T_{opt} + T_{min} - 2T)\right]} \qquad (19.74)$$

$$\gamma(a_w) = \frac{(a_w - a_{w,max})(a_w - a_{w,min})^2}{(a_{w,opt} - a_{w,min})\left[(a_{w,opt} - a_{w,min})(a_w - a_{w,opt}) - (a_{w,opt} - a_{w,max})(a_{w,opt} + a_{w,min} - 2a_w)\right]} \qquad (19.75)$$

$$\gamma(pH) = \frac{(pH - pH_{min})(pH_{max} - pH)}{(pH_{opt} - pH_{min})(pH_{max} - pH_{opt})} \qquad (19.76)$$

Note that the function $\gamma(pH)$ is different with respect to the other ones, since only this one has a symmetry around the optimum $pH_{opt} = C^{te}$.

19.3.6.6 Limiting Step

It is necessary to well apply the tools described in this chapter before going further. It should be recalled in this respect that good approaches need to start from a simple basis, to which complexity is added if necessary. On the other hand, even if the biological phenomena are complex, it is useless to model everything. The procedure should highlight important points or keys, which are most often represented by the limiting steps. We have already seen the Monod law; even though several media components are necessary for growth, only one substrate (usually the carbon source or nitrogen source) is limiting. Similarly, during a complex degradation by a community of microorganisms, it seems necessary to show what stage of the transformation is limiting in the process and which microorganisms are involved.

In the context of microbial ecology, the consideration of a single population, considered homogeneous, is usually performed under controlled conditions, to understand the environmental factors that may affect (positively or negatively) the saprophytic competence of a strain of particular interest. The nature of the planktonic unicellular or bacterial cells could partly justify a greater number of examples of models of population dynamics for these microorganisms (Ponciano et al. 2005). But it is mainly the role of bacteria in biotechnological processes (Patwardhan and Srivastava 2004), food production (Fujikawa et al. 2006), and human (Tam et al. 2005) or animal diseases (Wood et al. 2006) which led to the development and use of mathematical models, some of which being of predictive purposes (Van Impe et al. 2005; Li et al. 2007), in order to anticipate and control the microbial population development. Nevertheless, the population dynamics of yeast-like or filamentous fungi related to biotechnology (Lejeune and Baron 1998), agriculture, or food processes (Dantigny 2004) has also been modeled. This is the case for fungi, potential biological control agents that we want to determine environmental requirements in order to bring them into an environment where they can settle and exercise antagonist activity that prevailed in their selection (Couteaudier and Steinberg 1990). This is also the case for plant pathogens for which we want to know the growth characteristics (Steinberg et al. 1999) or determine conditions which favor their propagation in the soil (Bailey et al. 2000; Boswell et al. 2003). However, it is necessary to reconsider population dynamics studied in isolation, to a more natural context in which they will be interacting with other populations belonging to the same or different species or genera or even branches. It is therefore necessary to extend the procedures described in this chapter to structured models that firstly consider at least two types of non-structured population models, each type being characterized by different metabolisms, and also take into account the diversity of environments where biotic interactions take place (Middelboe et al. 2001; Stelling 2004). Depending on environmental conditions, secondary models explain how one type of population can grow to the detriment or benefit of another. Indeed, if there are phenomena of interaction between these populations, they may be of antagonistic order (antibiosis by producing harmful metabolites such as antibiotics, predation, or parasitism) or of competition type for a given resource or of synergistic type, a metabolite produced by a type of population being a substrate for the other.

To represent a microbial community, it is necessary to include the interactions between microbial populations in the model. For the sake of clarity, we present the usual approaches where only two populations interact. We then show how the ideas developed here extend to any number of populations. In particular, we show an example with more than two populations interact, the type of interaction between the two of them may depend on the abundance of the third population.

19.4 Different Types of Interactions and Their Representation

To understand how to represent the interactions between different populations in a model, it must be emphasized that what is expressed in the mathematical models is the effect of the density of one population on another one. The first point is to determine the sign of this effect: if one species has a positive effect on the growth of the two species, the interaction term in the equations will be assigned a "plus" sign. If the effect is negative, the corresponding interaction term is affected by a "minus" sign. In the case of two interacting populations, there will be interaction type +/+ (mutualism), type +/− (predation, parasitism), or type −/− (competition). Other types of interactions (commensalism, amensalism or neutralism) will not be discussed here for brevity, but the principles that we develop can be applied.

The usual way to proceed is to write the specific growth rate of population i as a function of the present population densities. In a community with n species, denote by N_i the population density i, the associated equation is

$$\frac{dN_i}{dt} = N_i F_i(N_1, \ldots, N_n) \quad (19.77)$$

where F_i is the specific growth rate of species i.

The influence of species j on species i is measured by the effect of an increase of N_j on the growth rate F_i. The mathematical tool allowing to know if, when a variable N_j increases, the function F_i, which depends on N_j, also increases or, on the contrary, decreases is the derivative of F_i with respect to N_j. Since F_i may depend on several variables, the previous derivative is called a *partial derivative*, denoted by $\partial F_i / \partial N_j$. The positive effect of species j on species i is translated by a positive derivative. If the derivative is negative, then species j has a negative effect on species i. Note that the notation ∂ in the derivative means that the other variables are assumed to be constant.

Note that in a community, the type of interaction between two populations is not always the same. For example, imagine two populations competing for a resource and add a generalist predator that can consume the previous two populations. Suppose that the consumption rate of each prey population by the predator is positively correlated with the proportion of the corresponding prey population. When the predator density is low, the two prey populations are in competition (by assumption). However, when the density of the predator is large and when one of the prey populations is low, the proportion of this prey is small. According to the assumptions, it will almost not be consumed, and the predator will focus and feed on the other prey population. The consequence is that at high density of predators, the prey populations are indirectly mutualistic. A simple small model based on the principles described here can quantitatively demonstrate our heuristic reasoning (e.g., Brauer and Castillo-Chavez 2000). This example illustrates the complex set of interactions in a community; the types of interactions between populations are the result of the structure and composition of the community as a whole. Therefore, it must be recalled that the following models are intended only to try to understand the effect of interactions in small frames, under specific assumptions, and their direct transfer to the natural environment without precaution would obviously be adventurous.

19.4.1 Competition

19.4.1.1 Lotka–Volterra Model

The first type of interaction that we consider here concerns the competition. We begin with a historical model, which has mainly a theoretical interest, the Lotka–Volterra. The principle of the model is to treat all individuals of each population as identical, randomly and uniformly in a homogeneous medium. The interaction between individuals of the two populations is then proportional to the number of encounters, which are themselves proportional to the product of the densities of the two populations. Note $N_i(t)$ is the density of population i at time t, where $i = 1, 2$. The number of encounters between individuals of population 1 and individuals of population 2 during a given period of time is proportional to

$$N_1(t)N_2(t)$$

This claim is rather intuitive since, as previously mentioned, the environment is assumed to be homogeneous. Indeed, if the number of individuals of a given population is doubled, then the number of encounters per unit of time is also doubled.

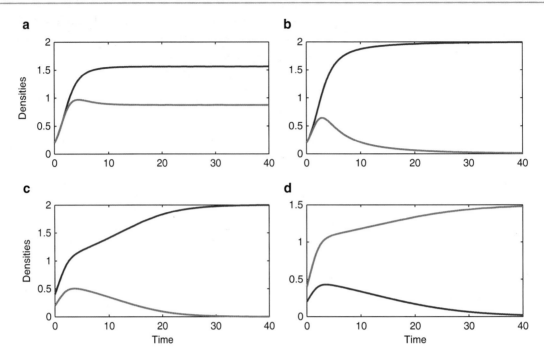

Fig. 19.11 This figure illustrates the different situations obtained with the Lotka–Volterra competition model. The intrinsic growth rates r_1 and r_2 are, respectively, set to 1 and 1.1. The carrying capacities K_1 and K_2 are, respectively, 2 and 1.5. In the *top-left panel* (**a**), the weak competition is illustrated, with coexistence of the species ($b_{12} = 0.5$ and $b_{21} = 0.4$). The *top-right panel* (**b**) shows an example with a weak competition for species 1 (*blue*) and a strong competition for species 2 (*green*); this one is excluded. *Bottom panels* (**c**) and (**d**) illustrate the case where the competition is strong for both species, with $b_{12} = 1.5$ and $b_{21} = 0.9$. In the graph (**c**), the species 2 (*green*) is outcompeted, whereas it is the species 1 in graph (**d**). The difference between graph (**c**) and graph (**d**) comes from the initial condition

In the case of competition, the interaction is characterized by a negative effect of each population on the other one. The competition is thus assigned with a negative sign. The Lotka–Volterra model then reads

$$\begin{cases} \dfrac{dN_1}{dt} = r_1 N_1 \left(1 - \dfrac{N_1}{K_1}\right) - r_1 N_1 b_{12} \dfrac{N_2}{K_1} \\ \dfrac{dN_2}{dt} = r_2 N_2 \left(1 - \dfrac{N_2}{K_2}\right) - r_2 N_2 b_{21} \dfrac{N_1}{K_2} \end{cases} \quad (19.78)$$

Again, here, dN_i/dt is the variation rate of the population density, $N_i(t)$, r_i is the intrinsic growth rate of population, i, K_i is the carrying capacity of the environment for the population i, and i, b_{ij} is the competition rate, which measures the interspecific competition of species j on species i. The terms $r_1 N_1 b_{12} \frac{N_2}{K_1}$ and $r_2 N_2 b_{21} \frac{N_1}{K_2}$ are the contribution of the interspecific competition of 2 to 1 and of 1 to 2, respectively, in the growth rate for populations 1 and 2, respectively. Let $C_{ij} = b_{ij} K_j / K_i$; this is a dimensionless number, which provides a measure of the competition strength. When $C_{ij} < 1$, species j exerts a weak competition on species i; when $C_{ij} > 1$, species j exerts a weak competition on species i. A mathematical analysis of the Lotka, according to the parameters c_{ij} (see illustration in Fig. 19.11), is as follows:

- If $C_{12} < 1$ and $C_{21} < 1$, it is a weak competition case since each species exerts a weak competition on the other one. It follows that both species coexist at a steady state provided by

$$\lim_{t \to +\infty} N_i(t) = \frac{1 - c_{ij}}{1 - c_{12} c_{21}}$$

- If $c_{12} < 1$ and $c_{21} > 1$, it is a strong competition case since the species 1 exerts a strong competition on species 2; this one is outcompete and

$$\lim_{t \to +\infty} N_1(t) = K_1$$

- If $c_{12} > 1$ and $c_{21} < 1$, it is a strong competition case since the species 2 exerts a strong competition on species 1; this one is outcompete and

$$\lim_{t \to +\infty} N_2(t) = K_2$$

- If $c_{12} > 1$ and $c_{21} > 1$, it is a strong competition since each species exerts a strong competition on the other one. One of the species is outcompeted, depending on the initial conditions. Either

$$\lim_{t \to +\infty} N_1(t) = K_1$$

$$\lim_{t \to +\infty} N_2(t) = 0$$

or
$$\lim_{t \to +\infty} N_1(t) = 0$$

$$\lim_{t \to +\infty} N_2(t) = K_2$$

Some theoretical works in ecology have shown how the interspecific competition parameters of this model may be estimated and how they can be linked to ecological niche overlapping.

This model has been widely used in theoretical ecology, despite the different weaknesses due to its simplicity. Among these problems, the competition is explicitly formulated by the model, while the resource is not explicit.

19.4.1.2 The Competitive Exclusion Principle (CEP)

In this section, a limiting resource is explicitly described to analyze the result of the competition. This model will be used to provide a mechanism to the previous one.

Consider two populations competing for a limiting resource. Denote by $N_1(t)$ and $N_2(t)$ the biomass of species 1 and 2, respectively. Let $S(t)$ be the amount of limiting resource for which the species compete. For instance, consider two bacteria species in competition for carbon substrate in a closed environment, like batch. A simple model reads

$$\begin{cases} \dfrac{dS}{dt} = -a_1 S N_1 - a_2 S N_2 \\ \dfrac{dN_1}{dt} = e_1 a_1 S N_1 - m_1 N_1 \\ \dfrac{dN_2}{dt} = e_2 a_2 S N_2 - m_2 N_2 \end{cases} \quad (19.79)$$

where a_i is the consumption rate for species i and e_i is the yield coefficient for species i, sometimes also called bacterial growth efficiency (BGE). The total amount of carbon in the system is

$$C(t) = S(t) + \frac{N_1(t)}{e_1} + \frac{N_2(t)}{e_2}$$

This total amount is constant in a closed environment, and this can be checked as follows:

$$\frac{dC(t)}{dt} = \frac{dS(t)}{dt} + \frac{1}{e_1}\frac{dN_1(t)}{dt} + \frac{1}{e_2}\frac{dN_2(t)}{dt} = -a_1 N_1 S - a_2 N_2 S + a_1 N_1 S + a_2 N_2 S = 0$$

Since the total amount of carbon is constant during the time course, $C(t) = C(0)$, it is then possible to replace in the model the state variable for the substrate by

$$S(t) = C(0) - \frac{N_1(t)}{e_1} - \frac{N_2(t)}{e_2}$$

A direct consequence is that, in the previous model, the state variable $S(t)$ can be eliminated in the bacterial population equations. Following this idea, the previous model (19.79) is equivalent to

$$\frac{dN_1}{dt} = e_1 a_1 N_1 \left(C(0) - \frac{N_1(t)}{e_1} - \frac{N_2(t)}{e_2} \right)$$

$$\frac{dN_2}{dt} = e_2 a_2 N_2 \left(C(0) - \frac{N_1(t)}{e_1} - \frac{N_2(t)}{e_2} \right)$$

This model is thus again a Lotka–Volterra model with $r_i = e_i a_i C(0)$, $K_i = e_i C(0)$, and $b_{ij} = e_i/e_j$.

The previous models are very simple and do not take into account processes like maintenance or death. Here is an example of a more general model to describe the competition between two species in a closed environment:

$$\begin{cases} \dfrac{dS}{dt} = -A_1(S) N_1 - A_2(S) N_2 \\ \dfrac{dN_1}{dt} = e_1 A_1(S) N_1 - m_1 N_1 \\ \dfrac{dN_2}{dt} = e_2 A_2(S) N_2 - m_2 N_2 \end{cases} \quad (19.80)$$

where A_i is the substrate absorption rate by species i, e_i is the yield coefficient (or BGE) for species i, and m_i is the death rate for species i. To determine the dynamics of this model, consider first the steady-state values for each species, which is obtained by vanishing the equations: $dN_i/dt = 0$

The equation $dN_1/dt = 0$ implies that $N_1 = 0$ or $e_1 A_1(S) = m_1$, and the equation $dN_2/dt = 0$ implies that $N_2 = 0$ or $e_2 A_2(S) = m_2$. These equations must be satisfied simultaneously to get the equilibrium. However, it is impossible that S satisfies simultaneously the equalities $e_1 A_1(S) = m_1$ and $e_2 A_2(S) = m_2$ and as a consequence either $N_1 = 0$ or $N_2 = 0$. This argument shows that from model (19.80), in a homogeneous environment, the competition of two species for one resource leads to extinction of one of these species, namely, the species which needs more substrate at steady state. This is called the competitive exclusion principle (CEP), which was first discussed by Gause (1935).

As it has been mentioned, the species which needs less substrate to grow is more efficient to compete. To illustrate this point, consider a particular formulation of the absorption rate, like the Monod (or Michaëlis–Menten) formulation:

$$A_i(S) = V_{\max,i} \frac{S}{K_{s,i} + S}$$

where $V_{\max,i}$ is the maximum uptake rate for one unit of biomass of species i and $K_{s,i}$ is the half-saturation constant. In this case, the solution of the equation $e_i A_i(S) = m_i$ is solution of

$$e_i V_{\max,i} \frac{S}{K_{s,i} + S} = m_i$$

and thus it reads

$$S_i^* = \frac{K_{s,i} m_i}{e_i V_{\max,i} - m_i}$$

The species for which the value S_i^* is lower is more efficient to compete; the other one is excluded.

19.4.2 Trophic Interactions

Predation plays a fundamental role in ecology, and its mathematical representation is a formidable challenge modeling (Mitra et al. 2003). Predation is the means by which the material passes through the different trophic levels of the community. A key concept to address predation in theoretical ecology is the functional response (Holling 1965) which is the amount of prey consumed by predator unit and per unit time. This response has an undeniable interest since it contains basic information on the amount of material that passes from one trophic level to the next level (Fu et al. 2000; Coleman et al. 2002). However, it is extremely difficult to quantify in practice for several reasons: the first reason is that it is not clear, in terms of manipulations, whether an individual disappears by consumption or by another process (exclusive intraspecific or interspecific competition, washout in a chemostat, etc.). The second reason is a little more difficult to identify and it concerns the level of observation: must this response be observed at the individual or at population level? This question arises in general ecology but even more in microbial ecology. Predation is basically an individual process, but the response of the population of the prey predation makes sense at the population level; it follows that the interaction between these two levels of organization does not simplify its evaluation. Modeling provides an interesting way to get an estimate.

The standard form (but not exclusive one) of a predator–prey model is

$$\begin{cases} \dfrac{dN}{dt} = f(N)N - g(N)P \\ \dfrac{dP}{dt} = eg(N)P - \mu P \end{cases} \quad (19.81)$$

where $f(N)$ is the growth rate of the prey population, $g(N)$ is the amount of prey consumed per unit predator biomass per unit time, e is the coefficient of conversion of biomass of prey into biomass of predator, and μ is the mortality rate of the predator population. The quantity $g(N)$ is the functional response (Solomon 1949). A well-known model of this type of example is that of Rosenzweig and MacArthur (1963), which is written as

$$\begin{cases} \dfrac{dN}{dt} = rN\left(1 - \dfrac{N}{K}\right) - \dfrac{aN}{b+N}P \\ \dfrac{dP}{dt} = e\dfrac{aN}{b+N}P - \mu P \end{cases} \quad (19.82)$$

where the first equation represents the dynamics of prey and the second equation that of predators. N is the biomass of the prey population, P is the biomass of the predator population, r is the intrinsic growth rate of the prey population, K is the carrying capacity, a is the maximum rate of prey consumption per unit of predator biomass, b is the half-saturation constant, e is the rate of conversion of biomass of prey in predator biomass, and μ is the mortality rate of the predator population. This model can generate three types of dynamics, which can be classified according to the values of the carrying capacity parameter. This parameter represents the nutritional richness of the environment because it reflects the amount of resources of the prey. The carrying capacity of the environment for a population A may be different from that for a population B in the same medium, since the species use different resources in different ways. When the carrying capacity parameter is low, it is a resource-poor environment (e.g., oligotrophic). When it is large, the environment is rich in nutritional resource (e.g., eutrophic). In the first case, the model shows an extinction of the predator population: the poor environment can sustain a small amount of prey, which is insufficient for the survival of the predator population. When the carrying capacity of the environment is higher, an equilibrium with coexisting prey and predator population is established. Finally, above a threshold of nutrient richness, predator–prey oscillations appear (Fig. 19.12). When the medium is further enriched, the amplitude of oscillations increases, so that the abundances can be very close to zero at some times. In this case, we can consider that such values in a real situation may lead to the extinction of one (or more) population. In other words,

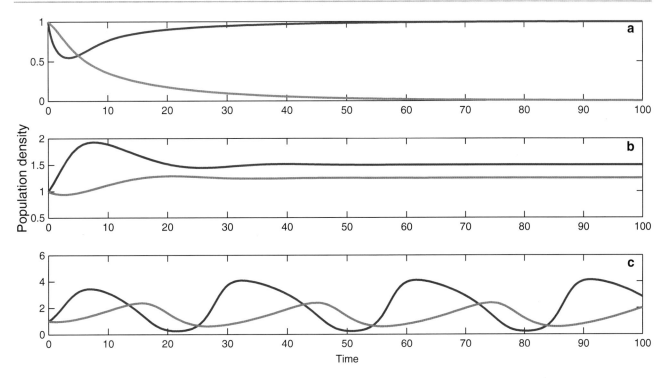

Fig. 19.12 This figure shows the different dynamics which can result from the Rosenzweig–MacArthur model (19.82). The prey density is represented in *blue*, and the predator density is in *green*. The *top panel* (**a**) shows the dynamics for a poor nutrient system; the predator population is exhausted. The *bottom panel* (**c**) shows predator–prey oscillations due to a rich environment. The panel (**b**) is an intermediate situation, where the densities reach an equilibrium value

increasing the nutritional richness of the environment, the model leads to a reduction of biodiversity, which has led to what the literature commonly called the paradox of enrichment (Rosenzweig 1971). This paradox is still the source of many publications.

In an open environment, whose resource richness is ensured by regular and abundant input medium, these oscillations can be maintained as long as resources allow prey to grow and reach a density close to the carrying capacity of the environment for their population. However, if the enrichment is not enough, the oscillations are damped, and the densities of prey and predators stabilize at an equilibrium value.

On the same principle, a trophic chain model with n levels can be derived. For instance, below is an example of a tritrophic chain representing interaction between a prey population (density N), a predator population (density P), and a top-predator population (density C):

$$\begin{cases} \dfrac{dN}{dt} = rN\left(1 - \dfrac{N}{K}\right) - \dfrac{a_1 N}{b_1 + N} P \\ \dfrac{dP}{dt} = e_1 \dfrac{a_1 N}{b_1 + N} P - \mu_1 P - \dfrac{a_2 P}{b_2 + P} C \\ \dfrac{dC}{dt} = e_2 \dfrac{a_2 P}{b_2 + P} C - \mu_2 C \end{cases} \quad (19.83)$$

Table 19.2 Names, meaning and units for the parameters used in model (19.4.7)

Parameter	Name/meaning	Unit
r	Intrinsic prey growth rate	h^{-1}
K	Carrying capacity	$[N]$
a_1	Maximum catch rate of prey per predator	$h^{-1}.[P]^{-1}$
b_1	Half-saturation constant of the functional response	$[N]$
e_1	Conversion coefficient	$[P].[N]^{-1}$
μ_1	Death rate of the predator	h^{-1}
a_2	Maximum catch rate of predators per top-predator	$h^{-1}.[C]^{-1}$
b_2	Half-saturation constant of the functional response	$[P]$
e_2	Conversion coefficient	$[C].[P]^{-1}$
μ_2	Death rate of the top — predator	h^{-1}

Table 19.2 summarizes the parameters, their meaning, and their units.

The mathematical analysis of this type of model is usually difficult and gives rise to numerous publications; the interested reader could refer to the article from de Boer et al. (1998) and the references therein. One of the main results is that this type of model can generate either simple dynamic with variables that tend toward equilibrium or very complex dynamic (chaotic), as illustrated in Fig. 19.13. In this situation, as shown in Fig. 19.14, two

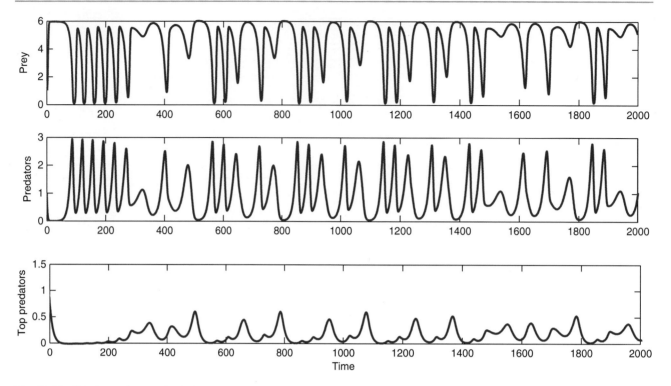

Fig. 19.13 This figure shows a simulation made with the tri-trophic food chain

Fig. 19.14 This figure represents two solutions of the third equation of system (19.83), obtained with two very close initial conditions. The curves are very close at the very beginning but then separate and have finally very different dynamics. It emphasizes that the prediction with a model must be made very cautiously

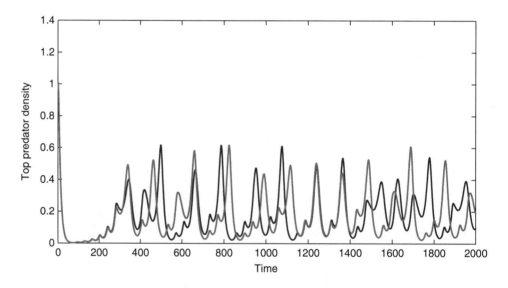

different, but very close, initial conditions lead rapidly to very different time series. It can be difficult to validate such a model in a quantitative way for many reasons. However, some works for the validation of such models are now emerging. For example, Fussmann et al. (2000) proposed a series of experiments on a planktonic food chain in chemostats. By varying the dilution rate of chemostat, they reproduce the dynamics provided by a model of the food chain in a chemostat. The authors have placed calyciflorus rotifers *Brachionus* and *Chlorella vulgaris* algal cells in a nitrogen-limited chemostat. The simplest model describing this system, presented in the article from Fussmann et al. (2000), is as follows:

$$\begin{cases} \dfrac{dN}{dt} = D(N_0 - N) - \dfrac{a_1 N}{b_1 + N} C \\ \dfrac{dC}{dt} = e_1 \dfrac{a_1 N}{b_1 + N} C - DC - \dfrac{a_2 C}{b_2 + C} B \\ \dfrac{dB}{dt} = e_2 \dfrac{a_2 C}{b_2 + C} B - \mu B \end{cases} \quad (19.84)$$

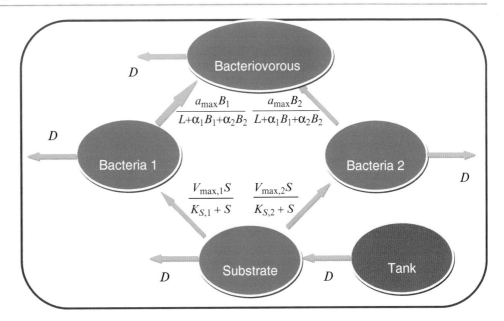

Fig. 19.15 Scheme of the competition–predation model (19.85). The mathematical expressions close to the *arrows* quantify the flux between the state variables

where the state variables N, C, and P are the nitrogen concentration in the chemostat, the concentration of *C. vulgaris*, and the concentration of *B. calyciflorus*, respectively. The parameters are the nitrogen concentration in the reservoir N_0, the dilution rate D, the maximum specific uptake rate a_1, the half-saturation constant b_1, the yield coefficient e_1, the maximum consumption rate a_2, the half-saturation constant of the functional response b_2, the conversion efficiency e_2, and the death rate μ. The mathematical analysis of these models shows that when D is low, only the phytoplankton population survives, and it reaches a steady-state value. When D is increased and cross a threshold value, the phytoplankton and the zooplankton coexist and reach equilibrium. After a further increase of the dilution rate, another threshold value is crossed and oscillations appear. Finally, a too large dilution rate leads to wash out the chemostat. The oscillations observed for large dilution rates are again related to the enrichment paradox. Fussmann et al. (2000) realized some experiments to illustrate the equilibrium destabilization and the occurrence of oscillations.

Biotic interactions of competition and predation type were presented through simple examples. Mathematical works have provided the analysis of increasing complexity, starting to 2 interacting species and complexifying the webs progressively. Among them is an interesting example in terms of the interaction between modeling and experimentation. This example is taken from the article in Becks et al. (2005). It is initially based on a work of a mathematical model representing two bacterial populations growing on a limiting resource and bacterivores consuming two bacterial populations. This type of model has been studied since the 1980s such as in Takeuchi and Adachi (1983) or Vayenas and Pavlou (1999). The dynamics exhibited by these models can be complex. For the sake of clarity, the example provided by Vayenas and Pavlou (1999) is discussed. It is a model of chemostat composed of four state variables: the limiting carbon substrate used by bacterial populations, biomasses of two bacterial populations, and phage biomass. The model is based on the diagram in Fig. 19.15.

The model reads

$$\begin{cases} \dfrac{dS}{dt} = D(S_0 - S) - \dfrac{V_{max,1}S}{K_{S,1}+S}B_1 - \dfrac{V_{max,2}S}{K_{S,2}+S}B_2 \\ \dfrac{dB_1}{dt} = e_1 \dfrac{V_{max,1}S}{K_{S,1}+S}B_1 - \dfrac{a_{max}B_1}{L+\alpha_1 B_1 + \alpha_2 B_2}P - DB_1 \\ \dfrac{dB_2}{dt} = e_2 \dfrac{V_{max,2}S}{K_{S,2}+S}B_2 - \dfrac{a_{max}B_2}{L+\alpha_1 B_1 + \alpha_2 B_2}P - DB_2 \\ \dfrac{dP}{dt} = \left(\dfrac{a_{max}B_1}{L+\alpha_1 B_1 + \alpha_2 B_2} + \dfrac{a_{max}B_2}{L+\alpha_1 B_1 + \alpha_2 B_2} \right)P - DP \end{cases}$$

(19.85)

Selective predation of bacteria by protozoa, illustrated by the previous example (see Fig. 19.16 for a model simulation and Fig. 19.17 for data obtained in a set of experiments), results in maintaining a population which would have been eliminated by competitive exclusion principle. More generally, this mechanism directly affects the functioning of the ecosystem by stimulating specific microbial activities as the mineralization of nitrogen (Bonkowski 2004) or structure (morphological and taxonomic composition) of prey communities (Becks et al. 2005; Ronn et al. 2002) and contributing to the evolution of the biodiversity of the ecosystem.

19.4.3 Virus

The population dynamics of the virus has been a great deal of work especially in the field of marine ecology, human health, and veterinary medicine and plant pathology. These

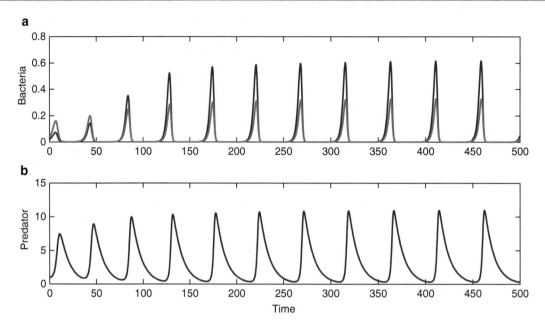

Fig. 19.16 Dynamics of bacteria biomass (**a**) and phage (**b**). The predator allows the coexistence between the competitors; otherwise, the CEP would have lead to one bacteria population outcompeting the other

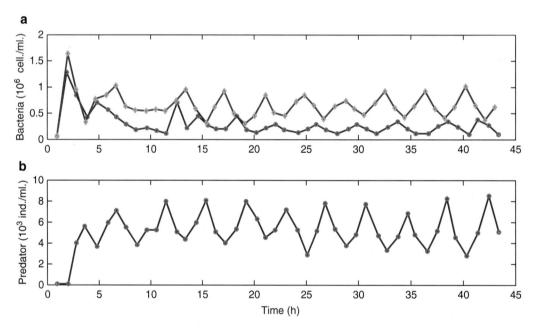

Fig. 19.17 Example of dynamics obtained with a set of experiments for a fixed value of the dilution rate in the chemostat (Data from Becks et al. 2005). The *top panel* (**a**) represents the bacteria populations, and the *bottom panel* (**b**) shows the predator population. This example, namely, shows that the predator allows the coexistence, whereas those bacteria species obey the CEP when they are cultivated together but without any predator. Not that the dynamics is not regular

works aim to understand, interpret, simulate, and anticipate the transmission of viruses. In ecology, it is important to understand the role of viruses in the functioning of ecosystems (Fuhrman 1999). In the medical field, studies aimed at understanding how to control the virus by appropriate treatment (Zhang et al. 2000; Ghosh et al. 2007; Monti et al. 2007).

As an example of representation of virus, a model of interaction between a population of virus and bacterial population (Beretta and Kuang 1998) is presented. This model has three state variables: $S(t)$ is the amount of susceptible bacteria, which is likely to be contaminated by the virus at time t; $I(t)$ is the amount of infected bacteria at time t; and $V(t)$ is the amount of virus at time t. The model reads

$$\begin{cases} \dfrac{dS}{dt} = rS\left(1 - \dfrac{S}{K}\right) - kSV \\ \dfrac{dI}{dt} = kSV - \lambda I \\ \dfrac{dV}{dt} = -kSV + \beta\lambda I \end{cases} \quad (19.86)$$

where k is the contact rate; kSV is the amount of susceptible bacteria contaminated by the virus per unit of time (amount proportional to the number of encounters); λI is the number of lysed bacteria per time unit, due to the infection; and $\beta\lambda I$ is the number of virus released in the environment. The parameter β is called burst coefficient; it is the number of virus released by one lysed bacteria. It plays an important role in the dynamics. Beretta and Kuang (1998) show for instance that if the burst coefficient is large enough, then oscillations occur. Some more realistic models (but more complex also), compared to experiment data sets, have been proposed in the literature in order to understand how the virus modifies the growth of population, the biotic interactions in microbial webs, and the functioning of these webs (Middelboe 2000; Middelboe et al. 2001).

19.5 The Formulation of a Process: A Problem of Scale

The following two examples raise the question of the formulation of a process. They were chosen because they illustrate two major problems currently seamless of the representation of processes in models of biogeochemistry and the serious consequences that a bad choice may have.

Example 1: The Co-limitation. This example concerns the representation of a phytoplankton cell population limited by two resources (two nutrients for instance). This is obviously a common phenomenon in nature, where the number of limiting factors may be even greater because the availability of resources can vary over time. There are many formulations, often based on empirical criteria. However, they have a common point on which we will build in this example: in any case, when one of the two limiting factors is not limiting, the model behaves like the Monod model describing the processes of absorption/growth with one limiting factor. There are many ways to reproduce this property with mathematical expressions; the two most common are

$$\mu = \mu_{\max} \dfrac{S_1}{k_{S_1} + S_1} \dfrac{S_2}{k_{S_2} + S_2} \quad (19.87)$$

called the *product law* and

$$\mu = \mu_{\max} \min\left(\dfrac{S_1}{k_{S_1} + S_1}; \dfrac{S_2}{k_{S_2} + S_2}\right) \quad (19.88)$$

called the Liebig law (or min law).

In the two previous examples, when of the nutrient (say, S_2) becomes abundant, the specific growth rate follows the Monod model:

$$\mu = \mu_{\max} \dfrac{S_1}{k_{S_1} + S_1}$$

as a function of the remaining limiting factor.

The second formulation expresses the fact that only one factor is limiting because the growth rate is given by the minimum growth rate that would take place if the limiting factors were considered separately. It is possible to interpret the first expression by considering the maximum specific growth rate when the limiting factor is S_1 as a function of the second limiting factor: this function tends to the maximum rate $\mu_{\max}$ when the availability of the factor S_2 is important. Both formulations have advantages and disadvantages: the latter seems to better represent the observations in general, but the instantaneous switching from one to another limiting factor when the minimum of the expression (19.88) changes can be a problem. This problem may have important consequences with numerical errors in simulations or in the case of parameter estimation from a data set. Finally, in both cases, these empirical models generalize the model of Monod, without repeating the mechanistic basis. Indeed, the Monod growth kinetics can be explained with a simplified reaction scheme, as follows: suppose that the absorption of a limiting factor S is done through the presence of an enzyme E and that each encounter between a molecule of S and an enzyme E contributes to the formation of a complex C_{ES}, which results in the synthesis of biomass. This mechanism can be represented by the scheme:

$$S + E \xrightarrow{k} C_{ES} \xrightarrow{k'} X + E$$

where the numbers k and k' are reactivity rates; they characterize the speed of enzymatic reactions. The reaction scheme above leads to the biomass-specific growth rate $\mu = k'\, C_{ES}$. Assume that enzyme reactions are fast with respect to the biomass growth and apply the quasi-steady-state assumption. This assumption means that the dynamics of E and C_{ES} are fast enough, and thus these variables reach their equilibrium value very fast. The complex C_{ES} dynamics is governed by

$$\dfrac{dC_{ES}}{dt} = kSE - k'C_{ES} = kS(\Theta - C_{ES}) - k'C_{ES}$$

where Θ is the total amount of enzymes, which is assumed to be constant here. Setting this equation equal to zero (equilibrium condition) leads to

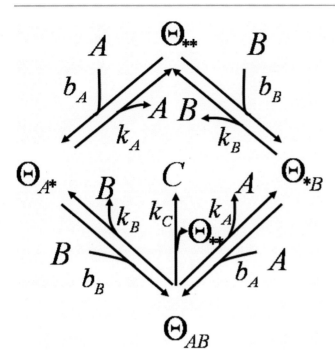

Fig. 19.18 Reaction scheme of two limiting factors A and B. The symbol Θ_{XY} is an enzyme (if $X = *$ and $Y = *$) or a complex associated with a molecule A or B when one of the symbols X and Y is replaced by A or B. C is a product, like biomass for instance

$$C_{ES} = \frac{kS\Theta}{(k' + kS)}$$

Replacing C_{ES} by this expression in the specific growth rate leads to the Monod formula:

$$\mu = \frac{k'k\Theta S}{k' + kS} = \frac{k'\Theta S}{\frac{k'}{k} + S}$$

which corresponds to $\mu_{\max} = k'\Theta$ and $k_S = k'/k$.

An interesting approach for formulating the co-limitation process consists in replicating the previous scheme diagram in the case of two limiting factors (see Kooijman 2010). Several types of interactions between the limiting factors may be considered in the absorption; Fig. 19.18 provides an example, and the interested reader can refer to Poggiale et al. (2010) and the reference therein for more details and applications.

This approach is an extension of the method used to build the model of Monod, when there are several limiting factors. It allows the formulation of mechanism-based models.

Example 2: The Functional Response. In models of biogeochemistry, primary production is often detailed, but the terms of zooplankton grazing, yet considered as crucial because this process controls the primary production, are often poorly represented for various reasons. The term grazing (which is specific) is typically what we have called the functional response in a more general situation (number of prey consumed per unit of predator). In the examples seen previously, in the trophic interaction section, the functional response was a hyperbolic function: $g(N) = a_H N/(1 + b_H N)$, which is the so-called Holling type II (or disc equation) functional response (Holling 1959). In general, it is quite natural to assume that this function satisfies the following three hypotheses:

- (H1) It vanishes when they is no prey.
- (H2) It increases when the number of prey increases.
- (H3) When the number of prey is large, the function saturates (it reaches a bounded value).

These three properties are fulfilled by the Holling function. But many other mathematical functions satisfy these properties. Among these are the following two functions that have been used in models of biotic interactions in microbial ecology: the *Ivlev* functional response $g(N) = a_I(1 - \exp(-b_I N))$ and the *trigonometric* functional response $g(N) = a_T \tanh(b_T N)$. Each of these functional responses involves two parameters, making them comparable in this respect. They therefore represent three different formulations of the same satisfactory phenomenological assumption set (H1–3). However, as shown by Fussmann and Blasius (2005), the choice of the formula in this example is fundamental. Indeed, consider the model (19.81) with one of the previous three functional responses. Assume that the parameters are selected such that the functions are quantitatively similar enough that the difference is smaller than the measurement error (Fig. 19.19).

Using these three responses in the model, a simulation provides the result shown in Fig. 19.20. This figure clearly shows that the dynamics achieved depends qualitatively and quantitatively on the choice of the functional response, even if it is set precisely by a data set. This example illustrates the importance of choosing the formulation of a process.

This example illustrates the importance of choosing the formulation of a process not only at the scale of the process itself but also at the scale of the population dynamics and communities.

19.6 Space

In the same way that has been done for the representation of time, the representation of space can be discrete or continuous, defining two different formalisms for spatially structured models. Many models in chemostat ignore the spatial structure assuming a good homogeneity of the medium. This assumption is obviously too strong to be applied to the natural environment. In addition, including bioreactors, more models take into account the spatial structure to understand, for instance, the formation and the role of biofilms.

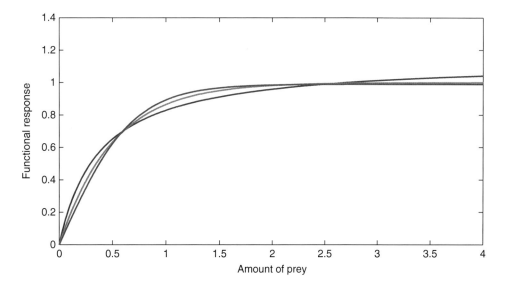

Fig. 19.19 Comparison of three functional responses (Holling, Ivlev trigonometric). They are very close; it is thus not possible to determine which one is the best model by means of a data set

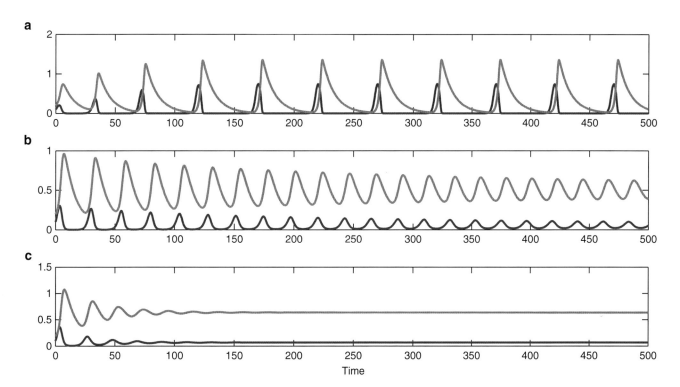

Fig. 19.20 Numerical simulations obtained with model (19.81), with different functional responses: (**a**) Holling functional response, (**b**) Ivlev functional response, and (**c**) trigonometric functional response. The dynamics are quantitatively and qualitatively different. In (**a**), periodic fluctuations take place, (**b**) damped oscillations are resulting, and (**c**) an equilibrium is reached

19.6.1 Discrete Spatial Structure

A simple way to represent the spatial variability of the variable of interest is to subdivide the study area into different sites. To keep it simple and concise, we provide an example with two sites, which may be, for example, a liquid phase and a biofilm or the euphotic zone and the aphotic layer in the water column or sediment and water. In both areas, the state variables must be described: $N_i(t)$ denote the population density on patch i at time t. For the case of two patches, the model reads

$$\begin{cases} \dfrac{dN_1}{dt} = m_{12}N_2 - m_{21}N_1 + f_1(N_1) \\ \dfrac{dN_2}{dt} = m_{21}N_1 - m_{12}N_2 + f_2(N_2) \end{cases} \quad (19.89)$$

where the parameters m_{ij} are the displacement rates from patch j to patch i and the functions f_i describe the set of processes influencing the state variable $N_i(t)$ on patch i. This approach can easily be extended to situations with more patches. The application of this type of models allows for instance to test assumptions on the role of spatial heterogeneity on community dynamics (Poggiale et al. 2005; Mchich et al. 2007).

19.6.2 Continuous Spatial Structure

This representation is more realistic and allows taking into account explicitly the geometry of the environment, which can be useful in the case of some realistic applications. However, the formulation is more complex and uses mathematical concepts that fall well outside the scope of this chapter. In order to not unnecessarily burden the chapter, we will only describe these formulations that are found in many models in microbial ecology, especially in models of marine biogeochemistry and ecology of soil, for example (Dassonville and Renault 2002). This brief description allows introducing biogeochemical models in the next section. The displacements can be classified into two types: the diffusion and advection.

The diffusion is a phenomenon where the individuals exhibit a random movement, so that the mean flow at a given point is oriented perpendicularly to the contour lines of the state variable, from the less concentrated to the most concentrated. To work with simple notations, we will consider one spatial dimension (e.g., depth, distance from the coast), which is denoted by x.

The state variable depends now on the time and on the space; it is denoted by $N(t, x)$ and is the population density at time t at point x. The population density gradient at time t and at point x is $\frac{\partial N}{\partial x}(t,x)$ where, again, the symbol ∂ means that the derivation is done with respect to the x variable, for a fixed value of t (partial derivative). Fick's law expresses that the flux at time t at point x, $\Phi(t, x)$, is proportional to the gradient of population density and is oriented from the highest density to the lowest one:

$$\Phi(t,x) = -D(x)\frac{\partial N}{\partial x}(t,x) \qquad (19.90)$$

where $D(x)$ is called *diffusion coefficient*, which may depend on the spatial position x. The time variation at the point x of the population density follows

$$\frac{\partial N}{\partial t}(t,x) = -\frac{\partial F}{\partial x}(t,x) = \frac{\partial}{\partial x}\left(-D(x)\frac{\partial N}{\partial x}(t,x)\right) \qquad (19.91)$$

If the diffusion coefficient is a constant (i.e., does not depend on x), the previous model reads

$$\frac{\partial N}{\partial t} = D\frac{\partial^2 N}{\partial x^2} \qquad (19.92)$$

The advection is a displacement of all the individuals in a direction and at a given speed, for example, in the case of a current which moves a mass of water and the individuals of a population inside. The advection term is

$$\frac{\partial N}{\partial t} = -\frac{\partial}{\partial x}(v(x)N) \qquad (19.93)$$

where $v(x)$ is the speed of the movement.

In summary, a population that undergoes a process of diffusion (e.g., with a constant diffusion coefficient D), a process of advection (e.g., with a constant speed v), and other processes (demography, epidemiology, etc.), represented by a function f, can be described by the model:

$$\frac{\partial N}{\partial t} = D\frac{\partial^2 N}{\partial x^2} - v\frac{\partial N}{\partial x} + f(N,x) \qquad (19.94)$$

More specific examples will be given in the section on biogeochemical models. It may be useful to get used to the notation in order to identify the processes in reading mathematical expressions. The generalization of this model to a three-dimensional space is simple in concept but a bit unwieldy with respect to the notation. It is why physicists have invented specific notations. A work of modeling, in which nonmathematician readers can find a lot of information on this type of models, including their implementation, is the book by Boudreau (1997). If the applications in this book are mainly oriented to early diagenesis in aquatic sediments, many ideas about modeling and spatial extension are beyond this framework.

19.7 Biofilms

Developments in modeling biofilms, during the last three decades, allow to illustrate the use of the different approaches previously described, for the consideration of the spatial structure in ecological systems. A biofilm is defined as a combination of microorganisms adhering to a surface and usually characterized by a significant presence of extracellular polymers called EPS abbreviation for "extracellular polymeric substances." Thus, it is an ecological system that develops an interface; the most studied and modeled are biofilms at the solid–water interface, but also biofilms develop at interfaces solid–liquid–air and air. Biofilms are ubiquitous in nature and in many human infrastructure, and their study is therefore of great interest in ecology, human health, fluid transport, and industry. Some biofilms are beneficial to humans, for example, by their action of purifying water, such as river biofilms and those used in the wastewater treatment processes. Many other biofilms cause problems for humans, such as dental plates, hull fouling of vessels, and biofilms that develop on medical implants and within the pipeline transport of fluids.

All kinds of microorganisms can grow in biofilms as eubacteria, archaebacteria, fungi, ciliates, and other unicellular eukaryotes and heterotrophic microalgae. Almost all biofilms in nature are composed of several species, and monospecific biofilms are considered laboratory artifacts that can be developed and maintained only in isolated enclosure, thanks to processes of work in aseptic microbiology. For the naked eye, biofilms often appear as a homogeneous high coverage that covers a surface. However, their microscopic observation often reveals significant spatial heterogeneity at the microscopic level, and it is thanks to the use of confocal microscopy that the study of the morphology has greatly advanced. Thus, observations of their morphology and complex spatial structure have driven a significant development of spatial modeling as we will describe below. The different types of morphologies of biofilms include (**i**) a network of relatively scattered microcolonies; (**ii**) fairly homogeneous covers with one or more cell layers; (**iii**) structures with several multispecies cell layers laminated, as for instance for microbial mats; and (**iv**) three-dimensional structures where islands of cellular biomass often "mushroom shaped" are interspersed with empty channels known as water channels.

The growth of a biofilm at the solid–water interface depends on the exchange of solutes between the water column and biofilm. Thus, the microorganisms in the biofilm get their growth substrates from the water column. In addition, their metabolic products, potentially inhibitory to the growth and spread, are diluted in water. Biomass in a biofilm may also suffer losses primarily by detachment and senescence. The first models of biofilms developed from the 1970s thus involved just two compartments: the biofilm and the water column (Fig. 19.21). This simplistic approach means we admit that the biofilm covers an area known homogeneously and, inside, the biomass distribution is uniform. The transfer of the growing media water column is coupled to the biofilm growth of biomass in the biofilm:

$$\frac{dX}{dt} = \text{Yield} \times (\text{Transfer of the dissolved substrate from the water to the biofilm})$$

For recent biofilms with a low density of biomass, biomass growth is determined by its biomass ($dX/dt = \mu X$), and an exponential growth phase of the biofilm requires an exponential increase in the transfer of growth substrates. However, this transfer is limited by a maximum depending on the physical conditions, because it depends on the molecular diffusion through a boundary layer. In this boundary layer, the velocity is zero, and the molecular diffusion remains the only mechanism for generating the stream what engineers call a mass transfer. This phenomenon can be observed using microelectrodes; the boundary layer is often a thickness between about 100 and 400 μm, but it is highly dependent on hydrodynamic conditions in the water column. The maximum transfer rate (J_{max}), expressed per unit area, can be calculated using Fick's law using the assumption that within the biofilm, the substrate concentration is close to zero. The equation is:

$$J_{max} = \Phi_{max} = -D\frac{\partial C}{\partial x} \simeq -D\frac{C_{eau}}{\Delta_{DBL}}$$

where D represents the *diffusion coefficient* for the compound, C_{eau} is the compound concentration in the supernatant water, and Δ_{DBL} is the thickness of the limit layer. Thus, the J_{max} shall impose the maximal growth rate of the biofilm, and the growth becomes linear, according to $dX/dt = \text{Yield} \times J_{max}$.

This type of model was insufficient to describe the development of a multispecies biofilm and does not reflect the spatial heterogeneity. However, this approach is still useful to other rating scales, such as at the landscape scale, to predict how biofilms will change the quality of the river water.

From the 1980s, models have been developed to describe the stratification of biomass in a biofilm in a single dimension in space. They are used to study the interactions between different species within the biofilm (Fig. 19.21). Interactions between species include competition for multiple substrates and the possibility of co-metabolism phenomena ("cross-feeding") where the metabolic product of species A can serve as a growth substrate for the species B. Similarly, this approach is appropriate when dealing with a biofilm composed of aerobic and anaerobic bacteria in a stratified oxycline. This approach requires the calculation of diffusion gradients of these solutes within the biofilm. Models in two and three dimensions were later developed mainly from the 1990s to better describe the phenomena of metabolism, growth, and spatial organization within heterogeneous biofilm structures, such as three-dimensional islets of cell biomass often "mushroom shaped," scattered, and separated by voids (Fig. 19.21). These models are mainly of two types: (**i**) biomass density or (**ii**) focusing on the individual.

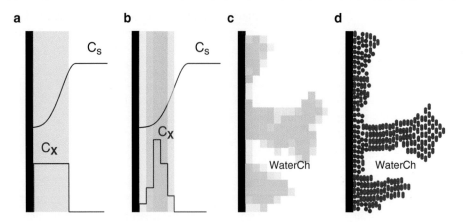

Fig. 19.21 Evolution of different models of biofilms: evolution from (**a**) simple models with two compartments (water and biofilm) in the 1970s to (**b**) models with a one-dimensional stratification of biomass within the biofilm and ability to model interactions between different species and several growth substrates, in the 1980s, to (**c**) modeling of multidimensional distributions of biomass and substrates developed from the late 1990s. C_s represents the substrate concentration, C_x is the biomass density in the biofilm, and *WaterCh* is the water channel (From Picioreanu 1999)

The so-called "biomass density" models offer digital simulations that will show a picture of the distribution of biomass density for the different species in the area from the initial conditions imposed. Thus, biofilm formation is considered an emergent property of the model, calculated from a set of equations describing the phenomena of reaction (conversion of substrate biomass production of metabolic product) and diffusion (solutes and biomass). These equations involve not only derivatives with respect to time (dt) but also derivatives with respect to space (dx, dy, dz depending on the model) to describe the diffusion process. Thus, the numerical scheme should be extended to account for the area concerned, introducing new sources of approximations and risk of *numerical errors*.

Typically, the area concerned, considered as a continuous space, is operationally divided into cells that are organized in a stack (1D) or a grid (2D and 3D). Within each cell, the equations describe the metabolic process similar to Fig. 19.21, as if they were micro-fermenters, and the inputs and outputs are determined by interactions with adjacent cells. For nonmobile organisms, once "inoculated," it can be admitted that the cells remain trapped in their cells as has been done for modeling microbial mats developing within the sediment (see below).

However, this approximation is not satisfactory for biofilms that project into the water from a settled area and become thicker as they grow. We must then consider the use of space and admit that biomass density within a biofilm is limited by a maximum value. Using cellular automata, Cristian Picioreanu and colleagues (1998) at the Technical University of Delft (the Netherlands) introduced rules for dividing and moving biomass according to an algorithm. At time t, where biomass density within a cell reaches a maximum ($B_{(t, x, y, z)} \geq B_{max}$), it is divided into two equal parts. One on both sides remains in its original cell, while the other is moved in a neighboring cell, preferably selected randomly from the still empty cells. If there are no empty adjacent cells, the other is moved randomly in a neighboring cell and pushes the people of this cell in another direction. The algorithm is repeated as many times as necessary to resolve all conflicts of occupation of space. In this way, a stochastic element is introduced into the simulation and the model became hybrid (deterministic/stochastic). Another element of randomness is represented by the initial conditions when the microcolonies that will initiate the formation of the biofilm were randomly distributed on the surface. Fig. 19.22 shows that the model produces the formation of a biofilm with a high spatial heterogeneity, with patches of cell biomass exhibiting "fingerlike and mushroom shape" (Picioreanu et al. 1998).

Furthermore, Fig. 19.22 shows the simulation of a monospecific development at different times (t) of the biofilm with isoclines of substrate concentrations in the boundary layer above the biofilm. After 2 days ($t = 2$), some irregularity reflects the colonization of the surface from the microcolonies, but the structure is a little more even homogeneous until the sixth day. From the 12th day, the structures "fingerlike and fungus" appear and become more pronounced. This structure is very similar to the structures of some natural biofilms observed using confocal microscopy. According to Fick's law, the vectors ($\Phi_{(t, x, y)}$) describing the direction of the diffusion of the growth substrate to the biofilm are perpendicular to the isoclines, and the slope of the gradient is proportional to their values. From the graphical analysis, we can conclude that the front structures "fingerlike and mushroom shape" receive substrate intake greater than the cells around the vacuum. This conclusion is contrary to the popular idea that the vacuum function as

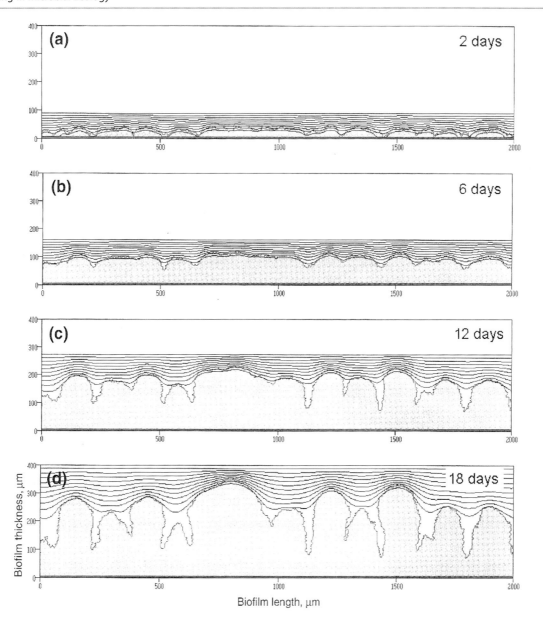

Fig. 19.22 Spatial distribution of biofilm and isoclines of substrate concentrations in the boundary layer, for a simulation of a model of "biomass density" of two spatial dimensions at different times. The isoclines were set at intervals of 10 % of the concentration in the water body (According to Picioreanu et al. 1998)

water channels allows "irrigating" the biofilm with substrates (Costerton et al. 1994).

The so-called models, "individual-based models" (IBMs), really consider microbial cells that are localized in space according to their coordinates (x, y, z). The growth of each cell is calculated, and functions come into play to describe the biovolume of each cell based on its growth and the time of cell division when the biovolume of cells is divided into two equal parts for forming the new cells (daughters). Numerical scheme again includes a spatial grid to calculate the diffusive transport of substrates for growth, and cellular automata rules determine how the cells will occupy space during their growth and division. Kreft et al. (2001) compared simulations of nitrifying biofilm composed of nitrifying bacteria ($NH_4^+ \rightarrow NO_2^-$) and nitrating bacteria ($NO_2^- \rightarrow NO_3^-$) using the so-called "biomass density" and IBM approach. Both approaches gave similar strong results and are well aware of the spatial structure "fingerlike and fungus." The nitrating bacteria ($NO_2^- \rightarrow NO_3^-$) formed colonies at the base of nitrifying bacteria ($NH_4^+ \rightarrow NO_2^-$). The IBM gave rise to more rounded structures that are probably closer to the reality of the biofilm. However, this small difference is not the main advantage of IBMs. The great advantage of these models is that they allow to better integrate important aspects of the biology of organisms. For example, this type of model can describe the genotypic and phenotypic variation in a population and thus can address the theoretical study of evolution. Note, however, that IBM may be based on computer algorithms or mathematics (differential equations, as in the theory "DEB" Kooijman 2010). The study of Jan Kreft and colleagues (Kreft 2004) is a good example in which they describe how the way of life in a biofilm facilitates the evolution of life history traits, enabling collaboration between cells of the same species. However, in a perfectly homogeneous medium, the evolution of such a kind of altruism is not possible because the survival of altruism is threatened by the emergence of "cheater" selfish.

The use of hybrid (stochastic/deterministic) model implies that each simulation is a unique event since random occurs. Therefore, researchers are forced to increase the number of simulations to verify whether their observations are generalizable. Different starting conditions must be taken into account, especially if the initial distribution of the biofilm-forming cells is random. Another problem is that these simulations generate a large number of graphical representations difficult to publish in a journal. To overcome this problem, the journals are increasingly called the "supplementary material online" that readers can visit on the web. This additional equipment often includes representations of simulations in the form of film. The Department of Environmental Biotechnology at the Technical University of Delft (the Netherlands) offers a site with access to a large number of publications with these cinematic representations as additional equipment: http://www.biofilms.bt.tudelft.nl/publications.html and http://www.biofilms.bt.tudelft.nl/material.html.

To date, few studies have used the numerical simulation of metabolic processes and growth in microbial mats, despite the fact that for these systems, we have a large number of measures of microgradient compounds (O_2, H_2S, pH, etc.) and light (cf also Chap. 17). Microbial mats can be considered as a specific example of photosynthetic biofilm that develops on the surface sediment. The light penetration is a structuring factor for photosynthetic biofilms. In microbial mats, microorganisms interact not only with the overlying environment (water column) but also with the underlying sedimentary layers that are often reduced and rich in H_2S and NH_4^+. Current microbial mats are restricted to the so-called extreme environments (hot, cold, and hypersaline environments undergoing flood and irregular exondation cycles). These systems are considered as an ideal model for the study of stromatolites (biosedimentary structures), and their study is of great importance in geology.

For example, a model reaction–diffusion of "biomass density" (1D) type that describes the growth of cyanobacteria and phototrophic sulfur-oxidizing bacteria (purple) and chemolithoautotrophic (without color) in a microbial mat was published by De Wit et al. (1995). This model consists of a stack of 25 sedimentary compartments 0.4 mm thick each, representing the first centimeter of the sediment column. It interacts with an upper compartment (supernatant water) and a lower compartment which simulates the supply of H_2S by sulfate reduction. The model calculates the molecular diffusion of O_2 and H_2S and light penetration separately for the visible fields (PAR 400–700 nm) used by cyanobacteria and near infrared (NIR 700–900 nm) used by bacteria sulfur-oxidizing phototrophic purple. The growth of cyanobacteria, bacteria, and sulfur-oxidizing phototrophic and chemolithoautotrophic bacteria has been set from ecophysiological studies of species *Microcoleus chthonoplastes*, *Thiocapsa roseopersicina*, and *Thiobacillus thioparus*, respectively. The model is fully deterministic, and from simulations, each compartment contains a low biomass of different groups (inoculum). The biomass formed remains confined in its original compartment.

Figure 19.23 shows the results of a typical simulation. The O_2 and H_2S profiles are similar to those observed using microelectrodes in microbial mats, and stratification of cyanobacteria and purple bacteria is very realistic. The spatial distribution of chemolithoautotrophic sulfur-oxidizing bacteria was still unknown, and the model predicts that they stratify between cyanobacteria and purple bacteria. The model has been used to further explore the potential interactions between bacteria and phototrophic sulfur-oxidizing chemolithoautotrophic by forcing changes in several factors. The predicted patterns were very clear: either only chemolithoautotrophic bacteria grow or these bacteria coexist with purple bacteria. However, when conditions are favorable, the purple bacteria are highly favored, and integrated biomass density per unit area may outweigh the chemolithoautotrophic bacteria by a factor of 17. This example shows how, after verification of the model with respect to known phenomena, it is possible to use it to get ideas on unknown process (generation of new hypotheses). However, it was emphasized that the predictive value of the model was probably limited, and it seems important to take into account the intermediate metabolic products, coupled with the recycling of carbon and active mobility of certain microorganisms, including chemolithoautotrophic bacteria (De Wit et al. 1995).

Decker et al. (2005) have adapted this model, particularly for comparing their observations with the recycling of compounds O, S, and C in the salt marsh microbial mats of Baja California (Mexico). Compared to the previous model, they included sulfate-reducing bacteria and coupled their activity of H_2S production with recycling of organic matter in the mat. Multiple observations have revealed that the

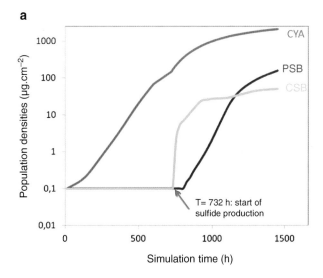
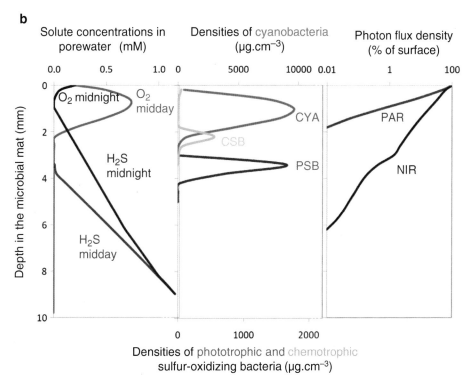

Fig. 19.23 Simulation of a model of a microbial mat with a reaction–diffusion model of "biomass density" type, according to a one-dimensional sediment gradient. (**a**) Simulation according to specific conditions showing integration on the sediment column (first cm) of cyanobacteria (CYA) biomass, bacteria, sulfur-oxidizing phototrophic purple (*PSB*), and chemolithoautotrophic bacteria (*CSB*). Note that the PSB and CSB did not develop during the first 720 h, because the production of H_2S is 0 during this period and starts from 720 h. This simulates the colonization of a virgin sediment by cyanobacteria before the installation of sulfate-reducing bacteria. (**b**) Results for $t = 1,452$ h (zenith) and $t = 1,464$ h (midnight) of the simulation described in (a) demonstrating the profiles of O_2 and H_2S and light penetration PAR (400–700 nm) and NIR (700–900 nm) during the day and the density distributions of biomass of CYA, PSB, and CSB (After De Wit et al. 1995)

sulfate-reducing bacteria are not confined to the deeper layers of the sediment, as is the case in the model of de Wit et al. (1995), but spread everywhere.

However, the new model is unable to predict the growth of sulfate-reducing bacteria, and therefore the authors have chosen to inject a known distribution. As with the previous model, the distributions calculated for cyanobacteria and purple bacteria are realistic and are very similar to that observed in the studied mat.

Their simulations (Decker et al. 2005) were also forced by solar radiance data and temperatures obtained by field monitoring, and the results were compared in detail with changes in patterns of O_2 and H_2S measured in situ during the same day. These measured profiles corresponded well with the simulated profiles, except during the afternoon where there were differences. A more important difference concerned the measured and simulated carbon flux through the water–sediment interface. Indeed, although the direction was consistent, the simulations overestimated flows. This indicates a lack of information at the coupling between the C cycle and the S cycle.

19.8 Biogeochemical Models

Biogeochemical models attempt to describe the dynamics of one or more elements in specific ecosystems. The number of publications on biogeochemical models and their applications increases significantly over the past fifteen years. Initially, they simply describe some global state variables such as biomass carbon global primary production in a given ecosystem. Today, a significant improvement has been made in the description of the process and the definition of functional groups for functions of interest in the course of one (or more) element set (Gabrielle et al. 2006). These models are very complex because they describe a large number of processes and their description is too long to be done properly in this chapter. However, we can note that these models classify living organisms into functional groups, each group being represented in the model with state variables representing the biomass of various group elements (carbon, nitrogen, phosphorus, etc.). Flows between these groups are then represented by mathematical formulas that attach to quantify the exchanges of material between the different functional groups. The influence of environmental parameters (temperature, light, natural forcings, etc.) is included in these formulas. Note that if the primary production was quickly taken into account in these models, the microbial loop has been introduced more recently and much remains to be done.

19.9 Conclusion

The environment, from marine fossa to the glaciers of the Himalayas, through the seas and oceans, rivers, arable land, nonagricultural land, forest and deserts, air, and all interfaces, is a huge reservoir of microorganisms. These microorganisms include on the one hand archaea and bacteria, prokaryotes which constitute two of the three largest biological domains in the phylogenetic tree of life (Woese 1987), and other fungi and protists belonging to the third domain, that of eukaryotes including, in addition to protists and fungi, plants, and animals. The numbers of species and individuals for each of these groups of microorganisms make the head spin (5×10^{30} cells, more than $1.5 \; 10^6$ species), especially when we know that we have access to less than 0.1 % of these microorganisms (Torsvik and Ovreas 2002), whereas their role in the functioning of ecosystems is fundamental. It thus includes the challenge of microbial ecology who wishes to understand the mechanisms that regulate the growth of these microorganisms, their intra- or interspecific interactions or intergeneric interactions with higher organisms, their activities, their role in biogeochemical cycles, and the constitution of food webs, genetic diversity, functional diversity, and evolution.

The spatial and temporal scales are a major problem. Observations and descriptions are carried out on samples whose representativeness of microbial communities and their environment is often discussed and makes generalization difficult. The operation of microbial systems is too complex to be understood without the development and use of tools to simplify, think and theorize, propose and formalize assumptions and mechanisms, simulate, acquire information on data virtually inaccessible, extrapolate, predict, and generalize.

Experimental studies can be conducted under controlled conditions in which simple biological models and increasing complexity are used. Mathematical modeling appears as an additional and exciting discipline providing tools for addressing complexity issues they face up.

This chapter aimed to explain the process of mathematical modeling in microbial ecology. It is not exhaustive, neither in the description of the methods nor in the various aspects of the applications discussed. We have tried to show how fairly general modeling allows to improve our knowledge at both the individual processes and the ecosystem functioning. This chapter is an open door on the scientific literature, which in the field of ecology – and microbial ecology in particular – relies more on the mathematical tool that models it. In addition, the specificity of microbial ecology is new in modeling in ecology. For example, the high biodiversity in a microbial sample or timeliness of genetic variation compared to population dynamics makes it a subject that conventional models of ecology cannot generally address. It is therefore in this area a challenge for modeling. Another exciting challenge is provided by the development of the "omic" technologies. Indeed, the new generations of high-speed sequencing now give better access to the taxonomic diversity and functional diversity present in complex ecosystems, but they remain global and descriptive and do not provide quantitative values or accurate information about the multi-trophic interactions between the various biotic components of the environment studied (Morales and Holben 2011). However, molecular techniques continue to improve and their cost to slightly decrease, which allows to consider a temporal dimension to their application.

Moreover, many efforts are being made in statistics and modeling to try to integrate the extensive data provided by these approaches to identify and assess the links between specific microbial populations and processes of ecosystems, as to predict the response of ecosystems to disturbances of both natural and anthropogenic origins.

Beyond the quantitative aspects that represent counts or measures of activity discussed in this chapter, mathematical modeling related to biology (and not only microbial ecology) evolves with the knowledge in this area and raises new questions thanks to the theoretical aspects which constitute the second aspect of this discipline (the first, as we have seen, being turned to practical answers). Indeed, by combining the mathematical models with the new tools offered by molecular biology, complex processes can be addressed, but also more fundamental issues can be addressed, from the behavior of an operon in a cell functionality expressed in a natural environment to the diversity of uncultured microorganisms (Sloan et al. 2007) to stimulate discussion and propose hypotheses about the evolution of this diversity (Pavé 2006, 2007; Maynard-Smith 1978).

References

Bailey DJ, Otten W, Gilligan CA (2000) Saprotrophic invasion by the soil-borne fungal plant pathogen Rhizoctonia solani and percolation thresholds. New Phytol 146:535–544

Becks L, Hilker FM, Malchow H, Jürgens K, Arndt H (2005) Experimental demonstration of chaos in a microbial food web. Nature 435:1226–1229

Beretta E, Kuang Y (1998) Modeling and analysis of a marine bacteriophage infection. Math Biosci 149:57–76

Boer MP, Kooi BW, Kooijman SALM (1998) Food chain dynamics in the Chemostat. Math Biosci 150:43–62

Bonkowski M (2004) Protozoa and plant growth: the microbial loop in soil revisited. New Phytol 162:617–631

Boswell GP, Jacobs H, Davidson FA, Gadd GM, Ritz K (2003) Growth and function of fungal mycelia in heterogeneous environments. Bull Math Biol 65:447–477

Boudreau PB (1997) Diagenetic models and their implementation: modelling transport and reactions in aquatic sediments. Springer, Berlin/Heidelberg/New York

Brauer F, Castillo-Chavez C (2000) Mathematical models in population biology and epidemiology, vol 40, Texts in applied mathematics. Springer-Verlag, New York

Coleman D, Fu SL, Hendrix P, Crossley D (2002) Soil foodwebs in agroecosystems: impacts of herbivory and tillage management. Eur J Soil Biol 38:21–28

Costerton JW, Lewandowski Z, De Beer D, Caldwell D, Korber D, James G (1994) Biofilms, the customized microniche. J Bacteriol 176:2137–2142

Couteaudier Y, Steinberg C (1990) Biological and mathematical description of growth pattern of *Fusarium oxysporum* in sterilized soil. FEMS Microbiol Ecol 74:253–259

Cunningham A, Maas P (1978) Time lag and nutrient storage effects in the transient growth response of *Chlamydomonas reinhardtii* in nitrogen limited batch and continuous culture. J Gen Microbiol 104:227–231

Dantigny P (2004) Predictive mycology. In: McKellar RC, Lu X (eds) Modeling microbial responses in food. CRC Press, Boca Raton, pp 313–320

Dassonville F, Renault P (2002) Interactions between microbial processes and geochemical transformations under anaerobic conditions: a review. Agronomie 22:51–68

De Wit R, Van den Ende FP, Van Gemerden H (1995) Mathematical simulation of the interactions among cyanobacteria, purple sulfur bacteria and chemotrophic sulfur bacteria in microbial mat communities. FEMS Microbiol Ecol 17:117–136

Decker KLM et al (2005) Mathematical simulation of the diel O, S, and C biogeochemistry of a hypersaline microbial mat. FEMS Microbiol Ecol 52:377–395

Droop MR (1968) Vitamin B12 and marine ecology. IV. The kinetics of uptake, growth and inhibition in Monochrysis lutheri. J Mar Biol Assoc UK 48:689–733

Fu SL, Cabrera ML, Coleman DC, Kisselle KW, Garrett CJ, Hendrix PF, Crossley DA (2000) Soil carbon dynamics of conventional tillage and no-till agroecosystems at Georgia Piedmont – HSB-C models. Ecol Model 131:229–248

Fuhrman JA (1999) Marine viruses and their biogeochemical and ecological effects. Nature 399:541–548

Fujikawa H, Yano K, Morozumi S (2006) Characteristics and modeling of Escherichia coli growth in pouched food. Shokuhin Eiseigaku Zasshi 47:95–98

Fussmann GF, Blasius B (2005) Community response to enrichment is highly sensitive to model structure. Biol Lett 1:9–12

Fussmann GF, Ellner SP, Shertzer KW, Hairston NG Jr (2000) Crossing the Hopf bifurcation in a live predator-prey system. Science 290:1358–1360

Gabrielle B, Laville P, Henault C, Nicoullaud B, Germon JC (2006) Simulation of nitrous oxide emissions from wheat-cropped soils using CERES. Nut Cycl Agroecosyst 74:133–146

Gause GF (1935) Vérifications expérimentales de la théorie mathématique de la lutte pour la vie. Actual Scient Ind 277

Ghosh S, Bhattacharyya S, Bhattacharya DK (2007) The role of viral infection in pest control: a mathematical study. Bull Math Biol 69:2649–2691

Holling CS (1959) Some characteristics of simple types of predation and parasitism. Can Entomol 80:274–287

Holling CS (1965) The functional response of predators to prey density and its role in mimicry and population regulation. Mem Entomol Soc Can 45:1–60

Kooijman SALM (2010) Dynamic energy and mass budgets in biological systems, 3rd edn. Cambridge University Press, Cambridge

Kreft JU (2004) Biofilms promote altruism. Microbiology 150:2751–2760

Kreft JU, Picioreanu C, Wimpenny JWT, Van Loosdrecht MCM (2001) Individual-based modelling of biofilms. Microbiology 147:2897–2912

Lejeune R, Baron GV (1998) Modelling the exponential growth of filamentous fungi during batch cultivation. Biotech Bioeng 60:169–179

Li H, Xie GH, Edmondson A (2007) Evolution and limitations of primary mathematical models in predictive microbiology. Br Food J 109:608–626

Maynard-Smith J (1978) Models in ecology. Cambridge University Press, Cambridge

Mchich R, Auger P, Poggiale J-C (2007) Effect of predator density dependent dispersal of prey on stability of a predator-prey system. Math Biosci 206:343–356

Middelboe M (2000) Bacterial growth rate and marine virus-host dynamics. Microb Ecol 40:114–124

Middelboe M, Hagstrom A, Blackburn N, Sinn B, Fischer U, Borch NH, Pinhassi J, Simu K, Lorenz MG (2001) Effects of bacteriophages on the population dynamics of four strains of pelagic marine bacteria. Microb Ecol 42:395–406

Mitra A, Davidson K, Flynn KJ (2003) The influence of changes in predation rates on marine microbial predator/prey interactions: a modelling study. Acta Oecol Int J Ecol 24:S359–S367

Monti GE, Frankena K, De Jong MCM (2007) Transmission of bovine leukaemia virus within dairy herds by simulation modelling. Epidemiol Infect 135:722–732

Morales SE, Holben WE (2011) Linking bacterial identities and ecosystem processes: can 'omic' analyses be more than the sum of their parts? FEMS Microbiol Ecol 75:2–16

Mulder MM (1988) Energetic aspects of bacterial growth: a mosaic non – equilibrium thermodynamic approach. PhD thesis, Amsterdam Universiteit

Patwardhan PR, Srivastava AK (2004) Model-based fed-batch cultivation of *R. eutropha* for enhanced biopolymer production. Biochem Eng J 20:21–28

Pavé A (2006) By way of introduction: modelling living systems, their diversity and their complexity: some methodological and theoretical problems. C R Biol 329:3–12

Pavé A (2007) Necessity of chance: biological roulettes and biodiversity. C R Biol 330:189–198

Picioreanu C (1999) Multidimensional modeling of biofilm structure. PhD thesis, Department of Biotechnology, TU Delft. ISBN 90-90133110-0

Picioreanu C, Van Loosdrecht MCM, Heijnen J (1998) Mathematical modelling of biofilm structure with a hybrid differential-discrete cellular automaton approach. Biotechnol Bioeng 58:101–116

Poggiale J-C, Auger P, Nérini D, Manté C, Gilbert F (2005) Global production increased by spatial heterogeneity in a population dynamics model. Acta Biotheor 53:359–370

Poggiale J-C, Baklouti M, Queguiner B, Kooijman SALM (2010) How far details are important in ecosystem modelling: the case of multi-limiting nutrients in phytoplankton – zooplankton interactions. Philos Trans R Soc Lond B Biol Sci 365:3495–3507

Ponciano JM, Vandecasteele FPJ, Hess TF, Forney LJ, Crawford RL, Joyce P (2005) Use of stochastic models to assess the effect of environmental factors on microbial growth. Appl Environ Microbiol 71:2355–2364

Ronn R, McCaig AE, Griffiths BS, Prosser JI (2002) Impact of protozoan grazing on bacterial community structure in soil microcosms. Appl Environ Microbiol 68:6094–6105

Rosenzweig ML (1971) The paradox of enrichment: destabilization of exploitation ecosystems in ecological time. Science 171:385–387

Rosenzweig ML, MacArthur RH (1963) Graphical representation and stability conditions of predator-prey interactions. Am Nat 97:209–223

Rosso L, Lobry JR, Bajard S, Flandrois JP (1993) An unexpected correlation between cardinal temperatures of microbial growth highlighted by a new model. J Theor Biol 162:447–463

Sloan WT, Woodcock S, Lunn M, Head IM, Curtis TP (2007) Modeling taxa-abundance distributions in microbial communities using environmental sequence data. Microb Ecol 53:443–455

Smith LH (1997) The periodically forced Droop model for phytoplankton growth in a chemostat. J Math Biol 35:545–556

Solomon E (1949) The natural control of animal populations. J Anim Ecol 18:1–35

Steinberg C, Whipps JM, Wood DA, Fenlon J, Alabouvette C (1999) Effects of nutritional sources on growth of one non-pathogenic strain and four strains of *Fusarium oxysporum* pathogenic on tomato. Mycol Res 103:1210–1216

Stelling J (2004) Mathematical models in microbial systems biology. Curr Opin Microbiol 7:513–518

Takeuchi Y, Adachi N (1983) Existence and bifurcation of stable equilibrium in two-prey-one-predator communities. Bull Math Biol 45:877–900

Tam VH, Schilling AN, Nikolaou M (2005) Modelling time-kill studies to discern the pharmacodynamics of meropenem. J Antimicrob Chemother 55:699–706

Torsvik V, Ovreas L (2002) Microbial diversity and function in soil: from genes to ecosystems. Curr Opin Microbiol 5:240–245

van den Berg HA, Kiselev YN, Kooijman SALM, Orlov MV (1998) Optimal allocation between nutrient uptake and growth in a microbial trichome. J Math Biol 37:28–48

Van Impe JF, Poschet F, Geeraerd AH, Vereecken KM (2005) Towards a novel class of predictive microbial growth models. Int J Food Microbiol 100:97–105

Vayenas DV, Pavlou S (1999) Chaotic dynamics of a food web in a chemostat. Math Biosci 162:69–84

Woese CR (1987) Bacterial evolution. Microbiol Rev 51:221–271

Wood JC, McKendrick IJ, Gettinby G (2006) A simulation model for the study of the within-animal infection dynamics of *E. coli* O157. Prev Vet Med 74:180–193

Zhang XS, Holt J, Colvin J (2000) A general model of plant-virus disease infection incorporating vector aggregation. Plant Pathol 49:435–444

Zwietering MH, Witjes T, de Wit JC, van't Riet K (1992) A decision support system for prediction of the microbial spoilage in foods. J Food Protect 12:973–979

Glossary

Abiotic That which is not biological.

Acetoclastic methanoarchaea Methanoarchaea which produce methane and carbon dioxide from acetate are called acetoclastic.

Acidophily Ability of an organism to grow in an acidic environment (pH below 5.5), thus containing high concentrations of protons (H^+).

Activated sludge Flocculating microbial biomass maintained in suspension by stirring in a tank, which transforms and digests the organic matter in wastewater and forms many settleable flocs.

Activator A small protein that interacts with a sequence upstream of a gene and activates its transcription except when physicochemical conditions (i.e., presence of a metabolite) change its conformation.

Adaptation All morphological and physiological characteristics resulting from selection that allow organisms to multiply in a given biotope.

Adhesin A cell structure that allows its adhesion to a support.

Advection In the field of oceanography, advection is referring mainly to the vertical or horizontal transport of water masses under the effects of current and wind.

Aerobe This term characterizes biological processes or organisms requiring molecular oxygen for their functioning.

Aerobic Capacity of an organism to grow in an environment exposed to air and therefore to dioxygen.

Air tolerant See "Anaerobic."

Akinete It is a thick-walled dormant cyanobacteria cell derived from the enlargement of a vegetative one. It serves as a survival structure under unfavorable condition.

Alkaliphilic Ability of an organism to grow under alkaline conditions (pH over 8.0), thus containing high concentrations of hydroxide ions (OH^-).

Amensalism (synonym of antagonism) An interaction that has a negative effect on one partner but without any effect on the other partner.

Amoeba Amoebae are cellular forms encountered in different unrelated Eukaryotic phyla. Two major phyla are entirely composed of amoeboid taxa: the Amoebozoa and the Rhizaria. Amoebae are characterized by temporary cytoplasmic expansions or projections called pseudopods of different forms, either bulbous, slender, or reticulated. Some amoebae can be united through their pseudopods to form colonies.

Anaerobe Term describing biological processes or organisms unable to use molecular oxygen; thus, molecular oxygen is not the final electron acceptor. Two types of anaerobic organisms can be defined: (1) the facultative anaerobes developing either in the presence of dioxygen (aerobic respiratory metabolism) or in the absence of dioxygen (anaerobic respiratory metabolism or fermentation) and (2) strict anaerobes unable to grow in an oxic environment or killed in the presence of dioxygen. Microorganisms known as air tolerant accept the presence of molecular oxygen while maintaining their anaerobic metabolic activity.

Anaerobic Capacity of an organism to develop in an environment devoid of air and thus that does not contain dioxygen.

Anaerobic oxidation of methane (AOM) Important process accounting for the removal of a large quantity of methane from anoxic environments and generally mediated by a syntrophic consortium of yet uncultivated *Archaea* and *Bacteria*.

Anoxic This term refers to environments in which molecular oxygen is absent. The term "anaerobic" characterizes favorable environmental conditions (anoxic environments) for the development of anaerobic microorganisms.

Anthropization Refers to the effects of man on the natural environment.

Antibiosis Ability of a microorganism to synthesize an antibiotic molecule that is detrimental to another microorganism, enabling it to defend its ecological niche.

Antibiotic (or antimicrobial) Toxic substance generally produced by certain fungi, bacteria, and other organisms that can kill other microorganisms and destroy or inhibit their growth.

Aphotic zone Portion of a lake or ocean, located below the photic zone, where there is little or no sunlight to sustain photosynthesis.

Apoptosis Also referred to as programmed cell death, apoptosis is a physiological process used by many organisms that eliminates cells during their embryonic or postembryonic development or during their morphogenesis.

Aposymbiotic State of an organism temporarily or permanently deprived of a symbiont.

Appressorium Fungal cell, usually bulbous and surrounded by a thick cell wall, which forms at the surface of a plant organ or of a plant cell and which allows the fungus to penetrate the corresponding organ or cell.

Arrhenius law In chemical kinetics, the Arrhenius law describes the change in speed of a chemical reaction as a function of temperature.

Autonomous sanitation system Still called on-site sanitation or individual sanitation. This treatment system includes a water tank that collects the whole domestic wastewater (kitchen, bathroom, washing machine, etc.), a purification device (i.e., filter bed, manure pipes, soil infiltration), and effluent discharge (generally in the ground). It is set in the remote areas of a community sanitation network (rural or dispersed housing, some camping sites, and holiday parks). Maintenance is done by the owner.

Autotrophs Organisms synthesizing organic compounds from CO_2 as sole carbon source.

Auxotrophs Organisms lacking or losing their capacity to synthesize one or more compounds essential for growth and requiring the presence of this or these compounds (e.g., growth factors) in the external environment.

Axenic Property of a habitat that is entirely free of all "contaminating" organisms. It is also said for a living organism containing no saprophytic or pathogenic microorganism or for a pure bacterial culture, free of other microorganisms. Axenic culture of many multicellular organisms is also possible.

Axopodia Long and thin pseudopod with an axial microtubule. They are used for phagocytosis.

Bacteremia Bacteremia is the presence of viable bacteria in the circulating blood. Usually, bacteremia does not cause infections because bacteria typically are present only in small numbers and are rapidly removed from the bloodstream by the immune system. However, if bacteria are present long enough and in large enough numbers, particularly in people who have a weakened immune system, bacteremia can lead to septicemia.

Bacteria beds Biological reactors with non-immersed fixed cultures, which form surface biofilms on natural materials (volcanic rocks, stones, etc.) or plastics through which the water to treat is filtered. The oxygen necessary to the metabolic activity of bacteria is brought either by the natural dissolution of oxygen from the air or by a forced ventilation.

Bacterial production The amount of carbon produced by a bacterial community per unit of time and sample volume.

Bacteriophage A virus that infects a prokaryotic host cell.

Bacteriorhodopsin Transmembrane protein of carotenoid pigment type that converts light energy into chemical energy; it is present in the cytoplasmic membrane of certain microorganisms, including extreme halophilic archaea, and involved in the synthesis of ATP via a proton transfer membrane. It is similar to the pigment rhodopsin present in visual cells of the retina.

Bacterivorous Feeding on bacteria.

Barophile See "Piezophile."

Bentonite Bentonite is a colloidal clay which consists primarily of montmorillonite: (Na, Ca) 0.33 (Al, Mg) $2Si_4O_{10}$ $(OH)_2 \cdot (H_2O)n$ (80 %) and clay, which explains its ability to retain water. Bentonite deposits have a volcanic and hydrothermal origin. Bentonite has various properties, including those of absorbing protein and reducing the activity of enzymes.

Bioaccumulation Accumulation of mineral or organic compound, such as copper or pesticide, in a living organism.

Biodegradation Biodegradation of an organic compound is a partial or total destruction of the molecular structure under the action of biochemical reactions catalyzed mostly by microorganisms. When biodegradation is complete (also referred to as ultimate degradation), it leads to the formation of water and carbon dioxide and/or other inorganic substances, in which case the term mineralization is then used.

Biodiversity Generic term to describe the presence of multiple types of organisms (genotypes, species, other taxonomic groupings) in a habitat or ecosystem. Diversity: a quantitative measure expressing the structure and/or composition of a population or community. Many diversity indices are used in microbiology. The essential property of these indices is that their average and variance can be calculated, which can be the basis for statistical testing of hypotheses. Can be described as "alpha diversity" (small spatial scale or within a community) or "beta diversity" (at large spatial scales or between communities).

Biofilm A biofilm is defined as an agglomeration of microorganisms adhering to the surface of a hard substrate or which develops at an interface between two phases (e.g., liquid-solid, liquid-liquid, liquid-gas, solid-gas). Biofilm formation is a natural phenomenon and biofilms are ubiquitous. They will occur on many occasions, when a non-sterile liquid enters into contact with a solid surface; many of the prokaryotes occurring in the liquid will progressively adhere to the surface. During the adhesion, many of these prokaryotes will express their capacities to synthesize and excrete biopolymers. As a result, bacteria in biofilms are surrounded by these polymers, which often confer protection and enhance their adherence to the surface. These biopolymers are also known as extracellular polymeric substances (EPS), which mainly comprise polysaccharides (polymers of sugar molecules) and proteins.

Biogeochemical tracer Compound which can be identified and quantified and is altered by a biogeochemical process. It can be a natural product involved in a biological activity (e.g., chlorophyll) or a compound deliberately introduced which interacts with a biogeochemical process (e.g., stable or radioactive isotope, fluorescent chemical).

Bioleaching Extraction of metals by dissolution mainly by the use of aerobic and acidophilic *Acidithiobacillus* bacteria. These bacteria transform sulfur of mineral compounds into sulfuric acid.

Biological pump Ocean absorbs atmospheric CO_2 according to 3 processes called CO_2 pumps. The *chemical pump* consists in dissolution and diffusion processes in the upper layer. The *physical pump* mixes surface, intermediary, and deep waters, drawing down CO_2 to the deep ocean. The *biological pump* consists in CO_2 conversion into organic matter by phytoplankton photosynthetic activity.

Biomagnification Process leading to the increase in the concentration of a natural or anthropic compound in the tissues of living organisms during its transfer at successively higher levels in a food chain.

Biomarker An organic molecule synthesized by specific (micro)organisms (or classes of (micro)organisms) as a function of (inter alia) environmental conditions (e.g., presence or absence of oxygen, temperature) and providing information about (1) its biological source (taxonomy, metabolism), (2) the origin and fate of organic matter (e.g., degradation processes), and (3) the involvement of (micro)organisms in biogeochemical processes. In petroleum geochemistry, the biomarker concept refers to the "molecular fossils" derived from natural molecules whose carbon skeletons have been preserved during diagenesis (e.g., hopanes and steranes formed from hopanoids and steroids, respectively). It is thus possible to link a molecular fossil to its biological precursor by structural analogy. Being present in both oils and source rocks, biomarkers allow inference of the conditions under which an oil has formed, as well as its age, maturity, and biodegradation state. In environmental geochemistry, biomarkers can be used to infer (inter alia) a pollutant's source and evolution in the environment (extent of degradation).

Biomass In ecology, biomass is the total mass from living organisms present at a given time in a particular biotope.

Bio-oxidation Oxidation of mineral or metal compounds by microorganisms, i.e., oxidation of sulfur compounds (metal sulfides) in sulfuric acid by *Thiobacillus*.

Biotic Which is biological in nature.

Biotic potential The maximum value that a population can reach in a given habitat, i.e., the biotic component.

Biotope Biotope is an area of uniform physicochemical environmental conditions providing a living place for a specific assemblage of living organisms.

Biotransformation Biochemical step during which an organic molecule (precursor) is transformed into another organic molecule (product). In metals the biotransformation corresponds to a change of valence or structure (i.e., mercury demethylation).

Biotroph Qualifies a parasitic organism that feeds on another living organism, without the latter's death.

Bloom Rapid increase or accumulation of microorganisms (typically phytoplankton) in an aquatic system.

Brackish Aquatic environment with salinity fluctuating between that of freshwater and seawater.

Candidatus Candidate species, describes a species name proposed but not yet accepted by the International Committee of bacterial nomenclature, given in general to noncultivated microorganisms.

Capsid Shell of protein that protects the nucleic acid of a virus; it is composed of structural units or capsomeres.

Capsule Extracellular polysaccharide and/or protein envelop providing cell protection toward external factors.

Carbon sink See "Link or sink."

Carotenoids Red-, orange-, and yellow-colored lipophilic pigments, which occur in different types of pigments. Carotenoids are involved in (1) providing protection against oxidation as antioxidant compound, (2) photocatalysis, and (3) light harvesting in photosynthesis as accessory pigments.

Carpospore Reproductive cell similar to conidia as it produces a clone. Different from conidia, carpospores are not facultative reproductive structures; they represent rather an obligate step in a reproductive cycle which allows the transition from one stage to the next one.

CD4 receptor Surface protein characteristic of lymphocytes that is involved in the modulation of the immune response.

Cell tropism This term can be used to describe the phenomenon by which pathogens will move toward or target particular host cell types.

Cellulolytic Metabolic capacity to catabolize cellulose.

Central metabolic pathways All metabolic changes common to all living beings. These include particularly glycolysis, gluconeogenesis, and the Krebs cycle. The carbon skeletons of the main monomers are derived from metabolic intermediates of these pathways.

Chaperone proteins A chaperone protein is a protein whose function is to help other proteins to fold into their specific conformation. Chaperone proteins guide newly synthesized polypeptides into their functional three-dimensional shape or can restore the original conformation of partly denatured proteins. A lot of chaperone proteins are heat shock proteins (Hsp), i.e., protein expressed in response to increasing temperature or other cellular stresses. The three-dimensional conformation of protein is sensitive to heat, and the proteins are denatured and lose their activity. The biological role of chaperone protein is to prevent the putative damage generated by the loss of biological function of protein resulting from misconformation.

Chemocline The chemocline is the zone characterized by a shape chemical vertical gradient between two water masses. For example, a double gradient of sulfide and oxygen can separate the oxygenated surface waters of a lake from the deep anoxic water containing sulfides. Very often chemocline and thermocline coincide.

Chemolithotrophs Microorganisms obtaining their energy from the oxidation of reduced mineral compounds (H_2, NH_3, HS^-, etc.).

Chemoorganoheterotrophic Microorganisms using an organic compound as a source of carbon and energy.

Chemoorganotrophic Microorganisms getting their energy from the oxidation of organic compounds.

Chemotaxis Property of a moving cell to respond to a chemical stimulus by moving to or away from the perceived substance.

Chemotropism Property of a nonmotile multicellular organism to respond to a chemical stimulus by growing toward or away from a perceived substance.

Chimeric gene, chimera Artificial DNA sequence composed of two or more natural sequences that associated themselves during genetic amplification such as PCR. Corresponds also to organisms that have two or more distinct genotypes.

Chlorophyll Green pigment from plants necessary for photosynthesis. This molecule is fundamental for photophosphorylation, a process by which light energy is transformed into chemical energy available for cell, allowing sugar photosynthesis from atmospheric CO_2 and H_2O. There are many kinds of chlorophylls (*a, b, c*, etc.) characterized by differing excitation wavelengths.

Chronic infection Persistent infection due to pathogenic microorganism, with episodes of reactivation.

Clade Monophyletic group of organisms consisting of an ancestor and all its descendants.

Class In systematics, set of orders that shares common attributes.

Clathrate A clathrate is an inclusion compound consisting of host molecules forming cage-like structures in which small guest molecules are physically trapped. In methane hydrate, a large amount of methane is enclosed within a host lattice of water ice.

Clonal Group of cells originating from the multiplication of a single cell and forming a genetically homogenous population.

Clonal reproduction Mode of asexual reproduction. In the prokaryotes and most unicellular eukaryotes, it results from the division (more rarely budding) of a mother cell followed by the separation of the resulting daughter cells. In multicellular organisms (e.g., *Trichodesmium*), fragmentation of the vegetative apparatus produces propagules, which give rise to new individuals. In plants, taking cuttings is a way for clonal reproduction. In all cases, the resulting cells or individuals are genetically identical to the parental cell or individual.

Clone A group of individuals derived from one organism by asexual reproduction, without genetic exchange. It represents an ideal notion as usually, in a clone, individuals are identical at the beginning, and then mutations could appear and accumulate through generations. Typically, a bacterial colony is a clone.

Cloning Technique for obtaining a bacterial strain with genetic properties modified by the addition of a DNA fragment, which can then be characterized, based on a cloning vector such as a plasmid or a bacteriophage.

Coastal zone Area where ocean, continents, and atmosphere interact with each other. There is no precisely defined limit between the coastal zone and the open ocean. However, it is sometimes assigned to the 200 m isobath.

Codon Group of three nucleotides coding for a nucleic acid or causing a stop in elongation.

Coenocytic A cellular structure that contains several nuclei in a common cytoplasm.

Coevolution Concerted evolution of two or more organisms that interact transiently or permanently.

Colimitation Term used when several resources needed for a species growth are limiting factors.

Colitis Colon inflammation.

Collection Institution funded for maintaining pure strains of microorganisms in viable conditions and involved in the distribution of a large number of strains of microorganisms; there are national collections, the main ones being the ATCC (USA, American Type Culture Collection), the DSMZ (Germany, Deutsche Sammlung von Mikroorganismen and Zellkulturen), the UKNCC (England, UK National Culture Collection), the JCM (Japan, Japan Collection of Microorganisms), the NCAIM (Sweden, National Collection of Agricultural and Industrial Microorganisms), the CIP (France, Collection of the Institut Pasteur), etc.

Colony A colony is defined as a visible cluster of cells, often resulting from an initial cell, on solid medium.

Cometabolism Degradation mechanism of a chemical compound by one or more microorganism(s) which do not possess all the enzymatic equipment for using this compound as the sole source of carbon and energy; degradation of the compound is only possible in the presence of an organic substrate easily used (cosubstrate) which can cause growth of microorganism(s). Generally, the chemical compound is partially degraded if one strain is involved in the degradation process.

Commensal A species which establishes biological interactions with other organisms to exploit their resources, without affecting them.

Commensalism Symbiotic relationship that is beneficial to one species without affecting the other interacting species (the opposite of parasitism).

Community All organisms, belonging to different taxa and living in a given habitat.

Community composition List and proportion of the different organisms (genotypes or phenotypes) shaping the community.

Competence Developmental state of the bacterium in which it is able to take up extracellular DNA and to recombine this DNA into the chromosome, thereby undergoing natural transformation.

Competition An interaction in which the two partners use the same resource (growth nutrients, water, space).

Competitive exclusion The displacement or extinction of one partner (less competitive) by the other (more competitive) when both compete for exactly the same resources.

Complex of species Group of species with similar characteristics, undergoing differentiation.

Compound-specific isotope analysis Study of the natural stable isotopic composition (e.g., $^{13}C/^{12}C$; D/H) of individual cellular components (e.g., specific lipids) biosynthesized by (micro)organisms or of individual organic pollutants.

Congener Refers to variants of the same chemical structure.

Conidia A disseminated mono- or multicellular spore which gives rise upon germination to an individual genetically identical to the one it originates from. Conidia give therefore rise to clones. They can be either haploid or diploid and are dispensable for completion of a reproductive sexual cycle.

Conjugation Genetic modification of a prokaryotic cell by transfer of genetic material from a donor cell to a recipient cell with a direct cell-to-cell contact. In eukaryotes, conjugation refers to the reproduction type (fusion of gamete cells) that is observed, particularly, in *Spirogyra* (Viridiplantae) and named cystogamy here.

Conserved core genes Set of genes common to a phylum.

Consortium Group of organisms belonging to two taxa or more whose physiological functions are complementary.

Cooperation Facultative interaction with reciprocal benefits.

Cosmid A hybrid plasmid often used as a cloning vector for large DNA fragments (up to 52 kb). It contains the specific sequence motifs (cos sites) that allow its packaging into a lambda bacteriophage capsid.

Cryoprotectant Substance that is used to protect cells from freezing damage.

Cyst Resting resistant cell of variable nature (spore, conidia, or zygote). A cyst is surrounded by a protective cell wall that can be almost indestructible as in the case of cysts from Dinophyceae. It allows organisms to survive for months or years in adverse conditions, waiting for favorable environmental ones.

Cystitis Inflammation of the bladder (urinary tract infection).

Deamination Elimination of an amine group catalyzed by a deaminase.

Detoxification (mechanisms, reactions) Process, generally biotic, by which a toxic compound is transformed into a nontoxic compound.

Dew point of a gas At constant pressure, it is the temperature at which appears the first drop of liquid during the cooling of a pure body vapors.

DGGE (denaturing gradient gel electrophoresis) DNA fingerprint method that discriminates DNA fragments of same length through their base composition by decreased electrophoretic mobility of partially melted double-stranded DNA molecules in polyacrylamide gels containing a linear gradient of denaturants (urea and formamide).

Diffusion coefficient (or molecular diffusivity) This coefficient is a proportionality constant between the flow of a species in the process of diffusion and the concentration gradient of this species. It measures the displacement of particles in the diffusion process. It is expressed in units of length squared per unit time (m^2/s, for instance, in the International System of Units).

Dikaryotic The dikaryotic phase of a reproductive cycle corresponds to the phase where each of the two haploid nuclei from the genitors (gametes) are not fused but instead coexist in each binucleate cell of the organism. It is a characteristic feature of many Basidiomycota species.

Dinitrogen fixation Metabolic capability of production of ammonium with dinitrogen as substrate by nitrogenase. This capability is restricted to some prokaryotes.

Disease incidence Number of cases (disease, infection) for a population in a specified period of time.

Disproportionation (or dismutation) Oxidation-reduction process of a chemical compound where part of the compound is oxidized while the other part is reduced; it is an energy-yielding reaction.

Dissolved organic matter (DOM) The definition to divide line between dissolved and particulate organic matter is closely dependent of the methodology used. In aquatic habitats, DOM is defined as whatever passes through a filter with pore sizes about 0.6 μm corresponding to the porosity of glass fiber filters (GF/F). Such GF/F filter can be cleaned easily of contamination by burning ("combusting") the filter at about 500 °C to remove all organic compounds. Particulate organic matter (POM) is the part retained on the GF/F filter.

DNA microarrays Commonly known as DNA chips or biochips are a technology in which thousands of nucleic acids (probes) are bound to a solid surface and are used to measure the relative concentration of nucleic acid sequences in a mixture via hybridization and subsequent detection of the hybridization events.

DNA sequencing Method for determining nucleotide sequences of DNA.

Domain The highest level of the classification of organisms. A domain is a monophyletic group of phyla. Three domains of life are currently recognized, two within prokaryotes (*Bacteria* and *Archaea*) and *Eukarya* grouping all eukaryotic organisms. They have been initially discovered through comparisons of ribosomal RNA sequences across all organisms. This term is also used in the context of protein sequences. A protein domain is a generally contiguous section of a protein chain that folds in a precise 3-dimensional structure and evolves independently from the rest of the protein. Many proteins consist of several domains. One domain may appear in a variety of different proteins.

Domain of attachment Region of a protein having the function of attachment to a substrate or to another protein.

Dormancy Cellular resting period during which the cell greatly reduces its metabolic activity to only maintain the minimum functions necessary for its survival.

Dystrophic crisis Physicochemical and biological disturbance and imbalance of an aquatic environment resulting from a massive influx of nutrients causing anoxic waters and their invasion by reduced compounds, including sulfides. It follows an intense eutrophication.

eratio Ratio between exported production and total primary production.

Ecological niche A set of resources and physicochemical conditions suitable for a specific biological species. This term can help to describe the way of life of a species.

Ecological succession Process by which, in a given habitat, population evolves naturally from a pioneer stage (initial stage) to a final stage (climax), through intermediary stages. The climax could theoretically be maintained indefinitely.

Ecosystem engineers Ecosystem engineers are organisms that directly or indirectly modulate the availability of resources to other species, by causing physical state changes in biotic or abiotic materials. In so doing they modify, maintain, and create habitats. Autogenic engineers (e.g., corals or trees) change the environment via their own physical structures (i.e., their living and dead tissues). Allogenic engineers (e.g., earthworms, beavers) change the environment by transforming living or nonliving materials from one physical state to another, via mechanical or other means.

Ectosymbiosis See "Symbiosis."

Effluent Discharge of domestic urban or industrial wastewater.

ELISA assay (enzyme-linked immunosorbent assay) Immunoenzymatic test that uses an antibody specific for a desired antigen. Antibodies are coupled to an enzyme, which specific substrate allows the colorimetric detection of the antibody-antigen complex.

Emerging disease An infectious disease that has newly appeared in a population or that has been known for some time but is rapidly increasing in incidence or geographic range.

Endergonic Chemical reaction that requires a supply of energy.

Endogenous internal standard of biodegradation Molecule naturally occurring in the environment, very resistant ("recalcitrant") to biodegradation and which can be used as a reference to calculate the rate of biodegradation of other molecules in the environment.

Endogenous respiration A situation in which living organisms oxidize some of their own cellular mass instead of new organic matter they adsorb or absorb from their environment.

Endolithic Pertaining to an organism that colonizes a rock. Endoliths can inhabit fractures, cracks, and pore spaces in a rock. They can also corrode tunnels in the rock through the production of organic acids. These kinds of habitats protect the microorganisms particularly in extreme environments, such as hot or cold deserts where insolation and UV radiation are high. Elements in the surrounding minerals are useful nutrients for these microorganisms, for example, iron. Corrosion due to this microbial activity produces typical exfoliation, for instance, in sandstones and granites, as well as on monuments constructed in stone.

Endospore See "Spore."

Endosymbiosis See "Symbiosis."

Endotoxin See "Toxin."

Enteric Adj. affecting the intestines or being within the intestines.

Enterotoxin An exotoxin produced by microbes and causing gastrointestinal diseases.

Enzymes Proteins with biochemical activities; some are anchored into membranes or soluble, linked to cellular structures; others are soluble, free in the cytoplasm (not linked to cellular structures).

Epidemy Disease (infectious, in particular) that affects a growing proportion of the population in a more or less extensive area and over a given period.

Epigenetic regulation Nonheritable regulation mode, based on changes in DNA bases, typically methylation.

Epilimnion The epilimnion is the topmost layer in a thermally stratified lake, occurring above the thermocline and the deeper zone, the hypolimnion. It is warmer and typically has a higher pH and higher dissolved oxygen concentration than the hypolimnion.

Episome Circular extrachromosomal DNA molecule showing autonomous replication (such as plasmid). It codes for the enzymes needed for its transient integration in the chromosome through episomal recombination. The episome remains autonomous with respect to the chromosome.

Epizootic Epidemic in animals.

EPS See "Exopolysaccharides" or "Extracellular polymeric substances."

Euphotic layer Layer of a lake or a sea between the surface and the maximum depth from which light is still sufficient to support positively the photosynthesis. This maximum depth depends on the turbidity of the water (from a few meters or less in estuaries to about 200 m in highly transparent areas of the open ocean). The illumination of its lower limit is of the order of 1 % of the surface illumination. Other regions can be defined below the euphotic layer, such as the mesopelagic (or twilight zone, 200–1,000 m) and the bathypelagic (1,000–4,000 m) layers. Beyond, the ocean waters are defined as the abyss.

Eutrophic State of an environment rich in minerals (nitrogen, phosphorus, etc.), allowing the maintenance of a high primary production.

Eutrophication Enrichment of an ecosystem in organic and inorganic nutrients (minerals) resulting in a significant development of primary production. The intake of nutrients may be natural or anthropogenic. The anthropogenic eutrophication, if it is rapid and extensive, can lead to an imbalance in the functioning of the ecosystem, can favor the proliferation of a taxon or a functional group, and can lead to a dystrophic crisis.

Evolution All processes by which organisms change gradually their genotypic and phenotypic characteristics in order to adapt to habitats.

Evolutionary convergence Character (or function) present in two different lineages but which does not derive from an ancestral character (or function) present in a common ancestor shared by the two lineages. This implies that the character (or the function) appeared independently in the lineages and thus that the molecular bases of the character (or function) are not homologous.

Exergonic A chemical reaction that produces energy.

Exonuclease See "Nuclease."

Exopolysaccharides Polymers of high molecular weight built from sugar monomers that are excreted by the microorganisms into the environment. Because of their high chemical diversity, the exopolysaccharides have been used for many different applications in agro-alimentary and pharmaceutical and cosmetic industries.

Export production Fraction of net primary production that can be exported from one ecosystem to another ecosystem. In oceanic ecosystems, it is the fraction of the net primary production that leaves the euphotic zone and is exported to deeper layers in particulate as well as dissolved forms. This is a central process in marine biogeochemical cycling.

Extinction Local or total disappearance of a taxon.

Extracellular polymeric substances or exopolymeric substances (EPS) Polymeric substances that are excreted into the environment by the microorganisms forming a matrix into which the microbial cells are embedded. The EPS contain a large amount of polysaccharides, which are built from sugar monomers forming polymeric and often branched chains. In addition, EPS contain other polymers as, e.g., peptides. The EPS matrix may provide protection to cells against environmental stresses.

f-ratio In oceanic biogeochemistry, the f-ratio is calculated by dividing the new production by the total primary production. This fraction is significant because it is assumed to be directly related to the export flux of organic matter from the surface ocean to the deep seafloor. The ratio was originally defined by Richard Eppley and Bruce Peterson in one of the first papers estimating global oceanic production.

Family A group of related objects or organisms. In the taxonomic hierarchy, a family is a subdivision of an order which encompasses different genera. A gene family groups homologous sequences which share a common ancestor and whose protein products usually share a common function. Different members of a gene family can be found in a single genome.

Fermentation Biochemical reaction of conversion of the chemical energy of a carbon compound (e.g., glucose) in an energy form used by the cell. This reaction takes place in the cytoplasm in the absence of dioxygen, by electron transfer between an exogenous electron donor (organic substrate) and an endogenous electron acceptor (intracellular organic compound, metabolism product) without involvement of transmembrane electron carriers. ATP production is obtained by substrate level phosphorylation.

Filopodia Long and thin pseudopodia devoid of a central axial microtubule (as opposed to axopodia).

Fitness Refers to the ability of an organism to reproduce and transmit its genes. This parameter is quantified by measuring the number of descendants and therefore by the rate of reproduction. It is a selective advantage conferred by a genetic property or feature of life, which translates into the ability of individuals with a genotype giving offspring to the next generation.

Food chain The food chain is a succession of organisms in an ecosystem that constitutes a transfer of food energy from one organism to another, as each consumes a lower member of the food chain and in turn is preyed upon by a higher member.

Food web All trophic links in a given ecosystem, from primary producers to top predators.

Fresh organic matter New organic matter introduced to the soil before any biotic or abiotic transformation or degradation.

Fumaroles Phenomenon associated with volcanism which corresponds to gas or vapor escape.

Functional group Group of organisms that may or may not be phylogenetically related, whose phenotypes, such as metabolic capacities, are very close permitting to colonize one particular ecological niche. For example, nitrogen-fixing bacteria or photosynthetic bacteria share a common feature but are largely distributed across the phylogenetic spectrum. Some eukaryotes belonging to distinct groups may have a common function in the ecosystem (e.g., bioturbation activity in sediments).

Fungal infection Parasitic disease caused by a fungus.

Gametogenic In a reproductive cycle the gametogenic stage corresponds to the one which produces the gametes. Botanists traditionally call it the gametophyte.

Gamy Fusion of two gametes which results in the formation of a zygote. Normally gametes are haploid and zygotes diploid. Fusion of cytoplasm and of nuclei occurs generally simultaneously. When this is not the case, we distinguish plasmogamy (fusion of the cytoplasm) from karyogamy (fusion of the nuclei).

Gas permeation Technique used for the separation and purification of gases.

Gene Molecular unit of heredity; deoxyribonucleic acid (DNA) sequence that codes for a polypeptide (amino acid chain) or a functional ribonucleic acid (RNA); an *allele* is one of the alternative forms of a gene that differs through mutations.

Gene cassette Gene cassettes are non-replicative mobile elements that are integrated into a genetic system (plasmid or chromosomal) for horizontal transfer of genes, called integrons, which promotes the adaptation of bacteria to their environment. Cassettes generally consist of a coding sequence (e.g., antibiotic resistance gene) and a recombination site (attC). They are integrated or excised in integrons through a system of site-specific recombination that involves integrase (tyrosine recombinase). Integrons constitute a unique system within the bacterial world, for capturing genes and thus effectively promoting the dissemination of certain genes such as genes for antibiotic resistance or degradation of contaminants.

Genetic diversity Correspond to variability of gene in a species or a population; it is characterized by quantification of difference between individual pair.

Genetically modified organism (GMO) An organism is genetically modified when exogenous genetic material has been introduced within its genome using genetic engineering techniques.

Genome All the genes and intergenes that constitute the genetic setup of an organism.

Genomic hybridization Method aiming at quantifying the differences between two genomes. It is based on denaturation, labeling, and renaturation of the DNA strands of two microorganisms under study. The percentage of DNA renaturation is used to estimate the proportion of common sequences between the two genomes.

Genomic species or genomospecies Prokaryotic species defined only with molecular criteria and in particular with the sequencing of rDNA 16S.

Genomics Study of all of the genetic information contained in a cell or a virus.

Genotype increase Operation consisting in introducing in a contaminated site, a microbial vector containing, naturally or after genetic modification, one or more degradation genes brought on a medium with high transfer rate to facilitate dispersal, survival, and expression of these genes in indigenous microbial community.

Genus (plural: genera) Low-level taxonomic rank below the family and above the species. It is designated by a Latin or Latinized capitalized singular noun.

Geothermalism Relative to internal thermal phenomena on Earth.

Gnotobiotic Status of environmental samples or organisms devoid of their natural microbial community and inoculated with specific microorganism(s).

Gray waters Wastewater from domestic activities except effluents from the toilets.

Gross annual flux Flux of mass or energy from one reservoir to another reservoir.

Gross photosynthesis The energy assimilated by primary producers through the process of photosynthesis. Losses due to processes such as respiration are not taken into account.

Gross primary production Overall energy assimilated by primary producers' photosynthetic activity. Losses due to processes such as respiration are not taken into account.

Halocline Interface characterized by a strong vertical salinity gradient in aquatic areas, separating surface waters, less salted and thus less dense, from deep waters with higher salinity.

Halophilic Characteristic of an organism which cannot grow in the absence of salt (NaCl) in its growth medium.

Halotolerant Ability of an organism to grow in the presence of high quantity of salt (NaCl) without requiring it for growth.

Haustorium A derived, usually highly branched fungal hypha, which forms inside living cells of infected hosts (as diverse as plants or cyanobacteria). It is specialized in nutrient uptake.

Helotism Symbiotic interaction in which one species is dominant and commands members of another species to perform the functions required for their mutual survival.

Heterotrophs Organisms using organic compounds as carbon source to synthesize their cellular constituents. The heterotrophy opposes the notion of autotrophy.

Histones Basic protein associating with DNA in eukaryotes to form the basic structure of chromatin. Histones play an important role in packaging and folding of DNA.

Homeostasis Maintaining the functional balance by physiological (individual) or environmental (ecosystem) regulations that ensure the consistency of the essential conditions characterizing the internal environment of an organism or an ecosystem, despite changes in the physicochemical habitat.

Homologous gene Two genes are homologous if they both derive from the same ancestral gene. For example, two genes in two species which are derived from a gene that was present in the genome of an ancestral species which is common to these two species are homologous. Another situation: two genes of the same species resulting from the duplication of a single ancestral gene are homologous.

Hopanoids Pentacyclic lipids biosynthesized by bacteria; these molecules have a similar role to that of sterols in eukaryotic cells. They are present in the cytoplasmic membrane and are derived from the metabolism of terpenes.

Horizontal (or lateral) gene transfer Mechanism allowing the transfer of genetic material among cells independently of the reproduction process. The transmission of genetic material from the mother cell to the daughter cells during the reproduction process (or cell division) is called vertical gene transfer. Three mechanisms of horizontal gene transfer have been described: the natural transformation, the conjugation, and the transduction. More recently a new type of horizontal gene transfer involving virus-like particles encoded by the cells has been described in *Rhodobacterales* (*Alphaproteobacteria*).

Housekeeping gene Gene needed for basic functioning of the cell, as opposed to function gene that allows organisms to occupy various niches.

Humic substances Humic substances are a set of organic macromolecules aliphatic and aromatic in nature from the random and highly heterogeneous mixture of compounds resulting from the breakdown of fresh organic material, in particular of plant lignin. They are brown and more or less resistant to biodegradation based on their degree of humification and condensation. Indeed, acidic in nature, humic substances exhibit high molecular weights ranging from a few hundred to several thousand daltons. These are the most abundant organic compounds in nature and they are found in terrestrial and aquatic ecosystems, thus they are the most important part of the stabilized soil material. Humic substances are often divided into three different families, defined by experimental practices: fulvic acids (the most important part of humus which are insoluble in water regardless of the pH and have a moderate molecular weight), humic acids (soluble only in a basic medium; they precipitate at pH below 2, exhibit higher molecular weight, and contain more aromatic compounds than fulvic acids), and humins (insoluble in a basic medium). The evolution of humic substances can be outlined as follows: fulvic acids → humic acids → humins.

Hydraulic retention time (HRT) Length of time of the liquid phase in a reactor, which can be assimilated to the flow rate (to = volume/flow).

Hydrogenosome Organelle present in some anaerobic single-celled eukaryotes taxa, phylogenetically distant: ciliates, Parabasalia, Microsporidia, Stramenopiles, etc. Hydrogenosomes are about 1 μm high in diameter. They are surrounded by a double membrane; they do not have genome and derive probably from mitochondria by loss or reduction of genome. This transformation would be produced several times separately along the evolution. Hydrogenosomes oxidize pyruvate to molecular hydrogen (H_2), CO_2, and acetate and synthesize ATP which is exported in the cytoplasm.

Hydrothermal springs Hydrothermal springs may be terrestrial or marine. Regarding the marine ones (e.g., hydrothermal vents or smokers), they send out hot fluids and are generally located close to the mid-ocean ridges. Seawater enters through cracks within these ridges, warms up at depth during contact with the magma, and leaves as hot (about 350 °C) and pressurized water. This water contains both heavy metals and gases of geochemical origin (H_2S, CO, CO_2, H_2). In contact with cold water, many mineral compounds precipitate and form a chimney diffusing hot fluids which grows slowly, reaching up to 20 m in height.

Hydrothermalism Water circulation in the Earth's crust; during its transfer within hot areas, water can be converted into steam or can be thermalyzed into hydrogen and dioxygen.

Hyphae Term used to designate filamentous structures found in various eukaryotic and prokaryotic taxa, especially in fungi. In the latter, the vegetative apparatus, composed of a mass of filaments (hyphae), is named mycelium.

Hypolimnion The hypolimnion is the bottom layer of water in a thermally stratified lake. It is the zone that lies below the thermocline. This layer is generally cold and with low dissolved oxygen concentrations.

Ice accretion Ice formed by freezing water in contact with existing ice.

IEF ("isoelectric focusing") Technique for separating different proteins by differences in their isoelectric point.

Immigration Transfer of a taxon from its biotope to another one where it will develop.

Immunocompromised A person who has an immunodeficiency of any kind. Immunodeficiency (or immune deficiency) is a state in which the immune system's ability to fight infectious disease is compromised or entirely absent.

Incertae sedis ("whose seat is uncertain") Said of a taxon whose position in the classification is uncertain.

Index Number (dimensionless) to summarize and characterize the relative variation of a simple or complex variation between two situations, one of which serves as a base (biological condition, time or place of reference).

Individual Living organism or microorganism characterized by its biological integrity that allows it to fulfill its own functions. Among prokaryotes, individuals, or more precisely the strains or clones sharing very similar characteristics, are grouped within a same species. Cells can be considered as individuals and, within a species, strains can also be considered as individuals if the genetic distances among the cells are small.

Inductive effect Effect related to differences in electronegativity of atoms and groups of a molecule which leads to polarization of bonds and an electronic moving along bonds which may affect their stability.

Inputs All mineral and organic elements imported in an ecosystem. It is a flux of matter and energy coming into the ecosystem.

Insertion sequences Also known as insertion sequence (IS) elements, these are segments of DNA that can move from one position on a chromosome to a different position on the same chromosome or on a different chromosome.

Integron A mobile DNA system of capture and gene expression. It encodes an integrase that mediates integration, by site-specific recombination, of genes contained in structures called gene cassettes. The integron/gene cassette system seems to have a broad role in adaptation (antibiotic resistance, development of multiple resistance phenotypes, degradation of organic contaminants).

Interference competition A type of competition in which one partner interferes directly with the other, typically via the release of toxic compounds in the environment.

Isotopes Atoms of the same element with different atomic masses (e.g., carbon-12, carbon-13, and carbon-14), bearing the same number of protons but different numbers of neutrons. Some isotopes are stable (e.g., ^{13}C), while others are unstable and radioactive (e.g., ^{14}C), disintegrating over time in radiogenic isotopes (or radionuclides).

Isotopic fractionation Discrimination between the light and heavy isotopes of an element, occurring during (bio)chemical reactions and physicochemical processes.

Isotopic labeling Study of the isotopic composition (e.g., $^{13}C/^{12}C$; D/H) of a (micro)organism (or of its cellular components) induced by the uptake of a labeled substrate (enriched in stable isotopes).

Kerogen Kerogen is the fraction of organic matter in sedimentary rocks and ancient sediments which is insoluble in organic solvents and basic aqueous solvents.

Kingdom Taxonomic rank, which is either the highest rank or in the more recent three-domain system, the rank below domain. Kingdoms are divided into smaller groups called phyla (singular: phylum).

Link or sink Theory – also called trophic link or carbon sink – indicating that the organic matter is either consumed or breathed by the organisms living in the water column.

Lipopolysaccharide The lipopolysaccharides (or LPS) are essential components of the cell wall of Gram-negative bacteria. In Gram-positive bacteria the equivalent structures are called lipoglycans.

Log K_{ow} Parameter used to measure the differential solubility (partition coefficient) of chemical compounds in two solvents ("o" for octanol and "w" for water); it allows to assess the molecule hydrophobic or hydrophilic nature.

Lyse Disintegration (decomposition) of cellular membrane.

Lysogenic cycle (or lysogeny) One of the ways of virus reproduction (other one: lytic cycle) in which the phage's nucleic acid is integrated into the host cell chromosome, so that genetic information of the virus, known as provirus, is transmitted through daughter cells; the host cell is not killed until the lytic cycle is activated.

Lysogenic infection Refer to *lysogenic cycle*.

Lytic cycle One of the ways of virus reproduction (other one: lysogenic cycle), which is usually considered as the main method of viral reproduction because it ends in the lysis of the infected cell releasing the progeny viruses that will in turn spread and infect other cells. Some viruses, though, can leave the infected cell not through lysis but rather by budding off from the cell taking a portion of the membrane with them. The latter is often known as the chronic cycle.

Macrocosm Set of organisms and substrates belonging to a wider structure, experimentally controlled, that can provide a representative image of the ecosystem.

Maximum likelihood Approach for reconstructing the phylogeny of a set of organisms based on the creation of trees that minimizes errors.

Mesophilic Relative to a microorganism with optimal growth temperature ranging from 20 to 40 °C.

Mesotrophic Refers to an environment moderately rich in minerals (intermediary between oligotrophic and eutrophic), at the origin of a moderate primary production.

Metabolism Whole biochemical processes occurring within a living cell or organism.

Metabolite Any substance produced by metabolism within an organism.

Metabolomics Study of all metabolites present in an organism.

Metagenomics The term metagenomics or environmental genomics is the study of metagenomes that represent the whole genome communities sampled in the environment; these communities are largely composed of noncultivable organisms. The method used is to study genes collectively, without detailing individual by individual.

Methanogen Microorganisms that produce methane as the major product of their energy-generating metabolism.

Methylase An enzyme that methylates DNA, either to protect it from restriction endonucleases or to mark the reference strand to help the cell checking the integrity of replication.

Microaerobiosis Capability of an organism to grow in a medium that is little exposed to air so presenting few quantity of dissolved dioxygen.

Microaerophilic Organism which optimal growth is in a medium presenting dissolved dioxygen pressure much lower than atmospheric one.

Microbial loop The microbial loop is a trophic pathway in the aquatic food web where dissolved organic carbon is returned to higher trophic levels via its incorporation into the bacterial biomass. The microbial loop is coupled with the classic food chain based on the trophic interactions between phytoplankton, zooplankton, and nekton.

Microbial mat A stratified microbial community comprising different layers of photosynthetic and non-photosynthetic microorganisms, which are embedded in an EPS matrix excreted by the mat-building organisms often cyanobacteria. Microbial mats can be considered as a specific case of biofilms which are particularly thick and comprise many different species. The thickness of the microbial mats ranges from one to several mm thickness; these systems develop at the surface of shallow sediments often in marine or lagoon environments and represent the origin of stromatolites.

Microbiome Area of life of the microbiota.

Microbiota Set of microorganisms, culturable or not, associated with an individual or an ecosystem. Term proposed to replace "microflora".

Microcosm Small enclosure designed to reproduce a biotope to control certain parameters and thus enabling the study of chemical or biological processes.

Mineralization In biology, mineralization is the process by which organic matter is converted in inorganic (mineral) compounds.

Miocene A geological span of time dating from 24 to 5 million years (My). It is the first epoch of the Neogene period (28–3.6 My), the latter being the second period of the Cenozoic era (72–0 My).

Mitosome Organelle present in some anaerobic eukaryotic cells that is evolutionarily related to mitochondria but that does not produce ATP or hydrogen and appears to lack most other metabolic pathways associated with mitochondria. Mitosomes have no genomic DNA.

Mixotrophic A mixotrophic organism is an organism that can use different sources of energy and carbon for sustaining its growth and maintenance.

MLST ("multilocus sequence typing") An approach that allows a detailed and stable characterization of microbial strains.

Molecular marker Conserved gene or genomic region, including variable zones, that can be used to classify an individual among taxonomic groups and estimate and assess their relationship.

Monoculture Production process of a single biological variety.

Monoxenic See "Gnotobiotic."

Mull A mull is characteristic of hardwood forests, deciduous forests, or grasslands in warm, humid climates. The porous, crumbly humus rapidly decomposes and becomes well mixed into the mineral soil, so that distinct layers are not apparent (absence of litter layer). Bacteria, earthworms, and larger insects are abundant.

Mutagens Factors or products inducing mutations in DNA. These include ionizing radiations and various chemical compounds such as polyaromatic hydrocarbons.

Mutation Alteration of genetic material (DNA or RNA) of a cell or virus that causes a lasting change in some characters because of the inheritance of this genetic material from generation to generation. The mechanisms are of several types (substitution, deletion, or insertion of one or more nucleic bases) and are triggered either spontaneously during cell division or under the influence of external agents called mutagens.

Mutualism Interaction (obligatory) between two or more organisms where each partner takes advantage of the interaction.

Mycelium Vegetative part of the fungi and the *Actinobacteria*, consisting of a set of filaments; these filaments are traditionally referred to as hyphae.

***N*-acyl homoserine lactone** Small lipidic metabolites composed of a homoserine lactone ring and an acyl chain and used as quorum-sensing signals by certain bacteria.

Natural attenuation Degradation that occurs in the natural environment without human intervention. Biodegradation by microorganisms in the environment plays a major role in the natural attenuation. This approach is preferred when the pollution does not pose an immediate threat and will not present during the time required for the removal of pollutants.

Necrotroph A microorganism feeding only on dead organic tissue.

Net flux Balance of the flux of mass or energy between two reservoirs.

Net photosynthesis The quantity of energy stored as biomass by primary producers. It corresponds to the energy assimilated by primary producers through the process of photosynthesis minus the energy used for respiration of these producers. Net community production takes into account respiration of primary producers and those of heterotrophic.

Net primary production Refers to the amount of energy stored as biomass by primary producers. It corresponds to the energy accumulated by the process of photosynthesis minus the energy lost by respiration.

Neutrophil Organism's ability to grow in a neutral medium, i.e., balanced in proton and hydroxide ions.

New primary production In aquatic environments, this notion is used to define the primary production fraction sustained by assimilable nitrogen supplied to the considered area.

Nosocomial infections Hospital-acquired infections that can be defined as those occurring within 48 h of hospital admission, 3 days of discharge, or 30 days of an operation.

Nuclease Enzyme able to cleave the phosphodiester bonds between nucleotides of nucleic acids (DNA or RNA). Nucleases that cleave the phosphodiester bond at either $5'$ or $3'$ end of nucleic acids are called exonucleases, whereas endonucleases cleave the phosphodiester bonds with nucleic acid molecules.

Nucleomorph An "additional" nucleus of a reduced size compared to the "main" nucleus. The nucleomorph results from a secondary or tertiary endosymbiosis where a non-photosynthetic organism has engulfed a eukaryotic photosynthetic alga, retaining its chloroplast and its nucleus. In the course of evolution, the algal nucleus can be lost following transfer of different genes to the "host" main nucleus. In several taxa (mainly the Cryptophyta and Chlorarachniophyta) degeneration is incomplete and a so-called nucleomorph remains associated to the chloroplast.

Offshore area Opposed to the coastal zone, the offshore area has no interaction with the continents. There is no real precise definition, an operational definition sets the limit between coastal areas and offshore at a depth of 200 m.

Oligotrophic Refers to a nutrient-deficient aquatic environment offering little nutrients to sustain life and characterized by a very low primary production.

Operon DNA sequence containing one or a cluster of genes under the control of a single regulatory signal.

Opportunistic pathogen Microorganism usually non-pathogenic, able to cause disease on an immuno-suppressed person, but not on healthy patient.

Order A taxonomic rank used to classify organisms composed of families sharing a set of similar characteristics.

Organelle A term in cellular biology that designates the structures contained in the cytoplasm that are delimited by a membrane and that have specific functions.

Orthologous gene Two homologous genes are termed orthologous if they have retained the same function.

Osmoregulation Regulation of the intracellular concentration of osmolytes during changes of the ionic concentration outside the cell, by accumulation or release of organic compounds (glycine betaine, ectoine, etc.) or minerals (potassium, etc.) for achieving an osmotic pressure balance between the internal and the external part of cell. This balance is achieved either through assimilation via synthesis of membrane transporters or by biosynthesis of osmoregulatory organic compounds.

Osmotrophy Relative to an osmotrophic organism able to assimilate dissolved nutrients (ions, dissolved organic matter) by diffusion through cell membrane and to use them as a significant food source.

Outgroup An outgroup is defined over a set of species of interest, the internal group. An outgroup is then a set of species that diverged from the common ancestor of all the compared species before that species of the internal group diverged. As an example, relative to *Rhizobiaceae*, any other bacterial family such as *Enterobacteriaceae* or *Pseudomonadaceae* is an outgroup. An outgroup is often used to anchor the phylogeny of species of interest because the root of the whole tree (internal and external) is necessarily placed outside the inner group.

Oxic Concerns environments in which the dioxygen is present. The term "aerobic" is sometimes used to refer environmental conditions (oxic environments) that are favorable for the development of aerobic organisms.

Oxidative phosphorylation Phosphorylation of ADP to ATP following electron transfer via a membrane respiratory transporter chain.

P450 cytochrome Family of cytochromes (hemoproteins) having monooxygenase enzymatic activity which maximal absorption, in the case of an in vitro reduction under carbon monoxide atmosphere, is observed at 450 nm. Cytochromes are catalyzing the insertion of oxygen in different substrates, among which hydrocarbons: the organic substrate (e.g., n-alkane) is hydroxylated (leading to the formation of corresponding alkanol in the case of n-alkanes), the second atom of oxygen being constitutive of a water molecule.

Pandemic An epidemic that affects a large proportion of the world population on a very large geographic area.

Paradigm Model permitting to explain a set of phenomena, based on a set of consistent scientific hypotheses and constituting a worldview.

Paralogous gene Various molecular phenomena can lead to the duplication of a gene within a genome: Initially, a genome contains two identical copies of the same gene. Most frequently, one of the two duplicated copies eventually disappears; however, when this is not the case, two distinct genes, similar in sequence, are found in the genomes. Such genes are called paralogs.

Paraprimary producer Refers to all prokaryotes that produce biomass and serve as food for bacterivorous predators (usually protists); these prokaryotes are the first link in a detrital food chain based on recycling dissolved organic or inorganic detritus. They are the equivalent of primary producers in a classic food chain. They include chemoorganotrophic bacteria and archaea (called decomposers) that consume dissolved organic matter, anoxygenic phototrophic bacteria (purple and green bacteria) and chemolithotrophic prokaryotes, which use electron donors whose origin is biological from the biodegradation of organic matter or excretory products of prokaryotic metabolic pathways (anaerobic respirations, fermentations).

Parasitism Biological interaction with a negative effect on one partner (the host) and a positive effect on the other partner (the parasite), the latter being physically associated with its host. Relationship between two species is designated as parasitism when one species favors a temporal variation to another species while the latter negatively affects the temporal variation of the former. In addition, parasitism implies the use of body or energy from the host for the growth and development of the parasite, resulting or not in death of the host.

Particulate organic matter (POM) See definition of dissolved organic matter.

Pathogen Microorganism that causes disease in its infected host.

Pathogenicity island A set of genes involved in virulence and harbored on a DNA region that can be acquired through a horizontal gene transfer event. This genomic island can encode an integrase favoring its recombination at specific DNA sites which are often located in tRNA genes.

Periphyton Biological communities colonizing surfaces of abiotic or biotic supports (e.g., plant or animal surfaces living in aquatic environments).

Permafrost The permafrost corresponds to a frozen soil for which the temperature is permanently at 0 °C or colder.

Permease The permease is membrane protein or protein complex that facilitates diffusion of a specific molecule across membrane.

Peroxisome Organelles found in all eukaryotic cells. They measure between 0.2 and 1.7 µm in diameter, are surrounded by a single membrane, and do not contain their own genetic material. They have been proposed to have an endosymbiotic origin (as mitochondria and chloroplasts), but they are most likely of autogenous origin by differentiation from the endoplasmic reticulum. Peroxisomes contain three enzymes (D-amino acid oxidase, urate oxidase, catalase) involved in the degradation of hydrogen peroxide (produced by the mitochondria and highly toxic), the oxidation of long-chain fatty acids (produced by phospholipid degradation in the lysosomes), fatty acid synthesis, and other functions.

PGPR (plant growth-promoting rhizobacteria) Rhizosphere bacteria that stimulate growth of the plant. PGPR can have a direct effect on the plant and/or indirect phytostimulatory effects when it inhibits pests.

Phage A virus capable of infecting a bacterial cell or bacteriophage and may cause lysis to its host cell.

Phagotrophy Nutritional mode of a microorganism based on the ingestion and intracellular digestion of particulate organic matter and its use as a source of energy and carbon.

Phase change A process of reversible, high-frequency phenotypic switching (i.e., mobility, antigenicity, or virulence) that is mediated by homologous recombination, insertion, and deletion or epigenetic regulation.

Phospholipid fatty acid (or PLFA) Aliphatic compound constituting the nonpolar (hydrophobic) part of a phospholipid molecule occurring in bacterial cytoplasmic membranes and walls. In intact phospholipids, a PLFA is linked via an ester group to a glycerophosphatidic polar head group (hydrophilic part).

Photolithotrophs Phototrophic microorganisms using reduced minerals (H_2O, H_2S, H_2) as electron donors for CO_2 reduction.

Photoorganotrophs Phototrophic microorganisms using organic compounds as carbon sources and electron donors.

Photophosphorylation Phosphorylation of ADP to ATP related to the transport of electrons produced during the use of light as an energy source.

Photosynthetic (or phototrophic) organisms Organisms able to convert light energy, normally from the sun, into chemical energy that can be used to fuel the organisms activities and to synthesize their cellular constituents.

Phototrophs Organisms obtaining the energy necessary for all cell activities (synthesis of cell constituents, motility, chemotaxis, etc.) through light.

Phototrophy Ability to use light as energy source.

Phylogenetic group A group of organisms having a recent common ancestor.

Phylogeny/phylogenetics Study of the evolutionary relationships among organisms or lineages. These relationships are represented by a phylogeny. Molecular phylogenetics: Study of the evolutionary relationships among molecules carrying genetic information (DNA, RNA, and proteins).

Phylum Wide monophyletic group of taxa. In *Archaea* and *Bacteria*, the phylum constitutes the highest hierarchical level. In eukaryotes, higher hierarchical levels can be found: subkingdom and kingdom. For example, the Nematoda phylum and the Echinodermata phylum belong to the subkingdom of metazoans and to the kingdom of opisthokonts.

Phylum In systematics, it is a set of classes with similar characteristics.

Phytoplankton All planktonic photosynthetic organisms, which can be unicellular (e.g., diatoms, cyanobacteria) or multicellular (e.g., Phaeophyceae of the genus *Sargassum* and cyanobacteria of the genus *Trichodesmium*). Phytoplankton is at the basis of the food webs in aquatic ecosystems.

Piezophile Current terminology for pressure-adapted microorganisms either as piezotolerant (similar growth rate at atmospheric pressure and high pressure), piezophilic (more rapid growth at high pressure than atmospheric pressure), or hyperpiezophilic (growth only at high pressure), with pressure maxima increasing in rank order (highest for hyperpiezophiles). Organisms that grow best at atmospheric pressure, with little to no growth at increased pressure, are termed piezosensitive. "Piezo" (from the Greek meaning pressure) is now currently used instead of "baro" (from the Greek meaning weight).

Pioneer species Species able to colonize a bare ecosystem and able to initiate an ecological succession. They are the first organisms to start the chain of events leading to a sustainable ecosystem.

Pioneer stage Initial stage in an ecological succession. The start of a succession usually takes place either when a new habitat becomes available (temporary pond, new island, lava flow) or following a natural or anthropogenic disturbance which "returns" a succession to its initial stage (e.g., wild fires).

Plankton The plankton is the collection of microscopic organisms, including bacteria, archaea, algae, protozoans, and small zooplankton that float in the water column of aquatic ecosystems and are incapable of swimming against a current.

Plasmid Nonessential extrachromosomal DNA elements (circular or linear).

Plate tectonics Process by which continental drift can be explained. The Earth's crust is divided into rigid plates that slide over the mantle, transported by convection within the mantle. When they diverge, the mantle rises up to the surface to become newly formed crust, as at mid-ocean ridges. When they collide with each other and if they transport continents, the resulting continental collision gives rise to mountain ranges, such as the Alps in the collision between the African and the European plates. Plates may also slide past each other. If the plate boundaries occur on land, this results in a zone of frequent earthquakes, such as the San Andreas fault in California, between the Pacific and the North American plates.

Pleiotropic Non-limiting nutrient conditions (as opposed to oligotrophic).

Pleiotropy Describes the multiple effects (e.g., phenotypic traits) of a single entity (gene, organism).

Polymerase Enzyme that catalyzes the synthesis of nucleic acids on preexisting nucleic acid templates (DNA polymerase or RNA polymerase).

Population The population is a collection of individuals of the same species.

Population composition List and proportion of the different physiological or cellular states present in a population.

Posttranslational regulation Regulation which is carried out by a chemical modification (activation) of proteins already present.

Predation Interaction in which the predator captures and consumes a prey.

Prevalence Percentage of population interacting with other organisms (parasite, pathogen, etc.).

Prevalence of a disease Number of cases (diseases, infections) in a given population.

Primary producer Organisms that produce biomass from inorganic compounds (autotrophs). In almost all cases, these are organisms performing oxygenic or anoxygenic photosynthesis (plants, cyanobacteria, and a number of other unicellular organisms). There are prokaryotes that produce biomass from the oxidation of inorganic chemical compounds such as NO_2^-, NH_4^+, H_2S, S^0, H_2, and Fe^{++} (chemoautotrophs).

Primary production Organic matter production from atmospheric or aquatic mineral carbon through photosynthesis (light energy). Organic matter production from mineral carbon by chemosynthesis (chemical energy) is often considered as primary production; it should rather be considered as a paraprimary production (referred to as paraprimary producer). The organic matter production through photosynthesis is processed by oxygenic phototrophic organisms (by CO_2 reduction with H_2O as electron donor) or by anoxygenic phototrophic bacteria able to reduce CO_2 with reduced mineral compounds produced by geochemical processes. Life on Earth directly or indirectly depends on primary production.

Probiotics In 2001, WHO and FAO gave the official definition of probiotics as "live microorganisms which, when administered in adequate amounts, confer a health benefit on the host." Etymologically, the term associates the Latin preposition *pro* ("for") and the Greek adjective βιωτικός (biotic), derived from βίος (bios, "life").

Promoter Promoter is a region of DNA from which initiation of transcription of a particular gene occurs. Promoters are located upstream and on the same strand as the genes they help transcribe.

Proteomics Proteomics is the large-scale study of protein set of an organism or a tissue. It serves to monitor the protein content during physiological changes.

Proterozoic The Proterozoic eon dating from 2.5 billion years to 542 million years is the third eon of the Precambrian supereon (4.56 billion years to 542 million years). It is during this eon that multicellular life appeared. The end of the Proterozoic coincides with the beginning of the Cambrian, the age of invertebrate organisms with hard carapaces.

Protists Set of polyphyletic eukaryotic microorganisms which includes autotrophic, heterotrophic, and mixotrophic microorganisms.

Protozoa Protozoa constitute a group of unicellular eukaryotic microorganisms which are mostly motile and heterotrophic.

Psychrophilic An organism that can live and grow at low temperatures (below 20 °C).

Pycnocline Area of the ocean (or lake) where the density of water changes rapidly with depth in relation to gradients of salinity (halocline) and/or of temperature (thermocline). Pycnocline acts as a density boundary that limits exchanges between the water surface layers (less dense) and deep layers (denser).

Pyrosequencing Next-generation sequencing (NGS) that is based on the detection of released pyrophosphate (PPi) during DNA synthesis.

Quorum sensing A cell-cell communication process based on production and recognition of small molecular signals and used to coordinate gene expression according to cell density and environmental conditions.

Recombination Genome modification involving two alleles or two similar DNA fragments carried out by specific enzymes, the recombinases.

Reducing power Capacity of a chemical compound to achieve reductions, i.e., to provide electrons, by extension, all the reducing equivalents necessary for the operation of a metabolic pathway.

Reductive dissimilation Reduction of a chemical compound, mainly minerals (NO_3^-, SO_4^{2-}, etc.), more rarely organic (fumarate reduced to succinate), that serves as terminal electron acceptor. In most of the cases, this compound is the terminal electron acceptor of a respiratory chain. The resulting reduced compound is released to the environment and not used for cell material synthesis.

Reemerging disease Disease that has resurged in incidence, after having been brought under control through effective health-care policy and improved living conditions. Examples are malaria, diphtheria, tuberculosis, and dengue.

Regenerated primary production In aquatic environments, this notion is used to define the primary production fraction sustained by assimilable nitrogen produced by regeneration within the considered area.

Regulon Regulatory network linking a collection of genes or operons for a coordinated physiological response to a modification of the environment.

Repressor Regulatory protein that prevents the transcription of genes by binding to the operator region and blocking the fixation of RNA polymerase to the promoter.

Resilience Characterizes the ability and the degree of restoration of an ecosystem, a community, or a population following a disturbance. It corresponds to the maximum amplitude of the disturbance for which the biological system is able to return to its historical average or reach a new state of equilibrium.

Respiration The metabolic process by which an organism obtains energy by reacting an electron acceptor (for instance, oxygen in aerobic respiration or nitrate and sulfate in anaerobic respirations) with organic compounds or reduced minerals (electron donor) to give energy (ATP) by oxidative phosphorylation.

Retrovirus A virus with an RNA genome (RNA virus) that replicates its genome by reverse transcription of its RNA into DNA. Some retroviruses are integrated and duplicated in a host cell while others, called endogenous retroviruses, are integrated into the genome of the host and inherited across generations.

Reverse flow of electrons Electrons move normally from a low redox potential compound to a higher redox potential compound; this electronic movement is accompanied by a loss of energy. In a reverse flow of electrons, the electron flow is reversed and requires an input of energy (movement of electrons against a thermodynamic gradient).

Reverse pathway of acetyl-CoA During acetogenesis, CO_2 is reduced by taking the acetate Wood pathway. The reverse reaction leading to the release of CO_2 following oxidation of acetate is called reverse pathway of acetyl-CoA.

Rhizobacteria Bacteria of the root zone of a plant; bacteria living in the area corresponding to the soil surrounding the roots of a plant and in contact with them.

Rhodamine Red fluorescent dye that can be used to trace the path of the water or of a water body.

RISA (ribosomal intergenic spacer analysis) RISA is a method to characterize a microbial community without requiring microbial culture. RISA involves PCR amplification of a hypervariable region of the rRNA gene operon located between genes for the small (16S) and large (23S) rRNAs called the intergenic spacer region.

Sampling Statistical technique to representatively sample biota (soil, air, water, etc.).

Saprophyte Organism that feeds on dead or decaying organic matter; synonym of saprotroph.

Schiff base Also called azomethine, is a functional group, e.g., bacteriorhodopsin, which contains a nitrogen-carbon double bond with the nitrogen atom connected to an alkyl group, called to honor the German chemist Hugo Schiff (1834–1915).

SDS-PAGE (sodium dodecyl sulfate polyacrylamide gel electrophoresis) A widely used biochemical technique to separate proteins according to their electrophoretic mobility, which is a function of the length and thus the charge of a polypeptide chain.

Secondary producer Organisms metabolizing exclusively organic energy sources (heterotrophs).

Selection The process by which some organisms survive and leave descendants while others are eliminated.

Semantide Term proposed by Linus Pauling and Emile Zuckerkandl in 1965 to design molecules carrying information that can be interpreted in terms of evolution. Genes correspond to primary semantides, messenger RNA to secondary semantides, and polypeptides to tertiary semantides.

Sepsis A serious and generalized infection of an animal body by a pathogenic microorganism which propagates around the site of infection and through the bloodstream in the case of generalized infection.

Serogroup Within a species, a group of closely microorganisms having the same serotype (serological group), that is to say, the same set of antigens characterized by serotyping.

Sewage Urban or industrial wastewater, mostly assimilated to liquid effluents contaminated by feces.

Sex pilus A hair-like extracellular protein complex involved in cell attachment and DNA transfer during conjugation.

Shotgun Cloning or sequencing technique used for sequencing long strands of DNA. The first step consists in broking randomly the DNA strand into short pieces. The fragmentation step is repeated several times leading to multiple overlapping fragments. The resulting DNA fragments are sequenced and assembled, allowing reconstituting the whole DNA strand.

Siderophore Membrane molecule that is used to chelate the iron present at low concentrations in the external environment and participate to its transport through plasma membrane.

Sigma (factor) RNA polymerase subunit that recognizes specifically the promoter region of a gene ensuring the initiation of the transcription; anti-sigma factors are proteins that interact with the sigma factor and prevent the initiation of transcription.

Signal A signal is a physicochemical modification; signaling concerns all mechanisms by which this signal is transmitted from sensor proteins to other proteins in order to modify cellular metabolism.

Signal transduction Mechanism by which a physicochemical modification of the environment is perceived by a sensor protein, which in turn modifies proteins of the overall regulatory network of the cell, in order to alter the cellular metabolism.

Skin appendages Skin-associated structures such as hair and nails that serve a particular function including sensation, contractility, lubrication, and heat loss.

Soil respiration Soil respiration refers to the production of carbon dioxide at the soil surface. This includes respiration of plant roots, the rhizosphere, microbes, and fauna.

Solfatara Crater of an ancient volcano in which fumaroles made of stream water (100–300 °C) may escape from the soil vents. These fumaroles contain hydrogen sulfide whose oxidation in the presence of dioxygen in the atmosphere leads to sulfur deposits.

Species Set of organisms which are present in different populations and can exchange genetic material. Within prokaryotes, the reproduction is essentially clonal; the species is defined by a set of individuals which share close characteristics and have a threshold of genetic similarity. In eukaryotes, the species is a group of individuals sharing close characteristics and defined by a fertile reproductive capacity.

Spore In the prokaryotes, a spore is a form of resistance to desiccation and heat which can participate in dissemination. It has a thick cell wall and its metabolic activity is reduced or even stopped. Its water content is as low as

10–20 % (compared to 80 % for an active vegetative cell) and accounts for thermotolerance. When environmental conditions become favorable, it "germinates" to produce an active cell. We can distinguish endospores produced within a mother cell (as in *Clostridium* or *Bacillus*) from exospores produced outside the mother cells in the form of sporangia or clusters of spores (e.g., *Frankia*, *Streptomyces*). In eukaryotes, the word spore has a broader meaning; it generally designates a unicellular or multicellular structure, usually in a physiologically resting stage, which participates to the dissemination of the species. We can distinguish vegetative spores, like conidia in the Fungi, which participate to the clonal dissemination of an individual from meiospores which contain haploid nuclei produced following meiosis.

Sporocarp The fungal sporocarp (also called carpophore or "fruiting body") is formed by agglomerated hyphal filaments, a pseudo-tissue called plectenchyma. The popular names for sporocarp are mushroom or toadstool. Sporocarps are reproductive organs where meiosis takes place followed by the formation of meiospores. Numerous Basidiomycota species form sporocarps.

SSCP (single-strand conformational polymorphism) Experimental approach for the detection of conformational polymorphism in single-stranded DNA. After denaturation of double-stranded DNA into single strands, intra-strand secondary structures form that result in different conformations that permit to separate two closely related DNA fragments. This technique is similar to DGGE or TGGE.

Steric effect Effect related to the presence on a molecule of high bulky substituents that prevent or hinder the approach of a reagent or a reaction site such as the active site of an enzyme.

Strain Pure cultures of microorganisms obtained from a single isolate.

Stranded Nucleic acids can be single or double stranded if they are made up of a unique or of two complementary strands.

Strategist of type K A species is considered to be a K strategist when it is characterized by slow growth, a strong allocation of energy in maintenance and defense of individuals, a low allocation of energy to reproduction, and a high life expectancy and generally large size; the term refers to the biological carrying capacity of a biotope.

Strategist of type r A species is considered to be an r strategist when it is characterized by rapid growth, a low allocation of energy in maintenance and defense of individuals, a high allocation of energy to reproduction, a short life expectancy, and a generally reduced size; the term refers to the maximum rate of growth of the population.

Stromatolite Stromatolites are finely laminated, lithified microbial mats of photosynthetic origin (mostly cyanobacteria), generally exhibiting a vertical structure (although flat-lying, lithified mats are referred to as tabular stromatolites). The structures grow vertically through trapping of detrital particles and/or by the in situ precipitation of carbonates within the microbial mats.

Substrate-level phosphorylation Phosphorylation of ADP to ATP directly coupled to enzymatic oxidation of substrate; it takes place in the cytoplasm.

Surfactant A surfactant is a compound which modifies the surface tension between two surfaces. Surfactant compounds are amphiphilic molecules, that is to say, they have two parts of different polarity, one apolar, lipophilic and the other polar, hydrophilic.

Symbiosis Sustainable interaction between two or more species. The species involved are referred to as symbionts. Currently, this interaction is most often considered, especially in microbial ecology, as being mutualistic, that is to say, with mutual benefits. This is clarified by the use of the term mutualistic symbiosis. Nevertheless, the word symbiosis can still be used in its historical sense that includes mutualism and parasitism, but this can lead to confusion. When only a mutually beneficial interaction is not required, the term associative symbiosis is preferentially used. The term *ectosymbiosis* is used in the case where the two organisms are outside one another; in the case of *endosymbiosis*, the smallest organism lives in cells or intercellular spaces in the larger host organism.

Synapse In the Rhodophyceae it designates a form of communication between two cells formed by a relatively large perforation in the transversal cell wall partially filled by a plug. A unique continuous plasmalemma surrounds the two contiguous cells. In the Metazoa, a synapse is a functional contact zone between either two neurons or a neuron and another cell (muscular, sensory cell, etc.).

Syncytial An organism is said to be syncytial or coenocytic when its vegetative apparatus is constituted of a single cytoplasmic mass encompassing numerous nuclei. Examples are *Caulerpa* (Plantae) or the oomycetes (Stramenopiles).

Syncytium Mass of cytoplasm containing a wide number of nuclei.

Synteny Gene order conservation between different loci or organisms.

Syntrophic Term used to describe organisms which are associated or mutually dependent upon one another with reference to food supply.

Systemic infection Infection caused by pathogenic organisms that are widespread throughout the body. An infection that is in the bloodstream is called a systemic infection.

Taxon (pl. taxa) A taxon is an element of a systematic classification considered without prejudice of rank it occupies. A taxon is a group of one (or more) populations of organism(s), which a taxonomist adjudges to be a unit; thus, it can correspond to a strain, variety, species, genus, family, order, class, phylum, etc. For example, genus *Anabaena* (cyanobacteria) and *Cyanobacteria* are taxa. A taxon can be also constituted of groups of strains with close phenotypic and genotypic characteristics.

TGGE (temperature gradient gel electrophoresis) DNA fingerprint method that discriminates DNA fragments of same length through their base composition by decreased electrophoretic mobility of partially melted double-stranded DNA molecules in polyacrylamide gels containing a linear temperature gradient.

Thermocline In an aquatic environment, the thermocline is the zone in which temperature changes more rapidly with depth than it does in the layers above (often warmer) or below (often colder and denser). The thermocline may be seasonal (summer) in temperate environments or permanent in tropical environments.

Thermophile Organisms that grow only at high temperature. It is unable (or with difficulty) to develop at temperatures below 40 °C.

Thylakoid A thylakoid is an organelle found within chloroplasts and cyanobacteria, which contains chlorophyll and other pigments and is involved in photosynthesis.

Toxin Toxic substance produced by a living microorganism to which it gives its pathogenicity. It can be produced outside the cell or be excreted (exotoxin). Some toxins are produced inside the cell and are liberated when cell is lysed (endotoxin).

Transcript A fragment of RNA (messenger RNA) generated, with the help of an RNA polymerase, from the transcription of a gene and that will eventually be translated into a protein.

Transcription Mechanism by which the genetic information encoded in DNA is transformed into messenger RNA and intermediary in the ultimate transformation of this genetic information into a protein.

Transcriptional regulation Regulation that occurs at the level of gene transcription into messenger RNAs. This control is activated by a small protein that interacts with a sequence upstream of a gene and activates its transcription except when physicochemical conditions (e.g., in the presence of a metabolite) change its conformation. It is then termed an activator. The control is repressed by a repressor, a small protein that interacts with a sequence upstream of a gene and prevents its transcription except when physicochemical conditions (a metabolite) change its conformation.

Transcriptomics The study of the ensemble of transcripts (messenger RNA) during a physiological process in an organism.

Transduction Horizontal DNA transfer and genomic modification made by bacteriophage.

Transformation Genetic alteration of a microbial strain resulting from the integration of exogenous genetic material (such as plasmid DNA).

Translation Process in which a messenger RNA is decoded by the ribosomes to produce proteins.

Translation regulation Regulation which is carried out at the level of translation of the messenger RNA already synthesized.

Transposon DNA sequence that can change its position within a genome, often creating mutations. Transposition functions are encoded by the transposon itself.

T-RFLP Terminal restriction fragment length polymorphism. This technique allows the characterization of microbial communities by the amplification of a conserved genome region followed by a restriction analysis that produces fragments of different lengths at the extremity of the amplified region according to the specific position of the restriction sites in the different species within the community.

Trophic link Transfer of organic matter between two organisms in a food web, for example, between predator and its prey, or between two organisms that exchange metabolites, for example, fermentative bacteria degrading macromolecules and producing substrates for anaerobic bacteria such as denitrifiers, sulfate reducers, and methanogens.

Two-component system Adaptive response of a microorganism to variations in environmental parameters. It consists of a sensor, typically a transmembrane protein of histidine kinase type because it is phosphorylated on a histidine residue when stimulated, and an intracellular regulator; the latter is activated by the transfer of a phosphoryl group to an aspartate residue, changing its conformation, allowing it to bind to DNA and modulate transcription, thus allowing the organism to adapt physiologically to external environmental changes.

Tyndallization Moderate sterilization method which comprises heating a mixture at a temperature of 60–70 °C for 30–60 min, 3 times at 24 h intervals. This process can destroy bacteria and their resistance forms without altering the organoleptic properties of the medium. Process developed by John Tyndall (1820–1893), a famous Irish physicist.

Typing by multilocus sequencing See "MLST."

Undulipodium In the literature it is often called "flagellum" and designates the whip-shaped structures that emerge outside of the cells that is surrounded by the cell membrane and which includes different internal structures allowing it to move. Lynn Margulis has proposed to restrict the use of the word flagellum to the prokaryotes and to use undulipodium to designate the completely different structures found in the Eukarya.

Indeed, the flagellum of prokaryotes is a hollow cylinder 10–25 nm in diameter which rotates on itself guided by two rings at its base. The eukaryotic undulipodium is a thicker structure of about 200 nm in diameter and including several hundreds of proteins such as tubulins, dyneins, or nexins. In section it displays 9 external doublets of microtubules and 2 axial ones. It does not turn on itself but rather produces waves resulting from the alternating contraction of the outer microtubules on one side and on the other.

Untreated waters Untreated urban or industrial wastewater.

Virus Is a small infectious agent that can only replicate inside a cell. Viruses are obligatory intracellular parasites; they cannot grow or multiply on their own and need to enter a cell and take over the cell metabolism to help them multiply.

Weeds Pioneer plants which thrive in abandoned or cultivated land at the expense of crops (e.g., stinging nettle, couch grass). They are often eliminated using herbicides.

Xenobiotics Chemical compounds resulting from a chemical synthesis that do not exist naturally on Earth.

Zoonosis Animal infectious disease transmissible to human.

Zooplankton The zooplankton is the animal component of the plankton (see Plankton).

Zoospore Motile flagellated asexual spore.

Subject Index

Note: Pages in bold refer to pages where substantial, heavy information are provided to the readers.

A
ABC transporters, **28**, 428
Abiotic factors, 552, 620, 642, 645, **647**, 848
Abortive transduction, 450
Abyssopelagic, 532
Acetoclastic, 364, 558, 560, 883
Acetogenesis, 46, 49, **50**, 390, 561, 585
Acetogenic, 50, 55, 62, 64, 498, 797
Acetyl-CoA reductive pathway, **62–63**
Acetylen inhibition technique (AIT), **570–571**, 576, 870
Acidic environments, 333, 379–380, **381–385**, 606
Acidic pools, 354
Acid mine drainage (AMD), 196, 271, **285**, 380, 383, 741, 742
Acidophiles/Acidophilic, 6, 44, 164, 166, 167, 170, 173, 185, **333–334**, 365, 366, **379–385**, 606, 607, 741, 822
Acid rains, 581, 612
Activated sludge, 643, 644, 664, 665, **670–674**, 684, 842, 883
Active transporters, **28–29**
Acyl-CoA: transacyclase, 358
Adaptability, 4, 28, 70, 294, 326, 346, 488, 630, 656, 699
Adaptation, 6, 22, 70, 76, 82, 100, 147, 149, 152, 173, 196, 228, 264, 281, **293–346**, 355, 358–360, 367, 372, 377, 385, 390, 399, 402, 422, 423, 430, 432, 437, 445, 453, 456–473, 478, 479, 506, 521, 561, 589, 635–637, 684–685, 691, 707, 768–770, 820, 822, 838
Adenosine phosphosulfate reductase (APS reductase), 48, 275, 587, 588
Adhesin, 310, 318, **428**, 429, 634
Adhesion, 18, 32, 312, 319, 341–345, 409, 413, 421, 428, 487
Advection, 498, 506, **522**, 874
Aerobe, 164, 166, 167, 173, 367, 372, 561, 573, **817–818**
Aerobic bacteria, 44, 168, 182, 184, 188, 358, 410, 514, 549, 586, 598, 672, 710, 711, 796
Aerobic biodegradation, 103, **561–562**, 661, 683, **711–715**, **723–728**, 734, 737
Aerobic microorganisms, 65, 68, 513, 553, 583, 593, 679, 745
Aerobic prokaryotes, 40, 539
Aerobic respiration, **37–45**, 47, 70, 83, 96, **103–104**, 164, 166, **327–328**, 328–329, 363, 376, 476, 486, 513, 516, 517, 532, **539–540**, 547, 558, 573–575, 593, 599, 611, 789, 792, 793
Aerobic sulfur-oxidizing bacteria, **43–44**, 62, 345, 597
Aerobic biodegradation, 103, **561–562**, 661, 683, **711–715**, **723–728**, 734, 737
Aerobiosis, 574, 673, 713, 722, **816–817**, 820
AFLP. *See* Amplified fragment length polymorphism (AFLP)
Agar, 13, 317, 327, 329, 367, 649, 650, 692, 805, **816**, 818, 820–822
Agriculture (nitrogen cycle), **579**
Agronomy, 6, 274, 405, 421
Air sparging, **663**

Air tolerants, 328, 329, 817
AIT. *See* Acetylen inhibition technique (AIT)
Akinetes, **31**, **569**, 625
Alginic, 232, 251
Alkaline, 166, 167, 169, 171, 315, 333–335, 361, 362, 368, 372, 374, **385–391**, 546, 601, 608, 623, 740
 enzymes, 335, 391
 lakes, 333, 368, 372, 389, 390
Alkaliphiles/Alkaliphilic, 170–172, **333–335**, 373, **385–391**
Alkalitolerant, 387, 390
Alkanes, 89, 661, 716–719, 722, **723–727**, 728, 729, 733
Allelopathy, 397
Alpha diversity, 266, 276, 884
AMD. *See* Acid mine drainage (AMD)
Amelioration process, 464
Amensalism, 123, **397**, 417, 863, 883
Ammonia dissimilative reduction, 46
Ammonia monooxygenase (AMO), 42, 175, **573**, 842
Ammonia-oxidizing bacteria, 42, 404, 544
Ammonification, 312, 387, 567, **569–570**
Ammonium assimilation, 565, 567
Ammonium oxidation, 571, 573, 574, 612
Ammonium oxidizing bacteria (AOB), 573, 578, 842
Amplified fragment length polymorphism (AFLP), 147, 157, 158, 814
Anabolism, 33–35, 61, 151
Anaerobe, 12, 22, 47, 49, 166, 167, 176, 182, 185, 364–367, 372, 391, 404, 471, 593, 599, 639, 796, 817
Anaerobic (chamber, jar), 557, **818–820**
Anaerobic air tolerant, 53, 817
Anaerobic archaea, 66, 167, 383, 557
Anaerobic bacteria, 20, 49, 65, 70, 183–185, 188, 328, 329, 334, 357, 358, 372, 373, 432, 514, 549, 586, 639, 799, 875
Anaerobic biodegradation, 268, **557**, 559, 661, **710–711**, 715, **728**
Anaerobic degradation, 42, 389, 390, **562–564**, 596, 600, 664, 679, 681, 715, 729
Anaerobic metabolism, 99, 303, 343, 498, 515, 539, 575, 664, 791
Anaerobic microorganisms, 12, 47, 50, 64, 67–69, 354, 513, 582, 591, 594, 599, 612, 680, 789, **817–819**
Anaerobic prokaryotes, 63, 96–98
Anaerobic respiration, 6, 27, 37, **45–51**, 67, 103, 176, 182, 185, 303, 315, 328, 329, 336, 363, 380, 383, 384, 410, 474, 516, **540**, 541, 557, 575, 583, 584, 598, 607, 817
Anaerobiosis, 49, 710, 711, 715
Anammox, 46, 47, 184, 564, 567, 568, 571, 572, 574, **576–578**, 580, 672, 842
Anammoxosome, 184, **576–578**
Anaplerotic sequences, **66**

Anoxia, 595, 626
Anoxic, 7, 19, **44**, 46, 51, 69–70, 362, 369–371, 384, 492, 513, 515–516, 539–542, 546, 554, 557–558, 562, 564, 580–584, 597, 599, 665, 672–673, 789, 798
Anoxygenic aerobic photosynthesis (AnAP), 526
Anoxygenic anaerobic photosynthesis (AnAnP), 526
Anoxygenic photosynthesis, 19, 51, **57–60**, 83, 102, 177, 179, 182, 512, 588, 589, 607
Anoxygenic phototrophic, 14, 35, 55–57, 59, 60, 70, 184, 275, 280, 326, 389, 491, 495, 526, 583, 584, **588–595**, 598, 599
Antagonism, 397, 401, 411, 417
Antennas, 56, 57, 172
Anthrax, 12, 185, 186, 632
Antibiosis, 13, 247, 294, **320–324**, 397, **401–402**, 417, 862
Antibiotic resistance, 121, 152, **320–324**, 404, **427**, 432, 448, 466–468, **635–638**, 805, 848
Antibody, 100, 622, 623, 630, 652, 653, 780
Antifreeze proteins, 360
Antigen O, 29
Anti-port transporters, 29, 47, 575
AOB. *See* Ammonium oxidizing bacteria (AOB)
API system, 152
Apicoplast, 101, 132, 221, 227, 251, 629
Apoptosis, **414**, 429, **433–436**
Aquifers, 104, 176, 285, 376, 379, 607, 743, 746, 758
Archaeal domain, 364
Aroclor, 708–713, 715
Aromatic hydrocarbons, 285, 420, 608, 661, 681, 697, 716, 717, 720, 722–724, 738
Arrhenius law, 355, 358, 884
Arsenate, 46, 70, 167, 176, 389, 601, 744
Arsenic, 44, 49, 285, 346, 389, 420, 512, 601, 611, 685, 740, 742, 744, 745, 837, 839, 841, 842
Arsenic cycle, 389
Assimilatory nitrate reduction, **67**, **565–566**
Assimilatory sulfate reduction, **69**, 581
Athalassohalines, 171, 333, 368, 372
Atmospheric CO_2, 382, 518–521, 552, 603
ATP synthase, 37, 39, 40, 57, 58, 70, 101, 171, 212, 390, 531, 578
Atrazine, 404, 686, 688, 691, 692, **693–697**, **699–704**, 746
Autoinducer, 313, 314, 316, 317, 633, 634
Autonomous sewerage system, **666**
Autotrophic microorganisms, **34**, 35, 45, **61–63**, 506, 591, 600, 625, 671, 783
Autotrophic organisms, 101, 239, 512, 514, 515, 530, 532, 534, 797
Autotrophic respiration, **519**
Autotrophy, 113, 137, 166, 372, 431, 492, 512, 516, 611, 798
Auxotrophs, **35**, 197, 449
Average nucleotide identity (ANI), 147, 148
Axenic cultures, 16, 816
Axopods, **207**, 208, 216, 229, 251
Azores, 361, 362

B
Bacteria beds, **670**
Bacterial artificial chromosome (BAC), 805, 833, 834
Bacterial carbon demand (BCD), 786
Bacterial domain, 55, 381
Bacteria exoenzymatic activity, **787**
Bacterial floc, 670
Bacterial growth efficiency (BGE), 786, 865
Bacterial production, 417, 494, 503, 531, **785**, 786
Bacterial respiration, 503, 533, **785–786**
Bacterial systematic, 113, 152
Bacteriochlorophyll (BChl), 55, **56**, 59, 60, 102, 161, 177, 179, 268, 589, 595, 813

BChl a, 56, 58–60, 179, 181, 591–593, 599, 813
BChl b, 56, 591, 592, 599, 813
BChl c, 56, 177, 184, 591–593, 813, 814
BChl d, 56, 184, 592, 813, 814
BChl e, 56, 184, 591–593, 813
BChl g, 56, 185, 592, 813
Bacteriocyte, 423, 424, 430, 600
Bacteriohopanepolyols, **93**, 95
Bacteriophage, 14, 108, 121, 401, 429, 449, 450, 501, 503, 634, 635, 640, 788
Bacteriorhodopsin, 27, **60–61**, 171, 172, 375, 391, 468, 841
Bacteriovory, 494
Banded iron formation (BIF), 103, **607**
Barophile, 376
Basic Local Alignment Search Tool (BLAST), 157, 454, 455, 806
Bathypelagic, 516, 531, 532, 771, 772
BChl. *See* Bacteriochlorophyll (BChl)
Beer-Lambert (law), **813**, 814
Benthic chamber, 786
Benthic ecosystems, **486–487**, **497–499**, 766, 799
Benthic microorganisms, 487, **490–492**, 505, 526, 587, 597–598
Benthic organims, 486, 487, 526, 537, 541
Bergey's Manual for Systematic Bacteriology, 113, 140, 160, 175, 187
Beta diversity, 266, 884
Betaine, 172, 330, 331, 333, 372, 373, 375, 390
Beta-keto-acyl-CoA synthase, 359
Beta (β)-oxidation, **38**, **40**, 65, 718, 719, 722, 725
Bioaccumulation, 227, 605, **660**, 709, 718, 733, **744–745**
Bioamplification, 605, 709, 713
Bioaugmentation, **661**, 662, 698–703, 708, **734–736**, 738, 746
Biochemical Oxygen Demand (BOD_5), **667**
Bioconstructions, 85
Biocycle, 224, 246
Biodiffusors, 537
Biodiversity, 9, 93, 96, 104, 108, 155, 193, 194, **261–286**, **355–358**, 368
Biofilm, 4, 5, 7, 18, 29, 87, 120, 215, 294, **341–345**, 359, 403, 487, 586, 631, 670, **719–722**, 758, **792–795**, 837, **839**, 847, **874–880**
Biofilters, 644, 662, **683–684**
Biogas production, 367, 561, 679–681
Biogeochemical cycles, 4, 6, 7, 10, 16, 19, 21, 22, 108, 164, 264, 485, 487, 496, 506, **511–612**, 660, 729, 730, 743, 824, 841, 848, 880
Biogeochemical tracers, **520**
Biohopanoids, **93**
Biohydrometallurgical processes, 382
Bioimmobilization, 660–661, **662**
Bioinformatic (sequences analysis), **806–808**
Bioleaching, 7, 380, 383, 384, 589, 606, 660, **661**, 666, **740–742**
Biological control, 243, 404, 419, 641
Biological Oxygen Demand (BOD), 595, 643, 644, **667**, 671
Biological pump, **519**, 521, 529, 533, 534, 536, 601, 603
Biological weapons, 185
BIOLOG method, **152**, **786–787**
Biomagnification, 885
Biomarker, **5**, **85**, **93–96**, 104, 122, 192, 268, 490, 499, 562, 717, 733, 776, 781, **795–800**, 811, 813, 814, **815**, 816
Biomass, 367, 391, 430, 488, 490, 493, 506, 516, 526, 529, 542, 549, 600, 625–626, 670–671, 758, **781–783**, 784–788, 796–797, 820, 840, 853, **854–855**, 856–858, 861, 865–866, 868–869, 871–872, 874–880
Biomineralization, 338, 842
Biooxidation, 380, **740–742**
Biopiles, **662–663**, 699, 715, 737, 738
Bioprecipitation, 745
Bioreactor, 279, 346, 367, 378, 382, 506, 557–559, 564, 661, **662**, 674, **675–676**, 679, **683–684**, 699, 710, 711, 715, 738, 743, 745, **821**, 872
Bioreduction (metals), **742–744**

Bioremediation, 5, 7, 21, 611, **661–663**, **715**, **734–740**
Bioscrubbers, **683**, 684
Biosignatures, 85, 87–89, 92
Biostimulation, 7, **661**, 662, 679, 699, 700, 713, 734, 735, **736–737**, 738
Biosurfactants, 699, 710, 713, **719–722**, 732, 734, 736
Biotic factors, 410, 412, **505–506**, 647
Biotransformation, 7, 16, 21, 152, 183, 611, **660**, 799, 820
Bioturbation, 498, 504, 505, 537, **541–542**, 544, 581, 734
Bioventing, **663**
Biovolume, 488, **781–782**, 877
Black smokers, 102, 166, 275, 362
BLAST. *See* Basic Local Alignment Search Tool (BLAST)
Bloom, 19, 111, 177, 224, 227, 237, 247, 264, **281–282**, 346, 490, 493, 523, 531, 534, 580, 596, **624–626**, 628, 648
BOD. *See* Biological oxygen demand (BOD)
Bootstrap method, 148
Botulism, 185, 633
Branched fatty acids, 358, 730, 815
Bromine, 211, 250
Bulking sludge, **670**

C

Calcium carbonate ($CaCO_3$), 89, 219, 333, 372, 386, 409, 515, 518, 519, 522, **533**, 552, 743
Calvin cycle, **61–62**, 64, 177, 573, 589, 606, 797
Canonical correspondence analysis, 265, 281
Capsid, 450, 451, 501, 503, **620**
Carbon-14 (method), 86, **784**, 840
Carbon cycle, **515–565**
    in marine sediments, **537–542**
    in soils, **542–551**
    in the lake ecosystems, **551–554**
    in the oceanic water column, **522–536**
Carbon sink, 495, 519, 521, 580, 885
Carboxysomes, 30, 62
Carotenoid(s), 56, 60, 129, 160, 177, 212, 234, 268, 282, 327, 345, 389, 390, 472, 491, 589, **591**, 592–594, 624, 813, 814
Carpoconidia, **198–199**, 202, 204, 230, 250
Carpoconidiogenous, **204**
Carpospores, 198, 199
Caryogamy, 205, 231, 249
Catalase, 49, 103, 328, 359, 603, 605, 609, 639
Cell lysis, 172, 450, **628**, 632, 633, 730, 807
Cellulose, 15, 16, 20, 65, 138, 184, 188, 208, 209, 212, 217, 224, 230, 242, 245, 247–252, 283, 306, 308, 309, 341, 402, 409, 410, 413, 423, 431, 436, 539, 542, 547, **548**, 559, 560, 679, 765, 840
Cenozoic Era, 369, 893
Central metabolic pathways, **65–66**, 685
Centroplast, 207, 250
Centrosome, 26, 33, 125, 213, 245
C4 pathway, **63–64**
Chemical gradients, 21, 343, 537, 599
Chemical microsensors, **788–792**
Chemical oxygen demand (COD), 641, **667**, 671
Chemical pump, 519
Chemiosmotic hypothesis, 104
Chemiosmotic theory, 40, 101, 531
Chemoautolithotrophs, 381
Chemoautotroph/chemoautotrophic, 396, 593, 595, **785**, 799, 896
Chemocline, 369, **593–594**
Chemolithoautotrophs/Chemolithoautotrophic, 41, 42, 44, 166, 170, 173, 377, 382, 383, 491, 497, 589, 600, 878, 879
Chemolithoheterotrophs, 41, 44, 589
Chemolithotrophic/Chemolithotrophs, 15–17, 34, 35, 37, **41–45**, 47, 48, 50, 61, 70, 91, 152, 166, 173, 175, 177, 181–183, 282, 285, 326, 332, 334, 341, 345, 362, **363**, 364, 366, 367, 380–385, 389, 404, **491**, 497, 512–513, 516, 544, 582–584, 588–591, 593–595, 599–601, 607, 611, 671, 741, 745, 791, 841
Chemolithotrophy, 27, 101, 179, 385, 512, 513, 558, 791
Chemoorganoheterotrophs, 44, 589, 792
Chemoorganotrophs/Chemoorganotrophic, **33–35**, **37–41**, 43, 47–50, 70, 152, 171, 173, 175–177, 179, 181–188, 326, 332–334, 336, 341, 343–344, **363**, 364–366, 390, 547, 575, 591, 593, 611, 745
Chemosynthesis/Chemosynthetic processes, 19, 431, 552, 600, 607
Chemotaxis, 6, 299, 300, 302, **303–308**, 340, 343, 408, 410, 504, **839**
Chimeras, **272**, 808
Chitinases, 65, 320, 341, 376, 414, 417, 539
Chitin, 32, 218, 242–243, 245, 252, 320, 372, 413, 436, 539, 548, 565, 840
Chlorination, 624, 644, 675, 708, 710
Chlorophyll (Chl), **55–56**, 58, 162, 177, 211, 231, 268, 356, 490, 529, 530, 601, 602, 624, 716, 732, 774, 778, **782**, 783, 813, 814, 841
    Chl *a*, **56**, 58, 123, 124, 126, 178, 208, 210, 212, 217, 221, 225, 234, 236, 239, 240, 250–252, 490, 523, 718, **782**, 783
    Chl *b*, **56**, 123, 126, 127, 212, 217, 225, 240, 250–252, 526
    Chl *c*, **56**, 123, 212, 221, 225, 234, 236, 239, 251, 252
Chloroplast, 26, 33, 37, 56, 62, 97–98, 101, 104, **122–136**, 177–178, 210, 212, 238–239, 477, 782
Chlorosis, 782
Chlorosomes, 30, 56, 57, 177, 184, 592–594
Cholera, 12, 13, 182, 308, 309, 424, 504, 620, 623, 630, 632–634, 649
Circadian cycle, 280, 599
Citric acid cycle, 38, 797
Classification, 108, 113–115, 135, 145, 151, 152, 154, 155, 158–160, 164–175, 221, 250, 354–356, 361, 368, 486, 488, 491, 492, 624, 655, 676–677, 740
Clathrate, 358, **554–556**, 886
Clonal reproduction, **146**, 886
Cloning sequencing, 269–270, **271**, 273, 337, **805–806**
Cluster analysis, 265
C:N:P ratio, 528, 529, 736
C:N ratio, 549, 573–574, 580, 734
Coals, 173, 363, 380, 382, 513, 514, 517, 520, 554, 556, 718
Coastal area/zone, 111, 280, 501, 513, 526, 529, 532, 537, 539, 540, 580, 597, 645–647, 732, 734, 782
Coastal environment, 183, 595, 645–646, 665, 796
Coastal ocean, 531
Coastal water, 182, 529, 530, 579, 645, 647
Codon usage, **461–465**, 835
Coenocytic, 92, 121, 192, 193, 211, 230, 245, 249, 250, 252
Coenzyme M, 50, 51, 557, 581
Co-evolution, 104, 414, 415, 430, 433, 634, **835**
Cold-acclimation proteins (Cap), 360
Cold deep-sea, 354, 768, 770
Cold-shock proteins (Csp), 311, 312, 359, **360**
Collections, **150–151**, 267, 325, 389, **822–824**
Colonization, 149, 263, 275, 311, 316, 318, 356, 380, 396, 410, 413, 414, 417, 418, 420, 421, 432, 436, 470, 473, 499, 505, 506, 544, 599, 630, **631**, 638, 642. 651, 758, **822**, 876, 879
Colony forming unit (CFU), 820
Colorless sulfo-oxidizing, **588–589**
CO_2 mechanism of concentration, 63
Cometabolism, **35**, **403**, 410, **660**, 684, 691, 692, 713, 738
Commensalism, 6, 123, 396, **397**, 403, **404**, 411, **432–433**
Community structure, 265, **271–275**, 277, 280, 391, 544, 738, 771, 848
Comparative genomics, **81–82**, 100, 149, 174, 175, 299, 464, 473–475, 809, 811, 812
Compatible solutes, 172, **330**, 332, 366, **373**, 375, 390
Competence, 303, 313, 316, 404, **446**, 634, 685, 703, 862, 887
Competition, 16, 81, 89, 123, 149, 283, 294, 307, 396–398, 401, **402**, 407, 410, 411, 417, 418, 493, 494, 497, 500, 505, 506, 534, 546, 548, 549, 573, 574, 580, 625, 626, 647, 648, 661, 665, 713, 849, **863–866**, 869, 875

Complete genome sequences, 20, 79, 154, 165, 454, 455, 478, 833, 843
Compost/Composting, 185, 268, 325, 334, 403, 404, 548, 631, 660–663, 674, **677–679**, 680, 683, 684, 700, 715, 740, 816, 843
Confocal scanning laser microscopy, **774–775**
Congener, 708–715, 717
Conidia, 121, 198, **199**, 200–201, 204, 206, 211, 216, 217, 220, 230, 234, 250, 629
Conidiophors, 187, 629
Conjugation, 32, 121, 202, 222, 269, 272, 346, 402, 404, 405, 432, **447–449**, 452, 466, 468, 634, 685, 688, 692, 699, **835**
Conjugative plasmid, 404, 410, 448, 449, 634, 696
Consortium, 87, 403, 558, 562, 563, 597, **661**, 703, 712, 713, 832, 887
Contamination, 5, 70, 87, 96, 172, 281, 285, 345, 358, 380, 408, 409, 425, 570, 605, 620, 622, 623, 629, 631, 632, 637–640, **642–645**, 663, 698–700, 708, 713, 718, 737, 738, 741, 743, 745, 758–761, 765, 770, 801, 853
Continuous cultures, 19, 817, **818–821**, 853
Conveyors, 537, 542
Cooperation, 6, 23, 313, 345, 397, 398, 402–403, **404**, 405, 411, **417–418**, 432, 559, 812
Core genes, 456, 474, 887
Core genome, 149, 463, 464, **474–475**, 634, 835
CO_2 respiration, 46, **50–51**, 516, 558
Coring tube, 767
Corrosion, 18, 20, 22, 45, 79, 85, 344, 367, 584, **586–587**, 588–589
Cosmid, 805, 887
Co-substrate, 691, 701, 702, 712, 713
Cryopreservation, **823–824**
Cryoprotectants, 356, 823, 824
Cultural approach, **267**, 281
Cultivated and uncultivated strains, 839–841
Current genome mispairing, 147
Cuttings, **198–200**, 204, 320, 409, 707, 708, 759
Cyanotoxins, 177, **626**, 628
Cycle of 3-hydroxypropionate. *See* hydroxypropionate
Cyclic photophosphorylation, 57–60
Cyclodextrin glycotransferases, 391
Cyclodextrins, 391, 713
Cyst, 4, 31, 93, 199, **201**, 217, 218, 222, 224–226, 229–231, 240, 242–244, 249, 251, 336, 338, 488, 493, 561, 621, 624, 629, 630, 640
Cystogamy, **202**, 203, 214, 222, 234, 250, 251
Cytochrome P450, **688**, 722, 724
Cytometry, 5, 7, 268, 488, 526, 649–651, 653, 758, **773–779**, 780–781, 787, 822
Cytoskeleton, 30, 33, 93, 125, 197, 209, 213, 225, 233, 236, 238, 241, 431

D

DAPI. *See* 4′,6′-Diamidino-2-phenylindole (DAPI)
DDT. *See* Dichlorodiphenyltrichloroethane (DDT)
Dead Sea, 171, **333**, **372**, **374**, **375**
Deamination, 40, 359, 417, 569, 704
Deep agar media, **818**
Deep aquifers, **376**, 758
Deep hypersaline anoxic basins (DHABs), **369–371**
Deep ocean, 5, 18, 392, 398, 518, 519, 522, 524, **528–531**, 533, 534, 555, 566, 589, **599–601**, 769, 771, 772, 816
Deep-sea hydrothermal vents, 170, 172, 174, 354, **361–365**, 367, **377–379**, **599**, 768
Deep-sea piezopsychrophiles, **376–377**

Deep waters, 89, **356**, 493, 519, 521, 523, 529, 552, 595, 597, 623, 764, 771, 784
Defense, 211, 222, 225, 234, 283, 314, 316, 328, 359, 397, 406, 408–410, 413–415, 417, 421, 429, 433–436, 620, 626, 631, 638, 650, 837
Degradation of organic matter in soils, 499, **546–551**
Denaturing gradient gel electrophoresis (DGGE), **269–272**, 276, 281, 383, 503, 802, **804–805**
Dendrogram, **154–158**
Denitrification, **46**, **47**, 171, 273, 279, 388, 389, 423, 498, 514, 515, 521, 539, 541, 564, 567, 568, **570–576**, 578–581, 612, 641, 644, 664, 672, **673**, **675**, 842
Denitrifying activity, 42, 545, **570–572**, 575, 576, 580
Denitrifying bacteria, 47, 181, 183, 573–575, 599, 611, 790, 792, 817
Desaturase, **358**, **359**
Detoxification, 328, 329, 403, 433, 487, 573, 589, 600, 638, **691**, **742**, **745**, 837, 839
Dew point, **683**
DGGE. *See* Denaturing gradient gel electrophoresis (DGGE)
Diagenetic hydrocarbons, **718**
4′,6′-Diamidino-2-phenylindole (DAPI), 194, 424, 491, 536, 780, 781, 798
Diazotrophs, 274, 566, 568
Diazotrophy, **67**, **69**, 521, 529
Dikaryotic, 201
Dichlorodiphenyltrichloroethane (DDT), **685**, **690**
2,4-Dichlorophenoxyacetic acid, 658, **747–749**
Differential interference contrast microscope, **774**
Diffuse source, **642**, **645**, **646**
Digenetic, **203–206**, 214, 218, 220, 224, 227, 234, 237, 239, 249, 251
Digestive tracts, 5, 19, 22, 53, 120, 130, 184, 188, 230, 242, 248, **283–284**, 283–285, 334, 346, 376, 430, 436, 541, 542, 544, 557, **559–561**, 566, 622, 629, 638, 639, 647
Dilution to extinction, **882**
Dimethylsulfide (DMS), 167, 237, 558, **581–583**, 681
Dimethylsulfoniopropionate (DMSP), 237, 251, **581**, 583
Dimethylsulfoxide (DMSO), 171, 583, 611, 823, 839
Dimictic lakes, **551–552**, 594
Dinitrogenase, 67, **566–569**
Dinitrogen fixation, **67–68**, 283, 389, **566–569**, 601
Dinitrogen-fixing bacteria, **579**
Dioxygen, **817–818**
Dioxygenase, **43**, 693, 694, 697, 706, 714, 724, 726, 727
Diploid, 198–201, **204–206**, 214, 215, 217, 218, 220, 222–224, 226, 228–231, 234, 235, 242, 245, 249, 252
Direct-light microscope, **773–774**
Discharge, 222, 225, 525, 537, 554, 557, 622, 642, 644, 645, 647, 660, 664, 666, 667, 674, 676, **680–681**, 732
Dissimilatory reduction of nitrate to ammonium, 46, 47, 494, 567, **572–573**, 575
Dismutation, **564**, **587–588**
Dispersant, **735**
Disproportionation, **49–50**, 101, 364, 558, **564**, **582**, 612
Dissemination, 7, 186, 187, 201, 242, 264, 338, 409, 448, 453, 466–468, 504, 634, 635, 641–642, 657, 697, 811
Dissimilatory reduction, **46–47**, 494, 570, 572–575, 583, 588, 600, 603, 607
Dissimilatory sulfate reduction, **47–49**, 564, **582–588**
Dissolved inorganic carbon (DIC), **519**, **521**, **522**, 784
Dissolved organic matter (DOM), 356, 494, 495, 497, 498, 503, **519**, **527**, **529**, 531, 533, 552, 553, 887
Diversity, 4, 14, 53, 93, 108, 146, 192, **262–268**, 307, 363, 409, 445, 488, 526, **625**, 660, 758
Diversity indices, **265–268**, 271, **275–277**
Divinyl chlorophyll, 56, **177**
DMS. *See* Dimethylsulfide (DMS)

DMSO. *See* Dimethylsulfoxide (DMSO)
DMSP. *See* Dimethylsulfoniopropionate (DMSP)
DNA
　arrays, 273, **808–810**
　chips, **808**, **837**
　gyrase, **366**
　microarray, 273, **808–812**, 888
　polymerase, 110, 264, **448**, 503, **800**, **805**, **843**
　reassociation, 146, **269**, 630
　repair, 4, 82, 177, 319, 322, 324, 337, 338, 346, 446
DNA-DNA (hybridization reassociation), **152–153**, **155**, **157–158**, 160, 655
Docosahexaenoic acid, **356**
DOM. *See* Dissolved organic matter (DOM)
Domains, 25, 27, 46, 77, 79, 80, **82–85**, 101, **107**, 113, **115–117**, **160–162**, 196, 298–302, 304–310, 319, 321, 323, 327, 346, 354, 363, 367, 372, 381, 428, 433–435, 449, 470, 486, 488, 503, 546, 607, 660, 707, 779, **844**, 880
Dormancy, 5, 336, 357, 629, 649, 656
Dystrophic crises, **595**

E
Ecological species, **148–150**
Ecosystem functioning, **262–264**, 274, **278–285**, 458, 500, 541, 729, **842**, 848
Ecotype, **148–149**
Ecto-enzymes, **539**
Ectoine, **330–333**, **373–375**, 390
Ectosymbiosis, 242, 398, 430, **597**
Effectors, 297, **302–305**, 317, 413, 414, 417, 428, 429, 433, 434, 436, 832, 837
EGT. *See* Endosymbiotic gene transfer (EGT)
Eicosapentaenoic acid, **356**
Electronic microscope cro-droplets (GMDs), **882**
Electron microscope, 14, 88, 110, 122, 124, 181, 244, 338, 340, 406, 527, 607, 641, **775**, 788
Electron reverse flow, **45**
Electroporation, 121, **446**, 805
Elemental sulfur disproportionation (Elemental sulfur dismutation), **49–50**, 101, 364, 558, 564, **582**, **612**
Emerging disease, **425**
Endocytosis, **32**, **33**, **65**
Endolithic, **87**, 211, 282
Endospore, 4, 11, **31–32**, 120, 198, 211, 243, 244, 252, 336, 338, 382, 622
Endosymbiosis, 6, **97–100**, 104, **122–136**, 169, 198, 205, 207, 208, 217, 221, 224, 225, 227, 234, 236, **238–241**, **395–398**, **407–408**, 430, 437, **476–478**
Endosymbiotic bacteria, **81**, **130**, 242, **403**
Endosymbiotic gene transfer (EGT), **476–478**
Endosymbiotic theory, **98**
Endotoxin, 29, **631–633**
Energy decoupling, **36**
Enrichment cultures, **15**, 174, 378, 383, 576, 596, 702, 712
Enteric, 429, 621, 622, 638, 640, 642, **644–649**, 651, 656, 669
Enterotoxin, **623**, **633**
Enumeration (heterotrophic bacteria), 60, 70, 154, 183, 277, 331, 344, 356, 488, 491, 494, 495, 499, 536, 560, 576, 599, 626, 639, 640, 652, 653, 655, 672, 717, 773, 776, 778, 785, 786, 799, **820–821**
Environment, 4, 9, 16, 26, 81, 109, 150, 193, 262, 293–346, 353, 445, 486, 620, 659–746, **757–824**, 848
Environmental microscope, **776**
Epidemics, 424, **425**, 465, 490, 620, **630**, **649**
Epilimnion, **551**, **594**, **595**
Epipelagic, **531–532**

Episome, **448–449**
EPS. *See* Extracellular polymeric substances (EPS)
E ratio, **529**
Ergosterol, **245**, **252**, 815
Ester bond, 27, **83**, **161**, 320, 690
Ether linkage, 27, **83**
Eukaryotes (origin), 6, 10, 25, 83, **108**, **122–135**, 145, 191–253, 294, 357, 395, 449, 486, 523, 629, 662, 778, 833, 875
Eukaryotic biomarkers, **94–96**, 122
Eukaryotic fossils, **92–94**
Euphotic, 178, 488, 494, 501, 522, **527–530**, **533**, **534**, **537**, **551**, 553, 580, 766, 768, 771, 772, 841, 873
Eurypsychrophilic (psychrotolerant), **171**, **355**, 357
Eutectophiles, **357**
Eutrophic, 389, 488, 489, **530**, 552, **580**, **584**, **596**, 625, 866
Eutrophication, 264, 346, 524, **580**, **593**, 842
Evaporites, **332–333**, **369**, 372
Evaporitic deposits, **368–369**
Evolution, 4, 6, 55, **76–84**, 89, 91, 92, 96–98, **100–104**, **107–141**, **146–150**, 175, 187, 189, 197, 198, 205, 209, 242, 262, 268, 273, 280, 294, 295, 324, 333, 354, 367, 385, 390, 395–397, 399, 400, 417, 420, 421, 424, 430, 433, 437, **455–457**, 459, 460, 464, 467, 468, 470, **473–476**, 503, 504, 514, 520, 521, 526, 570, 624, 626, 638, 664, 675, 678, 680, 684, 694, 695, 704, 707, 711, 734, 746, 773, 834, 835, 843, 848, 869, 876, 878, 880, 881
Evolutionary convergence, **80**
Exaptation, **402**
Exobiology, 6, 104, **353–392**
Exocytosis, **32–33**
Exoenzyme, **65**, 317, 320
Exopolymer, 29, 87, 120, 185, **341–345**, 599, **670**
Exopolysaccharide, 29, 89, 308, 310, 338, **341–343**, 356, **359–360**, 403, 409, 410, 417, **631**
Experimental paleobiochemistry, **83**
Exported primary production, **529**
Extracellular enzyme, 65, 341, 356, 413, **548**, **579**, 631, 634, **680**, 787
Extracellular polymeric substances (EPS), 29, 87, 338, **341**, **343–345**, **359**, **360**, 375, 487, 491, **497–499**, 874
Extracellular sulfur globules, **589**, **592**
Extreme environments, 175, 189, 196, 264, 271, 296, 334, 337, 346, **353–354**, 375, 379, 385, 423, 492, 505, 575, 719, 841, 843, 878
Extreme halophilic/halophiles, 27, 55, 60, 161, 162, 182, **331–333**, **367–375**, 592, 815
Extremophiles, 6, 168, **353**, **354**, **377**, 385, **391–392**, 770, 843
Exudate, 12, 68, 316, 405, **407–410**, 420, **542–544**, 701, 738, 739, 744

F
Facultative (aerobes/anaerobes), 166, 173, 182, **328–329**, 365, 366, 573, 599, 639, **817–818**
Facultative alkaliphiles, **385**
Faeces, 522, 537, 560
Fatty acid synthase, **358**
Fecal, 281, 560, 620, 622, 623, 629, **637–645**, 647, 651, 655, 765, 766, 771, 820
Fecal pellets, 221, 494–495, 500, 581, 765, **771**
Fermentation, 6, **10–12**, 29, 36, 45, **51–55**, 70, 99, 101, 152, 166, 167, 172, 182, 185, 187, 247, 325, 328, 329, 332, 334, 343, 344, 363, 373, 375, 392, 410, 436, 513, 518, 540, **541**, **547**, **557**, **559–561**, 584, 587, 588, 593, 595, 664, **678–680**, 701, **852–853**, **855–858**, 860
Fermentative bacteria, 20, **52**, 182, 184, 185, 187, 188, 331, 333, 344, 364, 373, 387, **513**, **611**
Fermentative metabolism, **53**, 404, 588
Fermentative microorganisms, **52**, **53**, **541**, 588
Fermentative pathway, **52**, **547**

Fermentative prokaryotes, 541, 573
Fermentors, 746, **852–858**
Ferric iron respiration, **49**, **607–608**
Ferri-reduction, iron-reducing bacteria, **49**, 183, 184, **607–608**
Ferrooxidant bacteria, **43–45**, 380–384, **606–607**, 741
Ferrous iron (Fe^{2+}), **37**, **44**, **55**, 70, 103, 382, 383, 588, 589, 603, **606–607**, 741
Fertilization, 198, 199, **201–203**, 206, 211, 213, 217, 220, 227, 229, 234, 245, **250–252**, 279, 419, 420, 547, 579, **601–603**
Filamentous algae, **383**
Filamentous cyanobacteria, 67, **89**, **177–178**, 389, 486, 491, **624–628**
Filtration (techniques), **499**
Fingerprints, **268–272**, 277, 491, 730, **802–805**, 838
Fischer-Tropsch (reaction), **89**
FISH. *See* Fluorescent in situ hybridization (FISH)
Fitness, 399, 404, 424, 430, 626, 697, 705, 707
Flagella, **31–33**, 80, 151, 152, 194, 201, 308, 317, 319, 330, 331, 335, **339–340**, 342, 400, 489, 629, 773
Flow cytometry (FCM), 5, 268, 488, 526, **649–651**, 773, **776–781**, 787, 822
Fluorescence microscope, **774**, **777**
Fluorescent in situ hybridization (FISH), 194–196, 371, 383, 562, 652, 653, **780**, **797–798**
Fluorescent probes (molecular, taxonomic), 649, 653, 775, **776–780**, 787
Fluorochrome, 273, 491, 652, 720, **774–780**, 792, **803–804**, 834, **837**
Fluorophore, **774**, **778**, **798**
Food chain, 35, 55, 227, 237, 281, 345, 383, 396, 404, **486**, **490**, **495–496**, 500, 513, 514, 541, 565, 580, 595, 600, 625, 636, 783, 784, 796, **799**, 868
Food web, 4, 6, 167, 224, 248, 262, 362, 396, 437, 485–506, 513, **515**, 522, 552, 554, 572, 598, 604, 605, 626, 762, **880**
Fossil
  carbon, **515**, **518–520**
  fuels, 513, **517–520**, 612
Fossil eukaryotic cells, **92–93**
F ratio, **529**
Freeze-drying, **823–824**
Fresh organic matter, **542–543**, **548–549**
Fruiting bodies, 306, **313**
Frustule, **523**, **529**
Fucogamy, **202**, **203**, 227, 234, 251
Fucosterol, **234**, 251
Fulvic acids, 547, **551**, 732
Fumarate respiration, **46**, **50**
Fumaroles, **361**
Functional diversity, 268, **280**, 704, 787, 880
Functional groups, 212, **262**, **264**, 268, 274, 285, 410, **491–492**, 535, **537**, 541, **597**, 611, 612, 684, 781, 814, 880
Functional microarray, **809**
Fungi, 7, 10, 33, 87, 108, **113**, **133**, **138**, 177, 192, **230**, **242–247**, 262, 299, 374, 397, 408, **411–412**, 416–430, 478, 486, 513, 519, **628**, 662, 815, 840, 852

G

Gametogen, **201–204**, 206, 214, 215, 220, 224, 230, 231, 234, 237, 246, **248–251**
Gaseous effluents, **663**, **680–684**
Gas vacuole, 26, 30, **339–340**, 372
Gas vesicles, 152, 186, **340–341**, 374, 594, 625
G + C content, **152**, 374, **461**
Gellan gum, **367**, **822**
Gene-augmentation, **704**
Gene bank, **801–803**, 833
Gene content, **81**, **463**, 474
Gene duplication, 118, 460, 474, 844
Gene loss, **80**, 84, 262, **457**, **458**, **460**, 474

Generalized transduction, **449–451**
Genetically modified microorganisms (GMMs), **661**, 737, 746
Genetic code, 79, 80, 110, 222, 396, **461**
Genetic criteria (prokaryotes), **152–154**, 160
Genetic drift, **148–150**, 399
Genome/genomics, 5, 20, 30, 76, **81**, 108, 145, **152**, 197, 262, 293, **355**, 400, 446, **454**, 493, 560, 625, **630**, 684, 780, **809–811**, **831–844**
Genomic hybridization, **146**
Genomic island, **429**, 468, 473, **474**, 619, **634–635**, 693
Genomic microarray, **811–812**
Genomic signature, **461**
Genomic species, **147–150**, 152
Genotype, 149, 160, **269–275**, 318, 411, 412, 414, 415, 420, 625, 822
Georespiration, **518**
Glaciers, 263, 345, **354–357**, 840, 880
Global carbon cycle, 247, **516–522**
Global production, **529–530**
Glycine betaine, 172, **330–331**, 333, 372, 375
Glycocalyx, **29**, 163
Glycolysis, 36, **38–40**, 65, 66, 81, 82, 573, 608
Glycoproteins, 29, **30**, 32, 33, **172**, 330, **360**, 413, 428, 477
Glycosylation, **312–313**
Glyoxylate cycle, 38, **65–67**
GMMs. *See* Genetically modified microorganisms (GMMs)
GO-FLO (sampling bottles), **761**, **764**, **765**
GPP. *See* Gross Primary Production (GPP)
Great Salt Lake, 171, **372**, 373
Green and red phyla, **526**
Green bacteria, 30, **56**, **58–59**, 60, 62, 340, 513, 588, **590–591**, 593
Greenhouse effect, 357, 516, 543, **553**, 556, **557**, 578, 612
Greenland, **76**, 356, 386, **557**
Green nonsulfur phototrophic, **56**, 63, **177**, **592**
Green phototrophic, 19, **589**
Green sulfur phototrophic, **56**, 59, **184**, 491, 590, **592–593**, 594, 600, 607, 797
Grey waters, 664
Gross photosynthesis, **794**
Gross Primary Production (GPP), 517, **519**, 523, 530, 532
Groundwater, 170, 358, 542, 546, 579, 580, 608, 620, 663, **686**, 700, 708, **737–740**, **744–746**, 812
GS-GOGAT system, **567**
Guaymas basin (Gulf of California), **362**

H

Hadopelagic, **532**
Haloalkaliphiles/Haloalkaliphilic, 171, 172, **385**, **387–391**
Haloalkaliphilic phototrophic bacteria, **389**
Halocline, **593**, 764, 891
Halophilic/Halophile, 6, 21, 27, 55, 60, 65, 70, 161, 162, 165, 168, 169, **171–172**, 182, 183, 187, 326, **329–334**, **367–375**, 377, 385, 390, 461, 473, 557, 575, 592, 815, 841
Halorhodopsin environment, **391**
Halotolerant, 169, **172**, 182, **329**, **356**, 357, 373, 375, **647**
Haploid, **198–201**, **203–206**, 214, 217, 218, 220, 222–224, **226–228**, 231, 234, 235, 245, 246, 249, 250, **466**
Haptonema, **235–237**, 251
Harmful, 131, **227**, 264, 304, 326, 360, 384, 586, 622, 624, 626, 698, 736, 862
Health, 5, 6, 13, 22, 108, 121, 160, 227, 228, 247, 264, 281, **283–286**, 320, 324, 401, 405, 415, 419, 420, 422, 423, 425, 427, 432, 436–437, 467, 468, 557, 605, 620–622, 624, 626, **628–631**, **636**, 637, 639, 649, 651, 656, 660, 663, 685, 703, 729, 735, 736, 816, 820, 869, 874
Heat-shock proteins, **100**
Heavy metal, 21, 152, 234, 345, 380, 383, 403, 420, 448, 468, 505, 634, 661, 699, 700, **740–746**, 762, 814

Helotism, 123, **127**, **130**, **131**
Hemicellulose, **212**, 242, 247, 402, 547, **548**, 559
Herbicide, 345, 404, **684–687**, 690, **691**, **693–696**, **699–701**, 744
Heterocyst, **67**, 120, **178**, 336, **569**, 625
Heterogeneous taxa, 158, 280, 455
Heterotrophic
    bacteria, 55, 60, **64–70**, 154, 182, 183, 331, 343, 344, 356, 488, **491–492**, 494, 495, 499, 536, 552, 599, 672, 717, 778, **785–786**, 796, 799, **820**
    denitrification, 388
    microorganisms, 47, **65–67**, 407, **494**, **496**, 499, 513, 526, 534, 549, 570
    nitrification, **573–547**
    prokaryotes, 4, **491**, 499, 500, 503, 534, **539–541**, 842
Heterotrophs, 7, 34, 37, 41, 43, 49, **67**, 111, 166, 167, 175, 215, 227, 235, 237, 240, 245, 381, 488, 489, **491**, 500, 529, 553, 574, 589, 593, 599
HGT. *See* Horizontal gene transfer (HGT)
Hierarchical classification, **113**
High frequency recombination (HFR), **449**
High Pressure Liquid Chromatography (HPLC), 268, 488, 490, 491, **782**, 813, 814, 838
High taxonomic ranks (prokaryotes), **160**, **196**
Histones, **33**, 100, **197**, 225, 251, **294**, 298, 366
History, 5, **9–23**, **76–78**, **80**, **83**, **87**, **96**, 101, 104, **108–118**, 123, 125, 140, 150, 194, 220, 262, **294–295**, 333, 338, 357, 376–377, 385
Holomictic, **593**, **595**
Homeostasis, 294, 297, 329, 335, 336, 346, **390–391**, 403, 433
Homoacetogenic (microorganisms), 184, **373**, 390
Homoacetogenic bacteria, **373**
Homologous genes, **79**, 80, 96, 274, **456–458**, 464, 465, 472, 474, 835
Homologous recombination, **148**, 150, **317**, 447, 449, **453**, 696, 697
Homologous sequences, 118, 157, **454**
Homology-facilitated illegitimate recombination (HFIR), **453**
Hopanes, 27, **93**, 95, **268**, 513, 716, 717, **727**, 730, 731, **734**, 737, 796
Horizontal gene transfer (HGT), 6, **77–79**, 98, 102, 121, 124, 242, 248, 266, **404–405**, 410, **445–479**, 630, 635, 636, 660, 685, 692–694, 696, 705, 707
Host, 13, 30, 76, 109, 162, 196, 266, 295, 356, **395–400**, 447, 490, 554, 620, 660, 775, 835
Hot environments, 82, 164, 177, 325, **361–363**, 366, 370, 470, 547
Hot spots, 285, 537, 542, **544**, 576, 835
Hot springs, 86, 164, 166, 167, 170, **173–175**, 177, 184, 185, 325, 333, 361, 362, **364–366**, 557, 560, 597, 624, 840
Housekeeping genes, 147, 148, **150**, **459**
HPLC. *See* High Pressure Liquid Chromatography (HPLC)
Humic acid, 267, 513, **542**, 800, 802
Humification, **551**, **661**
Humus, 185, 505, 547, **549–551**
Hungate tubes, 20, 557, **817**, 819
Hybridization, 111, **146**, **153**, **155**, **157**, 158, 194, 355, 424, 463, **652–653**, 655, **780**, 798, 800, 801, **808–811**, 837
Hydraulic detention time, **671**
Hydrazine, **577–578**
Hydrocarbon, 22, 27, 314, 358, 371, 554, 612, **716–724**, **728–729**, **731–734**, 736, 737, 740, 799
Hydrocarbonoclastic bacteria, 18, 718, 719, 734
Hydrocarbonoclasts, 18, **718–720**, 722, 727, **734–737**, 769, 815
Hydrogen
    consumption, **541**
    hypothesis, **98**, **99**
    production, **52**, **541**
Hydrogenogenes, **541**
Hydrogenosomes, **53**, **100**, 101, **124**, 125, 222, 241, 243, 252
Hydrogenotrophic/Hydrogenotrophs, 17, **54–55**, **62**, 167, 364, 373, 390, 403, **559–561**

Hydrogen peroxide, 103, **327**, **328**
Hydrolase, 367, **688–690**, 694, 696
Hydrolytic enzyme, 33, **539**
Hydrostatic pressure, 354, 356, 369, **375–379**, 650, 764, **768–770**, 772
Hydrothermal chimneys, **362**
Hydrothermal fluid, **361**, **362**, **377**
Hydrothermalism, **361**
Hydrothermal vents, 19, 102, 109, 172–174, 183, 196, 241, 334, 336, 356, **361–362**, **365–367**, **377–379**, 430, 525, 557, **599–600**, 607, 609
Hydroxylamine oxidoreductase, 42, **573**, 578
Hydroxylase, **688–689**, 693, 722, 724–727
Hydroxyl radical, 327, **328**, 568
Hydroxypropionate (3-hydroxypropionate cycle), 61, **63**, 177, 797
Hyperbaric, 5, **769**, **771**
Hypersaline, 19, 167, 168, **171–172**, 196, **331–333**, 363, **367–369**, **371–375**, 392, 492, 559, 597, 841, 878
Hypersaline environments, 171, **331–333**, **367–368**, **372–373**, **375**, 841, 878
Hyperthermophile, 21, **82–84**, 162, **164**, 166, 167, 171, 172, **325**, 354, **361**, **366**, **367**, 377–379, 381, 391, 392, 469, 470, 728
Hyperthermophilic bacteria, **82**, **176**, **325–326**, **361–367**, **469–471**
Hyperthermophily, **82–84**, **361**
Hyphae, 187, 188, **245**, 246, 297, 306, 339, 407, 416, 419, 430, 493, 499, **628**, 847
Hypolimnion, **551**, **552**, **594**

I
Iberian Pyritic Belt, 380, 384, 385
Iceland springs, 361
Identification, 5, 13, 19, 87–88, 92, 96, **108**, **113**, 150, 152–155, 157, 159, 265, 267, 268, 280–282, 385, 421, 437, 459, 463, 477, 534, 626, 651, 652, 655, 656, 697, 727, 731, 780, 796–800, 804, 806, 807, 811, 838, 840–842, 848
Immobilization, 582, 699, 740, 743
Immunity, 22, 413, 414
Immunofluorescence (IF), **652–653**, **780**
Incineration, 681, **683**, 740
Indicator, 196, 266, 285, 316, 543, 620, **638–640**, 644, 645, 651, 655, 656, 709, 733, 789, 792, 796, 799, 814, 818
Individual (taxonomy), 145, 147, 161, 162, 199, 200, 202, 203, 216, 217, 225, 228, 251
Infection, 182, 184, 187, 215, 242, 247, 281, 316, 344, 401, 404, 413, 414, 416, 417, 419, 421, 424, 425, 427, 428, 432–436, 450, 465, 467, 492, 503, 504, 620–624, 628–631, 633, 636, 638, 641, 644, 647, 788, 871
Infectious disease, 5, 12–14, 109, 152, 399, **423–427**, 435–436, 620, 635, 641, 648, 649, 656
Injectisome, 32
Insect, 10, 81, 182, 240, 247, 283, 324, 346, 408, 412, 413, 423, 424, 428, 430, 432–434, 436, 490, 505, 544, 552, 557, 579, 685, 689, 691, 708, 833, 843
Insertion sequences (IS), **121**, 319, 324, 404, 429, 449, 619, 693, 694, 705, 724
In situ colonizers, 822
Interaction, 4, 13, 77, 108, 181, 247, 266, 295, 358, **395**, 476, 485, 521, 620, 680, 761, 839, 849, **866**
Internal conserved biodegradation standards, 733
International Code, 113, 114, 192
Interspecies hydrogen transfer, **54–55**, 98, 183, 540, 559
Intestinal flora, 177, 185, 188
Intestinal tract, 169, 170, 185, 273, **283–284**, 334, 335, 404, 423, 432, 559, 560, 620, 642
Intracellular parasites, 109, 195, 218, 227, 243

Invertebrate, 224, 372, 374, 421, 423, 430, 487, 490, 497–499, 505, 542, 545, 547, 566
Ionophore, 789
Iron (respiration), 46, 49
Iron assimilation, 603
Iron cycle, 383, 384, **601–603**, 607
Iron in oceanic environments, **601–603**
Iron Mountain, 382
Iron oxidation/oxidizing, 42, 44, 181, 182, 334, 380, 382, 384, 599, **606**, 607, 612
Iron reduction/reducing, 183, 184, 365, 382, 540, **607–608**, 610
Isoelectric focusing (IEF), 838, 892
Isomerase, 39, 327, 359, 688, 690, 694, 726, 727
Isoprenoid hydrocarbons, 716
Isotope, **86–87**, 92, 104, 275, 499, 500, 532, 562, 565, 570–572, 581, 585, 784, 795–800
Isotope paring method, 5, 7, 75, 86–87, 101, 104, 491, 567, 570–572, 784, 796, 797, 799

J
Jaccard coefficient, 154–156
Jurassic, 369

K
Karyoklepty, **130**, 132, 133, 224
Kerogen, 89, 513, 542
Kinetic apparatus, 122, **125–127**, 192, 197, 201, 210, 213, 225, 232, 237, 242, 245, 250, 252
Kinetoplast, 240, 252
Kinetosome, 33, 125, 128, 211, 213, 217, 239, 240, 242, 245
Kleptoplasty, **128**, 130, 132–134, 136, 138, 224
Krebs cycle, **38–40**, 65–67, 311, 718
K strategists, 501, 549

L
Lactic acid bacteria, 187, 286, 329
Lactic acid fermentation, 53
Ladderane, 577
Lagoons, 90, 103, 156, 215, 220, 227, 282, 345, 386, 486, 488, 490, 494, 593, **595–597**, 637, 664, 665, 732, 741, 746, 760, 816
Lagoon treatment plant, **664–665**
Landfarming, 662–663, 738, 740
Last universal common ancestor (LUCA), 6, **75–85**, 101, 103, 104, 117–118, 175, 354
Leachate, 680, 741, 745
Leghemoglobin, 20, 68, 416
Leprosy, 188, 424, 635, 840
Leucine (radiolabeled), 785
Life cycles, 192, **198–206**, 209, 211, 214, 216–218, 220–222, 224, 226–228, 230, 231, 234, 237, 239, 240, 242, 244, 246, 247, 249–253, 353, 354, 397, 414, 426, 478, 505, 624
Light-harvesting molecules, 813
Lignin, 65, 212, 247, 250, 431, 499, 542, 544, 548–552, 679, 697, 701
Lindane, **278**, 346, 701, **705–707**, 746
Linkage disequilibrium analysis, 269
Lipid desaturases, 359
Lipoglycane, 892
Lipopolysaccharides (LPS), 29, 319, 322, 359, 413, 428, 433, 434, 626, 633, 722
Long branch attraction artefact, 174, 241, 459
LPS. *See* Lipopolysaccharides (LPS)
LUCA. *See* Last universal common ancestor (LUCA)
Luciferase, 834
Lumen, 127, 239, 560
Lyme desease, 184, 465, 648
Lysis, 87, 172, 322, 408, 409, 450, 451, 494, 501, 503, 534, **628**, 632, 647, 648, 650, 730, 807
Lysogenic, 449, 450, 504, 788
Lysozyme, 780
Lytic cycle, 449, 504, 788

M
Macrobenthic organisms, **537**, 541
Macrofauna, 492, 497–499, 504, 505, 537, 542, 544, 545, 576, 581, 767, 799
Macrofossils, 93
Macrophytic plant, 666
Magnetosomes, 30, 606
Magnetotactic bacteria, 606
Malolactic fermentation, 29, 52
Mammal, 118, 164, 169, 237, 244, 283, 324, 335, 342, 401, 408, 425, 428, 432–435, 514, 557, 629
Manganese oxidation, **608–609**
Manganese reduction, **608, 609**, 610
Marine snow, 766
Mars, 141, 336, 354, 357, 391, 392, 739
Mass spectrometry, 140, 499, 562, 570, 640, 781, 797, 814, 838
Matrix Assisted Laser Desorption Ionization-Time Of Flight (MALDI-TOF), 838
Matrix similarity, 154
Maximum likelihood, 457, 893
Meiosis, 198, 199, 217, 222, 224, 225, 230, 231, 237, 242, 245, 249, 252, 400
Membrane bioreactor (MBR), **674–676**
Membrane lipid(s), 27, **83–84**, 151, 152, 326, 366, 781, 824
Mercury, 279, 345, 346, 404, 600, 601, 604, 605, 637, 695, 745, 746, 774, 791
Meromictic lagoons, 595
Meromictic lakes, 551–553
Mesofauna, 492, 544
Mesopelagic, 516, 531, 532, 772
Mesophilic/Mesophile, **82–83**, 104, 162, 168–171, 173, 174, 176, 325–327, 354, 355, 358–361, 364, 366, 372, 374, 381, 390, 421, 469–472, 557, 560, 588, 677–679, 740, 840
Mesothermic, 358
Mesotrophic, **530–531**, 552, 625
Mesozoic, 92, 369
Messinian, 369
Metabolic capacities of community, 268–269
Metabolic plasticity, 612
Metabolism, 6, 9, 28, **33–61**, 77, **101–102**, 109, 151, 211, 280, 293, 365, 396, 492, 513, 626, 660, 769, 831, 853
Metabolites, 20, 22, 28, 68, 82, 140, 262, 267–268, 275, 284, 285, 295, 297, 298, 302, 310, 311, 315, 317, 320, 324, 346, 391, 397, 407, 410, 417, 498, 553, 564, 604, 606, 626, 638, 680, 681, 685, 686, 688, 692, 695, 696, 698, 700, 701, 714, 718, 723, 736, 781, 799, 831, 838, 839, 842–844, **858–859**, 863
Metabolomics, metabolome, 7, 831, **838–839**
Metagenome, 121, 296, 479, 503, 706, 707, 841–844
Metagenomic, 5, 20, 111, 121, 146, 175, 189, 275, 282, 283, 286, 355, 366, 468, 503, 504, 611, 707, 808, 811, 841–844
Metalimnion, 625
Metalloids, 600–611, 637, **740–746**
Metals
   bioprecipitation, 745
   bioreduction, **742–744**
Metapopulation, 149
Metatranscriptomic, 366, 843

Methane
  anaerobic oxidation, 49, 364, **562–564**, 797, 883
  degradation, **562–564**
  destruction, 561
  hydrate, 554–556
  oxidation, 364, 516, 553, 558, **562–564**, 798, 799
  production, 367, 516, 547, **554–565**
  removal, 561–562
  sources, 554–556
Methanization, 675, **679–680**
Methanoarchaea, 364, 372, 373, 390, 391, **557**, 559, 560
Methanogen, 50, **167–171**, 364, 541, **557**
Methanogenesis, 17, 27, 46, 49, 50, 55, 161, 162, 168–170, 174, 284, 371, 373, 378, 390, 516, 539, 540, 553, **557–562**, 584, 677, 711, 796, 843
Methanotrophic/Methanotroph, 49, 55, 63, 87, 99, 389, 431, 516, 547, 553, 558, **561–563**, 790, 797, 799, 815
Methanotrophy, 17, 431, 516
Methylamines, 50, 63, 64, 169, 335, 373, 374, 390, 559
Methylhopanes, 93, 95, 104
Methylmercury, 584, 604, 605
Methylotrophic/Methylotrophs, **63–64**, 68, 181–183, 390, 516, 553, 558, 561, 600, 796
Micro-aerophilic/aerophiles, 15, **44**, 104, 171, 175, 183, **329**, 364, 539, 541, 573, 623, 789, 817
Microautoradiography, 778–779
Microbial loop, 404, **494–496**, 501, 526, 553, 880
Microbial mat, 4, 19–21, 88–91, 93, 120, 121, 177, 179, 183, 266, 268, 270, 272, 275, **280–282, 341–345**, 403, 505, 537, 566, **597–600**, 722, 786, 789, 844, 875, 876, 878, 879
Microbiota, **421–423**, 431, 432, 436, 498, 500, 545, 560, 561, 650, 698
Microcolony, 408
Microcosm, 263, 280, 506, 596, 642, 693, 701, 711–713, 746, 799, 820
Microcystin, 177, 626–628
Microelectrodes, 5, 21, **788–794**, 875, 878
Microfauna, 492, 500, 506, 544, 670
Microfossils, 87–89, 92, 104
Micro-habitats, 149, 196, 280, 490, 498, 514, 546, 789
Micromanipulation, 122, 123, 774, **821**
Microniches, 49, 387, 514, 542, 581, 679
Micro-optodes, **791–792**
Microorganisms (study methods), **757–829**
Microorganisms (culture media), **816–817, 822**
Microorganisms (genetically modified organisms, GMO), **661**
Microplates, 152, 654, 655, 786, 787, 822
Microscope, Microscopy, 11, **773–776**
Micthaploid, 201, 204, 206, 245
Microsensors, 788–790, 792, **788–794**
Minamata disease, 584
Miniaturized biosensor, 791, 792
Minimal genome, 80, 82
Miocene, 357, 369
Mitochondria, 6, 26, 33, 38, 41, 53, 97–101, 104, **122–127**, 130, 161, 168, 197, 198, 207, 210, 213, 215, 217, 221, 222, 225, 227, 230, 234, 237–239, 241, 243, 248, 250–252, 397, 399, 433, 435, 476–478, 775
Mitosomes, **100**, 101, 124, 197, 241, 243, 248, 252
Mixotrophic/Mixotrophs, **37**, 41, 42, 166, 175, 196, 227, 235, 237, 381–383, 486, **488–490**, 492, **495**, **526**, 574, 589
Moderate halophiles, 185, 367
Modification/restriction system, 452
Molecular beacons, 780
Molecular clock, **117–118**, 158, 454, 806
Molecular fingerprint, 268, 503, 730, **802–805**
Molecular markers, 83, 84, 118, **150**, **157**, 588, 628, 806

Molecular paleontology, 116
Molecular phylogeny, 80, 100, 135, 150, 158, 173, 250, 376, **457**
Monogenetic life cycle, 234
Mono Lake, 386, 389
Monomictic lakes, 594
Monooxygenase, 42, 175, 275, 561, 562, 573, 574, 688, 693, 696, 699, 722, 842
Mono-unsaturated fatty acids, 814
Morphology, 6, 88, 89, 111, 118, 152, 179, 184, 187, 199, 200, 220, 250, 318, 339, 382, 391, 493, 500, 563, 576, 593, 628, 776, 875
Most probable number, 267, 268, 587, 649, 655, 817, **820**
Multi-locus approach, 157
Multi locus sequence typing (MLST), 150, 152
Multitrophic interactions, 407
Multivariate analysis, 265
Murein, 29, 162, 402
Mutualism, 6, 123, 127, 130, 227, 246, 396–398, 403, 411, 421, 424, 430–433, 437, 863
Mutualistic symbiosis, 98, 197, 215, 415, 433, 566, 589
Mycolic waxes, 188
Mycorrhizae, 398, 416, 815
Mycosporines, 813
Mycotoxin, 283, 408, 409, 411, 412

N
Natural selection, 297, 414, 453
Neighbor-joining, 110, 117, 157, 176
Nei index, 276–278
Nekton, 552, 893
Neoglucogenesis, 65, 66
Neritic, 526
Net community production, 786
Net primary production, 247, **519**, **526–530**
New production, 494, **529**, 784
Niskin (bottles), 761, 762, 764, 770, 771
Nitrate assimilation, **565–568**
Nitrate dissimilatory reduction, 572–575
Nitrate production, 576
Nitrate reductase, 46, 47, 67, 339, 468, 474, 566, 571, 573, 575
Nitrate respiration, **46**, **47**, 187, 468
Nitric oxide, 46, 47, 78, 407, 564, 565, 568, 573, **575–578**
Nitrification, 15, 20, 42, 175, 274, 279, 404, 410, 411, 498, 529, 547, 565, 567, 568, **570–576**, 578–581, 643, 644, 842
Nitrifying bacteria, 29, 42, 45, 181, 183, 334, 389, 407, 514, 573, 574, 599, 612, 808, 842, 877
Nitrite oxidizing bacteria, 42, 61, 274, 389, 404, 573
Nitrite reductase, 46, 47, 67, 274, 339, 571, 573, 575–577
Nitrogenase, 67, 120, 178, 274, 282, 415, 416
Nitrogen assimilation, 565–569
Nitrogen cycle, 175, 274, 282, 389, **565–581**
Nitrogen fixation, 15, 16, 20, 67, 303, 312, 336, 389, 410, 417, 419, 420, 514, **566–569**, 571, 840, 842
Nitrogen fixing bacteria, 14, 20, 404, 407, 418, 419
Nitrous oxide, 46, 47, 274, 420, 545–546, 565, 567, 568, 570, 571, 573–576, **578**, 580, 672, 790, 792
Nodules, 68, 298, 371, 398, 407, 415, 416, 420, 555, 556, 579, 608, 609, 797, 841
Nomenclature, 113, 114, 119, 136, 147, 150, 151, 160, 192, 196, 206, 708, 709, 815
Noncyclic photophosphorylation, 57
Non homologous recombination/illegitimate recombination, 452, 453
Non-parametric test, 265, 267
Nosocomial, 182, 344, 404, 427, 436, 622, 636, 641
NPC. *See* Nucleoplastidial complex (NPC)
Nucleating proteins, 360

Nucleic acids, 65, 67, 70, 157, 247, 269–273, 285, 324, 326, 328, 358, 395, 400, 504, 515, 534, 549, 565, 620, 635, 649, 650, 652, 655–656, 721, 774, 778, 780, 797, **800–802**, 806–809
Nucleoid, **30–31**, 198, 225
Nucleomorph, 127, 217, 238, 239, 251, 252
Nucleoplastidial complex (NPC), 127, 128, 234, 236, 238, 251, 252
Nucleus, 25, 31, 32, 83, 86, 92, 98, 110, 115, 118, 121–123, **125–127**, 130, 131, 133, 161, 197, 207, 215, 217, 221, 222, 224, 225, 230, 237–239, 242, 244, 249, 251, 252, 399, 414, 478, 622, 773, 775
Numerical taxonomy, 147, 154
Nutrient concentration, 490, 522, 527–529, 537, 734, 860

O

Obligate aerobes, 164, 817
Obligate alkaliphiles, 385
Obligate anaerobes, 817
Obligate parasites, 187, 218, 411, 620, 629
Ocean color, 529, 530, 782, 783
Oceanic phosphate cycle, 533–536
Oil, 21, 169, 170, 172, 173, 176, 183, 268, 280, 281, 283, 342, 344, 362, 363, 365, 373–376, 379, 514, 517, 520, 556, 584, 587, 607, 643, 662–664, 666, 668, 669, 685, 708, **716–719**, 721, 728–738, 740
 production, 286, 375
 reservoir, 170, 176, 362, 363, 365, 374, 376, 379, 607
Oilfield, 556, 758
Oligoheterotroph, 377
Oligonucleotide probes, 194, 809, 810, 812
Oligotroph/Oligotrophy/Oligotrophic, 5, 19, 44, 178, 267, 376, 390, 486, 488, 489, 494, 495, 523, **526–527**, 530, 531, 533, 552, 574, 602, 647, 692, 703, 785, 866
Olivine, 386
Oogamy, **201–203**, 213, 214, 234, 251
Open ocean, 178, 522, 526, 529–533, 537, 540, 566
Operational taxonomic units (OTUs), 195, 265, **275**, 276, 278, 282, 806
Operon, 27, 269, 297, 298, 308–311, 314, 318, 320, 333, 377, 450, 453, 468, 573, 575, 604, 685, 693–698, 707, 715, 724–727, 802, 803, **834**, 881
Opportunistic, 317, 343, 413, **427**, 428, 499, 622, 628, 630, 631, 633, 636–638, 641, 642, 656, 701, 822, 840
Optical microscope, 109, 773
Optical micro-sensors, 786, 788–795
Organelles, 25, 27, 33, 53, 92, 101, 111, 123, 124, 126–128, 130, 138, 197, 198, 208, 210, 213, 226, 233, 234, 236–238, 240, 241, 340, 476, 576, 593, 833, 835
Organic matter
 degradation, 4, **539–542**, 553, 559, 598, 665
 recalcitrant, 499, 552
 sequestration, **542**
Organic sulfur (biosynthesis, degradation), 582–583
Organochlorine, 49, 561, 691
Orthologous genes, 894
Osmoregulatory compound, 330, 581
Osmotic pressure, 172, 294, 302, 303, 322, 329, 330, 332, 336, 373, 375, 421
Osmotrophic/Osmotrophy, 478, 479, **489**, 497, 498
OTUs. *See* Operational taxonomic units (OTUs)
Outgroup, 84, 212, 232
Oxic-anoxic interfaces, 607, 609
Oxic conditions, **44**, 47, 70, 516, 542, 546, 548, 558, 607, 612, 734
Oxic environments, 70, 513, 894
Oxic zone, 539, 715, 734, 818
Oxidative phosphorylation, **36–39**, 49, 572
Oxido-reduction, 537, 539, 611

Oxygenic photosynthesis, 4, 35, **56–58**, 70, 83, **93–94**, 96, **102–104**, 139, 161, 177, 345, 476, 512, 526, 527, 598, 599, 608, 665, 789, 794
Ozonation, 644

P

PAI. *See* Pathogenicity island (PAI)
Paleomicrobiology, **285**
Pan-genome, **456**, 811
Paralogous genes, **84**, 85
Parametric test, **265**, 266–267
Para-primary producers, **35**, 491, 895, 897
Para-primary production, 595
Parasite, 12, 19, 53, 69, 81, 109, 111, 120, 126, 181, 184, 187, 195, 215, 218, 219, 221, 222, 224, 227, 228, 230, 231, 240, 242, 243, 247, 248, 317, 397, 399, 401, 405, 406, 408, 411–413, 415, 417, 423, 425–427, 432, 433, 435, 478, 479, 490, 620, 629, 644, 705, 795, 839
Parasitism, 6, 123, **130–131**, 227, **396–398**, 400, 403, 411–415, 421, 424, 432–433, 437, 501, 647, 863
Parasitoid, 400, 430
Particles, 5, 11, 14, 30, 32, 33, 49, 65, 88, 89, 101, 110, 219, 237, 241, 250, 341, 409, 424, 446, 449–451, 486–488, 490–492, 495, 498, 501, 514, 515, 522, 534, 537, 539, 542, 548, 551, 576, 602, 605, 621, 631, 634, 643, 645, 668, 669, 713, 715, 732, 744, 763, **765–767**, 769, 771–772, 774–778, 785, 788, 793, 820, 840
Particulate organic matter (POM), 494, 495, 497, 519, **528**, 531, 552, 771
Pathogen, 5, 12, 30, 117, 148, 224, 263, 312, 396, 467, 504, 545, 619, **653**, **655**, 660, 773, 840, 862
Pathogenic bacteria, 13, 30, 32, 81, 113, 131, 152, 182, 184, 284, 309, 317, 342, 413, 428, 435, 620, 622, 631, 634, 635, 641, 645, **648–650**, 838
Pathogenicity island (PAI), 428, 429, **634–635**
PCBs. *See* Polychlorobiphenyls (PCBs)
PCR. *See* Polymerase chain reaction (PCR)
Pelagic, 7, 211, 216, 221, 235, 485–486, 488–491, 493–498, 500–504, 506, 531, 532, **551–553**, 716, 732, 769
Peptidoglycan, **29**, **31**, 162, 184, 185, 187, 208, 303, 306, 321, 322, 324, 413, 428, 433, 548, 565, 744, 780
Permafrost, 354, **355**, **357**, 555, 556
Peroxidase, 103, 328, 548, 549, 603, 701
Peroxisome, 33, 124, 125, 895
Persistence, 432, 434, 446, 453, 465, 478, 561, 636, 638, 640, 642
Pesticides
 biodegradation, **686–699**
 bioremediation, **698–708**
Petroleum
 hydrocarbons, 268, **718**, 728, 729
 products, 18, 716–740
 reservoirs, 104, 362–364
PGPR. *See* Plant growth-promoting rhizobacteria (PGPR)
Phage, 281–283, 405, 449–451, 463, 501, 503, 504, 634, 640, 697, 788, **834**, 869, 870
Phagotrophy, 122, 130, 133, 198, 245, 478, **489**
Phase contrast microscope, 773, 774
Phase variation, **317–320**, 410
Phenotype, 121, **158–160**, 162, 267–269, 295, 308, 312–314, 316–319, 432–433, 447, 449, 501, 546, 630, 634, 638, 705, 722
Phenotypic criteria (prokaryotes), 151–152, 158
pH homeostasis, **403**
Phosphate removal, 672, 842
Phospholipid fatty acids (PLFA), 267, **781**, 795–799, 814–816
Phospholipids membrane, 28, 31–33, 83, 161, **366**, 781
Phosphotransferase system, **29**

Photoautotrophic, 55, 491, 497, 504–505, **530**
Photochemical, 561, 578, 605, 762
Photoheterotrophic, 55, **530**, 591
Photolithotrophic/Photolithotrophs, 34, **55, 56**, 513, 516, 530, 582, 591
Photoorganotrophs, 34, 56
Photophosphorylation, **36**, 57–60
Photosynthesis, 3, 6, 19, 25, 26, **55**, 75, 113, 161, 205, 275, 332, 376, 396, 476, 491, 512, 624, 665, 765, 841, 860
Photosynthetic bacteria, 19, 87, 91, 123, 130, 151, 152, 156, 177, 179, 181, 182, 282, 583, 596, 611, 795, 813
Photosynthetic microbial mats, 88, 91, **597–599**
Photosynthetic production, 35, **526**, 527, 598, 793
Photosynthetic sulfur oxidizing microorganisms, **589–593**
Photosystem, 17, 56–58, 60, 102, 177, 212, 588–590, 608, 815
Phototrophic bacteria, 14, 19, 35, 37, 55–57, 59, 60, 68, 70, 181, **275**, 280, 281, 326, 331, 372, 389, 491, 495, 526, 583, 584, 588–592, 594–596, 598, 599, 813
Phototrophic green bacteria, 30, 56, 58, 60, 62, 588, **593**
Phototrophic nonsulfur bacteria, 589
Phototrophic purple bacteria, 15, 29, **56**, 57, 61, 333, 591–593, 599
Phototrophic purple non-sulfur bacteria, 179, 181, 591, 593
Phototrophic sulfur bacteria, 589, 597–599
Phototrophy, **101–102**, 172, 179, 274–275, 841
Phycobilins, 123, 126, 127, 177, 178, 208, 210, 211, 225, 238, 239, 250, 252, 624
Phycobilisome, 123, 126, 127, 208, 210, 239, 250
Phylogenetic analysis, 110, 147, 153, 155, 157, 159, 458, **464**, 465
Phylogenetic diversity, 137, 215, 231, 526, 568, 607
Phylogenetic microarray, 809
Phylotype, 193–196, 261, 266, 281, 283–285, 357, 422, 562
Physiology, 5, 7, 13, 14, 16–22, 77, 101, 104, 108, 134, 152, 161, 182, 196, 285, **294**, 299, 312, 335, 365, 367, 379, 389, 410, 433, 563, 574, 685, 719, 722, 760, 767, 768, 788, 820, 835, 837, 839, 841–843, 847
Phytobenthos, 537
Phytoremediation, 420, 699, 703, **738–740**
Phytosphere, 408
Phytostimulation, 419, 661, 701
Picoeukaryote, 216, 553, 778
Picoplankton/picoplanktonic(s), 111, 215, 282, 486, **488–489**, 494–496, 501, 526
Piezomicrobiology, 376, 770
Piezophiles/Piezophilic, 6, **336**, 365, 375–379, 765, 768, 816, 840
Piezopsychrophiles, **376–377**
Piezosensitive, 377, 768, 769
Piezotolerant, 377, **768**, 769
Pigment, 27, 30, 33, 34, 48, 55–57, 59, 102, 111, 123, 129, 152, 161, 172, 192, 208, 210–212, 217, 225, 231, 234, 238, 268, 275, 282, 313, 327, 345, 391, 488, 490, 491, 526, 551, 588, 589, 591–595, 624, 765, 773, 774, 782, 783, 789, 792, **813–814**, 841
Pillow lava, 91, 362
Pilus, 32, 228, 312, 405, 446, **448**, 634
Plague, 182, 319, 424, 635
Planktonic, 60, 172, 220, 221, 227, 237, 307, 486, 488, 496, 506, 526, 600, 720–722, 786, 798, 799, 852, 862, 868
Planogamy, **201–203**, 213, 214, 220, 225, 234, 250, 251
Plant growth-promoting rhizobacteria (PGPR), 271, 406, **417–420**
Plasmid, 26, 30, 150, 171, 271, 295, 313, 314, 319, 324, 335, 346, 374, 402, 404, 405, 410, 412, 416, 420, 427, 429, 446–449, 452, 453, 634–637, 660, 661, 692–698, 704, 705, 713, 718, 724, 746, 802, **805**, 823
Plasmogamy, 201, 202, 206, 245, 249
Plastid, 130, 212, 217, 476–478
Pleiotrophic, 432, 628, 692
Pleiotropy, **432–433**
PLFA. *See* Phospholipid fatty acids (PLFA)

Pluricellular, 205, 491
Point source, **642–645**
Poisoning, 22, 177, 182, 187, 227, 425, 584, 605, 620, 623, 633
Pollution, 5, 280, 281, 284, 286, 419, 420, 579, 580, 628, 639, **640**, 659–746
Polychlorobiphenyls (PCBs), 345, 420, 661, 684, **708–715**
Polymerase chain reaction (PCR), 21, 141, 174, 177, 193, 195, 196, 264, 267, 269, 271, 273, 275, 277, 280, 367, 503, 655, 656, 707, 715, 738, 780, 797, **800–805**, 809, 810, 834, 842, 844
Polymetallic nodules, 608
Polymictic lake, 552
Polymorphic taxa, 158
Polysulfides, 70, 384, 583, 589
Polyunsaturated fatty acids, 356, 816
POM. *See* Particulate organic matter (POM)
Potential activity, 16, **571**
Potentiometric microelectrodes, **789–792**
Predation, 6, 123, 130, 201, 227, 281, 396, **397**, 400, 401, 407, 410, 411, 423, 431, 486, 494, 495, 497, 499–501, 505, 531, 553, 625, 640, 647, 787–788, 863, 866, 869
Predator, 7, 89, 130, 133, 140, 196, 224, 227–229, 235, 237, 240, 241, 383, **397**, 399, 401, 407, 431, 489, 492, 495–497, 500, 505, 523, 553, 599, 626, 647, 648, 782, 787, 788, 849, 863, 866, 867, 870, 872
Predatory bacteria, 183
Prevalence, 424, 432, 479, 504, 561, 621, 628, 636, 637, 692
Primary consumers, 396, **493**, 514, 519
Primary endosymbioses, **126–127**
Primary producers, 35, 167, 247, 263, 388–389, 396, 487, 491, 495, 497, 503, 506, 512, 514, 521, 526, 530, 552, 554, 589, 599, 782–784
Primary production, 35, 111, 194, 247, 282, 372, 389, 405, 437, 486, 494, 495, 499, 503, 505, 506, 513, 517, 519, 521, 523, **526–533**, 537, 542, 552, 594, 595, 598, 603, 605, 624, 625, 664, 665, **783–784**, 799, 860, 861, 872, 880
Primers, 111, 150, 174, 195, 196, 270–274, 276, 280, 780, 797, 800–807, 832, 834
Priming effect, **548–549**, 551
Primitive atmosphere, 4, **77–79**, **83**, **103**
Primitive Earth, 78, 86, 392
Principal component analysis (PCA), 265, 463
Production of ammonium, 569, 572, 578
Production of dinitrogen, 571, **574–578**
Production of sulfide, **582–588**, 597, 745
Programmed cell death, 429, 433
Prokaryote fossils, **87–92**
Propagule, 199, 267, 407, 413, 490
Prophage, 431, **450**, 504, 788
Protein modification, 295
Proteorhodopsin, **61**, 172, 468, 841
Proterozoic, 89, 93, 601
Proton-motive force, 38, 40, 45, 52, 56, 57, 60, 64, 339, 374, **390**
Protrichogamy, **201–203**, 211
Pseudofermentation, 558
Pseudopeptidoglycan, 162
Psychrophilic/Psychrophiles, 6, 169, **325–327**, 336, 354–361, 372, 376, 378, 557, 588, 679, 815, 840
Psychrophily, 19
Psychropiezophiles, 356
Psychrotolerant, 171, 355, 357
Psychrotrophic, 326, **355**, 356, 360
Purple anoxygenic bacteria, 387
Purple bacteria, 15, 18, 29, **56–61**, 102, 160, 179, 181, 280, 333, 334, 345, 389, 590–593, 599, 878, 879
Purple phototrophic (red phototrophic) bacteria, 59
Purple phototrophic sulfur bacteria, 59
Pycnocline, **529**

Pyrite, 87, 166, 173, 334, 363, 380, 382, 384, 581, 584, 585, 588, 605, 606, 741
Pyrolytic hydrocarbons, 718
Pyrosequencing, 140, 270, 272, 273, 285, 808, **834**
Pyroxene, 338, 386

Q
Quantitative PCR (qPCR), **656**, 738, 780, **802**
Quorum sensing, 293, 294, 303, 308, **313–315**, 317, 343, 404, 410, 411, 413, 626, 633, 634, 821, 822

R
Radiation resistance, **177**
Radionucleide, **86**, 468, 745
Rarefaction curve, 261, **272**
RBR. *See* Relative binding ratio (RBR)
Reaction centers, 56, **58–60**, 275, 280, 589, 591, 593
Real-time PCR (RT-PCR), 271, 656, **801–802**
Recalcitrant organic compounds, 403, 499, **513**, **548–550**, 552, 733, 749
Recombination, 82, 148, 150, 211, 242, 262, 272, 317, 319, 322, 400, 402, 405, 447, 449, 450, 452, 453, **465–467**, 504, 635, 693, 696, 697, **705–707**, 746
Red tide, **227**, 589, **595**
Regenerated primary production, **529**
Regenerated production, 494, 523, **529**, 784
Regenerators, **496**, 539, 542
Relative binding ratio (RBR), **146–148**
Reservoir
 Reservoir(s) (carbon cycle), 515, **516–518**, 519–521, 537
 Reservoir (s) (gas hydrates), 556
 Reservoir (s) (pathogens), 621–625, 629, 635, 638, 641, 645, 648, 656
 Reservoir (s) (petroleum), 104, **362–363**, 364
Resistome, 324, **638**
Respiration, **37–51**, 67, 100, 152, 279, 324, 403, 409, 495, 499, 503–504, 506, 517–522, 527, **530–533**, 542, 552–553, 581, 601, **785–786**, 792, 793, 817, 839–840. *See also* aerobic respiration, anaerobic respiration
Respiratory chain, **37–51**, 98, 101, 103, 152, 328, 330, 331, 334, 358, 468, 573, 574, 606
Restriction enzyme, 161, 270, 450, **803–805**
Reverse cycle of tricarboxylic acids, **61–62**
Reverse gyrase, **470–471**
Reverse methanogenesis, 49, 55, 553, **562**
Reverse transcriptase, 76, 656, **801–802**, 837
Reversion, 319, **447**
Rhizodegradation, 661, 699, **738–740**
Rhizodeposition, **407**, 408
Rhizosphere/ Rhizospheric, 68, 121, 271, 273, 315, 316, 320, 338, 339, 398, 404, 405, **407–411**, 417, 419, 420, 505, 544, 545, 561, 641, 661, 666, **701–703**, 713, **738–739**, 758
Rhizostimulation, 7, 659, **661**, 703, 737
Rhodamines, **645**
Ribosomal Intergenic Sequence Amplification (RISA), 261, **269–272**, 281, 282, **802–803**
Ribosomal 23S RNA, 30, 150, 652, **802**
Ribosomal 16S RNAs, 124, 146, 150, 151, **153–155**, **157**, 158, 161, 164, 165, 167, 170, 174–176, 269, 270, 272, 274, 275, 277, 280, 284, 358, 363, 366, 374, 392, 422, 463, 464, 492, 562, 578, 588, 652, 679, 798, **806**, **811**
Ribosome, 26, 27, **30**, **32**, **33**, 79, 80, 82, 97, 98, 124, 128, 146, 161, 198, 209, 210, 213, 223, 226, 233, 236, 238, 239, 241, 243, 244, 252, 295–297, 321, 324, 326, 330, 459, 467, 779, 780

Ribozymes, **76**
Ribulose 1,5-biphosphate carboxylase (RuBisCo), **61–62**, 128, 274, 531, 797
Rice field, 554, 557, 558, **561**
Richness, 18, 193–195, 227, 264, 271, 272, **275–277**, 279, 281, 422, 503, 506, 545, 599, 685, 741, 866, 867
Rift valley Eastern African, **386**
Rio Tinto, **383–385**, 741
RISA. *See* Ribosomal Intergenic Sequence Amplification (RISA)
RNA world, **75–76**, 77–85
Rolling circle replication, **448**
R strategists, 501, **549**
Rumen, 20, 169, 177, 182, 185, 283, 477, 478, 500, **557–560**, 822

S
Salt evaporation pools, 368
Salt-in strategy, **172**, 373
Salt lakes, 171, 333, 334, **368–372**, 373
Salt marsh, 170, 280, 333, 354, **368–374**, 375, 584, 878
Saprophytic, 188, 316, **398**, 410, **411**, **417**, 544, 862
Satellite imagery, 782
Saturated fatty acids, 326, 327, 377, 403, **814–815**
Saturated hydrocarbons, 513, **716**, 731
Saturation analysis, **271**
Sciaphilous, 211
SDS-PAGE. *See* Sodium Dodecyl Sulfate-Polyacrylamide Gel Electrophoresis (SDS-PAGE)
Sea ice, **354–357**, 360
Seawater, 128, 194, 221, 237, 280, 332, 333, 345, 354, 356, 361, 362, **367–372**, 386, 516, 517, 519, 521, 527, 534, 562, 595, 599, 601, 608, 647, 649, 655, 718, 719, 732, **761**, **764**, **765**, **767–768**, 770, 771, 785, 816
Secondary consumers, 396, **493**, 500, 514
Secondary endosymbioses, **127–128**, 132, 134, 476
Secondary metabolism, **20**, 306, 402
Secretion system, **32**, 310, 417, **421**, **428**, **429**, 436, 447, 448, 632, 843
Sediments (ecosystem), 86, 91, 93, 94, 216, 282, 341, 370, 389, 490, **497**, **499**, **537–542**
Sediment sampling, **767**
Sediment trap, 539, **766–767**, 771
Sedimentation (microorganisms), 580, 644–645, **668**, 766
Seeding, 674, 713, **735–736**, 818
Selection, 118, 148, 149, 262, **293–295**, 323, 324, 346, 356, **397**, **399**, 404, 410, 414, 423, 425, 446, 448, 453, **466**, 479, 503, 545, 625, 635, 637, 681, 692, 697, 703, 704, 708, 738, 766, 774, 811, 817, 822, **835**, 842, 862
Selective advantage, 317, 402, 446, **466**, 476–477, 635, 692, 703, 704
Semantides, **115**
Sequencing, 5, 79, 101, 104, 111, 113, 140, 148, 153, 154, 164, 174, 184, 193, 195, 196, 230, 267, **269–271**, 273, 275, 280, 282, 285, 286, 295, 298, 337, 355, 358, 359, 371, 391, 422, 433, 437, 473–475, 492, 503, 625, 693, 696, 697, 727, 780, **805**, 806, 808, 811, 812, **832**, 833–844, 880
Serogroup, **623**
Serpentine, **387**
Settler, 344, **669**, 670, 675
Sewage, 169, 170, 182, 281, 470, 621, 636, 640, **642–645**, 656, 660, **664–676**, 679, 681, 684, 685, 746, 842, 844, 853
Sexual pilus, 32, 312, **448**, 634
Shannon alpha diversity index, **276**
Shotgun, **833**
Siderophore, 70, 417, **602–606**, 631, 634, 744
Signaling, **299–304**, 305–307, 310, 312, 314–316, 340, 396, 403, 416, 417, 433, 822
Signal transduction, 294, 296, **299–304**

Silica, 88, 89, 216, 219, 232–234, 251, 515, 518, **522–524**, 608, 712, 745
Silicium cycle, 518, **522–524**
Similarity coefficient, **154**
Simpson index, **276–278**
Single strand conformational polymorphism (SSCP), **269–272**, 284, 802, **804**
Sinking organic carbon, **537**, 541
SIP. *See* Stable isotope probing (SIP)
Slightly halophiles, **367**
Smokers, 102, 166, 275, **362**, 378, 599, 600
Soda lakes, 333, 334, **385–391**
Sodium Dodecyl Sulfate-Polyacrylamide Gel Electrophoresis (SDS-PAGE), **838**
Soil
   bioremediation, **698**, 700, 703
   biota, **544**
   engineers, **487, 545, 546**
   respiration, **542**, 786
   sampling, **758**
Solfatara, **365**, 366, 383
Solid phase cytometry (SPC), 650–653, **777–779**
Solubility pump, **519**, 521
Somatogamy, **202**, 204, 245, 246, 252
Source, 4, 12, 33, 81, 139, 151, 198, 268, 303, 358, 402, 445, 487, 513, **516**, 621, **645**, 659, 758, 839, 851
Spandrel, **402**
SPC. *See* Solid phase cytometry (SPC)
Spearman correlation tests, **265**
Specialized transduction, **450**
Speciation, 146, 148, 150, 174, 377, 414, 433, 459, 460, 464, 504, **602–603**, 740, **742–745**
Species complex, **148**, **150**
Species richness index, **276**
Spontaneous generation, **10**, 12, 773
Sporangium, **188**
Spore, 10–12, **31**, **32**, 70, 113, 125, 138, 151, 152, 162, 166, 185, 186, 188, 198–202, 204, 205, 217, 218, 220, 224, 226, 229–231, 237, 243–246, **248–250**, 252, **294**, 336, 337, 339, 345, 354, 357, 374, 387, 390, 408, 413, 419, 436, 500, 609, 628, 641, 821
Sporocarp, **245**, 246, 250
Sporogenous generation, **204**, 206, 214
Sporophore, **188**, 336, 338
SSCP. *See* Single strand conformational polymorphism (SSCP)
Stable isotope probing (SIP), 275, **797–799**
Stenopsychrophilic (psychrophilic), **355**
Steranes, **94–95**, 104, 122, 197, 268, 716, **717, 727, 728**, 730, 731
Sterol, 32, 33, **93–96**, 122, 197, 211, 213, 230, 234, 245, 250–252, 409, 496, 513, 640, 730, 815
Stochastic, **452–453**, 876
Stratified lakes, **593–595**
Stromatolite, 4, 19, **85, 88–91**, 344, 597, 878
Subglacial, 104, 275, 355, **357–358**, 392
Suboxic, **539–540**
Substrate-level phosphorylation, **36**, 39, **50–52**, 101
Sulfate production, 588, **589**
Sulfate-reducer, **48–49**, 162, 173, 177, 364, 365, 390, 540, **541**, 547, **584**, 587, 588, **593–594**, 596, 610, 612, 798
Sulfate reducing bacteria, 19, 20, **48–49, 183**, 285, 344, **584**, 586, 588, 596–597, 605, 610, 743, 758, 799, 878–879
Sulfate respiration, **46**, 582, 583
Sulfhydrization, **582**
Sulfur (intracellular), 70, **590**, **592**
Sulfur (respiration), **46**, 164, 166

Sulfur cycle, 14, 19, 275, 345, 383–385, 391, **581–584, 593–596**, 612, 812
Sulfuretum, **584**, 590
Sulfur-oxidizing bactderia, 22, **43–45**, 182, 183, 380, 384, 389, 430, 491, 583, **588–590**, 595, 597, 599, 611, 680, 741, 790, 878
Sulfur oxidation, **43–44, 380**, 385, 582, 589
Sulfur reducers, **46**, 48, 364, 366
Sulfur reduction, **46**, 48, 610
Superoxide dismutase, 49, 103, 118, 298, **328**, 359, 651
Superoxide ion, 103, **327**, 568
Surface microlayer (sampling), **762–764**
Surfactants, 375, 639, 661, 663, 699, 713, **722, 735**, 736, 737
Survival, 5, 9, 16, 21, 31, 49, 81, 113, 130, 131, 227, 294, 295, 324, 336, 341, **357**, 360, 421, 433, 574, 620, 621, 629, 632, 638–642, **646–650**, 656, 692, 702–704, 746, 823, 841, 866, 878
Swallow hot springs, **365**
Swine erysipelas, **187**
Symbiont, 81, 98, 100, 127, 128, 132, 175, 181, 182, 188, 207, 215, 216, 220, 224, 230, 242, 298, 317, 408, **430**, 464, 476, 833, 835, 840, 843
Symbiosis, 6, 19, 20, **98, 123, 127**, 177, 197, 215, 227, 263, 313, 362, **396–398**, 403, 405, 411, 415–417, 420, 423, **430, 431, 433**, 476, 566, 579, 589, 840, 843
Symbiotic, 16, 20, 53, 68, **96–99, 128**, 131, 174, 188, 269, 275, 280, 314, 316, 328, **396–398**, 410, 411, 416, 418–420, 423, 424, 428, **430–432**, 435, 476, 544, 546, 566, 579, 600, 635, 739, 840, 842, 843
Synapse, **118**, 245
Syncytium, **33**, 120
Synteny, 474, **835**
Syntrophic, 20, **54–55**, 98, 99, 168, **403**, 541, 553, 562, 564, 611, 797
Syntrophy, 48, **54–55**, 183, 185, **403**, 562, 660
Syphilis, **184**, 635
Systematics, 6, 19, 81, 107–109, **111–113**, 135, 146, 151, 152, **160**, 175, 177, 187, 264, 280, 295, 306, 470, 476, 835, 844, 847

T

TAC. *See* Tricarboxylic acid cycle (TAC)
Taq polymerase, 21, 267, 271, 367, 843
Taxa, 6, 32, 33, 101, 107, 109, 113, **146**, 191, 261, 293, 398, 400–404, 410, 420, 421, 427, 432, 434, 475, 491, 493, 526, 544, 548, 589, 780, 835, 841, 843
Taxonomic diversity, 4, 196, 264, 274, 391, 456, 493, 611, 612, 880
Taxonomic groups (prokaryotes), 113, 154, 158, 161, 164, 173–175, 194–196, 207, 266, 268, 269, 273, 280, 317, 455, 468
Taxonomy, 6, 19, 108, **113**, 114, 117–118, 135, 145–147, 149–155, 157–159, 189, **191**, 198, 275, 377, 501, 806, 836, 843
TEM. *See* Transmission electron microscope (TEM)
Temperate phage, 449, 450, 504
Temperature gradient gel electrophoresis (TGGE), **269**, 270, 272, **804–805**
Terminal restriction fragment length polymorphism (T-RFLP), 269, **270**, 272, 280, 285, 802–804
Termites, 20, 169, 242, 247, 387, 423, 492, 500, 544, **545–546**, 554
Terrestrial geothermal ecosystems, 361
Terrestrial hot springs, 362, 364, 365
Tertiary endosymbioses, 6, **127–130**, 133, 135, 207
Tetanus, 185, 633
Tetrazolium, 787
TGGE. *See* Temperature gradient gel electrophoresis (TGGE)
Thalassohalines, 171, 333, 368
Thermal spring, 361, 386
Thermoacidophile/Thermoacidophilic, 44, 161, 162, 365, 472
Thermocline, 534, 551, **593**, **594**

Thermophilic/Thermophiles, 6, 21, 27, 44, 78, 82–84, 104, 151, 162, 164, 169–171, 173, 175–177, 185, 187, 268, 325, 326, 336–338, 354, 359, **361–367**, 374, 379–382, 390, 392, 403, 461, 470, 471, 557, 575, 588, 600, 611, 647, 677–679, 740, 741, 816, 822, 840, 843
Thiosulfate-reducing bacteria, thiosulfate reducers, 586, 588
Thiosulfate shunt, 582
Thiotrophy, 431
Thylakoid, 30, 33, 37, 56–58, 123, 124, 127, 128, 177, **208**, 209, 210, 212, 213, 217, 221, 225, 226, 233, 234, 236, 238–240, 250–252, 527
Thymidine, 346, 785, 788
Toga, 176, 364
Toxin, 29, 185, 227, 237, 282, 284, 295, 309, 403, 409, 410, 425, 428, 429, 436, 478, 504, 621, 624–626, 628, 631–634, 663
Transamidation, 69
Transcription, 76, 82, 96, 100, 164, 166, **294–299**, 302, 306, 308–312, 316, 318–320, 322, 324, 328, 358, 360, 377, 409, 604, 697, 727, 802
Transcriptomic microarray, **812**
Transducing particle, 450, 451
Transduction, 121, **294**, 296, **299–311**, 382, 405, 432, 447, **449–451**, 466, 504, 634, 685, 697, 724, 835
Transformasome, 446
Transformation, 402, 405, 406, 410, 430, 432, **445–447**, 448, 449, 452, 466, 503, 805, 835
Transforming DNA, 446, 447
Transmission electron microscope (TEM), 636, **775**
Transposable elements, 693
Transposon, 324, 396, 404, 604, **634**, 635, 637, 660, 693, 695, 704, **834**
Treatment plants, 7, 185, 280, 281, 345, 402, 626, 631, 636, 640, 642–644, 646, 661, 662, **664–668**, 670–672, 674, 677, 681, 684, 741, 745, 842
Tree of life, **82–85**, 110, 118, 121, **122**, **135–140**, 162, 194, 245, 392, 836, 880
T-RFLP. *See* Terminal restriction fragment length polymorphism (T-RFLP)
Triazine(s), 684, 688, 690, 691, **693–697**, 700–702, 704
Tricarboxylic acid cycle (TAC), 38, **39**, 49, 62, 65, 66, 718, 723, 726
Trichogamy, 201–203, 211, 245, 250, 252
Trichome, 198, 569
Trigenetic, 204, 206, 211, 237, 245, 250–253
Trophic link, 216, 493, 495–496, 498, 541
Trophic network, 431, **485–506**, 531, 553
Tuberculosis, 13, 14, 148, 188, 300, 306, 424, 425, 427, 435, 436, 623, 631, 635, 641, 838
Tyndallization, **11**
Type II secretion system, 447
Type IV pilus, 307, 317, 341, 447
Type IV secretion system, 447

U

Ultimate waste storage facilities (UWSF/landfills), 680
Undulipodium, 125, 126, 138, **201**, 202–203, 205, 207, 209, 211, 213, 214, 216–218, 220–222, 224–226, 229, 230, 232, 233, 236, 237, 239–245, 248–252, 629
Unicellularity, **118–121**, 209
Universal genes, 79
Unsaturated fatty acids, 327, 336, 356, 358–359, 377, 403, 814–816

Unsaturated hydrocarbons, 561, **716**
Uptake signal sequences (USS), 446
Upwellings, 486, 493, 519, 522–523, 529, 534–535, 733
Uranium, 49, 70, 346, 601, 607, 730, 742–744, 839
Urban wastewater (UWW), 561, 642, **664–676**

V

Vaccination, 12
Vaccine, 12, 426, 427, 436, 624
Vacuole, 26, 32, 33, 128, 138, 197, 209, 210, 213, 216, 220, 222, 233, 236, 238, 241, 243, 244, 773, 787
Variable shell, 474, 475
Viability, 373, 650, 651, 653, 656, 823
Viable non-culturable, 16, 21, 281, **649–651**
Viral production, **788**
Virulence, 30, 294, 295, 303, 306, 309, 310, 312–318, 320, 403, 413–415, 417, 424, **428–430**, 434, 436, 446, 463, 478, 504, 630, **633–635**, 641, 651, 656
Volatile organic compounds (VOCs), 663, 681, 682
Volcanic activity, 78, 361–362, 379
Volcanic underwater environments, 176
Volcanism, volcanic. 91, 101, 166, 173, 176, 263, 275, 333, 361, 365, 378, 380, 391, 419, 513, 518, 579
Voltametric micro-electrodes, **791–792**
Vostok, 275, **357–358**, 392

W

Wall (cell), 26–27, **29–31**, 32, 70, **162**, 164
Wastewater, **664–676**
Wastewater treatment plant (WWTP), 642, **643–644**, 645–646, 660, **667–676**
Waterborne, 620, 622–623, 640, 642
Watershed, 486, 487, 579, 642, 645
Wetlands, 498, 516, 554, 557, 558, **579–580**, 623, 629, 663, 701, 741
Windrow(s), **662–663**, 679
Winkler method, 532, **785**
Winograsky columns, 595, **596–597**
WWTP. *See* Waste water treatment plant (WWTP)

X

Xenobiotics, 4, 5, 19, 21, 30, 35, 152, 181, 183, 188, 189, 275, 345, 346, 561, 571, 612, **660**, 675, **684–715**, 746
Xylanases, 65, 391

Y

Yellowstone, 21, 174, 188, 361, 364, 557

Z

Zoobenthos, 537
Zoonosis, 187
Zoonotic, 424
Zooplankton, 224, 486, 490, 494, 495, 514, 522, 535, 537, 552, 553, 601, 625, 626, 644, 664–665, 709, 716, 717, 765, 766, 772, 782, 784, 869, 872
Zygote, **198–199**

Taxonomy Index

Note: Pages in bold refer to pages where substantial, heavy information are provided to the readers.

A

Abies, 263
Acanthamoeba polyphaga, 144
Acanthocystis turfacea, 207
Acanthopodina, 248
Acetabularia, 121
Acetobacter, 181, 423
Acetobacteraceae, 181
Acetohalobium arabaticum, 373
Acetonema longum, 186
Acetonoma, 185
Acholeplasma, 187
Acholeplasmataceae, 187
Acholeplasmatales, 187
Achromobacter, 711, 722
    A. xylosoxidans, 641
Acidaminococcaceae, 185
Acidaminococcus, 185
Acidianus, 44, 63, 109, 164, 166, 334, 363, 365, 366
    A. ambivalens, 383
    A. brierleyi, 163, 383
    A. convivator, 109
Acidilobales, 164, 167
Acidilobus, 166, 167
Acidimicrobiaceae, 187
Acidimicrobiales, 187
Acidimicrobidae, 187
Acidimicrobium, 44, 187, 381–383
    A. ferrooxidans, 382, 741
Acidiphilium, 381–383
Acidithiobacillus, 43, 44, 286, 334, 381, 382, 589
    A. caldus, 381, 382
    A. ferrooxidans, 43–45, 380–383, 606, 741
    A. thiooxidans, 381, 382, 741
Acidobacteria, 160, **184**, 280, 456
Acidobacteriaceae, 184
Acidobacteriales, **184**
Acidobacterium, 184
Acidothermaceae, 188
Acidothermus, 188
Acidovorax, 423
Aciduliprofundum boonei, 365
Acinetobacter, 46, 182, 432, 468, 622, 711, 722, 724
    A. baumannii, 636
    A. baylyi, 447, 453
Acrasiobionta, 239
Acrasiomycota, 239
Acritarchs, 93, 122

Actinobacteria, 14, 31, **100**, 120, 139, 140, 160, 161, 185, 186, **187–188**, 192, 243, 269, 270, 280, 283, 285, 298, 300, 321, 324, 328, 336–341, 356–358, 377, 387, 398, 401, 417, 419, 422, 423, 427, 436, 456, 464, 579, 631, 670, 678, 679, 816, 836
Actinobacteridae, 187
Actinobaculum, 187
Actinomyces, 187, 334
Actinomycetaceae, 187
Actinomycetales, 187, 387
Actinomycinae, **187**
Actinophryda, 207, 229, 243
Actinophyris, 383
Actinoplanes, 188
Actinosynnema, 188
Actinosynnemataceae, 188
Adenovirus, 621
Aedes, 423
    A. albopictus, 424
Aerococcaceae, 187
Aerococcus, 187
Aeromonadaceae, 182
Aeromonadales, 182
Aeromonas, 182, 387, 389, 423, 622, 642, 743
    A. hydrophila, 634, 642
    A. salmonicida, 182, 634
Aeropyrum, 163, 166, 363, 366
    A. pernix, 163
Aeschynomene, 407
Agaricus bisporus, 319, 401
Agrobacterium, 338, 404, 405, 420
    A. tumefaciens, 148, 149, 181, 300, 313, 315, 404, 412, 634, 704
Agromyces, 188
    A. ramosus, 401
Agropyron desertorum, 701
Aigarchaeota, 162, 174, 175
Alcaligenaceae, 181
Alcaligenes, 44, 46, 68, 181, 698, 711, 712, 722
    A. eutrophus, 692, 693
    A. paradoxus, 692
Alcalinovorax (syn: *Alcanivorax*) *borkumensis*, 300, 724
Alcanivorax, 182, 719, 722
    A. borkumensis, 303, 724
Algae, 19, 56, 62, 63, 93, 114, 115, 135, **138–141**, 171, 177, 192, 198, 199, 207, 209, 211, 221, 228, 229, 231, 234, 250, 263, 372, 374, 375, 383, 396, 404, 405, 476, 490, 498, 506, 513, 523, 526, 580, 594, 596, 598, 620, 624, 718, 765, 766, 782, 799, 815
Alicyclobacillaceae, 187

Alicyclobacillus, 187
Alkalibacillus, 387
Alkalibacter, 387
Alkalibacterium, 387
Alkaliflexus, 387
Alkalilimnicola, 388, 389
Alkalimonas, 388
Alkaliphilus, 387
Alkalispirillum, 388, 389
Allochromatium, 181
    A. vinosum, 181
Alnus, 263
Alphaproteobacteria, 68, 97–100, 124, 126, 160, 172, **179–181**, 280, 281, 300, 310, 388, 389, 415, 416, 456, 475, 476, 562, 722
Alternaria alternata, 316
Alteromonadaceae, 182, 388
Alteromonadales, 182, 388
Alteromonas, 182, 356, 359, 389
    A. infernus, 536
Alveolata, 120, 124, 127, 129, 130, 132, 137, 138, 196, 197, 205, 207, 208, **221–228**, 237–239, 243, 251, 478
Amaricoccus, 181
Ammonifex degensii, 84
Amoeba, 110, 131, 138, 192, 234, 239, 477, 486, 629–631, 647, 670, 883
Amoebidium, 243
Amoebobacter, 591
Amoebobionta, 121, 205, **247–250**, 252
Amoebophrya, 221
Amoebozoa, 137, 247
Amphibacillus, 387
Anabaena, 68, 96, 120, 177, 178, 338, 415, 420, 464, 569, 621, 628
    A. cylindrical, 178
    A. flosaquae, 340, 628
Anabaenopsis, 387, 389
Anaerobacter, 185
Anaerobaculum, 185
Anaerobiospirillum, 182
Anaerobranca, 387
Anaerococcus, 285
Anaerolinea, 177
Anaerolineae, 160, 177
Anaeromyces, 53
Anaerophilum, 185
Anaeroplasma, 187
Anaeroplasmataceae, 187
Anaeroplasmatales, 187
Anaerovirgula, 387
Anammoxoglobus *(Candidatus)*, 576
Anaplasma phagocytophilum, 435
Ancalomicrobium, 181
Ancylobacter aquaticus, 340
Andalucia, 478
Animalia, 115
Anopheles, 227, 423, 425, 426
Anoxybacillus, 387, 389
Anoxynatronum, 387
Antarctobacter, 181
Anterokonta, **205**
Antonospora locustae, 243
Aphanizomenon, 621, 628
    A. flosaquae, 628
Apicomplexa, 101, 132, 221, 227–228, 251, 629
Apis
    A. cerana, 243
    A. mellifera, 244

Aquabacter, 181, 340
Aquabacterium, 181
Aquaspirillum, 182
    A. magnetotacticum, 603
Aquifex, 44, 175, 363, 364
    A. aeolicus, 469, 470
    A. pyrophilus, 84
Aquificaceae, 175
Aquificae, 160, **175–176**, 456, 469
Aquificales, 82, 175, 364
Arabidopsis, 126, 416
    A. thaliana, 299, 316, 477, 699, 836
Archaea, **154, 161–175**, 300, 364–366, 388, 469, 560–561, 562–564, 568
Archaebacteria, 161, 162, 875
Archaeoglobales, 162, 163, 167, 169, **173–174**, 365
Archaeoglobus, 48, 62, 63, 66, 173, 363, 365
    A. fulgidus, 163, 173, 471, 728
Archaeopteris, 114
Archaeopteryx, 114
Archaeplastida, 114, 121, 124, 126, 127, 132, 134, 137–139, 192, 194, 197, 201, **204–215**, 218, 227, 229, 230, 234, 238, 243, 248, 250
Archamoeba, **248**, 252
Archezoa, 100
Armatimonadaceae, 185
Armatimonadales, 185
Armatimonadetes, 160, 185, 456
Armatimonadia, 160, 185
Armatimonas, 185
Arsenophonus, 423
Artemia salina, 372
Arthrobacter, 188, 338, 357, 387, 389, 692, 696, 713, 815
    A. aurescens, 697
    A. globiformis, 360
Arthrobotrys robusta, 401
Arthropods, 243, 398, 408, 423, 424, 432, 433, 436, 477, 500, 548, 565, 840
Arthrospira, 387, 389
Asaia, 423
Ascomycota, 115, 199, 202, 243–247
Asobara tabida, 433
Aspergillus, 284, 412, 427, 628, 722
    A. flavus, 284
    A. fumigatus, **628**
    A. nidulans, 299
Astasia, 240
Atopobium, 187
Atta, 430
Aureobasidium, 263
Aureococcus, 232
Aureoumbra, 232
Axiothella rubrocincta, 120, 121, 227
Azolla, 68, 177, 415
Azomonas, 68, 182
Azorhizobium, 68, 475
Azospirillum, 68, 181, 419, 420
    A. brasilense, 406, 417
    A. lipoferum, 319
Azotobacter, 31, 68, 338

B

Bacillaceae, 186, 187, 281, 387
Bacillales, 186, 387
Bacillariophyceae, 111, 197, 202, 232, 234, 235
Bacilli, 108, 160, 186, 387

Bacillus, 12, 13, 15, 44, 46, 63, 68, 120, 158, 160, 186, 315, 321, 325, 326, 335, 337, 387, 389–391, 403, 423, 432, 609, 641, 679, 711, 722, 743
    B. amyloliquefasciens, 359
    B. anthracis, 148, 185, 186, 641
    B. cereus, 150, 641
    B. megaterium, 326, 340
    B. stearothermophilus, 326
    B. subtilis, 11, 81, 150, 185, 294, 297, 299, 300, 303, 316, 326, 446, 447, 453, 461, 469, 833, 836, 839, 843
    B. thuringiensis, 404, 436
Bacteria, **154, 161, 175–188, 300, 364, 387–388, 469, 491–492, 568, 622–628**
Bacteriophages, 14, 108, 121, 401, 429, 449, 450, 501, 503, 634, 635, 640, 788, 884
Bacteriovorax, 401
Bacterium globiforme, 686
Bacteroidaceae, 183
Bacteroidales, 183, 242, 387
Bacteroides, 65, 160, **183**, 337, 357, 404, 423, 432, 640
    B. fragilis, 640
    B. thetaiotaomicron, 432
Bacteroidetes, 160, 172, 183–184, 283, 284, 356–358, 364, 422, 423, 456, 472, 473
Bacteroidia, 183, 387
Balaniger balticus, 237
Balantidium, 629
Balnearium, 363, 364
Bangiomorpha pubescens, 93, 122, 201
Bartonella, 181, 435
    B. henselae, 435
Bartonellaceae, 181
Basidiomycota, 115, 202, 243–247
Bathymodiolus, 431
Batrachochytrium dendrobatidis, 247
Bdellovibrio, 183, 400, 401
    B. bacteriovorus, 400
Bdellovibrionaceae, 183
Bdellovibrionales, 183
Beggiatoa, 14, 44, 109, **112**, 180, 183, 341, 344, 345, 491, 497, 588–590
Beijerinckia, 181
Beijerinckiaceae, 181
Bemisia tabaci, 423
Betaproteobacteria, 160, 175, 181–182, 280, 300, 388, 415, 416, 446, 456, 575, 672, 704, 722, 842
Beutenbergia, 188
Beutenbergiaceae, 188
Bifidobacteriaceae, 188
Bifidobacteriales, 187, 188
Bifidobacterium, 284, 334, 640
    B. bifidus, 188
Bigelowiella longifila, 217
Bikonts, **205**
Blastochloris, 181
    B. sulfoviridis, 728
Blattabacteriaceae, 183
Blattabacterium, 183
Bodo, 241
    B. caudatus, 489
Bogoriella, 387
Bogoriellaceae, 387
Boleophthalmus boddarti, 537
Bolidomonas, 231
    B. mediterraneus, 235
Bonamia, 114
Bordetella pertussis, 300, 303

Borrelia, 184
    B. burgdorferi, **465**, 469, 648
Bostrychia, 114
Botrytis cinerea, 411, 413
Brachionus, 868
    B. calyciflorus, 869
Bradyrhizobiaceae, 181, 388
Bradyrhizobium, 68, 160, 181, 407, 415, 476
    B. japonicum, 300
Brevibacillus, 421, 679
Brevibacteriaceae, 188
Brevibacterium, 188
Brocadia (Candidatus), 576
Brucella, 181
Brucellaceae, 181
Brugia malayi, 423, 432
Bryophyta, 115, 207, 212, 214
Buchnera, 81, 100, 109, 423, 430, 433, 464
    B. aphidicola, 464, 840
Burkholderia, 158, 181, 316, 403, 417, 430, 641, 650, 704, 711, 715, 722
    B. cenocepacia, 641
    B. cepacia, 315, 427, 634, 638, 641, 642, 694, 698
    B. pseudomallei, 313, 641, 642, 650, 840
    B. vietnamiensis, 641
    B. xenovorans, 715
Burkholderiaceae, 181
Burkholderiales, 181
Butyribacterium, 50, 53
Butyrivibrio, 185

C

Caenorhabditis, 401
Caldalkalibacillus, 387
Calderobacterium, 175
Caldilineae, 160, 177
Caldiserica, 160, 185, 456
Caldisericaceae, 185
Caldisericales, 185
Caldisericum exile, 185
Caldisphaera, 167
Caldivirga, 167
Callosobruchus chinensis, 477
Calonympha, 242
    C. grassii, 242
Camerina, 220
Caminicella, 363, 364
Campylobacter, 183, 428, 621–622, **623**, 642
    C. fetus, 317, 318
    C. jejuni, 312, 313
Campylobacteraceae, 183
Candida, 121, 628, 629, 722
    C. albicans, 247, 299, 430, 629
    C. glabrata, 629
    C. krusei, 629
    C. tropicalis, 629
Carboxydibrachium, 364
Cardiobacteriaceae, 183
Cardiobacteriales, 183
Cardiobacterium, 183
Carnobacteriaceae, 187, 387
Carnobacterium, 187, 387, 768
Carsonella ruddi, 81, 130, 833, 835
Caryophanaceae, 187
Caryophanon, 187
Casuarina equisetifolia, 407

Caulerpa, **120**, 121
    C. taxifolia, 120
Caulobacter, 181, 341
    C. crescentus, 300
Caulobacteraceae, 181
Caulobacterales, 181
Cellulomonadaceae, 188, 387
Cellulomonas, 188, 387
Cellulophaga, 183
Cenarchaeaceae, 175
Cenarchaeales, 175
Cenarchaeum symbiosum, 163, 175
Centrohelida, 138, 206, **207–208**, 229, 243, 250
Centroheliozoa, 195, 207
Centrohelomonada, 207
Cercomonads, 383
Chaetosphaeridium minus, 215
Chelatococcus, 338
Chilomonas, 239
Chironomidae, 491
Chitridia, 115
Chlamydia, 109, 184, 423, 435, 436
Chlamydiaceae, 184, 422
Chlamydiae, 160, **184**, 456
Chlamydiales, 184, 356
Chlamydomonas, 137, 213, 215, 383
    C. reinhardtii, 197, 215, 316, 478, 861
Chlorarachniobionta, 114, 120, 127, 128, 132, 215–217, 251
Chlorella, 207, 215, 383, 868
    C. vulgaris, 868, 869
Chlorobi, 160, 184, 446, 456, 590, 593
Chlorobia, 160, 184
Chlorobiaceae, 184, 491
Chlorobiales, 184
Chlorobionta, 92, 109, 111, 115, 118, 120, 127, 130, 136,
    197, 207, 211–215, 217, 220, 224, 240, 245
Chlorobium, 56, 62, 184, 607
    C. ferrooxidans, 606
Chloroflexaceae, 177
Chloroflexales, 177
Chloroflexi, 160, **177**, 456, 590, 593
Chloroflexus, 56, 63, 177, 364, 797
Chloroherpeton, 184
Chlorokybus atmophyticus, 215
Chloronema, 177
Chlorophyceae, 212, 214, 215, 245, 552
Chlorophyta, 383
Chondrilla, 114
Chondromyces, 183
Chromalveolata, 132, 134, 136, 137, 208, 221,
    225, 230, 238, 239
Chromatiaceae, 182, 388, 491, 591, 592
Chromatiales, 182, 388
Chromatium, 56, 154, 180–182
    C. okenii, 591
Chromera velia, 227
Chromista, 139
Chromobacterium, 182
    C. violaceum, 313
Chromobionta, 115, 121, 123, 128, 130, 132–134, 137, 138,
    215, 218, 225, 228–230, **231–235**, 236, 239, 251
Chroococcales, 177, 178
Chroococcus, 389
Chrysiogenaceae, 176
Chrysiogenales, 176
Chrysiogenes, 176
Chrysiogenetes, 160, **176**, 456

Chrysochromulina, 237
Chrysophyta, 115
Chrysopogom zizanioides, 699
Chthonomonadaceae, 185
Chthonomonadales, 185
Chthonomonadetes, 160, 185
Chthonomonas, 185
Chytridiales, 490
Ciliata, 222
Ciliophora, 221, **222–224**, 229, 251
Citrobacter, 182, 285, 622
Cladonia, 215
Cladosporium, 374
Clavibacter michiganense, 413
Claviceps purpurea, 284, 411, 412
Closterium, 215
Clostridia, 50, 160, **185**, 389
Clostridiaceae, 185, 387
Clostridiales, 185, 374, 387
Clostridium, 50, 53, 54, 65, 68, 120, 160, 185, 334, 357, 387,
    390, 471, 641, 899
    C. acetobutylicum, 53, 300
    C. botulinum, 185, 633, 641
    C. difficile, 641
    C. mayombei, 186
    C. pasteurianum, 15, 852
    C. perfringens, 631, 633, 640, 641, 838
    C. tetani, **185**, **633**, 641
Coccomyxa, 245
    C. astericola, 215
    C. ophiurae, 215
    C. parasitica, 215
Codium, 130
Coelodiceras spinosum, 216
Colacium, 240
Coliforms, 182, 281, 638–639, 820
Colwellia, 356, 359, 376, 768
    C. psychrerythraea, 359, 815
Comamonadaceae, 181
Comamonas, 181, 711
Coniothyrium minitans, 419
Conopodina, 348
Conosa, 348
Convoluta, 215
Coprococcus, 185
Corallochytrium, 243
Coriobacteriaceae, 187
Coriobacteriales, 187
Coriobacteridae, 187
Coriobacterineae, 285
Coriobacterium, 187
Corynebacteriaceae, 188
Corynebacterineae, **188**
Corynebacterium, 53, 188, 334, 387, 421, 711
    C. diphtheriae, 188, 633
    C. glutamicum, 188
Coxsackievirus, 621
Crambe, 114
Crenarchaeota, 160, 162–163, **164–167**, 364, **365–366**,
    403, 456, 472, 573, 588
Crenothrix, 183
Crenotrichaceae, 183
Cristispira, 184
Crocosphaera watsonii, 464
Crustomastix, 215
Cryptista, 238
Cryptobacterium, 187

Cryptobionta, 127, 128, 132, 139, 197, 205, **238–239**, 252
Cryptochlora perforans, 217
Cryptococcus, 427, 628
    C. neoformans, 629
Cryptomonas, 238
Cryptophyta, 127, 128, 197, 224, 237–239, 476, 478
Cryptosporidium, 478, 629, 647
    C. parvum, 227, 621, 629
Cryptotermes brevis, 242
Cupriavidus necator, 692, 693, 698
Cyanidiophytina, 210
Cyanidioschyzon, 840
    C. merolae, 197, 840
Cyanidium, 209, 211, 383
Cyanobacteria, 19–20, 30–31, 35, 56, 61–62, 67, 87, **89–91**, 96, 98, 103, 114–115, 118, 120, 126, 130, 137, 140, **177–178**, 281–282, 302, 336, 340, 344–345, 356, 364, 389, 396, 404, 420, 422, 446, 456, **464**, 473, 476, 486–488, **491**, 526–527, 542, 552, 566, 569, 558, 594, 598–599, 610, **624–628**, 642, 648, 716, 722, 745, 773, 777, 785, 794, 813, 815, 878
Cyanobium, 177
Cyanochlorobionta, 115
Cyanophora, 208, 209
Cyanospira, 387, 389
Cyanothece, 177, 178
Cycas, 177
Cyclobacteriaceae, 387
Cycloclasticus, 719
Cyclostephanos dubius, 489
Cyclotella, 233
    C. comta, 489
Cylindrocarpon tonkinense, 575
Cylindrospermopsis raciborskii, 625
Cystodinium, 138
Cytophaga, 183, 341, 815
Cytophagaceae, 387
Cytophagales, 171, 387
Cytophagia, 387

D

Dactylosporangium, 188
Dasycladales, 92, 118
Dasytricha, 53
Debaryomyces, 283
Dechloromonas aromatica, 300
Deferribacter, 176, 363
Deferribacteraceae, 176
Deferribacteres, 160, **176**, 364, 456
Deferribacteria, 422
Dehalococcoides, 710, 811, 812
    D. ethenogenes, 710, 711
Deinococcaceae, 176
Deinococcales, 176
Deinococcus, 160, **176–177**, 337, 338, 345, 346, 838
    D. radiodurans, 176
Deltaproteobacteria, 96, 99, 160, **183**, 281, 300, 364, 388, 390, 562, 566, 841
Dendrobaena octaedra, 506
Dermabacter, 188
Dermabacteraceae, 188
Dermatophilaceae, 188
Dermatophilus, 188
Dermatophytes, 628, 629
Dermocarpa, 178, 200
Desulfacinum, 48, 183

Desulfitobacterium hafniense, 471
Desulfoarculaceae, 183
Desulfoarculus, 48, 183
Desulfobacter, 48, 49, 62, 183, 814, 815
    D. latus, 181
Desulfobacteraceae, 183, 388
Desulfobacterales, 183, 388
Desulfobacterium, 48, 63, 183
    D. autotrophicum, 181
    D. vacuolatum, 181
Desulfobulbaceae, 183, 388
Desulfobulbus, 48, 183, 814, 815
    D. propionicus, 181
Desulfofrigus, 798
Desulfohalobiaceae, 183, 388
Desulfohalobium, 183
Desulfomicrobiaceae, 183
Desulfomicrobium, 48, 183, 586
Desulfomonadaceae, 183
Desulfomonadales, 183
Desulfomonas, 183
Desulfomonile, 48, 183
Desulfonatronaceae, 388
Desulfonatronobacter, 388
Desulfonatronospira, 388
Desulfonatronovibrio, 183, 388
    D. hydrogenovorans, 390
Desulfonatronum, 388
    D. lacustre, 390
Desulfonema ishimotoi, 181
Desulfopila, 388
Desulforomonas, 588
Desulforudis audaxviator *(Candidatus)*, 611
Desulfosporosinus, 383
    D. acidophilus, 383
Desulfotalea psychrophila, 359
Desulfotomaculum, 48, 185, 387, 584
    D. acetoxidans, 186, 798
    D. alkaliphilum, 390
Desulfovibrio, 48, 49, 55, 181, 183, 403, 585, 710, 743, 768, 814, 815
    D. desulfuricans, 181
    D. halophilus, 181
    D. sulfodismutans, 587
    D. vulgaris, 300
Desulfovibrionaceae, 183
Desulfovibrionales, 183, 388
Desulfurella, 48, 49, 183, 588
Desulfurellaceae, 183
Desulfurellales, 183
Desulfurivibrio, 388
Desulfurobacteriales, 364
Desulfurobacterium, 363, 364
Desulfurococcaceae, 166
Desulfurococcales, 163, 164, **166–167**, 174, 365, 366
Desulfurococcus, 48, 166, 363, 366
Desulfuromonadales, 388
Dethiobacter, 387
Dethiosulfovibrio peptidovorans, 586
Diatoms, **63**, **64**, **70**, 111, 128, 130, 137, 202, 216, 217, 219–221, 224, 227, **232–235**, 237, 247, 346, 374, 478, 486, 491, 493, 498, 499, 523, 524, 526, 529, 533, 534, 552, 601, 773, 799, 815
Dickeya dadantii, 298
Dictyoglomaceae, 185
Dictyoglomales, 185

Dictyoglomi, 160, 185
Dictyoglomus, 185
Dictyosphaeria, 93, 94
Dictyostelium, 121, 249, 252, 253
    D. discoideum, 249, 477
Dietzia, 387
    D. cinnamea, 641
Dietziaceae, 188, 387
Dinamoebium, 138
Dinobionta, 114, 120, 121, 128, 129, 132–134, 137, 138, 196,
    197, 201, 207, 215, 220, 221, **224–227**, 229, 237, 239, 251
Dinobryon, 235
    D. cylindricum, 234
Dinoflagellida, 224
Dinophysis norvegica, 239
Dinozoa, 196, 224
Diplomonadida, 241, **242**, 252, 478
Discicristates, 197, **239–242**, 252, 629, 630
Discorbis orbicularis, 220
Dissodinium pseudolunula, 227
Distigma, 240
    D. proteus, 489
Dolichomastix, 215
Drosophila melanogaster, 433, 434
Dryas, 263
Dunaliella, 171, 372, 375, 383
    D. salina, 372

E

Earthworms, 114, 492, 500, 506, 544–546, 888
Echovirus, 621
Ectothiorhodospira, 181, 182, 372, 374, 388, 389
Ectothiorhodospiraceae, 333, 388, 491, 591, 592
Ectrogella eurychasmoides, 230
Edwardsiella ictaluri, 335
Ehrlichia, 181
    E. chaffeensis, 435
Ehrlichiaceae, 181
Eisenia, 114
Ellobiopsidae, 221, 222, 243
Ellobiopsis, 221
Elusimicrobia, 160, 185, 456
Elusimicrobiaceae, 185
Elusimicrobiales, 185
Elusimicrobium, 185
Elysia
    E. chlorotica, **130**, **134**
    E. viridis, 130
Emiliania huxleyi, 237
Entamoeba, 53, 69, 629
    E. histolytica, 248, 621, **629**
Enterobacter, 53, 182
Enterobacteria, 152, 284, 334, 343, 423, 427, 471, 575, 743
Enterobacteriaceae, 46, 50, 147, 160, 182, 474, 622, 623,
    636, 639, 780, 894
Enterobacteriales, 182, 475
Enterococcaceae, 187
Enterococcus, 187, **636**
    E. avium, 639
    E. durans, 639
    E. faecalis, 287, 316, 468, 639
    E. faecium, 639, 650
Enterocytozoon bieneusi, 243
Enteroviruses, 621
Entomoplasma, 187

Entomoplasmataceae, 187
Entomoplasmatales, 187
Epidermophyton, 628, 629
Episeptum inversum, 244
Epsilonproteobacteria, 160, **183**, 300, 456
Erwinia, 32, 53, 313, 314, 780
    E. amylovora, 412, 413
    E. carotovora, 313, 315
    E. chrysanthemi, 298
Erysipelothrix, 187
Erysipelotrichaceae, 187
Erysiphe graminis, 411
Erythrobacter, 60
Erythrobionta, 115
Erythrolobus, 209, 211
    E. coxiae, 211
Erythromicrobium, 60
Escherichia, 53, 160, 464
    E. coli, 81, 161, 189, 271, 295, 299, 302–304, 306–309, 315, 316,
    318, 326, 328, 339, 360, 402, 404, 423, 428, 432, 447, **449–450**,
    453, 461, **463**, 462–467, 473, 474, **479**, 575, 623, 631, 635, 636,
    639, 642, 644, 645, 650, 651, **653–656**, 703, 707, 708, 780, 781,
    788, 805, 806, 836, 839–841, 843
Eubacteria, 150, 161, 162, 534, 635, 875
Eubacteriaceae, 185, 387
Eubacterium, 50, 54, 185, 357
Eucarya, 77, 79, 80, 83, 85, 100, **122–135**, 161, 162, 448, **478–479**
Euglena, 46, 114, 240, 252, 383
Euglena gracilis, 124
Euglenoidea, 197, 240–241, 252
Euglenozoa, 195, 476, 478
Euplotes, 383
Euprymna scolopes, 313, 423, 431
Euryarchaeota, 160, 162–164, 167, 169, 172, 174, 175, 364–365,
    388, 390, 456, 471, 472, 557, 563, 588, 841
Eurychasma, 231
Excavates, 205, 217, 239, **241–242**, 252, 629
Exiguobacterium, 387

F

Fabaceae, 411, 415–418, 420, 581
Fagus sylvatica, 505
Ferribacterium, 49
Ferrimicrobium, 382, 383
Ferroglobus, 46, 173, 363, 365
    F. placidus, 163, 607
Ferroplasma, 173, 383, 841
    F. acidarmanus, 383
    F. acidiphilum, 383
Ferroplasmaceae, 173, 381
Fervidicoccales, 164, 167
Fervidicoccus, 167
Fervidobacterium, 363, 364, 588
Festuca arundinacea, 701
Fibrobacter, 184, 284
    F. succinogenes, 284
Fibrobacteraceae, 184
Fibrobacterales, 184
Fibrobacteres, 160, **184**, 456
Firmicutes, 148, 160, 161, 166, **185–187**, 192, 283–285, 300, 316,
    317, 336–340, 357, 358, 364, 387, 417, 419, 422, 446, 447,
    456, 461, 469, 473–475, 560, 590, 609, 722, 836
Fischerella, 178
Flagellates, 170
Flammeovirga, 183

Flammeovirgaceae, 183
Flavivirus, 649
Flavobacteria, 160, 183, 356
Flavobacteriaceae, 183
Flavobacteriales, 183
Flavobacterium, 183, 341, 343, 622, 698, 815
    F. johnsoniae, 342
Flexibacteaceae, 183
Flexibacter, 183
Flexithrix, 183
Foraminifera, 46, 130, 195, 215, **219–221**, 227, 251, 486, 490, 790
Francisellaceae, 183
Francisella tularensis, 631, 650
Frankia, 20, 46, 68, 120, 186, 188, 263, 298, 398, 407, 415, **463–464**, 579, 835
    F. alni, 188, 300, 321, 464, 840
Frankiaceae, 188
Frankineae, **188**
Fuchsiella, 388
    F. alkaliacetigena, 373, 390
Fucus, 234
Fuligo, 253
    F. septica, 248, 249
Fungi, 7, 10, 33, 87, 108, **138**, 177, 192, **243–247**, 262, 299, 374, 397, 478, 486, 513, 620, **628–629**, 662, 815, 840, 852
Fungia, 230
Fusarium, 76, 266, 408, 409, 411, 412, 414, 417, 419, 722
    F. graminearum, 852
    F. oxysporum, 266, 406, 412–415, 417–419, 575
Fusibacter paucivorans, 186
Fusobacteria, 160, 184, 283, 422, 456
Fusobacteriaceae, 184
Fusobacteriales, 184
Fusobacterium, 184, 284

G

Gaeumannomyces tritici, 419
Galdieria, 209, 211
Gallionella, 44, 182, 606, 609
    G. ferruginea, 606
Gallionellaceae, 182
Gambierdiscus toxicus, 227
Gammaproteobacteria, 96, 100, 160, 172, 175, **182–183**, 296, 300, 356, 388–390, 446, 447, 456, 461, 463–465, 469, 562, 573, 590, 591, 722, 822, 833, 839, 840, 842
Gardnerella, 285
Gemmata, 184
    G. obscuriglobus, 96
Gemmatimonadaceae, 184
Gemmatimonadales, 184
Gemmatimonadetes, 160, 184, 456
Gemmatimonas aurantiaca, 184
Geoalkalibacter, 388
Geobacillus, 679, 722
Geobacter, 49, 183, 605, 607, 608, 742, 743
    G. metallireducens, 49, 607, 610, 743
    G. sulfurreducens, 300
Geobacteraceae, 183, 388, 608
Geodermatophilaceae, 188
Geodermatophilus, 188, 345
Geogemma, 166
Geoglobus, 173, 365
Georgenia, 387
Geosiphon pyriformis, 130, **132**
Geospirillum, 607

Geothrix, 49, 184
    G. fermentans, 607
Geotoga, 176, 364
Geotrichum candidum, 283
Geovibrio, 176
Giardia, 242, **629**
    G. intestinalis, 100
    G. lamblia, 242, 478, 621, 629
Glaciecola, 356
Glaucocystis, 208, 209
Glaucocystobionta, 126, 127, 197, 207–209, 250
Gliocladium, 417–419
Gloeobacter violaceus, 176, 464
Gloeocapsa, 179
Gloeochaete, 208, 209
Gloeothece, 177, 569
    G. membranacea, 178
Glomeromycota, 130, 132, 246, 252, 416
Glossina, 240
Gluconacetobacter, 68
    G. diazotrophicus, 420
    G. xylinus, 306, 309
Gluconobacter, 181, 423
Glycomyces, 188
Glycomycineae, 188
Goniomonas, 239
Gordonia, 44
Gordoniaceae, 188
Gracilaria, 114
Grypania spiralis, 92
Gymnodinium, 128, 138

H

Haemophilus, 82, 182, 319, 446, 464, 836
    H. influenzae, 82, 182, 319, 446, 464, 836
Hafnia, 182
Haladaptatus, 171
Halalkalicoccus, 171
Halanaerobiaceae, 373
Halanaerobiales, 373, 374, 388, 390
Halanaerobium
    H. acetethylicum, 374
    H. congolense, 374
    H. kuschneri, 374
    H. lacusrosei, 374
    H. salsuginis, 374
Halanaerocella petrolearia, 374
Halarchaeum, 171
Haloanaerobacter, 185
Haloanaerobiaceae, 333
Haloanaerobiales, 172, 185
Haloanaerobium, 185, 186
Haloarchaea, 333, 374, 390, 391
Haloarchaeobius, 171
Haloarcula, 46, 171, 575
    H. marismortui, 300
Halobacteria, 27, 60, 160, 161, 333, 371, 388, 390, 473
Halobacteriaceae, 171, 373, 374, 388, 390
Halobacteriales, 162–164, 167, 171–172, 388
Halobacterium, 25, 46, 66, 163, 171, 330, 333, 374, 471, 575
    H. halobium, 340
    H. salinarum, 60, 163, 171, 313
    H. volcanii, 124
Halobacteroidaceae, 185, 373, 374, 388, 390
Halobacteroides, 185

Halobaculum, 171
Halobellus, 171
Halobiforma, 171
Halocella cellulolytica, 373
Halochromatium, 374
Halococcus, 171, 374
  H. morrhuae, 162, 163
Haloferax, 46, 171, 340, 575
  H. volcanii, 163, 171, 391
Halogeometricum, 171
Halogranum, 171
Halomarina, 171
Halomethanococcus, 170
Halomicrobium, 171
Halomicronema, 179
  H. excentricum, 280
Halomonadaceae, 182, 388
Halomonas, 182, 388, 389
  H. campisalis, 390
  H. eurihalina, 375
Halonatronum, 388
  H. saccharophilum, 390
Halonotius, 171
Halopelagius, 171
Halopenitus, 171
Halopiger, 171
Haloplanus, 171
Haloplasma, 187
Haloplasmatales, 187
Haloquadratum, 162, 171
Halorhabdus, 171
Halorhodospira, 182, 333, 374, 388, 389
  H. abdelmalekii, 592
Halorhodospiraceae, 591
Halorientalis, 171
Halorubellus, 171
Halorubrobacterium, 388
Halorubrum, 171, 388, 390
Halorussus, 171
Halosarcina, 171
Halosimplex, 171
Halosphaera, 215
Halostagnicola, 171
Haloterrigena, 171
Halothermothrix, 185
  H. orenii, 186, 374
Halothiocapsa, 182
  H. halophila, 180
Halovivax, 171
Hamiltonella, 423
Hansenula, 722
Haplozoon axiothellae, 120, 121, 224
Haptobionta, 127, 128, 132, 139, 197, 205, 207, 221, 235–238, 251
Helicobacter, 283, 312
  H. pylori, 183, 283, 335, 437, 447, 452, 650, 838
Helicobacteraceae, 183
Heliobacillus, 185
Heliobacteria, 102, 389, 590
Heliobacteriaceae, 185, 387, 592
Heliobacterium, 185
Heliorestis, 387, 389
Heliothrix, 177
Heliozoa, 207, 229
Helminthosporium oryzae, 168, 412
Hepatitis A virus (HAV), 621–622

Hepatitis E virus (HEV), 621–622
Herminiimonas arsenicoxydans, 838, 839
Herpetosiphonales, 177
Heterorhabditis, 423, 428
Heterosigma nitzschia, 235
Hexamita, 53
Hippasteria phrygiana, 215
Hirsutella minnesotensis, 401
Histoplasma, 427
Hodotermopsis sjoestedti, 242
Holophaga, 184
Holosporaceae, 181
Holothurians, 376
Hoplonympha, 242
Human immunodeficiency virus (HIV), 285, 427, 631
Hyalophacus, 240
Hydrogenobacter, 44, 62, 175
Hydrogenophaga, 44, 182
Hydrogenophilaceae, 182
Hydrogenophilales, 182
Hydrogenophilus, 182
Hydrogenothermus, 44
Hypermastigids, 242
Hyperthermus, 166, 366
Hyphomicrobiaceae, 181
Hyphomicrobium, 181, 583

I
Ichthyophthirius multifilis, 224
Ictalurus punctatus, 335
Idiomarina loihiensis, 840
Ignicoccus, 112, 166, 363, 366, 403
  I. hospitalis, 162, 163, 174
Ignisphaera, 166
Indibacter, 387
Intrasporangiaceae, 188
Intrasporangium, 188
Isotricha, 53
Ixodes, 648

J
Jettenia *(Candidatus)*, 576
Jonesia, 188
Jonesiaceae, 188
Jonquetella, 177
Juniperus, 263

K
Karlodinium armiger, 196, 227
Katablepharida, 238, 239
Kinetoplastida, 239, **240–241**, 252
Klebsiella, 68, 182, 622
  K. aeruginosa, 404
  K. pneumoniae, 315, 427, 568, 696
Klebsormidium, 383
Kluyveromyces, 283
Kochia scoparia, 701
Korarchaeota, 160, 162, 163, **174**, 175, 366, 456
Korarchaeum cryptofilum (Candidatus), 163, 174
Kosmotoga, 176
Kuenenia *(Candidatus)*, 46, 576
Kuenenia stuttgartiensis *(Candidatus)*, 46, 570, 577

L

Labyrinthulobionta, 118, 120, 221, 228, **230**, 243, 252
Lachnospira, 285
Lachnospiraceae, 185
Lactobacillaceae, 187
Lactobacillales, 187, 387
Lactobacilli, 29, 334, 432
Lactobacillus, 53, 54, 187, 283, 285, 325, 329, 334, 402
    L. acidophilus, 187
    L. bulgaricus, 187, 398
    L. sakei, 360
Lactococcus, 53, 187, 283, 402, 423
Laminaria, 234
Lamprobacter, 182
Lebetimonas, 364
Legionella, 182, 189, 428, 620–622, 624, 642, 647
    L. pneumophila, 182, 319, 320, **623–624**, 631, 633, 647, 650, 651
Legionellaceae, 182
Legionellales, 182
Legumes, 68, 316, 338, 412, 476, 566, 739
Leishmania, 46, 69, 240, 241, 435
Lentisphaera araneosa, 185
Lentisphaeraceae, 185
Lentisphaerae, 160, 185, 456
Lentisphaerales, 185
Lentisphaeria, 160, 185
Lepidodinium, 128
Lepiota, 430
Leptospira, 184
Leptospiraceae, 184
Leptospirillum, 177, 381, 382, 841
    L. ferrodiazotrophum, 842
    L. ferrooxidans, 44, 380, 382, 383, 606, 741
Leptothrix, 181, 182
    L. discophora, 609
    L. ochracea, 606
Leptotrichia, 184
Leucoagaricus, 430
Leucobacter, 188
Leuconostoc, 53, 54, 187, 283, 334
Leuconostocaceae, 187
Leucothrix, 341, 670
Lichens, 115, 177, 245, 246
Licmophora, 231
Listeria, 187, 402, 428
    L. ivanovii, 635
    L. monocytogenes, 187, 294, 326, 635, 642, 650
Listeriaceae, 187
Litoribacter, 387
Lobelia, 580
Loftusia, 220
Lolium perenne, 701
Luteococcus, 188
Lyngbia, 179
Lyromonas, 239

M

Macrococcus, 186
Macromonas, 44, 589
Macrophytes, 486, 487, 497, 498, 530, 552, 598, 666, 717
Macrotermes gilvus, 423
Magnaporthe grisea, 413
Magnetobacterium, 177
Magnetospirillum, 181
    M. magnetotacticum, 603
Magnetotacticum bavaricum, 603
Magnoliophyta, 114, 207, 212, 214, 215, 218, 230, 487
Malassezia, 285
Mamiella, 215
Marichromatium, 388
Marinilabiliaceae, 387
Marinilactibacillus, 387
Marinithermus, 363, 364
Marinitoga, 176, 363, 364, 378, 768
    M. piezophila, 378, 379
Marinobacter, 46, 182, 719, 722
    M. hydrocarbonoclasticus, 719, 720
Marinobacterium, 182
Marinomonas primoryensis, 360
Marinospirillum, 182, 388
Mariprofundaceae, 183
Mariprofundales, 183
Mariprofundus, 44
Mariprofundus ferrooxidans, 183
Maritimibacter, 388
Maullinia, 230
Medicago, 316
    M. sativa, 699
    M. truncatula, 316, 415
Mediophyceae, 232, 234
Megasphaera, 285
Meiothermus, 364
Melosira nummuloides, 235
Mentha spicata, 713
Mesorhizobium, 181
Mesostigma viride, 211
Metallosphaera, 44, 63, 164, 166, 363, 383
Methanimicrococcus, 170
Methanoarchaea, 364, 372, 373, 390, 391, 557, 559, 560, 883
Methanobacteria, 160, 388
Methanobacteriaceae, 170, 388
Methanobacteriales, 162, 163, 167–169, **170**, 171, 173, 388
Methanobacterium, 168, 170, 334, 388, 390, 557, 558
    M. alcaliphilum, 170
    M. bryantii, 168
    M. espanolense, 170
    M. formicicum, 124
    M. oryzae, 168
Methanobrevibacter smithii, 170, 500
Methanocalculus halophilus, 168
Methanocaldococcaceae, 170
Methanocaldococcus, 170, 363, 364
    M. indicus, 365
    M. infernus, 171
    M. jannaschii, 163, 471
Methanocella, 170
Methanocellaceae, 170
Methanocellales, 162, 167–170, 172, 173
Methanococcaceae, 170, 171
Methanococcales, 162–164, 167, 168, **170–171**, 173, 364
Methanococcoides, 170, 563
    M. burtonii, 838
Methanococcus, 169, 170, 364, 557, 558
    M. vannielii, 163, 171
Methanocorpusculaceae, 169
Methanocorpusculum, 169
Methanoculleus, 169
    M. thermophilus, 169
Methanofollis, 169
Methanogenium, 169
    M. frigidum, 168, 169

Methanohalobium, 168, 170, 374
Methanohalophilus, 170, 374, 388, 390
Methanolacinia, 169
Methanolinea, 169
Methanolobus, 170, 558
Methanomethylovorans, 170, 558
Methanomicrobia, 168, 388
Methanomicrobiaceae, 169
Methanomicrobiales, 162, 163, 167–168, **169**, 170, 172, 563
Methanomicrobium, 169
Methanoplanus, 169
    M. endosymbiosus, 169
Methanoplasmatales, 162, 167, 169, 173
Methanopyraceae, 170
Methanopyrales, 162, 163, 167–169, **170**, 171, 173
Methanopyrus, 168, 170, 363, 366, 557
    M. kandleri, 163, 170, 364, 365
Methanoregula, 169
Methanoregulaceae, 169
Methanosaeta, 162, 169, 558
    M. thermophila, 169
Methanosalsum, 170, 334, 388
Methanosalsus, 390
Methanosarcina, 50, 66, 162, 169, 170, 558
    M. acetivorans, 163, 169, 464
    M. mazei, 163, 168, 169, 470, 471
Methanosarcinaceae, 169, 374, 388, 390
Methanosarcinales, 162–164, 167–168, **169–170**, 172, 173, 364, 388, 471, 563, 815
Methanosphaera, 170, 557
    M. stadtmaniae, 560
Methanosphaerula, 169
Methanospirillaceae, 169
Methanospirillum, 169
    M. hungatei, 162, 163, 169, 299, 300, 403
Methanothermaceae, 170
Methanothermobacter, 168, 170
    M. thermautotrophicus, 471
Methanothermococcus, 171, 364
    M. okinawensis, 171
    M. thermolithotrophicus, 171
Methanothermus, 170, 363
Methanothrix, 50
Methanotorris, 170, 171, 364
    M. igneus, 171
Methermicoccaceae, 169, 170
Methermicoccus, 170
Methylobacillus, 63, 182
Methylobacter, 63, 183, 561
Methylobacteriaceae, 181
Methylobacterium, 63
Methylococcaceae, 183, 388, 389
Methylococcales, 183, 388
Methylococcus, 183, 561, 562, 815
Methylocystaceae, 181
Methylocystis, 63, 561, 562, 815
Methylohalobius, 388
Methylomicrobium, 388
    M. alcaliphilum, 170, 389
    M. buryatense, 389
    M. kenyense, 389
Methylomonas, 561
Methylophaga, 388
Methylophilaceae, 182
Methylophilales, 182
Methylosinus, 561, 562

Metopus contortus, 169
Micrasterias, 215
Microalgae, 108, 115, 420, 487, 490, **491**, 497–499, 504, 625, 665, 793, 875
Microbacteriaceae, 188
Microbacterium, 188
Micrococcaceae, 188, 387, 421
Micrococcinae **188**
Micrococcus, 188
    M. agilis, 814
    M. cryophilus, 358
    M. halobius, 814
Microcoleus, 178
    M. chthonoplastes, 280, 281, 568, 878
    M. chtonoplastes, 179
Microcystis, 177, 282, 621, 628
    M. aeruginosa, 625, 626, **628**
Micromonas, 215
Micromonospora, 188
Micromonosporaceae, 188
Micromonosporineae, **188**
Microsporum, 628, 629
Microvirgula aerodenitrificans, 575
Mimiviridae, 144
Mimivirus, **109–112**
Ministeria, 243
Mollicutes, 160, 187
Molluscs, 552, 589, 600
Monera, 115
Mongoliicoccus, 387
Moorella, 50
Moraxella, 182
Moraxellaceae, 182
Moritella, 336, 356, 376, 768
Moritellaceae, 388
Mosquitoes, 423, 425, 426, 432–434, 436, 648, 649
Mycena
    M. androsaceus, 500
    M. galopus, 500
Mycetobionta, **248–250**, 252, 253
Mycobacteria, 306, 310, 428, 430, 435, 436
Mycobacteriaceae, 188
Mycobacterium, 153, 188, 284, 434, 435, 622, 641, 707, 722, 738
    M. avium, 153
    M. bovis, 306, 641
    M. leprae, 153, 188, 840
    M. tuberculosis, 13, 14, 148, 188, 300, 306, 427, 435, 436, 631, 641, 838
Mycoplasma, 27, 81, 100, 161, 173, 187, 423
    M. capricolum, 153
    M. genitalium, 81, 82, 100, 187, 465
    M. pneumoniae, 153, 300
Mycoplasmataceae, 187
Mycoplasmatales, 187
Myovirus, 501, 502
Myrionecta rubra, 224
Mytilus edulis, 215
Myxobacteria, 31, 96, 99, 183, 833
Myxobiota, 115
Myxococcaceae, 183
Myxococcales, 183
Myxococcus, 183
    M. xanthus, 306, 313, 316
Myxomycetes, 248
Myxomycota, 115

Taxonomy Index

N

Naegleria, 477, 621, 629
    N. fowleri, 239, **629–630**, 631, 647
Nannochloris, 215
Nannochloropsis, 231
Nanoarchaeota, **174**, 403, 456
Nanoarchaeum equitans, 81, 109, **112**, 162, 163, 166, 174, 366, 403
Nanoflagellates, 489, 647, 785
Natranaerobiaceae, 388
Natranaerobiales, 388
Natranaerobius, 388
Natrialba, 171, 388, 390
Natrinema, 171, 388
Natroniella, 388, 390
    N. acetigena, 390
Natronincola, 387
    N. histidinovorans, 390
Natronoarchaea, 334
Natronoarchaeum, 171, 388
Natronobacillus, 387, 389
Natronobacter, 334
Natronobacterium, 171, 374, 388–390
Natronococcus, 171, 333, 334, 388–391
Natronolimnobius, 171, 388
Natronomonas, 374, 388, 390
Natronorubrum, 171, 388
Nautilia, 364
Navicula radiosa, 137
Neidium, 489
Neisseria, 46, 312, 435
    N. gonorrhoeae, 182, 313, 317, 319, **446**, 467
    N. meningitidis, 182, 319
    N. pneumoniae, 182
Neisseriaceae, 182
Neisseriales, 182
Nematodes, 182, 230, 246, 401, 408, 410, 411, 413, 417, 419, 423, 428, 432, 433, 477, 490, 493, 498–500, 505, 544, 670
Neocallimastix, 53
Neomura, 119
Nereis diversicolor, 538
Nesterenkonia, 387
Neurospora crassa, 299
Neusina, 220
Nicotiana tabacum, 699
Nitriliruptor, 387
Nitriliruptoraceae, 387
Nitrincola, 388
Nitritalea, 387
Nitrobacter, 42, 46, 61, 181, 274, 388, 404, 573
    N. alkalicus, 389
    N. hamburgensis, 181
Nitrococcus, 182, 573, 574
Nitrosoarchaeum *(Candidatus)*, 175
Nitrosobacter, 574
Nitrosococcus, 42, 573
Nitrosolobus, 573
Nitrosomonadaceae, 388
Nitrosomonadales, 182, 388
Nitrosomonas, 42, 46, 182, 388, 573, 574
    N. europaea, 42, 573, 839
Nitrosopumilaceae, 175
Nitrosopumilales, 175
Nitrosopumilis maritimus *(Candidatus)*, 573
Nitrosopumilus, 175
    N. maritimus, 163, 175
Nitrososphaera, 175
Nitrososphaeraceae, 175
Nitrososphaerales, 175
Nitrosovibrio, 573
Nitrospina, 573
Nitrospira, 42, 177, 274, 573
Nitrospiraceae, 177
Nitrospirae, 160, **177**, 456
Nitrospirales, 177
Nocardia, 44, 46, 187, 188, 622, 631, 641, 679, 722, 815
    N. asteroides, 188, 641
    N. cyriacigeorgica, 188
Nocardiaceae, 188
Nocardioidaceae, 188
Nocardioides, 188, 696, 702
Nocardiopsaceae, 188
Noctiluca, 215, 227
Nodularia, 628
Noegleria, 383
Nonionella stella, 220
Norrisiella sphaerica, 217
Norwalk virus, 621, 622
Nosema
    N. bombycis, 243
    N. ceranae, 243
    N. locustae, 125
Nostoc, 120, 132, 178, 299, 300, 338, 628
    N. muscorum, 340
    N. punctiforme, 464
    N. symbioticum, 130, 132
Nostocales, **178**, 569
Nuclearia, 243
    N. simplex, 245
Nyctotherus ovalis, 124, 170

O

Oceanithermus, 363, 364
Oceanobacillus, 387, 389
Oceanospirillaceae, 182, 388
Oceanospirillales, 182, 388
Oceanospirillum, 182
Odontella sinensis, 221
Oenococcus, 187
Olavius algarvensis, 841
Oleispira, 719
Oligotropha, 44
Olpidium, 413
Onchocerca volvulus, 423, 432
Oobiontes, **230–231**, 493
Opalinida, 230
Ophiura texturata, 215
Opisthokonta, 4, 121, 130, 138, 191, 192, 196–199, 201, 202, 205–207, 209, 211, 241, **242–248**, 252
Orenia, 373
Orpinomyces, 53
Orthoseira gremmenii, 235
Oscillatoria, **178–179**, 200, 338, 569, 628
    O. limnetica, 613
    O. sancta, 178
Oscillatoriales, 178
Oscillochloridaceae, 177
Oscillochloris, 177
Ostreococcus, 92, 213, 215, 776
    O. tauri, 109, 215, 478, 488, 775
Oxalobacter, 182
Oxalobacteraceae, 182
Oxymonadida, 241
Oxytricha nova, 124

P
Paenibacillaceae, 187
Paenibacillus, 187, 387, 422, 679
Paleococcus, 172
Parabasalia, 124, 125, 134, 136, 139, 193, 241, **242**, 252
Parachlamydia, 184
Parachlamydiaceae, 184
Paracoccus, 44, 46, 63, 181, 387
    P. denitrificans, 41, 47, 574, 575
Parafusulina, 220
Paramoritella, 388
Paranema, 114
Pasteurella, 182
Pasteurellaceae, 182
Pasteurellales, 182
Pasteuria, 187
Pavlovophyceae, 235
Pectinobacterium carotovorum, 313, 412, 417
Pedinomonas, 215
Pediococcus, 187, 334, 402
Pelagibacter ubique, 841
Pelagomonas, 232
Pelobacter, 183
    P. acetylenicus, 728
Pelobacteraceae, 183
Pelodyction, 184
Penicillium, 283, 284, 412
    P. camembertii, 283
    P. notatum, 13
    P. roqueforti, 283
Pennisetum clandestinum, 701
Peptococcaceae, 185, 387
Peptococcus, 185
Peptostreptococcaceae, 185
Peptostreptococcus, 185
Percolobionta, 239, 241
Percolomonas, 239
Peredibacter starrii, 401
Peridinium, 128, 138
Perkinsus, 490
Peronosporales, 230
Persephonella, 363, 364
Petrotoga, 176, 363, 364
Pfiesteria, 227
Phacus, 489
Phaeocystis, 237, 783
    p. antarctica, 237
Phaeophyceae, 114, 197, 202, 231, 232, 234
Phaerochete chrysosporium, 701, 711
Phage, 281, 283, 405, 449–451, 463, 501, 503, 504, **634**, 640, 697, 788, **834**, 869, 870, 895
Phagomyxa, 217
Phagomyxida, 217
Phormidium, 179, 338
Photobacterium, 182, 356, 376, 377, 768
    P. fischeri, 431
    P. profundum, 840
Photorhabdus, 182, 423, 428
Phyllobacteriaceae, 181
Phyllobacterium, 181
Phytomonas, 413
Phytomyxea, 192, 193, 215, **217–219**, 243, 251
Phytophthora, 230, 413
    p. infestans, 230, 412, 835
    P. parasitica, 411
Picobiliphytes, 194

Picochlorum, 215
Picocyanobacteria, 488
Pico-eucaryotes, 109, 111–112, 194, 231
Picophagus, 232
    P. flagellatus, 235
Picrophilaceae, 173
Picrophilus, 66, 173, 365, 383
    P. torridus, 163, 472
Pinguiophyceae, 232, 234
Pinnularia mesolepta, 489
Pirellula, 184
Piscirickettsiaceae, 183, 388
Placopecten magellanicus, 215
Planctomyces, 184
Planctomycetaceae, 184, 576
Planctomycetales, 96, 184, 377, 577
Planctomycetes, 27, 47, 114, 139, 140, 160, **184**, 243, 337, 456, 576
Planktothrix, 282
    P. agardhii, 625–628
    P. rubescens, 178, 625
Planococcaceae, 186
Planococcus, 186
Plantae, 4, 115, 138, 139, 191, 205, 476
Plants, 4, 10, 33, 81, 108, 145, 192, 262, 298, 363, 396, 449, 487, 512, 620, 661, 766, 833, 862
Plasmodiophora, 217
Plasmodiophora brassicae, 218
Plasmodiophorida, 217, 218
Plasmodium, 101, 397, 423, 425, **426**, 434
    P. falciparum, 101, 132, 227, 228
    p. ovale, 426
    P. vivax, 426
Plasmopara viticola, 230, 406, 411
Pleurocapsa, 178, 389
Pleurocapsales, **177–178**
Pneumococci, 22, 467
Podoviridae, 501
Polaribacter, 183, 356
Polaromonas, 357
Poliovirus, 421
Polyangiaceae, 183
Polymyxa, 217
    P. betae, 218, 413
    P. graminis, 218
    P. palustris, 248
Poriferae, 115
Porphyra purpurea, 126
Porphyridiophyceae, 209, 211
Porphyridium, 209, 211
Porphyrobacter, 60
Porphyromonadaceae, 183
Porphyromonas, 183
Portiera, 423
Posidonia, 114, 207
Potamogeton, 218
Prasinophyceae, 111, 211–215, 217
Prevotella, 183, 285
Prevotellaceae, 183
Primoplantae, 205
Prochlorococcus, 178, 216, 235, 282, 526, 527, 773, 776, 778
    P. marinus, 464
Prochloron, **124, 178**
    P. didemni, 123
Prochlorophyta, 115, **124, 178**
Prochlorothrix, **178**

Promicromonospora, 188
Promicromonosporaceae, 188
Propionibacter, 181
Propionibacteriaceae, 188
Propionibacterineae, **188**
Propionibacterium, 52–54, 283
  P. acnes, 285
Propioniferax, 188
Prorocentrum micans, 124
Prosthecochloris, 156, 600
Prosthecomicrobium, 181
  P. pneumaticum, 340
Proteobacteria, 102, 148, 160, **179–183**, 280, 283, 285, 302, 313, 317, 336–338, 357, 358, 364, 371, 377, 388, 398, 417, 419, 420, 422, 423, 430, 566, 591, 609, 633, 722, 728, 836, 840, 841
Proteomonas sulcata, 239
Proteus, 53, 182, 622
Protista, 115
Prototheca, 215
Protozoa, 10, 53, 65, 67, 108, 113–115, 135, **138–140**, 221, 250, 252, 374, 397, 401, 405, 407, 410, 411, 413, 421, 436, 488, 492, 493, 505, 506, 620, 621, 624, 640, 643, 645, 647, 648, 651, 766, 787–788, 815, 842, 869, 897
Providencia, 622
Prymnesiophyceae, 111, 235, 237
Prymnesium parvum, 237
Psalteriomonas, 239
Pseudoalteromonas, 356, 359, 360
Pseudoaminobacter, 702
Pseudociliata, 239
Pseudomonadaceae, 388
Pseudomonadales, 182, 388
Pseudomonas, 44, 46, 182, 314, 315, 320, 321, 330, 338, 342, 345, 346, 357, 387–389, 401, 419, 423, 432, 468, 661, 693, 696–699, 701, 702, 704, 711–713, 722, 726, 727, 738, 743
  P. aeruginosa, 32, 182, 276, 294, 300, 307, 308, 310, 312–316, 344, 404, 423, 427, 464, 631, 633–638, 641, 642, 656, 713, 726, 727
  P. aureofaciens, 313
  P. chlororaphis, 284
  P. fluorescens, 272, 276, 310, 314, 737, 780
  P. fragi, 360
  P. methanica, 691
  P. nautica, 576
  P. oleovorans, 724
  P. paucimobilis, 713
  P. pickettii, 694
  P. putida, 360, 609, 694, 698, 704, **724–727**
  P. syringae, 276, 412, 413
  P. testosteroni, 124
  P. tolaasii, 319
Pseudonocardia, 188
Pseudonocardiaceae, 188
Pseudonocardineae, **188**
Pseudoparanema, 114
Pseudovahlkampfia, 239
Psychromonas, 182, 356, 768
Puccinia, 411
  P. graminis, 218
Pyramimonas, 215
Pyrobacterium islandicum, 163, 607
Pyrobaculum, 48, 49, 62, 167, 363
  P. aerophilum, 167, 471, 575
Pyrococcus, 46, 48, 62, 66, 172, 363, 365, 379
  P. abyssi, 367, 379, 471
  P. furiosus, 163, 367, 472
  P. yayanosii, 379

Pyrodictiaceae, 166
Pyrodictium, 166, 363, 366
Pyrolobus, 166, 363, 366
  P. fumarii, 163, 166, 365, 366
Pythium, 230, 412
  P. insidiosum, 230

R
Radiolaria, 196, **215–216**, 229, 251
Ralstonia, 641, 704, 711, 726, 727
  R. eutropha, 692, 693, 713
  R. metallidurans, 181
  R. solanacearum, 316, 319, 320, 411, 412
Ralstoniaceae, 181
Ramlibacter, 337, 338
  R. tataouinensis, 338
Raphidiophrys pallida, 208
Raphidophyceae, 232, 234
Reovirus, 621
Retaria, 134–136, 139–141, 193
Reticulitermes speratus, 423
Retortamonadida, 241, 253
Rhabdomonas, 240
Rhizaria, 120, 127, 130, 191, 194, 205, 207, **215–221**
Rhizellobiopsis, 221
Rhizobacter, 182
Rhizobiaceae, 181, 566
Rhizobiales, 179, 181, 388
Rhizobium, 46, 68, 158, 181, 314, 338, 398, 415, 416, 419, 566, 579
  R. etli, 314
  R. leguminosarum, 314
  R. loti, 443
  R. meliloti, 276, 303
Rhizoctonia solani, 412, 414, 417
Rhizopus, 403
Rhodella, 209, 211
Rhodellophyceae, 209, 210
Rhodobaca, 388, 389
Rhodobacter, 156, 179, 181, 591, 593
  R. sphaeroides, 25
Rhodobacteraceae, 388
Rhodobacterales, 179, 181, 281, 388, 891
Rhodobacteria, 119
Rhodobiaceae, 181
Rhodobionta, 114, 121–123, 125–128, 132, 134, 136, 197, 201, 204, **209–211**, 221, 225, 227, 230, 238, 239, 242, 245, 250, 253
Rhodobium, 156, 179, 181, 593
Rhodococcus, 188, 696, 703, 711, 715, 722, 724, 726, 727
  R. corallinus, 695, 703
  R. rhodochrous, 284
Rhodocyclaceae, 181
Rhodocyclales, 181
Rhodocyclus, 181
Rhodoferax, 181, 591
Rhodomicrobium, 181
Rhodomonas, 128
Rhodonellum, 387
Rhodophytina, 210
Rhodoplantae, 209
Rhodopseudomonas, 46, 156, 179, 181, 300, 568, 593
Rhodopseudomonas palustris, 181, 300
Rhodosorus, 209, 211
Rhodospira, 181
Rhodospirillaceae, 181

Rhodospirillales, 179, 181
Rhodospirillum, 56, 179, 181, 591
Rhodovibrio, 181
Rhodovulum, 156, 179, 181, 593
Rhodovulum iodosum, 600, 607
Rhodovulum robiginosum, 600, 607
Rhoicosphaenia abbreviata, 137
Rickenella, 183
Rickenellaceae, 183
Rickettsia, 97, 109, 111, 181, 423, 435, 475
Rickettsiaceae, 181
Rickettsia conorii, 181
Rickettsiales, 181, 432, 435
Rickettsia prowazekii, 126, 300
Rickettsia rickettsii, 435
Riftia pachyptila, 19, 430, 431, 600
Rissoella, 114
Roseibaca, 388
Roseinatronobacter, 388
Roseobacter, 60, 181
Roseococcus, 60
Roseovivax, 181
Rotavirus, **622**
Rotifers, 230, 374, 401, 490, 495, 552, 670, 868
Rubrivivax, 181
Rubrobacter, 187, 338
Rubrobacteraceae, 187
Rubrobacterales, 187
Rubrobacteridae, 187
Ruminants, 284, 431, 477, 546, 554, 557
Ruminobacter, 182
Ruminococcus albus, 284
Ruminococcus flavefaciens, 284

S
Saccharococcus, 186
Saccharomyces, 283
    S. cerevisiae, 81, 121, 245, 247, 299, 477, 478, 833, 852
Salarchaeum, 171
Salinarchaeum, 171
Salinibacter ruber, 171, 172, **472–473**
Salinicoccus, 387
Salinimonas, 388
Salinivibrio, 182, 388
Salix, 263
Salmonella, 53, 182, 307, 428, 429, 436, **463–464**, 620–621, **622**, 642, 656
    S. enterica, 300, 307, 435, 463–465, 631, 642
    S. paratyphi, 182, 621, 622
    S. typhi, 182, 463, 465, 621, 622, 630, 631
    S. typhimurium, 303, 307, 308, 320, **449**, 463, 464, 642
Saprolegnia, 230, 231
Saprolegniales, 230
Saprospira, 183
Saprospiraceae, 183
Sarcina ventriculi, 334
Satka favosa, 93, 94
Scalindua *(Candidatus)*, 576
Scenedesmus, 215
Schistosoma
    S. haematobium, 425
    S. japonicum, 425
    S. mansoni, 425
Schizocladiophyceae, 232
Schizosaccharomyces pombe, 299

Scytomonas, 240
Scytonema, 178
Serpulina, 184
Serpulinaceae, 184
Serratia, 53, 316, 622
    S. liquefaciens, 316
    S. marcescens, 313, 314, 638
Shewanella, 46, 70, 356, 359, 376, 607, 611, 768, 839, 840
    S. oneidensis, 49, 120, 839, 840
    S. putrefaciens, 600, 610, 743
Shigella, 53, 160, 429, 434, 436, 620–621, **622**
    S. flexneri, 160, 319, 435
    S. sonnei, 182
Simkania, 184
Simkaniaceae, 184
Simuliidae, 498
Sinorhizobium, 148, 401
    S. meliloti, 181, 314, 405, 841
Siphovirus, 501, 502
Sitophilus, 424
Skeletonema costatum, 221
Sneathia, 285
Sorangium cellulosum, 833
Sorghum vulgare, 701
Sphaerobacter, 187
Sphaerobacteraceae, 187
Sphaerobacterales, 187
Sphaerobacteridae, 187
Sphaerotilus, 181, 182
    S. natans, 606
Sphingobacteria, 160, 183, 371
Sphingobacteriaceae, 183
Sphingobacteriales, 183
Sphingobacterium, 183
Sphingobium, 404, **705–707**
    S. chlorophenolicum, 705
    S. francense, 705–707
    S. japonicum, 425, 705, 706
Sphingomonadaceae, 181
Sphingomonadales, 181
Sphingomonas, 181, 346, 423, 703, 711, 722
Spirillaceae, 182
Spirillum, 182, 356, 376
Spirochaeta, 184, 388, 422
    S. smaragdinae, 168
Spirochaetaceae, 184, 388
Spirochaetales, 184, 388
Spirochaetes, 125, 160, **184**, 300, 317, 388, 456, 469
Spirogyra, 137
Spironucleus salmonicida, 478
Spiroplasma, 187, 423
    S. citri, 187
Spiroplasmataceae, 187
Spirulina, 178, 334, 387, 389, 391
    S. subsalsa, 178
Spongospora, 217
    S. subterranea, 218
Sporichaetaceae, 188
Sporobacter, 185
    S. termitidis, 186
Sporohalobacter, 373
Sporolactobacillaceae, 187
Sporolactobacillus, 187
Sporomusa, 50, 185
Spumella, 235
Staphylococcaceae, 387

Staphylococcus, 12, 46, 186, 432
　S. aureus, 187, 294, 300, 303, 314, 316, 318, 319, 425, 427, 461, **462**, 468, 631, 633–636
　S. epidermidis, 285
　S. lugdunensis, 285
Staphylothermus, 163, 166, 363, 366
Steinernema, 423, 428
Stenotrophomonas, 389, 423
　S. maltophilia, 316, 636, 638, 641
Stephanodiscus astrea, 489
Steptomyces kanamyceticus, 322
Stetteria, 166
Stigonema, 178
Stigonematales, **178**
Stomatococcus, 188, 344
Stramenopiles, 4, 114, 118, 121, 127, 130, 138, 139, 192, 194, 197, 199, 202, 203, 205, 207, 218, 221, **228–235**, 237–239, 243, 247, 251, 383
Streptobacillus, 184
Streptobionta, 211–214
Streptococcaceae, 187, 639
Streptococcus, 53, 187, 283, 423, 446
　S. agalactiae, 474, 475
　S. bovis, 639
　S. equinus, 639
　S. mitis, 285
　S. oralis, 344
　S. pneumoniae, 313, 316, 319, 446, 447, 453
　S. pyogenes, 187, 633
　S. thermophilus, 398
Streptomyces, 14, 20, 188, 316, 341, 387, 402, 419, 436, 638, 679, 722
　S. coelicolor, 20, 187, 297, 300, 306, 321
　S. griseus, 313
　S. scabies, 188, 412
Streptomycetaceae, 188
Streptomycineae, **188**
Streptosporangiaceae, 188
Streptosporangineae, **188**
Stygiolobus, 164, 166, 363
Stylonematophyceae, 209
Succinivibrio, 182
Succinivibrionaceae, 182
Sulfobacillus, 381, 382
　S. acidophilus, 382
Sulfolobales, 161, 163, **164–166**, 174, 365, 383, 472
Sulfolobus, 44, 63, 66, 164, 166, 334, 363, 365, 366, 589
　S. acidocaldarius, 163, 300, 383
　S. metallicus, 383
　S. solfataricus, 163, 383, 471, 472
Sulfophobococcus, 166
Sulfurimonas denitrificans, 589
Sulfurisphaera, 164, 166
Sulfurococcus, 164, 166, 366
Symbiobacterium, 185
　S. thermophilum, 403
Symbiodinium, 227
Synchroma grande, 234, 235
Synchromophyceae, 232, 234, 235
Syndiniales, **221–222**
Synechococcus, 177, 282, 334, 364, 464, 526, **527**, 773, 776, 778
Synechocystis, 126, 300, 464, 469
Synergistes, 177
Synergistetes, 160, **177**, 456
Syntrophaceae, 183
Syntrophobacter, 54, 55, 183, 815
Syntrophobacteraceae, 183

Syntrophobacterales, 183
Syntrophomonadaceae, 185, 387
Syntrophomonas, 54, 55, 185
Syntrophus, 183
　S. aciditrophicus, 403
Synurophyceae, 231, 232, 234

T
Tappania plana, 93, 94
Telonemia, 207, 237
Tenericutes, 160, **187**, 456
Tepidibacter, 363, 364
Termites, 169, 242, 247, 387, 423, 492, 500, 544–546, 554
Terrabacter, 389
Tetraselmis, 215, 217
Thalassia, 207
Thalassolituus, 719
Thalassomyces, 221, 222
Thauera mechernichensis, 575
Thaumarchaeota, 162, 163, 174–175, 364, 456, 471
Thermaceae, 177
Thermales, 176, 177
Thermoactinomyces, 187, 679
Thermoactinomycetaceae, 187
Thermoanaerobacter, 185
　T. brockii, 186
Thermoanaerobacteriaceae, 185
Thermoanaerobacteriales, 185
Thermoanaerobacterium, 185, 364
Thermocladium, 167
Thermococcaceae, 172
Thermococcales, 162–164, 167, **172–173**, 365, 367, 472
Thermococcus, 48, 62, 172, 365, 379
　T. barophilus, 378
Thermocrinis, 175, 363
　T. ruber, 364
Thermodesulfatator, 363, 364
Thermodesulfobacteria, 160, 176, 364, 456
Thermodesulfobacteriaceae, 176
Thermodesulfobacterium, 176, 363, 364
　T. commune, 84, 607
Thermodesulfovibrio, 177
Thermodiscus, 166
Thermofilaceae, 167
Thermofilum, 163, 167, 363
Thermogladius, 166
Thermohalobacter berrensis, 186
Thermomicrobia, 160, **176**
Thermomicrobiaceae, 176
Thermomicrobium, 176
　T. roseum, 124
Thermomonospora, 150, 806
Thermomonosporaceae, 188
Thermoplasma, 66, 122, 161, 173, 334, 363, 365, 383, 399
　T. acidophilum, 163, 383, 472
Thermoplasmataceae, 173
Thermoplasmatales, 162–164, 167, 169, **173**, 365, 383, 472, 841
Thermoproteaceae, 167
Thermoproteales, 163, 164, **167**, 365
Thermoproteus, 48, 62, 167, 363
　T. tenax, 163
Thermosipho, 363, 364, 588
Thermosphaera, 166
Thermosulfurimonas dismutans, 364
Thermosynechococcus elongatus, 464

Thermotoga, 49, 176, 325, 363, 364, 588
   T. hypogea, 168
   T. maritima, **469–471**, 607
Thermotogaceae, 176
Thermotogae, 160, 175, **176**, 456, 469
Thermotogales, 82, 176, 364, 365
Thermovibrio, 364
Thermus, 49, **176–177**, 357, 364
   T. aquaticus, 21, 264, **367**
   T. thermophilus, **468**, 474
Thielaviopsis basicola, 412, 419
Thioalkalicoccus, 388, 389
Thioalkalimicrobium, 388, 389
Thioalkalispira, 388, 389
Thioalkalispiraceae, 388
Thioalkalivibrio, 388, 389
   T. nitratireducens, 390
Thiobacillus, 44, 46, 68, 182, 334, 589, 600, 885
   T. denitrificans, 589, 611
   T. ferrooxidans, 380
   T. thioparus, 878
Thiobacteria, 119
Thiobacterium, 44, 182, 589
Thiocapsa, 70, 156, 182, 340, 591, 595
   T. roseopersicina, **611**, 878
Thiococcus, 182
   T. pfennigii, 591
Thiocystis, 156, 182
Thiodictyon, 182, 340
Thiohalospira, 388
Thiomargarita, **112**, 182, 344, 491
   T. namibiensis, 26, 92, 109, 112, 497
Thiomicrospira, 44, 589
   T. denitrificans, 47
Thiomonas, 182, 286, 840
Thiopedia, 182
Thioploca, 109, 112, 182, 491, 497
Thiorhodospira, 388, 389
Thiospira, 44, 183
Thiospirillum jenense, 591
Thiospora pantotropha, 574
Thiothrix, 44, 182, 341, 344, 589, 670
Thiotricaceae, 183
Thiotrichales, 388
Thiovulum, 183
Tilapia, 372
Tindallia, 387
   T. magadiensis, 390
Toxoplasma, 101, 397, 435
   T. gondii, 101, 227
Toxoplasmes, 101, 228
Toxothrix, 183
Tracheophyta, 115
Trebouxia, 215, 245
Trebouxiophyceae, 212, 214, 215, 245
Trentepohlia, 215, 245
Trentepohliophyceae, 211, 214, 245
Treponema, 110
   T. denticola, 300
   T. pallidum, 184
Trichoderma, 417–419
Trichodesmium erythraeum, 464
Trichomonads, 100, 242
Trichomonas, 53, 242
   T. vaginalis, 100, 242
Trichonympha, 53
Trichophyton, 628, 629

Trimyema, 53
Tritrichomonas, 242
Tropheryma whipplei, 840
Trypanosoma, 240, 424, 435
   T. brucei, 240
   T. cruzi, 424
Trypanosomatids, 478
Tsukamurellaceae, 188
Tubulina, 248
Turbinaria, 114

U

Ulva, 316
Ulvophyceae, 211, 213, 214
Uncinula necator, 411, 413
Unikonts, 205
Ureibacillus, 679
Uromyces, 411
Ustilago maydis, 411

V

Vahlkampfia, 239
Valeria lophostriata, 93, 94
Valhkampfia, 383
Vampirovibrio, 183
Vannellina, 248
Variovorax, 694, 696, 704
   V. paradoxus, 349
Vaucheria litorea, 130, **134**
Venturia inaequalis, 263, 413
Verrucomicrobia, 160, **184**, 456
Verrucomicrobiaceae, 184
Verrucomicrobiales, 184, 356
Verrucomicrobiota, 422
Verrucomicrobium, 184
Verticillium, 412, 413
Vibrio, 158, 182, 314, 317, 334, 342, 359, 389, 402, 622, **623**
   V. alginolyticus, 623
   V. cholerae, 13, 182, 294, 300, 307–311, 317, 340, 429, 464, 504, 621, 623, 630, 633, 634, 649
   V. fischeri, 313, 314, 431, 634
   V. harveyi, 317
   V. marinus, 377
   V. mimicus, 623
   V. parahaemolyticus, 623
   V. vulnificus, 623
Vibrionaceae, 182, 388
Vibrionales, 182, 388
Victivallales, 185
Victivallis vadensis, 185
Viridiplantae, 92, 109, 114, 118, 120, 121, 123, 124, 126–128, 130, 132–134, 137, 177, 178, 207, 208, **211–215**, 218, 225, 234, 247, 250, 316
Virus, 12, 14, **109–111**, 218, 237, 397, 413, 424, 425, 437, 450, 470, 486, **501–504**, 526, **620–622**, 624, 631, 640, 775, 782, **788**, 822, **869–871**
Volvox, 215
Vulcanisaeta, 167
Vulcanithermus, 363, 364

W

Waddlia, 184
Waddliaceae, 184
Westiella, 200

Williamsiaceae, 188
Wolbachia, 181, 269, 298, 300, 398, 423, 424, **432–433**, 435, 477
Wollinella succinogenes, 50, 300
Wuchereria bancrofti, 432

X
Xanthomonadaceae, 182
Xanthomonadales, 182
Xanthomonas campestris, 310, 316, 406, 412, 413, 465
Xanthophyceae, 197, 231, 232, 234
Xanthoria, 215
Xenopus laevis, 247
Xenorhabdus, 423, 428
Xiphinema, 413

Y
Yersinia, 428, 429, 622, 623
   Y. enterocolitica, 313, 335, 621, **623**, 642
   Y. pestis, 148, 182, 308, 319, 326, 465, 635
   Y. pseudotuberculosis, **623**

Z
Zea mays, 124, 126
Zetaproteobacteria, 160, 183
Zooglea, 181, 182
Zoogloea ramigera, 670
Zostera, 207, 580
Zygnematophyceae, 212, 214, 215
Zygnemopsis, 383
Zygomycota, 115, 243, 245, 246